pK$_a$ values for various compounds and compound classes

Compound or Class	pK$_a$
HI	~-9.5
HBr	~-9
HCl	~-7
H$_2$SO$_4$	~-5
H$_3$O$^+$	-1.7
C$_6$H$_5$SO$_3$H	~-0.6
H$_3$PO$_4$	2.12
HF	3.45
CH$_3$CO$_2$H	4.75
RCO$_2$H	3–6
H$_2$CO$_3$	6.37
HCN	9.31
$\overset{\text{O}}{\overset{\|}{\text{CH}_3\text{C}}}\text{CH}_2\overset{\text{O}}{\overset{\|}{\text{C}}}\text{CH}_3$	9
NH$_4{}^+$	9.4
C$_6$H$_5$OH	10.00
CH$_3$NO$_2$	10.21
$\overset{\text{O}}{\overset{\|}{\text{CH}_3\text{C}}}\text{CH}_2\overset{\text{O}}{\overset{\|}{\text{C}}}\text{OCH}_2\text{CH}_3$	11
$\text{CH}_3\text{CH}_2\text{O}\overset{\text{O}}{\overset{\|}{\text{C}}}\text{CH}_2\overset{\text{O}}{\overset{\|}{\text{C}}}\text{OCH}_2\text{CH}_3$	13
H$_2$O	15.74
CH$_3$CH$_2$OH	15.9
ROH	15–19
$\text{CH}_3\overset{\text{O}}{\overset{\|}{\text{C}}}\text{CH}_3$	20
$\text{CH}_3\overset{\text{O}}{\overset{\|}{\text{C}}}\text{OCH}_2\text{CH}_3$	25
RC≡CH	~25
CH≡CH	~26
NH$_3$	~35
C$_6$H$_6$	~43
CH$_2$=CH$_2$	~45
CH$_3$CH$_3$	~50

Editor: *Keith Dodson*
Marketing Team: *Caroline Croley, Michelle Mootz*
Editorial Assistant: *Georgia Jurickovich*
Project Development Editor: *Beth Wilbur*
Production Editor: *Nancy L. Shammas*
Production Service Manager: *Donna King*
Production Service: *Progressive Publishing Alternatives*
Manuscript Editor: *Peter Fong*
Interior Design: *Progressive Publishing Alternatives*

Interior Illustration: *Progressive Information Technologies*
Cover Design: *Jamie Dagdigian*
Cover Photo: *Dennis Kunkel / Phototake*
Design Editor: *Roy R. Neuhaus*
Indexer: *James Minkin*
Typesetting: *Progressive Information Technologies*
Cover Printing: *Phoenix Color Corporation*
Printing and Binding: *R. R. Donnelley & Sons / Williard*

COPYRIGHT © 1998 by Brooks/Cole Publishing Company
A Division of International Thomson Publishing Inc.
I(T)P The ITP logo is a registered trademark under license.

For more information, contact:

BROOKS/COLE PUBLISHING COMPANY
511 Forest Lodge Road
Pacific Grove, CA 93950
USA

International Thomson Publishing Europe
Berkshire House 168–173
High Holborn
London WC1V 7AA
England

Thomas Nelson Australia
102 Dodds Street
South Melbourne, 3205
Victoria, Australia

Nelson Canada
1120 Birchmount Road
Scarborough, Ontario
Canada M1K 5G4

International Thomson Editores
Seneca 53
Col. Polanco
11560 México, D. F., México

International Thomson Publishing GmbH
Königswinterer Strasse 418
53227 Bonn
Germany

International Thomson Publishing Asia
221 Henderson Road
#05–10 Henderson Building
Singapore 0315

International Thomson Publishing Japan
Hirakawacho Kyowa Building, 3F
2-2-1 Hirakawacho
Chiyoda-ku, Tokyo 102
Japan

Printed in the United States of America.

10   9   8   7   6   5

Library of Congress Cataloging-in-Publication Data
Fessenden, Ralph J., [date]
    Organic chemistry.—6th ed. / Ralph J. Fessenden, Joan S. Fessenden, Marshall Logue.
        p.   cm.
    Includes index.
    ISBN 0-534-35199-9
    1. Chemistry, Organic.   I. Fessenden, Joan S.   II. Logue, Marshall.   III. Title.
QD251.2.F49 1998
547—dc21   97-43828
                                                   CIP

ORGANIC CHEMISTRY
6TH EDITION

Ralph J. Fessenden
University of Montana

Joan S. Fessenden

with contributions by
Marshall W. Logue
Michigan Technological University

Brooks/Cole Publishing Company

I T P® An International Thomson Publishing Company

Pacific Grove • Albany • Belmont • Bonn • Boston • Cincinnati • Detroit • Johannesburg
London • Madrid • Melbourne • Mexico City • New York • Paris • Singapore
Tokyo • Toronto • Washington

amines by catalytic hydrogenation as an alternative to metal/acid reductions has been included in Chapter 12 (Substituted Benzenes). A significant number of the end-of-chapter problems for Chapters 5 through 17 have been revised.

Organization

The text is organized into three parts; introductory material (concepts of structure and bonding); organic reactions and mechanism (sigma-bond chemistry followed by pi-bond chemistry); and topics of more specialized interest.

Introductory material   Chapters 1 and 2 are primarily reviews of atomic and molecular structure, electronegativity, hydrogen bonding, acid-base reactions, and molecular orbitals (presented for the most part in a pictorial way). Bonding in some simple nitrogen and oxygen compounds is included as a way of introducing functional groups. A brief introduction to resonance theory is also presented here as well as its application to the acidity of phenol and carboxylic acids.

The student's introduction to structural isomerism and nomenclature is in Chapter 3. In addition to discussing the nomenclature of alkanes, we introduce a few other compound classes that are used early in the book.

Chapter 4 on stereochemistry discusses structure: geometric isomerism, conformation, chirality, and resolution. The terms *stereogenic center* and *stereocenter* are defined, but we will continue to use the terms chiral carbon and chiral atom. Fischer projections are also introduced in this chapter, although they are not used again until Chapter 23 (Carbohydrates).

Organic reactions and mechanisms   Mechanisms are introduced in Chapter 5 with substitution and elimination reactions of alkyl halides. We have several reasons for using the reactions of alkyl halides for our introduction to mechanisms. First, the typical S_N2 reaction path is a concerted reaction with a single transition state and thus is ideal for introducing transition-state diagrams and reaction kinetics. Second, the S_N1 path follows logically from the S_N2 mechanism and allows introduction of steric hindrance and carbocations early in the course. Finally, ionic reactions allow us to apply the stereochemical principles just covered in Chapter 4.

A discussion of the E2 reaction path of alkyl halides follows the discussion of substitution reactions. Although the E1 mechanism is mentioned here, the principal discussion of this reaction is postponed until Chapter 7, where the reactions of alcohols are presented.

Chapter 6 covers free-radical chemistry, principally halogenation and polymerization. Chapter 7 (Alcohols) includes the introduction to Grignard reagents and lithium reagents. The differences between alcohols and phenols are emphasized. Chapter 8 discusses the chemistry of ethers, epoxides, and sulfides.

Infrared and NMR spectroscopy, presented in Chapter 9, provides a break in the discussion of organic reactions. The utility of high-field spectrometers has been added as well as a number of 300 MHz spectra.

Chapter 10 (Alkenes and Alkynes) is an introduction to pi-bond chemistry including a discussion of line-bond formulas and a discussion of organoborane

PREFACE

For the first edition of *Organic Chemistry,* our goal was to produce a manageable text for the one-year introductory organic chemistry course. Our intent was to *guide* students in their studies and not simply provide a compilation of organic chemical knowledge. The response from users of our earlier editions has been gratifying and, we believe, shows that we have been successful in achieving our goals.

The basic organization of the sixth edition is the same as that of previous editions. Most changes have been made to improve the logical flow or to help the student correlate the material. The major changes are outlined below.

Changes in the Sixth Edition

Stereochemistry The topic of conformational enantiomers (dynamic enantiomers) using the *gauche* conformations of butane has been added to Chapter 4 (Stereochemistry). Previously, this concept was informally introduced with substituted cyclohexanes in an in-chapter problem. The acyclic example should be easier for the student to visualize and will be reinforced by the more complicated cyclohexane example. A method for the assignment of (*R*) and (*S*) configurations directly from Fischer projections also been added to Chapter 4. The concept of diastereotopic hydrogens has been included in the section on prochirality in Chapter 4.

NMR spectroscopy The discussion of high-field NMR spectroscopy has been expanded in Chapter 9 (Spectroscopy). At various places in the remainder of the text 60 MHz spectra have been replaced with 300 MHz spectra, but only when such a change facilitates interpretation. Because spin-spin splitting caused by allylic coupling is so prevalent in the spectra of alkenes and alkynes, a brief discussion of this topic has been included in Chapter 10 (Alkenes and Alkynes). Thus the student will understand the origin of these splittings that are apparent in many of the NMR spectra of alkenes and alkynes that appear in subsequent chapters, and not be misled.

Nucleophilic substitution In previous editions, the back-side displacement by nucleophiles in S_N2 reactions has been invoked to explain the observed inversion of configuration, but the reason for the back-side attack of nucleophiles has never been presented. A molecular orbital explanation for back-side attack by nucleophiles using antibonding sigma orbitals on the substrate has been added to Chapter 5 (Alkyl Halides).

Miscellaneous changes In response to comments by reviewers and users of the previous editions, the segment on the assignment of formal charge has been reinstated in Chapter 1 (Atoms and Molecules—A Review), as has the appendix on organic nomenclature. The reduction of nitroaromatics to aryl-

Each chapter contains an essay problem. These essay problems consist of short, descriptive passages that discuss chemistry not specifically covered in the chapter and are followed by questions on that material. The essay problems are intended to accomplish three goals. The first is to provide a bridge between the subject material found in the chapter and the current chemical literature. Almost all of the essay problems emphasize modern organic synthesis and stereochemistry. Whenever possible, the essay problem is based on and referenced to an *Organic Syntheses* procedure or a research literature article.

The second goal of the essay problems is to challenge the student to think in terms of organic chemistry. Because each essay discusses material not found in the chapter, a knowledge of the chemistry presented in that chapter is required to understand the essay and answer the questions. However, merely memorizing the material will not be satisfactory. The student will need to understand the material and use it in a new context to answer the essay questions. To help accomplish this goal, the essays increase in difficulty as the student progresses through the text.

The third goal is to provide practice answering this type of question. Essay problems are now used in national examinations. This text's problems are slightly different from those. National exams use multiple choice questions while the questions in this text use an open-ended format requiring a written response. Our answers to the questions are contained in the *Student's Solutions Manual* that accompanies this text.

Ancillaries

The *Solutions Manual* includes answers and full solutions to end-of-chapter problems and the essay problems, as well as the *Organic Chemistry Toolbox 1998* on CD-ROM. This software works as an electronic study guide and provides tools that help students visualize and analyze the chemistry they need to solve approximately 200 problems from the text. The *Study Guide with Additional Drill Problems* includes a brief discussion of each chapter and additional drill problems with answers. Test items are available in both print and electronic formats. Many figures from the text, especially spectra, have been included in a set of overhead transparencies, available from the publisher.

Also available is *Organic Chemistry Online,* a hybrid CD-ROM that includes independent modules on key topics, numerous interactive problems, and state-of-the-art three-dimensional graphics.

Acknowledgments

We are indebted to users of previous editions for their suggestions. Many changes in this edition are the result of their comments. We also wish to thank the reviewers of previous editions. Their suggestions helped us firm our ideas and approach some topics with a new light.

Finally, we thank the editorial and production staff of Brooks/Cole Publishing Company, who have encouraged and helped us in this edition: Keith Dod-

oxidation. Chapter 11 covers benzene and electrophilic aromatic substitution. Chapter 12 includes the chemistry of alkylbenzenes, phenols, arylamines (including aryldiazonium salts), and aryl halides.

Carbonyl chemistry is introduced in Chapter 13 with the chemistry of aldehydes and ketones. Chapter 14 continues carbonyl chemistry with the chemistry of carboxylic acids. The reasons for differences in acid strength of many types of compounds are discussed here. Chapter 15 describes the chemistry of carboxylic acid derivatives.

Chapter 16 contains conjugate addition, including the Diels–Alder reaction. Chapter 17 (Enolates and Carbanions) introduces alkylation and condensation reactions. Finally, Chapter 18 (Amines) completes the basic discussion of organic reactions and mechanisms.

Topics of specialized interest Chapters 19 through 22 cover the chemistry of polycyclic and heterocyclic compounds, natural products and some modern synthetic techniques, pericyclic reactions, and other spectroscopic techniques (UV and visible spectroscopy and mass spectrometry). Chapters 23 through 26 include bio-organic topics: carbohydrates, lipids, amino acids and proteins, and nucleic acids.

Spectroscopy Infrared and nuclear magnetic resonance spectroscopy are discussed as early as we think feasible—Chapter 9. By this time, a student has a working knowledge of structure, a few functional groups, and a few reactions. However, those who wish to do so may cover the spectroscopy chapter right after Chapter 4, as soon as students are familiar with organic structures.

Ultraviolet and mass spectra are described in Chapter 22. These topics are designed to stand alone; therefore, either or both can be presented along with infrared and NMR spectra.

Synthesis Our formal presentation of synthesis begins in Chapter 7 with a discussion on approaching multistep synthesis problems and retrosynthetic analysis. Thereafter, we have placed our discussions of synthesis in separate sections at the end of each appropriate chapter so they may be emphasized or deemphasized in the lecture presentations.

Bio-organic material Many students in the introductory organic course are majoring in medical or biological fields. Therefore, numerous sections and problems that are biological in nature are included.

Problems We are firm believers in problem solving as an important part of learning organic chemistry, and we have included more than 1100 unsolved problems. Within the chapters, worked-out sample problems are included to illustrate the approach to problem solving and to provide further information. Often these sample problems are followed directly by study problems with answers at the end of the book. Some of these study problems are designed to relate previous material to the present discussion. Others are designed to test students on their mastery of new material.

In the chapter-end problem sets, the first study problems follow the order of the chapter. The *Additional Problems* do not follow chapter order, but are in order of increasing difficulty.

son, senior developmental editor; Beth Wilbur, project development editor; and Nancy Shammas, senior production editor. We are also grateful to Donna King, project manager at Progressive Publishing Alternatives, and to Peter Fong, our copy editor.

Ralph J. Fessenden
Marshall W. Logue

PREFACE TO THE STUDENT

Studying organic chemistry is similar to studying a foreign language. The vocabulary is new as are the "rules of grammar," such as predicting the course of an organic reaction.

When approaching new material, you may find it helpful to scan (1) the introductory paragraphs of a chapter or section, (2) the chapter summary, and (3) the chapter itself. If you do this *before the lecture* on that material, you will find the lecture more interesting and understandable. After the lecture, study the chapter sections and your lecture notes carefully, outlining the material wherever possible.

As a check on your understanding, you should work the problems within the chapter as you go along. (Cover the answers to Sample Problems with a sheet of paper, and try to work them as well.)

It is important to *write* your answers to the problems rather than simply to envision the answers in your mind. There are two reasons for writing out the answers: (1) for practice in the mechanical art of writing organic formulas, and (2) for reinforcing the subject material.

The chapter-end *Study Problems* are organized so that you can work them as you proceed through the text or you can work them when you have completed the entire chapter. The *Additional Problems* which draw on subject material more randomly should not be attempted until you have finished the chapter. These problems are gradated according to difficulty, with the early ones being drill problems and the later ones being thought problems.

See if you can answer the problems yourself, and *then* check the answers and the text. (If you try to take shortcuts, you are only deceiving yourself.) If your answers are correct, you have a reasonable mastery of the material. If you have several incorrect answers, go back and restudy the chapter, your lecture notes, and hints in the *Study Guide* which accompanies this text.

Organic molecules are three dimensional. Besides a pencil and paper, you will need a molecular model kit. At the start of the course, especially in discussions of stereochemistry, you will find it helpful to construct models of organic molecules and compare them with their representations on paper.

For your convenience, this text contains an appendix covering the rules of organic nomenclature. (Nomenclature is also included where appropriate within the text.) A glossary of prefix symbols is included at the end of the appendix. The in-chapter problems are answered at the end of the text. The answers to the text's chapter-end problems are available in the *Solutions Manual*.

The *Study Guide* also contains additional hints and suggestions for studying organic chemistry, as well as a brief chapter-by-chapter listing of important points, additional drill problems, and their answers.

BRIEF CONTENTS

CONTENTS

Chapter 10 Alkenes and Alkynes **413**

Chapter 11 Aromaticity and Benzene; Electrophilic Aromatic Substitution **473**

Chapter 24 Lipids 961

Chapter 25 Amino Acids and Proteins 985

Chapter 26 Nucleic Acids 1019

ATOMS AND MOLECULES—A REVIEW

Jöns Berzelius (1779–1848), a Swedish scientist, began using the term *organic* to refer to substances isolated from living systems. At that time, the vital force theory was in vogue. This theory held that a special force, or life force, was necessary to produce organic compounds. It was believed that since life itself was not understandable, organic compounds also were not understandable.

Organic chemistry became a distinct branch of science in the early 1800s. However, organic chemicals and the tools to separate mixtures of organic compounds were known long before this time. Cane sugar (sucrose) had been crystallized as a pure substance in northwestern India about 300 A.D. and distillation had been described in detail in the fifth century. But in the 1800s a change occurred in the approach used by chemists. They recognized that compounds could be characterized by their chemical properties and began to assign specific properties to specific compounds. With this concept, the structure of organic compounds came into prominence.

The demise of the vital force theory began with an experiment by Friedrich Wöhler (1800–1882). He showed in 1828 that urea, an organic compound, could be made from ammonium cyanate, an inorganic, nonliving material. However, even today we find remnants of the vital force theory. Some believe that organic compounds (like vitamins) that have been isolated from a living system are different—and more healthful—than the same compounds synthesized in a laboratory.

Today, organic chemistry is defined as **the chemistry of the compounds of carbon.** This definition is not entirely correct, because a few carbon compounds, such as carbon dioxide, sodium carbonate, and potassium cyanide, are considered to be inorganic. We accept this definition, however, because all organic compounds do contain carbon.

Carbon is but one element among many in the periodic table. What is so unique about carbon that its compounds justify a major subdivision in the study of chemistry? The answer is that carbon atoms can be covalently bonded to other carbon atoms and to atoms of other elements in a wide variety of ways, leading to an almost infinite number of different compounds. These compounds range in complexity from the simple compound methane (CH_4), the major component of natural gas and marsh gas, to the quite complex nucleic acids, the carriers of the genetic code in living systems.

A knowledge of organic chemistry is indispensable to many scientists. For example, because living systems are composed primarily of water and organic compounds, almost any area of study concerned with plants, animals, or microorganisms depends on the principles of organic chemistry. These areas of study include medicine and the medical sciences, biochemistry, microbiology, agriculture, and many others. However, these are not the only fields that depend on organic chemistry. Plastics and synthetic fibers are also organic compounds. Petroleum and natural gas consist mostly of compounds of carbon and hydrogen. Coal is a mixture of elemental carbon combined with compounds of carbon and hydrogen.

Where do we start? The cornerstone of organic chemistry is the covalent bond. Before we discuss the structure, nomenclature, and reactions of organic compounds in detail, we will first review some aspects of atomic structure and bonding (Chapter 1) and then molecular orbitals (Chapter 2) as these topics apply to organic compounds.

Section 1.1

Electron Structure of the Atom

The most important elements to organic chemists are carbon, hydrogen, oxygen, and nitrogen. These four elements are in the first two periods of the periodic table, and their electrons are all found in the two electron shells closest to the nucleus. Consequently, our discussion of the electron structures of atoms will center mainly on elements with electrons only in these two electron shells.

Each electron shell is associated with a certain amount of energy. Electrons close to the nucleus are more attracted by the protons in the nucleus than are electrons farther away. Therefore, the closer an electron is to the nucleus, the lower its energy is and the more difficult it is to remove the electron in a chemical reaction.

The electron shell closest to the nucleus is the shell of lowest energy, and an electron in this shell is said to be at the **first energy level.** Electrons in the second shell, at the **second energy level,** are of higher energy than those in the first shell. Electrons in the third shell, at the **third energy level,** are of higher energy yet.

A. Atomic Orbitals

We cannot accurately determine the position of an electron relative to the nucleus of an atom. Instead, we must rely upon quantum theory to describe the most likely location of an electron. Each electron shell of an atom is subdivided into **atomic orbitals,** regions in space where the probability of finding electrons of a specific energy content is high (90–95%). **Electron density** is another term used to describe the probability of finding an electron in a particular spot; a higher electron density means a greater probability, while a lower electron density means a lesser probability.

The first electron shell contains only the spherical $1s$ orbital. The probability of finding a $1s$ electron is highest in this sphere. The second shell, which is slightly farther from the nucleus than the first shell, contains one $2s$ orbital and three $2p$ orbitals. The $2s$ orbital, like the $1s$ orbital, is spherical.

Figure 1.1 shows a graph of electron density in the $1s$ and $2s$ orbitals as a function of distance from the nucleus. It may be seen from the graph that the $1s$ and $2s$ orbitals do not have sharply defined surfaces, but rather the electron density increases and decreases over a range of distances from the nucleus. The result is that the $1s$ and $2s$ orbitals overlap each other.

The electron-density–distance curve for the $2s$ orbital reveals two areas of high electron density separated by a zero point. This zero point, called a **node,** represents a region in space where the probability of finding an electron (the $2s$ electron in this case) is very small. All orbitals except the $1s$ orbital have nodes. Pictorial representations of the $1s$ and $2s$ orbitals are shown in Figure 1.2.

The second energy level also contains three $2p$ atomic orbitals. The $2p$ orbitals are at a slightly greater distance from the nucleus than the $2s$ orbital and are of slightly higher energy. The p orbitals are shaped rather like dumb-

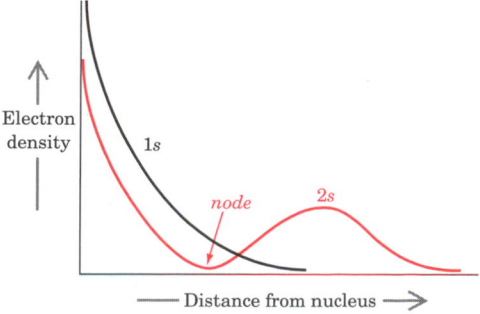

Figure 1.1   Graphic relationship between the $1s$ and $2s$ atomic orbitals.

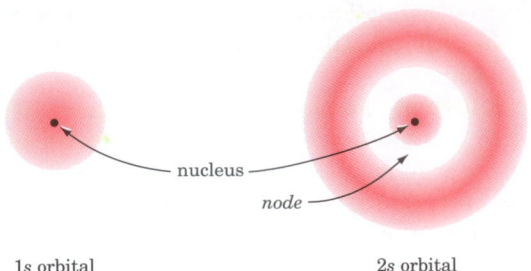

1s orbital   2s orbital

Figure 1.2   Pictorial representations of the 1s and 2s atomic orbitals.

bells; each *p* orbital has two lobes separated by a node (a nodal plane in this case) at the nucleus (see Figure 1.3).

A sphere (an *s* orbital) is nondirectional; that is, it appears the same when viewed from any direction. This is not the case with a *p* orbital, which can assume different orientations about the nucleus. The three 2*p* orbitals are at *right angles* to one another. This orientation allows maximum distance between the electrons in the three *p* orbitals and thus minimizes repulsions between electrons in different *p* orbitals. The mutually perpendicular *p* orbitals are sometimes designated p_x, p_y, and p_z. The subscript letters refer to the $x, y,$ and z axes that can be drawn through pictures of these *p* orbitals, as in Figure 1.3.

Since the three 2*p* orbitals are equivalent in shape and in distance from the nucleus, they have equal energies. Orbitals that have the same energy, such as the three 2*p* orbitals, are said to be **degenerate.**

The third electron shell contains one 3*s* orbital, three 3*p* orbitals, and also five 3*d* orbitals. The numbers of atomic orbitals at each of the first three energy levels are summarized in Table 1.1.

B. Filling the Orbitals

Electrons have spin, which can be either "clockwise" or "counterclockwise" ($+\frac{1}{2}$ or $-\frac{1}{2}$). The spin of a charged particle gives rise to a small magnetic field, or **magnetic moment,** and two electrons with opposite sign have *opposite magnetic moments*. The repulsion between the negative charges of two electrons with opposite spin is minimized by the opposite magnetic moments, allowing two such electrons to become *paired* within an orbital. For this reason, any orbital can hold a maximum of two electrons, and these electrons must be of opposite spin. Because of the number of orbitals at each energy level (one

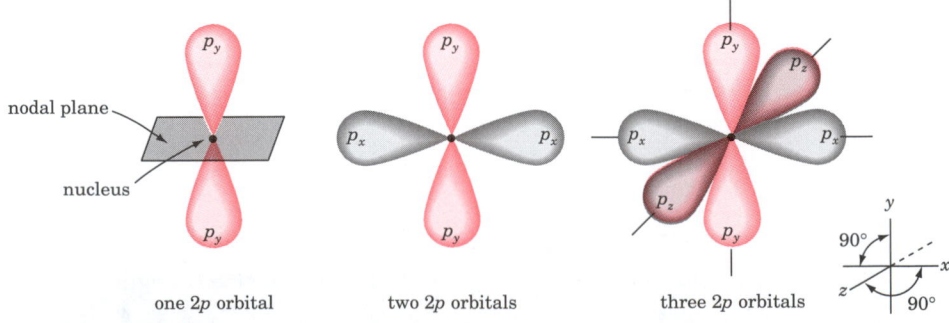

one 2*p* orbital   two 2*p* orbitals   three 2*p* orbitals

Figure 1.3   The shapes and orientations of the 2*p* orbitals.

Table 1.1	Atomic orbitals in the first three energy levels

Energy Level	Atomic Orbitals
1	$1s$
2	$2s\ 2p_x\ 2p_y\ 2p_z$
3	$3s\ 3p_x\ 3p_y\ 3p_z$ plus five $3d$

at the first energy level, four at the second, and nine at the third), the first three energy levels can hold up to 2, 8, and 18 electrons, respectively.

The **aufbau principle** (German, "building up") states that as we progress from hydrogen (atomic number 1) to atoms of successively higher atomic number, orbitals become filled with electrons in such a way that the *lowest-energy orbitals are filled first*. A hydrogen atom has its single electron in a $1s$ orbital. The next element, helium (atomic number 2), has its second electron also in the $1s$ orbital. The two electrons in this orbital are paired.

A description of the electron structure for an element is called its **electron configuration.** The electron configuration for H is $1s^1$, which means one electron (superscript 1) in the $1s$ orbital. For He, the electron configuration is $1s^2$, meaning *two* electrons (superscript 2) in the $1s$ orbital. Lithium (atomic number 3) has two electrons in the $1s$ orbital and one electron in the $2s$ orbital; its electron configuration is $1s^2 2s^1$.

The electron configurations for the first- and second-period elements are shown in Table 1.2. In carbon and the succeeding elements, each $2p$ orbital receives one electron before any $2p$ orbital receives a second electron. This is an example of **Hund's rule:** In filling atomic orbitals, pairing of two electrons in degenerate orbitals does not occur until each degenerate orbital contains one electron. Therefore, an atom of carbon has an electron configuration of $1s^2 2s^2 2p_x{}^1 2p_y{}^1$, abbreviated $1s^2 2s^2 2p^2$ in Table 1.2.

Table 1.2	Electron configurations of the elements in periods 1 and 2

Element	Atomic Number	Electron Configuration
H	1	$1s^1$
He	2	$1s^2$
Li	3	$1s^2\ 2s^1$
Be	4	$1s^2\ 2s^2$
B	5	$1s^2\ 2s^2\ 2p^1$
C	6	$1s^2\ 2s^2\ 2p^2$
N	7	$1s^2\ 2s^2\ 2p^3$
O	8	$1s^2\ 2s^2\ 2p^4$
F	9	$1s^2\ 2s^2\ 2p^5$
Ne	10	$1s^2\ 2s^2\ 2p^6$

STUDY PROBLEMS

1.1 Write the electron configuration of:

   (a) Mg   **(b)** Si

1.2 Which element corresponds with each of the following configurations?

   (a) $1s^2 2s^2 2p^1$ **(b)** $1s^2 2s^2 2p^4$

Section 1.2
Atomic Radius

The **radius of an atom** is the distance from the center of the nucleus to the outermost electrons, which are called the **valence electrons.** The atomic radius is determined by measuring the **bond length** (the distance between nuclei) in a symmetrical covalent compound such as Cl—Cl or H—H and dividing by two. Therefore, atomic radii are often called **covalent radii.** Values for atomic radii are usually given in nanometers, where 1 nm $= 10^{-9}$ m, or in angstroms, where 1 Å $= 10^{-10}$ m.

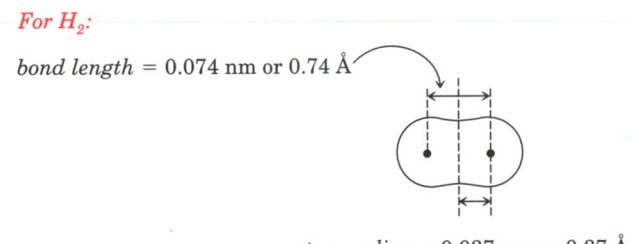

For H₂:

bond length = 0.074 nm or 0.74 Å

atom radius = 0.037 nm or 0.37 Å

Atomic radii vary depending on the extent of attraction between the nucleus and its electrons. The greater the attraction, the smaller the atomic radius. What factors affect this attraction? The most important factors are *the number of protons in the nucleus* and *the number of shells containing electrons.*

A nucleus with a greater number of protons has a greater attraction for its electrons, including the outermost electrons. Consider the elements of the second row of the periodic table (lithium to fluorine). An atom of any of these elements has electrons in only the first two electron shells. As we progress stepwise from lithium to fluorine, a proton is added to the nucleus. At each step, the nucleus has a greater attraction for the electrons, and the atomic radius decreases (refer to Figure 1.4).

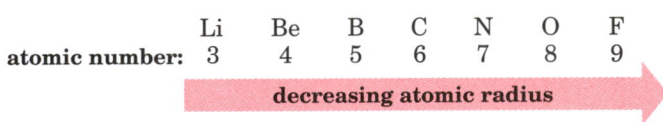

	Li	Be	B	C	N	O	F
atomic number:	3	4	5	6	7	8	9

decreasing atomic radius

As we proceed from top to bottom within a group in the periodic table, the number of electron shells increases and, therefore, so does the atomic radius.

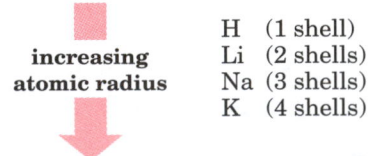

increasing atomic radius

H   (1 shell)
Li   (2 shells)
Na   (3 shells)
K   (4 shells)

In organic chemistry, atoms are bonded together in close proximity to one another by covalent bonds. We will find the concept of atomic radii useful in

H 0.37						
Li 1.225	Be 0.889	B 0.80	C 0.771	N 0.74	O 0.74	F 0.72
Na 1.572	Mg 1.364	Al 1.248	Si 1.173	P 1.10	S 1.04	Cl 0.994
						Br 1.142
						I 1.334

Figure 1.4 Atomic radii of some of the elements (in angstroms, Å, where $1 \text{ Å} = 10^{-8}$ cm).

estimating the attractions and repulsions between atoms and in discussing covalent bond strengths.

Section 1.3
Electronegativity

Electronegativity is a measure of an atom's attraction for its outer bonding electrons, especially in covalently bonded compounds. The concept of electronegativity is used for predicting and explaining chemical reactivity. Like the atomic radius, electronegativity is affected by the number of protons in the nucleus and by the number of shells containing electrons. A greater number of protons means a greater positive nuclear charge and thus an increased attraction for the bonding electrons. Therefore, electronegativity increases as we go from left to right in a given period of the periodic table.

Li   Be   B   C   N   O   F

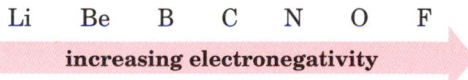

increasing electronegativity

Attractions between oppositely charged particles increase with decreasing distance between the particles. Thus, electronegativity increases as we proceed from bottom to top in a given group of the periodic table because the valence electrons are closer to the nucleus.

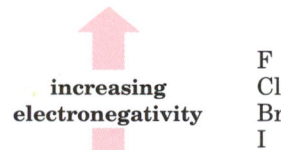

increasing
electronegativity

F
Cl
Br
I

The **Pauling scale** (Figure 1.5) is a numerical scale of electronegativities. This scale is derived from bond-energy calculations for different elements joined by covalent bonds. In the Pauling scale, fluorine, the most electronega-

H 2.1						
Li 1.0	Be 1.5	B 2.0	C 2.5	N 3.0	O 3.5	F 4.0
Na 0.9	Mg 1.2	Al 1.5	Si 1.8	P 2.1	S 2.5	Cl 3.0
						Br 2.8
						I 2.5

Figure 1.5   Electronegativities of some elements (Pauling scale).

tive element, has an electronegativity value of 4. Lithium, which has a low electronegativity, has a value of 1. An element with a very low electronegativity (such as lithium) is sometimes called an **electropositive** element. Carbon has an intermediate electronegativity value of 2.5.

> Linus Carl Pauling (1901–1995) received the forty-sixth Nobel Prize in chemistry in 1954 for his research on the nature of the chemical bond. Pauling is further distinguished by being the first person to receive two unshared Nobel prizes in two different fields. He was awarded the 1962 Nobel Peace Prize for his writing and lecturing against the danger of radioactive fallout.

Section 1.4
Introduction to the Chemical Bond

Because of their different electron structures, atoms can become bonded together in different ways. In 1916, the German physicist Walter Kössel noted that stable ions tend to form when atoms gain or lose electrons in sufficient number so that they attain the nearest noble-gas configuration. For example, sodium loses one electron to form a positive ion with a neon configuration, while fluorine gains one electron to form a negative ion, also with the neon configuration. One month after Kössel's work appeared in print, G. N. Lewis, an American chemist, published a paper outlining the concept of shared electron pairs (the covalent bond). The insights of Kössel and Lewis have been expanded by others, and subsequent proposals have resulted in the following theories.

1. An **ionic bond** results from the transfer of electrons from one atom to another.

2. A **covalent bond** results from the sharing of a pair of electrons by two atoms.

3. Atoms transfer or share electrons so as to gain a **noble-gas electron configuration.** This configuration is usually eight electrons in the outer shell, corresponding to the electron configuration of neon and argon. This theory is called the **octet rule.**

An ionic bond is formed by electron transfer. One atom donates one or more of its outermost electrons to another atom or atoms. The atom that loses electrons becomes a positive ion, or **cation.** The atom that gains the electrons becomes a negative ion, or **anion.** The ionic bond results from the electrostatic attraction between these oppositely charged ions. We can illustrate electron transfer using dots to represent the valence electrons.

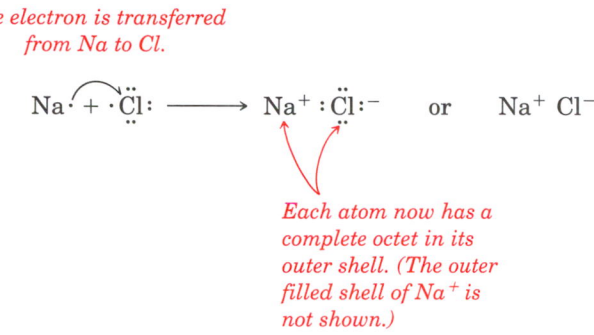

One electron is transferred from Na to Cl.

Each atom now has a complete octet in its outer shell. (The outer filled shell of Na⁺ is not shown.)

A covalent bond is produced by the sharing of a pair of valence electrons between two atoms. Shared electrons result from the merging of the atomic orbitals into shared orbitals called **molecular orbitals,** a topic that we will discuss in Chapter 2. For now, we will use dots to represent valence electrons. With dot formulas, called **Lewis formulas** or **Lewis structures,** we can easily count electrons and see that the atoms attain noble-gas configurations: two electrons (helium configuration) for hydrogen and eight electrons for most other atoms. Note that both the bonding electrons and the unshared electrons of an outermost electron shell are shown as dots in Lewis formulas.

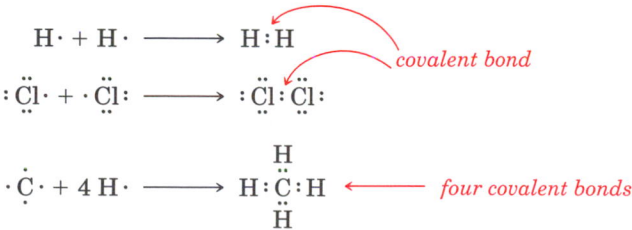

covalent bond

four covalent bonds

The sharing of one pair of electrons between two atoms is called a **single bond.** Two atoms can share two pairs or even three pairs of electrons; these multiple bonds are called **double bonds** and **triple bonds,** respectively.

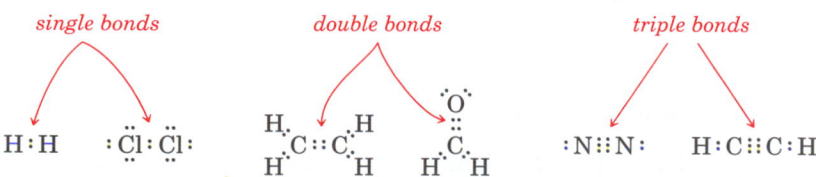

single bonds   *double bonds*   *triple bonds*

SAMPLE PROBLEMS

Circle the eight bonding electrons associated with the carbon atom in each of the following structures:

$$\ddot{O}\colon$$
$$H\colon\ddot{C}\colon H \qquad \ddot{O}\colon\colon C\colon\colon\ddot{O}\colon \qquad H\colon C\colon\colon\colon N\colon$$

Solution

For the structures in the preceding problem, circle the two electrons associated with each hydrogen atom and the eight electrons associated with each oxygen or nitrogen atom.

Solution

STUDY PROBLEM

1.3   Circle the eight bonding electrons for the nitrogen and the oxygen atoms in the following structures.

(a)
$$\begin{array}{c} H.\\ \ddot{C}\colon\colon\ddot{C}\colon N \colon\\ H^{.} \end{array} \begin{array}{c} H\colon\ddot{O}\colon\quad H\\ \ddot{C}\colon\quad \ddot{C}\colon H\\ H\colon\ddot{C}\colon H\ H\\ H \end{array}$$

(b)
$$\begin{array}{c} H\colon\ddot{O}\colon\\ H\colon\ddot{C}\colon\ddot{C}\colon\ddot{N}\colon H\\ \ddot{H}\quad \ddot{H} \end{array}$$

When do atoms form ionic bonds and when do they form covalent bonds? Ionic bonds are formed when the electronegativity difference between two atoms is large (greater than about 1.7). For example, a sodium atom (electronegativity 0.9), with little attraction for its outer electron, readily loses this electron to a chlorine atom (electronegativity 3.0). On the other hand, the electronegativity difference between two carbon atoms is 0; between carbon and hydrogen, only 0.4; and between carbon and chlorine, 0.5. Because carbon has an electronegativity of 2.5, intermediate between the extremes of high and low electronegativity, it almost never forms ionic bonds with other elements. Instead, *carbon forms covalent bonds with other carbon atoms and with atoms of other elements.*

A. Number of Covalent Bonds

The number of covalent bonds that an atom forms depends upon the number of additional electrons needed for the atom to attain a noble-gas configuration. For example, a neutral atom of hydrogen needs to share one electron to attain the electron configuration of helium; therefore, hydrogen forms one covalent bond. A neutral atom of chlorine has seven electrons in its outer shell and needs to share one electron to attain the electron configuration of argon; therefore, chlorine also forms one covalent bond. A neutral atom of carbon has four electrons in its outermost shell. Carbon needs to share four electrons to attain the electron configuration of neon; therefore, carbon forms four covalent bonds.

$$\cdot \overset{\cdot}{\underset{\cdot}{C}} \cdot \quad + \quad 4\,H\cdot \quad \longrightarrow \quad H : \overset{\cdot\cdot}{\underset{\cdot\cdot}{C}} : H \quad \overset{H}{\underset{H}{}}$$

forms 4   forms 1
covalent bonds   covalent bond

$$\cdot \overset{\cdot}{\underset{\cdot}{C}} \cdot \quad + \quad 4 : \overset{\cdot\cdot}{\underset{\cdot\cdot}{Cl}} \cdot \quad \longrightarrow \quad : \overset{\cdot\cdot}{\underset{\cdot\cdot}{Cl}} : \overset{:\overset{\cdot\cdot}{Cl}:}{\underset{:\overset{\cdot\cdot}{Cl}:}{C}} : \overset{\cdot\cdot}{\underset{\cdot\cdot}{Cl}} :$$

forms 4   forms 1
covalent bonds   covalent bond

Table 1.3 lists typical numbers of covalent bonds formed by elements commonly found in organic compounds. Table 1.3 also lists the groups in the periodic table where these elements are found. For the neutral atoms of elements in Groups IA–VIIA, the group number is the same as the number of outer electrons.

For simple structures, we can often deduce the Lewis formula for a compound of known composition by considering the usual number of covalent bonds an element forms.

Table 1.3   Numbers of covalent bonds usually formed by some elements typically encountered in organic compounds

Element	Number of Covalent Bonds	Group Number in Periodic Table
H	1	IA
C	4	IVA
N	3[a]	VA
O	2	VIA
Cl	1	VIIA
I	1	VIIA
Br	1	VIIA

[a] N can form four covalent bonds in some cases, such as in the NH_4^+ ion.

SAMPLE PROBLEMS

Write the Lewis formulas for H_2O and C_2H_6.

Solution
1. Determine the number of valence electrons of each atom: H = 1; O = 6; C = 4.

2. Draw the skeleton of the molecule, keeping in mind that H can form one covalent bond, O can form two, and C can form four.

$$
\begin{array}{cc}
& \text{H H} \\
\text{H O H} & \text{H C C H} \\
& \text{H H}
\end{array}
$$

3. Distribute the valence electrons in such a way that each H has two electrons and each other atom has an octet.

$$
\begin{array}{cc}
& \text{H H} \\
\text{H:\ddot{O}:H} & \text{H:\ddot{C}:\ddot{C}:H} \\
& \text{H H}
\end{array}
$$

Write the Lewis formula for C_3H_8 and two Lewis formulas for C_2H_6O.

Solution

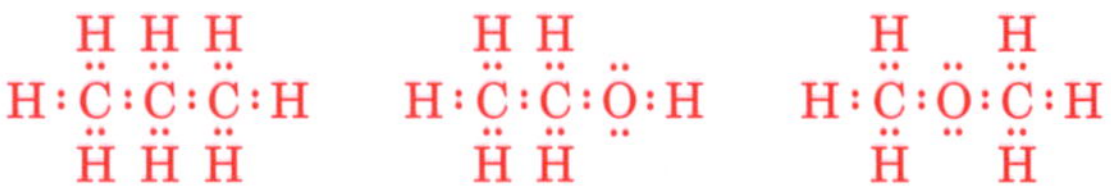

$$
\begin{array}{ccc}
\text{H H H} & \text{H H} & \text{H\quad H} \\
\text{H:\ddot{C}:\ddot{C}:\ddot{C}:H} & \text{H:\ddot{C}:\ddot{C}:\ddot{O}:H} & \text{H:\ddot{C}:\ddot{O}:\ddot{C}:H} \\
\text{H H H} & \text{H H} & \text{H\quad H}
\end{array}
$$

STUDY PROBLEM

1.4   Each of the following structures contains a double or triple bond. Write a Lewis formula for each one.

(a) Cl_2CO (b) C_2Cl_4 (c) C_3H_4O (d) $C_2H_4O_2$

B. Formal Charge

In some formulas, a few elements appear to form an unusual number of covalent bonds. We find that drawing completely correct Lewis formulas of these compounds is not possible unless we assign electrostatic charges, called **formal charges,** to some of the elements in these structures. For example, let us consider the Lewis structure of nitric acid.

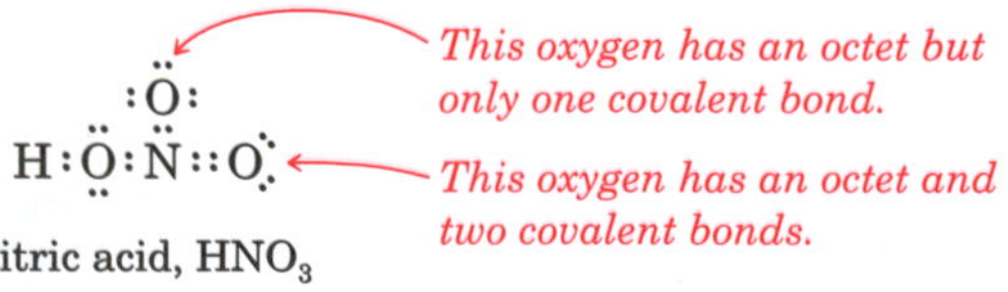

$$
\begin{array}{c}
:\ddot{O}: \\
\text{H:\ddot{O}:N::\ddot{O}:}
\end{array}
$$

nitric acid, HNO_3

This oxygen has an octet but only one covalent bond.

This oxygen has an octet and two covalent bonds.

The formula for nitric acid shows three oxygens bonded to a nitrogen atom. Each oxygen and nitrogen atom has a complete octet, but one oxygen atom is bonded by only one covalent bond instead of by the usual two. If the two electrons in the N—O single bond were divided between the N and the O (one electron to each atom), the oxygen would have seven valence electrons—one more electron than a neutral oxygen atom has. Therefore, this oxygen is assigned an electrostatic charge, or formal charge, of -1. Similarly, the nitrogen atom has only four valence electrons—one fewer electron than a neutral nitrogen atom has. This electron-deficient nitrogen atom has a formal charge of $+1$. The other atoms in nitric acid all have the same number of electrons as would their neutral atoms; their formal charges are all 0.

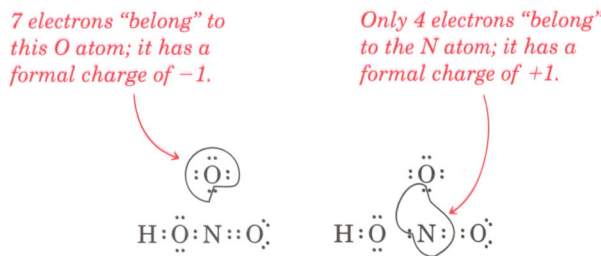

In nitric acid, or any other electrically neutral molecule, the formal charges must cancel, or sum to 0.

$$H:\ddot{O}:\overset{..}{\underset{..}{N}}\,^{+}::\ddot{O}: \qquad \begin{array}{l} O\ has\ formal\ charge\ of\ -1 \\ N\ has\ formal\ charge\ of\ +1 \end{array} \Bigg\} +1 + (-1) = 0$$

To determine if formal charges must be assigned to atoms within a Lewis structure, we use the following equation for each atom in the structure:

formal charge = (number of valence e^- in a neutral atom)
$-\frac{1}{2}$(number of shared e^-) − (number of unshared valence e^-)

Example
Using the formula, we can calculate the formal charges on N and each O in HNO_3. (H does not carry a formal charge in covalent molecules.)

$$\begin{array}{c} :\ddot{O}: \\ H:\ddot{O}:\ddot{N}::\ddot{O}: \end{array}$$

no. of valence e^- for neutral N $= 5$

$\frac{1}{2}$(no. of shared e^- for N) $= 4$

no. of unshared e^- for N $= 0$

formal charge for N $= 5 - 4 - 0 = +1$

For each O, the same technique is used.

formal charge:
$6 - 2 - 4 = 0$

$$\begin{array}{c} :\ddot{O}: \\ H:\ddot{O}:\ddot{N}::\ddot{O}: \end{array}$$

formal charge:
$6 - 1 - 6 = -1$

formal charge:
$6 - 2 - 4 = 0$

Example

Let us calculate the formal charges, if any, in the following molecule.

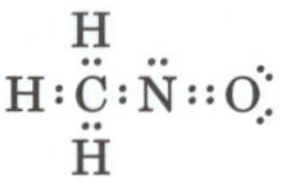

$$
\begin{array}{l}
\quad\ \text{H} \\
\text{H} : \overset{..}{\text{C}} : \text{N} :: \overset{..}{\text{O}} : \\
\quad\ \text{H}
\end{array}
\qquad
\begin{array}{l}
\text{for C:} \quad 4 - 4 - 0 = 0 \\
\text{for N:} \quad 5 - 3 - 2 = 0 \\
\text{for O:} \quad 6 - 2 - 4 = 0
\end{array}
$$

Note that none of the atoms in this molecule has a formal charge.

   If the formal charges within a structure do not sum to zero, then the structure is an ion. The sum of the formal charges represents the ionic charge.

Example

Let us calculate the formal charges in the following formula to determine the ionic charge.

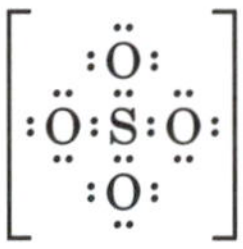

$$
\left[
\begin{array}{c}
\quad\ : \overset{..}{\text{O}} : \\
: \overset{..}{\text{O}} : \overset{..}{\text{S}} : \overset{..}{\text{O}} : \\
\quad\ : \overset{..}{\text{O}} :
\end{array}
\right]
\qquad
\begin{array}{l}
\text{for S:} \quad 6 - 4 - 0 = +2 \\
\text{for each O:} \quad 6 - 1 - 6 = -1
\end{array}
$$

$$
\begin{aligned}
\text{ionic charge} &= \text{(formal charge of sulfur)} + 4\text{(formal charge of each oxygen)} \\
&= (+2) + 4(-1) \\
&= -2
\end{aligned}
$$

The formula, of course, represents the sulfate ion, $SO_4{}^{2-}$.

STUDY PROBLEMS

1.5 Calculate the formal charges on C, N, and O in the following formulas and determine whether each represents a neutral molecule or an ion.

$$
\textbf{(a)}\ \begin{array}{l} \text{H}\ \text{H} \\ \text{H} : \overset{..}{\text{C}} : \text{N} : \text{H} \\ \text{H} \end{array}
\qquad
\textbf{(b)}\ \begin{array}{l} \text{H}\ \text{H} \\ \text{H} : \overset{..}{\text{C}} : \overset{..}{\text{N}} : \text{H} \\ \text{H}\ \text{H} \end{array}
\qquad
\textbf{(c)}\ : \text{C} :: \text{O} :
$$

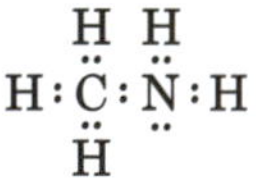

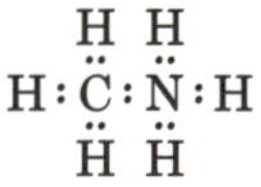

1.6 Calculate the formal charge on each atom in the following structures:

$$
\textbf{(a)}\ \text{H} : \overset{..}{\text{O}} : \overset{..}{\text{N}} :: \overset{..}{\text{O}} :
\qquad
\textbf{(b)}\ \begin{array}{l} \quad\ \text{H} \\ \text{H} : \overset{..}{\text{C}} :: \text{N} :: \overset{.}{\text{N}} : \end{array}
\qquad
\textbf{(c)}\ \begin{array}{l} \text{H}\quad : \overset{..}{\text{O}} : \\ \text{H} : \overset{..}{\text{C}} : \overset{..}{\text{O}} : \overset{..}{\text{S}} : \overset{..}{\text{O}} : \text{H} \\ \text{H}\quad : \overset{..}{\text{O}} : \end{array}
$$

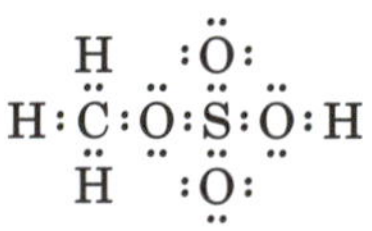

C. Polar Covalent Bonds

Atoms with equal or nearly equal electronegativities form covalent bonds in which both atoms exert equal or nearly equal attractions for the bonding electrons. This type of covalent bond is called a **nonpolar bond.** In organic molecules, carbon–carbon bonds and carbon–hydrogen bonds are the most common types of nonpolar bonds.

Some compounds containing relatively nonpolar covalent bonds:

$$H-H \qquad N \equiv N \qquad \underset{\displaystyle H}{\overset{\displaystyle H}{H-\underset{|}{\overset{|}{C}}-H}} \qquad \underset{\displaystyle H \quad H}{\overset{\displaystyle H \quad H}{H-\underset{|}{\overset{|}{C}}-\underset{|}{\overset{|}{C}}-H}} \qquad \underset{H}{\overset{H}{}}C=C\underset{H}{\overset{H}{}}$$

In covalent compounds like H_2O, HCl, CH_3OH, or $H_2C{=}O$, one atom has a substantially greater electronegativity than the others. The more electronegative atom has a greater attraction for the bonding electrons—not enough of an attraction for the atom to break off as an ion, but enough so that this atom takes the larger share of electron density. The result is a **polar covalent bond,** a bond with an uneven distribution of electron density.

Besides electronegativity, another factor determining the degree of polarity of a bond is the *polarizability* of the atoms, which is the ability of the electron cloud to be distorted to induce polarity. The outer electrons of larger atoms are farther from the nucleus and are thus less tightly held than those of smaller atoms. Therefore, larger atoms are more polarizable than smaller ones. This means that the C—I bond, for example, can act as if it is polar even though the electronegativity difference between C and I is negligible.

The result of electronegativity differences and polarizability differences is a variety of bond types. We may think of chemical bonds as a continuum from nonpolar covalent bonds to ionic bonds. Within this continuum, we speak of the increasing **ionic character** of the bonds.

$$H-H \qquad CH_3-O-CH_3 \qquad H-O-H \qquad H-Cl \qquad Na^+ \; Cl^-$$

increasing ionic character of bonds →

The distribution of electron density in a polar bond may be symbolized by **partial charges:** $\delta+$ (partial positive) and $\delta-$ (partial negative). Another way of representing different electron densities within a bond is by a crossed arrow ($\longmapsto$) that points from the partially positive end to the partially negative end of the bond.

$$\overset{\delta+ \quad \delta-}{H-Cl} \quad \text{or} \quad \overset{\longmapsto}{H-Cl} \qquad \underset{\delta+ \quad\;\; \delta+}{\overset{\displaystyle \overset{\delta-}{O}}{CH_3 \;\; CH_3}} \quad \text{or} \quad \underset{CH_3 \;\; CH_3}{\overset{\displaystyle O}{}}$$

STUDY PROBLEM

1.7 Using partial charges, indicate the polarity of the carbon–chlorine and carbon–oxygen bonds in the following compounds.

(a) CH_3CH_2Cl **(b)** $\overset{\displaystyle \overset{O}{\parallel}}{CH_3CCH_3}$

D. Bond Lengths and Bond Angles

The distance that separates the nuclei of two covalently bonded atoms is called the **bond length.** Covalent bond lengths, which can be measured experimentally, range from 0.74 Å to 2 Å.

If there are more than two atoms in a molecule, the bonds form an angle, called the **bond angle.** Bond angles vary from about 60° up to 180°.

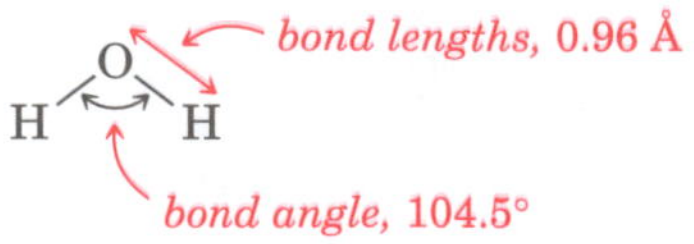
bond lengths, 0.96 Å
bond angle, 104.5°

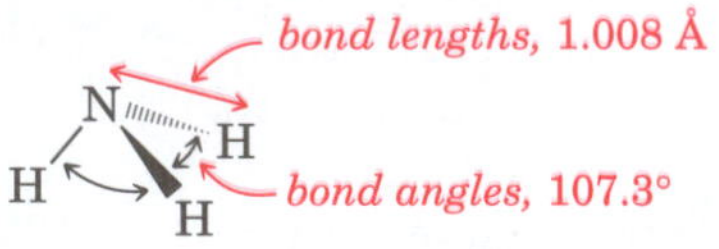
bond lengths, 1.008 Å
bond angles, 107.3°

Most organic structures contain more than three atoms and are three-dimensional rather than two-dimensional. The preceding formula for ammonia (NH_3) illustrates one technique for representing a three-dimensional structure. A line bond (—) represents a bond in the plane of the paper. The solid wedge (◀) represents a bond coming out of the paper toward the viewer; the H at the wide end of the solid wedge is in front of the paper. The broken wedge (⦙⦙⦙) represents a bond pointing back into the paper; the H at the small end of the broken wedge is behind the paper.

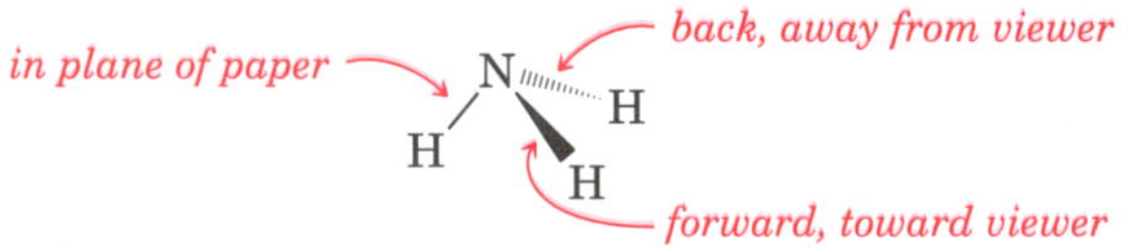
in plane of paper
back, away from viewer
forward, toward viewer

E. Bond Dissociation Energy

There are two ways a bond can dissociate, or break apart. One way is by **heterolytic cleavage** (Greek *heteros,* "different"), in which both bonding electrons are retained by one of the atoms. Heterolytic cleavage usually results in the formation of ionic species.

Heterolytic cleavage:

$$H{-}H \longrightarrow H^+ + H{:}^-$$

$$H{-}\ddot{\underset{..}{Cl}}{:} \longrightarrow H^+ + {:}\ddot{\underset{..}{Cl}}{:}^-$$

$$H{-}\ddot{\underset{..}{O}}H \longrightarrow H^+ + {}^-{:}\ddot{\underset{..}{O}}H$$

We use a curved arrow (↷) in these equations to show the direction in which the pair of bonding electrons moves during bond breakage. In the heterolytic cleavage of HCl or H_2O, the bonding electrons are transferred to the more electronegative Cl or O.

The other process by which a bond may dissociate is **homolytic cleavage** (Greek *homos,* "same"). In this case, each atom involved in the covalent bond receives one electron from the original shared pair. Electrically neutral atoms or groups of atoms usually result.

Homolytic cleavage:

$$H{-}H \longrightarrow H\cdot + H\cdot$$

$$H{-}\ddot{\underset{..}{Cl}}{:} \longrightarrow H\cdot + \cdot\ddot{\underset{..}{Cl}}{:}$$

$$H_3C{-}H \longrightarrow H_3C\cdot + H\cdot$$

Note that the curved arrows in these equations have only half an arrowhead. This type of arrow (⌒), called a fishhook, is used to show the direction of shift of *one* electron, whereas the curved arrow with a complete head ($\frown$) is used to show the direction of shift of a *pair* of electrons.

Homolytic cleavage results in atoms or groups of atoms having unpaired electrons. An atom such as H· or a group of atoms such as $H_3C·$ that contains an unpaired electron is called a **free radical** or, simply a **radical.** Radicals are usually electrically neutral; therefore, there are no electrostatic interactions between radicals as there are between ions. Because a radical contains an atom with an incomplete octet, most radicals are unstable and very reactive. We usually represent a radical by including a single dot in its formula, without showing the remaining valence electrons. Thus, we represent $·\ddot{\underset{..}{Cl}}:$ as ·Cl or $H\ddot{O}·$ as HO·.

Homolytic cleavage is more useful than heterolytic cleavage in determining the energies required for bond dissociations because calculations are not complicated by ionic attractions between the products. From measurements of the components of dissociating gases at high temperatures, the **change in enthalpy** ΔH (change in heat content, or energy) has been calculated for a large number of bond dissociations. For the reaction $CH_4 \rightarrow CH_3· + H·$, ΔH equals 104 kcal/mol. In other words, to cleave one hydrogen atom from each carbon atom in 1 mol of CH_4 requires 104 kcal. This value (104 kcal/mol) is the **bond dissociation energy** for the H_3C-H bond. The bond dissociation energies for several types of bonds are listed in Table 1.4.

To break a more stable bond requires a higher energy input. For example, cleavage of HF to H· and F· (135 kcal/mol) is difficult compared with cleavage of the $O-O$ bond in hydrogen peroxide, HOOH (35 kcal/mol).

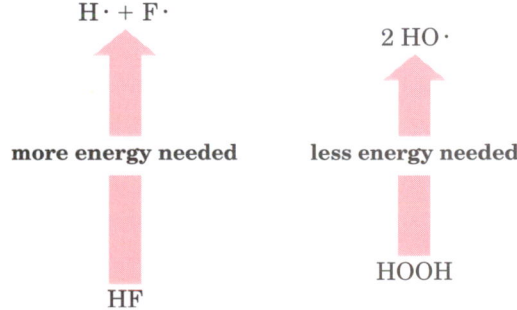

	Table 1.4 Selected bond dissociation energies (in kcal/mol)		
Miscellaneous Bonds	C—H Bonds	C—X Bonds[a]	C—C Bonds
H—H 104	CH_3—H 104	CH_3—Cl 83.5	CH_3—CH_3 88
N≡N 226	CH_3CH_2—H 98	CH_3CH_2—Cl 81.5	CH_2=CH_2 163
F—F 37	$(CH_3)_2$CH—H 94.5	$(CH_3)_2$CH—Cl 81	CH≡CH 230
Cl—Cl 58	$(CH_3)_3$C—H 91	$(CH_3)_3$C—Cl 78.5	
Br—Br 46	CH_2=CH—H 108	CH_2=CH—Cl 84	
I—I 36			
H—F 135			
H—OH 110			
H—Cl 103		CH_3—Br 70	
H—Br 87		CH_3CH_2—Br 68	
H—I 71		$(CH_3)_2$CH—Br 68	
HO—OH 35		$(CH_3)_3$C—Br 67	

[a] X refers to a halogen.

Note in Table 1.4 that atoms joined by multiple bonds require more energy for dissociation than the same atoms joined by single bonds ($CH \equiv CH$, 230 kcal/mol, versus $CH_3 - CH_3$, 88 kcal/mol). Also note that other parts of a molecule may affect the bond dissociation energy:

$$H_3\overset{\frown\ \frown}{C \quad H} + 104 \text{ kcal/mol} \xrightarrow{\text{more difficult}} H_3C \cdot + H \cdot$$

$$(CH_3)_3\overset{\frown\ \frown}{C \quad H} + 91 \text{ kcal/mol} \xrightarrow{\text{easier}} (CH_3)_3C \cdot + H \cdot$$

STUDY PROBLEM

1.8 Write equations for the heterolytic and homolytic cleavage of the C—Cl bonds in the following structures.

$$
\begin{array}{c}
CH_3 \\
| \\
\textbf{(a)} \ CH_3C{-}Cl \qquad \textbf{(b)} \ CH_3CH_2CH_2{-}Cl \\
| \\
CH_3
\end{array}
$$

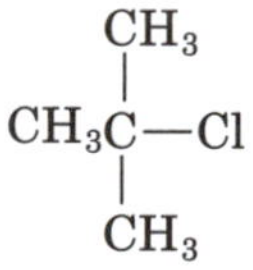

Section 1.5
Chemical Formulas in Organic Chemistry

The formulas that we use today to represent organic compounds were introduced in the latter half of the nineteenth century. Historically, three types of chemical formulas have emerged. These are the empirical formula, the molecular formula, and the structural formula. An **empirical formula** tells us the types of atoms and their numerical ratio in a molecule. For example, a molecule of ethane contains carbon and hydrogen atoms in a ratio of 1 to 3; the empirical formula is CH_3. A **molecular formula** tells us the actual number of each type of atom in a molecule, not simply the ratio. The molecular formula for ethane is C_2H_6. A **structural formula** shows the *structure* of a molecule, that is, the order of attachment of the atoms. In order to explain or predict chemical reactivity, we need to know the structure of a molecule. Therefore, structural formulas are the most useful of the different types of formulas.

$$
CH_3 \qquad\qquad C_2H_6 \qquad\qquad
\begin{array}{c}
\ \ \ H \ \ H \\
\ \ \ | \ \ \ \ | \\
H{-}C{-}C{-}H \\
\ \ \ | \ \ \ \ | \\
\ \ \ H \ \ H
\end{array}
$$

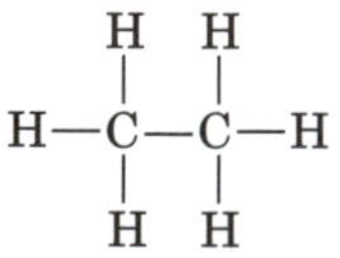

empirical formula   molecular formula   structural formula
for ethane   for ethane   for ethane

Of key importance to the development of the structural formula concept was the idea that carbon can bond with itself in chains. This notion was first published in 1858 by Friedrich August Kekulé (1829–1896), a German chemist. His publication was followed almost immediately by that of Archibald Couper (1831–1892), a Scottish chemist. Many other concepts of valence were proposed, but those of Kekulé and Couper became the basis of our modern theory. Kekule's eloquence, however, propelled him into eminence, whereas Couper faded into retirement. A few years later, Kekulé proposed that carbon could form rings as well as chains. In his words:

> I was sitting writing at my textbook, but the work did not progress: my thoughts were elsewhere. I turned my chair to the fire, and dozed. Again the atoms were gamboling before my eyes. This time the smaller groups kept modestly in the background. My mental eye, rendered more acute by repeated visions of this kind, could now distinguish larger structures of manifold conformations; long rows, sometimes more closely fitted together; all twisting and turning in snake-like motion. But look! What was that? One of the snakes had seized hold of its own tail, and the form whirled mockingly before my eyes. As if by a flash of lightning I woke. . . . Let us learn to dream, gentlemen, and then perhaps we shall learn the truth. (1890).

Kekulé was also the first to propose a workable structure for benzene (see Section 2.8).

A. Structural Formulas

Lewis formulas are one type of structural formula. However, chemists usually represent a covalent structure by using a dash for each shared pair of electrons, and rarely show unshared pairs of valence electrons. Formulas with dashes for bonds are called **line-bond formulas.** In this text, we will also refer to them as **complete structural formulas.**

$$H:H \quad \text{becomes} \quad H-H$$

$$:\ddot{C}l:\ddot{C}l: \quad \text{becomes} \quad :\ddot{C}l-\ddot{C}l: \quad \text{or} \quad Cl-Cl$$

$$H:\overset{\overset{H}{\cdot\cdot}}{\underset{\cdot\cdot}{C}}:H \quad \text{becomes} \quad H-\overset{\overset{\displaystyle H}{|}}{\underset{\underset{\displaystyle H}{|}}{C}}-H$$

$$\overset{H}{\underset{H}{\cdot}}C::\overset{H}{\underset{H}{\cdot}}\ \text{becomes}\ \overset{H}{\underset{H}{\diagup}}C=C\overset{H}{\underset{H}{\diagdown}}$$

Although unshared pairs of valence electrons are not usually shown in line-bond formulas, we will sometimes show these electrons when we want to emphasize their role in a chemical reaction.

All represent the same molecule:

$$H:\overset{\overset{\cdot\cdot}{N}}{\underset{\displaystyle H}{}}:H \quad \text{or} \quad H-\overset{\displaystyle |}{\underset{\displaystyle H}{N}}-H \quad \text{or} \quad H-\overset{\overset{\cdot\cdot}{N}}{\underset{\displaystyle H}{}}-H$$

— unshared pair of valence e⁻

SAMPLE PROBLEM

Write the complete structural formula, showing each bond as a line, for each of the following Lewis formulas:

$$H:\overset{\overset{\displaystyle H}{\cdot\cdot}}{\underset{\underset{\displaystyle H}{\cdot\cdot}}{C}}:\ddot{O}:\overset{\overset{\displaystyle H}{}}{\underset{\underset{\displaystyle H}{}}{C}}:H \qquad H:\overset{\overset{\displaystyle H}{}}{\underset{\underset{\displaystyle H}{}}{C}}:\overset{\overset{\displaystyle H}{}}{\underset{\underset{\displaystyle H}{}}{C}}:\ddot{O}:H \qquad H:\overset{\overset{\displaystyle \cdot\cdot}{\ddot{O}}}{C}:H$$

Solution

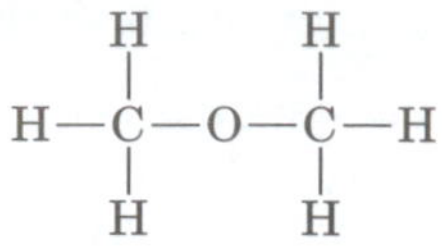

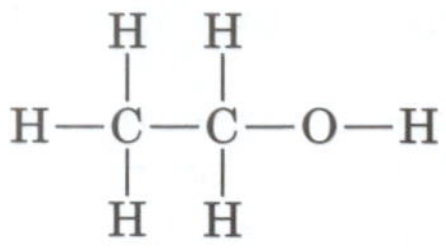

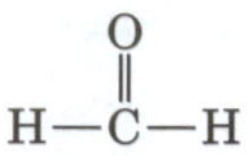

$$\begin{array}{ccc}
\overset{\displaystyle H}{\underset{\displaystyle H}{\overset{|}{\underset{|}{H-C}}}}-O-\overset{\displaystyle H}{\underset{\displaystyle H}{\overset{|}{\underset{|}{C-H}}}} & \overset{\displaystyle H}{\underset{\displaystyle H}{\overset{|}{\underset{|}{H-C}}}}\overset{\displaystyle H}{\underset{\displaystyle H}{\overset{|}{\underset{|}{-C}}}}-O-H & \overset{\displaystyle O}{\overset{\|}{H-C-H}}
\end{array}$$

B. Condensed Structural Formulas

Complete structural formulas are frequently condensed to shorter, more convenient formulas. In **condensed structural formulas,** bonds are not always shown and atoms of the same type bonded to one other atom are grouped together. The structure of a molecule is still evident from a condensed structural formula.

CH_3CH_3 is a condensed structural formula for $\overset{\displaystyle H}{\underset{\displaystyle H}{\overset{|}{\underset{|}{H-C}}}}\overset{\displaystyle H}{\underset{\displaystyle H}{\overset{|}{\underset{|}{-C}}}}-H$

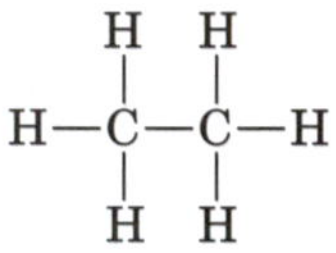

CH_3CH_2OH is a condensed structural formula for $\overset{\displaystyle H}{\underset{\displaystyle H}{\overset{|}{\underset{|}{H-C}}}}\overset{\displaystyle H}{\underset{\displaystyle H}{\overset{|}{\underset{|}{-C}}}}-O-H$

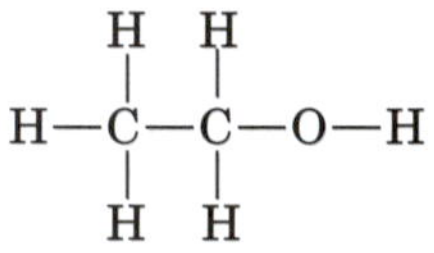

SAMPLE PROBLEM

Write **(a)** the complete structural formula (showing all bonds as dashes), and **(b)** a condensed structural formula, for each of the following Lewis formulas:

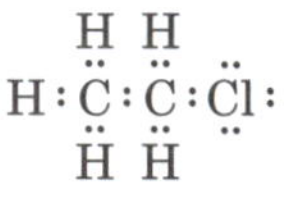

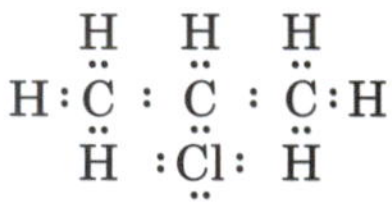

$$\begin{array}{cc}
\overset{\displaystyle H\ \ H}{\underset{\displaystyle H\ \ H}{\overset{..}{H:C:C:\overset{..}{\underset{..}{Cl}}:}}} &
\overset{\displaystyle H\ \ \ H\ \ \ H}{\underset{\displaystyle H\ \ :\overset{..}{\underset{..}{Cl}}:\ \ H}{H:\overset{..}{C}:\ \overset{..}{C}\ :\ \overset{..}{C}:H}}
\end{array}$$

Solution

(a) 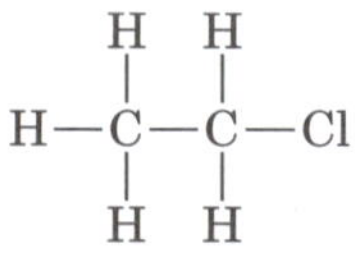 $H-\overset{\displaystyle H}{\underset{\displaystyle H}{\overset{|}{\underset{|}{C}}}}-\overset{\displaystyle H}{\underset{\displaystyle H}{\overset{|}{\underset{|}{C}}}}-Cl$ $H-\overset{\displaystyle H}{\underset{\displaystyle H}{\overset{|}{\underset{|}{C}}}}-\overset{\displaystyle H}{\underset{\displaystyle Cl}{\overset{|}{\underset{|}{C}}}}-\overset{\displaystyle H}{\underset{\displaystyle H}{\overset{|}{\underset{|}{C}}}}-H$

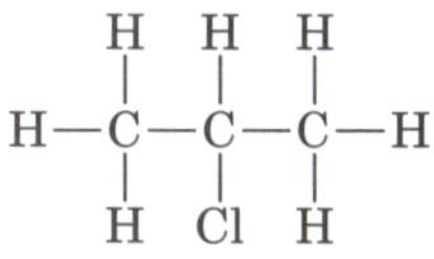

(b) CH_3CH_2Cl $CH_3CHClCH_3$

STUDY PROBLEM

1.9 Write the complete structural formula, and a condensed structural formula, for each of the following Lewis formulas:

(a) 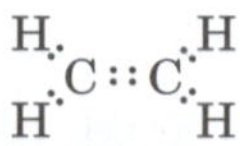 $\overset{\displaystyle H.}{\underset{\displaystyle H}{\ .C::C.}}\overset{\displaystyle .H}{\underset{\displaystyle H}{.}}$ **(b)** 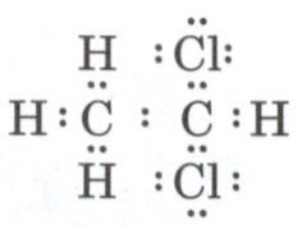 $\overset{\displaystyle H\ :\overset{..}{\underset{..}{Cl}}:}{\underset{\displaystyle H\ :\overset{..}{\underset{..}{Cl}}:}{H:\overset{..}{C}\ :\ \overset{..}{C}:H}}$

Structural formulas may be condensed even further if a molecule has two or more identical groups of atoms. In these cases, parentheses are used to enclose a repetitive group of atoms. The subscript following the second parenthesis indicates the number of times the entire group is found at that position in the molecule.

$$(CH_3)_2CHOH \quad \text{is the same as} \quad CH_3-\underset{\underset{\displaystyle H}{|}}{\overset{\overset{\displaystyle CH_3}{|}}{C}}-OH$$

$$(CH_3)_3CCl \quad \text{is the same as} \quad CH_3-\underset{\underset{\displaystyle CH_3}{|}}{\overset{\overset{\displaystyle CH_3}{|}}{C}}-Cl$$

$$CH_3(CH_2)_3CH_3 \quad \text{is the same as} \quad CH_3CH_2CH_2CH_2CH_3$$

For the sake of clarity, double or triple bonds are usually shown in a condensed structural formula.

$$CH_3CH{=}CH_2 \quad \text{is the same as} \quad$$

$$CH_3C{\equiv}CH \quad \text{is the same as} \quad H-\underset{\underset{\displaystyle H}{|}}{\overset{\overset{\displaystyle H}{|}}{C}}-C{\equiv}C-H$$

$$\overset{\displaystyle O}{\underset{\displaystyle \|}{}}$$
$$CH_3CCH_2CH_3 \quad \text{is the same as} \quad H-\overset{\overset{\displaystyle H}{|}}{\underset{\underset{\displaystyle H}{|}}{C}}-\overset{\overset{\displaystyle O}{\|}}{C}-\overset{\overset{\displaystyle H}{|}}{\underset{\underset{\displaystyle H}{|}}{C}}-\overset{\overset{\displaystyle H}{|}}{\underset{\underset{\displaystyle H}{|}}{C}}-H$$

STUDY PROBLEM

1.10   For each of the following formulas, write a more condensed formula:

(a) $CH_3\underset{\overset{\displaystyle |}{\displaystyle CH_3}}{\overset{\overset{\displaystyle CH_3}{|}}{CH}}CH_2Cl$ (b) $CH_3\overset{\overset{\displaystyle Cl}{|}}{CH}Cl$ (c) $CH_3CH_2CH_2CH_2\overset{\overset{\displaystyle Cl}{|}}{CH}CH_2Cl$

(d) $\underset{\displaystyle CH_3}{\overset{\displaystyle CH_3}{}}{\Large C}{=}{\Large C}\underset{\displaystyle CH_3}{\overset{\displaystyle CH_3}{}}$ (e) $N{\equiv}C-CH_2-C{\equiv}N$

C. Cyclic Compounds and Polygon Formulas

A compound such as $CH_3CH_2CH_2CH_3$ is said to have its carbon atoms connected in a chain. Carbon atoms can be joined together in rings as well as in chains; a compound containing one or more rings is called a **cyclic compound.**

Cyclic structures are usually represented by **polygon formulas,** which are another type of condensed structural formula. For example, a triangle is used to represent a three-membered ring, while a hexagon is used for a six-membered ring.

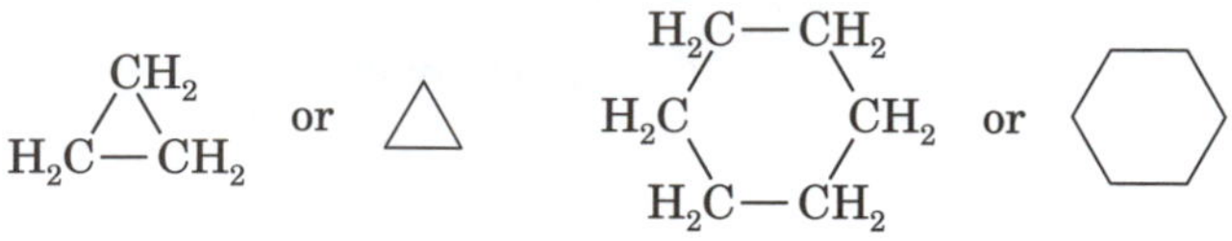

In polygon formulas, a corner represents a carbon atom along with its hydrogens; the sides of the polygon represent the bonds joining the carbons. If an atom or group other than hydrogen is attached to a carbon of the ring, the number of hydrogens at that position is reduced accordingly.

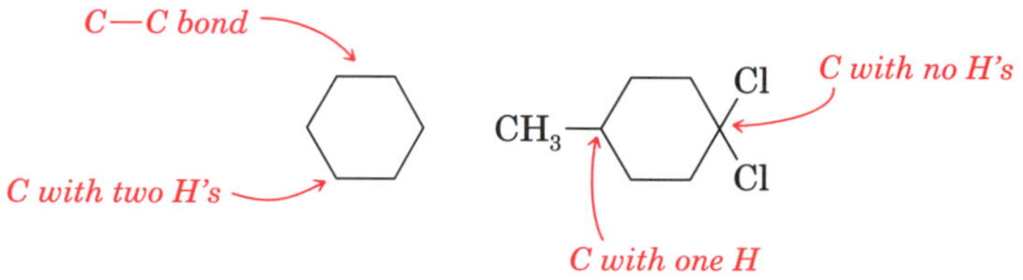

Rings can contain atoms other than carbon. These atoms and any hydrogens attached to them must be indicated in the polygon formula. Double bonds must also be indicated.

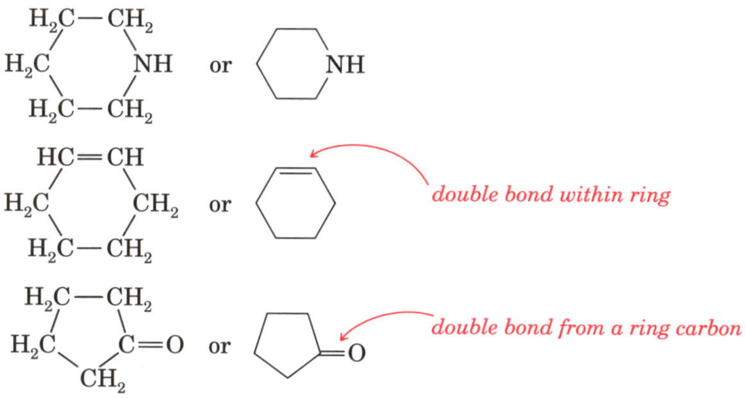

STUDY PROBLEMS

1.11 Draw complete structural formulas for the following structures, showing each C, each H, and each bond.

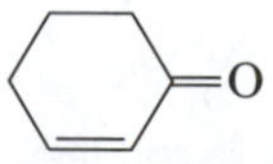

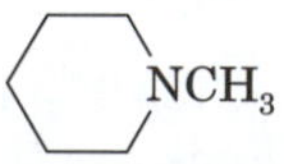

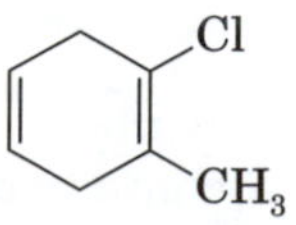

1.12 Draw polygon formulas for the following structures:

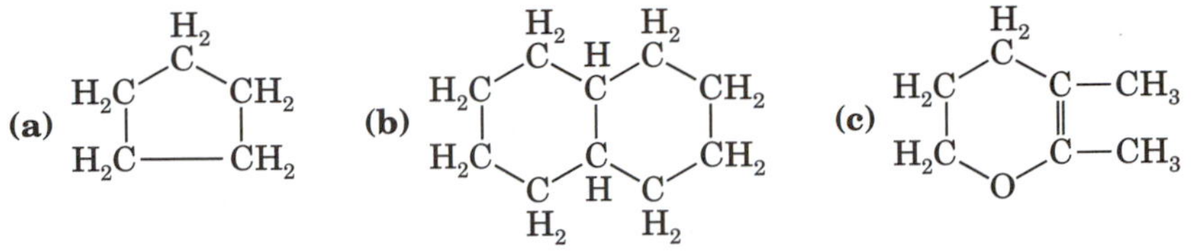

(a) H_2C—CH_2 ... H_2C—CH_2 ... C (H_2) (cyclopentane structure)

(b) H_2C—CH_2 ... (fused bicyclic structure)

(c) H_2C—C—CH_3 ... H_2C—C—CH_3 ... O (oxygen-containing ring structure)

Section 1.6
Attractions Between Molecules

A. Dipole–Dipole Interactions

Except in a dispersed gas, molecules attract and repel one another. These attractions and repulsions arise primarily from dipole–dipole interactions: attractions between opposite charges and repulsions between like charges. For example, in liquid CH_3Cl the partially negative chlorine atom in one molecule is attracted to the partially positive carbon in another molecule.

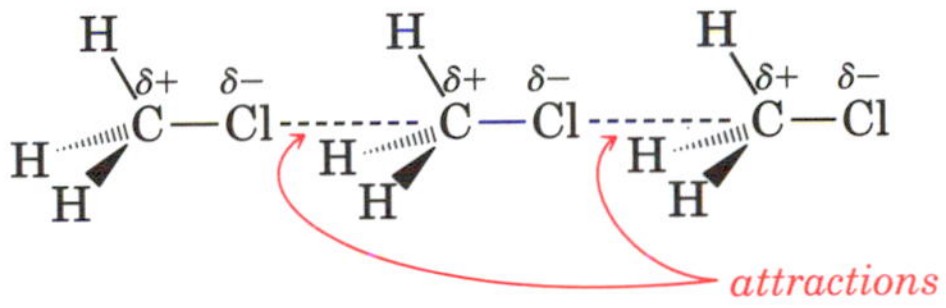

attractions

When the chlorine ends of two CH_3Cl molecules approach closely, the two molecules repel each other.

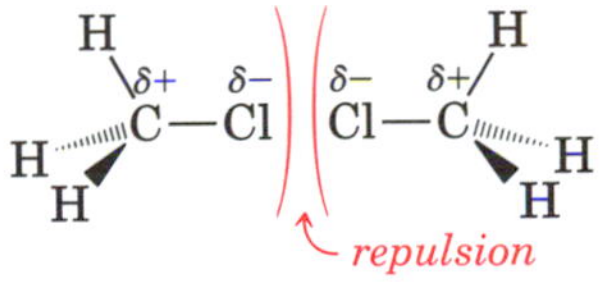

repulsion

Nonpolar molecules are attracted to one another by weak dipole–dipole interactions called **London forces.** London forces arise from dipoles *induced* in one molecule by another. In this case, electrons of one molecule are weakly attracted to a nucleus of a second molecule; then the electrons of the second molecule are repelled by the electrons of the first. The result is an uneven distribution of electrons and an induced dipole. Figure 1.6 depicts how an induced dipole can arise when two molecules approach each other.

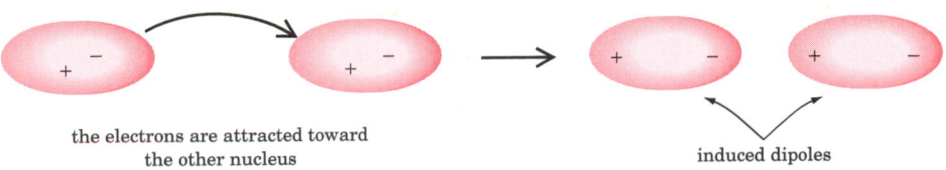

the electrons are attracted toward
the other nucleus

induced dipoles

Figure 1.6 Nonpolar molecules can induce dipoles in each other.

The various dipole–dipole interactions (attractive and repulsive) are collectively called **van der Waals forces.** The distance between molecules has an important effect on the strength of van der Waals forces. The distance at which attraction is greatest is called the **van der Waals radius.** If two atoms approach each other more closely than this distance, repulsions develop between the two nuclei and between the two sets of electrons. When the distance between two molecules becomes larger than the van der Waals radius, the attractive forces between the molecules decrease.

Continuous-chain molecules, such as $CH_3CH_2CH_2CH_2CH_3$, can align themselves in zigzag chains, enabling the atoms of different molecules to assume positions that match the van der Waals radii. Maximal van der Waals attractions can develop between such long-chain molecules. Branched molecules cannot approach one another closely enough for all the atoms to assume optimal van der Waals distances. Because more energy is necessary to overcome van der Waals attractions and to free molecules from the liquid state, continuous-chain compounds have higher boiling points than branched compounds of the same molecular weight and otherwise similar structures.

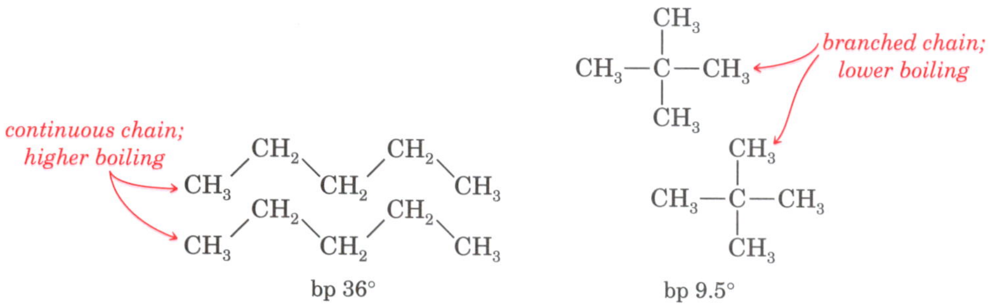

continuous chain;
higher boiling

bp 36°

branched chain;
lower boiling

bp 9.5°

B. Hydrogen Bonding

An especially strong type of dipole–dipole interaction occurs between molecules containing a hydrogen atom bonded to nitrogen, oxygen, or fluorine. Each of these latter elements is electronegative and has unshared valence electrons. Some typical compounds that contain an NH, OH, or FH bond are:

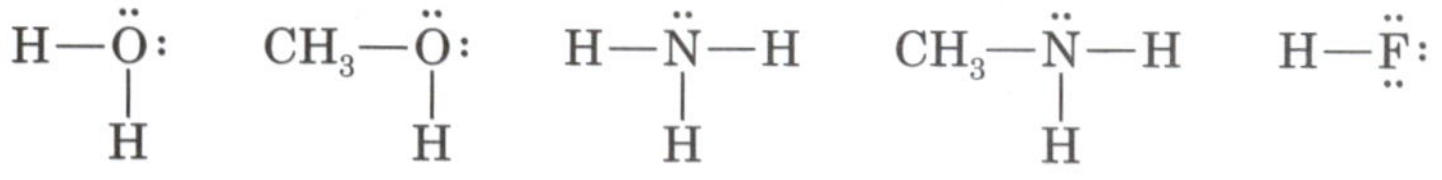

In the liquid state, the molecules of these compounds have strong attractions for one another. A partially positive hydrogen atom of one molecule is attracted to the unshared pair of electrons of the electronegative atom of another molecule. This attraction is called a **hydrogen bond.**

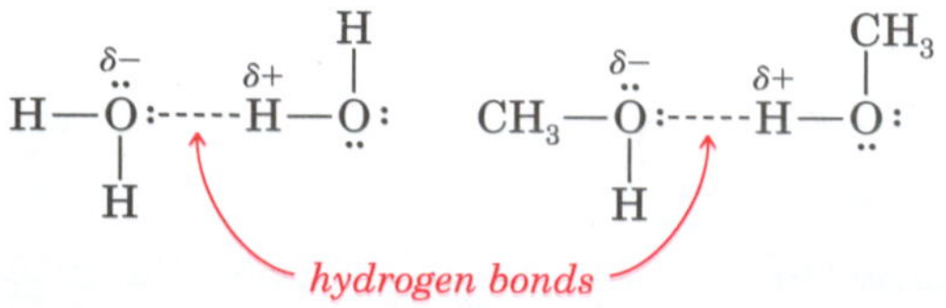

hydrogen bonds

Compounds or groups containing only carbon and hydrogen cannot undergo hydrogen bonding. As an example, consider methane, CH_4. Methane cannot undergo hydrogen bonding for two reasons:

1. Because the CH bond is relatively nonpolar, a CH_4 molecule does not have a partially positive H.

2. The carbon atom in CH_4 has no unshared electrons to attract a hydrogen atom.

The dissociation energy of a hydrogen bond is only 5–10 kcal/mol, much lower than the bond dissociation energy of a typical covalent bond (80–100 kcal/mol) but substantially stronger than most dipole–dipole attractions. The reason for this difference is the size of the atoms involved. A hydrogen atom is small compared to other atoms and can occupy a position very close to the unshared electrons of an electronegative atom. A strong electrostatic attraction results. Atoms larger than hydrogen cannot occupy positions so near each other; consequently, dipole–dipole attractions between other atoms are weaker.

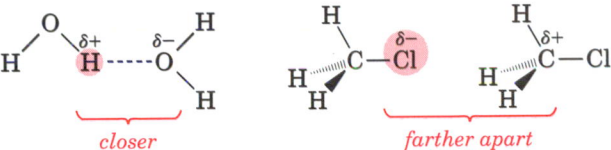

SAMPLE PROBLEM

Show the hydrogen bonding between two molecules of $CH_3CH_2NH_2$.

Solution

1. Look for one or more partially positive hydrogens, which are those bonded to a relatively electronegative atom.

$$CH_3CH_2-N-H$$

2. Look for an electronegative atom (N, O, or F) with unshared electrons.

$$CH_3CH_2-N-H$$

3. Draw two molecules with a hydrogen bond between one partially positive H and the N of the other molecule.

$$CH_3CH_2-N:\text{-----}H-N-CH_2CH_3$$

Hydrogen bonds are not all the same strength. An O---HO hydrogen bond is stronger than an N---HN hydrogen bond. Why is this true? Oxygen is more electronegative than nitrogen; therefore, the O—H group is more polar and has a more positive H. This more positive H is more strongly attracted by a negative center.

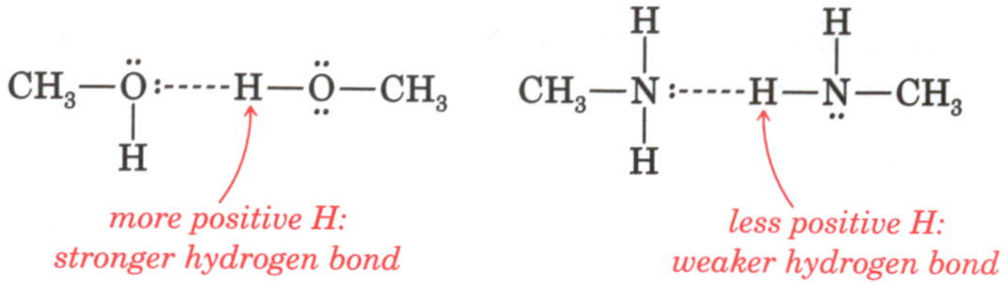

$$CH_3—\ddot{O}{:}\text{-----}H—\ddot{O}—CH_3 \qquad CH_3—N{:}\text{-----}H—N—CH_3$$

more positive H:
stronger hydrogen bond
less positive H:
weaker hydrogen bond

Hydrogen bonds may form between two different compounds, such as between CH_3OH and H_2O or between CH_3NH_2 and H_2O. In these cases, there is often more than one possibility for hydrogen bonding. The following structures show two types of hydrogen bonds between CH_3NH_2 and H_2O. (In a mixture of these two compounds, hydrogen bonds can also form between two molecules of H_2O and between two molecules of CH_3NH_2.)

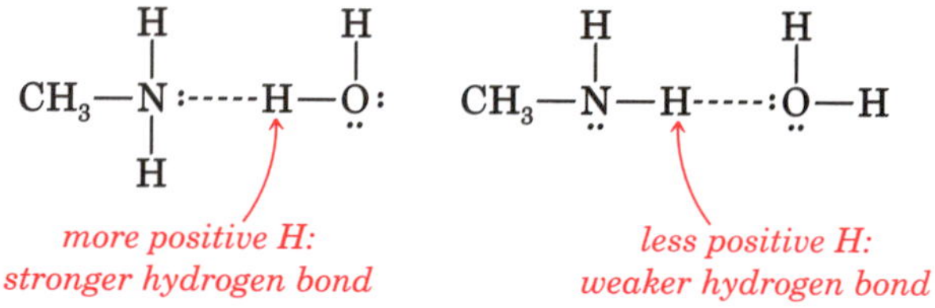

$$CH_3—N{:}\text{-----}H—\ddot{O}{:} \qquad CH_3—N—H\text{-----}{:}\ddot{O}—H$$

more positive H:
stronger hydrogen bond
less positive H:
weaker hydrogen bond

Table 1.5 shows the amount of energy needed to break some different types of hydrogen bonds. Note that the OH---N hydrogen bond is the strongest of the group. Because nitrogen is less electronegative than oxygen, its electrons are more loosely held and more easily attracted by another atom. The combination of nitrogen's loose electrons and the more positive hydrogen of an OH group leads to a quite strong hydrogen bond.

C. Effects of Hydrogen Bonding

Hydrogen bonds act rather like glue between molecules. Although a single hydrogen bond by itself is weak, all the molecules of a substance together may form a great many hydrogen bonds.

Table 1.5   Approximate dissociation energies of some hydrogen bonds

Type of Hydrogen Bond	Approximate Dissociation Energy (kcal/mol)
$—O—H\text{---}{:}N—$	7
$—O—H\text{---}{:}\ddot{O}—$	5
$—N—H\text{---}{:}N—$	3
$—N—H\text{---}{:}\ddot{O}—$	2

Boiling points generally increase with higher molecular weight because of increased van der Waals attractions. However, a hydrogen-bonded compound has a *higher boiling point* than would be predicted from molecular weight considerations alone. For a hydrogen-bonded liquid to be volatilized, additional energy must be supplied for breaking all the intermolecular hydrogen bonds.

Ethanol (CH_3CH_2OH) and dimethyl ether (CH_3OCH_3) have the same molecular weight. However, ethanol has a much higher boiling point than dimethyl ether—ethanol is a liquid at room temperature while dimethyl ether is a gas. The difference in boiling points between these two compounds can be directly attributed to the fact that ethanol molecules form hydrogen bonds, while dimethyl ether molecules cannot form hydrogen bonds among themselves. Note that hydrogen bonding affects the boiling point to a much greater extent than branching does.

hydrogen bond

no H to form a hydrogen bond

$$CH_3CH_2\ddot{O} \text{-----} H\text{---}\ddot{O} \quad\quad CH_3$$

ethanol
bp 78.5°

dimethyl ether
bp −23.6°

Solubility of covalent compounds in water is another property affected by hydrogen bonding. A compound that can form hydrogen bonds with water tends to be far more soluble in water than a compound that cannot. Sugars, such as glucose, contain many —OH groups and are quite soluble in water. Cyclohexane, however, cannot form hydrogen bonds and cannot break the existing hydrogen bonds in water; therefore, cyclohexane is insoluble in water.

glucose

soluble in water

cyclohexane

insoluble in water

The *shapes of large biomolecules* are also determined principally by hydrogen bonding. For example, proteins are held in specific orderly shapes by hydrogen bonds within each molecule or between molecules. Molecules of DNA—the genetic material in plant and animal cells—are held in double spirals partly because of hydrogen bonds between pairs of molecules.

STUDY PROBLEM

1.13   Show all the types of hydrogen bonds (if any) that would be found in:

(a) liquid $CH_3CH_2CH_2NH_2$ (b) a solution of CH_3OH in H_2O

(c) liquid ⬡—OH   (d) ⬡=O in H_2O

Section 1.7

Acids and Bases

A. Brønsted–Lowry Theory of Acidity

According to the Brønsted–Lowry concept of acids and bases, an **acid** is a substance that can donate a hydrogen ion (H^+) to a base. A hydrogen ion is also called a **proton.** A **base** is defined as a substance that can accept a hydrogen ion from an acid. Hydrogen ions are positively charged and bases are either negatively charged or have an unshared pair of electrons that can form a single bond with the hydrogen ion. Hydroxide ion (^-OH) and ammonia (NH_3) are typical bases.

An acid–base reaction is one in which a hydrogen ion is transferred from the acid to the base. Although we speak of hydrogen ion transfer, in organic chemical equations we emphasize the direction of electron movement rather than hydrogen ion movement, using curved arrows.

$$HÖ:^- + H-\overset{..}{\underset{..}{Cl}}: \longrightarrow H-\overset{..}{O}H + :\overset{..}{\underset{..}{Cl}}:^-$$

direction of electron movement

In 1923, J. N. Brønsted (1879–1947), director of the Fysisk-Kemiske Institute of Copenhagen, Denmark, and Thomas M. Lowry (1878–1936), physical chemist at Cambridge University, England, published papers nearly simultaneously on the theory of acidity. Brønsted's paper clearly distinguished an acid from a base. While Lowry did not explicitly define an acid or a base, he emphasized the importance of the hydrogen ion transfer in acid–base reactions.

B. Strong and Weak Acids and Bases

Recall from your general chemistry course that a **strong acid** is an acid that undergoes essentially complete ionization in water. Representative strong acids are HCl, HNO_3, and H_2SO_4. The ionization of these strong acids is a typical acid–base reaction. The acid (HCl, for example) loses a proton to the base (H_2O). The equilibrium lies far toward the right (complete ionization of HCl) because H_2O is a stronger base than Cl^- and HCl is a stronger acid than H_3O^+.

$$H-\overset{..}{O}-H + H-\overset{..}{\underset{..}{Cl}}: \longrightarrow H-\overset{H}{\underset{..}{\overset{|}{O^+}}}H + :\overset{..}{\underset{..}{Cl}}:^-$$

stronger base   *stronger acid*
than Cl^-   *than H_3O^+*

A **weak acid,** by contrast, has only a small proportion of its molecules broken into ions in water. Carbonic acid is a typical weak acid. The equilibrium lies toward the nonionized side of the equation because H_3O^+ is the stronger acid and HCO_3^- is the stronger base.

$$H-\overset{..}{O}-H + H-\overset{O}{\overset{\|}{\underset{..}{O}COH}} \rightleftharpoons H-\overset{H}{\underset{..}{\overset{|}{O^+}}}H + \ ^-:\overset{O}{\overset{\|}{\underset{..}{O}COH}}$$

carbonic acid   bicarbonate ion

stronger acid   *stronger base*
than H_2CO_3   *than H_2O*

Also, recall that *bases* are classified as strong (such as OH^-) or weak (NH_3), depending on their affinity for hydrogen ions.

$$:NH_3 \; + H\overset{\curvearrowleft}{-}\ddot{\overset{..}{O}}-H \; \rightleftharpoons \; \overset{+}{N}H_4 \; + \; {}^-:\ddot{\overset{..}{O}}H$$

ammonia	ammonium ion	hydroxide ion
a weak base	*stronger acid* *than H$_2$O*	*stronger base* *than NH$_3$*

Let us now consider some organic compounds that act as acids and bases. **Amines** are a class of organic compounds structurally similar to ammonia; an amine contains a nitrogen atom that is covalently bonded to one or more carbon atoms and that has an unshared pair of electrons.

Some common amines:

$$CH_3-\overset{..}{N}-H \qquad CH_3-\overset{..}{N}-H \qquad CH_3-\overset{..}{N}-CH_3$$
$$\quad\;\; | \qquad\qquad\qquad | \qquad\qquad\qquad\quad |$$
$$\quad\;\; H \qquad\qquad\qquad CH_3 \qquad\qquad\qquad CH_3$$

Amines, like ammonia, are weak bases and undergo reversible reactions with water or other weak acids. (The use of a strong acid drives the reaction to completion.)

In a reversible reaction, the reaction arrows are shown as equilibrium arrows ($\rightleftharpoons$) with the longer arrow used to indicate the direction of the equilibrium. In a nonreversible reaction, the reaction arrow is shown as a single one ($\rightarrow$).

$$CH_3\ddot{N}H_2 + H\overset{\curvearrowleft}{-}\ddot{\overset{..}{O}}H \; \rightleftharpoons \; CH_3\overset{+}{N}H_3 + {}^-:\ddot{\overset{..}{O}}H$$

$$CH_3\ddot{N}H_2 + H\overset{\curvearrowleft}{-}\ddot{\overset{..}{C}}l: \; \longrightarrow \; CH_3\overset{+}{N}H_3 + :\ddot{\overset{..}{C}}l:^-$$

An organic compound containing a **carboxyl group** ($-CO_2H$) is a weak acid. Compounds that contain carboxyl groups are called **carboxylic acids.** Acetic acid, CH_3CO_2H, is an example. One of the reasons for the acidity of carboxylic acids is the polarity of the O—H bond. (Another reason for this acidity will be discussed in Section 2.9E.)

The carboxyl group:

$$\overset{\displaystyle \ddot{\overset{..}{O}}}{\underset{}{\overset{\|}{-C}-\underset{..}{\overset{\delta-}{\ddot{O}}}-\overset{\delta+}{H}}} \quad \text{usually written} \quad -\overset{\displaystyle O}{\overset{\|}{C}}OH, \; -COOH, \; \text{or} \; -CO_2H$$

Example

$$CH_3\overset{\displaystyle O}{\overset{\|}{C}}OH \quad \text{or} \quad CH_3COOH \quad \text{or} \quad CH_3CO_2H$$

A carboxyl group in a carboxylic acid donates a hydrogen ion to a base. In doing so, the carboxyl group becomes a carboxylate ion, an anion. Because carboxylic acid compounds are only weakly acidic, acid–base reactions do not proceed to completion. In order to drive the reaction to completion, a base stronger than water, such as hydroxide (^-OH) must be used.

$$CH_3\overset{\overset{\displaystyle :\!O\!:}{\|}}{C}-\overset{..}{\underset{..}{O}}:-H + H-\overset{..}{\underset{..}{O}}-H \rightleftharpoons CH_3\overset{\overset{\displaystyle :\!O\!:}{\|}}{C}-\overset{..}{\underset{..}{O}}:^- + H-\overset{\overset{\displaystyle H}{|}}{\underset{..}{O}}\!\!^+\!\!-H$$

acetic acid	acetate ion
a carboxylic acid	*a carboxylate ion*

$$CH_3\overset{\overset{\displaystyle :\!O\!:}{\|}}{C}-\overset{..}{\underset{..}{O}}:-H + :NH_3 \rightleftharpoons CH_3\overset{\overset{\displaystyle :\!O\!:}{\|}}{C}-\overset{..}{\underset{..}{O}}:^- + \overset{+}{N}H_4$$

$$CH_3\overset{\overset{\displaystyle :\!O\!:}{\|}}{C}-\overset{..}{\underset{..}{O}}:-H + {}^-:\overset{..}{\underset{..}{O}}H \longrightarrow CH_3\overset{\overset{\displaystyle :\!O\!:}{\|}}{C}-\overset{..}{\underset{..}{O}}:^- + H\overset{..}{\underset{..}{O}}H$$

STUDY PROBLEMS

1.14 Which of the following compounds or ions act as acids and which act as bases in H_2O?

(a) $^-:\overset{..}{N}H_2$ (b) $CH_3CH_2CH_2\overset{\overset{\displaystyle :\!O\!:}{\|}}{C}\overset{..}{O}H$ (c) $^-:\overset{..}{O}CH_2CH_3$

(d) [piperidine ring with $\overset{\overset{\displaystyle H}{\diagup}}{\underset{\diagdown H}{\overset{+}{N}}}$]   (e) [cyclohexane ring with :NH]

1.15 Rewrite and complete the following equations for acid–base reactions. Include in your answer: (1) electron dots; (2) curved electron-shift arrows; and (3) reaction arrows that show the direction of the equilibrium. (If the reaction proceeds essentially to completion, use a single reaction arrow.)

(a) $CH_3CH_2CO_2H + H_2O$ (b) [benzene ring]$-CO_2H + OH^-$

(c) $(CH_3)_2NH + H_2O$ (d) [cyclohexane ring]$NH + CH_3CO_2H$

C. Conjugate Acids and Bases

The concept of conjugate acids and bases is useful for comparisons of acidities and basicities. The **conjugate base** of an acid is the ion or molecule that is formed when an acid loses its hydrogen ion. For example, the chloride ion is the conjugate base of HCl. The **conjugate acid** of a base is the product of the reaction of the base and a hydrogen ion. We say the base has been protonated. Thus, the conjugate acid of NH_3 is NH_4^+.

conjugate base
of H_3O^+

conjugate acid
of H_2O

$$HCl \quad + H_2O \longrightarrow H_3O^+ + \quad Cl^-$$

conjugate acid
of Cl^-

conjugate base
of HCl

conjugate base
of $^+NH_4$

conjugate acid
of NH_3

$$CH_3\overset{O}{\overset{\|}{C}}OH \ + NH_3 \rightleftharpoons \ CH_3\overset{O}{\overset{\|}{C}}O^- \ + \ \overset{+}{N}H_4$$

conjugate acid
of $CH_3CO_2^-$

conjugate base
of CH_3CO_2H

If an acid is a strong acid, its conjugate base is a weak base:

$$HCl \quad + H_2O \longrightarrow H_3O^+ + \quad Cl^-$$

strong acid
(loses H^+ readily)

very weak base
(has little attraction for H^+)

On the other hand, if an acid is weak or very weak, its conjugate base is a moderately strong base or a strong base, depending on the affinity of the conjugate base for H^+.

$$CH_3\overset{O}{\overset{\|}{C}}OH + H_2O \rightleftharpoons \ CH_3\overset{O}{\overset{\|}{C}}O^- \ + H_3O^+$$

weak acid

moderate base

$$2\,H_2O \quad \rightleftharpoons \ H_3O^+ + \quad OH^-$$

very weak acid

strong base

Thus, as the acid strengths of a series of compounds increase, the base strengths of their conjugate bases decrease.

Conjugate acids

$$H_2O \quad HCN \quad CH_3CO_2H \quad H_3PO_4 \quad HCl$$

increasing acidity

Corresponding conjugate bases

$$^-OH \quad ^-CN \quad CH_3CO_2^- \quad H_2PO_4^- \quad Cl^-$$

decreasing basicity

STUDY PROBLEM

1.16 Write the conjugate base of structures **(a)–(c)** and the conjugate acid of structures **(d)–(f)**:

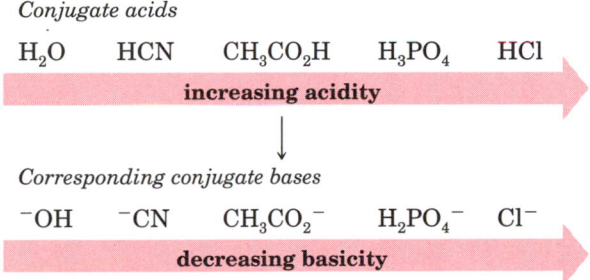

(d) ... S—NH   **(e)** ... —O⁻   **(f)** ...

D. Some Factors Affecting Acidity

The various factors that affect the strength of acids and bases are discussed in detail in Sections 14.4 and 18.5. Here, we will briefly describe only two of these factors: the size and the electronegativity of atoms.

In any equilibrium, such as the ionization of a weak acid, the thermodynamically more stable reactant or product predominates in the equilibrium mixture. We say that these reversible reactions are under **equilibrium control,** or **thermodynamic control.** The strength of an acid or base is thus a function of the stability of the protonated acidic form (H:B) versus that of the nonprotonated base (:B⁻), where H:B represents any Brønsted acid and :B⁻, its conjugate base.

$$\text{H:B} \quad \rightleftharpoons \quad \text{H}^+ \quad + \quad \text{:B}^-$$

A more stable protonated structure shifts the equilibrium toward the conjugate acid.

A more stable base shifts the equilibrium toward the conjugate base.

The ease with which a molecule or ion loses a proton depends on the strength of the bond to the proton. Two factors that affect bond strength are the size and the electronegativity of the atom bonded to the proton.

In the same group of the periodic table, the **size** of an atom bonded to a proton affects the acid strength of a compound. A larger atom can disperse the negative charge over a larger region and thus add stability to the conjugate base. As the stability of the conjugate base increases, the greater is the acid strength of the acid. For example,

$$\text{H—F} \quad \text{H—Cl} \quad \text{H—Br} \quad \text{H—I}$$

increasing size of halogen; increasing acid strength →

In the same row of the periodic table, the **electronegativity** of the atom bonded to the proton also affects the acidity. A more electronegative atom can carry a negative charge more readily than can a less electronegative atom. A proton is more readily removed from the more electronegative atom. For example,

$$\text{CH}_3\text{—}\underset{\underset{\text{CH}_3}{|}}{\overset{\overset{\text{CH}_3}{|}}{\text{C}}}\text{—H} \quad \underset{\underset{\text{CH}_3}{\diagup}}{\overset{\overset{\text{CH}_3}{\diagdown}}{\text{N}}}\text{—H} \quad \overset{\overset{\text{CH}_3}{\diagdown}}{\text{O}}\text{—H} \quad \text{F—H}$$

increasing electronegativity of circled atom; increasing acid strength →

The proximity of an electronegative atom to an acidic hydrogen can have a major influence on the acidity of a compound. Consider the following compound:

electronegative chloro group   O   *acidic hydrogen*
                                          ‖
                    ⟶ (Cl)CH₂CH₂CH₂CO—(H) ⟵

4-chlorobutanoic acid

The acidic hydrogen is bonded to the oxygen in the carboxyl group and is far removed from the electronegative chlorine atom. Even so, 4-chlorobutanoic is slightly more acidic than butanoic acid—the same compound without the chlorine atom. The electronegative chlorine increases the acidity of the 4-chloro compound by passing its electronegative effect through the single bonds to the carboxyl group. The closer the chlorine is to the carboxyl group, the more acidic is the carboxylic acid.

$$
\underset{\mathrm{Cl}}{\mathrm{CH_2CH_2CH_2\overset{\displaystyle O}{\overset{\|}{C}}OH}}
\qquad
\underset{\mathrm{Cl}}{\mathrm{CH_3CHCH_2\overset{\displaystyle O}{\overset{\|}{C}}OH}}
\qquad
\underset{\mathrm{Cl}}{\mathrm{CH_3CH_2CH\overset{\displaystyle O}{\overset{\|}{C}}OH}}
$$

increasing acidity ⟶

STUDY PROBLEMS

1.17 Predict which compound in each of the following pairs would be the stronger acid. Explain your choices.

(a) CH_4 or H_2O (b) CH_4 or NH_3 (c) H_2O or H_2S

1.18 Which member of each of the following pairs would be the strongest acid? Explain your answers.

(a) $ICH_2\overset{\displaystyle O}{\overset{\|}{C}}OH$, $FCH_2\overset{\displaystyle O}{\overset{\|}{C}}OH$

(b) $\underset{\mathrm{Cl}}{CH_3\overset{\mathrm{Cl}}{\overset{|}{C}}{-}\overset{\displaystyle O}{\overset{\|}{C}}OH}$, $\underset{\mathrm{Cl}}{CH_3CH\overset{\displaystyle O}{\overset{\|}{C}}OH}$

(c) $\underset{\mathrm{Cl\ \ Cl}}{CH_2CH\overset{\displaystyle O}{\overset{\|}{C}}OH}$, $\underset{\mathrm{Cl}}{CH_3\overset{\mathrm{Cl}}{\overset{|}{C}}{-}\overset{\displaystyle O}{\overset{\|}{C}}OH}$

E. Lewis Acids and Bases

The Brønsted—Lowry theory of acidity and basicity focuses on the hydrogen ion. A more general theory of acidity and basicity is the Lewis theory, which focuses on electron pairs. In the Lewis theory, a **Lewis acid** is a substance that can *accept a pair of electrons* and a **Lewis base** is a substance that can *donate a pair of electrons.*

A Lewis acid–base reaction

$$FeBr_3 \quad + \quad :\overset{..}{\underset{..}{Br}}\!-\!\overset{..}{\underset{..}{Br}}: \quad \rightleftharpoons \quad FeBr_4^- + \overset{..}{\underset{..}{Br}}:^+$$

a Lewis acid	*Lewis base*
(electron-pair acceptor)	*(electron-pair donor)*

The Brønsted–Lowry theory is actually a limited version of the Lewis theory, since in the Brønsted–Lowry theory there is only one acid, the hydrogen ion (H^+). In the Lewis theory, not only is the hydrogen ion an acid, but so are anhydrous metal salts such as $ZnCl_2$, $FeCl_3$, and $AlCl_3$ and compounds like BF_3.

In general, when organic chemists speak of an acid they are referring to the one Brønsted–Lowry acid, the hydrogen ion; when they speak of Lewis acids, they mean the metal salts, such as $AlCl_3$.

Many organic reactions that we will consider in this text can be considered to be reactions of Lewis acids or bases. Two examples follow.

Lewis acid–base reactions:

$$H^+ + CH_3CH_2\overset{..}{\underset{..}{O}}H \quad \rightleftharpoons \quad CH_3CH_2\overset{+}{\underset{..}{O}}\overset{\textstyle H}{\overset{|}{}}H$$

$$CH_3\overset{\displaystyle :\overset{..}{O}:}{\overset{\|}{C}}CH_3 + {}^-\!:C\!\equiv\!N: \quad \rightleftharpoons \quad CH_3\underset{\underset{C\equiv N:}{|}}{\overset{\overset{:\overset{..}{O}:^-}{|}}{C}}CH_3$$

SAMPLE PROBLEM

Methylamine (CH_3NH_2) undergoes a Lewis acid–base reaction with boron trifluoride (BF_3) to yield CH_3NH_2—BF_3. (a) Write the equation for this reaction, showing the complete structural formula for the product and the charges on N and B. (b) Identify each reactant as a Lewis acid or Lewis base.

Solution

$$CH_3\overset{..}{N}H_2 \quad + \quad BF_3 \quad \longrightarrow \quad H-\overset{\overset{\textstyle H}{|}}{\underset{\underset{\textstyle H}{|}}{C}}-\overset{\overset{\textstyle H}{|}}{\underset{\underset{\textstyle H}{|}}{N^+}}\!-\!\overset{\overset{\textstyle F}{|}}{\underset{\underset{\textstyle F}{|}}{B^-}}\!-\!F$$

Lewis base	*Lewis acid*

STUDY PROBLEM

1.19   Identify the reactants in the following equations as Lewis acids or Lewis bases:

$$\text{(a)} \quad CH_3\overset{\overset{\textstyle :\overset{..}{O}:}{\|}}{C}CH_3 + H^+ \quad \rightleftharpoons \quad CH_3\overset{\overset{\textstyle :\overset{+}{O}H}{\|}}{C}CH_3$$

(b) $(CH_3)_3C^+ + :\overset{..}{\underset{..}{Cl}}:^- \longrightarrow (CH_3)_3C\overset{..}{\underset{..}{Cl}}:$

(c) $CH_3\overset{\displaystyle O}{\overset{\|}{C}}OCH_3 + {}^-:\overset{..}{O}CH_3 \rightleftharpoons {}^-:CH_2\overset{\displaystyle O}{\overset{\|}{C}}OCH_3 + H\overset{..}{O}CH_3$

F. Acidity Constants

A chemical reaction has an **equilibrium constant** K that reflects how far the reaction proceeds toward completion. For the ionization of an acid in water, this constant is called an **acidity constant** K_a. An equilibrium constant is determined by the following general equation, with concentration values given in molarity, M:

$$K = \frac{\text{concentrations of products in } M}{\text{concentrations of reactants in } M}$$

For acetic acid:　$CH_3CO_2H \rightleftharpoons CH_3CO_2^- + H^+$

$$K_a = \frac{[CH_3CO_2^-][H^+]^*}{[CH_3CO_2H]}$$

where $[H^+]$ = molar concentration of H^+

$[CH_3CO_2^-]$ = molar concentration of $CH_3CO_2^-$

$[CH_3CO_2H]$ = molar concentration of CH_3CO_2H

The more ionized an acid is, the larger is the value for K_a because the values in the numerator are larger. *A stronger acid has a larger K_a value.* Any acid with a $K_a > 10$ is considered a strong acid. (For HCl, $K_a \cong 10^7$.) By contrast, typical carboxylic acids, such as acetic acid, have K_a values much smaller than 1. (For CH_3CO_2H, $Ka = 1.75 \times 10^{-5}$.)

$$K_a = \frac{[H^+][\text{anion}]}{[\text{un-ionized acid}]} \quad \xleftarrow{\quad} \quad \textit{as numerator increases, } K_a \textit{ increases}$$

---

* More correctly, the *activity,* or *effective concentration,* should be used instead of molarity. Because activities of ions approach their molarities in dilute solution, molarity is used for the sake of simplicity. In addition, the equilibrium expression should contain the hydrogen acceptor, water:

$$CH_3CO_2H + H_2O \rightleftharpoons CH_3CO_2^- + H_3O^+$$

$$K_a' = \frac{[CH_3CO_2^-][H_3O^+]}{[CH_3CO_2H][H_2O]}$$

In dilute solutions, the molar concentration of water is about 55.5. This constant factor is generally grouped with the equilibrium constant K_a, and the $[H_3O^+]$ term is simplified to $[H^+]$. Even though "naked" protons do not exist as such in solution, we will frequently use the symbol H^+ instead of H_3O^+ in this book to represent an aqueous acid.

$$K_a = K_a'[H_2O] = \frac{[CH_3CO_2^-][H_3O^+]}{[CH_3CO_2H]} = \frac{[CH_3CO_2^-][H^+]}{[CH_3CO_2H]}$$

Table 1.6 Acidity constants and pK_a values for some acids		
Formula	K_a	pK_a
Strong:		
HCl	$\sim 10^7$	~ -7
H_2SO_4	$\sim 10^5$	~ -5
Moderately Strong:		
H_3PO_4	7.52×10^{-3}	2.12
Weak:		
HF	35.3×10^{-5}	3.45
HCO_2H	17.5×10^{-5}	3.75
CH_3CO_2H	1.75×10^{-5}	4.75
$CH_3CH_2CO_2H$	1.34×10^{-5}	4.87
Very Weak:		
HCN	4.93×10^{-10}	9.31
H_2O	1.80×10^{-16}	15.74

Just as pH is defined as the negative logarithm of hydrogen ion concentration, pK_a is defined as the negative logarithm of K_a. We will use pK_a values in this text for comparison of acid strengths. (The K_a values and pK_a values for some carboxylic acids are listed in Table 1.6. The pK_a table inside the front cover lists some other types of compounds as well.)

$$pH = -\log[H^+]$$

$$pK_a = -\log K_a$$

Examples

$$\text{If} \quad K_a = 10^{-3}, \quad \text{then} \quad pK_a = 3$$

$$\text{If} \quad K_a = 10^2, \quad \text{then} \quad pK_a = -2$$

As K_a gets larger (stronger acid), pK_a gets smaller. *The smaller the value for* pK_a, *the stronger the acid.*

$$
\begin{array}{ccccc}
K_a: & 10^{-10} & 10^{-5} & 10^{-1} & 10^2 \\
pK_a: & 10 & 5 & 1 & -2
\end{array}
$$

increasing acid strength →

SAMPLE PROBLEM

Calculate the pK_a of an acid with K_a equal to 136×10^{-5}.

Solution

$$
\begin{aligned}
pK_a &= -\log K_a \\
&= -\log(136 \times 10^{-5}) \\
&= -\log(1.36 \times 10^{-3}) \\
&= -(\log 1.36 - 3) \\
&= 3 - \log 1.36 \\
&= 3 - 0.133 \quad \longleftarrow \text{\textit{from log table or calculator}} \\
&= 2.87
\end{aligned}
$$

When the concentration of the undissociated carboxylic acid equals the concentration of the dissociated acid, the pH of the solution equals the pK_a of the carboxylic acid. This useful relationship can be expressed by the **Henderson–Hasselbalch equation,** a reexpression of the mass law equation.

Mass law equation:

$$K_a = \frac{[H^+][A^-]}{[HA]}$$

Henderson–Hasselbalch equation:

$$pH = pK_a + \log \frac{[A^-]}{[HA]}$$

When $[A^-] = [HA]$, $pH = pK_a$

STUDY PROBLEM

1.20 Determine the pK_a of chloroacetic acid ($ClCH_2CO_2H$) if a solution containing 0.100 mol of undissociated acid and 0.100 mol of its sodium salt ($ClCH_2CO_2^- Na^+$) has a pH of 2.85.

G. Basicity Constants

The reversible reaction of a weak base with water, like the reaction of a weak acid with water, results in a small but constant concentration of ions at equilibrium. The **basicity constant K_b** is the equilibrium constant for this reaction. As in the case of K_a, the value for $[H_2O]$ is included in K_b in the equilibrium expression.

$$NH_3 + H_2O \rightleftharpoons NH_4^+ + OH^-$$

$$K_b = \frac{[NH_4^+][OH^-]}{[NH_3]}$$

$$pK_b = -\log K_b$$

With an increase in base strength, the value for K_b increases and the pK_b value decreases. *The smaller the value for pK_b, the stronger the base.*

K_b: 10^{-10} 10^{-7} 10^{-5}

pK_b: 10   7   5

increasing base strength

The basicity of an amine (R_3N) is often expressed as the pK_a of its conjugate acid (R_3NH^+). Conversions between pK_b and pK_a values are discussed in Section 18.5.

STUDY PROBLEMS

1.21 List the following compounds in order of increasing basicity (weakest first). See Table 1.7 for pK_b values.

(a) NH_3 **(b)** CH_3NH_2 **(c)** $(CH_3)_2NH$

1.22   List the following anions in order of increasing basicity (weakest first):

   (a) CH_3O^-, $pK_b = -1.5$ **(b)** $CH_3CO_2^-$, $pK_b = 9.25$ **(c)** Cl^-, $pK_b = 21$

Table 1.7   Basicity constants and pK_b values for ammonia and some amines

Formula	K_b	pK_b
NH_3	1.79×10^{-5}	4.75
CH_3NH_2	45×10^{-5}	3.34
$(CH_3)_2NH$	54×10^{-5}	3.27
$(CH_3)_3N$	6.5×10^{-5}	4.19

Summary

The probable location (relative to the nucleus) of an electron with a particular energy is called an **atomic orbital.** The first electron shell (closest to the nucleus, lowest energy) contains only the spherical 1s orbital. The second shell (higher energy) contains a spherical 2s orbital and three mutually perpendicular, two-lobed 2p orbitals. Any orbital can hold a maximum of two paired (opposite spin) electrons.

The **atomic radius** equals half the distance between nuclei bonded by a nonpolar covalent bond, such as in H—H. The atomic radius increases as we go down any group in the periodic table and decreases as we go from left to right across a period. **Electronegativity** is a measure of the pull of the nucleus on the outer electrons of an atom in a covalent compound. Electronegativity decreases as we go down any group and increases as we go from left to right in the periodic table.

A chemical bond results from electron transfer **(ionic bond)** or electron sharing **(covalent bond).** The number of bonds an atom can form is determined by the number of valence electrons. Carbon has four valence electrons and forms four covalent bonds. A **polar covalent bond** is a covalent bond with a charge separation.

An atom may share two, four, or six electrons with another atom—two atoms may be joined by a **single bond,** a **double bond,** or a **triple bond.**

single bond ⟶   *triple bond* ⟶ $H-C\equiv C-\overset{\overset{\displaystyle H}{|}}{C}=CH_2$ ⟵ *double bond*

The **bond length** is the distance between nuclei of covalently bonded atoms. The **bond angle** is the angle between two covalent bonds in a molecule. The **bond dissociation energy** (ΔH in the following equation) is the amount of energy needed to effect **homolytic cleavage** of a covalent bond. A greater bond dissociation energy means a more stable bond.

$$H_3C-H \longrightarrow H_3C\cdot + H\cdot \qquad \Delta H = +104 \text{ kcal/mol}$$

An **empirical formula** tells us the *relative number* of different atoms in a molecule, and the **molecular formula** tells us the *actual number* of each kind of atom in a molecule.

$$C_2H_5 \qquad C_4H_{10}$$

empirical   *molecular*
formula   *formula*

In **structural formulas,** which depict the structures of molecules, pairs of electrons may be represented by dots or by lines. Unshared valence electrons are not always shown in structural formulas.

$$
\begin{array}{ccc}
\overset{\displaystyle H}{\underset{\displaystyle \overset{..}{\underset{..}{H}} \ \overset{..}{\underset{..}{H}}}{H : \overset{..}{C} : \overset{..}{N} : H}}
&
\overset{\displaystyle H}{\underset{\displaystyle \underset{H}{\overset{|}{}} \ \underset{H}{\overset{|}{}}}{H - \underset{|}{\overset{|}{C}} - \overset{..}{N} - H}
&
CH_3NH_2
\end{array}
$$

Lewis formula   *complete*   *condensed*
                 structural formula   *structural formula*

Dipole–dipole attractions between molecules **(van der Waals attractions)** are generally well under 5 kcal/mol except for **hydrogen bonds** (attractions between a partially positive H and an unshared pair of electrons of N, O, or F), which require 5–10 kcal/mol for their dissociation. Hydrogen bonding leads to an increase in boiling point and water solubility of a compound.

A **Brønsted–Lowry acid** is a substance that can lose H^+; a **Brønsted–Lowry base** is a substance that can accept H^+. The strength of an acid or base is reported as K_a (or pK_a) or as K_b (or pK_b), respectively. A stronger acid has a larger value for K_a (and a smaller pK_a); a stronger base has a larger K_b (and a smaller pK_b). (See Tables 1.6 and 1.7.)

anion

$$K_a = \frac{[H^+][A^-]}{[HA]} \quad \text{and} \quad pK_a = -\log K_a$$

$$K_b = \frac{[\overset{+}{B}H][OH^-]}{[B:]} \quad \text{and} \quad pK_b = -\log K_b$$

base

strong acids ($pK_a < -1$): HCl, HNO_3, H_2SO_4

weak acids ($pK_a > 3$): CH_3CO_2H, HCN, H_2O

strong bases: $^-OH, \ ^-OCH_3$

weak bases: NH_3, CH_3NH_2

The **conjugate base** of a strong acid is a weak base, while the conjugate base of a very weak acid is a strong base. In the following equation, as HA decreases in acid strength, A^- increases in base strength.

$$H - A \quad \rightleftharpoons \quad H^+ + \quad :A^-$$

conjugate acid of A^- *conjugate base of HA*

A **Lewis acid** is a substance that can *accept* a pair of electrons, while a **Lewis base** is a substance that can *donate* a pair of electrons.

$$CH_3\overset{..}{N}H_2 \; + \quad H^+ \quad \longrightarrow \quad CH_3\overset{\overset{\displaystyle H}{\overset{\displaystyle |}{+|}}}{N}H_2$$

Lewis base   *Lewis acid*

Essay Problem for Chapter 1

Amphoteric Reactions of Amines

Although amines are generally considered to be organic bases rather than acids, they are actually amphoteric. Amines can react as acids or as bases depending upon the reagent used for the reaction. For example, when an amine (such as propylamine) is treated with a mineral acid (such as HCl) the amine accepts a hydrogen ion from the HCl. In this reaction, the amine acts as a Brønsted base and becomes an amine salt, a cation.

An amine acting as a base

$$CH_3CH_2CH_2NH_2 + H^+Cl^- \longrightarrow CH_3CH_2CH_2\overset{+}{N}H_3 + Cl^-$$

propylamine   *an amine salt*

However, when an amine is treated with an extremely strong base, such as potassium hydride (KH), the amine loses a hydrogen ion from its nitrogen and forms an anion called an amide. Since in this reaction the amine is a proton donor, it acts as a Brønsted acid.

An amine acting as an acid

$$CH_3CH_2CH_2NH_2 + \quad KH \quad \longrightarrow CH_3CH_2CH_2NH^-K^+$$

propylamine   potassium   *an amide*
                      hydride

Questions*

1. What is the ionic charge on the hydrogen atom of potassium hydride?
2. For the reaction of KH with water: (a) Write the ionic equation for the reaction. (b) Show all unshared electrons of the reactants. (c) Use an arrow to show what happens to the unshared electrons of the base. (d) Label each reactant in the equation as a Brønsted acid or as a Brønsted base.
3. Only one of the two products formed in the reaction of potassium hydride and propylamine is shown above. What is the other product?
4. Write the equation for the reaction of KH with propylamine. Label each hydrogen in the reactants as acidic, basic, or neither.
5. Write the equation for the reaction between water and the amide formed by the reaction between KH and propylamine. Label each reactant as a Brønsted acid or as a Brønsted base.

* Answers to the *Essay Problem* questions are found in the *Solutions Manual*

Study Problems

Note to student: The following *Study Problems* follow the order of the chapter presentation. The *Additional Problems* that come after them *do not* follow the chapter order; these problems should be worked after you have covered the entire chapter. The *Additional Problems* are gradated in order of difficulty—the first problems are routine drill problems, while the last are more challenging. In each chapter in this book, the problems are similarly arranged. Solutions to all chapter-end problems are given in the *Solutions Manual*.

1.23 If an atom used two *p* atomic orbitals to form single covalent bonds with two hydrogen atoms, what would be the expected bond angle?

1.24 Give the *Lewis formula* (electron-dot formula) for each of the following structures:

(a)

$$H-\overset{\overset{\displaystyle H}{|}}{\underset{\underset{\displaystyle H}{|}}{C}}-\overset{\overset{\displaystyle H}{|}}{\underset{\underset{\displaystyle H}{|}}{C}}-\overset{\overset{\displaystyle H}{|}}{\underset{\underset{\displaystyle H}{|}}{C}}-H$$

(b)

(c) $CH_3CH_2CH_2Br$

(d) H_2O_2

1.25 Calculate the formal charges of all atoms except H in each of the following structures:

(a) $CH_3-\overset{\overset{\displaystyle :\ddot{O}:}{|}}{\underset{\underset{\displaystyle :O:}{|}}{S}}-\ddot{\underset{\cdot\cdot}{C}l:$

(b) $CH_3-C{\equiv}N:$

(c) $CH_3\overset{\overset{\displaystyle :O:}{\|}}{C}-\ddot{\underset{\cdot\cdot}{O}}:^-$

(d) $CH_3-\overset{\overset{\displaystyle :\ddot{O}:}{|}}{\underset{\underset{\displaystyle \cdot\cdot}{S}}{}}-CH_3$

(e) $CH_3-\overset{\overset{\displaystyle CH_3}{|}}{\underset{\underset{\displaystyle CH_3}{|}}{C}}{}^+$

(f) $\left[\, CH_3-\overset{\overset{\displaystyle H}{|}}{\underset{\underset{\displaystyle \cdot\cdot}{O}}{}}-H \,\right]^+$

1.26 Give the *complete structural formula* (showing each atom and using lines for bonds) for each of the following condensed formulas:

(a) $(CH_3)_3CCl$

(b) $CH_3(C{\equiv}C)_3CH_3$

(c)

(d) $(CH_3)_2CHO_2CC(CH_3)_3$

1.27 Write a *condensed structural formula* for each of the following structures:

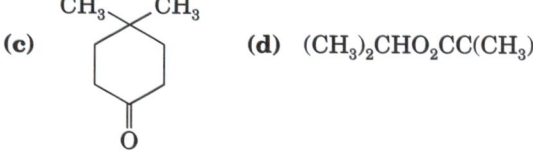

(a) $H-\overset{\overset{\displaystyle H}{|}}{\underset{\underset{\displaystyle H}{|}}{C}}-\ddot{\underset{\cdot\cdot}{O}}-H$

(b) $H-\overset{\overset{\displaystyle H}{|}}{\underset{\underset{\displaystyle H}{|}}{C}}-\overset{\overset{\displaystyle H}{|}}{\underset{\underset{\displaystyle H}{|}}{C}}-\overset{\overset{\displaystyle H}{|}}{\underset{\underset{\displaystyle H}{|}}{N}}:$

(c)
$$\begin{array}{ccc} H & H & H \\ H\!:\!\ddot{C}\!:\!\ddot{N}\!:\!\ddot{C}\!:\!H \\ H & & H \end{array}$$

(d)
$$CH_3\!\!\underset{\underset{CH_3}{|}}{\overset{\overset{CH_3}{|}}{C}}\!\!-CH_2CH_2CH_2\underset{\searrow Cl}{\overset{\nearrow Cl}{CH}}$$

1.28 Write the *molecular formula* for each of the following condensed structural formulas:

(a) $CH_3CH_2CH_2CH_2OH$ **(b)** $CH_3CH_2\overset{\overset{\textstyle OH}{|}}{C}HCH_3$ **(c)** $(CH_3)_3COH$

1.29 Give the complete structural formula (showing each atom and using lines for bonds) for each of the following compounds:

(a) $H_2C{=}CHCH{=}CHCN$ **(b)** CH_3COCH_3

(c) CH_3CO_2H **(d)** $OHCCH_2CHO$

(*Hint:* Each structure contains at least one double or triple bond that is not shown.)

1.30 Show *the unshared pairs of valence electrons* (if any) in the following formulas:

(a) CH_3NH_2 **(b)** $(CH_3)_3N$ **(c)** $(CH_3)_3NH^+$ **(d)** CH_3OH

(e) $(CH_3)_3COH$ **(f)** $CH_2{=}CH_2$ **(g)** $H_2C{=}O$

1.31 Draw a polygon formula for each of the following cyclic structures:

(a)
$$\begin{array}{c} \overset{\nearrow CH_2 \quad \nwarrow CH_3}{} \\ CH_2 \quad C{\searrow} \\ | \qquad | {>}O \\ CH_2{-}CH \end{array}$$

(b)
$$\begin{array}{c} HOCH_2 \quad O \qquad OH \\ \overset{\nwarrow}{C} \qquad \overset{\nearrow}{C} \\ H \quad CH{-}CH \quad H \\ \qquad OH \quad OH \end{array}$$

(c)
$$\begin{array}{c} H \quad H \\ \overset{\nwarrow}{N}\overset{\nearrow}{} \\ H \quad C \\ \overset{\nwarrow}{C} \quad {\searrow} N \\ | \qquad || \\ C \qquad C \\ \overset{\nearrow}{H} \quad N \quad {\searrow} O \\ \qquad | \\ \qquad H \end{array}$$

(d)
$$\begin{array}{c} H \quad H \\ H \quad {\searrow}\overset{/H}{C} \quad H \\ H{-}C \quad C{-}C{-}H \\ H{-}C \quad | \quad C{-}H \\ H{-}C \quad C \quad C{\searrow}H \\ \quad C \quad N \\ H \quad H \quad H \end{array}$$

1.32 Convert each of the following polygon formulas to a complete structural formula, showing each atom and each bond. Show also any unshared pairs of valence electrons.

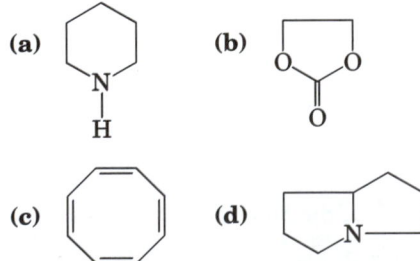

1.33 Draw the polygon and the structural formulas for a carbon ring system containing:
(a) six ring carbons and a double bond;
(b) five ring carbons, one of which is part of a carbonyl group (C=O).

1.34 Write chemical equations for (1) the *homolytic cleavage*, and (2) the *heterolytic cleavage*, of each of the following compounds at the indicated bond. (Apply your knowledge of electronegativities in the heterolytic cleavages and show by curved arrows the movement of the electrons.)

(a) CH_3CH_2—Cl   (b) H—NH_2

(c) CH_3—OH   (d) CH_3O—H

1.35 Which is the positive end and which is the negative end of the dipole in each of the following bonds? (Use $\delta+$ and $\delta-$.)

(a) C—Mg   (b) C—Br   (c) C—O

(d) C—Cl   (e) C—H   (f) C—B

1.36 Circle the most electronegative element in each of the following structures, and show the direction of polarization of its bond(s):

(a) CH_3OH (b) $CH_3\overset{\displaystyle O}{\overset{\|}{C}}CH_3$

(c) FCH_2CH_2OH (d) $(CH_3)_2NCH_2CH_2OH$

1.37 Draw structures to show the hydrogen bonding (if any) you would expect in the following compounds in their pure liquid states:

(a) $(CH_3)_2NH$ (b) $CH_3CH_2OCH_3$ (c) CH_3CH_2F

(d) $(CH_3)_3N$ (e) $(CH_3)_2C{=}O$ (f) $CH_3OCH_2CH_2OH$

1.38 Show all types of hydrogen bonds in an aqueous solution of $(CH_3)_2NH$. Which is the strongest hydrogen bond?

1.39 Write an equation for the acid—base reaction that would take place when each of the following Brønsted bases is treated with concentrated HCl:

(a) $CH_3\overset{\displaystyle O}{\overset{\|}{C}}O^-K^+$ (b) $CH_3CH_2NH_2$ (c) $(CH_3)_3COH$ (d) $Na^+HCO_3^-$

(e) 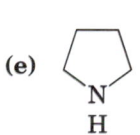   (f) ⬡—NH_2 (g) $H_3\overset{+}{N}CH_2\overset{\displaystyle O}{\overset{\|}{C}}O^-$ (h) pyridine

1.40 Write an equation for the acid—base reaction when each of the following Brønsted acids is treated with an aqueous solution of NaOH.

(a) $H\overset{\displaystyle O}{\overset{\|}{C}}OH$ (b) $CH_3\overset{\displaystyle O}{\overset{\|}{C}}OH$ (c) ⬡—$\overset{\displaystyle O}{\underset{\displaystyle O}{\overset{\|}{\underset{\|}{S}}}}OH$ (d) $CH_3CH_2\overset{+}{O}H_2\ Cl^-$

(e) $H_3\overset{+}{N}CH_2\overset{\displaystyle O}{\overset{\|}{C}}O^-$

1.41 Write the formulas for (a) the conjugate acid of $(CH_3)_2NH$, and (b) the conjugate base of $CH_3CH_2CH_2CO_2H$.

1.42 From the pK_a values for the conjugate acids in Table 1.6, list the following anions in order of increasing base strength:

(a) $CH_3CO_2^-$ (b) HCO_2^- (c) Cl^- (d) $H_2PO_4^-$

1.43 Calculate the pK_a of each of the following compounds, and arrange in order of increasing acidity (weakest acid first):

Structure	K_a
(a) CH_3CO_2H	1.75×10^{-5}
(b) ⬡—OH	1.0×10^{-10}
(c) ⬡—CH_2CO_2H	5.2×10^{-5}

1.44 Calculate the pK_b of the following bases, and arrange in order of increasing base strength:

Structure	K_b
(a) $CH_3\overset{O}{\overset{\|}{C}}NH_2$	4.3×10^{-14}
(b) ⬡—NH_2	4.3×10^{-10}
(c) $(CH_3)_3CNH_2$	6.8×10^{-4}

Additional Problems

1.45 Indicate the most likely position of attack by a Lewis acid on each of the following structures. Explain your answers.

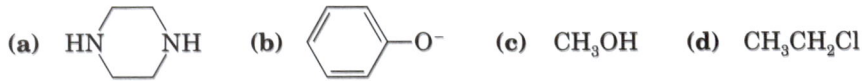

(a) $HN\underset{\smile}{\frown}NH$ **(b)** ⬡—O^- **(c)** CH_3OH **(d)** CH_3CH_2Cl

1.46 Assign the proper ionic charge to each of the following ions:

(a) $:C{\equiv}N:$ **(b)** $:C{\equiv}CH$ **(c)** $CH_3\overset{..}{O}H_2$

(d) $CH_3\overset{..}{\underset{..}{O}}:$ **(e)** $:\overset{..}{N}H_2$ **(f)** $CH_3\overset{\overset{..}{O}}{\overset{\|}{C}}\overset{..}{\underset{..}{O}}:$

1.47 Assume that none of the following compounds contains double bonds. Can you devise a condensed structural formula for each? (More than one answer may be possible.)

(a) C_3H_6 **(b)** C_2H_4O **(c)** C_4H_8O

1.48 Assuming that each of the following compounds contains a double bond, write a condensed structural formula for each. (There may be more than one correct answer.)

(a) C_3H_6 **(b)** C_2H_4O **(c)** C_4H_8O

1.49 Write an equation to show the heterolytic cleavage (at the most likely position) of each of the following molecules. (Remember to take electronegativities into consideration.)

(a) $(CH_3)_2CHBr$ **(b)** CH_3CH_2Li

(c) $(CH_3)_2CHOCH(CH_3)_2$ **(d)** 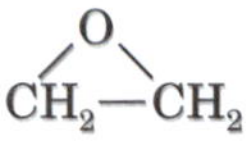 $\overset{\displaystyle O}{\overset{\diagup\diagdown}{CH_2-CH_2}}$

1.50 Arrange the following compounds according to increasing solubility in water, least soluble first:

(a) CH_3CO_2H **(b)** CH_3CH_3 **(c)** $CH_3CH_2OCH_2CH_3$

1.51 Why does compound **A** boil at a much lower temperature than compound **B**? (Use formulas in your answer.)

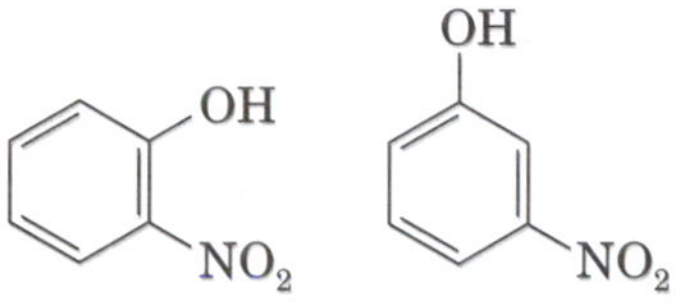

A   **B**

1.52 Diethyl ether, $CH_3CH_2OCH_2CH_3$, and 1-butanol, $CH_3CH_2CH_2CH_2OH$, are equally soluble in water, but the boiling point of 1-butanol is 83° higher than that of diethyl ether. What explanation can you give for these observations? (Use formulas in your answer.)

1.53 Complete the following equations:

(a) 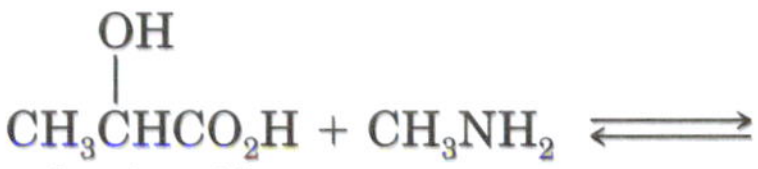 $\underset{\text{lactic acid}}{CH_3\overset{\displaystyle OH}{\overset{|}{C}}HCO_2H}$ + CH_3NH_2 $\rightleftharpoons$

(b) $\underset{\text{limestone}}{CaCO_3}$ + $2\ \underset{\text{acetic acid}}{CH_3CO_2H}$ $\longrightarrow$

(c) 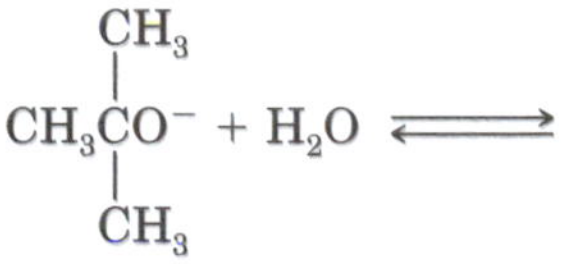 $CH_3\overset{\displaystyle CH_3}{\underset{\displaystyle CH_3}{\overset{|}{\underset{|}{C}}}}O^- + H_2O$ $\rightleftharpoons$

1.54 When a person is exercising strenuously, some of the energy used by the muscles comes from the anaerobic (without oxygen) conversion of glucose to lactic acid (see Problem 1.53a). Within the body, lactic acid undergoes a reaction with either **(a)** bicarbonate ions (HCO_3^-) or **(b)** hydrogen phosphate ions (HPO_4^{2-}). Write equations for each of these reactions.

1.55 Amino acids can undergo an acid–base reaction in their pure state to yield dipolar ions. Such an acid–base reaction of glycine is given below:

glycine   the dipolar ion
                       of glycine

For each of the following amino acids, write an equation that illustrates dipolar-ion formation.

(a) 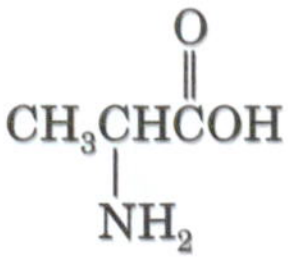 $CH_3\overset{\displaystyle O}{\overset{\|}{\underset{\displaystyle NH_2}{\underset{|}{C}H}}COH}$ **(b)** $HSCH_2\overset{\displaystyle O}{\overset{\|}{\underset{\displaystyle NH_2}{\underset{|}{C}H}}COH}$ **(c)** $CH_3CH_2\overset{\displaystyle CH_3}{\underset{|}{C}}H\overset{\displaystyle O}{\overset{\|}{\underset{\displaystyle NH_2}{\underset{|}{C}H}}COH}$

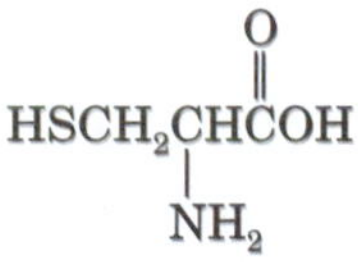

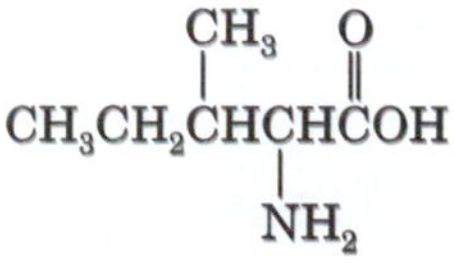

2 ORBITALS AND THEIR ROLE IN COVALENT BONDING

In Chapter 1, we took a brief look at atomic orbitals and at covalent bonding. In this chapter, we will discuss how covalent bonds are produced by the formation of molecular orbitals. Various approaches to molecular orbitals have been developed. The **molecular orbital (MO) theory** provides mathematical descriptions of orbitals, their energies, and their interactions. The **valence–shell electron–pair repulsion (VSEPR) theory** is based on the premise that valence electrons or electron pairs of an atom repel each other. These repulsions can be used to explain observed bond angles and molecular geometry. In the **valence–bond theory,** line-bond formulas are used to describe covalent bonds and their interactions.

Within their limitations, all of these theories are successful and are often in agreement. However, no single one of them is practical to use in all discussions of organic compounds and their reactions. For this reason, we will present some of the highlights of each theory without necessarily attempting to differentiate among them.

You will find that a knowledge of molecular orbitals and molecular shapes is invaluable in a study of organic reactions—whether they occur in a flask in the laboratory or in the cells of an animal.

Properties of Waves

Until 1923, chemists assumed that electrons were nothing more than negatively charged particles whirling about the atomic nuclei. In 1923, Louis de Broglie, a French graduate student, proposed the revolutionary idea that electrons have properties of waves as well as properties of particles. De Broglie's proposal met with skepticism at first, but his idea was the seed that grew into today's quantum-mechanical concept of electron motion and the molecular orbital theory.

Quantum mechanics is a mathematical subject. For our understanding of covalent bonds, we need only the results of quantum-mechanical studies, rather than the mathematical equations themselves. With this in mind, let us survey some of the basic concepts of wave motion as they pertain to the current theories of covalent bonds.

We will begin with some simple **standing waves** (Figure 2.1), the type of wave that results when you pluck a string, like a guitar string, that is fixed at both ends. This type of wave exhibits motion in only one direction. By contrast, the standing waves caused by beating the head of a drum are two-dimensional, and the wave system of an electron is three-dimensional. The height of a standing wave is its **amplitude,** which may be up (positive value) or down (negative value) in relation to the resting position of the string. (Note that the + or − sign of amplitude is a mathematical sign, not an electrical charge.) A position on the wave at which the amplitude is zero is called a **node** and corresponds to a position on the guitar string that does not move as the string vibrates.

Two standing waves can be either **in phase** or **out of phase** in reference to each other. Intermediate states in which waves are only partially in phase are also possible. We can illustrate these terms with two wave systems on two identical vibrating strings. If the positive and negative amplitudes of the two waves correspond to each other, the two waves are *in phase*. If the mathematical signs of the amplitudes are opposite, the waves are *out of phase* (see Figure 2.2).

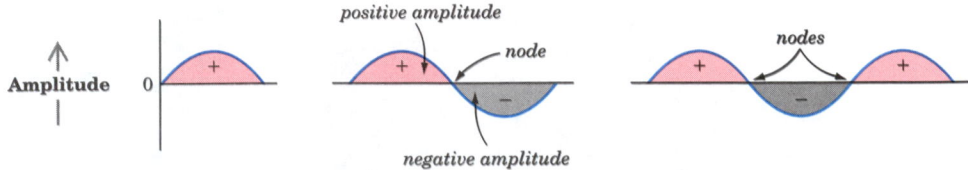

Figure 2.1   Some standing waves of vibrating strings with fixed ends (positive amplitudes shaded red).

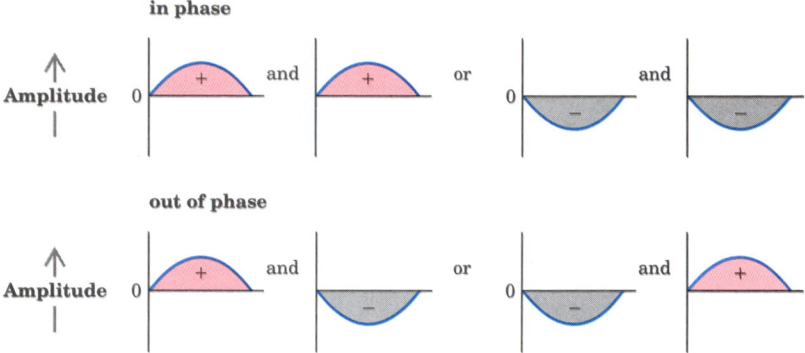

Figure 2.2   Two standing waves may be either in phase or out of phase.

If two in-phase waves on the same string overlap, they **reinforce** each other. The reinforcement is expressed by addition of the mathematical functions of the same sign describing the waves. Conversely, a pair of overlapping waves that are out of phase **interfere** with each other. The process of interference is represented by addition of two mathematical functions of opposite sign. Complete interference results in the cancelling of one wave by another. Partial overlap of two out-of-phase waves gives rise to a node. Figure 2.3 illustrates reinforcement and interference.

Although a three-dimensional electron wave system is more complicated than a one-dimensional string system, the principles are similar. Each atomic orbital of an atom behaves as a wave function and may have a positive or negative amplitude. If the orbital has both positive and negative amplitudes, it has a node. Figure 2.4 depicts the $1s$, $2s$, and $2p$ orbitals, including their signs of amplitude and their nodes.

One atomic orbital can overlap an atomic orbital of another atom. (Mathematically, the wave functions describing each overlapping orbital are added. These calculations are referred to as the **linear combination of atomic orbitals, or LCAO, theory.**) When the overlapping orbitals are in phase, the result is reinforcement and a **bonding molecular orbital.** On the other hand, interaction between atomic orbitals that are out of phase results in in-

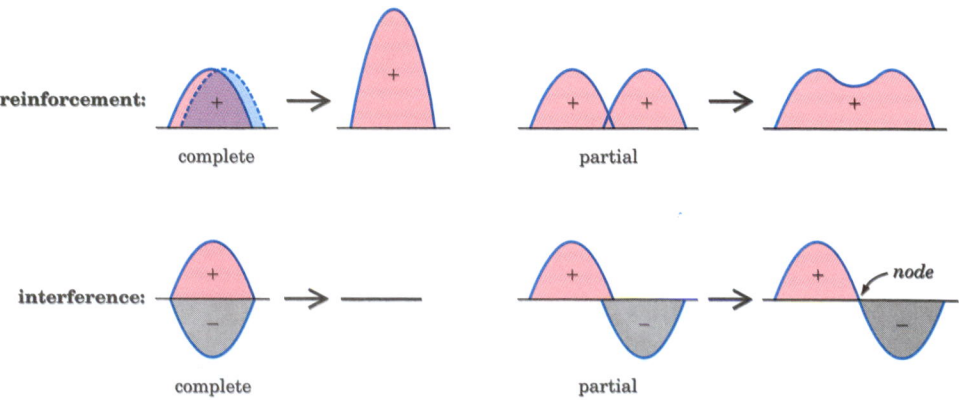

Figure 2.3   Reinforcement and interference of waves.

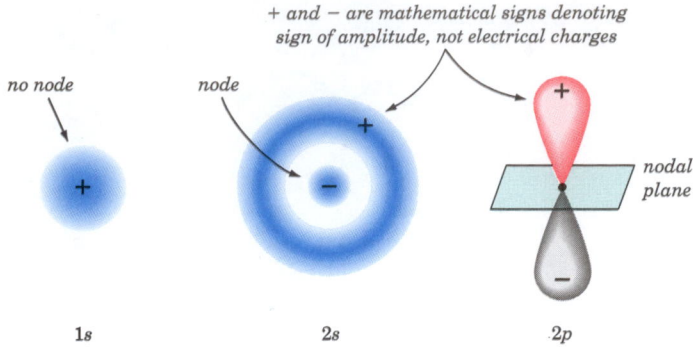

Figure 2.4   The 1s, 2s, and 2p orbitals with their signs of amplitude.

terference, creating a node between the two nuclei. Interference leads to an **antibonding molecular orbital.** We will expand upon these definitions of bonding and antibonding orbitals in Section 2.2B.

Section 2.2
Bonding in Hydrogen

Hydrogen (H_2) is the simplest molecule. We will look at the covalent bond of H_2 in some detail because many features of this bond are similar to those of more complex covalent bonds.

Let us consider two isolated hydrogen atoms, each with one electron in a 1s atomic orbital. As these two atoms begin bond formation, the electron of each atom becomes attracted by the nucleus of the other atom, as well as by its own nucleus. When the nuclei are at a certain distance from each other (the bond length, 0.74 Å for H_2), the atomic orbitals merge, or overlap, to reinforce each other and form a bonding molecular orbital. This molecular orbital encompasses both hydrogen nuclei and contains two paired electrons (one from each H). Both electrons are now equally attracted to both nuclei. Because a large portion of the negatively charged electron density of this new orbital is located between the two positively charged nuclei, repulsions between the nuclei are minimized. This molecular orbital results in the covalent bond between the two hydrogen atoms in H_2 (see Figure 2.5).

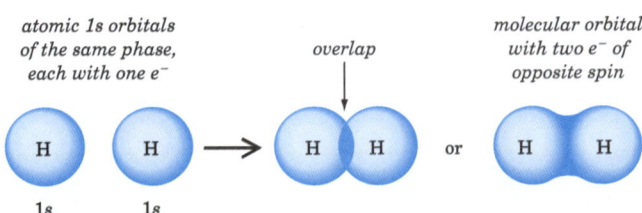

Figure 2.5   The formation of the bonding molecular orbital in H_2 from in-phase 1s orbitals.

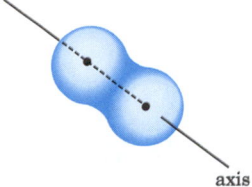

axis

Figure 2.6 The sigma bond of hydrogen is symmetrical about the axis joining the two nuclei.

A. The Sigma Bond

The molecular orbital that bonds two hydrogen atoms together is *cylindrically symmetrical*—that is, symmetrical about a line, or axis, joining the two nuclei. Think of the axis as an axle, and rotate the orbital around this axis. If the appearance of the orbital is not changed by the rotation, the orbital is symmetrical around that axis (see Figure 2.6).

Any molecular orbital that is symmetrical about the axis connecting the nuclei is called a **sigma (σ) molecular orbital; the bond is a sigma bond.** The bond in H_2 is only one of many sigma bonds we will encounter. (We will also encounter molecular orbitals that are not sigma orbitals—orbitals that are not symmetrical about their nuclear axes.)

B. The Bonding Orbital and the Antibonding Orbital

When a pair of waves overlap, they either reinforce each other or interfere with each other. Addition of two in-phase $1s$ atomic orbitals of two H atoms results in reinforcement and produces the σ bonding molecular orbital with a high electron density between the bonded nuclei.

If two waves are out of phase, they interfere with each other. Interference of two out-of-phase atomic orbitals of two hydrogen atoms gives a molecular orbital with a *node between the nuclei.* In this molecular orbital, the probability of finding an electron between the nuclei is *very low.* Therefore, this particular molecular orbital gives rise to a system in which the two nuclei are not shielded by the pair of electrons, and the nuclei repel each other. Principally because of the nuclear repulsion, this system is of *higher energy* than the system of two independent H atoms. This higher-energy orbital is the **antibonding orbital,** in this case, a "sigma star," or $\sigma*$, orbital (the * meaning "antibonding"). Figure 2.7 compares the shapes of the σ and $\sigma*$ orbitals for H_2.

The energy of the H_2 molecule with two electrons in the σ bonding orbital is *lower* by 104 kcal/mol than the combined energy of two separate hydrogen atoms. The energy of the hydrogen molecule with electrons in the $\sigma*$ antibonding orbital, on the other hand, is *higher* than that of two separate hydrogen atoms. These relative energies may be represented by the following diagram:

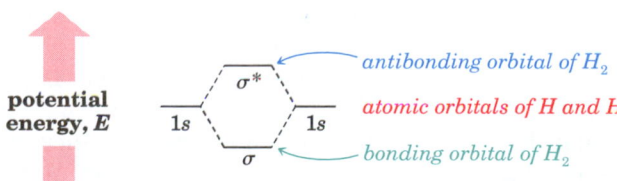

low e⁻ density

σ* orbital
(antibonding)

potential
energy, *E*

interference

out of phase

reinforcement

in phase

high e⁻ density

σ orbital
(bonding)

Figure 2.7   Reinforcement and interference of two 1s orbitals. (The + and − refer to phases of the wave functions, not electrical charges.)

A molecular orbital, like an atomic orbital, can hold no electrons, one electron, or two paired electrons. The two electrons in a hydrogen molecule go into the lowest-energy orbital available, the σ bonding orbital. In the following diagram, we use a pair of arrows (one pointing up and one pointing down) to represent a pair of electrons of opposite spin.

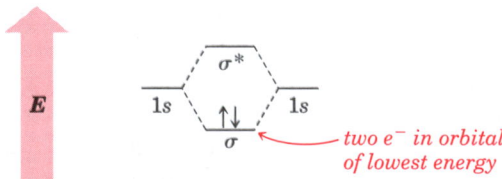

E   1s   σ*   1s   *two e⁻ in orbital of lowest energy*   σ

We said in Chapter 1 that electrons in different atomic orbitals differ in energy because of the various distances of these electrons from the nucleus. The higher energy of a molecule with electrons in an antibonding orbital (compared to the energy of the molecule with electrons in a bonding orbital) does not arise from the electrons being farther from the nuclei. Instead, the higher energy arises from the presence of the node between the nuclei.

For the hydrogen molecule, the σ and σ* orbitals are in the same general region of space. Although two particles of matter cannot occupy the same space at the same time, two orbitals may. Remember, orbitals are not matter, but are simply regions of space where the probability of finding an electron with a particular energy is high.

All molecular orbitals are formed from atomic orbitals. The number of molecular orbitals formed equals the number of atomic orbitals used for their formation. Usually two atomic orbitals are used to form two molecular orbitals. One of these molecular orbitals is called the bonding molecular orbital and the other is called the antibonding molecular orbital. Because antibonding orbitals are of high energy, the electrons are not generally found there. Almost all the chemistry in this text will deal with molecules in the **ground state,** the state in which the electrons are in the lowest-energy orbitals. However, we will encounter a few situations in which energy absorbed by a molecule is used to promote an electron from a low-energy orbital to a higher-energy orbital. A molecule is said to be in an **excited state** when one or more electrons are not in the orbital of lowest energy.

Section 2.3
Some General Features of Bonding and Antibonding Orbitals

Let us summarize some general rules that apply to all molecular orbitals, not only the molecular orbitals of H_2:

1. Any orbital (molecular or atomic) can hold a maximum of two electrons, which must be of opposite spin.

2. The number of molecular orbitals equals the number of atomic orbitals that went into their formation. (For H_2, two $1s$ atomic orbitals yield two molecular orbitals: σ and σ^*.)

3. In the filling of molecular orbitals with electrons, the lowest-energy orbitals are filled first. If two orbitals are degenerate (of equal energies), each gets one electron before either is filled.

Section 2.4
Hybrid Orbitals of Carbon

When a hydrogen atom becomes part of a molecule, it uses its $1s$ atomic orbital for bonding. The situation with the carbon atom is somewhat different. Carbon has two electrons in the $1s$ orbital; consequently, the $1s$ orbital is a filled orbital that is not used for bonding. The four electrons at the *second energy level* of carbon are the bonding electrons.

There are four atomic orbitals at the second energy level: one $2s$ and three $2p$ orbitals. However, carbon does not use these four orbitals in their pure states for bonding. Instead, carbon blends, or **hybridizes,** its four second-level atomic orbitals in one of three different ways for bonding:

1. *sp^3* **hybridization,** used when carbon forms four single bonds.

2. *sp^2* **hybridization,** used when carbon forms one double bond.

3. *sp* hybridization, used when carbon forms a triple bond or *cumulated* double bonds (two double bonds to a single carbon atom).

$$H-C{\equiv}C-H \qquad \underset{sp\ carbons}{\overset{}{}} \qquad \underset{H}{\overset{H}{\diagdown}}C{=}C{=}C\underset{H}{\overset{H}{\diagup}}$$

Why does a carbon atom form compounds with hybrid orbitals rather than with unhybridized atomic orbitals? The answer is that hybridization gives stronger bonds because of greater overlap and therefore results in more-stable, lower-energy molecules. As we discuss each type of hybridization, note that the *shape* of each hybrid orbital is favorable for maximum overlap with an orbital of another atom. Also note that the *geometries* of the three types of hybrid orbitals allow attached groups to be as far from each other as possible, thus minimizing their repulsions for each other.

A. *sp³* Hybridization

In methane (CH_4), the carbon atom has four equivalent bonds to hydrogen. Each C—H bond has a bond length of 1.09 Å and a bond dissociation energy of 104 kcal/mol. The bond angle between each C—H bond is 109.5°. From this experimental evidence alone, it is evident that carbon does not form bonds by means of one *s* atomic orbital and three *p* atomic orbitals. If that were the case, the four C—H bonds would not all be equivalent.

According to present-day theory, these four equivalent bonds arise from complete hybridization of the four atomic orbitals (one *2s* orbital and three *2p* orbitals) to yield four equivalent sp^3 orbitals. For this to be accomplished, one of the *2s* electrons must be promoted to the empty *2p* orbital. This promotion requires energy (about 96 kcal/mol), but this energy is more than regained by the concurrent formation of chemical bonds. The four sp^3 orbitals have equal energies—slightly higher than that of the *2s* orbital, but slightly lower than that of the *2p* orbitals. Each of the sp^3 orbitals contains one electron for bonding.

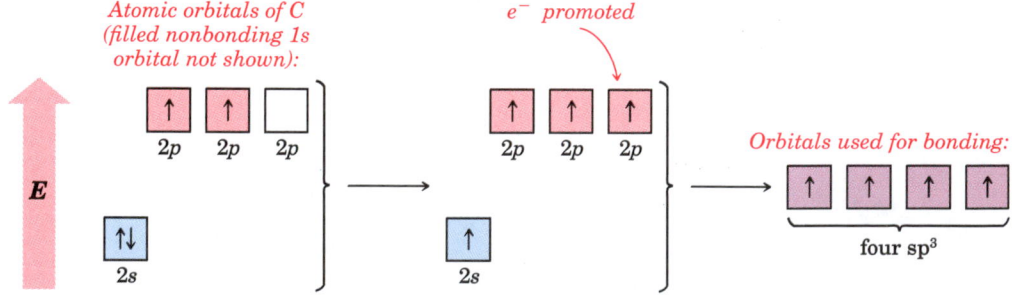

The preceding diagram is called an **orbital diagram.** Each box in the diagram represents an orbital. The relative energies of the various orbitals are signified by the vertical positions of the boxes within the diagram. Electrons are represented by arrows, and the direction of electron spin is indicated by the direction of the arrow.

The sp^3 orbital, which results from a blend of the *2s* and *2p* orbitals, is shaped rather like a bowling pin: it has a large lobe and a small lobe (of opposite amplitude) with a node near the nucleus. Figure 2.8 shows one isolated sp^3 orbital. The small lobe of the hybrid orbital is not used for bonding

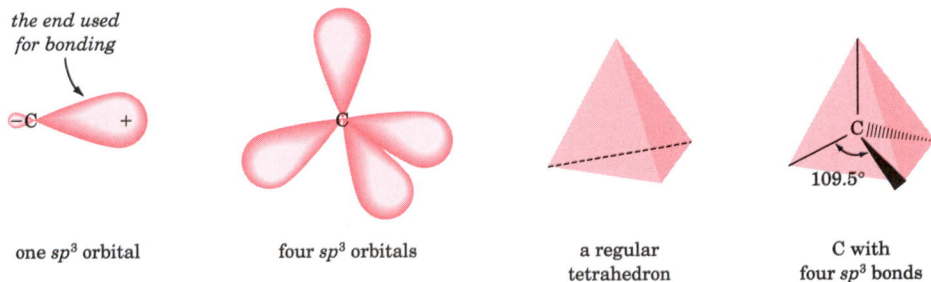

Figure 2.8   The four sp^3-hybrid orbitals of carbon point toward the corners of a regular tetrahedron.

because overlap of the large lobe with another orbital gives more complete overlap and results in a stronger bond.

Four sp^3-hybrid orbitals surround the carbon nucleus. Because of repulsions between electrons in different orbitals, these sp^3 orbitals lie as far apart from each other as possible while still extending away from the same carbon nucleus—the four orbitals point toward the corners of a regular tetrahedron (Figure 2.8). This geometry gives idealized bond angles of 109.5°. An sp^3-hybridized carbon atom is often referred to as a **tetrahedral carbon atom** because of the geometry of its bonds.

When an sp^3-hybridized carbon atom forms bonds, it does so by overlapping each of its four sp^3 orbitals (each with one electron) with orbitals from four other atoms (each orbital in turn containing one electron). In methane (Figures 2.9 and 2.10), each sp^3 orbital of carbon overlaps with a $1s$ orbital of

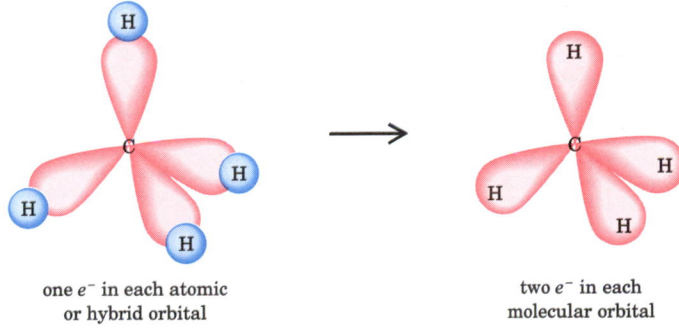

Figure 2.9   Formation of C—H sigma bonds in methane, CH_4. (The small lobes of the sp^3 orbitals are not shown.)

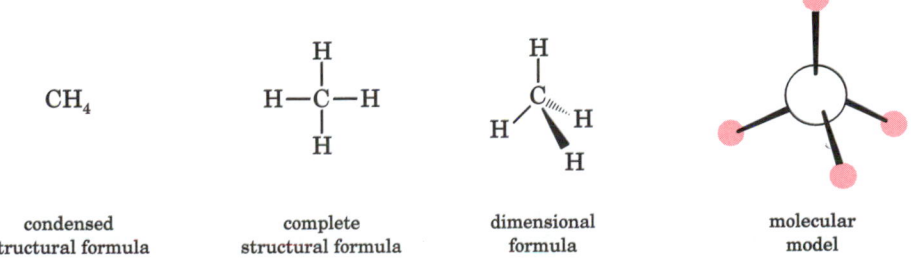

Figure 2.10   Some different ways of representing methane.

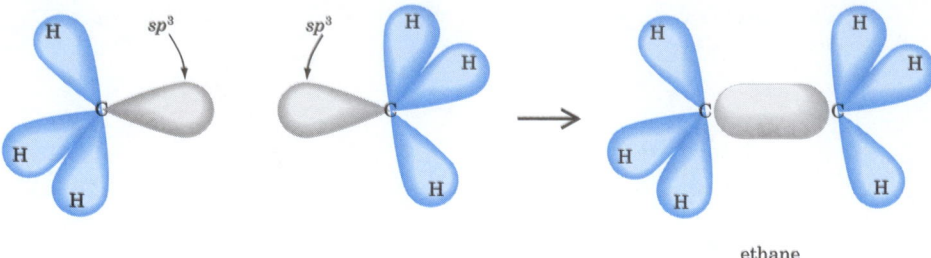

Figure 2.11 Formation of the sp^3–sp^3 sigma bond in ethane, CH_3CH_3.

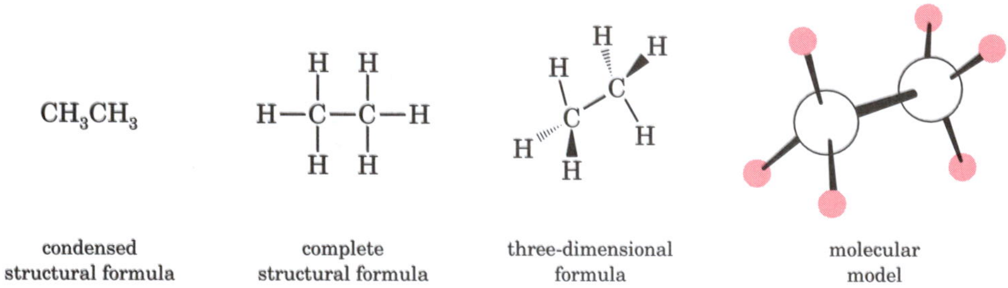

CH_3CH_3

| condensed structural formula | complete structural formula | three-dimensional formula | molecular model |

Figure 2.12 Some different ways of representing ethane.

hydrogen. Each of the resultant sp^3–s molecular orbitals is symmetrical around the axis passing through the nuclei of the carbon and the hydrogen. The covalent bonds between C and H in methane, like the H—H covalent bond, are sigma bonds.

Ethane (CH_3CH_3) contains two sp^3-hybridized carbon atoms. These two carbon atoms form a C—C sigma bond by the overlap of one sp^3 orbital from each carbon (sp^3–sp^3 sigma bond). Each carbon atom has three remaining sp^3 orbitals, and each of these overlaps with a 1s orbital of a hydrogen atom to form a C—H sigma bond. Each carbon atom in ethane is tetrahedral (see Figures 2.11 and 2.12).

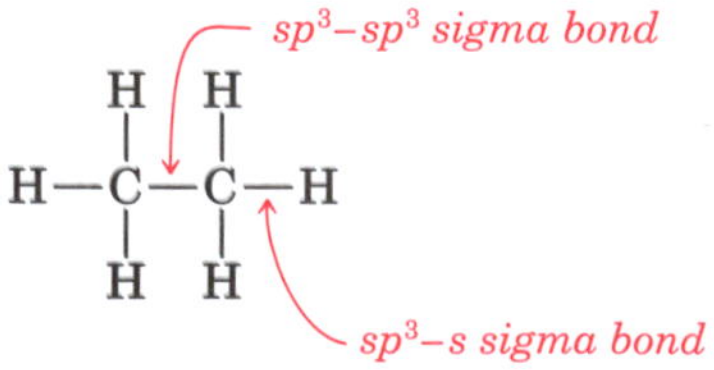

sp³–sp³ sigma bond

sp³–s sigma bond

In any molecule, any carbon atom bonded to four other atoms is in the sp^3-hybrid state, and the four bonds from that carbon are sigma bonds. When carbon is bonded to four other atoms, the sp^3 hybridization allows maximal overlap and places the four attached atoms at the maximum distances from each other. If possible, the sp^3 bond angles are 109.5°. However, other factors, such as dipole–dipole repulsions or the geometry of a cyclic compound, can cause deviations from this ideal bond angle.

Examples of structures with sp³-hybridized carbons (each C has four sigma bonds):

$$H-\underset{\underset{\displaystyle H}{|}}{\overset{\overset{\displaystyle H}{|}}{C}}-H \qquad H-\underset{\underset{\displaystyle H}{|}}{\overset{\overset{\displaystyle H}{|}}{C}}-\underset{\underset{\displaystyle H}{|}}{\overset{\overset{\displaystyle H}{|}}{C}}-H \qquad H-\underset{\underset{\displaystyle H}{|}}{\overset{\overset{\displaystyle H}{|}}{C}}-O-H$$

SAMPLE PROBLEM

Give the complete structural formula (showing all atoms and bonds) for propane ($CH_3CH_2CH_3$). Which types of orbitals overlap to form each bond?

Solution

each C—H bond is sp³–s

$$H-\underset{\underset{\displaystyle H}{|}}{\overset{\overset{\displaystyle H}{|}}{C}}-\underset{\underset{\displaystyle H}{|}}{\overset{\overset{\displaystyle H}{|}}{C}}-\underset{\underset{\displaystyle H}{|}}{\overset{\overset{\displaystyle H}{|}}{C}}-H$$

sp³–sp³

STUDY PROBLEM

2.1 Write the complete structural formula for each of the following compounds, showing each bond. Which types of orbitals overlap to form each bond?

(a) $CH_3CH_2CH_3$ **(b)**

B. *sp²* Hybridization

When carbon is bonded to another atom by one double bond, the carbon atom is in the *sp²*-hybrid state.

Examples of compounds with sp²-hybridized carbons:

$$\underset{H}{\overset{H}{>}}C=C\underset{H}{\overset{H}{<}} \qquad \underset{H}{\overset{H}{>}}C=O$$

ethylene   formaldehyde

To form *sp²* bonding orbitals, carbon hybridizes its 2s orbital with only two of its 2p orbitals. One *p* orbital remains unhybridized on the carbon atom. Because three atomic orbitals are used to form the *sp²* orbitals, three *sp²*-hybrid orbitals result. Each *sp²* orbital has a shape similar to that of an *sp³* orbital and contains one electron that can be used for bonding.

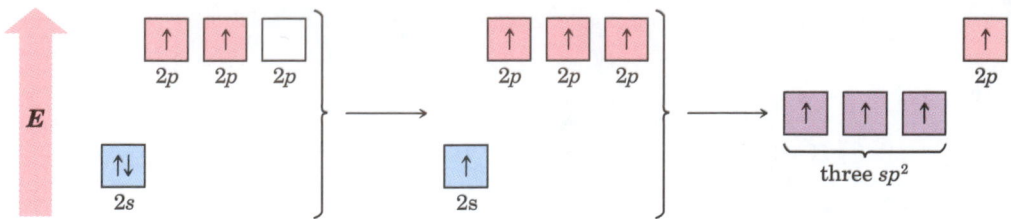

The three sp^2 orbitals around a carbon nucleus lie as far apart from one another as possible, that is, the sp^2 orbitals lie in a plane with angles of 120° (ideally) between them. An sp^2-hybridized carbon atom is said to be a **trigonal** (triangular) carbon. Figure 2.13 shows a carbon atom with three sp^2 orbitals and the one unhybridized p orbital, which is perpendicular to the sp^2 plane.

In ethylene (CH_2=CH_2), two sp^2-hybridized carbons are joined by a sigma bond formed by the overlap of one sp^2 orbital from each carbon atom. (This sigma bond is one of the bonds of the double bond.) Each carbon atom still has two sp^2 orbitals left for bonding with hydrogen. (Each carbon atom also has a p orbital, which is not shown in the following structure.)

Planar sigma-bond structure of ethylene (p orbitals not shown):

$$\begin{array}{ccc} H & 121° & H \\ & \diagdown \quad \diagup & \\ & C-C & 118° \\ & \diagup \quad \diagdown & \\ H & & H \end{array}$$

What of the remaining p orbital on each carbon? Each p orbital has two lobes, one above the plane of the sigma bonds and the other (of equal amplitude, but opposite phase) below the plane. Each p orbital contains one electron. If these p electrons become paired in a bonding molecular orbital, then the energy of the system is lowered. Because the p orbitals lie side by side in the ethylene molecule, the *ends* of the orbitals cannot overlap, as they do in sigma-bond formation. Instead, the two p orbitals overlap their *sides* (see Figure 2.14). The result of this side-to-side overlap is the **pi (π) bond**—a bond joining the two carbons and located above *and* below the plane of the sigma bonds. The pi bond is the second bond of the double bond.

Any carbon atom that is bonded to three other atoms is in the sp^2-hybrid state. In stable compounds, the p orbital on the sp^2-hybridized carbon must

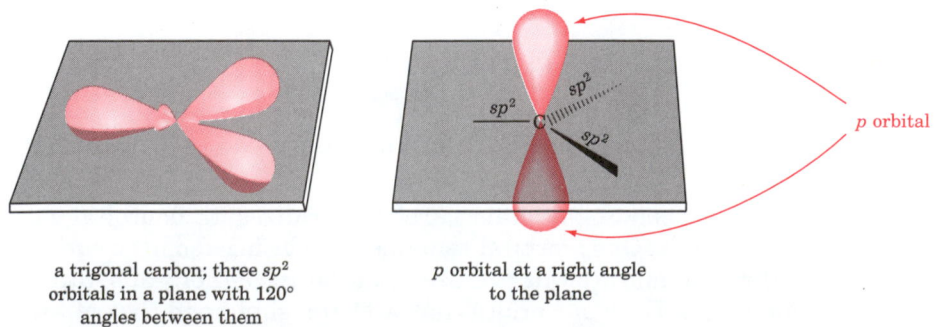

a trigonal carbon; three sp^2 p orbital at a right angle
orbitals in a plane with 120°   to the plane
angles between them

Figure 2.13 Carbon in the sp^2-hybrid state.

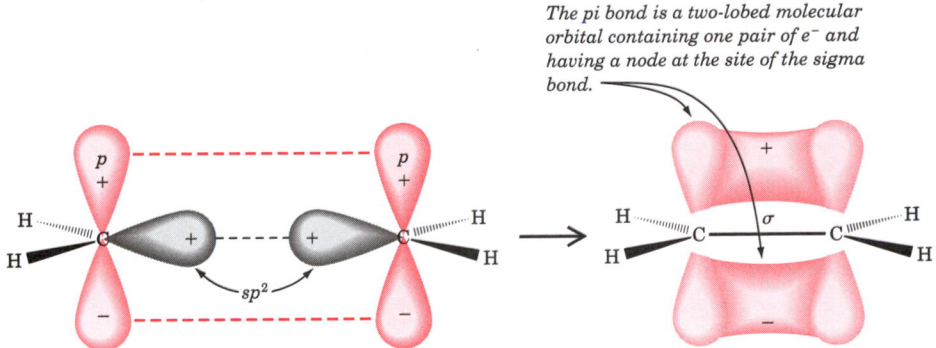

The pi bond is a two-lobed molecular orbital containing one pair of e⁻ and having a node at the site of the sigma bond.

Figure 2.14   Formation of the sp^2–sp^2 sigma bond and the p–p pi bond in ethylene, $CH_2{=}CH_2$. (The + and − refer to phases of the wave functions, not electrical charges.)

overlap with a p orbital of an adjacent atom, which can be another carbon atom or an atom of some other element.

$$H_2C{=}CH_2 \qquad H_2C{=}O \qquad H{-}C\!\!\begin{array}{c} O \\[-2pt] OH \end{array}$$

sp^2-hybridized carbons

C. Some Features of the Pi Bond

Each p orbital contributing to a pi bond has two lobes and has a node at the nucleus. It is not surprising that the pi orbital also is two-lobed and has a node. Unlike a sigma orbital, a pi orbital is not cylindrically symmetrical. However, just like any molecular orbital, a pi orbital can hold a maximum of two paired electrons.

A $2p$ orbital of carbon is of slightly higher energy than an sp^2 orbital. For this reason, a pi bond, which is formed from two $2p$ orbitals, has slightly higher energy and is slightly less stable than an sp^2–sp^2 sigma bond. The bond dissociation energy of the sigma bond of ethylene's carbon–carbon double bond is estimated to be 95 kcal/mol, while that of the pi bond is estimated to be only 68 kcal/mol.

The more-exposed pi electrons are more vulnerable to external effects than are electrons in sigma bonds. The pi bond is polarized more easily—we might say that the pi electrons are more mobile. The pi electrons are more easily promoted to a higher-energy (antibonding) orbital. Also, they are more readily attacked by an outside atom or molecule. What does this vulnerability mean in terms of the chemistry of pi-bonded compounds? In a molecule, the pi bond is a site of chemical reactivity.

Another property of the pi bond is that it holds the doubly bonded carbon atoms, and the atoms to which these carbons are bonded, in a rigid shape. Rotation of the doubly bonded carbon atoms and their substituent groups around the double bond would require that the pi bond first be broken (see Figure 2.15). In chemical reactions, molecules may have sufficient energy (about 68 kcal/mol) for this bond to break. In a flask at room temperature,

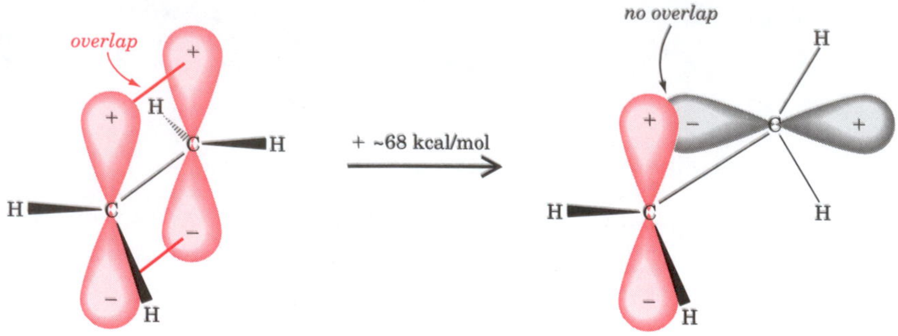

Figure 2.15 The portion of a molecule surrounding a pi bond is held in a planar structure unless enough energy is supplied to break the pi bond.

however, molecules do not have enough energy for this bond breakage to occur. (Approximately 20 kcal/mol is the maximum energy available to molecules at room temperature.) The significance of the rigidity of pi bonds will be discussed in Chapter 4.

In a structural formula, a double bond is indicated by two identical lines. Keep in mind that the double bond is not simply two identical bonds, but that the double line represents one relatively stable sigma bond and one relatively reactive pi bond.

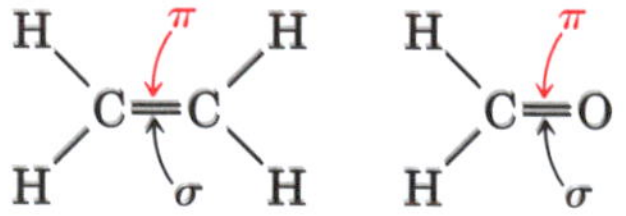

SAMPLE PROBLEM

What type of overlap (sp^3–s, for example) is present in each bond of $CH_3CH{=}CH_2$? What is each bond angle (approximately)?

Solution

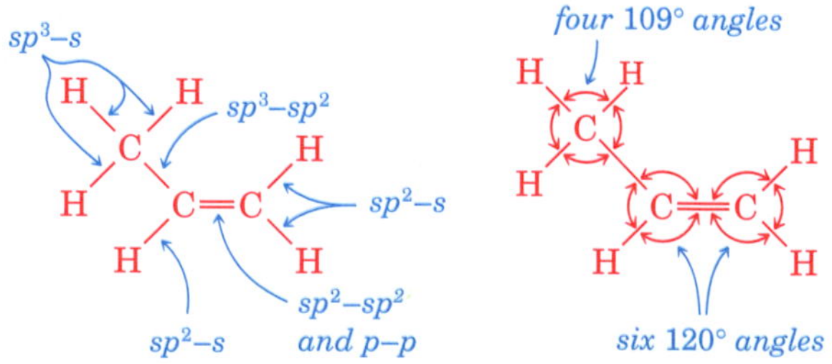

STUDY PROBLEMS

2.2 Give the complete structural formula for each of the following compounds. Indicate which types of orbitals are used to form each bond.

(a) $CH_3CH_2CH{=}CH_2$ **(b)**

2.3 For compound (b) in Problem 2.2, draw the structure showing the pi bonds with their correct geometry in reference to the sigma bonds. (Use lines to represent the sigma bonds.)

D. The Bonding and Antibonding Orbitals of Ethylene

The carbon–carbon sigma bond in ethylene results from overlap of two sp^2 orbitals. Because *two* sp^2 orbitals form this bond, *two* molecular orbitals result. The other molecular orbital, arising from interference between the two sp^2 orbitals, is the antibonding σ^* orbital. This antibonding orbital is similar to the σ^* orbital of hydrogen: it has a node between the two carbon nuclei and is of high energy. The two electrons of the C—C sigma bond in ethylene are usually found in the lower-energy σ orbital.

The π molecular orbital for the pi bond is formed by the side-to-side overlap of the two atomic p orbitals, one from each carbon atom. Because two atomic p orbitals are used for molecular orbital formation, two molecular orbitals result. One of these is the π-bonding orbital that arises from the overlap of two in-phase p orbitals. The other orbital is the π^* antibonding orbital, which arises from interference between two p orbitals of opposite phase. These two orbitals are frequently designated π_1 for the bonding orbital and π_2^* for the antibonding orbital. Some molecules contain several π orbitals, which we number in order of increasing energy; the subscript numbers are useful for differentiating the π orbitals. (We do not usually use subscript numbers for σ orbitals because σ^* orbitals are usually of minor importance to the organic chemist.)

Figure 2.16 shows the orbital representations of the π_1 and π_2^* orbitals of ethylene. Note that besides the node at the σ-bond site, the π_2^* orbital has an additional node *between the two carbon nuclei*. A minimum of the pi electron density is located between the nuclei in this orbital; thus, the π_2^* orbital is of higher energy than the π_1 orbital.

Recall that lower energy means greater stability. Therefore, in the most stable state, the ground state, the electrons occupy the lowest-energy orbital. In the ground state of ethylene, two of the four electrons of the C—C double bond are in the σ molecular orbital and the other two electrons are in the π_1 molecular orbital. The σ^* and π_2^* molecular orbitals are empty.

The following diagram comparing energies of the σ, σ^*, π_1, and π_2^* orbitals shows that the σ^* orbital is of higher energy than the π_2^* orbital. The amount of energy required to promote an electron from the σ orbital to the σ^* orbital is therefore greater than the energy required to promote a π electron to the π^* orbital.

Ground state of C—C in ethylene:

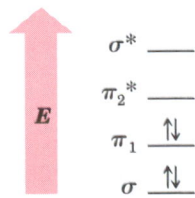

Why is more energy required to promote a sigma electron than a pi electron? Sigma-bonding electrons are found close to the nucleus and, for the

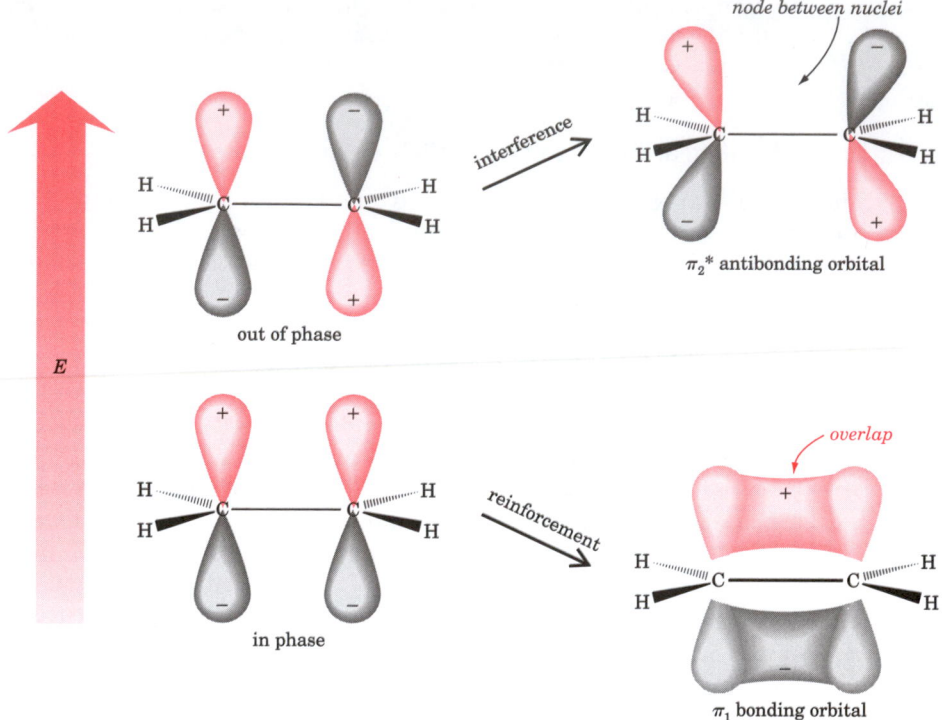

Figure 2.16 The π_1 bonding orbital and the $\pi_2{}^*$ antibonding orbital of ethylene.

most part, directly between the nuclei. Promotion of one of these sigma electrons results in severe nuclear repulsion. However, the pi-bonding electrons are farther from the nucleus. Promotion of one of these pi electrons does not result in such severe nuclear repulsions. Also, when a pi electron is promoted, the nuclei are still shielded from each other by sigma-bonding electrons.

Because of the large amount of energy required to promote a sigma electron, electron transitions of the $\sigma \rightarrow \sigma^*$ type are rare and relatively unimportant to the organic chemist. However, $\pi \rightarrow \pi^*$ electron transitions, which require less energy, are important. For example, $\pi \rightarrow \pi^*$ transitions are responsible for vision, a topic that will be mentioned in Chapter 22, and for the energy capture needed for photosynthesis.

Two possible excited states of C=C in ethylene:

$$\sigma \rightarrow \sigma^*, \text{ higher } \Delta E \left\{ \begin{array}{cc} \sigma^* \ \underline{\ \downarrow\ } & \sigma^* \ \underline{\ \ \ } \\ \pi_2{}^* \ \underline{\ \ \ } & \pi_2{}^* \ \underline{\ \downarrow\ } \\ \pi_1 \ \underline{\ \uparrow\downarrow\ } & \pi_1 \ \underline{\ \uparrow\ } \\ \sigma \ \underline{\ \uparrow\ } & \sigma \ \underline{\ \uparrow\downarrow\ } \end{array} \right\} \pi_1 \rightarrow \pi_2{}^*, \text{ lower } \Delta E$$

E. *sp* Hybridization

When a carbon atom is joined to only two other atoms, as in acetylene (CH≡CH), its hybridization state is *sp*. One 2s orbital blends with only one

$2p$ orbital to form two sp-hybrid orbitals. Two unhybridized $2p$ orbitals remain, each with one electron.

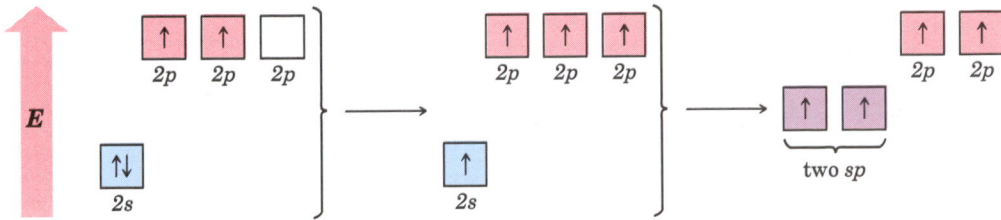

The two sp orbitals lie as far apart as possible, in a straight line with an angle of 180° between them. The p orbitals are perpendicular to each other and to the line of the sp orbitals (see Figure 2.17).

In CH≡CH, the two carbon atoms are joined by an sp–sp sigma bond. Each carbon is also bonded to a hydrogen atom by an sp–s sigma bond. The two p orbitals of one carbon then overlap with the two p orbitals of the other carbon to form *two* pi bonds. One pi bond is above and below the line of the sigma bonds, as shown in Figure 2.18; the other pi bond is located in front and back.

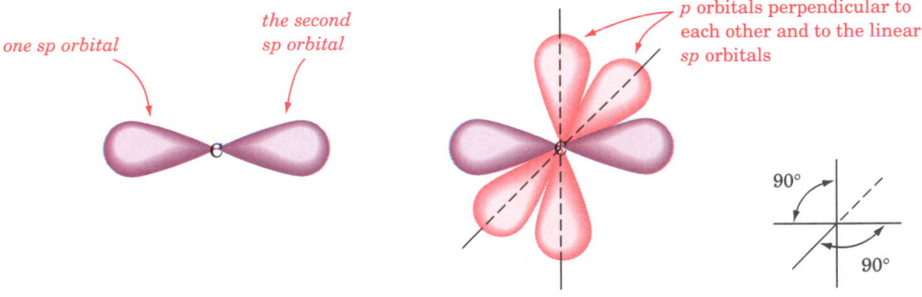

Figure 2.17   Carbon in the sp-hybrid state.

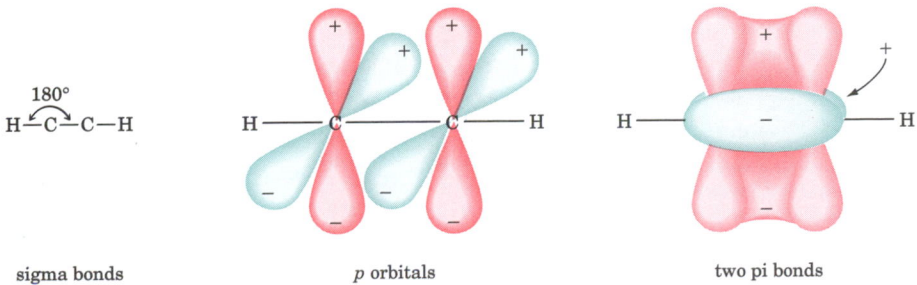

Figure 2.18   Bonding in acetylene, CH≡CH.

As you might guess, the chemical reactions of a compound containing a triple bond are not too different from those of a compound containing a double bond. Instead of one pi bond, there are two.

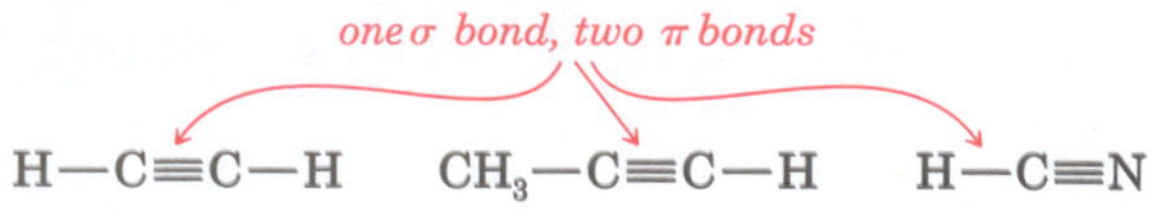

one σ bond, two π bonds

$$H-C\equiv C-H \quad CH_3-C\equiv C-H \quad H-C\equiv N$$

STUDY PROBLEM

2.4   What type of overlap is present in each carbon–carbon bond of $CH_3-C\equiv CH$? What are the approximate bond angles around each carbon?

F. Effects of Hybridization on Bond Lengths

A $2s$ orbital is of slightly lower energy than a $2p$ orbital. On the average, $2s$ electrons are found closer to the nucleus than $2p$ electrons. For this reason, a hybrid orbital with a greater proportion of s character is of lower energy and is closer to the nucleus than a hybrid orbital with less s character.

An sp-hybrid orbital is one-half s and one-half p; we may say that the sp orbital has 50% s character and 50% p character. At the other extreme is the sp^3 orbital, which has only one-fourth, or 25%, s character.

	Hybridization of Carbon	Percent s Character
$CH\equiv CH$	sp	50
$CH_2=CH_2$	sp^2	$33\frac{1}{3}$
CH_3CH_3	sp^3	25

Because the sp orbital contains more s character, it is closer to its nucleus; it forms shorter and stronger bonds than the sp^3 orbital. The sp^2 orbital is intermediate between sp and sp^3 in its s character and in the length and strength of the bonds it forms.

Table 2.1 shows the differences in bond lengths among the three C—C and C—H bond types. Note that the sp–s CH bond in $CH\equiv CH$ is the shortest, while the sp^3–s CH bond is the longest. We see an even wider variation in the C—C bonds because these bond lengths are affected by the *number* of bonds joining the carbon atoms as well as by the hybridization of the carbon atoms.

Table 2.1   Effect of hybridization on bond length

longest bonds from C   *shortest bonds from C*

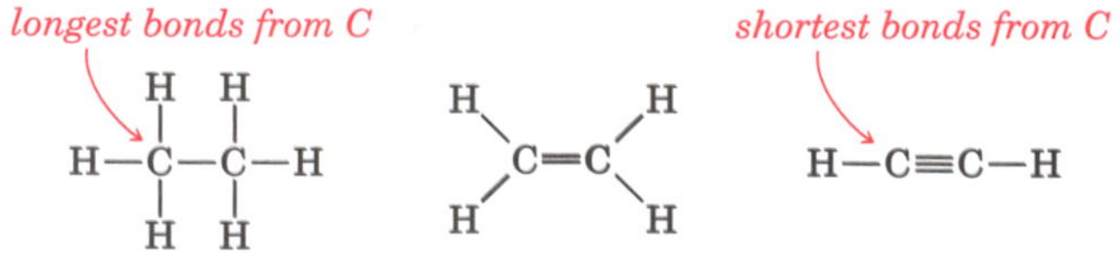

percent s character:	25	$33\frac{1}{3}$	50
C—C bond length:	1.54 Å	1.34 Å	1.20 Å
C—H bond length:	1.09 Å	1.08 Å	1.06 Å

STUDY PROBLEM

2.5   For each of the following structures, list the numbered bonds in order of increasing bond length (shortest bond first):

(a) $CH_3$①$-CH_2CH$②$=CH_2$ **(b)** $H-\overset{\overset{\displaystyle H}{|}}{\underset{\underset{\displaystyle H}{|}}{C}}-CH_2\overset{\overset{\displaystyle H}{|}}{C}=CH_2$ **(c)**

G. Summary of the Hybrid Orbitals of Carbon

1. When a carbon atom is bonded to *four other atoms,* the bonds from the carbon atoms are formed from four equivalent sp^3 orbitals. The sp^3-hybridized carbon is **tetrahedral.**

Examples: CH_4, $CHCl_3$,

tetrahedral

2. When carbon bonds to *three other atoms,* the bonds from the carbon atom are formed from three equivalent sp^2 orbitals with one p orbital remaining. The sp^2 orbitals form three sigma bonds; the remaining p orbital forms a pi bond. The sp^2-hybridized carbon is **trigonal.**

Examples: $CH_3CH=CH_2$, $H_2C=O$,

trigonal

3. When carbon bonds to *two other atoms,* the bonds from the carbon atom are formed from two equivalent sp orbitals, with two p orbitals remaining. The two p orbitals overlap with two p orbitals of another atom to form two pi bonds. The sp orbitals form two equivalent and **linear** sigma bonds.

one σ, two π

$H-C\equiv C-H$ *Examples:* $CH_3C\equiv CH$, $HC\equiv N$

linear   *sp carbons*

Section 2.5
Functional Groups

Although sp^3–sp^3 carbon–carbon bonds and sp^3–s carbon–hydrogen bonds are common to almost all organic compounds, surprisingly, these bonds do not usually play a major role in organic reactions. For the most part, it is the presence of either pi bonds or other atoms in an organic structure that confers reactivity. A site of chemical reactivity in a molecule is called a **functional group.** A pi bond or an electronegative (or electropositive) atom in an organic molecule can lead to chemical reaction; either one of these is considered a functional group or part of a functional group.

Some functional groups (marked with color):

Br CH$_2$ CH=C H$_2$ CH$_3$CH$_2$NH$_2$ CH$_3$CH$_2$OH

STUDY PROBLEM

2.6   Circle the functional groups in the following structures:

    (a)  CH$_2$=CHCH$_2$CH (with O double bonded above the last CH)   (b) 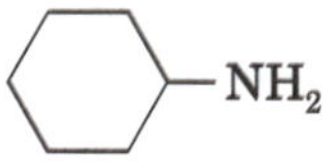 (cyclohexane ring)—NH$_2$ (c)  (cyclohexenone ring with O at top and OH at bottom)

Compounds with the same functional group tend to undergo similar chemical reactions. For example, each of the following series of compounds contains a **hydroxyl group** (—OH). These compounds all belong to the class of compounds called **alcohols,** and all undergo similar reactions.

Some alcohols:

    CH$_3$CH$_2$OH   (CH$_3$)$_3$COH   (cyclohexane ring)—OH

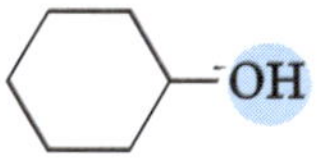

Because of the similarities in reactivity among compounds with the same functional group, it is frequently convenient to use a general formula for a series of these compounds. We usually use R to represent an **alkyl group,** a group that contains only sp^3-hybridized carbon atoms plus hydrogens. By this technique, we can represent an alcohol as ROH. Table 2.2 shows some functional groups and some classes of compounds with generalized formulas. A more extensive table is shown inside the front cover of this book.

R—   means an alkyl group, such as CH$_3$—, CH$_3$CH$_2$—, or (cyclohexane ring)—

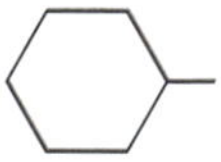

ROH   means an alcohol, such as CH$_3$OH, CH$_3$CH$_2$OH, or (cyclohexane ring)—OH

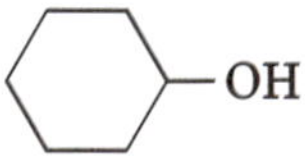

Table 2.2   Some functional groups and compound classes

Functional Group		Class of Compound	
Structure	Name	General Formula	Class Name
C=C	double bond	R$_2$C=CR$_2$	alkene
C≡C	triple bond	RC≡CR′[a]	alkyne
—NH$_2$	amino group	RNH$_2$	amine
—OH	hydroxyl group	ROH	alcohol
—OR	alkoxyl group	R′OR[a]	ether

[a] R′ refers to an alkyl group that may be the same as or different from R.

SAMPLE PROBLEM

The following compounds are all **amines.** Write a general formula for an amine.

$$CH_3NH_2 \quad CH_3CH_2NH_2 \quad (CH_3)_2CHCH_2NH_2$$

Solution
RNH_2

STUDY PROBLEMS

2.7 The following compounds are all **carboxylic acids.** Write the general formula for a carboxylic acid.

$$CH_3CO_2H \quad CH_3CH_2CH_2CO_2H \quad \langle \hexagon \rangle\!-\!CO_2H$$

2.8 The symbol $R\overset{\overset{\textstyle O}{\|}}{C}H$ is a general formula for an aldehyde. Write the structure for **(a)** a one-carbon aldehyde, **(b)** a five-carbon aldehyde, and **(c)** an aldehyde that is bonded to a six-membered ring.

Section 2.6

Hybrid Orbitals of Nitrogen and Oxygen

A. Amines

Many important functional groups in organic compounds contain nitrogen or oxygen. Because both nitrogen and carbon have only s and p valence electrons, the atomic orbitals of nitrogen hybridize in a manner very similar to those of carbon:

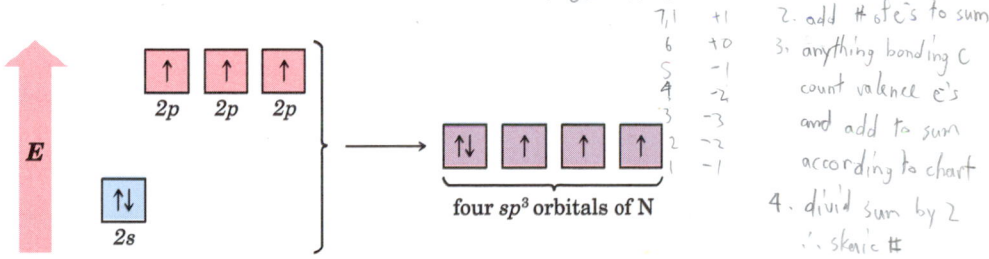

1. count valence e's   if 9   +2   1. find central atom
                            7,1   +1   2. add # of e's to sum
                            6   +0   3. anything bonding C
                            5   -1   count valence e's
                            4   -2   and add to sum
                            3   -3   according to chart
                            2   -2   4. divid sum by 2
                            1   -1   ∴ steric #

As these orbital diagrams show, nitrogen can hybridize its four second-level atomic orbitals to four equivalent sp^3 bonding orbitals. However, note one important difference between nitrogen and carbon. While carbon has four electrons to distribute in four sp^3 orbitals, nitrogen has *five* electrons to distribute in four sp^3 orbitals. One of the sp^3 orbitals of nitrogen is filled with a pair of electrons, and nitrogen can form compounds with only three covalent bonds to other atoms.

A molecule of ammonia contains an sp^3-hybridized nitrogen atom bonded to three hydrogen atoms. An **amine** molecule has a similar structure: an sp^3 nitrogen atom bonded to one or more carbon atoms. In either ammonia or an amine, the nitrogen has one orbital filled with a pair of unshared valence electrons. Figure 2.19 shows the geometry and the filled orbitals of ammonia and two amines: the similarities in structure are evident in this figure.

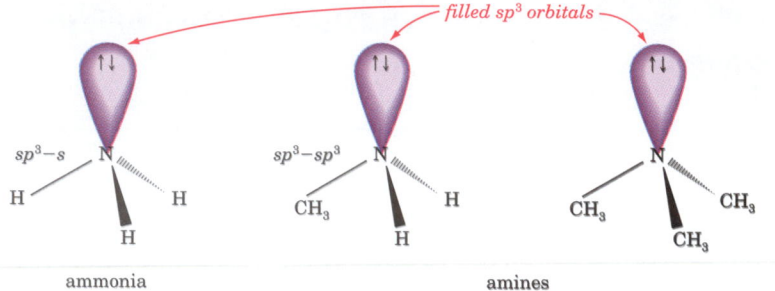

Figure 2.19   Bonding in ammonia and in two amines.

The unshared pair of electrons in the filled orbital on the nitrogen of ammonia and amines allows these compounds to act as bases. When an amine is treated with an acid, these unshared electrons are used to form a sigma bond with the acid. The product is an **amine salt.**

$$CH_3-\overset{\cdot\cdot}{N}-CH_3 + H-\overset{\cdot\cdot}{\underset{\cdot\cdot}{Cl}}: \longrightarrow CH_3-\overset{\overset{H}{|}}{\underset{\underset{CH_3}{|}}{N^+}}-CH_3 \quad :\overset{\cdot\cdot}{\underset{\cdot\cdot}{Cl}}:^-$$

an amine   an amine salt

Like carbon, nitrogen is also found in organic compounds in the sp^2- and sp-hybrid states. Again, the important difference between nitrogen and carbon is that one of the orbitals of nitrogen is filled with a pair of unshared electrons.

sp² hybridized   *sp hybridized*

$$CH_3CH=\overset{\cdot\cdot}{\underset{CH_3}{N}} \qquad CH_3CH_2C\equiv N:$$

STUDY PROBLEM

2.9 Give the complete structural formula for each of the following compounds and tell which types of orbitals overlap to form each bond:

(a) $(CH_3)_2NCH_2NH_2$ (b) $H_2N\overset{\overset{NH}{\|}}{C}NHCH_2CN$ (c) N⬡NH

B. Water, Alcohols, and Ethers

Like carbon and nitrogen, oxygen forms bonds with sp^3-hybrid orbitals. Because oxygen has six bonding electrons, it forms two covalent bonds and has two filled orbitals.

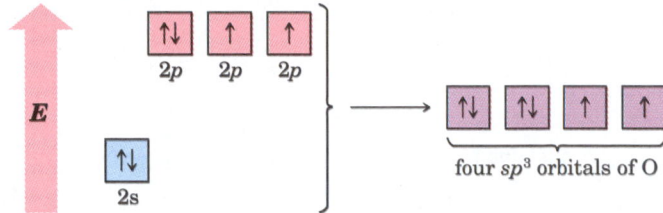

There are a number of classes of organic compounds that contain sp^3-hybridized oxygen atoms. For the present, let us consider just two, alcohols and ethers: ROH and ROR′. The bonding to the oxygen in alcohols and ethers is directly analogous to the bonding in water. In each case, the oxygen is sp^3 hybridized and has two pairs of unshared valence electrons, as is shown in Figure 2.20.

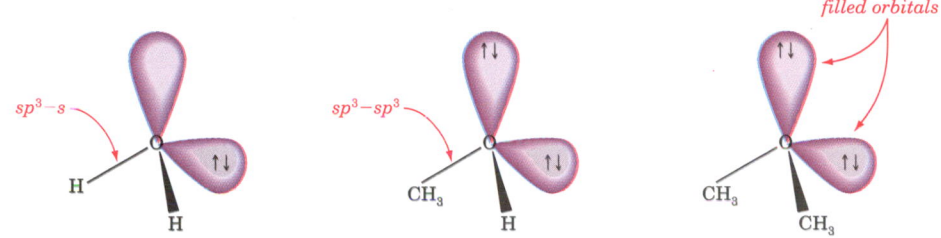

Figure 2.20   Bonding in water, in the alcohol CH_3OH, and in the ether CH_3OCH_3.

C. Carbonyl Compounds

The **carbonyl group** (C=O) contains an sp^2-hybridized carbon atom connected to an oxygen atom by a double bond. It is tempting to think that a carbonyl oxygen is in the sp^2-hybrid state just as the carbonyl carbon is; however, chemists are not truly sure of the hybridization of a carbonyl oxygen because there is no bond angle to measure.

The geometry of a carbonyl group is determined by the sp^2-hybridized carbon. The carbonyl group is *planar* around the trigonal sp^2 carbon. The carbon–oxygen bond contains a pair of *exposed pi electrons*. The oxygen also has *two pairs of unshared electrons*. Figure 2.21 shows the geometry of a carbonyl group.

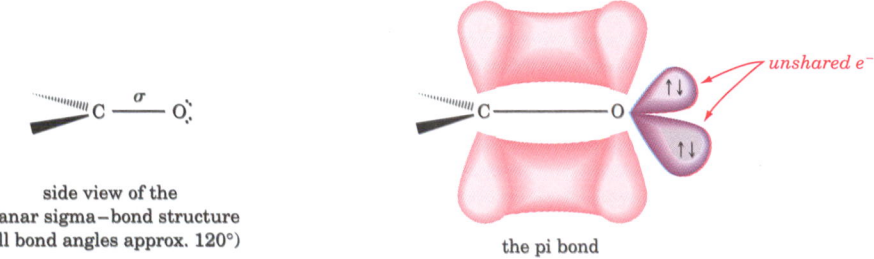

side view of the
planar sigma–bond structure
(all bond angles approx. 120°)

the pi bond

Figure 2.21   Bonding in the carbonyl group.

The carbonyl group:

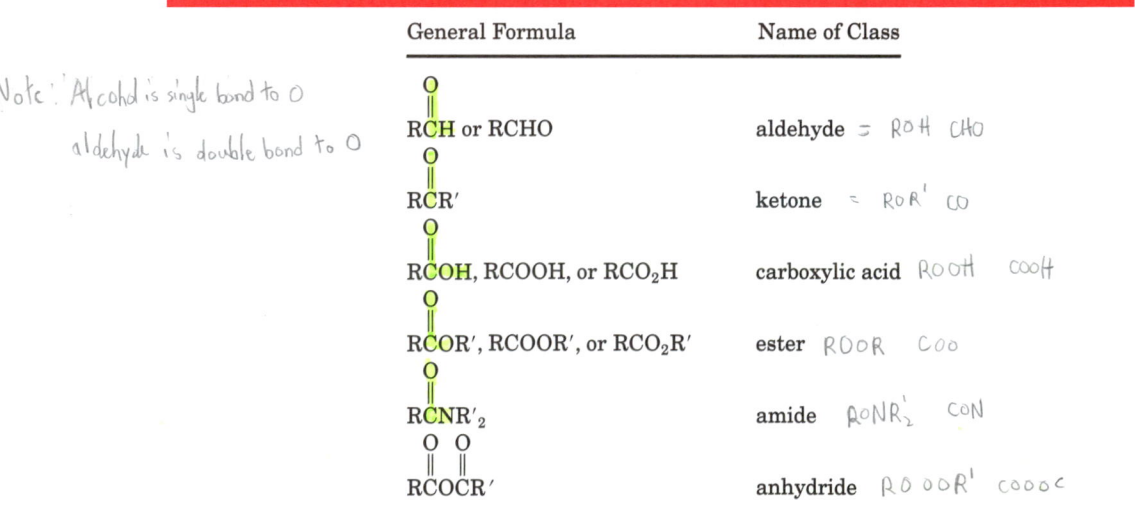

trigonal and planar   unshared pairs of e⁻   polar

The carbonyl group is more polar than the C—O group in an alcohol or ether. The probable reason for this enhanced polarity is that the pi electrons are more easily drawn toward the electronegative oxygen than are the C—O sigma electrons.

The carbonyl group is part of a variety of functional groups. The functional group and class of compound are determined by the other atoms bonded to the carbonyl carbon. If one of the atoms bonded to the carbonyl carbon is a hydrogen, then the compound is an **aldehyde.** If two carbons are bonded to the carbonyl carbon, then the compound is a **ketone.** In Chapter 1 you already encountered **carboxylic acids,** in which an OH group is bonded to the carbonyl group. Table 2.3 lists a few classes of compounds containing carbonyl groups.

Table 2.3	Some compound classes that contain carbonyl groups
General Formula	Name of Class
$\overset{O}{\overset{\|}{R C H}}$ or RCHO	aldehyde = ROH CHO
$\overset{O}{\overset{\|}{R C R'}}$	ketone = ROR' CO
$\overset{O}{\overset{\|}{R C O H}}$, RCOOH, or RCO_2H	carboxylic acid ROOH COOH
$\overset{O}{\overset{\|}{R C O R'}}$, RCOOR', or RCO_2R'	ester ROOR COO
$\overset{O}{\overset{\|}{R C N R'_2}}$	amide RONR'₂ CON
$\overset{O\ \ O}{\overset{\|\ \|}{R C O C R'}}$	anhydride RO OOR' COOOC

(handwritten margin note:) Note: Alcohol is single bond to O
aldehyde is double bond to O

SAMPLE PROBLEM

Classify each of the following compounds as an aldehyde, a ketone, or a carboxylic acid:

(handwritten labels: carboxilylic acid, aldehyde, ketone, aldehyde)

(a) 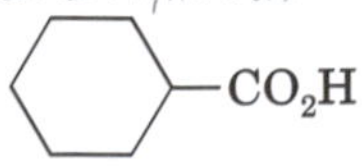 ⬡—CO_2H **(b)** $CH_3CH_2\overset{O}{\overset{\|}{C}}H$ **(c)** 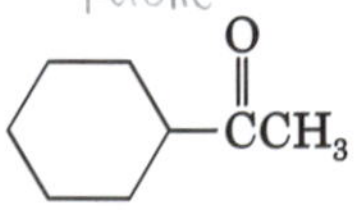 ⬡—$\overset{O}{\overset{\|}{C}}CH_3$ **(d)** HCHO

Solution
(a) carboxylic acid   **(b)** aldehyde   **(c)** ketone   **(d)** aldehyde

2.10   Write condensed structural formulas for four-carbon compounds that exemplify **(a)** an aldehyde; **(b)** a ketone; **(c)** a carboxylic acid; and **(d)** a carboxylic acid that is also a ketone. (There may be more than one correct answer for each part.)

2.11   Circle and identify each carboxylic acid, aldehyde, and ketone in the following structure.

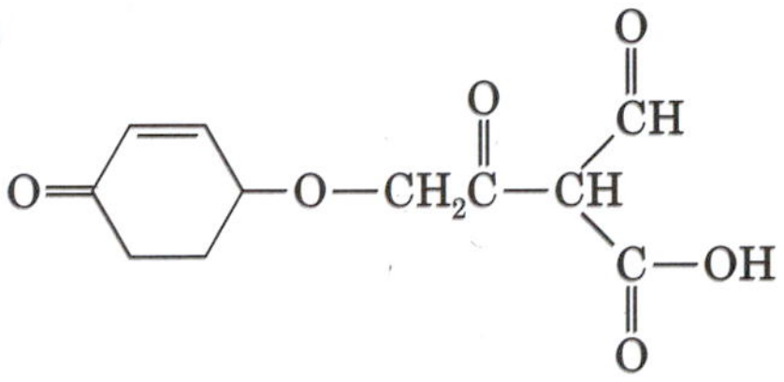

Section 2.7
Conjugated Double Bonds

An organic molecule may contain more than one functional group. In most polyfunctional compounds, each functional group is independent of another; however, this is not always the case. Let us consider some compounds with more than one carbon–carbon double bond.

There are two principal ways double bonds can be positioned in an organic molecule. Two double bonds originating at adjacent atoms are called **conjugated double bonds.**

Conjugated double bonds:

$$CH_2{=}CH{-}CH{=}CH_2$$

adjacent carbon atoms

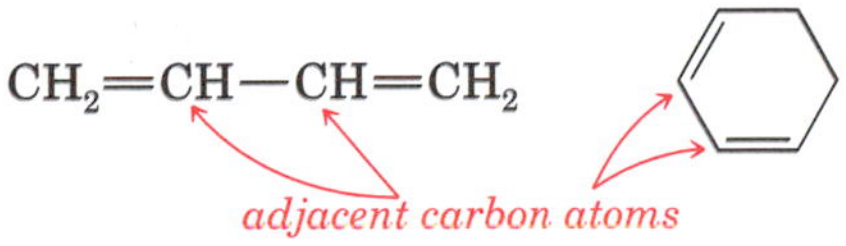

Double bonds joining atoms that are not adjacent are called **isolated,** or **nonconjugated,** double bonds.

Isolated double bonds:

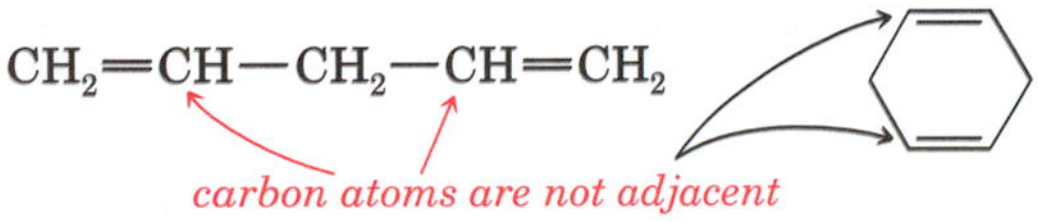

$$CH_2{=}CH{-}CH_2{-}CH{=}CH_2$$

carbon atoms are not adjacent

2.12   Vitamin A_1 has the structure that follows. How many conjugated double bonds does it contain? How many isolated double bonds?

$$H_3C \quad CH_3 \quad H \atop{C} \quad \overset{CH_3}{\underset{|}{C}} \quad H \atop{C} \quad \overset{CH_3}{\underset{|}{C}} \quad H \atop{C} \quad \overset{CH_3}{\underset{|}{C}} \quad CH_2OH$$

2.13 Draw the condensed structural formula of an eight-carbon compound that has **(a)** three conjugated double bonds; **(b)** two conjugated double bonds and one isolated double bond; **(c)** three isolated double bonds. (There may be more than one correct answer.)

Isolated double bonds behave independently; each double bond undergoes reaction as if the other were not present. Conjugated double bonds, on the other hand, are not independent of each other; there is electronic interaction between them. Let us choose the simplest of the conjugated systems, $CH_2=CH-CH=CH_2$, called 1,3-butadiene, to discuss this phenomenon. Figure 2.22 illustrates the p-orbital overlap in 1,3-butadiene.

We have numbered the carbon atoms of 1,3-butadiene in Figure 2.22 for reference. There are two pairs of p orbitals that form two pi bonds: one pi bond between carbons 1 and 2, and one pi bond between carbons 3 and 4. However, the p orbitals of carbons 2 and 3 are also adjacent, and *partial overlap* of these p orbitals occurs. Although most of the pi-electron density is located between carbons 1–2 and 3–4, some pi-electron density is also found between carbons 2–3. (Figure 2.22 shows a composite of the bonding orbitals in 1,3-butadiene. For a more-detailed discussion of the π molecular orbitals of this compound, see Section 21.1. Also, refer to Figure 21.1, page 854, for a representation of the four π molecular orbitals.)

We use a number of terms to describe this pi-bond interaction in conjugated systems. We may say that there is **partial p-orbital overlap** between the central carbons. We may also say that the bond between carbons 2 and 3 in 1,3-butadiene has **partial double-bond character.** Yet another way of

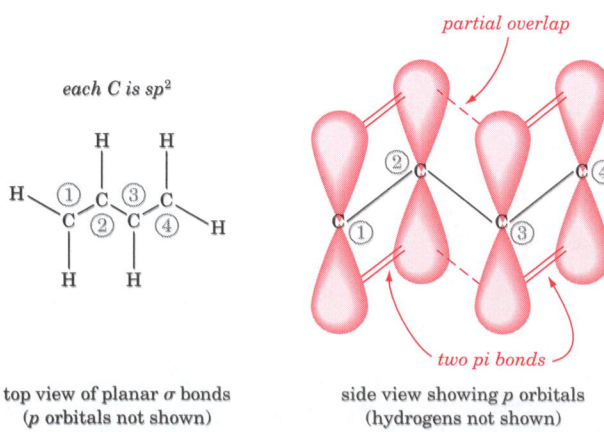

each C is sp^2

top view of planar σ bonds
(p orbitals not shown)

partial overlap

two pi bonds

side view showing p orbitals
(hydrogens not shown)

Figure 2.22   The bonding in 1,3-butadiene, $CH_2=CH-CH=CH_2$.

describing the pi bonding in the molecule is to say that the pi electrons are **delocalized,** which means that the pi-electron density is distributed over a somewhat larger region within the molecule. By contrast, **localized electrons** are restricted to two nuclei; nonconjugated double bonds contain localized pi electrons.

STUDY PROBLEMS

2.14 Draw the *p* orbitals of each of the following compounds, using lines to show pi-bond overlap and dotted lines to show partial overlap, as we have done in Figure 2.22.

(a) CH_2=CHCH$_2$CH=CHCH=CH$_2$ (b)

2.15 Write the formula for an open-chain four-carbon ketone in which the carbonyl group is in conjugation with a carbon–carbon double bond.

Section 2.8

Benzene

Benzene (C_6H_6) is a cyclic compound with six carbon atoms joined in a ring. Each carbon atom is sp^2 hybridized, and the ring is planar. Each carbon atom has one hydrogen atom attached to it, and each carbon atom also has an unhybridized *p* orbital perpendicular to the plane of the sigma bonds of the ring. Each of these six *p* orbitals can contribute one electron for pi bonding (see Figure 2.23).

With six *p* electrons, benzene should contain three pi bonds. We could draw three pi bonds in the ring one way (formula **A**), or we could draw them another way (formula **B**). However, we might also wonder—could this type of *p*-orbital system lead to *complete* delocalization of all six *p* electrons (formula **C**) instead of just partial delocalization?

sigma-bond structure
(Each C is sp^2 and
has one *p* electron.)

placement of pi bonds
(The circle represents complete delocalization.)

It is known that all carbon–carbon bond lengths in benzene are the same, 1.40 Å. All six bonds are longer than C—C double bonds, but shorter than C—C single bonds. If the benzene ring contained three localized double bonds

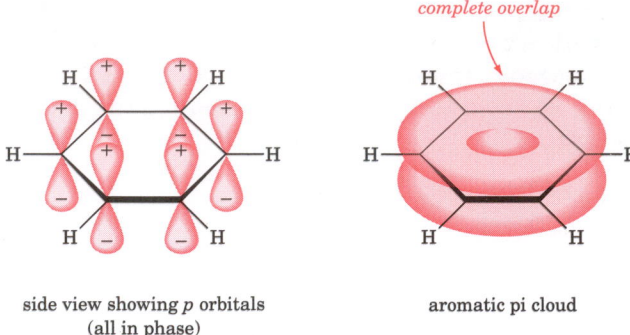

side view showing *p* orbitals
(all in phase)

aromatic pi cloud

Figure 2.23   The bonding in benzene, C_6H_6.

separated by three single bonds, the bonds would be of different lengths. The fact that all the carbon–carbon bonds in the benzene ring have the same length suggests that the benzene ring does not contain alternate single and double bonds.

From the bond lengths plus a body of other evidence that will be presented in later chapters, chemists have concluded that benzene is a symmetrical molecule and that each of the six ring bonds is like each of the other ring bonds. Instead of alternate double and single bonds, the six pi electrons are *completely delocalized* in a cloud of electronic charge shaped rather like a pair of donuts. This cloud of pi electrons is called the **aromatic pi cloud** of benzene. Figure 2.23 depicts the lowest-energy bonding orbital of benzene, the one commonly used to represent the aromatic pi cloud. (Figure 11.5, page 483, shows the six π orbitals of benzene.) For most purposes in this text, we will use a hexagon with a circle inside it to represent benzene; the circle represents the aromatic pi cloud.

Benzene is just one member of a class of **aromatic compounds,** compounds that contain aromatic pi clouds. Historically, the term "aromatic" derives from the fact that many of these compounds have distinctive odors. We will discuss aromatic compounds in detail in Chapters 11, 12, and 19.

Section 2.9
Resonance

Methane (CH_4) and ethylene (CH_2=CH_2) are examples of organic compounds with structures that may reasonably be described using single line-bond formulas. In each case, a line joining two atomic symbols represents a covalent bond between two atoms.

Benzene is a compound that cannot be accurately represented by a single line-bond formula. In benzene, the electron delocalization causes the pi electrons to encompass more than two atoms. (The circle in the hexagon representing the aromatic pi cloud in benzene is a fairly recent addition to organic symbolism.) In order to describe the pi-electron distribution in benzene using classical line-bond formulas, we must use *two* formulas.

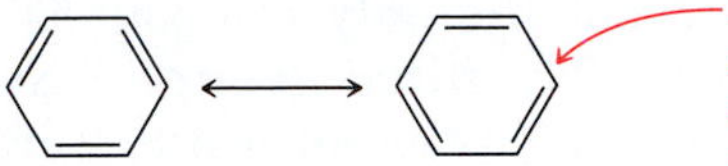

resonance structures for benzene
(Kekulé formulas)

Neither structure actually exists—the real structure is a composite of these two resonance structures.

These two line-bond formulas for benzene are called **Kekulé formulas** in honor of Friedrich August Kekulé, who first proposed them in 1872. Kekulé's original proposal was brilliant for its time, but unfortunately incorrect. His idea was that the two structures for benzene shifted back and forth so fast that neither could be isolated independently of the other. We now know that benzene does not shift between two different structures; the real structure of benzene is a composite of the two structures. We say that benzene is a **resonance hybrid** of the two structures, which are called **resonance structures.**

Whenever we can describe a molecular structure by two or more line-bond formulas that differ *only in the positions of the electrons* (usually pi electrons), none of these formulas will be in complete accord with the chemical and physical properties of the compound. If different resonance structures can be written for a compound, we can assume a delocalization of electron density. These statements are true for all aromatic structures, as well as for some other structures we will mention shortly.

The real structure of naphthalene is a composite of the resonance structures:

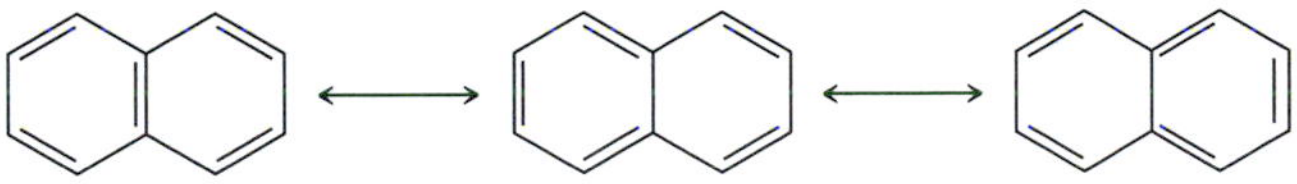

The real structure of pyridine is a composite of the resonance structures:

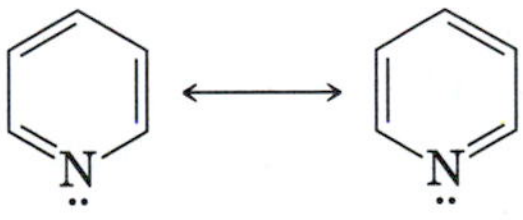

The important thing to keep in mind is that resonance symbols are not *true* structures; the true structure is a composite of all the resonance symbols. Many chemists have made the analogy that a rhinoceros (a real animal) might be described as a resonance hybrid of a unicorn (imaginary) and a dragon (imaginary). A rhinoceros does not shift back and forth from unicorn to dragon, but is simply an animal with characteristics of both a unicorn and a dragon.

To show that two or more formulas represent resonance structures (imaginary) and not real structures in equilibrium, we use a double-headed arrow (↔). By contrast, we indicate an equilibrium by *two* arrows (⇌).

Resonance: rhinoceros: unicorn ⟷ dragon
                              (real)   *(imaginary)*

benzene: ⟷ 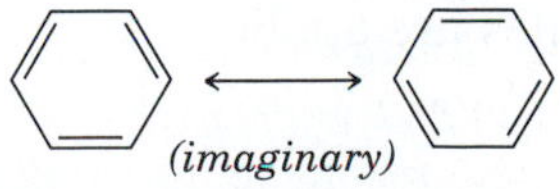
            (real)   *(imaginary)*

Equilibrium: $2\,H_2O \rightleftharpoons H_3O^+ + OH^-$
                  (real)   *(real)*   *(real)*

Aromatic compounds are not the only compounds for which single line-bond formulas are inadequate. The **nitro group** ($-NO_2$) is a group of atoms that we can best describe using resonance structures. A single line-bond structure for the nitro group shows two types of $N-O$ bond. However, it is known that the two $N-O$ bonds are the same length, intermediate between the lengths of an $N-O$ single bond and an $N=O$ double bond. We must use two line-bond resonance structures to describe this fact for a compound such as CH_3NO_2. To show that the bonds from nitrogen to the oxygen in the NO_2 group are the same, some chemists represent the nitro group with dotted lines for the partial double bonds.

$$\left[CH_3-\overset{+}{N}\begin{smallmatrix}\ddot{O}:\\ \\ \ddot{O}:^-\end{smallmatrix} \longleftrightarrow CH_3-\overset{+}{N}\begin{smallmatrix}\ddot{O}:^-\\ \\ \ddot{O}:\end{smallmatrix} \right] \quad \text{or} \quad CH_3-\overset{+}{N}\begin{smallmatrix}O^{\delta-}\\ \\ O^{\delta-}\end{smallmatrix}$$

The **carbonate ion** ($CO_3{}^{2-}$) is an ion that cannot be represented by a single line-bond structure. Each $C-O$ bond in the carbonate ion is the same length. We must use *three* resonance structures to describe the real structure.

$$\underset{:\ddot{O}:^- \quad :\ddot{O}:^-}{\overset{\ddot{O}}{C}} \longleftrightarrow \underset{:\ddot{O}: \quad :\ddot{O}:}{\overset{:\ddot{O}:^-}{C}} \longleftrightarrow \underset{:O \quad :\ddot{O}:^-}{\overset{:\ddot{O}:^-}{C}}$$

A. Electron Shifts

When writing resonance structures, keep in mind that nuclei of the atoms in a molecule cannot change positions—only electrons are delocalized. Also keep in mind that, with rare exceptions, only pi electrons or unshared valence electrons can be delocalized.

In a series of resonance symbols, we often show electron delocalization by small curved arrows that allow us to progress systematically from one resonance symbol to another. These electron shifts are purely artificial because the electrons do not truly shift, but are delocalized. Electron-shift arrows may be drawn only in the following ways:

from a pi bond to an adjacent bond position: $=X- \leftrightarrow -X=$

from a pi bond to an adjacent atom: $=X- \leftrightarrow -\ddot{X}-$

from an atom to an adjacent bond position: $-\ddot{X}- \leftrightarrow -X=$

Check each arrow in the following series of resonance structures to see how these rules are applied.

$$\text{(benzene)} \longleftrightarrow \text{(benzene)}$$

$$CH_2=CH-\overset{+}{C}H_2 \longleftrightarrow H_2\overset{+}{C}-CH=CH_2$$

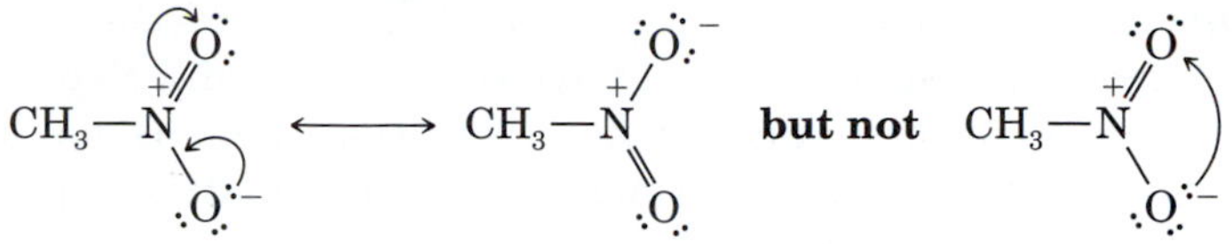

$$CH_3-\overset{+}{N}\overset{\nwarrow\ddot{O}\colon}{\diagdown_{\ddot{O}\colon^-}} \longleftrightarrow CH_3-\overset{+}{N}\overset{\ddot{O}\colon^-}{\diagdown_{\ddot{O}\colon}} \quad \textbf{but not} \quad CH_3-\overset{+}{N}\overset{\ddot{O}\colon}{\diagdown_{\ddot{O}\colon^-}}$$

$$\underset{^-\colon\ddot{O}\colon \quad \colon\ddot{O}\colon^-}{\overset{\ddot{O}\colon}{C}} \longleftrightarrow \underset{\colon\ddot{O} \quad \colon\ddot{O}\colon^-}{\overset{\colon\ddot{O}\colon^-}{C}} \longleftrightarrow \underset{^-\colon\ddot{O} \quad \colon\ddot{O}\colon}{\overset{\colon\ddot{O}\colon^-}{C}}$$

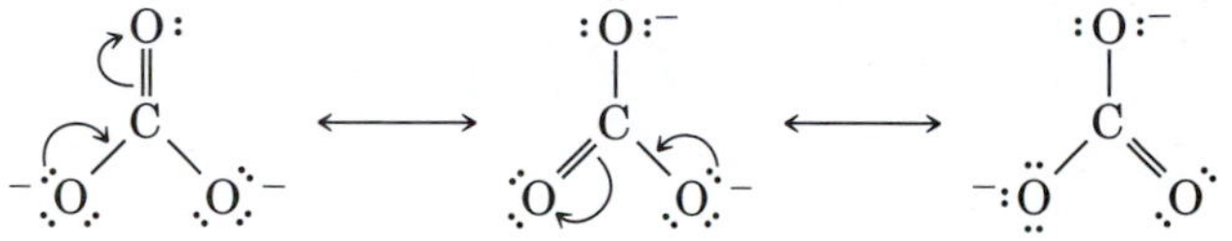

When using electron-shift arrows, we must pay special attention to how many electrons an atom can accommodate. An atom of the period-2 elements can accommodate a maximum of eight valence electrons. For purposes of keeping track of these electrons, it is advisable to show the unshared electrons in resonance structures.

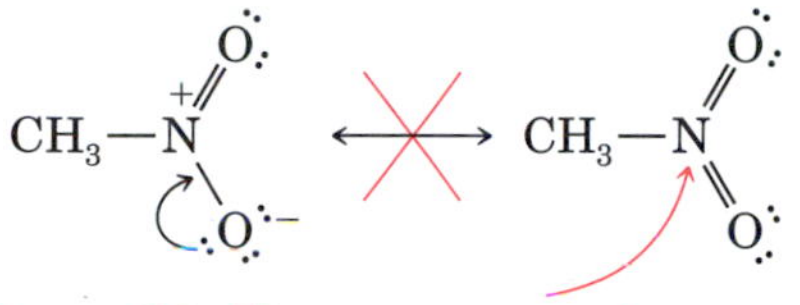

$$CH_3-\overset{+}{N}\overset{\ddot{O}\colon}{\diagdown_{\ddot{O}\colon^-}} \quad\xcancel{\longleftrightarrow}\quad CH_3-\overset{\ddot{O}\colon}{N}\diagdown_{\ddot{O}\colon}$$

Not possible: N cannot accommodate ten e⁻

In this book, we show the series of resonance structures to be read from left to right. For example, the curved arrows in the following left-hand resonance structure for nitromethane show how the electrons are "shifted" to get to the right-hand resonance structure.

Electron-shift arrows showing how to proceed from the left-hand structure to the right-hand structure:

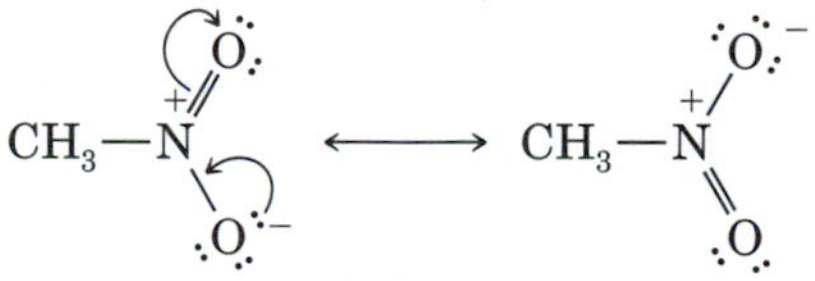

$$CH_3-\overset{+}{N}\overset{\ddot{O}\colon}{\diagdown_{\ddot{O}\colon^-}} \longleftrightarrow CH_3-\overset{+}{N}\overset{\ddot{O}\colon^-}{\diagdown_{\ddot{O}\colon}}$$

Some instructors prefer that curved arrows be shown on all resonance structures so that the series can be read in either direction—from right to left or from left to right.

Electron-shift arrows showing how to proceed in either direction:

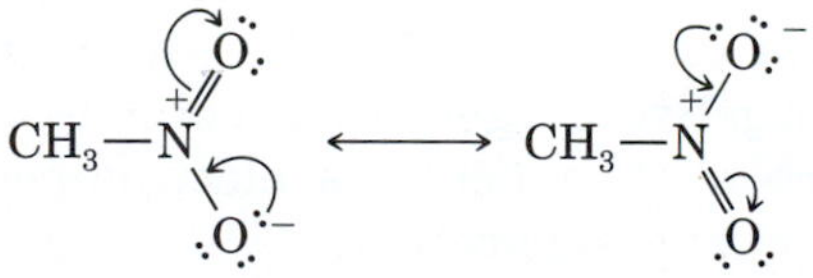

$$CH_3-\overset{+}{N}\overset{\ddot{O}\colon}{\diagdown_{\ddot{O}\colon^-}} \longleftrightarrow CH_3-\overset{+}{N}\overset{\ddot{O}\colon^-}{\diagdown_{\ddot{O}\colon}}$$

SAMPLE PROBLEM

Fill in the electron-shift arrows for the resonance structures of (a) pyrimidine and (b) the acetate ion.

(a) 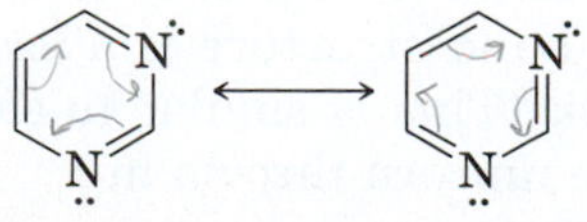 [pyrimidine ring structure] ⟷ [pyrimidine resonance structure]

(b) 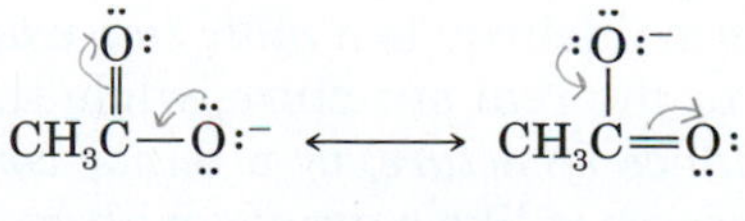 $CH_3\overset{\ddot{O}\colon}{\underset{}{C}}-\ddot{O}\colon^- \longleftrightarrow CH_3\overset{\colon\ddot{O}\colon^-}{\underset{}{C}}=\ddot{O}\colon$

Solution

(a) Write the first structure. To arrive at the second structure, shift each double bond to the adjacent bond position, just as for benzene. Draw the curved arrows one at a time. (It does not matter where you begin in this structure.)

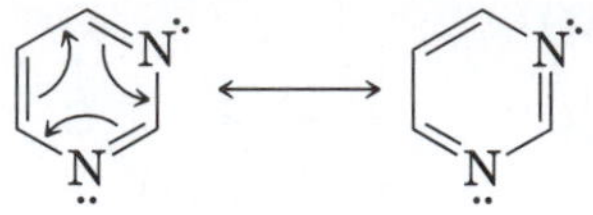

(b) Write the first structure. Draw the arrows one at a time. Then draw the second structure.

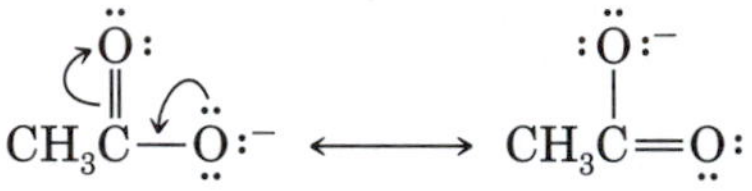

STUDY PROBLEM

2.16   The nitrate ion, NO_3^-, contains three equivalent N—O bonds. Write resonance structures for the nitrate ion.

B. Major and Minor Contributors

In each example we have shown so far, the bonding has been the same in each resonance symbol. When the bonding is the same, the resonance structures are of equal energy and are equivalent to each other. *Equivalent resonance structures contribute equally to the real structure.*

Dissimilar resonance structures for a compound or ion may not all contribute equally to the real structure. Consider the following example:

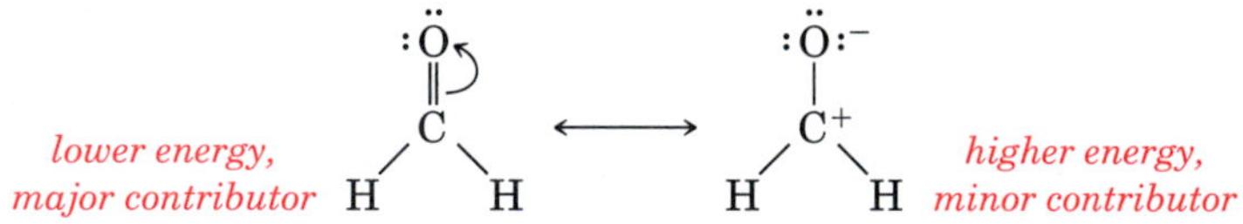

lower energy, major contributor   *higher energy, minor contributor*

formaldehyde

This pair of resonance structures shows that electron shifts can lead to an atom with an incomplete octet (the C in the second resonance structure). Here, the more electronegative oxygen atom becomes electron-rich and the less electronegative carbon atom becomes electron-poor—the second resonance structure shows a charge separation.

In general, resonance structures in which all period-2 elements have octets of valence electrons make a greater contribution to the real composite structure than a resonance structure with incomplete octets. Also, a resonance structure with a lesser amount of charge separation makes a greater contribution to the real structure. Thus, we say that the left-hand structure shown for formaldehyde is a *more important resonance structure,* or a *major contributor* to the real structure, while the right-hand structure is a *less important resonance structure,* or a *minor contributor.* (This is similar to saying that a rhinoceros is like a dragon with just a little unicorn thrown in.)

Formaldehyde is like $H-\overset{\overset{\displaystyle \cdot\cdot}{O}}{\underset{\displaystyle \|}{C}}-H$ with a small amount of $H-\overset{\overset{\displaystyle :\ddot{O}:^-}{|}}{\underset{\displaystyle +}{C}}-H$ thrown in.

C. Summary of Rules for Writing Resonance Structures

We will encounter resonance structures frequently in the coming chapters. Let us summarize the rules for writing resonance structures:

1. *Only pi electrons or unshared pairs of valence electrons (not sigma electrons or nuclei) may be shifted,* and they may be shifted only to adjacent atoms or bond positions.

2. Resonance structures in which an atom carries more than its quota of electrons (eight for the period-2 elements) are not contributors to the real structure.

3. The more important resonance structures show each atom with a complete octet and as little charge separation as possible.

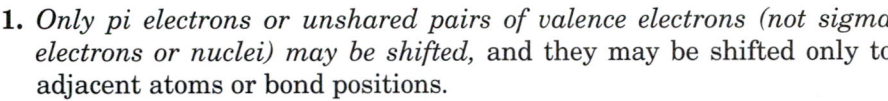

more important because
each atom has an octet

less important because N has
only six electrons and because
a greater charge separation exists

SAMPLE PROBLEM

Which of the following resonance structures is the major contributor to the real structure?

$$CH_3\overset{\overset{\displaystyle \ddot{O}:}{\|}}{C}-\ddot{O}CH_3 \longleftrightarrow CH_3\overset{\overset{\displaystyle :\ddot{O}:^-}{|}}{\underset{\displaystyle +}{C}}-\ddot{O}CH_3 \longleftrightarrow CH_3\overset{\overset{\displaystyle :\ddot{O}:^-}{|}}{C}=\overset{\displaystyle +}{\ddot{O}}CH_3$$

Solution
The first structure is the major contributor because each C and O atom has an octet and no C or O atom has a charge. The second and third structures are less important contributors because they contain charge separations and the second structure contains a C atom with an incomplete octet.

STUDY PROBLEM

2.17 Which of the following resonance structures is the major contributor to the real structure? Explain your answer.

$$CH_3\overset{\overset{\displaystyle :\ddot{O}:^-}{|}}{C}=CH-\overset{\displaystyle +}{C}H_2 \longleftrightarrow CH_3\overset{\overset{\displaystyle :\ddot{O}:^-}{|}}{\underset{\displaystyle +}{C}}-CH=CH_2 \longleftrightarrow CH_3\overset{\overset{\displaystyle :\ddot{O}}{\|}}{C}-CH=CH_2$$

D. Resonance Stabilization

If a structure is a resonance hybrid of two or more resonance structures, *the energy of the real structure is lower than that of any single resonance struc-ture.* The real structure is said to be **resonance-stabilized.** In most cases, the energy difference is slight, but for aromatic systems, like benzene or naphthalene, the energy difference is substantial. (The reasons for aromatic stabilization will be discussed in Chapter 11.)

The hypothetical structure [structure] is of higher energy than [structure]

The hypothetical structure [structure] is of substantially higher energy than [structure]

Resonance stabilization confers significant stability to a structure if two or more resonance structures of a compound are *equivalent or nearly equivalent in energy.* A high-energy, minor-contributing resonance structure adds little stabilization.

H—C—H   gains only a small amount of stabilization from   H—C—H.

The reason for the energy differences between hypothetical resonance structures and the real structure of a compound is not entirely understood. Certainly, part of the reason is that a delocalized electron is attracted to more than one nucleus. It is generally true that a system with delocalization of electrons or of electronic charge has lower energy and greater stability than a similar system with localized electrons or electronic charge.

SAMPLE PROBLEM

Which of the following compounds are resonance-stabilized? (*Note:* Because the valence electrons of sulfur are in the third energy level, a sulfur atom can accommodate more than eight electrons.)

(a) $CH_3CH_2CH_3$ (b) [structure: H,H on left, CH_3,H on right of C=C]   (c) [structure: CH_3O-S-O^- with O double bonds]

Solution

(a) is not resonance-stabilized because it contains no pi electrons.

(b) is not resonance-stabilized because there is no position in the molecule where the pi electrons can be delocalized.

(c) is resonance-stabilized:

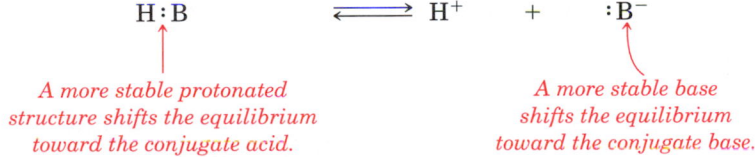

STUDY PROBLEM

2.18   (1) Write resonance structures for the following ions. (2) Indicate which, if any, of the resonance structures is the major contributor or whether the structures are of similar energy. [Use Kekulé formulas in (b).]

(a) $CH_3CH = \overset{+}{C}HCHCH_3$ **(b)** ⬡—$\overset{+}{C}H_2$

(c) $CH_3\overset{\overset{\ddot{O}\cdot}{\|}}{C}\overset{-}{C}H_2$ **(d)** $:\overset{..}{C}H_2C \equiv N:$

E. Resonance and Acidity

In Section 1.7D, we briefly discussed some of the factors that affect acidity. In that section, we mentioned that acid–base reactions are reversible and under equilibrium or thermodynamic control with the more stable reactant predominating in the equilibrium mixture. Consequently, the relative stabilities of the undissociated acid and its conjugate base determine their relative concentrations at equilibrium.

$$H:B \rightleftharpoons H^+ \; + \; :B^-$$

A more stable protonated structure shifts the equilibrium toward the conjugate acid.   *A more stable base shifts the equilibrium toward the conjugate base.*

As we also mentioned in Section 1.7D, the conjugate base can be stabilized if an electronegative atom carries the negative charge or if an electronegative atom is located near the atom that carries the negative charge.

Oxygen is more electronegative than nitrogen

$$CH_3\overset{..}{\underset{..}{O}}{-}H + H_2O \rightleftharpoons CH_3\overset{..}{\underset{..}{O}}{:}^- + H_3O^+$$

more acidic

$$\begin{array}{c} CH_3 \\ \\ CH_3 \end{array}\!\!\!\!\overset{..}{N}{-}H + H_2O \rightleftharpoons \begin{array}{c} CH_3 \\ \\ CH_3 \end{array}\!\!\!\!\overset{..}{N}{:}^- + H_3O^+$$

less acidic

The electronegative Cl is closer to the acidic carboxyl group

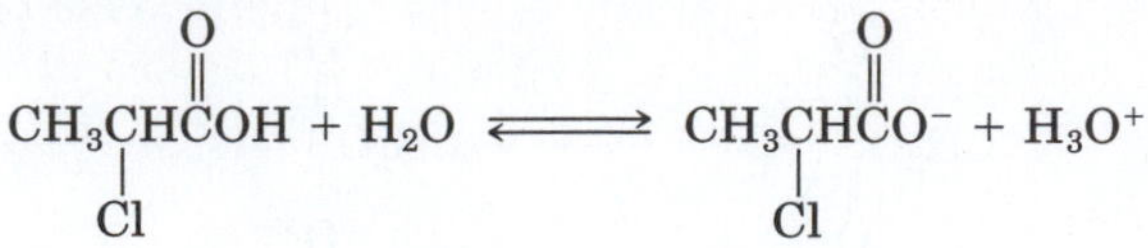

$$CH_3\overset{\overset{\displaystyle O}{\|}}{C}HCOH + H_2O \;\rightleftharpoons\; CH_3\overset{\overset{\displaystyle O}{\|}}{C}HCO^- + H_3O^+$$
$$\underset{Cl}{|} \qquad\qquad\qquad \underset{Cl}{|}$$

more acidic

$$\overset{}{C}H_2CH_2\overset{\overset{\displaystyle O}{\|}}{C}OH + H_2O \;\rightleftharpoons\; \overset{}{C}H_2CH_2\overset{\overset{\displaystyle O}{\|}}{C}O^- + H_3O^+$$
$$\underset{Cl}{|} \qquad\qquad\qquad\qquad \underset{Cl}{|}$$

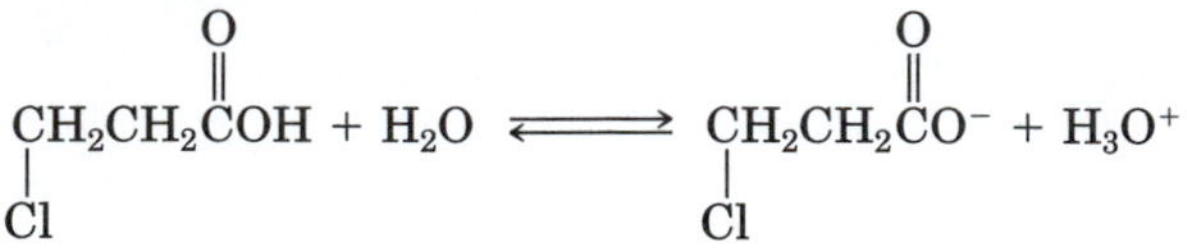

less acidic

Delocalization of pi electrons stabilizes a structure. Will delocalization of electrons in a conjugate base increase the stability of that anion and, therefore, the acidity of its conjugate acid? The answer is yes. Consider the relative acidities of the following three conjugate acids:

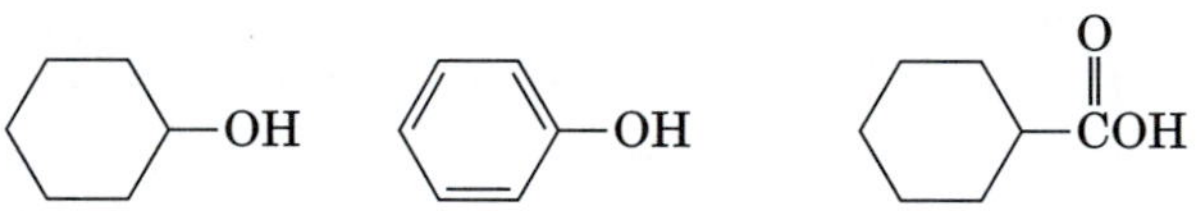

cyclohexanol; $pK_a = 18$ phenol; $pK_a = 10$ cyclohexanecarboxylic acid; $pK_a = 4.9$

Of the three, cyclohexanol is the weakest acid because the negative charge in its conjugate base must reside on only one oxygen atom. Consequently, the conjugate base of cyclohexanol is not resonance-stabilized.

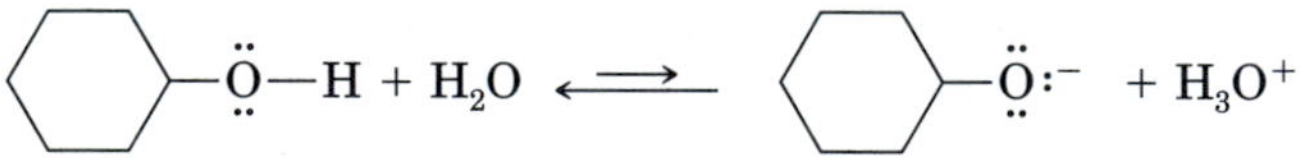

cyclohexanol

not resonance-stabilized; weakly acidic

The conjugate base of phenol, however, is resonance-stabilized. The aromatic pi cloud delocalizes the negative charge. As a result, phenol is many powers of 10 more acidic than cyclohexanol.

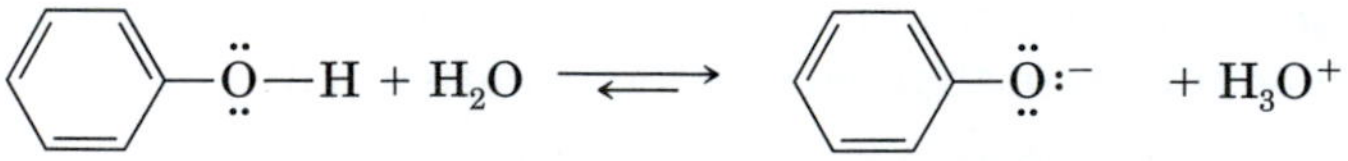

phenol

resonance-stabilized

Resonance structures for the conjugate base of phenol

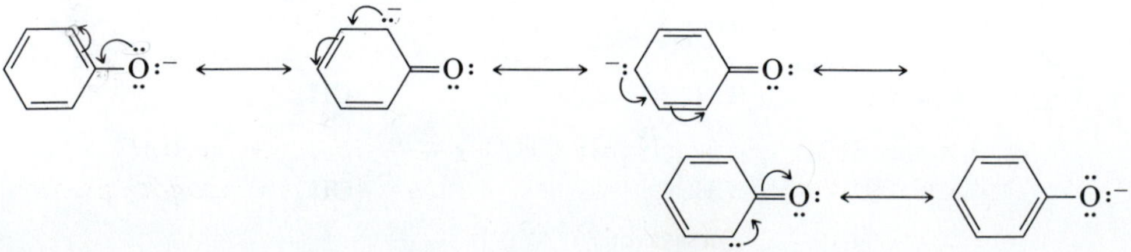

The third acid in the series, cyclohexanecarboxylic acid, is many powers of 10 more acidic than phenol. Like the conjugate base of phenol, the conjugate base of cyclohexanecarboxylic acid is resonance-stabilized.

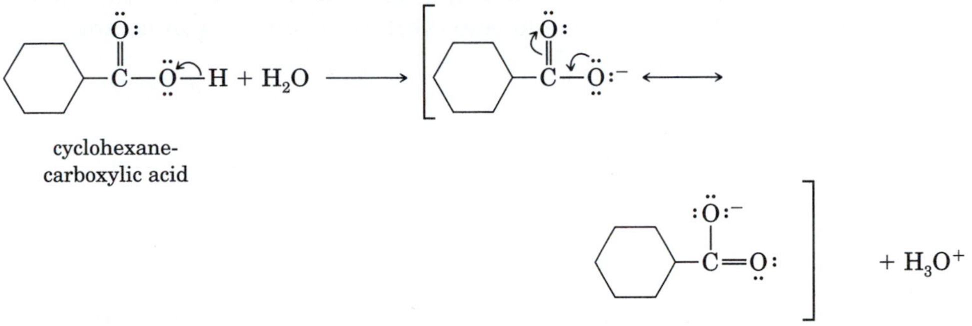

cyclohexane-
carboxylic acid

The negative charge is shared by two electronegative oxygen atoms (two equivalent resonance structures).

Why is cyclohexanecarboxylic acid more acidic than phenol? The reason is that oxygen is more electronegative than carbon. The conjugate bases of both acids are resonance-stabilized. They differ in the number of atoms which share the negative charge. In the conjugate base of cyclohexanecarboxylic acid, two electronegative oxygen atoms share the negative charge. In the conjugate base of phenol, only one electronegative oxygen atom shares the negative charge. The other atoms sharing the charge are carbon. Since carbon atoms are far less electronegative than oxygen atoms, they are far less able to help delocalize the charge.

STUDY PROBLEM

2.19 Which of the following phenols is more acidic? Use resonance structures of the conjugate base to defend your answer.

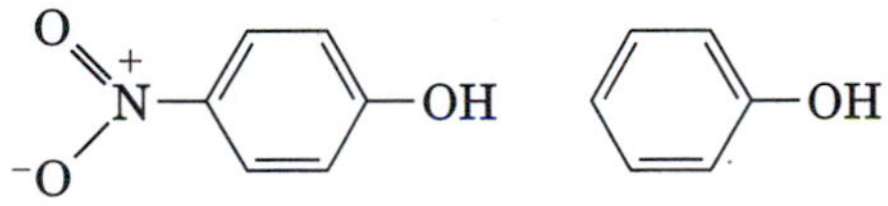

p-nitrophenol   phenol

Summary

A covalent bond is the result of two atomic orbitals overlapping to form a lower-energy **bonding molecular orbital** in which two electrons are paired.

The four atomic orbitals of C undergo **hybridization** in bond formation in three different ways:

***sp*^3 hybridization:** four equivalent single bonds *(tetrahedral)*.
***sp*^2 hybridization:** three equivalent single bonds *(trigonal)* plus one unhybridized *p* orbital.

sp hybridization: two equivalent single bonds *(linear)* plus two unhybridized *p* orbitals.

Overlap of a hybrid orbital with an orbital of another atom (end-to-end overlap) results in a **sigma bond.** Overlap of a *p* orbital with a parallel *p* orbital of another atom (side-to-side overlap) results in a **pi bond.**

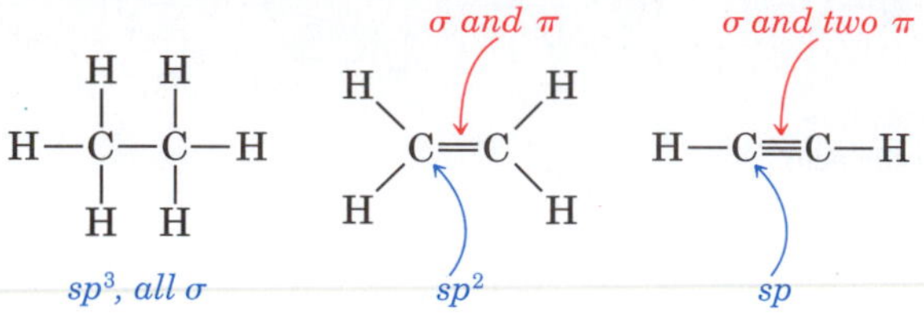

Each bonding molecular orbital has an **antibonding molecular orbital** associated with it. Antibonding orbitals are of higher energy than either bonding orbitals or the atomic orbitals that went into their formation. In the ground state, molecules have their electrons in the lowest-energy orbitals, usually the bonding orbitals.

A **functional group** is a *site of chemical reactivity* in a molecule and arises from a *pi bond* or from *differences in electronegativity* between bonded atoms. A double bond or a triple bond is a functional group. The following general formulas show examples of other functional groups:

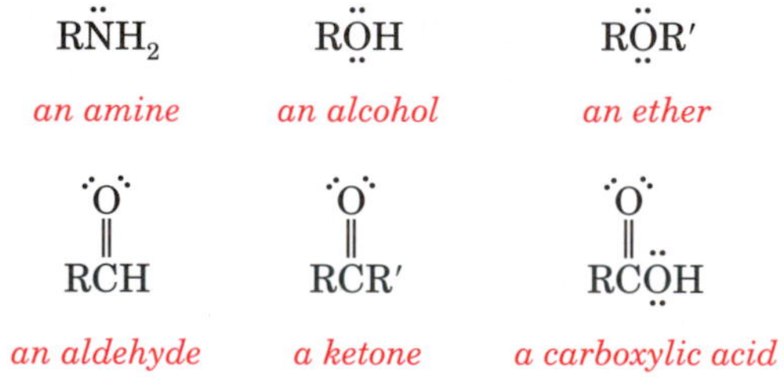

Pi bonds extending from adjacent carbon atoms are called **conjugated double bonds.** In the case of benzene and other aromatic compounds, complete delocalization of pi electrons results in the **aromatic pi cloud.**

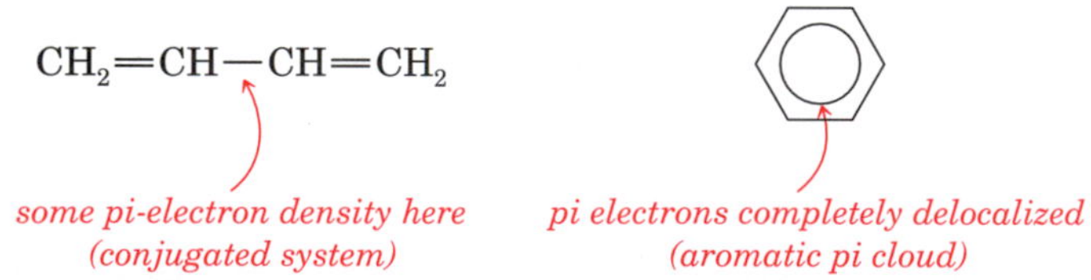

Resonance structures can be used to show delocalization of *pi* electrons.

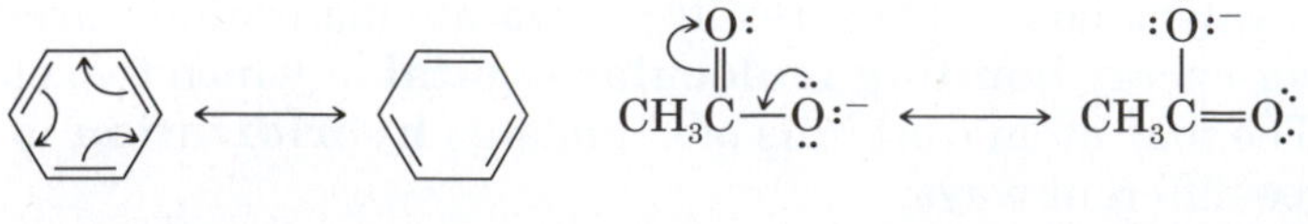

If all resonance structures are not equivalent, the lowest-energy resonance structure is the **major contributor.**

$$CH_3-\overset{\overset{\displaystyle \ddot{O}:}{\|}}{C}-CH_3 \longleftrightarrow CH_3-\overset{\overset{\displaystyle :\ddot{O}:^-}{|}}{\underset{+}{C}}-CH_3$$

major contributor

Delocalization of pi electrons in a molecule, or of electronic charge in an ion, results in a slight increase in the stability of the system. However, for aromatic compounds, the increase in stability is substantial. Delocalization of electrons in a conjugate base increases the stability of the anion and, consequently, the acidity of the conjugate acid.

Essay Problem for Chapter 2

Isomerization of Allenes to 1,3-Dienes

1,3-Dienes and allenes (1,2-dienes) are isomers. Of the two, 1,3-dienes are more stable; that is, they have less energy. Consequently, when allenes and 1,3-dienes are equilibrated, the equilibrium lies on the side of the 1,3-diene. For example, when ethyl 3,4-decadienoate, an allene, is heated to reflux in contact with activated aluminum oxide (Al_2O_3), ethyl 2,4-decadienoate (the isomeric 1,3-diene) is formed. Starting with the pure allene, the 1,3-diene can be isolated from the equilibrium mixture in a 75–88% yield. (S. Tsuboi, T. Masuda, S. Mimura, and A. Takeda, *Org. Syn.* **1988,** *66,* 22.)

ethyl 3,4-decadienoate

an allene
(a 1,2-diene)

ethyl 2,4-decadienoate

(a 1,3-diene)

Questions

1. Which carbon in the allene is *sp*-hybridized?
2. Are the two carbon–carbon double bonds in the allene conjugated? Draw an orbital picture of the pi bonds in the allene to explain your answer.
3. How many pi bonds in the 1,3-diene ethyl 2,4-decadienoate are conjugated? Draw an orbital picture of the pi bonds in the diene to defend your answer.
4. Why is the 1,3-diene more stable than the allene?
5. Alkynes (compounds with a carbon–carbon triple bond) are also isomeric with allenes and dienes. What is a possible alkyne isomer of ethyl 3,4-decadienoate? Would the isomeric alkyne be more or less stable than the 1,3-diene? Explain your answer.

2.20 Do the following structures represent the same or different compounds? Explain your answer.

(a) 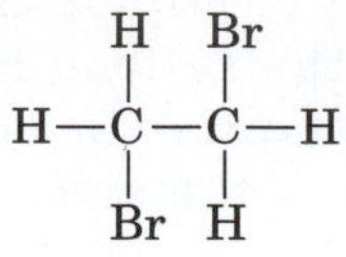
$$
\begin{array}{c}
\text{H} \quad \text{Br} \\
| \quad\quad | \\
\text{H—C—C—H} \\
| \quad\quad | \\
\text{Br} \quad \text{H}
\end{array}
$$

(b)
$$
\begin{array}{c}
\text{H} \quad \text{H} \\
| \quad\quad | \\
\text{Br—C—C—Br} \\
| \quad\quad | \\
\text{H} \quad \text{H}
\end{array}
$$

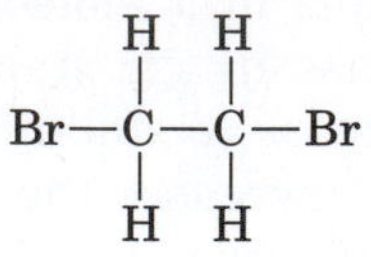

(c)
$$
\begin{array}{c}
\text{H} \quad \text{H} \\
| \quad\quad | \\
\text{H—C—C—H} \\
| \quad\quad | \\
\text{Br} \quad \text{Br}
\end{array}
$$

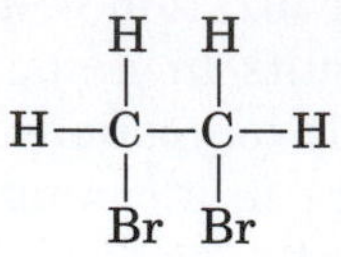

2.21 Draw a structure that shows the *p*-orbital overlap forming the pi bonds in each of the following compounds. (Use lines for sigma bonds.)

(a) $(CH_3)_2C{=}C(CH_3)_2$ (b) $CH_3C{\equiv}CCH_3$

2.22 Indicate the hybridization of each carbon in the following structures:

(a) $CH_3OCH_2CH_3$ (b) 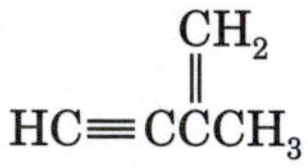 $HC{\equiv}C\overset{\displaystyle CH_2}{\underset{\|}{C}}CH_3$ (c)

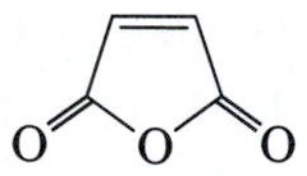

2.23 In each of the following structures, do the two indicated carbons lie in the same plane as each other and the double-bonded carbons?

(a) 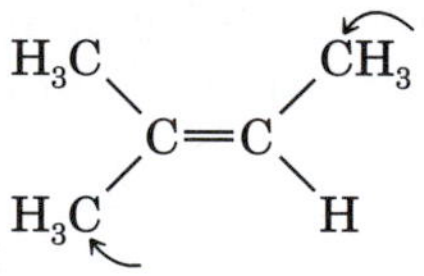
$$
\begin{array}{c}
H_3C \\
 \\
H_3C
\end{array}
C{=}C
\begin{array}{c}
CH_3 \\
 \\
H
\end{array}
$$

(b) 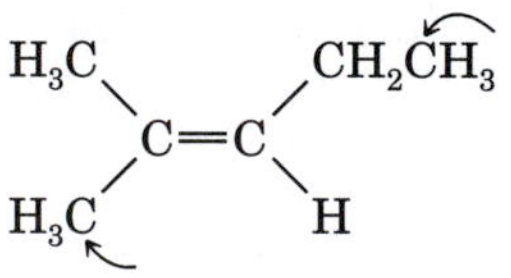
$$
\begin{array}{c}
H_3C \\
 \\
H_3C
\end{array}
C{=}C
\begin{array}{c}
CH_2CH_3 \\
 \\
H
\end{array}
$$

(c)

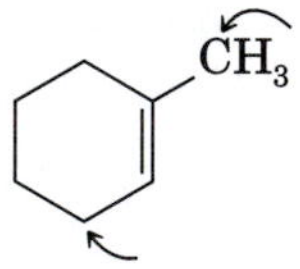

2.24 What would be the expected *sigma-bond angles* in the following structures?

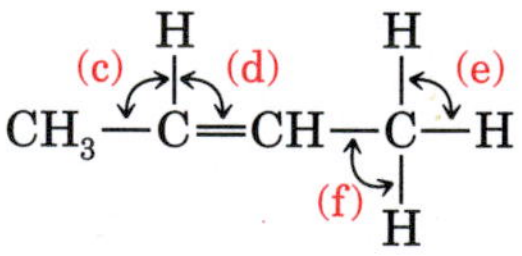

(a) (b) $C{\equiv}C{-}H$

(c) (d) (e) (f) $CH_3{-}\overset{H}{C}{=}CH{-}\overset{H}{\underset{H}{C}}{-}H$

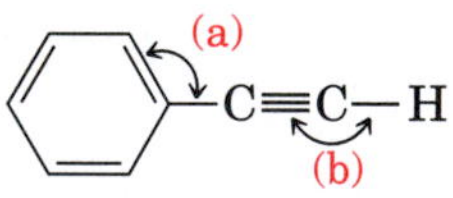

2.25 Show the structure of each of the following molecules or ions. Use lines to represent sigma bonds and dumbbell shapes to represent *p* orbitals. Show *p*-orbital overlap with dashed lines. For example, $CH_2{=}CH_2$:

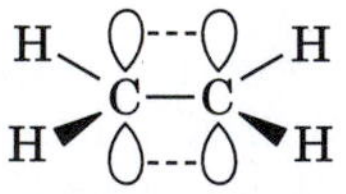

$$
\begin{array}{c}
H \\
 \\
H
\end{array}
C{-}C
\begin{array}{c}
H \\
 \\
H
\end{array}
$$

(a) H_3O^+ (b)   HCN   (c) OH^- (d) NH_2NH_2

(e) $(CH_3)_4N^+$ (f) $^-OCH_3$ (g) $CH_2{=}CHC{\equiv}CH$

2.26 The molecular-orbital diagram that follows is the for the carbon–carbon double bond in cyclohexene.

---

* For information concerning the organization of the *Study Problems* and *Additional Problems*, see the *Note to student* on page 41.

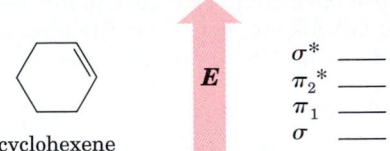

cyclohexene

$$\sigma^* \underline{\quad}$$
$$\pi_2^* \underline{\quad}$$
$$\pi_1 \underline{\quad}$$
$$\sigma \underline{\quad}$$

E

(a) Indicate by arrows which orbitals contain the double-bond electrons when cyclo-hexene is in the ground state.

(b) Absorption of ultraviolet light by a double bond results in the promotion of an electron to a higher-energy orbital. Which electron transition requires the least energy when cyclohexene goes to an excited state?

2.27 In each of the following structures or pairs of structures, which of the indicated bonds (1 or 2) is the shorter one?

(a) Cl①—CH$_3$,   ②—Cl

(b)
$$\begin{array}{c} H \\ \searrow \\ H \end{array} C{=}CH{-}C{\equiv}C②{-}H$$ with ① on top H

(c)   ①—CH$_3$,   ②—CH$_3$

(d) CH$_3$①—OCH$_3$,   CH$_3$C②CH$_3$ (with O double bonded)

2.28 What hybrid orbitals or atomic orbitals are used to form the indicated covalent bonds?

(a) H—(1)C≡C—(2)CH=CH$_2$

(b) H$_3$C—(1)CH=N—(2)OH

(c) H—(benzene ring with H's)—H (1) and (2)

(d) CH$_3$CH$_2$—CH$_3$

(e) H$_3$C—CH (with C=O)

(f) N≡C—CH$_2$—C≡N

2.29 Circle and name the functional groups in the following structures:

(a)   (CH$_3$)$_2$C=CHCH$_2$CH$_2$CHCH$_2$CH
with CH$_3$ branch and O (double bond)

citronellal

*chief constituent of citronella oil;
used in perfumes and as a mosquito repellent*

(b)   HOCH$_2$CCH$_2$OH (with C=O)

dihydroxyacetone

*formed in the metabolism
of carbohydrates; also
used in the tanning industry*

(c)

estradiol

a female hormone

(d)

cortisone

*an adrenal steroid
used to treat arthritis*

2.30 Draw a three-carbon open-chain structure to illustrate each of the following compound types: (**a**) alkene; (**b**) alkyne; (**c**) ether; (**d**) alcohol; (**e**) amine; (**f**) ketone; (**g**) aldehyde; (**h**) carboxylic acid. (There may be more than one correct answer in each case.)

2.31 Rewrite the formulas of the following compounds to emphasize the carbonyl groups:

(**a**) H_2CO

(**b**) $CH_3CO_2CH_3$

(**c**) CH_3CH_2CHO

(**d**) $(CH_3)_2CHCOCH_3$

(**e**) $OHCCH_2CHO$

(**f**) $(CH_3CO)_2O$

2.32 Identify and name the functional groups in the following compounds:

(**a**) 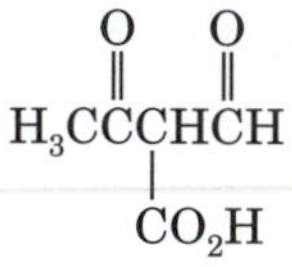 $H_3CCCHCH$ with two O double bonds and CO_2H substituent

(**b**) $H_3CNHCH_2CH_2OCH_2CCO_2H$ with O double bond

(**c**) HCH with O double bond

(**d**) $HOCCH_2CCH_3$ with two O double bonds

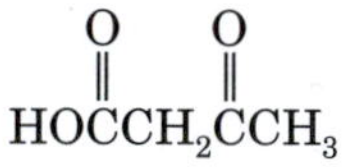

(**e**)

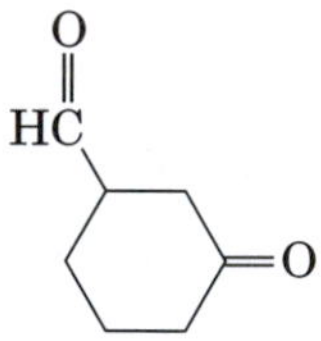

(**f**)

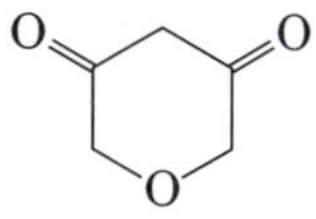

2.33 For each of the following pairs, which is the more important resonance structure? Explain your answers.

(**a**)

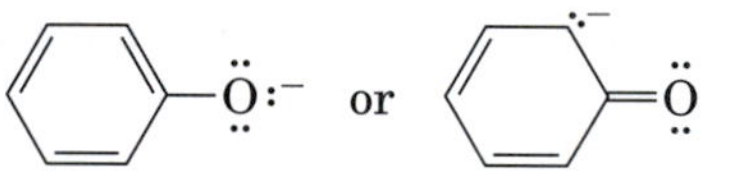

(**b**) 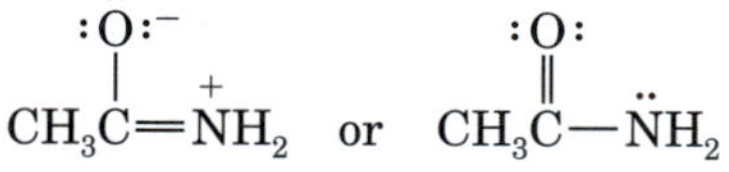 $CH_3C{=}NH_2$ or $CH_3C{-}NH_2$

(**c**)

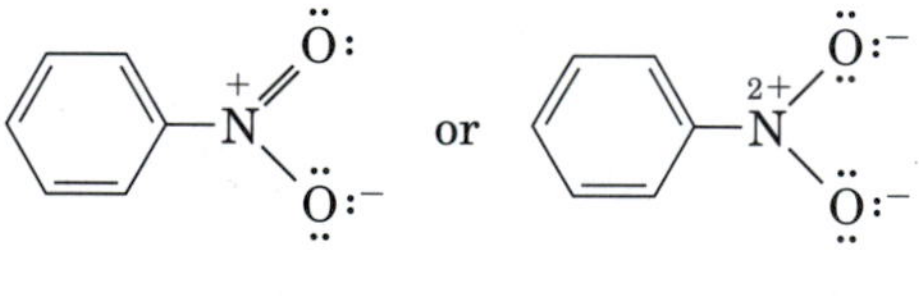

2.34 Write a condensed structural formula for each of the following compounds. (There may be more than one answer.)

(**a**)   a ketone, C_4H_8O

(**b**)   a cyclic alcohol, $C_6H_{12}O$

(**c**)   an alkene, C_5H_{10}

(**d**)   an aldehyde, C_3H_6O

2.35 For each of the following structures, indicate the single bond that has some partial double-bond character.

(**a**)

(**b**) $CH_2{=}CH{-}CH_2{-}CH{=}CH{-}C{\equiv}CH$

(**c**)

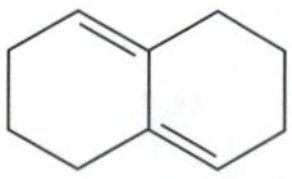

2.36 Draw the structure of an open-chain five-carbon compound with: (**a**) two double bonds in conjugation; (**b**) two isolated double bonds. (More than one answer may be possible.)

2.37 Which of the following pairs of structures represent resonance symbols for a molecule or ion?

(a) $(CH_3)_2CHCO^-$, $(CH_3)_2CHC=O$ (with C=O double bond and O^-)

(b) CH_3CCH_3 (with ^+OH), CH_3CCH_3 (with OH and $+$)

(c) CH_3CCH_3 (with O), $CH_3C=CH_2$ (with OH)

(d) (cyclohexene ring)=O, (cyclohexene ring with $+$)—O^-

(e) $CH_2=C=O$, $CH\equiv C-OH$

2.38 Draw the important resonance contributors for each of the following structures:

(a) HCO^- (with C=O)

(b) $HOCO^-$ (with C=O)

(c) (ring)—CH_3

(d) $CH_3\overset{+}{C}HCH=CH_2$

(e) $CH_2=CHCH=\overset{+}{C}HCH_2$

(f) $CH_3\overset{O}{C}\overset{..}{C}H_2^-$

(g) $:N\equiv C\overset{-}{C}HC\equiv N:$

(h) ^-O—(ring)—$\overset{O}{C}H$

(i) (benzene ring with $\overset{+}{C}H_2$ and OCH_3)

2.39 Insert electron-shift arrows and unshared pairs of valence electrons in the following structures:

(a) (series of six resonance structures of CCH_3 with O top and O^- bottom, connected by resonance arrows)

(b) (series of six resonance structures of NH_2 with nitro group, connected by resonance arrows)

2.40 Write the important Kekulé resonance-contributing structures for each of the following aromatic compounds.

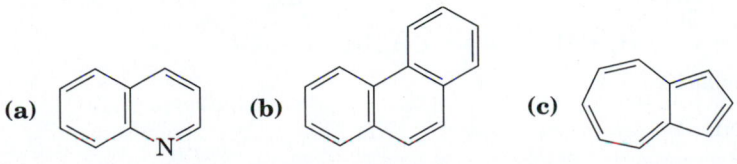

(a) (quinoline with N)   (b) (phenanthrene)   (c) (azulene)

2.41 Which ion in each of the following pairs would be more resonance-stabilized? Explain your answers by drawing resonance structures.

(a) 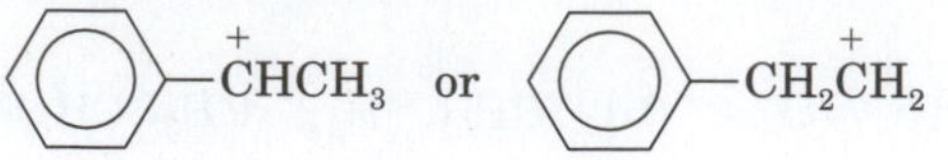 $\overset{+}{C}HCH_3$ or $CH_2\overset{+}{C}H_2$

(b) $CH_3CH_2\overset{+}{C}HCH_3$ or $CH_2{=}CH\overset{+}{C}HCH_3$

(c) $:\overline{C}H_2CH_3$ or $:\overline{C}H_2NO_2$

Additional Problems

2.42 *Allene* has the following structure: $CH_2{=}C{=}CH_2$. **(a)** What is the hybridization state of each carbon? **(b)** What would be the expected angle between the two pi bonds? **(c)** Are the two pi bonds conjugated? **(d)** Can electronic charge be delocalized throughout both pi bonds?

2.43 The O—C—O bond angle in CO_2 is 180°. Write the Lewis structure and draw the *p*-orbital picture. What is the hybridization state of the carbon?

2.44 What are the electron distributions in the bonding and antibonding orbitals of H_2^+ and H_2^-?

2.45 *Triplet methylene* is a very unstable molecule that has the formula $:CH_2$. Assume that the H—C—H bond angle is 180°. Draw the orbital picture (using lines for sigma bonds). What is the hybridization state of the carbon atoms?

2.46 Insert electron dots and draw resonance structures for each of the following ions:

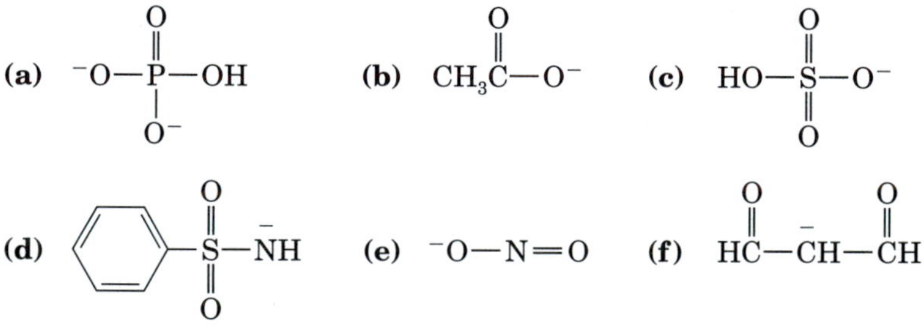

(a) $^-O{-}\overset{\overset{O}{\|}}{\underset{\underset{O^-}{|}}{P}}{-}OH$ (b) $CH_3\overset{\overset{O}{\|}}{C}{-}O^-$ (c) $HO{-}\overset{\overset{O}{\|}}{\underset{\underset{O}{\|}}{S}}{-}O^-$

(d) $\overset{\overset{O}{\|}}{\underset{\underset{O}{\|}}{S}}{-}\overset{-}{N}H$ (e) $^-O{-}N{=}O$ (f) $H\overset{\overset{O}{\|}}{C}{-}\overset{-}{C}H{-}\overset{\overset{O}{\|}}{C}H$

2.47 Which of the following pairs of structures are resonance structures? Explain your answers.

(a) $H{-}C{\equiv}\overset{+}{N}H$; $H{-}\overset{+}{C}{=}\overset{\cdot\cdot}{N}H$ (b) $CH_3CH{=}CHCH_3$; $CH_3CH_2CH{=}CH_2$

(c) 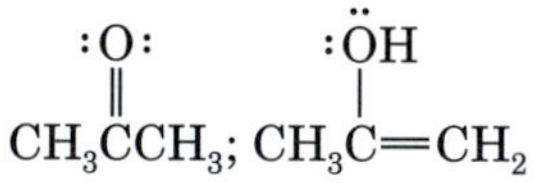 $CH_3\overset{\overset{:O:}{\|}}{C}CH_3$; $CH_3\overset{\overset{:\overset{\cdot\cdot}{O}H}{|}}{C}{=}CH_2$ (d)

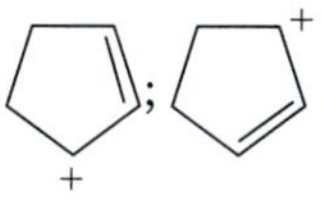

2.48 *Guanidines* and *amidines* are classes of compounds that react as bases. Draw the resonance structures for the conjugate acids—that is, the cations.

(a) 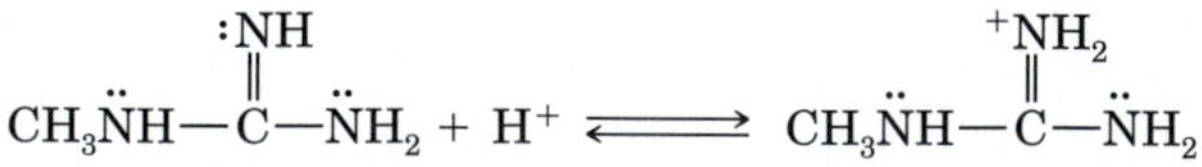 $CH_3\overset{\cdot\cdot}{N}H{-}\overset{\overset{:NH}{\|}}{C}{-}\overset{\cdot\cdot}{N}H_2 + H^+ \rightleftharpoons CH_3\overset{\cdot\cdot}{N}H{-}\overset{\overset{+NH_2}{\|}}{C}{-}\overset{\cdot\cdot}{N}H_2$

    a guanidine

(b) 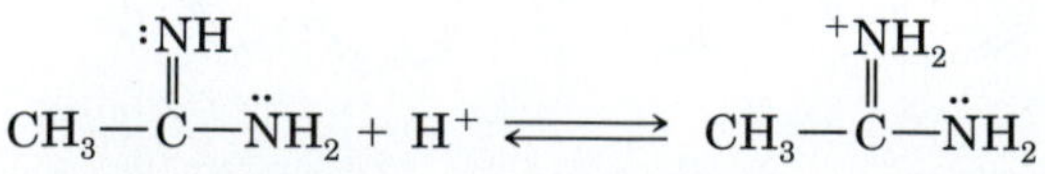 $CH_3{-}\overset{\overset{:NH}{\|}}{C}{-}\overset{\cdot\cdot}{N}H_2 + H^+ \rightleftharpoons CH_3{-}\overset{\overset{+NH_2}{\|}}{C}{-}\overset{\cdot\cdot}{N}H_2$

    an amidine

CHAPTER 3

STRUCTURAL ISOMERISM, NOMENCLATURE, AND ALKANES

In Chapter 2, we discussed the bonding of several compounds that contain only carbon and hydrogen. A compound containing only these two elements is called a **hydrocarbon.** Methane (CH_4), ethylene ($CH_2\!\!=\!\!CH_2$), and benzene (C_6H_6) are all examples of hydrocarbons.

Hydrocarbons with only sp^3 carbon atoms (that is, only single bonds) are called **alkanes** (or **cycloalkanes** if the carbon atoms are joined in rings). Some typical alkanes are methane, ethane (CH_3CH_3), propane ($CH_3CH_2CH_3$), and butane ($CH_3CH_2CH_2CH_3$). All these alkanes are gases found in petroleum deposits and used as fuels. Gasoline is primarily a mixture of alkanes. At the end of this chapter, we will discuss some of these aspects of hydrocarbon chemistry.

Alkanes and cycloalkanes are said to be **saturated hydrocarbons**, meaning "saturated with hydrogen." These compounds do not undergo reaction with hydrogen. Compounds containing pi bonds are said to be **unsaturated;** under the proper reaction conditions, they undergo reaction with hydrogen to yield saturated products.

Saturated hydrocarbons:

$$CH_4 \; + \; H_2 \; \xrightarrow{\text{catalyst}} \; \text{no reaction}$$

methane

an alkane

$+ \; H_2 \; \xrightarrow{\text{catalyst}} \;$ no reaction

cyclohexane

a cycloalkane

Unsaturated hydrocarbons:

$$CH_2{=}CH_2 + H_2 \; \xrightarrow{\text{Ni catalyst}} \; CH_3CH_3$$

ethylene   ethane

$$CH_3C{\equiv}CH + 2\,H_2 \; \xrightarrow{\text{Ni catalyst}} \; CH_3CH_2CH_3$$

propyne   propane

$+ \; 3\,H_2 \; \xrightarrow[\text{heat, pressure}]{\text{Ni catalyst}} \;$

benzene   cyclohexane

In this chapter, we will discuss primarily alkanes and cycloalkanes—the saturated hydrocarbons. Because saturated hydrocarbons lack a true functional group, their chemistry is not typical of most organic compounds; however, these compounds provide the carbon skeletons of organic compounds that do contain functional groups. Therefore, the discussion of saturated hydrocarbons provides an excellent opportunity for introducing the variations in the structures of organic compounds and the naming of organic compounds.

Section 3.1

Structural Isomers

Variations in the structures of organic compounds can arise from different numbers of atoms or types of atoms in molecules. However, variations in structure can also arise from *the order in which the atoms are bonded to one another in a molecule.* For example, we can write two different structural formulas for the molecular formula C_2H_6O. These two structural formulas represent two different compounds: dimethyl ether (bp $-23.6°$), a gas that has been used as a refrigerant and as an aerosol propellant, and ethanol (bp $78.5°$), a liquid that is used as a solvent and in alcoholic beverages.

$$CH_3OCH_3 \qquad\qquad CH_3CH_2OH$$

dimethyl ether   ethanol

a gas at room temperature   *a liquid at room temperature*

Two or more different compounds that have the same molecular formula are called **isomers** of each other. If the compounds with the same molecular formula have their atoms bonded in different orders, they are said to be **structural isomers** of each other. Structural isomers are also called **constitutional isomers.** (We will encounter other types of isomerism later.) Dimethyl ether and ethanol are examples of a pair of structural isomers.

Alkanes that contain three or fewer carbons have no isomers. In each case, there is only one possible way in which the atoms can be arranged.

No isomers: $\qquad$ CH_4 $\qquad$ CH_3CH_3 $\qquad$ $CH_3CH_2CH_3$

methane $\qquad$ ethane $\qquad\qquad$ propane

The four-carbon alkane (C_4H_{10}) has two possibilities for arrangement of the carbon atoms. As the number of carbon atoms increases, so does the number of possible isomers. The molecular formula C_5H_{12} represents three structural isomers and C_6H_{14} represents five structural isomers. As the number of carbon atoms increases, the number of possible isomers increases astronomically. There are more possible structural isomers of $C_{167}H_{336}$ (calculated to be 1.6×10^{80}) than there are particles in the universe (estimated to be about 10^{80}).

Structural isomers for C_4H_{10}:

$$\overset{\displaystyle CH_3}{\underset{\displaystyle |}{}}$$

$$CH_3CH_2CH_2CH_3 \qquad\qquad CH_3CHCH_3$$

butane, bp $-0.5°$ $\qquad$ methylpropane, bp $-12°$

a continuous-chain alkane   *a branched-chain alkane*

SAMPLE PROBLEM

Write formulas for the three structural isomers of C_5H_{12}.

Solution

(a) $CH_3CH_2CH_2CH_2CH_3$ $\quad$ **(b)** $CH_3CH_2\underset{|}{CH}CH_3$ $\quad$ **(c)** $CH_3\underset{|}{\overset{|}{C}}CH_3$

$\qquad\qquad\qquad\qquad\qquad\qquad\qquad\qquad CH_3 \qquad\qquad\qquad CH_3$

You might have listed $CH_3\underset{|}{CH}CH_2CH_3$. This structure is the same as (b).

$\qquad\qquad\qquad\qquad\qquad CH_3$

Different positions of a functional group in a molecule also lead to structural isomerism. The alcohols 1-propanol and 2-propanol are structural isomers with slightly different properties. The alkenes 1-butene and 2-butene are also structural isomers with different properties.

$$CH_3CH_2CH_2OH \qquad CH_3\overset{\overset{\displaystyle OH}{|}}{C}HCH_3$$

1-propanol   2-propanol

bp 97°   bp 82°

$$CH_2{=}CHCH_2CH_3 \qquad CH_3CH{=}CHCH_3$$

1-butene   2-butene

bp −6.3°   bp 3.7°

STUDY PROBLEMS

3.1 Draw the formula of a structural isomer for each of the following compounds. (There is more than one answer for each.)

(a) $CH_3\overset{\overset{\displaystyle O}{\|}}{C}H\underset{\underset{\displaystyle CH_3}{|}}{C}H$ (b) $(CH_3)_3COH$ (c)

3.2 For each of the following compounds, write a structural formula of an isomer that has a *different functional group*. (There may be more than one correct answer.)

(a) $CH_3\overset{\overset{\displaystyle O}{\|}}{C}OH$ (b) ⟨ ⟩—OCH_3 (c) $CH_3CH_2\overset{\overset{\displaystyle O}{\|}}{C}CH_2CH_3$

Example

An isomer of $CH_3CH_2CH_2OH$ that has a different functional group is $CH_3OCH_2CH_3$, but not $(CH_3)_2CHOH$, which is an isomer with the *same* functional group.

A. Isomers or Not?

Molecules can move around in space and twist and turn in "snake-like motion," as Kekulé once described it. (Kekulé, you will recall, was the chemist who proposed a structure for benzene.) We may write the same structure on paper in a number of ways. The *order of attachment* of the atoms is the factor that determines if two structural formulas represent isomers or the same compound. For example, all the following formulas show the same order of attachment of atoms; they all represent the same compound and do not represent isomers.

All represent the same compound:

$$CH_3CH_2\overset{\overset{\displaystyle OH}{|}}{C}HCH_3 \qquad \overset{\displaystyle CH_3CH_2}{CH_3\overset{|}{C}HOH} \qquad \overset{\displaystyle CH_3CH_2}{\underset{\displaystyle CH_3}{\diagdown CHOH\diagup}} \qquad CH_3\overset{\overset{\displaystyle OH}{|}}{C}HCH_2CH_3$$

SAMPLE PROBLEM

Which of the following pairs of formulas represent isomers and which represent the same compound?

(a) $CH_3\overset{\overset{\displaystyle Cl}{|}}{C}HCHCl_2$ and $Cl_2CH\overset{\overset{\displaystyle Cl}{|}}{C}HCH_3$ (b) ⬡ and ⬡

(c) Cl—⬡—Cl and ⬡ (with Cl substituents)

(d) $(CH_3)_3CCHCH_3$ and $(CH_3)_3CCH_2CH_2OH$
(with OH group)

Solution
The formulas in (c) and (d) represent pairs of structural isomers. In (a) and (b) the structures are oriented differently, but the order of attachment of the atoms is the same; therefore, the formulas in (a) and (b) represent the same compound.

B. A Ring or Unsaturation?

Given a molecular formula for a hydrocarbon, we can often deduce a reasonable amount of information about its structure. For example, all acyclic (noncyclic) alkanes have the general formula C_nH_{2n+2}, where n equals the number of carbon atoms in the molecule. Propane ($CH_3CH_2CH_3$, or C_3H_8) has three carbon atoms ($n = 3$). The number of hydrogen atoms in propane is $2n + 2$, or 8. Try the formula on the following alkanes:

$$CH_4 \qquad CH_3\overset{\overset{\displaystyle CH_3}{|}}{C}HCH_3 \qquad CH_3CH_2CH_2\overset{\overset{\displaystyle CH_3}{|}}{C}HCH_3$$

The presence of a ring or a double bond reduces the number of hydrogens in the formula by two for each double bond or ring; that is, a compound with the general formula C_nH_{2n} contains one **double-bond equivalent**—either one double bond or one ring.

$$CH_3CH_2CH_2CH_3 \qquad CH_3CH_2CH=CH_2 \qquad \square$$
$$C_4H_{10} \qquad\qquad C_4H_8 \qquad\qquad C_4H_8$$
$$C_nH_{2n+2} \qquad\qquad C_nH_{2n} \qquad\qquad C_nH_{2n}$$

an alkane   *an alkene*   *an cycloalkane*

A compound with the general formula C_nH_{2n-2} might have one triple bond, two rings, two double bonds, or one ring plus one double bond. We can say that the formula C_nH_{2n-2} represents a structure with two double-bond equivalents.

Three of many possible structural isomers for C_8H_{14}:

$$CH_3(CH_2)_5C{\equiv}CH \qquad \text{[two fused rings]} \qquad \text{[ring]}-CH_2CH_3$$

one triple bond   *two rings*   *one double bond and one ring*

3.3   How many double-bond equivalents does each of the following compounds have?

(a) 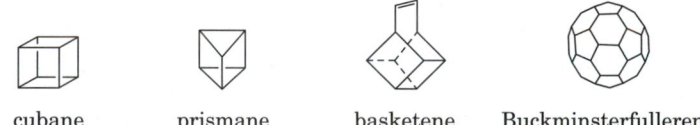   (b) $CH_2\!=\!CH\!-\!C\!\equiv\!CH$

Section 3.2
How Organic Nomenclature Developed

In the middle of the nineteenth century, the structures of many organic compounds were unknown. At that time, compounds were given names that were illustrative of their origins or their properties. Some compounds were named after friends or relatives of the chemists who first discovered them. For example, the name *barbituric acid* (and hence the common drug classification *barbiturates*) comes from the woman's name Barbara. At one time, the carboxylic acid HCO_2H was obtained from the distillation of red ants. This acid was given the name *formic acid* from the Latin *formica,* which means "ants." These names are both called **trivial names** or **common names.** In many respects, these trivial names are like nicknames; both these compounds have more formal (but seldom-used) names.

Even today, trivial names are coined for new compounds, especially for compounds with unwieldy formal names. Three examples of compounds with descriptive trivial names follow.

cubane   prismane   basketene   Buckminsterfullerene

Faced with the specter of an unlimited number of organic compounds, each with its own quaint name, organic chemists in the late nineteenth century decided to systematize organic nomenclature to correlate the names of compounds with their structures. The system of nomenclature that has been developed is called the **IUPAC system,** for the *International Union of Pure and Applied Chemistry,* the organization responsible for the continued development of organic nomenclature. Other systems of nomenclature, related to the IUPAC system, are now being developed to ease the computer-indexing of organic names. However, all the formal systems of nomenclature are very similar.

In the next section, we will present a brief survey of the IUPAC nomenclature system, along with some frequently used trivial names. In future chapters we will cover the nomenclature of each class of compounds in more detail. A more complete outline of organic nomenclature is presented in the appendix.

Section 3.3
A Survey of Organic Nomenclature

A. Continuous-Chain Alkanes

The IUPAC system of nomenclature is based upon the idea that the structure of an organic compound can be used to derive its name and, in turn, that a

Table 3.1	The first ten continuous-chain alkanes	

Number of Carbons	Structure	Name
1	CH_4	methane
2	CH_3CH_3	ethane
3	$CH_3CH_2CH_3$	propane
4	$CH_3(CH_2)_2CH_3$	butane
5	$CH_3(CH_2)_3CH_3$	pentane
6	$CH_3(CH_2)_4CH_3$	hexane
7	$CH_3(CH_2)_5CH_3$	heptane
8	$CH_3(CH_2)_6CH_3$	octane
9	$CH_3(CH_2)_7CH_3$	nonane
10	$CH_3(CH_2)_8CH_3$	decane

unique structure can be drawn for each name. The foundations of the IUPAC system are the names of the continuous-chain alkanes. The structures and names of the first ten continuous-chain alkanes are shown in Table 3.1. (Alkanes with a greater number of carbons are listed in Table A1 in the appendix.)

The compounds in Table 3.1 are arranged so that each compound differs from its neighbors by only a methylene (CH_2) group. Such a grouping of compounds is called a **homologous series,** and the compounds in such a list are called **homologs.**

Homologs generally undergo similar reactions. With knowledge of the chemistry of one member of a homologous series, you can predict the chemistry of its homologs with reasonable success. This fact was recognized in 1845 by the French chemist Charles Gerhardt (1816–1856). He wrote that "These [related] substances undergo reactions according to the same equation, and it is only necessary to know the reactions of one in order to predict the reactions of the others."

STUDY PROBLEM

3.4 Draw formulas for the first five members of each homologous series beginning with the following compounds:

(a) $H-CO_2H$ **(b)** $CH_3-\overset{\overset{\textstyle O}{\|}}{C}CH_3$ **(c)** $H-CH=CH_2$

From Table 3.1, you can see that all the alkane names end in **-ane,** which is the IUPAC ending denoting a saturated hydrocarbon. The first parts of the names of the first four alkanes (methane through butane) are derived from the traditional trivial names. The higher alkane names are derived from Greek or Latin numbers; for example, pentane is from the Greek *penta,* "five."

Let us briefly consider the derivations of the names for the first four alkanes. *Methane* (CH_4) is named after methyl alcohol (CH_3OH). *Methyl,* in turn, is a combination of the Greek words *methy* (wine) and *hyle* (wood). Methyl alcohol can be prepared by heating wood in the absence of air. Even today, this alcohol is sometimes referred to as wood alcohol.

The name *ethane* (CH_3CH_3) is derived from the Greek word *aithein,* which means "to kindle or blaze." Ethane is quite flammable. The name for *propane*

($CH_3CH_2CH_3$) is taken from the trivial name for the three-carbon carboxylic acid, propionic acid ($CH_3CH_2CO_2H$). *Propionic* is a combination of the Greek *proto* (first) and *pion* (fat). Propionic acid is the first (or lowest-molecular-weight) carboxylic acid to exhibit properties of fatty acids, which are acids that can be obtained from fats. *Butane* ($CH_3CH_2CH_2CH_3$) is named after butyric acid—the odorous component of rancid butter (Latin *butyrum,* "butter").

Starting with pentane, the continuous-chain alkanes have been more systematically named, with Greek prefixes generally used to indicate the number of carbon atoms in the molecule. The use of Greek prefixes is not totally consistent, however. In a few cases, such as nonane, the prefix is Latin rather than Greek.

B. Cycloalkanes

Cycloalkanes are named according to the number of carbon atoms in the ring, with the prefix **cyclo-** added.

cyclopentane   cyclohexane   cyclooctane

C. Side Chains

When alkyl groups or functional groups are attached to an alkane chain, the continuous chain is called the **root,** or **parent.** The groups are then designated in the name of the compound by prefixes and suffixes on the name of the parent.

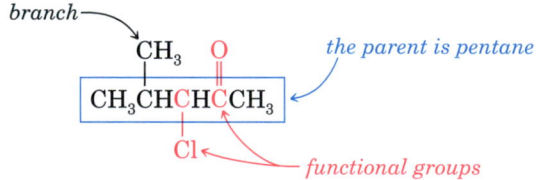

A **side chain,** or **branch,** is an alkyl group branching from a parent chain. A continuous-chain alkyl group is named after its own alkane parent with the **-ane** ending changed to **-yl.** (CH_4 is methane; therefore, the —CH_3 group is the *methyl group.* CH_3CH_3 is ethane; therefore, the —CH_3CH_2 group is the *ethyl group.*) The names for the first five continuous-chain alkyl groups are listed in Table 3.2.

Table 3.2 The first five continuous-chain alkyl groups	
Structure	**Name**
CH_3—	methyl
CH_3CH_2—	ethyl
$CH_3CH_2CH_2$—	propyl
$CH_3(CH_2)_2CH_2$—	butyl
$CH_3(CH_2)_3CH_2$—	pentyl

How is the name of a side chain incorporated into the name of a compound? To illustrate the technique, we will use hexane as the parent. If there is a methyl group on the second carbon of the hexane chain, the compound is named 2-methylhexane; *2-* for the position of attachment on the parent, *methyl* for the branch at this position, *hexane* for the parent. Methylhexane is one word.

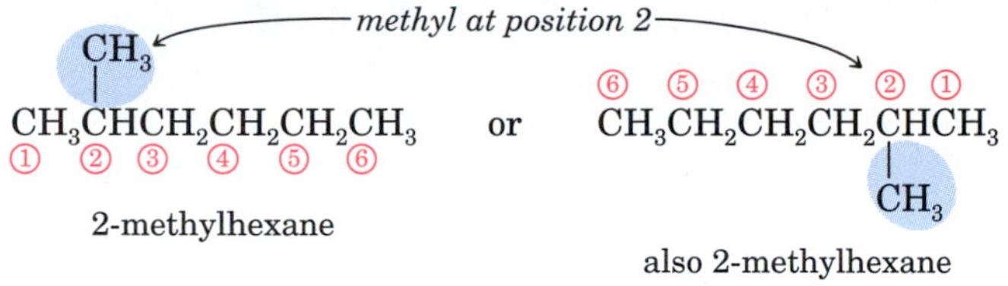

$$\text{methyl at position 2}$$

$$\underset{①②③④⑤⑥}{CH_3CHCH_2CH_2CH_2CH_3} \quad \text{or} \quad \underset{⑥⑤④③②①}{CH_3CH_2CH_2CH_2CHCH_3}$$

2-methylhexane

also 2-methylhexane

SAMPLE PROBLEM

Why is the following compound *not* named 1-methylhexane? What is its correct name?

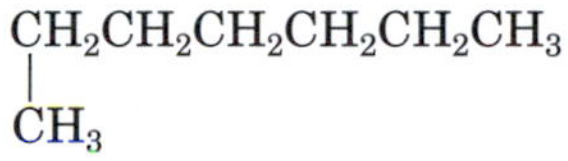

$$CH_2CH_2CH_2CH_2CH_2CH_3$$
$$|$$
$$CH_3$$

Solution
The structure contains a continuous chain of seven carbons; it is called *heptane.*

The general procedure to be followed in naming branched alkanes follows:

1. Find the longest continuous chain (the parent chain), which may or may not be shown in a straight line, and name this chain.

2. Number the parent chain, starting at the end closer to the branch.

3. Identify the branch and its position.

4. Attach the number and the name of the branch to the name of the parent.

SAMPLE PROBLEMS

Name the following compound:

$$CH_3CH_2CHCH_2CH_2CH_2CH_3$$
$$|$$
$$CH_2CH_2CH_3$$

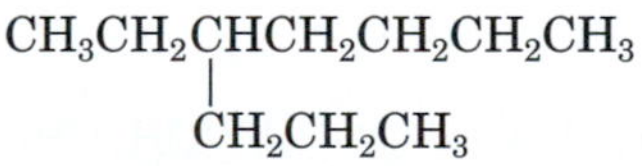

Solution
The longest continuous chain is eight carbons. The parent is *octane.*

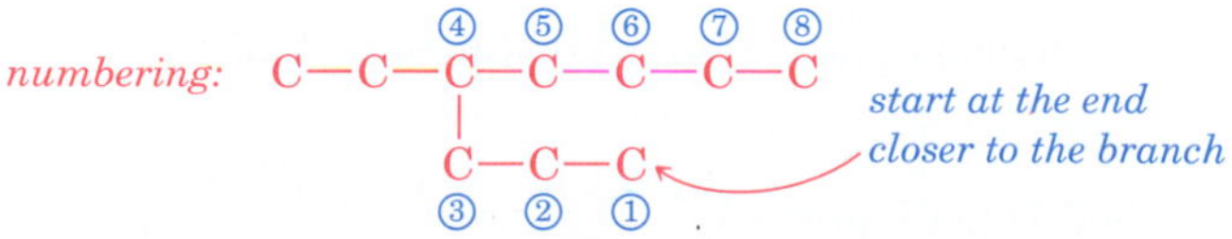

numbering: $\underset{④\;⑤\;⑥\;⑦\;⑧}{C-C-C-C-C-C-C}$

$\underset{③\quad②\quad①}{C-C-C}$

start at the end closer to the branch

alkyl group: ethyl at carbon 4

name: 4-ethyloctane

Write the structure for 3-methylpentane.

Solution

the parent chain: $C-C-C-C-C$ (pentane)

numbering: $\underset{①\ \ ②\ \ ③\ \ ④\ \ ⑤}{C-C-C-C-C}$

the alkyl group: $\overset{\overset{\displaystyle CH_3}{|}}{C-C-C-C-C}$ (3-methyl)

add the hydrogens (each carbon must have four bonds):

$$\overset{\overset{\displaystyle CH_3}{|}}{CH_3CH_2CHCH_2CH_3}$$

Name the following structure. In this name, a prefix number is not needed. Why not?

$$(CH_3)_2CHCH_2CH_3$$

Solution
Methylbutane. A prefix number is not needed because there is only one methylbutane:

$$\overset{\overset{\displaystyle CH_3}{|}}{CH_3CHCH_2CH_3} \quad \text{is the same as} \quad \overset{\overset{\displaystyle CH_3}{|}}{CH_3CH_2CHCH_3}$$

The only other place to attach the methyl group would be on an end carbon; however, this compound is pentane, not methylbutane:

$$\underset{\underset{CH_3}{|}}{CH_2CH_2CH_2CH_3}$$

STUDY PROBLEMS

3.5 Name the following compounds:

(a) $\underset{\underset{CH_2CH_2CH_3}{|}}{CH_3CH_2CHCH_2CH_3}$ **(b)** $\overset{\overset{\displaystyle CH_3}{|}}{CH_3CHCH_3}$

3.6 Write the structure of the following compounds:

(a) 3-ethylheptane   **(b)** 4-ethyl-5-methylnonane

D. Branched Side Chains

An alkyl group may be branched, rather than a continuous chain. The following examples show branched side chains on a cyclohexane ring and on a heptane chain, respectively.

$$CH_3CH_2CH_3 \quad \text{(structures)}$$

Common branched groups have specific trivial names. For example, the two propyl groups are called the **propyl group** and the **isopropyl group.**

$$CH_3CH_2CH_2-$$

propyl (or *n*-propyl)

$$CH_3\overset{\displaystyle CH_3}{\underset{\displaystyle |}{CH}}-$$

isopropyl

To emphasize that a side chain is *not* branched, a prefix *n*- (for *normal*-) is sometimes used. (The *n*- is redundant since the absence of a prefix indicates a continuous chain.) The prefix iso- (for *isomeric*) is used to indicate a methyl branch at the end of the alkyl chain.

A four-carbon side chain has four structural possibilities. The **butyl** group is the continuous-chain group. The **isobutyl group** is the four-carbon group with a methyl branch at the end of the chain. These two butyl groups are named similarly to the propyl group.

methyl branch at end

Note! "iso" is only used when substituents off the main chain have branches themselves.

$$CH_3CH_2CH_2CH_2-$$

butyl

$$CH_3\overset{\displaystyle CH_3}{\underset{\displaystyle |}{CH}}CH_2-$$

isobutyl

The ***secondary*-butyl group** (abbreviated *sec*-butyl) has two carbons bonded to the head (attaching) carbon. The ***tertiary*-butyl group** (abbreviated *tert*-butyl or *t*-butyl) has three carbons attached to the head carbon.

two C's on head carbon

three C's on head carbon

$$CH_3CH_2\overset{\displaystyle CH_3}{\underset{\displaystyle |}{CH}}- \quad \quad CH_3-\overset{\displaystyle CH_3}{\underset{\displaystyle |}{\underset{\displaystyle CH_3}{\overset{\displaystyle |}{C}}}}-NH_2$$

sec-butyl   *tert*-butyl

SAMPLE PROBLEM

Name the following compounds:

(a) (cyclohexane with $-\overset{\displaystyle CH_3}{\underset{\displaystyle |}{CH}}CH_2CH_3$)

(b) $CH_3CH_2CH_2\overset{\displaystyle CH_3CHCH_3}{\underset{\displaystyle |}{CH}}CH_2CH_2CH_3$

Solution

(a) *sec*-butylcyclohexane;   (b) 4-isopropylheptane.

> **3.7** Give structures for **(a)** propylcyclohexane; **(b)** isobutylcyclohexane; **(c)** 4-*tert*-butyloctane.

E. Multiple Branches

If two or more branches are attached to a parent chain, more prefixes are added to the parent name. The prefixes are placed alphabetically, each with its number indicating the position of attachment to the parent.

$$
\begin{array}{cc}
\text{CH}_3 & \text{CH}_3 \\
| & | \\
\text{CH}_3\text{CH}_2\text{CHCHCH}_2\text{CH}_2\text{CH}_3 & \text{CH}_3\text{CH}_2\text{CCH}_2\text{CH}_2\text{CH}_3 \\
| & | \\
\text{CH}_2\text{CH}_3 & \text{CH}_2\text{CH}_3
\end{array}
$$

4-ethyl-3-methylheptane   3-ethyl-3-methylhexane

If two or more substituents on a parent are the same (such as two methyl groups or three ethyl groups), these groups are consolidated in the name. For example, *dimethyl* means "two methyl groups" and *triethyl* means "three ethyl groups." The prefixes (*di-, tri-,* etc.) that denote number are listed in Table 3.3. In a name, the *di-* or *tri-* prefix is preceded by position numbers. If *di* is used, two numbers are required; if *tri* is used, three numbers must be present. Note the use of commas and hyphens in the following examples:

2,2-dimethylpentane   1,3,5-triethylcyclohexane

Name the following compounds:

Solution:

(a) 1,1-diisopropylcyclohexane   **(b)** 4-methyl-1,2-dipropylcyclopentane

(c) 4,5-diisopropylnonane. (We number a ring so that the prefix numbers are as low as possible. We alphabetize *dipropyl* as *propyl.*)

Table 3.3	Prefixes for naming multiple substituents

Number	Prefix
2	di-
3	tri-
4	tetra-
5	penta-
6	hexa-

F. Other Prefix Substituents

Like alkyl branches, some functional groups are named as prefixes to the parent name. Some of these groups and their prefix names are listed in Table 3.4. The rules governing the use of these prefixes are identical to those for the alkyl groups except that the parent is the longest continuous chain *containing the functional group*. The position of the functional group is specified by a number (as low as possible), and identical groups are preceded by *di-* or *tri-*.

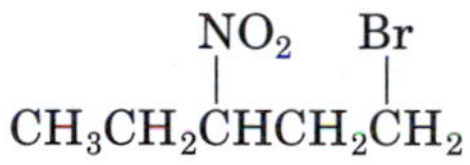

$$CH_3CH_2\overset{\overset{\displaystyle NO_2}{|}}{C}H\overset{\overset{\displaystyle Br}{|}}{C}H_2CH_2 \qquad CH_3-\overset{\overset{\displaystyle Cl}{|}}{\underset{\underset{\displaystyle Cl}{|}}{C}}-\overset{\overset{\displaystyle Cl}{|}}{\underset{\underset{\displaystyle Cl}{|}}{C}}-CH_3$$

1-bromo-3-nitropentane 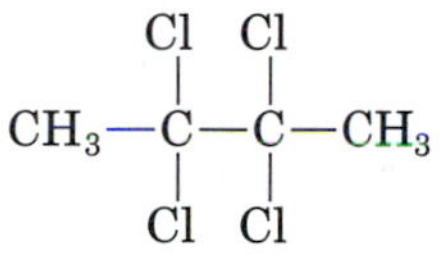   2,2,3,3-tetrachlorobutane

Table 3.4	Some substituents named as prefixes

Substituent	Prefix Name
—NO_2	nitro-
—F	fluoro-
—Cl	chloro-
—Br	bromo-
—I	iodo-

STUDY PROBLEMS

3.8   Give structures for:

   (a) 1,1,2-trichloroethane   **(b)** 1,2-dichloro-4-nitrocyclohexane

3.9   Name the following compounds:

(a) 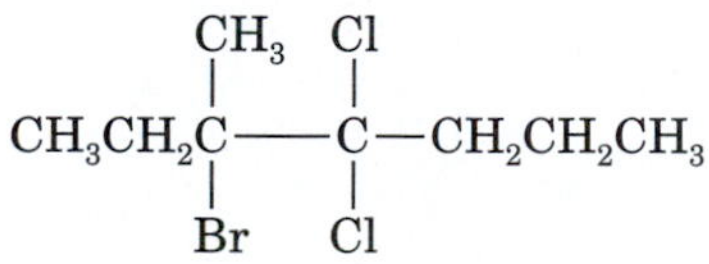 $CH_3CH_2\overset{\overset{\displaystyle CH_3}{|}}{\underset{\underset{\displaystyle Br}{|}}{C}}-\overset{\overset{\displaystyle Cl}{|}}{\underset{\underset{\displaystyle Cl}{|}}{C}}-CH_2CH_2CH_3$

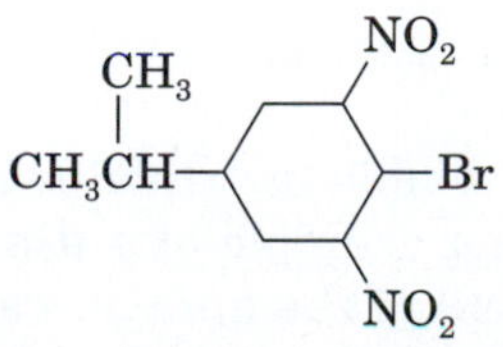

Note: when alphatizing "iso" counts ie. if methyl ; isopropyl isopropyl is first

(b) $CH_3\overset{\overset{\displaystyle CH_3}{|}}{C}H-$ [cyclohexane ring with NO_2 at top, Br at right, NO_2 at bottom]

G. Alkenes and Alkynes

In IUPAC nomenclature, carbon–carbon unsaturation is always designated by a change in the *ending* of the parent name. As we have indicated, if the parent hydrocarbon contains no double or triple bonds, the suffix **-ane** is used. If a double bond is present, the **-ane** ending is changed to **-ene**; the general name for a hydrocarbon with a double bond is **alkene.** A triple bond is indicated by **-yne;** a hydrocarbon containing this group is an **alkyne.**

<table>
<tr><td align="center">CH_3CH_3</td><td align="center">$CH_2\!=\!CH_2$</td><td align="center">$HC\!\equiv\!CH$</td></tr>
<tr><td align="center">ethane</td><td align="center">ethene</td><td align="center">ethyne</td></tr>
<tr><td align="center">*an alkane*</td><td align="center">(*trivial name:* ethylene)
an alkene</td><td align="center">(*trivial name:* acetylene)
an alkyne</td></tr>
</table>

When the parent contains four or more carbons, a prefix number must be used to indicate the position of the double or triple bond. The chain is numbered so as to *give the double or triple bond as low a number as possible,* even if a prefix group must then receive a higher number. Only a single number is used for each double and triple bond; it is understood that the double or triple bond *begins at this numbered position and goes to the carbon with the next higher number.* Thus, a prefix number 2 means the double or triple bond is between carbons 2 and 3, not between carbons 2 and 1.

C=C starts at carbon 1   *C=C starts at carbon 2*

$CH_2\!=\!CHCH_2CH_3$ $CH_3CH\!=\!CHCH_3$

1-butene   2-butene

a branch precedes the name

CH_3
|
$CH_3CHCH_2CH_2C\!\equiv\!CH$

5-methyl-1-hexyne

In some names, the prefix 1 is implied, but not written, for a carbon–carbon double bond or other principal functional group. The number can be omitted only in cases where there is no doubt about the position of a functional group.

carbon 1

—Cl

1-chlorocyclohexene
instead of 1-chloro-1-cyclohexene

If a structure contains more than one double or triple bond, the name becomes slightly more complex. The following example of a **diene,** a compound with two double bonds, shows the placement of the numbers and the *di.* (Note that an *a* is inserted before *di* to ease the pronunciation.)

$$CH_2\!\!=\!\!CHCH\!\!=\!\!CH_2$$

1,3-butadiene

— *two double bonds*

four C's

The nomenclature of alkenes, as well as that of other compounds, is discussed in greater detail in later chapters.

SAMPLE PROBLEM

Name the following compounds:

(a) $(CH_3)_2CHCH_2CH_2CH\!\!=\!\!CH_2$ **(b)** $CH_2\!\!=\!\!CHCH\!\!=\!\!CHCH\!\!=\!\!CH_2$

(c) $CH\!\!\equiv\!\!CC\!\!\equiv\!\!CH$

Solution
(a) 5-methyl-1-hexene **(b)** 1,3,5-hexatriene **(c)** 1,3-butadiyne, or simply butadiyne (the numbers are understood because there is only one possible way to position two triple bonds in a four-carbon chain)

STUDY PROBLEMS

3.10 Name the following compounds:

    (a) $CH_3CH\!\!=\!\!CHCH_3$ **(b)**   **(c)** $CH_3CH_2CCl\!\!=\!\!CHCl$

3.11 Give structures for:

    (a) cyclooctene   **(b)** 1,3-hexadiyne   **(c)** 1-isopropylcyclopropene

H. Alcohols

In the IUPAC system, the name of an alcohol (ROH) is the name of the parent hydrocarbon with the final **-e** changed to **-ol.** Prefix numbers are used when necessary; the hydroxyl group (—OH) receives the lowest prefix number possible.

$CH_3CH_2CH_3$	$CH_3CH_2CH_2OH$	$\overset{\displaystyle OH}{\underset{\displaystyle}{|}}$ CH_3CHCH_3
propane	1-propanol	2-propanol

$\overset{\displaystyle CH_3}{\underset{\displaystyle}{|}}$ $CH_3CHCH_2CH_3$	$\overset{\displaystyle CH_3}{\underset{\displaystyle}{|}}$ $HOCH_2CHCH_2CH_3$	$\overset{\displaystyle CH_3}{\underset{\displaystyle}{|}}$ $CH_3CHCH_2CH_2OH$
methylbutane	2-methyl-1-butanol	3-methyl-1-butanol

cyclopentane   cyclopentanol—OH

I. Aldehydes and Ketones

Because an aldehyde (RCHO) contains a carbonyl group bonded to a hydrogen atom, the aldehyde group must be at the beginning of a carbon chain. The aldehyde carbon is considered carbon 1; therefore, no number is used in the name to indicate the position. The ending of an aldehyde name is **-al.**

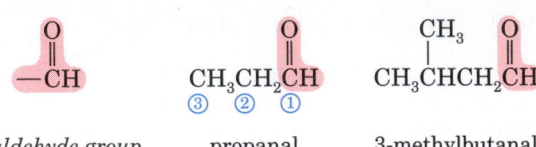

the aldehyde group	propanal	3-methylbutanal

   A keto group, by definition, cannot be at the beginning of a carbon chain. Therefore, except for propanone and a few other simple ketones, a prefix number is necessary. The chain should be numbered to give the carbonyl group the lowest possible number. The ending for a ketone name is **-one.**

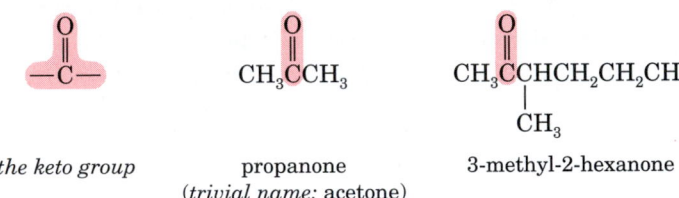

the keto group	propanone (*trivial name:* acetone)	3-methyl-2-hexanone

J. Carboxylic Acids

A carboxyl group, like an aldehyde group, must be at the beginning of a carbon chain and again contains the first carbon atom (carbon 1). Again, a number is not needed in the name. The ending for a carboxylic acid name is **-oic acid.**

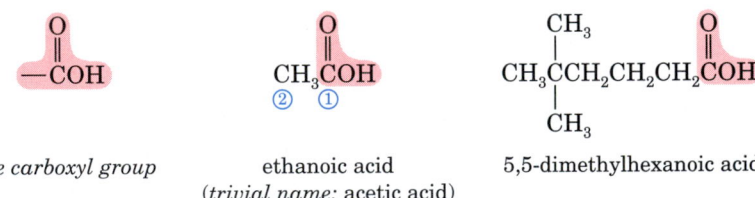

the carboxyl group	ethanoic acid (*trivial name:* acetic acid)	5,5-dimethylhexanoic acid

STUDY PROBLEM

3.12   Write structures and names for an aldehyde, a ketone, and a carboxylic acid, each containing five carbons in a continuous chain.

K. Benzene Compounds

The benzene ring is considered to be a parent in the same way that a continuous-chain alkane is. Alkyl groups, halogens, and the nitro group are named as prefixes to benzene.

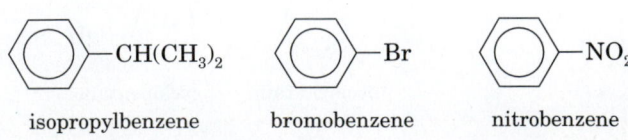

isopropylbenzene	bromobenzene	nitrobenzene

When a benzene ring is attached to an alkane chain with a functional group or to an alkane chain of seven or more carbon atoms, benzene is considered a substituent instead of a parent. The name for benzene as a substituent is **phenyl.**

the phenyl group

$(C_6H_5—)$

2-phenylethanol

2-phenyloctane

L. Conflicts in Numbering

A structure that has more than one type of substituent can sometimes be numbered in more than one way. Should the name for $ClCH_2CH=CH_2$ be 1-chloro-2-propene or 3-chloro-1-propene? To cover such situations, a *system of nomenclature priorities* for prefix numbers has been developed (Table 3.5); the higher-priority substituent receives the lower number.

From Table 3.5, we can see that a double bond is higher in priority than Cl. When numbering a carbon chain, we give the double bond the lowest possible number. The name of $ClCH_2CH=CH_2$ is therefore 3-chloro-1-propene, and not 1-chloro-2-propene. Similarly, the following compound is numbered to give the higher-priority group the lower prefix number.

start at end closer to OH

Cl_2CHCH_2OH
② ①

2,2-dichloroethanol

In a molecule containing both a double bond and a triple bond, the longest continuous chain containing these groups is numbered to give the lower prefix numbers. If there is a choice, the double bond receives the lower number. In the name, the suffix for the double bond precedes that for the triple bond.

Table 3.5	Nomenclature priorities of selected functional groups	
	Partial Structure	Name
	$—CO_2H$ carboxylic acid	-oic acid
	$—\overset{O}{\overset{\|}{C}}H$ Aldehyde	-al
increasing priority	$—\overset{O}{\overset{\|}{C}}—$ Ketone	-one
	$—OH$ Alcohol	-ol
	$—NR_2$ Amine	-amine
	$>C=C<$ Alkene	-ene
	$—C\equiv C—$ Alkyne	-yne
	$R—, C_6H_5—, Cl—, Br—, —NO_2$, etc.	prefix substituents

$$CH_2{=}CH{-}C{\equiv}C{-}CH_3 \qquad CH{\equiv}CCH{=}CHCH_3$$

1-penten-3-yne 3-penten-1-yne ←

*lower prefix numbers
than 2-penten-4-yne*

Other examples illustrating the use of nomenclature priority will be discussed in subsequent chapters.

STUDY PROBLEM

3.13   Name each of the following compounds:

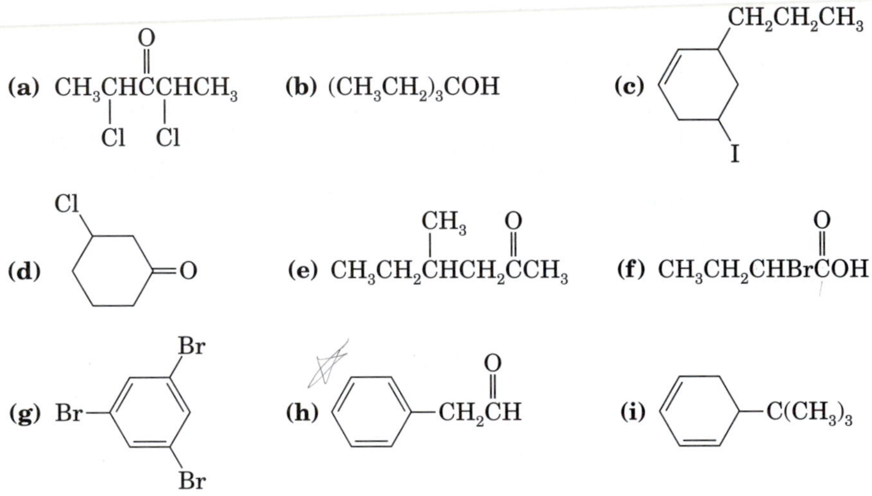

(a) $\overset{\displaystyle O}{\underset{\underset{Cl}{|}}{CH_3CH}}\overset{\displaystyle \parallel}{C}\underset{\underset{Cl}{|}}{CHCH_3}$

(b) $(CH_3CH_2)_3COH$

(c)

(d)

(e) $CH_3CH_2\underset{\underset{CH_3}{|}}{CH}CH_2\overset{\overset{O}{\parallel}}{C}CH_3$

(f) $CH_3CH_2CHBr\overset{\overset{O}{\parallel}}{C}OH$

(g) Br—

(h) —$CH_2\overset{\overset{O}{\parallel}}{C}H$

(i) —$C(CH_3)_3$

Section 3.4
Alkanes

Most organic compounds have a portion of their structures composed of carbon atoms and hydrogen atoms. A fat is one example of an organic compound with ester groups and with long hydrocarbon chains, which may be alkyl or alkenyl (containing a carbon–carbon double bond).

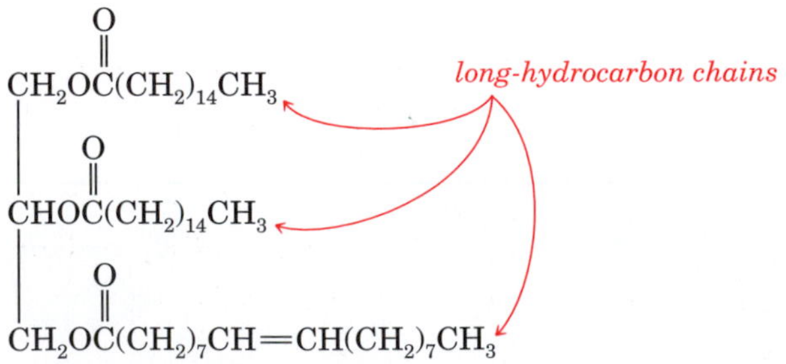

$$CH_2O\overset{\overset{O}{\parallel}}{C}(CH_2)_{14}CH_3$$
$$CHO\overset{\overset{O}{\parallel}}{C}(CH_2)_{14}CH_3$$
$$CH_2O\overset{\overset{O}{\parallel}}{C}(CH_2)_7CH{=}CH(CH_2)_7CH_3$$

long-hydrocarbon chains

a typical animal fat

Early chemists did not know the molecular structure of a fat, but they did know that many compounds containing long hydrocarbon chains behave similarly to fats—most of these compounds are insoluble in water and are less dense than water. For this reason, compounds with hydrocarbon chains are referred to as **aliphatic compounds** (Greek *aleiphatos,* "fat"). The term *aliphatic compound* is usually contrasted to *aromatic compound,* such as benzene or a substituted benzene.

Some aliphatic compounds:

$$CH_3CHCH_2CH_3 \qquad CH_3CH_2CH_2OH \qquad CH_3CH_2NHCH_2CH_3$$
with a $\overset{\displaystyle CH_3}{\underset{|}{}}$ on the second carbon

Some of the physical and chemical properties of an aliphatic compound arise from the alkyl part of its molecules. Therefore, much of what we have to say about alkanes and cycloalkanes is true for other organic compounds as well. Of course, the properties of a compound are also greatly determined by any functional groups that may be present. For example, a hydroxyl group in a molecule leads to a hydrogen bonding and a large change in physical properties. Ethane (CH_3CH_3) is a gas at room temperature, while ethanol (CH_3CH_2OH) is a liquid.

A. Chemical Properties of Alkanes

Alkanes and cycloalkanes are relatively unreactive compared with organic compounds containing functional groups. For example, many organic compounds undergo reaction with strong acids, bases, oxidizing agents, or reducing agents. Alkanes and cycloalkanes generally do not react with these reagents. Because of their lack of reactivity, alkanes are sometimes referred to as **paraffins** (Latin *parum affinis,* "slight affinity"), a term first used by the Austrian chemist K. Reichenback (1788–1869) to describe a wax he had isolated from the destructive distillate of beechwood.

We will discuss two principal reactions of alkanes in this text. One is the *reaction with halogens,* such as chlorine gas. We will present this reaction in detail in Chapter 6. The other important reaction of alkanes is *combustion.* The remainder of this chapter will be concerned primarily with the combustion of alkanes and their use as a source of energy.

Halogenation:

$$CH_4 + Cl_2 \xrightarrow{\text{light}} CH_3Cl + HCl + \text{other products}$$
methane $\qquad\qquad\qquad$ chloromethane

$$CH_3CH_2CH_3 + Br_2 \xrightarrow{\text{light}} CH_3\overset{\displaystyle Br}{\underset{|}{C}}HCH_3 + HBr + \text{other products}$$
propane $\qquad\qquad\qquad$ 2-bromopropane

Combustion:

$$CH_4 + 2\,O_2 \xrightarrow{\text{spark}} CO_2 + 2\,H_2O$$
carbon dioxide

$$CH_3CH_2CH_3 + 5\,O_2 \xrightarrow{\text{spark}} 3\,CO_2 + 4\,H_2O$$

B. Combustion

Combustion is the process of burning—the rapid reaction of a compound with oxygen. Combustion is accompanied by the release of light and heat, two forms of energy that humans have sought since they first built a fire and found that it kept them warm. Although we will present the subject of combustion under the heading of alkanes, keep in mind that almost all organic compounds can burn.

Combustion of organic mixtures, such as wood, is not always a simple conversion to CO_2 and H_2O. Instead, combustion is the result of a large number of complex reactions. One type of reaction that occurs is **pyrolysis,** the thermal fragmentation of large molecules into smaller molecules in the absence of oxygen. Pyrolysis of large molecules in wood, for example, yields smaller gaseous molecules that then react with oxygen above the surface of the wood. This reaction with oxygen gives rise to the flames. On the surface of the wood, a slow, but very hot, oxidation of the carbonaceous residue takes place. Most of the heat from a wood or coal fire results from this slow oxidation, rather than from the actual flames.

Complete combustion is the conversion of a compound to CO_2 and H_2O. If the oxygen supply is insufficient for complete combustion, **incomplete combustion** occurs. Incomplete combustion leads to carbon monoxide, or sometimes carbon as carbon black or soot.

Incomplete combustion:

$$2\ CH_3CH_2CH_3 + 7\ O_2 \longrightarrow 6\ CO + 8\ H_2O$$
carbon
monoxide

$$CH_3CH_2CH_3 + 2\ O_2 \longrightarrow 3\ C + 4\ H_2O$$
carbon

Section 3.5
The Hydrocarbon Resources

A. Natural Gas and Petroleum

Petroleum and natural gas are major energy sources. In 1994, world use of petroleum was approximately 2.5×10^{10} barrels (42 US gallons per barrel). In the same year, about 2.1×10^{13} cubic feet of natural gas were consumed in the United States alone.

Natural gas contains 60–90% methane. Its other components are ethane and propane along with nitrogen and carbon dioxide. The natural gas found in the Texas Panhandle and in Oklahoma also contains helium. Deposits of natural gas are usually found with petroleum deposits.

Petroleum, also called *crude oil,* is a complex mixture of aliphatic and aromatic compounds, including sulfur and nitrogen compounds (1–6%). In fact, over 500 compounds have been detected in a single sample of petroleum. The actual composition varies from deposit to deposit. Although the origin of petroleum is still subject to debate, it was probably formed by the anaerobic decay of microorganisms or other organic material.

Because crude oil is a mixture of many compounds, it is not very useful in the form removed from the ground. Separating the crude oil into useful com-

Table 3.6	Fractions of straight-run distillation		
Boiling Range, °C	Number of Carbons	Name	Use
under 30	1–4	gas fraction	heating fuel
30–180	5–10	gasoline	automobile fuel
180–230	11–12	kerosene	jet fuel
230–305	13–17	gas oil	diesel fuel, heating fuel
305–405	18–25	heavy gas oil	heating fuel

Residue: (1) Volatile oils: lubricating oils, paraffin wax, and petroleum jelly. (2) Nonvolatile material: asphalt and petroleum coke.

ponents is called **refining.** The first step in refining is a fractional distillation, called **straight-run distillation.** The fractions that are collected are listed in Table 3.6.

The gasoline fraction of straight-run distillation is too scanty for the needs of our automobile-oriented society, and straight-run gasoline has poor burning characteristics in automobile engines. To increase both quantity and quality of the gasoline fraction, the higher-boiling fractions are subjected to *cracking* and *reforming.*

Catalytic cracking is the process of heating the high-boiling material under pressure in the presence of a catalyst (usually finely divided, acid-washed aluminum silicate clay). Under these conditions, large molecules are cracked, or broken, into smaller fragments, and continuous-chain alkanes are isomerized to branched-chain alkanes. Catalytic cracking was invented by Eugene Houdry (1892–1962), a French mechanical engineer. The first successful commercial plant using catalytic cracking was brought on line in the United States just before World War II. The process played a major role in the production of aviation gasoline during that war and after.

Catalytic reforming converts alkanes to aromatic compounds. In automobile engines, continuous-chain hydrocarbons burn unevenly and cause knocking, which is the ticking or rattling noise heard when a car accelerates uphill. To prevent knocking, branched alkanes and aromatic compounds, which burn more evenly than continuous-chain alkanes, are added to automobile fuels.

Steam cracking is a technique for converting alkanes to alkenes. Alkenes and aromatic compounds are used to make plastics and other synthetic organic materials.

At one time, isooctane (a trivial name for 2,2,4-trimethylpentane) was the alkane with the best antiknock characteristics for automobile engines, and heptane was the poorest. These two compounds were used to develop an octane rating for petroleum fuels.

$$CH_3CCH_2CHCH_3 \qquad CH_3(CH_2)_5CH_3$$

2,2,4-trimethylpentane   heptane
("isooctane")

A gasoline's **octane number,** which is a rating of the gasoline's quality in automobile engines, is assigned by comparing the gasoline's burning characteristics with those of isooctane–heptane mixtures. An octane number of 100 means that the gasoline is equivalent in burning characteristics to pure isooctane. Gasoline with an octane number of 0 is equivalent to pure heptane. An octane number of 75 is given to gasoline that is equivalent to a mixture of 75% isooctane and 25% heptane. Straight-run gasoline has an octane number of about 70. Some compounds have better burning characteristics than isooctane; therefore, octane numbers of greater than 100 are possible.

Since 1973, octane numbers at the gasoline pump have been quoted using the (R + M)/2 method. R stands for **research octane number** (RON) and M for **motor octane number** (MON). RON values are obtained by testing the fuel under rather mild conditions. MON values are obtained by testing the fuel under higher engine speeds and temperatures. The average of the RON and MON values is a better measure of the actual performance of the fuel.

STUDY PROBLEM

3.14   A gasoline fuel has an octane rating of 85. What percentages of heptane and isooctane must a test fuel contain to have the same burning characteristics?

Additives are also mixed with gasoline to increase octane ratings. Some common additives with octane numbers over 100 are benzene, ethanol, *tert*-butyl alcohol [$(CH_3)_3COH$], and *tert*-butyl methyl ether [$(CH_3)_3COCH_3$]. The additive mixture used in leaded gasoline, called *Ethyl fluid,* contains approximately 65% tetraethyllead, 25% 1,2-dibromoethane, and 10% 1,2-dichloroethane. The halogenated hydrocarbons are essential for conversion of the lead to volatile lead halides, which are removed from the cylinder in the exhaust.

$(CH_3CH_2)_4Pb$	$BrCH_2CH_2Br$	$ClCH_2CH_2Cl$
tetraethyllead	1,2-dibromoethane (ethylene dibromide, EDB)	1,2-dichloroethane (ethylene dichloride)

A gasoline engine puts forth a variety of pollutants: unburned hydrocarbons, carbon monoxide, and nitrogen oxides. Catalytic converters are now installed in automobiles to convert nonoxidized and partly oxidized compounds to more highly oxidized and acceptable forms of exhaust. The platinum catalyst used in these converters is "poisoned" (made nonfunctional) by lead compounds; therefore, leaded gasoline cannot be used in cars equipped with catalytic converters without destroying the catalyst.

B. Coal

Coal is formed by the bacterial decomposition of plants under varying degrees of pressure. The final product is a complex, heterogeneous mixture of organic compounds and inorganic material. Coal contains aromatic hydrocarbons bonded together by CH_2 groups. It has no definitive organic structure. The inorganic content of coal can be more clearly defined. It consists of kaolinite (aluminum silicates), pyrite (FeS_2), and calcite ($CaCO_3$). Since coal is used primarily as a fuel, the mineral content is a nuisance because it gives rise to air and water pollution and creates a waste disposal problem. The sulfur content of some coal is particularly troublesome because it leads to severe air pollution and acid rain.

Type of Coal	% Carbon	BTU/lb	Comments
Table 3.7		**Properties of Coal**	
anthracite	> 86	> 15,000	reflective surface
bituminous	70–86	9900–15,000	plant structure partly recognizable
lignite	60–70	7200–9900	plant structure recognizable but no free cellulose
peat	< 60	< 7200	free cellulose and recognizable plant material

Coal is classified by its carbon content: *anthracite,* or hard coal (highest carbon content), followed by *bituminous* (soft) coal, *lignite,* and, finally, *peat.* See Table 3.7 for a comparison of these types of coals.

When coal is subjected to heat and distillation in the absence of air, a process called **destructive distillation,** three crude products result: *coal gas* (primarily CH_4 and H_2), *coal tar* (the condensable distillate), and *coke* (the residue). Both coal gas and coke are useful fuels. (Today, coke is used primarily in the manufacture of steel.) Coal tar is rich in aromatic compounds, which are formed in the destructive distillation.

Until petroleum became plentiful and cheap in the 1940s, coal was a primary source of synthetic organic compounds. Currently, however, over 90% of the organic chemicals produced in the United States are synthesized from petroleum. Petroleum-refining processes are less costly and give rise to less pollution than most coal-refining processes. Unfortunately, the world's petroleum resources are rapidly being depleted and becoming more expensive, while the coal reserves of the world are still very extensive. Economical ways to convert coal to useful fuels and chemicals (with a minimum of air pollution) are under investigation at the present time.

The conversion of coal to gaseous or liquid fuels, called synthetic fuels or "syn fuels," is referred to as **coal gasification** or **coal liquefaction,** respectively. Many gasification plants use the German-developed **Lurgi process,** or modifications of this technique, in which a bed of coal is treated with steam at high temperatures to yield *synthesis gas* ($CO + H_2$). Synthesis gas itself is only a moderately efficient fuel, and carbon monoxide is highly toxic. Thus, synthesis gas is treated with additional hydrogen to yield methane.

Coal gasification:

$$C + H_2O \xrightarrow{\text{heat}} \underbrace{CO + H_2}_{\text{``synthesis gas''}} \xrightarrow[\text{Ni catalyst}]{2\,H_2} CH_4 + H_2O$$

coal steam   methane

The liquefaction of coal is its conversion to liquid alkanes. The classical process for accomplishing this conversion is the **Fischer–Tropsch synthesis,** which was developed in Germany during World War II.

Summary

Compounds with the same molecular formula, but different structures (order of attachment of the atoms), are **structural isomers** of each other.

The **IUPAC system of nomenclature** is based on the names of the *continuous-chain alkanes* as parents. If a hydrocarbon chain forms a ring, the prefix *cyclo-* is added to the alkane name. Branches and functional groups are indicated in a name by prefixes or suffixes.

The longest continuous chain containing the functional group (if any) is the parent. The chain is numbered from the end nearer to the branches or functional groups. (The functional group of highest priority, as listed in Table 3.5, receives the lowest number.) Positions of substitution on the chain are then specified by these numbers.

$$
\begin{array}{c}
\quad\quad\quad\quad\quad\quad\quad \overset{\text{Cl}}{\underset{|}{\quad}} \\[-2pt]
\text{1,4-dichloro-4-methyl-2-pentene:}\quad \overset{⑤}{\text{CH}_3}\overset{④}{\text{C}}\overset{③}{\text{CH}}{=}\overset{②}{\text{CH}}\overset{①}{\text{CH}_2}\text{Cl} \\[-2pt]
\quad\quad\quad\quad\quad\quad\quad\quad\quad \underset{|}{\quad} \\[-4pt]
\quad\quad\quad\quad\quad\quad\quad\quad\quad \text{CH}_3
\end{array}
$$

The principal reactions of alkanes are **halogenation** and **combustion.**

$$CH_4 + Cl_2 \xrightarrow{\text{light}} CH_3Cl + HCl + \text{other products}$$

$$CH_4 + 2O_2 \xrightarrow{\text{spark}} CO_2 + 2H_2O$$

Petroleum is the world's principal source of gasoline and organic chemicals today. Alkanes, alkenes, and aromatic compounds are obtained by **refining:** straight-run distillation, cracking, and reforming. In the future, **coal-gasification** or **coal-liquefaction** processes may be the principal sources of methane and other alkanes.

Essay Problem for Chapter 3

Nomenclature of Fused-Ring Hydrocarbons

In addition to the standard ring systems (such as cyclohexane), cyclic compounds can also be bicyclic, tricyclic, etc., or they can be spirocyclic. A bicyclic compound has two rings that share atoms. The rings in a bicyclic structure are called fused rings, and the atoms at their points of attachment are called the bridgehead atoms. Some bicyclic compounds, such as camphor, are commonly found in plants. Others, like norbornane, can be synthesized in the laboratory.

camphor

a naturally occurring bicyclic compound

norbornane

a synthetic bicyclic compound

Norbornane and other bicyclic compounds are commonly drawn using three-dimensional line formulas. The three-dimensional formula of norbornane shows the two fused five-membered rings and the two bridgehead carbons more clearly.

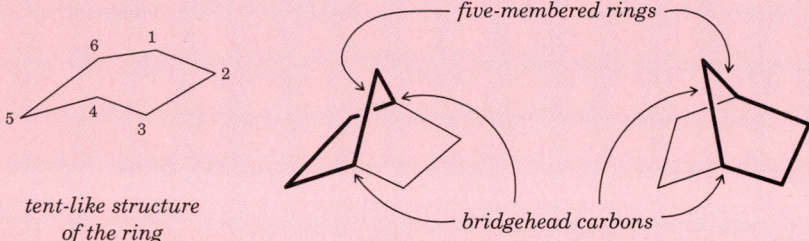

tent-like structure of the ring

five-membered rings

bridgehead carbons

The formal names of bicyclic and related ring systems are based on: (1) the total number of atoms in the molecule; and (2) the number of atoms in each bridge connecting the bridgehead atoms. Norbornane has a total of seven carbon atoms, with three bridges connecting the two bridgehead carbon atoms. Two of the bridges have two carbons and the third has one carbon. The formal name for norbornane is bicyclo[2.2.1]heptane.

Spirocyclic compounds have two fused rings, but only one bridgehead atom. Spirocyclic compounds are named like bicyclic compounds, but have the prefix *spirocyclo-*.

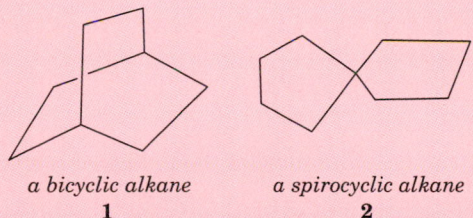

a bicyclic alkane
1

a spirocyclic alkane
2

Questions

1. Bicyclic hydrocarbon **1** has a total of how many carbon atoms? How many carbon atoms are in each bridge? What is its name?
2. What is the name of the spirocyclic alkane **2**?
3. Draw the structure of bicyclo[4.1.0]heptane.
4. *Oxa-* is a prefix indicating the substitution of an oxygen atom for a carbon. Draw the structure of 2-oxabicyclo[2.2.1]heptane.
5. The name for norbornane with one more carbon–carbon bond is tricyclo[2.2.1.0^{2,6}]heptane. Draw its structure. (Hint: the superscript numbers are nomenclature numbers.)

Study Problems*

3.15 Which of the following compounds are unsaturated? (We have used trivial names for these compounds because these names are encountered in industry and medicine.)

* For information concerning the organization of the *Study Problems* and *Additional Problems,* see the *Note to student* on page 41.

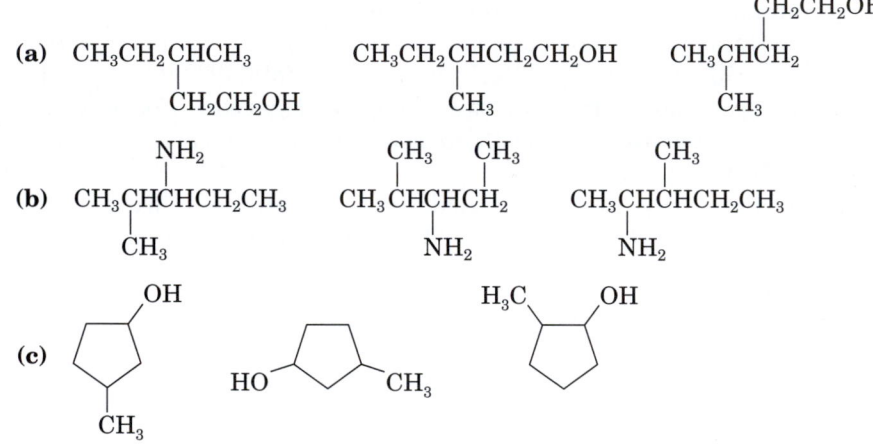

(a)   phenylephrine

a nasal decongestant

(b) $CH_2{=}CHCN$

acrylonitrile

a carcinogenic compound
used in the plastics industry

(c) $(CH_3)_2CHOH$

isopropyl alcohol

rubbing alcohol

(d) HCO_2H

formic acid

used in tanning and dyeing

3.16 In each of the following sets of three structures, two represent the same compound while the third is a structural isomer. Identify the structural isomer.

(a) $CH_3CH_2\underset{\underset{\displaystyle CH_2CH_2OH}{|}}{C}HCH_3$ $CH_3CH_2\underset{\underset{\displaystyle CH_3}{|}}{C}HCH_2CH_2OH$ $CH_3\underset{\underset{\displaystyle CH_3}{|}}{\overset{\overset{\displaystyle CH_2CH_2OH}{|}}{C}}HCH_2$

(b) $CH_3\underset{\underset{\displaystyle CH_3}{|}}{\overset{\overset{\displaystyle NH_2}{|}}{C}}HCHCH_2CH_3$ $CH_3\underset{\underset{\displaystyle NH_2}{|}}{\overset{\overset{\displaystyle CH_3}{|}}{C}}H\overset{\overset{\displaystyle CH_3}{|}}{C}HCH_2$ $CH_3\underset{\underset{\displaystyle NH_2}{|}}{\overset{\overset{\displaystyle CH_3}{|}}{C}}HCHCH_2CH_3$

(c)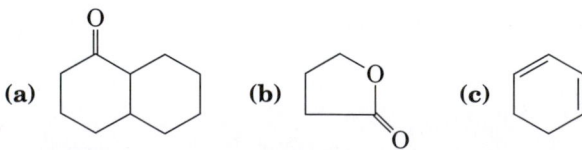

3.17 Write condensed structural formulas for the indicated compounds:

(a)   five structural isomers for C_6H_{14}

(b)   all the isomeric alcohols for $C_4H_{10}O$

(c)   all the isomeric amines for $C_4H_{11}N$

(d)   all the structural isomers for C_3H_6BrCl

3.18 What is the molecular formula and the general formula (e.g., C_nH_{2n}) for each of the following compounds?

(a) $CH_2{=}CHCH{=}CHCH_2CH(CH_3)_2$ (b)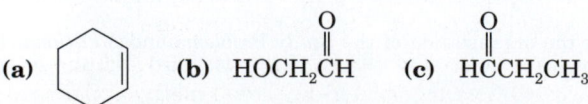

3.19 Write the formula for one structural isomer of each of the following compounds:

(a)   (b)   (c)

3.20 For each of the following compounds, write the structural formula of an isomer that has a *different functional group* (see Problem 3.2, page 94).

(a)   (b) $HOCH_2\overset{\overset{\displaystyle O}{\|}}{C}H$ (c) $H\overset{\overset{\displaystyle O}{\|}}{C}CH_2CH_3$

3.21 The antimicrobial activity of a homologous series of 4-alkylresorcinols against *Eberthella typhosa* increases with increasing chain length up to six carbons and then falls off dramatically. Use formulas of three of the active compounds to illustrate the term "homologus series."

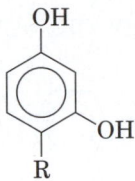

a 4-alkylresorcinol

3.22 Give the condensed structural (or polygon) formula of each of the following compounds:

    (a) 2,2-dimethyloctane   **(b)** 3,4-diethylheptane

    (c) 4-ethyl-2,4-dimethylnonane   **(d)** 1,3-diisopropylcyclohexane

    (e) *sec*-butylcyclopentane   **(f)** *tert*-butylbenzene

    (g) isobutylcycloheptane   **(h)** 1-methyl-3-pentylcyclohexane

    (i) 4-isopropylheptane

3.23 Name the following hydrocarbons by the IUPAC system:

    (a) CH_3CH_2 $CH_2CH(CH_3)_2$ (cyclopentane ring)

    (b) $(CH_3)_3C-$ (cyclohexane ring) $-C(CH_3)_3$

    (c) $CH_3CHCHCH=CCHCH_3$ with CH_3 substituents

    (d) $CH_3C{\equiv}C(CH_2)_4CH=CHCH_3$

3.24 Draw a condensed structural formula for each of the following:

    (a) 2,3,4-trimethyldecane   **(b)** methylcyclononane

    (c) 3-ethyl-2,5-dimethylheptane   **(d)** 4-isopropylcyclohexene

    (e) 2,3,4-trimethyl-1-pentene   **(f)** 1,5-hexadiene

3.25 Each of the following names is reasonable, but incorrect. State the reason for the name being incorrect and write the correct name.

    (a) 1-ethyl-5-methyl-3-cyclopentene

    (b) 3,4,5-trimethyl-4-heptene

    (c) 2,2-dimethylcyclobutane

    (d) 2-isopropylhexane

    (e) 2,4-hexaene

    (f) 2-ethyl-4,4-dimethyl-2-pentene

3.26 Give structures for all the monochloro isomers of the following alkanes: **(a)** pentane; **(b)** cyclopentane; **(c)** 2,2-dimethylbutane; **(d)** 2,2-dimethylpropane.

3.27 Write the formula for each of the following names: **(a)** 1-bromo-1,2-diphenylpropane; **(b)** hexachloroethane; **(c)** 2-iodo-1-octanol; **(d)** 1,1-dichloro-3-methylcyclohexane.

3.28 Write IUPAC names for the following compounds:

(a) $(CH_3)_2C{=}CHCH_3$ (b) $Cl_2C{=}CHCl$

(c) [pentagon structure]   (d) $CH_2{=}CHC{=}CHCH{=}CH_2$ with CH_3 above the central carbon

(e) [cyclohexene ring]—Br   (f)

3.29 Write the formulas and IUPAC names for (a) all the isomeric cyclopentanones that contain one methyl substituent on a ring carbon atom; (b) all the isomeric cyclopentenes with one isopropyl branch.

3.30 Write the IUPAC names for the following compounds:

(a) Cl_2CHCO_2H

a corrosive compound used as a topical astringent

(b) $ClCH_2\overset{\displaystyle O}{\overset{\|}{C}}H$

a strong irritant used to facilitate tree bark removal

(c) $BrCH_2\overset{\displaystyle O}{\overset{\|}{C}}CH_3$

an irritant that has been used as a chemical war gas

3.31 Draw condensed structural formulas for the following compounds (the IUPAC names are given in parentheses): (a) *erythritol* (1,2,3,4-butanetetrol), a naturally occurring alcohol isolated from algae and lichens; (b) *ethohexadiol* (2-ethyl-1,3-hexanediol), an insect repellent that can cause irritation of the mucous membranes and eyes; (c) *ethchlorvynol* (1-chloro-3-ethyl-1-penten-4-yn-3-ol), an addictive or habituating hypnotic with a pungent odor; (d) *citral* (3,7-dimethyl-2,6-octadienal), which is found in lemon-grass oil and is used as a citrus flavoring; (e) *piperitone* (6-isopropyl-3-methyl-2-cyclohexenone), which is found in the oils of Sitka spruce, eucalyptus, and peppermint.

Additional Problems

3.32 An alcohol with no carbon−carbon double bonds has the formula C_4H_8O. What are its possible structures?

3.33 A four-carbon compound contains an aldehyde group and a carboxyl group. What are two possible structures for this compound?

3.34 *Biacetyl,* $C_4H_6O_2$, is a ketone used to flavor margarine. This compound contains no rings or carbon−carbon double bonds. What is its structure?

3.35 *Oxirane,* also called ethylene oxide, C_2H_4O, is used as a sterilizing agent and a fumigant. It is extremely flammable and is also a strong irritant. This compound does not contain a double or triple bond. What is its structure?

3.36 When a continuous-chain alkane is subjected to catalytic cracking, one reaction that occurs is the cleaving of the alkane into a smaller alkane and an alkene. Write a chemical equation showing the conversion of heptane to propene and an alkane.

3.37 When petroleum alkanes of six or more carbons are heated in the presence of platinum and rhenium on aluminosilicates, they can undergo cyclization and dehydrogenation (loss of H_2) to yield aromatic hydrocarbons. What aromatic hydrocarbons could be formed when a mixture of hexane, heptane, and octane is subjected to these conditions? (Use balanced equations in your answer.)

4 STEREOCHEMISTRY

Stereochemistry is the study of molecules in space, that is, how atoms in a molecule are arranged in space relative to one another. The three aspects of stereochemistry that will be covered in this chapter are:

1. **Geometric isomers:** how rigidity in a molecule can lead to isomerism

2. **Conformations of molecules:** the shapes of molecules and how they can change

3. **Chirality of molecules:** how the right- or left-handed arrangement of atoms around a carbon atom can lead to isomerism.

It is often difficult to visualize a three-dimensional molecule from a two-dimensional illustration. Therefore, in our discussions of stereochemistry here and in subsequent chapters, we strongly urge you to use a set of molecular models.

Geometric Isomerism in Alkenes

In Chapter 3, we defined structural isomers as compounds with the same molecular formula but with different orders of attachment of their atoms. Structural isomerism is only one type of isomerism. A second type of isomerism is **geometric isomerism,** which results from rigidity in molecules and occurs in only two classes of compounds: *alkenes* and *cyclic compounds*.

Molecules are not quiet, static particles. They move, spin, rotate, and flex. Atoms and groups attached only by sigma bonds can rotate so that the overall shape of a molecule is in a state of continuous change. However, groups attached by a double bond cannot rotate around the double bond without the pi bond being broken. The amount of energy needed to break a carbon–carbon pi bond (68 kcal/mol in ethylene) is not available to molecules at room temperature. Because of the rigidity of a pi bond, groups attached to pi-bonded carbons are fixed in space relative to one another.

We usually write the structure for an alkene as if the sp^2-hybridized carbon atoms and the atoms attached to them are all in the plane of the paper. In this representation, we can visualize one lobe of the pi bond as being in front of the paper and the other lobe of the pi bond as being underneath the paper, behind the front lobe (see Figure 4.1).

In Figure 4.1, we show a structure with two Cl atoms (one on each sp^2-hybridized carbon) on one side of the pi bond and two H atoms on the other side. Because the double bond is rigid, this molecule is not readily interconvertible with the compound in which the Cl atoms are on opposite sides of the pi bond.

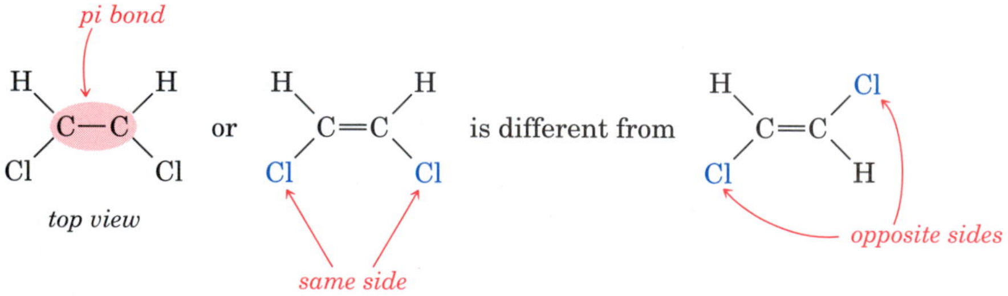

Figure 4.1   The groups attached to sp^2-hybridized carbons are fixed in relation to one another.

Two groups on the *same side of the pi bond* are said to be **cis** (Latin, "on the side"). Groups on the *opposite sides* are said to be **trans** (Latin, "across"). Note how the *cis* or *trans* designation is incorporated into the name.

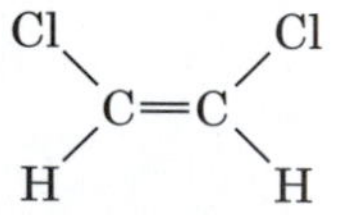

cis-1,2-dichloroethene   *trans*-1,2-dichloroethene
bp 60°   bp 48°

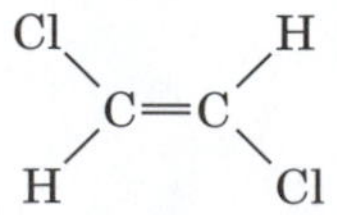

The *cis*- and *trans*-1,2-dichloroethenes have different physical properties (such as boiling points); they are different compounds. However, these two compounds are not structural isomers because the sequence of atoms and the location of the double bond are the same in each compound. This pair of isomers falls into the general category of **stereoisomers**: different compounds that have the same structure, differing only in the *arrangement of the atoms in space*. This pair of isomers falls into a more specific category of **geometric isomers** (also called *cis–trans* **isomers**): stereoisomers that differ by groups being on the same side or on opposite sides of a site of rigidity in a molecule.

The requirement for geometric isomerism in alkenes is that each carbon atom involved in the pi bond have two different groups attached to it, such as H and Cl, or CH_3 and Cl. If one of the carbons of the double bond has two identical groups, such as two H atoms or two CH_3 groups, then geometric isomerism is not possible. (We urge you to make molecular models and verify for yourself this requirement of geometric isomerism.)

Geometric isomers:

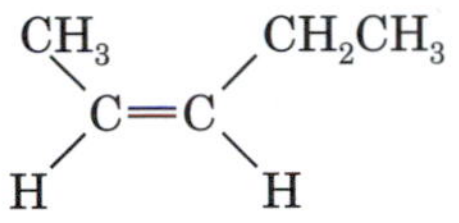

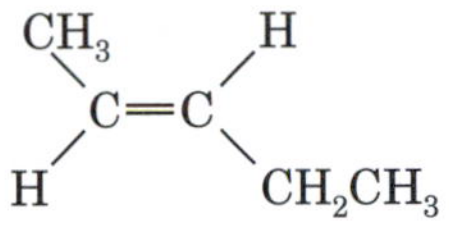

cis-2-pentene   *trans*-2-pentene

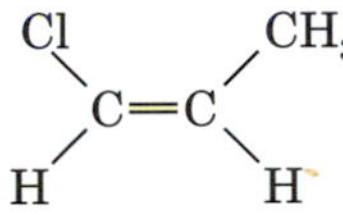

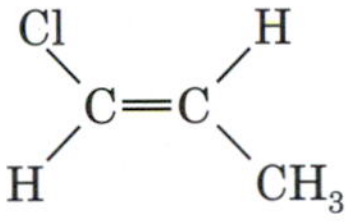

cis-1-chloropropene   *trans*-1-chloropropene

Not geometric isomers:

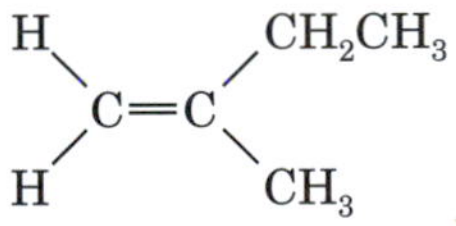

is the same as

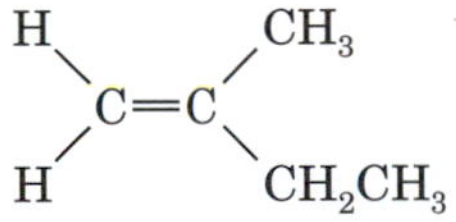

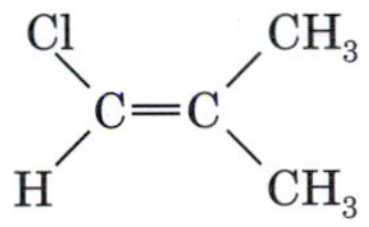

is the same as

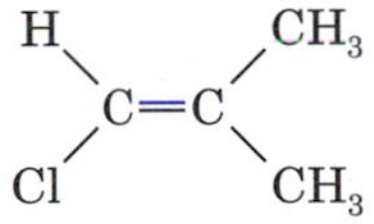

SAMPLE PROBLEM

Label each of the following pairs of structures as *structural isomers* of each other, as *geometric isomers* of each other, or as the *same compound*:

(a)

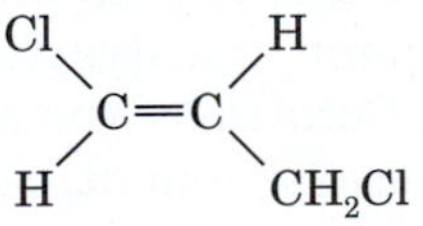

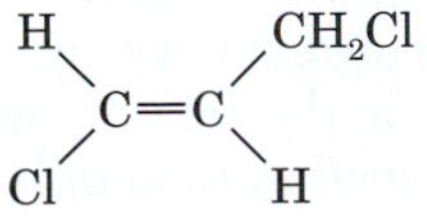

(b)

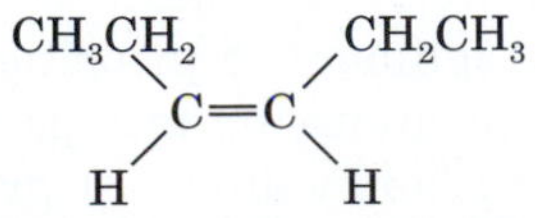

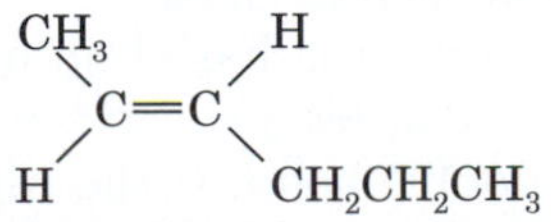

(c)

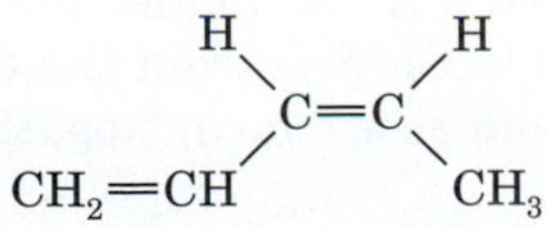

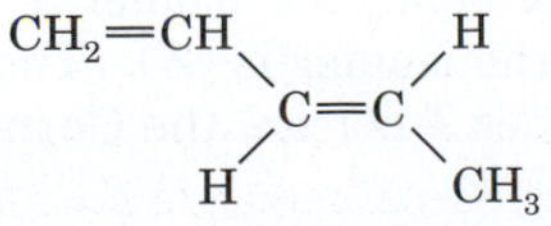

Solution

(a) *same compound* (H's *trans* to each other in each);

(b) *structural isomers* (position of double bond is different; each structural isomer has a geometric isomer);

(c) *geometric isomers* (the first is *cis;* the second is *trans*).

STUDY PROBLEM

4.1   Match the following compounds as pairs of:

   (a) same compounds   **(b)** structural isomers

   (c) geometrical isomers

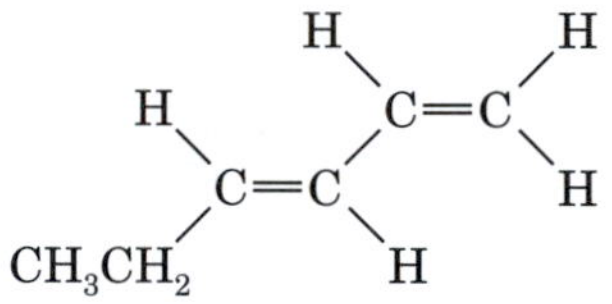

A. *(E)* and *(Z)* System of Nomenclature

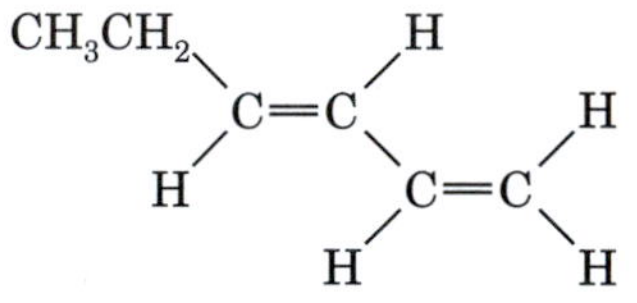

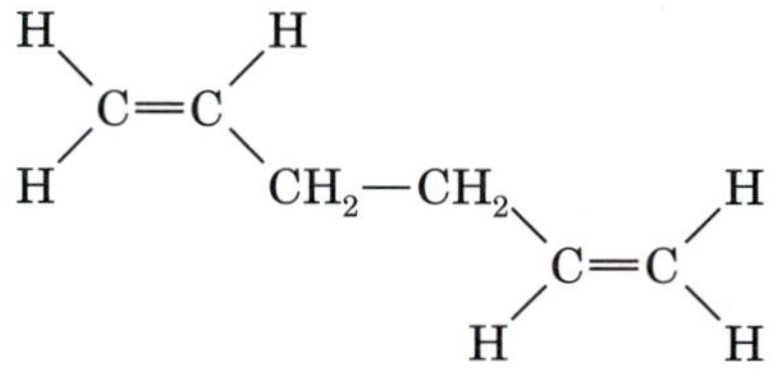

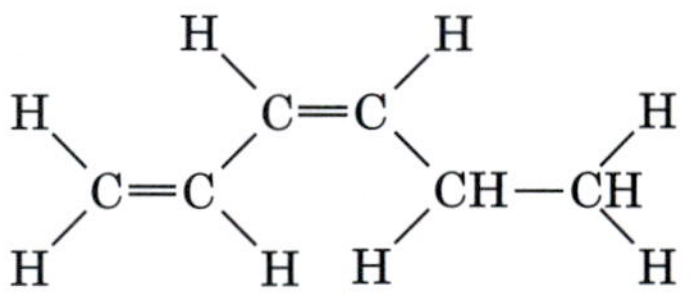

When there are three or four different groups attached to the carbon atoms of a double bond, a pair of geometric isomers exists, but it is sometimes difficult to assign *cis* or *trans* designations to the isomers.

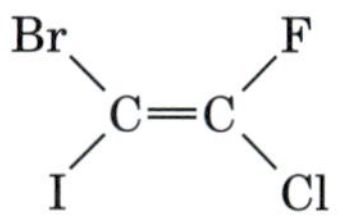

cis or *trans?*

In our example, we can say that Br and Cl are *trans* to each other or that I and Cl are *cis* to each other. However, we cannot name the structure in its entirety as being either the *cis* or the *trans* isomer. Because of the ambiguity in cases of this type, a more general system of isomer assignment, called the *(E)* **and** *(Z)* **system,** has been devised.

   The *(E)* and *(Z)* system is based on an assignment of priorities (not to be confused with nomenclature priorities) to the atoms or groups attached to each carbon of the double bond. If the higher-priority atoms or groups are on *opposite sides* of the pi bond, the isomer is *(E)*. If the higher-priority groups are on the *same side,* the isomer is *(Z)*. (The letter *E* is from the German *entgegen,* "across"; the letter *Z* is from the German *zusammen,* "together.")

If the two atoms on each double-bond carbon are different, priority is based on the atomic numbers of the single atoms directly attached to the double-bond carbons. *The atom with the higher atomic number receives a higher priority.* In our example, I has a larger atomic number than does Br; I is of higher priority. On the other carbon of the double bond, Cl is of higher priority than F.

$$
\begin{array}{ccccc}
 & F & Cl & Br & I \\
\textbf{atomic number:} & 9 & 17 & 35 & 53
\end{array}
$$

increasing priority →

I higher priority than Br

Cl higher priority than F

$$
\underset{\text{I}}{\overset{\text{Br}}{\diagdown}} C = C \underset{\text{Cl}}{\overset{\text{F}}{\diagup}}
$$

(*Z*)-1-bromo-2-chloro-
2-fluoro-1-iodoethene

$$
\underset{\text{I}}{\overset{\text{Br}}{\diagdown}} C = C \underset{\text{F}}{\overset{\text{Cl}}{\diagup}}
$$

(*E*)-1-bromo-2-chloro-
2-fluoro-1-iodoethene

SAMPLE PROBLEM

Name the following compound by the (*E*) and (*Z*) system:

$$
\underset{H_3C}{\overset{H}{\diagdown}} C = C \underset{Cl}{\overset{CH_2CH_2CH_3}{\diagup}}
$$

Solution

The left-hand carbon of the double bond has an H and a C attached to it; the C has the higher priority. The right-hand carbon has a Cl and a C attached to it; the Cl has a higher priority. (Look at the *single atoms* directly attached to the double-bond carbon: C, and not the entire $-CH_2CH_2CH_3$ group.) The higher-priority atoms are on the same side. The compound is named (*Z*)-3-chloro-2-hexene.

STUDY PROBLEMS

4.2 Identify the following compounds as (*E*) or (*Z*).

(a)
$$
\underset{H_3Si}{\overset{CH_3}{\diagdown}} C = C \underset{Cl}{\overset{H}{\diagup}}
$$

(b)
$$
\underset{Cl}{\overset{Br}{\diagdown}} C = C \underset{SCH_3}{\overset{CH_3}{\diagup}}
$$

4.3 The formula for maleic acid, which is a toxic dicarboxylic acid, is (*Z*)-$HO_2CCH = CHCO_2H$. Its (*E*) isomer, fumaric acid, is not toxic and is, in fact, an intermediate in some cellular reactions. Write the structures of maleic acid and fumaric acid, showing their stereochemistry.

B. Sequence Rules

Determination of priorities by atomic number alone cannot handle all cases. For example, how would we name the following compound by the (E) and (Z) system?

$$H_3C \diagdown \hspace{3cm} CH_2CH_3$$
$$C=C$$
$$H \diagup \hspace{3cm} CH_3$$

To handle such a case, and others like it, a set of *sequence rules* to determine order of priority has been developed. These priority rules form the basis of the **Cahn–Ingold–Prelog nomenclature system,** named in honor of the chemists who developed the system. (See also Section 4.8.)

Sequence Rules for Order of Priority:

1. If the atoms in question are different, the sequence order is by atomic number, with the atom of highest atomic number receiving the highest priority.

F   Cl   Br   I

increasing priority →

2. If two isotopes of the same element are present, the *isotope of higher mass* receives the higher priority.

^1_1H, or H ^2_1H, or D
hydrogen   deuterium

increasing priority →

3. If two atoms are identical, the atomic numbers of the *next atoms* are used for priority assignment. If these atoms also have identical atoms attached to them, priority is determined at the first point of difference along the chain. The atom that has attached to it an atom of higher priority has the higher priority. (Do not use the sums of the atomic numbers, but look for the single atom of highest priority.)

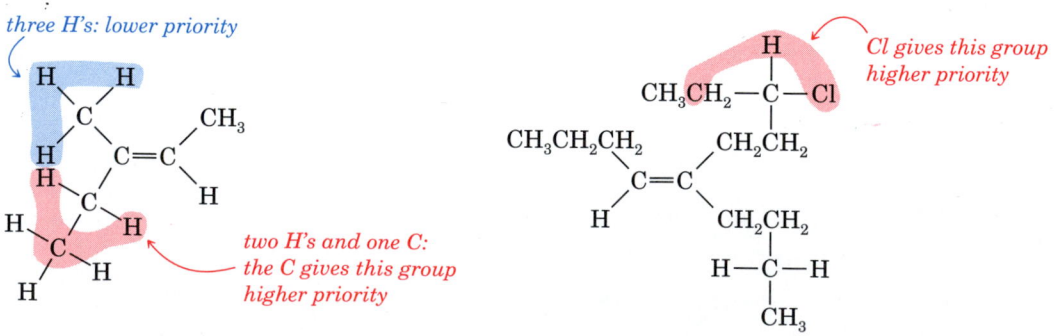

three H's: lower priority

two H's and one C: the C gives this group higher priority

Cl gives this group higher priority

(E)-3-methyl-2-pentene   (Z)-5-butyl-8-chloro-4-decene

4. Atoms attached by double or triple bonds are given single-bond *equivalencies* so that they can be treated like single-bonded groups in determining priority. Each doubly bonded atom is duplicated (or triplicated for triple bonds), a process better seen in examples.

Structure   *Equivalent for priority determination*

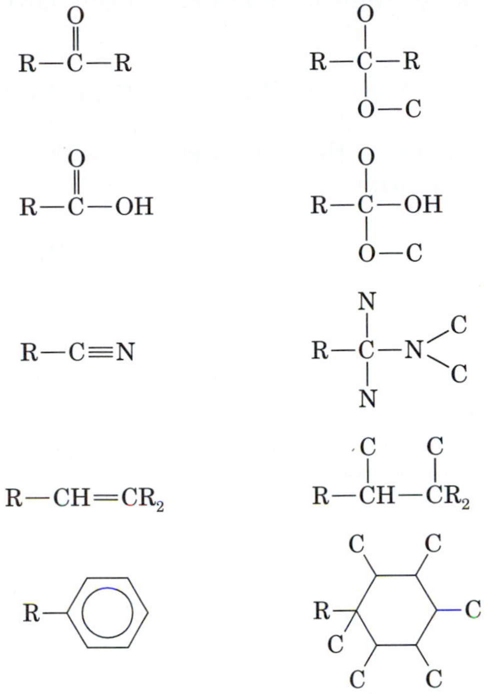

$$R-\overset{\overset{\displaystyle O}{\|}}{C}-R \qquad R-\overset{\overset{\displaystyle O}{|}}{\underset{\underset{\displaystyle O-C}{|}}{C}}-R$$

$$R-\overset{\overset{\displaystyle O}{\|}}{C}-OH \qquad R-\overset{\overset{\displaystyle O}{|}}{\underset{\underset{\displaystyle O-C}{|}}{C}}-OH$$

$$R-C{\equiv}N \qquad R-\overset{\overset{\displaystyle N}{|}}{\underset{\underset{\displaystyle N}{|}}{C}}-N{\overset{\diagup C}{\diagdown C}}$$

$$R-CH{=}CR_2 \qquad R-\overset{\overset{\displaystyle C}{|}}{CH}-\overset{\overset{\displaystyle C}{|}}{CR_2}$$

R—⬡   R-⬡-C (with C substituents)

By this rule, we obtain the following priority sequence:

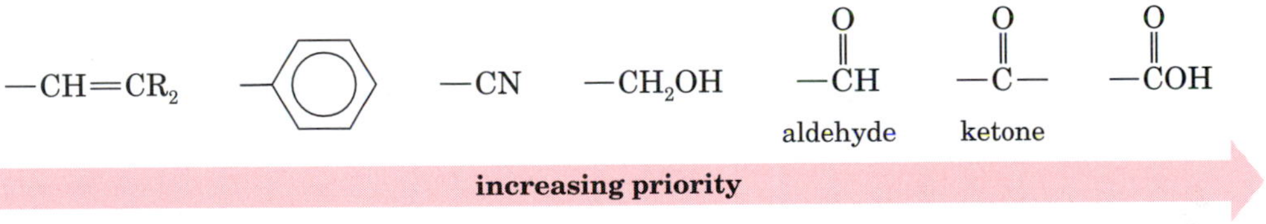

$-CH{=}CR_2$ ⬡ $-CN$ $-CH_2OH$ $-\overset{\overset{\displaystyle O}{\|}}{C}H$ (aldehyde) $-\overset{\overset{\displaystyle O}{\|}}{C}-$ (ketone) $-\overset{\overset{\displaystyle O}{\|}}{C}OH$

increasing priority →

SAMPLE PROBLEMS

List the following atoms or groups in order of increasing priority (lowest priority first):

$$-NH_2 \qquad -H \qquad -CH_3 \qquad -Cl$$

Solution
Increasing atomic number of the attached atom (N, H, C, Cl) gives increasing priority: H, CH_3, NH_2, Cl.

List the order of priority of the following groups (lowest to highest):

$$-CO_2H \qquad -CO_2CH_3 \qquad -CH_2OH \qquad -OH \qquad -H$$

Solution
Following the sequence rules:

$$-H \qquad -CH_2OH \qquad -CO_2H \qquad -CO_2CH_3 \qquad -OH$$

Which group is of higher priority, the butyl group or the isobutyl group?

Solution
The first carbon is the same in each (two H's and one C attached).

$$-\overset{\downarrow}{C}H_2CH_2CH_2CH_3 \qquad -\overset{\downarrow}{C}H_2CH(CH_3)_2$$

Therefore, we proceed to the second carbon, and we find that the isobutyl group is of higher priority.

$$-CH_2\overset{\overset{\textstyle H}{|}}{\underset{\underset{\textstyle H}{|}}{C}}CH_2CH_3 \qquad -CH_2\overset{\overset{\textstyle H}{|}}{\underset{\underset{\textstyle CH_3}{|}}{C}}CH_3$$

The butyl group has one C and two H's at the first point of difference.   *The isobutyl group has two C's and one H at the first point of difference.*

STUDY PROBLEMS

4.4   Tell whether each of the following compounds is (*E*) or (*Z*):

(a)
$$\begin{array}{c} D \\ \diagdown \\ H \diagup \end{array} C=C \begin{array}{c} H \\ \diagup \\ \diagdown D \end{array}$$

(b)
$$\begin{array}{c} H_3C \\ \diagdown \\ H \diagup \end{array} C=C \begin{array}{c} CH_2CH_3 \\ \diagup \\ \diagdown CH_3 \end{array}$$

(c)
$$\begin{array}{c} Cl \\ \diagdown \\ H \diagup \end{array} C=C \begin{array}{c} CH_2CH_3 \\ \diagup \\ \diagdown CH(CH_3)_2 \end{array}$$

(d)
$$\begin{array}{c} \overset{O}{\overset{||}{C}}H_3C \\ \diagdown \\ ClCH_2 \diagup \end{array} C=C \begin{array}{c} \overset{O}{\overset{||}{C}}CH_2Cl \\ \diagup \\ \diagdown Cl \end{array}$$

4.5   Draw the structures for the following names.

(a)   (*E*)-1-chloro-2-ethyl-1-fluoro-1,3-butadiene

(b)   (*Z*)-2-chloro-2-butenal

Section 4.2

Geometric Isomerism in Cyclic Compounds

We have described how restricted rotation around a double bond can lead to geometric isomerism. Let us now consider restricted rotation in cyclic compounds.

Atoms joined in a small ring are not free to rotate around the sigma bonds of the ring. Rotation around the sigma bonds forming the ring would require that attached atoms or groups pass through the center of the ring. Van der Waals repulsions prevent this from happening unless the ring contains eight or more carbon atoms, depending on the size of the substituent. The most common rings in organic compounds are five- and six-membered rings. Therefore, we will concentrate our discussion on rings of six and fewer carbon atoms.

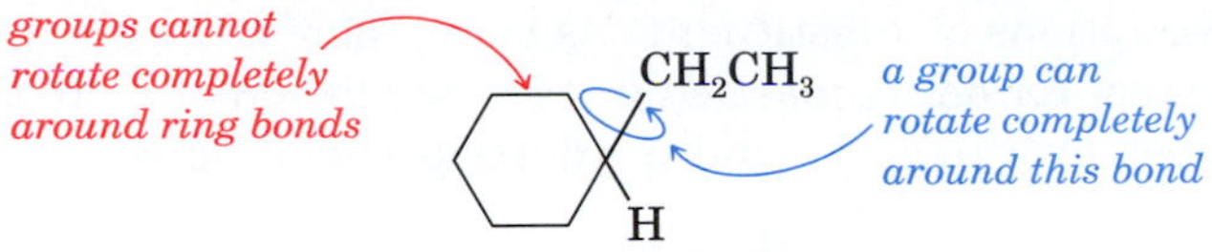

groups cannot rotate completely around ring bonds

CH$_2$CH$_3$

a group can rotate completely around this bond

H

For the moment, we will assume that the carbon atoms of a cyclic structure such as cyclohexane form a plane. (While this is not strictly correct, as we will show later in this chapter, it is often convenient to assume that they do lie in a plane.) For the present discussion, we will view the plane of the ring as being almost horizontal. The edge of the ring projected toward us is shaded more heavily.

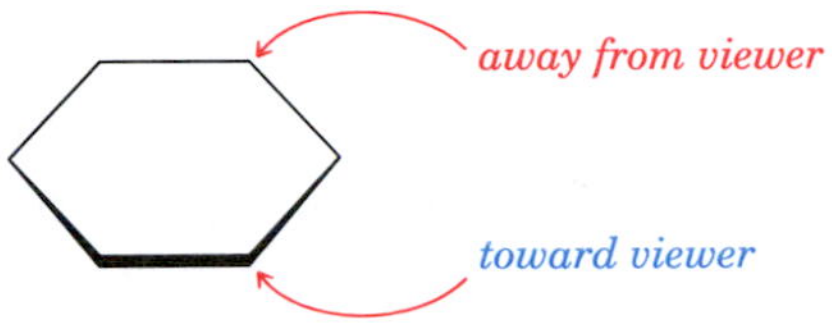

away from viewer

toward viewer

Each carbon atom in the cyclohexane ring is joined to its neighboring ring carbon atoms and also to two other atoms or groups. The bonds to these two other groups are represented by vertical lines. A group attached to the top of a vertical line is said to be *above the plane of the ring,* and the group attached to the bottom of a vertical line is said to be *below the plane of the ring.*

H   H

H   CH$_3$ *above the plane*

H   H

H   H

H   H

H   OH   *below the plane*

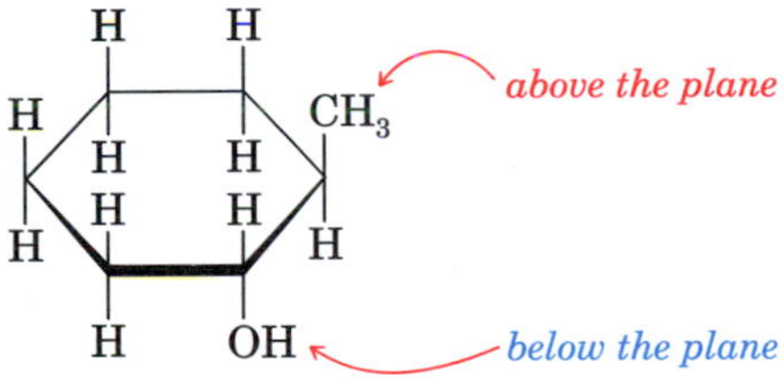

In this symbolism, hydrogen atoms attached to the ring and their bonds are not always shown.

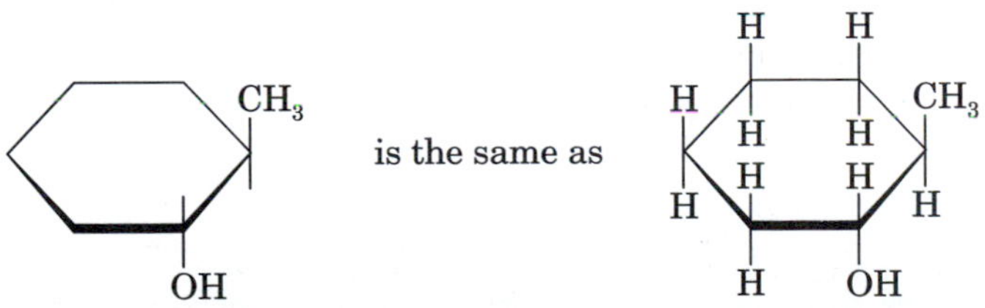

CH$_3$

OH   is the same as

H   H

H   CH$_3$

H   H

H   H

H   H

H   OH

Another way of showing how groups are attached to the ring is by using a broken wedge to indicate a group below the plane of the ring and a solid line bond or a wedge to represent a group above the plane.

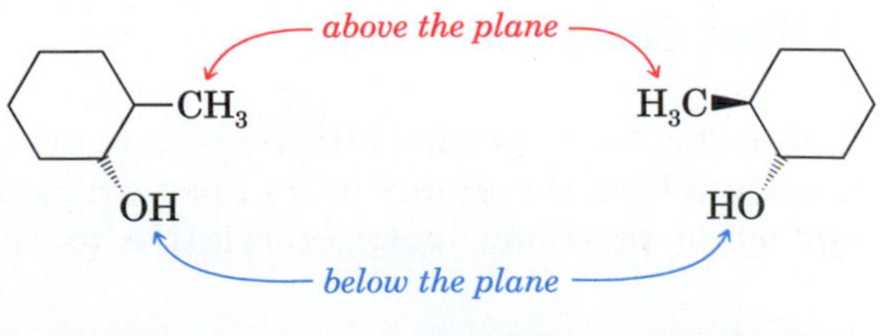

—— *above the plane* ——

—CH$_3$ H$_3$C◄

OH   HO

—— *below the plane* ——

Descriptions of substituents as being "above the plane" or "below the plane" are correct for only a particular representation of a structure. A molecule can be flipped over in space and the descriptions reversed.

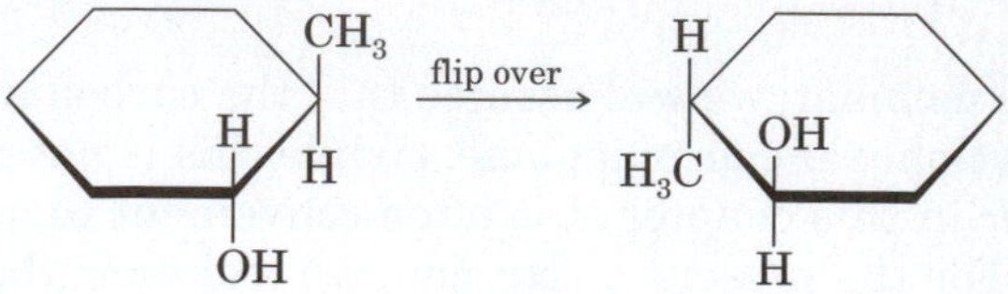

The important point is that, in all the preceding formulas, the methyl group and the hydroxyl group are on *opposite sides* of the plane of the ring. When the two groups are on opposite sides of the ring, they are *trans;* when they are on the same side, they are *cis.* These designations are directly analogous to *cis* and *trans* in alkenes. The *cis-* and *trans-*compounds are geometric isomers of each other, just as *cis-* and *trans-*alkenes are.

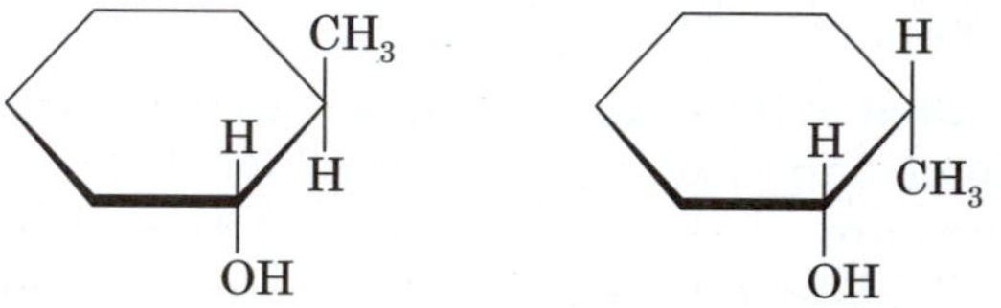

trans-2-methylcyclohexanol   *cis*-2-methylcyclohexanol

SAMPLE PROBLEM

Tell whether each of the following compounds is *cis, trans,* or *neither:*

(a)   **(b)**

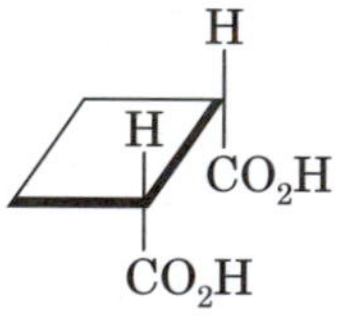

Solution

(a) *trans* because the hydroxyl groups are on opposite sides of the ring; **(b)** *cis* because the carboxyl groups are on the same side.

STUDY PROBLEM

4.6   Draw formulas for the geometric isomers of 2-isopropyl-5-methylcyclohexanol. (*Menthol,* used in throat lozenges, is one of these compounds.)

Section 4.3

Conformations of Open-Chain Compounds

In open-chain compounds, groups attached by sigma bonds can rotate around these bonds. Therefore, the atoms in an open-chain molecule can assume an infinite number of positions in space relative to one another. Ethane is a

three-dimensional formula   *ball-and-stick formula*   *Newman projection*

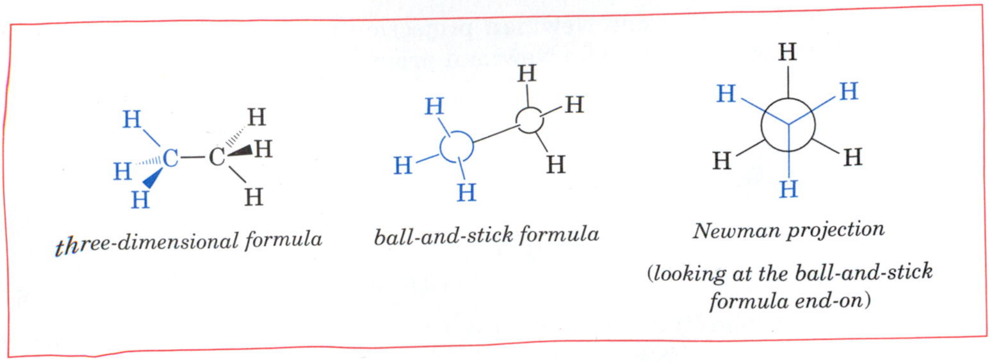

*(looking at the ball-and-stick
formula end-on)*

Figure 4.2   A dimensional formula, a ball-and-stick formula, and a Newman projection of ethane.

small molecule, but even ethane can assume different arrangements in space,
called **conformations.**

In recent years, computers have been applied to conformational analysis to
create a new field—**computational chemistry.** The shapes of complex molecules, such as enzymes, can be calculated and then graphically displayed in
their most stable conformations. This technique makes it possible, for example, to design medicinal agents that will interact with enzymes in a more systematic manner.

There are two general approaches to computerized molecular modeling.
The simplest is to assume that each bond in the molecule is a spring with
force and torsional constants. The computation involves calculation of the energy of the molecule in different conformations and then selection of those
that have the lowest energy. The second approach is similar to the first except
that the bonds are assumed to be molecular orbitals and, consequently, the
calculations are considerably more complicated and time-consuming.

To represent conformations, we will use three types of formulas: **three-dimensional formulas** (which we will call simply **dimensional form** as)
ball-and-stick formulas, and **Newman projections.** (We suggest you
also use models to compare different conformations.) A ball-and-stick he mo-
and a dimensional formula are three-dimensional representation *is an*
lecular model of a compound (see Figure 4.2). A Newman pring these
end-on view of only *two carbon atoms* in the molecule. The bor appear to
two carbons is hidden. The three bonds attached to the front carbon are
go to the center of the projection, and the three bonds of th
only partially shown.

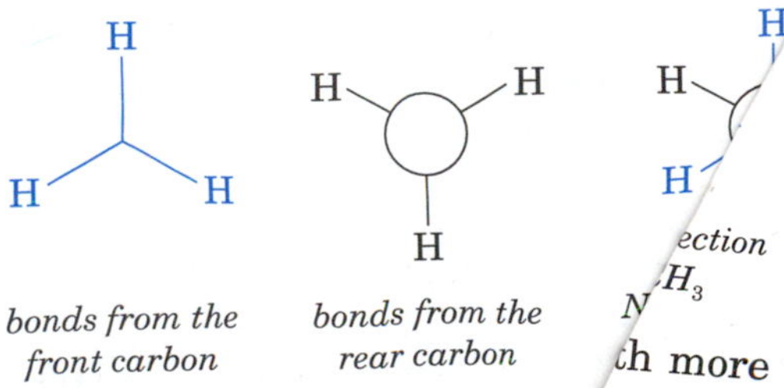

*bonds from the
front carbon*   *bonds from the
rear carbon*

Newman projections can be drawn for mo
bon atoms. Because only two carbon atoms

jection, more than one Newman projection can be drawn for a molecule. For example, we can show two Newman projections for 3-chloro-1-propanol.

Looking at carbons 1 and 2:

OH
①CH₂
②CH₂
③CH₂Cl

3-chloro-1-propanol

dimensional

ball-and-stick

Newman

Looking at carbons 2 and 3:

OH
①CH₂
②CH₂
③CH₂Cl

dimensional

ball-and-stick

Newman

Because of rotations around sigma bonds, a molecule can assume any number of conformations. The different structures formed by the bond rotations are called **conformers** (from "conformational isomers"). Because conformers are easily interconverted, they cannot usually be isolated from one another, as can structural isomers.

STUDY PROBLEM

4.7   Draw Newman projections of the conformers shown for the following compounds.

(a)

front carbon

O
‖
CH

HO
H—C—C⸻OH
HOCH₂   H

back carbon

(b)

back carbon

O
‖
CH

HO
H—C—C⸻OH
HOCH₂   H

front carbon

ur formulas of ethane and 3-chloro-1-propanol, we have shown **stag-** **conformers,** in which the hydrogen atoms or the attached groups are de part from one another as possible. Because the C—C bond can un- sib tion, the hydrogen atoms might also be **eclipsed,** or as close as pos- quit hind the other in the Newman projection. We will show them not d so they can be seen.

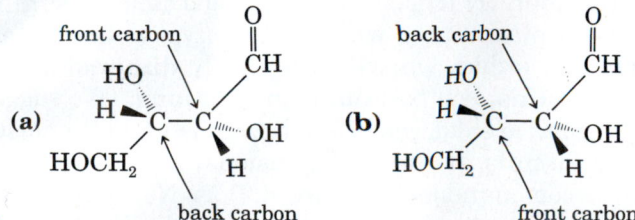

repulsion

ggered

eclipsed

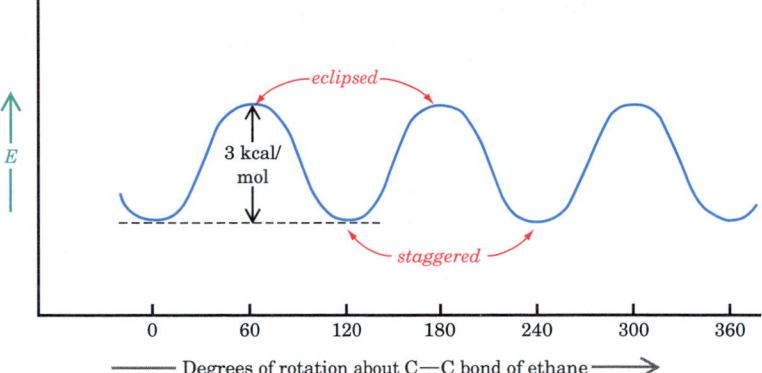

Figure 4.3   Energy changes involved in rotation around the carbon–carbon sigma bond of ethane.

The rotation around sigma bonds is often called **free rotation,** but it is not entirely free. The eclipsed conformation of ethane is about 3 kcal/mol less stable (of higher energy) than the staggered conformer because of minor repulsions between the bonding electrons to the hydrogen atoms. To undergo rotation from a staggered conformation to an eclipsed conformation, a mole of ethane molecules would require 3 kcal. Since this amount of energy is readily available to molecules at room temperature, the rotation can occur easily. However, even though the conformations of ethane are interconvertible at room temperature, at any given time we would expect a greater percentage of ethane molecules to be in the staggered conformation because of its lower energy. A diagram showing the changes in potential energy with rotation around the C—C bond in ethane is presented in Figure 4.3.

Butane ($CH_3CH_2CH_2CH_3$), like ethane, can exist in eclipsed and staggered conformations. In butane, there are two methyl groups, which are relatively large, attached to the center two carbons. If butane is viewed from the center two carbons, the presence of these methyl groups gives rise to two types of staggered conformations that differ in the positions of the methyl groups in relation to each other. The staggered conformation in which the methyl groups are at the maximum distance apart is called the ***anti*** conformer (Greek *anti,* "against"). The staggered conformations in which the methyl groups are closer are called ***gauche*** conformers (French *gauche,* "left" or "crooked"). Newman projections for one-half a complete rotation follow.

Partial rotation around the carbon 2–carbon 3 bond of butane (rear carbon rotating):

anti
(lowest energy)   *eclipsed*   *gauche*   *eclipsed methyls*
(highest energy)

The larger the groups attached to two carbon atoms, the greater is the energy difference between the molecule's conformations. It takes more energy to push together two bulky groups than two small groups. While it takes only

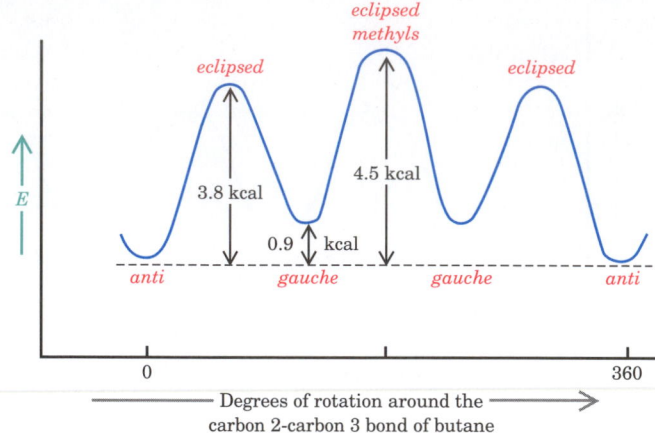

Figure 4.4 Energy relations (in kcal/mol) of the different conformations of butane.

3 kcal/mol for ethane to rotate from staggered to eclipsed, it takes 4.5 kcal/mol for butane to rotate from *anti* to the conformation in which the methyls are eclipsed. The energy relationships of the complete rotation around the carbon 2–carbon 3 bond of butane are shown in Figure 4.4.

STUDY PROBLEM

4.8 Draw Newman projections for the *anti* and *gauche* conformations of **(a)** 1-bromo-2-chloroethane, and **(b)** 3-hydroxypropanoic acid, $HOCH_2CH_2CO_2H$.

Section 4.4

Shapes of Cyclic Compounds

A. Ring Strain

In 1885, German chemist Adolf von Baeyer (1835–1917; Nobel Prize, 1905) theorized that cyclic compounds form planar rings. Baeyer further theorized that all cyclic compounds except for cyclopentane would be "strained" because their bond angles are not close to the tetrahedral angle of 109.5°. He proposed that, because of the abnormally small bond angles of the ring, cyclopropane and cyclobutane would be more reactive than an open-chain alkane. According to Baeyer, cyclopentane would be the most stable ring system (because its bond angles are closest to tetrahedral), and then reactivity would increase again starting with cyclohexane.

Bond angles according to Baeyer:

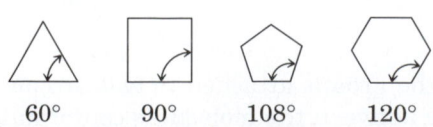

60°   90°   108°   120°

Baeyer's theory was not entirely correct. Cyclohexane and larger-sized rings are not more reactive than cyclopentane. We now know that cyclohexane is not a flat ring with bond angles of 120°, but rather a puckered ring with bond angles close to 109°, the normal sp^3 bond angles.

However, there is indeed what we call **ring strain** in the smaller ring systems. Cyclopropane is the most reactive of the cycloalkanes. Its heat of combustion is higher per CH_2 group than that of other alkanes (Table 4.1).

$+ 9\, O_2 \longrightarrow 6\, CO_2 + 6\, H_2O + 944.5$ kcal/mol (157.4 kcal/mol for each CH_2)

cyclohexane

2 $+ 9\, O_2 \longrightarrow 6\, CO_2 + 6\, H_2O + 499.8$ kcal/mol (166.6 kcal/mol for each CH_2)

cyclopropane

When treated with hydrogen gas at moderate temperatures, unstrained cycloalkanes such as cyclopentane do not react, but cyclopropane undergoes ring opening.

$+ H_2 \xrightarrow{\text{catalyst}}$ no reaction

cyclopentane

$$\underset{\text{cyclopropane}}{\overset{\displaystyle CH_2}{H_2C - CH_2}} + H_2 \xrightarrow{\text{catalyst}} \underset{\text{propane}}{CH_3CH_2CH_3}$$

Today we would say that the sp^3 orbitals of the carbon atoms in cyclopropane cannot undergo complete overlap with each other because the angles between the carbon atoms of cyclopropane are geometrically required to be

Table 4.1	Strain energies from heat of combustion data			
	$-\Delta H$ (kcal/mol)	$-\Delta H$ per CH_2 (kcal/mol)[a]	Strain Energy per CH_2 (kcal/mol)[b]	Strain Energy, Total (kcal/mol)[c]
cyclopropane	499.8	166.6	9.2	27.6
cyclobutane	655.9	164.0	6.6	26.4
cyclopentane	793.5	158.7	1.3	6.5
cyclohexane	944.5	157.4	0	0

[a] $-\Delta H$/mol divided by number of CH_2 groups.
[b] The difference between (1) the value of $-\Delta H/CH_2$ for that compound, and (2) the value for cyclohexane, the assumption being that cyclohexane is not strained.
[c] Strain energy/CH_2 × number of CH_2 groups.

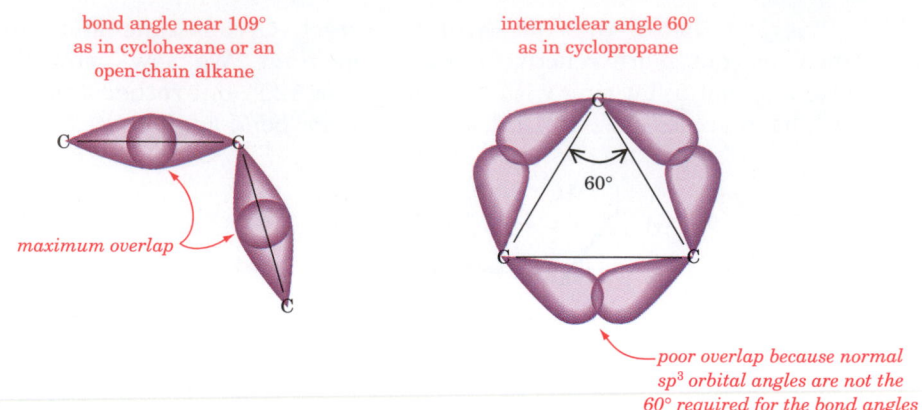

Figure 4.5   Maximum overlap cannot be achieved between the ring carbon atoms in cyclopropane.

60° (see Figure 4.5). The ring sigma bonds of cyclopropane are of higher energy than sp^3 sigma bonds that have the normal tetrahedral angle. The cyclopropane bonds are more easily broken than most other C—C sigma bonds and, in comparable reactions, more energy is released.

Cyclobutane is less reactive than cyclopropane but more reactive than cyclopentane. Following along reasonably with Baeyer's theory, the cyclopentane ring is stable and is far less reactive than the three- and four-membered rings.

With cyclohexane and larger rings, Baeyer's predictions fail. Cyclohexane and rings larger than cyclohexane are found in puckered conformations rather than as flat rings and are not particularly reactive. Larger rings are not as common in naturally occurring compounds as five- and six-membered rings. Baeyer felt the rarity was due to ring strain. We now realize that the rarity of larger rings is not due primarily to unusually high bond energies. Instead, the scarcity of these compounds arises from the decreasing probability that the ends of longer molecules will find each other to undergo a ring-forming reaction.

SAMPLE PROBLEM

Would you expect the benzene ring to be flat or puckered?

Solution
The benzene ring is flat because the carbon atoms are sp^2 (not sp^3) hybridized. The normal positioning of sp^2 bonds is planar and with angles of 120° between them, corresponding to a regular planar hexagon.

B. Ring Puckering and Hydrogen–Hydrogen Repulsions

If the cyclohexane ring were flat, all the hydrogen atoms on the ring carbons would be eclipsed. In the puckered conformer that we have shown, however, all the hydrogens are staggered.

flat   *puckered*

The idea of puckered rings was proposed by Herman Sachse in 1890, five years after Baeyer proposed his strain theory. Sachse's idea was not accepted, however, because he predicted there would be two stable isomers of *trans*-1,4-dimethylcyclohexane. (We now know there are two conformers of this compound, but not two isomers. See page 142.) Sachse's concept of puckered rings lay fallow for almost 60 years. In 1950, D. H. R. Barton, an English chemist, proposed what we now call conformational theory. In 1969, Barton and Odd Hassel (who used X-ray crystallography to study chlorocyclohexanes) shared the Nobel Prize for their work on conformational theory.

The energy of the puckered conformer of cyclohexane is lower than the energy of flat cyclohexane, both because of more-favorable sp^3 bond angles and also because of fewer hydrogen–hydrogen repulsions.

What of the other cyclic compounds? Cyclopentane would have near-optimal bond angles (108°) if it were flat, but cyclopentane also is slightly puckered so that the hydrogen atoms attached to the ring carbons are staggered. Cyclobutane (flat bond angles of 90°) also is puckered, even though the puckering causes more-strained bond angles. Cyclopropane must be planar; geometrically, three points (or three carbon atoms) define a plane. The hydrogen atoms in cyclopropane necessarily are eclipsed.

*envelope form
of cyclopentane*

*butterfly form
of cyclobutane*

Section 4.5

The Conformers of Cyclohexane

The cyclohexane ring, either alone or in fused-ring systems (ring systems that jointly share carbon atoms), is the most important of all the ring systems. In this section, we will study the conformations of cyclohexane and substituted cyclohexanes. In Chapter 24, we will discuss the conformations of fused-ring systems.

There are many shapes that a cyclohexane ring can assume, and any single cyclohexane molecule is in a continuous state of flexing into different shapes. (Molecular models are invaluable for showing the relationships among the

various conformations.) So far, we have shown the **chair form** of cyclohexane. This and some other shapes the cyclohexane molecule can assume
are as follows:

chair   *half-chair*   *twist-boat*   *boat*

None of these other conformations has the favorable staggered-hydrogen
structure of the chair form. The eclipsing of hydrogens, as in the boat form,
adds to the energy of the molecule. Figure 4.6 shows models and Newman
projections of the chair form and boat form; the staggered and eclipsed hydrogens are apparent in these representations.

The energy requirements for the interconversion of the different conformations of cyclohexane are shown in Figure 4.7. We can see that the chair form
has the lowest energy, while the half-chair (which has an almost-planar structure) has the highest energy. At any given time, we would expect most cyclo-

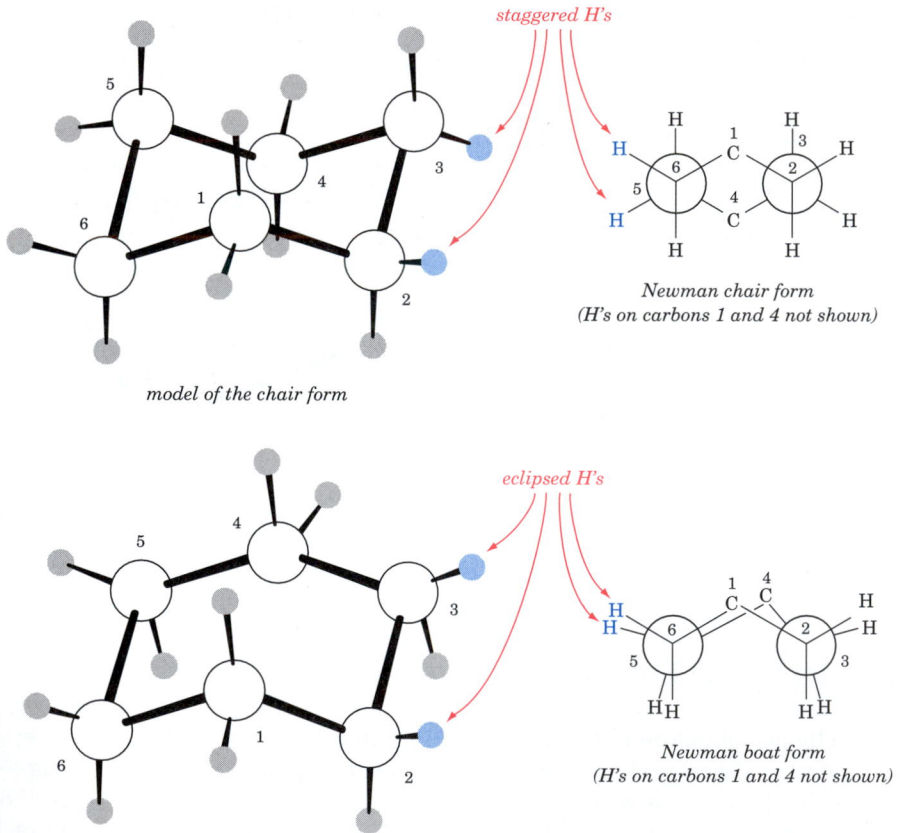

model of the chair form

staggered H's

*Newman chair form
(H's on carbons 1 and 4 not shown)*

eclipsed H's

model of the boat form

*Newman boat form
(H's on carbons 1 and 4 not shown)*

Figure 4.6   Molecular models and Newman projections of the chair and boat forms
of cyclohexane.

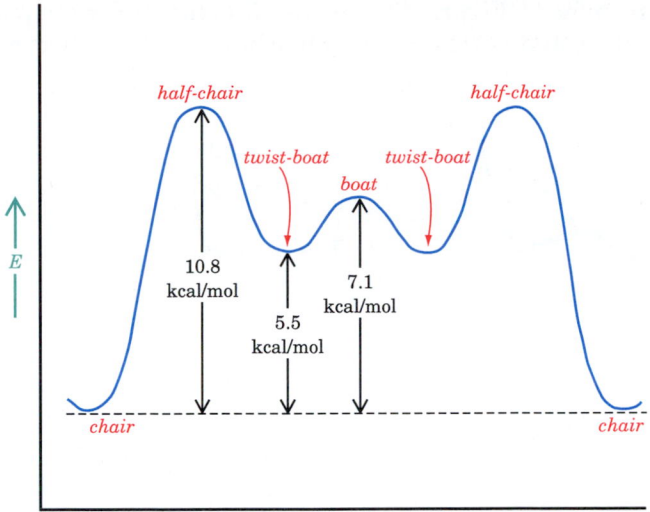

Figure 4.7   Relative potential energies of the conformations of cyclohexane.

hexane molecules to be in the chair form. Indeed, it has been calculated that about 99.9% of cyclohexane molecules are in the chair form at any one time.

A. Equatorial and Axial Substituents

The carbon atoms of the chair form of cyclohexane roughly form a plane. For purposes of discussion, an axis may be drawn perpendicularly to this plane. These operations are shown in Figure 4.8.

   Each ring carbon of cyclohexane is bonded to two hydrogen atoms. The bond to one of these hydrogens is in the rough plane of the ring; this hydrogen atom is called an **equatorial hydrogen.** The bond to the other hydrogen atom is parallel to the axis; this is an **axial hydrogen.** Each of the six carbon atoms of cyclohexane has one equatorial and one axial hydrogen atom.

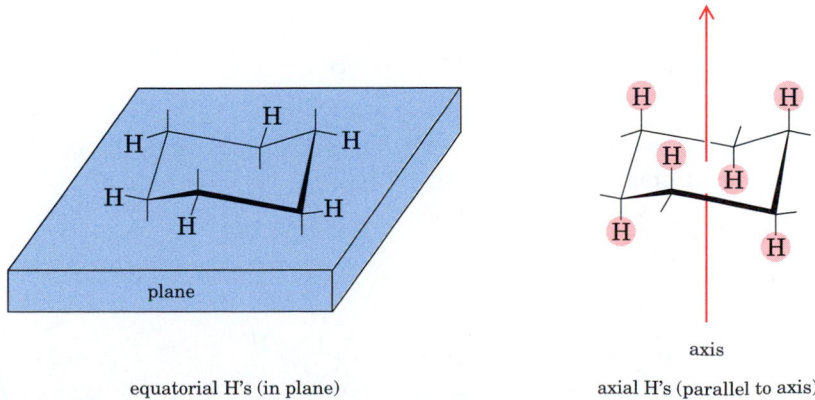

equatorial H's (in plane)   axial H's (parallel to axis)

Figure 4.8   The equatorial and axial hydrogens of cyclohexane.

(Again, refer to Figure 4.8.) In the flipping and reflipping between the conformers, axial becomes equatorial, while equatorial becomes axial.

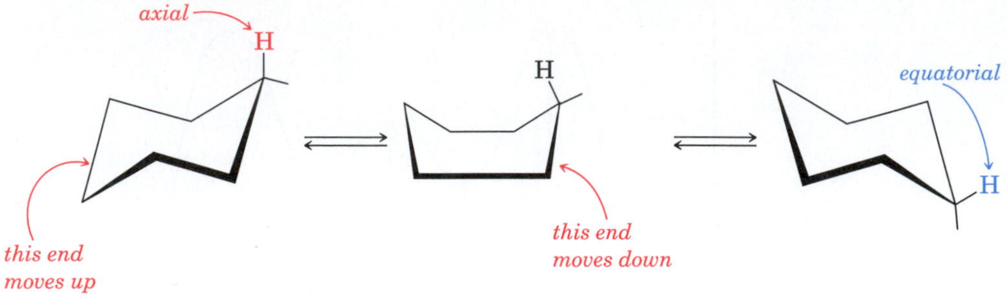

A methyl group is larger than a hydrogen atom. When the methyl group in methylcyclohexane is in the axial position, the methyl group and axial hydrogens on the same side of the ring repel each other. Interactions between axial groups are called **1,3-diaxial interactions.** When the methyl group is in the equatorial position, the repulsions are minimized. Thus, the energy of the conformer with an equatorial methyl is lower. At room temperature, about 95% of methylcyclohexane molecules are in the conformation in which the methyl group is equatorial.

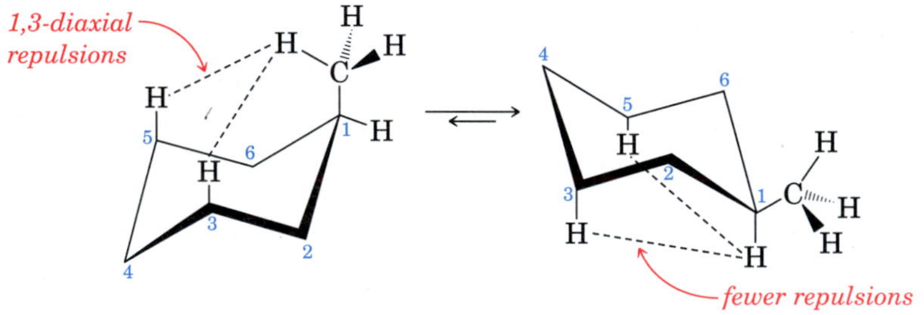

When the methyl group is in the axial position, the conformation of the ring is similar to that of butane with *gauche* methyl groups.

Gauche relationship in axial methylcyclohexane:

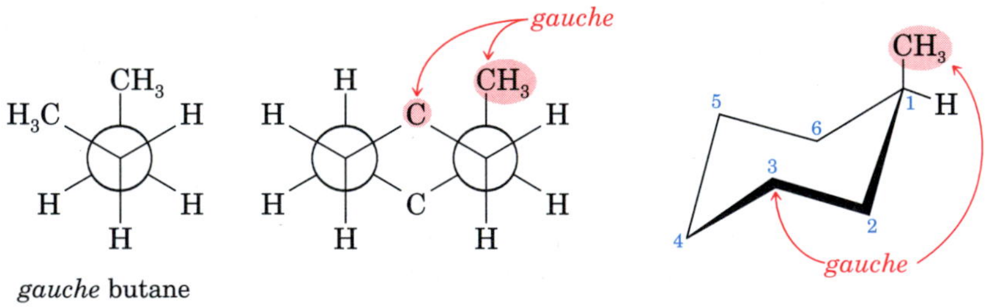

gauche butane

When the methyl group is in the equatorial position, the conformation of the ring is similar to that of butane with *anti* methyl groups.

Anti relationship in equatorial methylcyclohexane:

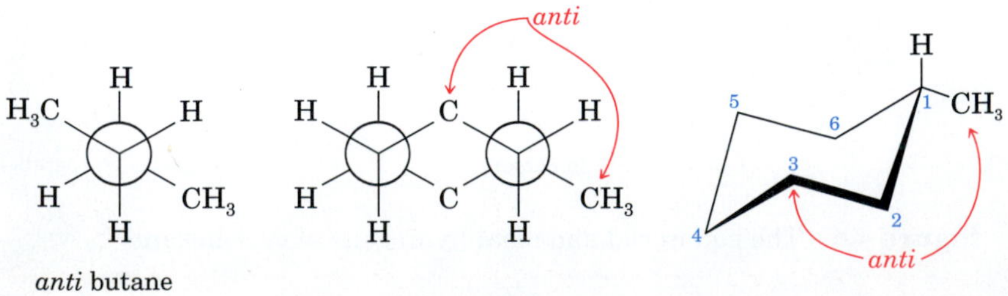

anti butane

Gauche butane is 0.9 kcal/mol less stable than *anti* butane. Methylcyclohexane with an axial methyl group is 1.8 kcal/mol less stable than the conformation with an equatorial methyl group. The energy is twice that for butane (2 × 0.9 kcal/mol) because the axial methyl group on the ring interacts with *two* axial hydrogen atoms (one at carbon 3 and one at carbon 5).

The bulkier the group, the greater is the energy difference between axial and equatorial conformers. In other words, a cyclohexane ring with a bulky substituent is more likely to have that group in the equatorial position. When the size of the substituent group reaches *tert*-butyl, the difference in energies between the conformers becomes quite large. *tert*-Butylcyclohexane is often said to be "frozen" in the conformation in which the *tert*-butyl group is equatorial. The ring is not truly frozen, but the energy difference (5.6 kcal/mol) between the equatorial and the axial positions of the *tert*-butyl group means that only 1 in 10,000 molecules has the *tert*-butyl group in an axial position at any given time.

Table 4.2 lists equilibrium constants for selected groups along with the percent equatorial at equilibrium.

Table 4.2   Conformational equilibria

Group	Equilibrium constant (K) at 25°	% Equatorial
—F	1.52	60.3
—Cl	2.32	69.8
—Br	2.32	69.8
—OH	5.41	84.4
—CH_3	17.6	94.6
—CH_2CH_3	20.9	95.4
—$CH(CH_3)_2$	34.6	97.2
—$C(CH_3)_3$	5,000–25,000	>99.9

B. Disubstituted Cyclohexanes

Two groups substituted on a cyclohexane ring may be either *cis* or *trans*. The *cis-* and *trans*-disubstituted rings are geometric isomers and are not interconvertible at room temperature; however, either isomer may assume a variety of conformations. For example, consider some chair forms of *cis*-1,2-dimethylcyclohexane.

Some different representations of cis-1,2-dimethylcyclohexane,

| both "down" | both "down" | both "up" |

Because this is the *cis*-isomer, the methyl groups must both be on the same side of the ring regardless of the conformation. In each chair conformation that we can draw, one methyl is axial and the other is equatorial. For any *cis*-1,2-disubstituted cyclohexane, one substituent must be axial and the other substituent must be equatorial. (Refer back to Figure 4.8 or use molecular models and verify this statement for yourself.)

When *cis*-1,2-dimethylcyclohexane changes from one chair conformer to the other, the two methyl groups reverse their equatorial–axial status. The energies of the two conformers are equal because their structures and bonding are equivalent. Therefore, this compound exists primarily as a 50:50 mixture of these two chair-form conformers.

Conformers of cis-1,2-dimethylcyclohexane:

axial, equatorial (or *a,e*)   *equatorial, axial* (or *e,a*)

In *trans*-1,2-dimethylcyclohexane, the methyl groups are on *opposite* sides of the ring. In the chair form of the *trans* isomer, one group must be attached to an "uppermost" bond, while the other is attached to a "lowermost" bond.

Some different representations of trans-1,2-dimethylcyclohexane,

No matter how the two adjacent *trans* groups are shown, they are *both axial* (*a,a*) or they are *both equatorial* (*e,e*). There is no way for two groups to be *trans* and 1,2 on the chair form of cyclohexane without assuming either the *a,a* or the *e,e* conformation.

Conformers of trans-1,2-dimethylcyclohexane:

a,a   e,e
favored

A single methyl group on a cyclohexane ring assumes the equatorial position preferentially. *Two* methyl groups on a cyclohexane ring also assume the equatorial positions preferentially, if possible. In *trans*-1,2-dimethylcyclohexane, the *e,e* conformer is the preferred conformer and is of lower energy than the *a,a* conformation. The *trans e,e* conformer is also of lower energy (by 1.87 kcal/mol) than either conformer of the *cis* compound, which must be *a,e* or *e,a*.

In summary,

For the 1,2-dimethylcyclohexanes:

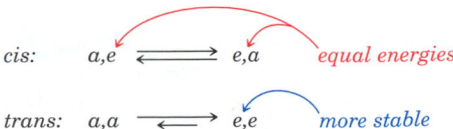

cis: a,e ⇌ e,a   *equal energies*

trans: a,a ⇌ e,e   *more stable*

In the case of a 1,2-disubstituted cyclohexane, the *trans* isomer is more stable than the *cis* isomer because both substituents can be equatorial. However, when the two substituents are 1,3 to each other on a cyclohexane ring, the *cis* isomer is more stable than the *trans* isomer. The reason is that both substituents in the *cis*-1,3 isomer can be equatorial. In the *trans*-1,3 isomer, one group must be axial.

cis-1,3-dimethylcyclohexane:

a,a

e,e
more stable and favored

trans-1,3-dimethylcyclohexane:

e,a

a,e

SAMPLE PROBLEM

What are the possible equatorial–axial relationships for the *cis-* and *trans-*1,4-dimethylcyclohexanes? In each case, which conformer is of lower energy?

Solution
cis-1,4: *a,e; e,a*
trans-1,4: *a,a; e,e*

The *cis* conformers are of equal energy. The *e,e trans* conformer is of lower energy than the *a,a trans* conformer (and of lower energy than either *cis* conformer).

STUDY PROBLEM

4.9 Label each of the following disubstituted rings as *cis* or *trans* and as *a,a; e,e;* or *a,e*.

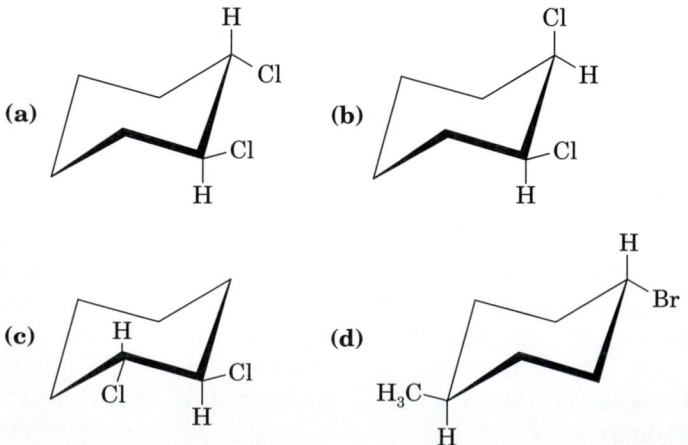

(a)

(b)

(c)

(d)

Section 4.6
Chirality

A. Chirality of Objects and Molecules

Consider your left hand. Your hand cannot be superposed on its mirror image, or placed over the mirror image so that all parts coincide. If you hold your left hand up to a mirror, the image looks like a right hand. If you do not have a mirror available, hold your hands together with the palms facing each other. You can see that they are mirror images. Try to superpose your hands (both palms down). You cannot do it (see Figure 4.9). This right- and left-handedness is also encountered in shoes and gloves. (Try wearing a left-handed glove on a right hand!)

Any object that *cannot be superposed on its mirror image* is said to be **chiral** (Greek *cheir,* "the hand"). Hands, gloves, and shoes are all chiral. Conversely, a plain cup or a cube is **achiral** (not chiral); these can be superposed on their mirror images. Figure 4.10 shows a cup being superposed on its mirror image. The same principles of right- and left-handedness also apply to molecules. A molecule that can be superposed upon its mirror image is *achiral.* A molecule that *cannot* be superposed on its mirror image is *chiral.* Figures 4.11 and 4.12 show an achiral molecule and a chiral molecule along with their mirror images.

An achiral molecule and its superposable mirror-image molecule are the same compound; they are not isomers. But a chiral molecule is *not* superposable on its mirror image; this molecule and its mirror-image molecule are different compounds, and represent a pair of stereoisomers called **enantiomers.** A pair of enantiomers is simply a pair of isomers that are *nonsuperposable mirror images.*

B. Chiral Carbon Atoms

The most common structural feature (but not the only one) that gives rise to chirality in molecules is that the molecule contains an sp^3-hybridized carbon

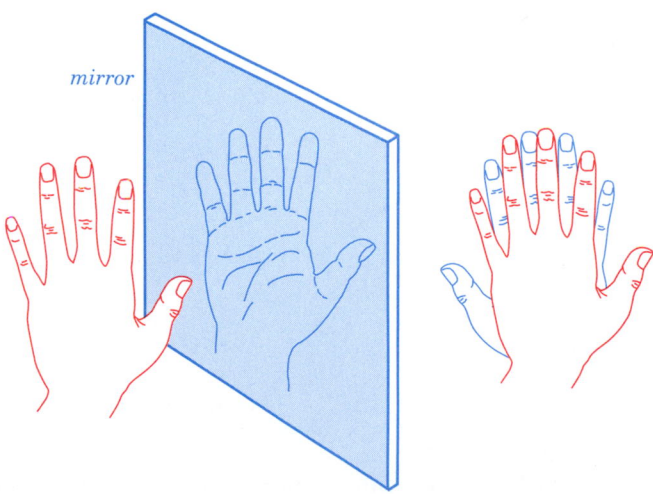

Figure 4.9   A *chiral* object *cannot* be superposed, or made to coincide completely, on its mirror image.

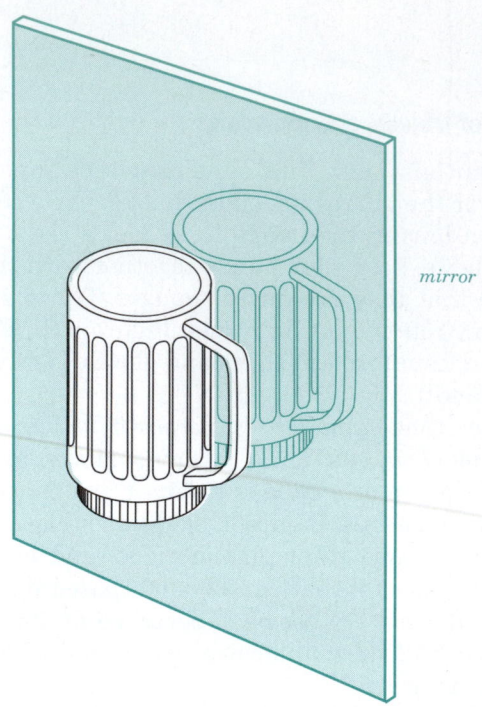

Figure 4.10 An *achiral* object is superposable on its mirror image.

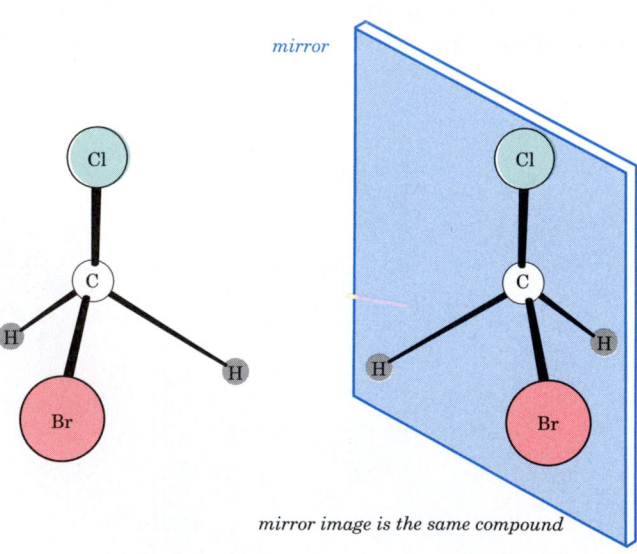

Figure 4.11 A molecule with a single carbon atom that has two identical substituents (H in this case) is achiral and can be superposed upon its mirror image. (**Try it with models.**)

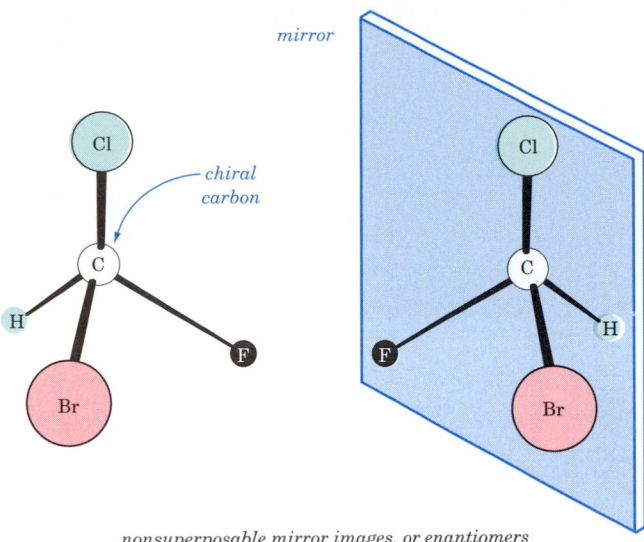

nonsuperposable mirror images, or enantiomers

Figure 4.12   A molecule that has four different groups attached to one single carbon atom is chiral and is not superposable on its mirror image. (Try it with models.)

atom with *four different groups* attached to it (Figure 4.12). Such a molecule is chiral and exists as a pair of enantiomers. A carbon atom with four different groups attached is often called a **chiral carbon** or chiral center, although, technically, it is the molecule and not the carbon atom that is chiral. It is also called an *asymmetric carbon, asymmetric center, stereocenter,* or *stereogenic center.* Until the nomenclature in this area clarifies, we will use the term *chiral carbon* rather than the alternative terms because (1) it avoids the introduction of a new word, (2) nearly all of the chiral molecules we will encounter will have carbon atoms as the chiral centers, and (3) it is consistent with the term *prochiral carbon* that will be introduced in Section 4.10. By learning to recognize chiral carbons within a formula, we can greatly simplify the problem of identifying structures that can exist as enantiomers.

In order to identify a chiral carbon, we must determine that all four groups attached to the sp^3-hybridized carbon are different. In many cases, the problem is trivial; for example, if the carbon is attached to two or more H atoms ($-CH_2-$ or $-CH_3$), then the carbon cannot be chiral. However, in a few cases the problem can be more challenging. In these cases we must inspect each *entire group* attached to the carbon in question, not just the atoms bonded directly to that carbon.

$$CH_3CH_2CH_2CH_2-\overset{\overset{\textstyle CH_2CH_3}{|}}{\underset{\underset{\textstyle CH_3}{|}}{C}}-CH_2CH_2CH_3$$

chiral carbon

SAMPLE PROBLEM

Chiral carbons are often starred for emphasis. Star the chiral carbons in the following structures:

$$\underset{\overset{|}{Br}}{\overset{\overset{Cl}{|}}{CH_3CCH_2CH}}=CHCH_3 \qquad \underset{\overset{|}{CH_2CH_3}}{CH_3CHCOCH_3}\overset{\overset{O}{\|}}{}$$

Solution

$$\underset{\overset{|}{Br}}{\overset{\overset{Cl}{|}}{CH_3\overset{*}{C}CH_2CH}}=CHCH_3 \qquad \underset{\overset{|}{CH_2CH_3}}{CH_3\overset{*}{C}HCOCH_3}\overset{\overset{O}{\|}}{}$$

STUDY PROBLEM

4.10 Star the chiral carbons (if any) in the following formulas:

(a) $CH_3\overset{*}{C}HCOH$ with $\overset{O}{\|}$ above C and NH_2 below

(b) structure with Cl-substituted benzene ring, $-CH-CH_2CH_3$, and a second benzene ring with Cl

(c) $ClCH_2\overset{*}{C}HCH_2OH$ with CH_3 below

(d) $HO\overset{*}{C}H$ with $\overset{\overset{O}{\|}}{CH}$ above and CH_2OH below

Drawing structures of a pair of enantiomers for a molecule with one chiral carbon is relatively easy. The *interchange of any two groups* around the chiral carbon results in the enantiomer. The following examples show two ways to convert the formula for one compound (**A**) to the formula for its enantiomer (**B**). (Use models to verify that the two formulas shown for **B** actually do represent the same molecule.)

interchange

$$H\blacktriangleright \underset{\overset{|}{CH_2CH_3}}{\overset{\overset{CH_3}{|}}{C}}\blacktriangleleft Cl \qquad Cl\blacktriangleright \underset{\overset{|}{CH_2CH_3}}{\overset{\overset{CH_3}{|}}{C}}\blacktriangleleft H \qquad H\blacktriangleright \underset{\overset{|}{CH_2CH_3}}{\overset{\overset{CH_3}{|}}{C}}\blacktriangleleft Cl \qquad H\blacktriangleright \underset{\overset{|}{Cl}}{\overset{\overset{CH_3}{|}}{C}}\blacktriangleleft CH_2CH_3$$

     A   B   *interchange*   A   B

        enantiomers   *enantiomers*

STUDY PROBLEM

4.11 Draw the formulas (either dimensional or ball-and-stick) for the two enantiomers of each of the following compounds:

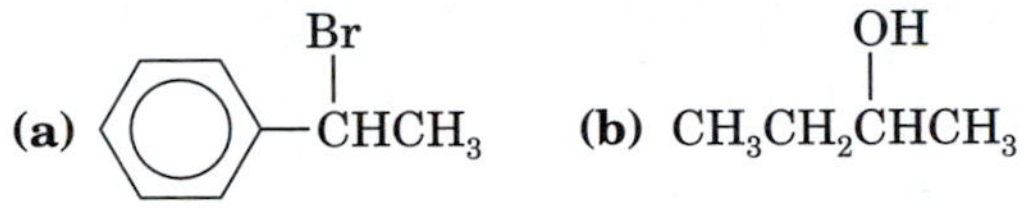

$$\text{(a)} \quad \text{C}_6\text{H}_5\!-\!\overset{\overset{\displaystyle Br}{|}}{\text{CHCH}_3} \qquad \text{(b)} \quad CH_3CH_2\overset{\overset{\displaystyle OH}{|}}{CHCH_3}$$

C. Fischer Projections

In 1891, the German chemist Emil Fischer introduced formulas showing the spatial arrangement of groups around chiral carbon atoms. Today these formulas are called **Fischer projections.** Fischer used his formulas to help determine the configurations of the four chiral carbons in glucose. He was awarded the 1902 Nobel Prize for this brilliant piece of research. We will illustrate the type of Fischer projection commonly used today with the simplest of sugars: 2,3-dihydroxypropanal (usually called glyceraldehyde) and 2,3,4-trihydroxybutanal (erythrose). Glyceraldehyde has one chiral carbon atom (carbon 2), while erythrose has *two* chiral carbons (carbons 2 and 3). As you can see from the following representations for these two compounds, a Fischer projection is simply a shorthand way to represent a ball-and-stick or dimensional formula.

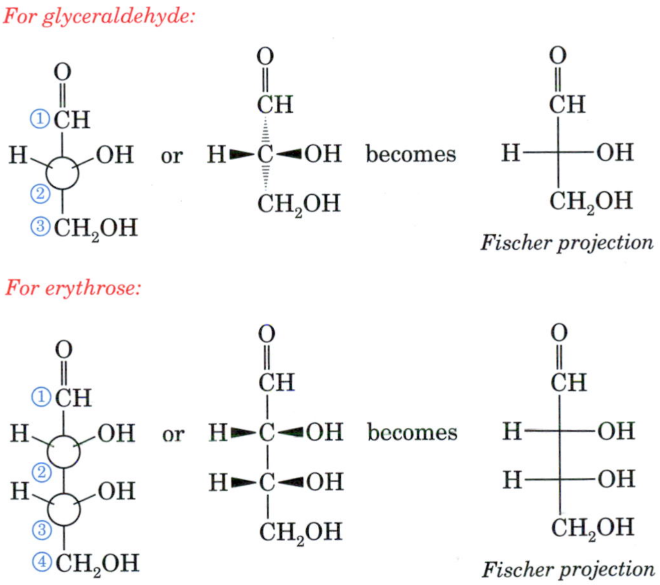

For glyceraldehyde:

For erythrose:

In drawing a Fischer projection, we assume that the molecule is *completely stretched out in the plane of the paper with all its substituents eclipsed,* regardless of any preferred conformation. The preceding formulas for erythrose show the conformation used for the Fischer projection. By convention, the carbonyl group (or group of highest nomenclature priority) is placed at or near the top. Thus, *the top carbon is carbon 1.* Each intersection of the horizontal and vertical lines represents a chiral carbon. *Each horizontal line represents a bond coming toward the viewer, while the vertical line represents bonds going back, away from the viewer.*

away from viewer

```
        CHO
         |
H————————OH ———— toward viewer
         |
      CH₂OH
```

A pair of enantiomers is easily recognized when Fischer projections are used.

```
    ¹CHO                         CHO
      |                           |
H —²— OH                    HO —— H
      |                           |
H —³— OH                    HO —— H
      |                           |
HO —⁴— H                     H —— OH
    ⁵ |                           |
    CH₂OH                       CH₂OH
```

Carbons 2, 3, and 4 are chiral.   *The enantiomer shows all groups or atoms on chiral carbons transposed from left to right.*

A Fischer projection may be rotated 180° in the plane of the paper, but it may not be flipped over or rotated by any other angle. Either of these last two operations would take the formula out of the Fischer projection and lead to an incorrect structure.

Correct:

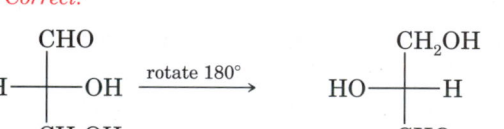

```
     CHO                          CH₂OH
      |          rotate 180°        |
H ——— OH      —————————→     HO ——— H
      |                            |
    CH₂OH                         CHO
```

same as left-hand structure

Incorrect:

```
     CHO                             H
      |          rotate 90°          |
H ——— OH      —————✗———→    HOCH₂ ——— CHO
      |                             |
    CH₂OH                          OH
```

NOT THE SAME

(To verify the difference, draw this projection as a dimensional formula with the horizontal bonds coming toward you.)

Fischer projections are a convenient shorthand way of representing chiral molecules. Because of the limitations placed upon them, such as the preceding one of rotation, Fischer projections must be used carefully. We suggest that you convert Fischer projections to ball-and-stick or dimensional formulas (or use models) when performing any spatial manipulations. In future chapters, we will stress dimensional formulas rather than Fischer projections until we discuss carbohydrates (Chapter 23).

STUDY PROBLEMS

4.12 The following Fischer projections represent two amino acids that are formed when proteins are hydrolyzed (cleaved with water). Convert these projections to dimensional formulas.

(a)
$$
\begin{array}{c}
CO_2H \\
H_2N\!-\!\!\!\!-\!\!|\!-\!\!\!\!-H \\
CH(CH_3)_2
\end{array}
$$
valine

(b)
$$
\begin{array}{c}
CO_2H \\
H_2N\!-\!\!\!\!-\!\!|\!-\!\!\!\!-H \\
CH_2CO_2H
\end{array}
$$
aspartic acid

4.13 Convert the following dimensional formulas for amino acids to Fischer projections.

(a)
$$
\begin{array}{c}
NH_2 \\
| \\
CH_3\!\cdots C\!\diagup\!\diagdown CO_2H \\
H
\end{array}
$$
alanine

(b)
$$
\begin{array}{c}
CH_2SH \\
| \\
H\!\cdots C\!\diagup\!\diagdown CO_2H \\
H_2N
\end{array}
$$
cysteine

Section 4.7
Rotation of Plane-Polarized Light

With the exception of chirality, the structures of a pair of enantiomers are the same. Therefore, almost all of their physical and chemical properties are the same. For example, each pure enantiomer of a pair has the same melting point and the same boiling point. Only two sets of properties are different for a single pair of enantiomers: *Note: rotation happens with enantiomers only:*

1. interactions with other chiral substances

2. interactions with polarized light.

Ordinary light travels in waves, and the waves are at right angles to the direction of travel. **Plane-polarized light** is light in which all wave vibrations have been filtered out except for those in one plane. The plane polarization can be effected by passing ordinary light through a pair of calcite crystals ($CaCO_3$) or a polarizing lens. (The same principle is used in Polaroid sunglasses.) Figure 4.13 shows a simplified diagram of the plane polarization of light.

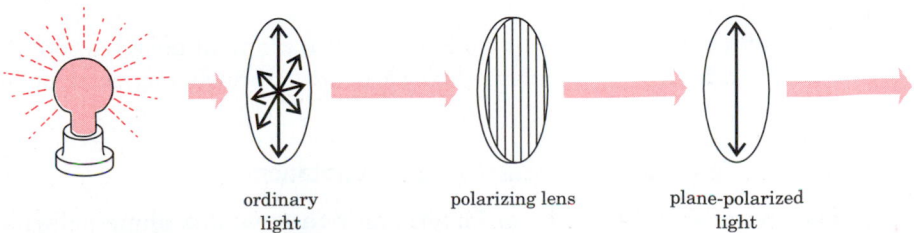

ordinary
light

polarizing lens

plane-polarized
light

Figure 4.13   The plane polarization of light.

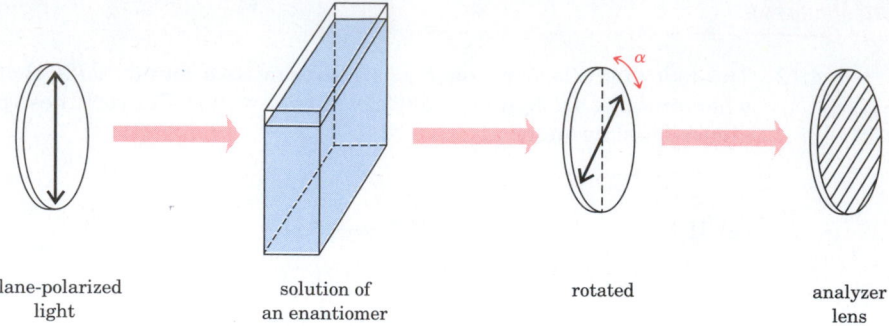

| plane-polarized light | solution of an enantiomer | rotated | analyzer lens |

Figure 4.14 The plane of polarization of plane-polarized light is rotated by a single enantiomer.

If plane-polarized light is passed through a solution containing a single enantiomer, the plane of polarization of the light is *rotated* either to the right or to the left (see Figure 4.14). The rotation of plane-polarized light is referred to as **optical rotation.** A compound that rotates the plane of polarization of plane-polarized light is said to be **optically active.** For this reason, enantiomers are sometimes referred to as **optical isomers.**

A **polarimeter** is an instrument designed for polarizing light and then showing the angle of rotation of the plane of polarization of the light by an optically active compound. The amount of rotation depends on (1) the structure of the molecules; (2) the temperature; (3) the wavelength; (4) the number of molecules in the path of the light; and, in some cases, (5) the solvent.

Specific rotation is the amount of rotation by 1.00 g of sample in 1.00 mL of solution in a tube with a path length of 1.00 decimeter (dm), at a specified temperature, wavelength, and solvent. The commonly used wavelength is 589.3 nm (the D line of sodium), where 1.0 nm = 10^{-9} m. The specific rotation for a compound (at 20°, for example) may be calculated from the observed rotation by the following formula:

$$[\alpha]_D^{20} = \frac{\alpha}{lc}$$

where $[\alpha]_D^{20}$ = specific rotation of sodium D line at 20°

α = observed rotation at 20°

l = cell length in dm (1.0 dm = 10 cm)

c = concentration of sample solution in g/mL.

The fact that some compounds rotate the plane of polarization of plane-polarized light was discovered in 1815 by the French physicist Jean-Baptiste Biot.

A. Some Terms Used in Discussing Optical Rotation

The enantiomer of any enantiomeric pair that rotates plane-polarized light to the *right* is said to be **dextrorotatory** (Latin *dexter,* "right"). Its mirror image, which rotates plane-polarized light to the *left,* is said to be **levorotatory**

(Latin *laevus,* "left"). The direction of rotation is specified in the name by (+) for dextrorotatory and (−) for levorotatory. (In older literature, *d* for dextrorotatory and *l* for levorotatory are sometimes encountered.)

$$\begin{array}{cccc}
O & O & O & O \\
\| & \| & \| & \| \\
CH & CH & CH & CH
\end{array}$$

$$H\blacktriangleright \!\!\! \overset{}{C} \!\!\! \blacktriangleleft OH \quad or \quad H\!-\!\!\!-\!\!\!-OH \qquad HO\blacktriangleright \!\!\! \overset{}{C} \!\!\! \blacktriangleleft H \quad or \quad HO\!-\!\!\!-\!\!\!-H$$

$$\begin{array}{cccc}
CH_2OH & CH_2OH & CH_2OH & CH_2OH
\end{array}$$

$$\begin{array}{cc}
\text{(+)-glyceraldehyde} & \text{(−)-glyceraldehyde} \\
[\alpha]_D^{20} = +8.7° & [\alpha]_D^{20} = -8.7°
\end{array}$$

A mixture of equal parts of any pair of enantiomers is called a **racemic mixture,** or **racemic modification.** A racemic mixture may be indicated in the name by the prefix (±). (In older literature, *dl* is used to signify a racemic mixture.) Thus, racemic glyceraldehyde is called (±)-glyceraldehyde. (The term *racemic* comes from the Latin *racemis,* "a bunch of grapes." The reason for this unusual derivation is that *racemic* was first used to describe racemic tartaric acid, which was isolated as a by-product of wine-making.)

$$50\% \text{ (+)-glyceraldehyde} + 50\% \text{ (−)-glyceraldehyde} = \text{(±)-glyceraldehyde}$$

$$[\alpha]_D^{20} = +8.7° \qquad\qquad [\alpha]_D^{20} = -8.7° \qquad\qquad [\alpha]_D^{20} = 0$$

a racemic mixture

A racemic mixture does not rotate the plane of polarization of plane-polarized light because the rotation by each enantiomer is cancelled by the equal and opposite rotation by the other. A solution of either a racemic mixture or of an achiral compound is said to be **optically inactive,** but the causes of the optical inactivity are different.

No Optical Activity
(1) racemic mixtures
(2) achiral compound

STUDY PROBLEM

4.14   (*S*)-2-Iodobutane has an $[\alpha]_D^{24}$ of +15.9°. **(a)** What is the observed rotation at 24°C of an equimolar mixture of (*R*)- and (*S*)-2-iodobutane? **(b)** What is the observed rotation (at 24°C, 1-dm sample tube) of a solution (1.0 g/mL) of a mixture that is 25% (*R*)- and 75% (*S*)-2-iodobutane?

It was Louis Pasteur who in 1848 made the momentous discovery that there are two types of sodium ammonium tartrate crystals and that these two types are mirror images of each other. (We will discuss the structure of tartaric acid in Section 4.9C.) Louis Pasteur was born in 1822. He attended college at Besançon in France, where his performance was creditable but not brilliant. At the age of 24, Pasteur completed his doctoral work at the École Normale in Paris. His first postdoctoral position was that of laboratory assistant at his graduate school, a job that allowed him freedom and opportunity for individual research.

To further his laboratory skills, Pasteur decided to repeat some previous studies on the crystals of tartaric acid—a project that led to his first major scientific discovery.

At that time, only the dextrorotatory (+) isomer (called tartaric acid) and the optically inactive racemic mixture (called "paratartaric acid" and thought to be a single compound) were known. The levorotatory (−) and *meso* (see Section 4.9C) isomers of tartaric acid, as well as the concept of a racemic mixture, were unknown. The year was 1848, many years before even the acceptance of a tetrahedral carbon.

Previous workers suggested that (+)-tartaric acid and paratartaric acid had the same structure, but Pasteur was convinced that there must be a correlation between molecular structure and the ability of a compound to rotate the plane of polarization of plane-polarized light.

He carefully crystallized the sodium ammonium salts of the two known tartaric acids and, to his surprise, detected two types of crystals. Paratartaric acid had formed mirror-image crystals, which Pasteur called right-handed and left-handed. However, (+)-tartaric acid formed only right-handed crystals.

The idea that paratartaric acid was actually a mixture of equal parts of (+)-tartaric acid and the unknown (−) isomer must have come to Pasteur in a flash of intuition. All of his subsequent efforts were aimed at obtaining physical evidence for this hypothesis. Using a hand lens and tweezers, he separated the right-handed crystals from the left-handed ones. Then he determined that a solution of right-handed crystals rotated the plane of polarization of plane-polarized light to the right—just like the salt of (+)-tartaric acid. The solution of left-handed crystals rotated the plane to the left by the same number of degrees. Pasteur completed the experiment by dissolving equal weights of the two crystals and observing that this solution did not rotate the plane of polarization.

At this point, Pasteur is said to have jumped up from the polarimeter and dashed into the hall shouting, "I have it! I have made a great discovery." Indeed, he had. At the age of 26, Pasteur had just completed one of the classic experiments in chemistry—the first resolution of a racemic mixture.

The scientific world of Paris was soon buzzing with talk about Pasteur's experiment. The news reached Biot, who at 74 was a "grand old man" of French science. He summoned Pasteur and, by voice and gesture, made it known that he was skeptical of Pasteur's claim that he could separate paratartaric acid into optically active substances. To personally verify this claim, Biot provided the chemicals Pasteur would need to repeat the experiment, including paratartaric acid previously determined to be optically inactive. Pasteur mixed the ingredients to make a solution of sodium ammonium tartrate in an evaporating dish. Biot then took the dish from Pasteur and dismissed him. Two days later, after the crystals had formed, Biot sent for Pasteur.

With the older man looking on, Pasteur separated the crystals.

"So you affirm," said Biot, "that your right-handed crystals will deviate to the right the plane of polarization, and your left-handed ones will deviate it to the left?"

"Yes," said Pasteur.

"Well, let me do the rest."

Biot dismissed Pasteur again and prepared solutions containing weighted amounts of the crystals for the polarimeter. When all was ready, he summoned Pasteur a second time.

Biot selected the solution that, according to Pasteur, contained the heretofore unknown levorotatory isomer of sodium ammonium tartrate. He peered into the polarimeter and instantly saw that the solution was indeed strongly levorotatory. He turned and grabbed Pasteur by the arm. "My dear child," he said, "I have loved science so much throughout my life that this makes my heart throb."

Luck had smiled on Pasteur that spring of 1848. From all possible salts, he had chosen to study sodium ammonium tartrate. It was Pasteur's good fortune that he did. The spontaneous crystallization of enantiomers from a racemic mixture to yield mirror-image crystals is an extremely rare occurrence. Only nine compounds are known to form enantiomeric crystals large enough to be separated with tweezers.

Luck smiled more than once on Pasteur. Sodium ammonium tartrate forms enantiomeric crystals only when crystallized at a temperature below 26°C (79°F). Had Biot's laboratory been warmer, the demonstration would have failed and Pasteur's reputation tarnished. Instead, Biot became Pasteur's sponsor, publicly lauding him and launching him on a brilliant scientific career.

Section 4.8
Assignment of Configuration: the (*R*) and (*S*) System

The order of arrangement of four groups around a chiral carbon atom is called the **absolute configuration** around that atom. (Do not confuse configurations with *conformations,* which are shapes arising from rotation around bonds.) A pair of enantiomers have opposite configurations. For example, (+)- and (−)-glyceraldehyde have opposite configurations. But which formula represents the dextrorotatory enantiomer and which represents the levorotatory one? Until 1951, chemists did not know! Prior to that time, it was known that (+)-glyceraldehyde and (−)-glyceric acid (2,3,-dihydroxypropanoic acid) have the same configuration around carbon 2, even though they rotate plane-polarized light in opposite directions. But it was not known whether the OH at carbon 2 was to the right or to the left in the formulas as shown:

same configuration, but is the —OH on the right (as shown) or on the left?

(−)-glyceric acid   (+)-glyceraldehyde

To make formulas easier to work with, chemists decided in the late nineteenth century to assume that (+)-glyceraldehyde has the absolute configuration with the OH on carbon 2 to the right, as shown in the preceding formula. In 1951, X-ray diffraction studies by J. M. Bijvoet at the University of Utrecht in Holland showed that the original assumption was correct. (+)-Glyceraldehyde does indeed have the absolute configuration that chemists had been using for 60 years. If the early chemists had guessed wrong, the chemical literature would have been in a state of confusion—all pre-1951 articles would show configurations the *reverse* of those in more modern articles. The original assumption was indeed a lucky guess!

The direction of rotation of plane-polarized light by a particular enantiomer is a physical property. The absolute configuration of a particular enantiomer is a characteristic of its molecular structure. *There is no simple relationship between the absolute configuration of a particular enantiomer and its direction of rotation of plane-polarized light.* As we have said, the enantiomer of glyceric acid with the same absolute configuration as (+)-glyceraldehyde is levorotatory, not dextrorotatory.

We have shown how the direction of rotation of plane-polarized light can be indicated by (+) or (−). However, a system is also needed to indicate the absolute configuration—that is, the actual arrangement of groups around a chiral carbon. This system of nomenclature is called the **(R) and (S) system** and was first introduced by R. S. Cahn and Sir Christopher Ingold of University College, London, in 1951. It was later modified in collaboration with Vlado Prelog of the Swiss Federal Institute of Technology. The letter (R) comes from the Latin *rectus*, "straight," or "right," while (S) comes from the Latin *sinister*, "left." Any chiral carbon atom has either an (R) configuration or an (S) configuration; therefore, one enantiomer is (R) and the other is (S). A racemic mixture can be designated (RS), meaning a mixture of the two.

In the (R) and (S) system, groups are assigned a priority ranking using the same set of rules as are used in the (E) and (Z) system; however, the priority ranking is used in a slightly different manner. To assign an (R) or (S) configuration to a chiral carbon:

---

1. Rank the four groups (or atoms) attached to the chiral carbon in order of priority by the Cahn–Ingold–Prelog sequence rules (pages 124–125).

2. Project the molecule with the group of *lowest priority* to the rear.

3. Select the group of *highest priority and draw a curved arrow to the group of next highest priority.*

4. If this arrow is *clockwise,* the configuration is (R). If the arrow is *counterclockwise,* the configuration is (S).

Let us illustrate the use of this procedure by assigning (R) and (S) to the enantiomers of 1-bromo-1-chloroethane.

enantiomers of 1-bromo-1-chloroethane

1. Rank the four groups. In the case at hand, the order of priority of the four atoms by atomic number is Br (highest), Cl, C, H (lowest).

2. Draw projections with the lowest-priority atom (H) in the rear. (This atom is hidden behind the carbon atom in the following projections.)

3. Draw an arrow from highest priority (Br) to second highest priority (Cl).

4. Assign (R) and (S). Note how the (R) and (S) designation is incorporated into the name.

clockwise = (R)   *counterclockwise = (S)*

(R)-1-bromo-1-chloroethane   (S)-1-bromo-1-chloroethane

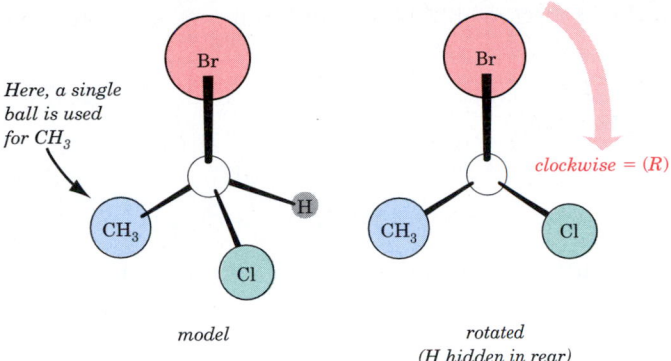

Figure 4.15   Using a molecular model to determine (R) or (S) configuration.

An easy way to remember which enantiomer is (R) and which is (S) is to compare the curved arrow to the steering wheel on an automobile. Turning the steering wheel to the *right* is equivalent to (R); turning it to the *left* is equivalent to (S).

Placing a structure in the correct position for an (R) or (S) assignment is easily done with a molecular model. Construct the model, grasp the group of lowest priority with one hand, and turn the model so that the remaining three groups face you. Determine if the structure is (R) or (S) in the usual way. (See Figure 4.15.)

SAMPLE PROBLEMS

Name the following compound, including an (R) or (S) designation.

$$\text{HO} \diagdown \underset{\underset{\text{CH}_2\text{CH}_3}{|}}{\overset{\overset{\text{H}}{|}}{\text{C}}} \diagup \text{CH}_3 \quad \text{or} \quad \text{HO} \blacktriangleright \underset{\underset{\text{CH}_2\text{CH}_3}{}}{\overset{\overset{\text{H}}{\vdots}}{\text{C}}} \blacktriangleleft \text{CH}_3$$

Solution

1. order of priority: OH (highest), CH_2CH_3, CH_3, H (lowest)

2. projection with H in rear:

$$\text{HO} \diagdown \underset{\text{CH}_2\text{CH}_3}{\overset{}{\text{C}}} \diagup \text{CH}_3 \quad \text{or} \quad \text{HO} \diagup \underset{\underset{\text{CH}_2\text{CH}_3}{|}}{\text{C}} \diagdown \text{CH}_3$$

3. counterclockwise = (S)

4. name: (S)-2-butanol

Draw the structure of (R)-$(-)$-2-butanol.

Solution

1. Write the formula of the compound without regard to configuration of the chiral carbon.

$$\overset{\text{OH}}{\underset{|}{\text{CH}_3\text{*CHCH}_2\text{CH}_3}}$$

2. Assign priorities around the chiral carbon: OH (highest), CH_2CH_3, CH_3, H (lowest).

3. Draw the projection with the lowest-priority group (H) in the rear. Place the remaining groups in such a way that from OH to CH_2CH_3 is clockwise for the (R) configuration.

4. Redraw the structure as a ball-and-stick or dimensional formula. (Models make this task easier.)

Note that the sign of rotation in the name is not used in the solution to this problem. The $(-)$ is the indication of direction of rotation of plane-polarized light, a physical property of (R)-2-butanol, and is not used in determining its configuration.

The method we have used for making assignments of configurations requires dimensional formulas. But how do we proceed if we are presented with a Fischer projection? One way is to first convert the Fischer projection to a dimensional formula and then apply the procedure we have just outlined. However, with a slight modification, the procedure can be used for configurational assignments directly from Fischer projections. In using this modified procedure, we must remember the following:

1. In Fischer projections, vertical bonds go away from the viewer, while horizontal bonds come toward the viewer.

2. Interchanging any two groups attached to a chiral carbon produces its mirror image (opposite configuration). Two consecutive two-group interchanges maintains the original configuration.

Thus, if the group with the lowest priority is on a vertical bond (either the top or bottom) of the Fischer projection, the molecule is in the proper orientation for assigning (R) or (S). If this group is not already on a vertical bond, interchange it with a group that is. Then interchange any other two groups to restore the original configuration. The molecule is now in the proper orientation.

Let us illustrate this procedure by assigning (R) or (S) to the Fischer projection of lactic acid.

1. Assign priorities. The order of priority is OH (highest), CO_2H, CH_3, H (lowest).

2. Interchange the H and CH_3 groups. Now interchange the OH and CH_3 groups to restore the original configuration.

lactic acid   opposite configuration   original configuration
                                                        (S)-lactic acid

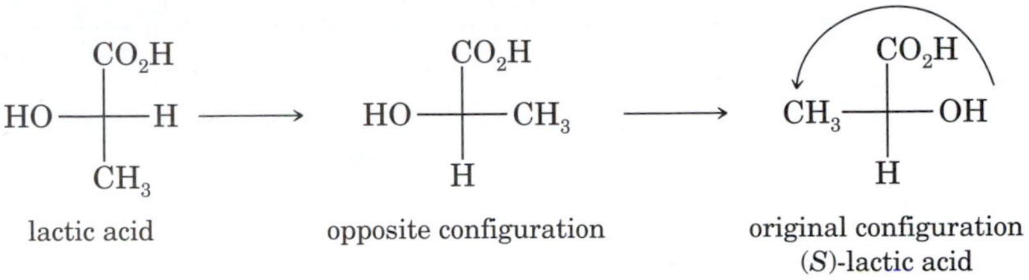

3. Going from OH to CO_2H to CH_3 is counterclockwise; the configuration is (S).

STUDY PROBLEMS

4.15   Assign each of the following molecules an (R) or (S) configuration:

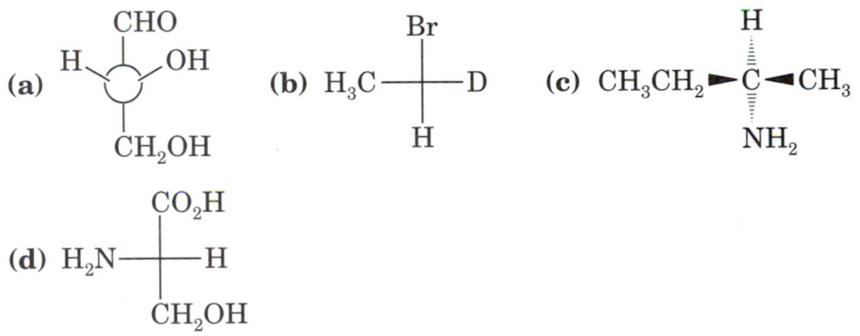

4.16   Draw formulas for **(a)** (R)-3-bromoheptane and **(b)** (S)-2-pentanol that show the absolute configurations around the chiral carbons.

Section 4.9
More Than One Chiral Carbon

Our discussion so far has concentrated primarily on compounds that contain only one chiral carbon, but compounds can have more than one chiral atom. Consider a compound with two different chiral carbons. Each of these two chiral carbons can be either (R) or (S); consequently, there are four different ways in which these configurations can be arranged in a molecule. A molecule with two different chiral carbons can therefore have four different stereoisomers.

chiral carbon 1	chiral carbon 2	total molecular configuration
(R)	(R)	$(1R,2R)$
(R)	(S)	$(1R,2S)$
(S)	(R)	$(1S,2R)$
(S)	(S)	$(1S,2S)$

SAMPLE PROBLEM

How many stereoisomers could exist for a compound that has three different chiral carbons?

Solution
eight: $(1R,2R,3R)$; $(1R,2R,3S)$; $(1R,2S,3R)$; $(1R,2S,3S)$; $(1S,2R,3R)$; $(1S,2R,3S)$; $(1S,2S,3R)$; $(1S,2S,3S)$.

The *maximum number of optical isomers* for a compound is 2^n, where n is the number of chiral atoms. If there are two chiral carbons, then there can be up to four stereoisomers ($2^2 = 4$); when there are three such carbon atoms, there can be up to eight stereoisomers ($2^3 = 8$).

STUDY PROBLEM

4.17   What is the maximum number of stereoisomers for each of the following compounds?

(a) 1,2-dibromo-1-phenylpropane;

(b) 1,2-dibromo-2-methyl-1-phenylpropane;

(c) 2,3,4,5-tetrahydroxypentanal

A. (*R*) and (*S*) System for a Compound with Two Chiral Carbon Atoms

To assign (R) or (S) configurations to two chiral carbons in one molecule, we consider each chiral carbon in turn. We will use the sugar erthyrose to illustrate the technique. (For convenience, we do not usually use broken wedges in a formula that shows more than one chiral carbon.)

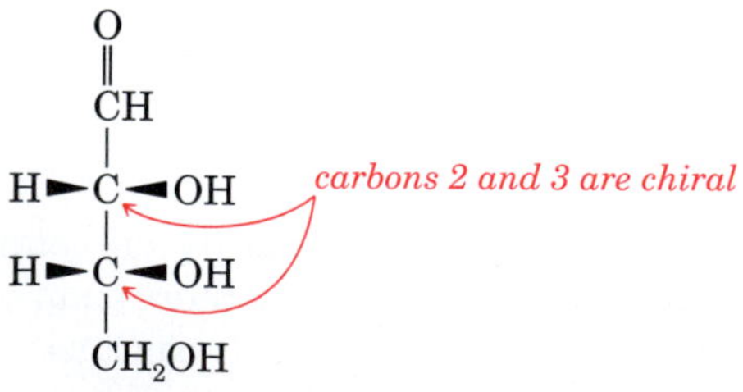

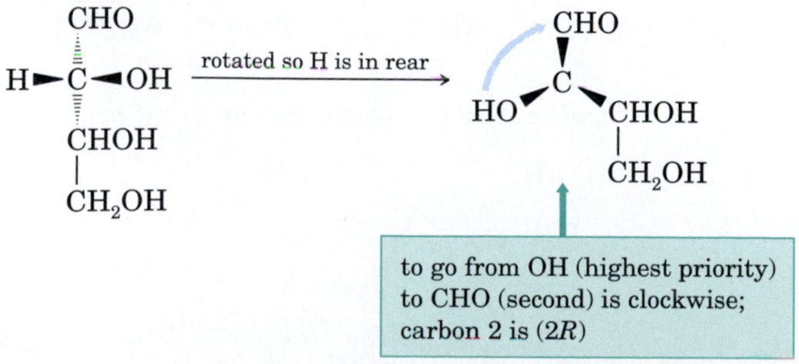

For carbon 3:

CHO
|
CHOH
|
H►C◄OH
|
CH$_2$OH

$\xrightarrow{\text{rotated}}$

CHO
|
CHOH
|
C
HO◄ ◄CH$_2$OH

to go from OH to —CH(OH)CHO
is clockwise; carbon 3 is (3*R*)

The IUPAC name for this stereoisomer is therefore (2*R*,3*R*)-2,3,4-trihydroxybutanal. Note that the numbers and letters in the single set of parentheses refer to the configurations around two different chiral carbons in one molecule. Contrast this notation with (2*RS*,3*RS*)-2,3,4-trihydroxybutanal, which means a racemic mixture. (Note that we must use the racemic symbol (*RS*) for each chiral carbon.)

Similarly, we may apply the previously outlined procedure for assigning configurations from Fischer projections to projections containing two or more chiral carbons. In fact, this procedure is more convenient than the one using dimensional formulas when two or more chiral carbons are present.

B. Diastereomers

When a molecule has more than one chiral carbon, not all of the optical isomers are enantiomers. By definition, enantiomers (mirror images) come in pairs.

The four stereoisomers of 2,3,4-trihydroxybutanal:

CHO	CHO		CHO	CHO
H►C◄OH	HO►C◄H		H►C◄OH	HO►C◄H
H►C◄OH	HO►C◄H		HO►C◄H	H►C◄OH
CH$_2$OH	CH$_2$OH		CH$_2$OH	CH$_2$OH
(2*R*,3*R*)	(2*S*,3*S*)		(2*R*,3*S*)	(2*S*,3*R*)

enantiomers   *enantiomers*

Note that the (2*R*,3*R*)- and (2*S*,3*S*)-stereoisomers are enantiomers. The (2*R*,3*S*)- and (2*S*,3*R*)-stereoisomers are also enantiomers. However, the (2*S*,3*S*)- and (2*R*,3*S*)-stereoisomers are *not* enantiomers. (What other nonenantiomeric pairings can be arranged?) Any pair of stereoisomers that are not enantiomers are called **diastereomers,** or **diastereoisomers.** Thus, the (2*S*,3*S*)- and (2*R*,3*S*)-stereoisomers are diastereomers. (*Geometric isomers* are also diastereomers by this definition.)

CHO   CHO
| |
HO►C◄H   H►C◄OH
| |
HO►C◄H   HO►C◄H
| |
CH$_2$OH   CH$_2$OH

a pair of diastereomers:
stereoisomers that are not enantiomers

A pair of enantiomers have identical physical and chemical properties except for interactions with other chiral molecules and with polarized light. Diastereomers, however, are chemically and physically different. They have different melting points and different solubilities and often undergo chemical reactions in a different fashion.

SAMPLE PROBLEM

Identify each pair of molecules as *structural isomers, enantiomers,* or *diastereomers.*

(a)
$$\begin{array}{cc} CO_2H & CO_2H \\ H_2N{\blacktriangleright}C{\blacktriangleleft}H & H{\blacktriangleright}C{\blacktriangleleft}NH_2 \\ CH_2OH & CH_2OH \end{array}$$

(b)
$$\begin{array}{cc} CH_3 \quad CH_3 & CH_3 \quad H \\ C{=}C & C{=}C \\ H \qquad H & H \qquad CH_3 \end{array}$$

(c)
$$\begin{array}{cc} CHO & CHO \\ H \quad OH & H \quad OH \\ H \quad OH & HO \quad H \\ CH_2OH & CH_2OH \end{array}$$

(d)
$$\begin{array}{cc} CHO & CHO \\ H{\blacktriangleright}C{\blacktriangleleft}OH & HO{\blacktriangleright}C{\blacktriangleleft}H \\ H{\blacktriangleright}C{\blacktriangleleft}OH & HO{\blacktriangleright}C{\blacktriangleleft}H \\ H{\blacktriangleright}C{\blacktriangleleft}OH & HO{\blacktriangleright}C{\blacktriangleleft}H \\ CH_2OH & CH_2OH \end{array}$$

Solution

The pairs in (a) and (d) are enantiomers; the pairs in (b) and (c) are diastereomers.

C. *Meso* Compounds

A compound with n chiral carbon atoms can have a maximum of 2^n stereoisomers, but it may not have that many. Consider a pair of structures (**A** and **B**) with two-chiral carbon atoms. At first glance, we might assume that **A** and **B** are enantiomers.

$$\begin{array}{cc} CO_2H & CO_2H \\ H{\blacktriangleright}C{\blacktriangleleft}OH & HO{\blacktriangleright}C{\blacktriangleleft}H \\ H{\blacktriangleright}C{\blacktriangleleft}OH & HO{\blacktriangleright}C{\blacktriangleleft}H \\ CO_2H & CO_2H \\ \textbf{A} & \textbf{B} \end{array}$$

Let us take compound **B** and rotate it 180° in the plane of the paper. We can see that compound **B** is identical to compound **A**! Indeed, **A** and **B** are mirror images, but the mirror images are superposable; therefore, **A** and **B** are the same compound.

$$\begin{array}{ccc} CO_2H & & CO_2H \\ HO{\blacktriangleright}C{\blacktriangleleft}H & \xrightarrow{\text{rotate 180°}} & H{\blacktriangleright}C{\blacktriangleleft}OH \\ HO{\blacktriangleright}C{\blacktriangleleft}H & & H{\blacktriangleright}C{\blacktriangleleft}OH \\ CO_2H & & CO_2H \\ \textbf{B} & & \textbf{B} \text{ superposable on } \textbf{A} \end{array}$$

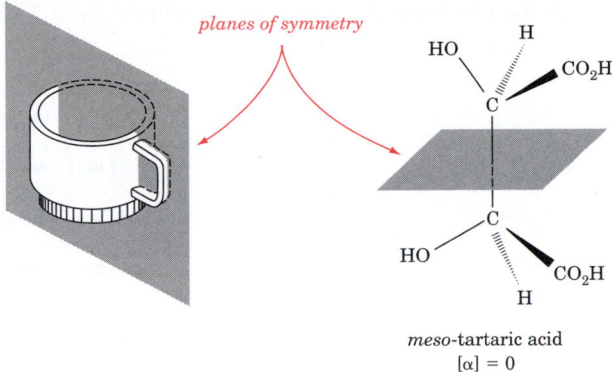

planes of symmetry

meso-tartaric acid
$[\alpha] = 0$

Figure 4.16 An internal plane of symmetry is a hypothetical plane that bisects an object into mirror-reflective halves. An object with an internal plane of symmetry is achiral (can be superposed on its mirror image).

How is it possible that a molecule with two chiral carbons is superposable on its mirror image? The answer is that, in at least one conformation, this molecule has an **internal plane of symmetry.** The "top" half of the molecule is the mirror image of the "bottom" half. We might say that the two halves of the molecule cancel each other so far as chirality is concerned. Therefore, the molecule as a whole is achiral and does not cause a rotation of plane-polarized light. Figure 4.16 illustrates an internal plane of symmetry.

A stereoisomer that contains chiral carbons but can be superposed on its mirror image is called a ***meso* form.** The compound we have been discussing is *meso*-tartaric acid.

A compound with two chiral carbons can have up to four stereoisomers. For tartaric acid, we have looked at two possibilities, and these two added up to only one isomer, ***meso*-tartaric acid. What about the other two stereoisomers of tartaric acid? Do they have an internal plane of symmetry? No, they do not. The top halves are not mirror images of the bottom halves. Rotation of either structure by 180° does not result in the other structure or in the *meso*-isomer. The following two stereoisomers of tartaric acid are enantiomers. They are both optically active and rotate the plane of polarization of plane-polarized light equally, but in opposite directions.

$$CO_2H$$
$$H\blacktriangleright C\blacktriangleleft OH$$
$$HO\blacktriangleright C\blacktriangleleft H$$
$$CO_2H$$

$$CO_2H$$
$$HO\blacktriangleright C\blacktriangleleft H$$
$$H\blacktriangleright C\blacktriangleleft OH$$
$$CO_2H$$

NOT a plane of symmetry

(2*R*,3*R*)-(+)-tartaric acid   (2*S*,3*S*)-(−)-tartaric acid

What are our conclusions about tartaric acid? Because of the symmetry in ***meso*-tartaric acid, there are only *three* stereoisomers for tartaric acid, rather than the four stereoisomers predicted by the 2^n rule. These three stereoisomers are a pair of enantiomers and the diastereomeric *meso* form.

The *meso* forms of some other compounds follow:

$$\begin{array}{c}
\text{Cl} \\
| \\
\text{H}_3\text{C} \blacktriangleright \text{C} \blacktriangleleft \text{H} \\
\text{------------} | \text{------------} \\
\text{H}_3\text{C} \blacktriangleright \text{C} \blacktriangleleft \text{H} \\
| \\
\text{Cl}
\end{array}
\qquad
\begin{array}{c}
\text{CHO} \\
| \\
\text{H} \blacktriangleright \text{C} \blacktriangleleft \text{OH} \\
| \\
\text{----HO} \blacktriangleright \text{C} \blacktriangleleft \text{H----} \\
| \\
\text{H} \blacktriangleright \text{C} \blacktriangleleft \text{OH} \\
| \\
\text{CHO}
\end{array}
\qquad
\begin{array}{c}
\text{CH}_2\text{OH} \\
| \\
\text{H} \blacktriangleright \text{C} \blacktriangleleft \text{OH} \\
| \\
\text{H} \blacktriangleright \text{C} \blacktriangleleft \text{OH} \\
\text{------------} | \text{------------} \\
\text{H} \blacktriangleright \text{C} \blacktriangleleft \text{OH} \\
| \\
\text{H} \blacktriangleright \text{C} \blacktriangleleft \text{OH} \\
| \\
\text{CH}_2\text{OH}
\end{array}$$

plane of symmetry

STUDY PROBLEM

4.18 Draw all possible stereoisomers of 2,3-dichlorobutane. Indicate any enantiomeric pairs.

D. Conformational Enantiomers

Butane contains no chiral carbons and is achiral. But the *gauche* conformation of butane is chiral. It is nonsuperposable on its mirror image.

   If butane can exist as chiral conformations, why is butane achiral? The rapid rotation about the carbon–carbon single bond interconverts one chiral *gauche* form into its enantiomer. (See Figure 4.17.) Interconverting chiral conformations are often called **conformational enantiomers,** or **dynamic enantiomers.** In contrast to normal enantiomers, conformational enantiomers cannot be isolated or separated.

   This fact that achiral compounds may exist as interconverting chiral conformations is more the rule than an exception. We will encounter this phenomenon again in the next section on chiral cyclic compounds.

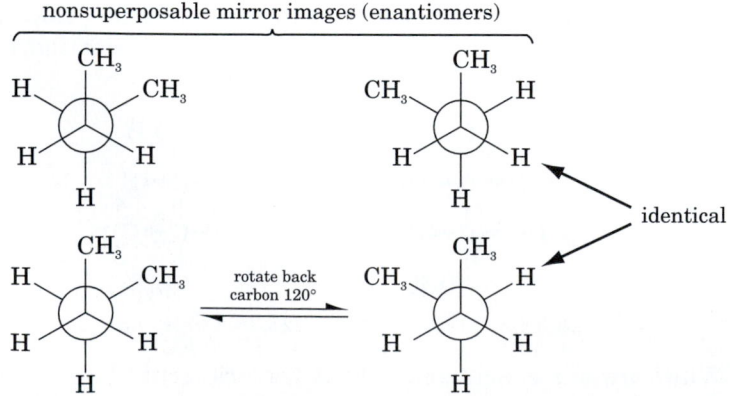

Figure 4.17   Chiral *gauche* conformations of butane and their interconversion.

E. Chiral Cyclic Compounds

Some substituted cycloalkanes are chiral. For example, both *cis*- and *trans*-dimethylcyclopropane contain two chiral carbons.

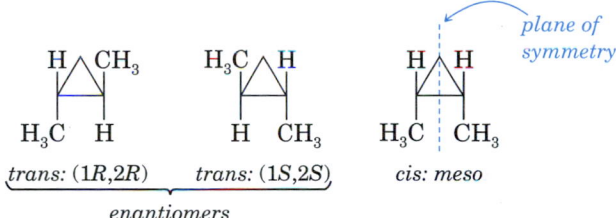

Carbon 1 has four different groups attached:

$-H, -CH_3, -CH_2-, and -CH-$
$\quad\quad\quad\quad\quad\quad\quad\quad\quad\quad\quad\quad |$
$\quad\quad\quad\quad\quad\quad\quad\quad\quad\quad\quad\quad CH_3$

Carbon 2 also has four different groups.

The *trans* isomer has no internal plane of symmetry and therefore exists as a pair of enantiomers—(1*R*,2*R*) and (1*S*,2*S*). The *cis* isomer does have an internal plane of symmetry; therefore, the *cis* isomer is a *meso* form.

The three stereoisomers of 1,2-dimethylcyclopropane:

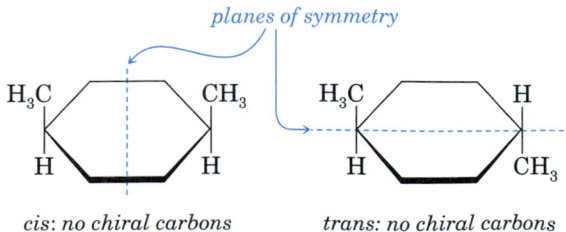

trans: (1*R*,2*R*)   *trans:* (1*S*,2*S*)   *cis: meso*

enantiomers

Substituted cycloalkanes other than the substituted cyclopropanes are more complex. For example, the dimethylcyclohexanes include *cis* and *trans* isomers of 1,2-; 1,3-; and 1,4-disubstituted rings. Unlike the planar cyclopropane ring, these cyclohexane rings can exist in a variety of conformations, primarily chair forms. Fortunately, when considering the chirality of substituted cyclohexanes, we can generally use the simple polygon formulas. (If you wish to analyze the actual conformers of these compounds, use models.)

Neither the *cis*- nor the *trans*-1,4-dimethylcyclohexane is chiral because they lack chiral carbons. Note that both possess an internal plane of symmetry. Thus, neither isomer exists as a pair of enantiomers.

The two stereoisomers of 1,4-dimethylcyclohexane:

planes of symmetry

cis: no chiral carbons   *trans: no chiral carbons*

The *cis*- and *trans*-1,3-dimethylcyclohexanes have two chiral carbons each. The *cis* isomer has an internal plane of symmetry (in either the polygon formula, as shown, or in its chair form) and is a *meso* form. The *trans* isomer has no plane of symmetry and exists as a pair of enantiomers.

The three stereoisomers of 1,3-dimethylcyclohexane:

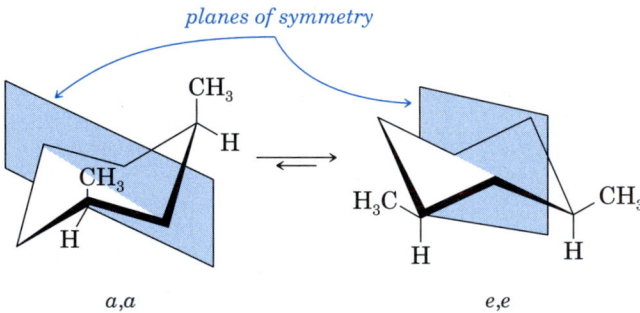

cis: meso   *trans: (1R,3R)*   *trans: (1S,3S)*

enantiomers

SAMPLE PROBLEM

Draw formulas for the chair–chair equilibrium for *cis*-1,3-dimethylcyclo-hexane and identify the internal plane of symmetry in each conformation.

Solution

planes of symmetry

a,a   *e,e*

STUDY PROBLEM

4.19   Draw formulas for the chair–chair equilibria for the two enantiomers of *trans*-1,3-dimethylcyclohexane and verify the mirror-image relationship of the enantiomers. (*Hint:* Make a model for each enantiomer.)

The 1,2-dimethylcyclohexanes also contain two chiral carbons. The *trans*-1,2-dimethylcyclohexane exists as a pair of enantiomers, while the *cis* isomer is a *meso* form.

The three stereoisomers of 1,2-dimethylcyclohexane:

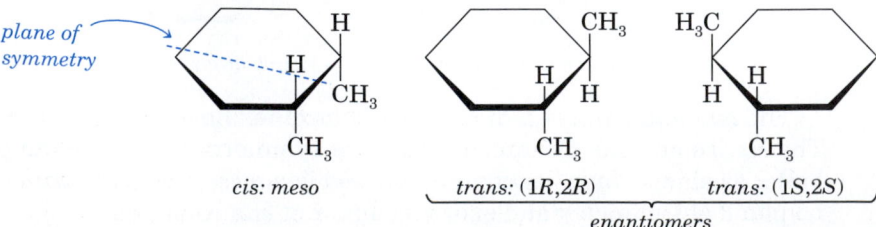

cis: meso   *trans: (1R,2R)*   *trans: (1S,2S)*

enantiomers

STUDY PROBLEMS

4.20 *cis*-1,2-Dimethylcyclohexane actually exists as a pair of chair conformations that are nonsuperposable mirror images. These two chair conformations, however, are interconvertible by way of a *meso* conformation.

(a) Verify that the 1*e*,2*a* chair conformation is indeed the enantiomer of the 1*a*,2*e* chair conformation. (Use models.)

(b) What conformation must the molecule assume in order to have an internal plane of symmetry?

4.21 Draw the stereoisomers of the following compounds. Indicate enantiomeric pairs and *meso* forms, if any.

(a)

(b)

proline

an amino acid

Section 4.10
Prochirality

Ethanol is not a chiral compound. Both carbons of ethanol have at least two groups (hydrogens) that are identical; therefore, neither carbon atom is chiral. However, the two carbons of ethanol are different—carbon 1 is prochiral and carbon 2 is not.

Prochirality means that substitution of an atom or group **could** lead to a chiral molecule. It does not mean the compound is chiral, nor does it mean that any reaction will produce a chiral compound. But it does mean that the compound has the potential for conversion to a chiral compound by a single substitution of a group. The atom at which substitution must take place for chirality to be generated is called the **prochiral atom.** In ethanol, carbon 1 is the prochiral carbon.

To illustrate why carbon 1 is prochiral and carbon 2 is not, we will substitute some group Y for one of the hydrogens at both carbon 1 and carbon 2.

Substitution for a hydrogen atom at the prochiral carbon atom

$$\overset{2}{C}H_3\overset{1}{C}H_2OH \xrightarrow[-H]{+Y} CH_3\overset{Y}{\underset{|}{C}}HOH$$

achiral — — *chiral*

Substitution for either hydrogen at carbon 1 generates a chiral carbon. A chiral carbon is not generated when the same substitution takes place at carbon 2. Substitution for any one of the three hydrogens with Y still leaves carbon 2 with two hydrogens. Therefore, it remains achiral.

Substitution for a hydrogen atom at the nonprochiral carbon

$$\overset{2}{C}H_3\overset{1}{C}H_2OH \xrightarrow[-H]{+Y} \overset{Y}{\underset{|}{C}}H_2CH_2OH$$

achiral   *achiral*

Because carbon 1 is prochiral, its two hydrogens are not equivalent. Replacement of one of them will produce an (R) enantiomer and replacement of the other will produce an (S) enantiomer. To emphasize this fact, the two hydrogens can be labeled as H_R (for pro-R) and H_S (for pro-S). In designating hydrogens as pro-R or pro-S, it is assumed that the priority of the incoming group will be 3. (In our example, if H_R is replaced by Y, then the priority at the new chiral carbon will be: OH, 1; CH_3, 2; Y, 3; and H, 4. The configuration will be (R).)

The hydrogens on the prochiral carbon are said to be **enantiotopic,** meaning that their replacement has the potential of forming enantiomers. Ethanol is an example of a prochiral compound that has two enantiotopic hydrogens.

$$H_S \underset{C}{\overset{H_R}{\diagup}} \quad \textit{enantiotopic hydrogens}$$

$$CH_3 \qquad OH$$

prochiral carbon atom

Enantiotopic hydrogens are identical in all respects except for their reactivity with a chiral reagent, such as an enzyme. For example, it has been shown that when ethanol is oxidized to acetaldehyde (ethanal) by the enzyme liver alcohol dehydrogenase, only H_R is removed from the prochiral carbon in the oxidation reaction. H_S remains bonded to the carbon.

SAMPLE PROBLEM

Identify the carbon(s) that are *not* prochiral in 1,3-propanediol. Label the enantiotopic hydrogens on the prochiral carbon(s) as H_R and H_S.

$$\overset{1}{H}OCH_2\overset{2}{C}H_2\overset{3}{C}H_2OH$$

Solution

Carbon 2 is not prochiral. Since the molecule will still have two identical groups (—CH_2OH) bonded to carbon 2 after replacement, an achiral compound results regardless of which hydrogen is replaced.

$$H \qquad H$$
$$C$$
$$HOCH_2 \qquad CH_2OH$$

achiral   *achiral*

$$Y \qquad H \qquad\qquad H \qquad Y$$
$$C \qquad\qquad\qquad C$$
$$HOCH_2 \quad CH_2OH \qquad HOCH_2 \quad CH_2OH$$

Both carbons 1 and 3 are prochiral. A chiral carbon will be produced if any one of the four hydrogens is replaced by another group. We will show replacement at carbon 3 to illustrate this fact.

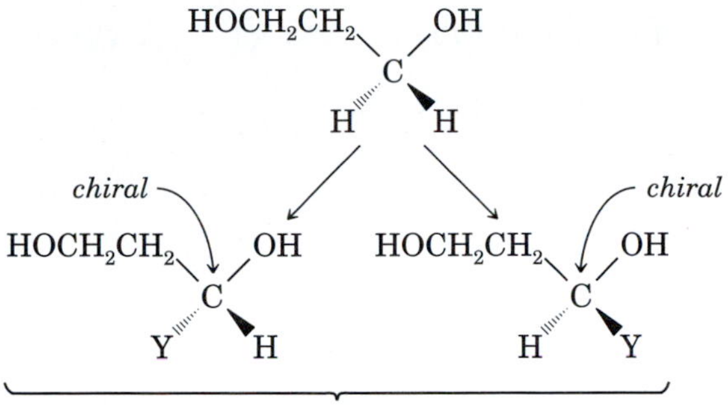

enantiomers

The labeling of the four enantiotopic hydrogens as pro-*R* and pro-*S* follows. (Again, we assume the incoming group will have a priority of 3.)

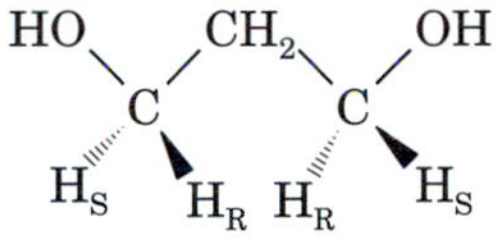

Carbon 3 of the chiral (*S*)-2-bromobutane is a prochiral carbon. But, because carbon 2 is chiral, replacement of the hydrogens on carbon 3 produces diastereomers instead of enantiomers as was done with ethanol. Thus, the hydrogens on the prochiral carbon are said to be **diastereotopic.**

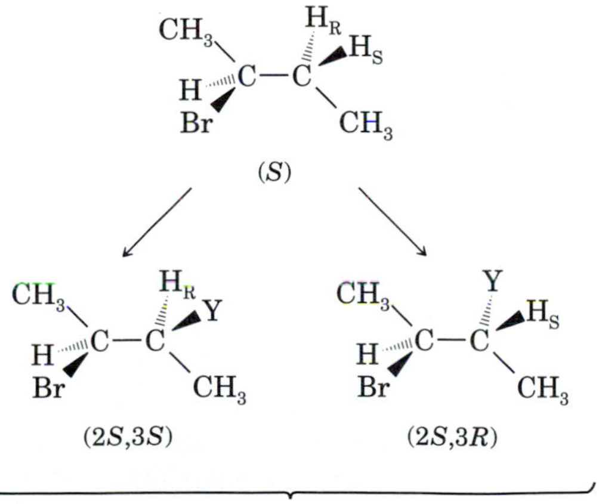

diastereomers

Unlike enantiotopic hydrogens, diastereotopic hydrogens are different in all respects—diastereotopic hydrogens have different reactivities toward even achiral reagents.

STUDY PROBLEM

4.22   Identify the prochiral carbon(s) in the following structures. Identify the hydrogens on the prochiral carbons as either enantiotopic or diastereotopic. Label each hydrogen on the prochiral carbons as H_R or as H_S assuming the incoming group will be priority 3.

(a)   HOCCH$_2$C—CH$_2$COH   (b) [benzene ring]—CH$_2$OH   (c) CH$_3$CHCH$_2$C≡N

Section 4.11
Resolution of a Racemic Mixture

In most laboratory reactions, a chemist uses achiral or racemic starting materials and obtains achiral or racemic products. Therefore, often we will ignore the chirality (or lack of it) of reactants or products in future chapters. On the other hand, we will also discuss many reactions in which stereochemistry is quite important. Because it is a very rare occurrence for enantiomers to crystallize separately, the method that Pasteur used cannot be considered a general technique. Because a pair of enantiomers exhibit the same chemical and physical properties, they cannot be separated by ordinary chemical or physical means. Instead, chemists must rely on chiral reagents or chiral catalysts (which, one way or another, almost always have their origins in living organisms).

One way to resolve a racemic mixture, or at least to isolate one pure enantiomer, is to treat the mixture with a microorganism that will consume only one of the two enantiomers. For example, pure (R)-nicotine can be obtained from (RS)-nicotine by incubating the racemic mixture with the bacterium *Pseudomonas putida,* which oxidizes (S)-nicotine, but not the (R)-enantiomer.

P. putida
metabolizes (S)-nicotine

(RS)-nicotine

racemic

(R)-(+)-nicotine (78%)

Unlike most laboratory reactions, most biological reactions start with chiral or achiral reactants and lead to chiral products. These biological reactions are possible because of biological catalysts called **enzymes,** which are chiral themselves. Recall that a pair of enantiomers have the same chemical properties *except* for their interactions with other chiral substances (Section 4.7). Because enzymes are chiral, they can act very selectively in their catalytic action. For example, when an organism ingests a racemic mixture of alanine, only the (S)-alanine becomes incorporated into protein structures. The (R)-alanine is not used in proteins; instead, it is oxidized to a keto acid with the aid of other enzymes and enters other metabolic schemes.

CO$_2$H   CO$_2$H

H$_2$N►C◄H   H►C◄NH$_2$

CH$_3$ CH$_3$

(S)-alanine   (R)-alanine

Some enzymes can be separated from their living organisms and still retain their unique activity. In certain reactions, enzymes can be used to generate chiral products from achiral starting materials.

$$\begin{array}{c} CH_3 \\ | \\ C=O \\ | \\ CH_2CH_3 \end{array} \xrightarrow{\text{enzymatic reduction}} \begin{array}{c} CH_3 \\ | \\ HO\blacktriangleright C\blacktriangleleft H \\ | \\ CH_2CH_3 \end{array}$$

2-butanone

achiral

(R)-2-butanol
$[\alpha]_D^{25} = -13°$

In the laboratory, the physical separation of a racemic mixture into its pure enantiomers is called the **resolution,** or **resolving,** of the racemic mixture. Pasteur's separation of racemic sodium ammonium tartrate was a resolution of that mixture.

The most general technique for resolving a pair of enantiomers is to subject them to a reaction with a chiral reagent to obtain a pair of **diastereomeric products.** Diastereomers, you will remember, are different compounds with different physical properties. Thus, a pair of diastereomers may be separated by ordinary physical means, such as crystallization.

Let us illustrate a general procedure for the laboratory resolution of (RS)-RCO_2H, a racemic mixture of a carboxylic acid, where (R)-RCO_2H and (S)-RCO_2H represent the two enantiomers. A carboxylic acid will react with an amine to yield a salt (Section 1.7B):

$$\underset{\text{a carboxylic acid}}{RC\overset{..}{\underset{..}{O}}{-}H} + \underset{\text{an amine}}{R'\overset{..}{N}H_2} \longrightarrow \underset{\text{a salt}}{RC\overset{..}{\underset{..}{O}}{:}^{-}\ R'\overset{+}{N}H_3}$$

Reaction of the (RS) carboxylic acid with an amine that is a pure enantiomer results in a pair of diastereomeric salts: the amine salt of the (R) acid and the amine salt of the (S) acid.

(R)-RCO_2H

and   + (S)-$R'NH_2$ $\longrightarrow$

(S)-RCO_2H *a pure enantiomer*

racemic mixture

$$\left\{ \begin{array}{c} (R)\text{-}RCO_2^-\ (S)\text{-}R'NH_3^+ \\ \text{and} \\ (S)\text{-}RCO_2^-\ (S)\text{-}R'NH_3^+ \end{array} \right\}$$

The (R,S)-salt and the (S,S)-salt are not enantiomers: they are diastereomers and can be separated.

In this reaction, the only possible products are the (R,S)-salt and the (S,S)-salt, which are not enantiomers of each other. The enantiomers of these two salts are the (S,R)-salt and the (R,R)-salt, respectively. Neither of these latter enantiomers can be formed in this reaction because only the (S)-amine was used as a reactant.

After separation, each diastereomeric salt is treated with a strong base to regenerate the amine. The amine and the carboxylate ion can be separated by extraction with a solvent such as diethyl ether (in which the amine is soluble, but the carboxylate salt is not). Acidification of the aqueous phase yields the free enantiomeric carboxylic acid.

$$(R)\text{-}RCO_2^-\ (S)\text{-}R'NH_3^+ \xrightarrow{\ OH^-\ } (R)\text{-}RCO_2^- + (S)\text{-}R'NH_2$$

one of the pure
diastereomeric salts

$$\xrightarrow{\ HCl\ } (R)\text{-}RCO_2H$$

as a pure enantiomer

This resolution of a racemic acid depends on salt formation with a single enantiomer of a chiral amine. Commonly used amines are amphetamine, which is commercially available as pure enantiomers, and the naturally occurring strychnine (page 761).

$$\underset{\text{amphetamine}}{\overset{\overset{\displaystyle CH_3}{|}}{\bigcirc\!\!\!\!\bigcirc\!-CH_2CHNH_2}}$$

amphetamine

STUDY PROBLEM

4.23 Describe the procedure by which $HO_2CCH_2\overset{\overset{\displaystyle OH}{|}}{C}HCO_2H$ can be resolved with an optically active amine, such as $(R)\text{-}C_6H_5\overset{\overset{\displaystyle }{|}}{C}HCH_3$. Use equations in your answer.
$\underset{NH_2}{}$

Summary

Stereoisomerism is isomerism resulting from different spatial arrangements of atoms in molecules. **Geometric isomerism,** one form of stereoisomerism, results from groups being *cis* (same side) or *trans* (opposite sides) around a pi bond or on a ring. Geometric isomers of alkenes may also be differentiated by the letter (*E*), opposite sides, or (*Z*), same side.

Rotation of groups around sigma bonds results in different **conformations,** such as the **eclipsed,** *gauche,* **staggered,** and *anti* conformations. Lower-energy conformers predominate. Conformers are interconvertible at room temperature and therefore are not usually isolable isomers. A cyclic compound assumes puckered conformations to relieve strain of unfavorable bond angles and, more important, to minimize repulsions of substituents. For the cyclohexane ring, the chair-form conformer with substituents **equatorial** instead of **axial** is favored.

A **chiral molecule** is a molecule that is nonsuperposable on its mirror image. A pair of nonsuperposable mirror images are called **enantiomers** and represent another type of stereoisomerism. Each member of a pair of enantiomers rotates the plane of polarization of plane-polarized light an equal amount, but in opposite directions. An equimolar mixture of enantiomers, called a **racemic mixture,** is optically inactive.

Chirality usually arises from the presence of a carbon with four different atoms or groups attached to it. The arrangement of these groups around the chiral carbon is called the **absolute configuration** and may be described as (*R*) or (*S*). **Fischer projections** are often used to depict chiral molecules.

A molecule with more than one chiral carbon has more stereoisomers than a single enantiomeric pair. Stereoisomers that are not enantiomers are **diastereomers.** If a molecule has more than one chiral carbon and has a plane of symmetry in at least one conformation, then it can be superposed on its mirror image. Such a compound is optically inactive and is called a *meso* **form.**

Achiral compounds that do not contain any chiral carbons may exist as rapidly interconverting chiral conformations. Such conformations are called **conformational enantiomers.**

The different types of isomerism may be summarized as follows:

A. Structural isomers differ in order of attachment of atoms.

$$(CH_3)_2CHCH_3 \quad \text{and} \quad CH_3CH_2CH_2CH_3$$

B. Stereoisomers differ in arrangement of atoms in space.

1. Enantiomers: nonsuperposable mirror images

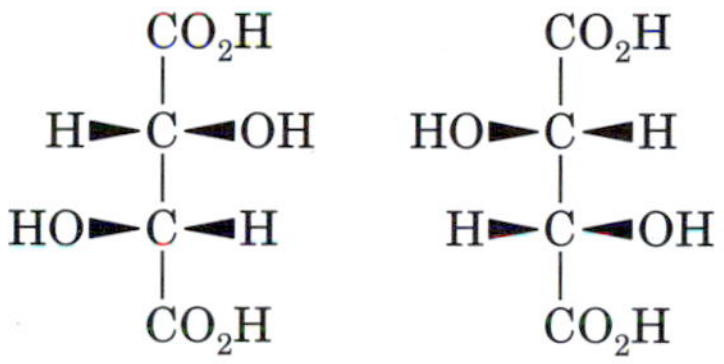

(2*R*,3*R*) (2*S*,3*S*)

2. Diastereomers: nonenantiomeric stereoisomers

Containing chiral carbons:

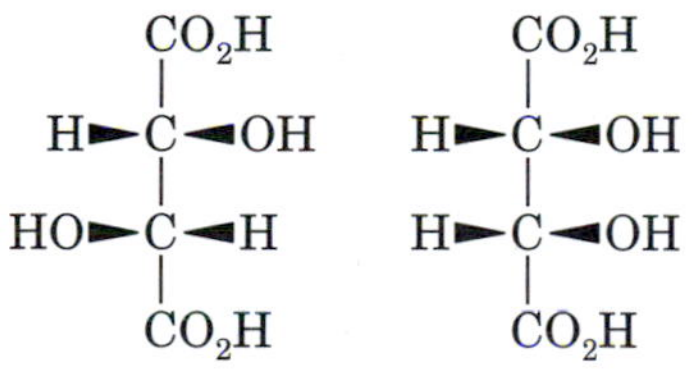

(2*R*,3*R*) *meso*

Achiral:

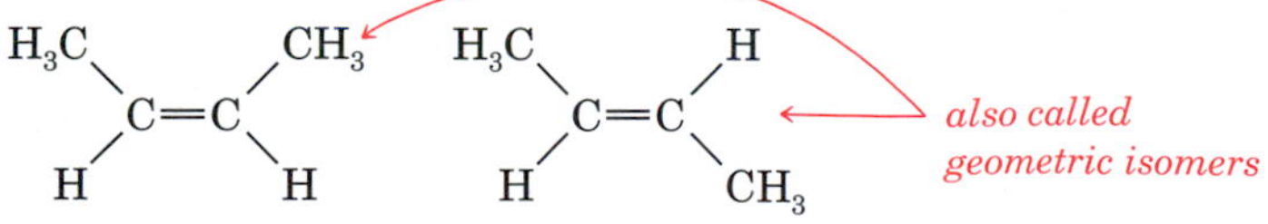

cis, or (*Z*) *trans*, or (*E*) *also called geometric isomers*

A pair of enantiomers have the same physical and chemical properties except for their interactions with other chiral substances and with polarized light. Enantiomers can be resolved by (1) treatment with a chiral reagent to yield a pair of diastereomers; (2) separation of the diastereomers, which do *not* have the same properties; and (3) regeneration of the separated enantiomers.

Prochirality means that substitution of an atom or group could lead to a chiral molecule. The atom at which substitution must take place for chirality to be generated is called the **prochiral atom.** The hydrogens on the prochiral carbon are said to be **enantiotopic,** when their replacement has the potential of forming enantiomers, or **diastereotopic,** when their replacement has the potential of forming diastereomers.

Essay Problem for Chapter 4

Enzymatic Resolution of Alcohols

When racemic *trans*-2-phenylcyclohexyl chloroacetate (**1**) is heated with lipase at pH 7 and 45–50°, optically active *trans*-2-phenylcyclohexanol (**2**) is formed in high yield. Lipase is an enzyme that catalyzes the hydrolysis of the carbonyl carbon–oxygen ester bond. The phenylcyclohexanol is levorotatory and the configuration of the chiral carbon at the hydroxyl group (carbon 1) is (*R*).

1		2
trans and racemic		*trans and optically active*

Further workup of the reaction mixture yields unreacted ester (**3**), which can be converted to another optically active *trans*-2-phenylcyclohexanol (**4**) by an alkaline hydrolysis reaction. Compounds **2** and **4** are not identical. (A. Schwartz, P. Madan, J. K. Whitesell, and R. M. Lawrence, *Org. Syn.* **1990**, *69*, 1.)

3		4
trans *ester remaining after lipase hydrolysis reaction*		*trans and optically active*

Questions

1. How many chiral carbons does the chloroacetate ester (**1**) have?
2. Draw all optical isomers of *cis*- and *trans*-phenylcyclohexyl chloroacetate.
3. Is carbon 2 of cyclohexanol (**2**) isolated from the lipase hydrolysis (*R*) or (*S*)?
4. Assign configurations to the chiral carbons of the cyclohexanol (**4**) isolated from the alkaline hydrolysis reaction.
5. What is the direction of optical rotation of the cyclohexanol (**4**) isolated from the alkaline hydrolysis reaction?

Study Problems*

4.24 *Cinnamic acid,* $C_6H_5CH{=}CHCO_2H$ (where C_6H_5 means a phenyl group), exists as a pair of stereoisomers. The *cis* isomer is a plant growth stimulant, while the *trans* isomer is not. Write formulas for the two isomers of cinnamic acid and label them.

4.25 Write structural formulas for the alkenes of molecular formula C_5H_{10} that exhibit geometric isomerism. Indicate the *cis* and *trans* structures.

4.26 Identify the group in each of the following pairs that has the higher priority by the Cahn–Ingold–Prelog nomenclature system.

    (a) —OH; —CH$_3$ **(b)** —CO$_2$H; —CH$_2$CO$_2$H

    (c) $-\overset{+}{N}(CH_3)_3$; $-\overset{+}{N}H_3$ **(d)** —CH=CH$_2$; —CH=CHCl

    (e) —Si(CH$_3$)$_3$; —C(CH$_3$)$_3$ **(f)** —C≡CH; —CH=CH$_2$

    (g) —Cl; —I   **(h)** —F; —Cl

    (i) —Br; —OH   **(j)** 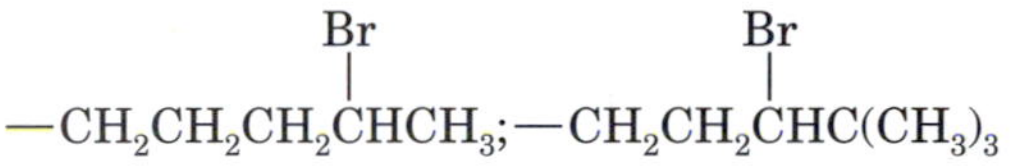 $-CH_2CH_2CH_2\overset{\overset{\displaystyle Br}{|}}{C}HCH_3$; $-CH_2CH_2\overset{\overset{\displaystyle Br}{|}}{C}HC(CH_3)_3$

    (k) 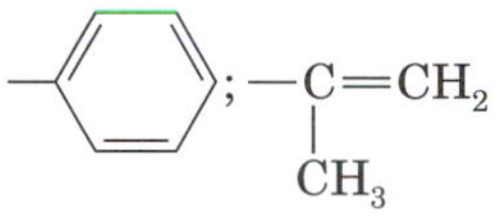 ⟨◯⟩; $-\overset{\underset{\displaystyle CH_3}{|}}{C}{=}CH_2$ **(l)** 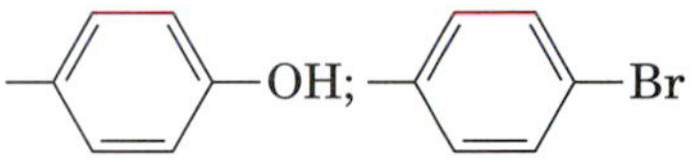 ⟨◯⟩—OH; ⟨◯⟩—Br

4.27 Identify each double bond in the following structures as (*E*) or (*Z*) where appropriate.

    (a) 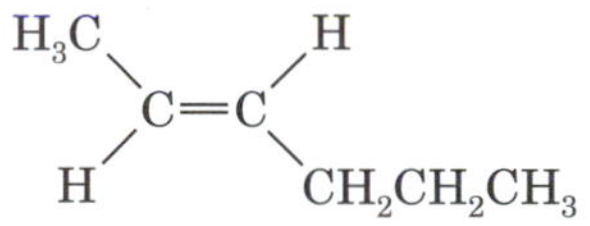 H$_3$C, H / C=C / H, CH$_2$CH$_2$CH$_3$

    (b) 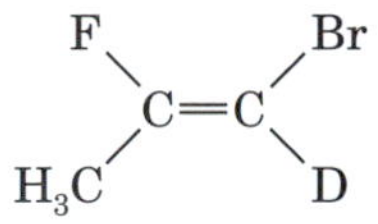 F, Br / C=C / H$_3$C, D

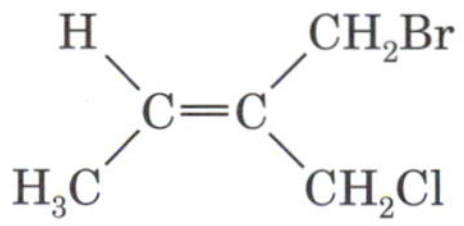

    (c) H, CH$_2$Br / C=C / H$_3$C, CH$_2$Cl

    (d) ClCH$_2$CH$_2$, H / C=C / ClCH$_2$, H

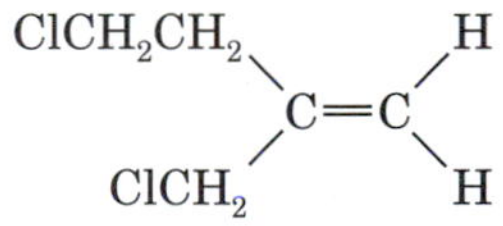

    (e) 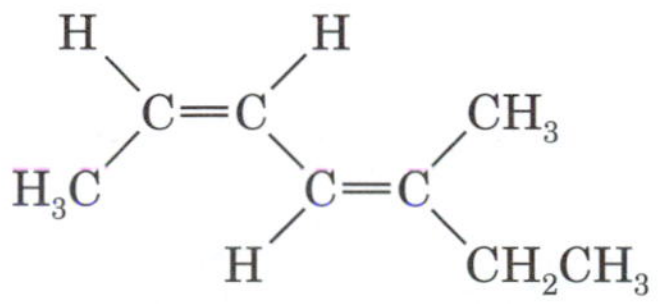 H, H / C=C / H$_3$C, (C=C) / H, CH$_3$ / CH$_2$CH$_3$

    (f) 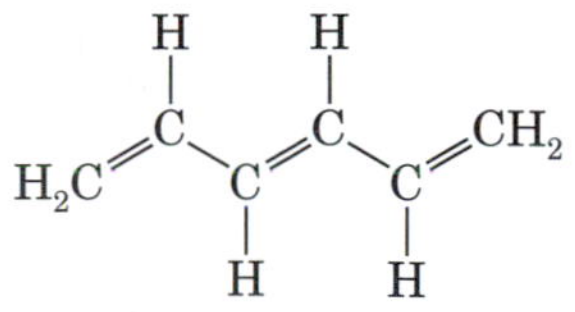 H$_2$C= C(H)= C(H)= C —CH$_2$ (H, H)

4.28 (a) One sex attractant compound of the female Douglas fir tussock moth has been identified as (*Z*)-1,6-heneicosadien-11-one. The (*E*) isomer shows little or no attractant power. Write formulas for these two compounds, identifying (*E*) and (*Z*) isomers. (*Hint:* Heneicosane is the 21-carbon continuous-chain alkane.)

    (b) The principal component of the sex attractant secretion of the female arctiid moth (*Utetheisa ornatrix*) is named (3*Z*,6*Z*,9*Z*)-heneicosatriene. What is the structure of this compound?

    (c) The principal component of the navel orangeworm (*Pamyelois transitella*) sex attractant is (11*Z*,13*Z*)-hexadecadienal. This sex attractant is unusual in that it

---

* For information concerning the organization of the *Study Problems* and *Additional Problems,* see the *Note to student* on page 41.

contains a conjugated (*Z,Z*) diene system. Draw the structure of this compound. (*Hint:* Hexadecane is the name of the 16-carbon continuous-chain alkane.)

4.29 Draw and label the *cis* and *trans* isomers of the following structures. (Ignore the possibility of optical isomers.)

(a) 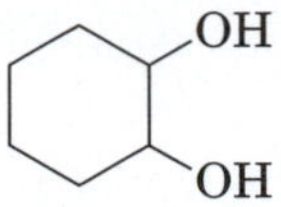 [structure: cyclohexane with OH and OH]   (b) 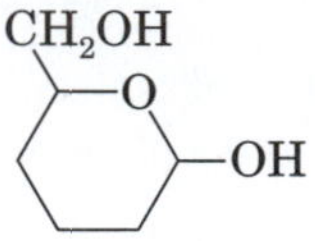 [structure with CH$_2$OH, O, —OH]   (c) 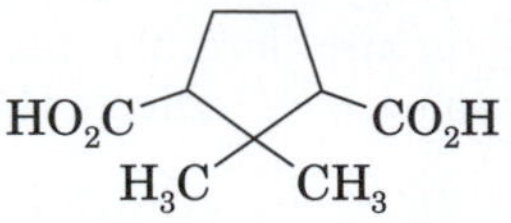 HO$_2$C [structure] CO$_2$H with H$_3$C and CH$_3$

4.30 Draw the Newman projections for the *anti* conformation and two types of eclipsed conformations of 1,2-diiodoethane. Of the two eclipsed conformations, which is of higher energy?

4.31 Draw the Newman projection for an *anti* conformer (if any) of each of the following compounds. Use the circled carbons as the center of the Newman projection.

(a) HO$_2$CCHCH$_2$CH$_2$CO$_2$H   (b) HO$_2$CCHCH$_2$CH$_2$CO$_2$H
      |
      CH$_3$ CH$_3$

(c) HO$_2$CCHCH$_2$CH$_2$CO$_2$H
      |
      CH$_3$

4.32 Draw equations illustrating the chair ⇌ boat ⇌ chair equilibria for each of the following substituted cyclohexanes:

(a)  H$_3$C— [cyclohexane] —CO$_2$H   (b) [cyclohexane with CH$_3$ and Cl]   (c) [cyclohexane with CHO and Br]

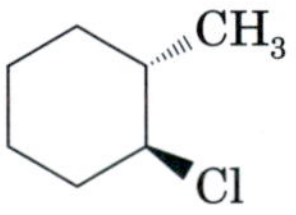

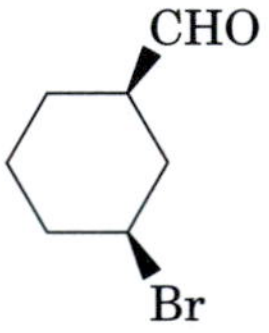

4.33 Which of the following conformations is more stable? Explain.

(a) 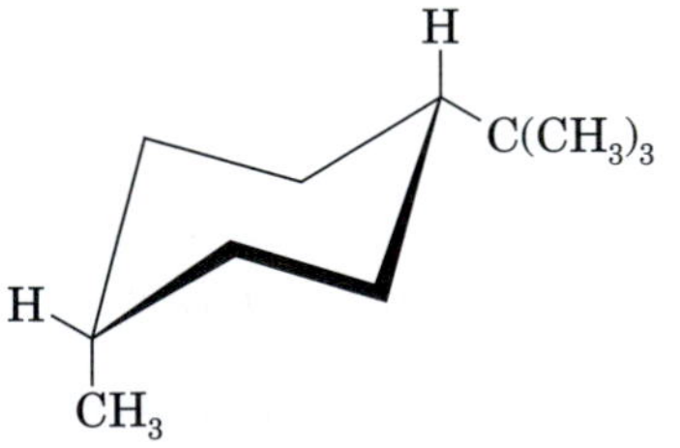 [chair structure with H, C(CH$_3$)$_3$, H., CH$_3$]   (b) 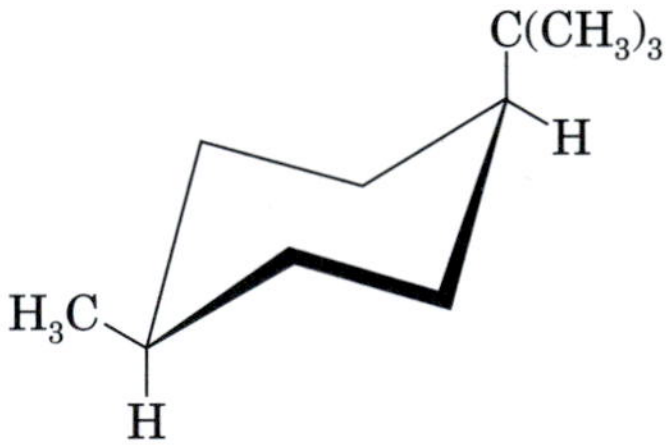 [chair structure with C(CH$_3$)$_3$, H, H$_3$C., H]

4.34 (a) Why is a *cis*-1,3-disubstituted cyclohexane more stable than the corresponding *trans*-structure?

(b) Is the *cis*-1,2-isomer more stable than the *trans*-1,2 disubstituted cyclohexane?

4.35 Draw the following compound in the chair form with all of the ring hydrogen atoms in axial positions. This compound is *glucose* (blood sugar). Suggest a reason why glucose is much more common in nature than any of its stereoisomers.

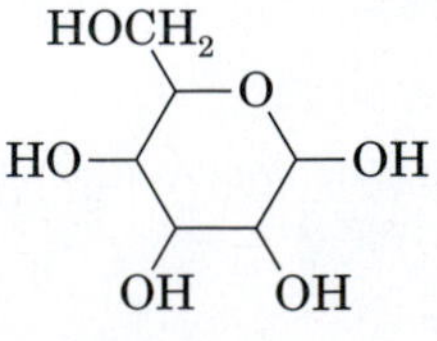

HOCH$_2$
HO— [pyranose ring with O] —OH
OH   OH

4.36 Draw the structure of the preferred conformation of **(a)** 1-methyl-1-propylcyclohexane; **(b)** *cis*-1-methyl-2-propylcyclohexane; **(c)** *trans*-1-methyl-3-propylcyclohexane.

4.37 Star the chiral carbons, if any, in the following structures:

(a) 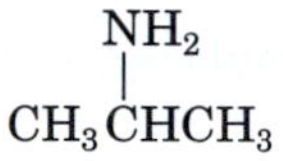 $\overset{\text{NH}_2}{\underset{|}{\text{CH}_3\text{CHCH}_3}}$ (b)  $\overset{\text{Cl}}{\underset{|}{\text{CH}_3\text{CHCH}_2\text{CH}_3}}$ (c) 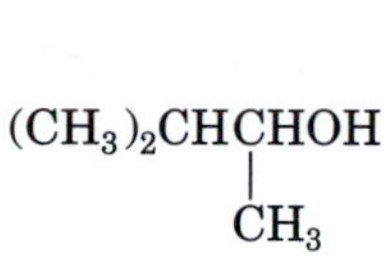 $(\text{CH}_3)_2\text{CHCHOH}$
$\underset{|}{\phantom{(\text{CH}_3)_2\text{CHCH}}}$
CH_3

(d) 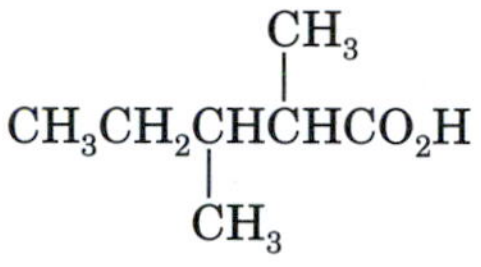 $\text{CH}_3\text{CH}_2\underset{|}{\overset{\text{CH}_3}{\text{CHCHCO}_2\text{H}}}$ (e) 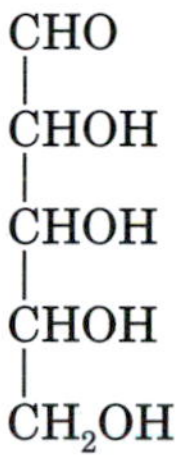 $\begin{array}{c}\text{CHO}\\|\\\text{CHOH}\\|\\\text{CHOH}\\|\\\text{CHOH}\\|\\\text{CH}_2\text{OH}\end{array}$ (f) 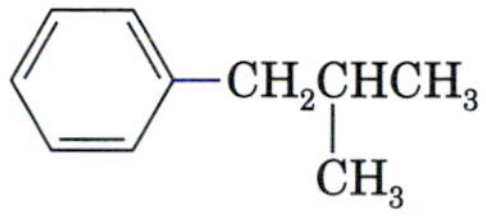 ⬡—$\text{CH}_2\underset{|}{\text{CHCH}_3}$
$\phantom{\text{CH}_3\text{CH}_2\text{CHC}}\text{CH}_3 \text{CH}_3$

(g)  $\text{CH}_2{=}\text{CHCH}{=}\text{CHC}\overset{\text{O}}{\overset{||}{\text{H}}}$ (h) $\text{CH}_3\overset{\text{Br}}{\underset{|}{\text{C}}}\text{FCH}_2\text{CH}_2\overset{\text{O}}{\overset{||}{\text{C}}}\text{H}$

(i) 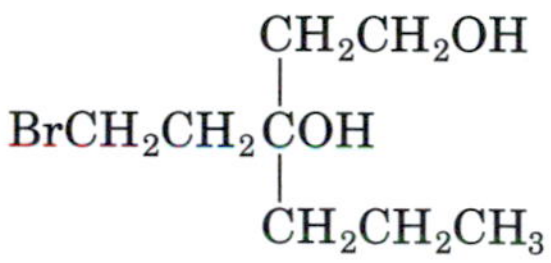 $\text{BrCH}_2\text{CH}_2\overset{\text{CH}_2\text{CH}_2\text{OH}}{\underset{\text{CH}_2\text{CH}_2\text{CH}_3}{||}}\text{COH}$

4.38 Show the configuration of the enantiomer of each of the following compounds:

(a) 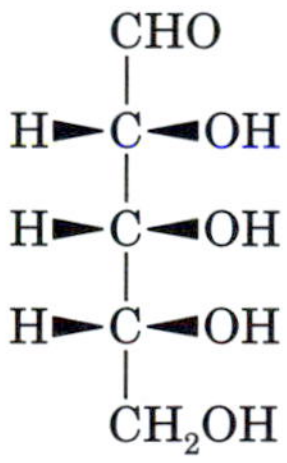 $\begin{array}{c}\text{CHO}\\|\\\text{H}{\blacktriangleright}\text{C}{\blacktriangleleft}\text{OH}\\|\\\text{H}{\blacktriangleright}\text{C}{\blacktriangleleft}\text{OH}\\|\\\text{H}{\blacktriangleright}\text{C}{\blacktriangleleft}\text{OH}\\|\\\text{CH}_2\text{OH}\end{array}$ (b) 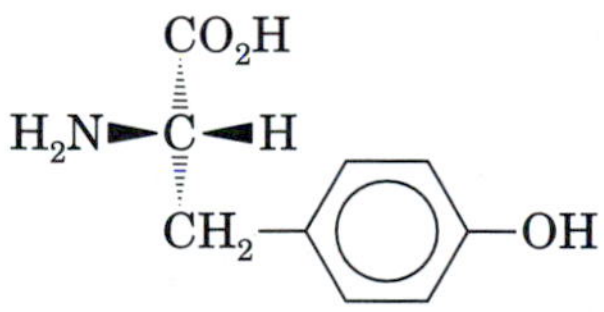 $\begin{array}{c}\text{CO}_2\text{H}\\|\\\text{H}_2\text{N}{\blacktriangleright}\text{C}{\blacktriangleleft}\text{H}\\|\\\text{CH}_2{-}⬡{-}\text{OH}\end{array}$

4.39 Give the structures and configurations for all the stereoisomers of saturated five-carbon alcohols (with one OH) that have stereoisomers, and indicate any enantiomeric pairs.

4.40 A carboxylic acid of the formula $C_3H_5O_2Br$ is optically active. What is its structure?

4.41 The compound 1,2,3-decanetriol contains two chiral carbon atoms. Using these chiral carbons as the center of the Newman projections, draw the Newman projections for all stereoisomers of this triol.

4.42 Convert the following dimensional formulas to Fischer projections. (Molecular models may be helpful.)

(a) 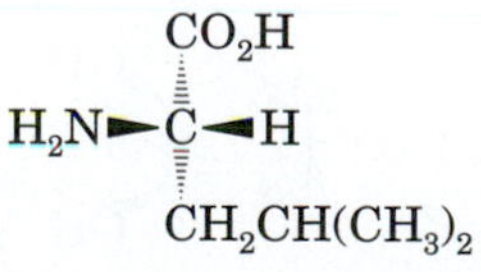 $\begin{array}{c}\text{CO}_2\text{H}\\|\\\text{H}_2\text{N}{\blacktriangleright}\text{C}{\blacktriangleleft}\text{H}\\|\\\text{CH}_2\text{CH}(\text{CH}_3)_2\end{array}$ (b) $\begin{array}{c}\text{CO}_2\text{H}\\|\\\text{C}\\\diagup\ \diagdown\\\text{H}\text{OH}\\\text{CH}_3\end{array}$ (c) $\begin{array}{c}\text{CH}_3\\|\\\text{C}{\cdots}\text{Br}\\\diagup\ \diagdown\\\text{CH}_3\text{CH}_2\text{CH}\\||\\\text{O}\end{array}$

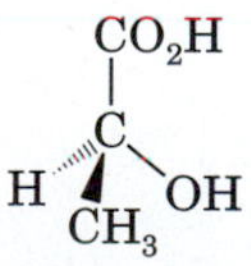

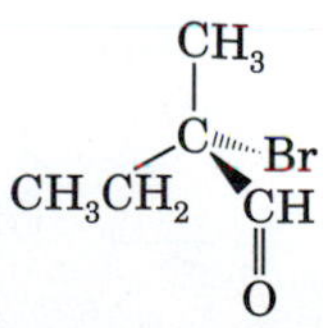

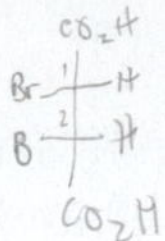

(d) 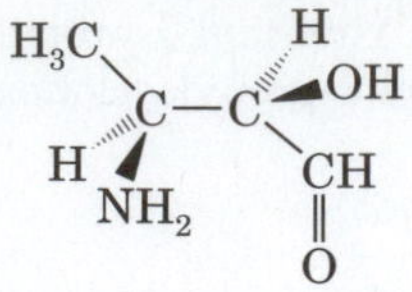 H_3C, ... $C-C$ ... OH, H, NH_2, CH, O (e) 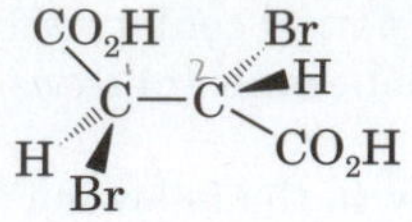 CO_2H, Br, $C-C$, H, Br, CO_2H

4.43 Convert the following Fischer projections to dimensional formulas (not necessarily the lowest-energy conformers):

(a) 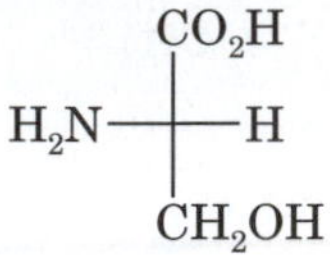 CO_2H
H$_2$N——H
CH$_2$OH

(b) 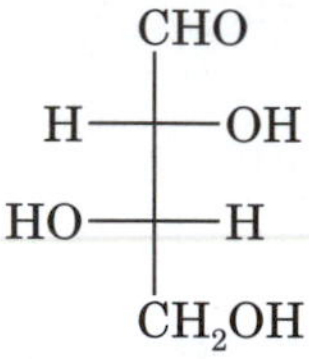 CHO
H——OH
HO——H
CH$_2$OH

(c) 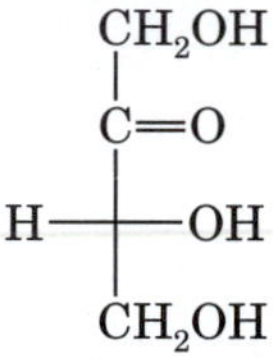 CH$_2$OH
C=O
H——OH
CH$_2$OH

4.44 For each Fischer projection in Problem 4.43, draw the Fischer projection of the enantiomer.

4.45 Which of the following pairs of formulas represent enantiomeric pairs?

(a) 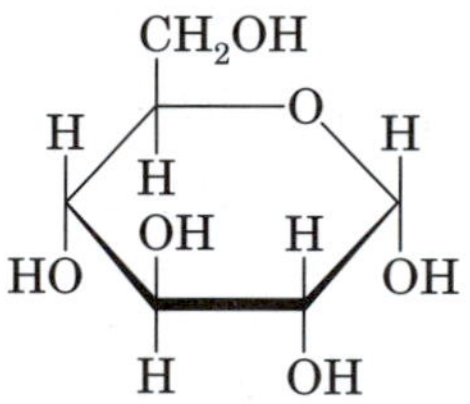 CH$_2$OH ... H ... O ... H ... OH ... H ... OH ... H ... OH   and   CH$_2$OH ... H ... O ... OH ... H ... OH ... H ... H ... OH

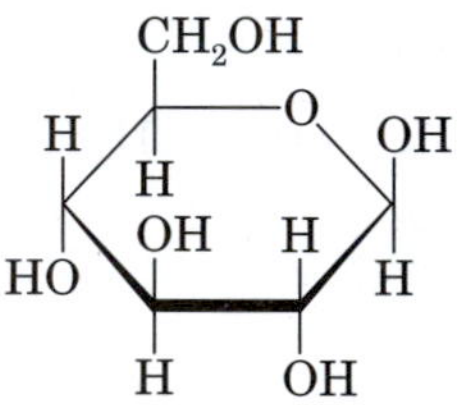

(b) 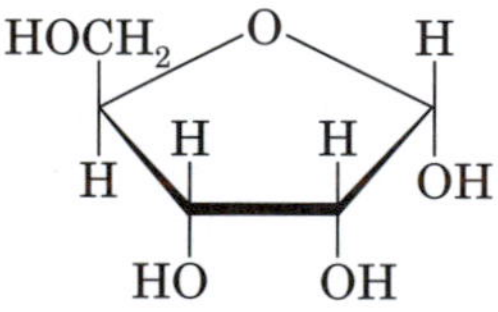 HOCH$_2$ ... O ... H ... H ... H ... OH ... HO ... OH   and   H ... O ... CH$_2$OH ... H ... H ... H ... HO ... OH

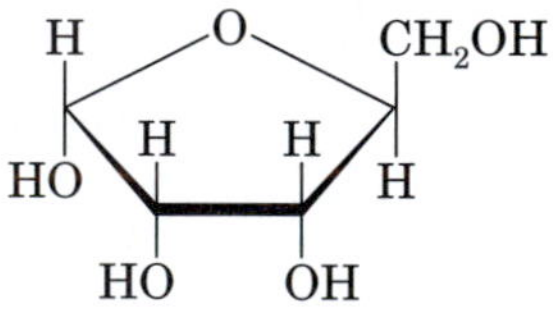

(c) 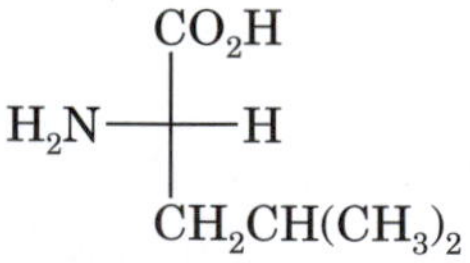 CO_2H
H$_2$N——H
CH$_2$CH(CH$_3$)$_2$ and CO_2H
H$_2$N——H
(CH$_3$)$_2$CHCH$_2$

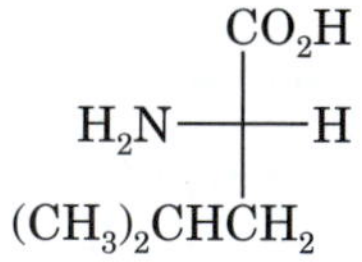

(d) 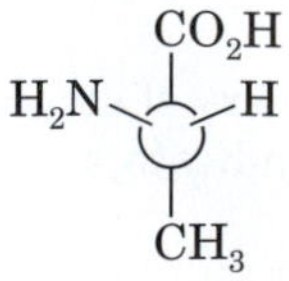 CO_2H
H$_2$N——H
CH$_3$ and   CH$_3$
H$_2$N——H
CO_2H

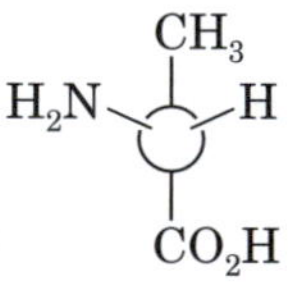

4.46 Star the chiral carbons, if any, in the following structures:

(a) 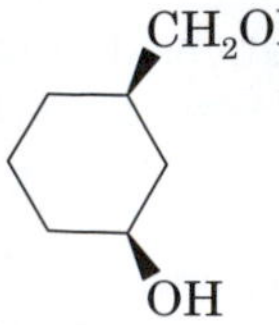 CH$_2$OH ... OH

(b) 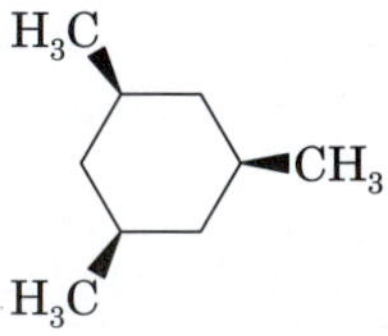 H$_3$C ... CH$_3$ ... H$_3$C

(c) OH ... O ... OH

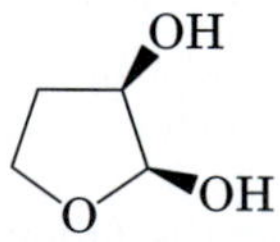

(d) 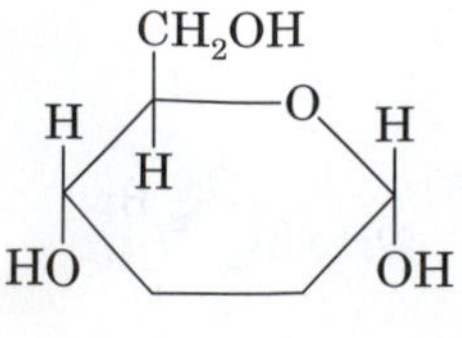 CH$_2$OH ... H ... O ... H ... HO ... OH

(e) 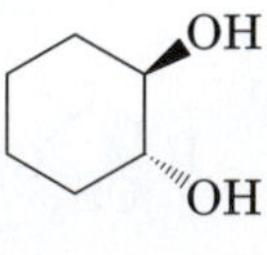 OH ... OH

(f) 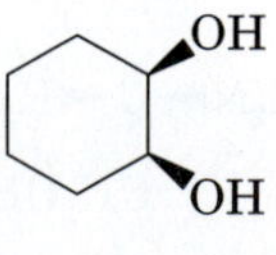 OH ... OH

4.47 Which of the following formulas represent *meso* compounds?

$$
\begin{array}{ccc}
\text{CH}_2\text{OH} & \text{CHO} & \text{CO}_2\text{H} \\
\text{H}\blacktriangleright\text{C}\blacktriangleleft\text{OH} & \text{H}\blacktriangleright\text{C}\blacktriangleleft\text{OH} & \text{H}\blacktriangleright\text{C}\blacktriangleleft\text{OH} \\
\text{(a)} \quad \text{HO}\blacktriangleright\text{C}\blacktriangleleft\text{H} & \text{(b)} \quad \text{HO}\blacktriangleright\text{C}\blacktriangleleft\text{H} & \text{(c)} \quad \text{HO}\blacktriangleright\text{C}\blacktriangleleft\text{H} \\
\text{H}\blacktriangleright\text{C}\blacktriangleleft\text{OH} & \text{HO}\blacktriangleright\text{C}\blacktriangleleft\text{H} & \text{HO}\blacktriangleright\text{C}\blacktriangleleft\text{H} \\
\text{H}\blacktriangleright\text{C}\blacktriangleleft\text{OH} & \text{H}\blacktriangleright\text{C}\blacktriangleleft\text{OH} & \text{H}\blacktriangleright\text{C}\blacktriangleleft\text{OH} \\
\text{CH}_2\text{OH} & \text{CH}_2\text{OH} & \text{CO}_2\text{H}
\end{array}
$$

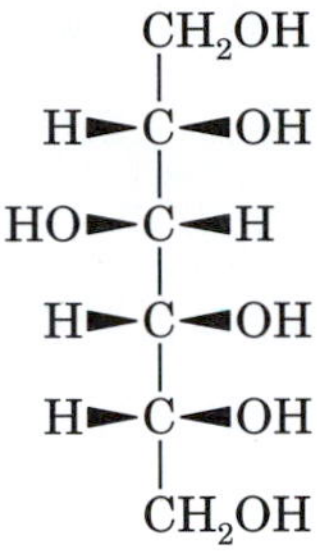

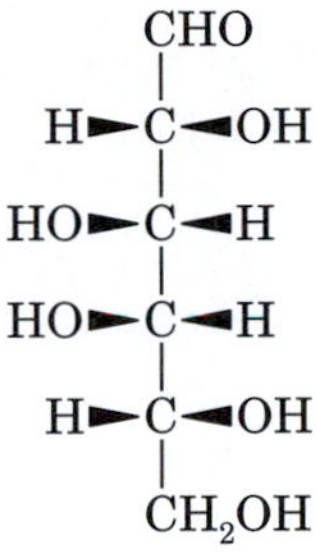

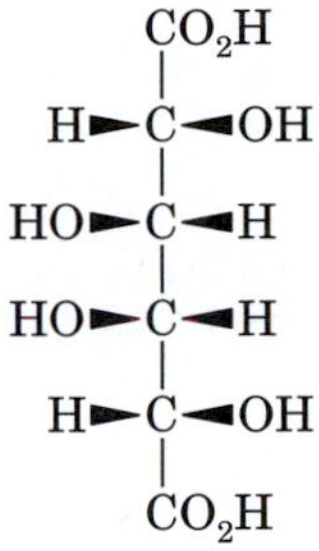

sorbitol   galactose   galactaric acid

an artificial sweetener   *a sugar*

4.48 Draw dimensional or ball-and-stick formulas for the following compounds. (Use models.)

(a) (*R*)-2-bromopropanoic acid

(b) ethyl (2*R*,3*S*)-3-amino-2-iodi-3-phenylpropanoate, $C_6H_5\overset{\overset{\displaystyle NH_2}{|}}{C}H\overset{}{C}HI\overset{\overset{\displaystyle O}{\|}}{C}OCH_2CH_3$

4.49 What is the *maximum number* of stereoisomers possible for **(a)** 3-hydroxy-2-methylbutanoic acid, and **(b)** 2,4-dimethyl-1-pentanol? Draw and correctly name each.

4.50 **(a)** Which of the following Fischer projections represent enantiomers?

(b) Which are diastereomers?

(c) Which is a *meso* form?

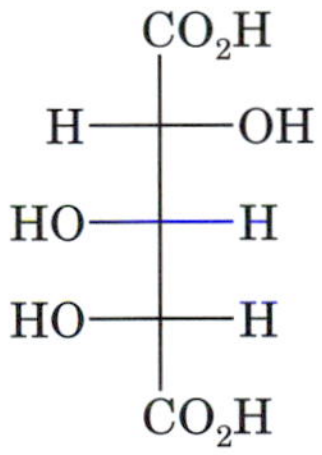

$$
\begin{array}{ccc}
\text{CO}_2\text{H} & \text{CO}_2\text{H} & \text{CO}_2\text{H} \\
\text{H}\!-\!\!-\!\text{OH} & \text{H}\!-\!\!-\!\text{OH} & \text{HO}\!-\!\!-\!\text{H} \\
\text{HO}\!-\!\!-\!\text{H} & \text{H}\!-\!\!-\!\text{OH} & \text{H}\!-\!\!-\!\text{OH} \\
\text{HO}\!-\!\!-\!\text{H} & \text{H}\!-\!\!-\!\text{OH} & \text{H}\!-\!\!-\!\text{OH} \\
\text{CO}_2\text{H} & \text{CO}_2\text{H} & \text{CO}_2\text{H} \\
\text{I} & \text{II} & \text{III}
\end{array}
$$

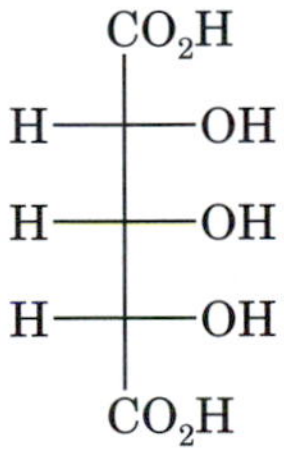

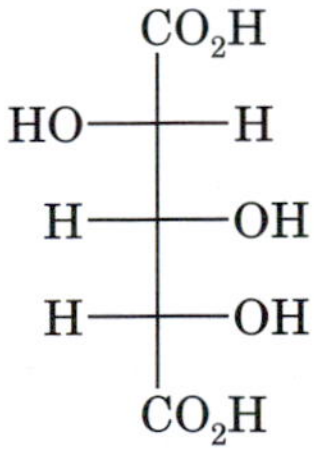

4.51 Write Fischer projections for all possible configurations of 2,3,4-pentanetriol. Indicate enantiomeric pairs. Are there any *meso* forms?

4.52 Draw dimensional or ball-and-stick formulas for

(a) any enantiomeric pairs,

(b) a *meso* form, and

(c) any diastereomeric pairs for $HO_2CCHBrCH_2CHBrCO_2H$.

4.53 Assign each chiral carbon in the following structures an (*R*) or (*S*) configuration:

(a) 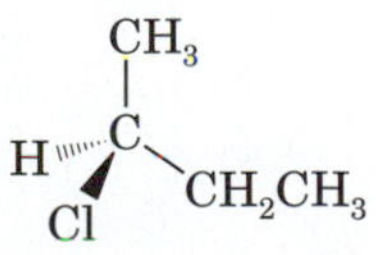 $H\overset{\overset{\displaystyle CH_3}{|}}{\underset{\displaystyle Cl}{C}}CH_2CH_3$

(b) $\overset{\overset{\displaystyle CO_2H}{|}}{\underset{\displaystyle H \quad OH}{C}}CH_3$

(c) $\overset{\overset{\displaystyle H}{|}}{\underset{\displaystyle D \quad CH_3}{C}}CH_2Cl$

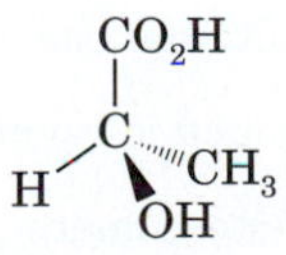

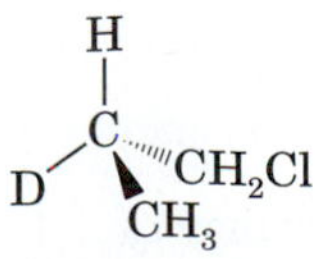

(d)

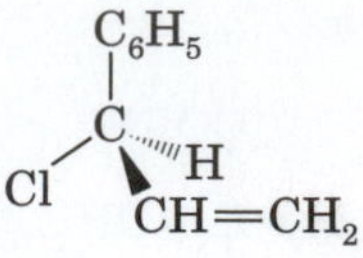

$$\begin{array}{c} C_6H_5 \\ | \\ Cl-C\cdots H \\ | \\ CH=CH_2 \end{array}$$

(e)

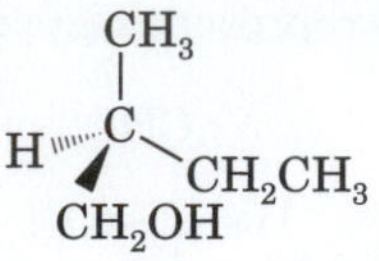

$$\begin{array}{c} CH_3 \\ | \\ H\cdots C \\ \diagdown CH_2CH_3 \\ CH_2OH \end{array}$$

(f) 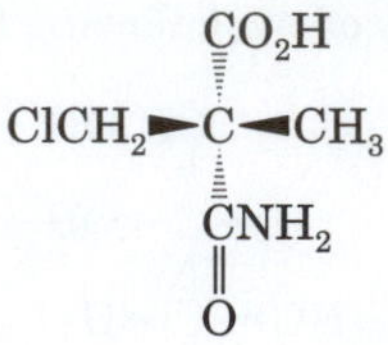 $ClCH_2-C\!-\!CH_3$

$$\begin{array}{c} CO_2H \\ | \\ ClCH_2\!-\!C\!-\!CH_3 \\ | \\ CNH_2 \\ \| \\ O \end{array}$$

4.54 Assign each chiral carbon in the following structures an (R) or (S) configuration:

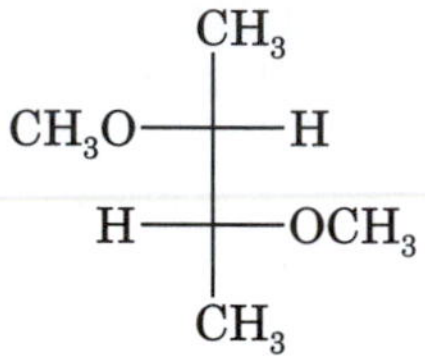

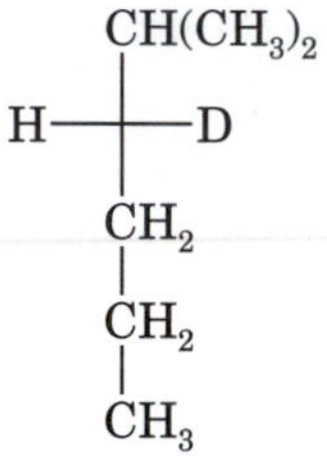

(a)

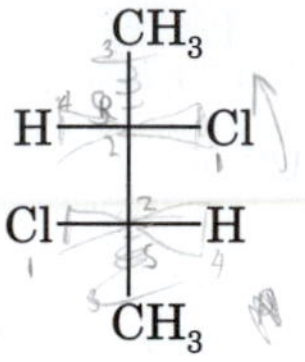

$$\begin{array}{c} CH_3 \\ | \\ H-\!\!-Cl \\ | \\ Cl-\!\!-H \\ | \\ CH_3 \end{array}$$

(b)

$$\begin{array}{c} CH_3 \\ | \\ CH_3O-\!\!-H \\ | \\ H-\!\!-OCH_3 \\ | \\ CH_3 \end{array}$$

(c)

$$\begin{array}{c} CH(CH_3)_2 \\ | \\ H-\!\!-D \\ | \\ CH_2 \\ | \\ CH_2 \\ | \\ CH_3 \end{array}$$

(d)

$$\begin{array}{c} CO_2H \\ | \\ H-\!\!-SH \\ | \\ HO-\!\!-OCH_3 \\ | \\ CH_2OH \end{array}$$

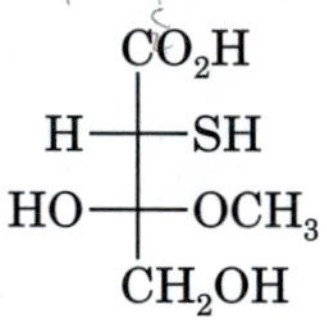

(e)

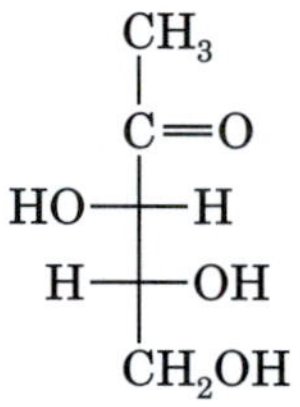

$$\begin{array}{c} CH_3 \\ | \\ C=O \\ | \\ HO-\!\!-H \\ | \\ H-\!\!-OH \\ | \\ CH_2OH \end{array}$$

(f)

$$\begin{array}{c} C\equiv CH \\ | \\ F-\!\!-CH=CHCH_3 \\ | \\ C_6H_5 \end{array}$$

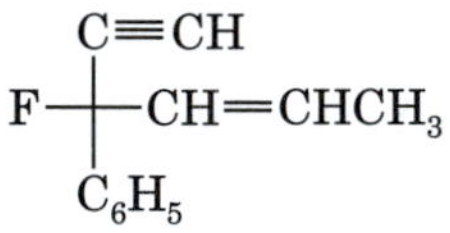

4.55 Which of the following formulas represents a *meso* form?

(a)

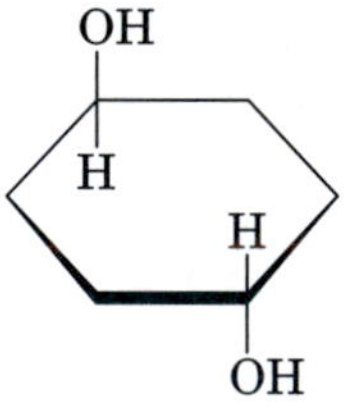

(b)

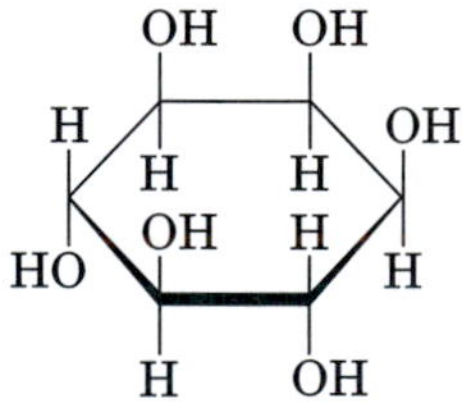

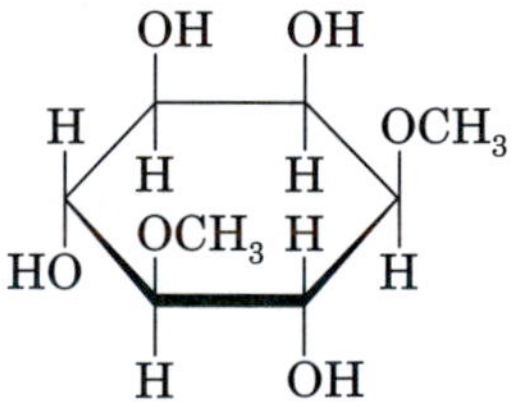

myo-inositol

a growth factor in animals

(c)

4.56 Draw the *meso* form of each of the following structures and indicate the plane of symmetry through the structure:

(a)

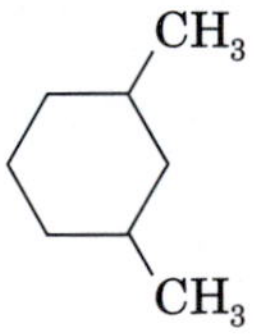

(b)

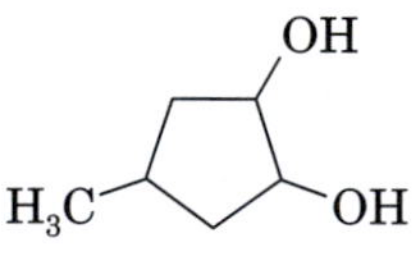

Additional Problems

4.57 Select one or more of the following seven structures to illustrate each term below.

    (a) structural isomers   **(b)** positional isomers

    (c) stereoisomers   **(d)** geometric isomers

    (e) enantiomers   **(f)** diastereomers

(g) chiral compound   **(h)** achiral compound

(i) *meso* compound   **(j)** racemic mixture

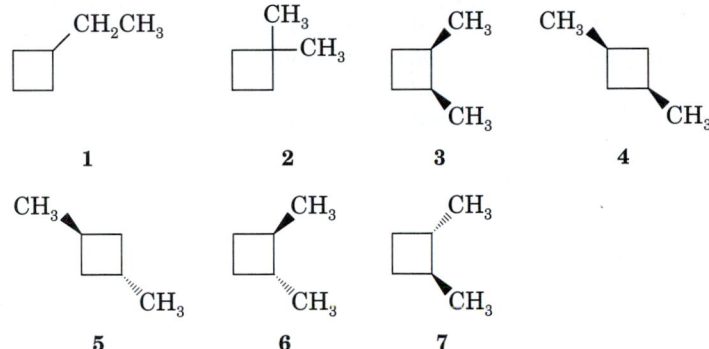

4.58 (a) Draw the structure for the most stable conformer of the following menthyl chloride:

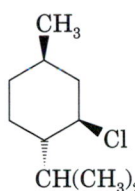

(b) Draw the structure of the chair-form conformer in which Cl is *anti* to an H on one of the adjacent carbon atoms. Would you expect a substantial percentage of the menthyl chloride molecules to be in this latter conformation? Explain.

4.59 *Inositols* are 1,2,3,4,5,6-hexahydroxycyclohexanes that are found in all cells.

(a) *scyllo*-Inositol is the most stable of all the inositols. Draw its structure in the chair form.

(b) Only one stereoisomeric pair of inositols is a pair of enantiomers. Draw the structures of this pair of enantiomers as polygon formulas.

4.60 Draw Newman projections for the most stable and the least stable conformations of **(a)** (1*S*,2*R*)-1,2-dibromo-1,2-diphenylethane and **(b)** *meso*-2,3-dichlorobutane.

4.61 Draw Newman projections for all stereoisomers of 1-bromo-1,2-diphenylpropane in which the H on carbon 2 and the Br are *anti*.

4.62 Assign (*R*) and (*S*) configurations to all stereoisomers of 1-bromo-1,2-diphenylpropane. Label enantiomeric pairs and diastereomers.

4.63 Write formulas for all the stereoisomers of 2,4-hexadiene. Name these isomers using the (*E*) and (*Z*) system.

4.64 What is the specific rotation of each of the following solutions at 20°, sodium D-line?

(a) 1.00 g of sample is diluted to 5.00 mL. A 3.00-mL aliquot is placed in a tube that is 1.0 cm long. The observed rotation is +0.45°.

(b) A 0.20-g sample is diluted to 2.0 mL and placed in a 10-cm tube. The observed rotation is −3.2°.

4.65 **(a)**   Although it contains no chiral carbons, 2,3-pentadiene can exist as a pair of enantiomers. Explain. (*Hint:* Consider the geometry of the molecule. Use models.)

(b)   Draw all possible stereoisomers for methyl 2,4,5-tetradecatrienoate, $CH_3(CH_2)_7CH=C=CHCH=CHCO_2CH_3$. One stereoisomer is the sex attractant produced by the male dried-bean weevil.

4.66   Which of the following compounds has an enantiomer? [Use models for **(b)** and **(c)**.]

(a)

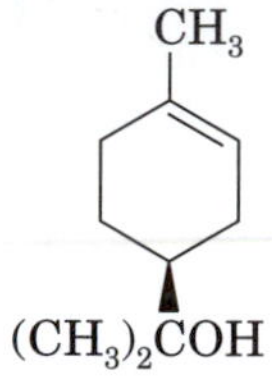

CH₃

$(CH_3)_2COH$

α-terpineol

found in many plant oils

(b)

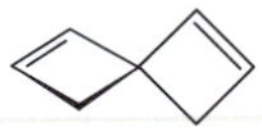

a spiroheptadiene

(c)

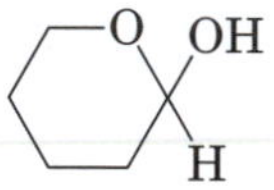

O   OH

H

(d)

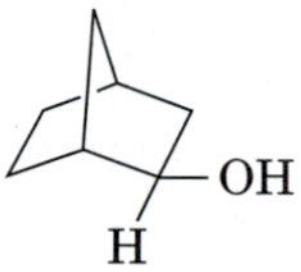

—OH

H

4.67   Draw polygon formulas for all isomeric dibromocyclobutanes. Label all enantiomeric pairs, *meso* forms, and achiral structures with no chiral atoms.

4.68   Draw polygon formulas for all the stereoisomers of **(a)** 1-chloro-2-methylcyclopentane, and **(b)** 1-chloro-3-methylcyclopentane.

4.69   Each of the following compounds can exist as a pair of enantiomers. Suggest why.

(a)

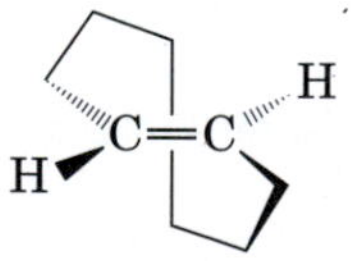

H

$C=C$

H

H

one enantiomer of
trans-cyclooctene

(b)

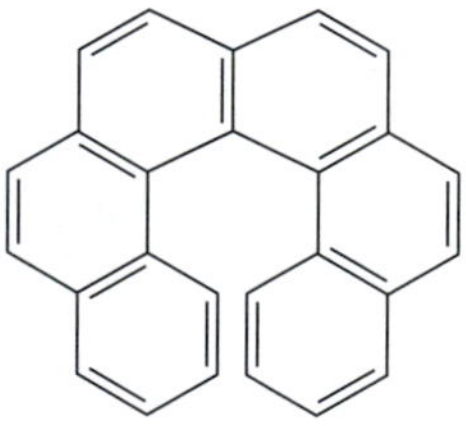

hexahelicene

4.70   Identify the prochiral carbons in the following compounds, and classify their hydrogens as either enantiotopic or diastereotopic.

(a) $CH_3CH_2\underset{\underset{Cl}{|}}{C}HCOH$ (with C=O)

(b) $\underset{\underset{Cl}{|}}{C}H_2CH_2CH_2COH$ (with C=O)

ALKYL HALIDES; SUBSTITUTION AND ELIMINATION REACTIONS

Organohalogen compounds are used extensively in modern society. Some are used as solvents, some as insecticides, and some as intermediates for the synthesis of other organic compounds. Naturally occurring organohalogen compounds are rather rare in land plants and animals. *Thyroxine*, a component of the human thyroid hormone thyroglobulin, is an example of a naturally occurring iodine compound. Chloramphenicol (Chloromycetin) is an antibiotic produced by *Streptomyces venezuelae*, a soil bacterium; it is very effective against typhoid fever.

thyroxine

chloramphenicol
Chloromycetin

Halogen compounds are somewhat more common in marine organisms, such as algae and seaweeds. The dye Tyrian purple is a bromine compound obtained in tiny amounts from a rare species of snail found in Crete. Tyrian purple was used as a dye by the Phoenician royalty and later by the Romans. (One may still hear the phrases "royal purple" and "born to the purple.") The structure of this compound is shown in Section 22.1.

Many organohalogen compounds are toxic and should be used with caution. For example, the solvents carbon tetrachloride (CCl_4) and chloroform ($CHCl_3$) both cause liver damage when inhaled in excess. Some halogen compounds, however, seem to be quite safe. A few of these are used as inhalation anesthetics, for example, halothane ($CF_3CHBrCl$) and methoxyflurane ($CH_3OCF_2CHCl_2$).

Section 5.1

Types of Organohalogen Compounds

Compounds containing only carbon, hydrogen, and a halogen atom generally fall into one of three categories: **alkyl halides**; **aryl halides** (in which a halogen is bonded to a carbon of an aromatic ring); and **vinylic halides** (in which a halogen is bonded to a double-bonded carbon). A few examples follow:

Alkyl halides (RX):

$$CH_3I$$

iodomethane

$$CH_3CH_2Cl$$

chloroethane

a topical anesthetic (bp 12°)
that numbs by chilling

Aryl halides (ArX):

bromobenzene

a polychlorinated biphenyl (PCB)

*one of a number of
toxic compounds that
have been used as cooling
fluids in transformers*

Vinylic halides:

$$CH_2{=}CHCl$$

chloroethene
(vinyl chloride)

*the starting material for
polyvinyl chloride (PVC),
a plastic used for piping,
house siding, phonograph
records, and trash bags*

$$CH_3CH{=}\overset{\overset{\displaystyle Br}{|}}{C}CH_3$$

2-bromo-2-butene

We have already defined R as the general symbol for an alkyl group. In a similar manner, Ar is the general symbol for an aromatic, or **aryl,** group. A halogen atom (F, Cl, Br, or I) may be represented by X. Using these general symbols, an alkyl halide is RX, and an aryl halide such as bromobenzene (C_6H_5Br) is ArX.

The carbon—halogen sigma bond is formed by the overlap of an orbital of the halogen atom and a hybrid orbital of the carbon atom. We cannot be sure of the hybridization (or lack of it) of the halogen atom in an organic halide because a halogen forms only one covalent bond and therefore has no bond angle around it. However, carbon uses the same type of hybrid orbital to bond to a halogen atom as it does to bond to a hydrogen atom or to another carbon atom.

An atom of F, Cl, or Br is electronegative with respect to a carbon atom. Although the electronegativity of iodine is close to that of carbon, iodine atoms are larger than carbon atoms or other halogens and are more readily polarized. Therefore, the C—X bonds of all the alkyl halides are polar.

A halogen atom in an organic compound is a functional group, and the C—X bond is a site of chemical reactivity. In this chapter, we will discuss reactions in only those cases where the halogen is bonded to an sp^3-hybridized carbon—that is, the alkyl halides. *Aryl halides and vinylic halides do not undergo these same types of reactions,* partly because a bond from an sp^2-hybridized carbon is stronger than a bond from an sp^3-hybridized carbon (Section 2.4F). Because this is the first chapter devoted to compounds containing functional groups, we will also use it as an introduction to organic chemical reactions.

Section 5.2

Nomenclature and Classification of Alkyl Halides

In the IUPAC system, an alkyl halide is named with a **halo-** prefix. Many common alkyl halides also have trivial **functional-group names.** In these names, the name of the alkyl group is given, followed by the name of the halide.

IUPAC:	2-chlorobutane	bromocyclohexane
trivial:	*sec*-butyl chloride	cyclohexyl bromide

In chemical reactions, the structure of the alkyl portion of an alkyl halide is important. Therefore, we need to differentiate the four types of alkyl halides: *methyl, primary, secondary,* and *tertiary.*

A **methyl halide** is a structure in which one hydrogen of methane has been replaced by a halogen.

The methyl halides:

CH_3F CH_3Cl CH_3Br CH_3I

fluoromethane chloromethane bromomethane iodomethane

The *"head" carbon* of an alkyl halide is the carbon atom bonded to the halogen. A **primary (1°) alkyl halide** (RCH_2X) has *one* alkyl group bonded to the head carbon. In the following examples, the head carbons and their hydrogens are circled.

Primary alkyl halides (one alkyl group attached to head):

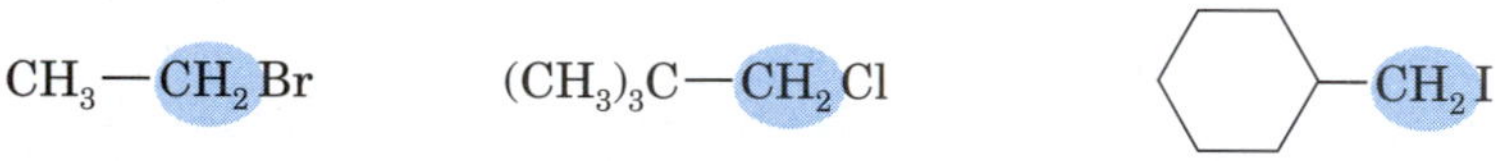

CH_3—CH_2Br $(CH_3)_3C$—CH_2Cl

bromoethane   1-chloro-2,2-dimethylpropane   (iodomethyl)cyclohexane
(ethyl bromide)   (neopentyl chloride)

A **secondary (2°) alkyl halide** (R_2CHX) has *two* alkyl groups attached to the head carbon, and a **tertiary (3°) alkyl halide** (R_3CX) has *three* alkyl groups attached to the head carbon. (Note that a halogen attached to a cycloalkane ring must be either secondary or tertiary.)

Secondary alkyl halides (two alkyl groups attached to head):

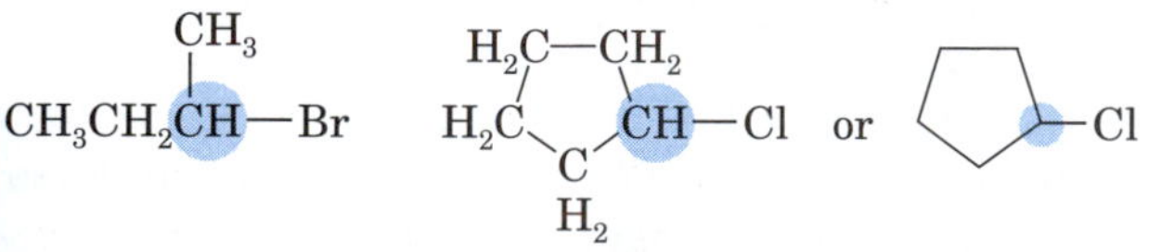

CH_3
|
CH_3CH_2CH—Br

H_2C—CH_2
H_2C CH—Cl or ⬠—Cl
C
$$$H_2$

2-bromobutane   chlorocyclopentane
(*sec*-butyl bromide)   (cyclopentyl chloride)

Tertiary alkyl halides (three alkyl groups attached to head):

CH_3
|
CH_3—C—Cl
|
CH_3

2-chloro-2-methylpropane   1-bromo-1-methylcyclopentane
(*tert*-butyl chloride)

SAMPLE PROBLEM

Classify each of the following alkyl halides as 1°, 2°, or 3°:

CH_3
|
(a) CH_3CHCH_2Cl **(b)**   **(c)** CH_3CHCl
$$$CH_2CH_3$
$$$Cl$
$$$CH_2CH_2CH_3$

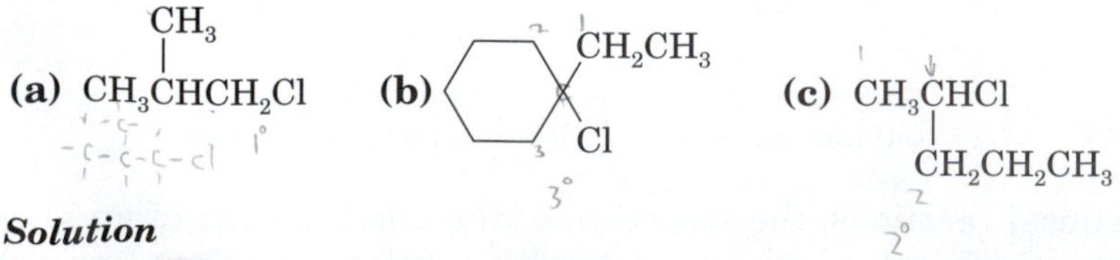

Solution
(a) 1°; **(b)** 3°; **(c)** 2°.

[Handwritten note in margin:]
Methyl Halide :- structure where 1 H replaced by 1 halogen
Primary Alkyl Halide : head carbon that has 1 alkyl group bonded to it
Secondary Alkyl Halide : head carbon bonded with 2 alkyl groups
Tertiary Alkyl Halide : head carbon bonded with 3 alkyl groups.
Note: (1) Halogen atoms are: F,Cl,Br,I
(2) Alkyl group is: CH_3, CH_2, CH, C
(3) Denoted 1°, 2°, 3°

STUDY PROBLEMS

5.1 Write an IUPAC name for each of the following structures. Classify each halogen as 1°, 2°, or 3°.

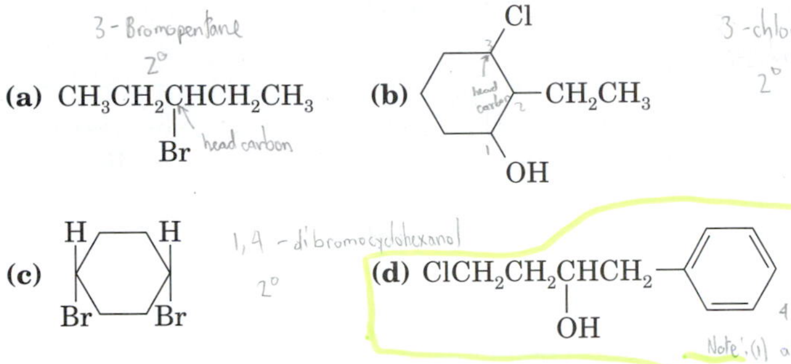

[handwritten: 3-Bromopentane]
[handwritten: 2°]

(a) $CH_3CH_2CHCH_2CH_3$
          |
          Br *[handwritten: head carbon]*

[handwritten: 3-chloro-2-ethylcyclohexanol]
[handwritten: 2°]

(b) *[cyclohexane ring with Cl at top, CH_2CH_3 on side, OH at bottom; handwritten "head carbon" labels]* —CH_2CH_3

[handwritten: 1,4-dibromocyclohexanol]
[handwritten: 2°]

(c) *[cyclohexane ring with H, H at top and Br, Br at bottom]*

(d) $ClCH_2CH_2CHCH_2$—*[benzene ring]*
                        |
                        OH
[handwritten: 1°]
[handwritten: 4-chloro-1-phenyl-2-butanol]
[handwritten: Note: (1) alcohol has priority]
[handwritten: (2) chunck structure (find longest c chain)]
[handwritten: (3) name main chain with ol then add in subst in alpha order.]

5.2 Draw structures for the following IUPAC names:

(a) bromobenzene   **(b)** 2,3-dichloro-1-pentanol

(c) 4-ethyl-3-iodo-4-methylheptane   **(d)** 5,5-dibromo-2-pentene

Section 5.3
A Preview of Substitution and Elimination Reactions

A. Substitution Reactions

A carbon–halogen bond is polar, and the halog[en] halide has a partial positive charge. This carbon is [open to] attack by an anion or by any other species that [has] electrons in the outer shell. The result is a **substitution reaction**—a reaction in which one atom, ion, or group is substituted for another.

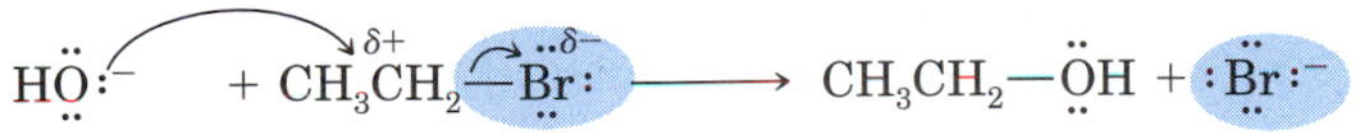

[blue note: Note: (1) carbon-halogen bond is polar
(2) carbon of alkyl halide is partially (+)
(3) $CH_3CH_2Br \rightarrow CH_3^{\delta+}CH_2^{\delta-}Br$
*   alkyl halide (leaving group)*
(4) halide ions good leaving groups ∴ weak base.
(5) substitution rx's of Alkyl Halides
*   I Br Cl increasing displacement]*

the sigma-bond electrons
leave with the halogen

$$HÖ:^- \quad + \quad CH_3CH_2 \overset{\delta+}{\frown} \overset{\delta-}{Br}: \quad \longrightarrow \quad CH_3CH_2—ÖH \; + :Br:^-$$

hydroxide ion   bromoethane   ethanol

$$CH_3Ö:^- \quad + \quad CH_3CH_2CH_2 \frown Cl: \quad \longrightarrow \quad CH_3CH_2CH_2—ÖCH_3 \; + :Cl:^-$$

methoxide ion   1-chloropropane   methyl propyl ether

In substitution reactions of the alkyl halides, the halide is called the **leaving group,** a term meaning any group that can be displaced from a carbon atom. Halide ions are *good* leaving groups because they are very weak bases. Strong bases, such as ⁻OH, are very poor leaving groups. In this chapter, we will discuss only the halides as leaving groups; we will introduce other leaving groups as we encounter them in subsequent chapters.

In substitution reactions of alkyl halides, the iodide ion is the halide most easily displaced, followed by the bromide ion and then the chloride ion. Because F⁻ is a stronger base than the other halide ions and because the C—F bond is distinctly stronger than other C—X bonds, fluorine is not as good a leaving group. From a practical standpoint, only Cl, Br, and I are good enough leaving groups to be useful in substitution reactions. For these reasons, when we refer to RX, we usually mean alkyl chlorides, bromides, and iodides.

$$RF \quad RCl \quad RBr \quad RI$$
increasing reactivity

The species that attacks an alkyl halide in a substitution reaction is called a **nucleophile** (literally, "nucleus lover"), often abbreviated Nu:$^-$. In the preceding equations, OH$^-$ and CH$_3$O$^-$ are the nucleophiles. Generally, a nucleophile is any species that is attracted to a positive center. Thus, a nucleophile is a Lewis base. Most nucleophiles are anions; however, some neutral polar molecules, such as H$_2$O, CH$_3$OH, and CH$_3$NH$_2$, can also act as nucleophiles. These neutral molecules all contain unshared electrons that can be used to form sigma bonds. Substitutions by nucleophiles are called **nucleophilic substitutions,** or **nucleophilic displacements.**

The opposite of a nucleophile is an **electrophile** ("electron lover"), often abbreviated E$^+$. An electrophile is any species that is attracted toward a negative center; that is, an electrophile is a Lewis acid, such as H$^+$ or ZnCl$_2$. Electrophilic reactions are common in organic chemistry. You will encounter many of these reactions in later chapters.

B. Elimination Reactions

When an alkyl halide is treated with a *strong base,* an **elimination reaction** can occur. An elimination reaction is one in which a molecule loses atoms or ions from its structure. The organic product of an elimination reaction of an alkyl halide is an alkene. In this type of elimination reaction, the elements H and X are lost from the alkyl halide. Therefore, these reactions are also called **dehydrohalogenation reactions.** (The prefix *de-* means "minus" or "loss of.")

$$CH_3CH-CH_2 + {}^-:\!\ddot{O}H \longrightarrow CH_3CH{=}CH_2 + H_2O + :\!\ddot{Br}:^-$$

2-bromopropane
(isopropyl bromide)

propene
(propylene)

$$CH_3-\underset{\underset{CH_3}{|}}{\overset{\overset{CH_3}{|}}{C}}-Cl + {}^-OH \longrightarrow CH_3-\underset{\underset{CH_3}{|}}{\overset{\overset{CH_2}{\|}}{C}} + H_2O + Cl^-$$

2-chloro-2-methylpropane
(*tert*-butyl chloride)

methylpropene
(isobutylene)

C. Competing Reactions

A hydroxide ion or an alkoxide ion (RO$^-$) can react as a *nucleophile* in a substitution reaction or as a *base* in an elimination reaction. Which type of reaction actually occurs depends on a number of factors, such as the structure of the alkyl halide (1°, 2°, or 3°), the strength of the base, the nature of the solvent, and the temperature.

Methyl and primary alkyl halides tend to yield substitution products, not elimination products. Under equivalent conditions, *tertiary alkyl halides* yield principally elimination products, and not substitution products. *Secondary alkyl halides* are intermediate in their behavior. The relative proportion of the substitution product to the elimination product depends to a large extent upon the experimental conditions.

1°: $CH_3CH_2Br + CH_3CH_2O^- \xrightarrow[25°]{CH_3CH_2OH} CH_3CH_2OCH_2CH_3$ + Br^-

(almost 100%)

2°: $(CH_3)_2CHBr + CH_3CH_2O^- \xrightarrow[25°]{CH_3CH_2OH} (CH_3)_2CHOCH_2CH_3 + CH_2{=}CHCH_3$

(20%)   (80%)

3°: $(CH_3)_3CBr + CH_3CH_2O^- \xrightarrow[25°]{CH_3CH_2OH} (CH_3)_3COCH_2CH_3 + CH_2{=}C(CH_3)_2$

(5%)   (95%)

Because more than one reaction can occur between an alkyl halide and a nucleophile or base, substitution reactions and elimination reactions are said to be **competing reactions.** Competing reactions are common in organic chemistry. Because mixtures of products are the rule rather than the exception when competing reactions occur, we will not balance most of the organic equations in this book.

In Sections 5.4–5.9, we will discuss two different types of substitution reactions (called S_N1 and S_N2 reactions) and two types of elimination reactions (E1 and E2). We will discuss each type individually. Then, we will summarize the factors that can help us predict which of these reactions will predominate in a given case.

The two different pathways for substitution reactions were postulated by several investigators during the 1920s and 1930s. However, between 1933 and 1935, Sir Christopher Ingold, Edward Hughes, and their collaborators at University College, London, solidified the theories into what we now call the S_N1 and S_N2 pathways. [Ingold, as you will recall, was also one of the originators of the sequence rules used for (*R*) and (*S*) assignments.]

STUDY PROBLEM

5.3   For each of the following reactants, write an equation showing competitive substitution and elimination.

(a)   (ring)—$CHCH_3$ + ^-OH
         |
         Br

(b)   (ring)—Br + CH_3O^-

(c) $CH_3CH_2CHCH_2CH_3$ + ^-OH
                |
                Br

(d)   (ring with Br) + $CH_3CH_2O^-$

D. Nucleophilicity versus Basicity

Before proceeding with the details of substitution and elimination reactions, let us briefly consider the similarities and the differences between bases and nucleophiles. Under the proper circumstances, all bases can act as nucleophiles. Conversely, all nucleophiles can act as bases. In either case, the reagent reacts by donating a pair of electrons to form a new sigma bond.

Basicity is a measure of a reagent's ability to accept a proton in an acid–base reaction. Therefore, the relative base strengths of a series of reagents are determined by comparing the relative positions of their equilibria in an acid–base reaction, such as the degree of ionization in water.

OH⁻ acting as a base:

$$HO\colon^- \quad H-Cl\colon \longrightarrow H_2O\colon + \colon Cl\colon^-$$

$$CH_3CH-CH_2 + {}^-\colon OH \longrightarrow CH_3CH=CH_2 + H_2O\colon + \colon Br\colon^-$$

strong bases

$$\colon I\colon^- \quad \colon Br\colon^- \quad \colon Cl\colon^- \quad ROH \quad H_2O\colon \quad {}^-\colon C\equiv N\colon \quad {}^-\colon OH \quad {}^-\colon OR$$

increasing basicity

In contrast to basicity, **nucleophilicity** is a measure of a reagent's ability to cause a substitution reaction. The relative nucleophilicities of a series of reagents are determined by their relative rates of reaction in a substitution reaction, such as a substitution reaction with bromoethane.

OH⁻ acting as a nucleophile:

$$HO\colon^- \quad CH_3CH_2-Br\colon \longrightarrow CH_3CH_2-OH + \colon Br\colon^-$$

$$H_2O\colon \quad ROH \quad \colon Cl\colon^- \quad \colon Br\colon^- \quad {}^-\colon OH \quad {}^-\colon OR \quad \colon I\colon^- \quad {}^-\colon C\equiv N\colon$$

increasing nucleophilicity

A list of relative nucleophilicities does not exactly parallel a list of base strengths; however, a stronger base is usually a better nucleophile than a weaker base. For example, OH⁻ (a strong base) is a better nucleophile than Cl⁻ or H_2O (weak bases). We will discuss the factors affecting nucleophilicity in greater detail in Section 5.9

Because some alkyl halides can undergo competing substitution and elimination reactions, a reagent such as OH⁻ can act both as a nucleophile and as a base in one reaction vessel.

$$(CH_3)_2CHBr + {}^-OH \xrightarrow{H_2O} (CH_3)_2CHOH + CH_2=CHCH_3$$

formed by ⁻OH acting as a nucleophile   *formed by ⁻OH acting as a base*

Section 5.4
The S_N2 Reaction

The reaction of bromoethane with hydroxide ion to yield ethanol and bromide ion (page 185) is a typical **S_N2 reaction.** (S_N2 means "substitution, nucleophilic, bimolecular." The term *bimolecular* will be defined in Section 5.4B.) Virtually any methyl or primary alkyl halide undergoes an S$_N$2 reaction with any relatively strong nucleophile: $^-$OH, $^-$OR, $^-$CN, and others that we have not yet mentioned. Methyl or primary alkyl halides also undergo reaction with weak nucleophiles, such as H$_2$O, but these reactions are too slow to be of practical value. Secondary alkyl halides can also undergo S$_N$2 reactions. However, tertiary alkyl halides do not.

A. Reaction Mechanism

The detailed description of how a reaction occurs is called a **reaction mechanism.** A reaction mechanism must take into account all known facts. For some reactions, the number of facts known is considerable, and the particular reaction mechanisms are accepted by most chemists. The mechanisms of some other reactions are still quite speculative. The S$_N$2 reaction is one that has been studied extensively. There is a large amount of experimental data supporting the mechanism that we will present.

For molecules to undergo a chemical reaction, they must first collide. Most collisions between molecules do not result in a reaction. Instead, the molecules simply rebound. To undergo reaction, the colliding molecules must contain enough *potential energy* for bond breakage to occur. Also, the *orientation* of the molecules relative to each other is often an important factor in determining whether a reaction will occur. This is particularly true in an S$_N$2 reaction. In this section, we will first discuss the stereochemistry of the S$_N$2 reaction, then we will discuss the energy requirements.

B. Stereochemistry of an S$_N$2 Reaction

In the S$_N$2 reaction between bromoethane and hydroxide ion, the oxygen of the hydroxide ion collides with the rear of the head carbon relative to the bromine and displaces the bromide ion.

Overall S$_N$2 reaction:

rear attack

When a nucleophile collides with the back side of a tetrahedral carbon atom bonded to a halogen, two things occur simultaneously: (1) a new bond begins to form, and (2) the C—X bond begins to break. The process is said to be a one-step, or **concerted,** process. If the potential energy of the two colliding species is high enough, a point is reached where it is energetically more favorable for the new bond to form and the old C—X bond to break. As the reactants are converted to products, they must pass through an in-between state that has a high potential energy relative to the reactants and the products. This state is called the **transition state,** or the **activated complex.** Because the transition state involves two particles (Nu:$^-$ and RX), the S$_N$2 reaction is said to be **bimolecular.** (The "2" in S$_N$2 indicates bimolecular.)

partial bonds

$$HO^- + \overset{\overset{\displaystyle H}{|}\;CH_3}{\underset{\overset{|}{H}}{C}}\!\!-\!Br \longrightarrow \left[HO\text{---}\overset{\overset{\displaystyle H}{}\;\;CH_3}{\underset{\overset{|}{H}}{C}}\text{---}Br \right]^- \longrightarrow HO\text{---}\overset{\overset{\displaystyle H_3C}{}\;\overset{\displaystyle H}{}}{\underset{\overset{|}{H}}{C}} + Br^-$$

reactants   *transition state:*   *products*
high potential energy,
equally able to go to reactants
or products

A transition state in any reaction is the fleeting high-energy arrangement of the reactants as they go to products. We cannot isolate a transition state and put it in a flask. The transition state is simply a description of "molecules in a state of transition." We will often use square brackets in an equation to show any temporary, nonisolable structure in a reaction. Here, we use brackets to enclose the structure of a transition state. Later, we will sometimes use brackets to indicate unstable products that undergo further reaction. In this text, we will label the transition states as such to help you differentiate them from intermediates.

For the S_N2 reaction, the transition state involves a temporary rehybridization of the head carbon from sp^3 to sp^2 and finally back to sp^3 again. In the transition state, the carbon atom has three planar sp^2 bonds, plus two half-bonds using the p orbital.

p orbital

sp² carbon

As the nucleophile attacks the rear of the molecule, relative to the halogen atom, the other three groups attached to the carbon flatten out in the transition state, then flip to the other side of the carbon atom, much as an umbrella being blown inside out. (Models would be useful to help visualize this.) The flipping is called **inversion of configuration,** or **Walden inversion,** after the Latvian and German chemist Paul Walden (1863–1957). He discovered the phenomenon of inversion in 1896 while at the Polytechnicum in Riga, Latvia.

The existence of inversion as part of the mechanism of an S_N2 reaction has been beautifully demonstrated by reactions of pure enantiomers of chiral secondary alkyl halides. For example, the S_N2 reaction of (R)-2-bromooctane with $^-$OH yields almost exclusively (S)-2-octanol.

$$HO^- + \overset{\overset{\displaystyle H}{}\;CH_2(CH_2)_4CH_3}{\underset{\overset{|}{CH_3}}{C}}\!\!-\!Br \xrightarrow{\;S_N2\;} HO\text{---}\overset{\overset{\displaystyle CH_2(CH_2)_4CH_3}{}\;\overset{\displaystyle H}{}}{\underset{\overset{|}{CH_3}}{C}} + Br^-$$

(R)-2-bromooctane (S)-2-octanol

Most reactions involving chiral molecules are carried out with racemic mixtures, that is, equal mixtures of (R) and (S) reactants. In these cases, the products also are racemic mixtures. Even though inversion occurs, we cannot observe the effects because half the molecules go one way and half go the other way.

$$(R)\text{-RX} + (S)\text{-RX} \xrightarrow[\text{S}_N2]{\text{OH}^-} (S)\text{-ROH} + (R)\text{-ROH}$$

racemic   *racemic*

The experimentally demonstrated inversion of configuration of chiral carbons during S$_N$2 reactions clearly requires that the nucleophile attack the tetrahedral carbon from the back side of the carbon–halogen bond. Is there some underlying reason for this preferred back-side displacement by the nucleophile? Indeed there is, and it comes from molecular orbital theory. When the nucleophile and the alkyl halide (RX) interact to begin forming a new bond, a filled nonbonding orbital (**n orbital**) on the nucleophile must overlap with an empty orbital on RX. The only empty orbitals available on RX are antibonding σ^* orbitals (σ^* orbitals were introduced in Section 2.2). And if the R—X bond is to be broken, the orbital interaction should occur between the C—X σ^* orbital and the nucleophile n orbital. What is the shape of the C—X σ^* orbital, and how does it control the geometry of an S$_N$2 displacement? The orbital interaction that occurs in an S$_N$2 reaction is displayed in Figure 5.1. Note that suitable orbital overlap to form a bond between the carbon and the nucleophile occurs only with the large lobe of the σ^* orbital that projects from the carbon opposite from the halogen. Overlap of the σ^* orbital lobes between the carbon and the halogen, as required for front-side displacement, is net nonbonding. Thus inversion, not retention, of configuration occurs in S$_N$2 reactions.

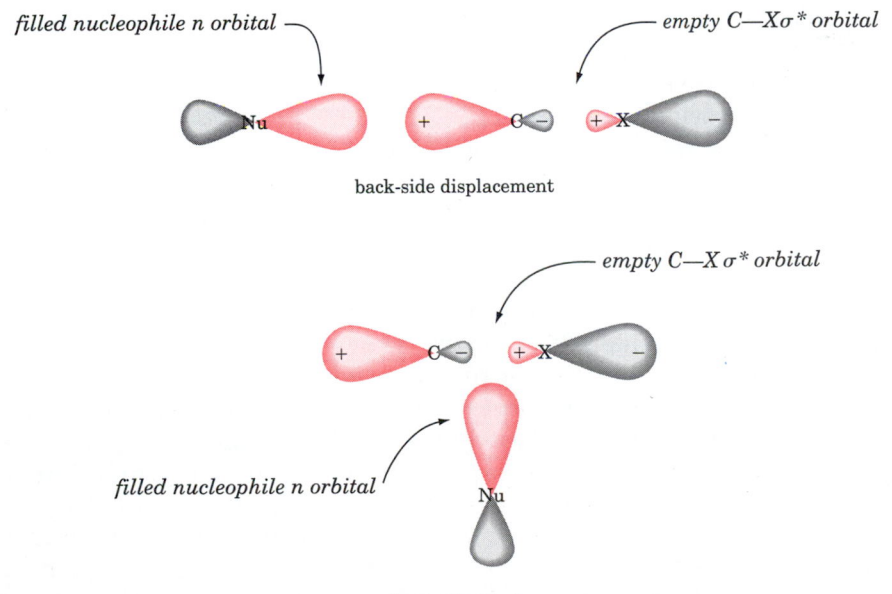

Figure 5.1 Orbital depiction of back-side versus front-side displacement in an S$_N$2 reaction.

5.4 Write an equation (showing the stereochemistry by using dimensional formulas) for the S_N2 reaction of (S)-2-bromobutane with ^-CN and designate the configuration of the product by the (R) and (S) system.

5.5 Write the S_N2-transition state structure for each of the following reactions.

(a) $+ \text{CH}_3\text{O}^- \xrightarrow{\text{S}_N2}$

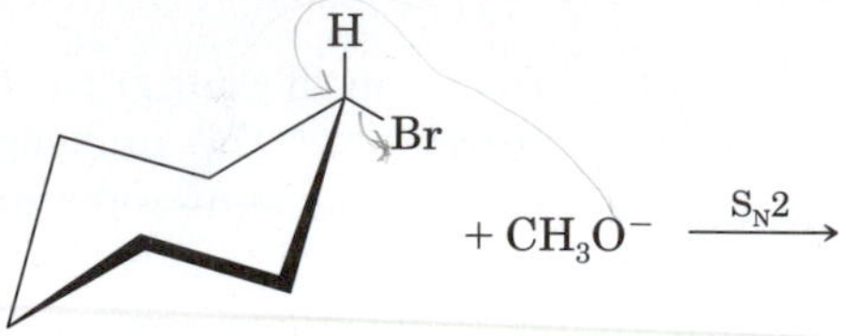

(b) $+ \ ^-\text{OH} \xrightarrow{\text{S}_N2}$

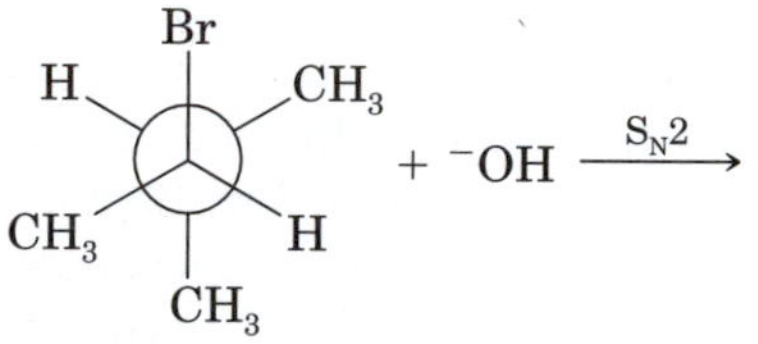

C. Energy in an S$_N$2 Reaction

We have mentioned that colliding molecules need energy to undergo reaction. We will now look at these energy requirements in more detail.

Molecules moving around in a solution contain a certain amount of potential energy in their bonds and a certain amount of kinetic energy from their movement. Not all molecules in solution have exactly the same amount of potential or kinetic energy. However, we may speak of the *average energy* of the molecules. The total energy of the reaction mixture may be increased, usually by heating the solution. When heated, the molecules gain kinetic energy, collide more frequently and more energetically, and exchange some kinetic energy for potential energy.

Before a reaction can begin to occur, some of the colliding molecules and ions in the flask must contain enough energy to reach the transition state upon collision. Reaching the potential-energy level of the transition state is rather like driving an old car to a mountain pass. Does the car have enough energy to make the top? Or, will it stall and slide back down the mountain? Once you reach the top, which way do you go—back the way you came or on down the other side? Once you are descending the far side, the choice is easy; you can relax and let the car roll to the bottom.

Figure 5.2 shows an energy diagram for the progress of an S_N2 reaction. The potential energy required to reach the transition state forms an energy barrier; it is the point of maximum energy on the graph. For a colliding alkyl halide and nucleophile to reach the transition state, they need a certain minimum amount of energy called the **energy of activation E_{act}**. At the transition state, the molecules find it just as easy to go back to reactants or on to products. But, once over the top, the path of least resistance is that of going to products. The difference between the average potential energy of the reactants and that of the products is the change in enthalpy ΔH for the reaction.

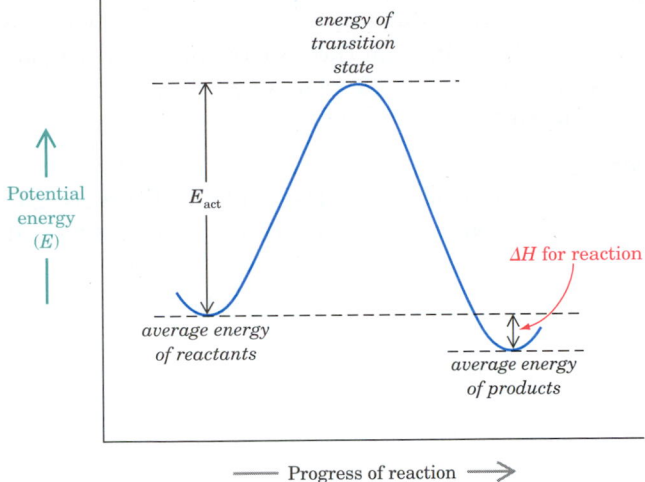

Figure 5.2   Energy diagram of an S$_N$2 reaction.

D. Rate of an S$_N$2 Reaction

Each molecule that undergoes reaction to yield product must pass through the transition state, both structurally and energetically. Since the energies of all the molecules are not the same, a certain amount of time is required for all the molecules present to react. This time requirement gives rise to the **rate of a reaction.** The rate of a chemical reaction is a measure of how fast the reaction proceeds, that is, how fast reactants are consumed and products are formed. **Reaction kinetics** is the term used to describe the study and measurement of reaction rates.

The rate of a reaction depends on many variables, some of which may be held constant for a given experiment (temperature and solvent, for example). In this chapter, we will be concerned primarily with two variables: (1) the concentrations of the reactants, and (2) the structures of the reactants.

Increasing the concentration of reactants undergoing an S$_N$2 reaction increases the rate at which products are formed because it increases the frequency of molecular collisions. Typically, the rate of an S$_N$2 reaction is proportional to the concentrations of both reactants. If all other variables are held constant and the concentration of either the alkyl halide or the nucleophile is doubled, the rate of product formation is doubled. If either concentration is tripled, the rate is tripled.

$$\text{Nu:}^- + \text{R}\ddot{\text{X}}: \longrightarrow \text{RNu} + :\ddot{\text{X}}:^-$$

$$\text{S}_N2 \text{ rate} = k[\text{RX}][\text{Nu:}^-]$$

In this equation, [RX] and [Nu:$^-$] represent the concentrations in moles per liter of the alkyl halide and the nucleophile, respectively. The term k is the proportionality constant, called the **rate constant,** between these concentrations and the measured rate of product formation. The value for k is constant for the same reaction under identical experimental conditions (solvent, temperature, etc.).

What would be the effect on the rate of the S_N2 reaction of CH_3I with CH_3O^- if the concentrations of *both* reactants were doubled and all other variables were held constant?

Solution

If the concentrations of both CH_3I and CH_3O^- were doubled, the rate would quadruple—the reaction would proceed four times as fast.

Because the rate of an S_N2 reaction depends on the concentrations of two particles (RX and Nu:$^-$), the rate is said to be **second order.** The S_N2 reaction is said to follow **second-order kinetics.** (Although the S_N2 reaction is also bimolecular, not all bimolecular reactions are second-order and not all second-order reactions are bimolecular. For example, see Problem 5.55, page 237).

E. Effect of E_{act} on Rate and on Products

The effect of the energies of activation on the relative rates of reaction may be stated simply: *Under the same conditions, the reaction with the lower E_{act} has a faster rate.* The reason for this relationship is that, if less energy is required for reaction, a greater number of molecules have enough energy to react.

Let us consider a case in which one starting material can undergo two different irreversible reactions leading to two different products. (A reaction in which the E_{act} of the reverse reaction is substantially greater than the E_{act} of the forward reaction is exothermic and essentially irreversible.) When the starting material can undergo two such reactions, *the product of the faster reaction (the one with the lower E_{act}) predominates.* We say that the irreversible reaction is under **kinetic control** (contrasted to the equilibrium control of a reversible reaction). Figure 5.3 shows energy curves for two such reactions of one starting material.

The E_{act} is the energy of the transition state relative to that of the reactants. Therefore, relative rates of reaction are related to the energies of the transition states. In competing reactions of the same starting material, *the*

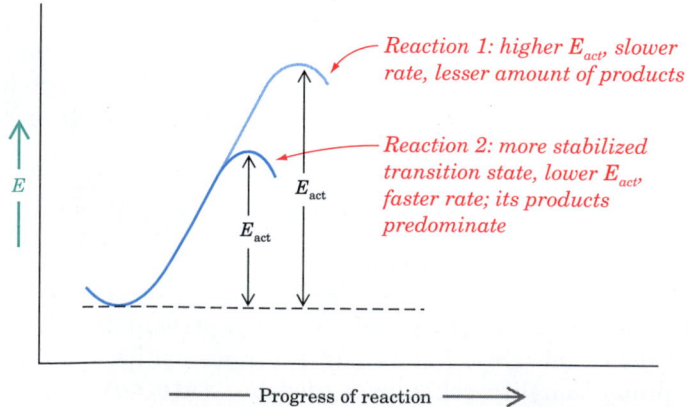

Figure 5.3   In competing reactions of a single starting material, the reaction with the lower E_{act} is the faster one. If the reactions are essentially irreversible, the products of the faster reaction predominate.

reaction with the lower-energy transition state is the faster reaction. From Figure 5.3, it is evident that the reaction with the lower-energy transition state has the lower E_{act}.

A species of low potential energy is more stable than one of high energy. Therefore, we can also say that *the reaction with a more stabilized transition-state structure is the faster reaction.* This concept is useful in analyzing competing reactions to determine which reaction predominates.

F. Effect of Structure on Rate

Reaction kinetics provide a valuable tool for exploring the effects of structure upon reactivity. Consider the following two reactions:

(1) $OH^- +$ CH_3Br $\longrightarrow$ $CH_3OH + Br^-$

bromomethane   methanol

a methyl halide

(2) $OH^- +$ CH_3CH_2Br $\longrightarrow$ $CH_3CH_2OH + Br^-$

bromoethane   ethanol

a 1° alkyl halide

Both are S_N2 reactions and both yield alcohols. The two reactions differ only in the alkyl portions of the alkyl halides. We may ask the following question: "Does the difference in the size of the alkyl group have an effect upon the rate of the S_N2 reaction?" To answer the question, we compare the rates of the two reactions measured under identical conditions. More commonly, however, we use the *relative rates* of the reactions to determine an answer.

$$CH_3Br \xrightarrow{OH^-} CH_3OH \qquad rate_1 = k_1[CH_3Br][OH^-]$$

$$CH_3CH_2Br \xrightarrow{OH^-} CH_3CH_2OH \qquad rate_2 = k_2[CH_3CH_2Br][OH^-]$$

relative rates of reaction of CH_3Br compared to CH_3CH_2Br $= \dfrac{rate_1}{rate_2}$

Under the experimental conditions used in one study, bromomethane undergoes reaction 30 times faster than bromoethane. (If it takes one hour for the bromoethane reaction to reach 50% completion, the bromomethane reaction would take about 1/30 as long, or only two minutes, to reach 50% completion!) We conclude that there is indeed a big difference in how the methyl and ethyl groups affect the rate of the reaction.

In a similar fashion, the relative rates of a variety of S_N2 reactions of alkyl halides have been determined. Table 5.1 shows some average relative rates (compared with the rates of ethyl halides) of S_N2 reactions of alkyl halides.

G. Steric Hindrance in S_N2 Reactions

In S_N2 reactions of the alkyl halides listed in Table 5.1, methyl halides show the fastest rate, followed by primary alkyl halides and then secondary alkyl halides. Tertiary alkyl halides do not undergo S_N2 reactions.

3°RX   2°RX   1°RX CH_3X

increasing rate of S_N2 reaction ➡

Table 5.1 Average relative reaction rates of some alkyl halides under typical S$_N$2 conditions

Alkyl Halide	Relative Rate
CH_3X	30
CH_3CH_2X	1
$CH_3CH_2CH_2X$	0.4
$CH_3CH_2CH_2CH_2X$	0.4
$(CH_3)_2CHX$	0.025
$(CH_3)_3CX$	~0

As the number of alkyl groups attached to the head carbon increases ($CH_3X \rightarrow 1° \rightarrow 2° \rightarrow 3°$), the transition state becomes increasingly crowded with atoms. Consider the example of reactions of alkyl bromides with the methoxide ion (CH_3O^-) as the nucleophile ($CH_3O^- + RBr \rightarrow CH_3OR + Br^-$), shown in Figure 5.4.

Spatial crowding in structures is called **steric hindrance.** When large groups are crowded in a small space, repulsions between groups become severe and therefore the energy of the system is high. In an S$_N$2 reaction, the energy of a crowded transition state is higher than that of a transition state with less steric hindrance. For this reason, the rates of reaction become progressively slower in the series methyl, primary, secondary, and tertiary (see Figure 5.5). The energy of the S$_N$2 transition state of a tertiary alkyl halide is so high relative to other possible reaction paths that the S$_N$2 reaction does not proceed.

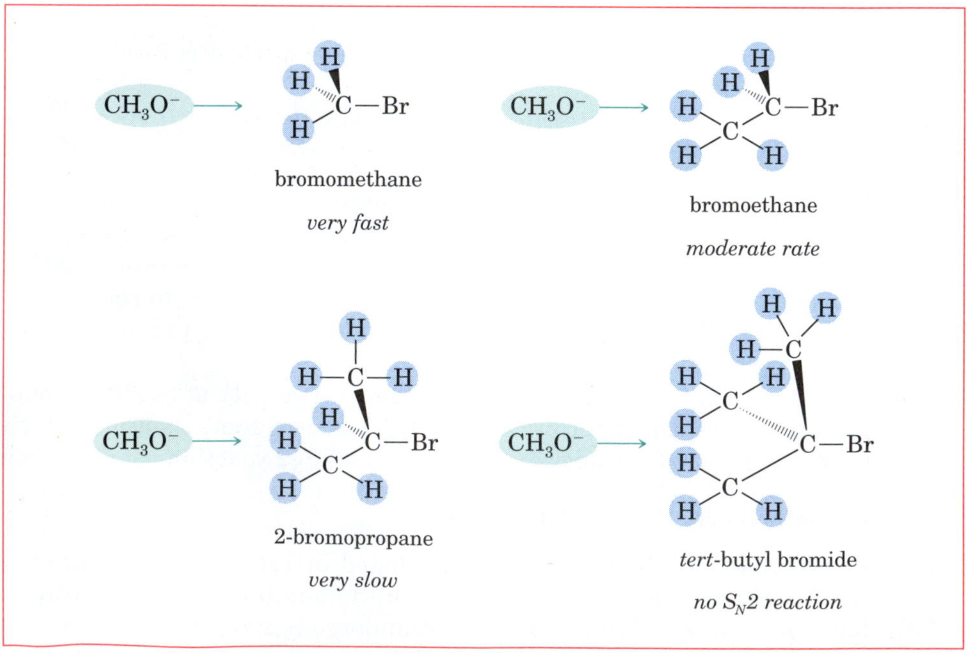

Figure 5.4 Steric hindrance in S$_N$2 reactions.

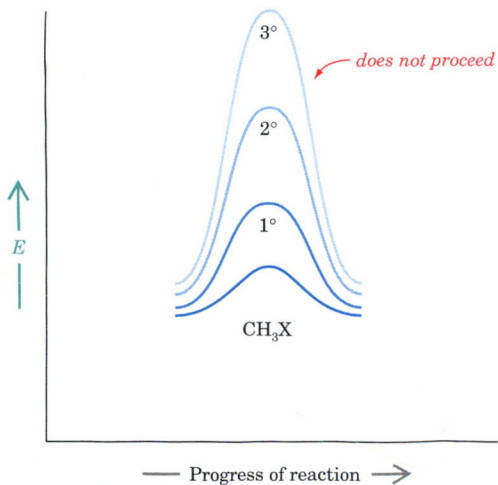

Figure 5.5   Energy diagram for S_N2 reactions of different types of alkyl halides.

F. Kehrmann in 1888 was the first to use the concept of steric hindrance to explain the ratio of products arising from competing reactions. The idea was used again by Victor Meyer in 1894 to explain the relative reactivity of carboxylic acids and alcohols in esterification reactions (see Section 14.6). The term *steric hindrance* was introduced by Rudolf Wegscheider in 1895.

SAMPLE PROBLEM

The S_N2 reaction of neopentyl bromide, $(CH_3)_3CCH_2Br$, with sodium ethoxide, $Na^+ \ ^-OCH_2CH_3$, proceeds about 0.00001 times as fast as the reaction of bromoethane. Explain.

Solution
Although neopentyl bromide is a primary alkyl halide, the alkyl group attached to the head carbon atom is very bulky. The steric hindrance in the transition state is considerable. Therefore, the E_{act} is high and the rate is slow.

STUDY PROBLEM

5.6   Which member of the following pairs of compounds will react more rapidly with a nucleophile in an S_N2 reaction? Explain.

(a) ⟨benzene ring⟩—Br   or   ⟨benzene ring⟩—CH_2CH_2Br

(b) $CH_3CH_2CH_2Cl$ or CH_3CHCH_3 1-chloropentane is faster ∵ less steric hindrance
                               |
                               Cl

(c) ⟨cyclohexane ring⟩—Cl   or   ⟨cyclohexane ring⟩—I   1-iodocyclohexane is faster ∵ I is better nu ∵ more polar.

$$CH_3$$
$$|$$
(d) $CH_3CH_2CH_2CH_2Cl$ or $CH_3C\!-\!Cl$
$$|$$
$$CH_3$$

1-chlorobutane is faster ∴ primary alkyl halide and 3° don't undergo S_N2 rx's

---

Section 5.5
The S$_N$1 Reaction

Because of steric hindrance, *tert*-butyl bromide and other tertiary alkyl halides do not undergo S$_N$2 reactions. Yet, if *tert*-butyl bromide is treated with a nucleophile that is a *very weak base* (such as H_2O or CH_3CH_2OH), substitution products are formed, along with elimination products. Because H_2O or CH_3CH_2OH is also used as the solvent, this type of substitution reaction is sometimes called a **solvolysis reaction** (from *solvent* and *-lysis,* "breaking down" or "loosening").

$$\xrightarrow[25°]{CH_3CH_2OH} (CH_3)_3COCH_2CH_3 + CH_2\!=\!C(CH_3)_2$$

tert-butyl ethyl ether   methylpropene
(80%)   (20%)

$(CH_3)_3CBr$

$$\xrightarrow[25°]{H_2O} (CH_3)_3COH + CH_2\!=\!C(CH_3)_2$$

tert-butyl alcohol   methylpropene
(70%)   (30%)

If tertiary alkyl halides cannot undergo S$_N$2 reactions, how are the substitution products formed? The answer is that tertiary alkyl halides undergo substitution by a different mechanism, called the **S$_N$1 reaction** (substitution, nucleophilic, unimolecular).

The experimental results obtained in S$_N$1 reactions are considerably different from those obtained in S$_N$2 reactions. Typically, if a pure enantiomer of an alkyl halide containing a chiral C—X carbon undergoes an S$_N$1 reaction, predominantly *racemic* substitution products are obtained (not the inverted products observed in an S$_N$2 reaction). It has also been determined that the concentration of nucleophile generally has little effect upon the overall rate of an S$_N$1 reaction. (In contrast, the rate of an S$_N$2 reaction is directly proportional to the concentration of the nucleophile.) To explain these experimental results, we will discuss the mechanism of the S$_N$1 reaction of *tert*-butyl bromide and water. For the moment, we will ignore the elimination product.

as the hydronium ion, H_3O^+

$$(CH_3)_3CBr + H_2O \longrightarrow (CH_3)_3COH + H^+ + Br^-$$

tert-butyl bromide   *tert*-butyl alcohol

A. The S$_N$1 Mechanism

An S$_N$1 reaction is an ionic reaction. The mechanism is complex because of interactions among solvent molecules, RX molecules, and the interm ions formed. For this reason, the mechanism we present here is somewhat idealized.

The S$_N$1 reaction of a tertiary alkyl halide is a *stepwise reaction*. Step 1 is the cleavage of the alkyl halide into a pair of ions: the halide ion and a **carbocation,** an ion in which a carbon atom carries a positive charge. Because S$_N$1 reactions involve ionization, these reactions are aided by polar solvents, such as H$_2$O, that can solvate, and thus stabilize, ions.

Step 1:

$$(CH_3)_3C—\ddot{B}r\colon \longrightarrow \left[(CH_3)_3C\overset{\delta+}{-}\!-\!-\overset{\delta-}{\ddot{B}r}\colon \right] \longrightarrow [(CH_3)_3C^+] \quad + \colon\!\ddot{B}r\colon^-$$

transition state 1   *unstable carbocation*
                                intermediate

Step 2 is the combining of the carbocation with the nucleophile (H$_2$O) to yield the initial product, a protonated alcohol.

Step 2:

$$[(CH_3)_3C^+] + \ H_2\ddot{O}\colon \longrightarrow \left[(CH_3)_3\overset{\delta+}{C}\!-\!-\overset{\delta+}{\underset{\cdot\cdot}{O}}\overset{H}{\underset{}{H}} \right] \longrightarrow (CH_3)_3C\!-\!\overset{H}{\underset{\cdot\cdot}{\overset{|}{O}}}\!\overset{+}{H}$$

nucleophile   *transition state 2*   protonated
                                                        tert-butyl alcohol

The final step in the sequence is the loss of H$^+$ by the protonated alcohol in a rapid, reversible acid–base reaction with the solvent.

Step 3:

$$(CH_3)_3C\overset{H\leftarrow}{\underset{}{\overset{|+}{O}H}} + H_2\ddot{O}\colon \ \rightleftharpoons \ (CH_3)_3C\ddot{O}H \ + H_3\overset{+}{O}\colon$$

excess   *tert*-butyl alcohol

The overall reaction of *tert*-butyl bromide with water is thus actually composed of two separate reactions: the S$_N$1 reaction (ionization followed by combination with the nucleophile) and an acid–base reaction. The steps can be summarized:

$$(CH_3)_3CBr \ \xrightarrow[\text{slow}]{-Br^-} \ [(CH_3)_3C^+] \ \xrightarrow[\text{fast}]{H_2O} \ (CH_3)_3C\overset{+}{O}H_2 \ \xrightarrow[\text{fast}]{-H^+} \ (CH_3)_3COH$$

—means loss of Br$^-$

$\underbrace{\qquad\qquad\qquad\qquad}_{S_N1 \ reaction}$ $\underbrace{\qquad\qquad\qquad}_{acid\text{-}base \ reaction}$

Let us consider the energy diagram for an S$_N$1 reaction (Figure 5.6). Typically, Step 1 (the ionization) has a high E_{act}. It is the *slow* step in the overall process. Enough energy must be supplied to the tertiary alkyl halide to break the C—X sigma bond and yield the carbocation and the halide ion.

The carbocation is an **intermediate** in this reaction, a structure that is formed during the reaction and then undergoes further reaction to products. An intermediate is not a transition state. An intermediate has a finite lifetime; a transition state does not. At the transition state, molecules are undergoing bond-breaking and bond-making. The potential energy of a transition state is a high point on a potential-energy curve. By contrast, an intermediate is a short-lived, reactive product. No bond-breaking or bond-making is occur-

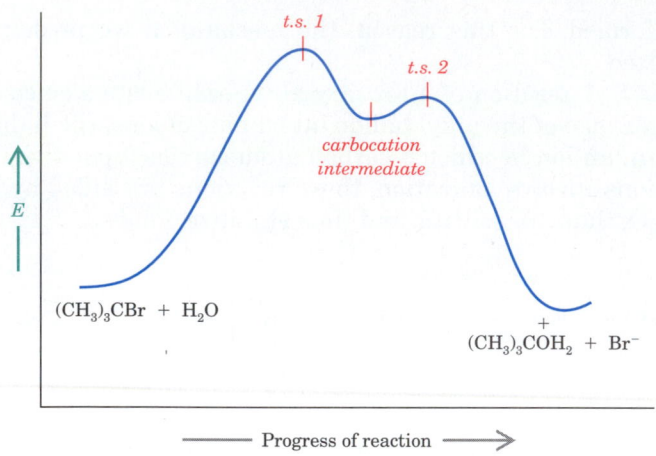

Figure 5.6   Energy diagram for a typical S$_N$1 reaction.

ring in an intermediate. An intermediate is of lower energy than the transition states that surround it (and is at a lower point on the potential-energy curve), although it is of higher energy than the final products. The energy diagram in Figure 5.6 shows a dip for the carbocation formation. The dip is not a large one because the carbocation is a high-energy, reactive species.

Carbocations are extremely important intermediates in chemical reactions. The existence of a carbocation as an intermediate was first postulated in the 1920s. However, general acceptance of carbocations is attributed to the work of three chemists: Sir Christopher Ingold (England), Hans Meerwein (Germany), and Frank Whitmore (United States). Throughout this text, we will explain many reactions using a carbocation intermediate.

Step 2 in the sequence of an S$_N$1 reaction is the reaction of the carbocation with a nucleophile. The two combine in a reaction having a low E_{act} (a fast reaction).

SAMPLE PROBLEM

Write equations for the steps in the S$_N$1 reaction of 2-chloro-2-methylbutane with methanol. (Include the deprotonation step in your answer.)

Solution

Step 1 (ionization):

$$(CH_3)_2\overset{\overset{\displaystyle :\ddot{C}l:}{|}}{C}CH_2CH_3 \longrightarrow [(CH_3)_2\overset{+}{C}CH_2CH_3] + :\ddot{C}l:^-$$

a carbocation

Step 2 (combination with Nu:⁻):

$$[(CH_3)_2\overset{+}{C}CH_2CH_3] + CH_3\ddot{O}H \longrightarrow (CH_3)_2\overset{\overset{\displaystyle \overset{+}{H}\ddot{O}CH_3}{|}}{C}CH_2CH_3$$

a protonated ether

Step 3 (loss of H$^+$ to solvent):

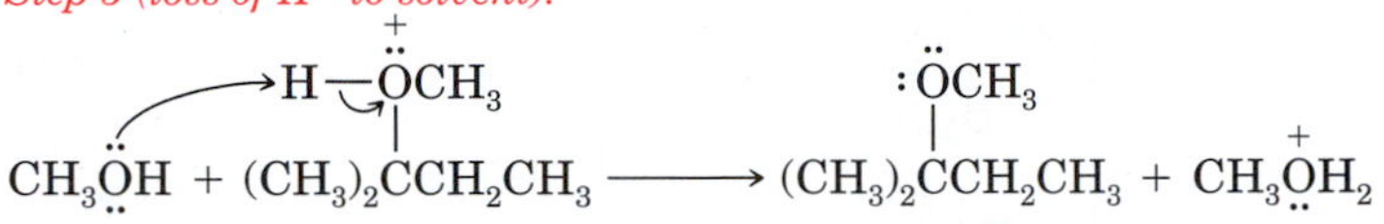

$$CH_3\overset{..}{\overset{..}{O}}H + (CH_3)_2\overset{+}{C}CH_2CH_3 \longrightarrow (CH_3)_2\overset{:\overset{..}{O}CH_3}{\underset{|}{C}}CH_2CH_3 + CH_3\overset{+}{\overset{..}{O}}H_2$$

$$\longrightarrow H\overset{+}{\underset{|}{\overset{..}{O}}}CH_3$$

an ether

STUDY PROBLEM

5.7   Complete the following equations to show the expected solvolysis products.

(a) $(CH_3CH_2)_3CI + H_2O \longrightarrow$

(b) 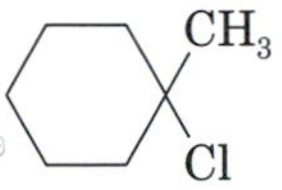 [cyclohexane with CH$_3$ and Cl substituents] $+ CH_3OH \longrightarrow$

(handwritten notes)
1) $(CH_3CH_2)_3C$:I: ⊖ 3) $(CH_3CH_2)_3C-\overset{\oplus}{\overset{H}{N}}_H$ +H$_2$O → $(CH_3CH_2)_3C-OH + H_3O^\oplus$
2) $(CH_3CH_2)_3C-\overset{\oplus}{\overset{H}{O}}H$

B. Stereochemistry of an S$_N$1 Reaction

A carbocation (or **carbonium ion,** as it is also called) is a carbon atom with only three groups attached, instead of the usual four. Because there are only three groups, the bonds to these groups lie in a plane, and the angles between the bonds from the positive carbon are approximately 120°. To attain this geometry, the positive carbon is sp^2-hybridized and has an empty p orbital.

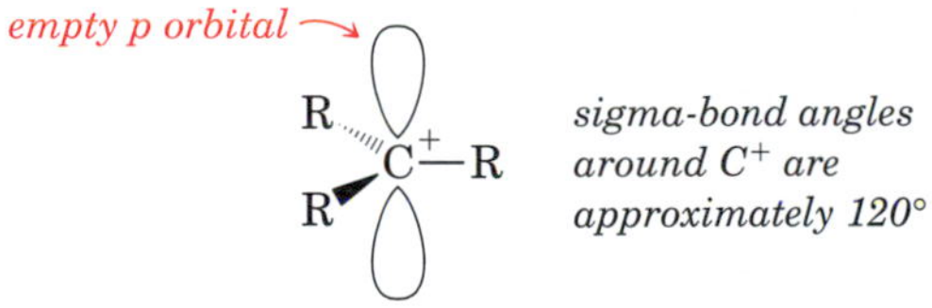

empty p orbital →

$$R\overset{\cdots}{\underset{R}{\overset{|}{C}}}\overset{+}{-}R$$

sigma-bond angles around C$^+$ are approximately 120°

Let us consider the S$_N$1 reaction of a chiral alkyl halide. When one enantiomer of 3-chloro-3,7-dimethyloctane is warmed with aqueous acetone, the halide undergoes solvolysis to 3,7-dimethyl-3-octanol that has been mostly racemized. The S$_N$1 mechanism can be used to explain this racemization. The first step is ionization to a carbocation and a halide ion.

Step 1:

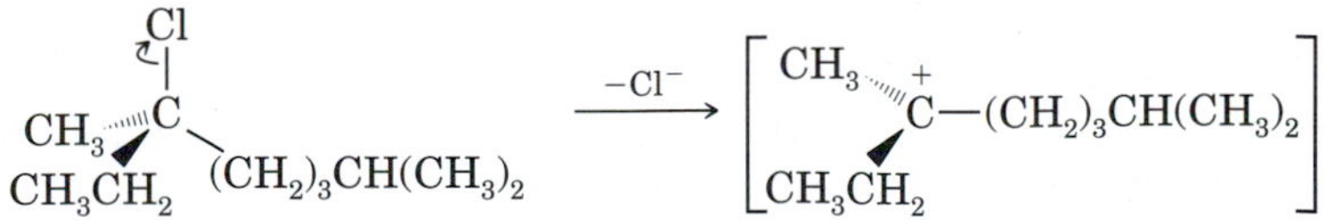

$$CH_3\overset{Cl}{\underset{CH_3CH_2}{\overset{|}{C}}}(CH_2)_3CH(CH_3)_2 \xrightarrow{-Cl^-} \left[\overset{CH_3}{\underset{CH_3CH_2}{C}}\overset{+}{-}(CH_2)_3CH(CH_3)_2 \right]$$

(R)-3-chloro-3,7-dimethyloctane   *planar carbocation*

In the second step, H$_2$O attacks the carbocation to form *two* protonated alcohols. (We show the alcohols *after* deprotonation in the following equations.) If the ionization of the alkyl halide molecule is complete, the H$_2$O molecule can attack the empty p orbital from the top side, as we have shown the structure, and the carbocation yields the *(R)*-enantiomer of the alcohol. However, the H$_2$O molecule can also attack from the bottom side, yielding the *(S)*-enantiomer. Because attack can be from either side, racemization can occur.

Steps 2 and 3 (attack of H_2O to yield ROH_2^+, followed by loss of H^+)

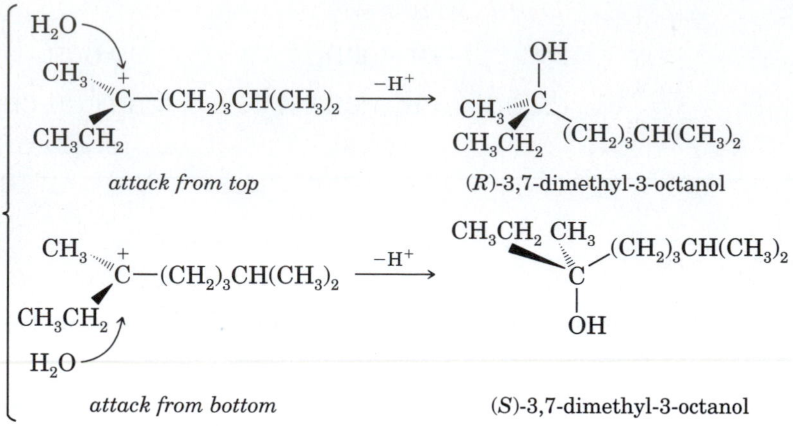

H₂O

CH₃ ⁺C—(CH₂)₃CH(CH₃)₂
CH₃CH₂

attack from top

$\xrightarrow{-H^+}$

OH
CH₃ C
CH₃CH₂ (CH₂)₃CH(CH₃)₂

(*R*)-3,7-dimethyl-3-octanol

CH₃ ⁺C—(CH₂)₃CH(CH₃)₂
CH₃CH₂
H₂O

attack from bottom

$\xrightarrow{-H^+}$

CH₃CH₂ CH₃
C (CH₂)₃CH(CH₃)₂
OH

(*S*)-3,7-dimethyl-3-octanol

STUDY PROBLEM

5.8 Draw structures of the intermediate carbocation and the expected S_N1 product, with stereochemistry where appropriate, for each of the following reactions:

(a)

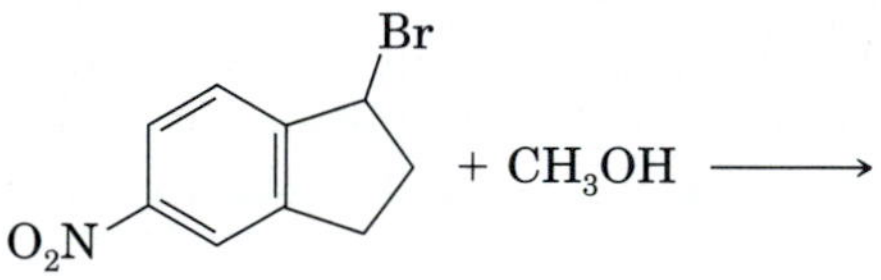

Br

O₂N

+ CH₃OH $\longrightarrow$

(b)

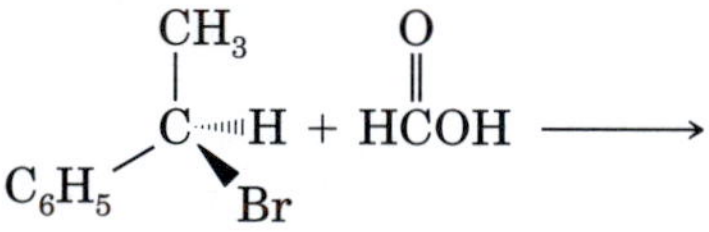

CH₃
C—H
C₆H₅ Br

+ HCOH (O) $\longrightarrow$

(c)

CH₃
Cl

(CH₃)₃C
H

+ H₂O $\longrightarrow$

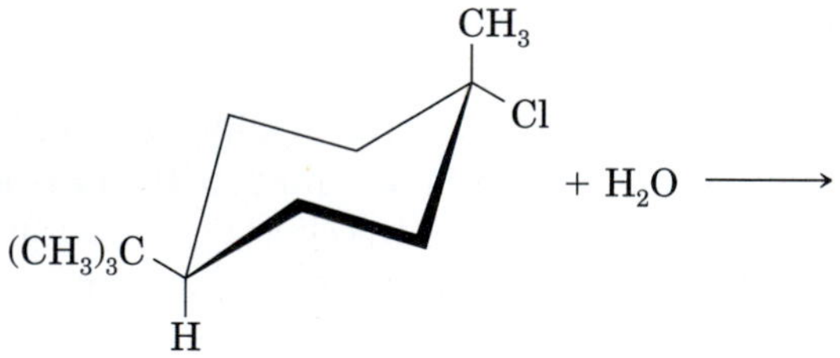

C. Rate of an S_N1 Reaction

We mentioned earlier that the rate of a typical S_N1 reaction does not depend on the concentration of the nucleophile, but depends only on the concentration of alkyl halide.

$$S_N1 \text{ rate} = k[RX]$$

The reason for this behavior is that the reaction between R^+ and $Nu:^-$ is very fast, but the concentration of R^+ is very low. The fast combination of R^+ and $Nu:^-$ occurs only when a carbocation is formed. Therefore, the rate of the overall reaction is determined entirely by how fast RX can ionize and form R^+ carbocations. This ionization step (Step 1 of the overall reaction) is called the **rate-determining step,** or the **rate-limiting step.** In most stepwise reactions, the slowest step in the entire sequence is the rate-determining step because a reaction cannot ordinarily proceed faster than its slowest step does.

An S$_N$1 reaction is characteristically **first order** in rate because the rate is proportional to the concentration of only one reactant (RX). It is a **unimolecular reaction** because only one particle (RX) is involved in the *transition state of the rate-determining step.* (The "1" in S$_N$1 refers to unimolecular.)

Rate-determining step:

$$R \overset{\frown}{-} X \longrightarrow \underset{\substack{\text{transition state 1} \\ \text{(from one particle)}}}{[\overset{\delta+}{R} \text{---} \overset{\delta-}{X}]} \longrightarrow R^+ + X^-$$

D.   Relative Reactivities in S$_N$1 Reactions

Table 5.2 lists the relative rates of reaction of some alkyl bromides under typical S$_N$1 conditions (solvolysis in water). Note that a secondary alkyl halide undergoes substitution 11.6 times faster than a primary alkyl halide under these conditions, while a tertiary alkyl halide undergoes reaction a million times faster than a primary halide!

Methyl:

$$CH_3Br \quad + H_2O \xrightarrow[S_N1]{\text{negligible rate}} CH_3OH \quad + Br^- + H^+$$

1°:

$$CH_3CH_2Br \quad + H_2O \xrightarrow[S_N1]{\text{negligible rate}} CH_3CH_2OH \quad + Br^- + H^+$$

2°:

$$(CH_3)_2CHBr + H_2O \xrightarrow[S_N1]{\text{very slow}} (CH_3)_2CHOH + Br^- + H^+$$

3°:

$$(CH_3)_3CBr \quad + H_2O \xrightarrow[S_N1]{\text{fast}} (CH_3)_3COH \quad + Br^- + H^+$$

The rates at which different alkyl halides undergo S$_N$1 reaction depend on the relative energies of activation leading to the different carbocations. In this reaction, the energy of the transition state leading to the carbocation is largely determined by the stability of the carbocation, which is partially formed in the transition state. We say that the transition state has **carbocation character.** Therefore, the reaction leading to a more stabilized, lower-energy carbocation has the faster rate. A tertiary alkyl halide yields a carbocation that is more stabilized than the carbocation from a methyl halide, a

Table 5.2	Relative reaction rates of some alkyl bromides under typical S$_N$1 conditions	
	CH$_3$Br	1.00^a
	CH$_3$CH$_2$Br	1.00^a
	(CH$_3$)$_2$CHBr	11.6
	(CH$_3$)$_3$CBr	1.2×10^6

^{a} The observed reaction of the methyl or primary bromide probably occurs by a different path (S$_N$2, not S$_N$1).

primary alkyl halide, or a secondary alkyl halide. Consequently, the reaction of a tertiary halide has the fastest rate.

E. Stability of Carbocations

A carbocation is unstable and quickly undergoes further reaction. However, we may still speak of *relative stabilities* of carbocations. The different types of carbocations that concern us here are the **methyl cation** (the carbocation resulting from ionization of a methyl halide), **primary carbocations** (from 1° alkyl halides, **secondary carbocations** (from 2° alkyl halides), and **tertiary carbocations** (from 3° alkyl halides). Some examples follow:

$$^{+}CH_3 \qquad CH_3\overset{+}{C}H_2 \qquad (CH_3)_2\overset{+}{C}H \qquad (CH_3)_3C^{+}$$

$$\text{methyl} \qquad\quad 1° \qquad\qquad\quad 2° \qquad\qquad\quad 3°$$

What increases the stability of a positively charged carbon atom? The answer is: anything that can *disperse the positive charge*. In alkyl cations, the principal phenomenon that disperses the positive charge is the **inductive effect**, a term used to describe the polarization of a bond by a nearby electronegative or electropositive atom. In a carbocation, the positively charged carbon is an electropositive center. The electron density of the sigma bonds is shifted toward the positive carbon. We will use arrows in place of line bonds to show the direction of this attraction.

Sigma-bond e^- are drawn toward the positive charge.

This shift of electron density creates a partial positive charge on the adjacent atoms. These partial positive charges, in turn, polarize the next sigma bonds. In this way, the positive charge of the carbocation is somewhat dispersed, and the carbocation is stabilized to some extent.

All the atoms help disperse the positive charge.

An alkyl group contains more atoms and electrons than does a hydrogen atom. When there are more alkyl groups attached to a positively charged carbon atom, there are more atoms that can help share the positive charge and help stabilize the carbocation. Because of the lack of stabilization, methyl and primary halides do not normally form carbocations.

$$^{+}CH_3 \qquad CH_3\overset{+}{C}H_2 \qquad (CH_3)_2\overset{+}{C}H \qquad (CH_3)_3C^{+}$$

$$\text{methyl} \qquad\quad 1° \qquad\qquad\quad 2° \qquad\qquad\quad 3°$$

increasing carbocation stability; increasing S_N1 rate of RX →

STUDY PROBLEM

5.9 List the following carbocations in order of increasing stability (least stable first):

(a) ⬡ +   **(b)** ⬡—$\overset{+}{C}H_2$ **(c)** ⬡—$\overset{+}{C}(CH_3)_2$

Another theory proposed to explain the relative stabilities of carbocations is **hyperconjugation,** the partial overlap of an sp^3–s orbital (a C—H bond) with the empty p orbital of the positively charged carbon.

An ethyl cation has only three C—H bonds that can overlap the empty p orbital, but the *tert*-butyl cation has *nine* C—H bonds that can help disperse the charge in this fashion. Therefore, a tertiary carbocation is stabilized by a greater dispersal of the positive charge.

F. Rearrangements of Carbocations

The following secondary alkyl bromide can undergo solvolysis with methanol. However, in addition to the expected product, a *second* substitution product is observed. (The total yield of substitution products is not 100% because alkenes are also formed.)

Let us examine the intermediate carbocation in this reaction more carefully. The expected carbocation is a *secondary* carbocation.

A secondary carbocation is of higher energy than a tertiary carbocation by about 11 kcal/mol. The energy of this particular carbocation can be lowered by the shift of a methyl group, with its bonding electrons, from the adjacent carbon atom. The result is the rearrangement of the secondary carbocation to a more stable tertiary carbocation.

1,2-Shift of a methyl group with its bonding electrons:

$$CH_3-\underset{\underset{CH_3}{|}}{\overset{\overset{CH_3}{|}}{C}}-\overset{+}{C}HCH_3 \longrightarrow CH_3-\underset{\underset{CH_3}{|}}{\overset{\overset{CH_3}{|}}{\overset{+}{C}}}-CHCH_3$$

a 2° carbocation   *a more stable 3° carbocation*

The shift of an atom or of a group from an adjacent carbon is called a **1,2-shift.** (The numbers 1,2 used in this context have nothing to do with nomenclature numbers, but refer to the positive carbon and the adjacent atom.) The 1,2-shift of a methyl group is called a **methyl shift,** or a **methide shift.** (The *-ide* suffix is sometimes used because ¯:CH_3 is an anion. However, the 1,2-shift is a concerted reaction step, and no methide anion is actually formed.) The current electronic concepts of alkyl and hydrogen atom rearrangements were proposed by Frank Whitmore of Pennsylvania State College in 1932.

The presence of both secondary and tertiary carbocations in solution leads to the two observed products, the so-called "normal" product and the **rearrangement product,** a product in which the skeleton or the position of the functional group is different from that of the starting material.

$$[(CH_3)_3\overset{+}{C}CHCH_3] \xrightarrow[-H^+]{CH_3OH} (CH_3)_3C\underset{\underset{CH_3}{|}}{\overset{\overset{OCH_3}{|}}{C}}HCH_3$$

2° carbocation   2-methoxy-3,3-dimethylbutane

$$[(CH_3)_2\overset{+}{C}CH(CH_3)_2] \xrightarrow[-H^+]{CH_3OH} (CH_3)_2\underset{\underset{CH_3}{}}{\overset{\overset{OCH_3}{|}}{C}}CH(CH_3)_2$$

3° carbocation   2-methoxy-2,3-dimethylbutane

If an alkyl group, an aryl group, or a hydrogen atom (each with its bonding electrons) on an adjacent carbon atom can shift and thereby create a more stable carbocation, rearrangement can occur. Rearrangement can also occur when a pair of carbocations are of equivalent stabilities. The extent of rearrangement that will be observed in a reaction is often hard to predict and depends on a number of factors, including the relative stabilities of the carbocations in question and the reaction conditions (solvent, etc.). The following rearrangements exemplify 1,2-shifts and the formation of more-stable carbocations.

A methide shift:

$$CH_3-\underset{\underset{CH_3}{|}}{\overset{\overset{CH_3}{|}}{C}}-\overset{+}{C}HCH_2CH_3 \longrightarrow CH_3-\underset{\underset{CH_3}{|}}{\overset{\overset{CH_3}{|}}{\overset{+}{C}}}-CHCH_2CH_3$$

a 2° carbocation   *a more stable 3° carbocation*

A hydride (H :¯) shift:

$$CH_3-\underset{\underset{CH_3}{|}}{\overset{\overset{H}{|}}{C}}-\overset{+}{C}HCH_3 \longrightarrow CH_3-\underset{\underset{CH_3}{|}}{\overset{+}{C}}-CH_2CH_3$$

a 2° carbocation   *a more stable 3° carbocation*

STUDY PROBLEMS

5.10 Write a formula for the expected rearranged carbocation, if any, from each of the following carbocations:

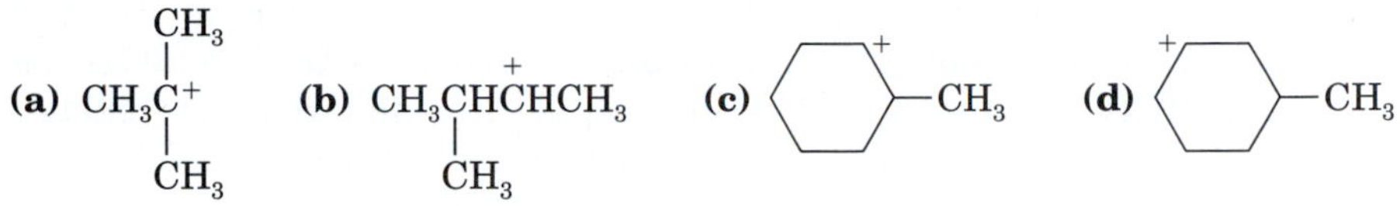

(a) $CH_3\overset{CH_3}{\underset{CH_3}{\overset{|}{\underset{|}{C^+}}}}$ (b) $CH_3\overset{+}{C}HCHCH_3$ with CH_3 below   (c) [cyclohexane ring with + and $-CH_3$]   (d) [cyclohexane ring with + and $-CH_3$]

5.11 Write flow equations for the solvolysis mechanism of the following alkyl halides in aqueous acetic acid. (Show intermediate carbocations in your answers.)

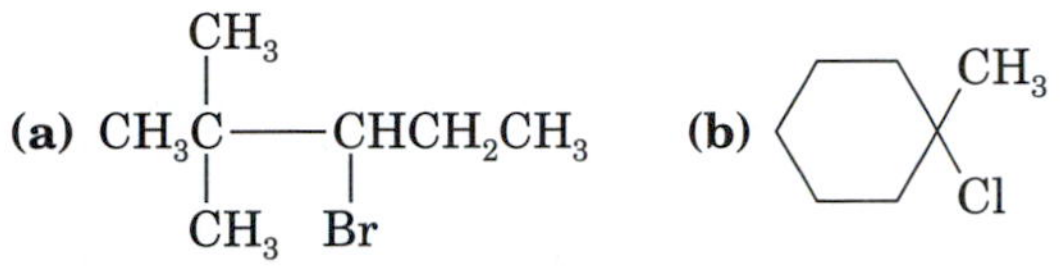

(a) $CH_3\overset{CH_3}{\underset{CH_3}{\overset{|}{\underset{|}{C}}}}\!\!-\!\!\overset{}{\underset{Br}{\overset{|}{CH}}}CH_2CH_3$ (b) [cyclohexane ring with CH_3 and Cl]

G. Summary of S_N1 and S_N2 Mechanisms

Substitution reactions of alkyl halides with nucleophiles can occur by an S_N1 path or by an S_N2 path. *Methyl, primary,* and *secondary alkyl halides* react principally by an S_N2 path, as shown in Figure 5.7. The rate of an S_N2 reac-

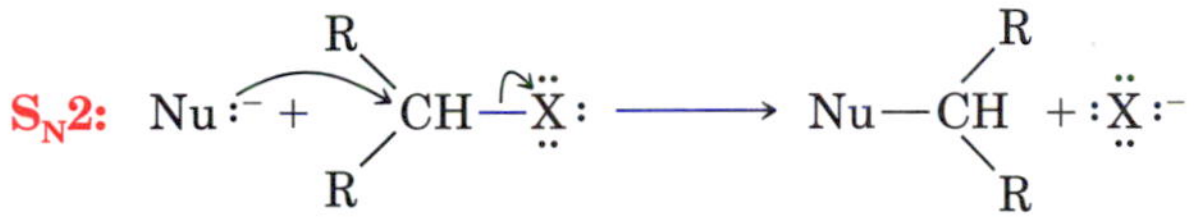

$$S_N2:\quad Nu\!:^- + \overset{R}{\underset{R}{>}}CH\!-\!\ddot{X}\!: \longrightarrow Nu\!-\!\overset{R}{\underset{R}{CH}} + :\ddot{X}:^-$$

A concerted back-side attack
A bimolecular reaction: $Nu:^-$ and R_2CHX both involved in the transition state
Second order in rate: rate is proportional to the concentrations of both $Nu:^-$ and R_2CHX
Stereochemistry: inversion of configuration
Relative rates: $CH_3X > 1° RX > 2° RX$

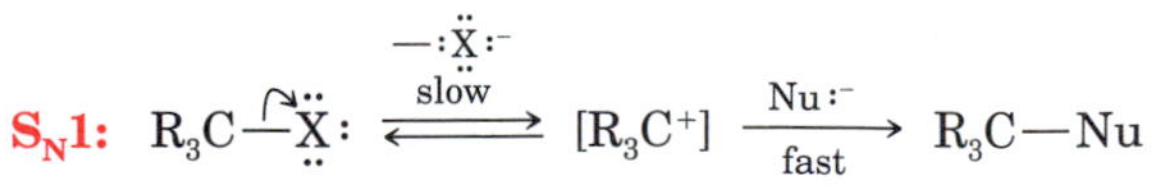

$$S_N1:\quad R_3C\!-\!\ddot{X}: \overset{\overset{-:\ddot{X}:^-}{slow}}{\underset{}{\rightleftharpoons}} [R_3C^+] \xrightarrow[fast]{Nu:^-} R_3C\!-\!Nu$$

An ionization reaction: sometimes accompanied by rearrangement, followed by combination with $Nu:^-$
A unimolecular reaction: only R_3CX is involved in the transition state of the rate-determining step
First order in rate: rate is proportional to the concentration of R_3CX only, because its ionization is the slow, rate-determining step
Stereochemistry: racemization
Relative rates: $3° RX > 2° RX$

Figure 5.7 Summary of S_N1 and S_N2 mechanisms for substitution reactions of alkyl halides.

tion is enhanced by increasing nucleophilicity of the attacking species. Typical good nucleophiles are $^-$OH, $^-$OR, and $^-$CN.

Increasing hindrance around the halogenated carbon decreases the rate of S_N2 reaction. *Tertiary alkyl halides* are too hindered to undergo reaction by an S_N2 path, but can undergo reaction by an S_N1 path (by way of a carbocation intermediate) with a nucleophile such as H_2O or ROH (see Figure 5.7). Methyl and primary alkyl halides do not undergo S_N1 reactions at all. Secondary alkyl halides react slowly by this pathway.

Section 5.6

Substitution Reactions of Allylic Halides and Benzylic Halides

Two special types of halides behave differently in S_N1 and S_N2 reactions from the alkyl halides we have been discussing. These are the **allylic halides** and the **benzylic halides.**

$CH_2=CH-CH_2-$ $CH_2=CHCH_2Cl$

the allyl group   3-chloro-1-propene
                      (allyl chloride)

〈benzene ring〉$-CH_2-$ 〈benzene ring〉$-CH_2Br$

the benzyl group   benzyl bromide

An atom or group that is attached to the carbon atom *adjacent to one of the sp^2-hybridized carbon atoms* is said to be in the **allylic position** or the **benzylic position.** The halogen atoms in our previous examples and also in the following examples are in allylic or benzylic positions.

CH_3 $CH=CHCH$ CH_3 〈benzene ring〉$-CH$ $CH_2CH_2CH(CH_3)_2$
              | |
              Cl   Br

4-chloro-2-pentene   1-bromo-4-methyl-1-phenylpentane

an allylic chloride   *a benzylic bromide*

STUDY PROBLEM

5.12 Classify each of the following compounds as a vinylic halide, an allylic halide, a benzylic halide, or an aryl halide:

(a) CH_3-〈benzene ring〉$-CH=CHCH_2Br$

(b) CH_3-〈benzene ring〉$-CH=CCH_3$ with Br on the C

(c) CH_3-〈benzene ring with Br〉$-CH=CHCH_3$

(d) $BrCH_2-$〈benzene ring〉$-CH=CHCH_3$

A. S_N1 Reactions

Most primary alkyl halides undergo substitution by the S_N2 path exclusively and do not undergo S_N1 reactions. However, a primary allylic halide or benzylic halide is very reactive in both S_N1 and S_N2 reactions. Table 5.3 lists the relative reactivities of some halides under typical S_N1 conditions. You can see from the table that an allyl halide is more than 30 times more reactive than an ethyl halide, and a benzyl halide is almost 400 times as reactive. If *two* phenyl groups are present, the halide is 100,000 times as reactive!

increasing rate of S_N1 reaction

$$CH_2{=}CHCH_2Cl + H_2O \longrightarrow CH_2{=}CHCH_2OH + Cl^- + H^+$$

allyl chloride

2-propen-1-ol
(allyl alcohol)

$$\langle\bigcirc\rangle{-}CH_2Cl \quad + H_2O \longrightarrow \langle\bigcirc\rangle{-}CH_2OH + Cl^- + H^+$$

benzyl chloride

benzyl alcohol

a lachrymator:
a compound that causes
tears to flow

The reason for the enhanced reactivity of these two types of halides in an S_N1 reaction lies in the *resonance stabilization of the carbocation and of the transition state leading to the carbocation*. Carbocations are stabilized by dispersal of the positive charge. *Inductive stabilization* involves dispersal of the positive charge through *sigma bonds*. We have used the inductive effect to explain the relative stabilities of 1°, 2°, and 3° carbocations. *Resonance stabilization* involves the dispersal of the positive charge by *pi bonds*.

Let us consider the S_N1 reaction of allyl chloride with H_2O:

$$CH_2{=}CHCH_2{-}\ddot{C}l\colon \xrightarrow[S_N1]{} [CH_2{=}CHCH_2]^+ \xrightarrow[-H^+]{H_2\ddot{O}} CH_2{=}CHCH_2\ddot{O}H$$

allyl cation

Recall from Section 2.9 that structures differing only in the position of pi electrons are resonance structures. If resonance structures can be drawn for a molecule or ion, the resonance hybrid (the real structure) is more stable than if delocalization of electrons or electrical charges could not occur. The two resonance structures for the allyl cation are equivalent in structure and bonding;

Table 5.3   Relative reaction rates of some organic halides under typical S_N1 conditions

Halide	Relative Rate
CH_3CH_2X	1.0^a
$CH_2{=}CHCH_2X$	33
$C_6H_5CH_2X$	380
$(C_6H_5)_2CHX$	$\sim10^5$

a The observed reaction probably proceeds by an S_N2 path.

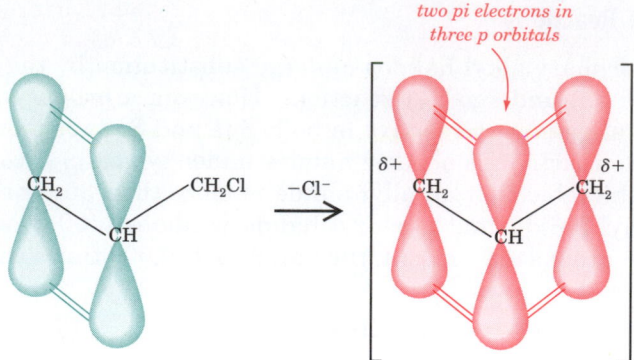

Figure 5.8   Formation of the allyl cation from allyl chloride in an S_N1 reaction.

therefore, they have the same energy content and contribute equally to the structure of the actual allyl cation. Because the allyl cation is resonance-stabilized, the energy of the transition state leading to its formation is relatively low. Consequently, the rate of its S_N1 reaction is fairly fast. Figure 5.8 shows the formation of the allyl cation and its bonding.

$$[CH_2\!\!=\!\!CH\!\!-\!\!\overset{+}{C}H_2 \longleftrightarrow \overset{+}{C}H_2\!\!-\!\!CH\!\!=\!\!CH_2]$$

resonance structures for the allyl cation
(equal contributors)

Both terminal (end) carbons in the allyl cation have an equal amount of positive charge. Which atom is attacked by the nucleophile? The answer is: either of them! Let us use another allylic system to illustrate this fact. The S_N1 reaction of 1-chloro-2-butene with H_2O in aqueous acetone leads to two products. These two products arise from attack of H_2O at either of the two partially positive carbon atoms. We say that the rearranged product is the result of **allylic rearrangement.**

$$CH_3CH\!\!=\!\!CHCH_2Cl \xrightarrow[S_N1]{-Cl^-} [CH_3CH\!\!=\!\!CH\!\!-\!\!\overset{+}{C}H_2 \longleftrightarrow CH_3\overset{+}{C}H\!\!-\!\!CH\!\!=\!\!CH_2]$$

1-chloro-2-butene

$$\Big\downarrow {\substack{H_2O \\ -H^+}}$$

$$CH_3CH\!\!=\!\!CHCH_2 \qquad\qquad CH_3CHCH\!\!=\!\!CH_2$$
$$\ \ \ \ \ |\ |$$
$$\ \ \ \ \ OH\ OH$$

2-buten-1-ol (56%)   3-buten-2-ol (44%)

Benzyl halides also show enhancement of the S_N1 rate because of resonance stabilization of the transition state leading to the carbocation. In this case, the pi electrons in the aromatic pi cloud of the benzene ring help disperse the charge.

$$\bigcirc\!\!-\!\!CH_2Cl \xrightarrow[S_N1]{-Cl^-} \Big[\bigcirc\!\!-\!\!\overset{+}{C}H_2\Big] \xrightarrow[-H^+]{H_2O} \bigcirc\!\!-\!\!CH_2OH$$

benzyl chloride   benzyl cation   benzyl alcohol

We usually symbolize the aromatic pi cloud of benzene by a circle in the ring. However, in discussions of delocalization of pi electrons, Kekulé formulas are more convenient. With Kekulé formulas, we can count the pi electrons in the ring and readily see which atoms are electron-deficient. Note the similarity between the resonance structures for the benzyl cation and those for the allyl cation. The benzyl cation has five resonance structures similar to allylic resonance structures.

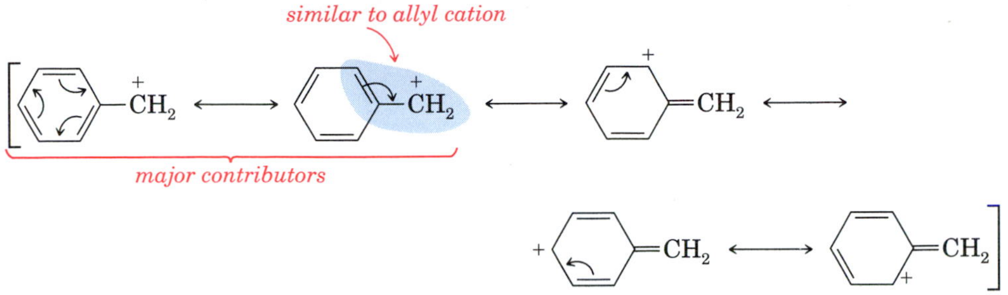

The first two resonance structures shown are the major contributors because these structures have aromatic stabilization. Therefore, the most positive carbon in the intermediate is the benzyl carbon. This is the carbon attacked by the nucleophile.

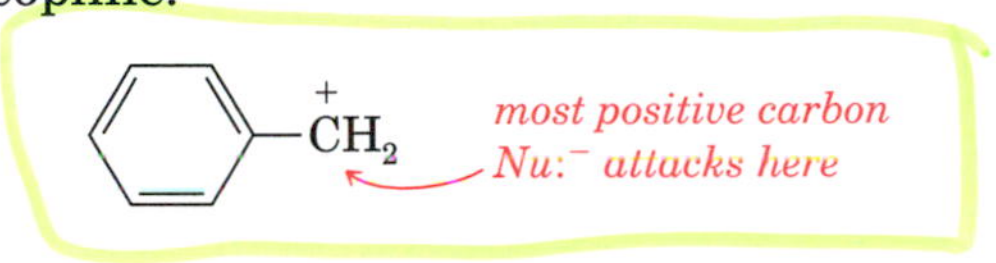

STUDY PROBLEM

5.13   Predict the products of the solvolysis in water of:

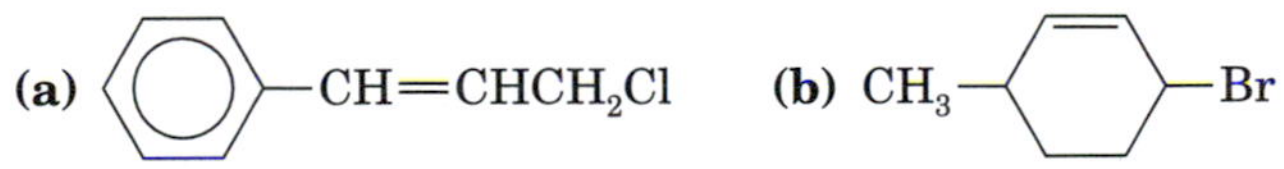

(a) C_6H_5—CH=CHCH$_2$Cl   (b) CH$_3$—⟨ ⟩—Br

B. S$_N$2 Reactions

Allylic halides and benzylic halides also undergo S$_N$2 reactions at faster rates than primary alkyl halides or even methyl halides. Table 5.4 lists the average relative rates of some halides in a typical S$_N$2 reaction.

The reason for the greater S$_N$2 reactivity of allylic and benzylic halides is that the allylic pi bond or the aromatic pi cloud reduces the energy of the

Table 5.4 Average relative S$_N$2 rates for some organic halides	
Halide	Relative Rate
CH$_3$X	30
CH$_3$CH$_2$X	1
(CH$_3$)$_2$CHX	0.025
CH$_2$=CHCH$_2$X	40
C$_6$H$_5$CH$_2$X	120

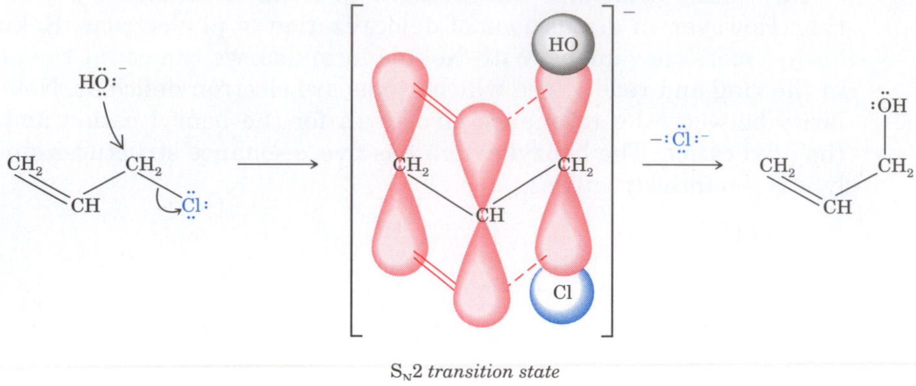

S_N2 *transition state*

Figure 5.9   Stabilization of the transition state in an S_N2 reaction of allyl chloride.

transition state of an S_N2 reaction. In the transition state, the carbon undergoing reaction changes from the sp^3-hybrid state to the sp^2-hybrid state and has a p orbital. This p orbital forms partial bonds with both the incoming nucleophile and the leaving group. The entire grouping of atoms carries a negative charge. Adjacent p orbitals, as in an allylic or benzylic group, undergo partial overlap with the transitional p orbital. In this way, adjacent p orbitals help delocalize the negative charge and thus lower the energy of the transition state. Figure 5.9 shows the p orbitals of the allylic case; the benzylic case is similar.

For increased stabilization to occur in either S_N1 or S_N2 reactions of compounds with pi systems, the pi system must be *adjacent* to the reacting carbon. If it is farther away, it cannot overlap and cannot help stabilize the transition state. In the S_N2 reaction, the pi system must be adjacent to, and align its p orbital with the orbital of the leaving group and nucleophile.

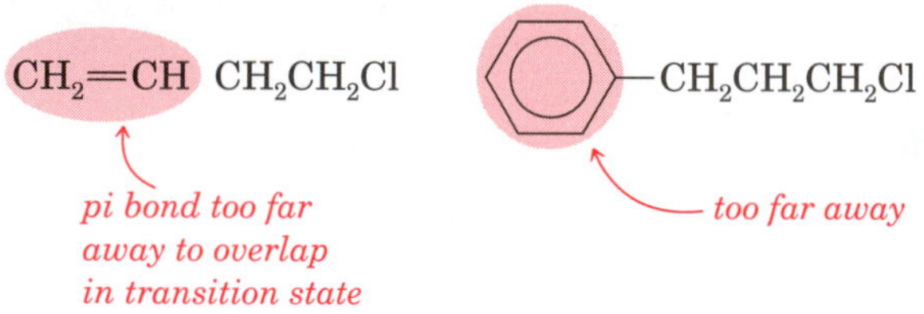

$$CH_2{=}CH \quad CH_2CH_2Cl \qquad \text{(benzene ring)}{-}CH_2CH_2CH_2Cl$$

pi bond too far away to overlap in transition state   *too far away*

SAMPLE PROBLEM

Which of the following compounds exhibit enhanced reactivities in S_N1 and S_N2 reactions because of resonance-stabilization or partial p-orbital overlap?

(a) $CH_3CH{=}CHCHBrCH{=}CH_2$ **(b)** $CH_3CH{=}CHCH_2CHBrCH_2CH_3$

(c) $C_6H_5CH{=}CHCH_2I$ **(d)** $CH_2{=}CHCH_2CHBrCH{=}CH_2$

Solution

Compounds **(a)**, **(c)**, and **(d)**, in which the halogen atom is allylic. In **(a)**, the Br atom is allylic with respect to *two* carbon–carbon double bonds; consequently, this halide is particularly reactive.

Section 5.7
The E1 Reaction

A carbocation is a high-energy, unstable intermediate that quickly undergoes further reaction. One way a carbocation can reach a stable product is by combining with a nucleophile. This, of course, is the S_N1 reaction. However, there is an alternative: the carbocation can *lose a proton to a base in an elimination reaction,* an **E1 reaction** in this case, and become an alkene.

Substitution (S_N1):

$$(CH_3)_3CBr \xrightarrow{-Br^-} [(CH_3)_3C^+] \xrightarrow[-H^+]{H_2O} (CH_3)_3COH$$

tert-butyl bromide   *tert*-butyl cation   *tert*-butyl alcohol

Elimination (E1):

$$(CH_3)_3CBr \xrightarrow{-Br^-} \left[(CH_3)_2\overset{+}{C}-CH_2\right] \longrightarrow (CH_3)_2C{=}CH_2 + H_3\overset{+}{O}{:}$$

tert-butyl cation   methylpropene

The first step in an **E1 reaction** is identical to the first step in an S_N1 reaction: the ionization of the alkyl halide. This is the slow step and thus the rate-determining step of the overall reaction. Like an S_N1 reaction, a typical E1 reaction shows first-order kinetics, with the rate of the reaction dependent on the concentration of only the alkyl halide. Since only one reactant is involved in the transition state of the rate-determining step, the **E1 reaction**, like the S_N1 reaction, is unimolecular.

Step 1 (slow):

$$(CH_3)_3C\overset{..}{\underset{..}{Br}}{:} \longrightarrow \left[(CH_3)_3\overset{\delta+}{C}{-}{-}{-}{-}\overset{\delta-}{\overset{..}{\underset{..}{Br}}}{:}\right] \longrightarrow \left[\begin{matrix}H_3C\\H_3C\end{matrix}\overset{+}{C}{-}CH_3\right] + {:}\overset{..}{\underset{..}{Br}}{:}^-$$

transition state 1   *carbocation intermediate*

In the second step of an elimination reaction, the base removes a proton from a carbon that is *adjacent to the positive carbon.* The electrons of that carbon–hydrogen sigma bond shift toward the positive charge, the adjacent carbon rehybridizes from the sp^3 state to the sp^2 state, and an alkene is formed.

Step 2 (fast):

$$\left[\begin{matrix}H_3C\\H_3C\end{matrix}\overset{+}{C}{\overset{H}{\underset{|}{-}}}CH_2\right] + H_2\overset{..}{O}{:} \longrightarrow \left[\begin{matrix}H_3C\\H_3C\end{matrix}\overset{\delta+}{C}{=}{=}{=}CH_2 \quad \overset{\delta+}{H}{-}{-}{-}\overset{..}{O}H_2\right] \longrightarrow \begin{matrix}H_3C\\H_3C\end{matrix}C{=}CH_2 + H_3\overset{+}{O}{:}$$

base   *transition state 2*   *an alkene*

Because an E1 reaction, like an S_N1 reaction, proceeds through a carbocation intermediate, it is not surprising that tertiary alkyl halides undergo this reaction far more rapidly than other alkyl halides. E1 reactions of alkyl halides occur under the same conditions as S_N1 reactions (polar solvent, very weak base, etc.). Therefore, S_N1 and E1 reactions are competing reactions. For this reason, E1 reactions of alkyl halides are relatively unimportant. You will learn in Chapter 7 that E1 reactions of alcohols, however, are quite important. We will discuss these reactions in greater detail at that time.

STUDY PROBLEM

5.14 Some tertiary alkyl halides yield mixtures of alkenes as well as a substitution product when they are subjected to S_N1 conditions. Predict all likely products from the reactions of 2-bromo-2-methylbutane with ethanol under S_N1 conditions.

Section 5.8
The E2 Reaction

The most useful elimination reaction of alkyl halides is the **E2 reaction** (bi-molecular elimination). E2 reactions of alkyl halides are favored by the use of strong bases, such as ^-OH or ^-OR, and high temperatures. Typically, an E2 reaction is carried out by heating the alkyl halide with K^+ ^-OH or Na^+ $^-OCH_2CH_3$ in ethanol.

$$\underset{\substack{\text{2-bromopropane}\\\text{(isopropyl bromide)}}}{\overset{\overset{\displaystyle Br}{|}}{CH_3CHCH_3}} + CH_3CH_2O^- \xrightarrow[\text{E2}]{\overset{CH_3CH_2OH}{\text{heat}}} CH_3CH=CH_2 + CH_3CH_2OH + Br^-$$

$$\underset{\text{propene}}{}$$

The E2 reaction does not proceed by way of a carbocation intermediate, but is a **concerted reaction**—that is, it occurs in one step, just as an S_N2 reaction does.

$$R\ddot{O}:^- + H-CH_2-CHCH_3 \overset{:\ddot{Br}:}{\longrightarrow} R\ddot{O}H + CH_2=CHCH_3 + :\ddot{Br}:^-$$

(1) The base is forming a bond with the hydrogen.
(2) The C—H electrons are forming the pi bond.
(3) The bromine is departing with the pair of electrons from the carbon–bromine sigma bond.

The preceding equation shows the mechanism with arrows representing "electron-pushing." The structure of the transition state in this one-step reaction follows:

$$\left[\begin{array}{c} R\ddot{O}:^{\delta-} \\ \vdots \\ H \\ | \\ CH_2 \cdots CHCH_3 \\ | \\ :\ddot{Br}:^{\delta-} \end{array}\right]$$

E2 transition state

In E2 reactions, as in E1 reactions, tertiary alkyl halides undergo reaction fastest, and primary alkyl halides react slowest. (When treated with a base, primary alkyl halides usually undergo substitution so readily that little alkene is formed unless the base is very strong and bulky.)

$$1°RX \qquad 2°RX \qquad 3°RX$$

increasing rate of E2 →

A. Kinetic Isotope Effect in E2 Reactions

One piece of experimental evidence that supports our understanding of the E2 mechanism is the difference in the rates of elimination of deuterated and nondeuteriated alkyl halides. A difference in the rates of reaction between compounds containing different isotopes is called a **kinetic isotope effect.**

Deuterium (^2_1H, or D) is an isotope of hydrogen with a nucleus containing one proton and one neutron. More energy (1.2 kcal/mol) is required to break a C—D bond than to break a C—H bond. We have postulated that the breaking of the C—H bond is an integral part of the rate-determining step (the *only* step) of an E2 reaction. What happens when the H that is eliminated is replaced by D? The stronger CD bond requires more energy to be broken. For this reason, the E_{act} is greater and the rate for the elimination reaction is slower.

When the following 2-bromopropanes are subjected to an E2 reaction with $CH_3CH_2O^-$ as the base, it has been observed that the deuterated compound undergoes reaction at *one-seventh* the rate that ordinary 2-bromopropane does, a fact that supports the E2 reaction mechanism we have described.

$$CH_3CH_2O^- + CH_3\overset{\overset{\displaystyle Br}{|}}{C}HCH_3 \xrightarrow{\text{fast}} CH_3CH_2OH + CH_2{=}CHCH_3 + Br^-$$

$$CH_3CH_2O^- + CD_3\overset{\overset{\displaystyle Br}{|}}{C}HCD_3 \xrightarrow{\text{slow}} CH_3CH_2OD + CD_2{=}CHCD_3 + Br^-$$

B. Mixtures of Alkenes

Often, E1 and E2 reactions are referred to as **beta (β) eliminations.** This term reflects which hydrogen atom is lost in the reaction. Different types of carbon and hydrogen atoms in a molecule may be labeled as α, β, and so forth, according to the Greek alphabet. The carbon atom *bonded to the principal functional group* in a molecule is called the **alpha (α) carbon,** and the adjacent carbon is the **beta (β) carbon.** The hydrogens bonded to the α carbon are called α hydrogens, while those bonded to the β carbon are β hydrogens. In a β elimination, a β hydrogen is lost when the alkene is formed. (Of course, an alkyl halide with no β hydrogen cannot undergo a β elimination.)

β carbons and hydrogens circled:

If 2-bromopropane or *tert*-butyl bromide undergoes elimination, there is only one possible alkene product. However, if the alkyl groups around the α carbon are different and there is more than one type of β hydrogen, then more than one alkene can result. The E2 reaction of 2-bromobutane yields two alkenes because two types of hydrogen atoms can be eliminated: a hydrogen from a CH_3 group or a hydrogen from a CH_2 group. (Both geometrical isomers of the 2-butene are formed. We will discuss this facet of the reaction in Section 5.8C.)

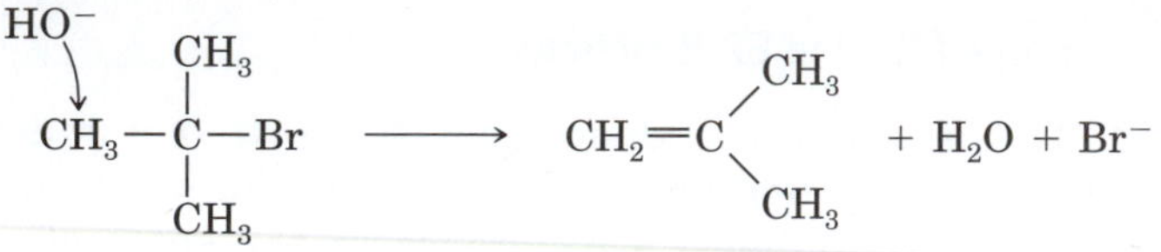

tert-butyl bromide · methylpropene

only one type of β H · *only possible alkene*

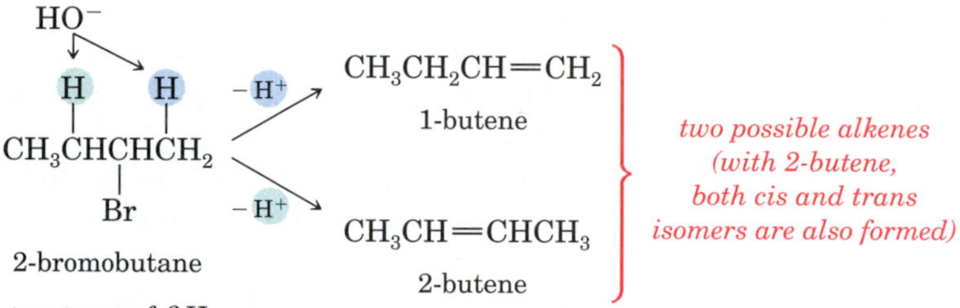

CH$_3$CH$_2$CH=CH$_2$
1-butene

CH$_3$CH=CHCH$_3$
2-butene

2-bromobutane

two types of β H

two possible alkenes (with 2-butene, both cis and trans isomers are also formed)

SAMPLE PROBLEMS

Circle the β carbons and hydrogens in the following structures:

(a) $CH_3CH_2CHCH_2CH_2CH_3$
      |
      Br

(b)

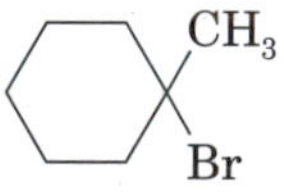

Solution

(a) CH$_3$ CH$_2$ CH CH$_2$ CH$_2$CH$_3$
            |
            Br

(b)

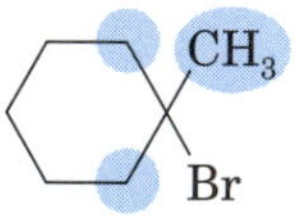

In the preceding problem, tell how many different *types* of β hydrogens are in each structure.

Solution
(a) two types; **(b)** two types (the ring CH_2 groups are equivalent to each other; the CH_3 is different).

STUDY PROBLEM

5.15 Show all possible alkene products in the following elimination reactions.

        Cl
        |
(a) CH$_3$CHCH$_2$CH$_2$CH$_3$ + $^-$OH $\xrightarrow{\text{heat}}$

(b) 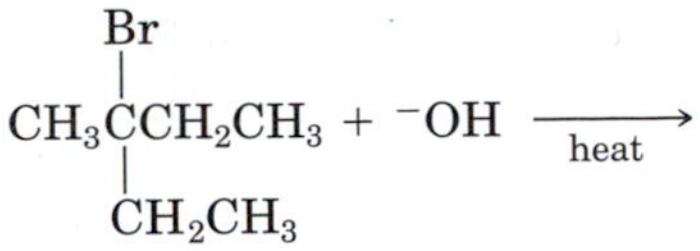 $CH_3\overset{\overset{\displaystyle Br}{|}}{\underset{\underset{\displaystyle CH_2CH_3}{|}}{C}}CH_2CH_3 + {}^-OH \xrightarrow{\text{heat}}$

C. Which Alkene Is Formed?

In 1875, the Russian chemist Alexander Saytzeff (also spelled Zaitsev; 1841–1910) formulated the following rule, now called the **Saytzeff rule.** *In elimination reactions, the alkene with the greatest number of alkyl groups on the doubly bonded carbon atoms predominates in the product mixture.* We will refer to this alkene as the *more highly substituted alkene.*

The Saytzeff rule predicts that 2-butene would predominate over 1-butene as a product in the E2 reaction of 2-bromobutane. This indeed is what occurs. In the following reaction, the mixture of alkene products consists of 80% 2-butene and only 20% 1-butene.

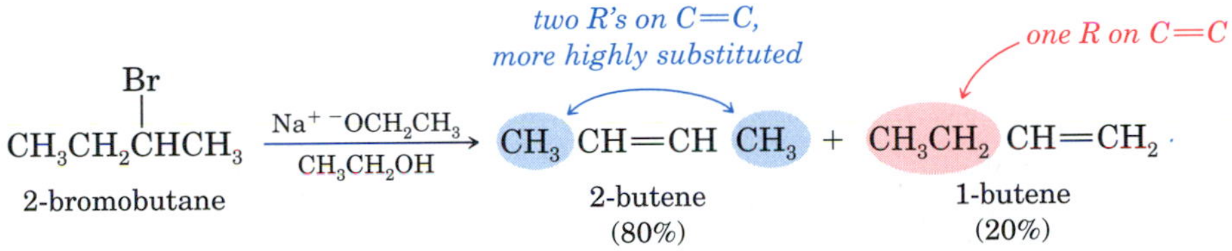

$$CH_3CH_2\overset{\overset{\displaystyle Br}{|}}{C}HCH_3 \xrightarrow[CH_3CH_2OH]{Na^+ \; {}^-OCH_2CH_3} CH_3CH{=}CHCH_3 + CH_3CH_2CH{=}CH_2 .$$

two R's on C=C, more highly substituted

one R on C=C

2-bromobutane   2-butene (80%)   1-butene (20%)

It has been determined that *more highly substituted alkenes are more stable than less substituted alkenes* (this will be discussed further in Chapter 10). Therefore, an E2 elimination leads to the *more stable alkene.* (In Chapter 7, we will describe how the E1 reaction also leads to the more stable alkene.)

$$CH_2{=}CH_2 \qquad CH_3CH{=}CH_2 \qquad CH_3CH{=}CHCH_3 \qquad (CH_3)_2C{=}C(CH_3)_2$$

increasing stability

To see why the more stable alkene (2-butene) is formed in preference to the less stable alkene (1-butene), let us consider the transition states leading to these two butenes. In either transition state, the base is removing a proton, and a double bond is being formed. We say that this transition state has some **double-bond character,** which we represent as a dotted line in the formula.

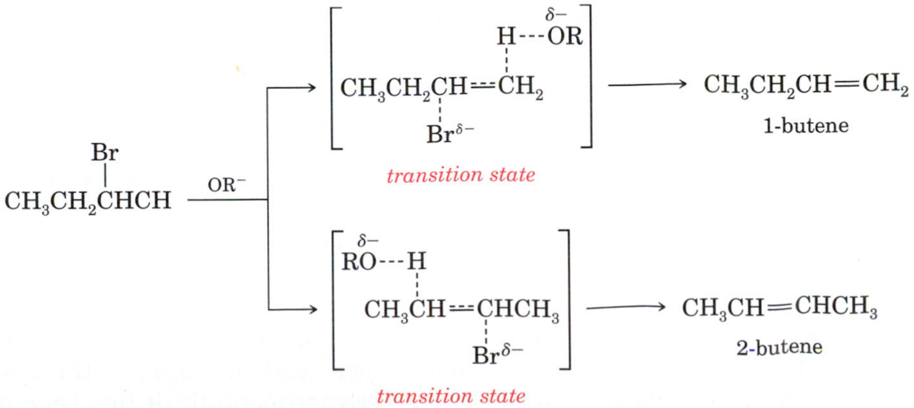

$$CH_3CH_2\overset{\overset{\displaystyle Br}{|}}{C}HCH \xrightarrow{OR^-}$$

$$\left[\begin{array}{c} \overset{\delta-}{H{-}{-}{-}OR} \\ | \\ CH_3CH_2CH{\cdots}CH_2 \\ | \\ Br^{\delta-} \end{array}\right] \longrightarrow CH_3CH_2CH{=}CH_2$$

transition state

1-butene

$$\left[\begin{array}{c} \overset{\delta-}{RO{-}{-}{-}H} \\ | \\ CH_3CH{\cdots}CHCH_3 \\ | \\ Br^{\delta-} \end{array}\right] \longrightarrow CH_3CH{=}CHCH_3$$

transition state

2-butene

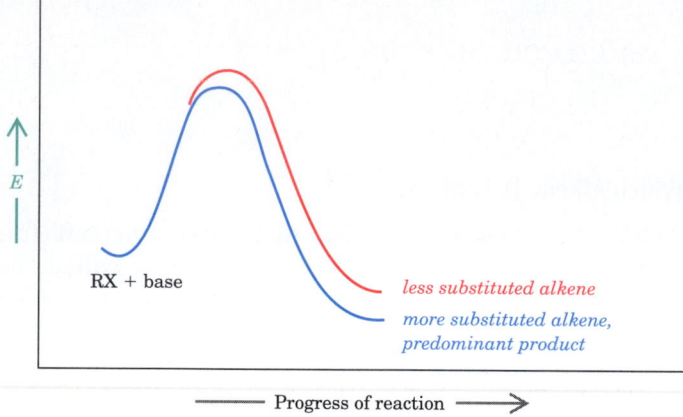

Figure 5.10   Energy diagram for a typical E2 reaction, showing why the more substituted alkene predominates.

Because both transition states leading to the alkenes have some double-bond character, the transition state leading to the more stable alkene is itself more stabilized and of lower energy. The reaction with the lower-energy transition state proceeds at a faster rate. Therefore, the more stable alkene is the predominant product (see Figure 5.10).

SAMPLE PROBLEMS

Which is the more stable alkene, **(a)** $(CH_3)_3CCH\!=\!CHCH_3$, or **(b)** $CH_3CH\!=\!C(CH_3)_2$?

Solution
Alkene **(b)** with three R's is more stable than **(a)** with only two R's.

Predict the major alkene product of the E2 dehydrohalogenation of $CH_3CH_2C(CH_3)_2Cl$.

Solution
The two possible alkenes are $CH_3CH\!=\!C(CH_3)_2$ and $CH_3CH_2\underset{\underset{\textstyle CH_3}{|}}{C}\!=\!CH_2$. The first predominates (3 R's versus 2 R's).

STUDY PROBLEM

5.16   Predict the major alkene product of the E2 dehydrohalogenation of 1-chloro-1-methylcyclohexane.

Part of our question about which alkene would be formed in dehydrohalogenation has been answered: the most highly substituted alkene predominates in the product mixture. The most highly substituted alkene can often exist as *cis* and *trans* diastereomers (geometric isomers). Is there a selectivity as to which diastereomer is formed? Experimentally, it has been determined that *trans* alkenes are generally more stable than their *cis* isomers, presumably because of less steric hindrance in the *trans* isomers. Therefore, it is not surprising that *trans* alkenes predominate as products of E2 reactions. Again, the reason is a more stabilized transition state. The following equation shows the results of an E2 reaction of 2-bromopentane.

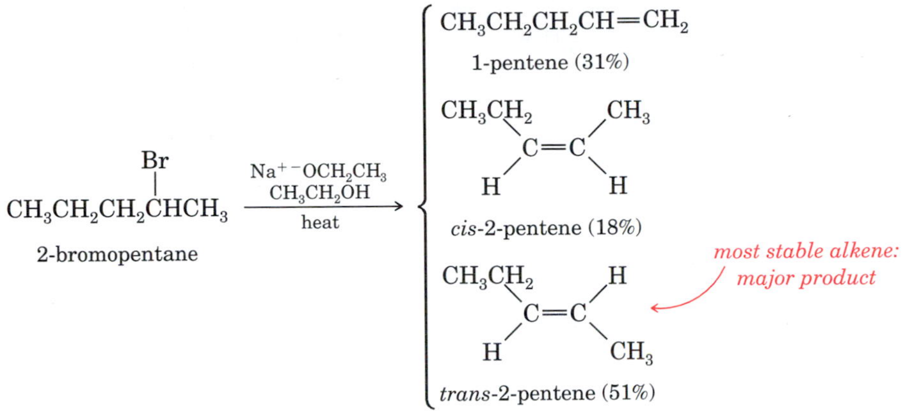

1-pentene (31%)

cis-2-pentene (18%)

trans-2-pentene (51%)

most stable alkene: major product

D. Stereochemistry of an E2 Reaction

In the transition state of an E2 elimination, the attacking base and the leaving group are generally as far apart as possible, or *anti*. For this reason, the E2 elimination is often referred to as ***anti*-elimination.**

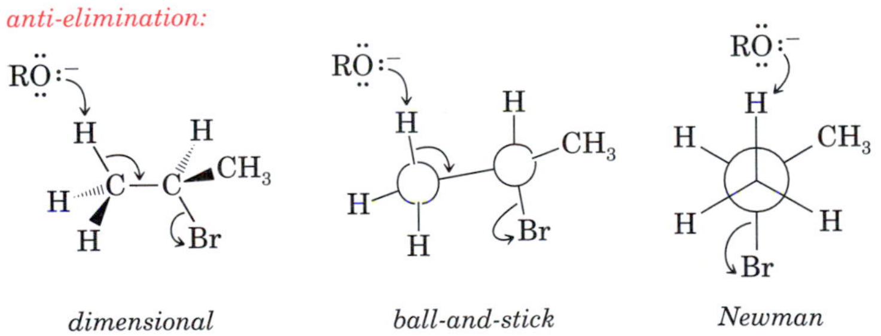

anti-elimination:

dimensional   *ball-and-stick*   *Newman*

The interesting feature about *anti*-elimination is that the *anti* positioning of the H and Br that are lost determines the stereochemistry of the product alkene. To see how this happens, let us look at the E2 reactions of some stereoisomeric halides. The compound 1-bromo-1,2-diphenylpropane has two chiral carbon atoms (carbons 1 and 2) and four stereoisomers.

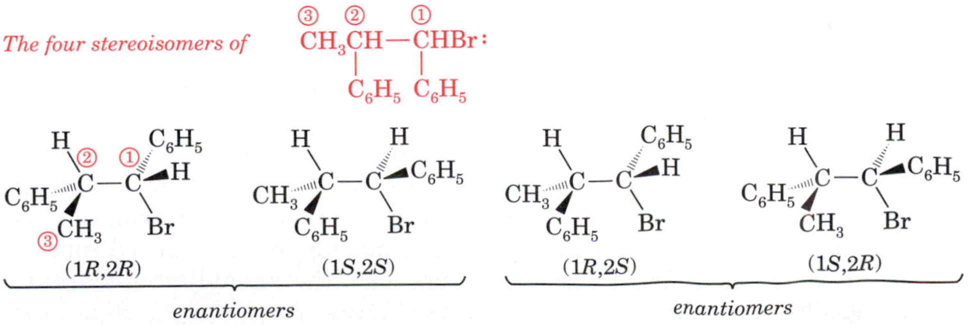

The four stereoisomers of

$$\overset{③}{C}H_3\overset{②}{C}H-\overset{①}{C}HBr:$$
$$\quad\quad\;\; | \quad\; |$$
$$\quad\quad\; C_6H_5 \; C_6H_5$$

(1R,2R)   (1S,2S)   (1R,2S)   (1S,2R)

enantiomers   *enantiomers*

Because there is only one β hydrogen in the starting halide, any of these stereoisomers yields $C_6H_5(CH_3)C{=}CHC_6H_5$. However, geometric isomerism is possible in this product.

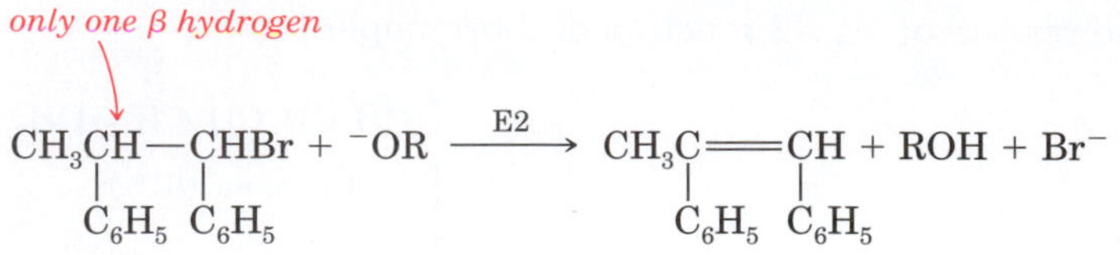

only one β hydrogen

$$CH_3CH{-}CHBr + {}^-OR \xrightarrow{E2} CH_3C{=\!=}CH + ROH + Br^-$$
$$\underset{C_6H_5\ \ C_6H_5}{} \qquad \underset{C_6H_5\ \ C_6H_5}{}$$

When either (1R,2R)-1-bromo-1,2-diphenylpropane or its (1S,2S)-enantiomer undergoes E2 reaction, the (Z) alkene is formed exclusively; no (E)-alkene is formed.

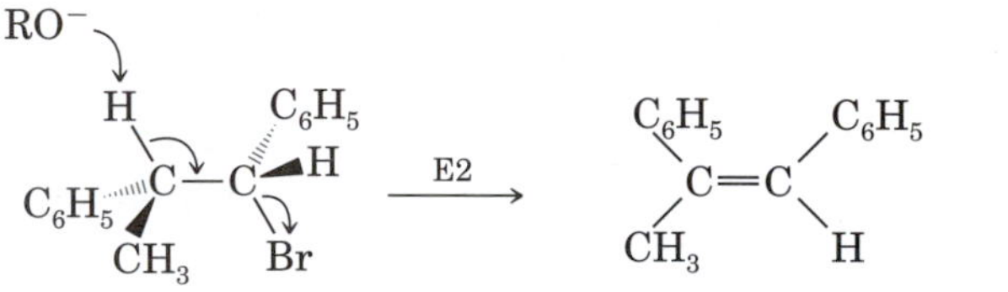

(1R,2R)   (Z)-1,2-diphenyl-1-propene

The reason for all (Z) and no (E) product is that there is only one conformation of either of these enantiomers in which the Br and the beta hydrogen are *anti*. In either the (1R,2R)- or the (1S,2S)-enantiomer, the *anti* alignment of H and Br puts the phenyl groups on the same side of the molecule, and the (Z)-alkene results. If the elimination could occur regardless of the conformation of the enantiomers, then some (E)-alkene would also be observed.

anti H and Br means cis-phenyls

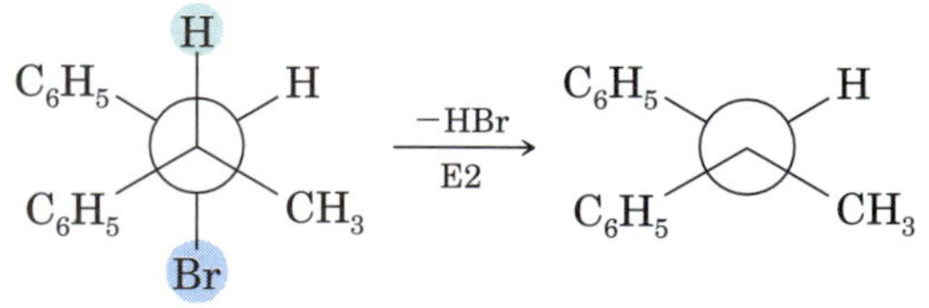

(1R,2R)   (Z)-alkene

STUDY PROBLEM

5.17 Write equations for the *anti*-elimination of the (1S,2S)-enantiomer, as we have done for the (1R,2R)-enantiomer.

Just the opposite situation prevails with the (1R,2S)- or (1S,2R)-enantiomers. Either of these isomers yields the (E)-alkene and not the (Z)-alkene. The reason, once again, is that there is only one conformation for each of these enantiomers in which the Br and the single beta hydrogen are in an *anti* relationship. In these conformations, the phenyl groups are on opposite sides of the molecule.

anti H and Br means trans-phenyls

(1R,2S)   (1S,2R)   (E)-alkene

A reaction in which different stereoisomers of the reactant yield stereoisomerically different products is said to be a **stereospecific reaction.** The E2 reaction is an example of a stereospecific reaction.

Halocycloalkanes, such as chlorocyclohexane, can also undergo E2 reactions. In these cases, the conformations of the ring play an important role in the course of the reaction. In order to be *anti* on a cyclohexane ring, the leaving group (such as chlorine) and a β hydrogen must be 1,2-*trans* and *diaxial*. No other conformation places the H and Cl *anti* to each other. (Try it with models.) Even though this conformation is not the favored one, a certain percentage of halocycloalkane molecules are in this conformation at any given time and can thus undergo elimination.

Cl is equatorial and not anti to any β hydrogens   *Cl is axial and anti to two β hydrogens*

OH⁻ could attack either H shown   cyclohexene

(from attack at either H)

STUDY PROBLEM

5.18   1,2-Dibromo-1,2-diphenylethane contains two chiral carbon atoms and has a pair of enantiomers plus a *meso* diastereomer. Any of the stereoisomers of this compound can undergo an E2 reaction to yield 1-bromo-1,2-diphenylethene. The *meso* form yields one geometric isomer of the alkene, while the racemic mixture of enantiomers yields the other geometric isomer. Predict the stereochemistry of the products of these two reactions.

E. Hofmann Products

Most dehydrohalogenations follow the Saytzeff rule, and the more substituted alkene predominates. However, under some circumstances, the major product of an E2 dehydrohalogenation is the *less substituted, less stable alkene.* When the less substituted alkene is the predominant product, we say that the reaction yields the **Hofmann product,** after the German chemist, August W. von Hofmann (1818–1892).

When is the less substituted alkene likely to be the predominant product? A common phenomenon leading to the less substituted alkene is **steric hindrance** in that transition state leading to the most substituted alkene. Steric hindrance can raise the energy of this transition state so substantially that the reaction follows a different course and yields the less substituted alkene. The steric hindrance may be caused by any one of three factors.

First of all, it may be caused by the **size of the attacking base.** The effect of increasing size on alkene product distribution in the E2 of 2-iodobutane is given in Table 5.5. The data in Table 5.5 show that as the size of the base increases, the percentage of the least-substituted alkene products increases.

Reaction with tricyclohexylmethoxide ion

attack at carbon 3: more steric hindrance than with attack at carbon 1

attack at carbon 1: less steric hindrance than with attack at carbon 3

Second, steric hindrance might be caused by the **bulkiness of groups surrounding the leaving group** in the alkyl halide. The hindered 2-bromo-2,4,4-trimethylpentane yields the less substituted alkene in an E2 reaction, even with a small base like the ethoxide ion.

Table 5.5	Butene product distribution from the E2 reaction of 2-iodobutane with alkoxide bases in dimethyl sulfoxide (DMSO)		
	Percent Products		
Base	1-butene	*trans*-2-butene	*cis*-2-butene
$NaOCH_3$	17.0	62.9	20.1
$NaOCH_2CH_2CH_3$	18.5	62.7	18.8
$KOC(CH_3)_3$	19.7	60.7	19.6
$KOC(C_6H_{11})_3$	27.2	54.8	18.0

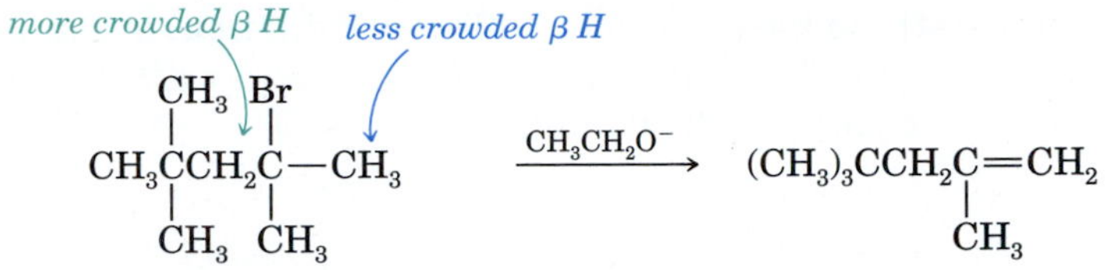

more crowded β H   *less crowded β H*

$$CH_3CCH_2C{-}CH_3 \xrightarrow{CH_3CH_2O^-} (CH_3)_3CCH_2C{=}CH_2$$

2-bromo-2,4,4-trimethylpentane   2,4,4-trimethyl-1-pentene

Third, if the **leaving group itself is large and bulky,** the Hofmann product may predominate. This type of reaction is discussed in Section 18.9.

5.19 Write formulas for both the Hofmann and Saytzeff products of the E2 reactions of **(a)** 3-bromo-2-methylpentane and **(b)** 1-chloro-1-methylcyclohexane.

F. Summary of E1 and E2 Mechanisms

Elimination reactions of alkyl halides can proceed by an E1 path or by an E2 path. *Tertiary alkyl halides* can undergo E1 elimination as a side reaction in solvolysis when water or an alcohol acts as a very weak base. Figure 5.11 summarizes some important features of the E1 pathway.

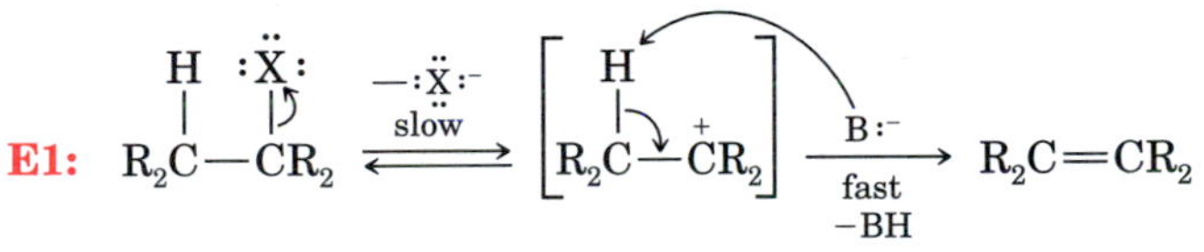

$$\textbf{E1:} \quad R_2C{-}CR_2 \underset{slow}{\overset{-:\ddot{X}:^-}{\rightleftharpoons}} \left[R_2C{\overset{+}{\cdots}}CR_2 \right] \xrightarrow[\substack{fast \\ -BH}]{B:^-} R_2C{=}CR_2$$

An ionization reaction is sometimes accompanied by rearrangement, followed by
  loss of H^+ to the base
Generally produces the most stable alkene (Saytzeff rule)
Stereochemistry: not stereospecific
A unimolecular reaction
First order in rate
Relative rates: 3° RX > 2° RX

$$\textbf{E2:} \quad B:^- + R_2C{\overset{H}{\cdots}}CR_2 \xrightarrow[{-:\ddot{X}:^-}]{-BH} R_2C{=}CR_2$$

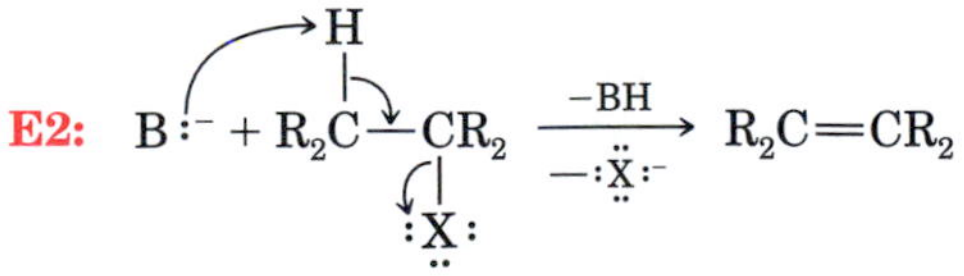

A concerted, bimolecular reaction
Produces the most stable alkene (with exceptions)
Stereochemistry: anti elimination of H and X
Second order in rate
Relative rates: 3° RX > 2° RX > 1° RX

Figure 5.11 Summary of E1 and E2 mechanisms for elimination reactions of alkyl halides. (The symbol $B:^-$ represents a weak base for E1 reactions or a strong base for E2 reactions of alkyl halides.)

When a strong base is used to effect elimination, *tertiary alkyl halides, secondary alkyl halides,* and, in some cases, *primary alkyl halides* undergo reaction by an E2 path, as shown in Figure 5.11.

Section 5.9

Factors Governing Substitution and Elimination Reactions

At the start of this chapter, we mentioned that S_N1, S_N2, E1, and E2 are *competing reactions.* A single alkyl halide could be undergoing substitution, elimination, and rearrangements all in the same reaction flask. If this happens, a mixture of a large number of products can result. However, a chemist can control the products of the reaction to a certain extent by a proper choice of the reagents and reaction conditions. We have already mentioned some of the factors that affect the course of substitution and elimination reactions. These and other major factors are:

1. the structure of the alkyl halide

2. the nature of the nucleophile or base

3. the nature of the solvent

4. the concentration of the nucleophile or base

5. the temperature.

A. The Alkyl Halide

We have mentioned that the type of alkyl halide affects the mechanism of the reaction. Now that we have looked at the four principal mechanisms by which an alkyl halide can undergo reaction with a nucleophile or base, we can summarize how the different alkyl halides act. (See also Figure 5.12.)

Methyl and primary alkyl halides tend to undergo S_N2 reactions. They do not form carbocations and thus cannot undergo S_N1 or E1 reactions. Primary alkyl halides undergo E2 reactions less readily than other alkyl halides.

Secondary alkyl halides can undergo reaction by any path, but S_N2 and E2 are more common than E1 or S_N1. The reactions of secondary alkyl halides are more subject to control by conditions in the reaction flask (concentration of nucleophile, solvent, etc.) than are reactions of other alkyl halides.

Tertiary alkyl halides undergo primarily E2 reactions with a strong base (such as $^-$OH or $^-$OR), but undergo the S_N1 reaction and some E1 reaction with a very weak base (such as H_2O or ROH).

Allylic and benzylic alkyl halides undergo substitution reactions readily: generally S_N1 with weak nucleophiles and S_N2 with moderately strong nucleophiles. With appropriate structures, elimination products can also be formed.

B. The Nucleophile or Base

The difference between nucleophilicity and basicity was discussed in Section 5.3D. As we mentioned in that section, a strong base is generally also a good nucleophile. Two other factors can affect the relative nucleophilicities of reac-

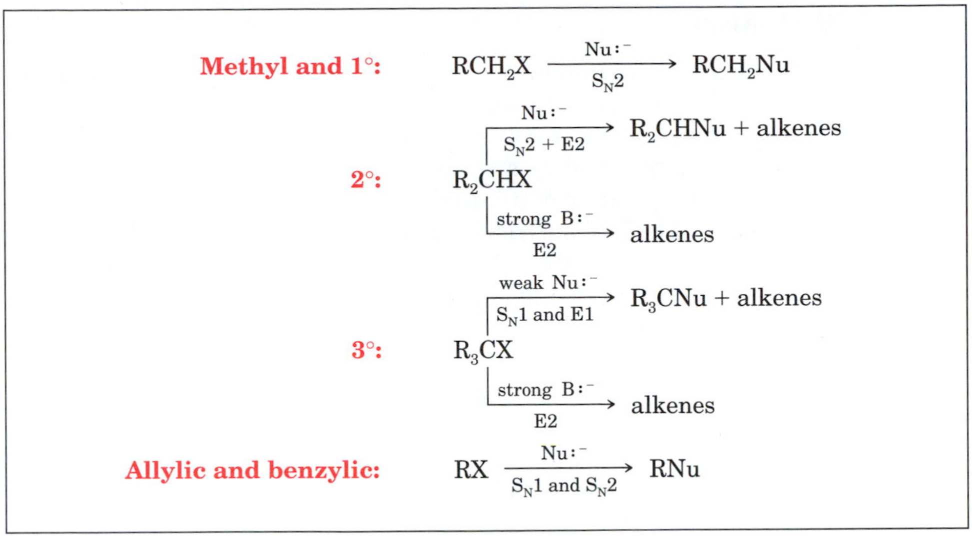

$$\textbf{Methyl and 1°:}\quad RCH_2X \xrightarrow[\;S_N2\;]{Nu:^-} RCH_2Nu$$

$$\textbf{2°:}\quad R_2CHX$$

$$\xrightarrow[\;S_N2\,+\,E2\;]{Nu:^-} R_2CHNu + alkenes$$

$$\xrightarrow[\;E2\;]{strong\ B:^-} alkenes$$

$$\textbf{3°:}\quad R_3CX$$

$$\xrightarrow[\;S_N1\ and\ E1\;]{weak\ Nu:^-} R_3CNu + alkenes$$

$$\xrightarrow[\;E2\;]{strong\ B:^-} alkenes$$

$$\textbf{Allylic and benzylic:}\quad RX \xrightarrow[\;S_N1\ and\ S_N2\;]{Nu:^-} RNu$$

Figure 5.12   The principal reactions that different classes of alkyl halides undergo. Typical nucleophiles ($Nu:^-$) are ^-OH and ^-CN. Typical weak nucleophiles are H_2O and ROH. Typical strong bases ($B:^-$) are ^-OH and $^-OCH_2CH_3$.

tants, sometimes dramatically. One of these factors is the *solvent* used for the reaction. Solvent effects will be discussed in Section 5.9C.

The *polarizability* of an ion or molecule is another factor that affects its nucleophilicity. The outer electrons of larger atoms are farther from the nucleus and less tightly held than those of smaller atoms. The outer electrons of larger atoms are therefore more easily distorted by attraction to a positive center and can attack a partially positive carbon atom more readily. For example, an iodide ion is usually a better nucleophile than a chloride ion.

The degree of nucleophilicity versus basicity can affect the course of a reaction. The reaction of a *primary alkyl halide* with a strong nucleophile (see page 187) follows an S_N2 path, even if the nucleophile is also a strong base. However, for a *tertiary alkyl halide,* any moderately strong base favors E2 reactions. Only the weakest bases (H_2O, ROH) result in substitution (by an S_N1 path).

For *secondary alkyl halides,* strong nucleophiles (such as ^-CN) favor S_N2 reactions, while weak nucleophiles (such as H_2O) favor carbocation reactions, primarily S_N1 with some E1. Strong bases (such as ^-OH or ^-OR) favor E2 reactions.

strong nucleophile:	S_N2
weak nucleophile:	S_N1
strong base:	E2

STUDY PROBLEM

5.20   Predict which is generally the better nucleophile: CH_3S^- or CH_3O^-. Explain the reason for your answer.

C. The Solvent

The solvent exerts its influence on substitution and elimination reactions by its ability or inability to solvate ions: carbocations, nucleophiles or bases, and leaving groups. The ability of a solvent to solvate ions is determined by its polarity, which is usually reported as a **dielectric constant.** The dielectric constant is a measure of the polarity of a liquid. A highly polar solvent has a large dielectric constant. Table 5.6 lists some common organic solvents, their dielectric constants, and the relative rates of a typical S_N1 reaction in that solvent.

While dielectric constants can provide a guide for solvent selection, there are no firm rules about how to predict the best solvent for a given reaction. (The solubilities of the reactants must be considered too!) In general, a *very polar solvent* (such as water) encourages S_N1 reactions by helping stabilize the carbocation through solvation.

S_N2 reactions are also favored by polar solvents, but only by solvents that cannot solvate the nucleophile. A solvent that can solvate (and thus stabilize) an anion reduces its nucleophilicity. By contrast, a solvent that cannot solvate an anion enhances its nucleophilicity. The chloride ion is a far better nucleophile in dimethylformamide (DMF) or dimethyl sulfoxide (DMSO), where it is not solvated, than in ethanol, where it is solvated.

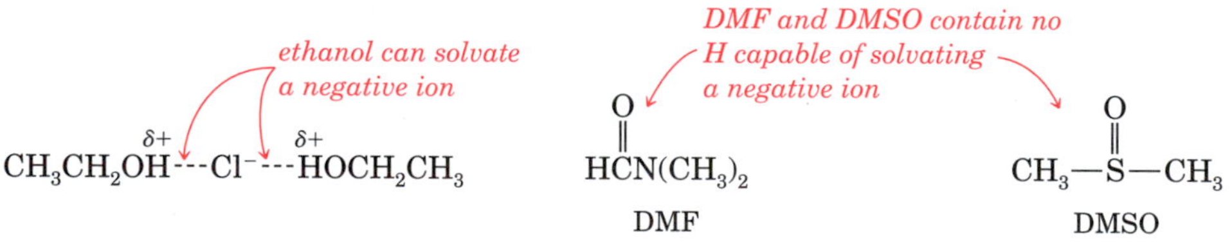

ethanol can solvate a negative ion

DMF and DMSO contain no H capable of solvating a negative ion

$$\overset{\delta+}{CH_3CH_2OH}\text{---}Cl^-\text{---}\overset{\delta+}{HOCH_2CH_3} \qquad\qquad HCN(CH_3)_2 \qquad\qquad CH_3\text{—}\overset{O}{\overset{\|}{S}}\text{—}CH_3$$

$$\text{DMF} \qquad\qquad\qquad \text{DMSO}$$

D. Concentration of the Nucleophile or Base

By controlling the concentration of nucleophile or base, a chemist has direct control over the rates of S_N2 and E2 reactions. Increasing the concentration of nucleophile generally has no effect on the rates of S_N1 or E1 reactions, but increases S_N2 or E2 reaction rates proportionally.

high contentrations of Nu:⁻ or base: S_N2 or E2

low concentration of Nu:⁻: S_N1 or E1

Table 5.6 Relative rates of typical S_N1 reactions in various solvents			
Solvent	Formula	Dielectric Constant	Approximate Relative Rate
formic Acid	HCO_2H	58	15,000
water	H_2O	78.5	4,000
80% aqueous ethanol	$CH_3CH_2OH—H_2O$	67	185
ethanol	CH_3CH_2OH	24	37
acetone	$CH_3\overset{O}{\overset{\|}{C}}CH_3$	21	0.5
tetrahydrofuran (THF)		7	0.05

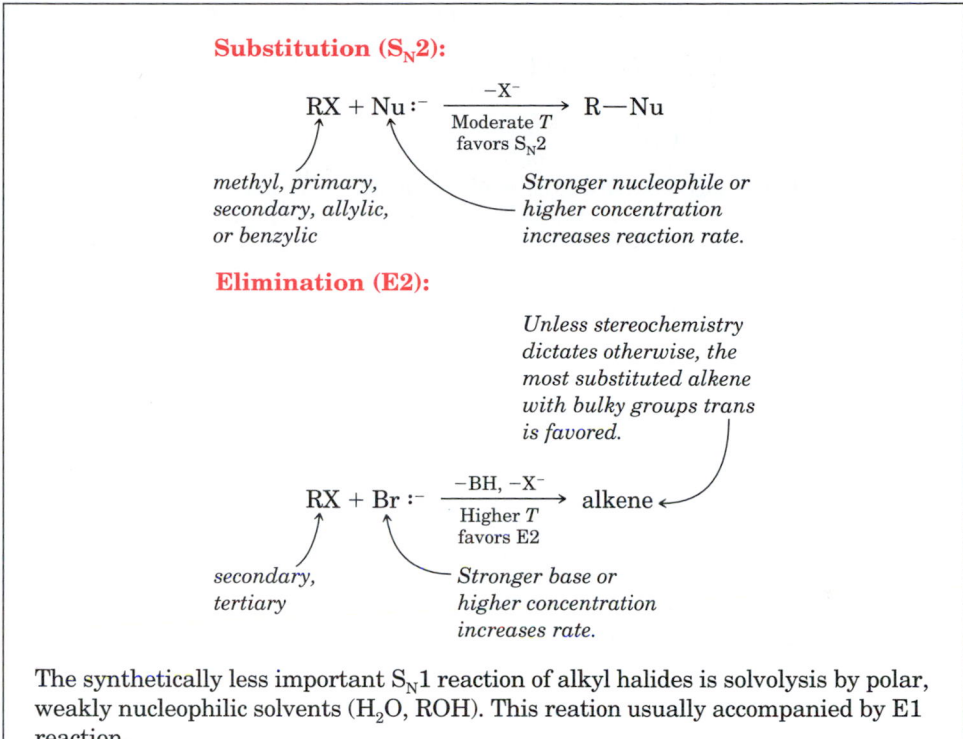

Figure 5.13 Summary of principal factors affecting important substitution and elimination reactions of alkyl halides.

E. Temperature

An increase in temperature increases the rates of all substitution and elimination reactions. However, an increase in temperature usually leads to a greater increase in elimination products. (The reason for this is that elimination reactions usually have higher E_{act}'s than do substitution reactions, and higher temperatures enable a greater number of molecules to reach the elimination transition state.)

The principal factors governing substitution (S_N2) and elimination (E2) are summarized in Figure 5.13.

SAMPLE PROBLEM

Predict the products of the following reactions, all of which generate one substitution or elimination product in good yield, and tell which mechanism(s) would be followed.

(a) $BrCH_2CH_2CH_2Br + 2\,Na^+\ ^-CN \xrightarrow[\text{heat}]{\text{aqueous } CH_3CH_2OH}$

(b) $C_6H_5S^-Na^+ + CH_3\overset{\displaystyle Br}{\underset{\displaystyle |}{C}}H-\overset{\displaystyle O}{\overset{\displaystyle \|}{C}}OCH_2CH_3 \xrightarrow[25°]{\text{THF}}$

(c) CH_3-⟨○⟩$-CH_2Cl + Na^+ \ {}^-\overset{\overset{\textstyle O}{\textstyle \|}}{O}CCH_3 \xrightarrow[\text{heat}]{CH_3CO_2H}$

(d) $CH_3CH_2CH_2\overset{\overset{\textstyle CH_3}{\textstyle |}}{C}{=}CHCH_2Cl + Na^+ \ {}^-OCH_3 \xrightarrow[25\text{–}50°]{CH_3OH}$

(e)

$+ K^+ \ {}^-OC(CH_3)_3 \xrightarrow[90°]{(CH_3)_2S{=}O}$

Solution

(a) Even though a polar solvent mixture is used (to dissolve the NaCN), the combination of primary halide and strong nucleophile yields the S_N2 products:

$$NCCH_2CH_2CH_2CN + 2Na^+ \ Br^-$$

(b) A strong nucleophile, a secondary halide, moderately low temperature, and nonpolar solvent lead to S_N2 products:

$$CH_3\overset{\overset{\textstyle C_6H_5S}{\textstyle |}}{C}H-\overset{\overset{\textstyle O}{\textstyle \|}}{C}OCH_2CH_3 + Na^+ \ Br^-$$

(c) The acetate ion is a weak nucleophile, but a benzylic halide is a reactive halide. From the material presented in this chapter, we would predict the substitution products to arise by either an S_N1 path or an S_N2 path.

$$CH_3-⟨○⟩-CH_2-\overset{\overset{\textstyle O}{\textstyle \|}}{O}CCH_3 + Na^+ \ Cl^-$$

(d) This is a reaction of an allylic halide. The conditions (strong nucleophile) indicate an S_N2 mechanism, not an S_N1 mechanism. For this reason, and because of the steric hindrance at carbon 3, we would not expect an allylic rearrangement. (In fact, only one substitution product could be isolated when this reaction was carried out.)

$$CH_3CH_2CH_2\overset{\overset{\textstyle CH_3}{\textstyle |}}{C}{=}CHCH_2OCH_3 + Na^+ \ Cl^-$$

(e) Do not worry about the complexity of a structure. This compound, which contains a tertiary halogen, is being subjected to E2 conditions—strong bulky base and heat. We would expect the Hofmann product because of the bulky base and because of the geometry of the substrate molecule.

(The formation of the most substituted alkene here would be geometrically impossible.)

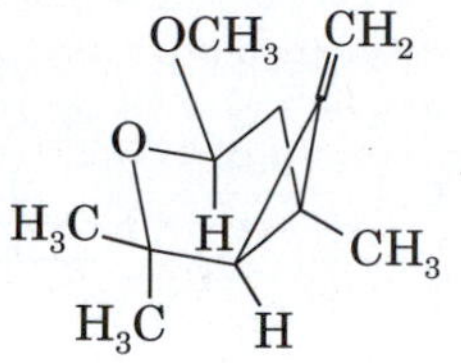

$$+ \text{ K}^+\text{Cl}^- + \text{HOC(CH}_3)_3$$

STUDY PROBLEM

5.21 Predict the products and probable mechanistic paths of the following reactions. Explain your answers.

(a) $\overset{\displaystyle \text{I}}{\underset{\displaystyle |}{\text{CH}_3(\text{CH}_2)_5\text{CHCH}_3}} \xrightarrow[\text{100}^\circ]{\text{aqueous CH}_3\text{CH}_2\text{OH}}$

(b) $\text{C}_6\text{H}_5\text{O}^-\text{ Na}^+ + \overset{\displaystyle \text{OH}}{\underset{\displaystyle |}{\text{HOCH}_2\text{CHCH}_2\text{Cl}}} \xrightarrow[\text{heat}]{\text{CH}_3\text{CH}_2\text{OH}}$

(c) $\text{CH}_3\text{CH}_2\text{OCH}_2\text{CH}_2\text{Br} + \text{Na}^+\ ^-\text{CN} \longrightarrow$

(d) $\text{CH}_3\text{CHClCH}=\text{CH}_2 + \text{CH}_3\text{CH}_2\text{OH} \xrightarrow{\text{heat}}$

Section 5.10
Synthesizing Other Compounds from Alkyl Halides

From a practical standpoint, only S_N2 and E2 reactions are useful for synthesizing other compounds from alkyl halides. S_N1 and E1 reactions usually yield mixtures of products.

A large number of compound classes can be prepared from alkyl halides and other halogen compounds by S_N2 reactions. We have presented only a few nucleophiles in the chapter discussions, but a variety of nucleophilic reagents are useful (see Table 5.7). Of special interest are the reactions of RX with an alkoxide (RO^-) or a phenoxide (ArO^-) to yield ethers. You will encounter these ether preparations in more detail in Section 8.2.

$$\text{RCH}_2\text{X} + \text{R'O}^- \longrightarrow \text{RCH}_2\text{OR'} + \text{X}^-$$

$$\text{HOCH}_2(\text{CH}_2)_{10}\text{CH}_2\text{Br} + \text{Na}^+\ ^-\text{OCH}_3 \xrightarrow{\text{CH}_3\text{OH}} \text{HOCH}_2(\text{CH}_2)_{10}\text{CH}_2\text{OCH}_3 + \text{NaBr}$$

$$(\sim 100\%)$$

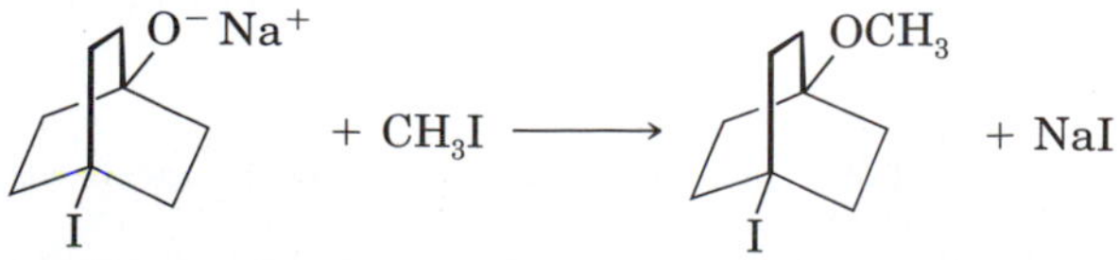

$+ \text{ CH}_3\text{I} \longrightarrow$ ... $+ \text{ NaI}$

$$(96\%)$$

The cyanide ion is a useful strong nucleophile that leads to nitriles (RCN). Nitriles can, in turn, be converted to amines or carboxylic acids. (See Section 15.11D.)

$$RCH_2X \quad \text{or} \quad R_2CHX + {}^-CN \longrightarrow RCH_2CN \quad \text{or} \quad R_2CHCN + X^-$$

(90%)

Ammonia and amines, too, can be used as nucleophiles. These reactions are discussed in Section 18.4A.

$$RCH_2X \quad \text{or} \quad R_2CHX + R_3N\colon \longrightarrow RCH_2\overset{+}{X}R_3 X^- \quad \text{or} \quad R_2CH\overset{+}{X}R_3 X^-$$

$$CH_3(CH_2)_3Br + \text{excess} \colon NH_3 \longrightarrow CH_3(CH_2)_3\overset{+}{N}H_3 Br^-$$

(45%)

Reactive alkyl halides (CH_3I, an allylic halide, or a benzylic halide), but not ordinary primary or secondary alkyl halides, react with the weakly nucleophilic carboxylate ion ($RCO_2{}^-$) to yield esters.

an allylic bromide   (96%)

An *alkene* can be prepared by heating a secondary or tertiary alkyl halide with a *strong base* such as potassium hydroxide or the alkali metal salt of an alcohol in an alcohol solvent. Generally, the more highly substituted, *trans*

Table 5.7   Some types of compounds that can be synthesized from alkyl halides

Reactants[a]			Principal Product	Typical Reagents
1° RX	+ ${}^-OR'$	$\longrightarrow$ ROR'	an ether	$Na^+ \ {}^-OCH_2CH_3$, $Na^+ \ {}^-OC_6H_5$
1° RX	+ ${}^-OH$	$\longrightarrow$ ROH	an alcohol	$Na^+ \ {}^-OH$, $K^+ \ {}^-OH$
1° or 2° RX	+ ${}^-CN$	$\longrightarrow$ RCN	a nitrile	$Na^+ \ {}^-CN$
1° or 2° RX	+ ${}^-SR'$	$\longrightarrow$ RSR'	a sulfide, or thioether	$Na^+ \ {}^-SCH_2CH_3$
1° or 2° RX	+ ${}^-O\overset{\text{O}}{\overset{\|}{C}}R'$	$\longrightarrow$ $RO\overset{\text{O}}{\overset{\|}{C}}R'$	an ester[b]	$Na^+ \ {}^-O_2CCH_3$
1° or 2° RX[c]	+ I^-	$\longrightarrow$ RI	an alkyl iodide	$Na^+ \ {}^-I$
1° or 2° RX	+ NR_3'	$\longrightarrow$ $RNR_3'X^-$	an ammonium salt	$(CH_3)_3N$
2° or 3° R_2CHCXR_2	+ ${}^-OR'$	$\longrightarrow$ $R_2C{=}CR_2$	an alkene	$K^+ \ {}^-OH$, $Na^+ \ {}^-OCH_2CH_3$, $K^+ \ {}^-OC(CH_3)_3$

[a] Where 1° RX is specified, methyl halides, allylic halides, and benzylic halides may also be used.
[b] A reactive halide must be used.
[c] Where X = Cl or Br.

alkene is the product. The less substituted alkene can sometimes be prepared if a bulky base, such as K^+ $^-OC(CH_3)_3$, is used.

Table 5.7 summarizes the types of products that can be obtained by S_N2 and E2 reactions of alkyl halides.

Study Problems

5.22 Insert the correct nucleophile in each of the following equations:

(a) C_6H_5CHBr + ? $\longrightarrow$ C_6H_5CHOH + Br^-
         | |
         CH_3 CH_3

(b) $C_6H_5CH_2Cl$ + ? $\longrightarrow$ $C_6H_5CH_2O\overset{\overset{\displaystyle O}{\|}}{C}CH_3$ + Cl^-

(c) $CH_3CH{=}CHCH_2Br$ + ? $\longrightarrow$ $CH_3CH{=}CHCH_2SCH_3$ + Br^-

(d) (furan)$\sim$$CH_2Cl$ + ? $\longrightarrow$ (furan)$\sim$$CH_2OC_6H_5$ + Cl^-

5.23 Write equations to show how you would synthesize the following compounds from organic halides and other appropriate reagents. If there are two routes to the compound, choose the better one. If two routes are equivalent, show both.

(a) $CH_3CH_2CH_2O{-}$(cyclohexyl)   (b) (phenyl)${-}CH_2SCH_3$

(c) $CH_3CH_2OCH_2CH_2CH_3$ (d) (phenyl)${-}\overset{\overset{\displaystyle O}{\|}}{C}OCH_2CH{=}CH_2$

(e) (phenyl)${-}OCH_2CH_3$ (f) $(CH_3)_2CHCH{=}CHCH_3$

Summary

An alkyl halide contains a good **leaving group** (X^-) and is readily attacked by **nucleophiles** ($Nu{:}^-$). Reaction occurs by one or more of four possible paths: **S_N1, S_N2, E1, or E2.**

An S_N1 or E1 reaction proceeds through a **carbocation intermediate**:

$$RX \xrightarrow{\;-X^-\;} [R^+] \begin{cases} \xrightarrow[S_N1]{\;Nu{:}^-\;} RNu \\ \xrightarrow[E1]{\;-H^+\;} alkene \end{cases}$$

A carbocation intermediate usually leads to a mixture of products: *a substitution product, an alkene,* and also *rearrangement products.* Rearrangement products occur if the carbocation can form a more stable carbocation by a 1,2-shift of H, Ar, or R. If RX is optically active, racemization can occur in an S_N1 reaction.

The rate of a typical S_N1 or E1 reaction depends on the concentration of only RX. Thus, these reactions are said to be **first order.** The **rate-determining step** (slow step) in an S_N1 or E1 reaction is the formation of R^+. The stability of R^+ determines the **energy of the transition state** (E_{act}) in this step because the transition state has carbocation character. The *order of stability* of carbocations is $3° > 2° \gg 1° \gg CH_3^+$. For this reason, the likelihood of RX to undergo S_N1 or E1 reaction is $3° > 2° \gg 1° \gg CH_3X$. Allylic and benzylic halides undergo S_N1 reactions readily because of resonance stabilization of the intermediate carbocation.

An S_N2 reaction is a **concerted reaction** that leads to *inversion of configuration.* Inversion can be observed if RX is optically active. An E2 reaction is also a concerted reaction that results by *anti*-elimination of H^+ and X^-.

$$Nu\!:^- + R\!-\!X \xrightarrow{S_N2} NuR + X^-$$

$$RO^- \searrow$$
$$\underset{\underset{X}{|}}{\overset{\overset{H}{|}}{R_2C\!-\!CR_2}} \xrightarrow{E2} ROH + R_2C\!=\!CR_2 + X^-$$

Both S_N2 and E2 reactions follow **second-order kinetics.** The rate is dependent on the concentrations of both RX and $Nu\!:^-$ because both are involved in the transition state. Because of steric hindrance, the order of reactivity of RX in S_N2 reactions is $CH_3X > 1° > 2° \gg 3°$.

Because the transition state has double-bond character, the order of reactivity of RX in E2 reactions is $3° > 2° \gg 1°$, the same order as in the E1 reaction. The *most substituted alkene* usually predominates in E2 reactions **(Saytzeff rule).** The *trans* alkene usually predominates over the *cis* alkene. If steric hindrance inhibits the formation of the most substituted alkene, then the *least substituted alkene* predominates **(Hofmann product).**

Essay Problem for Chapter 5

Stereoselectivity in Substitution Reactions

Under classical S_N2 reaction conditions, the carbon at the site of reaction undergoes inversion of configuration. For example, if racemic *cis*-4-chloro-1-methylcyclohexane is treated with a nucleophile in an S_N2 reaction, the *trans* racemic product predominates. Under S_N1 reaction conditions, the carbon at the site of reaction undergoes both retention and inversion, yielding a mixture of products. Solvolysis of *cis*-4-chloro-1-methylcyclohexane yields a mixture of *cis* and *trans* products. Classical substitution reactions do not give retention of configuration at the carbon undergoing reaction.

However, the stereochemistry of some substitution reactions can be controlled by using a catalyst, such as palladium acetate-triphenylphosphine. For example, when racemic *cis*-1-acetoxy-4-chloro-2-cyclohexene, an allylic halide, is treated with dimethyl sodiomalonate using a palladium acetate-triphenylphosphine catalyst, the *cis* product predominates.

In this palladium-catalyzed reaction, a palladium π-allylic complex is initially formed. This complex undergoes reaction with the sodiomalonate to yield the observed product.

When the same reaction is carried out under classical reaction conditions, only the *trans* product is formed. (J.-E. Backvall and J. O. Vagberg, *Org. Syn.* **1990**, *69*, 38.)

Questions

1. Are the products in these reactions optically active? Draw the enantiomers of the final products.
2. Could the palladium-catalyzed reaction have been used to synthesize the following compound? Explain your answer.

3. What would have been the product(s) if the classical substitution reaction between *cis*-1-acetoxy-4-chloro-2-cyclohexene and sodiomalonate had been carried out using CH_3OH as solvent? What type of reaction pathway would have been involved?
4. If the *cis* product were desired and the palladium-catalyst method was not available, what would be the required starting material?
5. Explain why the reaction with the palladium catalyst involves retention of configuration.

Study Problems*

5.24 Name each of the following compounds by the IUPAC system:

(a) $CCl_3CH{=}CH_2$ (b) $CH_3\overset{\overset{\displaystyle Br}{|}}{C}HCH_2CH_2Br$ (c) [cyclohexane with Br and CH_3]

(d) [cyclopentane with OH, H, H, Cl] (e) $H-\overset{\overset{\displaystyle I}{|}}{\underset{\displaystyle CH_3}{C}}\!\!\!\!\ldots CO_2H$

5.25 Give the structure for each of the following compounds:

(a) isobutyl iodide
(b) 1-iodo-2-methylpropane
(c) *cis*-1,3-dichlorocyclohexane
(d) 2-bromo-3-methyl-1-butanol
(e) (2*R*,3*R*)-2-bromo-3-chlorobutane

5.26 Classify the following organohalogen compounds as methyl, 1°, 2°, or 3°:

(a) $(CH_3)_3CCH_2Cl$ (b) $(CH_3CH_2)_3CCl$ (c) [cyclohexane with CH_3 and Cl]

5.27 Which compound in each of the following pairs would undergo more rapid S_N2 reaction?

(a) $(CH_3)_3CI$ or $(CH_3CH_2)_2CHI$ (b) $(CH_3)_2CHI$ or $(CH_3)_2CHCl$

(c) [cyclohexyl]—Cl or [cyclohexyl]—CH_2Cl (d) [cyclohexyl]—Cl or [cyclohexane with CH_3 and Cl]

5.28 What is the effect on the rate of the S_N2 reaction of CH_3I and OH^- when:

(a) the concentration of CH_3I is tripled and that of OH^- is doubled?

(b) the concentration of OH^- is halved?

(c) the temperature is increased?

(d) the ratio of solvent to reactants is doubled?

5.29 Which of the following two syntheses for *tert*-butyl ethyl ether would be preferred and why?

(a) $(CH_3)_3CO^- + CH_3CH_2Br$ (b) $(CH_3)_3CBr + CH_3CH_2O^-$

5.30 Write equations showing the structures of the transition state and the product for the S_N2 reaction of $^-OCH_3$ with each of the following alkyl halides:

(a) (*R*)-2-bromobutane
(b) *trans*-1-chloromethyl-4-methylcyclohexane
(c) (*R*)-2-bromo-3-methylbutane
(d) (*S*)-2-bromo-3-methylbutane

5.31 Complete the following equations for S_N2 reactions.

(a) [cyclohexyl]—Cl + $NaSCH_3$ $\longrightarrow$ (b) [ring]N: + CH_3I $\longrightarrow$

* For information concerning the organization of the *Study Problems* and *Additional Problems*, see the *Note to student* on page 41.

(c) $(CH_3CH_2O\overset{\displaystyle O}{\overset{\|}{C}})_2\overset{\displaystyle ..}{C}H^- + CH_3I \longrightarrow$

(d) 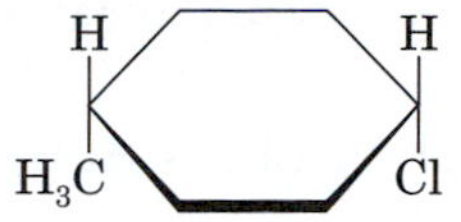 $+ OH^- \longrightarrow$

(e) $CH_3CH_2CH_2Br + {}^-SCH_2CH_3 \longrightarrow$

(f) $(S)-CH_3CH(CH_2)_3CH_3 + (R)-CH_3CH_2\overset{\displaystyle O^-}{\overset{\|}{C}}HCH_2CH_2CH_3$

with I below the first structure.

5.32 Suggest reagents for the preparation of each of the following compounds by an S_N2-type reaction of an organohalogen compound with a nucleophile:

(a) $(CH_3)_2CHOC_6H_5$ **(b)**    **(c)** 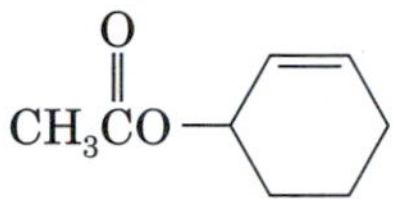 $CH_3\overset{\displaystyle O}{\overset{\|}{C}}O-$

5.33 Rank the following alkyl bromides in order of increasing rate of S_N2 reaction with sodium azide ($Na^+N_3^-$).

(a) $(CH_3)_2CHCH_2Br$ **(b)** $CH_3CH_2CH{=}CHBr$

(c) $CH_3CH_2CH_2CH_2Br$ **(d)** $CH_3CH_2CHBrCH_3$

5.34 Which of the following carbocations is the most stable? Which is the least stable? Explain.

(a) 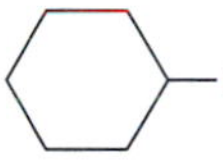 $-CH_2{}^+$ **(b)** $(CH_3CH_2)_2CH^+$ **(c)** $(CH_3CH_2)_3C^+$

5.35 **(1)** Show the products of the solvolysis of the following compounds in *aqueous ethanol*. **(2)** Which halide would react the fastest?

(a) $(CH_3)_2CHBr$ **(b)** $(CH_3)_3CBr$ **(c)** $(CH_3)_3CI$

5.36 Predict the products of the aqueous solvolysis of the following halides:

(a) *cis*-1-iodo-3-methylcyclohexane   **(b)** (*R*)-2-iodo-4-ethylhexane

(c) (3*S*,5*R*)-3-iodo-3,5-dimethylheptane   **(d)**

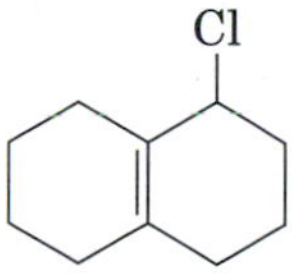

5.37 Each of the following carbocations is capable of undergoing rearrangement to a more stable carbocation. Suggest a structure for the rearranged carbocation.

(a) $(CH_3)_3C\overset{+}{C}HCH_2CH_3$ **(b)** $CH_2{=}CHCH_2\overset{+}{C}HCH_3$

(c) $(CH_3)_2CH\overset{+}{C}HCH_2CH(CH_3)_2$ **(d)**  $-\overset{+}{C}HCH_3$

5.38 The following reaction can proceed with rearrangement. Show the initial carbocation, the rearranged carbocations and all possible rearranged substitution products.

$$(CH_3)_2\overset{\displaystyle CH_2CH_3}{\overset{|}{C}}{-}CHICH_3 \xrightarrow[\text{heat}]{\text{aqueous } CH_3OH}$$

5.39 Give the important resonance structures of each of the following cations:

(a) $CH_2{=}CHCH{=}\overset{+}{C}HCH_2$ (b)

(c) $\overset{+}{-}CH{=}CH_2$ (d) $-\overset{+}{C}HCH_3$

5.40 (a)   Write the equations for the steps of the E1 reaction of 2-iodo-2-phenylpentane with H_2O, showing only the major alkene product.

   (b)   Which step determines the rate of reaction?

   (c)   What other alkene products could be formed?

5.41 Each of the following alkyl halides can undergo rearrangement in an E1 reaction. For each, show the initial carbocation, the rearranged carbocation, and the anticipated major rearrangement product.

   (a) $(CH_3)_2CHCHClCH(CH_3)_2$ **(b)** $(CH_3)_3CCHBrCH_2CH_2CH_3$

5.42 Which compound in each of the following pairs undergoes E2 reaction more rapidly? Why?

   (a) $(CH_3)_2CHCHBrCH_2CH_3$ or $(CH_3)_2CBrCH_2CH_2CH_3$

   (b) $(CH_3)_2CHCHICH_3$ or $(CH_3)_2CHCH_2CH_2I$

   (c) CH_3CHICH_3 or $CH_3CHBrCH_3$

5.43 Each of the following compounds yields two alkenes upon dehydrohalogenation. Predict the structures of the alkenes and predict which alkene would predominate in each product mixture.

(a) $CH_3(CH_2)_8\overset{\overset{\displaystyle CH_3}{|}}{\underset{\underset{\displaystyle Br}{|}}{C}}CO_2H$ (b)

5.44 Which hydrogen(s) must be replaced by deuterium if a maximum kinetic isotope effect is to be observed in an E2 reaction of each of the following compounds?

(a) $(CH_3)_2CClCH_3$ (b)   (c)

5.45 Predict the E2 products of the reaction of each of the following compounds with $Na^{+-}OCH_3$. If more than one alkene is possible, which is the major product?

   (a)   2-bromopentane   **(b)**   *trans*-1-chloro-2-methylcyclohexane

   (c)   (1*S*,2*S*)-1-bromo-1,2-diphenylbutane   **(d)**   (*S*)-1-chloro-1-cyclohexylethane

5.46 Which of the following E2 reactions will produce the most 1-alkene (Hofmann product)? Why?

   (a) $CH_3CH_2CH_2CHBrCH_3 + {}^-OC)CH_3)_3 \longrightarrow$

   (b) $CH_3CH_2CHBrCH_3 + OH^- \longrightarrow$

5.47 Complete the following equations, showing only the major organic product, and predict which reaction mechanism (S_N1, S_N2, E1, E2) is the most likely:

(a) $(CH_3)_2CHBr + KI \xrightarrow{\text{acetone}}$

(b) $(CH_3)_2CHBr + KOH \xrightarrow[\text{heat}]{CH_3CH_2OH}$

(c) $(C_6H_5)_2CHBr + CH_3CH_2OH \xrightarrow{\text{heat}}$

(d) $(CH_3)_2CClCH_2CH_3 + Na^+ {}^-OCH_3 \xrightarrow{CH_3OH}$

(e) $CH_3CHBrCH_2CH_2CH_3 + KOH \xrightarrow[\text{heat}]{CH_3CH_2OH}$

(f) $(CH_3)_2CBrCH_2CH_3 + K^+ {}^-OC(CH_3)_3 \xrightarrow{(CH_3)_3COH}$

(g) $CH_2{=}CHCH_2Cl + Na^+ C_6H_5Se^- \xrightarrow{CH_3CH_2OH}$

(h) $ClCH_2CO_2H + 2Na^+ {}^-OCH_2CH_3 \longrightarrow$
   chloroacetic acid
      a herbicide

Additional Problems

5.48 Herbicides can be selective or nonselective. 2,4-D (the sodium salt of 2,4-dichloro-phenoxyacetic acid) is an example of a selective herbicide. It overstimulates the growth of broadleaf plants (but not grasses), causing them to die. *Paraquat* is an example of a nonselective herbicide. Suggest the starting materials that would be needed to prepare each herbicide by an S_N2 reaction.

2,4-D   paraquat diiodide

5.49 2,3-Dibromopropene reacts with an excess of ethylamine ($CH_3CH_2NH_2$) to yield $CH_3CH_2\overset{+}{N}H_3\,Br^-$ and another product.

(a)   What is the structure of the initial substitution product?

(b)   What is the structure of the final product, which is the product of an acid-base reaction?

5.50 The reaction of 2-chloro-2-methyl-1-phenylpropane with methanol yields a mixture composed of a substitution product A (54%) and two elimination products, B (27%) and C (19%). What are the structures of A, B, and C?

5.51 Draw dimensional formulas for the conformers of (1R,2S)-1-bromo-1,2-diphenyl-propane and (1S,2R)-1-bromo-1,2-diphenylpropane that undergo E2 reaction.

5.52 A compound with the formula C_4H_9Cl, upon treatment with a strong base, yields three isomeric alkenes. What is the structure of this alkyl halide?

5.53 (2R,3S)-2-Bromo-3-deuteriobutane undergoes an E2 reaction when treated with $NaOCH_2CH_3$ in ethanol. What would be the principal product(s)?

5.54 (2S,3S)-3-Bromo-2-methoxybutane undergoes an S_N2 reaction with CH_3O^- to yield an optically inactive product. Explain. (Use an equation in your answer.)

5.55 Consider the following hypothetical two-step reaction:

$$A + B \underset{}{\overset{fast}{\rightleftharpoons}} [AB] \xrightarrow{slow} products$$

(a)   Would this reaction show overall *first-order* or *second-order* kinetics? Explain.

(b)   Is this reaction *unimolecular* or *bimolecular*? Explain.

5.56 Is the stereochemical result given for the following reaction consistent with an S_N2 mechanism? Explain.

$$\overset{\overset{\text{Br}}{|}}{(R)\text{-CH}_3\text{CHCH}_2\text{OCH}_3} + \text{NaCN} \xrightarrow[\text{heat}]{\text{H}_2\text{O}} \overset{\overset{\text{CN}}{|}}{(R)\text{-CH}_3\text{CHCH}_2\text{OCH}_3}$$

5.57 Predict the structure of the organohalide that would be needed to prepare the following structures by a substitution or elimination reaction.

(a) 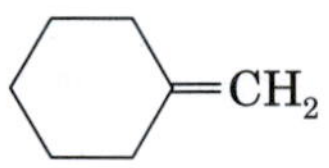 [ring]=CH_2 (b) 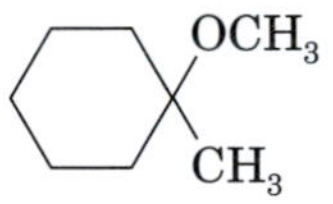 [ring with OCH_3 and CH_3]

5.58 Suggest a mechanism for each of the following reactions:

(a) $(CH_3)_2CHCl + AgNO_3 + H_2O \longrightarrow (CH_3)_2CHOH + AgCl$

(b) $\overset{\overset{\text{Br}}{|}}{\text{CH}_2{=}\text{CHCHCH}_3} \xrightarrow[\text{H}_2\text{O}]{\text{LiBr}} \text{CH}_3\text{CH}{=}\text{CHCH}_2\text{Br}$

(c) $CH_3CH_2CH_2Cl + NO_2^- \longrightarrow CH_3CH_2CH_2NO_2 + CH_3CH_2CH_2ONO + Cl^-$

(d) $\overset{\overset{\text{Cl}}{|}}{\text{CH}_2{=}\text{CHC(CH}_3)_2} + H_2O \xrightarrow{Ag_2O} \overset{\overset{\text{OH}}{|}}{\text{CH}_2{=}\text{CHC(CH}_3)_2} + HOCH_2CH{=}C(CH_3)_2$

   (86%)   (14%)

(e) $BrCH_2(CH_2)_8CH_2CO_2H \xrightarrow[100°]{\overset{K_2CO_3}{(CH_3)_2S{=}O}}$ [macrocyclic lactone structure]

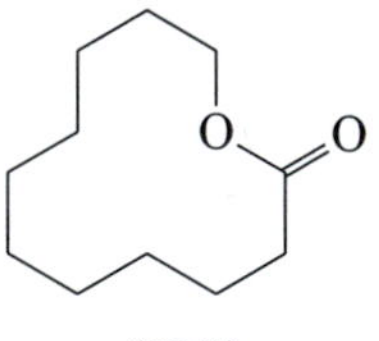

(83%)

5.59 Explain the following observation:

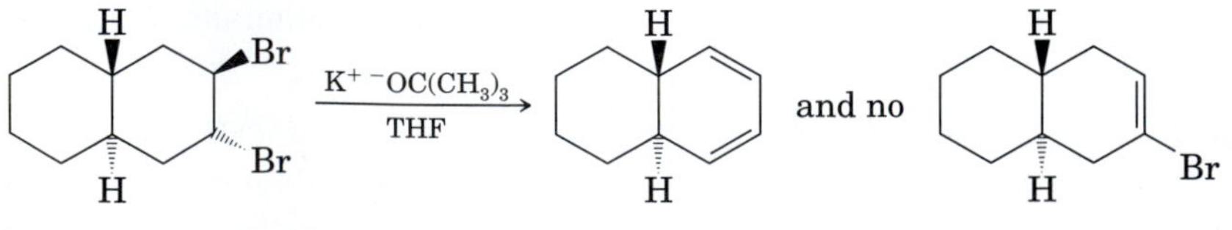

[bicyclic decalin structures] $\xrightarrow[\text{THF}]{K^+ {}^-OC(CH_3)_3}$ [diene structure] and no [bromo-alkene structure]

(76%)

5.60 Formulas for two of the stereoisomers of menthyl chloride follow. One of these menthyl chlorides undergoes rapid E2 reaction when treated with a base to yield two alkenes:

76% A and 25% B. The other menthyl chloride undergoes a slow E2 reaction and yields only one alkene.

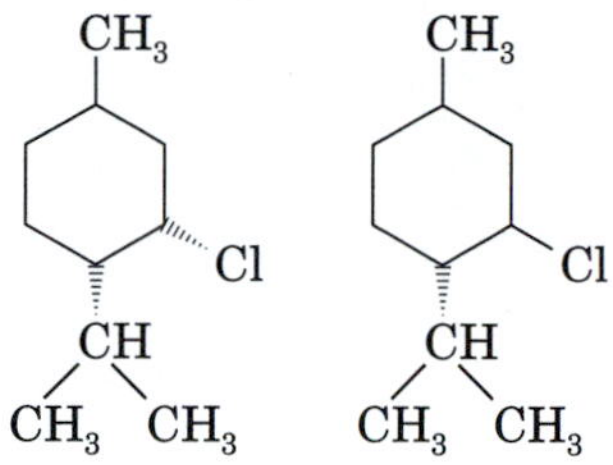

(a) Which menthyl chloride undergoes the more rapid reaction? Why?

(b) What two alkenes are formed in this reaction, and which one predominates? Explain.

(c) Which menthyl chloride undergoes the slow reaction, and what is the structure of the single alkene product? Explain why the reaction is slow and yields only one alkene.

5.61 Which nucleophile will produce the greatest amount of substitution versus elimination in the following reaction? Which will produce the greatest amount of elimination versus substitution?

$$\underset{\text{Br}}{\text{CH}_3\overset{|}{\text{C}}\text{HCH}_2\text{CH}_3} + \text{Nu}:^- \quad \xrightarrow[\text{heat}]{\text{DMF}}$$

where $\text{Nu}:^-$ = **(a)** $\text{CH}_3\text{CH}_2\text{O}^-$,   **(b)** $(\text{CH}_3)_3\text{CO}^-$, or   **(c)** CH_3CO_2^-

5.62 A beginning graduate student needed some (+)-2-iodobutane for a research project. Although none of the (+)-isomer was available, several kilograms of (−)-2-iodobutane ($[\alpha]_D = -32°$) were on hand. Remembering that S_N2 reactions occur with inversion of configuration, the student heated the (−)-isomer with excess sodium iodide in acetone. When the optical rotation of the isolated 2-iodobutane was measured, the student was surprised to find it had an $[\alpha]_D = 0°$ instead of the expected $[\alpha]_D = +32°$. What happened?

6

FREE-RADICAL REACTIONS

Many organohalogen compounds are prepared industrially by the reaction of hydrocarbons and halogens, two relatively inexpensive starting materials. Direct halogenation reactions often proceed explosively and, as a general rule, give mixtures of products. For these reasons, direct halogenation is used only occasionally in the laboratory.

$$CH_3CH_3 + Cl_2 \xrightarrow{\text{light}} CH_3CH_2Cl$$
ethane

$$+ ClCH_2CH_2Cl + HCl + \text{other products}$$

Direct halogenation reactions proceed by a **free-radical**, or **radical, mechanism**, which is different from the mechanisms discussed in Chapter 5 for nucleophilic substitutions and eliminations. Radical reactions have biological and practical importance. For example, organisms utilize atmospheric oxygen by a sequence of reactions that begins with a radical oxidation–reduction. Butter and other fats become rancid partly by radical reactions with oxygen.

Polymerization reactions are reactions that yield giant molecules from the joining of small molecules. Many polymerization reactions are also radical reactions.

$$\text{many units of } CH_2{=}CH_2 \xrightarrow{\text{initiator}} \{-CH_2CH_2-CH_2CH_2-CH_2CH_2-\}$$

ethylene

a portion of a polyethylene molecule,
a polymer used for plastic bags and
many other items

Section 6.1

A Typical Radical Reaction: Chlorination of Methane

The term **free radical** or **radical** refers to any atom or group of atoms that has *one or more unpaired electrons.* Although a radical usually has no positive or negative charge, such a species is highly reactive because of its unpaired electron and incomplete octet. A radical is usually encountered as a high-energy, highly reactive, short-lived, nonisolable reaction intermediate.

We usually symbolize a radical with a single dot representing the unpaired electron.

Lewis formulas for typical radicals:

$$:\ddot{C}l\cdot \qquad :\ddot{B}r\cdot \qquad H{:}\overset{\displaystyle H}{\underset{\displaystyle H}{\ddot{C}}}\cdot$$

Usual formulas for radicals (note the omission of the paired electrons):

$$Cl\cdot \qquad Br\cdot \qquad H_3C\cdot \quad \text{or} \quad CH_3\cdot \quad \text{or} \quad \cdot CH_3$$

Moses Gomberg (1866–1947), a Russian-born chemist working at the University of Michigan, was the first to propose the existence of a free radical. In 1900, Gomberg was attempting to prepare hexaphenylethane by the reaction of triphenylchloromethane with zinc. The reaction did yield a product but its properties were inconsistent with hexaphenylethane. Gomberg wrote that "The experimental evidence . . . forces me to the conclusion that we have to deal here with a free radical, triphenylmethyl $(C_6H_5)_3C$. . . . [T]he fourth valence of the methane is bound either to take up the complicated group $(C_6H_5)_3C-$ or remains as such, with carbon as trivalent. Apparently the latter is what happens." It was ten years before his proposal found acceptance.

STUDY PROBLEM

6.1   Write Lewis formulas for the following radicals:

(a) $HS\cdot$ (b) $CH_3CH_2O\cdot$ (c) $\dot{C}H_2Cl$

The chlorination of methane in the presence of light (symbolized $h\nu$; see Section 9.1) is a classical example of a radical reaction. The result of the reaction of Cl_2 with CH_4 is the *substitution* of one or more chlorine atoms for hydrogen atoms on the carbon.

$$CH_4 \; + Cl_2 \; \xrightarrow{h\nu} \; CH_3Cl \; + \; CH_2Cl_2 \; + \; CHCl_3$$

methane   chloromethane   dichloromethane   trichloromethane
                          (methyl chloride)   (methylene chloride)   (chloroform)

$$+ \quad CCl_4 \qquad + HCl$$

tetrachloromethane
(carbon tetrachloride)

Although methane is the simplest alkane, four organic products can be formed in its chlorination. Small amounts of higher alkanes, such as ethane, and their chlorinated products may also be formed. First, we will discuss the reactions leading to CH_3Cl. Then, we will expand the discussion to the formation of other products.

The mechanism of this type of radical reaction is best thought of as a series of stepwise reactions. Each step falls into one of the following categories: (1) **initiation** of the radical reaction; (2) **propagation** of the radical reaction; and (3) **termination** of the radical reaction.

A. Initiation

As the term implies, the initiation step is the initial formation of radicals. In the chlorination of methane, the initiation step is the homolytic cleavage of Cl_2 into two chlorine radicals. (You may wish to review the discussion of bond dissociation in Section 1.4E). The energy for this reaction step is provided by light or by heating the mixture to a very high temperature.

Step 1 (initiation):

$$Cl\overbrace{}Cl + 58 \text{ kcal/mol} \xrightarrow{h\nu \text{ or heat}} 2 \text{ Cl} \cdot$$

B. Propagation

After its formation, the chlorine radical starts a series of reactions in which new radicals are formed. Collectively, these reactions are called the **propagation steps** of the radical reaction. In effect, the initial formation of a few radicals results in the propagation of new radicals in a self-perpetuating reaction called a **chain reaction.**

As the first propagation step, the reactive chlorine radical abstracts a hydrogen atom from methane to yield a methyl radical and HCl.

$$Cl \cdot \quad H \colon CH_3 + 1 \text{ kcal/mol} \longrightarrow H \colon Cl + \cdot CH_3$$

The methyl radical is also reactive. In the second propagation step, the methyl radical abstracts a chlorine atom from Cl_2.

$$\cdot CH_3 \quad Cl \colon Cl \longrightarrow CH_3Cl \quad + Cl \cdot + 25.5 \text{ kcal/mol}$$

chloromethane

This step yields one of the products of the overall reactions, chloromethane. This step also regenerates a chlorine radical that can abstract a hydrogen atom from another methane molecule and start the propagation sequence over again.

The overall sequence so far is:

Initiation:

$$Cl_2 \xrightarrow{\text{$h\nu$ or heat}} 2\,Cl\cdot$$

Propagation:

$$CH_4 + Cl\cdot \longrightarrow \cdot CH_3 + HCl$$

$$\cdot CH_3 + Cl_2 \longrightarrow CH_3Cl + Cl\cdot \quad \textit{can undergo reaction with CH}_4$$

The reaction is self-perpetuating because a single Cl· starts a reaction sequence in which a new Cl· (along with products) is formed. However, as you might imagine, the reaction does not continue indefinitely. The *number of cycles* (that is, the number of passes through the propagation steps) is called the **chain length.** The chain length of a radical reaction depends partly upon the energies of the radicals involved in the propagation. (We will discuss this subject shortly.) For chlorination of a hydrocarbon, the chain length is about 10,000.

C. Termination

The propagation cycle is broken by **termination reactions.** Any reaction that results in the destruction of radicals or in the formation of stable, nonreactive radicals can terminate the propagation cycle. The chlorination of methane is terminated principally by radicals combining with other radicals. This is a process of destruction of radicals. In Section 6.7, we will mention termination by formation of stable, nonreactive radicals.

Termination steps:

$$Cl\cdot + \cdot CH_3 \longrightarrow CH_3Cl$$

$$\cdot CH_3 + \cdot CH_3 \longrightarrow CH_3CH_3$$

The second termination step shown is an example of a **coupling reaction**: the joining together of two alkyl groups.

Termination reactions are often difficult to determine. In contrast to propagation reactions, termination reactions rarely occur and, consequently, they form only a small amount of product. Most of the products arise from propagation reactions, not termination reactions.

In summary,

1. **initiation**: increases the net number of radicals

2. **propagation**: leaves the net number of radicals unchanged

3. **termination**: decreases the net number of radicals

STUDY PROBLEM

6.2   Write equations for the initiation, propagation, and termination reactions leading to the formation of chlorocyclohexane from cyclohexane and chlorine.

D. Why Radical Reactions Yield Mixtures of Products

Radical reactions are often characterized by a multitude of products. For example, the chlorination of methane can yield four organic products. The reason for the formation of these mixtures is that the high-energy chlorine radical is not particularly selective about which hydrogen it abstracts during the propagation step.

While chlorine is undergoing reaction with methane, chloromethane is being formed. In time, the chlorine free radicals are more likely to collide with chloromethane molecules than with methane molecules, and a new propagation cycle is started. In this new cycle, chloromethyl radicals ($\cdot CH_2Cl$) are formed. These undergo reaction with chlorine molecules to yield dichloromethane (CH_2Cl_2). As in the previous cycle leading to CH_3Cl, another chlorine radical is regenerated in the process.

Propagation steps leading to dichloromethane:

$$Cl\cdot + CH_3Cl \longrightarrow HCl + \cdot CH_2Cl$$

$$\cdot CH_2Cl + Cl_2 \longrightarrow CH_2Cl_2 + Cl\cdot$$

dichloromethane

SAMPLE PROBLEM

Write the propagation steps leading to the formation of trichloromethane (chloroform) from dichloromethane.

Solution

$$\cdot Cl + CH_2Cl_2 \longrightarrow HCl + \cdot CHCl_2$$

$$\cdot CHCl_2 + Cl_2 \longrightarrow Cl\cdot + CHCl_3$$

STUDY PROBLEM

6.3 Write the propagation steps leading to the formation of tetrachloromethane (carbon tetrachloride) from trichloromethane.

We've seen that the radical chlorination of methane can yield four organic products (or more, if coupling products are considered). Because higher alkanes have several additional kinds of hydrogens available, their propagation reactions can produce even larger numbers of products.

STUDY PROBLEM

6.4 How many chloroalkanes could be produced in the chlorination of ethane? Write formulas for their structures.

SAMPLE PROBLEM

A chemist wishes to make chloroethane from chlorine and ethane. If higher chlorination products are to be avoided would the chemist use: (a) an equimolar mixture of CH_3CH_3 and Cl_2; (b) an excess of Cl_2; or (c) an excess of CH_3CH_3?

Solution
(c) By using an excess of CH_3CH_3, the chemist increases the probability of collisions between $Cl \cdot$ and CH_3CH_3 and decreases the probability of collisions between $Cl \cdot$ and CH_3CH_2Cl.

Section 6.2
Relative Reactivities of the Halogens

A. Bond Dissociation Energies

We discussed bond dissociation energies in Section 1.4E. Besides providing a quantitative measure of bond strength, bond dissociation energies allow a chemist to calculate whether a reaction is **endothermic** (heat absorbing) or **exothermic** (heat releasing). For example, is the chlorination of methane, CH_4, exothermic or endothermic? To answer this question, we break the bonds undergoing reaction homolytically to obtain free radicals, and assign the bond dissociation energy to each reaction. We then sum the bond dissociation energies to determine whether energy is released or absorbed. (The term "$+\Delta H$" means energy is absorbed; the term "$-\Delta H$" means energy is released.)

		ΔH	
(1) $Cl-Cl + 58\,kcal/mol \longrightarrow Cl\cdot + Cl\cdot$		$+58$	kcal/mol
(2) $H_3C-H + 104\,kcal/mol \longrightarrow H_3C\cdot + H\cdot$		$+104$	kcal/mol
(3) $H_3C\cdot + Cl\cdot \longrightarrow H_3C-Cl + 83.5\,kcal/mol$		-83.5	kcal/mol
(4) $H\cdot + Cl\cdot \longrightarrow H-Cl + 103\,kcal/mol$		-103	kcal/mol

Net reaction:

$$Cl_2 + CH_4 \longrightarrow CH_3Cl + HCl + 24.5\,kcal/mol \qquad net\ \Delta H = -24.5\,kcal/mol$$

We calculate that the chlorination of methane releases energy. If we ran this reaction in the laboratory, we would find that it is indeed exothermic.

For most purposes, we can simplify the above calculation and still obtain the correct answer. If we combine equations (2) with (3) and (1) with (4), we obtain:

Combination of equations (2) and (3)

(2) $H_3C-H + 104\,kcal/mol \longrightarrow H_3C\cdot + \cdot H$
(3) $H\cdot + Cl\cdot \longrightarrow HCl + 103\,kcal/mol$

(a) $H_3C-H + Cl\cdot + 1\,kcal/mol \longrightarrow HCl + \cdot CH_3 \quad \Delta H = +1\,kcal/mol$

Combination of equations (1) and (4)

(1) $Cl-Cl + 58\,kcal/mol \longrightarrow Cl\cdot + Cl\cdot$
(4) $\cdot CH_3 + Cl\cdot \longrightarrow CH_3Cl + 83.5\,kcal/mol$

(b) $Cl-Cl + \cdot CH_3 \longrightarrow CH_3Cl + Cl\cdot + 25.5\,kcal/mol \quad \Delta H = -25.5\,kcal/mol$

Using the combined equations (a) and (b), we capture the essential information in an abbreviated equation, (c).

(a) H_3C—H + $Cl\cdot$ + 1 kcal/mol $\longrightarrow$ HCl + $CH_3\cdot$ ΔH = +1 kcal/mol
(b) Cl—Cl + $\cdot CH_3$ $\longrightarrow$ CH_3Cl + $Cl\cdot$ + 25.5 kcal/mol ΔH = −25.5 kcal/mol

(c) CH_4 + Cl_2 $\longrightarrow$ CH_3Cl + HCl + 24.5 kcal/mol ΔH = −24.5 kcal/mol

STUDY PROBLEM

6.5 Using bond dissociation energies from Table 1.4 (page 17), predict which of the following reactions liberates more energy:

  (a) $(CH_3)_3CH$ + Cl_2 $\longrightarrow$ $(CH_3)_3CCl$ + HCl

  (b) CH_4 + Cl_2 $\longrightarrow$ CH_3Cl + HCl

B. Halogen Reactivity

The halogens vary dramatically in their reactivity toward alkanes in radical reactions. Fluorine undergoes explosive reactions with hydrocarbons. Chlorine is next in terms of reactivity, followed by bromine. Iodine is nonreactive toward alkanes.

$$I_2 \quad Br_2 \quad Cl_2 \quad F_2$$

increasing reactivity as free-radical agents ➤

The relative reactivity of the halogens toward alkanes is *not* due to the ease with which X_2 molecules are cleaved into radicals. From the bond dissociation energies for the halogens, we can see that the relative ease of homolytic cleavage is almost the reverse of their reactivity in halogenation reactions.

	F_2	Cl_2	Br_2	I_2
bond dissociation energy (kcal/mol):	37	58	46	36

The order of reactivity of the halogens in radical halogenations (relative rates of reaction under similar conditions) depends on the energy of activation for the rate-determining step. In radical halogenations, we can approximate the order of reactivity by examining the heats of reaction. The propagation steps of fluorination are highly exothermic. Because fluorination has a low E_{act}, the heat produced by the exothermic reaction increases the rate of reaction by increasing the temperature. The result is an extremely rapid, explosive reaction.

$F\cdot$ + CH_4 $\longrightarrow$ HF + $CH_3\cdot$ + 31 kcal/mol ΔH = −31 kcal/mol
$CH_3\cdot$ + F_2 $\longrightarrow$ CH_3F + $F\cdot$ + 71 kcal/mol ΔH = −71 kcal/mol

CH_4 + F_2 $\longrightarrow$ CH_3F + HF + 102 kcal/mol   net ΔH = −102 kcal/mol

Just the reverse situation is encountered with iodine: taken together the two propagation steps are *endothermic*—that is, the products are of higher energy than the reactants. Most important, the energy required by $I\cdot$ to abstract hydrogen from a C—H bond is *substantially* endothermic. The result is that the iodine radical does not enter into a chain reaction. $I\cdot$ is an example of a *stable radical,* a radical that does not abstract hydrogens.

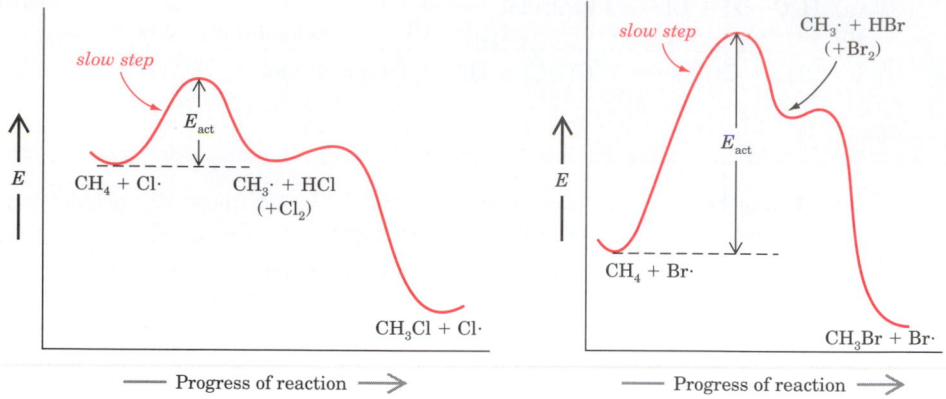

Figure 6.1 Energy diagrams for the radical chlorination and bromination of methane.

$$\begin{array}{lll} \text{I}\cdot + \text{CH}_4 + 33\,\text{kcal/mol} \longrightarrow \text{HI} + \text{CH}_3\cdot & \Delta H = +33\,\text{kcal/mol} \\ \text{CH}_3\cdot + \text{I}_2 \longrightarrow \text{CH}_3\text{I} + \text{I}\cdot + 20\,\text{kcal/mol} & \Delta H = -20\,\text{kcal/mol} \\ \hline \text{CH}_4 + \text{I}_2 + 13\,\text{kcal/mol} \longrightarrow \text{CH}_3\text{I} + \text{HI} & \text{net } \Delta H = +13\,\text{kcal/mol} \end{array}$$

Chlorine and bromine are intermediate between fluorine and iodine in their overall ΔH of the propagation steps and therefore are also intermediate in reactivity. Figure 6.1 shows energy diagrams for the reactions of Cl_2 and Br_2 with methane.

$$\begin{array}{lll} \text{Cl}\cdot + \text{CH}_4 + 1\,\text{kcal/mol} \longrightarrow \text{HCl} + \text{CH}_3\cdot & \Delta H = +1 \quad\,\text{kcal/mol} \\ \text{CH}_3\cdot + \text{Cl}_2 \longrightarrow \text{CH}_3\text{Cl} + \text{Cl}\cdot + 25.5\,\text{kcal/mol} & \Delta H = -25.5\,\text{kcal/mol} \\ \hline \text{CH}_4 + \text{Cl}_2 \longrightarrow \text{CH}_3\text{Cl} + \text{HCl} + 24.5\,\text{kcal/mol} & \text{net } \Delta H = -24.5\,\text{kcal/mol} \end{array}$$

$$\begin{array}{lll} \text{Br}\cdot + \text{CH}_4 + 17\,\text{kcal/mol} \longrightarrow \text{HBr} + \text{CH}_3\cdot & \Delta H = +17\,\text{kcal/mol} \\ \text{CH}_3\cdot + \text{Br}_2 \longrightarrow \text{CH}_3\text{Br} + \text{Br}\cdot + 24\,\text{kcal/mol} & \Delta H = -24\,\text{kcal/mol} \\ \hline \text{CH}_4 + \text{Br}_2 \longrightarrow \text{CH}_3\text{Br} + \text{HBr} + 7\,\text{kcal/mol} & \text{net } \Delta H = -7 \quad\,\text{kcal/mol} \end{array}$$

In summary, we find that chlorine and bromine are the most useful radical halogenating agents.

Section 6.3

Stereochemistry of Radical Halogenation

An alkyl radical is a species in which a carbon atom has three atoms or groups bonded to it and a single, unpaired electron. We will look at the structure of the methyl radical. Other alkyl radicals have similar bonding around the radical carbon.

A radical carbon is in the sp^2-hybrid state. The three sp^2 orbitals are planar in the methyl radical and very nearly planar in most other radicals. The unpaired electron of the radical is in the p orbital. The structure is very similar to that of a carbocation, except that the p orbital of a carbocation is empty.

methyl radical, $CH_3\cdot$

When a pure enantiomer of a chiral alkyl halide undergoes S_N1 reaction at the chiral carbon, racemization is observed. As we discussed in Chapter 5, racemization arises from the nucleophile being able to attack either lobe of the empty p orbital of the carbocation. If a hydrogen is abstracted from the chiral carbon of a pure enantiomer in a radical reaction, racemization also occurs.

(S)-1-chloro-2-methylbutane   (RS)-1,2-dichloro-2-methylbutane

(racemic)

The preceding reaction can lead to a number of products. There are five carbon atoms in the molecule that can lose a hydrogen and gain a chlorine. We would also expect to find as products trichlorinated alkanes, tetrachlorinated alkanes, and so forth. But we are interested only in the one product that has been chlorinated at the chiral carbon. When we isolate this specific product, we find that it is a racemic mixture of the (R) and (S) enantiomers. Just as in an S_N1 reaction, this evidence leads us to believe that the bonds around the radical carbon are planar (sp^2-hybrid orbitals) and that a chlorine atom can add to either lobe of the p orbital.

STUDY PROBLEM

6.6   Another dichloroalkane formed in the chlorination of (S)-1-chloro-2-methylbutane is $CH_3CHClCH(CH_3)CH_2Cl$. Is this dichloroalkane racemic or not?

Section 6.4
Hydrogen Abstraction: The Rate-Determining Step

Unlike the kinetics of substitution and elimination reactions, the kinetics of a radical reaction are quite complex. Simple rate expressions, such as first-order or second-order, are not encountered in radical chemistry. The reason for this complexity is that the steps in a radical reaction are enmeshed in a cyclical process of varying chain lengths. However, evidence does point to the **hydrogen-abstraction step** as the step governing the overall rate at which products are formed. For example, methane (CH_4) undergoes chlorination 12 times faster than perdeuteriomethane (CD_4), indicating that the CH bond is broken in the rate-determining step of the reaction.

H (or D) abstraction is the rate-determining step:

$$CH_4 + Cl\cdot \xrightarrow{\text{faster}} CH_3\cdot + HCl$$

$$CD_4 + Cl\cdot \xrightarrow{\text{slower}} CD_3\cdot + DCl$$

A. Which Hydrogen Is Abstracted?

The hydrogen atoms in organic compounds can be classified as **methyl** (CH_4), **primary** (bonded to a 1° carbon), **secondary** (bonded to a 2° carbon), **tertiary** (bonded to a 3° carbon), **allylic** (on a carbon adjacent to a double bond), or **benzylic** (on a carbon adjacent to an aromatic ring.

2° hydrogens

$CH_3CH_2CH_3$ $CH_2{=}CHCH_3$ ⬡—CH_3

1° hydrogens   *allylic hydrogens*   *benzylic hydrogens*

These different types of hydrogen atoms are not abstracted at identical rates by radicals. Instead, there is a degree of selectivity in hydrogen abstraction. The reaction of propane with a small amount of chlorine under radical conditions yields two monochlorinated products, 1-chloropropane and 2-chloropropane, with 2-chloropropane predominating.

$$CH_3CH_2CH_3 + Cl_2 \xrightarrow{h\nu} \underset{\substack{\text{2-chloropropane (55\%)}\\ \text{(isopropyl chloride)}}}{CH_3\overset{\displaystyle Cl}{\underset{\displaystyle |}{C}}HCH_3} + \underset{\substack{\text{1-chloropropane (45\%)}\\ \text{(propyl chloride)}}}{CH_3CH_2CH_2Cl}$$

propane

There are *six* primary hydrogens and *two* secondary hydrogens in propane. The ratio of primary to secondary hydrogens in propane is 6/2, or 3/1. If all the hydrogens underwent abstraction at equal rates, we would observe three times more 1-chloropropane than 2-chloropropane in the product mixture. This is *not* what is observed when the reaction is carried out. Instead, slightly more 2-chloropropane is formed. We conclude that secondary hydrogens are abstracted at a faster rate than are primary hydrogens.

Another example of how the relative rates of hydrogen abstraction affect the product ratio follows. (We will discuss the greater selectivity of Br_2 in Section 6.5.)

$$CH_3\underset{\substack{\displaystyle |\\ \displaystyle CH_3}}{C}HCH_3$$

methylpropane
(isobutane)

$\xrightarrow{\underset{h\nu}{Cl_2}}$ $\underset{\substack{\text{1-chloro-2-methylpropane}\\ \text{(isobutyl chloride)}\\ \text{(50\%)}}}{(CH_3)_2CHCH_2Cl}$ + $\underset{\substack{\textit{tert}\text{-butyl chloride}\\ \text{(30\%)}}}{(CH_3)_3CCl}$ + other products (20%)

$\xrightarrow{\underset{h\nu}{Br_2}}$ $\underset{\substack{\textit{tert}\text{-butyl bromide}\\ \text{(almost 100\%)}}}{(CH_3)_3CBr}$

Through these and similar experiments, the order of reactivity of hydrogens toward radical halogenation has been determined. The relative rates for halogenation reactions of a few compounds are given in Table 6.1.

Table 6.1 Average relative rates of hydrogen abstraction		
	Reagent[a]	
Hydrocarbon	Br_2	Cl_2
CH_3—H	0.0007	0.004
CH_3CH_2—H	1	1
$(CH_3)_2CH$—H	220	4.3
$(CH_3)_3C$—H	19,400	6.0
$C_6H_5CH_2$—H	64,000	1.3
$(C_6H_5)_2CH$—H	6.2×10^5	2.6
$(C_6H_5)_3C$—H	1.14×10^6	9.5

[a] The two columns contain data from two separate studies of relative rates. The chlorination of ethane proceeds much more rapidly than bromination under the same conditions.

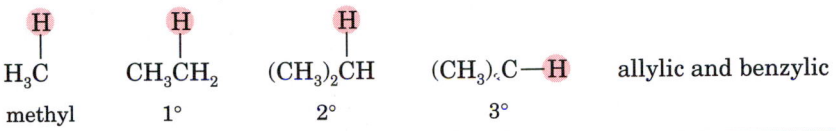

H_3C CH_3CH_2 $(CH_3)_2CH$ $(CH_3)_3C$—H   allylic and benzylic

methyl   1°   2°   3°

increasing rate of reaction toward Br_2

STUDY PROBLEM

6.7 For each of the following structures, label those hydrogen atoms that are (1) 2° and allylic and those that are (2) 3° and benzylic.

(a) H_3C — [benzene ring] —$CH(CH_3)_2$

(b) [benzene ring] —$CH_2CH=CH_2$

(c) [benzene ring] —$\overset{\displaystyle CH_3}{\underset{\displaystyle CH_3}{C}}CH=CHCH_2CH_3$

(d) [bicyclic structure with CH_3 groups]

B. Relative Stabilities of Alkyl Radicals

To understand why some hydrogens are abstracted more easily than others, we must look at the transition states of the hydrogen-abstraction steps. The following equations show the hydrogen-abstraction steps in the chlorination of methane and methylpropane. (The symbol $\delta\cdot$ is used to show that both the chlorine atoms and the carbon atoms have partial radical character in the transition states.)

$$Cl\cdot + H-C \longrightarrow \left[\overset{\delta\cdot}{Cl}----H----\overset{\delta\cdot}{C} \right] \longrightarrow Cl-H + \ \ C\cdot$$

methane   *transition state*   *planar methyl radical* methylpropane   transition state   planar tert-butyl radical

The reactivity sequence $3° > 2° > 1° > CH_4$ arises from the stabilities of the transition states leading to the radicals. Because the transition states have radical character, their stabilities parallel the stabilities of the radicals themselves. These stabilities can be related to the bond dissociation energies of the C—H bonds being broken. A bond breakage leading to a more stable radical requires less energy than one leading to a less stable (higher-energy) radical.

bond dissociation energy (kcal/mol):	CH_3—H	CH_3CH_2—H	$(CH_3)_2CH$—H	$(CH_3)_3C$—H
	104	98	94.5	91

decreasing bond strength →

The order of radical stability, just like that of carbocation stability, increases as we proceed from methyl to tertiary. It is thought that the radical intermediates are stabilized by interaction with neighboring sigma bonds.

$\cdot CH_3$	$\cdot CH_2CH_3$	$(CH_3)_2\dot{C}H$	$(CH_3)_3C\cdot$	allylic and benzylic
methyl	1°	2°	3°	

increasing stability →

The energies of formation of free radicals and carbocations are given in Table 6.2. A lower energy of formation means a more stable intermediate. As seen in Table 6.2, there are larger differences in energy of formation between carbocations than between free radicals.

As in the case of carbocation reactions, we find enhanced radical reactivity at allylic and benzylic positions because of resonance stabilization of the intermediate.

Table 6.2   Energies of formation of carbocations and radicals[a,b]

Carbocation[c]		Free Radical[d]	
Structure	Energy (kcal/mol)	Structure	Energy (kcal/mol)
CH_3^+	217.7	$CH_3\cdot$	104
$CH_3CH_2^+$	181.9	$CH_3CH_2\cdot$	98
$(CH_3)_2CH^+$	162.9	$(CH_3)_2CH\cdot$	95
$(CH_3)_3C^+$	148	$(CH_3)_3C\cdot$	92

[a] Taken from Jerry March, *Advanced Organic Chemistry, 3rd ed.* (John Wiley and Sons, New York, 1985).
[b] Stability is in reverse order. It takes less energy to form the more stable species.
[c] R—Br → R^+Br^- dissociation in the gas phase. Dissociation energies in solution are substantially lower.
[d] R—H → R·H· dissociation.

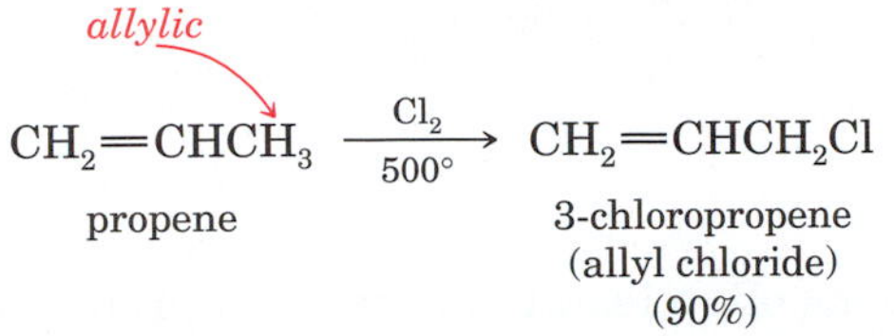
allylic
CH_2=CHCH$_3$
Cl_2
500°
CH_2=CHCH$_2$Cl
propene
3-chloropropene
(allyl chloride)
(90%)

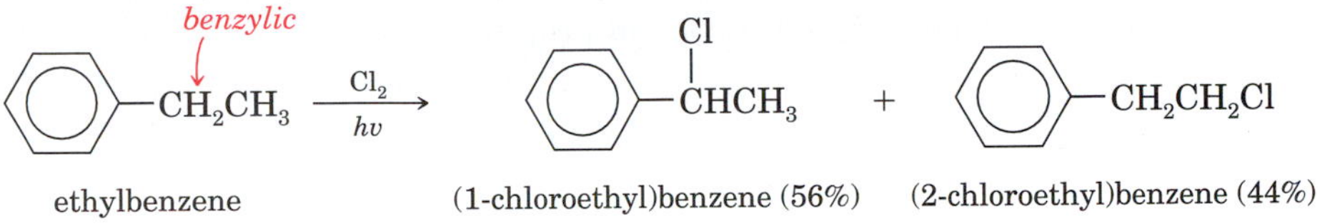
benzylic
CH$_2$CH$_3$
Cl_2
hv
Cl
CHCH$_3$
+
CH$_2$CH$_2$Cl
ethylbenzene
(1-chloroethyl)benzene (56%)
(2-chloroethyl)benzene (44%)

Study Problems

6.8 State which of the two indicated hydrogen atoms would be abstracted at the more rapid rate by Br_2 in the presence of light.

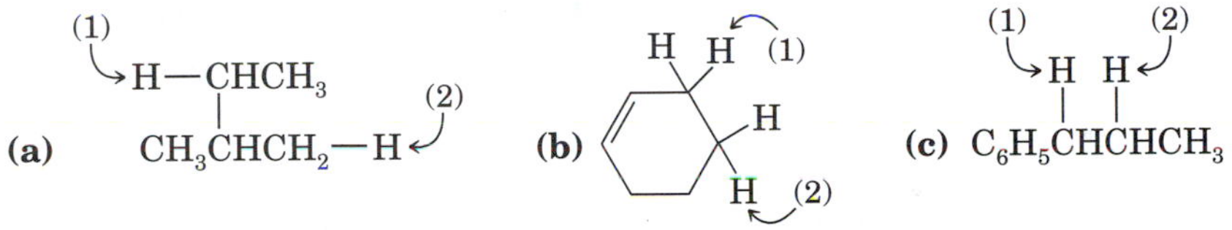
(1)
H—CHCH$_3$
(2)
(a)
CH$_3$CHCH$_2$—H
(b)
H H (1)
H
H (2)
(c) C$_6$H$_5$CHCHCH$_3$

6.9 Write resonance structures (using curved arrows) for the following radicals.

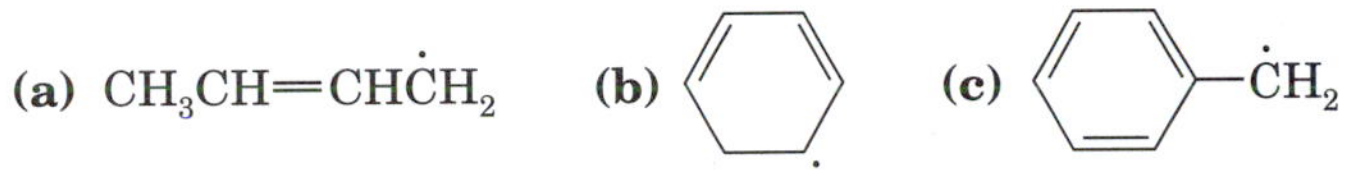
(a) CH$_3$CH=CHĊH$_2$
(b)
(c)
ĊH$_2$

C. Rearrangement of Radicals

Alkyl radicals are, in many respects, similar to carbocations. Both are sp^2 hybrids; both undergo racemization if reaction occurs at a resolved chiral carbon; and both show the same order of stability in terms of structure. Carbocations tend to undergo rearrangement to more-stable carbocations. Does the same hold true for radicals? No, this is one of the differences between free radicals and carbocations. While radical rearrangements are not unknown, they are not common.

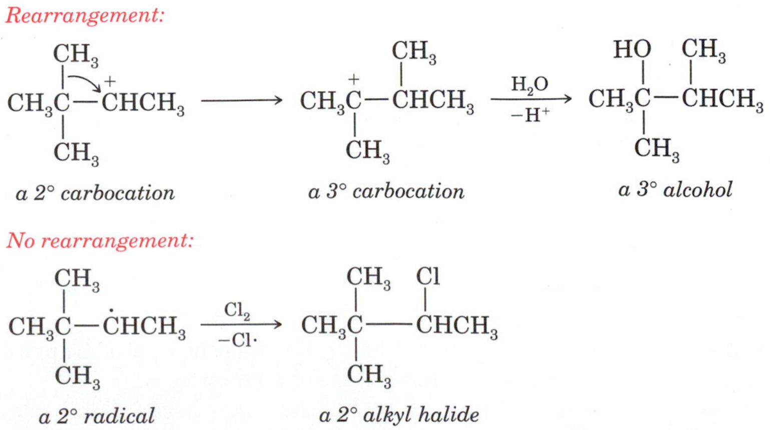
Rearrangement:
CH$_3$
CH$_3$C—CHCH$_3$
CH$_3$
a 2° carbocation
a 3° carbocation
H$_2$O
−H$^+$
HO CH$_3$
a 3° alcohol
No rearrangement:
Cl_2
−Cl·
CH$_3$ Cl
a 2° radical
a 2° alkyl halide

Section 6.5

Selective Radical Halogenations

A. Bromine versus Chlorine

Although radical halogenations often lead to mixtures of products, good yields of single products can be obtained in some cases. Compare the product ratios of chlorination and bromination of propane:

$$
CH_3CH_2CH_3 \quad
\begin{cases}
\xrightarrow[hv]{Cl_2} & \underset{\substack{\text{2-chloropropane} \\ (55\%)}}{CH_3\overset{\displaystyle Cl}{\overset{|}{C}}HCH_3} + \underset{\substack{\text{1-chloropropane} \\ (45\%)}}{CH_3CH_2CH_2Cl} \\
\\
\xrightarrow[hv]{Br_2} & \underset{\substack{\text{2-bromopropane} \\ (98\%)}}{CH_3\overset{\displaystyle Br}{\overset{|}{C}}HCH_3} + \underset{\substack{\text{1-bromopropane} \\ (2\%)}}{CH_3CH_2CH_2Br}
\end{cases}
$$

We can see that bromine, which yields 98% 2-bromopropane, is more selective about abstracting a secondary hydrogen than is chlorine. The selectivity of bromine arises from the fact that bromine is less reactive than chlorine in radical halogenations. To see why this is so, we will consider a pair of hypothetical energy diagrams (Figure 6.2).

Reaction 1 in Figure 6.2 is an *exothermic reaction with a low* E_{act}. The structure of the transition state in Reaction 1 is *very close to that of the reactants*. Like Reaction 1, the hydrogen-abstraction step in the chlorination of propane is exothermic and has a low E_{act}. Therefore, the transition state in this reaction step resembles the reactants more than it does the products.

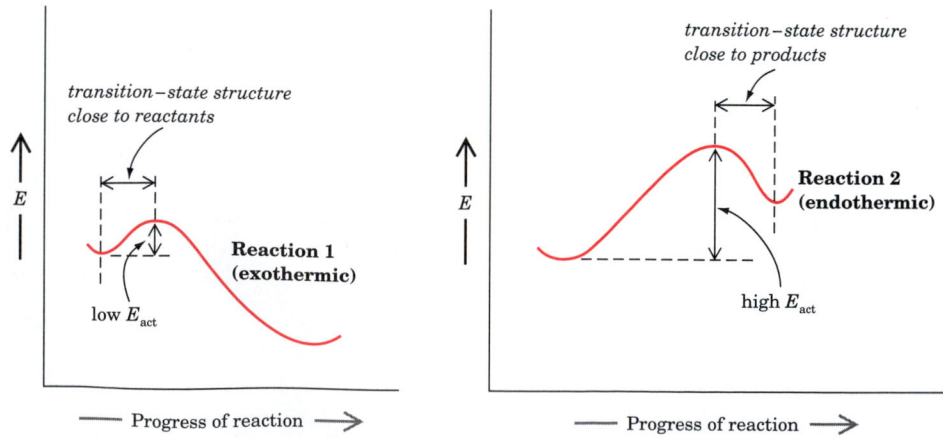

Figure 6.2   Energy diagrams showing the relationship of transition-state structure to the exothermic or endothermic nature of a reaction.

In this transition state, very little bond breaking or bond forming has occurred.

$$Cl\cdot + H-\underset{CH_3}{\overset{H\quad CH_3}{C}} \longrightarrow \left[\overset{\delta\cdot}{Cl}\text{-------}H\text{---}\underset{CH_3}{\overset{H\quad CH_3}{C}}{}^{\delta\cdot} \right] \longrightarrow Cl-H + \underset{CH_3}{\overset{H\quad CH_3}{C}}\cdot$$

propane	transition state resembles reactants	isopropyl free radical

Now let us look at Reaction 2 in Figure 6.2. This is an *endothermic reaction with a high* E_{act}. The structure of the transition state in Reaction 2 is *close to that of the products*. The hydrogen-abstraction step in radical bromination is more endothermic and has a higher E_{act} than in chlorination. The structure of the transition state in bromination has a greater resemblance to the product alkyl radical because bond breaking and bond forming are practically complete.

$$Br\cdot + H-\underset{CH_3}{\overset{H\quad CH_3}{C}} \longrightarrow \left[\overset{\delta\cdot}{Br}\text{---}H\text{-------}\underset{CH_3}{\overset{H\quad CH_3}{C}}{}^{\delta\cdot} \right] \longrightarrow Br-H + \underset{CH_3}{\overset{H\quad CH_3}{C}}\cdot$$

*transition state
resembles products*

Because this transition state in bromination resembles the alkyl radical, it is highly influenced by the stability of the alkyl radical. The reaction proceeds through the lower-energy transition state to yield the lower-energy, more stable radical. $CH_3\dot{C}HCH_3$ is highly favored over $CH_3CH_2\dot{C}H_2$. By contrast, the transition state in chlorination is less influenced by the stability of the alkyl radical. $CH_3\dot{C}HCH_3$ is only slightly favored over $CH_3CH_2\dot{C}H_2$. Therefore, chlorine is more likely to yield product mixtures.

In 1955, the American chemist George F. Hammond first proposed what we now call the **Hammond postulate,** a working rule for the relationships of different states in a reaction. According to Hammond, in a single reaction path, reaction states that differ little in energy also differ little in geometry. What we have just described concerning chlorination and bromination is part of this working rule. In an exothermic reaction with a low E_{act}, the geometry of the transition state resembles that of the reactant because their energies are similar. In the endothermic reaction with a high E_{act}, the energy difference between the reactants and the transition state is large. Therefore, the geometry of the transition state is more similar to that of the product than to that of the reactants.

B. Benzylic and Allylic Halogenations

Toluene can be selectively halogenated at the benzylic position with either chlorine or bromine. If more than one alkyl position on a benzene side chain is open to attack, as in ethylbenzene, the more selective bromine is the reagent of choice for halogenation at the benzylic position.

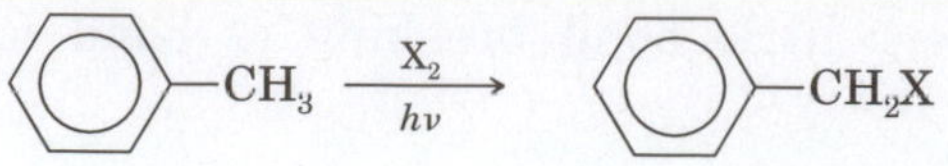

toluene → a benzyl halide (100%) where X = Cl or Br

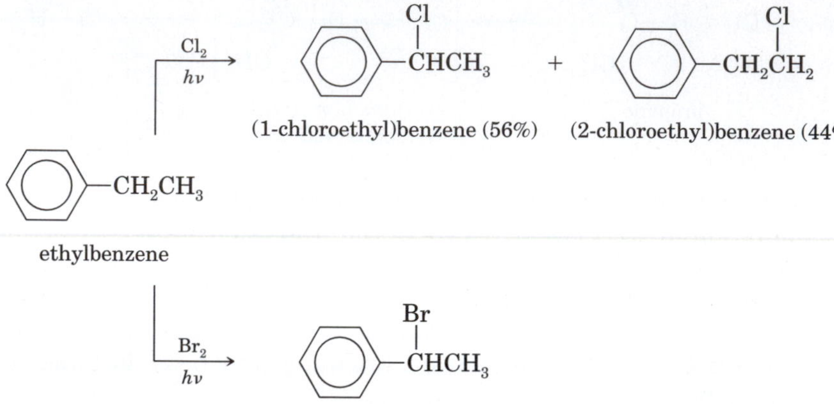

(1-chloroethyl)benzene (56%) + (2-chloroethyl)benzene (44%)

ethylbenzene

(1-bromoethyl)benzene (100%)

Alkenes can be directly halogenated in the allylic position, but very high temperatures and a low concentration of halogen must be used to prevent reaction at the double bond. A reagent that is more specific than Br_2 for allylic and benzylic halogenations is **N-bromosuccinimide (NBS).** An NBS reaction can be initiated by light or by some source of radicals, such as a peroxide (ROOR).

General:

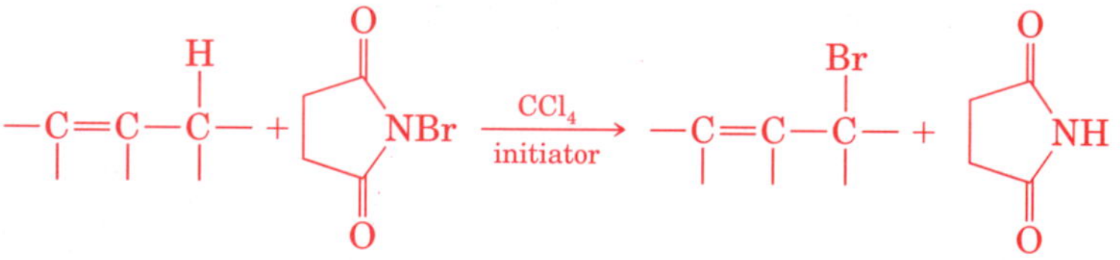

NBS succinimide

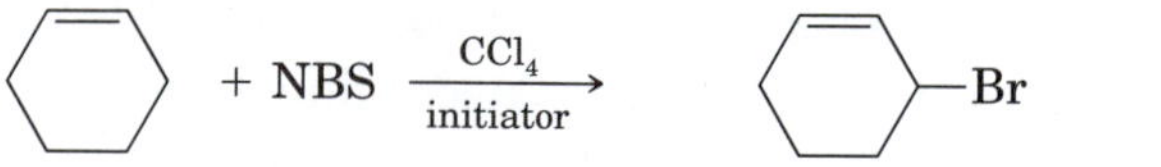

cyclohexene + NBS → 3-bromocyclohexene (87%)

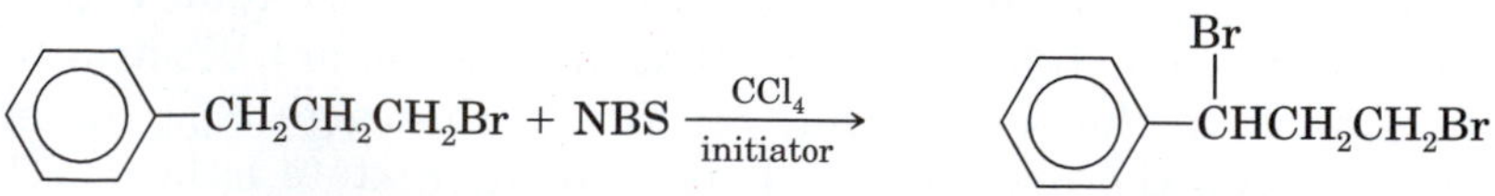

3-bromo-1-phenylpropane + NBS → 1,3-dibromo-1-phenylpropane (~100%)

NBS is not the halogenating agent; it is simply a carrier for the Br_2. NBS acts by providing a low, but constant, concentration of Br_2. The radical reaction can be envisioned as beginning with a small amount of Br_2. The propagation steps of the free-radical reactions are shown below.

$$CH_2{=}CHCH_3 + Br\cdot \longrightarrow CH_2{=}CHCH_2\cdot + HBr$$

$$CH_2{=}CHCH_2\cdot + Br_2 \longrightarrow CH_2{=}CHCH_2Br + Br\cdot$$

One product of the free-radical reaction is HBr. It reacts with NBS by an ionic mechanism to generate Br_2. The process is cyclic—as Br_2 is consumed by the radical reaction, more is generated by the ionic reaction of HBr with NBS.

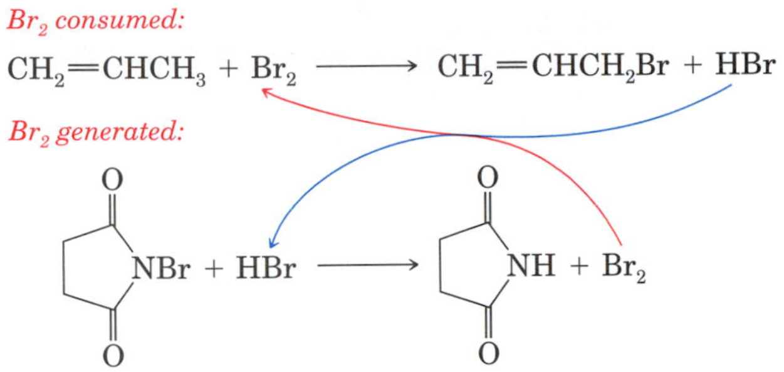

Br₂ consumed:

$$CH_2{=}CHCH_3 + Br_2 \longrightarrow CH_2{=}CHCH_2Br + HBr$$

Br₂ generated:

The following diagram shows the interrelationship of HBr and Br_2 in this reaction.

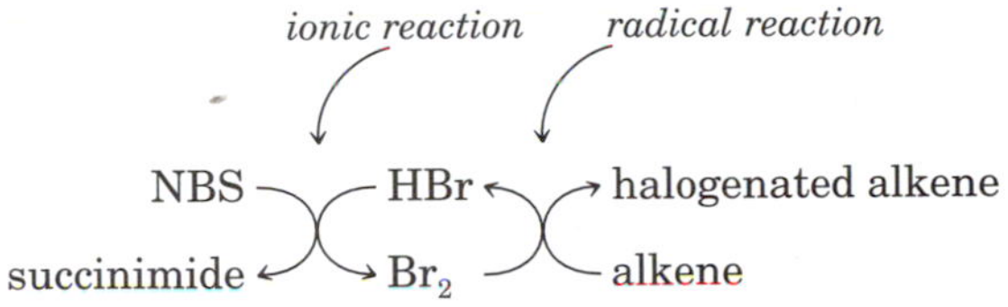

Bromine can also add to a double bond in an ionic reaction. (We will discuss this in more detail in Chapter 10.)

Addition reaction of bromine

$$CH_3CH{=}CH_2 + Br_2 \longrightarrow CH_3\underset{\underset{Br}{|}}{C}H{-}\underset{\underset{Br}{|}}{C}H_2$$

Allylic bromination and the double-bond addition appear to take place under the same reaction conditions. Why does Br_2 from NBS attack the allylic hydrogen and not the double bond? Because of the low concentration of Br_2. At low concentration, the radical reaction has a faster rate than the addition reaction. For addition to the double bond to occur, the bromine concentration must be high. At high concentration, the ionic addition reaction will have the faster rate.

SAMPLE PROBLEM

When an unsymmetrical alkene is brominated with NBS, a mixture of allylic bromides is formed. For example, bromination of 1-butene yields two products. Explain this observation.

$$CH_3CH_2CH{=}CH_2 \xrightarrow{\text{NBS}} CH_3\underset{\underset{Br}{|}}{C}H{-}CH{=}CH_2 + CH_3CH{=}CH\underset{\underset{Br}{|}}{C}H_2$$

Solution

In the propagation cycle, a resonance-stabilized radical is formed. This radical reacts with Br_2 at both of the free-radical sites.

$$CH_3CH_2CH{=}CH_2 + Br\cdot \longrightarrow$$

$$\left[CH_3\overset{.}{C}H{-}CH{=}CH_2 \longleftrightarrow CH_3CH{=}CH\overset{.}{C}H_2 \right]$$

$$
\begin{array}{ccc}
Br_2 & & Br_2 \\
\downarrow & & \downarrow \\
CH_3CH{-}CH{=}CH_2 + Br\cdot & & CH_3CH{=}CHCH_2 + Br\cdot \\
\mid & & \mid \\
Br & & Br
\end{array}
$$

STUDY PROBLEM

6.10   Predict the major monohalogenation products of NBS halogenation of the following compounds: **(a)** butylbenzene; **(b)** 1-phenylpropene.

Section 6.6
Other Radical Reactions

As we have mentioned, radical reactions are not limited to the halogenation of hydrocarbons, but are encountered in many areas of organic chemistry. Here, we will limit our discussion to sigma-bond radical reactions and consider only a few of the many processes known to proceed by radical mechanisms.

A. Pyrolysis

In Section 3.4B, we defined **pyrolysis** as the *thermal decomposition of organic compounds in the absence of oxygen.* When organic molecules are heated to high temperatures, carbon–carbon sigma bonds rupture and the molecules are broken into radical fragments. (The temperature required depends on the bond dissociation energies.) This fragmentation step, called **thermally induced homolysis** (homolytic cleavage caused by heat), is the initiation step for a series of radical reactions. The following equations illustrate some possible pyrolysis reactions of pentane. (There are other possible positions of cleavage and subsequent reaction.)

Initiation (homolysis):

$$CH_3CH_2CH_2{-}CH_2CH_3 \xrightarrow{heat} CH_3CH_2CH_2\cdot + CH_3CH_2\cdot$$

Propagation (hydrogen abstraction):

$$CH_3CH_2\cdot \quad \overset{\overset{\displaystyle H}{|}}{CH_3CH_2CHCH_2CH_3} \longrightarrow CH_3CH_3 + CH_3CH_2\overset{.}{C}HCH_2CH_3$$

Termination by coupling (joining together):

$$CH_3CH_2 \cdot \quad CH_3CH_2 \cdot \longrightarrow CH_3CH_2CH_2CH_3$$

Termination by disproportionation (oxidation–reduction of two equivalent species):

$$CH_3CH_2 \cdot \quad \overset{H}{\underset{}{\ddot{C}H_2} } CH_2 \longrightarrow CH_3CH_3 + CH_2{=}CH_2$$

two ethyl radicals   ethane   ethylene

                                   reduced   *oxidized*

Controlled pyrolysis has been used industrially for the cracking of high-molecular-weight compounds into lower-molecular-weight compounds. Until about 1925, pyrolysis of wood was the major source of methanol (wood alcohol). Thermal cracking was once the only petroleum cracking technique available. Now, the cracking of petroleum is accomplished using catalysts (Section 3.5), and methanol is largely produced by the catalytic hydrogenation of carbon monoxide ($CO + 2H_2 \rightarrow CH_3OH$).

B. Oxygen as a Radical Reagent

Molecular oxygen is different from the compounds we have been studying so far because a stable molecule of O_2 in the ground state has two unpaired electrons. Oxygen is said to be a **diradical.** The structure of O_2 cannot be adequately explained by line-bond formulas because two $2p$ electrons are in antibonding orbitals. For our purposes, we will represent molecular oxygen as $\cdot O{-}O\cdot$ or simply O_2.

Oxygen is a stable diradical and therefore is a selective radical agent. A compound that contains double bonds, allylic or benzylic hydrogens, or tertiary hydrogens is susceptible to **air oxidation,** also called **autoxidation.** Compounds with only primary and secondary hydrogens are not as susceptible. (From our discussion of radical halogenation reactions, the relative reactivities of these hydrogens should not be surprising.)

(57%)

Fats and vegetable oils often contain double bonds. Autoxidation of a fat yields a mixture of products that includes low-molecular-weight (and foul-smelling) carboxylic acids. For example, rancid butter contains the odorous butanoic acid.

Linseed oil and other vegetable oils, which contain many double bonds, are used as drying oils in paint and varnish. These compounds are purposely allowed to undergo air oxidation because the molecules join together, or *polymerize,* into a tough film on the painted surface.

The radical reactions in autoxidation are initiation, propagation, and termination. The initiation reaction is a two-step process. The first step is the abstraction of a reactive hydrogen to form hydroperoxy and hydrocarbon radicals.

The second step, which is also a reaction in the propagation cycle, is the addition of molecular oxygen to the hydrocarbon radical to form a peroxy radical.

Initiation:

$$R{-}H + O_2 \longrightarrow R\cdot + \quad \cdot OOH$$

hydroperoxide radical

$$R\cdot + \cdot O{-}O\cdot \longrightarrow \quad ROO\cdot$$

an alkyl peroxy radical

Hydroperoxy and hydrocarbon radicals can abstract hydrogens or add to double bonds to generate new free radicals. Together, these reactions constitute the propagation cycle. We will illustrate part of the propagation cycle with the reactions of the hydrocarbon radical.

Two reactions of the propagation cycle:

$$ROO\cdot + R{-}H \longrightarrow ROOH + R\cdot$$

$$R\cdot + \cdot O{-}O\cdot \longrightarrow ROO\cdot$$

The chain length of these two radical-reaction steps is about 100; one R· can produce about 100 ROOH. Contrast this chain length with that of chlorination—near 10,000. Chlorination is a more energetic radical reaction.

Termination occurs by radical–radical reactions. A typical termination would be a disproportionation reaction.

Termination by disproportionation:

$$2\ \underset{\underset{\displaystyle H}{|}}{\overset{\overset{\displaystyle O{-}O\cdot}{|}}{R{-}C{-}R}} \longrightarrow \underset{\underset{\displaystyle H}{|}}{\overset{\overset{\displaystyle OH}{|}}{R{-}C{-}R}} + \overset{\overset{\displaystyle O}{\|}}{R{-}C{-}R} + O_2$$

Ethers and aldehydes are quite susceptible to autoxidation. With ethers, the carbon adjacent to the oxygen is the position of attack. Peroxides of ethers explode when heated. For example, diethyl ether is a common laboratory solvent that is purified by distillation. Unless the peroxides have been removed prior to distillation (by a reducing agent, for example), they will become concentrated in the distilling flask as the ether is boiled away. The result could easily be an explosion.

$$\underset{\text{diethyl ether}}{CH_3CH_2OCH_2CH_3} + O_2 \longrightarrow \underset{\text{a hydroperoxide}}{\overset{\overset{\displaystyle OOH}{|}}{CH_3CH_2OCHCH_3}}$$

The product of aldehyde autoxidation is a carboxylic acid, which is formed by reaction of the intermediate peroxy acid with another molecule of the aldehyde.

$$\underset{\substack{\text{ethanal}\\\text{(acetaldehyde)}}}{\overset{\overset{\displaystyle O}{\|}}{CH_3CH}} \xrightarrow{\ O_2\ } \underset{\text{peroxyacetic acid}}{\overset{\overset{\displaystyle O}{\|}}{CH_3COOH}} \xrightarrow{\overset{\displaystyle \overset{O}{\|}}{CH_3CH}} \underset{\text{acetic acid}}{2\ \overset{\overset{\displaystyle O}{\|}}{CH_3COH}}$$

In this series of reactions, isopropylbenzene (trivial name, cumene) is heated in air to yield a hydroperoxide. Oxygen attacks at the reactive benzylic position, just as we would predict. Treatment with aqueous sulfuric acid causes the hydroperoxide to undergo rearrangement to yield the products.

Section 6.7
Free-Radical Initiators and Inhibitors

A **free-radical initiator** is anything that can initiate a radical reaction. The action of light to bring about radical halogenation is the action of an initiator. There are several types of compounds that may be added to a reaction mixture to initiate radical reactions. These compounds are sometimes erroneously called *free-radical catalysts*. They are not truly catalysts because they are often consumed in the reaction.

Any compound that can easily decompose into free radicals can act as an initiator. **Peroxides** (ROOR) are one example. They form radicals easily because the RO—OR bond dissociation energy is only about 35 kcal/mol, lower than that for most other bonds. Benzoyl peroxide and peroxybenzoic acid are two peroxides that are commonly used in conjunction with NBS brominations.

benzoyl peroxide

peroxybenzoic acid

As the name implies, a **free-radical inhibitor** inhibits a radical reaction. An inhibitor is sometimes referred to as a **free-radical "trap."** The usual action of a free-radical inhibitor is to undergo reaction with reactive radicals to form relatively stable and nonreactive radicals.

An inhibitor used to control autoxidation is called an **antioxidant** or, in the food industry, a **preservative.** *Phenols,* compounds with an —OH group attached to an aromatic ring carbon, are effective antioxidants because the radical products of these compounds are resonance-stabilized and thus nonreactive compared with most other radicals.

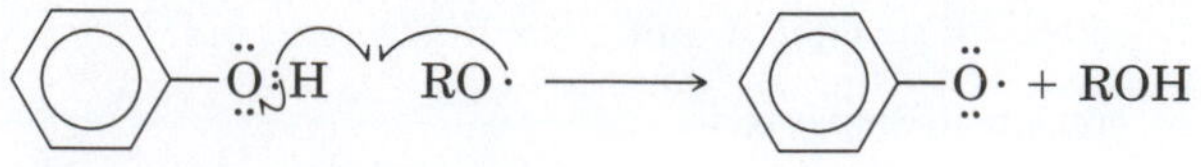

phenol

"trapped"

*the simplest of
the phenols*

Resonance structures of the phenol radical:

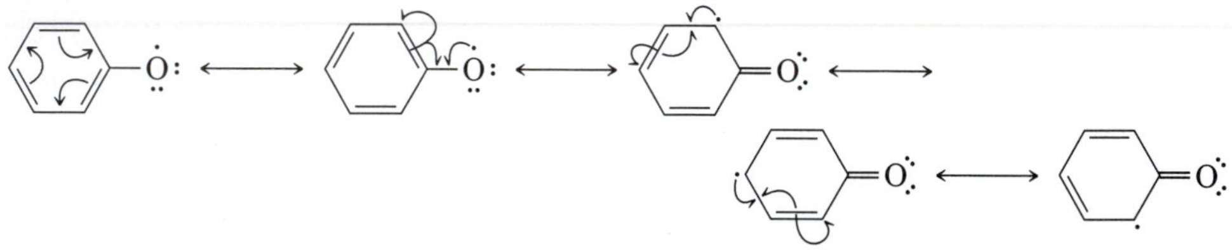

The food preservative BHT is a synthetic phenol. (BHA, another preservative, is closely related to BHT. Instead of a methyl group on the ring, BHA has an —OCH$_3$ group.) A naturally occurring preservative found in vegetable oils, especially wheat germ oil, is α-tocopherol, or vitamin E.

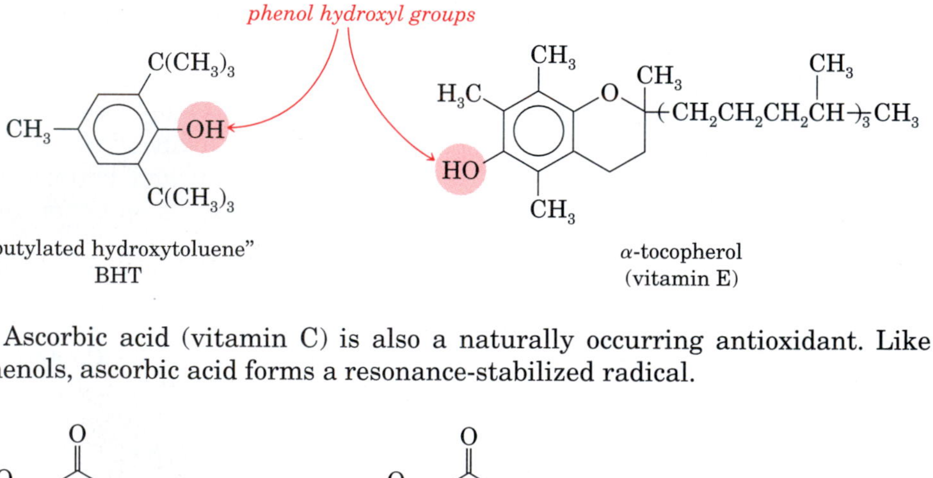

phenol hydroxyl groups

"butylated hydroxytoluene"
BHT

α-tocopherol
(vitamin E)

Ascorbic acid (vitamin C) is also a naturally occurring antioxidant. Like phenols, ascorbic acid forms a resonance-stabilized radical.

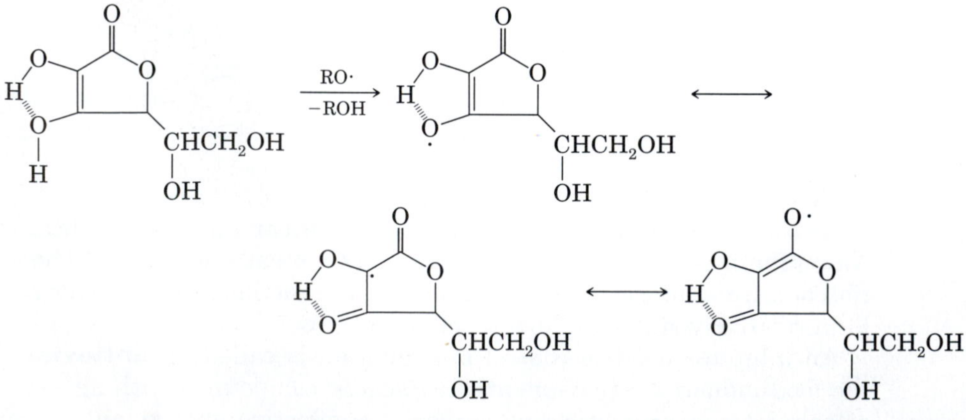

STUDY PROBLEM

6.11 Arylamines, like phenols, can act as antioxidants. For example, *N*-phenyl-2-naphthylamine is added to rubber articles to prevent radical degradation of the rubber. Write an equation that shows how this amine can function as an antioxidant.

$$NHC_6H_5$$

N-phenyl-2-naphthylamine

Section 6.8
Polymers

Polymers are giant molecules, or **macromolecules.** Natural polymers include proteins (such as silk, muscle fibers, and enzymes), polysaccharides (starch and cellulose), and nucleic acids. Synthetic polymers are about as diverse as nature's polymers. The original synthetic polymers were devised to mimic natural polymers. For example, the DuPont Company introduced synthetic rubber and nylon ("artificial silk") in the 1930s. Today, we wear polyester clothes, sit on vinyl chairs, and write on Formica table tops. Our rugs may be made of polyester, polyacrylic, or polypropylene. Sky divers use nylon parachutes. We paint walls with latex paint and protect wood floors with polyurethane. Automobiles may have synthetic rubber tires and vinyl upholstery. Dishes may be melamine. Other common polymeric products include food wrap, Teflon coating for frying pans, hairbrushes, toothbrushes, epoxy glue, electrical insulators, plastic jugs, heart valves, airplane windshields—the list could continue! The technology of macromolecules has become a giant in the world of industry.

Polymers fall into three general classifications. *Elastomers* are those polymers with elastic properties, like rubber. *Fibers* are threadlike polymers, such as cotton, silk, or nylon. *Plastics* can be thin sheets (kitchen wrap), hard and moldable solids (piping, children's toys), or coatings (car finishes, varnishes). The multiplicity of properties depends on the variety of structures that are possible in polymers.

Many useful polymers can be prepared by radical reactions, and we will discuss some of these here. Other types of polymers will be covered in appropriate sections elsewhere in this text.

A polymer (Greek *poly-* and *meros,* "many parts") is made up of thousands of repeating units of small parts, the **monomers** ("one part"). In a polymerization reaction, the first products are **dimers** ("two parts"), then **trimers, tetramers,** and finally, after a series of reaction steps, the polymer molecules, The polymers that we will discuss here are called **addition polymers** because they are formed by the addition of monomers to each other without the loss of atoms or groups.

A synthetic polymer is usually named from the name of its monomer prefixed with **poly-**. For example, ethylene forms the simple polymer *polyethylene,* which is used for things like cleaners' bags and plastic piping.

$$CH_2{=}CH_2 \xrightarrow[\text{or catalyst}]{\text{O}_2,\text{ heat, pressure}} -\overbrace{CH_2CH_2}-\overbrace{CH_2CH_2}-\overbrace{CH_2CH_2}-$$

repeating monomeric units

ethylene   polyethylene

the monomer   *the polymer*

The equations for polymerization are conveniently represented in the following format, where x is used to mean "a large number."

$$x\,CH_2{=}CH_2 \xrightarrow{\text{catalyst}} {+}CH_2CH_2{\overline{)}_x}$$

Frequently, the end groups of macromolecules are unknown—they may arise from impurities in the reaction mixture. In some cases, the end groups can be controlled. The properties of a polymer are governed almost entirely by the bulk of the polymer molecule rather than by the end groups. To emphasize the basic structure of the polymer, it is customary not to include the end groups in the formula unless they are specifically known.

A. Mechanism of Radical Polymerization

A radical polymerization is started by a catalyst or an initiator such as O_2 or a peroxide. The resulting polymer is formed by a chain-propagation process. Let us use the polymerization of styrene as our example.

Overall reaction:

$$x\,CH_2{=}CH \xrightarrow{\text{initiator}} {+}CH_2{-}CH{\overline{)}_x}$$

styrene   polystyrene

used in toothbrush handles and in the manufacture of Styrofoam

In this reaction, the initiation step is the cleavage of the initiator into radicals. An initiator radical attacks the pi electrons of the carbon–carbon double bond. The results are a new sigma bond (between RO and C) and a new radical, as shown in the following equations. The new radical can then attack another carbon–carbon double bond to add a second monomer to the growing chain.

Initiation:

$$ROOR \longrightarrow 2\,RO\cdot$$

Propagation:

$$RO: \overset{\frown}{CH_2} = \overset{\curlywedge}{CH} \longrightarrow RO-CH_2\dot{C}H$$
$$\overset{|}{C_6H_5} \qquad\qquad\qquad\qquad \overset{|}{C_6H_5}$$

$$ROCH_2\dot{C}H \overset{\curlyvee}{\frown} \overset{\frown}{CH_2} = \overset{\curlywedge}{CH} \longrightarrow ROCH_2CH-CH_2\dot{C}H \quad etc.$$
$$\overset{|}{C_6H_5} \qquad \overset{|}{C_6H_5} \qquad\qquad\qquad \overset{|}{C_6H_5} \qquad \overset{|}{C_6H_5}$$

Theoretically, the chain growth could go on indefinitely, which of course does not happen. The termination steps for polymerization are typical radical termination steps. Two radicals may meet and join, or two radicals might undergo **disproportionation,** one being oxidized to an alkene and one being reduced to an alkane. (The squiggles in the following formulas are used to indicate that a much larger molecule than shown is involved in the reaction.)

Coupling:

$$\xi-CH_2\dot{C}H \; + \; \dot{C}HCH_2-\xi \longrightarrow \xi-CH_2CH \text{———} CHCH_2-\xi$$
$$\overset{|}{C_6H_5} \quad \overset{|}{C_6H_5} \qquad\qquad\qquad\qquad \overset{|}{C_6H_5} \quad \overset{|}{C_6H_5}$$
$$\textit{terminated}$$

Disproportionation:

$$\xi-CH_2\dot{C}H \; + \; \dot{C}HCH_2-\xi \longrightarrow \xi-CH=CH \; + \; CH_2CH_2-\xi$$
$$\overset{|}{C_6H_5} \quad \overset{|}{C_6H_5} \qquad\qquad\qquad\qquad \overset{|}{C_6H_5} \quad \overset{|}{C_6H_5}$$
$$\textit{terminated}$$

There are two ways in which styrene molecules could join together to form polystyrene: (1) *head-to-tail,* or (2) *head-to-head and tail-to-tail.*

tail — *head*

$$CH_2 = CH$$
$$\overset{|}{C_6H_5}$$

head-to-tail

$$\xi-CH_2CH-CH_2CH-CH_2CH-\xi$$
$$\overset{|}{C_6H_5} \quad \overset{|}{C_6H_5} \quad \overset{|}{C_6H_5}$$

head-to-head   *tail-to-tail*

$$\xi-CH_2CH \text{———} CHCH_2-CH_2CH-\xi$$
$$\overset{|}{C_6H_5} \quad \overset{|}{C_6H_5} \qquad \overset{|}{C_6H_5}$$

Polystyrene is an example of a *head-to-tail polymer.* Let us discuss the reason for this orientation of the monomers. A more stable radical intermediate means a lower-energy transition state and a faster rate of reaction. Of the two possible modes of radical attack, one leads to a less stable primary radical, while the other leads to the more stable secondary benzylic radical. The repetitive formation of benzylic radicals leads to a head-to-tail joining of styrene monomers.

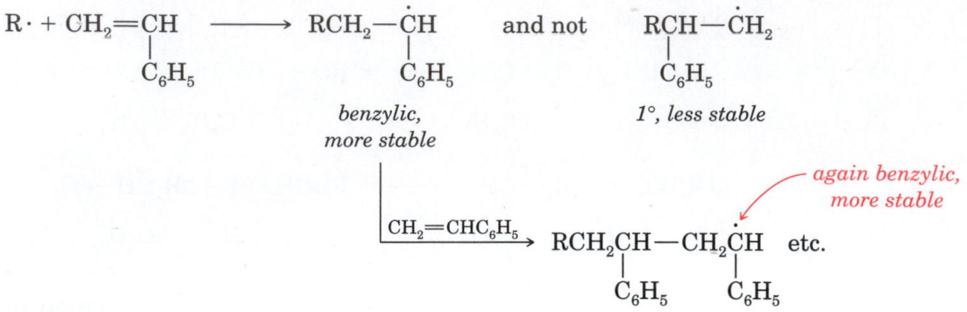

$$R\cdot + CH_2\!\!=\!\!\underset{\underset{C_6H_5}{|}}{CH} \longrightarrow RCH_2\!-\!\underset{\underset{C_6H_5}{|}}{\overset{\cdot}{CH}} \quad \text{and not} \quad RCH\!-\!\overset{\cdot}{CH_2}$$

benzylic,
more stable

1°, less stable

— *again benzylic,*
more stable

$$\underset{CH_2=CHC_6H_5}{\big\downarrow} \quad RCH_2CH\!-\!CH_2\overset{\cdot}{CH} \quad \text{etc.}$$

Resonance structures of a benzylic radical:

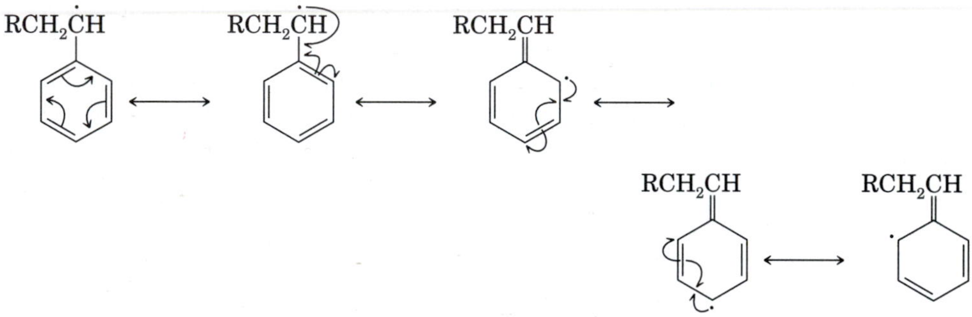

The polymerization of vinyl chloride and methyl methacrylate are other examples of radical reactions leading to head-to-tail products.

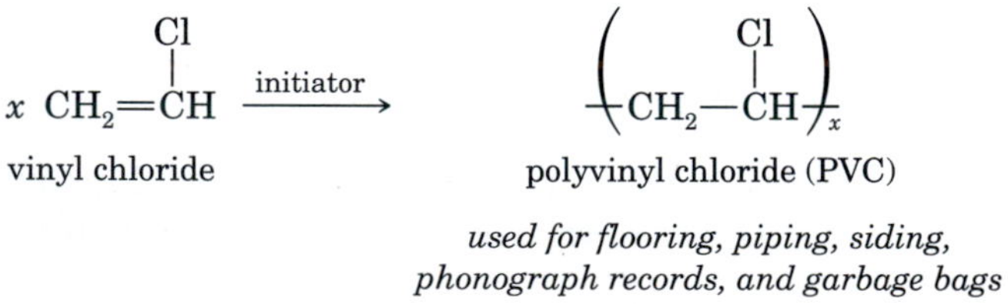

$$x\ CH_2\!\!=\!\!\underset{\underset{Cl}{|}}{CH} \xrightarrow{\text{initiator}} \left(\!\!CH_2\!-\!\underset{\underset{Cl}{|}}{CH}\!\!\right)_{\!\!x}$$

vinyl chloride

polyvinyl chloride (PVC)

used for flooring, piping, siding,
phonograph records, and garbage bags

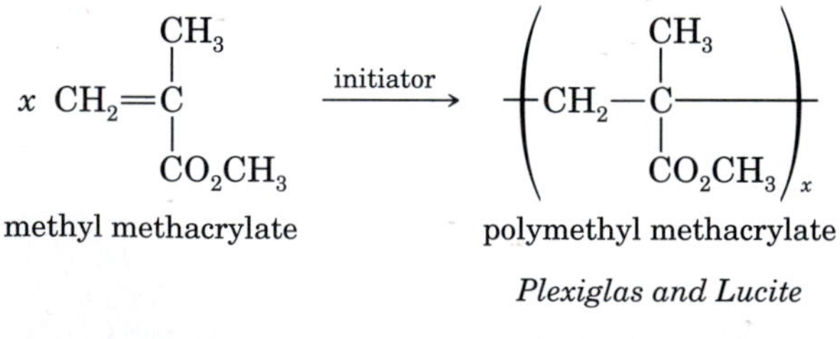

$$x\ CH_2\!\!=\!\!\underset{\underset{CO_2CH_3}{|}}{\overset{\overset{CH_3}{|}}{C}} \xrightarrow{\text{initiator}} \left(\!\!CH_2\!-\!\underset{\underset{CO_2CH_3}{|}}{\overset{\overset{CH_3}{|}}{C}}\!\!\right)_{\!\!x}$$

methyl methacrylate

polymethyl methacrylate

Plexiglas and Lucite

A chemist is not limited to using a single monomer in the production of a polymer. To achieve desired properties, a chemist might use a mixture of two, three, or even more monomers. A mixture of two different monomers results in a **copolymer,** such as Saran (used in kitchen wrap).

$$x\ CH_2\!\!=\!\!CHCl + x\ CH_2\!\!=\!\!CCl_2 \xrightarrow{\text{initiator}} (CH_2CHCl\!-\!CH_2CCl_2)_x$$

vinyl chloride   1,1-dichloroethene   Saran
(vinylidene chloride)

a copolymer
(not necessarily alternating)

STUDY PROBLEMS

6.12   The monomer for *Teflon* is $CF_2=CF_2$. What is the structure of Teflon?

6.13   *Orlon* has the formula $\left(CH_2CH \underset{CN}{}\right)_x$. What is the structure of its monomer?

B. Structure and Stereochemistry of Polymers

Polymers, like any other organic compound, may have functional groups and chiral carbons. They can undergo hydrogen bonding and dipole–dipole inter-actions. The chemical composition of a polymer chain is referred to as its **primary structure.** How the chain is arranged in relation to itself and to other chains is called the **secondary structure.** This secondary structure may be as important to the properties of a polymer as its chemical composition.

A polymer may be a tangled mass of continuous chains or branched chains. The result is a soft amorphous solid such as soft rubber. On the other hand, a polymer may be composed of continuous chains held together by hydrogen bonds or by other dipole–dipole attractions. This type of polymer structure lends itself to fibers or hard, moldable plastics. A more-ordered polymer is said to have a higher degree of **crystallinity** than the amorphous, or non-crystalline, polymer.

a noncrystalline polymer   a crystalline polymer

Let us consider the polymerization of propylene.

$$x\ CH_2\!\!=\!\!\underset{\underset{CH_3}{|}}{CH} \xrightarrow{\text{catalyst}} \left(CH_2\!-\!\underset{\underset{CH_3}{|}}{CH}\right)_x$$

polypropylene

used for rugs and upholstery

Three types of products could result from the head-to-tail polymerization of propylene. (1) The methyl groups at the newly formed chiral carbons could be protruding from the chain in a random fashion. This is an **atactic polymer** (a soft, amorphous product). (2) The methyl groups could alternate from one side of the chain to the other. This is a **syndiotactic polymer.** (3) The methyl groups could all be on the same side. Then, the polymer is said to be **isotactic** (see Figure 6.3). Because of their orderly arrangements, the chains of the latter two polymers can lie closer together and the polymers are more crystalline.

Until 1955, most addition polymers were made by radical paths, which pro-duce atactic polymers. In that year, however, Karl Ziegler and Giulio Natta in-

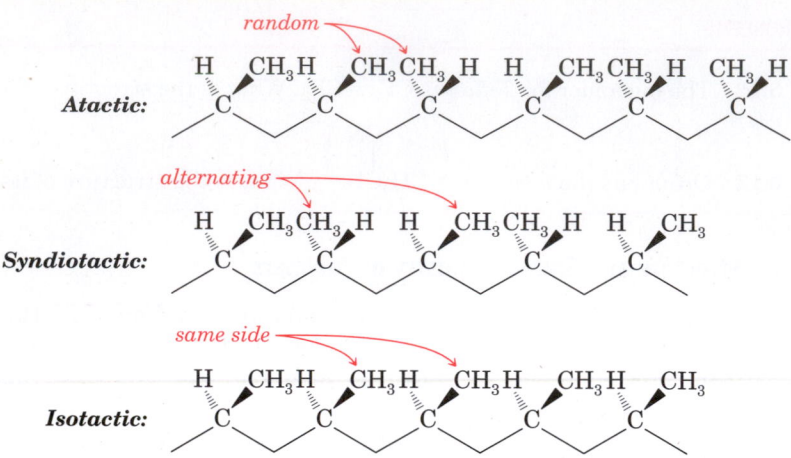

Figure 6.3 The three types of polypropylene molecules.

troduced a new technique for polymerization. These two chemists received the 1963 Nobel Prize for their discovery, a type of catalyst that permits control of the stereochemistry of a polymer during its formation. [A commonly used Ziegler–Natta catalyst is $(CH_3CH_2)_3Al$ complexed $TiCl_4$.] Ziegler–Natta catalysts function by undergoing reaction with the monomeric alkene. Then, new monomers are *inserted* between the catalyst and the growing polymer.

$$\boxed{\text{catalyst}}\!\leftharpoonup\!CH_2CH_3 + CH_2\!=\!CH_2 \longrightarrow \boxed{\text{catalyst}}\!-\!CH_2CH_2\!-\!CH_2CH_3 \quad \text{etc.}$$

In the commercial synthesis of polypropylene and other polymers, Ziegler–Natta catalysts are used in order to produce the more desirable syndiotactic or isotactic polymers.

Summary

A **free radical,** or **radical,** is an atom or group of atoms with an unpaired electron. Radical reactions are **chain reactions** that involve **initiation** (formation of radicals); **propagation** (reactions in which new radicals are consumed and formed); and **termination** (coupling, disproportionation, or the formation of stable radicals). Racemization of a reacting chiral carbon occurs during a radical reaction.

The order of reactivity of H toward radical abstraction is $CH_4 < 1° < 2° < 3° < $ allyic or benzylic. The order of reactivity is the result of the relative stabilities of the radical intermediates.

Cl_2 is more reactive and less selective than Br_2 in radical halogenations. *N*-**Bromosuccinimide** (NBS) is a selective brominating agent for *allylic* and *benzylic positions.*

$$\text{CH}_3\text{CH}_2\text{CH}_3$$

$$\xrightarrow[hv]{\text{Cl}_2} \text{CH}_3\text{CHClCH}_3 + \text{CH}_3\text{CH}_2\text{CH}_2\text{Cl} + \text{HCl}$$

$$\xrightarrow[hv]{\text{Br}_2} \text{CH}_3\text{CHBrCH}_3 + \text{HBr}$$

$$\text{CH}_2\text{=CHCH}_3 \xrightarrow{\text{NBS}} \text{CH}_2\text{=CHCH}_2\text{Br}$$

$$\langle\bigcirc\rangle\text{—CH}_2\text{R} \xrightarrow{\text{NBS}} \langle\bigcirc\rangle\text{—CHBrR}$$

Other radical reactions besides halogenation include **pyrolysis,** a thermal free-radical decomposition of organic compounds, and **autoxidation,** a free-radical oxidation by O_2 that results in decomposition of fats, oils, rubber, ethers, and aldehydes. The **cumene process** is a useful autoxidation.

Free-radical initiators are substances that cause the formation of radicals. Light and peroxides (which contain the easily broken —O—O— bond) are examples. **Free-radical inhibitors,** such as phenols, are substances that form nonreactive radicals.

Polymers are giant molecules composed of repetitive units of *monomers.* Some polymers can be formed by radical addition reactions of alkenes.

$$x\ \text{CH}_2\text{=CH}\underset{\overset{|}{\text{R}}}{} \xrightarrow[\text{catalyst}]{\overset{\text{initiator}}{\text{or}}} \left(\text{CH}_2\text{CH}\underset{\overset{|}{\text{R}}}{}\right)_{\!\!x}$$

The stereochemistry of a polymer can sometimes be controlled by the use of an appropriate catalyst in the polymerization reaction.

Free-Radical Reducing Agents

Free-radical reactions are generally assumed to be nonselective and, consequently, yield a mixture of products. This assumption is true when high-energy radicals, such as chlorine radicals, are generated. But when less energetic radicals are generated, only the most reactive bonds are broken. Under these conditions, radical reactions can be quite selective and are useful as synthetic tools. Bromination of allylic and benzylic positions with *N*-bromosuccinimide (NBS) is an example of a selective radical reaction.

Tributylstannane is a selective free-radical reducing reagent. The bond dissociation energy of the Sn—H is low (74 kcal/mol); therefore, radical reactions can occur preferentially at this site. In free-radical reactions, tributylstannane is a hydrogen atom donor. For example, treatment of 1-bromooctane with tributylstannane yields octane in 80% yield. The other product of the reaction is tributyltin bromide, which is formed in 90% yield. The net result of this reaction is the exchange of a hydrogen atom and a bromine atom between 1-bromooctane and the stannane. (H. G. Kuivila, L. W. Menapace, and C. R. Warner, *J. Am. Chem. Soc.* **1962,** *8*, 3584.)

$$CH_3(CH_2)_6CH_2Br + (C_4H_9)_3Sn-H \xrightarrow{AIBN} CH_3(CH_2)_6CH_3 + (C_4H_9)_3Sn-Br$$

1-bromooctane	tributyl- stannane	octane (80%)	tributyltin bromide (90%)

Reactions with tributylstannane are generally initiated by a free-radical catalyst (initiator). Usually, the thermal decomposition of azobisisobutyronitrile (AIBN) is used to start the radical chain reaction.

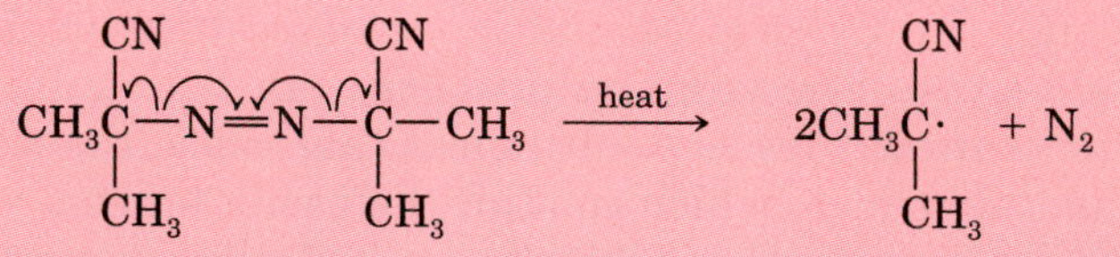

$$\underset{\substack{\text{azobisisobutyronitrile (AIBN)}}}{CH_3\overset{\overset{\displaystyle CN}{|}}{\underset{\underset{\displaystyle CH_3}{|}}{C}}-N{=}N-\overset{\overset{\displaystyle CN}{|}}{\underset{\underset{\displaystyle CH_3}{|}}{C}}-CH_3} \xrightarrow{\text{heat}} \underset{\substack{\textit{isobutyronitrile radical}}}{2CH_3\overset{\overset{\displaystyle CN}{|}}{\underset{\underset{\displaystyle CH_3}{|}}{C}}\cdot\ +\ N_2}$$

Questions

1. Write a reaction in which the isobutyronitrile radical abstracts a hydrogen atom from the Sn—H bond.
2. What are the two steps in the propagation cycle in the free-radical reduction of 1-bromooctane by tributylstannane?
3. In these radical-reduction reactions, less alkane is formed than tin halide. What step(s) in a radical mechanism can be used to explain this fact?
4. Is the stannane reduction of 1-bromooctane endothermic or exothermic? [Bond dissociation energies (kcal/mol): C—Br, 68; Sn—Br, 47; Sn—H, 74; C—H, 98.]
5. Tris(trimethylsilyl)silane, $[(CH_3)_3Si]_3SiH$, like tributylstannane, is a free-radical reducing agent. Write a mechanism for the radical reaction of the silane with bromocyclohexane using AIBN as the radical initiator.

Study Problems*

6.14 Label the following reactions as *initiation, propagation,* or *termination* steps:

(a) $(CH_3)_3C\cdot + CH_2{=}CH_2 \longrightarrow (CH_3)_3C-CH_2CH_2\cdot$

(b) $C_6H_5\overset{\overset{\displaystyle O}{\|}}{C}OOH \longrightarrow C_6H_5\overset{\overset{\displaystyle O}{\|}}{C}O\cdot\ +\ \cdot OH$

(c) $2\ CH_3CH_2CH_2\cdot \longrightarrow CH_3CH_2CH_3 + CH_3CH{=}CH_2$

(d) $2\ CH_3CH_2\cdot \longrightarrow CH_3CH_2CH_2CH_3$

(e) $Br\cdot + CH_2{=}CH_2 \longrightarrow \cdot CH_2CH_2Br$

* For information concerning the organization of the *Study Problems* and *Additional Problems,* see the *Note to student* on page 41.

$$\begin{matrix} & \overset{\text{CN}}{\underset{|}{}} & \overset{\text{CN}}{\underset{|}{}} & & \overset{\text{CN}}{\underset{|}{}} \\ \textbf{(f)} & (CH_3)_2 C & - N = N - C(CH_3)_2 & \longrightarrow & 2\,(CH_3)_2 C \cdot\; + N_2 \end{matrix}$$

6.15 Write equations for the steps in the radical dichlorination of cyclopentane to yield 1,2-dichlorocyclopentane.

6.16 If all H's were abstracted at equivalent rates, what would be the relative percentages of monochlorination products of an equimolar mixture of $CH_3CH_2CH_3$ and cyclohexane?

6.17 Write formulas for all possible dibromo products from the monobromination of the following compound. Which of the dibromo products would be optically active?

$$\overset{\text{Br}}{\underset{|}{}}$$
$$(R)-C_6H_5\overset{|}{C}HCH_2C_6H_5$$

6.18 Only one monochlorination product is obtained from an alkane with the molecular formula C_5H_{12}. What is the structure of the alkane?

6.19 Rank the following radicals in order of increasing stability (least stable first):

(a) $C_6H_5CH\!=\!CHCH_2\overset{.}{C}HCH_3$ **(b)** $C_6H_5\overset{\overset{\displaystyle CH_3}{|}}{C}CH\!=\!CHCH_2CH_3$

(c) $C_6H_5CH_2CH\!=\!CHCH_2\overset{.}{C}H_2$ **(d)** $C_6H_5CH_2CH\!=\!CH\overset{.}{C}HCH_3$

(e) $C_6H_5\overset{\overset{\displaystyle CH_3}{|}}{\underset{.}{C}}CH_2CH\!=\!CH_2$

6.20 Rank the following hydrocarbons in order of increasing ease of radical bromination:

(a) $CH_3CH_2CH(CH_3)_2$ **(b)** $CH_3CH_2CH_3$ **(c)** $C_6H_5CH_2CH_3$

6.21 Each of the following compounds is treated with NBS and a peroxide in CCl_4. Predict the monohalogenation product.

(a) $(CH_3)_2C\!=\!CHC_6H_5$ **(b)** 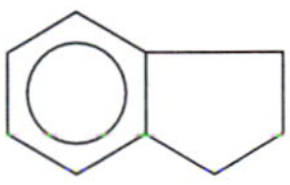   **(c)**

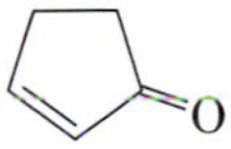

6.22 Predict the major products, including stereochemistry where appropriate:

(a) $C_6H_5CH_2CH_3 + Br_2 \xrightarrow{\;h\nu\;}$

(b) 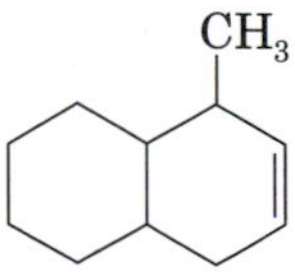 $+ \text{NBS} \xrightarrow{\;CCl_4\;}$

(c) $CH_3\overset{\overset{\displaystyle CH_3}{|}}{C}HCH_3 + Cl_2 \xrightarrow{\;350°\;}$

(d) $CH_3\overset{\overset{\displaystyle CH_3}{|}}{C}H - \overset{\overset{\displaystyle CH_3}{|}}{C}HCH_3 + 2\,Br_2 \xrightarrow{\;h\nu\;}$

(e) $C_6H_5\overset{\overset{\displaystyle CH_3}{|}}{C}HCH_2CH_3 + Br_2 \xrightarrow{\;h\nu\;}$

(f) 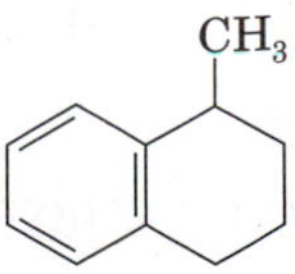 $\xrightarrow{\;O_2\;}$

(g) $(R)\text{-}CH_3CHBrCH_2CH_3 + Cl_2 \xrightarrow{h\nu}$

6.23 For the molecule shown below, write equations to illustrate the following reactions:

$$\begin{array}{cc} CH_3 & CH_3 \\ | & | \\ CH_3-CH-CH-C_6H_5 \end{array}$$

(a)   preferred bond homolysis yielding two radicals

(b)   all possible disproportionations of these radicals

(c)   all possible recombinations of these radicals

(d)   most favored hydrogen abstraction from the hydrocarbon by one of the radicals

6.24 Predict the coupling and disporportionation products of the following radicals:

(a) $(CH_3)_2\overset{\cdot}{C}CH_2CH_2CH_3$ **(b)**

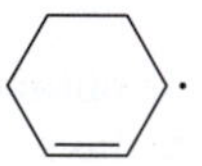

6.25 Which of the following compounds would form a hydroperoxide (ROOH) readily upon exposure to air?

(a) 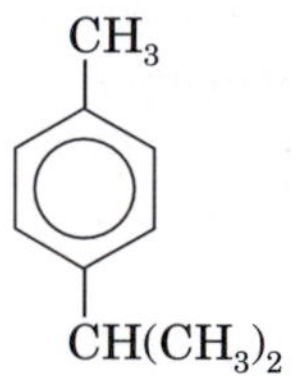   **(b)**   **(c)**

p-cymene

found in plant oils

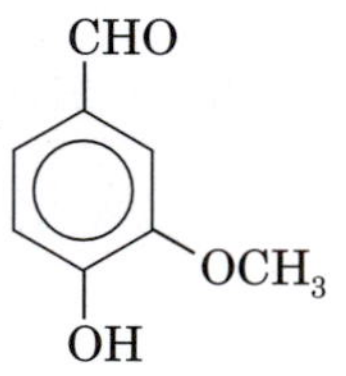

vanillin

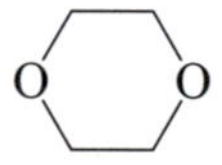

dioxane

a useful solvent, but carcinogenic

6.26 Write the formulas for the monomers of the following polymers:

(a)

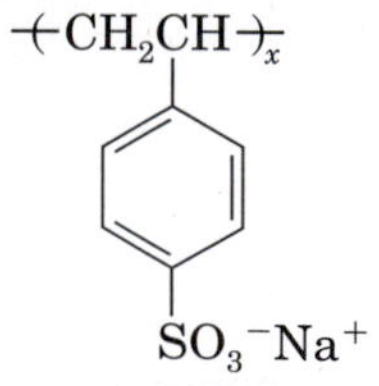

(b) $\begin{array}{cc} CH_3 & Cl \\ | & | \\ +CH_2CHCH_2CH+_x \end{array}$

(c) $\begin{array}{c} +CH_2CH+_x \\ | \\ CH=CH_2 \end{array}$

(d) $\begin{array}{ccc} CH_3 & CH_3 & CH_3 \\ | & | & | \\ +CH_2C-CH_2C-CH_2C+_x \\ | & | & | \\ CO_2CH_3 & CO_2CH_3 & CO_2CH_3 \end{array}$

6.27 Write the formula for the head-to-tail trimer that would be formed by each of the following compounds. Do not show end groups.

(a) $CH_2{=}CHCN$

(b) $\begin{array}{c} Cl \\ | \\ CH_3C{=}CH_2 \end{array}$

(c) $CH_2{=}CHO\overset{\overset{\textstyle O}{||}}{C}CH_3$

(d) $CH_2{=}CHC\overset{\overset{\textstyle O}{||}}{O}CH_3$

Additional Problems

6.28 Suggest a synthesis by which the hydrocarbon on the left can be converted to the product on the right.

(a) $C_6H_5{-}CH_3 \longrightarrow C_6H_5{-}CH_2OCH_3$

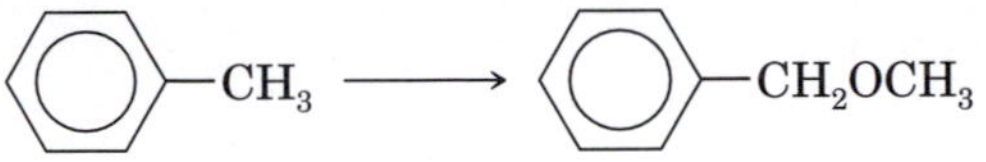

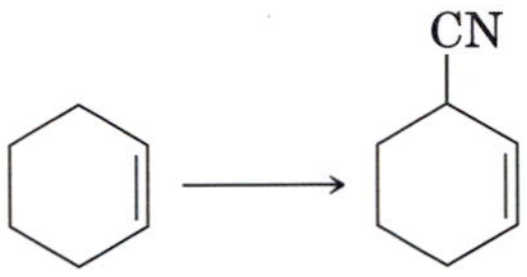

(b) (cyclohexene) $\longrightarrow$ (1-cyano-cyclohexene with CN)

(c) $C_6H_5{-}CH_2CH_2{-}C_6H_5 \longrightarrow C_6H_5{-}CH{=}CH{-}C_6H_5$

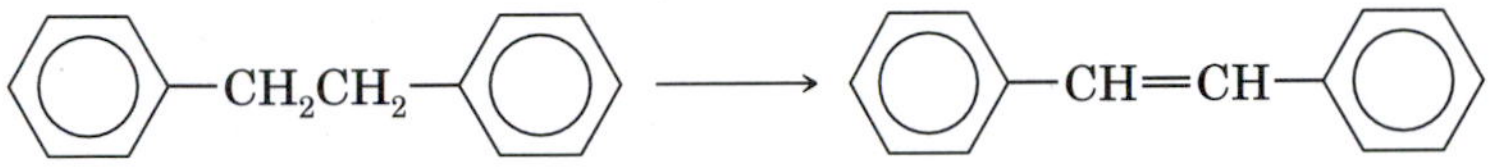

6.29 The aqueous electrolysis of salts of carboxylic acids (called the **Kolbe electrolysis**) yields carbon dioxide and hydrocarbons by a radical path. What would be the products of the electrolysis of sodium acetate (CH_3CO_2Na)?

6.30 Upon radical bromination, pentane yields almost exclusively two monobromo compounds, A and B. Upon treatment with $NaOCH_3$ under E2 conditions, A and B yield predominantly the same product, C. What are the structures of A, B, and C?

6.31 Write equations for each step in the following free-radical reaction:

$$CH_2{=}CHCH_3 + (CH_3)_3C{-}O{-}Cl \xrightarrow{h\nu} CH_3{=}CHCH_2Cl + (CH_3)_3COH$$

6.32 If not prepared with an antioxidant, polypropylene undergoes rapid deterioration when exposed to the air. Polyethylene does not exhibit the same rate of deterioration. Explain why these two polymers differ in their air stabilities.

6.33 When an excess of radical initiator is used in the polymerization of an alkene, the average polymeric molecular weight is often less than that when just a trace of initiator is used. Explain why this is true.

6.34 Some addition polymers can be depolymerized by heat. **(a)** Write equations for the depolymerization of (1) polystyrene and (2) Teflon (polytetrafluoroethene). **(b)** What danger would reactions such as these pose to electronic technicians soldering insulated wiring or to firefighters?

6.35 A chemist attempted to polymerize the alkene shown below by heating it with a small amount of AIBN. The polymerization failed to occur. Explain why?

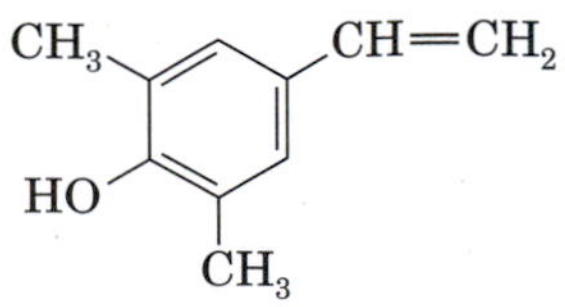

(structure: benzene ring with CH_3, $CH{=}CH_2$, HO, and CH_3 substituents)

7 ALCOHOLS

Alcohols (ROH) are compounds containing a hydroxyl group (—OH) bonded to an sp^3-hybridized carbon. The word *alcohol* is derived from the Arabic word *alkuhl*, meaning finely powdered antimony, used as a cosmetic to darken eyelids. This unusual word derivation arose in the eighteenth century, when *alkuhl* was taken to mean "essence of." Ethanol, the essence of wine, was called *alcool vini*. By the start of the nineteenth century, the *vini* had been dropped and the *alcool* had become *alcohol*.

Many alcohols are commonly encountered in everyday life. Ethanol (ethyl alcohol, grain alcohol, or just alcohol), CH_3CH_2OH, is used in beverages. 2-Propanol (isopropyl alcohol, or rubbing alcohol), $(CH_3)_2CHOH$, is used as a bactericidal agent. Methanol (methyl alcohol, or wood alcohol), CH_3OH, is used as an automobile gas-line antifreeze. In the laboratory and in industry, all these compounds are used as solvents and reagents.

Nomenclature and Classification of Alcohols

A. IUPAC Names

The IUPAC names of alcohols are taken from the names of the parent alkanes, but with the ending **-ol**. A prefix number, chosen to be as low as possible, is used if necessary.

$$CH_3OH \qquad CH_3CH_2CH_2OH \qquad CH_3\overset{\displaystyle OH}{\underset{\displaystyle |}{C}}HCH_3$$

| IUPAC: | methanol | 1-propanol | 2-propanol |

More than one hydroxyl group is designated by *di-, tri-*, etc., just before the *-ol* ending. The final *-e* of the parent alkane name is retained before the consonant "d" of the *-diol* ending.

$$CH_3\overset{\displaystyle OH}{\underset{\displaystyle |}{C}}HCH_2CH_2OH \qquad CH_3CH_2CH_2\underset{\displaystyle OH}{\overset{\displaystyle |}{C}}H-\underset{\displaystyle CH_2CH_3}{\overset{\displaystyle |}{C}}H-CH_2OH$$

1,3-butanediol   2-ethyl-1,3-hexanediol

a diol   *a popular insect repellent*

B. Trivial Names

Just as CH_3I may be called methyl iodide, CH_3OH may be called methyl alcohol. This type of name is a popular way of naming alcohols with common alkyl groups.

$$(CH_3)_3COH \qquad (CH_3)_2CHOH$$

tert-butyl alcohol   isopropyl alcohol

A diol (especially a 1,2-diol) is often referred to as a **glycol**. The trivial name for a 1,2-diol is that of the corresponding **alkene** followed by the word **glycol**. Epoxides and 1,2-dihalides are often named similarly. The naming of a saturated compound as a derivative of an alkene is unfortunate. However, the practice arose quite innocently in the early years of organic chemistry, because all these compounds can be prepared from alkenes.

	$CH_2{=}CH_2$	$\underset{\displaystyle CH_2}{\overset{\displaystyle OH}{\underset{|}{\vphantom{x}}}}{-}\underset{\displaystyle CH_2}{\overset{\displaystyle OH}{\underset{|}{\vphantom{x}}}}$	$\underset{\displaystyle CH_2}{\overset{\displaystyle Br}{\underset{|}{\vphantom{x}}}}{-}\underset{\displaystyle CH_2}{\overset{\displaystyle Br}{\underset{|}{\vphantom{x}}}}$	$CH_2{-}CH_2$ (epoxide O)
IUPAC:	ethene	1,2-ethanediol	1,2-dibromoethane	oxirane
trivial:	ethylene	ethylene glycol	ethylene dibromide	ethylene oxide

an epoxide

STUDY PROBLEMS

7.1 Name the following compounds:

(a) $(CH_3)_3CCH_2OH$ (b) 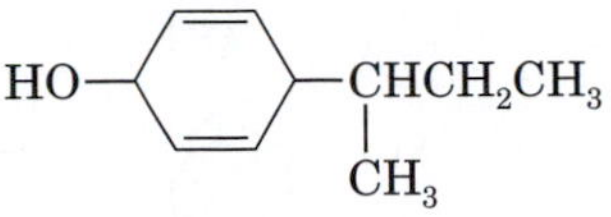 $HO-\!\!\bigcirc\!\!-CHCH_2CH_3$ (c) 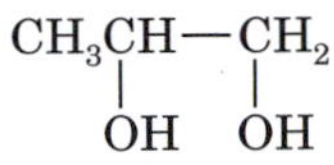 CH_3CH-CH_2
$\qquad\qquad\qquad\qquad\qquad\qquad\qquad | \qquad\qquad\qquad\qquad\quad | \quad\;\; |$
$\qquad\qquad\qquad\qquad\qquad\qquad\;\; CH_3 \qquad\qquad\qquad\quad OH \;\; OH$

7.2 Write structures for:

(a) 2-methyl-2-butanol   (b) 2-cyclohexenol

(c) *trans*-4-ethylcyclohexanol   (d) 4-cyclohexyl-1,6-heptadien-4-ol

C. Classification of Alcohols

Alcohols, like alkyl halides, can be classified as **methyl, primary, secondary,** or **tertiary,** as well as **allylic** or **benzylic.**

$$CH_3OH \qquad CH_3CH_2OH \qquad (CH_3)_2CHOH \qquad (CH_3)_3COH$$
$$\text{methyl} \qquad\quad 1° \qquad\qquad\quad 2° \qquad\qquad\qquad 3°$$

$$CH_3CH\!=\!CHCH_2OH \qquad\qquad \bigcirc\!\!-\overset{\displaystyle OH}{\underset{\displaystyle |}{CHCH_3}}$$

an allylic alcohol   *a benzylic alcohol*
(and 1°)   *(and 2°)*

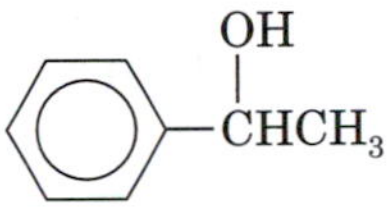

SAMPLE PROBLEM

Name the following structures (1) by the IUPAC system, and (2) by a trivial name. Classify each as 1°, 2°, or 3°.

(a) $CH_3(CH_2)_6CH_2OH$ (b) $CH_2\!=\!CHCH_2OH$

Solution
Structure (a) is 1-octanol by the IUPAC system and its trivial name is octyl alcohol. Structure (b) is 2-propen-1-ol by the IUPAC system and its trivial name is allyl alcohol. Both structures are 1° alcohols.

STUDY PROBLEM

7.3 Classify each of the following alcohols as 1°, 2°, 3°, allylic, or benzylic (all classifications that apply):

(a) 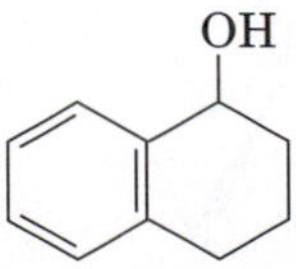   (b) 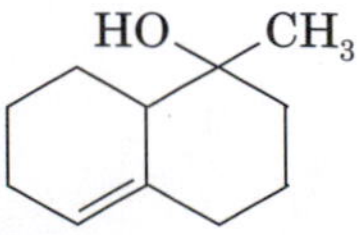   (c)

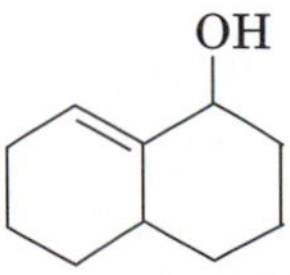

When a hydroxyl group is bonded to a carbon of an aromatic ring, the compound belongs to the class of phenols, not alcohols. The chemistry of phenols is very different from that of alcohols. We will discuss phenols in Chapter 12.

OH   OH

phenol   1-naphthol

not an alcohol   *not an alcohol*

Section 7.2
Basicity and Acidity of Alcohols

In acidic solution, alcohols are protonated. This reaction is an acid–base equilibrium with the alcohol acting as a base. It is the same type of reaction that occurs between water and a proton.

$$H-\overset{\overset{\displaystyle H}{|}}{\underset{..}{O}}: + HCl \rightleftharpoons H-\overset{\overset{\displaystyle H}{|}}{\underset{..}{O}}^+{-}H + Cl^-$$

hydronium ion

$$R-\overset{\overset{\displaystyle H}{|}}{\underset{..}{O}}: + HCl \rightleftharpoons R-\overset{\overset{\displaystyle H}{|}}{\underset{..}{O}}^+{-}H + Cl^-$$

an alcohol   *an oxonium ion*

In each case, an empty $1s$ orbital of H^+ overlaps with one of the filled valence orbitals of the oxygen, and an O—H sigma bond is formed (see Figure 7.1). The product of the reaction with water is the protonated water molecule, or the **hydronium ion.** A protonated alcohol molecule is called an **oxonium ion.**

An alcohol can also *lose* a proton to a strong base to yield an **alkoxide ion,** RO^-. Alkoxides are *strong bases,* generally stronger than hydroxides. To prepare an alkoxide from an alcohol, we need a base stronger than the alkoxide itself, such as an alkali metal hydride (NaH, KH).

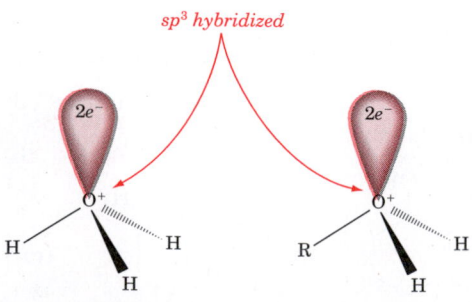

Figure 7.1   Bonding in the hydronium ion (H_3O^+) and in an oxonium ion (ROH_2^+).

$$\text{⬡—OH} + \text{Na}^+ \text{ H}^- \longrightarrow \text{⬡—O}^- \text{ Na}^+ + \text{H}_2 \uparrow$$

Another convenient method for the preparation of alkoxides is the treatment of an alcohol with an alkali metal such as sodium or potassium. The alkali metal is oxidized to a cation, and the hydrogens of the OH groups are reduced to hydrogen gas.

$$\text{CH}_3\text{CH}_2\text{OH} + \text{Na} \longrightarrow \text{CH}_3\text{CH}_2\text{O}^- \text{ Na}^+ + \tfrac{1}{2}\text{H}_2 \uparrow$$

$$(\text{CH}_3)_3\text{COH} + \text{K} \longrightarrow (\text{CH}_3)_3\text{CO}^- \text{ K}^+ + \tfrac{1}{2}\text{H}_2 \uparrow$$

Methanol and ethanol undergo fairly vigorous reaction with sodium metal. As the size of the R group is increased, the vigor of the reaction decreases. Sodium and water react explosively. Sodium and ethanol undergo reaction at a very controllable rate. Sodium and 1-butanol undergo a very sluggish reaction. With alcohols of four or more carbons, the more reactive potassium metal is generally used to prepare the alkoxide.

$$\text{H}_2\text{O} \quad \text{CH}_3\text{OH} \quad \text{CH}_3\text{CH}_2\text{OH} \quad \text{CH}_3\text{CH}_2\text{CH}_2\text{OH}$$

decreasing reactivity toward Na or K

Like water, a pure alcohol can undergo ionization. However, in their pure states, alcohols ionize to a lesser extent than does water.

$$\text{CH}_3\ddot{\text{O}}\text{—H} + \text{CH}_3\ddot{\text{O}}\text{H} \rightleftharpoons \text{CH}_3\ddot{\text{O}}\text{:}^- + \text{CH}_3\overset{+}{\ddot{\text{O}}}\text{H}_2$$

One reason for the lesser amount of ionization of pure alcohols is that alcohols have lower dielectric constants. Because they are less polar, alcohols are less able to support ions in solution than are water molecules.

In dilute aqueous solution, alcohols have approximately the same pK_a values as water (see Table 7.1). For example, the pK_a of methanol in water is 15.5, while that of pure water is 15.74 (not 14, which is the pK_w).

Water (pK_a = 15.74):

$$\text{H}\ddot{\text{O}}\text{—H} + \text{H}_2\ddot{\text{O}}\text{:} \rightleftharpoons \text{H}\ddot{\text{O}}\text{:}^- + \text{H}_3\text{O}^+$$

Methanol in H_2O (pK_a = 15.5):

$$\text{CH}_3\ddot{\text{O}}\text{—H} + \text{H}_2\ddot{\text{O}}\text{:} \rightleftharpoons \text{CH}_3\ddot{\text{O}}\text{:}^- + \text{H}_3\text{O}^+$$

Table 7.1 pK_a values for water and some alcohols in dilute aqueous solution

Compound	pK_a
H_2O	15.74
CH_3OH	15.5
CH_3CH_2OH	15.9
$(CH_3)_3COH$	~18

7.4 Write equations for the acid–base reactions of cyclopentanol with **(a)** $NaNH_2$ and **(b)** H_2SO_4. Label each reactant as an acid or base.

Section 7.3
Preparation of Alcohols

The ethanol used in beverages is obtained by the enzyme-catalyzed **fermentation of carbohydrates** (sugars and starches). Fermentation is the decomposition of organic compounds into simpler compounds by the action of the biological catalysts known as enzymes. In the fermentation of carbohydrates, one type of enzyme converts carbohydrates to glucose and then to ethanol. Another type leads to vinegar (acetic acid), with ethanol as an intermediate.

$$C_6H_{12}O_6 \xrightarrow{\text{enzymes}} CH_3CH_2OH$$

glucose   ethanol

a sugar

The source of the carbohydrates used for fermentation depends on availability and on the purposes of the alcohol. In the United States, carbohydrates are obtained primarily from corn and from the molasses residue of sugar refining. However, potatoes, rice, rye, or fruit (grapes, blackberries, etc.) may also be used.

Fermentation of any of these fruits, vegetables, or grains ceases when alcoholic content reaches 14–16%. If a higher concentration of alcohol is desired, the mixture is distilled. The distillate is an azeotrope of 95% ethanol–5% water. (An azeotrope is a mixture that boils at a constant boiling point as if it were a pure compound.) This distillate may then be used to fortify the fermentation mixture, or it may be diluted with water to the desired strength.

Because alcoholic beverages are taxed in almost all countries of the world, most ethanol sold for laboratory or industrial purposes (and not taxed as a liquor) is **denatured,** that is, small amounts of toxic impurities are added so that the ethanol cannot be diverted from the laboratory or factory into illegal beverages.

A. Nucleophilic Substitution Reactions

The reaction of an alkyl halide with hydroxide ion is a nucleophilic substitution reaction. When primary alkyl halides are heated with aqueous sodium hydroxide, reaction occurs by an S_N2 path. Primary alcohols can be prepared in good yields by this technique. Because secondary and tertiary alkyl halides are likely to give elimination products, they are not generally as useful for preparing alcohols.

$$CH_3CH_2CH_2Br + OH^- \xrightarrow{\text{heat}} CH_3CH_2CH_2\ OH + Br^-$$

1-bromopropane   1-propanol

a 1° alkyl halide   *a 1° alcohol*

B. Reduction of Carbonyl Compounds

Alcohols can be prepared from carbonyl compounds by **reduction reactions** in which hydrogen atoms are added to the carbonyl group. For example, *reduction of a ketone* by catalytic hydrogenation or with a metal hydride yields a *secondary alcohol*. Yields are often 90–100%. These reactions will be discussed in more detail in Section 13.6.

$$
\underset{\text{acetone}}{CH_3\overset{\overset{\displaystyle O}{\|}}{C}CH_3} \xrightarrow[\text{(2) } H_2O, H^+]{\text{(1) } NaBH_4} \underset{\text{2-propanol}}{CH_3\overset{\overset{\displaystyle OH}{|}}{C}HCH_3}
$$

$$
\underset{\text{cyclohexanone}}{\bigcirc=O} \xrightarrow[\text{heat, pressure}]{H_2,\ Ni\ catalyst} \underset{\text{cyclohexanol}}{\bigcirc-OH}
$$

C. Hydration of Alkenes

When an alkene is treated with water plus a strong acid, which acts as a catalyst, water (H^+ and OH^-) adds to the double bond in a **hydration reaction.** The product is an alcohol. Many alcohols, such as laboratory ethanol, are made commercially by the hydration of alkenes. Limitations and variations of hydration reactions, as well as the mechanism, will be discussed in Chapter 10.

$$
\underset{\text{ethylene}}{CH_2{=}CH_2} + H_2O \xrightarrow{H^+} \underset{\text{ethanol}}{CH_3CH_2OH}
$$

$$
\underset{\text{cyclohexene}}{\bigcirc} + H_2O \xrightarrow{H^+} \underset{\text{cyclohexanol}}{\bigcirc-OH}
$$

STUDY PROBLEMS

7.5 Write equations that show how each of the following alcohols can be prepared from (1) an alkene, and (2) a ketone:

(a) 2-butanol   **(b)** 2,4-dimethylcyclopentanol

7.6 What alkyl halide would be required to prepare each of the following alcohols by an S_N2 reaction?

(a) $\bigcirc-\overset{\overset{\displaystyle OH}{|}}{C}HCH_3$ **(b)** $CH_3CH_2\overset{\overset{\displaystyle OH}{|}}{C}HCH_2CH_3$

(c) $(S)\text{-}\bigcirc-CH_2\overset{\underset{\displaystyle OH}{|}}{C}HCH_3$ **(d)** $(CH_3)_3C-\bigcirc\overset{\displaystyle H}{\underset{\displaystyle OH}{}}$

D. Grignard Reactions

An **organometallic compound** is defined as a compound in which *carbon is bonded directly to a metallic atom* (such as mercury, zinc, lead, magnesium, or lithium) or to certain *metalloids* (such as silicon, arsenic, or selenium).

$$CH_3CH_2CH_2CH_2Li \qquad\qquad (CH_3)_4Si \qquad\qquad CH_3ONa$$

butyllithium   tetramethylsilane (TMS)   sodium methoxide

organometallic   *organometallic*   *not considered organometallic (no carbon-metal bond)*

Some of the most useful reagents in organic synthesis are the organomagnesium halides (RMgX). An organomagnesium halide is the product of a reaction between magnesium metal and an organohalogen compound in an ether solvent. An indirect preparation of organomagnesium compounds was first reported in 1899 by François Philippe Antoine Barbier (1848–1922), a professor at the University of Lyon, France. However, organomagnesium compounds are called **Grignard reagents** after Victor Grignard (1871–1935), a student of Barbier's and later a professor at the University of Nancy, France. In 1900, Grignard reported that alkylmagnesium halides could be prepared by the simple and direct reaction of magnesium metal and an organohalogen compound using diethyl ether as the solvent. He also reported the results of the reaction of organomagnesium compounds with a variety of other compounds. Grignard was awarded the 1912 Nobel Prize in chemistry for his work in this area.

$$R{-}X + Mg \xrightarrow{\text{diethyl ether}} R{-}Mg{-}X$$

a Grignard reagent

Grignard reagents exist in solution as a mixture of several components in equilibrium. The major components are shown in the following equation, called the *Schlenk equilibrium* after its discoverer.

The Schlenk equilibrium of a Grignard reagent

$$2\,RMgX \rightleftharpoons R_2Mg + MgX_2$$

Dimers and trimers also exist in the equilibrium mixture, but they make a minor contribution to its composition.

Dimer structures in a Grignard solution

Because the major organic species in the mixture (RMgX and R_2Mg) yield the same products in reactions, we will, for simplicity, represent the structure of the Grignard reagent as RMgX.

The most used solvent for a Grignard reagent is diethyl ether ($CH_3CH_2OCH_2CH_3$), which is nonreactive toward Grignard reagents, but can donate unshared electrons to the empty orbitals of Mg and help dissolve the

organomagnesium compound. The ethyl groups provide a hydrocarbon environment that acts as the solvent for the alkyl portion of a Grignard reagent.

$$CH_3CH_2 \quad CH_2CH_3$$
$$\ddot{O}$$

$$CH_3—Mg—I$$

$$\ddot{O}$$
$$CH_3CH_2 \quad CH_2CH_3$$

[handwritten note on blue sticky:] Grignard Rx : Organometallic Compound - Carbon bonded to metallic ion • ethyl ether used ∴ can dissolve organometallic • Rx not depended on R group ; all can form Grignard • Grignard Reagent ⎫ ① strong Base   R-C̄-M̄g-X ⎬ ② acts as nucleophile.

The reaction is general and does not depend to any great extent upon the nature of the R group. Primary, secondary, and tertiary alkyl halides, as well as allylic and benzylic halides, all form Grignard reagents.

$$(CH_3)_3CBr \quad + \quad Mg \quad \longrightarrow \quad (CH_3)_3CMgBr$$

tert-butyl bromide   *tert*-butylmagnesium bromide

$$\langle\bigcirc\rangle—CH_2Cl + Mg \longrightarrow \langle\bigcirc\rangle—CH_2MgCl$$

benzyl chloride   benzylmagnesium chloride

Aryl and vinylic halides (X on the doubly bonded carbon) are generally quite inert toward nucleophilic substitution and elimination. These compounds are not as reactive as alkyl halides toward magnesium, but their Grignard reagents can still be prepared.

$$\langle\bigcirc\rangle—Br + Mg \longrightarrow \langle\bigcirc\rangle—MgBr$$

bromobenzene   phenylmagnesium bromide

an aryl Grignard reagent

$$CH_2{=}CHI + Mg \longrightarrow CH_2{=}CHMgI$$

iodoethene   ethenylmagnesium iodide
(vinyl iodide)   (vinylmagnesium iodide)

a vinylic Grignard reagent

In most organic compounds, carbon carries either no partial charge or a partial positive charge. In a Grignard reagent, carbon is bonded to an electropositive element and consequently carries a *partial negative charge*.

$$\overset{\delta+}{CH_3CH_2}—\overset{\delta-}{Br} + Mg \xrightarrow{\text{diethyl ether}} \overset{\delta-}{CH_3CH_2}—\overset{\delta+}{Mg}—Br$$

carbon is δ−

Because a Grignard reagent has a partially negative carbon, (1) it is an *extremely strong base,* and (2) the alkyl or aryl portion of the Grignard reagent can act as a *nucleophile.*

Grignard reagents as bases   A hydrogen that can be abstracted from a compound by a Grignard reagent is said to be an *acidic hydrogen* in relation to the Grignard reagent. Grignard reagents undergo rapid reaction with compounds that have acidic hydrogens such as water, alcohols (ROH), amines (R_2NH), carboxylic acids (RCO_2H), and terminal alkynes ($RC{\equiv}CH$), to yield hydrocarbons and metal salts. For example,

$$\overset{\delta-}{R}{-}MgX + H{-}\ddot{O}H \longrightarrow RH + XMg\ddot{O}H$$

Because of the basicity of the carbon in a Grignard reagent, water and other compounds containing acidic hydrogens must be excluded from a Grignard reaction mixture unless a hydrocarbon is the desired product.

Grignard reagents as nucleophiles   The most important reactions of Grignard reagents are those with carbonyl compounds. In a carbonyl group (C=O), the electrons in the carbon–oxygen bonds (sigma and pi) are drawn toward the electronegative oxygen. The carbon of the carbonyl group, which has a partial positive charge, is attacked by the nucleophilic carbon of the Grignard reagent. The following equations show how Grignard reagents undergo reactions with ketones.

Reaction of RMgX with ketones:

General:

$$R{-}\overset{\overset{\displaystyle \ddot{O}:\,\delta-}{\|}}{\underset{\delta+}{C}}{-}R + R'{-}MgX \longrightarrow R{-}\overset{\overset{\displaystyle :\ddot{O}:^- \ ^+MgX}{|}}{\underset{\underset{\displaystyle R'}{|}}{C}}{-}R$$

$$CH_3\overset{\overset{\displaystyle O}{\|}}{C}CH_3 + CH_3CH_2CH_2CH_2{-}MgBr \longrightarrow CH_3\overset{\overset{\displaystyle O^- \ ^+MgBr}{|}}{\underset{\underset{\displaystyle CH_2CH_2CH_2CH_3}{|}}{C}}CH_3$$

The product of the reaction of RMgX with a ketone is the magnesium salt of an alcohol. When treated with water or aqueous acid, this magnesium salt yields the alcohol and a mixed inorganic magnesium salt.

$$\underset{a\ strong\ base}{R{-}\overset{\overset{\displaystyle :\ddot{O}:^-\ ^+MgX}{|}}{\underset{\underset{\displaystyle R'}{|}}{C}}{-}R} + \underset{an\ acid}{H^+} \longrightarrow \underset{a\ 3°\ alcohol}{R{-}\overset{\overset{\displaystyle :\ddot{O}H}{|}}{\underset{\underset{\displaystyle R'}{|}}{C}}{-}R} + Mg^{2+} + X^-$$

Grignard reactions provide an excellent route to alcohols with complex carbon skeletons. A Grignard reaction:

1. with formaldehyde yields a *primary alcohol*

2. with any other aldehyde yields a *secondary alcohol*

3. with a ketone yields a *tertiary alcohol.*

Reaction of RMgX with formaldehyde, H₂C=O:

$$\underset{\substack{\text{methanal}\\ \text{(formaldehyde)}}}{\overset{\overset{\displaystyle O}{\overset{\|}{}}}{\text{HCH}}} \xrightarrow[\text{(2) H}_2\text{O, H}^+]{\text{(1) RMgX}} \underset{\substack{a\ 1°\ alcohol}}{\text{RCH}_2\text{OH}}$$

Reaction of R′MgX with other aldehydes:

$$\underset{\substack{an\ aldehyde}}{\overset{\overset{\displaystyle O}{\overset{\|}{}}}{\text{RCH}}} \xrightarrow[\text{(2) H}_2\text{O, H}^+]{\text{(1) R'MgX}} \underset{\substack{a\ 2°\ alcohol}}{\overset{\overset{\displaystyle OH}{\overset{|}{}}}{\text{RCHR'}}}$$

Some other Grignard reactions also lead to alcohols. The reaction of a Grignard reagent with *ethylene oxide* yields a *primary alcohol*. The reaction of a Grignard reagent with an *ester* leads to a *tertiary alcohol*. (If a formate ester is used, a secondary alcohol is the product.)

1° Alcohols from ethylene oxide (Section 8.4A):

$$\underset{\text{ethylene oxide}}{\overset{\overset{\displaystyle O}{\diagdown\ \diagup}}{\text{CH}_2 - \text{CH}_2}} \xrightarrow[\text{(2) H}_2\text{O, H}^+]{\text{(1) C}_6\text{H}_5\text{MgBr}} \underset{\text{2-phenylethanol}}{\bigcirc\!\!-\text{CH}_2\text{CH}_2\text{OH}}$$

2° Alcohols from formate esters (Section 15.5C):

$$\underset{\text{methyl formate}}{\overset{\overset{\displaystyle O}{\overset{\|}{}}}{\text{HCOCH}_3}} \xrightarrow[\text{(2) H}_2\text{O, H}^+]{\text{(1) 2 CH}_3\text{CH}_2\text{MgBr}} \underset{\text{3-pentanol}}{\overset{\overset{\displaystyle OH}{\overset{|}{}}}{\text{HC}-\text{CH}_2\text{CH}_3}}$$

from RMgX

3° Alcohols from other esters (Section 15.5C):

$$\underset{\text{ethyl acetate}}{\overset{\overset{\displaystyle O}{\overset{\|}{}}}{\text{CH}_3\text{COCH}_2\text{CH}_3}} \xrightarrow[\text{(2) H}_2\text{O, H}^+]{\text{(1) 2 CH}_3\text{CH}_2\text{MgBr}} \underset{\text{3-methyl-3-pentanol}}{\overset{\overset{\displaystyle OH}{\overset{|}{}}}{\text{CH}_3\text{C}-\text{CH}_2\text{CH}_3}}$$

from RMgX

Lithium reagents   Lithium reagents (RLi) are closely related to Grignard reagents. These reagents are prepared similarly to Grignard reagents, by the reaction of lithium metal with an organic halide. They undergo the same types of reactions as Grignard reagents do. Because Li is more electropositive than Mg, lithium reagents are more reactive than Grignard reagents. Therefore, lithium reagents can be used for the synthesis of even very hindered alcohols. Also, lithium reactions can be run at very low temperatures.

General:

$$\text{RX} + 2\ \text{Li} \longrightarrow \text{RLi} + \text{LiX}$$

$$\underset{}{\overset{\overset{\displaystyle O}{\overset{\|}{}}}{\text{R}'-\text{C}-\text{R}''}} \xrightarrow[\text{(2) H}_2\text{O, H}^+]{\text{(1) RLi}} \underset{}{\overset{\overset{\displaystyle OH}{\overset{|}{}}}{\underset{\underset{\displaystyle R}{\overset{|}{}}}{\text{R}'-\text{C}-\text{R}''}}}$$

Specific example:

$$(CH_3)_3CCC(CH_3)_3 + (CH_3)_3C{-}Li \xrightarrow{-70°} (CH_3)_3CCC(CH_3)_3 \xrightarrow{H_2O,\ H^+} (CH_3)_3CCC(CH_3)_3$$

$$\overset{\overset{\displaystyle O}{\|}}{} \qquad \qquad \qquad \overset{O^-\ {}^+Li}{|} \qquad \qquad \overset{OH}{|}$$

$$\underset{C(CH_3)_3}{} \qquad \qquad \underset{C(CH_3)_3}{}$$

(81%)

Table 7.2 summarizes the preparations of alcohols.

Table 7.2	Summary of laboratory syntheses of alcohols[a]
Reaction	**Section Reference**
Primary Alcohols:	
$RCH_2X + {}^-H \xrightarrow{S_N2} RCH_2OH$	5.4
$HCH \xrightarrow[\text{(2) }H_2O,\ H^+]{\text{(1) RMgX}} RCH_2OH$	7.3D, 13.4D
$CH_2CH_2 \xrightarrow[\text{(2) }H_2O,\ H^+]{\text{(1) RMgX}} RCH_2CH_2OH$	8.4A
$R_2C{=}CH_2 \xrightarrow[\text{(2) }H_2O_2,\ OH^-]{\text{(1) }BH_3} R_2CHCH_2OH$	10.9
$RCH \xrightarrow{[H]^b} RCH_2OH$	13.6
$RCO_2R' \xrightarrow{[H]^b} RCH_2OH + R'OH$	15.5C
$RCO_2H \xrightarrow[\text{(2) }H_2O,\ H^+]{\text{(1) }LiAlH_4} RCH_2OH$	14.7
Secondary Alcohols:	
$RCH \xrightarrow[\text{(2) }H_2O,\ H^+]{\text{(1) }R'MgX} RCHR'\ (OH)$	7.3D, 13.4D
$HCOR \xrightarrow[\text{(2) }H_2O,\ H^+]{\text{(1) 2 }R'MgX} R'CHR'\ (OH)$	15.5C
$RCR' \xrightarrow{[H]^b} RCHR'\ (OH)$	7.3D, 13.4D
$RCH{=}CHR^c \xrightarrow[]{H_2O,\ H^+} RCHCH_2R\ (OH)$	10.7
Tertiary Alcohols:	
$RCR' \xrightarrow[\text{(2) }H_2O,\ H^+]{\text{(1) }R''MgX} RCR'\ (OH)(R'')$	7.3D, 13.4D
$RCCl\ \text{or}\ RCOR' \xrightarrow[\text{(2) }H_2O,\ H^+]{\text{(1) }R''MgX} RCR''_2\ (OH)$	15.3C, 15.5C

[a] The synthesis of 1,2-diols is discussed in Section 8.2C.

[b] The symbol [H] means a reducing agent such as H_2 + catalyst or a metal hydride (see page 299).

[c] Other preparations of 2° alcohols from alkenes include oxymercuration-demercuration (Section 10.8) and hydroboration-oxidation (Section 10.9).

SAMPLE PROBLEMS

A chemist (a) treats iodobenzene with magnesium metal and diethyl ether; (b) adds acetone; and finally, (c) adds a dilute solution of HCl. Write an equation to represent each reaction.

Solution

(a) $C_6H_5I + Mg \xrightarrow{\text{diethyl ether}} C_6H_5MgI$

(b)

$$CH_3\overset{O}{\overset{\|}{C}}CH_3 + C_6H_5MgI \longrightarrow CH_3\underset{\underset{C_6H_5}{|}}{\overset{\overset{OMgI}{|}}{C}}CH_3$$

(c)

$$(CH_3)_2\overset{\overset{OMgI}{|}}{C}C_6H_5 + H^+ \longrightarrow (CH_3)_2\overset{\overset{OH}{|}}{C}C_6H_5 + Mg^{2+} + I^-$$

A student maintains that a Grignard reagent undergoes reaction with water to yield an alcohol. What is wrong with this statement?

Solution

The carbon of the Grignard reagent is partially *negative* and undergoes reaction with a *positive group* (such as H^+), not with a negative group (such as ^-OH).

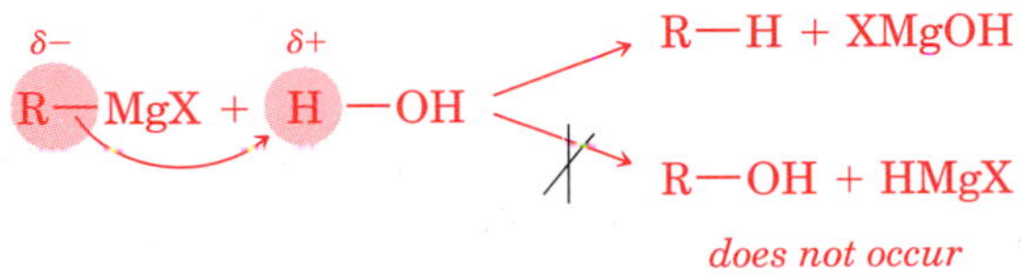

$$\overset{\delta-}{R}-MgX + \overset{\delta+}{H}-OH$$

$$\longrightarrow R-H + XMgOH$$

$$\xrightarrow{\times} R-OH + HMgX$$
does not occur

STUDY PROBLEMS

7.7 Which of the following compounds could *not* be used to prepare a Grignard reagent? Explain.

(a) $CH_3\overset{\overset{NH_2}{|}}{C}HCH_2CH_2Br$ (b) 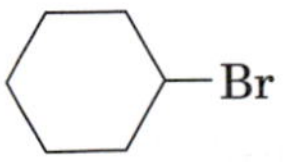 ⬡—Br (c) 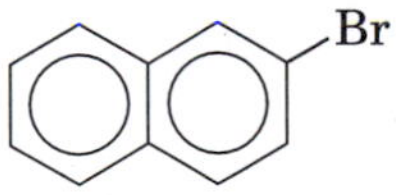 ⬡⬡—Br

(d) $BrCH_2\overset{O}{\overset{\|}{C}}OH$ (e) $CH_3C{\equiv}CCH_2CH_2I$ (f) $HOCH_2CH_2CH_2Br$

7.8 Write equations to show how the following conversions could be made:

(a) 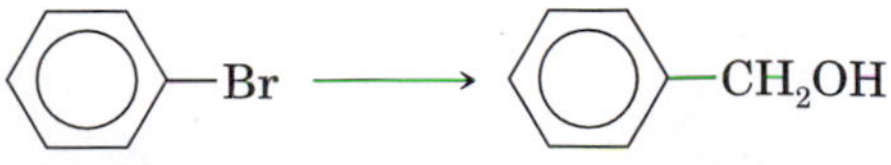 ⬡—Br $\longrightarrow$ ⬡—CH_2OH

(b) $CH_3CH_2CH_2CH_2Cl \longrightarrow CH_3CH_2CH_2CH_2OH$

(c) $CH_3CH_2Br \longrightarrow$ ⬡—$\underset{\underset{OH}{|}}{C}HCH_2CH_3$

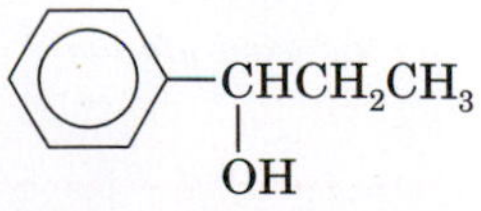

7.9 Suggest three different Grignard syntheses for each of the following alcohols:

$$\text{(a)} \quad \langle \text{C}_6\text{H}_5 \rangle - \underset{\underset{\text{CH}_3}{|}}{\overset{\overset{\text{OH}}{|}}{\text{CH}_2\text{CH}_2\text{CCH}_2\text{CH}_3}}$$

$$\text{(b)} \quad \text{CH}_3\text{CH}_2\underset{\underset{\text{C}_6\text{H}_5}{|}}{\overset{\overset{\text{OH}}{|}}{\text{C}}}\text{CH}_3$$

Section 7.4
Substitution Reactions of Alcohols

In acidic solution, alcohols can undergo substitution reactions.

$$\text{CH}_3\text{CH}_2\text{CH}_2\text{CH}_2-\underset{\text{OH}}{} + \text{H}-\text{Br} \xrightarrow[\text{heat}]{\text{H}_2\text{SO}_4} \text{CH}_3\text{CH}_2\text{CH}_2\text{CH}_2-\text{Br} + \text{H}_2\text{O}$$

1-butanol   1-bromobutane (95%)

$$\underset{\text{2-butanol}}{\text{CH}_3\text{CH}_2\overset{\overset{\text{CH}_3}{|}}{\text{CH}}-\text{OH}} + \text{H}-\text{Cl} \xrightarrow{\text{ZnCl}_2} \underset{\text{2-chlorobutane (66\%)}}{\text{CH}_3\text{CH}_2\overset{\overset{\text{CH}_3}{|}}{\text{CH}}-\text{Cl}} + \text{H}_2\text{O}$$

$$\underset{\textit{tert}\text{-butyl alcohol}}{(\text{CH}_3)_3\text{C}-\text{OH}} + \text{H}-\text{Cl} \longrightarrow \underset{\textit{tert}\text{-butyl chloride (88\%)}}{(\text{CH}_3)_3\text{C}-\text{Cl}} + \text{H}_2\text{O}$$

Unlike alkyl halides, alcohols do not undergo substitution in neutral or alkaline solution. Why not? The reason is that, in general, a leaving group must be a fairly weak base. In Chapter 5, we saw that Cl^-, Br^-, and I^- are good leaving groups and are readily displaced from alkyl halides. These ions are very weak bases. But ^-OH, which would be the leaving group of an alcohol in neutral or alkaline solution, is a *strong base* and thus is a very poor leaving group.

good leaving group   *poor leaving group*

$$\text{CH}_3\text{CH}_2-\text{Br} + {}^-\text{OH} \longrightarrow \text{CH}_3\text{CH}_2\text{OH} + \text{Br}^- \qquad \text{CH}_3\text{CH}_2-\text{OH} + \text{Br}^- \longrightarrow \text{no reaction}$$

In acidic solution, alcohols are protonated. Although $-\text{OH}$ is a poor leaving group, $-\text{OH}_2^+$ is a good leaving group because it is lost as water, a very weak base. A weak nucleophile such as a halide ion can displace the water molecule to yield an alkyl halide, as shown in the equations at the start of this section.

a good leaving group

$$\text{R}-\overset{..}{\underset{|}{\text{O}}:} \underset{}{\overset{\text{H}^+}{\rightleftharpoons}} \text{R}-\overset{+}{\underset{\underset{\text{H}}{|}}{\overset{..}{\text{O}}\text{H}}} \xrightarrow{:\overset{..}{\underset{..}{\text{X}}}:^-} \text{R}-\overset{..}{\underset{..}{\text{X}}}: + :\overset{..}{\underset{\underset{\text{H}}{|}}{\text{O}}}\text{H}$$

an alcohol   *an oxonium ion*   *an alkyl halide*

A. Reactivity of the Hydrogen Halides

In alcohol substitution reactions, the reactivity of the hydrogen halides is as follows:

	HF	HCl	HBr	HI
pK_a:	3.45	-7	-9	-9.5

increasing acid strength and increasing nucleophilicity of anion; increasing reactivity toward ROH

Although HI, HBr, and HCl are all considered strong acids (almost completely ionized in water), HI is the strongest acid of the group. HF is a weak acid (Section 1.7D). Also, I^- is the strongest nucleophile of the series and F^- is the weakest (see Section 5.3D). The order of reactivity of hydrogen halides toward alcohols parallels the relative acid strengths and the relative nucleophilicities.

increasing rate of reaction

$$ROH + HI \longrightarrow RI + H_2O$$

$$ROH + HBr \longrightarrow RBr + H_2O$$

$$ROH + HCl \longrightarrow RCl + H_2O$$

B. Reactivity of Alcohols Toward Hydrogen Halides

The order of reactivity of alcohols toward the hydrogen halides follows:

methyl   1°   2°   3°   benzylic and allylic

increasing reactivity of ROH toward HX

All alcohols undergo reaction readily with concentrated aqueous HBr or HI to yield alkyl bromides and iodides. Tertiary alcohols, benzylic alcohols, and allylic alcohols also undergo reaction with concentrated aqueous HCl. However, primary and secondary alcohols are less reactive and require the help of anhydrous $ZnCl_2$ or a similar catalyst before they can undergo reaction with the less reactive HCl in a reasonable period of time.

increasing reactivity toward HX

$$3°: \quad (CH_3)_3COH + HCl \xrightarrow{25°} (CH_3)_3CCl + H_2O$$

$$2°: \quad (CH_3)_2CHOH + HCl \xrightarrow{ZnCl_2} (CH_3)_2CHCl + H_2O$$

$$1°: \quad CH_3CH_2OH + HCl \xrightarrow[\text{heat}]{ZnCl_2} CH_3CH_2Cl + H_2O$$

The function of the zinc chloride is similar to that of H^+. Anhydrous zinc chloride is a powerful Lewis acid with empty orbitals that can accept electrons from the oxygen. The formation of a complex of $ZnCl_2$ with the alcohol oxygen weakens the C—O bond and thus enhances the leaving ability of the oxygen group.

C. S$_N$1 or S$_N$2?

It has been observed that secondary alcohols and tertiary alcohols sometimes undergo rearrangement when treated with HX. Most primary alcohols do not. The conclusion is that secondary and tertiary alcohols undergo reaction with hydrogen halides by the S$_N$1 path (through a carbocation), while primary alcohols undergo reaction by the S$_N$2 path (back-side displacement).

Methyl and primary alcohols, S$_N$2:

$$CH_3CH_2\overset{..}{\underset{..}{O}}H \; \underset{\longleftarrow}{\overset{H^+}{\longrightarrow}} \; CH_3CH_2-\overset{+}{\underset{..}{O}}H_2 \; \overset{X^-}{\longrightarrow}$$

protonated

$$\left[\overset{\delta-}{:\overset{..}{\underset{..}{X}}} ---\underset{\underset{CH_3}{|}}{CH_2}--- \overset{\delta+}{\underset{..}{O}H_2} \right] \longrightarrow :\overset{..}{\underset{..}{X}}-\underset{\underset{CH_3}{|}}{CH_2} + H_2\overset{..}{\underset{..}{O}}$$

S$_N$2 transition state

Other alcohols, S$_N$1:

$$(CH_3)_2CH\overset{..}{\underset{..}{O}}H \; \underset{\longleftarrow}{\overset{H^+}{\longrightarrow}} \; (CH_3)_2CH\overset{+}{\underset{..}{O}}H_2 \; \underset{\longleftarrow}{\overset{-H_2\overset{..}{O}}{\longrightarrow}} \; [(CH_3)_2\overset{+}{CH}] \; \overset{X^-}{\longrightarrow} \; (CH_3)_2CHX$$

protonated   *carbocation intermediate*

SAMPLE PROBLEM

The predominant alkyl halide formed in the reaction of 3-methyl-2-butanol with HBr is a product of rearrangement. Give the structure of the product and show how it is formed.

Solution

$$\underset{\underset{\text{H}}{\underset{|}{}}\overset{\overset{CH_3}{|}}{CH_3C}-\underset{\underset{\text{OH}}{|}}{CHCH_3}} \quad \overset{H^+}{\underset{-H_2O}{\longrightarrow}} \quad \left[CH_3\overset{\overset{CH_3}{|}}{\underset{\underset{H}{|}}{C}}-\overset{+}{C}HCH_3 \right] \longrightarrow$$

a 2° carbocation

$$\left[CH_3\overset{\overset{CH_3}{|}}{\underset{+}{C}}-CH_2CH_3 \right] \quad \overset{Br^-}{\longrightarrow} \quad CH_3\overset{\overset{CH_3}{|}}{\underset{\underset{Br}{|}}{C}}-CH_2CH_3$$

a 3° carbocation   2-bromo-2-methylbutane

Section 7.5

Other Reagents Used to Convert Alcohols to Alkyl Halides

Other halogenating reagents besides hydrogen halides can convert alcohols to alkyl halides. Two common halogenating reagents are thionyl chloride (SOCl$_2$) and phosphorus tribromide (PBr$_3$). Both these reagents undergo reaction with alcohols to form intermediate *inorganic esters,* discussed in Section 7.7B. The resulting inorganic ester groups are good leaving groups that can be displaced by halide ions. These reactions do not proceed by way of carbocations and thus no rearrangement occurs.

The following reactions are excellent methods for the preparation of alkyl halides from primary and secondary alcohols. (Tertiary alcohols generally undergo elimination reactions under these conditions.)

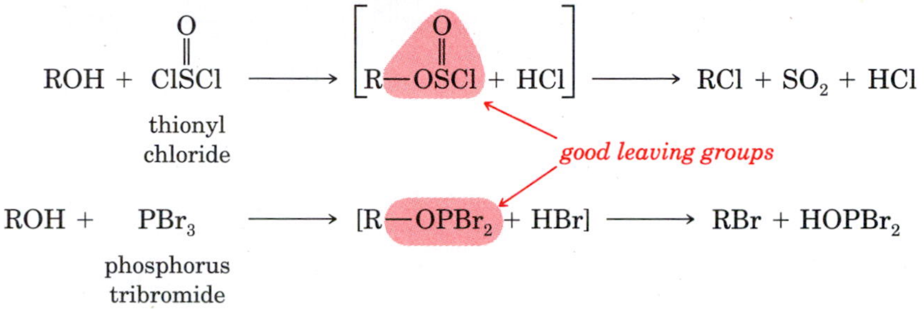

The reaction of $SOCl_2$ with an alcohol is often carried out in the presence of pyridine (page 75) or a tertiary amine (R_3N:), which removes H^+ from the reaction mixture. In the presence of these compounds, inversion of configuration is observed if the starting alcohol is chiral. The following mechanism explains why inversion occurs.

Step 1:

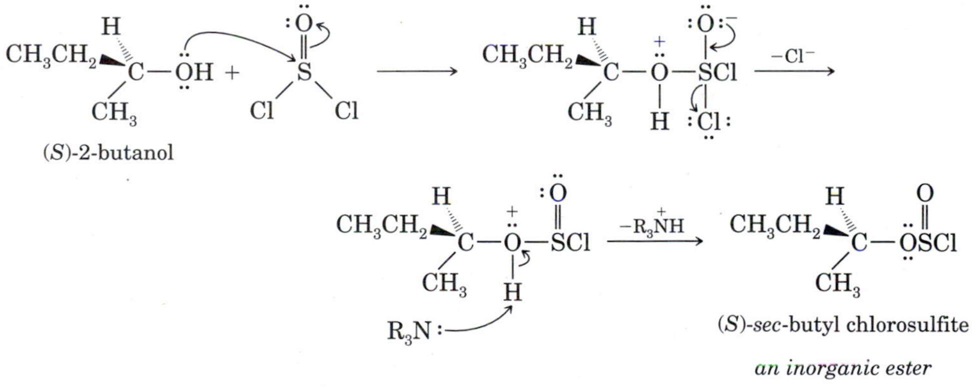

Step 2 in an amine solvent, an S_N2 reaction:

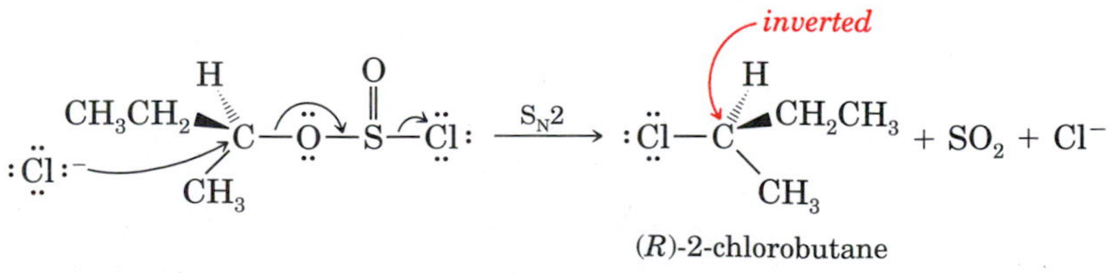

Phosphorus tribromide reacts with alcohols by a similar path. However, each PBr_3 molecule can brominate three ROH molecules.

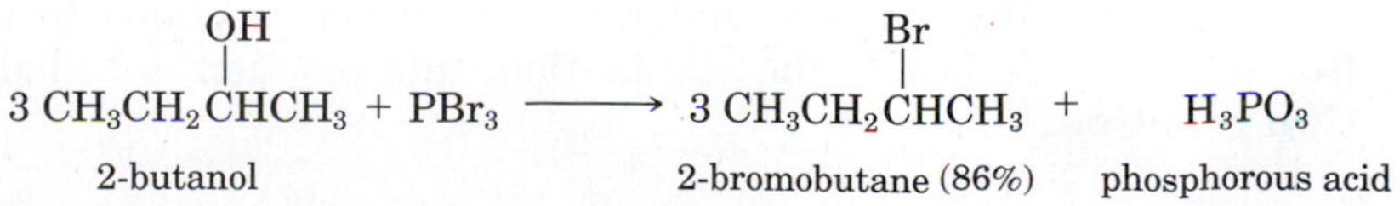

Figure 7.2 summarizes the reactions for converting alcohols to alkyl halides by substitution reactions.

$$
\begin{array}{ccc}
 & \xrightarrow[\text{S}_\text{N}2]{\text{HCl} + \text{ZnCl}_2 \text{ or HBr or HI}} & \text{RCH}_2\text{X} \\
\mathbf{1°:}\quad \text{RCH}_2\text{OH} & & \\
 & \xrightarrow[\text{S}_\text{N}2,\text{ by way of inorganic ester}]{\text{PBr}_3 \text{ or SOCl}_2} & \text{RCH}_2\text{X} \\
 & \xrightarrow[\text{S}_\text{N}1]{\text{HCl} + \text{ZnCl}_2 \text{ or HBr or HI}} & \text{R}_2\text{CHX} \\
\mathbf{2°:}\quad \text{R}_2\text{CHOH} & & \\
 & \xrightarrow[\text{S}_\text{N}2,\text{ by way of inorganic ester}]{\text{PBr}_3 \text{ or SOCl}_2} & \text{R}_2\text{CHX} \\
\mathbf{3°:}\quad \text{R}_3\text{COH} & \xrightarrow[\text{S}_\text{N}1]{\text{HCl, HBr, or HI}} & \text{R}_3\text{CX}
\end{array}
$$

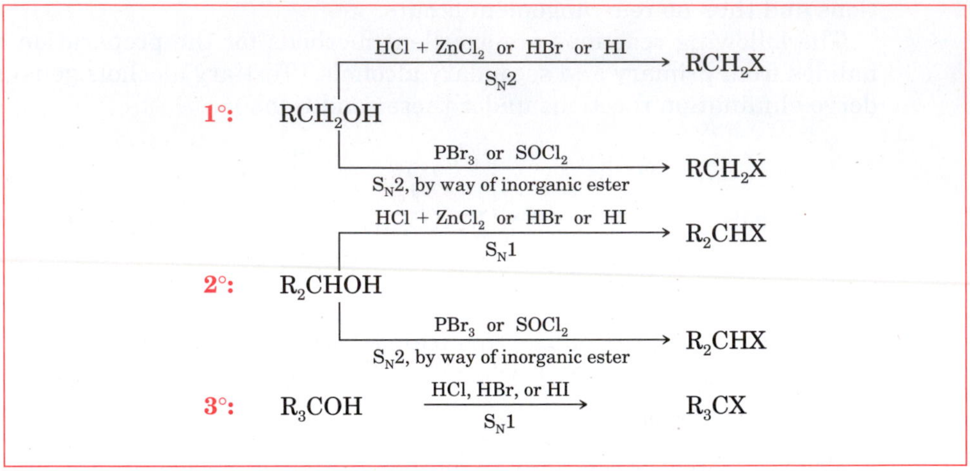

Figure 7.2 Summary of substitution reactions of alcohols leading to alkyl halides.

STUDY PROBLEMS

7.10 Write equations for the following conversions:

(a) ⟨tetrahydrofuranyl⟩–CH₂OH ⟶ ⟨tetrahydrofuranyl⟩–CH₂Br

(b) ⟨tetrahydrofuranyl⟩–CH₂OH ⟶ ⟨tetrahydrofuranyl⟩–CH₂Cl

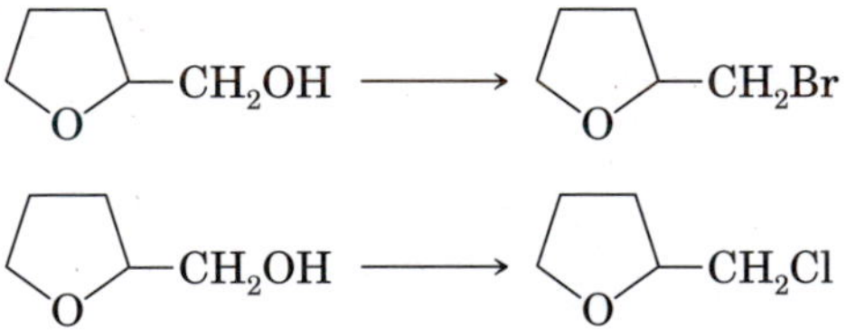

7.11 Would you expect deuterium and chlorine to be *cis* or *trans* in the product of the following reaction?

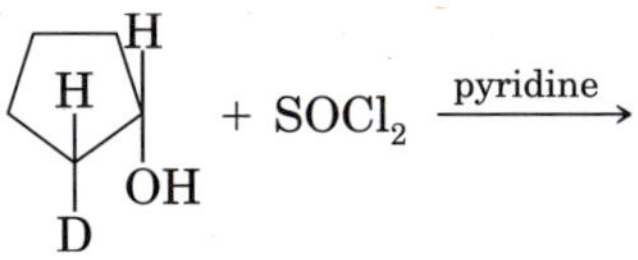

$$\text{(cyclopentane with H, H, OH, D substituents)} + \text{SOCl}_2 \xrightarrow{\text{pyridine}}$$

Section 7.6

Elimination Reactions of Alcohols

Alcohols, like alkyl halides, undergo elimination reactions to yield alkenes. Because water is lost in the elimination, this reaction is called a **dehydration reaction.**

increasing ease of dehydration

3°: $(CH_3)_3COH$ $\xrightarrow[60°]{\text{conc. } H_2SO_4}$ $(CH_3)_2C{=}CH_2 + H_2O$

tert-butyl alcohol   methylpropene
                                        (isobutylene)

2°: $(CH_3)_2CHOH$ $\xrightarrow[100°]{\text{conc. } H_2SO_4}$ $CH_3CH{=}CH_2 + H_2O$

2-propanol   propene
                                        (propylene)

1°: CH_3CH_2OH $\xrightarrow[180°]{\text{conc. } H_2SO_4}$ $CH_2{=}CH_2 + H_2O$

ethanol   ethene
                                        (ethylene)

Although sulfuric acid is often the acid of choice for a dehydration catalyst, any strong acid can cause dehydration of an alcohol. Note the comparative ease with which a tertiary alcohol undergoes elimination. Simply warming it with concentrated H_2SO_4 leads to the alkene. Elimination is a prevalent side reaction in substitution reactions of tertiary alcohols with HX.

For secondary and tertiary alcohols, dehydration follows an **E1 path.** The hydroxyl group is protonated, a carbocation is formed with loss of a water molecule, and then a proton is eliminated to yield the alkene. (Primary alcohols probably undergo dehydration by an E2 path.) A secondary or tertiary alcohol yields little, if any, substitution product in hot sulfuric acid. However, primary alcohols may yield ethers or sulfates. These competing reactions are discussed in Section 8.2.

Let us consider the dehydration of 2-pentanol, a secondary alcohol that undergoes a typical E1 reaction.

Step 1 (protonation and loss of water):

$$CH_3CH_2CH_2\overset{\displaystyle :\ddot{O}H}{\underset{|}{C}}HCH_3 \;\; \underset{H^+}{\rightleftharpoons} \;\; CH_3CH_2CH_2\overset{\displaystyle {}^+\ddot{O}H_2}{\underset{|}{C}}HCH_3 \;\; \underset{-H_2\ddot{O}:}{\rightleftharpoons} \;\; [CH_3CH_2CH_2\overset{+}{C}HCH_3]$$

protonated   *a carbocation*

Step 2 (loss of H⁺):

$$\overset{\displaystyle H}{\underset{|}{CH_3CH_2C}}H\overset{+}{-}CHCH_3 \;\; \rightleftharpoons \;\; \left[CH_3CH_2\overset{H^{\delta+}}{\underset{|}{C}}H{\cdots}\overset{\delta+}{C}HCH_3 \right] \;\; \underset{-H^+}{\rightleftharpoons} \;\; CH_3CH_2CH{=}CHCH_3$$

transition state   2-pentene

In the second step, the carbocation loses H^+ (to H_2O, HSO_4^-, or another molecule of alcohol). In this second step, the double bond is partially formed in the transition state. For this reason, when more than one alkene can be formed, a typical E1 reaction yields predominantly the *more substituted, more stable alkene* (Saytzeff rule). Recall from Chapter 5 that E2 reactions also usually yield the more stable alkene.

$$
\underset{\text{2-pentanol}}{CH_3CH_2CH_2\overset{\overset{\displaystyle OH}{|}}{C}HCH_3} \xrightarrow[\substack{\text{E1}}]{\substack{H_2SO_4 \\ \text{heat}}} \underset{\text{2-pentene (80\%)}}{CH_3CH_2CH=CHCH_3} + \underset{\text{1-pentene (5\%)}}{CH_3CH_2CH_2CH=CH_2}
$$

— more stable alkene

In any elimination reaction in which the double bond can be in *conjugation with a benzene ring,* the conjugated product is formed in preference to the nonconjugated product. The conjugated alkene is of lower energy, as is the transition state leading to its formation. In fact, it is often difficult to isolate alcohols in which the hydroxyl group is one or two carbons away from a benzene ring. Dehydration is usually spontaneous under acidic conditions.

$$
\underset{\text{1-phenyl-2-butanol}}{C_6H_5-CH_2\overset{\overset{\displaystyle OH}{|}}{C}HCH_2CH_3} \xrightarrow[-H_2O]{H^+}
$$

$$
\underset{\text{1-phenyl-1-butene}}{C_6H_5-CH=CHCH_2CH_3} \quad \text{and not} \quad \underset{\text{1-phenyl-2-butene}}{C_6H_5-CH_2CH=CHCH_3}
$$

Recall from Section 5.5F that carbocations can undergo rearrangements. Are rearrangements observed in the E1 reactions of alcohols? Yes, they are. If a carbocation can undergo a 1,2-shift to a more stable carbocation, rearrangement products will be observed.

Rearrangements:

$$
\underset{\text{3,3-dimethyl-2-butanol}}{CH_3-\overset{\overset{\displaystyle CH_3}{|}}{\underset{\underset{\displaystyle CH_3}{|}}{C}}-\overset{\overset{\displaystyle OH}{|}}{C}HCH_3} \xrightarrow[95°]{H_2SO_4} \underset{\text{2,3-dimethyl-2-butene}}{\overset{CH_3}{\underset{CH_3}{}}C=C\overset{CH_3}{\underset{CH_3}{}}}
$$

$$
\underset{\text{2,3-dimethyl-1-butanol}}{CH_3\overset{\overset{\displaystyle CH_3}{|}}{C}H-\overset{\overset{\displaystyle H}{|}}{\underset{\underset{\displaystyle CH_3}{|}}{C}}-CH_2OH} \xrightarrow[140°]{H_2SO_4} \underset{\text{2,3-dimethyl-2-butene}}{\overset{CH_3}{\underset{CH_3}{}}C=C\overset{CH_3}{\underset{CH_3}{}}}
$$

Because rearrangements can occur in the dehydration of alcohols and because the dehydrations of primary alcohols are slow, the dehydration of an alcohol is not always the best method for preparing an alkene. In many cases, it is preferable to convert the alcohol to the alkyl halide and then subject the alkyl halide to an E2 reaction.

Figure 7.3 summarizes the dehydration reactions of alcohols.

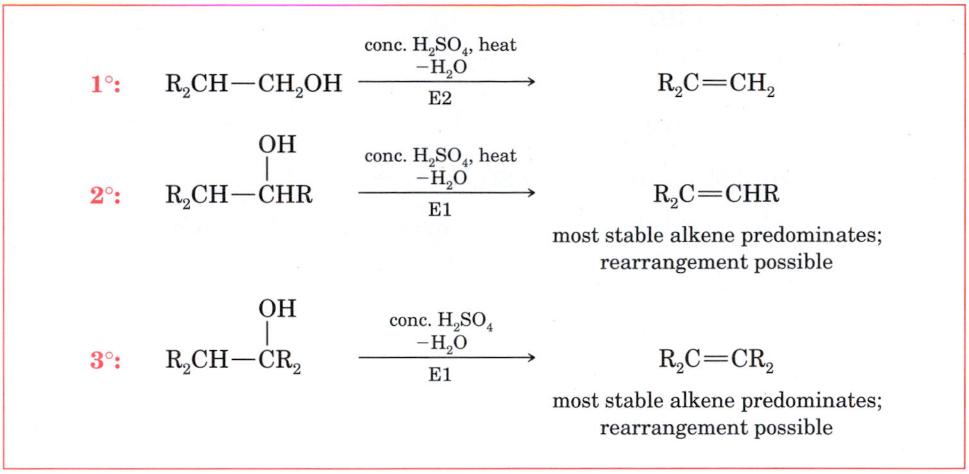

1°: R_2CH-CH_2OH $\xrightarrow[\text{E2}]{\substack{\text{conc. H}_2\text{SO}_4, \text{ heat} \\ -\text{H}_2\text{O}}}$ $R_2C=CH_2$

2°: $R_2CH-\overset{\overset{\displaystyle OH}{|}}{C}HR$ $\xrightarrow[\text{E1}]{\substack{\text{conc. H}_2\text{SO}_4, \text{ heat} \\ -\text{H}_2\text{O}}}$ $R_2C=CHR$

most stable alkene predominates;
rearrangement possible

3°: $R_2CH-\overset{\overset{\displaystyle OH}{|}}{C}R_2$ $\xrightarrow[\text{E1}]{\substack{\text{conc. H}_2\text{SO}_4 \\ -\text{H}_2\text{O}}}$ $R_2C=CR_2$

most stable alkene predominates;
rearrangement possible

Figure 7.3 Summary of dehydration reactions of alcohols.

SAMPLE PROBLEM

Predict the major products of dehydration of (a) 3-pentanol and (b) neopentyl alcohol (2,2-dimethyl-1-propanol). (*Hint:* Neopentyl alcohol undergoes rearrangement when treated with H_2SO_4.)

Solution

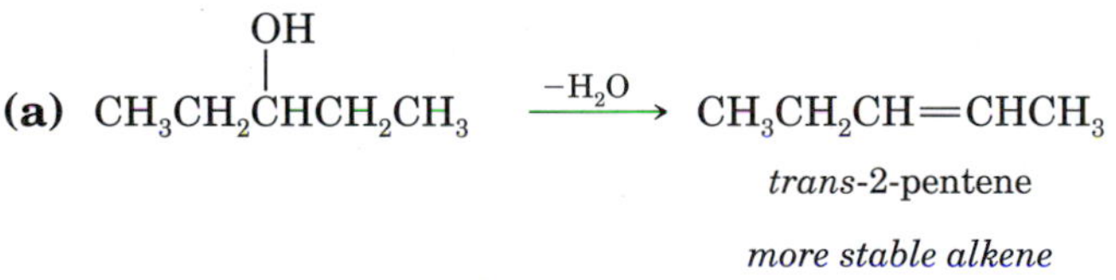

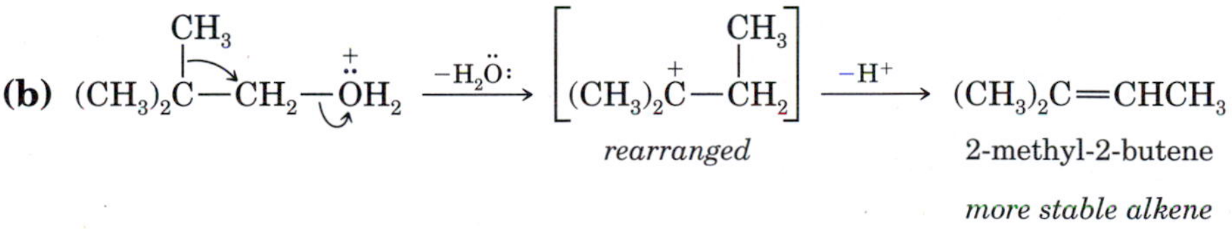

(a) $CH_3CH_2\overset{\overset{\displaystyle OH}{|}}{C}HCH_2CH_3$ $\xrightarrow{-\text{H}_2\text{O}}$ $CH_3CH_2CH=CHCH_3$

trans-2-pentene

more stable alkene

(b) $(CH_3)_2\overset{\overset{\displaystyle CH_3}{|}}{C}-CH_2-\overset{+}{\ddot{O}}H_2$ $\xrightarrow{-\text{H}_2\ddot{\text{O}}:}$ $\left[(CH_3)_2\overset{+}{C}-\overset{\overset{\displaystyle CH_3}{|}}{C}H_2\right]$ $\xrightarrow{-\text{H}^+}$ $(CH_3)_2C=CHCH_3$

rearranged   2-methyl-2-butene

more stable alkene

STUDY PROBLEMS

7.12 Write formulas for the expected elimination product(s) for each of the following reactions. If more than one alkene can be formed, indicate which alkene would predominate in the reaction mixture.

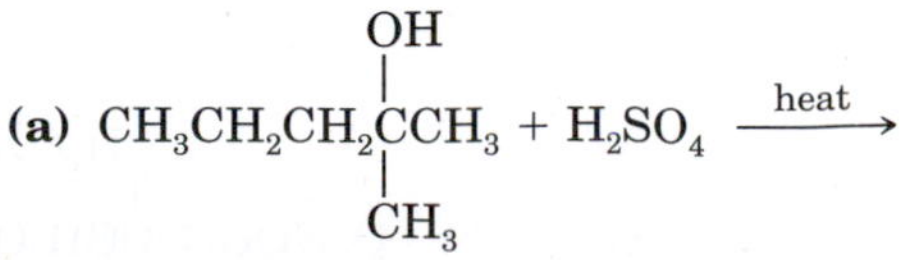

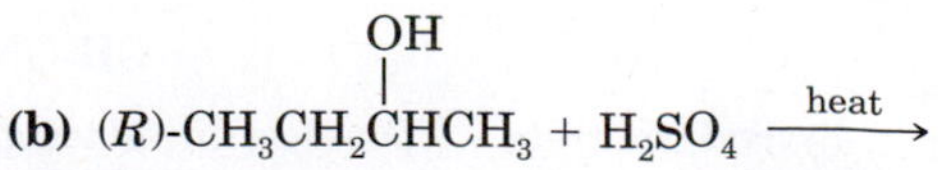

(a) $CH_3CH_2CH_2\overset{\overset{\displaystyle OH}{|}}{\underset{\underset{\displaystyle CH_3}{|}}{C}}CH_3 + H_2SO_4$ $\xrightarrow{\text{heat}}$

(b) $(R)\text{-}CH_3CH_2\overset{\overset{\displaystyle OH}{|}}{C}HCH_3 + H_2SO_4$ $\xrightarrow{\text{heat}}$

(c) + H_2SO_4 $\xrightarrow{\text{heat}}$

(d) $(C_6H_5)_2\overset{\overset{\displaystyle OH}{|}}{C}CH_2CH_3 + H_2SO_4 \longrightarrow$

7.13 Write an equation showing the mechanism of the conversion of 3,3-dimethyl-2-butanol to 2,3-dimethyl-2-butene.

Section 7.7

Esters of Alcohols

A. Carboxylic Esters

Alcohols undergo reaction with carboxylic acids and carboxylic acid derivatives to yield **esters of carboxylic acids.** These reactions, called *esterification reactions,* their mechanisms, and the product esters will be covered in detail in Chapters 14 and 15.

$$\underset{\substack{\text{acetic acid}}}{CH_3\overset{\overset{\displaystyle O}{\|}}{C}OH} + \underset{\substack{\text{ethanol}}}{H\,OCH_2CH_3} \underset{\xleftarrow{\hspace{1em}}}{\overset{H^+,\,heat}{\rightleftharpoons}} \underset{\substack{\text{ethyl acetate}}}{CH_3\overset{\overset{\displaystyle O}{\|}}{C}OCH_2CH_3} + H_2O$$

a carboxylic acid   *an ester*

$$\underset{\substack{\text{benzoic acid}}}{\overset{\overset{\displaystyle O}{\|}}{-C}OH} + \underset{\substack{\text{1-propanol}}}{H\,OCH_2CH_2CH_3} \overset{H^+,\,heat}{\rightleftharpoons} \underset{\substack{\text{propyl benzoate}}}{\overset{\overset{\displaystyle O}{\|}}{-C}OCH_2CH_2CH_3} + H_2O$$

a carboxylic acid   *an ester*

B. Inorganic Esters

Inorganic esters of alcohols are compounds prepared by the reaction of alcohols and either mineral acids (such as HNO_3 or H_2SO_4) or acid halides of mineral acids (such as $SOCl_2$).

Nitrate esters Nitrate esters, $RONO_2$ (for example, nitroglycerin and PETN), are explosives. When detonated, these compounds undergo fast intramolecular oxidation–reduction reactions to yield large volumes of gases (N_2, CO_2, H_2O, O_2). Organic nitrates and nitrites, $RONO$, are also used as *vasodilators* (substances that dilate blood vessels) in the treatment of certain types of heart disease.

$$\begin{array}{cc} \overset{\displaystyle CH_2ONO_2}{|} & \overset{\displaystyle CH_2ONO_2}{|} \\ \overset{\displaystyle CHONO_2}{|} & O_2NOCH_2\overset{\displaystyle C}{\underset{|}{}}CH_2ONO_2 \\ CH_2ONO_2 & CH_2ONO_2 \end{array}$$

glyceryl trinitrate   pentaerythritol tetranitrate
(nitroglycerin)   (PETN)

Phosphate esters   Phosphate esters of phosphoric acid and its anhydrides are extremely important in biochemistry. In living systems, diphosphate groups are common leaving groups in substitution reactions.

good leaving group

$$\underset{\textit{an alkyl phosphate}}{R-O\overset{\displaystyle O}{\underset{\displaystyle OH}{\|}}P-OH} \qquad \underset{\textit{an alkyl diphosphate}}{R-O\overset{\displaystyle O}{\underset{\displaystyle OH}{\|}}P-O\overset{\displaystyle O}{\underset{\displaystyle OH}{\|}}P-OH} \qquad \underset{\textit{an alkyl triphosphate}}{R-O\overset{\displaystyle O}{\underset{\displaystyle OH}{\|}}P-O\overset{\displaystyle O}{\underset{\displaystyle OH}{\|}}P-O\overset{\displaystyle O}{\underset{\displaystyle OH}{\|}}P-OH}$$

Sulfates   The reaction of concentrated sulfuric acid with alcohols can lead to monoalkyl or dialkyl sulfate esters. The monoesters are named as **alkyl hydrogen sulfates, alkylsulfuric acids,** or **alkyl bisulfates.** The three terms are synonymous. Alkyl hydrogen sulfates are strong acids, but dialkyl sulfates are not acidic.

acidic

$$\underset{\substack{\text{methyl hydrogen}\\\text{sulfate}}}{CH_3OSOH} \qquad \underset{\text{dimethyl sulfate}}{CH_3OSOCH_3} \qquad \underset{\text{methyl ethyl sulfate}}{CH_3OSOCH_2CH_3}$$

Sulfonates   Sulfonates are inorganic esters with the general formula RSO_2OR. (Do not confuse the sulfonate structure with the sulfate structure. A sulfonate has an alkyl or aryl group attached *directly to the sulfur atom.*)

acidic

benzenesulfonic acid   methyl benzenesulfonate

a sulfonic acid   *a sulfonate*
(a strong acid)

The *p*-toluenesulfonates (4-methylbenzenesulfonates) are usually called **tosylates** and abbreviated ROTs. The tosylates are prepared by the reaction of an alcohol with *p*-toluenesulfonyl chloride (tosyl chloride). An amine, such as pyridine (page 75), is often added to the reaction mixture to "trap" the HCl as it is formed ($R_3N\colon + HCl \rightarrow R_3NH^+ Cl^-$).

ROH   + ClS———CH$_3$ ⟶ ROS———CH$_3$ + HCl

an alcohol

tosyl chloride   *an alkyl tosylate*
(TsCl)   (ROTs)

The tosylate anion (as well as other sulfonate anions) is resonance-stabilized and is a very weak base.

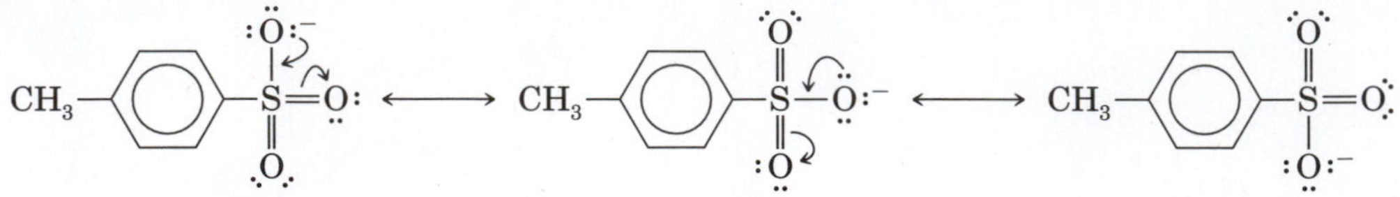

For this reason, the tosylate group is a far better leaving group than an OH group. The tosylate group can be displaced in S_N2 reactions by such weak nucleophiles as halide ions or alcohols. (No acidic catalyst is necessary.)

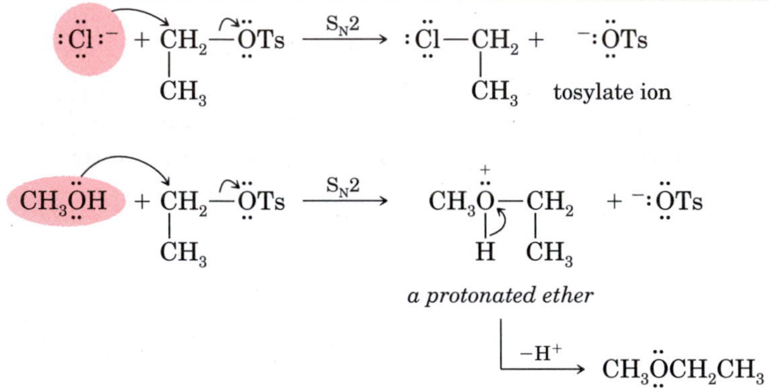

Figure 7.4 summarizes the types of esters of alcohols.

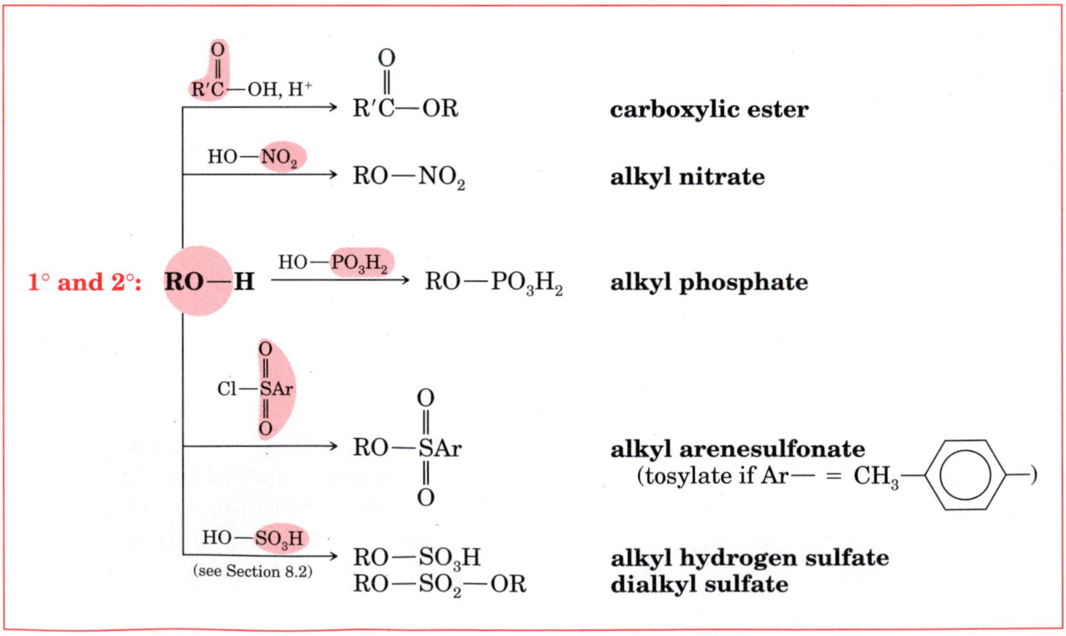

Figure 7.4 Summary of the types of esters of alcohols. (Esters of carboxylic acids are discussed in more detail in Sections 14.6 and 15.5.) Tertiary alcohols are more likely to yield alkenes in these reactions. Secondary alcohols yield alkenes when treated with concentrated H_2SO_4.

STUDY PROBLEMS

7.14 Write resonance structures for the methanesulfonate ion.

7.15 Predict the S_N2 product of the reaction of water with (R)-1-methylheptyl tosylate $[(R)\text{-}CH_3(CH_2)_5\underset{\underset{\displaystyle CH_3}{|}}{CH}\text{—OTs}]$.

Section 7.8
Oxidation of Alcohols

A. Oxidation of Organic Compounds

Oxidation is defined as the *loss of electrons* by an atom, while **reduction** is the *gain of electrons* by an atom.

$$
\begin{array}{ll}
\textit{Oxidation:} & \textit{Reduction:}\\[4pt]
Na^0 \xrightarrow{\;-e^-\;} Na^+ & Fe^{3+} \xrightarrow{\;+e^-\;} Fe^{2+}\\[8pt]
Mg^0 \xrightarrow{\;-2e^-\;} Mg^{2+} & Cu^{2+} \xrightarrow{\;+2e^-\;} Cu^0
\end{array}
$$

In organic reactions, it is not always easy to determine whether a carbon atom "gains" or "loses" electrons. However, oxidation and reduction of organic compounds are common reactions. Good rules of thumb to determine if an organic compound has been oxidized or reduced follow:

If a molecule gains oxygen or loses hydrogen, it is oxidized:

The symbol [O] represents an oxidizing agent.

$$CH_3CH_2OH \xrightarrow{\;[O]\;} CH_3CO_2H$$

$$CH_3\overset{\overset{\displaystyle OH}{|}}{C}HCH_3 \xrightarrow{\;[O]\;} CH_3\overset{\overset{\displaystyle O}{\|}}{C}CH_3$$

If a molecule loses oxygen or gains hydrogen, it is reduced:

The symbol [H] represents a reducing agent.

$$CH_3CO_2H \xrightarrow{\;[H]\;} CH_3CH_2OH$$

$$CH_3\overset{\overset{\displaystyle O}{\|}}{C}CH_3 \xrightarrow{\;[H]\;} CH_3\overset{\overset{\displaystyle OH}{|}}{C}HCH_3$$

We can list a series of compounds according to the increasing oxidation state of carbon:

$$
\begin{array}{ccccc}
 & CH_2{=}CH_2 & CH{\equiv}CH & \overset{\overset{\displaystyle O}{\|}}{} & \\
CH_3CH_3 & CH_3CH_2OH & CH_3CHO & CH_3COH & CO_2\\
 & CH_3CH_2Cl & CH_3CHCl_2 & \text{and its derivatives} &
\end{array}
$$

increasing oxidation state of C →

Note that $CH_2{=}CH_2$ and CH_3CH_2OH are at the same oxidation level. This is not surprising because the difference between the two molecules is only a molecule of water. No oxidation–reduction reaction takes place in the interconversion of ethylene and ethanol.

$$CH_2{=}CH_2 \underset{-H_2O}{\overset{H_2O}{\rightleftharpoons}} CH_3CH_2OH$$

STUDY PROBLEMS

7.16   List the following compounds in order of increasing oxidation state.

(a) $HO_2C(CH_2)_4CO_2H$ (b) ⬡—OH   (c) ⬡=O

7.17   For each of the following conversions, would an oxidizing agent, a reducing agent, or neither be required?

(a) $(CH_3)_3CH \longrightarrow (CH_3)_3CBr$ (b) [cyclohexane with Br and Cl] $\longrightarrow$ [cyclohexene]

(c) $CH_3CH_2Cl \longrightarrow CH_3CH_2I$

(d) $CH_3CH_2CH_2Br \longrightarrow CH_3CH_2CH_2MgBr$

(e) ⬡=CH_2 $\longrightarrow$ ⬡—CH_2Br

(f) $CH_3CH{=}CH_2 \longrightarrow \overset{O}{\overset{\|}{CH_3CH}} + \overset{O}{\overset{\|}{HCH}}$

Alcohols can be oxidized to ketones, aldehydes, or carboxylic acids. These oxidations are widely used in the laboratory and in industry, and they also occur in biological systems.

$$RCH_2OH \underset{a\ 1°\ alcohol}{} \overset{[O]}{\underset{[O]}{\Big\langle}} \quad \begin{array}{l} \overset{O}{\overset{\|}{RCH}} \\ an\ aldehyde \\[1em] RCO_2H \\ a\ carboxylic\ acid \end{array}$$

$$\underset{a\ 2°\ alcohol}{\overset{OH}{\overset{|}{RCHR}}} \overset{[O]}{\longrightarrow} \underset{a\ ketone}{\overset{O}{\overset{\|}{RCR}}}$$

B. Biological Oxidation of Ethanol

Ingested ethanol is oxidized primarily in the liver with the aid of an enzyme called *alcohol dehydrogenase*. The product of this dehydrogenation is acetaldehyde, CH_3CHO. (The biological oxidation of methanol leads to formaldehyde, HCHO, which is toxic.) The acetaldehyde from ethanol is further oxidized enzymatically to the acetate ion, $CH_3CO_2^-$, which undergoes esterification with the thiol **coenzyme A** (often abbreviated HSCoA). The product of the esterification is **acetylcoenzyme A.** (The complete structure of acetyl-coenzyme A is shown on page 736.) The acetyl group (CH_3CO-) in acetyl-coenzyme A can be converted to CO_2, H_2O, and energy, or it can be converted to other compounds, such as fat.

$$CH_3CH_2OH \xrightarrow[\text{dehydrogenase}]{\text{alcohol}} CH_3\overset{\overset{O}{\|}}{C}H \xrightarrow{[O]} CH_3\overset{\overset{O}{\|}}{C}O^- \xrightarrow{\text{HSCoA}}$$

$$CH_3\overset{\overset{O}{\|}}{C}-SCoA \underset{\text{acetylcoenzyme A}}{} {\displaystyle \begin{array}{l} \nearrow CO_2 + H_2O + \text{energy} \\ \searrow \text{fat, etc.} \end{array}}$$

C. Laboratory Oxidation of Alcohols

In general, laboratory oxidizing agents oxidize primary alcohols to carboxylic acids and secondary alcohols to ketones.

$$RCH_2OH \xrightarrow{[O]} R\overset{\overset{O}{\|}}{C}OH$$

$$\textit{a 1° alcohol} \qquad\qquad \textit{a carboxylic acid}$$

$$\overset{\overset{OH}{|}}{R}CHR \xrightarrow{[O]} R\overset{\overset{O}{\|}}{C}R$$

$$\textit{a 2° alcohol} \qquad\qquad \textit{a ketone}$$

Some typical oxidizing agents used for these oxidations are:

1. alkaline potassium permanganate solution: $KMnO_4 + {}^-OH$

2. hot, concentrated HNO_3

3. chromic acid: H_2CrO_4 prepared *in situ* from CrO_3 with aqueous H_2SO_4 (Jones' reagent) or $Na_2Cr_2O_7$ with aqueous H_2SO_4

4. chromium trioxide (CrO_3) complexed with pyridine or with pyridine and HCl.

$$CrO_3 \cdot 2 : N{\bigcirc} \overset{\text{pyridine}}{}$$

chromium trioxide–
pyridine complex

pyridinium chlorochromate (PCC)

Primary alcohols are oxidized first to aldehydes. In aqueous solution, aldehydes are more easily oxidized than alcohols. Therefore, the oxidation usually continues until the carboxylic acid (or, in alkaline solution, its anion) is formed.

$$\text{CH}_3(\text{CH}_2)_8\text{CH}_2\text{OH} \xrightarrow[\text{H}^+]{\text{H}_2\text{CrO}_4} \text{CH}_3(\text{CH}_2)_8\text{CO}_2\text{H}$$

1-decanol   decanoic acid (93%)

$$\underset{\substack{\text{2-ethyl-1-hexanol}}}{\text{CH}_3(\text{CH}_2)_3\overset{\overset{\displaystyle\text{CH}_2\text{CH}_3}{|}}{\text{CH}}\text{CH}_2\text{OH}} \xrightarrow[\text{(2) H}^+]{\text{(1) KMnO}_4,\ \text{OH}^-} \underset{\substack{\text{2-ethylhexanoic acid (74\%)}}}{\text{CH}_3(\text{CH}_2)_3\overset{\overset{\displaystyle\text{CH}_2\text{CH}_3}{|}}{\text{CH}}\text{CO}_2\text{H}}$$

If the intermediate aldehyde has a low boiling point, it can be distilled from the reaction mixture before it is oxidized to the carboxylic acid. Yields of aldehydes by this method are usually low. Therefore, this technique is of limited synthetic value. Better reagents for oxidizing a primary alcohol to an aldehyde are the chromium trioxide–pyridine complex or pyridinium chlorochromate. These reagents are soluble in nonaqueous solvents, such as CH_2Cl_2, and thus do not oxidize aldehydes further to carboxylic acids.

$$\underset{\text{1-heptanol}}{\text{CH}_3(\text{CH}_2)_5\text{CH}_2\text{OH}} + \text{CrO}_3 \cdot 2\,\text{N} \bigcirc \quad \text{or} \quad \text{PCC} \xrightarrow{\text{CH}_2\text{Cl}_2} \underset{\text{heptanal (72–84\%)}}{\text{CH}_3(\text{CH}_2)_5\overset{\overset{\displaystyle\text{O}}{\|}}{\text{CH}}}$$

Secondary alcohols are oxidized to ketones in excellent yields by standard oxidizing agents. (Acidic conditions are usually used because ketones can be oxidized further in alkaline solution.)

$$\underset{\text{2-octanol}}{\text{CH}_3(\text{CH}_2)_5\overset{\overset{\displaystyle\text{OH}}{|}}{\text{CH}}\text{CH}_3} \xrightarrow[\text{H}^+]{\text{H}_2\text{CrO}_4} \underset{\text{2-octanone (96\%)}}{\text{CH}_3(\text{CH}_2)_5\overset{\overset{\displaystyle\text{O}}{\|}}{\text{C}}\text{CH}_3}$$

menthol

in mint

menthone (84%)

Tertiary alcohols are not oxidized under alkaline conditions. If the oxidation is attempted in acidic solution, the tertiary alcohol undergoes dehydration and then the alkene is oxidized. Alkene oxidation will be discussed in Chapter 10.

$$\text{R}_3\text{COH} \underset{\text{a 3° alcohol}}{} \quad \overset{\text{[O]}}{\underset{\text{OH}^-}{\nearrow}} \text{no reaction}$$

$$\overset{\text{H}^+}{\searrow} \text{alkenes} \xrightarrow{\text{[O]}} \text{alkene oxidation products}$$

STUDY PROBLEM

7.18 Predict the organic products of H_2CrO_4 oxidation of: **(a)** cyclopentanol, and **(b)** benzyl alcohol. (*Hint:* the benzene ring is not affected by H_2CrO_4.)

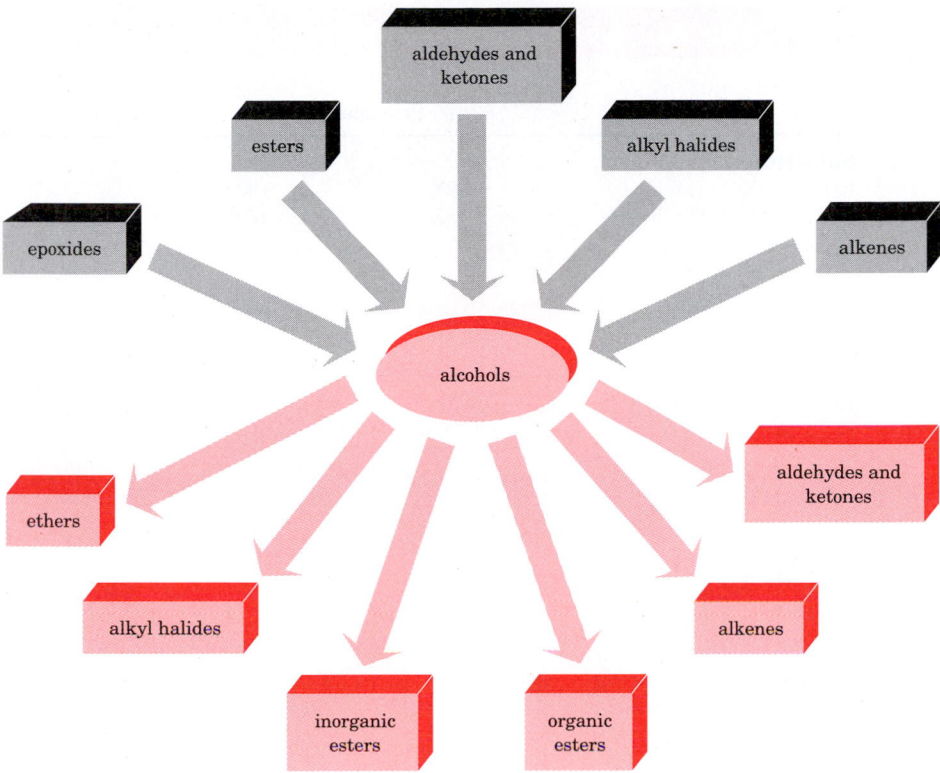

Figure 7.5 The synthetic relationship of alcohols to some other organic compounds.

Section 7.9
Use of Alcohols in Synthesis

Alcohols are versatile starting materials for the preparation of alkyl halides, alkenes, carbonyl compounds, and ethers. The types of compounds that can be obtained from alcohols are shown in Figure 7.5 and Table 7.3. From the reactions presented in this chapter, together with those presented previously, many types of compounds can be prepared from a variety of starting materials.

Section 7.10
Synthesis Problems

Organic chemists often synthesize compounds in the laboratory. The syntheses may be simple and straightforward (for example, the preparation of a particular simple alcohol for a rate study), or they may be very involved (for example, the laboratory synthesis of a complex biological molecule). Even if you do not become a laboratory chemist, designing synthetic schemes on paper is a valuable way to learn to think in the language of organic chemistry.

In this text, you will encounter many **synthesis problems**: problems in which you are asked to show by equations how you would prepare a particu-

Table 7.3   Types of compounds that can be obtained from alcohols

Reaction		Principal Product		Section Reference
Substitution:				
ROH + HX	$\longrightarrow$ RX		alkyl halide	7.4
ROH + PX_3 or $SOCl_2$	$\longrightarrow$ RX		alkyl halide	7.5
Elimination:				
$\underset{\underset{\displaystyle R_2CCHR_2}{\displaystyle |}{OH}}{}$ + H_2SO_4	$\xrightarrow{\text{heat}}$ $R_2C{=}CR_2$		alkene	7.6
Alkoxide Formation:[a]				
ROH + Na	$\longrightarrow$ $RO^- Na^+$		alkoxide	7.2
ROH + NaH	$\longrightarrow$ $RO^- Na^+$		alkoxide	7.2
Esterification:				
ROH + $R'CO_2H$	$\xrightarrow{H^+}$ $R'CO_2R$		ester	7.7a, 14.6, 15.5
ROH + TsCl	$\longrightarrow$ ROTs		inorganic ester	7.7B
Oxidation:				
RCH_2OH + $CrO_3 \cdot 2$ pyridine	$\longrightarrow$ RCHO		aldehyde	7.8
RCH_2OH + $[O]$[b]	$\longrightarrow$ RCO_2H		carboxylic acid	7.8
R_2CHOH + $[O]$	$\longrightarrow$ $R_2C{=}O$		ketone	7.8

[a] RO^- is used with methyl and 1° halides to synthesize ethers.
[b] Typical oxidizing agents used synthetically are $KMnO_4$ + ^-OH and H_2CrO_4.

lar compound. In some of these problems (but not all), a starting material will be specified. Often, there will be more than one correct solution to a synthesis problem. However, only one correct answer to each synthesis problem is usually given in this text or the accompanying study guide. If your answer is significantly different from the one given, you should verify its correctness with your instructor.

The synthesis problems posed in this text are not intended to be valid in a laboratory sense. The solution to a true laboratory synthesis problem includes a complete search of the chemical literature to see if a particular compound or sequence of reactions has already been studied by other chemists. Then, various possible pathways to the desired compound are drawn up. Each pathway is evaluated from a practical laboratory standpoint (likelihood of success, cost in terms of reagents and time, availability of starting materials, hazards, etc.). Of the various proposed pathways, one is selected. Finally, the synthetic sequence is tested in the laboratory. In solving synthesis problems in this text, you need consider only reactions or reaction sequences that are reasonable, based on the information previously presented in the text.

A. Solving Synthesis Problems

The following suggestions may help you answer synthesis problems correctly.

1. A typical study problem gives the reactants and asks for the product or products:

$$A + B \longrightarrow ?$$

In synthesis problems, the reverse question is asked: Given the product, what are the reactants?

$$? + ? \longrightarrow C$$

Thus, as you study organic reactions, you must learn them both ways. You must be able to answer such questions as "What reactions yield alcohols?" as well as "The reaction of a Grignard reagent with formaldehyde yields what?"

2. Use only reactions that give the product in reasonable yield. If one reaction gives 100% of the desired product, use that reaction. However, if no such reaction is available, use a reaction that gives 50 or 60% yield. Do not use reactions that give very low yields (under 25%).

3. It is acceptable to use flow equations, with reagents and reaction conditions above or below the arrows. This technique, while not necessary with a one-step synthesis, saves considerable time in writing the solutions to multistep syntheses.

$$A \xrightarrow{\text{xy}} B \xrightarrow[100°]{\text{yz}} C$$

4. Unless you are asked to do so, you need not balance equations nor indicate minor products.

5. If the text shows a reaction of a simple compound, you can usually extrapolate the reaction to more-complex, but similar, structures.

Example

$$\underset{\text{CH}_3\overset{\displaystyle O}{\overset{\displaystyle \|}{C}}\text{CH}_3}{} \xrightarrow[\text{(2) H}_2\text{O, H}^+]{\text{(1) CH}_3\text{MgI}} (\text{CH}_3)_3\text{COH}$$

This reaction can be extrapolated to other ketones and Grignard reagents.

$$\text{CH}_3\text{CH}_2\overset{\displaystyle O}{\overset{\displaystyle \|}{C}}\text{CH}_2\text{CH}_3 \xrightarrow[\text{(2) H}_2\text{O, H}^+]{\text{(1)} \bigcirc\!-\text{MgI}} \text{CH}_3\text{CH}_2\overset{\displaystyle OH}{\overset{\displaystyle |}{C}}\text{CH}_2\text{CH}_3$$

6. In a synthesis problem, do not be intimidated by the complexity of a structure, but focus your attention on the functional groups. Inspect the structure for its important features and worry only about the small portion that undergoes reaction. When you have a larger vocabulary of organic reactions, you will want to inspect a complex structure for other functional groups that might also undergo reaction under the same conditions.

B. Multistep Synthetic Problems

Most syntheses in the laboratory (and many synthesis problems in this book) require more than a single step from commercially available starting materials to desired products. When you are confronted with a synthesis that will require two or more steps and a reaction sequence is not immediately evident, do *not* choose a likely starting material and try to convert it to the product. Instead, *start with the product and work backwards, one step at a time, to the starting material.* This procedure, called **retrosynthetic analysis,** is best explained with an example.

Example
Show by flow equations how you would prepare 3-deuteriopropene from a nondeuteriated hydrocarbon, standard inorganic reagents, and appropriate solvents.

1. Write the structure of the product.

$$CH_2\!=\!CHCH_2D$$

2. Consider not a possible starting hydrocarbon, but a reaction that leads directly to this given product. A Grignard reagent treated with D_2O is a way to introduce deuterium into a structure. Write the equation for this reaction.

$$CH_2\!=\!CHCH_2MgBr \xrightarrow{D_2O} CH_2\!=\!CHCH_2D$$

3. Next, what reagents are needed to prepare the allyl Grignard reagent? (Again, you are working "backwards.")

$$CH_2\!=\!CHCH_2Br \xrightarrow[\text{diethyl ether}]{Mg} CH_2\!=\!CHCH_2MgBr$$

4. Finally, what reaction could be used to prepare allyl bromide?

$$CH_2\!=\!CHCH_3 \xrightarrow{NBS} CH_2\!=\!CHCH_2Br$$

5. Because we have worked our way backwards to a nondeuteriated hydrocarbon, we have solved the problem. The answer is now written forwards, rather than backwards.

$$CH_2\!=\!CHCH_3 \xrightarrow{NBS} CH_2\!=\!CHCH_2Br \xrightarrow[\text{diethyl ether}]{Mg}$$

$$CH_2\!=\!CHCH_2MgBr \xrightarrow{D_2O} CH_2\!=\!CHCH_2D$$

SAMPLE PROBLEM

Suggest a synthesis for 3-methyl-3-hexanol from alcohols of four carbons or fewer.

Solution
1. Write the formula for 3-methyl-3-hexanol:

$$CH_3CH_2CH_2\overset{\displaystyle OH}{\underset{\displaystyle CH_3}{\overset{|}{\underset{|}{C}}}}CH_2CH_3$$

2. Decide on the reactants. This is a 3° alcohol, so it can be prepared by a Grignard reaction.

$$CH_3\overset{\overset{O}{\|}}{C}CH_2CH_3 \xrightarrow[\text{(2) H}_2\text{O, H}^+]{\text{(1) CH}_3\text{CH}_2\text{CH}_2\text{MgBr}} \text{product}$$

3. The organic reactants in the preceding step can be obtained from alcohols.

$$CH_3\overset{\overset{OH}{|}}{C}HCH_2CH_3 \xrightarrow{\text{H}_2\text{CrO}_4} CH_3\overset{\overset{O}{\|}}{C}CH_2CH_3$$

$$CH_3CH_2CH_2OH \xrightarrow{\text{HBr}} CH_3CH_2CH_2Br \xrightarrow[\text{diethyl ether}]{\text{Mg}} CH_3CH_2CH_2MgBr$$

4. Write the entire synthetic sequence.

$$CH_3CH_2CH_2OH \xrightarrow{\text{HBr}} CH_3CH_2CH_2Br \xrightarrow[\text{diethyl ether}]{\text{Mg}} CH_3CH_2CH_2MgBr$$

$$CH_3\overset{\overset{OH}{|}}{C}HCH_2CH_3 \xrightarrow{\text{H}_2\text{CrO}_4} CH_3\overset{\overset{O}{\|}}{C}CH_2CH_3 \xrightarrow[\text{(2) H}_2\text{O, H}^+]{\text{(1) CH}_3\text{CH}_2\text{CH}_2\text{MgBr}} \text{product}$$

STUDY PROBLEMS

7.19 Suggest a method for preparing ethyl acetate ($CH_3CO_2CH_2CH_3$) from ethanol and *no other organic reagent*.

7.20 Suggest syntheses for the following compounds from organic compounds containing six or fewer carbon atoms and any other required reagents:

(a) 3,5-dimethyl-3-hexanol **(b)** cyclohexylmethanol

(c) (E)-5-methyl-1,5-heptadien-4-ol

7.21 Suggest a synthesis for each of the following compounds from an alcohol of four carbons or less and any other reagents needed:

(a) $CH_2\overset{\overset{CH_2}{\diagup}}{-}CHCH_2CHClCH_2CH_3$ **(b)** $(CH_3)_2C{=}CHCH_3$

(c) $(CH_3)_2CHCH_2CH_2CH_2OH$ **(d)** $(CH_3)_2CHCH_2\overset{\overset{}{\underset{\underset{CN}{|}}{C}}}HCH_3$

Summary

An **alcohol** (ROH) is a compound containing a hydroxyl group bonded to an sp^3-hybridized carbon. The molecules of an alcohol are polar and can form hydrogen bonds with other alcohol molecules, with water molecules, or with any molecules containing NH or OH.

An **alkoxide** (a compound containing RO^-) can be prepared from the reaction of an alcohol with a strong base, such as NaH, or with an alkali metal, such as Na.

In the laboratory, alcohols can be prepared by the S_N2 reaction of primary alkyl halides with OH^-; by the hydration of alkenes; by the reduction of carbonyl compounds; or by Grignard reactions of carbonyl compounds or epoxides. These reactions are summarized in Table 7.2.

Alcohols undergo *substitution reactions* with HX (1° alcohols, S_N2; 2° and 3° alcohols, S_N1). Alcohols undergo *elimination reactions* with H_2SO_4 or other strong acids. In either case, the order of reactivity of alcohols is 3° > 2° > 1°. Alkyl halides can be prepared from alcohols without rearrangement with $SOCl_2$ or PBr_3.

Substitution:

$$ROH + HX \longrightarrow RX + H_2O$$

Elimination:

$$R_2CH{-}\overset{\overset{\displaystyle OH}{|}}{C}R_2 \xrightarrow[\text{heat}]{H_2SO_4} R_2C{=}CR_2 + H_2O$$

An alcohol can undergo reaction with an acid or an acid derivative to yield an **ester of a carboxylic acid** (RCO_2R) or an **inorganic ester,** such as $RONO_2$ or $ROSO_3H$. Most inorganic ester groups are good leaving groups.

The **oxidation** of a primary alcohol results in a carboxylic acid (or aldehyde), while the oxidation of a secondary alcohol yields a ketone (see Table 7.3).

Essay Problem for Chapter 7

Yeast Reduction of 2,2-Dimethyl-1,3-cyclohexanedione

Enzymatic reaction at a prochiral center generally yields only one enantiomer. Enzymes are chiral. In an enzyme-catalyzed reaction, the substrate (compound to undergo reaction) forms a complex with the chiral enzyme. In the complex, the substrate is bound onto a chiral surface. Consequently, reaction on only one side or at only one of two atoms at the prochiral center can occur.

Enzymatic reduction of prochiral 2,2-dimethyl-1,3-cyclohexanedione **(1)** to (*S*)-(+)-3-hydroxy-2,2-dimethylcyclohexanone **(2)** is accomplished by stirring it in an aqueous solution of sucrose and baker's yeast at 30°C. The yield of the reduced product **2** is 47–52%. The optical purity of the product is estimated to be 96–98%. (K. Mori and H. Mori, *Org. Syn.* **1990,** *68,* 56.) The optical purity of a compound is the percent excess of one enantiomer compared to the amount of that enantiomer in the racemic mixture. A pure enantiomer is 100% optically pure, while a racemic mixture is 0% optically pure.

1		**2**
2,2-dimethyl-1,3-cyclohexanedione		(*S*)-(+)-3-hydroxy-2,2-dimethylcyclohexanone (47–52%)

If the reduction had occurred without the aid of a chiral catalyst (in this case, the enzyme in baker's yeast), both enantiomers of **2** would have been formed because there would have been equal probability of attack on either face (side) of each equivalent carbonyl group. However, once the prochiral dione has been bound onto the enzyme's chiral surface, attack on one face of one of the carbonyl groups would yield only one enantiomer. Attack on the other face of that same carbonyl group would yield its mirror image. Because of symmetry in **1**, if these same reactions were carried out at the other carbonyl group of the bound dione, the mirror stereoisomers would have been observed.

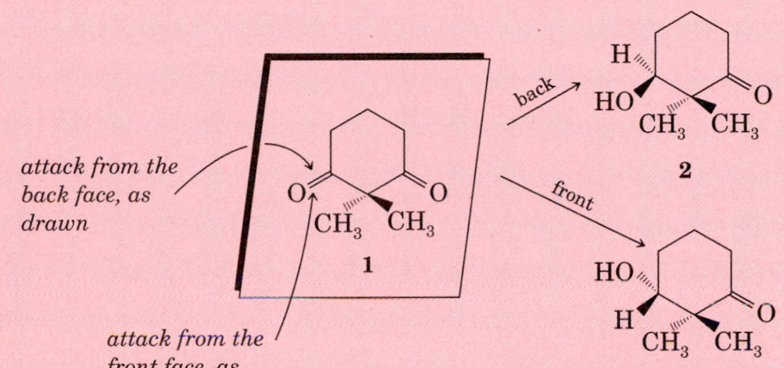

Questions

1. Which carbons in 2,2-dimethyl-1,3-cyclohexanedione are prochiral? Write equations that show the formation of both enantiomers at the prochiral carbons to explain your answer.

2. Draw the geometrical isomers that would result if both keto groups are reduced to hydroxyl groups. Which of these geometrical isomers is *meso?* Which is optically active? Draw the enantiomers of the optically active isomer.

3. As the yeast-reduction equation of **1** to **2** is written in the essay, the keto group on the left of dione **1** has been reduced. If the dione had been bound onto an enzyme surface as shown, what would be the product if the keto group on the right had been reduced?

4. Ethyl acetoacetate ($CH_3\overset{\text{O}}{\overset{\|}{C}}CH_2\overset{\text{O}}{\overset{\|}{C}}OCH_2CH_3$) can be reduced using a similar yeast-reduction reaction. The products are 93% of the (S) enantiomer and 7% of the (R) enantiomer. Write an equation for this reaction.

5. Write a reduction reaction for 1-deuterioacetaldehyde in which 1-deuterioethanol is formed. Assume the aldehyde is bound onto the surface of a chiral enzyme in the configuration shown below. Attack on which face of the carbonyl group would yield (S)-1-deuterioethanol? (R)-1-deuterioethanol?

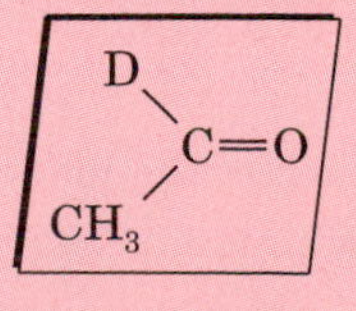

1-deuterioacetaldehyde

7.22 Draw formulas for the following compounds:

(a) 3-phenyl-1-propanol   (b) cyclopentanol

(c) isobutyl alcohol   (d) *cis*-2-buten-1-ol

(e) (*E*)-4-bromo-2,3-dimethyl-2-penten-1-ol   (f) propylene glycol

(g) *trans*-3-methylcyclohexanol   (h) (*R*)-2-butanol

7.23 Write a trivial name for each of the following compounds:

(a) $CH_3CH_2CH_2OH$ (b) $CH_2\!=\!CHCH_2OH$

(c) [benzene ring]$-CH_2OH$ (d) $CH_3\overset{\displaystyle OH}{\underset{|}{C}}HCH_3$ (e) [cyclobutane ring]$-OH$

7.24 Complete the following equations for acid-base reactions:

(a) $CH_3CH_2OH + H^+ \rightleftharpoons$ (b) [cyclohexane ring]$-OH + K \longrightarrow$

(c) $(CH_3)_3CO^- + H_2O \rightleftharpoons$ (d) $CH_3CH_2CH_2CH_2OH + NaH \longrightarrow$

(e) $CH_3OH + conc.\ H_2SO_4 \rightleftharpoons$ (f) $CH_3CH_2\overset{+}{O}H_2 + NH_3 \rightleftharpoons$

7.25 Write equations to show how the following alcohols could be prepared by S_N2 reactions.

(a) [benzene ring]$-CH_2CH_2OH$ (b) (*R*)-$CH_3CH_2\overset{\displaystyle OH}{\underset{|}{C}}HCH_3$ (c) $(CH_3)_3C$[cyclohexane ring, H and OH]

7.26 Write equations that show how each of the following alcohols can be prepared (1) by the reduction of a carbonyl compound, and (2) by a Grignard reaction:

(a) CH_3[benzene ring]$-CH_2OH$ (b) $(CH_3)_2CHCH_2\overset{\displaystyle OH}{\underset{|}{C}}HCH_3$

7.27 (1) Which of the following compounds contain acidic hydrogens that would be removed by CH_2MgI? (2) Write equations showing the products (if any) of these reactions, assuming that an excess of CH_3MgI is used.

(a) $C_6H_5C\!\equiv\!CCH_3$ (b) $HOCH_2CH_2OH$

(c) $(CH_3CH_2)_2NH$ (d) $HO_2C\!-\!\overset{\displaystyle CO_2H}{\underset{|}{C}}H\!-\!CO_2H$

7.28 How could you prepare each of the following compounds, starting with 2-bromo-propane, magnesium, and other appropriate reagents? **(a)** 2,3-dimethyl-2-butanol **(b)** 3-methyl-2-butanol

7.29 Write equations showing how each of the following alcohols could be prepared by (1) a nucleophilic substitution reaction; (2) a Grignard synthesis; and (3) the reduction of a carbonyl compound.

* For information concerning the organization of the *Study Problems* and *Additional Problems*, see the *Note to student* on page 41.

(a) $CH_3CH_2\overset{\underset{|}{OH}}{C}HCH_3$

(b) 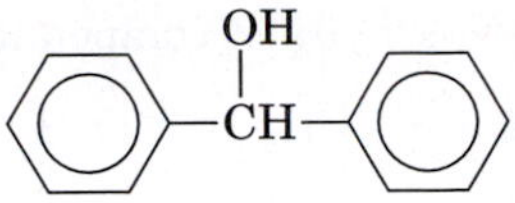 (ring)—$\overset{\underset{|}{OH}}{C}H$—(ring)

(c) $CH_3CH_2CH_2CH_2OH$

(d) 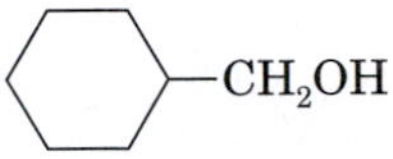 (ring)—CH_2OH

7.30 Predict the product of the reaction of methylmagnesium iodide with **(a)** H_2O, **(b)** HCHO, **(c)** $(CH_3)_2C{=}O$, **(d)** CH_3CHO, **(e)** CH_3OH

7.31 Write the equation for the reaction that occurs when each of the following alcohols is treated with HI (show the mechanisms): **(a)** 2-propanol **(b)** 1-butanol

7.32 Give the structure of the expected rearranged halide from each of the following reactions:

(a) 3,3-dimethyl-2-butanol + HCl $\xrightarrow{\text{ZnCl}_2}$

(b) 2,2-diphenyl-1-ethanol + HI $\longrightarrow$

7.33 For each of the following alcohols, would you expect a substitution reaction with HBr to proceed by an S_N1 path or by an S_N2 path? Explain.

(a) CH_3OH **(b)** 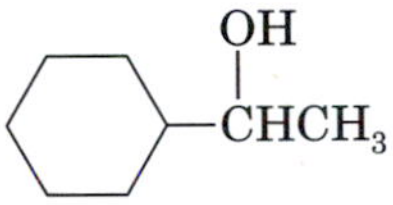 (ring)—$\overset{\underset{|}{OH}}{C}HCH_3$ **(c)** (decalin ring with OH) **(d)** 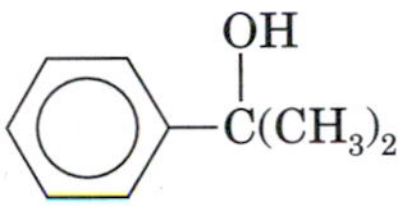 (ring)—$\overset{\underset{|}{OH}}{C}(CH_3)_2$

7.34 Complete the following equations. Show only the substitution product and ignore any possible rearrangements.

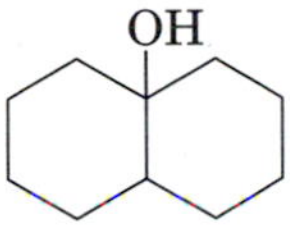

(a) $(CH_3)_2CHCH_2OH + HI \longrightarrow$

(b) 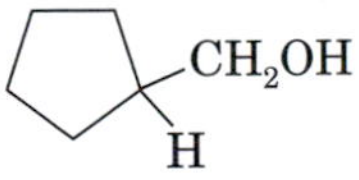 (cyclopentane ring)$\overset{CH_2OH}{\underset{H}{}}$ + HBr $\longrightarrow$

(c) 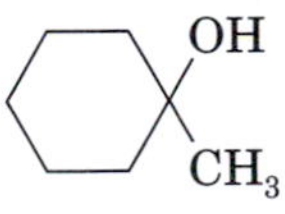 (cyclohexane ring)$\overset{OH}{\underset{CH_3}{}}$ + HCl $\longrightarrow$

(d) 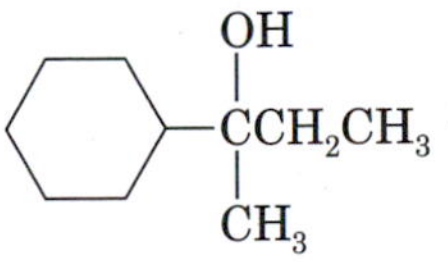 (ring)—$\overset{\underset{|}{OH}}{\underset{\underset{|}{CH_3}}{C}}CH_2CH_3$ + HCl $\longrightarrow$

7.35 Predict the organic products:

(a) $CH_3CH_2OCH_2CH_2OH \xrightarrow{\text{PBr}_3}$

(b) 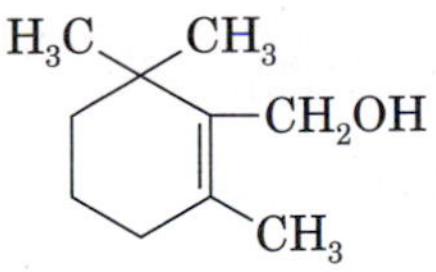 (cyclohexene ring with H_3C, CH_3, CH_3)$-CH_2OH \xrightarrow{\text{PBr}_3}$

(c) $(CH_3)_2NCH_2CH_2OH \xrightarrow{\text{SOCl}_2}$

(d) (S)-$CH_3CH_2\overset{\underset{|}{OH}}{C}HCH_3 \xrightarrow[\text{acetone}]{\text{NaI}}$

7.36 Write equations for the following reactions: **(a)** (S)-2-pentanol and tosyl chloride; **(b)** (R)-2-butanol and H_2SO_4 at 180°; **(c)** (R)-2-butanol and $ClSO_3H$; **(d)** (R)-sec-butyl tosylate and ethanol under S_N1 conditions.

7.37 Identify the inorganic ester groups in the following compounds: **(a)** a *lecithin*, found in egg yolks (page 969); **(b)** *sphingomyelin*, a component of nerve sheaves (page 969); **(c)** a *nucleotide*, a hydrolysis product of nucleic acids (page 1027)

7.38 In which of the following pairs of compounds is carbon in the higher oxidation state?

(a) $CH_3CH_2C\equiv CH$ or $CH_3CH_2CO_2H$

(b) CH_3CH_2CHO or $CH_3CH_2CH_2OH$

(c) 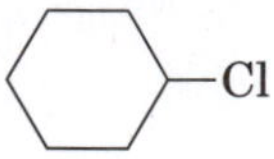   [cyclohexane ring]—Cl   or   [cyclohexane ring with two Cl on same carbon]

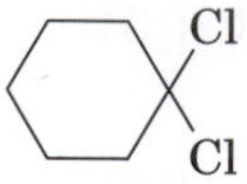

7.39 For each of the following conversions, would an oxidizing agent, a reducing agent, or neither be required?

(a) $CH_4 \longrightarrow CH_3Cl$

(b) 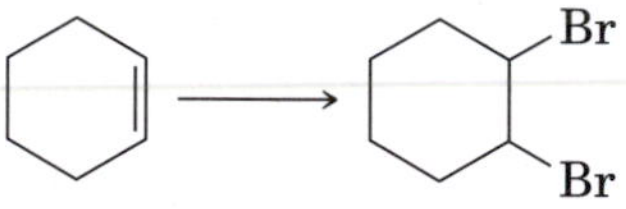   [cyclohexene] $\longrightarrow$ [cyclohexane with two Br]

(c) $CH_3CH_2Br \longrightarrow CH_3CH_2OH$

(d) 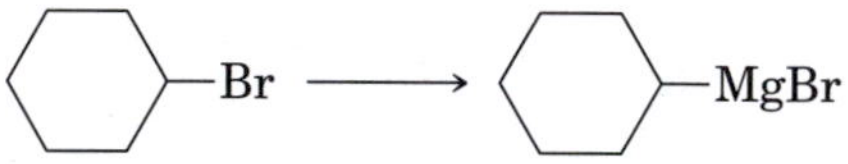   [cyclohexane]—Br $\longrightarrow$ [cyclohexane]—MgBr

(e) $CH_3CH=CH_2 \longrightarrow CH_3CH_2CH_2Br$

7.40 Suggest an alcohol and an oxidizing agent for preparing:

(a)   2-butanone   (b)   butanal   (c)   butanoic acid

(d)   1,3-dichloropropanone   (e)   cyclooctanone   (f)   decanal

7.41 Suggest reagents for the following conversions. Other organic reactants may also be used.

(a)   3,5-dimethylcyclohexanol to 1,3,5-trimethylcyclohexanol

(b)   butyl bromide to pentyl alcohol

(c)   cyclopentanol to cyclopentanone

(d)   cyclopentanone to cyclopentanol

(e)   benzaldehyde (C_6H_5CHO) to 1-phenyl-3-buten-1-ol

(f)   (R)-2-butanol to (S)-2-iodobutane

(g)   benzaldehyde (C_6H_5CHO) to styrene (phenylethene)

(h)   cycloheptanone to cycloheptyl bromide

(i)   propene to 1-hepten-4-ol

Additional Problems

7.42 Predict the major organic products of each of the following reactions.

(a) $(CH_3)_2CHCH_2OH + HCl \xrightarrow{ZnCl_2}$

(b) $CH_3CH_2CH_2CH_2O^- K^+ + H_2O \longrightarrow$

(c)   (S)- [cyclopentene ring with H_3C and OH substituents] $+ SOCl_2 \xrightarrow{pyridine}$

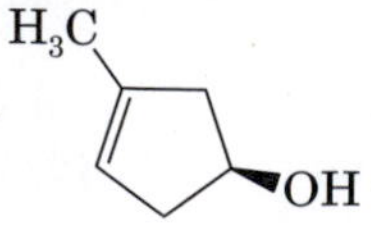

(d) 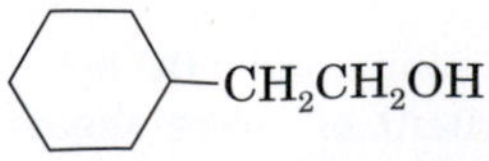   [cyclohexane]—$CH_2CH_2OH + Na \longrightarrow$

(e) $CH_3CH_2CH_2CH_2MgBr + CH_3OH \longrightarrow$

(f) $(C_6H_5)_2\overset{\overset{\displaystyle OH}{|}}{C}CH_2CH_3 \xrightarrow{H^+}$

7.43 Suggest reagents for the following conversions. Other organic reactants may also be used.

(a) $(CH_3)_2CHCH_2O\overset{\overset{\displaystyle CH_3}{|}}{C}HCH_2CH_3$ from alcohols containing four or fewer carbon atoms

(b) cycloheptanol from cycloheptanone

(c) cycloheptanone from cycloheptanol

(d) (S)-2-chlorobutane from (R)-2-butanol

(e) benzaldehyde from benzyl alcohol

(f) 1-butanol from 1-bromopropane

(g) benzyl alcohol from bromobenzene

(h) 1-phenylethanol from benzaldehyde (C_6H_5CHO)

(i) $CH_3\overset{\overset{\displaystyle O}{\|}}{C}CH_2CH_2\overset{\overset{\displaystyle O}{\|}}{C}H$ from $CH_3\overset{\overset{\displaystyle OH}{|}}{C}HCH_2CH_2\overset{\overset{\displaystyle O}{\|}}{C}H$

7.44 Compound **A**, $C_4H_{10}O$, reacts with potassium metal to liberate hydrogen gas. **A** is oxidized by chromium trioxide in pyridine to give compound **B**, C_4H_8O. **B** reacts with methylmagnesium iodide to give, after an acidic workup, compound **C**, $C_5H_{12}O$. **C** readily undergoes a reaction with concentrated hydrochloric acid. What are the structures of **A, B,** and **C**? Write equations showing what occurs in each reaction.

7.45 When $(2R,3S)$-3-methyl-2-pentanol is treated with aqueous HBr, both rearranged and unrearranged alkyl bromides are formed. Write equations for mechanisms of these reactions that would predict the products along with their stereochemistry.

7.46 Predict the major organic product of the reaction (if any) of (R)-2-heptanol with each of the following reagents: **(a)** H_2CrO_4; **(b)** HI; **(c)** Li metal; **(d)** hot, conc. H_2SO_4; **(e)** CH_3MgI; **(f)** aqueous NaCl; **(g)** aqueous NaOH; **(h)** $SOCl_2$ in pyridine.

7.47 Suggest one synthetic route (or more, if possible) to each of the following alcohols, starting with an organic halide and other needed organic reagents:

(a) $(CH_3)_3C\overset{\overset{\displaystyle OH}{|}}{C}HCH_3$

(b) [bicyclic structure with CH_2OH substituent]

(c) [bicyclic structure with HO and CH_3 substituents]

(d) $CH_3CH_2\overset{\overset{\displaystyle OH}{|}}{\underset{\underset{\displaystyle CH_3}{|}}{C}}$—[cyclohexane ring]

(e) $(CH_3)_2CH\overset{\overset{\displaystyle OH}{|}}{C}H$—[cyclopentane ring]

(f) $CH_2{=}CHC\overset{\overset{\displaystyle CH_2}{\|}}{\underset{\underset{\displaystyle OH}{|}}{C}}(CH_3)_2$

7.48 A chemist treated $CH_3CH{=}CHCH_2Cl$ with magnesium in anhydrous ether and then added acetone (propanone). After hydrolysis, instead of a single alkenyl alcohol as a product, the chemist obtained *two* alcohols. What are the structures of the two alcohols? Explain your answer.

7.49 Upon dehydration, 2,2-dimethylcyclohexanol yields two alkenes, both the result of re-arrangements. One contains a five-membered ring. What are these alkenes? Write a mechanism for the formation of each alkene.

7.50 Inorganic reagents can often be complexed with an organic polymer to provide a conve-nient insoluble source of the reagent, which can be easily removed from the reaction mix-ture in the workup. For example, when poly(vinylpyridine) is treated with CrO_3/HCl, poly(vinylpyridinium chlorochromate) is obtained. In one study, this reagent was used to oxidize the following alcohols, resulting in product yields of 76–100%. What were the organic products? **(a)** 2-octanol; **(b)** 1-phenylethanol; **(c)** 3-phenyl-2-propen-1-ol (cin-namyl alcohol).

7.51 Suggest a synthesis for each of the following compounds, starting with any organic compounds of three or fewer carbon atoms: **(a)** 1-bromo-2-butene **(b)** 3-ethyl-3-pen-tanol

7.52 Propose a reaction sequence for each of the following conversions. Use any inorganic or organic reagents you wish.

(a) any alcohol to CH_3O—⟨benzene⟩—CH_2CN

(b) benzyl bromide to stilbene (*trans*-1,2-diphenylethene).

(c) $C_6H_5CH_3$ to $C_6H_5CH_2\overset{\underset{\displaystyle |}{OH}}{C}HCH_3$

(d) ⟨cyclopentyl⟩—CH_2CH_2OH to   ⟨cyclopentyl⟩—$CH_2CH_2CH_2\overset{\underset{\displaystyle |}{OH}}{C}H$—⟨benzene⟩

7.53 Suggest ways to carry out the following conversions. More than one step is required.

(a) $CH_3CH_2\overset{\overset{\displaystyle O}{\|}}{C}CH_2CH_3$ from alcohols of three or fewer carbons

(b) $CH_3\overset{\underset{\displaystyle |}{\overset{\displaystyle CH_3}{}}}{C}DCH_3$ from $CH_3\overset{\underset{\displaystyle |}{\overset{\displaystyle CH_3}{}}}{C}HCH_3$ (*Hint:* consider a Grignard reagent.)

7.54 When 1,2-diols are heated in acidic solution, they undergo a rearrangement called a **pinacol rearrangement.** The following conversion is just one example of this reac-tion. Suggest a mechanism.

$$(CH_3)_2\overset{\underset{\displaystyle |}{\overset{\displaystyle OH}{}}}{C}—\overset{\underset{\displaystyle |}{\overset{\displaystyle OH}{}}}{C}(CH_3)_2 \xrightarrow{\text{H}^+} (CH_3)_3\overset{\overset{\displaystyle O}{\|}}{C}CCH_3 \quad + H_2O$$

2,3-dimethyl-2,3-butanediol   3,3-dimethyl-2-butanone (72%)
(pinacol)   (pinacolone)

7.55 2-Butanol can be converted into the corresponding nitrile by either of the two methods below. If the 2-butanol were the (*R*)-isomer, what would be the stereochemistry of the nitrile obtained from each method? Explain.

(a) $CH_3CH_2\overset{\underset{\displaystyle |}{\overset{\displaystyle OH}{}}}{C}HCH_3 \xrightarrow[\text{pyridine}]{\text{TsCl}} \xrightarrow{\text{KCN}} CH_3CH_2\overset{\underset{\displaystyle |}{\overset{\displaystyle CN}{}}}{C}HCH_3$

(b) $CH_3CH_2\overset{\underset{\displaystyle |}{\overset{\displaystyle OH}{}}}{C}HCH_3 \xrightarrow{\text{HBr}} \xrightarrow{\text{KCN}} CH_3CH_2\overset{\underset{\displaystyle |}{\overset{\displaystyle CN}{}}}{C}HCH_3$

ETHERS, EPOXIDES, AND SULFIDES

Like water and alcohols, an ether (ROR′) contains an sp^3-hybridized oxygen. But in an ether, the oxygen is bonded to two carbon atoms. The groups bonded to the ether oxygen can be alkyl, aryl, vinylic, or other carbon groups. The ether oxygen, like other sp^3-hybridized oxygens, contains two pairs of unshared valence electrons.

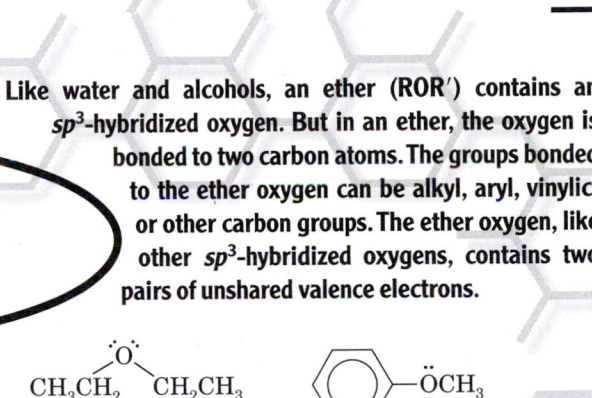

diethyl ether
("ether")

a solvent and
inhalation anesthetic

methyl phenyl ether
(anisole)

used in perfumery

divinyl ether

an inhalation
anesthetic

Ethers can be either open-chain or cyclic. When the ring size (including the oxygen) is five or greater, the chemistry of the ether is similar to that of an open-chain ether. (There are some differences in rates of reactions because the oxygen atom in a cyclic ether is less sterically hindered than the oxygen atom in a comparable open-chain ether, as the alkyl substituents are tied back in the ring.)

Epoxides are three-membered ring ethers. Because of their small ring size, epoxides are more reactive than other ethers. Large ring systems with repeating $-OCH_2CH_2-$ units are called **crown ethers.** These compounds are valuable reagents that can be used to help dissolve inorganic salts in organic solvents.

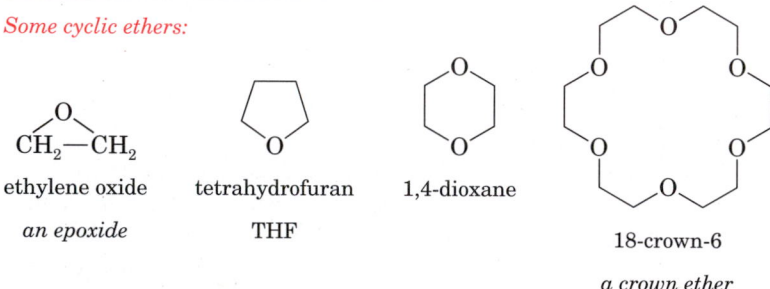

Some cyclic ethers:

ethylene oxide   tetrahydrofuran   1,4-dioxane

an epoxide   THF

18-crown-6

a crown ether

We will discuss first the open-chain ethers and then the epoxides and crown ethers. At the end of the chapter, we will briefly discuss the sulfur analogs of alcohols and ethers—thiols (RSH) and sulfides (RSR').

Section 8.1

Nomenclature of Ethers and Epoxides

Simple open-chain ethers are named almost exclusively by their trivial names, as **alkyl ethers.**

$$CH_3CH_2OCH_2CH_3 \qquad (CH_3)_2CHOCH(CH_3)_2 \qquad CH_3OCH_2CH_3$$

diethyl ether   diisopropyl ether   ethyl methyl ether
(or ethyl ether, or
simply "ether")

The names of more-complex ethers follow systematic nomenclature rules. An **alkoxy-** prefix is used when there is more than one alkoxyl (RO—) group or when there is a functional group of higher priority. (Note that a hydroxyl group has priority over an alkoxyl group.)

$$\overset{\overset{\textstyle OH}{|}}{CH_3CH_2OCH_2CH_2CH_2CHCH_3} \qquad CH_3OCH_2CH_2OCH_3$$

5-ethoxy-2-pentanol   1,2-dimethoxyethane
(DME, or "glyme")

a common solvent

In the IUPAC system, epoxides may be named as either **epoxyalkanes** or **oxiranes.** To name epoxides as epoxyalkanes, simply affix the appropriate numbers for the two carbons bridged by the oxygen to the prefix **epoxy-.** The prefix *epoxy-* always comes just before the alkane name, not in alphabetical order with any other substituents. When named as oxiranes, the epoxide ring is always numbered starting with the oxygen as position 1.

① ② O ③ ④ ⑤
CH_3CH—$CHCH_2CH_3$

2,3-epoxypentane

① O ②
③
CH_3CH—$CHCH_2CH_3$

2-ethyl-3-methyloxirane

Because an epoxide is cyclic, a substituted epoxide may be capable of geometric isomerism.

H_3C O CH_3
  C—C
H H

cis-2,3-epoxybutane
(*cis*-2,3-dimethyloxirane)

H_3C O H
  C—C
H CH_3

trans-2,3-epoxybutane
(*trans*-2,3-dimethyloxirane)

Epoxide rings may be part of fused-ring systems. In such cases, the epoxide must be *cis* on the other ring unless the other ring is eight-membered or larger. (The required bond angles for the three-membered ring makes the *trans* configuration impossible for smaller rings.) The prefix *cis* is not used in the names of these compounds. These fused-ring systems are most conveniently named epoxycycloalkanes.

H

H

O *cis*

1,2-epoxycyclopentane
(also called cyclopentene oxide)

a fused-ring system

Sample Problem

Name the following compounds:

(a) $(CH_3)_3COCH_3$

(b)

H
H
OCH_2CH_3
OCH_2CH_3

(c)

H_3C O H
  C—C
H $CH_2CH_2CH_3$

Solution

(a) *tert*-butyl methyl ether; **(b)** *cis*-1,2-diethoxycyclohexane; **(c)** *trans*-2,3-epoxyhexane or *trans*-2-methyl-3-propyloxirane.

STUDY PROBLEMS

8.1 Write formulas for:

(a) cyclohexyl isopropyl ether   (b) 2,2-diethyloxirane

(c) 1,2-epoxycyclooctane   (d) 2-phenoxyethanol

8.2 Write acceptable names for the following ethers:

(a) $(CH_3)_3C-O-$⟨benzene ring⟩$-OC(CH_3)_3$ (b) ⟨cyclohexene oxide structure⟩O

Section 8.2

Preparation of Ethers and Epoxides

A. Diethyl Ether

Diethyl ether is undoubtedly the most popular organic laboratory solvent. Of historical interest is its introduction in the 1800s, along with chloroform and nitrous oxide (N_2O, laughing gas), as a general anesthetic. Diethyl ether and nitrous oxide are still used as anesthetics. (Chloroform, however, has a narrow margin of safety and leads to liver damage, as do many of the chlorinated hydrocarbons.) Diethyl ether is volatile, its vapors are explosive, and it has a tendency to cause nausea. Despite these drawbacks, it is physiologically a relatively safe anesthetic. Other ethers that are used as anesthetics are methyl propyl ether ($CH_3OCH_2CH_2CH_3$) and ethyl vinyl ether ($CH_3CH_2OCH=CH_2$).

Industrially, diethyl ether is prepared from ethanol in the presence of sulfuric acid. This reaction was first reported in the 1500s! Until 1800, it was thought that diethyl ether contained sulfur in its structure when, in fact, sulfur was an impurity arising from the sulfuric acid.

$$CH_3CH_2OH \xrightarrow{\text{H}_2\text{SO}_4} CH_3CH_2OSO_3H \xrightarrow{\text{CH}_3\text{CH}_2\text{OH}} CH_3CH_2OCH_2CH_3$$

ethanol   ethyl hydrogen sulfate   diethyl ether

In Chapter 7, we described the conversion of alcohols to alkenes in the presence of sulfuric acid. The reactions that actually occur depend on the reaction conditions.

When an alcohol is mixed with H_2SO_4, a series of reversible reactions occurs. (The scheme that follows is simplified; alkenes can also go to sulfates, sulfates can go to ethers, etc.) Which reaction product predominates depends on the structure of the alcohol, the relative concentrations of reactants, and the temperature of the reaction mixture. In general, *primary alcohols* give sulfate esters at low temperatures, ethers at moderate temperatures, and alkenes at high temperatures. (In all cases, mixtures would be expected.) *Tertiary alcohols* and, to a large extent, *secondary alcohols* yield alkene products.

1°: $ROH + H_2SO_4$ ⇌ (0°) $ROSO_2OH + ROSO_2OR + H_2O$

(140°) $ROR + H_2O$

(170°) alkenes $+ H_2O$

2° and 3°: $ROH + H_2SO_4$ ⇌ alkenes $+ H_2O$

Because of the large numbers of possible products, this type of dehydration reaction is seldom used in the laboratory for the synthesis of ethers.

B. Williamson Ether Synthesis

The **Williamson ether synthesis** is a versatile laboratory procedure for synthesizing ethers. Alexander William Williamson (1824–1904) was a professor at University College, London. His synthesis in 1850 of diethyl ether from ethyl iodide and sodium ethoxide showed that the structures of ethyl alcohol and diethyl ether are related. This was a major step in the development of organic structure theory.

The Williamson synthesis is the S_N2 reaction of an alkyl halide with an alkoxide or phenoxide, a reaction we discussed in detail in Chapter 5.

$$R' = CH_3 \text{ or } 1°$$

$$RO^- + R'X \xrightarrow{\ S_N2\ } ROR' + X^-$$

$$R = CH_3, 1°, 2°, 3°, \text{ or } Ar$$

Best yields are obtained when the alkyl halide is methyl or primary. (Secondary and tertiary alkyl halides lead to alkenes, while aryl and vinyl halides do not undergo S_N2 reactions.) The alkoxide that can be used in a Williamson synthesis has fewer limitations. It may be methyl, primary, secondary, tertiary, allylic, or benzylic. Usually, either the sodium or potassium alkoxide or phenoxide is used.

Synthesis of dialkyl ethers:

$$CH_3O^- \ + CH_3CH_2CH_2{-}Cl \xrightarrow{\ S_N2\ } CH_3OCH_2CH_2CH_3 + Cl^-$$

methoxide ion   1-chloropropane   methyl propyl ether

$$CH_3CH_2CH_2O^- + \quad CH_3I \xrightarrow{\ S_N2\ } CH_3OCH_2CH_2CH_3 + I^-$$

propoxide ion   iodomethane   methyl propyl ether

$$(CH_3)_3CO^- \ + CH_3I \xrightarrow{\ S_N2\ } \quad (CH_3)_3COCH_3 \quad + I^-$$

tert-butoxide ion   *tert*-butyl methyl ether

but not $CH_3O^- + (CH_3)_3CCl$ (Why not?)

Synthesis of an alkyl aryl ether:

$$\langle\bigcirc\rangle{-}O^- + CH_3CH_2Br \xrightarrow{\ S_N2\ } CH_3CH_2O{-}\langle\bigcirc\rangle + Br^-$$

phenoxide ion   bromoethane   ethyl phenyl ether

but not $CH_3CH_2O^- + \langle\bigcirc\rangle{-}Br$ (Why not?)

STUDY PROBLEM

8.3 Write equations for Williamson syntheses of the following ethers:

(a) ethyl isopropyl ether

(b) *tert*-butyl propyl ether

(c) cyclohexyl 3,5-dimethylphenyl ether

(d) benzyl phenyl ether

C. Epoxides

Treatment of an alkene with a peroxycarboxylic acid (RCO_3H or $ArCO_3H$) in an inert solvent, such as $CHCl_3$ or CCl_4, yields an epoxide, or epoxyalkane. Peroxybenzoic acid ($C_6H_5CO_3H$) and *m*-chloroperoxybenzoic acid are commonly used peroxy acids.

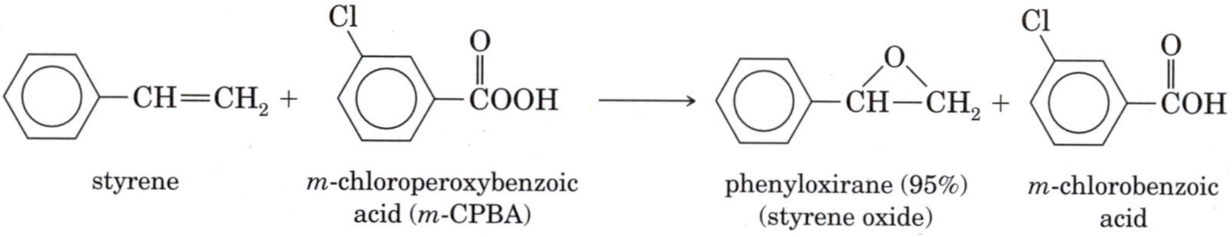

styrene   *m*-chloroperoxybenzoic acid (*m*-CPBA)   phenyloxirane (95%) (styrene oxide)   *m*-chlorobenzoic acid

The reaction path involves transfer of an oxygen from the peroxy acid directly to the alkene in one step.

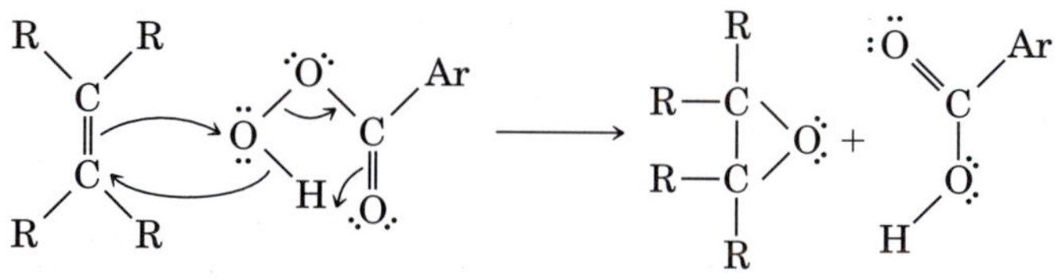

A second way to synthesize an epoxide is by an intramolecular Williamson synthesis using a **1,2-halohydrin,** a compound with a halogen and a hydroxyl group on adjacent (1,2) carbons.

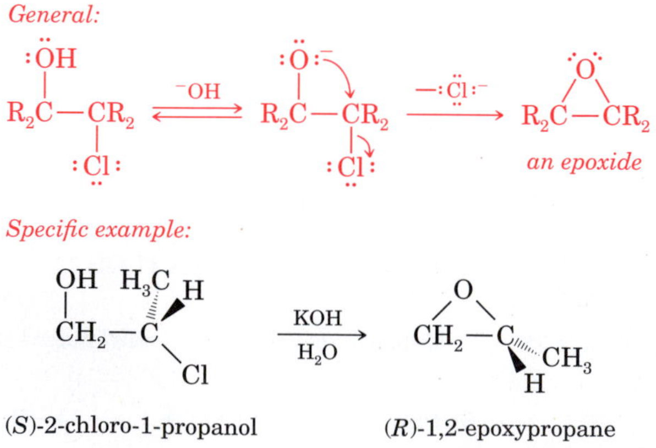

General:

an epoxide

Specific example:

(*S*)-2-chloro-1-propanol   (*R*)-1,2-epoxypropane

STUDY PROBLEM

8.4   Write equations for two syntheses of *cis*-2,3-epoxybutane.

Section 8.3
Substitution Reactions of Ethers

Ethers are quite unreactive and behave more like alkanes than like organic compounds containing functional groups. Ethers undergo autoxidation (Section 6.6B) and combustion (which occurs readily), but they are not oxidized by laboratory reagents; nor do ethers undergo reduction, elimination, or reactions with bases. When they are heated with strong acids, ethers do undergo substitution reactions. For example, when heated with HI or HBr, an ether yields an alcohol and an alkyl halide. Under these conditions, the alcohol can react further with the HI or HBr to yield additional alkyl iodide or bromide.

$$CH_3CH_2OCH_2CH_3 + HI \xrightarrow{heat} CH_3CH_2I + HOCH_2CH_3$$

diethyl ether   iodoethane   ethanol

$$\xrightarrow{HI} CH_3CH_2I$$

Ether cleavage with HI or HBr proceeds by almost the same path as the reaction of an alcohol with HX: protonation of the oxygen, followed by S_N1 or S_N2 reaction. (Protonation is necessary because RO^- is a poor leaving group, while ROH, like H_2O, is easily displaced.)

$$CH_3CH_2 - \overset{..}{\underset{..}{O}} - CH_2CH_3 \xrightleftharpoons{H^+} CH_3CH_2 - \overset{\overset{H}{|}}{\underset{..}{O}}{}^+ - CH_2CH_3 \xrightarrow{I^-}$$

protonated

$$\left[\overset{\delta-}{:}\overset{..}{\underset{..}{I}} - - - \overset{\overset{CH_3}{|}}{CH_2} - - - \overset{\overset{H}{|}}{\underset{\delta+}{O}} - CH_2CH_3 \right] \longrightarrow CH_3CH_2\overset{..}{\underset{..}{I}}: + H\overset{..}{\underset{..}{O}}CH_2CH_3$$

S_N2 *transition state*

An alkyl phenyl ether, such as anisole, yields the alkyl iodide and phenol (not iodobenzene and methanol) because sp^2-hybridized carbons do not undergo reaction by an S_N1 or S_N2 path.

not cleaved

$$\text{C}_6\text{H}_5{-}O{-}CH_3 + HI \xrightarrow{heat} \text{C}_6\text{H}_5{-}OH + CH_3I$$

methyl phenyl ether   phenol   iodomethane
(anisole)

SAMPLE PROBLEM

Give the steps for the cleavage of diisopropyl ether by HI (an S_N1 reaction).

Solution

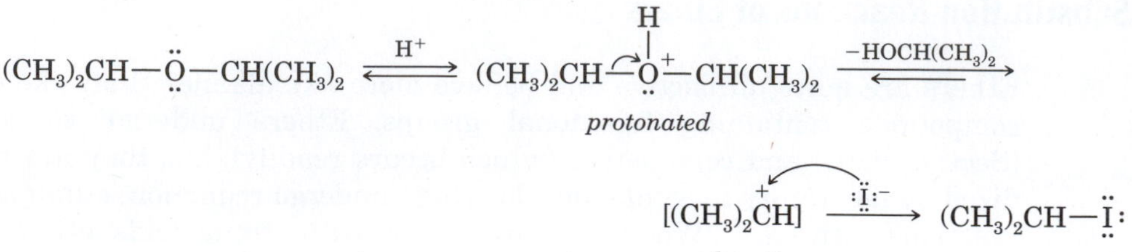

$$(CH_3)_2CH-\overset{\cdot\cdot}{\underset{\cdot\cdot}{O}}-CH(CH_3)_2 \underset{}{\overset{H^+}{\rightleftharpoons}} (CH_3)_2CH\overset{\overset{\overset{H}{|}}{\nearrow}}{\underset{}{O}}\overset{+}{\underset{}{}}-CH(CH_3)_2 \overset{-HOCH(CH_3)_2}{\underset{}{\rightleftharpoons}}$$

protonated

$$[(CH_3)_2\overset{+}{C}H] \overset{:\overset{\cdot\cdot}{\underset{\cdot\cdot}{I}}:^-}{\longrightarrow} (CH_3)_2CH-\overset{\cdot\cdot}{\underset{\cdot\cdot}{I}}:$$

a carbocation

STUDY PROBLEM

8.5   Write equations to show the mechanisms of the following reactions:

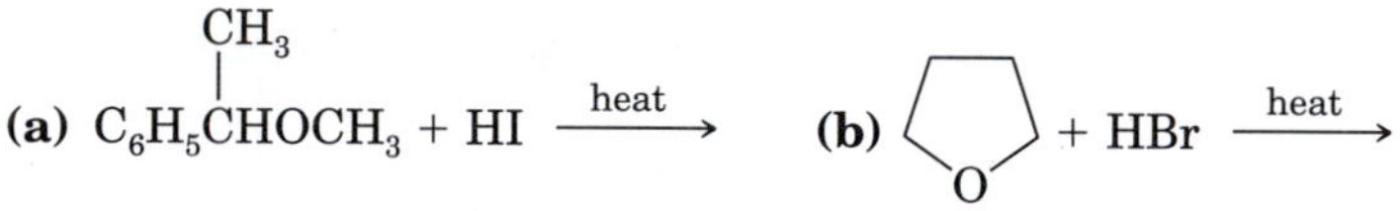

(a) $\underset{\underset{C_6H_5CHOCH_3}{}}{\overset{\overset{CH_3}{|}}{}}$ + HI $\overset{heat}{\longrightarrow}$ (b) ⬠(O) + HBr $\overset{heat}{\longrightarrow}$

Section 8.4
Substitution Reactions of Epoxides

An epoxide ring, like a cyclopropane ring, cannot have normal sp^3 bond angles of 109°; instead, the internuclear angles are 60°, a geometric requirement of the three-membered ring. The orbitals forming the ring bonds are incapable of maximum overlap. Therefore, epoxide rings are strained. The polarity of the C—O bonds, along with this ring strain, contributes to the high reactivity of epoxides compared to the reactivity of other ethers.

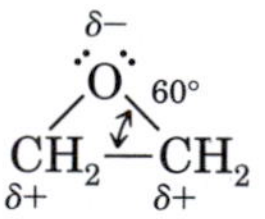

$$\underset{\delta+ \qquad \delta+}{\underset{CH_2-CH_2}{\overset{\overset{\delta-}{\overset{\cdot\cdot}{\underset{\cdot\cdot}{O}}}}{\diagup\diagdown}}} \quad 60°$$ *polar and strained*

Opening of the strained three-membered ring results in a lower-energy, more stable product. The characteristic reaction of epoxides is ring opening, which can occur under either alkaline or acidic reaction conditions. These reactions of epoxides are referred to as **base-catalyzed** or **acid-catalyzed cleavage reactions.**

In base:

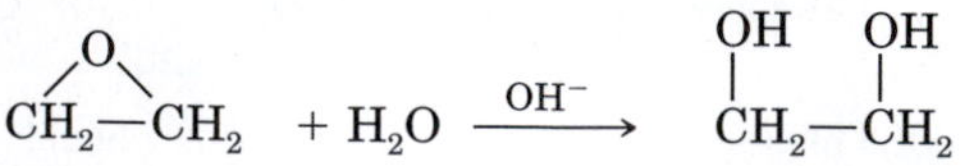

$$\underset{\substack{epoxyethane \\ (ethylene\ oxide)}}{\overset{O}{\overset{\diagup\diagdown}{CH_2-CH_2}}} + H_2O \overset{OH^-}{\longrightarrow} \underset{\substack{1,2\text{-}ethanediol \\ (ethylene\ glycol)}}{\overset{\overset{OH \quad OH}{|\qquad|}}{CH_2-CH_2}}$$

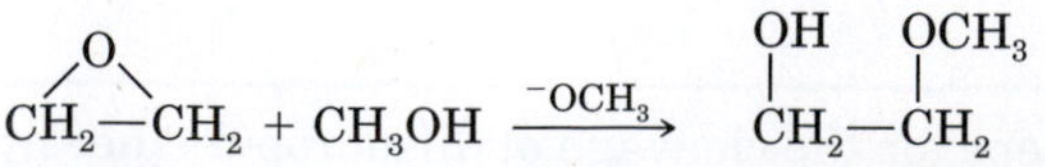

$$\overset{O}{\overset{\diagup\diagdown}{CH_2-CH_2}} + CH_3OH \overset{-OCH_3}{\longrightarrow} \underset{\substack{2\text{-}methoxyethanol}}{\overset{\overset{OH \quad OCH_3}{|\qquad|}}{CH_2-CH_2}}$$

In acid:

$$CH_2-CH_2 + H_2O \xrightarrow{H^+} \overset{\overset{OH}{|}}{CH_2}-\overset{\overset{OH}{|}}{CH_2}$$

$$CH_2-CH_2 + HCl \longrightarrow \overset{\overset{OH}{|}}{CH_2}-\overset{\overset{Cl}{|}}{CH_2}$$

2-chloroethanol
(ethylene chlorohydrin)

A. Base-Catalyzed Cleavage

Epoxides undergo S_N2 attack by nucleophiles such as the hydroxide ion or alkoxides. The steps in the reactions of ethylene oxide with hydroxide ion (NaOH or KOH in water) and with methoxide ion (NaOCH$_3$ in methanol) follow:

$$CH_2-CH_2 + {}^-:\ddot{O}H \xrightarrow{S_N2} CH_2-CH_2OH \rightleftharpoons CH_2CH_2OH + {}^-:\ddot{O}H$$

*abstracting a
proton from H$_2$O*

1,2-ethanediol

$$CH_2-CH_2 + {}^-:\ddot{O}CH_3 \xrightarrow{S_N2} CH_2-CH_2OCH_3 \rightleftharpoons CH_2CH_2OCH_3 + CH_3\ddot{O}:^-$$

2-methoxyethanol

In base-catalyzed cleavage, the nucleophile attacks the *less hindered carbon,* just as we would expect from an S_N2 attack (1° > 2° > 3°).

$$CH_3-\overset{\overset{O}{\diagup\,\diagdown}}{\underset{\underset{H}{|}}{C}}-CH_2 \quad :Nu^-$$

*in base, attack at
less hindered carbon*

$$CH_3CH-CH_2 \xrightarrow[CH_3CH_2OH]{Na^+ \, {}^-OCH_2CH_3} \overset{\overset{OH}{|}}{CH_3CHCH_2OCH_2CH_3}$$

1,2-epoxypropane
(propylene oxide)

1-ethoxy-2-propanol (83%)

A Grignard reagent contains a partially negative carbon atom and attacks an epoxide ring in the same manner as other nucleophiles. The product is the magnesium salt of an alcohol. The alcohol can be obtained by hydrolysis. The reaction of a Grignard reagent with ethylene oxide is a method by which the hydrocarbon chain of the Grignard reagent may be *extended by two carbons.*

$$CH_3(CH_2)_3MgBr + CH_2-CH_2 \longrightarrow CH_3(CH_2)_3CH_2\overset{\overset{O^- \, {}^+MgBr}{|}}{CH_2}$$

butylmagnesium
bromide

two carbons added

$$\xrightarrow{H_2O, \, H^+} CH_3(CH_2)_3CH_2CH_2OH$$

1-hexanol (62%)

SAMPLE PROBLEM

Suggest a synthesis for 2-phenylethanol starting with bromobenzene.

Solution

1. Convert bromobenzene to a Grignard reagent.

$$C_6H_5Br + Mg \xrightarrow{\text{ether}} C_6H_5MgBr$$

2. Treat the Grignard reagent with ethylene oxide.

$$C_6H_5MgBr + CH_2\overset{O}{-}CH_2 \longrightarrow C_6H_5CH_2CH_2OMgBr$$

3. Hydrolyze the magnesium alkoxide.

$$C_6H_5CH_2CH_2OMgBr + H^+ \xrightarrow{H_2O} C_6H_5CH_2CH_2OH + Mg^{2+} + Br^-$$

STUDY PROBLEM

8.6 Show how the following conversions can be carried out.

(a) $(CH_3)_3CCH_2CH_2OH \longrightarrow (CH_3)_3CCH_2CH_2CH_2CH_2OH$

(b) $CH_3CH_2CH{=}CHCH_2CH_3 \longrightarrow (CH_3CH_2)_2CH\underset{\underset{OH}{|}}{-}CHCH_2CH_3$

(c) $CH_3CH{=}CH_2 \longrightarrow CH_3\underset{\underset{OH}{|}}{CH}CH_2OCH_2CH_3$

B. Acid-Catalyzed Cleavage

In acidic solution, the oxygen of an epoxide is protonated. A protonated epoxide can be attacked by weak nucleophiles such as water, alcohols, or halide ions.

General:

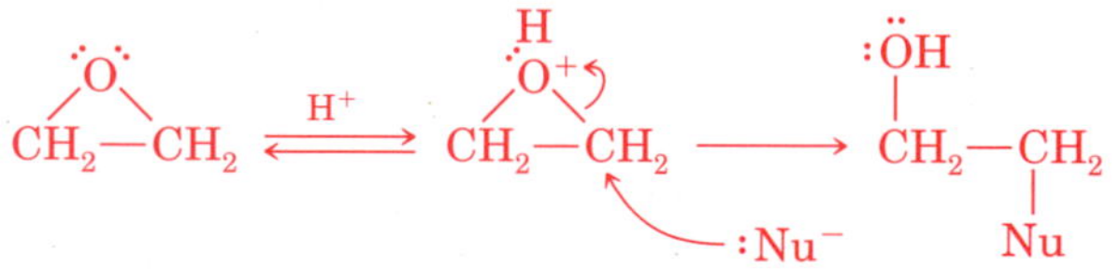

protonated

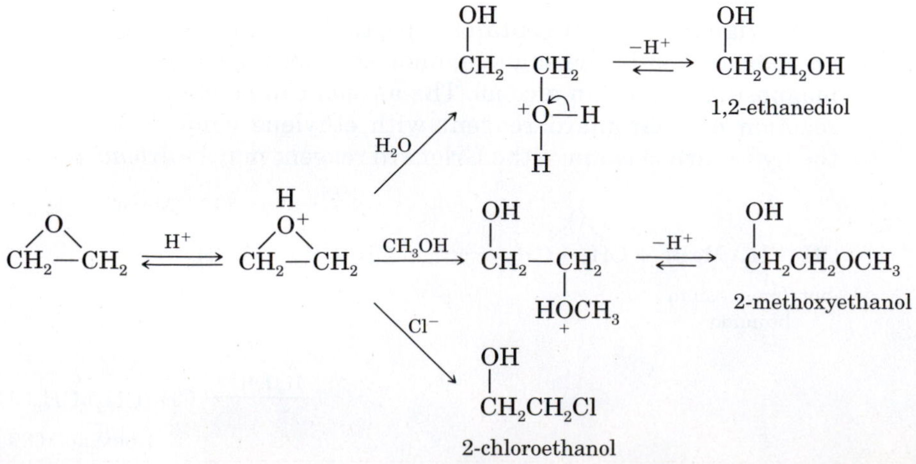

1,2-ethanediol

2-methoxyethanol

2-chloroethanol

As contrasted with base-catalyzed cleavage, attack under acidic conditions occurs at the more hindered carbon.

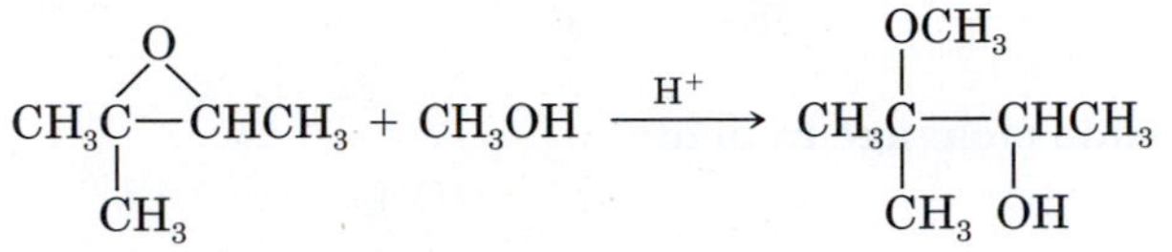

$$CH_3\overset{\displaystyle O}{\overset{\displaystyle \diagdown\!\!\diagup}{C}}\!-\!\underset{\displaystyle CH_3}{\overset{\displaystyle |}{C}}HCH_3 + CH_3OH \xrightarrow{H^+} CH_3\underset{\displaystyle CH_3}{\overset{\displaystyle OCH_3}{\overset{\displaystyle |}{C}}}\!-\!\underset{\displaystyle OH}{\overset{\displaystyle |}{C}}HCH_3$$

We conclude that if the protonated epoxide has a substantial amount of carbocation character, attack will occur at that site. This is the case when there are two alkyl groups on one carbon of the epoxide, because it is similar to a 3° carbocation.

similar to a 3° carbocation; most attack occurs here

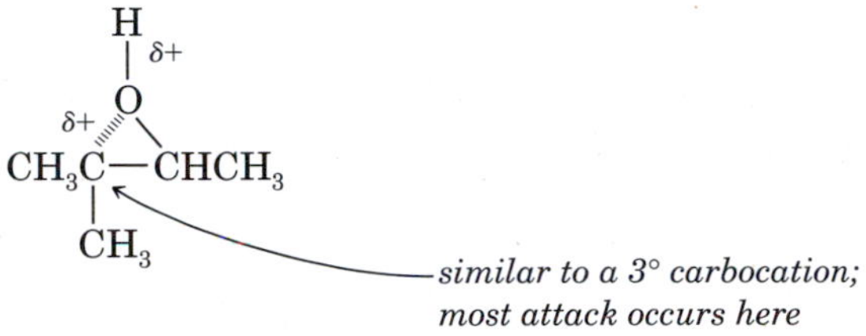

However, when the carbon undergoing attack is similar to a 2° carbocation, there is little selectivity in the reaction.

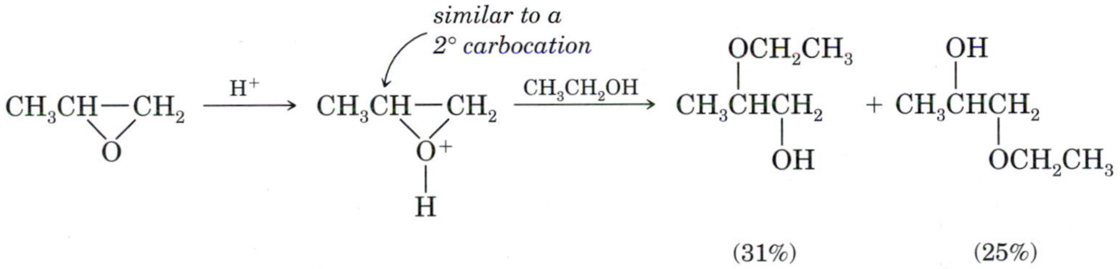

similar to a 2° carbocation

(31%)   (25%)

If the attacking nucleophile is a halide, then attack at the least hindered carbon is favored.

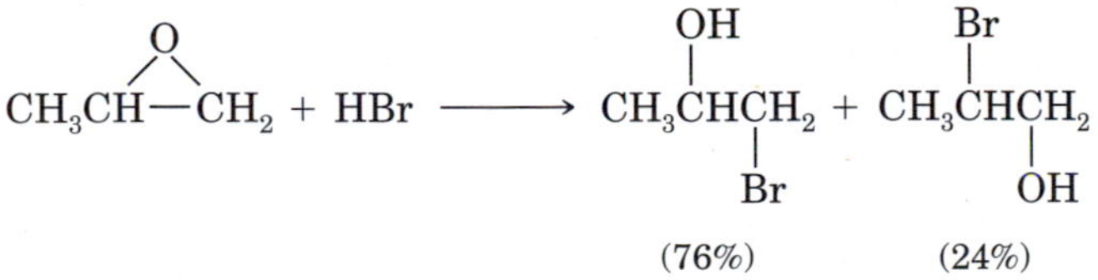

$$CH_3\overset{\displaystyle O}{\overset{\displaystyle \diagdown\!\!\diagup}{C}}H\!-\!CH_2 + HBr \longrightarrow CH_3\underset{\displaystyle Br}{\overset{\displaystyle OH}{\overset{\displaystyle |}{C}}}HCH_2 + CH_3\underset{\displaystyle OH}{\overset{\displaystyle Br}{\overset{\displaystyle |}{C}}}HCH_2$$

(76%)   (24%)

Figure 8.1 summarizes the substitution reactions of ethers.

STUDY PROBLEM

8.7 Predict the products of reaction of the following epoxide with **(a)** sodium methoxide in methanol, and **(b)** concentrated aqueous HCl:

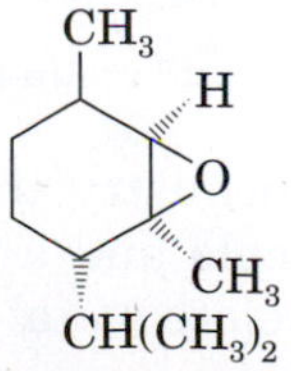

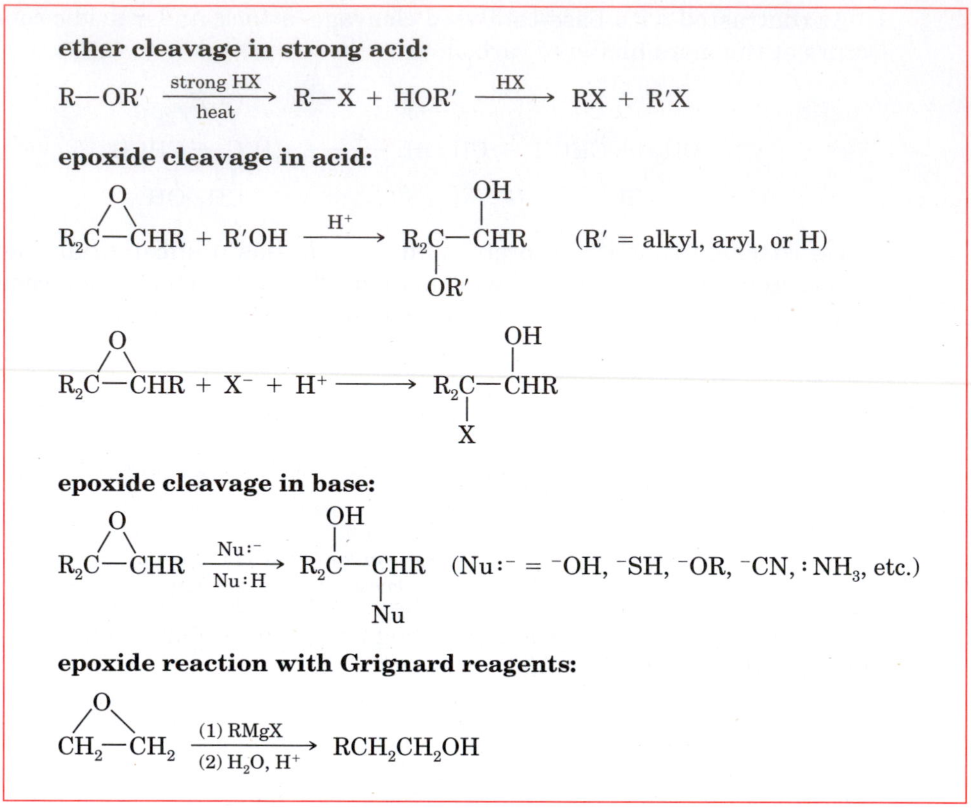

ether cleavage in strong acid:

$$R—OR' \xrightarrow[\text{heat}]{\text{strong HX}} R—X + HOR' \xrightarrow{\text{HX}} RX + R'X$$

epoxide cleavage in acid:

$$R_2C—CHR + R'OH \xrightarrow{H^+} R_2C—CHR \quad (R' = \text{alkyl, aryl, or H})$$

$$R_2C—CHR + X^- + H^+ \longrightarrow R_2C—CHR$$

epoxide cleavage in base:

$$R_2C—CHR \xrightarrow[\text{Nu:H}]{\text{Nu:}^-} R_2C—CHR \quad (Nu:^- = {}^-OH, {}^-SH, {}^-OR, {}^-CN, :NH_3, \text{etc.})$$

epoxide reaction with Grignard reagents:

$$CH_2—CH_2 \xrightarrow[\text{(2) H}_2\text{O, H}^+]{\text{(1) RMgX}} RCH_2CH_2OH$$

Figure 8.1   Summary of substitution reactions of ethers.

Section 8.5
Crown Ethers

Crown ethers are cyclic ethers with structures consisting of repeating —OCH$_2$CH$_2$— units, derived from 1,2-ethanediol. These compounds are named as *x*-crown-*y*, where *x* is the total number of atoms in the ring and *y* is the number of oxygen atoms in the ring.

one unit of

O(H)
CH$_2$
CH$_2$
(OH)

18 atoms in the ring; 6 oxygen atoms in the ring

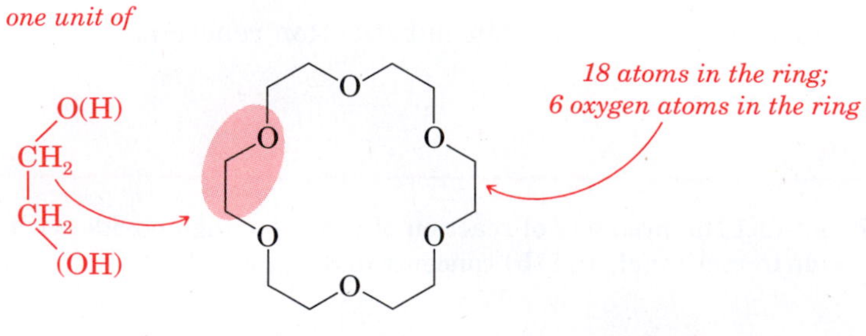

18-crown-6

Recall from general chemistry that a **chelate** (Greek *chele,* "claw") is a cyclic complex of a molecule and a metal ion. The unique feature of crown ethers is that they can chelate metal ions. Different-sized cyclic ethers chelate different-sized metal ions. Thus, 18-crown-6 chelates K$^+$, 15-crown-5 chelates the smaller Na$^+$, and 12-crown-4 chelates Li$^+$. In these complexes, the crown ether is often referred to as the *host,* while the metal ion is called the *guest.* The ring closures leading to crown ethers are easily accomplished if the appropriate metal ion is in solution. The chelation occurs even before the ring is closed and thus the two end groups can find each other readily.

An ionic compound in which the metal ion is chelated by a crown ether is soluble in nonpolar organic solvents. For example, "purple benzene" is a reagent in which KMnO$_4$, complexed by 18-crown-6, is dissolved in benzene.

One advantage of having an ionic compound dissolved in a nonpolar solvent is fairly evident: an ionic reagent can be dissolved in an organic phase, where it can react with a water-insoluble organic compound. A second advantage is that the nucleophilicity of an anion such as $^-$CN or CH$_3$CO$_2$$^-$ is greatly enhanced in nonpolar solvents, where the anion is poorly solvated, or "naked." An example of how a crown ether increases the rate of a substitution reaction in CH$_3$CN, a polar solvent that does not dissolve ionic compounds, follows:

5% yield with no crown ether;
100% yield with 18-crown-6

Nonactin and similar cyclic ethers are ionophorous antibiotics. Healthy cells must maintain a specific ratio of K$^+$ to Na$^+$ within the cell. Ionophorous antibiotics allow K$^+$ to escape rapidly from a cell, thus upsetting the delicate balance. They surround and bond to K$^+$, forming a complex much like the crown ethers do. The hydrocarbon-like portion of the antibiotic allows the complex to pass through the hydrocarbon lipid barrier of the cell.

nonactin

an ionophorous antibiotic

Section 8.6
Thiols and Sulfides

Sulfur is just below oxygen in the periodic table. Many organic compounds containing oxygen have sulfur analogs. The sulfur analog of an alcohol is called an **alkanethiol,** or simply **thiol,** or by its older name **mercaptan.** The —SH group is called a **thiol group** or a **sulfhydryl group.**

$$CH_3\overset{..}{\underset{..}{S}}H \qquad CH_3CH_2\overset{\overset{\displaystyle :\overset{..}{S}H}{|}}{C}HCH_3$$

methanethiol   2-butanethiol

The most characteristic property of a thiol is its odor! The human nose is very sensitive to these compounds and can detect their presence at levels of about 0.02 parts thiol to one billion parts air. The odor of a skunk's spray is due primarily to a few simple thiols.

Hydrogen sulfide (H_2S, $pK_a = 7.04$) is a stronger acid than water ($pK_a = 15.7$). Thiols ($pK_a = \sim 8$) are also substantially stronger acids than alcohols ($pK_a = \sim 16$).

$$CH_3CH_2SH + \quad OH^- \quad \rightleftharpoons \quad CH_3CH_2S^- + H_2O$$

stronger acid *stronger base*
than H_2O   *than RS^-*

Sulfur is less electronegative than oxygen, and its outer electrons are more diffuse; therefore, sulfur atoms form weaker hydrogen bonds than oxygen atoms. For this reason, H_2S (bp $-61°$) is more volatile than water (bp $100°$), and thiols are more volatile than their analogous alcohols.

Treatment of an alkyl halide with the hydrogen sulfide ion (HS^-) leads to thiols. Good yields are obtained only if an excess of inorganic hydrogen sulfide is used because the resulting thiol (which is acidic) can ionize to form the RS^- ion, also a good nucleophile. The subsequent reaction of RS^- with the alkyl halide yields the **sulfide,** R_2S, the sulfur analog of an ether.

$$CH_3I + {}^-SH \longrightarrow CH_3SH + I^-$$

methanethiol

$$CH_3SH \underset{}{\overset{-H^+}{\rightleftharpoons}} CH_3S^- \underset{-I^-}{\overset{CH_3I}{\longrightarrow}} CH_3SCH_3$$

dimethyl sulfide

When a thiol is treated with a mild oxidizing agent such as I_2 or $K_3Fe(CN)_6$, it undergoes coupling to form a **disulfide,** a compound containing the S—S linkage. This reaction can be reversed by treatment of the disulfide with a reducing agent such as lithium metal in liquid NH_3.

$$2\ CH_3CH_2SH \underset{[H]}{\overset{[O]}{\rightleftharpoons}} CH_3CH_2S{-}SCH_2CH_3$$

ethanethiol   diethyl disulfide

This disulfide link is an important structural feature of some proteins (Section 25.1B). The disulfide bond helps hold protein chains together in their proper shapes. The locations of the disulfide bonds determine, for example, whether hair (a protein) is curly or straight.

A sulfide can be oxidized to a **sulfoxide** or a **sulfone,** depending on the reaction conditions. For example, 30% hydrogen peroxide in the presence of an acidic catalyst oxidizes a sulfide to a sulfoxide at 25° or to a sulfone at 100°. With other oxidants, such as potassium permanganate and potassium dichromate, oxidations cannot be stopped at the sulfoxide stage.

$$CH_3SCH_3 \quad + \quad H_2O_2 \xrightarrow[\text{H}^+, 100°]{\text{H}^+, 25°}$$

dimethyl sulfide

$$\overset{\text{O}}{\underset{\|}{CH_3SCH_3}}$$

dimethyl sulfoxide

$$\underset{\overset{\|}{O}}{\overset{\overset{O}{\|}}{CH_3SCH_3}}$$

dimethyl sulfone

Dimethyl sulfoxide (DMSO) is prepared industrially by the air oxidation of dimethyl sulfide, which is a by-product of the paper industry. DMSO is a unique and versatile solvent. It has a high dielectric constant (49 D), but does not form hydrogen bonds in the pure state. (Why not?) It is a powerful solvent for both inorganic ions and organic compounds (see Section 5.9C). Reactants often have enhanced reactivities in DMSO compared to that in alcohol solvents. DMSO readily penetrates the skin and has been used to promote the dermal absorption of drugs; however, DMSO can also cause the absorption of dirt and poisons. A common complaint of people working with DMSO is that, when it is spilled on their hands, they can taste it!

Sulfoxides can be chiral. The unshared electrons of the sulfur atom occupy one corner of the tetrahedron.

Enantiomers of methyl p-tolyl sulfoxide

(S)   (R)

STUDY PROBLEMS

8.8   Explain with formulas why dimethyl sulfoxide is miscible with water.

8.9   What is the priority ranking of the unshared electrons on the sulfur in methyl *p*-tolyl sulfoxide?

Summary

Ethers can be prepared by the reaction of an alkoxide (RO^-) or a phenoxide (ArO^-), both good nucleophiles, with a methyl, primary, allylic, or benzylic halide **(Williamson ether synthesis).**

$$RO^- + R'X \longrightarrow ROR' + X^-$$

Epoxides ($R_2\overset{\displaystyle O}{\overset{\displaystyle \diagdown}{C}}\!\!-\!\!CR_2$) are prepared by the reaction of a peroxy acid (RCO_3H) with an alkene or by a Williamson ether synthesis with a 1,2-halohydrin.

Ethers do not undergo elimination reactions, but can undergo *substitution reactions* when heated with HBr or HI.

$$ROR + HI \xrightarrow{\text{heat}} RI + ROH \xrightarrow{\text{HI}} RI$$

Epoxides are more reactive than other ethers and undergo ring opening with nucleophiles (in either alkaline or acidic solution) or with Grignard reagents. These reactions are summarized in Figure 8.1. **Crown ethers** are cyclic ethers that are used to chelate metal ions.

Thiols (RSH) and **sulfides** (RSR') are the sulfur analogs of alcohols and ethers. Thiols can be oxidized to disulfides (RSSR). Sulfides can be oxidized to sulfoxides and sulfones.

Essay Problem for Chapter 8

Enantioselective Epoxidation

Epoxidation of an alkene using a peroxide is a stereospecific reaction. For example, when *trans*-stilbene is treated with peroxybenzoic acid, *trans*-stilbene oxide is obtained. When *cis*-stilbene is used, *cis*-stilbene oxide is the product. In each case, the stereochemistry of the starting alkene is retained in the product.

trans-stilbene + $C_6H_5\overset{\displaystyle O}{\overset{\displaystyle \|}{C}}OOH$ → *trans*-stilbene oxide (55%)

cis-stilbene + $C_6H_5\overset{\displaystyle O}{\overset{\displaystyle \|}{C}}OOH$ → *cis*-stilbene oxide (52%)

With allyl alcohols, *enantioselective epoxidation* (Sharpless epoxidation) can also be carried out. With this procedure, only one of two possible enantiomeric epoxides will be formed. For example, when (*E*)-2-hexen-1-ol is treated with *tert*-butylhydroperoxide, titanium(IV) isopropoxide, and a chiral compound, diethyl (2*R*,3*R*)-tartrate, only the (2*S*,3*S*) epoxide is formed. The yield is 80–81%. (J. G. Hill, K. B. Sharpless, C. M. Exon, and R. Regenye, *Org. Syn.* **1985,** *63,* 66.)

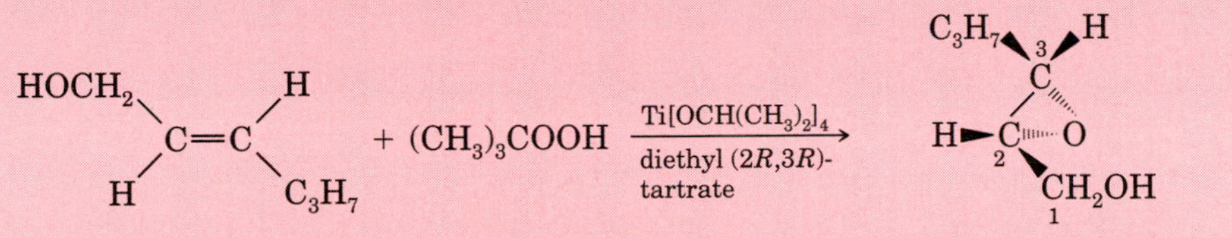

(2*S*,3*S*)-2-hydroxymethyl-3-propyloxirane

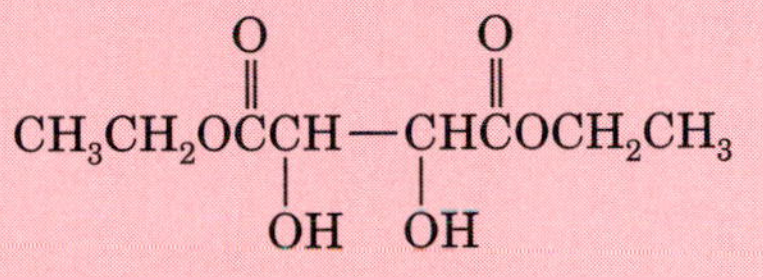

$$CH_3CH_2O\overset{O}{\overset{\|}{C}}\underset{OH}{\overset{|}{C}}H - \underset{OH}{\overset{|}{C}}H\overset{O}{\overset{\|}{C}}OCH_2CH_3$$

diethyl tartrate

The reaction can be envisioned as proceeding by the following scheme (T. Katsuki and K. B. Sharpless, *J. Am. Chem. Soc.* **1980,** *102,* 5976).

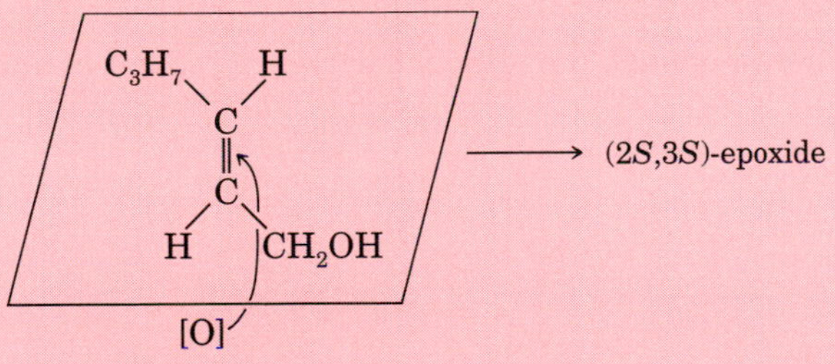

$\longrightarrow$ (2*S*,3*S*)-epoxide

Questions

1. What is the stereochemical relationship of *cis*- and *trans*-stilbene oxide?
2. Write a mechanism showing how *tert*-butylhydroperoxide transfers one of its oxygens to the pi bond of the alkene.
3. If the (*Z*) isomer of the following allyl alcohol yields the (2*S*,3*R*) enantiomer upon enantioselective epoxidation with diethyl (2*R*,3*R*)-tartrate as the chiral agent, what will be the product when the (*E*) isomer is subjected to the same reaction conditions?

$$C_9H_{19}CH_2CH{=}CHCH_2OH$$

4. Draw the structure of diethyl (2*R*,3*R*)-tartrate and diethyl (2*S*,3*S*)-tartrate.
5. If diethyl (2*S*,3*S*)-tartrate were used as the chiral agent for the enantioselective epoxidation of the (*E*) alkene whose structure was given in Problem 3, what would be the expected product? Draw its structure.

Study Problems*

8.10 Identify the ether, thiol, and sulfide groups in the following structures:

   (a)   cysteine, an amino acid found in proteins (page 987)

   (b)   scopolamine (page 840), a preoperative anesthetic

   (c)

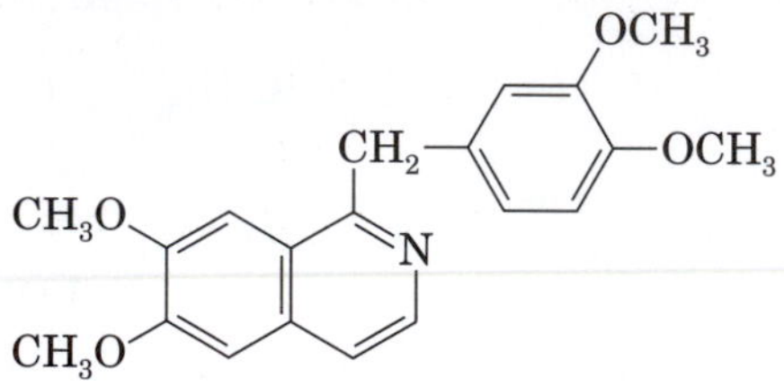

papaverine

a component of opium

   (d)

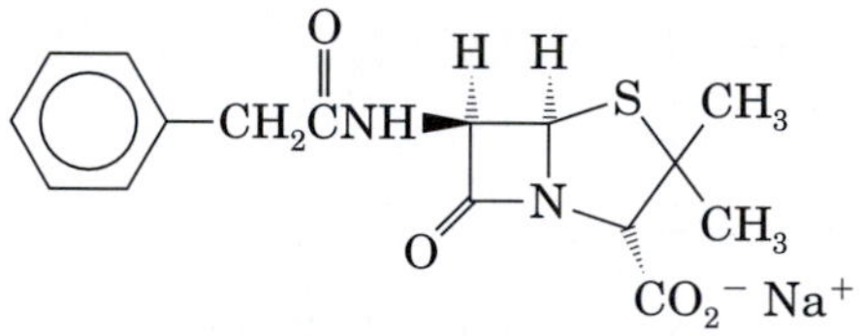

benzylpenicillin sodium

   (e)   the drug codeine (page 839)

   (f)   tetrahydrocannabinol, the principal active ingredient in marijuana

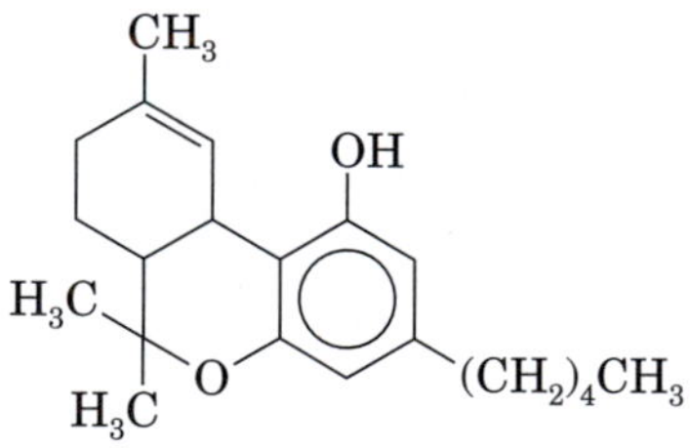

tetrahydrocannabinol

the principal active ingredient in marijuana

8.11 Draw formulas for the (R) and (S) enantiomers of 1-chloro-2,3-epoxypropane (epichlorohydrin).

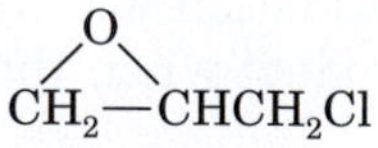

$$CH_2-CHCH_2Cl$$

epichlorohydrin

8.12 Write formulas for the following ethers and epoxides:

   (a)   *sec*-butyl phenyl ether

   (b)   3-isobutoxyhexane

---

* For information concerning the organization of the *Study Problems* and *Additional Problems,*
  see the *Note to student* on page 41.

(c)   *cis*-1-cyclohexyl-3-methyl-1,2-epoxybutane

(d)   styrene oxide (Styrene is the trivial name for phenylethene.)

8.13 Complete the following equations.

(a) $CH_3OH + (C_6H_5)_2CHOH \xrightarrow[\text{heat}]{H_2SO_4}$

(b) $CH_3CH_2\overset{\overset{\displaystyle OH}{|}}{C}HCH_2SCH_3 \xrightarrow[\text{CH}_3\text{I}]{\text{NaH}}$

(c)   [benzene ring]$\!\!-O^-Na^+$ $+ CH_3CH_2O\overset{\overset{\displaystyle O}{\|}}{\underset{\underset{\displaystyle O}{\|}}{S}}-OCH_2CH_3 \longrightarrow$

(d) $HOCH_2CH_2CH_2CH_2Br + NaH \longrightarrow$

(e) $(C_6H_5)_2CHCl + Na^+ {}^-OCH_2CH_2N(CH_3)_2 \longrightarrow$

8.14 Complete the following equations:

(a)   [cyclohexane ring with Cl and OH] $\xrightarrow[\text{H}_2\text{O}]{\text{KOH}}$

(b)   [cyclooctene] $+$ [benzene ring]$-\overset{\overset{\displaystyle O}{\|}}{C}OOH \longrightarrow$

(c) $(2S,3S)$-$CH_3CH_2\overset{\overset{\displaystyle CH_3}{|}}{C}H\underset{\underset{\displaystyle Cl}{|}}{C}HCH_2OH \xrightarrow[\text{H}_2\text{O}]{\text{KOH}}$

8.15 Complete the following equations:

(a)   [diphenyl ether with CH$_3$ substituent] $+$ excess HBr $\xrightarrow[\text{heat}]{H_2SO_4}$

(b)   [benzene ring]$-OCH_2CH_2CH_2OCH_3$ $+$ excess HI $\xrightarrow{\text{heat}}$

(c) $(CH_3)_2CHCH_2OCH_3 + 1\ HBr \xrightarrow[\text{heat}]{H_2SO_4}$

(d)   [tetrahydrofuran ring with two CH$_3$ groups] $+ 1\ HCl \xrightarrow{\text{heat}}$

(e)   [bicyclic ether] $+$ excess HBr $\xrightarrow[\text{heat}]{H_2SO_4}$

8.16 Triphenylmethyl (trityl) ethers are often used as protecting groups for alcohol (—OH) functions because, unlike most ethers, they readily undergo hydrolysis when treated with aqueous CF_3CO_2H at room temperature. Write a mechanism that will explain why the trityl ether below undergoes cleavage with aqueous CF_3CO_2H, whereas the corresponding ethyl ether does not.

$$\text{(cyclohexyl)}CH_2OC(C_6H_5)_3 \xrightarrow[CF_3CO_2H]{H_2O} \text{(cyclohexyl)}CH_2OH + (C_6H_5)_3COH$$

$$\text{(cyclohexyl)}CH_2OCH_2CH_3 \xrightarrow[CF_3CO_2H]{H_2O} \text{no reaction}$$

8.17 Predict the major organic products when 2-methyl-2,3-epoxybutane is treated with the following reagents: **(a)** 1-pentanol and HCl; **(b)** $CH_3CH_2NH_2$; **(c)** phenyllithium, then dilute HCl; **(d)** a solution of phenol and NaOH.

8.18 Complete the following equations:

(a) $\overset{O}{\overset{\displaystyle\triangle}{CH_2CH_2}}$ + HBr $\longrightarrow$

(b) $\overset{O}{\overset{\displaystyle\triangle}{CH_2CHCH_2N(CH_2CH_3)_2}}$ + $Na^+{}^-CN$ $\xrightarrow{H_2O}$

(c) $\overset{O}{\overset{\displaystyle\triangle}{CH_2CHCH_2Cl}}$ + excess CH_3SH $\xrightarrow[CH_3OH]{OH^-}$

(d) (cyclopentane epoxide)O $\xrightarrow[\text{(2) } H_2O,\ H^+]{\text{(1) } CH_3CH_2CH_2Li}$

(e) (cyclohexane with CH_3 and O epoxide) + HCl $\longrightarrow$

(f) (naphthalene with $OCH_2\overset{O}{\overset{\displaystyle\triangle}{C}{-}CH_2}$, H) $\xrightarrow{(CH_3)_2CHNH_2}$

8.19 Complete the following equations:

(a) $CH_3CH_2SCH_2CH_3 \xrightarrow[25°]{H_2O_2}$

(b) (cyclohexyl)—SH $\xrightarrow[\text{(2) } CH_3CH_2I]{\text{(1) NaOH}}$

(c) $CH_3CH_2CH_2SH + I_2 \longrightarrow$

(d) $CH_3CH_2\overset{\overset{\displaystyle CH_3}{|}}{\underset{\underset{\displaystyle CH_2CH_3}{|}}{S}}CSCH_2CH_3 \xrightarrow[\text{warm}]{KMnO_4}$

8.20 Suggest reagents for the following conversions, which may require more than one step.

(a) 1-methoxy-2-propanol from an epoxide

(b) 1-butanol from bromoethane

(c) 3,3-dimethyl-1-butanol from *tert*-butyl alcohol

(d) *cis*-1,2-dichlorocyclohexane from 1,2-epoxycyclohexane

(e) *trans*-1,2-dichlorocyclohexane from 1,2-epoxycyclohexane

Additional Problems

8.21 When *trans*-2-chloro-1-cyclohexanol is treated with base, 1,2-epoxycyclohexane is the product. However, when *cis*-2-chloro-1-cyclohexanol is treated with base, the product is cyclohexanone.

(a) Why doesn't the *cis* isomer yield the epoxide?

(b) Write a mechanism for each reaction.

8.22 Allyl disulfide (CH_2=$CHCH_2SSCH_2CH$=CH_2) is a contributor to the odor of garlic. How would you prepare this compound from allyl alcohol?

8.23 The red fox *(Vulpes vulpes)* uses a chemical communication system based partly upon odorous compounds in its urine. The characteristic "skunky" odor of fox urine is due to two compounds, the structures of which are shown below. A synthetic mixture of these compounds induces characteristic marking behavior in wild red foxes. Suggest a synthesis for each compound, starting with a nonsulfur-containing alcohol.

(a) CH_2=$\overset{\overset{\displaystyle CH_3}{|}}{C}CH_2CH_2SCH_3$

3-methyl-3-butenyl methyl sulfide

(b) 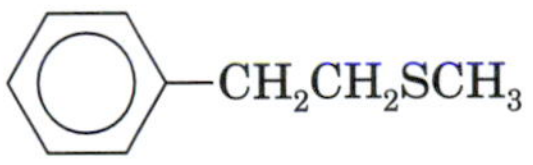   —$CH_2CH_2SCH_3$

2-phenylethyl methyl sulfide

8.24 Write flow equations to show how you would make the following conversions:

(a) 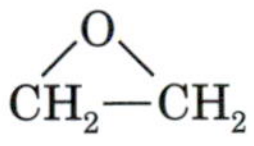 $\overset{\displaystyle O}{\overset{\diagup\diagdown}{CH_2-CH_2}}$ to $C_6H_5OCH_2CH_2OC_6H_5$

(b) 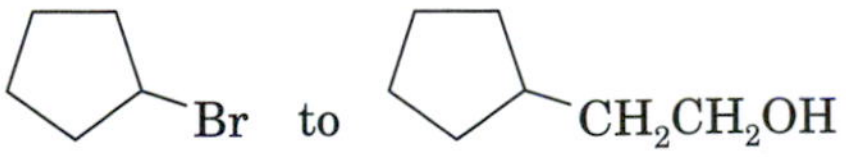   to

(c) $ClCH_2CH_2SCH_2CH_2Cl$ to

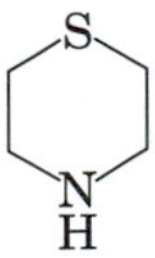

(d) 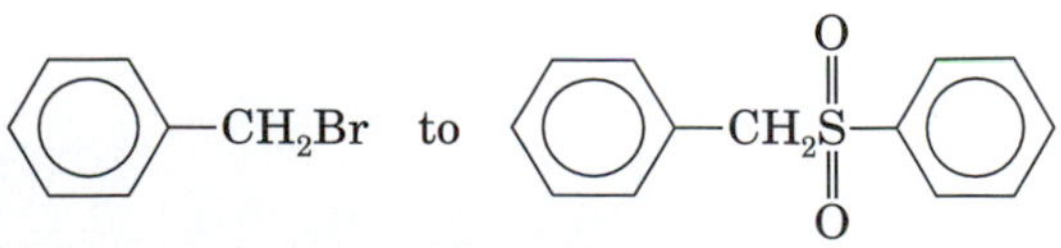   —CH_2Br to

8.25 Predict if the following equilibrium results in a racemic mixture. Explain your answer.

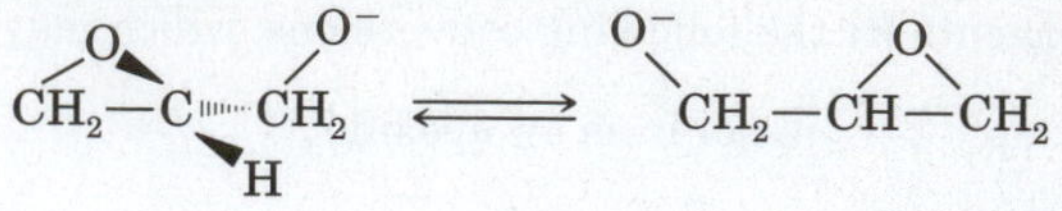

$$\underset{\underset{\text{(2R)}}{\underset{H}{\overset{O}{\text{CH}_2-\text{C}\text{\small \textbar\textbar\textbar}\cdot\cdot\cdot\text{CH}_2}}}} \rightleftharpoons \underset{\underset{\textit{optically active?}}{}}{\overset{O^-}{\text{CH}_2-\text{CH}-\text{CH}_2}}$$

8.26 In a study of naphthalene metabolism in rats, the following diepoxide A was impli-cated as a biochemical intermediate:

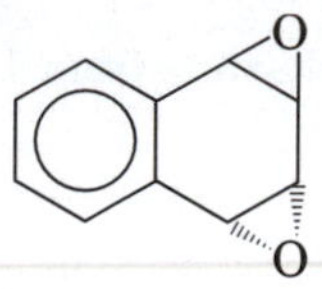

A

When A was treated with methanol at room temperature, it was converted to com-pound B in 96% yield. Compound B was subsequently treated with an excess of sodium methoxide in methanol to yield compound C in 80% yield. Compound A was also converted to compound C in 76% yield by treatment with *p*-toluenesulfonic acid in methanol. Write an equation for each reaction, showing the structures and stereo-chemistry of B and C.

8.27 The following reaction is postulated to involve the displacement of the chlorine to form an epoxy group at that carbon. Write a plausible mechanism for the reaction with such an intermediate.

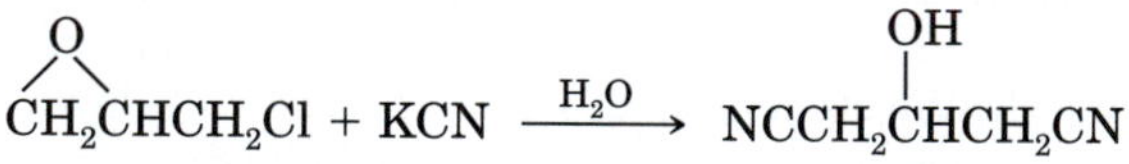

$$\overset{O}{\overset{/\backslash}{\text{CH}_2\text{CHCH}_2\text{Cl}}} + \text{KCN} \xrightarrow{\text{H}_2\text{O}} \text{NCCH}_2\overset{\text{OH}}{\overset{|}{\text{CHCH}_2\text{CN}}}$$

epichlorohydrin (54–62%)

9

SPECTROSCOPY I: INFRARED AND NUCLEAR MAGNETIC RESONANCE

Spectroscopy is the study of the interactions between radiant energy and matter. The colors that we see and the fact that we can see at all are consequences of energy absorption by organic and inorganic compounds. The capture of the sun's energy by plants in the process of photosynthesis is another aspect of the interaction of organic compounds with radiant energy. Of primary interest to the organic chemist is the fact that the wavelengths at which an organic compound absorbs radiant energy are *dependent upon the structure of the compound*. Therefore, we can use spectroscopic techniques to determine the structures of unknown compounds and to study the bonding characteristics of known compounds.

In this chapter, our emphasis will be on **infrared (IR) spectroscopy** and **nuclear magnetic resonance spectroscopy (NMR)**, both of which are used extensively in organic chemistry. In Chapter 22, we will broaden our discussion to include some other types of spectroscopy.

The discovery of the IR region of the spectrum was made by W. Herschel in 1800. Nearly a century passed, however, before an IR spectrum of an organic compound was determined. The first IR absorption spectra of organic compounds were reported in 1882 by W. Abnery and E. R. Festing. They were also the first to correlate functional groups with the presence of absorption bands in the spectrum. The early use of IR spectroscopy was limited to investigators who could build their own instruments. In the late 1930s, Richard Perkin, an investment banker, and Charles Elmer, a court stenographer, formed the Perkin–Elmer group to make precision optical components. Perkin and Elmer shared a love of astronomy. During World War II, their firm built one of the first operating IR spectrometers. (The Beckman Instrument group also built an IR device about the same time.) These IR instruments were used to analyze hydrocarbons needed for the production of synthetic rubber.

The theoretical basis for NMR was first postulated by Wolfgang Pauli in 1924. He recognized that some nuclei have magnetic moment and can absorb radiofrequency radiation when in an external magnetic field. Experimental evidence for this postulate was obtained in 1946 by Felix Bloch at Stanford and Edward Purcell at Harvard. Bloch and Purcell were jointly awarded the 1952 Nobel Prize for their discoveries. The first commercial NMR instrument, a 30 MHz machine, was built in 1953 by Varian Associates in Palo Alto, California. By 1962, the Varian 60 MHz instrument was in widespread use.

Section 9.1
Electromagnetic Radiation

Electromagnetic radiation is energy that is transmitted through space in the form of waves. Each type of electromagnetic radiation (radio waves, ultraviolet, infrared, visible, and so forth) is characterized by its **wavelength** (λ), the distance from the crest of one wave to the crest of the next wave. (See Figure 9.1.)

The entire spectrum of electromagnetic radiation is represented in Figure 9.2. The wavelengths that lead to vision range from 400 nm to 750 nm (1 nm $= 10^{-9}$ m or 10^{-7} cm). However, the visible region is a very small part of the entire electromagnetic spectrum. Wavelengths slightly shorter than those of the visible region fall into the ultraviolet region, while slightly longer wavelengths fall into the infrared region.

In addition to being characterized by its wavelength, radiation can be characterized by its **frequency** (ν), which is defined as the number of complete cycles per second (cps), also called *Hertz* (Hz). (See Figure 9.3.) Radiation of a

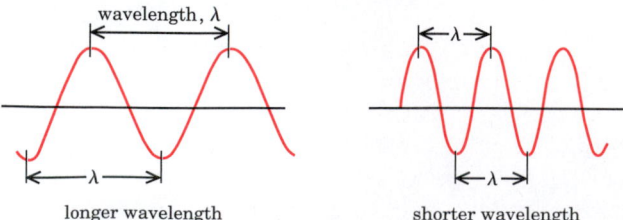

longer wavelength   shorter wavelength

Figure 9.1 Wavelength of electromagnetic radiation.

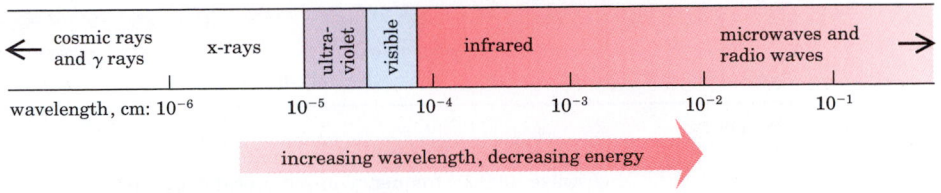

Figure 9.2 The electromagnetic spectrum.

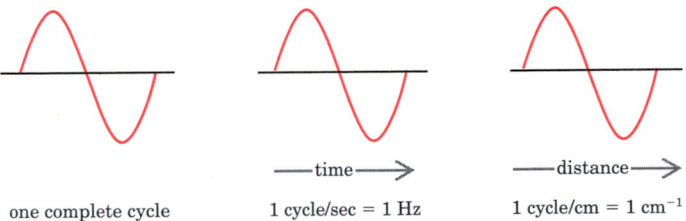

Figure 9.3 Frequency of electromagnetic radiation.

higher frequency contains more waves per second; therefore, the wavelength must be shorter. By their definitions, wavelength and frequency are *inversely proportional*. This relationship can be expressed mathematically:

$$\nu = \frac{c}{\lambda}$$

where ν = frequency in Hz,

$c = 3 \times 10^{10}$ cm/sec (the speed of light), and

λ = wavelength in cm

In infrared spectroscopy, frequency is expressed as **wavenumbers**: the number of cycles per centimeter. Wavenumbers have units of *reciprocal centimeters* (1/cm, or cm^{-1}). The unit used for wavelength in infrared spectroscopy is the *micrometer,* μm (or *micron,* μ), where 1.0 $\mu m = 10^{-6}$ m or 10^{-4} cm. Wavenumbers and wavelengths can be interconverted by the following equation:

$$\text{wavenumber in } cm^{-1} = \frac{1}{\lambda \text{ in cm}} = \frac{1}{\lambda \text{ in } \mu m} \times 10^4$$

A few common symbols and units used in spectroscopy are listed in Table 9.1.

Electromagnetic radiation is transmitted in particle-like packets of energy called **photons** or **quanta.** The energy of a photon is *inversely proportional to the wavelength.* (Mathematically, $E = hc/\lambda$, where h = Planck's constant.) Radiation of shorter wavelength has a higher energy. Therefore, a photon of ultraviolet light has more energy than a photon of visible light and has substantially more energy than a photon of radio waves.

Conversely, the energy of a photon of radiation is *directly proportional to the frequency* (more waves per unit time mean higher energy). The

Table 9.1 Symbols commonly encountered in spectroscopy	
Symbol or Unit	Definition
Frequency:	
ν	frequency in Hz (cycles per second)
cm^{-1}	wavenumber: frequency in reciprocal cm, or $1/\lambda$
Wavelength:	
λ	wavelength, usually expressed in units of meters (μm, nm, etc.)
μm	micrometer, same as micron (μ), 10^{-6} m
nm	nanometer, same as millimicron (mμ), 10^{-9} m
Å	angstrom, 10^{-10} m or 10^{-1} nm

relationship is expressed in the equation $E = h\nu$. (The symbol $h\nu$ is often used in chemical equations to represent electromagnetic radiation.)

ultraviolet   visible   infrared   radio

increasing λ (or decreasing ν) means decreasing energy

Molecules absorb specific wavelengths of electromagnetic radiation. Absorption of ultraviolet light (high-energy radiation) results in the promotion of an electron to a higher-energy orbital. Infrared radiation does not contain enough energy to promote an electron. Absorption of infrared radiation results in increased amplitudes of vibration of bonded atoms.

The magnitude of energy of electromagnetic radiation is determined by the wavelength (or frequency) of that radiation. The amount of energy of that magnitude is determined by the intensity or brightness of the radiation. The **intensity** of radiation is proportional to the number of photons. When a sample absorbs photons from a beam of transmitted radiation, whether of infrared or ultraviolet radiation, the number of transmitted photons in the beam must decrease. Thus, the intensity of the transmitted radiation decreases. It is this change in intensity that is measured in absorption spectroscopy.

Section 9.2

Features of a Spectrum

An infrared, visible, or ultraviolet spectrum of a compound is a graph of either *wavelength* or *frequency,* continuously changing over a small portion of the electromagnetic spectrum, versus either *percent transmission (%T)* or *absorbance (A).*

$$\%T = \frac{\text{intensity}}{\text{original intensity}} \times 100 \qquad A = \log\left(\frac{\text{original intensity}}{\text{intensity}}\right)$$

Most infrared spectra record wavelength or frequency versus $\%T$. The absence of absorption by a compound at a specific wavelength is recorded as $100\%T$ (ideally). When a compound absorbs radiation at a specific wavelength, the intensity of radiation being transmitted decreases. This results in

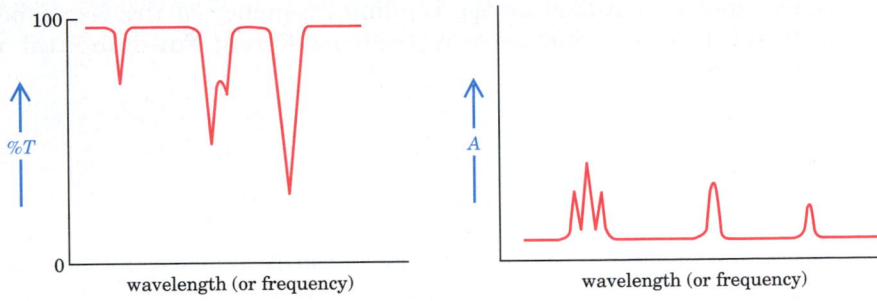

Figure 9.4 Spectra are graphs of percent transmission ($\%T$) or absorbance (A) by a sample versus wavelength or frequency of radiation.

a *decrease* in $\%T$ and appears in the spectrum as a dip, called an **absorption peak,** or **absorption band.** The portion of the spectrum where $\%T$ measures 100 (or near 100) is called the **base line,** which is recorded at the top of an infrared spectrum.

Visible and ultraviolet spectra (Chapter 22) are usually presented as graphs of A versus wavelength. In these cases, the base line (zero absorbance) runs along the bottom of the spectrum, and absorption is recorded as an *increase* of the signal. The general appearance of spectra using $\%T$ and A is shown in Figure 9.4.

Although the physical appearance of a nuclear magnetic resonance spectrum is similar to that of a visible or ultraviolet spectrum (base line at the bottom), the physical principles of nuclear magnetic resonance spectroscopy are different from those of other types of spectroscopy. Therefore, we will defer a discussion of these spectra until Section 9.6.

Section 9.3
Absorption of Infrared Radiation

Nuclei of atoms bonded by covalent bonds undergo vibrations, or oscillations, in a manner similar to two balls attached by a spring. When molecules absorb infrared radiation, the absorbed energy causes an increase in the amplitude of the vibrations of the bonded atoms. The molecule is then in an **excited vibrational state.** The absorbed energy is subsequently dissipated as heat when the molecule returns to the ground state.

The vibrational states of bonds occur at fixed, or quantized, energy levels. The particular wavelength of absorption by a given type of bond depends on the energy difference between the ground state and the excited state. Therefore, different types of bonds, such as C—H or C=O, etc., absorb infrared radiation at different and characteristic wavelengths.

Because a bond within a molecule may undergo different types of oscillations, a specific bond may absorb energy at more than one wavelength. For example, an O—H bond absorbs energy near 3330 cm^{-1} (3.0 μm). Energy of this wavelength causes increasing **stretching vibrations** of the O—H bond. Therefore, this band is identified as the stretching frequency of the O—H bond. An O—H bond also absorbs energy near 1250 cm^{-1} (8.0 μm); energy of this wavelength causes increased **bending vibrations.** This band is consequently identified as the bending frequency of the O—H bond. These different types of vibrations are called different **fundamental modes of vibration.**

stretching   bending

The amount of energy that a bond absorbs depends on the change in the bond moment as the bonded atoms vibrate—a greater change in bond moment results in the absorption of a larger amount of energy. Nonpolar bonds do not absorb infrared radiation because there is no change in bond moment when the atoms oscillate. Relatively nonpolar bonds (most C—C and C—H bonds in organic molecules) give rise to weak absorption. By contrast, polar bonds (such as O—H and C=O) exhibit strong absorption.

STUDY PROBLEM

9.1 Arrange the following compounds in order of increasing intensity of absorption by the double bond (least intense first) assuming the same concentrations and other conditions. Explain your answer.

            O
            ‖
(a) $CH_3CCH_2CH_3$ **(b)** $(CH_3)_2C=C(CH_3)_2$ **(c)** $CH_3CH_2CH=CH_2$

Section 9.4

The Infrared Spectrum

The instrument used to measure absorption of infrared radiation is called an **infrared spectrophotometer.** A diagram of a typical instrument is shown in Figure 9.5. At one end of the instrument is the light source, which emits all wavelengths of infrared radiation. The light from this source is split by mirrors (not shown) into two beams, the reference beam and the sample beam. After passing through the reference cell (which contains solvent, if used in the sample, or nothing if the sample is pure) and the sample cell, the two beams are combined in the chopper (another mirror system) into one beam that alternates from reference beam to sample beam. This alternating beam is diffracted by a grating that separates the beam into its different wave-

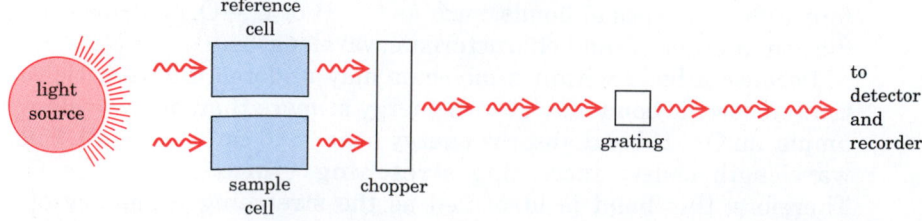

Figure 9.5 The infrared spectrophotometer.

lengths. The detector measures the difference in intensities of the two segments of the beam at each wavelength and passes this information on to the recorder, which produces the spectrum.

Modern infrared instruments employ a fast Fourier analysis to obtain the spectrum and are called **Fourier transform infrared (FTIR) spectrometers.** Infrared radiation is split into two beams. One beam remains static, while the other is directed to a moving mirror to provide a moving beam. The two beams are then combined to obtain a modulated beam. In the modulated beam, there will be either more light energy (constructive interference) or less light energy (destructive interference) for a given wavelength. The modulated beam is passed through the sample, digitized, then Fourier-transformed by a computer into the infrared spectrum.

Figure 9.6 shows two infrared spectra of 1-hexanol. The top spectrum is a typical example of the type of spectrum obtained directly from a spectrophotometer. The lower spectrum is an artist's rendition of the real spectrum. In the artist-rendered spectrum, many small extraneous peaks arising from impurities and electronic "noise" have been deleted. For the sake of clarity, we will use this second type of spectrum in this text.

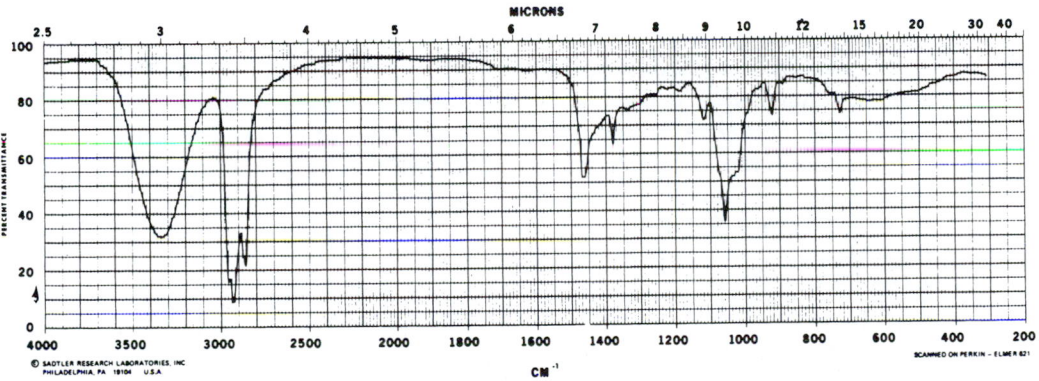

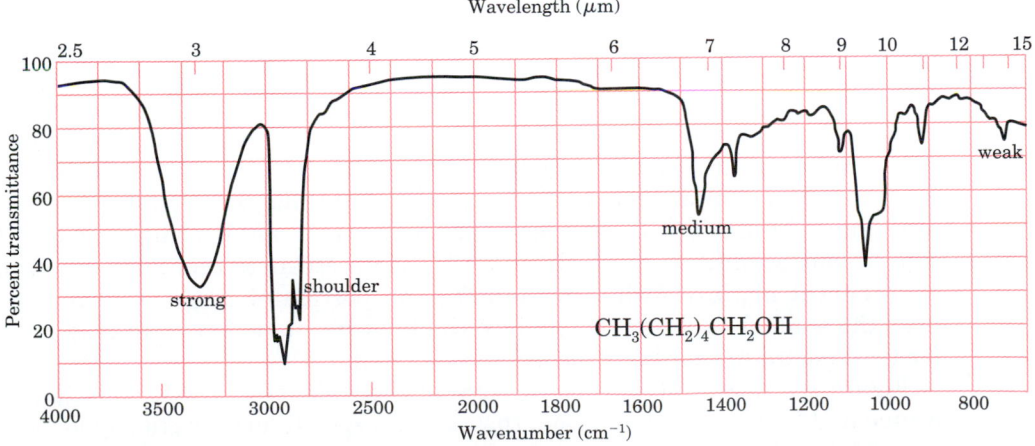

Figure 9.6 Infrared spectra of 1-hexanol. The upper spectrum is a reproduction of an actual spectrum, while the lower one is an artist's rendition. Upper spectrum © Sadtler Research Laboratories, Division of Bio-Rad Laboratories, Inc., 1986.

The scales at the bottoms of these spectra are in wavenumbers, decreasing from 4000 cm^{-1} to about 670 cm^{-1} or lower. The corresponding wavelengths in μm (or μ) are shown at the top. The wavelength or frequency of the *minimum point* (maximum absorption) of an absorption band is used to identify each band. This point is more reproducible than the range of a wide band, which may vary with concentration or with the sensitivity of the instrument.

The infrared bands in a spectrum can be classified by intensity: *strong (s), medium (m),* and *weak (w).* A weaker band overlapping a stronger band is called a *shoulder (sh).* These terms are relative, and the assignment of any given band as *s, m, w,* or *sh* is qualitative at best. In Figure 9.6, a few bands in the lower spectrum are labeled according to this classification method.

The number of identical groups in a molecule alters the relative strengths of the absorption bands in a spectrum. For example, a single OH group in a molecule produces a relatively strong absorption, while a single CH absorption is relatively weak. However, if a compound has many CH bonds, the collective effect of the CH absorption gives a peak that is medium or even strong.

Section 9.5
Interpretation of Infrared Spectra

Chemists have studied thousands of infrared spectra and have determined typical ranges of the wavelengths of absorption for each of the functional groups. **Correlation charts** provide summaries of this information.

A typical correlation chart for the stretching and bending frequencies of various groups is shown inside the back cover of this book. From the chart we see that OH and NH stretching bands are found between 3000 and 3700 cm^{-1} (2.7–3.3 μm). (Figure 9.6, the infrared spectrum of 1-hexanol, shows absorption in this region.) If the infrared spectrum of a compound of unknown structure shows absorption in this region, then we suspect that the compound contains either an OH or an NH group in its structure. If this region does not contain an absorption band, we conclude that the structure probably does not have an OH group or an NH group.

The region from 1400 to 4000 cm^{-1} (2.5 μm to about 7.1 μm), to the left in the infrared spectrum, is especially useful for identification of the various functional groups. This region shows absorption arising from stretching modes. The region to the right of 1400 cm^{-1} is often quite complex because both stretching and bending modes give rise to absorption here. In this region, correlation of an individual band with a specific functional group usually cannot be made with accuracy. However, each organic compound has its own unique absorption here. This part of the spectrum is therefore called the **fingerprint region.** Although the left-hand portion of a spectrum may appear the same for similar compounds, the fingerprint region must also match for two spectra to represent the same compound.

Figure 9.7 shows the infrared spectra of two alkanes of the formula C_8H_{18}, octane and 2-methylheptane. Note that the two spectra are practically identical from 1400 to 4000 cm^{-1}, but that the fingerprint regions are slightly different.

In the following sections, we will discuss the characteristic infrared absorption of compounds containing aliphatic C—C and C—H bonds and a few functional groups. Our intent is to develop familiarity with the features of typical infrared spectra. As we discuss the various functional groups in fu-

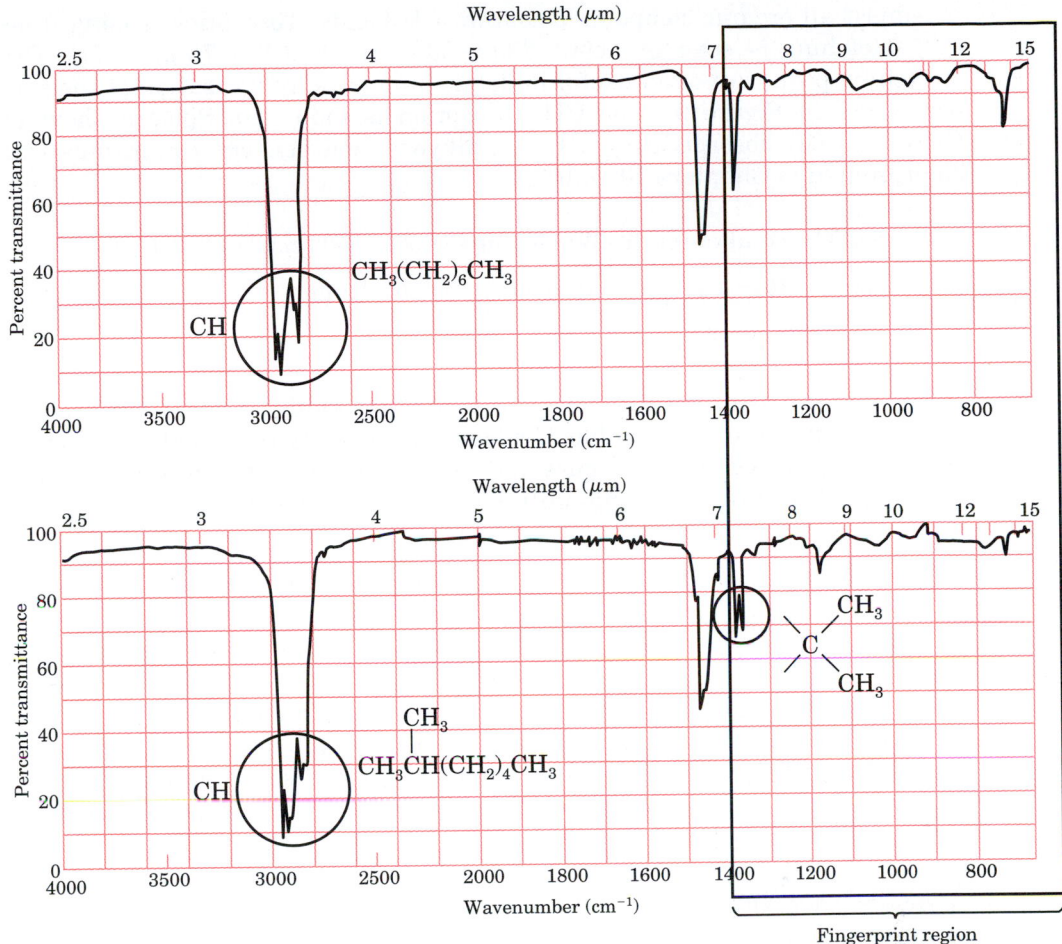

Figure 9.7 Infrared spectra of octane and 2-methylheptane, showing slight differences in the fingerprint regions. Also note the CH absorption peaks.

ture chapters, we will also include discussions of their infrared spectral characteristics.

A. Carbon–Carbon and Carbon–Hydrogen

Bonds between sp^3-hybridized carbon atoms (C—C single bonds) give rise to weak absorption bands in the infrared spectrum. In general, these absorption bands are not very useful for structure identification. Bonds between sp^2-hybridized carbons (C=C) often exhibit characteristic absorption (variable in strength) around $1600-1700$ cm^{-1} ($5.8-6.2$ μm). Aryl carbon–carbon bonds show absorption at slightly lower frequencies (to the right in the spectrum). Bonds between sp-hybridized carbons (C≡C) show weak, but extremely characteristic, absorption at $2100-2250$ cm^{-1} ($4.4-4.8$ μm), a region of the spectrum where most other groups show no absorption.

$$sp^2 \ \textbf{C=C: } 1600-1700 \text{ cm}^{-1} \ (5.8-6.2 \ \mu\text{m})$$

$$sp^2 \ \textbf{C—C (aryl): } 1450-1600 \text{ cm}^{-1} \ (6.25-6.9 \ \mu\text{m})$$

$$sp \ \textbf{C≡C: } 2100-2250 \text{ cm}^{-1} \ (4.4-4.8 \ \mu\text{m})$$

Almost all organic compounds contain CH bonds. Absorption arising from CH stretching is seen at about 2800–3300 cm^{-1} (3.1–3.75 μm). The CH stretching peaks are often useful in determining the hybridization of the carbon atom. In Figure 9.7 the CH absorption is indicated. Spectra showing C—C and CH absorption by alkenes, alkynes, and benzene compounds will be presented in Chapters 10 and 11.

sp^3 **C—H (alkanes or alkyl groups):** 2800–3000 cm^{-1} (3.3–3.6 μm)

sp^2 **C—H (=CH—):** 3000–3300 cm^{-1} (3.0–3.3 μm)

sp **C—H (≡CH):** ~3300 cm^{-1} (3.0 μm)

The ***geminal-*** or ***gem*-dimethyl** grouping (two methyl groups on the same carbon) often exhibits a double CH bending peak in the 1360–1385 cm^{-1} (7.22–7.35 μm) region (see Figure 9.7). Unfortunately, the two peaks are not visible in all spectra; sometimes only a single peak is observed.

C(CH$_3$)$_2$: 1360–1385 cm^{-1} (7.22–7.35 μm) (two peaks)

B. Haloalkanes

The stretching absorption of the CX bond of a haloalkane falls in the fingerprint region of the infrared spectrum, from 500 to 1430 cm^{-1} (7–20 μm). (See Figure 9.8.) Without additional information, the presence or absence of a band in this region cannot be used for verifying the presence of a halogen in an organic compound.

C. Alcohols and Amines

Alcohols and amines exhibit distinctive OH and NH stretching absorption at 3000–3700 cm^{-1} (2.7–3.3 μm), to the *left* of CH absorption. If there are two hydrogens on an amine nitrogen (—NH$_2$), the NH absorption appears as a double peak. If there is only one H on the N, only one peak is observed. Of

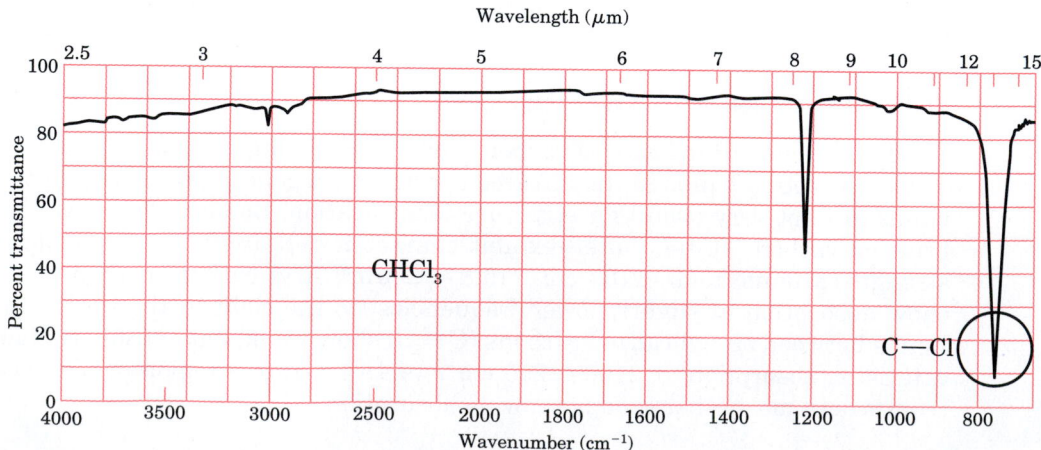

Figure 9.8 Infrared spectrum of chloroform, a trihaloalkane.

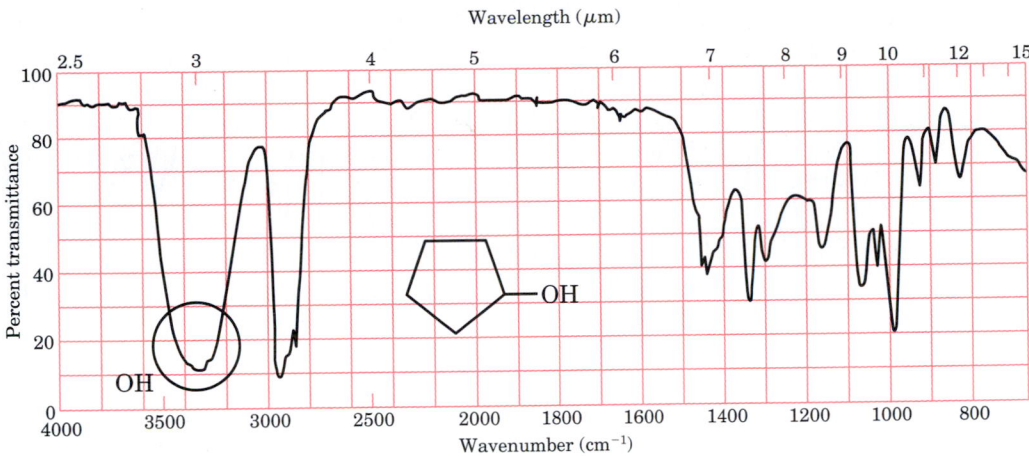

Figure 9.9 Infrared spectrum of cyclopentanol, an alcohol.

course, if there is no NH (as in the case of a tertiary amine, R_3N), there is no absorption in this region. Figure 9.9 shows the spectrum of an alcohol, while Figure 9.10 shows spectra of the three types of amines.

Alcohols and amines also exhibit C—O and C—N absorption in the fingerprint region. These bands are not always easy to identify because this region of the spectrum often contains a large number of peaks.

OH or NH: $3000-3700$ cm^{-1} $(2.7-3.3 \ \mu\text{m})$

C—O or C—N: $900-1300$ cm^{-1} $(8-11 \ \mu\text{m})$

Hydrogen bonding changes the position and appearance of an infrared absorption band. The spectra in Figures 9.9 and 9.10 are those of pure liquids in which hydrogen bonding is extensive. The OH absorption in Figure 9.9 is a wide band at about 3330 cm^{-1} (3.0 μm). When hydrogen bonding is less extensive, a sharper, less intense OH peak is observed. Figure 9.11 shows two partial spectra of an alcohol. One spectrum is of a pure liquid (hydrogen bonded); the other spectrum is of the alcohol in the vapor phase (not hydrogen bonded). The differences in OH absorption are apparent in this figure.

Absorption by NH bonds is less intense than OH absorption, partly because of weaker hydrogen bonds in amines and partly because NH bonds are less polar.

STUDY PROBLEM

9.2 Figure 9.12 (page 349) gives the infrared spectra of two compounds, I and II, both of which appear in the following list. Which is I and which is II?

(a) $CH_3(CH_2)_6CH_3$ (b) $CH_3(CH_2)_6CH_2OH$

(c) $CH_3(CH_2)_5N(CH_3)_2$ (d) $CH_3CH_2CH_2CH_2NH_2$

(e) $CH_3(CH_2)_6CH_2I$ (f) $CH_3CH_2CH_2NHCH_3$

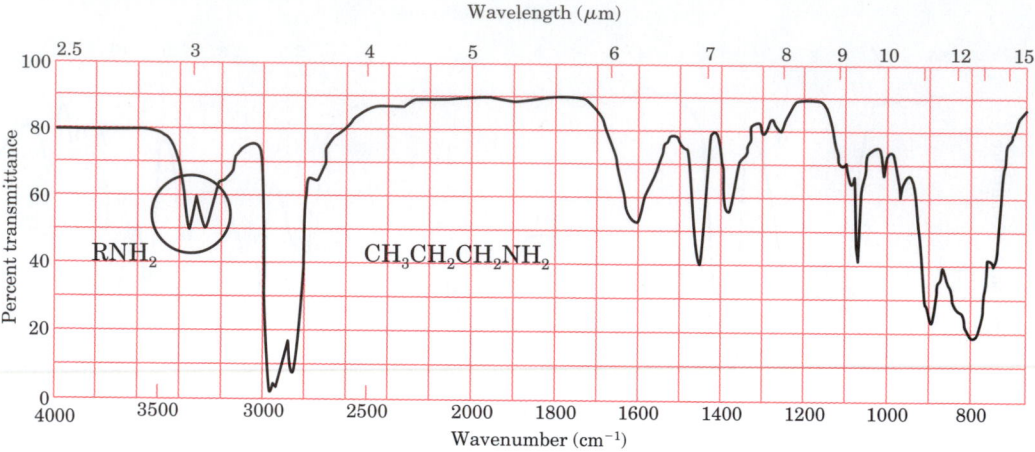

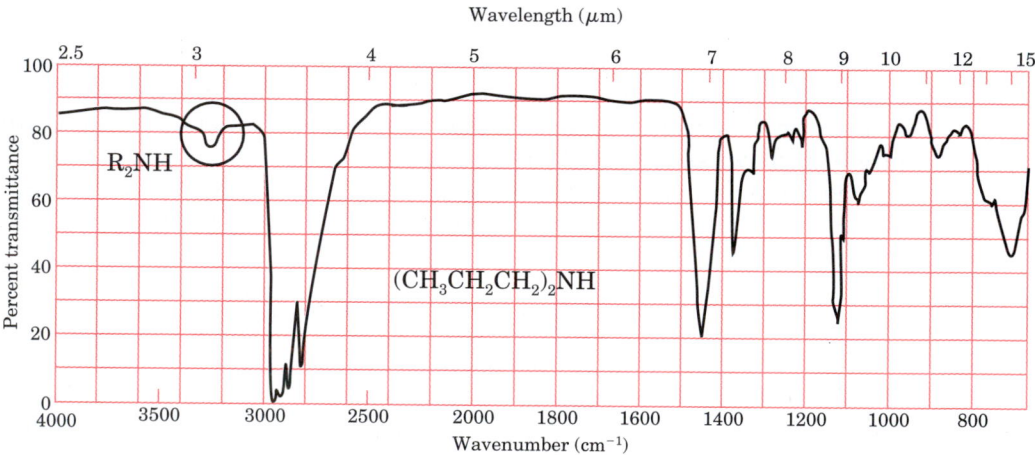

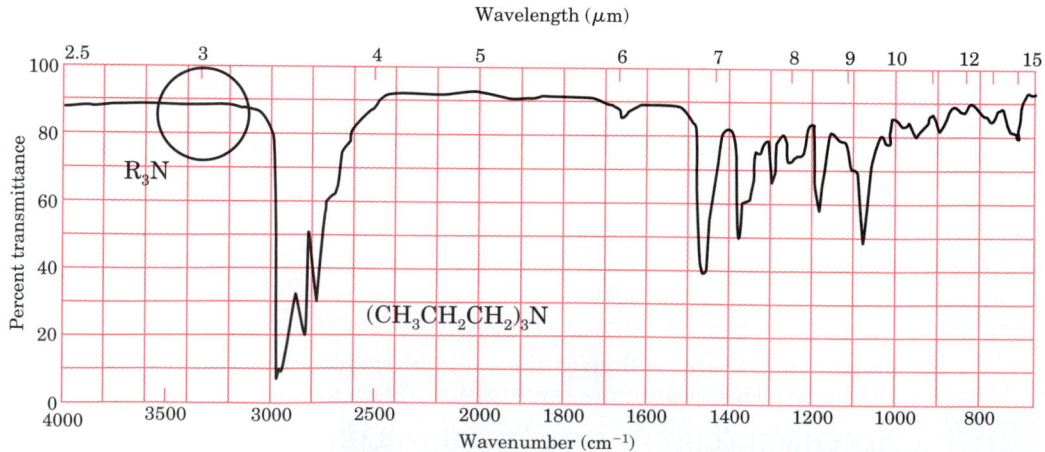

Figure 9.10 Infrared spectra of a primary amine, propylamine (top); a secondary amine, dipropylamine (center); and a tertiary amine, tripropylamine (bottom).

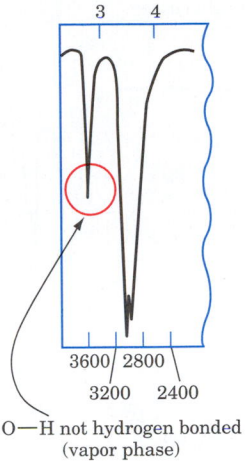
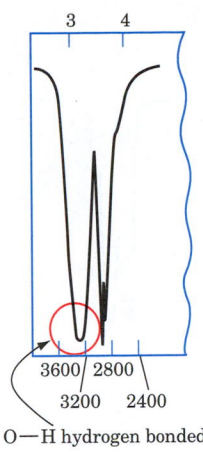

O—H not hydrogen bonded
(vapor phase)

O—H hydrogen bonded

Figure 9.11 Partial infrared spectra of an alcohol in the vapor and liquid phases, showing hydrogen-bonded and nonhydrogen-bonded OH absorption.

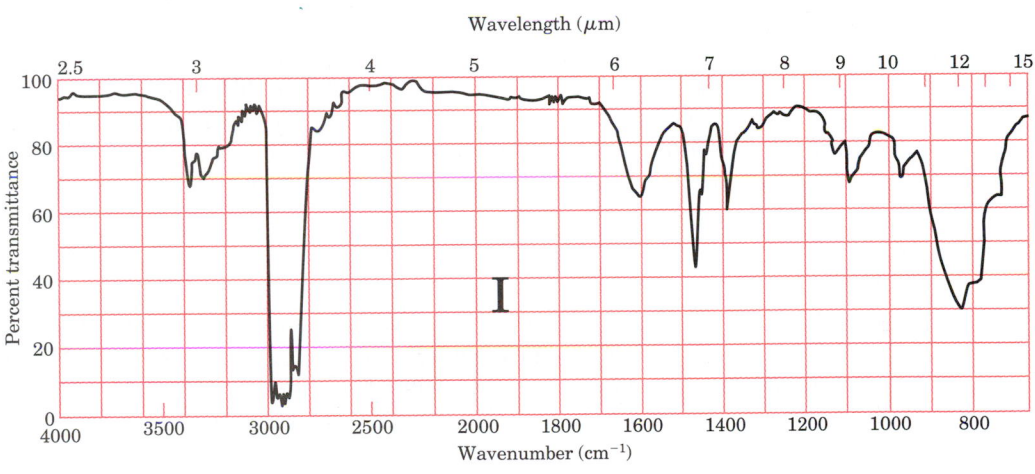

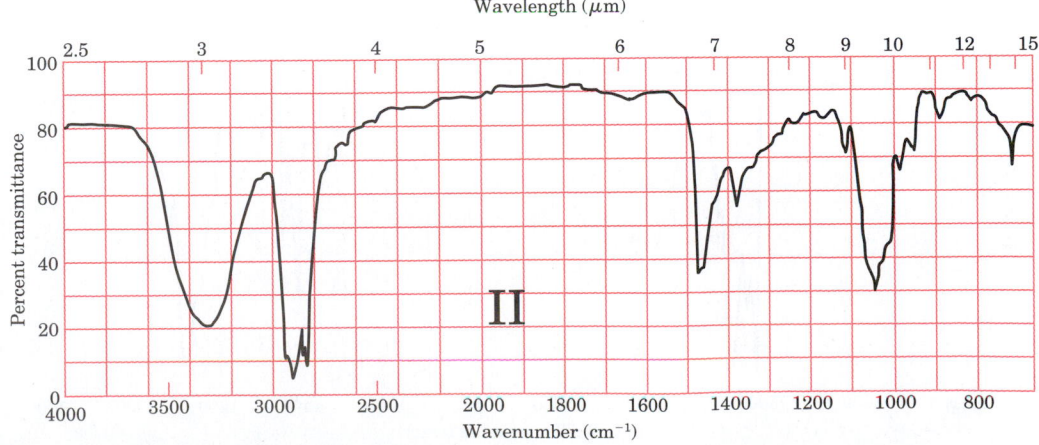

Figure 9.12 Infrared spectra for Problem 9.2.

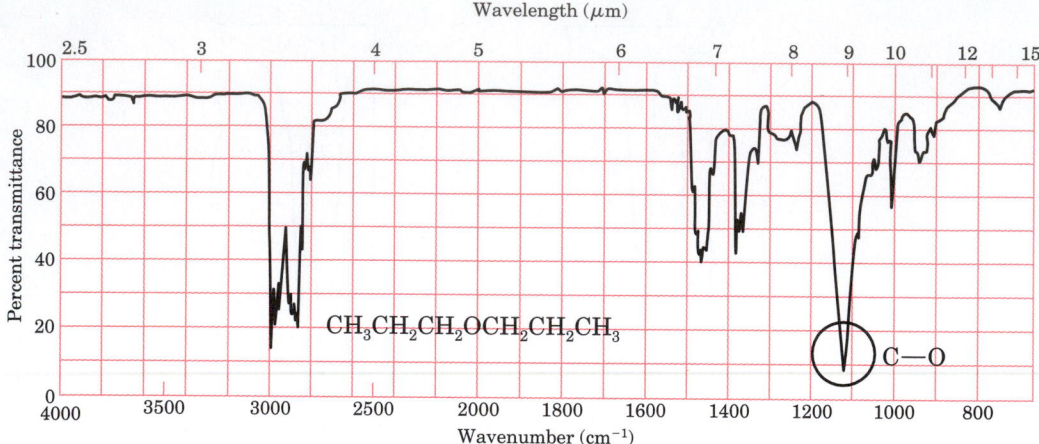

Figure 9.13   Infrared spectrum of dipropyl ether.

D. Ethers

Ethers have a C—O stretching band in the fingerprint region at 1050–1260 cm^{-1} (7.9–9.5 μm). Because oxygen is electronegative, the stretching causes a large change in the C—O bond moments. Therefore, the C—O absorption is usually strong (see Figure 9.13). The absorption, unfortunately, is not unique with ethers. Any compound with a C—O single bond (alcohols, esters, etc.) will also have a strong absorption band in this region.

STUDY PROBLEM

9.3   A compound yielded only iodoethane as a product when heated with HI. The infrared spectrum of the original compound is shown in Figure 9.14. What is its structure? Make assignments to the prominent spectral bands.

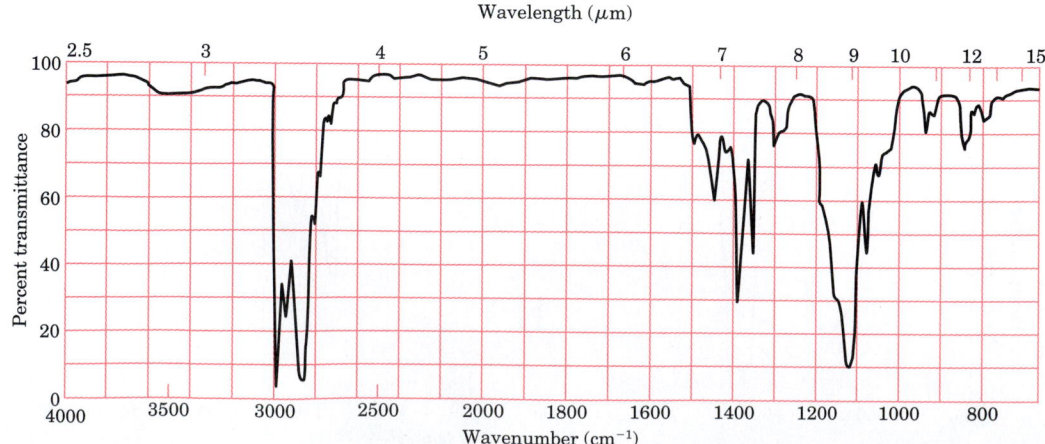

Figure 9.14   Infrared spectrum of unknown compound for Problem 9.3.

Table 9.2	Stretching vibrations for some carbonyl compounds[a]		

		Position of Absorption	
Type of Compound		cm^{-1}	μm
aldehyde, $R\overset{\overset{\displaystyle O}{\|}}{C}H$		1720–1740	5.75–5.81
ketone, $R\overset{\overset{\displaystyle O}{\|}}{C}R$		1705–1750	5.71–5.87
carboxylic acid, $R\overset{\overset{\displaystyle O}{\|}}{C}OH$		1700–1725	5.80–5.88
ester, $R\overset{\overset{\displaystyle O}{\|}}{C}OR$		1735–1750	5.71–5.76

[a] In each case, R is saturated and aliphatic. Compounds with ring strain, conjugation, and other substituents may fall outside the listed absorption range.

E. Carbonyl Compounds

One of the most distinctive bands in an infrared spectrum is the one arising from the carbonyl stretching mode. This is a strong peak observed between 1640 and 1820 cm^{-1} (5.5–6.1 μm).

The carbonyl group is part of a number of functional groups. The exact position of the carbonyl absorption, the positions of other absorption bands in the infrared spectrum, and other spectral techniques (particularly NMR) may be needed to identify the functional group. The positions of C=O absorption for aldehydes, ketones, carboxylic acids, and esters are listed in Table 9.2.

Ketones give the simplest spectra of the carbonyl compounds. If a compound is an aliphatic ketone, all strong stretching absorption bands will arise from C=O, C—H or C—C groups. The infrared spectrum of a typical aliphatic ketone is shown in Figure 9.15.

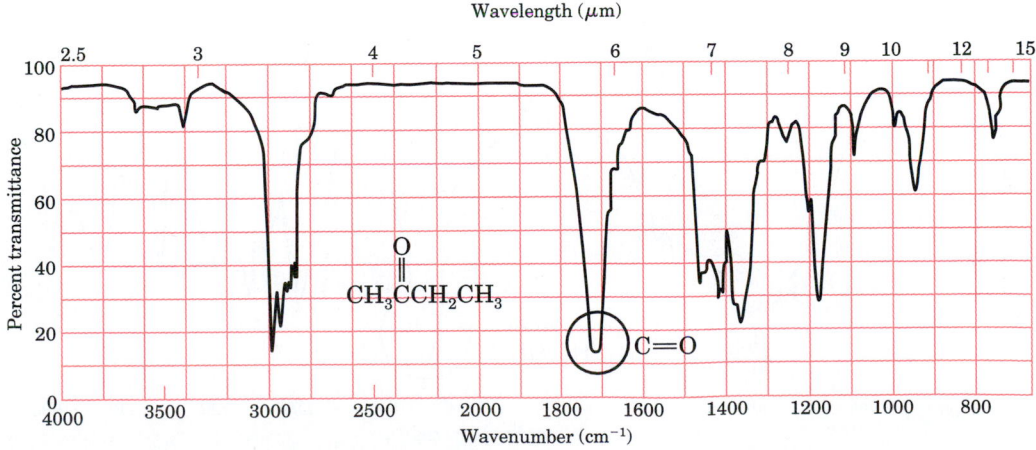

Figure 9.15 Infrared spectrum of butanone, a ketone.

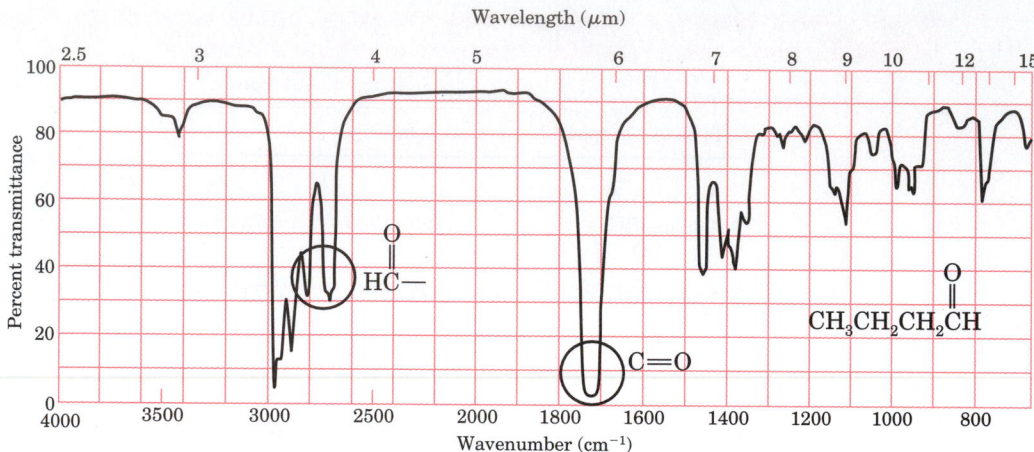

Figure 9.16 Infrared spectrum of butanal, an aldehyde.

Aldehydes give infrared spectra that are similar to those of ketones. The important structural difference between an aldehyde and a ketone is that an aldehyde has an H bonded to the carbonyl carbon. This particular CH bond shows two characteristic stretching bands (just to the right of the aliphatic CH band) at 2820–2900 cm^{-1} (3.45–3.55 μm) and 2700–2780 cm^{-1} (3.60–3.70 μm). Both these CH peaks are sharp, but weak, and the peak at 2900 cm^{-1} (3.45 μm) may be obscured by overlapping absorption of other CH bonds (see Figure 9.16). The aldehyde CH also has a very characteristic NMR absorption. (See Section 9.7.) If the infrared spectrum of a compound suggests that the structure is an aldehyde, the NMR spectrum should be checked.

Carboxylic acids exhibit typical C=O absorption and also show a very distinctive O—H band, which begins at about 3330 cm^{-1} (3.0 μm) and slopes into the aliphatic CH absorption band (see Figure 9.17). The reason that a carboxyl OH band is different from an OH band of an alcohol is because carboxylic acids form hydrogen-bonded dimers. The strong hydrogen bonding broadens the OH stretching band dramatically.

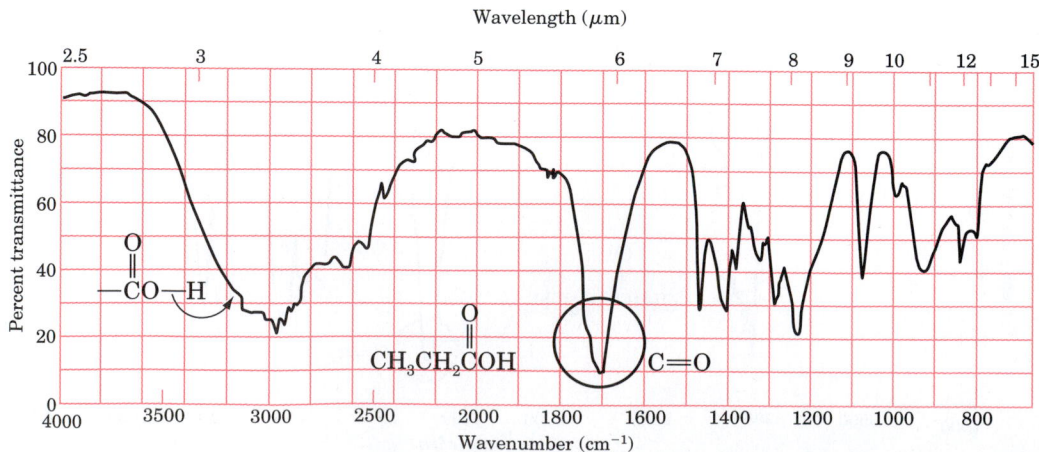

Figure 9.17 Infrared spectrum of propanoic acid, a carboxylic acid.

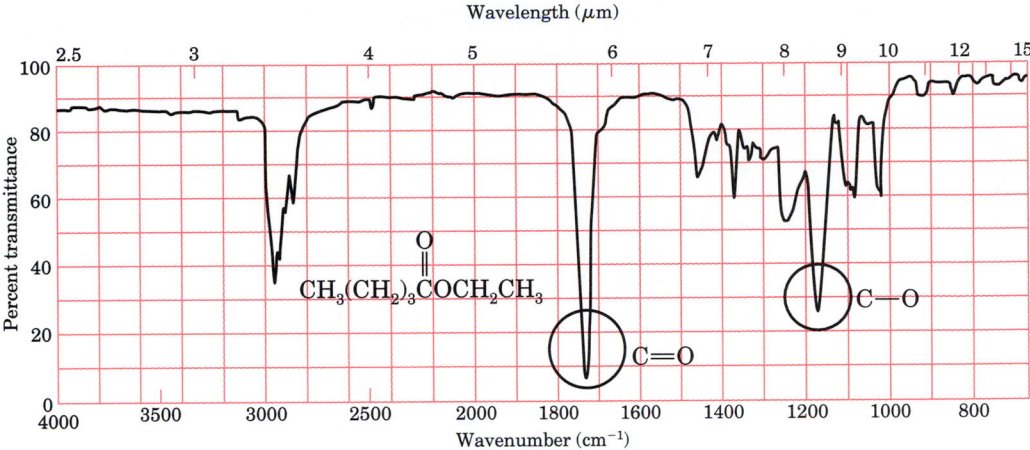

Figure 9.18 Infrared spectrum of ethyl pentanoate, an ester.

$$R-C \overset{\ddot{O}:\text{---}H-\ddot{O}:}{\underset{:\ddot{O}-H\text{---}:\ddot{O}}{}} C-R$$

Esters exhibit typical C=O bands and C—O single-bond bands. The C—O band, like that in ethers, is observed in the fingerprint region of the spectrum, 1100–1300 cm^{-1} (7.7–9.1 μm), and is sometimes difficult to identify. However, when the C—O band is strong, it can be used to distinguish between esters and ketones. See Figure 9.18 for the infrared spectrum of a typical ester.

STUDY PROBLEM

9.4   One of the two spectra in Figure 9.19 is that of a ketone; the other is of an ester. Which is which? Explain your answer.

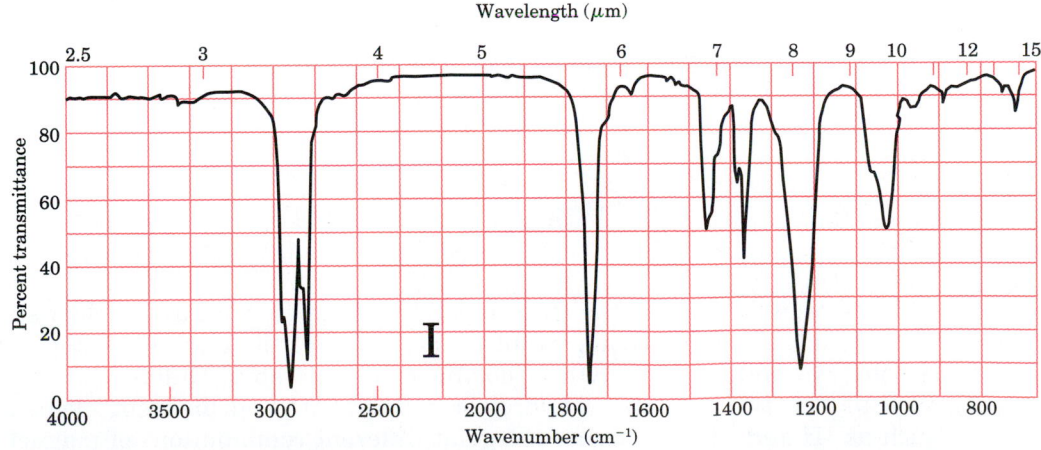

Figure 9.19 Infrared spectra of unknown compounds for Problem 9.4.

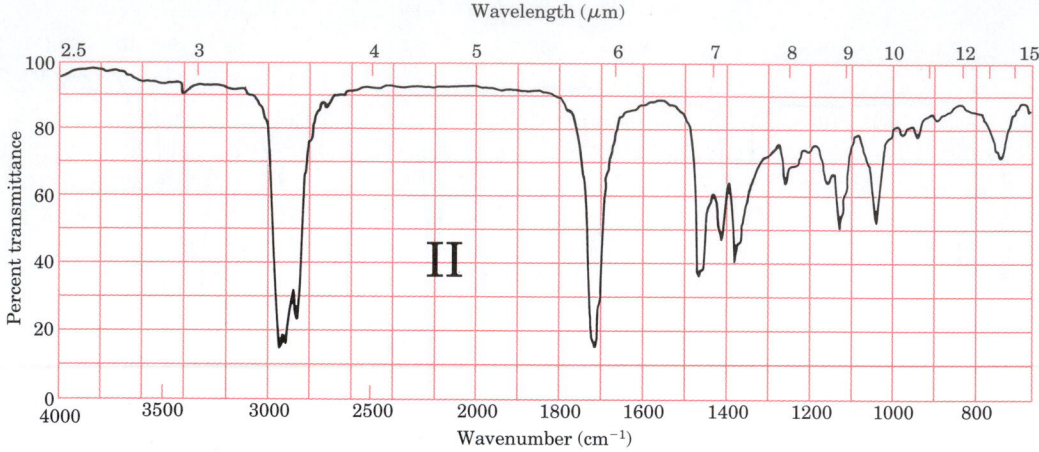

Figure 9.19 *(continued)* Infrared spectra for Problem 9.4.

Section 9.6

Nuclear Magnetic Resonance Spectroscopy

The infrared spectrum of a compound gives a picture of the different functional groups in an organic molecule, but gives only meager clues about the hydrocarbon portion of the molecule. **Nuclear magnetic resonance (NMR) spectroscopy** fills this gap by providing information about the carbon and hydrogen atoms in the molecule.

NMR spectroscopy is based upon the absorption of radio waves by certain nuclei in organic molecules when they are in a strong magnetic field. Before proceeding with a discussion of NMR spectra and their use in organic chemistry, let us consider the physical principles that give rise to the NMR phenomenon.

A. Origin of the NMR Phenomenon

The nuclei of atoms of all elements can be classified as either *having spin* or *not having spin.* A nucleus with spin gives rise to a small magnetic field, which is described by a **nuclear magnetic moment,** a vector.

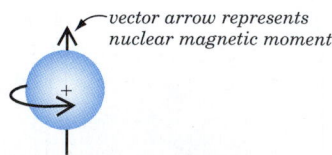

To the organic chemist, the most important isotopes that have nuclear spin are ^1_1H and $^{13}_6C$. Equally important is the fact that the common isotopes of carbon (^{12}C) and oxygen (^{16}O) do not have nuclear spin. Different isotopes (such as the two isotopes of hydrogen, ^1H and ^2H) and different elements (such as ^1H and ^{13}C) all absorb energy at different combinations of magnetic field strength and radiofrequency. Consequently, we can obtain an NMR spectrum of each isotope or element without interference from the others. We will first describe NMR spectroscopy using ^{1}H, or proton, NMR spectra, then discuss ^{13}C spectroscopy in Section 9.12.

In NMR spectroscopy, an **external magnetic field** is generated by a permanent magnet or an electromagnet. The strength of this external field is symbolized by H_0, and its direction is represented by an arrow. The symbol H_0 has the units of gauss. It is important that H_0, the symbol for magnetic strength, not be confused with H, the symbol for hydrogen (proton). In the discussions that follow, both symbols will be used — and both may even appear in the same figure. (For example, see Figure 9.24 on page 360.)

Symbol representing the external magnetic field: $\uparrow$
H_0

A spinning proton with its nuclear magnetic moment is similar, in many respects, to a tiny bar magnet. When molecules containing hydrogen atoms are placed in an external magnetic field, the magnetic moment of each hydrogen nucleus, or proton, aligns itself in one of two different orientations with respect to the external magnetic field. (Keep in mind that it is only the magnetic moments of *hydrogen nuclei,* not molecules, that become aligned.) The two orientations that the nuclear magnetic moment may assume are **parallel** and **antiparallel** to the external field. In the parallel state, the magnetic moment of the proton points in the *same direction* as that of the external field. In the antiparallel state, the magnetic moment of the proton *opposes* the external field. At any given time, approximately half the protons in a sample are in the parallel state and half are in the antiparallel state.

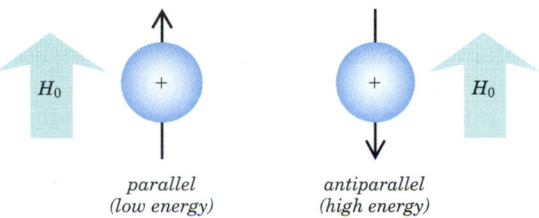

parallel
(low energy)

antiparallel
(high energy)

The parallel state of a proton is slightly more stable (lower energy) than the antiparallel state. When exposed to the proper frequency of radio waves, the magnetic moments of a small fraction of the parallel protons absorb energy and turn around, or *flip,* to the higher-energy antiparallel state. After being converted to the higher-energy antiparallel state, a nucleus can lose the absorbed energy to its surroundings as heat and thereby return to the lower-energy parallel state. When the sample is scanned with a properly adjusted instrument, the absorption and loss of energy by the nuclei are sufficiently rapid that a slight excess of nuclei in the parallel state can be maintained. Therefore, the sample continues to absorb energy, and the signal in the spectrum is observed.

The amount of energy required to flip the magnetic moment of a proton from parallel to antiparallel depends, in part, upon the strength of H_0. If H_0 is increased, the energy difference between the parallel and antiparallel states increases. Thus, if H_0 is increased, the nucleus is more resistant to being flipped and higher-energy, higher-frequency radiation is required.

The relationship of external field strength with the energy difference between parallel and antiparallel spin states is illustrated in Figure 9.20.

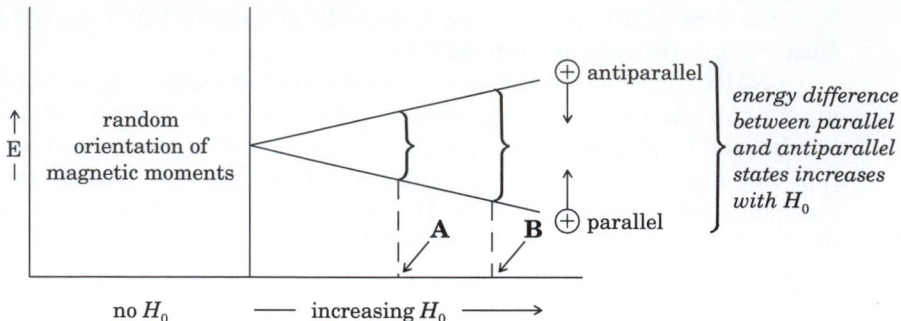

Figure 9.20 The energy difference between parallel and antiparallel spin states increases with the strength of the external field, H_0. A flip at the field strength labeled **A** on the graph requires lower frequency, less energetic radiowaves than a flip at the field strength labeled **B.**

When the particular combination of external magnetic field strength and radiofrequency causes a proton to flip from parallel to antiparallel, the proton is said to be in **resonance** (a different type of resonance from that of "resonance" structures of benzene). The term **nuclear magnetic resonance** means "nuclei in resonance in a magnetic field."

It would seem that all protons should come into resonance at the same combination of H_0 and radiofrequency. However, this is not the case. The magnetic field actually observed by a proton in a particular molecule is a combination of two fields: (1) the applied external magnetic field (H_0) and (2) **induced molecular magnetic fields,** small magnetic fields induced in the bonds of a molecule by H_0. Thus, protons may be exposed to magnetic fields of different intensities. The magnetic field observed by a proton is also modified by the spin states of nearby protons, a topic we will discuss later in this chapter.

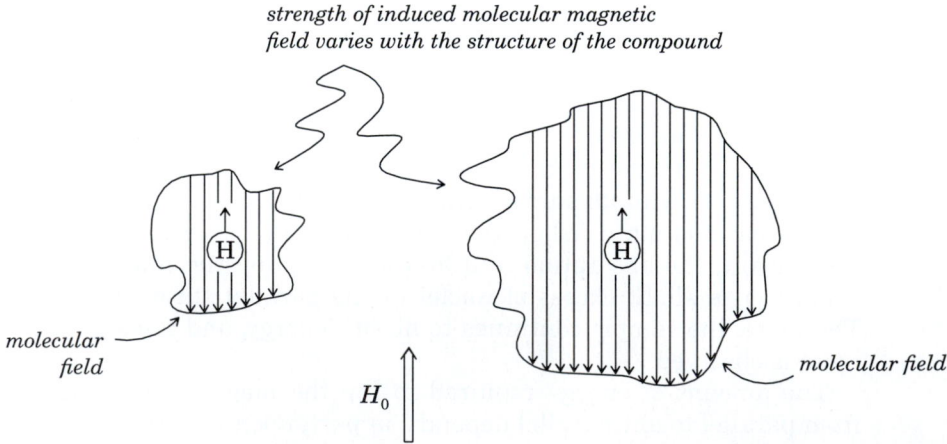

Different protons in an organic compound are surrounded by molecular fields of different strengths. It takes a stronger or a weaker H_0 to overcome the influence of these molecular fields. Consequently, different protons come into resonance at different positions in the spectrum. With knowledge of the relationships between absorption pattern and structure, we will be able to identify the types of hydrogens in an organic structure.

B. The NMR Spectrum

A diagram of an NMR spectrometer is shown in Figure 9.21. The sample is placed between the poles of a magnet and is irradiated with radio waves. When the protons flip from the parallel to the antiparallel state, the absorption of energy is detected by a power indicator.

In one type of NMR spectrometer, the radiofrequency is held constant at 60 MHz (60 megaHertz, or 60×10^6 Hz), H_0 is varied over a small range, and the frequency of energy absorption is recorded at the various values for H_0. Thus, the NMR spectrum is a graph of the amount of energy absorbed versus magnetic field strength.

Modern NMR spectrometers employ fast Fourier analysis to obtain the spectrum. These instruments are called **Fourier transform NMR (FT NMR) spectrometers.** A Fourier transform spectrometer can generate a ^{1}H NMR spectrum in a few seconds and store a large number of repetitive scans in computer memory. The stored scans are computer-averaged, a process that averages the noise to zero and increases the intensities of the real signals. Then, the data are converted by a mathematical technique called **Fourier transform analysis** to provide the spectrum that we see.

One feature that makes the new spectrometers practical is their speed compared to that of the conventional continuous-wave spectrometers. In a Fourier transform NMR spectrometer, the sample is irradiated with a brief intense pulse of a range of radiofrequencies that brings all the nuclei under study into resonance simultaneously. As the nuclei return to their normal distribution, the frequencies of energy losses are measured simultaneously and stored in computer memory. Fourier transform analysis makes interpretation of the stored signals possible.

While we will emphasize spectra obtained using a 60 MHz instrument, most NMR spectrometers in use today employ a much stronger magnetic field

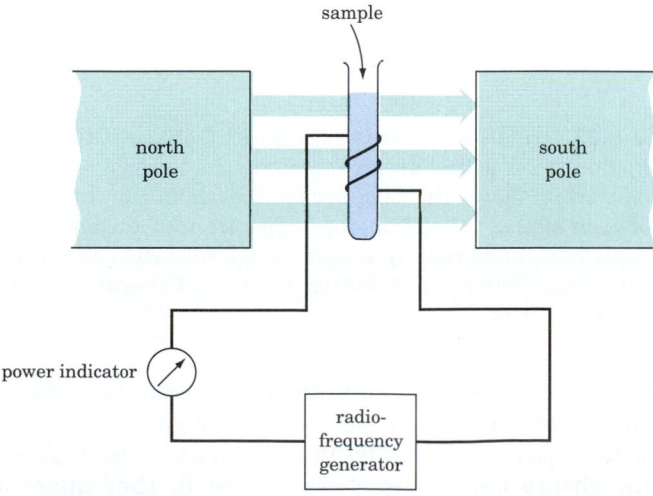

Figure 9.21 Schematic diagram of an NMR spectrometer.

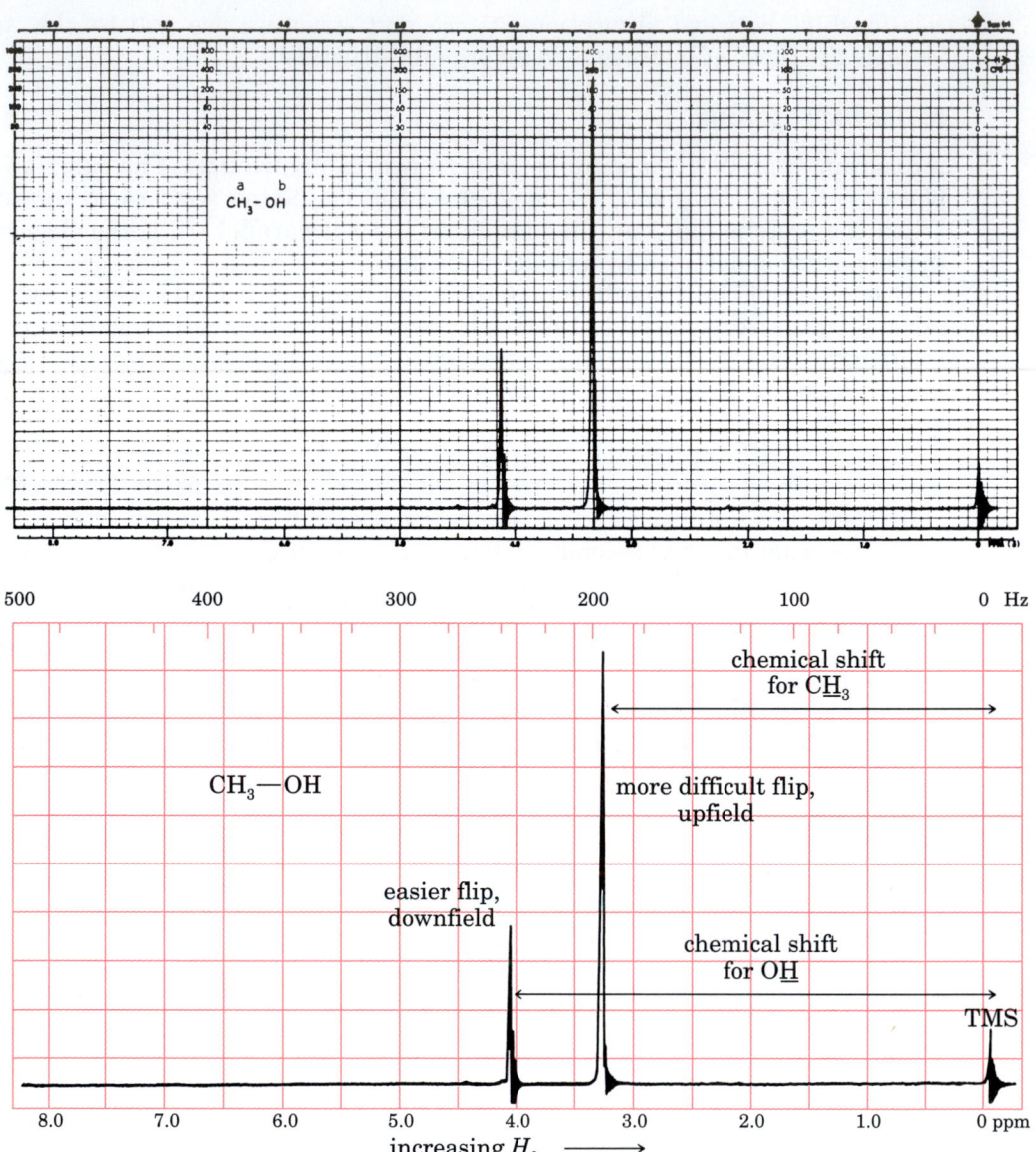

Figure 9.22 ^{1}H NMR spectra of methanol (solvent, CCl_4). The upper spectrum is a reproduction of an actual spectrum, while the lower one is an artist's rendition. The lower spectrum shows the field sweep and chemical shifts. (The fluctuations to the right of each signal, called "ringing," arise from distortion of the magnetic field when the sample is scanned rapidly. Ringing is an indication that the instrument is properly adjusted.) Upper spectrum © Sadtler Research Laboratories, Division of Bio-Rad Laboratories, Inc., 1986.

and, therefore, use higher radiofrequency radiation. For example, a 200 MHz instrument uses a field strength of 46,972 gauss, while a 60 MHz instrument uses a field strength of only 14,092 gauss. With higher field strengths, the protons' absorption patterns are spaced further apart and the spectra are much easier to interpret. We will use high-field spectra in place of 60 MHz spectra only when they facilitate interpretation. You may assume that NMR spectra were obtained at 60 MHz unless otherwise noted.

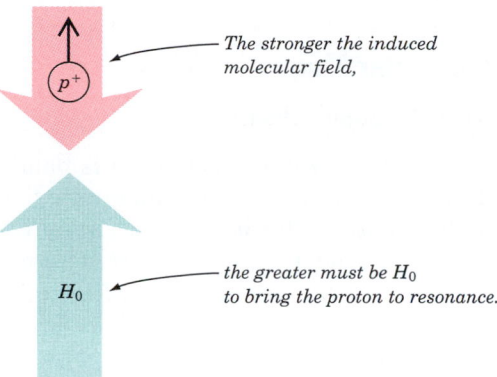

The stronger the induced molecular field,

the greater must be H_0 to bring the proton to resonance.

Figure 9.23 Induced molecular fields oppose H_0, the external applied magnetic field. The strength of H_0 needed to flip the spin states of protons depends partly on how strong the induced molecular magnetic field is.

Figure 9.22 shows two NMR spectra of methanol: a reproduction of an actual spectrum and an artist's copy. On the artist's spectrum, we have indicated the direction of field sweep from weak H_0 to strong H_0. You can see two principal peaks in these spectra. The left-hand peak arises from the OH proton; the right-hand peak from the CH_3 protons. The protons that flip more easily absorb energy at a lower H_0. They give rise to an absorption peak **downfield** (farther to the left). The protons that flip with greater difficulty absorb energy at a higher H_0 and give a peak that is **upfield** (to the right).

In an NMR spectrum, the position of absorption by a proton depends on the net strength of the magnetic field that surrounds the proton. This field is a sum of the applied field H_0 and the induced molecular field. If the induced field is strong, more H_0 is required to bring the proton into resonance. A proton surrounded by a strong molecular field is said to be **shielded** and we observe its signal *upfield* in the spectrum. On the other hand, if the induced molecular field is weak, then less applied field is needed to bring it into resonance. Such a proton is said to be **deshielded** and its signal appears *downfield* (see Figure 9.23).

Shielding and deshielding are relative terms. In order to obtain quantitative measurements, we need a reference point. The compound that has been chosen for this reference point is tetramethylsilane (TMS), $(CH_3)_4Si$, the protons of which absorb to the far right in the NMR spectrum. The absorption for most other protons is observed downfield from that of TMS. In practice, a small amount of TMS is added directly to the sample, and the peak for TMS is observed on the spectrum along with any absorption peaks from the sample compound. The difference between the position of absorption of TMS and that of a particular proton is called the **chemical shift.** You can see the TMS peak in the spectra in Figure 9.22, where we have also indicated the chemical shifts.

Chemical shifts are reported in δ **values,** which are expressed as *parts per million (ppm) of the applied radiofrequency.* At 60 MHz, 1.0 ppm is 60 Hz. Therefore, a δ value of 1.0 ppm is 60 Hz downfield from the position of absorption of TMS, which is set at 0 ppm (δ 0). At 200 MHz, a value of 1.0 is 200 Hz. As may be seen in Figure 9.22, the two types of protons in CH_3OH have δ values of 3.4 ppm and 4.15 ppm, respectively. (Note that the lower scale on an NMR spectrum shows value in ppm. The use of the Hz scale at the top of an NMR spectrum will be discussed in Section 9.10.)

Section 9.7
Types of Induced Molecular Magnetic Fields

A. Fields Induced by Sigma Electrons

A hydrogen atom in an organic compound is bonded to carbon, oxygen, or another atom by a σ bond. The external magnetic field causes these σ-bond electrons to circulate within the bond. The circulating electrons create a weak molecular magnetic field that opposes H_0 (Figure 9.24). It takes a slightly stronger external field to overcome the effect of the induced molecular field and bring the proton into resonance. Therefore, the proton absorbs upfield compared to a hypothetical naked proton. The strength of the induced field depends upon the electron density near the hydrogen atom in the sigma bond. The higher the electron density, the greater the induced field (and the farther upfield the observed absorption).

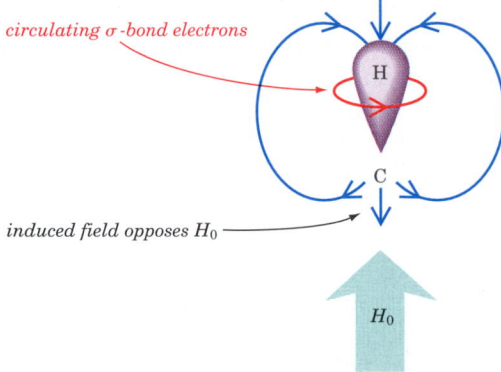

Figure 9.24   The induced field from circulating σ-bond electrons opposes H_0 in the vicinity of the proton. (For simplicity, the figure shows only one CH bond of an organic molecule.)

The electron density of a carbon–hydrogen covalent bond is affected by the electronegativity of the other atoms bonded to the carbon. Let us consider a specific example. The CF bond in CH_3F is polar: the fluorine atom carries a partial negative charge and the carbon atom carries a partial positive charge. Because the carbon has a partial positive charge, the electrons in each C—H σ bond are drawn toward the carbon and away from the hydrogen atom. Recall from our discussion of carbocation stability that the polarization of bonds by positive or negative centers is called the **inductive effect.** This shift in electron density toward an electronegative element (F) is another example of the inductive effect. In this case, the electron-withdrawing effect of F results in a greater electron density around F and a lesser electron density around each hydrogen. The protons of CH_3F are deshielded and absorb downfield compared to the protons of CH_4.

$$
\begin{array}{c}
H \\
\downarrow \\
H \rightarrow C \rightarrow F \\
\uparrow \\
H
\end{array}
\qquad
\begin{array}{l}
\textit{F causes a decrease in} \\
\textit{e}^{-} \textit{ density around each H}
\end{array}
$$

The following series of methyl halides show increasing shielding of the hydrogen nuclei with decreasing electronegativity of the atom attached to $-CH_3$.

$$H_3C-F \qquad H_3C-Cl \qquad H_3C-Br \qquad H_3C-I$$
$$\delta 4.3 \qquad\qquad \delta 3.0 \qquad\qquad \delta 2.7 \qquad\qquad \delta 2.1$$

increased shielding of H →

SAMPLE PROBLEM

Which of the circled protons is the most shielded? The most deshielded? What are the relative positions of NMR absorption?

(a) CH_3CH_2Cl **(b)** CH_3CH_3 **(c)** CH_3CH_2I

Solution

Chlorine is more electronegative than iodine, which in turn is more electronegative than hydrogen. Therefore, (a) is the most deshielded and absorbs downfield; (b) is the most shielded and absorbs upfield, closer to TMS; and (c) is intermediate.

In a single molecule, a proton that is bonded to the same carbon as an electronegative atom is more deshielded than protons on other carbons. Figure 9.32 shows the spectrum of chloroethane. The peaks at $\delta 1.5$ arise from the CH_3 protons, while the deshielded CH_2Cl protons absorb farther downfield, at $\delta 3.55$. (The complexity of these peaks will be discussed later in this chapter.)

The inductive effect of an electronegative atom falls off rapidly when it is passed through a number of sigma bonds. In NMR spectra, the inductive effect is negligible three carbons away from the electronegative atom.

effect of X is of little importance in decreasing e^- density around this proton

$$\begin{array}{ccc} & H & H & H \\ & | & | & | \\ H-&C-&C-&C-X \\ & | & | & | \\ & H & H & H \end{array}$$

effect of X is important in decreasing e^- density around this proton

Most elements encountered in organic compounds are more electronegative than carbon. Their inductive effect is one of electron-withdrawal, and protons affected by them are deshielded. However, silicon is *less* electronegative than carbon. The silicon–carbon bond is polarized such that carbon carries a partial negative charge. The electrons in the CH bond of an $SiCH_3$ group are repelled from the negative carbon and pushed toward the hydrogen atoms. In this case, the inductive effect is *electron-releasing*. The protons of an $SiCH_3$ group are highly shielded because of the increased electron density around them. This is the reason that the protons of TMS absorb upfield and that TMS provides a good reference peak in an NMR spectrum.

greater e^- density on H; highly shielded

$$\begin{array}{c} CH_3 \\ \uparrow \\ H_3C \leftarrow Si \rightarrow CH_3 \\ \downarrow \\ CH_3 \end{array}$$

B. Fields Induced by Pi Electrons

The molecular magnetic fields generated by the π *electrons* are *directional* (that is, unsymmetrical). Any measurement that varies depending on the direction in which the measurement is taken is said to be **anisotropic.** Because the effects of molecular fields generated by π electrons are direction-dependent, they are called **anisotropic effects.** (Anisotropic effects are contrasted to inductive effects, which are symmetrical around the proton.) Anisotropic effects occur in addition to the ever-present molecular fields generated by σ-bond electrons.

In benzene, the π electrons are delocalized around the ring. Under the influence of an external magnetic field, these π electrons *circulate around the ring*. This circulation, called **ring current,** generates a molecular magnetic field with the geometry shown in Figure 9.25. (All the benzene rings, of course, do not line up as we have represented the ring in this figure. This figure shows the net effect of the vector forces.)

Referring to Figure 9.25, you can see that the field generated by the ring current is in the same direction as the applied external field H_0 near the benzene protons. Therefore, less external field (H_0) is needed to bring these protons into resonance compared to that needed for the alkyl protons. Aryl protons, such as the benzene hydrogens, are deshielded and absorb farther downfield than alkyl protons.

Vinylic and aldehyde hydrogens are also deshielded by an anisotropic effect. The π electrons are set in motion within the bond and generate a molecular field that adds to the applied field near the $=$CH proton (see Figure 9.25).

Figure 9.26 shows the NMR spectrum of a compound with aryl protons, vinylic protons and alkyl protons. In this spectrum, the alkyl protons are farthest upfield, the aryl protons are the farthest downfield, and the vinylic protons are intermediate.

Figure 9.27 shows the spectrum of an aldehyde. The aldehyde proton is shifted downfield both by anisotropic effects and by electron-withdrawal by the carbonyl oxygen. The combination of effects results in absorption that is far downfield ($\delta 9 - 10$), outside the normal scan for some NMR spectrometers. Instrument design allows scanning at field strengths that are lower than normal; the scan of this region ($\delta 8 - 20$), traced above the standard NMR scan, is called an **offset scan.**

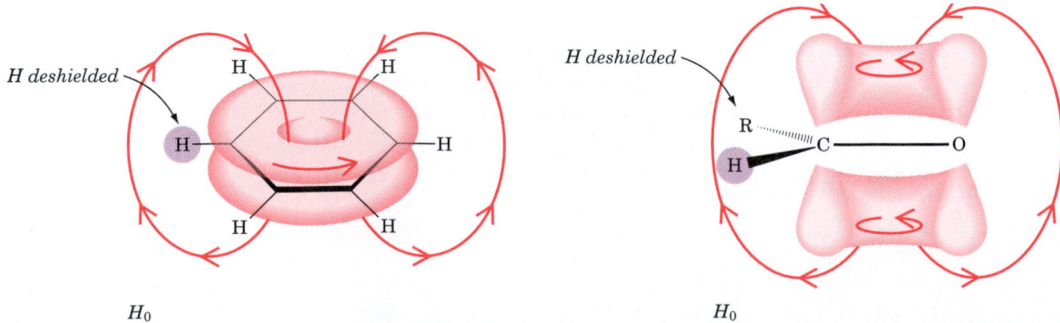

Figure 9.25 Circulating π electrons in benzene or aldehydes induce a magnetic field that deshields the adjacent protons. (All six benzene protons are deshielded.)

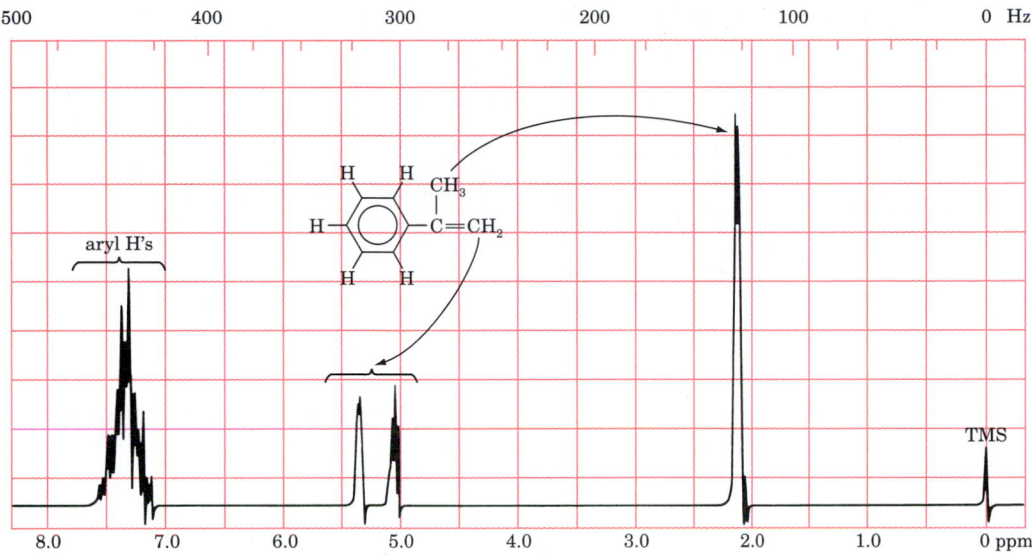

Figure 9.26 ^{1}H NMR spectrum of 2-phenylpropene, showing absorption by aryl, vinylic, and methyl protons.

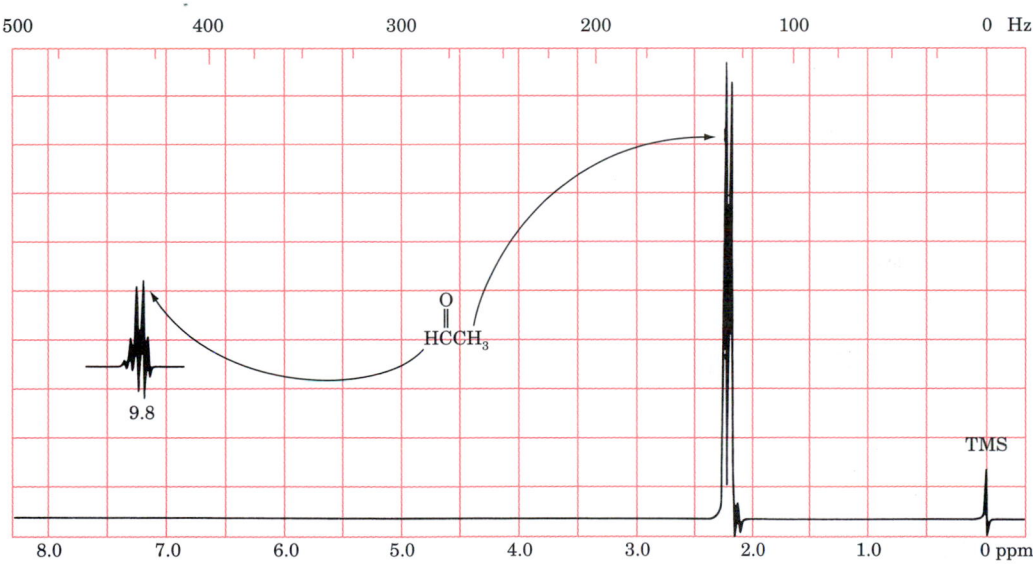

Figure 9.27 ^{1}H NMR spectrum of ethanal (acetaldehyde), showing the offset scan for the aldehyde proton.

C. Summary of Induced-Field Effects

The presence of an electronegative atom causes a decrease in electron density around a proton by the **inductive effect.** Such a proton is deshielded and absorbs downfield. In aromatic compounds, alkenes, and aldehydes, a proton bonded to the sp^2-hybridized carbon is deshielded by **anisotropic effects** and absorbs even farther downfield. The absorption positions are summarized in Figure 9.28. A chart inside the back cover lists the δ values for a number of types of protons.

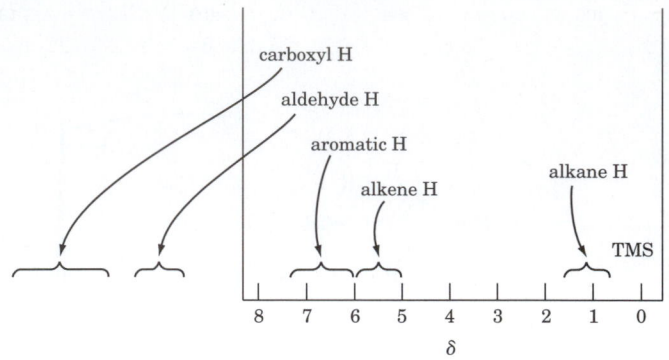

Figure 9.28 Relative positions of proton absorption in the ^1H NMR spectrum. (See also the chart inside the back cover of this book.)

STUDY PROBLEM

9.5 For each of the following, which group of protons, 1 or 2, will have the greater chemical shift?

(a) [structure: benzene ring with "1: the aryl protons"] $\quad$ CH$_3$CH with O double bond, labeled 2

1: the aryl protons

(b) $1 \searrow$ $2 \swarrow$ $Cl-CH_2CH=CH_2$

(c) CH_3CO-H $\quad$ $CH_3CH_2NH_2$ (labels 1, 2)

(d) CH_3OCCH_3 (labels 1, O, 2)

(e) [benzene]$-CH_3$ (1) $\quad$ [cyclohexane]$-CH_3$ (2)

Section 9.8
Counting the Protons

A. Equivalent and Nonequivalent Protons

Protons that are in the same magnetic environment in a molecule have the same chemical shift in an NMR spectrum. Such protons are said to be **magnetically equivalent protons.** Protons that are in different magnetic environments have different chemical shifts and are said to be **nonequivalent protons.**

Magnetically equivalent protons in NMR spectroscopy are generally the same as *chemically equivalent protons*. In chloroethane, the three methyl protons are magnetically equivalent and are also chemically equivalent. To see that they are chemically equivalent, imagine a chemical reaction in which one of these three protons is replaced by another atom, such as bromine. If only one product can result, regardless of which proton is replaced, then the protons are chemically equivalent. In the following example, note that replacement of any of the methyl protons by Br leads to the same compound, 1-bromo-2-chloroethane, and not to isomers.

equivalent

If any H is replaced by Br, only one bromo compound is obtained.

$$H—C—CH_2Cl \longrightarrow Br—C—CH_2Cl$$

The two protons of the CH_2Cl group are also magnetically and chemically equivalent to each other. However, the three protons of the CH_3 group are not equivalent to the two CH_2Cl protons.

three equivalent protons (but nonequivalent to CH_2)   *two equivalent protons (but nonequivalent to CH_3)*

$$CH_3CH_2Cl$$

The three CH_3 protons have the same chemical shift and absorb at the same position in the NMR spectrum. The two CH_2 protons are deshielded compared to the methyl protons and have a greater chemical shift. A low-resolution spectrum of chloroethane would look like the stylized spectrum shown in Figure 9.29.

Equivalent protons need not necessarily be on the same carbon atom. For example, diethyl ether contains only two types of equivalent protons: the CH_3 protons and the CH_2 protons.

six equivalent protons

$$CH_3CH_2OCH_2CH_3$$

four equivalent protons

Restricted rotation, the geometry of the molecule, and the presence of chiral carbons can affect the equivalence of protons. For example, rotation about a carbon–carbon double bond is restricted. Therefore, in chloroethene, the proton *cis* to the Cl atom is in a different magnetic environment from that of the *trans* proton. Both these protons are in different magnetic environments

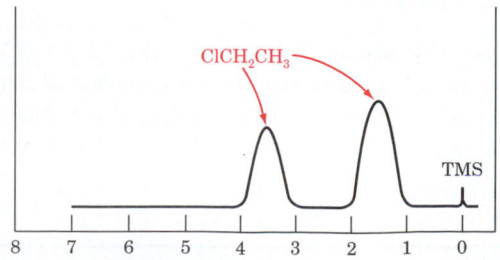

Figure 9.29   Stylized low-resolution ^1H NMR spectrum of CH_3CH_2Cl.

from that of the proton on the C—Cl carbon. In chloroethene, all three protons are nonequivalent.

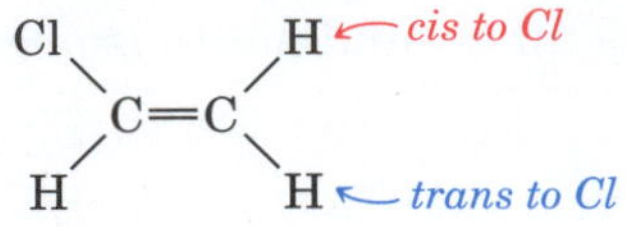

chloroethene
(vinyl chloride)

three nonequivalent protons

SAMPLE PROBLEM

Which protons in *p*-chlorotoluene (1-chloro-4-methylbenzene) are chemically equivalent and which are nonequivalent?

Solution

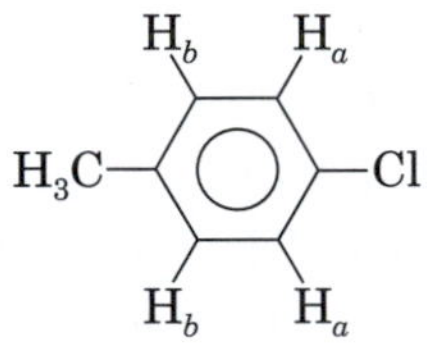

the three methyl protons are equivalent;
the two H_a's are equivalent;
the two H_b's are equivalent;
the methyl protons, H_a and H_b, are nonequivalent

Replacement of either proton labeled H_a by Br would result in the same compound. Similarly, replacement of either H_b would yield the same compound. However, replacement of H_a by Br would result in a different compound from that obtained by replacing H_b by Br.

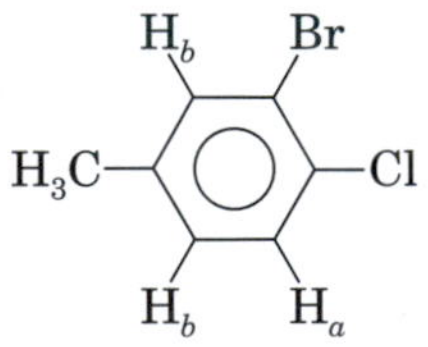 is the same as 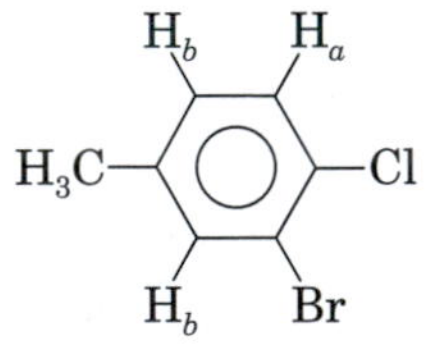

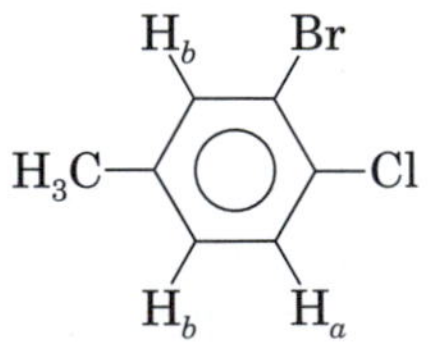 is *not* the same as 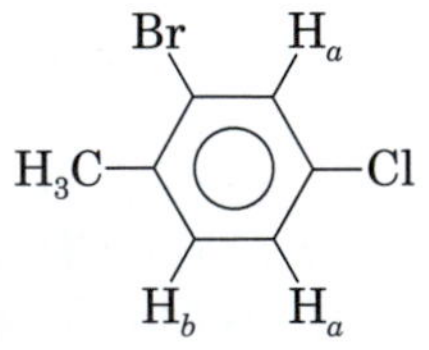

B. Areas Under the Peaks

If we measure the areas under the peaks in an NMR spectrum, we find that *the areas are in the same ratio as the number of protons that give rise to each signal.* With chloroethane (Figure 9.30), the ratio is 2:3. (Note that the area under the peak is the important feature, not the height of the peak.)

SAMPLE PROBLEM

How many types of equivalent protons are there in each of the following structures? What would be the relative areas under the NMR absorption bonds?

(a) $(CH_3)_2CHCl$ **(b)** $CH_3CH_2OCH_2CH_3$ **(c)** Cl—⟨O⟩—OCH_3

Solution

(a) two, $6:1$ **(b)** two, $3:2$, **(c)** three, $3:2:2$ (3 for CH_3, 2 for aryl protons adjacent to the oxygen, 2 for aryl protons adjacent to Cl).

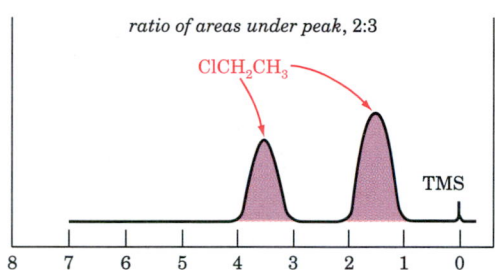

Figure 9.30 Stylized low-resolution ^1H NMR spectrum of CH_3CH_2Cl, showing areas under the peaks.

STUDY PROBLEM

9.6 For each of the following structures, circle the proton(s), if any, that are chemically equivalent to the indicated proton(s).

(a)
$$\begin{array}{c} CH_2—CHCH_3 \\ | \qquad | \\ CH_2—\underline{CH}_2 \end{array}$$

(b) $\underline{CH}_3OCH_2OCH_2OCH_3$

(c)
$$\begin{array}{ccc} \underline{CH}_3 & OH & CH_3 \\ | & | & | \\ CH_3CCH_2 & CHCH_2 & CCH_3 \\ | & & | \\ CH_3 & & CH_3 \end{array}$$

Most NMR spectrometers are equipped with **integrators,** which give a signal that shows the relative areas under the peaks in a spectrum. The integration appears as a series of steps superimposed upon the NMR spectrum. The height of the step over each absorption peak is proportional to the area under that peak. From the relative heights of the steps on the integration curve, the relative areas under the peaks may be determined. In Figure 9.31, the heights of the steps of the integration curve were measured with a ruler and found to be 33 mm, 100 mm, and 50 mm. For determining the relative numbers of equivalent protons, these values are converted to ratios of small whole numbers, $2:6:3$. (In the laboratory, the numbers rarely turn out to be exactly whole numbers; therefore, some rounding of numbers must be done.) We can compare these values with the numbers of protons in the known structure of 1-bromo-2,4,6-trimethylbenzene and, indeed, the values agree.

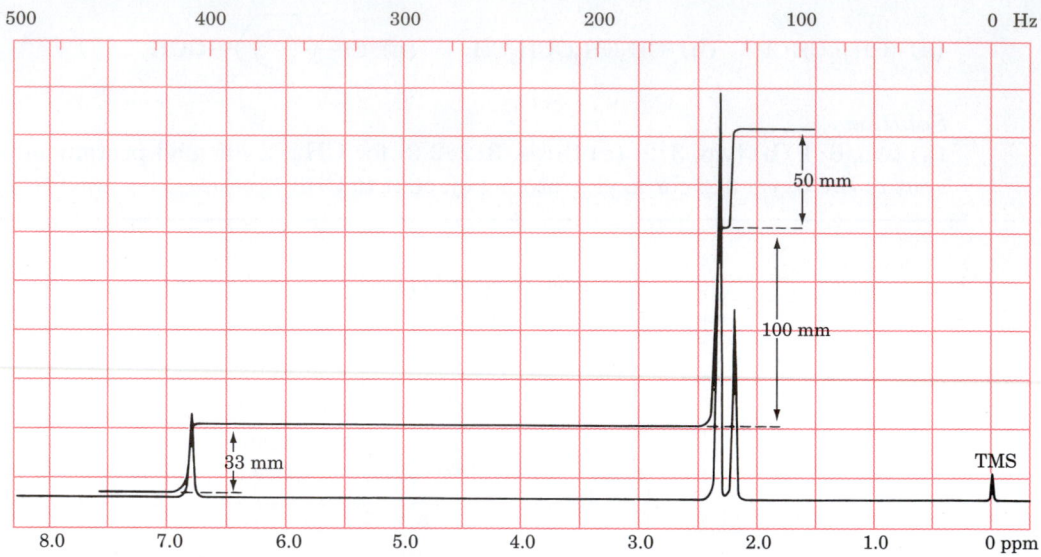

Figure 9.31 ^{1}H NMR spectrum of 1-bromo-2,4,6-trimethylbenzene, showing an integration curve.

In this text, we will not show the integration lines in the spectra, but will indicate relative areas as numbers directly above the proton absorption peaks, as shown in Figure 9.32.

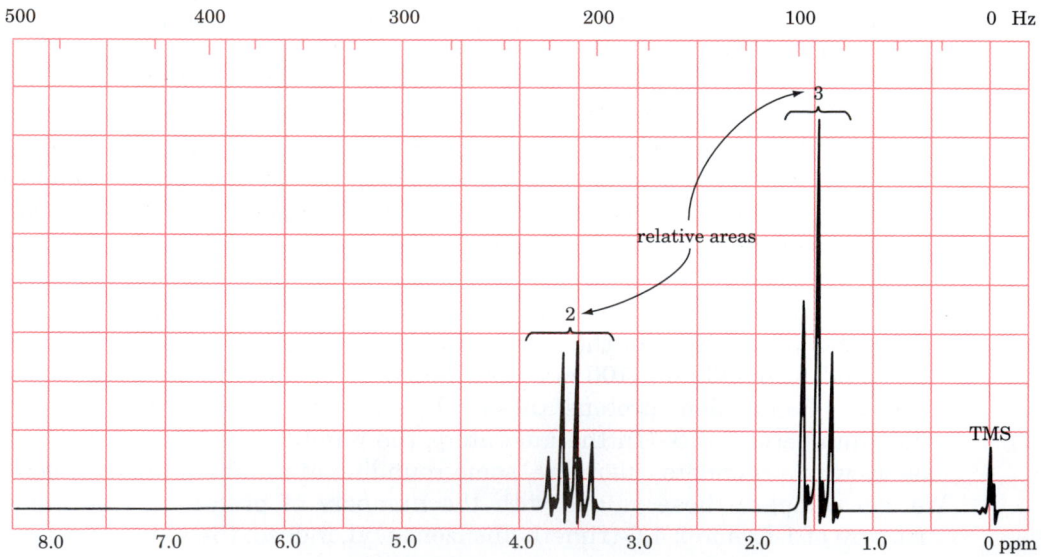

Figure 9.32 ^{1}H NMR spectrum of chloroethane, CH_3CH_2Cl.

SAMPLE PROBLEM

The heights of four steps of an integration curve were measured to be 90 mm, 36 mm, 37 mm, and 54 mm. Calculate the ratio of the different types of protons in the sample compound.

Solution

1. Divide each height by the smallest one (36 mm in this case):

$$\frac{90 \text{ mm}}{36 \text{ mm}} = 2.5 \qquad \frac{36 \text{ mm}}{36 \text{ mm}} = 1.0 \qquad \frac{37 \text{ mm}}{36 \text{ mm}} = 1.0 \qquad \frac{54 \text{ mm}}{36 \text{ mm}} = 1.5$$

2. Multiply the quotients by the integer that will convert them to the smallest whole numbers possible (rounding off, if necessary). In this case, multiplying by 2 gives the ratio of 5:2:2:3. Therefore, the ratio of the numbers of nonequivalent protons is 5:2:2:3.

STUDY PROBLEMS

9.7 Predict the relative areas under the principal ^{1}H NMR signals for the following compounds:

                                             Cl
                                             |

(a) $CH_3CH_2CH_2CH_2Cl$ (b) $CH_3CH_2CHCH_2CH_3$

9.8 Two ^{1}H NMR integration curves had the following step heights. Calculate the relative areas in each case.

(a) 14.0, 13.9, 125.0 mm   (b) 50.6, 8.3, 50.2 mm

Section 9.9
Spin–Spin Coupling

The spectrum of chloroethane in Figures 9.29 and 9.30 is a stylized low-resolution spectrum. If we increase the resolution (that is, the sensitivity), the peaks are resolved, or split, into groups of peaks. (See Figure 9.32.) This type of splitting is called **spin–spin splitting** and is caused by the presence of *vicinal, or neighboring, protons (protons on an adjacent carbon) that are nonequivalent.* Protons that split each other's signals are said to have undergone **spin–spin coupling.**

these protons split   *these protons split*
the signal for   *the signal for*
the CH$_2$ protons   *the CH$_3$ protons*

$$CH_3CH_2Cl$$

Why do protons undergo spin–spin coupling? The splitting of the signal arises from the two spin states (parallel and antiparallel) of the neighboring protons.

Signal for H$_a$ split by the two spin states of H$_b$:

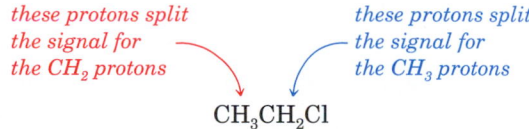

H$_b$ *is parallel;*   H$_b$ *is antiparallel;*
its moment is   *its moment is*
added to H$_0$ *subtracted from H$_0$*

The spin of a proton generates a magnetic moment. If the spin of the neighboring proton (H_b in the preceding formula) is parallel, its magnetic moment *adds* to the applied magnetic field. Consequently, the first proton (H_a in the preceding formula) sees a slightly stronger field and comes into resonance at a slightly lower applied field strength. If the neighboring proton is in the antiparallel state, its magnetic moment *decreases* the magnetic field around the first proton. It takes slightly more applied H_0 for H_a to come into resonance. Because about half the H_b nuclei are always parallel and half are always antiparallel, there are two types of H_a in the sample: those with a neighboring H_b in a parallel spin state and those with a neighboring H_b in an antiparallel spin state. Consequently, we observe two peaks for H_a.

For many compounds, we can predict the number of spin−spin splitting peaks in the NMR absorption of a particular proton (or a group of equivalent protons) by counting *the number (n) of neighboring protons nonequivalent to the proton in question and adding 1*. This is called the **n + 1 rule.**

These three equivalent protons see *two* neighboring, nonequivalent protons. Their NMR band is split into 2 + 1, or 3, peaks.	These two equivalent protons see *three* neighboring, nonequivalent protons. Their NMR band is split into 3 + 1, or 4, peaks.

$$CH_3CH_2Cl$$

Protons that have the same chemical shift do not split each other's signals. Only neighboring protons that have different chemical shifts cause splitting. Some examples follow:

no neighboring H: not split

one neighboring, nonequivalent H: H_b split into two

one neighboring, nonequivalent H: H_a split into two

twelve equivalent H's: no splitting

six equivalent H's: no splitting

In some cases, the magnetic environments of chemically nonequivalent protons are so similar that the protons exhibit identical chemical shifts. In this case, no splitting is observed. For example, toluene has four groups of chemically nonequivalent protons, yet the NMR spectrum shows only two absorption peaks (one for the CH_3 protons and one for the ring protons).

one peak

one peak

In summary, the **chemical shift** for a proton depends on its molecular field. The **area** under an absorption band is determined by the number of equivalent protons giving that signal. **Spin–spin splitting** of a signal is determined by the number of neighboring, nonequivalent protons.

SAMPLE PROBLEM

Predict the spin–spin splitting patterns for the protons in 2-chloropropane.

Solution

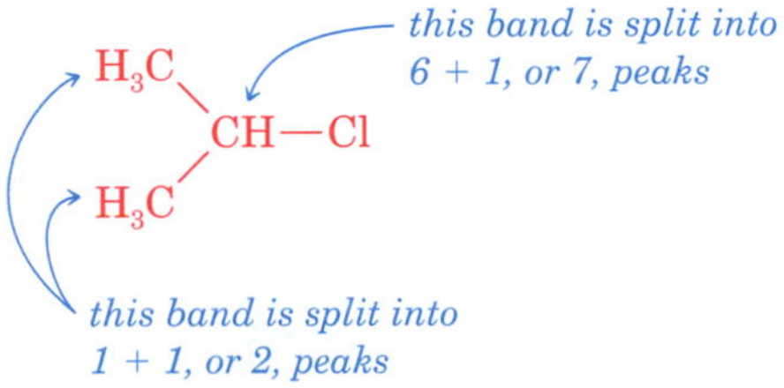

H_3C, H_3C CH—Cl

this band is split into 6 + 1, or 7, peaks

this band is split into 1 + 1, or 2, peaks

STUDY PROBLEM

9.9 Using the $n + 1$ rule, predict the number of NMR peaks for each of the indicated protons:

(a) $CH_3C\underline{H}_2CH_3$ (b) $(CH_3)_3C\underline{H}$ (c) $(C\underline{H}_3)_3CH$ (d) 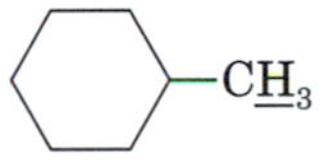 ⬡—$C\underline{H}_3$

Section 9.10
Splitting Patterns

A. The Singlet

A proton with no neighboring, nonequivalent protons shows a single peak, called a **singlet,** in the NMR spectrum. Figure 9.33, the NMR spectrum of *p*-methoxybenzaldehyde, shows two singlets: one for the CH_3O— protons and one for the —CHO proton.

STUDY PROBLEMS

9.10 Which of the following compounds would show at least one singlet in the NMR spectrum?

(a) CH_3CH_3 (b) $(CH_3)_2CHCH(CH_3)_2$ (c) $(CH_3)_3CCl$

(d) Cl_3CCHCl_2 (e) $ClSi(CH_3)_3$

9.11 How many peaks would you expect to observe in the NMR spectra of (a) cyclohexane, and (b) benzene?

B. The Doublet

A proton with one neighboring, nonequivalent proton gives a signal that is split into a double peak, or **doublet.** In the following example, a pair of doublets is produced, one for each proton.

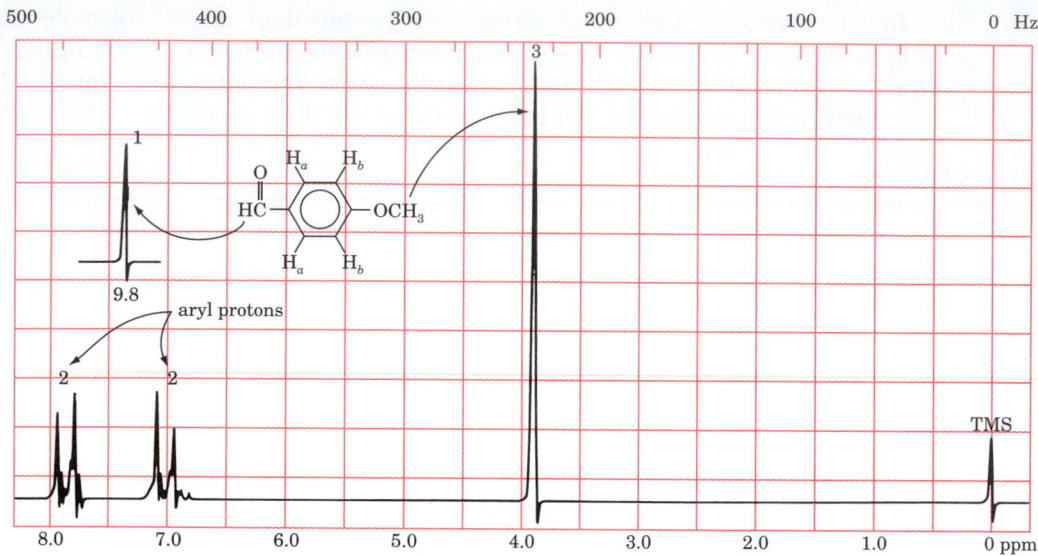

Figure 9.33 ^{1}H NMR spectrum of *p*-methoxybenzaldehyde. (Note the offset signal for the aldehyde proton; see Section 9.7B.)

H_a gives a doublet — *H_b also gives a doublet*

In the NMR spectrum of the preceding hypothetical structure, the δ value for each proton is the value at the *center* of the doublet. (See Figure 9.34.) The relative areas under the entire doublets in this case are 1:1, reflecting the fact that each doublet arises from the absorption by one proton. (The two peaks *within* any doublet also have an area ratio that is ideally 1:1, but may be slightly different, as may be seen for the aryl protons in Figure 9.33.)

The separation between the two peaks of a doublet is called the **coupling constant J,** and varies with the environment of the protons and their geometric relationship to each other. (Because splitting patterns are caused by

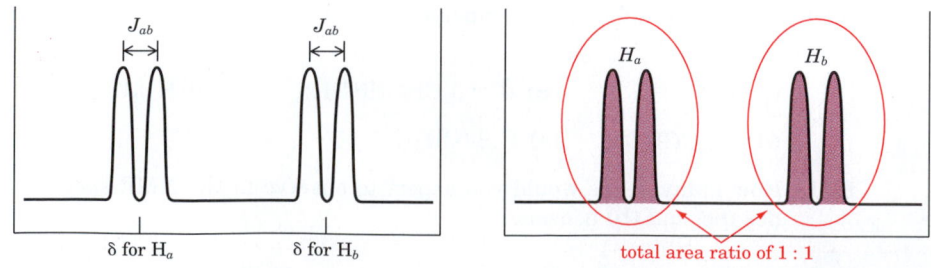

Figure 9.34 Spin–spin splitting pattern for $-\overset{\overset{\displaystyle H_a}{|}}{C}-\overset{\overset{\displaystyle H_b}{|}}{C}-$.

internal forces, coupling constants are independent of the strength of H_0.)
The symbol J_{ab} means the coupling constant for H_a split by H_b or for H_b split
by H_a. For any pair of coupled protons, the J value is the same in each of the
two doublets. By convention, J values are reported in Hz. Therefore, the Hz
scale at the top of an NMR spectrum is used to determine coupling constants.
For a pair of neighboring, nonequivalent protons attached to freely rotating
carbons, the J value is about 7 Hz.

Figure 9.33, the NMR spectrum of p-methoxybenzaldehyde, shows a pair of
doublets in the aromatic region. The two protons labeled H_a are equivalent,
as are the two H_b protons. H_a and H_b are nonequivalent to each other and are
neighboring. Their signals are split into a pair of doublets. Note that the
peaks within the doublets are not perfectly symmetrical. The inside peaks of
a doublet are taller. This phenomenon is called **leaning.** Leaning occurs
whenever coupled protons have similar chemical shifts. As the chemical shift
difference becomes smaller, leaning becomes more pronounced. This leaning
of multiplets can be used to tell where to look in the spectrum for the proton
that is causing the splitting. The multiplet always leans toward the proton to
which it is coupled. If the multiplet leans upfield, the proton to which it is
coupled resides upfield. If the multiplet leans downfield, the proton to which
it is coupled resides downfield.

STUDY PROBLEM

9.12   Assuming that the NMR spectrum of each of the following partial structures
would exhibit two groups of peaks, tell how many peaks would be in each group.
(Use the $n + 1$ rule.)

(a) $-\text{CH}-\text{CH}-$ (b) $-\text{CH}_2-\text{CH}-$ (c) $\text{CH}_3-\text{CH}-$

SAMPLE PROBLEMS

Which of the following compounds would show a doublet (as well as other signals) in its NMR spectrum?

(a) CH_3CHCl_2 (b) $\text{CH}_3\text{CH}_2\text{CH}_2\text{Cl}$ (c) $\text{CH}_3\text{CHClCH}_3$

Solution
Underlined protons would appear as doublets: (a) $\underline{\text{CH}_3}\text{CHCl}_2$: (b) no doublet:
(c) $\underline{\text{CH}_3}\text{CHCl}\underline{\text{CH}}_3$ as one doublet and not as a pair of doublets. (Why?)

Give the total relative areas under all of the absorption bands for the compounds in the preceding problem.

Solution
(a) 3:1   (b) 3:2:2   (c) 6:1

Tell how many singlets and how many doublets would appear in the NMR
spectrum of each of the following substituted benzenes:

(a) $\text{CH}_3-\!\!\left\langle\bigcirc\right\rangle\!\!-\text{OCH}_3$ (b) $\text{Cl}-\!\!\left\langle\bigcirc\right\rangle\!\!-\text{NH}_2$

Solution

(a) two singlets (CH_3 and OCH_3) and two doublets for the aryl protons; **(b)** one singlet (NH_2) and two doublets. (A pair of doublets in the aryl region is typical of 1,4-disubstituted benzenes.)

STUDY PROBLEM

9.13 The NMR spectrum of an aryl ketone ($Ar\overset{O}{\overset{\|}{C}}R$), C_8H_7ClO, is shown in Figure 9.35. What is the structure of this compound?

C. The Triplet

If a proton (H_a) sees two neighboring protons equivalent to each other, but not equivalent to itself, the NMR signal of H_a is a **triplet** ($2 + 1 = 3$). If the two protons labeled H_b are equivalent, they give a signal that is split into a doublet by H_a.

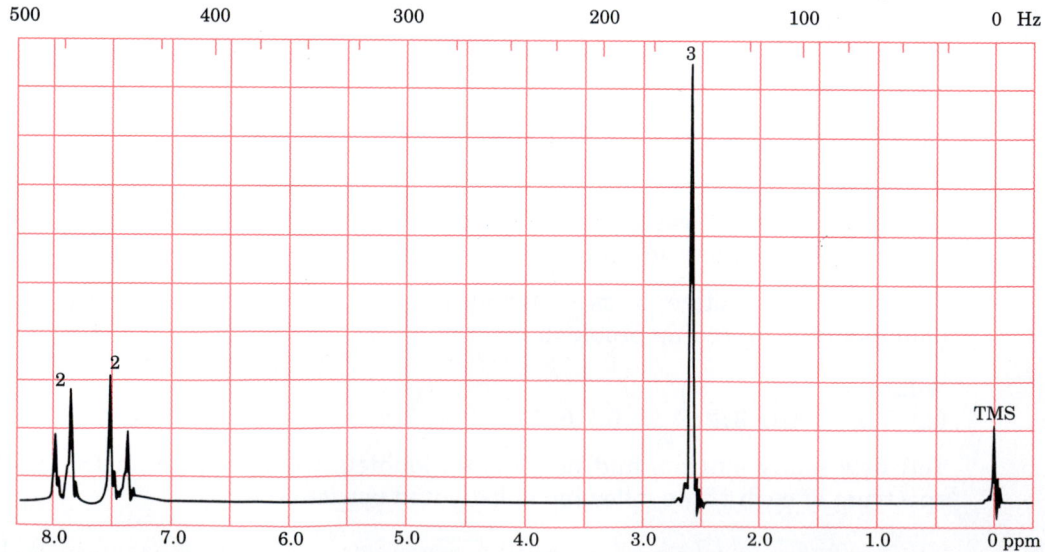

H_a *sees two neighboring H's, signal is a triplet*

H_b *sees one neighboring H, signal is a doublet*

The NMR absorption pattern for all three protons in the preceding partial structure consists of a doublet and a triplet. The peaks in the triplet each are separated by the *same J value as that for the doublet.* The total width of the triplet (from side peak to side peak) is therefore 2*J* (see Figure 9.36). The areas under the entire triplet and the entire doublet in our example are in

Figure 9.35 ^1H NMR spectrum for unknown aryl ketone in Problem 9.13.

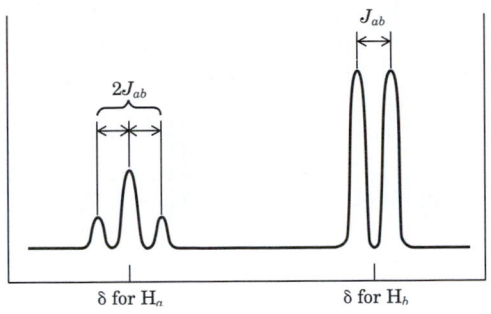

Figure 9.36 Spin–spin splitting pattern for

$$\begin{array}{cc} H_a & H_b \\ | & | \\ -C-C & -H_b. \\ | & | \end{array}$$

the ratio of 1 for H_a to 2 for H_b. However, the relative areas within the triplet are in a ratio of $1:2:1$. The spins of both of the neighboring H_b protons will affect the magnetic field felt by the H_a proton. Let us label the neighboring protons as H_b and $H_{b'}$ so that we may distinguish them. What are the possible combinations of spins (parallel or antiparallel to H_0) for H_b and $H_{b'}$ that H_a will see? At one extreme, both may be parallel to the applied field and increase the field felt by H_a. At the other extreme, both may be antiparallel to the applied field and decrease the field felt by H_a. There are two combinations with one parallel and one antiparallel that have no net effect upon the field felt by H_a. Because the parallel–antiparallel combinations are twice as probable to occur as either extreme combination, we obtain a ratio of $1:2:1$ for the three peaks in the triplet. (Another way to arrive at the $1:2:1$ ratio for the triplet peaks will be discussed in Section 9.11.)

The NMR spectrum of 1,1,2-trichloroethane, which has a doublet and a triplet, is shown in Figure 9.37.

more shielded: upfield
sees one H: doublet
relative total area: 2

less shielded: downfield
sees two H's: triplet
relative total area: 1

9.14 Suggest a reason for the fact that the proton in the Cl_2CH group in 1,1,2-trichloroethane is less shielded than the protons in the CH_2Cl group.

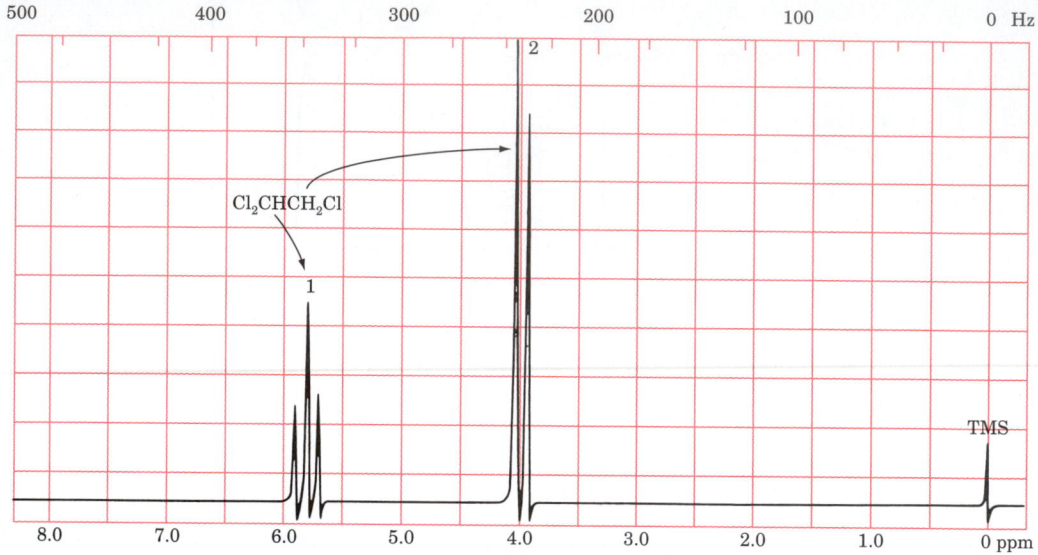

Figure 9.37 ^{1}H NMR spectrum of 1,1,2-trichloroethane.

SAMPLE PROBLEM

Which of the following compounds shows a triplet (among other signals) in the NMR? How many triplets will each compound show?

(a)
$$\begin{array}{cc} Cl & OCH_3 \\ | & | \\ CH_2 & CH_2 \end{array}$$

(b) CH_3CH_2— ⬡

(c) $Cl_2CHCH_2CHCl_2$

Solution

(a) two triplets, one for each CH_2; (b) one triplet for CH_3; (c) two triplets. H_a split by two H_b's and H_b split by two H_a's.

$$\begin{array}{ccc} H_a & H_b & H_a \\ | & | & | \\ Cl_2C & —C— & CCl_2 \\ & | & \\ & H_b & \end{array}$$

STUDY PROBLEM

9.15   Figure 9.38 shows the NMR spectrum of 2-phenylethyl acetate. Assign each signal to the proper protons.

D. The Quartet

Consider a compound with a methyl group and one nonequivalent proton on an adjacent carbon.

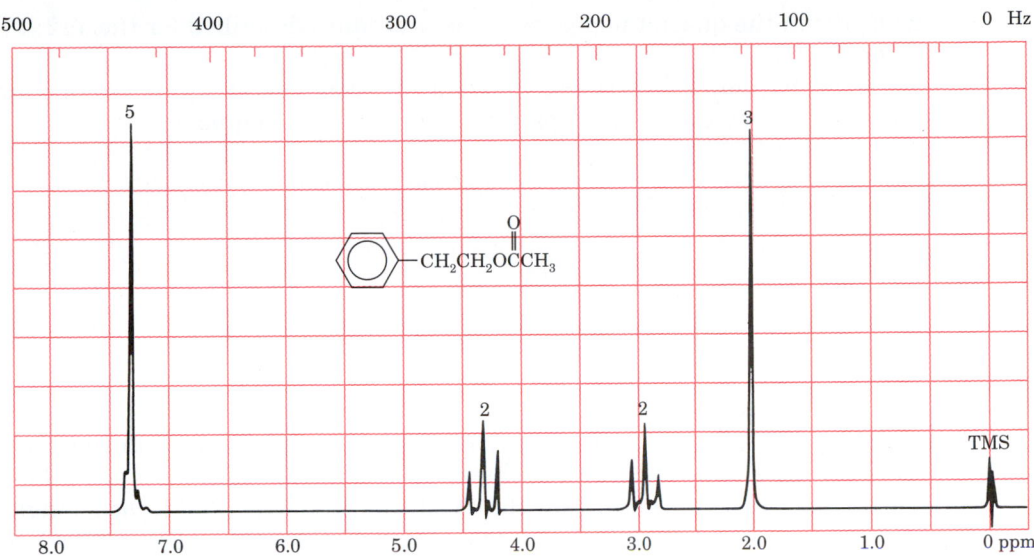

Figure 9.38 ^{1}H NMR spectrum of 2-phenylethyl acetate.

quartet for H$_a$ H$_a$ H$_b$ *doublet for CH$_3$(H$_b$)*

—C—C—H$_b$

H$_b$

The three equivalent methyl protons (H$_b$) see one neighboring proton and appear in the spectrum as a doublet, the total relative area of which is 3 (for three protons).

The signal arising from H$_a$ is observed as a **quartet** (3 + 1) because it sees three neighboring protons. The J values between each pair of peaks in the quartet is the same as the J value between the peaks in the doublet. In our example, the total area under the quartet for H$_a$ is 1. (The area ratio *within* a quartet is 1:3:3:1; see Figure 9.39. See if you can derive the 1:3:3:1 ratio of

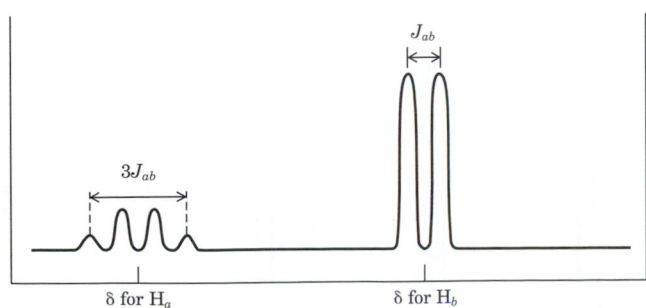

Figure 9.39 Spin–spin splitting pattern for —C—C—H$_b$.

the peaks in the quartet using the same procedure described for the 1:2:1 ratio for the triplet.)

The ethyl group (CH_3CH_2—), which is very common in organic compounds, exhibits a characteristic NMR pattern—a triplet and a quartet.

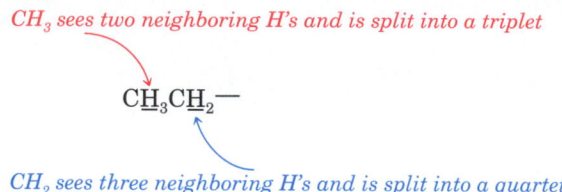

The chemical shifts of an ethyl group are also characteristic. The CH_2 is often bonded to an electronegative atom, such as oxygen, which deshields the CH_2 protons. The quartet is thus observed downfield, while the triplet for the more shielded CH_3 group is observed upfield. The NMR spectrum of chloroethane (Figure 9.32) shows typical ethyl absorption. Figure 9.40 contains another example of an NMR spectrum that shows an ethyl pattern: an upfield triplet and a downfield quartet.

STUDY PROBLEM

9.16   Match each NMR spectrum in Figure 9.41 with a compound in the following list:

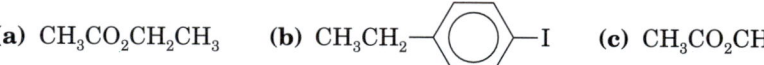

(a) $CH_3CO_2CH_2CH_3$ (b) CH_3CH_2—⟨○⟩—I   (c) $CH_3CO_2CH(CH_3)_2$

(d) $CH_3CH_2CH_2NO_2$ (e) CH_3CH_2I (f) $(CH_3)_2CHNO_2$

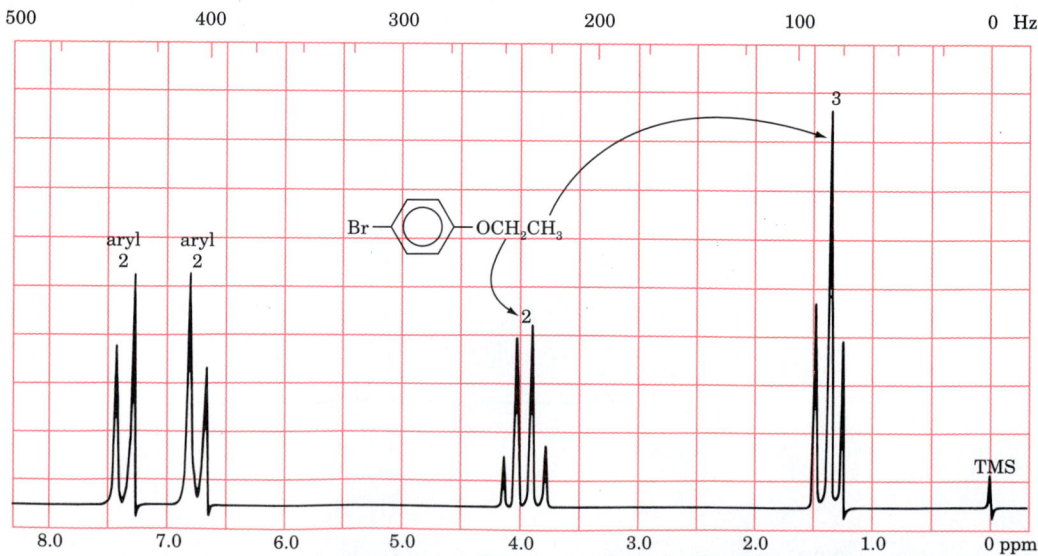

Figure 9.40 ^1H NMR spectrum showing a typical ethyl pattern.

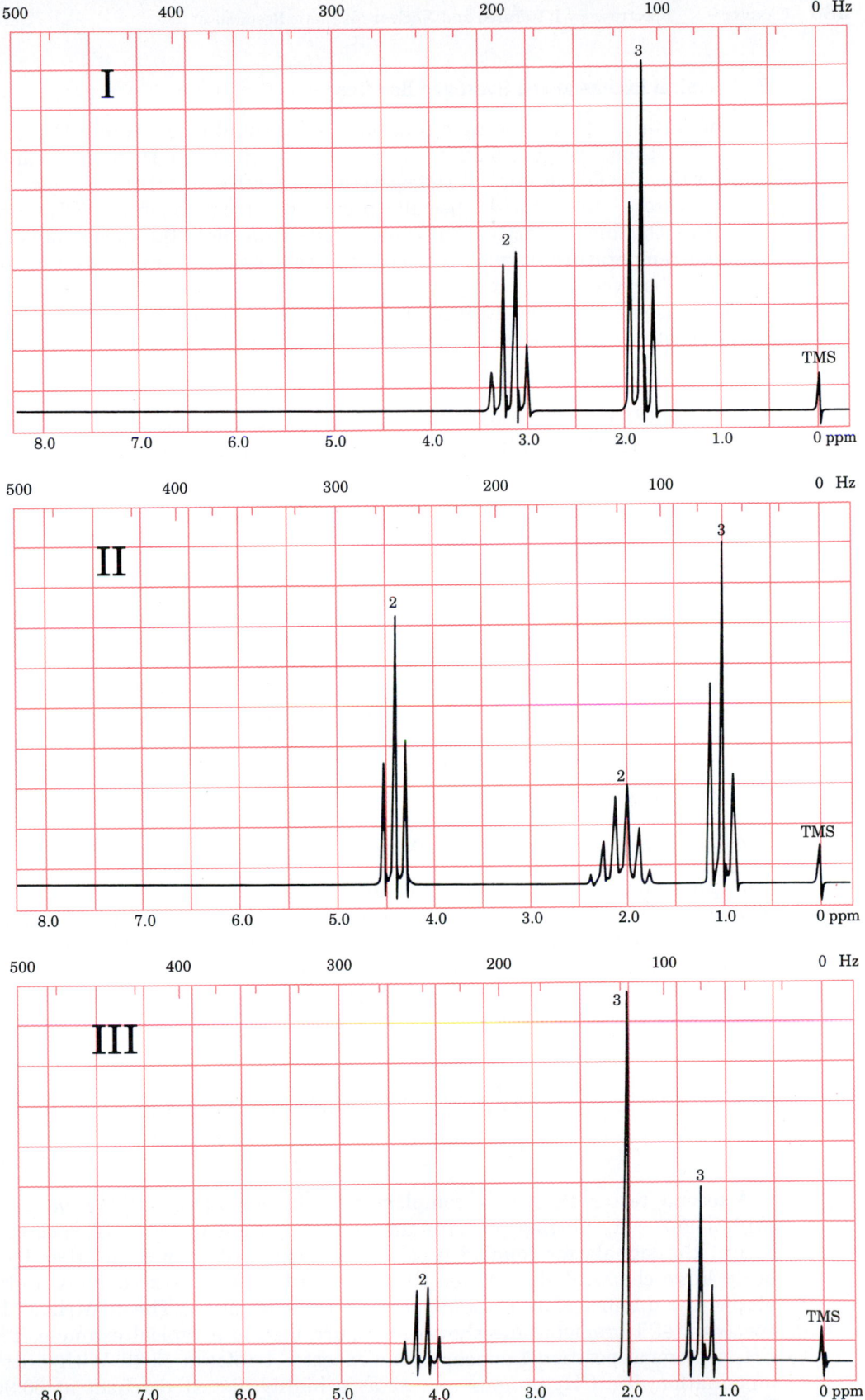

Figure 9.41 ¹H NMR spectra for Problem 9.16.

E. Chemical Exchange and Hydrogen Bonding

On the basis of the preceding discussions, one would expect the NMR spectrum of methanol (Figure 9.22) to show a doublet (for the CH_3 protons) and a quartet (for the OH proton). If the spectrum of methanol is run at a very low temperature ($-40°$) or with specially prepared solvents such as CCl_4 stored over Na_2CO_3, this is exactly what is observed. However, if the spectrum is run at room temperature or with ordinary spectroscopic solvents, only two singlets are observed.

The reason for this behavior of methanol is that alcohol molecules undergo rapid reaction with one another at room temperature in the presence of a trace of acid, exchanging OH protons in a process called **chemical exchange.** This exchange is so rapid that neighboring protons cannot distinguish any differences in spin states and thus see only the average value of zero. Amines (RNH_2 and R_2NH) also undergo chemical exchange.

$$CH_3\ddot{O}: + H'^{+} \rightleftharpoons CH_3\overset{+}{\ddot{O}}H' \rightleftharpoons CH_3\ddot{O}H' + H^{+}$$
$$\quad\; |\qquad\qquad\qquad\qquad |$$
$$\quad\; H\qquad\qquad\qquad\qquad H$$

Of more practical importance to most organic chemists is the fact that the chemical shifts of OH and NH protons are *solvent-, temperature-,* and *concentration-dependent* because of hydrogen bonding. In a non-hydrogen-bonding solvent (such as CCl_4) and at low concentrations (1% or less), OH proton absorption is observed at a δ value of about 0.5 ppm. At the more usual, higher concentrations, the absorption is observed in the 4–5.5 ppm region because of hydrogen bonding between CH_3OH molecules. In hydrogen-bonding solvents, the OH proton absorption can be shifted even farther downfield.

F. Other Factors That Affect Splitting Patterns

The spin–spin splitting patterns we have been discussing in this chapter are idealized cases. Most NMR spectra are more complex than those we have been using as examples. The complexity arises from a number of factors, of which we will mention only three. One factor that adds complexity to an NMR spectrum is the *nonequivalence of coupling constants of neighboring protons:*

$$\begin{array}{c} O \\ \parallel \\ ClCH_2C\;H\;CH \\ | \\ CHCl_2 \end{array}$$

signal is split by three types of neighboring H's with nonequivalent coupling constants

A second factor that adds complexity is the *magnitude of the chemical shift.* The $n + 1$ splitting patterns are truly apparent in an NMR spectrum only if the signals for coupled protons are separated from each other by a fairly large chemical shift. When chemical shifts are close, a complex multiplet with a number of overlapping peaks is often observed. The 60 MHz NMR spectrum of 1-bromobutane shown in Figure 9.42 is a typical example. The CH_2 protons at position 1 appear as the expected triplet at $\delta 3.40$. With a little imagination, the CH_3 protons appear as a highly distorted triplet at $\delta 0.90$, leaning toward the signal of the remaining CH_2 groups. The CH_2 protons at

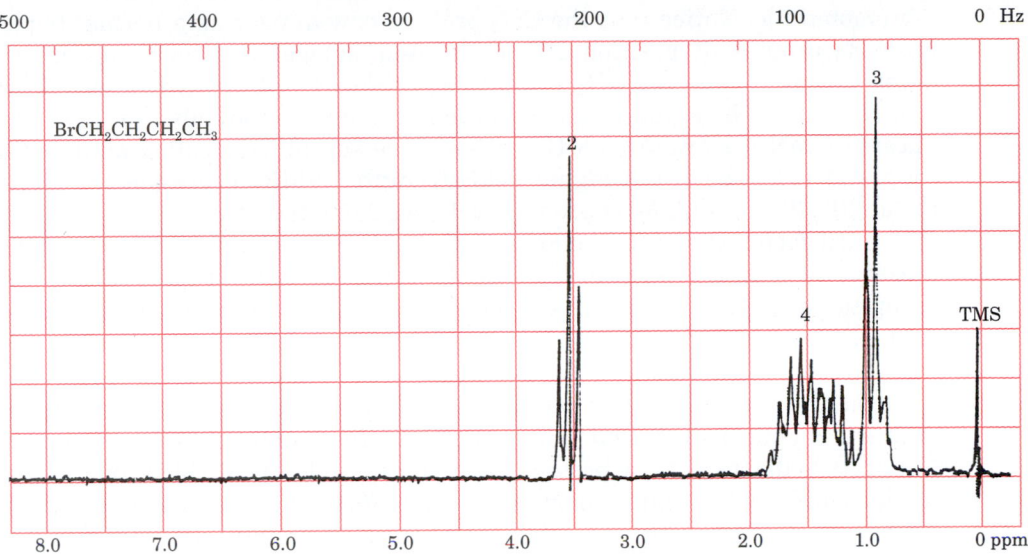

Figure 9.42 60 MHz ^{1}H NMR spectrum of 1-bromobutane.

positions 2 and 3 appear as an indecipherable multiplet between $\delta 1.1$ and $\delta 2.1$ because the difference in chemical shifts is so small that their splitting patterns overlap. This situation is not entirely hopeless. Remember that as the operating field strength increases, the frequency difference in chemical shifts increases (even though the δ value remains the same). Thus if the spectrum is obtained at a high field strength, the resonance frequencies for the protons on the two CH_2 groups become far enough apart to give separate, distinct $n + 1$ multiplets. Figure 9.43 shows the 300 MHz NMR spectrum of

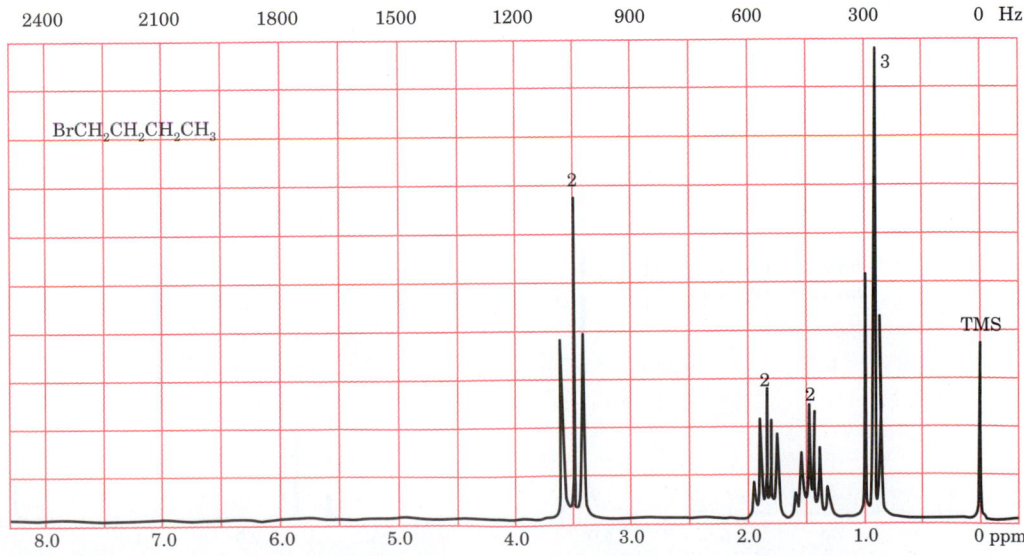

Figure 9.43 300 MHz ^{1}H NMR spectrum of 1-bromobutane.

1-bromobutane. Notice that the CH_3 protons now appear as a normal triplet, the CH_2 protons at position 2 (split by four adjacent protons) appear as a quintet at $\delta 1.77$, and the CH_2 protons at position 3 (split by five adjacent protons) appear as a sextet at $\delta 1.45$. Similarly, many compounds that give complicated NMR spectra at 60 MHz will produce simplified (readily analyzed by the $n + 1$ rule) spectra at higher field strengths. NMR spectrometers operating at 200, 300, or 400 MHz are routinely available today.

A third factor that adds complexity is the *presence of a chiral atom*. The presence of a chiral atom in a molecule will cause the protons of any CH_2 group to be diastereotopic. (Diastereotopic protons were discussed in Section 4.10.) Recall that diastereotopic protons are chemically nonequivalent and will have different chemical shifts. Figure 9.44 shows the NMR spectrum of the disodium salt of (R)-citramalic acid. The diastereotopic CH_2 protons have different chemical shifts and split each other into doublets at $\delta 2.45$ and $\delta 2.76$. Not only do diastereotopic protons have different chemical shifts, but they often have different coupling constants with neighboring protons which leads to further complexities in the NMR spectra. Occasionally, the difference in chemical shifts of diastereotopic protons is so small that they have practically the same chemical shift and do not split each other. The main point of all this is to watch for the presence of diastereotopic protons and to remember that when they are present, the NMR spectrum may be more complex than expected from a first glance at the chemical formula.

Many organic molecules have all three of these factors operating simultaneously. (We will encounter an example in Problem 9.41.) Thus, you can imagine the complexities in the splitting patterns that can arise. These are situations in which high-field NMR spectrometers are very beneficial. They often simplify spectra such that they are easily analyzed by application of the $n + 1$ rule.

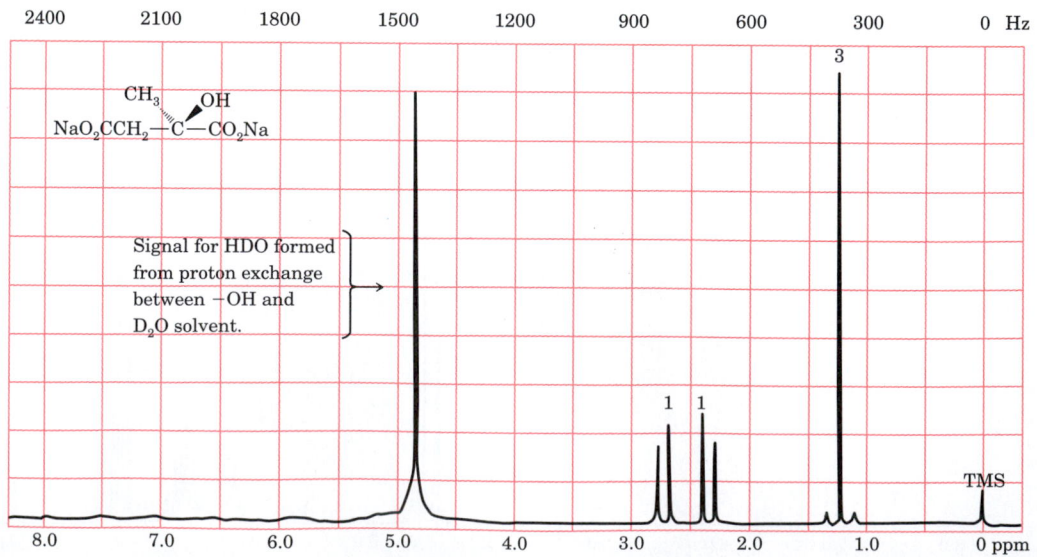

Figure 9.44 300 MHz ^{1}H NMR spectrum of (R)-citramalic acid, disodium salt.

Section 9.11

Spin–Spin Splitting Diagrams

A **spin–spin splitting diagram,** also called a **tree diagram,** is a convenient technique for the analysis of splitting patterns. Let us consider H_a in the simple partial structure $\text{>CH}_a\text{—CH}_b\text{<}$. The splitting of the signal for H_a into a doublet by H_b can be symbolized by the following tree diagram, which we "read" starting at the top and moving toward the bottom.

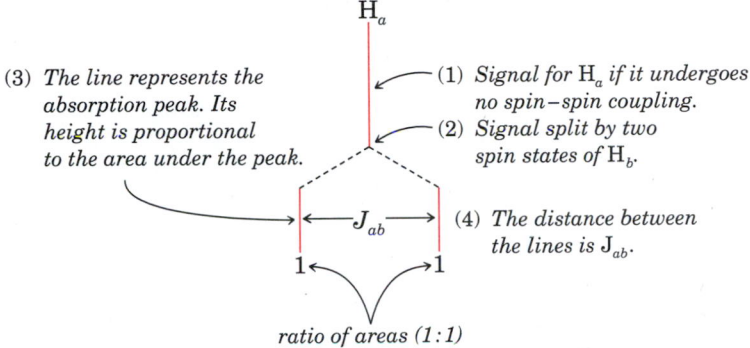

(3) The line represents the absorption peak. Its height is proportional to the area under the peak.

(1) Signal for H_a if it undergoes no spin–spin coupling.

(2) Signal split by two spin states of H_b.

(4) The distance between the lines is J_{ab}.

ratio of areas (1:1)

The splitting of the signal for H_b by the spin states of H_a may be represented by a similar tree diagram. These two diagrams can then be superimposed on an NMR spectrum (see Figure 9.45).

The tree diagram describing a triplet is a direct extension of that for a doublet. Consider the absorption pattern for H_a in the following grouping:

$$-\underset{|}{\overset{|}{C}}_{H_a}-\underset{|}{\overset{|}{C}}_{H_b}-H_b$$

In this case, H_a sees two neighboring protons and is split into a triplet. The coupling constant is J_{ab}. The triplet comes about because the peak for H_a is

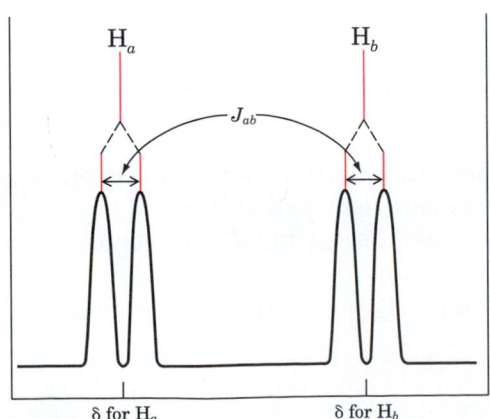

Figure 9.45 Spin–spin splitting pattern of two nonequivalent neighboring H's.

split twice, once for each H_b. Using a tree diagram, we can see the result of each of the two splits. H_a is first split into a pair of doublets, and then the resulting two peaks are split again. We observe a triplet because the center two peaks absorb at the same position in the NMR spectrum. Consequently, the area of the center peak of the triplet is twice that of the two outside peaks.

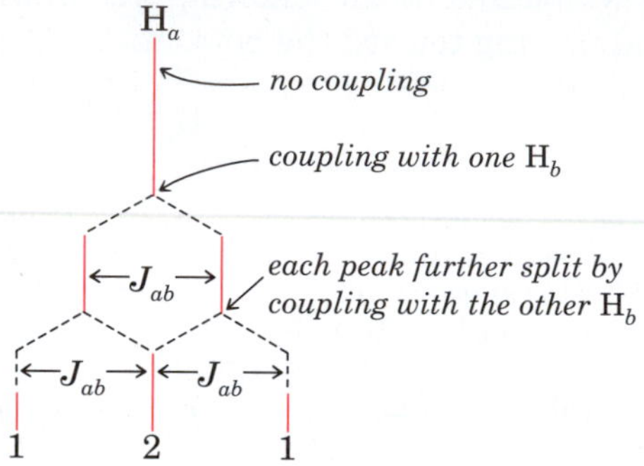

H_a

no coupling

coupling with one H_b

$\leftarrow J_{ab} \rightarrow$ *each peak further split by coupling with the other H_b*

$\leftarrow J_{ab} \rightarrow \leftarrow J_{ab} \rightarrow$

1   2   1

SAMPLE PROBLEMS

Draw a spin–spin splitting diagram for H_m of the following system, where $J_{am} = 10$ Hz and $J_{mx} = 5$ Hz:

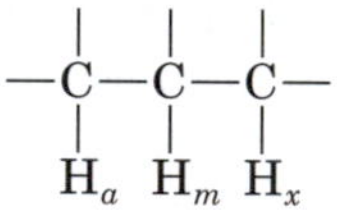

$$-\overset{|}{C}-\overset{|}{C}-\overset{|}{C}-$$
$$\quad H_a \ H_m \ H_x$$

Solution

H_m

$\leftarrow J_{am} \rightarrow$

$\leftarrow J_{mx} \rightarrow$ $\leftarrow J_{mx} \rightarrow$

1   1   1   1

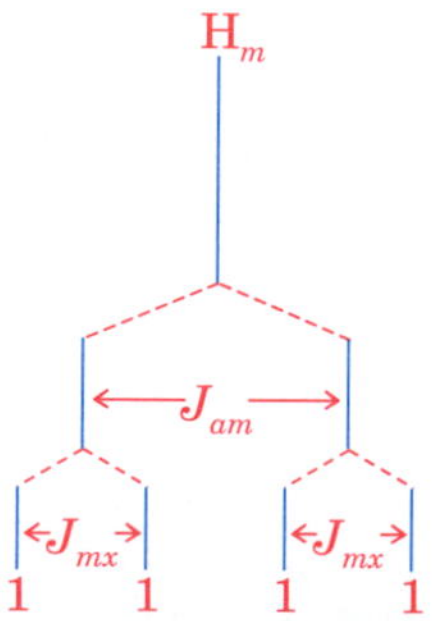

Here the absorption pattern is not a triplet, as would be predicted by the $n + 1$ rule. Instead, four distinct lines of equal heights are observed because the coupling constants J_{am} and J_{mx} are not equal.

Draw a splitting diagram for the absorption of H_a in the partial structure.

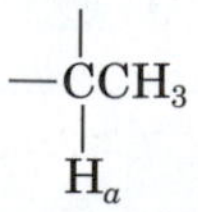

$$-\overset{|}{C}CH_3$$
$$\quad H_a$$

Solution

H_a

all J values $= J_{ax}$

1   3   3   1

The original signal is split three times. The $1:3:3:1$ ratio of the areas arises from the fact that all the protons have the same coupling constant and, consequently, the protons have superimposed absorption positions.

STUDY PROBLEMS

9.17   Construct a tree diagram for each of the indicated protons:

   (a) $ClCH_2\underline{CH}_2CH_2Cl$ **(b)** $Cl_2CH\underline{CH}_3$ **(c)** $CH_3\underline{CH}_2OCH_3$

9.18   Draw a tree diagram for H_x in the following partial structure, where $J_{ax} = 10$ Hz and $J_{bx} = 5$ Hz. (Be careful to relate the height of each line to the relative area of each peak.)

$$-\underset{H_a}{\overset{|}{C}}-\underset{H_x}{\overset{|}{C}}-\underset{H_b}{\overset{|}{C}}-H_b$$

Section 9.12
Carbon-13 NMR Spectroscopy

Proton, or ^1H, NMR spectroscopy provides structural information about the hydrogen atoms in an organic molecule. Carbon-13, or ^{13}C, NMR spectroscopy produces structural information about the *carbons* in an organic molecule.

In ^1H NMR spectroscopy, we deal with the common natural isotope of hydrogen; 99.985% of natural hydrogen atoms are ^1_1H. However, 98.9% of the carbon atoms in nature are $^{12}_6C$, an isotope with nuclei that have no spin. Carbon-13 constitutes only 1.1% of naturally occurring carbon atoms. Also, the transition of a ^{13}C nucleus from parallel to antiparallel is a low-energy transition. Consequently, ^{13}C NMR spectra can be obtained only on very sensitive spectrometers. These spectrometers have become widely available in recent years, and ^{13}C NMR spectroscopy has become an important tool in organic laboratories.

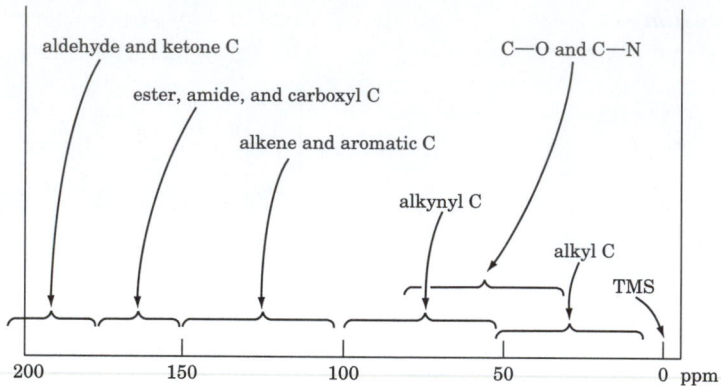

Figure 9.46 Relative positions of ^{13}C NMR absorption.

Continuous-wave NMR spectrometers could be used to obtain ^{13}C NMR spectra, but many hours, or even days, would be required to generate a single ^{13}C NMR spectrum with this technique. FT NMR spectrometers can generate a ^{13}C NMR spectrum in a few seconds. Without the widespread availability of FT NMR spectrometers, ^{13}C NMR spectroscopy would never have become such an important and commonplace technique.

While the low abundance of ^{13}C nuclei complicates instrument design, it also reduces the complexity of ^{13}C spectra compared to ^{1}H spectra. Although adjacent ^{13}C nuclei in a molecule will split each other's signals, the chances of finding adjacent ^{13}C nuclei are very small. For this reason, no $^{13}C-^{13}C$ splitting patterns are observed in ^{13}C spectra.

In a typical ^{13}C spectrum, the areas under the peaks are not proportional to the numbers of equivalent carbon atoms. Consequently, ^{13}C NMR spectra are not integrated.

There are two principal types of ^{13}C spectra: those that show $^{13}C-^{1}H$ spin–spin splitting patterns and those that do not. These two types of spectra are frequently used in conjunction with each other. In both types of spectra, TMS is used as an internal standard, and chemical shifts are measured downfield from the TMS peak. The chemical shifts in ^{13}C NMR are far greater than those observed in ^{1}H NMR. Most protons in ^{1}H NMR spectra show absorption between δ values of $0-10$ ppm downfield from TMS. Only a few, such as aldehyde or carboxyl protons, show peaks outside this range. Carbon-13 absorption is observed over a range of $0-200$ ppm downfield from TMS. This large range of chemical shifts is another factor that simplifies ^{13}C spectra compared with ^{1}H spectra. In ^{13}C spectra we are less likely to observe overlapping absorption.

The relative shifts in ^{13}C NMR spectroscopy are roughly parallel to those in ^{1}H NMR spectroscopy. TMS absorbs upfield, while aldehyde and carboxyl carbons absorb far downfield. Figure 9.46 shows the general locations of absorption by different types of carbon atoms.

A. Proton Decoupling

A ^{13}C **proton-decoupled spectrum** is, as the name implies, a spectrum in which ^{13}C is not coupled with ^{1}H and thus shows no spin–spin splitting. Decoupling is accomplished electronically by the application of a second radiofrequency to the sample. This extra energy causes rapid interconversions between the parallel and antiparallel spin states of the protons. As a result, a ^{13}C nucleus sees only an average of the two spin states of the protons, and its signal is not split.

Because there is no splitting in a proton-decoupled spectrum, the signal for each group of magnetically equivalent carbon atoms appears as a *singlet*. By simply counting the number of peaks in the spectrum, we can determine the number of different types of carbon atoms in a molecule of the sample.

Figure 9.47 shows the proton-decoupled ^{13}C NMR spectrum of octane, a compound containing four types of nonequivalent carbons (carbons 1 and 8, carbons 2 and 7, carbons 3 and 6, carbons 4 and 5). The spectrum shows four singlets. Compare this spectrum with the ^{1}H spectrum of octane in Figure 9.48. In the

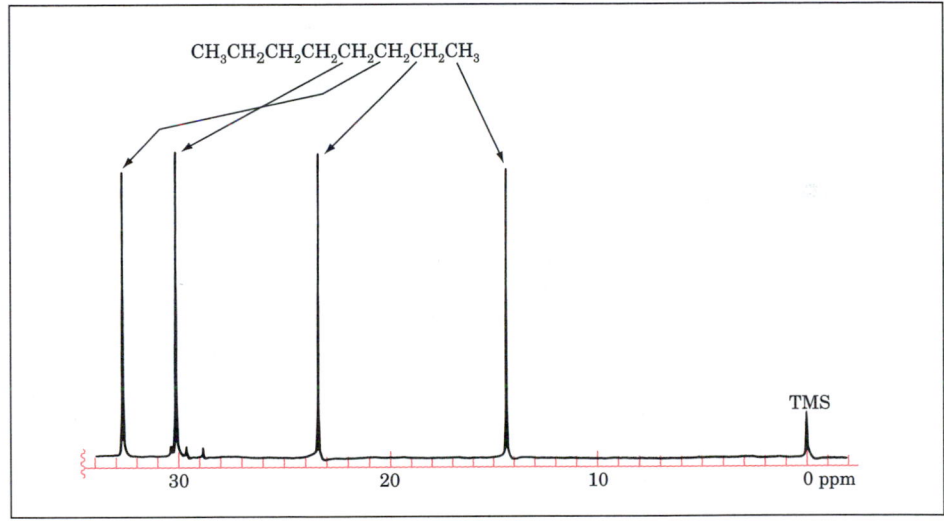

Figure 9.47 The proton-decoupled ^{13}C NMR spectrum of octane.

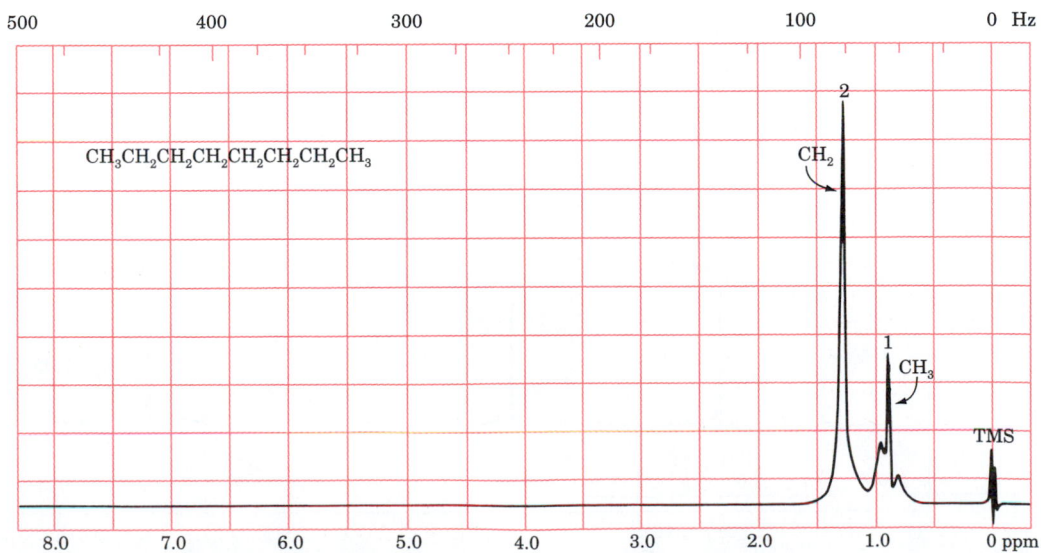

Figure 9.48 ^{1}H NMR spectrum of octane.

^{1}H spectrum, the proton signals are crowded together in the $\delta 0-1.5$ area, and analysis of the peaks is impossible. In the ^{13}C spectrum, the signals are farther apart, in the $\delta 0-35$ range, and are easily distinguished.

B. Off-Resonance Decoupling

The second common type of ^{13}C NMR spectra, called **off-resonance-decoupled spectra,** is the type in which the ^{13}C$-^1$H splitting is not suppressed. In these spectra, the signal for each carbon is split by *protons bonded directly to it.* As in ^{1}H NMR spectroscopy, the $n + 1$ rule is followed. In ^{13}C NMR, n is the number of hydrogen atoms bonded to the carbon in question.

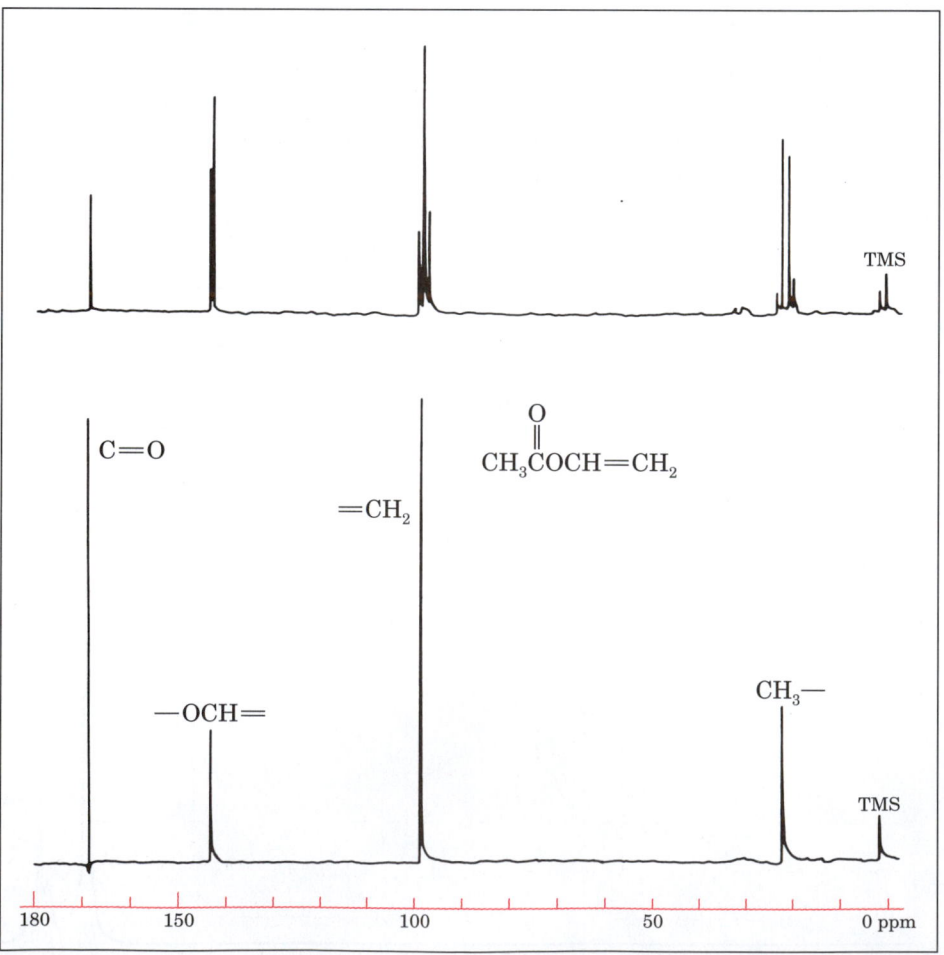

Figure 9.49 The off-resonance-decoupled (top) and the proton-decoupled (bottom) ^{13}C NMR spectra of vinyl acetate.

An impressive amount of information can be deduced about the structure of a compound when the off-resonance-decoupled and the proton-decoupled spectra of the compound are compared. Consider the two spectra of vinyl acetate shown in Figure 9.49. Inspection of the spectra reveals that the carbonyl carbon at $\delta169$ is a singlet in both spectra; therefore, this carbon is not bonded to any hydrogens. The signal for the $CH_3CO_2\underline{C}H=CH_2$ carbon is split into a doublet in the coupled spectrum, indicating one bonded H. Similarly, the $\underline{C}H_2$ signal at $\delta98$ is a triplet, and the $\underline{C}H_3$ signal at $\delta20$ is a quartet.

Newer and faster techniques than off-resonance-decoupled spectra are available for determining the number of protons on each carbon. One of the most useful of these techniques involves what are called **DEPT** spectra (DEPT stands for distortionless enhancement by polarization transfer). The theory behind these newer techniques requires a more in-depth study of NMR spectroscopy than we have time for in this text.

We will use off-resonance-decoupled spectra for the determination of the number of attached protons because they are readily interpreted with the $n + 1$ rule, which has already been learned for interpreting splitting patterns in ^{1}H NMR spectra.

STUDY PROBLEM

9.19   Figure 9.50 shows the ^{13}C NMR spectra of an ester with the molecular formula $C_5H_7O_2Br$. What is the structure of this compound?

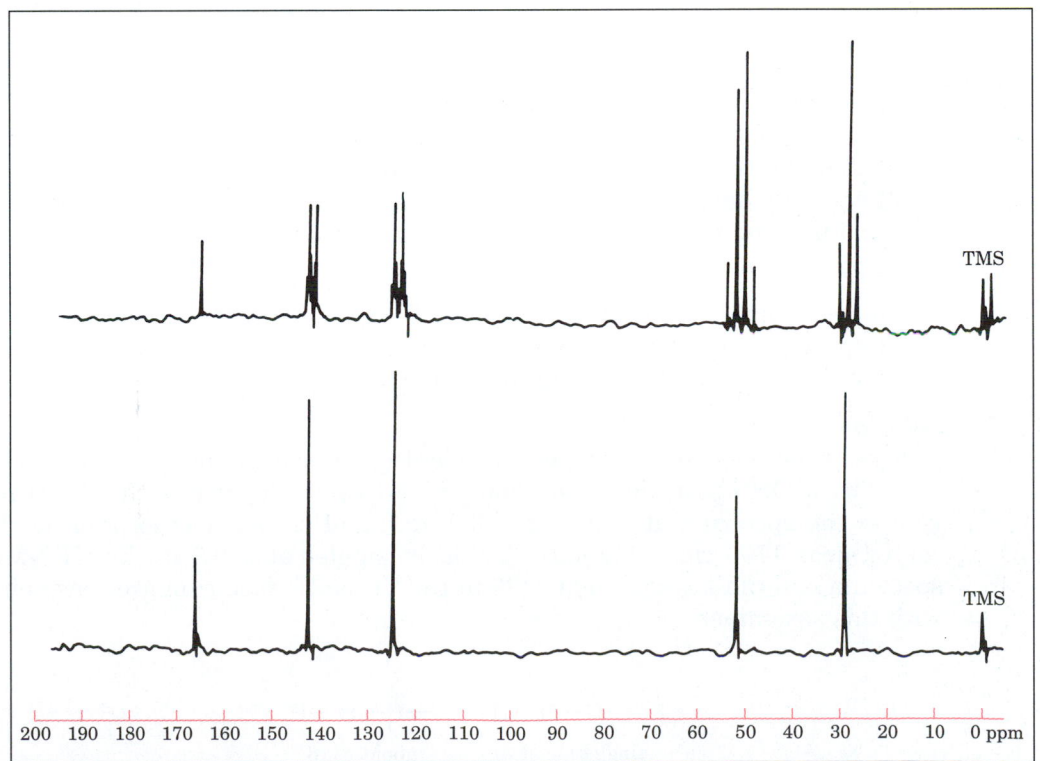

Figure 9.50 ^{13}C NMR spectra for Problem 9.19.

Section 9.13

Using Infrared and NMR Spectra for the Identification of Organic Structures

From an infrared spectrum, we can deduce the identities of functional groups. From a ^{1}H NMR spectrum, we can often deduce the structure of the hydrocarbon portion of a molecule. Sometimes it is possible to deduce the complete structure of a compound from just the infrared and ^{1}H NMR spectra. More commonly, additional information is needed (such as chemical reactivity, elemental analysis, or other spectra). In this text, we provide additional information generally in the form of a molecular formula. From the molecular formula, we can calculate the number of rings or double bonds. (For example, C_6H_{12} is C_nH_{2n}; the structure contains either one double bond or a ring.) We can determine fragments of the structure from the spectra. Then, we try to match the fragments with the molecular formula.

SAMPLE PROBLEMS

A compound has a molecular formula of C_3H_6O. Its infrared and ^{1}H NMR spectra are given in Figure 9.51. What is the structure of this compound?

Solution
From the molecular formula, we know that the compound has one oxygen; therefore, the compound must be an alcohol, an ether, an aldehyde, or a ketone. Because the molecular formula is of the type $C_nH_{2n}O$, we know that the structure contains either one double bond (C=C or C=O) or one ring. To distinguish between the possible functional groups, we use the infrared spectrum. We see strong absorption in the C=O region of 1750 cm^{-1} (5.8 μm). We conclude that the oxygen is not in a hydroxyl group or ether group, but rather is in a carbonyl group (aldehyde or ketone). This means, then, that no ring is present.

In the ^{1}H NMR spectrum, we see no downfield, offset absorption; we conclude that the compound is a ketone instead of an aldehyde. The rest of the NMR spectrum shows only a singlet; therefore, all six hydrogens are equivalent and have no neighboring, nonequivalent hydrogens. The compound must be propanone (acetone): $(CH_3)_2C$=O.

A compound has a molecular formula of $C_8H_8O_2$. Its infrared spectrum, ^{1}H NMR spectrum, and proton-decoupled ^{13}C NMR spectrum are shown in Figure 9.52. What is the structure of the compound?

Solution
Using the infrared spectrum, we can identify the compound as a carboxylic acid. The —OH absorption near 3200 cm^{-1} (3.1 μm) slopes into the C—H region of the spectrum at 3000 cm^{-1} (3.3 μm), and carbonyl absorption is observed near 1700 cm^{-1} (5.9 μm). The offset singlet at δ10.9 in the ^{1}H NMR spectrum and the absorption at δ178 in the ^{13}C NMR spectrum are consistent with this assignment.

$$\underset{\text{singlet about }\delta 11}{-\overset{\overset{\textstyle O}{\|}}{C}-O-H} \qquad \underset{\text{about }\delta 178}{-\overset{\overset{\textstyle O}{\|}}{C}-O-H}$$

The total proton count in the ^{1}H NMR spectrum is eight. This value is the same as the number of hydrogens in the molecular formula. Thus, the ratio of

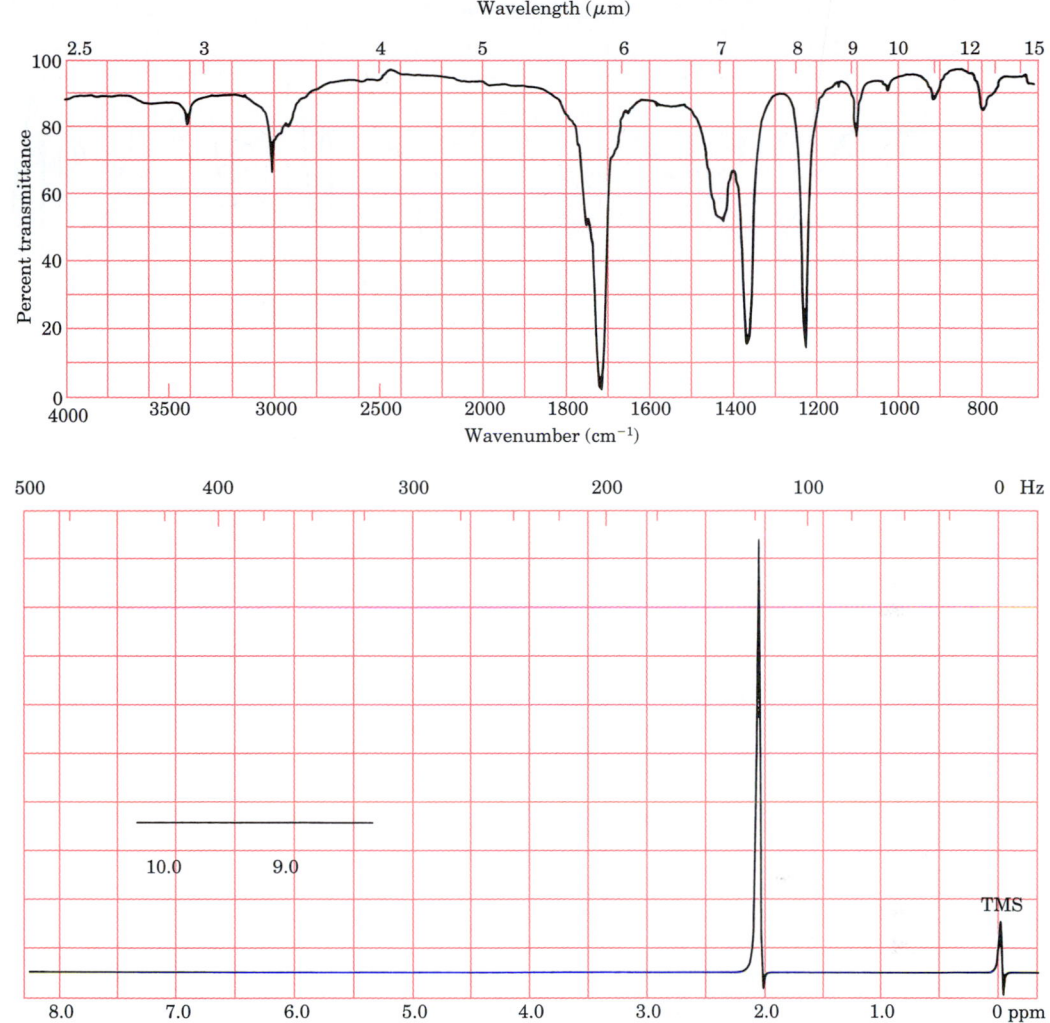

Figure 9.51 Spectra for C_3H_6O in Sample Problem, page 390.

proton signals in the ^{1}H NMR spectrum reflects the actual number of protons in the structure.

The ^{1}H NMR signal at δ7.2 (area 5) suggests a phenyl group (C_6H_5—). This information is not corroborated by the infrared spectrum because the regions where aryl absorption is expected (3100 cm^{-1} and about 1600 cm^{-1}) are masked by wide absorption bands. However, corroborating evidence is found in the ^{13}C NMR spectrum at about δ130. The appearance of four peaks in this region is consistent with phenyl absorption. There are only four absorption peaks for the six carbons of the phenyl group because of symmetry. The four bands are assigned as follows:

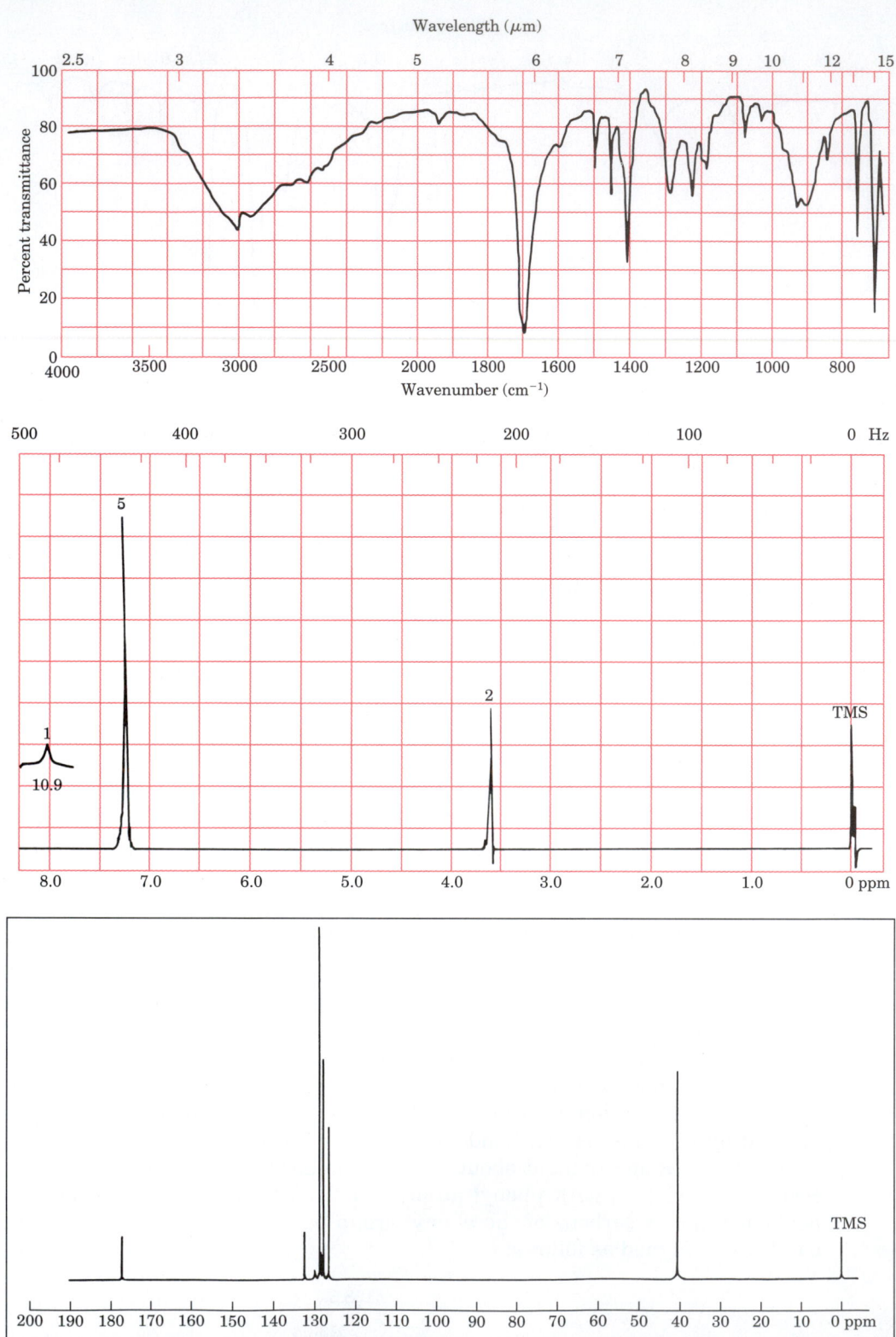

Figure 9.52 Infrared spectrum, ^{1}H NMR spectrum, and proton-decoupled ^{13}C NMR spectrum for $C_8H_8O_2$ in Sample Problem, page 390.

The phenyl group and the carbonyl group account for seven of the eight carbons in the unknown. The remaining carbon must be bonded to both the phenyl group and the carboxyl group. This is evidenced by (1) the benzene ring being monosubstituted (phenyl group) and (2) a two-proton absorption in the ^{1}H NMR spectrum (not three as it would be for $—CH_3$).

Therefore, the structure of the unknown is as follows.

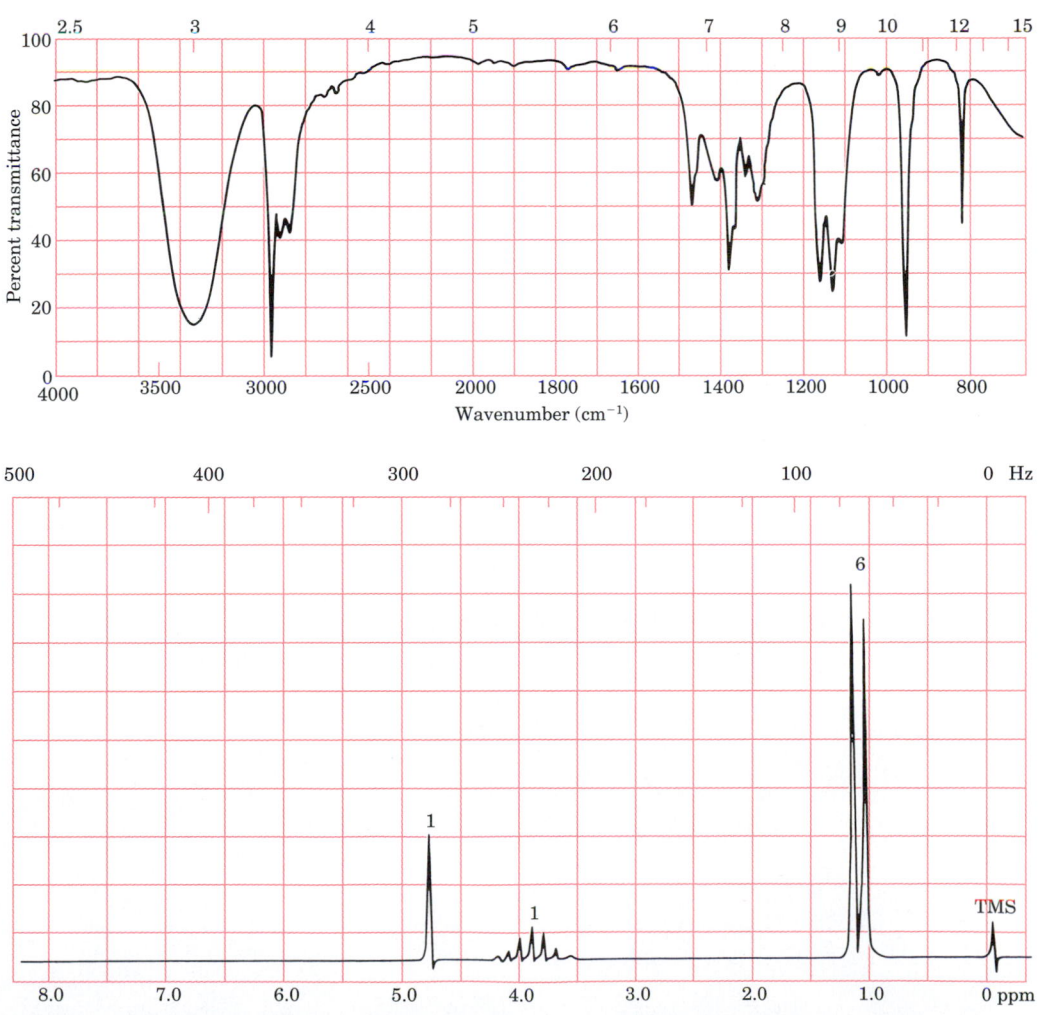

phenylacetic acid

STUDY PROBLEMS

9.20   A compound has the formula C_3H_8O. Its infrared and ^{1}H NMR spectra are shown in Figure 9.53. What is the structure of this compound?

Figure 9.53 Spectra for C_3H_8O in Problem 9.20.

9.21 Figure 9.54 shows the infrared and ^1H NMR spectra for a compound with the molecular formula C_7H_8O. What is the structure of this compound?

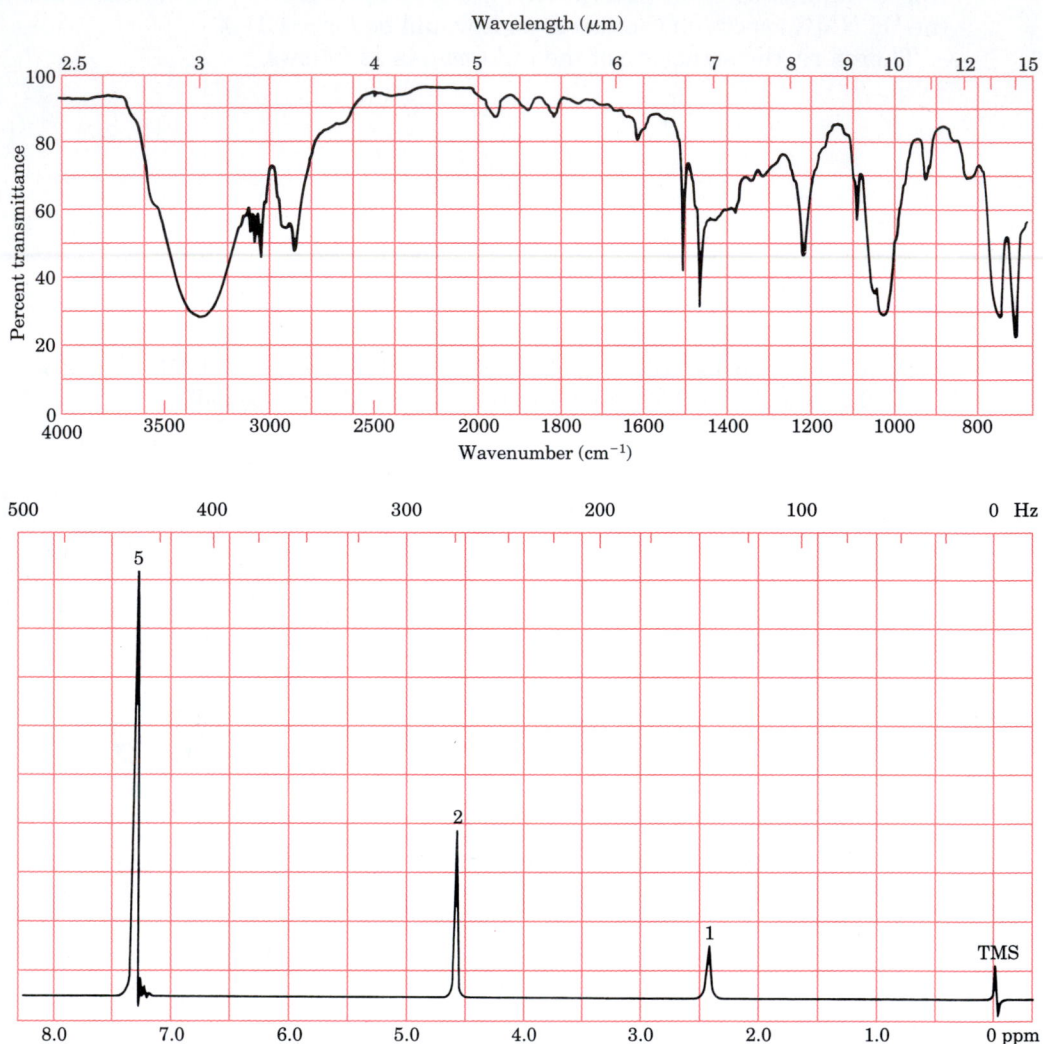

Figure 9.54 Spectra for C_7H_8O in Problem 9.21.

9.22   Figure 9.55 shows ^{1}H NMR and infrared spectra for a compound with the molecular formula $C_4H_7BrO_2$. What is the structure of the compound?

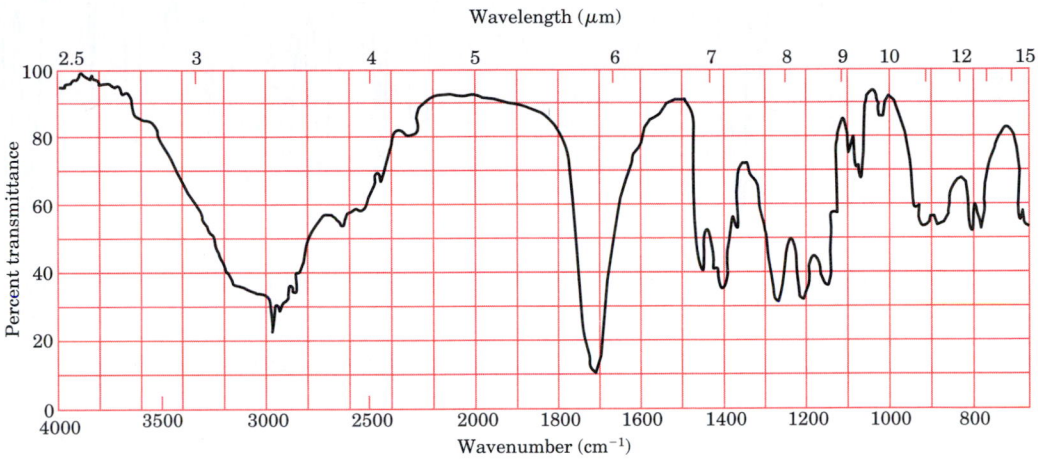

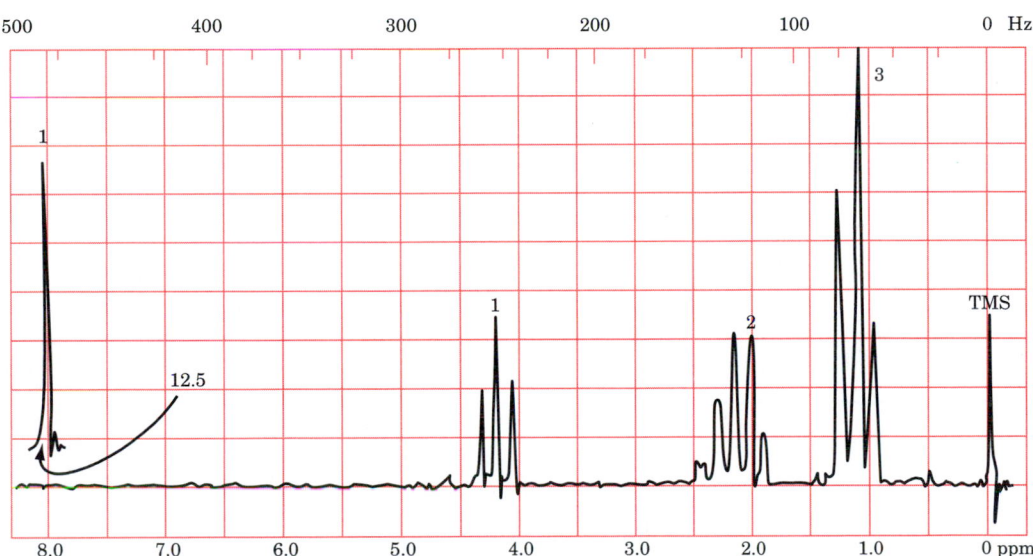

Figure 9.55   Spectra for $C_4H_7BrO_2$ in Problem 9.22.

9.23   Figure 9.56 shows the infrared spectrum, ^{1}H NMR spectrum, and proton-decoupled ^{13}C NMR spectrum for a compound with the molecular formula $C_{10}H_{12}O$. What is the structure of the compound?

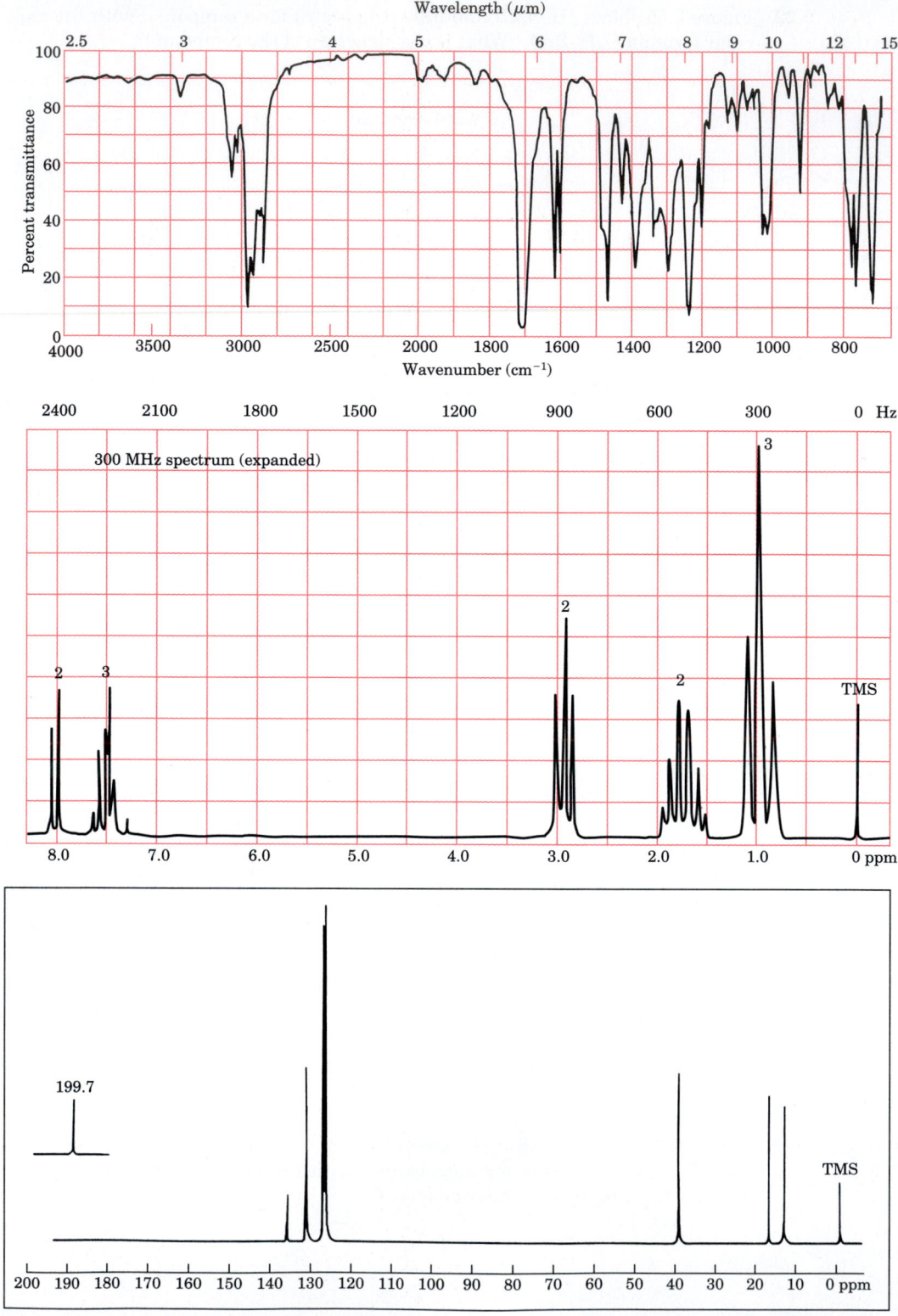

Figure 9.56 Infrared spectrum, ^{1}H NMR spectrum, and proton-decoupled ^{13}C NMR spectrum for $C_{10}H_{12}O$ in Problem 9.23.

Summary

Organic compounds can absorb **electromagnetic radiation** of various wavelengths. Absorption in the **infrared region** results in **vibrational excitations** of bonds. Different types of bonds require differing amounts of energy for vibrational excitations. In an infrared spectrum, the region of $1400-4000$ cm^{-1} ($2.5-7.1$ μm) is useful for determination of functional groups, while the region beyond is the **fingerprint region.**

Nuclear magnetic resonance is the result of certain nuclei in a magnetic field (H_0) absorbing electromagnetic radiation in the radiofrequency region and flipping from the *parallel* to the *antiparallel spin state*. An **induced molecular magnetic field** can *shield* protons (oppose H_0) or *deshield* protons ("augment" H_0) and cause a *chemical shift* (δ) of the absorption band. The induced field is a result of **anisotropic effects** and **inductive effects.** In ^{1}H NMR, a shielded proton absorbs *upfield,* close to the reference TMS, while a deshielded proton absorbs *downfield.*

$$\underset{RCO\underline{H}}{\overset{O}{\overset{\|}{}}} \quad \underset{RC\underline{H}}{\overset{O}{\overset{\|}{}}} \quad \underset{}{\bigcirc}\!-\!H \quad R_2C\!=\!C\underline{H}R \quad \underset{R_2C\underline{H}}{\overset{Cl}{\overset{|}{}}} \quad R\underline{H} \quad TMS$$

increasing shielding →

Spin–spin splitting of an absorption band in ^{1}H NMR results from the spin states of neighboring nonequivalent protons. The signal of a particular proton (or group of equivalent protons) is split into $n + 1$ peaks, where n is the number of neighboring protons equivalent to each other, but nonequivalent to the proton in question.

The distance (in Hz) between any two peaks in a split band is the **coupling constant J.** For protons that are *coupled* (splitting each other's signals), the J values are the same.

The *area* under an entire absorption band is proportional to the relative number of protons giving rise to that signal.

Carbon-13 NMR spectra show the absorption frequencies of ^{13}C atoms instead of protons. Spectra may be **proton-decoupled,** in which case no splitting is observed, or they may be **off-resonance-decoupled,** in which case the ^{13}C signals are split by protons bonded to them.

Essay Problem for Chapter 9

NMR Spectra of Enantiotopic and Diastereotopic Hydrogens

A prochiral carbon is an achiral carbon that can become chiral upon substitution. Groups on a prochiral carbon, such as hydrogen, are either enantiotopic or diastereotopic. Substitution of an enantiotopic hydrogen results in an enantiomer. Substitution of a diastereotopic hydrogen results in a pair of diastereomers.

chiral carbon

$$CH_3CH_2OH \qquad\qquad CH_3CH_2CHOH$$
$$\underset{\textit{prochiral carbon}}{} \qquad\qquad\qquad CH_3$$

Ethanol has no chiral carbons and is, therefore, an achiral compound. However, replacement of either hydrogen bonded to ethanol's prochiral carbon will result in a chiral compound. Replacement of one hydrogen will result in an (R) enantiomer; replacement of the other hydrogen will result in an (S) enantiomer. Therefore, the hydrogens on the prochiral carbon of ethanol are said to be *enantiotopic*.

substitute Y on the prochiral carbon

$$CH_3CH_2OH \longrightarrow \quad \overset{OH}{\underset{CH_3}{H-C-Y}} \quad \overset{OH}{\underset{CH_3}{Y-C-H}}$$

enantiomers

2-Butanol has a chiral carbon and a prochiral carbon. The hydrogens on the prochiral carbon of 2-butanol are *diastereotopic*. Replacement of one hydrogen will yield a pair of diastereomers. Replacement of the other diastereotopic hydrogen will result in another pair of diastereomers, the mirror images of the first pair. The following equation shows the result of the replacement of both diastereotopic hydrogens for one enantiomer.

$$\overset{CH_3}{\underset{CH_3}{\overset{|}{HO-C-H}\atop \overset{|}{H-C-H}}} \longrightarrow \overset{CH_3}{\underset{CH_3}{\overset{|}{HO-C-H}\atop \overset{|}{H-C-Y}}} + \overset{CH_3}{\underset{CH_3}{\overset{|}{HO-C-H}\atop \overset{|}{Y-C-H}}}$$

(R)-2-butanol $\qquad$ $(2R,3R)$-2-butanol $\quad$ $(2R,3S)$-2-butanol

(assume Y to be priority 3)
a pair of diastereomers

Enantiotopic hydrogens are chemically equivalent. In NMR spectra, they have identical chemical shifts. Diastereotopic hydrogens are not chemically equivalent. In NMR spectra, they have different chemical shifts. Therefore, they split each other and appear in the spectrum as multiplets. NMR splitting patterns of the benzylic–methylene hydrogens of *cis-* and *trans-N*-benzyl-2,6-dimethylpiperidine illustrate the differences between enantiotopic and diastereotopic hydrogens (see Figure 9.57).

the benzylic-methylene hydrogens

N-benzyl-2-6-dimethylpiperidine

Due to the low inversion barrier of the nitrogen, the benzyl group in *N*-benzyl-2,6-dimethylpiperidine is not *cis* or *trans* to the methyl groups. Consequently, there are only two geometric isomers of this piperidine. In one geometric isomer, the benzylic–methylene hydrogens are enantiotopic. They appear in the spectrum as a singlet. The hydrogens of the other geometric isomer are diastereotopic. They split each other and appear in the spectrum as a four-line pattern (a pair of doublets). (R. K. Hill and T.-H Chan, *Tetrahedron* **1965,** *21,* 2015.)

Spectrum A

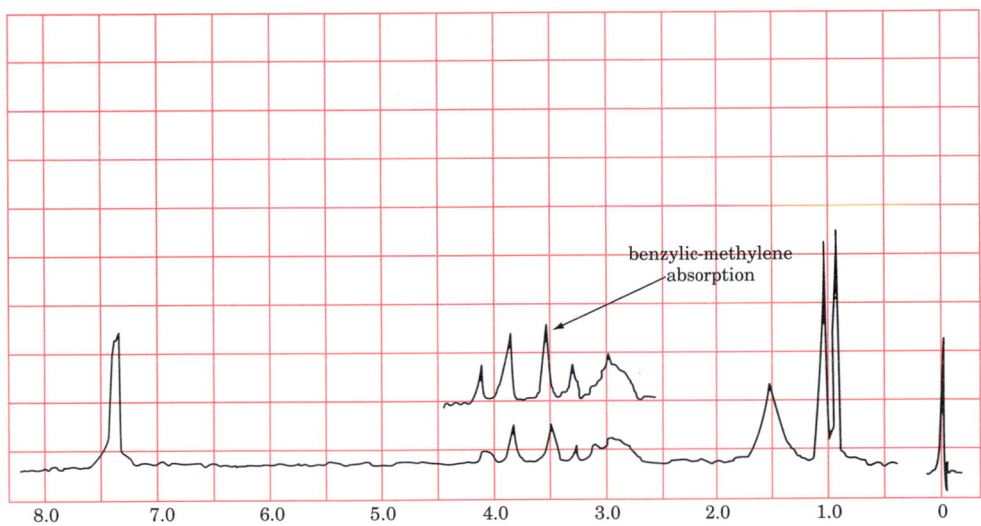

Spectrum B

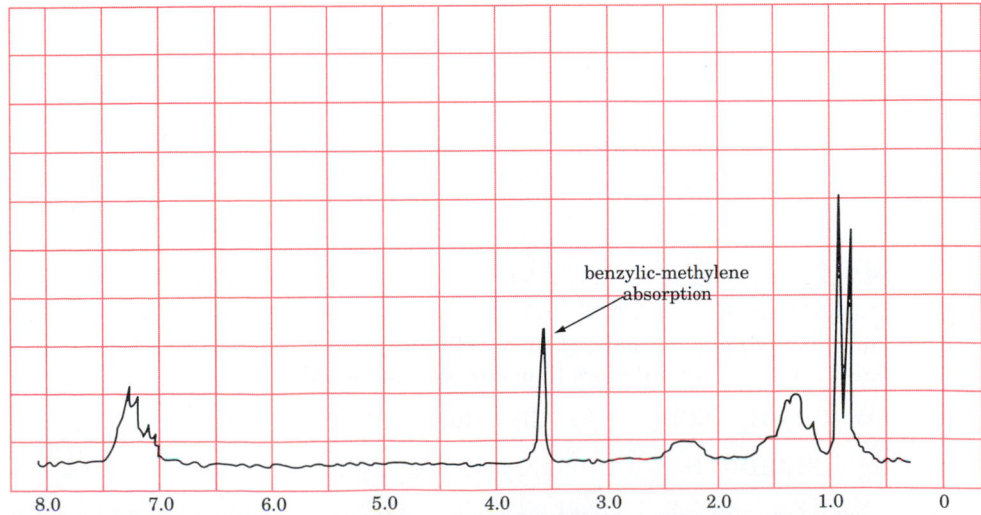

Figure 9.57   NMR spectra of the geometric isomers of *N*-benzyl-2,6-dimethylpiperidine. (R. K. Hill and T.-H. Chan, *Tetrahedron* **1965,** *21,* 2015.)

Questions

1. Draw the structures of the two geometric isomers of *N*-benzyl-2,6-dimethylpiperidine.

2. Two ring carbons of *N*-benzyl-2,6-dimethylpiperidine are chiral. Therefore, there is the possibility of four optical isomers. Draw these four structures.

3. Of the four possible optical isomers of *N*-benzyl-2,6-dimethylpiperidine, two have a special stereochemical relationship. What is this relationship?

4. Identify the benzylic–methylene hydrogens in *cis*- and in *trans-N*-benzyl-2,6-dimethylpiperidine as enantiotopic or diastereotopic.

5. Assign the structures of *cis*- and *trans-N*-benzyl-2,6-dimethylpiperidine to the two spectra in Figure 9.57.

Study Problems*

9.24 In otherwise similar compounds, which one of each of the following pairs of partial structures would give stronger infrared absorption, and why?

   (a) $C=O$ or $C=C$ **(b)** $C=C-Cl$ or $C=C-H$

   (c) $O-H$ or $N-H$

9.25 Tell how you could distinguish each of the following pairs of compounds by their infrared spectra:

   (a) $CH_3CH_2CH_2N(CH_3)_2$ and $CH_3CH_2CH_2NH_2$

   (b) $CH_3CH_2CH_2CO_2H$ and $CH_3CH_2CO_2CH_3$

   (c) $CH_3CH_2\overset{\overset{\textstyle O}{\|}}{C}CH_3$ and $CH_3CH_2CO_2CH_3$

   (d) $CH_3C\equiv CCH_3$ and $CH_3CH_2C\equiv CH$

9.26 Indicate which underlined proton in each of the following groups of compounds will absorb farther upfield:

   (a) $CH_3CH_2C\underline{H}_2Cl$, $CH_3CHClC\underline{H}_3$, $CH_2ClCH_2C\underline{H}_3$

   (b) $CH_3C\underline{H}_2Cl$, $CH_3C\underline{H}Cl_2$ **(c)** 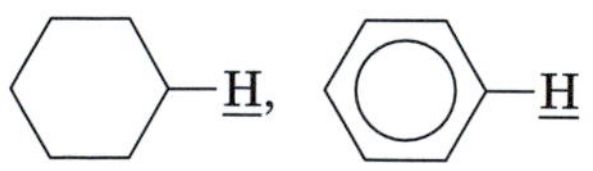 ⬡—$\underline{H}$, ⬡—$\underline{H}$

   (d) 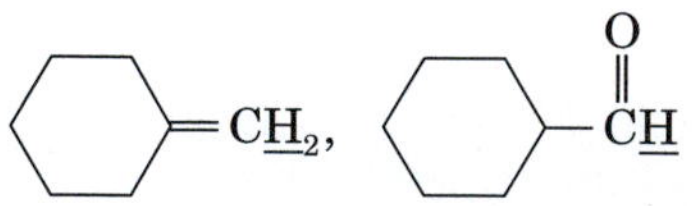 ⬡=$C\underline{H}_2$, ⬡—$\overset{\overset{\textstyle O}{\|}}{C}\underline{H}$

9.27 How many different groups of chemically equivalent protons are present in each of the following structures? If more than one, indicate each group.

   (a) $CH_3CH_2CH_2CH_3$ **(b)** $CH_3CH_2OCH_2CH_3$

   (c) $CH_3CH=CH_2$ **(d)** *trans*-$CHCl=CHCl$

   * For information concerning the organization of the *Study Problems* and *Additional Problems*, see the *Note to student* on page 41.

(e) $cis\text{-CHCl}=\text{CHCl}$

(f) $CH_3CH_2CO_2CH_2CH_3$

(g) (R)-2-chlorobutane

(h) $(2R,3R)$-2,3-dichlorobutane

(i) $(CH_3)_2CHN(CH_3)CH(CH_3)_2$

(j) $BrCH_2CH_2CH(CH_3)_2$

(k) 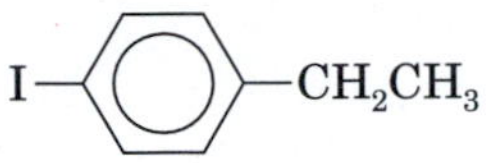 $I-\!\!\left\langle \bigcirc \right\rangle\!\!-CH_2CH_3$

(l)  $CH_3O-\!\!\left\langle \bigcirc \right\rangle\!\!-OCH_3$

(m) 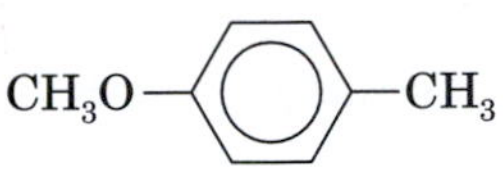 $CH_3O-\!\!\left\langle \bigcirc \right\rangle\!\!-CH_3$

(n)   18-crown-6 ether (Section 8.5)

9.28 In Problem 9.27, tell how many principal signals would probably be observed in the ^1H NMR spectrum of each compound, and predict the relative areas under the signals.

9.29 An NMR spectrum shows two principal signals with an area ratio of 3:1. Based on this information only, which of the following structures are possibilities?

(a) $CH_2=CHCH_3$

(b) $CH_2=CHCO_2CH_3$

(c) $CH_3CH_2CH_2CH_2CH_2CH_3$

(d) $CH_3CH_2CH_3$

(e) $CH_2=C(CH_3)_2$

(f) $CH_3CO_2CH_3$

9.30 Calculate the ratios of the different types of hydrogen atoms in a sample when the steps of the integration curve measure 81.5, 28, 55, and 80 mm.

9.31 A ^1H NMR spectrum shows four signals with integration curve heights of 4.0, 3.5, 5.4, and 5.5 cm. What is the ratio of the types of protons in this sample?

9.32 For each of the following compounds, predict the *multiplicity* (number of peaks arising from spin-spin coupling) and the *total relative area* under the signal of each set of equivalent protons in the ^1H NMR spectrum.

(a) $CH_3CH_2CO_2CH_3$

(b) $CH_3OCH_2CH_2OCH_3$

(c) $CH_3\overset{\displaystyle O}{\overset{\|}{C}}CH_2CH_3$

(d) 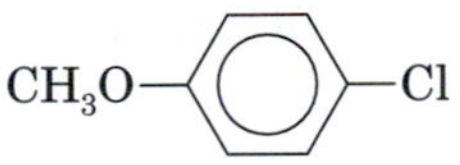 $CH_3O-\!\!\left\langle \bigcirc \right\rangle\!\!-Cl$

(e) Cl_2CHCH_2Br

9.33 How would you distinguish between each of the following pairs of compounds by ^1H NMR spectroscopy?

(a) $CH_3CH_2CH_2CH_2CH=CH_2$ and $(CH_3)_2C=C(CH_3)_2$

(b) CH_3CH_2CHO and $CH_3\overset{\displaystyle O}{\overset{\|}{C}}CH_3$

(c) $CH_3O\overset{\displaystyle O}{\overset{\|}{C}}CH_2CH_3$ and $CH_3\overset{\displaystyle O}{\overset{\|}{C}}OCH_2CH_3$

(d) 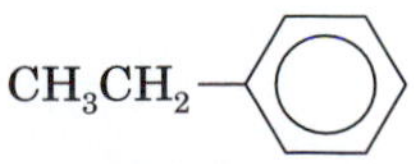 $CH_3CH_2-\!\!\left\langle \bigcirc \right\rangle$ and $CH_3-\!\!\left\langle \bigcirc \right\rangle\!\!-CH_3$

9.34 Using graph paper and a ruler, construct a spin–spin splitting diagram for the indicated proton(s) in the following partial structures:

(a)

$$H_a \underset{}{\diagdown} C = C \underset{}{\diagup} \overset{H_b}{\underset{H_c}{\diagup}}$$

H_a where $J_{ab} = 10$ Hz and $J_{ac} = 16$ Hz

(b)

$$-CH_2-\overset{O}{\overset{\|}{C}}-H_b$$
$\uparrow$
a

H_b where $J_{ab} = 2$ Hz

9.35 Draw a tree diagram for H_m in the following partial structure, where $J_{am} = 8$ Hz and $J_{mx} = 2$ Hz. Then for H_m where $J_{am} = 2$ Hz and $J_{mx} = 8$ Hz.

$$H_a-\overset{|}{\underset{\underset{H_a}{|}}{C}}-\overset{|}{\underset{\underset{H_m}{|}}{C}}-\overset{|}{\underset{\underset{H_x}{|}}{C}}-$$

9.36 Predict **(a)** the number of principal peaks in the ^{1}H NMR spectrum of the local anesthetic *xylocaine* (following); **(b)** the splitting patterns of these peaks; **(c)** the characteristic absorption peaks in the infrared spectrum.

$$\text{(aromatic ring)} \begin{array}{c} CH_3 \\ \\ -NH\overset{O}{\overset{\|}{C}}CH_2N(CH_2CH_3)_2 \\ \\ CH_3 \end{array}$$

xylocaine

9.37 Sketch the expected ^{1}H NMR spectrum for 1,1,3,3-tetrachloropropane. Be sure to include *(qualitatively)* anticipated chemical shifts, splitting patterns, and appropriate areas.

9.38 Sketch the expected ^{1}H NMR spectrum for each of the following compounds:

(a) $ClCH_2CH_2\overset{O}{\overset{\|}{C}}OCH_2CH_3$

(b) $CH_3O-\text{(aromatic ring)}-CH_2O\overset{O}{\overset{\|}{C}}CH_2CH_3$

9.39 A compound (molecular weight 72) containing only C, H, and O shows four signals in its proton-decoupled, ^{13}C NMR spectrum: $\delta 8$, 30, 37, and 208. The off-resonance-decoupled spectrum shows these signals as a quartet, a quartet, a triplet, and a singlet. What is the structure of the compound?

Additional Problems

9.40 Under the influence of H_0, [18]annulene has an induced ring current not too different from that of benzene. Predict the shielding and deshielding of the protons of this ring system.

[18]annulene

9.41 Figure 9.58 shows a partial ^1H NMR spectrum (60 MHz) of the disodium salt of (S)-malic acid. It contains only the signals for the CH_2 and CH groups. Give an interpretation of their spin–spin splitting patterns.

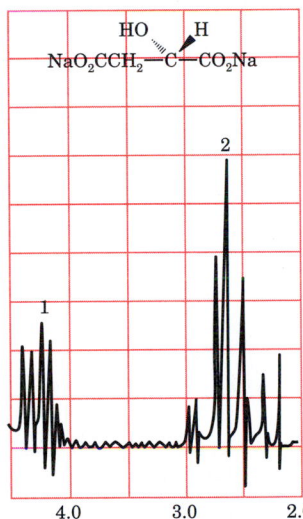

Figure 9.58 Partial 60 MHz ^1H NMR spectrum of (S)-malic acid, disodium salt.

9.42 Predict the multiplicity of each absorption band in the off-resonance-decoupled ^{13}C NMR spectra of the following compounds:

(a) $(CH_3)_2\overset{OH}{\underset{|}{C}}C\equiv CH$ (b)

9.43 Determine the structures of the following compounds:

(a) The ^1H NMR spectrum of an alcohol ($C_5H_{12}O$) shows the following absorption: one singlet (relative area 1); two doublets (areas 3 and 6); and two multiplets (areas both 1). When treated with HBr, the alcohol yields an alkyl bromide ($C_5H_{11}Br$). Its NMR spectrum shows only a singlet (area 6); a triplet (area 3); and a quartet (area 2). What are the structures of the alcohol and the alkyl bromide?

(b) A compound (C_4H_7NO) shows moderately strong infrared absorption at 3000 cm^{-1} (3.33 μm) and weak absorption at 2240 cm^{-1} (4.46 μm). The only other absorption is in the fingerprint region. The ^1H NMR spectrum shows a triplet (δ2.6), a singlet (δ3.3), and a triplet (δ3.5). The area ratio of these signals is 2 (triplet): 5 (singlet and second triplet combined).

(c) A compound (molecular weight 122) containing only C, H, and O shows the following ^{1}H NMR absorption: singlet (δ2.15), singlet (δ3.45), and a group of four peaks (centered around δ6.7); area ratio of 3:3:4.

(d) A compound with the molecular formula $C_5H_{10}O_2$ shows strong infrared absorption at 1750 cm^{-1} (5.71 μm). The ^{1}H NMR spectrum shows a doublet (δ1.2), a singlet (δ2.0), and a septet (δ4.95); area ratio of 6:3:1.

(e) A compound ($C_8H_{12}O_2$) shows its principal infrared absorption at 2900–3050 cm^{-1} (3.28–3.45 μm), 1760 cm^{-1} (5.68 μm), 1460 cm^{-1} (6.85 μm), and 1380 cm^{-1} (7.25 μm). Its ^{1}H NMR spectrum shows only a singlet at δ1.3.

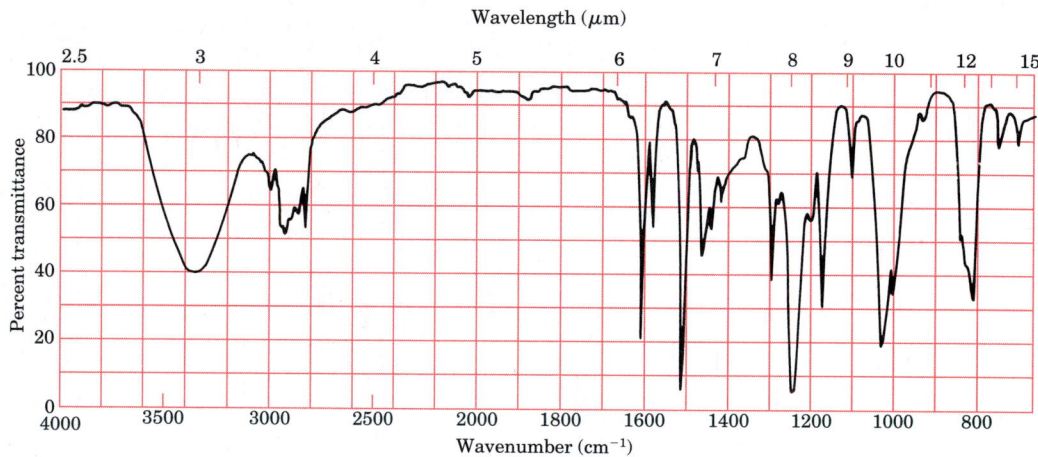

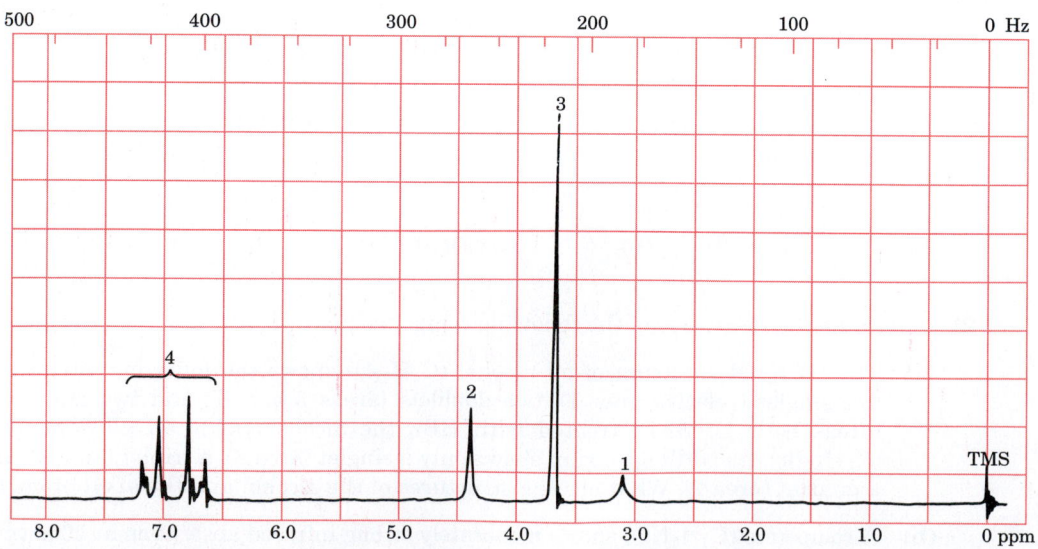

Figure 9.59 Spectra for Problem 9.44(a): $C_8H_{10}O_2$.

9.44 Figures 9.59 through 9.65 each give a molecular formula, an infrared spectrum, and a ^{1}H NMR spectrum for an unknown compound. What is the structure of each compound?

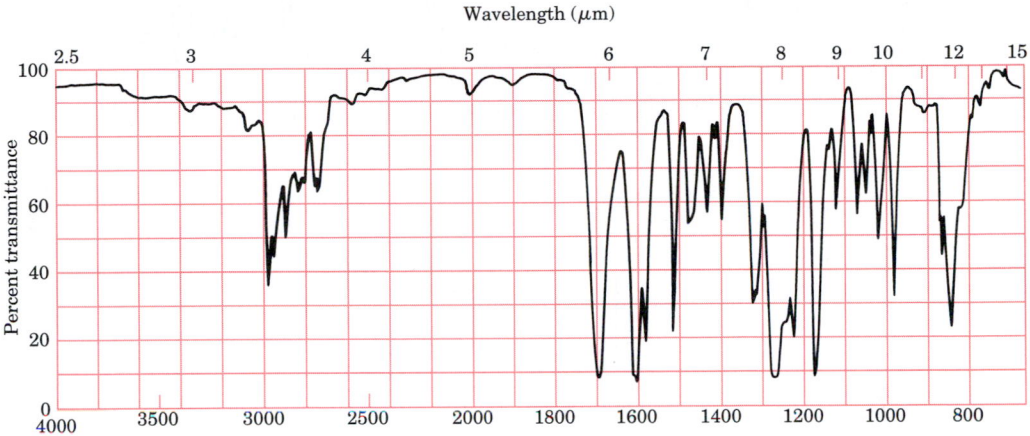

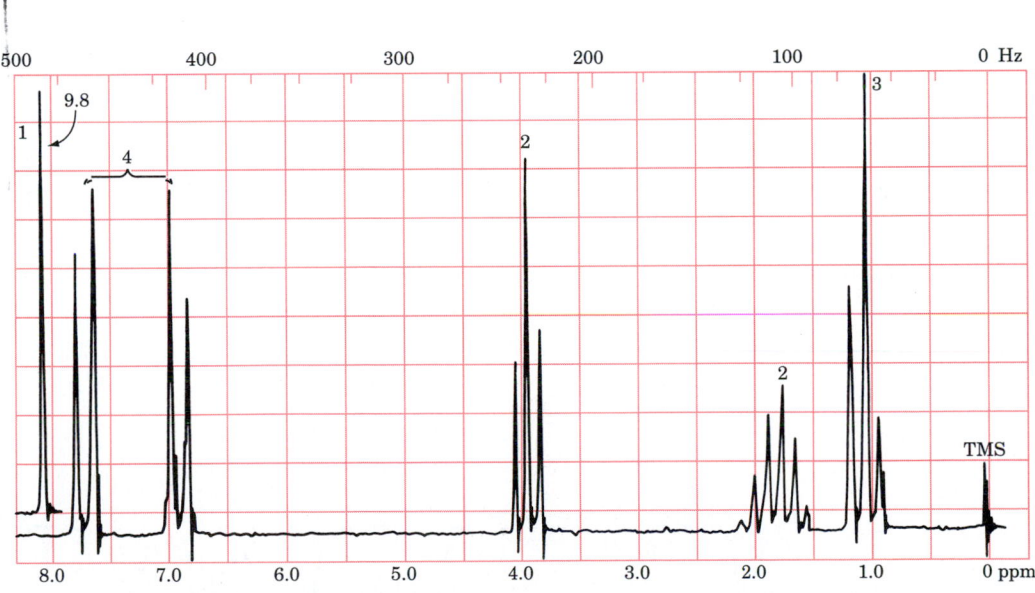

Figure 9.60 Spectra for Problem 9.44(b): $C_{10}H_{12}O_2$.

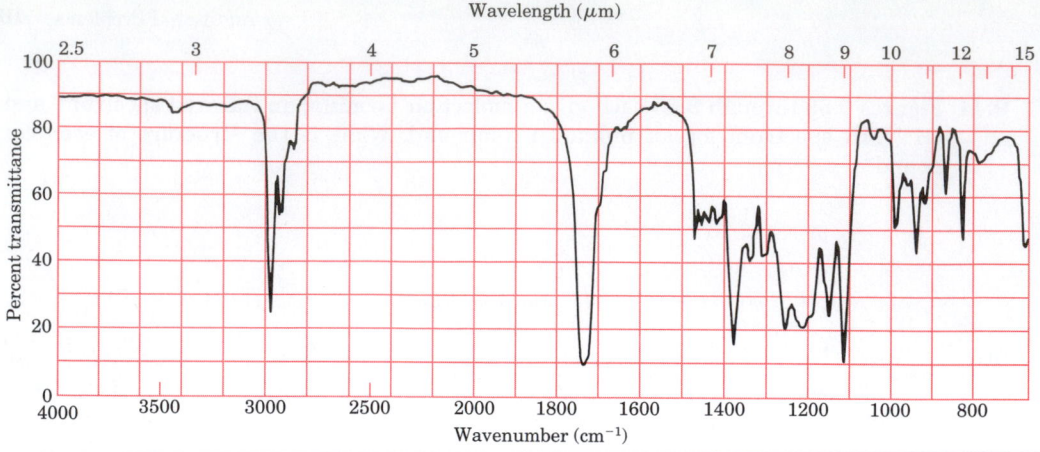

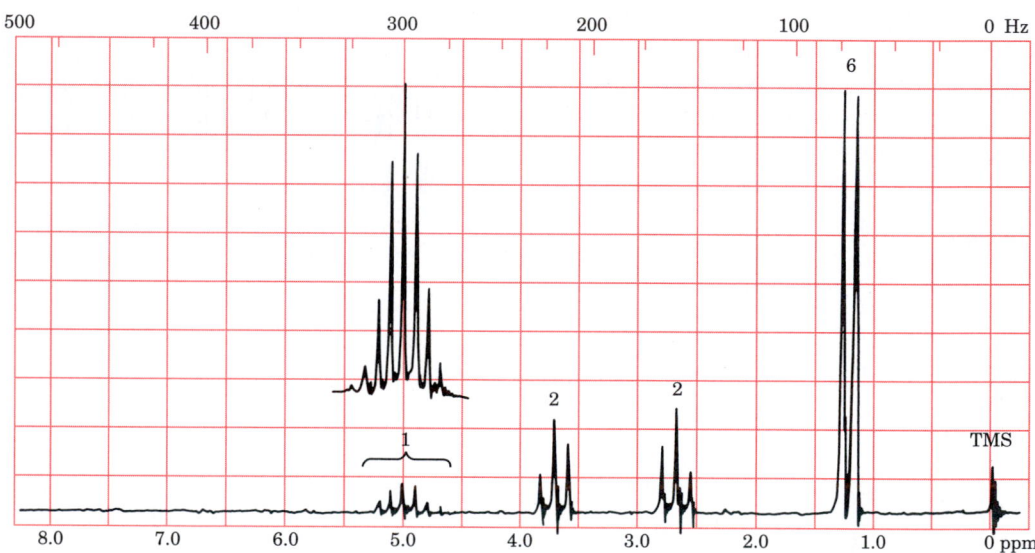

Figure 9.61 Spectra for Problem 9.44(c): $C_6H_{11}O_2Cl$.

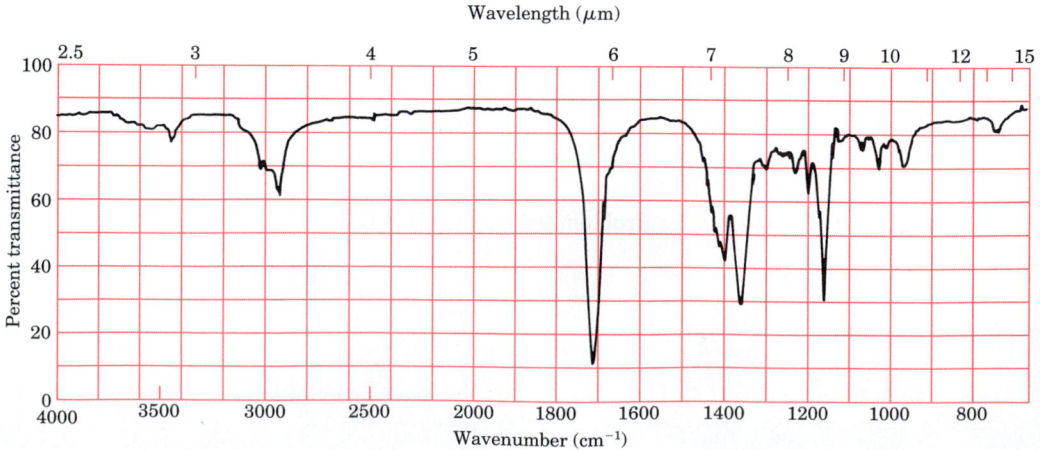

Figure 9.62 Spectra for Problem 9.44(d): $C_6H_{10}O_2$.

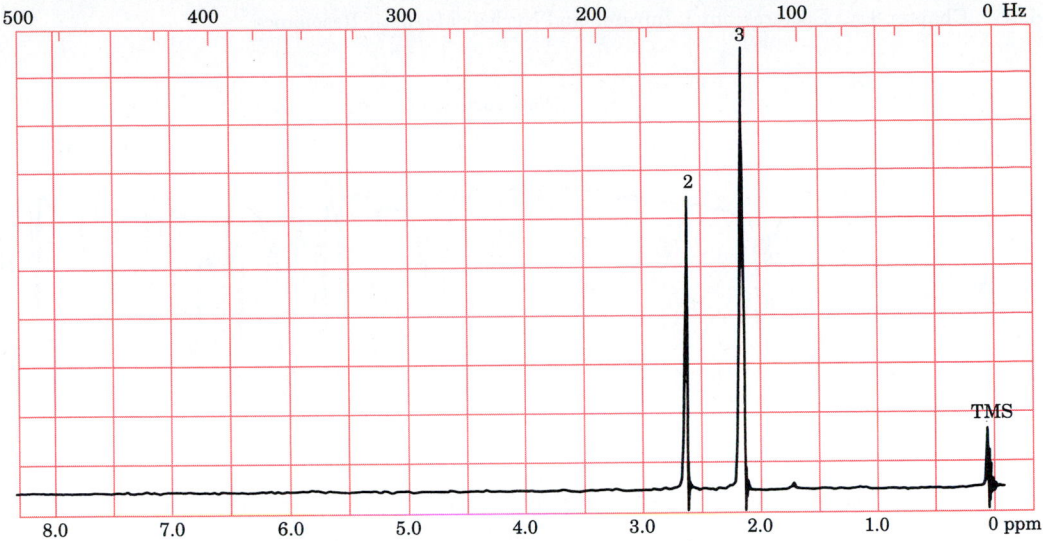

Figure 9.62 *(continued)* Spectra for Problem 9.44(d): $C_6H_{10}O_2$.

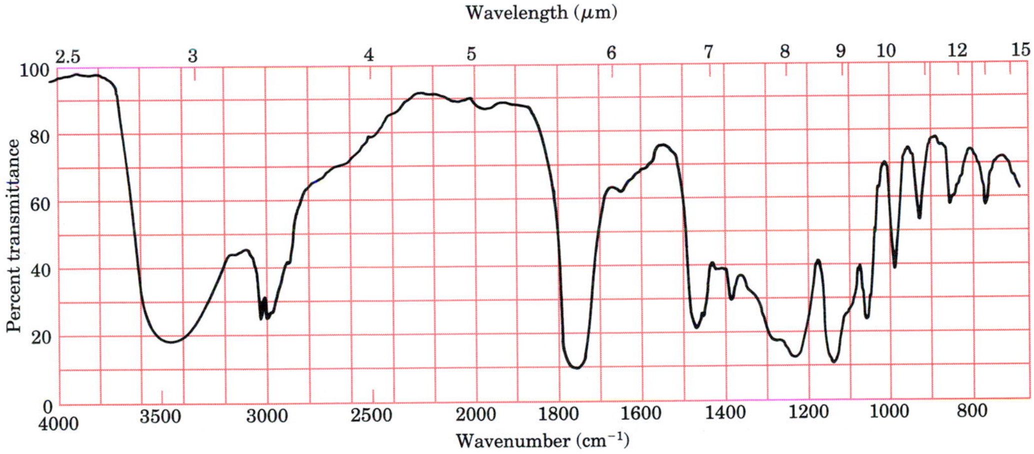

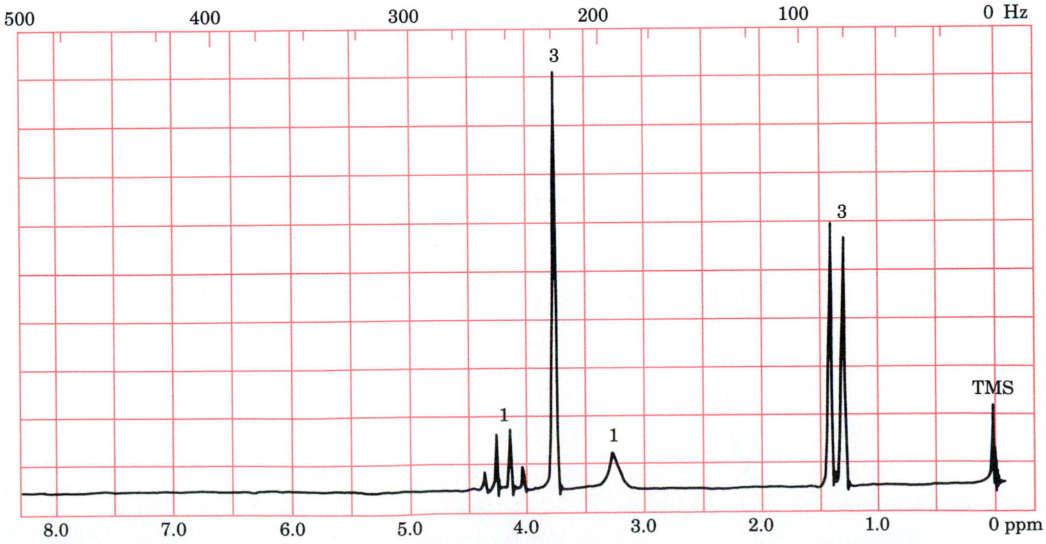

Figure 9.63 Spectra for Problem 9.44(e): $C_4H_8O_3$.

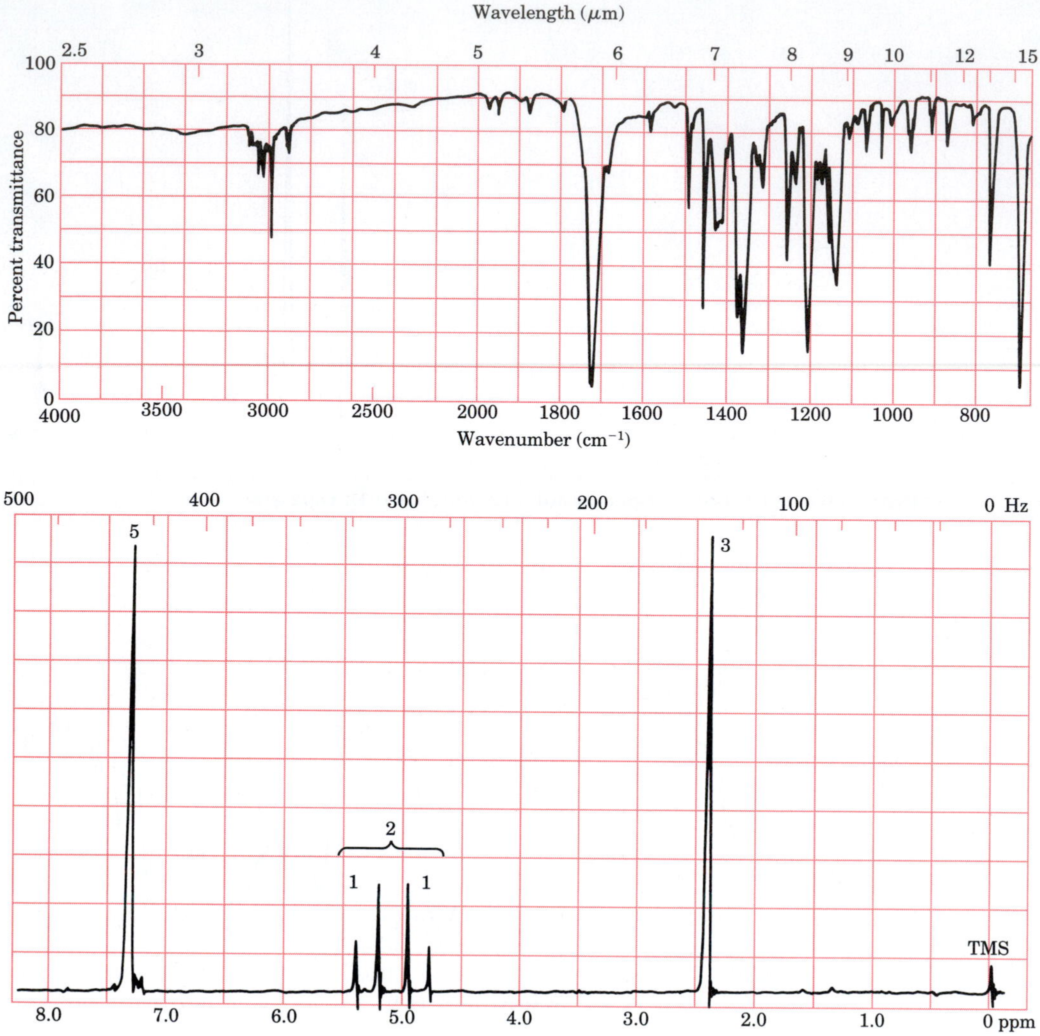

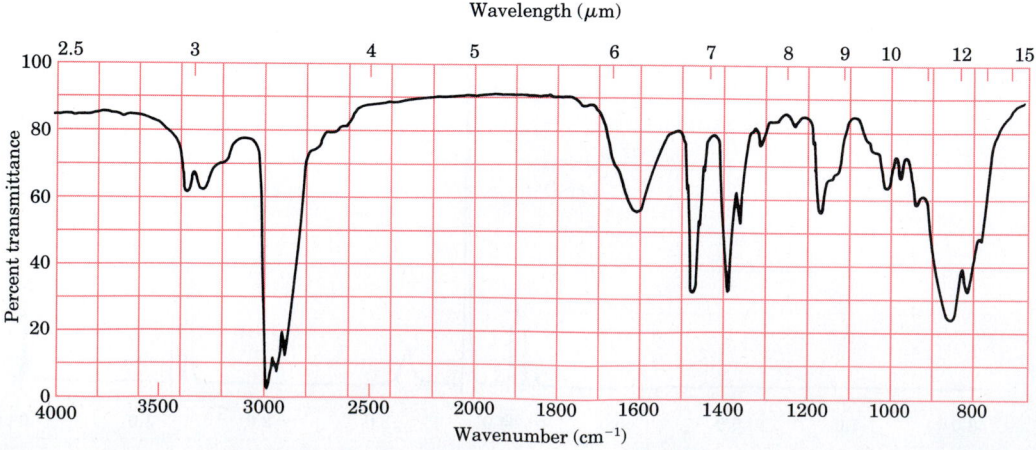

Figure 9.64   Spectra for Problem 9.44(f): $C_{10}H_{10}Br_2O$.

Figure 9.65   Spectra for Problem 9.44(g): $C_4H_{11}N$.

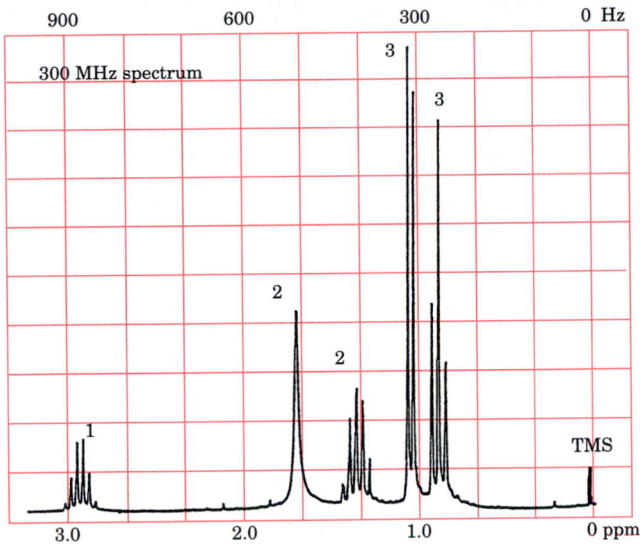

Figure 9.65 *(continued)* Spectra for Problem 9.44(g): $C_4H_{11}N$.

9.45 Figures 9.66 through 9.68 show ^{13}C NMR spectra, along with additional information concerning compounds of unknown structure. What are the structures of these compounds?

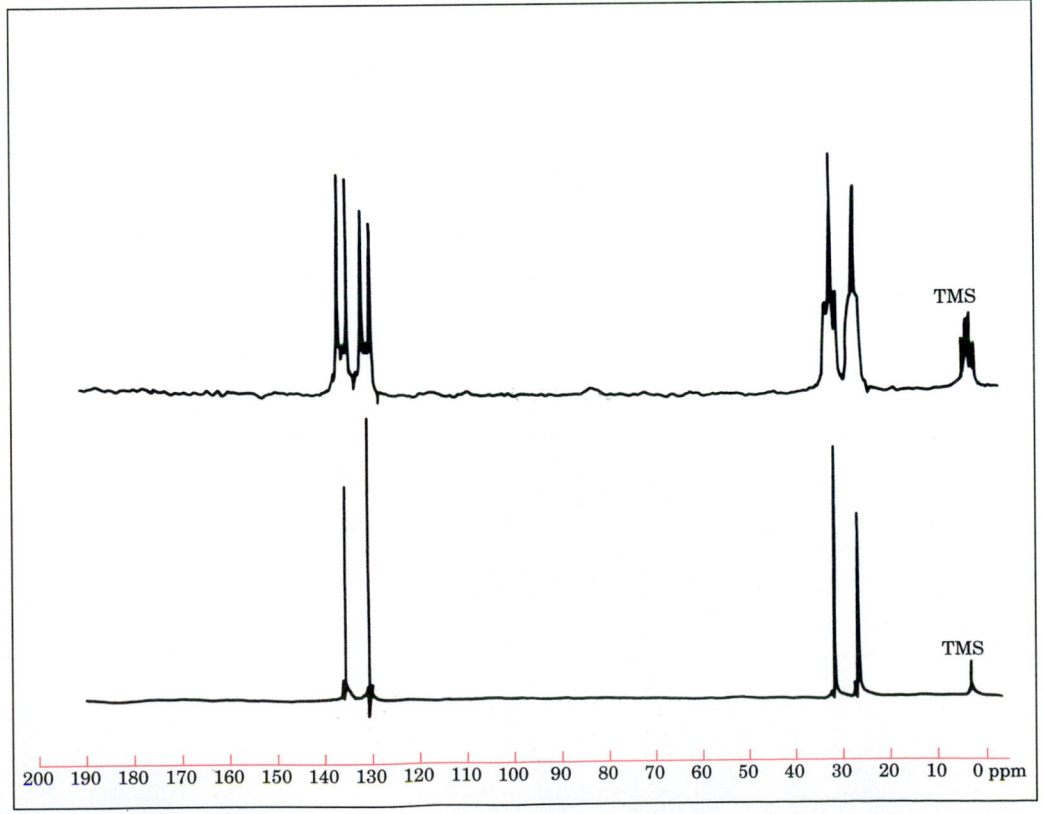

Figure 9.66 ^{13}C NMR spectra for Problem 9.45(a): C_8H_{12}.

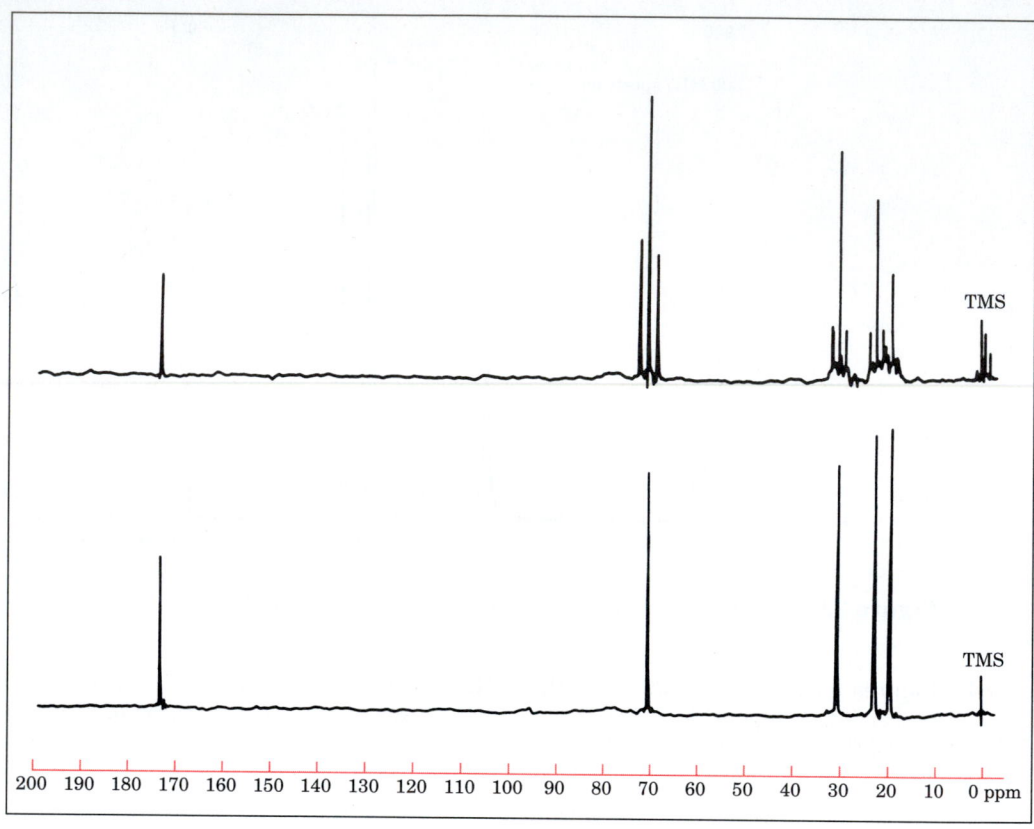

Figure 9.67 ^{13}C NMR spectra for Problem 9.45(b): an ester with the molecular formula $C_5H_8O_2$.

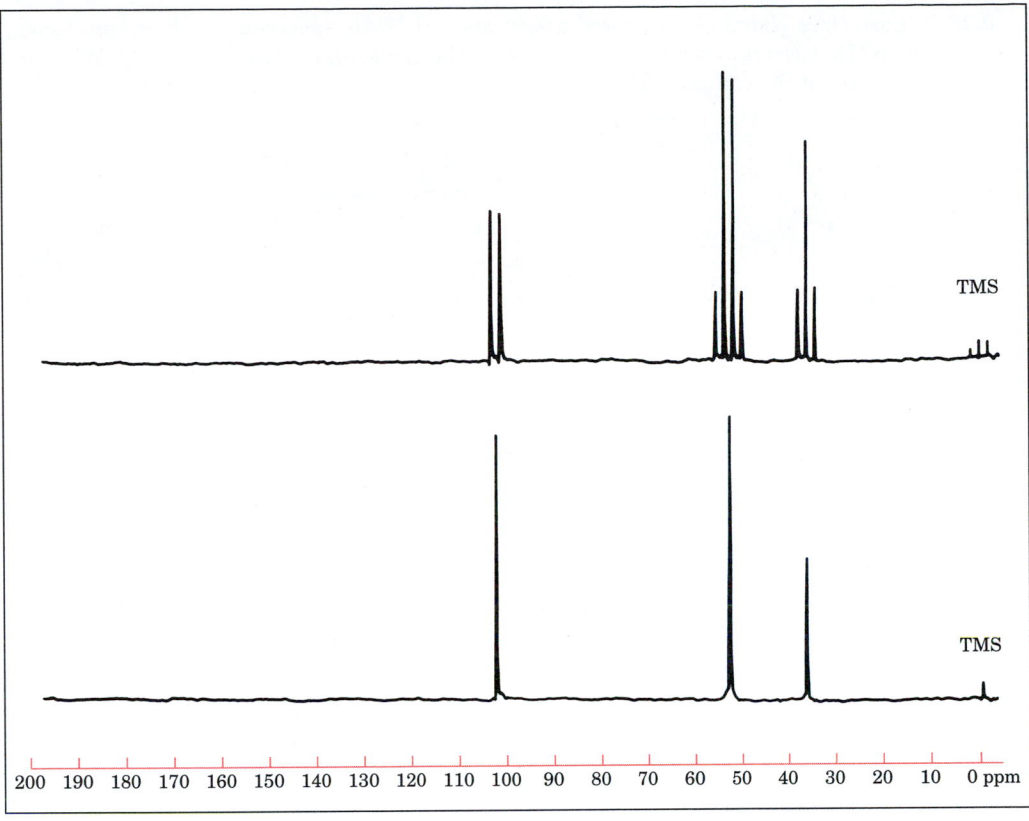

Figure 9.68 ^{13}C NMR spectra for Problem 9.45(c): a compound with the molecular formula $C_7H_{16}O_4$. The ^{1}H NMR spectrum of this compound shows only a triplet ($\delta 1.95$), a singlet ($\delta 3.3$), and a triplet ($\delta 4.5$), with relative areas of $1:6:1$.

9.46 Figure 9.69 shows the infrared spectrum, ^{1}H NMR spectrum, and proton-decoupled ^{13}C NMR spectrum for a compound with the molecular formula $C_7H_{14}O$. What is the structure of the compound?

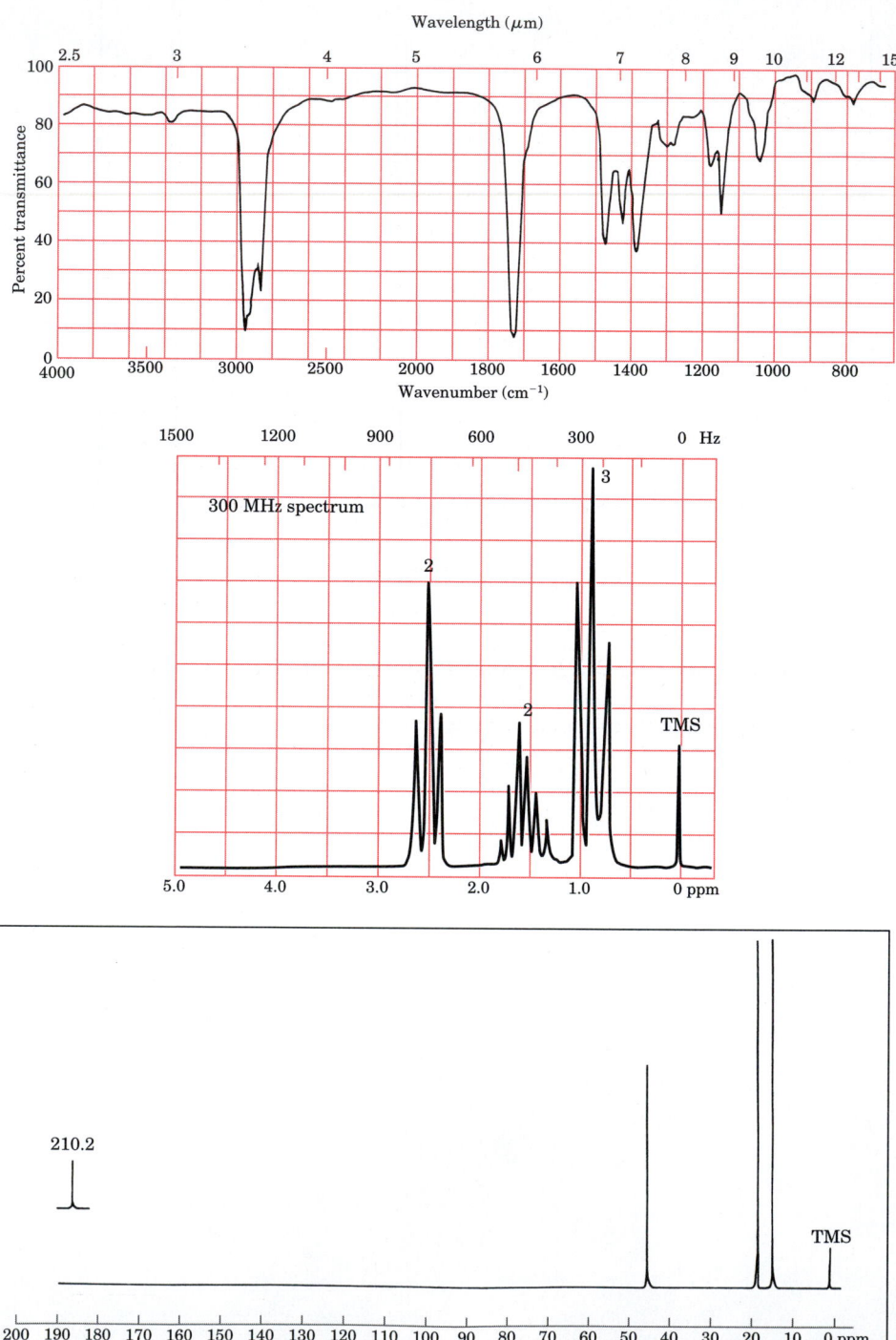

Figure 9.69 Infrared spectrum, ^{1}H NMR spectrum, and proton-decoupled ^{13}C NMR spectrum for $C_7H_{14}O$, Problem 9.46.

10 ALKENES AND ALKYNES

An **alkene** is a hydrocarbon with one double bond. Alkenes are sometimes called **olefins,** from *olefiant gas* ("oil-forming gas"), an old name for ethylene ($CH_2{=}CH_2$). An **alkyne** is a hydrocarbon with one triple bond. Acetylene ($CH{\equiv}CH$) is the simplest alkyne.

A carbon–carbon double bond is a common functional group in natural products. Most frequently, the double bond occurs in conjunction with other functional groups. However, alkenes with no other functionality are not at all rare and are often found as plant products and in petroleum. Two interesting examples of naturally occurring compounds containing carbon–carbon double bonds follow. Both these compounds are examples of *insect pheromones,* compounds emitted by one insect to transmit information to another insect of the same species (see Section 20.2).

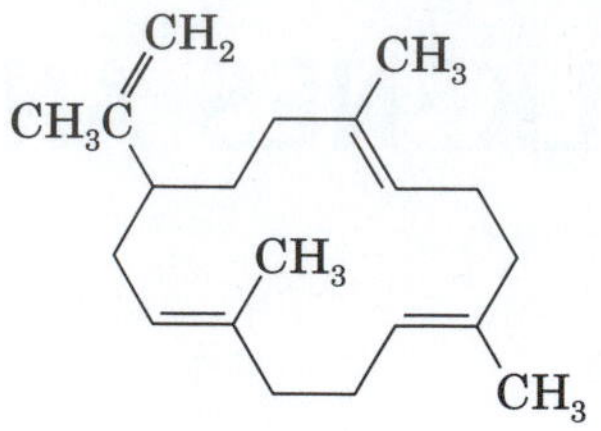

neocembrene

a termite trail marker

$$\text{(CH}_3\text{)}_2\text{C}=\text{CHCH}_2\text{CH}_2\overset{\overset{\displaystyle \text{CH}_3}{|}}{\text{C}}=\text{CHCH}_2\text{CH}_2\overset{\overset{\displaystyle \text{CH}_2}{\|}}{\text{C}}\text{CH}=\text{CH}_2$$

3-methylene-7,11-dimethyl-1,6,10-dodecatriene

an aphid alarm pheromone

Section 10.1

Bonding in Alkenes and Alkynes; Acidity of Alkynes

The bonding in ethylene and acetylene was discussed in detail in Chapter 2. Recall that the two carbon atoms in ethylene are in the sp^2-hybrid state. The three sp^2 bonds from each carbon atom lie in the same plane, with bond angles of approximately 120°. The π bond joining the two sp^2-hybridized carbons lies above and below the plane of the σ bonds (see Figure 2.14, page 59).

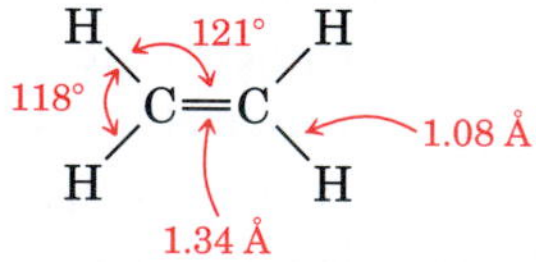

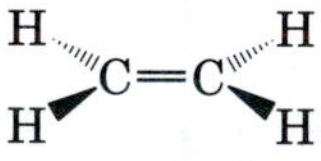

side view

All the atoms in the ethylene molecule lie in the same plane. However, in a molecule that also has sp^3-hybridized carbons, only those atoms bonded to the double-bond carbons lie in the same plane.

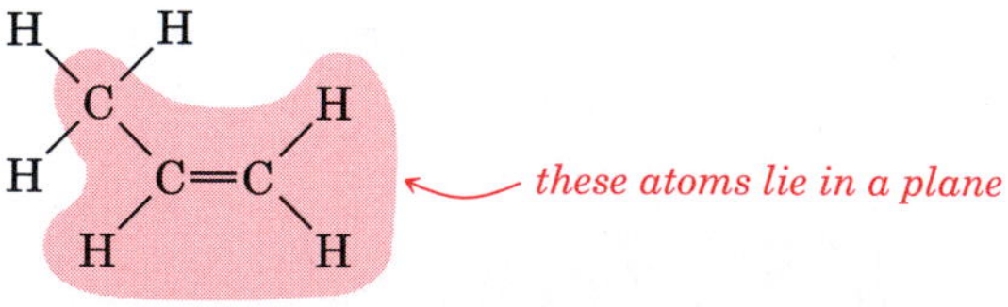

← *these atoms lie in a plane*

The two carbon atoms in acetylene are in the sp-hybrid state. The σ bonds formed by the sp-hybridized carbon atoms are linear. In acetylene, *two π* bonds join the sp-hybridized carbon atoms (Figure 2.18, page 63).

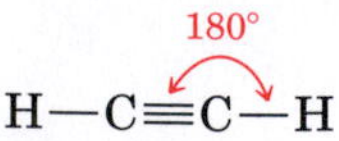

$$\text{H}-\text{C}\overset{180°}{\equiv}\text{C}-\text{H}$$

acetylene

A triple-bond carbon is in the sp-hybrid state. The sp orbital is one-half s, while an sp^2 orbital is one-third s, and an sp^3 orbital is only one-fourth s. Because an *sp* orbital has more *s* character, the electrons in this orbital are closer to the carbon nucleus than are electrons in an sp^2 or sp^3 orbital. (See Section 2.4F). In an alkyne, the *sp*-hybridized carbon is therefore *more electronegative* than most other carbon atoms. Thus, an alkynyl CH bond is *more polar* than an alkane CH bond or an alkene CH bond.

$$\overset{\delta+}{R} - \overset{\delta-}{C} \equiv \overset{\delta-}{C} - \overset{\delta+}{H}$$

sp-hybridized carbons electron-withdrawing

One of the most important results of the polarity of the alkynyl carbon–hydrogen bond is that $RC \equiv CH$ can lose a hydrogen ion to a strong base. The resulting anion ($RC \equiv C:^-$) is called an **acetylide ion.** With a pK_a of 25, alkynes are not strong acids. They are weaker acids than water (pK_a 15.7), but stronger acids than ammonia ($pK_a \sim 35$). Alkynes undergo reaction with a strong base like sodamide ($NaNH_2$) or a Grignard reagent. Alkanes and alkenes are not sufficiently acidic to react under these conditions.

$$CH_3C \equiv C - H + :NH_2^- \xrightarrow{\text{liq. } NH_3} CH_3C \equiv C:^- + :NH_3$$

propyne   *an acetylide ion*

$$CH_3C \equiv CH + CH_3MgI \longrightarrow CH_3C \equiv CMgI + CH_4$$

Section 10.2
Nomenclature of Alkenes and Alkynes

A. Nomenclature

In the IUPAC system, the continuous-chain alkenes are named after their alkane parents, but with the **-ane** ending changed to **-ene.** For example, CH_3CH_3 is ethane and $CH_2 = CH_2$ is ethene (trivial name, ethylene).

$$CH_2 = CH_2 \qquad CH_3CH = CH_2$$

IUPAC: ethene   propene   cyclohexene

A hydrocarbon with two double bonds is called a **diene,** while one with three double bonds is called a **triene.** The following examples illustrate diene and triene nomenclature:

$$CH_2 = CHCH = CH_2 \qquad CH_2 = \overset{\overset{\textstyle CH_3}{|}}{C}CH = CH_2 \qquad CH_2 = CHCH = CHCH = CH_2$$

1,3-butadiene   2-methyl-1,3-butadiene   1,3,5-hexatriene

a diene   *a diene*   *a triene*

In the names of most alkenes, we need a prefix number to show the position of the double bond. Unless there is functionality of higher nomenclature priority, we number the chain from the end that gives the lower number to the double bond. The prefix number specifies the carbon atom in the chain where the double bond begins.

$$CH_3C{=}CHCH_2CH_3 \qquad CH_2{=}CHCH_2CH_2OH \qquad CH_3CH{=}CHCOH$$

$$\overset{CH_3}{|}$$

2-methyl-2-pentene   3-buten-1-ol   2-butenoic acid

Some alkenyl groups have trivial names that are in common use. A few of these are shown in Table 10.1.

A π bond prevents free rotation of groups around a double bond. Consequently, some alkenes exhibit geometric isomerism. This topic and the nomenclature of geometric isomers were covered in Section 4.1. (Because alkynes are linear molecules, they do not exhibit geometric isomerism.)

cis-2-butene
or (Z)-2-butene

trans-2-butene
or (E)-2-butene

The IUPAC nomenclature of alkynes is directly analogous to that of the alkenes. The suffix for an alkyne is **-yne,** and a position number is used to signify the position of the triple bond in the parent hydrocarbon chain. Unless there is functionality of higher nomenclature priority in the molecule, the chain is numbered to give the triple bond the lower number.

In an older, trivial system of nomenclature for the simple alkynes, acetylene ($CH{\equiv}CH$) is considered the parent. Groups attached to the sp-hybridized carbons are named as substituents on acetylene. In this text we will use the IUPAC nomenclature system for the alkynes except for acetylene itself.

$$C{\equiv}CH \qquad CH_3C{\equiv}CCH_2CH_3$$

IUPAC: phenylethyne   2-pentyne
trivial: phenylacetylene   ethylmethylacetylene

Table 10.1   Trivial names of some alkenyl groups

Structure	Name	Example
$CH_2{=}$	methylene[a]	$=CH_2$ methylenecyclohexane
$CH_2{=}CH{-}$	vinyl	$CH_2{=}CHCl$ vinyl chloride
$CH_2{=}CHCH_2{-}$	allyl	$CH_2{=}CHCH_2Br$ allyl bromide

[a] The term *methylene* is also used to refer to a disubstituted sp^3-hybridized carbon ($-CH_2-$); for example, CH_2Cl_2 is called methylene chloride.

B. Line Formulas

Line formulas provide a very convenient representation for many open-chain structures. Line formulas are directly analogous to the polygon formulas for cyclic structures that we presented in Section 1.5C. In line formulas, the line represents a single bond. An intersection or endpoint represents a carbon atom with its complement of hydrogens. Double and triple bonds are shown as double and triple lines. All functional groups are drawn out.

$\bigwedge\!\!\bigvee$ or $\bigwedge\!\!\bigvee$ means $CH_3(CH_2)_3CH_3$

$\parallel$ or $=$ means $CH_2{=}CH_2$

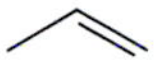

$\bigwedge\!\!\diagup$ or $\bigvee\!\!\diagup$ means $CH_3CH{=}CH_2$

$\bigwedge\!\!\!\diagup$ or $\parallel\diagdown$ means $CH_2{=}CHCH{=}CH_2$

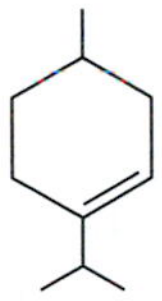

means

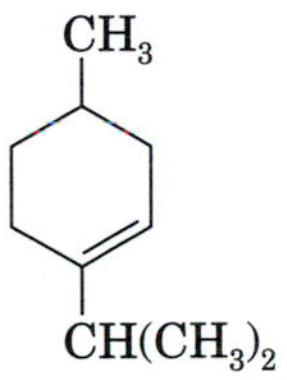

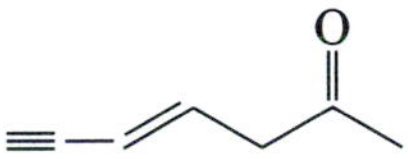

$$\underset{CH(CH_3)_2}{\overset{CH_3}{\bigcirc}}$$

$={-}\diagup\!\!\diagup\overset{O}{\underset{\parallel}{C}}\diagdown$ means $HC{\equiv}C{-}CH{=}CH{-}CH_2\overset{O}{\underset{\parallel}{C}}CH_3$

STUDY PROBLEMS

10.1   Give structures for:

  (a) (*E*)-4-methyl-2-hexene

  (b) *trans*-2-chloro-3-hexene

  (c) *cis*-1,3-pentadiene

  (d) (1*R*,2*S*)-2-methyl-3-cyclopentenol

10.2   Write line formulas for **(a)**, **(b)**, and **(c)** of Problem 10.1.

10.3   Write the IUPAC names of the following alkynes:

  (a) $CH_3\overset{Cl}{\underset{|}{C}}HC{\equiv}CCH_2CH_3$ **(b)** $HOCH_2C{\equiv}CCH_2OH$ **(c)** 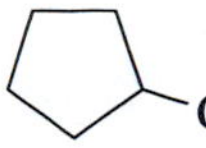 $\bigpentagon\!\!-C{\equiv}CH$

Section 10.3
Spectra of Alkenes and Alkynes

A. Infrared Spectra

Alkenes   Ethylene, tetrachloroethylene, and other alkenes containing a nonpolar C=C double bond do not absorb infrared radiation in the C=C range.

Table 10.2   Infrared absorption characteristics of alkenes and alkynes

	Position of Absorption	
Type of Vibration	cm^{-1}	μm
Alkenes:		
=C—H stretching	3000–3100	3.2–3.3
=C—H bending	800–1000	10.0–12.5
=CH$_2$ bending	855–995	10.0–11.7
C=C stretching	1600–1700	5.8–6.2
Alkynes:		
≡C—H stretching	~3300	~3.0
C≡C stretching	2100–2250	4.4–4.8

Alkenes such as RR′C=CHR and other unsymmetrical alkenes contain a polar C=C double bond and thus do absorb infrared radiation. Stretching of the C=C double bond gives rise to absorption at 1600–1700 cm^{-1} (5.8–6.2 μm). Because the stretching usually results in only a small change in bond moment, the absorption is weak, 10 to 100 times less intense than that of a carbonyl group. The absorption due to the stretching of the alkenyl, or vinylic, carbon–hydrogen bond (=C—H) at about 3000–3100 cm^{-1} (3.2–3.6 μm) is also weak. Alkenyl carbon–hydrogen bonds exhibit bending absorption in the fingerprint region of the infrared spectrum. (See Table 10.2.) Figure 10.1 shows the spectra of heptane and 1-heptene. The differences between an alkane and an alkene are evident in these spectra.

Alkynes   The C≡C stretching frequency of alkynes is at 2100–2250 cm^{-1} (4.4–4.8 μm). This absorption is quite weak and can easily be lost in the background noise of the spectrum. However, with the exception of C≡N and Si—H, no other commonly encountered groups show absorption in this region. The ≡C—H stretching frequency is about 3300 cm^{-1} (3.0 μm) as a sharp peak (see Figure 10.2).

B. ^{1}H NMR Spectra

Alkenes   The chemical shift for a vinylic proton is at an approximate δ value of 5.0 ppm. The exact position of the absorption depends on the location of the double bond in the hydrocarbon chain. In general, protons on terminal alkenyl carbons absorb near δ4.7, while the protons on nonterminal carbons absorb slightly farther downfield, at δ values of about 5.3 ppm.

$$CH_3CH_2\underset{\delta5.3}{CH}=\underset{\delta4.7}{CH_2}$$

The splitting patterns for vinylic protons are more complex than those for alkyl protons. The complexity arises from the lack of rotation around the double bond. Let us look at a general example:

$$\begin{array}{c} H_x \diagdown \qquad \diagup H_a \\ C=C \\ R \diagup \qquad \diagdown H_b \end{array} \qquad \begin{array}{l} J_{ab} = 0\text{–}3.5 \text{ Hz} \\ J_{ax} = 6\text{–}14 \text{ Hz} \\ J_{bx} = 11\text{–}18 \text{ Hz} \end{array}$$

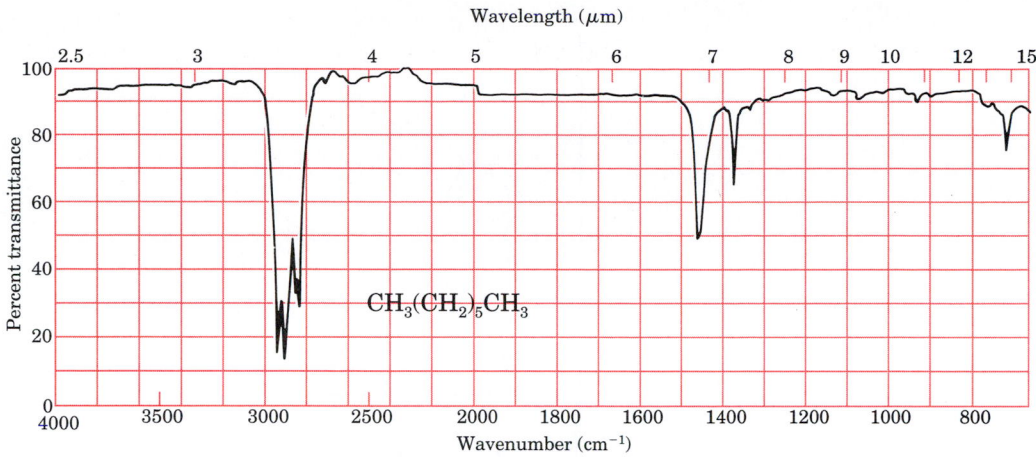

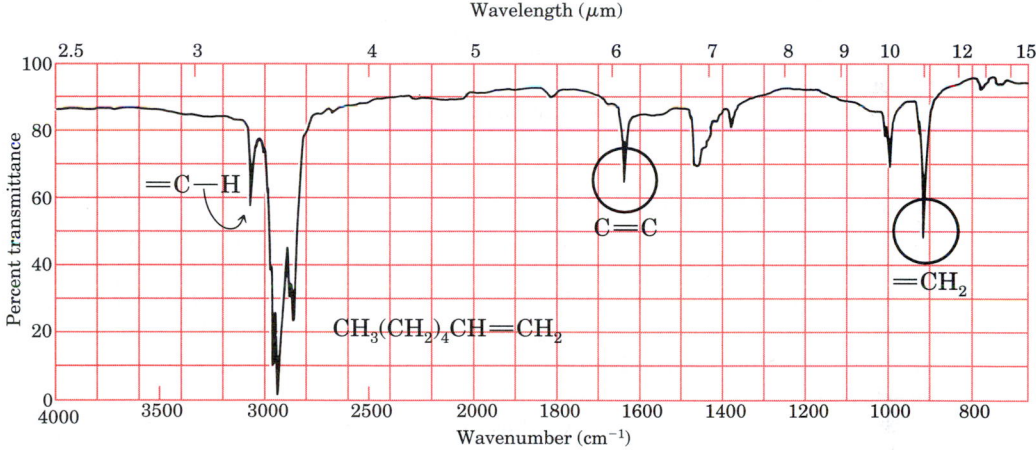

Figure 10.1 Infrared spectra of heptane and 1-heptene.

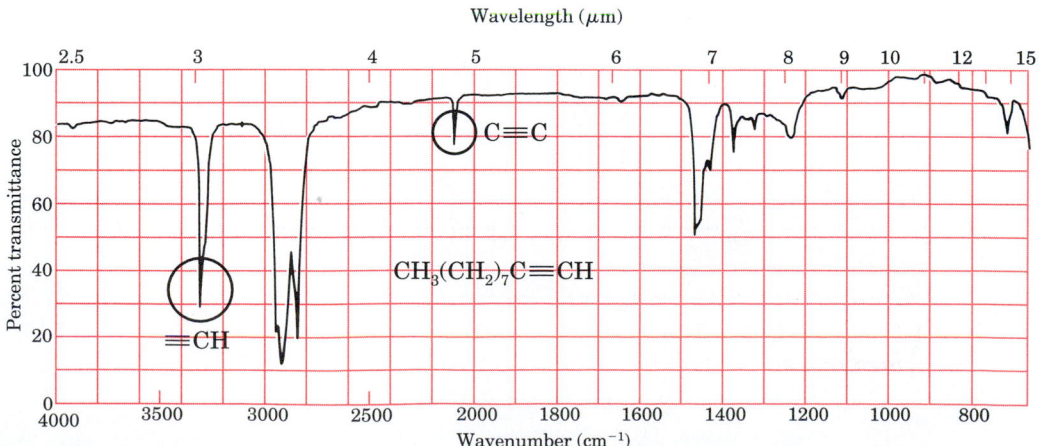

Figure 10.2 Infrared spectrum of 1-decyne.

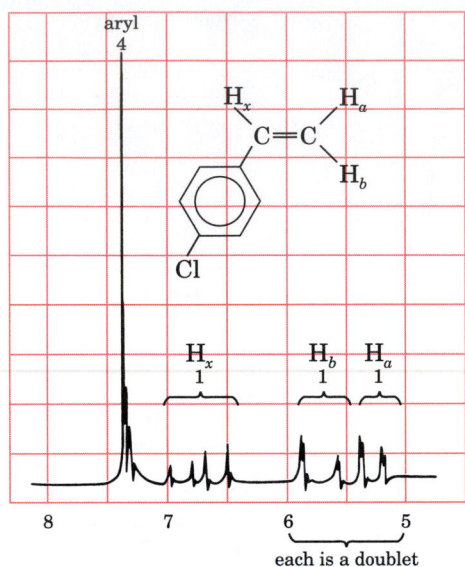

Figure 10.3 Partial ^{1}H NMR spectrum of *p*-chlorostyrene.

In this example, all three vinylic protons (H_a, H_b, and H_x) are nonequivalent. Therefore, they exhibit different chemical shifts and give three separate signals. In addition, the coupling constants between any two of the protons (J_{ax}, J_{bx}, and J_{ab}) are different. Each of the three signals is therefore split into four peaks. (For example, the signal for H_x is split into two by H_b and again into two by H_a.) A total of twelve peaks may be observed in the NMR spectrum for these three protons.

The ^{1}H NMR spectrum of *p*-chlorostyrene (Figure 10.3) shows this alkene pattern of twelve peaks. Figure 10.4 shows tree diagrams for the splitting patterns of the double-bond protons in *p*-chlorostyrene. With the aid of the

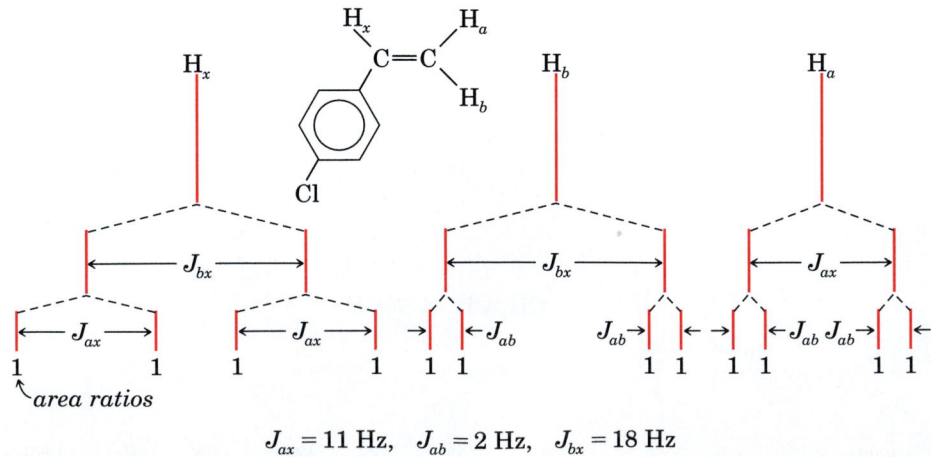

$J_{ax} = 11$ Hz, $J_{ab} = 2$ Hz, $J_{bx} = 18$ Hz

Figure 10.4 Tree diagrams for the splitting patterns of the three alkenyl protons in *p*-chlorostyrene.

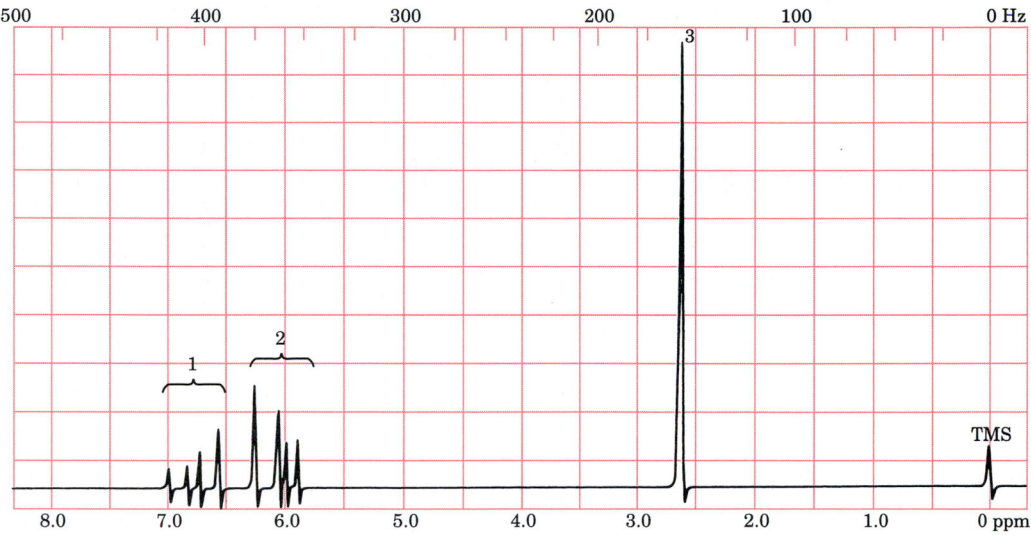

Figure 10.5 ^{1}H NMR spectrum of $CH_2{=}CHSO_2CH_3$ for Problem 10.4.

tree diagram, we can see that none of the absorption peaks are superimposed because of the differences in J values.

In the spectrum of *p*-chlorostyrene (Figure 10.3), the chemical shift for H_a is at $\delta 5.3$, while the chemical shift for H_b, which is *cis* (and closer) to the benzene ring, is at $\delta 5.7$. The signal for H_b is downfield because H_b is somewhat deshielded by the induced field of the benzene ring. The signal for H_x is even farther downfield because H_x is more deshielded by the induced field of the ring.

Although any terminal vinylic group of the type $RCH{=}CH_2$ should give a spectrum with twelve vinyl-proton peaks, the twelve peaks are not always evident. (For example, see Figure 10.5 for Problem 10.4). Whether or not twelve peaks will be seen in the spectrum depends upon the ability of the NMR spectrometer to resolve the signals.

The splitting patterns of vinylic protons can be even more complex than we have just seen above. When allylic protons are present, they are spin–spin coupled to nonadjacent vinylic protons and splitting of their signals may be seen. This type of spin–spin coupling is called **allylic coupling.** Because the allylic coupling constant is fairly small (0.5–3 Hz), splitting may not always be apparent. Figure 10.6 shows the ^{1}H NMR spectrum of 3-chlorocrotonic acid, which shows a typical splitting pattern caused by allylic coupling. The CH_3 protons are split into a doublet by the vinylic proton, which in turn is split into a quartet.

allylic proton

$$H_a{-}\overset{|}{\underset{|}{C}}{-}\overset{|}{C}{=}C\diagdown$$

$$H_b \quad \longleftarrow \textit{vinylic proton}$$

either cis or trans

$J_{ab}{=}0.5{-}3$ Hz (typically ~ 2 Hz)

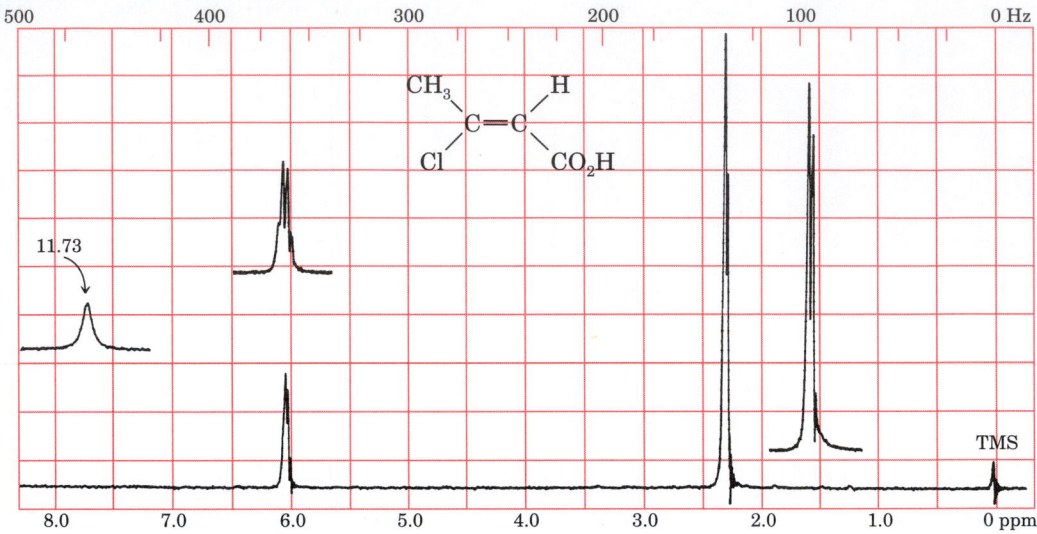

Figure 10.6 ^{1}H NMR spectrum of 3-chlorocrotonic acid.

STUDY PROBLEMS

10.4 Figure 10.5 is the ^{1}H NMR spectrum of methyl vinyl sulfone, $CH_2\!\!=\!\!CHSO_2CH_3$. The spectrum contains 8 peaks (not 12) in the alkene region. Construct a tree diagram that would explain the "missing" peaks.

10.5 Construct a tree diagram for each proton in the NMR spectrum of the following partial structure:

$$
\begin{array}{cc}
H_x & H_a \\
\!\!\!\!\!\!\!\diagdown & \!\!\!\!\diagup \\
C\!\!=\!\!C \\
\!\!\!\!\diagup & \!\!\!\!\diagdown \\
Y & H_b
\end{array}
\qquad
\begin{aligned}
J_{ab} &= 2.0 \text{ Hz} \\
J_{ax} &= 8.0 \text{ Hz} \\
J_{bx} &= 13.0 \text{ Hz}
\end{aligned}
$$

Alkynes A disubstituted alkyne ($RC\!\!\equiv\!\!CR$) has no alkynyl proton. Therefore, a disubstituted alkyne has no characteristic ^{1}H NMR absorption. (The rest of the molecule, however, will probably have NMR absorption.) A monosubstituted alkyne, $RC\!\!\equiv\!\!CH$, shows absorption for the alkynyl proton at a δ value of about 3 ppm. This absorption is not nearly as far downfield as that for a vinylic or aryl proton because the alkynyl proton is *shielded* by the induced field of the triple bond. Figure 10.7 shows how the circulation of the π electrons results in this field. Note the difference between this anisotropic effect and the effect for a vinylic proton, $\!\!=\!\!CHR$ (Section 9.7B). In the case of an alkyne, the induced field opposes, rather than adds to, H_0.

As with vinyl protons, alkynyl protons on terminal alkynes show spin–spin splitting caused by allylic coupling. The J values (2–3 Hz) are similar to those for vinylic protons.

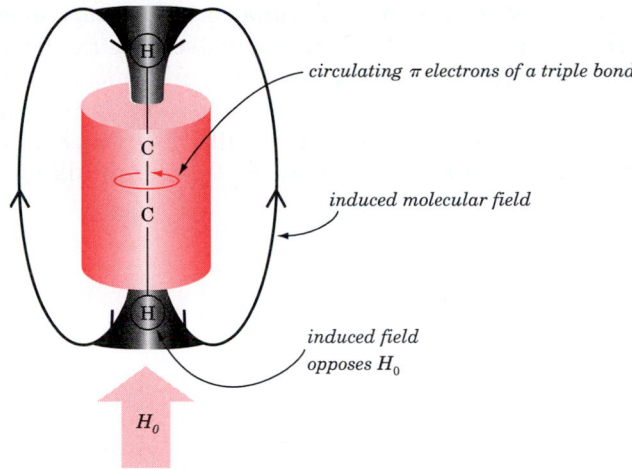

circulating π electrons of a triple bond

induced molecular field

induced field opposes H_0

H_0

Figure 10.7 An alkynyl proton is shielded by the induced magnetic field of the triple bond.

Section 10.4

Preparation of Alkenes and Alkynes

Alkenes can be prepared by the elimination reactions of alcohols (in strong acid) or alkyl halides (in base). Table 10.3 lists the sections where these reactions are discussed in detail.

Table 10.3   Summary of laboratory syntheses of alkenes and alkynes

Reaction	Section Reference
Alkenes:	
$R_2\overset{\displaystyle X}{\underset{\displaystyle \|}{C}}CHR_2 + OH^- \xrightarrow{\text{heat}} R_2C{=}CR_2$	5.7, 5.8
$R_2\overset{\displaystyle OH}{\underset{\displaystyle \|}{C}}CHR_2 + H_2SO_4 \xrightarrow{\text{heat}} R_2C{=}CR_2$	7.6
$R_2C{=}O + (C_6H_5)_3P{=}CR'_2 \longrightarrow R_2C{=}CR'_2$	13.5D
$RC{\equiv}CR + H_2 \xrightarrow{\text{catalyst}} RCH{=}CHR \ (cis)$	10.12B
$RC{\equiv}CR \xrightarrow[\text{liq. NH}_3]{\text{Na or Li}} RCH{=}CHR \ (trans)$	10.12B
Alkynes:	
$R\overset{\displaystyle X\ \ X}{\underset{\displaystyle \|\ \ \ \|}{C}HCH_2} \xrightarrow[\text{(2) H}_2\text{O}]{\text{(1) 3 NaNH}_2,\ \text{liq. NH}_3} RC{\equiv}CH$	10.4
$RC{\equiv}CH \xrightarrow[\text{liq. NH}_3]{\text{NaNH}_2} RC{\equiv}C{:}^-\ Na^+ \xrightarrow{R'X} RC{\equiv}CR'$	10.4
$RC{\equiv}CH \xrightarrow{\text{CH}_3\text{MgI}} RC{\equiv}CMgI \xrightarrow[\text{(2) H}_2\text{O, H}^+]{\text{(1) R}'_2\text{C}{=}\text{O}} RC{\equiv}C\overset{\displaystyle OH}{\underset{\displaystyle \|}{C}}R'_2$	10.4

Primary alcohols undergo elimination reactions slowly. In hot, concentrated H_2SO_4, the product alkene may also undergo isomerization and other reactions. Therefore, primary alcohols are not usually useful in alkene preparation. *Primary alkyl halides* also undergo elimination reactions slowly by an E2 path. However, if a bulky base such as the *tert*-butoxide ion is used to minimize reaction by an S_N2 path, an alkene may be obtained in good yield.

1° RX (E2 and S_N2):

$$CH_3CH_2CH_2CH_2CH_2Br \xrightarrow[\text{warm}]{K^+ \ ^-OC(CH_3)_3}$$

1-bromopentane

$$\begin{cases} CH_3CH_2CH_2CH=CH_2 \\ \text{1-pentene (85\%)} \\ + \\ CH_3CH_2CH_2CH_2CH_2OC(CH_3)_3 \\ \textit{tert}\text{-butyl pentyl ether (12\%)} \end{cases}$$

Secondary alcohols undergo elimination by an E1 path when heated with a strong acid, and rearrangement of the intermediate carbocation may occur. Except in simple cases, secondary alcohols are therefore not useful intermediates for the preparation of alkenes. *Secondary alkyl halides* can undergo E2 reactions. Although product mixtures might be expected, the predominant product is usually the more highly substituted *trans*-alkene.

2° RX (E2):

$$\overset{\overset{\displaystyle Br}{|}}{CH_3CH_2CH_2CHCH_3} \xrightarrow[\text{warm}]{Na^+ \ ^-OCH_2CH_3}$$

2-bromopentane

$$\begin{cases} \begin{array}{c} CH_3CH_2 \\ \diagdown \\ H \diagup \end{array} C=C \begin{array}{c} H \\ \diagdown \\ \diagup CH_3 \end{array} \\ \textit{trans}\text{-2-pentene (51\%)} \\ + \\ \begin{array}{c} CH_3CH_2 \\ \diagdown \\ H \diagup \end{array} C=C \begin{array}{c} CH_3 \\ \diagup \\ \diagdown H \end{array} \\ \textit{cis}\text{-2-pentene (18\%)} \\ + \\ CH_3CH_2CH_2CH=CH_2 \\ \text{1-pentene (31\%)} \end{cases}$$

Tertiary alcohols undergo elimination readily through carbocations (E1) when treated with a strong acid. *Tertiary alkyl halides* undergo elimination with base principally by the E2 reaction. In both cases, excellent yields can be obtained if all three R groups of R_3CX or R_3COH are the same. Otherwise, mixtures might be obtained.

Alkynes can also be prepared by elimination reactions, but stronger conditions are required to remove the second halide ion. Under these conditions, the triple bond may migrate from one position to another.

General:

$$\underset{\underset{RCH-CHR}{\overset{\overset{Br\quad\ \ Br}{|\qquad\ |}}{}}}{} \xrightarrow[-HBr]{(1)} \underset{\underset{RCH=CR}{\overset{\overset{Br}{|}}{}}}{} \xrightarrow[-HBr]{(2)} RC\equiv CR$$

(1) OH^- or OR^- in CH_3CH_2OH
(2) $NaNH_2$, molten KOH, or very concentrated KOH + heat

$$\text{Ph}-\underset{\underset{CH}{\overset{\overset{Br}{|}}{}}}{}-\underset{\underset{CH}{\overset{\overset{Br}{|}}{}}}{}-\text{Ph} \xrightarrow[-2\,HBr]{2\,KOH,\ CH_3CH_2OH} \text{Ph}-C\equiv C-\text{Ph}$$

1,2-dibromo-1,2-diphenylethane   diphenylethyne (85%)

$$CH_3(CH_2)_{13}\underset{\underset{CH}{\overset{\overset{Br}{|}}{}}}{}-\underset{\underset{CH_2}{\overset{\overset{Br}{|}}{}}}{} \xrightarrow[-2\,HBr]{2\,NaNH_2} [CH_3(CH_2)_{13}C\equiv CH] \xrightarrow{NaNH_2}$$

1,2-dibromohexadecane

$$CH_3(CH_2)_{13}C\equiv C{:}^- \ Na^+ \xrightarrow{H_2O} CH_3(CH_2)_{13}C\equiv CH$$

1-hexadecyne (65%)

In Section 10.1, we mentioned that treatment of an alkyne with a strong base yields an acetylide. An acetylide ion may be used as a nucleophile in S_N2 reactions with primary alkyl halides. (Secondary and tertiary alkyl halides are more likely to give elimination products.) This reaction provides a synthetic route to substituted or more complex alkynes from simpler ones.

Preparation of acetylide:

$$CH_3C\equiv CH + NaNH_2 \xrightarrow[-30°]{liq.\ NH_3} CH_3C\equiv C{:}^- \ Na^+ + NH_3$$

Reaction with an alkyl halide:

$$CH_3C\equiv C{:}^- + CH_3CH_2CH_2-Cl \xrightarrow{S_N2} CH_3CH_2CH_2C\equiv CCH_3 + Cl^-$$

1-chloropropane   2-hexyne

As we also mentioned in Section 10.1, alkynyl Grignard reagents can be prepared by the reaction of a Grignard reagent and a 1-alkyne. In this reaction, the Grignard reagent acts as a base while the alkyne acts as an acid.

Preparation of $RC\equiv CMgX$:

$$CH_3C\equiv C\overset{\delta+}{H} + \overset{\delta-}{C}H_3MgI \longrightarrow CH_3C\equiv CMgI + CH_4$$

propyne

As with other Grignard reagents, the nucleophilic carbon of an alkynyl Grignard reagent attacks partially positive centers, such as the carbon of a carbonyl group. The advantage of this type of Grignard synthesis is that more-complex alkynes can be prepared more easily this way than by S_N2 reactions.

Reaction with a ketone:

$$CH_3C\overset{\delta-}{\equiv}\overset{\delta+}{C}-MgI + CH_3\overset{\overset{\ddot{O}{:}^{\delta-}}{\|}}{\underset{\delta+}{C}}CH_3 \longrightarrow CH_3\underset{\underset{C\equiv CCH_3}{\overset{\overset{{:}\ddot{O}MgI}{|}}{}}}{C}CH_3 \xrightarrow[H_2O]{H^+} CH_3\underset{\underset{C\equiv CCH_3}{\overset{\overset{{:}\ddot{O}H}{|}}{}}}{C}CH_3$$

2-methyl-3-pentyn-2-ol

Table 10.3 summarizes the methods of synthesizing alkenes and alkynes.

STUDY PROBLEMS

10.6 Predict the major organic product in each of the following reactions. Include stereochemistry, where appropriate.

(a) $(CH_3)_2\overset{\underset{\displaystyle |}{OH}}{C}CH_2CH_3 + H_2SO_4 \xrightarrow{\text{warm}}$

(b) $C_6H_5\overset{\underset{\displaystyle |}{CH_2CH_3}}{\underset{\underset{\displaystyle |}{Br}}{C}}CH_3 + OH^- \longrightarrow$

(c) 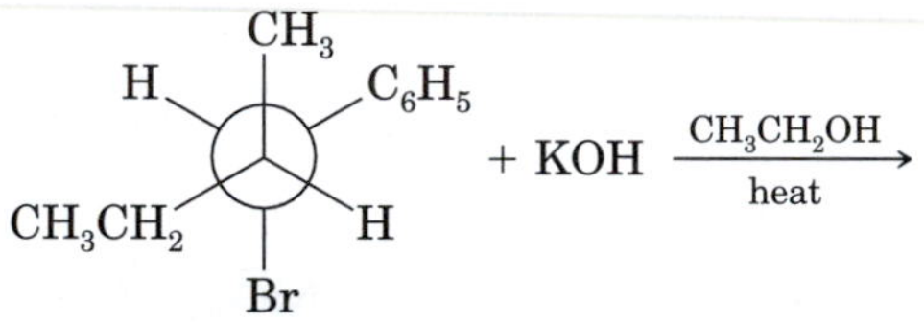 $+ \text{ KOH} \xrightarrow[\text{heat}]{CH_3CH_2OH}$

(d) 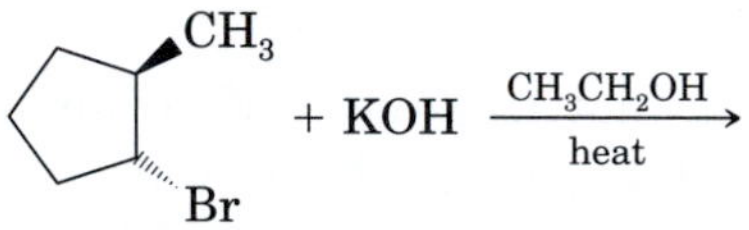 $+ \text{ KOH} \xrightarrow[\text{heat}]{CH_3CH_2OH}$

10.7 Show how the following conversions could be carried out.

(a) $CH_3C{\equiv}CH \longrightarrow CH_3C{\equiv}CCH_2CH_3$

(b) $CH_3CH_2\overset{\underset{\displaystyle |}{O}}{\overset{\displaystyle \|}{C}}H \longrightarrow CH_3CH_2\overset{\underset{\displaystyle |}{OH}}{C}H{-}C{\equiv}CH$

Section 10.5
Preview of Addition Reactions

Three typical reactions of alkenes are the reactions with hydrogen, with chlorine, and with a hydrogen halide:

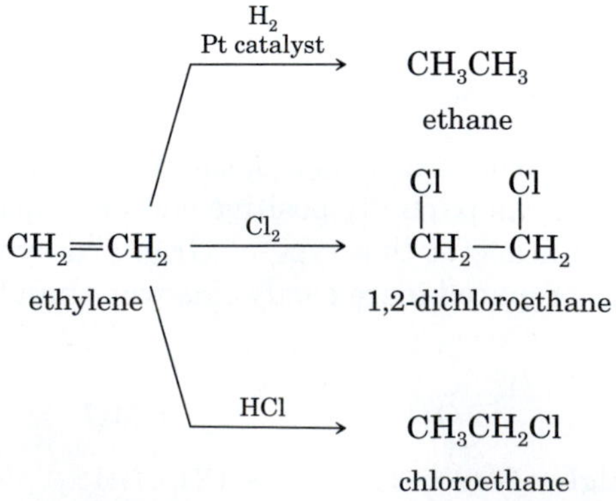

Each of these reactions is an **addition reaction.** In each case, a reagent adds to the alkene without the loss of any other atoms. We will find that the characteristic reaction of unsaturated compounds is the *addition of reagents to π bonds*.

In an addition reaction of an alkene, the π bond is broken and its pair of electrons is used in the formation of two new σ bonds. In each case, the sp^2-hybridized carbon atoms are rehybridized to sp^3. Compounds containing π bonds are usually of higher energy than comparable compounds containing only σ bonds. Consequently, an addition reaction is usually exothermic.

In general, carbon–carbon double bonds are not attacked by nucleophiles because there is no partially positive carbon atom. However, the exposed π electrons in the carbon–carbon double bond are attractive to *electrophiles* (E^+), such as H^+. Many reactions of alkenes and alkynes are, therefore, initiated by an *electrophilic attack,* a reaction step that results in a carbocation. The carbocation can then be attacked by a nucleophile to yield the product. We will discuss this type of addition reaction first and then proceed to other types of alkene reactions.

an alkene · *a carbocation*

Section 10.6

Addition of Hydrogen Halides to Alkenes and Alkynes

Hydrogen halides add to the π bonds of alkenes to yield alkyl halides. Alkynes react in an analogous manner and yield either vinylic halides or 1,1-dihaloalkanes, depending on the amount of HX used. We will not stress alkynes in our discussions, however, because alkenes are more important both in the laboratory and in nature.

$$CH_2{=}CH_2 + HX \longrightarrow CH_3CH_2X$$

ethylene · *an ethyl halide*

$$CH{\equiv}CH \xrightarrow{\text{HX}} CH_2{=}CHX \xrightarrow{\text{HX}} CH_3CHX_2$$

acetylene · *a vinyl halide* · *a 1,1-dihaloethane*

The addition of hydrogen halides to alkenes is often used as a synthetic re-action. Usually, the gaseous HX is bubbled through a solution of the alkene. (Concentrated aqueous solutions of hydrogen halides give mixtures of prod-ucts because water can also add to double bonds.) The relative reactivity of HX in this reaction is HI > HBr > HCl > HF. The strongest acid (HI) is the most reactive toward alkenes, while the weakest acid (HF) is the least reactive.

A hydrogen halide contains a highly polar H—X bond and can easily lose H^+ to the π bond of an alkene. The result of the attack of H^+ is an intermedi-ate carbocation, which quickly undergoes reaction with a negative halide ion to yield an alkyl halide. Because the initial attack is by an electrophile, the addition of HX to an alkene is called an **electrophilic addition reaction.**

Step 1 (slow):

$$H-\ddot{C}l\colon$$

$$CH_3CH \!=\! CHCH_3 \longrightarrow \left[\begin{array}{c} H \\ | \\ CH_3\overset{+}{C}HCHCH_3 \end{array}\right] + \colon\!\ddot{C}l\colon^-$$

2-butene   *intermediate carbocation*

Step 2 (fast):

$$\overset{..}{C}l\colon$$

$$[CH_3\overset{+}{C}HCH_2CH_3] + \colon\!\ddot{C}l\colon^- \longrightarrow CH_3\overset{|}{C}HCH_2CH_3$$

2-chlorobutane

A. Markovnikov's Rule

If an alkene is *unsymmetrical* (that is, if the groups attached to the two sp^2-hybridized carbons differ), there is the possibility of two different products from HX addition.

$$CH_3CH \!=\! CHCH_3 \qquad\qquad CH_3CH \!=\! CH_2$$

symmetrical alkenes   *unsymmetrical alkenes*

$$CH_3CH=CHCH_3 \xrightarrow{HCl} \begin{array}{cc} H & Cl \\ | & | \\ CH_3CH\!-\!CHCH_3 \end{array}$$

2-butene   2-chlorobutane

symmetrical   *only one possible product*

$$CH_3CH=CH_2 \xrightarrow{HCl}$$

propene

unsymmetrical

$$CH_3CH_2\!-\!CH_2Cl$$
1-chloropropane

$$\begin{array}{c} Cl \\ | \\ CH_3CH\!-\!CH_3 \end{array}$$
2-chloropropane

two possible products

In an electrophilic addition that can lead to two products, one product usu-ally predominates over the other. In 1869, the Russian chemist Vladimir

Markovnikov formulated the following empirical rule. *In additions of HX to unsymmetrical alkenes, the H^+ of HX goes to the double-bonded carbon that already has the greatest number of hydrogens.*

Vladimir Vasil'evich Markovnikov (also spelled Markownikoff; 1838–1904) formulated his rule in his 1869 doctoral dissertation and published it in *Annalen der Chemie* in 1870. As he stated it, the rule emphasized the direction of substitution of the halogen (rather than the hydrogen) of a hydrogen halide. Over time, the wording of Markovnikov's rule has been modified and today emphasizes the direction of substitution of the hydrogen (rather than the halogen) of the hydrogen halide. This modern version of Markovnikov's rule correlates with our current mechanistic concepts of the reaction.

Markovnikov was appointed to the chair of chemistry at the University of Moscow in 1873. The previous holder of the chair had died and no chemistry had been taught in the intervening two years. Upon his arrival, Markovnikov was required to teach chemistry to all science, medical, and pharmacy students who had not taken chemistry during the last two years as well as to students of the current year.

In 1881 Markovnikov was summarily dismissed from his academic position because he refused to sign an apology required of the faculty by a political official who had been insulted by a student.

By Markovnikov's rule, we would predict that the reaction of HCl with propene yields 2-chloropropane (and not the 1-chloro isomer). Examples of reactions that obey Markovnikov's rule follow:

$$CH_3CH{=}CH_2 \xrightarrow{\text{HCl}} CH_3\overset{\overset{\displaystyle Cl}{|}}{C}H{-}CH_3$$

H goes here

propene   2-chloropropane

$$(CH_3)_2C{=}CHCH_3 \xrightarrow{\text{HBr}} (CH_3)_2\overset{\overset{\displaystyle Br}{|}}{C}{-}CH_2CH_3$$

H goes here

2-methyl-2-butene   2-bromo-2-methylbutane

H goes here

1-methylcyclohexene $\xrightarrow{\text{HI}}$ 1-iodo-1-methylcyclohexane

The addition of HX to an alkene is referred to as a **regioselective reaction** (from the Latin *regio*, "direction"), a reaction in which one direction of addition to an unsymmetrical alkene predominates over the other direction. Therefore, a regioselective addition reaction is a reaction in which two isomeric products could be obtained, but one predominates. The addition of HX to an alkene is regioselective because the H^+ of HX becomes bonded to the alkenyl carbon that already has the greater number of hydrogens. The reason for this selectivity is that this path of addition yields the more stable of the two possible intermediate carbocations, as explained in the next section.

B. The Reason for Markovnikov's Rule

Markovnikov formulated his rule because of experimental observations. Why is this empirical rule followed? To answer this question, let us return to the

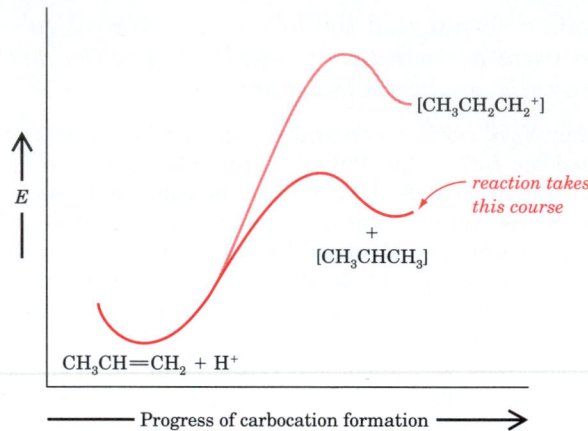

Figure 10.8 Energy diagram for the protonation of propene.

mechanism of HX addition. *Step 1* is the formation of a carbocation. For propene, two possible carbocations could be formed:

$$CH_3CH\overset{H^+}{=\!\!\!=}CH_2 \longrightarrow \left[\begin{matrix} H^{\delta+} \\ \vdots \\ CH_3CH\overset{\delta+}{=\!=\!=}CH_2 \end{matrix} \right] \longrightarrow CH_3CH_2\overset{+}{C}H_2$$

transition state   *1°, less stable*

$$CH_3CH\overset{H^+}{=\!\!\!=}CH_2 \longrightarrow \left[\begin{matrix} H^{\delta+} \\ \vdots \\ CH_3\overset{\delta+}{CH}=\!=\!=\overset{}{C}H_2 \end{matrix} \right] \longrightarrow CH_3\overset{+}{C}HCH_3$$

transition state   *2°, more stable*

The order of stability of carbocations is 3° > 2° > 1°. For propene, the two positions of H^+ addition lead to (1) a high-energy, unstable, primary carbocation, or (2) a lower-energy, more-stable, secondary carbocation. The transition states leading to these intermediates have carbocation character. Therefore, the secondary carbocation has a lower-energy transition state and a faster rate of formation. (See Figure 10.8.)

Addition of a reagent to an unsymmetrical alkene proceeds by way of the more stable carbocation. This is the reason that Markovnikov's rule is followed.

$$CH_3CH\overset{H^+}{=\!\!\!=}CH_2 \xrightarrow{H^+} [CH_3\overset{+}{C}HCH_3] \xrightarrow{Cl^-} \overset{\displaystyle Cl}{\underset{\displaystyle |}{CH_3CHCH_3}}$$

$$(CH_3)_2C\overset{H^+}{=\!\!\!=}CHCH_3 \xrightarrow{H^+} [(CH_3)_2\overset{+}{C}CH_2CH_3] \xrightarrow{Br^-} \overset{\displaystyle Br}{\underset{\displaystyle |}{(CH_3)_2CCH_2CH_3}}$$

Predict the relative rates of reaction of the following alkenes toward HBr (lowest rate first):

(a) $CH_3CH_2CH\!=\!CH_2$ **(b)** $CH_2\!=\!CH_2$ **(c)** $(CH_3)_2C\!=\!CHCH_3$

Solution

The alkene that can form the most stable carbocation has the lowest E_{act} and the fastest reaction.

$$CH_2{=}CH_2 \qquad CH_3CH_2CH{=}CH_2 \qquad (CH_3)_2C{=}CHCH_3$$

increasing rate of reaction →

STUDY PROBLEMS

10.8 For the alkenes in (a) and (c) in the preceding sample problem, give the structures of the carbocation intermediate and the major product of the reaction with HBr.

10.9 Because the addition of HX to an alkene proceeds by way of a carbocation, rearrangements can occur (see Section 5.5F). For each of the following reactions, predict the products, including the rearranged product (if any).

(a) 3,3-dimethyl-1-butene + HI $\longrightarrow$

(b) styrene ($C_6H_5CH{=}CH_2$) + HCl $\longrightarrow$

(c) 3-methyl-1-butene + HCl $\longrightarrow$

(d) 4,4-dimethyl-2-pentene + HCl $\longrightarrow$

C. Anti-Markovnikov Addition of HBr

The addition of HBr to alkenes sometimes proceeds by Markovnikov's rule, but sometimes does not. (This effect is not observed with HCl or HI.)

$$CH_3CH{=}CH_2 \xrightarrow{\text{HBr}} \underset{\text{2-bromopropane}}{CH_3\overset{\text{Br}}{\underset{|}{C}}HCH_3} \quad \text{but sometimes} \quad \underset{\substack{\text{1-bromopropane} \\ \textit{anti-Markovnikov product}}}{CH_3CH_2CH_2Br}$$

It has been observed that the primary alkyl bromide is obtained only when a peroxide or O_2 is present in the reaction mixture. Oxygen is a stable diradical (see Section 6.6B), and peroxides (ROOR) are easily cleaved into radicals. When O_2 or peroxides are present, HBr addition proceeds through a *free-radical* mechanism instead of an ionic one.

Formation of Br·:

$$ROOR \longrightarrow 2\,RO\cdot$$

$$RO\cdot + HBr \longrightarrow ROH + Br\cdot$$

Addition of Br· to alkene:

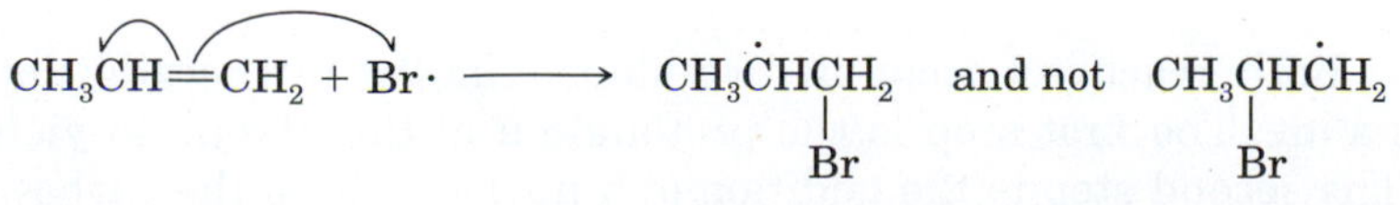

$$CH_3CH{=}CH_2 + Br\cdot \longrightarrow CH_3\overset{\cdot}{C}HCH_2Br \quad \text{and not} \quad CH_3CH\overset{\cdot}{C}H_2Br$$

$$\underset{2°,\ \textit{more stable}}{} \qquad \underset{1°,\ \textit{less stable}}{}$$

Formation of product:

$$CH_3\overset{\cdot}{C}HCH_2Br + H\!-\!Br \longrightarrow CH_3CH_2CH_2Br + Br\cdot$$

When Br· attacks the alkene, the more stable radical is formed. (The stability of radicals, like that of carbocations, is in the order 3° > 2° > 1°.) The result of the free-radical addition in our example is 1-bromopropane.

Hydrogen chloride does not undergo free-radical addition to alkenes because of the relative slowness of the homolytic cleavage of HCl into radicals. Hydrogen iodide does not undergo this reaction because the addition of I· to alkenes is endothermic and too slow to sustain a chain reaction.

STUDY PROBLEM

10.10   (a) Predict the major product, and (b) write a complete reaction mechanism for each of the following reactions.

(a)   [structure] + HBr $\xrightarrow{\text{ROOR}}$

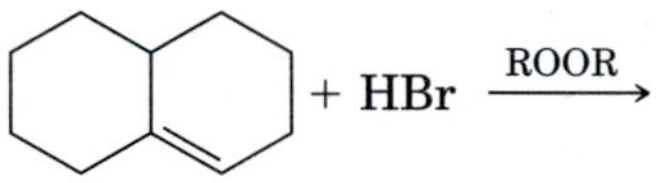

(b)   [structure with CH$_2$] + HBr $\xrightarrow{\text{no ROOR}}$

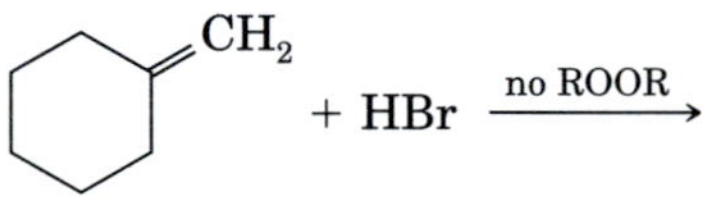

Section 10.7
Addition of H₂SO₄ and H₂O to Alkenes and Alkynes

Sulfuric acid undergoes addition to an alkene just as a hydrogen halide does. The product is an alkyl hydrogen sulfate, which can be used to synthesize alcohols or ethers. (See Section 8.2A.)

$$CH_3CH\!=\!CH_2 + H\!-\!OSO_3H \longrightarrow \overset{\displaystyle OSO_3H}{\underset{\displaystyle CH_3\overset{|}{C}H\!-\!CH_3}{}}$$

propene   1-methylethyl hydrogen sulfate

In strongly acidic solution (such as aqueous sulfuric acid), water adds to a double bond to yield an alcohol. This reaction is called the **hydration of an alkene.**

$$CH_3CH\!=\!CH_2 + H_2O \xrightarrow{\text{H}^+} \overset{\displaystyle OH}{\underset{\displaystyle CH_3\overset{|}{C}H\!-\!CH_3}{}}$$

propene   2-propanol (60%)

Both reactions occur in two steps, similar to the addition of a hydrogen halide. The first step is the protonation of the alkene to yield a carbocation. The second step is the addition of a nucleophile to the carbocation. Because a carbocation is formed initially, both reactions follow Markovnikov's rule. Rearrangements are to be expected if the carbocation can undergo a 1,2-shift of H or R to yield a more stable carbocation.

Step 1:

$$R_2C{=}CHR + H^+ \rightleftharpoons [R_2\overset{+}{C}{-}CH_2R]$$

Step 2:

$$[R_2\overset{+}{C}{-}CH_2R] + H_2\ddot{O}: \rightleftharpoons \begin{matrix} H{-}\overset{+}{\ddot{O}}H \\ | \\ R_2C{-}CH_2R \end{matrix} \rightleftharpoons \begin{matrix} :\ddot{O}H \\ | \\ R_2C{-}CH_2R + H^+ \end{matrix}$$

a protonated alcohol   an alcohol

Hydration reactions of alkenes are the reverse of dehydration reactions of alcohols. The products depend on the conditions used. (See Sections 7.6 and 8.2A.)

Alkynes also undergo hydration, but the initial product is a vinylic alcohol, or **enol.** An enol is in equilibrium with an aldehyde or ketone. (See Section 13.9.) The equilibrium almost always favors the carbonyl compound. Therefore, hydration of an alkyne actually results in an aldehyde or ketone. (The hydration of alkynes proceeds more smoothly when a mercuric salt is added to catalyze the reaction.

$$CH_3(CH_2)_3C{\equiv}CH + H_2O \xrightarrow[H_2SO_4]{HgSO_4} \left[\begin{matrix} OH \\ | \\ CH_3(CH_2)_3C{=}CH_2 \end{matrix}\right] \rightleftharpoons \begin{matrix} O \\ \| \\ CH_3(CH_2)_3CCH_3 \end{matrix}$$

an enol   2-hexanone (80%)

Acid-catalyzed hydration of an alkene or alkyne is seldom used in the laboratory because of the relatively low yields and possibilities of rearrangements and polymerization. However, many simple alcohols, such as ethanol and 2-propanol, are synthesized industrially by this technique.

Section 10.8
Hydration Using Mercuric Acetate

Mercuric acetate, $Hg(O_2CCH_3)_2$, and water add to alkenes in a reaction called **oxymercuration.** Unlike the addition reactions we have discussed so far, oxymercuration proceeds *without rearrangement*. The product of oxymercuration is usually reduced with sodium borohydride ($NaBH_4$) in a subsequent reaction called **demercuration** to yield an alcohol, the same alcohol that would be formed if water had been added across the double bond. Oxymercuration–demercuration reactions give better yields of alcohols than the addition of water with H_2SO_4 because the alkene cannot undergo acid-catalyzed rearrangement and polymerization.

Oxymercuration:

$$CH_3CH_2CH_2CH{=}CH_2 \xrightarrow[H_2O]{Hg\left(\begin{smallmatrix} O \\ \| \\ OCCH_3 \end{smallmatrix}\right)_2} \begin{matrix} OH \\ | \\ CH_3CH_2CH_2CHCH_2 \\ | \\ HgO_2CCH_3 \end{matrix}$$

1-pentene

Demercuration:

$$\underset{\substack{| \\ \text{CH}_3\text{CH}_2\text{CH}_2\,\text{CHCH}_2\text{—HgO}_2\text{CCH}_3}}{\overset{\text{OH}}{}} \xrightarrow{\text{NaBH}_4} \underset{\substack{| \\ \text{CH}_3\text{CH}_2\text{CH}_2\,\text{CHCH}_3 + \text{Hg}}}{\overset{\text{OH}}{}}$$

2-pentanol
(90% overall)

Like the addition of other reagents to alkenes, oxymercuration is a two-step process. The addition proceeds by electrophilic attack of $^+\text{HgO}_2\text{CCH}_3$ followed by nucleophilic attack of H_2O. Because rearrangements do not occur, the intermediate formed by electrophilic attack cannot be a true carbocation. On the other hand, since Markovnikov's rule is followed, the intermediate must have some carbocation character. Both these facts are explained by postulating a **bridged ion,** or **cyclic ion,** as the intermediate.

Dissociation of mercuric acetate:

$$\text{Hg(O}_2\text{CCH}_3)_2 \rightleftharpoons \ ^+\text{HgO}_2\text{CCH}_3 + \ ^-\text{O}_2\text{CCH}_3$$

Electrophilic attack:

$$\underset{\searrow \ ^+\text{HgO}_2\text{CCH}_3}{\text{R}_2\text{C}=\text{CHR}} \longrightarrow \left[\ \overset{\delta+}{\text{R}_2\text{C}}\cdots\underset{\substack{| \\ \underset{\delta+}{\text{HgO}_2\text{CCH}_3}}}{\text{CHR}}\ \right]$$

a bridged intermediate

Attack of H₂O and proton loss:

$$\left[\ \overset{\delta+}{\text{R}_2\text{C}}\cdots\underset{\substack{| \\ \underset{\delta+}{\text{HgO}_2\text{CCH}_3}}}{\text{CHR}}\ \right] \xrightarrow{\text{H}_2\ddot{\text{O}}:} \left[\ \underset{\substack{| \\ \text{HgO}_2\text{CCH}_3}}{\overset{^+:\text{OH}_2}{\text{R}_2\text{C}}}\text{—CHR}\ \right] \xrightarrow{-\text{H}^+} \underset{\substack{| \\ \text{HgO}_2\text{CCH}_3}}{\overset{:\ddot{\text{O}}\text{H}}{\text{R}_2\text{C}}}\text{—CHR}$$

The formation of a bridged intermediate is not very different from the formation of a carbocation. The reaction of this intermediate is also very similar to that of a carbocation. The difference between this bridged intermediate and a true carbocation is that Hg is partially bonded to *each* double-bond carbon and rearrangements cannot occur. The more positive carbon in the bridged intermediate (the carbon attacked by H_2O) can be predicted from carbocation stabilities ($3° > 2° > 1°$). We can compare the reaction of this bridged ion to the reaction of the protonated intermediate formed during the acid-catalyzed substitution reaction of an epoxide (Section 8.4B).

more positive carbon;
H₂O attacks here

$$\underset{\substack{| \\ \underset{\delta+}{\text{HgO}_2\text{CCH}_3}}}{\overset{\delta+}{\text{R}_2\text{C}}}\text{—CHR} \qquad \text{similar to} \qquad \underset{\substack{| \\ \text{HgO}_2\text{CCH}_3}}{\overset{+}{\text{R}_2\text{C}}}\text{—CHR}$$

bridged intermediate   *true carbocation*

The reducing agent in the demercuration reaction, sodium borohydride, is an important reducing agent in organic chemistry. It forms stable solutions in aqueous base, but decomposes and releases H_2 in acidic solution. We will encounter this reagent again as a reducing agent for aldehydes and ketones (Section 13.6B).

$$Na^+ \quad H-\overset{\overset{\displaystyle H}{|}}{\underset{\underset{\displaystyle H}{|}}{B}}{}^-\!\!-H$$

sodium borohydride

STUDY PROBLEMS

10.11 **(a)** Write equations for the steps in the oxymercuration–demercuration of 3,3-dimethyl-1-butene.

(b) Compare the product of this reaction sequence with the product from the reaction of 3,3-dimethyl-1-butene and dilute aqueous HCl.

10.12 Oxymercuration–demercuration results in alcohols as products. If the reaction is carried out in an alcohol solvent instead of in water (a sequence called **solvomercuration–demercuration**), an *ether* is obtained as a product. Write equations to show how you would prepare 2-methoxy-2-methylbutane from 2-methyl-1-butene by this technique.

Section 10.9
Addition of Borane to Alkenes

A. Hydroboration

Diborane (B_2H_6) is a toxic gas prepared by the reaction of sodium borohydride and boron trifluoride ($3\,NaBH_4 + 4\,BF_3 \rightarrow 2\,B_2H_6 + 3\,NaBF_4$). In diethyl ether solution, diborane dissociates into borane (BH_3) solvated by an ether molecule: $(CH_3CH_2)_2O\!:\!\text{----}BH_3$. Borane undergoes rapid and quantitative reaction with most alkenes to form **organoboranes** (R_3B). The overall reaction is the result of three separate reaction steps. In each step, one alkyl group is added to borane until all three hydrogen atoms have been replaced by alkyl groups. This sequence of reactions is called **hydroboration.**

Step 1:

$$CH_2{=}CH_2 + \overset{\overset{\displaystyle H}{|}}{\underset{\underset{\displaystyle H}{|}}{B}}{-}H \longrightarrow CH_3-\overset{\overset{\displaystyle BH_2}{|}}{CH_2}$$

Step 2:

$$CH_2{=}CH_2 + CH_3CH_2BH_2 \longrightarrow (CH_3CH_2)_2BH$$

Step 3:

$$CH_2{=}CH_2 + (CH_3CH_2)_2BH \longrightarrow (CH_3CH_2)_3B$$

triethylborane

an organoborane

Organoboranes were discovered in the 1950s by Herbert C. Brown at Purdue University, who was awarded a Nobel prize in 1979 for his work with organoboron compounds. The value of these compounds arises from the variety of other compounds that can be synthesized from them. Let us first consider the addition of BH_3 to alkenes and then look at some of the products that can be obtained from the organoboranes.

Borane is different from the other addition reagents we have mentioned, because H is the *electronegative* portion of the molecule instead of the electropositive portion as it is in HCl or H_2O. When borane adds to a double bond, the hydrogen (as a hydride ion, H^-) becomes bonded to the *more substituted carbon*. The result is what appears to be anti-Markovnikov addition.

$$CH_3CH{=}CH_2 \longrightarrow CH_3CH{-}CH_2$$

H on more substituted carbon

Steric hindrance also plays a role in the course of this reaction. Best yields of the anti-Markovnikov organoborane are obtained when one carbon of the double bond is substantially more hindered than the other.

less hindered

$$CH_3CH_2CH{=}CH_2$$

94% yield of anti-Markovnikov product

more hindered

$$CH_3CH_2C{=}CH_2$$

99% yield of anti-Markovnikov product

Organoboranes are easily oxidized to alcohols by alkaline hydrogen peroxide. The result of borane addition, followed by H_2O_2 oxidation, appears as if water had been added to the double bond in an anti-Markovnikov manner.

$$3\ CH_3CH_2CH_2CH_2CH{=}CH_2 \xrightarrow{BH_3} (CH_3CH_2CH_2CH_2CH_2CH_2)_3B$$

1-hexene

$$\xrightarrow{H_2O_2,\ OH^-} 3\ CH_3(CH_2)_4CH_2OH$$

1-hexanol (90%)

Besides being oxidized to alcohols, organoboranes can be converted to alkanes, alkyl halides, or other products. In each case, the newly introduced atom or group becomes bonded to the less-substituted carbon of the double bond.

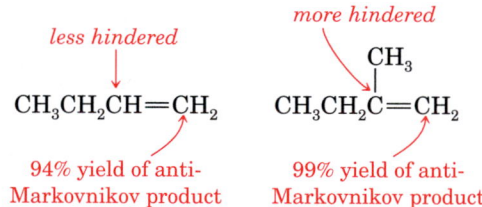

$$3\ CH_3CH{=}CH_2 \xrightarrow{BH_3} (CH_3CH_2CH_2)_3B$$

$\xrightarrow{3\ CH_3CO_2D}$ $3\ CH_3CH_2CH_2D$

1-deuteriopropane

a deuteriated alkane

$\xrightarrow{3\ Br_2,\ CH_3O^-}$ $3\ CH_3CH_2CH_2Br$

1-bromopropane

an alkyl halide

B. Stereochemistry of Hydroboration

When borane adds to a double bond, the boron atom and the hydride ion add to the two carbon atoms of the double bond *simultaneously*. The result is that the B and H must add to the same side of the double bond. An addition reaction in which two species add to the same side is called a **syn-addition.** (*Syn,* like *cis,* means "on the same side, or face.")

A syn-addition:

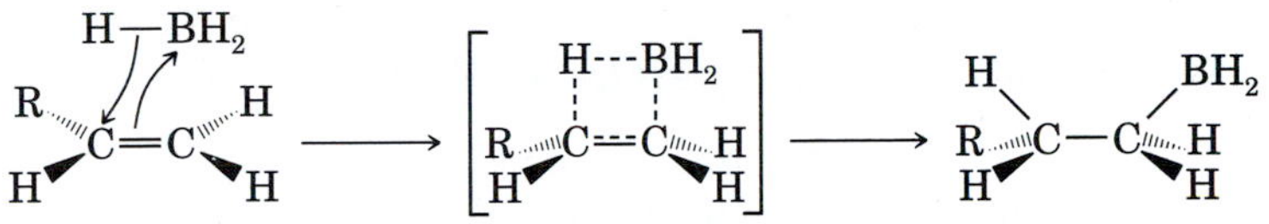

transition state

If the addition product is capable of geometric isomerism, such as the addition product of 1-methylcyclopentene, the B and the H are thus *cis* to each other in the product.

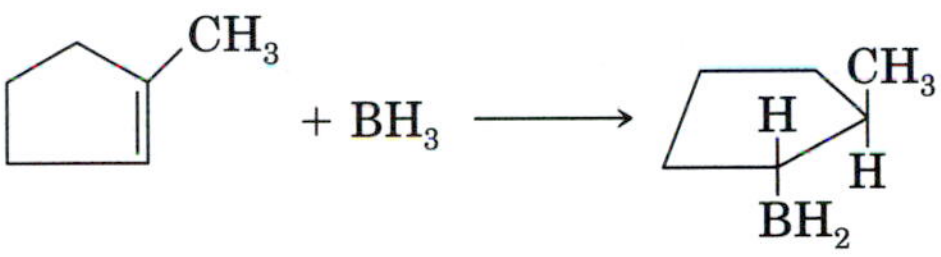

When an organoborane is oxidized to an alcohol, the hydroxyl group that replaces the boron atom does so with **retention of configuration at the carbon.**

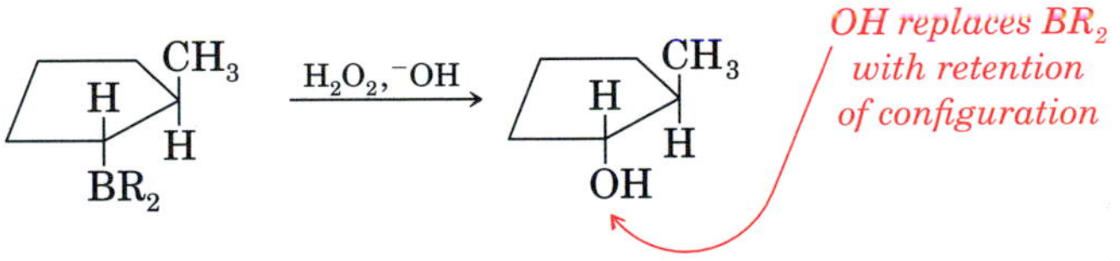

OH replaces BR$_2$ with retention of configuration

trans isomer   *trans*-2-methylcyclopentanol (87%)

The configuration is retained because the oxidation occurs by a 1,2-shift. Subsequent hydrolysis (not an oxidation) of the boron−oxygen groups does not affect stereochemistry of the carbon−oxygen group.

Retention of configuration in oxidation of a borane:

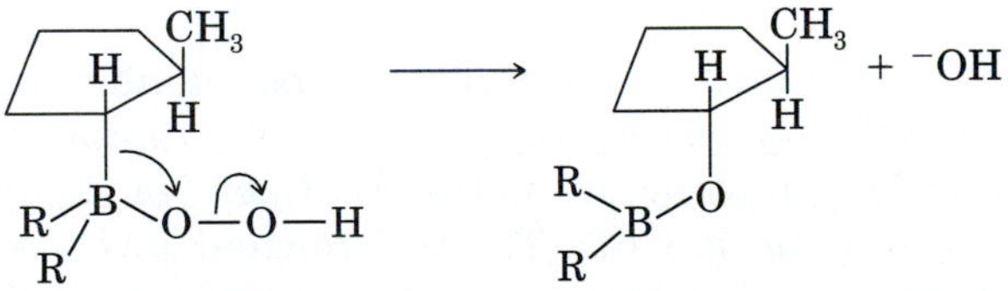

General equation for the oxidation and hydrolysis of boranes:

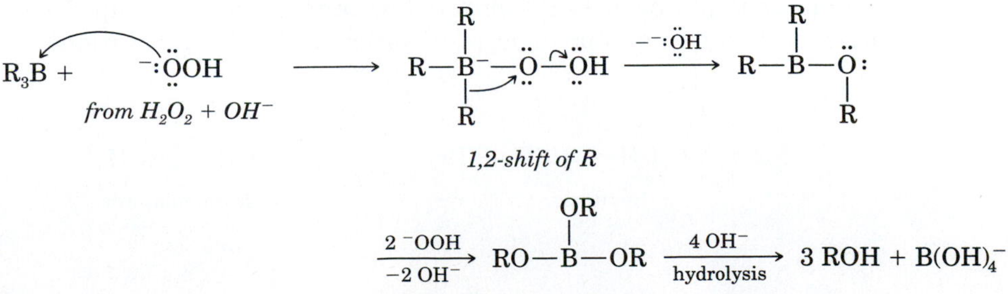

from H$_2$O$_2$ + OH$^-$

1,2-shift of R

$$\xrightarrow[-2\,^-\text{OH}]{2\,^-\text{OOH}} \quad RO-\underset{\underset{OR}{|}}{\overset{\overset{OR}{|}}{B}}-OR \quad \xrightarrow[\text{hydrolysis}]{4\,OH^-} \quad 3\;ROH + B(OH)_4^-$$

STUDY PROBLEMS

10.13   Use a transition-state structure to show why *trans*-2-methylcyclohexylborane cannot yield *cis*-2-methylcyclohexanol when treated with alkaline hydrogen peroxide.

10.14   Predict the major organic products (including stereochemistry) of the hydroboration and oxidation of **(a)** 1-ethylcyclopentene: **(b)** (*Z*)-3-methyl-2-pentene.

C. Summary of Alcohol Synthesis from Alkenes

Now, we have discussed three routes to alcohols from alkenes. These are summarized in the following flow diagram:

$$
R_2C{=}CHR
\begin{cases}
\xrightarrow{\text{H}_2\text{O, H}^+} & R_2\overset{\displaystyle OH}{\underset{\displaystyle |}{C}}CH_2R & \textit{low yields, possible rearrangement} \\[2em]
\xrightarrow[\text{(2) NaBH}_4]{\text{(1) Hg(O}_2\text{CCH}_3)_2,\ \text{H}_2\text{O}} & R_2\overset{\displaystyle OH}{\underset{\displaystyle |}{C}}CH_2R & \textit{excellent yields, Markovnikov product, no rearrangement} \\[2em]
\xrightarrow[\text{(2) H}_2\text{O}_2,\ \text{OH}^-]{\text{(1) BH}_3} & R_2CH\overset{\displaystyle OH}{\underset{\displaystyle |}{C}}HR & \textit{excellent yields, anti-Markovnikov, no rearrangement}
\end{cases}
$$

STUDY PROBLEM

10.15   Write equations showing the product(s) for the conversion of 1-methyl-cyclohexene to an alcohol using the three different synthetic methods discussed in this section.

Section 10.10

Addition of Halogens to Alkenes and Alkynes

Chlorine and bromine add to carbon–carbon double bonds and triple bonds. A common laboratory test for the presence of a double or triple bond in a compound of unknown structure is the treatment of the compound with a dilute solution of bromine in CCl_4. The test reagent has the reddish-brown color of Br_2; disappearance of this color is a positive test. The decolorization of a Br_2/CCl_4 solution by an unknown is suggestive, but not definitive proof, that a double or triple bond is present. A few other types of compounds, such as aldehydes, ketones, and phenols, also decolorize Br_2/CCl_4 solution.

$$
\underset{\text{2-butene}}{CH_3CH{=}CHCH_3} + \underset{\textit{red}}{Br_2} \xrightarrow{CCl_4} \underset{\substack{\text{2,3-dibromobutane} \\[0.3em] \textit{colorless}}}{CH_3\overset{\displaystyle Br}{\underset{\displaystyle |}{C}}H{-}\overset{\displaystyle Br}{\underset{\displaystyle |}{C}}HCH_3}
$$

$$\underset{\underset{red}{\text{2-butyne}}}{CH_3C{\equiv}CCH_3 + 2\ Br_2} \xrightarrow{CCl_4} \underset{\underset{\text{colorless}}{\text{2,2,3,3-tetrabromobutane}}}{CH_3\overset{\displaystyle Br}{\underset{\displaystyle Br}{C}}-\overset{\displaystyle Br}{\underset{\displaystyle Br}{C}}CH_3}$$

Neither F_2 nor I_2 is a useful reagent for alkene addition reactions. Fluorine undergoes an explosive free-radical reaction with organic compounds (see Section 6.2). Iodine adds to a double bond, but the reaction is reversible and the equilibrium lies on the alkene–iodine side.

$$R_2C{=}CR_2 + I_2 \;\rightleftharpoons\; R_2\overset{\displaystyle |}{\underset{\displaystyle I}{C}}-\overset{\displaystyle |}{\underset{\displaystyle I}{C}}R_2$$

A more substituted alkene is more reactive toward X_2 (Br_2 or Cl_2) than a less substituted alkene (see Table 10.4). This is the same general order of reactivity as that toward HX.

$$CH_2{=}CH_2 \quad RCH{=}CH_2 \quad R_2C{=}CH_2 \quad R_2C{=}CHR \quad R_2C{=}CR_2$$

increasing reactivity toward X_2 or HX addition →

A. Electrophilic Attack of Br_2

Halogens undergo addition reactions with alkenes. But, what is the source of the electrophile in Br_2? When Br_2 approaches the π-bond electrons, polarity is induced in the Br_2 molecule by repulsion between the π electrons and the electrons in the Br_2 molecule.

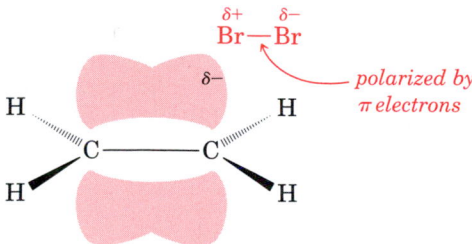

As the Br—Br bond becomes more polarized, it becomes progressively weaker until it finally breaks. The result is a halide ion and a positively charged organohalogen ion, called a **halonium ion.** The halonium ion from

Table 10.4	Relative reactivities of some alkenes toward Br_2 in methanol
Compound	Relative Rate
$CH_2{=}CH_2$	1.0
$CH_3CH_2CH{=}CH_2$	97
$cis\text{-}CH_3CH_2CH{=}CHCH_3$	4,300
$(CH_3)_2C{=}C(CH_3)_2$	930,000

bromine, the **bromonium ion,** is bridged, similar to the intermediate in oxymercuration. (Chloronium ions are also sometimes bridged.) In the addition of Br_2 to ethylene or some other symmetrical alkene, the bridged bromonium ion is symmetrical, with Br bonded equally to each carbon.

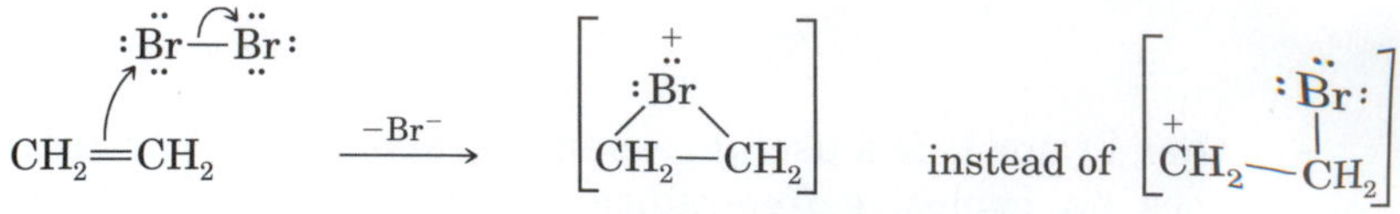

a symmetrical bridged bromonium ion

If the alkene is unsymmetrical, most of the positive charge is carried on the more substituted carbon. Carbocation stability is followed.

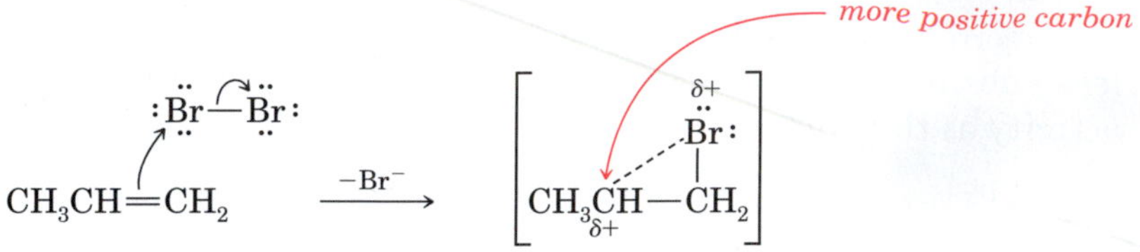

more positive carbon

propene

an unsymmetrical bridged bromonium ion

B. *anti*-Attack of Br⁻

The bridged intermediate ion is positively charged and of high energy. Like a carbocation, it exists only momentarily in solution. Reaction is completed by attack by a nucleophile (in this case, Br^-). Br^- cannot attack a carbon of the bridged intermediate from the top (as we have shown the structure); that path is blocked by the Br bridge. Therefore, Br^- attacks from the opposite side of the intermediate. The result is ***anti*-addition** of Br_2 to the double bond. (Contrast this mode of addition to the *syn*-addition of BH_3, Section 10.9B.)

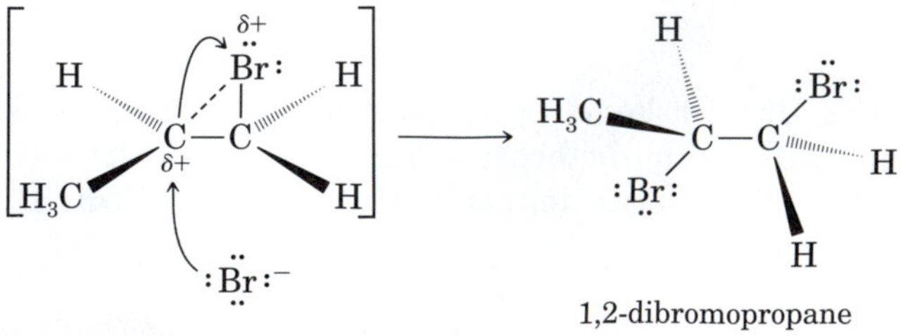

1,2-dibromopropane

General mechanism:

Step 1 (slow):

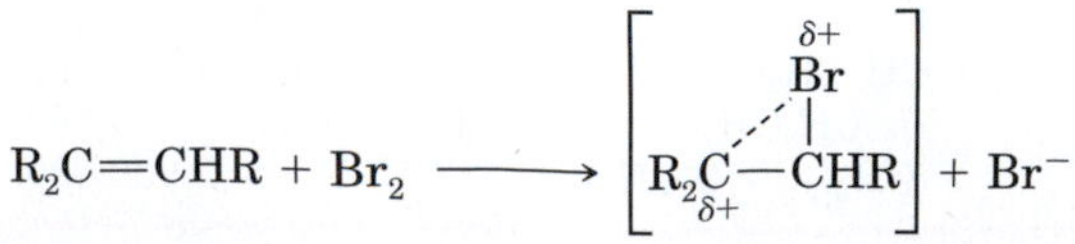

$$R_2C{=}CHR + Br_2 \longrightarrow \left[\underset{\delta+}{R_2C}{-}CHR \right] + Br^-$$

Step 2 (fast):

$$\left[\begin{array}{c} \overset{\delta+}{Br} \\ | \\ R_2\overset{}{C}\text{---}CHR \\ {\scriptstyle\delta+} \end{array} \right] + Br^- \longrightarrow \begin{array}{c} Br \\ | \\ R_2C\text{---}CHR \\ | \\ Br \end{array}$$

C. Evidence for *anti*-Addition

Considerable experimental evidence supports a bromination mechanism involving a bridged-ion intermediate (*anti*-addition) instead of a mechanism involving a simple carbocation intermediate. Bromination reactions of alkenes and alkynes are **stereoselective**—they produce predominantly one stereoisomeric product in reactions where two or more products might be expected. For example, when cyclohexene is brominated, the product is *trans*-1,2-dibromocyclohexane. If the intermediate were a simple carbocation, both *cis*- and *trans*-1,2-dibromocyclohexane should be formed.

cyclohexene

bridge on top

bridged intermediate

trans-1,2-dibromocyclohexane (95%)

and not a carbocation intermediate

Attack of Br⁻ here would lead to the cis isomer.

Attack here would lead to the trans isomer.

STUDY PROBLEM

10.16   2,3-Dibromobutane exists as three stereoisomeric forms: a pair of enantiomers, (2*R*,3*R*) and (2*S*,3*S*), and a *meso* form, (2*R*,3*S*). Use an equation to show which isomers are formed when *cis*-2-butene is brominated.

D. Mixed Addition

Bromination reactions of alkenes proceed by way of a bromonium-ion intermediate, followed by attack by a bromide ion to yield the dibromide. Is the second step limited to attack only by bromide ion? Can other nucleophiles compete with bromide ion in the second step to yield other products? Consider the case of a bromination reaction carried out with Br_2 in a solution containing Cl^- (from, say, NaCl). In this reaction, two nucleophiles (Br^- and Cl^-) are present. In such a case, *mixed dihalide products* are observed—along with the dibromoalkane, we find some bromochloroalkane.

$$CH_2=CH_2 + Br_2 \longrightarrow \left[\underset{CH_2-CH_2}{\overset{\overset{+}{Br}}{\diagdown}} \right] \begin{array}{l} \overset{Br^-}{\nearrow} \quad \underset{CH_2-CH_2Br}{\overset{\overset{Br}{\mid}}{}} \quad \text{1,2-dibromoethane} \\[2em] \underset{Cl^-}{\searrow} \quad \underset{CH_2-CH_2Cl}{\overset{\overset{Br}{\mid}}{}} \end{array}$$

1-bromo-2-chloroethane

STUDY PROBLEM

10.17 Would you expect to find 1,2-dichloroethane as a product in the preceding example? Explain.

SAMPLE PROBLEM

When propene is treated with $Br_2 + Cl^-$, only one bromochloropropane is isolated as a product. What is its structure? Show, by equations, its formation.

Solution

Step 1:

$$CH_3CH=CH_2 + Br_2 \xrightarrow{-Br^-} \left[\underset{\delta+}{CH_3}\overset{\overset{\delta+}{\cdots Br}}{\underset{CH}{\mid}}-CH_2 \right], \quad \text{not} \quad \left[CH_3\overset{\overset{\delta+}{\cdots Br}}{CH}-\underset{\delta+}{CH_2} \right]$$

Step 2:

$$\left[\underset{\delta+}{CH_3}\overset{\overset{\delta+}{\cdots Br}}{\underset{CH}{\mid}}-CH_2 \right] + Cl^- \longrightarrow \underset{\underset{Cl}{\mid}}{CH_3}\overset{\overset{Br}{\mid}}{CHCH_2}, \quad \text{not} \quad CH_3\overset{\overset{Br}{\mid}}{CHCH_2Cl}$$

1-bromo-2-chloropropane

E. Addition of Halogens and Water

When an alkene is treated with a mixture of Cl_2 or Br_2 in water, a 1,2-halohydrin (a compound with a halogen and OH on adjacent carbon atoms) is formed. Recall from Section 8.2C that 1,2-halohydrins can be used to synthesize epoxides. The path is similar to that for mixed halogen addition:

Step 1:

$$(CH_3)_2C=CH_2 + Br_2 \longrightarrow \left[(CH_3)_2\underset{\delta+}{C}\overset{\overset{\delta+}{Br}}{-}CH_2 \right] + Br^-$$

Step 2:

$$\left[(CH_3)_2\underset{\delta+}{C} \overset{\overset{\displaystyle \delta+}{\overset{\displaystyle Br}{\,}}}{-} CH_2 \right] + H_2\ddot{O}: \longrightarrow$$

$$(CH_3)_2\underset{\underset{+}{:\ddot{O}H_2}}{\overset{\overset{Br}{|}}{C}} -CH_2 \underset{\xrightarrow{-H^+}}{\rightleftharpoons} (CH_3)_2\underset{OH}{\overset{\overset{Br}{|}}{C}} -CH_2$$

1-bromo-2-methyl-2-propanol (73%)

a 1,2-bromohydrin

STUDY PROBLEM

10.18 When cyclohexene is treated with aqueous Cl_2, one product is obtained in 70% yield. Write equations for the steps in this reaction (complete with stereochemistry).

Section 10.11

Addition of Carbenes to Alkenes

If a student were asked whether a compound with the structure CH_2 exists, the answer might be, "No, because the carbon only has two bonds." However, in 1959 spectroscopic evidence showed that such a species does have a fleeting existence. The structure $:CH_2$ is called *methylene* and belongs to a class of highly reactive intermediates called **carbenes** ($R_2C:$). There are two types of methylene: *singlet methylene,* with an sp^2-hybridized carbon, and *triplet methylene,* with an sp-hybridized carbon (see Figure 10.9). Singlet methylene is more useful in organic reactions and thus we will confine our discussion to this form of methylene.

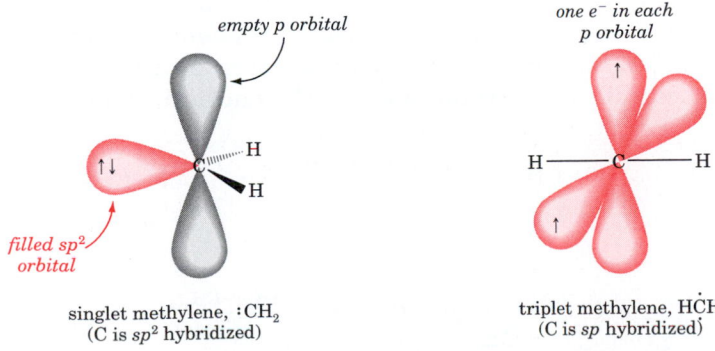

singlet methylene, $:CH_2$
(C is sp^2 hybridized)

triplet methylene, $H\overset{.}{C}H$
(C is sp hybridized)

Figure 10.9 The bonding in singlet methylene and triplet methylene.

Singlet methylene can be prepared by the photolysis (cleavage by light) of diazomethane (CH_2N_2), an unusual and reactive compound itself.

$$[:\bar{C}H_2-\ddot{N}=\overset{+}{N}: \longleftrightarrow :\widehat{C}H_2-\overset{+}{N}\equiv N: \longleftrightarrow CH_2=\overset{+}{N}=\ddot{N}:^- \longleftrightarrow {}^+CH_2-\ddot{N}=\ddot{N}:^-]$$

resonance structures for diazomethane

$$:\bar{C}H_2-\underset{+}{\overset{}{N}}\equiv N: \xrightarrow{h\nu} \quad :CH_2 \quad + :N\equiv N:$$

singlet
methylene

The carbon in $:CH_2$, with only six valence electrons, is electrophilic and adds to the double bond of an alkene to yield a substituted cyclopropane. This reaction is a stereospecific *syn*-addition. Thus, *cis* alkenes yield *cis* cyclopropanes, and *trans* alkenes yield *trans* cyclopropanes.

cis + $:CH_2$ singlet methylene ⟶ a *cis*-dialkylcyclopropane *(meso)*

trans + $:CH_2$ singlet methylene ⟶ a *trans*-dialkylcyclopropane *(racemic)*

STUDY PROBLEM

10.19 Reaction of $CHCl_3$ with *tert*-butoxide ion [$^-OC(CH_3)_3$] yields dichlorocarbene ($:CCl_2$). The reaction proceeds by an initial formation of a carbanion, followed by an α-elimination. **(a)** Write equations for the two-step elimination reaction. **(b)** Write the equation for the reaction of dichlorocarbene with cyclohexene. The product of this reaction is formed in a 60% yield.

Another way to prepare cyclopropanes is with a zinc–copper alloy and diiodomethane, which react to yield a **carbenoid** ("like a carbene"). The carbenoid, in turn, reacts with alkenes just as a carbene would. This reaction sequence, called the **Simmons–Smith reaction,** is fairly general.

$$CH_2I_2 + Zn(Cu) \xrightarrow{\text{diethyl ether}} I-CH_2-ZnI$$

a carbenoid,
which reacts like a carbene

(67%)

The electrophilic addition reactions of alkenes are outlined in Figure 10.10.

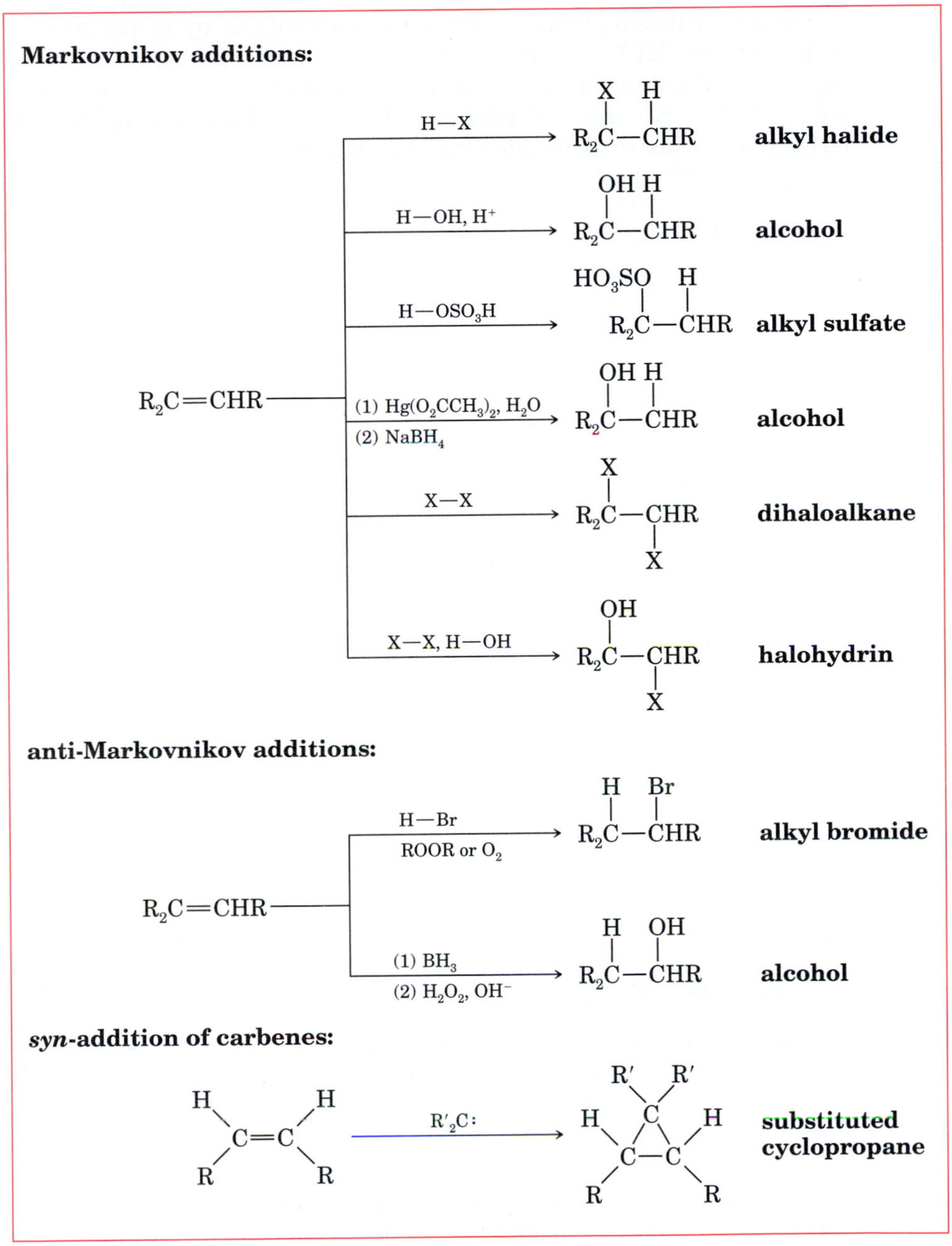

Markovnikov additions:

$$
\begin{array}{ccc}
& & \overset{X}{\underset{|}{}}\ \overset{H}{\underset{|}{}} \\
\xrightarrow{\text{H—X}} & R_2C—CHR & \textbf{alkyl halide}
\end{array}
$$

$$
\xrightarrow{\text{H—OH, H}^+} \quad R_2C—CHR \quad \textbf{alcohol} \quad (OH\ H)
$$

$$
\xrightarrow{\text{H—OSO}_3\text{H}} \quad R_2C—CHR \quad \textbf{alkyl sulfate} \quad (HO_3SO\ H)
$$

$$
\xrightarrow[\text{(2) NaBH}_4]{\text{(1) Hg(O}_2\text{CCH}_3)_2\text{, H}_2\text{O}} \quad R_2C—CHR \quad \textbf{alcohol} \quad (OH\ H)
$$

$$
\xrightarrow{\text{X—X}} \quad R_2C—CHR \quad \textbf{dihaloalkane} \quad (X\ \ldots\ X)
$$

$$
\xrightarrow{\text{X—X, H—OH}} \quad R_2C—CHR \quad \textbf{halohydrin} \quad (OH\ \ldots\ X)
$$

$R_2C{=}CHR$

anti-Markovnikov additions:

$R_2C{=}CHR$

$$
\xrightarrow[\text{ROOR or O}_2]{\text{H—Br}} \quad R_2C—CHR \quad \textbf{alkyl bromide} \quad (H\ Br)
$$

$$
\xrightarrow[\text{(2) H}_2\text{O}_2\text{, OH}^-]{\text{(1) BH}_3} \quad R_2C—CHR \quad \textbf{alcohol} \quad (H\ OH)
$$

syn-addition of carbenes:

$$
\underset{R}{\overset{H}{\diagdown}}C{=}C\underset{R}{\overset{H}{\diagup}} \quad \xrightarrow{\text{R}'_2\text{C:}} \quad \text{substituted cyclopropane}
$$

Figure 10.10   Summary of the electrophilic addition reactions of alkenes.

Catalytic Hydrogenation

The catalytic addition of hydrogen gas to an alkene or alkyne is a reduction of the π-bonded compound. The reaction is general for alkenes, alkynes, and other compounds with π bonds.

The first catalytic hydrogenation was carried out by H. Debus in 1863. He hydrogenated HCN using Pt as catalyst and obtained CH_3NH_2. Systematic investigation of hydrogenation was begun in 1897 by Paul Sabatier (1854–1941). He was awarded the 1912 Nobel Prize (jointly with Victor Grignard) for his work in this area of chemistry.

Alkenes and alkynes:

$$CH_3CH=CH_2 + H_2 \xrightarrow{Pt} CH_3CH_2CH_3$$

propene   propane

$$CH_3C\equiv CH + 2\,H_2 \xrightarrow{Pt} CH_3CH_2CH_3$$

propyne   propane

Other π systems:

$$\overset{O}{\overset{\|}{CH_3CCH_3}} + H_2 \xrightarrow{Pt} \overset{OH}{\overset{|}{CH_3CHCH_3}}$$

acetone   2-propanol

$$CH_3C\equiv N + 2\,H_2 \xrightarrow{Pt} CH_3CH_2NH_2$$

ethanenitrile   ethylamine
(acetonitrile)

$$\text{benzene} + 3\,H_2 \xrightarrow[\text{heat, pressure}]{Pt} \text{cyclohexane}$$

benzene   cyclohexane

A. Action of the Catalyst

Hydrogenation reactions are exothermic, but they do not proceed spontaneously because the energies of activation are extremely high. Heating cannot supply the energy needed to get the molecules to the transition state. However, reaction proceeds smoothly when a catalyst is added.

A finely divided metal or a metal adsorbed onto an insoluble, inert carrier (such as elemental carbon or barium carbonate) is often used as a hydrogenation catalyst. The metal chosen depends on the compound to be reduced and the conditions of the hydrogenation. For example, platinum, palladium, nickel, rhenium, and copper are all suitable for the reduction of alkenes. For esters, which are far more difficult to reduce, a copper–chromium catalyst (plus heat and pressure) is usually used.

A **poisoned catalyst** (one that is partly deactivated) is used for hydrogenation of an alkyne to an alkene instead of to an alkane. Palladium that has been treated with quinoline (page 474) is a typical poisoned catalyst.

$$CH_3C\equiv CH + H_2 \xrightarrow{\text{deactivated Pd}} CH_3CH=CH_2$$

propyne   propene

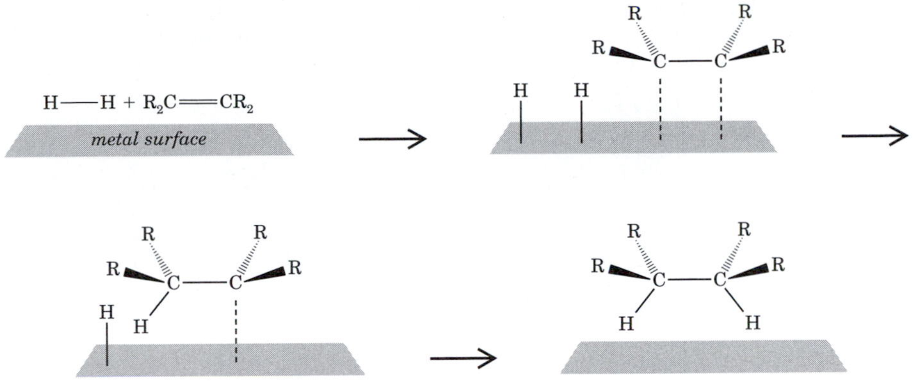

Figure 10.11 Hydrogenation of an alkene.

How does a hydrogenation catalyst ease the course of a hydrogenation reaction? Experimental evidence supports the theory that first the hydrogen
molecules are adsorbed onto the metallic surface; then the H_2 σ bonds are
broken and metal–H bonds are formed. The alkene is also adsorbed onto the
metallic surface, with its π bond interacting with the empty orbitals of the
metal. The alkene molecule moves around on the surface until it collides with
a metal-bonded hydrogen atom, undergoes reaction, and then leaves as the
hydrogenated product (see Figure 10.11).

The overall effect of the catalyst is to provide a surface on which the reaction can occur and to weaken the bonds of both H_2 and the alkene. The result
is a *lowering of the energy of activation for the reaction.* Figure 10.12 shows
energy diagrams for a hydrogenation reaction. Note that *the catalyst does not
affect the energies of reactants or products.* The ΔH for the reaction is not
changed by catalytic action; only the E_{act} is changed.

In recent years, soluble catalysts have been developed so that hydrogenation can occur in a homogeneous solution, rather than on a surface. These catalysts, organic metal complexes such as $[(C_6H_5)_3P]_3RhCl$, selectively catalyze
the reduction of nonhindered double bonds. For example, if two double bonds
of a diene differ in steric hindrance, the less hindered double bond can be selectively hydrogenated.

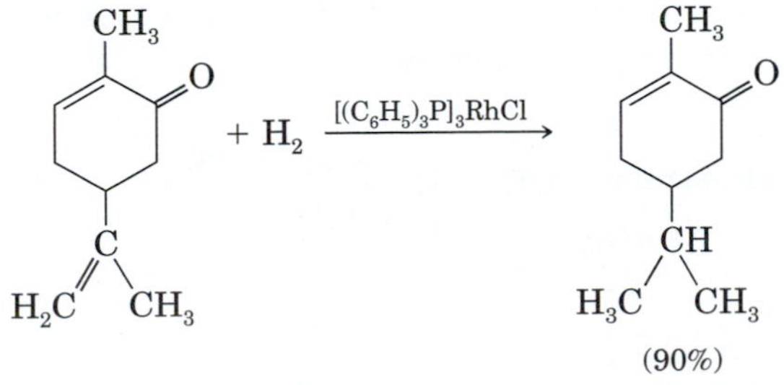

If the soluble catalyst is complexed with an optically active chelating
agent, asymmetry can be introduced by the hydrogenation reaction.

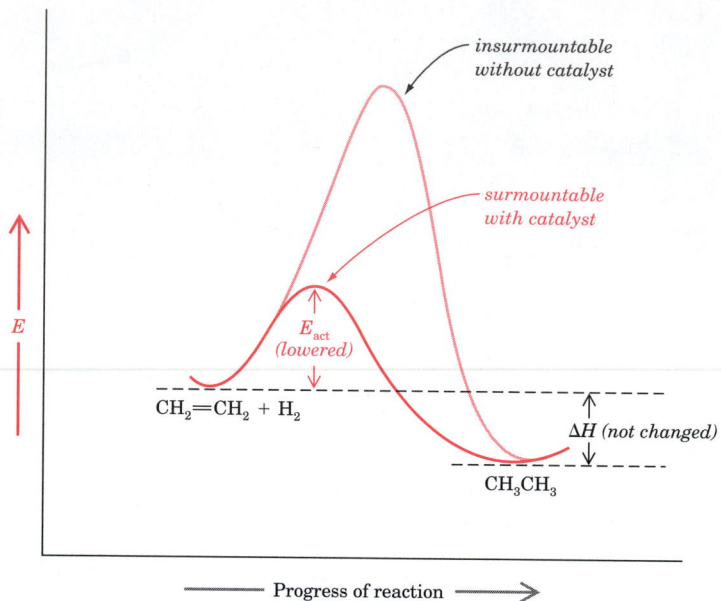

Figure 10.12 Idealized energy diagrams for a hydrogenation reaction. (The reaction actually occurs stepwise.)

B. Stereochemistry of Hydrogenation

Evidence shows that the two hydrogen atoms add to the same side of, or *syn* to, the double bond when a solid catalyst is used. The *syn*-addition arises from the reaction occurring on a surface. Access by H_2 to only one side of the π bond is more favorable than access to both sides. If the hydrogenation products are capable of geometric isomerism, the *cis* product is usually observed as the predominant product. (In some cases, however, isomerization to the more stable *trans* product occurs.)

SAMPLE PROBLEM

The *cis* isomers of alkenes can be prepared by the catalytic hydrogenation of alkynes. The *trans* isomers of alkenes can be obtained in excellent yields by a **dissolving-metal reduction** of the alkynes: a reduction using a solution of sodium or lithium metal in liquid ammonia at $-33°$. In a dissolving-metal reduction, the alkali metal is oxidized to a cation, while protons for the reduction are obtained from NH_3. Write balanced equations showing the reduction of 3-octyne by (a) hydrogenation and (b) Na in liquid NH_3. Name the products.

Solution

(a) $CH_3CH_2C \equiv C(CH_2)_3CH_3 + H_2 \xrightarrow{\text{catalyst}}$

$$\underset{\substack{\\ \textit{cis-}3\text{-octene}}}{\overset{\displaystyle CH_3CH_2 \diagdown \diagup (CH_2)_3CH_3}{\underset{\displaystyle H \diagup \diagdown H}{C=C}}}$$

(b) $CH_3CH_2C \equiv C(CH_2)_3CH_3 + 2\,Na\cdot + 2:NH_3 \xrightarrow[-33°]{\text{liq. } NH_3}$

$$\underset{\substack{\\ \textit{trans-}3\text{-octene}}}{\overset{\displaystyle CH_3CH_2 \diagdown \diagup H}{\underset{\displaystyle H \diagup \diagdown (CH_2)_3CH_3}{C=C}}} + 2\,Na^+ + 2:\overset{..}{N}H_2^-$$

C. How Heats of Hydrogenation Show Alkene Stability

The **heat of hydrogenation** of an alkene is the energy difference between the starting alkene and the product alkane. It is calculated from the amount of heat released in a hydrogenation reaction. Table 10.5 lists the heats of hydrogenation of a few alkenes.

Let us consider the three alkenes that can be reduced to butane:

$$CH_3CH_2CH=CH_2 \xrightarrow{H_2,\,Pt}$$
$$\textit{cis-}CH_3CH=CHCH_3 \xrightarrow{H_2,\,Pt} CH_3CH_2CH_2CH_3$$
$$\textit{trans-}CH_3CH=CHCH_3 \xrightarrow{H_2,\,Pt} \quad \text{butane}$$

The product butane has the same energy regardless of the starting alkene. Any differences in ΔH for the three reactions reflect *differences in the energies of the starting alkenes*. The greater the value of the ΔH of hydrogenation, the higher is the energy of the starting alkene (see Figure 10.13).

From the differences in ΔH, we can see that 1-butene contains 1.7 kcal/mol more energy than does *cis*-2-butene. This means that 1-butene is 1.7 kcal/mol less stable than *cis*-2-butene. On the other hand, cis-2-butene contains 1.0 kcal/mol more energy than *trans*-2-butene. The relative heats of hydrogenation thus show that *trans*-2-butene is the most stable butene and that 1-butene is the least stable.

$$CH_2=CHCH_2CH_3 \qquad \textit{cis-}CH_3CH=CHCH_3 \qquad \textit{trans-}CH_3CH=CHCH_3$$

decreasing energy content; increasing stability $\longrightarrow$

Table 10.5 Heats of hydrogenation for some alkenes and dienes		
Name	Structure	$-\Delta H$, kcal/mol
ethene (ethylene)	$CH_2{=}CH_2$	32.8
propene (propylene)	$CH_3CH{=}CH_2$	30.1
1-butene	$CH_3CH_2CH{=}CH_2$	30.3
cis-2-butene	*cis*-$CH_3CH{=}CHCH_3$	28.6
trans-2-butene	*trans*-$CH_3CH{=}CHCH_3$	27.6
2-methyl-2-butene	$\overset{\displaystyle CH_3}{\overset{\displaystyle \|}{CH_3C}}{=}CHCH_3$	26.9
3-methyl-1-butene	$\overset{\displaystyle CH_3}{\overset{\displaystyle \|}{CH_3CHCH}}{=}CH_2$	30.3
1,3-pentadiene	$CH_2{=}CHCH{=}CHCH_3$	54.1
1,4-pentadiene	$CH_2{=}CHCH_2CH{=}CH_2$	60.8

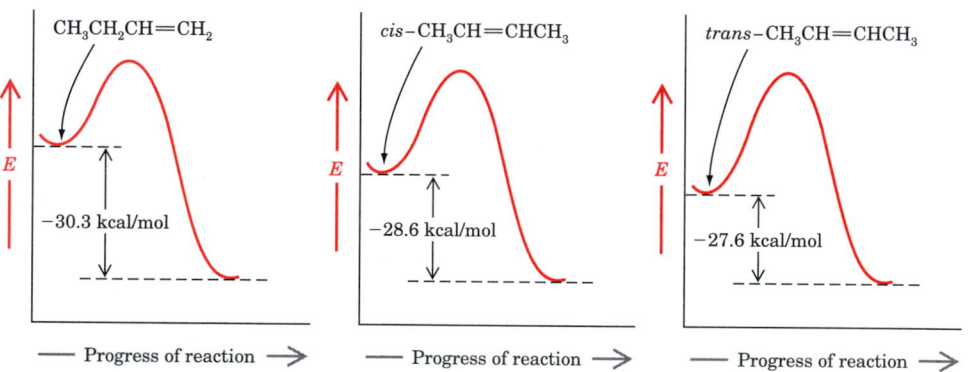

Figure 10.13 Comparison of the heats of hydrogenation of the three butenes that yield butane upon reduction.

From just such ΔH comparisons, the relative stabilities of a large number of alkenes have been determined. The following statements summarize what we have learned about alkene stabilities:

1. Increasing the alkyl substitution on the carbons of a double bond stabilizes an alkene. The reasons for this trend are not well understood.

$CH_2{=}CH_2$ $RCH{=}CH_2$ $RCH{=}CHR$ $R_2C{=}CH_2$ $R_2C{=}CHR$ $R_2C{=}CR_2$

increasing stability →

2. Conjugated dienes are more stable than dienes with isolated double bonds (because of delocalization of the π-electron density in the conjugated system).

3. *trans*-Alkenes are more stable than *cis* alkenes (because there are fewer steric repulsions in *trans* isomers).

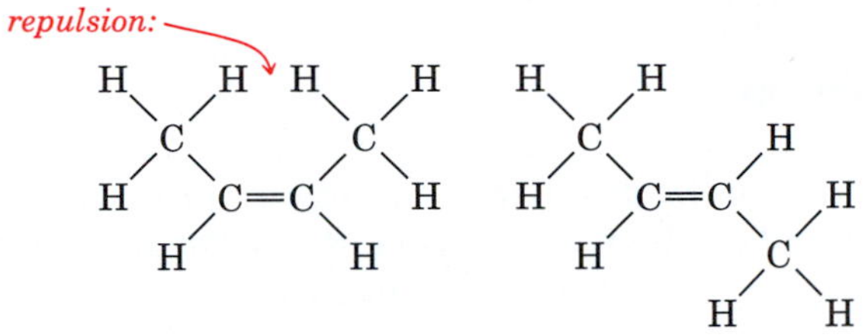

repulsion:

less stable   *more stable*

10.20 Which member of each of the following pairs would release the greater amount of energy per mole upon hydrogenation of the alkene double bonds? Explain.

(a) $(CH_3)_2C$=CH_2 or $(CH_3)_2C$=$CHCH_3$

(b)

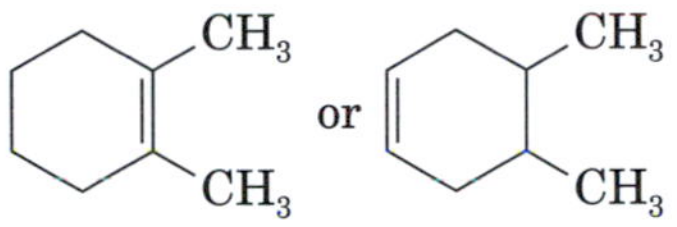

or

(c) (E)-C_6H_5CH=$CHCH_3$ or (Z)-C_6H_5CH=$CHCH_3$

D. Hydrogenation of Fats and Oils

The molecules of animal fats and vegetable oils contain long hydrocarbon chains. In vegetable oils, these chains are **polyunsaturated** (have several double bonds). Solid fats, on the other hand, usually contain few, if any, double bonds.

A vegetable oil can be converted to a substance of more solid consistency by partial hydrogenation of the carbon–carbon double bonds. The process of converting liquid oils to solid fats by this technique is called **hardening.** Although the polyunsaturates may be more healthful, the hydrogenated products are generally more palatable. Partially hydrogenated peanut oil is used to make peanut butter, and partially hydrogenated corn oil or safflower oil is used in margarine. Note that the carbonyl groups in the vegetable oil do not undergo hydrogenation under these conditions because they are more difficult to hydrogenate. Also, any or all of the carbon–carbon double bonds could be hydrogenated. Therefore, a mixture of partially hydrogenated products would result.

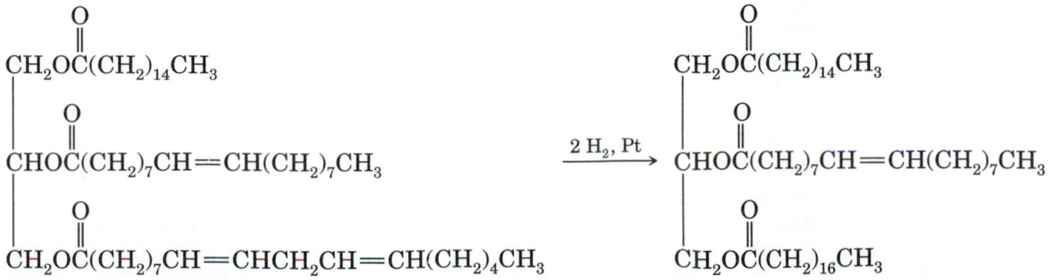

a typical vegetable oil   *a typical fat*

Section 10.13
Oxidation of Alkenes

Alkenes can be oxidized to a variety of products depending on the alkene structure and on the reagent used. Reactions involving oxidation of a carbon–carbon bond can be classified into two general groups: (1) oxidation of the π bond *without cleavage of the σ bond,* and (2) oxidation of the π bond *with cleavage of the σ bond.*

The products of oxidation without cleavage are generally either 1,2-diols or epoxides.

Without cleavage:

an epoxide   a 1,2-diol, or glycol

When both the σ bond and the π bond of an alkene are cleaved in an oxidation, the products are ketones, aldehydes, or carboxylic acids.

With cleavage:

ketones   aldehydes   carboxylic acids

A variety of reagents are used to oxidize alkenes. Some of the more common ones are listed in Table 10.6.

A. Oxidation Without Cleavage

Diol formation   One reagent used to oxidize an alkene to a 1,2-diol is a cold, alkaline, aqueous solution of potassium permanganate ($KMnO_4$). Unfortunately, this reagent usually gives low yields. Osmium tetroxide (OsO_4), followed by reduction with a reagent such as Na_2SO_3 or $NaHSO_3$, gives better

Table 10.6 Common reagents for oxidation of alkenes	
Reagent	Products
Oxidation Without Cleavage:	
$KMnO_4$ with OH^- (cold)	1,2-diols
OsO_4 followed by Na_2SO_3	1,2-diols
$C_6H_5CO_3H$	epoxides
O_2 (catalyst)	methyl ketones
Oxidation With Cleavage:	
$KMnO_4$ (hot)	carboxylic acids and ketones
O_3 followed by oxidation	carboxylic acids and ketones
O_3 followed by reduction	aldehydes and ketones

yields of diols. If OsO_4 is used together with hydrogen peroxide, only a catalytic amount of this toxic and expensive reagent is needed.

Both MnO_4^- and OsO_4 add *syn (cis)* to the π bond of a double bond to form an intermediate cyclic ester in the oxidation step of the reaction. Hydrolysis of the cyclic ester results in the observed products. The *cis* diol is formed if the product is capable of geometric isomerism.

The **Baeyer test** for unsaturation consists of treating a solution containing an unknown compound with a cold, aqueous solution of potassium permanganate. A positive test consists of the loss of the purple color of the test solution with the concurrent formation of a brown MnO_2 precipitate. The Baeyer test is really a test for the presence of an easily oxidizable group. Aldehydes, alkenes, alkynes, some alcohols, and other similar groups give a positive test.

STUDY PROBLEM

10.21   The following compounds are treated with OsO_4, followed by Na_2SO_3. What products would you expect? (Indicate any stereoisomerism.)

(a) $CH_3CH{=}C(CH_3)_2$ (b) —CH_3

Epoxide formation   In Section 8.2C, we described how the oxidation of an alkene by a peroxycarboxylic acid yields the epoxide.

In Chapter 8, we discussed the S_N2 cleavage of epoxides of cycloalkanes, which results in *trans*-1,2-diols. By contrast, oxidation of an alkene with OsO_4 or cold $KMnO_4$ yields *cis*-1,2-diols. Thus, either type of diol may be prepared from an alkene, depending on the choice of reagents.

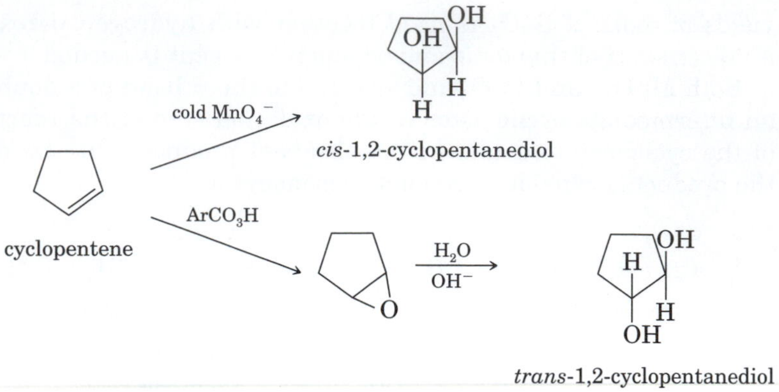

cyclopentene

cold MnO_4^- → *cis*-1,2-cyclopentanediol

ArCO$_3$H → (epoxide) $\xrightarrow[\text{OH}^-]{\text{H}_2\text{O}}$ *trans*-1,2-cyclopentanediol

Methyl ketone formation When treated with oxygen and a catalyst, alkenes with a $-CH{=}CH_2$ group can be converted to methyl ketones.

$$n\text{-}C_8H_{17}CH{=}CH_2 + O_2 \xrightarrow[\text{CuCl}_2]{\text{PdCl}_2} C_8H_{17}\overset{\displaystyle O}{\overset{\|}{C}}CH_3$$

This general synthetic method yields the same products as the hydration of the corresponding terminal alkyne (Section 10.7).

STUDY PROBLEM

10.22 Write equations showing how the following compound can be converted to **(a)** an epoxide, **(b)** a diol, and **(c)** a ketone.

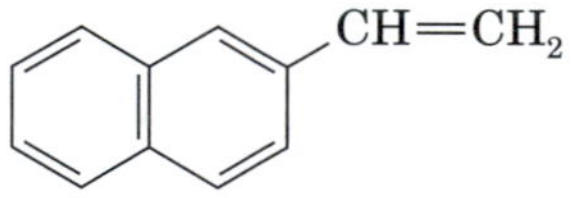

2-vinylnaphthalene

B. Oxidation with Cleavage

The products of oxidation with cleavage depend both upon the oxidizing conditions and upon the structure of the alkene. Let us consider first the structure of the alkene.

The structural feature of the alkene that determines the products of oxidative cleavage is the *presence or absence of hydrogen atoms on the sp^2 carbons*. If neither alkene carbon is bonded to a hydrogen atom (that is, each alkene carbon is disubstituted), then oxidative cleavage results in a pair of *ketone* molecules.

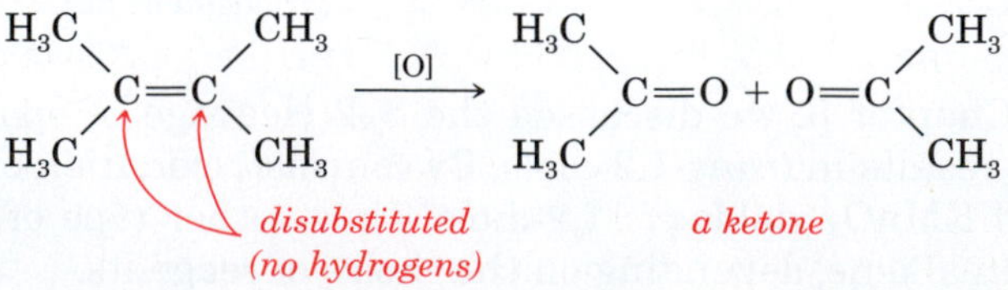

*disubstituted
(no hydrogens)*

a ketone

If, on the other hand, each alkene carbon has a hydrogen bonded to it, then the products of oxidative cleavage are either *aldehydes* or *carboxylic acids,* depending upon the reaction conditions.

$$
\underset{\substack{\textit{monosubstituted}\\ \textit{(one hydrogen)}}}{\underset{H}{\overset{H_3C}{\diagdown}}C=C\underset{H}{\overset{CH_3}{\diagup}}}
\quad
\begin{array}{c}
\xrightarrow{[O]} \\
\\
\xrightarrow{[O]}
\end{array}
\quad
\begin{array}{c}
\underset{H}{\overset{H_3C}{\diagdown}}C=O \; + \; O=C\underset{H}{\overset{CH_3}{\diagup}} \qquad \textit{an aldehyde}\\
\\
\underset{HO}{\overset{H_3C}{\diagdown}}C=O \; + \; O=C\underset{OH}{\overset{CH_3}{\diagup}} \qquad \textit{a carboxylic acid}
\end{array}
$$

Cleavage with KMnO₄ A hot solution of $KMnO_4$ is a vigorous oxidizing agent that leads to only ketones and carboxylic acids. (Aldehydes cannot be isolated from $KMnO_4$ solutions. They would be oxidized promptly to carboxylic acids.)

$$
\underset{\text{2-methyl-2-butene}}{\underset{H}{\overset{H_3C}{\diagdown}}C=C\underset{CH_3}{\overset{CH_3}{\diagup}}}
\xrightarrow[\text{heat}]{MnO_4^-}
\underset{\text{acetic acid}}{CH_3\overset{O}{\overset{\|}{C}}OH} \; + \; \underset{\text{acetone}}{CH_3\overset{O}{\overset{\|}{C}}CH_3}
$$

Under these vigorous oxidizing conditions, the carbon of a terminal double bond is oxidized to CO_2.

$$
\overset{\text{methylene-}}{\underset{\text{cyclopentane}}{\bigcirc\!\!=\!\!CH_2}}
\xrightarrow[\text{heat}]{MnO_4^-}
\underset{\text{cyclopentanone}}{\bigcirc\!\!=\!\!O} \; + \; CO_2 + H_2O
$$

to CO₂

The reason for CO_2 formation is that the methylene group is first oxidized to formic acid, which is further oxidized to carbonic acid. The latter undergoes spontaneous decomposition to CO_2 and H_2O.

$$
\underset{R}{\overset{R}{\diagdown}}C=CH_2
\xrightarrow[-R_2C=O]{[O]}
\left[\underset{\text{formic acid}}{H\overset{O}{\overset{\|}{O}C}H}\right]
\xrightarrow{[O]}
\left[\underset{\text{carbonic acid}}{HO\overset{O}{\overset{\|}{C}}OH}\right]
\longrightarrow H_2O + CO_2
$$

Ozonolysis Ozonolysis (cleavage by ozone) has been used for determining the structures of unsaturated compounds because this reaction results in the degradation of large molecules into smaller, identifiable fragments. Today, spectroscopy has largely replaced ozonolysis as an analytical technique.

An ozone molecule consists of three oxygen atoms bonded in a chain. The two O—O bonds are of equal length (1.29 Å) with a bond angle of 116°. The structure is best represented as a resonance hybrid.

resonance structures of ozone

Ozonolysis consists of two separate reactions: (1) oxidation of the alkene by ozone to an **ozonide,** and (2) either oxidation or reduction of the ozonide to the final products. The initial oxidation is usually carried out by bubbling ozone through a solution of the alkene in an inert solvent such as CCl_4. Ozone attacks the π bond to yield an unstable intermediate called a 1,2,3-tri-oxolane. This intermediate then goes through a series of transformations in which the carbon–carbon σ bond is cleaved. The product is an ozonide (a 1,2,4-trioxolane), which is rarely isolated, but is carried on to the second step.

2-methyl-2-butene · ozone · *a 1,2,3-trioxolane*

an ozonide (a 1,2,4-trioxolane)

The second reaction in ozonolysis is either the oxidation or the reduction of the ozonide. If the ozonide is subjected to a *reductive work-up,* the part of the original alkene with a monosubstituted carbon yields an aldehyde. If an *oxidative work-up* is used, the part of the alkene with a monosubstituted carbon yields a carboxylic acid. In either case, the portion of an alkene containing a disubstituted carbon yields a ketone. Common reducing agents are zinc metal in aqueous acid and dimethyl sulfide. Hydrogen peroxide is a common oxidizing agent.

Reductive work-up to aldehydes and ketones:

Oxidative work-up to carboxylic acids and ketones:

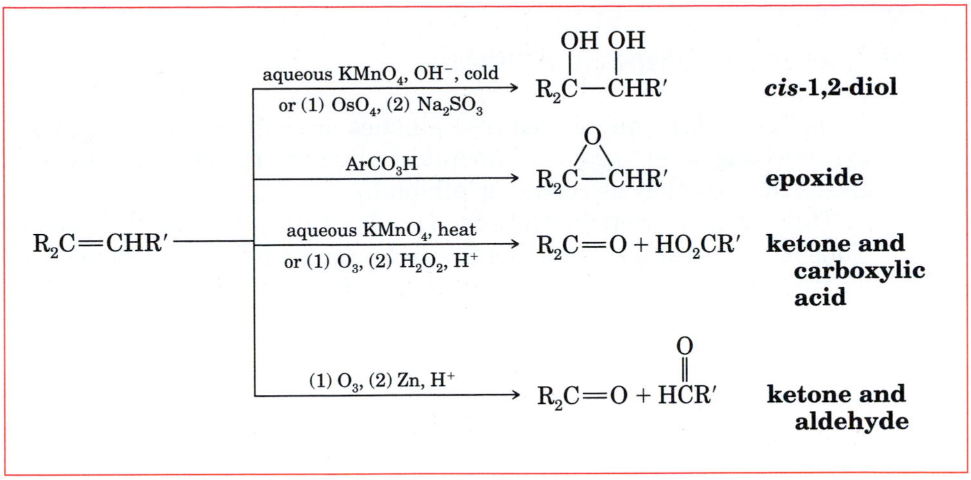

$$R_2C\!=\!CHR' \longrightarrow$$

aqueous $KMnO_4$, OH^-, cold
or (1) OsO_4, (2) Na_2SO_3
$\longrightarrow$ $R_2\overset{\underset{OH}{|}}{C}\!-\!\overset{\underset{OH}{|}}{C}HR'$ ***cis*-1,2-diol**

$ArCO_3H$
$\longrightarrow$ $R_2\overset{O}{\overset{/\backslash}{C}}\!-\!CHR'$ **epoxide**

aqueous $KMnO_4$, heat
or (1) O_3, (2) H_2O_2, H^+
$\longrightarrow$ $R_2C\!=\!O + HO_2CR'$ **ketone and carboxylic acid**

(1) O_3, (2) Zn, H^+
$\longrightarrow$ $R_2C\!=\!O + H\overset{\overset{O}{\|}}{C}R'$ **ketone and aldehyde**

Figure 10.14 Summary of the oxidation of alkenes. (Alkenes with terminal double bonds can also be oxidized to methyl ketones.)

Figure 10.14 summarizes the oxidation reactions of alkenes.

SAMPLE PROBLEM

Predict the products of reductive ozonolysis of γ-terpinene, a compound found in coriander oil:

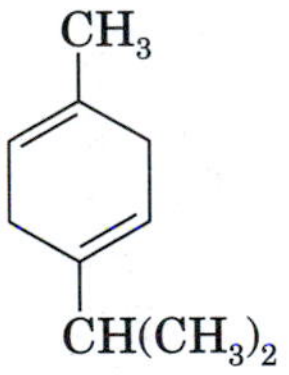

CH₃ ... CH(CH₃)₂

Solution

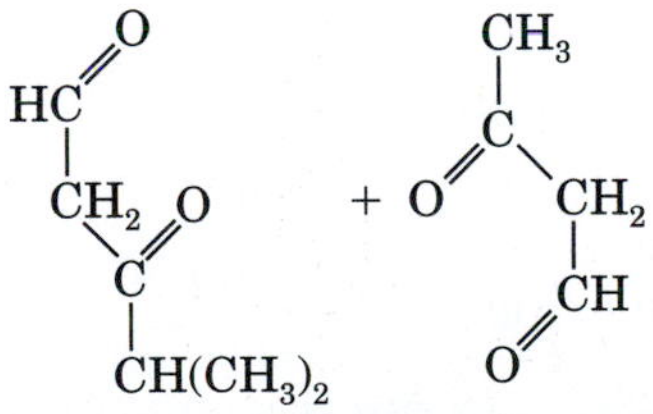

STUDY PROBLEM

10.23 What would be the products of both reductive and oxidative ozonolyses of each of the following alkenes?

(a) $CH_3(CH_2)_7CH\!=\!CH(CH_2)_7\overset{\overset{O}{\|}}{C}OH$ (b) 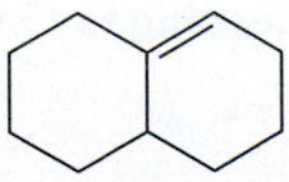 ⬡   (c) 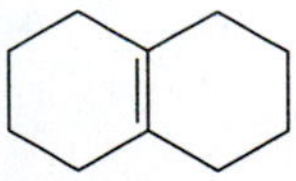 ⬡⬡

Section 10.14
Use of Alkenes and Alkynes in Synthesis

From Table 10.7, you can see that alkenes are valuable starting materials for synthesizing other organic compounds. Alkynes are not as widely used in synthesis (nor are they as readily available).

The addition reactions that use H^+ as a catalyst proceed through carbocations and yield Markovnikov products (and possible rearrangement products). These reactions include those with HX and with H_2O (H^+).

Table 10.7 Types of compounds that can be obtained from alkenes		
Reaction	Product	Section Reference
Markovnikov Addition:		
$R_2C=CHR + HX \longrightarrow R_2\overset{X}{\underset{|}{C}}CH_2R$	**alkyl halide**	10.6A–B
$+ H_2O \xrightarrow{H^+} R_2\overset{OH}{\underset{|}{C}}CH_2R$	**alcohol**	10.7
$\xrightarrow[\text{(2) NaBH}_4]{\text{(1) Hg(O}_2\text{CCH}_3)_2, \text{H}_2\text{O}} R_2\overset{OH}{\underset{|}{C}}CH_2R$	**alcohol**	10.8
$\xrightarrow[\text{(2) NaBH}_4]{\text{(1) Hg(O}_2\text{CCH}_3)_2, \text{R'OH}} R_2\overset{OR'}{\underset{|}{C}}CH_2R$	**ether**	10.8
$+ X_2 \longrightarrow R_2\overset{X}{\underset{\underset{X}{|}}{C}}-CHR$	**dihaloalkane**	10.10
$+X_2 \xrightarrow{H_2O} R_2\overset{OH}{\underset{\underset{X}{|}}{C}}CHR$	**1,2-halohydrin**	10.10D
"Anti-Markovnikov" Addition:		
$R_2C=CHR \xrightarrow[\text{(2) H}_2\text{O}_2, \text{OH}^-]{\text{(1) BH}_3} R_2CH\overset{OH}{\underset{|}{C}}HR$	**alcohol**	10.9
$\xrightarrow[\text{(2) RCO}_2\text{H}]{\text{(1) BH}_3} R_2CHCH_2R$	**alkane**	10.9
$\xrightarrow[\text{(2) Br}_2, \text{CH}_3\text{O}^-]{\text{(1) BH}_3} R_2CH\overset{Br}{\underset{|}{C}}HR$	**alkyl bromide**	10.9
$+ HBr \xrightarrow[\text{peroxide}]{\text{O}_2 \text{ or}} R_2CH\overset{Br}{\underset{|}{C}}HR$	**alkyl bromide**	10.6C
Addition Leading to Cyclic Products:		
$R_2C=CHR + :CH_2 \longrightarrow R_2C\overset{\overset{CH_2}{\diagup\diagdown}}{-}CHR$	**a cyclopropane**	10.11

Table 10.7 *(continued)*   Types of compounds that can be obtained from alkenes

Reaction	Products	Section Reference
Reduction:		
$R_2C{=}CHR + H_2 \xrightarrow{Pt} R_2CHCH_2R$	**alkane**	10.12
Oxidation:		
$R_2C{=}CHR + MnO_4^- \xrightarrow{25°} R_2\overset{\displaystyle OH}{\underset{\displaystyle OH}{C}}CHR$	**1,2-diol**[a]	10.13A
$+ C_6H_5CO_3H \longrightarrow R_2\overset{\displaystyle O}{\overset{\displaystyle /\backslash}{C{-}CHR}}$	**expoxide**[a]	10.13A
$+ MnO_4^- \xrightarrow{heat} R_2C{=}O + HO\overset{\displaystyle O}{\overset{\displaystyle \|}{C}}R$	**ketone, carboxylic acid**	10.13B
$\xrightarrow[\text{(2) Zn, H}^+\text{, H}_2\text{O}]{\text{(1) O}_3} R_2C{=}O + H\overset{\displaystyle O}{\overset{\displaystyle \|}{C}}R$	**ketone, aldehyde**[b]	10.13B
$RCH{=}CH_2 \xrightarrow[\text{catalyst}]{O_2} R\overset{\displaystyle O}{\overset{\displaystyle \|}{C}}CH_3$	**methyl ketone**	10.13A

[a] A *cis*-1,2-diol can be prepared from an alkene and cold MnO_4^-; the *trans*-1,2-diol can be prepared from hydrolysis of the epoxide.
[b] Ozonolysis with an oxidative work-up yields ketones and carboxylic acids.

Markovnikov products:

$$R_2C{=}CHR \xrightarrow{H^+} [R_2\overset{+}{C}{-}CH_2R] \underset{\underset{-H^+}{H_2O}}{\overset{Cl^-}{\rightleftarrows}} \begin{matrix} R_2\overset{\displaystyle Cl}{\underset{\displaystyle}{C}}CH_2R \\ R_2\overset{\displaystyle OH}{\underset{\displaystyle}{C}}CH_2R \end{matrix}$$

Alcohols (or ethers) can be prepared *without rearrangement,* in excellent yield, by oxymercuration–demercuration reactions, a sequence that also leads to Markovnikov products.

$$R_2C{=}CHR \xrightarrow[-CH_3CO_2H]{\underset{H_2O}{Hg(O_2CCH_3)_2}} R_2\overset{\displaystyle OH}{\underset{\displaystyle HgO_2CCH_3}{C{-}CHR}} \xrightarrow{NaBH_4} R_2\overset{\displaystyle OH}{\underset{\displaystyle}{C}}CH_2R$$

"Anti-Markovnikov" products, with the functional group added to the less substituted carbon of the alkene, can be prepared with HBr (and O_2 or a peroxide catalyst) or by hydroboration.

"Anti-Markovnikov" products:

$$3\,R_2C{=}CHR \xrightarrow{BH_3} (R_2CH{-}\overset{\displaystyle R}{\underset{\displaystyle}{CH}})_3B \underset{\overset{\displaystyle Br_2, CH_3O^-}{\nearrow}}{\overset{H_2O_2, OH^-}{\searrow}} \begin{matrix} 3\,R_2CH\overset{\displaystyle OH}{\underset{\displaystyle}{C}}HR \\ 3\,R_2CH\overset{\displaystyle Br}{\underset{\displaystyle}{C}}HR \end{matrix}$$

Difunctional compounds arise from the addition of X_2 (or $X_2 + H_2O$) to a double bond; or by oxidation of an alkene to a 1,2-diol. For example,

$$R_2C{=}CHR \xrightarrow{\ X^+\ } R_2\underset{\delta+}{C}{\cdots}\overset{\overset{\displaystyle \delta+}{X}}{CHR} \ \substack{\xrightarrow{\ X^-\ }\\ \\ \xrightarrow[-H^+]{H_2O}} \ \substack{R_2\overset{\displaystyle X}{C}\overset{\displaystyle X}{CHR}\\ \\ R_2\overset{\displaystyle X}{C}{-}\underset{\displaystyle OH}{CHR}}$$

Oxidation of alkenes is a way to introduce other functionality, such as an epoxide ring. Vigorous oxidizing conditions result in cleavage of the double bond and yield carbonyl compounds, as shown in Table 10.6.

SAMPLE PROBLEM

Suggest a synthesis for 3-methyl-2-pentanone from 3-bromo-3-methylpentane.

Solution

1. Write the structures:

$$\underset{\underset{\displaystyle CH_3}{|}}{CH_3CH_2\overset{\overset{\displaystyle Br}{|}}{C}CH_2CH_3} \longrightarrow \underset{\underset{\displaystyle CH_3}{|}}{CH_3\overset{\overset{\displaystyle O}{\|}}{C}CHCH_2CH_3}$$

2. There is no one-step reaction from RX to a ketone. Therefore, working backwards, ask yourself what reactions lead to ketones (without cleavage of the carbon skeleton). Oxidation of a 2° alcohol is a standard reaction leading to a ketone.

$$\underset{\underset{\displaystyle CH_3}{|}}{CH_3\overset{\overset{\displaystyle OH}{|}}{C}HCHCH_2CH_3} \xrightarrow{\ H_2CrO_4\ } \underset{\underset{\displaystyle CH_3}{|}}{CH_3\overset{\overset{\displaystyle O}{\|}}{C}CHCH_2CH_3}$$

3. The hydroxyl group in the alcohol is not in the same position as the Br in the starting material. Therefore, a simple substitution reaction (RX + OH⁻) would not be a route to the alcohol; however, an alcohol can be obtained from an alkene.

$$\underset{\underset{\displaystyle CH_3}{|}}{CH_3CH{=}CCH_2CH_3} \longrightarrow \underset{\underset{\displaystyle CH_3}{|}}{CH_3\overset{\overset{\displaystyle OH}{|}}{C}H{-}CHCH_2CH_3}$$

Because the alcohol is the *anti-Markovnikov product,* a reaction with $H_2O + H^+$ or oxymercuration–demercuration will not yield the proper al-

Table 10.7 (continued)   Types of compounds that can be obtained from alkenes

Reaction	Products	Section Reference
Reduction:		
$R_2C{=}CHR + H_2 \xrightarrow{Pt} R_2CHCH_2R$	**alkane**	10.12
Oxidation:		
$R_2C{=}CHR + MnO_4^- \xrightarrow{25°} R_2\overset{\overset{\displaystyle OH}{\vert}}{C}\underset{\underset{\displaystyle OH}{\vert}}{C}HR$	**1,2-diol**[a]	10.13A
$+\ C_6H_5CO_3H \longrightarrow R_2\overset{O}{\overset{\triangle}{C{-}CHR}}$	**expoxide**[a]	10.13A
$+\ MnO_4^- \xrightarrow{heat} R_2C{=}O + HO\overset{O}{\overset{\Vert}{C}}R$	**ketone, carboxylic acid**	10.13B
$\xrightarrow[\text{(2) Zn, H}^+\text{, H}_2\text{O}]{\text{(1) O}_3} R_2C{=}O + H\overset{O}{\overset{\Vert}{C}}R$	**ketone, aldehyde**[b]	10.13B
$RCH{=}CH_2 \xrightarrow[\text{catalyst}]{O_2} R\overset{O}{\overset{\Vert}{C}}CH_3$	**methyl ketone**	10.13A

[a] A *cis*-1,2-diol can be prepared from an alkene and cold MnO_4^-; the *trans*-1,2-diol can be prepared from hydrolysis of the epoxide.

[b] Ozonolysis with an oxidative work-up yields ketones and carboxylic acids.

Markovnikov products:

$$R_2C{=}CHR \xrightarrow{H^+} [R_2\overset{+}{C}{-}CH_2R] \begin{array}{c} \xrightarrow{Cl^-} R_2\overset{\overset{\displaystyle Cl}{\vert}}{C}CH_2R \\ \\ \xrightarrow[-H^+]{H_2O} R_2\overset{\overset{\displaystyle OH}{\vert}}{C}CH_2R \end{array}$$

Alcohols (or ethers) can be prepared *without rearrangement,* in excellent yield, by oxymercuration–demercuration reactions, a sequence that also leads to Markovnikov products.

$$R_2C{=}CHR \xrightarrow[-CH_3CO_2H]{\substack{Hg(O_2CCH_3)_2 \\ H_2O}} R_2\overset{\overset{\displaystyle OH}{\vert}}{\underset{\underset{\displaystyle HgO_2CCH_3}{\vert}}{C}}{-}CHR \xrightarrow{NaBH_4} R_2\overset{\overset{\displaystyle OH}{\vert}}{C}CH_2R$$

"Anti-Markovnikov" products, with the functional group added to the less substituted carbon of the alkene, can be prepared with HBr (and O_2 or a peroxide catalyst) or by hydroboration.

"Anti-Markovnikov" products:

$$3\ R_2C{=}CHR \xrightarrow{BH_3} (R_2CH{-}\overset{\overset{\displaystyle R}{\vert}}{CH})_3B \begin{array}{c} \xrightarrow{H_2O_2,\ OH^-} 3\ R_2CH\overset{\overset{\displaystyle OH}{\vert}}{C}HR \\ \\ \xrightarrow{Br_2,\ CH_3O^-} 3\ R_2CH\overset{\overset{\displaystyle Br}{\vert}}{C}HR \end{array}$$

Difunctional compounds arise from the addition of X_2 (or X_2 + H_2O) to a double bond; or by oxidation of an alkene to a 1,2-diol. For example,

$$R_2C\!\!=\!\!CHR \xrightarrow{X^+} \underset{\delta+}{R_2C}\!\!\overset{\overset{\delta+}{\overset{X}{\vdots}}}{-}\!\!CHR$$

with branches leading to:

$$\xrightarrow{X^-} \underset{\overset{|}{X}}{\overset{\overset{X}{|}}{R_2C\!\!-\!\!CHR}}$$

$$\xrightarrow[-H^+]{H_2O} \underset{\overset{|}{OH}}{\overset{\overset{X}{|}}{R_2C\!\!-\!\!CHR}}$$

Oxidation of alkenes is a way to introduce other functionality, such as an epoxide ring. Vigorous oxidizing conditions result in cleavage of the double bond and yield carbonyl compounds, as shown in Table 10.6.

SAMPLE PROBLEM

Suggest a synthesis for 3-methyl-2-pentanone from 3-bromo-3-methylpentane.

Solution

1. Write the structures:

$$\underset{\overset{|}{CH_3}}{\overset{\overset{Br}{|}}{CH_3CH_2CCH_2CH_3}} \longrightarrow \underset{\overset{|}{CH_3}}{\overset{\overset{O}{\|}}{CH_3CCHCH_2CH_3}}$$

2. There is no one-step reaction from RX to a ketone. Therefore, working backwards, ask yourself what reactions lead to ketones (without cleavage of the carbon skeleton). Oxidation of a 2° alcohol is a standard reaction leading to a ketone.

$$\underset{\overset{|}{CH_3}}{\overset{\overset{OH}{|}}{CH_3CHCHCH_2CH_3}} \xrightarrow{H_2CrO_4} \underset{\overset{|}{CH_3}}{\overset{\overset{O}{\|}}{CH_3CCHCH_2CH_3}}$$

3. The hydroxyl group in the alcohol is not in the same position as the Br in the starting material. Therefore, a simple substitution reaction (RX + OH⁻) would not be a route to the alcohol; however, an alcohol can be obtained from an alkene.

$$\underset{\overset{|}{CH_3}}{CH_3CH\!\!=\!\!CCH_2CH_3} \longrightarrow \underset{\overset{|}{CH_3}}{\overset{\overset{OH}{|}}{CH_3CH\!\!-\!\!CHCH_2CH_3}}$$

Because the alcohol is the *anti-Markovnikov product,* a reaction with H_2O + H^+ or oxymercuration–demercuration will not yield the proper alcohol, but hydroboration–oxidation will. The reagents for the preceding conversion are therefore (1) BH_3 (or, more correctly, B_2H_6); (2) H_2O_2, ^-OH.

4. Can the preceding alkene be prepared from the starting alkyl bromide? Yes, this is an elimination reaction that proceeds by an E2 path.

$$CH_3CH_2\underset{\underset{CH_3}{|}}{\overset{\overset{Br}{|}}{C}}CH_2CH_3 \xrightarrow[\text{E2}]{Na^+ \; {}^-OCH_3} CH_3CH{=}\underset{\underset{CH_3}{|}}{C}CH_2CH_3$$

5. Now, the entire sequence can be written.

$$CH_3CH_2\underset{\underset{CH_3}{|}}{\overset{\overset{Br}{|}}{C}}CH_2CH_3 \xrightarrow{Na^+ \; {}^-OCH_3} CH_3CH{=}\underset{\underset{CH_3}{|}}{C}CH_2CH_3 \xrightarrow[(2)\,H_2O_2,\,OH^-]{(1)\,BH_3}$$

$$CH_3\underset{\underset{CH_3}{|}}{\overset{\overset{OH}{|}}{C}}HCHCH_2CH_3 \xrightarrow{H_2CrO_4} CH_3\overset{\overset{O}{\|}}{C}\underset{\underset{CH_3}{|}}{C}HCH_2CH_3$$

STUDY PROBLEMS

10.24 Suggest syntheses for the following compounds:

(a) $(CH_3)_2C{=}\underset{\underset{}{}}{\overset{\overset{CH_3}{|}}{C}}CH_2CH_3$ from organic compounds containing four or fewer carbon atoms

(b) $CH_3CH_2C{\equiv}CH$ from an alkene

10.25 Write flow diagrams for the following conversions:

(a) cyclohexene $\longrightarrow$ [cyclohexyl–C(=O)–cyclohexyl]

(b) cyclopentanol $\longrightarrow$ *cis*-1,2-cyclopentanediol

(c) *cis*-1-bromo-2-methylcyclohexane $\longrightarrow$ *trans*-2-methylcyclohexanol

(d) propyne $\longrightarrow$ 2-methyl-3-penten-2-ol

Summary

The reactivities of alkenes and alkynes arise from the weakness and exposure of the π bond, as well as from the exposure of the trigonal linear carbon. An alkyne with a $—C{\equiv}CH$ group is more *acidic* than other hydrocarbons.

Alkenes are susceptible to **electrophilic attack.** If the alkene and reagent are unsymmetrical, reaction goes through the most stable carbocation.

(**Markovnikov's rule**: H^+ adds to the carbon that already has the most H's.) As reagents, hydrogen halides, H_2SO_4, or $H_2O + H^+$ can be used. Reactions with halogens, halogens plus water, and $Hg(O_2CCH_3)_2$ proceed through bridged ions and thus are *stereoselective*. (These reactions are summarized in Table 10.7.) The addition of BH_3 yields products that appear to be anti-Markovnikov.

Some addition reactions proceed by **syn-additions** instead of by way of carbocations or bridged ions. Hydrogenation, hydroboration, and carbene-addition reactions are in this category. *syn*-Additions are usually stereospecific and often lead to *cis* products. (Note that hydroboration also leads to "anti-Markovnikov" products.)

Hydrogenation reactions can be used to determine relative *stabilities of alkenes*. The most highly substituted alkenes are the most stable. Acyclic *trans* alkenes are usually more stable than *cis* alkenes. Conjugated dienes are more stable than nonconjugated dienes.

Oxidation of alkenes can lead to *cis*-1,2-diols, epoxides, ketones, aldehydes, or carboxylic acids. These reactions are summarized in Figure 10.14 and Table 10.7.

Essay Problem for Chapter 10

Stereochemistry of Osmium Tetroxide Hydroxylation Reactions

Hydroxylation of an alkene by osmium tetroxide (OsO_4) is a *syn*-addition reaction. For example, if *trans*-stilbene is hydroxylated with OsO_4, an enantiomeric mixture of diols is obtained.

trans-stilbene

enantiomers of 1,2-diphenyl-1,2-ethanediol

There are two experimental methods for carrying out a hydroxylation reaction with OsO_4. The first and classical method is to treat the alkene with OsO_4 to form an osmate ester, which is then reduced with a reagent such as Na_2SO_3. The second and modern method is to treat the alkene with a catalytic amount of OsO_4 along with an oxidizing agent, such as H_2O_2 or an amine oxide. With this procedure, the osmate ester that forms in the mixture is oxidized to yield the diol. The oxidation reaction also regenerates OsO_4, which forms more osmate ester. Consequently, only a catalytic amount of osmium tetroxide, a toxic and expensive reagent, is needed.

Asymmetric induction can be employed with the catalytic method. In asymmetric induction, a chiral agent is added to the reaction mixture. Under these conditions the reaction produces only one enantiomer. Therefore, when *trans*-stilbene is hydroxylated with dihydroquinidine 4-chlorobenzoate as the chiral agent, the (*R*,*R*)-diol is obtained in a 72–75% yield. The (*S*,*S*) enantiomer is not produced. (B. H. McKee, D. G. Gilheany, and K. B. Sharpless, *Org. Syn.* **1992**, *70*, 47.)

trans-stilbene

an amine oxide

(1*R*,2*R*)-1,2-diphenyl-1,2-ethanediol
(72–75%)

dihydroquinidine 4-chlorobenzoate

a chiral agent

Questions

1. What are the products of the hydroxylation of *cis*-stilbene?
2. If hydroxylation were an *anti*-addition reaction, what would be the structures of the pair of compounds obtained from *trans*-stilbene? from *cis*-stilbene?
3. What happens to the osmium atom in the classical procedure?
4. What is the structure of the intermediate osmate ester that leads to the (*R*,*R*) enantiomer?
5. Using asymmetric hydroxylation, how would you obtain the (*S*,*S*) enantiomer?

Study Problems*

10.26 Write the IUPAC name for each of the following compounds:

(a) $BrCH_2CH{=}CH_2$

(b) $(CH_3)_2CHC{\equiv}CH$

(c)

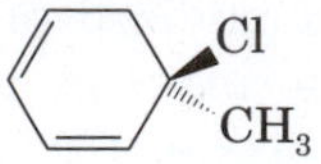

(d) 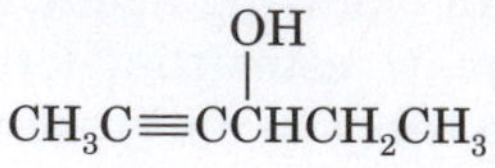 $CH_3C{\equiv}CCHCH_2CH_3$
  with OH on the CH

(e) 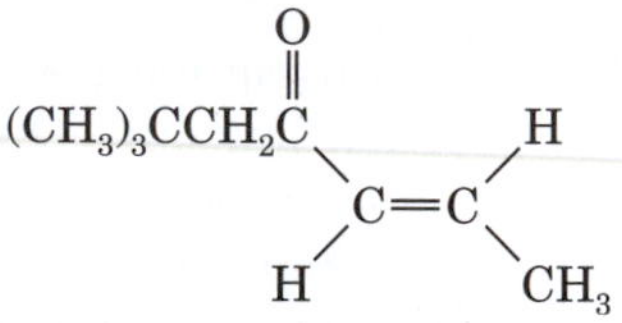 $(CH_3)_3CCH_2C$ (with =O) attached to $C{=}C$ bearing H, H, and CH_3

10.27 Give the structure for each of the following compounds:

(a) 2-methyl-1-butene

(b) 2-pentyne

(c) 2-cyclopentenol

(d) *cis*-2-hexene

(e) *trans*-1,2-diphenylethene

(f) (*Z*)-3-methoxy-2-pentene

(g) (2*E*,4*Z*)-2,4-hexadiene

(h) (*E*)-3-bromo-2-butenoic acid

10.28 Draw a tree diagram for the alkene protons H_x and H_y in the following structure (J_{xy} = 10 Hz, J_{ay} = 2 Hz, J_{ax} = 16 Hz).

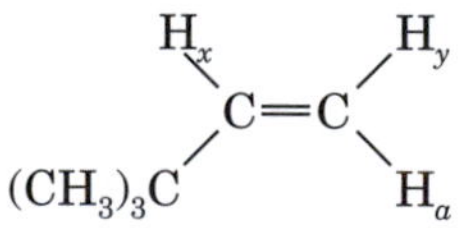

$(CH_3)_3C$ — $C{=}C$ with H_x, H_y, H_a

10.29 Write equations for the synthesis of the following compounds from alcohols or alkyl halides.

(a) 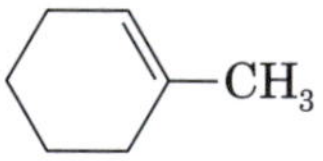 cyclohexene —CH_3

(b) 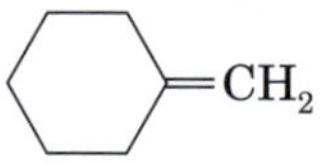 cyclohexane ${=}CH_2$

(c) C_6H_5, H on one carbon; C_6H_5, CH_3 on the other of $C{=}C$

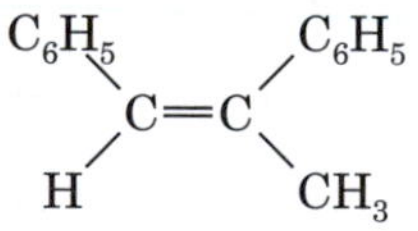

(d) 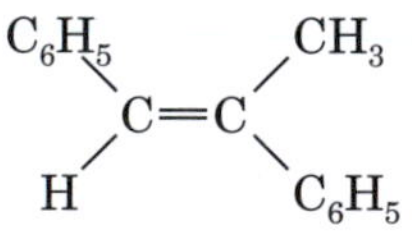 C_6H_5, H on one carbon; CH_3, C_6H_5 on the other of $C{=}C$

10.30 Predict the products:

(a) $(CH_3)_2C{=}CHCH_3 + HCl \xrightarrow{\text{diethyl ether}}$

(b) 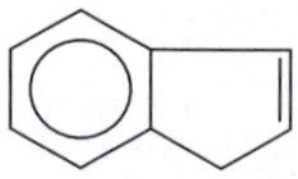 (indene structure) + excess gaseous HCl $\xrightarrow{5-10°}$

(c) $CH_3CH_2C{\equiv}CH + 1\ HBr \xrightarrow[\text{no peroxides}]{15°,\ \text{dark}}$

---

* For information concerning the organization of the *Study Problems* and *Additional Problems*, see the *Note to student* on page 41.

(d) 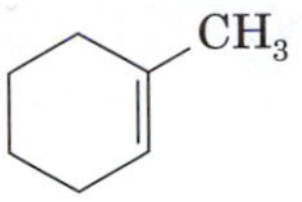 [structure: methylcyclohexene with CH₃] + HBr $\xrightarrow{\text{no peroxides}}$

(e) 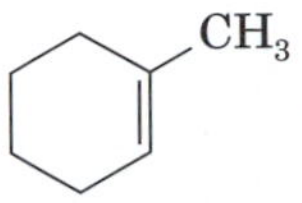 [structure: methylcyclohexene with CH₃] + HBr $\xrightarrow{\overset{\text{O}\quad\text{O}}{\overset{\|\quad\|}{C_6H_5COOCC_6H_5}}}$

10.31 Predict the major organic products of the following reactions:

  (a) 3-methyl-2-pentene with aqueous H_2SO_4

  (b) 2-methylpropene with H_2SO_4 in ethanol

  (c) 2,2-dimethyl-3-hexene with aqueous H_2SO_4

  (d) 1-butene with $0.1M$ aqueous HI

  (e) $(CH_3)_3CCH{=}CH_2$ $\xrightarrow[\text{(2) NaBH}_4]{\text{(1) Hg(O}_2\text{CCH}_3)_2,\ \text{H}_2\text{O}}$

  (f) 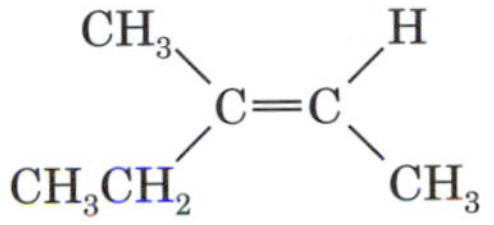 [structure:
$$\underset{CH_3CH_2}{\overset{CH_3}{\diagdown}}C{=}C\underset{CH_3}{\overset{H}{\diagup}}$$] $\xrightarrow[\text{(2) H}_2\text{O}_2,\ \text{OH}^-]{\text{(1) BH}_3}$

  (g) $C_6H_5C{\equiv}C{-}CH_3$ $\xrightarrow{\text{H}_2\text{SO}_4,\ \text{H}_2\text{O}}$

10.32 Write equations for the mechanism to predict the stereochemistry of the solvomercuration product of 1-methylcyclohexene in methanol.

10.33 Predict the products. Show stereochemistry where applicable.

  (a) $(E)\text{-}C_6H_5CH{=}CHCOCH_2CH_3$ + Br$_2$ $\longrightarrow$
  (with O double-bonded above the C)

  (b) $(Z)\text{-}CH_3(CH_2)_7CH{=}CH(CH_2)_7COCH_3$ + Br$_2$ $\longrightarrow$
  (with O double-bonded above the C)

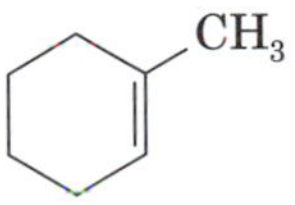

  (c) [structure: methylcyclohexene with CH₃] + Br$_2$ $\longrightarrow$

  (d) 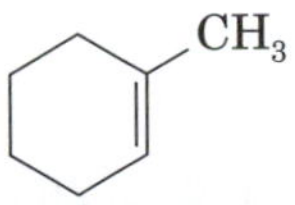 [structure: methylcyclohexene with CH₃] + Br$_2$ $\xrightarrow{\text{CH}_3\text{OH}}$

10.34 When 5-hexen-2-ol is treated with a dilute solution of bromine in chloroform, the isolated product contains only one bromine atom and has the formula $C_6H_{11}BrO$. What is the structure of the product? Write a mechanism to show how it is formed.

10.35 Fumaric acid (*trans*-$HO_2CCH{=}CHCO_2H$), which is essential for cell respiration, is found in all plant and animal tissues. The *Boletaceae* family of mushrooms is especially rich in this acid. (One trivial name of fumaric acid is *boletic acid*.) When fumaric acid is treated with bromine, an 84% yield of a dibrominated product is obtained. Predict the stereochemistry of the product.

10.36 Styrene, $C_6H_5CH{=}CH_2$, was mixed with bromine and anhydrous acetic acid. Two products were obtained. Write equations to show the identities of these products, and write a mechanism that shows how they are formed.

10.37 Predict the major organic products (showing stereochemistry):

(a) 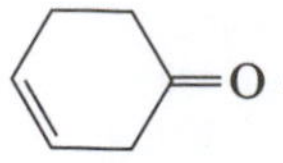 $=O + H_2 \xrightarrow{Pt}$

(b) $CH_3CH_2C{\equiv}CCH(CH_3)_2 \xrightarrow[\text{liq. NH}_3]{\text{Na}}$

(c) $CH_3C{\equiv}C\overset{\overset{\text{O}}{\|}}{C}OCH_2CH_3 + 1\,H_2 \xrightarrow{\text{deactivated Pd}}$

(d) 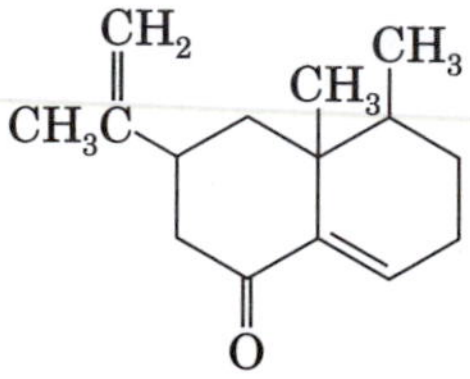 $+\ 1\,H_2 \xrightarrow{[(C_6H_5)_3P]_3RhCl}$

10.38 Predict the products when (RS)-limonene (1-methyl-4-isopropenylcyclohexene) is treated with an excess of the following reagents: **(a)** $C_6H_5CO_3H$; **(b)** hot $KMnO_4$ solution; **(c)** cold $KMnO_4$ solution.

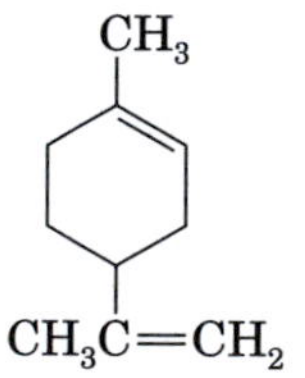

limonene

found in citrus oils

10.39 During an investigation of naturally occurring antitumor agents, the following compound was subjected to OsO_4 oxidation, followed by treatment with $NaHSO_3$. A single stereoisomeric product was obtained in 93% yield. What was this product?

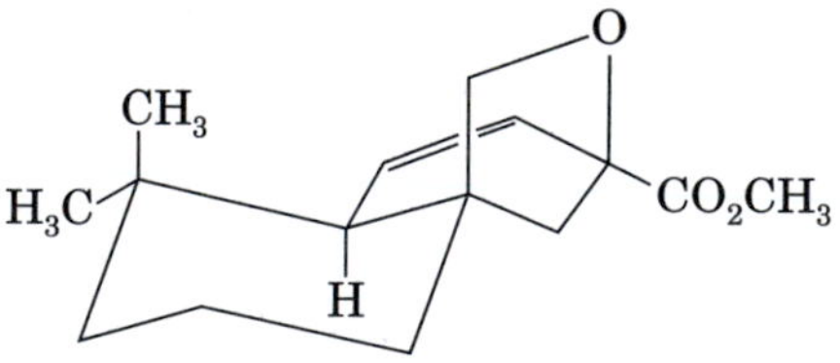

10.40 Each of a series of unknown alkenes is treated with (1) O_3, and (2) Zn, H^+, H_2O. The following products are obtained. Give the structure (or structures, if there is more than one possibility) of each unknown alkene.

(a) $H\overset{\overset{\text{O}}{\|}}{C}CH_2CH_2CH_2\overset{\overset{\text{O}}{\|}}{C}H$ only

(b) $CH_3CH_2\overset{\overset{\text{O}}{\|}}{C}H + CH_3\overset{\overset{\text{O}}{\|}}{C}CH_3$

(c) $CH_3\overset{\overset{\text{O}}{\|}}{C}H$ only

(d)

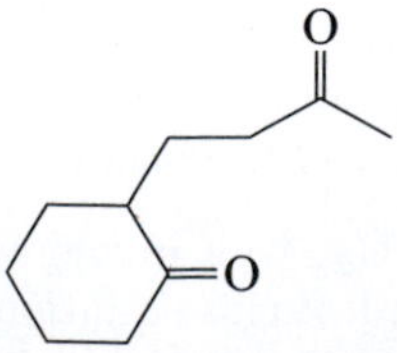

(e) $CH_3\overset{O}{\overset{\|}{C}}-\overset{O}{\overset{\|}{C}}H + 2\ CH_3\overset{O}{\overset{\|}{C}}CH_3$ (f) $H\overset{O}{\overset{\|}{C}}CH_2\overset{O}{\overset{\|}{C}}CH_2\overset{O}{\overset{\|}{C}}H + CH_3\overset{O}{\overset{\|}{C}}CH_3$

10.41 Write flow diagrams for the following conversions:

(a) $CH_3\underset{\underset{Br}{|}}{C}HCH_3 \longrightarrow CH_3CH_2CH_2Br$ (b) $CH_3CH_2CH_2Br \longrightarrow CH_3\underset{\underset{Br}{|}}{C}HCH_3$

(c) 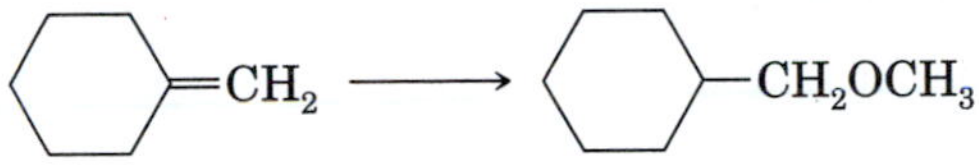 [cyclohexane ring]$=CH_2 \longrightarrow$ [cyclohexane ring]$-CH_2OCH_3$

10.42 Write flow equations showing how each of the following conversions might be carried out:

(a) cyclohexanol $\longrightarrow$ *trans*-1,2-dibromocyclohexane

(b) propyne $\longrightarrow$ 1-phenyl-2-butyne

(c) 2-bromopropane $\longrightarrow$ 1,2-propanediol

(d) 1-butyne $\longrightarrow$ *cis*-3-hexene

(e) $C_6H_5CH{=}CHC_6H_5 \longrightarrow C_6H_5C{\equiv}CC_6H_5$

Additional Problems

10.43 Predict the organic product of the reaction of *cis*-2-butene with each of the following reagents. (Show the stereochemistry where appropriate.) **(a)** HI; **(b)** OsO_4 (cat), H_2O_2; **(c)** $C_6H_5CO_3H$, followed by $H_2O + HCl$; **(d)** H_2SO_4; **(e)** O_3, followed by $Zn + HCl$; **(f)** $CHCl_3$ and $K^+\ {}^-OC(CH_3)_3$.

10.44 Give the structure of the product formed when 1-methylcyclopentene is treated with each of the following reagents: **(a)** H_2 (Pt catalyst); **(b)** Cl_2 in H_2O; **(c)** Cl_2 in CCl_4; **(d)** cold alkaline $KMnO_4$; **(e)** hot $KMnO_4$ solution; **(f)** dilute solution of Br_2 in CH_3OH; **(g)** HCl; **(h)** HBr with H_2O_2; **(i)** mercuric acetate in water followed by $NaBH_4$; **(j)** OsO_4 (cat), H_2O_2.

10.45 Three different methods for converting alkenes into alcohols were discussed in this chapter. Write equations showing how these three methods could be used to convert 1-methyl-1-vinylcyclopentane into the following three alcohols:

(a) 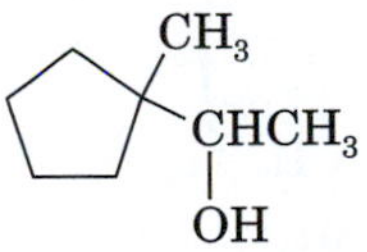 [cyclopentane ring with CH_3 and $\underset{\underset{OH}{|}}{C}HCH_3$ groups]   (b) [cyclopentane ring with CH_3 and CH_2CH_2OH groups]   (c) [cyclohexane ring with CH_3, OH, and CH_3 groups]

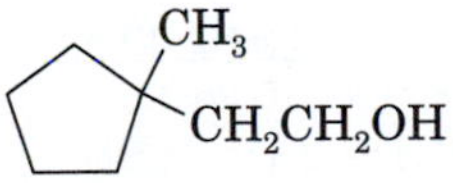

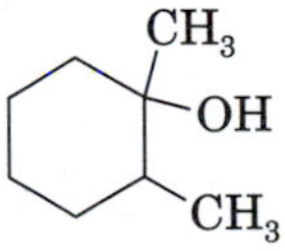

10.46 Show how you could synthesize the following compounds from propyne:

(a) 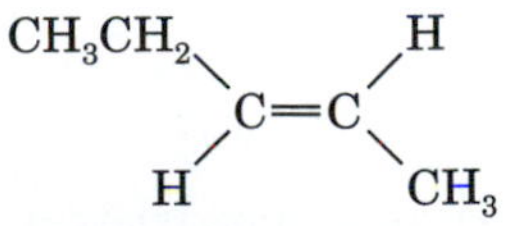 $\underset{H}{\overset{CH_3CH_2}{\diagdown}}C{=}C\underset{CH_3}{\overset{H}{\diagup}}$ (b) $CH_3\overset{O}{\overset{\|}{C}}CH_2CH_3$

(c) $CH_3\underset{\underset{OCH_3}{|}}{C}HC{\equiv}CCH_3$ (d) $CH_2\underset{\underset{Br}{|}}{C}H\underset{\underset{Br}{|}}{C}H_2Br$

10.47 Under acidic conditions, alkenes undergo additions of alcohols in much the same way that they undergo hydration. The addition of methanol to 1-methoxypropene in the presence of H_2SO_4 is highly regioselective and produces only 1,1-dimethoxypropane. Furthermore, this reaction occurs much faster with 1-methoxypropene than it does with 2-butene. Write a mechanism for this reaction and use it to explain the regioselectivity and the greater reactivity of 1-methoxypropene.

10.48 Starting with only acetylene, ethylene, iodomethane, and any appropriate inorganic reagents, show how you would prepare the following compounds:

(a) *cis*-4-octene; **(b)** 1-butene; **(c)** 1-chloro-2-butanol;

(d) 2-methyl-2-butanol; **(e)** *meso*-2,3-butanediol; **(f)** racemic 2,3-butanediol.

10.49 Suggest a mechanism showing all steps for each of the following reactions:

(a) $(CH_3)_2C=CH(CH_2)_2CH=C(CH_3)_2$ $\xrightarrow{H^+}$

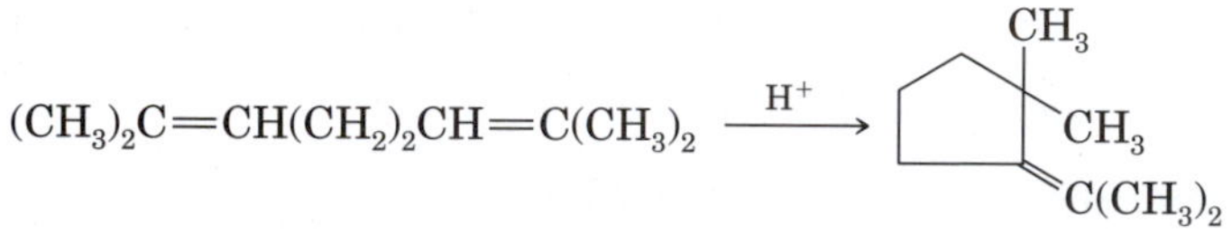

(b)

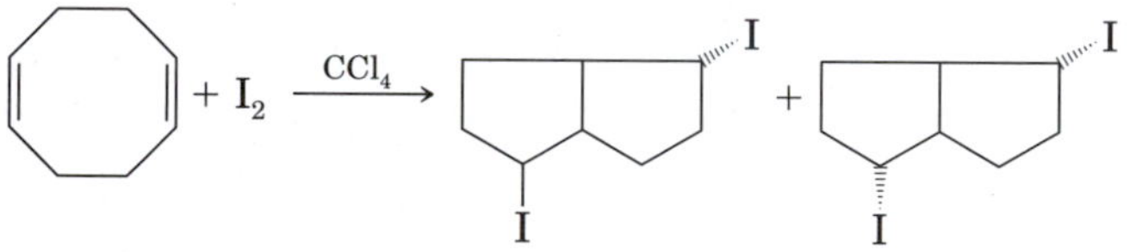

10.50 How would you make the following conversions?

(a)

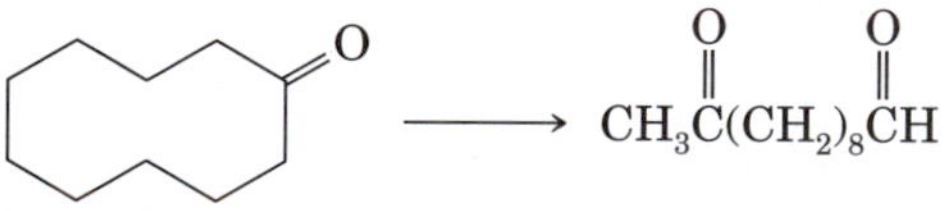

(b)

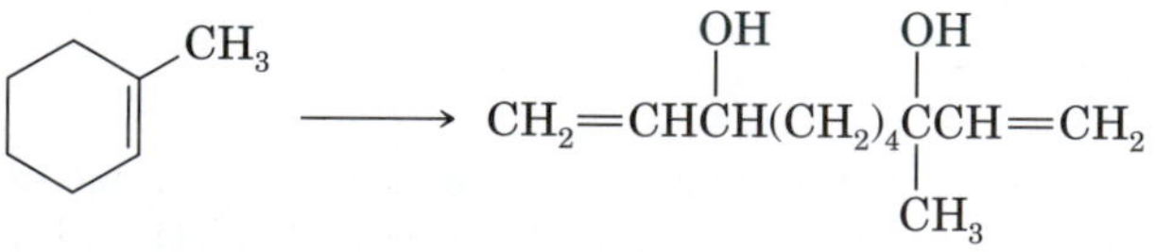

10.51 We saw in Section 6.8 that many alkenes readily undergo radical-catalyzed polymerization. Some alkenes also undergo **cationic polymerization,** an electrophilic addition that proceeds through a carbocation intermediate. The Lewis acid BF_3, with a trace of water, is an effective catalyst for cationic polymerizations. For example, isobutylene readily polymerizes to polyisobutylene under these conditions:

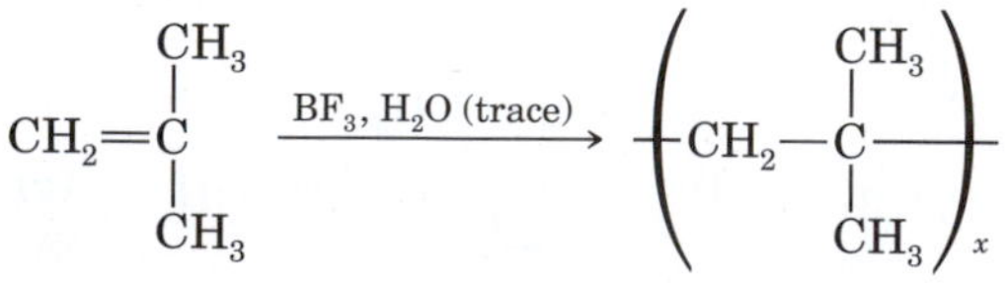

isobutylene   polyisobutylene

(a) What is the function of the trace of water in the catalyst system?

(b) Using two monomer units, write an equation to show why the polymer forms in a head-to-tail manner.

(c) Based on your mechanism, would you expect styrene (phenylethene) and acrylic acid (propenoic acid) to undergo cationic polymerization? Explain why or why not.

10.52 *Muscalure*, the sex attractant of the common housefly *(Musca domestica)*, is a hydrocarbon with the formula $C_{23}H_{46}$. Catalytic hydrogenation of this compound yields $C_{23}H_{48}$, while oxidation with a hot, alkaline solution of $KMnO_4$ followed by acidification yields $CH_3(CH_2)_{12}CO_2H$ and $CH_3(CH_2)_7CO_2H$. Addition of bromine to the hydrocarbon yields one pair of enantiomeric dibromides $(C_{23}H_{46}Br_2)$. What is the structure of the housefly sex attractant?

10.53 Compound **A** has the formula C_6H_{10}. Catalytic hydrogenation of **A** over deactivated palladium gives compound **B** (C_6H_{12}). **A** does not react with methylmagnesium bromide to liberate methane gas. **B** reacts with Simmons–Smith reagent to produce compound **C** (a cyclopropane, C_7H_{14}) that is optically inactive and cannot be resolved into enantiomers. What are the structures of **A**, **B**, and **C**?

10.54 The three deuterated methylcyclopentanes shown below can all be prepared from 1-methylcyclopentene using hydroboration reactions. Show how you would convert 1-methylcyclopentene into each of the three. Can you think of an alternate route to convert 1-methylcyclopentene into the deuterated compound in **(c)** that does not use hydroboration?

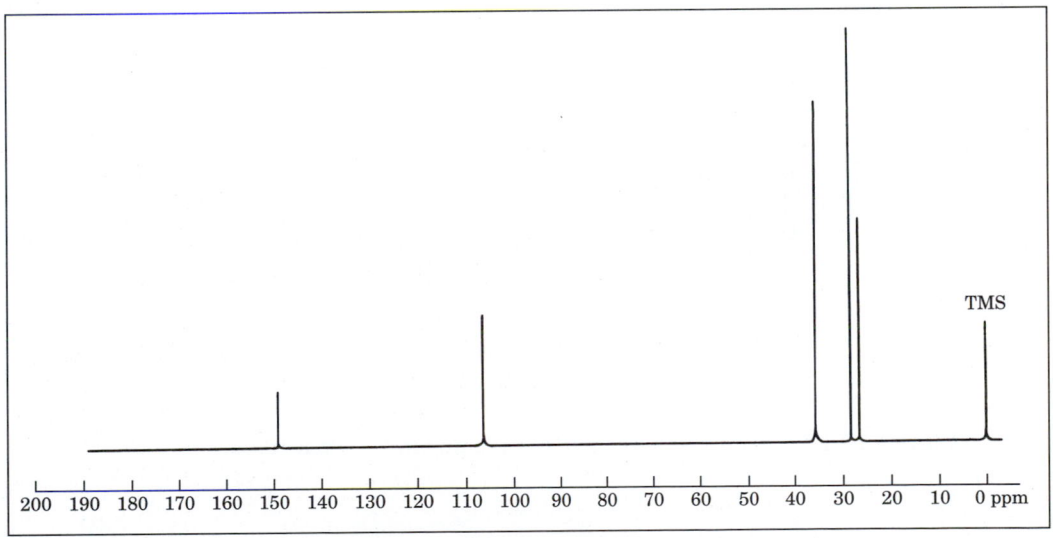

10.55 The ^{13}C NMR spectrum in Figure 10.15 is of a hydrocarbon containing a methylcyclohexane skeleton and one double bond. Locate the position of the double bond within the structure.

Figure 10.15 ^{13}C NMR spectrum for Problem 10.55.

10.56 A six-carbon compound, A, can be converted to B by hydrogenation. The ^{1}H NMR spectrum of A and the infrared spectrum of B are shown in Figure 10.16. What are the structures of A and B?

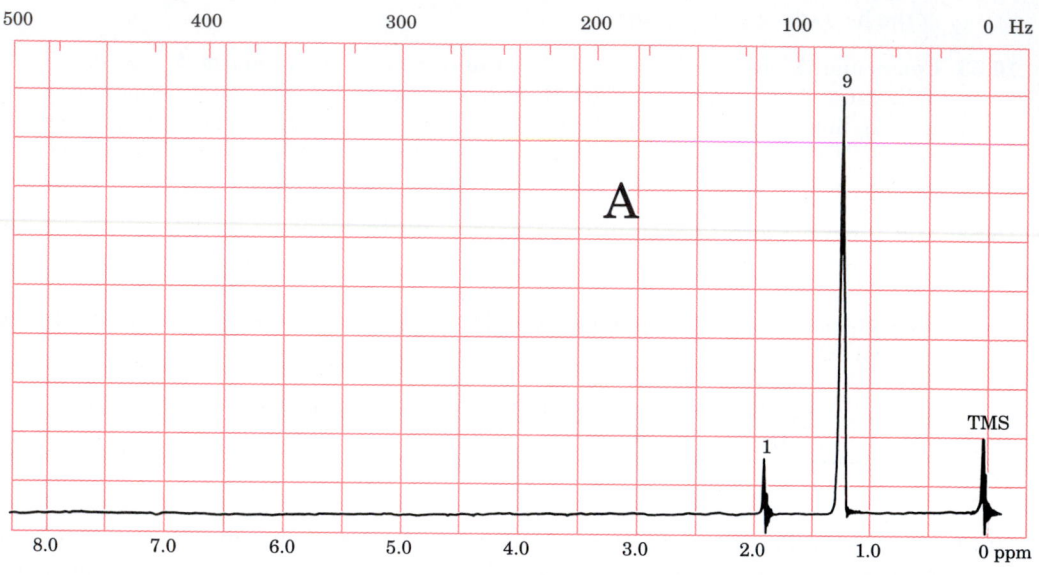

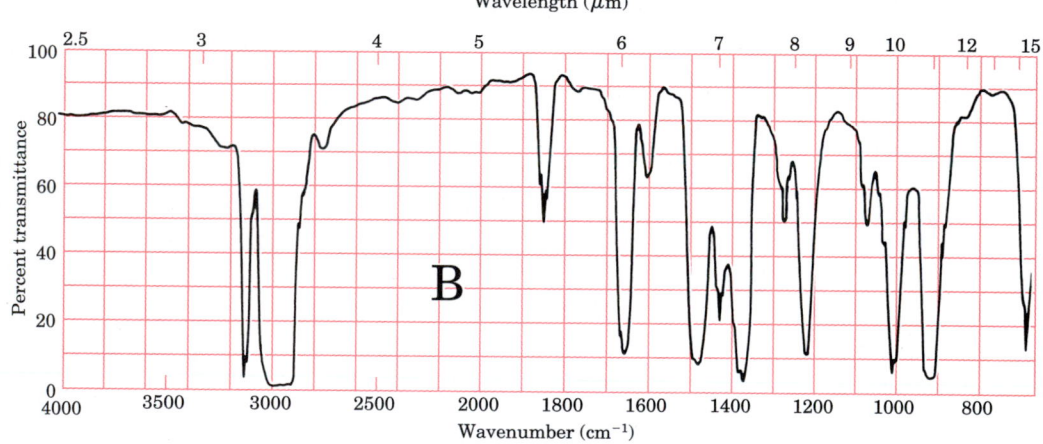

Figure 10.16 Spectra for unknown compounds in Problem 10.56.

10.57 Suggest a structure for each spectrum given in Figure 10.17.

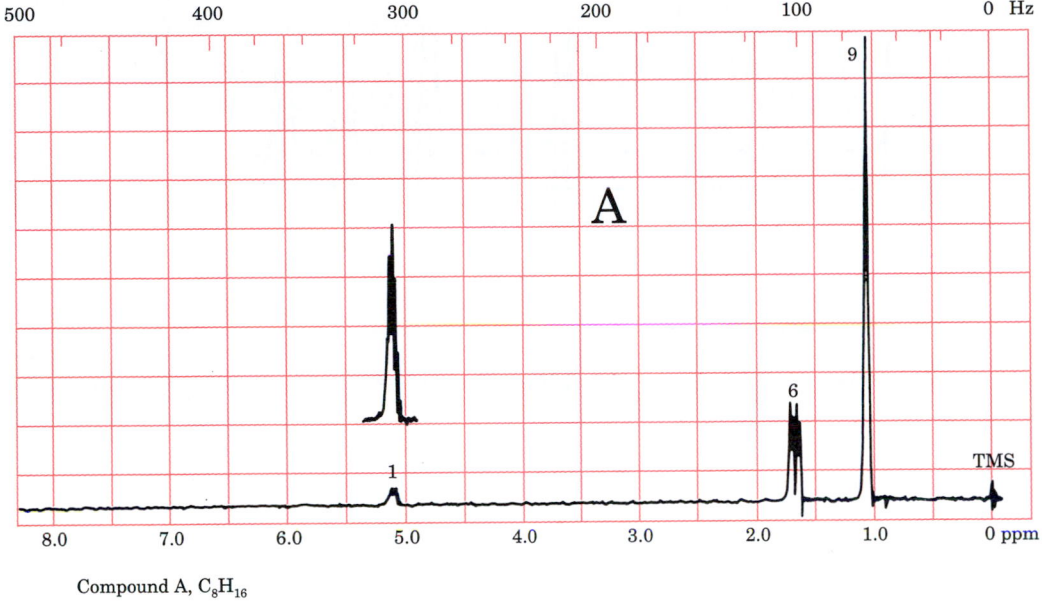

Compound A, C_8H_{16}

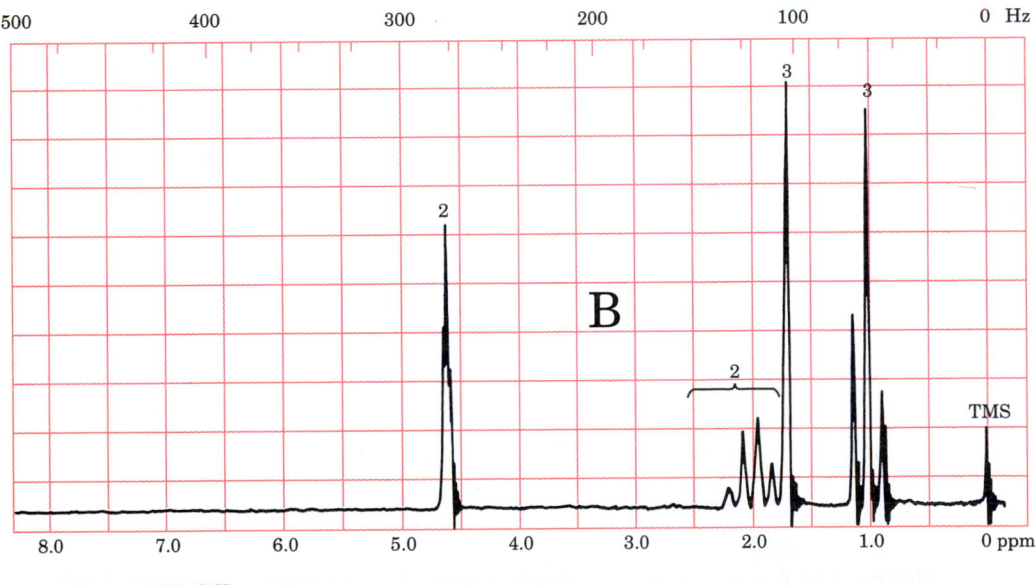

Compound B, C_5H_{10}

Figure 10.17 ^{1}H NMR spectra for unknown compounds in Problem 10.57.

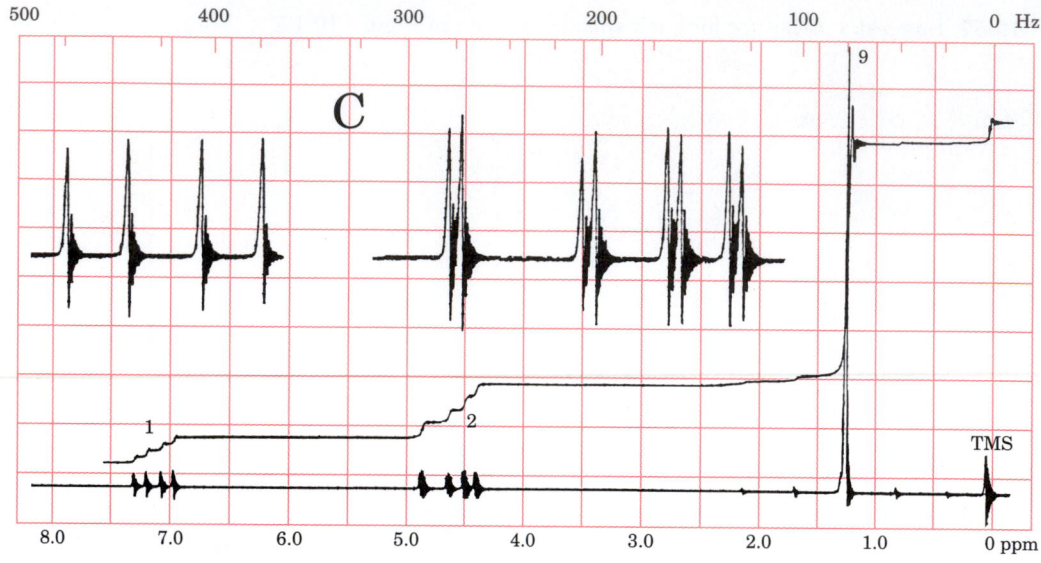

Compound C, $C_7H_{12}O_2$

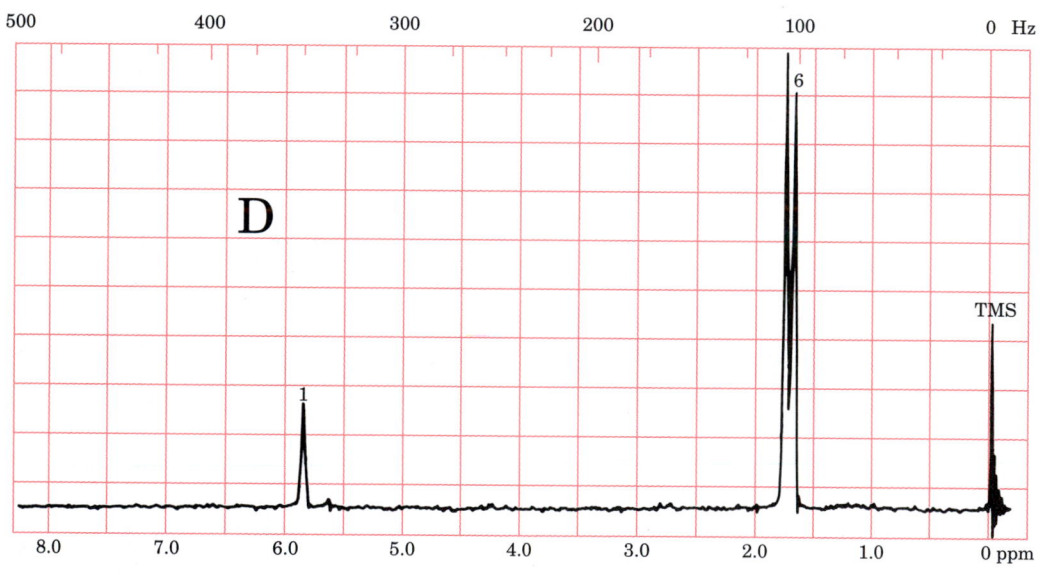

Compound D, C_8H_{14}

Figure 10.17 *(continued)* ^{1}H NMR spectra for unknown compounds in Problem 10.57.

AROMATICITY AND BENZENE; ELECTROPHILIC AROMATIC SUBSTITUTION

Benzene is the simplest of the aromatic compounds and one that we have already encountered many times. In this chapter and in Chapter 12, we will formalize the definition of aromaticity, and we will discuss the properties and reactions of benzene and substituted benzenes. In Chapter 19, we will discuss the chemistry of polycyclic and heterocyclic aromatic compounds. (*Heterocyclic* aromatic compounds have at least two different kinds of ring atoms, such as carbon and nitrogen.)

Benzene was first isolated in 1825 by Michael Faraday from oily residues that had accumulated in London gas mains. Today the main source of benzene, substituted benzenes, and other aromatic compounds is petroleum. Until the 1940s, coal tar was the principal source. The types of aromatic compounds obtained from these sources are hydrocarbons, phenols, and aromatic heterocycles.

Aromatic hydrocarbons:

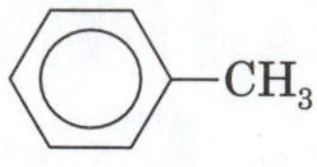

toluene 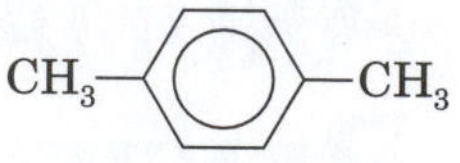   *p*-xylene

substituted benzenes

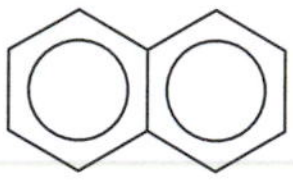

naphthalene 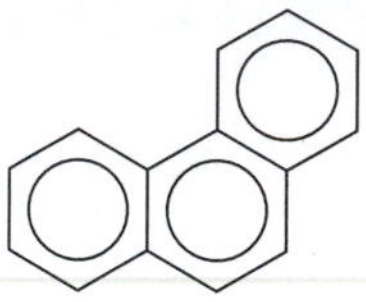   phenanthrene

polycyclic compounds

Aromatic nitrogen heterocycles:

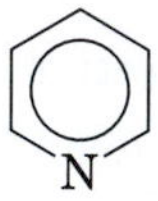

pyridine 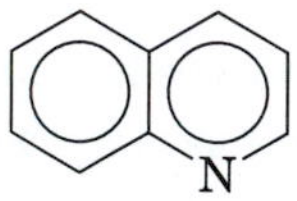   quinoline

Compounds containing benzene rings and aromatic heterocyclic rings are also exceedingly common in biological systems.

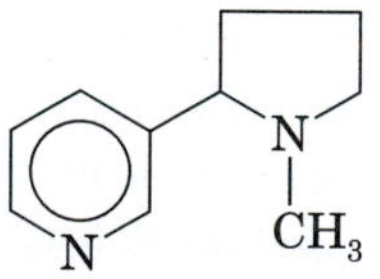

nicotine

in tobacco

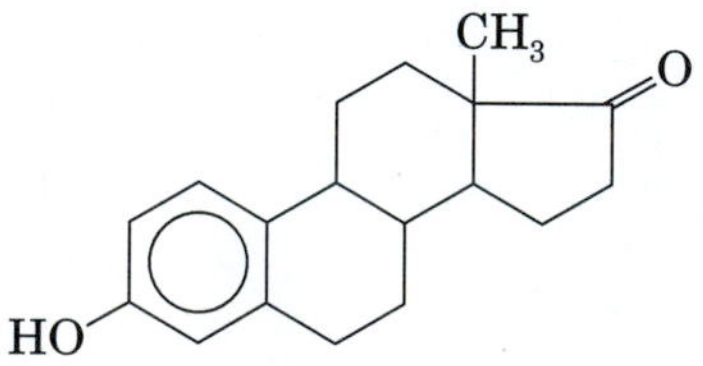

estrone

an estrogen, or female hormone

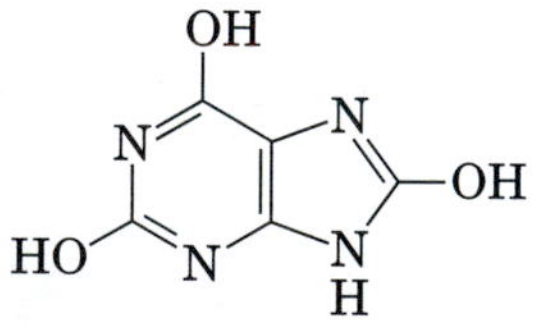

uric acid

associated with gout

The first recognition of chemical carcinogenesis was made in 1775 by the English surgeon Percivall Potts. He correlated the high incidence of scrotal cancer of chimney sweeps with the lodgment of soot and coal tar in the scrotum. By the 1700s, burning of coal had become widespread in Europe and England. Chimney sweeps cleaned the long English chimneys to prevent chimney fires. In the practice of that day, parish orphanages offered "master" chimney sweeps a fee to accept children as "apprentices." These pathetic "apprentices" were described by a poet of that day as "little black things . . . clothed . . . in the clothes of death." Indifference, neglect, and the lack of personal hygiene forced the sweeps to be in continual contact with coal tar. Studies have shown that many of the compounds found in coal tar have structures with four or more fused benzene rings. Such compounds are carcinogens (cancer-causing chemicals). Benzo[*a*]pyrene, a potent carcinogen, was first isolated from coal tar in 1933. Benzo[*a*]pyrene is also found in cigarette tar and provides the chemical link between smoking and lung cancer. Benzene itself is toxic and somewhat carcinogenic; therefore, it should be used in the laboratory only when necessary. (In many cases, toluene, which is less toxic, may be used as a substitute.)

Two highly carcinogenic hydrocarbons:

benzo[*a*]pyrene   benzanthracene

Section 11.1
Nomenclature of Substituted Benzenes

In 1833, E. Mitscherlich (1794–1855) isolated benzene by heating a mixture of benzoic acid and lime. He named the compound he isolated *benzin*. The editor of Mitscherlich's publication, Justus von Liebig (1803–1873), appended a note suggesting that benzin be renamed *benzol*. In 1837, Auguste Laurent (1807–1853) proposed yet a third name, *pheno* (Greek, "I bear light"), in honor of Faraday's discovery of the compound from illuminating gas. All three of these names have found their way into current usage. *Benzol* is the modern German name for benzene. In England and France, the suffix *ol* of benzol was dropped because it signified a hydroxyl group. The suffix for a double bond, *ene,* was added, generating the name *benzene*. The name *pheno* is found in *phenyl,* which is the name for benzene as a substituent.

The naming of monosubstituted benzenes was introduced in Section 3.3K. Many common benzene compounds have their own names, names that are not necessarily systematic. Some of the more commonly used names are listed in Table 11.1

Table 11.1   Structures and names of some common substituted benzenes

Structure	Name	Structure	Name
⬡–CH$_3$	toluene	⬡–OH	phenol
CH$_3$–⬡–CH$_3$	*p*-xylene	⬡–CO$_2$H	benzoic acid
⬡–CH=CH$_2$	styrene	⬡–CH$_2$OH	benzyl alcohol
⬡–NH$_2$	aniline	CH$_3$–⬡–SO$_2$Cl	*p*-toluenesulfonyl chloride (tosyl chloride)
⬡–NHCCH$_3$ (O)	acetanilide	CH$_3$C–⬡ (O)	acetophenone
		⬡–C–⬡ (O)	benzophenone

Disubstituted benzenes are named either with position numbers or with the prefixes **ortho-, meta-,** and **para-**. The prefix *ortho-* signifies that two substituents are 1,2 to each other on a benzene ring. *Meta-* signifies a 1,3-relationship, and *para-* means a 1,4-relationship. The use of *ortho-, meta-,* and *para-* in lieu of position numbers is reserved exclusively for disubstituted benzenes. The system is never used with cyclohexanes or other ring systems.

ortho-, or *o-*   *meta-*, or *m-*   *para-*, or *p-*

The use of these prefixes in the naming of some disubstituted benzenes follows:

o-dibromobenzene   *m*-chloroaniline   *p*-chlorophenol

In reactions of benzene compounds, we will speak of *ortho-, meta-,* or *para-*substitution. A monosubstituted benzene has two *ortho* and two *meta* positions, but only one *para* position.

ortho   *meta*   *para*

If there are three or more substituents on a benzene ring, the *o-, m-,* and *p-* system is no longer applicable. In this case, numbers must be used. As in numbering any compound, we number the benzene ring in such a way as to keep the prefix numbers as low as possible and give preference to the group of highest nomenclature priority. If a substituted benzene, such as aniline or toluene, is used as the parent, that substituent is understood to be at position 1 on the ring.

1,2,4-tribromobenzene   2-chloro-4-nitroaniline   2,4,6-trinitrotoluene (TNT)

Benzene as a substituent is called a **phenyl group.** How a *toluene* substituent is named depends on the point of attachment.

phenyl   benzyl   *p*-tolyl   *o*-tolyl

SAMPLE PROBLEM

Name the following substituted benzenes:

(a)   (b) Br—⟨ ⟩—CH₂Cl   (c)

Solution
(a) 2,6-dimethylaniline; (b) *p*-bromobenzyl chloride; (c) *m*-tolylcyclohexane
(or *m*-cyclohexyltoluene).

STUDY PROBLEMS

11.1 Write formulas for: **(a)** isopropylbenzene, **(b)** *p*-bromoaniline, **(c)** triphenylethene, **(d)** 3,5-dinitrophenol

11.2 Name the following compounds:

(a)   (b)   (c)   (d)

Section 11.2

Spectra of Substituted Benzenes

Both infrared and NMR spectra provide data useful in the structure determination of substituted benzenes. The ^1H NMR spectrum provides the most clear-cut answer to the absence or presence of aryl protons (and thus an aromatic ring).

A. Infrared Spectra

The infrared absorption bands of substituted benzenes are summarized in Table 11.2. The presence of a benzene ring in a compound of unknown structure can often be determined by inspection of two regions of the infrared

Table 11.2 Infrared absorption characteristics of benzene compounds		
	Position of Absorption	
Type of Vibration	cm^{-1}	μm
aryl C—H	3000–3300	3.0–3.3
aryl C—C (four peaks)	1450–1600	6.25–6.9

spectrum. The absorption for aryl CH stretching, which is generally weak, falls near 3030 cm^{-1} (3.3 μm), just to the left of aliphatic CH absorption. Absorption for aryl C—C vibrations gives a series of four peaks, generally between 1450 and 1600 cm^{-1} (about 6–7 μm). However, all four peaks are not always apparent. In the spectrum of chlorobenzene (Figure 11.1), the first peak is visible, but the second peak is a barely visible shoulder. In this spectrum, the third and fourth peaks are quite evident. Yet, in some spectra, the fourth peak at 1450 cm^{-1} (6.90 μm) is obscured by aliphatic CH$_2$ bending absorption.

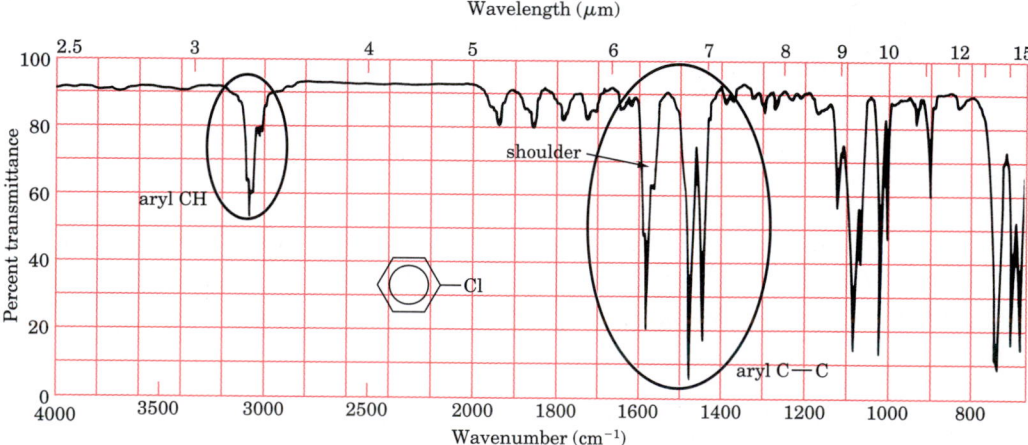

Figure 11.1 Infrared spectrum of chlorobenzene.

Table 11.3 The C—H bending absorption of substituted benzenes				
			Position of Absorption	
Substitution		Appearance	cm^{-1}	μm
monosubstituted,		two peaks	730–770	12.9–13.7
			690–710	14.0–14.5
o-disubstituted,		one peak[a]	735–770	12.9–13.6
m-disubstituted,		three peaks	860–900	11.1–11.6
			750–810	12.3–13.3
			680–725	13.7–14.7
p-disubstituted,		one peak	800–860	11.6–12.5

[a] An additional band is often observed at about 680 cm^{-1} (14.7 μm).

We can sometimes determine the positions of substitution on a benzene ring from the infrared spectrum. Differently substituted benzene rings often give characteristic absorption at about 680–900 cm^{-1} (11–15 μm). The patterns observed are summarized in Table 11.3. Figure 11.2 shows infrared spectra of the three isomeric chlorotoluenes. By comparison of these spectra

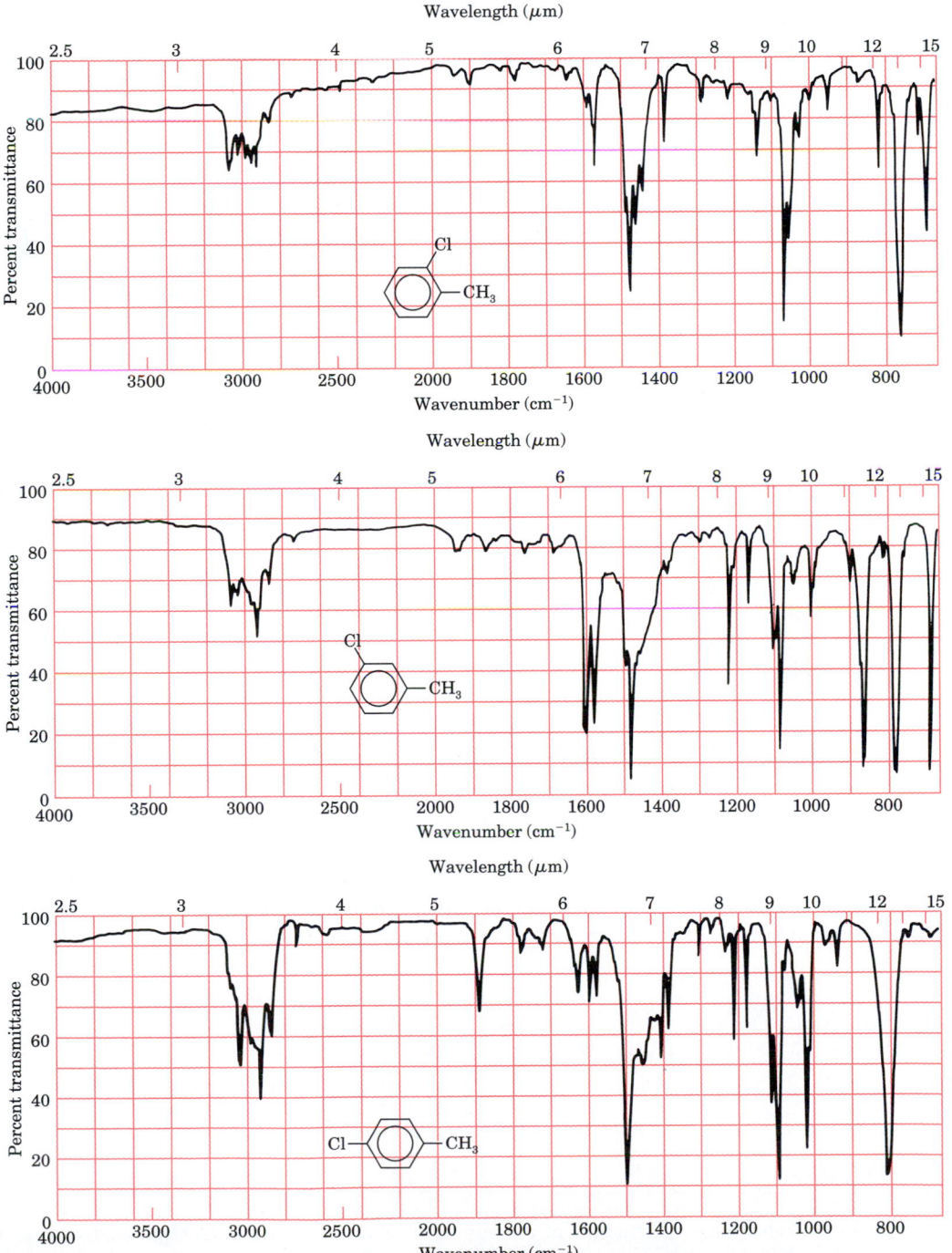

Figure 11.2 Infrared spectra for *o*-, *m*-, and *p*-chlorotoluene.

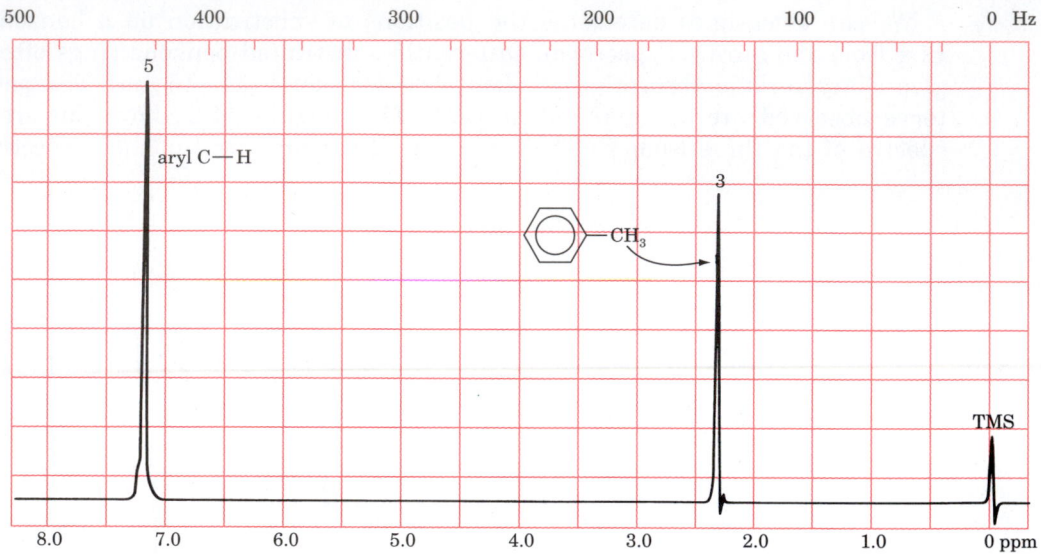

Figure 11.3 ^{1}H NMR spectrum of toluene, showing aryl and benzylic CH absorption.

with Table 11.3, you can see how the absorption in the fingerprint region may be used. Also compare these spectra with that of chlorobenzene in Figure 11.1. Unfortunately, the absorption in this region is not always so clear-cut.

B. ^{1}H NMR Spectra

The ^{1}H NMR spectra of aromatic compounds are distinctive. Protons on an aromatic ring absorb downfield, with δ values between 6.5 ppm and 8 ppm. This downfield absorption is due to the ring current, which gives rise to a molecular magnetic field that deshields protons bonded to the ring. The chemical shift for the protons in benzene itself is at $\delta 7.27$. Electronegative substituents on the ring shift the absorption of adjacent protons farther *downfield,* while electron-releasing groups shift absorption *upfield* from that of unsubstituted benzene. Simple splitting patterns of aryl protons are sometimes observed. In many cases, however, the splitting patterns are very complex.

Benzylic protons are not as affected by the aromatic ring current as are ring protons; their absorption is observed farther upfield, in the region of $\delta 2.3$. (See the NMR spectrum of toluene in Figure 11.3.

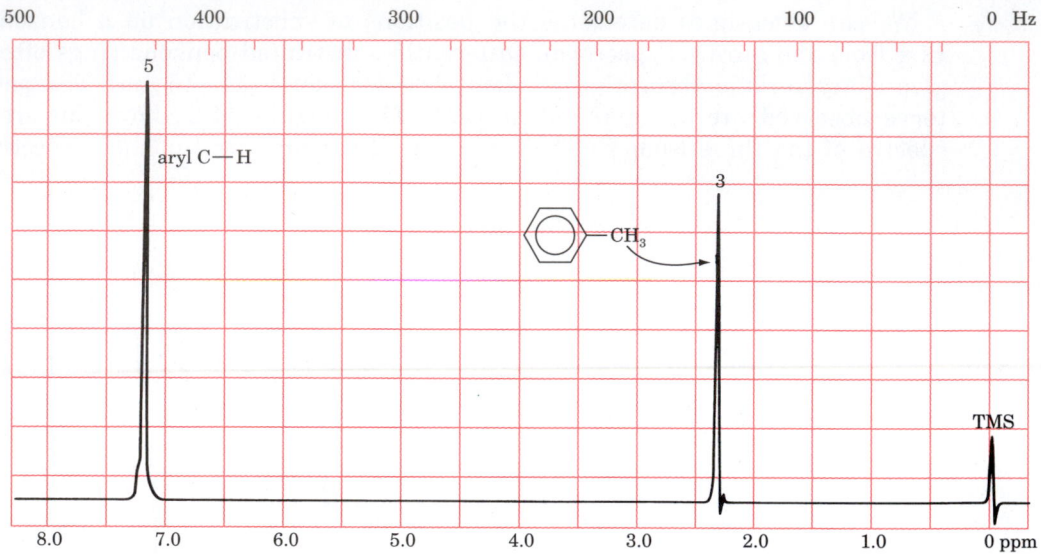

11.3 What is the structure of a compound whose formula is $C_8H_{10}O$ and whose ^{1}H NMR spectrum has singlets at $\delta 2.1$ and 3.7 and a "quartet" centered at $\delta 7.0$?

Section 11.3
Stability of the Benzene Ring

The heat of hydrogenation of cyclohexene is 28.6 kcal/mol. If benzene contained three alternate single and double bonds without any π-electron delocalization, we would expect its heat of hydrogenation to be 3×28.6, or 85.8, kcal/mol. However, benzene liberates only 49.8 kcal/mol of energy when it is hydrogenated.

cyclohexene $+ H_2$ $\xrightarrow{\text{Pt}}$ cyclohexane $+ 28.6$ kcal/mol

benzene $+ 3H_2$ $\xrightarrow[\substack{225°, \\ 35 \text{ atm}}]{\text{Pt}}$ cyclohexane $+ 49.8$ kcal/mol

The hydrogenation of benzene liberates 36 kcal/mol less energy than would be liberated by the hydrogenation of the hypothetical cyclohexatriene. Therefore, benzene, with delocalized π electrons, contains 36 kcal/mol less energy than it would contain if the π electrons were localized in three isolated double bonds. This difference in energy between benzene and the imaginary cyclohexatriene is called the **resonance energy** of benzene. The resonance energy is the energy lost (stability gained) by the complete delocalization of electrons in the π system. It is a measure of the added stability of the aromatic system compared to that of the localized system. Figure 11.4 shows the energy diagrams for these hydrogenations.

What does the resonance energy of benzene mean in terms of chemical reactivity? It means that more energy is required for a reaction in which the aromatic character of the ring is lost. Hydrogenation is one example of such a

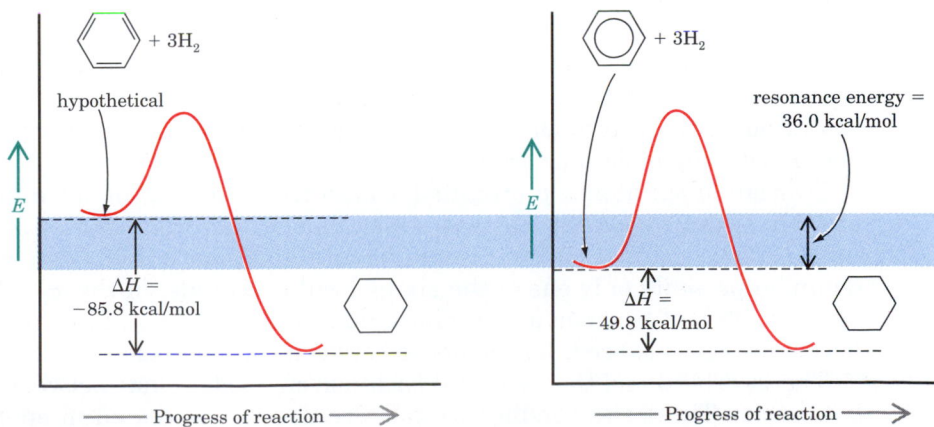

Figure 11.4 Energy diagrams for the hydrogenation of cyclohexatriene (hypothetical) and benzene.

reaction. Whereas an alkene can be hydrogenated at room temperature under atmospheric pressure, benzene requires high temperature and high pressure. Also, benzene does not undergo most reactions that are typical of alkenes. For example, benzene does not undergo addition of HX or X_2, nor does it undergo oxidation with $KMnO_4$ solution.

Section 11.4
The Bonding in Benzene

Although the molecular formula of benzene (C_6H_6) was determined shortly after its discovery in 1825, 40 years elapsed before Kekulé's brilliant proposal for the structure of benzene—a hexagon with three double bonds. To explain the existence of only three (not four) isomeric disubstituted benzenes, Kekulé later suggested that the benzene ring is in rapid equilibrium with the structure in which the double bonds are in the alternative positions. This idea survived for more than 50 years before it was replaced by the theories of resonance and molecular orbitals.

benzene in 1865   benzene in 1872   benzene in 1940

The Kekulé formulas for benzene, showing three double bonds instead of a circle in the ring, do not explain the unique stability of the benzene ring. However, these formulas have one advantage—they allow us to count the number of π electrons at a glance. In conjunction with resonance theory, these formulas are quite useful. Therefore, we will use Kekulé formulas when we discuss the reactions of benzene.

Let us now consider the π molecular orbitals of benzene. Benzene has six sp^2-hybridized carbons in a ring. The ring is planar, and each carbon atom has a p orbital perpendicular to this plane. Figure 2.23 (page 74) shows a representation of the p orbitals of benzene and how they overlap in the lowest-energy bonding molecular orbital.

Overlap of six atomic p orbitals leads to the formation of six π molecular orbitals. When we look at all six possible π molecular orbitals of benzene (Figure 11.5), we see that our representation of the aromatic π cloud as a "double donut" represents only one of the six molecular orbitals. In the π_1 orbital, all six p orbitals of benzene are in phase and overlap equally. This orbital is of lowest energy because it has no nodes between carbon nuclei.

The π_2 orbital and the π_3 orbital each has one nodal plane between the carbon nuclei. These two bonding orbitals are degenerate (equal in energy) and of higher energy than the π_1 molecular orbital. Benzene, with six p electrons, has the π_1, π_2, and π_3 orbitals each filled with a pair of electrons. These are the bonding orbitals of benzene.

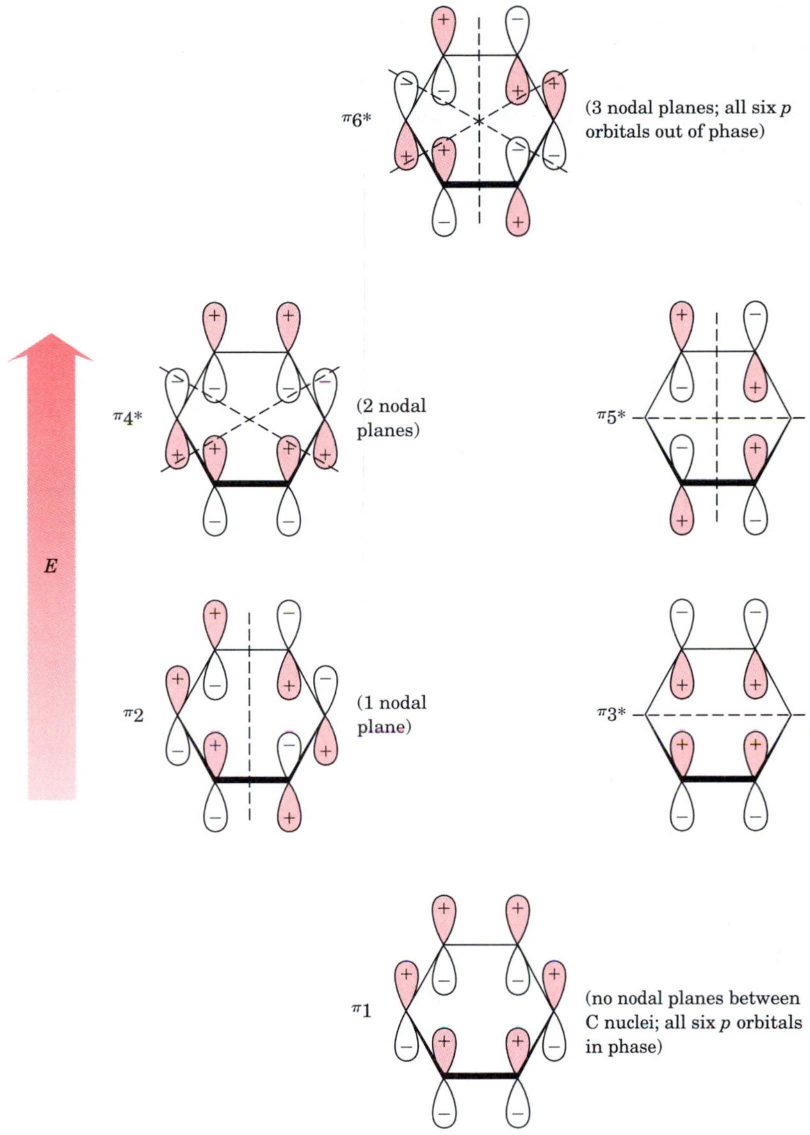

π_6^* (3 nodal planes; all six p orbitals out of phase)

E

π_4^* (2 nodal planes)

π_5^* (2 nodal planes)

π_2 (1 nodal plane)

π_3^* (1 nodal plane)

π_1 (no nodal planes between C nuclei; all six p orbitals in phase)

Figure 11.5 The π orbitals in benzene. Nodal planes are represented by dashed lines; the "missing p orbitals" in π_3 and π_5^* are a result of nodes at these positions. (The + and − signs are mathematical signs of phase, not electrical charges.)

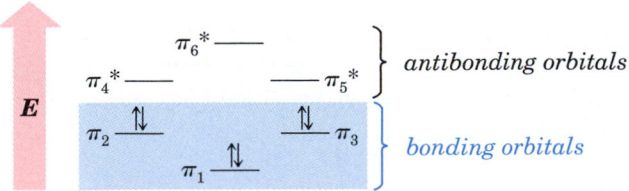

Along with the three bonding orbitals in benzene are three antibonding orbitals. Two of these antibonding orbitals (π_4^* and π_5^*) have two nodal planes each and the highest-energy orbital (π_6^*) has three nodal planes between carbon nuclei. Recall that a node or a nodal plane is a region of very low electron density. A molecular orbital with a nodal plane between nuclei is of higher energy than a molecular orbital without a nodal plane between nuclei. As we progress from π_1 to π_6^*, the number of nodal planes increases; this is the reason that the energy associated with these molecular orbitals increases.

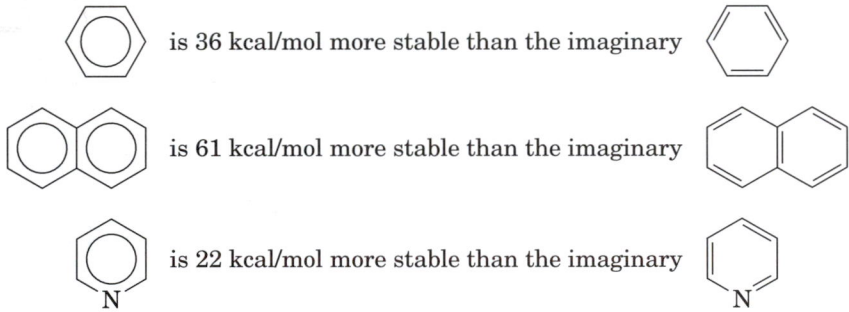

Section 11.5
What Is an Aromatic Compound?

Benzene is one member of the large class of aromatic compounds—compounds that are *substantially stabilized by π-electron delocalization*. The resonance energy of an aromatic compound is a measure of its gain in stability. (The structural features that give rise to aromaticity will be discussed shortly.)

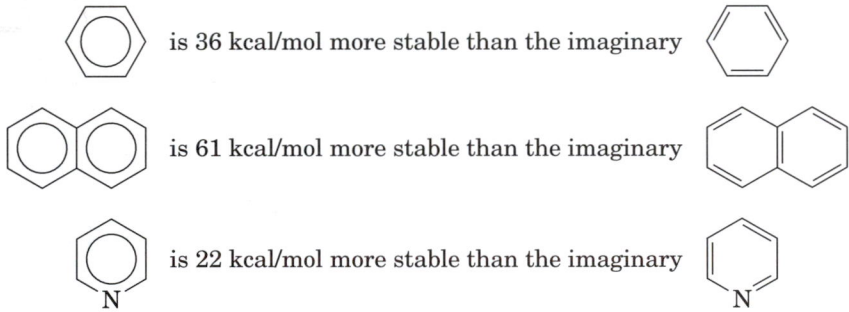

The most convenient way to determine if a compound is aromatic is by observing the chemical shifts of hydrogens bonded to the ring atoms. Hydrogens on the outer periphery of an aromatic ring are highly deshielded and absorb far downfield from most other hydrogens, usually beyond $\delta 7$. (See Section 9.7B.)

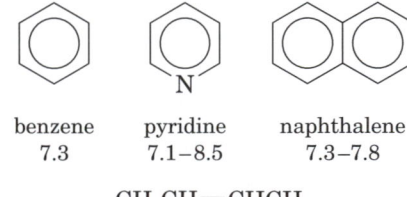

	benzene	pyridine	naphthalene
δ of aromatic CH, ppm:	7.3	7.1–8.5	7.3–7.8

$$CH_3\underline{CH}\!=\!\underline{CH}CH_3$$

δ of nonaromatic sp^2 CH, ppm:	5.3

11.4 The ^{1}H NMR spectrum of cyclooctatetraene shows only a singlet at $\delta 5.7$. On the basis of this chemical shift, would you say that cyclooctatetraene is aromatic or not?

cyclooctatetraene

Section 11.6
Requirements for Aromaticity

What structural features are necessary for a molecule to be aromatic? The molecule must be *cyclic* and *planar*, and each atom of the ring or rings must have a *p orbital perpendicular to the plane of the ring*. If a system does not fit these criteria, there cannot be complete delocalization of the π electrons. Whether these three criteria are met can often be deduced from inspection of the formula of an organic compound. The line-bond formula of an aromatic compound usually shows a ring with alternate single and double bonds. There are cases, however, of cyclic organic compounds with alternate single and double bonds that are *not* aromatic. Cyclooctatetraene is such a compound. Cyclooctatetraene undergoes addition reactions with the hydrogen halides and with the halogens. These reactions are typical of alkenes, but are not typical of benzene and other aromatic compounds. Cyclooctatetraene is not planar, but has been shown to be shaped like a tub.

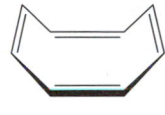

cyclooctatetraene

a cyclic tetraene; not aromatic

Why is cyclooctatetraene not aromatic? To answer this question, we must proceed to a fourth criterion for aromaticity, a criterion usually called **the Hückel rule.**

A. The Hückel Rule

In 1931, the German chemist Erich Hückel proposed that, to be aromatic, a monocyclic (one ring), planar compound must have **$(4n + 2)$ π electrons,** where n is an integer. According to the Hückel rule, a ring with 2, 6, 10, or 14 π electrons may be aromatic, but a ring with 8 or 12 π electrons may not. Thus, cyclooctatetraene (with 8 π electrons) would not be aromatic even if it were planar.

six πe^-   *eight πe^-*

$(4n + 2)$ $4n$
$n = 1$ $n = 2$

aromatic   *not aromatic*

Why can a monocyclic compound with six or ten π electrons be aromatic, but not a compound with eight π electrons? The answer is found in the number of π electrons versus the number of π orbitals available. To be aromatic, a molecule must have *all π electrons paired and all bonding orbitals filled*. This system provides the maximum and complete overlap required for aromatic stabilization. If some π orbitals are *not* filled (that is, there are unpaired π electrons), overlap is not maximized, and the compound is not aromatic.

Benzene has six π electrons and three bonding π orbitals. The three bonding π orbitals are filled to capacity, all π electrons are paired, and benzene is aromatic.

Let us look at the π molecular orbitals for cyclooctatetraene. This compound has eight p orbitals on the ring. Overlap of eight p orbitals would result in eight π molecular orbitals.

If cyclooctatetraene were planar and had a π system similar to that in benzene, the π_1, π_2, and π_3 orbitals would be filled with six of the π electrons. The remaining two electrons would be found, one each, in the degenerate π_4 and π_5 orbitals (Hund's rule; see Section 1.1B). The π electrons of cyclooctatetraene would not all be paired, and overlap would not be maximal. Thus, cyclooctatetraene cannot be aromatic.

B. The Ions of Cyclopentadiene

Cyclopentadiene is a conjugated diene and is not aromatic. The principal reason that cyclopentadiene is not aromatic is that one of the carbon atoms is sp^3 hybridized. This sp^3-hybridized carbon has no p orbital available for bonding. However, removal of a hydrogen ion from this carbon changes the hybridization to sp^2. This carbon now has a p orbital containing a pair of electrons.

The cyclopentadiene cation also has all its carbon atoms in the sp^2 state. (The p orbital pictures for the ions of cyclopentadiene are shown in Figure 11.6.)

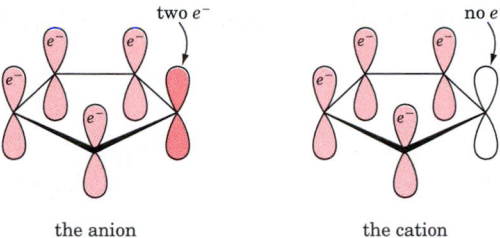

the anion   the cation

Figure 11.6 Orbital pictures of the anion and the cation of cyclopentadiene. (Each carbon is sp^2 hybridized and is bonded to one H.)

Would either or both of these ions be aromatic? Either ion would have five π molecular orbitals (from five p orbitals, one per carbon). The cyclopentadienyl anion, with six π electrons ($4n + 2$), has three π orbitals filled and all π electrons paired. *The anion is aromatic.* The cation, however, would have only four π electrons ($4n$) in the three orbitals. The π electrons would *not* all be paired, and there would *not* be maximum overlap. *The cation is not aromatic.*

Cyclopentadiene is a stronger acid than most other hydrocarbons because its anion is aromatic and thus is resonance-stabilized. Although cyclopentadiene is not nearly as strong an acid as a carboxylic acid, it is a proton donor in the presence of a strong base. (The pK_a of cyclopentadiene is 16, similar to that of an alcohol. By contrast, the pK_a of cyclopentane is about 50.)

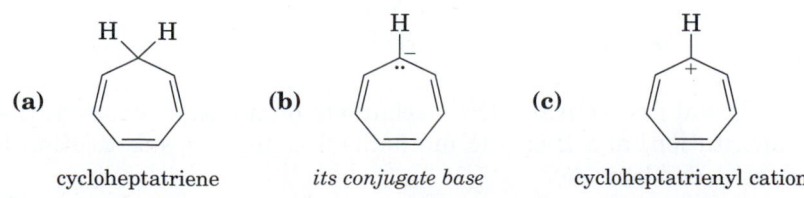

STUDY PROBLEMS

11.5 Analyze cycloheptatriene as we have just done for cyclopentadiene. **(a)** Is cycloheptatriene aromatic? **(b)** Is its conjugate base aromatic? **(c)** Is the cycloheptatrienyl cation aromatic? Explain your answers.

(a)   **(b)**   **(c)**

cycloheptatriene   *its conjugate base*   cycloheptatrienyl cation

11.6   Which of the following structures would you expect to be aromatic?

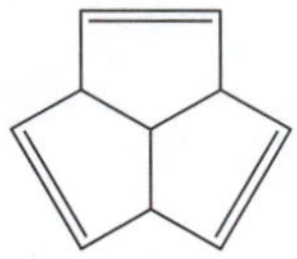

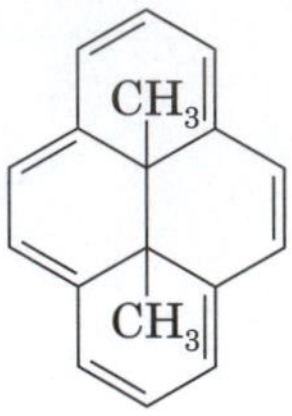

Section 11.7
Electrophilic Aromatic Substitution

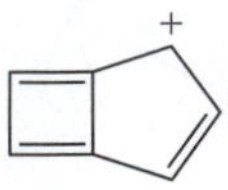

The aromaticity of benzene confers a unique stability to the π system, and benzene does not undergo most of the reactions typical of alkenes. However, benzene is far from inert. Under the proper conditions, benzene readily undergoes **electrophilic aromatic substitution reactions**—reactions in which an electrophile is substituted for one of the hydrogen atoms on the aromatic ring. Two examples of this type of substitution reaction follow. Note that the aromaticity of the ring is retained in each product.

Halogenation:

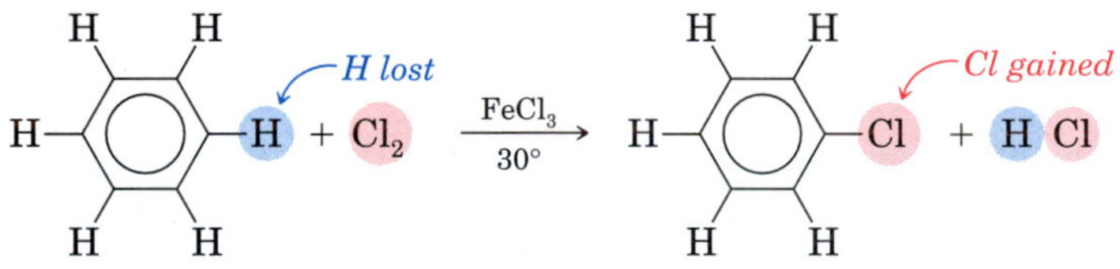

chlorobenzene (90%)

Nitration:

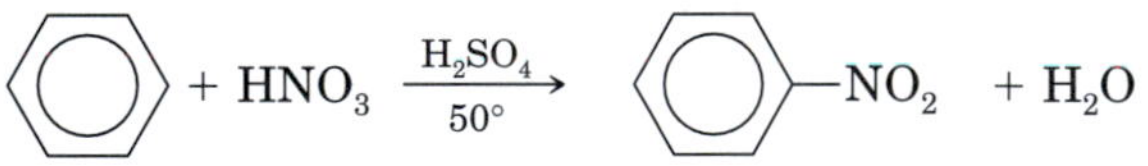

nitrobenzene (85%)

The preceding examples show **monosubstitution** of the benzene ring. Further substitution is possible:

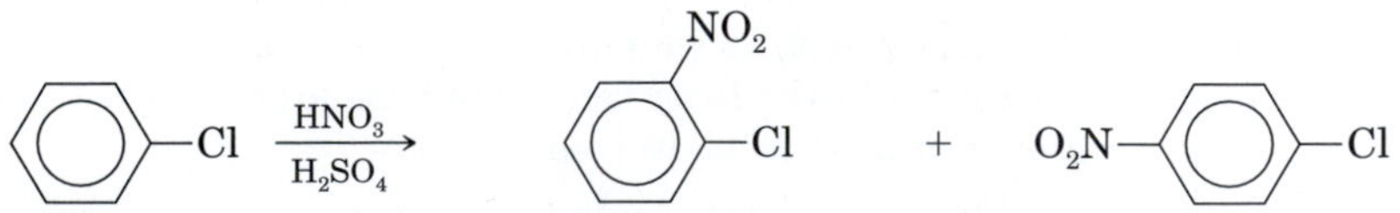

chlorobenzene   *o*-chloronitrobenzene (30%)   *p*-chloronitrobenzene (70%)

We will first consider the mechanism of monosubstitution (that is, the first substitution) and then the mechanism of further substitution leading to disubstituted benzenes.

Section 11.8
The First Substitution

In each of the two monosubstitution reactions shown, a Lewis acid is used as a catalyst. The Lewis acid undergoes reaction with the reagent (such as X_2 or HNO_3) to generate an electrophile, which is the actual agent of substitution. For example, H_2SO_4 (a very strong acid) can remove a hydroxyl group from nitric acid to yield the nitronium ion, $^+NO_2$.

Formation of an electrophile by a Lewis acid:

$$\ddot{H\ddot{O}}-NO_2 + H_2SO_4 \xrightleftharpoons[]{-HSO_4^-} H_2\overset{+}{\ddot{O}}-NO_2 \rightleftharpoons H_2\ddot{O}: + \quad ^+NO_2$$

an electrophile

An electrophile can attack the π electrons of a benzene ring to yield a resonance-stabilized carbocation intermediate called a *benzenonium ion,* or an *arenium ion.* Like other carbocations, this carbocation reacts further. A hydrogen ion is removed from the intermediate (by HSO_4^-, for example) to yield the substitution product. In showing these structures, we use Kekulé formulas, which allow us to keep track of the number of π electrons. In the following equation, we also show the hydrogen atoms bonded to the ring carbons so you can see how they are affected (or not) in the reaction.

benzene   *intermediate*   *product*
                     carbocation

As we discuss the various types of electrophilic aromatic substitution reactions, you will see that the mechanisms are all simply variations of this general one.

A. Halogenation

Aromatic halogenation is typified by the bromination of benzene. The catalyst in aromatic bromination is $FeBr_3$ (often generated *in situ* from Fe and Br_2). The function of the catalyst is to generate the electrophile Br^+. This may occur by direct reaction and fission of the Br—Br bond. More likely, Br_2 is not completely cleaved upon reaction with the $FeBr_3$ catalyst, but is polarized. For the sake of simplicity, we will show Br^+ as the electrophile.

$$:\ddot{Br}-\ddot{Br}: + FeBr_3 \rightleftharpoons \overset{\delta+}{:\ddot{Br}}-\overset{\delta-}{\ddot{Br}}\text{:}\cdots FeBr_3 \rightleftharpoons :\ddot{Br}^+ \quad + FeBr_4^-$$

electrophilic

polarized   *cleaved*

When an electrophile such as Br^+ collides with the electrons of the aromatic π cloud, a pair of the π electrons forms a σ bond with the electrophile. This step is the slow step, and therefore the rate-determining step, in the reaction.

Step 1 (slow):

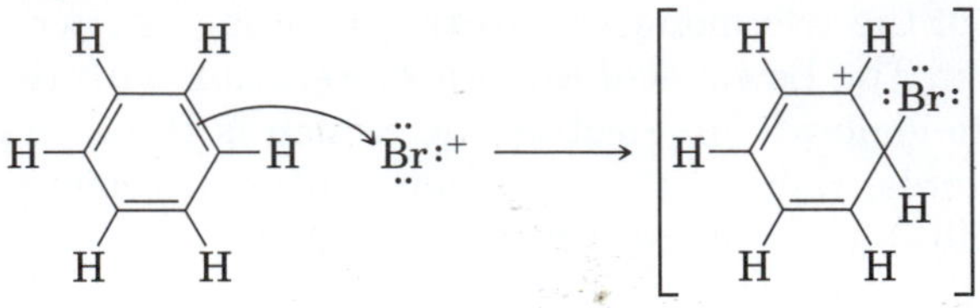

The intermediate carbocation loses a proton to a base in the reaction mixture. The product is bromobenzene, a product in which the aromatic character of the ring has been restored.

Step 2 (fast):

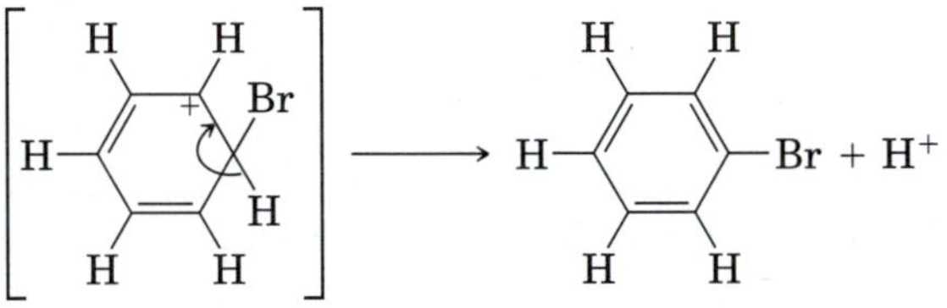

bromobenzene

The third step in the reaction mechanism (which is concurrent with the loss of H⁺) is the regeneration of the Lewis acid catalyst. The proton released in Step 2 undergoes reaction with the $FeBr_4^-$ ion to yield HBr and $FeBr_3$.

Step 3 (fast):

$$H^+ + FeBr_4^- \rightleftharpoons FeBr_3 + HBr$$

Ignoring the function of the catalyst, we can write an equation for the overall reaction of the aromatic bromination of benzene:

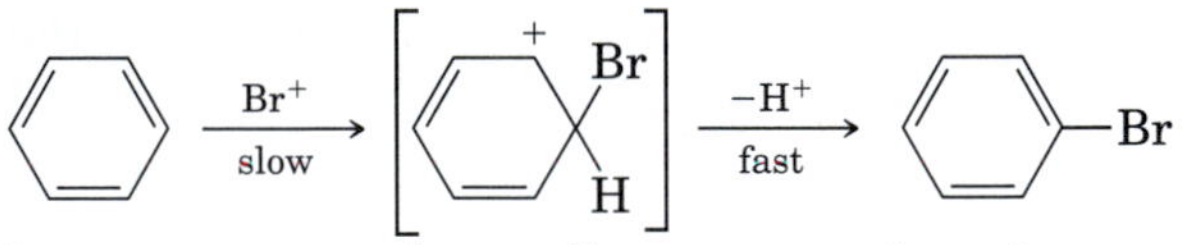

benzene   *intermediate*   bromobenzene

SAMPLE PROBLEM

Write resonance structures for the intermediate carbocation in the aromatic bromination of benzene.

Solution

The intermediate is often represented as

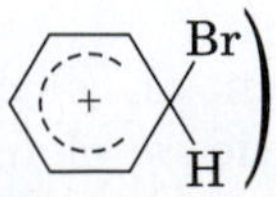

Note the similarity between an electrophilic aromatic substitution reaction and an E1 reaction. In an E1 reaction, an intermediate alkyl carbocation eliminates a proton to form an alkene.

$$\left[-\overset{+}{\underset{|}{C}}\overset{H}{\underset{|}{C}}- \right] \xrightarrow[-H^+]{E1} \overset{\diagdown}{\diagup}C=C\overset{\diagup}{\diagdown}$$

An alkyl carbocation also can undergo reaction with a nucleophile in an S_N1 reaction. However, the intermediate carbocation in aromatic substitution reaction does *not* undergo this reaction with a nucleophile. The addition of the nucleophile would destroy the aromatic stabilization of the benzene ring, whereas loss of a proton restores the aromatic stabilization.

B. Isotope Effect in Aromatic Substitution

Recall from Section 5.8A that a C—D bond is stronger than a C—H bond. If the breaking of a C—H bond is part of the rate-determining step of a reaction, the rate of reaction for a C—D compound is slower than the rate for the corresponding C—H compound.

If, as we have said, the rate-determining step of electrophilic aromatic substitution is the formation of the intermediate carbocation, then reaction of deuteriated benzene would proceed at the *same rate* as the reaction of normal benzene. Experiments have shown this to be true. Benzene and perdeuteriobenzene (C_6D_6) undergo electrophilic bromination at the same rate; no kinetic isotope effect is observed.

Step 2 in the reaction mechanism, loss of H^+ or D^+, does involve breaking of the C—H or C—D bond. Undoubtedly, the rate of elimination of D^+ is slower than that of H^+, but in either case the second step is so fast compared to Step 1 that no change in overall rate of reaction is observed.

C. Nitration

Benzene undergoes nitration when treated with concentrated HNO_3. The Lewis acid catalyst in this reaction is concentrated H_2SO_4. Like halogenation, aromatic nitration is a two-step reaction. The first step (the slow step) is electrophilic attack. In nitration, the electrophile is $^+NO_2$; the equation for its formation is shown on page 489. The result of the attack is a carbocation, which undergoes rapid loss of H^+ in the second step. This H^+ combines with HSO_4^- to regenerate the catalyst, H_2SO_4.

benzene   nitrobenzene

D. Alkylation

The alkylation of benzene is the substitution of an alkyl group for a hydrogen on the ring. Alkylation with an alkyl halide and a trace of $AlCl_3$ as catalyst is often referred to as a **Friedel–Crafts alkylation,** after Charles Friedel (1832–1899), a French chemist, and James Crafts (1839–1917), an American chemist, who developed this reaction in 1877.

> The story of their meeting is an interesting one. In 1874, due to ill health, Crafts resigned his position as professor of chemistry at the Massachusetts Institute of Technology (MIT), so that he could recuperate and travel. On one of his stops, he visited the laboratory at the Sorbonne, France, where he met Friedel. Crafts extended his visit for seventeen years, and did not return to MIT until 1891. He became head of the chemistry department in 1895, and president of MIT in 1897.

The reaction of 2-chloropropane with benzene in the presence of $AlCl_3$ is a typical Friedel–Crafts alkylation reaction.

2-chloropropane   isopropylbenzene
(isopropyl chloride)   (cumene)

The first step in alkylation is the generation of the electrophile: an alkyl cation.

$$R-\overset{..}{\underset{..}{Cl}}: + AlCl_3 \rightleftharpoons R^+ + AlCl_4^-$$

The second step is electrophilic attack on benzene, while the third step is elimination of a hydrogen ion. The product is an alkylbenzene.

an alkylbenzene

One problem in Friedel–Crafts alkylations is that the substitution of an alkyl group on the benzene ring activates the ring so that a *second* substitution may also occur. (We will discuss ring activation and second substitutions later in this chapter.) To suppress this second reaction, an excess of the aromatic compound is commonly used.

excess

Another problem in Friedel–Crafts alkylations is that the attacking electrophile may undergo rearrangement by 1,2-shifts of H or R.

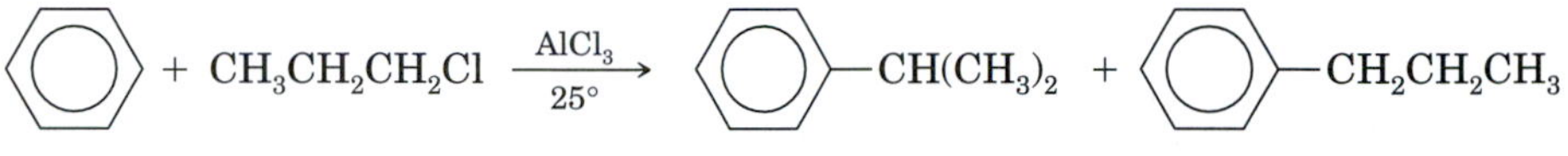

1-chloropropane   isopropylbenzene (70%)   propylbenzene (30%)

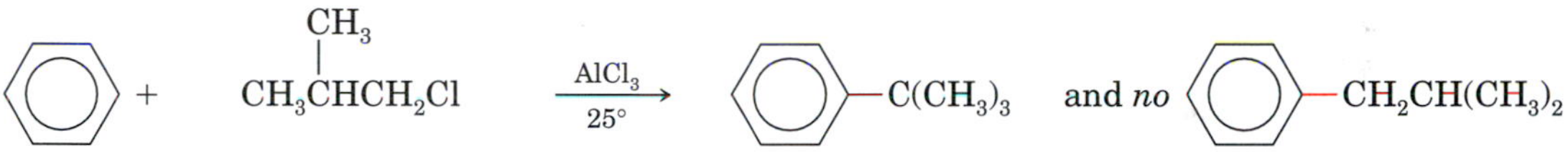

1-chloro-2-methylpropane   *tert*-butylbenzene (100%)   isobutylbenzene

The rearrangements shown are of primary alkyl halides, which do not readily form carbocations. In these cases, reaction probably proceeds through an RX—$AlCl_3$ complex.

$$CH_3CH_2CH_2-\overset{..}{\underset{..}{Cl}}: + AlCl_3 \rightleftharpoons CH_3CH_2\overset{\delta+}{CH_2}-\overset{\delta-}{\overset{..}{Cl}}\text{------}AlCl_3$$

a complex

This complex may (1) undergo reaction with benzene to yield the nonrearranged product, or (2) undergo rearrangement to a secondary or tertiary carbocation, which leads to the rearranged product.

No rearrangement:

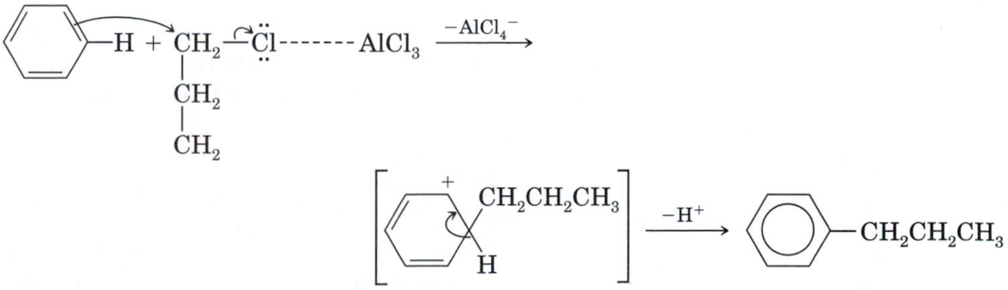

propylbenzene

Rearrangement:

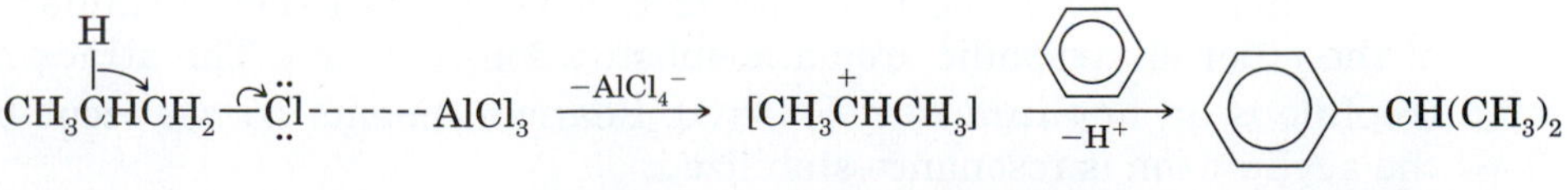

a 2° carbocation   isopropylbenzene

Alkylations can also be accomplished with alkenes in the presence of HCl and $AlCl_3$. The mechanism is similar to alkylation with an alkyl halide and proceeds by way of the more stable alkyl carbocation.

$$CH_3CH{=}CH_2 + HCl + AlCl_3 \longrightarrow [CH_3\overset{+}{C}HCH_3] + AlCl_4{}^-$$

propene

*more stable
2° carbocation*

isopropylbenzene (85%)

STUDY PROBLEM

11.7 Predict the *major* organic product of the reaction of benzene in the presence of $AlCl_3$ (and HCl in the case of an alkene) with: **(a)** 1-chlorobutane; **(b)** methylpropene; **(c)** neopentyl chloride (1-chloro-2,2-dimethylpropane); **(d)** dichloromethane (using an excess of benzene).

E. Acylation

The $R\overset{O}{\overset{\|}{C}}{-}$ group or the $Ar\overset{O}{\overset{\|}{C}}{-}$ group is called an **acyl group.** Substitution of an acyl group on an aromatic ring by reaction with an acid halide is called an **aromatic acylation reaction,** or a **Friedel–Crafts** acylation.

acetyl chloride

an acid halide

acetophenone
(97%)

This reaction is often the method of choice for preparing an aryl ketone. The carbonyl group of the aryl ketone can be reduced to a CH_2 group (Section 13.6C). By the combination of a Friedel–Crafts acylation and reduction, an alkylbenzene may be prepared without rearrangement of the alkyl group.

propanoyl chloride

phenyl ethyl ketone
(90%)

propylbenzene
(80%)

The mechanism of the Friedel–Crafts acylation reaction is similar to that of the other electrophilic aromatic substitution reactions. The attacking electrophile is an **acylium ion** ($R\overset{+}{C}{=}O$). Rearrangements do not occur because the acylium ion is resonance-stabilized.

the electrophilic carbon
↓

$$RC\overset{O}{\underset{||}{C}}Cl: + AlCl_3 \longrightarrow RC\overset{O}{\underset{||}{\underset{\delta+}{C}}} \cdots \underset{\delta-}{Cl} \cdots AlCl_3 \xrightarrow{-AlCl_4^-} \left[R-\overset{+}{C} = \overset{..}{O}: \longleftrightarrow R-C \equiv \overset{+}{O}: \right]$$

polarized   *resonance structures for*
                                     an acylium ion

Because $AlCl_3$ forms complexes with carbonyl groups ($C=\overset{..}{\underset{..}{O}}$:---$AlCl_3$), a slight excess of $AlCl_3$, such as 1.1 mol of $AlCl_3$ to 1.0 mol of RCOCl, is used in a Friedel–Crafts acylation. The product ketone is also complexed with $AlCl_3$, but treatment with water during the work-up of the reaction mixture liberates the ketone.

STUDY PROBLEMS

11.8 Write a mechanism for the reaction of an acylium ion with benzene to yield a ketone.

11.9 Predict the major organic products of the reaction of benzene with each of the following compounds and $AlCl_3$ as the catalyst:

$$\overset{CH_3}{\underset{|}{\text{(a)}\ CH_3CH_2CHCH_2Cl}}$$

$$\overset{CH_3\ \ O}{\underset{|\ \ \ ||}{\text{(b)}\ CH_3CH_2CH-CCl,\ \text{followed by Zn/Hg, HCl}}}$$

F. Sulfonation

The sulfonation of benzene with fuming sulfuric acid ($H_2SO_4 + SO_3$) yields benzenesulfonic acid.

benzenesulfonic acid (50%)

Unlike the other electrophilic substitution reactions of benzene, sulfonation is a readily reversible reaction if water is present, and it shows a moderate kinetic isotope effect. Perdeuteriobenzene undergoes sulfonation at about half the rate of ordinary benzene. From these data, we conclude that the intermediate benzenonium ion in sulfonation can revert to benzene or go on to benzenesulfonic acid with almost equal ease (or the reaction would not be readily reversible). Also, the rates of reaction of Steps 1 and 2 must be more nearly equal than for other electrophilic aromatic substitution reactions (or the reaction would not show a kinetic isotope effect). Figure 11.7 shows an energy diagram for the sulfonation of benzene. Note that the energies of the transition states of Steps 1 and 2 are roughly the same.

The sulfonic acid group is easily displaced by a variety of other groups; therefore, arylsulfonic acids are useful synthetic intermediates. We will introduce a few of these reactions of arylsulfonic acids in Section 12.5.

Figure 11.8 summarizes the monosubstitution reactions of benzene.

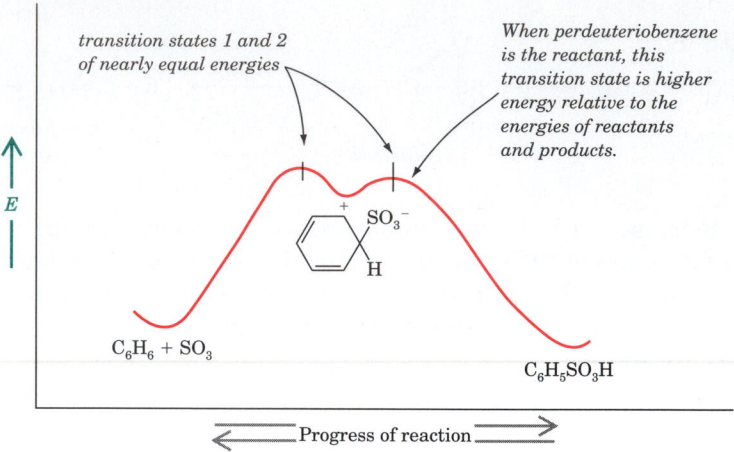

Figure 11.7 Energy diagram for the sulfonation of benzene.

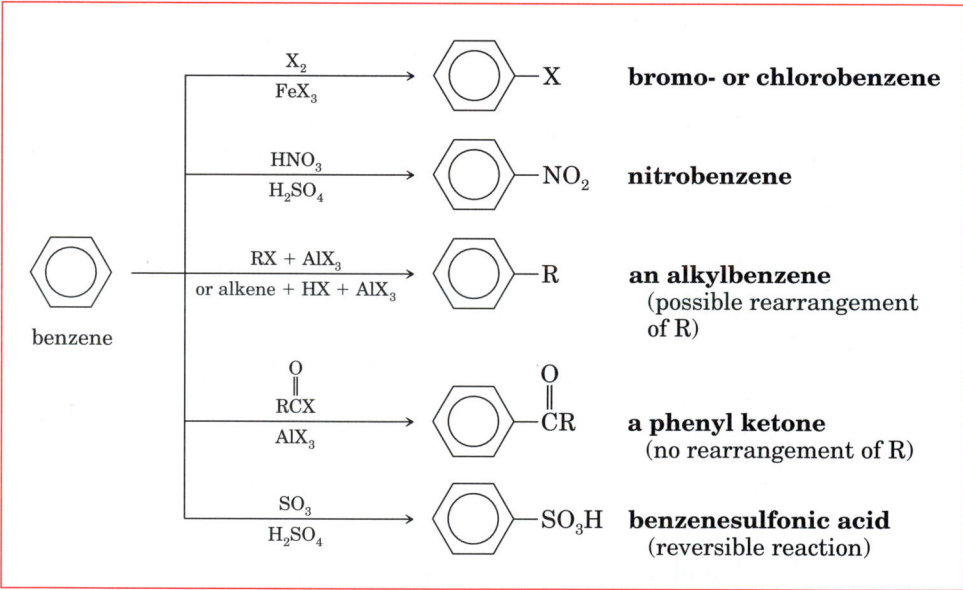

Figure 11.8 Summary of the monosubstitution reactions of benzene.

The Second Substitution

A substituted benzene may undergo substitution of a second group. Some substituted benzenes undergo reaction *more easily* than benzene itself, while other substituted benzenes undergo reaction *less easily*. For example, aniline undergoes electrophilic substitution a million times faster than benzene. Nitrobenzene, on the other hand, undergoes reaction at approximately one-millionth the rate of benzene! (Rather than allowing a reaction to proceed over a much longer period of time, a chemist commonly uses stronger reagents and higher temperatures for a less reactive compound.)

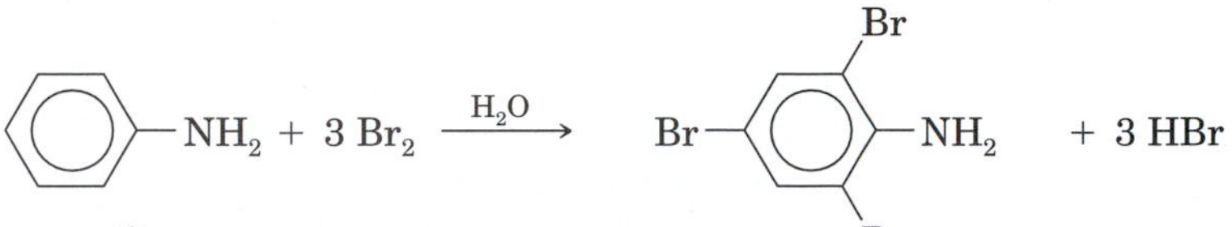

aniline — 2,4,6-tribromoaniline (100%)

no catalyst needed as it would be for benzene

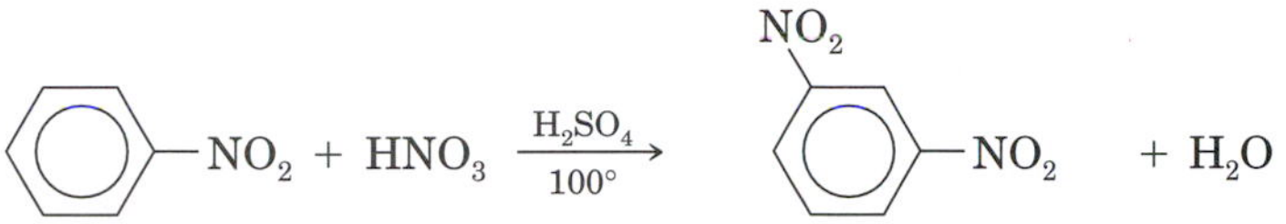

nitrobenzene — *m*-dinitrobenzene (93%)

much slower to nitrate than benzene, unless stronger reagents and higher temperatures are used

In these examples, we would say that NH_2 is an **activating group.** Its presence causes the ring to be *more* susceptible to further electrophilic substitution. On the other hand, the NO_2 group is a **deactivating group.** Its presence causes the ring to be *less* susceptible than benzene to substitution.

Besides the differences in reaction rates of substituted benzenes, the *position* of the second attack varies:

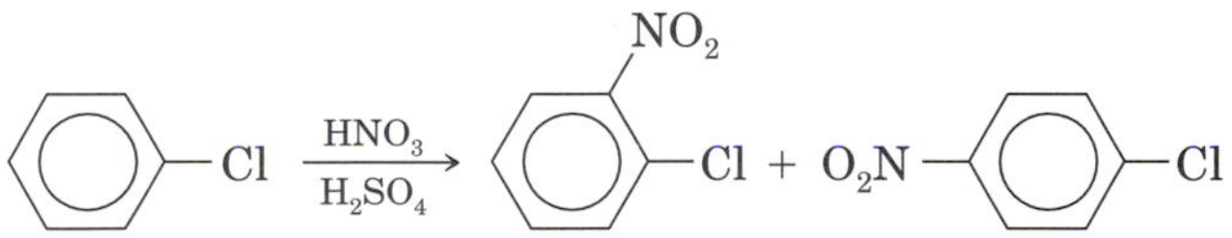

but *no meta*

chlorobenzene — *ortho* (30%) — *para* (70%)

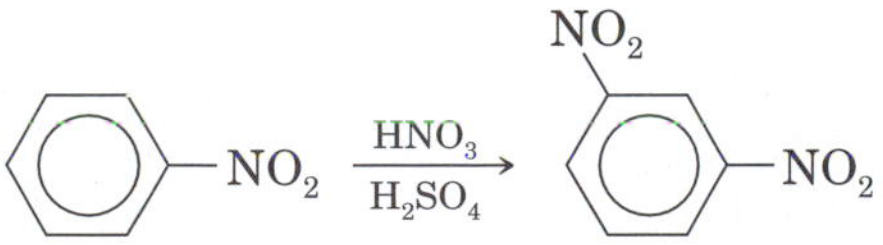

and very little *ortho* or *para*

nitrobenzene — *meta* (93%)

Study Problem

11.10 If chlorobenzene were nitrated at equal rates at each possible position of substitution, what would be the ratio of *o*-, *m*-, and *p*- products?

Chlorobenzene is nitrated in the *ortho* and *para* positions, but not in the *meta* position. However, nitrobenzene undergoes a second nitration in the *meta* position; very little substitution at the *ortho* or *para* positions occurs. These examples show that the nature of the incoming group has little effect on its own positioning on the ring. The position of second substitution is determined by the group that is already on the ring.

Table 11.4 Orientation of the nitro group in aromatic nitration

Reactant	Approximate Percents of Products		
	o	p	m
C_6H_5OH	50	50	—
$C_6H_5CH_3$	60	40	—
C_6H_5Cl	30	70	—
C_6H_5Br	40	60	—
$C_6H_5NO_2$	7	—	93
$C_6H_5CO_2H$	20	—	80

Table 11.5 Effect of the first substituent on the second substitution

	o,p-Directors	m-Directors (All Deactivating)	
increasing activation	$-\ddot{N}H_2, -\ddot{N}HR, -\ddot{N}R_2$	$\overset{O}{\overset{\|}{-CR}}$	**increasing deactivation**
	$-\ddot{O}H$	$-CO_2R$	
	$-\ddot{O}R$	$-SO_3H$	
	$\overset{O}{\overset{\|}{-\ddot{N}HCR}}$	$-CHO$	
	$-C_6H_5$ (aryl)	$-CO_2H$	
	$-R$ (alkyl)	$-CN$	
	$-\ddot{\underset{..}{X}}$: (deactivating)	$-NO_2$	
		$-NR_3{}^+$	

To differentiate between these two types of substituent, Cl is called an *or-tho-, para-director,* while NO_2 is called a **meta-director.** Any substituent on a benzene ring is either an o,p-director or an m-director, although to varying degrees, as is shown in Table 11.4.

Table 11.5 contains a summary of commonly encountered benzene substituents classified as *activating* or *deactivating* and as *o,p-directors* or *m-directors*. Note that all o,p-directors except the halogens are also activating groups. All m-directors are deactivating. Also note that all o,p-directors except aryl and alkyl groups have an unshared pair of electrons on the atom bonded to the ring. None of the m-directors has an unshared pair of electrons on the atom bonded to the ring.

unshared e^-

$\ddot{O}H$

o,p-director,
activating

no unshared e^-

m-director,
deactivating

STUDY PROBLEM

11.11   Predict the products:

(a) $C_6H_5Br + SO_3 \xrightarrow{H_2SO_4}$

(b) $C_6H_5CH_3 + CH_3\overset{\overset{\displaystyle O}{\|}}{C}Cl \xrightarrow{AlCl_3}$

(c) $C_6H_5NO_2 + Br_2 \xrightarrow{FeBr_3}$

(d) $C_6H_5\overset{\overset{\displaystyle O}{\|}}{C}CH_3 + CH_3\overset{\overset{\displaystyle O}{\|}}{C}Cl \xrightarrow{AlCl_3}$

(e) $C_6H_5CO_2CH_2CH_3 + Cl_2 \xrightarrow{FeCl_3}$

(f) $C_6H_5NHCH_3 + HCl \longrightarrow$

(g) $C_6H_5NH\overset{\overset{\displaystyle O}{\|}}{C}CH_3 + CH_2{=}CH_2 + HCl \xrightarrow{AlCl_3}$

A. Mechanism of the Second Substitution with an *o,p*-Director

Why are most *o,p*-directors activating groups? Why do they direct incoming groups to the *ortho* and *para* positions? To answer these questions, let us consider aniline, a compound with the *o,p*-directing NH_2 group on the ring.

The amino group in aniline activates the benzene ring toward substitution to such an extent that (1) no Lewis acid catalyst is needed, and (2) it is very difficult to obtain a monobromoaniline. Aniline quickly undergoes reaction to form the 2,4,6-tribromoaniline (both *ortho* positions and the *para* position brominated).

The mechanism of the bromination of aniline is similar to the mechanism of the bromination of benzene itself.

aniline   p-bromoaniline

The difference between the mechanism of the bromination of aniline and that of benzene lies in the stabilization of the intermediate carbocation. The carbocation from aniline is resonance-stabilized just as is the carbocation from benzene but, in the case of aniline, the amino group can increase the stabilization. Increased stabilization of the intermediate means a transition state of lower energy in Step 1 (see Figure 11.9) and therefore a faster reaction.

Resonance structures for the p-intermediate:

added stabilization

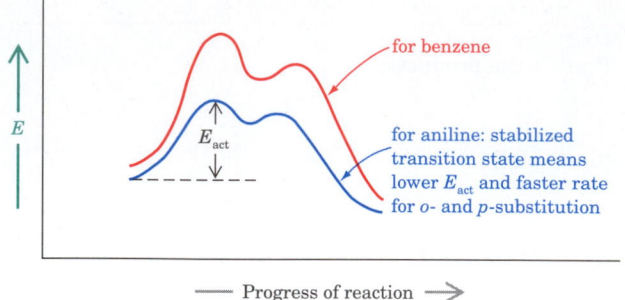

Figure 11.9   Energy diagrams for the bromination of aniline and benzene.

In the intermediate for *m*-substitution, the nitrogen of the amino group cannot help share the positive charge. (Draw the Kekulé structures for the *m*-substituted intermediate and verify this statement for yourself.) Therefore, the intermediate for *m*-substitution is of higher energy than are the intermediates leading to either *o*- or *p*-substituted products. Because this intermediate is of higher energy, its transition state also has a higher energy, and the rate of reaction at the *meta* position is lower.

The amino group, along with —OH, —OR, —NHCOR, and phenyl groups, activates the benzene ring toward electrophilic substitution by donating a pair of electrons to the ring through resonance. Substitution occurs at the *ortho* and *para* positions because the group helps share the positive charge in these intermediates.

Although the amino group is an *o,p*-director and a ring activator, its character changes in a reaction mixture containing a Lewis acid such as H_2SO_4, HNO_3, or $AlCl_3$. The reason for this is that the amino group reacts with a Lewis acid to yield a *m*-directing, deactivating ammonium ion group.

<table>
<tr><td>*o,p-director,*
activating</td><td></td><td>*m-director,*
deactivating</td></tr>
</table>

$$\text{C}_6\text{H}_5-\overset{\displaystyle ..}{\text{N}}\text{H}_2 + \text{H}_2\text{SO}_4 \longrightarrow \text{C}_6\text{H}_5-\overset{+}{\text{N}}\text{H}_3 + \text{HSO}_4^-$$

STUDY PROBLEMS

11.12   Draw the resonance structures for the intermediate in the nitration of phenol to yield *o*-nitrophenol.

11.13   The benzene ring in acetanilide (Table 11.1) is *less* reactive toward electrophilic substitution than the ring in aniline. Suggest a reason for this lesser activation by the —NHCCH$_3$ group.
$$\overset{\displaystyle \|}{\text{O}}$$

Halogens are different from the other *o,p*-directors. They direct an incoming group *ortho* or *para,* but they *deactivate* the ring to electrophilic substitution. A halogen substituent on the benzene ring directs an incoming group to the *ortho* or *para* position for the same reason the amino or hydroxyl group does. The halogen can donate electrons and help share the positive charge in the intermediate.

But why does a halogen deactivate the ring? A halogen, oxygen, or nitrogen withdraws electronic charge from the ring by the inductive effect. We would expect that any electronegative group would decrease the electron density of the ring and would make the ring less attractive to an incoming electrophile.

e^- withdrawal should deactivate
the ring (and does deactivate the ring
when X is Cl, Br, I)

In phenol or aniline, the effect of ring deactivation by electron withdrawal is counterbalanced by the release of electrons by resonance. Why is the same effect not observed in halobenzenes? In phenol or aniline, the resonance structures of the intermediate that confer added stability arise from overlap of $2p$ orbitals of carbon and $2p$ orbitals of N or O. These $2p$ orbitals are about the same size, and overlap is maximal.

In chlorobenzene, bromobenzene, or iodobenzene, the overlap in the intermediate is $2p$–$3p$, $2p$–$4p$, or $2p$–$5p$, respectively. The overlap is between orbitals of different size and is not so effective. The intermediate is less stabilized, the transition-state energy is higher, and the rate of reaction is lower. (Fluorobenzene, with $2p$–$2p$ overlap in the intermediate, is more reactive than the other halobenzenes toward electrophilic substitution in spite of its greater electronegativity.)

In summary, OH, NH_2, or a halogen determines the orientation of an incoming group by *donating electrons* by resonance and by adding resonance stabilization to *o*- and *p*-intermediates. For OH and NH_2, electron release by resonance activates the ring toward electrophilic substitution. Electron release by resonance is less effective for Cl, Br, or I than it is for OH or NH_2. Chlorobenzene, bromobenzene, and iodobenzene contain deactivated rings because the inductive electron-withdrawal by these substituents is relatively more effective.

An *alkyl group* does not have unshared electrons to donate for resonance stabilization. However, an alkyl group is electron-releasing by the inductive effect, a topic we discussed in Section 5.5E. Because an alkyl group releases electrons to the benzene ring, the ring gains electron density and becomes more attractive toward an incoming electrophile.

R groups activate
because they are
e^- releasing

To see why alkyl groups direct incoming electrophiles to *ortho* and *para* positions, we must again look at the resonance structures of the intermediates.

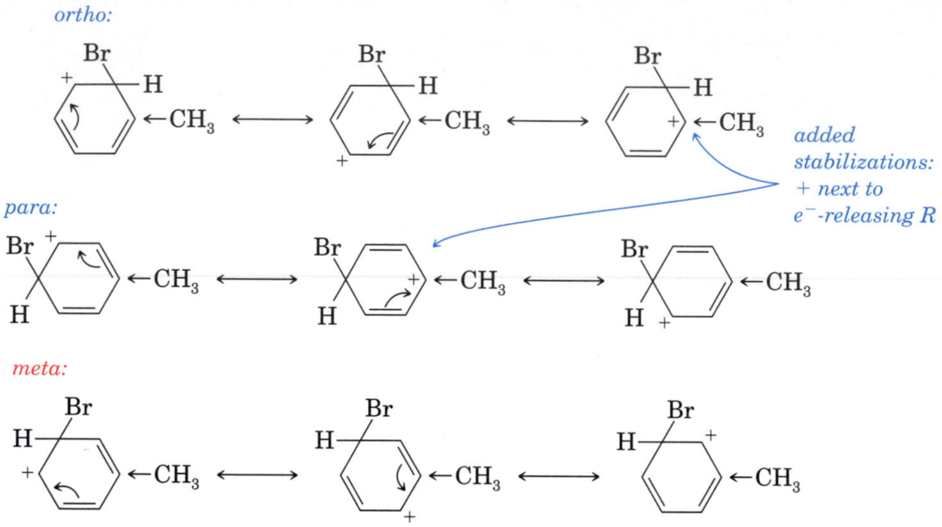

ortho:

*added
stabilizations:
+ next to
e⁻-releasing R*

para:

meta:

The intermediates for *o*- or *p*-substitution both have resonance structures in which the positive charge is adjacent to the R group. These structures are especially important contributors to resonance stabilization because the R group can help delocalize the positive charge by electron release and lower the energy of the transition state leading to these intermediates. The situation is directly analogous to that of an R group stabilizing a carbocation. The resonance structures for the intermediate in *m*-substitution have no such contributor. The *m*-intermediate is of higher energy. Attack on an alkylbenzene occurs *ortho* and *para* at a rate that is much faster than attack at a *meta* position.

SAMPLE PROBLEM

Which compound would you expect to undergo aromatic nitration more readily, $C_6H_5CH_3$ or $C_6H_5CCl_3$?

Solution
While the CH_3 group is electron-releasing and activates the ring, the CCl_3 group is strongly *electron-withdrawing* because of the influence of the electronegative chlorines. $C_6H_5CH_3$ has an activated ring; $C_6H_5CCl_3$ has a *deactivated* ring and undergoes substitution more slowly.

STUDY PROBLEM

11.14 Predict the major organic products of the reaction of each of the following compounds with 2-chloropropane and $AlCl_3$. Indicate the relative order of the reaction rates.

(a) bromobenzene   **(b)** phenol   **(c)** toluene   **(d)** benzene

B. Mechanism of the Second Substitution with a *m*-Director

In benzene substituted with a *m*-director (such as NO_2 or CO_2H), the atom bonded to the benzene ring has no unshared pair of electrons and carries a positive or a partial positive charge. It is easy to see why the *m*-directors are deactivating. Each one is *electron-withdrawing* and cannot donate electrons by resonance. Each one decreases the electron density of the ring and makes it less attractive to an incoming electrophile. The energy of the Step-1 transition state is higher than it would be for unsubstituted benzene.

A *m*-director does not activate the *meta* position toward electrophilic substitution. A *m*-director *deactivates all positions in the ring,* but it deactivates the *meta* position less than the other positions. The resonance structures of the intermediates resulting from attack at the various positions show that the *o*- and *p*-intermediates are destabilized by the nearness of two positive charges. The *m*-intermediate has no such destabilizing resonance structure. The following resonance structures are for the intermediate in the bromination of nitrobenzene:

Figure 11.10 summarizes how the three main types of substituents affect the course of a second substitution.

11.15 Identify each of the following compounds or ions as containing an *o,p*-directing or *m*-directing substituent:

(a) $C_6H_5CH_2CH_3$ (b) $C_6H_5\overset{\overset{\displaystyle O}{\|}}{C}CH_2CH_3$ (c) C_6H_5Cl (d) C_6H_5F

(e) $C_6H_5CCl_3$ (f) $C_6H_5\overset{+}{S}(CH_3)_2$ (g) $C_6H_5\overset{+}{N}(CH_3)_3$

Electron-releasing substituent: an *o,p*-director; activates the ring toward E$^+$.

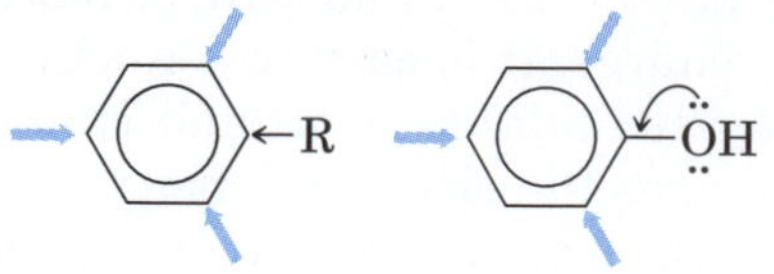

Halogen substituent: an *o,p*-director; deactivates the ring toward E$^+$.

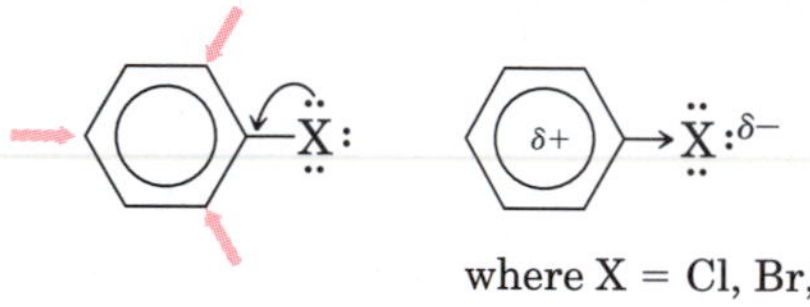

where X = Cl, Br, or I

Electron-withdrawing substituent: a *m*-director, deactivates the ring toward E$^+$.

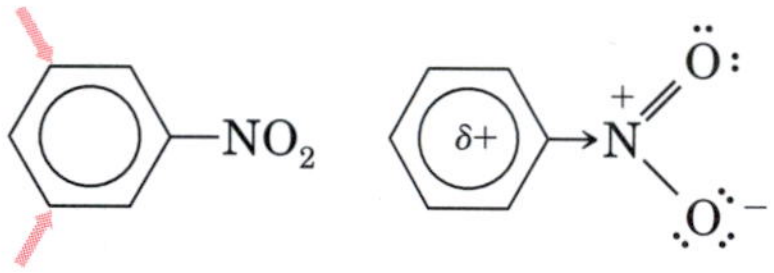

Figure 11.10 Summary of substituent effects in the second substitution. (Table 11.5 lists some common substituents.

Section 11.10
The Third Substitution

What if a benzene ring has two substituents already? Where does a third substituent go? A few general rules cover the majority of cases.

1. If two substituents direct an incoming group to the same position, that will be the principal position of the third substitution.

ortho to CH$_3$ and meta to NO$_2$

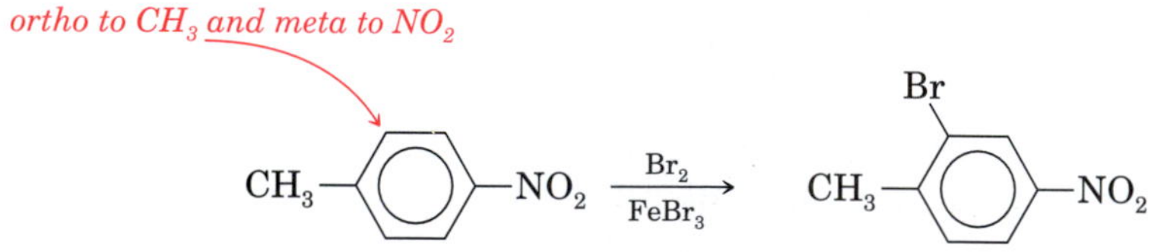

p-nitrotoluene   2-bromo-4-nitrotoluene (90%)

2. If two groups conflict in their directive effects, the more powerful activator (Table 11.5) will exert the predominant directive effect.

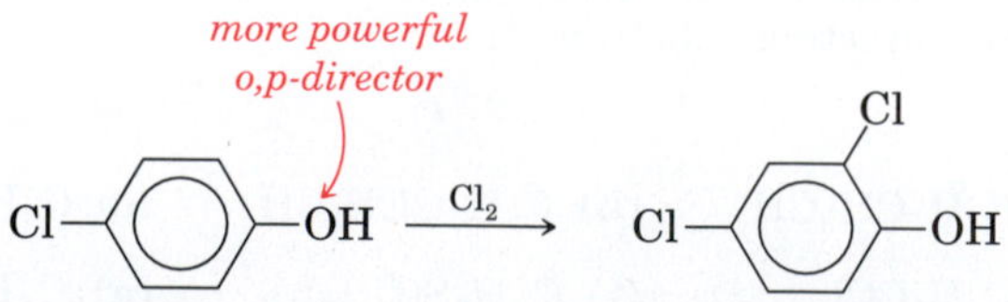

*more powerful
o,p-director*

p-chlorophenyl   2,4-dichlorophenol (94%)

3. If two deactivating groups are on the ring, regardless of their positions, it may be difficult to effect a third substitution.

4. If two groups on a ring are *meta* to each other, the ring does not usually undergo substitution at the position between them, even if the ring is activated. The lack of reactivity at this position is probably due to steric hindrance.

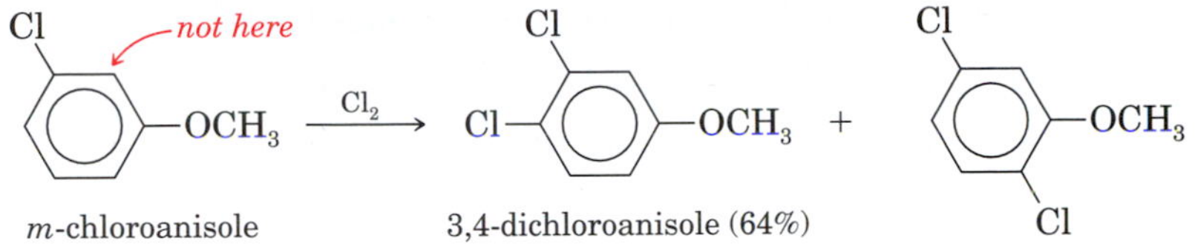
Cl
not here
OCH3
Cl2
Cl
Cl
OCH3
+
Cl
OCH3
Cl

m-chloroanisole — 3,4-dichloroanisole (64%) — 2,5-dichloroanisole (18%)

Study Problem

11.16 Predict the products of the next substitution:

(a)

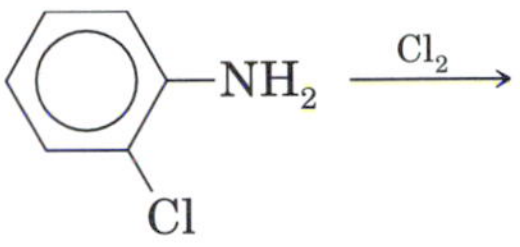
NH2
Cl2
Cl

(b)

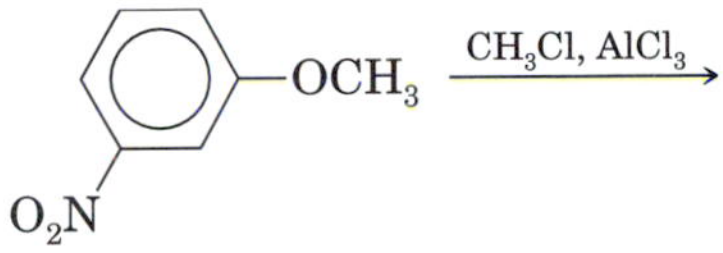
OCH3
CH3Cl, AlCl3
O2N

(c)

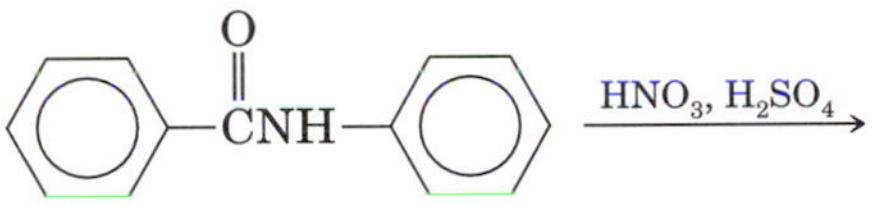
O
CNH
HNO3, H2SO4

Summary

An **aromatic compound** is a type of compound that gains substantial stabilization by π-electron delocalization. To be aromatic, a compound must be cyclic and planar. Each ring atom must have a *p* orbital perpendicular to the plane of the ring, and the *p* orbitals must contain $(4n + 2)$ π electrons **(Hückel rule).**

Benzene and other aromatics undergo **electrophilic aromatic substitution reactions.** These reactions are summarized in Table 11.6.

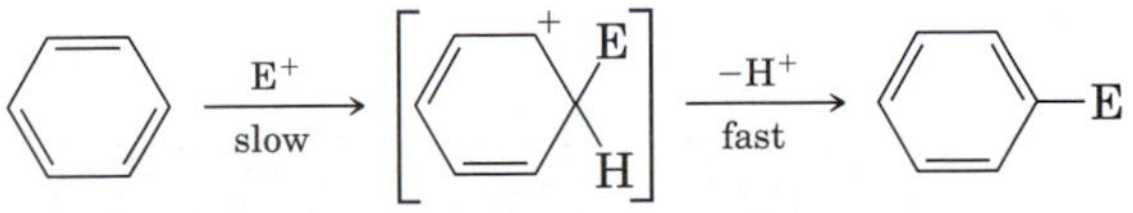
E+
slow
E
H
+
−H+
fast
E

A second substitution results in *o*- and *p*-isomers or else *m*-isomers, depending on the first substituent (see Table 11.5). The *o,p*-directors (except R) have electrons that can be donated to the ring by resonance.

Electron release by resonance:

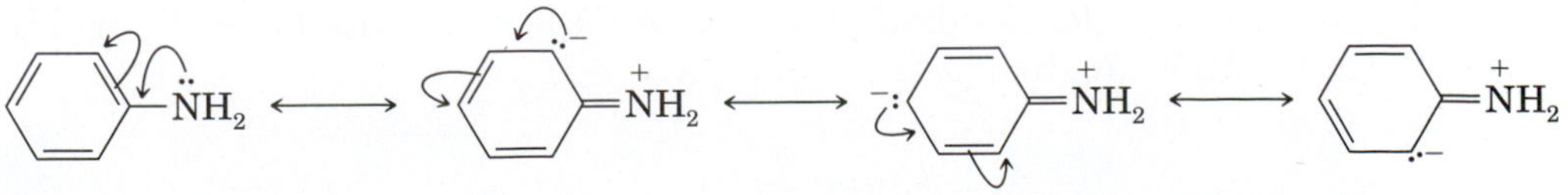
NH2
NH2
NH2
NH2

Table 11.6   Summary of electrophilic substitution reactions of benzene and its derivatives

Reaction	Product	Section Reference
$C_6H_6 + X_2 \xrightarrow{FeX_3} C_6H_5X$	**halobenzene**	11.8A
$+ HNO_3 \xrightarrow{H_2SO_4} C_6H_5NO_2$	**nitrobenzene**[a]	11.8C
$+ RX \xrightarrow{AlX_3} C_6H_5R$	**alkylbenzene**	11.8D
$+ R_2C{=}CHR \xrightarrow[HX]{AlX_3} C_6H_5CR_2CH_2R$	**alkylbenzene**	11.8D
$+ R\overset{\overset{O}{\|}}{C}X \xrightarrow{AlX_3} C_6H_5\overset{\overset{O}{\|}}{C}R$	**aryl ketone**[b]	11.8E
$+ SO_3 \xrightarrow{H_2SO_4} C_6H_5SO_3H$	**benzenesulfonic acid**[c]	11.8F
$C_6H_5{-}\ddot{Y} + E^+ \longrightarrow$ where Y is OH, NH$_2$, X, R, etc.	***o*- and *p*-products**	11.9
$C_6H_5{-}\overset{\delta+}{Z} + E^+ \longrightarrow$	***m*-products**	11.9

[a] Can be reduced to aniline (Section 12.3).
[b] Can be reduced to an alkylbenzene (Section 13.6C).
[c] Can be converted to C_6H_6 or C_6H_5OH (Section 12.5).

Electron release by inductive effect:

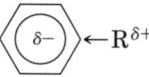

All *o,p*-directors except X activate the entire ring toward electrophilic substitution. The *ortho*- and *para*- positions are the preferred positions of substitution because of added resonance stabilization of their intermediates.

All *m*-directors and X *deactivate* the ring toward further electrophilic substitution by electron withdrawal.

deactivated toward E⁺

Essay Problem for Chapter 11

Metallocenes

Metallocenes are aromatic compounds that contain a transition metal bonded within the π system. Ferrocene (dicyclopentadienyliron), the first metallocene discovered, was isolated in 1951 from the reaction of cyclopentadienylmagnesium bromide and ferrous chloride. Ferrocene can be more easily obtained by the reaction of ferrous chloride, diethylamine, and 1,3-cyclopentadiene (G. Wilkinson, *Org. Syn.* **1956,** *36,* 34).

$$FeCl_2 + 2 \quad \text{(cyclopentadiene)} \quad + 2\,(C_2H_5)_2NH \longrightarrow \text{ferrocene} + 2\,(C_2H_5)_2NH \cdot HCl$$

ferrocene (73–84%)

Since the discovery of ferrocene, other transition metals have been found to form aromatic metallocenes. Typical metallocenes are listed below.

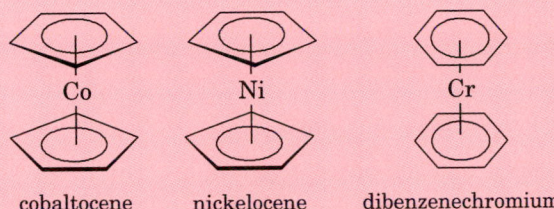

cobaltocene   nickelocene   dibenzenechromium

The bonding in these aromatic compounds is typified by ferrocene. The iron atom is sandwiched between two cyclopentadienyl rings. To form an aromatic bond, the two rings and the iron atom each contribute six electrons for a total of 18 electrons. This number of electrons agrees with Hückel's $(4n + 2)$ rule, where n is 4. The rings are not rigidly joined by carbon–iron bonds but are rotating freely on the iron atom.

The metallocenes undergo typical aromatic substitution reactions. For example, ferrocene when treated with acetic anhydride and phosphoric acid gives acetylferrocene in a Friedel–Crafts acylation.

$$\text{ferrocene} + CH_3\overset{O}{\underset{\|}{C}}O\overset{O}{\underset{\|}{C}}CH_3 \xrightarrow{H_3PO_4} \text{acetylferrocene}$$

acetylferrocene

Questions

1. The method shown for the preparation of ferrocene can also be used for the synthesis of nickelocene in an 80% yield. Write the equation for this reaction using the bromide salt of the metal rather than the chloride.
2. Where would you expect to observe **(a)** the IR C—H absorption, and **(b)** the approximate position of ^{1}H NMR absorption in the spectra of ferrocene?
3. If the rings of ferrocene were *not* freely rotating, how many isomers of diacetylferrocene would be possible if each ring had one acetyl group bonded to it? How many isomers of diacetylferrocene actually exist?

4. Ferrocene can be oxidized to a ferricinium ion. Write an equation for this reaction using nitric acid as the oxidant. Would you expect the series of ferrocenes substituted with $RC\overset{\overset{\displaystyle O}{\|}}{}$—, H—, R—, —OH to be easier or more difficult to oxidize as we proceed from $RC\overset{\overset{\displaystyle O}{\|}}{}$— to —OH? Explain your choice.

5. Cobaltocene and nickelocene are unstable and readily form cations. Predict the charge of these cations. Explain your answer.

Study Problems*

11.17 Draw formulas for the following compounds:

    (a) *o*-phenylphenol, an agricultural fungicide

    (b) estragole (1-allyl-4-methoxybenzene), main constituent of tarragon oil used in food flavorings

    (c) *p*-phenetidine (4-ethoxyaniline), used in the synthesis of an analgesic and an artificial sweetener

    (d) DDT [1,1,1-trichloro-2,2-bis(*p*-chlorophenyl)ethane], an insecticide

    (e) Chloramben (2,5-dichloro-3-aminobenzoic acid), a selective pre-emergence herbicide

    (f) *m*-cresol (*m*-methylphenol), used in photographic developers

11.18 Name the following structures:

    (a) 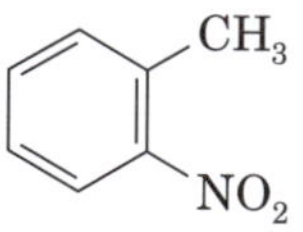 [structure: benzene ring with CH₃ and NO₂ ortho substituents]

    (b) 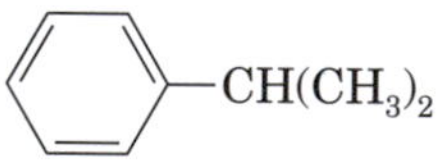 [structure: benzene ring with —CH(CH₃)₂]

    (c) 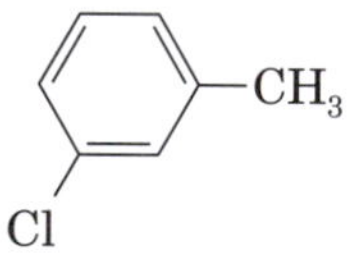 [structure: benzene ring with —CH₃ and Cl]

    (d) 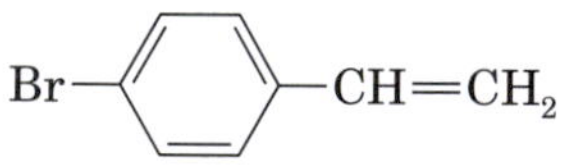 [structure: Br— benzene ring —CH=CH₂]

    (e) 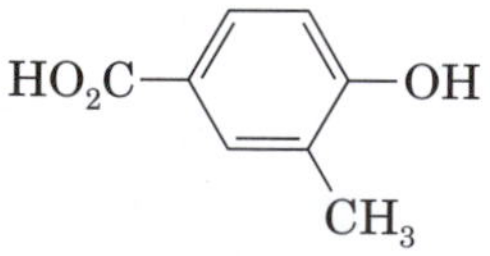 [structure: HO₂C— benzene ring —OH with CH₃]

    (f) 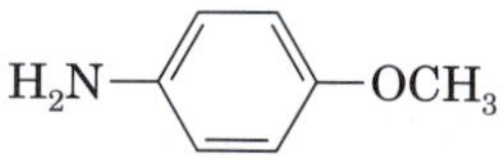 [structure: H₂N— benzene ring —OCH₃]

11.19 Give the structures and the names of all the isomeric: **(a)** mononitroanilines; **(b)** dibromotoluenes.

11.20 Which of the following compounds could be aromatic? (*Hint:* In many cases, the Hückel rule may be extrapolated to structures with more than one ring.)

---

* For information concerning the organization of the *Study Problems* and *Additional Problems*, see the *Note to student* on page 41.

(a) 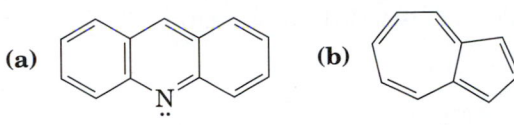   **(b)**

azulene

(c)   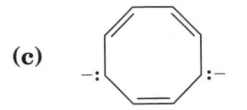   **(d)**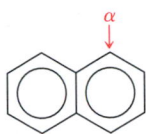

cyclooctatetraene
dianion

11.21 Naphthalene undergoes electrophilic aromatic substitution at the α-position:

α

(1) Predict the organic products of the monosubstitution reactions of naphthalene with each of the following sets of reagents. (2) Write equations to show how the electrophile is generated in each case.

(a) Br_2, $FeBr_3$ **(b)** HNO_3, H_2SO_4

(c) $ClCH_2CH_3$, $AlCl_3$ **(d)** $CH_3CH{=}CH_2$, $AlCl_3$, HCl

(e) $CH_3\overset{O}{\overset{\|}{C}}Cl$, $AlCl_3$ **(f)** $(CH_3)_2CHOH$, H_2SO_4

11.22 Write equations showing the preparation of the following compounds from benzene:

(a) nitrobenzene   **(b)** chlorobenzene

(c) 1-phenyl-1-pentanone   **(d)** benzenesulfonic acid

(e) butylbenzene   **(f)** bromobenzene

(g) isopropylbenzene (two routes)

11.23 Write a flow equation for the mechanism of the nitration of benzene.

11.24 Predict the product (including stereochemistry) of the following reaction:

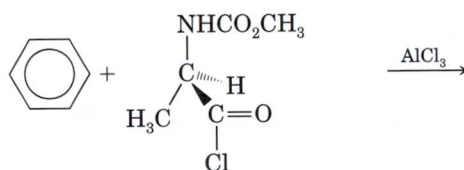

11.25 Which one of each of the following pairs is more reactive toward aromatic bromination?

(a) acetanilide ($C_6H_5NH\overset{O}{\overset{\|}{C}}CH_3$) or benzene

(b) acetanilide or aniline

(c) *p*-methoxytoluene or *p*-toluic acid (*p*-methylbenzoic acid)

(d) *m*-dinitrobenzene or *m*-nitrotoluene

(e) chlorobenzene or *m*-dichlorobenzene

11.26 Label each of the following groups as an *o,p*-director or as an *m*-director:

(a) $-N(CH_3)_2$ (b) $-\overset{+}{H}N(CH_3)_2$ (c) $-OC_6H_5$

(d) $-CF_3$ (e) $-\overset{\overset{\displaystyle O}{\|}}{C}C_6H_5$ (f) [naphthalene with methyl substituent structure]

(g) $-O\overset{\overset{\displaystyle O}{\|}}{C}CH_3$ (h) $-\overset{\overset{\displaystyle O}{\|}}{C}OCH_3$ (i) [cyclohexyl structure]

11.27 Draw resonance structures for the following carbocation intermediates:

(a) [structure with H, Br, and CCH₃ (O) group]   (b) HO–[structure with NO₂, H]   (c) [structure with Br, H, CH₃]   (d) [structure with NO₂, H, F]

11.28 Ring monobromination of a diethylbenzene yields three isomeric bromodiethylbenzenes (two formed in minor amounts). Give the structures of the diethylbenzene and the brominated products.

11.29 Predict the major products of aromatic monochlorination of: (a) chlorobenzene; (b) *o*-dichlorobenzene; (c) *m*-bromochlorobenzene; (d) *m*-xylene; (e) acetophenone (see Table 11.1); (f) (trichloromethyl)benzene ($C_6H_5CCl_3$).

11.30 Write equations for the mononitration of the following compounds:

(a) *p*-bromophenol   (b) *p*-nitrotoluene

(c) *m*-methoxybenzoic acid   (d) 1-bromo-3-methoxybenzene

11.31 Complete the following equations:

(a) [benzene]$-CH_3 + SO_3 \xrightarrow{H_2SO_4}$

(b) Cl–[benzene]$-OCH_3 + (CH_3)_2CHCH_2Cl \xrightarrow{AlCl_3}$

(c) [benzene with CH₃]$-CH_3 + CH_3CH_2\overset{\overset{\displaystyle}{}}{\underset{\underset{\displaystyle CH_3}{|}}{C}}H\overset{\overset{\displaystyle O}{\|}}{C}Cl \xrightarrow{AlCl_3}$

(d) CH_3O–[benzene]$-C(CH_3)_3 + HNO_3 \longrightarrow$

(e)

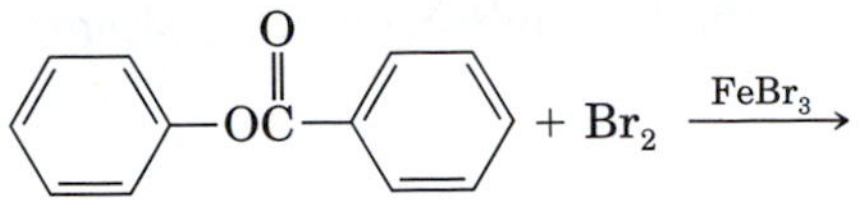

(f)

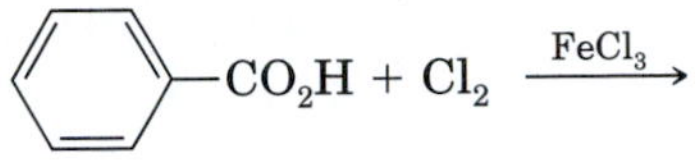

(g)

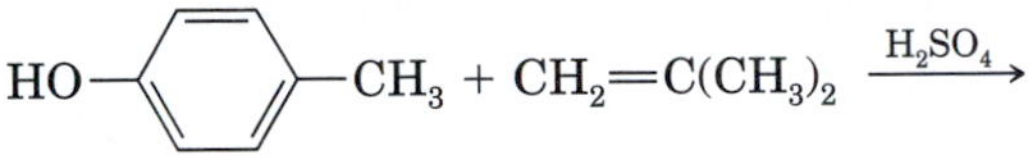

(h)

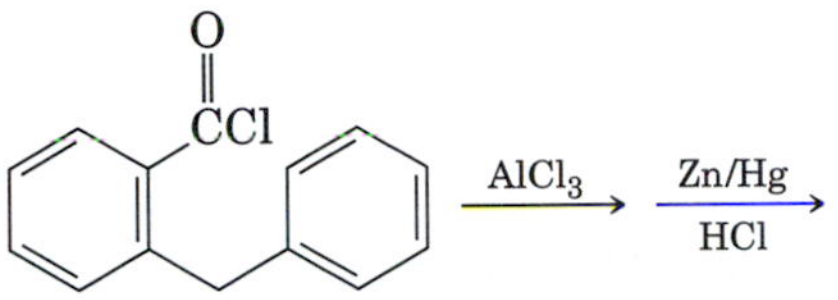

(i)

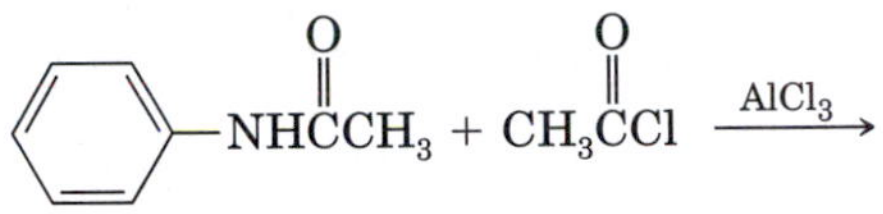

Additional Problems

11.32 A fluorine atom is more electronegative than a chlorine atom, yet fluorobenzene is more reactive toward electrophilic substitutions than is chlorobenzene. Explain. (*Hint:* recall the reason for halogens being *o,p*-directors.)

11.33 The basicity of 2,6-dimethylpyridine is much greater than the basicity of 2,5-dimethylpyrrole; the conjugate acid of 2,6-dimethylpyridine has a $pK_a = 6.43$, while the conjugate acid of 2,5-dimethylpyrrole has a $pK_a = -0.58$. Explain.

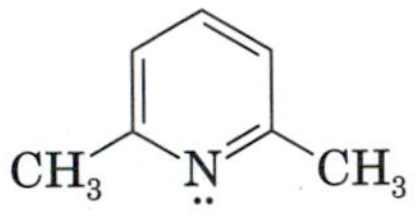

2,6-dimethylpyridine

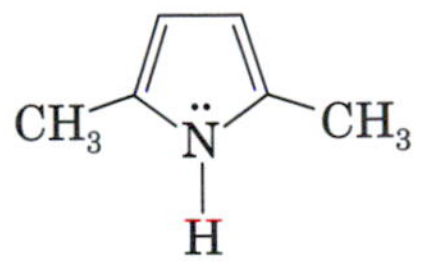

2,5-dimethylpyrrole

11.34 Would you expect the following structures to be aromatic? Explain.

(a)

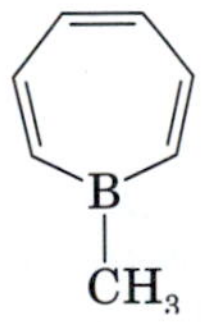

(b)

11.35 When 2,3-di-*tert*-butylcyclopropenone is treated with *tert*-butyllithium and the resulting intermediate subjected to dehydrating conditions, a stable cation is formed. Why is this cation stable?

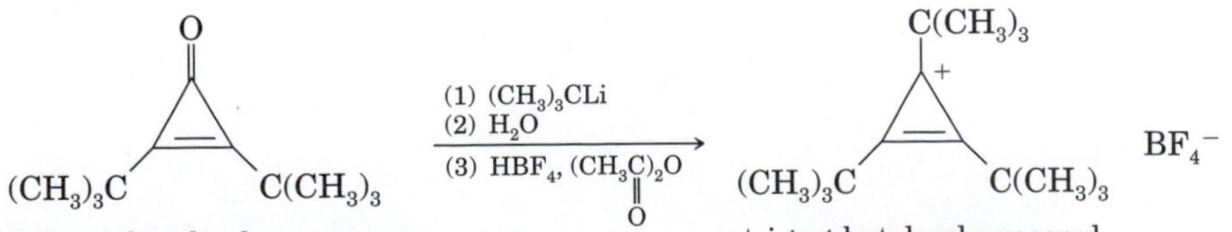

2,3-di-*tert*-butylcyclopropenone

tri-*tert*-butylcyclopropenyl

tetrafluoroborate

11.36 Friedel–Crafts acylation reactions can be accomplished with acid anhydrides as well as with acid halides:

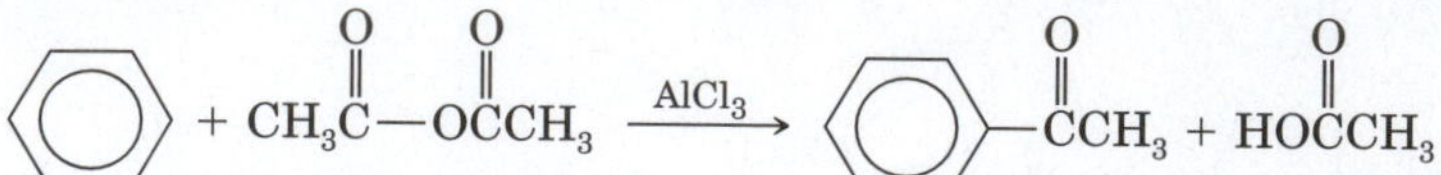

an acid anhydride

Predict the product of the AlCl$_3$-catalyzed reaction of benzene with:

(a) 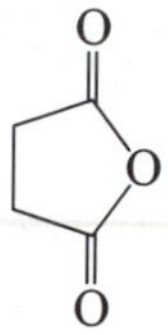   **(b)**

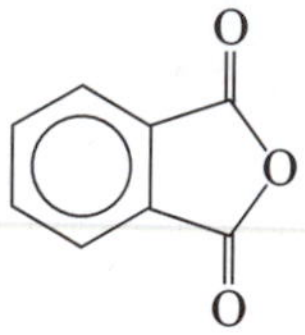

11.37 Explain the following observations: **(a)** Treatment of benzene with an excess of fuming nitric acid and sulfuric acid yields *m*-dinitrobenzene, but not the trinitrated product. **(b)** Under the same conditions, toluene yields 2,4,6-trinitrotoluene (TNT).

11.38 In acidic solution, nitrous acid (HONO) reacts with some aromatic compounds in a manner similar to that of nitric acid.

(a)   Write an equation showing the formation of the electrophile.

(b)   Write the equation showing the electrophilic substitution of phenol by this species. (Include the structure of the intermediate.)

11.39 Show how the following conversion could be carried out:

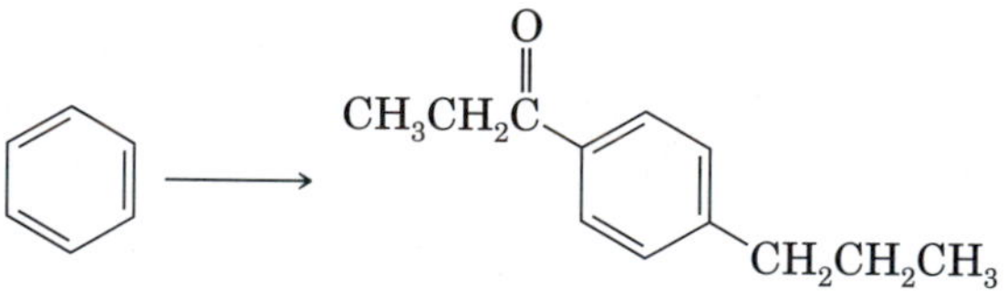

11.40 In the ^{1}H NMR spectrum of 1,6-methano[10]annulene, the ring protons absorb at low field ($\delta 6.8 - 7.5$), while the —CH$_2$— protons absorb far upfield ($\delta - 0.5$, to the *right* of TMS). Explain.

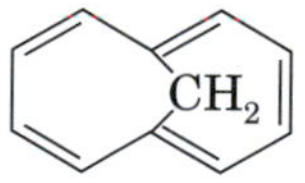

1,6-methano[10]annulene

11.41 The chemical shift of the proton bonded to carbon 5 of 4-ethynylphenanthrene is 1.71 ppm downfield from the chemical shift of the absorption of the same proton on phenanthrene itself. Explain why this is true.

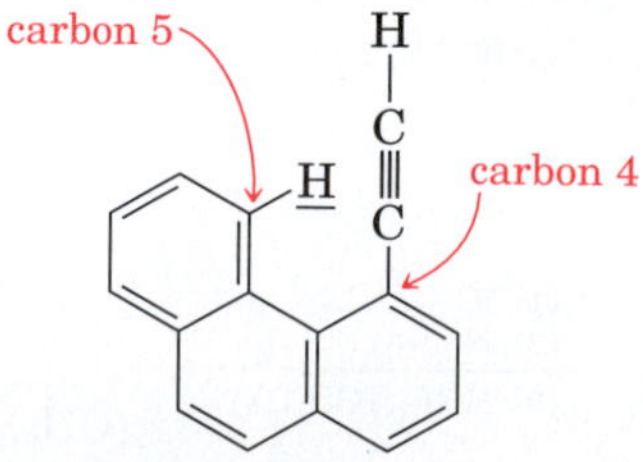

4-ethynylphenanthrene 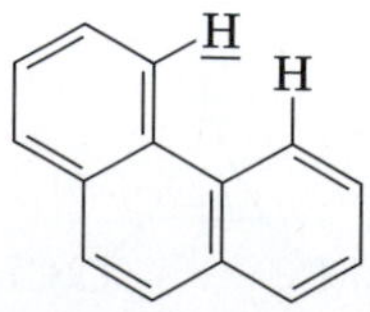   phenanthrene

11.42 A compound with the formula $C_{15}H_{14}O$ gives the ^1H NMR spectrum shown in Figure 11.11. Its infrared spectrum shows strong absorption about 1750 cm^{-1}(5.75 μm). What is the structure of the compound?

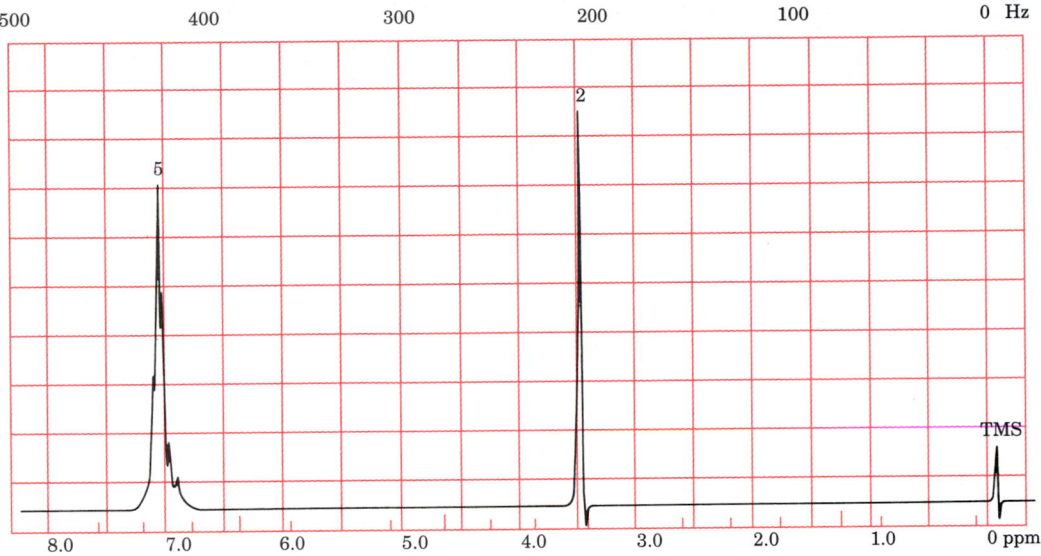

Figure 11.11 ^1H NMR spectrum for Problem 11.42.

SUBSTITUTED BENZENES

In Chapter 11, we described electrophilic aromatic substitution reactions of substituted and unsubstituted benzene rings. We discussed how a substituent on the ring can change the reactivity of the ring. However, the reverse is also true. A substituent on an aromatic ring often exhibits different types of reactivity than the same substituent in other compounds. For example, the methyl group in toluene reacts differently from the methyl group in methylpropane. A phenol hydroxyl group reacts differently from an alcohol hydroxyl group.

In this chapter we will discuss the reactions of the substituents on aromatic rings. Then, we will summarize the different types of syntheses that can be carried out both with electrophilic aromatic substitution reactions and with substituent reactions.

Section 12.1

Alkylbenzenes

Although alkyl groups in general are relatively nonreactive, alkyl groups bonded to an aromatic ring show a unique reactivity. The carbon adjacent to an aromatic ring is a *benzylic carbon*.

A benzyl cation, a benzyl radical, and a benzyl carbanion are all resonance-stabilized by the benzene ring. Consequently, the benzylic position is a site of attack in many reactions.

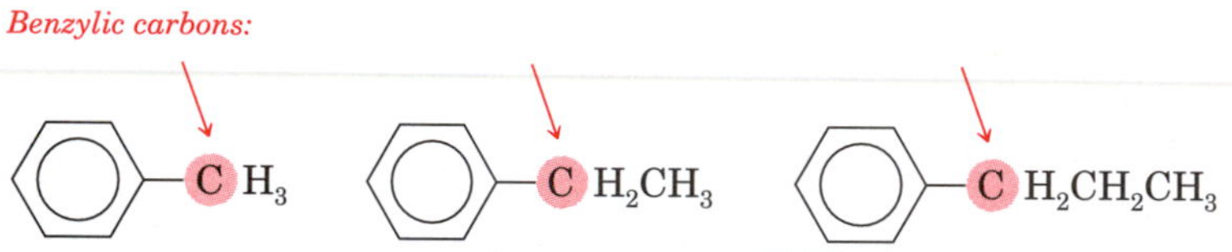

Benzylic carbons:

Although an alkene is readily oxidized by such reagents as hot $KMnO_4$ solution, a benzene ring is not oxidized under these laboratory conditions. However, the *alkyl group* of an alkylbenzene can be oxidized. Because of the reactivity of the benzylic position, alkylbenzenes with a benzylic H atom all yield the same product, benzoic acid, upon vigorous oxidation.

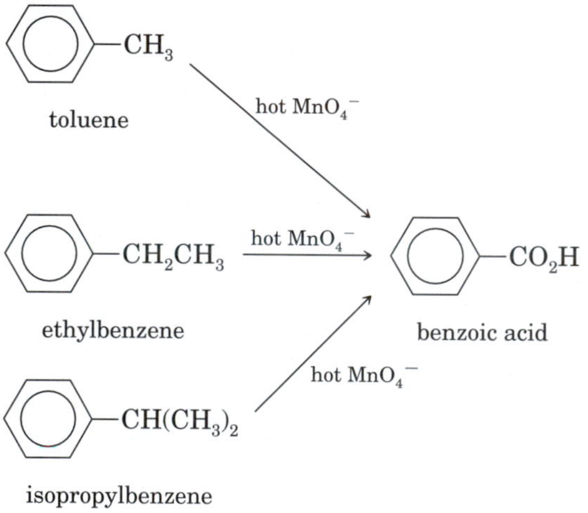

Free-radical halogenation is another reaction that takes place preferentially at the benzylic position (Section 6.5B). (Note that the conditions of radical halogenation are different from those for electrophilic aromatic halogenation, which is an ionic reaction, not a radical reaction.)

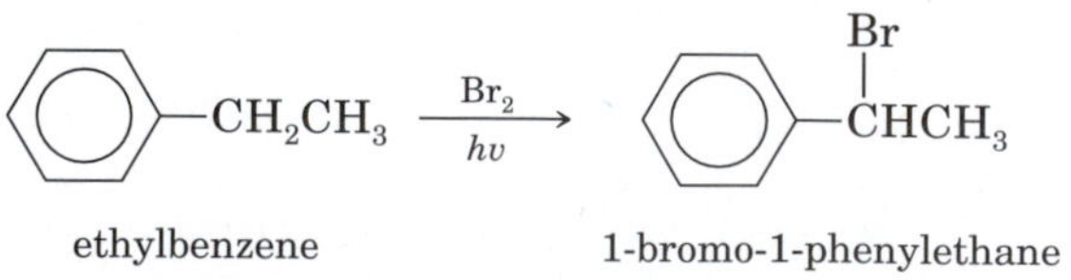

STUDY PROBLEMS

12.1   Draw resonance structures of the intermediates in **(a)** ring bromination and **(b)** side-chain bromination of *p*-xylene (1,4-dimethylbenzene).

12.2 Write equations showing **(a)** the oxidation, **(b)** the free-radical bromination, and **(c)** the ionic bromination of tetralin.

tetralin

Section 12.2
Phenols

A phenol (ArOH) is a compound with a hydroxyl group bonded to an aromatic ring. As we mentioned in Section 11.9, the OH group is a powerful activator in electrophilic aromatic substitution reactions.

Because a bond from an sp^2-hybridized carbon is stronger than a bond from an sp^3-hybridized carbon (Section 2.4F) and because of geometric restraints, the C—O bond of a phenol is not easily broken. Phenols do not undergo S_N1 or S_N2 reactions or elimination reactions as alcohols do.

$$R—OH + HBr \xrightarrow{S_N1 \text{ or } S_N2} RBr + H_2O$$

an alcohol

$$Ar—OH + HBr \longrightarrow \text{no reaction}$$

a phenol

A. Acidity of Phenols

Although the C—O bond of a phenol is not easily broken, the OH bond is readily broken. Phenol, with a pK_a of 10.00, is a stronger acid than an alcohol or water. Because a phenoxide ion (ArO$^-$) is a weaker base than OH$^-$, a phenoxide can be prepared by treatment of a phenol with aqueous NaOH. This reactivity is in direct contrast with alcohols. The reason for the difference in acidity between alcohols and phenols was discussed in Section 2.9E.

— not favored

$$CH_3CH_2OH + NaOH \rightleftharpoons CH_3CH_2O^-\ Na^+ + H_2O$$

ethanol, $pK_a = 15.9$ sodium ethoxide $pK_a = 15.74$

— favored

$$\text{C}_6\text{H}_5—OH + NaOH \rightleftharpoons \text{C}_6\text{H}_5—O^-\ Na^+ + H_2O$$

phenol, $pK_a = 10.00$ sodium phenoxide $pK_a = 15.74$

Phenoxides are useful in the preparation of aryl alkyl ethers (see Section 8.2B).

$$ArO^- + RX \xrightarrow{S_N2} ArOR + X^-$$

a phenoxide   a 1° alkyl   an ether
ion   halide

Because of its acidity, phenol was originally called *carbolic acid.* In the 1800s, the British surgeon Joseph Lister urged that phenol be used as a hospital antiseptic. Prior to that time, no antiseptics were used because it was thought that odors, not microorganisms, were the cause of infection. As an antiseptic, phenol itself has been replaced by less irritating compounds. Interestingly, many modern antiseptics still contain phenolic groups.

hexachlorophene

banned for most uses
because it can be
absorbed through the skin

hexylresorcinol

STUDY PROBLEM

12.3   What are the principal ions in solution when the following reagents are mixed?

(a) sodium ethoxide and phenol **(b)** sodium phenoxide and ethanol

B. Esterification of Phenols

Esterification of phenol does not involve cleavage of the strong C—O bond of the phenol, but depends on the cleavage of the OH bond. (We will discuss the mechanism of esterification reactions in Chapters 14 and 15.) Therefore, esters of phenols can be synthesized by the same reactions that lead to alkyl esters.

Although a carboxylic acid could be used in the esterification of a phenol, yields are usually low. Better results are obtained if a more reactive derivative of a carboxylic acid is used instead of the acid itself. Acetic anhydride is a reactive derivative that leads to acetate esters (Section 15.4C).

acetic anhydride   phenol   phenyl acetate

an ester

C. The Kolbe Reaction

Because the OH group is such a powerful ring activator in electrophilic substitution reactions, phenols and phenoxides react with weak electrophiles. The **Kolbe reaction** is the reaction of sodium phenoxide with CO_2 to yield sodium salicylate. This salt yields salicylic acid when acidified.

salicylate ion   salicylic acid

Salicylic acid is used to synthesize acetylsalicylic acid, commonly called *aspirin.*

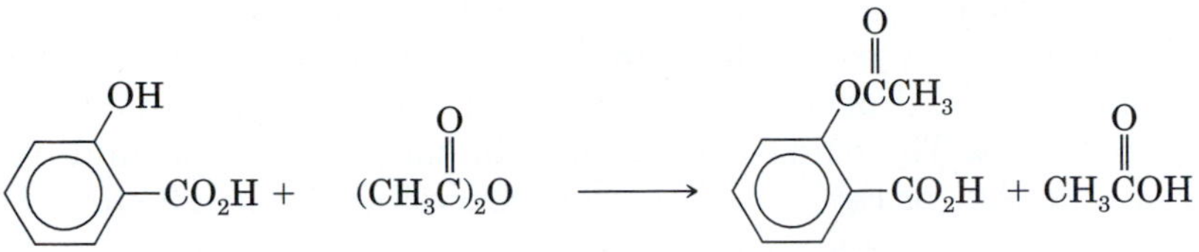

acetic anhydride   acetylsalicylic acid
                                          (aspirin)

D. The Reimer–Tiemann Reaction

The reaction of phenol with chloroform in an aqueous base, followed by acidification of the mixture, yields an aldehyde, salicyladehyde. This reaction is called the **Reimer–Tiemann reaction.**

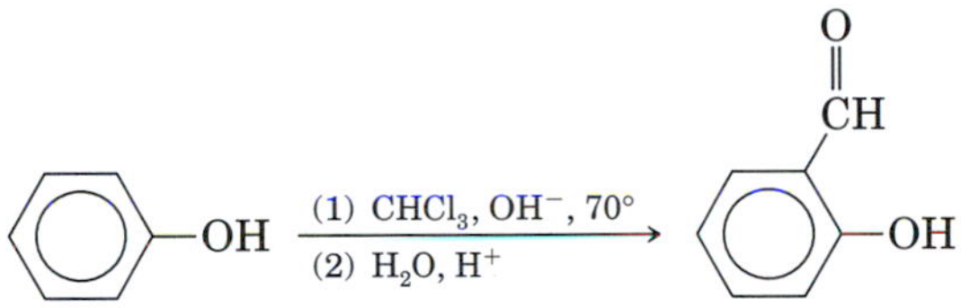

salicylaldehyde

The first step in the Reimer–Tiemann reaction is the formation of the reactive intermediate dichlorocarbene.

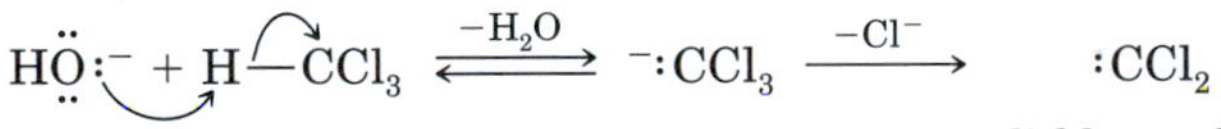

dichlorocarbene

The carbon in $:CCl_2$ contains only six valence electrons; thus, $:CCl_2$ can act as an electrophile in aromatic substitution.

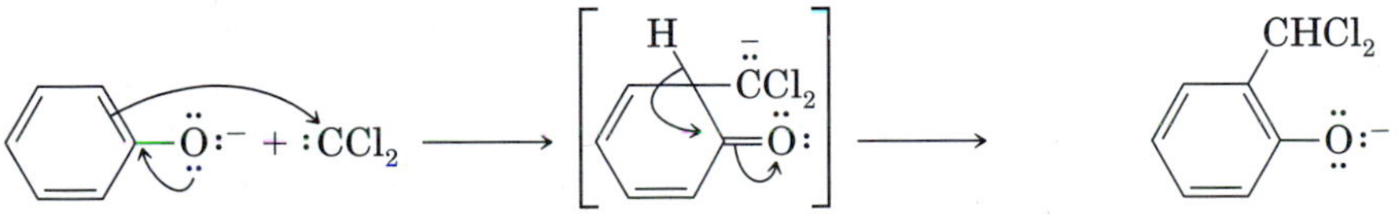

o-(dichloromethyl)phenoxide ion

The (dichloromethyl)phenoxide ion contains two chlorides in a benzylic position. Benzylic halides undergo rapid S_N1 and S_N2 reactions with OH^-. The dichloro compound undergoes just such a substitution to yield an unstable 1,1-chlorohydrin that subsequently loses H^+ (to the base) and Cl^-. Acidification of the alkaline solution yields salicylaldehyde.

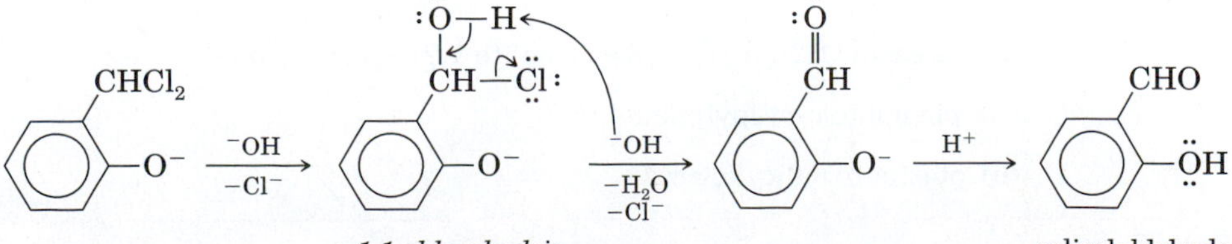

a 1,1-chlorohydrin   salicylaldehyde

E. Oxidation of Phenols

Phenol itself resists oxidation because formation of a carbonyl group at a ring carbon would lead to loss of aromatic stabilization. However, 1,2- and 1,4-dihydroxybenzenes, called **hydroquinones,** can be oxidized to **quinones.** The oxidation proceeds in the presence of mild oxidizing agents, such as Ag^+ or Fe^{3+}, and is readily reversible. Although simple hydroquinones are colorless, quinones are colored.

hydroquinone   1,4-benzoquinone
                          (quinone)

catechol   1,2-benzoquinone

The ability of hydroquinone to reduce silver ions to silver metal is the chemical basis of photography. Silver ions in a silver halide crystal that has been exposed to light are more easily reduced than the silver ions of an unexposed crystal. Hydroquinone in the developer fluid reduces these light-activated silver ions at a faster rate than the nonexposed silver ions. In the fixing process, unreacted silver halide is converted to a water-soluble silver complex by sodium thiosulfate, $Na_2S_2O_3$ (called *hypo*), and washed from the film. The result is the familiar photographic negative.

Bombardier beetles rely on the easy oxidation of hydroquinone as a defensive mechanism against spiders, mice and frogs. These beetles have glands that store hydrogen peroxide as well as glands that store hydroquinones. When the beetle is threatened, secretions from the two types of glands are mixed, along with enzymes that catalyze the oxidation of the hydroquinones. The end products are irritating quinones and O_2 (a propellant), which are squirted at the predator.

STUDY PROBLEM

12.4   Suggest a procedure by which each of the following conversions could be carried out:

(a) catechol (1,2-dihydroxybenzene) to 1,2-benzoquinone

(b) phenol to *o*-propylphenol

(c) phenol to salicylic acid

Benzenediazonium Salts

A. Preparation and Reactivity of Benzenediazonium Chloride

Arylamines such as aniline can be prepared in the laboratory by the nitration of aromatic compounds followed by reduction of the nitro group. A mixture of iron filings and concentrated HCl is a common reducing agent for this reaction. Because the reaction is run in acid, a *protonated* amine is the product. Subsequent treatment with base generates the amine itself. Alternatively, the nitro group can be readily reduced by catalytic hydrogenation.

Aniline and other arylamines undergo reaction with cold nitrous acid (HONO) in HCl solution to yield **aryldiazonium chlorides** ($ArN_2^+ Cl^-$). The nitrous acid is usually generated *in situ* by the reaction of sodium nitrite ($Na^+ \, {}^-ONO$) with HCl. Diazonium salts are very reactive. Therefore, the reaction mixture must be chilled. (Alkyldiazonium salts, $RN_2^+ X^-$, are not stable even in the cold; see Section 18.8.)

aniline

benzenediazonium
chloride

The name *diazonium* is based upon the French word for nitrogen *(azot)*. Diazonium salts were discovered in 1858 by Johan Peter Griess (1829–1888), investigated extensively, and within five years of their discovery used to commercially produce azo dyes for cloth (Section 12.3B). Like Grignard reagents, diazonium salts are versatile intermediates and can be used to synthesize a variety of aromatic compounds.

The high reactivity of diazonium salts arises from the excellent leaving ability of nitrogen gas, N_2. Because of this leaving ability, the diazonium group may be displaced by a variety of nucleophiles, such as I^-. Some of the substitution reactions that we will present are thought to proceed by a radical mechanism. In other cases, the substitution reaction proceeds through an aryl cation, by a mechanism similar to that of an S_N1 reaction.

In these displacement reactions, the diazonium salt is generally prepared (but not isolated), the nucleophilic reagent is added, and the mixture is allowed to warm or is heated. Yields of substitution products are generally good to excellent: 70–95% from the starting arylamine.

B. Reactions of Benzenediazonium Salts

Displacement by halide ions and cyanide ions The diazonium group is readily displaced as N_2 by halide ions (F^-, Cl^-, Br^-, or I^-) or the cyanide ion (^-CN). These reactions provide synthetic routes to aryl fluorides, iodides, and nitriles (ArCN), none of which can be obtained by direct electrophilic substitution. Although aryl bromides and chlorides can be synthesized by electrophilic substitution reactions, the products are often contaminated with disubstituted by-products. Diazonium displacement yields pure monochloro or monobromo compounds, uncontaminated by disubstituted products.

For the substitution of —Cl, —Br, or —CN, a copper(I) salt is used as the source of the nucleophile, and the reaction mixture is warmed to 50–100°. (The Cu^+ ion acts as a catalyst in these reactions.) This reaction, using the copper(I) salt, is called the **Sandmeyer reaction** after the Swiss chemist Traugott Sandmeyer, who first reported it in 1884.

Aryl iodides and fluorides are prepared with KI (no catalyst needed) or fluoroboric acid (HBF_4), respectively.

One use of these diazonium reactions is the preparation of compounds that cannot be readily synthesized by other routes. For example, consider the synthesis of *p*-toluic acid (*p*-methylbenzoic acid). The usual route to an aryl carboxylic acid is the oxidation of an alkylarene (R—Ar). Oxidation cannot be used for the preparation of *p*-toluic acid from *p*-xylene (*p*-dimethylbenzene) because *both* methyl groups would be oxidized to carboxyl groups. However, *p*-toluic acid can be obtained by way of a diazonium intermediate, followed by treatment with CuCN–KCN and subsequent hydrolysis of the —CN group to —CO_2H, a reaction that will be discussed in Section 15.11.

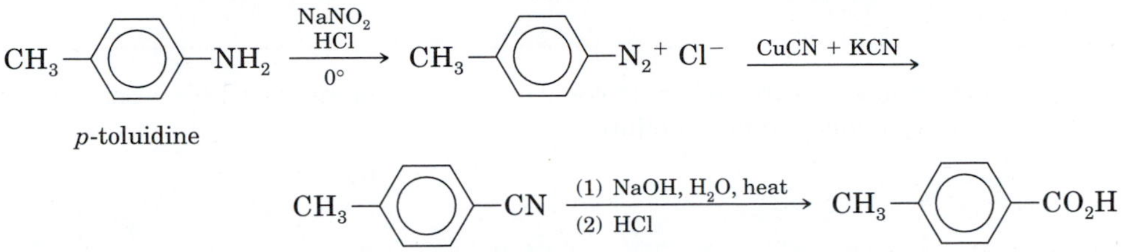

CH$_3$—⟨ ⟩—NH$_2$ $\xrightarrow[0°]{\substack{\text{NaNO}_2 \\ \text{HCl}}}$ CH$_3$—⟨ ⟩—N$_2{}^+$ Cl$^-$ $\xrightarrow{\text{CuCN + KCN}}$

p-toluidine

CH$_3$—⟨ ⟩—CN $\xrightarrow[\text{(2) HCl}]{\text{(1) NaOH, H}_2\text{O, heat}}$ CH$_3$—⟨ ⟩—CO$_2$H

p-tolunitrile

p-toluic acid
(59% overall)

Displacement by —OH   Phenols can be prepared from diazonium salts by reaction with hot aqueous acid. This reaction provides one of the few laboratory routes to phenols.

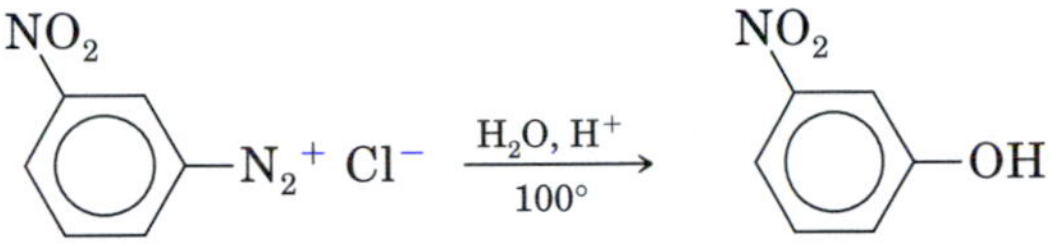

NO$_2$ ⟨ ⟩—N$_2{}^+$ Cl$^-$ $\xrightarrow[100°]{\text{H}_2\text{O, H}^+}$ NO$_2$ ⟨ ⟩—OH

m-nitrophenol (86%)

Displacement by —H   The —N$_2{}^+$ group may be replaced with —H by treatment of the diazonium salt with hypophosphorous acid, H$_3$PO$_2$. This sequence provides a method of *removal of an NH$_2$ group from an aromatic ring*.

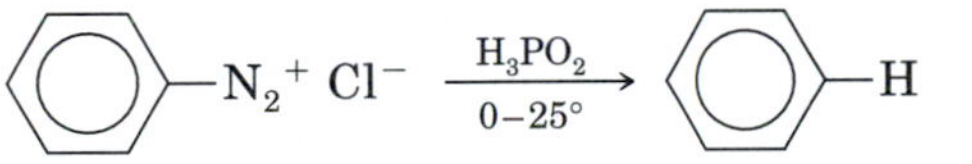

⟨ ⟩—N$_2{}^+$ Cl$^-$ $\xrightarrow[0-25°]{\text{H}_3\text{PO}_2}$ ⟨ ⟩—H

benzene

SAMPLE PROBLEM

Show how you would synthesize 1,3,5-tribromobenzene from aniline. (Remember that bromine is an *o,p*-director in electrophilic substitution reactions.)

Solution
Aniline undergoes tribromination readily (see Section 11.9). Therefore, the following reaction sequence will yield the desired product.

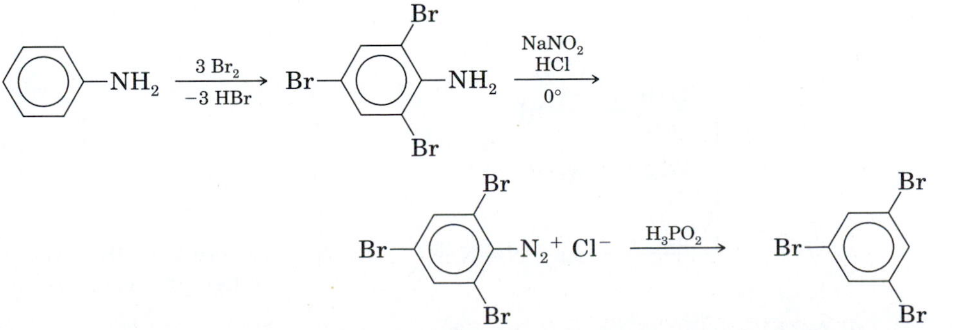

⟨ ⟩—NH$_2$ $\xrightarrow[-3 \text{ HBr}]{3 \text{ Br}_2}$ Br—⟨ ⟩—NH$_2$ (with Br at top and bottom) $\xrightarrow[0°]{\substack{\text{NaNO}_2 \\ \text{HCl}}}$

Br—⟨ ⟩—N$_2{}^+$ Cl$^-$ (with Br at top and bottom) $\xrightarrow{\text{H}_3\text{PO}_2}$ Br—⟨ ⟩ (with Br at top and bottom)

1,3,5-tribromobenzene

12.5 Show how the following compounds could be prepared from benzene or toluene by a diazonium procedure.

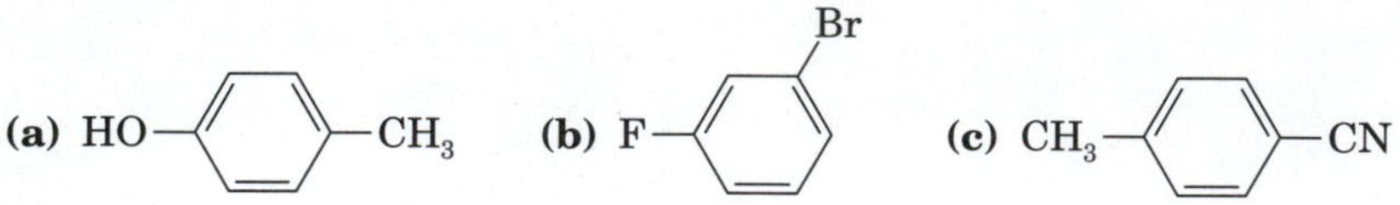

(a) HO—⟨ ⟩—CH₃   (b) F—⟨ ⟩ (with Br)   (c) CH₃—⟨ ⟩—CN

Coupling reactions Coupling reactions of aryldiazonium salts are used to prepare dyes from aniline and from substituted anilines. In these reactions, the diazonium ion acts as an *electrophile*. Resonance structures for the diazonium ion show that each nitrogen carries a partial positive charge.

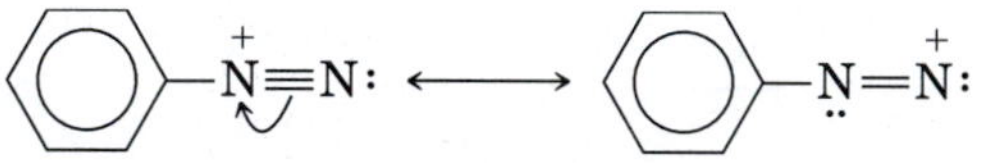

The terminal nitrogen attacks predominantly the *para* position of an activated benzene ring (one substituted with an electron-releasing group like NH_2 or OH). The coupling product contains an **azo group** (—N=N—) and is generally referred to as an **azo compound.** Many azo compounds are used as dyes.

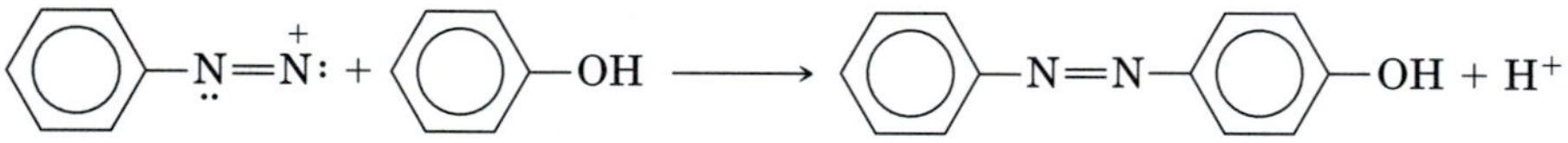

p-hydroxyazobenzene

orange

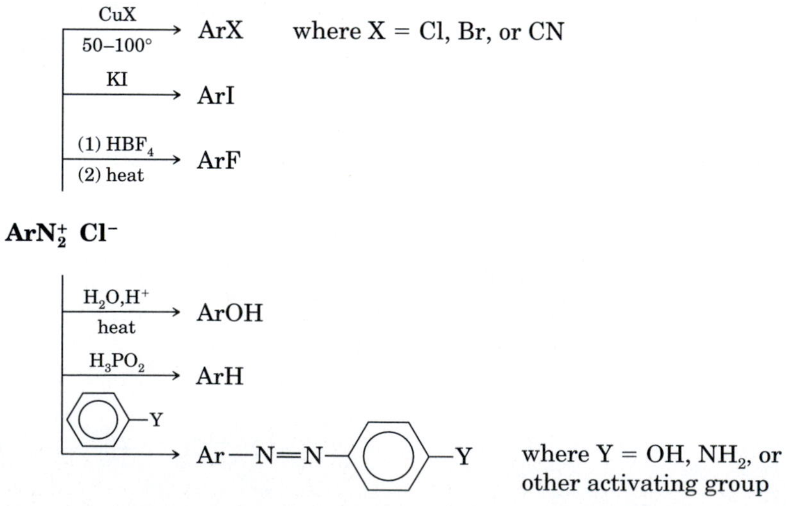

ArN_2^+ Cl^-

CuX, 50–100°	ArX where X = Cl, Br, or CN
KI	ArI
(1) HBF₄ (2) heat	ArF
H₂O,H⁺ heat	ArOH
H₃PO₂	ArH
⟨ ⟩—Y	Ar—N=N—⟨ ⟩—Y where Y = OH, NH₂, or other activating group

Figure 12.1 Reactions of an aryldiazonium chloride.

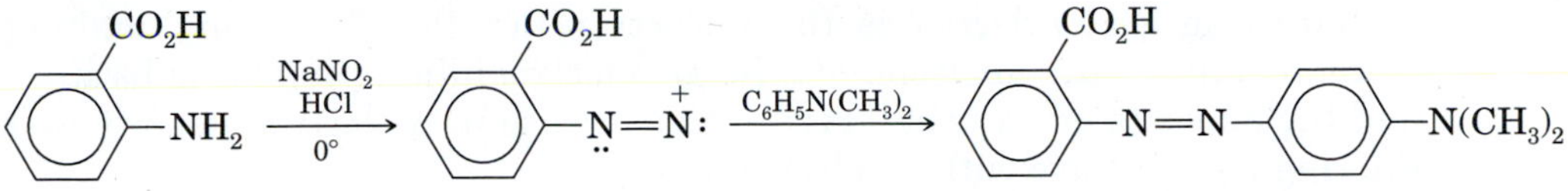

The reactions of aryldiazonium salts are summarized in Figure 12.1

12.6   How would you make the following conversions?

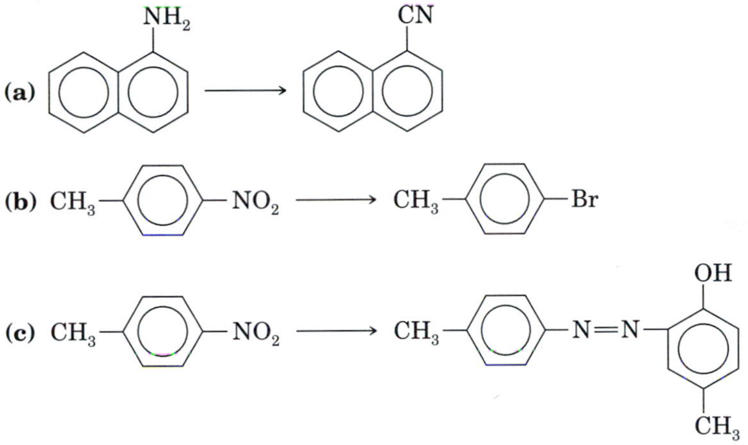

(a)

(b) CH_3—⟨○⟩—NO_2 ⟶ CH_3—⟨○⟩—Br

(c) CH_3—⟨○⟩—NO_2 ⟶ CH_3—⟨○⟩—N=N—⟨○⟩

Section 12.4
Halobenzenes and Nucleophilic Aromatic Substitution

In Chapter 5, we mentioned that aryl halides do not undergo the substitution and elimination reactions characteristic of alkyl halides.

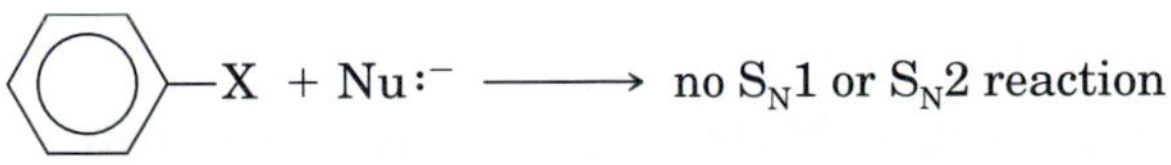

⟨○⟩—X + Nu:$^-$ ⟶ no S_N1 or S_N2 reaction

an aryl halide

Under certain circumstances, however, an aryl halide can undergo a **nucleophilic aromatic substitution reaction.**

⟨○⟩—X + Nu:$^-$ ⟶ ⟨○⟩—Nu + X$^-$

Although this reaction appears to be similar to an S_N1 or S_N2 reaction, it is quite different. It is also different from *electrophilic* substitution, which is initiated by E$^+$, not Nu:$^-$.

The reactivity of aryl halides toward nucleophiles is enhanced by the presence of electron-withdrawing substituents on the ring. An electron-withdrawing group decreases the π-electron density of the ring, making it more reactive with nucleophiles. In an electrophilic aromatic substitution, just the opposite effect is observed. The decrease in π-electron density makes the ring less reactive with an electrophile.

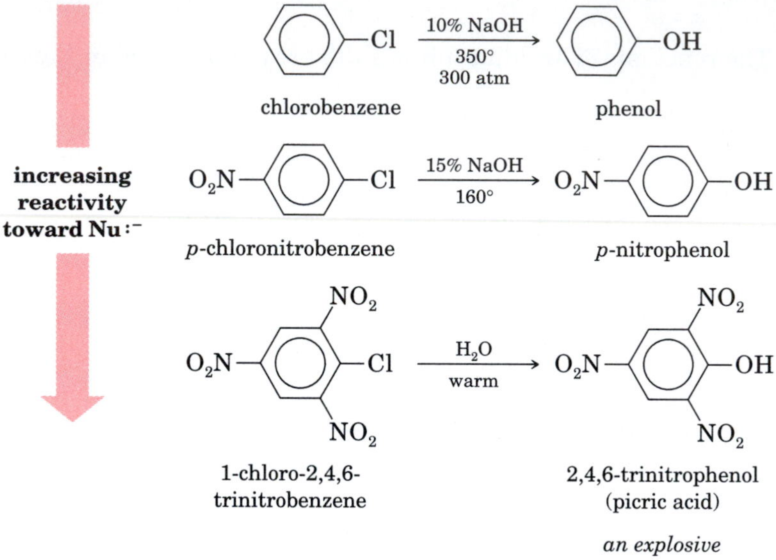

increasing reactivity toward Nu :⁻

chlorobenzene   phenol

p-chloronitrobenzene   p-nitrophenol

1-chloro-2,4,6-trinitrobenzene   2,4,6-trinitrophenol (picric acid)

an explosive

Nucleophilic aromatic substitution is not limited to ⁻OH and H_2O as nucleophiles. Other nucleophiles, such as NH_3, can also undergo reaction with activated aryl chlorides.

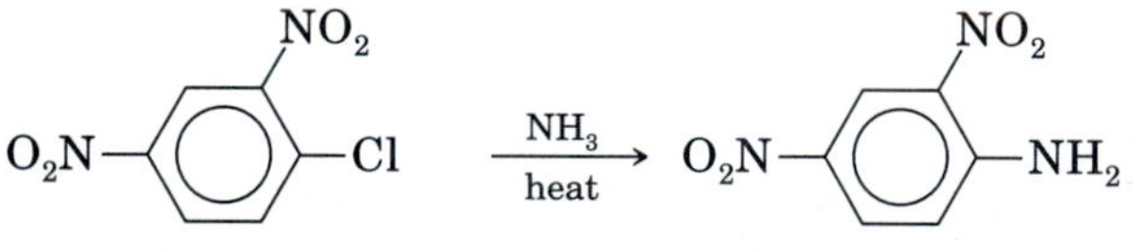

1-chloro-2,4-dinitrobenzene   2,4-dinitroaniline (76%)

The mechanism of nucleophilic aromatic substitution proceeds by one of two routes. The first is through a carbanion intermediate and the second is through a benzyne intermediate.

Carbanion mechanism   If the ring is activated toward nucleophilic substitution by an electron-withdrawing group, the reaction proceeds by a two-step mechanism: (1) addition of the nucleophile to form a carbanion (stabilized by the electron-withdrawing group), and (2) subsequent loss of the halide ion.

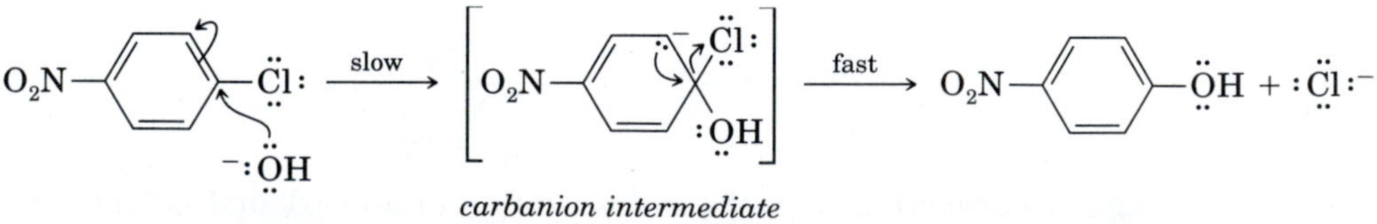

carbanion intermediate

Although the carbanion intermediate is unstable and reactive, it is stabilized to some extent by resonance and by dispersal of the negative charge by the electron-withdrawing group. Resonance structures of this intermediate show that an electron-withdrawing *o-* or *p*-substituent lends more stability to the carbanion intermediate than does a *m*-substituent. When the substituent is *ortho* or *para*, the negative charge is adjacent to the electron-withdrawing group in one resonance structure. In the case of the nitro group, added stability is gained because the nitro group helps disperse the negative charge by resonance.

Resonance structures for p-substituted intermediate:

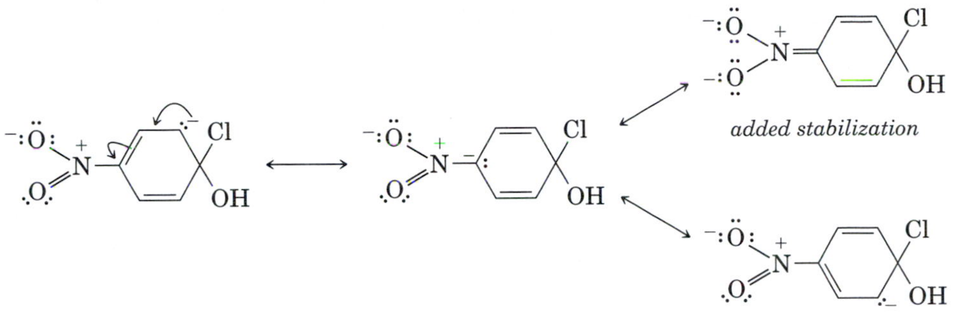

STUDY PROBLEMS

12.7   Write resonance structures that show why the following carbanion is not resonance-stabilized by the nitro groups.

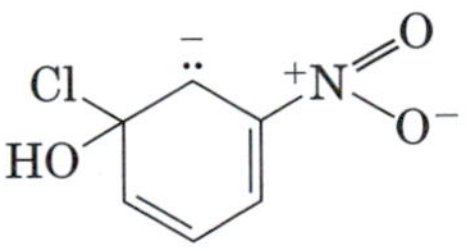

carbanion needed for an m-substituted product

12.8   Draw resonance structures for the carbanion intermediate in the reaction of sodium hydroxide and *o*-chloronitrobenzene.

Benzyne mechanism   If there is no electron-withdrawing substituent on the ring, nucleophilic aromatic substitution is very difficult and is thought to proceed by way of a **benzyne** intermediate.

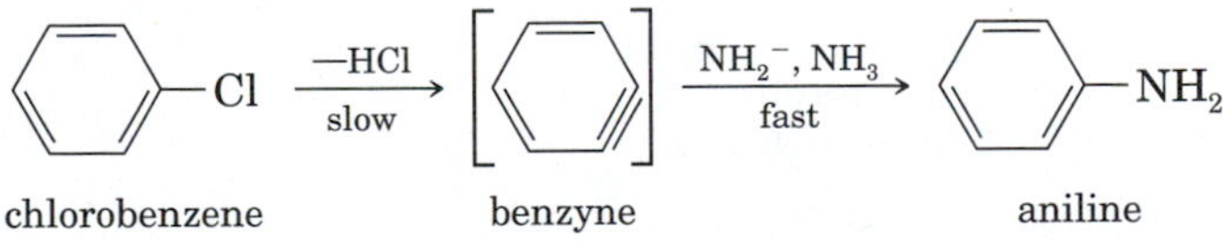

The structure of benzyne is not like that of an alkyne. The two ring carbons that have a triple bond between them are joined by an sp^2–sp^2 sigma bond and p–p overlap (the aromatic π cloud). These two bonds are the same as in benzene. The third bond is the side-to-side overlap of two sp^2 orbitals, the ones that originally were used in the bonds to H and X. Figure 12.2 shows this overlap. Because of the rigid geometry of the ring and the unfavorable angles of normal sp^2 orbitals, this overlap cannot be very good. The new bond is very weak, and benzyne is a highly reactive intermediate.

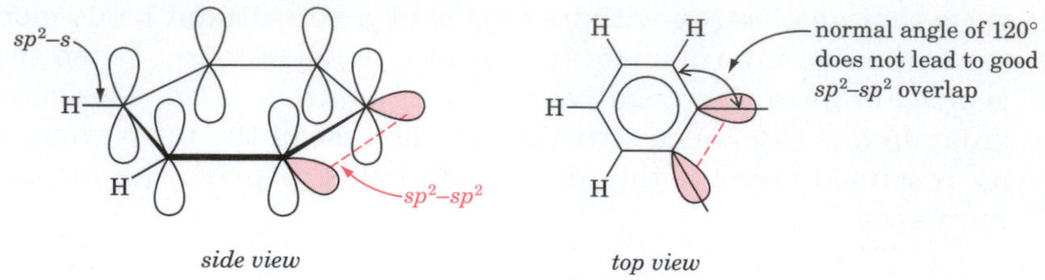

Figure 12.2   The bonding in the benzyne intermediate.

The formation of the benzyne intermediate in the reaction of chlorobenzene with $NaNH_2$ in liquid NH_3 is thought to occur by the following route:

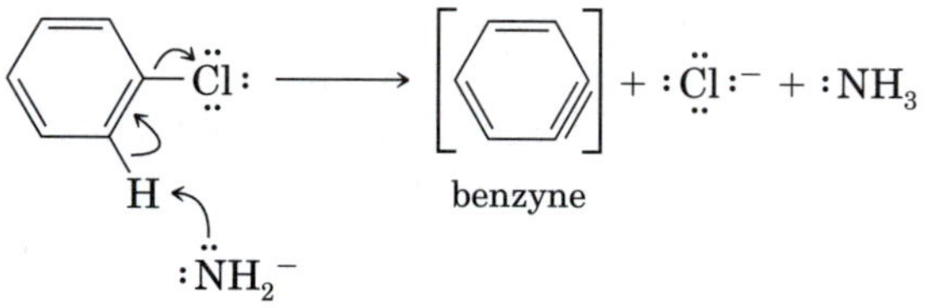

Benzyne undergoes rapid addition of a nucleophile to yield a carbanion that can abstract a proton from NH_3 to yield the product (aniline, here) and another NH_2^- ion.

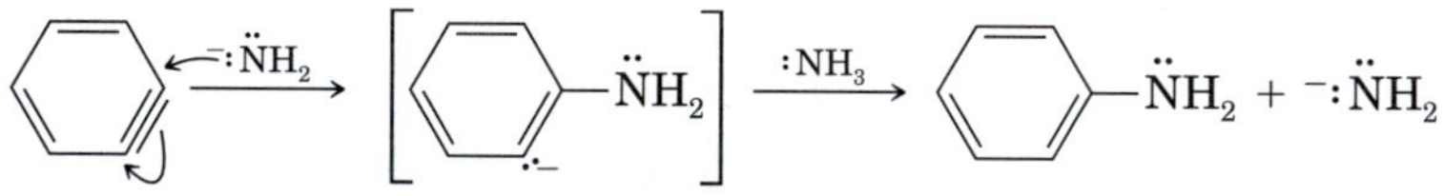

Figure 12.3 summarizes the aromatic nucleophilic substitution reactions of halobenzenes.

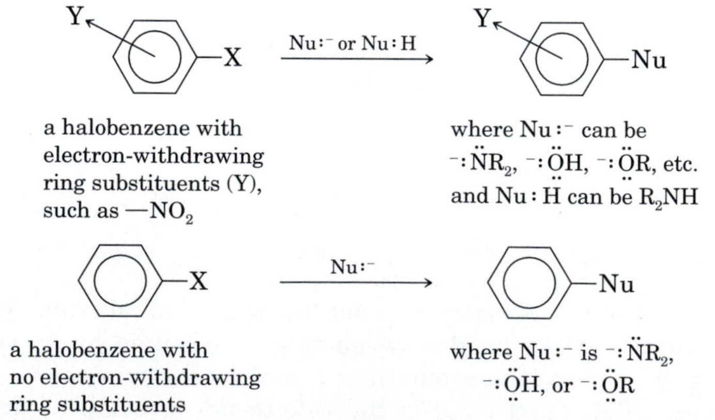

Figure 12.3   Summary of aromatic nucleophilic substitution reactions. (A bond to the center of a benzene ring denotes an unspecified position of substitution.)

Section 12.5
Syntheses Using Substituted Benzene Compounds

Table 11.6 (page 506) and Table 12.1 show some possible applications of benzene and its derivatives in synthesis. In the laboratory, electrophilic aromatic substitutions are more widely used as synthetic reactions than are nucleophilic aromatic substitution reactions because there are fewer limitations on the starting material. (For aromatic nucleophilic substitution, a laboratory chemist would use a ring activated by an electron-withdrawing group.)

When using electrophilic aromatic substitution reactions to prepare substituted benzene compounds, a chemist must keep in mind whether the substituents are *o,p*-directors or *m*-directors. For example, in the synthesis of *m*-chloronitrobenzene, chlorination would be a poor choice as the first step because this reaction places an *o,p*-director on the ring. Subsequent nitration would give *o*- and *p*-chloronitrobenzene, but not the desired *m*-chloronitrobenzene. The first substitution should be nitration because the nitro group is a

---

Table 12.1   Summary of substitution and substituent reactions of substituted benzenes

Reaction	Product	Section References
Electrophilic Substitution:		
$C_6H_5Y + E^+ \longrightarrow$ *o*- and *p*-products		11.9
$C_6H_5Z + E^+ \longrightarrow$ *m*-products		11.9
where Y is e^--releasing and Z is e^--withdrawing		
Nucleophilic Substitution:		
$C_6H_5X + OH^- \longrightarrow C_6H_5OH$	**phenol**[a]	12.4
$C_6H_5X + NH_3 \longrightarrow C_6H_5NH_2$	**aniline**[a]	12.4
Substituent Reactions:		
$C_6H_5R + MnO_4^- \xrightarrow{\text{heat}} C_6H_5CO_2H$	**carboxylic acid**	12.1
$C_6H_5CH_2R + Br_2$ or NBS $\xrightarrow{\text{catalyst}} C_6H_5CHBrR$	**benzylic bromide**	6.5B, 12.1
$C_6H_5OH + OH^- \longrightarrow C_6H_5O^-$	**phenoxide**	12.2A
$C_6H_5OH + (R\overset{\overset{\text{O}}{\|}}{C})_2O \longrightarrow C_6H_5O\overset{\overset{\text{O}}{\|}}{C}R$	**ester**	12.2B
$C_6H_5OH \xrightarrow[\text{(2) H}_2\text{O, H}^+]{\text{(1) CHCl}_3\text{, OH}^-}$ (CHO)(OH ring)	*o*-**hydroxy aryl aldehyde**	12.2D
$HO-\langle\text{ring}\rangle-OH + [O] \longrightarrow O=\langle\text{ring}\rangle=O$	**quinone**	12.2E
Reactions of Aryldiazonium Salts:[b]		
$C_6H_5N_2^+Cl^- + Nu:^- \longrightarrow C_6H_5Nu$	**substituted (or unsubstituted) benzene**	12.3B
$C_6H_5N_2^+Cl^- + C_6H_5Y \longrightarrow C_6H_5N{=}NC_6H_4Y$	**azo compound**	12.3B
where Y is OH, NH_2, etc.		

[a] Practical in the laboratory only if one or more other electron-withdrawing groups is present on the ring.
[b] The preparation of aryldiazonium salts is discussed in Section 12.3A. Some specific reactions of these salts are summarized in Figure 12.1.

m-director. Therefore, in synthetic work with substituted benzenes, the *order of substitution reactions* is important.

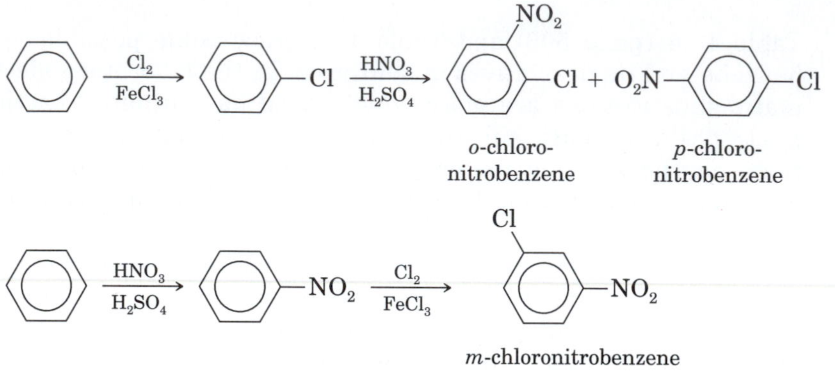

o-chloro-
nitrobenzene

p-chloro-
nitrobenzene

m-chloronitrobenzene

Conversion of one group to another may also be necessary. For instance, reduction of the nitro group to an amino group gives a route to *m*-substituted anilines. Benzene may first be nitrated, then subjected to *m*-substitution, and finally the nitro group may be reduced.

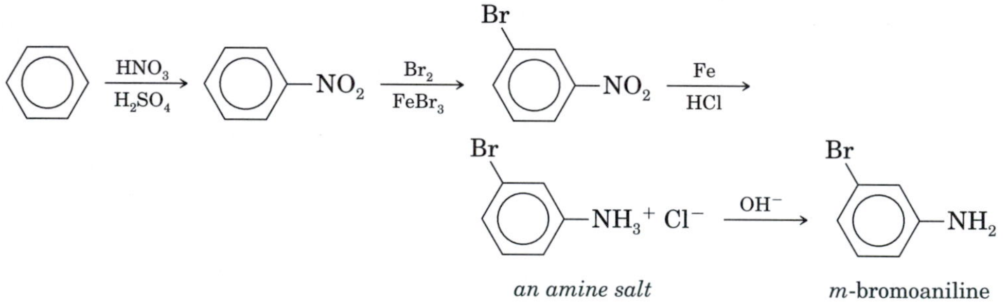

an amine salt

m-bromoaniline

Arylamines can also be converted to aryldiazonium salts, making possible a diversity of substitution products, as was shown in Figure 12.1. In viewing aromatic compounds from a synthetic standpoint, consider every nitro group as a potential diazonium group.

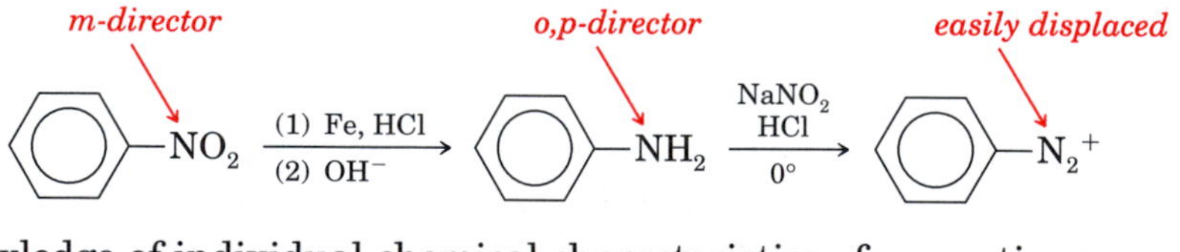

m-director   *o,p-director*   *easily displaced*

Knowledge of individual chemical characteristics of aromatic compounds is often required for successful solution of synthetic problems. For example, aniline does not undergo Friedel–Crafts reactions because an amino group (a basic group) undergoes reaction with Lewis acids to yield a strongly deactivating group.

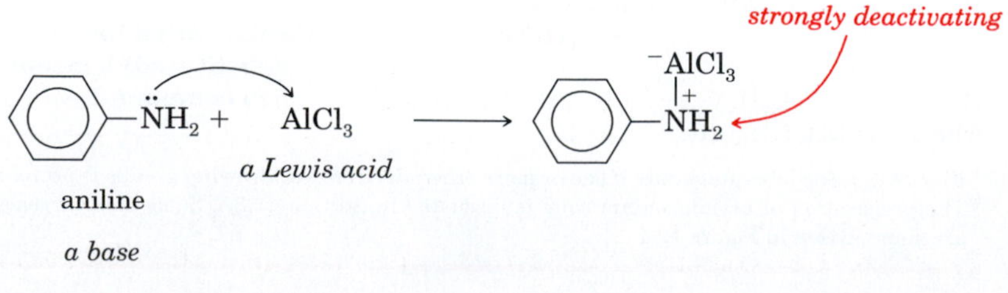

strongly deactivating

aniline

a base

a Lewis acid

An amino group is an *o,p*-director, but an ammonium group (—NR$_3^+$) is a *m*-director, and deactivating.

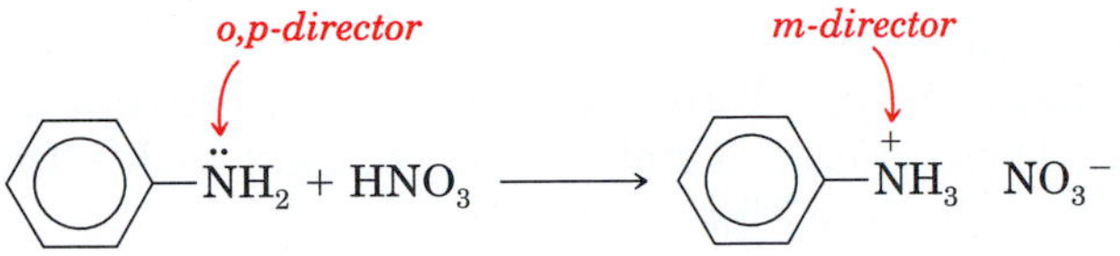

Like aniline, nitrobenzene also does not undergo Friedel–Crafts reactions. In this case, lack of reaction is due to deactivation of the ring by the electron-withdrawing nitro group.

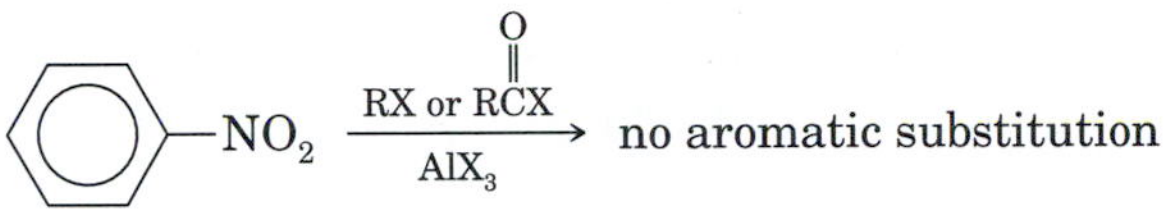

nitrobenzene

Like the diazonium group, the sulfonic acid group is easily removed from an aromatic ring, and therefore can be displaced by a variety of reagents. Phenols, for example, can be prepared from arylsulfonic acids as well as from diazonium salts.

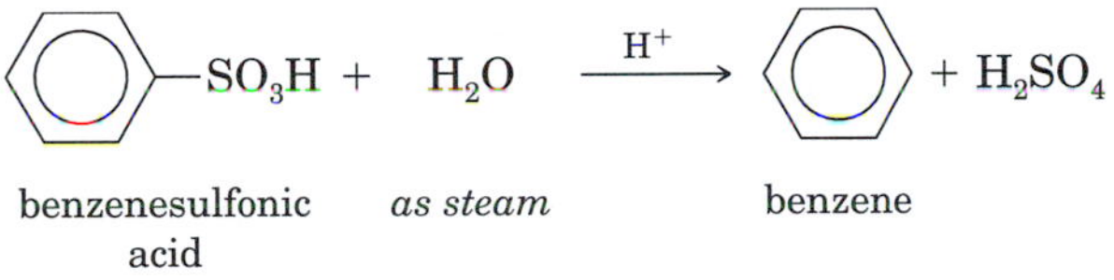

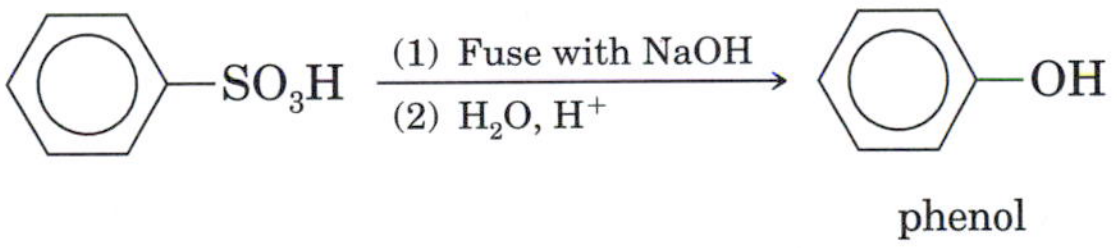

benzenesulfonic   *as steam*   benzene
acid

phenol

STUDY PROBLEM

12.9 Show how you would synthesize the following compounds from benzene: **(a)** *m*-chloroaniline; **(b)** pentylbenzene; **(c)** 1,2-diphenylethane; **(d)** triphenylmethane; **(e)** 2,4,6-tribromophenol.

Summary

Alkylbenzenes contain a benzylic position that is active toward many reagents, such as oxidizing agents and free-radical halogenating agents.

Phenols are more acidic than alcohols and react with aqueous sodium hydroxide to yield sodium phenoxides.

$$ArOH + Na^+ \ ^-OH \longrightarrow ArO^- \ Na^+ + H_2O$$

Phenols can be *esterified* with a reactive esterification agent, such as a carboxylic acid anhydride (see page 518).

A phenoxide can undergo reaction with CO_2 to yield a phenolic carboxylic acid **(Kolbe reaction).** Treatment of a phenol with chloroform ($CHCl_3$) and base, followed by acidification yields a phenol-aldehyde (**Reimer–Tiemann reaction**).

Hydroquinones (1,2- and 1,4-dihydroxybenzenes) undergo reversible oxidation to **quinones**.

Arylamines are converted to **aryldiazonium salts** by reaction with HONO. These salts are stable at 0°, but are highly reactive toward a variety of nucleophiles (see Figure 12.1).

$$\text{ArNH}_2 \xrightarrow[\substack{\text{HCl} \\ 0°}]{\text{NaNO}_2} \text{ArN}_2{}^+ \text{Cl}^- \xrightarrow{\text{Nu:}^-} \text{ArNu} + \text{N}_2 + \text{Cl}^-$$

Halobenzenes do not undergo S_N1 or S_N2 reactions, but X^- can be displaced in **nucleophilic aromatic substitution reactions** if the ring is activated by electron-withdrawing groups such as NO_2.

Essay Problem for Chapter 12

Reductive Arylation of Electron-Deficient Alkenes

Treatment of a benzenediazonium chloride with a Ti(III) salt results in a benzenediazonium radical. Benzenediazonium radicals are unstable and rapidly decompose to phenyl radicals and nitrogen gas.

a benzenediazonium ion   *a benzenediazonium radical*   *a phenyl radical*

Phenyl radicals generated by this sequence of reactions can be used to carry out arylation reactions (reactions involving the addition of a phenyl group to a double bond). In reductive arylation, the reaction appears to involve the addition of a benzene to a double bond. The reaction proceeds by a free-radical pathway.

Only electron-deficient alkenes can be used as reactants in these arylation reactions. Steric hindrance is also important. When the alkene is electron-deficient and not sterically hindered, yields are high. For example, reductive arylation of buten-2-one using the *p*-chlorophenyl radical yields 4-(4-chlorophenyl)-butan-2-one in a 65–75% yield. (A. Citterio, *Org. Syn.* **1984,** *62,* 67.)

4-(4-chlorophenyl)-butan-2-one (65–75%)

Questions

1. Is the diazonium group in benzenediazonium ion resonance-stabilized? What about the diazonium group in benzenediazonium radical? Write the appropriate resonance structures for the above structures.

2. Write an equation showing the formation of the benzenediazonium radical from benzenediazonium ion and Ti(III) that accounts for any electron transfers.

3. Compounds such as the one shown below are not formed by reaction in any significant amounts. How can this observation be explained by resonance, or lack of resonance, in the phenyl radical?

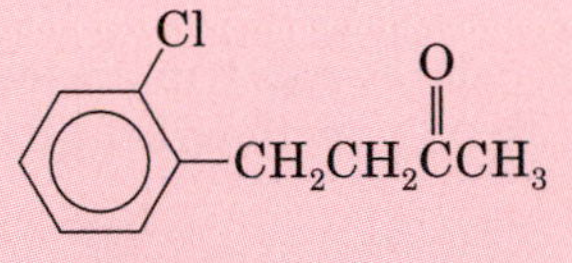

$$
\begin{array}{c}
\text{Cl} \\
\end{array}
\quad \text{-CH}_2\text{CH}_2\overset{\displaystyle O}{\overset{\displaystyle \|}{\text{C}}}\text{CH}_3
$$

4. Why are products such as the one shown below not formed in significant amounts by the reaction?

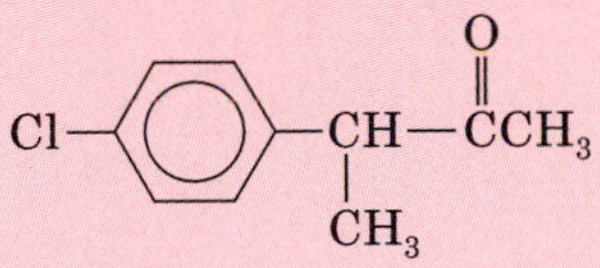

$$
\text{Cl}-\bigcirc-\text{CH}-\overset{\displaystyle O}{\overset{\displaystyle \|}{\text{C}}}\text{CH}_3 \\
\qquad\quad \overset{\displaystyle |}{\text{CH}_3}
$$

5. Which of the following are electron-deficient alkenes?

   (a) $CH_2{=}CHCN$ **(b)** $CH_2{=}CHNO_2$

   (c) $CH_2{=}CHCl$ **(d)** $CH_2{=}CHCH_3$

Study Problems*

12.10 Complete the following equations:

   (a) $C_6H_5CH_2CH_2CH_3 + KMnO_4 \xrightarrow[\text{H}_2\text{O}]{\text{H}^+,\ \text{heat}}$

   (b) $C_6H_5CH_2CH_3 + Br_2 \xrightarrow{h\nu}$

   (c) $C_6H_5CH_3 + CH_3\overset{\displaystyle O}{\overset{\displaystyle \|}{\text{C}}}Cl \xrightarrow{\text{AlCl}_3}$

   (d) $C_6H_5CH_2CH_2CH_3 + HNO_3 \xrightarrow{\text{H}_2\text{SO}_4}$

   (e) $C_6H_5CH(CH_3)_2 + O_2 \longrightarrow$

   (f) $C_6H_5CH_2CH_3 + Br_2 \xrightarrow{\text{FeBr}_3}$

12.11 TNT (2,4,6-trinitrotoluene) is treated with an excess of concentrated sulfuric acid and sodium dichromate. One product is obtained in 69% yield. What is the structure of this product?

---

* For information concerning the organization of the *Study Problems* and *Additional Problems*, see the *Note to student* on page 41.

12.12 Write equations for the expected reactions of 3,5-dimethylphenol with the following reagents. If little or no reaction is expected, write "no reaction."

(a) Na   (b) NaOH   (c) HI

(d) CH_3MgBr (e) NaH   (f) $(CH_3CO)_2O$

(g) NaCN   (h) PCl_3 (i) $(CH_3O)_2SO_2$, NaOH

12.13 Write formulas for the principal product(s) formed by the reaction of *p*-cresol (*p*-methylphenol) with each of the following reagents:

(a) CO_2, NaOH   (b) $CHCl_3$, OH^- (c) H_2SO_4, heat

(d) HNO_3, cold   (e) Br_2, H_2O

12.14 (a) Why is benzyl chloride more reactive in nucleophilic substitution (S_N1) reactions than heptyl chloride? (b) Would *p*-methoxybenzyl chloride show greater or less reactivity? Explain.

12.15 Complete the following equations. If no reaction occurs write "no reaction."

(a) 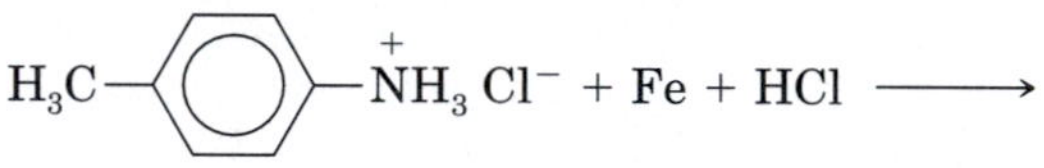 H_3C—⟨○⟩—$\overset{+}{N}H_3\ Cl^-$ + Fe + HCl $\longrightarrow$

(b) 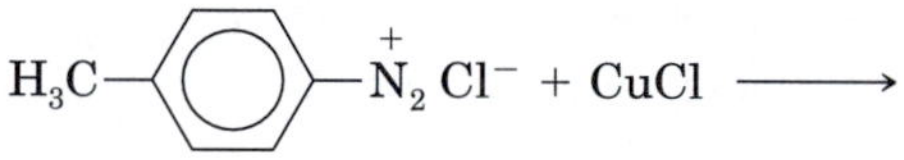 H_3C—⟨○⟩—$\overset{+}{N}_2\ Cl^-$ + CuCl $\longrightarrow$

(c) 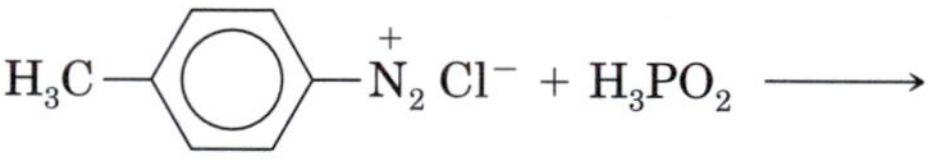 H_3C—⟨○⟩—$\overset{+}{N}_2\ Cl^-$ + H_3PO_2 $\longrightarrow$

(d) 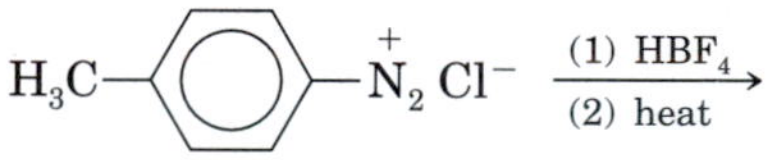 H_3C—⟨○⟩—$\overset{+}{N}_2\ Cl^-$ $\xrightarrow{\text{(1) } HBF_4}{\text{(2) heat}}$

12.16 Predict the major organic products:

(a) *p*-ethylaniline + $NaNO_2$ + HCl $\xrightarrow{0°}$

(b) β-naphthylamine + $NaNO_2$ + HCl$\xrightarrow{0°}$

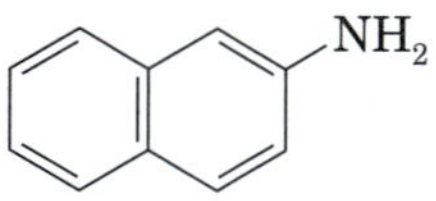

β-naphthylamine

a potent carcinogen

(c) [the product from (a)] + CuCN $\xrightarrow[\text{KCN}]{100°}$

(d) [the product from (b)] + phenol $\xrightarrow{0°}$

12.17 How would you make the following conversions?

(a)

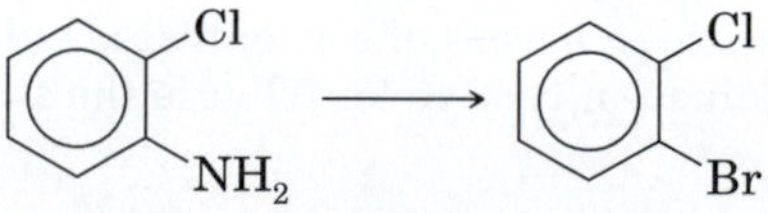

(b) 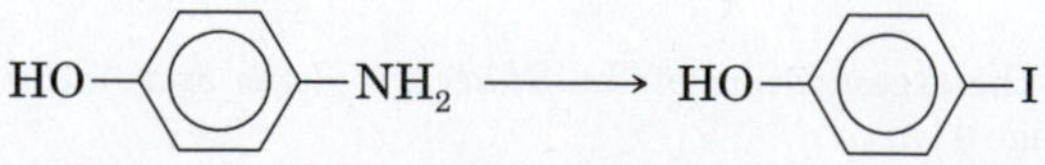 HO—⟨○⟩—NH_2 $\longrightarrow$ HO—⟨○⟩—I

(c) 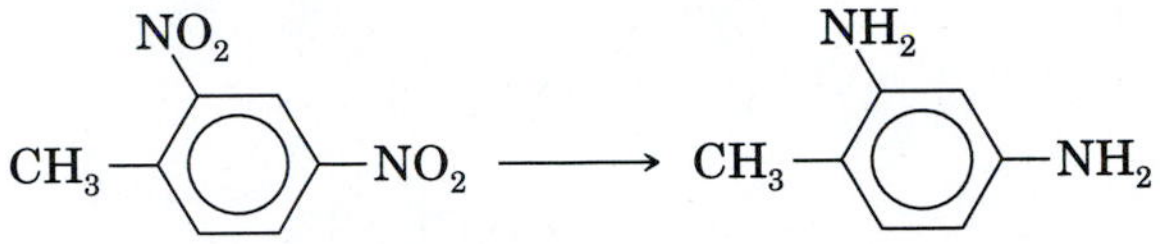

(d) 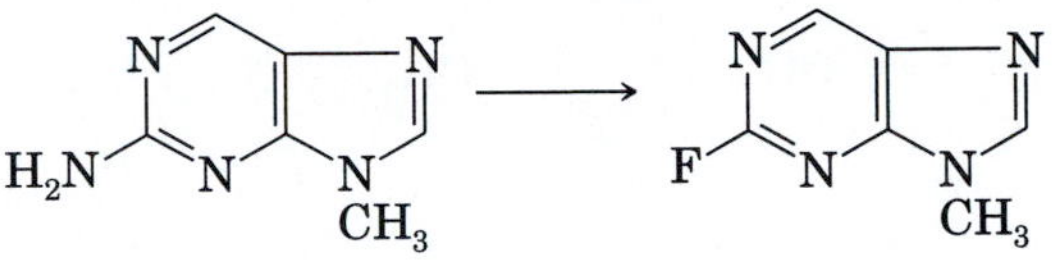

(e) 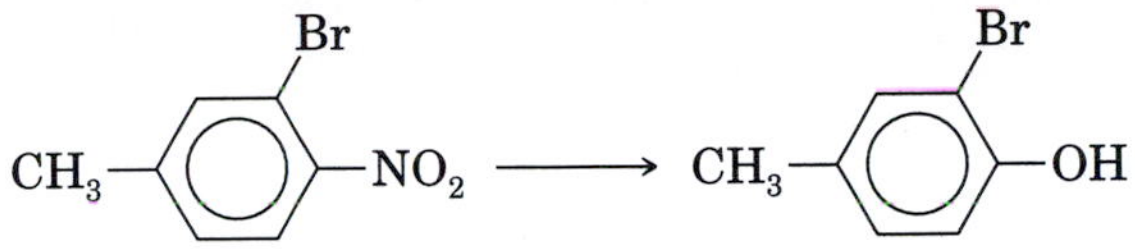

12.18 Complete the following equations:

(a) 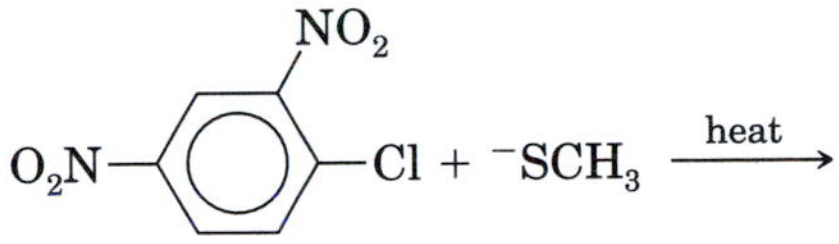

(b) 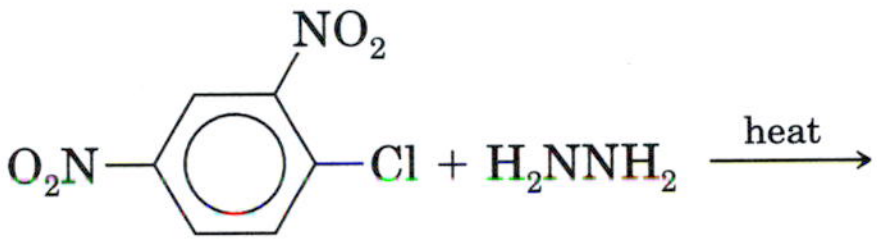

(c) 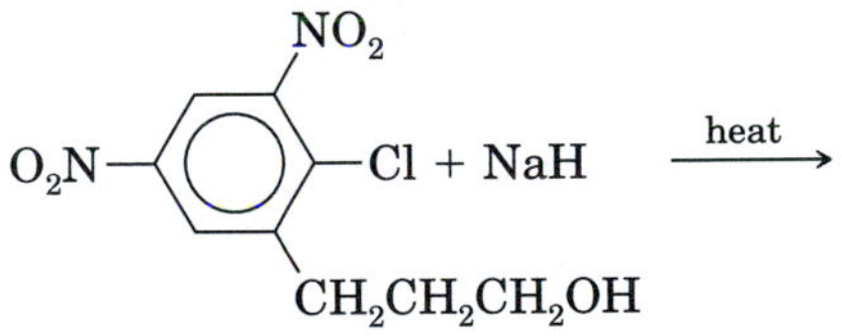

(d) 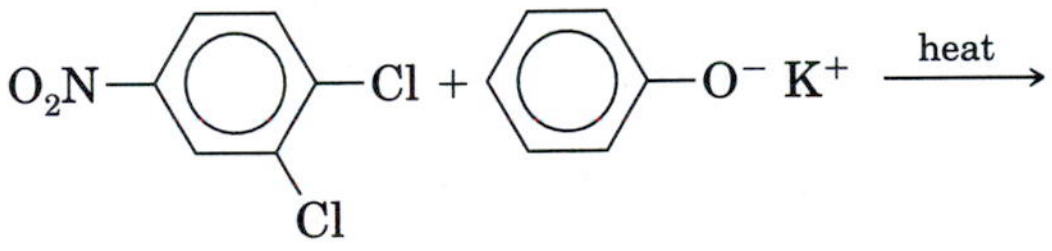

12.19 How would you make the following conversions?

(a) 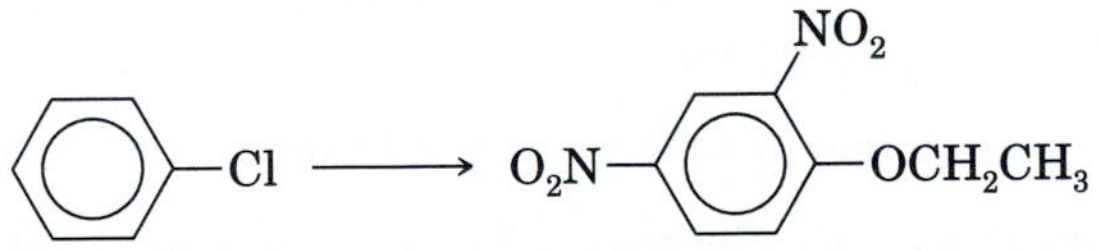

(b) 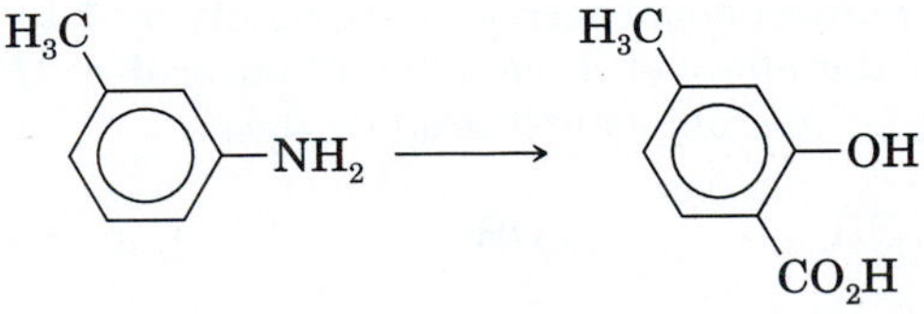

(c) 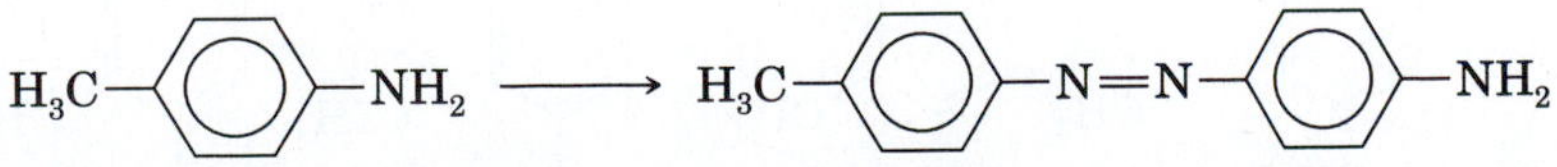

(d) 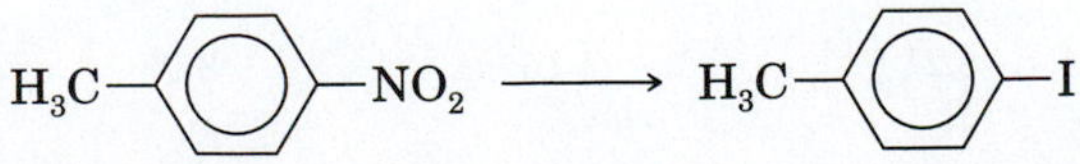

(e) $CH_3O-\!\!\bigcirc\!\!-\!\!\longrightarrow\ CH_3O-\!\!\bigcirc\!\!-CN$

(f) $CH_3-\!\!\overset{Br}{\bigcirc}\!\!-CH_3\ \longrightarrow\ CH_3-\!\!\overset{CH_2CH_2CH_3}{\bigcirc}\!\!-CH_3$

Additional Problems

12.20 Predict the major organic products:

(a) ethylbenzene + Cl_2 $\xrightarrow{\text{FeCl}_3}$

(b) styrene + Br_2 $\longrightarrow$

(c) *p-tert*-butyltoluene + MnO_4^- $\xrightarrow{\text{heat}}$

(d) toluene + 1-chloropropane $\xrightarrow{\text{AlCl}_3}$

(e) toluene + fuming H_2SO_4 $\longrightarrow$

(f) 3-phenylpropene $\xrightarrow[\text{(2) Zn, H}^+]{\text{(1) O}_3}$

12.21 Suggest syntheses for the three isomeric mononitrobenzoic acids from toluene.

12.22 How would you synthesize each of the following compounds from benzene (assuming that you can separate *o-*, *m-*, and *p-*isomers)?

(a) phenylcyclohexane   **(b)** 1-chloro-2-methyl-2-phenylpropane

(c) anisole ($C_6H_5OCH_3$)   **(d)** terephthalic acid ($p\text{-}HO_2C\text{—}C_6H_4\text{—}CO_2H$)

(e) *m*-dibromobenzene   **(f)** *m*-nitrophenol

(g) 2-phenyl-2-propanol   **(h)** 1,1-diphenylethane

(i) benzyl methyl ether

12.23 (a) Suggest a reason for the fact that aniline (with a pK_b of 9.37) is much less basic than cyclohexylamine, $C_6H_{11}NH_2$ ($pK_b = 3.34$).

(b) Would you expect *p*-nitroaniline to be more or less basic than aniline? Why?

12.24 The acidity of *p*-cyanophenol decreases by only 0.26 pK units upon 3,5-dimethylation, but the same substitution on *p*-nitrophenol decreases its acidity by 1.10 pK units. Explain this large difference in the effect of dimethylation on acidity. (*Hint:* Start by drawing resonance structures for the conjugate bases.)

OH	OH	OH	OH
	CH_3 CH_3		CH_3 CH_3
CN	CN	NO_2	NO_2
$pK_a = 7.95$	8.21	7.15	8.25

12.25 Write a mechanism to show how the following reaction might occur. (*Hint:* Remember that sulfonations are reversible.)

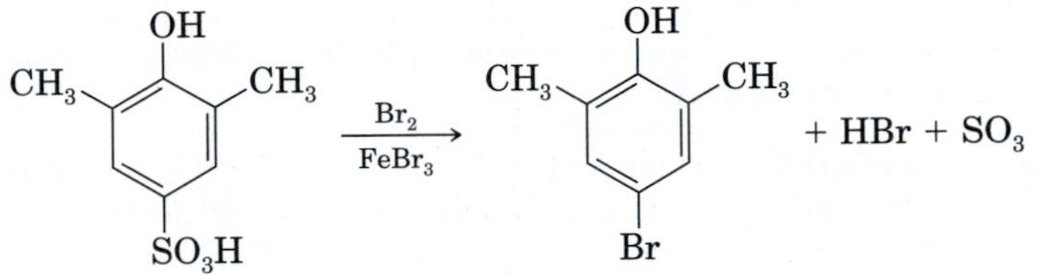

$$+ HBr + SO_3$$

12.26 1-fluoro-2,4-dinitrobenzene is used in structure determinations of polypeptides (small proteins). This compound undergoes reaction with the free amino group at one end of a peptide chain. Write a mechanism for its reaction with glycylglycine.

$$\overset{O}{\overset{\|}{H_2NCH_2CNHCH_2CO_2H}}$$

glycylglycine

12.27 Starting with aniline, benzene, or toluene, show how you could prepare the following compounds:

(a) 2-bromo-4-nitrotoluene   **(b)** *m*-iodobenzoic acid   **(c)** *m*-bromoaniline

12.28 Write a flow diagram showing how each of the following conversions could be carried out:

(a) toluene to *p*-fluorotoluene

(b) C_6H_5CHO to $C_6H_5CH_2CH_2Br$

(c) $C_6H_5NO_2$ to

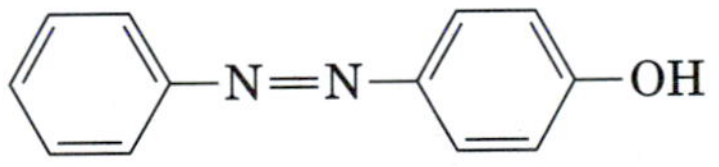

(d) $C_6H_5CH_3$ to

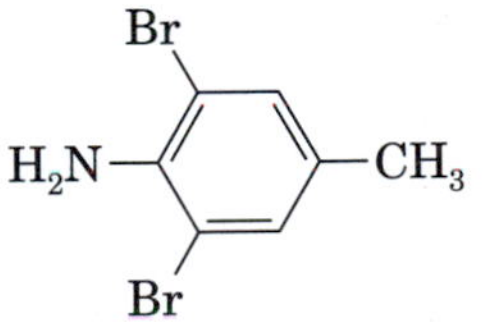

(e) $C_6H_5CH_2CH_2CH_3$ to

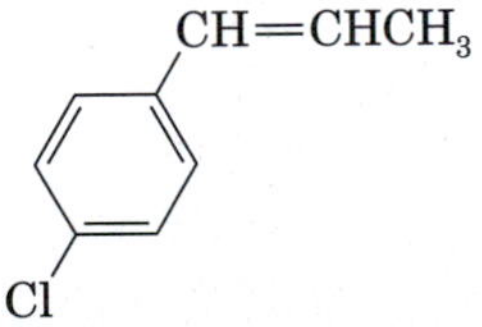

12.29 A nitrogen atom in an aromatic ring is analogous to a nitro group in its ability to aid nucleophilic aromatic substitution. 2-Chloropyrimidine, for example, reacts rapidly with dimethylamine to yield 2-(dimethylamino)pyrimidine. Why?

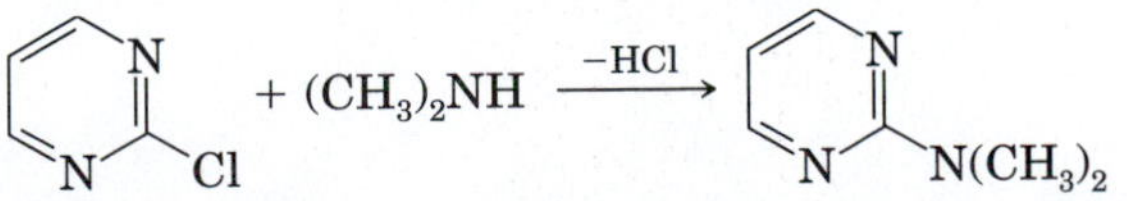

$+ (CH_3)_2NH \xrightarrow{-HCl}$

(86%)

12.30 The nitroso group (—NO) is an *o,p*-director and is also an activator for nucleophilic aromatic substitution. Write equations for **(a)** the *p*-bromination of nitrosobenzene and **(b)** the nucleophilic substitution reaction of *p*-bromonitrosobenzene with methoxide. Show resonance structures for each intermediate.

12.31 When *m*-bromotoluene is treated with $NaNH_2$ in liquid NH_3, a mixture of *o*-, *p*-, and *m*-toluidine (methylaniline) is produced. Write chemical equations showing how the three isomers are formed.

12.32 A compound C_6H_5Cl was treated with HNO_3 and H_2SO_4. The products of this reaction were then treated with Fe and HCl, followed by neutralization. The ^1H NMR spectrum of one of the products of this reaction is given in Figure 12.4. What is the structure of this component? What would be the structures of the other components of the product mixture?

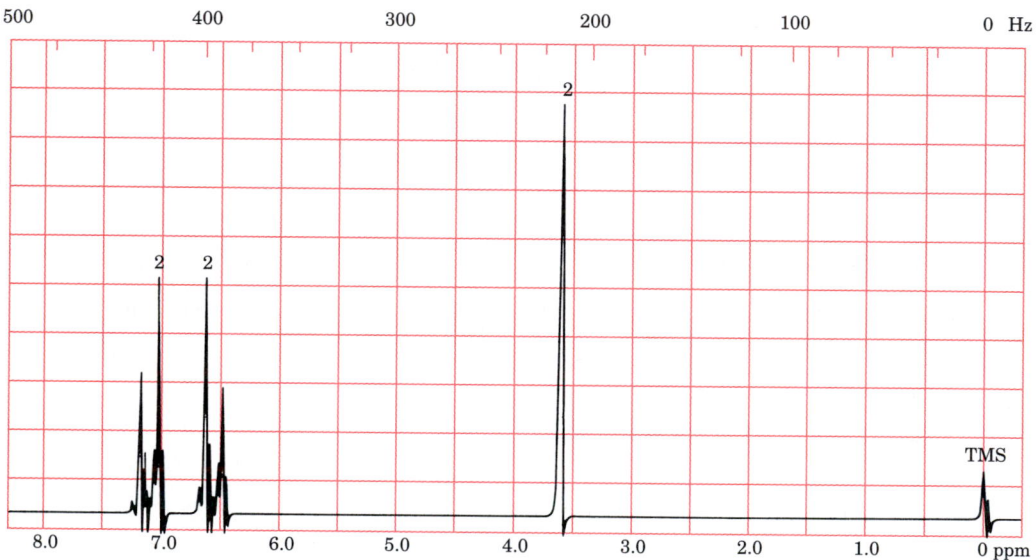

Figure 12.4 ^1H NMR spectrum for Problem 12.32.

12.33 One of the isomeric cresols (methylphenols) was treated with NaOH and iodomethane. The ^{1}H NMR spectrum of the product is shown in Figure 12.5. What are the structures of the cresol and the product?

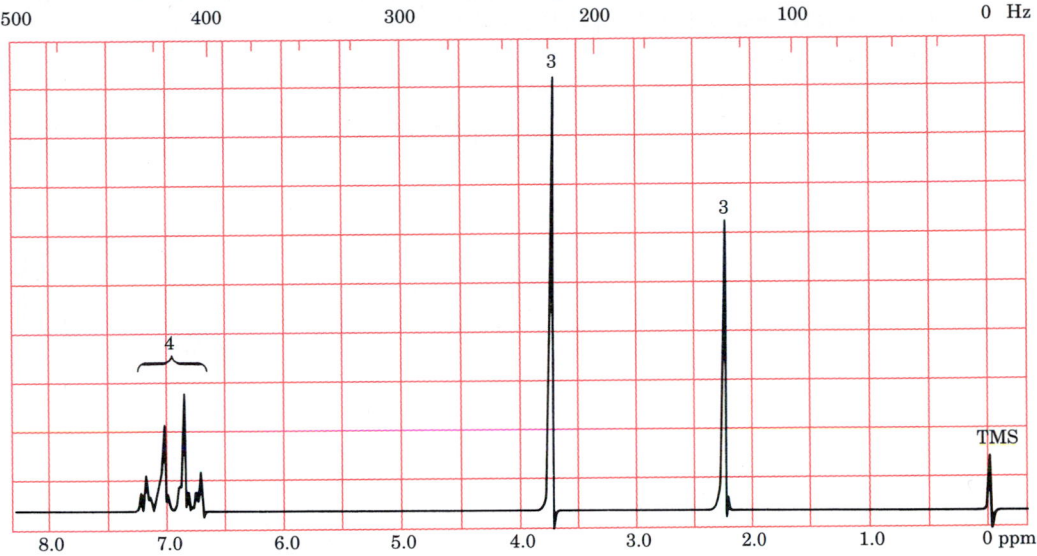

Figure 12.5 ^{1}H NMR spectrum for Problem 12.33.

12.34 Upon treatment with HNO_3, a monosubstituted aromatic compound yielded two isomeric products, A and B. Treatment of A with NaOH followed by CH_3I gave C. In an
identical manner, compound B yielded compound D. The infrared spectra of A, B, C,
and D are given in Figure 12.6. Reaction of C with Fe and HCl, followed by treatment
with base, yielded E. Following this same procedure, compound D yielded compound F.
The ^1H NMR spectra of E and F are given in Figure 12.7. What are the structures of
the original compound, A, B, C, D, E, and F?

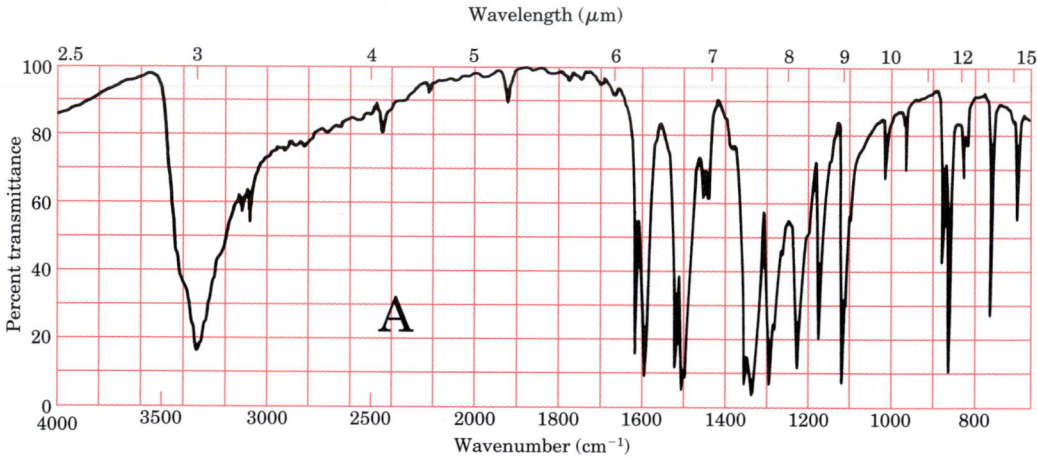

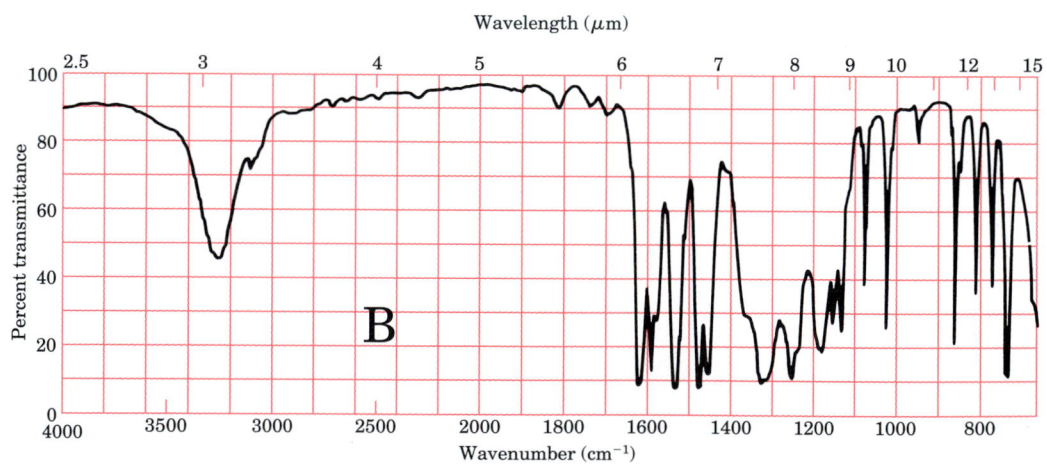

Figure 12.6 Infrared spectra for Problem 12.34.

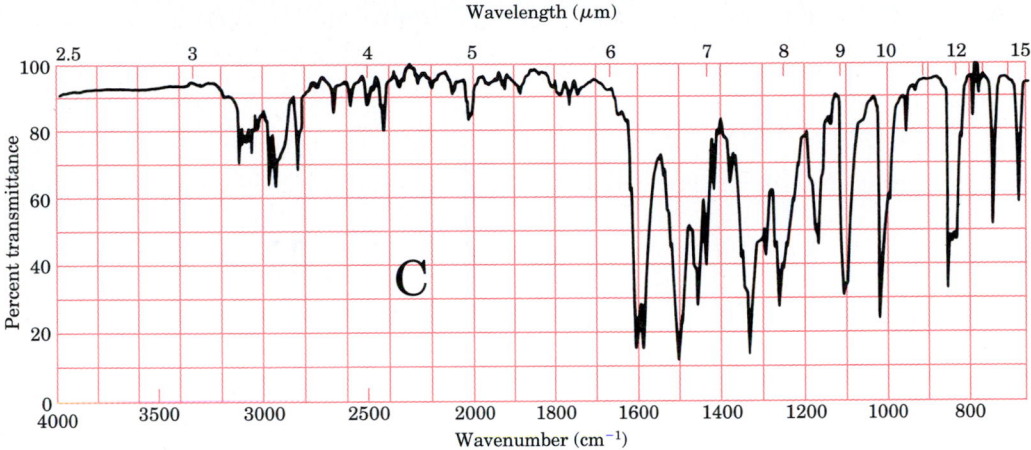

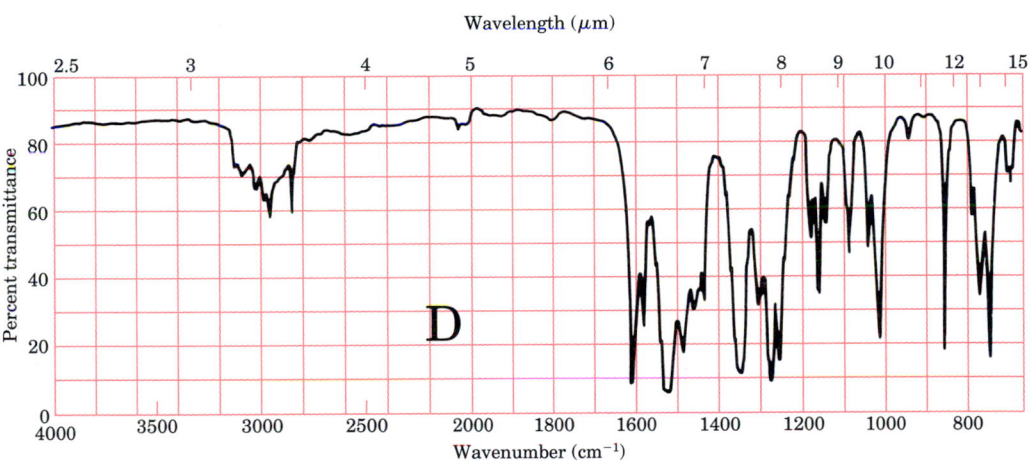

Figure 12.6 *(continued)*   Infrared spectra for Problem 12.34.

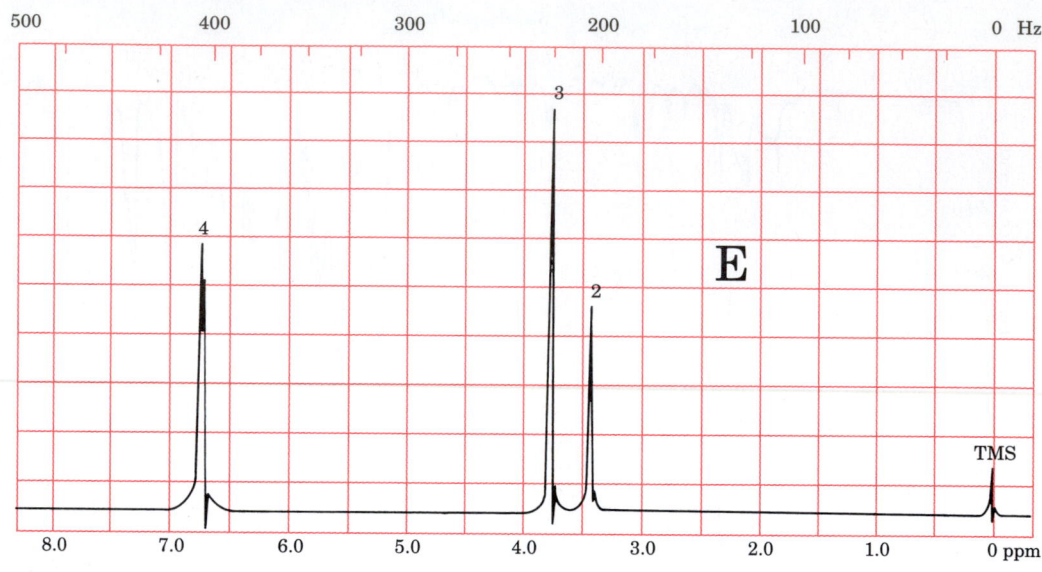

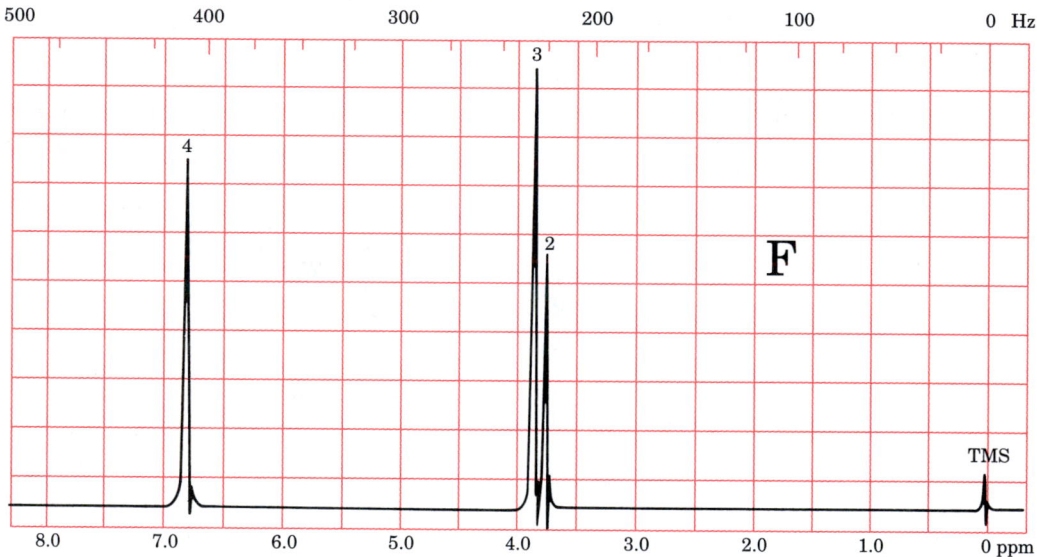

Figure 12.7 ¹H NMR spectra for Problem 12.34.

13 ALDEHYDES AND KETONES

Aldehydes and ketones are but two of many classes of organic compounds that contain carbonyl groups. A **ketone** has *two alkyl (or aryl) groups* attached to the carbonyl carbon, while an **aldehyde** has at least one hydrogen atom attached to the carbonyl carbon. The other group in an aldehyde (R in the following formulas) can be alkyl, aryl, or H.

$$
\begin{array}{c} O \\ \parallel \\ R-C-H \end{array} \quad \text{or} \quad RCHO
$$

an aldehyde

$$
\begin{array}{c} O \\ \parallel \\ R-C-R \end{array} \quad \text{or} \quad RCOR
$$

a ketone

Aldehydes and ketones are common in living systems. The sugar ribose and the female hormone progesterone are two examples of biologically important aldehydes and ketones.

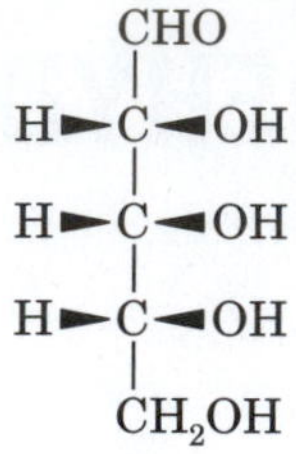

$$\begin{array}{c} CHO \\ | \\ H\blacktriangleright C\blacktriangleleft OH \\ | \\ H\blacktriangleright C\blacktriangleleft OH \\ | \\ H\blacktriangleright C\blacktriangleleft OH \\ | \\ CH_2OH \end{array}$$

ribose

a carbohydrate

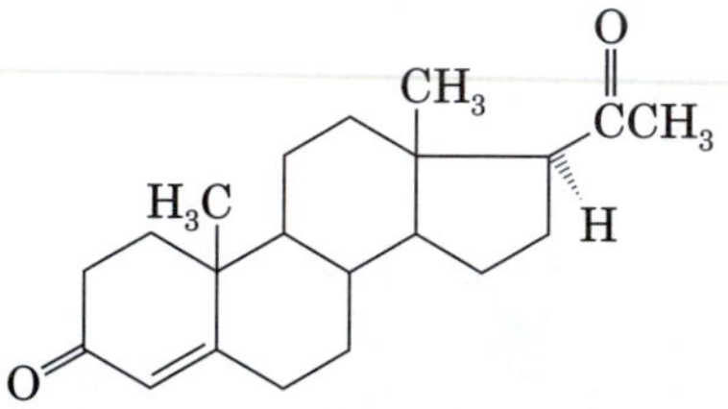

progesterone

a female hormone

Many aldehydes and ketones have distinctive odors. Aldehydes are generally pungent-smelling and ketones are sweet-smelling. For example, *trans*-cinnamaldehyde is the principal component of cinnamon oil, and the enantiomers of carvone are responsible for the odors of caraway seeds and spearmint.

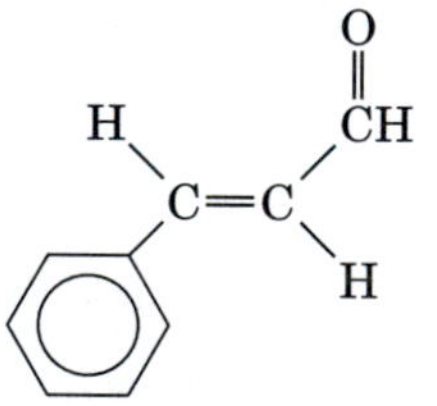

the chiral carbon

(*E*)-3-phenylpropenal
(*trans*-cinnamaldehyde)

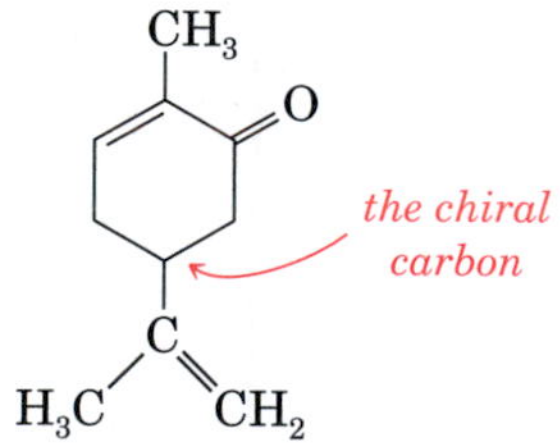

(+)-carvone: spearmint oil
(−)-carvone: caraway-seed oil

The carbonyl group of aldehydes and ketones consists of an sp^2-hybridized carbon atom joined to an oxygen atom by a σ bond and a π bond. (See Figure 2.21.) The σ bonds of the carbonyl carbon lie in a *plane,* with bond angles of approximately 120°. The π bond joining the C and the O lies above and below the plane of these σ bonds. The carbonyl group is *polar,* the electrons in the σ bond, and especially those in the π bond, being drawn toward the electronegative oxygen. The oxygen of the carbonyl group has *two pairs of unshared valence electrons.* All these structural features—the flatness, the π bond, the polarity, and the unshared electrons—contribute to the properties and the reactivity of the carbonyl group.

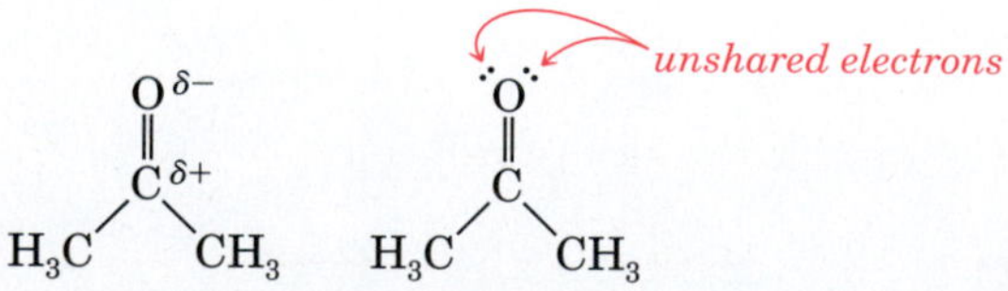

unshared electrons

Section 13.1
Nomenclature of Aldehydes and Ketones

In the IUPAC system, the name of an aldehyde is derived from the name of the parent alkane by changing the final **-e** to **-al.** No number is needed; the —CHO group always contains carbon 1.

$$
\underset{\text{CH}_3\overset{\displaystyle O}{\overset{\|}{\text{C}}}\text{H}}{} \qquad
\underset{\underset{\text{Cl}}{|}}{\text{CH}_3\text{CH}\overset{\displaystyle O}{\overset{\|}{\text{C}}}\text{H}} \qquad
\text{CH}_3\text{CH}{=}\text{CH}\overset{\displaystyle O}{\overset{\|}{\text{C}}}\text{H}
$$

IUPAC: ethanal   2-chloropropanal   2-butenal

Ketones are named by changing the **-e** of the alkane name to **-one.** The *-e* is retained in dione names. A number is used where necessary.

$$
\bigcirc{=}O \qquad
\text{CH}_3\overset{\displaystyle O}{\overset{\|}{\text{C}}}\text{CH}_2\text{CH}_2\text{CH}_3 \qquad
\text{CH}_3\overset{\displaystyle O}{\overset{\|}{\text{C}}}\text{CH}_2\overset{\displaystyle O}{\overset{\|}{\text{C}}}\text{CH}_3
$$

IUPAC: cyclohexanone   2-pentanone   2,4-pentanedione

Trivial names for the common aldehydes and ketones are widely used. Aldehydes are named after the parent carboxylic acids with the *-oic acid* or *-ic acid* ending changed to *-aldehyde.* Table 13.1 lists a few examples.

Propanone is usually called *acetone,* while the other simple ketones are sometimes named by a functional-group name. The alkyl or aryl groups attached to the carbonyl group are named, then the word *ketone* is added.

$$
\text{CH}_3\overset{\displaystyle O}{\overset{\|}{\text{C}}}\text{CH}_3 \qquad
\text{CH}_3\overset{\displaystyle O}{\overset{\|}{\text{C}}}\text{CH}_2\text{CH}_3 \qquad
(\text{CH}_3)_2\text{CH}\overset{\displaystyle O}{\overset{\|}{\text{C}}}\text{C}(\text{CH}_3)_3
$$

IUPAC: propanone   butanone   2,2,4-trimethyl-3-pentanone
trivial: acetone   methyl ethyl ketone   isopropyl *tert*-butyl ketone

Table 13.1   Trivial names for some carboxylic acids and aldehydes

Carboxylic acid		Aldehyde	
$\text{H}\overset{O}{\overset{\|}{\text{C}}}\text{OH}$	formic acid	$\text{H}\overset{O}{\overset{\|}{\text{C}}}\text{H}$	formaldehyde
$\text{CH}_3\overset{O}{\overset{\|}{\text{C}}}\text{OH}$	acetic acid	$\text{CH}_3\overset{O}{\overset{\|}{\text{C}}}\text{H}$	acetaldehyde
$\text{CH}_3\text{CH}_2\overset{O}{\overset{\|}{\text{C}}}\text{OH}$	propionic acid	$\text{CH}_3\text{CH}_2\overset{O}{\overset{\|}{\text{C}}}\text{H}$	propionaldehyde
$\text{CH}_3\text{CH}_2\text{CH}_2\overset{O}{\overset{\|}{\text{C}}}\text{OH}$	butyric acid	$\text{CH}_3\text{CH}_2\text{CH}_2\overset{O}{\overset{\|}{\text{C}}}\text{H}$	butyraldehyde
$\bigcirc{-}\overset{O}{\overset{\|}{\text{C}}}\text{OH}$	benzoic acid	$\bigcirc{-}\overset{O}{\overset{\|}{\text{C}}}\text{H}$	benzaldehyde

Other positions in a molecule in relation to the carbonyl group may be referred to by Greek letters. The carbon adjacent to the C=O group (or any other functional group for that matter) is called the **alpha** (α) carbon. The next carbon is **beta** (β), then **gamma** (γ), **delta** (δ), and so forth, according to the Greek alphabet. Occasionally, **omega** (ω), the last letter in the Greek alphabet, is used to designate the terminal carbon of a long chain, regardless of the actual length of the chain. The groups (or atoms) attached to an α carbon are called α groups; those attached to the β carbon are called β groups.

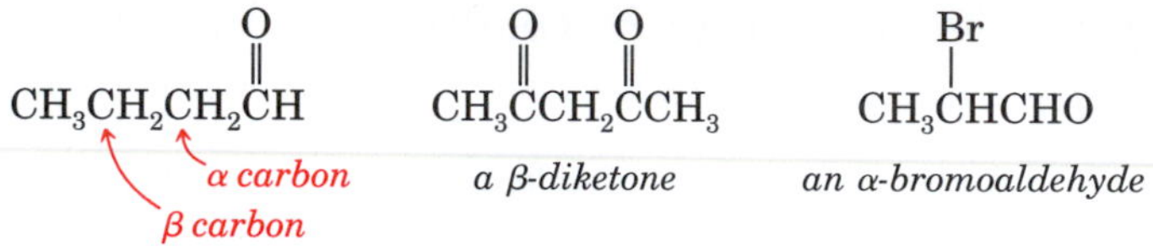

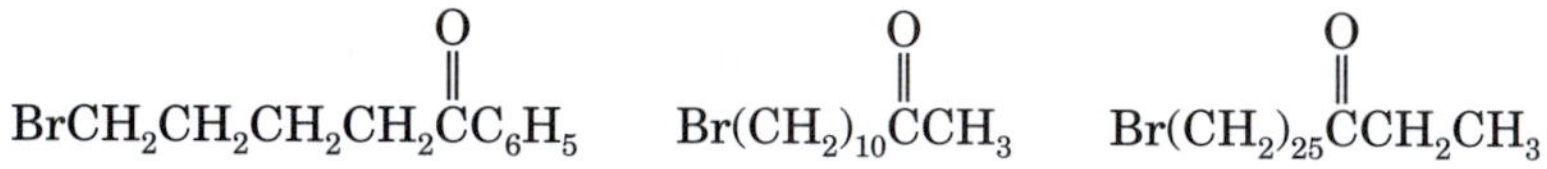

$$\underset{\substack{\\ \beta\ carbon \\ \alpha\ carbon}}{CH_3CH_2CH_2\overset{\displaystyle O}{\overset{\|}{C}}H} \qquad \underset{a\ \beta\text{-diketone}}{CH_3\overset{\displaystyle O}{\overset{\|}{C}}CH_2\overset{\displaystyle O}{\overset{\|}{C}}CH_3} \qquad \underset{an\ \alpha\text{-bromoaldehyde}}{CH_3\overset{\displaystyle Br}{\overset{|}{C}}HCHO}$$

$$\underset{all\ are\ \omega\text{-bromoketones}}{BrCH_2CH_2CH_2CH_2\overset{\displaystyle O}{\overset{\|}{C}}C_6H_5 \qquad Br(CH_2)_{10}\overset{\displaystyle O}{\overset{\|}{C}}CH_3 \qquad Br(CH_2)_{25}\overset{\displaystyle O}{\overset{\|}{C}}CH_2CH_3}$$

The Greek-letter designations may be used in the trivial names of carbonyl compounds, but *not in the IUPAC names* because this latter practice would be mixing two systems of nomenclature.

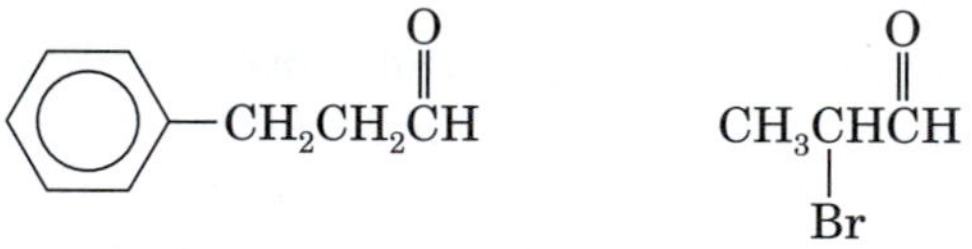

	$\langle\text{phenyl}\rangle-CH_2CH_2\overset{\displaystyle O}{\overset{\|}{C}}H$	$CH_3\overset{\displaystyle O}{\overset{\|}{C}}HCH$ with Br
IUPAC:	3-phenylpropanal	2-bromopropanal
trivial:	β-phenylpropionaldehyde	α-bromopropionaldehyde

STUDY PROBLEMS

13.1   Write formulas for:

(a) bromoacetone, a lachrymator that has been used as a war gas

(b) biacetyl (2,3-butanedione), used to give margarine a butter flavor

(c) 4-(4-hydroxyphenyl)-2-butanone, responsible for the flavor of raspberries

(d) geranial, (E)-3,7-dimethyl-2,6-octadienal, which is found in lemongrass oil and is used in perfumes and flavorings

13.2   Write an acceptable name for each of the following structures:

(a) 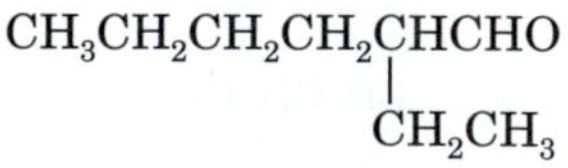 $CH_3CH_2CH_2CH_2\underset{\displaystyle CH_2CH_3}{CH}CHO$

(b) 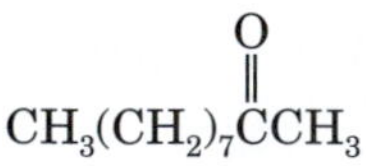 $CH_3(CH_2)_7\overset{\displaystyle O}{\overset{\|}{C}}CH_3$

(c) $O=\langle\text{ring}\rangle=O$ with CH_3

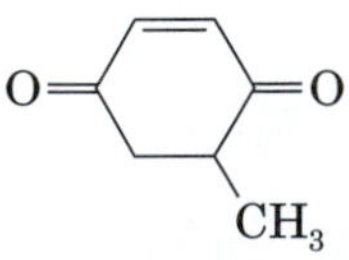

(d) 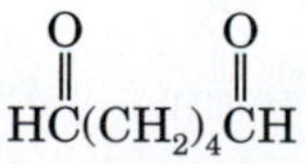 $H\overset{\displaystyle O}{\overset{\|}{C}}(CH_2)_4\overset{\displaystyle O}{\overset{\|}{C}}H$

(e) 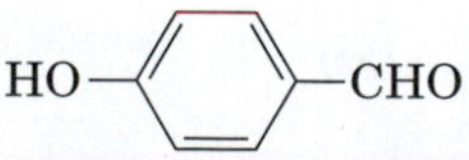 $HO-\langle\text{phenyl}\rangle-CHO$

(f) 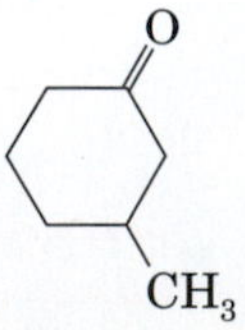 cyclohexanone ring with CH_3

13.3   Give formulas for: **(a)** a β-ketoaldehyde; **(b)** an α,β-unsaturated ketone; **(c)** an α-bromoaldehyde; **(d)** a β-hydroxyketone.

Section 13.2
Preparation of Aldehydes and Ketones

In the laboratory, the most common way of synthesizing a simple aldehyde or ketone is by the *oxidation of an alcohol*. Aryl ketones can be prepared by *Friedel–Crafts acylation reactions*. General equations for these reactions are shown in Table 13.2.

$$CH_3(CH_2)_5CH_2OH + CrO_3 \cdot 2\,N\langle\bigcirc\rangle \longrightarrow CH_3(CH_2)_5\overset{O}{\overset{\|}{C}}H$$

1-heptanol   heptanal (84%)

menthol   menthone (84%)

Table 13.2   Summary of laboratory syntheses of aldehydes and ketones[a]

Reaction	Section Reference
Aldehydes:	
$RCH_2OH \xrightarrow[\text{pyridine}]{\text{CrO}_3} RCHO$	7.8C
1° alcohol	
$R\overset{O}{\overset{\|}{C}}Cl \xrightarrow[(2)\ H_2O,\ H^+]{(1)\ LiAlH[OC(CH_3)_3]_3} RCHO$	15.3C
acid chloride	
Ketones:	
$R_2CHOH \xrightarrow[H^+]{H_2CrO_4} R_2C{=}O$	7.8C
2° alcohol	
$RC{\equiv}CR \xrightarrow[Hg^{2+}]{H_2O,\ H^+} RC\overset{O}{\overset{\|}{C}}CH_2R$	10.7
alkyne	
$R\overset{O}{\overset{\|}{C}}Cl \xrightarrow{R'_2CuLi} R\overset{O}{\overset{\|}{C}}R'$	15.3C
acid chloride	
Aryl Ketones:	11.8E
$\langle\bigcirc\rangle + R\overset{O}{\overset{\|}{C}}Cl \xrightarrow{AlCl_3} \langle\bigcirc\rangle{-}\overset{O}{\overset{\|}{C}}R$	

[a] Polyfunctional aldehydes and ketones can be prepared by condensation and alkylation reactions, which will be discussed in Chapter 17.

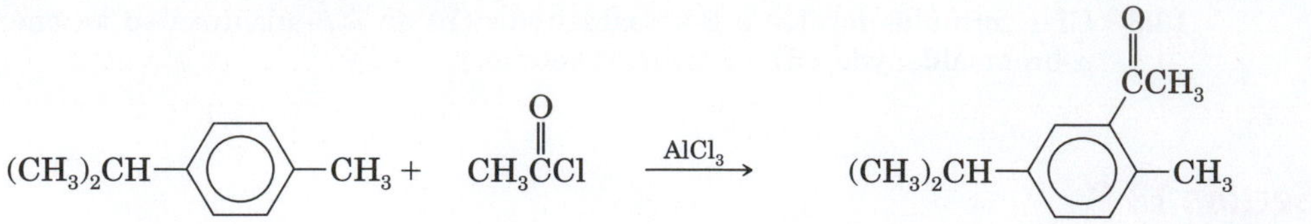

$$(CH_3)_2CH-\!\!\bigcirc\!\!-CH_3 + CH_3\overset{\overset{\displaystyle O}{\|}}{C}Cl \xrightarrow{\text{AlCl}_3} (CH_3)_2CH-\!\!\bigcirc\!\!-CH_3$$

p-isopropyltoluene   acetyl chloride   5-isopropyl-2-methylacetophenone (55%)

One of the more important aldehydes, *formaldehyde,* which is used as a reagent and as a preservative for biological specimens, is a gas. However, it is conveniently shipped or stored in water solution (formalin = 37% formaldehyde and 7–15% methanol in H_2O) or as a solid polymer or trimer. Heating any one of these preparations yields the gaseous formaldehyde.

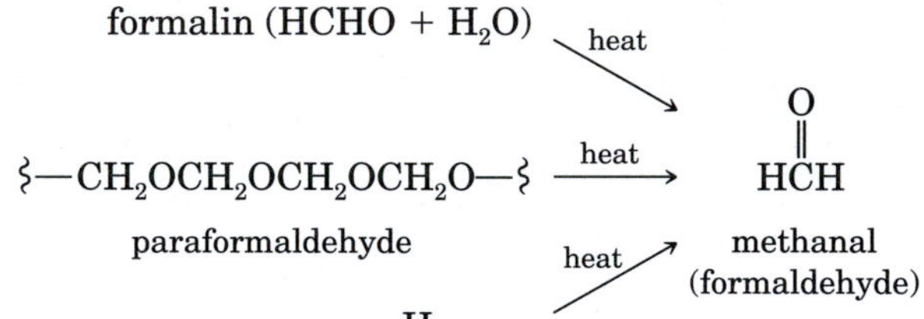

formalin ($HCHO + H_2O$)

$\{-CH_2OCH_2OCH_2OCH_2O-\}$

paraformaldehyde

$$\overset{\overset{\displaystyle O}{\|}}{H\overset{}{C}H}$$

methanal (formaldehyde)

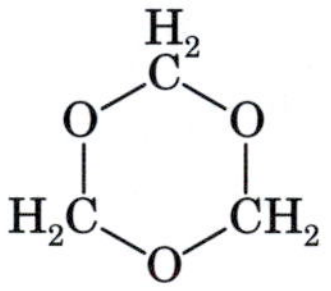

trioxane
mp 62°C

STUDY PROBLEM

13.4 Write equations showing how each of the following aldehydes or ketones could be prepared by an oxidation reaction:

(a) $CH_3CH_2CH_2CH_2\overset{\overset{\displaystyle O}{\|}}{C}H$ (b)

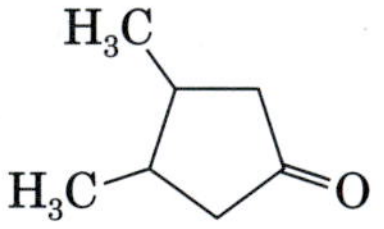

(c) 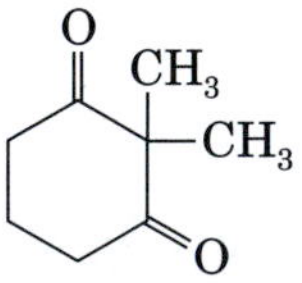   (d) $(CH_3)_2CHCHO$

Section 13.3

Spectral Properties of Aldehydes and Ketones

A. Infrared Spectra

The infrared spectrum is useful in the detection of a carbonyl group in a ketone or an aldehyde. (Characteristic absorption bands are listed in Table 13.3.) However, carbonyl groups are also found in other compounds (carboxylic acids, esters, and so forth). For this reason, the fact that a carbonyl

Table 13.3 **Characteristic infrared absorption of aldehydes and ketones**		
	Position of Absoprtion[a]	
Type of Vibration	cm^{-1}	μm
Aldehydes:		
C—H stretching of —CHO	2700–2900	3.45–3.7
C=O stretching	1700–1740	5.7–5.9
Ketones:		
C=O stretching	1660–1750	5.7–6.0

[a] Other substituents or ring strain cause the carbonyl absorption to fall outside the listed ranges.

group is present does not mean that an unknown is necessarily an aldehyde or a ketone.

For aldehydes, corroborating evidence can be found in both the infrared and NMR spectra because of the unique absorption by the aldehyde hydrogen. Ketones, unfortunately, cannot be positively identified by spectral methods. The usual procedure is to eliminate the other carbonyl compounds as possibilities. If a carbonyl compound is not an aldehyde, carboxylic acid, ester, amide, etc., it is probably a ketone.

The C=O absorption of both aldehydes and ketones appears somewhere near 1700 cm^{-1} (about 5.8 μm). If the carbonyl group is in conjugation with a double bond or benzene ring, the position of absorption is shifted to a slightly lower frequency (about 1675 cm^{-1}, or 6 μm, for ketones). Figure 13.1 shows the infrared spectra of cyclohexanone (nonconjugated) and 2-cyclohexenone (conjugated).

The CH stretching of the aldehyde group, which shows absorption just to the right of aliphatic CH absorption, is characteristic of an aldehyde. Usually two peaks appear in this region. The two CH peaks of the aldehyde are clearly evident in the spectrum of butanal in Figure 13.2, but the peak closer to the aliphatic CH absorption is often obscured by that absorption.

STUDY PROBLEM

13.5 Match the infrared spectrum in Figure 13.3 to one of the following structures:

$$\text{(a)} \ C_6H_5\overset{\overset{\displaystyle O}{\|}}{C}CH_2CH_3 \qquad \text{(b)} \ CH_3(CH_2)_2CH{=}CH\overset{\overset{\displaystyle O}{\|}}{C}H \qquad \text{(c)} \ C_6H_5CH_2\overset{\overset{\displaystyle O}{\|}}{C}H$$

B. ¹H NMR Spectra

The molecular magnetic field induced by the electrons in a carbonyl group has a profound effect on the NMR absorption of the aldehyde proton. You will recall from Section 9.7B that NMR absorption for an aldehyde proton is shifted far downfield ($\delta 9$–10). This large shift arises from the additive effects of both anisotropic deshielding by the π electrons and inductive deshielding by the electronegative oxygen of the carbonyl group.

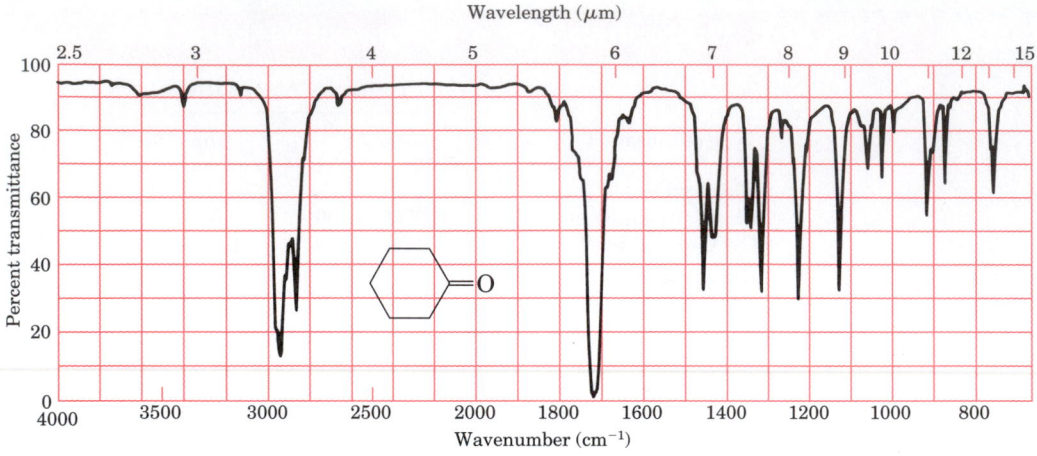

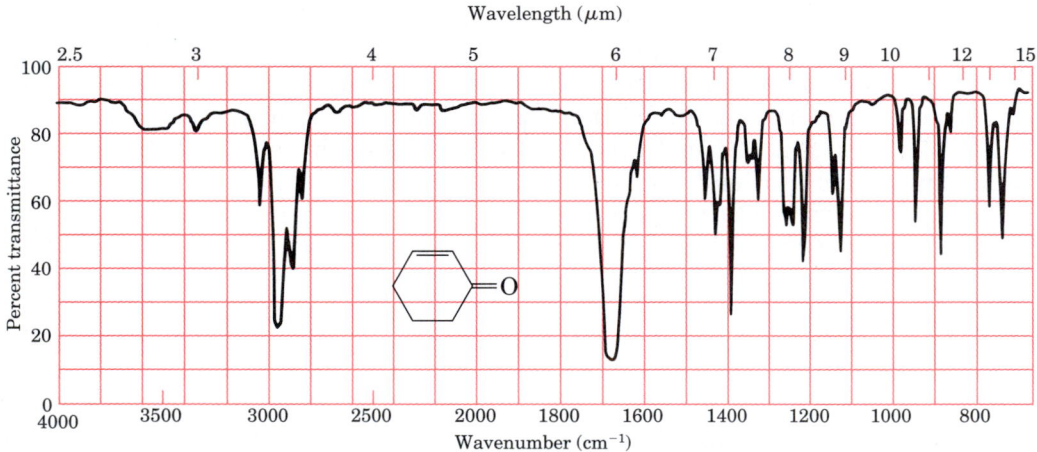

Figure 13.1 Infrared spectra of cyclohexanone and 2-cyclohexenone, illustrating the slight shift of C=O absorption to lower frequency (longer wavelength) by conjugation.

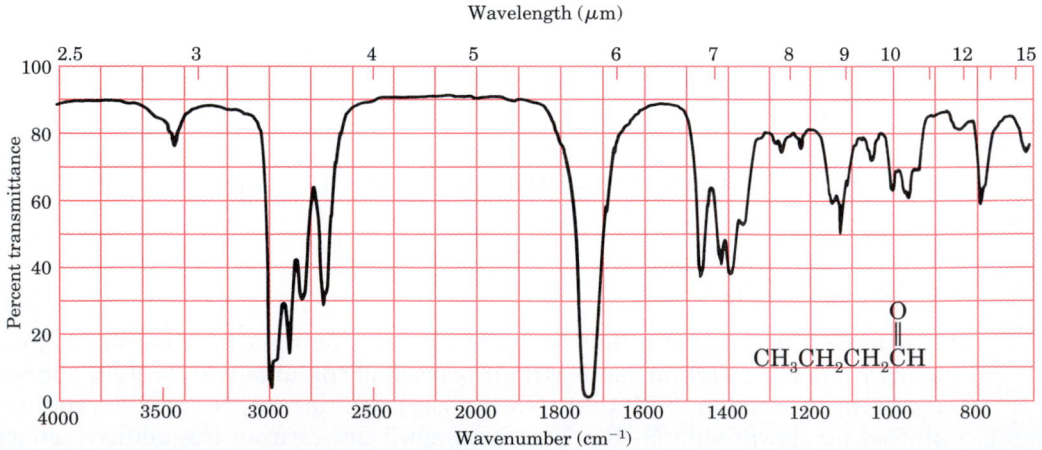

Figure 13.2 Infrared spectrum of butanal.

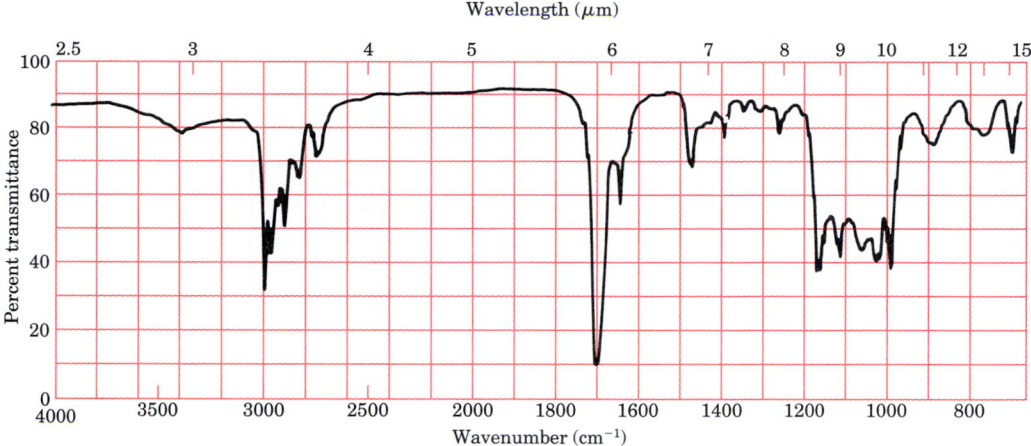

Figure 13.3 Infrared spectrum for Problem 13.5.

The α hydrogens of either aldehydes or ketones are not affected to such a large extent by the carbonyl group. The NMR absorption for the α protons ($\delta 2.1$–2.6) appears slightly downfield from that of other CH absorption (about $\delta 1.5$) because of electron withdrawal by the electronegative oxygen atom. The effects of this deshielding by the inductive effect are evident in the NMR spectra of butanal (Figure 13.4) and 1-phenyl-2-propanone (Figure 13.5). In an aldehyde, the splitting of the aldehyde proton can sometimes be used to determine the number of α hydrogens. The spectrum of butanal shows a triplet for the —CHO proton, an indication of two α hydrogens.

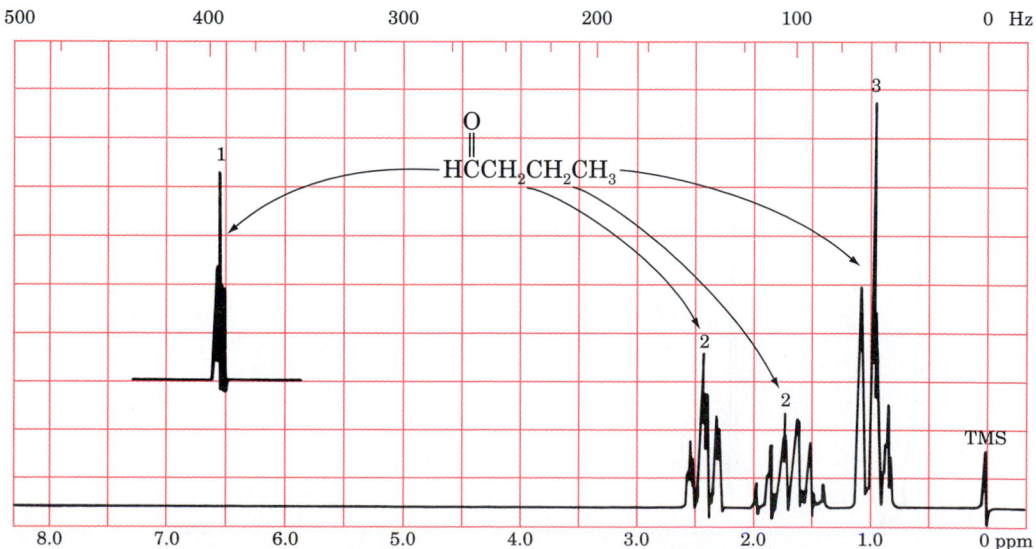

Figure 13.4 ^{1}H NMR spectrum of butanal, showing the relative chemical shifts of α, β, and γ protons and the aldehyde proton (at $\delta 9.74$).

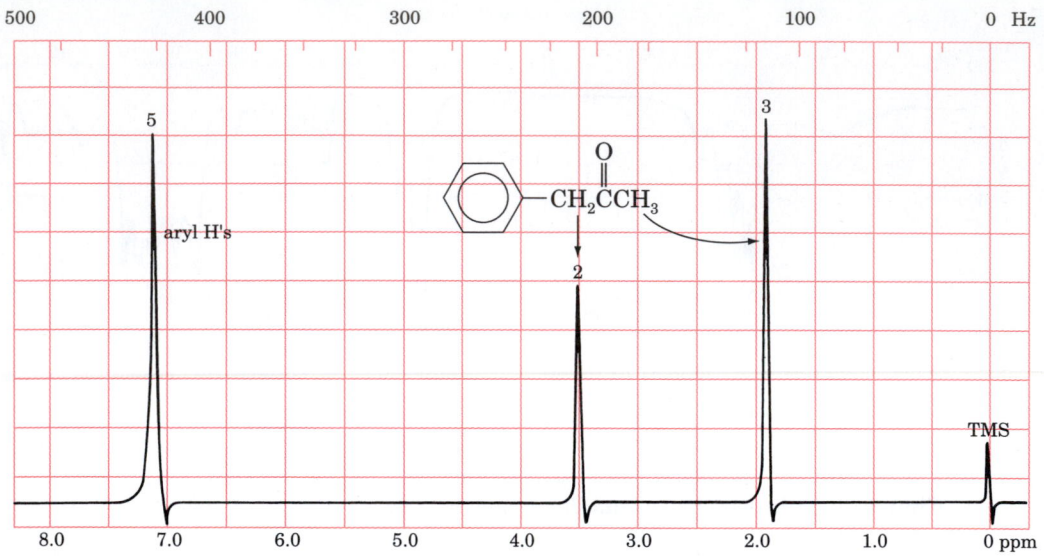

Figure 13.5 ¹H NMR spectrum of 1-phenyl-2-propanone.

STUDY PROBLEM

13.6 A compound with the molecular formula C_7H_6O has the ¹H NMR spectrum shown in Figure 13.6. What is the structure of this compound?

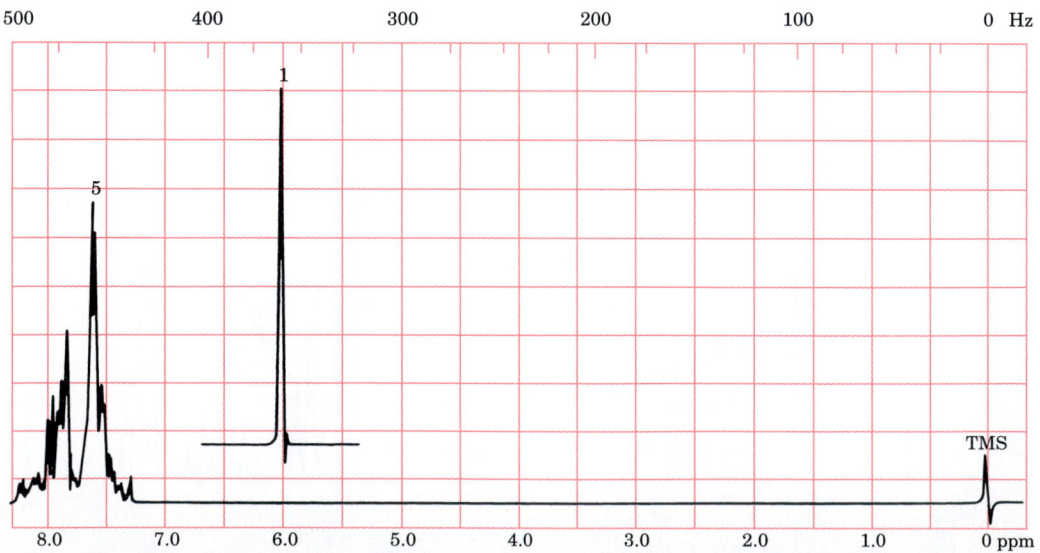

Figure 13.6 ¹H NMR spectrum for Problem 13.6. (The offset absorption is at δ10.0.)

Addition Reactions of Aldehydes and Ketones

Isolated carbon–carbon double bonds are nonpolar. For reaction, an electrophile is generally needed to attack the π-bond electrons. However, the carbon–oxygen double bond is polar even without electrophilic attack. A carbonyl compound may be attacked either by a nucleophile or by an electrophile.

Many reactions of carbonyl groups involve an initial protonation of the oxygen. This protonation enhances the positive charge of the carbonyl carbon so that this carbon is more easily attacked by weaker nucleophiles.

major contributor   *minor contributor*

resonance structures for a protonated carbonyl group

Like alkenes, aldehydes and ketones undergo addition of reagents to the π bond. Many of these addition reactions, especially those with weak nucleophiles, are acid-catalyzed for the foregoing reason.

The relative reactivities of aldehydes and ketones in addition reactions may be attributed partly to the *amount of positive charge on the carbonyl carbon*. A greater positive charge means a higher reactivity. If this partial positive charge is dispersed throughout the molecule, then the carbonyl compound is more stable and less reactive.

The carbonyl group is stabilized by adjacent alkyl groups, which are electron-releasing. A ketone, with two R groups, is about 7 kcal/mol more stable than an aldehyde, with only one R group. Formaldehyde, with no alkyl groups, is the most reactive of the aldehydes and ketones.

a ketone   *an aldehyde*   *formaldehyde*

increasing reactivity

SAMPLE PROBLEM

List the following aldehydes in terms of increasing reactivity: CH_3CHO, $ClCH_2CHO$, Cl_2CHCHO, Cl_3CCHO.

Solution

Cl is *electron-withdrawing*. The carbonyl carbon becomes *increasingly positive* and *increasingly reactive* as more Cl atoms are added to the α carbon. Therefore, the order of reactivity is as already shown, with CH_3CHO being the least reactive and Cl_3CCHO being the most reactive.

$$CH_3 \rightarrow \overset{\delta+}{CHO} \qquad Cl_3C \leftarrow \overset{\delta+}{CHO}$$

stabilized   *destabilized*

Steric factors also play a role in the relative reactivities of aldehydes and ketones. An addition reaction of the carbonyl group leads to an increase in steric hindrance around the carbonyl carbon.

Bulky groups around the carbonyl group lead to more steric hindrance in the product (and in the transition state). The product is of higher energy because of steric repulsions. A more hindered ketone is therefore less reactive than an aldehyde or a less hindered ketone. The lack of steric hindrance is another reason that formaldehyde is more reactive than other aldehydes.

$$\underset{\text{an ethyl ketone}}{CH_3CH_2CH_2\overset{O}{\overset{\|}{C}}CH_2CH_3} \qquad \underset{\text{a methyl ketone}}{CH_3CH_2CH_2\overset{O}{\overset{\|}{C}}CH_3} \qquad \underset{\text{an aldehyde}}{CH_3CH_2CH_2\overset{O}{\overset{\|}{C}}H}$$

increasing reactivity →

A. Reaction with Water

Water can add to a carbonyl group to form a 1,1-diol, called a **gem-diol,** or **hydrate.** The reaction is reversible, and the equilibrium generally lies on the carbonyl side.

General:

$$\underset{\substack{\text{an aldehyde} \\ \text{or ketone}}}{R-\overset{O}{\overset{\|}{C}}-R} + H_2O \underset{H^+}{\rightleftarrows} \underset{\substack{\text{a hydrate} \\ \text{(two OH's on C)}}}{R-\overset{OH}{\underset{OH}{\overset{|}{C}}}-R}$$

$$CH_3-\overset{\overset{\displaystyle O}{\|}}{C}-CH_3 + H_2O \underset{\xrightarrow{\hspace{1cm}}}{\overset{H^+}{\xleftarrow{\hspace{1cm}}}} CH_3-\overset{\overset{\displaystyle OH}{|}}{\underset{\underset{\displaystyle OH}{|}}{C}}-CH_3$$

acetone

2,2-propanediol
(acetone hydrate)

Stable hydrates are known, but they are the exception rather than the rule. *Chloral hydrate* (a hypnotic and chief ingredient of the "Mickey Finn," or knock-out drops) is an example of a stable solid hydrate. *Formalin* contains a stable aqueous hydrate of formaldehyde.

$$Cl_3\overset{\overset{\displaystyle O}{\|}}{C}H + H_2O \underset{\xrightarrow{\hspace{1cm}}}{\overset{H^+}{\xleftarrow{\hspace{1cm}}}} Cl_3CCH(OH)_2$$

chloral   chloral hydrate

$$H\overset{\overset{\displaystyle O}{\|}}{C}H \ + H_2O \underset{\xrightarrow{\hspace{1cm}}}{\overset{H^+}{\xleftarrow{\hspace{1cm}}}} HCH(OH)_2$$

formaldehyde   *in formalin*

Both formaldehyde and chloral are more reactive than most other aldehydes or ketones because the carbonyl carbon in each has a fairly large amount of positive charge. In formaldehyde, there are no alkyl groups to help disperse the positive charge. In chloral, the strongly electron-withdrawing Cl_3C- group enhances the positive charge by removing electron density.

Let us compare the equilibrium constants for the hydration reactions of chloral (with a more positive carbonyl carbon) and acetone (with the positive charge more dispersed). The equilibrium constants differ by a factor of 10^7!

$$Cl_3\overset{\overset{\displaystyle O}{\|}}{C}H + H_2O \underset{\xrightarrow{\hspace{1cm}}}{\overset{H^+}{\xleftarrow{\hspace{1cm}}}} Cl_3\overset{\overset{\displaystyle OH}{|}}{\underset{\underset{\displaystyle OH}{|}}{C}}CH \qquad K = \frac{[Cl_3CCH(OH)_2]}{[Cl_3CCHO][H_2O]} = 30{,}000$$

$$CH_3\overset{\overset{\displaystyle O}{\|}}{C}CH_3 + H_2O \underset{\xrightarrow{\hspace{1cm}}}{\overset{H^+}{\xleftarrow{\hspace{1cm}}}} CH_3\overset{\overset{\displaystyle OH}{|}}{\underset{\underset{\displaystyle OH}{|}}{C}}CH_3 \qquad K = \frac{[CH_3C(OH)_2CH_3]}{[CH_3COCH_3][H_2O]} = 0.002$$

STUDY PROBLEM

13.7 Write the equation for the formation of the hydrate of each of the following aldehydes. Which hydrate would be more stable?

(a) Br_2CHCHO **(b)** Br_2CHCH_2CHO

B. Reactions with Alcohols

Like water, an alcohol can add to a carbonyl group. In most cases, the equilibrium lies on the aldehyde or ketone side of the equation, just as in the reaction with water.

The product of addition of *one* molecule of an alcohol to an aldehyde is called a **hemiacetal,** while the product of addition of *two* molecules of alcohol (with the loss of H_2O) is called an **acetal.** (**Hemiketal** and **ketal** are the corresponding terms used for ketone products.) All these reactions are catalyzed by a trace of strong acid.

General:

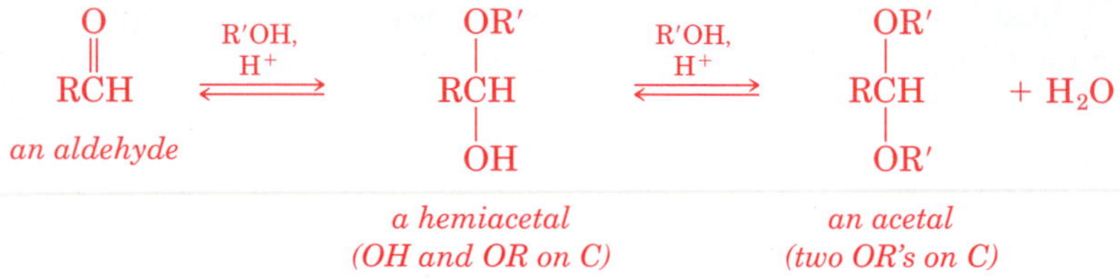
O
RCH
an aldehyde
R′OH, H+
OR′
RCH
OH
a hemiacetal (OH and OR on C)
R′OH, H+
OR′
RCH
OR′
+ H2O
an acetal (two OR's on C)

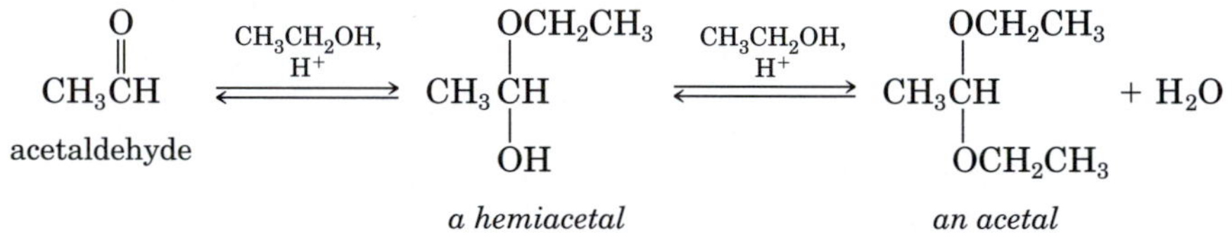
O
CH3CH
acetaldehyde
CH3CH2OH, H+
OCH2CH3
CH3 CH
OH
a hemiacetal
CH3CH2OH, H+
OCH2CH3
CH3CH
OCH2CH3
+ H2O
an acetal

Sample Problem

Give the structures of the organic compounds present in a methanol solution of cyclohexanone that contains a trace of HCl.

Solution

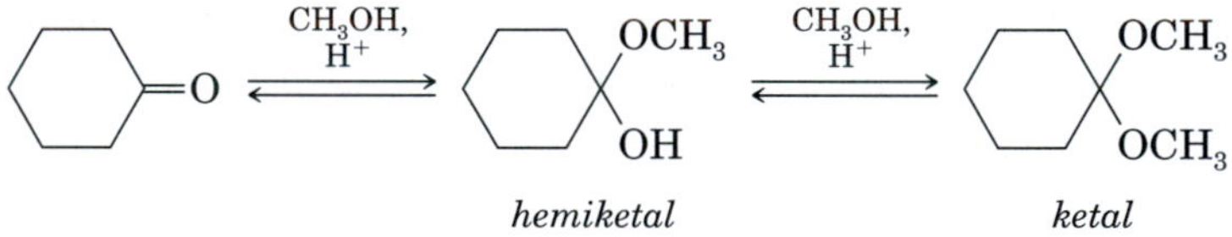
O
CH3OH, H+
OCH3
OH
hemiketal
CH3OH, H+
OCH3
OCH3
ketal

Study Problems

13.8 Which of the following structures contains a hemiacetal or hemiketal group, and which contains an acetal or ketal group? Circle and identify each group.

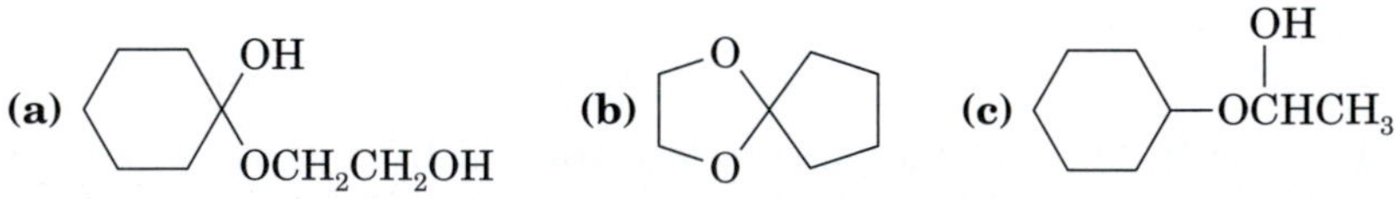
(a)
OH
OCH2CH2OH
(b)
O
O
(c)
OH
OCHCH3

13.9 Give the structures of the alcohol and the aldehyde or ketone that are needed to prepare each of the compounds shown in the preceding problem.

The mechanism for the reversible reaction of an aldehyde or a ketone with an alcohol is typical of the mechanisms for many acid-catalyzed addition reactions of carbonyl compounds: a series of protonations and deprotonations of oxygen-containing groups.

Protonation:

$$
\underset{\text{an aldehyde}}{\overset{\ddot{\text{O}}:}{\underset{\text{RCH}}{\parallel}}} \quad + \text{ H}^+ \quad \rightleftharpoons \quad \left[\underset{\text{a protonated aldehyde}}{\overset{+\atop\ddot{\text{O}}\text{H}}{\underset{\text{RCH}}{\parallel}}} \right]
$$

Attack of R'OH:

$$
\left[\underset{\text{R}-\overset{}{\underset{}{\text{C}}}-\text{H}}{\overset{+\atop\ddot{\text{O}}\text{H}}{}} \right] + \text{R}'\ddot{\text{O}}\text{H} \rightleftharpoons \left[\underset{\underset{+}{\overset{}{\text{H}\ddot{\text{O}}\text{R}'}}}{\overset{:\ddot{\text{O}}\text{H}}{\text{R}-\overset{}{\underset{}{\text{C}}}-\text{H}}} \right] \overset{-\text{H}^+}{\rightleftharpoons} \underset{:\ddot{\text{O}}\text{R}'}{\overset{:\ddot{\text{O}}\text{H}}{\text{R}-\overset{}{\underset{}{\text{C}}}-\text{H}}}
$$

a protonated   a hemiacetal
hemiacetal

In the mechanism for acetal formation from the hemiacetal, again protonation and deprotonation, along with loss of water, are the major reaction steps. Acetal formation from a hemiacetal is therefore a two-step *substitution* of an OR group for an OH group.

Protonation and loss of water:

$$
\underset{\text{OR}'}{\overset{:\ddot{\text{O}}\text{H}}{\text{R}-\overset{}{\underset{}{\text{C}}}-\text{H}}} \overset{\text{H}^+}{\rightleftharpoons} \left[\underset{:\ddot{\text{O}}\text{R}'}{\overset{+\ddot{\text{O}}\text{H}_2}{\text{R}-\overset{}{\underset{}{\text{C}}}-\text{H}}} \right] \overset{-\text{H}_2\text{O}}{\rightleftharpoons} \left[\underset{+\ddot{\text{O}}\text{R}'}{\overset{}{\text{R}-\overset{\parallel}{\text{C}}-\text{H}}} \right]
$$

a hemiacetal

Attack of R'OH:

$$
\left[\underset{+\ddot{\text{O}}\text{R}'}{\overset{}{\text{R}-\overset{\parallel}{\text{C}}-\text{H}}} \right] \overset{\text{R}'\ddot{\text{O}}\text{H}}{\rightleftharpoons} \left[\underset{:\ddot{\text{O}}\text{R}'}{\overset{+\atop\text{H}\ddot{\text{O}}\text{R}'}{\text{R}-\overset{}{\underset{}{\text{C}}}-\text{H}}} \right] \overset{-\text{H}^+}{\rightleftharpoons} \underset{:\ddot{\text{O}}\text{R}'}{\overset{:\ddot{\text{O}}\text{R}'}{\text{R}-\overset{}{\underset{}{\text{C}}}-\text{H}}}
$$

an acetal

In the equilibrium between an aldehyde, a hemiacetal, and an acetal, the aldehyde is generally favored. In an equilibrium mixture, we would usually find a large amount of aldehyde and only small amounts of hemiacetal and acetal. There is one important exception to this generality. A molecule that has an OH group γ or δ (1,4 or 1,5) to an aldehyde or ketone carbonyl group undergoes an intramolecular reaction to form a five- or six-membered hemiacetal ring. *These cyclic hemiacetals are favored over the open-chain aldehyde forms.*

$$
\begin{array}{c}
\overset{⑤}{\text{H}_2\text{C}} — \overset{\ddot{\text{O}}\text{H}}{\underset{①}{}} \quad \text{H} \\
\overset{④}{\text{CH}_2} \qquad \text{C} \\
\text{H}_2\text{C} — \underset{②}{\text{CH}_2} \quad \ddot{\text{O}}: \\
\overset{③}{}
\end{array}
\overset{\text{H}^+}{\rightleftharpoons}
\begin{array}{c}
\text{H}_2\text{C} — \text{O} \\
\text{CH}_2 \qquad \text{CHOH} \\
\text{H}_2\text{C} — \text{CH}_2
\end{array}
$$

the hemiacetal carbon has OH and OR

94% at equilibrium

The reason that cyclic hemiacetals are important is that glucose and other sugars contain hydroxyl groups γ and δ to carbonyl groups. Sugars, therefore, form cyclic hemiacetals in water solution. This topic will be discussed in Section 23.4.

glucose

a sugar

favored

the hemiacetal carbon

In most cases, a hemiacetal cannot be isolated. Acetals, however, are stable in nonacidic solution and can be isolated. (In acidic solution, of course, they are in equilibrium with their aldehydes.) If an acetal is the desired product from reaction of an aldehyde and an alcohol, an *excess of alcohol* is used to drive the series of reaction steps to that product. *Removing water* as it is formed also helps drive the reversible reactions to the acetal. Best results in this type of reaction are obtained when the acetal is *cyclic*:

benzaldehyde   2,3-butanediol   *a cyclic acetal* (96%)

If a desired reaction can be carried out under alkaline conditions, acetal and ketal groups are effective blocking groups for aldehydes and ketones. For example, after changing an aldehyde group to an acetal, we can oxidize a double bond in the same molecule without oxidizing the aldehyde to a carboxylic acid.

Blocking:

propenal
(acrolein)   *an acetal*

Oxidation of double bond and regeneration of aldehyde:

2,3 -dihydroxypropanal

C. Reaction with Hydrogen Cyanide

Hydrogen cyanide (bp 26°) can be considered to be either a gas or a liquid with a low boiling point. In typical laboratory operations, it is used as a gas, but with special apparatus it can be used as a liquid (and, in some cases, even as a solvent). Often, HCN is generated directly in a reaction mixture from KCN or NaCN and a strong acid. Hydrogen cyanide is toxic and is particularly insidious because some people can detect its odor only at levels that may be lethal.

Like water and alcohols, hydrogen cyanide can add to the carbonyl group of an aldehyde or a ketone. The product in either case is referred to as a **cyanohydrin.**

$$
\underset{\text{acetaldehyde}}{CH_3\overset{\displaystyle O}{\overset{\|}{C}}H} + HCN \underset{}{\overset{CN^-}{\rightleftarrows}} \underset{\text{acetaldehyde cyanohydrin (75\%)}}{CH_3\overset{\displaystyle OH}{\underset{}{\overset{|}{C}}}H-CN}
$$

General:

$$
\underset{\substack{\text{an aldehyde} \\ \text{or ketone}}}{R-\overset{\displaystyle O}{\overset{\|}{C}}-R} + HCN \overset{CN^-}{\rightleftarrows} \underset{\text{a cyanohydrin}}{R-\overset{\displaystyle OH}{\underset{\displaystyle CN}{\overset{|}{\underset{|}{C}}}}-R}
$$

Hydrogen cyanide does not add directly to a carbonyl group. Successful addition requires slightly alkaline reaction conditions such as those found in a NaCN–HCN buffer solution. In alkaline solution, the concentration of cyanide ion is increased, and addition proceeds by nucleophilic attack of CN⁻ on the carbonyl group. Although weak nucleophiles (such as H_2O and ROH) require acid catalysis for addition to the carbonyl group, the strongly nucleophilic CN⁻ does not require a catalyst.

$$
HCN: + {}^-\!:\!\ddot{O}H \rightleftarrows H_2\ddot{O}: + {}^-\!:\!CN:
$$

$$
R-\overset{\displaystyle \ddot{O}:}{\overset{\|}{C}}-H \rightleftarrows R-\overset{\displaystyle :\ddot{O}:^-}{\underset{\displaystyle CN:}{\overset{|}{\underset{|}{C}}}}-H \overset{HCN}{\rightleftarrows} R-\overset{\displaystyle :\ddot{O}H}{\underset{\displaystyle CN:}{\overset{|}{\underset{|}{C}}}}-H + {}^-\!:\!CN:
$$

Cyanohydrins are useful synthetic intermediates. For example, the CN group can be hydrolyzed to a carboxyl group or converted to an ester group. Also, the —OH group of a cyanohydrin is far more reactive than that of an ordinary alcohol and can be replaced by ammonia to yield an amino group. This latter property finds use in the synthesis of amino acids.

$$
\underset{}{CH_3\overset{\displaystyle O}{\overset{\|}{C}}CH_3} \xrightarrow[CN^-]{HCN} CH_3-\overset{\displaystyle OH}{\underset{\displaystyle CH_3}{\overset{|}{\underset{|}{C}}}}-CN \xrightarrow[H_2SO_4]{CH_3OH} CH_2=\overset{\displaystyle O}{\underset{\displaystyle CH_3}{\overset{}{\underset{|}{C}}}}-\overset{\|}{C}OCH_3
$$

methyl methacrylate

used to prepare methacrylate polymers

$$
\underset{}{CH_3\overset{\displaystyle O}{\overset{\|}{C}}H} \xrightarrow[NH_4Cl]{NaCN} \underset{\displaystyle OH}{CH_3\overset{}{\underset{|}{C}}HCN} \xrightarrow{NH_3} \underset{\displaystyle NH_2}{CH_3\overset{}{\underset{|}{C}}HCN} \xrightarrow[H_2O]{HCl} \underset{\displaystyle NH_2}{CH_3\overset{\displaystyle O}{\underset{}{C}}H\overset{\|}{C}OH}
$$

alanine (60%)

an amino acid

The millipede *Apheloria corrugata* carries its own poison-gas generator in the form of *mandelonitrile,* a cyanohydrin stored in its defensive glands. When the millipede is attacked, the cyanohydrin is mixed with an enzyme that causes a rapid dissociation to a mixture of benzaldehyde and HCN, which is squirted on the predator to ward off the attack. A single millipede can emit enough HCN to kill a mouse! It is interesting to note that mandelonitrile, benzaldehyde, and HCN all have the odor of bitter almonds, despite their disparity in structure.

$$\underset{\text{mandelonitrile}}{\overset{\overset{\text{OH}}{|}}{C_6H_5CHCN}} \xrightarrow{\text{enzyme}} \underset{\text{benzaldehyde}}{\overset{\overset{O}{\|}}{C_6H_5CH}} + HCN$$

In plants of the genus *Prunus* (which includes plums, apricots, cherries, and peaches), cyanohydrins are biosynthesized and stored as sugar derivatives in the kernels of the pits. Amygdalin and laetrile are the best known of these cyanohydrins. (These two compounds are closely related structurally; indeed, amygdalin is often sold as laetrile.) Because these cyanohydrins can be hydrolyzed enzymatically to HCN, the pits of cherries and other *Prunus* species should not be eaten in quantity.

amygdalin

laetrile

("*lae*vorotatory glycosidic ni*trile*")

D. Reaction with Grignard Reagents

The reaction of a Grignard reagent with a carbonyl compound is another example of a nucleophilic addition to the positive carbon of a carbonyl group. However, the addition of a Grignard reagent is *not* a reversible reaction. (Why not?)

Reaction of a Grignard reagent with an aldehyde or a ketone provides an excellent method for the synthesis of alcohols and was discussed in that context earlier (Section 7.3D). The reaction sequence consists of two separate steps: (1) the reaction of the Grignard reagent with the carbonyl compound, and (2) hydrolysis of the resulting magnesium alkoxide to yield the alcohol.

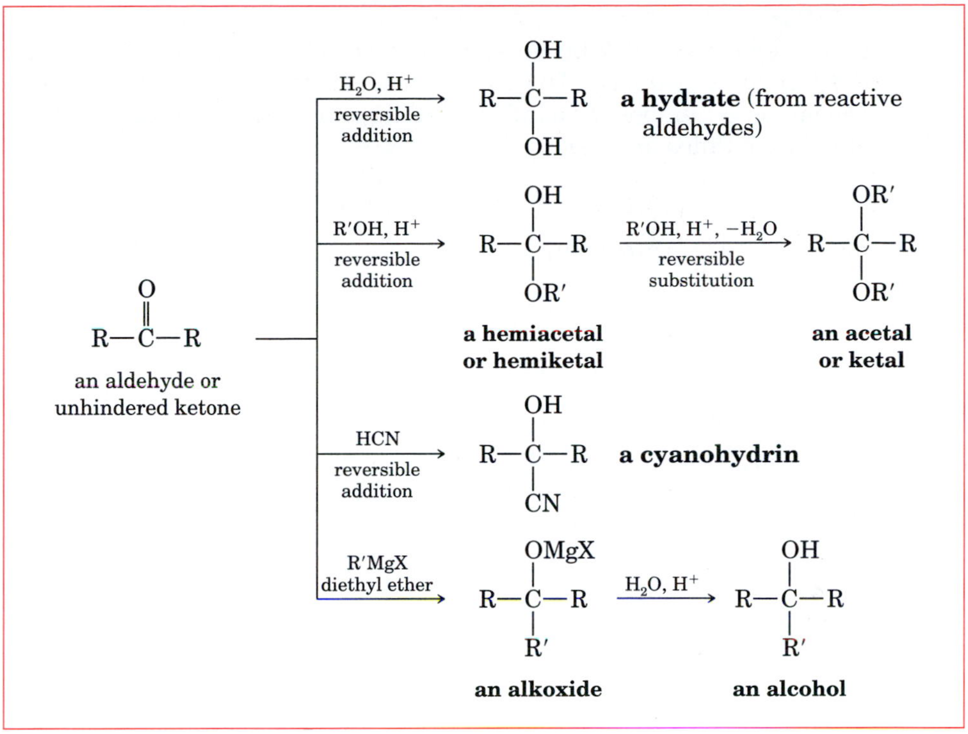

Figure 13.7 Summary of the addition reactions of aldehydes and ketones.

Recall that the Grignard reaction of formaldehyde yields a *primary alcohol*. Other aldehydes yield *secondary alcohols*. Ketones yield *tertiary alcohols*.

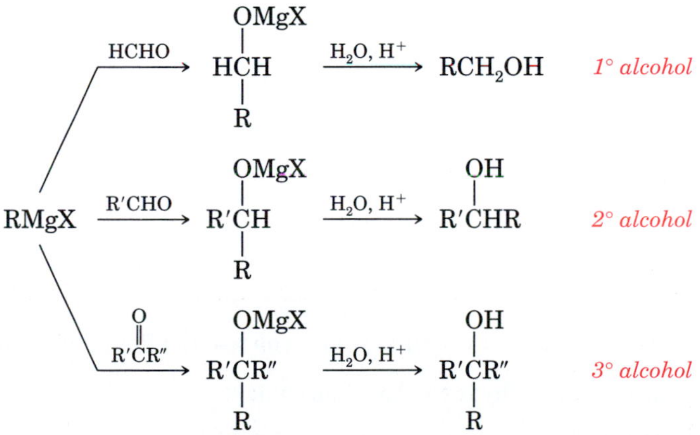

Figure 13.7 summarizes the addition reactions of aldehydes and ketones.

SAMPLE PROBLEMS

Suggest *two* synthetic routes to 2-butanol from an aldehyde or ketone and a Grignard reagent.

Solution

In the synthesis of 2-butanol, either (1) CH_3MgX and CH_3CH_2CHO or (2) CH_3CH_2MgX and CH_3CHO can be used. In the laboratory, the choice would depend on a number of factors, including availability and cost of the appropriate alkyl halides and aldehydes.

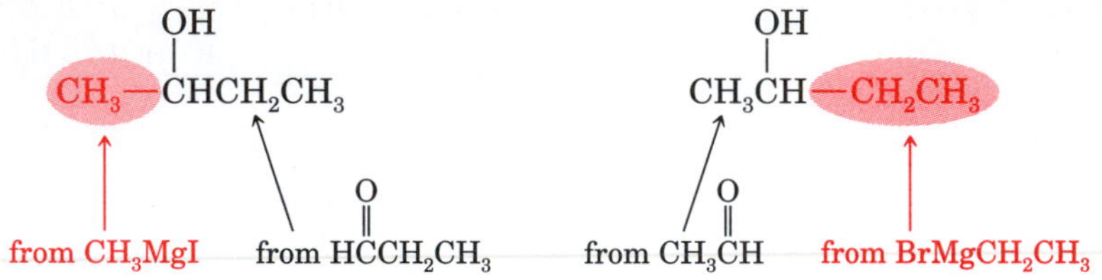

The two sequences to 2-butanol:

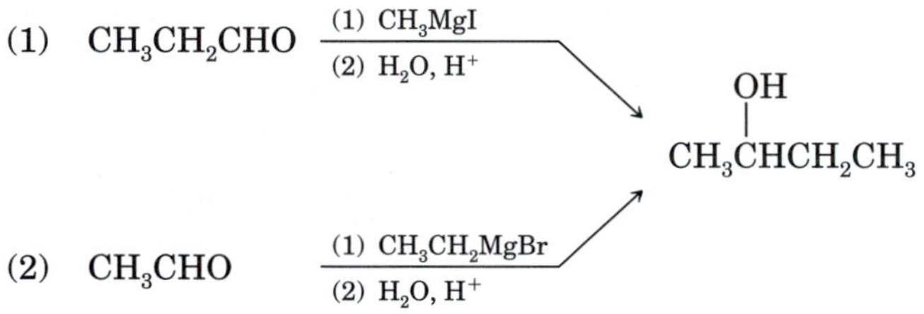

(1) CH_3CH_2CHO $\xrightarrow{\text{(1) } CH_3MgI}{\text{(2) } H_2O, H^+}$

(2) CH_3CHO $\xrightarrow{\text{(1) } CH_3CH_2MgBr}{\text{(2) } H_2O, H^+}$

Suggest three different Grignard reactions leading to 2-phenyl-2-butanol.

Solution

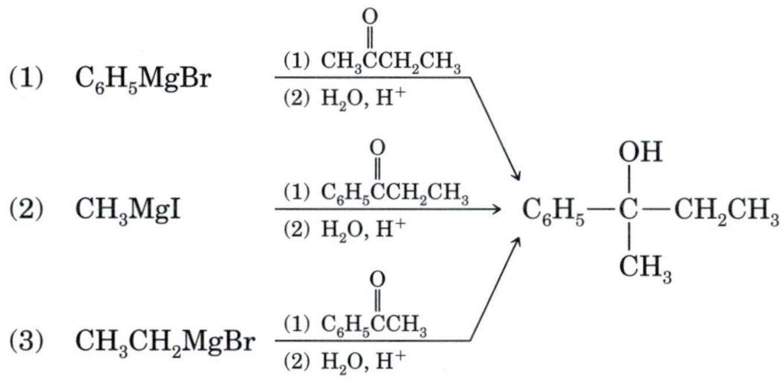

(1) C_6H_5MgBr $\xrightarrow{\text{(1) } CH_3CCH_2CH_3}{\text{(2) } H_2O, H^+}$

(2) CH_3MgI $\xrightarrow{\text{(1) } C_6H_5CCH_2CH_3}{\text{(2) } H_2O, H^+}$

(3) CH_3CH_2MgBr $\xrightarrow{\text{(1) } C_6H_5CCH_3}{\text{(2) } H_2O, H^+}$

Study Problems

13.10 What Grignard reagents would you use to effect the following conversions?

(a) formaldehyde to benzyl alcohol;

(b) cyclohexanone to 1-propylcyclohexanol.

13.11 Which of the following compounds could *not* be used as a carbonyl starting material in a Grignard synthesis? (*Hint:* See Section 7.3D.)

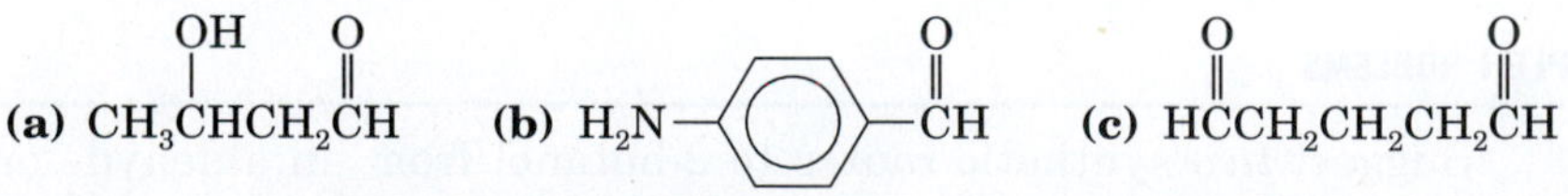

(a) CH_3CHCH_2CH (b) $H_2N-\!\!\bigcirc\!\!-CH$ (c) $HCCH_2CH_2CH_2CH$

Section 13.5
Addition–Elimination Reactions of Aldehydes and Ketones

In the preceding section, we discussed the addition reactions of aldehydes and ketones. Some reagents undergo addition to aldehydes and ketones followed by elimination of water or other small molecule to yield a product containing a double bond.

$$
\underset{\substack{\text{O}\\\|}}{R-C-R} + H\!-\!NuH \xrightarrow{\text{addition}} \left[\underset{\substack{|\\\text{NuH}}}{\overset{\overset{\text{OH}}{|}}{R-C-R}} \right] \xrightarrow[-H_2O]{\text{elimination}} \underset{\substack{\|\\\text{Nu}}}{R-C-R}
$$

*an unstable
addition product*

A. Reaction with Ammonia and Primary Amines

Ammonia is a nucleophile that can attack a carbonyl group of an aldehyde or a ketone in an addition–elimination reaction. The reaction is catalyzed by a trace of acid. The product is an **imine,** a compound that contains the C=N grouping.

$$
\underset{\substack{\text{O}\\\|}}{RCH} + H\!-\!NH_2 \rightleftharpoons[\text{}]{H^+} \left[\underset{\substack{|\\\text{}}}{\overset{\overset{\text{OH}}{|}}{RCH-NH_2}} \right] \xrightarrow{-H_2O} RCH{=}NH
$$

an imine

Unsubstituted imines formed from NH_3 are unstable and polymerize on standing. However, if a *primary amine* (RNH_2) is used instead of ammonia, a more stable, substituted imine (sometimes called a **Schiff base**) is formed. Aromatic aldehydes (such as benzaldehyde) or arylamines (such as aniline) give the most stable imines, but other aldehydes, ketones, or primary amines can be used.

$$
\text{benzaldehyde} \quad \text{methylamine} \qquad\qquad\qquad\qquad\qquad\qquad \textit{an imine (95\%)}
$$

$$
\text{benzaldehyde} \qquad\qquad \text{aniline} \qquad\qquad \textit{an imine (87\%)}
$$

The mechanism for imine formation is essentially a two-step process. The first step is the *addition* of the nucleophilic amine to the partially positive carbonyl carbon, followed by the loss of a proton from the nitrogen and the gain of a proton by the oxygen.

Step 1, addition:

$$
\underset{\substack{\|\\RCR}}{\overset{\overset{..}{O}:}{}} + R'NH_2 \rightleftharpoons[\text{}]{\text{fast}} \underset{\substack{|\\RCR\\|\\R'NH_2\\+}}{\overset{:\overset{..}{O}:^-}{}} \rightleftharpoons[\text{}]{\text{fast}} \underset{\substack{|\\RCR\\|\\R'NH\\..}}{\overset{:\overset{..}{O}H}{}}
$$

Step 2 is the protonation of the OH group, which then can be lost as water in an *elimination* reaction.

Step 2, elimination:

$$
\underset{R_2\overset{..}{C}NHR'}{\overset{:\overset{..}{O}H}{|}} \;\overset{H^+}{\underset{fast}{\rightleftarrows}}\; \underset{R_2C-NHR'}{\overset{:\overset{+}{O}H_2}{|}} \;\overset{-H_2O}{\underset{slow}{\rightleftarrows}}\; \underset{}{R_2C\overset{+}{=}NHR'} \;\overset{-H^+}{\underset{fast}{\rightleftarrows}}\; R_2C=\overset{..}{N}R'
$$

the imine

Imine formation is a reaction that is *pH-dependent*. Why? Consider the two steps in the mechanism. The first step is the addition of the free nonprotonated amine to the carbonyl group. If the solution is too acidic, the concentration of the amine becomes negligible. If this happens, the usually fast addition step becomes slow and actually becomes the rate-determining step in the sequence.

In acid:

— not nucleophilic

$$
R\overset{..}{N}H_2 + H^+ \;\rightleftarrows\; RNH_3{}^+
$$

The second step in the reaction is the elimination of the protonated OH group as water. Unlike the first step (amine addition), the rate of the second step increases with increasing acid concentration. (Remember, OH^- is a strong base and a poor leaving group, while $-OH_2{}^+$ can leave as the weak base and good leaving group H_2O.)

An increase in acidity causes Step 2 to go faster, but Step 1 to go slower, while decreasing acidity causes Step 1 to go faster, but Step 2 to go slower. Between these two extremes is the optimum pH (about pH 3–4), at which the rate of the overall reaction is greatest. At this pH, some of the amine is protonated, but some is free to initiate the nucleophilic addition. At this pH, too, enough acid is present so that elimination can proceed at a reasonable rate.

STUDY PROBLEM

13.12   How would you prepare each of the following imines from a carbonyl compound?

(a) ⬡—CH=NCH$_2$CH$_3$ **(b)** CH$_3$—◯—CH=N—◯

(c) (CH$_3$)$_2$C=N—◯—CH$_3$ **(d)**
$$\text{NH}$$
(fluorenone imine structure)

B. Reaction with Secondary Amines

With primary amines, aldehydes and ketones yield imines. With *secondary amines* (R_2NH), aldehydes and ketones yield **iminium ions,** which undergo further reaction to **enamines** (vinylamines). The enamine is formed by loss of a proton from a carbon atom β to the nitrogen, which results in a double bond between the α and β carbon atoms. Enamines are useful synthetic intermediates. You will encounter them again in Chapter 17.

double bond
between carbons
α and β to
the nitrogen

$$CH_3CH + (CH_3)_2NH \underset{\xrightarrow{\hspace{1cm}}}{\overset{H^+}{\underset{-H_2O}{\rightleftharpoons}}} \left[\overset{H}{\underset{}{CH_2}} - CH = \overset{+}{N}(CH_3)_2 \right] \overset{-H^+}{\rightleftharpoons} CH_2 = CH\ddot{N}(CH_3)_2$$

dimethylamine   *an iminium ion*   *an enamine*

a 2° amine

13.13   Predict the product of the reaction of cyclohexanone with:

   (a) CH_3NH_2 **(b)** $(CH_3)_2NH$ **(c)** NH

C. Reaction with Hydrazine and Related Compounds

Imines are easily hydrolyzed (cleaved by water). The initial step of hydrolysis is protonation of the imine nitrogen. If an *electronegative group* is attached to the imine nitrogen, the basicity of the nitrogen is reduced and the hydrolysis is suppressed.

$$\underset{H_3C}{\overset{H_3C}{\diagdown}}C = \ddot{N} \quad H^+ \quad \rightleftharpoons \quad \underset{H_3C}{\overset{H_3C}{\diagdown}}C = \overset{+}{N}\underset{OH}{\overset{H}{\diagup}}$$

electron-withdrawing   *not favored*
by the inductive effect

Imine-type products formed from aldehydes or ketones and a nitrogen compound of the type H_2N-NH_2 or H_2N-OH (reagents with an electronegative group attached to the N) are quite stable. Table 13.4 lists the variety of nitrogen compounds that react with aldehydes and most ketones to form stable imine-type products.

$$\underset{H_3C}{\overset{H_3C}{\diagdown}}C = O + H_2NNH_2 \overset{H^+}{\rightleftharpoons} \underset{H_3C}{\overset{H_3C}{\diagdown}}C = NNH_2 + H_2O$$

acetone   hydrazine   acetone hydrazone

$$\bigcirc = O \; + \; H_2NNHC_6H_5 \overset{H^+}{\rightleftharpoons} \bigcirc = NNHC_6H_5 + H_2O$$

cyclopentanone   phenylhydrazine   cyclopentanone
                                                 phenylhydrazone

The hydrazones and other products listed in Table 13.4, especially the high-molecular-weight 2,4-dinitrophenylhydrazones, or DNP's, are generally

Name	Structure	Product with RCHO
	Table 13.4 Some nitrogen compounds that form stable substitution products with aldehydes and ketones	
hydroxylamine	$HONH_2$	$RCH{=}NOH$ *an oxime*
hydrazine	H_2NNH_2	$RCH{=}NNH_2$ *a hydrazone*
phenylhydrazine	⬡—$NHNH_2$	$RCH{=}NNHC_6H_5$ *a phenylhydrazone*
2,4-dinitro- phenylhydrazine	O_2N—⬡—$NHNH_2$ with NO_2	$RCH{=}NNH$—⬡—NO_2 with NO_2 *a 2,4-dinitrophenylhydrazone*
semicarbazide	$\overset{\overset{\text{O}}{\|\|}}{H_2NNHCNH_2}$	$RCH{=}NNH\overset{\overset{\text{O}}{\|\|}}{C}NH_2$ *a semicarbazone*

solids. Before the wide use of spectrometers, these derivatives were used extensively for identification purposes. A liquid ketone of unknown structure could be converted to the solid DNP, purified by crystallization, and its melting point compared to those of DNP's of known structure.

STUDY PROBLEM

13.14 Predict the product of the reaction of: **(a)** butanone with semicarbazide; **(b)** cyclohexanone with 2,4-dinitrophenylhydrazine; and **(c)** acetophenone $(C_6H_5COCH_3)$ with hydrazine.

D. Reaction with Phosphonium Ylides (the Wittig Reaction)

In 1954, Georg Wittig reported a general synthesis of alkenes from carbonyl compounds using *phosphonium ylides*. This synthesis is called the **Wittig reaction.**

from the ylide

$$\underset{\substack{\text{an aldehyde}\\\text{or ketone}}}{\overset{R}{\underset{R}{>}}C{=}O} + \underset{\substack{\text{a phosphonium}\\\text{ylide}}}{(C_6H_5)_3P{=}C\overset{R'}{\underset{R'}{<}}} \longrightarrow \underset{\substack{\text{an alkene}\\\text{(where at least one}\\R \text{ or } R' = H)}}{\overset{R}{\underset{R}{>}}C{=}C\overset{R'}{\underset{R'}{<}}} + \underset{\substack{\text{triphenylphosphine}\\\text{oxide}}}{(C_6H_5)_3P{=}O}$$

An **ylide** is a molecule with *adjacent + and − charges* (see the following resonance structures). An ylide is formed by removal of a proton from the carbon adjacent to a positively charged heteroatom (such as P^+, S^+, or N^+). The phosphonium ylide for a Wittig reaction is prepared by (1) nucleophilic substitution (S_N2) of an alkyl halide with a tertiary phosphine, such as triphenylphosphine (a good nucleophile, a weak base) and (2) treatment with a strong base, such as butyllithium ($CH_3CH_2CH_2CH_2Li$), a reaction in which the intermediate phosphonium ion eliminates a proton to form the ylide.

$$(C_6H_5)_3P: \quad + \quad \overset{R'}{\underset{R'}{\overset{|}{C}H}} - X \quad \xrightarrow[S_N2]{-X^-} \quad (C_6H_5)_3\overset{+}{P} - \overset{H}{\underset{R'_2}{\overset{|}{C}}} \quad \xrightarrow{CH_3CH_2CH_2CH_2Li}$$

triphenylphosphine   *a phosphonium ion*

$$CH_3CH_2CH_2CH_3 + \underbrace{(C_6H_5)_3\overset{+}{P} - \overset{..}{C}R'_2 \quad \longleftrightarrow \quad (C_6H_5)_3P{=}CR'_2}$$

resonance structures for the ylide

The Wittig reaction is versatile. The alkyl halide used to prepare the ylide may be methyl, primary, or secondary, but not tertiary (why not?). The halide may also contain other functionality, such as double bonds or alkoxyl groups. The product of the Wittig reaction is an alkene with the double bond in the desired position, even if it is not the most stable alkene. Yields are generally good (about 70%) for mono-, di-, and trisubstituted alkenes. (The reaction fails for tetrasubstituted alkenes, probably because of steric hindrance.) Unfortunately, it is sometimes difficult to predict whether the *cis* or the *trans* product will predominate in a particular reaction.

cyclohexanone   methylenecyclohexane (84%)

benzaldehyde

1,4-diphenyl-1,3-butadiene (67%)

The mechanism of the Wittig reaction involves nucleophilic attack on the carbonyl group by the negative carbon of the ylide.

nucleophilic carbon

$$(C_6H_5)_3P{=}CR'_2 \quad \longleftrightarrow \quad (C_6H_5)_3\overset{+}{P} - \overset{-}{C}R'_2$$

resonance structures

Addition to carbonyl:

a betaine

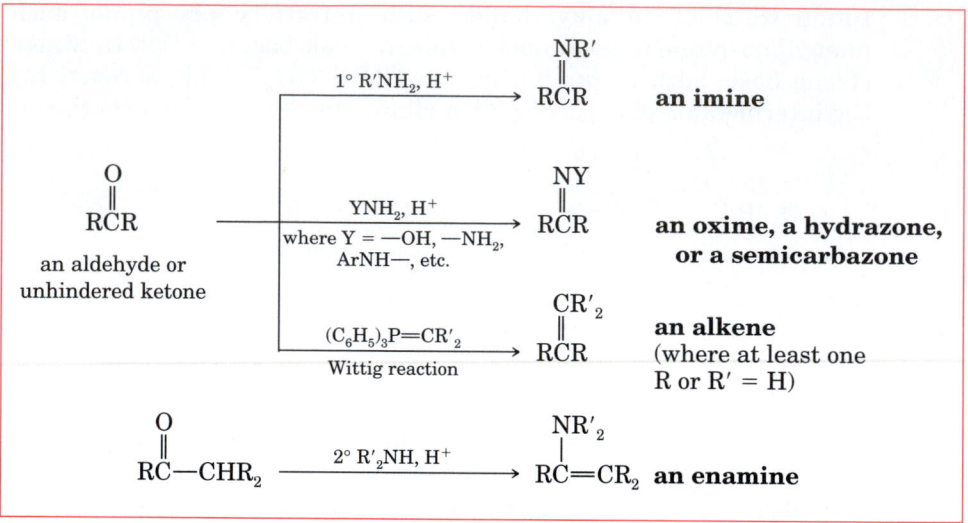

Figure 13.8   Summary of the addition–elimination reactions of aldehydes and ketones.

The addition product of the ylide and an aldehyde or ketone is a **betaine** (a molecule having *nonadjacent* opposite charges). The betaine undergoes cyclization and elimination of triphenylphosphine oxide to form the alkene.

Elimination to form alkene:

$$R_2C-CR'_2 \longrightarrow R_2C \stackrel{\frown}{\underset{}{CR'_2}} \longrightarrow R_2C{=}CR'_2 + {:}O{=}P(C_6H_5)_3$$
$$\underset{:O:}{\;} \quad \underset{P(C_6H_5)_3}{\;} \qquad :O-P(C_6H_5)_3$$

STUDY PROBLEM

13.15   Write equations to show how you would prepare the following compounds by Wittig reactions. (Begin with organic halides and carbonyl compounds.)

(a) $C_6H_5CH{=}CHCH_2CH_3$ (b) [cyclopentylidene]=CH–[cyclopentane ring]   (c) [benzo-fused cyclooctene ring structure]

Figure 13.8 summarizes the addition–elimination reactions of aldehydes and ketones.

Section 13.6
Reduction of Aldehydes and Ketones

An aldehyde or a ketone can be reduced to an alcohol, a hydrocarbon, or an amine. The product of the reduction depends on the reducing agent and on the structure of the carbonyl compound.

$$
\underset{\substack{an\ aldehyde \\ or\ ketone}}{\overset{\overset{\textstyle O}{\overset{\|}{}}}{RCR}} \xrightarrow{[H]} \underset{an\ alcohol}{\overset{\overset{\textstyle OH}{\overset{|}{}}}{RCHR}} \quad or \quad \underset{a\ hydrocarbon}{RCH_2R} \quad or \quad \underset{an\ amine}{\overset{\overset{\textstyle NR'_2}{\overset{|}{}}}{RCHR}}
$$

A. Hydrogenation

The π bond of a carbonyl group can undergo hydrogenation, just as can the π bond of an alkene. The reaction conditions depend on the compound being reduced and on the catalyst. Unhindered ketones, such as cyclohexanone, can be hydrogenated at room temperature and four atmospheres of pressure with a platinum or ruthenium catalyst. Other carbonyl compounds and other catalyst systems, such as copper chromite or Raney nickel, may require harsher reaction conditions.

$$
cyclohexanone + H_2 \xrightarrow{Pt} cyclohexanol\ (91\%)
$$

a ketone → a 2° alcohol

$$
\underset{\substack{acetaldehyde}}{\overset{\overset{\textstyle O}{\overset{\|}{}}}{CH_3CH}} + H_2 \xrightarrow{Pt} \underset{ethanol}{CH_3CH_2OH}
$$

an aldehyde → a 1° alcohol

STUDY PROBLEM

13.16 2-Heptanol can be obtained by two Grignard reactions as well as by a hydrogenation reaction. Write the equations for the three sets of reactions that would lead to this alcohol.

If both a double bond and a carbonyl group are present in a structure, the double bond may be hydrogenated, leaving the carbonyl intact, or both may be hydrogenated. However, the carbonyl group cannot be catalytically hydrogenated independently of the double bond. To reduce a carbonyl group while leaving a carbon–carbon double bond intact, a metal hydride reduction is the method of choice.

C=C reduced (but not C=O):

$$
\underset{\substack{3\text{-pentenal}}}{CH_3CH=CHCH_2\overset{\overset{\textstyle O}{\overset{\|}{}}}{CH}} + H_2 \xrightarrow[25°]{Ni} \underset{\substack{pentanal}}{CH_3CH_2CH_2CH_2\overset{\overset{\textstyle O}{\overset{\|}{}}}{CH}}
$$

C=C and C=O reduced:

$$
\underset{\substack{3\text{-pentenal}}}{CH_3CH=CHCH_2\overset{\overset{\textstyle O}{\overset{\|}{}}}{CH}} + 2H_2 \xrightarrow[\text{heat, pressure}]{Ni} \underset{\substack{1\text{-pentanol}}}{CH_3CH_2CH_2CH_2CH_2OH}
$$

If the hydrogenation of a keto group yields a chiral alcohol, a racemic mixture is formed unless the hydrogenation catalyst is chelated with one enantiomer of a chiral compound. Under these conditions, asymmetry is induced into the product, and only one enantiomer of the product alcohol is isolated. This type of reaction is called **asymmetric hydrogenation.**

Asymmetric hydrogenation of a ketone:

an achiral
keto lactone

a chiral hydroxy lactone (98%)
$[\alpha]_D^{20} = -49.87°$

B. Metal Hydrides

Hydrogen gas is inexpensive on a molar basis; however, a hydrogenation reaction is rather inconvenient. The apparatus usually consists of gas tanks and a metal pressure vessel. An alternative reduction procedure involves the use of a metal hydride. Two valuable reducing agents are *lithium aluminum hydride* (often abbreviated LAH) and *sodium borohydride.* Both reduce aldehydes and ketones to alcohols.

lithium aluminum hydride (LAH)   sodium borohydride

These two metal hydrides are quite different in their reactivities. $LiAlH_4$ is a powerful reducing agent that reduces not only aldehydes and ketones, but also carboxylic acids, esters, amides, and nitriles. $LiAlH_4$ undergoes violent reaction with water; reductions are usually carried out in a solvent such as anhydrous ether.

Sodium borohydride is a milder reducing agent than $LiAlH_4$. Its reactions can be carried out in water or aqueous alcohol as the solvent. For the reduction of an aldehyde or ketone, $NaBH_4$ is the preferred reagent; it is certainly more convenient to use because of its lower reactivity toward water. While $NaBH_4$ reduces aldehydes and ketones rapidly, it reduces esters very slowly. Therefore, an aldehyde or ketone carbonyl group can be reduced without the simultaneous reduction of an ester group in the same molecule. This selectivity is not possible with $LiAlH_4$.

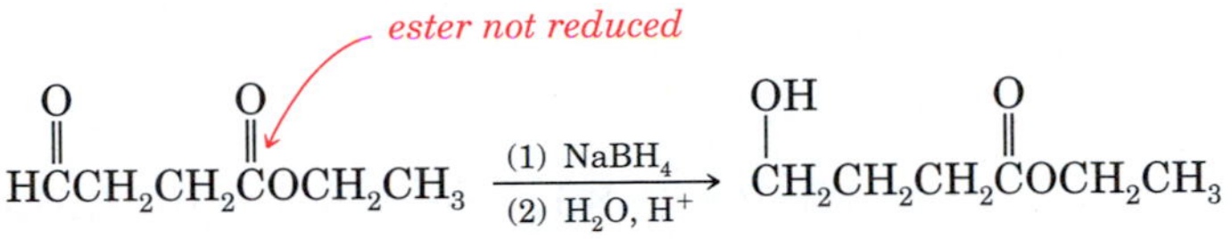

ester not reduced

$$\underset{\text{HCCH}_2\text{CH}_2\text{COCH}_2\text{CH}_3}{\overset{\text{O}\qquad\qquad\text{O}}{\|\qquad\qquad\|}} \xrightarrow[\text{(2) H}_2\text{O, H}^+]{\text{(1) NaBH}_4} \underset{\text{CH}_2\text{CH}_2\text{CH}_2\text{COCH}_2\text{CH}_3}{\overset{\text{OH}\qquad\quad\;\text{O}}{|\qquad\qquad\|}}$$

Neither $NaBH_4$ nor $LiAlH_4$ reduces isolated carbon–carbon double bonds, although C=C in conjugation with a carbonyl group is sometimes attacked. Consequently, a structure that contains both a double bond and a carbonyl group can often be reduced selectively at the carbonyl position. In this respect, the metal hydrides are complementary to the hydrogen gas as reducing agents.

C=O reduced (but not C=C):

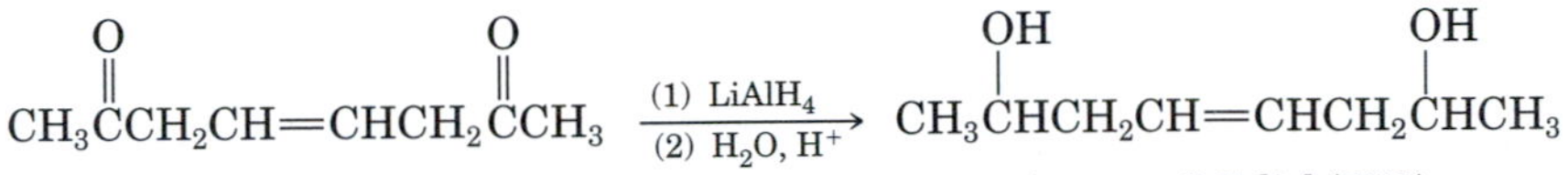

$$\underset{\underset{\text{4-octene-2,7-dione}}{\text{CH}_3\text{CCH}_2\text{CH}=\text{CHCH}_2\text{CCH}_3}}{\overset{\text{O}\qquad\qquad\qquad\qquad\quad\text{O}}{\|\qquad\qquad\qquad\qquad\quad\|}} \xrightarrow[\text{(2) H}_2\text{O, H}^+]{\text{(1) LiAlH}_4} \underset{\underset{\text{4-octene-2,7-diol (70\%)}}{\text{CH}_3\text{CHCH}_2\text{CH}=\text{CHCH}_2\text{CHCH}_3}}{\overset{\text{OH}\qquad\qquad\qquad\qquad\quad\text{OH}}{|\qquad\qquad\qquad\qquad\quad|}}$$

Diisobutylaluminum hydride (DIBAL-H), $[(CH_3)_2CHCH_2]_2AlH$, is a newer, but popular, metal hydride reducing agent that is similar to $LiAlH_4$ in reducing power. Besides reducing aldehydes or ketones to alcohols, DIBAL-H reduces carboxylic acids and esters to aldehydes or alcohols (depending on the reaction conditions). Other hydrides that have specialized reducing activity are also available; for example, see Section 15.3C.

STUDY PROBLEM

13.17   Show how each of the following alcohols could be prepared by the $NaBH_4$ reduction of an aldehyde or ketone:

(a) $CH_3CH_2CH_2OH$ **(b)** 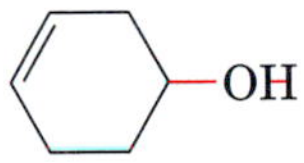 —OH   **(c)**

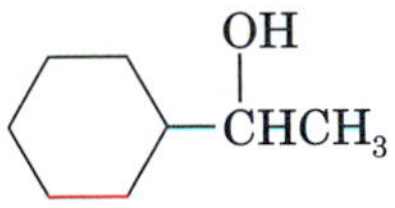

Metal hydrides react by transferring a negative hydride ion to the positive carbon of a carbonyl group, just as a Grignard reagent transfers R to the carbonyl group.

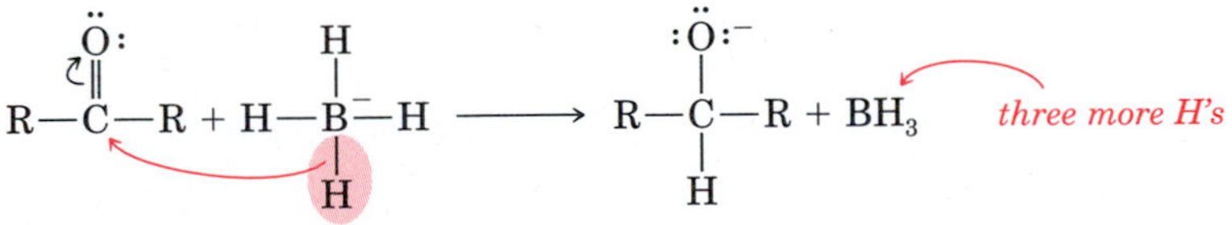

$$\underset{\text{R}-\overset{\overset{\displaystyle \ddot{\text{O}}:}{\|}}{\text{C}}-\text{R} + \underset{\overset{|}{\underset{\text{H}}{\text{H}}}}{\text{H}-\text{B}-\text{H}}} \longrightarrow \underset{\text{R}-\overset{\overset{\displaystyle :\ddot{\text{O}}:^-}{|}}{\underset{\underset{\text{H}}{|}}{\text{C}}}-\text{R} + \text{BH}_3} \qquad \textit{three more H's}$$

Each hydride ion can reduce one carbonyl group. Therefore, one mole of $NaBH_4$ can reduce *four* moles of aldehyde or ketone, theoretically. After the reaction is completed, treatment with water or aqueous acid liberates the alcohol from its salt. (Of course, if water, methanol, or ethanol is used as a solvent for a borohydride reduction, this step occurs spontaneously.) In the hydrolysis, the boron portion of the organoborate is converted to boric acid, H_3BO_3.

Step 1:

$$4 \overset{\overset{\text{O}}{\|}}{\text{RCR}} + \text{NaBH}_4 \longrightarrow \text{Na}^+ \ ^-\text{B}\left(\overset{\overset{\text{R}}{|}}{\text{OCHR}}\right)_4$$

Step 2:

$$\overset{\overset{\text{O}^-}{|}}{\text{RCHR}} + \text{H}^+ \longrightarrow \overset{\overset{\text{OH}}{|}}{\text{RCHR}}$$

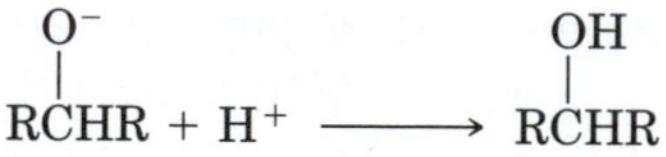

Camphor is a bridged, cyclic compound with a ketone group. Reduction of camphor with LiAlH$_4$ leads to 90% of the isomer in which the OH group is *cis* to the bridge. Why is this so? Let us look at the structure of camphor. Note that the bridge provides substantial steric hindrance on one side of the carbonyl group—on top, as it is shown here.

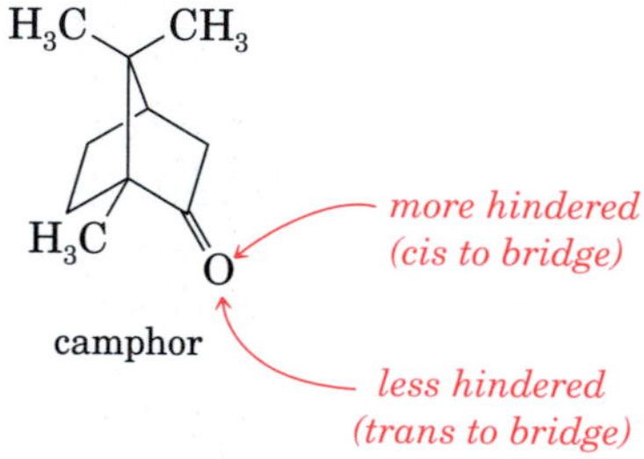

camphor

more hindered (cis to bridge)

less hindered (trans to bridge)

When a ketone is reduced by LiAlH$_4$, it is not just a tiny hydride ion attacking; it is the relatively bulky AlH$_4^-$ ion or an alkoxyaluminum hydride ion, such as $^-$AlH$_2$(OR)$_2$. There is evidence that the lithium ion forms a complex with the carbonyl oxygen while the hydride ion is transferred from the AlH$_4^-$ ion to the carbonyl carbon. A look at the likely structure of the transition state shows how AlH$_4^-$ attacks the *less hindered* side of the camphor structure; that is, AlH$_4^-$ attacks the carbonyl group from the side that is *trans* to the bridge. The resultant OH group is then formed *cis* to the bridge.

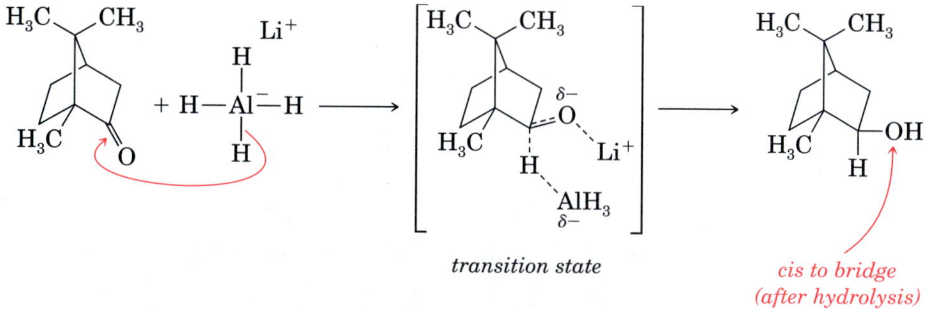

transition state

cis to bridge (after hydrolysis)

C. Wolff–Kishner and Clemmensen Reductions

The Clemmensen reduction and the Wolff–Kishner reduction are primarily used to reduce aryl ketones obtained from Friedel–Crafts reactions (Section 11.8E), but may sometimes be used to reduce other aldehydes and ketones. Both these methods of reduction result in the conversion of a C=O group to a CH$_2$ group.

In the Wolff–Kishner reduction, the aldehyde or ketone is first converted to a hydrazone by reaction with hydrazine. The hydrazone is then treated with a strong base, such as potassium hydroxide or potassium *tert*-butoxide in dimethyl sulfoxide (DMSO) as solvent. The reaction is therefore limited to carbonyl compounds that are stable in base.

Wolff-Kishner reduction:

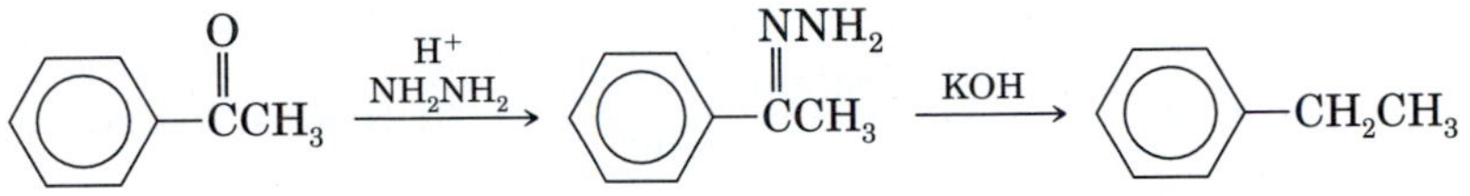

acetophenone   ethylbenzene (73%)

In the Clemmensen reduction, on the other hand, a zinc amalgam (an alloy of zinc and mercury) and concentrated HCl are used. These reagents would be the reagents of choice for a compound unstable in base but stable in acid.

Clemmensen reduction:

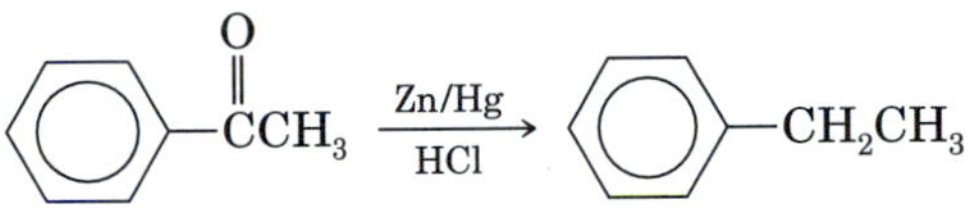

(45%)

D. Reductive Amination

If an amine is the desired reduction product, the carbonyl compound is treated with ammonia or a primary amine to form an imine in the presence of hydrogen and a catalyst. The imine C=N group then undergoes catalytic hydrogenation in the same way that a C=C or a C=O group does.

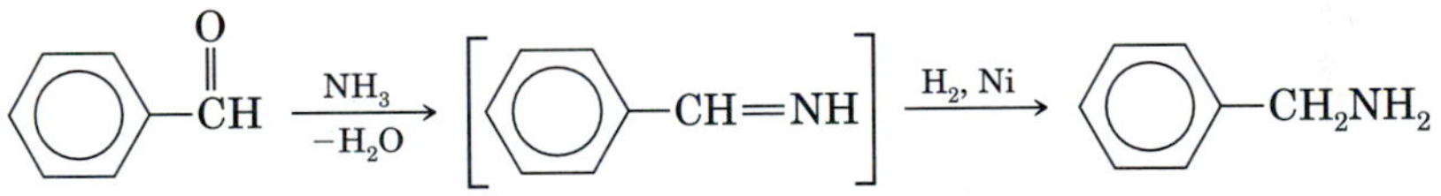

benzaldehyde   *an imine*   benzylamine (89%)

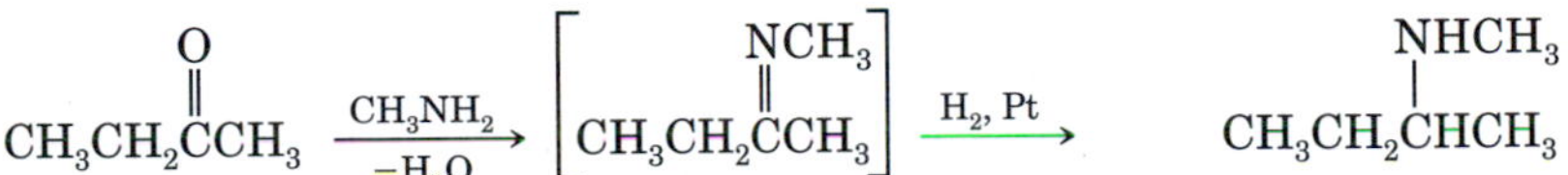

butanone   *an imine*   *N*-methyl-*sec*-butylamine (69%)

Reductive amination is a good method for the preparation of an amine with a secondary alkyl group: R_2CHNH_2. (Treating the secondary alkyl halide R_2CHX with NH_3 in an S_N2 reaction may result in elimination or in dialkyl-amines, a reaction we will discuss in Chapter 18.)

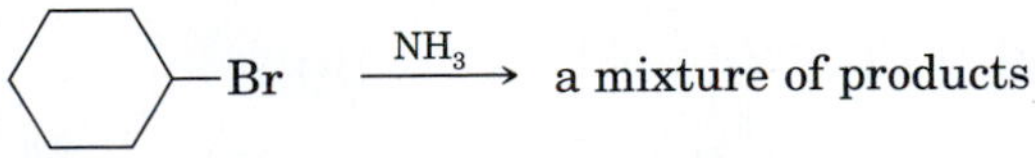

bromocyclohexane

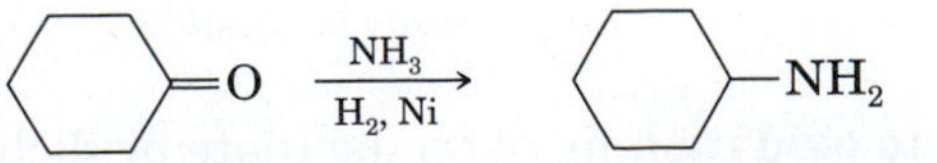

cyclohexanone   cyclohexylamine (80%)

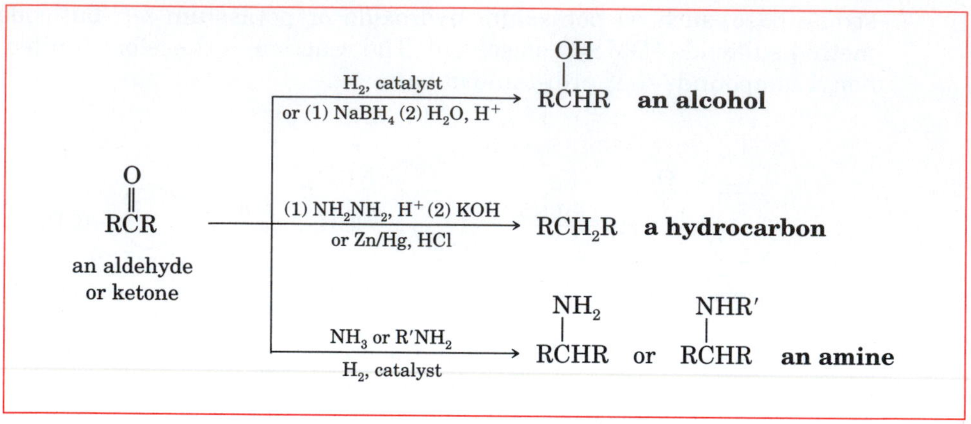

$$
\underset{\substack{\text{an aldehyde}\\\text{or ketone}}}{\overset{\overset{\displaystyle O}{\|}}{RCR}}
\begin{cases}
\xrightarrow[\text{or (1) NaBH}_4\text{ (2) H}_2\text{O, H}^+]{\text{H}_2\text{, catalyst}} \underset{}{\overset{\overset{\displaystyle OH}{|}}{RCHR}} \quad \textbf{an alcohol} \\[3em]
\xrightarrow[\text{or Zn/Hg, HCl}]{\text{(1) NH}_2\text{NH}_2\text{, H}^+\text{ (2) KOH}} RCH_2R \quad \textbf{a hydrocarbon} \\[3em]
\xrightarrow[\text{H}_2\text{, catalyst}]{\text{NH}_3\text{ or R'NH}_2} \underset{}{\overset{\overset{\displaystyle NH_2}{|}}{RCHR}} \text{ or } \underset{}{\overset{\overset{\displaystyle NHR'}{|}}{RCHR}} \quad \textbf{an amine}
\end{cases}
$$

Figure 13.9 Summary of reductions of aldehydes and ketones.

Figure 13.9 summarizes the reduction reactions of aldehydes and ketones.

Section 13.7
Oxidation of Aldehydes and Ketones

Ketones are not easily oxidized (see Sections 13.9 and 13.10 for exceptions), but aldehydes are very easily oxidized to carboxylic acids. Almost any reagent that oxidizes an alcohol also oxidizes an aldehyde (see Section 7.8C). Permanganate or dichromate salts are the most popular oxidizing agents, but are by no means the only reagents that can be used.

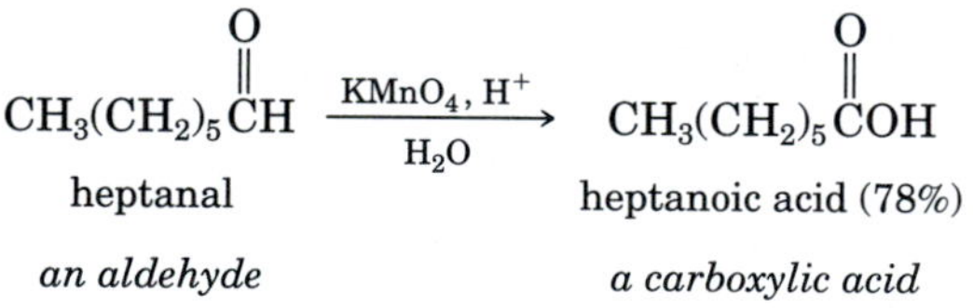

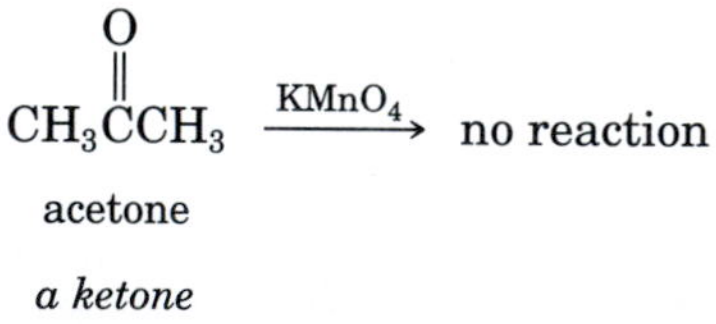

$$
\underset{\substack{\text{heptanal}\\ \textit{an aldehyde}}}{CH_3(CH_2)_5\overset{\overset{\displaystyle O}{\|}}{CH}} \xrightarrow[\text{H}_2\text{O}]{\text{KMnO}_4\text{, H}^+} \underset{\substack{\text{heptanoic acid (78\%)}\\ \textit{a carboxylic acid}}}{CH_3(CH_2)_5\overset{\overset{\displaystyle O}{\|}}{C}OH}
$$

$$
\underset{\substack{\text{acetone}\\ \textit{a ketone}}}{CH_3\overset{\overset{\displaystyle O}{\|}}{C}CH_3} \xrightarrow{\text{KMnO}_4} \text{no reaction}
$$

Hydrate formation is the reason that aldehydes are more readily oxidized to carboxylic acids in aqueous media than in nonaqueous media.

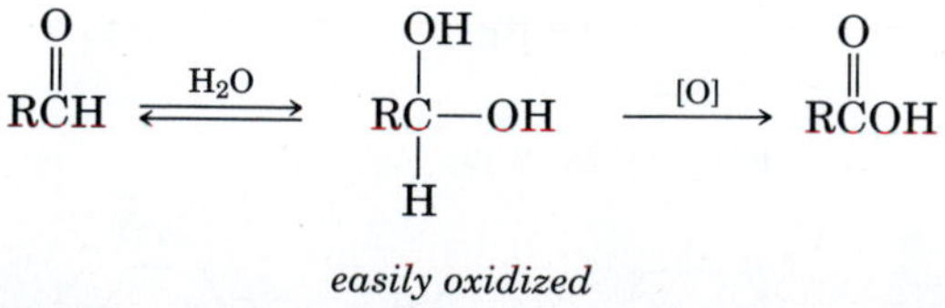

$$
\overset{\overset{\displaystyle O}{\|}}{RCH} \underset{}{\overset{\text{H}_2\text{O}}{\rightleftharpoons}} \underset{\substack{|\\H\\ \textit{easily oxidized}}}{\overset{\overset{\displaystyle OH}{|}}{RC-OH}} \xrightarrow{[O]} \overset{\overset{\displaystyle O}{\|}}{RCOH}
$$

In addition to oxidation by permanganate or dichromate, aldehydes are oxidized by very mild oxidizing agents, such as Ag^+ or Cu^{2+}. **Tollens reagent** (an alkaline solution of the silver–ammonia complex ion) is used as a test for aldehydes. The aldehyde is oxidized to a carboxylate anion; the Ag^+ in the Tollens reagent is reduced to Ag metal. A positive test is indicated by the formation of a silver mirror on the wall of the test tube. With the widespread use of spectroscopy, the Tollens test is no longer the test of choice for an aldehyde, but mirrors are sometimes still made this way.

$$\underset{\substack{\text{from Tollens reagent}}}{\overset{\displaystyle O \atop \displaystyle \|}{R\overset{}{C}H} + \quad Ag(NH_3)_2{}^+} \quad \xrightarrow{\ OH^-\ } \quad \underset{\substack{\text{mirror}}}{\overset{\displaystyle O \atop \displaystyle \|}{R\overset{}{C}O^-} + \quad Ag}$$

STUDY PROBLEM

13.18 Which of the following compounds would give a positive Tollens test?

(a) CH_3CHO **(b)** $CH_3\overset{\displaystyle O}{\overset{\|}{C}}CH_3$ **(c)** [structure: tetrahydrofuran ring with —OH at the 2-position]   **(d)** [structure: tetrahydrofuran ring with —OCH₃ at the 2-position]

Section 13.8
Reactivity of the Alpha Hydrogens

A carbon–hydrogen bond is usually stable, nonpolar, and certainly not acidic. But the presence of a carbonyl group results in an *acidic α hydrogen*. If a hydrogen is α to *two* carbonyl groups, it is acidic enough that salts can be formed by treatment with an alkoxide. The pK_a of ethyl acetoacetate ($CH_3COCH_2CO_2CH_2CH_3$) is 11; it is more acidic than ethanol ($pK_a = 16$) or water ($pK_a = 15.7$). Treatment of this β-dicarbonyl compound with sodium ethoxide (or any other strong base, such as NaH or $NaNH_2$) yields the sodium salt. (Sodium hydroxide is usually not used with keto esters because the ester groups undergo hydrolysis in NaOH and water. This reaction will be discussed in Section 15.5C.)

[reaction scheme:]

α to one C=O

$$\underset{\substack{\text{acetone}\\pK_a = 20}}{CH_3\overset{\displaystyle O}{\overset{\|}{C}}CH_3}$$

α to two C=O groups

$$\underset{\substack{\text{ethyl acetoacetate}\\pK_a = 11}}{CH_3\overset{\displaystyle O}{\overset{\|}{C}}CH_2\overset{\displaystyle O}{\overset{\|}{C}}OCH_2CH_3}$$

$$\underset{\substack{| \\ H}}{CH_2}\overset{\displaystyle O}{\overset{\|}{C}}CH_3 + Na^+\ {}^-{:}\ddot{O}CH_2CH_3 \quad \rightleftharpoons \quad Na^+\ {}^-{:}CH_2\overset{\displaystyle O}{\overset{\|}{C}}CH_3 + CH_3CH_2OH$$

not favored

$$\underset{\substack{| \\ H}}{CH_3\overset{\displaystyle O}{\overset{\|}{C}}CH}\overset{\displaystyle O}{\overset{\|}{C}}OCH_2CH_3 + Na^+\ {}^-{:}\ddot{O}CH_2CH_3 \quad \rightleftharpoons \quad \underset{\substack{\\ Na^+}}{CH_3\overset{\displaystyle O}{\overset{\|}{C}}\overset{-}{C}H}\overset{\displaystyle O}{\overset{\|}{C}}OCH_2CH_3 + CH_3CH_2OH$$

favored

Why is a hydrogen α to a carbonyl group acidic? The answer is two-fold. First, the α carbon is adjacent to one or more partially positive carbon atoms. The α carbon, too, partakes of some of this positive charge (inductive effect by electron-withdrawal), and C—H bonds are consequently weakened.

$$\underset{\|}{\overset{O}{\underset{}{}}} \quad \underset{\|}{\overset{O}{\underset{}{}}}$$
$$-\overset{\|}{C}\leftarrow\overset{\delta+}{CH_2}\rightarrow\overset{\|}{C}-$$

Second, and more important, is the resonance stabilization of the **enolate ion,** the anion formed when the proton is lost. From the resonance structures, we can see that the negative charge is carried by the carbonyl oxygens as well as by the α carbon. This delocalization of the charge stabilizes the enolate ion and favors its formation.

Adjacent to one carbonyl group:

$$\left[-\overset{..}{\underset{..}{C}}H-\overset{\overset{\displaystyle :\overset{..}{O}}{\|}}{C}- \quad \longleftrightarrow \quad -CH=\overset{\overset{\displaystyle :\overset{..}{O}:^-}{|}}{C}- \right] \quad or \quad -\overset{\delta-}{CH}\overset{\overset{\displaystyle O}{\|}}{\equiv}\overset{\delta-}{C}-$$

Adjacent to two carbonyl groups:

$$\left[-\overset{\overset{\displaystyle :\overset{..}{O}}{\|}}{C}-\overset{..}{C}H=\overset{\overset{\displaystyle :\overset{..}{O}:^-}{}}{C}- \quad \longleftrightarrow \quad -\overset{\overset{\displaystyle :\overset{..}{O}}{\|}}{C}-\overset{..}{C}H-\overset{\overset{\displaystyle \overset{..}{O}:}{\|}}{C}- \quad \longleftrightarrow \quad -\overset{\overset{\displaystyle :\overset{..}{O}:^-}{}}{C}=CH-\overset{\overset{\displaystyle \overset{..}{O}:}{\|}}{C}- \right] \quad or \quad -\overset{\overset{\displaystyle \overset{\delta-}{O}}{}}{C}\equiv CH\overset{\overset{\displaystyle \overset{\delta-}{O}}{}}{\underset{\delta-}{\equiv}}C-$$

STUDY PROBLEM

13.19 Give the resonance structures of the enolate ions formed when the following diones are treated with sodium ethoxide:

(a) $\text{C}_6\text{H}_5-\overset{\overset{\displaystyle O}{\|}}{C}\text{CH}_2\overset{\overset{\displaystyle O}{\|}}{C}\text{CH}_3$ (b) $\text{CH}_3\overset{\overset{\displaystyle O}{\|}}{C}\text{CH}_2\overset{\overset{\displaystyle O}{\|}}{C}\text{CH}_3$

Section 13.9

Tautomerism

Even if a strong base is not present, the acidity of an α hydrogen may be evident. A carbonyl compound with an acidic α hydrogen may exist in two forms called **tautomers:** a special type of interconvertible structural isomers that differ from each other only in the location of a double bond and a hydrogen atom relative to the oxygen. The two tautomers of a simple ketone are called the *keto tautomer* and the *enol tautomer.* The keto tautomer of a carbonyl compound has the expected carbonyl structure. The enol tautomer (from *-ene* + *-ol*), which is a vinylic alcohol, is formed by transfer of an acidic hydrogen from the α carbon to the carbonyl oxygen. Because a hydrogen atom is in different positions, the two tautomeric forms are not resonance structures, but are two different structures in equilibrium. (Remember that resonance structures vary only in the positions of *electrons*.)

$$\underset{\text{keto form}}{-\overset{\overset{\displaystyle :\ddot{O}}{\|}}{C}-\overset{\overset{\displaystyle H}{|}}{\underset{|}{C}}-} \; \rightleftharpoons \; \underset{\text{enol form}}{-\overset{\overset{\displaystyle :\ddot{O}-H}{|}}{C}=\overset{}{\underset{|}{C}}-}$$

$$\underset{\substack{\text{keto form} \\ \text{of acetone}}}{CH_3-\overset{\overset{\displaystyle O}{\|}}{C}-\overset{\overset{\displaystyle H}{|}}{\underset{|}{C}}H_2} \; \rightleftharpoons \; \underset{\substack{\text{enol form} \\ \text{of acetone}}}{CH_3-\overset{\overset{\displaystyle OH}{|}}{C}=CH_2}$$

The relative quantities of enol versus keto tautomer in a pure liquid can be estimated by infrared or NMR spectroscopy. Acetone exists primarily in the keto form (99.99%, as determined by a specialized titration procedure). Most other simple aldehydes and ketones also exist primarily in their keto forms. However, 2,4-pentanedione exists as 80% enol! How can this tremendous difference be explained? Let us consider the structures of the 2,4-pentanedione tautomers:

keto form (20%)   enol form (80%)

The enol form not only has conjugated double bonds, which add a small amount of stability, but it is also structurally arranged for internal hydrogen bonding, which helps stabilize this tautomer.

SAMPLE PROBLEM

Suggest reasons why 1,2-cyclohexanedione exists 100% in an enol form.

Solution

dipole–dipole   repulsions relieved;
repulsions   stabilized by hydrogen
bonding

STUDY PROBLEM

13.20 A trace of either acid or base will catalyze a keto-enol tautomerization. Consequently, it is difficult to isolate a pure tautomer free of the other form. In 1911, the German chemist Ludwig Knorr isolated the keto tautomer of ethyl acetoacetate by crystallization at $-78°$, a temperature at which the rate of equilibration is slow. The formula for the keto tautomer follows. Write the formula for the enol tautomer that illustrates its hydrogen bonding.

$$CH_3\overset{\overset{\displaystyle O}{\|}}{C}CH_2\overset{\overset{\displaystyle O}{\|}}{C}OCH_2CH_3$$

ethyl acetoacetate

Tautomerism can affect the reactivity of a compound. An exception to the generality that ketones are not easily oxidized is the oxidation of a ketone with at least one α hydrogen. A ketone that can undergo tautomerization can be oxidized by a strong oxidizing agent at the carbon–carbon double bond of the enol tautomer. Yields in this reaction are poor because, under these conditions, other C—C bonds may be cleaved. This reaction is not used in synthetic work, but is used often in structure determinations.

$$
\begin{array}{c}
\underset{\text{OH}}{\text{CH}_3\text{CH}_2\text{CH}\!=\!\overset{|}{\text{C}}\text{CH}_2\text{CH}_3} \xrightarrow[\text{heat}]{\text{conc. HNO}_3} 2\ \text{CH}_3\text{CH}_2\text{CO}_2\text{H} \\
\text{propanoic acid}
\end{array}
$$

$$
\underset{\text{3-hexanone}}{\text{CH}_3\text{CH}_2\text{CH}_2\!-\!\overset{\text{O}}{\overset{\|}{\text{C}}}\!-\!\text{CH}_2\text{CH}_3} \rightleftharpoons
$$

+

$$
\begin{array}{c}
\underset{\text{OH}}{\text{CH}_3\text{CH}_2\text{CH}_2\overset{|}{\text{C}}\!=\!\text{CHCH}_3} \xrightarrow[\text{heat}]{\text{conc. HNO}_3}
\left\{
\begin{array}{c}
\text{CH}_3\text{CH}_2\text{CH}_2\text{CO}_2\text{H} \\
\text{butanoic acid} \\
+ \\
\text{CH}_3\text{CO}_2\text{H} \\
\text{acetic acid}
\end{array}
\right.
\end{array}
$$

Alpha Halogenation

Ketones are readily halogenated at the α carbon. The reaction requires either alkaline conditions or an acidic catalyst. (Note that base is a *reactant*, whereas acid is a *catalyst*.)

In base:

$$
\underset{\text{acetone}}{\text{CH}_3\overset{\text{O}}{\overset{\|}{\text{C}}}\text{CH}_3} + \text{Br}_2 + \text{OH}^- \longrightarrow \underset{\text{bromoacetone}}{\text{BrCH}_2\overset{\text{O}}{\overset{\|}{\text{C}}}\text{CH}_3} + \text{Br}^- + \text{H}_2\text{O}
$$

$$
\underset{\text{cyclohexanone}}{\bigcirc\!\!=\!\text{O}} + \text{Br}_2 + \text{OH}^- \longrightarrow \underset{\substack{\text{2-bromo-}\\\text{cyclohexanone}}}{\overset{}{\bigcirc}\!\!=\!\text{O}} + \text{Br}^- + \text{H}_2\text{O}
$$

In acid:

$$
\text{CH}_3\overset{\text{O}}{\overset{\|}{\text{C}}}\text{CH}_3 + \text{Br}_2 \xrightarrow{\text{H}^+} \text{BrCH}_2\overset{\text{O}}{\overset{\|}{\text{C}}}\text{CH}_3 + \text{HBr}
$$

$$
\bigcirc\!\!=\!\text{O} + \text{Br}_2 \xrightarrow{\text{H}^+} \overset{}{\bigcirc}\!\!=\!\text{O} + \text{HBr}
$$

The first step (the slow step) in the reaction under alkaline conditions is the formation of the enolate ion. The anion of a ketone with only one carbonyl group is a much stronger base than the hydroxide ion. Therefore, the acid–base equilibrium favors the hydroxide ion rather than the enolate ion. Nonetheless, a *few* enolate ions exist in alkaline solution. As these few anions are used up, more are generated to go on to Step 2. In Step 2, the enolate ion quickly undergoes reaction with halogen to yield the α-halogenated ketone and a halide ion.

In base:

Step 1 (slow):

$$CH_3\overset{\overset{\displaystyle O}{\|}}{C}CH_3 + OH^- \underset{-H_2O}{\longleftarrow} \left[CH_3\overset{\overset{\displaystyle \ddot{O}:}{\|}}{C}\underset{\cdot}{\ddot{C}}H_2 \longleftrightarrow CH_3\overset{\overset{\displaystyle :\ddot{O}:^-}{|}}{C}=CH_2 \right]$$

resonance structures for the enolate ion

Step 2 (fast):

$$CH_3\overset{\overset{\displaystyle O}{\|}}{C}\ddot{C}H_2 + :\ddot{B}r\!-\!\ddot{B}r: \longrightarrow CH_3\overset{\overset{\displaystyle O}{\|}}{C}CH_2\ddot{B}r: + :\ddot{B}r:^-$$

Alpha halogenation in acid usually gives higher yields than the reaction in base. The acid-catalyzed reaction proceeds by way of the enol, the formation of which is the rate-determining step. The carbon–carbon double bond of the enol undergoes electrophilic addition, just like any carbon–carbon double bond, to form the more stable carbocation. In this case, the more stable carbocation is the one in which the positive charge is on the carbon of the carbonyl group (because this intermediate is resonance-stabilized). This carbocation intermediate quickly loses a proton and forms the ketone, which is now halogenated in the α position.

In acid:

Step 1 (fast):

$$CH_3\overset{\overset{\displaystyle \ddot{O}}{\|}}{C}CH_3 + H^+ \rightleftharpoons \left[CH_3\overset{\overset{\displaystyle ^+\ddot{O}H}{\|}}{C}CH_3 \right]$$

Step 2 (slow):

$$\left[CH_3\overset{\overset{\displaystyle ^+\ddot{O}H}{\|}}{C}\!-\!\underset{\underset{\displaystyle H}{|}}{C}H_2 \right] \rightleftharpoons \left[CH_3\overset{\overset{\displaystyle :\ddot{O}H}{|}}{C}=CH_2 \right] + H^+$$

the enol

Step 3 (fast):

$$\left[CH_3\overset{\overset{\displaystyle :\ddot{O}H}{|}}{C}=CH_2 \right] + Br\!-\!Br \longrightarrow \left[CH_3\overset{\overset{\displaystyle ^+\ddot{O}H}{\|}}{C}\!-\!CH_2Br \right] + Br^-$$

Step 4 (fast):

$$\left[CH_3\overset{\overset{\displaystyle ^+\ddot{O}H}{\|}}{C}CH_2Br \right] \rightleftharpoons CH_3\overset{\overset{\displaystyle \ddot{O}}{\|}}{C}CH_2Br + H^+$$

13.21 Predict the organic products:

(a) [structure: phenyl ring]—$\overset{\overset{\displaystyle O}{\|}}{C}CH_3$ + Cl_2 $\xrightarrow{H^+}$ (b) [cyclohexane-1,3-dione structure] + 2 Br_2 $\xrightarrow{-OH}$

A. Haloform Reaction

Alpha halogenation is the basis of the **iodoform test** for methyl ketones. The methyl group of a methyl ketone is iodinated stepwise until the yellow solid iodoform (CHI_3) is formed.

Iodoform test:

[reaction scheme]

$\overset{\overset{\displaystyle O}{\|}}{C}CH_3$ + 3 I_2 $\xrightarrow[H_2O]{OH^-}$ $\left[\overset{\overset{\displaystyle O}{\|}}{C}CI_3\right]$ $\xrightarrow{OH^-}$ $\overset{\overset{\displaystyle O}{\|}}{C}O^-$ + CHI_3

cyclohexyl
methyl ketone

*intermediate
triiodomethyl ketone*

cyclohexane-
carboxylate ion

iodoform

yellow solid

The intermediate is unstable in base because a trihalomethyl anion is a good leaving group.

[reaction scheme]

$\underset{\underset{:\ddot{O}H^-}{\overset{\displaystyle \curvearrowleft}{}}}{RC}\!-\!CI_3$ $\rightleftharpoons$ $\underset{:\ddot{O}H}{\overset{:\ddot{O}:^-}{RC}}\!-\!CI_3$ $\longrightarrow$ $\left[\underset{:\ddot{O}H}{\overset{:\ddot{O}\cdot}{RC}} + {}^-CI_3\right]$ $\xrightarrow[\text{transfer}]{\text{proton}}$ $\underset{:\ddot{O}:^-}{\overset{\overset{\displaystyle O}{\|}}{RC}}$ + CHI_3

The test is not uniquely specific for methyl ketones. Iodine is a mild oxidizing agent, and any compound that can be oxidized to a methyl carbonyl compound also gives a positive test.

$\underset{\text{2-propanol}}{\overset{\overset{\displaystyle OH}{|}}{CH_3CHCH_3}}$ $\xrightarrow[OH^-]{I_2}$ $\underset{\text{acetone}}{\overset{\overset{\displaystyle O}{\|}}{CH_3CCH_3}}$ $\xrightarrow[OH^-]{I_2}$ $\underset{\text{acetate ion}}{\overset{\overset{\displaystyle O}{\|}}{CH_3CO^-}}$ + CHI_3

Bromine and chlorine also react with methyl ketones to yield bromoform ($CHBr_3$) and chloroform ($CHCl_3$), respectively. "Haloform" is the general term used to describe CHX_3; hence this reaction is often referred to as the **haloform reaction.** Because bromoform and chloroform are nondistinctive liquids, their formations are not useful for test purposes. However, the reaction of a methyl ketone with any of these halogens provides a method for the conversion of these compounds to carboxylic acids.

$\underset{\text{3,3-dimethyl-2-butanone}}{(CH_3)_3C\overset{\overset{\displaystyle O}{\|}}{C}CH_3}$ $\xrightarrow[\text{(2) } H^+]{\text{(1) 3 } Br_2, OH^-}$ $\underset{\text{2,2-dimethylpropanoic acid (74\%)}}{(CH_3)_3C\overset{\overset{\displaystyle O}{\|}}{C}OH}$

STUDY PROBLEM

13.22 What methyl ketones can be used to prepare the following carboxylic acids by haloform reactions?

(a) [structure: bicyclic decalin ring with CO$_2$H substituent] (b) $(CH_3)_2CHCO_2H$ (c) HO_2C—[benzene ring]—CO_2H

Section 13.11
Use of Aldehydes and Ketones in Synthesis

Aldehydes and ketones are readily available by the oxidation of alcohols and can be converted to a variety of other types of compounds, as may be seen in Table 13.5. When viewed from a synthesis standpoint, the reactions in this table can be grouped into three major categories:

1. reactions in which the carbonyl group is retained (for example, α halogenation)

2. reactions in which the carbonyl group is converted to another functional group (for example, reduction or conversion to a hemiacetal)

3. reactions in which extension of the carbon skeleton occurs at the carbonyl group (Grignard and Wittig reactions).

In designing synthetic sequences, you should keep all three classes of reactions in mind.

SAMPLE PROBLEMS

How would you make the following conversion?

$$CH_3\overset{\overset{\displaystyle O}{\|}}{C}CH_3 \longrightarrow CH_3\overset{\overset{\displaystyle O}{\|}}{C}CH_2OH$$

Solution
In this conversion, the carbonyl group is retained, but a functional group is inserted α to the C=O group. Alpha halogenation is a way to insert an α functional group.

$$CH_3\overset{\overset{\displaystyle O}{\|}}{C}CH_3 \xrightarrow{\;Cl_2,\,H^+\;} CH_3\overset{\overset{\displaystyle O}{\|}}{C}CH_2Cl$$

The alcohol can then be obtained by treatment of the α-chloroketone with aqueous NaOH (S$_N$2 reaction).

$$CH_3\overset{\overset{\displaystyle O}{\|}}{C}CH_2Cl \xrightarrow{\;^-OH\;} CH_3\overset{\overset{\displaystyle O}{\|}}{C}CH_2OH$$

Table 13.5	Types of compounds that can be obtained from aldehydes and ketones[a]		

Reaction		Product	Section Reference

Addition:[b]

$$RCHO + 2\ R'OH \xrightarrow{H^+} RC\overset{\overset{\displaystyle OR'}{|}}{H}OR'$$ acetal 13.4B

$$RCHO + HCN \xrightarrow{CN^-} RC\overset{\overset{\displaystyle OH}{|}}{H}CN$$ cyanohydrin 13.4C

$$R\overset{\overset{\displaystyle O}{\|}}{C}R \xrightarrow[\text{(2) } H_2O,\ H^+]{\text{(1) } R'MgX} R\overset{\overset{\displaystyle OH}{|}}{\underset{\underset{\displaystyle R'}{|}}{C}}R$$ alcohol 7.3D, 13.4D

Addition–Elimination:[b]

$RCHO + R'NH_2 \longrightarrow RCH{=}NR'$ imine 13.5A
$RCH_2CHO + R'_2NH \longrightarrow RCH{=}CHNR'_2$ enamine 13.5B, 17.5
$RCHO + R'NHNH_2 \longrightarrow RCH{=}NNHR'$ hydrazone 13.5C
$R_2C{=}O + (C_6H_5)_3P{=}CR'_2 \longrightarrow R_2C{=}CR'_2$ alkene 13.5D

Reduction:

$$R_2C{=}O \xrightarrow[\text{or metal hydride}]{H_2,\ \text{catalyst}} R_2CHOH$$ alcohol 13.6A,B

$$R_2C{=}O \xrightarrow[\text{or Zn/Hg, HCl}]{\text{(1) } NH_2NH_2,\ \text{(2) } OH^-} R_2CH_2$$ alkane or alkylbenzene 13.6C

$$R_2C{=}O \xrightarrow[H_2,\ Ni]{R'_2NH} R_2CHNR'_2$$ amine 13.6D, 18.4B

α Halogenation:

$$RCH_2\overset{\overset{\displaystyle O}{\|}}{C}R + X_2 \xrightarrow{H^+} RC\overset{\overset{\displaystyle O}{\|}}{\underset{\underset{\displaystyle X}{|}}{H}}CR$$ α-halo carbonyl 13.10

$$R\overset{\overset{\displaystyle O}{\|}}{C}CH_3 + 3\ X_2 \xrightarrow[\text{(2) } H^+]{\text{(1) } OH^-} RCO_2H$$ carboxylic acid 13.10A

[a] The condensation and alkylation reactions of aldehydes and ketones will be discussed in Chapter 17.
[b] Nonhindered ketones, such as methyl ketones, can also be used.

How would you make the following conversion?

Solution
This conversion involves an extension of the carbon skeleton at the carbonyl group. The Wittig reaction could be used for this conversion.

$$C_6H_5\overset{\overset{\displaystyle O}{\|}}{C}H \xrightarrow{(C_6H_5)_3P{=}CHCH_2CH_3} \text{product}$$

The product could also be obtained from the alcohol $C_6H_5CH(OH)CH_2CH_2CH_3$ by dehydration (spontaneous in this case because the product C=C is in conjugation with a benzene ring). The alcohol, in turn, can be obtained by a Grignard reaction, another reaction that allows us to build up a carbon skeleton.

$$C_6H_5\overset{\overset{\displaystyle O}{\|}}{C}H \xrightarrow[\text{(2) H}_2\text{O, H}^+]{\text{(1) CH}_3\text{CH}_2\text{CH}_2\text{MgBr}} C_6H_5\overset{\overset{\displaystyle OH}{|}}{C}HCH_2CH_2CH_3 \xrightarrow{-\,\text{H}_2\text{O}} \text{product}$$

Suggest a synthesis for *N*-ethyl-3-hexanamine from organic compounds containing three or fewer carbon atoms.

Solution

1. Write the structure: $CH_3CH_2\overset{\overset{\displaystyle NHCH_2CH_3}{|}}{C}HCH_2CH_2CH_3$

2. It is apparent that more than one step will be needed to synthesize this compound. We must both build up a carbon skeleton and also insert the ethylamino group. Let us consider the amino group first. It could be placed in this structure by reductive amination of a ketone.

$$CH_3CH_2\overset{\overset{\displaystyle O}{\|}}{C}CH_2CH_2CH_3 \xrightarrow[\text{H}_2,\ \text{Ni}]{\text{CH}_3\text{CH}_2\text{NH}_2} \text{product}$$

3. Is there a one-step reaction to 3-hexanone from starting materials containing three or fewer carbons? No, but a Grignard reaction can provide us with an alcohol, which is readily oxidized to a ketone. Thus, we can write the initial series of reactions.

$$CH_3CH_2\overset{\overset{\displaystyle O}{\|}}{C}H \xrightarrow[\text{(2) H}_2\text{O, H}^+]{\text{(1) CH}_3\text{CH}_2\text{CH}_2\text{MgBr}} CH_3CH_2\overset{\overset{\displaystyle OH}{|}}{C}HCH_2CH_2CH_3 \xrightarrow{\text{H}_2\text{CrO}_4}$$

$$CH_3CH_2\overset{\overset{\displaystyle O}{\|}}{C}CH_2CH_2CH_3$$

The Grignard reagent can be prepared by the reaction of RX with Mg.

$$CH_3CH_2CH_2Br \xrightarrow[\text{diethyl ether}]{\text{Mg}} CH_3CH_2CH_2MgBr$$

The proposed synthesis is complete, and the series of reactions may be written forward.

STUDY PROBLEM

13.23   Write flow equations for syntheses of the following compounds:

   (a) 4-methoxy-2-phenyl-2-butene from compounds containing six or fewer carbons

   (b) 4-methyl-1,3-pentadiene from acetone

   (c) 3,5-heptanediol from compounds containing four or fewer carbons

(d) 2,3-dihydroxycyclohexane from cyclohexanone

(e) 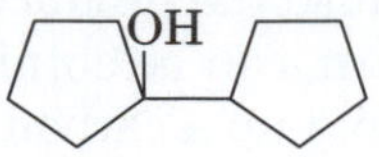 from cyclopentanol

(f) 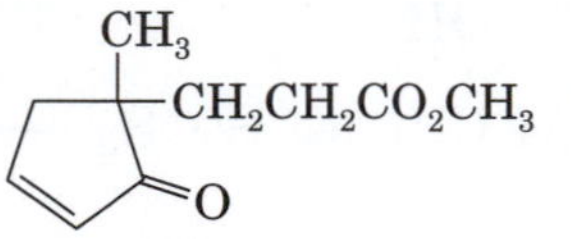 from

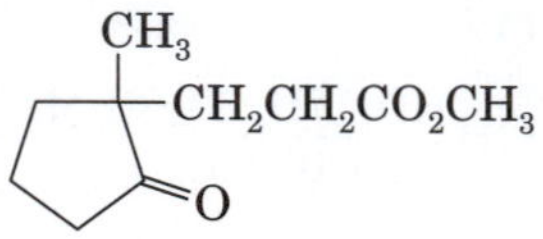

Summary

The **carbonyl group** is *planar* and *polar,* and the oxygen has *two pairs of unshared valence electrons.* The carbonyl group can undergo *electrophilic* or *nucleophilic* attack.

Because of (1) *inductive stabilization* of the partial positive charge on the carbonyl carbon, and (2) *steric hindrance,* ketones are less reactive than aldehydes.

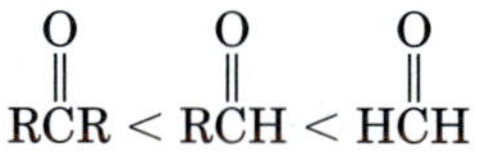

$$\overset{O}{\underset{\parallel}{RCR}} < \overset{O}{\underset{\parallel}{RCH}} < \overset{O}{\underset{\parallel}{HCH}}$$

Many reactions of aldehydes and ketones are simple **addition reactions** at the $C=O$ π bond. These reactions can lead to hydrates, hemiacetals (or hemiketals) and cyanohydrins.

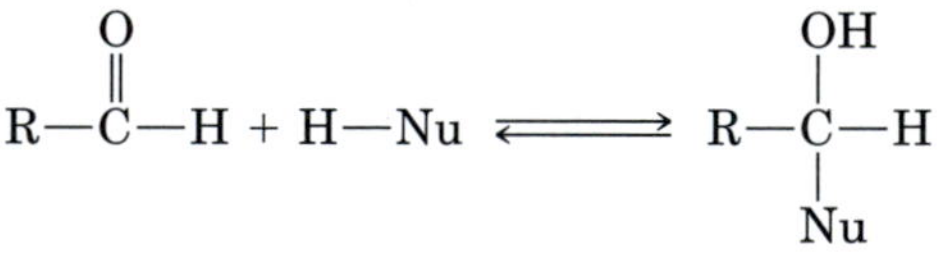

where HNu = H_2O, ROH, or HCN

Other addition reactions are **reduction** and **Grignard reactions.**

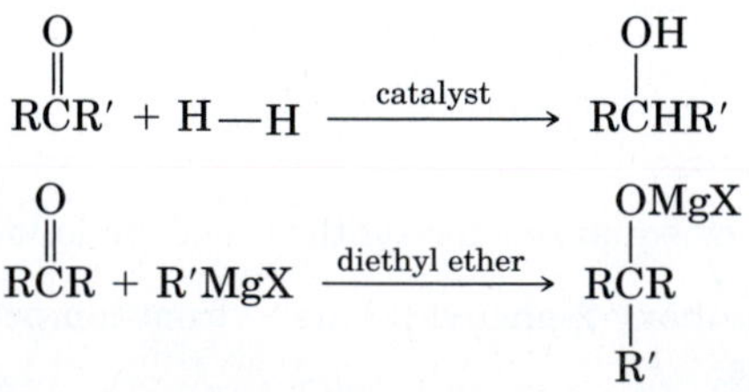

Substitution reactions of aldehydes and ketones are the result of initial addition reactions followed by elimination, or **addition–elimination reac-**

tions. The formation of imines, enamines, hydrazones, and alkenes (by the Wittig reaction) all fall into this category. These reactions are summarized in Table 13.5.

$$RCR \xrightarrow[\text{addition}]{R'NH_2} \left[\begin{array}{c} OH \\ | \\ RCR \\ | \\ NHR' \end{array} \right] \xrightarrow[\text{elimination}]{-H_2O} RCR$$

Because the carbonyl group is polar and because its π electrons can participate in resonance-stabilization, an *α hydrogen is acidic,* especially if it is α to two carbonyl groups. This acidity can give rise to **tautomerism.** Because of tautomerism, ketones can undergo *α **halogenation,*** as shown in Table 13.5, or **oxidative cleavage** between the carbonyl carbon and the α carbon.

acidic

$$CH_3CH + OH^- \;\rightleftharpoons\; {}^-{:}CH_2CH + H_2O$$

$$CH_3CCH_2CCH_3 + OH^- \;\rightleftharpoons\; CH_3CCHCCH_3 + H_2O$$

$$CH_3C-CH-CCH_3 \;\rightleftharpoons\; CH_3C=CH-CCH_3$$

keto tautomer   *enol tautomer*

Essay Problem for Chapter 13

Titanium-Induced Deoxygenations

When titanium(III) salts are reduced with an alkali metal such as potassium or lithium, a very reactive metallic titanium powder is formed. Active titanium metal is a potent oxygen scavenger and can be used to reductively remove oxygen from organic compounds. When the reactant is a glycol, the product is an alkene. Typical reactions follow:

$$\text{(glycol)} \xrightarrow{K/TiCl_3} \text{(alkene)} \quad (85\%)$$

$$\underset{C_4H_9}{\overset{OH}{H\cdots C}}-\underset{C_4H_9}{\overset{OH}{C\cdots H}} \xrightarrow{K/TiCl_3} \underset{C_4H_9}{\overset{H}{}}C=C\underset{H}{\overset{C_4H_9}{}} \quad (75\%)$$

(trans, 60%; *cis,* 40%)

The mechanism involves an initial formation of a five-membered ring. This step is the redox reaction in the mechanism. The alkene is formed by a nonconcerted collapse of the intermediate. The collapse is believed to occur by a free-radical pathway.

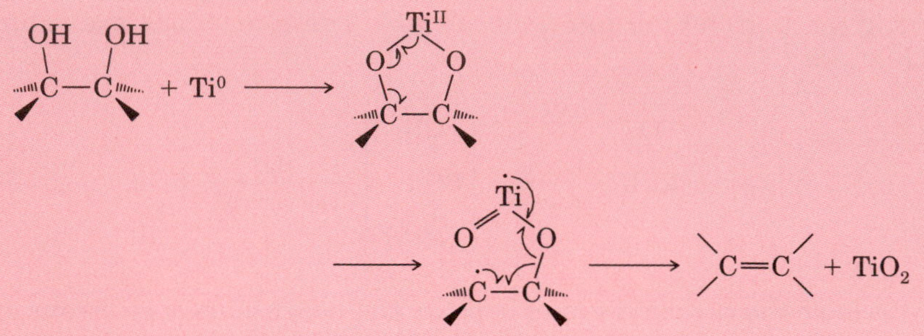

With aldehydes and ketones, the reaction involves an initial reductive coupling, which forms a glycol. Deoxygenation of the glycol gives an alkene. When aldehydes or unsymmetrical ketones are used as starting materials, a mixture of *cis* and *trans* isomers is produced. (M. P. Fleming and J. E. McMurry, *Org. Syn.* **1981**, *60,* 113.)

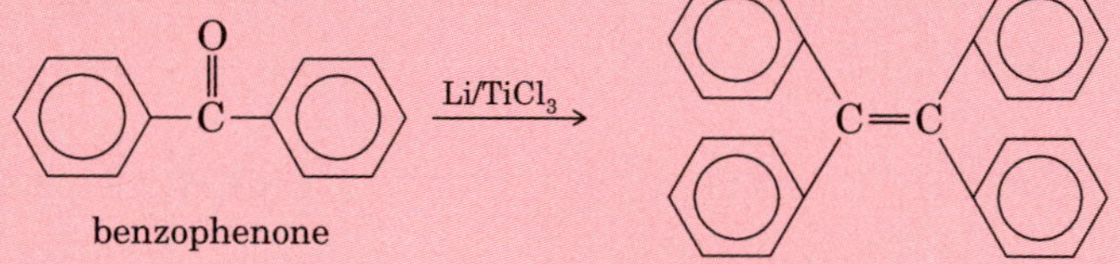

benzophenone

1,1,2,2,-tetraphenylethene (96%)

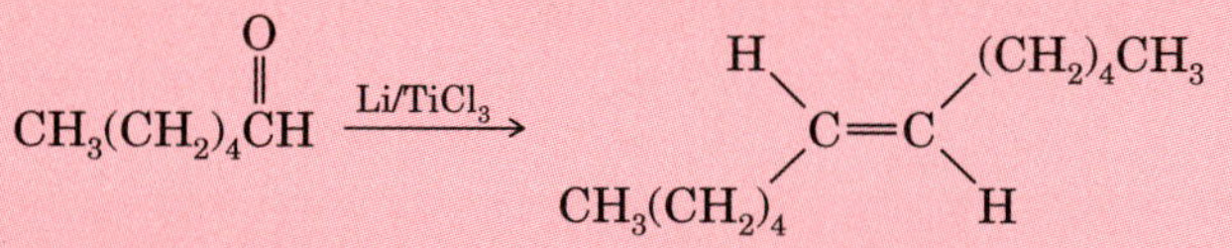

$CH_3(CH_2)_4CH$ → $\xrightarrow{Li/TiCl_3}$

(58%)

(*trans,* 72%; *cis,* 28%)

Questions

1. How many stereoisomers are there of $C_4H_9CH-CHC_4H_9$?

2. If (*S*),(*S*)- or (*R*),(*R*)-$C_4H_9CH-CHC_4H_9$ was used as the starting material for the reaction, would you expect the *cis* or the *trans* geometric isomer to predominate in the reaction mixture?

3. What is the evidence against the following concerted mechanism for the collapse of the intermediate?

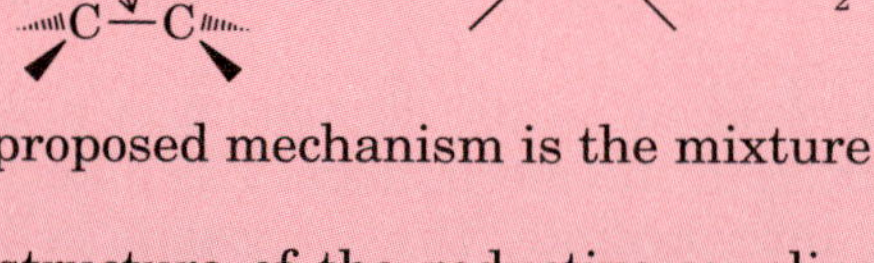

4. At which step in the proposed mechanism is the mixture of geometric isomers formed?

5. What would be the structure of the reductive-coupling intermediates from benzophenone?

13.24 Write the structure for each of the following compounds:

   (a) 2-phenylpropanal   **(b)** 1,3-cyclopentanedione

   (c) methyl vinyl ketone   **(d)** β-hydroxybutyraldehyde

   (e) 4-methyl-2-hexanone   **(f)** cycloheptanone

13.25 Each of the following compounds is dissolved in water to which a trace of HCl has been added. Give the structures of any other compounds (besides HCl, water, and the compound in question) that would be found in each solution.

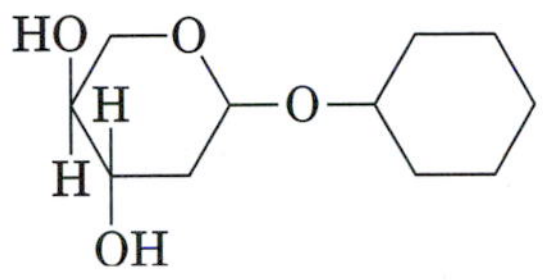

13.26 Write equations for the reactions of **(a)** cyclohexanone with 1,2-ethanethiol ($HSCH_2CH_2SH$) and a trace of acid; **(b)** formaldehyde ($HCHO$) with aqueous potassium cyanide and sulfuric acid; **(c)** *cis*- or *trans*-tetrahydropyran-2,3-diol with methanol and a trace of acid.

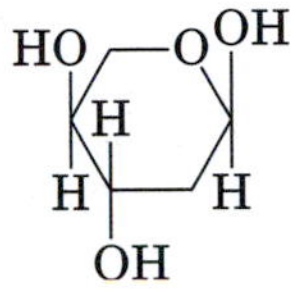

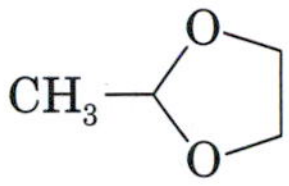

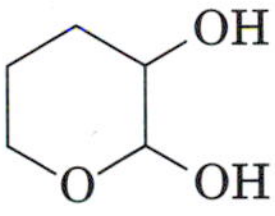

tetrahydropyran-2,3-diol

13.27 Predict the major organic products:

(a)

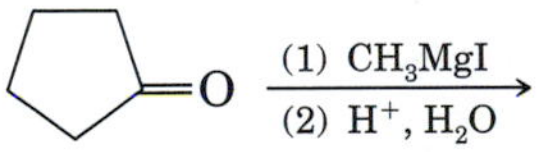

(b)

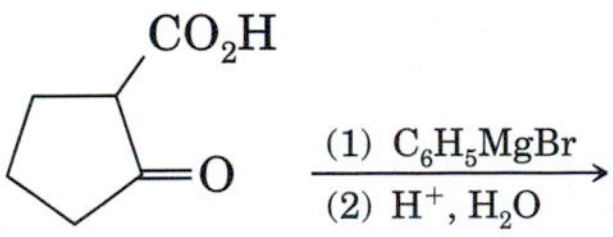

(c) 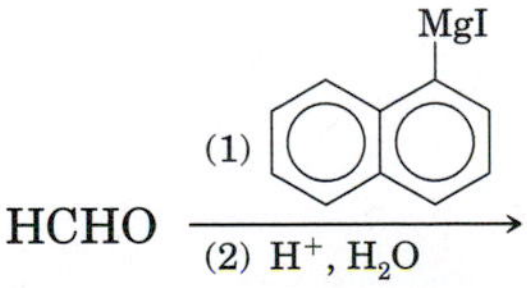   HCHO

13.28 Write equations for Grignard syntheses of the following alcohols from aldehydes or ketones. Give all possible correct answers.

(a)

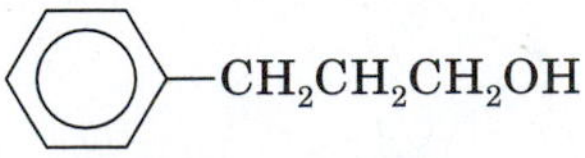

(b)

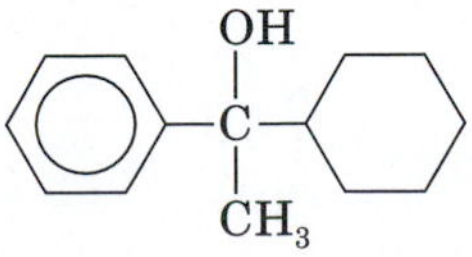

* For information concerning the organization of the *Study Problems* and *Additional Problems,* see the *Note to student* on page 41.

13.29 Show how you would prepare the following compounds from compounds containing six or fewer carbon atoms:

(a) $CH_3CH_2CH{=}CHN(CH_2CH_3)_2$ (b)

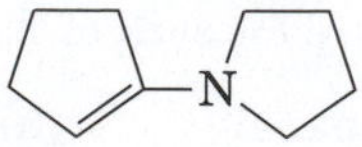

(c) $CH_3CH_2CH{=}NC_6H_5$ (d) $CH_3CH_2CH{=}NNHC_6H_5$

13.30 List the following compounds in terms of reactivity toward 2,4-dinitrophenylhydrazine (least reactive first): (a) 2-pentanone; (b) 3-pentanone; (c) pentanal.

13.31 Predict the organic products:

(a)

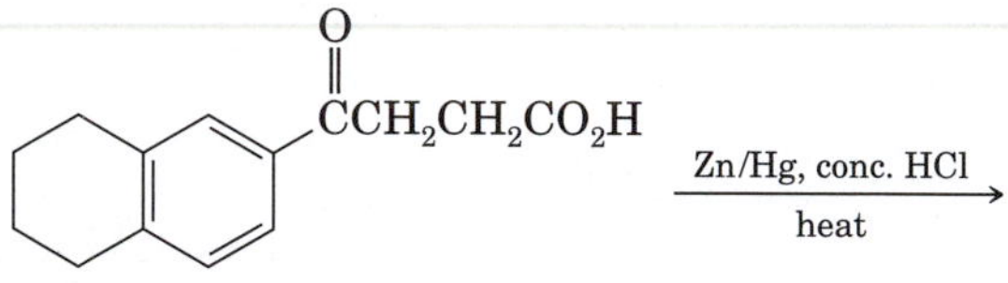

(b)

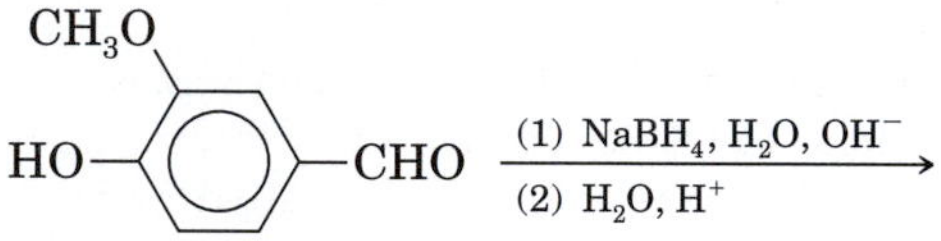

vanillin

(c)

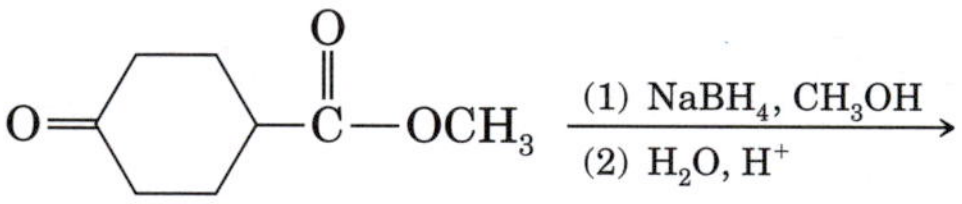

(d) $CH_3\overset{O}{\underset{\|}{C}}CH_3 + HOCH_2CH_2NH_2 \xrightarrow{H_2,\ Pt}$

(e)

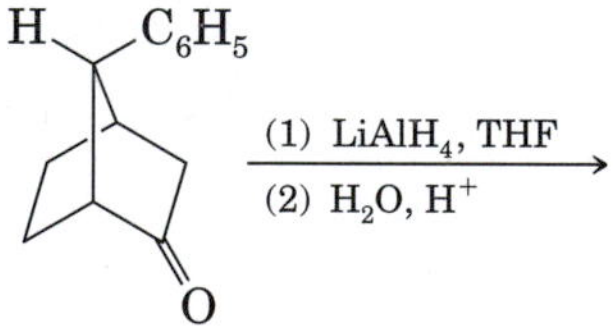

13.32 How would you effect the following conversions by Wittig reactions? (Start with organic halides.)

(a)

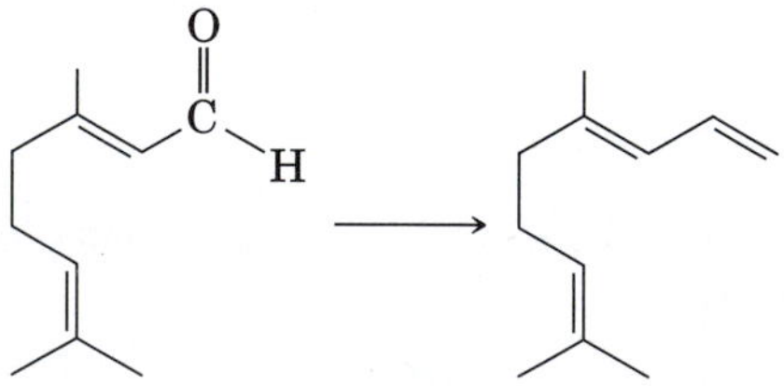

(b)

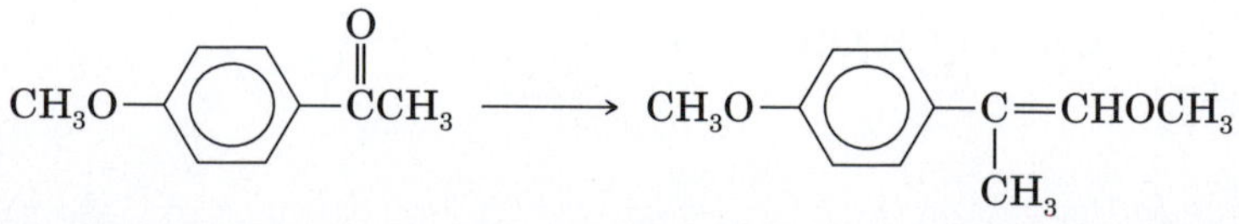

13.33 Suggest a synthetic scheme to convert 3-hydroxypropanal to HO_2CCH_2CHO.

13.34 Complete the following equations for acid–base reactions:

(a) $CH_3\overset{\overset{O}{\|}}{C}H + OH^- \;\rightleftharpoons$

(b)   (cyclopentane-1,3-dione structure) $+ CO_3{}^{2-} \;\rightleftharpoons$

(c) $(C_6H_5)_2CH\overset{\overset{O}{\|}}{C}OCH_2CH_3 + Na^+\; {}^-OCH_2CH_3 \;\rightleftharpoons$

(d) $(S)\text{-}CH_3CH_2CH_2\overset{\overset{CH_3}{|}}{C}HCHO + OH^- \;\rightleftharpoons$

13.35 Write equations that illustrate tautomerism of the following compounds:

(a) (cyclopentane-1,2-dione)   (b) (C$_6$H$_5$—CCH$_3$ with =O, acetophenone)   (c) (cyclic HN—C(=O)—NH)

(d) (cyclohexyl—N=C with =O)   (e) (anthracene ring system with CHOH and O)

13.36 When a *cis*-3-alkyl-4-*tert*-butylcyclohexanone is treated with LiAlH$_4$, the predominant product is that in which the OH is *cis* to the *tert*-butyl group and the alkyl group. Suggest a reason.

13.37 The LiAlH$_4$ reduction of a stereoisomeric 3-phenyl-2-pentanone yielded 75% of the $(2R,3S)$ alcohol and 25% of the $(2S,3S)$ alcohol.

(a)   What was the stereochemistry of the starting ketone?

(b)   Why was the $(2R,3S)$ alcohol the predomiant product? (*Hint:* Use models.)

13.38 What would be observed if each of the following compounds were placed in a test tube and a solution of I$_2$ in dilute aqueous NaOH were added? Write equations where appropriate.

(a) (phenyl—CCH$_3$ with =O)   (b) (phenyl—CH$_2$COCH$_3$ with =O)   (c) (phenyl—CHCH$_3$ with OH)

13.39 Write flow equations for syntheses of the following compounds:

(a) (bicyclic structure with CH$_3$, OH, CHCH=CH$_2$, CH$_3$, CH$_3$O, CH$_3$ substituents) from (bicyclic structure with CH$_3$, O, CH$_3$O, CH$_3$ substituents)

(b) [structure: 2,3-dimethylquinoxaline] from $CH_3\overset{O}{\overset{\|}{C}}-\overset{O}{\overset{\|}{C}}CH_3$

(c) CH_2=[cyclohexane ring]=$CHCH$=CH_2 from CH_2=[cyclohexane ring]=O

(d) CH_3CH=$CH-\overset{OH}{\overset{|}{C}}HC_6H_5$ from $CH_3CH_2CH_2\overset{O}{\overset{\|}{C}}C_6H_5$

(e) 1,2,3-decanetriol from octanal

(f) [bicyclic structure with H and CH_2OH] from [bicyclic structure with OH and H]

Additional Problems

13.40 Predict the product of the reaction of cyclopentanone with:

(a) Br_2 in acetic acid

(b) $NaBH_4$, followed by H^+, H_2O

(c) phenylhydrazine with H^+

(d) $(CH_3)_2CHMgBr$, followed by H^+, H_2O

(e) $(C_6H_5)_3P$=$CHCH_3$

(f) hydrazine, followed by heating in K^+ $^-OC(CH_3)_3$ solution

13.41 Treatment of one mole of $LiAlH_4$ with three moles of CH_3CH_2OH yields lithium tri-ethoxyaluminum hydride, $LiAlH(OCH_2CH_3)_3$. When 3,3,5-trimethylcyclohexanone is treated with this reducing agent, followed by hydrolysis, an 83% yield of one diastereomer of an alcohol is obtained. Write the formula for this diastereomer, and explain why it is the principal product.

13.42 Aldehydes and nonhindered ketones (primarily methyl ketones) yield water-soluble **bisulfite addition products** when treated with concentrated, aqueous sodium bisulfite.

$$R\overset{O}{\overset{\|}{C}}H + Na^+ \ ^-SO_3H \longrightarrow R\overset{OH}{\overset{|}{C}}H-SO_3^- \ Na^+$$

bisulfite addition product

This reaction is sometimes used to separate an aldehyde or ketone from water-insoluble organic compounds. (The aldehyde or ketone can be regenerated by treatment of the aqueous bisulfite solution with acid or base.)

(a) Suggest a mechanism for the formation of the bisulfite addition product of acetaldehyde.

(b) Explain how the following compounds (in a diethyl ether solution) could be separated from one another by a series of extractions: heptanoic acid, 4-heptanone, and heptanal.

13.43 In part of the synthesis of a fungitoxin found in brown seaweed, the following conversion was carried out. Suggest a reaction sequence for this conversion.

13.44 *Glutaraldehyde* forms a cyclic hydrate that contains one equivalent of water. What is its structure, and how is it formed? (*Hint:* If *one* aldehyde group is hydrated, what reaction might the molecule undergo?)

$$\overset{\text{O}}{\overset{\|}{\text{HC}}}(\text{CH}_2)_3\overset{\text{O}}{\overset{\|}{\text{CH}}}$$

glutaraldehyde

used as an antiseptic

13.45 *Civetone* is an active ingredient of *civet,* a mixture isolated from the scent glands of the African civet cat and used in perfumes. Civetone, which has the formula $C_{17}H_{30}O$, shows strong absorption in the infrared spectrum at 1700 cm^{-1} (5.8 μm) and shows no offset downfield absorption in the ^{1}H NMR spectrum. Treatment of civetone with Br_2 in CCl_4 yields a single dibromide A, $C_{17}H_{30}Br_2O$. Oxidation of civetone with $KMnO_4$ solution yields a diacid B, $C_{17}H_{30}O_5$. Oxidation of civetone with hot, concentrated HNO_3 yields principally $HO_2C(CH_2)_7CO_2H$ and $HO_2C(CH_2)_6CO_2H$. Hydrogenation of civetone (Pd catalyst, no heat or pressure), followed by oxidation with hot HNO_3, yields a diacid $HO_2C(CH_2)_{15}CO_2H$. What are the structures of civetone, A, and B?

13.46 When 1,3-diphenyl-1,3-propanedione reacts with alkaline Br_2, it gives bromoform and benzoate ion instead of 2,2-dibromo-1,3-diphenyl-1,3-propanedione, which might have been anticipated at first glance. Write chemical equations to show how this reaction could occur.

$$\text{C}_6\text{H}_5-\overset{\text{O}}{\overset{\|}{\text{C}}}\text{CH}_2\overset{\text{O}}{\overset{\|}{\text{C}}}-\text{C}_6\text{H}_5 \xrightarrow{\text{Br}_2, \text{OH}^-} 2\,\text{C}_6\text{H}_5\text{CO}_2^- + \text{CHBr}_3$$

13.47 The reaction of a cyanohydrin with NH_3 to give an aminonitrile (page 559) most likely does not occur by a direct displacement of the —OH group by NH_3. Propose a reasonable mechanism for this reaction. (*Hint:* Cyanohydrin formation is reversible under basic conditions.)

$$\underset{\underset{\text{OH}}{|}}{\text{CH}_3\text{CHCN}} \xrightarrow{\text{NH}_3} \underset{\underset{\text{NH}_2}{|}}{\text{CH}_3\text{CHCN}}$$

13.48 Vinyl ethers, such as dihydropyran, react with alcohols in the presence of an acid catalyst to form acetals. Because acetals readily undergo hydrolysis with aqueous acid, acetals derived from vinyl ethers can be used as protecting groups for —OH functions.

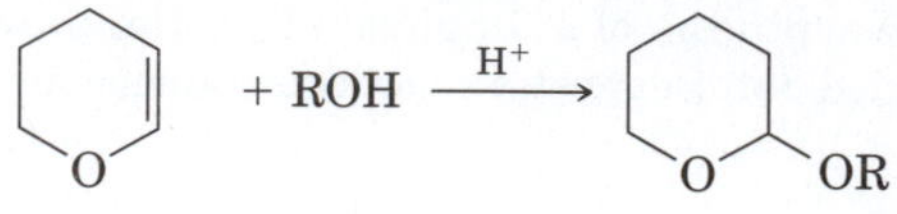

dihydropyran

Write flow equations to illustrate this use of dihydropyran in the following conversion:

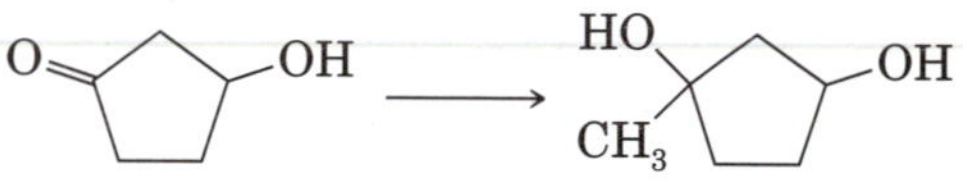

13.49 Show how the following conversions could be carried out.

(a)

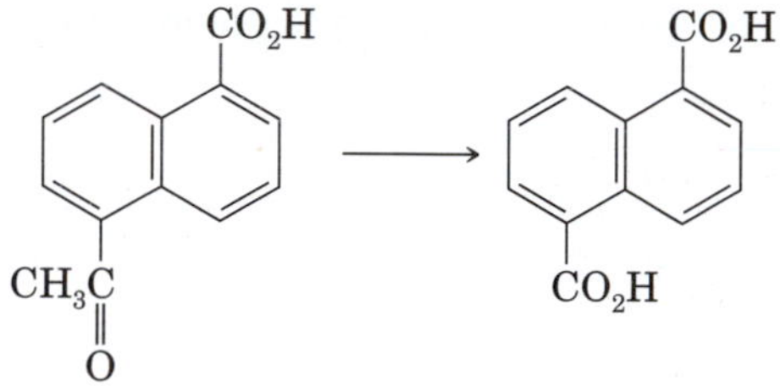

(b)

(c)

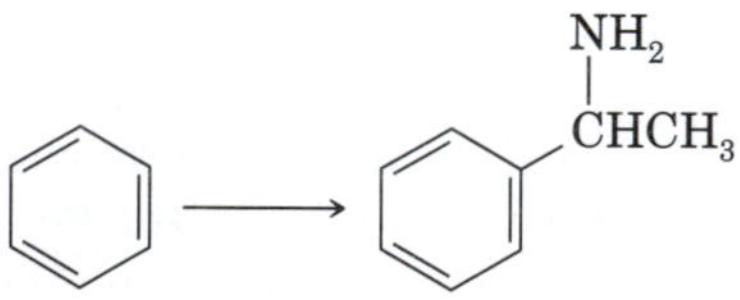

(d)

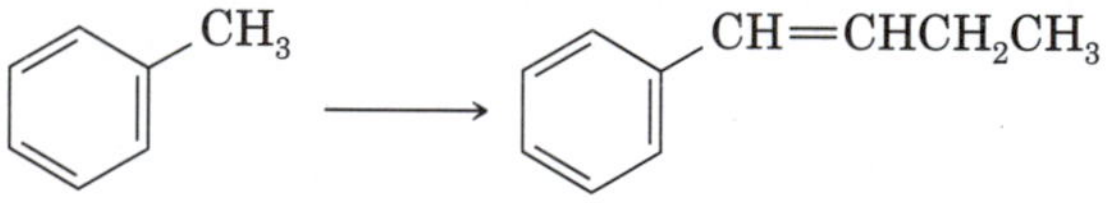

(e)   3-pentanone $\longrightarrow$ 1-penten-3-one

13.50 When the following iodo ketone was treated with KOH in CH_3OH, Compound A ($C_{10}H_{16}O_2$) was obtained in 93% yield. The infrared spectrum of A showed absorption at 2940, 2860, 1200, 1150, 1090, 1010, and 995 cm^{-1} (3.40, 3.50, 8.33, 8.70, 9.17, 9.90, and 10.05 μm). The ^{13}C NMR spectrum (off-resonance decoupled) showed the following peaks: δ95.9 (singlet), 71.6 (doublet), 48.2 (quartet), 38.9 (triplet), 35.1 (triplet), 35.0 (triplet), 29.2 (doublet). What is the structure of compound A?

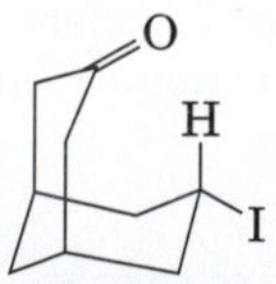

the iodo ketone

13.51 Match each of the spectra in Figure 13.10 with one of the following structures:

(a) $(CH_3)_3CCHO$

(b)

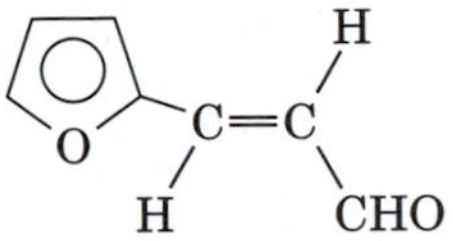

(c)
$$\underset{\underset{CH_3}{|}}{\overset{\overset{CH_3}{|}}{NCCH_2CH_2CCHO}}$$

(d) 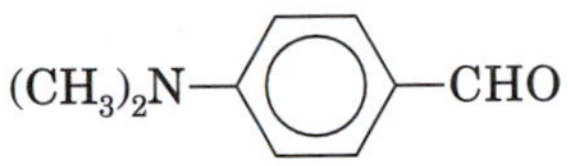 $(CH_3)_2N\!-\!\langle\bigcirc\rangle\!-\!CHO$

(e) 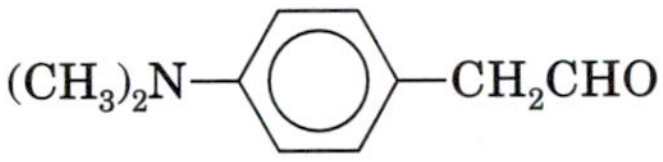 $(CH_3)_2N\!-\!\langle\bigcirc\rangle\!-\!CH_2CHO$

13.52 Compound A has a molecular weight of 132. Treatment of A with $NaBH_4$ in aqueous methanol yields Compound B. The ^1H NMR spectrum of A and the infrared spectrum of B are shown in Figure 13.11. What are the structures of A and B?

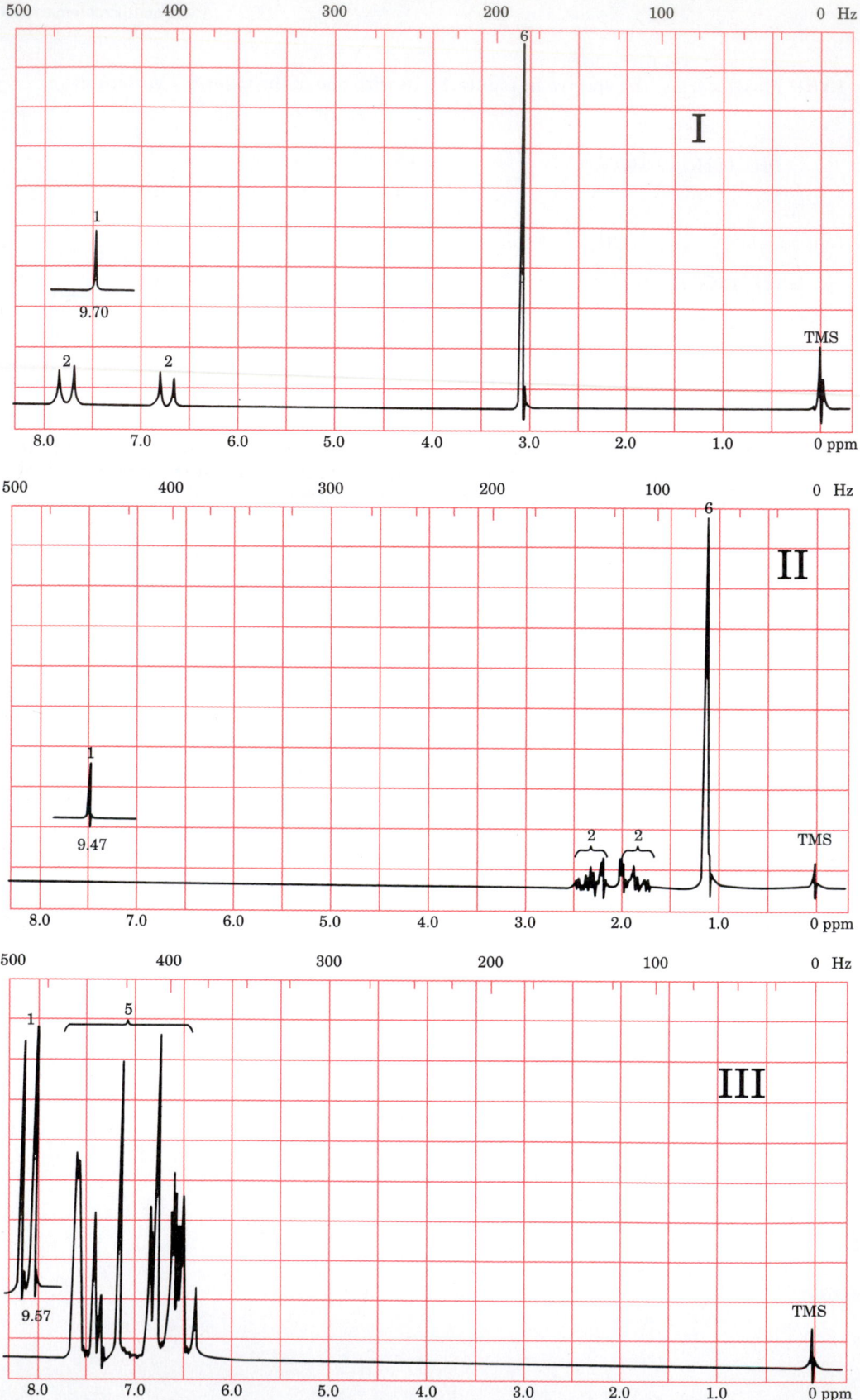

Figure 13.10 ^{1}H NMR spectra for Problem 13.51.

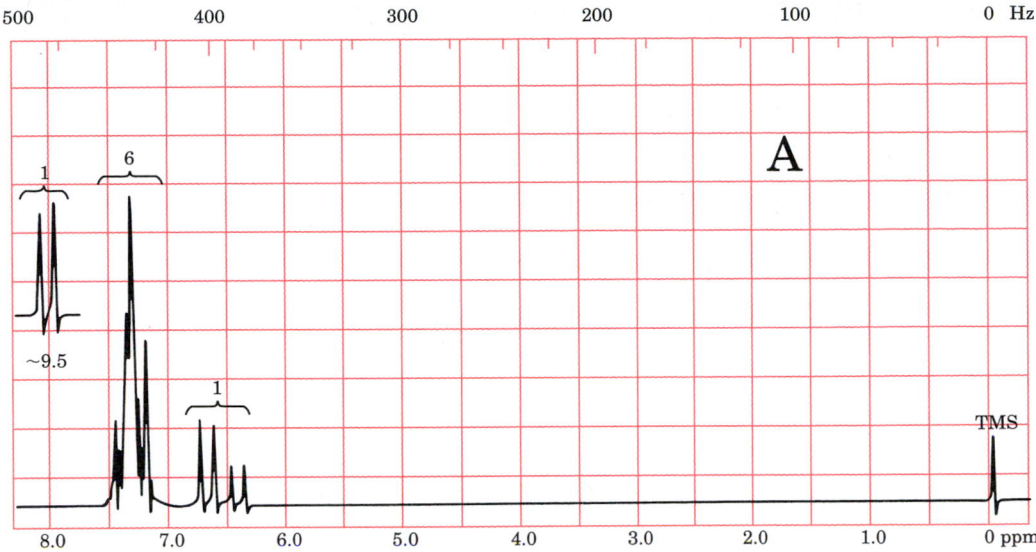

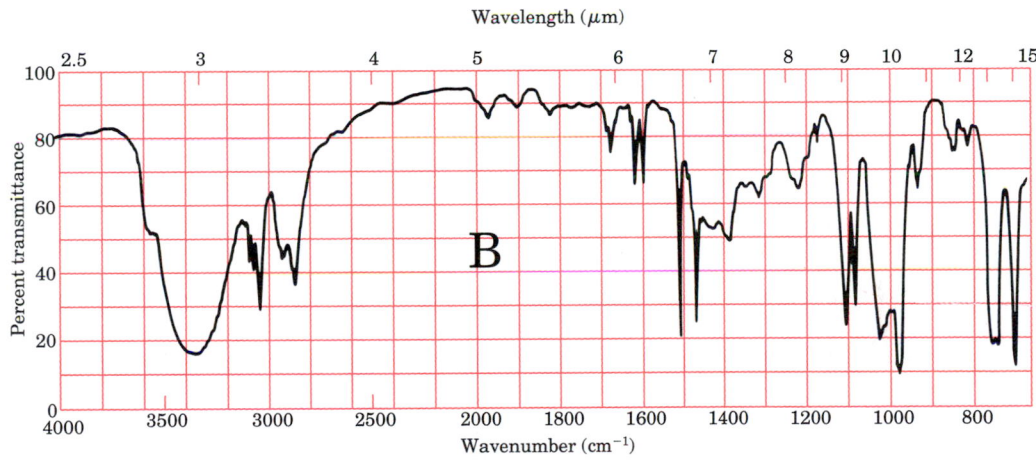

Figure 13.11 Spectra for Problem 13.52.

CARBOXYLIC ACIDS

A **carboxylic acid** is an organic compound containing the **carboxyl group,** $-CO_2H$. The carboxyl group contains a carbonyl group and a hydroxyl group. The interactions of these two groups lead to a chemical reactivity that is unique to carboxylic acids.

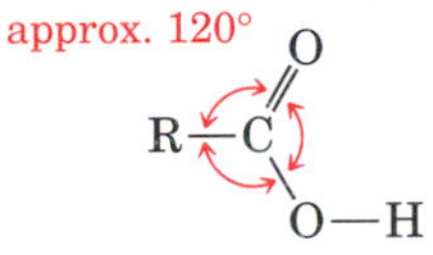

planar 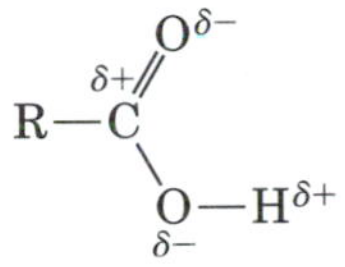 *polar*

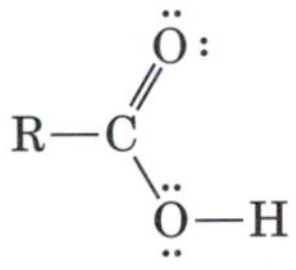

unshared electrons

Carboxylic acids are important biologically and commercially. Aspirin is a carboxylic acid, as are oleic acid (Section 24.1) and the prostaglandins (Section 24.3).

acetylsalicylic acid
(aspirin)

cis-$CH_3(CH_2)_7CH{=}CH(CH_2)_7CO_2H$

oleic acid

a fatty acid: a component of fats

prostaglandin E_1 (PGE_1)

a moderator of hormone activity

Because the carboxyl group is polar and generally nonhindered, its reactions are not usually affected to a great extent by the rest of the molecule. The carboxyl groups in aspirin, oleic acid, and other carboxylic acids undergo similar reactions.

The most important chemical property of carboxylic acids is their acidity. Ranked against mineral acids such as HCl or HNO_3 (pK_a values about -1 or smaller), carboxylic acids are weak acids (pK_a values typically about 5). However, carboxylic acids are more acidic than alcohols or phenols, primarily because of resonance stabilization of the carboxylate anion, RCO_2^-. A p-orbital picture of the carboxylate ion is shown in Figure 14.1.

$$CH_3\overset{\overset{\displaystyle O}{\|}}{C}OH + H_2O \rightleftharpoons \left[CH_3\overset{\overset{\displaystyle \ddot{O}:}{\|}}{C}{-}\ddot{\underset{\cdot\cdot}{O}}{:}^- \longleftrightarrow CH_3\overset{:\ddot{O}:^-}{\underset{}{C}}{=}\overset{}{\underset{\cdot\cdot}{O}}{:} \right] + H_3O^+$$

resonance-stabilized

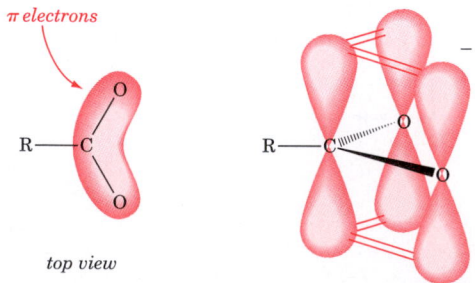

top view

side view showing p-orbital overlap
(unshared valence electrons not shown)

Figure 14.1 Bonding in the carboxylate ion, RCO_2^-.

Section 14.1
Nomenclature of Carboxylic Acids

The IUPAC name of an aliphatic carboxylic acid is that of the alkane parent with the **-e** changed to **-oic acid.** The carboxyl carbon is carbon 1, just as the aldehyde carbon is. (Note that the final *-e* of the alkane name is retained in the name of a dioic acid.)

$$\overset{\displaystyle Cl}{\underset{\displaystyle |}{}}$$

	HCO_2H	$CH_3CH_2CHCO_2H$	$HO_2CCH_2CO_2H$
IUPAC:	methanoic acid	2-chlorobutanoic acid	propanedioic acid

For the first four carboxylic acids, the trivial names are used more often than the IUPAC names. (See Table 14.1.) The name **formic acid** comes from *formica* (Latin for "ants"). In medieval times, alchemists obtained formic acid by distilling red ants! **Acetic acid** is from the Latin *acetum,* "vinegar." In its pure form, it is called *glacial* acetic acid. The term *glacial* arises from the fact that pure acetic acid is a viscous liquid that solidifies into an icy-looking solid. The name for **propionic acid** literally means "first fat." Propionic acid is the first carboxylic acid (the one of lowest molecular weight) to exhibit some of the properties of **fatty acids,** which are carboxylic acids obtained from the hydrolysis of fats (discussed in Section 24.1). **Butyric acid** is found in rancid butter (Latin, *butyrum*).

A notable property of the lower-molecular-weight carboxylic acids is their odor. Formic and acetic acids have pungent odors. Propionic acid has a pungent odor reminiscent of rancid fat. The odor of rancid butter arises in part from butyric acid. Caproic acid smells like a goat. (Goat sweat, incidentally, contains caproic acid). Valeric acid (from the Latin *valere,* "to be strong") is not a strong acid, but it does have a strong odor somewhere between that of rancid butter and goat sweat. (Interestingly, valeric acid is the sex attractant of the sugar beet wireworm.) Dogs can differentiate the odors of individual humans because of differing proportions of carboxylic acids in human sweat.

Table 14.1	Trivial names of the first ten carboxylic acids		

Number of Carbons	Structure	Trivial Name	Occurrence and Derivation of Name
1	HCO_2H	formic	ants (L. *formica*)
2	CH_3CO_2H	acetic	vinegar (L. *acetum*)
3	$CH_3CH_2CO_2H$	propionic	milk, butter, and cheese (Gr. *protos*, first; *pion*, fat)
4	$CH_3(CH_2)_2CO_2H$	butyric	butter (L. *butyrum*)
5	$CH_3(CH_2)_3CO_2H$	valeric	valerian root (L. *valere*, to be strong)
6	$CH_3(CH_2)_4CO_2H$	caproic	goat (L. *caper*)
7	$CH_3(CH_2)_5CO_2H$	enanthic	(Gr. *oenanthe*, vine blossom)
8	$CH_3(CH_2)_6CO_2H$	caprylic	goat
9	$CH_3(CH_2)_7CO_2H$	pelargonic	Its ester is found in *Pelargonium roseum*, a geranium.
10	$CH_3(CH_2)_8CO_2H$	capric	goat

The odors of the aliphatic carboxylic acids of ten and more carbons diminish, probably because of their lack of volatility.

Some other commonly encountered carboxylic acids and names follow:

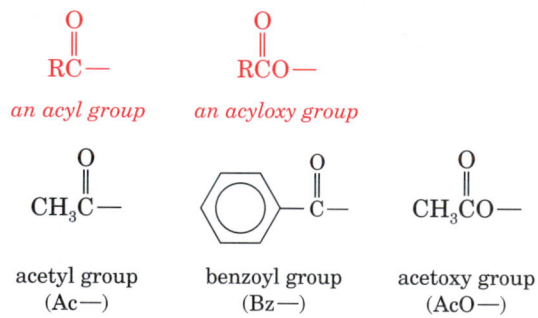

CH_3—⬡—CO_2H $\overset{CO_2H}{\underset{CO_2H}{⬡}}$ ⬡—CO_2H $CH_2{=}CHCO_2H$

p-toluic acid   *o*-phthalic acid   cyclohexanecarboxylic acid   acrylic acid

As with aldehydes and ketones, Greek letters may be used in the trivial names of carboxylic acids to refer to a position in the molecule relative to the carboxyl group.

$$CH_3CH_2CO_2H \qquad CH_3CH_2\overset{Br}{\underset{|}{C}}HCO_2H$$

β *carbon*   α *carbon*   α-bromobutyric acid
                                     or 2-bromobutanoic acid

It is sometimes convenient to refer to the RCO— group as an **acyl group** and to RCO$_2^-$ as an **acyloxy group.** For example, the *acylation* of benzene is the substitution of an RCO— group on the aromatic ring.

$$\overset{O}{\overset{\|}{R\overset{}{C}}}{-} \qquad\qquad \overset{O}{\overset{\|}{R\overset{}{C}O}}{-}$$

an acyl group   *an acyloxy group*

$$\overset{O}{\overset{\|}{CH_3C}}{-} \qquad ⬡{-}\overset{O}{\overset{\|}{C}}{-} \qquad \overset{O}{\overset{\|}{CH_3CO}}{-}$$

acetyl group   benzoyl group   acetoxy group
(Ac—)   (Bz—)   (AcO—)

STUDY PROBLEMS

14.1   Name the following compounds:

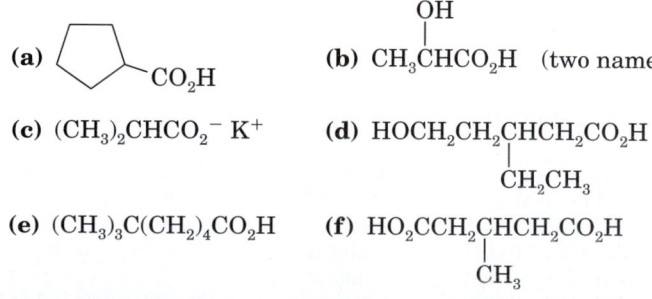

(a) ⬠—CO_2H

(b) $CH_3\overset{OH}{\underset{|}{C}}HCO_2H$ (two names)

(c) $(CH_3)_2CHCO_2^- K^+$

(d) $HOCH_2CH_2\overset{}{\underset{\underset{\textstyle CH_2CH_3}{|}}{C}}HCH_2CO_2H$

(e) $(CH_3)_3C(CH_2)_4CO_2H$

(f) $HO_2CCH_2\overset{}{\underset{\underset{\textstyle CH_3}{|}}{C}}HCH_2CO_2H$

14.2   Write the structure for each of the following compounds:

   (a) nonanoic acid   **(b)** potassium 5-methylnonanoate

   (c) octanedioic acid   **(d)** formic acid

(e) phenylacetic acid   **(f)** 4-benzyl-7-cycloheptylheptanoic acid

(g) β-methoxybutyric acid   **(h)** *trans*-2-pentenoic acid

Section 14.2
Spectral Properties of Carboxylic Acids

A. Infrared Spectra

Carboxylic acids, either as pure liquids or in solution at concentrations in excess of about 0.01 *M*, exist primarily as hydrogen-bonded dimers rather than as discrete monomers. The infrared spectrum of a carboxylic acid is therefore the spectrum of the dimer. Because of the hydrogen bonding, the OH stretching absorption of carboxylic acids is very broad and very intense. This OH absorption starts around 3300 cm^{-1} (3.0 μm) and slopes into the region of aliphatic carbon–hydrogen absorption. (See Figure 14.2.) The broadness of the carboxylic acid OH band can often obscure both aliphatic and aromatic CH absorption, as well as any other OH or NH absorption in the spectrum.

The carbonyl absorption is observed at about 1700–1725 cm^{-1} (5.8–5.88 μm) and is moderately strong. Conjugation shifts this absorption to lower frequencies: 1680–1700 cm^{-1} (5.9–5.95 μm).

The fingerprint region in an infrared spectrum of a carboxylic acid often shows C—O stretching and OH bending. (See Table 14.2.) Another OH bending vibration of the dimer results in a broad absorption near 925 cm^{-1} (10.8 μm).

B. ¹H NMR Spectra

In the NMR spectrum, the absorption of the acidic proton of a carboxylic acid is seen as a singlet far downfield (δ10–13).

The α protons are only slightly affected by the C=O group. Their absorption is slightly downfield (about δ2.2) because of the inductive effect of the partially positive carbonyl carbon. There is no unique splitting pattern associ-

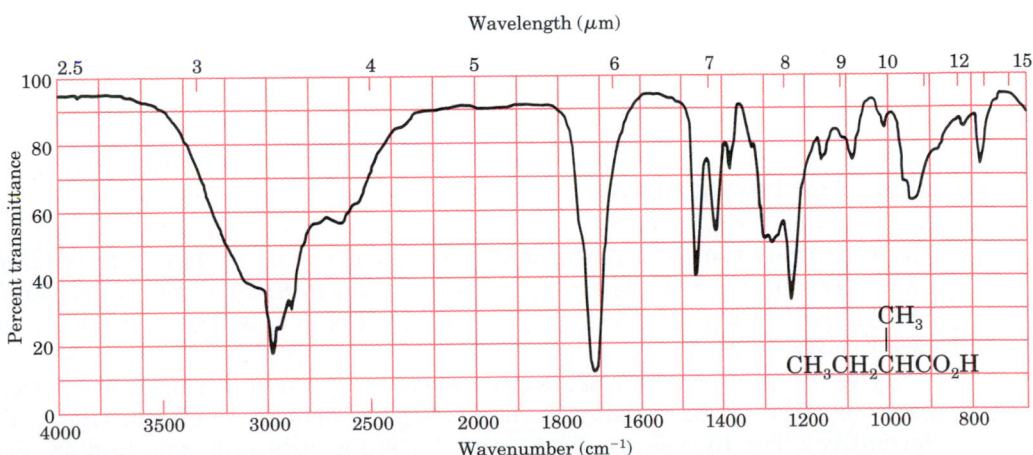

Figure 14.2 Infrared spectrum of 2-methylbutanoic acid.

Table 14.2 Characteristic infrared absorption for carboxylic acids		
	Position of Absorption	
Type of Vibration	cm^{-1}	μm
O—H stretching	2860–3300	3.0–3.5
C=O stretching	1700–1725	5.8–5.88
C—O stretching	1210–1330	7.5–8.26
O—H bending	1300–1440	6.94–7.71
O—H bending (dimer)	~925	~10.8

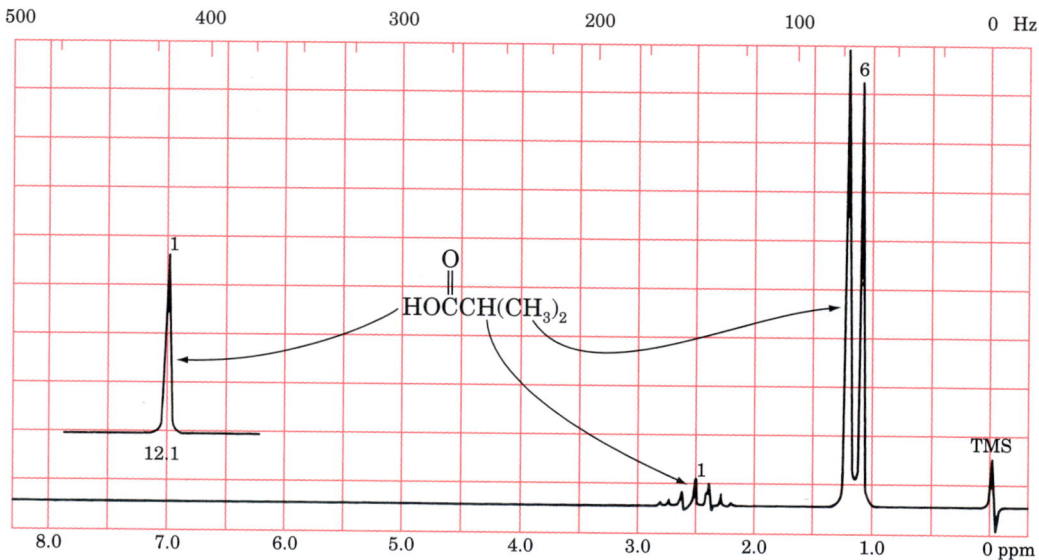

Figure 14.3 ^{1}H NMR spectrum of 2-methylpropanoic acid.

ated with the carboxylic acid group because the carboxyl proton has no neighboring protons. (See Figure 14.3.)

Section 14.3

Preparation of Carboxylic Acids

The numerous synthetic paths that lead to carboxylic acids can be grouped into three types of reactions: (1) *hydrolysis of the derivatives of the carboxylic acids;* (2) *oxidation reactions;* and (3) *Grignard reactions.* These reactions are summarized in Table 14.3.

Hydrolysis of carboxylic acid derivatives results from the attack of water or OH$^-$ on the carbonyl carbon (or the —CN carbon of a nitrile) of the derivatives. The hydrolysis of an ester to yield a carboxylic acid and an alcohol is typical of this class of reactions. We will discuss these reactions in detail in Chapter 15.

Table 14.3   Summary of laboratory syntheses of carboxylic acids

	Reaction	Section Reference
Hydrolysis:[a]		
ester:	$\overset{\text{O}}{\overset{\|}{\text{RC}}}\!-\!\text{OR}' \ + \text{H}_2\text{O} \xrightarrow{\text{H}^+ \text{ or OH}^-} \text{RCO}_2\text{H} + \text{HOR}'$	15.5C
amide:	$\overset{\text{O}}{\overset{\|}{\text{RC}}}\!-\!\text{NR}'_2 \ + \text{H}_2\text{O} \xrightarrow{\text{H}^+ \text{ or OH}^-} \text{RCO}_2\text{H} + \text{HNR}'_2$	15.8C
anhydride:	$\overset{\text{O}}{\overset{\|}{\text{RC}}}\!-\!\overset{\text{O}}{\overset{\|}{\text{OC}}}\text{R}' + \text{H}_2\text{O} \xrightarrow{\text{H}^+ \text{ or OH}^-} \text{RCO}_2\text{H} + \text{HO}_2\text{CR}'$	15.4C
acid halide:	$\overset{\text{O}}{\overset{\|}{\text{RC}}}\!-\!\text{X} \quad + \text{H}_2\text{O} \xrightarrow{\text{H}^+ \text{ or OH}^-} \text{RCO}_2\text{H} + \text{X}^-$	15.3C
nitrile:	$\text{RC}\equiv\text{N} \quad + \text{H}_2\text{O} \xrightarrow{\text{H}^+ \text{ or OH}^-} \text{RCO}_2\text{H} + \text{NH}_3$	15.11D
Oxidation:		
1° alcohol:	$\text{RCH}_2\text{OH} \quad + [\text{O}]^b \longrightarrow \text{RCO}_2\text{H}$	7.8C
aldehyde:	$\overset{\text{O}}{\overset{\|}{\text{RC}}}\text{H} \quad + [\text{O}] \longrightarrow \text{RCO}_2\text{H}$	13.7
alkene:	$\text{RCH}\!=\!\text{CR}_2 + [\text{O}] \longrightarrow \text{RCO}_2\text{H} + \text{R}_2\text{C}\!=\!\text{O}$	10.13B
alkylarene:	$\text{Ar}\!-\!\text{R} \quad + [\text{O}] \longrightarrow \text{ArCO}_2\text{H}$	12.1
methyl ketone:[a]	$\overset{\text{O}}{\overset{\|}{\text{RC}}}\text{CH}_3 \quad + \text{X}_2 \xrightarrow{\text{OH}^-} \text{RCO}_2\text{H} + \text{CHX}_3$	13.10A
Grignard Reaction:		
$\text{RX} \xrightarrow[\substack{(2)\ \text{CO}_2 \\ (3)\ \text{H}_2\text{O, H}^+}]{(1)\ \text{Mg, ether}} \text{RCO}_2\text{H}$		14.3

[a] In alkaline solution, the carboxylate is obtained. The acid can be generated by acidification:
$\text{RCO}_2^- + \text{H}^+ \rightarrow \text{RCO}_2\text{H}$.
[b] Typical oxidizing agents are KMnO_4 or H_2CrO_4 solutions.

$$\underset{\text{ethyl acetate}}{\underset{\text{an ester}}{\text{CH}_3\overset{\text{O}}{\overset{\|}{\text{C}}}\text{OCH}_2\text{CH}_3}} + \text{H}_2\text{O} \underset{\text{H}^+}{\rightleftharpoons} \underset{\underset{\text{a carboxylic acid}}{\text{acetic acid}}}{\text{CH}_3\overset{\text{O}}{\overset{\|}{\text{C}}}\text{OH}} + \underset{\underset{\text{an alcohol}}{\text{ethanol}}}{\text{HOCH}_2\text{CH}_3}$$

$$\underset{\underset{\text{a nitrile}}{\text{acetonitrile}}}{\text{CH}_3\text{CN}} + \text{H}_2\text{O} + \text{H}^+ \longrightarrow \underset{\text{acetic acid}}{\text{CH}_3\overset{\text{O}}{\overset{\|}{\text{C}}}\text{OH}} + \text{NH}_4^+$$

Oxidation of primary alcohols and aldehydes to yield carboxylic acids was discussed in Sections 7.8C and 13.7. The principal limitation of alcohol oxidation is that the necessary strength of the oxidizing agent precludes the presence of another oxidizable functional group in the molecule. Even with this limitation, the oxidation of primary alcohols is the most common oxidative procedure leading to carboxylic acids because alcohols are often available.

$$\xrightarrow{\hspace{1cm}} RCH_2OH \xrightarrow[\text{heat}]{H_2CrO_4} RCO_2H$$

no unblocked double bond,
aldehyde, benzyl
group, or other
—OH group

The oxidation of aldehydes proceeds with mild oxidizing agents (such as Ag^+) that do not oxidize other groups. However, aldehydes are not as readily available as primary alcohols.

Oxidation of alkenes is primarily an analytical technique, but also can be used in synthesis of carboxylic acids. Like alcohols, alkenes require vigorous oxidizing agents.

$$\xrightarrow[\text{heat}]{MnO_4^-,\, H^+} HO_2CCH_2CH_2CH_2CH_2CO_2H$$

cyclohexene

hexanedioic acid
(adipic acid)

Oxidation of substituted alkylbenzenes is an excellent route to substituted benzoic acids. A carboxyl group is a *m*-director, but an alkyl group is an *o,p*-director. Electrophilic substitution of an alkylbenzene, followed by oxidation, yields *o*- and *p*-substituted benzoic acids.

$$O_2N-\text{（ring）}-CH_3 \xrightarrow[\text{heat}]{H_2CrO_4} O_2N-\text{（ring）}-CO_2H$$

2,4,6-trinitrotoluene

2,4,6-trinitrobenzoic acid (69%)

A **Grignard reaction** between a Grignard reagent (1°, 2°, 3°, vinylic, or aryl) and carbon dioxide (as a gas or as dry ice) is often the method of choice for preparing a carboxylic acid.

This Grignard reaction is directly analogous to the Grignard reactions of ketones or other carbonyl compounds. However, the reaction of a Grignard reagent with carbon dioxide does not yield an alcohol but a *magnesium carboxylate salt*. The magnesium salt is insoluble in the ether solvent used in a Grignard reaction. Therefore, only one of the two π bonds of CO_2 reacts. Treatment of the insoluble magnesium salt with aqueous acid liberates the *carboxylic acid*.

Step 1:

$$:O=C=O: + R-MgX \longrightarrow :O=C-O:^- \; ^+MgX$$
$$\qquad\qquad\qquad\qquad\qquad\qquad | $$
$$\qquad\qquad\qquad\qquad\qquad\qquad R$$

a carboxylate
(insoluble in ethers)

Step 2:

$$
\underset{\text{a carboxylic acid}}{
\overset{\displaystyle \overset{O}{\underset{\|}{}}}{RCO^-}\ {}^+MgX + H^+ \longrightarrow \overset{\displaystyle \overset{O}{\underset{\|}{}}}{RCOH} + Mg^{2+} + X^-
}
$$

$$(CH_3)_3CMgCl \xrightarrow[\text{(2) } H_2O, H^+]{\text{(1) } CO_2} (CH_3)_3CCO_2H$$

(63%)

STUDY PROBLEMS

14.3 Show how propanoic acid can be obtained:

(a) from an alkyl halide that has only two carbons

(b) from a three-carbon alcohol

(c) from an alkene that has at least four carbons

14.4 Using only a nitrile or a Grignard method of synthesis, propose a feasible route for each of the following conversions.

(a) $HOCH_2CH_2Cl \longrightarrow HOCH_2CH_2CO_2H$

(b)

Section 14.4
How Structure Affects Acid Strength

Carboxylic acids, sulfonic acids (RSO_3H), and alkyl hydrogen sulfates ($ROSO_3H$) are the only three classes of organic compounds more acidic than carbonic acid (H_2CO_3). Of these three classes, carboxylic acids are by far the most common.

	RCH_3	RNH_2	$RC{\equiv}CH$	ROH	H_2O	$ArOH$	H_2CO_3	RCO_2H
approx. pK_a:	~45–50	~35	~25	15–19	15.7	10	6.4	5

increasing acid strength →

Acid strength is a term describing the extent of ionization of a Brønsted acid in water: the greater the amount of ionization, the stronger is the acid. The strength of an acid is expressed as its K_a or its pK_a. Table 14.4 lists some carboxylic acids and their pK_a values. (For a more extensive list of pK_a values see the table on the page facing the front cover. You may want to review the discussion of K_a and pK_a in Section 1.7F and the discussions of acidity found in Sections 1.7 and 2.9E before proceeding.) In this section, we will discuss the general structural features that affect the acid strength of an organic compound. Our emphasis will be on carboxylic acids, but we will not limit the discussion to just these compounds.

Table 14.4	pK_a values for some carboxylic acids	
Trivial Name	**Structure**	**pK_a**
formic	HCO_2H	3.75
acetic	CH_3CO_2H	4.75
propionic	$CH_3CH_2CO_2H$	4.87
butyric	$CH_3(CH_2)_2CO_2H$	4.81
trimethylacetic	$(CH_3)_3CCO_2H$	5.02
fluoroacetic	FCH_2CO_2H	2.66
chloroacetic	$ClCH_2CO_2H$	2.81
bromoacetic	$BrCH_2CO_2H$	2.87
iodoacetic	ICH_2CO_2H	3.13
dichloroacetic	Cl_2CHCO_2H	1.29
trichloroacetic	Cl_3CCO_2H	0.7
α-chloropropionic	$CH_3CHClCO_2H$	2.8
β-chloropropionic	$ClCH_2CH_2CO_2H$	4.1
lactic	$\overset{\displaystyle OH}{\underset{\displaystyle |}{CH_3CHCO_2H}}$	3.87
vinylacetic	$CH_2{=}CHCH_2CO_2H$	4.35

The reaction of a weak acid with water is reversible. The equilibrium favors the lower-energy side of the equation. Any structural feature that *stabilizes the anion* with respect to its conjugate acid *increases the acid strength* by driving the equilibrium to the H_3O^+ and anion (A^-) side.

$$HA + H_2O \rightleftharpoons H_3O^+ + A^-$$

lower energy means HA is stronger acid

The principal factors that affect the stability of A^-, and thus the acid strength of HA, are: (1) electronegativity of the atom carrying the negative charge in A^-; (2) size of A^-; (3) hybridization of the atom carrying the negative charge in A^-; (4) inductive effect of other atoms or groups attached to the negative atom in A^-; (5) resonance stabilization of A^-; and (6) solvation of A^-. We will discuss each of these features in turn, but keep in mind that they do not work independently of one another.

A. Electronegativity

We discussed the effect of electronegativity on acidity in Section 1.7D. As described in that section, a more electronegative atom holds its bonding electrons more tightly than does a less electronegative atom. In comparisons of anions, the anion with a more electronegative atom carrying the negative ionic charge is generally the more stable anion. Therefore, as we proceed from left to right in the periodic table, we find that the elements form progressively more stable anions and that the conjugate acids are progressively stronger acids.

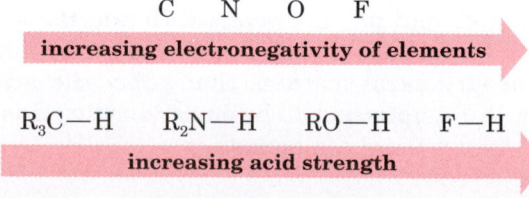

Just the reverse is true when we consider the base strengths of the conjugate bases. The anion of a very weak acid is a very strong base, while the anion of a stronger acid is a weaker base.

$$R_3C^- \quad R_2N^- \quad RO^- \quad F^-$$

increasing base strength

Consider, for example, the ionization reactions of ethanol and HF in water. As an element, F is more electronegative than O; thus, a fluoride ion is better able to carry a negative charge than is the alkoxide ion. Although HF is a weak acid, it is a much stronger acid than ethanol. Conversely, the fluoride ion is a weaker base than the ethoxide ion.

$$CH_3CH_2OH + H_2O \rightleftharpoons CH_3CH_2O^- + H_3O^+$$

weaker acid   *stronger base*
$pK_a = 15.9$

$$HF + H_2O \rightleftharpoons F^- + H_3O^+$$

stronger acid   *weaker base*
$pK_a = 3.45$

B. Size

A larger atom is better able to disperse a negative charge than is a smaller atom. Dispersal of a charge results in stabilization. Thus, as the size of an atom attached to H increases through a series of compounds in any group of the periodic table, the stability of the anion increases and so does the acid strength. The effect of size can be more important than the effect of electronegativity. Because of the small size of the fluorine atom, HF is a weaker acid than the other hydrogen halides, even though fluorine is more electronegative than the other halogens.

$$F^- \quad Cl^- \quad Br^- \quad I^-$$

increasing ionic radii

	HF	HCl	HBr	HI
pK_a:	3.45	-7	-9	-9.5

increasing acid strength

C. Hybridization

In Section 10.1, we discussed why an alkyne with a $\equiv CH$ group is weakly acidic. The increasing s character of the hybrid orbitals of carbon in the series $sp^3 \rightarrow sp^2 \rightarrow sp$ means increasing electronegativity of the carbon, and thus increasing polarity of the CH bond and increasing acid strength. A greater electronegativity of the atom bonded to H also enhances anion stability and thus the acidity of the compound. For these reasons, an alkyne $C\equiv CH$ loses a proton more readily than does an alkene $C=CH$. An alkene, in turn, can lose a proton more readily than an alkane.

$$\begin{array}{cccc} & CH_3CH_3 & CH_2{=}CH_2 & CH{\equiv}CH \\ \textit{approx. pK}_a: & {\sim}50 & {\sim}45 & {\sim}26 \end{array}$$

increasing acid strength →

Again, the strongest acid of the series yields the anion that is the weakest base.

$$\begin{array}{ccc} CH_3CH_2^- & CH_2{=}CH^- & CH{\equiv}C^- \end{array}$$

← **increasing base strength**

D. Inductive Effect

So far, we have discussed how the atom bonded directly to a hydrogen affects acid strength. However, other parts of a molecule can also affect acid strength. Compare the pK_a values for acetic acid and chloroacetic acid:

$$\begin{array}{cc} CH_3CO_2H & ClCH_2CO_2H \end{array}$$

acetic acid   chloroacetic acid
pK_a = 4.75   pK_a = 2.81

Chloroacetic acid is a much stronger acid than acetic acid (see Section 1.7D). This enhanced acidity arises from the inductive effect of the electronegative chlorine. In the un-ionized carboxylic acid, the electron-withdrawing Cl decreases the electron density at the α carbon. The result is a relatively high-energy structure with adjacent positive charges.

$$\overset{\delta-}{Cl}{\leftarrow}\underset{\delta+}{CH_2}{-}\overset{\overset{O}{\|}}{\underset{\delta+}{C}}OH \qquad \textit{adjacent } \delta\textit{+ charges} \\ \textit{destabilize acid}$$

The presence of chlorine, however, *reduces* the energy of the anion. The negative charge of the carboxylate group is partially dispersed by the nearby δ+ charge.

$$\underset{\delta-}{Cl}{\leftarrow}\underset{\delta+}{CH_2}{\leftarrow}\overset{\overset{O}{\|}}{C}{-}O^- \qquad \textit{nearby } \delta\textit{+ and } - \\ \textit{charges stabilize anion}$$

The effect of an electronegative group near the carboxyl group is to strengthen the acid by destabilizing the acid and stabilizing the anion relative to each other.

$$ClCH_2\overset{\overset{O}{\|}}{C}OH + H_2O \rightleftharpoons ClCH_2\overset{\overset{O}{\|}}{C}O^- + H_3O^+$$

less stable   *more stable*
than CH$_3$CO$_2$H   *than CH$_3$CO$_2^-$*

A list of groups in order of their electron-withdrawal power follows:

$$CH_3-   H-   CH_2=CH-   C_6H_5-   HO-   CH_3O-   I-   Br-   Cl-$$

increasing power of electron-withdrawal

The pK_a values of the following carboxylic acids reflect the differences in electron withdrawal by groups attached to $-CH_2CO_2H$:

$$CH_3CH_2CO_2H \quad CH_3CO_2H \quad CH_2=CHCH_2CO_2H \quad C_6H_5CH_2CO_2H \quad HOCH_2CO_2H \quad ClCH_2CO_2H$$

pK_a: 4.87   4.75   4.35   4.31   3.87   2.81

increasing acid strength

Additional electron-withdrawing groups amplify the inductive effect. Dichloroacetic acid is a stronger acid than is chloroacetic acid, and trichloroacetic acid is the strongest of the three.

$$ClCH_2CO_2H \qquad Cl_2CHCO_2H \qquad Cl_3CCO_2H$$

chloroacetic acid   dichloroacetic acid   trichloroacetic acid
pK_a = 2.81   pK_a = 1.29   pK_a = 0.7

The influence of the inductive effect upon acid strength diminishes with an increasing number of atoms between the carboxyl group and the electronegative group. 2-Chlorobutanoic acid is a substantially stronger acid than butanoic acid itself. However, 4-chlorobutanoic acid has a pK_a value very close to that of the unsubstituted acid.

$$CH_3CH_2CH_2CO_2H \quad \overset{\overset{\displaystyle Cl}{\displaystyle |}}{CH_2}CH_2CH_2CO_2H \quad CH_3\overset{\overset{\displaystyle Cl}{\displaystyle |}}{CH}CH_2CO_2H \quad CH_3CH_2\overset{\overset{\displaystyle Cl}{\displaystyle |}}{CH}CO_2H$$

pK_a: 4.8   4.5   4.0   2.9

increasing acid strength as a distance between $-$Cl and $-CO_2$H decreases

STUDY PROBLEM

14.5   Which is the stronger acid: **(a)** phenylacetic acid or bromoacetic acid? **(b)** dibromoacetic acid or bromoacetic acid? **(c)** 2-iodopropanoic acid or 3-iodopropanoic acid? Explain your answers.

E. Resonance Stabilization

Alcohols, phenols, and carboxylic acids all contain OH groups (see Section 2.9E). Yet these classes of compounds vary dramatically in acid strength. The differences may be attributed directly to the resonance stabilization (or lack of it) of the anion with respect to its conjugate acid.

$$ROH \qquad ArOH \qquad RCO_2H$$

approx. pK_a: 15–19   10   5

In the case of *alcohols,* the anion is not resonance-stabilized. The negative charge of an alkoxide ion resides entirely on the oxygen and is not delocalized. At the opposite end of the scale are the *carboxylic acids*. The negative charge of the carboxylate ion is equally shared by two electronegative oxygen atoms. *Phenols* are intermediate between carboxylic acids and alcohols in acidity. The oxygen of a phenoxide ion is adjacent to the aromatic ring and the negative charge is partially delocalized by the ring, as shown in Section 2.9E.

F. Solvation

The solvation of the anion can play a major role in the acidity of a compound. Solvent molecules can stabilize the anion by helping disperse the negative charge through dipole–dipole interactions. Any factor that increases the degree of solvation of the anion increases the acidity of that compound in solution. For example, water has a greater ability to solvate ions than does ethanol. A water solution of a carboxylic acid is more acidic than an ethanol solution by a factor of about 10^5!

G. Acid Strengths of Substituted Benzoic Acids

You might expect that resonance stabilization by the aromatic ring would play a large role in the relative acid strengths of benzoic acid and substituted benzoic acids; however, this is not the case. The negative charge of the carboxylate ion is shared by the two carboxylate oxygen atoms but cannot be effectively delocalized by the aromatic ring. (The oxygens of the carboxylate anion are not directly adjacent to the aromatic ring; resonance structures in which the negative charge is delocalized by the ring cannot be drawn.)

Even though the negative charge of the benzoate ion is not delocalized by the benzene ring, benzoic acid is a stronger acid than phenol. In the benzoate ion, the negative charge is equally shared by two electronegative oxygen atoms. In the phenoxide ion, however, most of the negative charge resides on the single oxygen atom.

Because the benzene ring does not participate in resonance stabilization of the carboxylate group, substituents on a benzene ring influence acidity primarily by the inductive effect. Thus, an electron-withdrawing group, such as the —NO_2 group, that is substituted either in the *meta* or the *para* position increases the acidity of a benzoic acid. An electron-releasing group in the same positions decreases acid strength.

Almost all *ortho*-substituents (whether electron-releasing or electron-withdrawing) increase acid strength of a benzoic acid. The reasons for this ***ortho-***

effect, as it is called, are probably a combination of both steric and electronic factors.

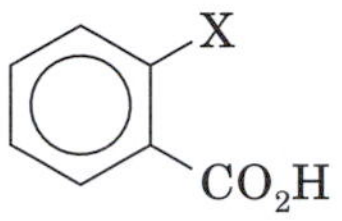

o-substituents increase acid strength

14.6   Which one of each of the following pairs of carboxylic acids is the stronger acid? **(a)** benzoic acid or *p*-bromobenzoic acid; **(b)** benzoic acid or *m*-bromobenzoic acid; **(c)** *m*-bromobenzoic acid or 3,5-dibromobenzoic acid. Explain.

Section 14.5
Reaction of Carboxylic Acids with Bases

The reaction of a carboxylic acid with a base results in a *salt*. An organic salt has many of the physical properties of its inorganic counterparts. Like NaCl or KNO_3, an organic salt is high melting, water soluble, and odorless.

cyclohexane-
carboxylic acid

sodium cyclohexane-
carboxylate

The carboxylate anion is named by changing the **-ic acid** ending of the carboxylic acid name to **-ate.** In the name of the salt, the name of the cation precedes the name of the anion as a separate word.

$$CH_3CO_2H \qquad CH_3CO_2^- \ Na^+$$

acetic acid   sodium acetate

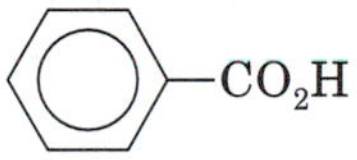

benzoic acid

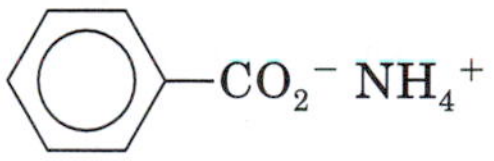

ammonium benzoate

$$\underset{\text{glutamic acid}}{HO_2CCH_2CH_2\overset{\displaystyle NH_2}{\overset{|}{C}HCO_2H}} \qquad \underset{\text{monosodium glutamate}}{HO_2CCH_2CH_2\overset{\displaystyle NH_2}{\overset{|}{C}HCO_2^- \ Na^+}}$$

an amino acid found in proteins   *a flavor enhancer (MSG)*

The carboxylate ion is a weak base and can act as a nucleophile. Esters, for example, are occasionally prepared by the reaction of especially reactive alkyl halides and carboxylates (see Section 5.10).

acetate ion   benzyl bromide

benzyl acetate

an ester

Because it is more acidic than carbonic acid, a carboxylic acid undergoes an acid–base reaction with sodium bicarbonate as well as with stronger bases such as NaOH.

$$CH_3\overset{\displaystyle O}{\overset{\|}{C}}OH + HCO_3^- \longrightarrow CH_3\overset{\displaystyle O}{\overset{\|}{C}}O^- + [H_2CO_3] \longrightarrow H_2O + CO_2$$

STUDY PROBLEM

14.7 Predict the major organic products:

(a) [cyclohexane ring with CO₂H and CO₂H substituents] + excess NaOH $\longrightarrow$

(b) $C_6H_5CH_2Cl + CH_3CH_2\overset{\displaystyle O}{\overset{\|}{C}}O^- \ Na^+ \longrightarrow$

(c) $CH_3CO_2H + CH_3O^- \longrightarrow$

(d) $CH_3CO_2H + CH_3NH_2 \longrightarrow$

(e) $CH_3CO_2^- + ClCH_2CO_2H \longrightarrow$

Section 14.6
Esterification of Carboxylic Acids

An **ester of a carboxylic acid** is a compound that contains the —CO₂R group, where R may be alkyl or aryl. An ester may be formed by the direct reaction of a carboxylic acid with an alcohol, a reaction called an **esterification reaction.** Esterification is acid-catalyzed and reversible.

General:

$$\underset{\text{a carboxylic acid}}{R\overset{\displaystyle O}{\overset{\|}{C}}OH} + \underset{\text{an alcohol}}{R'OH} \underset{}{\overset{H^+, \text{ heat}}{\rightleftharpoons}} \underset{\text{an ester}}{R\overset{\displaystyle O}{\overset{\|}{C}}OR'} + H_2O$$

$$\underset{\text{acetic acid}}{CH_3\overset{\displaystyle O}{\overset{\|}{C}}OH} + \underset{\text{ethanol}}{CH_3CH_2OH} \overset{H^+, \text{ heat}}{\rightleftharpoons} \underset{\text{ethyl acetate}}{CH_3\overset{\displaystyle O}{\overset{\|}{C}}OCH_2CH_3} + H_2O$$

[benzoic acid] —$\overset{\displaystyle O}{\overset{\|}{C}}$OH + HO—[cyclohexyl] $\overset{H^+, \text{ heat}}{\rightleftharpoons}$ [benzene]—$\overset{\displaystyle O}{\overset{\|}{C}}$O—[cyclohexyl] + H₂O

benzoic acid   cyclohexanol   cyclohexyl benzoate

The rate at which a carboxylic acid is esterified depends primarily upon the steric hindrance in the alcohol and the carboxylic acid. The acid strength of the carboxylic acid plays only a minor role in the rate at which the ester is formed.

Reactivity of alcohols toward esterification:

$$3° \text{ ROH} \qquad 2° \text{ ROH} \qquad 1° \text{ ROH} \qquad CH_3OH$$

increasing reactivity →

Reactivity of carboxylic acids toward esterification:

$$R_3CCO_2H \qquad R_2CHCO_2H \qquad RCH_2CO_2H \qquad CH_3CO_2H \qquad HCO_2H$$

increasing reactivity →

Like many reactions of aldehydes and ketones, esterification of a carboxylic acid proceeds through a series of protonation and deprotonation steps. The carbonyl oxygen is protonated, the nucleophilic alcohol attacks the positive carbon, and elimination of water yields the ester.

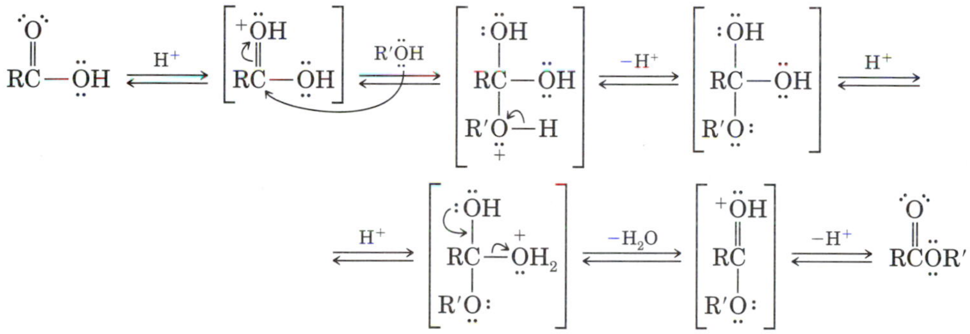

We can abbreviate this mechanism in the following way:

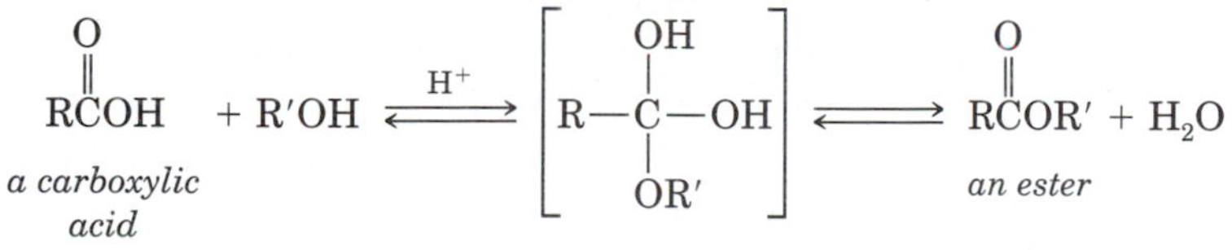

In an esterification reaction, it is the C—O bond of the carboxylic acid that is broken and not the O—H bond of the acid nor the C—O bond of the alcohol. Evidence for the mechanism is the reaction of a labeled alcohol such as $CH_3{}^{18}OH$ with a carboxylic acid. In this case, the ^{18}O stays with the methyl group.

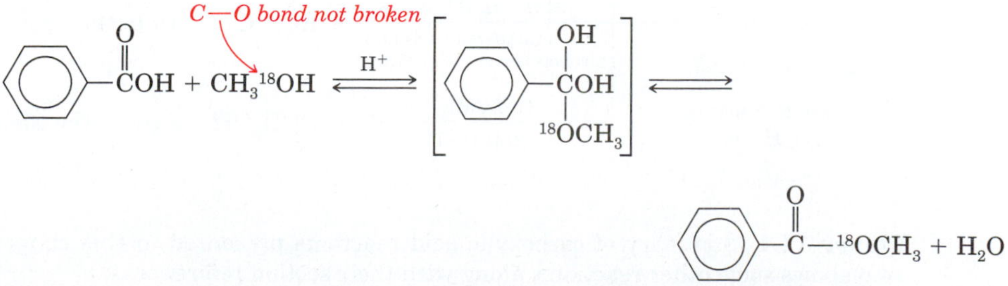

The esterification reaction is reversible. To obtain a high yield of an ester, we must shift the equilibrium to the ester side. One technique for accomplishing this is to use an excess of one of the reactants (the cheaper one). Another technique is to remove one of the products from the reaction mixture (for example, by the azeotropic distillation of water).

As the amount of steric hindrance in the intermediate increases, the yield of ester decreases. The reason is that esterification is a reversible reaction and the less hindered species (the reactants) are favored. If bulky esters are to be prepared, it is better to use another synthetic route, such as the reaction of an alcohol with an acid anhydride or an acid chloride, which are more reactive than the carboxylic acid (see Chapter 15) and undergo irreversible reactions with alcohols.

Phenyl esters ($RCO_2C_6H_5$) are not generally prepared directly from phenols and carboxylic acids because the equilibrium favors the acid–phenol side rather than the ester side. Phenyl esters, like bulky esters, can be obtained from phenol and the more reactive acid derivatives.

STUDY PROBLEMS

14.8 Predict the esterification products of: **(a)** *p*-toluic acid (page 600) and 2-propanol; **(b)** terephthalic acid (p-HO_2C—C_6H_4—CO_2H) and excess ethanol; **(c)** acetic acid and (*R*)-2-butanol.

14.9 4-Hydroxybutanoic acid spontaneously forms a cyclic ester, or *lactone*. What is the structure of this lactone?

Section 14.7
Reduction of Carboxylic Acids

The carbonyl carbon of a carboxylic acid is at the highest oxidation state it can attain and still be part of an organic molecule. (The next higher oxidation state is in CO_2.) Other than combustion or oxidation by very strong reagents, such as hot H_2SO_4–CrO_3 (cleaning solution), the carboxylic acid group is inert toward further oxidative reaction.

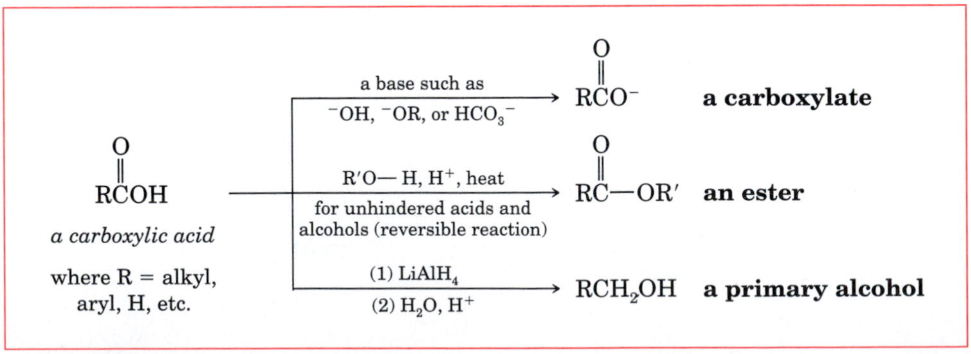

Figure 14.4 Summary of carboxylic acid reactions presented in this chapter. Table 14.6 shows some other reactions, along with their section references.

Surprisingly, the carboxylic acid group is also inert toward most common reducing agents (such as hydrogen plus catalyst). This inertness made necessary the development of alternative reduction methods, such as conversion of the carboxylic acid to an ester and then reduction of the ester. However, the introduction of lithium aluminum hydride in the late 1940s simplified the reduction because $LiAlH_4$ reduces a carboxyl group directly to a $-CH_2OH$ group. (Other carbonyl functionality in the molecule is, of course, reduced as well. See Section 13.6.)

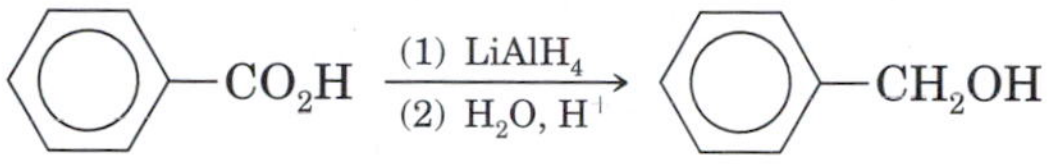

benzoic acid   benzyl alcohol

STUDY PROBLEM

14.10 Give the structures for the $LiAlH_4$ reduction products of:

(a) 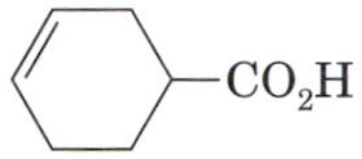   **(b)**

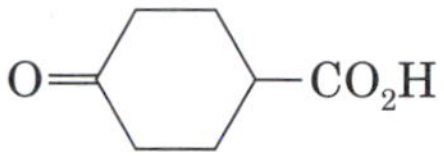

Figure 14.4 summarizes the reactions of carboxylic acids presented in Sections 14.5–14.7.

Section 14.8
Polyfunctional Carboxylic Acids

Dicarboxylic acids and carboxylic acids containing other functional groups often show unique chemical properties. In this section, we will consider a few of the more important polyfunctional carboxylic acids. Hydroxy acids will be mentioned in Chapter 15 and amino acids will be discussed in Chapter 25.

A. Acidity of Dibasic Acids

A **dibasic acid** is one that reacts with *two equivalents of base* (1.0 mol acid reacts with 2.0 mol base). The term **diprotic acid** (two acidic protons) would perhaps be a better term to describe such a compound. In general, dibasic carboxylic acids have a chemistry similar to that of monocarboxylic acids, but let us examine a few of the differences.

With any dibasic acid (inorganic or organic), the first hydrogen ion is removed more easily than the second. Thus, K_1 (the acidity constant for the ionization of the first H^+) is larger than K_2 (that for the ionization of the second H^+), and the value for pK_1 is smaller than that for pK_2. The difference between pK_1 and pK_2 decreases with increasing distance between the carboxyl groups. (Why?) The first and second pK_a values for a few diacids are listed in Table 14.5.

$$HO_2CCH_2CO_2H + H_2O \rightleftharpoons HO_2CCH_2CO_2^- + H_3O^+$$

malonic acid
$pK_1 = 2.83$

$$HO_2CCH_2CO_2^- + H_2O \rightleftharpoons {}^-O_2CCH_2CO_2^- + H_3O^+$$

$pK_2 = 5.69$

	Table 14.5 pK_a values for some diacids		
Trivial Name[a]	Structure	pK_1	pK_2
oxalic	HO_2C—CO_2H	1.2	4.2
malonic	$HO_2CCH_2CO_2H$	2.8	5.7
succinic	$HO_2C(CH_2)_2CO_2H$	4.2	5.6
glutaric	$HO_2C(CH_2)_3CO_2H$	4.3	5.4
adipic	$HO_2C(CH_2)_4CO_2H$	4.4	5.4
pimelic	$HO_2C(CH_2)_5CO_2H$	4.5	5.4
o-phthalic	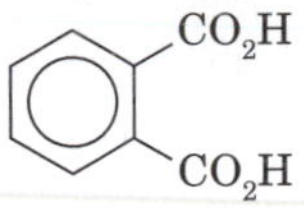 (structure shown)	2.9	5.5

[a] To memorize the trivial names of the diacids with two to seven carbons, note that the first letters (o, m, s, g, a, p) fit the phrase: "Oh my, such good apple pie." (o-Phthalic acid is, of course, not part of the homologous series.)

B. Anhydride Formation by Dibasic Acids

An **anhydride of a carboxylic acid** has the structure of two carboxylic acid molecules joined together with the loss of water.

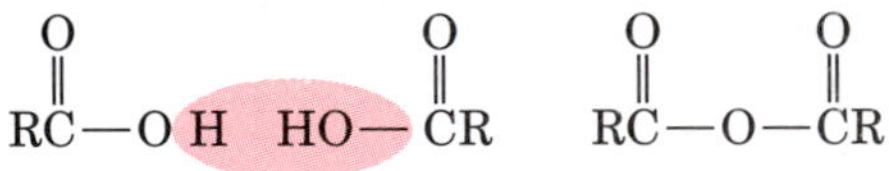

two carboxylic acids   an anhydride

Although it would seem reasonable at first glance, heating most carboxylic acids to drive off water does *not* result in the anhydride. Exceptions are dicarboxylic acids that can form five- or six-membered cyclic anhydrides. These diacids yield anhydrides when heated to 200–300°, but milder conditions can be used if acetic anhydride is added. Acetic anhydride reacts with the diacid to form the cyclic anhydride and acetic acid. Distillation of the acetic acid drives the reaction to completion. Anhydrides will be discussed further in Section 15.4.

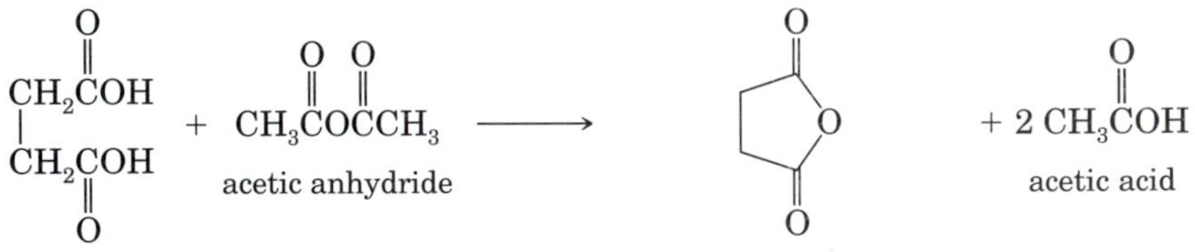

succinic acid   succinic anhydride (95%)

STUDY PROBLEM

14.11   What would be the product when each of the following diacids is heated with a dehydrating agent such as acetic anhydride?

(a) o-phthalic acid (Table 14.5)   (b) 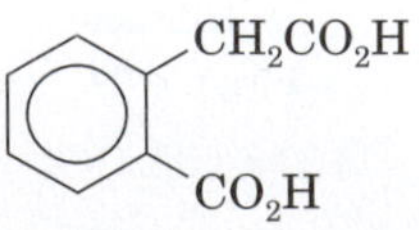 (structure shown)

C. Decarboxylation of β-Keto Acids and β-Diacids

Simply heating most carboxylic acids does not result in any chemical reaction. However, a carboxylic acid with a β carbonyl group undergoes **decarboxylation** (loss of CO_2) when heated. The temperature necessary depends on the individual compound.

General:

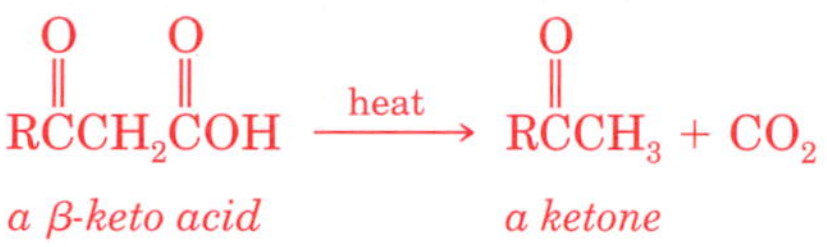

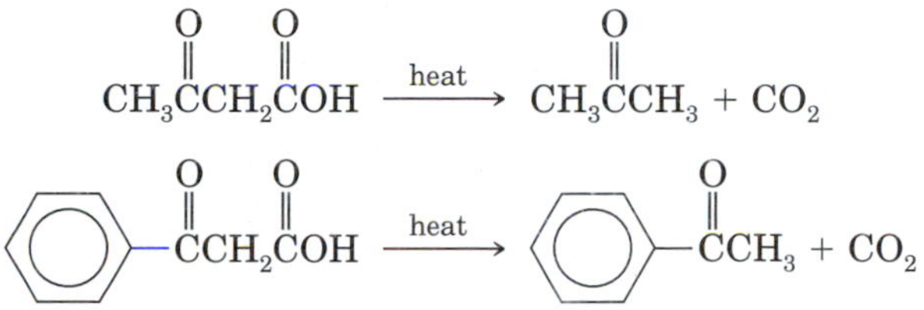

$$\underset{\text{a }\beta\text{-keto acid}}{\text{RCCH}_2\text{COH}} \xrightarrow{\text{heat}} \underset{\text{a ketone}}{\text{RCCH}_3} + CO_2$$

$$CH_3CCH_2COH \xrightarrow{\text{heat}} CH_3CCH_3 + CO_2$$

$$\text{C}_6\text{H}_5\text{—CCH}_2\text{COH} \xrightarrow{\text{heat}} \text{C}_6\text{H}_5\text{—CCH}_3 + CO_2$$

Decarboxylation takes place through a cyclic transition state:

transition state   *an enol*

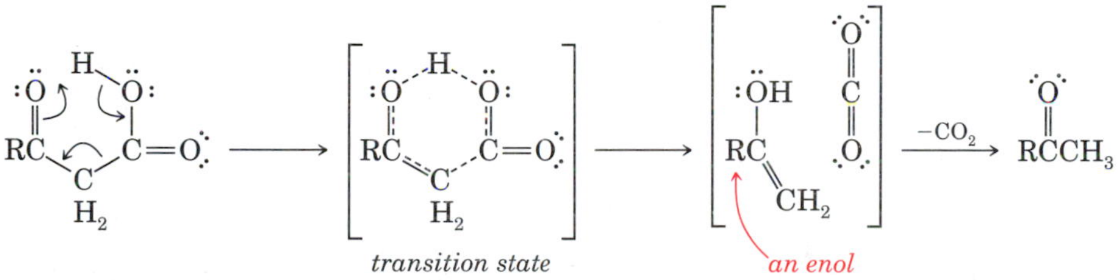

The cyclic transition state requires only a carbonyl group beta to the carboxyl group. This carbonyl group need not necessarily be a keto group. A β-diacid also undergoes decarboxylation when heated. Decarboxylation of substituted malonic acids is especially important in organic synthesis, and you will encounter this reaction again in Chapter 17.

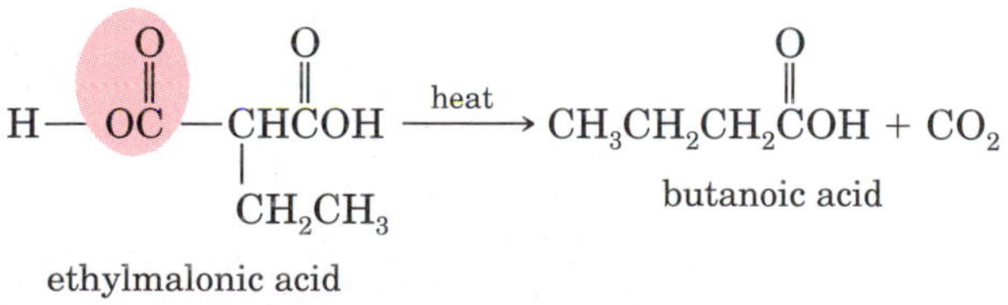

$$\underset{\text{ethylmalonic acid}}{H\!-\!OC\!-\!\underset{\underset{CH_2CH_3}{|}}{CHCOH}} \xrightarrow{\text{heat}} \underset{\text{butanoic acid}}{CH_3CH_2CH_2COH} + CO_2$$

STUDY PROBLEMS

14.12   Write an equation for the mechanism of the decarboxylation of ethylmalonic acid.

14.13   Predict the products (if any) when each of the following acids is heated:

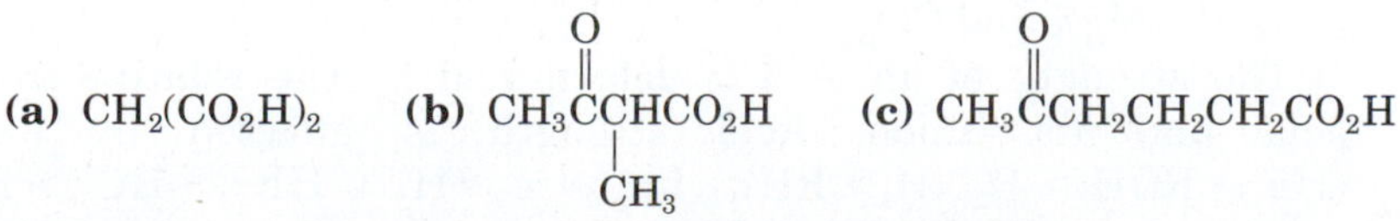

(a) $CH_2(CO_2H)_2$ (b) CH_3CCHCO_2H with CH_3 (c) $CH_3CCH_2CH_2CH_2CO_2H$

A few α-carbonyl acids, such as oxalic acid, can also undergo decarboxylation. The decarboxylation of α-keto acids is common in biological systems in which enzymes catalyze the reaction.

$$\underset{\text{oxalic acid}}{HOC\!-\!CO\,H} \xrightarrow{\;150°\;} \underset{\text{formic acid}}{HCO_2H} \;+\; CO_2$$

$$\underset{\substack{\text{pyruvic acid} \\ \textit{formed in glucose} \\ \textit{metabolism}}}{CH_3C\!-\!CO\,H} \xrightarrow{\;\text{enzymes}\;} \underset{\text{acetaldehyde}}{CH_3CH} \;+\; CO_2$$

Section 14.9

Use of Carboxylic Acids in Synthesis

The types of products that can be obtained from carboxylic acids are shown in Table 14.6. In many of these reactions, an acid halide or an acid anhydride gives better yields. Esterification is such a reaction. However, the acids themselves are usually more readily available.

STUDY PROBLEMS

14.14   Show two routes to methyl butanoate from butanoic acid.

14.15   Show by flow equations how you would make the following conversions:

    (a) propanoic acid $\longrightarrow$ butanoic acid

    (b) 3-chloropropanoic acid $\longrightarrow$ butanedioic acid

    (c) $\longrightarrow$ $\bigcirc\!\!-\!CH_2CO_2H$

Summary

Carboxylic acids (RCO_2H) undergo hydrogen bonding to form **dimers,** a structural feature that has an effect on their physical and spectral properties.
Carboxylic acids can be synthesized by: (1) the *hydrolysis of their derivatives* (esters, amides, anhydrides, acid halides, or nitriles); (2) *oxidation of primary alcohols, aldehydes, alkenes, or alkylbenzenes;* and (3) *Grignard reactions* of RMgX and CO_2.
The strength of an acid is determined by the relative stabilities of the acid and its anion. Acid strength is affected by *electronegativity* (HF > ROH > R_2NH > RH), by *size* (HI > HBr > HCl > HF), and by

Table 14.6 Compounds that can be obtained from carboxylic acids

Reaction	Product	Section Reference
Neutralization:[a]		
$RCO_2H + Na^+\,OH^- \xrightarrow{H_2O} RCO_2^-\,Na^+$	**carboxylate salt**	14.4, 14.5
Esterification:		
$RCO_2H + R'OH \xrightarrow{H^+} RCO_2R'$	**ester**	14.6
Reduction:		
$RCO_2H \xrightarrow[\text{(2) } H_2O,\,H^+]{\text{(1) } LiAlH_4^+} RCH_2OH$	**1° alcohol**	14.7
Anhydride Formation:		
$HO_2C(CH_2)_nCO_2H \xrightarrow[\text{heat}]{(CH_3C)_2O}$ where $n = 2$ or 3 → cyclic anhydride structure	**cyclic anhydride**	14.8B 15.4B
Decarboxylation:		
$RCCR'_2CO_2H \xrightarrow{\text{heat}} RCCHR'_2$	**ketone**	14.8C
$HO_2CCR_2CO_2H \xrightarrow{\text{heat}} R_2CHCO_2H$	**carboxylic acid**	14.8C
Acid Halide Formation:[b]		
$RCO_2H + SOCl_2 \text{ or } PCl_3 \longrightarrow RCCl$	**acid halide**	15.3B
α Halogenation:		
$RCH_2CO_2H + Cl_2 \xrightarrow{PCl_3} RCHClCO_2H$	**α-chloro acid**	15.3C

[a] Carboxylate salts can be used to make esters by S_N2 reactions with reactive halides:

$$RCO_2^- + R'X \longrightarrow RCO_2R' \quad \text{(See Section 5.10.)}$$

[b] Acid halides may be used to make a variety of compounds, including anhydrides and esters. These reactions are discussed in Chapter 15.

hybridization ($\equiv CH > {=}CH_2 > {-}CH_3$). The *inductive effect* of electron-withdrawing groups causes an increase in acid strength ($ClCH_2CO_2H > CH_3CO_2H$). *Resonance stabilization* of the anion also strengthens an acid ($RCO_2H > ArOH > ROH$). The anion of an acid can be partially stabilized by *solvation*—increased solvation of the anion strengthens an acid.

The strength of a benzoic acid is determined largely by inductive effects because the $-CO_2^-$ group does not enter into resonance with the aromatic ring. Electron-withdrawing substituents strengthen the acid, while electron-releasing groups weaken the acid. Substitution at the *o*-position almost always increases acid strength.

$$O_2N{\leftarrow}\langle\bigcirc\rangle{-}CO_2H > \langle\bigcirc\rangle{-}CO_2H > CH_3{\rightarrow}\langle\bigcirc\rangle{-}CO_2H$$

Carboxylic acids are one of the few general classes of organic compounds that are more acidic than H_2CO_3; carboxylic acids react with HCO_3^-. **Carboxylate salts** result from the reaction of a carboxylic acid with base.

$$RCO_2H \;\; + NaOH \longrightarrow RCO_2^- \, Na^+ + H_2O$$

$$\begin{array}{ccc} \textit{a carboxylic} & \textit{a base} & \textit{a sodium} \\ \textit{acid} & & \textit{carboxylate} \end{array}$$

The reactions of carboxylic acids are summarized in Table 14.6.

Essay Problem for Chapter 14

Quantitative Relationship Between Structure and Reactivity: The Hammett Equation

In the 1930s and 1940s, L. P. Hammett, working at Columbia University, developed an equation that provides a quantitative relationship between structure and reactivity. For correlations, Hammett used substituted benzoic acids as the reference group and the acid ionization of the benzoic acids as the reference reaction. The Hammett equation is:

$$\log K - \log K_0 = \rho\sigma$$

In this equation, K can be either the equilibrium or the rate constant for a reaction, and K_0 is the corresponding value for a benzoic acid. Rho (ρ) is defined as the reaction constant and sigma (σ) as the substituent constant.

A **sigma value** (σ) is a measure of a substituent's ability to donate or to withdraw electron density from a reaction site. The sigma value for a substituent is the difference between the pK_a of a substituted benzoic acid and that of the benzoic acid itself. For example, the sigma value for the nitro group is calculated from the pK_a of benzoic acid ($pK_a = 4.2$) and that of p-nitrobenzoic acid ($pK_a = 3.4$) using the Hammett equation as follows:

$$\log K_{p\text{-nitrobenzoic acid}} - \log K_{\text{benzoic acid}} = \sigma\rho$$

since $pK_a = -\log K_a$ and $\rho = 1.00$ for the benzoic acids,
$$-3.4 - (-4.2) = \sigma = 0.8 \text{ (literature value, 0.78)}$$

Sigma values vary with the position of substitution on the benzene ring. What we calculated above was the σ_{para}. The σ_{meta} for the nitro group is 0.71. The *ortho* values are not used because steric hindrance and other interactions complicate the effects of *ortho* substituents on acidity.

A positive value for sigma means the substituent withdraws electron density from the reaction site. A negative value means that the substituent donates electron density to the reaction. Typical σ_{para} values are given in Table 14.7.

A **reaction constant** or **rho value** (ρ) is a measure of a reaction's sensitivity to substituent effects. A rho value is calculated for a reaction by using compounds with substituents that have known sigma values. The rho value is the slope of a line obtained by plotting the equilibrium (or rate) constants of a series of substituted compounds versus benzoic acids substituted with the same substituents. For example, plotting the data given in Table 14.8 gives a line, shown in Figure 14.5, whose slope is 0.5. This is the rho value for the ionization of phenylacetic acid. (The literature rho value for this reaction is 0.489.)

Table 14.7	Selected *para* substituent constants (σ_p)		
Group	σ_p	Group	σ_p
—OH	− 0.37	—Cl	+ 0.23
—OCH_3	− 0.27	—CF_3	+ 0.55
—CH_3	− 0.17	—CN	+ 0.66
—C_6H_5	− 0.01	—NO_2	+ 0.78

Table 14.8	Acidity constants (pK_a) of some carboxylic acids		
Compound	pK_a	Compound	pK_a
CH_3—⬡—COH	4.4	CH_3—⬡—CH_2COH	4.4
⬡—C—OH	4.2	⬡—CH_2COH	4.3
Cl—⬡—C—OH	4.0	Cl—⬡—CH_2COH	4.2
O_2N—⬡—C—OH	3.4	O_2N—⬡—CH_2COH	3.9

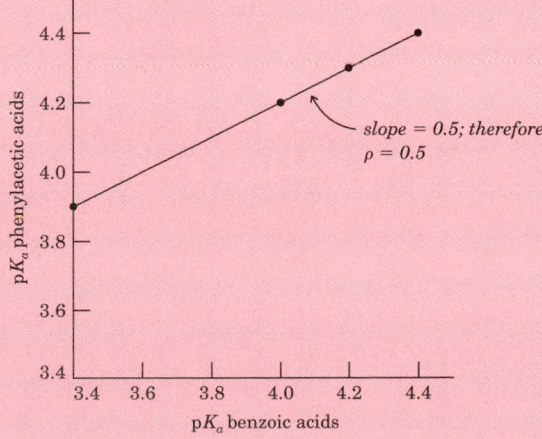

slope = 0.5; therefore,
ρ = 0.5

Figure 14.5 pK_a values for some substituted benzoic acids versus pK_a of substituted phenylacetic acids. Data taken from Table 14.8.

Table 14.9 Selected reaction constants (ρ)[a]

Reaction	Solvent	
	H_2O	1.000[b]
	H_2O	0.212
	60% acetone	2.229
	ethanol	−5.090

[a] Temperature, 25°C.
[b] Defined as 1.000 for benzoic acid.

A positive rho means that the reaction is aided by an electron-withdrawing group. A negative rho means that the reaction is aided by electron-donating substituents. The magnitude of the rho value tells us the sensitivity of the reaction to the effect of the substituent. With phenylacetic acid, we see that the rho is positive and has a magnitude of one-half that of benzoic acid. Typical reactions and their rho values are given in Table 14.9.

Questions

1. Explain why the ionization of phenylacetic acid is not as sensitive to substituents as is the ionization of benzoic acid.
2. Write resonance structures for the following two methoxy benzoic acids. How can these structures be used to explain why the σ_{para} for the —OCH_3 group is −0.27 while the σ_{meta} value is +0.12?

3. Identify each group in Table 14.7 as an o,p-director or as a m-director (see Table 11.5). Is the sigma value of the group in reasonable agreement with the activating influence of the group?
4. Write the mechanism for the solvolysis of diphenylchloromethane with ethanol. Identify the rate-determining step in the mechanism. Would you expect the transition state for this step to be stabilized or destabilized by a para—OCH_3? by a para—NO_2? Does your assessment agree with the ρ value for the reaction given in Table 14.9?

5. Predict the following pK_a's using the Hammett equation.

(a) NC—⬡—COH (with O double bonded)

(b) HO—⬡—COH (with O double bonded)

(c) CH_3O—⬡—CH_2COH (with O double bonded)

Study Problems*

14.16 Write an IUPAC name for each of the following carboxylic acids or salts.

(a) $(CH_3)_3CCO_2H$ (b) $CH_3CH_2CHBrCHBrCO_2H$

(c) ⬡ with OH and —CO_2H and HO substituents
an analgesic

(d) Cl—⬡—OCH_2CO_2H
a plant growth regulator and fruit thinner

14.17 Give the structures for: **(a)** 4-iodobutanoic acid; **(b)** potassium formate; **(c)** disodium *o*-phthalate; **(d)** sodium benzoate; **(e)** *m*-methylbenzoic acid.

14.18 Give the structure of each of the following groups: **(a)** propionyl group; **(b)** butyryl group; **(c)** *m*-nitrobenzoyl group.

14.19 Show the principal types of hydrogen bonding (there may be more than one type) that occur in each of the following systems: **(a)** an aqueous solution of propanoic acid; **(b)** an aqueous solution of lactic acid, $CH_3CH(OH)CO_2H$.

14.20 Show how you could synthesize butanoic acid from each of the following compounds:

(a) 1-bromopropane (b) 1-butanol

(c) butanal (d) 4-octene

(e) $CH_3CH_2CH_2CO_2CH_2CH_3$ (f) 1-pentyne

14.21 Usually, there are two optional methods for the transformation of alkyl halides into carboxylic acids: (1) By conversion of the alkyl halide into its Grignard reagent followed by carboxylation with CO_2, or (2) by conversion of the alkyl halide into its nitrile with cyanide ion followed by acidic hydrolysis. However, when the alkyl halide is a tertiary halide, one of the methods is unsatisfactory. Which method is unsatisfactory? Why?

---

* For information concerning the organization of the *Study Problems* and *Additional Problems*, see the *Note to student* on page 41.

14.22 List the following compounds in order of increasing acid strength (least acidic first):

(a) $CH_3CH_2CHBrCO_2H$ **(b)** $CH_3CHBrCH_2CO_2H$

(c) $CH_3CH_2CHClCH_2OH$ **(d)** $CH_3CH_2CH_2CH_2OH$

(e) C_6H_5OH **(f)** H_2CO_3

(g) Br_3CCO_2H **(h)** H_2O

14.23 Which compound in each pair is the stronger acid? Why?

(a) NH_4^+ or PH_4^+ **(b)** $CH_3CH{=}CH_2$ or $CH_3C{\equiv}CH$

(c) $HO_2CCH_2CO_2H$ or $^-O_2CCH_2CO_2H$ **(d)**

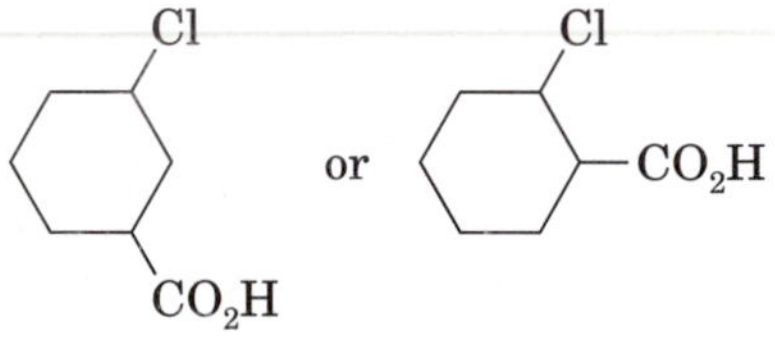

(e)

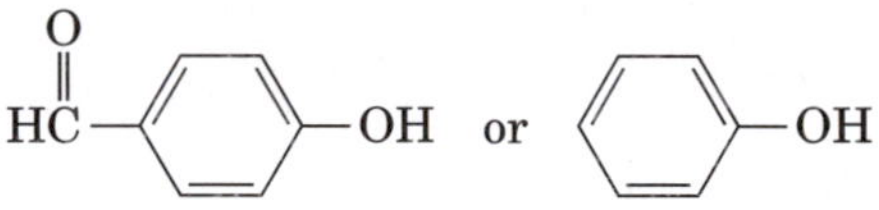

(f)

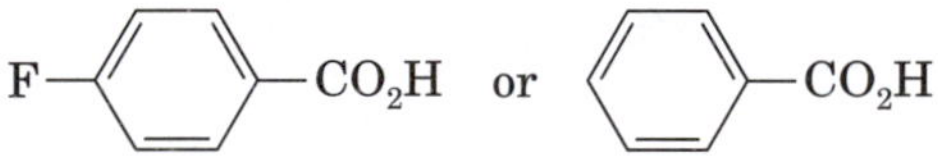

(g) $BrCH_2CO_2H$ or $ClCH_2CO_2H$

(h) $CH_3CH(Cl){-}CH(Cl)CO_2H$ or $CH_3C(Cl)_2CH_2CO_2H$

14.24 Arrange the following compounds in order of increasing acidity (weakest acid first). Give a reason for your answer.

p-methoxyphenol *p*-cyanophenol phenol

14.25 In each pair, which is the stronger base? Explain.

(a) $CH_3CH{=}CH^-$ or $CH_3C{\equiv}C^-$

(b) Cl^- or $CH_3CO_2^-$

(c) $ClCH_2CO_2^-$ or $Cl_2CHCO_2^-$

(d) $(CH_3)_3CO^-$ or $(CH_3)_3CCO_2^-$

(e) $CH_3CHClCO_2^-$ or $ClCH_2CH_2CO_2^-$

14.26 Complete the following equations, giving the organic products:

(a)

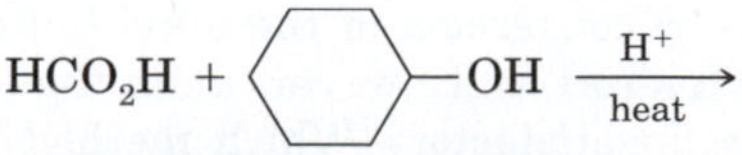

(b)

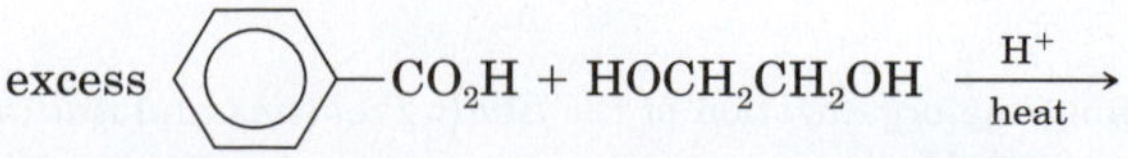

$$^{18}OH$$

(c) 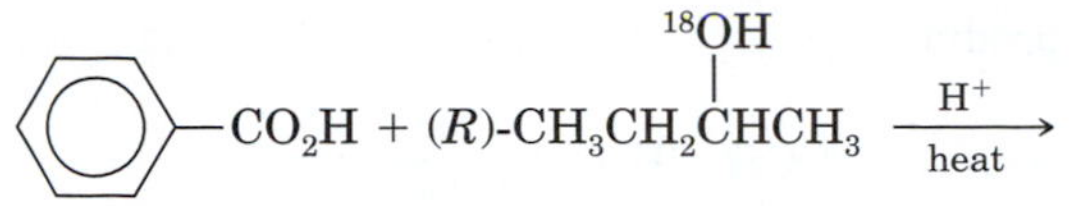 ⬡—CO$_2$H + (*R*)-CH$_3$CH$_2$CHCH$_3$ $\xrightarrow[\text{heat}]{\text{H}^+}$

14.27 List the following acids and alcohols in order of increasing rates of reaction leading to the ester (slowest first). Explain your answer.

(a) acetic acid with methanol and HCl

(b) 2-ethylbutanoic acid with 2,2-dimethyl-1-propanol and HCl

(c) 2-ethylbutanoic acid with ethanol and HCl

14.28 Racemic 2-bromopropanoic acid reacts with (*R*)-2-butanol to produce an ester mixture.

(a) What are the structures (stereochemistry included) of the esters?   **(b)** Are these esters enantiomers or diastereomers?

14.29 Predict the products when the following compounds are treated with LiAlH$_4$ followed by treatment with aqueous acid:

(a) 2,2-dimethylpropanoic acid

(b) 3-pentenoic acid

(c)

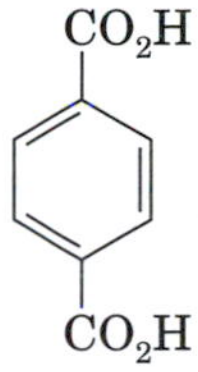

14.30 What are the major organic products when the following compounds are heated with acetic anhydride?

(a) 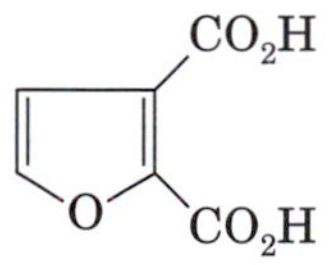   **(b)** HO$_2$CCH$_2$CH$_2$CH$_2$CO$_2$H

14.31 Predict the products of heating the following compounds:

(a) 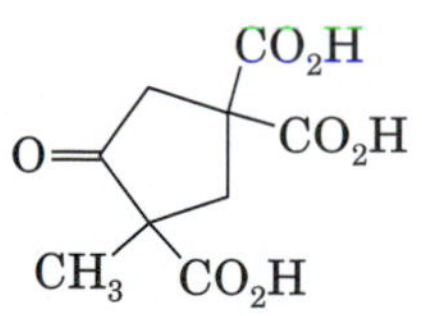   **(b)**

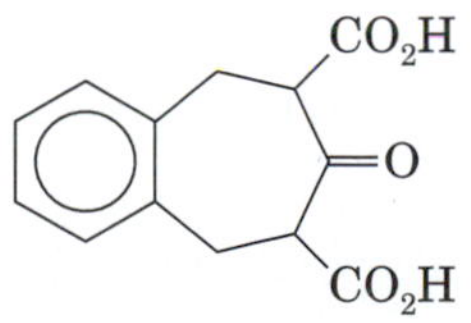

14.32 Write flow equations that would explain each of the following reactions.

(a) (HOCH$_2$)$_2$C(CO$_2$CH$_2$CH$_3$)$_2$ + HBr $\xrightarrow{\text{heat}}$ CH$_2$=CCO$_2$H (43%)

$$\qquad\qquad\qquad\qquad\qquad\qquad\qquad\qquad\qquad\qquad\qquad\qquad |$$
$$\qquad\qquad\qquad\qquad\qquad\qquad\qquad\qquad\qquad\qquad\qquad\text{CH}_2\text{Br}$$

(b) 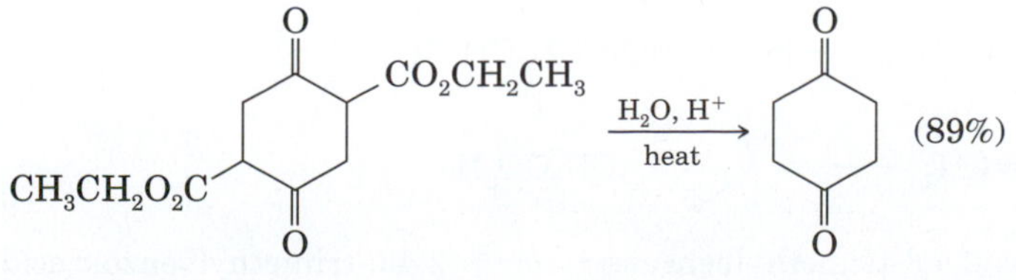 $\xrightarrow[\text{heat}]{\text{H}_2\text{O, H}^+}$ (89%)

14.33 Predict the major organic products:

(a) $CH_3CH_2CH_2CO_2H$ + excess $CH_3CH_2CH_2OH$ $\xrightarrow[\text{heat}]{H^+}$

(b) $CH_3CH_2CH_2CO_2H$ $\xrightarrow[\text{(2) } H_2O, \, H^+]{\text{(1) LiAlH}_4}$

(c) $CH_3CH_2CH_2CO_2H$ $\xrightarrow[\text{heat, pressure}]{H_2, \, Pt}$

(d) $CH_3CH_2CH_2CO_2C_6H_5$ + excess H_2O $\xrightarrow[\text{heat}]{H^+}$

(e) $CH_3\overset{O}{\overset{\|}{C}}CH_2\overset{O}{\overset{\|}{C}}OH$ $\xrightarrow[\text{(2) } H_2O, \, H^+]{\text{(1) LiAlH}_4}$

(f) $HO_2CCH_2CH_2\overset{Cl}{\overset{|}{C}H}CO_2H$ + 1 NaOH $\xrightarrow[\text{cold}]{H_2O}$

(g) $HO_2CCH_2CH_2\overset{Cl}{\overset{|}{C}H}CO_2H$ + 2 NaOH $\xrightarrow[\text{cold}]{H_2O}$

(h) $HO_2CCH_2CH_2\overset{Cl}{\overset{|}{C}H}CO_2H$ + excess $CH_3\overset{O}{\overset{\|}{C}}O\overset{O}{\overset{\|}{C}}CH_3$ $\xrightarrow{\text{warm}}$

(i) $HO_2CCH_2\overset{O}{\overset{\|}{C}}CH_2CO_2H$ $\xrightarrow{\text{heat}}$

Additional Problems

14.34 Predict the major organic products when $CH_3CH_2CH_2CH(CO_2H)_2$ is treated with an excess of each of the following reagents: **(a)** aqueous $NaHCO_3$; **(b)** LiAlH$_4$ followed by aqueous acid; **(c)** methanol + HCl; **(d)** methanol + NaOH; **(e)** a mixture of xylenes (dimethylbenzenes) and heat; **(f)** methylmagnesium iodide in diethyl ether; **(g)** aqueous NH_3.

14.35 Suggest reagents for the following conversions:

(a) $H_2NCH_2CH_2CH_2Br$ $\longrightarrow$ $^+H_3NCH_2CH_2CH_2CO_2^-$

(b)   2-bromobutane $\longrightarrow$ 2-methylbutanoic acid

(c) CH_3CH_2OH $\longrightarrow$ $CH_3\underset{OH}{\overset{|}{C}H}CO_2H$

(d) $BrCH_2CH_2CH_2Br$ $\longrightarrow$ $HO_2CCH_2CH_2CH_2CO_2H$

(e)   ⬠=CH_2 $\longrightarrow$ ⬠–CH_2CO_2H

(f)   1-bromo-2,4,6-trimethylbenzene $\longrightarrow$ 2,4,6-trimethylbenzoic acid

(g) [structure: cyclohexane ring with $CO_2CH_2CH_3$ and CCH_3 (with =O) substituents] $\longrightarrow$ [structure: cyclohexane ring with $C(=O)CH_3$ substituent]

(h) $CH_3CH_2CH_2Br \longrightarrow CH_3(CH_2)_3CO_2H$

14.36 *Nicotinic acid* (also called *niacin*), a B vitamin, can be obtained by the vigorous oxidation of nicotine (page 474). What is the structure of nicotinic acid?

14.37 *Dacron* is a synthetic polymer that can be made by the reaction of terephthalic acid and 1,2-ethanediol. What is the structure of Dacron?

[structure: benzene ring with $HOC(=O)$ and $C(=O)OH$ substituents]

terephthalic acid

14.38 The salt of a carboxylic acid does not show carbonyl absorption at $1660-2000$ cm^{-1} ($5-6$ μm) in the infrared spectrum. Explain.

14.39 Compound A, $C_4H_6O_4$, yielded Compound B, $C_4H_4O_3$, when heated. When A was treated with an excess of methanol and a trace of sulfuric acid, Compound C, $C_6H_{10}O_4$, was obtained. Upon treatment with LiAlH$_4$ followed by hydrolysis, Compound A yielded Compound D, $C_4H_{10}O_2$. What are the structures of Compounds A, B, C, and D?

14.40 Predict the product(s), including stereochemistry, when the following compound is heated under reflux with an excess of 1 *M* HCl. (Use equations in your answer.)

[structure: benzene ring with two CH_3O substituents, and a side chain with H_3C, H, C, $C(CO_2CH_2CH_3)_2$, NH_2]

14.41 Upon treatment with alkaline permanganate followed by neutralization, the geometric isomers A and B of 1,3-divinylcyclopentane produce isomeric compounds C and D, $(C_7H_{10}O_4)$, respectively. Isomer C is recovered unchanged when heated with acetic anhydride, but isomer D produces compound E $(C_7H_8O_3)$ under the same conditions. What are the stereochemical formulas for A, B, C, D, and E?

14.42 Suggest synthetic schemes for the following conversions:

(a) propyl 2-methylpropanoate from propene as the only organic starting material

(b) [structure: cyclohexane ring with OH and CO_2H substituents] from cyclohexanone

(c) [structure: cyclobutane ring with CO_2CH_2-(benzene)-$CH=CH_2$, H, H, and CO_2CH_2-(benzene)-$CH=CH_2$ substituents] from $ClCH_2$-(benzene)-$CH=CH_2$

(d) 1,4-butanediol $\longrightarrow$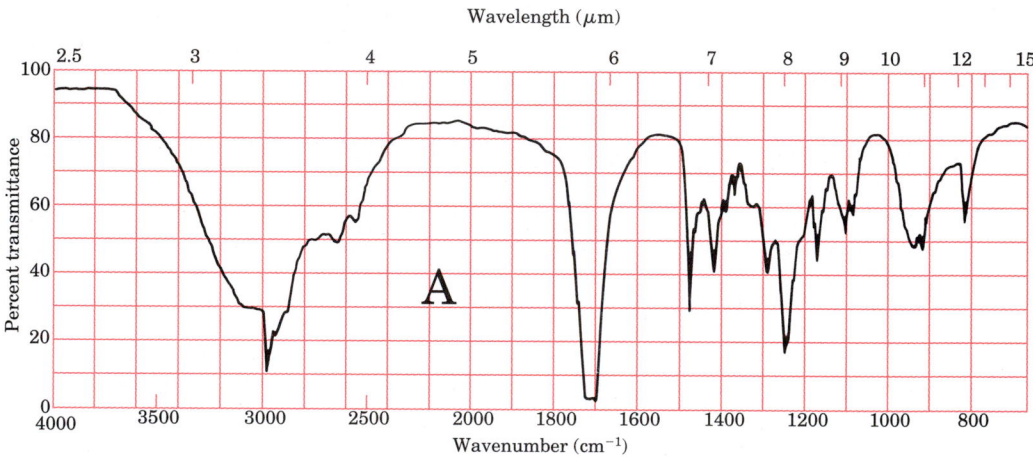

(e) iodoethene $\longrightarrow$ propenoic acid

(f) bromoethane $\longrightarrow$ ethyl butanoate

(g) 1-bromopropane $\longrightarrow$ ethyl butanoate

(h) 3-hydroxybutanoic acid $\longrightarrow$ acetone

14.43 Compound A ($C_4H_8O_2$) was treated with $LiAlH_4$. After hydrolysis, Compound B was isolated. The infrared spectrum of A and the ^{1}H NMR spectrum of B are shown in Figure 14.6. What are the structures of A and B?

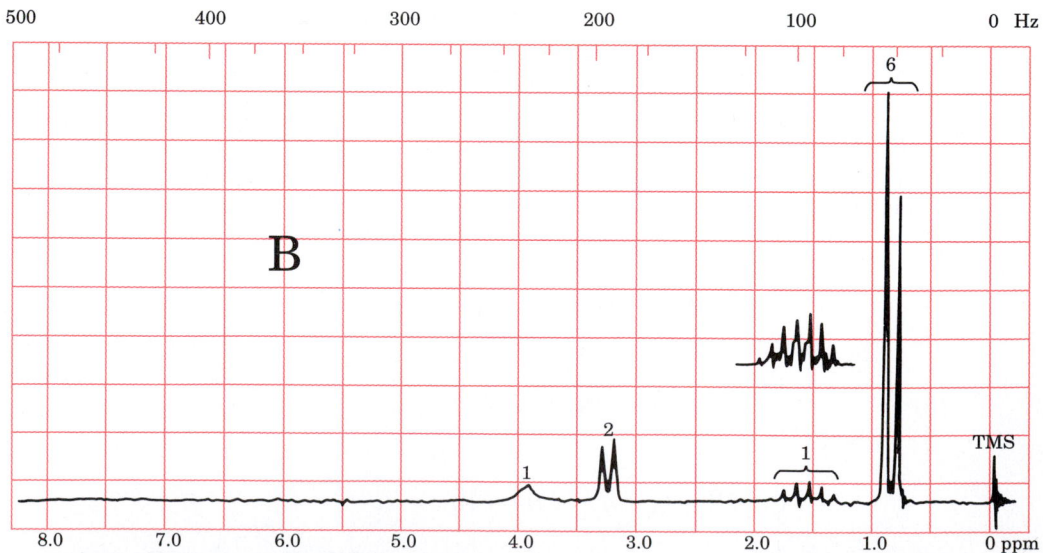

Figure 14.6 Spectra for Problem 14.43.

14.44 A chemist treated Compound A ($C_4H_7O_2Br$) with potassium *tert*-butoxide in *tert*-butyl alcohol. After acidification of the product mixture, the chemist isolated two isomeric products: B and C. The infrared spectrum of B and the ^1H NMR spectrum of C are shown in Figure 14.7. What are the possible structures of A, B, and C?

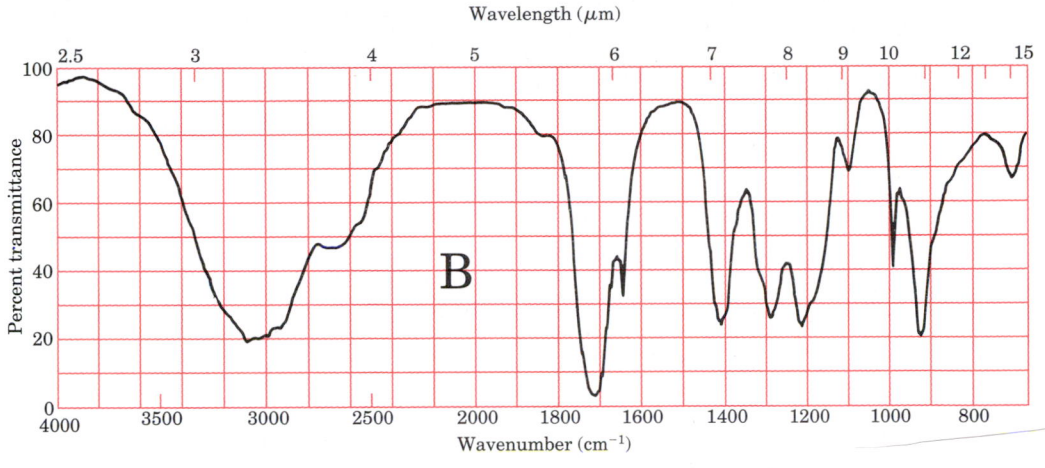

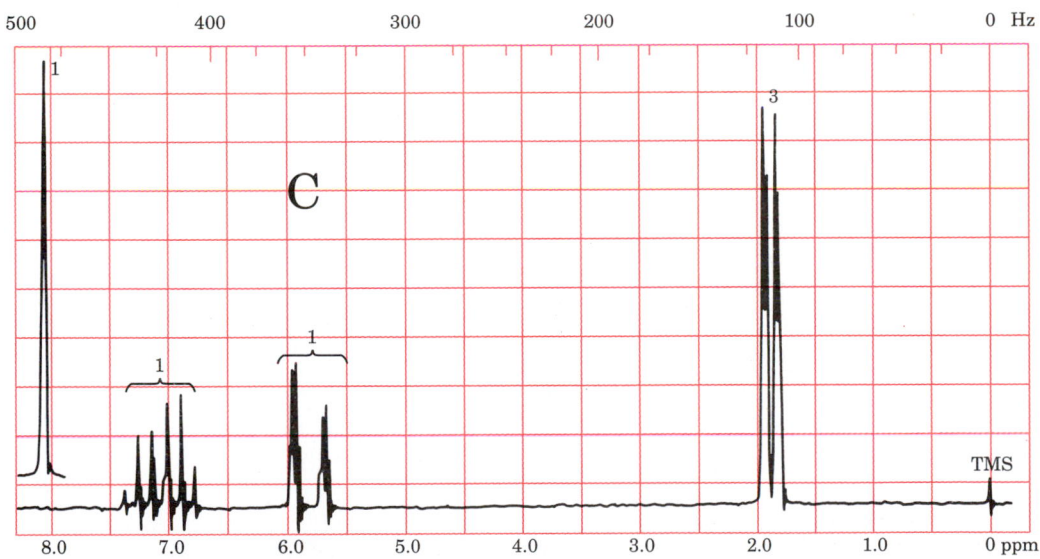

Figure 14.7 Spectra for Problem 14.44. (The offset absorption for C is at $\delta12.18$.)

DERIVATIVES OF CARBOXYLIC ACIDS

A derivative of a carboxylic acid is a compound that yields a carboxylic acid upon reaction with water.

$$\underset{\text{ethyl acetate}}{CH_3\overset{\displaystyle O}{\overset{\|}{C}}OCH_2CH_3} + H_2O \; \underset{}{\overset{H^+,\, \text{heat}}{\rightleftharpoons}}$$

$$\underset{\text{acetic acid}}{CH_3\overset{\displaystyle O}{\overset{\|}{C}}OH} + HOCH_2CH_3$$

$$\underset{\text{acetamide}}{CH_3\overset{\displaystyle O}{\overset{\|}{C}}NH_2} + H_2O + H^+ \; \overset{\text{heat}}{\longrightarrow}$$

$$CH_3\overset{\displaystyle O}{\overset{\|}{C}}OH + NH_4{}^+$$

In this chapter, we will discuss carboxylic acid halides, anhydrides, esters, amides, and nitriles. Table 15.1 shows some representative examples of these compounds. Note that all the derivatives except

Table 15.1	**Some derivatives of carboxylic acids**	
	Examples	
Class	Structure	Trivial Name
acid halide	CH₃COCl	acetyl chloride
	⬡—COCl	benzoyl chloride
acid anhydride	(CH₃CO)₂O	acetic anhydride
	(⬡—CO)₂O	benzoic anhydride
ester	CH₃CO₂CH₂CH₃	ethyl acetate
	⬡—CO₂CH₃	methyl benzoate
amide	CH₃CONH₂	acetamide
	⬡—CONH₂	benzamide
nitrile	CH₃CN	acetonitrile
	⬡—CN	benzonitrile

nitriles contain **acyl groups,** RCO—. In each case, an electronegative atom is attached to the carbonyl carbon of the acyl group. For this reason, the chemistry of each of these compound classes is similar.

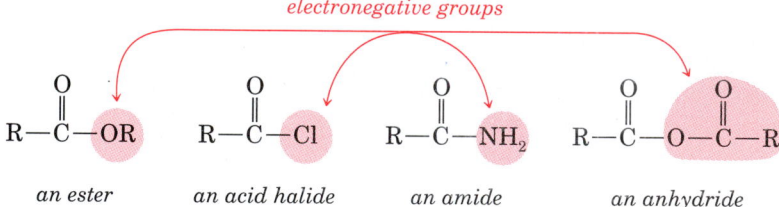

an ester   *an acid halide*   *an amide*   *an anhydride*

   Carboxylic acids and some of their derivatives are found in nature. Fats are triesters, waxes are monoesters, and proteins are polyamides, to name just a few. Acid halides are never found in nature, and anhydrides are found only rarely. One example of a naturally occurring anhydride is *cantharidin,* a cyclic anhydride found in "Spanish fly." Cantharidin is a toxic irritant of the urinary tract. The dried insects were used by the ancient Greeks and Romans as an aphrodisiac. Cantharidin is also reputed to remove warts.

cantharidin

Section 15.1
Reactivity of Carboxylic Acid Derivatives

Why are carboxylic acids, esters, and amides commonly found in nature, while acid halides and anhydrides are not? Why are the carboxylic acid derivatives different from aldehydes or ketones? We can answer these questions by considering the relative reactivities of carboxylic acid derivatives and how they undergo reaction.

The derivatives of carboxylic acids contain *leaving groups* bonded to the acyl carbons, whereas aldehydes and ketones do not. Reagents generally *add* to the carbonyl group of ketones and aldehydes, but *substitute* for the leaving groups of acid derivatives:

Addition:

$$\underset{\text{CH}_3\text{CCH}_3}{\overset{\overset{\displaystyle O}{\|}}{}} \quad \overset{\text{HCN}}{\underset{}{\rightleftharpoons}} \quad \text{CH}_3 - \underset{\underset{\displaystyle \text{CN}}{|}}{\overset{\overset{\displaystyle \text{OH}}{|}}{\text{C}}} - \text{CH}_3$$

Substitution:

$$\underset{\text{CH}_3\text{C}-\text{Cl}}{\overset{\overset{\displaystyle O}{\|}}{}} \quad \overset{\text{H}_2\text{O}}{\longrightarrow} \quad \underset{\text{CH}_3\text{C}-\text{OH}}{\overset{\overset{\displaystyle O}{\|}}{}} + \text{HCl}$$

good leaving group

In Chapter 5, we mentioned that a good leaving group is a weak base. Therefore, Cl^- is a good leaving group, but ^-OH and ^-OR are poor leaving groups. The reactivity of carbonyl compounds toward substitution at the carbonyl carbon may be directly attributed to the basicity of the leaving group:

$$^-\text{CH}_3 \quad ^-\text{NH}_2 \quad ^-\text{OR} \quad ^-\overset{\overset{\displaystyle O}{\|}}{\text{OCR}} \quad X^-$$

**decreasing basicity
(increasing ease of displacement)**

$$\underset{\text{R}-\text{C}-\text{R}}{\overset{\overset{\displaystyle O}{\|}}{}} \quad \underset{\text{RC}-\text{NH}_2}{\overset{\overset{\displaystyle O}{\|}}{}} \quad \underset{\text{RC}-\text{OR}'}{\overset{\overset{\displaystyle O}{\|}}{}} \quad \underset{\text{RC}-\text{OCR}}{\overset{\overset{\displaystyle O}{\|}}{\overset{\displaystyle O}{\|}}} \quad \underset{\text{RC}-\text{Cl}}{\overset{\overset{\displaystyle O}{\|}}{}}$$

increasing reactivity toward water

Acid chlorides and acid anhydrides, with good leaving groups, are readily attacked by water. Therefore, we would not expect to find these compounds in the cells of plants or animals. Because of their high reactivity, however, these acid derivatives are invaluable in the synthesis of other organic compounds. A relatively nonreactive carboxylic acid may be converted to one of these more reactive derivatives and then converted to a ketone, an ester, or an amide (Figure 15.1).

Esters and amides are relatively stable toward water. In the laboratory, these compounds require an acid or a base and, usually, heating to undergo

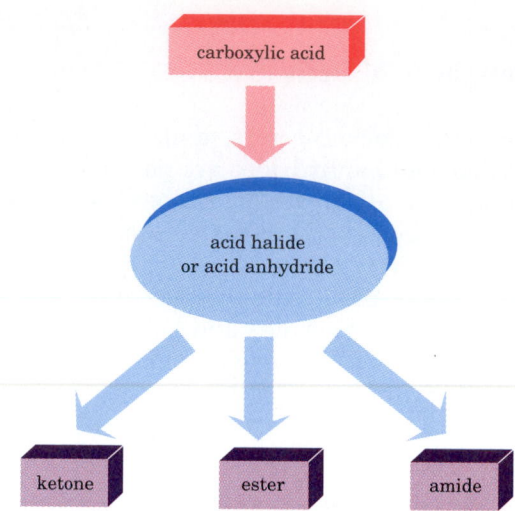

Figure 15.1   The synthetic relationship of acid halides and anhydrides to the other carboxylic acid derivatives and ketones.

hydrolysis. In nature, enzymes can perform the functions of acid or base and heat.

15.1   Arrange the following compounds in order of increasing reactivity toward methanol:

$$\text{(a)} \; CH_3\overset{\displaystyle O}{\overset{\|}{C}}OCH_2CH_3 \qquad \text{(b)} \; CH_3\overset{\displaystyle O}{\overset{\|}{C}}O\overset{\displaystyle O}{\overset{\|}{C}}CH_3 \qquad \text{(c)} \; CH_3\overset{\displaystyle O}{\overset{\|}{C}}Cl \qquad \text{(d)} \; CH_3\overset{\displaystyle O}{\overset{\|}{C}}NH_2$$

Section 15.2

Spectral Properties of Carboxylic Acid Derivatives

The ^1H NMR spectra of carboxylic acid derivatives provide little information about the functionality in these compounds. The signals for α hydrogens of these carbonyl compounds are shifted slightly downfield from the signals for ordinary aliphatic hydrogens because of deshielding by the partially positive carbonyl carbon atom. The α hydrogens of an acid chloride exhibit a greater chemical shift than those of the other acid derivatives. This large chemical shift arises from the greater ability of the Cl (compared to O or N) to withdraw electron density from nearby bonds.

$$CH_3CN \quad CH_3\overset{\displaystyle O}{\overset{\|}{C}}OCH_3 \quad CH_3\overset{\displaystyle O}{\overset{\|}{C}}NH_2 \quad CH_3\overset{\displaystyle O}{\overset{\|}{C}}OH \quad CH_3\overset{\displaystyle O}{\overset{\|}{C}}Cl$$

$\delta:$ 2.00   2.03   2.08   2.10   2.67

increasing chemical shift of α hydrogens

Table 15.2   Infrared carbonyl absorption for carboxylic acid derivatives

		Position of Absorption	
Class	Structure	cm^{-1}	μm
acid chloride	$\overset{\text{O}}{\overset{\|}{\text{RCCl}}}$	1785–1815	5.51–5.60
acid anhydride	$\overset{\text{O O}}{\overset{\|\ \ \|}{\text{RCOCR}}}$	1740–1840 (usually two peaks)	5.45–5.75
ester	$\overset{\text{O}}{\overset{\|}{\text{RCOR}}}$	1740	5.75
amide	$\overset{\text{O}}{\overset{\|}{\text{RCNH}_2}}$	1630–1700	5.9–6.0

The infrared spectra of acid derivatives provide more information about the type of functional group than do NMR spectra. With the exception of the nitriles, the principal distinguishing feature of the infrared spectra of all the carboxylic acid derivatives is the carbonyl absorption found at about 1630–1840 cm^{-1} (5.4–6 μm). The positions of carbonyl absorption for the various acid derivatives are summarized in Table 15.2. Anhydrides and esters also show C—O absorption in the region of 1050–1250 cm^{-1} (8–9.5 μm).

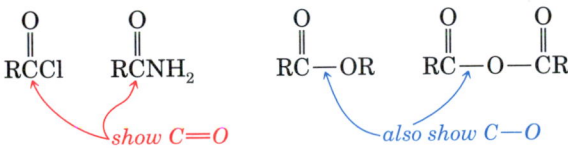

A. Acid Chlorides

The carbonyl infrared absorption of acid chlorides is observed at slightly higher frequencies than that of other acid derivatives. There is no other distinguishing feature in the infrared spectrum that signifies "This is an acid chloride." See Figure 15.2 for the infrared spectrum of a typical acid chloride.

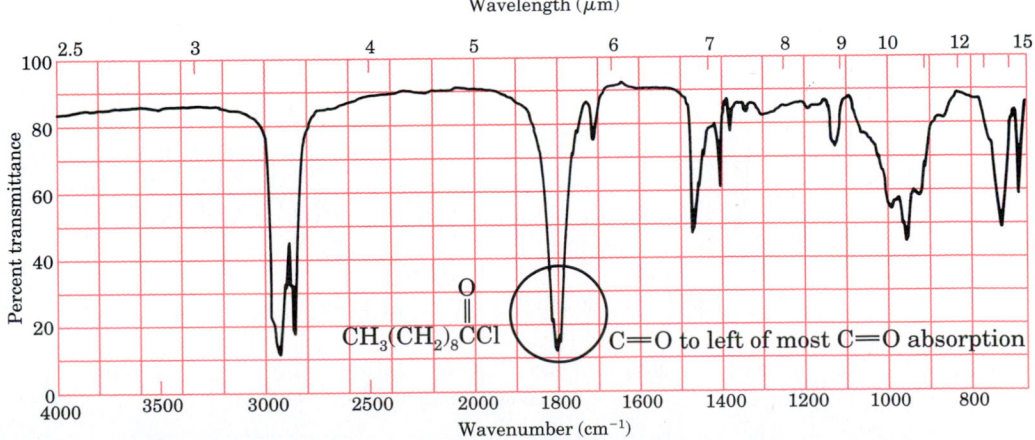

Figure 15.2   Infrared spectrum of decanoyl chloride.

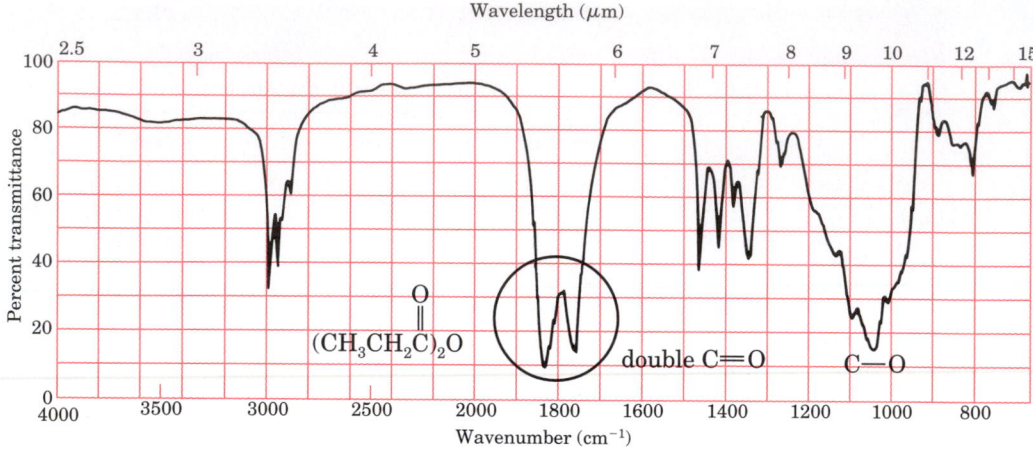

Figure 15.3 Infrared spectrum of propanoic anhydride.

B. Anhydrides

A carboxylic acid anhydride, which has two C=O groups, generally exhibits a *double carbonyl peak* in the infrared spectrum. Anhydrides also exhibit a C—O stretching band around 1100 cm^{-1} (9 μm). Figure 15.3 shows the infrared spectrum of a typical aliphatic anhydride.

C. Esters

The carbonyl infrared absorption of aliphatic esters is observed at about 1740 cm^{-1} (5.75 μm). However, conjugated esters (either α,β-unsaturated esters or α-aryl esters) absorb at slightly lower frequencies, about 1725 cm^{-1} (5.8 μm). Esters also exhibit C—O stretching absorption in the fingerprint region. See Figure 15.4 for the spectrum of a typical ester.

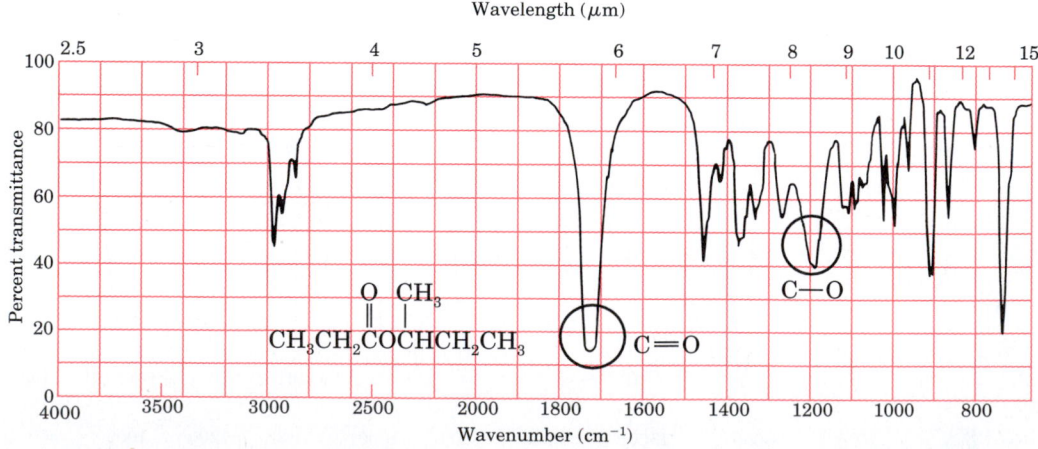

Figure 15.4 Infrared spectrum of *sec*-butyl propanoate.

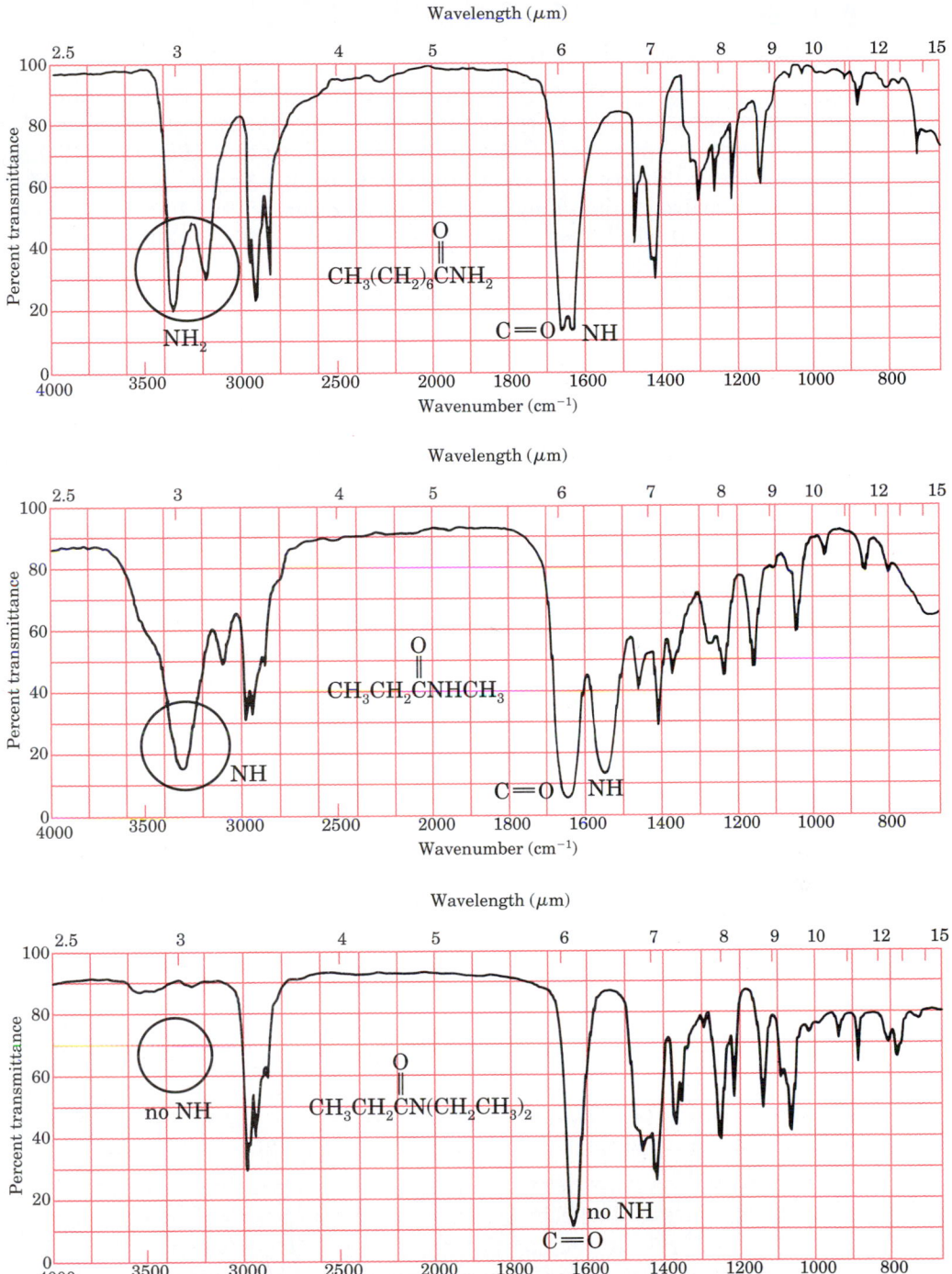

Figure 15.5 Infrared spectrum of a 1° amide (top), a 2° amide (center), and a 3° amide (bottom).

D. Amides

The position of absorption of the carbonyl group of an amide is variable and depends upon the extent of hydrogen bonding between molecules. The infrared spectrum of a pure liquid amide (maximum hydrogen bonding) shows a carbonyl peak called the **amide I band** around 1650 cm^{-1} (6.0 μm). (The spectra in Figure 15.5 show this C=O peak.) As the sample is diluted with a nonhydrogen-bonding solvent, the extent of the hydrogen bonding diminishes, and the C=O absorption is shifted to a higher frequency (1700 cm^{-1}; 5.88 μm).

The **amide II band** appears between 1515–1670 cm^{-1} (6.0–6.6 μm), just to the right of the C=O absorption. This absorption arises from NH bending. Therefore, a disubstituted, or tertiary, amide does not show an amide II band.

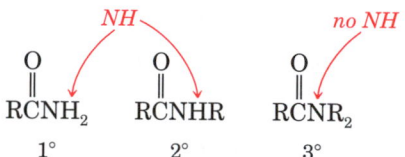

The NH stretching vibrations give rise to absorption to the left of aliphatic CH absorption at 3125–3570 cm^{-1} (2.8–3.2 μm). (This is about the same region where the NH of amines and OH absorb.) *Primary amides* (RCONH$_2$) show a double peak in this region. *Secondary amides* (RCONHR), with only one NH bond, show a single peak. *Tertiary amides* (RCONR$_2$), with no NH, show no absorption in this region. Figure 15.5 shows the infrared spectra of the three types of amides. Compare the NH stretching and bending absorptions of these three compounds.

E. Nitriles

The C≡N absorption is found in the triple-bond region of the infrared spectrum (2200–2300 cm^{-1}; 4.3–4.5 μm) and is medium to weak in intensity. (See Figure 15.6.)

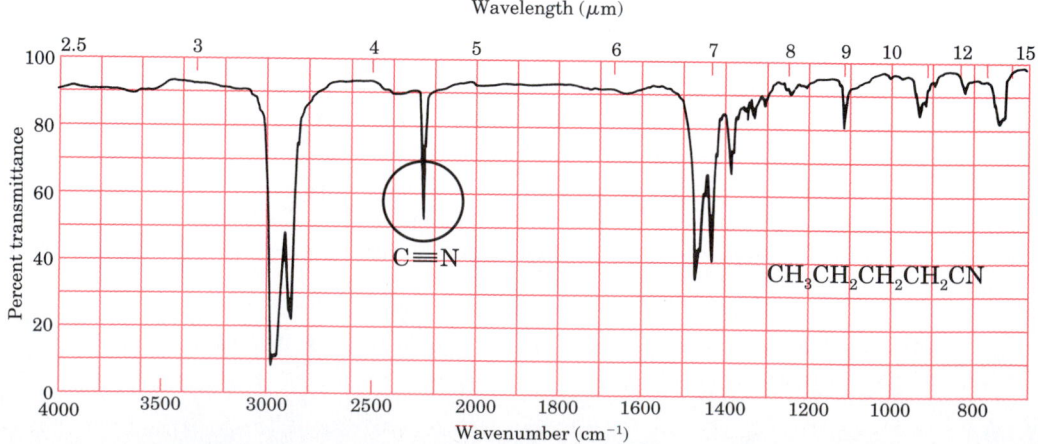

Figure 15.6   Infrared spectrum of pentanenitrile.

STUDY PROBLEM

15.2 An unknown has the molecular formula $C_5H_7O_2N$. Its infrared and ^1H NMR spectra are given in Figure 15.7. What is the structure of the unknown?

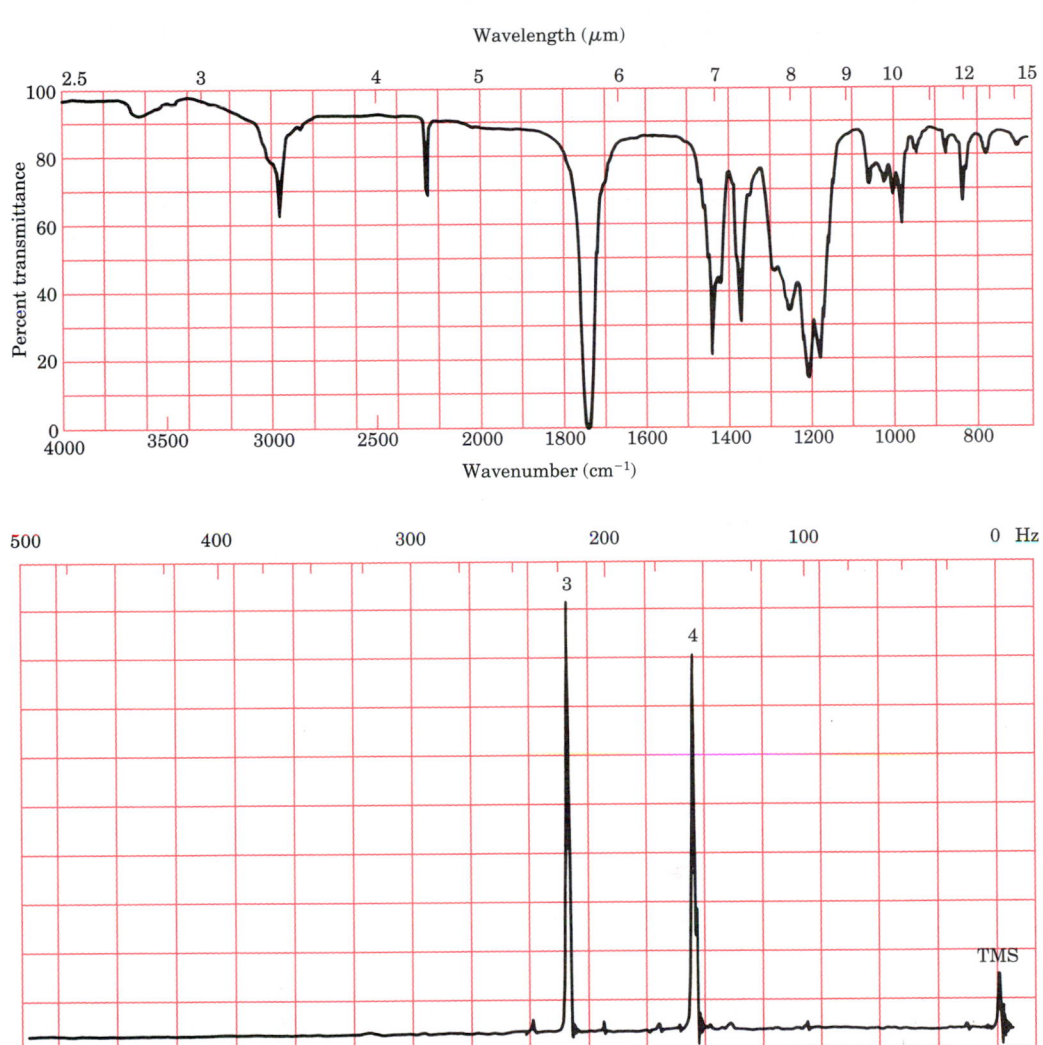

Figure 15.7 Spectra for Problem 15.2.

Section 15.3
Acid Halides

Acid fluorides, chlorides, bromides, and iodides all undergo similar reactions. Because acid chlorides are the most common of the acid halides, our discussions will emphasize these compounds.

A. Nomenclature of Acid Chlorides

Acid chlorides are named after the parent carboxylic acid with the **-ic acid** ending changed to **-yl chloride.**

$$\underset{\text{CH}_3\text{CCl}}{\overset{\overset{\displaystyle O}{\|}}{}} \qquad \underset{\text{CH}_3\text{CH}_2\text{CCl}}{\overset{\overset{\displaystyle O}{\|}}{}} \qquad \underset{\text{CH}_3\text{CH}_2\text{CH}_2\text{CCl}}{\overset{\overset{\displaystyle O}{\|}}{}}$$

IUPAC: ethanoyl chloride   propanoyl chloride   butanoyl chloride
trivial: acetyl chloride   propionyl chloride   butyryl chloride

---

STUDY PROBLEM

15.3 Name the following compounds.

$$\textbf{(a) } \text{CH}_3(\text{CH}_2)_8\overset{\overset{\displaystyle O}{\|}}{\text{C}}\text{Cl} \qquad \textbf{(b)}$$

(b) structure: 3,5-dimethoxybenzoyl chloride drawn with CH$_3$O groups and —CCl

---

B. Preparation of Acid Chlorides

Acid chlorides can be obtained directly from their parent carboxylic acids by reaction with thionyl chloride (SOCl$_2$) or some other inorganic acid chloride such as phosphorus trichloride (PCl$_3$):

General:

$$\underset{\substack{\text{a carboxylic}\\\text{acid}}}{\overset{\overset{\displaystyle O}{\|}}{\text{RCOH}}} + \text{SOCl}_2 \longrightarrow \underset{\substack{\text{an acid}\\\text{chloride}}}{\overset{\overset{\displaystyle O}{\|}}{\text{RCCl}}} + \text{SO}_2 + \text{HCl}$$

$$3\,\overset{\overset{\displaystyle O}{\|}}{\text{RCOH}} + \text{PCl}_3 \longrightarrow \overset{\overset{\displaystyle O}{\|}}{\text{RCCl}} + \text{H}_3\text{PO}_3$$

$$\underset{\text{butanoic acid}}{\text{CH}_3\text{CH}_2\text{CH}_2\overset{\overset{\displaystyle O}{\|}}{\text{C}}\text{OH}} + \text{SOCl}_2 \xrightarrow{\text{heat}} \underset{\text{butanoyl chloride (85\%)}}{\text{CH}_3\text{CH}_2\text{CH}_2\overset{\overset{\displaystyle O}{\|}}{\text{C}}\text{Cl}} + \text{SO}_2 + \text{HCl}$$

$$3\,\underset{\text{acetic acid}}{\text{CH}_3\overset{\overset{\displaystyle O}{\|}}{\text{C}}\text{OH}} + \text{PCl}_3 \xrightarrow{\text{heat}} 3\,\underset{\text{acetyl chloride (70\%)}}{\text{CH}_3\overset{\overset{\displaystyle O}{\|}}{\text{C}}\text{Cl}} + \text{H}_3\text{PO}_3$$

Note the similarity between these reactions and the corresponding reactions of alcohols (Section 7.5).

$$\text{ROH} + \text{SOCl}_2 \longrightarrow \text{RCl} + \text{SO}_2 + \text{HCl}$$

15.4 Write an equation for the preparation of each of the following acid chlorides:

(a) $CH_3(CH_2)_4\overset{O}{\overset{\|}{C}}Cl$

(b)

$\overset{O}{\overset{\|}{CCl}}$

(c) CH_3O- with CH_3O (top), CH_3O (bottom) $-\overset{O}{\overset{\|}{C}}-Cl$

C. Reactions of Acid Chlorides

The acid halides are the most reactive of all the derivatives of carboxylic acids. The halide ion is a good leaving group. Bonded to the positive carbon of a carbonyl group, it is displaced even more easily than when it is bonded to an alkyl carbon. In the following general mechanism for the reaction of an acid chloride with a nucleophile, note that displacement of Cl^- is not a simple displacement like an S_N2 reaction. Rather, the reaction consists of two steps: (1) *addition of the nucleophile to the carbonyl group,* followed by (2) *elimination of the chloride ion.* The result of this reaction is a **nucleophilic acyl substitution,** which means "nucleophilic substitution on an acyl (RCO—) carbon."

$$R-\overset{O}{\overset{\|}{C}}-Cl \xrightarrow{(1)\ addition} \left[R-\overset{O^-}{\overset{|}{\underset{Nu}{C}}}-Cl \right] \xrightarrow{(2)\ elimination} R-\overset{O}{\overset{\|}{\underset{Nu}{C}}} + :Cl:^-$$

intermediate

Hydrolysis Cleavage by water, called **hydrolysis,** is a typical reaction of an acid chloride with a nucleophile. (Because of hydrogen bonding and solvation effects, the actual mechanism is more complex than we show here.)

Nucleophilic attack and elimination of Cl^-:

$$CH_3\overset{O}{\overset{\|}{C}}Cl + H_2O: \xrightarrow{addition} \left[CH_3\overset{O^-}{\overset{|}{\underset{^+OH_2}{C}}}-Cl: \right] \xrightarrow{elimination} CH_3\overset{O}{\overset{\|}{\underset{^+OH_2}{C}}} + :Cl:^-$$

Loss of a proton:

$$CH_3\overset{O}{\overset{\|}{\underset{^+OH_2}{C}}} \rightleftharpoons CH_3\overset{O}{\overset{\|}{\underset{:OH}{C}}} + H^+$$

Overall reaction:

$$CH_3\overset{\overset{O}{\|}}{C}Cl \; + \; H_2O \; \longrightarrow \; CH_3\overset{\overset{O}{\|}}{C}OH \; + \; HCl$$

acetyl chloride   acetic acid

Although all acid chlorides undergo acidic hydrolysis to yield carboxylic acids and alkaline hydrolysis to yield carboxylic acid salts, their rates of reaction vary. An acid chloride with a bulky alkyl group bonded to the carbonyl group undergoes reaction more slowly than an acid chloride with a small alkyl group. For example, acetyl chloride reacts almost explosively with water, but butanoyl chloride requires gentle reflux.

The effect of the size of the alkyl group on the reaction rate is one of *solubility in water* instead of steric hindrance. An acid chloride with a small alkyl group is more soluble and undergoes reaction faster. An increase in the size of the alkyl portion of the molecule renders the acid chloride less water-soluble; the reaction is slower. If hydrolysis of a variety of acid chlorides is carried out in an inert solvent that dissolves both the acid chloride and water, the rates of hydrolysis are similar.

$$CH_3CH_2CH_2\overset{\overset{O}{\|}}{C}Cl \qquad CH_3CH_2\overset{\overset{O}{\|}}{C}Cl \qquad CH_3\overset{\overset{O}{\|}}{C}Cl$$

increasing rate of hydrolysis in pure H$_2$O $\longrightarrow$

Reaction with alcohols Acid chlorides react with alcohols to yield esters and HCl in a reaction that is directly analogous to hydrolysis. The cleavage of an organic compound with an alcohol is referred to as **alcoholysis.** Alcoholysis of acid chlorides is valuable for the synthesis of hindered esters or phenyl esters.

General:

$$R\overset{\overset{O}{\|}}{C}Cl \; + \; R'OH \; \longrightarrow \; R\overset{\overset{O}{\|}}{C}OR' \; + \; HCl$$

               an alcohol   *an ester*

It is usually wise to remove HCl from the reaction mixture as it is formed because HCl can undergo reaction with alcohols. A tertiary amine or pyridine is usually added as a scavenger for HCl.

$$CH_3\overset{\overset{O}{\|}}{C}Cl \; + \; (CH_3)_3COH \; + \; \langle\bigcirc\rangle\!-\!\ddot{N}(CH_3)_2 \; \longrightarrow$$

acetyl chloride   *tert*-butyl alcohol   *N,N*-dimethylaniline

$$CH_3\overset{\overset{O}{\|}}{C}OC(CH_3)_3 \; + \; \langle\bigcirc\rangle\!-\!\overset{+}{N}H(CH_3)_2Cl^-$$

         tert-butyl acetate (68%)   *N,N*-dimethylanilinium chloride

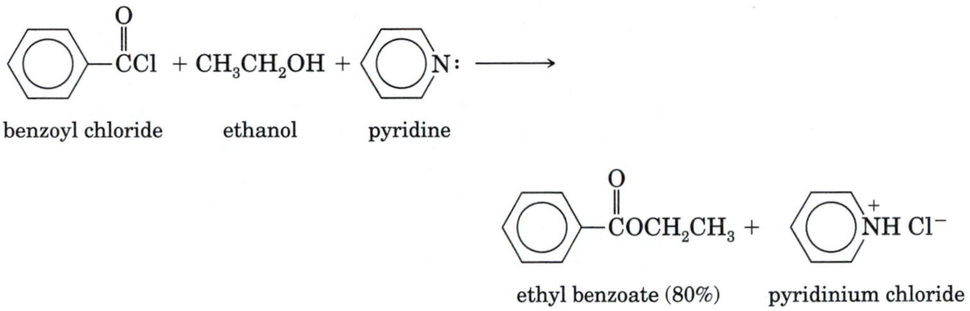

benzoyl chloride   ethanol   pyridine

ethyl benzoate (80%)   pyridinium chloride

Reaction with ammonia and amines   Ammonia and amines are good nucleophiles. Like other nucleophiles, they react with acid chlorides. The organic product of the reaction is an *amide*. As protons are lost in the deprotonation step, they react with the basic NH_3 or amine. For this reason, at least two equivalents of NH_3 or amine must be used.

$$CH_3CCl + NH_3 \longrightarrow CH_3CNH_2 + HCl \xrightarrow{NH_3} NH_4^+ Cl^-$$

ammonia   a 1° amide

$$CH_3CCl + 2\ CH_3NH_2 \longrightarrow CH_3CNHCH_3 + CH_3NH_3^+ Cl^-$$

methylamine   a 2° amide

a 1° amine

$$CH_3CCl + 2\ (CH_3)_2NH \longrightarrow CH_3CN(CH_3)_2 + (CH_3)_2NH_2^+ Cl^-$$

dimethylamine   a 3° amide

a 2° amine

If an amine is expensive, a chemist may not want to use an excess in the reaction with acid chloride. Only one mole of amine is needed to undergo reaction with acid chloride. A second mole is wasted when it is acting merely as a scavenger for HCl. In this case, a less expensive base can be used to remove the HCl. For example, an inexpensive tertiary amine might be used.

If the acid halide (such as benzoyl chloride) is not very reactive toward water, aqueous NaOH can be added to remove HX. The reactants and the aqueous NaOH form two layers. As HX is formed, it reacts with NaOH in the water layer. This reaction of an acid chloride and an amine in the presence of NaOH solution is called the **Schotten–Baumann reaction.**

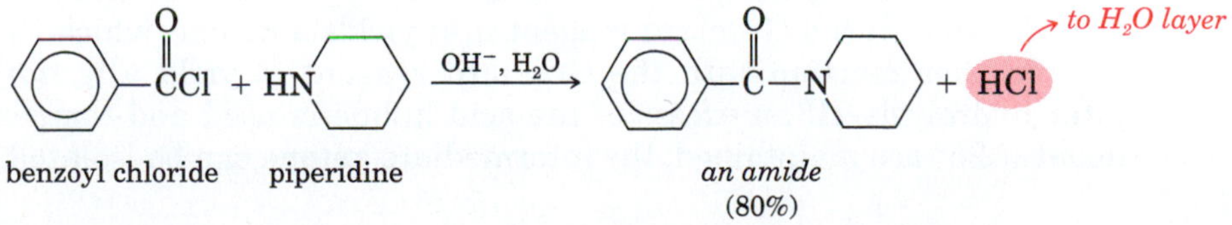

to H₂O layer

benzoyl chloride   piperidine   an amide
(80%)

STUDY PROBLEM

15.5 The Schotten–Baumann reaction is not applicable if the acid chloride and amine are water-soluble. What products would you expect from a mixture of acetyl chloride, methylamine, NaOH, and H_2O?

Conversion to anhydrides Carboxylate ions are nucleophiles, and carboxylate salts (RCO_2Na) can be used for displacement of the chloride of acid chlorides. The product of the reaction is an acid anhydride.

$$CH_3(CH_2)_5\overset{O}{\overset{\|}{C}}Cl + CH_3(CH_2)_5\overset{O}{\overset{\|}{C}}O^- \longrightarrow CH_3(CH_2)_5\overset{O}{\overset{\|}{C}}O\overset{O}{\overset{\|}{C}}(CH_2)_5CH_3 + Cl^-$$

heptanoyl chloride   heptanoate ion   heptanoic anhydride (60%)

STUDY PROBLEM

15.6 Show how the following compounds could be prepared using an acid chloride.

(a) $CH_3\!-\!\langle\!\bigcirc\!\rangle\!-\!\overset{O}{\overset{\|}{C}}\!-\!OCHCH_3$ with CH_3 branch

(b) $(CH_3)_2CH(CH_2)_8\overset{O}{\overset{\|}{C}}O\overset{O}{\overset{\|}{C}}CH_3$

(c) cyclohexyl$-\overset{O}{\overset{\|}{C}}\!-\!N\!\langle\text{morpholine}\rangle\!O$

(d) naphthyl$-O\overset{O}{\overset{\|}{C}}\!-\!$acenaphthene

Conversion to aryl ketones Acid halides are usually the reagents of choice for Friedel–Crafts acylation reactions (Section 11.8E). This reaction is a route to aryl alkyl ketones without rearrangement of the alkyl side chain. (Recall that similar *alkylations* go through alkyl carbocations and that rearrangements are common.)

$$\langle\!\bigcirc\!\rangle + CH_3CH_2\overset{O}{\overset{\|}{C}}Cl \xrightarrow{AlCl_3} \langle\!\bigcirc\!\rangle\!-\!\overset{O}{\overset{\|}{C}}CH_2CH_3 + HCl$$

benzene   propanoyl   1-phenyl-1-propanone
            chloride   (90%)

Reaction with organometallic compounds An acid chloride undergoes reaction with a variety of nucleophiles, including organometallic compounds. Reaction of an acid chloride with a Grignard reagent first yields a ketone, which then undergoes further reaction with the Grignard reagent to yield a *tertiary alcohol* after hydrolysis. (If an excess of the acid halide is used and temperatures of about $-25°$ are maintained, the intermediate ketone can be isolated.)

General:

$$R-\overset{\overset{\displaystyle O}{\|}}{C}-Cl \xrightarrow{R'MgX} R-\overset{\overset{\displaystyle O}{\|}}{C}-R' \xrightarrow{R'MgX} R-\overset{\overset{\displaystyle OMgX}{|}}{\underset{\underset{\displaystyle R'}{|}}{C}}-R' \xrightarrow{H_2O,\ H^+} R-\overset{\overset{\displaystyle OH}{|}}{\underset{\underset{\displaystyle R'}{|}}{C}}-R'$$

an acid chloride   *a ketone*   *a 3° alcohol*

benzoyl chloride $\xrightarrow[\text{(2) }H_2O,\ H^+]{\text{(1) }2\ CH_3MgI}$ 2-phenyl-2-propanol

a 3° alcohol

Lithium dialkylcopper reagents, also called **cuprates,** are synthesized from an alkyllithium and a copper(I) halide, such as CuI.

$$2\ CH_3Li + CuI \longrightarrow (CH_3)_2CuLi + LiI$$

a cuprate

Lithium dialkylcuprates are commonly used to convert acid halides to ketones, often in very high yields. These reactions are usually carried out in an ether solvent at low temperatures (0 to $-78°$). The other derivatives of carboxylic acids do not react with cuprates.

General:

$$RCCl + R'_2CuLi \longrightarrow RCR'$$

a cuprate   a ketone

$$(CH_3)_3CCCl \quad + \quad (CH_3)_2CuLi \xrightarrow[-78°]{\text{ether}} (CH_3)_3CCCH_3$$

2,2-dimethylpropanoyl chloride   *from CuI and CH₃Li*   3,3-dimethyl-2-butanone (60%)

STUDY PROBLEM

15.7 Show by equations how you would prepare the following ketones from alkyl halides and acid chlorides:

(a) phenyl—CCH(CH₃)₂   **(b)** cyclohexyl—CCH₃

Reduction Reduction of acid chlorides with lithium aluminum hydride yields primary alcohols. Because primary alcohols also can be obtained by LiAlH₄ reduction of the parent acids, this reaction finds little synthetic use.

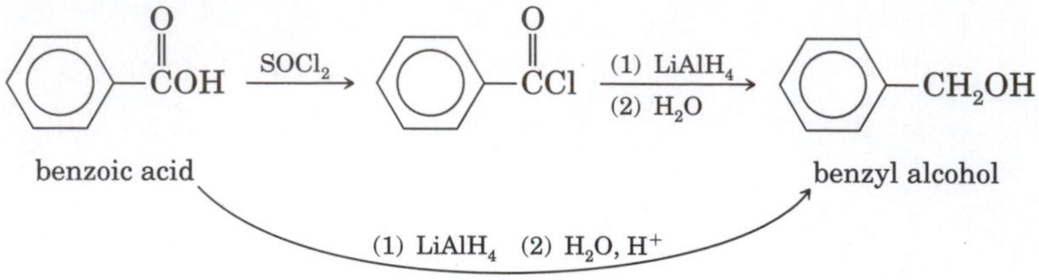

benzoic acid — benzyl alcohol

A much more valuable reaction is the partial reduction of an acid chloride to an aldehyde. (A carboxylic acid itself is not readily reduced to an aldehyde.) A milder reducing agent than $LiAlH_4$ is required to reduce $RCOCl$ to $RCHO$ instead of to RCH_2OH. A suitable reagent is **lithium tri-*tert*-butoxyaluminum hydride,** which is obtained from *tert*-butyl alcohol and $LiAlH_4$. This reducing reagent is less reactive than $LiAlH_4$ because of both steric hindrance and electron withdrawal by the oxygen atoms.

Preparation of reducing agent:

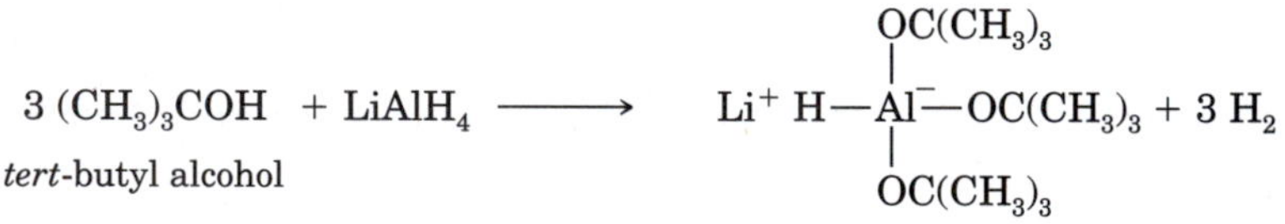

$$3\,(CH_3)_3COH \;+\; LiAlH_4 \longrightarrow Li^+\; H{-}\overset{\displaystyle OC(CH_3)_3}{\underset{\displaystyle OC(CH_3)_3}{|}}\!\!\!\overline{Al}{-}OC(CH_3)_3 + 3\,H_2$$

tert-butyl alcohol   lithium tri-*tert*-butoxyaluminum hydride

Reaction with RCOCl:

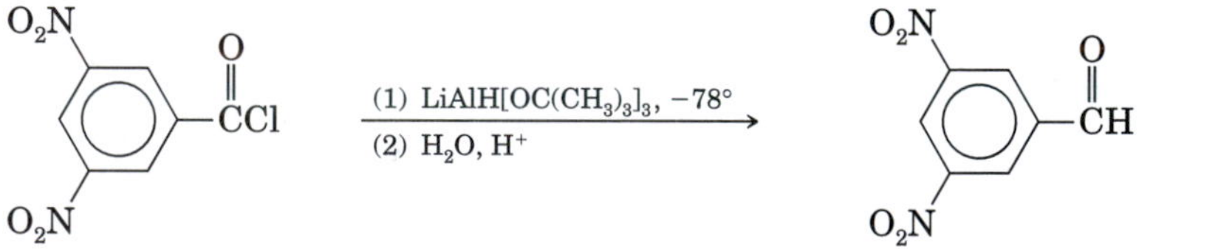

3,5-dinitrobenzoyl chloride   3,5-dinitrobenzaldehyde (63%)

Alpha halogenation   Ketones can be halogenated in the α position by treatment with X_2 and H^+ or OH^-. This reaction, which was discussed in Section 13.10, proceeds by way of the enol. Acid halides also undergo tautomerization and therefore undergo α halogenation.

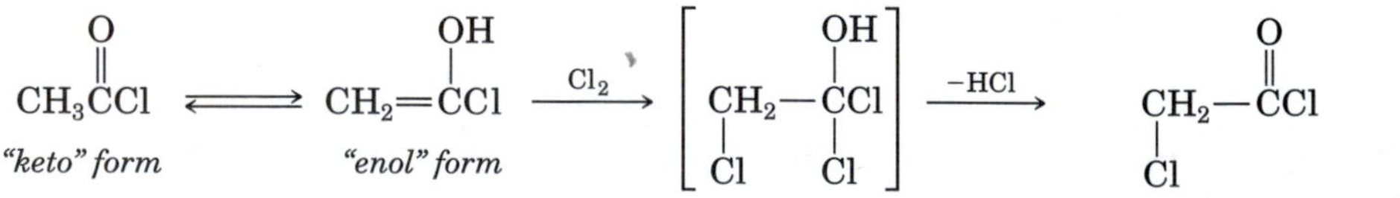

"keto" form   *"enol" form*

chloroacetyl chloride

an α-halo acid halide

Carboxylic acids do not tautomerize readily and thus do not undergo α halogenation. However, the halogenation of acid halides provides a technique by which we can obtain α-halocarboxylic acids. If a catalytic amount of PCl_3 or PBr_3 is added to a carboxylic acid along with the halogenating agent, the PCl_3 or PBr_3 converts a small proportion of the acid to the acid halide, which undergoes α halogenation.

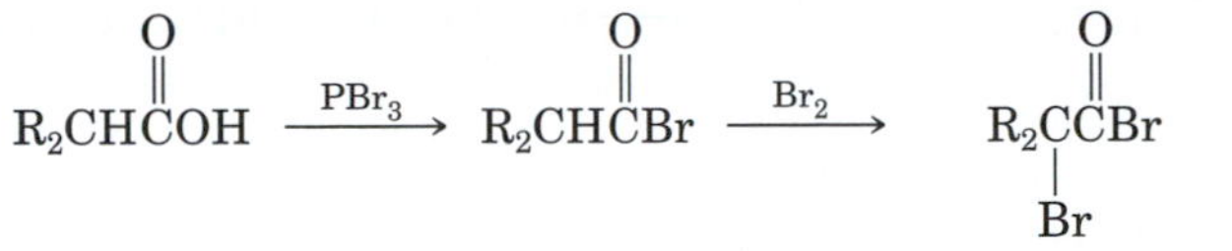

$$R_2CHCOH \xrightarrow{\ PBr_3\ } R_2CHCBr \xrightarrow{\ Br_2\ } R_2CCBr$$

$$\overset{Br}{|}$$

the halogenated acid halide

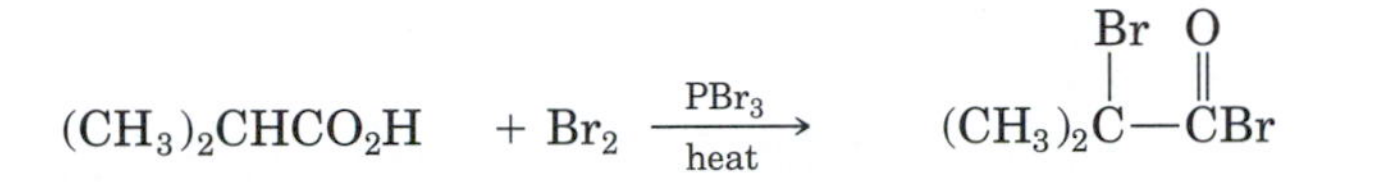

$$(CH_3)_2CHCO_2H \ + \ Br_2 \xrightarrow[heat]{\ PBr_3\ } (CH_3)_2C-CBr$$

2-methylpropanoic acid   2-bromo-2-methylpropanoyl
bromide (83%)

In the reaction mixture, the acid halide is in equilibrium with the carboxylic acid itself. Because the acid is present in excess, the α-halo acid halide is converted to the α-halo acid. This reaction is called the **Hell–Volhard–Zelinsky reaction** after the chemists who developed the technique.

General:

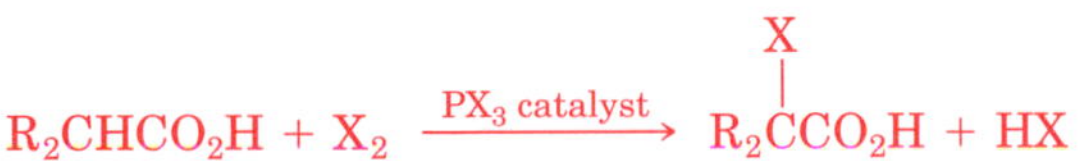

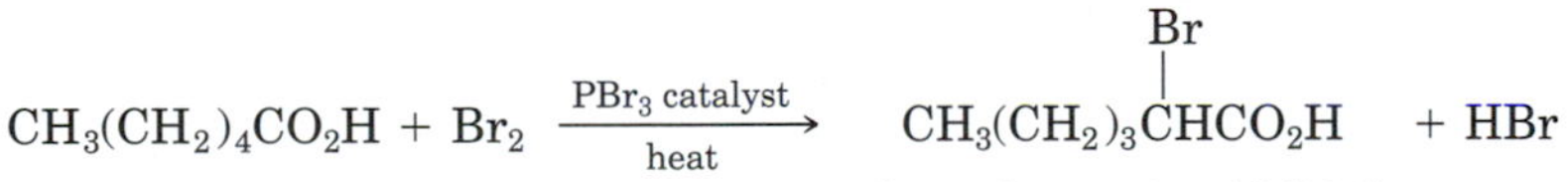

$$R_2CHCO_2H + X_2 \xrightarrow{\ PX_3\ catalyst\ } R_2\overset{X}{\underset{|}{C}}CO_2H + HX$$

$$CH_3(CH_2)_4CO_2H + Br_2 \xrightarrow[heat]{\ PBr_3\ catalyst\ } CH_3(CH_2)_3\overset{Br}{\underset{|}{C}}HCO_2H \ + \ HBr$$

hexanoic acid   2-bromohexanoic acid (89%)

Figure 15.8 summarizes the reactions of acid halides.

STUDY PROBLEMS

15.8 Predict the organic products of each of the following reactions:

(a) 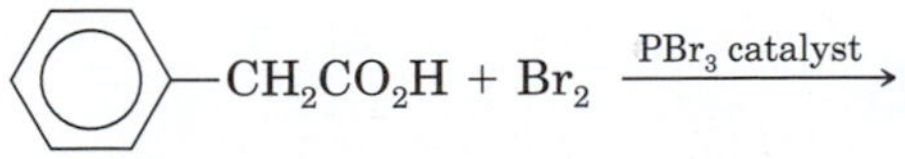 ⟨benzene ring⟩—$CH_2CO_2H + Br_2 \xrightarrow{\ PBr_3\ catalyst\ }$

(b) $ClCH_2CH_2CO_2H + Cl_2 \xrightarrow{\ PCl_3\ catalyst\ }$

15.9 Write an equation showing how 2-methylpropanoic acid can be converted to an acid halide. Then show by equations how the acid halide can be converted **(a)** to the ethyl ester, **(b)** to the aldehyde, and **(c)** to 2-methyl-3-pentanone.

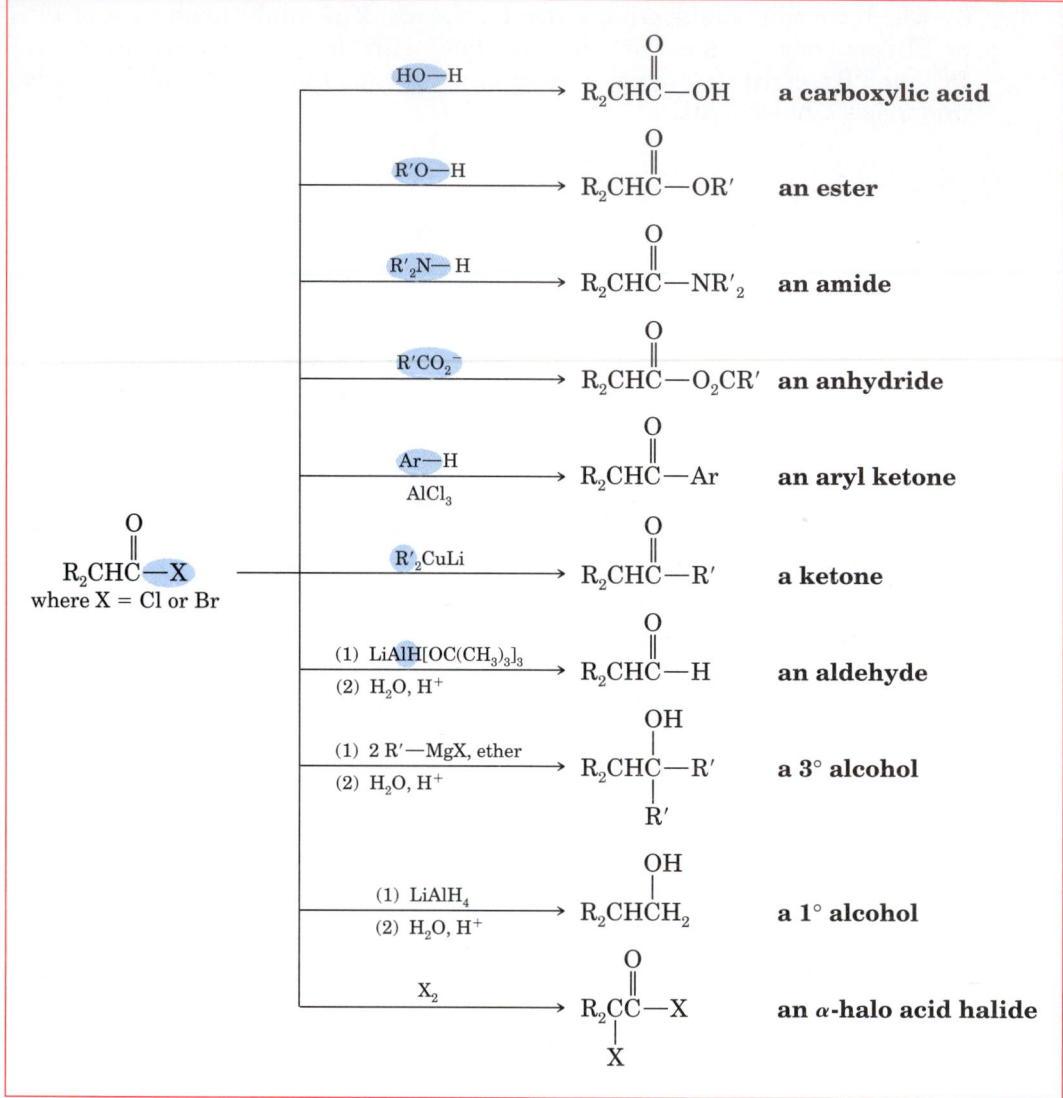

Figure 15.8 Summary of reactions of acid halides.

Section 15.4

Anhydrides of Carboxylic Acids

An **anhydride of a carboxylic acid** has the structure of two carboxylic acid molecules with a molecule of water removed. (*Anhydride* means "without water.")

$$CH_3CO \text{ H} \quad HO \text{ } CCH_3 \xrightarrow{-H_2O} CH_3COCCH_3$$

an anhydride

A. Nomenclature of Anhydrides

Symmetrical anhydrides are those in which the two acyl groups are the same. They are named after the parent carboxylic acid followed by the word **anhydride.**

$$\underset{\displaystyle \text{O} \quad \text{O}}{CH_3\overset{\parallel}{C}O\overset{\parallel}{C}CH_3} \qquad\qquad CH_3CH_2\overset{\parallel}{C}O\overset{\parallel}{C}CH_2CH_3$$

| **IUPAC:** | ethanoic anhydride | propanoic anhydride |
| **trivial:** | acetic anhydride | propionic anhydride |

Unsymmetrical anhydrides are named by the word **anhydride** preceded by the names of the *two* parent acids.

$$CH_3\overset{\parallel}{C}O\overset{\parallel}{C}CH_2CH_3$$

| **IUPAC:** | ethanoic propanoic anhydride |
| **trivial:** | acetic propionic anhydride |

STUDY PROBLEM

15.10 Write a suitable name for each of the following anhydrides:

(a) $CH_3CH_2CH_2CH_2\overset{\parallel}{C}O\overset{\parallel}{C}CH_2CH_2CH_2CH_3$ (b) $C_6H_5\overset{\parallel}{C}O\overset{\parallel}{C}CH_3$

B. Preparation of Anhydrides

With few exceptions, acid anhydrides cannot be formed directly from their parent carboxylic acids, but must be prepared from the more reactive derivatives of carboxylic acids. One route to an anhydride is from an acid chloride with a carboxylate, a reaction mentioned in Section 15.3C.

$$\underset{\text{an acid choloride}}{R\overset{\displaystyle O}{\overset{\parallel}{C}}-Cl} \quad + \quad \underset{\text{a carboxylate ion}}{{}^-O\overset{\displaystyle O}{\overset{\parallel}{C}}R'} \quad \longrightarrow \quad \underset{\text{an anhydride}}{R\overset{\displaystyle O}{\overset{\parallel}{C}}-O\overset{\displaystyle O}{\overset{\parallel}{C}}R'} + Cl^-$$

Another route to an anhydride is by treatment of the carboxylic acid with acetic anhydride. A reversible reaction occurs between a carboxylic acid and an anhydride. The equilibrium can be shifted in the desired direction by distilling the acetic acid as it is formed. This reaction has been mentioned previously (Section 14.8B).

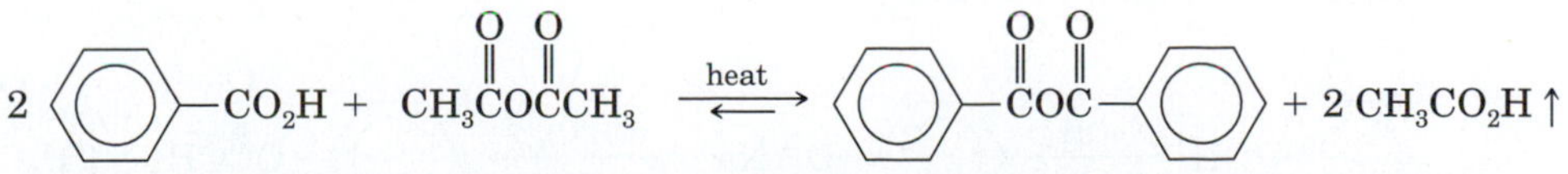

$$2\ \underset{\text{benzoic acid}}{\text{⬡}-CO_2H} + \underset{\text{acetic anhydride}}{CH_3\overset{\displaystyle O}{\overset{\parallel}{C}}O\overset{\displaystyle O}{\overset{\parallel}{C}}CH_3} \xrightarrow{\text{heat}} \underset{\text{benzoic anhydride (74\%)}}{\text{⬡}-\overset{\displaystyle O}{\overset{\parallel}{C}}O\overset{\displaystyle O}{\overset{\parallel}{C}}-\text{⬡}} + \underset{\text{acetic acid}}{2\ CH_3CO_2H\uparrow}$$

A five- or six-membered cyclic anhydride can be obtained by simply warming the appropriate diacid with acetic anhydride for a few minutes. See Section 14.8B.

C. Reactions of Anhydrides

Like acid halides, acid anhydrides are more reactive than carboxylic acids and can be used to synthesize ketones, esters, or amides. Acid anhydrides undergo reactions with the same nucleophiles that the acid chlorides react with. However, the rates of reaction are slower. (A carboxylate ion is not quite as good a leaving group as a halide ion.) The other product in these reactions is a carboxylic acid or, when the reaction mixture is alkaline, its anion.

General:

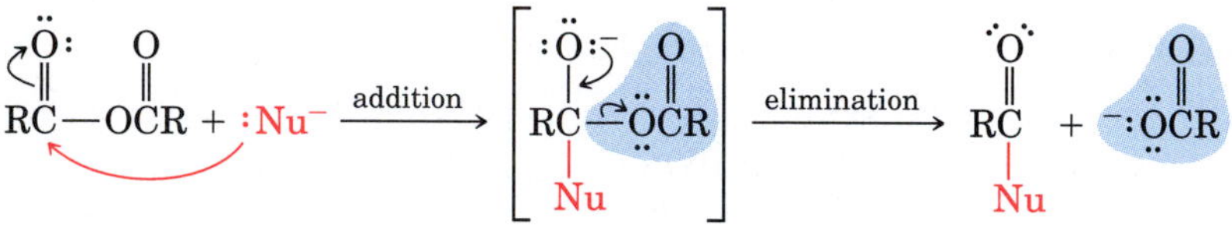

intermediate

Hydrolysis Anhydrides react with water to yield carboxylic acids. The rate of reaction, like the rate of hydrolysis of an acid chloride, depends on the solubility of the anhydride in water.

General:

$$R\overset{O}{\overset{\|}{C}}O\overset{O}{\overset{\|}{C}}R' + H_2O \longrightarrow RCO_2H + R'CO_2H$$

an anhydride → *carboxylic acids*

$$CH_3\overset{O}{\overset{\|}{C}}O\overset{O}{\overset{\|}{C}}CH_3 + H_2O \longrightarrow 2\ CH_3\overset{O}{\overset{\|}{C}}OH$$

acetic anhydride → acetic acid

Reaction with alcohols and phenols The acid-catalyzed reaction of an anhydride with an alcohol or a phenol yields an ester. The reaction is particularly useful with commercially available acetic anhydride.

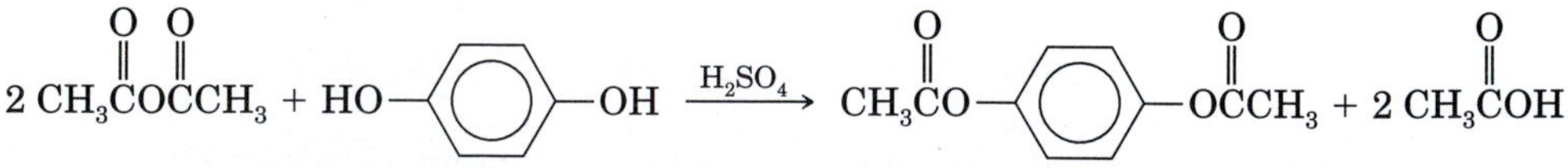

hydroquinone — hydroquinone diacetate (93%)

Phenyl esters can be prepared under either acidic or alkaline conditions. Under alkaline conditions, the sodium salt of the phenol is first prepared and then treated with the anhydride.

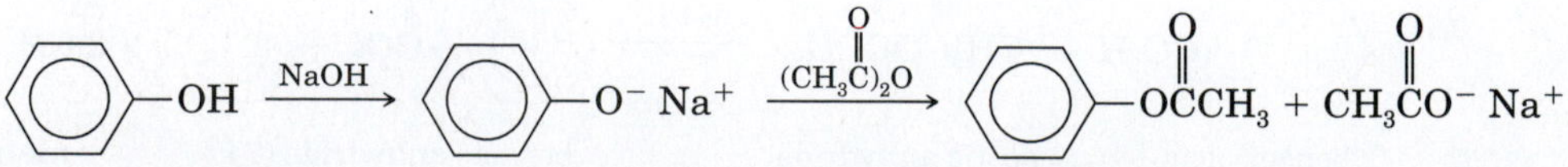

phenol — phenyl acetate

STUDY PROBLEM

15.11   Complete the following equations:

(a) 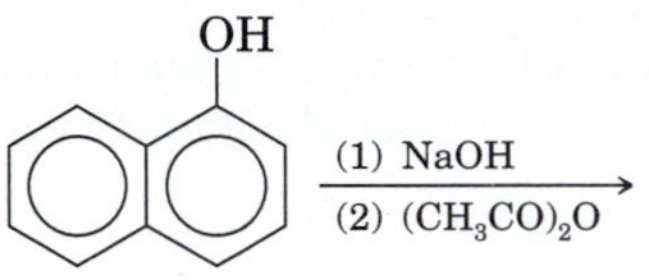 [naphthalene structure with OH] $\xrightarrow[\text{(2) }(CH_3CO)_2O]{\text{(1) NaOH}}$ (b) 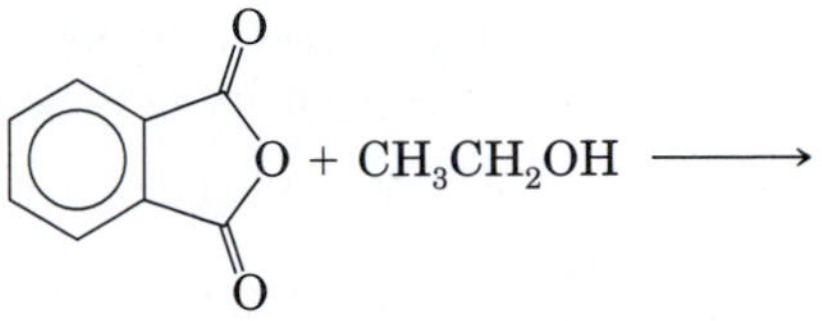 [phthalic anhydride structure] $+ CH_3CH_2OH \longrightarrow$

Reaction with ammonia and amines   Ammonia, primary amines, and secondary amines react with anhydrides to yield *amides*. For example, ammonia and acetic anhydride yield acetamide. Amines and acetic anhydride give substituted acetamides. With ammonia or the amines, one mole of the base is consumed in neutralizing the acetic acid formed in the reaction.

$$CH_3\overset{\overset{\displaystyle O}{\|}}{C}O\overset{\overset{\displaystyle O}{\|}}{C}CH_3 + 2NH_3 \longrightarrow CH_3\overset{\overset{\displaystyle O}{\|}}{C}NH_2 + CH_3CO_2^- NH_4^+$$

ammonia   acetamide

Figure 15.9 summarizes the reactions of anhydrides of carboxylic acids.

$$RC\overset{\overset{\displaystyle O}{\|}}{-}OCR$$

$\xrightarrow{HO-H}$ $RC\overset{\overset{\displaystyle O}{\|}}{-}OH + H-O\overset{\overset{\displaystyle O}{\|}}{C}R$
carboxylic acids

$\xrightarrow[\text{or ArO-H}]{R'O-H}$ $RC\overset{\overset{\displaystyle O}{\|}}{-}OR' +$ $H-O\overset{\overset{\displaystyle O}{\|}}{C}R$
an ester   **a carboxylic acid**

$\xrightarrow{R'_2N-H}$ $RC\overset{\overset{\displaystyle O}{\|}}{-}NR'_2 +$ $H-O\overset{\overset{\displaystyle O}{\|}}{C}R$
an amide   **a carboxylic acid**

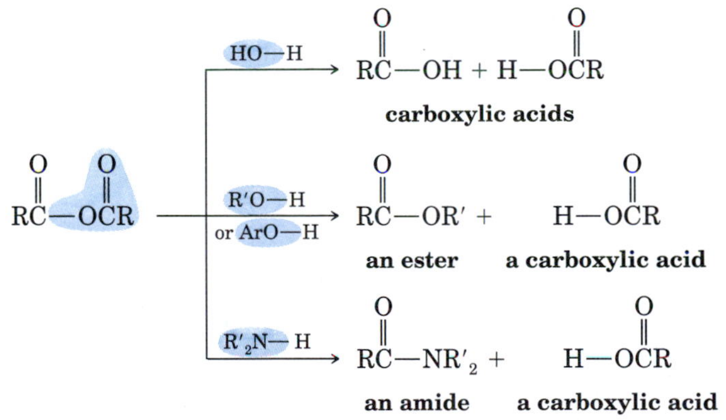

Figure 15.9   Summary of reactions of anhydrides of carboxylic acids.

SAMPLE PROBLEM

Predict the product of reaction of one mole of succinic anhydride with two moles of ammonia.

Solution

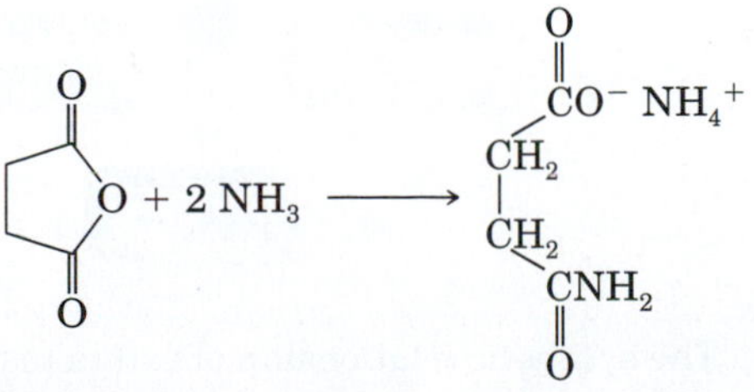

[succinic anhydride structure] $+ 2 NH_3 \longrightarrow$ product: $\overset{\overset{\displaystyle O}{\|}}{C}O^- NH_4^+$, CH_2, CH_2, $\overset{\overset{\displaystyle }{}}{C}NH_2$ with O

15.12 The acetyl group is often used as a blocking group to protect an amine function. For example, *o*-aminotoluene can be nitrated at carbon 3 (the carbon *ortho* to the nitrogen) in 55% overall yield by converting the amino group to an amide group, nitrating the amide, and hydrolyzing the amide back to an amine.

(a) Write equations for these reactions.

(b) What product probably constitutes the remaining 45% of the yield? Write equations for its formation.

(c) Predict what would happen if *o*-aminotoluene were treated with nitric acid directly.

Section 15.5

Esters of Carboxylic Acids

Esters, one of the most useful classes of organic compounds, can be converted to a variety of other compounds (Figure 15.10). Esters are common in nature. Fats and waxes are esters (Section 24.1). Esters are also used for synthetic polymers. For example, Dacron is a polyester (Section 15.7). Table 15.3 lists some representative esters.

Volatile esters lend pleasant aromas to many fruits and perfumes. (You might find it interesting to compare the odors of some of these esters to the odors of carboxylic acids, Section 14.1.) Natural fruit flavors are complex

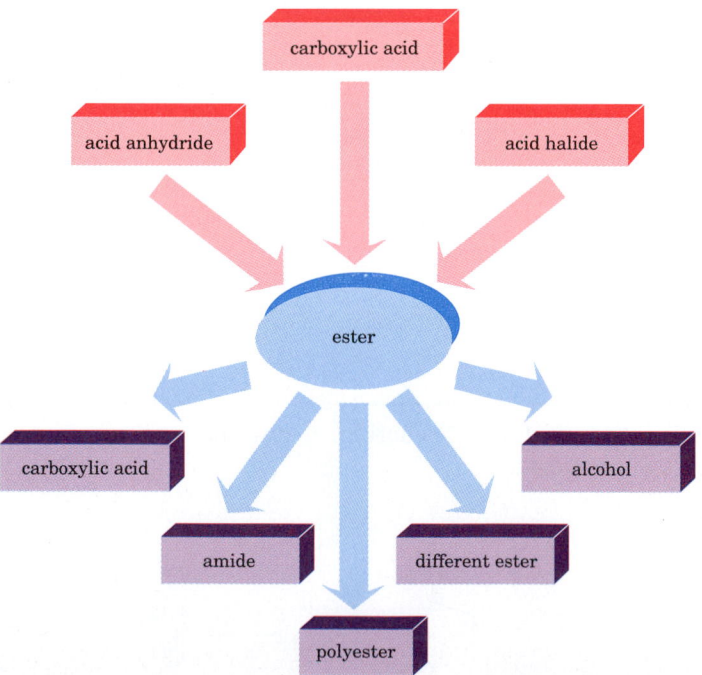

Figure 15.10 The synthetic relationship of esters to other compounds.

Table 15.3	Names and odors of selected esters	
Trivial Name	Structure	Odor
methyl acetate	$CH_3CO_2CH_3$	pleasant
ethyl acetate	$CH_3CO_2CH_2CH_3$	pleasant
propyl acetate	$CH_3CO_2CH_2CH_2CH_3$	like pears
ethyl butyrate	$CH_3(CH_2)_2CO_2CH_2CH_3$	like pineapple
isoamyl acetate	$CH_3CO_2(CH_2)_2CH(CH_3)_2$	like bananas
isobutyl propionate	$CH_3CH_2CO_2CH_2CH(CH_3)_2$	like rum
methyl salicylate		like wintergreen

blends of many esters with other organic compounds. Synthetic fruit flavors are usually simple blends of just a few esters with a few other substances. Therefore, these synthetic flavors seldom completely duplicate the fuller-bodied natural flavors.

A. Nomenclature of Esters

The name of an ester consists of two words. The first word is the name of the **alkyl group** attached to the ester oxygen. The second word is derived from the carboxylic acid name with **-ic acid** changed to **-ate.** Note the similarity between the name of an ester and that of a carboxylate salt.

$$CH_3CH_2\overset{\overset{\displaystyle O}{\|}}{C}O-H \qquad CH_3CH_2\overset{\overset{\displaystyle O}{\|}}{C}O^-\,Na^+ \qquad CH_3CH_2\overset{\overset{\displaystyle O}{\|}}{C}O-CH_3$$

	an acid	*a salt*	*an ester*
IUPAC:	propanoic acid	sodium propanoate	methyl propanoate
trivial:	propionic acid	sodium propionate	methyl propionate

STUDY PROBLEM

15.13   Write structures for the following esters:

(a) *tert*-butyl acetate   (b) phenyl butanoate

(c) benzyl benzoate   (d) ethenyl 4,4-dimethyl-2-pentenoate

B. Preparation of Esters

Most methods for ester synthesis have been covered elsewhere in this text. In this section, we will provide a summary of these methods.

From carboxylic acids and alcohols (Section 14.6):

$$\text{C}_6\text{H}_5-CH_2CH_2CH_2\overset{\overset{\displaystyle O}{\|}}{C}OH + CH_3CH_2OH \underset{}{\overset{H^+,\,heat}{\rightleftharpoons}} \text{C}_6\text{H}_5-CH_2CH_2CH_2\overset{\overset{\displaystyle O}{\|}}{C}OCH_2CH_3 + H_2O$$

4-phenylbutanoic acid   ethanol   ethyl 4-phenylbutanoate (85%)

From acid halides and alcohols (for hindered systems and for phenols) (Section 15.3C):

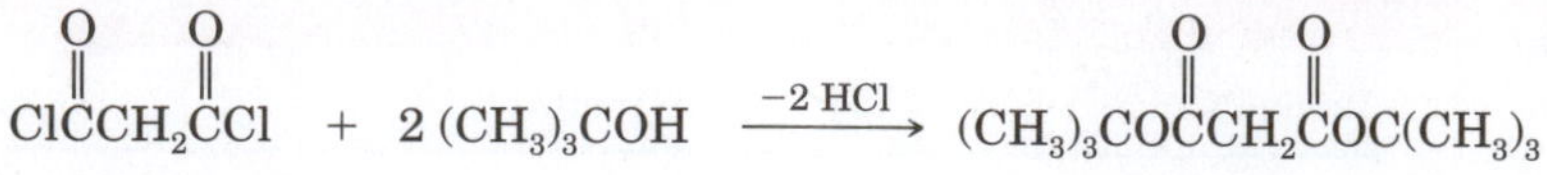

$$ClCCH_2CCl \quad + \quad 2\,(CH_3)_3COH \quad \xrightarrow{-2\ HCl} \quad (CH_3)_3COCCH_2COC(CH_3)_3$$

malonyl chloride   *tert*-butyl alcohol   di-*tert*-butyl malonate (80%)

From an anhydride and an alcohol or phenol (Section 15.4C):

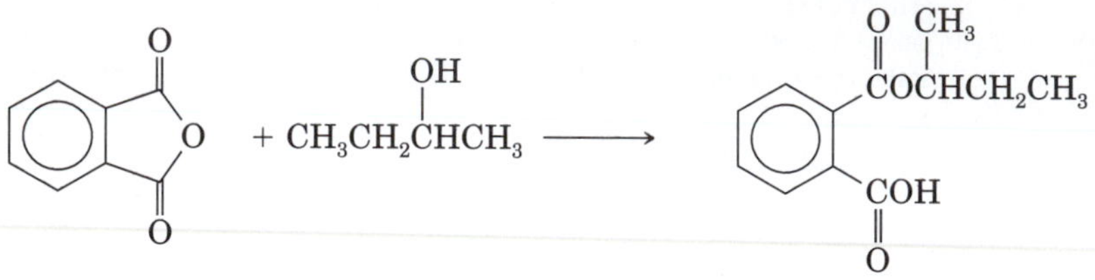

phthalic anhydride   2-butanol   *sec*-butyl hydrogen phthalate (97%)

From a carboxylate and a reactive alkyl halide (Section 14.5):

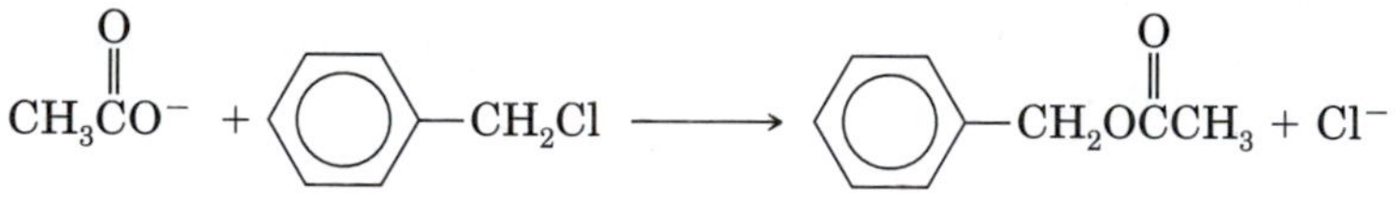

acetate ion   benzyl chloride   benzyl acetate (93%)

STUDY PROBLEM

15.14 Write equations to show two different preparations of: **(a)** methyl butanoate (odor of apples); and **(b)** 3-methyl-2-butenyl acetate ("Juicy fruit" odor).

C. Reactions of Esters

In *acidic solution,* the carbonyl oxygen of an ester can be protonated. The partially positive carbon can then be attacked by a weak nucleophile such as water.

Protonation:

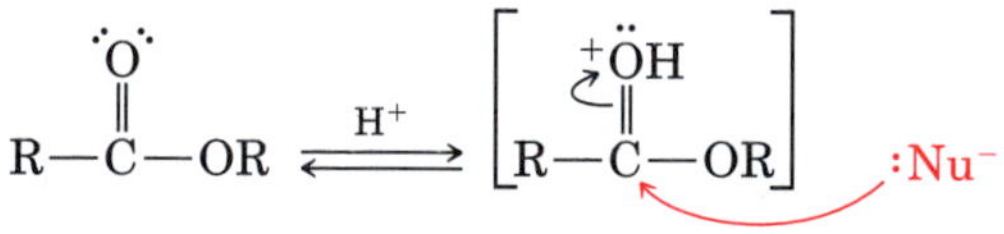

In *alkaline solution,* the carbonyl carbon of an ester can be attacked by a good nucleophile without prior protonation. This is the same addition–elimination path as for nucleophilic attack on acid chlorides or anhydrides.

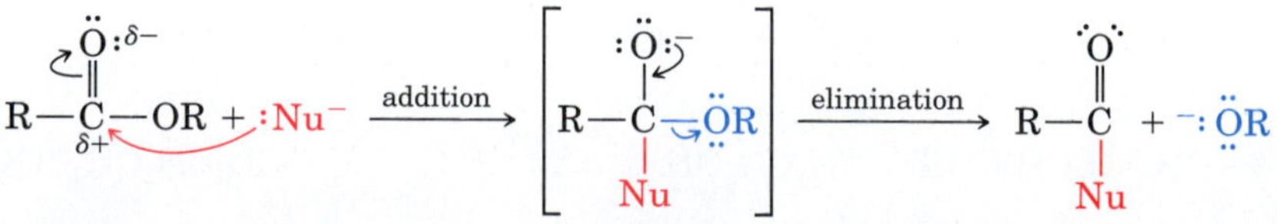

intermediate

Acid hydrolysis   The esterification of a carboxylic acid with an alcohol (Section 14.6) is a reversible reaction. When a carboxylic acid is esterified, an excess of the alcohol is used. To cause the reverse reaction—that is, *acid-catalyzed hydrolysis of an ester to a carboxylic acid*—an excess of water is used. The excess of water shifts the equilibrium to the carboxylic acid side of the equation.

Esterification:

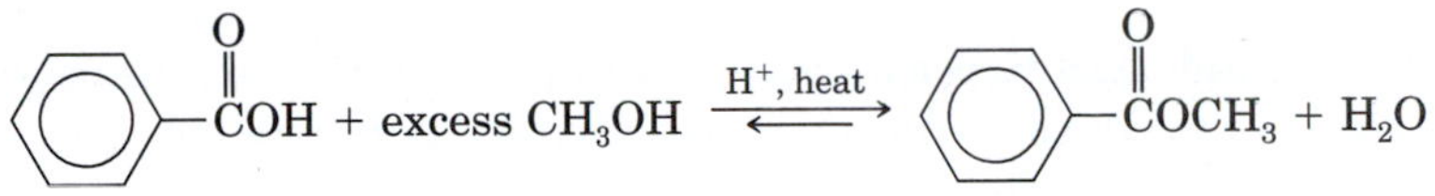

benzoic acid   methanol   methyl benzoate

Hydrolysis:

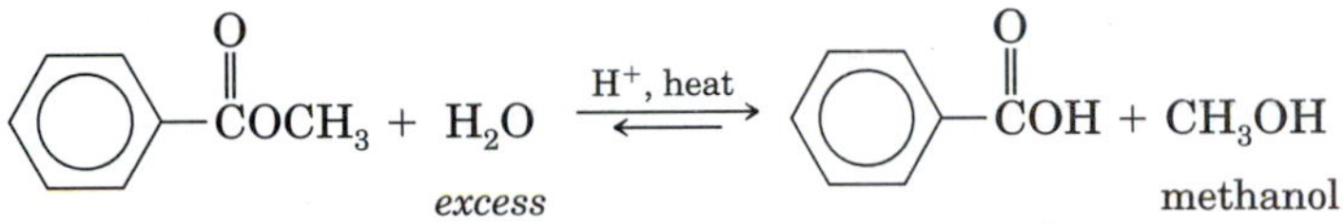

methyl benzoate   benzoic acid   methanol

If water labeled with oxygen-18 is used in the hydrolysis, the labeled oxygen ends up in the carboxylic acid.

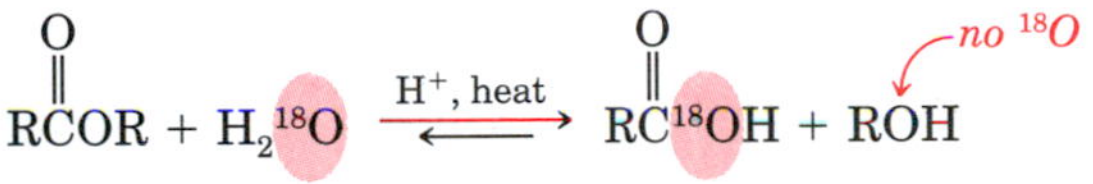

$$RCOR + H_2{}^{18}O \xrightleftharpoons{H^+,\ heat} RC^{18}OH + ROH \quad (no\ {}^{18}O)$$

The reason for this is that the water attacks the *carbonyl group*. The RO bond is not broken in the hydrolysis.

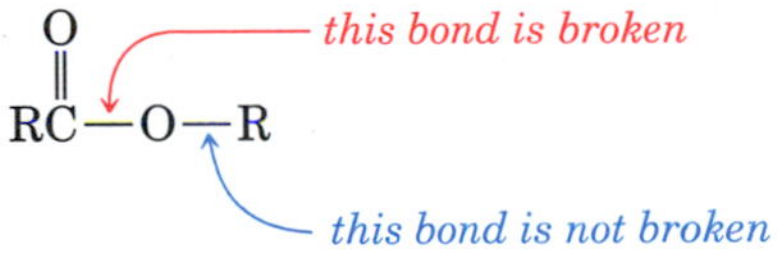

$$RC{-}O{-}R$$

—— *this bond is broken*

—— *this bond is not broken*

The following mechanism accounts for this observation. The first step is *protonation,* followed by *addition of H_2O.* A deprotonation, another protonation, and then *elimination of R'OH,* followed by a final deprotonation complete the mechanism.

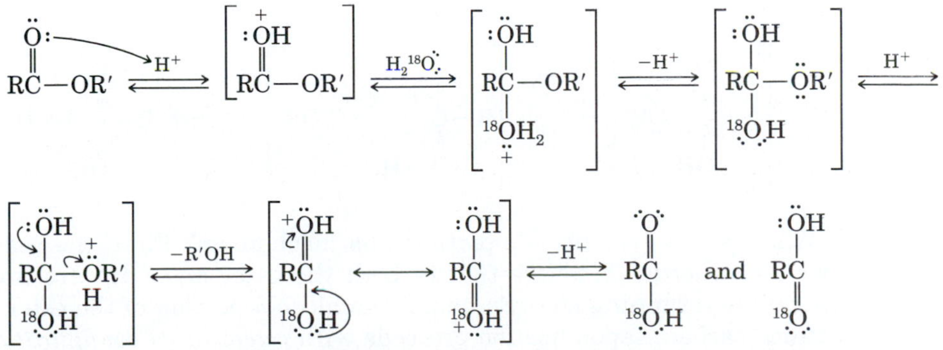

*resonance structures
for protonated acid*

An abbreviated mechanism for ester hydrolysis can be written:

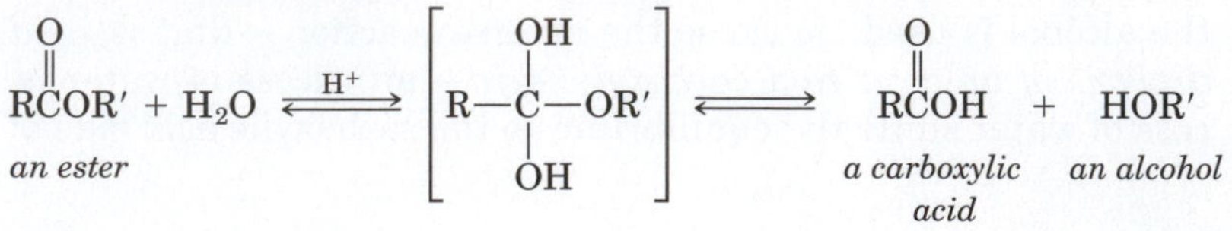

$$\underset{\text{an ester}}{\overset{\overset{\displaystyle O}{\parallel}}{R C O R'}} + H_2O \underset{}{\overset{H^+}{\rightleftharpoons}} \left[\underset{OH}{\overset{OH}{R - \underset{|}{\overset{|}{C}} - OR'}} \right] \rightleftharpoons \underset{\substack{\text{a carboxylic} \\ \text{acid}}}{\overset{\overset{\displaystyle O}{\parallel}}{R C O H}} + \underset{\text{an alcohol}}{H O R'}$$

Alkaline hydrolysis (saponification) Hydrolysis of an ester in base, or **saponification,** is an *irreversible reaction.* Because it is irreversible, saponification often gives better yields of carboxylic acid and alcohol than does acidic hydrolysis. Because the reaction occurs in base, the product of saponification is the carboxylate salt. The free acid is generated when the solution is acidified. Note that OH⁻ is a reactant, not a catalyst, in this reaction.

Saponification:

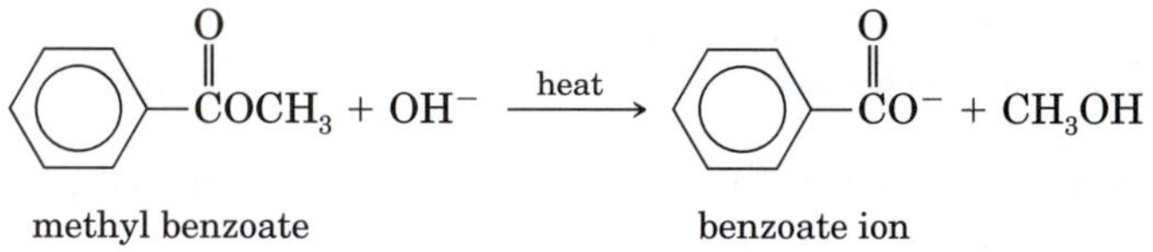

methyl benzoate   benzoate ion

Acidification:

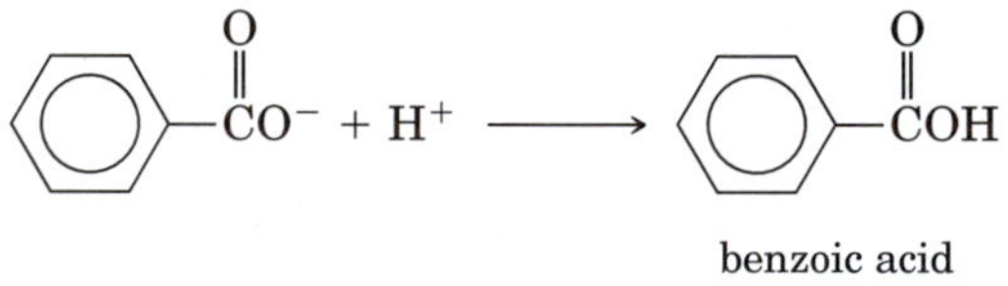

benzoic acid

A large body of evidence supports the following mechanistic scheme, which is typical of nucleophilic attack on a carboxylic acid derivative.

Step 1 (addition of OH⁻) (slow):

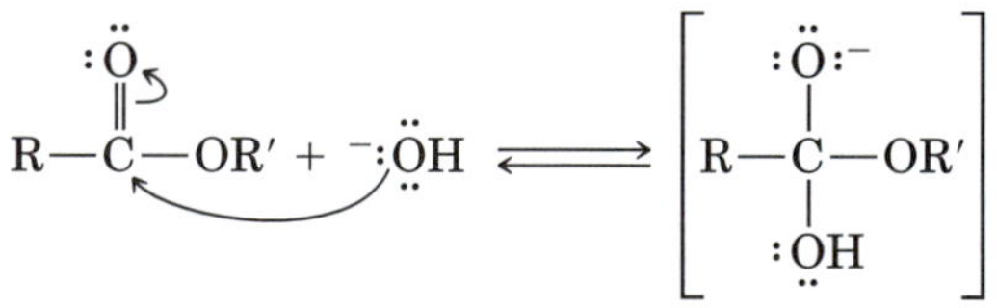

Step 2 (elimination of ⁻OR′ and proton transfer) (fast):

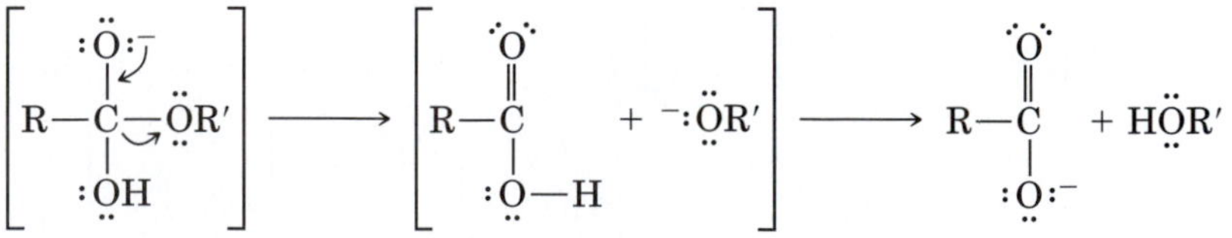

What is the evidence supporting this mechanism? First, the reaction follows *second-order kinetics*—that is, both the ester and OH⁻ are involved in the rate-determining step. Second, if the alcohol portion of the ester contains a chiral carbon, saponification proceeds with *retention of configuration* in the alcohol. This evidence supports the cleavage of the carbonyl–oxygen bond, not cleavage of the alkyl–oxygen bond.

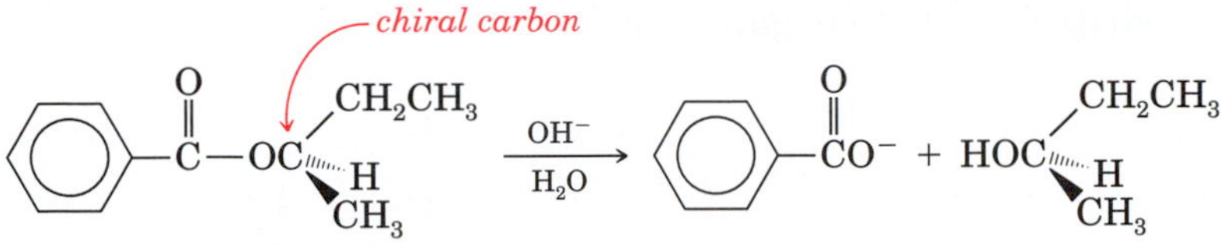
chiral carbon
OH⁻
H₂O

(*R*)-*sec*-butyl benzoate | benzoate ion | (*R*)-2-butanol

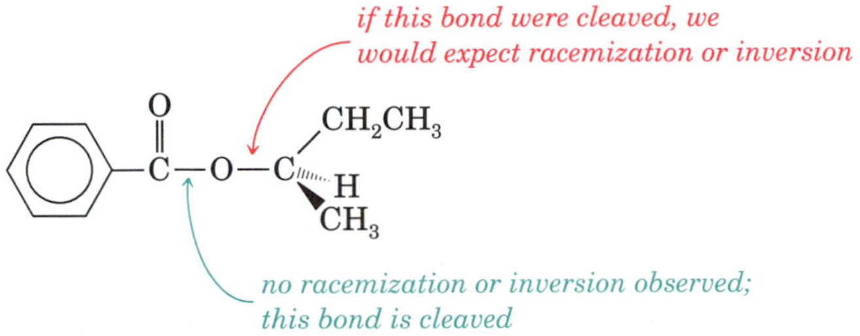
if this bond were cleaved, we would expect racemization or inversion
no racemization or inversion observed; this bond is cleaved

Study Problem

15.15 Write equations for the saponification of the following esters with aqueous NaOH:

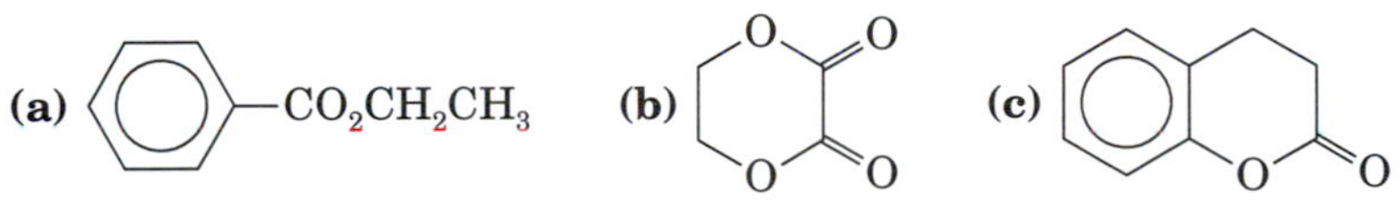
(a)
CO₂CH₂CH₃
(b)
(c)

The word *saponification* means "the making of soap." Soaps, which are synthesized by the saponification of fats, will be discussed at greater length in Chapter 24.

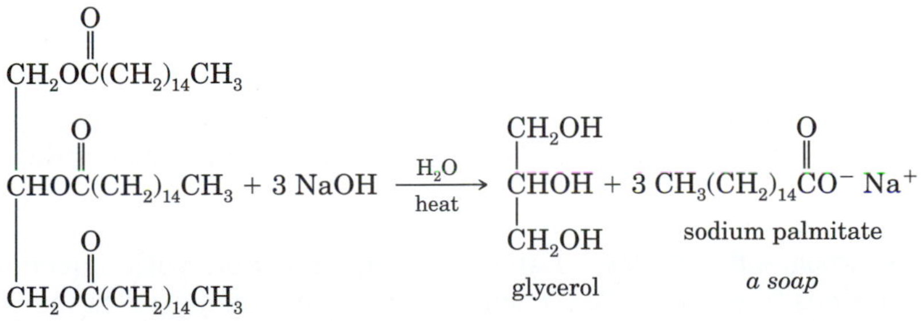
H₂O
heat

glyceryl tripalmitate
a fat, or triglyceride

glycerol

sodium palmitate
a soap

Sample Problem

Which would have a faster rate of saponification: **(a)** ethyl benzoate, or **(b)** ethyl *p*-nitrobenzoate? Why?

Solution

(b) has the faster rate because it has an electron-withdrawing nitro group. The transition state leading to the intermediate in Step 1 is more stabilized by dispersal of the negative charge:

$$O_2N \leftarrow \bigcirc \leftarrow \underset{\underset{OH}{|}}{\overset{\overset{O^-}{|}}{C}} - OCH_2CH_3$$

STUDY PROBLEM

15.16 With excess sodium hydroxide solution and identical reaction conditions, which member of the following pairs of esters would hydrolyze faster? Why?

(a) $CH_3CO_2CH_3$ or $CH_3CO_2C_6H_5$ (b) $CH_3CO_2CH_3$ or $C_6H_5CO_2CH_3$

(c) $CH_3CO_2CH_3$ or $CF_3CO_2CH_3$ (d) $CH_3CO_2CH_3$ or $(CH_3)_3CCO_2CH_3$

Transesterification Exchange of the alcohol portion of an ester can be accomplished in acidic or basic solution by a reversible reaction between the ester and an alcohol. These **transesterification reactions** are directly analogous to hydrolysis in acid or base. Because the reactions are reversible, an excess of the initial alcohol is generally used.

$$\underset{\substack{\text{methyl propenoate} \\ \text{(methyl acrylate)}}}{CH_2{=}CH\overset{\overset{O}{\|}}{C}OCH_3} + \underset{excess}{CH_3CH_2OH} \underset{}{\overset{H^+}{\rightleftharpoons}} \underset{\text{ethyl propenoate (99\%)}}{CH_2{=}CH\overset{\overset{O}{\|}}{C}OCH_2CH_3} + CH_3OH$$

STUDY PROBLEM

15.17 Suggest mechanisms for the transesterification reactions of ethyl acetate with (a) methanol and HCl, and (b) methanol and sodium methoxide.

Reaction with ammonia Esters undergo reaction with aqueous ammonia or amines to yield amides. The reaction is slow compared to the reactions of acid halides or anhydrides with ammonia. This slowness of the ester reaction can be an advantage because the reaction of an acid chloride with an amine can sometimes be violent.

$$\underset{\text{ethyl chloroacetate}}{ClCH_2\overset{\overset{O}{\|}}{C}OCH_2CH_3} + NH_3 \xrightarrow[\text{1 hr}]{0°} \underset{\text{chloroacetamide (80\%)}}{ClCH_2\overset{\overset{O}{\|}}{C}NH_2} + CH_3CH_2OH$$

The ester route to amides is also the reaction of choice for the synthesis of an amide with another functional group that would not be stable toward an acid chloride.

STUDY PROBLEM

15.18 Why is it not possible to convert 2-hydroxypropanoic acid to 2-hydroxy-propanamide by way of an acid chloride? How could this carboxylic acid be converted to its amide?

Reduction Esters can be reduced under pressure by catalytic hydrogenation, a reaction sometimes called **hydrogenolysis of esters,** or by lithium aluminum hydride. An older technique is the reaction of the ester with sodium metal in ethanol. Regardless of the reducing agent, a pair of alcohols (with at least one being primary) results from the reduction of an ester.

General:

$$
\underset{\substack{\textit{to a primary} \\ \textit{alcohol}}}{RC}\!\!-\!\!\underset{\textit{to the other alcohol}}{OR'} \xrightarrow{\text{[H]}} RCH_2OH + HOR'
$$

$$
CH_3(CH_2)_8COCH_2CH_3 \xrightarrow[\text{CH}_3\text{CH}_2\text{OH}]{\text{Na}} CH_3(CH_2)_8CH_2OH + HOCH_2CH_3
$$

ethyl decanoate   1-decanol (70%)   ethanol

$$
CH_3CH_2OC(CH_2)_4COCH_2CH_3 \xrightarrow[\text{255°, 200 atm}]{\substack{\text{H}_2 \\ \text{CuCr}_2\text{O}_4 \text{ catalyst}}} HOCH_2(CH_2)_4CH_2OH + 2\ HOCH_2CH_3
$$

diethyl adipate   1,6-hexanediol (90%)   ethanol

$$
\underset{\substack{\textit{a cyclic ester,} \\ \textit{or lactone}}}{H_3C} \xrightarrow[\text{(2) H}_2\text{O, H}^+]{\text{(1) LiAlH}_4} CH_3\overset{\overset{\displaystyle OH}{|}}{C}HCH_2CH_2CH_2OH
$$

1° hydroxyl from the carbonyl carbon

1,4-pentanediol (86%)

Reaction with Grignard reagents The reaction of esters with Grignard reagents is an excellent technique for the preparation of *tertiary alcohols with two identical R groups.*

General:

$$
\underset{\textit{an ester}}{RCOR'} \xrightarrow[\text{(2) H}_2\text{O, H}^+]{\text{(1) 2 R''MgX}} R\!-\!\underset{\underset{\displaystyle R''}{|}}{\overset{\overset{\displaystyle OH}{|}}{C}}\!-\!R''
$$

two R″ groups the same

a 3° alcohol

$$
CH_3COCH_2CH_3 \xrightarrow[\text{(2) H}_2\text{O, NH}_4\text{Cl}]{\text{(1) 2 C}_6\text{H}_5\text{MgBr}} CH_3\overset{\overset{\displaystyle OH}{|}}{C}(C_6H_5)_2 \xrightarrow{-\text{H}_2\text{O}} CH_2{=}C(C_6H_5)_2
$$

ethyl acetate   1,1-diphenylethanol   1,1-diphenylethene (70%)

If a formate ester is subjected to a Grignard reaction, a *secondary alcohol with two identical R groups* is obtained. Formates are a special case because the carbonyl carbon is attached to an H atom, not an alkyl or aryl group.

$$
\underset{\text{a formate ester}}{\overset{\overset{\displaystyle O}{\|}}{HCOR}} \xrightarrow[\text{(2) } H_2O,\, H^+]{\text{(1) } 2\ R'MgX} \underset{\underset{\displaystyle R'}{|}}{\overset{\overset{\displaystyle OH}{|}}{H-C-R'}} \quad \begin{array}{l}\textit{two R' groups}\\ \textit{the same}\end{array}
$$

a 2° alcohol

The mechanism of the reaction of a Grignard reagent with an ester is similar to that of the reaction of a Grignard reagent with an aldehyde or ketone; that is, the nucleophilic carbon of the Grignard reagent attacks the positive carbon of the carbonyl group. With esters, as with acid halides (Section 15.3C), *two* equivalents of Grignard reagent attack the carbonyl carbon atom. To see why this is so, let us consider the reaction stepwise. First, the negative carbon of the Grignard reagent attacks the carbon of the carbonyl group. The product of this step has a hemiketal-like structure that loses an alkoxyl group to yield a ketone.

Initial attack:

$$
\underset{}{\overset{\overset{\displaystyle \ddot{O}:}{\|}}{RC}}{-}OR' + \overset{\delta-\quad\delta+}{CH_3MgX} \longrightarrow \left[\underset{\underset{\displaystyle CH_3}{|}}{\overset{\overset{\displaystyle :\ddot{O}:^-\ ^+MgX}{|}}{RC{-}\ddot{O}R'}} \right] \longrightarrow \left[\underset{\underset{\displaystyle CH_3}{|}}{\overset{\overset{\displaystyle \ddot{O}:}{\|}}{RC}} \right] + R'\ddot{O}:^-\ ^+MgX
$$

a type of hemiketal   *a ketone*

The ketone then reacts with a *second* molecule of Grignard reagent. The second reaction is faster than the first; hence the ketone cannot be isolated.

Second attack and hydrolysis:

$$
\left[\overset{\overset{\displaystyle \ddot{O}:}{\|}}{RCCH_3} \right] + CH_3MgX \longrightarrow \underset{\underset{\displaystyle CH_3}{|}}{\overset{\overset{\displaystyle :\ddot{O}MgX}{|}}{RCCH_3}} \xrightarrow{H_2O,\, H^+} \underset{\underset{\displaystyle CH_3}{|}}{\overset{\overset{\displaystyle OH}{|}}{RCCH_3}} + Mg^{2+} + X^- + H_2O
$$

a 3° alcohol

STUDY PROBLEMS

15.19 Show by equations how you would prepare the following alcohols from methyl propanoate: **(a)** 2-methyl-2-butanol; **(b)** 3-ethyl-3-pentanol.

15.20 Predict the product of the reaction of ethyl formate with ethylmagnesium bromide, followed by work-up with aqueous acid.

Figure 15.11 summarizes the reactions of esters.

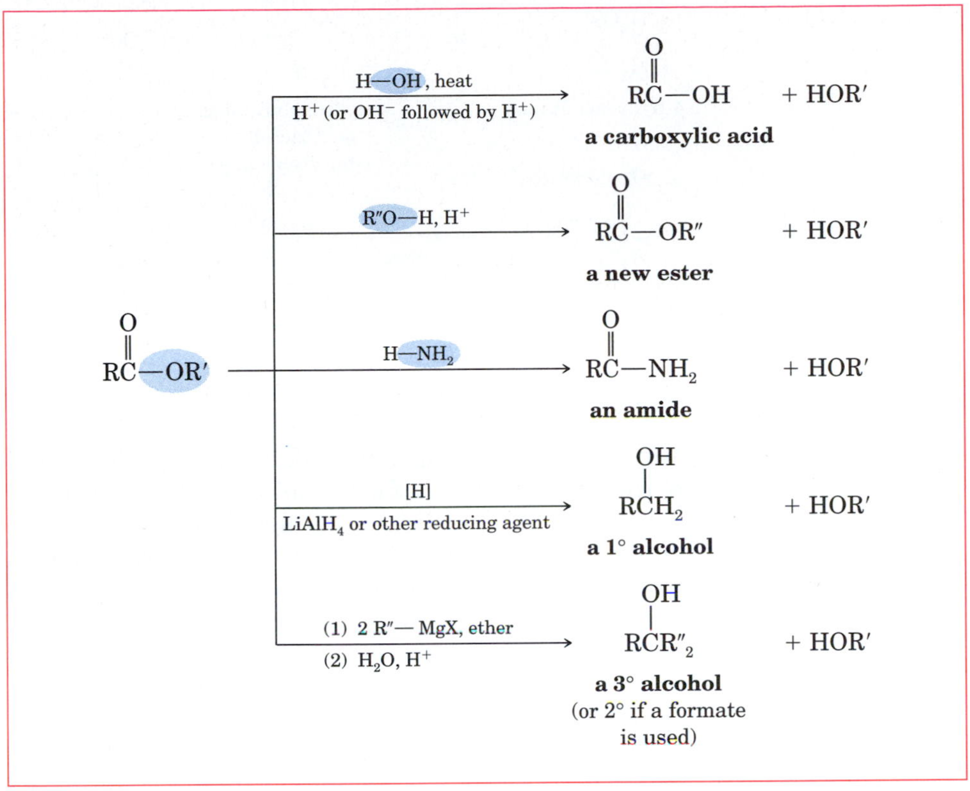

Figure 15.11   Summary of reactions of esters.

Section 15.6
Lactones

A **lactone** is a cyclic ester. Lactones are fairly common in natural sources. For example, vitamin C and nepetalactone, the cat attractant in catnip, are both lactones. It is interesting to note the close structural relationship of nepetalactone to iridomyrmecin, an odorous compound found in the *iridomyrmex* species of ants.

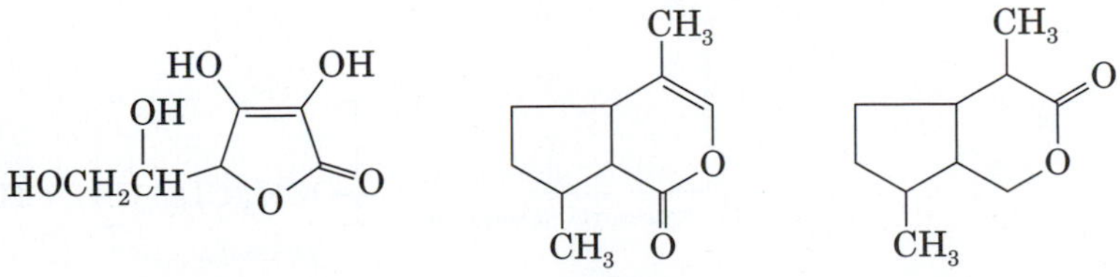

vitamin C
(ascorbic acid)

nepetalactone

cat attractant in catnip

iridomyrmecin

in ants

Lactones are formed from molecules that contain a carboxyl group and a hydroxyl group. These molecules can undergo an intramolecular esterification.

$$\underset{\substack{\text{4-hydroxybutanoic}\\\text{acid}}}{\underset{\substack{\text{CH}_2\text{OH}\\|\\\text{CH}_2\text{CH}_2}}{\quad\text{CO}_2\text{H}}} \; \underset{}{\overset{\text{H}^+}{\rightleftharpoons}} \; \underset{\substack{\text{4-hydroxybutanoic}\\\text{acid lactone}\\(\gamma\text{-butyrolactone})}}{\quad} + \text{H}_2\text{O}$$

a γ-hydroxy acid

$$\underset{\substack{\text{5-hydroxypentanoic}\\\text{acid}}}{\underset{\substack{\text{CH}_2\text{OH}\\\text{CH}_2\quad\quad\text{CO}_2\text{H}\\\text{CH}_2\text{CH}_2}}{}} \; \underset{}{\overset{\text{H}^+}{\rightleftharpoons}} \; \underset{\substack{\text{5-hydroxypentanoic}\\\text{acid lactone}\\(\delta\text{-valerolactone})}}{\quad} + \text{H}_2\text{O}$$

a δ-hydroxy acid

With γ- *or* δ-hydroxy acids, which form lactones that are five- or six-membered rings, the cyclization is so facile that the hydroxy acids often cannot be isolated. Although the reaction is catalyzed by acid or base, even a trace of acid from the glassware is sufficient to catalyze lactone formation if a five- or six-membered ring is the product.

Carboxylic acids with hydroxyl groups in the α or β position do not form ordinary cyclic lactones readily because of the small, strained rings that would result. Carboxylic acids with hydroxyl groups farther away than the γ or δ position do not form lactones spontaneously. However, the lactones of these hydroxy acids may be synthesized under the usual conditions for esterification. In these cases, a *dilute solution* of hydroxy acid in an inert solvent is used. An intramolecular reaction is favored by dilute solution because collisions between molecules are less apt to occur. If the solution is *concentrated,* the hydroxy acid molecules undergo reaction with one another to yield a **polyester.** In either case, a solvent such as benzene allows the product water to be distilled as an azeotrope and drives the reaction toward the lactone (or polyester).

$$\xrightarrow[\text{dilute solution}]{\text{H}^+,\text{ heat}} \quad \underset{\text{the lactone}}{\quad} + \text{H}_2\text{O}$$

$$\underset{\text{7-hydroxyheptanoic acid}}{\text{HOCH}_2(\text{CH}_2)_5\overset{\displaystyle\overset{\text{O}}{\|}}{\text{C}}\text{OH}}$$

$$\xrightarrow[\text{concentrated solution}]{\text{H}^+,\text{ heat}} \quad \underset{\text{the polyester}}{\left[\text{OCH}_2(\text{CH}_2)_5\overset{\displaystyle\overset{\text{O}}{\|}}{\text{C}}\right]_x} + \text{H}_2\text{O}$$

Section 15.7

Polyesters

The synthetic fiber **Dacron** (Figure 15.12) is a polyester made by the transesterification reaction of dimethyl terephthalate and ethylene glycol. The reason that polymer formation can occur is that the reactants are *bifunctional.* Thus, each reactant can undergo reaction with two other molecules.

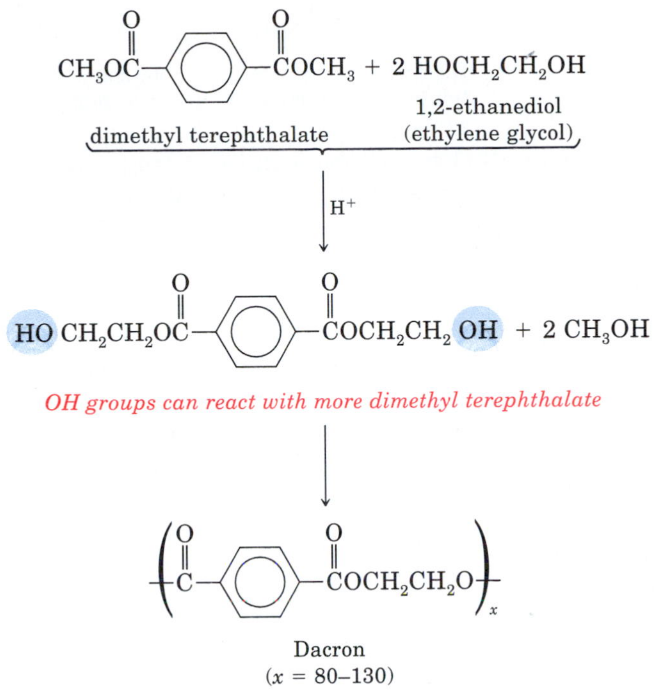

Figure 15.12   Synthesis of the polyester Dacron by a transesterification reaction.

When the monomers are bifunctional, such as dimethyl terephthalate and ethylene glycol, polymer growth must occur in a *linear* fashion. Linear polymers often make excellent textile fibers. If more than two reactive sites are present in one of the monomers, then the polymer can grow into a cross-linked network. **Glyptal** (a polymer of glycerol and phthalic anhydride) is an example of a cross-linked polyester.

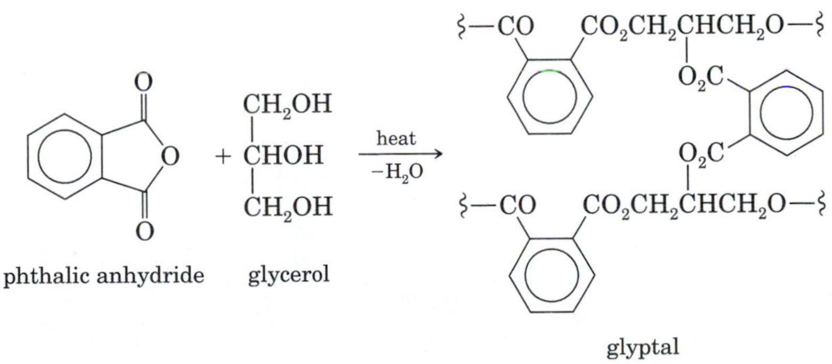

Section 15.8

Amides

A. Nomenclature of Amides

An **amide** is a compound that has a trivalent nitrogen bonded to a carbonyl group. An amide is named from the parent carboxylic acid with the **-oic** (or **-ic**) **acid** ending changed to **-amide**.

$$\text{CH}_3\overset{\overset{\text{O}}{\|}}{\text{C}}\text{NH}_2 \qquad \text{CH}_3\text{CH}_2\text{CH}_2\,\overset{\overset{\text{O}}{\|}}{\text{C}}\text{NH}_2$$

IUPAC: ethanamide   butanamide
trivial: acetamide   butyramide

Amides with alkyl substituents on the nitrogen have their names preceded by *N*-alkyl, where *N* refers to the nitrogen atom.

$$\overset{\overset{\text{O}}{\|}}{}\text{C}\text{NHCH}_3 \qquad \text{H}\overset{\overset{\text{O}}{\|}}{\text{C}}\text{N(CH}_3)_2$$

N-methylbenzamide   *N,N*-dimethylformamide

A few amides of interest follow; the amide groups are highlighted in color:

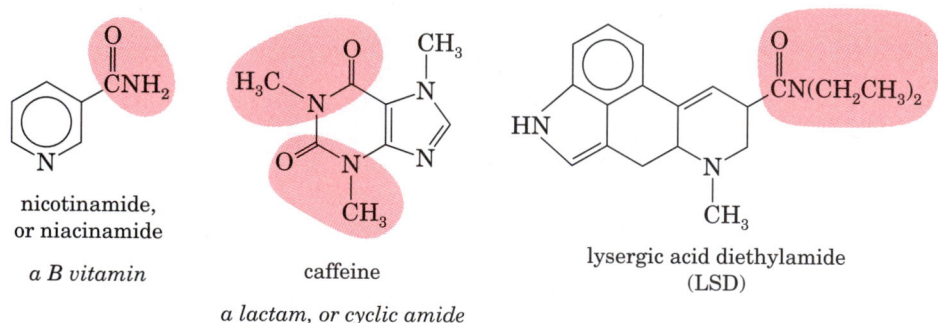

nicotinamide,
or niacinamide

a B vitamin

caffeine

a lactam, or cyclic amide

lysergic acid diethylamide
(LSD)

B. Preparation of Amides

Amides are synthesized from derivatives of carboxylic acids and ammonia or the appropriate amine. These reactions have been discussed previously in this chapter.

$$\begin{array}{c}
\overset{\overset{\text{O}}{\|}}{\text{R}\text{C}\text{Cl}} \\[4pt]
\overset{\overset{\text{O}\quad\text{O}}{\|\quad\|}}{\text{R}\text{C}\text{O}\text{C}\text{R}} \xrightarrow{\;\text{R}'_2\text{NH}\;} \overset{\overset{\text{O}}{\|}}{\text{R}\text{C}\text{N}\text{R}'_2} \\[4pt]
\overset{\overset{\text{O}}{\|}}{\text{R}\text{C}\text{O}\text{R}'}
\end{array}$$

C. Reactions of Amides

An amide contains a nitrogen with a pair of unshared valence electrons. It would be reasonable to expect amides to undergo reaction with acids, as do amines; however, they do not. Amides are *very* weak bases with pK_b values of 15–16. (By contrast, methylamine has a pK_b of 3.34.) The resonance structures for an amide show why the nitrogen of an amide is neither particularly basic nor nucleophilic.

$$CN_3\ddot{N}H_2 \quad + \text{ dilute HCl } \longrightarrow \quad CH_3NH_3{}^+Cl^-$$

methylamine   methylammonium
                                                  chloride

$$\overset{\displaystyle O}{\underset{\displaystyle \parallel}{}}$$
$$CH_3\overset{\parallel}{C}\ddot{N}H_2 + \text{ dilute HCl } \longrightarrow \text{ no appreciable salt formation}$$

acetamide

Resonance structures for an amide:

R—C—NH₂ ⟷ R—C=NH₂   *less basic than
                          an amine nitrogen*

The effect of the partial double-bond character of the bond between the carbonyl carbon and the nitrogen of an amide is evident in the ¹H NMR spectrum of *N,N*-dimethylformamide (Figure 15.13). The spectrum shows one peak for *each* methyl group. If the —N(CH₃)₂ group underwent free rotation around its bond to the carbonyl group, the methyl groups would be magnetically equivalent and would give rise to one singlet. Because of the restricted rotation around the carbonyl carbon–nitrogen bond, the two methyl groups are in different magnetic environments. (The energy barrier for rotations around the carbonyl carbon–nitrogen bond in an amide has been found to be about 18 kcal/mol at room temperature.)

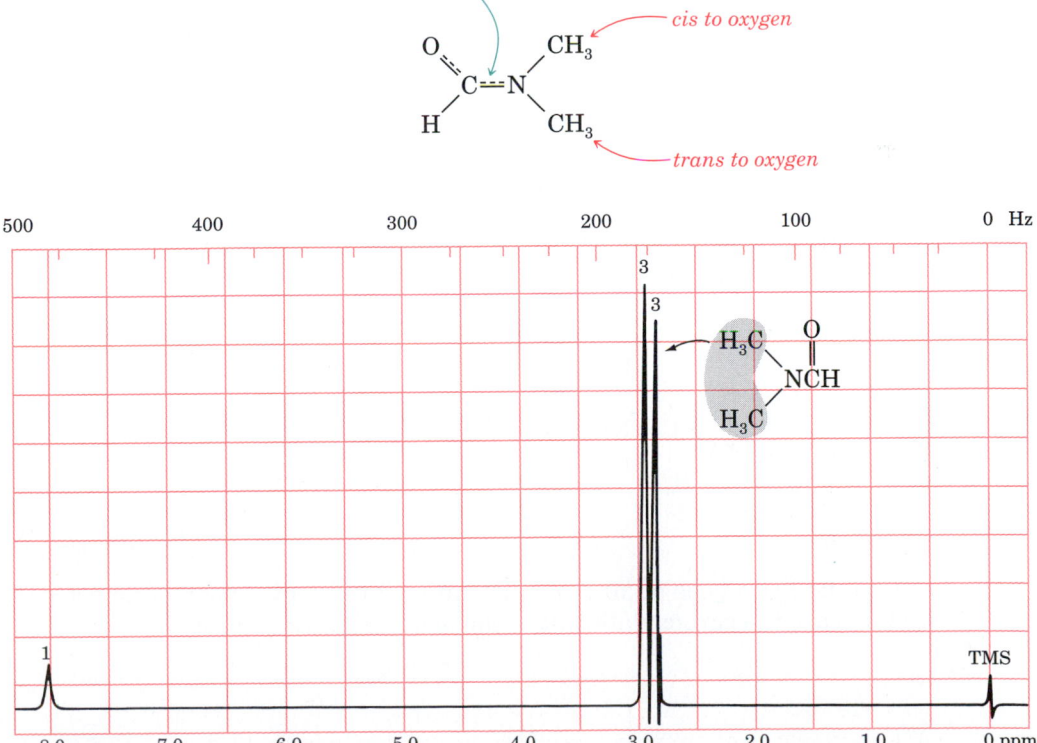

Figure 15.13 The ¹H NMR spectrum of *N,N*-dimethylformamide, showing a pair of peaks for the *N*-methyl groups.

Hydrolysis   Like esters, amides can be hydrolyzed in either acidic or alkaline solution. In either case, the acid or base is a reactant, not a catalyst, and must be used in a 1:1 molar ratio or in excess. Neither type of hydrolysis reaction is reversible.

$$CH_3CH_2\overset{\overset{\textstyle O}{\|}}{C}N(CH_3)_2 \quad + H_2O + H^+ \longrightarrow CH_3CH_2CO_2H + \; H_2\overset{+}{N}(CH_3)_2$$

N,N-dimethylpropanamide   propanoic   dimethyl-
                                                        acid   ammonium ion

In base:

$$CH_3CH_2\overset{\overset{\textstyle O}{\|}}{C}N(CH_3)_2 + OH^- \longrightarrow CH_3CH_2\overset{\overset{\textstyle O}{\|}}{C}O^- + \; HN(CH_3)_2$$

propanoate ion   dimethylamine

Hydrolysis of an amide in acidic solution proceeds in a fashion similar to hydrolysis of an ester. The carbonyl oxygen is protonated, the carbonyl carbon is attacked by H_2O, protons are transferred, and an amine is expelled. This amine then reacts with H^+ to yield the amine salt. The formation of the amine salt explains (1) why H^+ is a reactant, not a catalyst, and (2) why the reverse reaction does not proceed. (Although R_2NH is a nucleophile, $R_2NH_2^+$ is not; this ion cannot attack the carbonyl group.)

In acid:

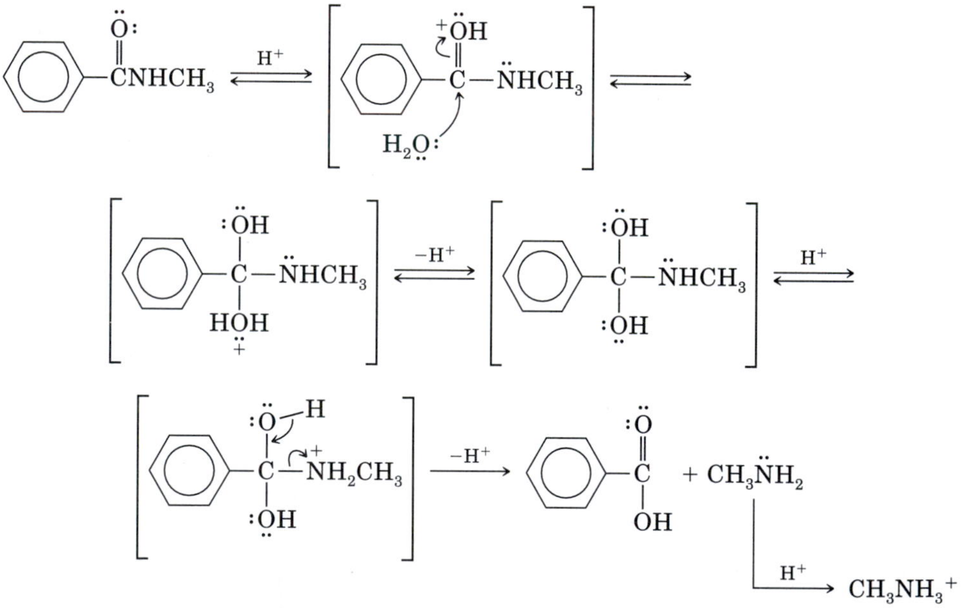

Alkaline hydrolysis of an amide is similar to saponification of an ester. The products are the carboxylate salt of the acid and a free amine or ammonia.

In base:

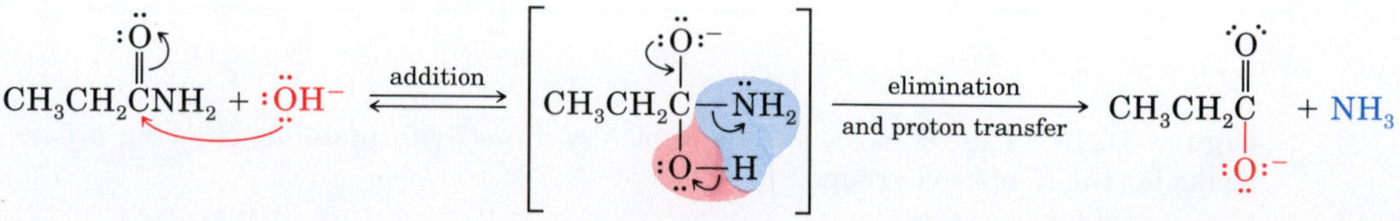

STUDY PROBLEM

15.21   *Aspartame* is a low-calorie synthetic sweetener introduced in 1983. This compound is a *dipeptide,* a small protein that yields two amino acid molecules upon hydrolysis. Predict the products of the hydrolysis of aspartame in **(a)** aqueous HCl, and **(b)** aqueous NaOH.

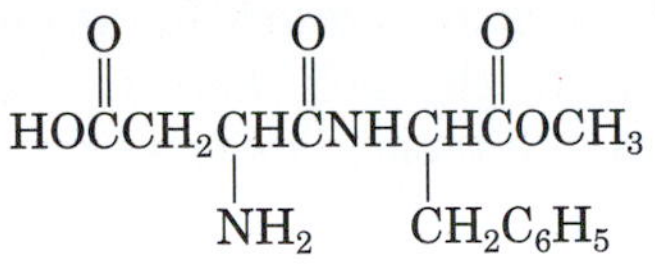

$$\underset{\substack{| \\ NH_2}}{HOCCH_2CHCNHCHCOCH_3}\underset{CH_2C_6H_5}{}$$

aspartame

Reduction   Reduction of amides with lithium aluminum hydride results in conversion of the carbonyl group to $—CH_2—$. The product is an amine. Sodium borohydride does not reduce amides.

General:

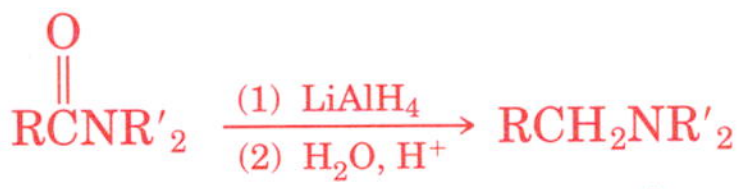

$$\underset{an\ amide}{RCNR'_2} \xrightarrow[\text{(2) } H_2O,\ H^+]{\text{(1) } LiAlH_4} \underset{an\ amine}{RCH_2NR'_2}$$

$$\underset{\textit{N-methyldodecanamide}}{CH_3(CH_2)_{10}\overset{O}{\overset{\|}{C}}NHCH_3} \xrightarrow[\text{(2) } H_2O,\ H^+]{\text{(1) } LiAlH_4} \underset{\textit{N-methyldodecylamine (95\%)}}{CH_3(CH_2)_{10}CH_2NHCH_3}$$

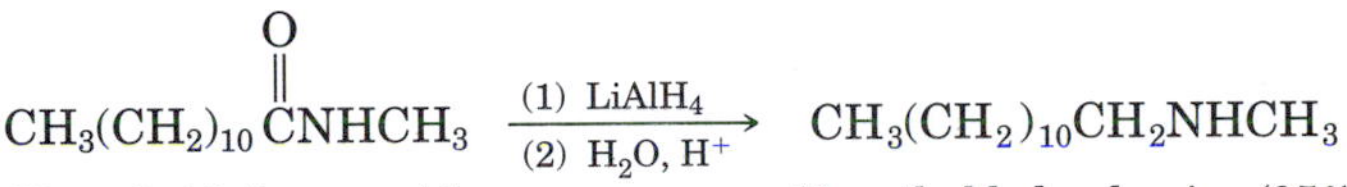

The first step in the reduction is the addition of $H:^-$ to the carbonyl carbon. Oxygen is then abstracted by the aluminum hydride to yield an imine or iminium ion, which in turn is reduced by another hydride transfer.

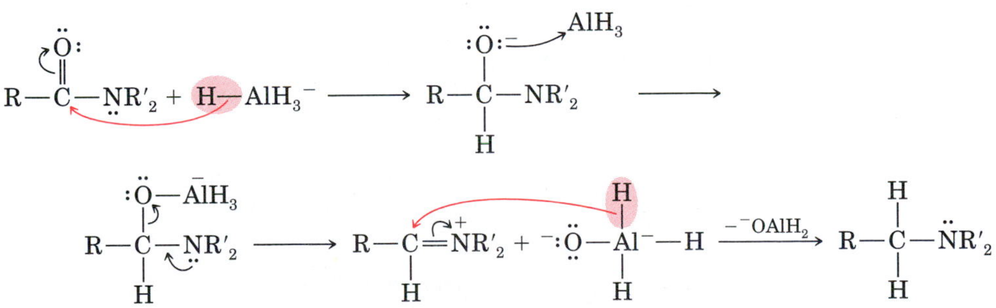

STUDY PROBLEM

15.22   What product is formed when 5,5-dimethyl-2-pyrrolidone is reduced with lithium aluminum hydride?

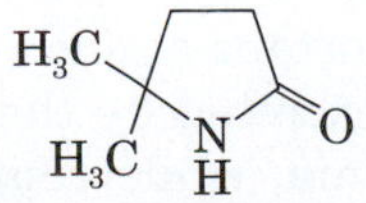

5,5-dimethyl-2-pyrrolidone

Section 15.9
Polyamides

There can be no question that the most important polyamides are the *proteins*. Chapter 25 is devoted to this subject. The most notable example of a man-made polyamide is the synthetic polyamide **nylon 6,6,** which is prepared from adipic acid (a diacid) and hexamethylenediamine (a diamine). As in the synthesis of the polyester Dacron, the result of the reaction of two types of bifunctional molecules is a linear polymer.

$$x\ HO_2C(CH_2)_4CO_2H\ +\ x\ H_2N(CH_2)_6NH_2 \xrightarrow[-H_2O]{heat}$$

$$\left[\begin{array}{c} O \quad\quad O \\ \| \quad\quad\quad \| \\ C(CH_2)_4C-NH(CH_2)_6NH \end{array}\right]_x$$

hexanedioic acid   1,6-hexanediamine   nylon 6,6
(adipic acid)   (hexamethylenediamine)

Nylon 6,6 is but one member of the family of synthetic nylons. Nylon 6,6 is made from a *six-carbon* diacid and *six-carbon* diamine. **Nylon 6,** on the other hand, is prepared from ε-caprolactam, a monomer that contains the acid and amine in the same molecule (with *six carbons*). In this reaction, ε-caprolactam undergoes ring opening with a small amount of water to produce some ε-aminocaproic acid. The ε-aminocaproic acid then ring opens another ε-caprolactam to produce a dimer with a new amino function, which can ring open yet another ε-caprolactam. Thus, one ring after another is opened to form the polymer chain.

ε-caprolactam   ε-aminocaproic acid   nylon 6

A newer fiber, "para-aramid or Kevlar," is so strong and indestructible that it is widely used in military protective (bulletproof) vests and helmets.

"para-aramid"
(Kevlar)

Section 15.10
Compounds Related to Amides

Some types of compounds related to amides are shown in Table 15.4. **Urea** is one of the most important amide relatives. Excess nitrogen from the metabolism of proteins is excreted by the higher animals as urea. Some lower animals excrete ammonia, while reptiles and birds excrete **guanidine.** Both guanidine and urea, as well as ammonia, are widely used as nitrogen fertilizers and as starting materials for synthetic polymers and drugs.

Table 15.4   Some types of compounds related to amides

Partial Structure	Class of Compound	Example
$-\overset{\overset{\displaystyle O}{\|}}{C}N\diagup$	amide	$CH_3\overset{\overset{\displaystyle O}{\|}}{C}NH_2$
$-\overset{\overset{\displaystyle O}{\|}}{C}N\diagup$ in ring	lactam	(six-membered ring lactam)
$-\overset{\overset{\displaystyle O}{\|}}{C}N\overset{\overset{\displaystyle O}{\|}}{C}-$	imide	(six-membered ring imide)
$\diagup N\overset{\overset{\displaystyle O}{\|}}{C}N\diagup$	urea	$H_2N\overset{\overset{\displaystyle O}{\|}}{C}NH_2$
$\diagup N\overset{\overset{\displaystyle O}{\|}}{C}O-$	carbamate, or urethane	$H_2N\overset{\overset{\displaystyle O}{\|}}{C}OCH_3$
$-\overset{\overset{\displaystyle O}{\|}}{\underset{\underset{\displaystyle O}{\|}}{S}}N\diagup$	sulfonamide	(phenyl)$-\overset{\overset{\displaystyle O}{\|}}{\underset{\underset{\displaystyle O}{\|}}{S}}NH_2$

$$H_2N-\overset{\overset{\displaystyle O}{\|}}{C}-NH_2 \qquad R_2N-\overset{\overset{\displaystyle O}{\|}}{C}-NR_2 \qquad H_2N-\overset{\overset{\displaystyle NH}{\|}}{C}-NH_2$$

urea   *a substituted urea*   guanidine

Urea is used for the synthesis of barbiturates (used as sedatives) by reaction with α-substituted diethyl malonates. This reaction is similar to the reaction of an ester with an amine to yield an amide.

diethyl malonate

not substituted

barbituric acid

A **carbamate,** or **urethane,** is a compound in which the $-NH_2$, $-NHR$, or $-NR_2$ group is bonded to an ester carbonyl group. A carbamate is related to a carbonate structure, with one O replaced by N.

$$RO-\overset{\overset{\displaystyle O}{\|}}{C}-OR \qquad H_2N-\overset{\overset{\displaystyle O}{\|}}{C}-OR$$

a carbonate   *a carbamate*

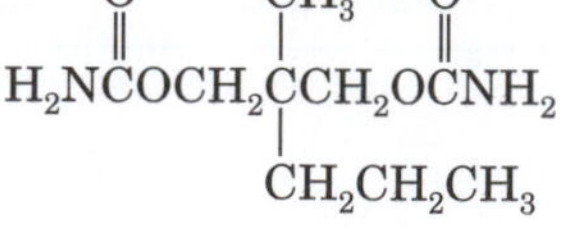

$$\underset{\substack{\text{meprobamate}}}{\text{H}_2\text{NCOCH}_2\overset{\displaystyle\text{CH}_3}{\underset{\displaystyle\text{CH}_2\text{CH}_2\text{CH}_3}{\text{C}}}\text{CH}_2\text{OCNH}_2}$$

meprobamate

*a dicarbamate used as
a tranquilizer (Miltown, Equanil)*

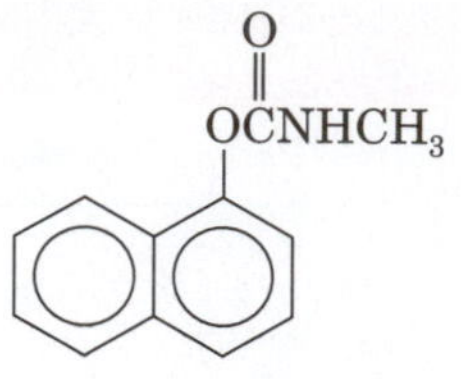

1-naphthyl-*N*-methylcarbamate
(Sevin)

a biodegradable insecticide

One way in which a carbamate may be prepared is by the action of an alcohol or phenol on an **isocyanate,** a compound containing the —N=C=O group.

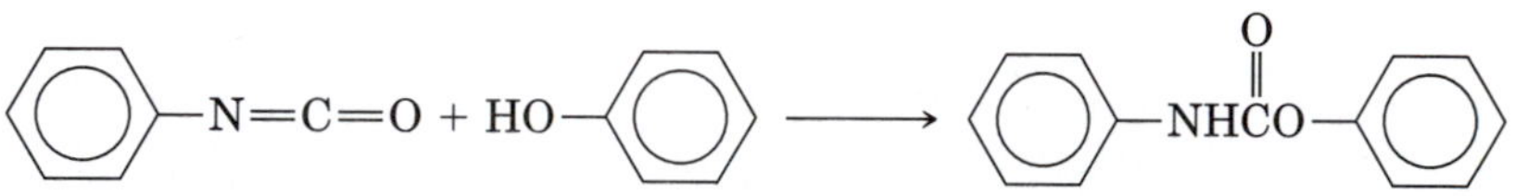

phenyl isocyanate   phenol   phenyl *N*-phenylcarbamate

An analogous reaction is used to make **polyurethanes,** such as that used for polyurethane foam insulation. The foaming effect in polyurethane foam is achieved by the addition of a low-boiling liquid, such as dichloromethane, which vaporizes during polymerization.

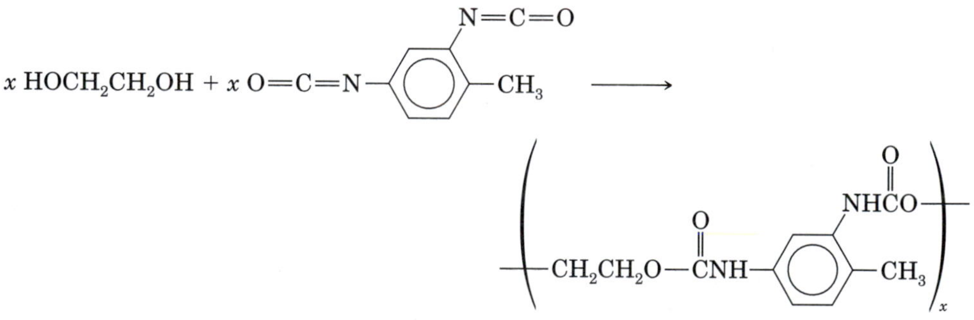

$$x\,\text{HOCH}_2\text{CH}_2\text{OH} + x\,\text{O}{=}\text{C}{=}\text{N}-$$

a polyurethane

Sulfa drugs are **sulfonamides,** compounds in which the nitrogen is attached to a sulfonyl group rather than to an acyl group. A sulfonamide is prepared by the action of an arylsulfonyl chloride on ammonia or on a primary or secondary amine.

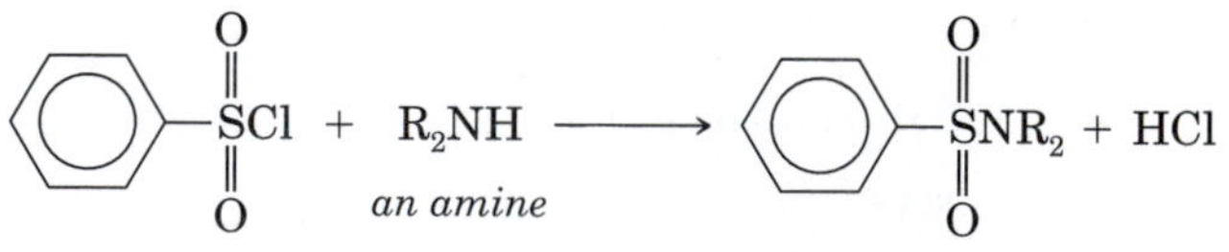

benzenesulfonyl
chloride

an amine

a sulfonamide

STUDY PROBLEM

15.23 Sodium cyclohexylsulfamate (a **cyclamate**) is an artificial sweetener that is thirty times sweeter than cane sugar. This compound can be prepared by the reaction of cyclohexylamine and chlorosulfonic acid, followed by treatment with sodium hydroxide. What is the structure of this cyclamate?

$$\text{cyclohexyl}-NH_2 \ + \ \underset{\substack{\text{chlorosulfonic} \\ \text{acid}}}{HO\overset{\overset{\displaystyle O}{\|}}{\underset{\underset{\displaystyle O}{\|}}{S}}Cl} \quad \xrightarrow{\quad\quad} \quad \xrightarrow{\ OH^- \ }$$

cyclohexylamine   chlorosulfonic
                            acid

Section 15.11
Nitriles

A. Nomenclature of Nitriles

Nitriles are organic compounds containing the C≡N group. They are also sometimes called *cyano compounds* or *cyanides*. In the IUPAC system, the number of carbon atoms, including that in the CN group, determines the alkane parent. The alkane name is suffixed with **-nitrile.** Some nitriles are named after the trivial names of their carboxylic acid parent with the **-ic acid** changed to **-nitrile,** or to **-onitrile** if the parent name lacks an o-.

$$CH_3C\equiv N \qquad\qquad \text{phenyl}-CN$$

IUPAC: ethanenitrile   benzenecarbonitrile
trivial: acetonitrile   benzonitrile

STUDY PROBLEM

15.24 Write formulas for **(a)** propanenitrile; **(b)** butyronitrile.

B. Bonding in Nitriles

The cyano group contains a triple bond—one σ bond and two π bonds (Figure 15.14). Although the nitrogen has a pair of unshared electrons, a nitrile is a very weak base. The pK_b of a nitrile is about 24, while the pK_b of NH_3 is about

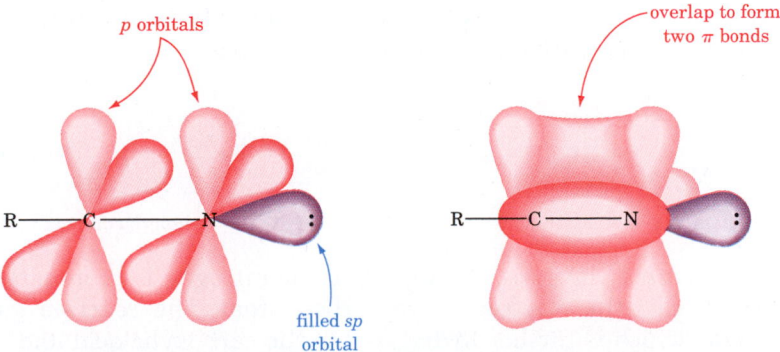

Figure 15.14   The bonding in a nitrile, RC≡N⊷.

4.5 (about 20 powers of ten difference). The lack of basicity of a —CN: group results from the unshared electrons being in an sp orbital. The greater amount of s character in an sp orbital (compared with that in an sp^2 or sp^3 orbital) means that these sp electrons are more tightly held and less available for bonding to a proton.

C. Preparation of Nitriles

The CN^- ion (from NaCN, for example) is a good nucleophile for S_N2 displacement of a halide ion from an alkyl halide. This reaction is the principal route to aliphatic nitriles. However, because of elimination reactions, high yields are obtained only with *primary alkyl halides* and, to a lesser extent, *secondary alkyl halides.*

$$CH_3CH_2CH_2CH_2Br + CN^- \xrightarrow{S_N2} CH_3CH_2CH_2CH_2CN + Br^-$$

1-bromobutane   pentanenitrile

a 1° alkyl halide   (90%)

Aryl nitriles are best obtained through the **diazonium salts,** compounds that were discussed in Section 12.3B.

$\langle\bigcirc\rangle$—NH$_2$ $\xrightarrow[\substack{0°C}]{\substack{NaNO_2 \\ HCl}}$ $\langle\bigcirc\rangle$—N$_2^+$ Cl$^-$ $\xrightarrow[100°]{CuCN + KCN}$ $\langle\bigcirc\rangle$—CN

aniline   benzenediazonium   benzonitrile
chloride

Another route to nitriles is the dehydration of amides with such dehydrating agents as $SOCl_2$, P_2O_5, or acetic anhydride.

$$\underset{\underset{CH_3CH_2}{|}}{CH_3(CH_2)_3CH\overset{\overset{O}{\|}}{C}NH_2} + SOCl_2 \longrightarrow \underset{\underset{CH_3CH_2}{|}}{CH_3(CH_2)_3CHCN}$$

2-ethylhexanamide   2-ethylhexanenitrile (94%)

D. Reactions of Nitriles

Hydrolysis   Nitriles are included as carboxylic acid derivatives because their hydrolysis yields carboxylic acids. The hydrolysis of a nitrile can be carried out by heating with either aqueous acid or base.

$\langle\bigcirc\rangle$—CH$_2$CN + 2 H$_2$O + H$^+$ $\xrightarrow[\text{4-hr reflux}]{42\% H_2SO_4}$ $\langle\bigcirc\rangle$—CH$_2\overset{\overset{O}{\|}}{C}$OH + NH$_4^+$

phenylacetonitrile   phenylacetic acid (78%)

In *acidic hydrolysis,* the weakly basic nitrogen is protonated and then water attacks the electropositive carbon atom. The reaction goes through an amide, which is further hydrolyzed to the carboxylic acid and ammonium ion. Because of the formation of the NH_4^+ ion (from NH_3 and H^+), an excess of acid must be used in nitrile hydrolysis.

In acid:

$$RC\equiv N: \rightleftharpoons [RC\overset{+}{\equiv}NH \longleftrightarrow RC\overset{+}{=}NH] \xrightarrow{H_2O} \left[\begin{array}{c} RC=NH \\ | \\ HOH \\ + \end{array} \right] \xrightarrow{-H^+}$$

$$\left[\begin{array}{c} RC=NH \\ | \\ :OH \end{array} \right] \xrightarrow{H^+} \left[\begin{array}{c} RC\overset{+}{=}NH_2 \\ || \\ H-O: \end{array} \right] \xrightarrow{-H^+} \begin{array}{c} RC-NH_2 \\ || \\ O: \end{array} \xrightarrow{H_2O,\,H^+} \begin{array}{c} OH \\ | \\ RC \\ || \\ O \end{array} + NH_4{}^+$$

intermediate amide

If comparatively mild conditions are used, the amide may be isolated as the product of the reaction.

$$\text{C}_6\text{H}_5-CH_2CN + H_2O \xrightarrow[40°,\,1\,hr]{35\%\,HCl} \text{C}_6\text{H}_5-CH_2\overset{O}{\overset{||}{C}}NH_2$$

phenylacetonitrile   2-phenylacetamide (86%)

Alkaline hydrolysis occurs by nucleophilic attack on the partially positive carbon of the nitrile group. The reaction again results in an amide, which is further hydrolyzed to the carboxylate and ammonia. The free acid is obtained when the solution is acidified.

In base:

$$RC\equiv N: + :\overset{..}{O}H^- \xrightarrow{heat} \left[\begin{array}{c} RC=\overset{..}{N}:^- \\ | \\ :OH \end{array} \right] \xrightarrow[-OH^-]{H-\overset{..}{O}H} \left[\begin{array}{c} RC=\overset{..}{N}H \\ | \\ :O-H \end{array} \right] \xrightarrow[-H_2O]{H-\overset{..}{O}H}$$

$$\begin{array}{c} RC-\overset{..}{N}H_2 \\ || \\ :O \end{array} \xrightarrow{OH^-} \begin{array}{c} O^- \\ | \\ RC \\ || \\ O \end{array} + NH_3$$

intermediate amide $\xrightarrow{H^+} RCO_2H$

Reduction   Nitriles can be reduced to *primary amines* of the type RCH_2NH_2 either by catalytic hydrogenation or by lithium aluminum hydride.

General:

$$RC\equiv N \xrightarrow{[H]} RCH_2NH_2$$

$$\text{C}_6\text{H}_5-CH_2C\equiv N \xrightarrow[\substack{heat\\pressure}]{2\,H_2,\,Ni} [\text{C}_6\text{H}_5-CH_2CH=NH] \longrightarrow \text{C}_6\text{H}_5-CH_2CH_2NH_2$$

phenylacetonitrile   (2-phenylethyl)amine (87%)

$$CH_3CH_2CH_2C\equiv N \xrightarrow[(2)\,H_2O]{(1)\,LiAlH_4} CH_3CH_2CH_2CH_2NH_2$$

butanenitrile   butylamine (85%)

Figure 15.15 summarizes the reactions of amides and nitriles.

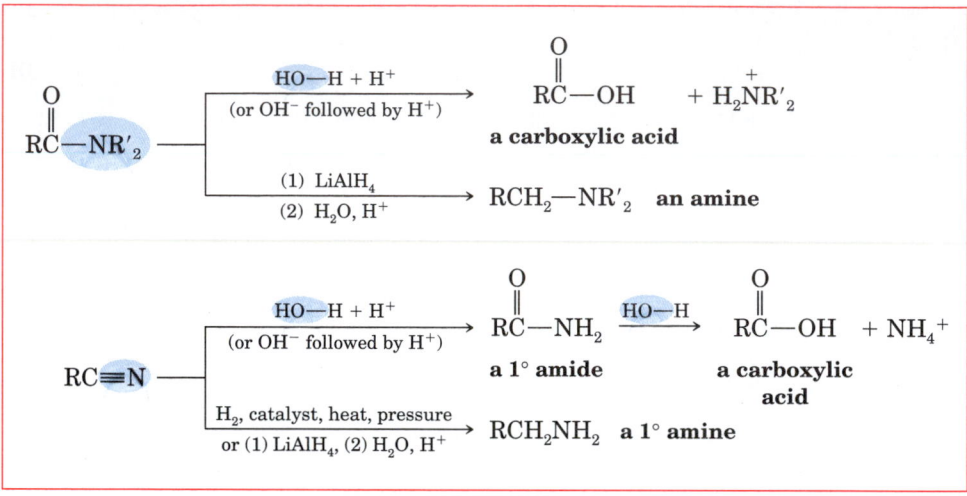

Figure 15.15 Summary of reactions of amides and nitriles.

Use of Carboxylic Acid Derivatives in Synthesis

Carboxylic acids and their derivatives are all synthetically interconvertible. However, of the carboxylic acid derivatives, the acid halides and anhydrides are probably the most versatile because they are more reactive than other carbonyl compounds. Either of these two reactants can be used to synthesize hindered esters or phenyl esters, which cannot be prepared in good yield by heating RCO_2H and R'OH with an acidic catalyst because of an unfavorable equilibrium. These two derivatives are also the most useful reagents for making N-substituted amides. By reduction with $LiAlH(OR)_3$, acid chlorides may be used to synthesize aldehydes as well.

Although esters are not as reactive as acid chlorides or anhydrides, they are useful for the synthesis of alcohols (by reduction or by Grignard reactions) and are valuable starting materials for the synthesis of complex molecules.

The nitrile group can be converted to a carboxylic acid by hydrolysis or it can be reduced to a primary amino group. In addition, the synthesis of nitriles by reaction of a primary alkyl halide with cyanide ion affords one of the most convenient techniques for extending an aliphatic carbon chain by one carbon.

The preparations and reactions of the carboxylic acid derivatives are summarized in Tables 15.5 and 15.6.

Table 15.5   Summary of laboratory syntheses of carboxylic acid derivatives

Reaction	Section Reference
Acid Chlorides:	
$RCO_2H + SOCl_2$ or $PCl_3 \longrightarrow R\overset{\displaystyle O}{\overset{\displaystyle \|}{C}}Cl$	15.3B
Acid Anhydrides:	
$R\overset{\displaystyle O}{\overset{\displaystyle \|}{C}}Cl + {}^-O\overset{\displaystyle O}{\overset{\displaystyle \|}{C}}R' \longrightarrow R\overset{\displaystyle O}{\overset{\displaystyle \|}{C}}O\overset{\displaystyle O}{\overset{\displaystyle \|}{C}}R'$	15.4B
$RCO_2H + \text{excess } (CH_3\overset{\displaystyle O}{\overset{\displaystyle \|}{C}})_2O \longrightarrow R\overset{\displaystyle O}{\overset{\displaystyle \|}{C}}O\overset{\displaystyle O}{\overset{\displaystyle \|}{C}}R$	15.4B
$HO_2C(CH_2)_nCO_2H \xrightarrow{\text{heat}}$ cyclic anhydride $(n = 2 \text{ or } 3)$	14.8B
Esters:[a]	
$RCO_2H + R'OH \xrightarrow{H^+} RCO_2R'$	14.6
$R\overset{\displaystyle O}{\overset{\displaystyle \|}{C}}Cl + R'OH \longrightarrow RCO_2R'$	15.3C
$(R\overset{\displaystyle O}{\overset{\displaystyle \|}{C}})_2O + R'OH \longrightarrow RCO_2R'$	15.4C
$RCO_2^- + R'X \longrightarrow RCO_2R'$[b]	5.10, 14.5
$RCO_2R' + R''OH \xrightarrow{H^+ \text{or } {}^-OR''} RCO_2R''$	15.5C
Amides:	
$R\overset{\displaystyle O}{\overset{\displaystyle \|}{C}}Cl + HNR'_2 \longrightarrow R\overset{\displaystyle O}{\overset{\displaystyle \|}{C}}NR'_2$	15.3C
$(R\overset{\displaystyle O}{\overset{\displaystyle \|}{C}})_2O + HNR'_2 \longrightarrow R\overset{\displaystyle O}{\overset{\displaystyle \|}{C}}NR'_2$	15.4C
$RCO_2R' + NH_3 \longrightarrow R\overset{\displaystyle O}{\overset{\displaystyle \|}{C}}NH_2$	15.5C
Nitriles:	
$RX + CN^- \longrightarrow RCN$	15.11C
$ArNH_2 \xrightarrow[0°]{\substack{NaNO_2 \\ HCl}} ArN_2^+ Cl^- \xrightarrow[\text{heat}]{CuCN+KCN} ArCN$	12.3B, 15.11C
$R\overset{\displaystyle O}{\overset{\displaystyle \|}{C}}NH_2 \xrightarrow[-H_2O]{\text{dehydrating agent}} RCN$	15.11C

[a] The syntheses of some complex esters will be discussed in Chapter 17.
[b] For this reaction to be successful, a reactive halide must be used.

Table 15.6   **Types of compounds that can be obtained from carboxylic acid derivatives**

Reaction		Product	Section Reference

Acid Chlorides:[a]

Reaction		Product	Section Reference
$\overset{\text{O}}{\overset{\|}{\text{RCCl}}} + H_2O \longrightarrow RCO_2H$		**carboxylic acid**	15.3C
$\overset{\text{O}}{\overset{\|}{\text{RCCl}}} + R'OH \longrightarrow RCO_2R'$		**ester**	15.3C
$\overset{\text{O}}{\overset{\|}{\text{RCCl}}} + R'_2NH \longrightarrow \overset{\text{O}}{\overset{\|}{\text{RCNR}'_2}}$		**amide**	15.3C
$\overset{\text{O}}{\overset{\|}{\text{RCCl}}} + R'CO_2^- \longrightarrow \overset{\text{O O}}{\overset{\|\ \|}{\text{RCOCR}'}}$		**anhydride**	15.3C
$\overset{\text{O}}{\overset{\|}{\text{RCCl}}} + C_6H_6 \xrightarrow{\text{AlCl}_3} \overset{\text{O}}{\overset{\|}{\text{RCC}_6\text{H}_5}}$		**aryl ketone**	11.8E, 15.3C
$\overset{\text{O}}{\overset{\|}{\text{RCCl}}} \xrightarrow[\text{(2) H}_2\text{O, H}^+]{\text{(1) 2 R'MgX}} \overset{\text{OH}}{\overset{\|}{\text{RCR}'_2}}$		**3° alcohol**	15.3C
$\overset{\text{O}}{\overset{\|}{\text{RCCl}}} + \text{LiCuR}'_2 \longrightarrow \overset{\text{O}}{\overset{\|}{\text{RCR}'}}$		**ketone**	15.3C
$\overset{\text{O}}{\overset{\|}{\text{RCCl}}} \xrightarrow[\text{(2) H}_2\text{O}]{\text{(1) LiAlH(OR')}_3} \overset{\text{O}}{\overset{\|}{\text{RCH}}}$		**aldehyde**	15.3C
$\overset{\text{O}}{\overset{\|}{\text{RCH}_2\text{CCl}}} + Cl_2 \longrightarrow \overset{\text{O}}{\overset{\|}{\text{RCHClCCl}}}$		**α-chloro acid chloride**	15.3C

Acid Anhydrides:

Reaction		Product	Section Reference
$\overset{\text{O O}}{\overset{\|\ \|}{\text{RCOCR}}} + H_2O \longrightarrow 2\ RCO_2H$		**carboxylic acid**	15.4C
$(RCO)_2O + R'OH \longrightarrow RCO_2R'$		**ester**	15.4C
$(RCO)_2O + R'_2NH \longrightarrow \overset{\text{O}}{\overset{\|}{\text{RCNR}'_2}}$		**amide**	15.4C

Esters:[a]

Reaction		Product	Section Reference
$RCO_2R' + H_2O \xrightarrow{\text{H}^+ \text{ or OH}^-} RCO_2H$		**carboxylic acid**	15.5C
$RCO_2R' + R''OH \xrightarrow{\text{H}^+ \text{ or }^-\text{OR}''} RCO_2R''$		**ester**	15.5C
$RCO_2R' + NH_3 \longrightarrow \overset{\text{O}}{\overset{\|}{\text{RCNH}_2}}$		**amide**	15.5C
$RCO_2R' + [H] \longrightarrow RCH_2OH + HOR'$		**alcohols**	15.5C
$RCO_2R' \xrightarrow[\text{(2) H}_2\text{O, H}^+]{\text{(1) 2 R''MgX}} \overset{\text{OH}}{\overset{\|}{\text{RCR}''_2}}$		**3° alcohol**	15.5C

(continued)

Table 15.6 *(continued)*	Types of compounds that can be obtained from carboxylic acid derivatives

Reaction		Product	Section Reference

Amides:

$$\underset{\overset{\displaystyle O}{\|}}{RCNR'_2} + H_2O \xrightarrow{H^+ \text{ or } OH^-} RCO_2H + HNR'_2$$ **carboxylic acid and amine**   15.8C

$$\underset{\overset{\displaystyle O}{\|}}{RCNR'_2} + [H] \longrightarrow RCH_2NR'_2$$ **amine**   15.8C, 18.4B

$$\underset{\overset{\displaystyle O}{\|}}{RCNH_2} + Br_2 + {}^-OH \longrightarrow RNH_2$$ **amine**   18.4C

Nitriles:[a]

$$RCN + H_2O \xrightarrow{H^+ \text{ or } OH^-} RCO_2H$$ **carboxylic acid**   15.11D

$$RCN + H_2O \xrightarrow{H^+ \text{ or } OH^-} \underset{\overset{\displaystyle O}{\|}}{RCNH_2}$$ **amide**   15.11D

$$RCN + [H] \longrightarrow RCH_2NH_2$$ **amine**   15.11D

[a] Acid chlorides, esters, and nitriles can be used to synthesize more-complex compounds. Some of these reactions will be discussed in Chapter 17.

STUDY PROBLEM

15.25 Suggest synthetic routes to the following compounds:

(a) $\langle\ \rangle$—$\underset{\overset{\displaystyle O}{\|}}{C}N(CH_3)_2$ from a carboxylic acid

(b) 2-methyl-4-propyl-4-heptanol   from compounds containing six or fewer carbon atoms

(c) $CH_3CH_2\underset{\overset{\displaystyle O}{\|}}{C}NH(CH_2)_3CH_3$ from 1-propanol and *no other organic reagents*

Summary

The derivatives of carboxylic acids are usually prepared from the carboxylic acids themselves or from other more reactive derivatives, as shown in Table 15.5

The reactions of the various carboxylic acid derivatives with nucleophiles are similar to one another. Differences arise from differences in reactivity of the various derivatives.

$$
\begin{array}{l}
\underset{\displaystyle \substack{\text{O} \\ \|}}{RC}-Cl \\[4pt]
\underset{\displaystyle \substack{\text{O} \;\;\;\; \text{O} \\ \| \;\;\;\; \|}}{RC}-OCR \\[4pt]
\underset{\displaystyle \substack{\text{O} \\ \|}}{RC}-OR \\[4pt]
\underset{\displaystyle \substack{\text{O} \\ \|}}{RC}-NH_2 \\[4pt]
RCN
\end{array}
\;\Bigg\}
\;\;\xrightarrow[\text{H}^+ \text{ or OH}^-]{\text{H}_2\text{O}}\;\;
\overset{\displaystyle \substack{\text{O} \\ \|}}{RC}-OH \quad \textit{a carboxylic acid}
$$

increasing reactivity (arrow pointing up)

$$
\begin{array}{l}
RCOCl \\
(RCO)_2O \\
RCO_2R
\end{array}
\;\Bigg\}
\;\;\xrightarrow{\text{NH}_3}\;\;
\overset{\displaystyle \substack{\text{O} \\ \|}}{RC}-NH_2 \quad \textit{an amide}
$$

increasing reactivity (arrow pointing up)

$$
\begin{array}{l}
RCOCl \\
(RCO)_2O \\
RCO_2R
\end{array}
\;\Bigg\}
\;\;\xrightarrow{\text{R'OH}}\;\;
\overset{\displaystyle \substack{\text{O} \\ \|}}{RC}-OR' \quad \textit{an ester}
$$

increasing reactivity (arrow pointing up)

In addition to these reactions, the more reactive acid halides undergo Friedel–Crafts reactions with aromatic compounds and also undergo reaction with lithium dialkylcuprates to yield ketones. (See Table 15.6.)

Esters react with Grignard reagents to yield tertiary alcohols.

$$
RCO_2R \;\xrightarrow[\text{(2) H}_2\text{O, H}^+]{\text{(1) 2R'MgX}}\; \overset{\displaystyle \text{OH}}{\underset{}{\underset{\displaystyle |}{RCR'_2}}} \quad \textit{a 3° alcohol}
$$

All the derivatives can be reduced. For example:

$$
RCO_2R' \;\xrightarrow[\text{(2) H}_2\text{O, H}^+]{\text{(1) LiAlH}_4}\; RCH_2OH + HOR' \quad \textit{alcohols}
$$

$$
\overset{\displaystyle \substack{\text{O} \\ \|}}{RC}NR'_2 \;\xrightarrow[\text{(2) H}_2\text{O, H}^+]{\text{(1) LiAlH}_4}\; RCH_2NR'_2 \quad \textit{an amine}
$$

$$
RCN \;\xrightarrow[\text{(2) H}_2\text{O, H}^+]{\text{(1) LiAlH}_4}\; RCH_2NH_2 \quad \textit{a 1° amine}
$$

Titanate-Mediated Transesterifications

Transesterification reactions are usually carried out using an acidic or a basic catalyst. These reactions are sensitive to steric hindrance and will fail if either the ester or the alcohol is highly hindered. To circumvent this limitation, a titanate-mediated transesterification procedure has been developed. In a typical reaction, dimethyl $(2S,3S)$-2,3-O-isopropylidenetartrate **(1)** is treated with absolute 2-propanol and a small amount of tetraisopropyltitanate at reflux. A transesterification reaction occurs. From the mixture, diisopropoxy $(2S,3S)$-2,3-O-isopropylidenetartrate **(2)** can be isolated in a 91–95% yield.

Titanate-mediated transesterifications can be used with an alcohol as solvent to exchange the alcohol component of an ester or to remove an acyl-protecting group.

Exchange of an alcohol group

Alternatively, by using an ester solvent system, hydroxyl groups can be protected by acylation or the acid component of an ester-blocking group can be exchanged. (R. Imwinkelried, M. Schiess, and D. Seebach, *Org. Syn.* **1987,** *65,* 230.)

Acylation of an alcohol

Questions

1. If dimethyl $(2S,3S)$-2,3-O-isopropylidenetartrate **(1)** were treated with aqueous isopropyl alcohol using an acid catalyst, what would be the expected product?

2. Why must an excess of isopropyl alcohol be used for the transesterification of **1**? What would happen if the two reactants had been used in a 1:1 molar ratio?
3. Why does the transesterification of **1** with isopropyl alcohol occur with retention of configuration? If the reaction had been carried out using isopropoxide ion [$(CH_3)_2CHO^-$] as an alkaline catalyst, would the configurations have been retained? Explain.
4. Write an equation showing how titanate-mediated transesterification can be used to prepare the following compound from its methyl ester.

5. Using the general formula $RCOR'$, write an equation showing how the titanate-mediated transesterification can be used to (a) exchange a carboxylic acid component of an ester and (b) remove an acyl-blocking group.

Study Problems*

15.26 Name the following compounds:

(a) $(CH_3)_2CHCH_2\overset{O}{\overset{\|}{C}}Cl$ (b) $Br-\!\!\bigcirc\!\!-\overset{O}{\overset{\|}{C}}Br$ (c) $CH_2\!\!=\!\!CHCH_2\overset{O}{\overset{\|}{C}}Cl$

15.27 Oleic acid, $cis\text{-}CH_3(CH_2)_7CH\!\!=\!\!CH(CH_2)_7CO_2H$, is a carboxylic acid that can be obtained from the hydrolysis of animal fats and vegetable oils. Oleoyl chloride can be synthesized from this acid in 99% yield. Write equations that illustrate two ways in which this synthesis could be carried out.

15.28 What would be the products of the reactions of propanoyl chloride with the following reagents?

(a) water

(b) cyclohexanol

(c) *p*-bromophenol

(d) piperidine (page 726)

(e) sodium ethoxide in ethanol

(f) $(CH_3CH_2)_2CuLi$

(g) sodium formate in tetrahydrofuran

(h) $LiAlH_4$ in ether, followed by aqueous acid

---

* For information concerning the organization of the *Study Problems* and *Additional Problems*, see the *Note to student* on page 41.

(i)   excess CH_3CH_2MgBr in ether, followed by acidification

(j)   bromine, followed by aqueous acid

15.29 Name the following compounds:

(a) $CH_3CH_2\overset{\overset{O}{\|}}{C}O\overset{\overset{O}{\|}}{C}CH_2CH_2CH_2CH_3$

(b) $Cl-\!\!\bigcirc\!\!-\overset{\overset{O}{\|}}{C}O\overset{\overset{O}{\|}}{C}CH_2-\!\!\bigcirc$

(c)

(d) $H\overset{\overset{O}{\|}}{C}O\overset{\overset{O}{\|}}{C}CH_3$

15.30 Write equations showing how the anhydrides in Problem 15.29 could be synthesized.

15.31 Predict the products when the following cyclic anhydride is treated with the following reagents:

(a)   aqueous NaOH   (b)   aqueous HCl   (c)   ethanol

(d)   sodium ethoxide in ethanol   (e)   methylamine

15.32 Write the IUPAC names of the following esters:

(a) $CH_3CH_2O_2CCH_2-\!\!\bigcirc$

odor of honey

(b) benzene ring with CO_2CH_3 and NH_2 substituents

odor of grapes

15.33 Write equations that show the syntheses of the esters in the preceding problem from carboxylic acids.

15.34 Show by equations how you would convert methyl benzoate to:

(a)   benzoic acid;   (b)   benzyl alcohol;

(c)   2-phenyl-2-propanol   (d)   acetophenone ($CH_3COC_6H_5$)

(e)   N-methylbenzamide   (f)   benzophenone ($C_6H_5COC_6H_5$)

15.35 Complete the following equations showing the major products:

(a) cyclic lactone $\xrightarrow[\text{(2) } H_2O,\ H^+]{\text{(1) } CH_3MgBr \text{ (excess), ether}}$

(b) $CH_3CH=\overset{\overset{\displaystyle CH_3}{|}}{C}Li \xrightarrow[\text{(2) } H_2O,\ H^+]{\text{(1) } CH_3CO_2CH_2CH_3}$

(c) $ClCH_2CO_2CH_2CH_3 + NH_3 \xrightarrow[\text{30 min}]{0°}$

(d) $CH_3\overset{\overset{O}{\|}}{C}CH_2O\overset{\overset{O}{\|}}{C}CH_3 + CH_3OH \xrightarrow{H^+}$

(e) $(S)\text{-}CH_3\overset{O}{\overset{\|}{C}}\ {}^{18}O\overset{CH_3}{\overset{|}{C}}HC_6H_5$ $\xrightarrow[\text{(2) }H_2O,\ H^+]{\text{(1) aqueous NaOH, heat}}$

(f) $CH_3CH_2CH_2\overset{O}{\overset{\|}{C}}OC_6H_5 + C_6H_5NH_2 \longrightarrow$

15.36 Name the following amides, using the IUPAC system:

(a) $CH_3(CH_2)_6\overset{O}{\overset{\|}{C}}NH_2$ **(b)** $CH_3\overset{O}{\overset{\|}{C}}NHCH_2C_6H_5$

(c) $(CH_3CH_2)_2CH\overset{O}{\overset{\|}{C}}NHC_6H_5$ **(d)** $C_6H_5\overset{O}{\overset{\|}{C}}N(CH_2CH_3)_2$

15.37 Write the equations that show *three* possible reactions leading to N-methylhexan-amide.

15.38 Name the following compounds by the IUPAC system:

(a) $(CH_3)_2CHCN$ **(b)** 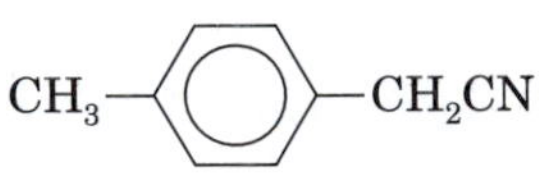 $CH_3\!-\!\!\bigcirc\!\!-\!CH_2CN$ **(c)** 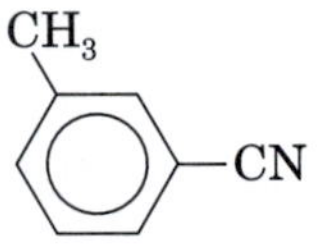 [benzene ring with CH_3 and $-CN$ substituents]

15.39 Write formulas for **(a)** 2-phenylpropanenitrile, and **(b)** β-chlorobutyronitrile.

15.40 Suggest syntheses for the following nitriles:

(a) 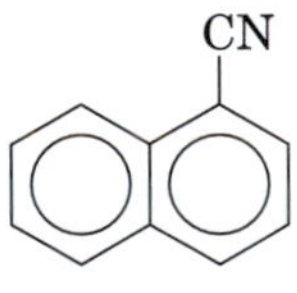 [naphthalene ring with CN substituent]   **(b)** $(CH_3)_2C{=}CHCH_2CN$

15.41 Predict the products of the following reactions:

(a) $NCCH_2CH_2CH_2CN + H_2O \xrightarrow[\text{heat}]{\text{HCl}}$

(b) 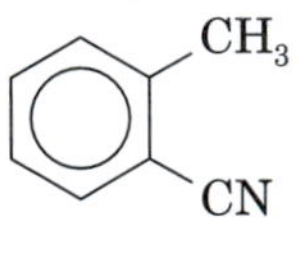 [benzene ring with CH_3 and CN substituents] $+ H_2O \xrightarrow[\text{heat}]{H_2SO_4}$

(c) 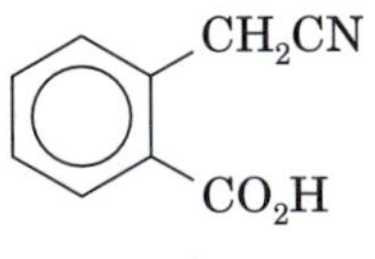 [benzene ring with CH_2CN and CO_2H substituents] $\xrightarrow[\text{(2) }H_2O,\ H^+]{\text{(1) LiAlH}_4}$

(d) $NCCH_2CH_2CH_2CH_2CO_2CH_3 \xrightarrow[\text{(2) heat}]{\text{(1) }H_2\text{ (excess), catalyst}}$

15.42 How would you make the following conversions?

(a) 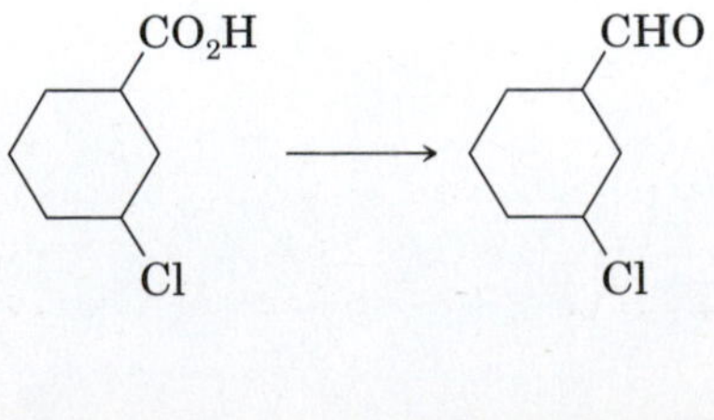 [cyclohexane ring with CO_2H and Cl substituents] $\longrightarrow$ [cyclohexane ring with CHO and Cl substituents]

(b) $(CH_3)_3CCH_2CO_2H \longrightarrow$ $(CH_3)_3CCH_2\overset{OH}{\underset{|}{C}}(CH_3)_2$

(c) [benzamide structure with CNH$_2$] $\longrightarrow$ [N-methylbenzamide structure with C—NHCH$_3$]

(d) $(CH_3)_3CCH_2CO_2H \longrightarrow [(CH_3)_3CCH_2\overset{O}{\overset{\|}{C}}]_2O$

(e) [bicyclic ring]—Br $\longrightarrow$ [bicyclic ring]—CH$_2$Br

(f) $CH_3CH_2CO_2H \longrightarrow CH_3CH_2CH_2NHCH_2CH_3$

(g) $CH_3CH_2CO_2H \longrightarrow CH_3CH_2CH_2CH_2NH_2$

Additional Problems

15.43 Predict the organic products of the LiAlH$_4$ reduction, followed by hydrolysis, of the following compounds: **(a)** dimethyl malonate, $CH_2(CO_2CH_3)_2$; **(b)** propanoyl bromide; **(c)** propanoic anhydride; **(d)** propanenitrile; **(e)** *N*-methylphenylacetamide.

15.44 Suggest a practical method for the synthesis of:

(a) hexanoic acid from 1-bromopentane

(b) 2-hydroxyhexanoic acid from pentanal

(c) β-phenylethylamine $(C_6H_5CH_2CH_2NH_2)$ from benzyl bromide

(d) *N*-cyclohexylacetamide from acetic acid

(e) acetic butanoic anhydride from carboxylic acids

(f) di-*tert*-butyl malonate from malonic acid, $CH_2(CO_2H)_2$

(g) 5-nonanol from compounds containing four or fewer carbons

(h) *trans*-$CH_3(CH_2)_8CH{=}CHCO_2H$ from $CH_3(CH_2)_8CH_2CH_2CO_2H$

(i) triphenylmethanol from benzoic acid

15.45 Compound A (C_3H_5ON) gives a negative Tollens test, but decolorizes a solution of Br$_2$ in CCl$_4$. When A is heated with aqueous NaOH, the fumes have a strong ammoniacal odor.

(a) Suggest a structure for A.

(b) How would you confirm your structure assignment by infrared or ^1H NMR spectroscopy? (Tell what absorption you would look for.)

15.46 2,6-Dimethyl-4-methoxy-1-bromobenzene was converted to a Grignard reagent and then was treated with 2,6-dimethyl-4-methoxybenzoyl chloride. Upon work-up, a tertiary alcohol was *not* obtained; however, another product was obtained in 48% yield.

Suggest a structure for the product and explain why this reaction proceeds in a different manner from that of the usual reaction of a Grignard reagent and an acid halide.

15.47 Suggest mechanisms for the following reactions:

(a) 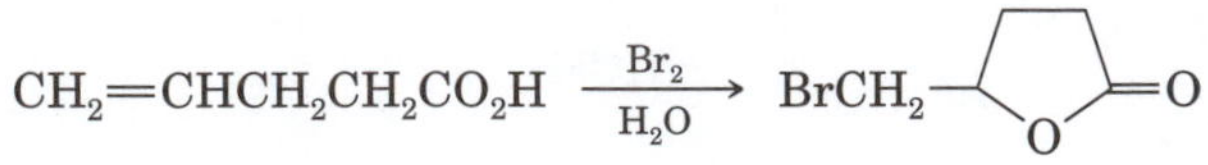 $CH_2{=}CHCH_2CH_2CO_2H \xrightarrow[\text{H}_2\text{O}]{\text{Br}_2}$

(b) 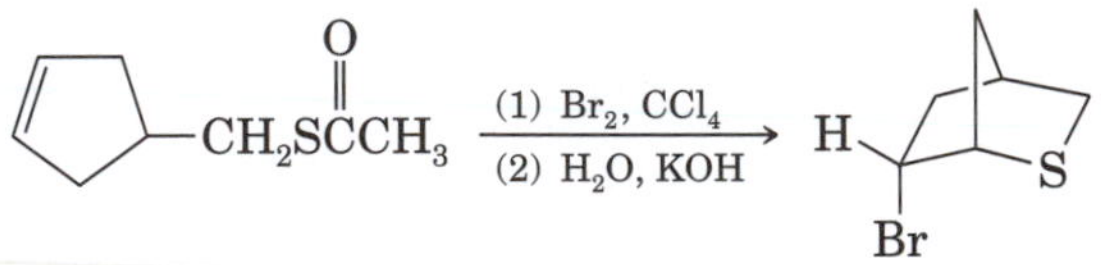 $\xrightarrow[\text{(2) H}_2\text{O, KOH}]{\text{(1) Br}_2\text{, CCl}_4}$

15.48 Predict the structures of the polymers that could be formed from the following monomers:

(a) $HO_2CCH_2CH_2CO_2H + H_2NCH_2CH_2CH_2CH_2NH_2 \xrightarrow{\text{heat}}$ nylon 4,4

(b) $\longrightarrow$ polyalanine

(c) 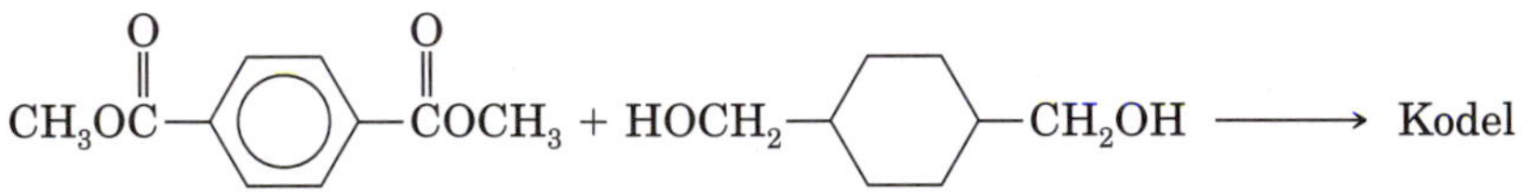 $\longrightarrow$ Kodel

15.49 Hindered esters and phenyl esters can be synthesized in high yield from acid chlorides and the lithium salt or sodium salt of the phenol. Write equations showing how the following phenyl esters can be synthesized from a carboxylic acid and a phenol using this technique.

(a) 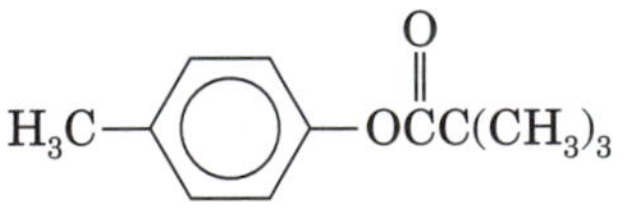   **(b)**

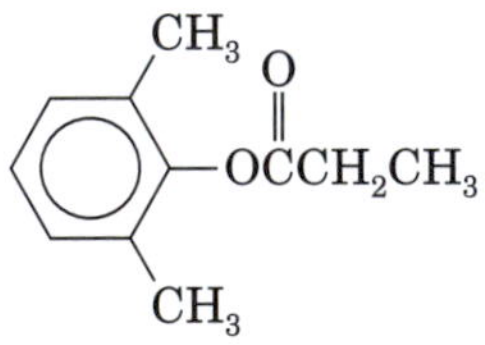

15.50 *Juvenile hormone* is associated with the pupal development of many insects. Predict the products of hydrolysis in aqueous NaOH of this compound.

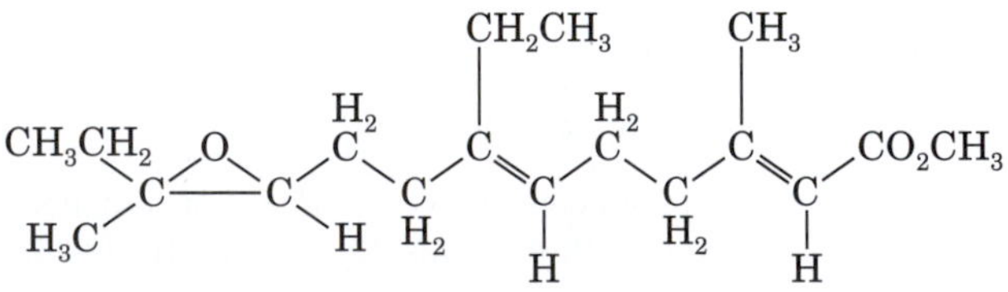

juvenile hormone

15.51 Suggest syntheses for the following compounds:

(a) butanal from compounds containing three or fewer carbons

(b) $NCCH_2CO_2CH_2CH_3$ from acetic acid and other appropriate reagents

(c) ⬡—$\overset{\overset{\text{O}}{\|}}{\text{C}}$N(CH$_3$)$_2$ from cyclohexanecarboxylic acid

(d)   ethyl 4-bromobutanoate   from a lactone

(e)   CH$_3$O$_2$CCH$_2\overset{\overset{\text{O}}{\|}}{\text{C}}CH_2CO_2$H   from a diacid

(f)   [cyclohexene with Br and CN(CH$_3$)$_2$ (C=O)]   from   [cyclohexene with Br and CHO]

15.52 Determine the structures of the following compounds:

(a) Compound A (C$_7$H$_{12}$O$_4$) shows absorption at 3000 cm^{-1} (3.33 μm) and 1750 cm^{-1} (5.71 μm) in the infrared spectrum. Its ^{1}H NMR spectrum shows a triplet (δ1.25), a singlet (δ3.3), and a quartet (δ4.1). The area ratio of these signals is 3:1:2.

(b) Compound B (C$_6$H$_{10}$O$_3$) shows its principal infrared absorption at 3000 cm^{-1} (3.33 μm), 1750 cm^{-1} (5.71 μm), and 1725 cm^{-1} (5.79 μm). The ^{1}H NMR spectrum shows the following peaks and relative areas: triplet (δ1.25), singlet (δ2.25), singlet (δ3.55), and quartet (δ4.15). The area ratios are 3:3:2:2.

(c) Compound C (C$_4$H$_7$NO) shows infrared absorption at 2920 cm^{-1} (3.42 μm), 2230 cm^{-1} (4.48 μm), 1370–1450 cm^{-1} (6.9–7.3 μm) and 1110 cm^{-1} (9.0 μm). The ^{1}H NMR spectrum shows two triplets (δ2.6 and 3.55) and a singlet (δ3.35), with area ratios of 2:2:3.

15.53 Explain why the ^{13}C NMR spectrum of *N*-ethylformamide (Figure 15.16) exhibits six peaks instead of three. (See Section 9.12.)

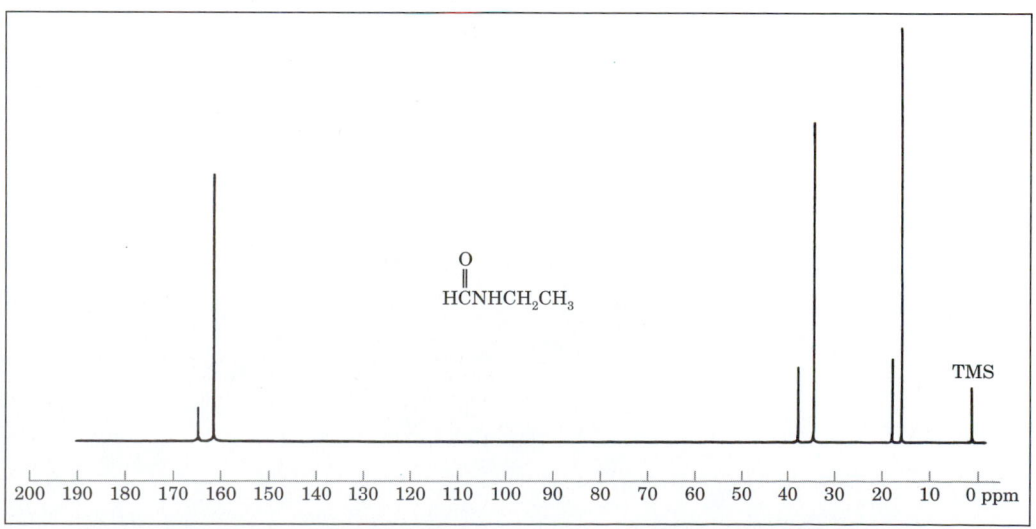

Figure 15.16 ^{13}C NMR spectrum of *N*-ethylformamide for Problem 15.53.

15.54 The ¹H NMR spectrum of Compound A is shown in Figure 15.17. When A is heated with aqueous acid, the products are acetic acid and acetaldehyde. What is the structure of A?

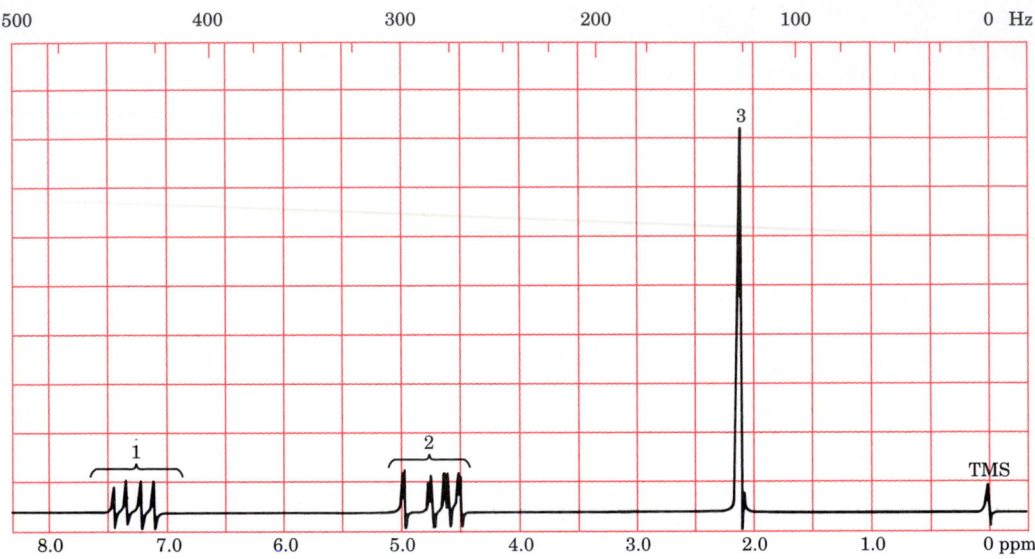

Figure 15.17 ¹H NMR spectrum for Problem 15.54.

15.55 Compound A ($C_{10}H_{18}O_3$) was treated with dilute aqueous acid to yield the single compound B. When A was heated with ethanol and a trace of H_2SO_4, C was obtained as the only product. The infrared spectra of A and B and the 300 MHz ¹H NMR spectrum of C are shown in Figure 15.18. What are the structures of A, B, and C?

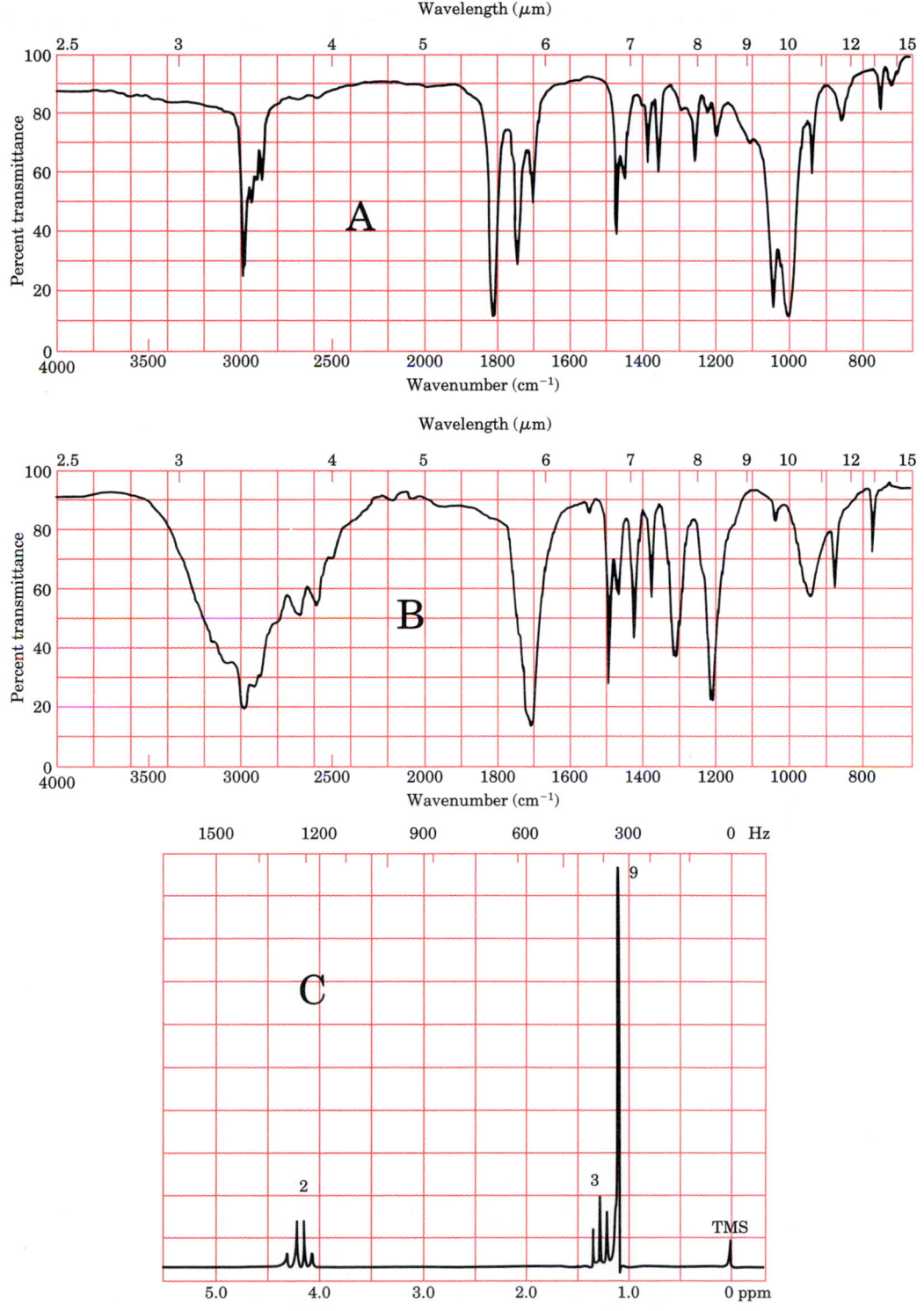

Figure 15.18 Spectra for Problem 15.55.

16 CONJUGATE ADDITION

In Chapter 2, we defined isolated and conjugated dienes (Section 2.7). We stated that conjugated dienes are resonance-stabilized and that they can react as a unit rather than just individually. In this chapter, we will expand the concept of conjugated dienes to conjugated π bonds and discuss their reactivity.

Section 16.1

Conjugated Pi Bonds

Conjugated carbon–carbon double bonds are defined as double bonds originating on adjacent carbon atoms. Isolated double bonds are defined as double bonds that do not originate on adjacent carbon atoms. Conjugated double bonds are separated by only one single bond while isolated double bonds are separated by two or more single bonds.

Conjugated double bonds

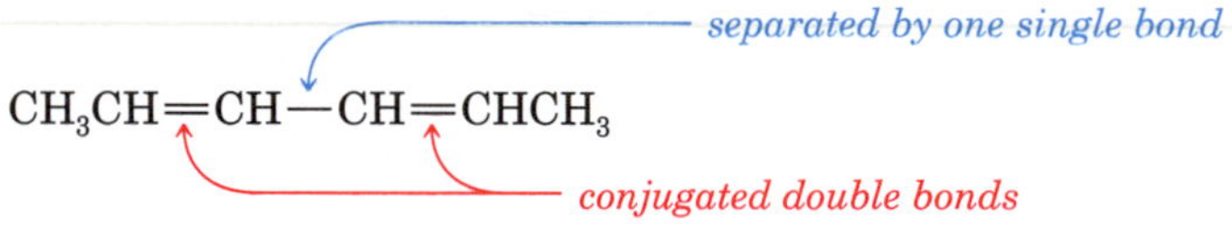

separated by one single bond

$$CH_3CH{=}CH{-}CH{=}CHCH_3$$

conjugated double bonds

Isolated double bonds

separated by two or more single bonds

$$CH_2{=}CH{-}CH_2{-}CH{=}CH_2$$

isolated double bonds

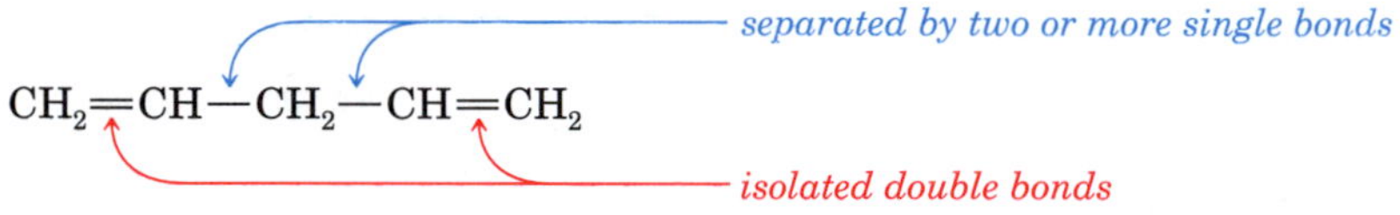

A compound can have any number of conjugated and isolated double bonds. γ-Carotene, a rare vitamin-A precursor, has eleven conjugated double bonds and one isolated double bond.

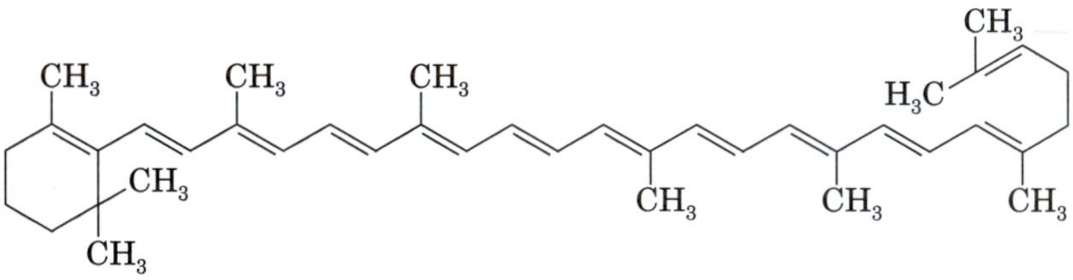

11 conjugated double bonds; 1 isolated double bond

A carbon–carbon double bond also can be conjugated with the π bond of a carbonyl group in an aldehyde, ketone, carboxylic acid, or carboxylic acid derivative.

Carbonyl compounds with conjugated π bonds

$$CH_2{=}CH{-}\overset{\overset{\textstyle O}{\|}}{C}{-}H \qquad CH_3CH{=}CH{-}\overset{\overset{\textstyle O}{\|}}{C}OCH_3$$

Isolated double bonds can only react individually. Each double bond of an isolated diene is independent of the other. Except in rare cases, there is no interaction between isolated double bonds.

Like isolated double bonds, each double bond of a conjugated diene can react by itself, independent of the other double bond. This type of addition is called **1,2-addition.** In this terminology, the 1 and 2 refer to the two adjacent carbons of one of the double bonds.

1,2-Addition

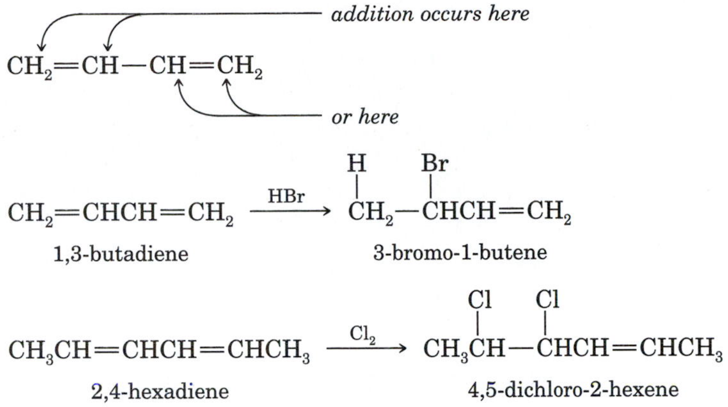

$$CH_2\!=\!CH\!-\!CH\!=\!CH_2$$

— addition occurs here

— or here

$$CH_2\!=\!CHCH\!=\!CH_2 \xrightarrow{\text{HBr}} \overset{\overset{\text{H}}{|}}{CH_2}\!-\!\overset{\overset{\text{Br}}{|}}{CH}CH\!=\!CH_2$$

1,3-butadiene   3-bromo-1-butene

$$CH_3CH\!=\!CHCH\!=\!CHCH_3 \xrightarrow{\text{Cl}_2} CH_3\overset{\overset{\text{Cl}}{|}}{CH}\!-\!\overset{\overset{\text{Cl}}{|}}{CH}CH\!=\!CHCH_3$$

2,4-hexadiene   4,5-dichloro-2-hexene

STUDY PROBLEM

16.1   Predict the product of 1,2-addition to the second double bond in each of the two preceding examples.

Conjugated π systems (dienes and conjugated carbonyl compounds) also can undergo reactions in which the entire π system reacts. This type of addition is called 1,4-addition. In this terminology, 1 and 4 refer to the first and fourth carbons of the conjugated π system. 1,4-Addition is also called **conjugate addition.**

1,4-Addition, also called conjugate addition

$$CH_2\!=\!CH\!-\!CH\!=\!CH_2$$

— addition occurs here

$$CH_2\!=\!CH\!-\!\overset{\overset{\text{O}}{\|}}{CH}$$

$$CH_2\!=\!CHCH\!=\!CH_2 \xrightarrow{\text{HBr}} \overset{\overset{\text{H}}{|}}{CH_2}\!-\!CH\!=\!CH\!-\!\overset{\overset{\text{Br}}{|}}{CH_2}$$

1,3-butadiene   1-bromo-2-butene

$$CH_3CH\!=\!CHCH\!=\!CHCH_3 \xrightarrow{\text{Cl}_2} CH_3\overset{\overset{\text{Cl}}{|}}{CH}\!-\!CH\!=\!CH\!-\!\overset{\overset{\text{Cl}}{|}}{CH}CH_3$$

2,4-hexadiene   2,5-dichloro-3-hexene

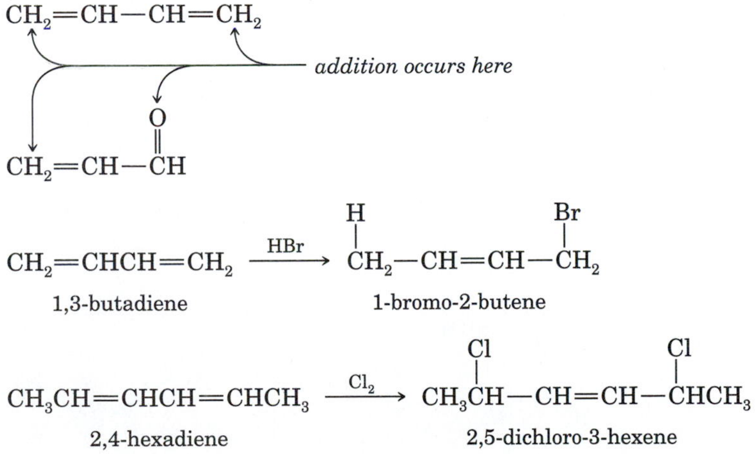

STUDY PROBLEMS

16.2   Each of the following structures contains conjugated π bonds. Identify the first and fourth atoms in each conjugated system.

(a)   (b)

(c)   (d)

16.3   A pair of conjugated π bonds has two 1,2-positions. In the following structures, are the two 1,2-positions equivalent?

(a) $CH_3CH{=}CH{-}CH{=}CHCH_3$ (b)

(c)   (d)

To see why conjugated double bonds have unique chemical properties, we must look at where the p orbitals are located in a conjugated diene, such as 1,3-butadiene.

From a pictorial standpoint (Figure 16.1), the inner two p orbitals are positioned on adjacent carbon atoms and are partially overlapped. The partial overlap allows electron density of the diene to be delocalized, thus conferring stability to the system.

From a molecular-orbital standpoint, conjugated dienes have four π electrons and four π molecular orbitals. Two of the four π molecular orbitals are

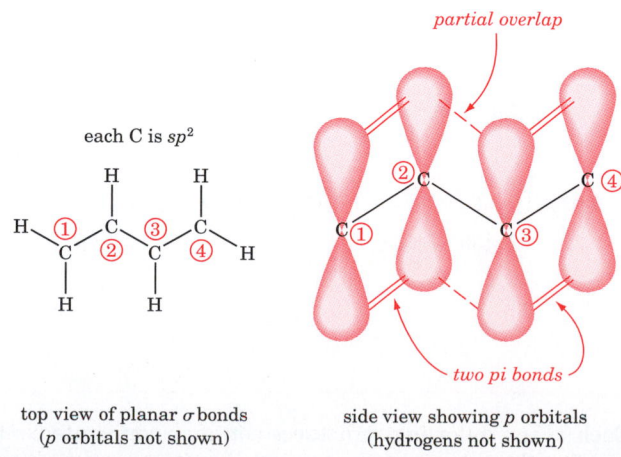

partial overlap

each C is sp^2

two pi bonds

top view of planar σ bonds
(p orbitals not shown)

side view showing p orbitals
(hydrogens not shown)

Figure 16.1 The bonding in 1,3-butadiene, $CH_2{=}CH{-}CH{=}CH_2$.

bonding orbitals and two are antibonding orbitals. One bonding molecular orbital encompasses the entire π system. The other bonding molecular orbital has an additional node and is higher in energy. The bonding and antibonding molecular orbitals of 1,3-butadiene are shown in Figure 16.2. The π-electron diagrams for a conjugated and an isolated diene follow.

π-electron diagrams for a conjugated and an isolated diene

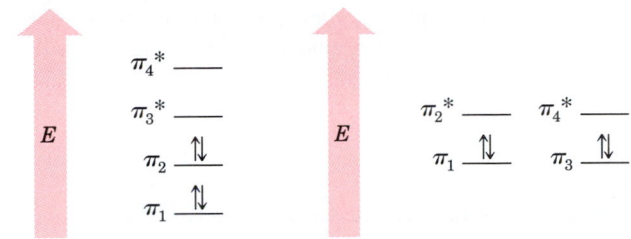

two conjugated double bonds   *two isolated double bonds*

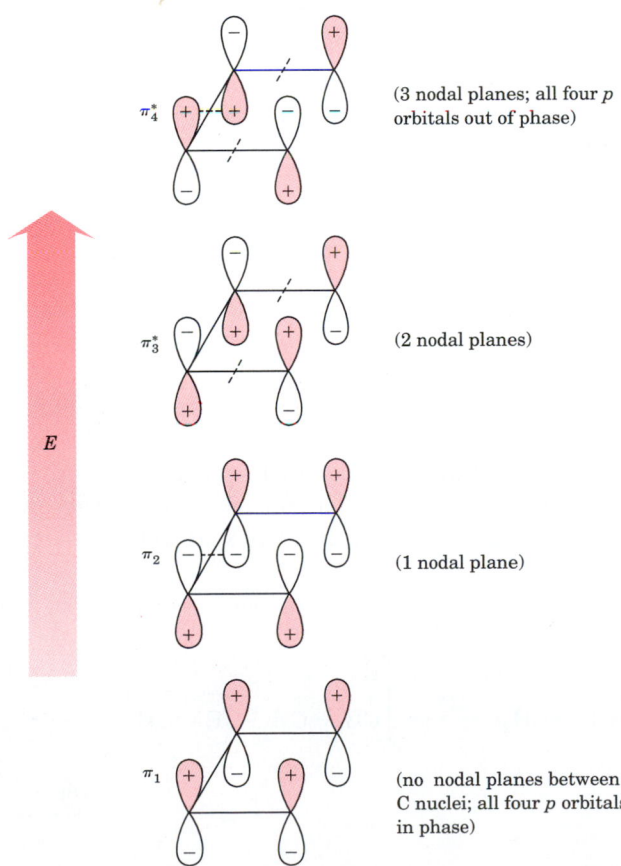

Figure 16.2 The bonding and antibonding π molecular orbitals of 1,3-butadiene, $CH_2{=}CHCH{=}CH_2$. The π_1 and π_2 orbitals are bonding orbitals; $\pi_3{}^*$ and $\pi_4{}^*$ are antibonding orbitals.

STUDY PROBLEMS

16.4   Draw the π-electron diagram for the following conjugated triene.

$$CH_2{=}CH{-}CH{=}CH{-}CH{=}CH_2$$

16.5   **(a)** Draw the π-electron diagram for $CH_2{=}CH{-}\overset{\overset{\textstyle O}{\textstyle \|}}{C}H$. **(b)** Draw a second diagram that includes the nonbonding (n) electrons on the oxygen.

Section 16.2

1,2-Addition and 1,4-Addition to Conjugated Dienes

Many of the reactions of conjugated dienes are the same as those of compounds with isolated double bonds. Acidic reagents and the halogens can add across one or both of the π bonds.

A. Mechanisms of 1,2- and 1,4-Addition

Let us look at the mechanism of each type of addition. The mechanism for 1,2-addition of **HX** is the same as that for addition to an isolated double bond. (The reaction of 1,3-butadiene goes through the more stable allylic carbocation and not through the less stable $^+CH_2CH_2CH{=}CH_2$.)

1,2-Addition:

$$CH_2{=}CHCH{=}CH_2 \xrightarrow{\;H^+\;} \left[\overset{\overset{\textstyle H}{\textstyle |}}{C}H_2{-}\overset{+}{C}HCH{=}CH_2 \right] \xrightarrow{\;:\ddot{Br}:^-\;} \overset{\overset{\textstyle H}{\textstyle |}}{C}H_2{-}\overset{\overset{\textstyle :\ddot{Br}:}{\textstyle |}}{C}HCH{=}CH_2$$

$$\text{a 2° allylic carbocation} \qquad\qquad \text{3-bromo-1-butene}$$

The mechanism for 1,4-addition is a direct extension of that for 1,2-addition. The carbocation in the preceding example is an *allylic cation* (Section 5.6A) and is resonance-stabilized. Because of the resonance stabilization of the allylic cation, there is a partial positive charge on carbon 4 of the diene system as well as on carbon 2. Attack at carbon 4 leads to the 1,4-addition product.

1,4-Addition:

$$CH_2{=}CHCH{=}CH_2 \xrightarrow{\;H^+\;} \left[\overset{\overset{\textstyle H}{\textstyle |}}{C}H_2{-}\overset{+}{C}H{-}CH{=}CH_2 \longleftrightarrow \overset{\overset{\textstyle H}{\textstyle |}}{C}H_2{-}CH{=}CH{-}\overset{+}{C}H_2 \right]$$

$$\xrightarrow[\;:\ddot{Br}:^-\;]{} \overset{\overset{\textstyle H}{\textstyle |}}{C}H_2{-}CH{=}CH{-}\overset{\overset{\textstyle :\ddot{Br}:}{\textstyle |}}{C}H_2$$

$$\text{1-bromo-2-butene}$$

If an equimolar quantity of reagent is added to 1,3-butadiene, a mixture of two products results: 3-bromo-1-butene from 1,2-addition, and 1-bromo-2-butene from 1,4-addition.

SAMPLE PROBLEM

(a) Write the structures of *all possible* carbocation intermediates in the addition of 1.0 mol of HI to 1.0 mol of 2,4-hexadiene.

(b) Which carbocation would you expect to be formed at the faster rate?

Solution

(a) $CH_3CH{=}CH{-}CH{=}CHCH_3$ $\xrightarrow{\ H^+\ }$

$[CH_3CH_2{-}\overset{+}{C}H{-}CH{=}CHCH_3]$ or $[CH_3\overset{+}{C}H{-}CH_2{-}CH{=}CHCH_3]$

(Addition of H^+ to the other double bond gives identical intermediates.)

(b) The first carbocation shown would be formed at the faster rate because it is a resonance-stabilized allylic carbocation.

STUDY PROBLEMS

16.6   Write the *resonance structures* for the principal intermediate in the preceding sample problem, and give the structures of the principal products.

16.7   Predict the products of the addition of an equimolar quantity of bromine to the following dienes. Write resonance structures for any intermediates formed.

(a)   **(b)**

In the reaction of 1,3-butadiene with an equimolar quantity of HBr, the ratio of 1,2-addition to 1,4-addition varies with the temperature at which the reaction is carried out. At $-80°$, the 1,2-addition product predominates. At $40°$, the 1,4-addition product predominates.

$$CH_2{=}CHCH{=}CH_2 + HBr$$
1,3-butadiene

$\xrightarrow{-80°}$

$\overset{\displaystyle Br}{\underset{\displaystyle |}{CH_3CHCH{=}CH_2}} + CH_3CH{=}CHCH_2Br$

1,2-product   *1,4-product*
(80%)   (20%)

$\xrightarrow{40°}$

$\overset{\displaystyle Br}{\underset{\displaystyle |}{CH_3CHCH{=}CH_2}} + CH_3CH{=}CHCH_2Br$

(20%)   (80%)

It has also been observed that warming 3-bromo-1-butene (the 1,2-addition product) to $40°$ with a trace of acid results in an equilibrium mixture that is predominantly 1-bromo-2-butene (the 1,4-addition product).

$$\overset{\displaystyle Br}{\underset{\displaystyle |}{CH_3CHCH{=}CH_2}} \underset{}{\overset{H^+,\, 40°}{\rightleftharpoons}} CH_3CH{=}CHCH_2Br$$

(20%)   (80%)

How can we explain these observations? At low temperature, the reaction yields predominantly the 1,2-addition product because the 1,2-addition has

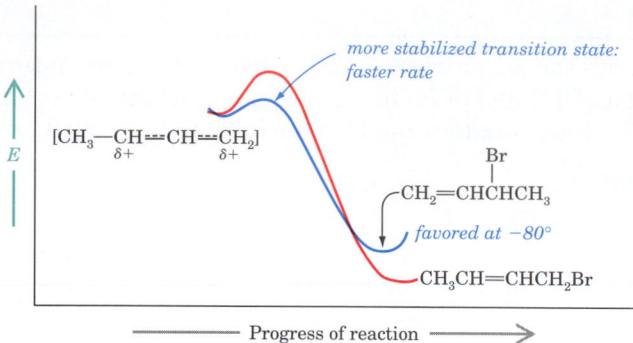

Figure 16.3 Partial energy diagram for the reaction of 1,3-butadiene with HBr. At −80°, the 1,2-addition product predominates.

the lower E_{act} (the 2° carbon carries more + charge than the 1° carbon) and thus the faster rate. The relative rates of the reactions control the product ratio. The reaction is under *kinetic control* at low temperatures. Figure 16.3 shows the energy diagram for the competing reactions.

At higher temperatures, a greater percentage of molecules can reach the higher-energy transition state, and the two products are in *equilibrium*. The more stable **1,4-product** (which is the more substituted alkene) predominates. At the higher temperatures, the relative stabilities of the products control the product ratios, and the reaction is under *thermodynamic control*. Figure 16.4 shows the energy diagram for the equilibrium.

STUDY PROBLEMS

16.8 Which *geometric isomer* of 1-bromo-2-butene would predominate at 40°?

16.9 1,3-Butadiene is treated with an equimolar quantity of Br_2 at −15°. Two structural isomers are obtained: 46% A and 54% B. When the reaction is carried out at 60°, the product mixture contains 90% A. What are the structures of A and B?

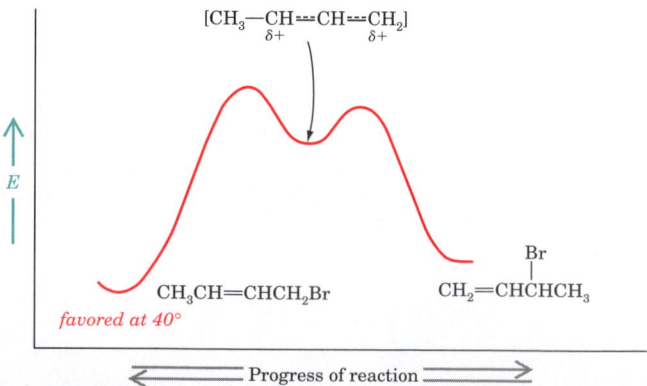

Figure 16.4 Energy diagram for the equilibrium between the 1,2- and 1,4-addition products. At 40°, the 1,4-addition product predominates.

B. 1,4-Addition Polymers

Conjugated dienes can be polymerized by 1,4-addition. The product still contains unsaturation; therefore, the polymer could contain all *cis* units, all *trans* units, or a mixture of *cis* and *trans* units. The following equation shows the 1,4-polymerization of isoprene.

$$x \; CH_2{=}\overset{\overset{\textstyle CH_3}{|}}{C}{-}CH{=}CH_2 \;\xrightarrow[\text{1,4-addition}]{\text{catalyst}}\; \left(CH_2\overset{\overset{\textstyle CH_3}{|}}{C}{=}CHCH_2{-}CH_2\overset{\overset{\textstyle CH_3}{|}}{C}{=}CHCH_2\right)_x$$

2-methyl-1,3-butadiene   polyisoprene
(isoprene)

Natural rubber is polyisoprene with *cis* double bonds. The *trans* polymer, called **gutta-percha,** is a hard polymer used as a covering for golf balls and in temporary dental fillings.

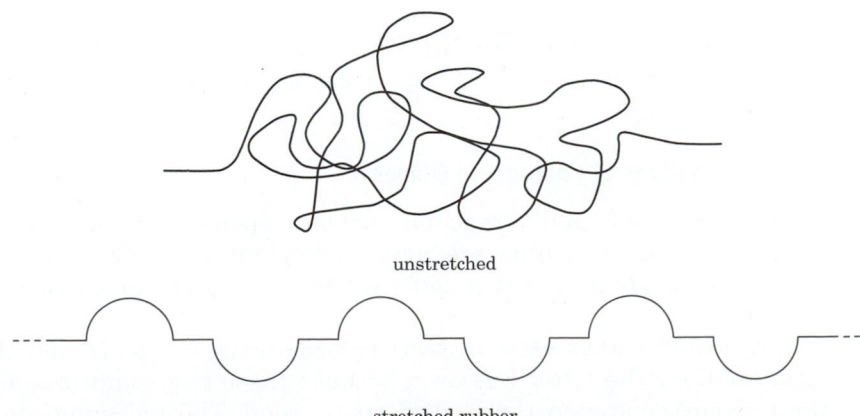

natural rubber (all *cis*)

gutta-percha (all *trans*)

The *cis* double bonds in natural rubber force the molecules into a distorted conformation that can be stretched and yet return to its original shape. The *trans* double bonds in gutta-percha allow the polymer molecules to pack more closely to one another and form a more crystalline, nonelastic substance.

Rubber is elastic because *cis*-polyisoprene molecules tend to coil, and stretching uncoils them. Unstretched rubber is a disorderly arrangement of coiled molecules (see Figure 16.5). Stretched rubber has a nearly linear and

unstretched

stretched rubber

Figure 16.5 The structure of unstretched and stretched rubber, polyisoprene.

orderly arrangement. The tendency for things to become disorderly is the reason stretched rubber spontaneously returns to its original unstretched condition.

Neither natural rubber nor gutta-percha is formed in nature by the direct polymerization of isoprene, as was once believed. Instead, they are formed by a biosynthetic path starting from acetylcoenzyme A. (See Section 24.5C.)

STUDY PROBLEM

16.10 *Neoprene,* developed in 1932, was the first synthetic rubber. It is used in washers, gaskets, and the like. Unlike natural rubber, neoprene does not swell when in contact with petroleum products. Therefore, it is also used extensively in the fabrication of petroleum containers, such as gasoline tanks. Neoprene is the all *trans,* head-to-tail polymer formed by a 1,4-addition of 2-chloro-1,3-butadiene. What is the structure of this polymer?

Section 16.3
The Diels–Alder Reaction

One very important type of 1,4-addition is represented by the **Diels–Alder reaction,** which is a route to cyclohexenes. The reaction is named after the German chemists Otto Diels and Kurt Alder, who jointly received the 1950 Nobel Prize for their work in this area. In a Diels–Alder reaction, a conjugated diene is heated with a second unsaturated compound, called the **dienophile** ("diene lover"), to yield a product containing a six-membered ring.

1,3-butadiene	propenal	3-cyclohexene-1-carbaldehyde
the diene	*the dienophile*	(100%)

The Diels–Alder reaction is but one example of a larger class of reactions called **pericyclic reactions.** In this section, we will limit our discussion to the Diels–Alder reaction. In Chapter 21, we will discuss the theory of pericyclic reactions.

A. Conformations of Conjugated Dienes

Because the Diels–Alder reaction converts open-chain compounds to cyclic compounds, line formulas are very convenient for representing the open-chain compounds in this reaction. (See Section 10.2B for a discussion of line formulas.)

Conjugated dienes exist in two conformations, the **s-cis** and the **s-trans** conformation. The letter *s* is used because these two conformations differ in their geometry around the central *single* bond. The following compounds illustrate the use of the terms. For open-chain compounds, these formulas represent interconvertible conformers because σ bond rotation (requiring about

4 kcal/mol for 1,3-butadiene) is all that is needed for the conversion from one to another.

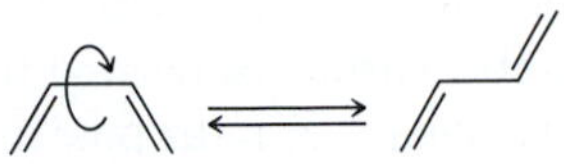

s-*cis*   s-*trans*

s-*cis*

even though the
stereochemistry at
each double bond is
(E), or trans

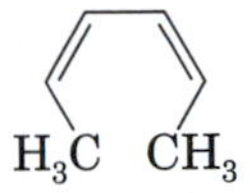

s-*cis*
and (Z), or
cis, at each
double bond

When the diene function is part of a cyclic system, the s-*cis* and s-*trans* structures represent different compounds. Interconversion cannot occur without bonds being broken.

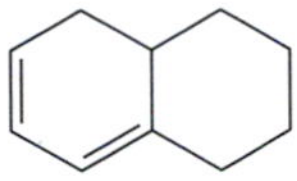

"s-*cis*" 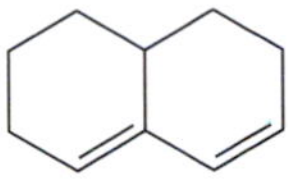   "s-*trans*"

STUDY PROBLEM

16.11 Classify each of the following dienes as s-*cis* or s-*trans*. Indicate which is interconvertible with the other form.

(a) 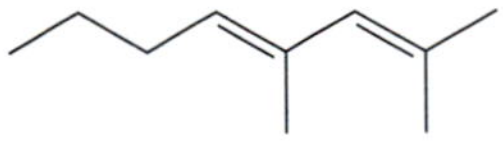   **(b)**    **(c)**

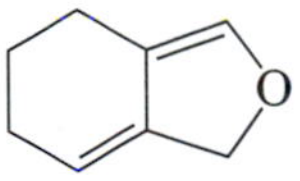

In a Diels–Alder reaction, the diene must have the s-*cis*, not the s-*trans*, conformation. In some structures (such as 1,3-butadiene), the s-*cis* and s-*trans* conformers are readily interconvertible. In other diene systems (such as in ring systems), the s-*trans* isomer does not undergo reaction.

Some s-cis dienes that can be used in a Diels–Alder reaction:

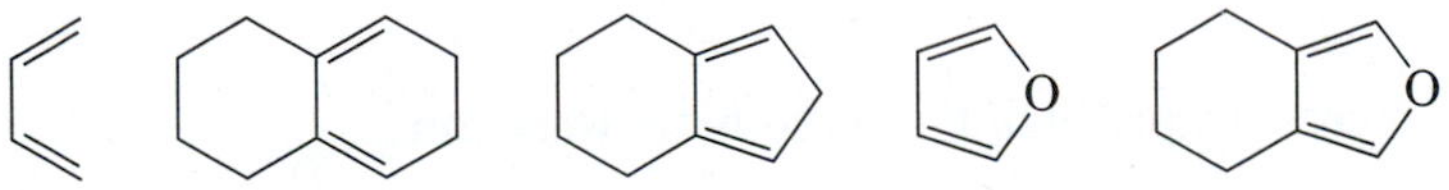

B. Examples of Diels–Alder Reactions

Using the line-formula symbolism, we can show the Diels–Alder reaction that was presented on page 698 as:

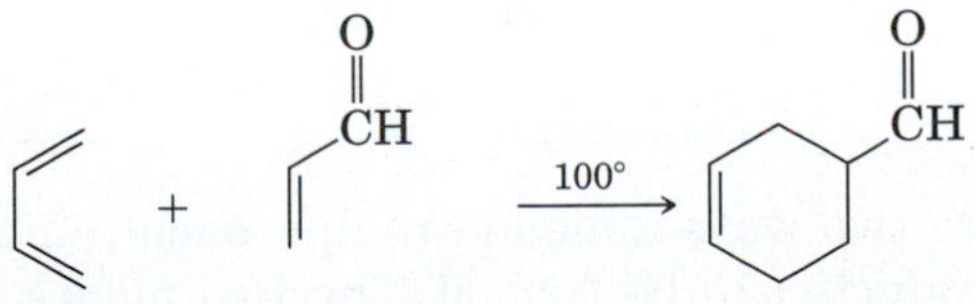

the diene   the dienophile   the product

The dienophile usually contains other unsaturation (an aldehyde group in the preceding example) that does not participate directly in the addition reaction. This group does, however, enhance the reactivity of the dienophile's carbon–carbon double bond (the site of reaction) by electron withdrawal. (Remember, the carbon in C—O carries a partial positive charge.)

You can see from the following examples the versatility of the Diels–Alder reaction for the synthesis of cyclic compounds containing cyclohexene rings. For practice, find the cyclohexene rings in the following examples.

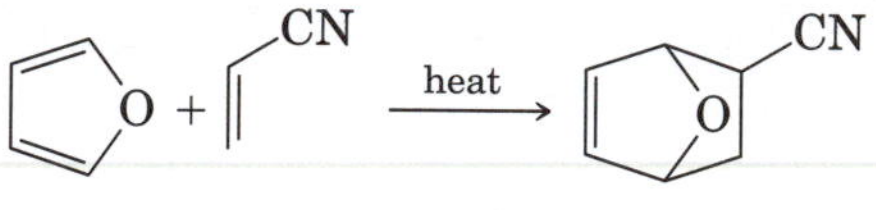

(39%)

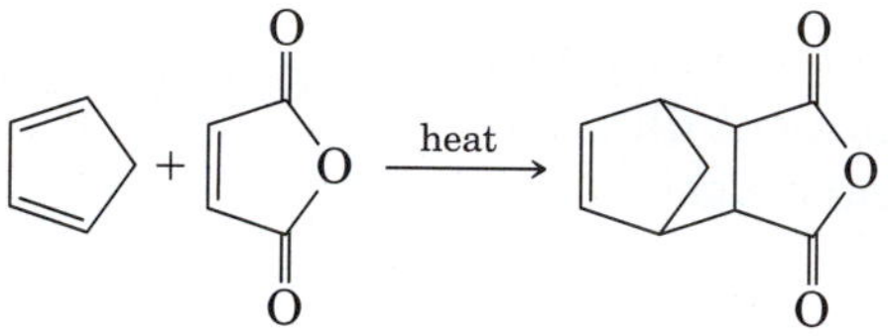

(100%)

STUDY PROBLEMS

16.12 Predict the Diels–Alder products:

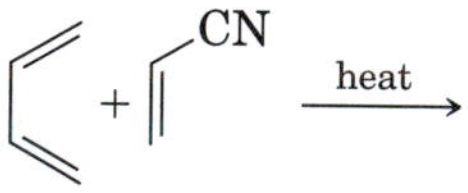

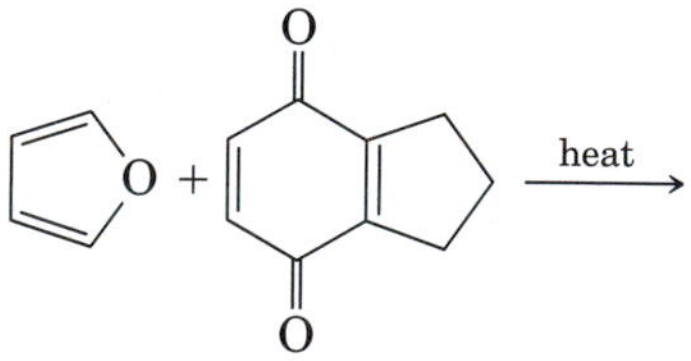

16.13 Suggest a synthesis for the following compound. (*Hint:* An alkyne may be used as a dienophile.)

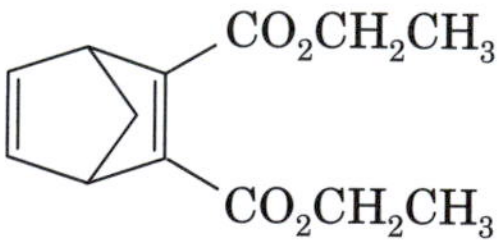

C. Stereochemistry of the Diels–Alder Reaction

A Diels–Alder reaction is a *cis,* or *syn*-addition, and is thus stereospecific.

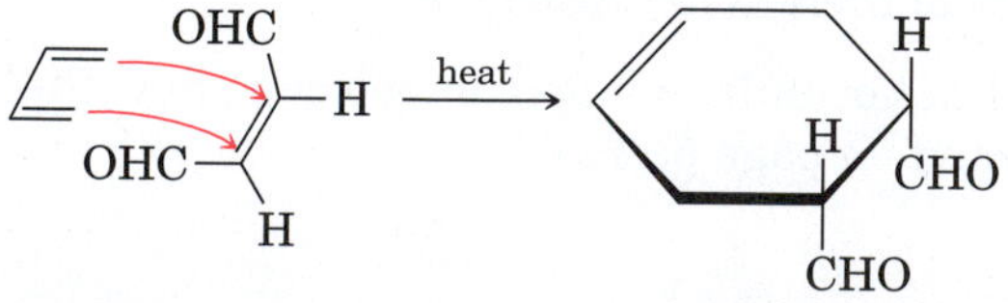

a *cis*-dienophile   a *cis*-product

There are two ways a diene can approach a dienophile. In some cases, two possible products can be formed. The two modes of addition are called ***endo*** (inside) and **exo** (outside), as the following examples show. The *endo* addition is usually favored, probably because of favorable interactions of the π orbitals of the developing double bond and the π orbitals in the unsaturated group. In the *endo* product of our example, the carbonyl group of the dienophile is *trans* to the bridge. In the *exo* product, the dienophile C=O is *cis* to the bridge.

STUDY PROBLEMS

16.14   Predict the product (including stereochemistry):

16.15   The following Diels–Alder reaction can yield two isomeric cyclohexenes. What are their structures?

Section 16.4
1,4-Addition to α,β-Unsaturated Carbonyl Compounds

A. Electrophilic 1,4-Addition

When an alkene reacts with HCl, the reaction proceeds by electrophilic attack of H^+ to yield the more stable carbocation, followed by attack of Cl^-.

An *α,β-unsaturated* aldehyde or ketone has a carbon–carbon double bond in conjugation with a carbonyl group. The carbon–carbon double bond in an alkene is nonpolar. However, a carbon–carbon double bond in conjugation with a carbonyl group is *polar*, as the following resonance structures indicate.

$$CH_2{=}CH{-}\overset{\overset{\ddot{O}:}{\|}}{C}H \longleftrightarrow CH_2{=}CH{-}\overset{\overset{:\ddot{O}:^-}{|}}{\underset{+}{C}}H \longleftrightarrow {^+}CH_2{-}CH{=}\overset{\overset{:\ddot{O}:^-}{|}}{C}H$$

The resonance structures show that the β carbon, as well as the carbonyl carbon, carries a partial positive charge, while the carbonyl oxygen carries a partial negative charge.

$$CH_2{=}CH{-}\overset{\overset{O^{\delta-}}{\|}}{C}H$$

β carbon is $\delta+$ ⟋   ⟍ *carbonyl carbon is $\delta+$*

Because the C=C grouping in an α,β-unsaturated carbonyl compound is polarized, the mechanism for electrophilic addition is somewhat different from that for electrophilic addition to an isolated, nonpolar, alkene double bond. Let us consider two examples of electrophilic addition reactions of α,β-unsaturated carbonyl compounds. Then we will discuss the mechanism.

$$CH_2{=}CH\overset{\overset{O}{\|}}{C}H + HCl \longrightarrow \overset{\overset{Cl}{|}}{C}H_2CH_2\overset{\overset{O}{\|}}{C}H$$

propenal   3-chloropropanal

$$CH_3CH{=}CH\overset{\overset{O}{\|}}{C}CH_3 + H_2O \xrightarrow{\ H^+\ } CH_3\overset{\overset{OH}{|}}{C}HCH_2\overset{\overset{O}{\|}}{C}CH_3$$

3-penten-2-one   4-hydroxy-2-pentanone

In each reaction, the nucleophilic part of the reagent (not H^+) becomes bonded to the β carbon. The reason for this is that the β carbon has a partial positive charge. The initial attack by H^+ occurs not at this positive carbon, but at the partially negative oxygen of the carbonyl group.

$$\overset{\delta+}{C}H_2{=}CH\overset{\overset{\delta-\ddot{O}:\ \curvearrowright H^+}{\|}}{C}H$$

The protonated intermediate is resonance-stabilized. In this intermediate, the β carbon still carries a partial positive charge and can be attacked by a nucleophile.

Protonation:

$$CH_2{=}CH\overset{\overset{\ddot{O}:}{\|}}{C}H \overset{H^+}{\rightleftharpoons} \left[CH_2{=}CH\overset{\overset{{^+}\ddot{O}H}{\|}}{C}H \longleftrightarrow CH_2{=}CH{-}\overset{\overset{:\ddot{O}H}{|}}{\underset{+}{C}}H \longleftrightarrow {^+}CH_2CH{=}\overset{\overset{:\ddot{O}H}{|}}{C}H \right]$$

resonance-stabilized

Attack of Nu:⁻:

$$\left[\text{⁺CH}_2\text{CH}{=}\overset{\overset{\text{OH}}{|}}{\text{CH}} \right] + \text{:}\overset{..}{\underset{..}{\text{Cl}}}\text{:}^- \longrightarrow \left[\text{:}\overset{..}{\underset{..}{\text{Cl}}}\text{CH}_2\text{CH}{=}\overset{\overset{\text{OH}}{|}}{\text{CH}} \right] \rightleftharpoons \text{ClCH}_2\text{CH}_2\overset{\overset{\text{O}}{\|}}{\text{CH}}$$

an enol of an aldehyde   an aldehyde

Note that this reaction is a **1,4-addition,** or **conjugate addition,** the same type of addition observed with conjugated dienes. The difference is that the initial addition product is an enol, which undergoes tautomerization to the final keto form of the aldehyde.

You might wonder why the nucleophile cannot attack the carbonyl carbon (which also carries a partial positive charge in the intermediate). This attack on the carbonyl carbon can occur, but the product is unstable and reverts to starting material. This is a concurrent, but nonproductive, side reaction.

$$\left[\text{CH}_2{=}\text{CH}\overset{\overset{\text{:O̤H}}{|}}{\underset{+}{\text{CH}}} \right] + \text{:}\overset{..}{\underset{..}{\text{Cl}}}\text{:}^- \rightleftharpoons \left[\text{CH}_2{=}\text{CHCH}{-}\overset{..}{\underset{..}{\text{Cl}}}\text{:} \right] \rightleftharpoons \text{CH}_2{=}\text{CH}\overset{\overset{\text{:O̤:}}{\|}}{\text{CH}} + \text{H}\overset{..}{\underset{..}{\text{Cl}}}\text{:}$$

B. Nucleophilic 1,4-Addition

The π bond of an alkene is not normally attacked by a nucleophile unless there has been prior attack by an electrophile. However, a double bond in conjugation with a carbonyl group is polarized. In this case, nucleophilic addition can occur *either at the C=C double bond or at the C=O double bond* (at either of the two partially positive carbons).

$$\text{CH}_2{=}\text{CH}\overset{\overset{\text{O}}{\|}}{\text{C}}\text{CH}_3 + {}^-\text{CN} \xrightarrow{\text{HCN}} \text{CH}_2{=}\text{CH}\overset{\overset{\text{OH}}{|}}{\underset{\underset{\text{CN}}{|}}{\text{C}}}\text{CH}_3 \text{ or } \overset{}{\underset{\underset{\text{CN}}{|}}{\text{CH}_2}}{-}\text{CH}_2\overset{\overset{\text{O}}{\|}}{\text{C}}\text{CH}_3$$

Let us look at the mechanism of each reaction. First, we will consider the attack of the cyanide ion (from HCN and dilute base) on the carbonyl group. In this case, the nucleophilic CN⁻ attacks the partially positive carbon of the carbonyl group. The reaction is no different from cyanohydrin formation by an ordinary ketone.

Attack of CN⁻ on the carbonyl carbon:

$$\text{CH}_2{=}\text{CH}\overset{\overset{\text{:O̤:}}{\|}}{\text{C}}\text{CH}_3 + {}^-\text{:CN:} \longrightarrow \left[\text{CH}_2{=}\text{CH}\overset{\overset{\text{:O̤:}^-}{|}}{\underset{\underset{\text{CN:}}{|}}{\text{C}}}\text{CH}_3 \right] \overset{\text{HCN}}{\rightleftharpoons} \text{CH}_2{=}\text{CH}\overset{\overset{\text{:O̤H}}{|}}{\underset{\underset{\text{CN}}{|}}{\text{C}}}\text{CH}_3 + \text{CN}^-$$

Now we will consider attack of the nucleophilic CN⁻ on the β carbon. This reaction is a 1,4-addition of CN⁻ and H⁺ to the conjugated system. The product of the 1,4-addition is an enol, which forms the product ketone.

Attack of CN⁻ on the β carbon:

$$CH_2\!\!=\!\!CH\!\!-\!\!\overset{\overset{\displaystyle ::\!O::}{\|}}{C}CH_3 \;\rightleftharpoons\; \left[CH_2\!\!-\!\!\overset{\displaystyle ::}{C}H\!\!-\!\!\overset{\overset{\displaystyle ::\!O::}{\|}}{C}CH_3 \;\longleftrightarrow\; CH_2\!\!-\!\!CH\!\!=\!\!\overset{\overset{\displaystyle ::\!O::^-}{\|}}{C}CH_3 \right] \overset{HCN}{\rightleftharpoons}$$

with the groups CN: on the lower carbons

$$CN^- + \left[\underset{CN}{\underset{|}{CH_2}}\!\!-\!\!CH\!\!=\!\!\overset{\overset{\displaystyle ::\!OH}{\|}}{C}CH_3 \right] \;\rightleftharpoons\; \underset{CN}{\underset{|}{CH_2}}CH_2\overset{\overset{\displaystyle O}{\|}}{C}CH_3$$

an enol

Which of the two addition reactions occurs? Sometimes both do, and a mixture of products results. In most cases, however, one product or the other predominates. Steric hindrance around the double bond or the carbonyl group may lead to preferred attack at the nonhindered position. Aldehydes, less hindered than ketones, usually undergo carbonyl attack.

$$(CH_3CH_2)_2C\!\!=\!\!CH\overset{\overset{\displaystyle O}{\|}}{C}H \qquad CH_2\!\!=\!\!\underset{CH_3}{\underset{|}{C}}\overset{\overset{\displaystyle O}{\|}}{C}CH_2CH_3$$

Nu:⁻ attacks here   *Nu:⁻ attacks here*

A highly basic nucleophile (such as RMgX or LiAlH$_4$) attacks preferentially at the carbonyl group, while a weaker base (such as CN⁻ or R$_2$NH) usually attacks at the carbon–carbon double bond.

Stronger bases attack 1,2 (at C=O):

(98%)

$$CH_3CH\!\!=\!\!CH\overset{\overset{\displaystyle O}{\|}}{C}CH_3 + CH_3MgI \;\xrightarrow{H_2O,\,H^+}\; CH_3CH\!\!=\!\!CH\underset{CH_3}{\underset{|}{\overset{\overset{\displaystyle OH}{|}}{C}}}CH_3$$

(72%)

Weaker bases attack 1,4 (at C=C):

$$(CH_3)_2C\!\!=\!\!CH\overset{\overset{\displaystyle O}{\|}}{C}CH_3 + :NH_3 \;\longrightarrow\; (CH_3)_2\underset{NH_2}{\underset{|}{C}}CH_2\overset{\overset{\displaystyle O}{\|}}{C}CH_3$$

(70%)

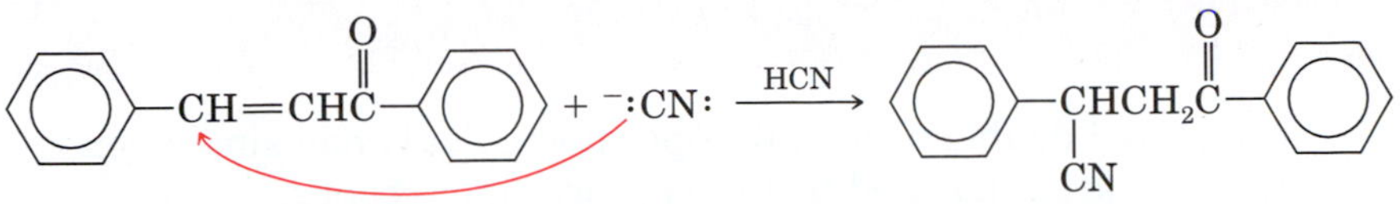

+ ⁻:CN:
HCN

(96%)

Lithium dialkylcuprates (R_2CuLi; Section 15.3C) attack an α,β-unsaturated carbonyl compound almost exclusively at the carbon–carbon double bond by 1,4-addition.

$$CH_2{=}CHCCH_3 \xrightarrow[(2)\ H_2O,\ H^+]{(1)\ (CH_3)_2CuLi} CH_3CH_2CH_2CCH_3$$

(97%)

C. α,β-Unsaturated Carboxylic Acids

An isolated double bond in an unsaturated carboxylic acid behaves independently of the carboxyl group. A *conjugated* double bond and carboxyl group, however, can undergo typical 1,4-addition reactions. These reactions take place in the same fashion as those of α,β-unsaturated aldehydes or ketones.

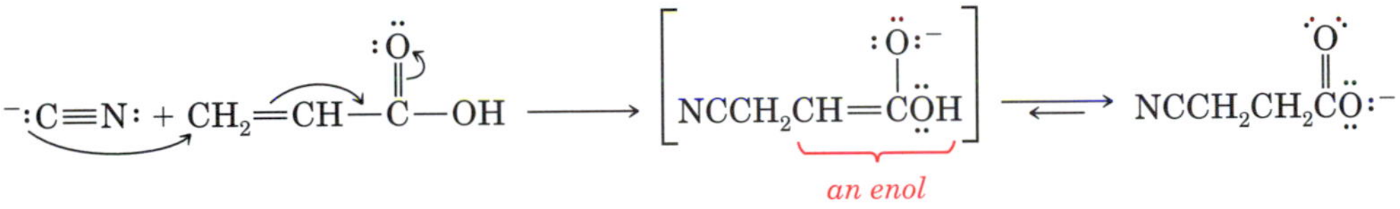

an enol

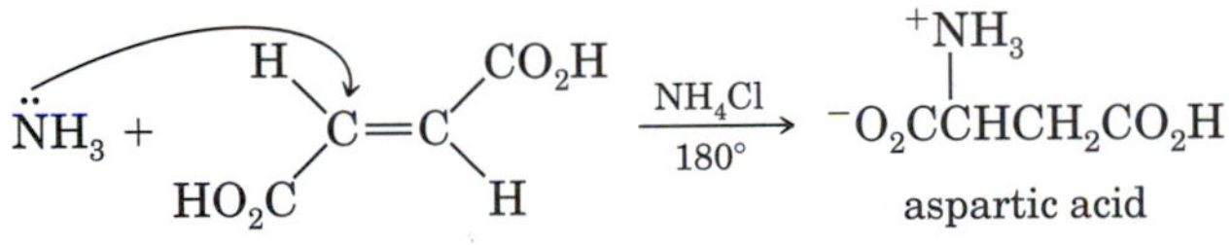

NH₄Cl
180°

fumaric acid

aspartic acid

an amino acid found in proteins

Study Problems

16.16 Predict the products:

(a) 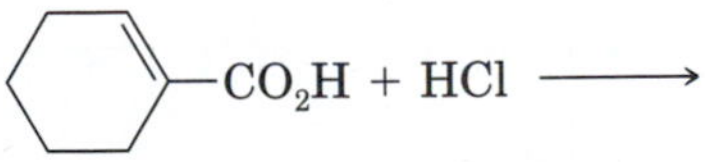

CO₂H + HCl

(b) 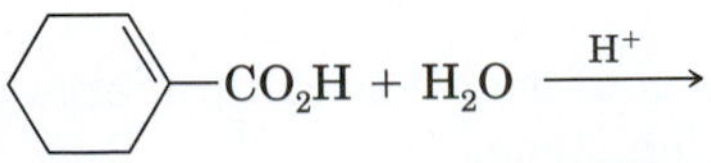

CO₂H + H₂O
H⁺

(c) $CH_3CH{=}CHCO_2H + 2\ CH_3NH_2 \longrightarrow$

16.17 Predict the major organic product of the reaction of each of the following reagents with 2-cyclohexenone: **(a)** CH_3MgI (followed by H^+, H_2O); **(b)** 1 equivalent of H_2 with Ni catalyst (25°); **(c)** $NaBH_4$ (followed by H^+, H_2O); **(d)** NH_3; **(e)** $(CH_2{=}CH)_2CuLi$ (followed by H^+, H_2O).

Summary

Conjugated double bonds are separated by only one single bond; isolated double bonds are separated by two or more single bonds.

$$CH_2\!\!=\!\!CH-CH_2-CH_2-CH\!\!=\!\!CH-CH\!\!=\!\!CH_2$$

an isolated double bond   *conjugated double bonds*

Conjugated dienes can undergo **1,2-addition** or **1,4-addition,** the latter by way of allylic carbocations. 1,4-Addition is also called **conjugate addition.**

$$
CH_2\!\!=\!\!CHCH\!\!=\!\!CH_2 \quad
\begin{cases}
\xrightarrow[1,2]{X_2} XCH_2CHXCH\!\!=\!\!CH_2 \\[2mm]
\xrightarrow[1,4]{X_2} XCH_2CH\!\!=\!\!CHCH_2X
\end{cases}
$$

When a C=C double bond is in conjugation with a C=O group, 1,2- and 1,4-additions also occur. With these reactions, both electrophilic and nucleophilic addition can take place. Under nucleophilic attack, 1,2-addition occurs when a strong base and a nonhindered carbonyl compound are used as reactants. When a weak base and a hindered carbonyl compound are used, 1,4-addition predominates.

Electrophilic:

$$
\underset{}{R_2C\!\!=\!\!CH-\overset{\overset{\displaystyle O}{\|}}{C}R} \xrightarrow[1,4]{HCl}
\left[R_2\overset{\overset{\displaystyle Cl}{|}}{\underset{}{C}}-CH\!\!=\!\!\overset{\overset{\displaystyle OH}{|}}{C}R \right]
\rightleftharpoons
R_2\overset{\overset{\displaystyle Cl}{|}}{\underset{}{C}}-CH_2-\overset{\overset{\displaystyle O}{\|}}{C}R
$$

Nucleophilic:

$$
\underset{\substack{nonhindered\ C=O}}{R_2C\!\!=\!\!CH\overset{\overset{\displaystyle O}{\|}}{C}H} \;+\; \underset{strong\ base}{R'MgX} \xrightarrow{1,2}
R_2C\!\!=\!\!CH\overset{\overset{\displaystyle OMgX}{|}}{\underset{\underset{\displaystyle R'}{|}}{C}}H
$$

$$
\underset{\substack{hindered\ C=O}}{CH_2\!\!=\!\!CH\overset{\overset{\displaystyle O}{\|}}{C}R} \;+\; \underset{weak\ base}{NH_3} \xrightarrow{1,4}
\left[\overset{\overset{\displaystyle OH}{|}}{CH_2}-CH\!\!=\!\!\underset{\underset{\displaystyle NH_2}{|}}{C}R \right]
\rightleftharpoons
CH_2CH_2\overset{\overset{\displaystyle O}{\|}}{\underset{\underset{\displaystyle NH_2}{|}}{C}}R
$$

Conjugated dienes in an s-*cis* conformation can undergo a **Diels–Alder** reaction to yield a substituted cyclohexene.

s-trans   *s-cis*   *a substituted cyclohexene*

Conjugate Allylation of α,β-Enones

In the presence of a Lewis acid, the allyl group of an allylsilane is nucleophilic. For example, it will react with the carbonyl group of an aldehyde or a ketone to give an addition product.

$$CH_3(CH_2)_2\overset{\overset{\displaystyle O}{\|}}{C}H + CH_3CH{=}CHCH_2Si(CH_3)_3 \xrightarrow[\text{2) } H_2O]{\text{1) } TiCl_4}$$

$$CH_3(CH_2)_2\overset{\overset{\displaystyle OH}{|}}{C}H{-}\underset{\underset{\displaystyle CH_3}{|}}{C}H{-}CH{=}CH_2 \quad (83\%)$$

Conjugate allylation using allylsilanes also can be carried out. Treatment of benzalacetone with allyltrimethylsilane in the presence of $TiCl_4$, a Lewis-acid catalyst, followed by hydrolysis, yields 4-phenyl-6-hepten-2-one in a 78–80% yield.

benzalacetone

$$\text{(78–80\%)}$$

The method can be used to stereoselectively introduce an allyl group at an angular position on a fused-ring system. (H. Sakurai, A. Hosomi, and J. Hayashi, *Org. Syn.* **1987,** *65,* 86.)

$$\text{(85\%)}$$

Questions

1. Which carbon of the allyl group becomes bonded to the beta carbon of benzalacetone in the conjugate allylation?

2. In the conjugate allylation reaction, is the allyl double bond acting as a nucleophile or as an electrophile? Explain your answer.

3. The experimental procedure for conjugate allylation of an aldehyde calls for the addition of $TiCl_4$ to the carbonyl compound prior to the addition of the allylsilane. (a) Write an equation for the reaction of the carbonyl group of hexanal with $TiCl_4$. (b) Write a second equation showing the reaction of $TiCl_4$ with

a conjugated carbonyl compound such as benzalacetone. Include the resonance structures for the benzalacetone–TiCl$_4$ complex.

4. Write a mechanism for the conjugate allylation reaction of benzalacetone using allyltrimethylsilane and TiCl$_4$.

5. Predict the fate of the trimethylsilyl group in these reactions before and after hydrolysis.

Study Problems*

16.18 Circle the conjugated π systems in the following structures.

(a)

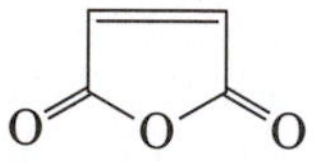

maleic anhydride

(b)

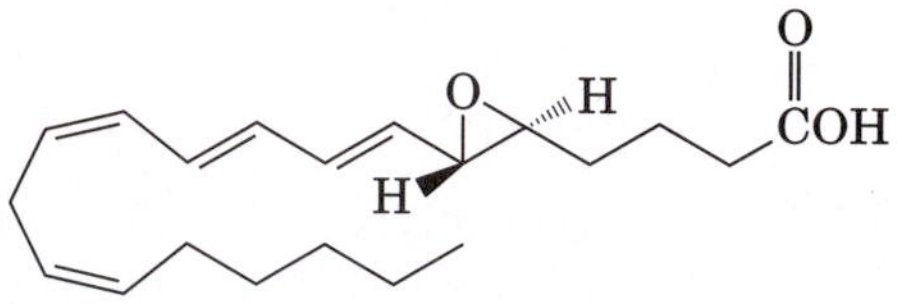

leukotriene A

(c)

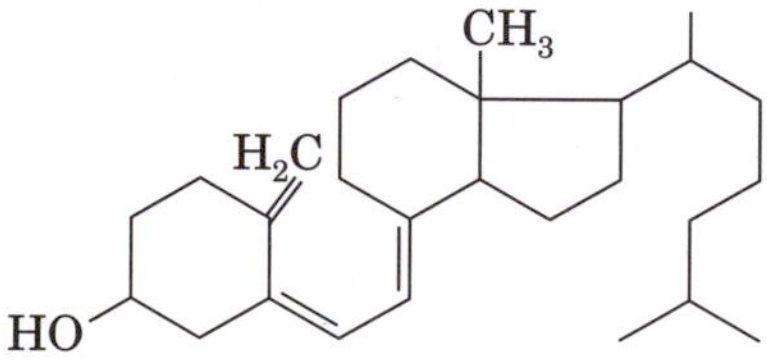

Vitamin D$_3$

(d)

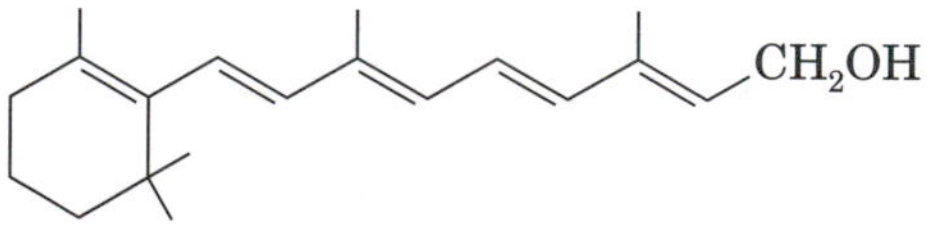

Vitamin A

16.19 Name the following compounds.

    (a) $CH_2{=}CH{-}CH{=}CH{-}CH_2CH_2CH_3$

    (b)

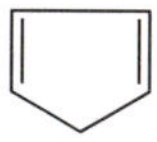

    (c) $CH{\equiv}C{-}CH{=}CH{-}CH{=}CH_2$

    (d)

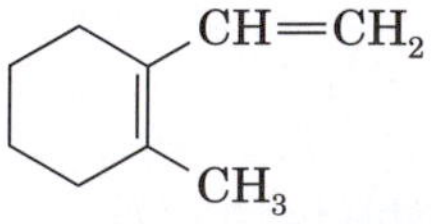

16.20 1,3,5-Cyclooctatriene undergoes a 1,6-addition in a 68% yield when treated with Br$_2$. Write the product of this reaction and the resonance structures of the intermediate.

16.21 (*E*)-1,2-Di-*tert*-butyl-1,3-butadiene does not undergo typical 1,4-addition reactions. Suggest a reason for this lack of reactivity.

   *For information concerning the organization of the *Study Problems* and *Additional Problems*, see the *Note to student* on page 41.

16.22 Predict the major product and write a plausible mechanism for each of the following reactions.

(a) + (1) CH$_2$=$\overset{\underset{\displaystyle CH_3}{|}}{C}$MgBr $\xrightarrow{\text{(2) H}_2\text{O}}$

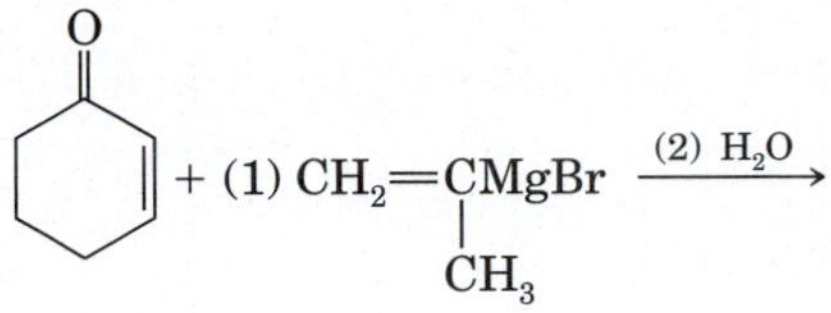

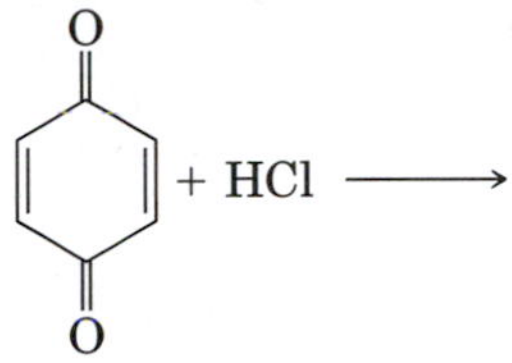

(b) + HCl $\longrightarrow$ (c) CH$_2$=CHCH$_2$$\overset{\overset{\displaystyle O}{\|}}{C}$H $\xrightarrow[\text{CN}^-]{\text{HCN}}$

16.23 Predict all likely organic products of the reaction of 4-phenyl-3-buten-2-one with:

(a) H$_2$O, H$^+$ (b) CH$_3$MgI   (c) HBr

(d) excess CH$_3$NH$_2$ (e) (CH$_3$)$_2$CuLi

16.24 Predict the product of the following reaction. What would have been the product of a 1,4-addition reaction?

$$\langle\!\!\!\bigcirc\!\!\!\rangle\text{-CH}=\text{CH}-\overset{\overset{\displaystyle O}{\|}}{\text{C}}\text{OCH}_2\text{CH}_3 + \text{Br}_2 \longrightarrow$$

16.25 An equimolar quantity of HCl is added to each of the following alkenes. Give the structure of the likely products in each case: **(a)** 1,3-pentadiene; **(b)** 2,2-dimethyl-3-heptene; **(c)** 1,3-cyclohexadiene.

16.26 Using flow equations, show how the following conversions can be carried out.

(a)

(b)

(c)

(d) CH$_3$CH=CH$\overset{\underset{\displaystyle OH}{|}}{C}HCH_3$ $\longrightarrow$ CH$_3$(CH$_2$)$_3$$\overset{\underset{\displaystyle CH_3}{|}}{C}HCH_2$$\overset{\overset{\displaystyle O}{\|}}{C}CH_3$

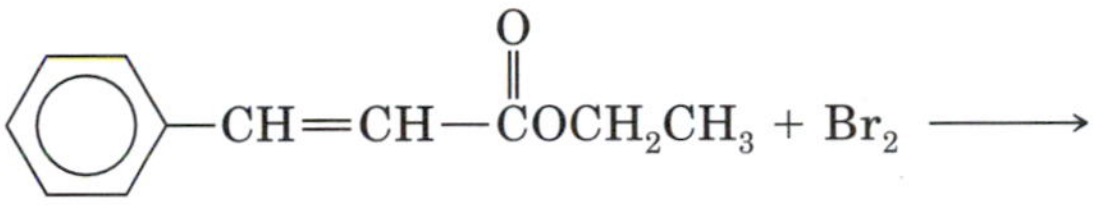

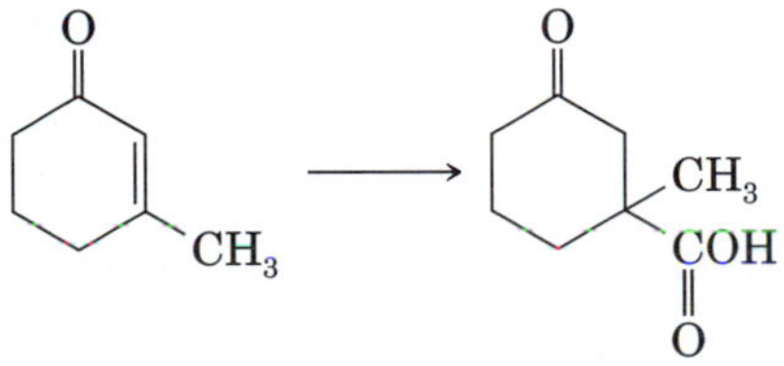

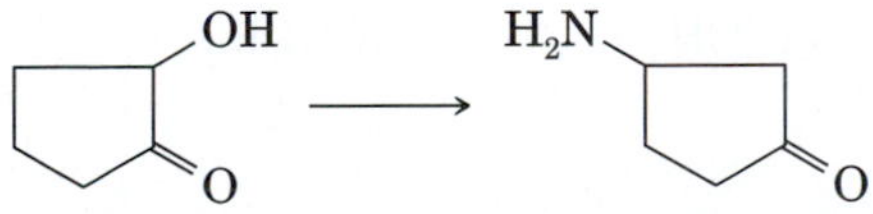

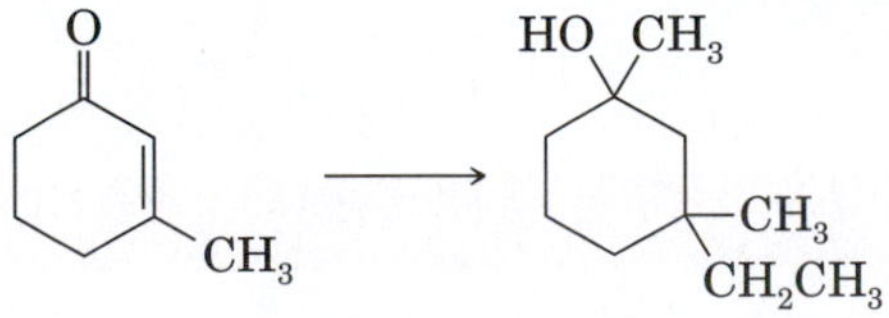

16.27 Complete the following equations.

(a) $CH_2{=}CH-\overset{\displaystyle O}{\overset{\|}{C}}CH_3$ + HBr $\longrightarrow$

(b) $CH_2{=}CHCH_2CH_2\overset{\displaystyle O}{\overset{\|}{C}}CH_3$ + HCl $\longrightarrow$

(c) 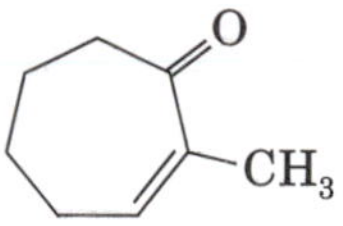   + CH_3OH $\xrightarrow{\;H^+\;}$

(d)   *trans*-$CH_3CH{=}CH\overset{\displaystyle O}{\overset{\|}{C}}OCH_3$ + BrCl $\longrightarrow$

16.28 Predict the products of the following Diels–Alder reactions, showing stereochemistry where appropriate:

(a)

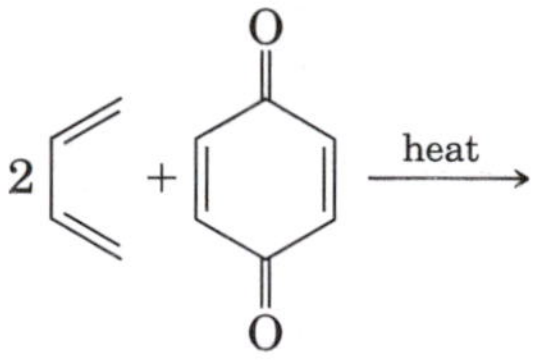

(b)

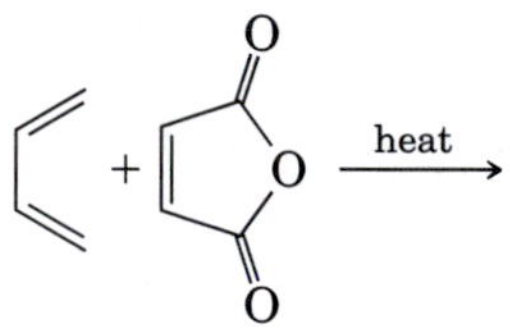

(c)

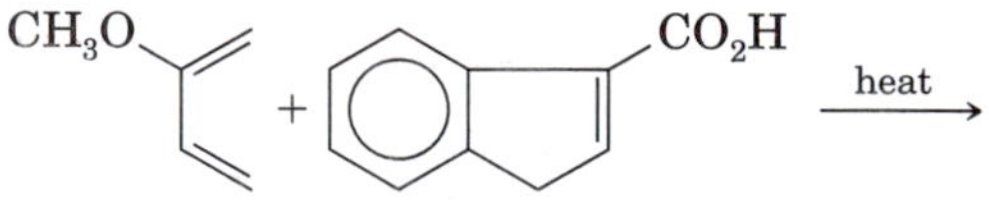

(d)

16.29 Predict the Diels–Alder reaction product of 1,3-butadiene and each of the following compounds:

(a) $CH_2{=}CH\overset{\displaystyle O}{\overset{\|}{C}}H$ (b) $(NC)_2C{=}C(CN)_2$ (c) $CH{\equiv}C\overset{\displaystyle O}{\overset{\|}{C}}OCH_2CH_3$

(d) $CH_2{=}CH\overset{\displaystyle O}{\overset{\|}{C}}OCH_2CH_3$ (e)   *cis*-$CH_3O\overset{\displaystyle O}{\overset{\|}{C}}CH{=}CH\overset{\displaystyle O}{\overset{\|}{C}}OCH_3$

(f)   *trans*-$CH_3O\overset{\displaystyle O}{\overset{\|}{C}}CH{=}CH\overset{\displaystyle O}{\overset{\|}{C}}OCH_3$

Additional Problems

16.30 Although styrene contains a conjugated π system, it does not undergo 1,4-addition. Explain.

16.31 Draw pictures of the bonding and nonbonding orbitals of a conjugated triene.

16.32 *Styrene-butadiene rubber* (SBR) is a synthetic rubber used in tires. It is a copolymer of three parts of 1,3-butadiene and one part styrene. About 80% of the polymer is formed by 1,4-addition, while the remaining 20% is formed by 1,2-addition. Most of the double bonds in SBR are *trans*. (These features of SBR are randomly incorporated in the polymer.) Write a formula for a hypothetical segment of this polymer that illustrates the preceding characteristics.

16.33 The reaction of 2-methyl-1,3-butadiene (isoprene), used in the manufacture of rubber and other polymers, with HCl at 25° yields compounds A and B in a ratio of 75:25. Write mechanisms for the formation of these compounds.

$$\underset{\text{A}}{(CH_3)_2\overset{\overset{\displaystyle Cl}{|}}{C}CH{=}CH_2} \qquad \underset{\text{B}}{(CH_3)_2C{=}CHCH_2Cl}$$

16.34 Suggest a Diels–Alder reaction that would lead to each of the following compounds.

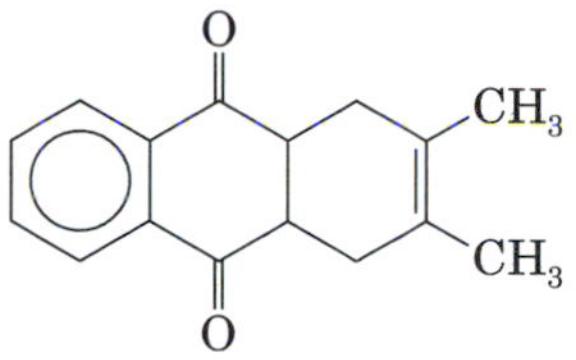

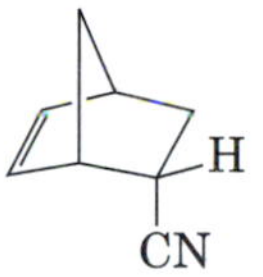

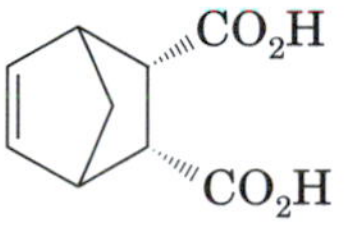

16.35 When 1,3-butadiene is treated with HCl in a solvent of acetic acid at room temperature, a mixture consisting of 78% 3-chloro-1-butene and 22% 1-chloro-2-butene is produced. When this mixture is treated with ferric chloride, its composition changes to 25% 3-chloro-1-butene and 75% 1-chloro-2-butene. How and why does the composition change with ferric chloride treatment?

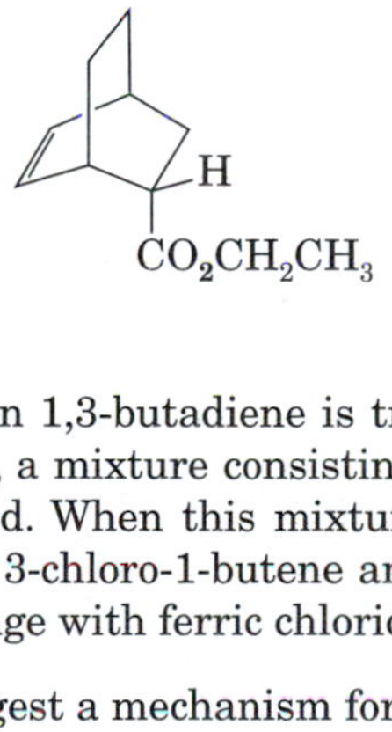

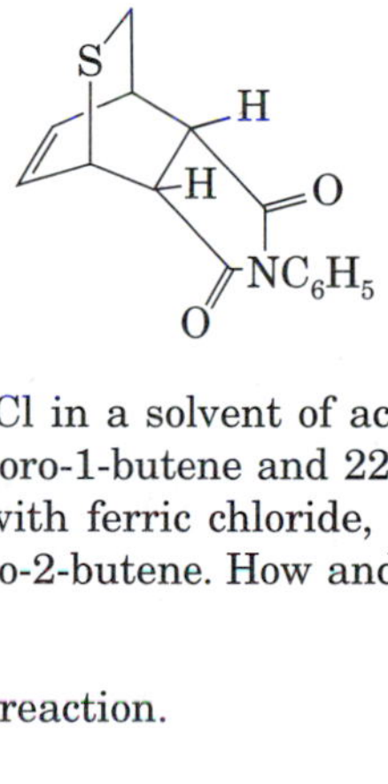

16.36 Suggest a mechanism for the following reaction.

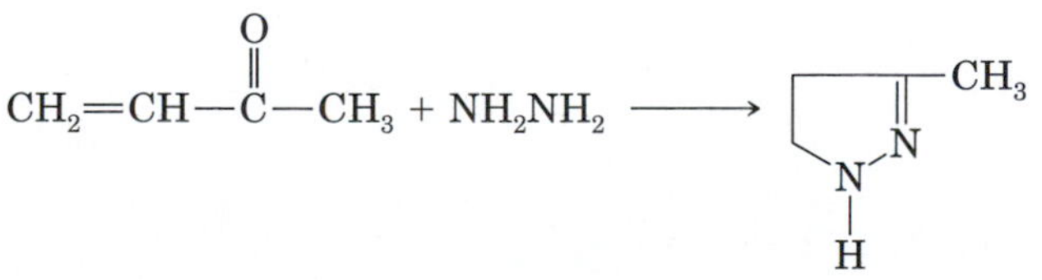

16.37 A mixture of methoxy chloro butenes is produced when 1,3-butadiene is treated with Cl_2 in methanol. Predict the products of this reaction and write a plausible mechanism for their formation.

16.38 The bicyclic diester below can be prepared by a Diels–Alder reaction between 1,3-cyclohexadiene and dimethyl acetylenedicarboxylate (dimethyl 2-butynedioate). When heated at a high temperature, the bicyclic ester undergoes a retro Diels–Alder reaction producing ethylene and a diene component.

(a) What is the structure of the diene component?

(b) Two other retro Diels–Alder pathways are available to the bicyclic diester: (1) cleavage to dimethyl acetylenedicarboxylate and a diene, or (2) cleavage to acetylene and a diene. Explain why the bicyclic ester undergoes a retro Diels–Alder reaction only by the ethylene-producing pathway. (*Hint:* Compare the structures of the diene components produced in the three pathways.)

ENOLATES AND CARBANIONS: BUILDING BLOCKS FOR ORGANIC SYNTHESIS

Nucleophilic reagents react with compounds that contain electron-deficient (partially positive) carbon atoms.

$$Nu{:}^- + \overset{\delta+}{R}{-}\overset{\delta-}{X} \longrightarrow Nu{-}R + X^-$$

$$Nu{:}^- + R{-}\underset{\delta+}{\overset{\overset{\displaystyle O^{\delta-}}{\|}}{C}}{-}R \longrightarrow R{-}\underset{Nu}{\overset{\overset{\displaystyle O^-}{|}}{C}}{-}R$$

Reagents that contain **nucleophilic carbon atoms**, carbon atoms with carbanion character, also attack partially positive carbon atoms. As an example, a **Grignard reagent**, which has a partially negative carbon atom, attacks carbonyl groups.

$$\overset{\delta-}{CH_3CH_2}{-}\overset{\delta+}{MgBr} + CH_3\overset{\overset{\displaystyle \ddot{O}{:}}{\|}}{C}CH_3 \longrightarrow$$

$$\underset{\underset{\displaystyle CH_2CH_3}{\overset{\displaystyle |}{}}}{\underset{\displaystyle CH_3\overset{\displaystyle |}{C}CH_3}{}} \overset{\displaystyle :\ddot{O}{:}^-\ {}^+MgBr}{}$$

The attack of one carbon upon another results in a new carbon–carbon bond. Reagents like Grignard reagents with nucleophilic carbon atoms allow us to synthesize compounds with complex carbon skeletons from simple compounds.

Grignard reagents are but one type of many reagents with nucleophilic carbons that are available to the organic chemist. Another versatile class of reagents for building complex molecules are the **enolates,** salts of carbonyl compounds, which also contain nucleophilic carbon atoms.

Section 17.1
Acidity of the Alpha Hydrogen

Recall from Section 13.8 that a hydrogen α to a carbonyl group is acidic and can be removed by a strong base. The α hydrogen is acidic primarily because of resonance stabilization of the product enolate ion.

$$CH_3\ddot{O}:^- + \overset{H}{\underset{}{C}}H_2\overset{O}{\underset{}{C}}CH_3 \rightleftharpoons CH_3OH + \left[{}^-CH_2-\overset{\ddot{O}:}{\underset{}{C}}CH_3 \longleftrightarrow CH_2=\overset{:\ddot{O}:^-}{\underset{}{C}}CH_3 \right]$$

acetone   resonance structures for
                                 the enolate ion of acetone

Because of resonance stabilization in the enolate ion of acetone, acetone is a far stronger acid than is an alkane. (However, acetone is still a weaker acid than ethanol.)

$$\underset{\substack{\text{ethane} \\ pK_a = 50}}{CH_3\overset{H}{\underset{}{C}}H_2} \qquad \underset{\substack{\text{acetone} \\ pK_a = 20}}{CH_2\overset{H}{\underset{}{\overset{|}{C}}}\overset{O}{\underset{}{C}}CH_3} \qquad \underset{\substack{\text{ethanol} \\ pK_a = 15.9}}{CH_3CH_2O-H}$$

In an ester, the carbonyl oxygen already carries a partial negative charge from delocalization of the electrons on the alkoxyl oxygen. Therefore, the carbonyl group is somewhat less able to delocalize the anionic negative charge of the enolate. The resonance structures for a typical ester, ethyl acetate, follow:

$$\underset{\text{major contributor}}{CH_3-\overset{\ddot{O}:}{\underset{}{C}}-\ddot{O}CH_2CH_3} \longleftrightarrow CH_3-\overset{:\ddot{O}:^-}{\underset{}{C}}=\overset{+}{\underset{}{\ddot{O}}}CH_2CH_3$$

Because the α hydrogen of an ester is less easily removed, a simple ester is less acidic than a ketone.

$$\underset{\substack{\text{ethyl acetate} \\ pK_a = 25}}{CH_2\overset{H}{\underset{}{\overset{|}{C}}}\overset{O}{\underset{}{C}}OCH_2CH_3}$$

A hydrogen α to a single carbonyl group is less acidic than a hydroxyl hydrogen. Therefore, a treatment of an aldehyde, a ketone, or an ester with an alkoxide results in a very low concentration of enolate ions. If we want a reasonably high concentration of the enolate, we must use a much stronger base, such as one of the following:

Some extremely strong bases:

$Na^+ {}^-NH_2$ $Na^+ H^-$

sodamide   sodium hydride

$CH_3CH_2CH_2CH_2Li$ $(CH_3)_2CH\overline{N}CH(CH_3)_2Li^+$

butyllithium   lithium diisopropylamide (LDA)

$$CH_3\overset{\overset{\text{O}}{\|}}{C}OCH_2CH_3 + {}^-:\ddot{O}CH_2CH_3 \rightleftharpoons {}^-:CH_2\overset{\overset{\text{O}}{\|}}{C}OCH_2CH_3 + H\ddot{O}CH_2CH_3$$

ethyl acetate   *not favored*   ethanol
$pK_a = 25$ $pK_a = 16$

$$CH_3\overset{\overset{\text{O}}{\|}}{C}OCH_2CH_3 + {}^-:\ddot{N}H_2 \rightleftharpoons {}^-:CH_2\overset{\overset{\text{O}}{\|}}{C}OCH_2CH_3 + :NH_3$$

$pK_a = 25$ *favored* $pK_a = 35$

If a hydrogen is α to two carbonyl groups, the negative charge on the anion can be delocalized by both C=O groups. Such a hydrogen is *more acidic than that of an alcohol*. A high concentration of enolate may be obtained by treatment of a β-dicarbonyl compound with an alkoxide. Table 17.1 lists the pK_a values for some compounds with hydrogens α to one and two carbonyl groups.

$$CH_3\overset{\overset{\text{O}}{\|}}{C}CH_2\overset{\overset{\text{O}}{\|}}{C}CH_3 + {}^-OCH_3 \rightleftharpoons CH_3\overset{\overset{\text{O}}{\|}}{C}\overset{-}{C}H\overset{\overset{\text{O}}{\|}}{C}CH_3 + CH_3OH$$

2,4-pentanedione   *favored* $pK_a = 15.5$
$pK_a = 9$

Resonance structures of the enolate ion:

$$CH_3\overset{\overset{\ddot{\text{O}}:}{\|}}{C}\overset{\curvearrowright}{-}\overset{:\ddot{\text{O}}}{\underset{\|}{C}}HCCH_3 \longleftrightarrow CH_3\overset{:\ddot{\text{O}}:{}^-}{\underset{\|}{C}}=CH\overset{:\ddot{\text{O}}}{\underset{\|}{C}}CH_3 \longleftrightarrow CH_3\overset{:\ddot{\text{O}}}{\underset{\|}{C}}CH=\overset{:\ddot{\text{O}}:{}^-}{\underset{\|}{C}}CH_3$$

If an enolate contains more than one α carbon atom with ionizable hydrogens, two or more enolate ions can be in equilibrium. For example, in solution, the preceding enolate is in equilibrium with the enolate formed at one of the terminal carbons. Because of the differences in acidity, however, only the one enolate would be present to any measurable extent.

Equilibrium mixture of enolates:

$$CH_3\overset{\overset{\text{O}}{\|}}{C}\overset{-}{C}HCCH_3 \rightleftharpoons {}^-:CH_2\overset{\overset{\text{O}}{\|}}{C}CH_2\overset{\overset{\text{O}}{\|}}{C}CH_3$$

more stable enolate
favored

Not only a carbonyl group, but any strongly electron-withdrawing group

	Table 17.1 pK_a values for some carbonyl compounds[a]	
Structure	Name	Approx. pK_a
$CH_3CCHCCH_3$ (O, O) H	2,4-pentanedione (acetoacetone)	9
$CH_3CCHCOCH_2CH_3$ (O, O) H	ethyl acetoacetate (acetoacetic ester)	11
$CH_3CCCOCH_2CH_3$ (O, O) R H	an alkylacetoacetic ester	13
$CH_3CH_2OCCHCOCH_2CH_3$ (O, O) H	diethyl malonate (malonic ester)	13
CH_2CCH_3 (O) H	acetone	20
$CH_2COCH_2CH_3$ (O) H	ethyl acetate	25

[a] For comparison, the pK_a of acetic acid is 4.75; ethanol, 15.9; water, 15.74; and ethane, about 50.

enhances the acidity of an α hydrogen. Some other compounds that are more acidic than ethanol are:

$$CH_3CH_2OCCHCN \quad NCCHCN \quad CH_2NO_2$$
$$\text{(O)} \quad \text{H} \qquad \text{H} \qquad \text{H}$$

STUDY PROBLEMS

17.1   Which hydrogen is the most acidic? Why?

(a) $CH_3CH_2CCH_2CCH_3$ (O, O)   (1) (2) (3)

(b) $CH_3COCH_2CH_3$ (O)   (1) (2) (3)

(c)

17.2 Write an equation for the reaction of each of the following compounds with sodium ethoxide:

(a) $CH_3CH_2\overset{O}{\overset{\|}{C}}OCH_2CH_3$ **(b)** $CH_3CH_2O\overset{O}{\overset{\|}{C}}CH_2CN$ **(c)** CH_3NO_2

(d) 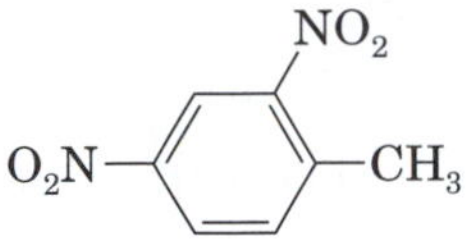 **(e)** 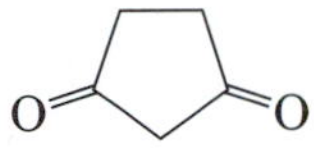

(f) $CH_3(CH_2)_4CH(CO_2C_2H_5)_2$

Section 17.2
Alkylation of Malonic Ester

One of the more powerful tools at the disposal of the synthetic organic chemist is the reaction of an enolate with an alkyl halide. In this section, we will emphasize the *alkylation of malonic ester*. In general, the end products from alkylation of malonic ester are *α-substituted acetic acids*. In the following example, the R group comes from RX. (In the discussions that follow, we will use $—C_2H_5$ to represent the ethyl group.)

General:

$$\underset{\text{diethyl malonate (malonic ester)}}{CH_2(CO_2C_2H_5)_2} \xrightarrow[(2)\ RX]{(1)\ Na^+\ ^-OC_2H_5} \underset{\textit{a diethyl alkylmalonate}}{RCH(CO_2C_2H_5)_2} \xrightarrow[\text{heat},\ -CO_2]{H_2O,\ H^+} \underset{\textit{an α-substituted acetic acid}}{\overset{\text{from RX}}{R}CH_2\overset{O}{\overset{\|}{C}}OH}$$

$$\underset{\text{diethyl malonate}}{CH_2(CO_2C_2H_5)_2} \xrightarrow[(2)\ CH_3CH_2Br]{(1)\ Na^+\ ^-OC_2H_5} \underset{\text{diethyl ethylmalonate}}{CH_3CH_2CH(CO_2C_2H_5)_2} \xrightarrow[115°,\ -CO_2]{H_2O,\ H^+} \underset{\text{butanoic acid (80\%)}}{\overbrace{CH_3CH_2}^{\text{from RX}}CH_2CO_2H}$$

A malonic ester alkylation consists of four separate reactions: (1) preparation of the enolate; (2) the actual alkylation; and (3) hydrolysis of the ester, followed by (4) decarboxylation of the resulting β-dicarboxylic acid.

Although there are many chemical reactions involved in this sequence, the laboratory procedure is quite simple because the same reaction vessel can be used for the entire sequence of reactions. (Sometimes the intermediate alkylated ester is purified prior to hydrolysis and decarboxylation in order to simplify the final purification.) Let us discuss each of the steps in this reaction in turn.

A. Formation of the Enolate

The enolate of malonic ester is usually prepared by treatment of the ester with sodium ethoxide, prepared by dissolving sodium metal in anhydrous ethanol (not the common 95% ethanol). (Why not?) Excess ethanol serves as the solvent for the reaction. Diethyl malonate is then added. The ethoxide ion is a stronger base than the enolate ion; therefore, the acid–base equilibrium lies on the side of the resonance-stabilized enolate anion.

$$Na + C_2H_5OH \longrightarrow Na^+ \ ^-:\ddot{O}C_2H_5 + \tfrac{1}{2}H_2 \uparrow$$

$$C_2H_5OCCH_2COC_2H_5 + \ ^-:\ddot{O}C_2H_5 \ \rightleftharpoons \ C_2H_5OCCHCOC_2H_5 + H\ddot{O}C_2H_5$$

The enolate ion takes # H here
more resonance stabilized
i.e. Oxygen's share charge

like OH⁻

favored
enolate

The enolate can also be prepared using NaH and an aprotic solvent, such as benzene, dimethylformamide (DMF), or 1,2-dimethoxyethane.

B. Alkylation

The alkylation reaction is a typical S_N2 displacement by a nucleophile. Methyl and primary alkyl halides give the best yields, with secondary alkyl halides giving lower yields because of competing elimination reactions. (Tertiary alkyl halides give exclusively elimination products, and aryl halides are nonreactive under S_N2 conditions.)

nucleophile

$$(C_2H_5O_2C)_2CH^- + CH_2\!-\!\ddot{B}r: \ \xrightarrow{\ S_N2\ } \ (C_2H_5O_2C)_2CHCH_2 + \ :\ddot{B}r:^-$$

enolate

CH_3

CH_3

17.3   Predict the products of the following reactions:

(a) $CH_2(CO_2C_2H_5)_2 \ \xrightarrow[\text{(2) } CH_3CH_2CH_2Br]{\text{(1) } Na^+ \ ^-OC_2H_5}$

(b) $CH_3CH(CO_2C_2H_5)_2 \ \xrightarrow[\text{(2) } CH_3CH_2I]{\text{(1) } Na^+ \ ^-OC_2H_5}$

The product of the alkylation still contains an acidic hydrogen:

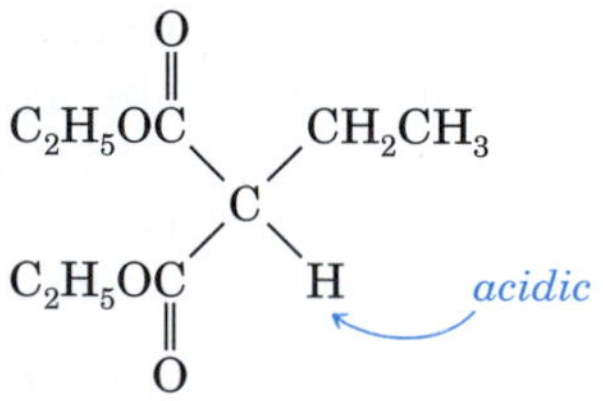

$$\begin{array}{c} \ \ \ \ \ \ \ \ \ \ O \\ \ \ \ \ \ \ \ \ \ \ \| \\ C_2H_5OC \ \ \ \ \ \ CH_2CH_3 \\ \diagdown \ \diagup \\ \ \ \ \ \ \ \ C \\ \diagup \ \diagdown \\ C_2H_5OC \ \ \ \ \ H \ \ \ \ \textit{acidic} \\ \| \\ O \end{array}$$

This second hydrogen can be removed by base, and a *second* R group can be substituted on the malonic ester. This second R group may be the same as, or different from, the first.

$$CH_3CH_2CH(CO_2C_2H_5)_2 \xrightarrow{Na^+ \ ^-OC_2H_5} CH_3CH_2\overset{..}{\overset{-}{C}}(CO_2C_2H_5)_2 \xrightarrow{CH_3I}$$

diethyl ethylmalonate   *an enolate*

$$\begin{array}{c} CH_3 \\ | \\ CH_3CH_2\,C(CO_2C_2H_5)_2 \end{array}$$

diethyl ethylmethylmalonate

The second hydrogen is not as acidic as the first hydrogen. Therefore, the second substitution requires more strenuous reaction conditions than does the first substitution. (For example, a stronger base such as NaH could be used.) Because of the difference in reactivity between the unsubstituted ester and the monoalkyl ester, diethyl malonate can be selectively mono- or dialkylated.

C. Hydrolysis and Decarboxylation

We have mentioned previously that a compound with a carboxyl group β to a carbonyl group undergoes decarboxylation when heated. The mechanism for decarboxylation was shown in Section 14.8C. If malonic ester (substituted or not) is hydrolyzed in hot acidic solution, the product β-diacid may undergo decarboxylation. (Sometimes decarboxylation does not occur until the diacid is distilled.)

$$\underset{\text{(structure)}}{\xi-\overset{O}{\overset{\|}{C}}CH_2\overset{O}{\overset{\|}{C}}OH} \xrightarrow{\text{heat}} \xi-\overset{O}{\overset{\|}{C}}CH_3 + CO_2$$

Hydrolysis and decarboxylation:

$$\underset{\substack{\text{an } \alpha,\alpha\text{-disubstituted} \\ \text{malonic ester}}}{\begin{array}{c} R \\ \diagdown \\ C \\ \diagup \\ R' \end{array}\begin{array}{c} O \\ \| \\ COC_2H_5 \\ \\ COC_2H_5 \\ \| \\ O \end{array}} \xrightarrow[\text{heat}]{H^+, H_2O} \underset{\substack{\text{a } \beta\text{-diacid}}}{\begin{array}{c} R \\ \diagdown \\ C \\ \diagup \\ R' \end{array}\begin{array}{c} O \\ \| \\ COH \\ \\ COH \\ \| \\ O \end{array}} \xrightarrow{-CO_2} \underset{\substack{\text{an } \alpha,\alpha\text{-disubstituted} \\ \text{acetic acid}}}{\begin{array}{c} R \\ \diagdown \\ CHCO_2H \\ \diagup \\ R' \end{array}}$$

A better yield of carboxylic acid is sometimes obtained when the diester is first saponified and the resulting disodium salt heated with aqueous acid.

$$R_2C(CO_2C_2H_5)_2 \xrightarrow[H_2O]{NaOH} R_2C(CO_2^- \ Na^+)_2 \xrightarrow[\text{heat}]{HCl} R_2CHCO_2H + CO_2$$

What if a chemist does not want a decarboxylation product, but wants a diacid? A diacid can be prepared by *saponification of the diester in base, followed by acidification of the cooled solution.* This way, the dicarboxylic acid itself is not subjected to heat and thus is less likely to undergo decarboxylation.

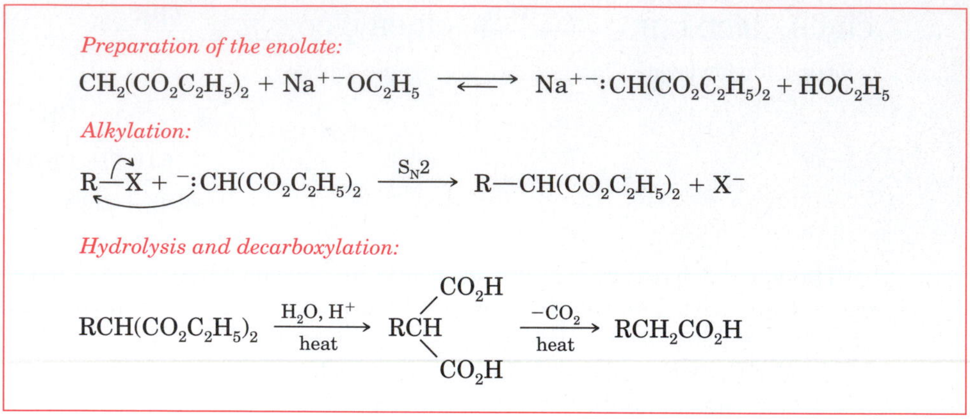

Preparation of the enolate:

$$CH_2(CO_2C_2H_5)_2 + Na^{+-}OC_2H_5 \rightleftharpoons Na^{+-}{:}CH(CO_2C_2H_5)_2 + HOC_2H_5$$

Alkylation:

$$R{\overset{\frown}{-}}X + {^-}{:}CH(CO_2C_2H_5)_2 \xrightarrow{S_N2} R{-}CH(CO_2C_2H_5)_2 + X^-$$

Hydrolysis and decarboxylation:

$$RCH(CO_2C_2H_5)_2 \xrightarrow[\text{heat}]{H_2O,\ H^+} RCH{\overset{\displaystyle CO_2H}{\underset{\displaystyle CO_2H}{\big\langle}}} \xrightarrow[\text{heat}]{-CO_2} RCH_2CO_2H$$

Figure 17.1 The steps in a malonic ester alkylation. The alkyl halide may be primary, secondary, allylic, or benzylic.

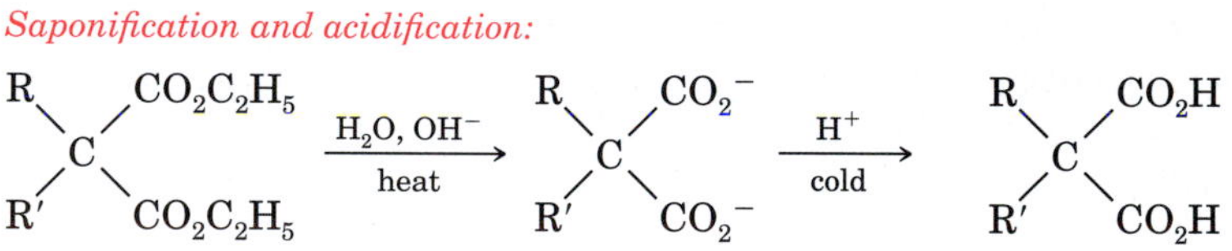

Saponification and acidification:

$$\underset{R'}{\overset{R}{>}}C\underset{CO_2C_2H_5}{\overset{CO_2C_2H_5}{<}} \xrightarrow[\text{heat}]{H_2O,\ OH^-} \underset{R'}{\overset{R}{>}}C\underset{CO_2{}^-}{\overset{CO_2{}^-}{<}} \xrightarrow[\text{cold}]{H^+} \underset{R'}{\overset{R}{>}}C\underset{CO_2H}{\overset{CO_2H}{<}}$$

an α,α-disubstituted
malonic acid

Figure 17.1 summarizes the steps in a malonic ester monoalkylation.

STUDY PROBLEMS

17.4 Write equations for the following reactions:

(a) saponification of diethyl propylmalonate, followed by treatment with cold HCl

(b) hydrolysis of diethyl dimethylmalonate in hot aqueous acid.

17.5 Give the mechanism for the decarboxylation of methylmalonic acid.

Section 17.3
Alkylation of Acetoacetic Ester

Alkylation reactions are not limited to the enolate of diethyl malonate. Other enolates also undergo S_N2 reaction with alkyl halides to yield alkylated products. Another commonly used enolate is that obtained from ethyl acetoacetate (acetoacetic ester). The end product of alkylation of acetoacetic ester is an *α-substituted acetone.*

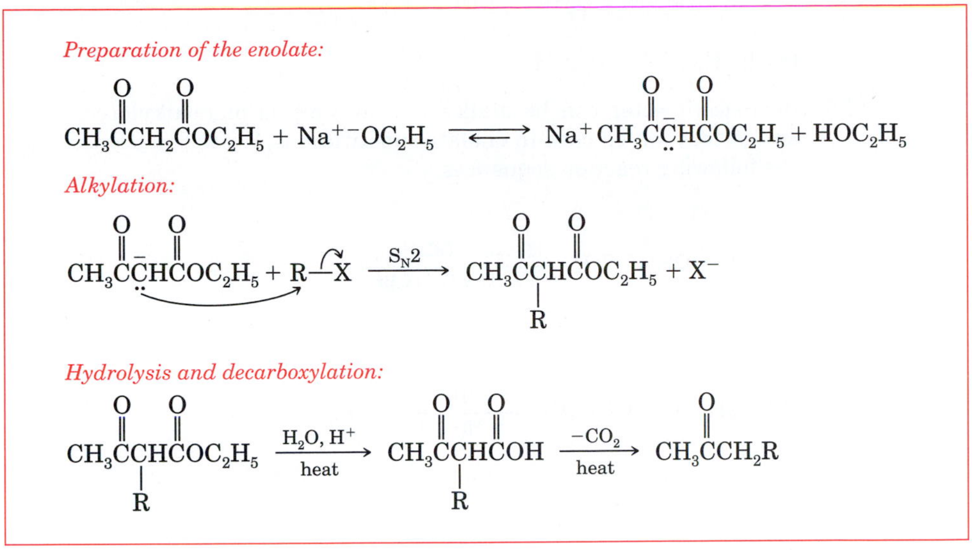

Preparation of the enolate:

$$CH_3CCH_2COC_2H_5 + Na^{+\;-}OC_2H_5 \;\rightleftharpoons\; Na^+ \; CH_3C\ddot{C}HCOC_2H_5 + HOC_2H_5$$

Alkylation:

$$CH_3C\ddot{C}HCOC_2H_5 + R\!-\!X \xrightarrow{S_N2} CH_3CCHCOC_2H_5 + X^-$$
$$\underset{R}{\big|}$$

Hydrolysis and decarboxylation:

$$CH_3CCHCOC_2H_5 \xrightarrow[\text{heat}]{H_2O,\,H^+} CH_3CCHCOH \xrightarrow[\text{heat}]{-CO_2} CH_3CCH_2R$$

Figure 17.2 The steps in an acetoacetic ester alkylation. The alkyl halide may be primary, secondary, allylic, or benzylic.

General:

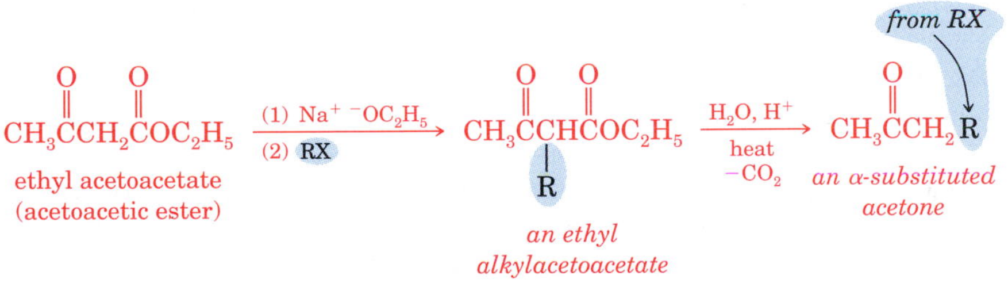

from RX

$$CH_3CCH_2COC_2H_5 \xrightarrow[\text{(2) RX}]{\text{(1) }Na^+\;^-OC_2H_5} CH_3CCHCOC_2H_5 \xrightarrow[\substack{\text{heat}\\-CO_2}]{H_2O,\,H^+} CH_3CCH_2\,R$$

ethyl acetoacetate
(acetoacetic ester)

an ethyl
alkylacetoacetate

*an α-substituted
acetone*

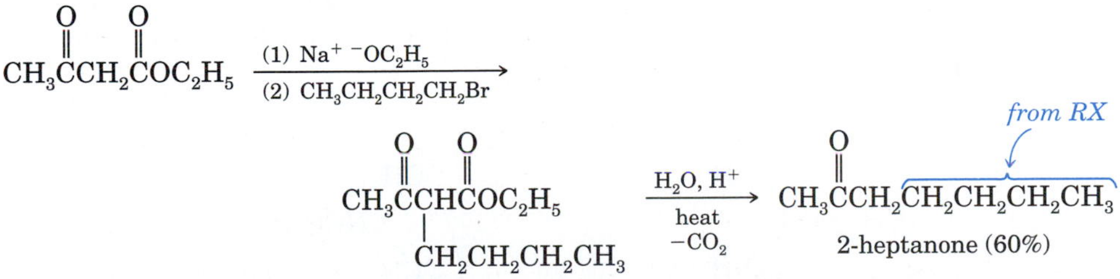

$$CH_3CCH_2COC_2H_5 \xrightarrow[\text{(2) }CH_3CH_2CH_2CH_2Br]{\text{(1) }Na^+\;^-OC_2H_5}$$

from RX

$$CH_3CCHCOC_2H_5 \xrightarrow[\substack{\text{heat}\\-CO_2}]{H_2O,\,H^+} CH_3CCH_2CH_2CH_2CH_2CH_3$$
$$\underset{CH_2CH_2CH_2CH_3}{\big|}$$

2-heptanone (60%)

The steps in an acetoacetic ester synthesis are similar to those for a malonic ester synthesis. (See Figure 17.2.)

STUDY PROBLEMS

17.6 State if the synthesis of each of the following compounds should start with diethyl malonate or ethyl acetoacetate:

(a) 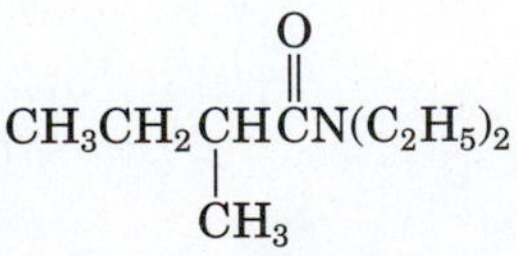 $CH_3CH_2\underset{\underset{CH_3}{|}}{CH}CN(C_2H_5)_2$ (with C=O)

(b) 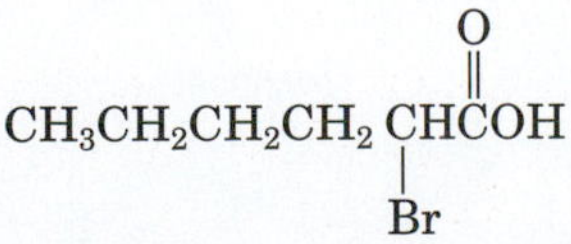 $CH_3CH_2CH_2CH_2\underset{\underset{Br}{|}}{CH}COH$ (with C=O)

$$\text{(c) } [(CH_3)_2CH]_2CHCCH_3 \overset{\displaystyle O}{\overset{\|}{}}$$

17.7   Acetoacetic ester can be dialkylated as well as monoalkylated. Also, other β-keto esters can be used in enolate substitutions. Write equations for the steps of the following reaction sequences:

(a) $CH_3\overset{O}{\overset{\|}{C}}\overset{O}{\underset{\underset{CH_3}{|}}{\overset{\|}{CH}}COC_2H_5 \xrightarrow[\text{(2) } CH_3CH_2CH_2Br]{\text{(1) } Na^+ \ ^-OC_2H_5}$

(b) $C_6H_5\overset{O}{\overset{\|}{C}}CH_2\overset{O}{\overset{\|}{C}}OC_2H_5 \xrightarrow[\text{(3) } H_2O, H^+, \text{ heat}]{\substack{\text{(1) } Na^+ \ ^-OC_2H_5 \\ \text{(2) } CH_3I}}$

Section 17.4

Syntheses Using Alkylation Reactions

A. Malonic Ester and Acetoacetic Ester Syntheses

In general, the products of alkylation reactions of malonic ester or acetoacetic ester are substituted acetic acids or substituted acetones. However, we can also obtain diacids, diesters, keto acids, and keto esters. The various products that can be obtained from alkylation of malonic ester or acetoacetic ester are summarized in Figure 17.3

It is comparatively easy to predict the products of a reaction when we are given the reactants. It is somewhat more difficult to decide upon specific reactants to use in a synthesis problem. Remember to work the problem backwards: if you are asked to synthesize a compound by an alkylation reaction, first decide what dicarbonyl compound you would need, then pick the alkyl halide.

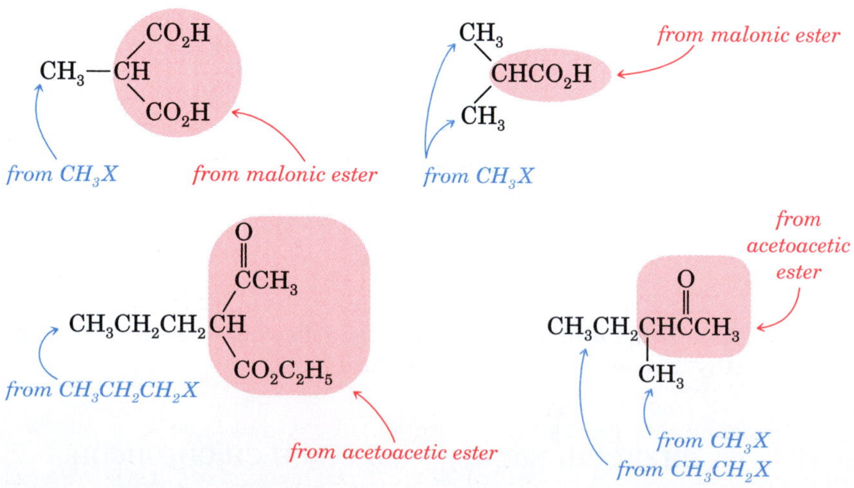

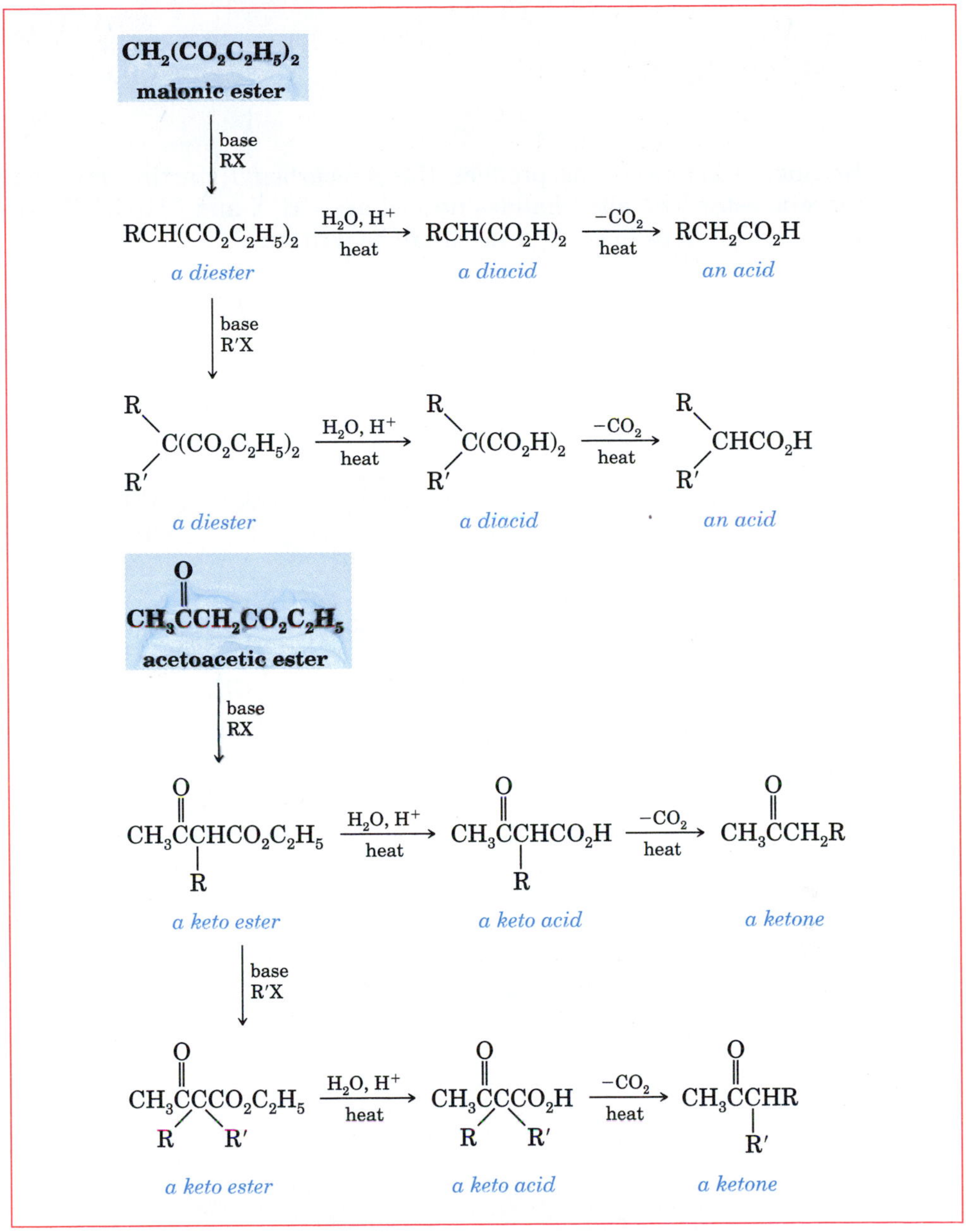

$$CH_2(CO_2C_2H_5)_2$$
malonic ester

$\downarrow$ base RX

$$RCH(CO_2C_2H_5)_2 \xrightarrow[\text{heat}]{H_2O, H^+} RCH(CO_2H)_2 \xrightarrow[\text{heat}]{-CO_2} RCH_2CO_2H$$

a diester   *a diacid*   *an acid*

$\downarrow$ base R'X

$$\begin{matrix} R \\ \diagdown \\ \diagup \\ R' \end{matrix} C(CO_2C_2H_5)_2 \xrightarrow[\text{heat}]{H_2O, H^+} \begin{matrix} R \\ \diagdown \\ \diagup \\ R' \end{matrix} C(CO_2H)_2 \xrightarrow[\text{heat}]{-CO_2} \begin{matrix} R \\ \diagdown \\ \diagup \\ R' \end{matrix} CHCO_2H$$

a diester   *a diacid*   *an acid*

$$\overset{O}{\overset{\|}{CH_3CCH_2CO_2C_2H_5}}$$
acetoacetic ester

$\downarrow$ base RX

$$\underset{R}{\overset{O}{\overset{\|}{CH_3CCHCO_2C_2H_5}}} \xrightarrow[\text{heat}]{H_2O, H^+} \underset{R}{\overset{O}{\overset{\|}{CH_3CCHCO_2H}}} \xrightarrow[\text{heat}]{-CO_2} \overset{O}{\overset{\|}{CH_3CCH_2R}}$$

a keto ester   *a keto acid*   *a ketone*

$\downarrow$ base R'X

$$\underset{R\quad R'}{\overset{O}{\overset{\|}{CH_3CCCO_2C_2H_5}}} \xrightarrow[\text{heat}]{H_2O, H^+} \underset{R\quad R'}{\overset{O}{\overset{\|}{CH_3CCCO_2H}}} \xrightarrow[\text{heat}]{-CO_2} \underset{R'}{\overset{O}{\overset{\|}{CH_3CCHR}}}$$

a keto ester   *a keto acid*   *a ketone*

Figure 17.3 Products from the alkylations of malonic ester and acetoacetic ester.

Example

If you were asked to write the equations for the synthesis of 3-methyl-2-pentanone, you would:

1. *write the structure*

2. *decide what β-dicarbonyl compound you would need*

3. *decide what alkyl halides would have to be used for the substitution.*

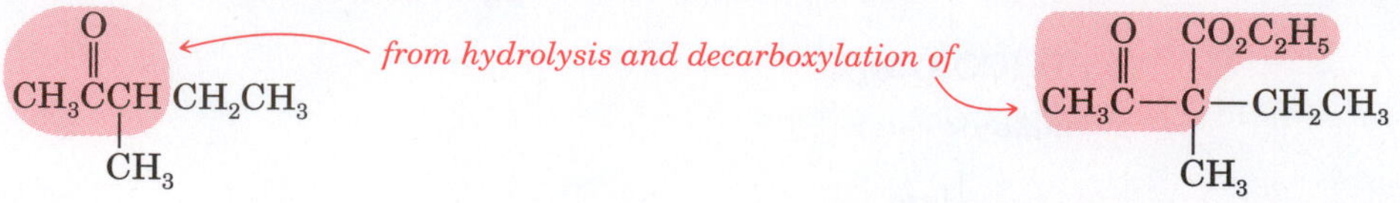

$$\underset{\underset{CH_3}{\overset{\overset{O}{\|}}{CH_3CCH}}CH_2CH_3} \quad \xleftarrow{\textit{from hydrolysis and decarboxylation of}} \quad \underset{\underset{CH_3}{CH_3C-\overset{CO_2C_2H_5}{\underset{|}{\overset{|}{C}}}-CH_2CH_3}}{\overset{O}{\|}}$$

Because a ketone is the product, the β-dicarbonyl starting material is acetoacetic ester. The alkyl halides needed are CH_3X and CH_3CH_2X. Now, equations for the steps in the synthesis can be written:

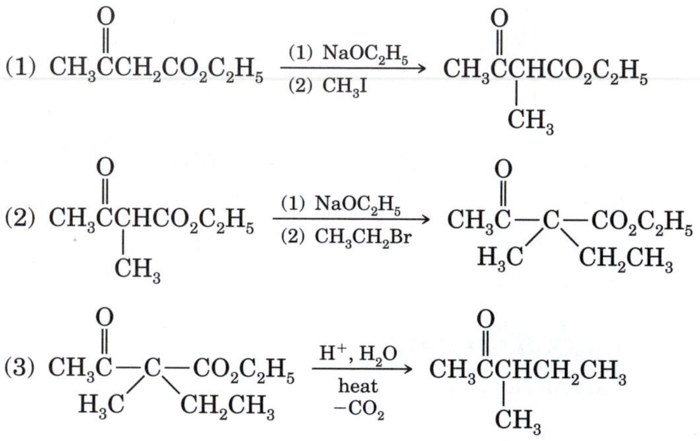

$$(1)\quad CH_3\overset{O}{\overset{\|}{C}}CH_2CO_2C_2H_5 \xrightarrow[\text{(2) } CH_3I]{\text{(1) } NaOC_2H_5} CH_3\overset{O}{\overset{\|}{C}}\underset{\underset{CH_3}{|}}{CH}CO_2C_2H_5$$

$$(2)\quad CH_3\overset{O}{\overset{\|}{C}}\underset{\underset{CH_3}{|}}{CH}CO_2C_2H_5 \xrightarrow[\text{(2) } CH_3CH_2Br]{\text{(1) } NaOC_2H_5} CH_3\overset{O}{\overset{\|}{C}}\underset{H_3C}{\overset{|}{\underset{CH_2CH_3}{\overset{|}{C}}}}CO_2C_2H_5$$

$$(3)\quad CH_3\overset{O}{\overset{\|}{C}}\underset{H_3C}{\overset{|}{\underset{CH_2CH_3}{\overset{|}{C}}}}CO_2C_2H_5 \xrightarrow[\substack{\text{heat}\\-CO_2}]{H^+,\,H_2O} CH_3\overset{O}{\overset{\|}{C}}\underset{\underset{CH_3}{|}}{CH}CH_2CH_3$$

SAMPLE PROBLEM

Suggest a reaction sequence leading to 3-phenylpropanoic acid.

Solution

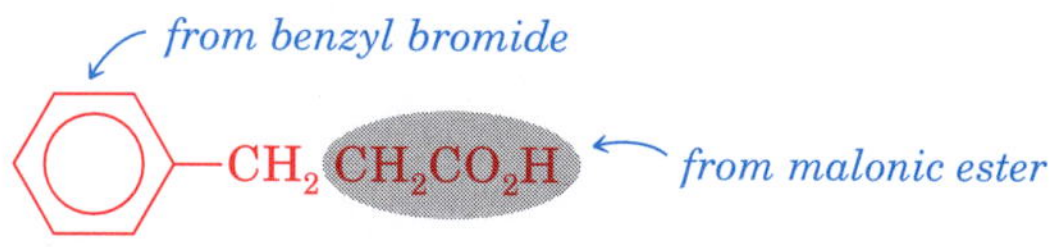

(structure: C_6H_5—CH_2—CH_2CO_2H, with "from benzyl bromide" pointing to the CH_2 and "from malonic ester" pointing to the CH_2CO_2H)

$$(1)\quad CH_2(CO_2C_2H_5)_2 \xrightarrow{Na^+\,{}^-OC_2H_5} {}^-CH(CO_2C_2H_5)_2$$

$$(2)\quad C_6H_5CH_2Br + {}^-CH(CO_2C_2H_5)_2 \xrightarrow{-Br^-} C_6H_5CH_2CH(CO_2C_2H_5)_2$$

$$(3)\quad C_6H_5CH_2CH(CO_2C_2H_5)_2 \xrightarrow[\substack{\text{heat}\\-CO_2}]{H_2O,\,H^+} C_6H_5CH_2CH_2CO_2H$$

STUDY PROBLEM

17.8 Show how you could synthesize the following compounds by alkylation reactions:

(a) $CH_3CH_2CH_2CH_2CH(CO_2H)_2$ **(b)** $(CH_3)_2CHCO_2H$

(c) 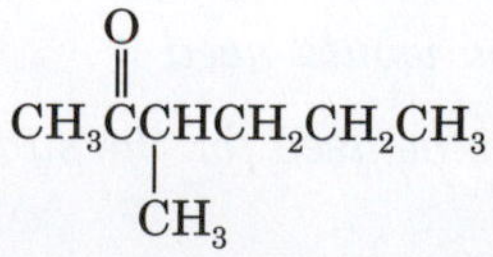 $CH_3\overset{O}{\overset{\|}{C}}\underset{\underset{CH_3}{|}}{CH}CH_2CH_2CH_3$ **(d)** 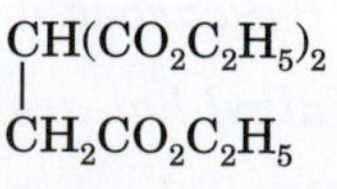 $\underset{CH_2CO_2C_2H_5}{\overset{CH(CO_2C_2H_5)_2}{\overset{|}{}}}$

(e) 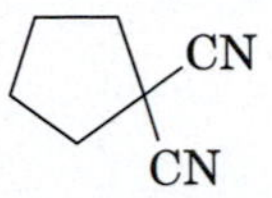 **(f)**

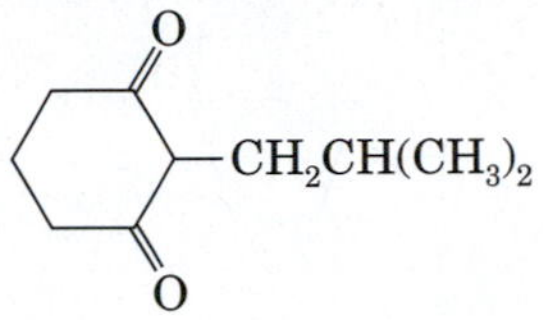

B. Alkylation Reactions of Other Carbonyl Compounds

Ketones and esters with only one carbonyl group and at least one α hydrogen can also be alkylated in the α position. Because only one activating group is present, a stronger base than an alkoxide is necessary. Lithium diisopropylamide (LDA) is commonly used.

One advantage to reactions with lithium enolates is that they can be carried out at very low temperatures—a reaction condition that minimizes equilibration among different enolates. Thus, these reactions are under kinetic (rate) control and not thermodynamic, or equilibrium, control as are many enolate reactions.

General:

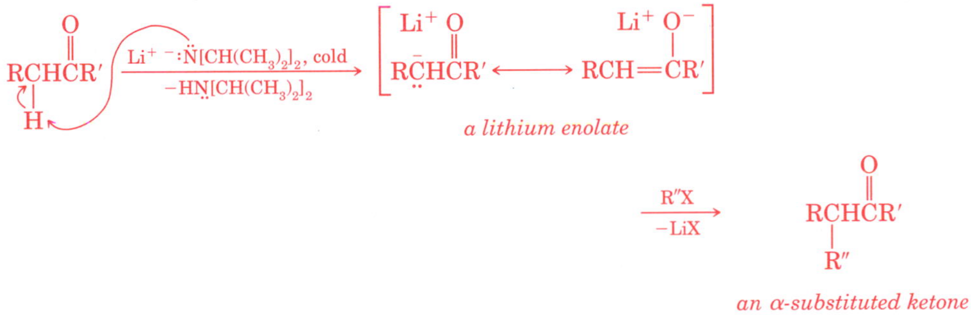

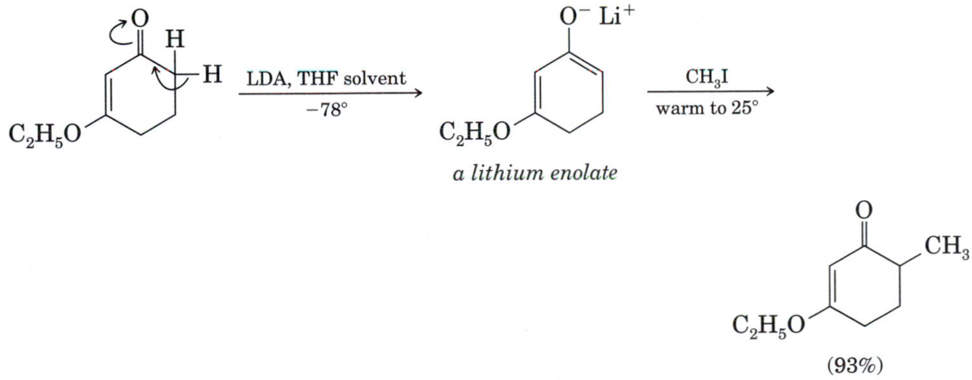

a lithium enolate

an α-substituted ketone

a lithium enolate

(93%)

17.9 Suggest alkylation reactions for the following conversions.

(a)

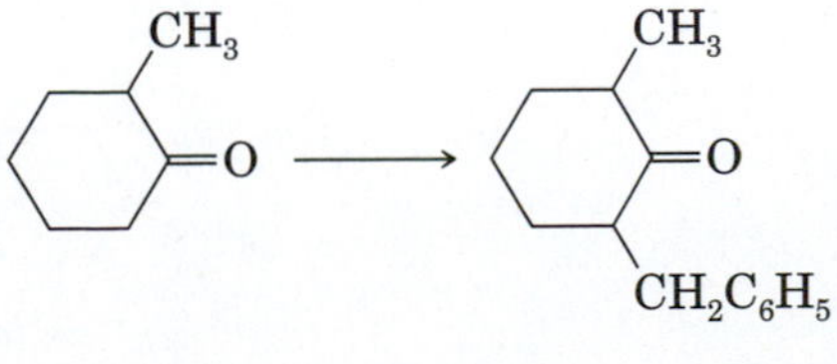

(b)

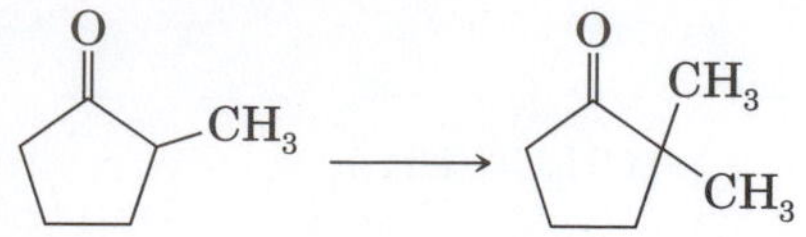

Section 17.5

Alkylation and Acylation of Enamines

Another type of organic compound containing a nucleophilic carbon that can undergo alkylation reactions is an **enamine.** In Section 13.5B, we discussed the formation of enamines from secondary amines and aldehydes or ketones.

Formation of an enamine:

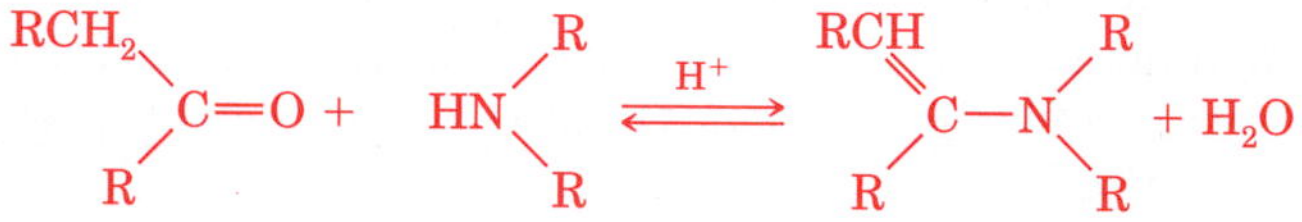

The nitrogen of an enamine has an unshared pair of electrons. These electrons are, in a sense, in an *allylic position* and consequently are in conjugation with the double bond. Resonance structures for the enamine show that the carbon β to the nitrogen has a partial negative charge.

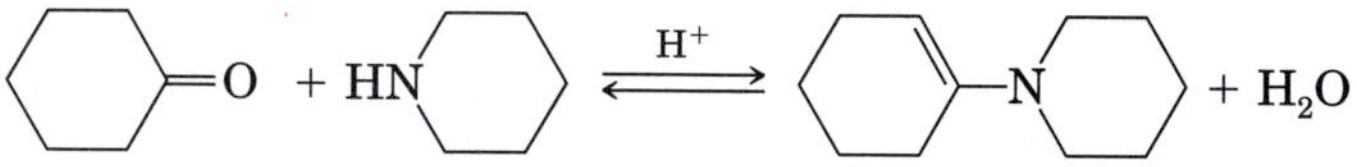

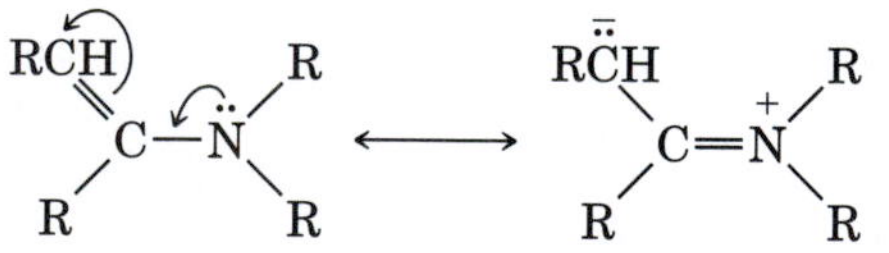

resonance structures for an enamine

This β carbon has carbanion character and can act as a nucleophile. For example, when an enamine is treated with an alkyl halide, such as CH_3I, the enamine displaces the halogen of the alkyl halide in an S_N2 reaction. The result is alkylation of the enamine at the position that is β to the nitrogen.

Alkylation:

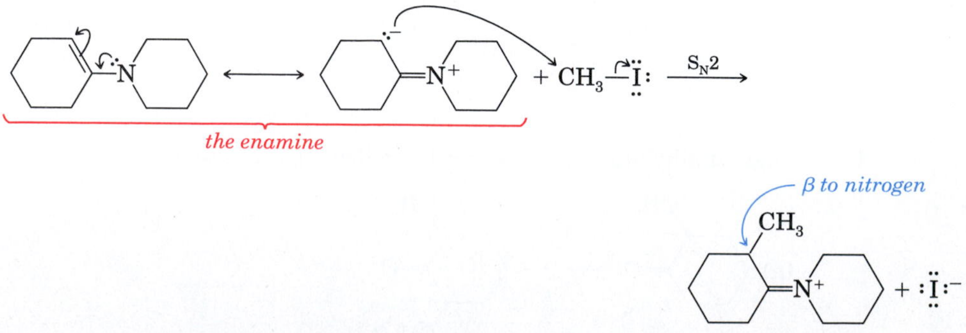

the enamine

β to nitrogen

an iminium ion

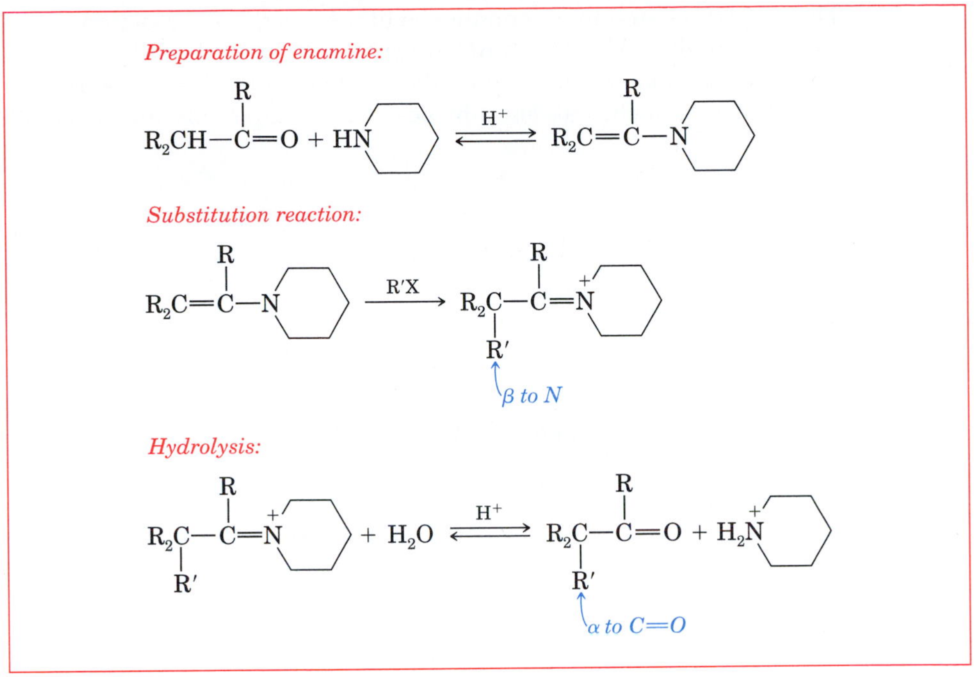

Preparation of enamine:

$$R_2CH-\overset{\overset{\displaystyle R}{|}}{C}=O + HN\bigcirc \overset{H^+}{\rightleftharpoons} R_2C=\overset{\overset{\displaystyle R}{|}}{C}-N\bigcirc$$

Substitution reaction:

$$R_2C=\overset{\overset{\displaystyle R}{|}}{C}-N\bigcirc \overset{R'X}{\longrightarrow} R_2C-\overset{\overset{\displaystyle R}{|}}{\underset{\underset{\displaystyle R'}{|}}{C}}=\overset{+}{N}\bigcirc$$

β *to N*

Hydrolysis:

$$R_2C-\overset{\overset{\displaystyle R}{|}}{\underset{\underset{\displaystyle R'}{|}}{C}}=\overset{+}{N}\bigcirc + H_2O \overset{H^+}{\rightleftharpoons} R_2C-\overset{\overset{\displaystyle R}{|}}{\underset{\underset{\displaystyle R'}{|}}{C}}=O + H_2\overset{+}{N}\bigcirc$$

α *to C=O*

Figure 17.4 The steps in an enamine alkylation or acylation. The alkyl halide must be a reactive one, such as methyl, benzylic, or allylic.

The product iminium ion is readily hydrolyzed to a ketone. The net result of the entire reaction sequence is alkylation of a ketone in the α position.

Hydrolysis:

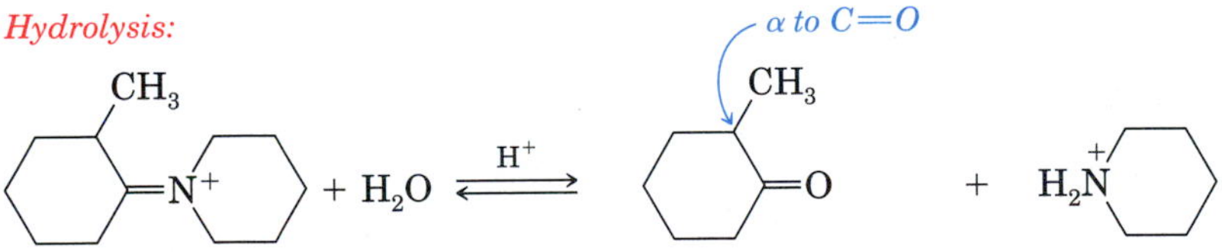

the iminium ion   2-methylcyclohexanone   piperidinium ion

a ketone

A generalized sequence for an enamine synthesis is shown in Figure 17.4.

SAMPLE PROBLEM

Give the steps in the enamine synthesis of the following ketone using piperidine as the amine.

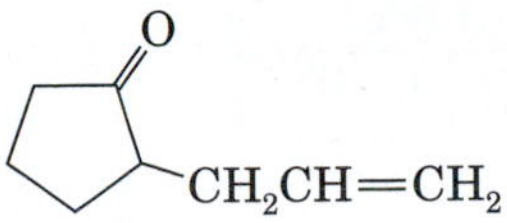

Solution

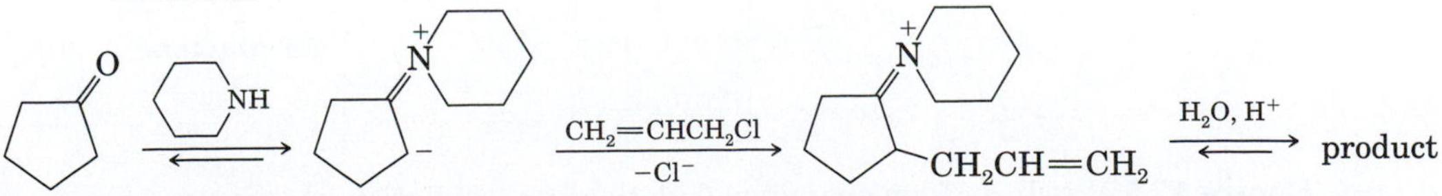

The alkylation step in an enamine synthesis is an S_N2 reaction with a rather weak nucleophile. (Why?) It is not surprising then that only the most reactive halogen compounds are suitable as alkylating agents. These compounds include allylic halides, benzylic halides, α-halocarbonyl compounds, and iodomethane.

Some reactive halides:

CH_3I	$C_6H_5CH_2Cl$	$CH_2{=}CHCH_2Cl$	$CH_3\overset{\displaystyle O}{\overset{\|}{C}}Cl$	$BrCH_2\overset{\displaystyle O}{\overset{\|}{C}}CH_3$
iodomethane	benzyl chloride	allyl chloride	acetyl chloride	bromoacetone

The reactions of enamines with α-halocarbonyl compounds and acid halides follow similar paths to that of alkylation. In each case, the final product (after hydrolysis) is a ketone substituted at the α position.

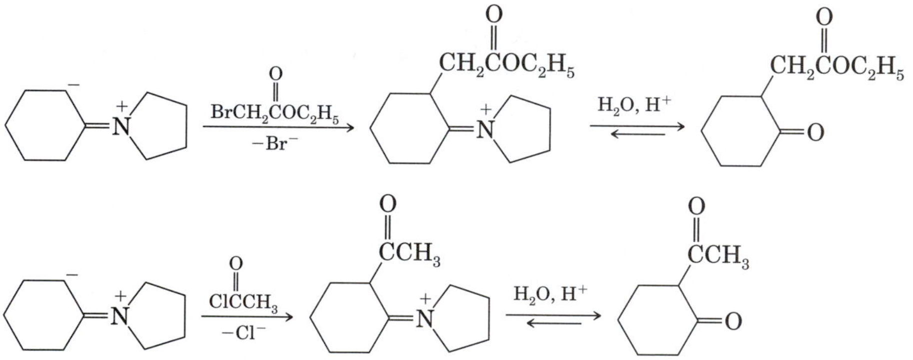

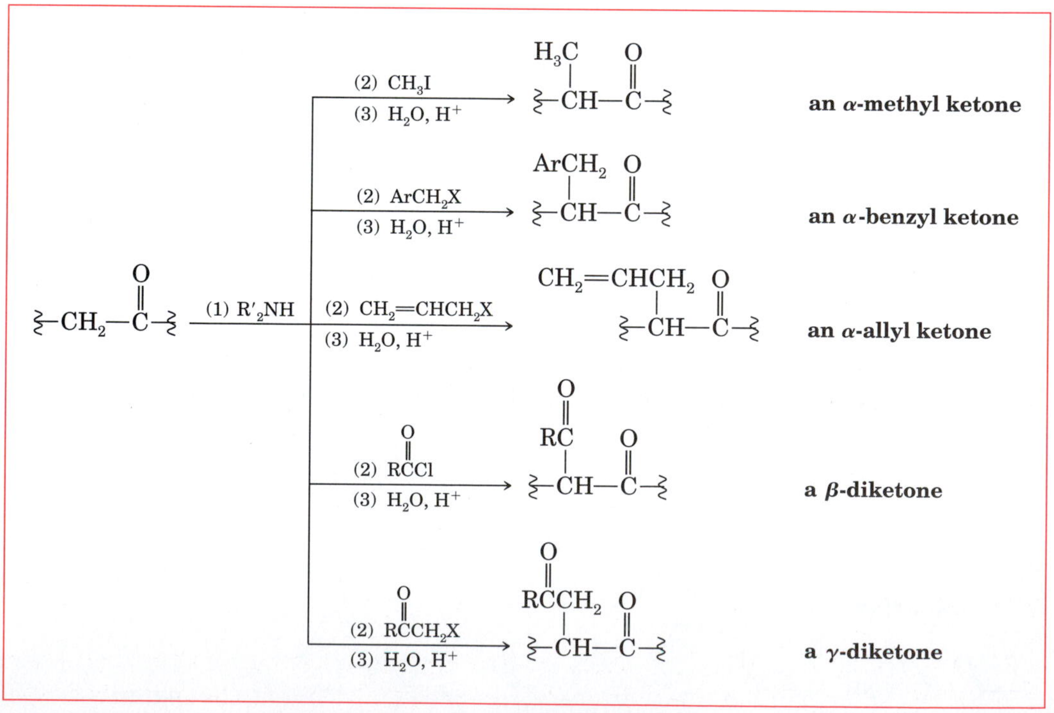

Figure 17.5 Products from enamine alkylations and acylations.

Figure 17.5 summarizes the products that can be obtained by enamine syntheses.

SAMPLE PROBLEM

How would you prepare the following compound by an enamine synthesis?

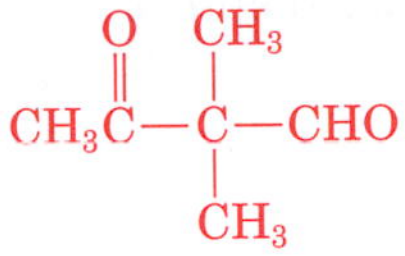

$$\underset{\underset{CH_3}{|}}{\overset{\overset{O}{\parallel}}{CH_3C}} -\overset{\overset{CH_3}{|}}{\underset{|}{C}} -CHO$$

Solution

from $CH_3\overset{O}{\overset{\parallel}{C}}Cl$ $\underset{\underset{CH_3}{|}}{\overset{\overset{O}{\parallel}}{CH_3C}}-\overset{\overset{CH_3}{|}}{\underset{|}{C}}-CHO$ from $(CH_3)_2CHCH\overset{O}{\overset{\parallel}{}}$

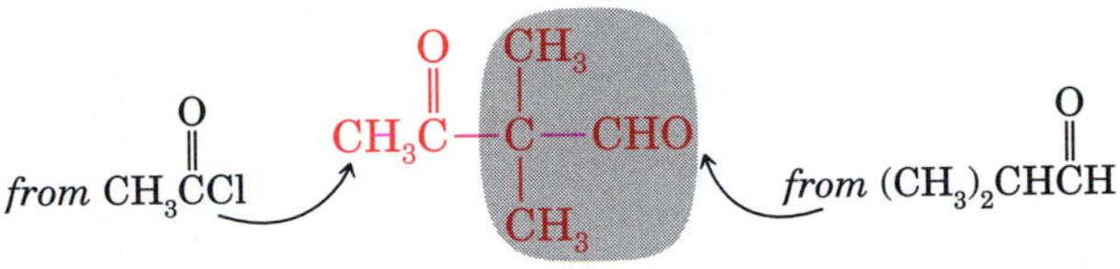

Reaction sequence:

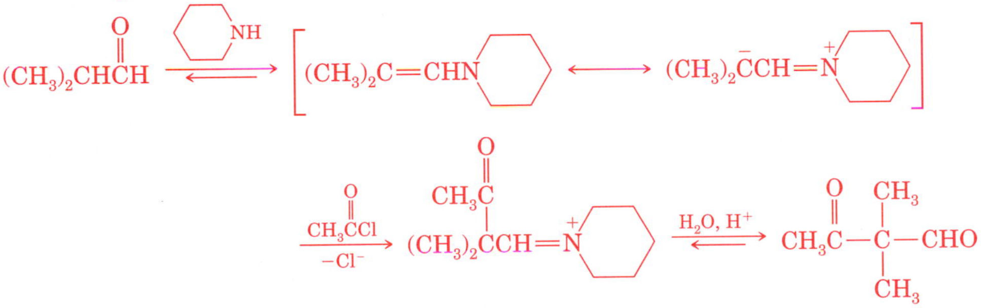

STUDY PROBLEM

17.10   Show how you would prepare the following compounds by enamine synthesis:

(a) 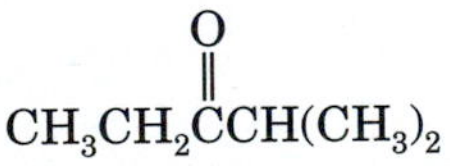 $CH_3CH_2\overset{\overset{O}{\parallel}}{C}CH(CH_3)_2$ (b)

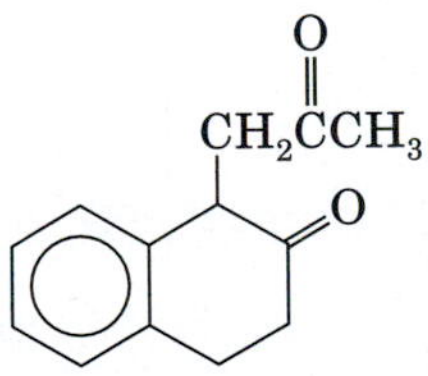

Section 17.6
Aldol Condensations

So far, we have been discussing the displacement of halide ions by nucleophiles. A reagent with a nucleophilic carbon atom can also attack the partially positive carbon of a carbonyl group. The rest of this chapter will be devoted to the reactions of enolates and related anions with carbonyl compounds.

$$Nu:^- + \overset{\overset{\displaystyle \ddot{O}:}{\|}}{\underset{|}{-C-}} \longrightarrow \overset{\overset{\displaystyle :\ddot{O}:^-}{|}}{\underset{\underset{\displaystyle Nu}{|}}{-C-}}$$

When an aldehyde is treated with a base such as aqueous NaOH, the resulting enolate ion can undergo reaction at the carbonyl group of another molecule of aldehyde. The result is the *addition of one molecule of aldehyde to another.*

from one aldehyde

$$2 \overset{\overset{\displaystyle O}{\|}}{CH_3CH} \xrightarrow{\ \ OH^-\ \ } \overset{\overset{\displaystyle OH}{|}}{CH_3CH} - \overset{\overset{\displaystyle O}{\|}}{CH_2CH}$$

acetaldehyde   3-hydroxybutanal (50%)
                                 (acetaldol or aldol)

This reaction is called an **aldol addition,** or **aldol condensation reaction.** The word "aldol," derived from *ald*ehyde and alcoh*ol*, describes the product, which is a *β-hydroxy aldehyde*. A **condensation reaction** is one in which two or more molecules combine into a larger molecule often with the loss of a small molecule (such as water).

How does an aldol condensation proceed? If acetaldehyde is treated with dilute aqueous sodium hydroxide, a low concentration of enolate ions is formed. The reaction is reversible—as enolate ions undergo reaction, more are formed.

$$\overset{\overset{\displaystyle O}{\|}}{CH_3CH} + OH^- \rightleftharpoons \left[\overset{\overset{\displaystyle \ddot{O}:}{\|}}{^-\ddot{C}H_2CH} \longleftrightarrow \overset{\overset{\displaystyle :\ddot{O}:^-}{|}}{CH_2{=}CH} \right] + H_2O$$

*resonance structures for
the enolate ion*

The enolate ion undergoes reaction with another acetaldehyde molecule by adding to the carbonyl carbon to form an alkoxide ion, which abstracts a proton from water to yield the product aldol.

α hydrogens

$$\overset{\overset{\displaystyle \ddot{O}:}{\|}}{CH_3CH} + \overset{\overset{\displaystyle O}{\|}}{^-CH_2CH} \rightleftharpoons \left[\overset{\overset{\displaystyle :\ddot{O}:^-}{|}}{CH_3CH} - \overset{\overset{\displaystyle O}{\|}}{CH_2CH} \right] \xrightarrow{\ H_2O\ } \overset{\overset{\displaystyle OH}{|}}{CH_3CHCH_2CH} \overset{\overset{\displaystyle O}{\|}}{} + OH^-$$

an alkoxide ion

The starting aldehyde in an aldol condensation must contain a hydrogen α to the carbonyl group so that it can form an enolate ion in base. The aldol product still has a carbonyl group with α hydrogens. Can it undergo further reaction to form trimers? tetramers? polymers? Yes, these materials are by-products of the reaction. For simplicity, we will show only the dimer products and ignore the fact that other higher-molecular-weight products may also be formed.

We have shown the aldol condensation for acetaldehyde. Other aldehydes also undergo this self-addition. Ketones undergo aldol condensations, but the equilibrium does not favor the ketone-condensation product. (Why not?) Although there are a number of laboratory procedures that can be used to induce ketone condensations of the aldol type, the reaction is not as useful with ketones as it is with aldehydes. Therefore, we will concentrate our present discussion on aldehydes. Two other examples of aldol condensations follow:

$$
\underset{}{CH_3CH_2CH_2\overset{\overset{\displaystyle O}{\|}}{C}H} + CH_3CH_2CH_2\overset{\overset{\displaystyle O}{\|}}{C}H \xrightleftharpoons{OH^-} CH_3CH_2CH_2\overset{\overset{\displaystyle OH}{|}}{C}H-\underset{\underset{\displaystyle CH_2CH_3}{|}}{\overset{\overset{\displaystyle O}{\|}}{C}H}CHCH
$$

an aldol (75%)

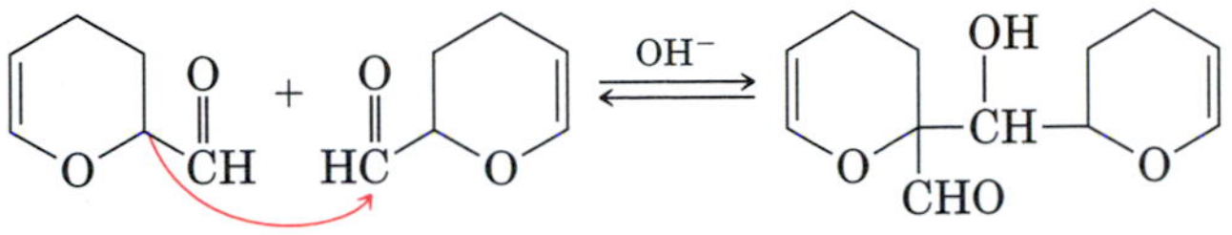

an aldol (76%)

SAMPLE PROBLEM

Show how you could prepare 3-hydroxy-2,2,4-trimethylpentanal by an aldol condensation.

Solution
Write the structure and indicate the carbon–carbon bond formed in the condensation.

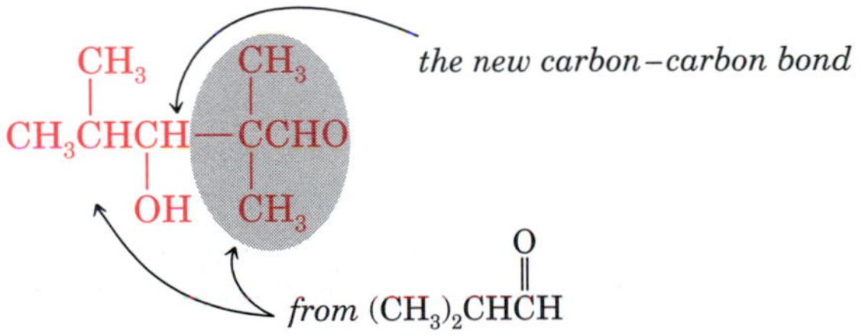

the new carbon–carbon bond

$$CH_3CHCH-CCHO$$

from $(CH_3)_2CHCH$

Write the equation.

$$
2\,(CH_3)_2CH\overset{\overset{\displaystyle O}{\|}}{C}H \xrightleftharpoons{OH^-} (CH_3)_2CH\overset{\overset{\displaystyle OH}{|}}{C}H-C(CH_3)_2\overset{\overset{\displaystyle O}{\|}}{C}H
$$

STUDY PROBLEM

17.11 Which of the following aldehydes can undergo self-condensations? Explain.

(a) 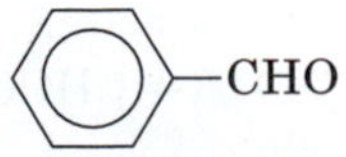 phenyl—CHO

(b) HCHO

(c) 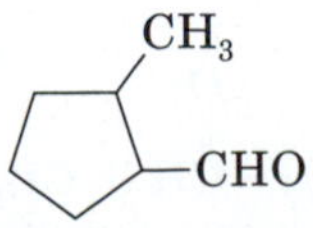 methylcyclopentyl—CHO

(d) $(CH_3)_3CCHO$

(e) $(CH_3CH_2)_2CHCHO$

A. Dehydration of Aldols

A β-hydroxy carbonyl compound, such as an aldol, undergoes dehydration readily because the double bond in the product is in conjugation with the carbonyl group. Therefore, an **α,β-unsaturated aldehyde** can be readily obtained as the product of an aldol condensation.

$$\underset{\text{3-hydroxybutanal}}{CH_3\overset{\underset{\displaystyle |}{OH}}{CH}-CH_2\overset{\underset{\displaystyle \|}{O}}{CH}} \xrightarrow[\text{warm}]{\text{dil. H}^+} \underset{\substack{\text{2-butenal} \\ \text{(crotonaldehyde)}}}{CH_3CH=CH\overset{\underset{\displaystyle \|}{O}}{CH}} + H_2O$$

When dehydration leads to a double bond in conjugation with an aromatic ring, dehydration is often spontaneous, even in alkaline solution.

$$\underset{\substack{\text{3-hydroxy-3-} \\ \text{phenylpropanal}}}{C_6H_5-\overset{\underset{\displaystyle |}{OH}}{CH}-CH_2\overset{\underset{\displaystyle \|}{O}}{CH}} \xrightarrow{\text{spontaneous}} \underset{\substack{\text{3-phenylpropenal} \\ \text{(cinnamaldehyde)}}}{C_6H_5-CH=CH\overset{\underset{\displaystyle \|}{O}}{CH}} + H_2O$$

B. Crossed Aldol Condensations

An aldehyde with no α hydrogens cannot form an enolate ion and thus cannot dimerize in an aldol condensation. However, if such an aldehyde is mixed with an aldehyde that *does* have an α hydrogen, a condensation between the two can occur. This reaction is called a **crossed aldol condensation.** A crossed aldol condensation is most useful when only one of the carbonyl compounds has an α hydrogen; otherwise, mixtures of products result. Methyl ketones can be used successfully in crossed aldol condensations with aldehydes that contain no α hydrogen, as in the second example that follows:

$$\underset{\substack{\text{benzaldehyde} \\ \textit{(no } \alpha \textit{ hydrogens)}}}{C_6H_5-\overset{\underset{\displaystyle \|}{O}}{CH}} + \underset{\text{acetaldehyde}}{CH_3\overset{\underset{\displaystyle \|}{O}}{CH}} \xrightleftharpoons{OH^-} C_6H_5-\overset{\underset{\displaystyle |}{OH}}{CH}CH_2\overset{\underset{\displaystyle \|}{O}}{CH} \xrightarrow{-H_2O} \underset{\text{cinnamaldehyde (90\%)}}{C_6H_5-CH=CH\overset{\underset{\displaystyle \|}{O}}{CH}}$$

$$C_6H_5-\overset{\underset{\displaystyle \|}{O}}{CH} + CH_3\overset{\underset{\displaystyle \|}{O}}{C}CH_3 \xrightarrow{OH^-} \underset{\substack{\text{4-phenyl-3-buten-2-one (90\%)} \\ \text{(benzalacetone)}}}{C_6H_5-CH=CH\overset{\underset{\displaystyle \|}{O}}{C}CH_3} + H_2O$$

$$\underset{\text{acetone}}{}$$

17.12   Predict the major products:

    (a) $C_6H_5CHO + CH_3CH_2CH_2CHO \xrightarrow{\text{OH}^-}$

    (b) $C_6H_5CHO + CH_3CH_2\overset{\displaystyle O}{\overset{\displaystyle \|}{C}}CH_2CH_3 \xrightarrow{\text{OH}^-}$

C.   Syntheses Using Aldol Condensations

In an aldol condensation, two types of product can result: (1) β-hydroxy alde-hydes or ketones, and (2) α,β-unsaturated aldehydes or ketones (Figure 17.6). In a synthesis problem, look for these functional groups and decide which aldehydes or ketones must be used for the starting materials.

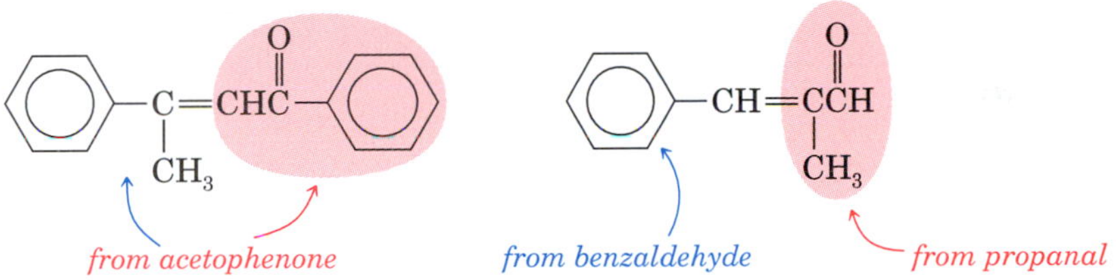

from acetophenone   *from benzaldehyde*   *from propanal*

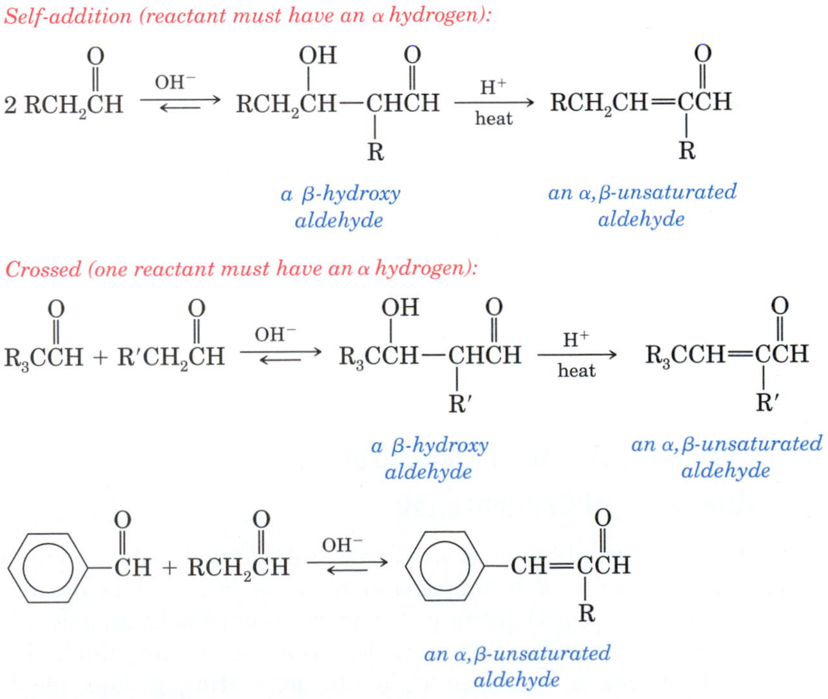

Figure 17.6   Products from aldol condensations.

SAMPLE PROBLEM

SAMPLE PROBLEM

The following ketone can be prepared in 90% yield by a crossed aldol condensation:

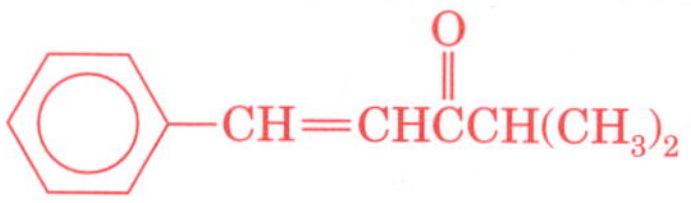

What organic reactants are needed?

Solution

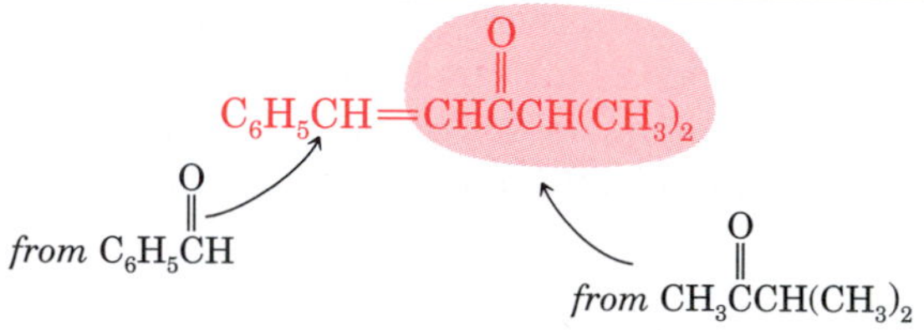

STUDY PROBLEM

17.13 Suggest syntheses for the following compounds from aldehydes or ketones:

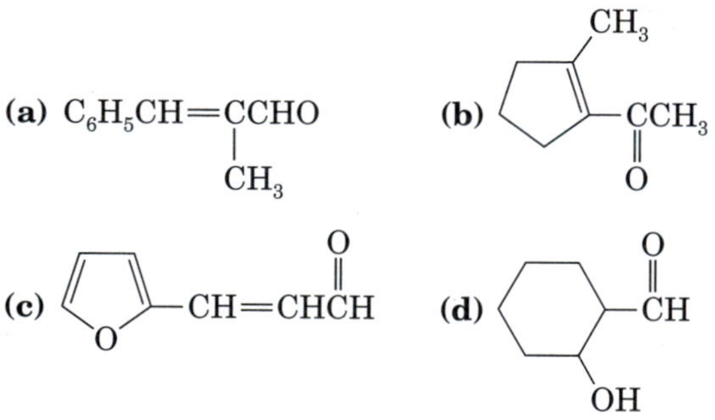

Section 17.7

Reactions Related to the Aldol Condensation

A. Knoevenagel Condensation

We have shown aldol condensations and crossed aldol condensations. However, for this type of condensation to occur, all that is needed is one compound with a carbonyl group plus one compound with an acidic hydrogen. The **Knoevenagel condensation** is the reaction of an aldehyde with a compound that has a hydrogen α to *two* activating groups (such as C=O or C≡N), usually with ammonia or an amine as the catalyst. Under these conditions, malonic acid itself can be used as a reactant.

Knoevenagel condensations:

$$(CH_3)_2CHCH_2\overset{\overset{\displaystyle O}{\|}}{C}H + CH_2(CO_2C_2H_5)_2 \xrightarrow[\text{heat}]{\substack{\text{piperidine} \\ \text{benzene}}} (CH_3)_2CHCH_2CH{=}C(CO_2C_2H_5)_2 + H_2O$$

3-methylbutanal   diethyl malonate   (78%)

benzaldehyde   malonic acid   3-phenylpropenoic acid (85%)
                                                          (cinnamic acid)

A variation of the Knoevenagel reaction allows the less reactive ketones to undergo condensation with the more acidic ethyl cyanoacetate ($pK_a = 9$, compared with $pK_a = 11$ for diethyl malonate).

ethyl cyanoacetate   (80%)

SAMPLE PROBLEM

How would you prepare the following compound?

Solution

from C_6H_5CHO *from* $C_6H_5CH_2CN$

$$C_6H_5\overset{\overset{\displaystyle O}{\|}}{C}H + \overset{\overset{\displaystyle CN}{|}}{C}H_2C_6H_5 \underset{}{\overset{OH^-}{\rightleftharpoons}} C_6H_5CH{=}\overset{\overset{\displaystyle CN}{|}}{C}C_6H_5 + H_2O$$

STUDY PROBLEM

17.14 Suggest synthetic routes to the following compounds:

(a) $(CH_3CH_2)_2C{=}\overset{\overset{\displaystyle O}{\|}}{\underset{\underset{\displaystyle CN}{|}}{C}}COC_2H_5$ (b) $=C(CN)_2$

B. A Biological Aldol-Type Condensation

Esters with the $\overset{\overset{\textstyle O}{\|}}{R C S}$— unit are called **thioesters.** Thioesters undergo reactions, such as hydrolysis, similar to those of ordinary esters. **Acetylcoenzyme A,** which is important in biological reactions, is a thioester.

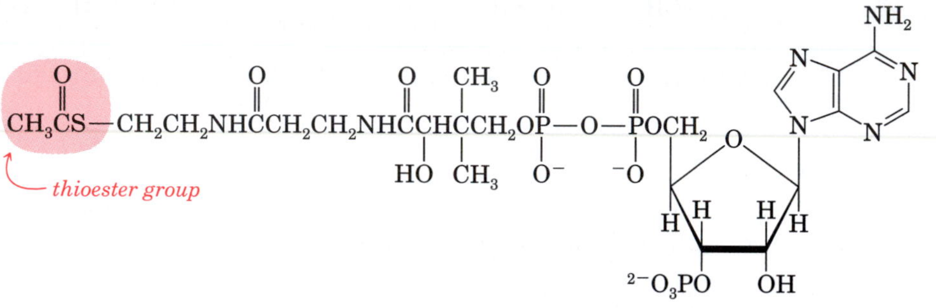

acetylcoenzyme A

A proton from the acetyl group (at the left in the preceding formula) of acetylcoenzyme A can be removed enzymatically. The resultant enolate ion is used in many biological transformations. For example, this enolate ion can undergo an aldol-type condensation by adding to other carbonyl compounds. The following equations show loss of H⁺ from acetylcoenzyme A, nucleophilic attack on the carbonyl group of the oxaloacetate ion, and finally hydrolysis of the thioester to yield the citrate ion and coenzyme A (abbreviated HSCoA).

Enzymatic loss of proton:

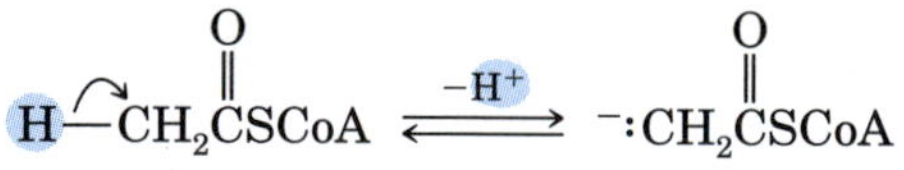

acetylcoenzyme A

Nucleophilic attack and thioester hydrolysis:

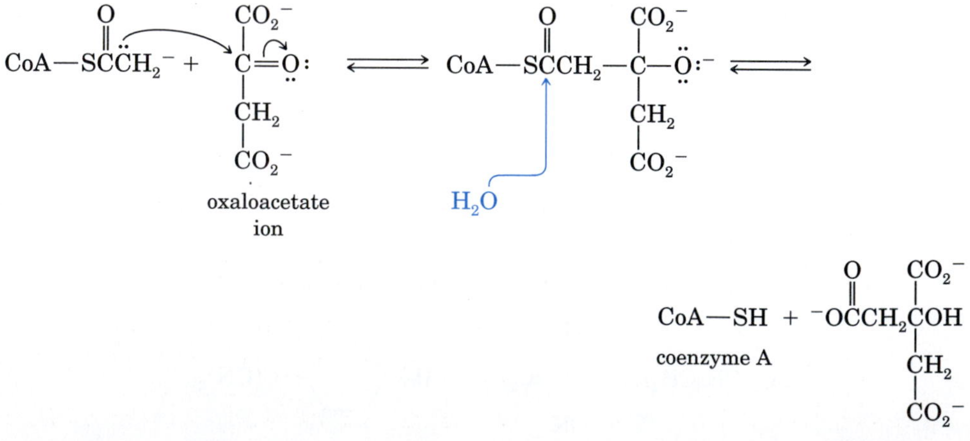

oxaloacetate
ion

H_2O

coenzyme A

citrate ion

Section 17.8
Ester Condensations

A. Claisen Condensations

Esters with α hydrogens can undergo self-condensation reactions to yield **β-keto esters.** An ester condensation is similar to an aldol condensation; the difference is that the —OR group of an ester can act as a leaving group. The result is therefore *substitution* (whereas aldol condensations are *additions*). Simple ester condensations, such as the following examples, are called **Claisen condensations.** (In the examples below, note the use of *oxo* as a prefix in the name of the β-keto ester to designate the position of the keto group.)

General:

$$RCH_2\overset{O}{\overset{\|}{C}}\!-\!OR' + H\!-\!\overset{}{\underset{R}{\overset{|}{C}H}}COR' \xrightarrow{\text{base}} RCH_2\overset{O}{\overset{\|}{C}}\!-\!\underset{R}{\overset{|}{C}H}\overset{O}{\overset{\|}{C}}OR' + R'OH$$

a β-keto ester

$$2\ CH_3\overset{O}{\overset{\|}{C}}OC_2H_5 \xrightarrow{Na^+\ {}^-OC_2H_5} CH_3\overset{O}{\overset{\|}{C}}\!-\!CH_2\overset{O}{\overset{\|}{C}}OC_2H_5\ +\ C_2H_5OH$$

ethyl acetate   ethyl 3-oxobutanoate (75%)
(ethyl acetoacetate)

$$2\ CH_3CH_2\overset{O}{\overset{\|}{C}}OC_2H_5 \xrightarrow{Na^+\ {}^-OC_2H_5} CH_3CH_2\overset{O}{\overset{\|}{C}}\!-\!\underset{CH_3}{\overset{|}{C}H}\overset{O}{\overset{\|}{C}}OC_2H_5\ +\ C_2H_5OH$$

ethyl propanoate

ethyl 2-methyl-3-oxopentanoate (45%)

Let us look at the stepwise reaction. First is the *formation of the enolate of the ester* by an acid–base reaction with the alkoxide ion. (An alkoxide, rather than a hydroxide, is used as the base to prevent saponification of the ester.) As in the aldol condensation, a low concentration of enolate is formed because the enolate (with only one carbonyl) is a stronger base than the alkoxide ion.

Enolate formation:

$$CH_3\overset{O}{\overset{\|}{C}}OC_2H_5 + {}^-\!\!:\!\overset{..}{O}C_2H_5 \rightleftharpoons \left[{}^-\!\!:\!CH_2\!-\!\overset{:\overset{..}{O}}{\overset{\|}{C}}OC_2H_5 \longleftrightarrow CH_2\!=\!\overset{:\overset{..}{O}:^-}{\overset{|}{C}}OC_2H_5 \right] + C_2H_5\overset{..}{O}H$$

ethyl acetate   *resonance structures for the enolate ion*

The nucleophilic carbon then attacks the carbonyl group in a typical carbonyl *addition* reaction. This addition of the enolate is followed by *elimination* of ROH. The entire sequence is thus a typical nucleophilic acyl substitution reaction of a carbonyl compound, similar to those you encountered in Chapter 15.

Attack on carbonyl group:

$$CH_3\overset{\overset{\textstyle \ddot{O}:}{\|}}{C}OC_2H_5 + {}^-CH_2CO_2C_2H_5 \underset{\text{addition}}{\rightleftharpoons} \left[CH_3\overset{\overset{\textstyle :\ddot{O}:^-}{|}}{\underset{\underset{\textstyle CH_2CO_2C_2H_5}{|}}{C}}OC_2H_5 \right]$$

Loss of ROH:

$$\left[CH_3\overset{\overset{\textstyle :\ddot{O}:^-}{|}}{\underset{\underset{\textstyle CH_2CO_2C_2H_5}{|}}{C}}{-}\ddot{O}C_2H_5 \right] \underset{\text{elimination}}{\rightleftharpoons} \left[CH_3\overset{\overset{\textstyle :\ddot{O}}{\|}}{\underset{\underset{\textstyle CH_2CO_2C_2H_5}{|}}{C}} + {}^-\!:\ddot{O}C_2H_5 \right] \rightleftharpoons$$

$$CH_3\overset{\overset{\textstyle O}{\|}}{C}\overset{-}{\ddot{C}}HCO_2C_2H_5 + H\ddot{O}C_2H_5$$

*the enolate of
ethyl acetoacetate*

The product β-keto ester is more acidic than an alcohol because it has hydrogens that are α to two carbonyl groups. Therefore, the product of the condensation is the enolate salt of the β-keto ester. In most Claisen condensations, the formation of this resonance-stabilized enolate ion is the single step that drives the reaction sequence to completion. The β-keto ester is produced when the reaction mixture is acidified with cold dilute mineral acid.

$$CH_3\overset{\overset{\textstyle O}{\|}}{C}\overset{-}{\ddot{C}}HCO_2C_2H_5 \xrightarrow{H^+} CH_3\overset{\overset{\textstyle O}{\|}}{C}CH_2CO_2C_2H_5$$

ethyl acetoacetate

a β-keto ester

A β-keto ester can be hydrolyzed by heating in acidic solution, in which case decarboxylation may occur.

Hydrolysis and decarboxylation:

$$CH_3\overset{\overset{\textstyle O}{\|}}{C}CH_2CO_2C_2H_5 \xrightarrow[-C_2H_5OH]{H^+,\ H_2O,\ \text{heat}} CH_3\overset{\overset{\textstyle O}{\|}}{C}CH_2CO_2H \xrightarrow[\text{heat}]{-CO_2} CH_3\overset{\overset{\textstyle O}{\|}}{C}CH_3$$

a β-keto ester   *a β-keto acid*   *a ketone*

SAMPLE PROBLEM

Predict the product of the ester condensation of methyl butanoate with sodium methoxide as the base, followed by acidification.

Solution

1. Write the structure of the starting ester and determine the structure of the enolate ion.

$$CH_3CH_2CH_2\overset{\displaystyle O}{\overset{\|}{C}}OCH_3 + {}^-OCH_3 \rightleftharpoons CH_3CH_2\overset{\displaystyle O}{\underset{..}{\overset{\|}{C}HC}}OCH_3 + CH_3OH$$

2. Write the overall equation for nucleophilic attack on the carbonyl group and loss of ROH.

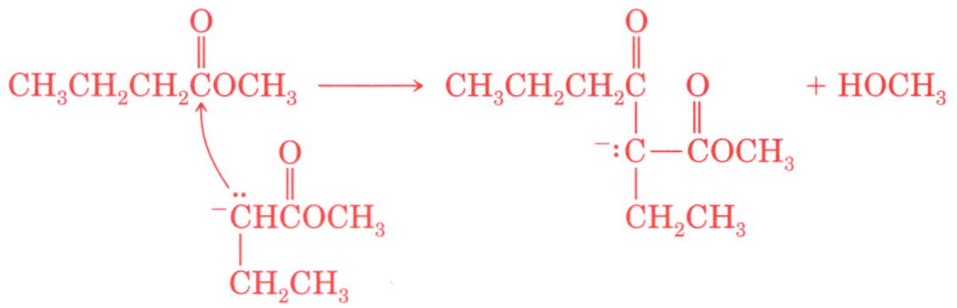

$$CH_3CH_2CH_2\overset{\displaystyle O}{\overset{\|}{C}}OCH_3 \longrightarrow CH_3CH_2CH_2C \quad O \quad + HOCH_3$$

$$\underset{\overset{\displaystyle |}{CH_2CH_3}}{\overset{\displaystyle O}{\underset{..}{\overset{\|}{\overset{\displaystyle -}{C}}HC}}OCH_3}$$

$$\overset{\displaystyle -:C-COCH_3}{\underset{CH_2CH_3}{}}$$

3. Acidify:

$$\xrightarrow{\ H^+\ } CH_3CH_2CH_2\overset{\displaystyle O \quad O}{\overset{\| \quad \|}{C}CHC}OCH_3$$
$$\underset{\overset{|}{CH_2CH_3}}{}$$

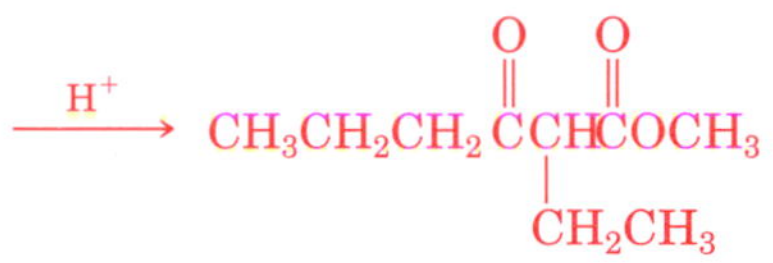

4. Write the equation for the combined reactions:

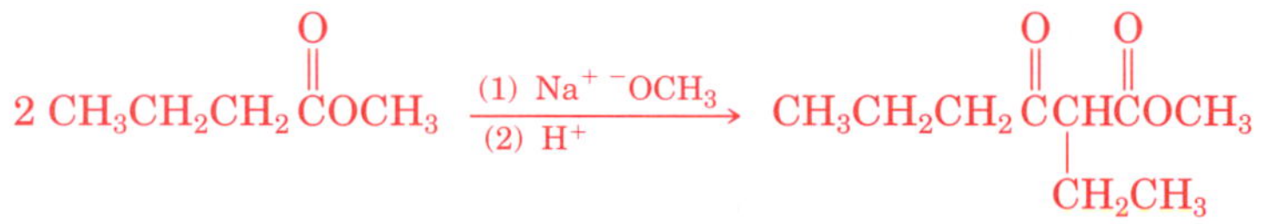

$$2\ CH_3CH_2CH_2\overset{\displaystyle O}{\overset{\|}{C}}OCH_3 \xrightarrow[\text{(2) } H^+]{\text{(1) } Na^+\ {}^-OCH_3} CH_3CH_2CH_2\overset{\displaystyle O \quad O}{\overset{\| \quad \|}{C}CHC}OCH_3$$
$$\underset{\overset{|}{CH_2CH_3}}{}$$

STUDY PROBLEM

17.15   Predict the major organic product:

(a) $C_6H_5CH_2CO_2C_2H_5 \xrightarrow[\text{(2) } H^+]{\text{(1) } Na^+\ {}^-OC_2H_5}$

(b) [the product from (a)] $\xrightarrow[\text{heat}]{H_2O,\ H^+}$

SAMPLE PROBLEM

A Claisen-like condensation, called a **Dieckmann ring closure,** is used to prepare the following cyclic ketone from a diester. What is the structure of the diester?

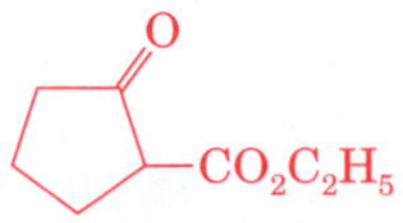

Solution
Because the keto group arises from attack of an α carbon, the ring must be closed at the following position:

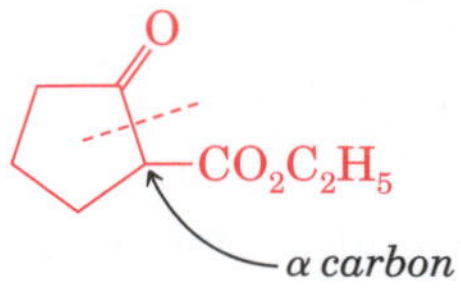

Therefore, the starting diester must be diethyl adipate.

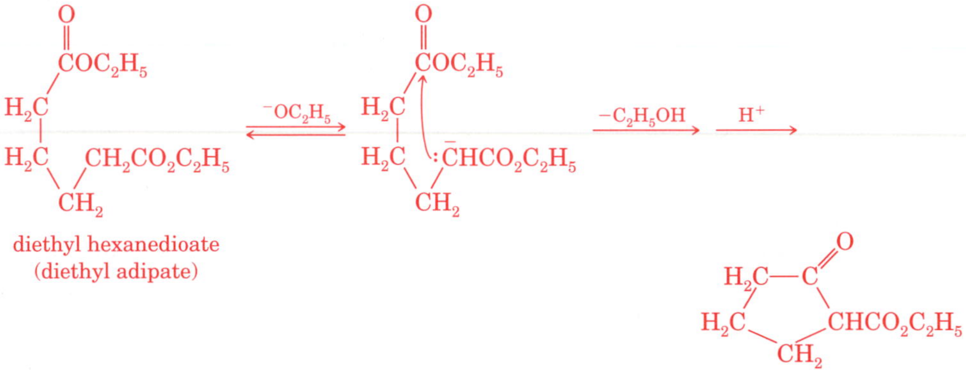

diethyl hexanedioate
(diethyl adipate)

STUDY PROBLEM

17.16 Write equations for the preparation of the following cyclic compounds from open-chain starting materials:

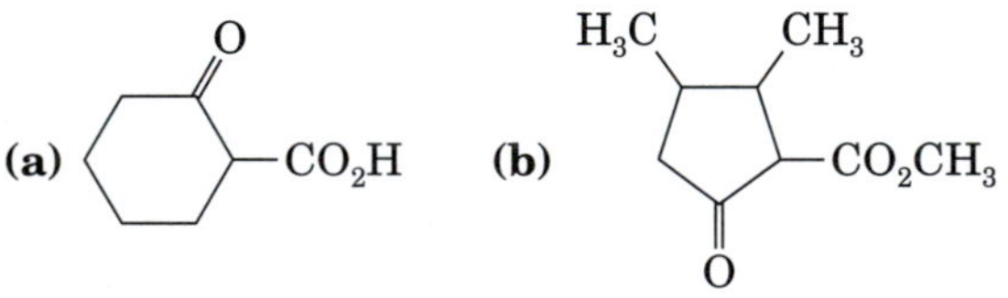

(a) $-CO_2H$ (b) $-CO_2CH_3$

B. Crossed Claisen Condensations

Two different esters can be used in a Claisen condensation. Best results (that is, avoidance of mixtures) are obtained if only one of the esters has an α hydrogen.

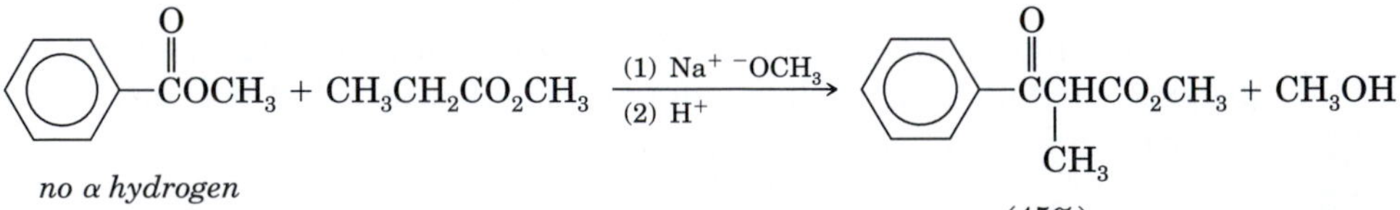

$$\text{C}_6\text{H}_5\text{—COCH}_3 + CH_3CH_2CO_2CH_3 \xrightarrow[\text{(2) H}^+]{\text{(1) Na}^+ \text{ }^-\text{OCH}_3} \text{C}_6\text{H}_5\text{—CCHCO}_2CH_3 + CH_3OH$$

no α hydrogen

(45%)

Crossed Claisen condensations can be successfully carried out between ketones and esters whether the ester contains an α hydrogen or not. The α hydrogen of the ketone is removed preferentially because ketones are more acidic than esters. For this reason, the crossed Claisen is favored over a self-Claisen condensation of the ester.

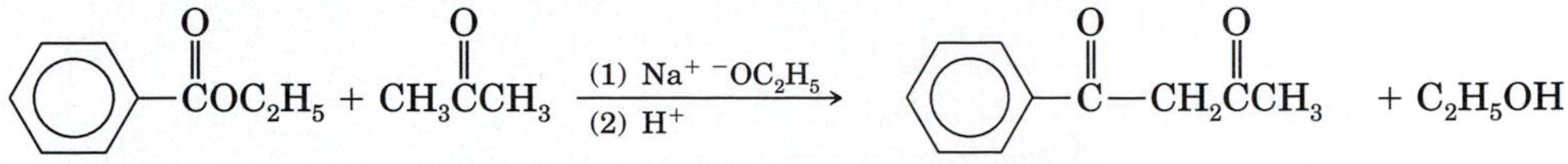

$$\text{C}_6\text{H}_5-\overset{\overset{\displaystyle O}{\|}}{\text{C}}\text{OC}_2\text{H}_5 + \text{CH}_3\overset{\overset{\displaystyle O}{\|}}{\text{C}}\text{CH}_3 \quad \xrightarrow[\text{(2) H}^+]{\text{(1) Na}^+ \ ^-\text{OC}_2\text{H}_5} \quad \text{C}_6\text{H}_5-\overset{\overset{\displaystyle O}{\|}}{\text{C}}-\text{CH}_2\overset{\overset{\displaystyle O}{\|}}{\text{C}}\text{CH}_3 \quad + \ \text{C}_2\text{H}_5\text{OH}$$

ethyl benzoate   acetone   1-phenyl-1,3-butanedione (40%)

$$\text{CH}_3(\text{CH}_2)_4\overset{\overset{\displaystyle O}{\|}}{\text{C}}\text{OC}_2\text{H}_5 + \text{CH}_3\overset{\overset{\displaystyle O}{\|}}{\text{C}}\text{CH}_3 \quad \xrightarrow[\text{(2) H}^+]{\text{(1) NaH}} \quad \text{CH}_3(\text{CH}_2)_4\overset{\overset{\displaystyle O}{\|}}{\text{C}}\text{CH}_2\overset{\overset{\displaystyle O}{\|}}{\text{C}}\text{CH}_3 + \text{C}_2\text{H}_5\text{OH}$$

ethyl hexanoate   acetone   2,4-nonanedione (65%)

SAMPLE PROBLEM

A mixture of acetone and diethyl oxalate ($C_2H_5O_2C-CO_2C_2H_5$) is added to a mixture of sodium ethoxide in ethanol. After the reaction is completed, the mixture is treated with cold, dilute HCl. A condensation product is isolated from the mixture in 60% yield. What is the product?

Solution

$$(1) \ \text{CH}_3\overset{\overset{\displaystyle O}{\|}}{\text{C}}\text{CH}_3 + \ ^-\text{OC}_2\text{H}_5 \ \rightleftharpoons \ ^-\text{CH}_2\overset{\overset{\displaystyle O}{\|}}{\text{C}}\text{CH}_3 + \text{HOC}_2\text{H}_5$$

$$(2) \ \text{C}_2\text{H}_5\text{O}\overset{\overset{\displaystyle O}{\|}}{\text{C}}-\overset{\overset{\displaystyle O}{\|}}{\text{C}}\text{OC}_2\text{H}_5 + \ ^-\text{CH}_2\overset{\overset{\displaystyle O}{\|}}{\text{C}}\text{CH}_3 \ \xrightarrow{\text{H}^+}$$

$$\underbrace{\text{C}_2\text{H}_5\text{O}\overset{\overset{\displaystyle O}{\|}}{\text{C}}-\overset{\overset{\displaystyle O}{\|}}{\text{C}}}_{\textit{from diethyl oxalate}}-\underbrace{\text{CH}_2\overset{\overset{\displaystyle O}{\|}}{\text{C}}\text{CH}_3}_{\textit{from acetone}} + \text{HOC}_2\text{H}_5$$

STUDY PROBLEM

17.17   Predict the products:

(a) $C_2H_5O_2C-CO_2C_2H_5 + CH_3CO_2C_2H_5 \xrightarrow[\text{(2) H}^+]{\text{(1) Na}^+ \ ^-\text{OC}_2\text{H}_5}$

(b)  (cyclopentylidene)$=O + CH_3CO_2C_2H_5 \xrightarrow[\text{(2) H}^+]{\text{(1) Na}^+ \ ^-\text{OC}_2\text{H}_5}$

C. Syntheses Using Ester Condensations

Because the product of an ester condensation between two esters is a β-keto ester (or a ketone after hydrolysis and decarboxylation), the decision of which starting materials to use is not difficult. The keto group comes from one starting ester; the ester group with its substituent comes from the other starting ester. The different types of products from ester condensations are summarized in Figure 17.7.

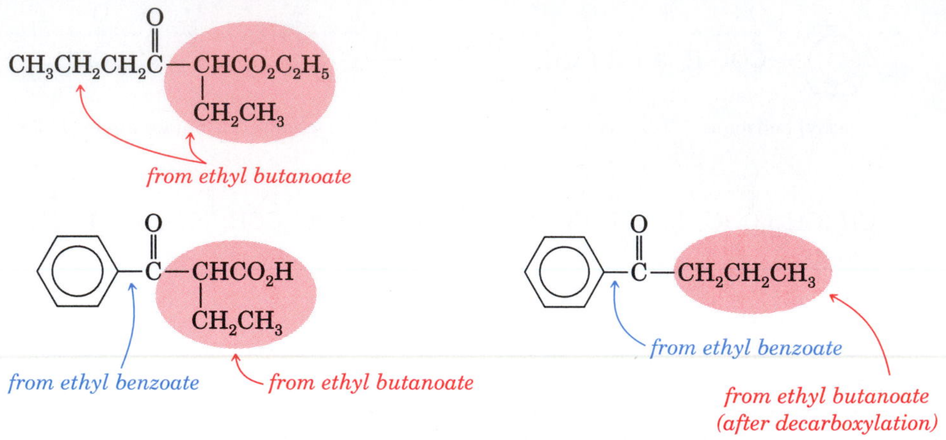

$$CH_3CH_2CH_2\overset{\overset{\displaystyle O}{\|}}{C}-\underset{\underset{\displaystyle CH_2CH_3}{|}}{CH}CO_2C_2H_5$$

from ethyl butanoate

$$\langle\!\!\!\bigcirc\!\!\!\rangle-\overset{\overset{\displaystyle O}{\|}}{C}-\underset{\underset{\displaystyle CH_2CH_3}{|}}{CH}CO_2H$$

from ethyl benzoate *from ethyl butanoate*

$$\langle\!\!\!\bigcirc\!\!\!\rangle-\overset{\overset{\displaystyle O}{\|}}{C}-CH_2CH_2CH_3$$

from ethyl benzoate

from ethyl butanoate
(after decarboxylation)

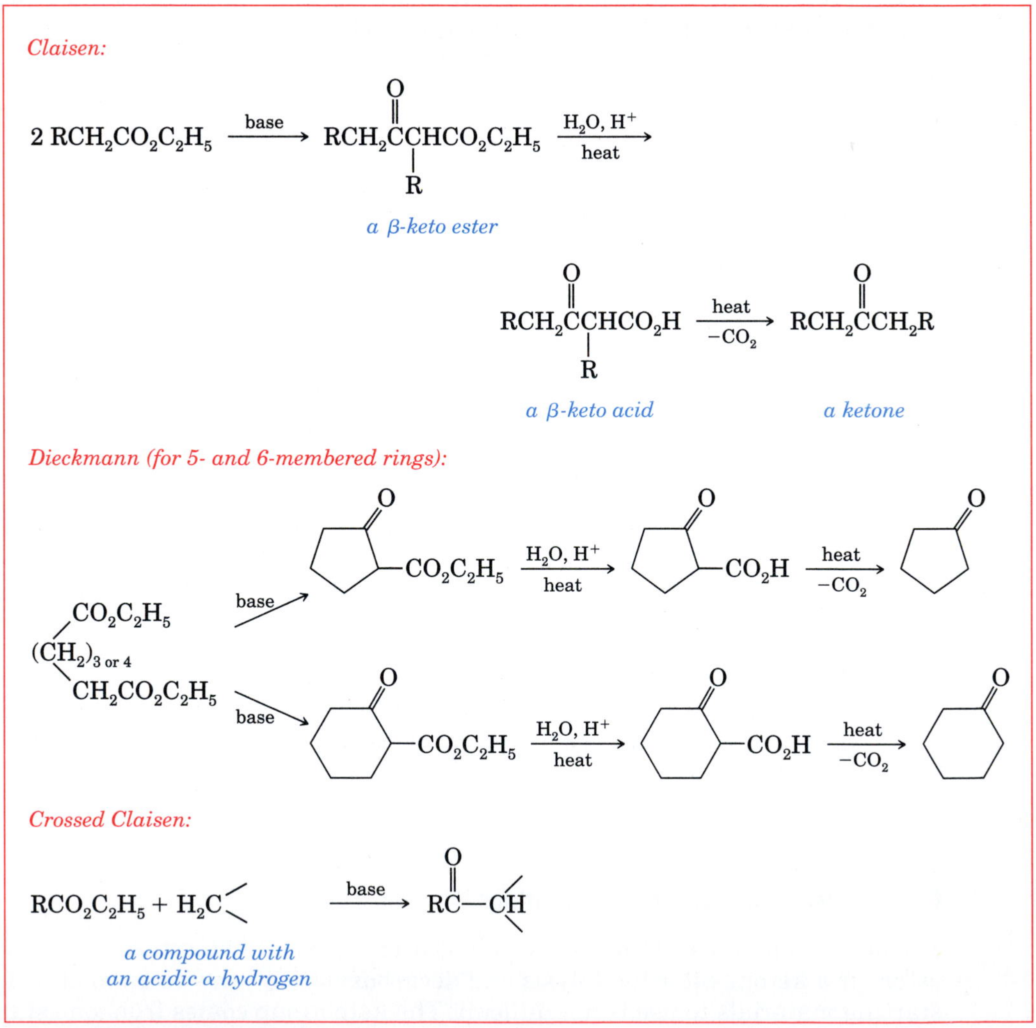

Claisen:

$$2\ RCH_2CO_2C_2H_5 \xrightarrow{\text{base}} RCH_2\overset{\overset{\displaystyle O}{\|}}{\underset{\underset{\displaystyle R}{|}}{C}}CHCO_2C_2H_5 \xrightarrow[\text{heat}]{H_2O,\ H^+}$$

a β-keto ester

$$RCH_2\overset{\overset{\displaystyle O}{\|}}{\underset{\underset{\displaystyle R}{|}}{C}}CHCO_2H \xrightarrow[-CO_2]{\text{heat}} RCH_2\overset{\overset{\displaystyle O}{\|}}{C}CH_2R$$

a β-keto acid *a ketone*

Dieckmann (for 5- and 6-membered rings):

$$(CH_2)_{3\ or\ 4}\overset{\overset{\displaystyle CO_2C_2H_5}{}}{\underset{\underset{\displaystyle CH_2CO_2C_2H_5}{}}{}}$$

base → [cyclopentanone with CO_2C_2H_5] $\xrightarrow[\text{heat}]{H_2O,\ H^+}$ [cyclopentanone with CO_2H] $\xrightarrow[-CO_2]{\text{heat}}$ [cyclopentanone]

base → [cyclohexanone with CO_2C_2H_5] $\xrightarrow[\text{heat}]{H_2O,\ H^+}$ [cyclohexanone with CO_2H] $\xrightarrow[-CO_2]{\text{heat}}$ [cyclohexanone]

Crossed Claisen:

$$RCO_2C_2H_5 + H_2C\!\!<\quad \xrightarrow{\text{base}}\quad R\overset{\overset{\displaystyle O}{\|}}{C}-CH\!\!<$$

a compound with
an acidic α hydrogen

Figure 17.7 Products of ester condensations.

SAMPLE PROBLEM

What reactants would you need to prepare the following compound by an ester condensation?

Solution

STUDY PROBLEMS

17.18   Each of the following β-keto esters can be prepared by an ester or a crossed-ester condensation. For each compound, write formulas for the ester or esters required.

(a) $C_6H_5\overset{O}{\overset{\|}{C}}CH_2\overset{O}{\overset{\|}{C}}OC_2H_5$ **(b)** $H\overset{O}{\overset{\|}{C}}CH_2\overset{O}{\overset{\|}{C}}OC_2H_5$

(c) $C_4H_9\overset{O}{\overset{\|}{C}}\underset{\underset{CH_2CH_2CH_3}{|}}{C}H\overset{O}{\overset{\|}{C}}OC_2H_5$ **(d)** $C_2H_5O\overset{O}{\overset{\|}{C}}-\overset{O}{\overset{\|}{C}}CH_2\overset{O}{\overset{\|}{C}}OC_2H_5$

17.19   In untreated *diabetes mellitus,* a lack of insulin or an inability to use insulin effectively leads to the poor utilization of glucose (blood sugar) by the cells. One result of this failure to metabolize glucose is that insufficient oxaloacetate ion is formed to break down acetylcoenzyme A (Section 17.7B). Consequently, the level of acetylcoenzyme A increases. In the liver, excess acetylcoenzyme A is converted to β-hydroxybutyrate ions and **ketone bodies:** acetoacetate ion and acetone. (A diabetic's breath and urine may smell of acetone.) Write equations showing how acetylcoenzyme A can be converted to acetoacetate ion and acetone.

Section 17.9

Nucleophilic Addition to α,β-Unsaturated Carbonyl Compounds

A double bond in conjugation with a carbonyl group is susceptible to nucleophilic attack in a 1,4-addition reaction (Section 16.4B).

$$\text{Nu:}^- + \text{CH}_2\!=\!\text{CH}\!-\!\overset{\overset{\displaystyle :\ddot{O}:}{\|}}{\text{C}}\text{CH}_3 \longrightarrow \left[\begin{array}{c} \overset{\overset{\displaystyle :\ddot{O}:^-}{\|}}{\text{CH}_2\!-\!\text{CH}\!=\!\text{C}\text{CH}_3} \\ | \\ \text{Nu} \end{array}\right] \xrightarrow{\text{H}^+} \overset{\overset{\displaystyle \cdot\ddot{O}\cdot}{\|}}{\text{CH}_2\text{CH}_2\text{C}\text{CH}_3} \\ \qquad\qquad\qquad\qquad\qquad\qquad\qquad\qquad\qquad\qquad\qquad\qquad\qquad\quad \text{Nu}$$

an enolate

If an α,β-unsaturated carbonyl compound can undergo nucleophilic attack, we might expect that an enolate ion could add to the double bond. Indeed, it does. The intermediate is the enolate of the 1,4-addition product. The final product is obtained by acidification. This useful synthetic reaction is called a **Michael addition.**

$$\text{CH}_2\!=\!\text{CH}\!-\!\overset{\overset{\displaystyle :\ddot{O}:}{\|}}{\text{C}}\text{H} + {}^-\text{CH}(\text{CO}_2\text{C}_2\text{H}_5)_2 \Longleftrightarrow \left[\begin{array}{c} \overset{\overset{\displaystyle :\ddot{O}:^-}{|}}{\text{CH}_2\!-\!\text{CH}\!=\!\text{CH}} \\ | \\ \text{CH}(\text{CO}_2\text{C}_2\text{H}_5)_2 \end{array}\right] \Longleftrightarrow$$

an enolate

$$\left[\begin{array}{c} \overset{\overset{\displaystyle O}{\|}}{\text{CH}_2\text{CH}_2\text{CH}} \\ | \\ {}^-\!:\text{C}(\text{CO}_2\text{C}_2\text{H}_5)_2 \end{array}\right] \xrightarrow{\text{H}^+} \overset{\overset{\displaystyle O}{\|}}{\text{CH}_2\text{CH}_2\text{CH}} \\ \qquad\qquad\qquad\qquad\qquad\qquad | \\ \qquad\qquad\qquad\qquad\qquad\qquad \text{CH}(\text{CO}_2\text{C}_2\text{H}_5)_2$$

more stable enolate   (50%)

$$\text{C}_6\text{H}_5\!-\!\text{CH}\!=\!\text{CH}\!-\!\overset{\overset{\displaystyle O}{\|}}{\text{C}}\text{OC}_2\text{H}_5 + {}^-\!:\text{CH}(\text{CO}_2\text{C}_2\text{H}_5)_2 \Longleftrightarrow \left[\begin{array}{c} \overset{\overset{\displaystyle O^-}{|}}{\text{C}_6\text{H}_5\!-\!\text{CH}\!-\!\text{CH}\!=\!\text{COC}_2\text{H}_5} \\ | \\ \text{CH}(\text{CO}_2\text{C}_2\text{H}_5)_2 \end{array}\right]$$

$$\Longleftrightarrow \left[\begin{array}{c} \overset{\overset{\displaystyle O}{\|}}{\text{C}_6\text{H}_5\!-\!\text{CHCH}_2\text{C}\text{OC}_2\text{H}_5} \\ | \\ {}^-\!:\text{C}(\text{CO}_2\text{C}_2\text{H}_5)_2 \end{array}\right] \xrightarrow{\text{H}^+} \overset{\overset{\displaystyle O}{\|}}{\text{C}_6\text{H}_5\!-\!\text{CHCH}_2\text{C}\text{OC}_2\text{H}_5} \\ \qquad\qquad\qquad\qquad\qquad\qquad\qquad\qquad | \\ \qquad\qquad\qquad\qquad\qquad\qquad\qquad\qquad \text{CH}(\text{CO}_2\text{C}_2\text{H}_5)_2 \\ \qquad\qquad\qquad\qquad\qquad\qquad\qquad\qquad\quad (98\%)$$

more stable enolate

The product of the last example shown is a triester. Saponification, followed by acidification, gives a triacid. Under the conditions of acid hydrolysis, however, decarboxylation may occur.

Acid hydrolysis and decarboxylation:

$$\text{C}_6\text{H}_5\!-\!\text{CHCH}_2\text{CO}_2\text{C}_2\text{H}_5 \underset{}{\overset{\text{H}_2\text{O, H}^+ \atop \text{heat}}{\rightleftarrows}} \text{C}_6\text{H}_5\!-\!\text{CHCH}_2\text{CO}_2\text{H} \xrightarrow[\text{heat}]{-\text{CO}_2} \\ \quad | \qquad\qquad\qquad\qquad\qquad\qquad\qquad\qquad | \\ \quad \text{CH}(\text{CO}_2\text{C}_2\text{H}_5)_2 \qquad\qquad\qquad\qquad \text{CH}(\text{CO}_2\text{H})_2$$

a triester   *a triacid*

$$\text{C}_6\text{H}_5\!-\!\text{CHCH}_2\text{CO}_2\text{H} \\ \qquad | \\ \qquad \text{CH}_2\text{CO}_2\text{H}$$

a diacid

In the decarboxylation of the triacid, it is the malonic acid grouping that loses CO_2. (Why?) This group is highlighted in color in the following equation.

The types of products that we can obtain from simple Michael additions are shown in Figure 17.8. Michael additions in combination with other condensations are exceedingly useful in laboratory syntheses of complex cyclic compounds, such as steroids (Section 24.5). A portion of one such synthesis is shown. This particular ring-forming sequence (Michael plus aldol) is called a **Robinson annulation.**

In an important variation of this sequence, one enantiomer of a chiral amine is used to catalyze the ring-closure reaction and thus to induce formation of a single enantiomer of the product.

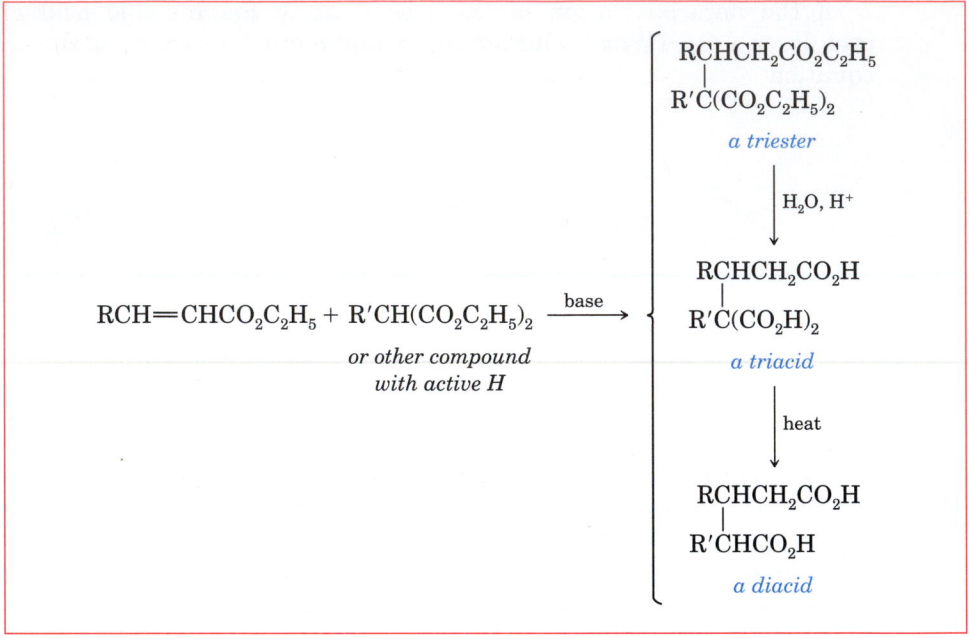

Figure 17.8 Products of Michael addition of a malonic ester with an α,β-unsaturated ester.

Show by equations how you would prepare the following keto acid by a Michael addition:

$$CH_3\overset{O}{\overset{\|}{C}}CH_2CH_2CH_2CO_2H$$

Solution

Addition occurs β to the keto group in a 1,4-addition.

β to ketone

$$CH_3\overset{O}{\overset{\|}{C}}CH_2CH_2 \quad \boxed{CH_2CO_2H}$$

from malonic ester

from $CH_3\overset{O}{\overset{\|}{C}}CH=CH_2$

1. $CH_2(CO_2C_2H_5)_2 \xrightarrow{NaOC_2H_5} \ ^-\!\!:CH(CO_2C_2H_5)_2$

2. $CH_3\overset{\overset{\ddot{O}:}{\|}}{C}-CH=CH_2 + \ ^-\!\!:CH(CO_2C_2H_5)_2 \rightleftharpoons CH_3\overset{:\ddot{O}:^-}{\overset{|}{C}}=CHCH_2CH(CO_2C_2H_5)_2$

$$\xrightarrow{H^+} CH_3\overset{O}{\overset{\|}{C}}CH_2CH_2CH(CO_2C_2H_5)_2$$

3. $\xrightarrow[\text{heat}]{H_2O, H^+} CH_3\overset{O}{\overset{\|}{C}}CH_2CH_2CH_2CO_2H$

A chemist decided to try the sequence of reactions in the preceding sample problem. After Step 2, the chemist discovered a *cyclic* product, along with the keto diester that had been predicted. What happened?

Solution

The initial product of Step 2 underwent a Dieckmann ring closure. (Note that this intermediate has a nucleophilic carbon that can attack a carbonyl group to yield a six-membered ring.)

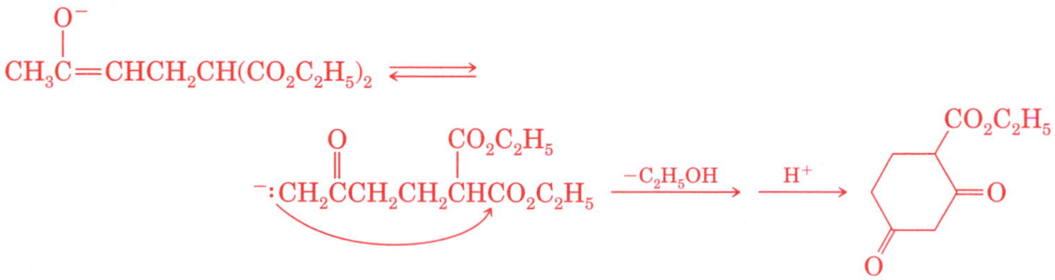

$$
\overset{\displaystyle O^-}{\underset{\displaystyle |}{CH_3C}}=CHCH_2CH(CO_2C_2H_5)_2 \;\rightleftharpoons\;
$$

$$
{}^-\!:CH_2\overset{\displaystyle O}{\overset{\displaystyle \|}{C}}CH_2CH_2\overset{\displaystyle CO_2C_2H_5}{\underset{\displaystyle |}{CH}}CO_2C_2H_5 \xrightarrow{-C_2H_5OH} \xrightarrow{H^+}
$$

How would you prepare the following keto ester by a Michael addition?

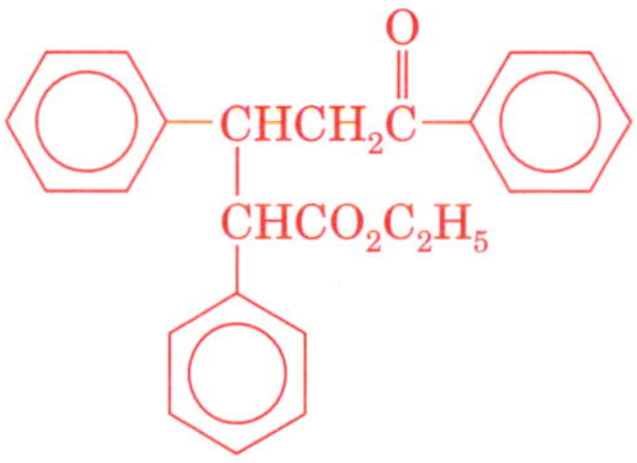

Solution

Addition occurs β to the ketone.

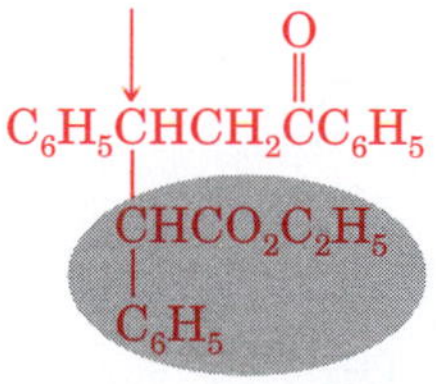

$$
C_6H_5\underset{\displaystyle |}{CH}CH_2\overset{\displaystyle O}{\overset{\displaystyle \|}{C}}C_6H_5
$$
$$
\underset{\displaystyle |}{CHCO_2C_2H_5}
$$
$$
C_6H_5
$$

$$
C_6H_5CH=CH\overset{\displaystyle O}{\overset{\displaystyle \|}{C}}C_6H_5 + \underset{\displaystyle |}{CH_2CO_2C_2H_5} \xrightarrow{Na^+\; {}^-OC_2H_5} \xrightarrow{H^+} product
$$
$$
\quad\quad\quad\quad\quad\quad\quad\quad\quad C_6H_5
$$

STUDY PROBLEMS

17.20 Show by equations how you would prepare the following carboxylic acids by Michael additions:

(a) $(HO_2C)_2\overset{\overset{\displaystyle CH_3}{|}}{C}\underset{\underset{\displaystyle CH_2CH_3}{|}}{C}HCH_2CO_2H$ **(b)** $HO_2CCH_2\underset{\underset{\displaystyle CH_3}{|}}{C}HCH_2CO_2H$

(c)

CO_2H / CH_2CO_2H / $C(CH_3)_3$

17.21 Write equations showing how you would prepare the following compounds by alkylation, addition, or condensation reactions.

(a) $(CH_3)_3\overset{\overset{\displaystyle O}{\|}}{C}C\overset{\overset{\displaystyle O}{\|}}{C}H_2\overset{\overset{\displaystyle O}{\|}}{C}OC_2H_5$ **(b)**

$CH_2CO_2C_2H_5$

(c) $CH_3CH_2\overset{\overset{\displaystyle OH}{|}}{C}H\overset{\overset{\displaystyle O}{\|}}{C}H\underset{\underset{\displaystyle CH_3}{|}}{C}H$ **(d)** $(CH_3)_2C{=}C(CN)_2$

(e) $C_6H_5\overset{\overset{\displaystyle O}{\|}}{C}CH_2\overset{\overset{\displaystyle O}{\|}}{C}{-}\overset{\overset{\displaystyle O}{\|}}{C}OC_2H_5$ **(f)** $H_3C{-}\bigcirc{-}CH{=}$

CH_3

Summary

In this chapter, we have looked at a variety of ways to synthesize compounds with complex carbon skeletons. Each of these reactions is caused by a species with carbanion character.

Alkylations:

$$\text{>}CH^- + RX \longrightarrow \text{>}CHR + X^-$$

Condensations:

$$\text{>}CH^- + R{-}\overset{\overset{\displaystyle O}{\|}}{C}{-}R \longrightarrow R{-}\underset{\underset{\displaystyle -CH-}{|}}{\overset{\overset{\displaystyle OH}{|}}{C}}{-}R$$

1,4-Additions:

$$\text{>}CH^- + CH_2{=}CH\overset{\overset{\displaystyle O}{\|}}{C}{-} \longrightarrow \underset{\underset{\displaystyle -CH-}{|}}{CH_2}{-}CH_2\overset{\overset{\displaystyle O}{\|}}{C}{-}$$

Table 17.2 gives an overview of the more important products available from these reactions.

Table 17.2   Major synthetic reactions involving enolates and carbanions

Reaction	Section Reference

Malonic Ester:

$$CH_2(CO_2C_2H_5)_2 \xrightarrow[\text{(2) } ^-OC_2H_5,\, R'X]{\text{(1) } ^-OC_2H_5,\, RX} \underset{R}{\overset{R}{>}}C(CO_2C_2H_5)_2$$

17.2

Acetoacetic Ester:

$$CH_3\overset{O}{\overset{\|}{C}}CH_2CO_2C_2H_5 \xrightarrow[\text{(2) } ^-OC_2H_5,\, R'X]{\text{(1) } ^-OC_2H_5,\, RX} CH_3\overset{O}{\overset{\|}{C}}C\underset{R\;\;R}{}CO_2C_2H_5$$

17.3

Enamine:

$$R_2CH\overset{O}{\overset{\|}{C}}R \xrightarrow[\text{(3) } H_2O,\, H^+]{\overset{\text{(1) } R_2NH,\, H^+}{\text{(2) } R'X}} R_2C\overset{O}{\overset{\|}{C}}\underset{R'}{}R$$

17.5

Aldol Condensation:

$$2\ RCH_2CHO \rightleftharpoons[\;OH^-\;] RCH_2\underset{R}{\overset{OH}{C}H}CHCHO$$

17.6

Crossed Aldol:

$$RCHO + R'CH_2CHO \rightleftharpoons[\;OH^-\;] R\overset{OH}{\underset{R'}{C}H}CHCHO$$

17.6B

Claisen Condensation:

$$2\ RCH_2CO_2C_2H_5 \xrightarrow[\text{(2) } H^+]{\text{(1) } ^-OC_2H_5} RCH_2\overset{O}{\overset{\|}{C}}\underset{R}{C}HCO_2C_2H_5$$

17.8

Crossed Claisen:

$$R\overset{O}{\overset{\|}{C}}OC_2H_5 + R'CH_2CO_2C_2H_5 \xrightarrow[\text{(2) } H^+]{\text{(1) } ^-OC_2H_5} R\overset{O}{\overset{\|}{C}}\underset{R'}{C}HCO_2C_2H_5$$

17.8B

Michael Addition:

$$RCH{=}CHCO_2C_2H_5 + R'CH(CO_2C_2H_5)_2 \xrightarrow[\text{(2) } H^+]{\text{(1) } ^-OC_2H_5} RCH{-}CH_2CO_2C_2H_5$$
$$R'C(CO_2C_2H_5)_2$$

17.9

Essay Problem for Chapter 17

Thiazolium Catalysts

In living systems, vitamin B_1 activates aldehydes for condensation reactions. The catalytic portion of vitamin B_1, a thiazolium ion, can be incorporated into synthetic compounds that are useful as catalysts for aldehyde condensation reactions.

Vitamin B$_1$
a biological thiazolium catalyst

a synthetic thiazolium catalyst

Thiazolium-catalyzed aldehyde reactions appear to involve the conjugate addition of an acyl group. For example, 2,5-hexanedione and related compounds can be prepared by the reaction of an aldehyde with an α,β-unsaturated ketone in the presence of a thiazolium catalyst. Without the thiazolium catalyst, an enolate-type addition of the aldehyde to the unsaturated ketone (Michael addition) would be expected.

$$CH_3\overset{O}{\overset{\|}{C}}H + CH_2=CHC\overset{O}{\overset{\|}{C}}CH_3 \xrightarrow[\text{catalyst}]{\text{thiazolium}} CH_3\overset{O}{\overset{\|}{C}}CH_2CH_2\overset{O}{\overset{\|}{C}}CH_3$$

61%

The catalytic activity of the thiazolium ion resides in the acidity of the carbon-2 hydrogen. The thiazolium ion is nucleophilic and will attack the carbonyl carbon of aldehydes. Rearrangement of the initial adduct gives a stabilized betaine. Attack of the betaine at the β carbon of the unsaturated ketone, followed by enolization and loss of the thiazolium ion, gives the final product. The mechanism for this reaction follows. [H. Setter, *Angew. Chem. Int. Ed. Engl.* **1976,** *15,* 639.]

(1) [thiazolium ion] $\rightleftharpoons$ [thiazolium ylide] $+ H^+$

(2) $CH_3\overset{O}{\overset{\|}{C}}H +$ [ylide] $\rightleftharpoons$ $CH_3\overset{O^-}{\underset{H}{C}}\text{—}[\text{thiazolium}]$ $\rightleftharpoons$ **2**

1
a betaine

(3) $CH_3\overset{O}{\overset{\|}{C}}CH=CH_2 + $ **2** $\longrightarrow$ $CH_3\overset{O^-}{C}=CHCH_2\overset{OH}{\underset{CH_3}{C}}\text{—}[\text{thiazolium}]$ $\longrightarrow$ **3**

(4) **3** $\longrightarrow$ $CH_3\overset{O}{\overset{\|}{C}}CH_2CH_2\overset{O}{\overset{\|}{C}}CH_3 +$ [thiazolium ylide]

Questions

1. Write the product of the enolate addition of acetaldehyde and 2-butenone.
2. Draw resonance structures that show why betaine **1** is resonance-stabilized.
3. If vitamin B$_1$ were dissolved in D$_2$O, which hydrogens of the vitamin would exchange with deuterium? Write the structure of the deuterated vitamin.
4. What are the structures of the intermediates **2** and **3** in the preceding mechanism?
5. Predict the products for the following reactions.

(a) 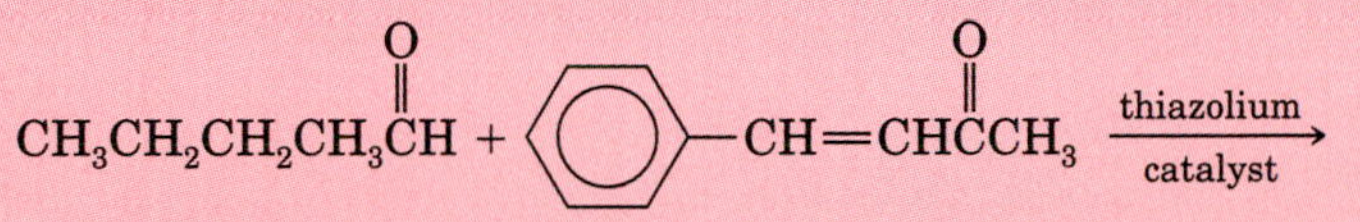 $CH_3CH_2CH_2CH_2\overset{\displaystyle O}{\overset{\|}{C}}H$ + (phenyl)$-CH=CHC\overset{\displaystyle O}{\overset{\|}{C}}CH_3$ $\xrightarrow[\text{catalyst}]{\text{thiazolium}}$

(b) 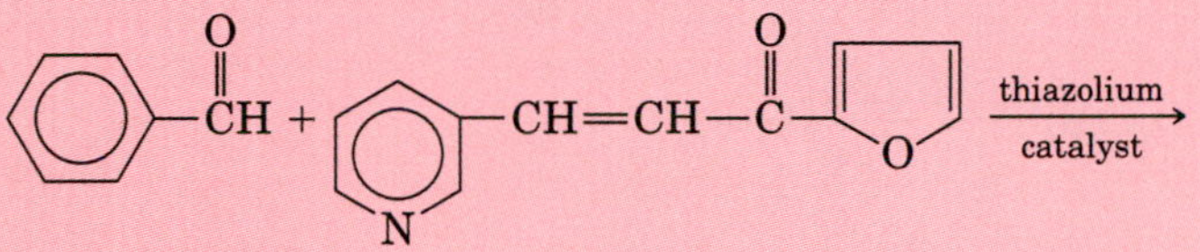 (phenyl)$-\overset{\displaystyle O}{\overset{\|}{C}}H$ + (pyridyl)$-CH=CH-\overset{\displaystyle O}{\overset{\|}{C}}-$(furyl) $\xrightarrow[\text{catalyst}]{\text{thiazolium}}$

(c) $(CH_3)_2C=CH\overset{\displaystyle O}{\overset{\|}{C}}H$ + $CH_2=CHCN$ $\xrightarrow[\text{catalyst}]{\text{thiazolium}}$

Study Problems*

17.22 Circle the most acidic hydrogen in each of the following compounds:

(a) $CH_3O\overset{\displaystyle O}{\overset{\|}{C}}\overset{\displaystyle}{\underset{\displaystyle CH_3}{C}}H\overset{\displaystyle O}{\overset{\|}{C}}OCH_3$

(b) $HO\overset{\displaystyle O}{\overset{\|}{C}}CH_2\overset{\displaystyle O}{\overset{\|}{C}}OCH_3$

(c) $HOCH_2\overset{\displaystyle O}{\overset{\|}{C}}CH_2\overset{\displaystyle O}{\overset{\|}{C}}OC_2H_5$

(d) 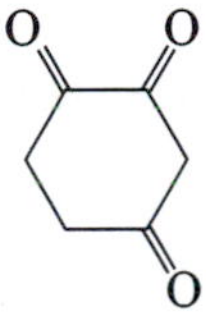 (cyclohexane-1,3,5-trione structure)

(e) 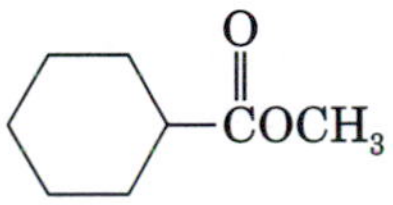 (cyclohexyl)$-\overset{\displaystyle O}{\overset{\|}{C}}OCH_3$

(f) 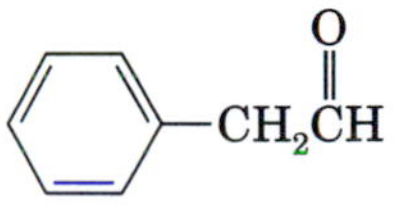 (phenyl)$-CH_2\overset{\displaystyle O}{\overset{\|}{C}}H$

17.23 Complete the following equations for acid–base reactions:

(a) $C_6H_5\underset{\displaystyle CH_2CH_3}{\overset{\displaystyle}{C}}HCO_2C_2H_5$ + $CH_3CH_2O^-\,Na^+$ $\rightleftharpoons$

(b) $CH_3CH_2NO_2$ + $NaNH_2$ $\rightleftharpoons$

(c) $CH_3\overset{\displaystyle O}{\overset{\|}{C}}CH_2CO_2C_2H_5$ + NaH $\rightleftharpoons$

(d) $CH_3\overset{\displaystyle O}{\overset{\|}{C}}CH_2\overset{\displaystyle O}{\overset{\|}{C}}CH_3$ + $Na^+\ :\!\bar{C}H(CO_2CH_2CH_3)_2$ $\rightleftharpoons$

* For information concerning the organization of the *Study Problems* and *Additional Problems*, see the *Note to student* on page 41.

17.24 Write resonance structures for the following anions. (Include electron dots.)

(a) $CH_3\bar{C}HCN$ (b) $CH_3\bar{C}(CO_2CH_3)_2$

(c) $CH_3\overset{\overset{\displaystyle S}{\|}}{C}CH_2^-$ (d) $CH_3\overset{\overset{\displaystyle O}{\|}}{C}-\underset{\underset{\displaystyle CH_3}{|}}{C}-\overset{\overset{\displaystyle O}{\|}}{C}OC_2H_5$

17.25 Predict the major organic product:

(a) 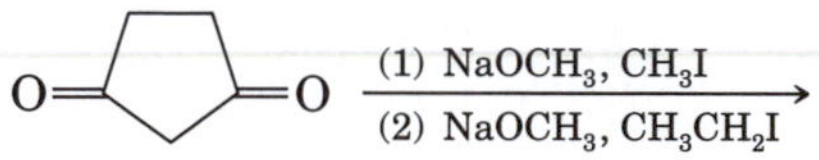 $O{=}\!\!\!\overset{\bigcirc}{}\!\!\!{=}O$ $\xrightarrow[\text{(2) NaOCH}_3,\ \text{CH}_3\text{CH}_2\text{I}]{\text{(1) NaOCH}_3,\ \text{CH}_3\text{I}}$

(b) 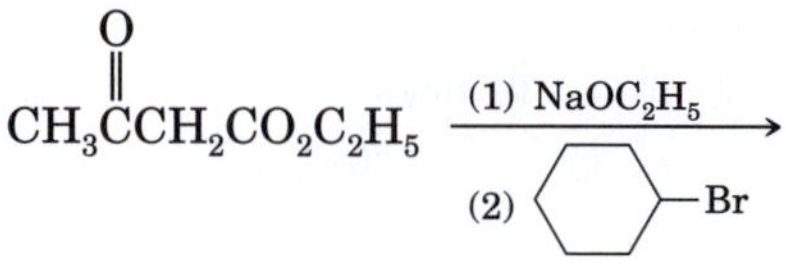 $CH_3\overset{\overset{\displaystyle O}{\|}}{C}CH_2CO_2C_2H_5$ $\xrightarrow[\text{(2)} \bigcirc\!\!-\!\text{Br}]{\text{(1) NaOC}_2\text{H}_5}$

(c)   [the product from (b)] $\xrightarrow[\text{heat}]{\text{H}_2\text{O, H}^+}$

(d) $CH_3\overset{\overset{\displaystyle O}{\|}}{C}CH_2CO_2C_2H_5$ $\xrightarrow[\text{(2) BrCH}_2\overset{\overset{\displaystyle O}{\|}}{C}CH_3]{\text{(1) NaOC}_2\text{H}_5}$

(e)   [the product from (d)] $\xrightarrow[\text{heat}]{\text{H}_2\text{O, H}^+}$

(f) 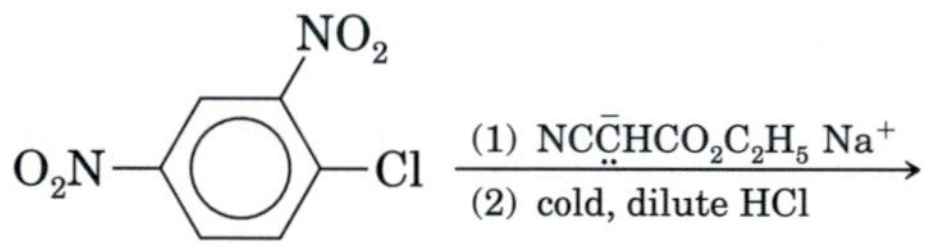 $O_2N\!-\!\!\overset{\overset{\displaystyle NO_2}{}}{\bigcirc}\!\!-\!Cl$ $\xrightarrow[\text{(2) cold, dilute HCl}]{\text{(1) NC}\bar{\text{C}}\text{HCO}_2\text{C}_2\text{H}_5\ \text{Na}^+}$

17.26 Using flow equations, show how each of the following compounds could be prepared starting with either malonic ester or acetoacetic ester:

(a) $C_6H_{13}CH_2\overset{\overset{\displaystyle O}{\|}}{C}CH_3$ (b) $CH_3CH(CH_2OH)_2$ (c) $CH_3\overset{\overset{\displaystyle O}{\|}}{C}\underset{\underset{\displaystyle CH_2CH_3}{|}}{C}HCH_2CH_2CH_3$

(d) $(CH_3CH_2)_2CHCO_2H$ (e) 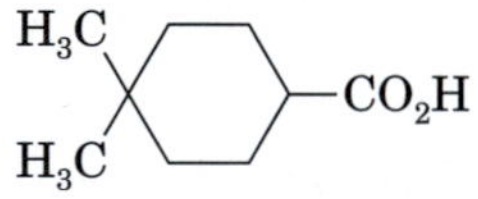 $\underset{H_3C}{\overset{H_3C}{}}\!\!\!>\!\!\!\bigcirc\!\!-\!CO_2H$

17.27 Predict the hydrolysis products of the following enamines.

(a) 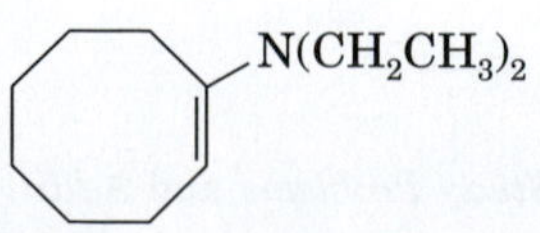 $\bigcirc\!\!\!-\!N(CH_2CH_3)_2$ (b) 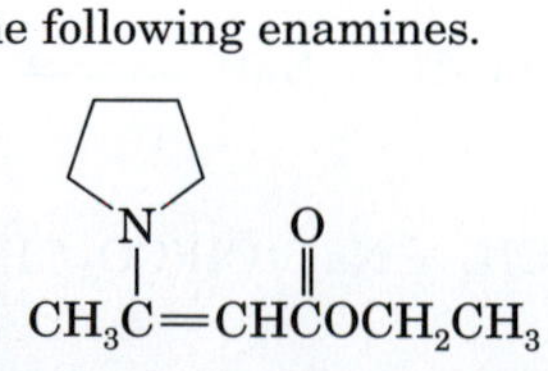 $CH_3C{=}CH\overset{\overset{\displaystyle O}{\|}}{C}OCH_2CH_3$ with $\overset{\bigcirc}{N}$ on the carbon

(c) 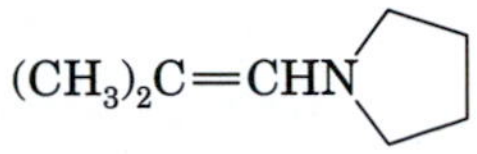 $(CH_3)_2C=CHN$⟨pyrrolidine⟩   (d) 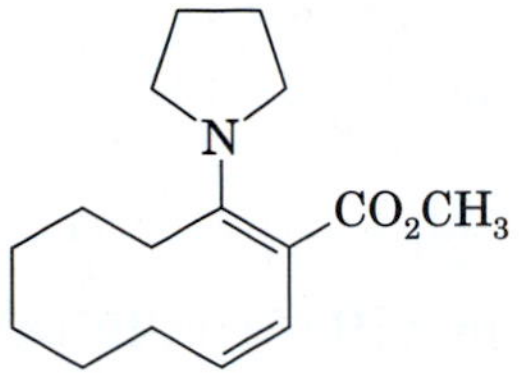 ⟨macrocyclic structure with N-pyrrolidine and CO_2CH_3 substituents⟩

17.28 Suggest a synthetic procedure for each of the following conversions:

(a) 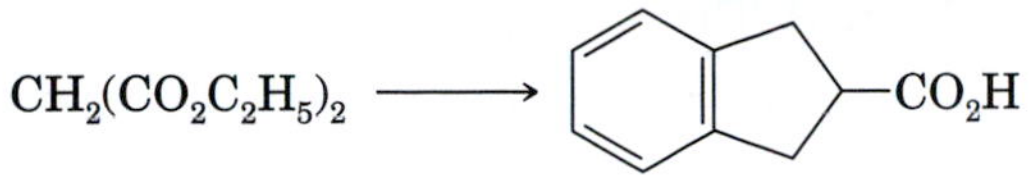 $CH_2(CO_2C_2H_5)_2$ ⟶ ⟨indane structure⟩$-CO_2H$

(b) 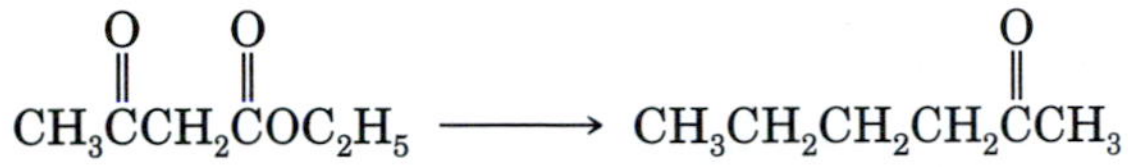 $CH_3\overset{O}{\overset{\|}{C}}CH_2\overset{O}{\overset{\|}{C}}OC_2H_5$ ⟶ $CH_3CH_2CH_2CH_2\overset{O}{\overset{\|}{C}}CH_3$

(c) ⟨cyclohexanone⟩$=O$ ⟶ ⟨cyclohexanone with $CH_2CH=CH_2$ substituent⟩$=O$ (by two different methods)

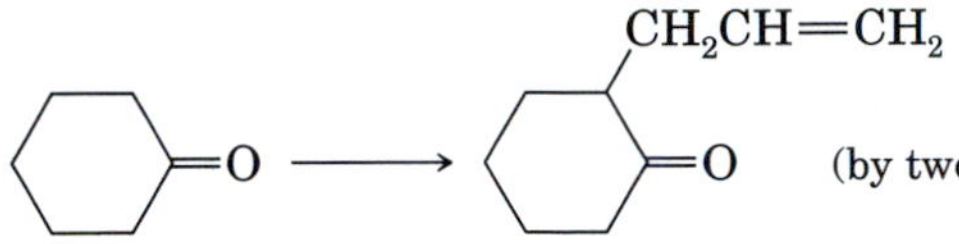

17.29 What aldehydes would be needed to prepare the following aldol condensation products?

(a) 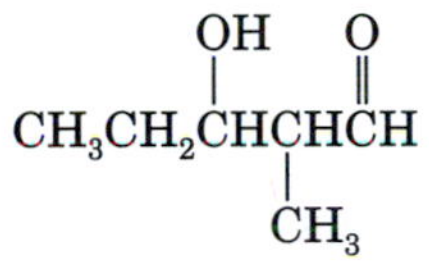 $CH_3CH_2\overset{OH}{\overset{|}{CH}}\overset{}{\underset{\underset{CH_3}{|}}{CH}}\overset{O}{\overset{\|}{CH}}$

(b) $C_6H_5CH_2\overset{OH}{\overset{|}{CH}}\overset{}{\underset{\underset{C_6H_5}{|}}{CH}}\overset{O}{\overset{\|}{CH}}$

(c) $C_6H_5\overset{OH}{\overset{|}{CH}}\overset{}{\underset{\underset{C_6H_5}{|}}{CH}}\overset{O}{\overset{\|}{CH}}$

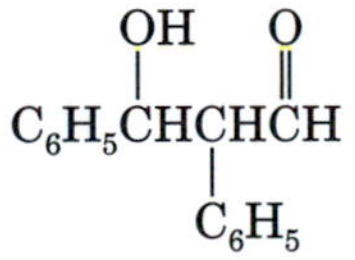

(d)  ⟨cyclooctane ring with $\overset{O}{\overset{\|}{C}}H$ and OH substituents⟩

17.30 Predict the major organic product:

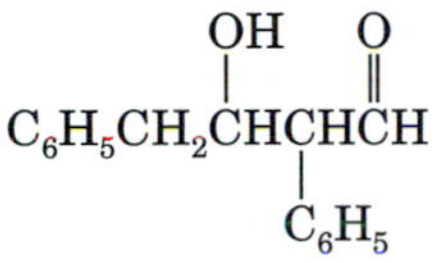

(a) 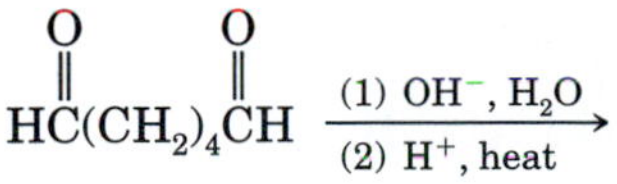 $H\overset{O}{\overset{\|}{C}}(CH_2)_4\overset{O}{\overset{\|}{C}}H$ $\xrightarrow[\text{(2) H}^+\text{, heat}]{\text{(1) OH}^-\text{, H}_2\text{O}}$

(b) $C_6H_5CHO +$ ⟨cyclopentanone⟩$=O$ $\xrightarrow{\text{OH}^-}$

(c) ⟨cyclodecane-1,6-dione⟩ $\xrightarrow[\text{(2) H}^+\text{, heat}]{\text{(1) OH}^-\text{, H}_2\text{O}}$

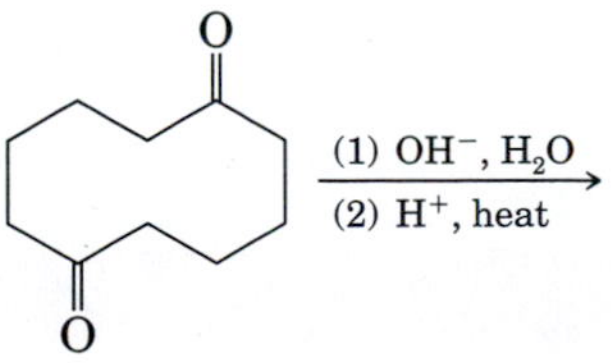

(d) 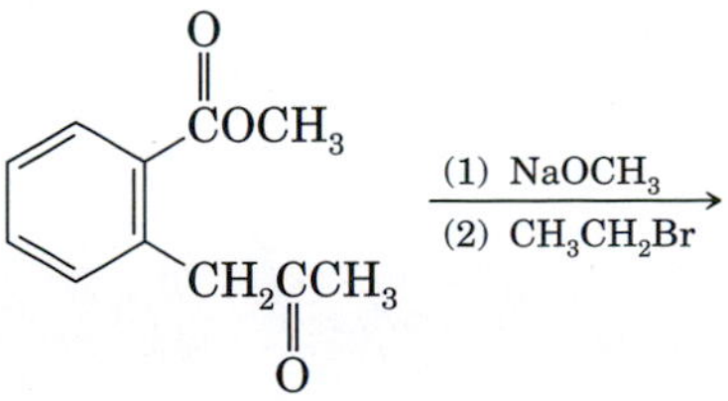 ⟨benzene ring with $\overset{O}{\overset{\|}{C}}OCH_3$ and $CH_2\overset{O}{\overset{\|}{C}}CH_3$ substituents⟩ $\xrightarrow[\text{(2) CH}_3\text{CH}_2\text{Br}]{\text{(1) NaOCH}_3}$

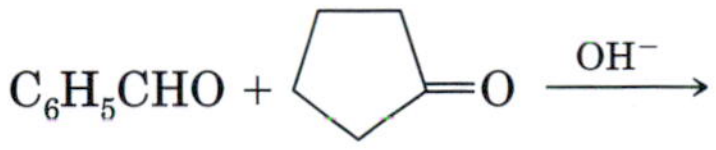

(e) 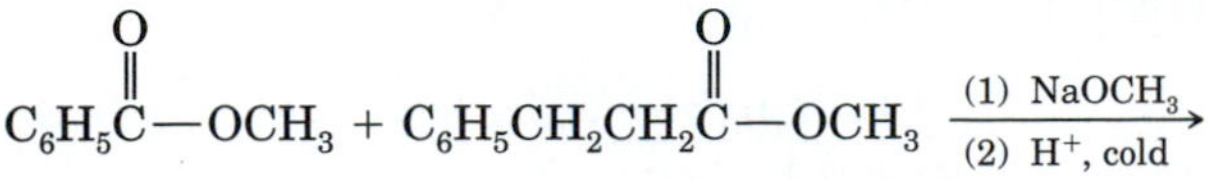 $C_6H_5\overset{O}{\overset{\|}{C}}-OCH_3 + C_6H_5CH_2CH_2\overset{O}{\overset{\|}{C}}-OCH_3$ $\xrightarrow[\text{(2) H}^+\text{, cold}]{\text{(1) NaOCH}_3}$

17.31 How would you make the following conversions?

(a) cyclopentanone to ⟨cyclopentanone with $\overset{O}{\overset{\|}{C}}OCH_2CH_3$ substituent⟩

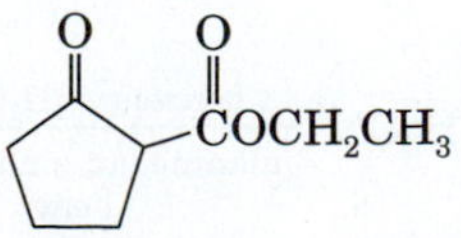

(b) 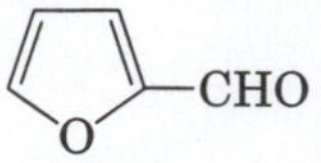 [furan]—CHO to [furan]—CH=CHNO$_2$

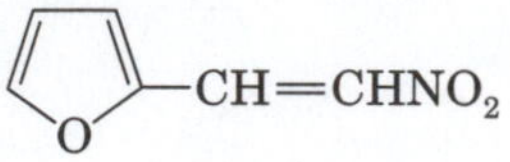

(c) benzaldehyde to $C_6H_5CH=CHCOCH_2CH_3$

(d) benzaldehyde to $C_6H_5CH=CHCCH=CHC_6H_5$ (with C=O)

17.32 Predict the major organic product of each of the following Michael addition reactions:

(a) 1 $CH_2=CHCCH_2CH_3$ + $CH_2(CO_2C_2H_5)_2$ $\xrightarrow[\text{(2) H}^+\text{, cold}]{\text{(1) NaOC}_2\text{H}_5}$

(b) 2 $CH_2=CHCCH_2CH_3$ + $CH_2(CO_2C_2H_5)_2$ $\xrightarrow[\text{(2) H}^+\text{, cold}]{\text{(1) NaOC}_2\text{H}_5}$

(c) [cyclohexenone]=O + $CH_3CCH_2CCH_3$ $\xrightarrow[\text{(2) H}^+\text{, cold}]{\text{(1) NaOC}_2\text{H}_5}$

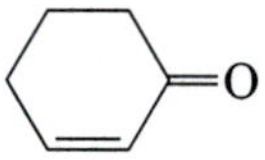

(d) $CH_3CH=CHCO_2C_2H_5$ + $C_6H_5CH_2CO_2C_2H_5$ $\xrightarrow[\text{(2) H}^+\text{, cold}]{\text{(1) NaOC}_2\text{H}_5}$

17.33 How would you prepare the following compounds by Michael additions?

(a) $CH_3CCHCH_2CH_2CN$ (with C=O; and $CO_2C_2H_5$ substituent)

(b) 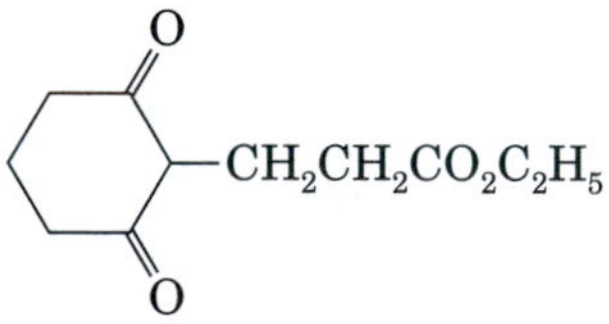 [cyclohexane-1,3-dione]—$CH_2CH_2CO_2C_2H_5$

(c) $(CH_3)_2CCH_2CH_2CN$ (with NO_2 substituent)

(d) 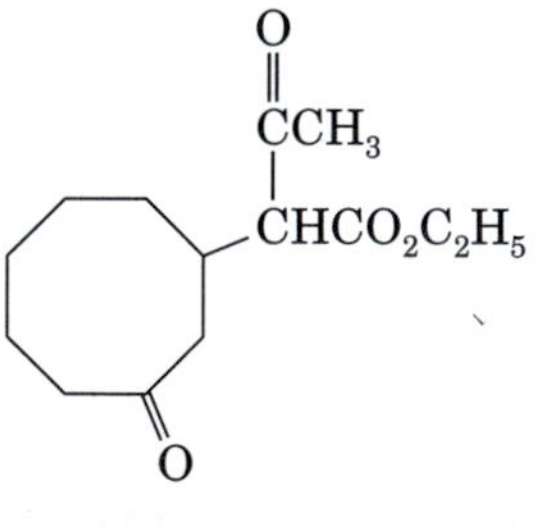 [cyclooctanone ring]—$CHCO_2C_2H_5$ (with CCH_3, C=O)

Additional Problems

17.34 Predict the major organic products:

(a) 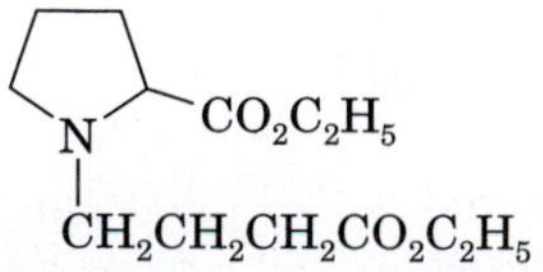 [pyrrolidine ring]N—$CO_2C_2H_5$ with $CH_2CH_2CH_2CO_2C_2H_5$ on N $\xrightarrow[\text{(2) H}_2\text{O, H}^+\text{, heat}]{\text{(1) NaOC}_2\text{H}_5}$

(b) CH_3CH_2CHO + $C_2H_5O_2CCH_2CN$ $\xrightarrow[\substack{\text{alanine (an amino acid)}\\ \text{heat}}]{\text{benzene, CH}_3\text{CO}_2\text{H}}$

(c)

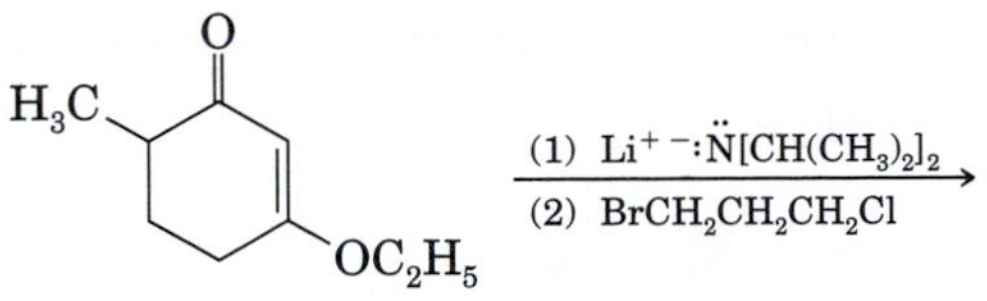

$$H_3C \quad \text{(cyclohexenone with } OC_2H_5\text{)} \xrightarrow[\text{(2) } BrCH_2CH_2CH_2Cl]{\text{(1) } Li^+ \; ^-\!:\!\ddot{N}[CH(CH_3)_2]_2}$$

(d)

$$\text{(cyclopentenone)}-CH_2\overset{O}{\overset{\|}{C}}CH_2CO_2C_2H_5 \xrightarrow[\text{(2) } H_2O, \, H^+]{\text{(1) } K_2CO_3}$$

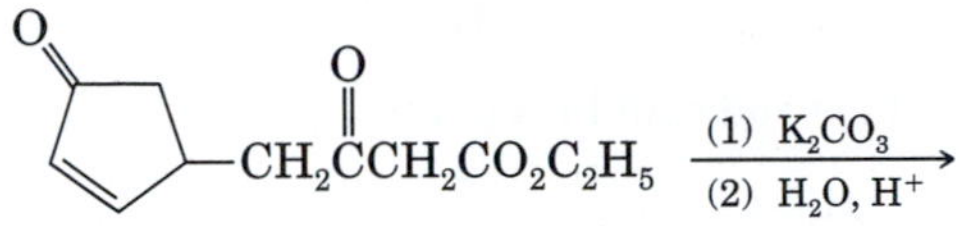

17.35 Suggest syntheses for the following compounds from readily available starting materials:

(a) 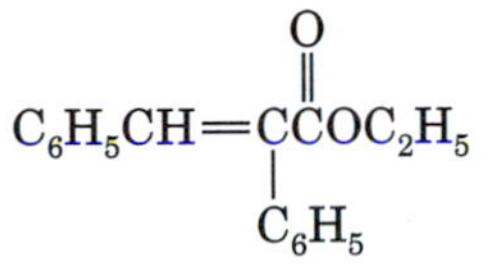 $C_6H_5CH\!=\!\overset{O}{\overset{\|}{C}}COC_2H_5$
            $\quad\quad\quad\quad\;\;\underset{C_6H_5}{|}$

(b) 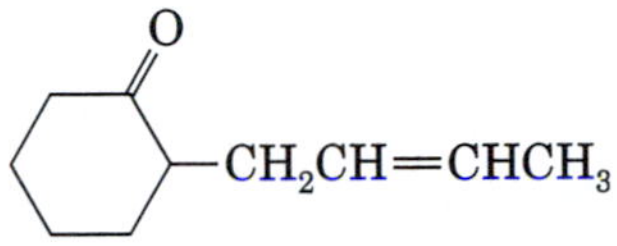 (cyclohexanone)$-CH_2CH\!=\!CHCH_3$

(c) $(CH_3)_2CHCH(CO_2CH_3)_2$

(d) $C_6H_5CH\!=\!C(CN)_2$

(e) 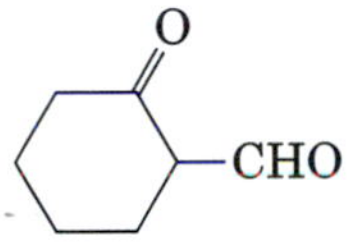 (cyclohexanone)$-CHO$

(f) 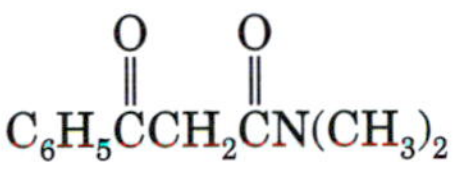 $C_6H_5\overset{O}{\overset{\|}{C}}CH_2\overset{O}{\overset{\|}{C}}N(CH_3)_2$

(g) 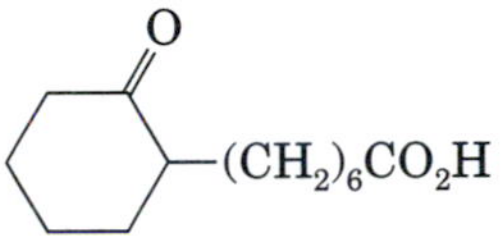 (cyclohexanone)$-(CH_2)_6CO_2H$

(h) 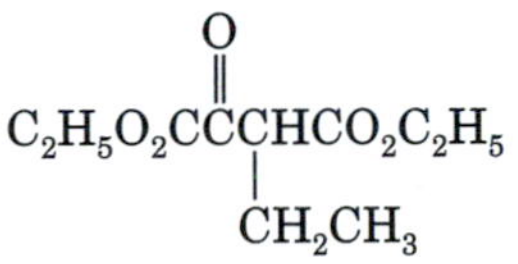 $C_2H_5O_2C\overset{O}{\overset{\|}{C}}CHCO_2C_2H_5$
            $\quad\quad\quad\quad\quad\;\underset{CH_2CH_3}{|}$

(i) 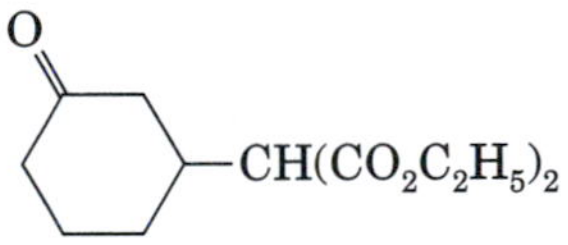 (cyclohexanone)$-CH(CO_2C_2H_5)_2$

17.36 Propose syntheses for the following compounds:

(a) 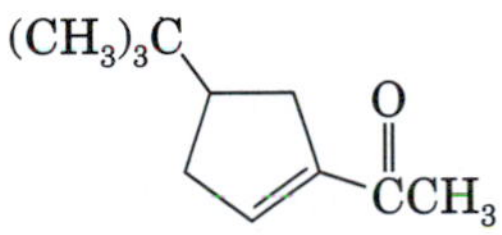 (tert-butyl cyclopentene with $\overset{O}{\overset{\|}{C}}CH_3$)   from an open-chain compound

(b) 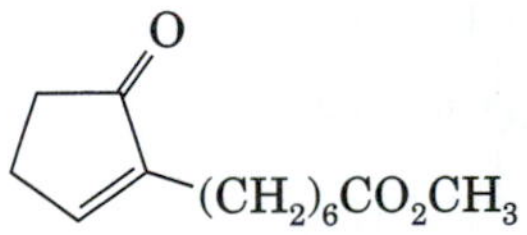 (cyclopentenone with $(CH_2)_6CO_2CH_3$)   a useful intermediate in prostaglandin syntheses (see Section 24.4), from an open-chain compound

(c) 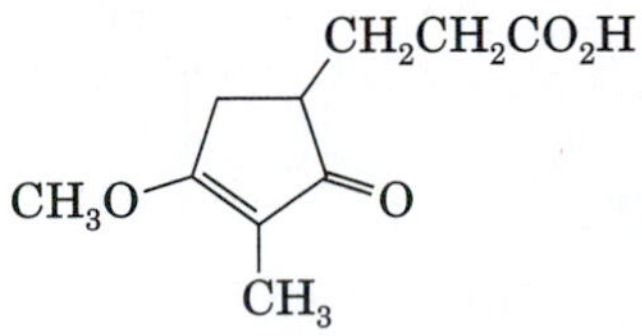 CH_3O-(cyclopentenone with $CH_2CH_2CO_2H$ and CH_3)   from 3-methoxy-2-methyl-2-cyclopentenone

(d) 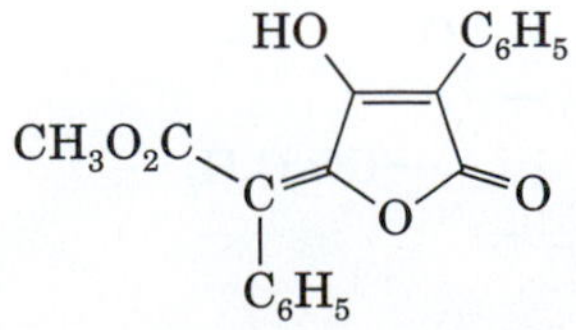 CH_3O_2C-(furanone with HO, C_6H_5, C_6H_5)   vulpinic acid, a yellow lichen metabolite with anti-inflammatory activity in humans, by an ester condensation of an open-chain compound

17.37 (a) How would you prepare the following compound from 4-methylcyclohexanone?

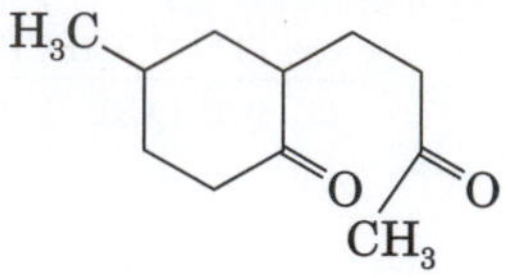

(b) From the dione in (a), show how you could prepare:

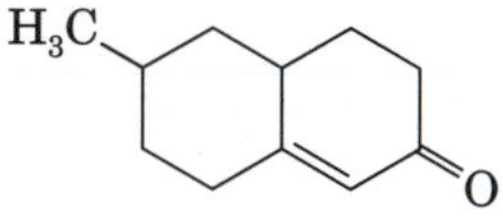

17.38 Show how the Robinson annulation can be used to prepare the following compounds:

(a)

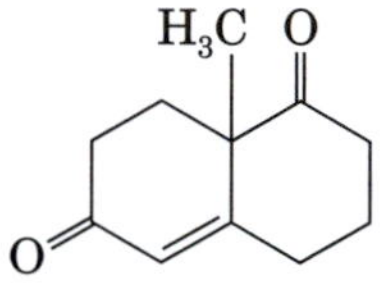

(b)

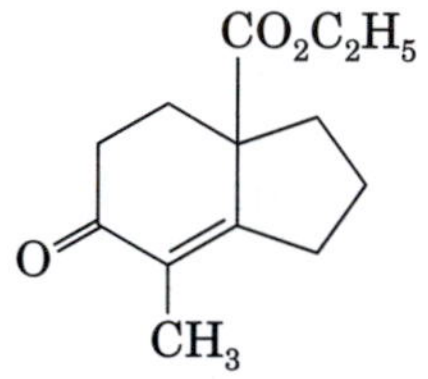

17.39 An aldehyde with no α hydrogen cannot undergo an aldol condensation when treated with aqueous base. However, if such an aldehyde is heated with concentrated NaOH or KOH solution, the aldehyde is converted to a 1:1 mixture of carboxylate and alcohol **(Cannizzaro reaction)**. Suggest a mechanism for the following Cannizzaro reaction. (*Hint:* Consider a hydride transfer as part of the mechanism.)

$$C_6H_5\overset{\overset{\displaystyle O}{\|}}{C}H \xrightarrow[\text{heat}]{\text{conc. KOH}} C_6H_5\overset{\overset{\displaystyle O}{\|}}{C}O^- K^+ + C_6H_5CH_2OH$$

17.40 Based upon the alkylation of malonic ester, outline a general synthesis for the preparation of:

(a) cycloalkanecarboxylic acids

(b) alkanedioic acids

17.41 Devise a general method for the synthesis of γ-keto acids from ethyl α-chloroacetate.

17.42 Write flow equations to show the steps in the following conversions:

(a) excess HCHO + CH₃CHO $\xrightarrow{\text{Ca(OH)}_2}$ HOCH₂CCH₂OH

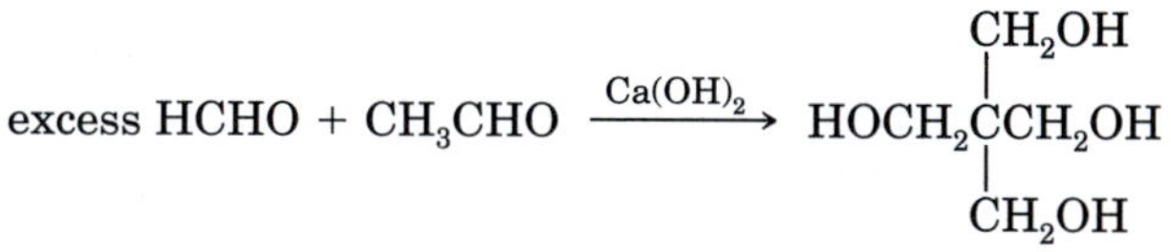

$$\begin{array}{c} CH_2OH \\ | \\ HOCH_2CCH_2OH \\ | \\ CH_2OH \end{array}$$

(b)

conc. HCl, heat
4 days

(c)

$\xrightarrow{\text{NaOC}_2\text{H}_5}$ $\xrightarrow{\text{H}^+}$

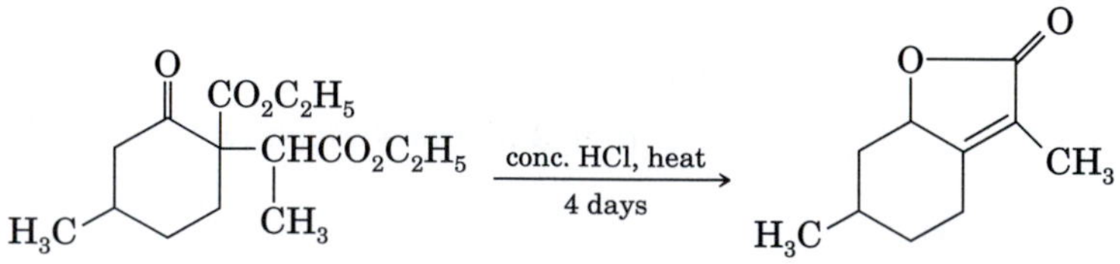

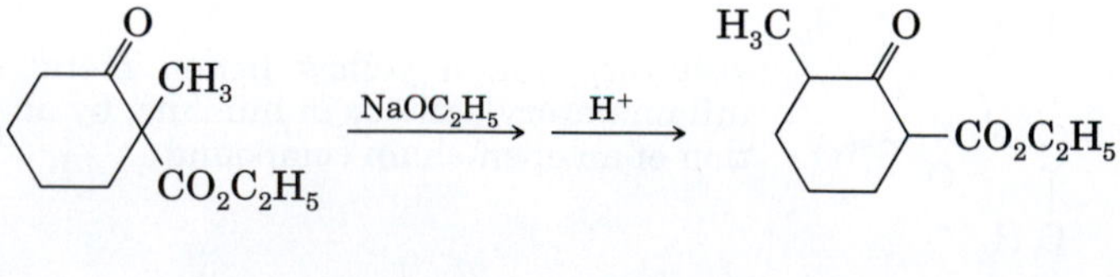

17.43 *Pulegone* is a fragrant component of oil of pennyroyal. When this compound is heated in aqueous base, acetone is formed. What other organic product or products could be isolated from the reaction mixture? Explain your answer.

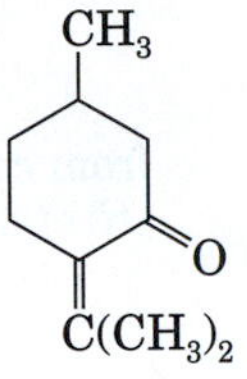

pulegone

17.44 Suggest syntheses for the following compounds:

(a)

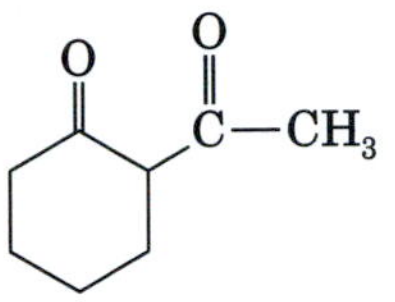

starting with cyclohexanone

(b)

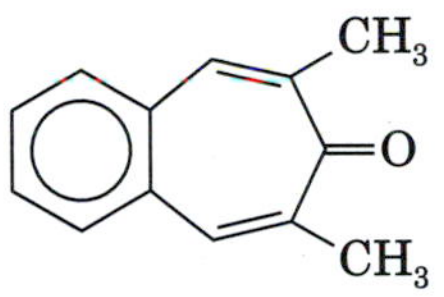

from monocyclic or acyclic starting materials

(c)

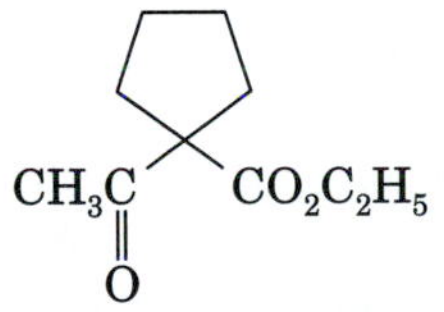

from acyclic starting materials

(d)

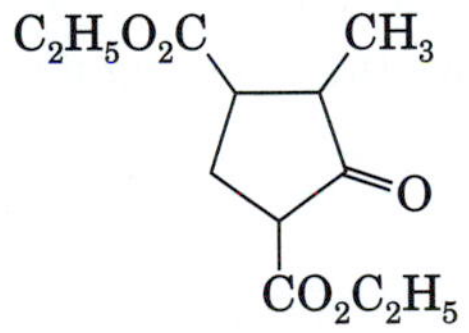

from acyclic starting materials

(e) 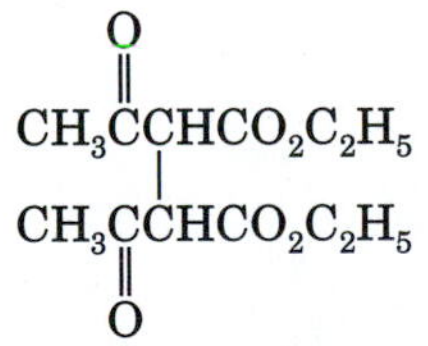 $CH_3\overset{O}{\overset{\|}{C}}CHCO_2C_2H_5$
$CH_3\overset{}{\underset{\overset{\|}{O}}{C}}CHCO_2C_2H_5$

ethyl acetoacetate as the only carbon source

(f) 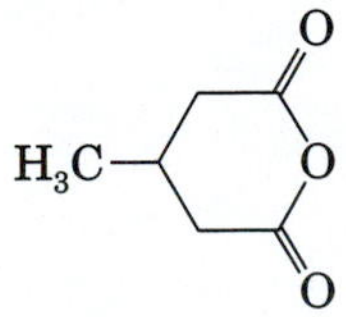 H_3C-

starting with diethyl malonate

(g)

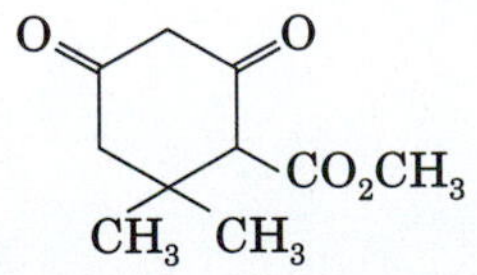

from 4-methyl-3-penten-2-one

(h) 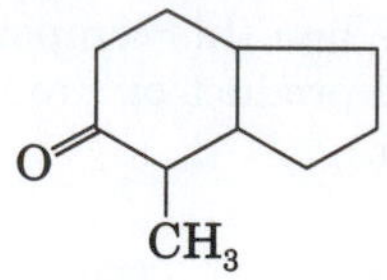   starting with cyclopentanone

(i) 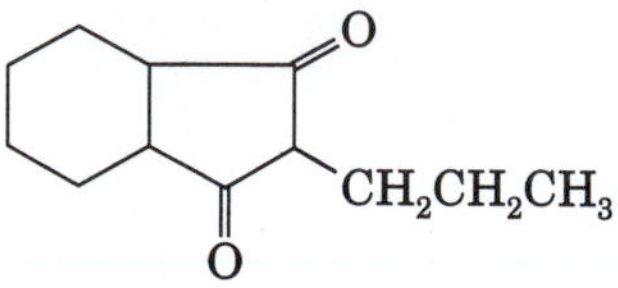   from compounds containing six or fewer carbons (*Hint:* Start with a Diels–Alder reaction.)

(j) 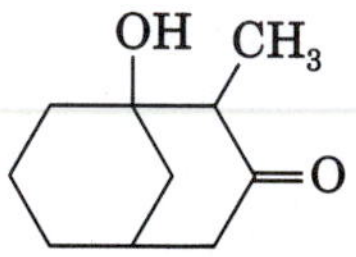   from 2-cyclohexenone and other appropriate reagents

(k) 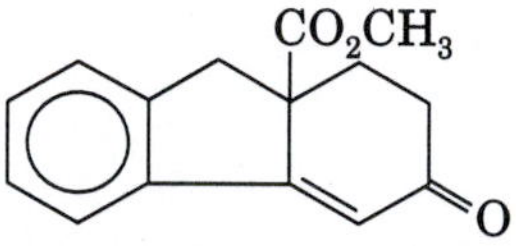   from

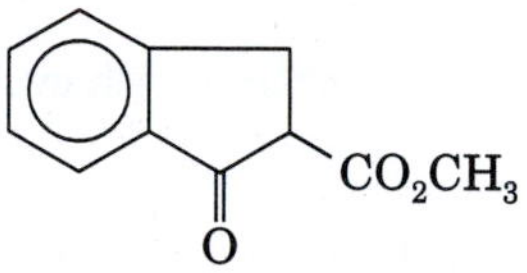

(l) 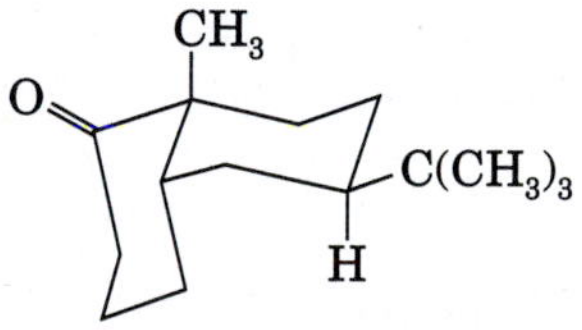   from

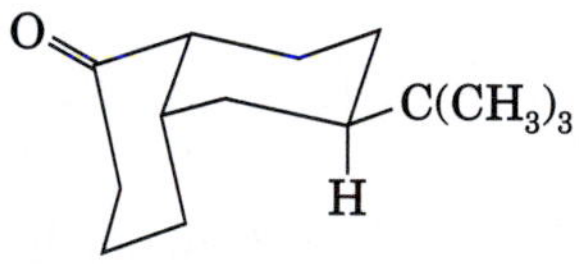

17.45   *cis*-Jasmone, a principal odorous component of jasmine oil, can be synthesized by heating the 11-carbon compound *cis*-8-undecen-2,5-dione with basic alumina (Al_2O_3) in benzene. The infrared spectrum of *cis*-jasmone shows principal peaks at 1645 cm^{-1} (6.08 μm) and 1700 cm^{-1} (5.88 μm). The molecular weight was determined by mass spectrometry to be 164. The ^{1}H NMR spectrum shows the following absorption. What is the structure of *cis*-jasmone?

δ	*splitting*	*relative area*
5.25	multiplet	2
2.85	doublet	2
1.9–2.7	multiplet	6
2.05	singlet	3
0.95	triplet	3

18 AMINES

Carbon, hydrogen, and oxygen are the three most common elements in living systems. Nitrogen is fourth. Nitrogen is found in proteins and nucleic acids, as well as in many other naturally occurring compounds of both plant and animal origin. In this chapter, we will discuss the **amines**, organic compounds containing trivalent nitrogen atoms bonded to one or more carbon atoms: RNH_2, R_2NH, or R_3N.

Amines are widely distributed in plants and animals, and many amines have physiological activity. For example, two of the body's natural stimulants of the sympathetic ("fight or flight") nervous system are norepinephrine and epinephrine (adrenaline).

$$\underset{\text{norepinephrine}}{\text{HO}\underset{\text{HO}}{\bigcirc}\overset{\overset{\displaystyle \text{OH}}{|}}{\text{CHCH}_2\text{NH}_2}}$$

norepinephrine

$$\underset{\text{epinephrine (adrenaline)}}{\text{HO}\underset{\text{HO}}{\bigcirc}\overset{\overset{\displaystyle \text{OH}}{|}}{\text{CHCH}_2\text{NHCH}_3}}$$

epinephrine (adrenaline)

Both norepinephrine and epinephrine are β-phenylethylamines (2-phenylethylamines). A number of other β-phenylethylamines act upon the sympathetic receptors. These compounds are referred to as *sympathomimetic amines* because they "mimic," to an extent, the physiological action of norepinephrine and epinephrine.

Well before the birth of Christ, the compound *ephedrine* was extracted from the *ma-huang* plant in China and used as a drug. Today, it is the active decongestant in nose drops and cold remedies. Ephedrine causes shrinkage of swollen nasal membranes and inhibition of nasal secretions. (Overdoses cause nervousness and sleeplessness.) *Mescaline,* a hallucinogen isolated from the peyote cactus, has been used for centuries by the Indians of the southwestern United States and Mexico in religious ceremonies. *Amphetamine* is a synthetic stimulant that causes sleeplessness and nervousness. Amphetamine is sometimes prescribed for obesity, because it is also an appetite depressant. Like many other sympathomimetic amines, amphetamine contains a chiral carbon and has a pair of enantiomers. The more active enantiomer of amphetamine (the dextrorotatory one) is called *dexedrine.*

ephedrine

a decongestant

mescaline

a hallucinogen

amphetamine

a stimulant

Single enantiomers of chiral amines are common in plants. Because of their basicity, some of these amines find use in the resolution of racemic carboxylic acids. (See Section 4.11.) Strychnine, isolated from the seeds of the Asiatic tree *Strychnos nux-vomica,* is an example of such an amine. Strychnine is a highly toxic stimulant of the central nervous system. Besides its laboratory use, it is used to kill rodents and predatory animals.

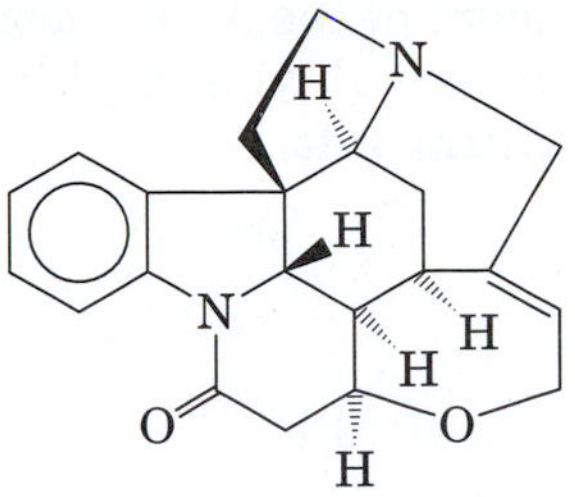

strychnine

Volatile amines have very distinctive and usually offensive odors. Methyl-amine has an odor similar to that of ammonia. Trimethylamine smells like dead saltwater salmon. Piperidine smells like dead freshwater fish. Aryl-amines are not as unpleasant smelling as alkylamines. However, arylamines such as aniline are toxic and are especially insidious because they can be absorbed through the skin. Some, like β-naphthylamine, are carcinogenic.

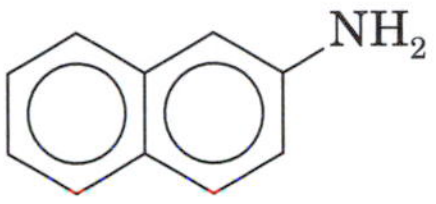

2-naphthylamine (β-naphthylamine)

a carcinogenic arylamine

Section 18.1
Classification and Nomenclature of Amines

Amines may be classified as **primary, secondary,** or **tertiary,** according to the number of alkyl or aryl substituents bonded to the nitrogen.

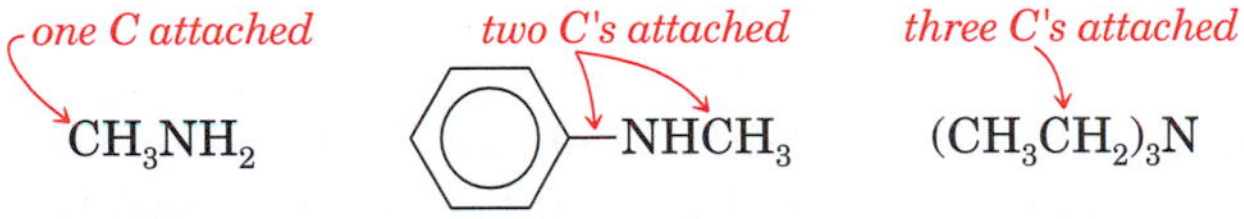

one C attached   *two C's attached*   *three C's attached*

CH_3NH_2 —NHCH$_3$ $(CH_3CH_2)_3N$

a 1° alkylamine   *a 2° arylalkylamine*   *a 3° trialkylamine*

This classification is different from that of alkyl halides or alcohols. Their classification is based on the number of groups attached to the carbon that has the halide or hydroxyl group.

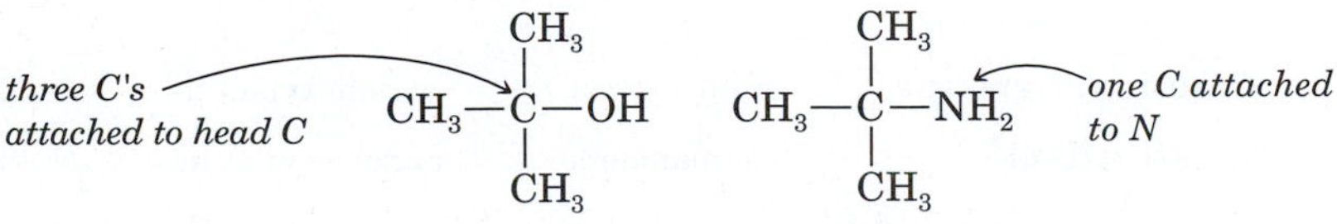

three C's attached to head C

CH_3
|
CH_3—C—OH
|
CH_3

CH_3
|
CH_3—C—NH_2
|
CH_3

one C attached to N

tert-butyl alcohol   *tert*-butylamine

a 3° alcohol   *a 1° amine*

An amine nitrogen can have *four* groups or atoms bonded to it, in which case the nitrogen is part of a positive ion. These ionic compounds fall into two categories. If one or more of the attachments is H, the compound is an **amine salt.** If all four groups are alkyl or aryl (no H's on the N), the compound is a **quaternary ammonium salt.**

Amine salts:

$$(CH_3)_2NH_2^+ \ Cl^-$$

dimethylammonium chloride

salt of a 2° amine

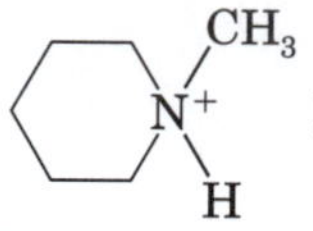

N-methylpiperidinium bromide

salt of a 3° amine

Quaternary ammonium salts:

$$(CH_3)_4N^+ \ Cl^-$$

tetramethylammonium
chloride

$$CH_3CO_2CH_2CH_2\overset{+}{N}(CH_3)_3 \ Cl^-$$

acetylcholine chloride

*involved in the transmission
of nerve impulses*

STUDY PROBLEM

18.1   Classify each of the following compounds as a 1°, 2°, or 3° amine; as a salt of a 1°, 2°, or 3° amine; or as a quaternary ammonium salt:

(a) $CH_3CH_2\overset{\overset{\displaystyle CH_3}{|}}{C}HNH_2$

(b)  ⬡NH

(c) $(CH_3)_3NH^+ \ NO_3^-$

(d) $(CH_3CH_2)_3CNH_2$

(e) 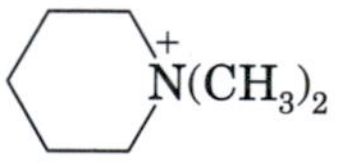 ⬡$\overset{+}{N}(CH_3)_2 \ Cl^-$

(f) 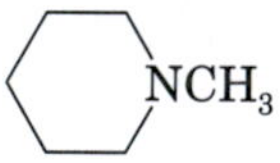 ⬡NCH_3

Simple amines are usually named by the functional-group system. The alkyl or aryl group is named, then the ending **-amine** is added. Amines may also be named by the substitutive system as derivatives of the parent hydrocarbon. The substitutive names are derived in the same manner as for alcohols, but with the ending **-amine.**

$$CH_3CH_2CH_2NH_2$$ ⬡—NH_2 $$(CH_3CH_2)_2NH$$

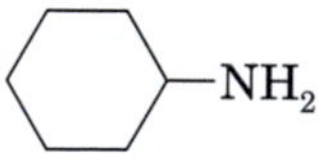

functional-group:	propylamine	cyclohexylamine	diethylamine
substitutive:	1-propanamine	cyclohexanamine	N-ethylethanamine

Diamines are named from the name of the parent alkane (with appropriate prefix numbers) followed by the ending **-diamine.**

$$H_2NCH_2CH_2CH_2NH_2$$

1,3-propanediamine

If more than one type of alkyl group is attached to the nitrogen, the largest alkyl group is considered the parent. A subsidiary alkyl group is designated by an **N-alkyl-** prefix.

$$CH_3CHNHCH_3 \quad\quad CH_3CHN(CH_3)_2$$

(with CH_3 substituent above each CH)

functional-group:	*N*-methylisopropylamine	*N,N*-dimethylisopropylamine
substitutive:	*N*-methyl-2-propanamine	*N,N*-dimethyl-2-propanamine

If a functional group of higher nomenclature priority is present, an **amino-** prefix is used.

$$H_2NCH_2CH_2OH \quad\quad CH_3CHCO_2H$$

(with $NHCH_3$ substituent above the CH)

2-aminoethanol	2-(methylamino)propanoic acid

The chemistry of the **nonaromatic heterocyclic amines** is similar to that of their open-chain counterparts. The commonly encountered compounds of this class generally have individual names:

pyrrolidine   piperidine   piperazine   morpholine

In the numbering of heterocyclic rings, the heteroatom is considered position 1. Oxygen has priority over nitrogen.

2-methylpyrrolidine   3,5-dimethylmorpholine

Study Problems

18.2 Name the following amines and ammonium ions:

(a) $C_6H_5CH_2NH_2$ **(b)** $(CH_3CH)_2NH$ (with CH_3 below)   **(c)** $(CH_3)_4N^+ Br^-$

(d) $CH_2{=}CHCH_2NH_2$ **(e)** $H_2N(CH_2)_{10}NH_2$

18.3 Write formulas for the following amines or ammonium ions:

(a) heptylamine   **(b)** 1,2-butanediamine

(c) ethanolamine   (d) (S)-1-phenyl-2-propanamine

(e) N-methyl-sec-butylamine   (f) dimethylammonium acetate

Section 18.2
Bonding in Amines

The bonding in an amine is directly analogous to that in ammonia: an sp^3-hybridized nitrogen atom bonded to three other atoms or groups (H or R) and with a pair of unshared valence electrons in the remaining sp^3 orbital (see Figure 2.19, page 68).

$$H-\ddot{N}-H \qquad CH_3-\ddot{N}-CH_3 \qquad \langle\rangle N\cdot\cdot$$

$$\qquad H \qquad\qquad\qquad CH_3 \qquad\qquad\qquad H$$

     ammonia   trimethylamine   piperidine

In an amine salt or a quaternary ammonium salt, the unshared pair of electrons forms the fourth σ bond. The cations are analogous to the ammonium ion.

$$H-\overset{+}{N}-H \; Cl^- \qquad CH_3-\overset{+}{N}-CH_3 \; Cl^- \qquad N^+ \; {}^-O_2CCH_3$$

    ammonium   tetramethylammonium   N-methylpiperidinium
    chloride   chloride   acetate

An amine molecule with three different groups bonded to the nitrogen is chiral. However, enantiomers of most amines cannot be isolated because rapid inversion between mirror images occurs at room temperature. The inversion proceeds by way of a planar transition state (sp^2 nitrogen). The result is that the nitrogen pyramid flips inside out, much as an umbrella in a strong wind. The energy required for this inversion is about 6 kcal/mol, about twice the energy required for rotation around a carbon–carbon σ bond.

The mirror images are interconvertible:

$$R_1-\ddot{N}\cdots R_3 \rightleftharpoons \left[R_1-\overset{sp^2}{N}{}_{R_2}^{R_3} \right] \rightleftharpoons R_1\diagdown \underset{\cdot\cdot}{N}{\overset{R_3}{\diagup}}R_2$$

*transition state with
two e⁻ in p orbital*

If an amine nitrogen has three different substituents and interconversion between the two mirror-image structures is restricted, then a pair of enantiomers can be isolated. **Tröger's base** is an example of such a compound.

The methylene bridge between the two nitrogens prevents interconversion between the mirror images; Tröger's base can be separated into a pair of enantiomers.

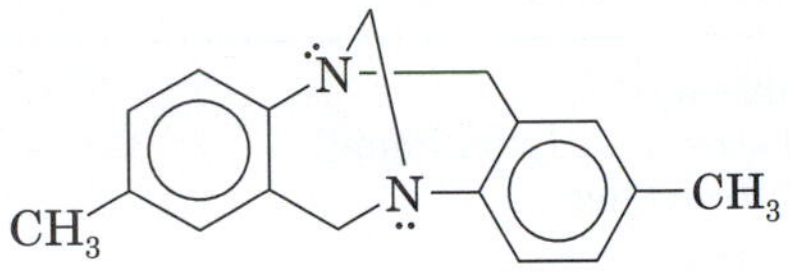

Tröger's base

Some quaternary ammonium salts can also exist as isolable enantiomers. These compounds are structurally similar to compounds containing sp^3-hybridized carbon atoms. If four different groups are bonded to the nitrogen, the ion is chiral and the salt can be separated into enantiomers.

A pair of enantiomers:

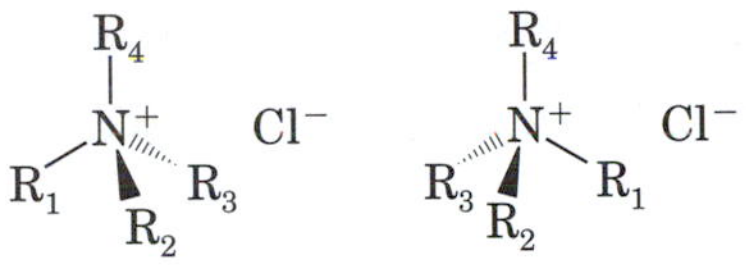

STUDY PROBLEM

18.4   Which of the following structures could exist as isolable enantiomers?

(a) $[(CH_3)_2CH]_2\overset{+}{N}(CH_2CH_2Cl)_2\ Cl^-$ (b) $CH_3NHCH_2CH_2Cl$

(c)

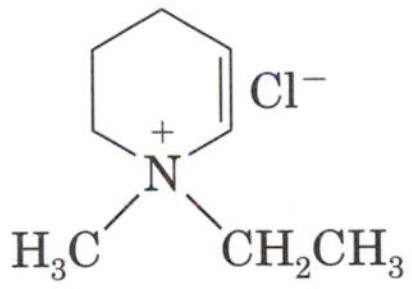

(d)

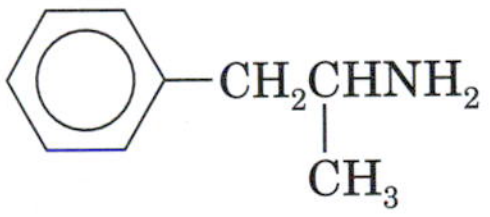

Section 18.3
Spectral Properties of Amines

A. Infrared Spectra

The bonds that give rise to infrared absorption characteristic of amines are the C—N and N—H bonds (Table 18.1). All aliphatic amines show C—N stretching in the fingerprint region. However, only primary and secondary amines show the distinctive NH stretching absorption, which is observed to the left of CH absorption in the spectrum. This is the same region where OH absorption is observed. However, the two can often be differentiated because the OH absorption is usually broader and stronger than NH absorption. The stronger absorption by an OH bond is due to the greater polarity and hydrogen bonding of this group.

Table 18.1	Characteristic infrared absorption for amines	

	Position of Absorption	
Type of Absorption	cm^{-1}	μm
1° amines:		
N—H stretching (pure liquid)	3250–3400 (2 peaks)	2.9–3.1
C—N stretching	1020–1250	8.0–9.8
2° amines:		
N—H stretching (pure liquid)	3330	3.0
C—N stretching	1020–1250	8.0–9.8
3° amines:		
C—N stretching	1020–1250	8.0–9.8

In Chapter 9, we described how primary amines show two NH absorption peaks, secondary amines show one NH peak, and tertiary amines show no absorption in this region. Spectra of the three types of amines are shown in Figure 9.10 (page 348).

B. ^{1}H NMR Spectra

The NH absorption in the ^{1}H NMR spectrum is generally a sharp singlet, not split by adjacent protons. In this respect, NH absorption is similar to OH absorption (Section 9.10E). Aliphatic amines show NH absorption at δ values of about 1.0–2.8 ppm, while arylamines absorb at about δ2.6–4.7. (The exact position depends upon the solvent used.) The α protons are somewhat deshielded by the electronegative nitrogen. The chemical shift for these protons is in the range of δ2.2–2.8. (See Figure 18.1.)

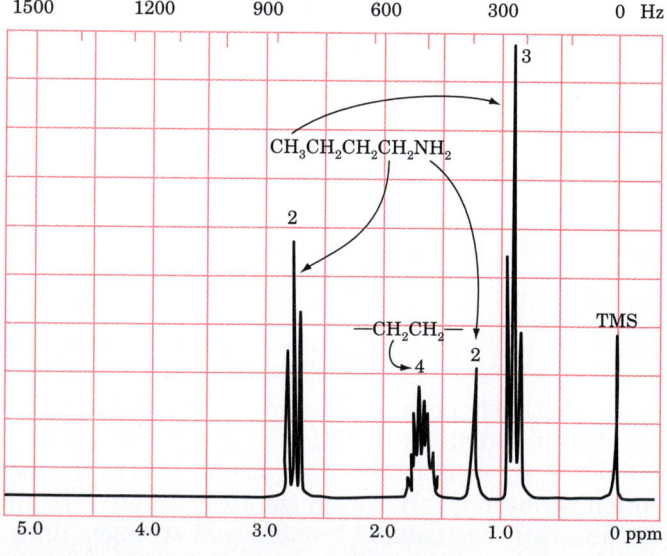

Figure 18.1 300 MHz ^{1}H NMR spectrum of butylamine.

Preparation of Amines

Techniques for the preparation of amines fall into three general categories. We will discuss each category in turn.

Nucleophilic substitution:

$$RX \quad + NH_3 \xrightarrow{\text{S}_N2} RNH_3^+ \, X^- \xrightarrow{OH^-} RNH_2$$

an alkyl halide

Reduction:

$$\underset{\text{an amide or nitrile}}{\overset{\displaystyle O}{\underset{\|}{R\overset{\|}{C}NH_2}} \quad or \quad RCN} \xrightarrow{[H]} RCH_2NH_2$$

Amide rearrangement:

$$\underset{\text{an amide}}{\overset{\displaystyle O}{R\overset{\|}{C}NH_2}} \xrightarrow{Br_2, OH^-} RNH_2$$

A. Synthesis by Substitution Reactions

Reaction of amines and alkyl halides Ammonia or an amine carries an unshared pair of electrons and can act as a nucleophile in a substitution reaction with an alkyl halide. The reaction of a nitrogen nucleophile is similar to the reaction of any other nucleophile with RX. The product of the reaction with ammonia or an amine is an amine salt. The free amine can be obtained by the treatment of this amine salt with a base such as NaOH.

S$_N$2 reaction:

$$\underset{\text{ammonia}}{H_3N:} + \underset{\text{bromoethane}}{\overset{CH_3}{\underset{\displaystyle |}{CH_2-\ddot{B}r:}}} \xrightarrow{\text{S}_N2} \underset{\underset{\text{\textit{an amine salt}}}{\text{ethylammonium bromide}}}{\overset{CH_3}{\underset{\displaystyle |}{H_3\overset{+}{N}-CH_2}} \, :\ddot{B}r:^-}$$

Treatment with base:

$$CH_3CH_2NH_3^+ \, Br^- + :\ddot{O}H^- \longrightarrow \underset{\text{ethylamine}}{CH_3CH_2\ddot{N}H_2} + H_2\ddot{O}: + Br^-$$

The order of reactivity of alkyl halides is typical for S$_N$2 reactions: $CH_3X >$ 1° > 2°. Tertiary alkyl halides do not undergo substitution reactions with ammonia or amines; elimination products are obtained.

The principal disadvantage of this route to amines is that the product amine salt can exchange a proton with the starting ammonia or amine.

$$CH_3CH_2NH_3^+ \, Br^- + NH_3 \rightleftharpoons CH_3CH_2NH_2 + NH_4^+ \, Br^-$$

also a nucleophile

This proton exchange results in two or more nucleophiles competing in the reaction with the alkyl halide. For this reason, a mixture of mono-, di-, and trialkylamines and the quaternary ammonium salt is frequently obtained from the reaction of ammonia with an alkyl halide.

$$NH_3 \xrightarrow{RX} RNH_2 \xrightarrow{RX} R_2NH \xrightarrow{RX} R_3N \xrightarrow{RX} R_4N^+ X^-$$

If the amine is very inexpensive or if ammonia is used, a large excess can be used to favor monoalkylation. In this case, RX is more likely to collide with the molecules of the desired reactant and less likely to collide with those of the alkylated product. In the following example, an excess of ammonia favors the primary amine product.

$$CH_3CH_2CH_2CH_2Br + \text{excess } NH_3 \longrightarrow \xrightarrow{OH^-} CH_3CH_2CH_2CH_2NH_2 + Br^-$$

1-bromobutane   butylamine (45%)

If the quaternary ammonium salt is desired, an excess of the alkyl halide would be used.

$$(CH_3CH_2)_2NH + \text{excess } CH_3CH_2I \longrightarrow (CH_3CH_2)_4N^+ I^-$$

Gabriel phthalimide synthesis   A reaction sequence that yields primary amines without secondary and tertiary amines is the Gabriel phthalimide synthesis. The first step in the sequence is an S_N2 reaction with the phthalimide anion as the nucleophile. Hydrolysis of the substituted phthalimide yields the amine.

potassium
phthalimide

N-ethylphthalimide

+ H$_2$NCH$_2$CH$_3$
ethylamine

free of 2° and 3° amines

Phthalimide is prepared by heating phthalic anhydride with ammonia. The potassium salt is made by treating phthalimide with KOH. Usually, a proton cannot be removed from an amide nitrogen so easily. However, like other β-dicarbonyl compounds, imides are acidic because the anion is resonance-stabilized.

Preparation of phthalimide anion:

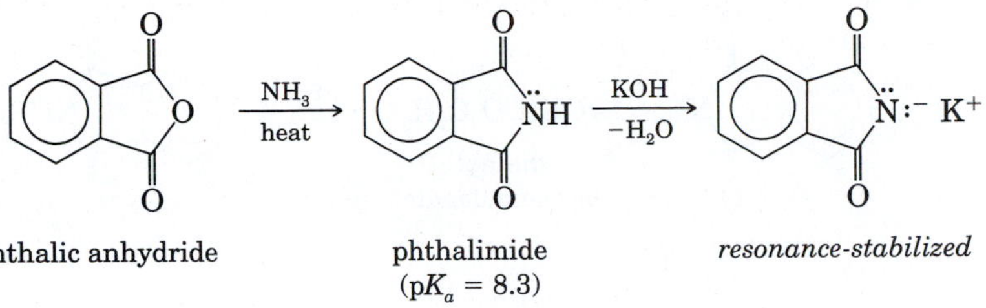

phthalic anhydride   phthalimide   *resonance-stabilized*
                             (pK_a = 8.3)

Treatment of the potassium phthalimide with an alkyl halide yields an *N*-alkylphthalimide.

Attack on RX:

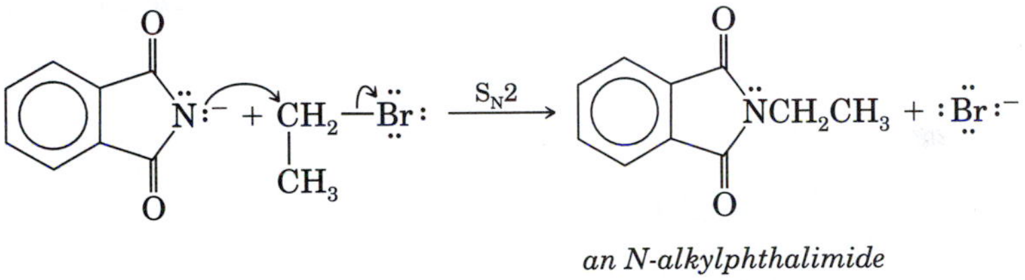

an N-alkylphthalimide

Finally, the alkylphthalimide is hydrolyzed. This reaction is simply the hydrolysis of an amide (Section 15.8C).

Hydrolysis:

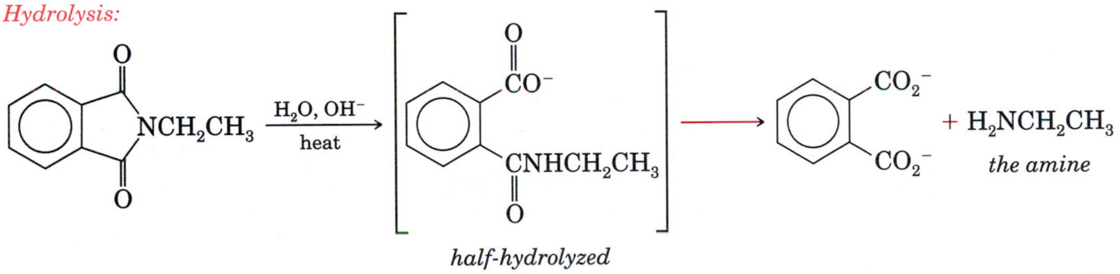

half-hydrolyzed   *the amine*

STUDY PROBLEM

18.5 List the sequence of reagents that you would add to potassium phthalimide to prepare: **(a)** propylamine; **(b)** allylamine; **(c)** benzylamine.

An ingenious variation of the Gabriel phthalimide synthesis is used to prepare α-amino acids, the building blocks of proteins. This sequence is: (1) treatment of potassium phthalimide with diethyl bromomalonate; (2) treatment of the imide-malonate with base to remove the α hydrogen; and (3) treatment with RX, which gives a typical malonic ester alkylation reaction.

1. Reaction with bromomalonic ester:

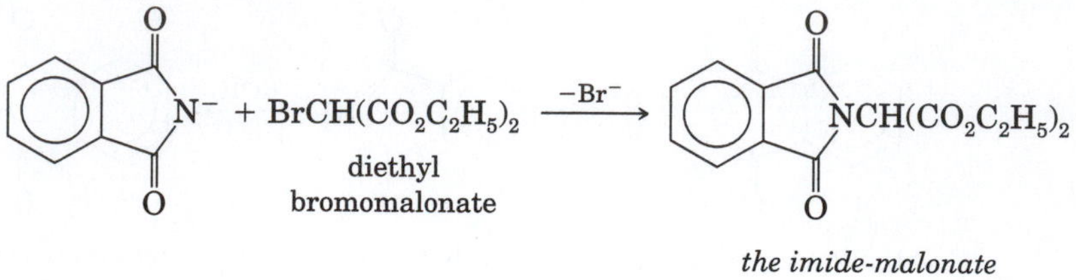

diethyl
bromomalonate

the imide-malonate

2. Treatment with base:

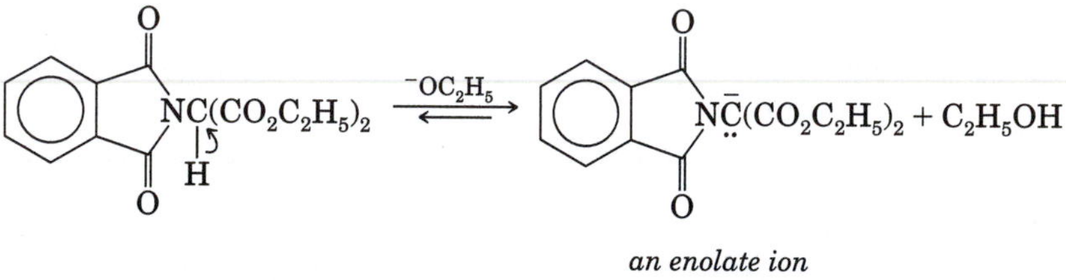

an enolate ion

3. Reaction with RX:

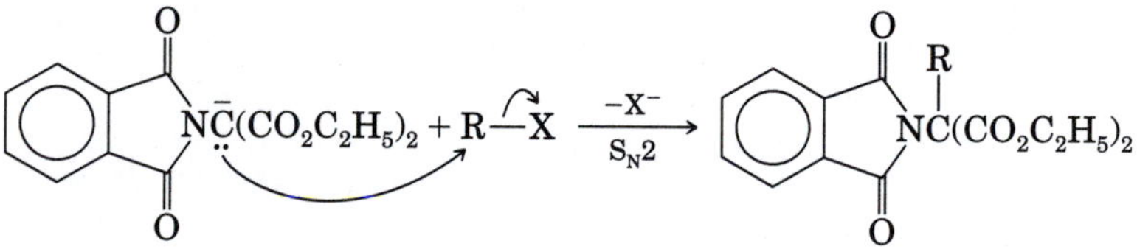

Acid hydrolysis results in hydrolysis of both imide and diester plus decarboxylation of the diacid. The product is the protonated amino acid.

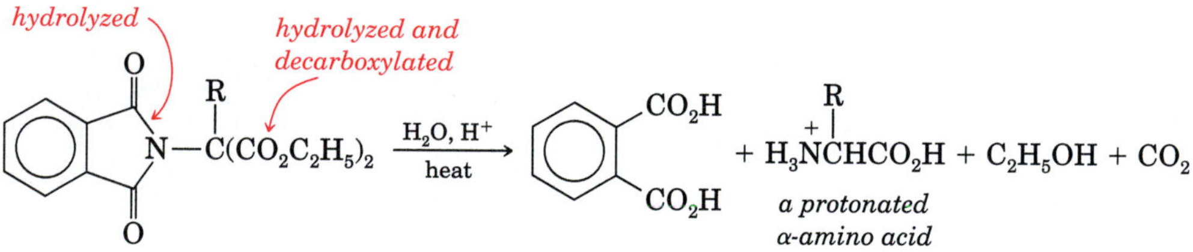

hydrolyzed

*hydrolyzed and
decarboxylated*

$+ H_3\overset{+}{N}CHCO_2H + C_2H_5OH + CO_2$

*a protonated
α-amino acid*

SAMPLE PROBLEM

How would you prepare phenylalanine, $C_6H_5CH_2CHCO_2H$, by a phthalimide synthesis?

$\underset{NH_2}{|}$

Solution

1. Determination of reagents needed:

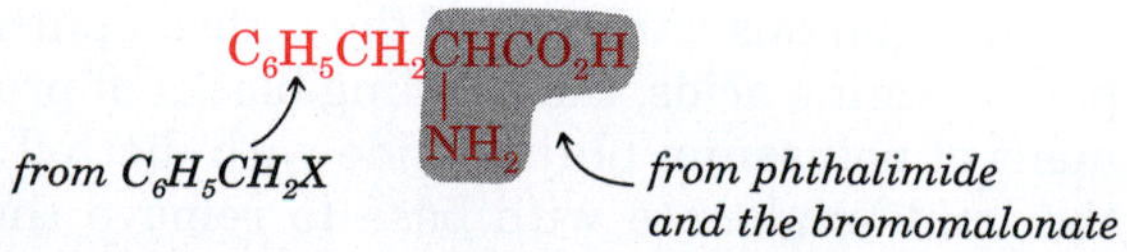

$C_6H_5CH_2CHCO_2H$
NH_2

from $C_6H_5CH_2X$

*from phthalimide
and the bromomalonate*

2. *Steps in the synthesis:*

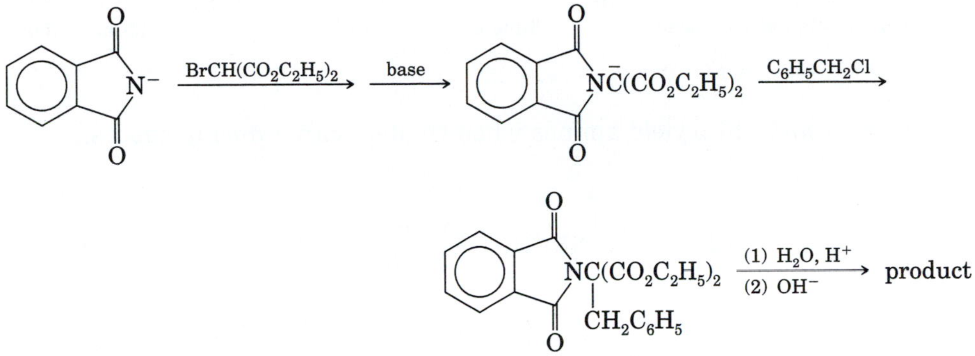

18.6   How would you prepare the following amino acid by a phthalimide synthesis?

$$(CH_3)_2CHCH_2CHCO_2H$$
$$\overset{|}{NH_2}$$

leucine

B. Synthesis by Reduction

Reduction reactions often provide convenient routes to amines. The reduction of *aromatic nitro compounds* to arylamines was discussed in Section 12.3A.

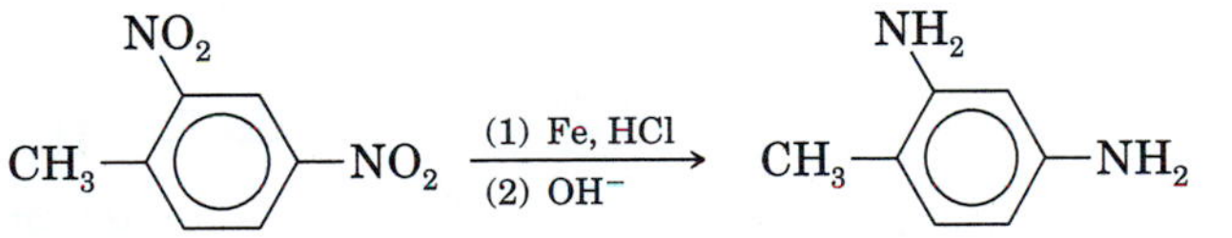

2,4-dinitrotoluene   2,4-toluenediamine (75%)

Some arylamines can be synthesized by the reaction of ammonia or amines with activated aryl halides (Section 12.4).

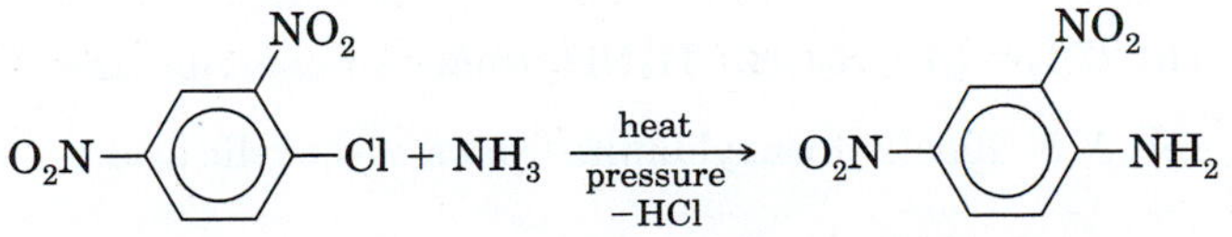

2,4-dinitroaniline (70%)

Nitriles undergo catalytic hydrogenation or reduction with $LiAlH_4$ to yield primary amines of the type RCH_2NH_2 in yields of approximately 70%. Nitriles are available from alkyl halides. Therefore, a nitrile synthesis is a technique for lengthening a carbon chain as well as for preparing an amine.

$$(CH_3)_2CHCH_2Br \xrightarrow[-Br^-]{CN^-} (CH_3)_2CHCH_2CN \xrightarrow[\text{(2) } H_2O, H^+]{\text{(1) LiAlH}_4} (CH_3)_2CHCH_2CH_2NH_2$$

1-bromo-2-methylpropane 3-methylbutanenitrile (3-methylbutyl)amine

a 1° alkyl halide

Amides also yield amines when treated with reducing agents:

$$CH_3(CH_2)_{10}\overset{\overset{\displaystyle O}{\|}}{C}NHCH_3 \xrightarrow[\text{(2) } H_2O, H^+]{\text{(1) LiAlH}_4} CH_3(CH_2)_{10}CH_2NHCH_3$$

N-methyldodecanamide *N*-methyldodecylamine (95%)

Reductive amination, a reaction that converts ketones or aldehydes to primary amines, was discussed in Section 13.6D. This reaction is much better for synthesizing an amine of the type R_2CHNH_2 than is the reaction of R_2CHBr and NH_3 because the latter reaction may lead to elimination products. Secondary and tertiary amines can also be synthesized by reductive amination if a primary and secondary amine is used instead of ammonia.

benzaldehyde *an imine* benzylamine (85%)

2-aminoethanol acetone *an imine*
(ethanolamine)

$$\xrightarrow{H_2, Pt} HOCH_2CH_2NHCH(CH_3)_2$$

2-(isopropylamino)ethanol (95%)

STUDY PROBLEM

18.7 Show with flow equations how you would carry out the following syntheses:

(a) cyclohexylamine from cyclohexanone

(b) $CH_2{=}CHCH_2CH_2CH_2NH_2$ from 4-bromo-1-butene

(c) *N,N*-dimethylbenzylamine from a carboxylic acid

C. Amide Rearrangement

When an unsubstituted amide ($RCONH_2$) is treated with an alkaline, aqueous solution of bromine, it undergoes rearrangement to yield an amine. This reaction is called the **Hofmann rearrangement.** The carbonyl group is lost as $CO_3{}^{2-}$. Therefore, the amine contains one less carbon than the starting amide.

$$\underset{\text{hexanamide}}{CH_3(CH_2)_4\overset{\overset{\displaystyle O}{\parallel}}{C}NH_2} + 4\ OH^- + Br_2 \xrightarrow{H_2O} \underset{\text{pentylamine (85\%)}}{CH_3(CH_2)_4NH_2 + CO_3^{2-} + 2\ H_2O + 2\ Br^-}$$

Because the carbonyl group appears to be lost from the interior of the molecule, let us consider the mechanism of the Hofmann rearrangement. The reaction proceeds by a series of discrete steps. *Step 1* is bromination at the nitrogen. *Step 2* is loss of a proton from the nitrogen and results in an unstable anion. The rearrangement step is *Step 3* of the sequence. Note that this rearrangement is a *1,2-shift* very similar to those in carbocation rearrangements (Section 5.5F). The product of the rearrangement is an isocyanate, stable under some conditions, but not in aqueous base. In aqueous base, the isocyanate undergoes hydrolysis *(Step 4)* to an amine and the carbonate ion.

Step 1 (bromination of N):

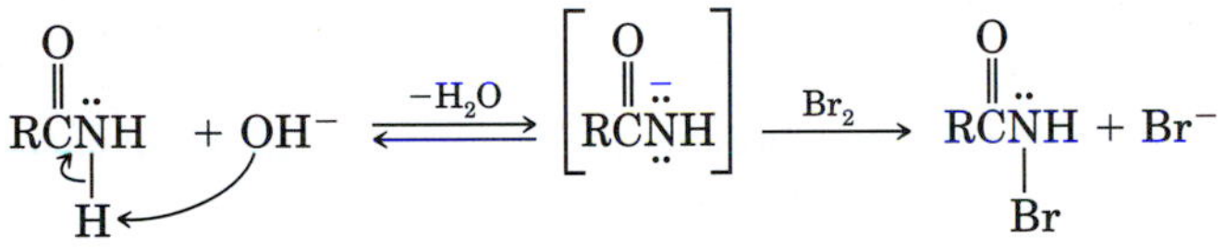

$pK_a = \sim 15$

Step 2 (extraction of H⁺ by OH⁻):

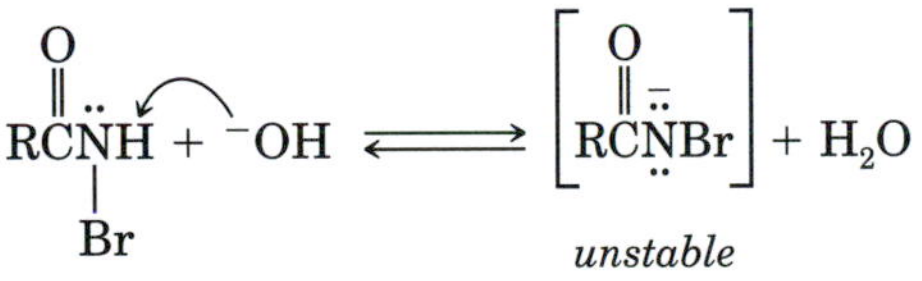

Step 3 (displacement of Br⁻ by R—, a 1,2 shift):

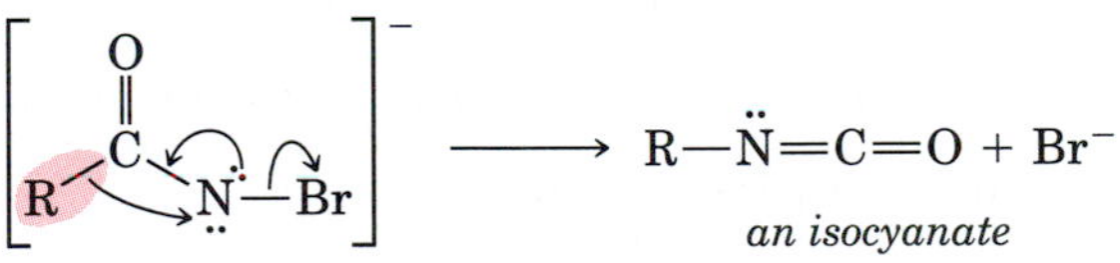

Step 4 (hydrolysis of the isocyanate):

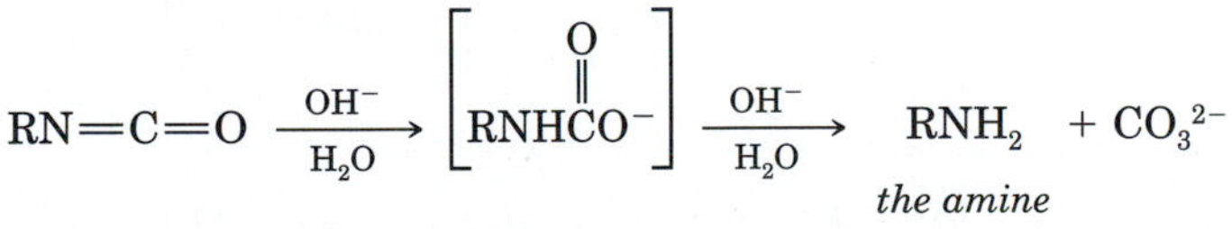

The Hofmann rearrangement has been found to proceed with *retention of configuration* at the α carbon of the amide. This evidence leads us to believe that the rearrangement step *(Step 3)* has a bridged transition state.

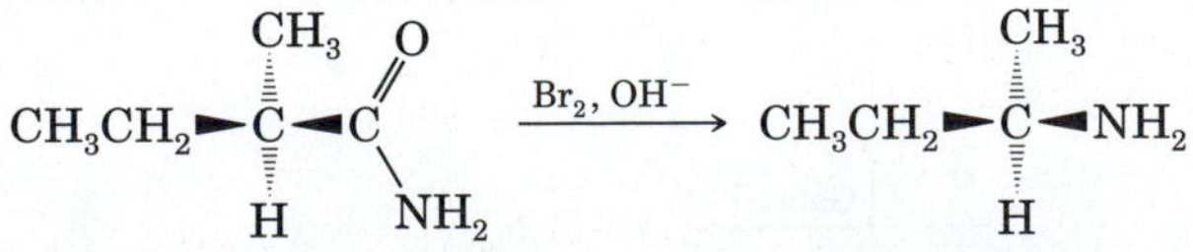

(*R*)-2-methylbutanamide   (*R*)-*sec*-butylamine

A bridged transition state in Step 3:

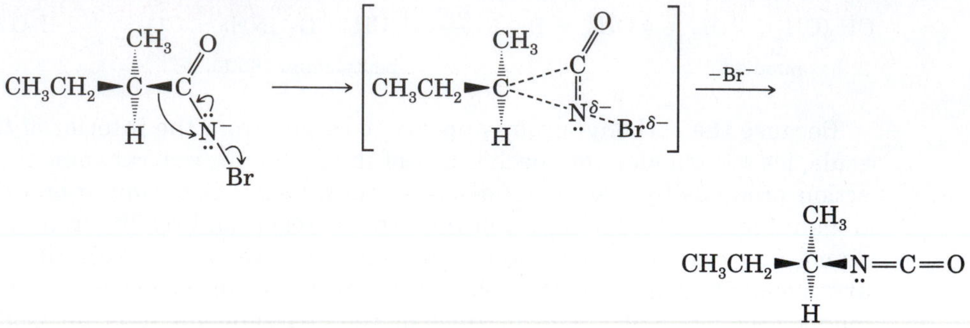

$$CH_3CH_2 \blacktriangleright \underset{\underset{H}{|}}{\overset{\overset{CH_3}{|}}{C}} \blacktriangleleft \overset{O}{\underset{\underset{Br}{\overset{|}{\ddot{N}^-}}}{\overset{\|}{C}}} \longrightarrow \left[CH_3CH_2 \blacktriangleright \underset{\underset{H}{|}}{\overset{\overset{CH_3}{|}}{C}} \cdots \overset{\overset{O}{\|}}{\underset{\underset{\ddot{N}^{\delta-} \cdots Br^{\delta-}}{}}{C}} \right] \xrightarrow{-Br^-}$$

$$CH_3CH_2 \blacktriangleright \underset{\underset{H}{|}}{\overset{\overset{CH_3}{|}}{C}} \blacktriangleleft \ddot{N}{=}C{=}O$$

The advantage of the Hofmann rearrangement is that yields of pure primary amines are good. This would be the best route to a primary amine containing a tertiary alkyl group, such as $(CH_3)_3CNH_2$.

STUDY PROBLEM

18.8 Predict the major organic products when the following compounds are treated with aqueous alkaline Br_2:

(a) 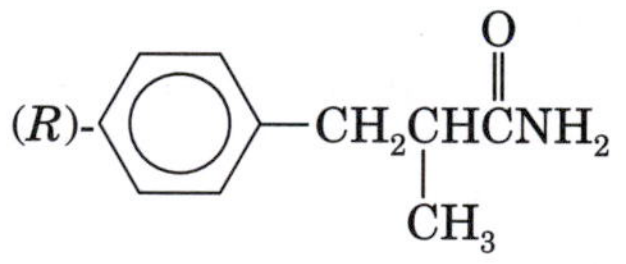 (R)- ⬡ $-CH_2\underset{\underset{CH_3}{|}}{C}H\overset{\overset{O}{\|}}{C}NH_2$ **(b)** 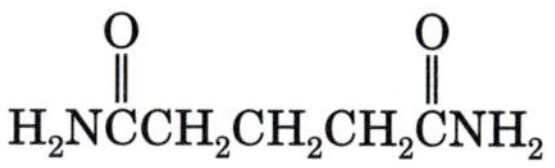 $H_2N\overset{\overset{O}{\|}}{C}CH_2CH_2CH_2\overset{\overset{O}{\|}}{C}NH_2$

D. Summary of Amine Syntheses

We have shown several routes to amines. By one or another of these routes, shown in the scheme below, a chemist may synthesize:

1. an amine with the same number of carbons as the starting material

2. an amine with one additional carbon

3. an amine with one less carbon.

The specific reactions are summarized in Table 18.2.

General routes to amines:

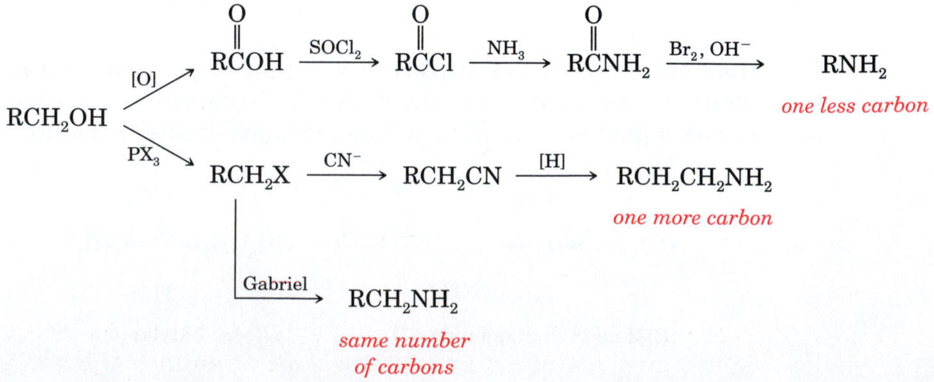

$$RCH_2OH \begin{cases} \xrightarrow{[O]} R\overset{\overset{O}{\|}}{C}OH \xrightarrow{SOCl_2} R\overset{\overset{O}{\|}}{C}Cl \xrightarrow{NH_3} R\overset{\overset{O}{\|}}{C}NH_2 \xrightarrow{Br_2,\ OH^-} RNH_2 \\ \qquad\qquad\qquad\qquad\qquad\qquad\qquad\qquad\qquad\qquad\quad \textit{one less carbon} \\ \\ \xrightarrow{PX_3} RCH_2X \xrightarrow{CN^-} RCH_2CN \xrightarrow{[H]} RCH_2CH_2NH_2 \\ \qquad\qquad\qquad\qquad\qquad\qquad\qquad\qquad\quad \textit{one more carbon} \end{cases}$$

$$RCH_2X \xrightarrow{Gabriel} RCH_2NH_2$$

same number of carbons

Table 18.2 Summary of laboratory syntheses of amines[a]		
Reaction		Section Reference

Primary Amines:

substitution

$$1° \text{ RX} \xrightarrow[\text{(2) OH}^-]{\text{(1) excess NH}_3} \text{RNH}_2 \qquad\qquad 18.4\text{A}$$

$$\text{RX} \xrightarrow[\substack{\text{(2) H}_2\text{O, H}^+ \\ \text{(3) OH}^-}]{\text{(1) K}^+ \text{ phthalimide}} \text{RNH}_2 \qquad\qquad 18.4\text{A}$$

reduction

$$\text{ArNO}_2 \xrightarrow[\text{(2) OH}^-]{\text{(1) Fe, HCl}} \text{ArNH}_2 \qquad\qquad 12.3\text{A}$$

$$\underset{\text{RCN or R\overset{\text{O}}{\overset{\|}{\text{C}}}NH}_2}{} \xrightarrow[\text{(2) H}_2\text{O, H}^+]{\text{(1) LiAlH}_4} \text{RCH}_2\text{NH}_2 \qquad\qquad 18.4\text{B}$$

$$\text{R}_2\text{C}{=}\text{O} \xrightarrow{\text{NH}_3, \text{H}_2, \text{Ni}} \text{R}_2\text{CHNH}_2 \qquad\qquad 13.6\text{C, } 18.4\text{B}$$

rearrangement

$$\text{R}\overset{\text{O}}{\overset{\|}{\text{C}}}\text{NH}_2 \xrightarrow{\text{X}_2, \text{OH}^-} \text{RNH}_2 \qquad\qquad 18.4\text{C}$$

Secondary and Tertiary Amines:

reduction

$$\text{R}\overset{\text{O}}{\overset{\|}{\text{C}}}\text{NR}'_2 \xrightarrow[\text{(2) H}_2\text{O}]{\text{(1) LiAlH}_4} \text{RCH}_2\text{NR}'_2 \qquad\qquad 18.4\text{B}$$

$$\text{R}_2\text{C}{=}\text{O} + \text{R}'_2\text{NH} \xrightarrow{\text{H}_2, \text{Ni}} \text{R}_2\text{CHNR}'_2 \qquad\qquad 13.6\text{D, } 18.4\text{C}$$

[a] Some arylamines can be synthesized by nucleophilic aromatic substitution. See Section 12.4.

STUDY PROBLEM

18.9   Write equations showing how each of the following amines could be prepared by (1) a Gabriel phthalimide synthesis and (2) a Hofmann rearrangement. (Different starting materials will be required.)

(a) 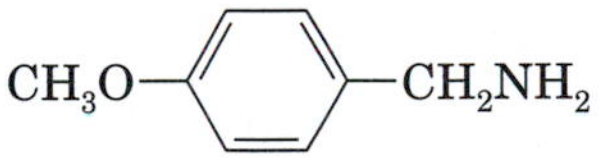 CH_3O—⟨ ⟩—CH_2NH_2 (b) $(S)\text{-}CH_3CH_2CHCH_3$ | NH_2

Section 18.5
Basicity of Amines

The unshared pair of valence electrons of ammonia or an amine can be donated to an electron-deficient atom, ion, or molecule. In water solution, an amine is a *weak base* and accepts a proton from water in a reversible acid–base reaction.

$$(CH_3)_3N\!: \;\; + H{-}\ddot{O}H \rightleftharpoons (CH_3)_3\overset{+}{N}H + :\ddot{O}H^-$$

trimethylamine

The calculation of basicity constants and pK_b values for weak bases was discussed in Section 1.7G. Table 18.3 lists a few amines along with their pK_b

Table 18.3 pK_b values for some amines and pK_a values for their conjugate acids

Structure	pK_b	Structure	pK_a
NH_3	4.75	NH_4^+	9.25
CH_3NH_2	3.34	$CH_3NH_3^+$	10.66
$(CH_3)_2NH$	3.27	$(CH_3)_2NH_2^+$	10.73
$(CH_3)_3N$	4.19	$(CH_3)_3NH^+$	9.81
⬡NH	2.88	⬡NH_2^+	11.12
⬡—NH_2	9.37	⬡—NH_3^+	4.63

values. (Recall that decreasing values for pK_b indicate increasing base strength.) We have also included the pK_a values for the conjugate acids of these amines. (Note that $pK_a = 14 - pK_b$.) An amine that is a stronger base has a *weaker* conjugate acid with a *higher* pK_a.

The same structural features that affect the relative acid strengths of carboxylic acids and phenols (Section 2.9E) affect the relative base strengths of amines.

1. *If the free amine is stabilized relative to the cation, the amine is less basic.*

2. *If the cation is stabilized relative to the free amine, the amine is a stronger base.*

if the free amine is stabilized, R_3N is a weaker base

if the cation is stabilized, R_3N is a stronger base

$$R_3N: + H_2O \rightleftharpoons R_3NH^+ + OH^-$$

An *electron-releasing group,* such as an alkyl group, on the nitrogen increases basicity by dispersing the positive charge in the cation. (This dispersal of positive charge is analogous to that in carbocations, Section 5.5E.) By dispersal of the positive charge, the cation is stabilized relative to the free amine. Therefore, base strength increases in the series NH_3, CH_3NH_2, and $(CH_3)_2NH$.

$$NH_3 \qquad CH_3NH_2 \qquad CH_3NHCH_3$$

ammonia   methylamine   dimethylamine

increasing basicity →

$$CH_3 \rightarrow \overset{..}{N}H_2 + H_2O \rightleftharpoons CH_3 \rightarrow NH_3^+ + OH^-$$

stabilized by dispersal of positive charge

The cation is also stabilized by *increasing solvation*. In this case, the solvent helps disperse the positive charge. Dimethylamine ($pK_b = 3.27$) is a slightly stronger base than methylamine. However, trimethylamine ($pK_b = 4.19$) is a *weaker base* than dimethylamine. The reason is that trimethylamine is more hindered, and the cation is less stabilized by solvation. These arguments explain why the nonaromatic heterocyclic amines (with their alkyl groups "tied back" away from the unshared electrons of the nitrogen) are more basic than comparable open-chain secondary amines.

$$CH_3-\underset{\underset{CH_3}{|}}{N}-CH_3 \quad \text{is a weaker base than} \quad CH_3-\underset{\underset{H}{|}}{N}-CH_3$$

trimethylamine
$pK_b = 4.19$

dimethylamine
$pK_b = 3.27$

$$CH_3CH_2NHCH_2CH_3 \quad \text{is a weaker base than}$$

diethylamine
$pK_b = 3.01$

pyrrolidine
$pK_b = 2.73$

Hybridization of the nitrogen atom in a nitrogen compound also affects the base strength. An sp^2 orbital contains more s character than an sp^3 orbital. A molecule with an sp^2-hybridized nitrogen is less basic because its unshared electrons are more tightly held and thus the free nitrogen compound is stabilized instead of the cation.

sp^2, *less basic*

sp^3, *more basic*

pyridine
$pK_b = 8.75$

piperidine
$pK_b = 2.88$

Resonance also affects the base strength of an amine. Cyclohexylamine is a far stronger base than is aniline.

aniline
$pK_b = 9.37$

cyclohexylamine
$pK_b = 3.3$

The reason for the low basicity of aniline is that the positive charge of the anilinium ion cannot be delocalized by the aromatic π cloud. However, the pair of electrons of the free amine are delocalized by the ring. The result is that the free amine is stabilized in comparison with the conjugate acid (the cation).

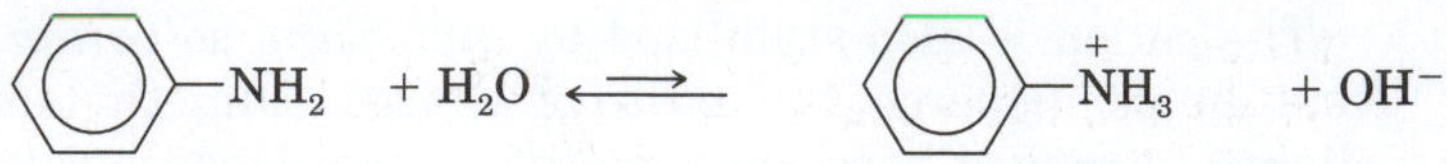

aniline

anilinium ion

resonance-stabilized
and favored

no resonance-stabilization
of positive charge

Resonance structures of aniline:

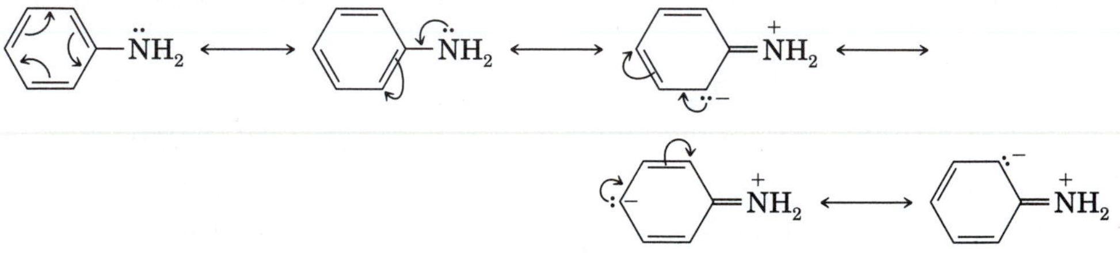

SAMPLE PROBLEM

Explain why piperidine is a stronger base than morpholine.

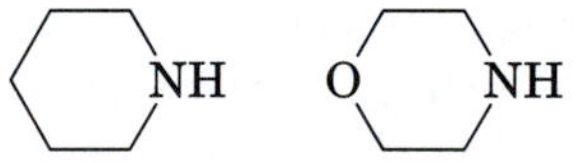

piperidine
$pK_b = 2.88$

morpholine
$pK_b = 5.67$

Solution

The oxygen atom in morpholine is *electron-withdrawing,* making the nitrogen more positive. The cation is not stabilized relative to the free morpholine, but is *destabilized:*

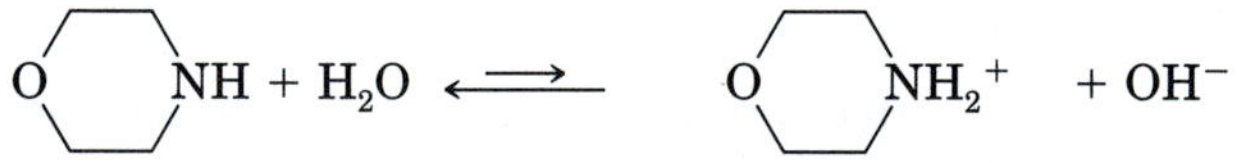

less stable
because of less
dispersal of + charge

Piperidine has no such destabilizing effect. Its cation is stabilized by electron release of the attached CH_2 groups:

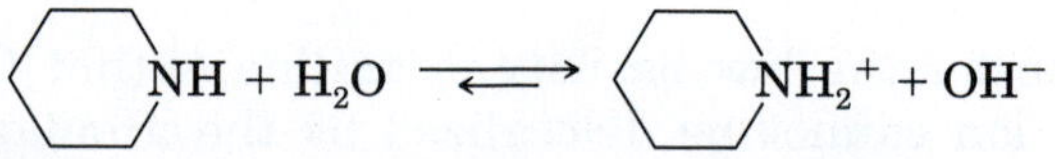

stabilized

STUDY PROBLEM

18.10   Explain the following trend in pK_b values:

$$O_2N-\text{⟨○⟩}-NH_2 \qquad \text{⟨○⟩}-NH_2 \qquad CH_3-\text{⟨○⟩}-NH_2$$

$$pK_b = 13.0 \qquad\qquad pK_b = 9.37 \qquad\qquad pK_b = 8.9$$

Section 18.6
Amine Salts

The reaction of an amine with a mineral acid (such as HCl) or a carboxylic acid yields an **amine salt.** The salts are commonly named in one of two ways: as **substituted ammonium salts** or, by an older system, as **amine-acid complexes.**

$$(CH_3)_3N\colon \;\; + HCl \longrightarrow \;\; (CH_3)_3NH^+ \, Cl^-$$

trimethylamine

trimethylammonium chloride
(trimethylamine hydrochloride)

$$CH_3CH_2\ddot{N}H_2 + CH_3CO_2H \longrightarrow CH_3CH_2\overset{+}{N}H_3 \;\; {}^-O_2CCH_3$$

ethylamine   acetic acid

ethylammonium acetate
(ethylamine acetate)

Because of its ability to form salts, an amine that is insoluble in water can be made soluble by treatment with dilute acid. In this fashion, compounds containing amino groups can be separated from water- and acid-insoluble materials. Naturally occurring amines in plants, called **alkaloids** (Section 20.3), can be extracted from their sources, such as leaves or bark, by aqueous acid. Many compounds containing amino groups are used as drugs. These drugs are often administered as their water-soluble salts rather than as water-insoluble amines.

$$(CH_3CH_2)_2\ddot{N}CH_2CH_2O\overset{O}{\overset{\|}{C}}-\text{⟨○⟩}-\ddot{N}H_2 \xrightarrow{\;HCl\;} (CH_3CH_2)_2\overset{Cl^-}{\overset{+}{N}H}CH_2CH_2O\overset{O}{\overset{\|}{C}}-\text{⟨○⟩}-\ddot{N}H_2$$

novocaine

water-insoluble

novocaine hydrochloride

water-soluble

STUDY PROBLEM

18.11   *Piperazine citrate* is a crystalline solid used in the treatment of pinworms and roundworms. Write an equation that shows the formation of the piperazine citrate formed from one molecule of each reactant.

$$\underset{\text{citric acid}}{\begin{array}{c} CH_2CO_2H \\ | \\ HOCCO_2H \\ | \\ CH_2CO_2H \end{array}} + \underset{\text{piperazine}}{HN\text{<}\bigcirc\text{>}NH} \longrightarrow$$

A free amine can be regenerated from one of its salts by treatment with a strong base, usually NaOH. Quaternary ammonium salts, which have no acidic protons, do not undergo this reaction.

$$\underset{\textit{an amine salt}}{RNH_3{}^+ \ Cl^- + OH^-} \longrightarrow \underset{\textit{an amine}}{R\overset{..}{N}H_2} + H_2O + Cl^-$$

$$\underset{\substack{\textit{a quaternary} \\ \textit{ammonium salt}}}{R_4N^+ \ Cl^-} + OH^- \longrightarrow \text{no reaction}$$

Because of the nitrogen's ionic charge, quaternary ammonium salts have some interesting applications. For example, quaternary ammonium salts with long hydrocarbon chains are used as detergents. The combination of a long, hydrophobic hydrocarbon tail with an ionic hydrophilic head results in two types of interactions with other substances. One part of the molecule is soluble in nonpolar organic solvents, fats, and oils, while the other part is soluble in water. Soaps (Section 24.2) exhibit similar behavior.

In Section 8.5, we discussed how crown ethers can act as *phase-transfer agents* by carrying inorganic ions into an organic solution. Quaternary ammonium salts can also act as phase-transfer agents. For example, let us say that we wish to carry out a substitution reaction using an alkyl halide and NaCN. When an aqueous NaCN solution is mixed with a solution of RX in a water-insoluble organic solvent, two layers result. Reaction can occur only at the interface of these layers. A quaternary ammonium salt can be used to transfer the CN$^-$ ions to the organic solution so that reaction can occur in the organic solution as well as at the interface. Also, recall that nucleophiles such as CN$^-$ are more nucleophilic and more reactive when they are not solvated by water. The result is that reaction is much faster.

The catalytic action of R$_4$N$^+$ X$^-$ arises from the fact that it is water-soluble and also slightly soluble in organic solvents. If R$_4$N$^+$ X$^-$ is dissolved in the aqueous phase of the two-phase reaction mixture, some of the salt also becomes dissolved in the organic layer. However, if the aqueous layer contains an excess of CN$^-$ ions, the salt that is transferred is mainly R$_4$N$^+$ CN$^-$, not R$_4$N$^+$ X$^-$.

Anion exchange in the aqueous phase:

$$\underset{\substack{\text{tetrabutylammonium} \\ \text{chloride}}}{(C_4H_9)_4N^+ \ Cl^-} + \underset{\textit{excess}}{CN^-} \rightleftarrows \underset{\substack{\textit{migrates to organic} \\ \textit{phase}}}{(C_4H_9)_4N^+ \ CN^-} + Cl^-$$

S$_N$2 reaction in the organic phase:

$$\underset{\text{1-chlorooctane}}{CH_3(CH_2)_7Cl} + (C_4H_9)_4N^+ \ CN^- \longrightarrow \underset{\substack{\text{nonanenitrile} \\ (90\%)}}{CH_3(CH_2)_7CN} + \underset{\substack{\textit{returns to aqueous phase} \\ \textit{for additional exchange}}}{(C_4H_9)_4N^+ \ Cl^-}$$

Substitution Reactions with Amines

We have already mentioned a variety of substitution reactions with amines. The problems of the reaction of an amine with an alkyl halide were discussed earlier in this chapter (Section 18.4A).

$$RNH_2 + R'Cl \xrightarrow{S_N2} R\overset{+}{N}H_2\,Cl^- \quad \text{and also} \quad R\overset{+}{N}HR'_2\,Cl^- \quad \text{and} \quad R\overset{+}{N}R'_3\,Cl^-$$
$$\underset{R'}{|}$$

In Chapter 15, we discussed the acylation of amines as a technique for the synthesis of amides. For example:

N-methylacetamide

The utility of this reaction is that amines can be used to synthesize other amines by conversion to the amide, followed by reduction.

| an acid chloride | an amine | | an amide | | a new amine |

In Section 15.10, the preparation of sulfonamides was discussed. The reaction of amines with benzenesulfonyl chloride is sometimes used to test whether an amine is primary, secondary, or tertiary. This test is called the **Hinsberg test.**

anion of a sulfonamide:
soluble in base

a sulfonamide:
insoluble in acid

a sulfonamide:
insoluble in acid or base

$$3° \ R_3N \xrightarrow[\text{OH}^-]{\text{Cl}-\text{SO}_2-\text{C}_6\text{H}_5} \left[\overset{+}{R_3N}-\text{SO}_2-\bigcirc \right] \longrightarrow$$

unstable in base:
no change observed

$$R_3N + {}^-\text{OSO}_2-\bigcirc \xrightarrow{\text{H}^+} \overset{+}{R_3NH} + \text{HOSO}_2-\bigcirc$$

soluble in acid

Amines also react with aldehydes and ketones to yield imines and enamines. (For the mechanisms, see Section 13.5.)

cyclohexanone

$$\xrightarrow[\text{H}^+]{\text{H}_2\text{NR}(1°)} \bigcirc =\text{NR} + \text{H}_2\text{O}$$

an imine

$$\xrightarrow[\text{H}^+]{\text{HNR}_2(2°)} \bigcirc -\text{NR}_2 + \text{H}_2\text{O}$$

an enamine

Section 18.8
Reactions of Amines with Nitrous Acid

In Section 12.3, we discussed the formation of benzenediazonium chloride ($C_6H_5N_2^+ \ Cl^-$) by the treatment of aniline with cold aqueous nitrous acid, HNO_2 (prepared *in situ* from $NaNO_2$ and HCl). Recall that the aryldiazonium salts are stable at 0° and are useful synthetic intermediates because of the excellent leaving ability of N_2.

The treatment of a *primary alkylamine* with $NaNO_2$ and HCl also results in a diazonium salt, but an alkyldiazonium salt is unstable and decomposes to a mixture of alcohol and alkene products along with N_2. The decomposition proceeds by way of a carbocation.

$$(CH_3)_2CHNH_2 \xrightarrow[0°]{\substack{\text{NaNO}_2 \\ \text{HCl}}} (CH_3)_2CH\overset{\frown}{-}N_2^+ \ Cl^- \xrightarrow[-\text{Cl}^-]{-\text{N}_2}$$

isopropylamine   isopropyldiazonium
                          chloride
a 1° amine

$$[(CH_3)_2CH^+] \xrightarrow{\text{H}_2\text{O}} (CH_3)_2CHOH + CH_3CH{=}CH_2$$

When treated with $NaNO_2$ and HCl, *secondary amines* (alkyl or aryl) yield **N-nitrosoamines,** compounds containing the N—N=O group. Many *N*-nitrosoamines are carcinogenic.

N-methylaniline

a 2° amine

an N-nitrosoamine (93%)

Tertiary amines are not entirely predictable in their reactions with nitrous acid. A tertiary arylamine usually undergoes ring substitution with —NO because of the ring activation by the —NR$_2$ group. A tertiary alkylamine (and sometimes tertiary arylamines, too) may lose an R group and form an *N*-nitroso derivative of a secondary amine.

STUDY PROBLEM

18.12 When butylamine is treated with cold aqueous solution of HCl and NaNO$_2$, the following products are obtained: 1-chlorobutane, 2-chlorobutane, 1-butanol, 2-butanol, 1-butene, 2-butene, and nitrogen gas. Suggest a mechanism that accounts for each of these products.

Section 18.9
Hofmann Elimination

Quaternary ammonium hydroxides (R$_4$N$^+$ OH$^-$) are amine derivatives that are used in structure-determination studies because they undergo elimination reactions to yield alkenes and amines. We will look briefly at how these compounds are prepared and at their elimination reactions.

A. Formation of Quaternary Ammonium Hydroxides

When a quaternary ammonium halide is treated with aqueous silver oxide, the quaternary ammonium hydroxide is obtained:

$$2\,R_4N^+ X^- \; + Ag_2O + H_2O \longrightarrow \; 2\,R_4N^+ OH^- \; + 2\,AgX\downarrow$$

a quarternary
ammonium halide

a quarternary
ammonium hydroxide

N,N-dimethylpiperidinium
chloride

N,N-dimethylpiperidinium
hydroxide

A quaternary ammonium hydroxide cannot be obtained by the ionic reaction of R$_4$N$^+$ X$^-$ with aqueous NaOH because the reactants and products are all water-soluble ionic compounds. If such a reaction were attempted, a mixture of R$_4$N$^+$ OH$^-$ and R$_4$N$^+$ Cl$^-$ (along with NaOH and NaCl) would result. However, silver hydroxide, which is formed *in situ* from moist silver oxide

($Ag_2O + H_2O \rightarrow 2AgOH$), removes the halide ion as an AgX precipitate. Removal of the AgX by filtration, followed by evaporation of the water, yields the pure quaternary ammonium hydroxide.

B. The Elimination

When a quaternary ammonium hydroxide (as a solid) is heated, an elimination reaction called a **Hofmann elimination** occurs. This reaction is an E2 reaction in which the leaving group is an amine.

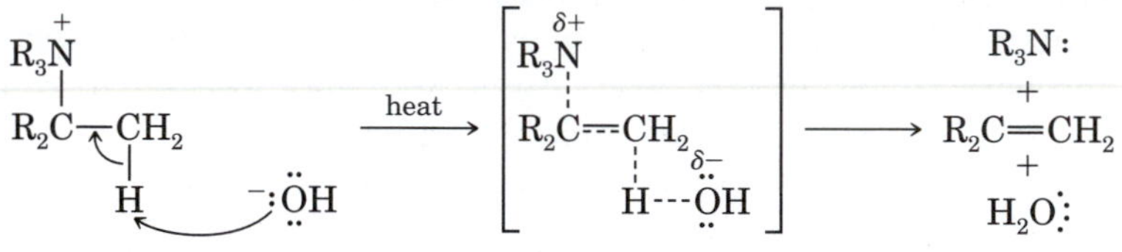

E2 transition state

This elimination generally yields the *Hofmann product,* the alkene with *fewer alkyl groups* on the π-bonded carbons. The formation of the less substituted, less stable alkene can be attributed to steric hindrance in the transition state (Section 5.8E) due to the bulky R_3N^+— group.

$$(CH_3)_3\overset{+}{N}-\underset{}{\overset{CH_3}{\overset{|}{C}}}HCH_2CH_3\ OH^- \xrightarrow{\text{heat}}$$

sec-butyltrimethylammonium hydroxide

$$(CH_3)_3N + \underset{\text{1-butene (95\%)}}{CH_2{=}CHCH_2CH_3} + \underset{\text{2-butene (5\%)}}{CH_3CH{=}CHCH_3} + H_2O$$

Study Problem

18.13 Predict the major organic products when the following compounds are heated:

(a) $CH_3CH_2CH_2CH_2\overset{+}{N}(CH_3)_2-CH_2CH_3\ OH^-$ **(b)** $C_6H_5CH_2CH_2\overset{+}{N}(CH_3)_2-CH_2CH_3\ OH^-$

Summary

An **amine** is a compound that contains a trivalent nitrogen that has from one to three alkyl or aryl groups attached: RNH_2, R_2NH, or R_3N. A compound with four groups attached to the nitrogen is an **amine salt** ($R_3NH^+ X^-$) or a **quaternary ammonium salt** ($R_4N^+ X^-$).

Amines can be prepared by *substitution reactions,* by *reduction reactions,* or by *rearrangement.* These synthetic reactions are summarized in Table 18.2.

Because the nitrogen of an amine has a pair of unshared valence electrons, amines are *weak bases.* The base strength is affected by *hybridization* (sp^3 >

$sp^2 > sp$), by *electron-withdrawing groups* (base-weakening), by *electron-releasing groups* (base-strengthening), and by *conjugation* (base-weakening).

hybridization   *electron-release*

$$CH_3CH_2NH_2 > HO \leftarrow CH_2CH_2NH_2$$

electron-withdrawal   *conjugation*

Amines undergo reaction with acids to yield amine salts:

$$R_3N \underset{OH^-}{\overset{HX}{\rightleftharpoons}} R_3NH^+ \ X^-$$

an amine salt

Most amines are nucleophiles and can displace good leaving groups or can attack carbonyl groups. When primary amines are treated with cold nitrous acid, diazonium salts are formed. Alkyldiazonium salts are unstable, but aryldiazonium salts can be used to prepare a variety of substituted aromatic compounds. The reactions of amines are summarized in Table 18.4.

Quaternary ammonium hydroxides, when heated, lose water and an amine. The least substituted alkene is usually formed. This reaction is called

Table 18.4 Summary of reactions of amines

Reaction		Product	Section Reference
$R_3N + R'X$	$\longrightarrow R_3\overset{+}{N}R' \ X^-$	**amine salt or quaternary ammonium salt**	18.4A
$\underset{\ }{\overset{O}{\overset{\|}{R_2NH + R'CCl}}}$	$\longrightarrow \underset{\ }{\overset{O}{\overset{\|}{R_2NCR'}}}$	**amide**[a]	15.3C, 18.7
1° $RNH_2 + R'_2C{=}O$	$\xrightarrow{H^+} RN{=}CR'_2$	**imine**	13.5A
$2°R_2NH + \underset{\ }{\overset{O}{\overset{\|}{R'_2CHCR'}}}$	$\xrightarrow{H^+} R'_2C{=}\overset{\overset{NR_2}{\|}}{C}R'$	**enamine**[b]	13.5B
$ArNH_2$	$\xrightarrow[0°]{\underset{HCl}{NaNO_2}} ArN_2^+ \ Cl^-$	**aryldiazonium salt**[c,d]	12.3, 18.8
$\underset{\ }{\overset{NR'_2}{\overset{\|}{R_2CHCR_2}}}$	$\xrightarrow[\text{(3) heat}]{\overset{\text{(1) CH}_3\text{I}}{\text{(2) Ag}_2\text{O, H}_2\text{O}}} R_2C{=}CR_2$	**alkene**	18.9

[a] Other routes to amides are the similar reactions of amines with acid anhydrides (Section 15.4C) and with esters (Section 15.5C).
[b] Enamines can be converted to α-substituted aldehydes or ketones (see Section 17.5).
[c] Aryldiazonium salts can be converted to aryl halides, nitriles, etc. (see Section 12.3B).
[d] Secondary amines (alkyl or aryl) yield N-nitrosoamines when treated with HNO_2.

the **Hofmann elimination.**

$$CH_3CH_2\overset{+}{\underset{\underset{\displaystyle CH_2CH_2CH_3}{|}}{N}}(CH_3)_2\ OH^- \xrightarrow{\text{heat}} CH_2{=}CH_2 + \underset{\underset{\displaystyle CH_2CH_2CH_3}{|}}{N}(CH_3)_2 \quad + H_2O$$

Essay Problem for Chapter 18

Lactonization with Retention of Configuration

When (S)-$(+)$-glutamic acid (**1**) is treated with nitrous acid, an unstable intermediate (**2**) is initially formed. This unstable intermediate undergoes reaction to give the alpha lactone (**3**), which is the second unstable intermediate in the reaction sequence. Nitrogen gas is released when **3** is formed. The α-lactone (**3**) spontaneously undergoes reaction to yield the observed product, (S)-$(+)$-γ-butyrolactone-γ-carboxylic acid (**4**) in a 70% yield. (O. H. Gringore and F. P. Rouessac, *Org. Syn.* **1985**, *63*, 121.)

Questions
1. What is the structure and stereochemistry of the intermediate product **2**?
2. Attack by which carboxyl group yields the α-lactone? Write a plausible mechanism for this step in the reaction sequence.
3. What is the configuration of the chiral carbon in the α-lactone? Is the α-lactone formed by retention or inversion of configuration of (S)-glutamic acid?
4. Does the conversion of **3** to **4** take place by retention or inversion of configuration? Write a plausible mechanism for this conversion.
5. Explain why the lactonization reaction presented in the essay occurs with retention of configuration.

Study Problems*

18.14   Write formulas for the following compounds:

   (a)   phenylalanine (2-amino-3-phenylpropanoic acid), an amino acid constituent of proteins

* For information concerning the organization of the *Study Problems* and *Additional Problems*, see the *Note to student* on page 41.

(b)   *N*-methyl-1-cyclohexyl-2-propanamine, a vasoconstrictor

(c)   *N,N*-dimethylaniline, a solvent and industrial intermediate

(d)   1,2-propanediamine, used with $CuSO_4$ to test for mercury

18.15 Name the following compounds:

(a) $C_6H_5CH_2NCH_3$
          $\quad\quad\quad\ |$
          $\quad\quad\quad CH_2CH_3$

(b) 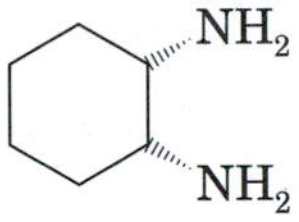   [cyclohexane with two NH_2 groups]

(c) $(C_6H_5)_2CHN(CH_2CH_3)_2$

(d) $CH_3CHCH_2CH_2CH$ with NH_2 substituent and $C{=}O$
$$CH_3\overset{\overset{\displaystyle NH_2}{|}}{C}HCH_2CH_2\overset{\overset{\displaystyle O}{\|}}{C}H$$

18.16 Which of the following structures has enantiomers, geometric isomers, both, or neither? [*Hint:* In (f), consider the hybridization of the N and the resultant geometry.]

(a) $C_6H_5CHNH_2$
          $\quad\quad\quad |$
          $\quad\quad CH_2CH_3$

(b) $CH_3NCH_2CH_3$
          $\quad\quad\ |$
          $\quad\quad C_6H_5$

(c) 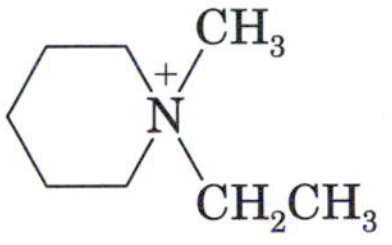   [piperidinium ring with CH_3 and CH_2CH_3 on N$^+$] Cl^-

(d)   [piperidinium ring with two CH_3 and CH_2CH_3 on N$^+$] Cl^-

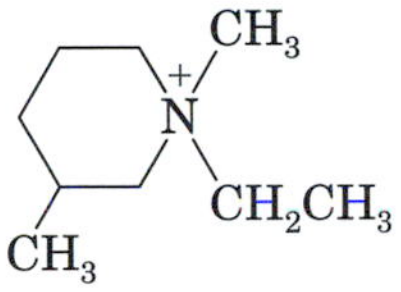

(e) $CH_3NCH_2CH_3$
          $\quad\quad\ \overset{+}{|}\ O^-$
          $\quad\quad C_6H_5$

(f) $C_6H_5CH{=}NOH$

18.17 Which of the following species can act as a nucleophile? Explain.

(a) $(CH_3)_2NH$ **(b)** $(CH_3)_3N$ **(c)** H_2N-NH_2

(d)    [pyridine ring with N]   **(e)** 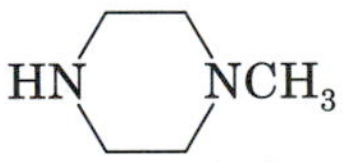 $HN\ \ NCH_3$ [piperazine ring]   **(f)** 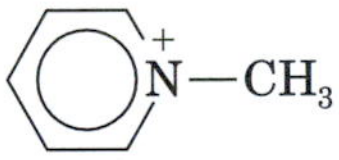   [pyridinium ring $\overset{+}{N}-CH_3$]

18.18 Suggest syntheses for the following compounds, starting with organohalogen compounds or alcohols:

(a) 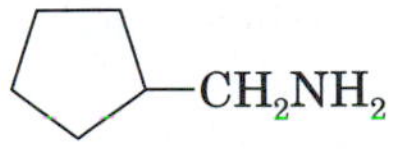   [cyclopentane]$-CH_2NH_2$

(b) 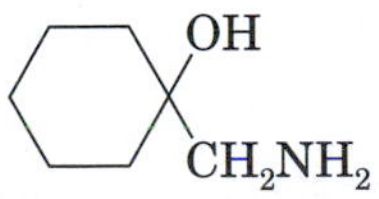   [cyclohexane with OH and CH_2NH_2]

(c) $HO_2CCH_2CHCO_2H$
          $\quad\quad\quad\quad |$
          $\quad\quad\quad\ NH_2$

(d) $(CH_3)_2CHNH_2$

(e)    [cyclohexane]$-NHCH_2CH_2CH_3$

(f) $CH_3CH_2CH_2CH_2N(CH_3)_2$

18.19 Suggest a way to make each of the following conversions: **(a)** benzene to aniline; **(b)** benzamide to aniline; **(c)** aniline to acetanilide ($C_6H_5NHCOCH_3$); **(d)** (*R*)-2 butanol to (*S*)-2-butanamine; **(e)** toluene to benzylamine; **(f)** acetic acid to acetamide.

18.20 Which is more basic? **(a)** aniline or *p*-bromoaniline; **(b)** trimethylamine or tetramethylammonium hydroxide; **(c)** *p*-nitroaniline or 2,4-dinitroaniline; **(d)** ethylamine or ethanolamine ($HOCH_2CH_2NH_2$); **(e)** *p*-toluidine (*p*-methylaniline) or *p*-(trichloromethyl)aniline. Explain your answers.

18.21 Complete the following equations:

(a) [ring]NH$^+$ + [ring]NH $\longrightarrow$ (b) [ring]NH$^+$ + OH$^-$ $\longrightarrow$

(c) [ring]—$\overset{+}{N}H_3$ Cl$^-$ + (CH$_3$)$_3$N $\longrightarrow$ (d) (CH$_3$)$_3$NH$^+$ + [ring]—NH$_2$ $\longrightarrow$

(e) [ring]NH + CH$_3$CO$_2$H $\longrightarrow$

18.22 List each of the following groups of cations in order of increasing acidity (weakest acid first):

(a) (1) [ring with Cl, NH$_2^+$]   (2) [ring with Cl, NH$_2^+$, Cl]   (3) [ring with NH$_2^+$]

(b) (1) [ring with O, NH$_2^+$]   (2) [ring with NH$_2^+$]   (3) [ring with O, NH$_2^+$]

(c) (1) [ring]—NH$_3^+$ (2) ([ring])$_2$—NH$_2^+$ (3) ([ring])$_3$—NH$^+$

(d) (1) H$_3$O$^+$ (2) [ring]NH$_2^+$

18.23 Predict the major product of the reaction of pyrrolidine (page 763) with: **(a)** benzoyl chloride; **(b)** acetic anhydride; **(c)** CH$_3$I (excess NaOH); **(d)** phthalic anhydride; **(e)** benzenesulfonyl chloride (C$_6$H$_5$SO$_2$Cl); **(f)** acetyl chloride, followed by LiAlH$_4$ (then hydrolysis); **(g)** cold nitrous acid; **(h)** dilute HCl; **(i)** acetone + H$^+$.

18.24 Explain, using equations, how each of the following amines would behave in a Hinsberg test.

(a) CH$_3$CH$_2$CH$_2$CH$_2$NH$_2$ (b) C$_6$H$_5$CH$_2$NHCH$_3$

(c) (CH$_3$CH$_2$CH$_2$)$_3$N   (d) (CH$_3$CH$_2$CH$_2$)$_4$N$^+$ Br$^-$

18.25 Write equations showing how each of the following conversions could be carried out:

(a) [ring] $\longrightarrow$ [ring]—N(CH$_3$)$_2$

(b) [ring]—CO$_2$H $\longrightarrow$ [ring]—NHCH$_2$CH$_3$

(c) CH$_3$CH$_2\underset{\underset{CH_3}{|}}{\overset{\overset{CH_3}{|}}{C}}$Br $\longrightarrow$ CH$_3$CH$_2\underset{\underset{CH_3}{|}}{\overset{\overset{CH_3}{|}}{C}}NH_2$

18.26 Predict the major organic products when each of the following compounds is heated:

(a) 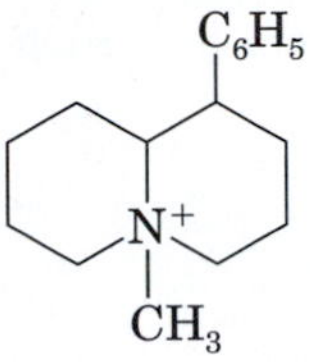   OH⁻   (b) 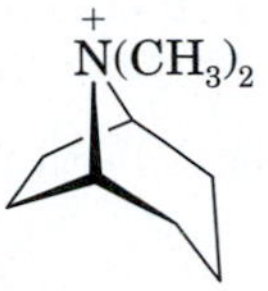   OH⁻

18.27 What would you expect the major organic products to be if *p*-methylbenzamide were treated with the following reagents?

(a) Br_2, NaOH   (b) Br_2, $FeBr_3$ (c) Br_2, light   (d) Br_2, CCl_4 (no light)

Additional Problems

18.28 Complete the following equations:

(a) $(CH_3CH_2)_2NH + ClCH_2CH_2OH \xrightarrow{\text{NaOH}}$

(b) $CH_3NH_2 \xrightarrow[\text{(2) H}^+]{\text{(1) excess ClCH}_2\text{CO}_2^- \text{ Na}^+, \text{ NaOH}}$

(c) 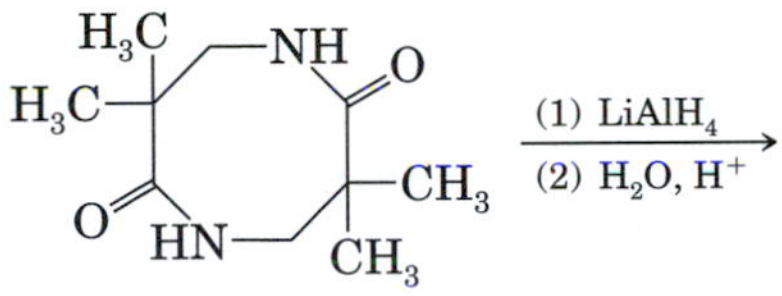 $\xrightarrow[\text{(2) H}_2\text{O, H}^+]{\text{(1) LiAlH}_4}$

(d) $CH_3NHCO_2C_2H_5 \xrightarrow[\text{HCl}]{\text{NaNO}_2}$

(e) $C_6H_5CH_2\overset{\overset{\displaystyle NH_2}{|}}{C}HCO_2CH_3 \xrightarrow[\text{CH}_3\text{CH}_2\text{OH}]{\text{C}_6\text{H}_5\text{CHO, (CH}_3\text{CH}_2)_3\text{N}}$

(f) 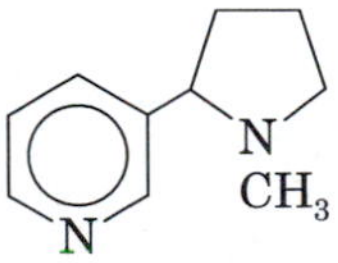 $+ 1 \text{ HCl} \longrightarrow$

nicotine

18.29 Suggest synthetic routes to the following compounds:

(a)   isoleucine (2-amino-3-methylpentanoic acid)

(b)   *β*-phenylethylamine from toluene

(c)   2-methyl-1-phenyl-1,3-pentanedione from benzoic acid

(d)   aniline from benzoic acid

(e)   the plant product *ubine* (2-dimethylamino-1-phenylethanol) from an alkene

(f)  H_2NCH_2—⬡—CH_2NH_2 from toluene

(g)   3,5-dimethoxyaniline from a nitrile

18.30   Suggest synthetic routes to the following compounds:

(a)   the pain reliever *phenacetin* (*p*-ethoxyacetanilide) from *p*-nitrophenol. (For the structure of acetanilide, see Problem 18.19 (c)).

(b)   methyl orange (page 894) from substituted benzenes

(c) 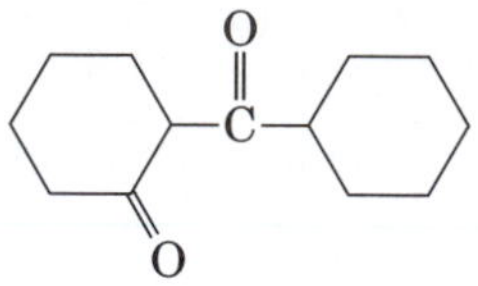   from compounds containing six or fewer carbon atoms

18.31   The treatment of (cyclopentylmethyl)amine with nitrous acid resulted in a 76% yield of cyclohexanol. Another alcohol and three alkenes were also present in the product mixture.

(a)   Give a plausible mechanism for the formation of cyclohexanol.

(b)   What are the likely structures of the other products?

18.32   A chemist attempted to carry out a Hofmann rearrangement of butanamide with bromine and potassium hydroxide in methanol, rather than in water. Instead of propylamine, the chemist obtained a carbamate, $CH_3CH_2CH_2NHCO_2CH_3$. Explain how this product was formed.

18.33   A tertiary amine is oxidized by peroxides such as H_2O_2 to an **amine oxide,** a compound containing the —NO group. An amine oxide with a β hydrogen undergoes elimination when heated. (This reaction is called a **Cope elimination.**) Suggest a mechanism for the following elimination reaction of an amine oxide.

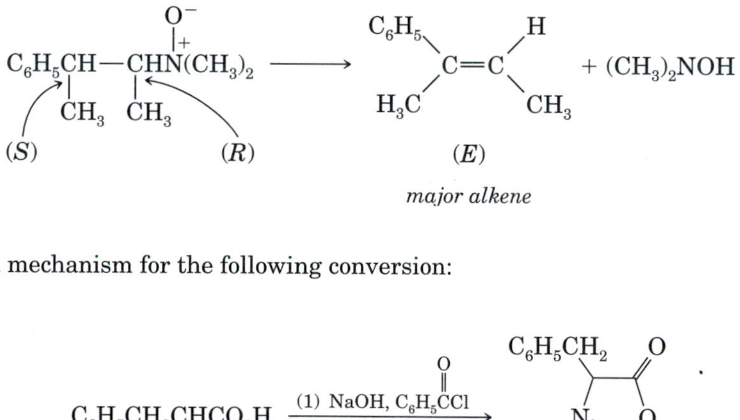

major alkene

18.34   Suggest a mechanism for the following conversion:

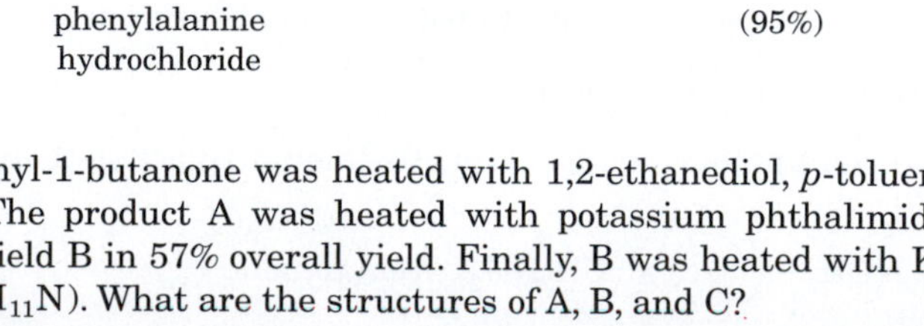

phenylalanine
hydrochloride

(95%)

18.35   4-Chloro-1-phenyl-1-butanone was heated with 1,2-ethanediol, *p*-toluenesulfonic acid, and benzene. The product A was heated with potassium phthalimide in dimethylformamide to yield B in 57% overall yield. Finally, B was heated with KOH in ethanol to yield C ($C_{10}H_{11}N$). What are the structures of A, B, and C?

18.36   A chemist treated 1-bromobutane with ammonia and isolated products A and B. When A was treated with acetic anhydride, C was obtained. When B was treated with acetic anhydride, D was obtained. The infrared spectra of C and D are shown in Figure 18.2. Identify A, B, C, and D.

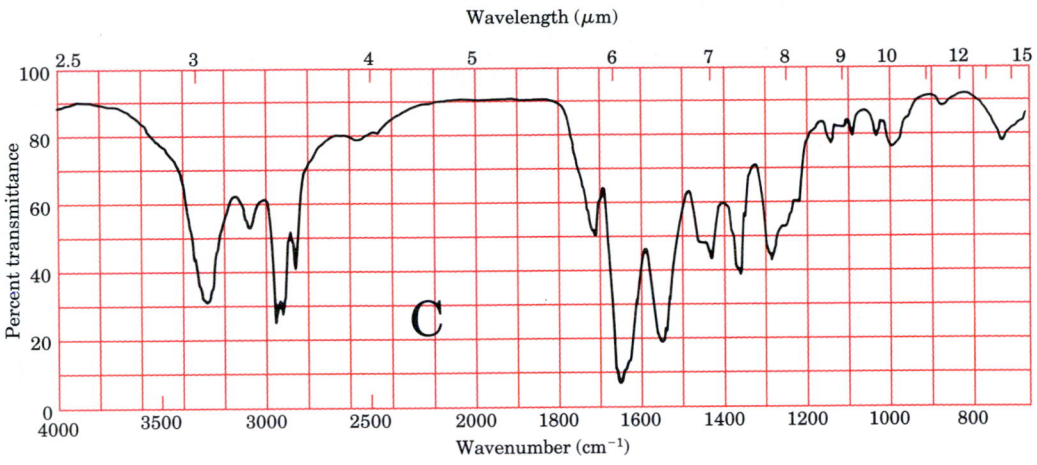

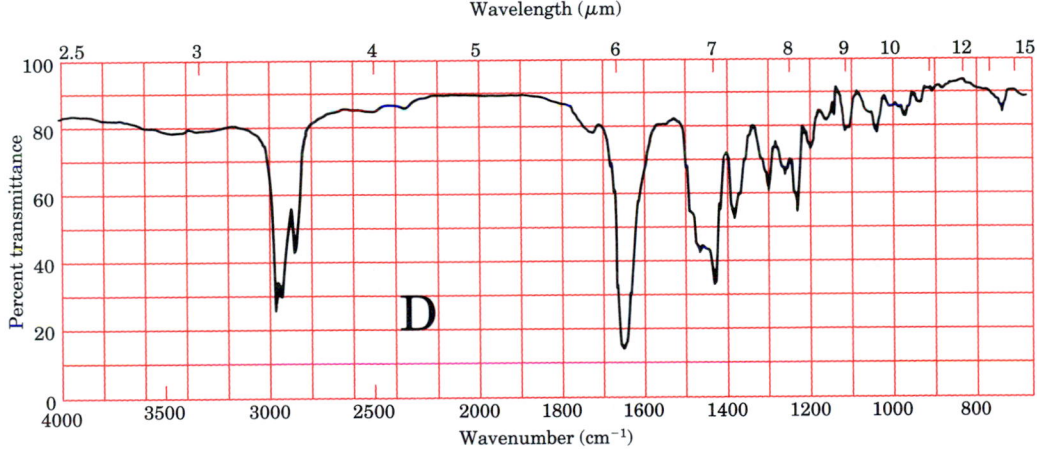

Figure 18.2 Infrared spectra for Problem 18.36.

18.37 The infrared spectrum for Compound A ($C_8H_{11}N$) is given in Figure 18.3. A is soluble in dilute acid. Oxidation of A with hot $KMnO_4$ yields benzoic acid. What are the two possible structures for A? How could you distinguish these two possibilities by ^1H NMR spectroscopy?

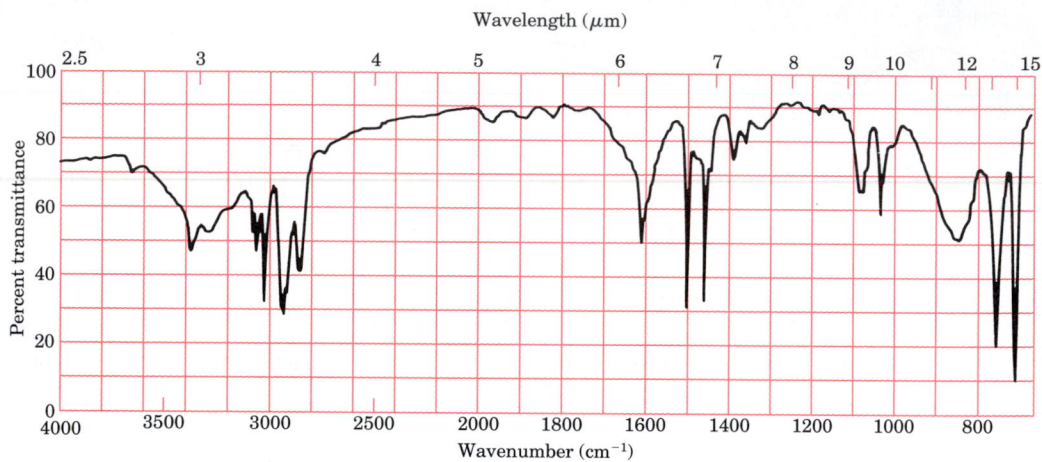

Figure 18.3 Infrared spectrum for Compound A in Problem 18.37.

POLYCYCLIC AND HETEROCYCLIC AROMATIC COMPOUNDS

In Chapters 11 and 12, we discussed benzene and substituted benzenes. However, benzene is but one member of a large number of aromatic compounds. Many other aromatic compounds can be grouped into two classes: **polycyclic compounds and heterocyclic compounds.** The polycyclic aromatic compounds are also referred to as *polynuclear, fused-ring,* or *condensed-ring* aromatic compounds. These aromatic compounds are characterized by rings that jointly share carbon atoms and by a common aromatic π cloud.

Some polycyclic aromatic compounds:

naphthalene   anthracene

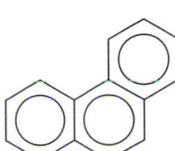

phenanthrene

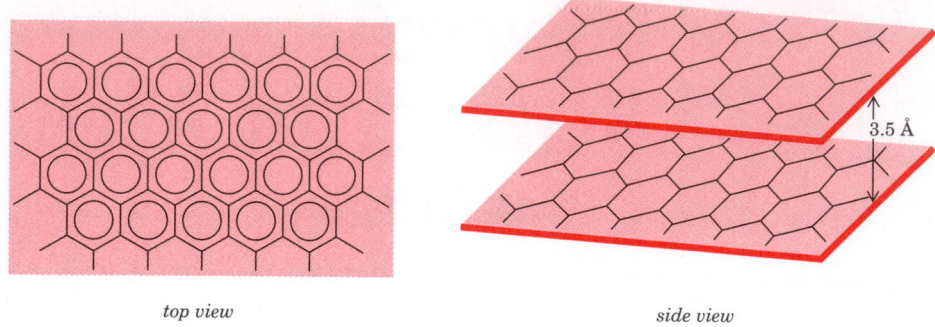

top view   *side view*

Figure 19.1 The structure of graphite.

The polycyclic aromatic hydrocarbons and most of their derivatives are solids. Naphthalene has been used as mothballs and flakes, and derivatives of naphthalene are used in motor fuels and lubricants. The structure for naphthalene was first proposed in 1866 by R. Erlenmeyer (1825–1909), who also invented the conical flask that bears his name. The most extensive use made of the polycyclic aromatics is that of synthetic intermediates, for example, in the manufacture of dyes. (See Section 22.4B.)

Graphite is one of the more interesting polycyclic compounds. The structure of graphite consists of planes of fused benzene rings (Figure 19.1). The distance (3.5 Å) between any pair of planes is believed to be the thickness of the π system of benzene. The "slipperiness" of graphite is due to the ability of these planes to slide across each other. Because of this property, graphite is a valuable lubricant that can be used even in outer space where ordinary oils and greases would solidify. Because of its mobile π electrons, graphite can conduct electricity and finds use where an inert electrode is needed. Flashlight batteries, for example, contain graphite electrodes.

A **heterocyclic compound** is a cyclic compound in which the ring atoms are of two more different elements. Heterocyclic rings can be aromatic, just as carbon rings can. Approximately a third of the organic chemical literature deals with heterocyclic compounds. Many **alkaloids,** such as morphine (Section 20.3A); the **nucleic acids,** the carriers of the genetic code; and many other compounds of biological importance are examples of heterocyclic compounds.

Two structurally similar heterocycles:

tryptophan

an amino acid

psilocin

a hallucinogen in Psilocybe mushrooms

Nomenclature of Polycyclic Aromatic Compounds

The ring systems of common polycyclic aromatic compounds have individual names. Unlike the numbering of benzene or a cycloalkane ring, which starts at the position of a substituent, the numbering of a polycyclic ring is fixed by convention and does not change with the position of a substituent.

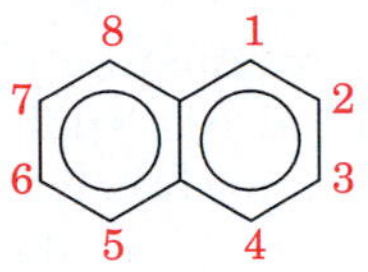

naphthalene 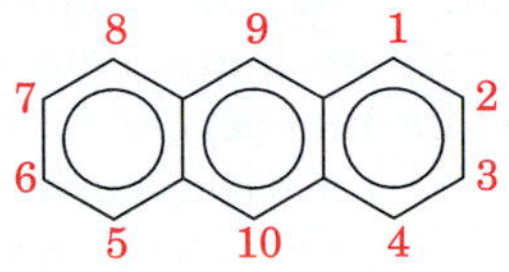   anthracene 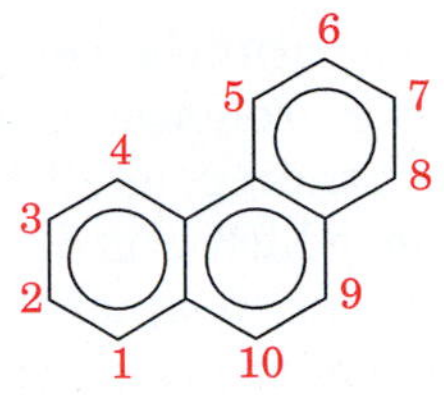   phenanthrene

The position of a substituent in a monosubstituted naphthalene is often designated by a Greek letter. The positions adjacent to the ring-junction carbons are called α positions, while the next positions are β positions. By this system, 1-nitronaphthalene is called α-nitronaphthalene, while 2-nitronaphthalene is called β-nitronaphthalene. Naphthalene itself has four equivalent α positions and four equivalent β positions. (Only number designations are used in the anthracene and phenanthrene systems.)

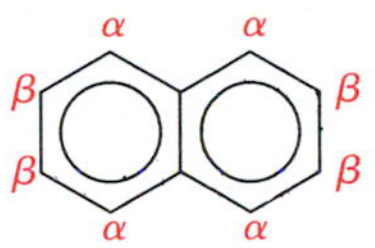

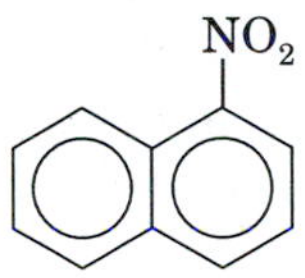

1-nitronaphthalene   2-nitronaphthalene
(α-nitronaphthalene)   (β-nitronaphthalene)

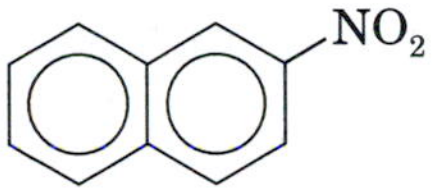

STUDY PROBLEM

19.1   Name the following compounds:

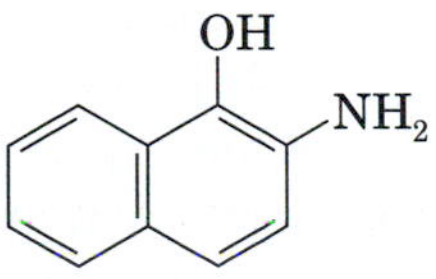

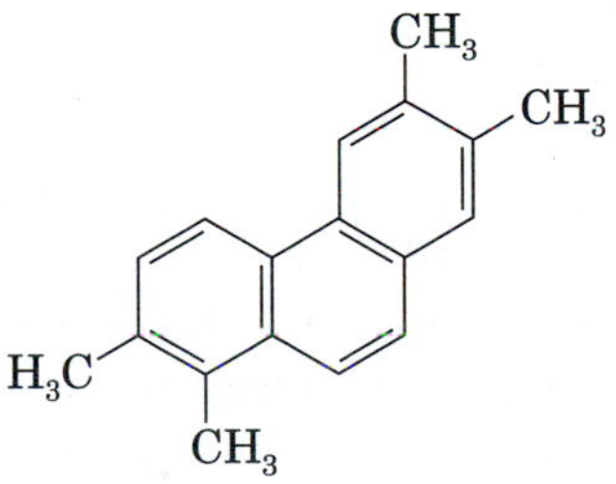

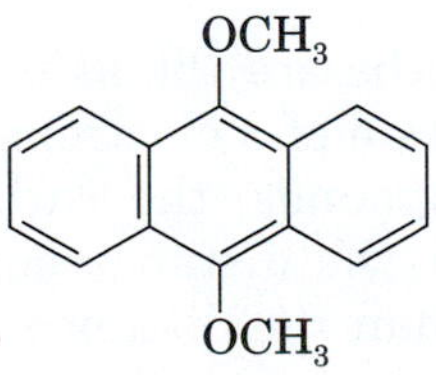

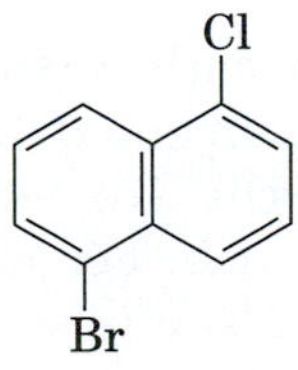

Section 19.2
Bonding in Polycyclic Aromatic Compounds

For a monocyclic ring system to be aromatic, it must meet three criteria:

1. Each atom in the ring system must have a p orbital available for bonding.

2. The ring system must be planar.

3. There must be $(4n + 2)$ π electrons in the ring system (Hückel rule).

These criteria were discussed in Section 11.6. The aromatic system of a poly-cyclic aromatic compound must also contain atoms with p orbitals available for bonding, and the entire aromatic ring system must be planar. The Hückel rule, which was devised for monocyclic systems, also is applicable to polycyclic systems in which the sp^2-hybridized carbons are peripheral, or on the outside edge of the ring system. In polycyclic compounds, the number of π electrons is easily counted when Kekulé formulas are used.

10 π electrons $(n = 2)$	*14 π electrons* $(n = 3)$	*14 π electrons* $(n = 3)$

Like benzene, the polycyclic aromatic systems are more stable than the corresponding hypothetical polyenes with localized π bonds. The energy dif-ferences between the hypothetical polyenes and the real compounds (that is, the *resonance energies*) have been calculated from heats of combustion and hydrogenation data.

resonance energy (kcal / mol):   36   61   84   92

The resonance energy of polycyclic compounds seems less per ring than that for benzene. For example, with two rings, naphthalene could be expected to have 72 kcal/mol—or twice as much resonance energy as benzene. But the resonance energy for naphthalene is only 61 kcal/mol. When viewed from the amount of resonance energy per p orbital however (about 6 kcal per orbital), we can see that all of the aromatic hydrocarbons are stabilized to approxi-mately the same extent.

In benzene, all carbon–carbon bond lengths are the same. This fact leads us to believe that there is an equal distribution of π electrons around the ben-zene ring. In the polycyclic aromatic compounds, the carbon–carbon bond lengths are *not* all the same. For example, the distance between carbons 1 and 2 (1.36 Å) in naphthalene is smaller than the distance between carbons 2 and 3 (1.40 Å).

C—C in ethane: 1.54 Å

C=C in ethylene: 1.34 Å

C—C in benzene: 1.40 Å

From these measurements, we conclude that there is *not* an equal distribution of π electrons around the naphthalene ring. From a comparison of bond lengths, we would say that the carbon 1–carbon 2 bond of naphthalene has more double-bond character than the carbon 2–carbon 3 bond. The resonance structures for naphthalene also indicate that the carbon 1–carbon 2 bond has more double-bond character.

two out of three resonance structures show a carbon 1– carbon 2 double bond

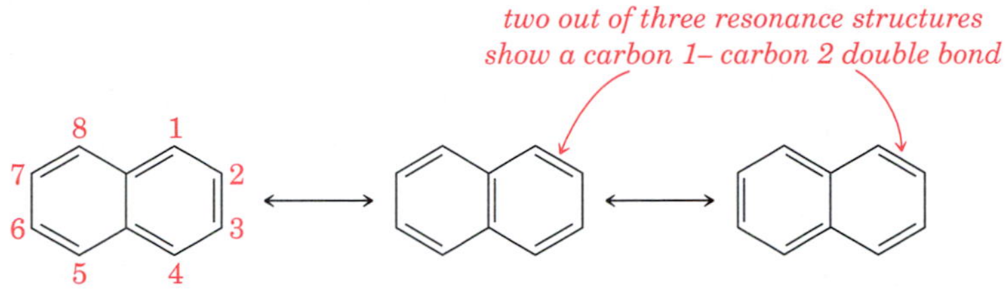

Because all the carbon–carbon bonds in naphthalene are not the same, many chemists prefer to use Kekulé-type formulas for this compound instead of using circles to represent the π cloud. We will use Kekulé-type formulas in our discussion of the reactions of naphthalene.

Phenanthrene shows similar differences among its bonds. The double-bond character of the 9,10-bond of phenanthrene is particularly evident in its chemical reactions. These positions of the phenanthrene ring system undergo addition reactions that are typical of alkenes but are not typical for benzene.

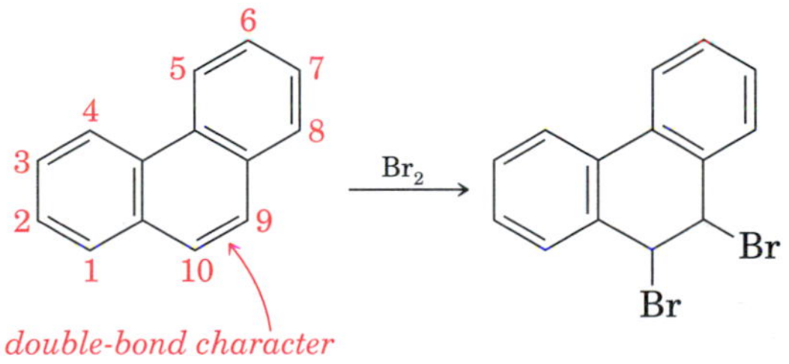

double-bond character

STUDY PROBLEM

19.2 What is the total number of π electrons in each of the following polynuclear hydrocarbons? Estimate the resonance energy of each structure.

(a) 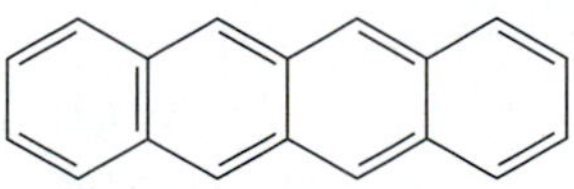   **(b)**

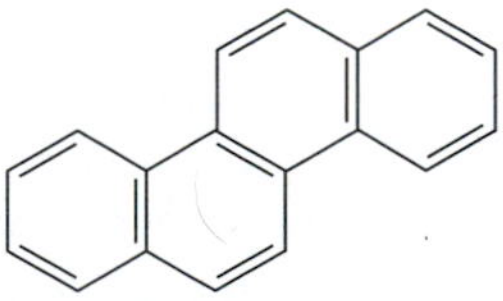

Section 19.3
Oxidation of Polycyclic Aromatic Compounds

The polycyclic aromatic compounds are more reactive toward oxidation, reduction, and electrophilic substitution than is benzene. The reason for the greater reactivity is that the polycyclic compounds can undergo reaction at one ring and still have one or more intact benzenoid rings in the intermediate and in the product. Less energy is required to overcome the aromatic character of a single ring of the polycyclic compounds than is required for benzene.

Benzene is not easily oxidized; however, naphthalene can be oxidized to products in which much of the aromaticity is retained. Phthalic anhydride is prepared commercially by the oxidation of naphthalene. This reaction probably proceeds by way of *o*-phthalic acid.

still has one benzenoid ring

naphthalene   *o*-phthalic acid   phthalic anhydride

Under controlled conditions, 1,4-naphthoquinone can be isolated from an oxidation of naphthalene (although yields are usually low).

naphthalene   1,4-naphthoquinone (22%)

STUDY PROBLEM

19.3   From the following observations, tell whether chromate oxidation of naphthalene derivatives involves an initial *electrophilic attack* or *nucleophilic attack*. Explain.

1-nitronaphthalene   3-nitro-1,2-phthalic acid

1-naphthylamine   *o*-phthalic acid

Section 19.4

Reduction of Polycyclic Aromatic Compounds

Unlike benzene, the polycyclic compounds can be partially hydrogenated without heat and pressure, or they can be reduced with sodium and ethanol.

$\xrightarrow[\text{heat}]{\text{Na, CH}_3\text{CH}_2\text{OH}}$ no reaction

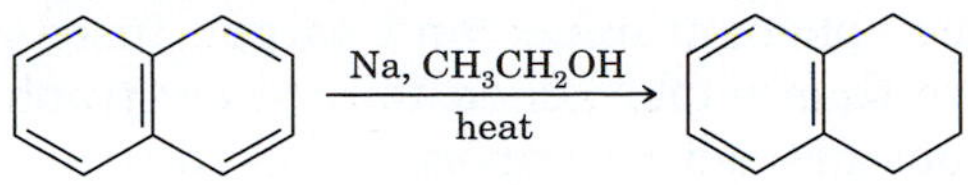
Na, CH_3CH_2OH
heat

tetralin

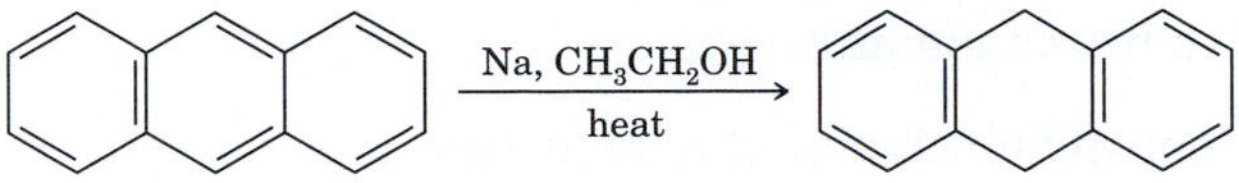
Na, CH_3CH_2OH
heat

9,10-dihydroanthracene

The partially reduced ring systems still contain one or more benzenoid rings. Most of the aromatic character of the original ring systems has been retained in these partially reduced products. To hydrogenate the polycyclic aromatics completely would, of course, require heat and pressure, just as it does for benzene.

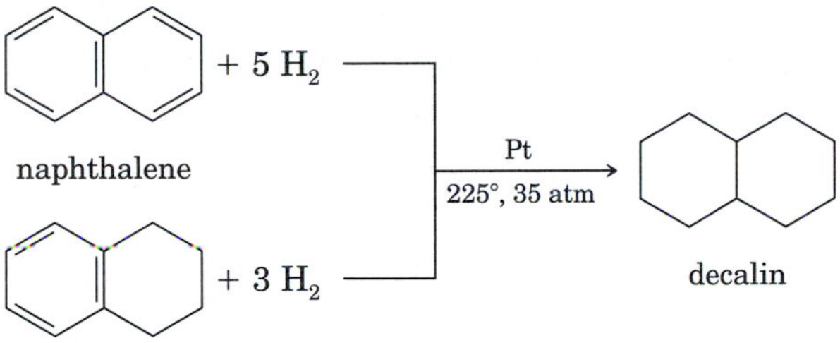
+ 5 H_2
naphthalene
Pt
225°, 35 atm
decalin
+ 3 H_2
tetralin

Section 19.5
Electrophilic Substitution Reactions of Naphthalene

The polycyclic aromatic ring systems are more reactive toward electrophilic attack than is benzene. Naphthalene undergoes electrophilic aromatic substitution reactions predominantly at the 1-position. The reasons for the enhanced reactivity and for this position of substitution will be discussed shortly.

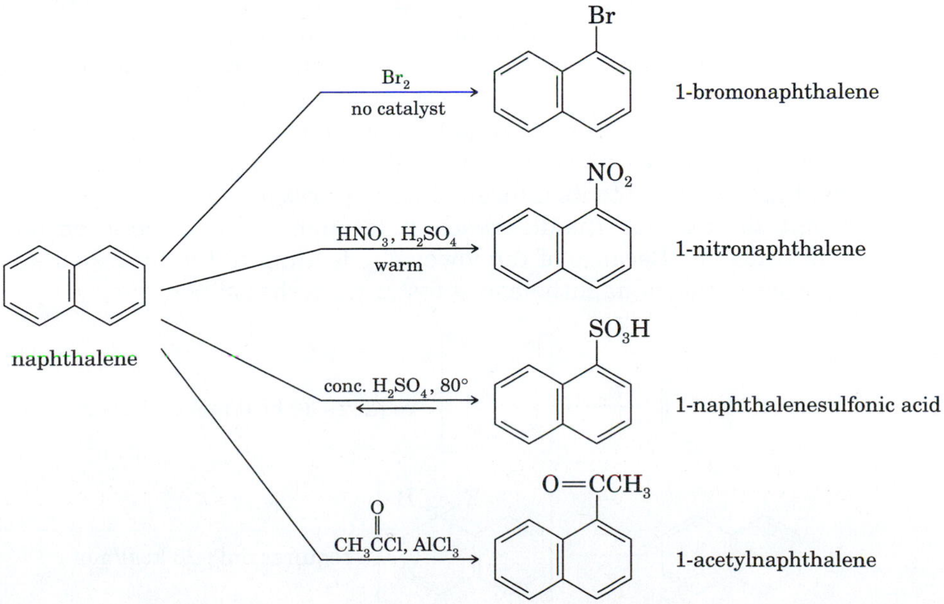
Br
Br_2
no catalyst
1-bromonaphthalene
NO_2
HNO_3, H_2SO_4
warm
1-nitronaphthalene
naphthalene
SO_3H
conc. H_2SO_4, 80°
1-naphthalenesulfonic acid
O=CCH_3
CH_3CCl, $AlCl_3$
1-acetylnaphthalene

Anthracene, phenanthrene, and larger fused-ring compounds are even more reactive than naphthalene toward electrophilic substitution. However, these reactions are not as important as those of naphthalene because mixtures of isomers (which are often difficult to separate) are obtained. Phenanthrene, for example, undergoes mononitration at each available position to yield five nitrophenanthrenes.

A. Position of Substitution of Naphthalene

The mechanism for naphthalene substitution is similar to that for benzene substitution. Let us look at the stepwise bromination reaction to see why substitution at the 1-position is favored and why this reaction occurs more readily than the bromination of benzene.

1-Substitution (favored):

intermediate

Resonance structures for the 1-substitution intermediate:

major contributors

The resonance structures of the intermediate for substitution at the 1-position show four contributors in which the benzene ring is intact. Because of aromatic resonance-stabilization, these structures are of lower energy than the other resonance structures and are major contributors to the real structure of the intermediate. This is the reason that naphthalene undergoes electrophilic substitution more readily than benzene. For benzene to go to a benzenonium ion requires the loss of aromaticity, about 36 kcal/mol. For naphthalene to go to its intermediate requires only partial loss of aromaticity, about 25 kcal/mol (the difference in resonance energy between naphthalene and benzene). Because of the lower E_{act} leading to the intermediate, the rate of bromination of naphthalene is faster than that of benzene.

requires 36 kcal/mol to destroy aromaticity

requires only 25 kcal/mol

Why is 1-substitution favored over 2-substitution for naphthalene? Inspect the resonance structures for the intermediate leading to 2-substitution:

2-Substitution (not favored):

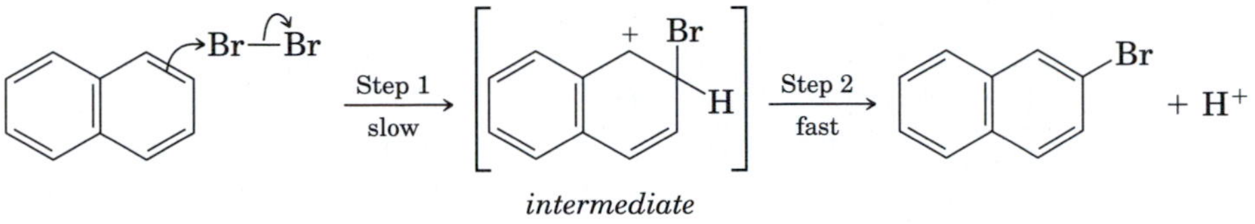

intermediate

Resonance structures for the 2-substitution intermediate:

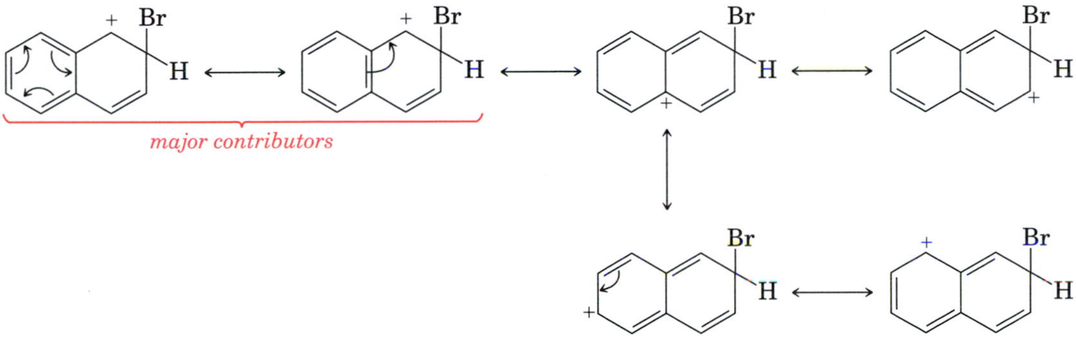

major contributors

The intermediate leading to the 2-substituted isomer has only two contributing resonance structures in which a benzenoid ring is intact, while the 1-intermediate shows four such resonance structures. The 1-intermediate is more stabilized by resonance, and its transition state is of lower energy. For this reason, the E_{act} is lower and the 1-intermediate is formed faster.

The sulfonation of naphthalene, which is a reversible reaction (Section 11.8F), is more complex than bromination. At 80°, the expected 1-naphthalenesulfonic acid is the product. However, at higher temperatures (160–180°), the product is 2-naphthalenesulfonic acid. At low temperatures, the reaction is under **kinetic control**—that is, the relative rates of reaction determine the product ratio. At high temperatures, the reaction is under **thermodynamic,** or **equilibrium, control**—the relative stabilities of the products determine the product ratio.

We have already seen why the 1-substitution of naphthalene is faster than 2-substitution. At less than 80°, the rate of formation of either naphthalenesulfonic acid is relatively slow. The reaction proceeds through the lower-energy 1-intermediate just as it does for bromination.

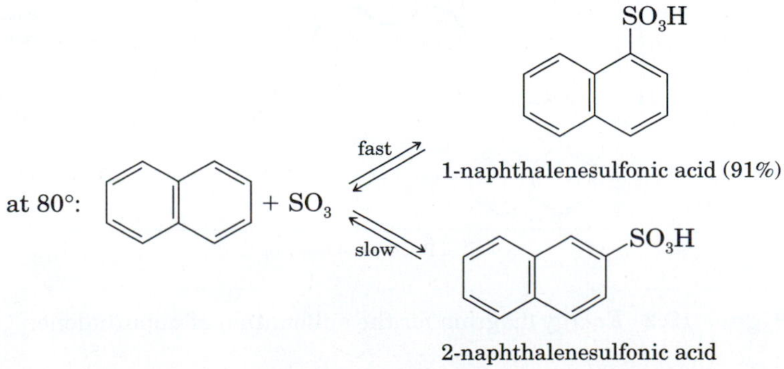

1-naphthalenesulfonic acid (91%)

at 80°:

2-naphthalenesulfonic acid

Even though 1-naphthalenesulfonic acid is formed at low temperatures, this isomer is less stable than the 2-isomer because of repulsions between the —SO$_3$H group and the hydrogen at position 8.

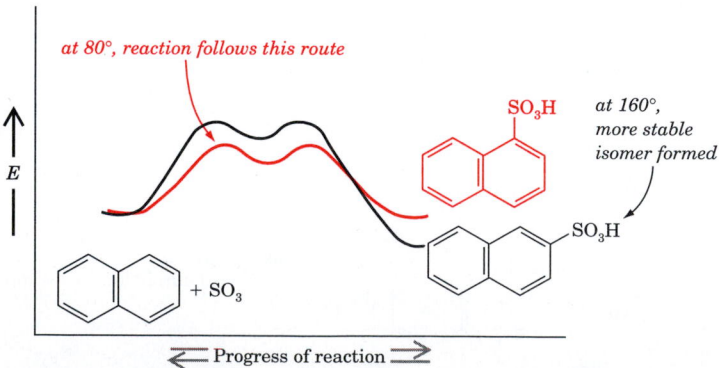

1-naphthalenesulfonic acid

less stable

2-naphthalenesulfonic acid

more stable

At a higher temperature, the rates of both forward reactions and the rates of both reverse reactions are all increased. Although the 1-product may be formed more readily, it can revert quickly to naphthalene. The 2-product is formed more slowly, but the rate of its reverse reaction is even slower because the 2-product is more stable and of lower energy. At higher temperatures, the 2-product accumulates in the reaction mixture and is the observed product. (See Figure 19.2.)

Figure 19.2   Energy diagram for the sulfonation of naphthalene.

STUDY PROBLEMS

19.4 On the basis of what you learned in Chapter 11 about activating and deactivating groups and the position(s) to which they direct an incoming substituent, predict the major organic products of aromatic nitration of the following compounds:

(a) [structure: naphthalene with CH_3 at position 2]

(b) [structure: naphthalene with CH_3 at position 1]

(c) [structure: naphthalene with SO_3H at position 2]

(d) [structure: naphthalene with NO_2 at position 1]

19.5 Write equations showing how you could prepare the following compounds from naphthalene.

(a) [structure: naphthalene with CN at position 1]

(b) [structure: naphthalene with $CH_2CH_2CH_3$ at position 1]

(c) [structure: naphthalene with Br at positions 1 and 3]

Section 19.6
Nomenclature of Aromatic Heterocyclic Compounds

Because of their widespread occurrence in nature, the aromatic heterocycles are of more general interest to chemists than are the polycyclic compounds containing only carbon atoms in their rings. Like the polycyclic aromatic compounds, the aromatic heterocycles generally have individual names. The names and structures of some of the more important members of this class of compounds are listed in Table 19.1.

The numbering of three representative heterocycles follows:

[structures of three heterocycles with ring numbering]

pyridine   thiazole*   imidazole*

When a heterocycle contains only one heteroatom, Greek letters may also be used to designate ring position. The carbon atom adjacent to the heteroatom is the α carbon. The next carbon is the β carbon. The next carbon in

---

* A circle in the ring is not a proper representation of an aromatic π cloud in a five-membered heterocycle. How the aromatic π cloud is formed in these compounds will be discussed in Section 19.9.

Structure	Name	Structure	Name
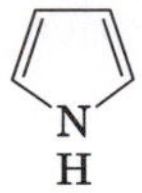	pyrrole	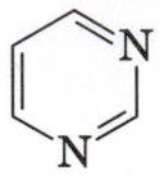	pyrimidine
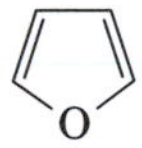	furan	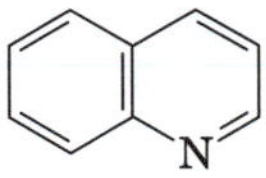	quinoline
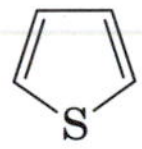	thiophene	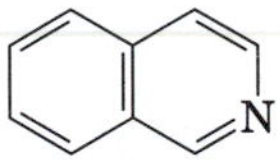	isoquinoline
	imidazole	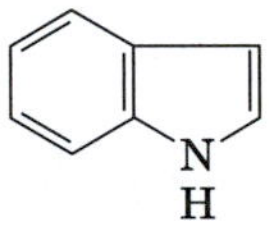	indole
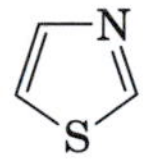	thiazole	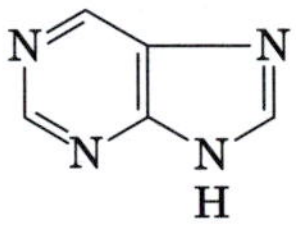	purine
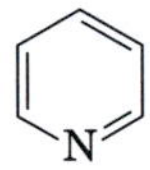	pyridine		

Table 19.1   Some important aromatic heterocyclic compounds

line, if any, is γ. Pyridine has two α positions, two β positions, and one γ position. Pyrrole has two α and two β positions.

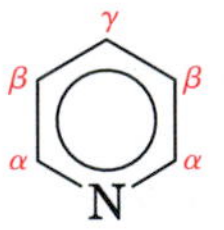

pyridine 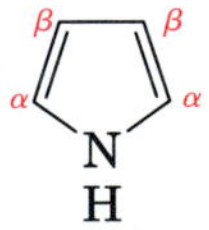   pyrrole

Section 19.7
Pyridine, a Six-Membered Aromatic Heterocycle

Of the common six-membered heterocycles, only the nitrogen heterocycles are stable aromatic compounds.

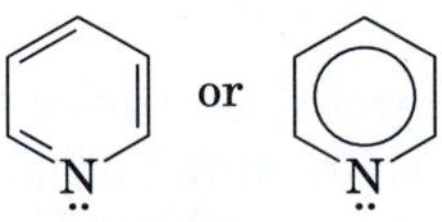

pyridine

aromatic

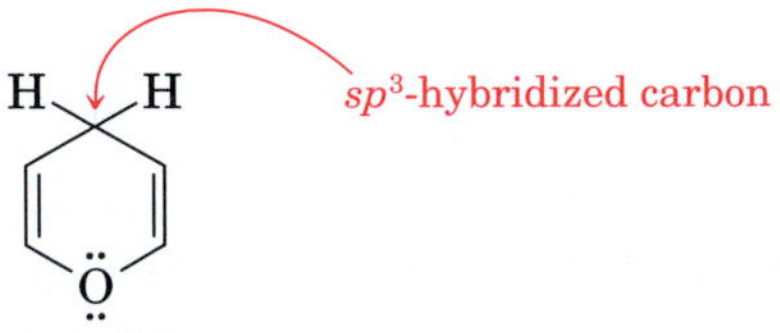

pyran

not aromatic

sp^3-hybridized carbon

Pyridine has a structure similar to that of benzene. Pyridine contains a planar, six-membered ring consisting of five carbons and one nitrogen. Each

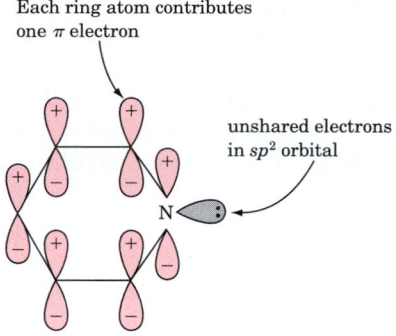

Each ring atom contributes
one π electron

unshared electrons
in sp^2 orbital

Figure 19.3 The lowest-energy π molecular orbital in pyridine.

of these ring atoms is sp^2-hybridized and has one electron in a p orbital that contributes to the aromatic π cloud (six π electrons). Figure 19.3 shows the lowest-energy π molecular orbital of pyridine.

Note the differences between benzene and pyridine. Benzene is symmetrical and nonpolar, but pyridine contains an electronegative nitrogen and therefore is *polar.*

E^+ *attack on ring
not favored*

Because the nitrogen is more electronegative than carbon, the rest of the pyridine ring is electron-deficient. An electron-deficient ring means that the carbon atoms in the ring carry a *partial positive charge.* A pyridine ring therefore has a low reactivity toward electrophilic substitution compared with benzene. Besides the electronegative nitrogen rendering the ring partially positive, pyridine forms a cation with many Lewis acids. Cation formation renders the ring even more electron-deficient.

e^--*deficient* ⟶   ⟵ *more* e^--*deficient*

Pyridine does not undergo Friedel–Crafts alkylations or acylations, nor does it undergo coupling with diazonium salts. Bromination proceeds only at high temperatures in the vapor phase and probably proceeds by a free-radical path. When substitution does occur, it occurs at the 3-position.

pyridine   3-bromopyridine (37%)   3,5-dibromopyridine (26%)

Another major difference between pyridine and benzene is that the nitrogen in pyridine contains an unshared pair of electrons in an sp^2 orbital. This pair of electrons can be donated to a hydrogen ion. Like amines, pyridine is basic. The basicity of pyridine ($pK_b = 8.75$) is less than that of aliphatic amines ($pK_b = \sim 4$) because the unshared electrons are in an sp^2 orbital instead of an sp^3 orbital. Nonetheless, pyridine undergoes many reactions typical of amines.

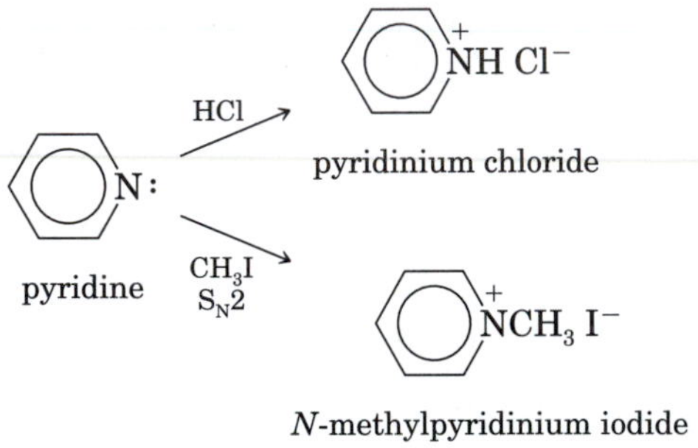

pyridine
HCl
CH_3I
S_N2
pyridinium chloride
N-methylpyridinium iodide

Like that of benzene, the aromatic ring of pyridine is resistant to oxidation. Side chains can be oxidized to carboxyl groups under conditions that leave the ring intact.

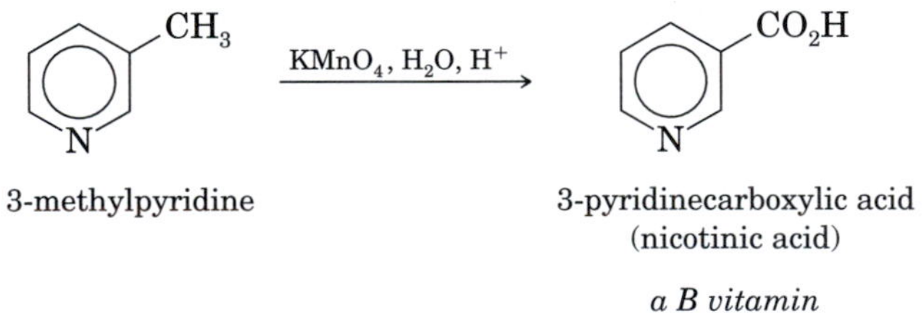

$KMnO_4$, H_2O, H^+
3-methylpyridine
3-pyridinecarboxylic acid
(nicotinic acid)
a B vitamin

A. Nucleophilic Substitution on the Pyridine Ring

When a benzene ring is substituted with electron-withdrawing groups such as —NO_2, *aromatic nucleophilic substitutions* can take place (Section 12.4).

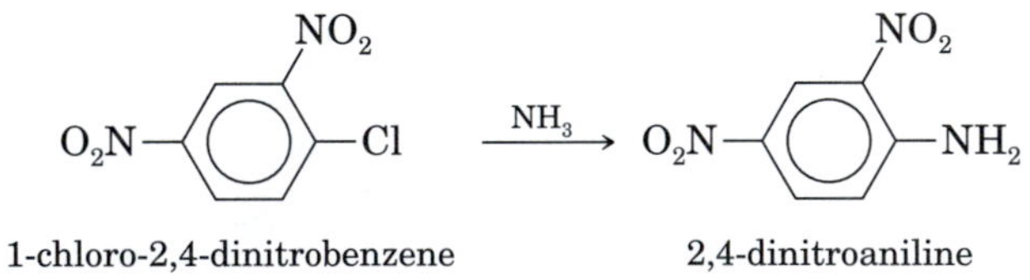

NH_3
1-chloro-2,4-dinitrobenzene
2,4-dinitroaniline

The nitrogen in pyridine withdraws electron density from the rest of the ring. It is not surprising, then, that nucleophilic substitution also occurs with pyridine. Substitution proceeds most readily at the 2-position, followed by the 4-position, but not at the 3-position.

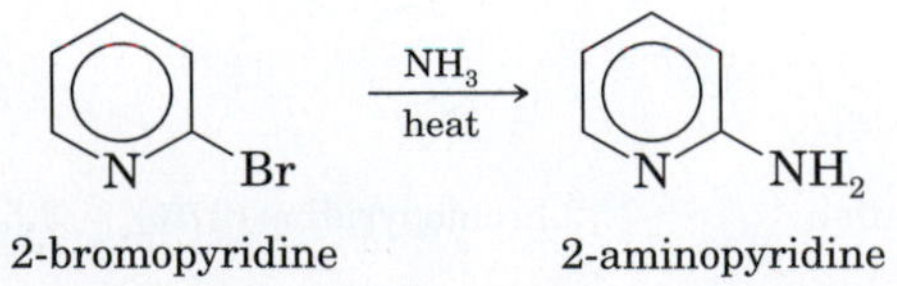

NH_3
heat
2-bromopyridine
2-aminopyridine

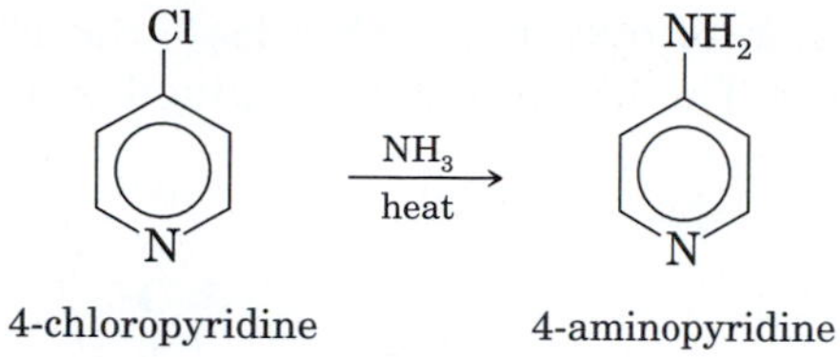

4-chloropyridine   4-aminopyridine

Let us look at the mechanisms for substitution at the 2- and 3-positions to see why the former reaction proceeds more readily.

2-Substitution (favored):

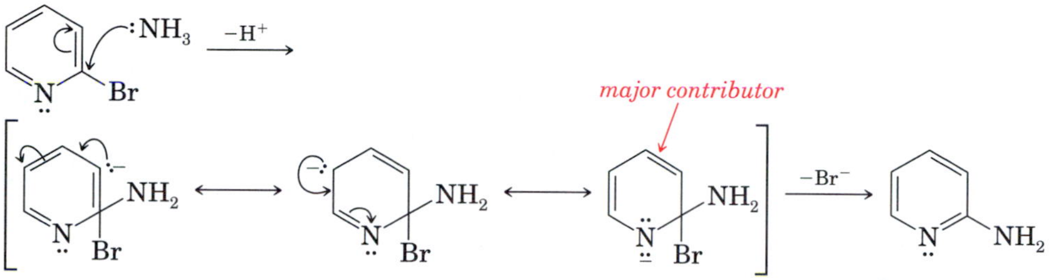

resonance structures for intermediate

The intermediate for 2-substitution is especially stabilized by the contribution of the resonance structure in which nitrogen carries the negative charge. Substitution at the 3-position goes through an intermediate in which the nitrogen cannot help stabilize the negative charge. The intermediate for 3-substitution is of higher energy; the rate of reaction going through this intermediate is slower.

STUDY PROBLEMS

19.6 Give the resonance structures for the intermediate in the reaction of 4-chloropyridine with ammonia.

19.7 Write the mechanism for substitution of 3-bromopyridine with ammonia. Include the resonance structures for the intermediate in your answer.

Benzene itself (with no substituents) does not undergo nucleophilic substitution. This reaction does occur with pyridine if an extremely strong base such as a lithium reagent or amide ion (NH_2^-) is used.

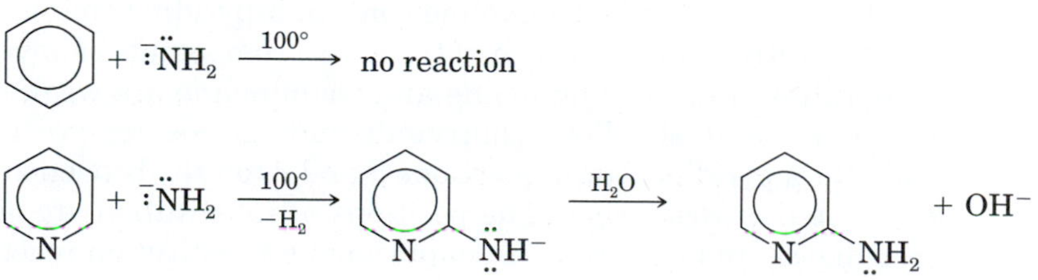

2-aminopyridine (70%)

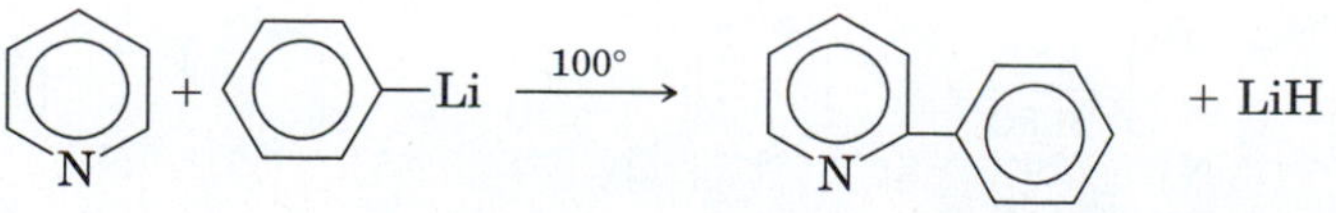

2-phenylpyridine (50%)

In the reaction of pyridine with NH_2^-, the initial product is the anion of 2-aminopyridine. The free amine is obtained by treatment with water.

Step 1 (attack of NH_2^- and loss of H_2):

resonance structures for intermediate

the anion of 2-aminopyridine

Step 2 (treatment with H_2O):

2-aminopyridine

Section 19.8

Quinoline and Isoquinoline

Quinoline is a fused-ring heterocycle that is similar in structure to naphthalene, but with a nitrogen at position 1. Isoquinoline is the 2-isomer. (Note that the numbering of isoquinoline starts at a carbon, not at the nitrogen.)

quinoline   isoquinoline

Both quinoline and isoquinoline contain a pyridine ring fused to a benzene ring. The nitrogen ring in each of these two compounds behaves somewhat like the pyridine ring. Both quinoline and isoquinoline are weak bases ($pK_b = 9.1$ and 8.6, respectively). Both compounds undergo *electrophilic substitution* more easily than pyridine, but in positions 5 and 8 (on the benzenoid ring, not on the deactivated nitrogen ring). The positions of substitution are determined by intermediates similar to those in naphthalene substitution reactions.

quinoline   5-nitroquinoline (52%)   8-nitroquinoline (48%)

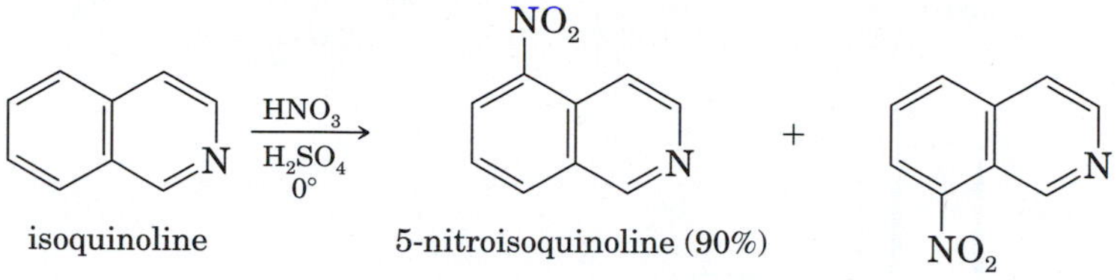

isoquinoline   5-nitroisoquinoline (90%)   8-nitroisoquinoline (10%)

STUDY PROBLEM

19.8 Draw resonance structures of the intermediates for nitration at the 5- and 6-positions of quinoline to show why 5-nitroquinoline is formed preferentially.

Like pyridine, the nitrogen-containing ring of either quinoline or isoquinoline can undergo **nucleophilic substitution.** The position of attack is α to the nitrogen in either ring system, just as it is in pyridine.

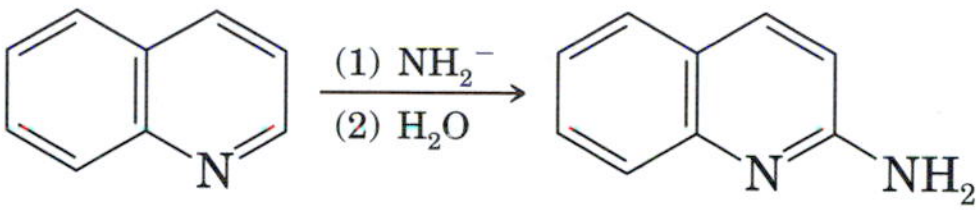

quinoline   2-aminoquinoline

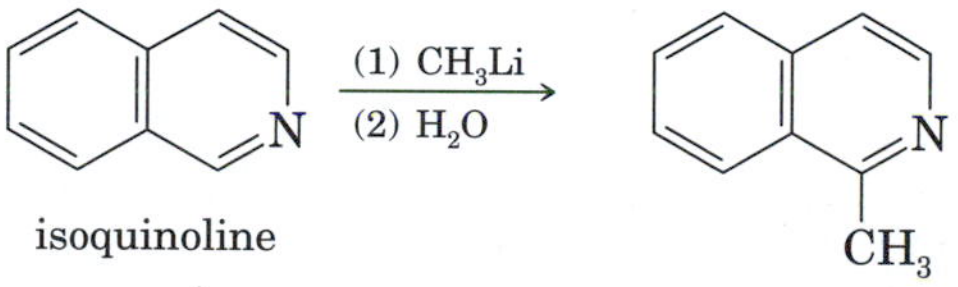

isoquinoline

1-methylisoquinoline

Section 19.9
Pyrrole, a Five-Membered Aromatic Heterocycle

For a five-membered-ring heterocycle to be aromatic, the heteroatom must have *two electrons* to donate to the aromatic π cloud. Pyrrole, furan, and thiophene all meet this criterion and therefore are aromatic. We will emphasize pyrrole in our discussion of five-membered aromatic heterocycles because it is typical in terms of both bonding and chemical reactivity.

pyrrole   furan   thiophene

Unlike pyridine and the amines, pyrrole ($pK_b = \sim 14$) is *not basic* under the usual conditions.

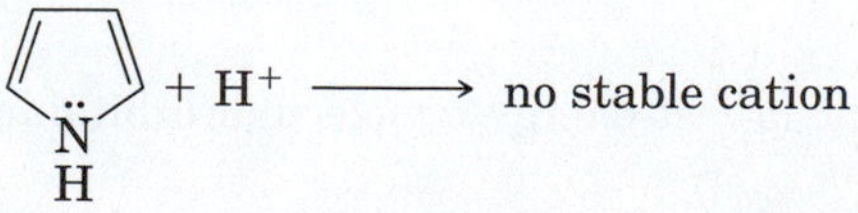

$+ \ H^+ \longrightarrow$ no stable cation

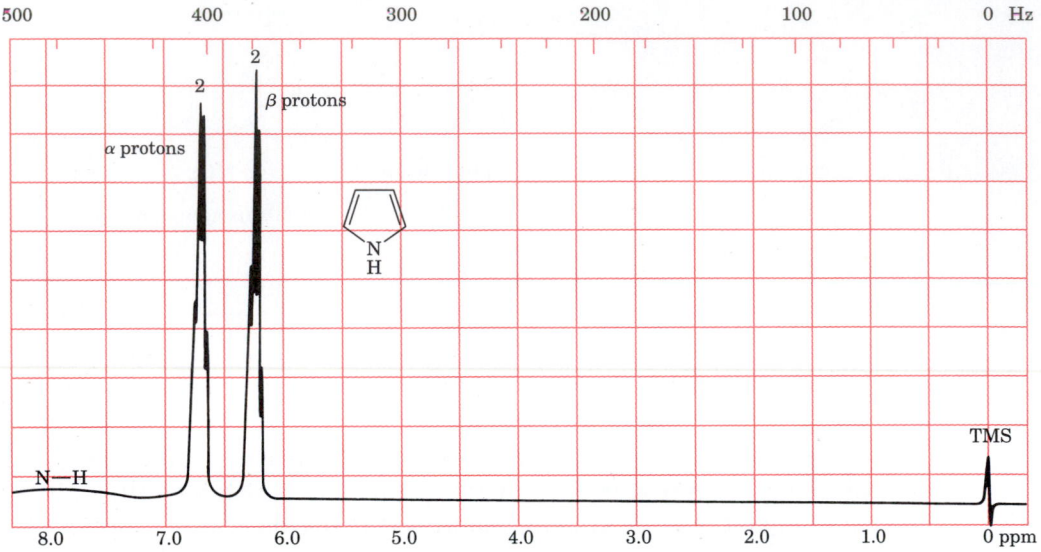

Figure 19.4 ^{1}H NMR spectrum of pyrrole, C_4H_5N. (The NH absorption is a low, broad band near 8 ppm.)

To see why pyrrole is not basic, we must consider the electronic structure of pyrrole. We know that pyrrole is aromatic because (1) its heat of combustion is about 25 kcal/mol less than that calculated for a diene structure; (2) pyrrole undergoes aromatic substitution reactions; and (3) the protons of pyrrole absorb in the aromatic region of the ^{1}H NMR spectrum (Figure 19.4). (Recall that aryl protons absorb downfield from most other protons because they are deshielded by the effects of the ring current. See Section 9.7B.)

In a five-membered ring, the minimum number of π electrons needed for aromaticity is six ($4n + 2$, where $n = 1$). The four carbons of pyrrole each contribute one electron; therefore, the nitrogen atom of pyrrole must contribute *two* electrons (not just one as it does in pyridine). In addition to contributing two electrons to the π molecular orbitals, the nitrogen in pyrrole shares three electrons in σ bonds to two ring carbons and to a hydrogen. Consequently, all five bonding electrons of the nitrogen are used in bonding. The pyrrole nitrogen does not have unshared electrons and is not basic. Figure 19.5 shows the orbital picture for the lowest-energy π molecular orbital.

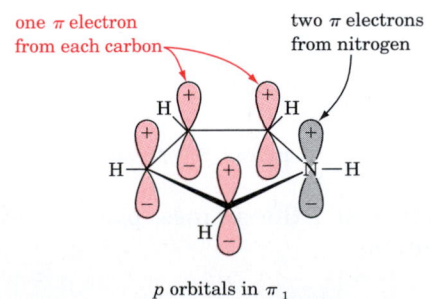

p orbitals in π_1

Figure 19.5 The lowest-energy π molecular orbital in pyrrole.

Because the nitrogen atom in pyrrole contributes two electrons to the aromatic π cloud, the nitrogen atom is electron-deficient and therefore not basic. The pyrrole ring, however, has six π electrons for only five ring atoms. The ring is electron-rich and therefore partially negative.

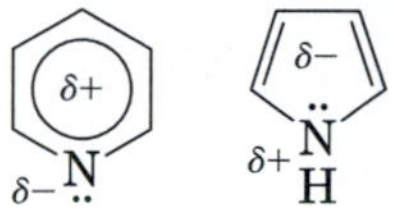

A. Electrophilic Substitution on the Pyrrole Ring

Because the ring carbons are the negative part of the pyrrole molecules, these carbons are *activated toward electrophilic attack, but deactivated to nucleophilic attack.* (This reactivity is opposite to that of pyridine.) The principal chemical characteristic of pyrrole and the other five-membered aromatic heterocycles is the ease with which they undergo electrophilic substitution.

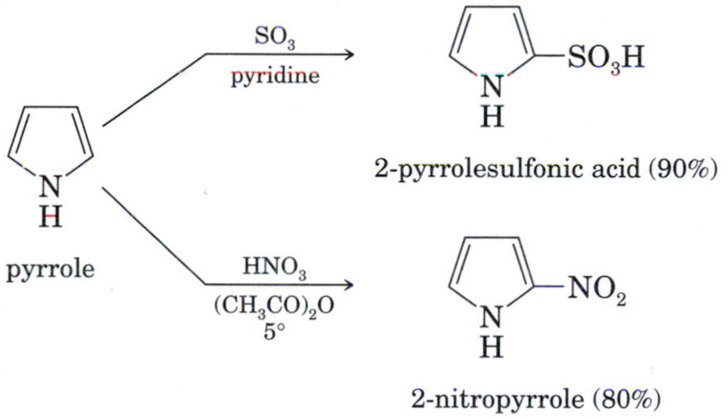

2-pyrrolesulfonic acid (90%)

2-nitropyrrole (80%)

Electrophilic substitution occurs principally at the 2-position of the pyrrole ring. A look at the resonance structures for the intermediates of 2- and 3-nitration shows why. Three resonance structures can be drawn for the 2-intermediate, while only two can be drawn for the 3-intermediate. There is greater delocalization of the positive charge in the intermediate leading to 2-nitration than there is in that leading to 3-nitration.

2-Nitration (favored):

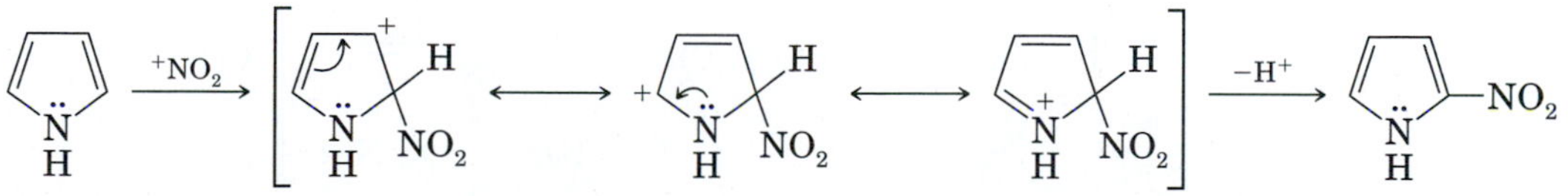

resonance structures for the intermediate

3-Nitration (not favored):

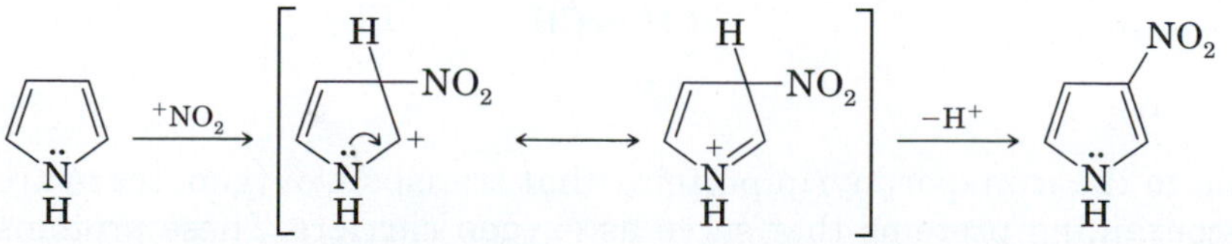

resonance structures for the intermediate

19.9   Predict the major organic monosubstitution products:

(a) furan + $(CH_3CO)_2O$ $\xrightarrow{BF_3}$

(b) thiophene + H_2SO_4 $\xrightarrow{25°}$

(c) pyrrole + $C_6H_5N_2^+$ Cl^- $\xrightarrow{25°}$

(d furan + Br_2 $\xrightarrow[25°]{dioxane}$

B. Porphyrins

The **porphyrin ring system** is a biologically important unit found in *heme,* the oxygen-carrying component of hemoglobin; in *chlorophyll,* a plant pigment; and in the *cytochromes,* compounds involved in utilization of O_2 by animals.

Note that the porphyrin ring system is composed of four pyrrole rings joined by $=CH-$ groups. The entire ring system is aromatic.

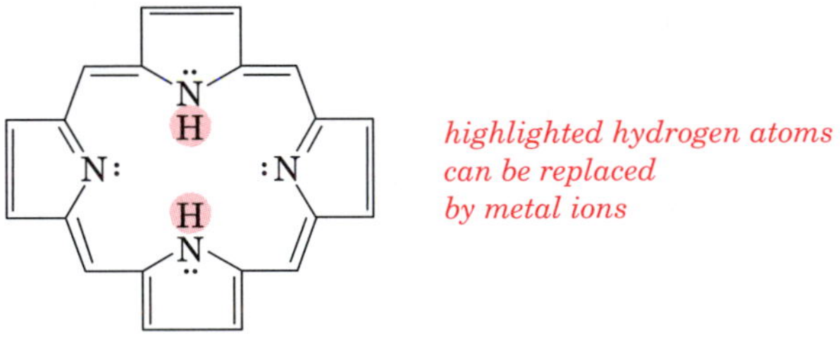

highlighted hydrogen atoms can be replaced by metal ions

porphyrin

The pyrrole hydrogens in the porphyrin ring system can be replaced by a variety of metal ions. The product is a **chelate** (Greek, *chele,* "crab's claw"), a compound or ion in which a metal ion is held by more than one bond from the original molecule. With porphyrin, the chelate is planar around the metal ion, and resonance results in four equivalent bonds from the nitrogen atoms to the metal.

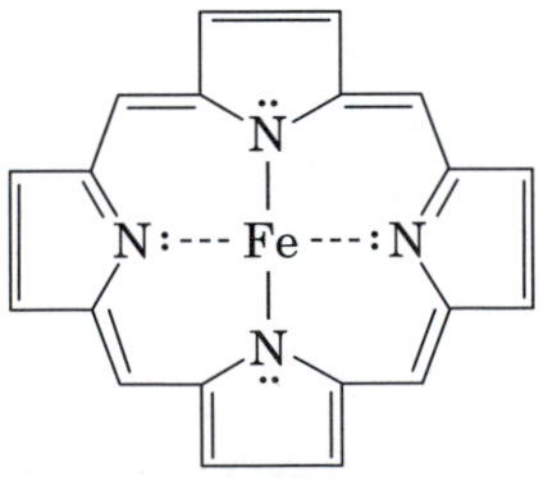

planar and resonance-stabilized

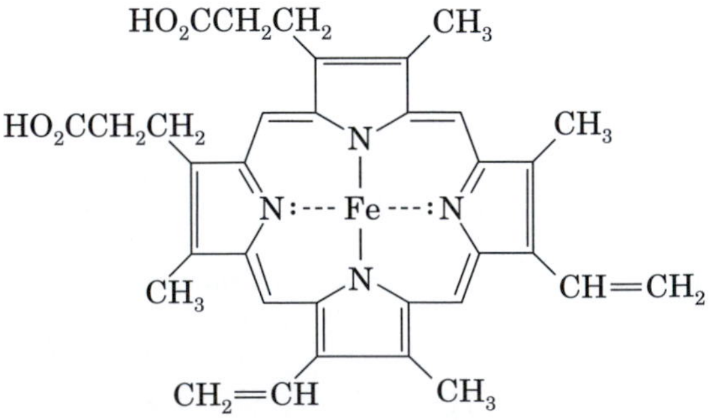

heme

In addition to the iron–porphyrin proteins that transport oxygen, there are also copper-containing proteins that serve as oxygen carriers. These proteins are found free in the blood of invertebrates (crabs, lobsters, snails, squid, and octopuses). The metal in these proteins is not bound into a porphyrin ring but is complexed directly with the protein. The oxygenated forms of these proteins are blue and the nonoxygenated forms are colorless.

STUDY PROBLEM

19.10   Predict the major organic products:

(a) [Cl-substituted pyridine] + Na⁺ ⁻OCH₂CH₃ $\xrightarrow{\text{CH}_3\text{CH}_2\text{OH}}$

(b) [naphthalene with NO₂ group] $\xrightarrow[160°]{\text{H}_2\text{SO}_4 + \text{SO}_3}$

(c) H₃C-[thiophene]-S + HNO₃ $\xrightarrow{\text{(CH}_3\text{CO)}_2\text{O}}$

(d) [methylnaphthalene, CH₃] + H₂SO₄ $\xrightarrow{40°}$

(e) [1-methylnaphthalene, CH₃] + Br₂ $\xrightarrow{h\nu}$

(f) [methylquinoline, CH₃] + HCl $\longrightarrow$

(g) [pyrrole, N-H] + CH₃CH₂MgBr $\longrightarrow$

(h) [furan derivative: H₃C, CH₃O₂C, O, C(CH₂)₃CH₃] $\xrightarrow[\text{(2) H}_2\text{O, H}^+]{\text{(1) NH}_2\text{NH}_2\text{, KOH,}\ \text{HOCH}_2\text{CH}_2\text{OH, heat}}$

Summary

Three common **polycyclic aromatic compounds** are **naphthalene, anthracene,** and **phenanthrene.** These compounds exhibit less resonance energy per ring than benzene, and certain C—C bonds have more double-bond character than others. These compounds may be oxidized to quinones or may be partially hydrogenated, as shown in Table 19.2.

Naphthalene undergoes electrophilic substitution at the 1-position.

[structure: naphthalene with E⁺]

The six-membered aromatic heterocycle **pyridine** is a weak base and has a partially positive ring. Compared to benzene, pyridine is deactivated toward electrophilic substitution, but activated toward nucleophilic substitution.

[structure: pyridine ring with $\delta+$ and N $\delta-$, :Nu⁻]

Table 19.2 Summary of the reactions of nephthalene, pyridine, and pyrrole

Reaction	Section Reference

Naphthalene:

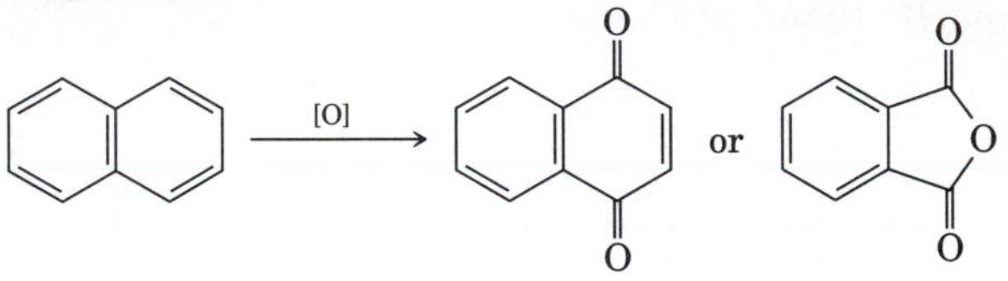

19.3

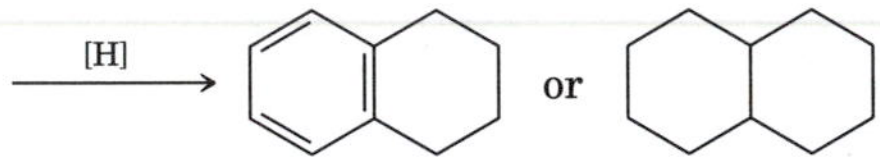

19.4

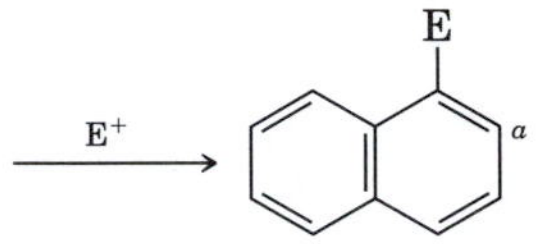

19.5

Pyridine:

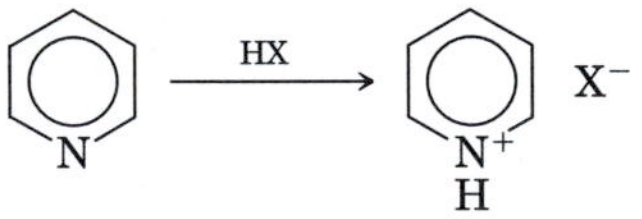

19.7

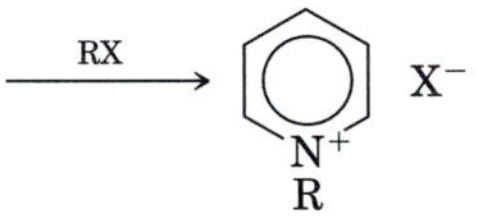

19.7

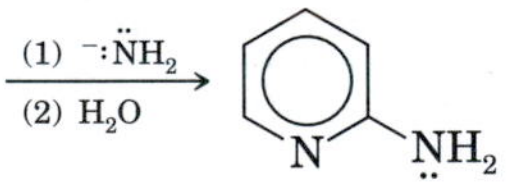

19.7A

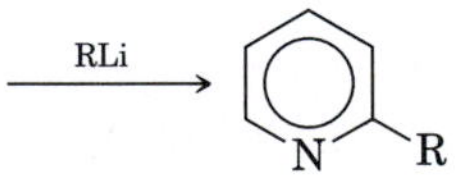

19.7A

Substituted Pyridine:

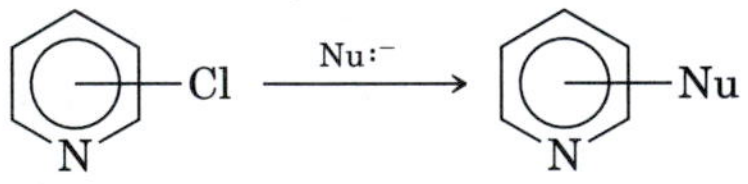

where Cl is 2- or 4-

19.7A

Quinoline:

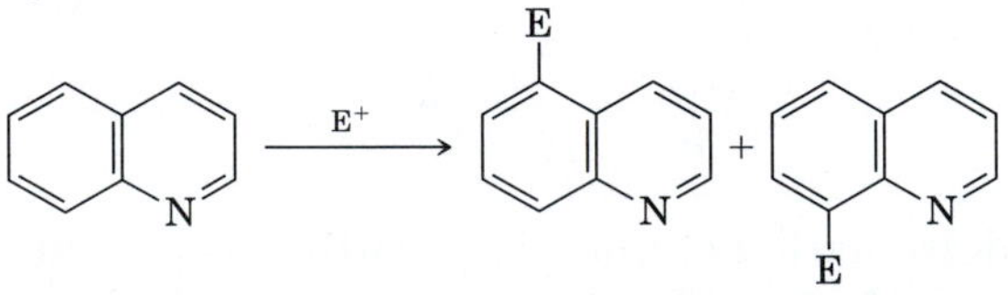

19.8

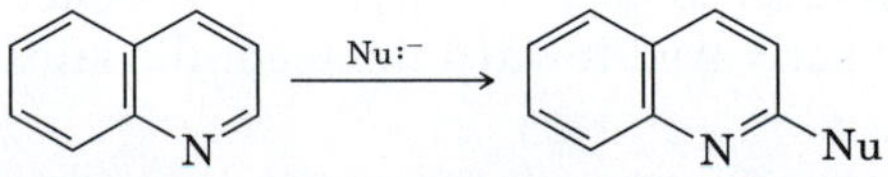

where Nu:⁻ is ⁻NH₂ or R⁻ from RLi

19.8

(continued)

Table 19.2 *(continued)*	Summary of the reactions of naphthalene, pyridine, and pyrrole

Reaction	Section Reference

Isoquinoline:

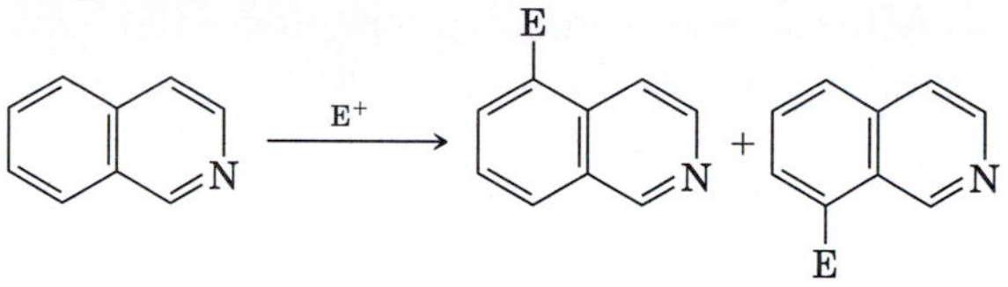

19.8

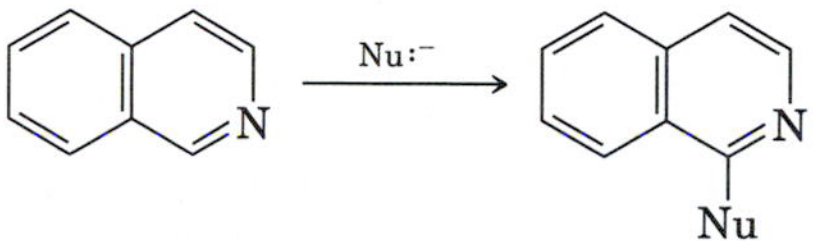

19.8

where Nu:$^-$ is $^-NH_2$ or R$^-$ from RLi

Pyrrole:

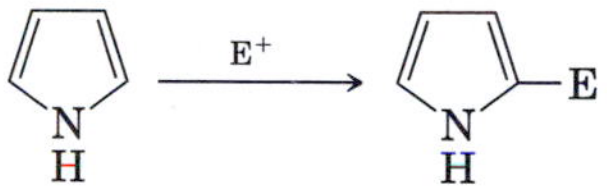

19.9A

^{a} At 160°, sulfonation yields 2-naphthalenesulfonic acid.

Quinoline and **isoquinoline** undergo electrophilic substitution on the benzenoid ring, but nucleophilic substitution on the nitrogen ring.

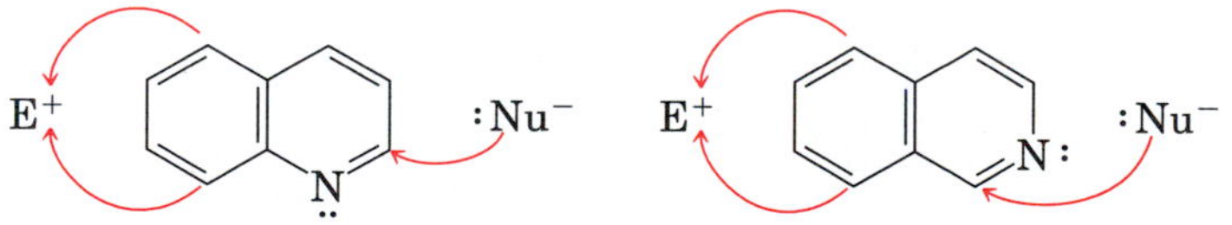

Pyrrole is an aromatic five-membered nitrogen heterocycle. It is not basic. Its ring is partially negative and is activated toward electrophilic substitution, but deactivated toward nucleophilic substitution.

electron pair in aromatic π system

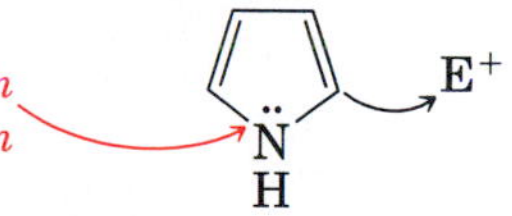

The order of reactivity of the heterocycles and benzene toward electrophilic and nucleophilic substitution follows:

Electrophilic substitution:

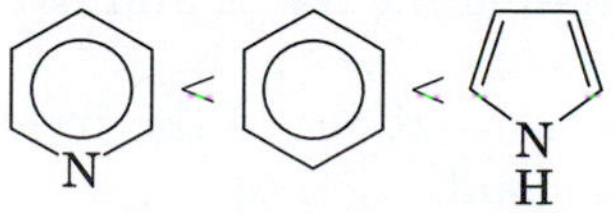

Nucleophilic substitution:

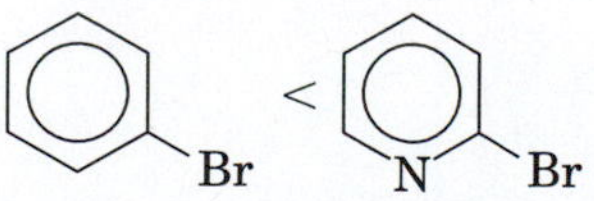

Imidazoles

The imidazole ring system is found in the amino acid histidine and in its decarboxylation product, histamine. Histamine is the compound responsible for many allergic symptoms. The over-the-counter remedies for colds and nasal congestion alleviate the effects of histamine activity.

imidazole   histidine   histamine

Imidazole is amphoteric. Its two acidity constants are 14.5 and 7.16. It also is aromatic. The ring is less reactive toward aromatic substitution than pyrrole but more reactive than pyridine. The aromatic chemistry of imidazole is complicated by the basicity of nitrogen 3. In strong acid, such as in a nitration or sulfonation reaction, the ring is protonated. Consequently, it is difficult to carry out an electrophilic substitution reaction. Under neutral or weak acidic conditions, however, the ring is reactive. For example, bromination or iodination of imidazole generates a 2,4,5-trihaloimidazole, rather than a monosubstituted derivative.

Imidazole is an excellent nucleophile and reacts readily with alkylating and acylating reagents. In these reactions, the two nitrogens of the ring are indistinguishable. For example, 4- and 5-methylimidazole are equivalent and often referred to as 4(5)-methylimidazole.

Questions

1. Write equations for the acid–base reactions of imidazole that are responsible for its two pK_a's.
2. Are 4- and 5-methylimidazole indistinguishable because they are resonance structures or because they are tautomers? Explain your answer.
3. Explain why 1-methylimidazole melts and boils at temperatures considerably lower than imidazole itself.
4. Identify the two π electrons of one of the imidazole's nitrogens that are used to form the aromatic π cloud.
5. 4(5)-Methylimidazole, when treated with CH_3I, undergoes an S_N2 substitution reaction. Write the equation for this reaction.

19.11 Name the following compounds:

(a)

NH₂ NH₂

*used as an antioxidant
in lubricating oils*

(b)

SO₃H

NH₂

1,6-Cleve's acid

used in the dye industry

(c)

CH₃
|
CHCO₂H

CH₃O

naproxen

an anti-inflammatory agent

(d)

OH   O

O

juglone

*a colored compound
in walnut shells*

19.12 Draw formulas for the following compounds:

(a) 1-methylnaphthalene

(b) 4,10-diethylphenanthrene

(c) 9,10-dibromophenanthrene

(d) 9,10-dihydroanthracene

(e) *N*-phenylpyrrole

(f) 5-hydroxyquinoline

(g) 5-hydroxyisoquinoline

19.13 Write resonance structures for the following polynuclear hydrocarbons:

(a)

(b)

19.14 Predict the major organic products. (If no reaction occurs, write *no reaction*.)

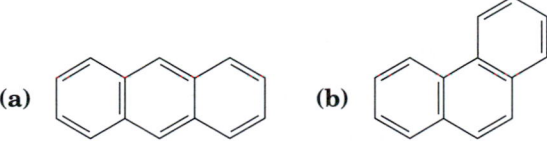

(a) $+ C_6H_5\overset{O}{\overset{\|}{C}}Cl \xrightarrow{AlCl_3}$

(b) $\xrightarrow[\text{(2) H}_2\text{O, H}^+]{\text{(1) C}_6\text{H}_5\text{MgBr}}$

(c)

OH

$+ HNO_3 \longrightarrow$

(d)

CO₂H

$+ HNO_3 \xrightarrow{H_2SO_4}$

(e)

NH₂

$\xrightarrow[\text{(2) HNO}_3]{\text{(1) CH}_3\overset{O}{\overset{\|}{C}}Cl}$

*For information concerning the organization of the *Study Problems* and *Additional Problems*,
see the *Note to student* on page 41.

(f) 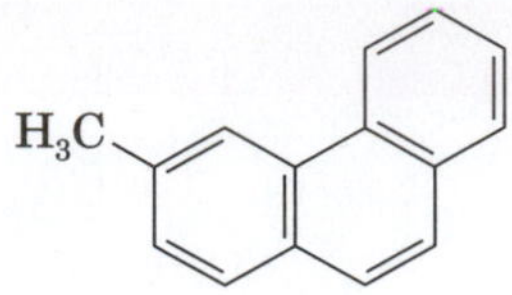 $\xrightarrow[\text{heat}]{\text{CrO}_3, \text{H}_2\text{SO}_4}$

(g) 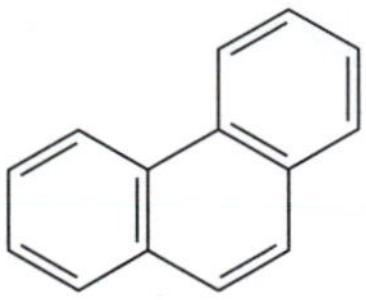 $+ \text{H}_2$ $\xrightarrow[\text{no heat or pressure}]{\text{CuCr}_2\text{O}_4}$

19.15 Propose syntheses for the following compounds from naphthalene:

(a)

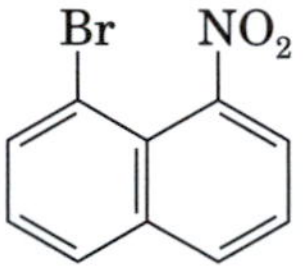

(b)

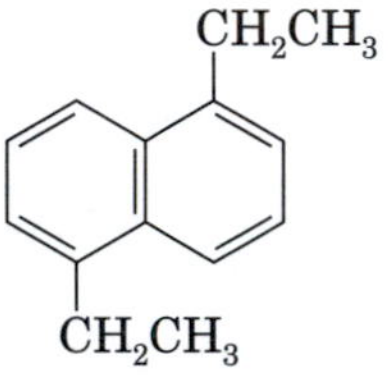

19.16 How would you explain the following observations? Aromatic nitration of 2-methyl-naphthalene yields 75% 2-methyl-1-nitronaphthalene, but aromatic sulfonation of 2-methylnaphthalene yields 80% 6-methylnaphthalene-2-sulfonic acid.

19.17 Name the following compounds:

(a)

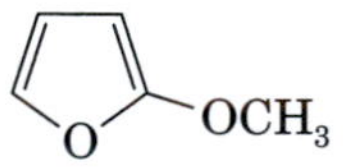

(b)

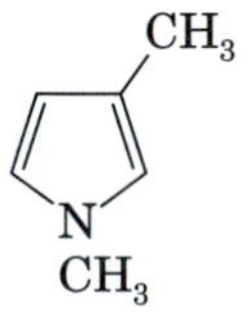

(c)

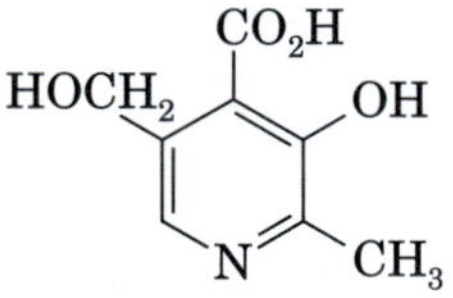

*a metabolic product
from pyridoxine*

(d)

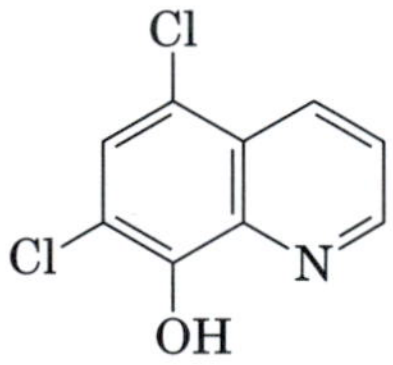

chloroxine

used in dandruff treatment

19.18 Draw structures for: **(a)** 2,4-dinitrothiophene; **(b)** *N*-phenylpyrrole; **(c)** furan-2-car-boxylic acid; **(d)** 1,8-dimethylisoquinoline.

19.19 Complete the following equations. (If no reaction occurs, write *no reaction.*)

(a) 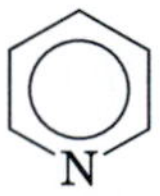 $+ \text{HBr} \longrightarrow$

(b) 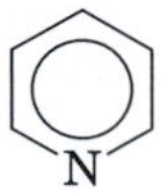 $+ \text{CH}_3\text{CH}_2\text{Br} \longrightarrow$

(c) 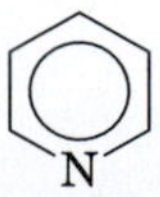 $+ (\text{CH}_3)_3\text{NH}^+ \text{Cl}^- \longrightarrow$

19.20 Which reaction would you expect to have the faster rate and why: **(a)** the reaction of pyridine and sodamide (NaNH_2) or **(b)** the reaction of 2-chloropyridine and sodamide?

19.21 Predict the major organic products:

(a) [isoquinoline structure] $\xrightarrow[\text{(2) H}_2\text{O, H}^+]{\text{(1) C}_6\text{H}_5\text{MgBr}}$

(b) [quinoline structure] $+ \text{Br}_2 \xrightarrow{\text{FeBr}_3}$

(c) [quinoline structure] $\xrightarrow[\text{(2) H}_2\text{O}]{\text{(1) CH}_3(\text{CH}_2)_3\text{Li}}$

19.22 Draw the *p*-orbital components in the lowest-energy π molecular orbitals for the following compounds (see Figure 19.3). Indicate the number of *p* electrons contributed by each atom and any unshared electrons.

(a) thiazole   (b) pyrimidine   (c) purine   (d) thiophene   (e) pyran

19.23 Predict the major organic products:

(a) [thiophene] $+ \text{CH}_3\overset{\overset{\text{O}}{\|}}{\text{C}}\text{Cl} \xrightarrow{\text{SnCl}_4}$

(b) [3-bromothiophene] $+ \text{HNO}_3 \xrightarrow{\text{CH}_3\text{CO}_2\text{H}}$

(c) [thiophene-3-carboxylic acid] $+ \text{Br}_2 \xrightarrow{\text{CH}_3\text{CO}_2\text{H}}$

19.24 Although pyrrole is not basic, thiazole is. Explain.

19.25 Pyrrole (pK_a = ~ 15) is a weak acid: it can lose a proton and form an anion. Suggest a reason for the stability of this anion compared to that of $(\text{CH}_3\text{CH}_2)_2\ddot{\text{N}}{:}^-$.

19.26 Complete the following equations.

(a) [furan] $+ \text{CH}_3\overset{\overset{\text{O}}{\|}}{\text{C}}\overset{\overset{\text{O}}{\|}}{\text{C}}\text{CH}_3 \xrightarrow{\text{BF}_3}$

(b) [pyrrole with N-H] $+ \text{CH}_3\text{MgBr} \longrightarrow$

(c) [quinoline] $+ \text{HNO}_3 \xrightarrow{\text{H}_2\text{SO}_4}$

(d) [nicotine-like structure with N-H] $+ 1 \text{ HCl} \longrightarrow$

(e) [furan] $+ \text{H}_3\text{C}-\text{C}_6\text{H}_4-\overset{+}{\text{N}}_2\,\text{Cl}^- \longrightarrow$

(f) [2-chloropyridine] $+ \text{NH}_3 \xrightarrow{\text{heat}}$

19.27 Draw formulas for the tautomer(s) of the following compounds:

(a) [thiazole-OH structure]   (b) [4-hydroxypyridine]   (c) [methyl-pyrimidine-OH structure]

19.28 Suggest syntheses of the following compounds by Friedel–Crafts reactions:

(a)

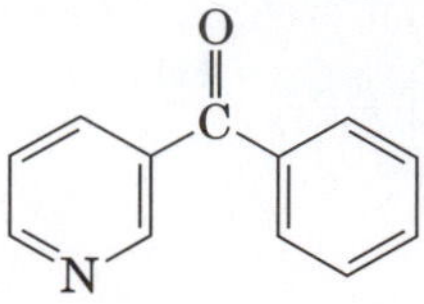

(b)

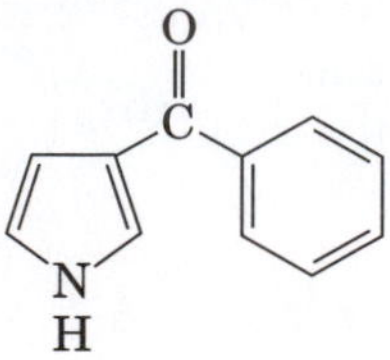

Additional Problems

19.29 Propose syntheses for the following compounds:

(a) 2-naphthoxyacetic acid, a plant growth hormone and a sedative, from 2-naphthol

(b) 9,10-phenanthrenediol from phenanthrene

(c) 4-amino-1-naphthalenesulfonic acid from naphthalene

(d) 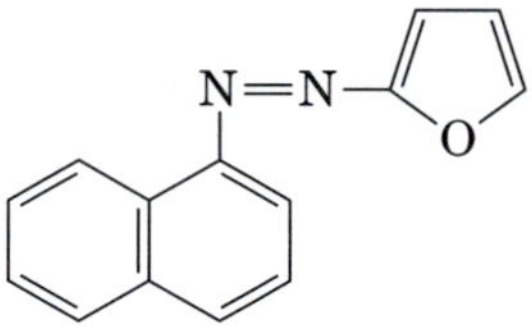 from naphthalene and furan

(e) 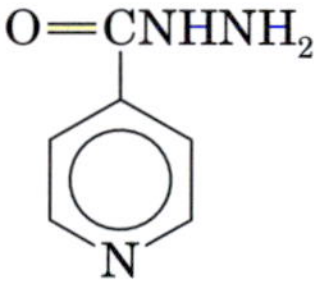 from 4-methylpyridine

isoniazid

used in the treatment of tuberculosis

(f) 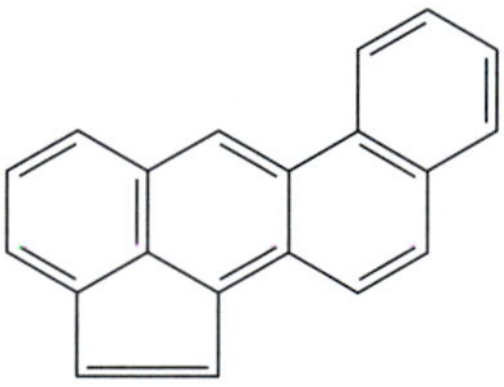 from

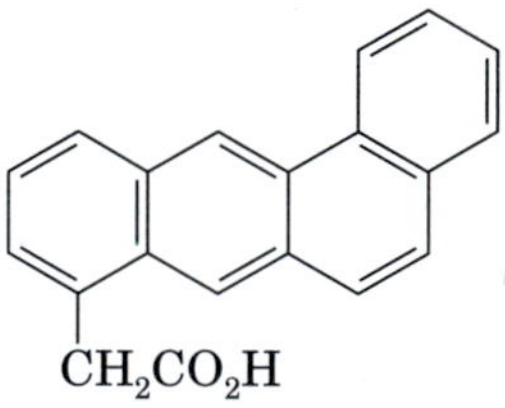

benz[*j*]aceanthrylene

a carcinogen

19.30 Only one of the three isomeric monohydroxypyridines (OH on a ring carbon) exhibits the chemical characteristics of a phenol. Explain.

19.31 Predict the position of aromatic electrophilic and nucleophilic substitution of pyrimidine (Table 19.1). Would you expect pyrimidine to be more or less reactive than pyridine toward electrophilic substitution? Toward nucleophilic substitution?

19.32 Substituted naphthalene compounds can be prepared by Friedel–Crafts acylations of substituted benzenes with succinic anhydride. Tell what intermediate and final products would be obtained in the following sequence. [*Hints:* HF catalyzes aromatic acylation with a carboxylic acid, and Pd/heat causes catalytic dehydrogenation (loss of H_2) leading to aromatization of nonaromatic rings.]

$$\text{toluene + succinic anhydride} \xrightarrow{\text{AlCl}_3} \textbf{(a)} \xrightarrow[\text{HCl}]{\text{Zn(Hg)}} \textbf{(b)} \xrightarrow{\text{HF}} \textbf{(c)} \xrightarrow[\text{heat}]{\text{Pd}} \textbf{(d)}$$

19.33 The aromatic dianion A of *sym*-dibenzocyclooctatetraene can be prepared by the treatment of the hydrocarbon with potassium metal. When A is treated with excess dry ice, followed by acidification, two products are formed: B ($C_{18}H_{12}O_3$) in 22% yield and C ($C_{18}H_{14}O_4$) in 51% yield. The ^{1}H NMR spectrum (in $CDCl_3$) of B showed a multiplet at $\delta 7.20$, a singlet at $\delta 6.98$, and a singlet at $\delta 4.64$ (area ratio 4:1:1). What are the structures of A, B, and C?

sym-dibenzocyclooctatetraene

19.34 Acid hydrolysis of the carbohydrates in oat hulls or corncobs gives Compound A, $C_5H_4O_2$, in almost 100% yield. Catalytic hydrogenation of A with $CuO—Cr_2O_3$ catalyst at 175° and 100 atm gives Compound B, $C_5H_6O_2$. The infrared spectrum of B and ^{1}H NMR spectra of A and B are shown in Figure 19.6. What are the structures of A and B?

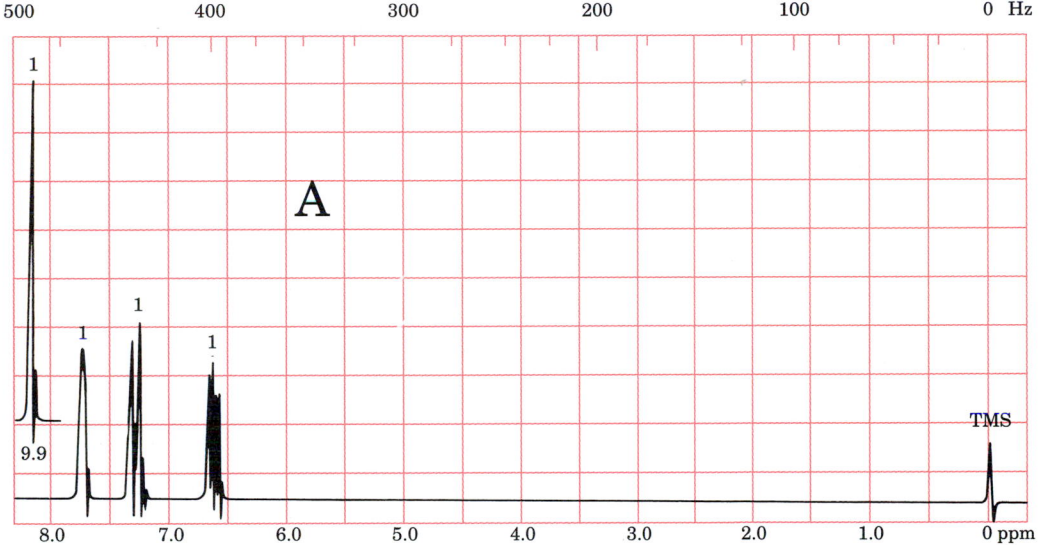

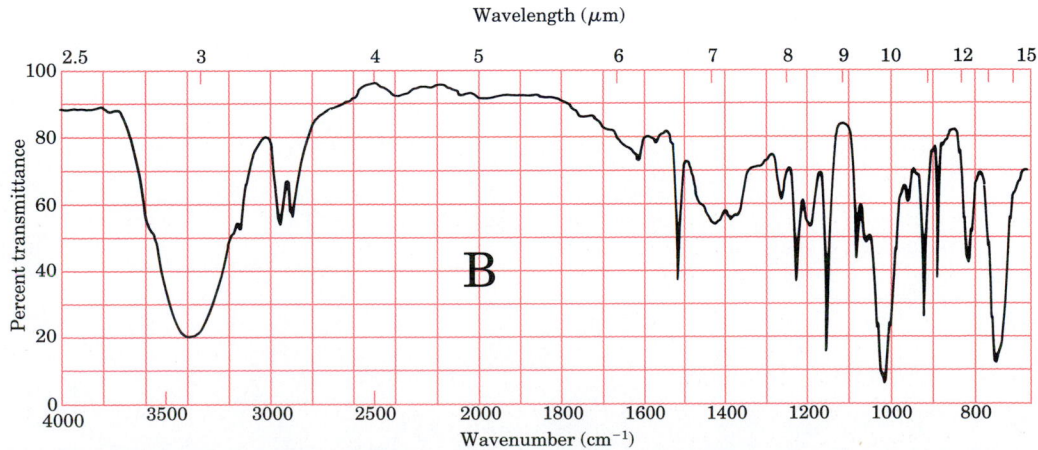

Figure 19.6 Spectra for Problem 19.34.

(continued)

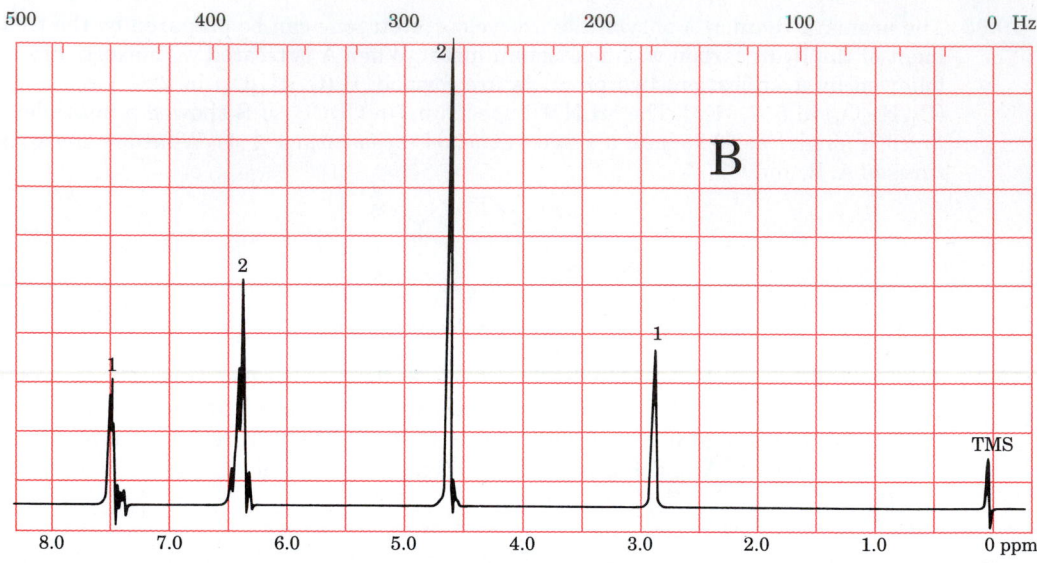

Figure 19.6 *(continued)* Spectra for Problem 19.34.

CHAPTER 20

NATURAL PRODUCTS: STUDIES IN ORGANIC SYNTHESIS

Identification of organic compounds and understanding of their chemical and physical properties are major goals of organic chemists. The synthesis of organic compounds is an equally important goal. In synthesis, the goal is to prepare a particular pure compound, with defined stereochemistry, by a short series of organic reactions.

Synthetic studies are frequently aimed at preparing natural products. Naturally occurring organic compounds often have unique structures as well as properties. They are often found in nature as a single enantiomer or diastereomer. The synthesis of one of these naturally occurring compounds can be a great challenge.

It is not uncommon for a synthesis of a natural product to begin as an intellectual challenge. However, such academic studies sometimes result in important chemical discoveries and new commercial products. Many natural products have potent biological properties and some have unusual physical properties. For example, plants biosynthesize a class of nitrogen-containing organic compounds called alkaloids. Morphine

(Section 20.3), a painkiller, is an alkaloid obtained from the poppy *Papaver somniferum*. Once an alkaloid with biological properties such as morphine is synthesized, variations on the synthesis can be carried out and structural analogs of the natural product prepared. Such studies have led to the development of numerous commercially important pharmaceuticals.

We will consider three syntheses and briefly look at the class of natural products to which the target compound belongs. The first synthesis is of abscisic acid, a terpene—a class of naturally occurring compounds composed of isoprene (2-methylbutyl) skeletal units. Abscisic acid is involved in the biological control of the dormancy of plants.

The second synthesis is of (S)-4-methyl-3-heptanone, an alarm pheromone of the leaf-cutting ant *(Atta texana)* and other insects. The (S) enantiomer is 400 times more active than the (R) enantiomer. Therefore, we will discuss methods by which a single stereoisomer of such a chiral compound can be synthesized.

The third synthesis is of (+)-retronecine, which is part of a pyrrolizidine alkaloid that is a cumulative plant poison of grazing animals.

(+)-abscisic acid

a terpenoid

(S)-4-methyl-3-heptanone

a pheromone

(+)-retronecine

an alkaloid component

Section 20.1

Terpenes

Essential oils are the odorous plant components that can be separated from other plant material by steam distillation. Many essential oils, such as those from flowers, are used in perfumes. Quite early in the history of organic chemistry, these essential oils, especially the oils of turpentine, attracted the attention of chemists. It was soon discovered that the ratio of the number of carbon atoms to the number of hydrogen atoms in turpentine is $5:8$. Other natural products with a carbon-to-hydrogen ratio of $5:8$ were then classified as **terpenes.** Later, chemists learned that the terpenes were composed of compounds containing a *head-to-tail* joining of isoprene skeletal units. (The *head* is the end closer to the methyl branch.)

The original definition of terpenes was then expanded to include all compounds containing isoprene skeletal units. To emphasize this relationship to isoprene, terpenes are also called **isoprenoids.** Terpenes may contain two, three, or more isoprene units. Their molecules may be open-chain or cyclic. They may contain double bonds, hydroxyl groups, carbonyl groups, or other functional groups. A terpene-like structure that contains elements other than C and H is called a **terpenoid.**

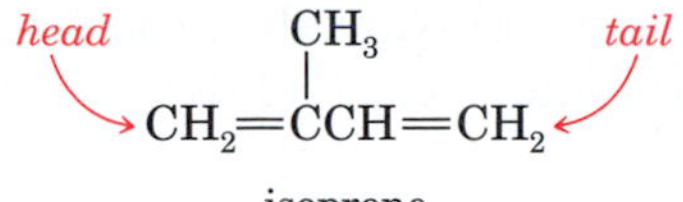

head   CH$_3$ tail

CH$_2$=CCH=CH$_2$

isoprene

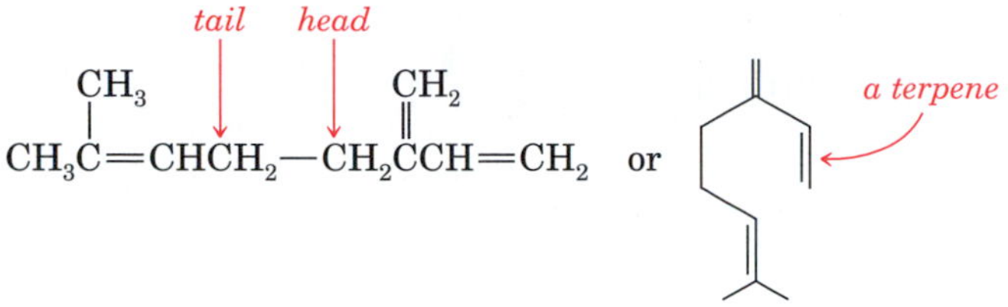

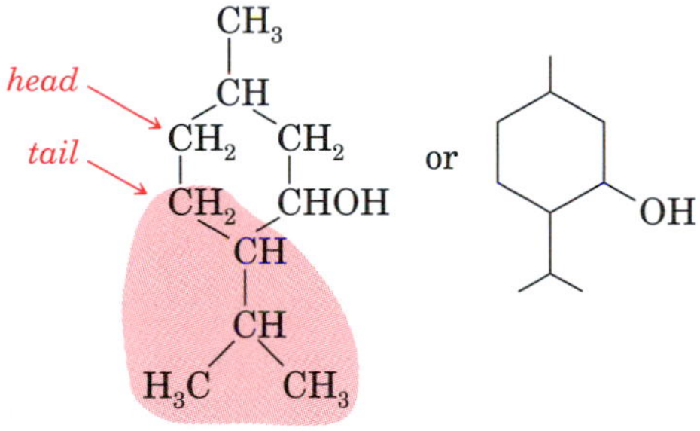

tail   head

CH$_3$ CH$_2$

CH$_3$C=CHCH$_2$—CH$_2$CCH=CH$_2$ or   *a terpene*

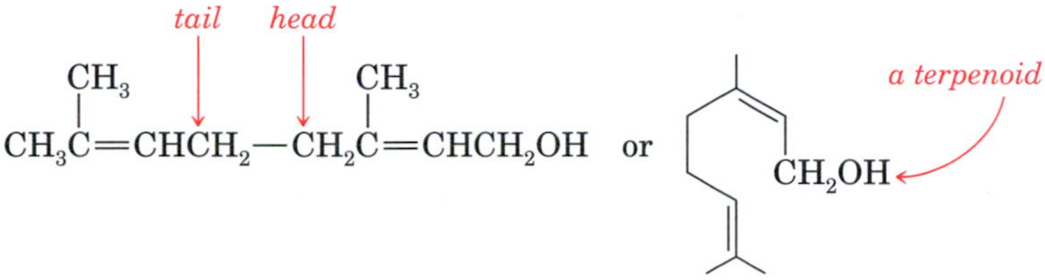

tail   head

CH$_3$ CH$_3$

CH$_3$C=CHCH$_2$—CH$_2$C=CHCH$_2$OH   or   CH$_2$OH   *a terpenoid*

CH$_3$
CH
head   CH$_2$ CH$_2$
tail   CH$_2$ CHOH   or   OH
CH
CH
H$_3$C   CH$_3$

Although the idea is appealing, naturally occurring terpenes do not arise from the polymerization of isoprene. The first step in terpene biosynthesis is an enzymatic ester condensation of the acetyl portions of acetylcoenzyme A. Intermediates in the formation of terpenes are the pyrophosphates (diphosphates) of mevalonic acid and a pair of isopentenyl alcohols. An abbreviated biosynthetic route to terpenes and steroids is shown in Section 24.5C.

A. Classification of Terpenes

Terpenes are classified by the *number of isoprene units* they contain:

monoterpenes:	two isoprene units (10 carbons)
sesquiterpenes:	three isoprene units (15 carbons)
diterpenes:	four isoprene units (20 carbons)
triterpenes:	six isoprene units (30 carbons)
tetraterpenes:	eight isoprene units (40 carbons)

Monoterpenes, with skeletons that contain only two isoprene units, are the simplest of the terpenes. Yet, even monoterpenes exhibit a variety of structures. (Although some terpenes and terpenoids, such as geraniol, are found in a variety of plants and animals, we have indicated sources especially rich in the following compounds.) Note that the trivial names of terpenes often are derived from the original sources of these compounds.

Acyclic monoterpenes:

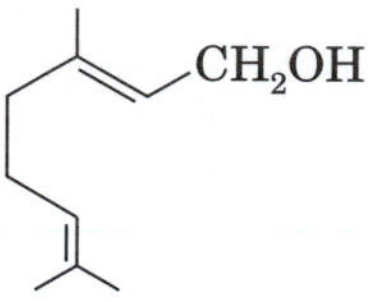

geraniol

in roses

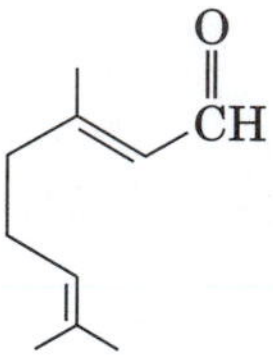

citral (geranial)

in lemongrass

Cyclic monoterpenes:

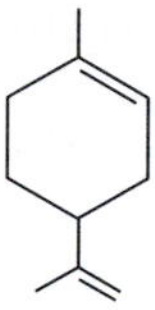

limonene	menthol	camphor	α-pinene
in citrus fruits	*in mint*	*from camphor trees*	*in turpentine (from pine trees)*

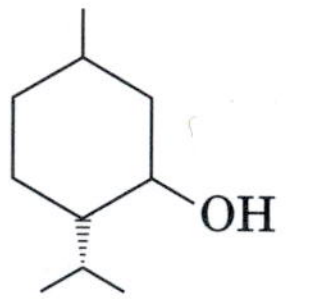

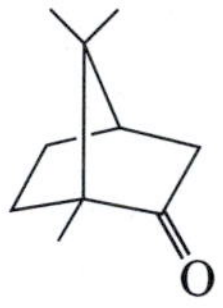

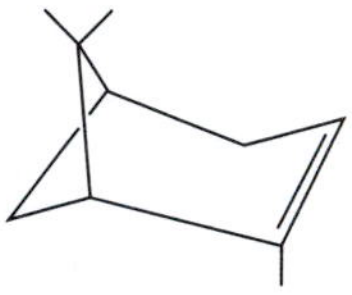

Carrots contain an orange-colored tetraterpene called *carotene*. (If a person eats too many carrots, the deposition of carotene will color his or her skin orange. However, time is a cure for this condition.) Carotene can be cleaved enzymatically into two units of vitamin A. (The role of vitamin A in vision will be discussed in Chapter 22.)

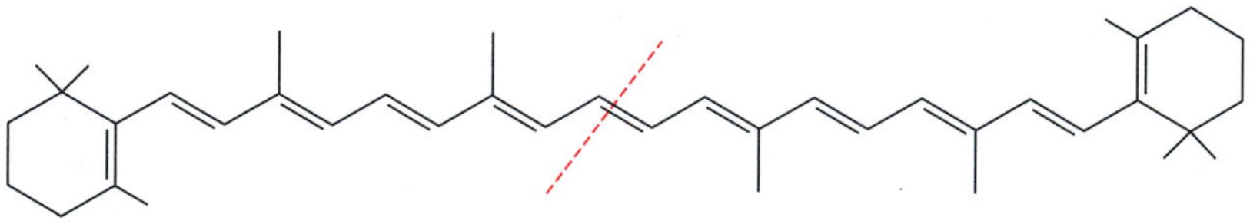

β-carotene

an orange tetraterpene

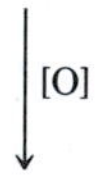

[O]

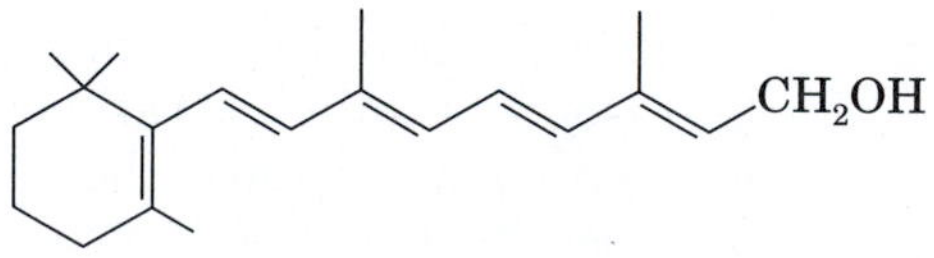

all-*trans*-vitamin A

STUDY PROBLEMS

20.1   Which of the following compounds belong to the class of terpenes or terpenoids?

(a) 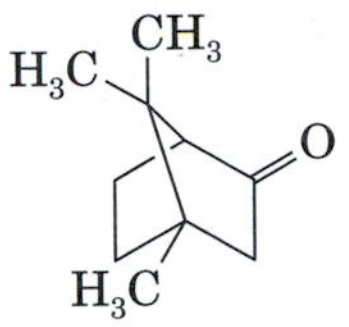

(b) 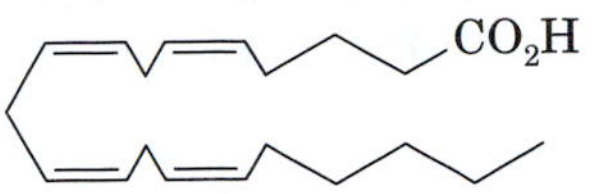

(c) $(CH_3)_2C{=}CH(CH_2)_2C(CH_3){=}CHCH_2OH$

(d) 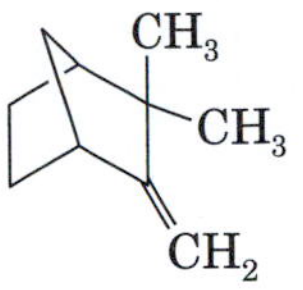

20.2 Circle the isoprene units in the following formulas:

(a) 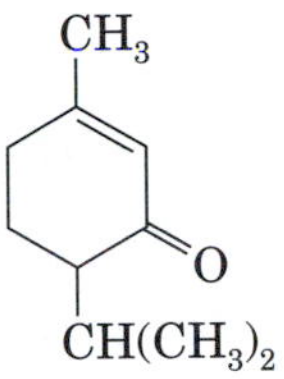

(b) 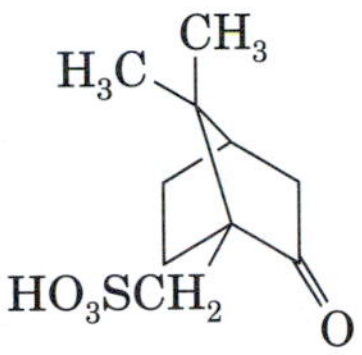

(c) 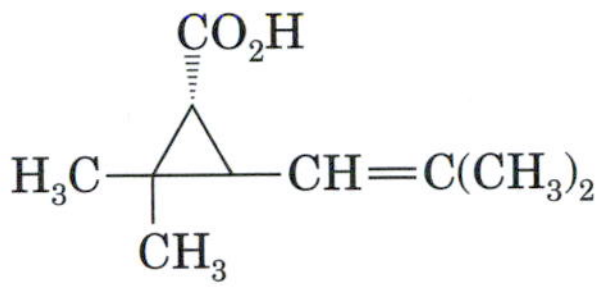

B. Synthesis of Abscisic Acid, a Terpenoid

Abscisic acid, a sesquiterpenoid, is widely distributed in the plant kingdom. It is involved in the regulation of the dormant state of plants, allowing them to survive under adverse environmental conditions.

Several different syntheses of abscisic acid have been reported. The one we describe (see Figure 20.1) begins with isophorone, a commercially available, achiral cyclohexenone synthesized by an aldol-dehydration reaction of acetone.

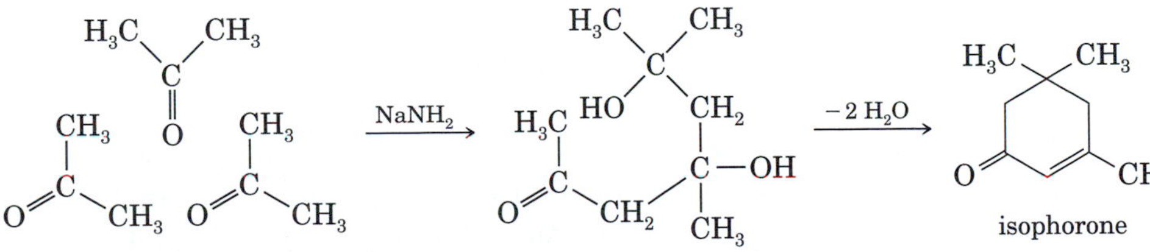

Comparison of the structures of abscisic acid and isophorone reveals that the keto group, the double bond, and the methyl groups needed are present in the starting isophorone. The synthetic work, therefore, involves building the allylic dienoic acid side chain and the insertion of the allylic tertiary hydroxyl group.

Retrosynthetic analysis (working backwards from the final, or target, compound) shows that one possible synthetic approach could be (1) the formation of the target tertiary alcohol and insertion of the side chain by the reaction of a Grignard reagent or related organometallic reagent with a ketone, preceded by (2) formation of a keto group at the key position.

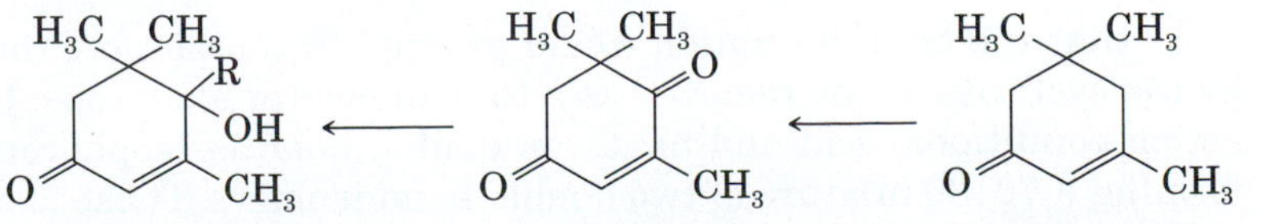

abscisic acid *a diketone* isophorone

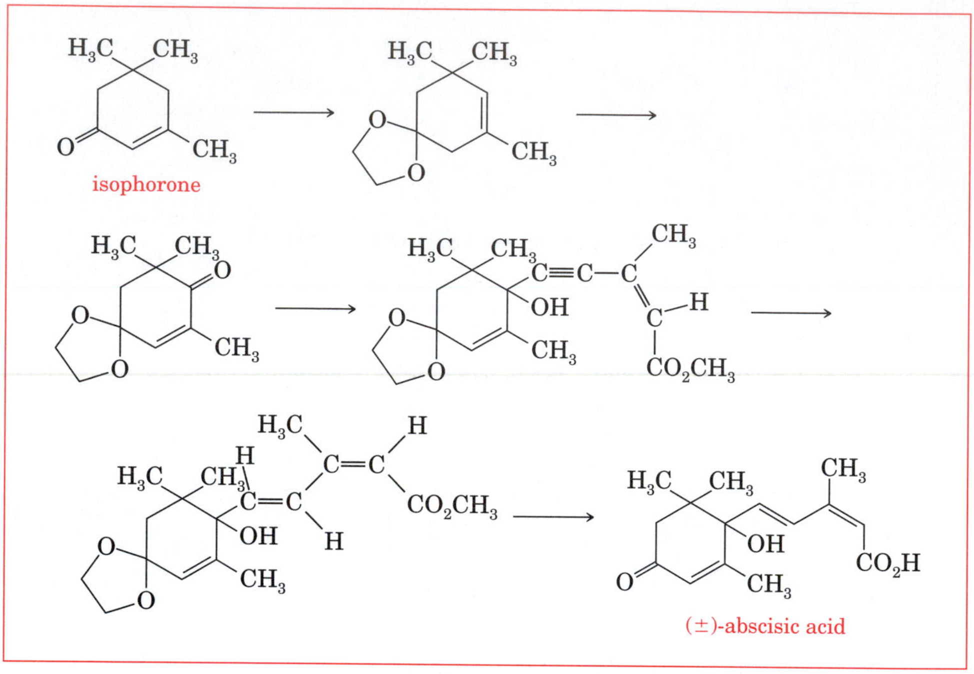

Figure 20.1 An outline of the synthesis of abscisic acid from isophorone. (M. G. Constantino, P. M. Donate, and N. Petragnani, *J. Org. Chem.* **1986,** *51,* 253.)

STUDY PROBLEM

20.3   Write an equation for the reaction of the intermediate diketone shown with excess CH_3MgI, followed by hydrolysis.

Part 1. Synthesis of the ketone   Because the synthesis hinges on the reaction of only one of two keto groups, the existing keto group in the starting isophorone must be blocked. The blocking group chosen was a cyclic ketal, a functional group inert to Grignard-type reactions, but readily reconverted to a keto group by acidic hydrolysis (Section 13.4B). Reaction of isophorone with ethylene glycol using *p*-toluenesulfonic acid as catalyst gave the ketal in 88.5% yield.

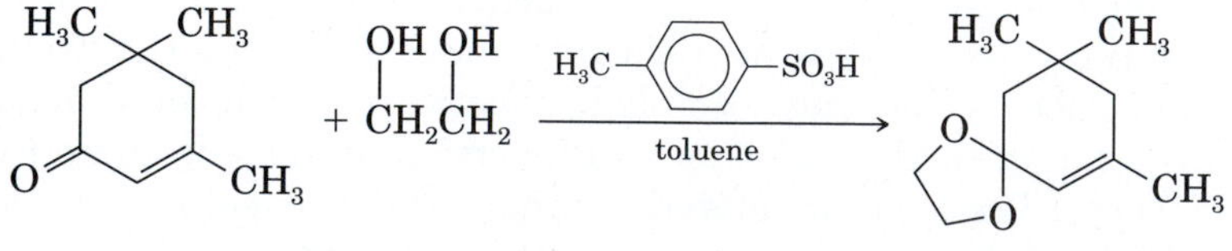

So that the ketal formation would proceed to completion, the water formed by the cyclization was removed as a toluene–water azeotrope. Under these reaction conditions, acid and heat, the double bond of isophorone equilibrates, forming a 70 : 30 mixture of two double-bond isomers. These isomers were separated by fractional distillation.

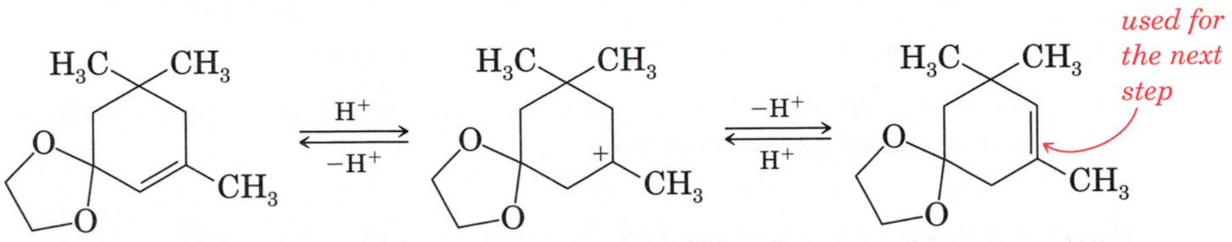

used for the next step

(30%) (70%)

Alkaline permanganate oxidation of the double bond of the ketal gave the α-hydroxy ketone in 72% yield. The alkene was *cis*-hydroxylated by permanganate to yield a diol with secondary and tertiary hydroxyl groups. Continued oxidation of this diol converted only the secondary—OH group to a carbonyl. Tertiary hydroxyl groups are inert to oxidation under alkaline conditions (Section 7.8C).

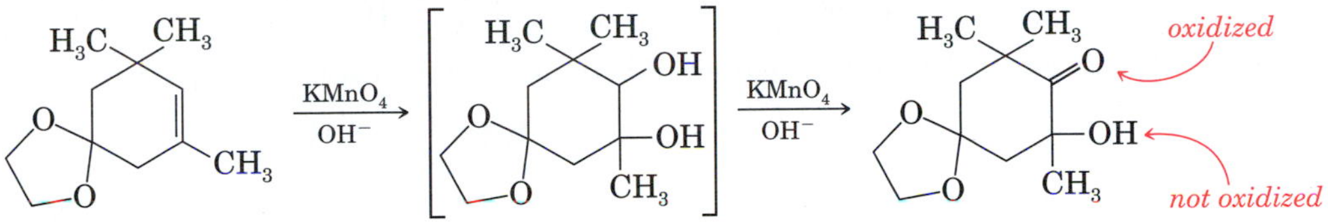

oxidized

not oxidized

(72%)

A mixture of methanesulfonyl chloride and pyridine was used to dehydrate the tertiary alcohol. Under these alkaline conditions, the ketal group remains intact.

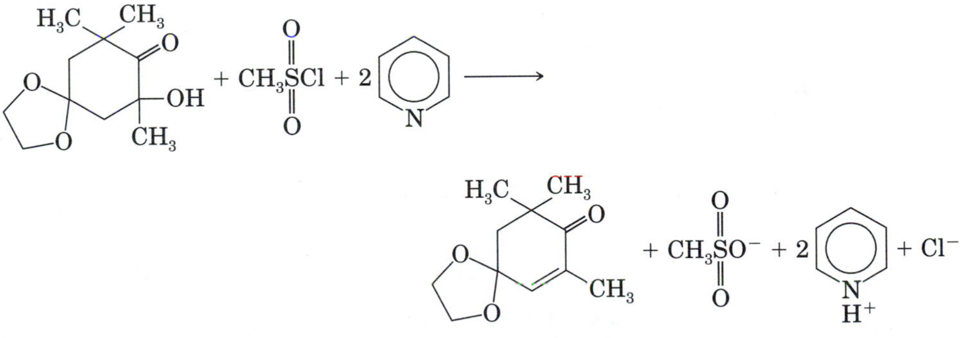

(63%)

A possible pathway for this alkaline dehydration is an initial formation of a methylsulfonate ester that decomposes into a carbocation. (Recall from Section 7.7B that sulfonate groups are good leaving groups.) The most substituted conjugated alkene forms from the carbocation.

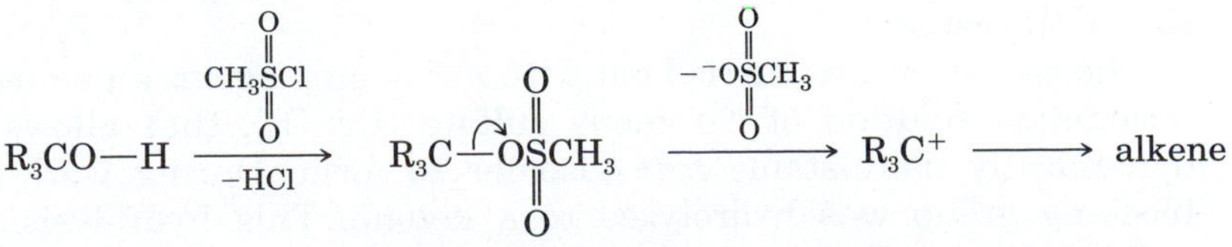

a methanesulfonate ester

20.4   Write an equation showing the dehydration of the preceding hydroxy–ketal–ketone under acidic conditions.

Part 2. Synthesis of the final product   Insertion of the side chain was accomplished by reaction of the ketone with an acetylenic lithium reagent at low temperature. The reagent was obtained by treating methyl 2-penten-4-ynoate with lithium diisopropylamide (LDA), a hindered (and thus non-nucleophilic) strong base, at $-78°C$. A rapid acid–base reaction between the lithium amide and the acetylenic proton yielded the organolithium reagent.

methyl (Z)-3-methyl-2-penten-4-ynoate   lithium diisopropylamide (LDA)

a lithium acetylide   diisopropylamine

Immediately following the formation of the lithium acetylide, the cyclic ketone was added to the reaction mixture and the mixture maintained at a low temperature ($-78°$ to $-30°$). (At these low temperatures, the ester group does not undergo reaction.) The product lithium salt of a tertiary alcohol was treated with water to yield the alcohol.

($\sim 100\%$)

The final steps in the synthesis were (1) reduction of the triple bond to a *trans* double bond, (2) removal of the ketal blocking group, and (3) saponification of the ester.

The reduction was carried out with a reducing agent, an aqueous dimethylformamide solution of chromous sulfate ($CrSO_4$), that allows the thermodynamically more-stable *trans* isomer to form. During work-up, the ketal blocking group was hydrolyzed to a ketone. This hydrolysis reaction was catalyzed by ammonium ions, which are sufficiently acidic to hydrolyze the ketal group.

Saponification, followed by acidification, provided the final abscisic acid in 80% yield from the methyl ester.

In designing and executing this synthesis of abscisic acid, the organic chemists had to consider the numerous possible side reactions that the three different oxygen-containing functional groups and the three differently substituted carbon–carbon double bonds could undergo. Note that this particular route to abscisic acid was not designed to be stereoselective. It produced the natural product as the racemic mixture.

STUDY PROBLEM

20.5 Suggest reasons why acidic hydrolysis of methyl abscisate would be a poor idea.

Section 20.2
Pheromones: Synthesis of Chiral Natural Products

Humans communicate by talking, using sign language, drawing pictures, or writing. Insects and some animals communicate, in part, by chemicals. A chemical or a mixture of chemicals secreted by one individual of a species that brings forth a response in another individual of the same species is called a **pheromone** (from the Greek *phero*, "carrier").

Insect pheromones are generally classified by the response they elicit. For example, *alarm pheromones* signify danger, and *sex attractants* help the different sexes of the same species to find one another. Other pheromones are used to recruit. For example, a pheromone secreted by one bee helps alert other bees to the location of a food source.

Extremely small quantities of a pheromone can elicit the desired response. A typical female insect may carry only 10^{-8} gram of sex attractant, yet that is enough to attract over a billion males from miles away! A male gypsy moth can "smell" a female at a distance of 7 miles.

The biological activity of a pheromone is usually highly dependent upon stereochemistry—one enantiomer or geometric isomer is active and all other stereoisomers inactive.

Many of the known insect pheromones are not complex in structure. Geraniol and citral, both terpenes, are recruiting pheromones for honeybees, while isoamyl acetate (not a terpene) is a bee alarm pheromone. (Isoamyl acetate is also the principal odorous component of banana oil.)

$$CH_3\overset{\overset{\textstyle O}{\|}}{C}OCH_2CH_2CH(CH_3)_2$$

geraniol   citral   3-methylbutyl acetate
                                  (isoamyl acetate)

The following compounds have sex-attractant activity for different species of insects.

$cis\text{-}(CH_3)_2CH(CH_2)_4\overset{\overset{\textstyle O}{\diagup\diagdown}}{C}HCH(CH_2)_9CH_3$ $cis\text{-}CH_3(CH_2)_3CH{=}CH(CH_2)_6\overset{\overset{\textstyle O}{\|}}{O}CCH_3$

gypsy moth   cabbage looper

$cis\text{-}CH_3(CH_2)_{12}CH{=}CH(CH_2)_7CH_3$

boll weevil   house fly

An example of a three-component pheromone mixture is the sex attractant from the male *Ips paraconfusus* (an engraver pine beetle). To elicit response, all three compounds (ipsenol, ipsdienol, and *cis*-verbenol) must be present. We say that the mixture is *synergistic*. It is postulated that the synergism has the effect of increasing specificity because only the correct combination functions as the pheromone. Thus, this synergism may allow the pheromone to be distinguished from the multitude of background chemicals found in a forest.

"ipsenol"   "ipsdienol"   *cis*-verbenol

Pheromone chemistry is an exciting and vigorous field of research that includes isolation, structure determination, synthesis, and biological studies. It is an area of study that also is of immediate practical value. For example, sex-attractant pheromones have been used for insect control. In some cases, male insects may be lured by a sex-attractant pheromone, trapped, and then sterilized and released to mate unproductively with females.

The research effort in this field has been directed toward the elucidation and synthesis of single compounds as pheromones. As more information becomes available, it may turn out that a single compound believed to be the active pheromone is actually only one component of a complex mixture of interacting substances, such as the three-component system shown.

A. Synthesis of Chiral Compounds

It is not possible to synthesize only one enantiomer of a chiral compound from an achiral starting material using achiral reagents and catalysts. When such a reaction is carried out, both configurations, (R) and (S), of the product are formed; that is, the product is always racemic.

To synthesize one stereoisomer of a chiral compound, chemists must use one of two synthetic approaches. The first is to synthesize the compound as a racemic mixture and then separate the enantiomers. The separation of enantiomers, called **resolution,** was discussed in Section 4.11. The second approach is to synthesize the compound with a chiral auxiliary—a chiral catalyst, reagent, solvent, or intermediate. This technique, called **asymmetric synthesis,** will be discussed here.

Because many asymmetric syntheses are not 100% efficient, the terms **optical purity** and **enantiomeric excess** are often used to describe the products.

$$\text{optical purity} = \frac{\text{observed } [\alpha]_D^T}{[\alpha]_D^T \text{ of pure enantiomer}} \times 100\%$$

$$\text{enantiomeric excess} = \frac{\text{excess of desired enantiomer}}{\text{total yield}} \times 100\%$$

$$\text{or} \quad \frac{A - B}{A + B} \times 100\%$$

Thus, a chemist might report a 61% yield with a 94% enantiomeric excess.

STUDY PROBLEM

20.6 **(a)** Calculate the enantiomeric excess of A in a mixture of enantiomers that contains 120 mg of A and 20 mg of B. **(b)** Calculate the enantiomeric excess of A in a mixture of 50 mg of A and 50 mg of B (a racemic mixture).

Chiral catalyst Enzymes are chiral catalysts. Because of their asymmetric surfaces, enzymes are able to hold an achiral substrate so that only one face of the molecule undergoes reaction. The reduction of ketones, hydroxy ketones, and keto esters with baker's yeast is an example of an enzymatic asymmetric synthesis.

$$\underset{\text{1-hydroxy-2-heptanone}}{CH_3(CH_2)_4\overset{\overset{\displaystyle O}{\|}}{C}CH_2OH} \xrightarrow[\substack{\text{sucrose} \\ H_2O}]{\text{baker's yeast}} \underset{\substack{(R)\text{-1,2-heptanediol (56\%)} \\ (\text{enantiomeric excess, 100\%})}}{\overset{\overset{\displaystyle OH}{|}}{\underset{CH_3(CH_2)_4 \quad CH_2OH}{C\cdots H}}}$$

Nonenzymatic chiral catalysts have also been developed. These catalysts function in the same manner as their enzymatic counterparts. An example is a chiral hydrogenation catalyst that can induce chirality during hydrogenation (Section 10.12A).

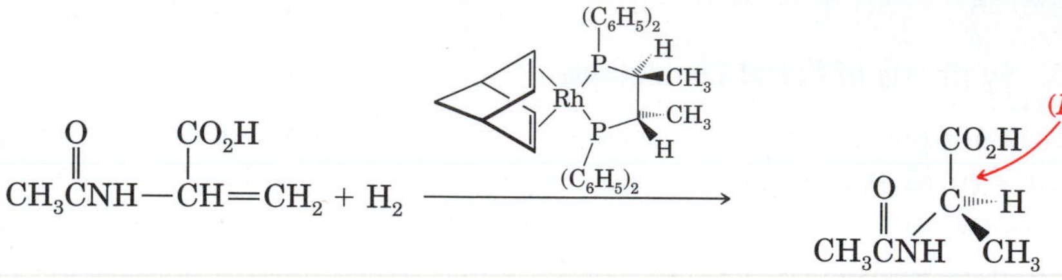

(enantiomeric excess, 90%)

Chiral reagent If a chiral reagent is used to form a chiral center in an achiral starting material, one enantiomer will usually be formed in excess. For example, a single enantiomer can be isolated from the hydroboration—oxidation of an achiral diene when a chiral dialkylborane is used in the hydroboration step.

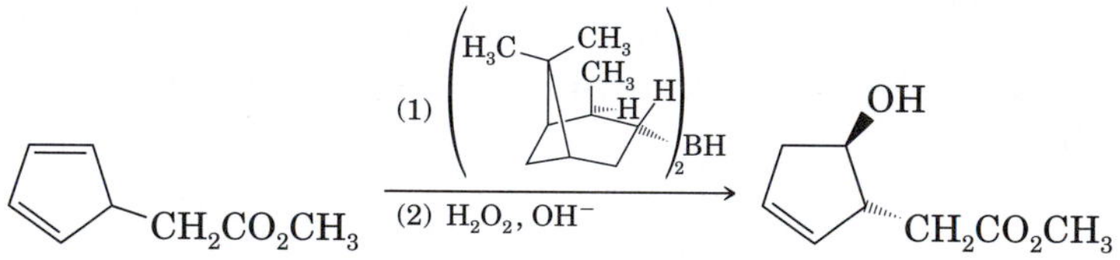

(21–31%)

Chiral solvent We can often enhance the formation of one enantiomer by using a chiral solvent for the reaction of two achiral reagents.

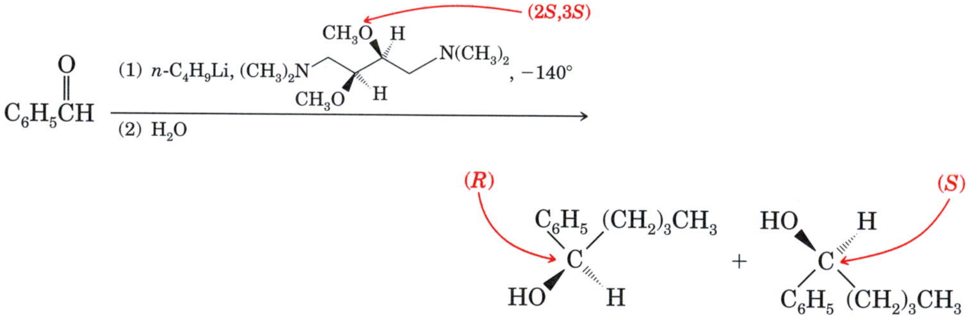

(enantiomeric excess, 33%)

Chiral intermediate Another approach to asymmetric syntheses is the use of chiral intermediates. In this procedure, a chiral auxiliary reagent is used to prepare a chiral intermediate from an achiral substrate. The intermediate is then subjected to a standard chemical reaction that forms a chiral center. The bonded chiral auxiliary allows only one face or side of the substrate to undergo reaction. After the reaction, the chiral auxiliary is removed and the chiral product is released. This approach is used in the pheromone synthesis that follows.

achiral substrate $\xrightarrow{\text{chiral auxiliary}}$ [chiral intermediate] $\xrightarrow[\text{creating chiral center}]{\text{chemical reaction}}$

[diastereomer of a new intermediate] $\xrightarrow{\text{remove chiral auxiliary}}$ chiral product

B. Synthesis of (*S*)-4-Methyl-3-heptanone, a Pheromone

The principal alarm pheromone of the leaf-cutting ant *(Atta texana)* is (*S*)-4-methyl-3-heptanone, a compound with a relatively simple structure containing one chiral carbon. The racemic form of this pheromone has been synthesized by a number of different routes, and the enantiomers have been separated by resolution.

We will discuss an asymmetric synthesis of the (*S*) enantiomer that starts with 3-pentanone, an achiral compound.

The synthetic pathway is outlined in Figure 20.2. The chiral auxiliary reagent is a hydrazine, (*S*)-1-amino-2-methoxymethylpyrrolidine (abbreviated SAMP). This optically active hydrazine is obtained from (*S*)-proline, a naturally occurring amino acid, by a multistep synthesis, which we will not discuss. However, it is important to recognize that the chirality of the chiral auxiliary reagent originates with a naturally occurring compound and is not created from achiral compounds by a laboratory procedure.

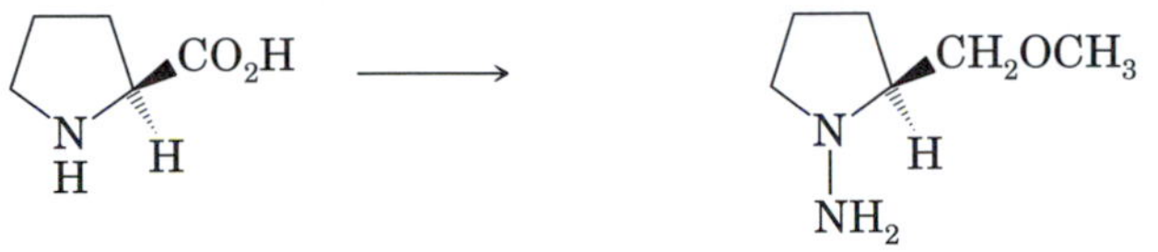

(*S*)-proline

naturally occurring

(*S*)-1-amino-2-methoxymethylpyrrolidine
(SAMP)
(50% overall)

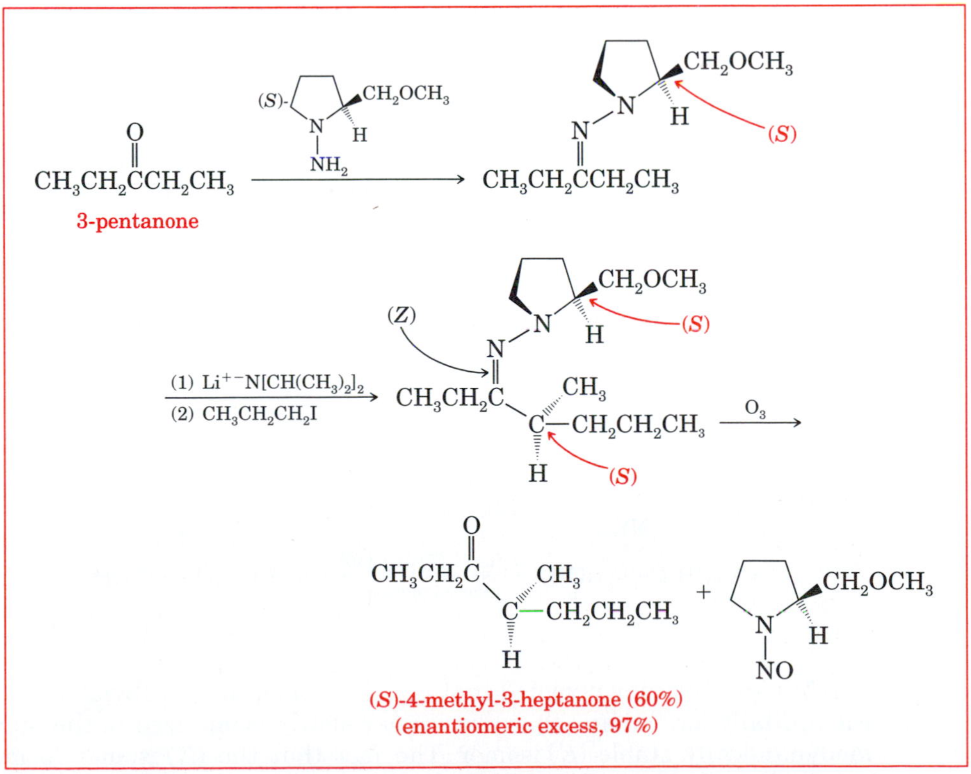

Figure 20.2 Synthesis of (*S*)-(+)-4-methyl-3-heptanone, a pheromone. (D. Enders, H. Kipphardt, and P. Fey, *Org. Syn.* **1987,** *65,* 183.)

The first step in the asymmetric synthesis of the ant pheromone was conversion of 3-pentanone to a chiral hydrazone. This step created the molecular chiral environment that allowed only one configuration at the new chiral center later in the synthesis.

hydrazine group

$$CH_3CH_2\overset{\overset{\displaystyle O}{\|}}{C}CH_2CH_3 + \begin{array}{c} \text{(pyrrolidine ring)} \\ CH_2OCH_3 \\ N \\ | \\ H \\ | \\ NH_2 \end{array} \longrightarrow \begin{array}{c} \text{(pyrrolidine ring)} \\ CH_2OCH_3 \\ H \\ N \\ \| \\ N \\ CH_3CH_2\overset{}{C}CH_2CH_3 \end{array}$$

a hydrazone

The α hydrogens of hydrazones, like those of aldehydes and ketones, are slightly acidic (Chapter 17). In this synthesis, lithium diisopropylamide (LDA) was used to convert the optically active hydrazone to the lithium salt, a lithium aza-enolate.

Enolate formation:

$$RC\overset{\overset{\displaystyle O}{\|}}{C}CH_2R + \text{base}{:}^- \longrightarrow RC\overset{\overset{\displaystyle O}{\|}}{\underset{\cdot\cdot}{C}}HR + \text{base—H}$$

$$CH_3CH_2\overset{\overset{\displaystyle NNR_2}{\|}}{C}CH_2CH_3 + Li^+ \ ^-{:}\overset{\cdot\cdot}{N}[CH(CH_3)_2]_2 \longrightarrow CH_3CH_2\overset{\overset{\displaystyle NNR_2}{\|}}{C}\overset{-}{\underset{\cdot\cdot}{C}}HCH_3 \ Li^+ + H\overset{\cdot\cdot}{N}[CH(CH_3)_2]_2$$

a lithium aza-enolate

The lithium aza-enolate is stabilized by forming a complex involving the lithium cation, the α carbanion, and the chiral auxiliary. This chiral chelated carbanion is a nucleophile that will react with alkyl halides in nucleophilic substitution reactions. Treatment of the complex with 1-iodopropane at −110° resulted in alkylation. The chiral pyrrolidine portion of the hydrazone allowed only one of two diastereomers, (*S*,*S*), and not (*R*,*S*), to form because of the chiral environment at the site of reaction. This type of reaction is called **asymmetric induction.**

achiral carbon   *chiral carbon, (S)*

$$(S) \quad CH_3CH_2\overset{\overset{\displaystyle NNR_2}{\|}}{C}{-}\overset{-}{\underset{\cdot\cdot}{C}}HCH_3 \xrightarrow[\text{(allowed to warm)}]{CH_3CH_2CH_2I, \ -110°} (S) \quad CH_3CH_2\overset{\overset{\displaystyle NNR_2}{\|}}{C}{-}\underset{\underset{\displaystyle CH_2CH_2CH_3}{|}}{C}HCH_3$$

NMR spectroscopy revealed that the (*Z*) isomer of the alkylated hydrazone was initially formed and that this isomer slowly isomerized to the more thermodynamically stable (*E*) isomer. The fact that the (*Z*) isomer formed first shows that the chelated face of the molecule is the one that undergoes alkylation. The equilibration between the (*Z*) and (*E*) hydrazones did not pose a problem in the overall synthesis because either hydrazone would yield the (S) ketone after hydrolysis.

Simple alkaline hydrolysis would convert the hydrazone group to the desired keto group, but would also cause racemization at the α carbon by way of the symmetrical enol tautomer. For this reason, ozone without a subsequent water work-up was used to oxidize the π bond of the hydrazone and to form the keto group. (In another study, a two-phase system of $3\ M$ HCl and pentane was used.)

an (S)-N-nitrosoamine

STUDY PROBLEMS

20.7 Write an equation showing the racemization of (S)-4-methyl-3-heptanone in aqueous sodium hydroxide. Include the structure of the intermediate in your answer.

20.8 Complete the following equations:

(a) $CH_3\overset{O}{\overset{\|}{C}}CH_2CH_3 + Li^+\ ^-CH_2CH_2CH_2CH_3 \longrightarrow$

(b) $CH_2{=}CHCH_2CH_2\overset{O}{\overset{\|}{C}}OCH_3 \xrightarrow[\text{(2) }H_2O_2,\ OH^-]{\text{(1) }R_2BH}$

$$\overset{\text{O}}{\overset{\|}{\text{(c) } CH_3CCH_3}} + C_6H_5NHNH_2 \xrightarrow{\text{H}^+}$$

(d) the product from (c) + $Li^+\ ^-N[CH(CH_3)_2]_2 \longrightarrow$

Section 20.3
Alkaloids

A. Some Common Alkaloids

Primitive people have often used extracts of roots, bark, leaves, flowers, berries, and seeds as drugs. This use of plants for medicinal purposes was not necessarily based on superstition or wishful thinking. Many plants contain compounds that have profound physiological impact. The active agents in many of these plant substances have been isolated and have been found to be heterocyclic nitrogen compounds.

Many of the nitrogen compounds in plants contain basic nitrogen atoms and thus can be extracted from the bulk of the plant material by dilute acid. These compounds are called **alkaloids,** which means "like an alkali." After extraction, the free alkaloids can be regenerated by subsequent treatment with aqueous base.

Extraction:

$$R_3N\colon + HCl \xrightarrow{\text{H}_2\text{O}} R_3NH^+\,Cl^-$$

Regeneration:

$$R_3NH^+\,Cl^- + OH^- \longrightarrow R_3N\colon + H_2O + Cl^-$$

Alkaloids vary from simple to complex in their structures. One of the simplest in structure, but not in its physiological effects, is *nicotine.*

nicotine

Nicotine is highly toxic and has been used as an insecticide. In very small doses (such as a smoker obtains from cigarettes), nicotine acts by stimulating the autonomic (involuntary) nervous system. If small doses are continued, nicotine can depress this same nervous system into less than normal activity.

The first isolation of an alkaloid in the pure state was reported in 1805. This alkaloid was *morphine* (from the Greek *Morpheus,* the god of dreams), one of many alkaloids to come from the gum and seeds of the opium poppy, *Papaver somniferum.*

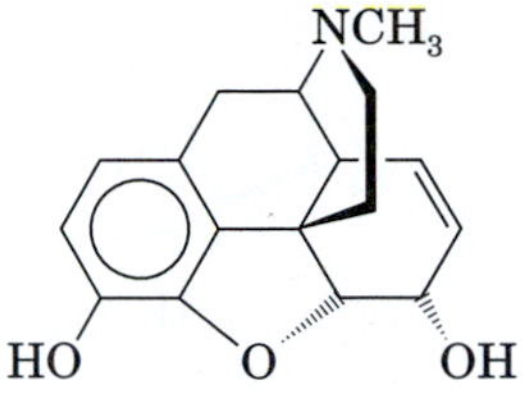

morphine 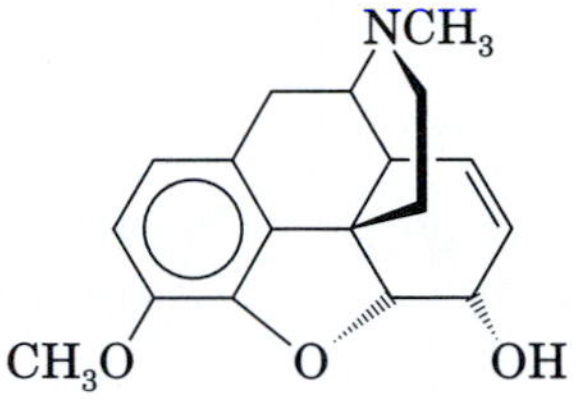   codeine 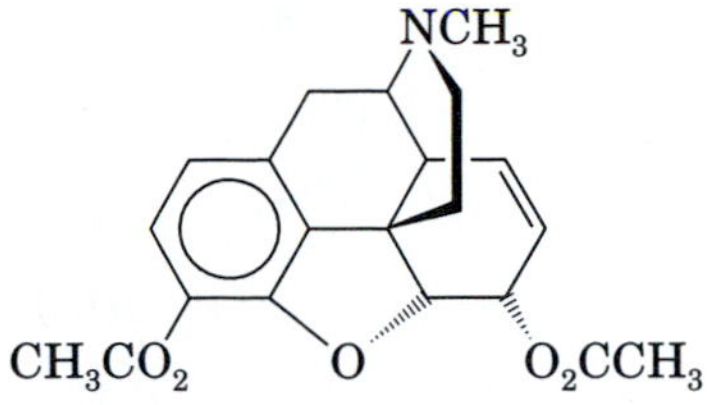   heroin

Codeine is the methyl derivative of morphine (at the phenolic group), while *heroin* is the diacetyl derivative. Codeine, like morphine, is a powerful analgesic and occurs naturally in the seeds of the opium poppy. Codeine is also an excellent cough suppressant that is sometimes used in prescription cough medicines. In recent years, it has been largely replaced by *dextromethorphan,* a nonaddictive, synthetic drug that is an equally effective cough suppressant. (Note the similarities in structure.)

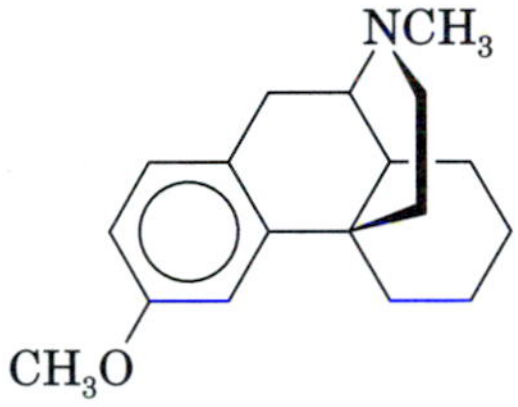

dextromethorphan

Heroin does not occur naturally, but may be synthesized from morphine in the laboratory. Heroin, like morphine and codeine, is a powerful analgesic. In some parts of the world, heroin is used to relieve pain in terminal cancer patients. Because it is even more addictive than morphine, its medicinal use is prohibited in the United States.

A large number of physiologically active alkaloids contain the **tropane ring system:**

tropane

One of the tropane alkaloids is *atropine,* found in *Atropa belladonna* and other members of the nightshade plant family. Atropine is used in eye drops to dilate the pupils. *Scopolamine* (a so-called "truth" serum) is used as a preoperative sedative. Chemically, it is the epoxide of atropine. *Cocaine,* an addictive stimulant and pain reliever, also contains the tropane ring system.

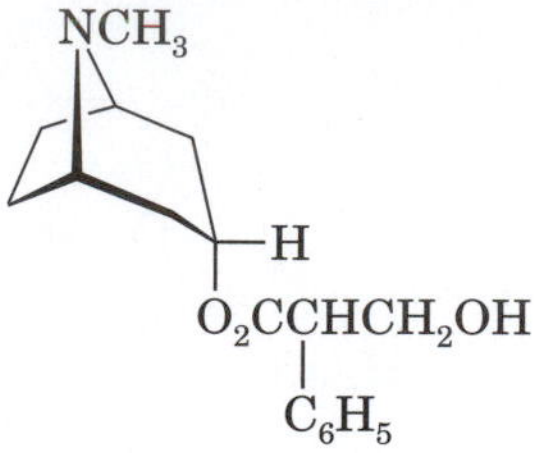

atropine

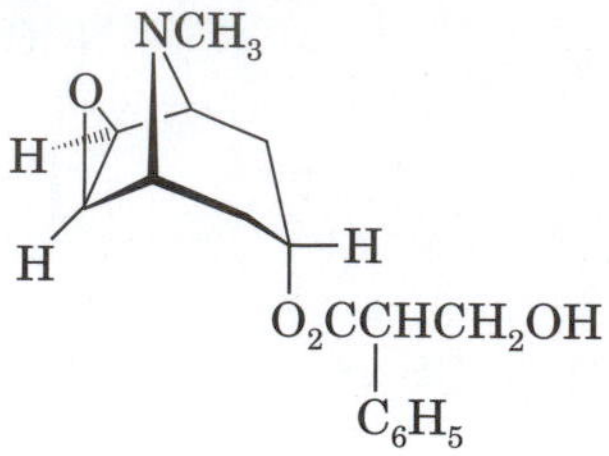

scopolamine

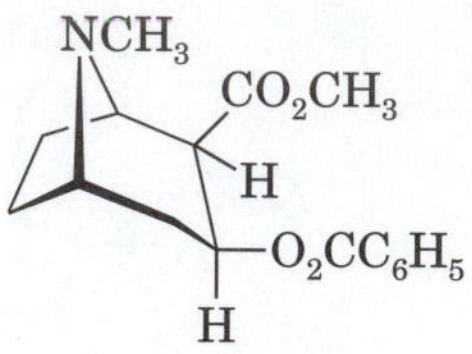

cocaine

20.9 When atropine is subjected to acid hydrolysis, two products can be isolated: *tropine,* which is not optically active, and *tropic acid,* which is obtained as a racemic mixture. What are the structures of these two compounds?

B. Synthesis of Retronecine, an Alkaloid

Alkaloids that contain a pyrrolizidine ring system are called **pyrrolizidine alkaloids.**

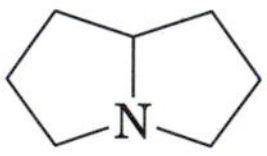

pyrrolizidine

These alkaloids are found worldwide in a variety of plants, such as asters and ragworts. They are of interest because they are cumulative poisons to grazing animals, causing liver cirrhosis (leading to a condition called "horse staggers"), liver tumors, and death.

About 100 different pyrrolizidine alkaloids have been identified. Most of these are found in nature as esters that can be hydrolyzed to a carboxylic acid, called a *necic acid,* and an amino alcohol (a substituted pyrrolizidine), called a *necine.*

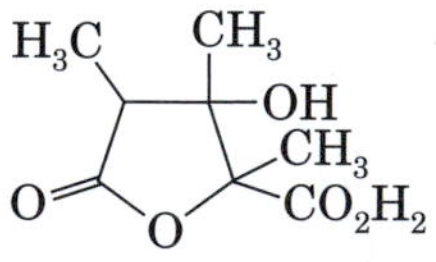

monocrotalic acid

retronecine

a typical necic acid   *the most widespread necine*

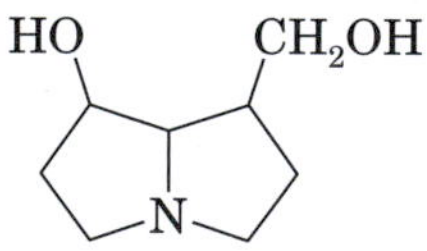

Although a large number of necines are known, the most common is retronecine, which was first isolated in 1909 and was synthesized in the laboratory in 1962. Figure 20.3 shows the reaction sequence used in this stereospecific synthesis.

The first step in the synthesis is a Michael addition (Section 17.9). Under the conditions of this addition, the product tetraethoxy compound **A** undergoes a Dieckmann ring closure (Section 17.8) to yield a β-keto ester **B**. Acid hydrolysis of this keto ester yields a β-keto acid (not isolated), which undergoes decarboxylation. The conditions of the hydrolysis and decarboxylation

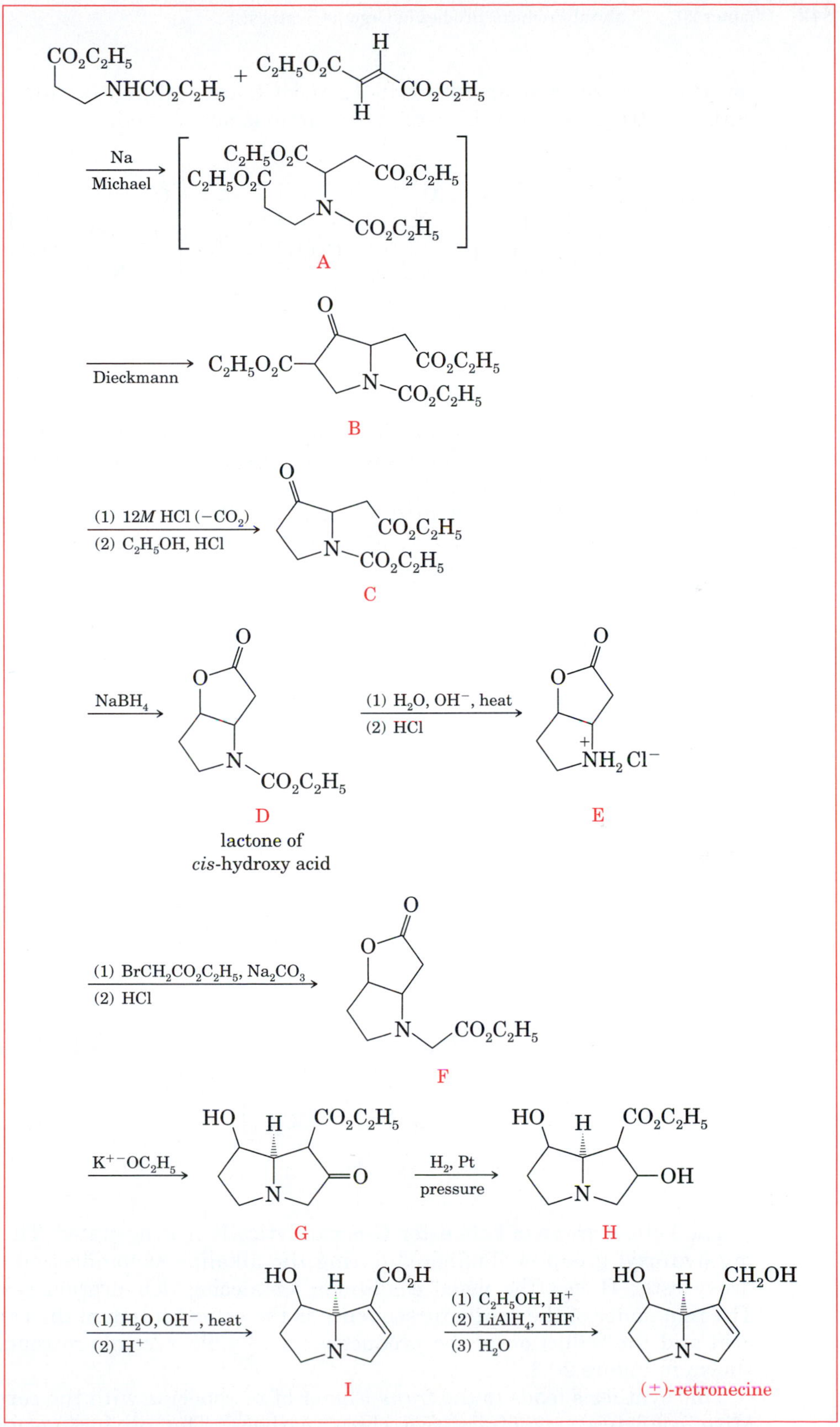

Figure 20.3 Synthesis of retronecine, a component of several widely spread plant toxins. (T. A. Geissman and A. C. Waiss, Jr., *J. Org. Chem.* **1962**, *27*, 139.)

(overnight at room temperature in 12 M HCl) are sufficiently mild that the carbon–nitrogen bond of the carbamate group is not cleaved.

The decarboxylated product is then re-esterified with ethanol to yield compound **C**. Reduction of the keto group in **C** with sodium borohydride yields a *cis* hydroxy acid, which undergoes lactonization to yield compound **D**. (The reason for the preferential formation of the *cis* acid instead of the *trans* isomer is discussed in Section 13.6B.)

Hydrolysis of the carbon–nitrogen bond in the carbamate group occurs when compound **D** is heated at reflux in aqueous hydroxide for 16 hours. These reaction conditions also open the lactone ring. Upon acidification, the lactone ring re-forms and the amino nitrogen atom is protonated to yield salt **E**. This salt is neutralized with sodium carbonate and alkylated (a nucleophilic substitution reaction between an amine and a reactive organohalogen compound). Treatment of the product ester **F** with potassium ethoxide results in transesterification of the lactone followed by another Dieckmann ring closure.

The ketone group of keto ester **G** is catalytically hydrogenated. The resulting hydroxyl group is eliminated during the alkaline saponification of dihydroxy ester **H** (not the usual conditions for alcohol dehydration reactions). The remainder of the synthetic sequence is the esterification of the carboxylic acid and the reduction of the product ester to yield racemic retronecine, as shown in Figure 20.3.

This synthesis leads to the *trans* isomer of retronecine with the correct relative configurations of its two chiral carbons. The product, however, is racemic. The natural product, (+)-retronecine, was separated from this racemic mixture by a resolution (see Section 4.11) using (+)-camphoric acid.

STUDY PROBLEM

20.10 Write equations showing the mechanism of the following conversion, shown at the start of the synthetic sequence in Figure 20.3.

$$C_2H_5O_2CCH_2CH_2NHCO_2C_2H_5 + \textit{trans-}C_2H_5O_2CCH{=}CHCO_2C_2H_5 \longrightarrow \mathbf{B}$$

Summary

Terpenes and **terpenoids,** found in both plants and animals, have carbon skeletons of diisoprene, triisoprene, etc.

A **pheromone** is a chemical excreted by an individual of a species (notably insects) that elicits a response in another individual of the same species. Generally of simple structure, pheromones are used to signify danger or food, or to act as sex attractants.

An **asymmetric synthesis** is the synthesis of a chiral compound from an achiral starting material. This type of synthesis requires a chiral catalyst, reagent, solvent, or auxiliary.

Alkaloids are acid-soluble, nitrogen-containing plant materials. Typical alkaloids are nicotine, morphine, and atropine.

Essay Problem for Chapter 20

Synthesis of Frontalin

Frontalin (**1**) is a component of the aggregation pheromone of southern and western pine beetles. Frontalin has two chiral carbons. The absolute configuration of the biologically active optical isomer is (1S,5R).

(1S,5R)-frontalin (**1**)   2,6-dimethyl-6-heptene-1,2-diol (**2**)

Two optical isomers of frontalin can be synthesized from the two enantiomers of 2,6-dimethyl-6-heptene-1,2-diol (**2**). Treatment of one enantiomer of **2** with ozone, followed by treatment of the ozonide with dimethyl sulfide yields an intermediate (**3**) that spontaneously cyclizes to (1S,5R)-frontalin (**1**). Treatment of the other enantiomer of **2** yields an enantiomer of **3** that yields the enantiomer of **1**. (J. K. Whitesell and C. M. Buchanan, *J. Org. Chem.* **1986,** *51,* 5443.)

$$\text{Compound } \mathbf{2} \xrightarrow[\text{(2) (CH}_3)_2\text{S}]{\text{(1) O}_3} \text{Compound } \mathbf{3} \longrightarrow (1S,5R)\text{-frontalin (}\mathbf{1}\text{)}$$

one optical isomer

Questions

1. Frontalin (1) has two chiral carbons and, therefore, should have four optical isomers. Can all four optical isomers exist? Explain your answer.
2. Carbon 5 of frontalin is bonded to how many oxygen atoms? What is the name of the functional group that has an sp^3 carbon bonded to two ether oxygens? What reactants do you need to synthesize this functional group?
3. What is the intermediate (3) that arises from the ozonolysis of (R)-2,6-dimethyl-6-heptene-1,2-diol (2)? What is the structure of its enantiomer? Draw both formulas showing their stereochemistry.
4. Draw a stereochemical equation that shows how 3 reacts to form (1S,5R)-frontalin.
5. Explain how, in this synthesis, it is possible to prepare one enantiomer of a compound that has two chiral carbons from a starting material that has only one chiral carbon—that is, why isn't a pair of diastereomers obtained?

Study Problems*

20.11 Identify each of the following as a monoterpene, sesquiterpene, or higher terpene:

(a)

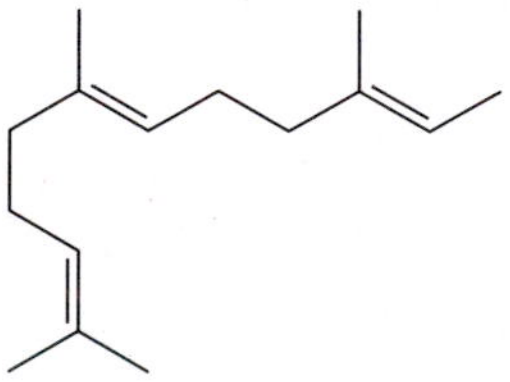

(b)

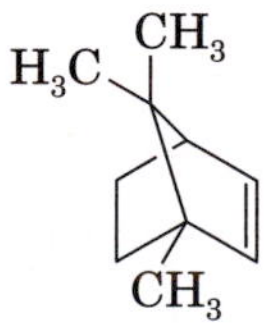

(c)

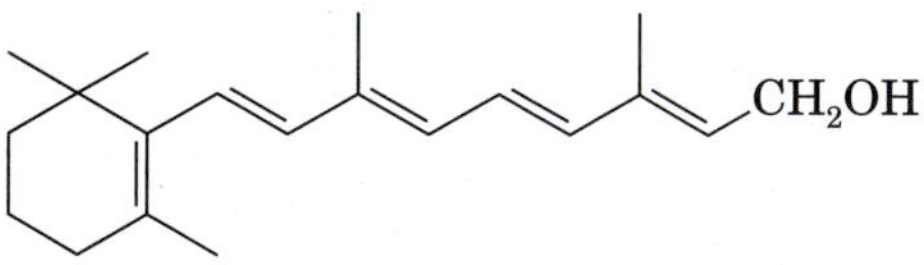

vitamin A

20.12 Tell whether the following compounds exemplify head-to-head, head-to-tail, or tail-to-tail isoprene units. Explain your answers.

(a) $CH_3CH_2CH(CH_2)_3CH(CH_3)_2$ with CH_3 branch

(b) $CH_3CH_2CH(CH_2)_2CHCH_2CH_3$ with CH_3, CH_3 branches

(c)

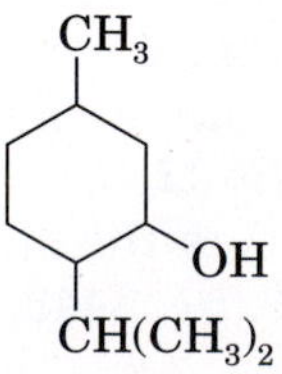

(d)

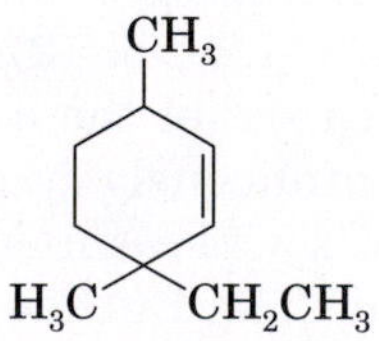

* For information concerning the organization of the *Study Problems* and *Additional Problems*, see the *Note to student* on page 41.

20.13 Draw the following terpenes in a chair or boat conformation:

(a)

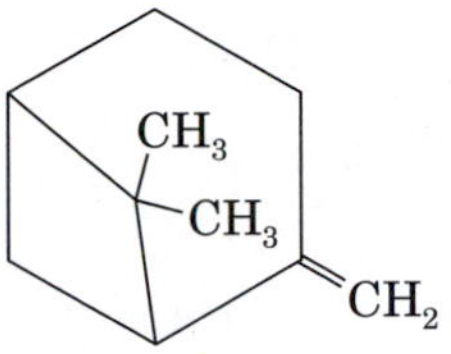

CH$_3$
—CH$_3$
CH$_2$

β-pinene

(b)

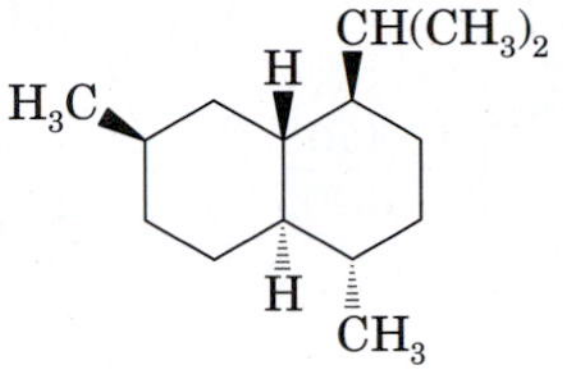

H$_3$C
H
CH(CH$_3$)$_2$
H
CH$_3$

(c)

CH$_3$
H$_3$C—CH$_3$
O

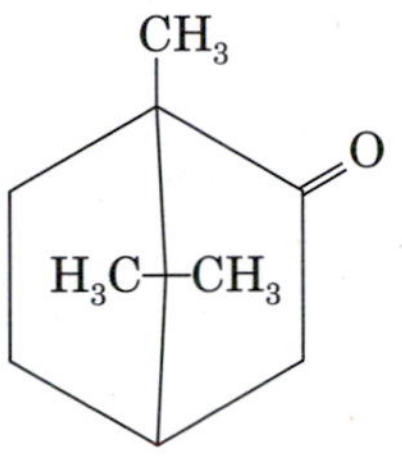

camphor

(d)

CH$_3$
O
O
CH(CH$_3$)$_2$

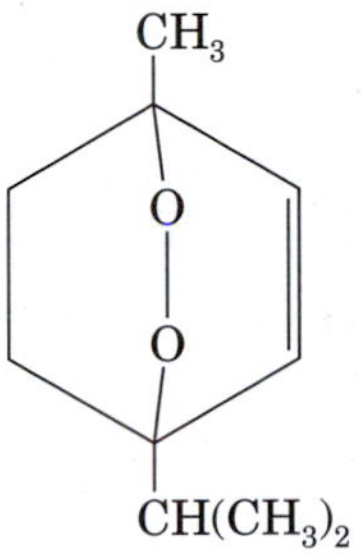

ascaridole

20.14 Complete the following equations:

(a) $CH_3CH_2C{\equiv}CH + Li^+\ {}^-N[CH(CH_3)_2]_2 \longrightarrow$

(b) $CH_3CH_2\overset{\displaystyle O}{\overset{\displaystyle \|}{C}}CH_3 + \overset{\displaystyle OH\ \ OH}{\underset{}{\overset{|\ \ \ \ |}{CH_2CH_2}}} \xrightarrow{H^+}$

(c) $CH_3CH{=}C(CH_3)_2 \xrightarrow[\text{cold}]{KMnO_4,\ {}^-OH}$

(d)   the product from **(c)** $\xrightarrow[\text{heat}]{KMnO_4,\ {}^-OH}$

(e)   the product from **(d)** $\xrightarrow[\text{heat}]{H^+}$

20.15 Write equations, showing starting materials and reagents, for the following conversions:

(a)   an aldehyde $\longrightarrow$ $CH_3CH_2CH{<}^{O-}_{O-}$

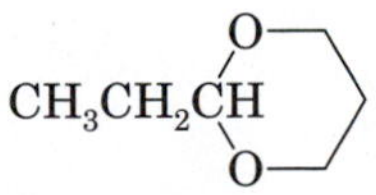

(b)   a hydrocarbon $\longrightarrow$

O
CH$_2$CH$_3$
CH$_3$

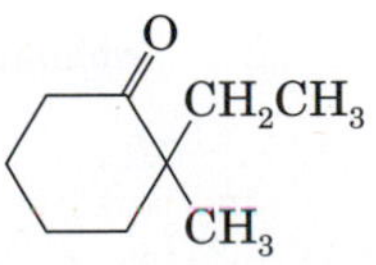

(c)   an alkyne $\longrightarrow$ $CH_3CH_2C{\equiv}C{:}^-\ Li^+$

20.16 Write flow equations to explain how the following conversions take place:

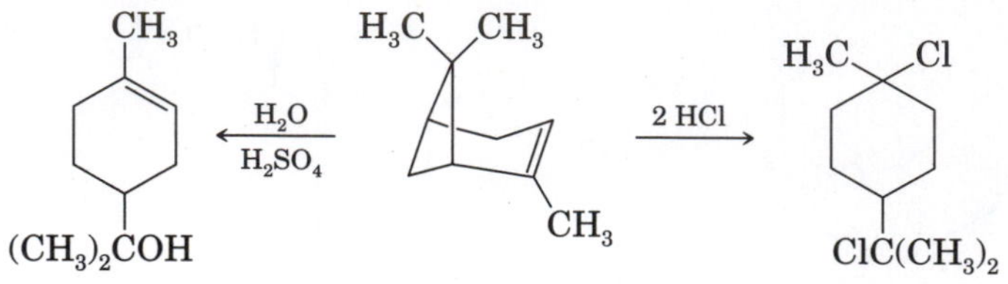

α-terpineol   α-pinene   limonene dihydrochloride

20.17 We have shown a synthesis in which chirality is induced by a reagent called SAMP (Section 20.2B). Another compound used to induce chirality is called RAMP. Suggest a complete name and formula for RAMP.

20.18 Give the structures of the products:

  (a) morphine + cold, dilute HCl ⟶

  (b) morphine + cold, dilute NaOH ⟶

  (c) codeine + cold, dilute NaOH ⟶

  (d) psilocin (page 794) + cold, dilute HCl ⟶

20.19 Circle and name the parent aromatic heterocyclic ring system in each of the following alkaloids.

  Example:

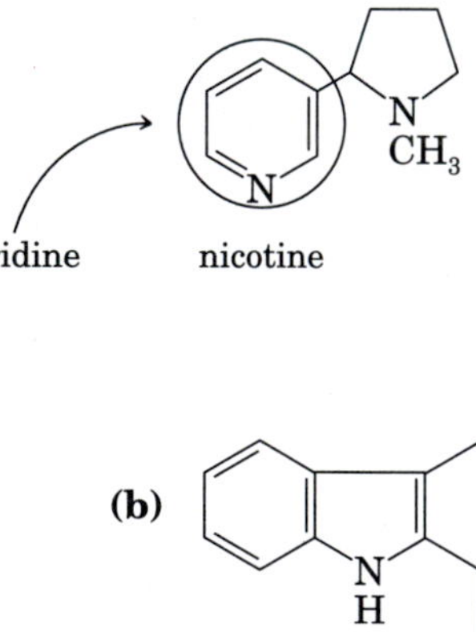

pyridine   nicotine

(a)

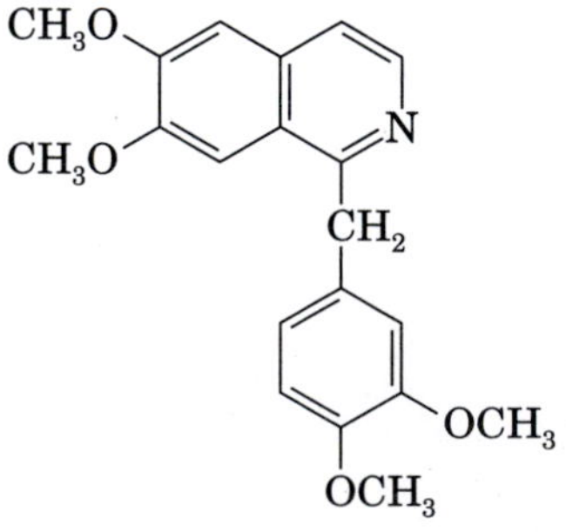

papaverine, found in opium

(b)

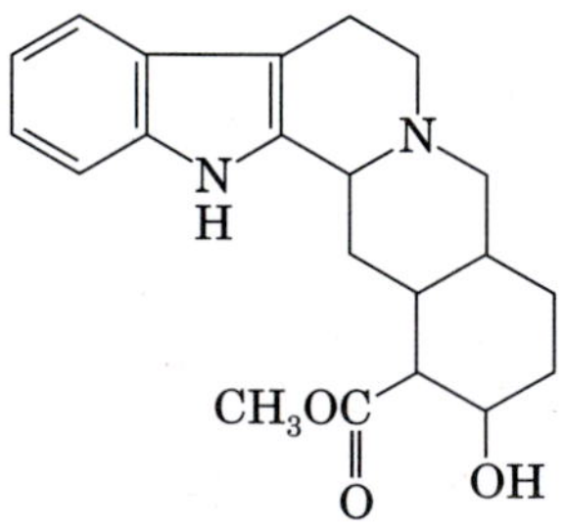

yohimbine, reputed to be an aphrodisiac

(c)

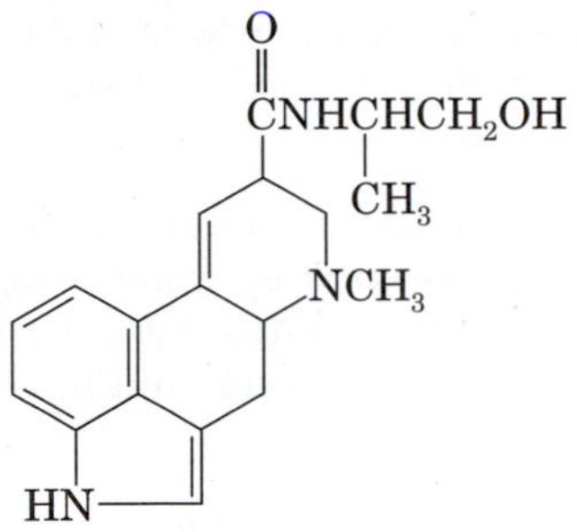

ergonovine, found in ergot

(d)

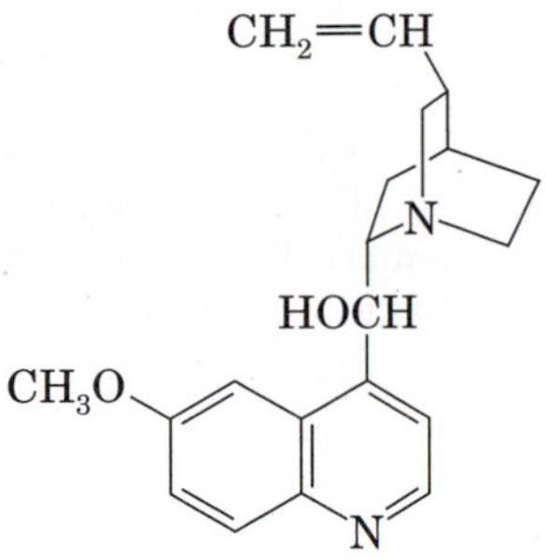

quinine, an antimalarial
from cinchona bark

Additional Problems

20.20 The structure of the aggregating pheromone of the bark beetle follows:

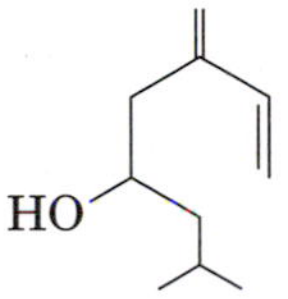

What are the organic products of its reaction with: **(a)** excess Br_2 in CCl_4; **(b)** excess gaseous HBr (no O_2 or peroxides); **(c)** hot $KMnO_4$ solution?

20.21 Many terpenes undergo rearrangements in the presence of acid. Suggest mechanisms for the following terpene rearrangements:

(a)

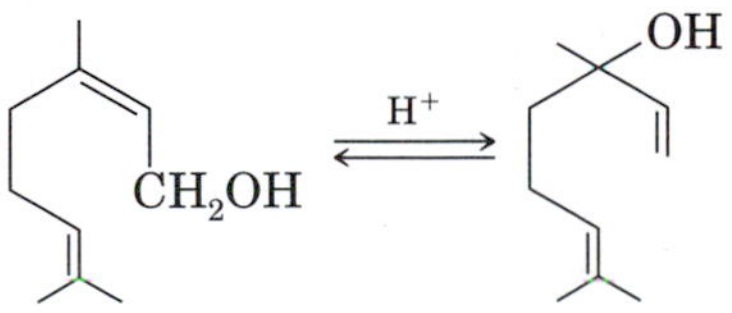

(b)

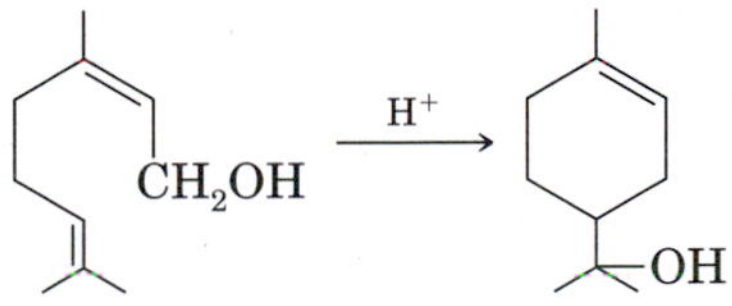

(c)

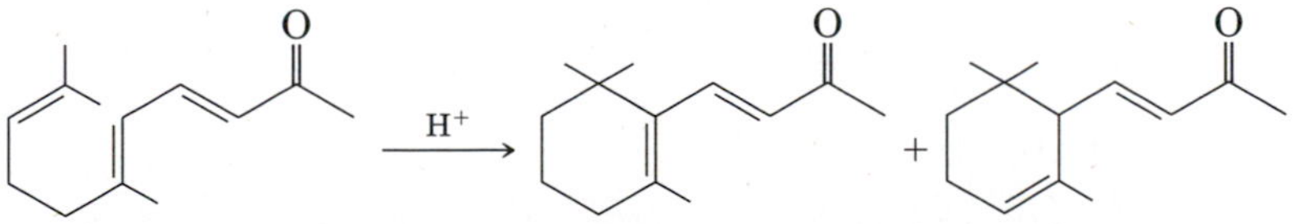

20.22 A sex attractant pheromone of the Douglas fir tussock moth was synthesized by the following route. The 11-carbon compound (Z)-5,10-undecadien-1-ol was oxidized to compound **A** with pyridinium chlorochromate, a reagent that oxidizes alcohols to aldehydes or ketones without oxidizing carbon–carbon double bonds. Compound **A** was treated with decylmagnesium bromide, followed by acidic hydrolysis to yield **B**. Compound **B** was subjected to pyridinium chlorochromate oxidation to yield the pheromone. Write a flow equation for the synthesis, showing the structures of **A**, **B**, and the pheromone.

20.23 *Citronellal* ($C_{10}H_{18}O$) is a terpenoid that undergoes reaction with Tollens reagent to yield citronellic acid ($C_{10}H_{18}O_2$). Chromate oxidation of citronellal yields acetone and $HO_2CCH_2CH(CH_3)CH_2CH_2CO_2H$. What is the structure of citronellal?

20.24 α-Terpinene ($C_{10}H_{16}$) is a lemon-scented compound that has been isolated from marjoram. Ozonolysis of α-terpinene, followed by oxidative work-up, yields oxalic acid and a $C_8H_{14}O_2$ compound. The latter compound can also be isolated from the reaction of 5-methyl-4-oxohexanoyl chloride (where oxo means a keto group) and lithium dimethylcuprate. What is the structure of α-terpinene?

20.25 The synthesis of the acyclic monoterpene *alloocimene* was accomplished by the following route. What is the structure of alloocimene? (Show the structures of the intermediate products in your answer.)

$$CH_3\overset{\overset{\displaystyle O}{\|}}{C}CH_2CH_3 \xrightarrow{NaC\equiv CH} \mathbf{A} \xrightarrow{C_2H_5MgBr} \mathbf{B} \xrightarrow[(2)\ H_2O]{(1)\ (CH_3)_2CHCHO}$$

$$\mathbf{C} \xrightarrow[\text{poisoned Pd}]{H_2} \mathbf{D} \xrightarrow[\text{heat}]{H^+} \text{alloocimene}$$

20.26 The terpenoid *pinol* ($C_{10}H_{16}O$) can be obtained by the treatment of the dibromide of α-terpineol with $NaOC_2H_5$. What is the structure of pinol? (*Hint:* Consider the stereochemistry of the intermediates.)

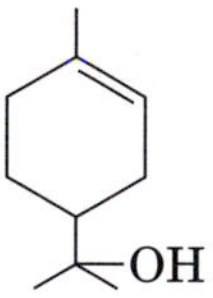

α-terpineol

20.27 Suggest a synthesis for α-terpineol from readily available, open-chain starting materials.

20.28 Complete the following equations. (Each of these reactions was carried out either in the synthesis or structure determination of an alkaloid.)

(a) 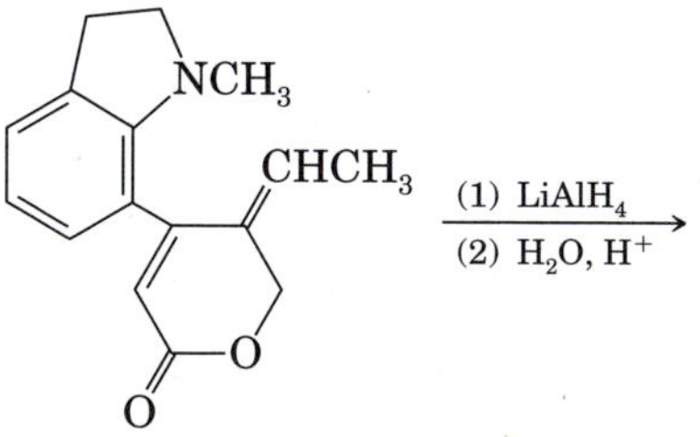

(b) 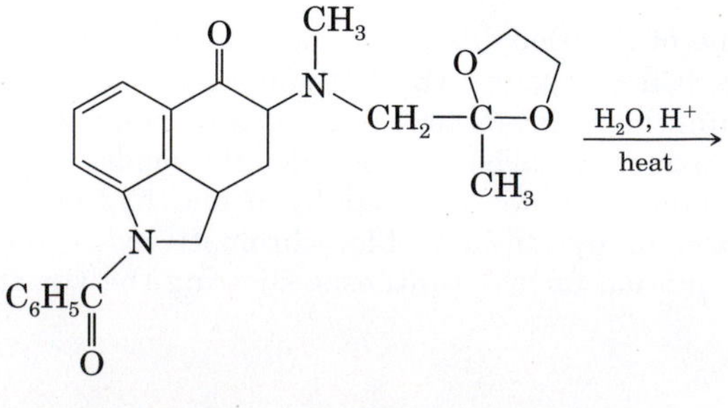

(c)

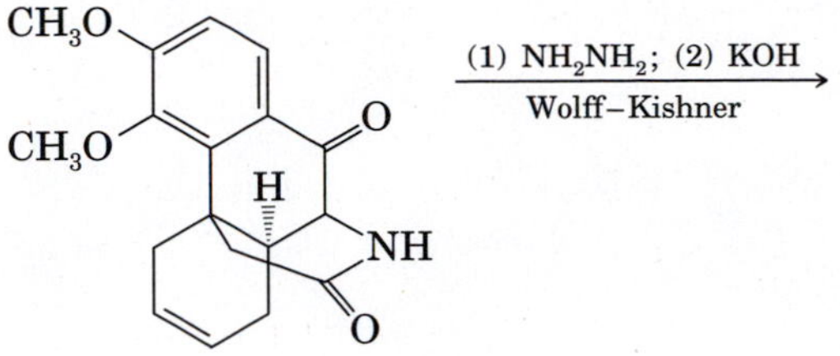

$$\xrightarrow[\text{Wolff–Kishner}]{\text{(1) NH}_2\text{NH}_2; \text{ (2) KOH}}$$

(d)

$$\xrightarrow[\text{(3) heat}]{\text{(1) CH}_3\text{I; (2) Ag}_2\text{O;}}$$

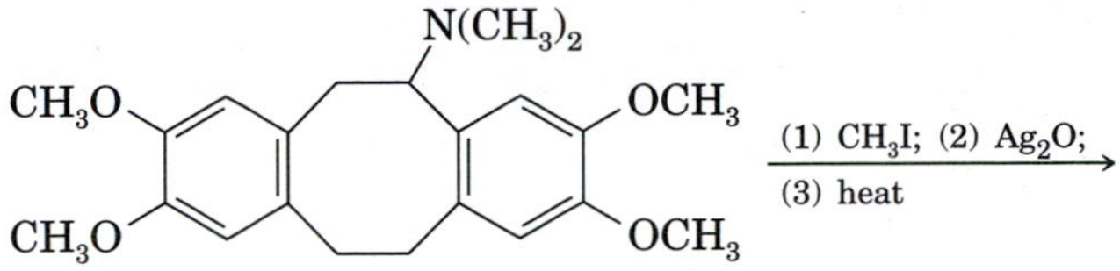

PERICYCLIC REACTIONS

Except for S_N2 and E2 reactions, most of the organic reactions we have discussed so far proceed *stepwise* by way of intermediates such as carbocations or free radicals. A large number of reactions of conjugated polyenes, called **pericyclic reactions** (from *peri*, "around" or "about"), proceed by concerted *(single-step)* mechanisms just as an S_N2 reaction does. That is, old bonds are broken as new bonds are formed, all in one step. Pericyclic reactions are characterized by a cyclic transition state involving the π bonds.

The energy of activation for pericyclic reactions is supplied by heat (**thermal induction**) or by ultraviolet light (**photo-induction**). (Solvents and electrophilic or nucleophilic reagents have little or no effect on the course of a pericyclic reaction.) Pericyclic reactions are generally stereospecific, and it is not uncommon that the two modes of induction yield products of opposite stereochemistry. For example, a thermally induced pericyclic reaction might yield a *cis* product, while the photo-induced reaction of the same reactant yields the *trans* product.

There are three principal types of pericyclic reactions:

1. **Cycloaddition reactions,** in which two molecules combine to form a ring. In these reactions two π bonds are converted to two σ bonds. The best-known example of a cycloaddition reaction is the Diels–Alder reaction, discussed in Section 16.3. "Line" formulas, such as those in the following equation, were discussed in Section 10.2B. Recall that a reactant must be in the s-*cis* (not s-*trans*) form to undergo cycloaddition.

1,3-butadiene	ethylene	cyclohexene

2. **Electrocylic reactions,** reversible reactions in which a compound with conjugated double bonds undergoes cyclization. In the cyclization, two π electrons are used to form a σ bond.

3. **Sigmatropic rearrangements,** concerted intramolecular rearrangements in which an atom or group of atoms shifts from one position to another.

For many years, a theoretical understanding of the mechanisms of pericyclic reactions eluded chemists. However, since 1960, several theories have been developed to rationalize these reactions. R. B. Woodward of Harvard University and R. Hoffmann of Cornell University have proposed explanations based upon the symmetry of the molecular orbitals of the reactants and products. Hoffmann received a Nobel prize in 1981 for this work. (Woodward received a Nobel prize in 1965 for his work on organic synthesis of complex compounds, including chlorophyll.) A similar treatment of pericyclic reactions has been developed by K. Fukui (Nobel prize, 1981) of Kyoto University in Japan. In this text, we will emphasize Fukui's approach, which is called the **frontier orbital method** of analyzing pericyclic reactions.

Before we discuss the mechanisms of pericyclic reactions, we will introduce some features of the molecular orbitals of conjugated systems. We suggest that you first review Sections 2.1–2.3 to refresh your understanding of bonding and antibonding molecular orbitals.

STUDY PROBLEM

21.1 Identify each of the following reactions as being (1) a cycloaddition; (2) an electrocyclic reaction; or (3) a sigmatropic rearrangement.

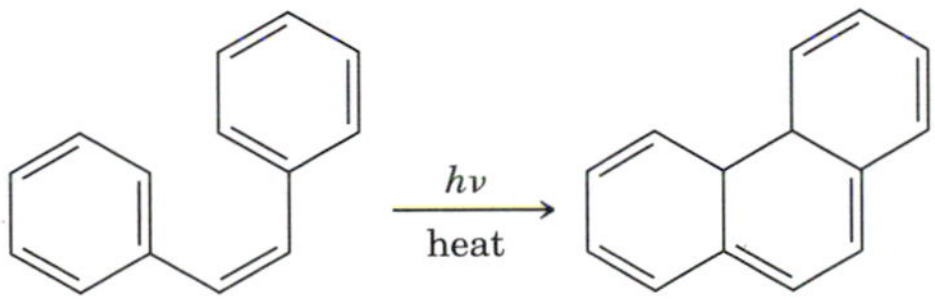

Section 21.1
Molecular Orbitals of Conjugated Polyenes

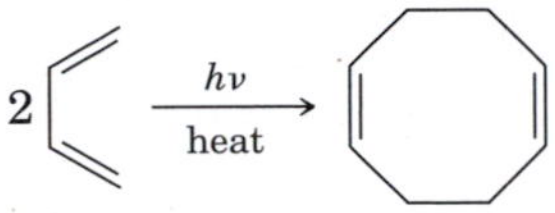

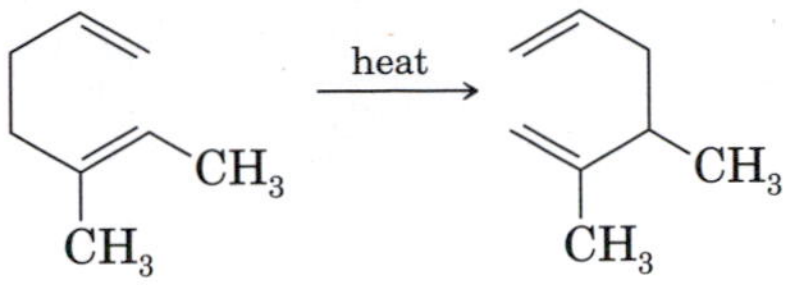

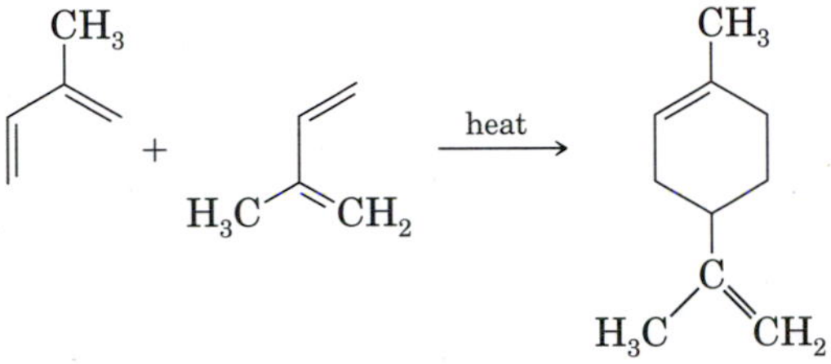

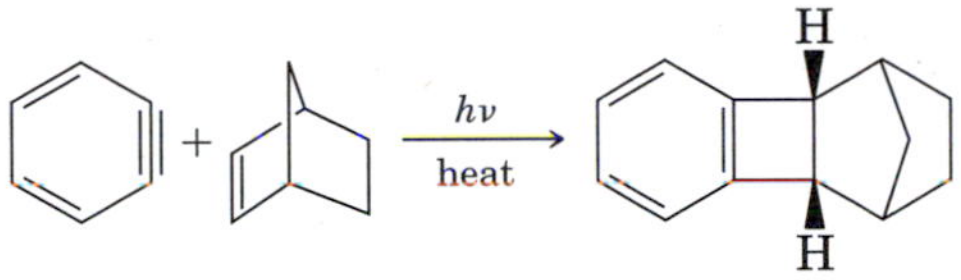

A conjugated polyene contains either $4n$ or $(4n + 2)$ π electrons, where n is an integer, in its conjugated system. The simplest $4n$ system is represented by 1,3-butadiene, where $n = 1$. Any conjugated diene contains π molecular orbitals similar to those of 1,3-butadiene. Therefore, we can use 1,3-butadiene as a model for all conjugated dienes.

In 1,3-butadiene, four p orbitals are used in the formation of the π molecular orbitals; thus, four π molecular orbitals result. In this system, π_1 and π_2 are the bonding orbitals and π_3^* and π_4^* are the antibonding orbitals. Figure 21.1 depicts these orbitals in terms of increasing energy. Note that the higher-energy molecular orbitals are those with a greater number of nodes between nuclei.

In the ground state, 1,3-butadiene has its four π electrons in the two orbitals of lowest energy: π_1 and π_2. In this case, π_2 is the **Highest Occupied Molecular Orbital, or HOMO,** and π_3^* is the **Lowest Unoccupied Molecular Orbital, or LUMO.** The HOMO and LUMO are referred to as **frontier orbitals** and are the orbitals used in the frontier orbital method of analyzing pericyclic reactions.

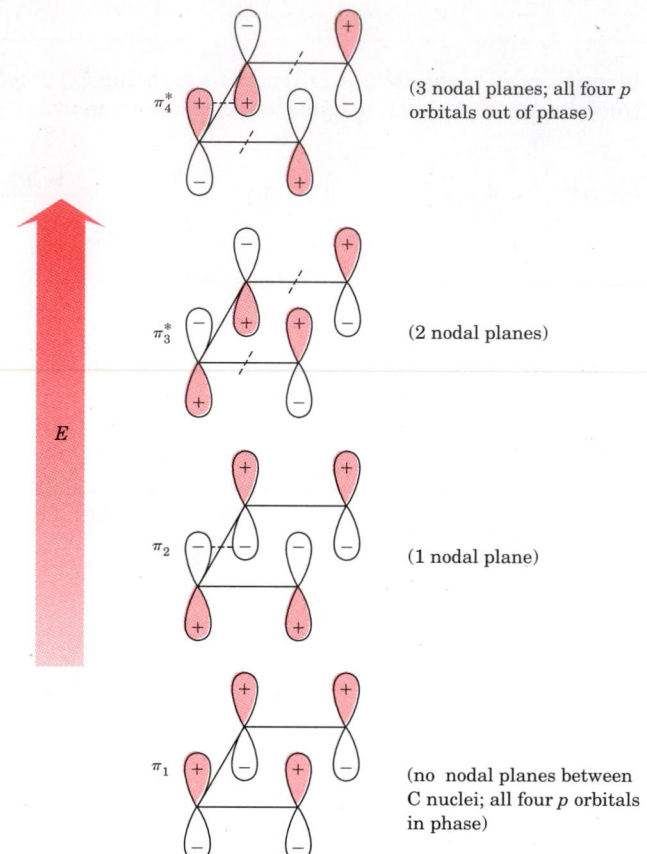

π_4^* (3 nodal planes; all four p orbitals out of phase)

π_3^* (2 nodal planes)

π_2 (1 nodal plane)

π_1 (no nodal planes between C nuclei; all four p orbitals in phase)

Figure 21.1 The bonding and antibonding π molecular orbitals of 1,3-butadiene, $CH_2=CHCH=CH_2$. The π_1 and π_2 orbitals are bonding orbitals; π_3^* and π_4^* are antibonding orbitals.

Ground state of 1,3-butadiene:

π_4^* ——

π_3^* —— ← *the LUMO*

π_2 ⇅ ← *the HOMO*

π_1 ⇅

When 1,3-butadiene absorbs a photon of the proper wavelength, an electron is promoted from the HOMO to the LUMO, which then becomes the new HOMO.

Excited state of 1,3-butadiene:

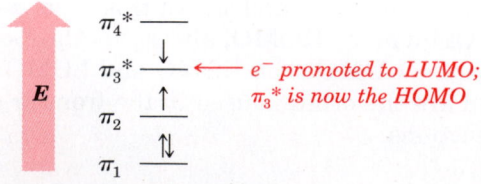

π_4^* ——

π_3^* ↓ ← *e⁻ promoted to LUMO; π_3^* is now the HOMO*

π_2 ↑

π_1 ⇅

Aside from ethylene ($n = 0$), the simplest ($4n + 2$) system is represented by a conjugated triene ($n = 1$), such as 1,3,5-hexatriene. Because a triene contains a π system formed from six p orbitals, a total of six π molecular orbitals results. These are depicted in Figure 21.2, along with the π orbital diagram of the ground state.

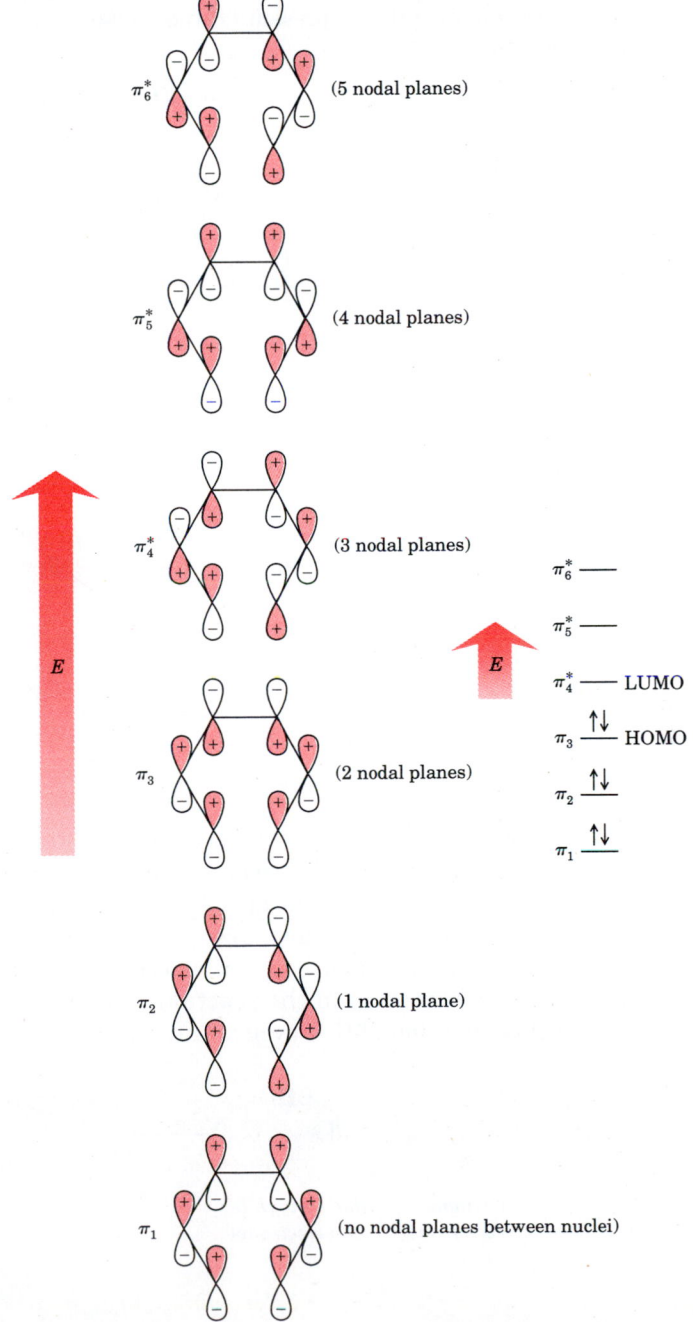

Figure 21.2   The bonding and antibonding π molecular orbitals of 1,3,5-hexatriene, $CH_2{=}CHCH{=}CHCH{=}CH_2$.

STUDY PROBLEMS

21.2 Draw the π orbital diagram for the lowest-energy *excited* state of 1,3,5-hexatriene.

21.3 Draw the π orbital diagram for the ground state of the nonconjugated diene 1,4-pentadiene.

21.4 Draw orbital pictures that represent (1) the HOMO and (2) the LUMO for each of the following π systems:

(a) $CH_2{=}CH_2$ (b) $CH_2{=}CHCH{=}CH_2$

(c) $CH_2{=}CHCH{=}CHCH{=}CH_2$ (d) $CH_2{=}CHCH{\overset{\displaystyle O}{\overset{\|}{}}}$

Section 21.2

Cycloaddition Reactions

A **cycloaddition reaction** is a reaction in which two unsaturated molecules undergo an addition reaction to yield a cyclic product. For example,

two π electrons   *two π electrons*

$$\underset{\text{ethylene}}{\begin{array}{cc} CH_2 & CH_2 \\ \| & \| \\ CH_2 & CH_2 \end{array}} \xrightarrow{h\nu} \underset{\text{cyclobutane}}{\begin{array}{c} CH_2{-}CH_2 \\ | \quad\quad | \\ CH_2{-}CH_2 \end{array}}$$

The cycloaddition of ethylene or any two simple alkenes is called a **[2 + 2] cycloaddition,** because *two π electrons + two π electrons* are involved. The Diels–Alder reaction is an example of a **[4 + 2] cycloaddition.** The diene contains four π electrons that are used in the cycloaddition, while the dienophile contains two. (The carbonyl π electrons in the following example are not used in bond formation in the reaction and therefore are not included in the number classification of this cycloaddition.)

the diene   the dienophile
(4 π electrons)   (2 π electrons)

$\text{(diene)} + \text{(dienophile, CHO)} \xrightarrow{\text{heat}} \text{(cyclohexene, CHO)}$

STUDY PROBLEM

21.5 Classify the following cycloaddition reaction by the number of π electrons involved:

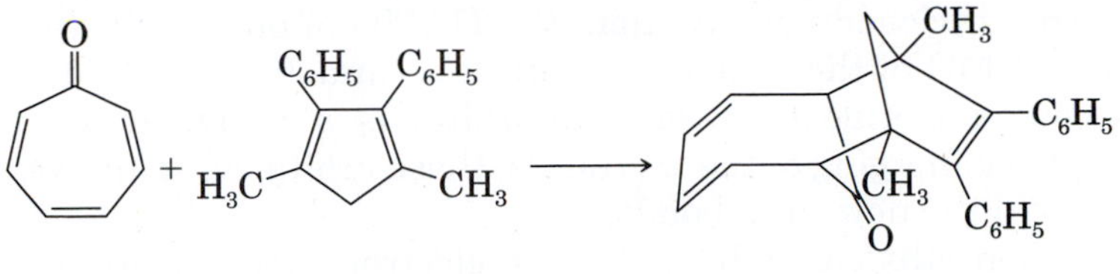

Cycloaddition reactions are concerted, stereospecific reactions. (See Section 16.3C for a discussion of the stereochemistry of the Diels–Alder reaction.) Also, any particular cycloaddition reaction is either thermally induced or photo-induced, but not both.

A. [2 + 2] Cycloaddition

Cycloaddition reactions of the [2 + 2] type proceed readily in the presence of light of the proper wavelength, but not when the reaction mixture is heated. This behavior is readily explained by the frontier orbital theory: by assuming that electrons "flow" from the HOMO of one molecule to the LUMO of the other.

Let us consider the [2 + 2] cycloaddition of ethylene to yield cyclobutane. Ethylene has two π molecular orbitals: π_1 and $\pi_2{}^*$. In the ground state, π_1 is the bonding orbital and the HOMO, while $\pi_2{}^*$ is the antibonding orbital and the LUMO.

Ethylene in the ground state:

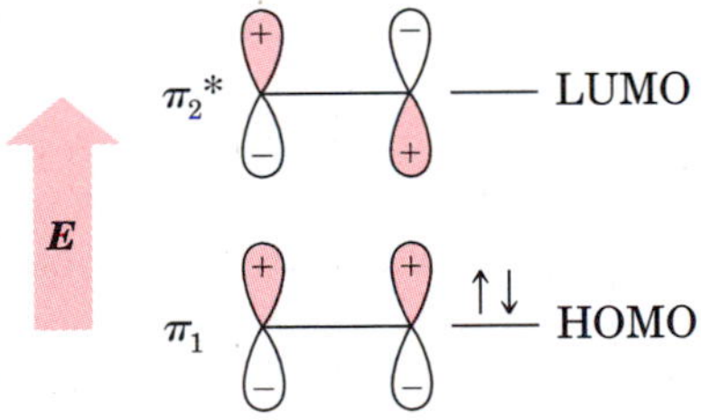

$\pi_2{}^*$ —— LUMO

E

π_1 $\uparrow\downarrow$ HOMO

SAMPLE PROBLEM

Draw the orbital diagram and the *p* orbitals for the lowest-energy *excited* state of ethylene, and indicate the HOMO.

Solution

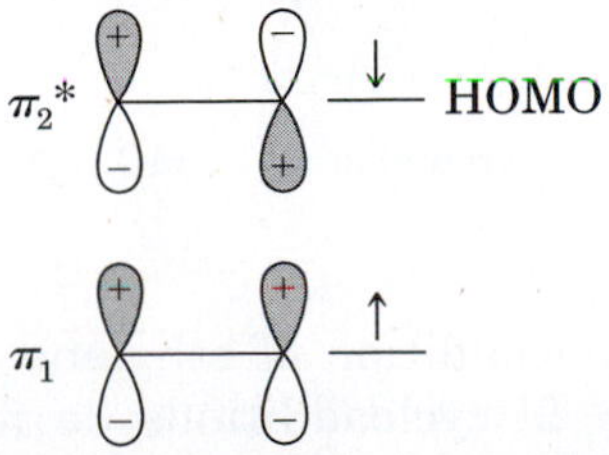

$\pi_2{}^*$ $\downarrow$ HOMO

π_1 $\uparrow$

In a cycloaddition reaction, the HOMO of one molecule must overlap with the LUMO of the second molecule. (It cannot overlap with the HOMO of the second molecule because that orbital is already occupied.) Simultaneously with the merging of the π orbitals, these orbitals also undergo rehybridization to yield the new sp^3 σ bonds.

When ethylene is heated, its π electrons are not promoted, but remain in the ground state, π_1. If we examine the phases of the ground-state HOMO of one ethylene molecule, and the LUMO of another ethylene molecule, we can see why cyclization does not occur by thermal induction.

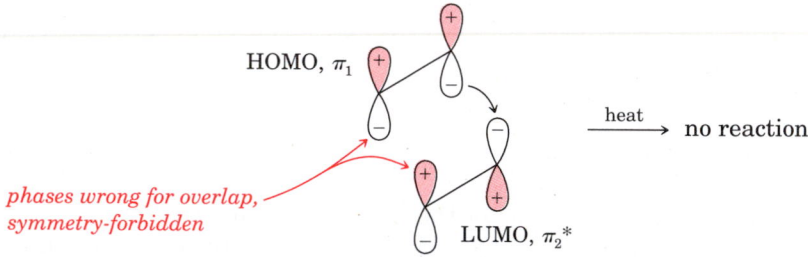

For bonding to occur, the phases of the overlapping orbitals must be the same. This is not the case for the ground-state HOMO and LUMO of two ethylene molecules or any other [2 + 2] system. Because the phases of the orbitals are incorrect for bonding, a thermally induced [2 + 2] cycloaddition is said to be a **symmetry-forbidden reaction.** A symmetry-forbidden reaction may occur under some circumstances, but the energy of activation would be very high—possibly so high that other reactions, such as free-radical reactions, would occur first.

When ethylene is irradiated with ultraviolet light, a π electron is promoted from the π_1 to the π_2^* orbital in some, but not all, of the molecules. The result is a mixture of the ground-state and excited-state ethylene molecules. If we examine the HOMO of an excited molecule (π_2^*) and the LUMO of a ground-state molecule (also π_2^*), we see that the phases are now correct for bonding. Such a reaction has a relatively low energy of activation, and is said to be **symmetry-allowed.**

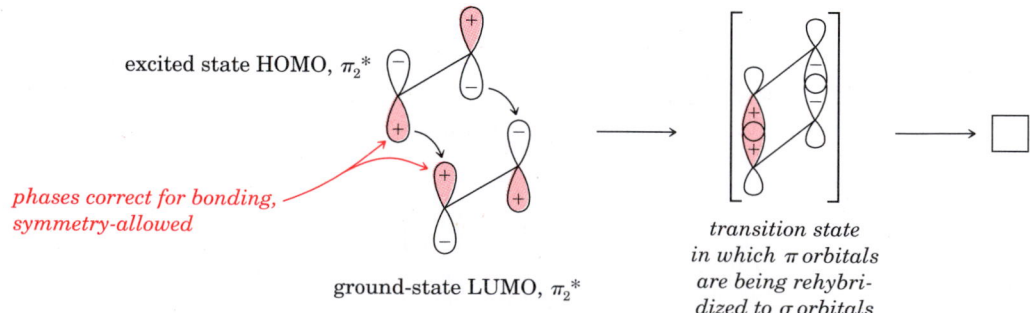

Although the cycloaddition of ethylene itself proceeds in poor yield, other photo-induced [2 + 2] cycloadditions do find synthetic utility. Probably the widest use of this type of reaction is with intramolecular cyclizations, which can yield some very unusual "cage" structures.

(74%)

(80%)

Study Problems

21.6 Suggest synthetic routes to the following compounds from open-chain starting materials.

(a)   **(b)**

21.7 When irradiated, tetrafluoroethene undergoes a cycloaddition reaction with 1-buten-3-yne to yield two products in almost equal amounts. What are the structures of these products?

B. [4 + 2] Cycloaddition

As we have mentioned, the Diels–Alder reaction is the best-known [4 + 2] cycloaddition. The examples on page 700 illustrate the versatility of this reaction. Note that the Diels–Alder reaction requires heat, not ultraviolet light, for its success. This experimental condition is different from that required for a [2 + 2] cycloaddition. To see why this is so, we will examine the HOMO–LUMO interactions of *only the p-orbital components that will form the new σ bonds* in a [4 + 2] cycloaddition. We will compare the HOMO–LUMO interactions for the ground state (for a thermally induced reaction) and those for the excited state (for an attempted photo-induced reaction). Based upon experimental observations, we would expect to find that the HOMO–LUMO interactions of the thermally induced reaction are symmetry-allowed and those of the photo-induced reaction are symmetry-forbidden.

We will use the simplest [4 + 2] system: the cycloaddition of 1,3-butadiene (the diene) and ethylene (the dienophile). The frontier orbital pictures may be extrapolated to other [4 + 2] cycloadditions. In the thermally induced reaction, we can visualize the π electrons "flowing" from the HOMO (π_2) of the diene (Figure 21.1) to the LUMO (π_2*) of the dienophile. Note the phases of the orbitals that lead to the thermally induced reaction. This reaction is symmetry-allowed.

When a diene is excited by light, its HOMO becomes the $\pi_3{}^*$ orbital, and this molecular orbital cannot overlap with the LUMO of the dienophile. The photo-induced [4 + 2] cyclization is therefore symmetry-forbidden.

HOMO, $\pi_3{}^*$

⟶ no reaction

symmetry-forbidden

LUMO, $\pi_2{}^*$

STUDY PROBLEMS

21.8 Predict whether a [4 + 2] cycloaddition could be photo-induced if the dienophile, instead of the diene, were the excited reactant. Explain your answer.

21.9 Is the cycloaddition reaction in Problem 21.5 a thermally induced or photo-induced reaction? Explain.

Section 21.3
Electrocyclic Reactions

An **electrocyclic reaction** is the concerted interconversion of a conjugated polyene and a cycloalkene. We will discuss primarily the cyclization. The reverse reaction, ring opening, proceeds by the same mechanism, but in reverse.

Electrocyclic reactions are induced either thermally or photochemically:

$$
\begin{array}{c}
\text{CH}_2 \\
\text{CH} \\
| \\
\text{CH} \\
\text{CH}_2
\end{array}
\quad \underset{}{\overset{\text{heat or } h\nu}{\rightleftharpoons}} \quad
\begin{array}{c}
\text{CH} - \text{CH}_2 \\
|| \quad\quad | \\
\text{CH} - \text{CH}_2
\end{array}
\qquad
\begin{array}{c}
\text{CH} \\
\text{CH} \quad \text{CH}_2 \\
|| \quad\quad | \\
\text{CH} \quad \text{CH}_2 \\
\text{CH}
\end{array}
\quad \underset{}{\overset{\text{heat or } h\nu}{\rightleftharpoons}} \quad
\begin{array}{c}
\text{CH} \\
\text{CH} \quad \text{CH}_2 \\
| \quad\quad | \\
\text{CH} \quad \text{CH}_2 \\
\text{CH}
\end{array}
$$

1,3-butadiene   cyclobutene   1,3,5-hexatriene   1,3-cyclohexadiene

An intriguing feature about electrocyclic reactions is that the stereochemistry of the product is dependent on whether the reaction is thermally induced or photo-induced. For example, when (2E,4Z,6E)-octatriene is heated, the *cis*-dimethylcyclohexadiene is the product. When the triene is irradiated with ultraviolet light, however, the *trans*-dimethylcyclohexadiene is formed. We will discuss the reasons for this in Section 21.3B.

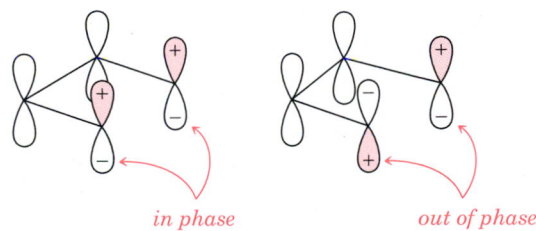

cis-5,6-dimethyl-1,3-cyclohexadiene

(2E,4Z,6E)-octatriene

trans-5,6-dimethyl-1,3-cyclohexadiene

A. Cyclization of 4n Systems

A conjugated polyene yields a cycloalkene by the end-to-end overlap of its p orbitals and the simultaneous rehybridization of the carbon atoms involved in bond formation. 1,3-Butadiene, which has $4n$ π electrons, is the simplest polyene. Therefore, we will introduce the mechanism with this compound. Because cyclobutanes are strained, the reverse reaction (ring opening) is usually favored. However, the mechanism for ring opening is simply the reverse of that for ring closing.

The two lobes of each p orbital that will form the new σ bond in cyclization are either *in phase* or *out of phase* with each other:

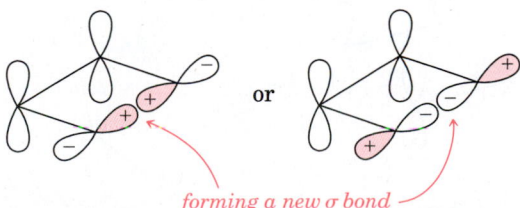

in phase   *out of phase*

To form a new σ bond, the existing C—C σ bonds must rotate so that the p orbitals can undergo end-to-end overlap. For this to occur, the existing π bonds must be broken. The energy for the π-bond breakage and the bond rotation is supplied by heat or ultraviolet light. To form a σ bond, the pair of overlapping lobes of the two p orbitals must be *in phase* after rotation.

forming a new σ bond

There are two different ways in which the existing C—C σ bonds can rotate in order to position the p orbitals for overlap. (1) The two C—C σ bonds can rotate in the *same direction* (either both clockwise or both counterclockwise). This type of rotation is referred to as **conrotatory motion.** (2) The two C—C σ bonds can rotate in *different* directions, one clockwise and one counterclockwise. This type of rotation is **disrotatory motion.**

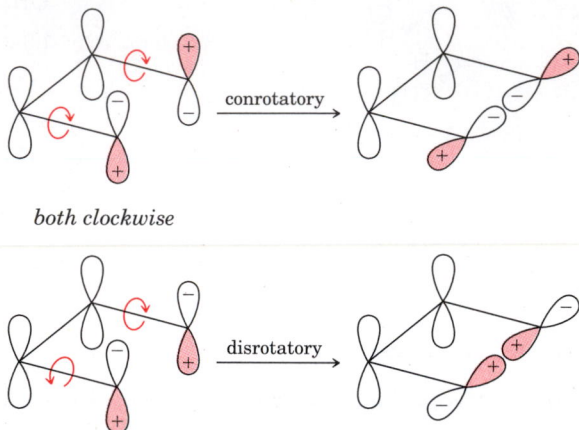

both clockwise

different directions

Note that in the two preceding equations, the phases of the p orbitals in the two starting dienes are different. Therefore, the direction of rotation for symmetry-allowed overlap depends upon the phases of the p orbitals just prior to cyclization. If the p orbitals are out of phase before rotation, then conrotatory motion brings them into phase after rotation. If the p orbitals are in phase before rotation, then disrotatory motion is required. To determine which diene system is present just prior to reaction, we must consider the phases of the p orbitals in the ground state and the excited state of the diene.

When 1,3-butadiene is *heated,* reaction takes place from the ground state. The electrons that are used for σ-bond formation are in the HOMO (π_2, in this case). In Figure 21.1, it can be seen that the pertinent p orbitals in this HOMO are out of phase with each other. For the new σ bond to form, rotation must be *conrotatory.* Only in this way are the in-phase lobes allowed to overlap. (Disrotatory motion would not place the in-phase lobes together.)

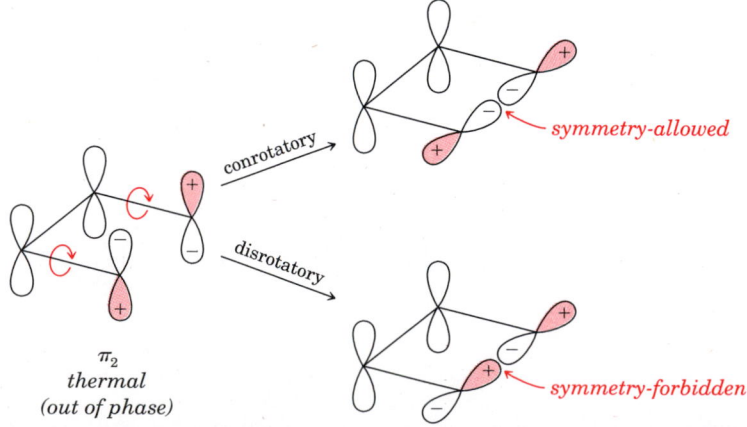

π_2
thermal
(out of phase)

In *photo-induced* cyclization, the phases of the p orbitals of the HOMO (now $\pi_3{}^*$) are the reverse of that in thermal cyclization (see Figure 21.1).

Therefore, the symmetry-allowed rotation is *disrotatory* instead of conrotatory.

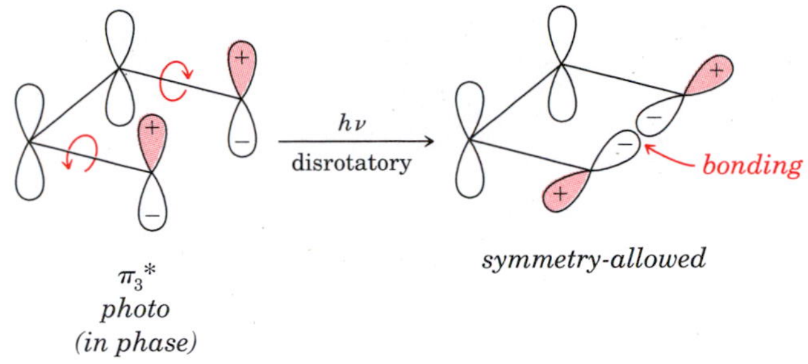

π_3^*
photo
(in phase)

symmetry-allowed

STUDY PROBLEM

21.10   Draw structures showing *conrotatory* motion of 1,3-butadiene in the excited state (π_3^*). Are the potential bonding p orbitals in a symmetry-allowed or symmetry-forbidden orientation?

B. Stereochemistry of a 4*n* Electrocyclization

If (2*E*,4*Z*)-hexadiene were cyclized, the *cis*-dimethylcyclobutane would result from heating. In thermal cyclization, conrotatory motion is required for bond formation. Both methyl groups rotate in the same direction. As a result, they end up on the same side of the ring, or *cis,* in the product.

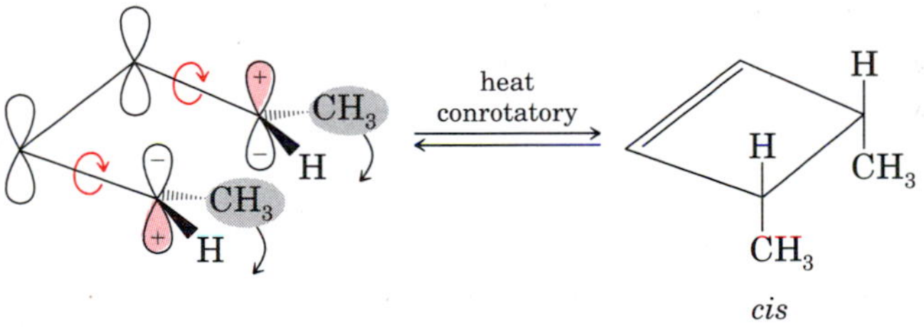

cis

Just the reverse would occur in photochemical cyclization. In the disrotatory motion, one of the methyl groups rotates up and the other rotates down. The result is that the two methyl groups are *trans* in the product.

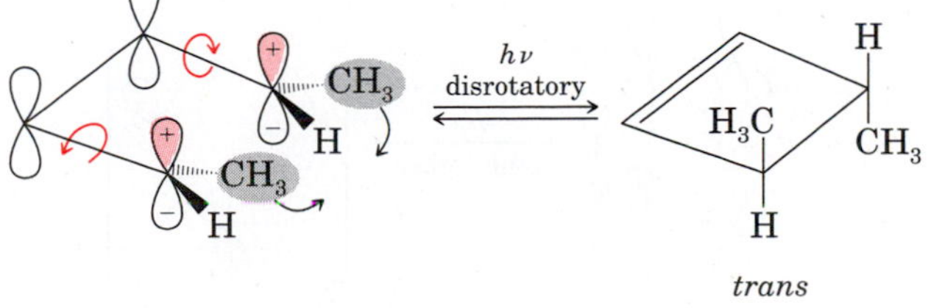

trans

SAMPLE PROBLEM

Would the photochemical electrocyclic reaction of (2*E*,4*E*)-hexadiene yield *cis*- or *trans*-3,4-dimethylcyclobutene?

Solution

2,4-Hexadiene is a $4n$ polyene. Therefore, the photochemical electrocyclic reaction takes place by disrotatory motion.

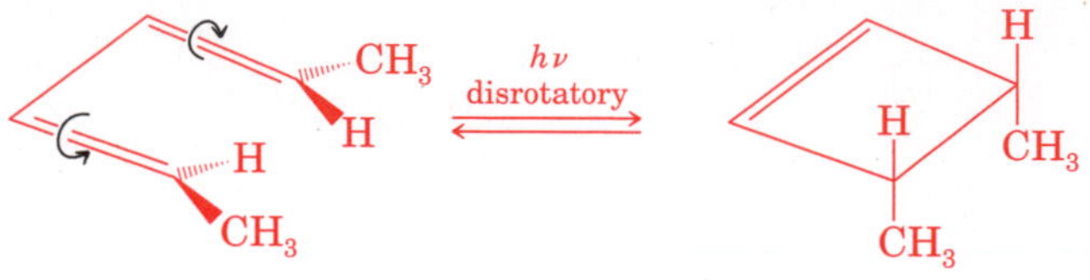

(2*E*,4*E*)-hexadiene   *cis*-3,4-dimethylcyclobutene

21.11   What is the structure and expected stereochemistry of the ring-opened product when *trans*-3,4-dimethylcyclobutene is heated?

C. Cyclization of (4n + 2) Systems

Figure 21.2 shows the π orbitals of 1,3,5-hexatriene, a $(4n + 2)$ polyene. In the HOMO of the ground state (π_3), the p orbitals that form the σ bond in the cyclization are in phase. Therefore, the thermal cyclization proceeds by *disrotatory motion*.

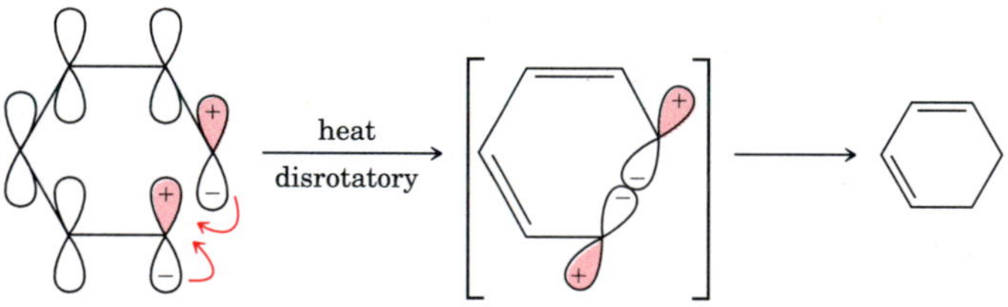

When an electron of 1,3,5-hexatriene is promoted by photon-absorption, π_4^* becomes the HOMO and thus the p orbitals in question become out of phase. Therefore, photo-induced cyclization proceeds by *conrotatory motion*. The symmetry-allowed reactions of this $(4n + 2)$ system are just the opposite of those for 1,3-butadiene, a $4n$ system.

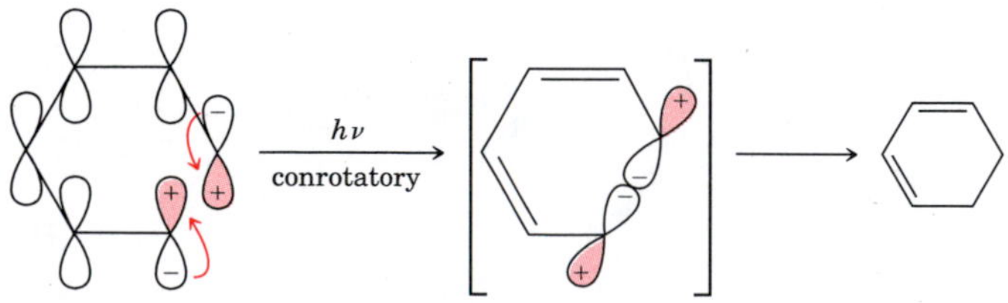

A summary of the types of motion to be expected from the different types of polyenes under the influence of heat and ultraviolet light is shown in Table 21.1.

Table 21.1	Types of electrocyclic reactions	
Number of π Electrons	Reaction	Motion
$4n$	thermal	conrotatory
$4n$	photochemical	disrotatory
$(4n + 2)$	thermal	disrotatory
$(4n + 2)$	photochemical	conrotatory

STUDY PROBLEM

21.12 Without referring to the start of this section, predict the stereochemistry of the products.

(a)

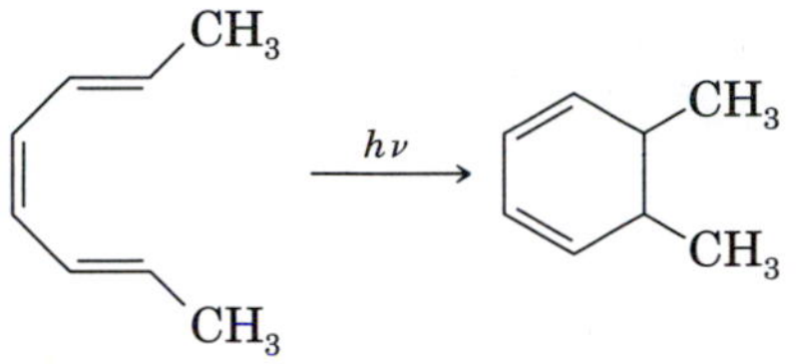

(b)

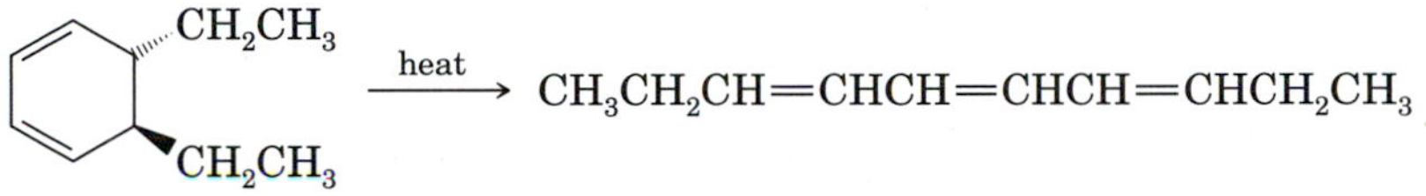

The thermally induced electrocyclic reactions of (2*E*,4*Z*,6*Z*,8*E*)-decatetraene provide elegant examples of electrocyclic reactions. The starting tetraene forms a cyclooctatriene near room temperature. The tetraene is a $4n$ polyene. Therefore, a conrotatory motion is the expected mode of cyclization. Indeed, the *trans*-dimethylcyclooctatriene is the product of this initial cyclization. When this cyclooctatriene is heated to a slightly higher temperature, another electrocyclic ring closure occurs. However, the cyclooctatriene is a $(4n + 2)$ polyene. Therefore, this thermally induced electrocyclic reaction proceeds with disrotatory motion, and a *cis* ring junction is formed.

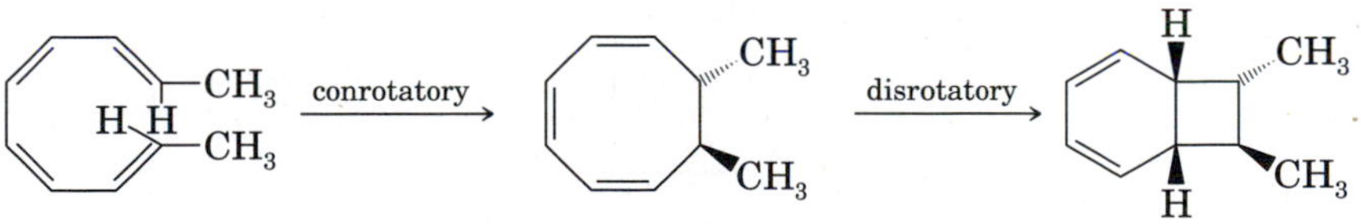

trans methyl groups   *cis ring junction*

Section 21.4
Sigmatropic Rearrangements

A **sigmatropic rearrangement** is a concerted intramolecular shift of an atom or a group of atoms. Two typical examples of sigmatropic rearrangements are:

Cope rearrangement:

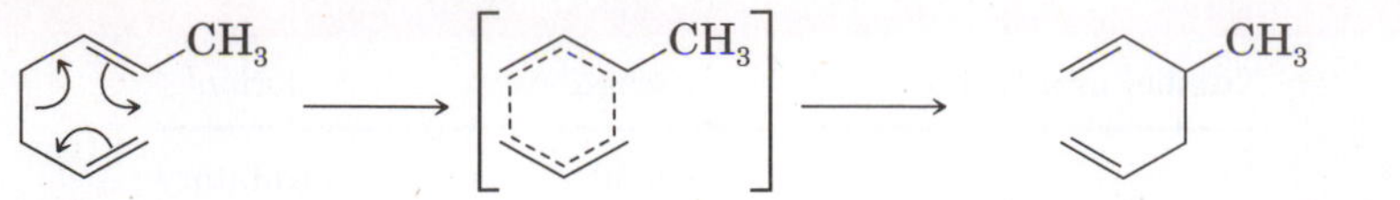

1,5-heptadiene   *transition state*   3-methyl-1,5-hexadiene

Claisen rearrangement:

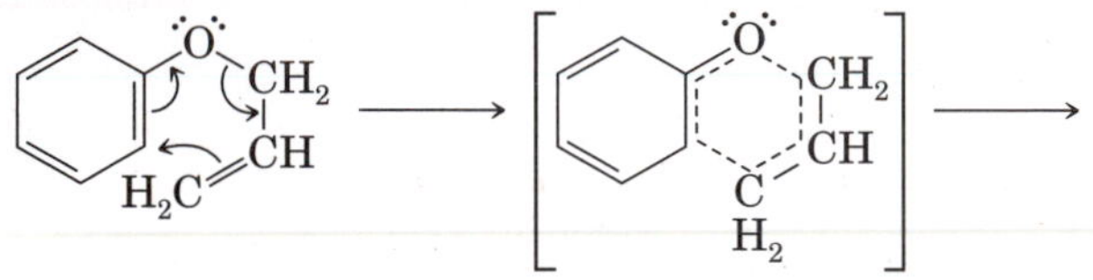

allyl phenyl ether   *transition state*

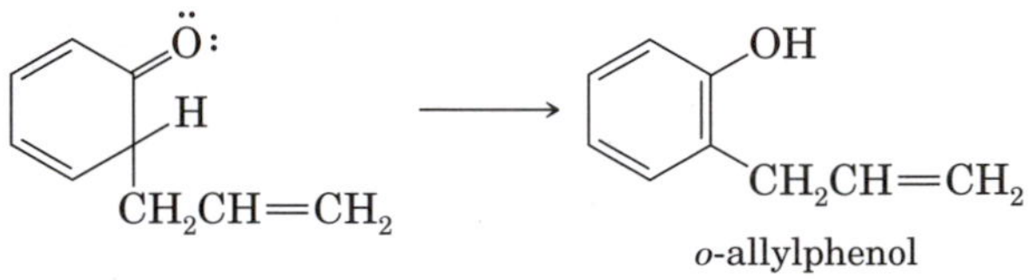

     keto form   *o*-allylphenol

                                   enol form

A. Classification of Sigmatropic Rearrangements

Sigmatropic rearrangements are classified by a double numbering system that refers to the relative positions of the atoms involved in the migration. This method of classification is different from those for cycloadditions or electrocyclic reactions, which are classified by the number of π electrons involved in the cyclic transition state.

The method used in classifying sigmatropic reactions is best explained by example. Consider the following rearrangement:

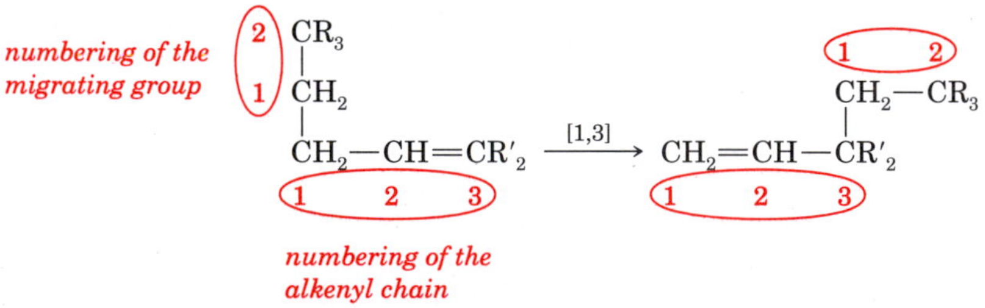

numbering of the migrating group

numbering of the alkenyl chain

Both the alkenyl chain and the migrating group are numbered *starting at the position of their original attachment,* not necessarily at a carbon atom. (Note that these numbers are not related to nomenclature numbers.) In the example, atom 1 of the migrating group ends up on atom 3 of the alkenyl chain. Therefore, this sigmatropic rearrangement would be classified as a [1,3] sigmatropic rearrangement.

In a similar manner, the following reaction would be classified as a [1,7] sigmatropic shift. (In this example, there is no atom 2 in the migrating group.)

$$\overset{\textcircled{1}H}{\underset{\textcircled{1}}{|}}CH_2CH=CHCH=CHCH=\underset{\textcircled{7}}{C}D_2 \xrightarrow{[1,7]} CH_2=CHCH=CHCH=CH\overset{H}{\underset{|}{C}}D_2$$

It is not always the first atom of the migrating group that becomes bonded to the alkenyl chain in the rearrangement. Consider the following example. In this case, atom 3 of the migrating group becomes bonded to atom 3 of the alkenyl chain. This is an example of a [3,3] sigmatropic rearrangement.

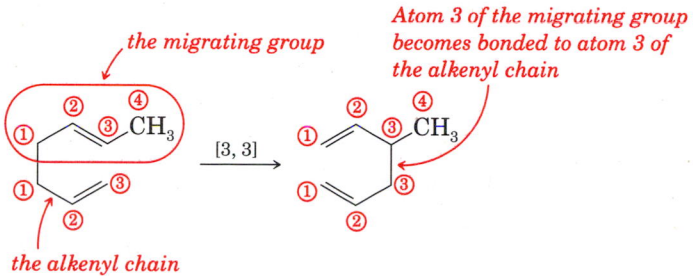

the migrating group

Atom 3 of the migrating group becomes bonded to atom 3 of the alkenyl chain

[3, 3]

the alkenyl chain

STUDY PROBLEM

21.13   Classify the following rearrangements by the preceding technique:

(a) the Claisen rearrangement

(b) $CH_3CH_2CH=CHCH=CDCH_3 \longrightarrow CH_3CH=CHCH=CHCHDCH_3$

B. Mechanism of Sigmatropic Rearrangements

Sigmatropic rearrangements of the [1,3] type are relatively rare, while [1,5] sigmatropic rearrangements are fairly common. We can use the frontier orbital approach to analyze these reactions and see why this is so. Let us first consider the following thermally induced sigmatropic rearrangement, which is a [1,3] shift:

$$\overset{H}{\underset{|}{C}}H_2CH=CD_2 \xrightarrow{\text{difficult}} CH_2=CH\overset{H}{\underset{|}{C}}D_2$$

For the purpose of analyzing the orbitals, it is assumed that the sigma bond connecting the migrating group to its original position (the CH bond in our example) undergoes homolytic cleavage to yield two radicals. This is *not* how the reaction takes place (the reaction is concerted), but this assumption does allow analysis of the molecular orbitals.

$$\overset{H}{\underset{|}{C}}H_2CH=CD_2 \xrightarrow[\text{homolytic cleavage}]{\text{hypothetical}} \overset{H\cdot}{} \cdot CH_2CH=CD_2$$

allyl radical

The products of the hypothetical cleavage are a hydrogen atom and an allyl radical, which contains three π electrons and thus three π molecular orbitals. The π molecular orbitals of the allyl radical are shown in Figure 21.3.

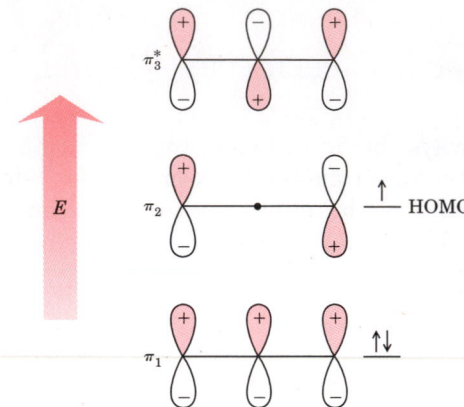

Figure 21.3 The three π molecular orbitals of an allyl radical. (Note that π_2 contains one node at carbon 2.)

The actual shift of the $H \cdot$ could take place in one of two directions. In the first case, the migrating group could remain on the same side of the π orbital system. Such a migration is termed a **suprafacial process.** As you can see, in this system a suprafacial migration is geometrically feasible but symmetry-forbidden.

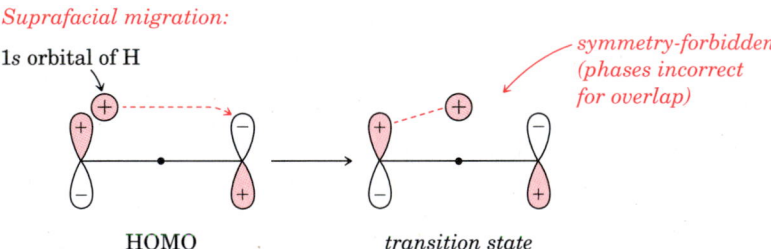

Let us consider the second mode of migration. For a symmetry-allowed [1,3] sigmatropic shift to occur, the migrating group ($H \cdot$ in our example) must shift by an **antarafacial process**—that is, it must migrate to the *opposite face* of the orbital system.

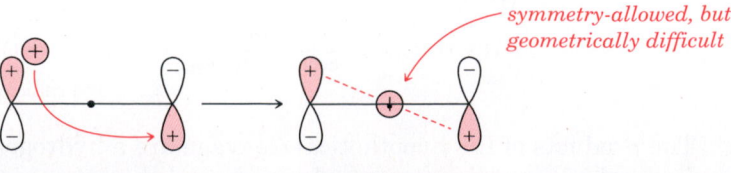

While symmetry-allowed, a [1,3] antarafacial sigmatropic rearrangement of H is not geometrically favorable. Our conclusion is that [1,3] sigmatropic shifts should not occur readily. This conclusion is in agreement with experimental facts. As we have mentioned, [1,3] sigmatropic rearrangements are rare.

By contrast, [1,5] sigmatropic shifts are quite common. A simple example follows:

$$\overset{\overset{\displaystyle H}{\displaystyle |}}{CH_2}CH=CHCH=CD_2 \xrightarrow{[1,5]} CH_2=CHCH=\overset{\overset{\displaystyle H}{\displaystyle |}}{CH}CD_2$$

If we again assume a homolytic bond cleavage for purposes of analysis, we must consider the π molecular orbitals of a pentadienyl radical, which contains five π electrons. These orbitals are depicted in Figure 21.4.

$$\overset{\overset{\displaystyle H}{\displaystyle |}}{CH_2}CH=CHCH=CH_2 \xrightarrow[\text{homolytic cleavage}]{\text{hypothetical}} \overset{\displaystyle H\cdot}{\cdot CH_2CH=CHCH=CH_2}$$

pentadienyl radical

Considering the HOMO of this radical and the orbital symmetry, we can see that the [1,5] shift is both symmetry-allowed and suprafacial.

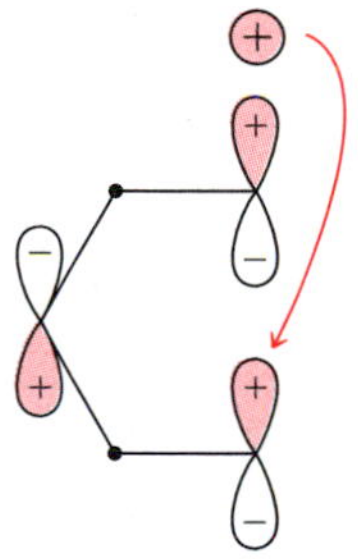

The [1.5] suprafacial shift is symmetry-allowed.

STUDY PROBLEM

21.14 Which of the following known sigmatropic rearrangements would proceed readily and which slowly? Explain your answers.

(a)

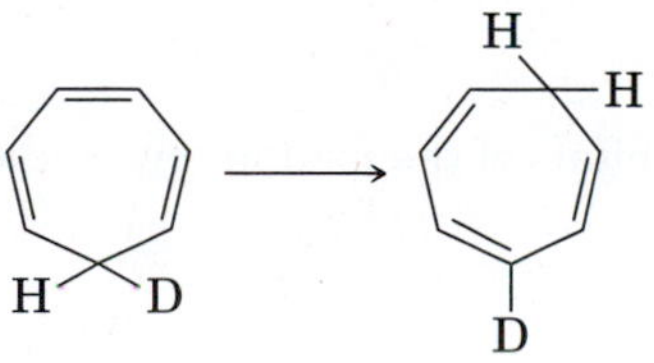

(b)

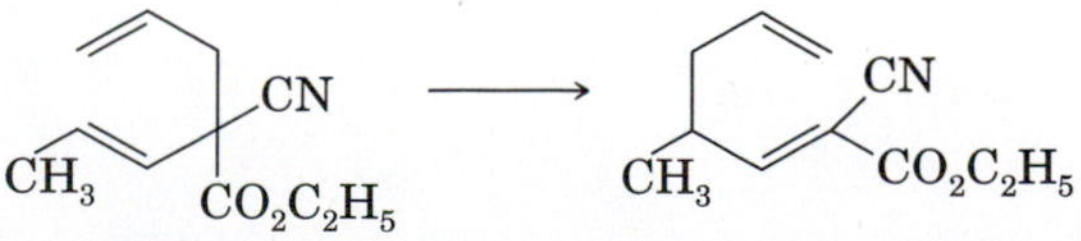

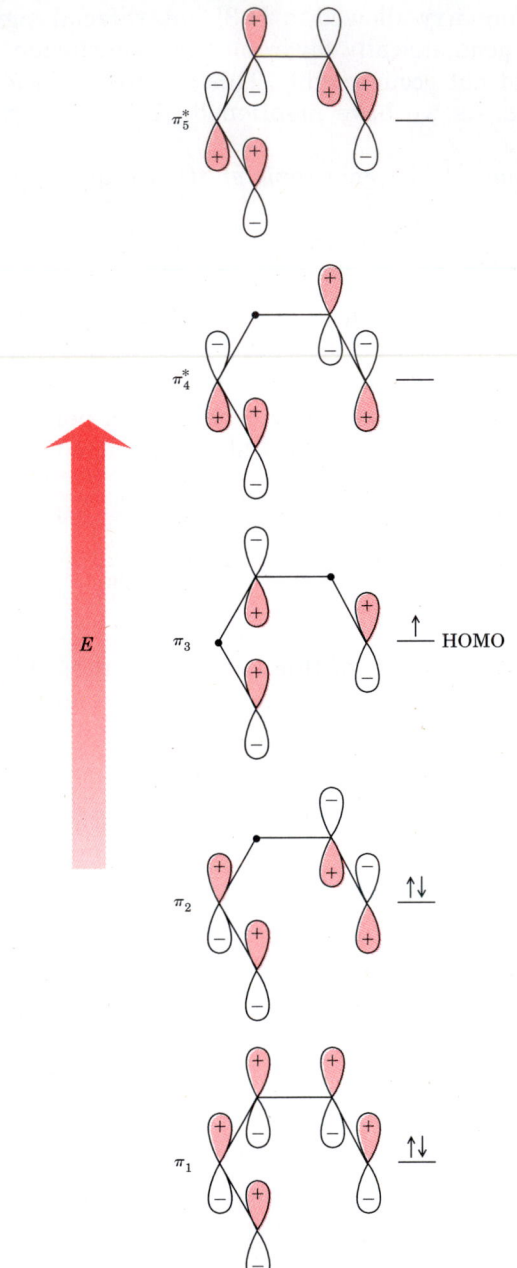

Figure 21.4 The five π molecular orbitals of the pentadienyl radical.

Section 21.5
Pericyclic Reactions Leading to Vitamin D

Pericyclic reactions are not merely laboratory curiosities; they are also observed in natural processes. As one example, let us briefly consider some of the transformations that occur in the vitamin D group of compounds.

Vitamin D is essential for the proper growth of bones. Lack of vitamin D results in defective bone growth, a condition called *rickets*. Humans can obtain vitamin D by a number of routes. One is by the action of sunlight on a particular steroid, 7-dehydrocholesterol, which is found in the skin. (Steroids are discussed in Section 24.5.) This steroid undergoes a photo-induced electrocyclic ring opening to yield a triene. The triene, previtamin D, then undergoes a thermally induced [1,7] sigmatropic shift to yield vitamin D_3. (The subscript 3 is used to differentiate this vitamin D from other structurally similar compounds with vitamin D activity.)

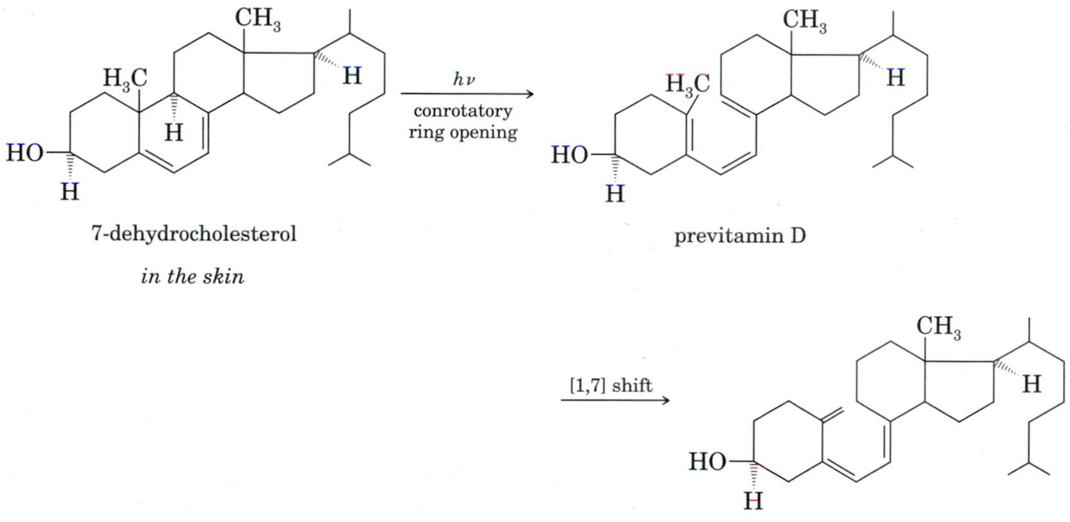

7-dehydrocholesterol

in the skin

previtamin D

vitamin D_3

Another source of vitamin D is *irradiated ergosterol,* which is commonly added to milk. The conversion of ergosterol to vitamin D_2 proceeds by the same series of reaction steps as the conversion of 7-dehydrocholesterol.

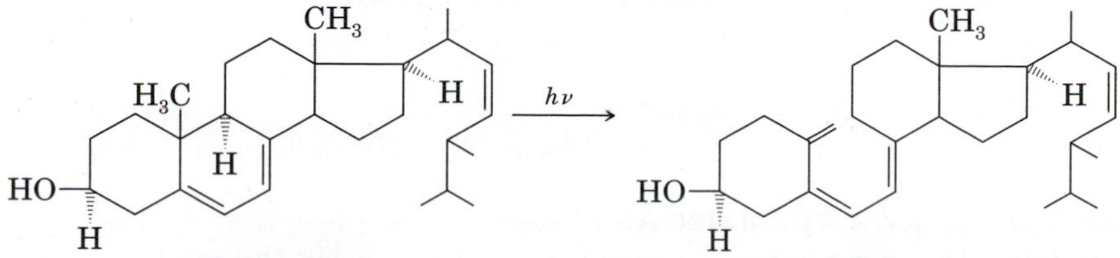

ergosterol

*in yeast, soybean oil,
and ergot (a fungus)*

vitamin D_2
(calciferol)

Summary

Pericyclic reactions are concerted, thermally induced or photo-induced reactions with cyclic transition states. Three types of pericyclic reactions are:

Cycloaddition:

Electrocyclization:

Sigmatropic rearrangement:

In the **frontier orbital method** of analyzing cycloaddition reactions, electrons are assumed to flow from the **HOMO** of one molecule to the **LUMO** of the other. If the phases of these orbitals are the same, the reaction is **symmetry-allowed.** If the orbital phases are opposite and show antibonding character, the reaction is **symmetry-forbidden.** Symmetry-allowed [2 + 2] cycloadditions are photo-induced, while [4 + 2] cycloadditions are thermally induced.

In electrocyclic reactions, *p*-orbital components of the HOMO undergo end-to-end overlap to form the new sigma bond. To do this, they must undergo **conrotatory** or **disrotatory** motion, which, in turn, determines the stereochemistry. A summary of the types of motion to be expected is shown in Table 21.1.

Sigmatropic rearrangements occur **suprafacially** or **antarafacially,** depending on the phases of the interacting orbitals in the HOMO of a hypothetical radical system. The geometry of the transition state determines whether the rearrangement proceeds readily or not. The classification of sigmatropic rearrangements is discussed in Section 21.4A.

Essay Problem for Chapter 21

Synthesis of γ,δ-Unsaturated Aldehydes

The following experimental procedure was given for the synthesis of 3-phenyl-4-pentenal. The procedure is general and can be used for the synthesis of a variety of γ,δ-unsaturated aldehydes. (D. E. Vogel and G. H. Buchi, *Org. Syn.* **1988,** *66,* 29.)

In a 100-mL round-bottomed flask equipped with a magnetic stirring bar is placed 20.1 g (0.098 mol) of (*E*)-3-[(*E*)-3-phenyl-2-propenoxy]acrylic acid. The flask is fitted with a distillation head for vacuum distillation and heated at 0.1 mm pressure. The oil-bath temperature is maintained between 160–165°C until all the material melts, while the mixture is stirred. Once the initial reaction is under control the oil bath is slowly heated to 180°C. The product is collected in a receiver flask cooled with a dry ice–acetone bath to give 13.3–14.3 g (84–91% yield).

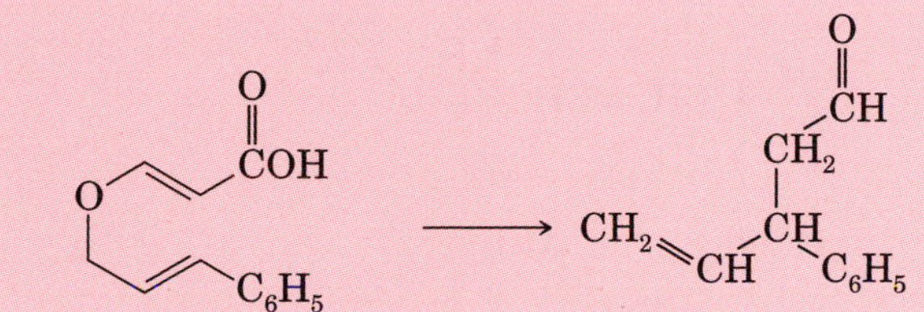

(*E*)-3-[(*E*)-3-phenyl-
2-propenoxy]acrylic acid

3-phenyl-4-pentenal

a γ,δ*-aldehyde*

Questions

1. Identify the type of pericyclic reaction involved in the formation of the unstable intermediate.
2. What elements are lost when the intermediate is converted to 3-phenyl-4-pentenal?
3. Identify the carbon atom in the starting acrylic acid that becomes the aldehyde carbon in the pentenal.
4. Write a complete mechanism for this transformation.
5. What starting material would be required to synthesize each of the following compounds?

(a)

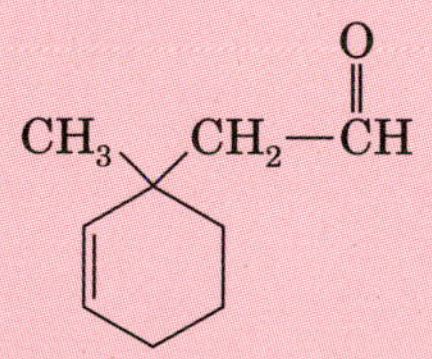

(b) $CH_2{=}CH{-}\overset{\displaystyle CH_3}{\underset{\displaystyle CH_3}{C}}{-}CH_2CH{=}O$

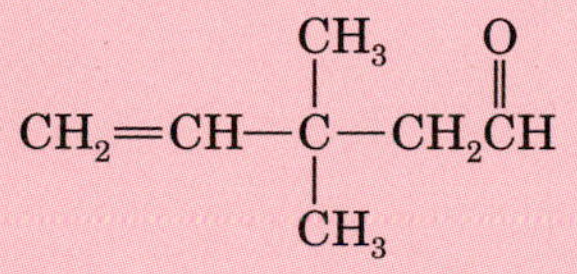

(c) $CH_3CH_2CH_2\underset{\displaystyle CH{=}CHCH_3}{CH}{-}CH{=}O$

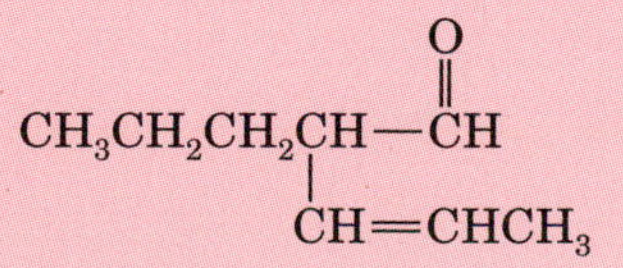

21.15 Identify each of the following pericyclic reactions as being (1) a cycloaddition; (2) an electrocyclic reaction; or (3) a sigmatropic rearrangement:

(a)

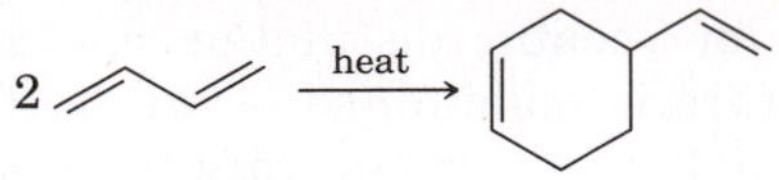

(b) 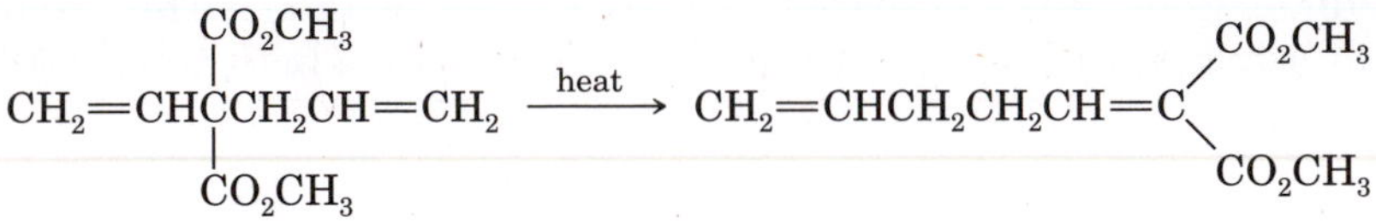 $CH_2{=}CHCCH_2CH{=}CH_2$ (with CO_2CH_3 above and CO_2CH_3 below) $\xrightarrow{\text{heat}}$ $CH_2{=}CHCH_2CH_2CH{=}C$ (with CO_2CH_3 and CO_2CH_3)

(c)

$\xrightarrow{\text{heat}}$ $CH_2{=}CHCH{=}C$ (with CH_3 and CH_3)

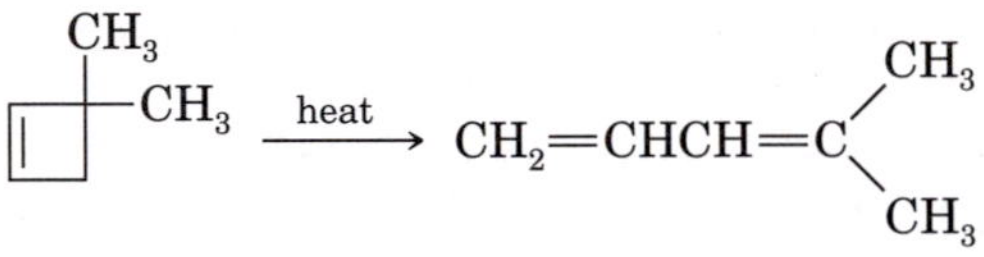

(d)

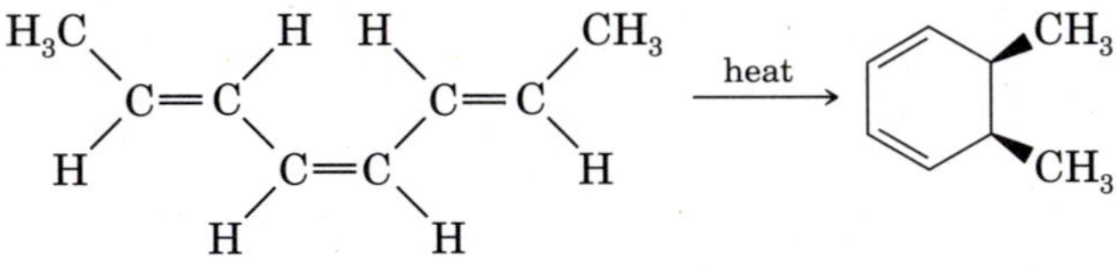

(e)

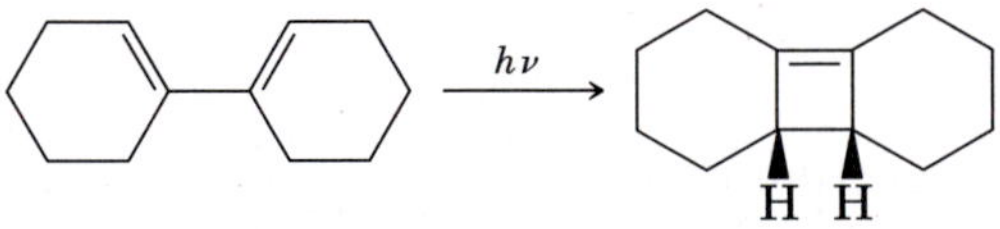

21.16 Construct a diagram of the array of bonding and antibonding π molecular orbitals (like those in Figure 21.2) for each of the following structures:

(a) *cis*-$CH_3CH{=}CH\bar{C}HCH_2CH_3$ **(b)** 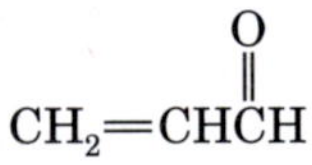 $CH_2{=}CHCH$ (with O double bonded)   **(c)**

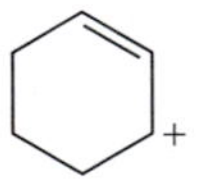

21.17 Classify each of the following cycloadditions as [2 + 2], [4 + 2], etc.:

(a) 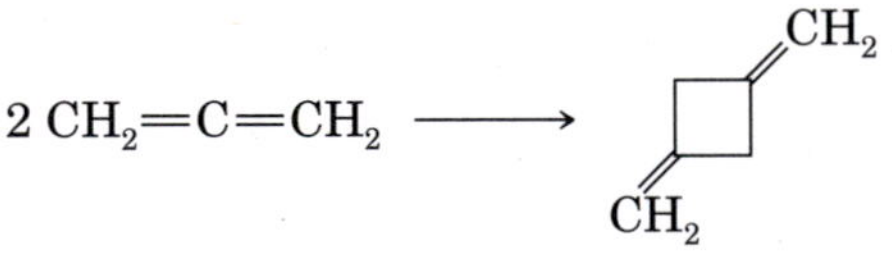 $2\ CH_2{=}C{=}CH_2 \longrightarrow$

(b)

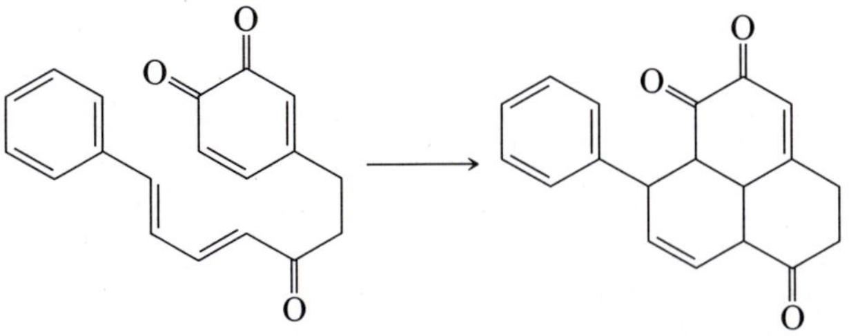

* For information concerning the organization of the *Study Problems* and *Additional Problems*, see the *Note to student* on page 41.

(c) 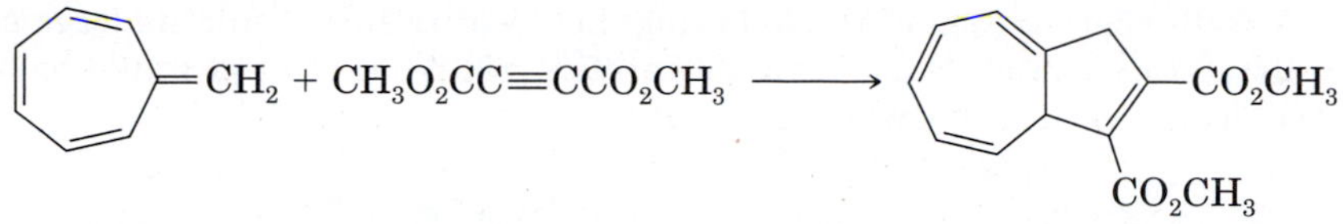 $=CH_2 + CH_3O_2CC\equiv CCO_2CH_3 \longrightarrow$ $-CO_2CH_3$

CO_2CH_3

21.18 Which of the following types of cycloadditions would you predict to proceed easily upon heating?

   (a) [6 + 2]   **(b)** [6 + 4]   **(c)** [8 + 2]   **(d)** [8 + 4]

21.19 1,2-Diphenylcyclobutene undergoes a photochemical dimerization. What is the structure of the dimer? (Include the stereochemistry in your answer.)

21.20 Would you expect the following conversions to require heat or light? Explain.

(a)

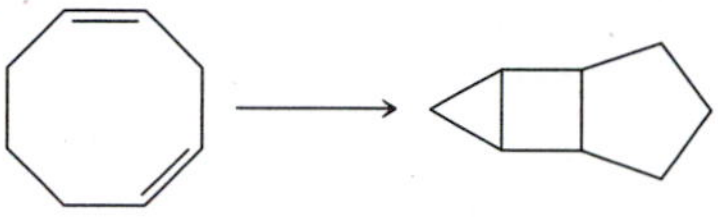

(b) 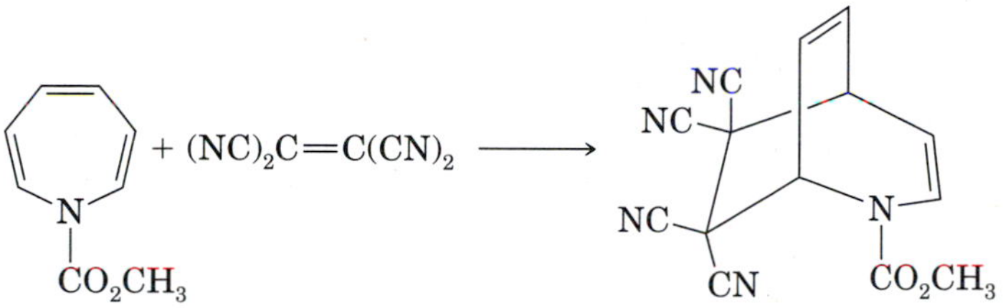 $+ (NC)_2C{=}C(CN)_2 \longrightarrow$

(c)

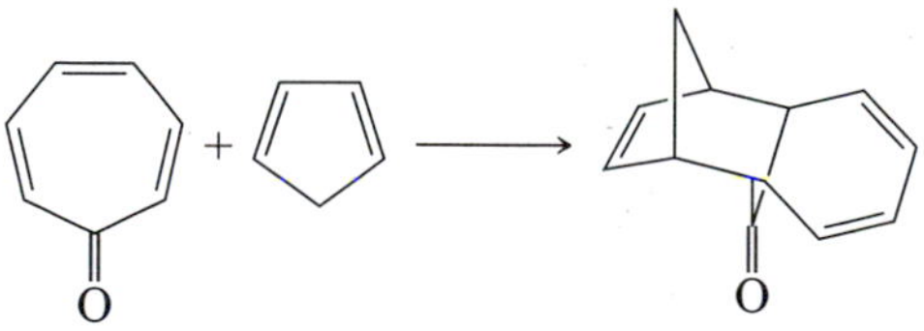

(d)

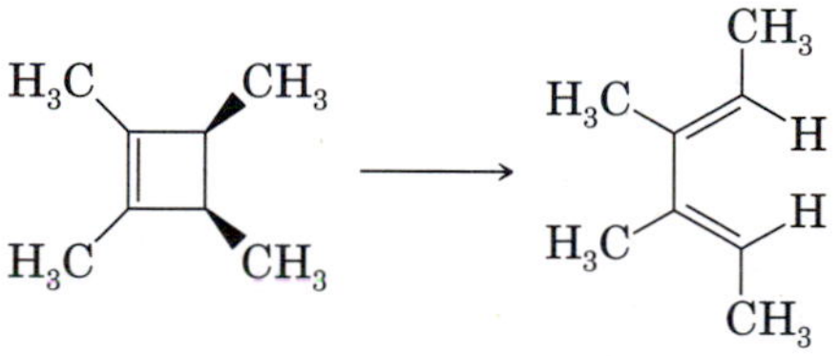

21.21 Suggest synthetic routes to the following compounds from monocyclic or acyclic starting materials:

(a)

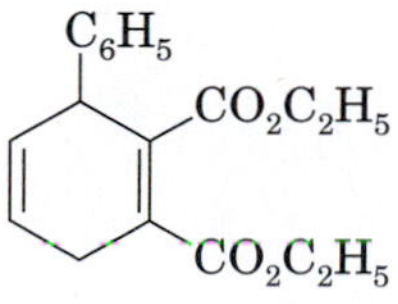

(b)

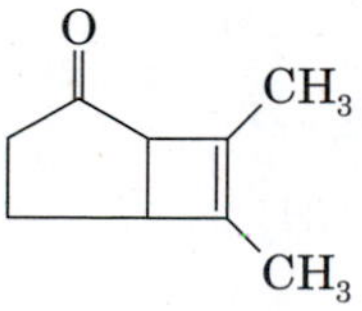

(c)

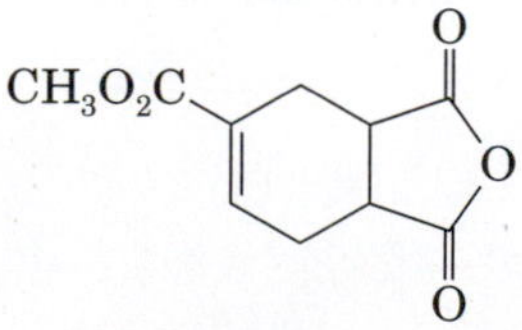

(d)

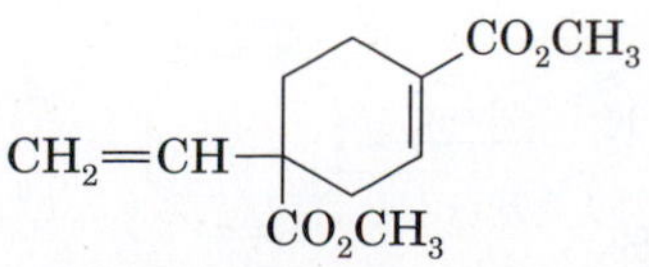

21.22 (1) Tell whether each of the following dienes or trienes would undergo conrotatory or disrotatory motion in a cyclization reaction. (2) What product would be observed? (Include stereochemistry where appropriate.)

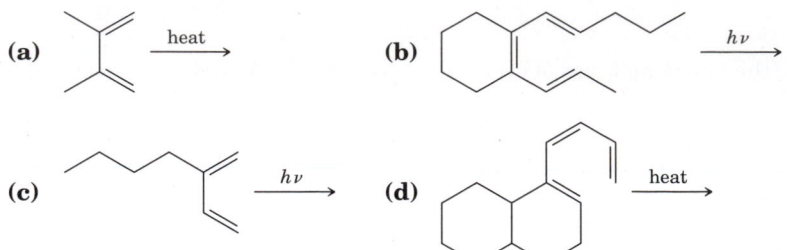

21.23 Predict the electrocyclic product and its stereochemistry:

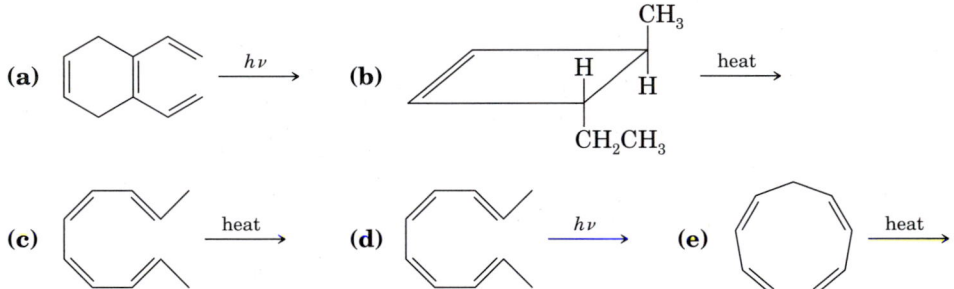

21.24 The following compounds both undergo photo-induced electrocyclic reactions. What are the structures and stereochemistry of the products?

(a)   (b)

21.25 How would you carry out the following conversions?

(a)

(b)

21.26 Classify each of the following sigmatropic rearrangements as [1,3], [3,3], etc.

(a)

(b)

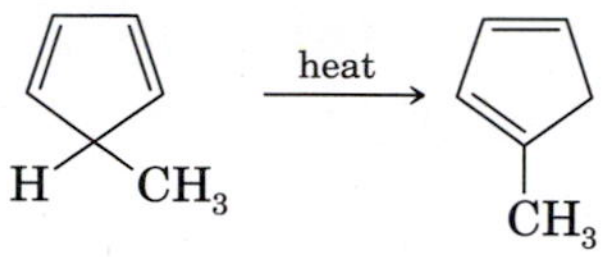

$$\xrightarrow{\text{heat}}$$

(c)

$$\xrightarrow{\text{heat}}$$

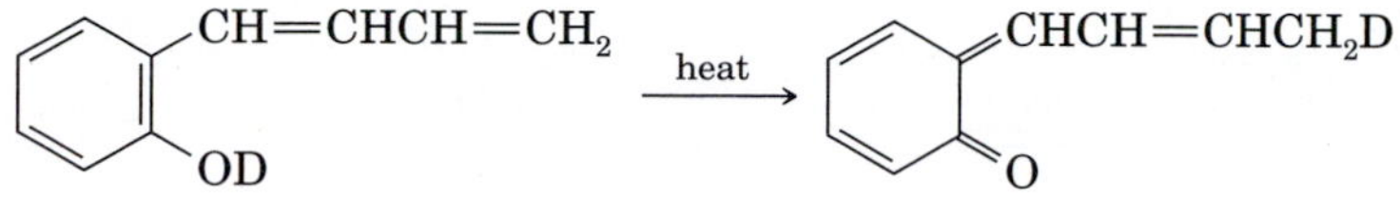

21.27 Predict the product or products of the sigmatropic rearrangement of 7,7-dideuterio-1,3,5-cycloheptatriene.

21.28 Explain the following observation:

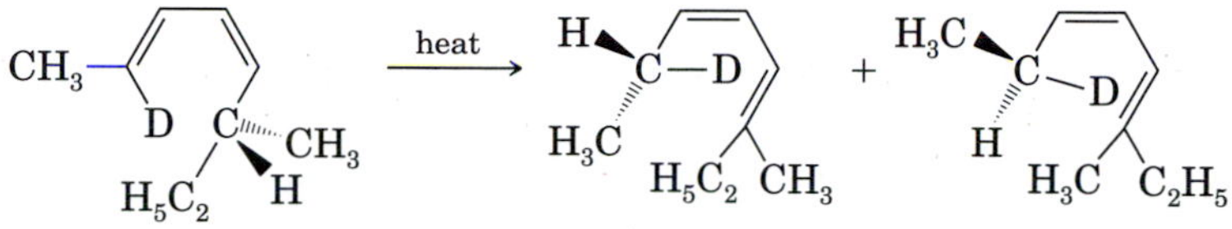

$$\xrightarrow{\text{heat}}$$

21.29 Should you use thermal induction or photo-induction to carry out the following reactions?

(a)

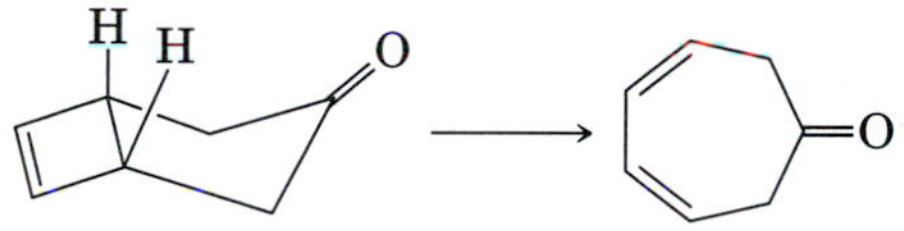

(b)

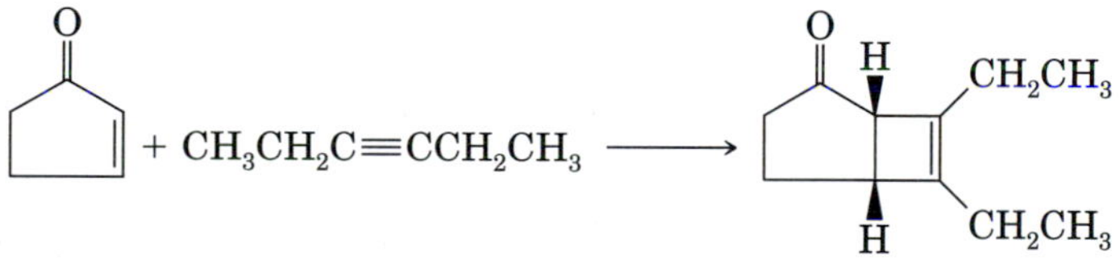

$+ \text{CH}_3\text{CH}_2\text{C}\equiv\text{CCH}_2\text{CH}_3 \longrightarrow$

(c) $\text{CH}_2{=}\text{CHCHCH}_2\text{CH}{=}\text{CH}_2 \longrightarrow \text{CH}_3\text{CH}{=}\text{CHCH}_2\text{CH}_2\text{CH}{=}\text{CH}_2$
with CH$_3$ substituent

21.30 When a mixture of quinone (page 520) and cyclopentadiene is heated, the two compounds undergo an addition reaction. When the product of this addition is exposed to sunlight, it undergoes isomerization. What are the structures of the addition product and its isomer?

Additional Problems

21.31 Complete the following equations, including stereochemistry where appropriate:

(a) 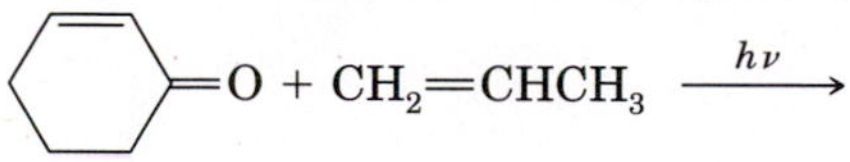 $={}O + \text{CH}_2{=}\text{CHCH}_3 \xrightarrow{h\nu}$

(b) 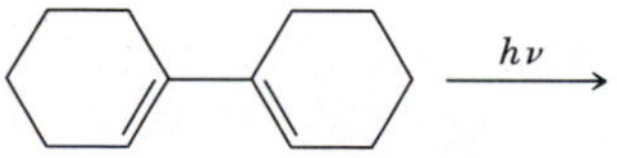 $\xrightarrow{h\nu}$

(c) 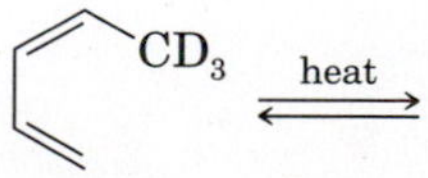 $\xrightleftharpoons{\text{heat}}$

(d)

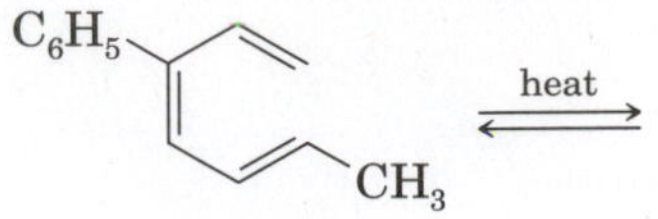

$$\text{heat} \rightleftharpoons$$

(e)

$$\text{heat} \rightleftharpoons$$

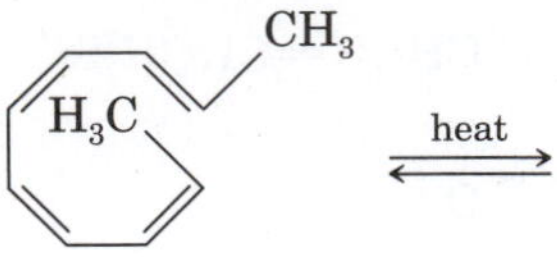

(f)   *cis*-2-butene $\overset{h\nu}{\rightleftharpoons}$

(g)   *trans*-2-butene $\overset{h\nu}{\rightleftharpoons}$

21.32   Explain the following observations:

(a)

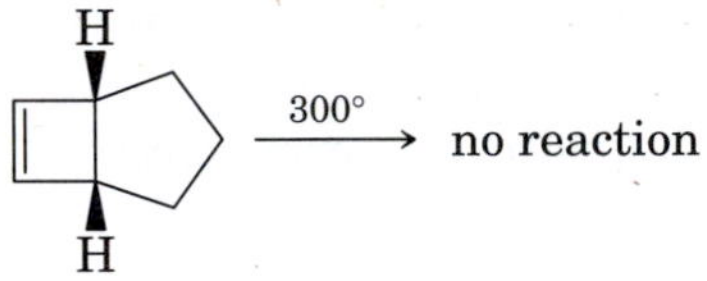

$$\xrightarrow{300°} \text{no reaction}$$

(b)

$$\xrightarrow{300°}$$

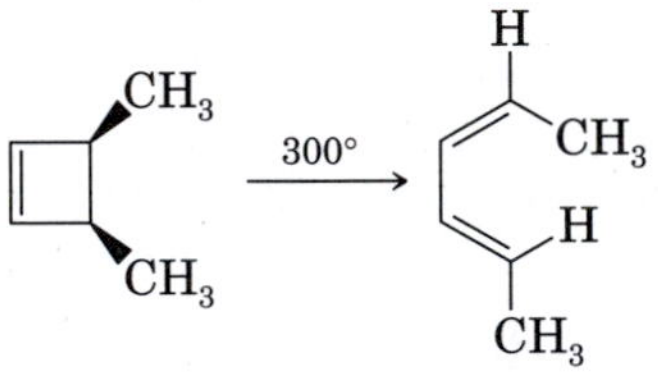

21.33   Did conrotatory or disrotatory motion take place in each of the following electrocyclic reactions?

(a)

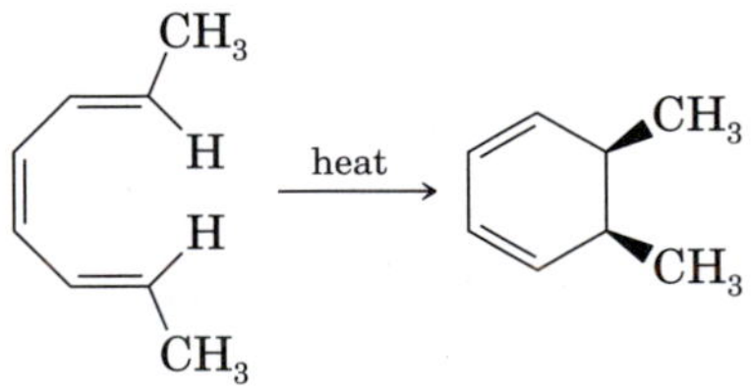

$$\xrightarrow{\text{heat}}$$

(b)

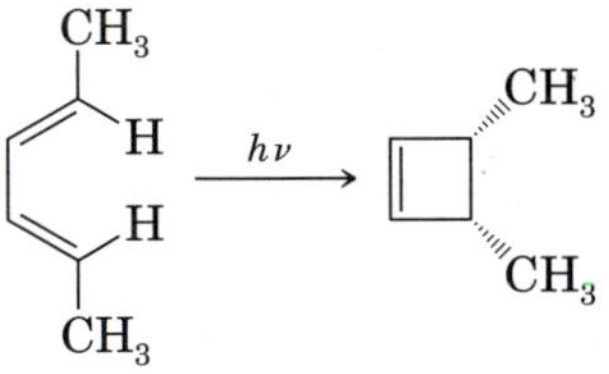

$$\xrightarrow{h\nu}$$

(c)

$$\xrightarrow{\text{heat}}$$

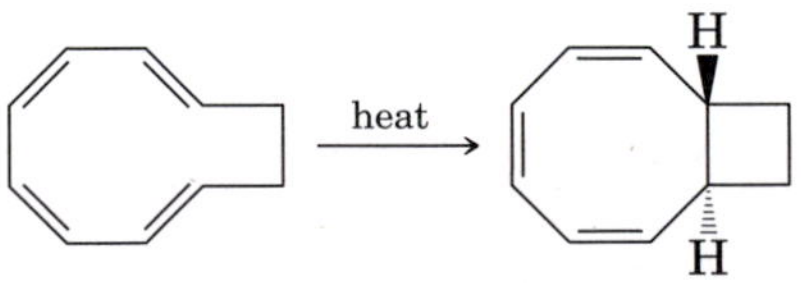

21.34   Predict whether the following sigmatropic rearrangements would proceed suprafacially or antarafacially:

(a)

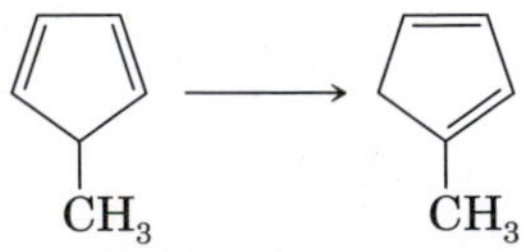

(b)

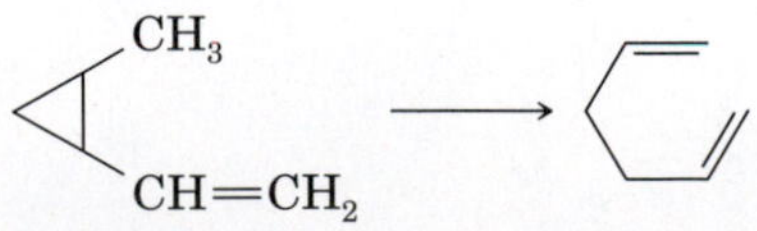

(c)

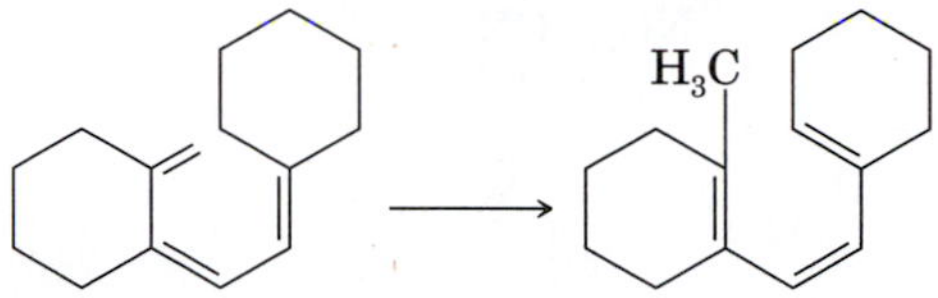

(d)

21.35 Previtamin D can undergo both thermally induced and photo-induced electrocyclic ring-closure reactions. What will be the stereochemistry at positions 9 and 10 of the resulting products? (See page 972 for the numbering of the steroid ring.)

21.36 Complete the following equations, showing the stereochemistry of the products:

(a) 2 $\xrightarrow{h\nu}$

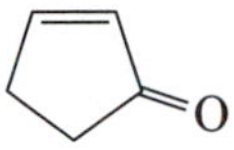

(b) 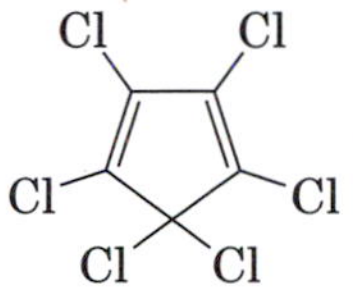 $+\ CH_2{=}C{=}CH_2 \xrightarrow{\text{heat}}$

(c) 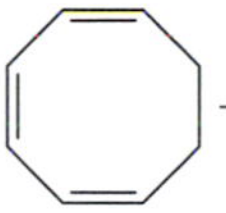 $+\ CH_2{=}CH_2 \xrightarrow{\text{heat}}$

(d) 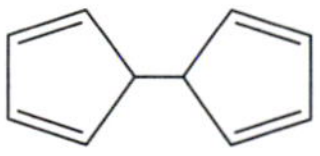 $+\ CH_3O_2CC{\equiv}CCO_2CH_3 \xrightarrow{\text{heat}}$

21.37 Suggest synthetic routes to the following compounds from acyclic or monocyclic starting materials:

(a)

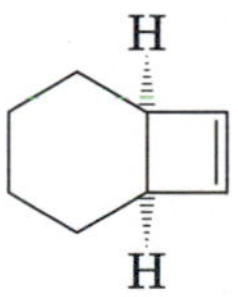

(b)

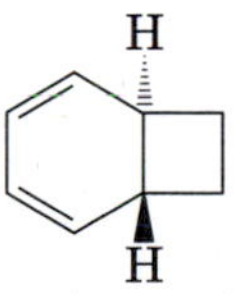

(c)

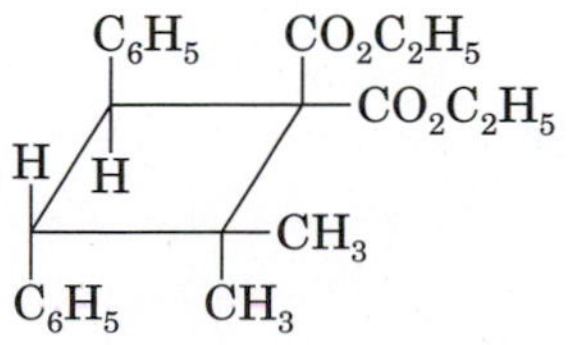

(d)

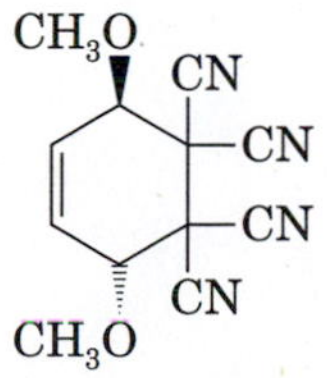

(e)

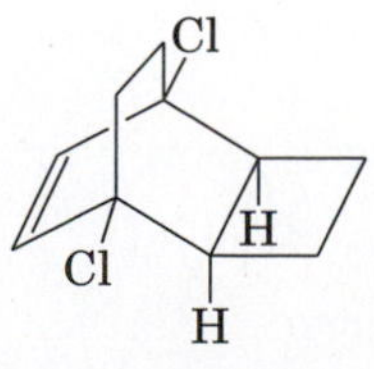

(f)

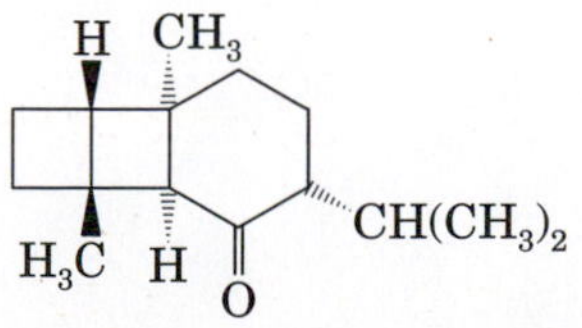

(g)

(h)

SPECTROSCOPY II: ULTRAVIOLET SPECTRA, COLOR AND VISION, MASS SPECTRA

In Chapter 9, we discussed the absorption of infrared and radiofrequency radiation by organic compounds and how the absorption can be used in structure identification. In this chapter, we will consider the absorption of **ultraviolet (UV) and visible light** by organic compounds. Ultraviolet and visible spectra are also used in structure determination. More important, the absorption of visible light results in vision. We will also discuss this topic, along with colors and dyes. Last, we will introduce **mass spectra**, which arise from fragmentation of molecules when they are bombarded with high-energy electrons.

Section 22.1
Ultraviolet and Visible Spectra

The wavelengths of UV and visible light are substantially shorter than the wavelengths of infrared radiation (see Figure 9.2, page 339). The unit we will use to describe these wavelengths is the *nanometer* (1 nm = 10^{-7} cm). The visible spectrum spans from about 400 nm (violet) to 750 nm (red), while the ultraviolet spectrum ranges from 100 to 400 nm.

The quantity of energy absorbed by a compound is inversely proportional to the wavelength of the radiation:

$$\Delta E = h\nu = \frac{hc}{\lambda}$$

where ΔE = energy absorbed, in ergs

h = Planck's constant, 6.6×10^{-27} erg-sec

ν = frequency, in Hz

c = speed of light, 3×10^{10} cm/sec

λ = wavelength, in cm

Infrared radiation is relatively low-energy radiation. Absorption of infrared radiation by a molecule leads to increased vibrations of covalent bonds. Molecular transitions from the ground state to an excited vibrational state require about 2–15 kcal/mol.

Both UV and visible radiation are of higher energy than infrared radiation. Absorption of ultraviolet or visible light results in **electronic transitions,** promotion of electrons from low-energy ground-state orbitals to higher-energy excited-state orbitals. These transitions require about 40–300 kcal/mol. The energy absorbed is subsequently dissipated as heat, as light, or in chemical reactions (such as isomerization or free-radical reactions).

The wavelength of UV or visible light absorbed depends on the ease of electron promotion. Molecules that require *more energy* for electron promotion absorb at *shorter wavelengths*. Molecules that require *less energy* absorb at *longer wavelengths*. Compounds that absorb light in the visible region (that is, colored compounds) have more-easily promoted electrons than compounds that absorb at shorter UV wavelengths.

absorption at 100 nm (UV) $\longrightarrow$ 750 nm (visible)

increasing ease of electronic transition

STUDY PROBLEM

22.1 Which would have the more-easily promoted electrons, anthracene (colorless) or Tyrian purple?

anthracene   Tyrian purple

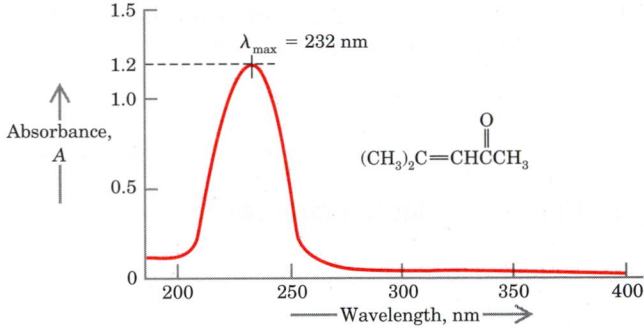

Figure 22.1   Ultraviolet spectrum of mesityl oxide, $9.2 \times 10^{-5}\,M$, 1.0-cm cell. (The $n \to \pi^*$ transition, at about 325 nm, is too weak to be detected at this solution concentration. The different types of transitions are described in Section 22.3.)

A UV or visible spectrophotometer has the same basic design as an infrared spectrophotometer (see Figure 9.5, page 342). Absorption of radiation by a sample is measured at various wavelengths and plotted by a recorder to give the spectrum. (Figure 22.1 shows a typical UV spectrum.)

Because energy absorption by a molecule is quantized, we might expect that the absorption for electronic transitions would be observed at discrete wavelengths as a spectrum of lines or sharp peaks. This is not the case. Instead, a UV or visible spectrum consists of broad absorption bands over a wide range of wavelengths. The reason for the broad absorption is that the energy levels of both the ground state and the excited state of a molecule are subdivided into *rotational and vibrational sublevels.* Electronic transitions may occur from any one of the sublevels of the ground state to any one of the sublevels of an excited state (Figure 22.2). Since these various transitions dif-

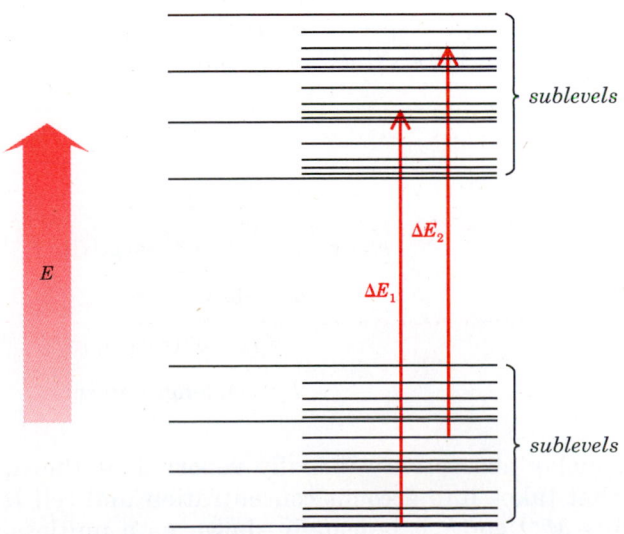

Figure 22.2   Schematic representation of electronic transitions from a low energy level to a high energy level.

fer slightly in energy, their wavelengths of absorption also differ slightly and give rise to the broad band observed in the spectrum.

Section 22.2

Expressions Used in Ultraviolet Spectroscopy

Figure 22.1 shows the UV spectrum of a dilute solution of mesityl oxide (4-methyl-3-penten-2-one). The spectrum shows the scan from 200 to 400 nm. (Because absorption by atmospheric carbon dioxide becomes significant below 200 nm, the 100–200 nm region is usually not scanned.) The wavelength of absorption is usually reported as λ_{max}, the wavelength at the highest point of the curve. The λ_{max} for mesityl oxide is 232 nm.

The absorption of energy is recorded as **absorbance** (not transmittance as in infrared spectra). The absorbance at a particular wavelength is defined by the equation:

$$A = \log \frac{I_0}{I}$$

where A = absorbance

I_0 = intensity of the reference beam

I = intensity of the sample beam

The absorbance by a compound at a particular wavelength increases with an increasing number of molecules undergoing transition. Therefore, the absorbance depends on the electronic structure of the compound and also upon the concentration of the sample and the length of the sample cell. For this reason, chemists report the energy absorption as **molar absorptivity** ϵ (sometimes called the *molar extinction coefficient*) rather than as the actual absorbance. Often, UV spectra are replotted to show ϵ or $\log \epsilon$ instead of A as the ordinate. The $\log \epsilon$ value is especially useful when values for ϵ are very large.

$$\epsilon = \frac{A}{cl}$$

where ϵ = molar absorptivity

A = absorbance

c = concentration, in M

l = cell length, in cm

The molar absorptivity (usually reported at the λ_{max}) is a reproducible value that takes into account concentration and cell length. Although ϵ has the units M^{-1} cm^{-1}, it is usually shown as a unitless quantity. For mesityl oxide, the ϵ_{max} is $1.2 \div (9.2 \times 10^{-5} \times 1.0)$, or 13,000 (values taken from Figure 22.1).

SAMPLE PROBLEM

A flask of cyclohexane is known to be contaminated with benzene. At 260 nm, benzene has a molar absorptivity of 230, and cyclohexane has a molar absorptivity of zero. A UV spectrum of the contaminated cyclohexane (1.0-cm cell length) shows an absorbance of 0.030. What is the concentration of benzene?

Solution

$$c = \frac{A}{\epsilon l} = \frac{0.030}{230 \times 1.0} = 0.00013M$$

STUDY PROBLEM

22.2 What is the molar absorptivity (ϵ_{max}) of a compound for which a $0.000100M$ solution in ethanol shows an absorbance of 1.05 at a λ_{max} of 237 nm in a 1.00-cm cell?

Section 22.3
Types of Electron Transitions

Let us consider the different types of electron transitions that give rise to ultraviolet or visible spectra. The ground state of an organic molecule contains valence electrons in three principal types of molecular orbitals: **sigma (σ) orbitals; pi (π) orbitals;** and **filled, but nonbonded, orbitals (n).**

$$\sigma\ electrons \qquad \pi\ electrons \qquad n\ electrons$$

$$H\!:\!CH_3 \qquad CH_2\!:\!:\!CH_2 \qquad CH_3\ddot{O}H$$

Both σ and π orbitals are formed from the overlap of two atomic or hybrid orbitals. Each of these molecular orbitals therefore has an antibonding σ^* or π^* orbital associated with it. An orbital containing n electrons does not have an antibonding orbital (because it was not formed from two orbitals). Electron transitions involve the promotion of an electron from one of the three ground state orbitals (σ, π, or n) to one of the two excited state orbitals (σ^* or π^*). There are six possible transitions; the four important transitions and their relative energies are shown in Figure 22.3.

The most useful region of the UV spectrum is at wavelengths longer than 200 nm. The following transitions give rise to absorption in the nonuseful 100–200 nm range: $\pi \rightarrow \pi^*$ for an isolated double bond and $\sigma \rightarrow \sigma^*$ for an ordinary carbon–carbon bond. The useful transitions (200–400 nm) are $\pi \rightarrow \pi^*$ for compounds with conjugated double bonds, and some $n \rightarrow \sigma^*$ and $n \rightarrow \pi^*$ transitions.

A. Absorption by Polyenes

Less energy is required to promote a π electron of 1,3-butadiene than is needed to promote a π electron of ethylene. The reason is that the energy dif-

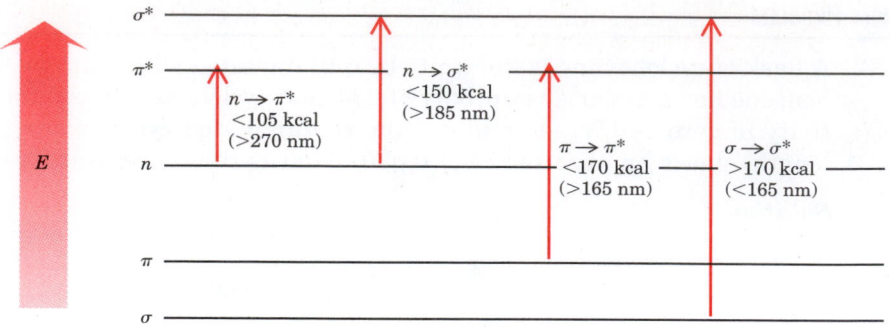

Figure 22.3 Energy requirements for important electronic transitions. (The corresponding wavelengths are in parentheses.)

ference between the HOMO (Highest Occupied Molecular Orbital) and the LUMO (Lowest Unoccupied Molecular Orbital) for conjugated double bonds is less than the energy difference for an isolated double bond.

For CH₂=CH₂:

$$\pi_2^* \; ———— \qquad\qquad \pi_2^* \; \downarrow$$

$$\text{greater } \Delta E \longrightarrow$$

$$\pi_1 \; \underline{\uparrow\downarrow} \qquad\qquad \pi_1 \; \underline{\uparrow}$$

For CH₂=CHCH=CH₂:

$$\pi_4^* \; ———— \qquad\qquad \pi_4^* \; ————$$

$$\pi_3^* \; ———— \qquad\qquad \pi_3^* \; \downarrow$$

$$\text{less } \Delta E \longrightarrow$$

$$\pi_2 \; \underline{\uparrow\downarrow} \qquad\qquad \pi_2 \; \underline{\uparrow}$$

$$\pi_1 \; \underline{\uparrow\downarrow} \qquad\qquad \pi_1 \; \underline{\uparrow\downarrow}$$

Because less energy is needed for a $\pi \rightarrow \pi^*$ transition of 1,3-butadiene, this diene absorbs UV radiation of longer wavelengths than does ethylene. As more conjugated double bonds are added to a molecule, the energy required to reach the first excited state decreases. Sufficient conjugation shifts the absorption to wavelengths that reach into the visible region of the spectrum; a compound with sufficient conjugation is colored. For example, lycopene, the compound responsible for the red color of tomatoes, has eleven conjugated double bonds.

lycopene
$\lambda_{max} = 505$ nm

Table 22.1 lists the λ_{max} values for $\pi \rightarrow \pi^*$ transitions of a series of aldehydes with increasing conjugation. Inspection of the table reveals that the position of absorption is shifted to longer wavelengths as the extent of the conjugation increases. Generally, this increase is about 30 nm per conjugated double bond in a series of polyenes.

Table 22.1 Ultraviolet absorption for some unsaturated aldehydes	
Structure	λ_{max}, nm
$CH_3CH{=}CHCHO$	217
$CH_3(CH{=}CH)_2CHO$	270
$CH_3(CH{=}CH)_3CHO$	312
$CH_3(CH{=}CH)_4CHO$	343
$CH_3(CH{=}CH)_5CHO$	370

STUDY PROBLEMS

22.3 List the following all-*trans*-polyenes in order of increasing λ_{max}:

   (a) $CH_3(CH{=}CH)_{10}CH_3$ **(b)** $CH_3(CH{=}CH)_9CH_3$ **(c)** $CH_3(CH{=}CH)_8CH_3$

22.4 The λ_{max} for Compound **(a)** in Problem 22.3 is 476 nm. Predict the λ_{max} values for Compounds **(b)** and **(c)**.

B. Absorption by Aromatic Systems

Benzene and other aromatic compounds exhibit more complex spectra than can be explained by simple $\pi \rightarrow \pi^*$ transitions. The complexity arises from the existence of *several* low-lying excited states. Benzene absorbs strongly at 184 nm ($\epsilon = 47{,}000$) and at 202 nm ($\epsilon = 7{,}000$) and has a series of absorption bands between 230–270 nm. A value of 260 nm is often reported as the λ_{max} for benzene because this is the position of strongest absorption above 200 nm. Solvents and substituents on the ring alter the UV spectra of benzene compounds.

The absorption of UV radiation by aromatic compounds composed of fused benzene rings is shifted to longer wavelengths as the number of rings is increased because of increasing conjugation and greater resonance stabilization of the excited state.

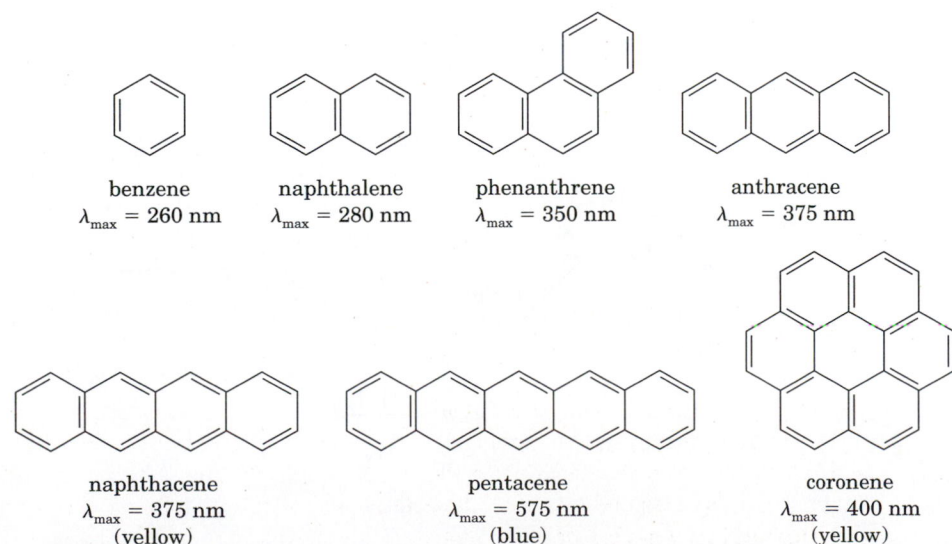

benzene
$\lambda_{max} = 260$ nm

naphthalene
$\lambda_{max} = 280$ nm

phenanthrene
$\lambda_{max} = 350$ nm

anthracene
$\lambda_{max} = 375$ nm

naphthacene
$\lambda_{max} = 375$ nm
(yellow)

pentacene
$\lambda_{max} = 575$ nm
(blue)

coronene
$\lambda_{max} = 400$ nm
(yellow)

Table 22.2	Ultraviolet absorption arising from $n \rightarrow \sigma^*$ transitions				
Structure	λ_{max}, nm	ϵ	Structure	λ_{max}, nm	ϵ
$CH_3\ddot{O}H$	177	200	$CH_3CH_2CH_2\ddot{B}\ddot{r}$	208	300
$(CH_3)_3\ddot{N}$	199	3950	$CH_3\ddot{I}$	259	400
$CH_3\ddot{C}\ddot{l}$	173	200			

C. Absorption Arising from Transitions of n Electrons

Compounds that contain nitrogen, oxygen, sulfur, phosphorus, or one of the halogens all have unshared n electrons. If the structure contains no π bonds, these n electrons can undergo only $n \rightarrow \sigma^*$ transitions. Because the n electrons are of higher energy than either the σ or π electrons, less energy is required to promote an n electron, and transitions occur at longer wavelengths than $\sigma \rightarrow \sigma^*$ transitions. Note that some of these values are within the usual UV spectral range of 200–400 nm (Table 22.2). The π^* orbital is of lower energy than the σ^* orbital. Consequently, $n \rightarrow \pi^*$ transitions require less energy than $n \rightarrow \sigma^*$ transitions and often are in the range of a normal instrument scan.

The n electrons are in a different region of space from σ^* and π^* orbitals, and the probability of an n transition is low. Since molar absorptivity depends on the number of electrons undergoing transition, ϵ values for n transitions are low, in the 10–100 range (compared to about 10,000 for a $\pi \rightarrow \pi^*$ transition).

A compound such as acetone that contains both a π bond and n electrons exhibits both $\pi \rightarrow \pi^*$ and $n \rightarrow \pi^*$ transitions. Acetone shows absorption at 187 nm ($\pi \rightarrow \pi^*$) and 270 nm ($n \rightarrow \pi^*$).

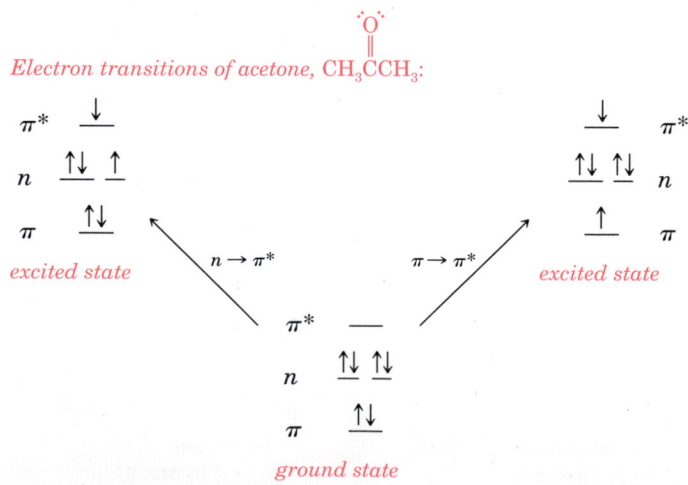

Electron transitions of acetone, $CH_3\overset{\overset{\displaystyle \ddot{O}}{\|}}{C}CH_3$:

Section 22.4
Color and Vision

Color has played a significant role in human society ever since people first learned to color clothes and other articles. Color is the result of a complex set of physiological and psychological responses to wavelengths of light of 400–750 nm striking the retina of the eye. If all wavelengths of visible light strike the retina, we perceive white. If none of them do, we perceive black or darkness. If a small range of wavelengths hits the eye, then we observe individual colors. Table 22.3 lists the wavelengths of the visible spectrum with their corresponding colors and complementary colors, which we will discuss shortly.

Our perception of color arises from a variety of physical processes. A few examples of how light of a particular wavelength may be directed to the eye follow. (1) The yellow-orange color of a sodium flame results from the **emission of light** with a wavelength of 589 nm. The emission is caused by excited electrons returning to lower-energy orbitals. (2) A prism causes a **diffraction of light** that varies with the wavelength. We observe the separated wavelengths as a rainbow pattern. (3) **Interference** results from light being reflected from two surfaces of a very thin film (e.g., soap bubbles or bird feathers). The light wave reflected from the farther surface is reflected out of phase with the reflection from the nearer surface, resulting in wave interference and cancellation of some wavelengths. Hence, we see color instead of white.

The fourth, and most common, process that leads to color is the **absorption of light of certain wavelengths** by a substance. Organic compounds with extensive conjugation absorb certain wavelengths of light because of $\pi \rightarrow \pi^*$ and $n \rightarrow \pi^*$ transitions. We do not observe the color absorbed, but we see its **complement,** which is reflected. A complementary color, sometimes called a **subtraction color,** is the result of subtraction of some of the visible wavelengths from the entire visual spectrum. For example, pentacene (page 887) absorbs at 575 nm, in the yellow portion of the visible spectrum. Thus, pentacene absorbs the yellow light (and, to a lesser extent, that of the surrounding wavelengths) and reflects the other wavelengths. Pentacene has a blue color, which is the complement of yellow.

Some compounds appear yellow even though their λ_{max} is in the ultraviolet range of the spectrum (for example, coronene, page 887). In such a case, the tail of the absorption band extends from the ultraviolet region into the visible region and absorbs the violet to blue wavelengths. Figure 22.4 depicts the spectrum of such a compound.

Table 22.3	Colors in the visible spectrum	
Wavelength, nm	Color	Complementary (Subtraction) Color
400–424	violet	green-yellow
424–491	blue	yellow
491–570	green	red
570–585	yellow	blue
585–647	orange	green-blue
647–700	red	green

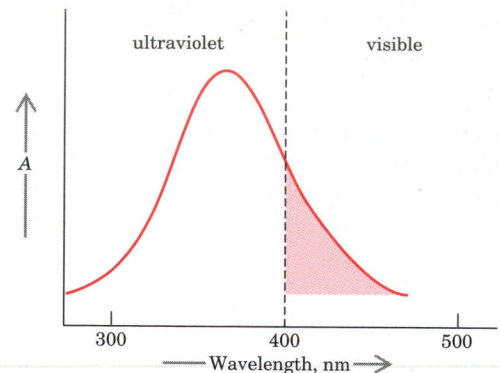

ultraviolet visible

300 400 500

← Wavelength, nm →

Figure 22.4 A compound with a λ_{max} in the UV region may also absorb light in the visible region.

A. Mechanism of Vision

The human eye is an amazingly intricate organ that converts photons of light into nerve pulses that travel to the brain and result in vision. The mechanism of the eye is remarkably sensitive. About one quantum of light energy is all that is necessary to trigger the mechanism resulting in a visual nerve pulse. We can detect as few as 100 photons of light. (For comparison, a typical flashlight bulb radiates about 2×10^{18} photons per second.)

The retina of the eye contains two types of photoreceptors—the **rods** and the **cones.** The cones are responsible for color vision and for vision in bright light. Animals that lack cones are colorblind. The rods are responsible for black and white photoreception and for vision in very dim light. While more is known about rods than cones, much is still to be learned, for example, about how the nerve impulse is generated.

In the rods, light is detected by a reddish-purple pigment called **rhodopsin,** or **visual purple** ($\lambda_{max} \cong 500$ nm). Rhodopsin is formed from an aldehyde called *11-cis-retinal* and a protein called *opsin.* These two components of rhodopsin are bonded together by a protonated imine link between the aldehyde group of 11-*cis*-retinal and an amino group of a lysine residue in opsin. As is frequently the case with protein complexes, opsin has a shape that fits around the 11-*cis*-retinal and holds it in a pocket. Compounds with other shapes do not fit into this pocket. In the combined form, the imine bond joining 11-*cis*-retinal and opsin is protected by the rest of the opsin molecule and is not readily hydrolyzed.

the 11-cis double bond

11
10
12
13

+ H_2N—opsin $\xrightarrow{-H_2O}$

CHO

11-*cis*-retinal

iminium ion link

CH=NH—opsin

rhodopsin

When a photon of light ($h\nu$) is absorbed by rhodopsin, the 11-*cis* double bond of the retinal portion is isomerized to a *trans* double bond. The product is a high-energy intermediate that undergoes a series of transformations. Finally, because all-*trans*-retinal does not fit into opsin's pocket, the now exposed iminium ion link is hydrolyzed and all-*trans*-retinal is released (see Figure 22.5).

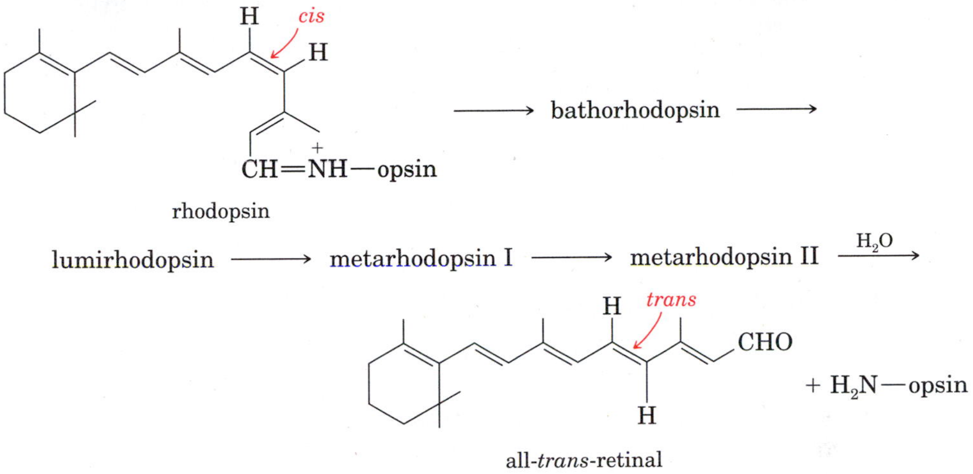

In the process of the hydrolysis, enzymes are activated that change the ionic permeability of the photoreceptor cell and thus change its electrical character. These changes are responsible for the generation of the nerve impulse.

Vitamin A is important in the human diet partly because it is the precursor of 11-*cis*-retinal.

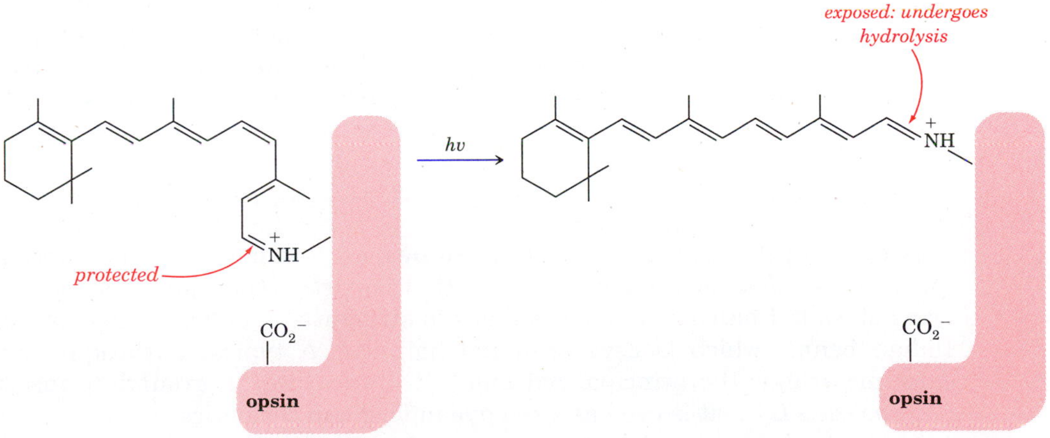

Figure 22.5 When 11-*cis*-retinal undergoes isomerization to all-*trans*-retinal, its shape changes so that it no longer fits into the pocket. Thus, the iminium ion link becomes exposed and can be hydrolyzed. Note that the isomerization also separates the iminium ion's positive charge from its balancing negative charge in opsin; this charge separation is one reason for the high energy contained in the photoisomerization product.

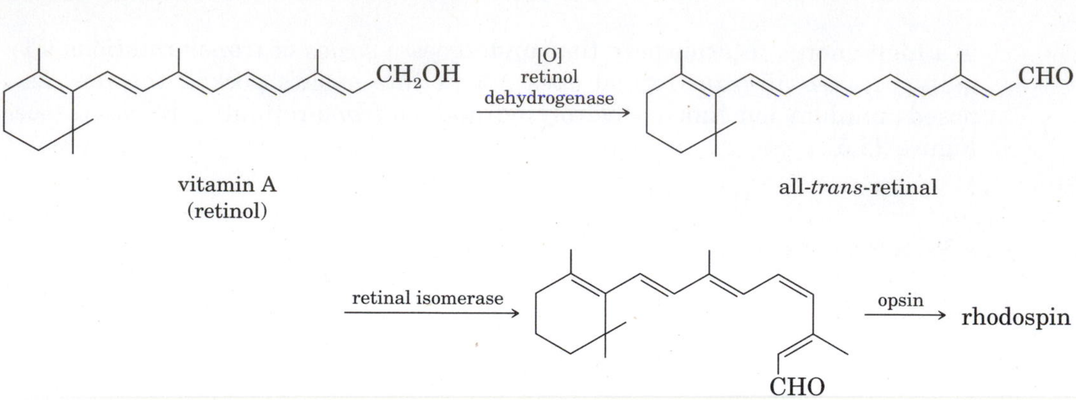

vitamin A
(retinol)

all-*trans*-retinal

11-*cis*-retinal

B. Colored Compounds, Dyes, and Indicators

Nature abounds with color. Some colors, such as those of hummingbird or peacock feathers, arise from light diffraction by the unique structure of the feathers. However, most of Nature's colors are due to the absorption of certain wavelengths of visible light by organic compounds.

Before the theories of electronic transition were developed, it was observed that some types of organic structures give rise to color, while others do not. The partial structures necessary for color (unsaturated groups that can undergo $\pi \to \pi^*$ and $n \to \pi^*$ transitions) were called **chromophores,** a term coined in 1876 (Greek *chroma,* "color," and *phoros,* "bearing").

Some chromophores:

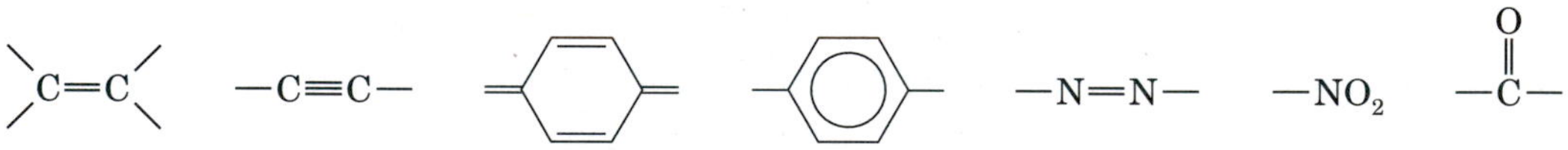

It was also observed that the presence of some other groups caused an intensification of color. These groups were called **auxochromes** (Greek *auxanein,* "to increase"). We now know that auxochromes are groups that cannot undergo $\pi \to \pi^*$ transitions, but can undergo transitions of n electrons.

Some auxochromes:

—OH   —OR   —NH$_2$ —NHR   —NR$_2$ —X

Naphthoquinones and **anthraquinones** are common natural coloring materials. *Juglone* is a naphthoquinone that is partly responsible for the coloring of walnut hulls. *Lawsone* is similar in structure to juglone; it is found in Indian henna, which is used as a red hair dye. A typical anthraquinone, carminic acid, is the principal red pigment of *cochineal,* a ground-up insect *(Coccus cacti L.)* that is used as a red dye in food and cosmetics.

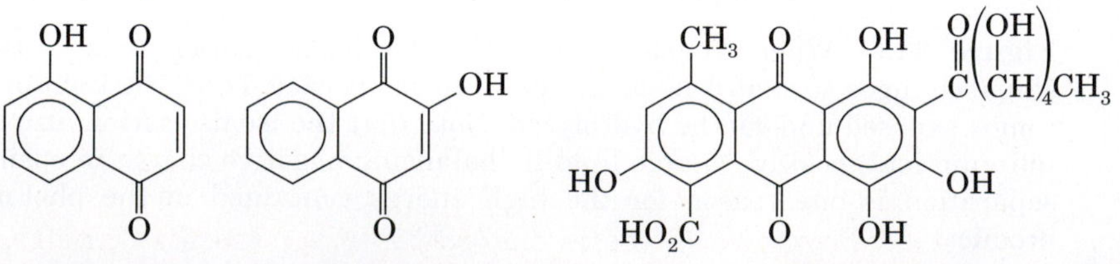

juglone   lawsone   carminic acid

A **dye** is a colored organic compound that is used to impart color to an object or a fabric. The history of dyes goes back to prehistoric times. *Indigo,* the oldest known dye, was used by the ancient Egyptians to dye mummy cloths. *Tyrian purple* (page 883), obtained from *Murex* snails found near the city of Tyre, was used by the Romans to dye the togas of the emperors.

To be useful as a dye, a compound must be *fast* (remain in the fabric during washing or cleaning). To be fast, a dye must, in one way or another, be bonded to the fabric. A fabric composed of fibers of polypropylene or a similar hydrocarbon is difficult to dye because it has no functional groups to attract dye molecules. Successful dyeing of these fabrics can be accomplished, however, by the incorporation of a metal–dye complex into the polymer. Dyeing of cotton (cellulose) is easier because hydrogen bonding between hydroxyl groups of the glucose units and groups of the dye molecule hold the dye to the cloth. Polypeptide fibers, such as wool or silk, are the easiest fabrics to dye because they contain numerous polar groups that can interact with dye molecules.

A **vat dye** is a dye that is applied to fabric (in a vat) in a soluble form and then is allowed to undergo reaction to an insoluble form. The blue coats supplied by the French to the Americans during the American Revolution were dyed with *indigo,* a typical vat dye. Indigo was obtained by a fermentation of the woad plant *(Isatis tinctoria)* of Western Europe or of plants of the *Indigofera* species, found in tropical countries. Both types of plants contain the glucoside *indican,* which can be hydrolyzed to glucose and *indoxyl,* a colorless precursor of indigo. Fabrics were soaked in the fermentation mixture containing indoxyl, then were allowed to air dry. Air oxidation of indoxyl yields the blue, insoluble indigo. Indigo is deposited in the *cis* form, which undergoes spontaneous isomerization to the *trans* isomer.

indoxyl   *cis*-indigo   *trans*-indigo

Azo dyes are the largest and most important class of dyestuffs, their numbers running into the thousands. In azo-dyeing, the fabric is first impregnated with an aromatic compound activated toward electrophilic substitution, then is treated with a diazonium salt to form the dye. (See Section 12.3B.)

Direct Blue 2B

an azo dye

An **acid–base indicator** is an organic compound that changes color with a change in pH. These compounds are most frequently encountered as titration endpoint indicators. Test papers, such as litmus paper, are impregnated with one or more of these substances.

Two typical indicators are *methyl orange* and *phenolphthalein*. Methyl orange is red in acidic solutions that have pH values less than 3.1. It is yellow in solutions with pH values greater than 4.4. Phenolphthalein, on the other hand, changes color on the alkaline side of the pH range. Up to pH 8.3, phenolphthalein is colorless. At pH 10, it is red. In strongly alkaline solutions, it again becomes colorless.

methyl orange:	red	orange		yellow	

pH: 2 3 4 5 6 7 8 9 10 11 12 13 14

phenolphthalein:	colorless		red	colorless

Indicators change color because the chromophoric system is changed by an acid–base reaction. In acidic solution, methyl orange exists as a resonance hybrid of a protonated azo structure; this resonance hybrid is red. The azo nitrogen is not strongly basic, and the protonated azo group loses the hydrogen ion at about pH 4.4. The loss of the proton changes the electronic structure of the compound, resulting in a change of color from red to yellow. Figure 22.6 shows spectra of methyl orange at two different pH values.

$$^-O_3S-\langle\bigcirc\rangle-\ddot{N}{=}\ddot{N}-\langle\bigcirc\rangle-\ddot{N}(CH_3)_2$$

methyl orange

yellow in base

$$OH^- \big\updownarrow\big| H^+$$

$$^-O_3S-\langle\bigcirc\rangle-\overset{+}{\underset{H}{N}}{=}\ddot{N}-\langle\bigcirc\rangle-\ddot{N}(CH_3)_2 \longleftrightarrow {}^-O_3S-\langle\bigcirc\rangle-\underset{H}{\ddot{N}}-\ddot{N}{=}\langle\bigcirc\rangle{=}\overset{+}{N}(CH_3)_2$$

red in acid

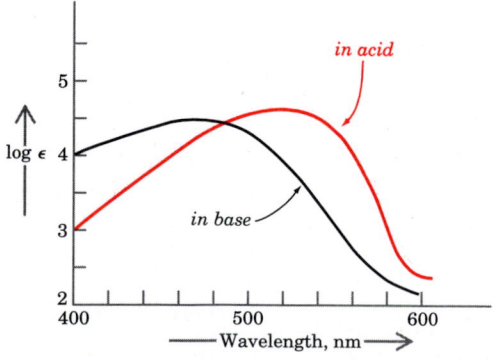

Figure 22.6 Visible spectra of methyl orange in acidic and alkaline solutions (pH 1 and 13, respectively).

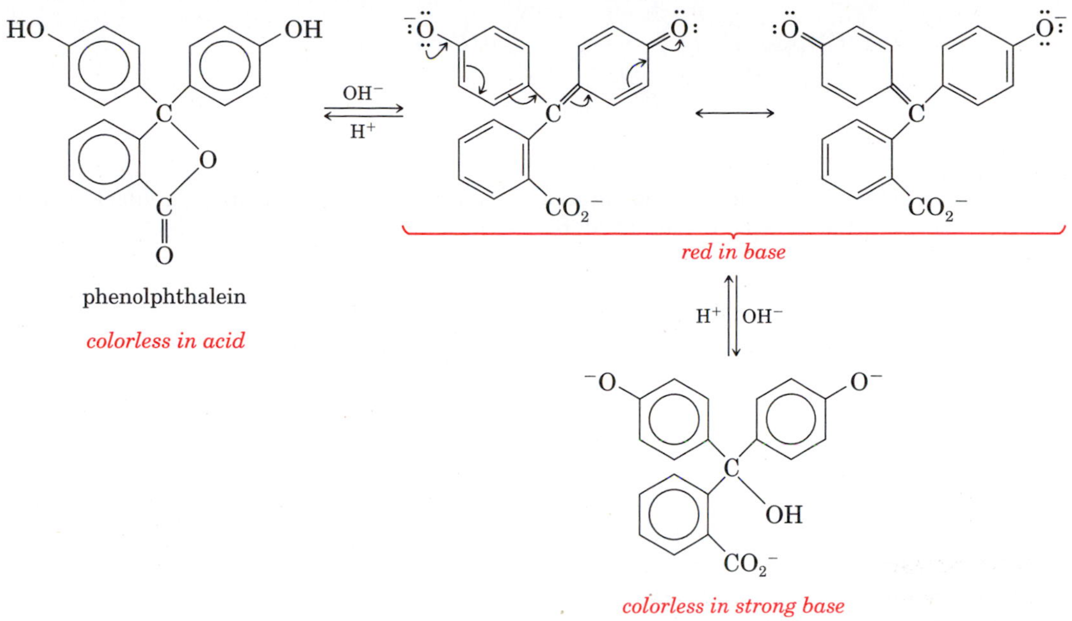

Figure 22.7   The acid–base reactions of phenolphthalein.

The commercial value of phenolphthalein is that it serves as the active ingredient in "candy" and "gum" laxatives. However, phenolphthalein is also one of the best-known titration indicators. In acidic solution, phenolphthalein is a colorless lactone. In the lactone, the center carbon is in the sp^3-hybrid state; consequently, the three benzene rings are isolated, not conjugated.

At pH values greater than 8.3 (alkaline solution), a phenolic hydrogen is removed from phenolphthalein, the lactone ring opens, and the center carbon becomes sp^2-hybridized. In this form, the benzene rings are in conjugation, and the extensive π system gives rise to the red color that is observed in mildly alkaline solution. Figure 22.7 shows these reactions.

In strongly alkaline solution, the center carbon of phenolphthalein is hydroxylated and is converted to the sp^3-hybrid state. This reaction isolates the three π systems again. At high pH values, phenolphthalein is colorless.

STUDY PROBLEMS

22.5   Identify (1) the chromophores and (2) the auxochromes in each of the following compounds:

(a)

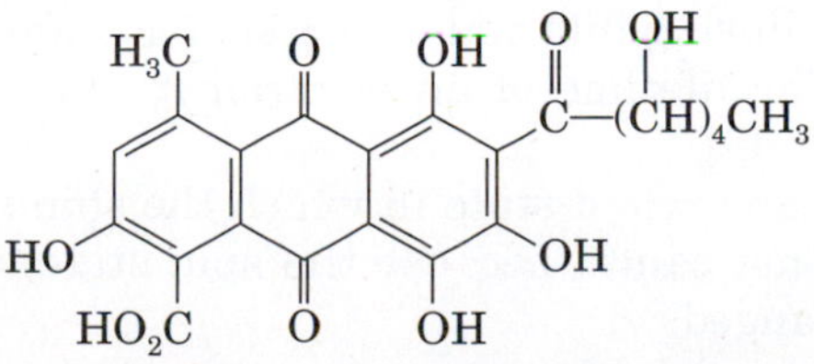

(b) $(CH_3)_2N$— —$C=$ $=\overset{+}{N}(CH_3)_2$ Cl^-

22.6 One of the following indicators is blue-green at pH 7; the other is violet. Which is which? Explain your answer.

(a)

(b)

Section 22.5
Photochemistry

The electrons of most stable compounds are paired and their spins (represented as $+1/2$ or $-1/2$) cancel each other. If the electron spins of a molecule cancel, regardless of whether the electrons are all paired in orbitals, the molecule is said to be in the **singlet state.** If the electrons are paired in their lowest-energy orbitals, the molecule is in the lowest-energy singlet state, or ground state, S_0.

When a molecule absorbs a photon of ultraviolet or visible light, an electron is promoted from a ground-state orbital to an excited-state orbital. This promotion of an electron is extremely rapid (about 10^{-15} second). The spin state of the electron that absorbs the energy does not change. Therefore, the molecule is still in the singlet state, now an **excited singlet state:** S_1, S_2, or higher excited state, depending on the energy of the excited state.

Immediately after promotion (in about 10^{-11} second), the electron drops to the lowest-energy excited singlet state (S_1) by a process called **internal conversion.** The energy lost during internal conversion is transformed into heat and molecular motion. The lifetime of an electron in the S_1 state is in the range of 10^{-8} to 10^{-7} second.

A **triplet state (T)** is an excited state in which the spin states of the electrons of the molecule do not cancel because the spin state of one electron in the molecule has been changed.

$$\pi_2^* \; \underline{\downarrow} \qquad\qquad\qquad \pi_2^* \; \underline{\uparrow}$$

$$\pi_1 \underline{\uparrow} \qquad\qquad\qquad \pi_1 \underline{\uparrow}$$

a π-bond excited singlet state;   *a π-bond triplet state;*
electrons of opposite spin   *electrons with the same spin*

The average energy of the lowest-energy triplet excited state T_1 is generally greater than that of S_0, but less than that of S_1. The process by which an electron in S_1 is transferred to T_1 is called **intersystem crossing.** An S_1 or T_1 electron must lose its excess energy and return to the ground state. The energy can be lost as heat, as light, or by a chemical reaction. In this section, we will briefly discuss two of these processes—photon emission (fluorescence) and photochemical reactions.

Figure 22.8 summarizes the relative energy relationships of the various energy states.

A. Fluorescence

Besides intersystem crossing, one way in which an excited molecule with electrons in the S_1 state can return to the ground state S_0 is to lose its energy as

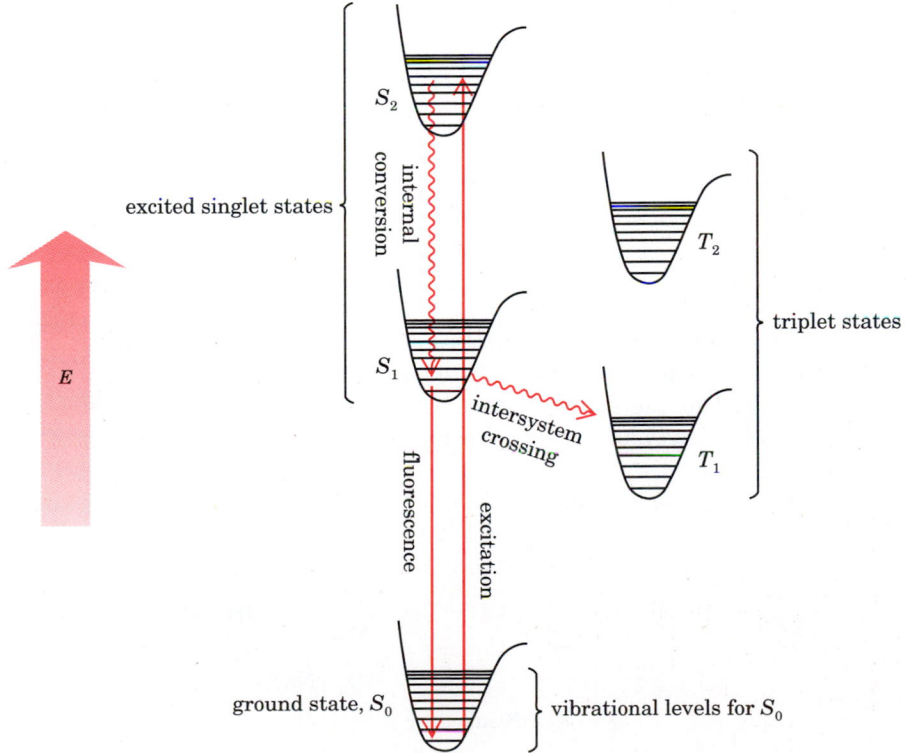

Figure 22.8 Simplified energy relationships of the ground state (S_0), the singlet excited states S_1 and S_2, and the triplet excited states T_1 and T_2. Each state contains various vibrational levels, as shown in the diagram and labeled for S_0. Energy can also be lost by excited species in chemical reactions and phosphorescence (emission of light when an electron in the T_1 state returns to the S_0 state).

light, a process called **fluorescence.** This process is fast (10^{-7} second). The energy lost by this emission of light is slightly *less* than the energy initially absorbed because of internal conversion. Consequently, the wavelength of light emitted is slightly longer than that absorbed.

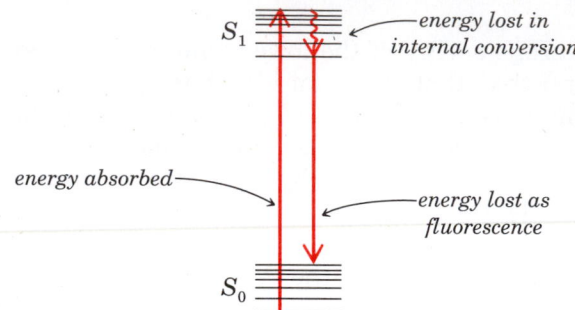

A compound that absorbs light in the visible range appears colored. When the same compound emits light of a different wavelength, it appears two-colored, or fluorescent. An example of a fluorescent compound is **fluorescein,** which has been used as a marker for airplanes downed at sea. In aqueous solution and in the presence of light, fluorescein appears red with an intense yellow-green fluorescence.

fluorescein

Some fluorescent compounds, called **optical bleaches,** are used as fabric whiteners. These are colorless compounds that absorb ultraviolet light just out of the visible range, then emit blue-violet light at the edge of the visible spectrum. This blue-violet color masks yellowing of the fabric.

Two optical bleaches:

Blankophor R

Calcofluor SD

B. Photochemical Reactions

We have already discussed some types of photochemical reactions. The photolysis, or cleavage caused by light, of Cl_2 into $2\ Cl\cdot$ as the initiation step in free-radical halogenation (Chapter 6) is one example. Many pericyclic reactions (Chapter 21) are also photochemical reactions. Let us introduce a few other types of these reactions.

Photoisomerization Isomerization of *cis* and *trans* isomers is possible under proper experimental conditions. A *cis*-to-*trans* isomerization is usually an exothermic process that occurs when the two isomers are allowed to equilibrate. The difference in energy between the *cis* and *trans* isomers determines the position of the equilibrium and, thus, the extent of the isomerization.

Isomerization of a *trans* isomer to a *cis* isomer is usually endothermic and, thus, equilibration does not favor the *cis* isomer. However, a *trans* isomer can be converted to *cis* by irradiation with a frequency that promotes a π electron of the *trans* system, but not a π electron of the *cis* system.

When a π electron is excited, the π bond of either the *cis* or *trans* isomer can assume a twisted shape. However, if the frequency of radiation is chosen properly, the *cis* isomer cannot absorb radiation, and its reaction leading back to the *trans* isomer cannot occur.

trans   twisted S_1 or T_1 *cis*
cannot absorb $h\nu$
and thus accumulates

An example of a photoisomerization reaction follows:

trans-dibenzoylethene   *cis*-dibenzoylethene

Photodimerization Irradiation of an α,β-unsaturated carbonyl compound or a conjugated polyene can lead to dimerization with the formation of four-membered rings (cycloaddition; see Section 21.2). For example, cyclopentenone undergoes dimerization to yield the following diones.

Photolysis A diatomic molecule, such as bromine or chlorine, can dissociate into a pair of radicals after the absorption of a photon of light. Ultraviolet light can also cause aldehydes and ketones to undergo photolytic reactions. Photolytic reactions take place by two principal pathways—the **Norrish type I** and the **Norrish type II,** named to honor R. G. W. Norrish, who received the 1967 Nobel Prize in Chemistry for his work on photochemical reactions.

Norrish type I reactions, which apply mainly to ketones in the gaseous state, involve the homolysis of the σ bond joining the carbonyl carbon to the α carbon and subsequent free-radical reactions. The products depend upon the structures of the radicals. An example of a Norrish type I reaction is the reaction caused by the irradiation of gaseous acetone at 313 nm. The primary photochemical reaction is the formation of a methyl radical ($\cdot CH_3$) and an ethanoyl radical ($CH_3CO\cdot$). The products result from the free-radical reactions of these two radicals.

Norrish type I fragmentation:

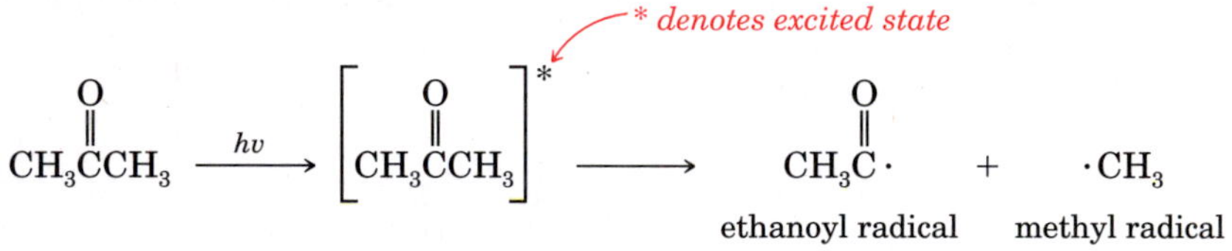

Subsequent free-radical reactions:

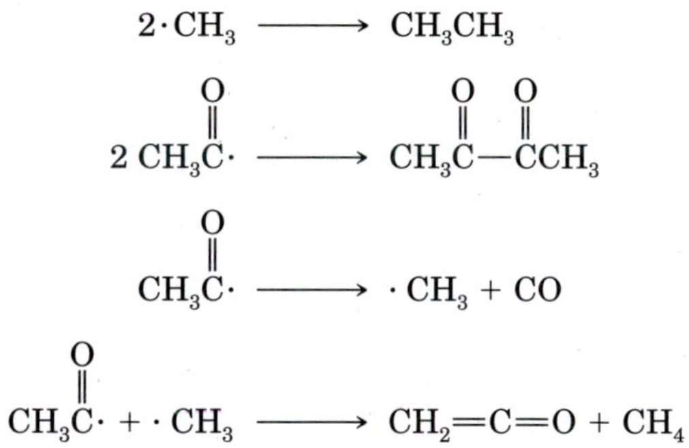

A *Norrish type II* pathway involves an initial excitation of an aldehyde or a ketone by ultraviolet light followed by an internal abstraction of a γ hydrogen. The resulting diradical can fragment, cyclize, or form other radical products.

A Norrish type II reaction:

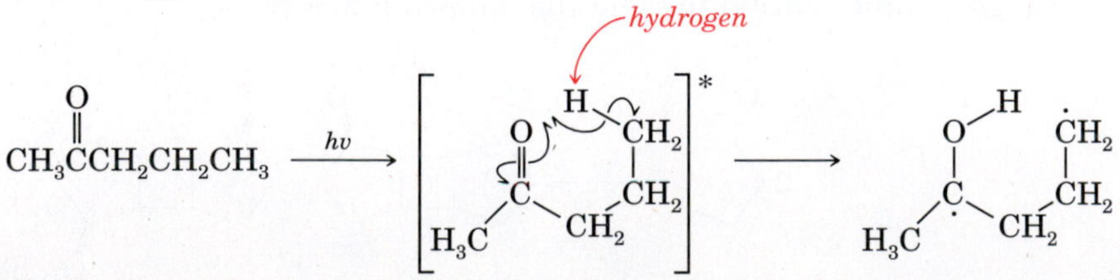

Subsequent free-radical reactions:

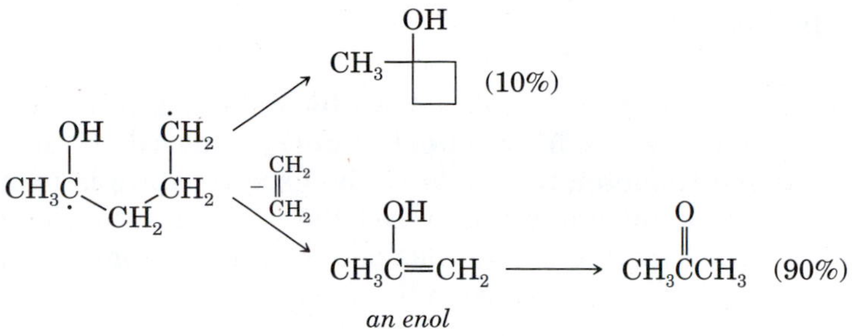

an enol

Photoreduction of aryl ketones

Irradiation of an aryl ketone, such as benzophenone, in the presence of a hydrogen donor, such as a secondary alcohol, causes a reductive coupling reaction.

The initial photo product is an excited singlet state, which then goes to a triplet state. The triplet contains two unpaired electrons and thus has diradical character. This radical can abstract a hydrogen atom from some other molecule in solution. In the photoreduction, the diradical abstracts a hydrogen atom from the secondary alcohol and forms two radicals that can dimerize to yield the observed product.

Step 1 (formation of T_1):

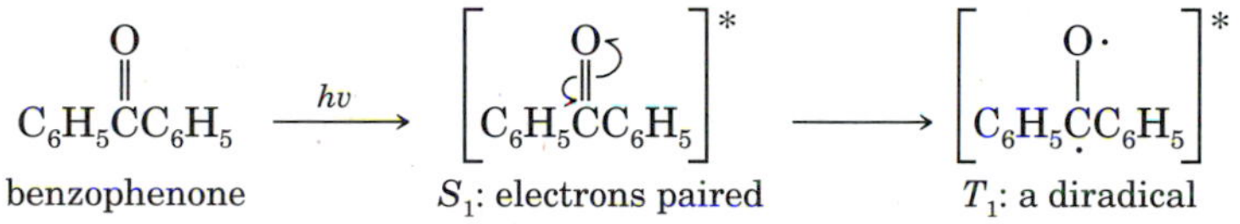

Step 2 (hydrogen abstraction):

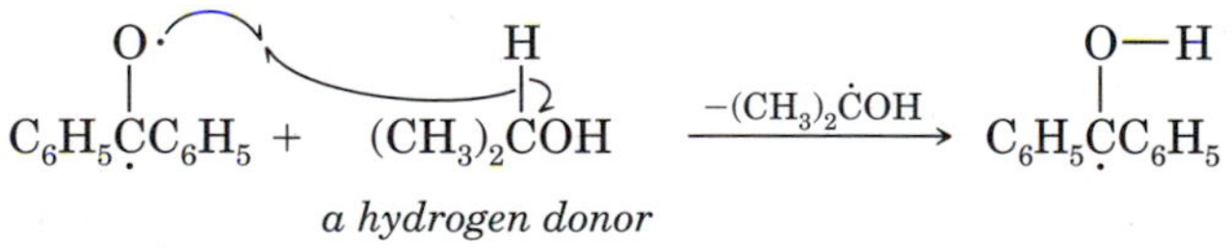

Step 3 (dimerization):

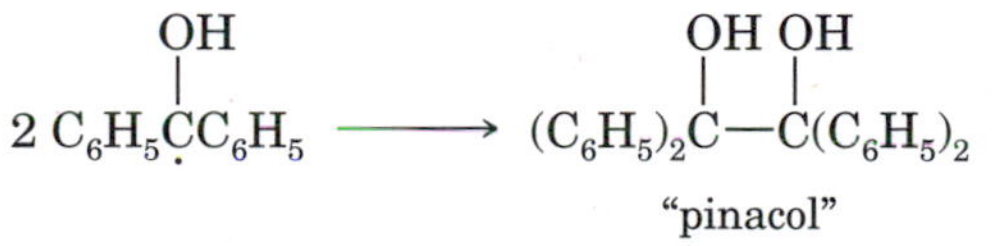

"pinacol"

STUDY PROBLEM

22.7 Large-ring ketones undergo a cyclization reaction upon photolysis. The mechanism is similar to a Norrish type II process. Suggest a mechanism for the following photochemical reaction.

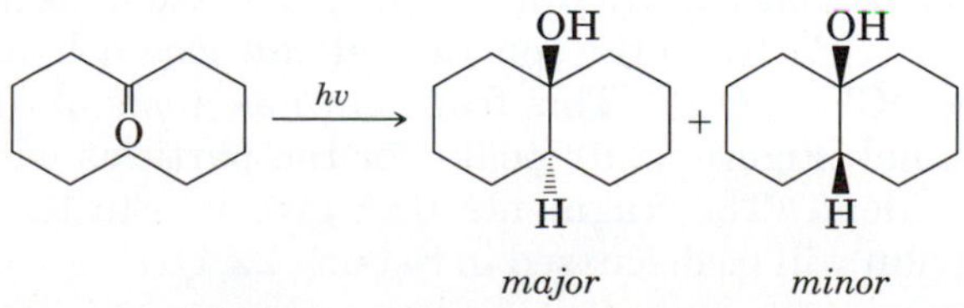

major   *minor*

Section 22.6

Mass Spectrometry

Most of the spectral techniques we have discussed arise from absorption of energy by molecules. **Mass spectrometry** is based on different principles. In a mass spectrometer, a sample in the gaseous state is bombarded with electrons of sufficient energy to exceed the first ionization potential of the compound. (The ionization potentials of most organic compounds are in the range of 185–300 kcal/mol.) Collision between an organic molecule and one of these high-energy electrons results in the loss of an electron from the molecule and the formation of an organic ion. The organic ions that result from this high-energy electron bombardment are unstable and fly apart into smaller fragments, both radicals and other ions. In a typical mass spectrometer, the *positively charged fragments* are detected. The **mass spectrum** is a plot of **abundance** (the relative amounts of the different positively charged fragments) versus the **mass-to-charge ratio** (*m/e*, or *m/z*) of the fragments. The ionic charge of most particles detected in a mass spectrometer is +1. The *m/e* value for such an ion is equal to its mass. Consequently, from a practical standpoint, the mass spectrum is a record of particle mass versus relative abundance of the particles.

How a molecule or ion breaks into fragments depends upon the carbon skeleton and functional groups present. Therefore, the structure and mass of the fragments give clues about the structure of the parent molecule. Also, it is frequently possible to determine the molecular weight of a compound from its mass spectrum.

Let us introduce mass spectrometry using methanol as an example. When methanol is bombarded with high-energy electrons, one of the valence electrons is lost. The result is an **ion radical,** a species with one unpaired electron and a charge of +1. An ion radical is symbolized by $^{+\cdot}$. The ion radical that results from abstraction of one electron of a molecule is called the **molecular ion** and is symbolized $M^{+\cdot}$. The mass of the molecular ion is the molecular weight of the compound. The molecular ion of methanol has a mass of 32 and a charge of +1. Its mass-to-charge ratio (*m/e*) is 32. (In the following example, the half-headed arrow signifies the loss of one electron from a methanol molecule.)

$$e^- + CH_3\ddot{O}H \xrightarrow{-2e^-} CH_3\dot{O}H^+, \quad \text{usually written} \quad [CH_3OH]^{+\cdot}$$

the molecular ion of methanol, m/e = 32

A molecular ion can undergo fragmentation after it has been formed. In the case of $[CH_3OH]^{+\cdot}$, the ion radical can lose a hydrogen atom and become a cation: $[CH_2{=}OH]^+$. This fragment has a *m/e* of 31. In the mass spectrum of methanol (Figure 22.9), peaks for the particles with *m/e* values of 31 and 32 are evident. (The fragments that give rise to the other peaks in this mass spectrum will be discussed in Section 22.11.)

$$\left[\begin{array}{c} \text{H} \\ \text{H:C:O:H} \\ \text{H} \end{array}\right]^{\ddagger} \longrightarrow \text{H}\cdot\; + \;\left[\begin{array}{c} \text{H} \\ \text{:C::O:H} \\ \text{H} \end{array}\right]^{+}$$

a cation, m/e = 31
(not an ion radical because
all electrons are paired)

As can be seen in Figure 22.9, a mass spectrum is presented as a bar graph. Each peak in the spectrum represents a fragment of the molecule. The fragments are scanned so that the peaks are arranged by increasing *m/e* from left to right in the spectrum. The intensities of the peaks are proportional to the relative abundance of the fragments, which in turn depends on their relative stabilities. By convention, the tallest peak in a spectrum, called the **base peak,** is given the intensity value of 100%. Lesser peaks are reported as 20%, 30%, or whatever their value is relative to the base peak. The base peak sometimes arises from the molecular ion, but often it arises from a smaller fragment.

STUDY PROBLEMS

22.8   Write the formula for each product and tell whether it is a cation, an ion radical, or a radical:

(a) CH_4 minus one e^- (b) $[CH_4]^{\ddagger}$ minus $H\cdot$

(c) $[CH_3CH_2]^+$ minus $H\cdot$ (d) $[CH_3CH_3]^{\ddagger}$ minus $H\cdot$

22.9   Write the formulas for the molecular ions of (a) CH_4 and (b) $CH_3CH_2CH_3$.

22.10   Give the *m/e* value for each of the following particles:

(a) $[CH_4]^{\ddagger}$ (b) $[(CH_3)_2CH]^+$ (c) $[O_2]^{\ddagger}$ (d) $[H_2O]^{\ddagger}$

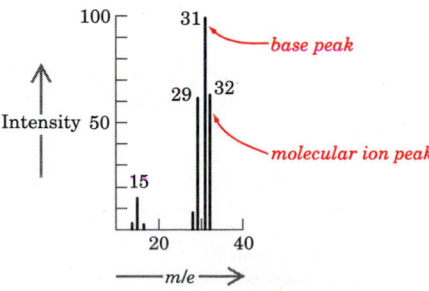

Figure 22.9   Mass spectrum of methanol, CH_3OH.

The Mass Spectrometer

A diagram of a common type of mass spectrometer is shown in Figure 22.10. The sample is introduced, vaporized, and allowed to feed in a continuous stream to the *ionization chamber.* The ionization chamber (as well as the entire instrument) is kept under vacuum to minimize collisions and reactions between radicals, air molecules, and so forth. In this chamber, the sample passes through a stream of high-energy electrons, which causes the ionization of some of the sample molecules into their molecular ions.

After its formation, a molecular ion can undergo fragmentation and rearrangement. These processes are extremely rapid ($10^{-10} - 10^{-6}$ sec). The longer-lived particles can be detected by the ion collector, but a shorter-lived particle may not have a sufficient lifetime to reach the ion collector. In some cases, the molecular ion is too short-lived to be detected, and only its fragmentation products exhibit peaks.

As the ion radicals and other particles are formed, they are fed past two electrodes, the *ion accelerator plates,* which accelerate the positively charged particles. (The neutral and negatively charged particles are not accelerated and are removed continuously by the vacuum pumps.) From the accelerator plates, the positively charged particles pass into the *analyzer tube,* where they are deflected into a curved path by a magnetic field.

The radius of the curved path depends upon the particle's velocity, which in turn is dependent on the magnetic field strength, the accelerating voltage, and the *m/e* of the particle. At the same field strength and voltage, the particles of higher *m/e* have a path with a wider radius, while the lower-*m/e* particles have a path of smaller radius (see Figure 22.11). The continuous flow of positively charged particles through the analyzer tube therefore forms a pattern: higher-*m/e* particles with a larger radius and lower-*m/e* particles with a smaller radius. If the accelerating voltage is slowly and continuously decreased, the velocities of all the particles decrease, and the radii of the paths of all the particles also decrease. By this technique, particles of successively higher *m/e* are allowed to strike the detector. Figure 22.11 shows the effects of decreasing the accelerating voltage on the paths of particles with three differ-

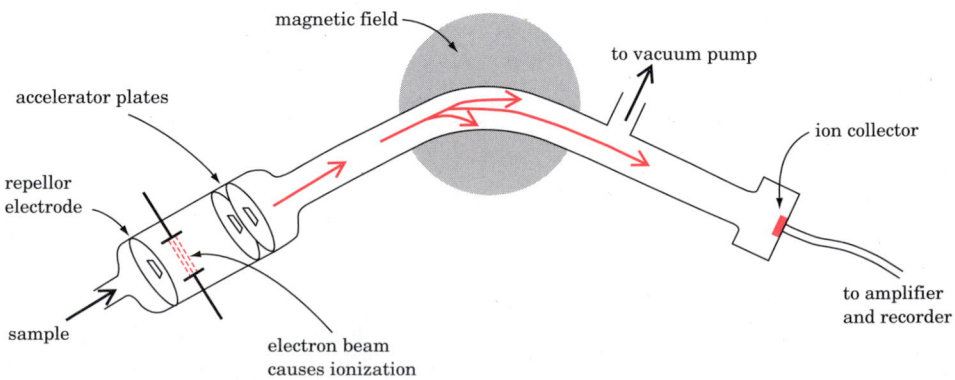

Figure 22.10 Diagram of a mass spectrometer.

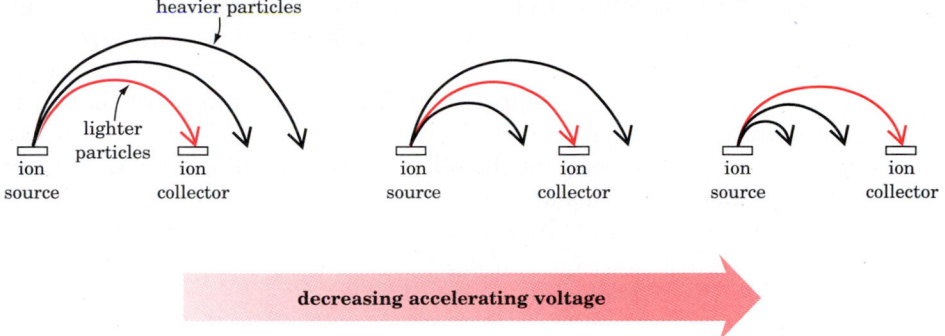

Figure 22.11 As the accelerating voltage is decreased, particles of successively higher *m/e* hit the ion collector.

ent *m/e* values. (The same effect can be obtained by increasing the magnetic field strength instead of decreasing the accelerating voltage.)

Section 22.8

Isotopes in Mass Spectra

A mass spectrometer is so sensitive that particles differing by 1.0 mass unit give separate signals. The molecular weight of CH_3Br is 94.9 (15.0 atomic mass units for CH_3 and 79.9 for Br). However, the mass spectrum of this compound (Figure 22.12) does not show one molecular ion peak at *m/e* = 94.9. Instead, two peaks are observed, one at *m/e* = 94 and the other at *m/e* = 96. The reason for the two peaks is that naturally occurring Br exists in two isotopic forms, one with atomic mass 79 and the other with atomic mass 81. When we calculate the molecular weight of a bromine compound, we use a weighted average of these two isotopic masses (79.9). Because a mass spectrometer detects particles containing each of these isotopes as individual species, we cannot use average atomic masses when dealing with mass spectra, as we do when calculating the stoichiometry of a chemical reaction.

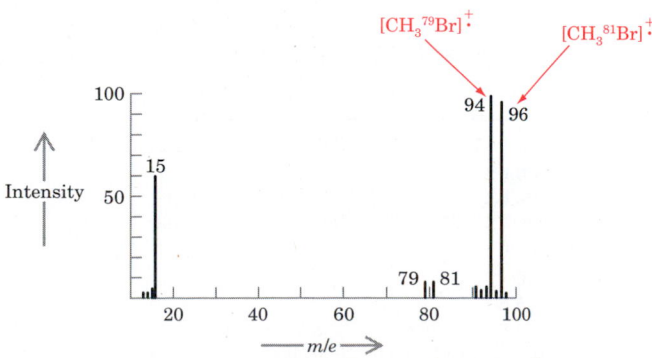

Figure 22.12 Mass spectrum of bromomethane, CH_3Br.

Table 22.4 lists the common natural isotopes and their relative abundance. Naturally occurring bromine exists as a 50.5%–49.5% mixture of bromine-79 and bromine-81, respectively. Particles of the same structure containing Br give a pair of peaks of approximately the same intensity that are 2.0 mass units apart. *The particle containing the lower-mass isotope is considered the molecular ion.* The peak for the other particle is called the **M + 2 peak** (molecular ion plus two mass units).

Naturally occurring chlorine is a mixture of 75.5% chlorine-35 and 24.5% chlorine-37. The particle containing chlorine-35 is considered the molecular ion, while the particle containing chlorine-37 gives rise to the M + 2 peak, which has an intensity approximately one-third that of the molecular ion peak.

Most elements common in organic compounds (except for Cl and Br) exist in nature as predominantly one isotope. For example, carbon is 98.89% carbon-12, and hydrogen is 99.985% hydrogen-1. For this reason, we generally assume for mass spectral purposes that all of C is carbon-12 and ignore the tiny proportion that is carbon-13. The presence of the isotopes of the common elements explains the multitude of small peaks surrounding a large peak in a mass spectrum (for example, the small peaks around the peaks at $m/e = 94$ and $m/e = 96$ in Figure 22.12.

STUDY PROBLEMS

22.11 Calculate the m/e value for the molecular ion of each of the following compounds:

(a) ethane; (b) 1,2-dichloroethane;

(c) ethanol; (d) *p*-bromophenol.

22.12 Figure 22.13 contains mass spectra of four compounds. Which of these compounds contains Br? Which contains Cl?

Table 22.4	Natural abundance of some isotopes				
Isotope	Abundance, %	Isotope	Abundance, %	Isotope	Abundance, %
^1H	99.985	^2H	0.015		
^{12}C	98.89	^{13}C	1.11		
^{14}N	99.63	^{15}N	0.37		
^{16}O	99.76	^{17}O	0.04	^{18}O	0.20
^{32}S	95.0	^{33}S	0.76	^{34}S	4.2
^{19}F	100				
^{35}Cl	75.5			^{37}Cl	24.5
^{79}Br	50.5			^{81}Br	49.5
^{127}I	100				

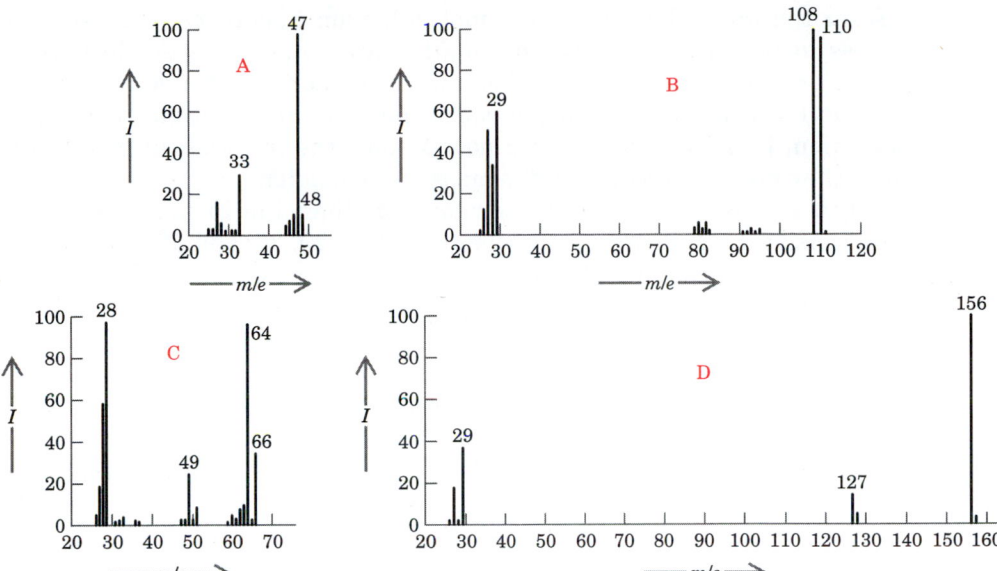

Figure 22.13 Mass spectra for Problem 22.12.

Section 22.9

Ionization and Fragmentation in Mass Spectra

In the mass spectrometer, the first reaction of a molecule is the initial ionization—the abstraction of a single electron. The loss of an electron gives rise to the molecular ion. From the peak for this ion radical, which is usually the peak that is farthest to the right in the spectrum, the molecular weight of the compound may be determined. (Remember, it is the exact molecular weight for a molecule containing single isotopes and not an average molecular weight.)

The question arises, "Which type of electron is lost from the molecule?" This question cannot be answered with accuracy. It is believed that the electron in the highest-energy orbital (the "loosest" electron) is the first to be lost. If a molecule has n (unshared) electrons, one of these is lost. If there are no n electrons, then a π electron is lost. If there are neither n electrons nor π electrons, the molecular ion is formed by loss of a σ electron.

$$CH_3\overset{\frown}{\underset{\cdot\cdot}{\ddot{O}}}H \longrightarrow [CH_3OH]^{\ddot{+}} + n\,e^-$$

$$CH_3CH\overset{\frown}{::}CH_2 \longrightarrow [CH_3CH-CH_2]^{\ddot{+}} + \pi\,e^-$$

$$H\overset{\frown}{:}CH_3 \longrightarrow [CH_4]^{\ddot{+}} + \sigma\,e^-$$

After the initial ionization, the molecular ion undergoes fragmentation, a process in which free radicals or small neutral molecules are lost from the molecular ion. A molecular ion does not fragment in a random fashion, but tends to form the *most stable fragments possible*. In equations showing fragmentation, it is common practice not to show the free-radical fragments because they are not detected by the mass spectrometer.

Let us reconsider the mass spectrum of methanol in Figure 22.9. The spectrum consists of three principal peaks at m/e = 29, 31, and 32. The structures of the fragments can often be deduced from their masses. The $M^{\ddot{+}}$ peak (at 32) of methanol arises from loss of one electron. The peak at 31 must arise from loss of an H atom (which has a mass of 1.0). The peak at 29 must arise from an ion that has lost two more H atoms. What about the minor peak at 15? This peak arises from the loss of $\cdot OH$ from the molecular ion.

$$[CH_2{=}OH]^+ \xrightarrow{\ -H_2\ } [CH{\equiv}O]^+$$
$$m/e = 31 \qquad\qquad m/e = 29$$

$$CH_3OH \xrightarrow{\ -e^-\ } [CH_3OH]^{\ddot{+}}$$
$$m/e = 32$$

$-H\cdot$ (upper branch) $\quad -\cdot OH$ (lower branch)

$$[CH_3]^+$$
$$m/e = 15$$

Could other fragmentation patterns occur? For example, could the molecular ion lose H^+ to become $[CH_3O]^{\cdot}$? It possibly could, but we do not know, because only the positively charged particles are accelerated and detected.

A. Effect of Branching

Branching in a hydrocarbon chain leads to fragmentation primarily at the branch because secondary ion radicals and secondary carbocations are more stable than primary ion radicals and primary carbocations. Carbocation stability is a more important factor than free-radical stability. For example, a methylpropane molecular ion yields predominantly an isopropyl cation and a methyl radical (not the reverse).

$$\left[\begin{array}{c} CH_3 \\ | \\ CH_3CH \frown CH_3 \end{array} \right]^{\ddot{+}} \longrightarrow \begin{array}{c} CH_3 \\ | \\ CH_3CH^+ \end{array} + \cdot CH_3 \quad\text{and very little}\quad \begin{array}{c} CH_3 \\ | \\ CH_3\underset{\cdot}{C}H \end{array} + {}^+CH_3$$

22.13 What would be the molecular ion and the principal positively charged fragments arising from ionization of:

(a) 2-methylpentane;   (b) 2,2-dimethylpropane;   (c) 1-pentene?

B. Effect of a Heteroatom or Carbonyl Group

Consider another spectrum, that of *N*-ethylpropylamine (Figure 22.14). The molecular ion has m/e = 87. Fragmentation of this molecular ion takes place

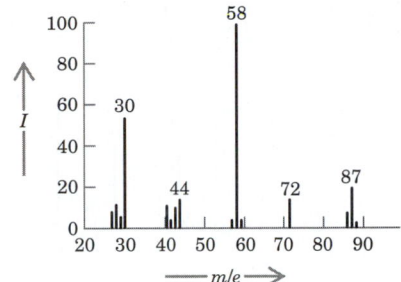

Figure 22.14 Mass spectrum of *N*-ethylpropylamine, $CH_3CH_2NHCH_2CH_2CH_3$.

α to the nitrogen atom and yields fragments with $m/e = 58$ (loss of an ethyl group) and 72 (loss of a methyl group). This type of fragmentation is called **α-fission** and is common in both amines and ethers.

$$[CH_3 - CH_2NHCH_2 - CH_2CH_3]^{\ddagger}$$
$$m/e = 87$$

$$\xrightarrow{-\cdot C_2H_5} [CH_3CH_2NH = CH_2]^+$$
$$m/e = 58$$

$$\xrightarrow{-\cdot CH_3} [CH_2 = NHCH_2CH_2CH_3]^+$$
$$m/e = 72$$

fission here

The reason for α-fission is that the cation formed in this reaction is resonance-stabilized:

$$[R - CH_2 - \overset{\cdot\cdot}{N}HR]^{\ddagger} \xrightarrow{-R\cdot} [CH_2 = \overset{+}{N}HR \longleftrightarrow \overset{+}{C}H_2 - \overset{\cdot\cdot}{N}HR]$$

A similar fragmentation occurs at a bond adjacent to a carbonyl group (or α to the oxygen). Again, the resultant cation is resonance-stabilized.

$$\left[\begin{array}{c} \overset{\cdot\cdot}{\underset{\cdot\cdot}{O}:} \\ \| \\ RC - R \end{array} \right]^{+} \xrightarrow{-R\cdot} [RC \equiv \overset{+}{O}: \longleftrightarrow R\overset{+}{C} = \overset{\cdot\cdot}{O}:]$$

STUDY PROBLEM

22.14 The mass spectrum of an ether is shown in Figure 22.15. What is the structure of the ether?

C. Loss of a Small Molecule

Small stable molecules, such as H_2O, CO_2, CO, and C_2H_4, can be lost from a molecular ion. An alcohol, for example, readily loses H_2O and shows a peak at 18 mass units less than the peak of the molecular ion. This peak is referred to

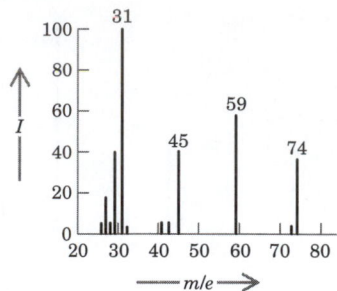

Figure 22.15 Mass spectrum for Problem 22.14.

as the **M–18 peak.** In many alcohols, elimination of H_2O is so facile that the molecular-ion peak is not even observed in the spectrum. The spectrum of 1-butanol (Figure 22.16) is a typical mass spectrum of an alcohol.

$$[CH_3CH_2CH_2CH_2OH]^{+\cdot} \xrightarrow{-H_2O} [CH_3CH_2CHCH_2]^{+\cdot}$$

$$M^{+\cdot},\ m/e = 74 \qquad\qquad M - 18,\ m/e = 56$$

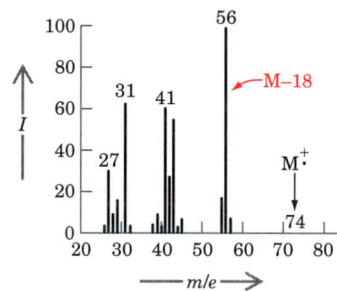

Figure 22.16 Mass spectrum of 1-butanol.

D. McLafferty Rearrangement

When there is a hydrogen atom γ to a carbonyl group in the molecular ion, a **McLafferty rearrangement** may occur. In this rearrangement, an alkene is lost from the molecular ion.

γ hydrogen

$$-\underset{|}{\overset{H}{\underset{|}{C}}}CH_2CH_2\overset{O}{\overset{\|}{C}}-$$

Rearrangement:

$$\left[\begin{array}{c} H_2C\ \overset{H}{\diagdown}\ \overset{\cdot\cdot}{O}\cdot \\ \| \\ H_2C\diagdown\ CH \\ CH_2 \end{array}\right]^{+\cdot} \longrightarrow \begin{array}{c} H_2C \\ \| \\ H_2C \end{array} + \left[\begin{array}{c} H\overset{\cdot\cdot}{O}\cdot \\ | \\ CH \\ \| \\ CH_2 \end{array}\right]^{+\cdot}$$

$$m/e = 72 \qquad\qquad m/e = 44$$

STUDY PROBLEM

22.15 Predict the *m/e* values for the products of the McLafferty rearrangement of the following compounds:

$$\text{(a)} \quad CH_3CH_2CH_2\overset{\displaystyle O}{\overset{\displaystyle \|}{C}}CH_3 \qquad \text{(b)} \quad CH_3\overset{\displaystyle CH_3}{\overset{\displaystyle |}{C}}HCH_2\overset{\displaystyle O}{\overset{\displaystyle \|}{C}}H$$

$$\text{(c)} \quad CH_3\overset{\displaystyle O}{\overset{\displaystyle \|}{C}}OCH_2CH_2CH_3 \qquad \text{(d)} \quad (CH_3)_2CHCH_2\overset{\displaystyle O}{\overset{\displaystyle \|}{C}}OCH_2CH_3$$

Summary

Absorption of ultraviolet (200–400 nm) or visible (400–750 nm) light results in **electronic transitions,** promotion of electrons from the ground-state orbitals to orbitals of higher energy. The wavelength λ of absorption is inversely proportional to the energy required. The UV or visible spectrum is a plot of **absorbance A** or **molar absorptivity ϵ** vs λ, where $\epsilon = A/cl$. The position of maximum absorption is reported as λ_{max}.

The important electronic transitions are $\pi \to \pi^*$ for conjugated systems and $n \to \pi^*$. Increasing amounts of conjugation result in shifts of λ_{max} toward longer wavelengths. Compounds that absorb at wavelengths longer than 400 nm are colored; the apparent color is the **complementary color** of the wavelength absorbed.

Table 22.5 Summary of some fragmentation patterns in mass spectra

Fragmentation	Reaction Type
Alkanes:	
$[R_2CH \!\!\dashv\!\! CH_3]^{\ddagger} \xrightarrow{-\cdot CH_3} R_2\overset{+}{C}H$	σ-bond fission to most stable carbocation
Amines and Ethers:	
$[R\!\!\dashv\!\!CH_2{-}NR_2']^{\ddagger} \xrightarrow{-R\cdot} CH_2{=}\overset{+}{N}R_2'$	α-fission
$[R\!\!\dashv\!\!CH_2{-}OR']^{\ddagger} \xrightarrow{-R\cdot} CH_2{=}\overset{+}{O}R'$	α-fission
Carbonyl Compounds:	
$\left[\begin{matrix} O \\ \| \\ RC\!\!\dashv\!\!Y \end{matrix}\right]^{\ddagger} \xrightarrow{-Y\cdot} R\overset{+}{C}{=}O$	α-fission
$\left[\begin{matrix} H & & O \\ \| & & \| \\ R_2CCH_2 \!\dashv\! CH_2CY \end{matrix}\right]^{\ddagger} \xrightarrow{-R_2C=CH_2} \left[\begin{matrix} OH \\ \| \\ CH_2{=}CY \end{matrix}\right]^{\ddagger}$ where Y = H, R', OH, OR', etc.	McLafferty rearrangement[a]
Alcohols:	
$\left[\begin{matrix} OH \\ \| \\ R_2CHCR_2 \end{matrix}\right]^{\ddagger} \xrightarrow{-H_2O} [R_2C{=}CR_2]^{\ddagger}$	loss of H_2O

[a]If a ketone or ester undergoes rearrangement and if Y contains a γ hydrogen, two types of rearrangements may be observed.

Vision is made possible by the conversion of 11-*cis*-retinal in rhodopsin to all-*trans*-retinal.

Dyes are colored compounds that adhere to fabric or other substance. An **acid–base indicator** is a compound that undergoes a color change in a reaction with acid or base. The color change arises from a change in the conjugated system and thus in the wavelength of absorption.

Photochemistry is the study of physical changes, such as fluorescence, and chemical changes caused by the absorption of ultraviolet or visible light. Photochemical reactions include *cis-trans* isomerization, dimerization of polyunsaturated compounds, photolysis of carbonyl compounds, and photoreduction of aryl ketones.

A **mass spectrum** is a graph of **abundance** versus **mass-to-charge** ratio (*m/e*) of positively charged particles that result from bombardment of a compound with high-energy electrons. Removal of one electron from a molecule of the compound results in the **molecular ion.** The molecular ion can lose atoms, ions, radicals, and small molecules to yield a variety of fragmentation products. Table 22.5 summarizes some of the fragmentation patterns.

Essay Problem for Chapter 22

Photocyclization of an Enone and an Alkene

A solution of 3-methyl-2-cyclohexenone in dichloromethane is placed in a Pyrex® reaction vessel and cooled with a dry ice–isopropyl alcohol bath. The solution is saturated with ethylene and then irradiated with ultraviolet light for about 8 hours. Concentration of the solution followed by distillation of the concentrate yields the product in an 86–90% yield. The spectral properties of the product follow: IR (CCl_4) cm^{-1}: 1700; ^1H NMR (CCl_4) δ: 1.21 (singlet, 3 H, methyl), 1.9 (multiplet, 11 H, all other protons); ^{13}C NMR δ (based on δ C_6D_6 128.00): 211.63, 51.34, 40.86, 39.45, 35.26, 31.20, 28.84, 21.45, 20.35. (R. L. Cargill, J. R. Dalton, G. H. Morton, and W. E. Caldwell, *Org. Syn.* **1984,** *62,* 118.)

Questions
1. Is this photocyclization a [2 + 2] or a [2 + 4] cycloaddition reaction?
2. How many different types of carbons are there in the product?
3. Predict the product of this photoaddition and its stereochemistry.
4. What is the reason for the use of a Pyrex® reaction vessel?
5. Predict the structures of the products when the following compounds are photocyclized with ethylene.

22.16 List the types of UV-useful electronic transitions that you would expect the following compounds to exhibit:

    (a) $CH_3\overset{\displaystyle O}{\overset{\|}{C}}H$ (b) $CH_2{=}CHBr$ (c) $CH_2{=}CH\overset{\displaystyle O}{\overset{\|}{C}}H$ (d) CH_3OH

22.17 Which of the following pairs of compounds could easily be differentiated by ultraviolet spectroscopy? Explain.

    (a) $CH_2{=}CHCH_2CH{=}CHCH_3$ and $CH_3CH{=}CHCH{=}CHCH_3$

    (b) $(CH_3)_2CHCH_2\overset{\displaystyle O}{\overset{\|}{C}}CH_3$ and $CH_3CH_2CH_2CH_2\overset{\displaystyle O}{\overset{\|}{C}}CH_3$

    (c) $BrCH_2CH_2CH_2\overset{\displaystyle O}{\overset{\|}{C}}CH{=}CH_2$ and $CH_3\overset{\displaystyle Br}{\overset{|}{C}}HCH_2\overset{\displaystyle O}{\overset{\|}{C}}CH{=}CH_2$

22.18 Calculate the molar absorptivities of the following compounds at the specified wavelengths:

    (a) adenine ($9.54 \times 10^{-5}\ M$ solution, 1.0-cm cell), absorbance of 1.25 at 263 nm

    (b) cyclohexanone (0.038 M solution, 1.0-cm cell), absorbance of 0.75 at 288 nm

22.19 3-Buten-2-one shows UV absorption maxima at 219 nm and 324 nm. **(a)** Why are there two maxima? **(b)** Which has the greater ϵ_{max}?

22.20 Predict which member of each of the following pairs of compounds would have a λ_{max} at the longer wavelength:

    (a) $CH_3CH{=}CHC\overset{\displaystyle O}{\overset{\|}{}}CH_3$ or $CH_2{=}CHCH_2\overset{\displaystyle O}{\overset{\|}{C}}CH_3$

    (b) 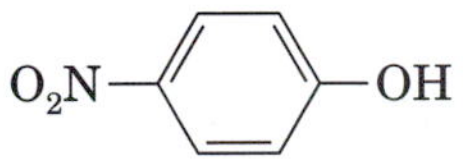 $O_2N{-}\langle\!\bigcirc\!\rangle{-}OH$ or (a benzene ring with NO_2 and OH substituents)

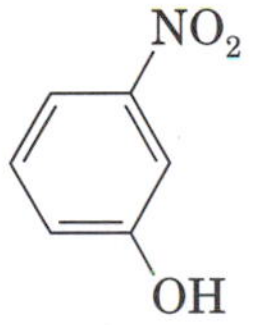

22.21 *Indophenol blue* is a vat dye that is oxidized to an insoluble blue dye after application to fabric. Give the structure of the oxidized form.

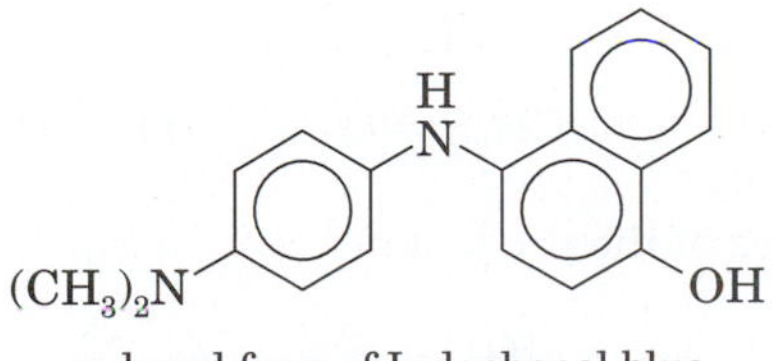

reduced form of Indophenol blue

22.22 *Phenol red* is an acid–base indicator that is yellow at pH 6 but red at pH 9. Draw the structures (and resonance structures) for the two forms of this compound.

---

* For information concerning the organization of the *Study Problems* and *Additional Problems*, see the *Note to student* on page 41.

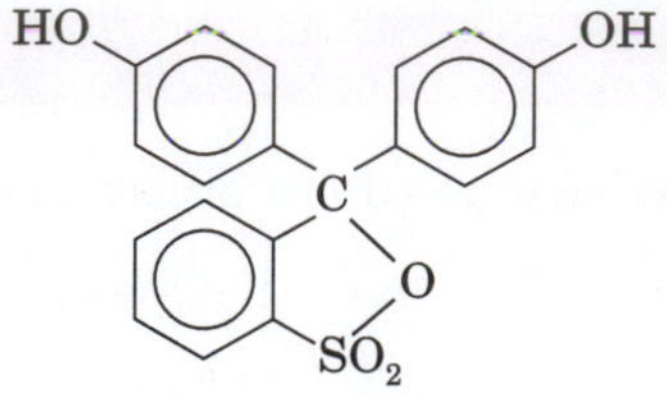

Phenol red

22.23 If an organic compound absorbs radiation of wavelength 575–580 nm, what color light is being absorbed? What color is the compound?

22.24 Each of the following reactions proceeds by either a Norrish type I path or a Norrish type II path. State which path, and suggest a mechanism for each.

(a)

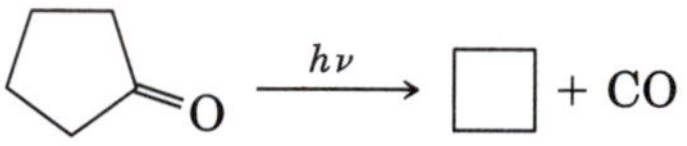

(b) 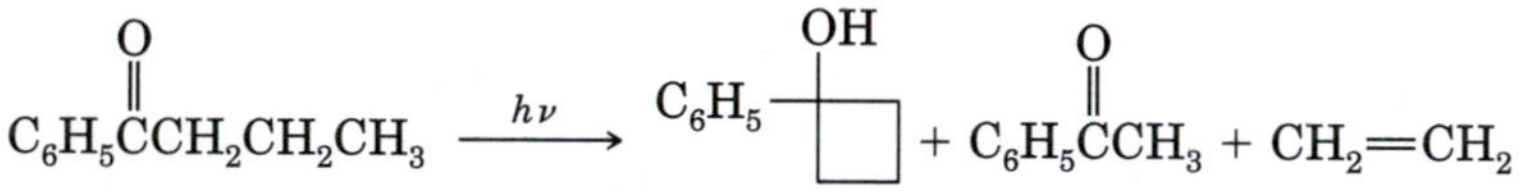 $C_6H_5\overset{O}{\overset{\|}{C}}CH_2CH_2CH_3 \xrightarrow{h\nu}$ $+ C_6H_5\overset{O}{\overset{\|}{C}}CH_3 + CH_2{=}CH_2$

(c) $CH_3\overset{O}{\overset{\|}{C}}CH_3 \xrightarrow{h\nu} CH_3\overset{O}{\overset{\|}{C}}{-}\overset{O}{\overset{\|}{C}}CH_3 + CH_3CH_3$

22.25 Predict the products of each of the following photo-induced reactions:

(a) 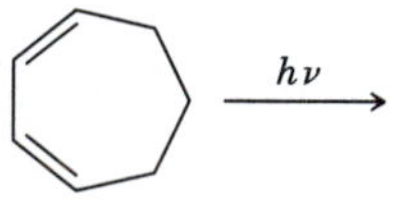 $\xrightarrow{h\nu}$ (b) 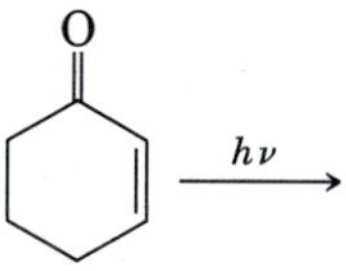 $\xrightarrow{h\nu}$ (c) $\underset{C_6H_5}{\overset{C_6H_5}{\diagdown}}N{=}N\underset{C_6H_5}{\diagup} \xrightarrow{h\nu}$

22.26 Write a mechanism for the following process:

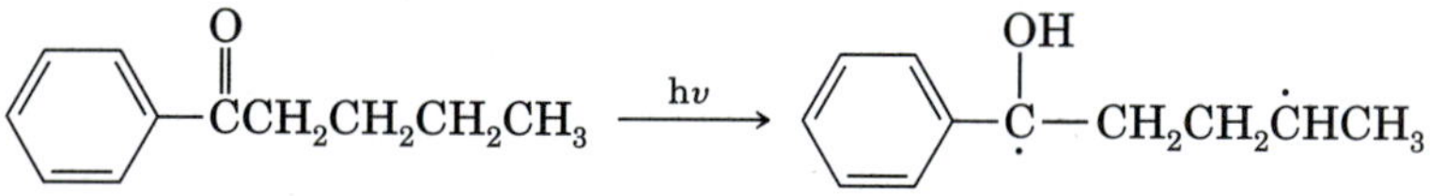

$\xrightarrow{h\nu}$

22.27 (1) Identify each of the following species as an ion, an ion radical, or a radical. (2) Assign an appropriate charge to each.

(a) $CH_3\overset{\cdot\overset{..}{O}H}{\underset{|}{C}H}CH_3$ (b) $CH_3\overset{\cdot}{C}{-}\overset{\cdot}{C}H_2 \atop \underset{H}{|}$ (c) $CH_3C{-}CH_3 \atop \underset{H}{|}$

(d) $H\overset{..}{\underset{..}{O}}\cdot$ (e) $CH_3C{=}\overset{..}{\underset{..}{O}}$ (f) $CH_2{=}CH{-}\overset{..}{\underset{..}{O}}H$

22.28 Give the structures of the products:

(a) $[CH_3CH_2CH_2OH]^{\dagger} \xrightarrow{-H_2O}$ (b) 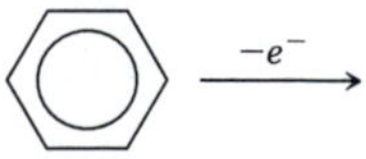 $\xrightarrow{-e^-}$

(c) 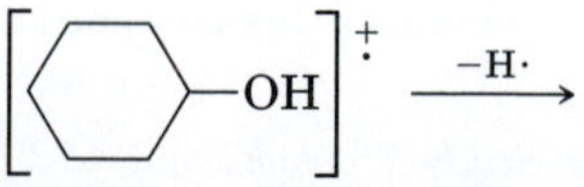 $\left[\text{}{-}OH \right]^{\overset{+}{\cdot}} \xrightarrow{-H\cdot}$ (d) $[(CH_3)_2CHCl]^{\dagger} \xrightarrow{-Cl\cdot}$

22.29 For each of the following compounds, predict the structures and m/e values for the molecular ion and likely positively charged fragmentation products:

(a) ethyl isopropyl ether   **(b)** ethyl isobutyl ether

(c) 2-chloropropane   **(d)** 2,5-dimethylhexane

(e) 2-propanol   **(f)** 4-cyclopentylbutanal (McLafferty fragment only)

22.30 Suggest structures and fragmentation patterns that account for the following observed peaks in the mass spectra:

(a) butane, $m/e = 58, 57, 43, 29, 15$

(b benzamide, $m/e = 121, 105, 77$

(c) 1-bromopropane, $m/e = 124, 122, 43, 29, 15$

(d) $C_6H_5CH_2OCH_3$, $m/e = 122, 121, 91, 77$

(e) 5-methyl-2-hexanone, $m/e = 71, 58, 43$

Additional Problems

22.31 What reactants would you need to prepare the yellow azo dye *Crysamine G?*

Crysamine G

22.32 *Phosphorescence* is the emission of light by a molecule returning from the T_1 excited state to the ground state. For the same compound, would you expect phosphorescence to be at lower or higher frequencies than fluorescence? Explain.

22.33 In UV spectra, the presence of an additional double bond in conjugation adds about 30 nm to the λ_{max}. From the following observed values for λ_{max}, state how the *degree of substitution* on the sp^2 carbons of a polyene affects the position of the λ_{max}.

Structure	λ_{max}, nm
$CH_2{=}CHCH{=}CH_2$	217
$CH_3CH{=}CHCH{=}CH_2$	223
$CH_3CH{=}CHCH{=}CHCH_3$	227

$CH_2{=}CCH{=}CHCH_3$
   |
   CH_3
	227

22.34 What would be the expected significant M + 1 and M + 2 peaks of each of the following compounds?

22.35 Which of the following pairs of compounds could be easily differentiated by mass spectrometry? Explain.

(a) $CH_3CH_2CH_2OH$ and $CH_3CH_2CH_2Br$

(b)

HO—⟨OH, H, H⟩ and H—⟨OH, H, OH⟩

(c) $CH_3OCH_2CH_2CH_3$ and $CH_3CH_2OCH_2CH_3$

(d) $CH_3CHCH_2CH_2CH_3$ and $CH_3CH_2CHCH_2CH_3$
         | |
         OH   OH

22.36 Predict the fragmentation products of the following molecular ions:

(a) $[CH_3CH_2OCH_2CH(CH_3)_2]^{+\cdot}$

(b) $\left[(CH_3)_3C{-}\bigcirc \right]^{+\cdot}$

(c) $\left[\triangleright{-}CH_2CH_2OH \right]^{+\cdot}$

(d) $\left[\text{(cyclopentanone with } CH_2CH_2CH_2CH_3 \text{ substituent)} \right]^{+\cdot}$

22.37 The following two ion radicals undergo McLafferty rearrangements to yield ethylene and a new ion radical. Write an equation showing the mechanism of each rearrangement.

(a) $[CH_3CH_2CH_2C{\equiv}N]^{+\cdot}$

(b) $[CH_3CH_2CH_2\overset{\displaystyle O}{\overset{\displaystyle \diagup\ \diagdown}{CH{-}CH_2}}]^{+\cdot}$

22.38 A compound contains only C, H, and O. The infrared spectrum shows strong absorption at 1724 cm^{-1} (5.8 μm), 1388 cm^{-1} (7.2 μm), and 1231 cm^{-1} (8.1 μm) (plus other minor absorption). The ^{1}H NMR spectrum shows only one singlet at $\delta 2.1$. The mass spectrum has principal peaks at 58 m/e and 43 m/e. What is the structure of the compound?

22.39 The infrared, ^{1}H NMR, and mass spectra for Compounds A through C are shown in Figures 22.17 through 22.19. From the spectra, deduce the structure of each compound.

22.40 Compound A is a terpenoid with one chiral carbon. The (S)-enantiomer is an essential oil in rose oil and geranium oil. Spectra for Compound A are shown in Figure 22.20. What is the structure of the (S)-enantiomer of Compound A? (*Hint:* What is the significance of the fact that Compound A is a terpenoid?)

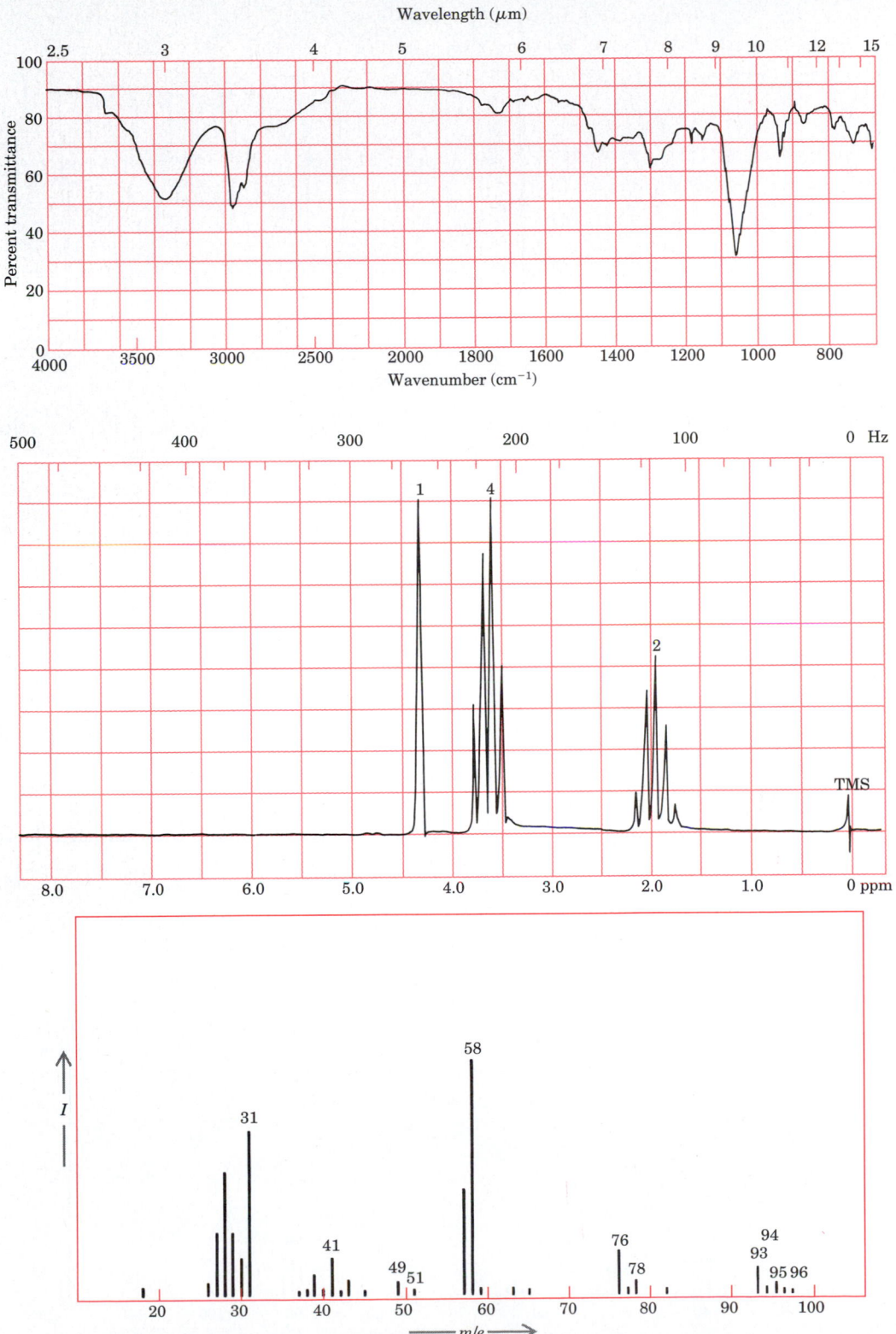

Figure 22.17 Spectra for Compound A, Problem 22.39.

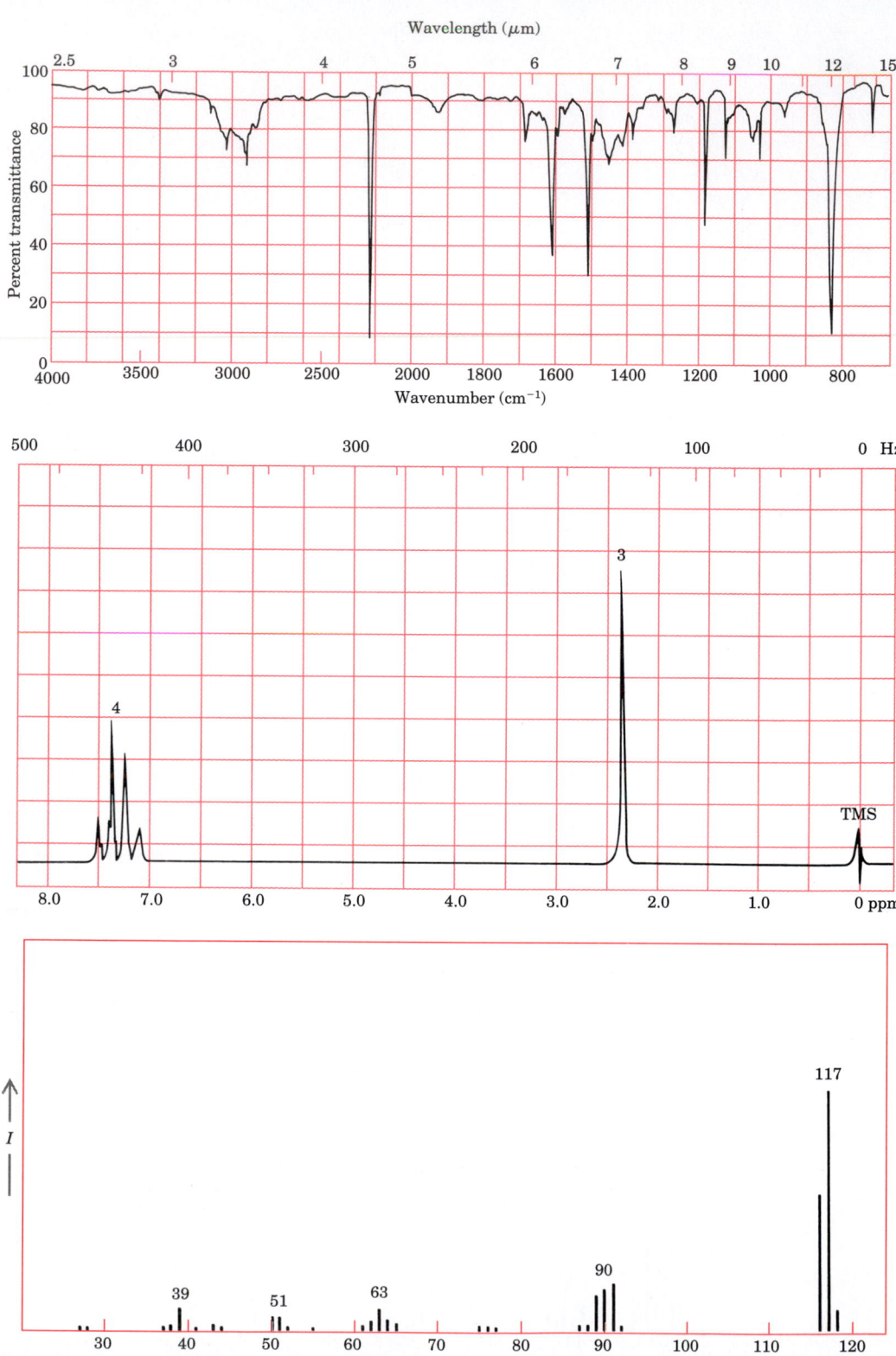

Figure 22.18 Spectra for Compound B, Problem 22.39.

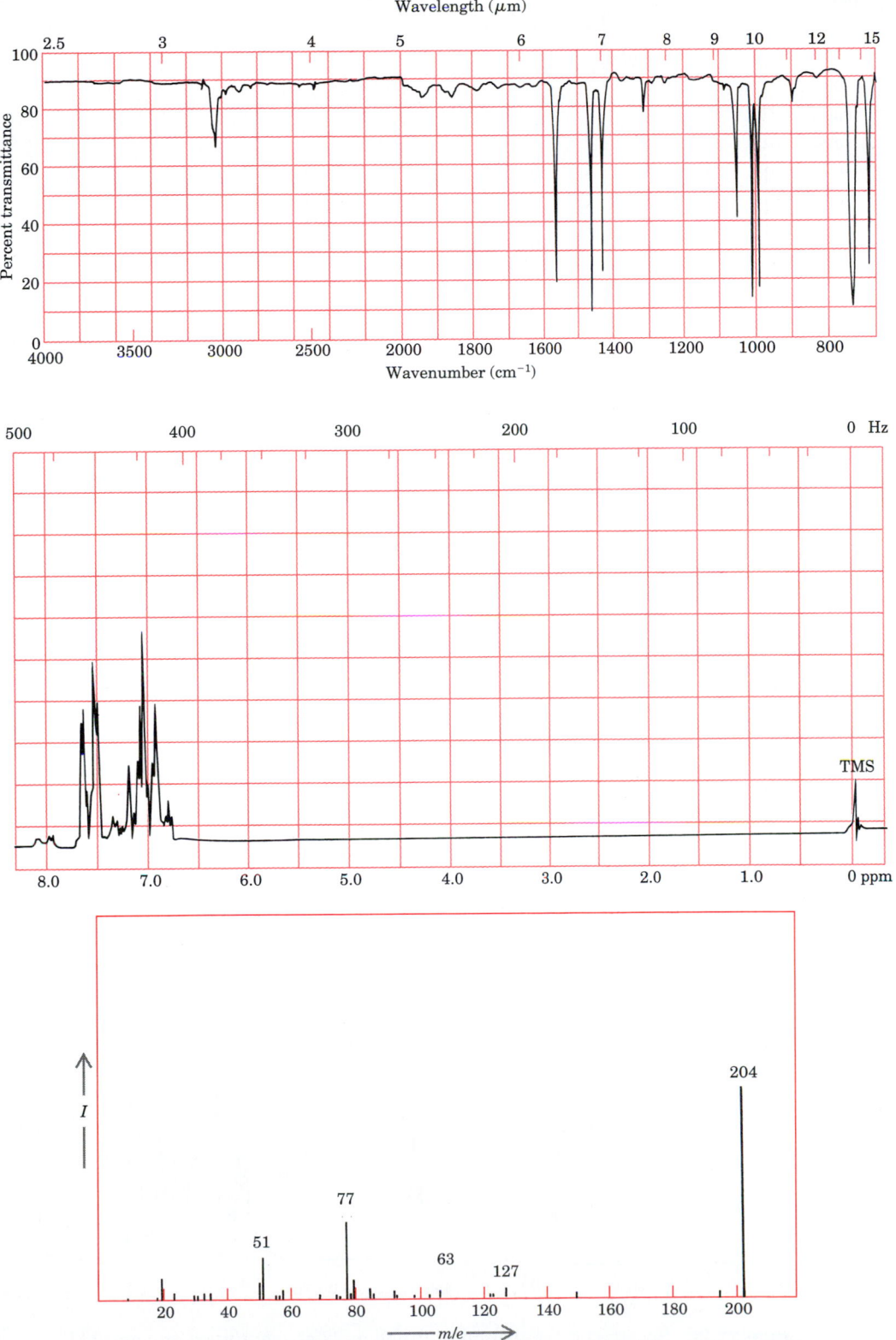

Figure 22.19 Spectra for Compound C, Problem 22.39.

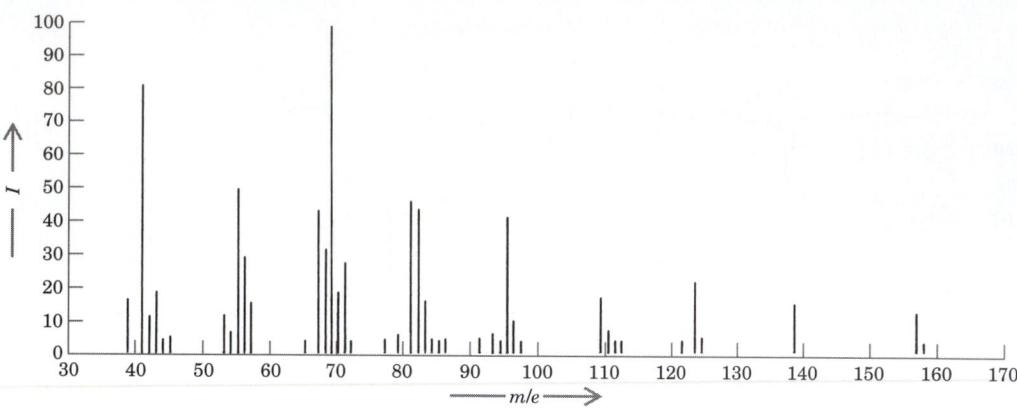

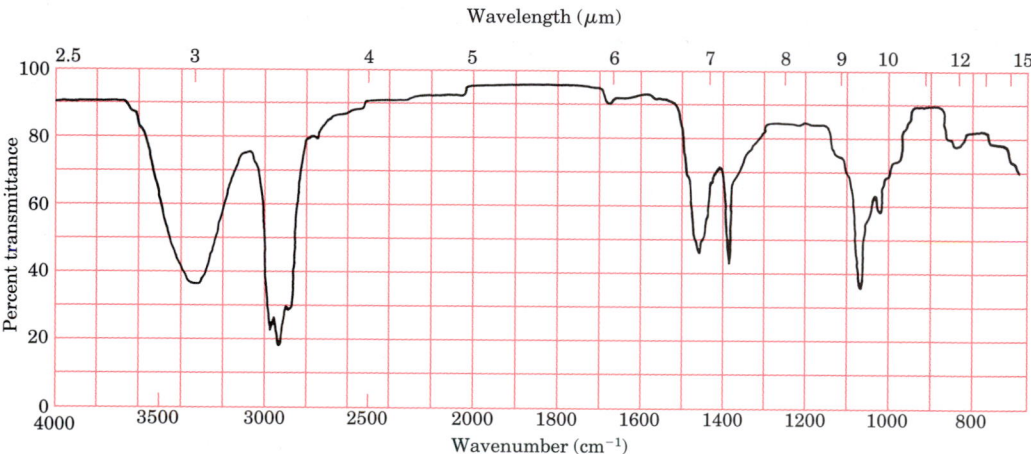

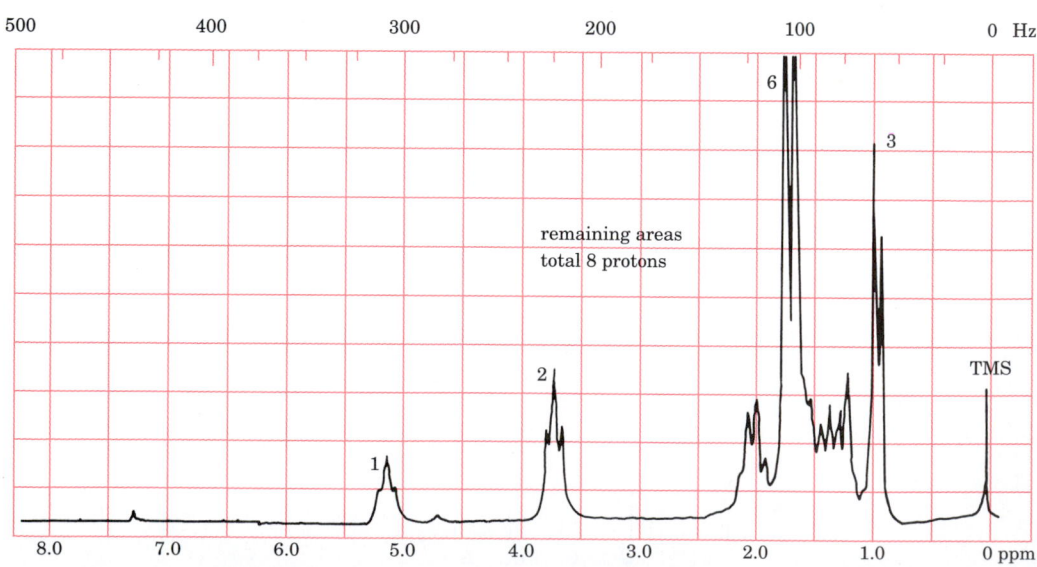

Figure 22.20 Spectra for Compound A, Problem 22.40 *(continued on page 921)*.

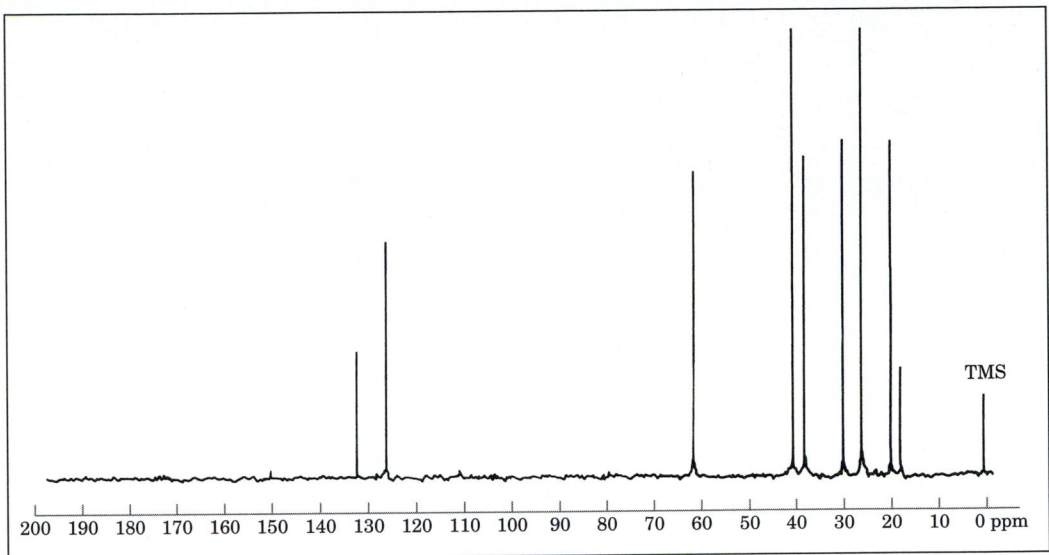

Figure 22.20 *(continued)* Spectra for Compound A, Problem 22.40.

23 CARBOHYDRATES

Carbohydrates are naturally occurring compounds of carbon, hydrogen, and oxygen. Many carbohydrates have the empirical formula CH_2O. For example, the molecular formula for glucose is $C_6H_{12}O_6$ (six times CH_2O). These compounds were once thought to be "hydrates of carbon," hence the name carbohydrates. In the 1880s, it was recognized that the idea was a misconception and that carbohydrates are actually polyhydroxy aldehydes and ketones or their derivatives.

Carbohydrates vary dramatically in their properties. For example, *sucrose* (table sugar) and *cotton* are both carbohydrates. One of the principal differences between various types of carbohydrates is the size of the molecules. The **monosaccharides** (often called *simple sugars*) are the simplest carbohydrate units. They cannot be hydrolyzed to smaller carbohydrate molecules. Figure 23.1 shows dimensional formulas and Fischer projections for five of the most important monosaccharides. (You may find it helpful to review Fischer projections in Section 4.6C.)

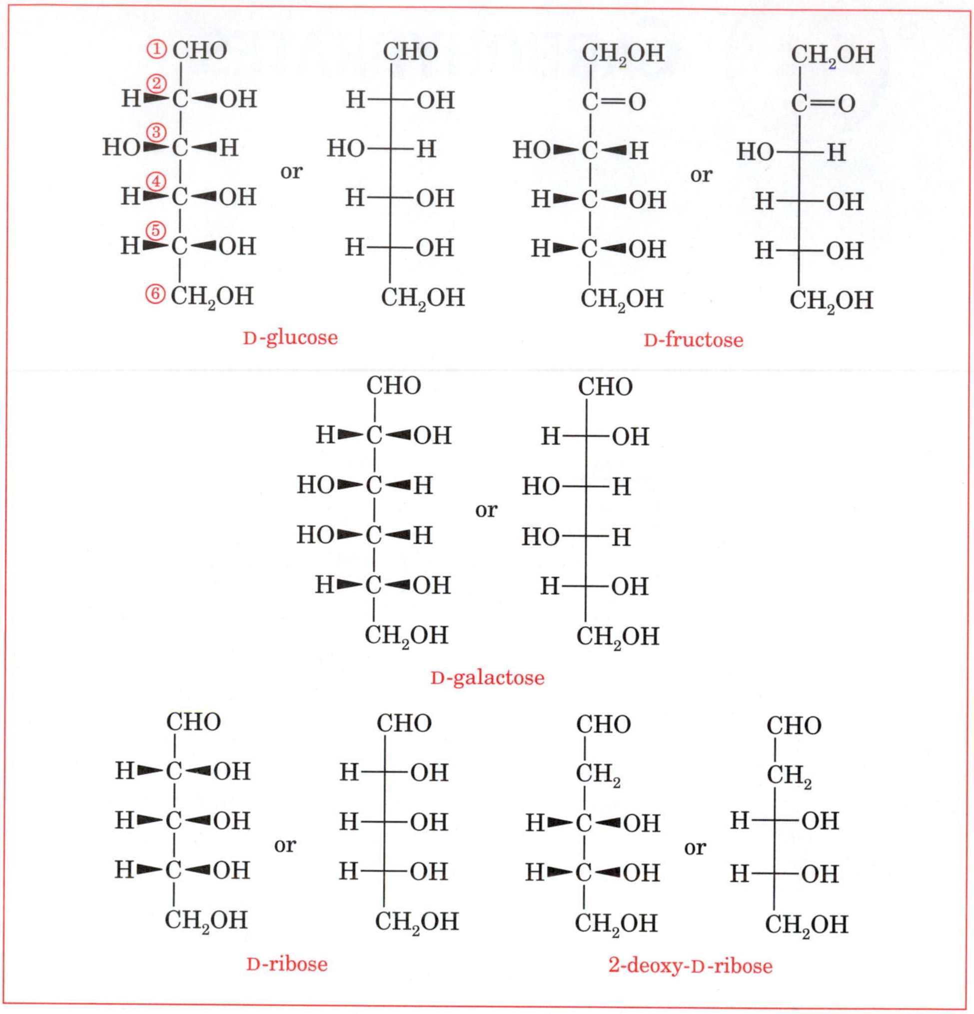

Figure 23.1 Some important monosaccharides.

Monosaccharides can be bonded together to form dimers, trimers, etc., and ultimately, polymers. The dimers are called **disaccharides.** Sucrose is a disaccharide that can be hydrolyzed to one unit of glucose plus one unit of fructose. The monosaccharides and disaccharides are soluble in water and are generally sweet-tasting.

$$1 \text{ sucrose } \xrightarrow[\text{heat}]{\text{H}_2\text{O, H}^+} 1 \text{ glucose} + 1 \text{ fructose}$$

a disaccharide

Carbohydrates composed of two to eight units of monosaccharide are referred to as **oligosaccharides** (Greek *oligo-,* "a few"). If more than eight units of monosaccharide result from hydrolysis, the carbohydrate is a **polysaccharide.** Examples of polysaccharides are *starch,* found in flour and cornstarch, and *cellulose,* a fibrous constituent of plants and the principal component of cotton.

$$\text{starch or cellulose} \xrightarrow[\text{heat}]{H_2O, H^+} \text{many units of glucose}$$
polysaccharides

In this chapter, we will consider first the monosaccharides and the conventions used by carbohydrate chemists. Then we will discuss some disaccharides, and finally, a few polysaccharides.

Section 23.1

Some Common Monosaccharides

Glucose, the most important monosaccharide, is sometimes called *blood sugar* (because it is found in the blood), *grape sugar* (because it is found in grapes), or *dextrose* (because it is dextrorotatory). Mammals can convert sucrose, lactose (milk sugar), maltose, and starch to glucose, which is then used by the organism for energy or stored as *glycogen* (a polysaccharide). When the organism needs energy, the glycogen is again converted to glucose. Excess carbohydrates can be converted to fat; therefore, a person can become obese on a fat-free diet. Carbohydrates can also be converted to steroids (such as cholesterol) and, to a limited extent, to protein. (A source of nitrogen is also needed in protein synthesis.) Conversely, an organism can convert proteins and fats to carbohydrates.

Fructose, also called *levulose* because it is levorotatory, is the sweetest-tasting sugar. It occurs in fruit and honey. Combined with glucose, it is also found in sucrose. **Galactose** is found, bonded to glucose, in the disaccharide lactose. **Ribose** and **deoxyribose** form part of the polymeric backbones of nucleic acids. The prefix *deoxy-* means "minus an oxygen"; the structures of ribose and deoxyribose (Figure 23.1) are the same except that deoxyribose lacks an oxygen at carbon 2.

Section 23.2

Classification of the Monosaccharides

The suffix **-ose** is used to designate a carbohydrate. All the monosaccharides and many oligosaccharides and polysaccharides have names ending in *-ose* (for example, sucrose and cellulose).

Monosaccharides that contain aldehyde groups are referred to as **aldoses** (*alde*hyde plus *-ose*). Glucose, galactose, ribose, and deoxyribose are all aldoses. Monosaccharides, such as fructose, with ketone groups are called **ketoses** (*ket*one plus *-ose*).

The number of carbon atoms in a monosaccharide (usually three to seven) may be designated by *tri-, tetr-,* etc. For example, a **triose** is a three-carbon monosaccharide, while a **hexose** is a six-carbon monosaccharide. Glucose is an example of a hexose. These terms may be combined. Glucose is an **aldohexose** (six-carbon aldose), while ribose is an **aldopentose** (five-carbon aldose). Ketoses are often given the ending **-ulose.** Fructose is an example of a **hexulose** (six-carbon ketose), or **ketohexose.**

STUDY PROBLEM

23.1   Classify each of the following monosaccharides by the preceding system:

(a)
$$
\begin{array}{c}
CH_2OH \\
| \\
C=O \\
HO \rule[0.5ex]{1em}{0.4pt} H \\
H \rule[0.5ex]{1em}{0.4pt} OH \\
H \rule[0.5ex]{1em}{0.4pt} OH \\
CH_2OH
\end{array}
$$

(b)
$$
\begin{array}{c}
CHO \\
H \rule[0.5ex]{1em}{0.4pt} OH \\
H \rule[0.5ex]{1em}{0.4pt} OH \\
H \rule[0.5ex]{1em}{0.4pt} OH \\
CH_2OH
\end{array}
$$

(c)
$$
\begin{array}{c}
CHO \\
H \rule[0.5ex]{1em}{0.4pt} OH \\
HO \rule[0.5ex]{1em}{0.4pt} H \\
HO \rule[0.5ex]{1em}{0.4pt} H \\
H \rule[0.5ex]{1em}{0.4pt} OH \\
CH_2OH
\end{array}
$$

Section 23.3

Configurations of the Monosaccharides

All monosaccharides with the same number of carbon atoms have similar structures. They differ from one another by being structural isomers or diastereomers. Glucose and fructose, for example, are hexoses that are structural isomers. Glucose is an aldehyde while fructose is a ketone. Diastereomers are nonenantiomeric stereoisomers that have two or more chiral centers and differ in the projection of at least one of them. Glucose, galactose, and talose are all diastereomers. If two diastereomers differ in the projection of only one chiral center, then they are called **epimers.** Glucose and galactose are epimers because they differ only in the projection of the hydroxyl group at carbon 4. Glucose and talose (as well as galactose and talose) are not epimers because they differ in the projections of more than one chiral center.

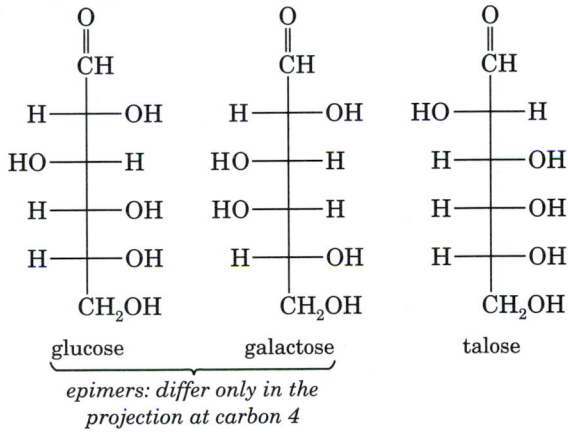

Two monosaccharides that are structural isomers

glucose   fructose

Three monosaccharides that are diastereomers

glucose   galactose   talose

*epimers: differ only in the
projection at carbon 4*

A. The D and L System

In the late nineteenth century, it was determined that the configuration of the last chiral carbon in each of the naturally occurring monosaccharides is the same as that of (+)-glyceraldehyde. Today we call that configuration the (*R*)-configuration, but chemists at that time had no way to determine the absolute configuration around a chiral carbon. Instead, chemists devised the D and L system for designating relative configurations. (Do not confuse D and L with *d* and *l*, sometimes used to refer to the direction of rotation of the plane of polarization of plane-polarized light; see Section 4.7A.)

In the D and L system, (+)-glyceraldehyde was arbitrarily assigned the configuration with its OH on carbon 2 to the right in the Fischer projection (an assumption later shown to be correct). A monosaccharide is a member of the **D-series** if the hydroxyl group on the chiral carbon farthest from carbon 1 is also *on the right* in the Fischer projection. (Almost all naturally occurring carbohydrates are members of the D-series.) In addition, each monosaccharide was given its own name. For example, the following two diastereomeric aldopentoses are named D-lyxose and D-ribose.

$$\begin{array}{ccc}
& \text{CHO} & \text{CHO} \\
\text{CHO} & \text{HO}-\!\!-\text{H} & \text{H}-\!\!-\text{OH} \\
\text{H}-\!\!-\text{OH} & \text{HO}-\!\!-\text{H} & \text{H}-\!\!-\text{OH} \\
\text{CH}_2\text{OH} & \text{H}-\!\!-\text{OH} & \text{H}-\!\!-\text{OH} \\
& \text{CH}_2\text{OH} & \text{CH}_2\text{OH}
\end{array}$$

D-(+)-glyceraldehyde   D-lyxose   D-ribose

If the OH on the last chiral carbon is projected to the *left,* then the compound is a member of the L-series. The following two examples are the enantiomers of D-lyxose and D-ribose.

$$\begin{array}{cc}
\text{CHO} & \text{CHO} \\
\text{H}-\!\!-\text{OH} & \text{HO}-\!\!-\text{H} \\
\text{H}-\!\!-\text{OH} & \text{HO}-\!\!-\text{H} \\
\text{HO}-\!\!-\text{H} & \text{HO}-\!\!-\text{H} \\
\text{CH}_2\text{OH} & \text{CH}_2\text{OH}
\end{array}$$

L-lyxose   L-ribose

STUDY PROBLEM

23.2 Identify each saccharide as D or L.

$$\begin{array}{lll}
\text{CHO} & \text{CHO} & \text{CHO} \\
\text{HO}-\!\!-\text{H} & \text{H}-\!\!-\text{OH} & \text{H}-\!\!-\text{OH} \\
\textbf{(a)}\quad\text{H}-\!\!-\text{OH} & \textbf{(b)}\quad\text{HO}-\!\!-\text{H} & \textbf{(c)}\quad\text{H}-\!\!-\text{OH} \\
\text{H}-\!\!-\text{OH} & \text{HO}-\!\!-\text{H} & \text{HO}-\!\!-\text{H} \\
\text{CH}_2\text{OH} & \text{CH}_2\text{OH} & \text{HO}-\!\!-\text{H} \\
& & \text{CH}_2\text{OH}
\end{array}$$

B. Relating Configurations

We have mentioned that early chemists could not determine the absolute configurations of chiral carbons. Instead, configurations relative to that of (+)-glyceraldehyde were determined. How are other compounds related to glyceraldehyde? One example of determining a relative configuration follows. If the aldehyde group of D-glyceraldehyde is oxidized to a carboxylic acid, the product, glyceric acid, necessarily has the same configuration around the chiral carbon as that in D-glyceraldehyde. The product, even though levorotatory, is still a member of the D-series.

$$\begin{array}{ccc}
\text{CHO} & & \text{CO}_2\text{H} \\
\text{H}\blacktriangleright\text{C}\blacktriangleleft\text{OH} & \xrightarrow{[O]} & \text{H}\blacktriangleright\text{C}\blacktriangleleft\text{OH} \\
\text{CH}_2\text{OH} & & \text{CH}_2\text{OH}
\end{array}$$

no change in configuration

D-(+)-glyceraldehyde   D-(−)-glyceric acid

The configurations of the tartaric acids relative to D-glyceraldehyde were established in 1917 by the sequence shown in Figure 23.2, which produced two of the three isomers of tartaric acid.

In the first step of the sequence, D-glyceraldehyde is treated with HCN to yield a mixture of cyanohydrins. A new site of chirality is introduced in this step, and both diastereomers are formed. The diastereomers are separated; then, in the second step, each diastereomeric cyanohydrin is hydrolyzed.

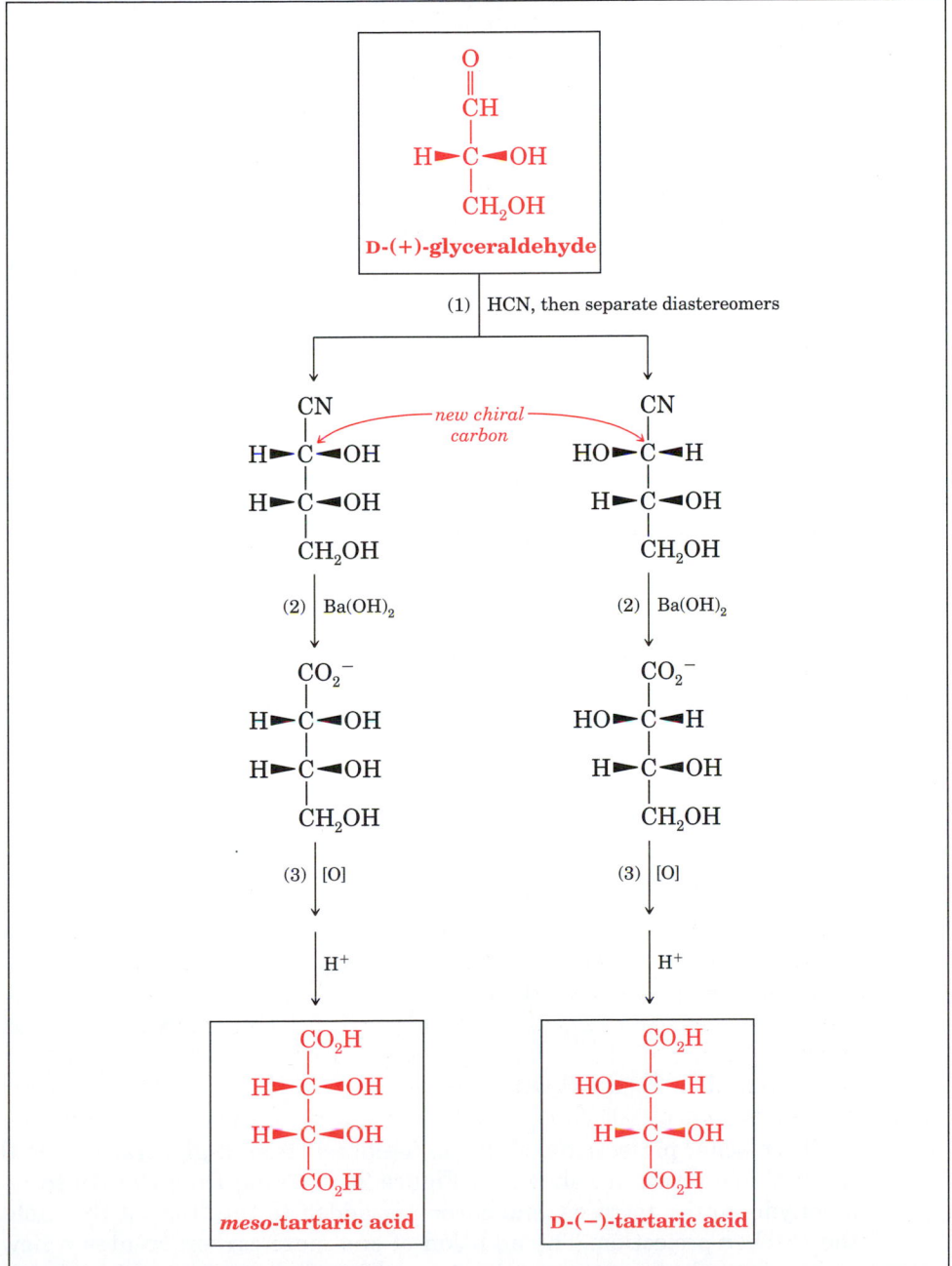

Figure 23.2 Determining the relative configurations of tartaric acids.

In the third step, the terminal CH_2OH group is oxidized to yield two tartaric acids. Carbon 3 of each of these tartaric acids has the same configuration as carbon 2 of D-glyceraldehyde because the series of reactions has not affected the configuration around that carbon. However, the configurations around carbon 2 in the tartaric acids are different. One of the tartaric acids obtained from this synthesis does not rotate plane-polarized light. This is the *meso* isomer, the one with an internal plane of symmetry. The other tartaric acid obtained from the synthesis rotates plane-polarized light to the left. It must have the second structure as follows:

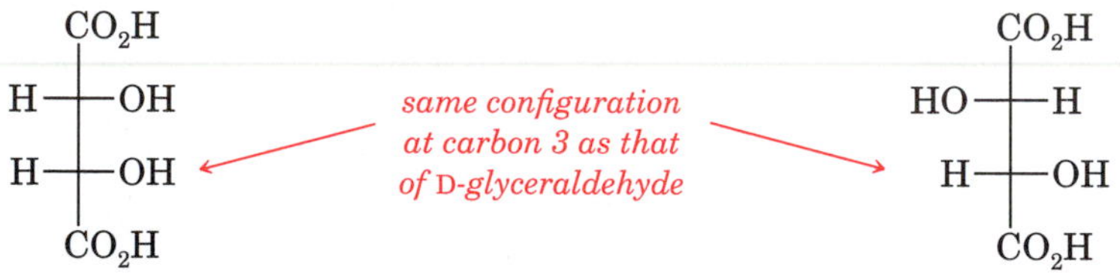

$$
\begin{array}{ccc}
& CO_2H && CO_2H \\
H & \!\!-\!\!|\!\!-\!\! & OH & \qquad HO & \!\!-\!\!|\!\!-\!\! & H \\
H & \!\!-\!\!|\!\!-\!\! & OH & \qquad H & \!\!-\!\!|\!\!-\!\! & OH \\
& CO_2H && CO_2H
\end{array}
$$

same configuration at carbon 3 as that of D-glyceraldehyde

meso-tartaric acid   D-(−)-tartaric acid

STUDY PROBLEM

23.3 Starting with L-(−)-glyceraldehyde, what tartaric acid(s) would be produced in the preceding sequence?

C. Configurations of the Aldohexoses

Glucose has six carbon atoms, four of which are chiral (carbons 2, 3, 4, and 5). Because the terminal carbon atoms of glucose have different functional groups, there can be no internal plane of symmetry; therefore, this compound has 2^4, or sixteen, stereoisomers. Only half of these sixteen stereoisomers belong to the D-series; of these, only D-glucose, D-galactose, and D-mannose occur in abundance.

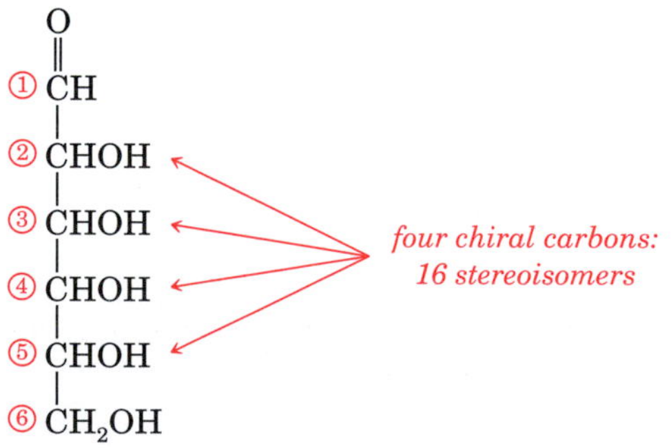

$$
\begin{array}{l}
\quad\;\; O \\
\quad\;\; \| \\
① \; CH \\
\quad | \\
② \; CHOH \\
\quad | \\
③ \; CHOH \\
\quad | \\
④ \; CHOH \\
\quad | \\
⑤ \; CHOH \\
\quad | \\
⑥ \; CH_2OH
\end{array}
$$

four chiral carbons: 16 stereoisomers

The Fischer projections of all the D-aldoses, from D-glyceraldehyde through the D-aldohexoses, are shown in Figure 23.3. Going from the triose, D-glyceraldehyde, to the tetroses, one carbon is added to the "top" of the molecule in the Fischer projection. The addition of one more carbon creates a new chiral carbon in each step down in the figure. Therefore, D-glyceraldehyde leads to a pair of tetroses, each tetrose leads to a pair of pentoses, and each pentose leads to a pair of hexoses.

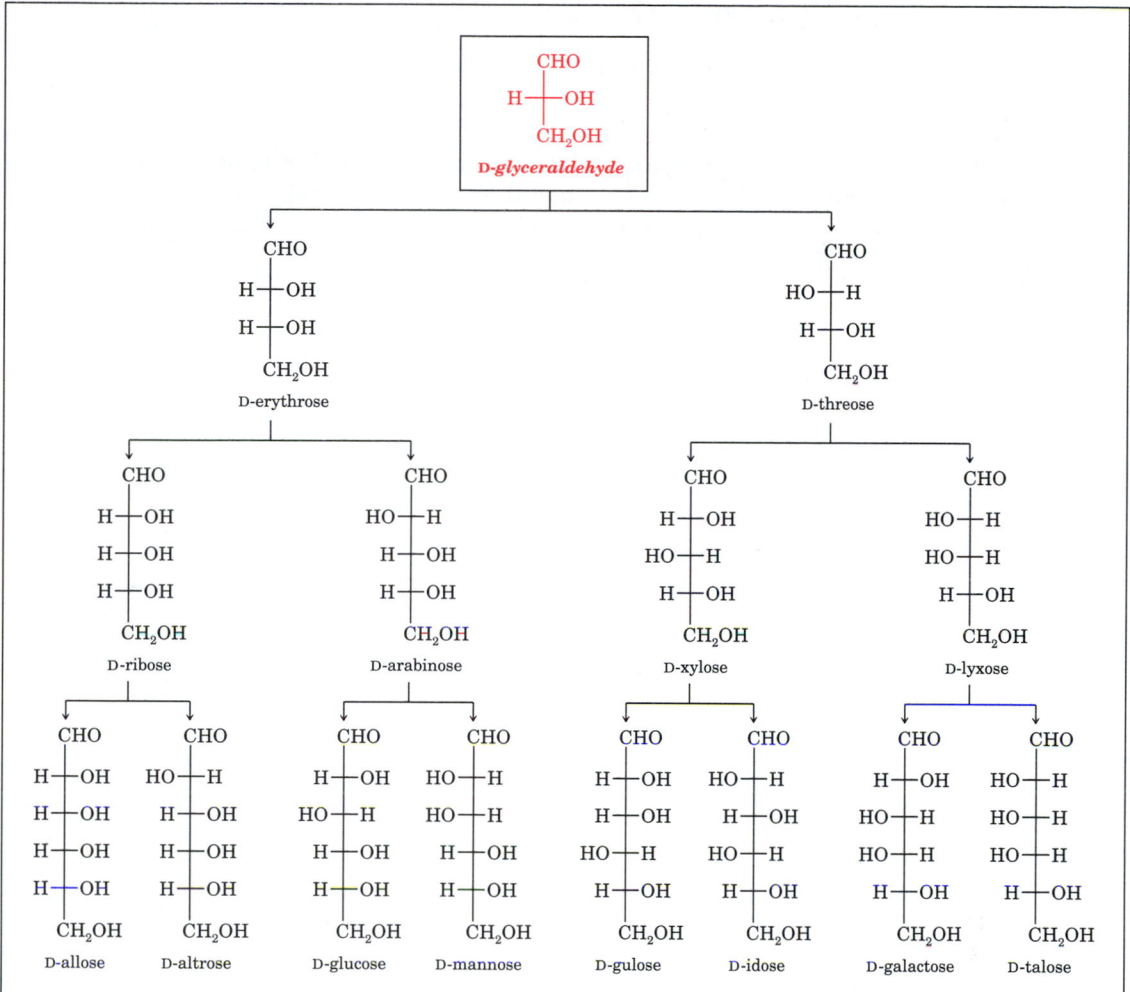

Figure 23.3 The D-aldoses.

23.4   Although most naturally occurring sugars belong to the D series, L-arabinose is more common in plants than D-arabinose is. Write the Fischer projection for L-arabinose.

23.5   What is the name of the aldohexose in which only the OH at carbon 5 has the opposite configuration from that of D-glucose?

Section 23.4
Cyclization of the Monosaccharides

Glucose has an aldehyde group at carbon 1 and hydroxyl groups at carbons 4 and 5 as well as at carbons 2, 3, and 6. A general reaction of alcohols and aldehydes is that of *hemiacetal formation* (see Section 13.4B).

$$\underset{O}{\overset{\displaystyle \parallel}{R C H}} + R'OH \underset{\overset{\displaystyle H^+}{\rightleftharpoons}}{} \underset{\text{\textit{a hemiacetal}}}{\overset{\displaystyle OH}{\overset{\displaystyle |}{R C H}}-OR'}$$

In water solution, glucose can undergo an intramolecular reaction to yield *cyclic hemiacetals*. Either five-membered ring hemiacetals (using the hydroxyl group at carbon 4) or six-membered ring hemiacetals (using the hydroxyl group at carbon 5) can be formed.

Although Fischer projections are useful in discussions of open-chain carbohydrates, they are awkward for cyclic compounds. Let us rewrite the formula for glucose to show the more important of the cyclization reactions, the reaction leading to the six-membered rings.

Carbon 1 (the aldehyde carbon), which is not chiral in the open-chain structure, becomes chiral in the cyclization. Therefore, a pair of diastereomers results from the cyclization. Because all the hemiacetal structures are in equilibrium with the aldehyde in water solution, they are also in equilibrium with each other.

$$\underset{\text{\textit{(two diastereomers)}}}{\text{5-membered cyclic hemiacetal}} \rightleftharpoons \underset{\text{\textit{(open-chain)}}}{\text{glucose}} \rightleftharpoons \underset{\text{\textit{(two diastereomers)}}}{\text{6-membered cyclic hemiacetal}}$$

A. Furanose and Pyranose Rings

A monosaccharide in the form of a five-membered ring hemiacetal is called a **furanose.** *Furan-* is from the name of the five-membered oxygen heterocycle *furan*. Similarly, the six-membered ring form is called a **pyranose** after *pyran*. The terms furanose and pyranose are often combined with the name of the monosaccharide. For example, **D-glucopyranose** is used for the six-membered ring of D-glucose, or **D-fructofuranose** for the five-membered ring of fructose.

furan   pyran

Of the two ring systems for glucose, the six-membered cyclic hemiacetal, or glucopyranose, is favored; we will emphasize this ring size in our discussion. Part of the reason that glucose preferentially forms the six-membered ring in solution is that the bond angles and staggering of attached groups are favorable in the chair form of this ring. Even though the pyranose ring of a monosaccharide may predominate in equilibrium in water, it may be the furanose ring that is incorporated enzymatically into natural products. For example, in ribonucleic acids, ribose is found as a furanose, and not as a pyranose.

B. Haworth and Conformational Formulas

To represent the cyclic structures of sugars, **Haworth perspective formulas** were developed. By convention, a Haworth formula is drawn with the ring oxygen on the far side of the ring and carbon 1 on the right. The terminal —CH_2OH group is positioned above the plane of the ring in the D-series and below the plane of the ring in the L-series. The hydrogen atoms on the ring carbons are not usually shown in Haworth projections.

Using Fischer and Haworth formulas, we can rewrite the equations showing the cyclization of glucose. *Note that any group that is to the right in the Fischer projection is down in the Haworth projection, and any group that is to the left in the Fisher projection is up in the Haworth formula.*

The flat Haworth formula is not an entirely correct representation of a pyranose ring (although it is fairly correct for the more planar furanose ring). A pyranose, like cyclohexane, exists primarily in the chair form, as the following conformational formula shows. In this chapter, we will use both Haworth formulas and conformational formulas.

α-D-glucopyranose

If an OH is down in a Haworth formula, it is also down (below the plane of the ring) in the conformational formula. Similarly, if an OH is up in the Haworth formula, it is also up in the conformational formula. As for any substituted six-membered ring, the ring assumes the conformation in which the majority of the groups are equatorial.

STUDY PROBLEM

23.6 Draw a conformational formula of α-D-glucopyranose in which the CH_2OH is in an axial position.

C. Anomers

To yield a pyranose, the hydroxyl group at carbon 5 of glucose attacks the aldehyde carbon, carbon 1. A hemiacetal group is formed. Two of the most important consequences of this cyclization reaction are that a new chiral carbon is formed (carbon 1) and that a pair of diastereomers results. These diastereomers, monosaccharides that differ only in the configuration at carbon 1, are called **anomers** of each other. The carbonyl carbon in any monosaccharide is the **anomeric carbon.** This is the carbon that becomes chiral in the cyclization reaction.

In a Haworth formula of a D-sugar, the structure in which the anomeric OH is projected *down* (*trans* to the terminal CH_2OH) is called the *α*-**anomer,** while the structure in which the anomeric OH is projected *up* (*cis* to the terminal CH_2OH) is called the *β*-**anomer.**

Thus, the two anomeric D-glucoses can be called *α*-**D-glucopyranose** (or simply *α*-D-glucose) and *β*-**D-glucopyranose** (or *β*-D-glucose).

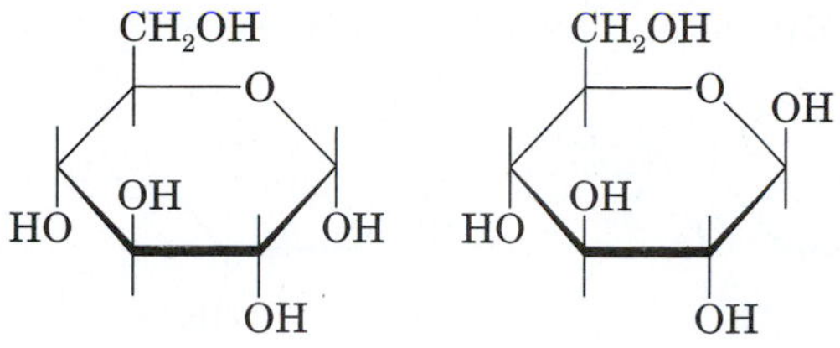

α-D-glucopyranose   β-D-glucopyranose

SAMPLE PROBLEM

Although fructose can form a six-membered cyclic hemiketal, in sucrose it is found in the furanose form. Draw Haworth formulas for α- and β-D-fructofuranose.

Solution

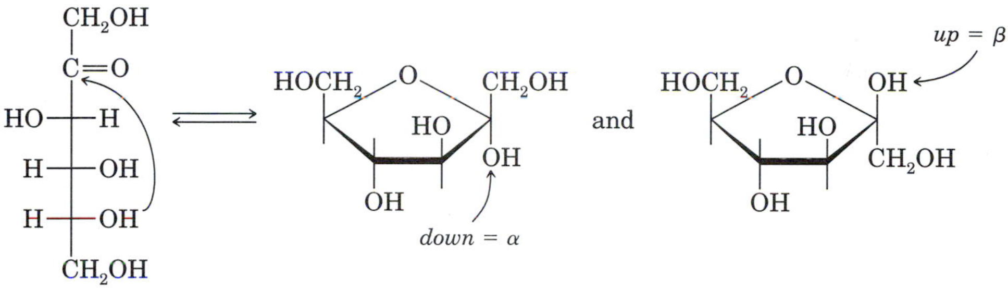

D-fructose

STUDY PROBLEM

23.7 Draw the Haworth and conformational formulas for the anomers of D-galactopyranose.

D. Mutarotation

Pure glucose exists in two crystalline forms: α-D-glucose and β-D-glucose. Pure α-D-glucose has a melting point of 146°C. The specific rotation of a freshly prepared solution is +112°. Pure β-D-glucose has a melting point of 150°C and a specific rotation of +18.7°. The specific rotation of a solution of either α- or β-D-glucose changes slowly until it reaches an equilibrium value of +52.6°. This slow spontaneous change in optical rotation, first observed in 1846, is called **mutarotation.**

   Mutarotation occurs because, in solution, either α- or β-D-glucose undergoes a slow equilibration with the open-chain form and with the other anomer. Regardless of which anomer is dissolved, the result is an equilibrium mixture of 64% β-D-glucose, 36% α-D-glucose, and 0.02% of the aldehyde form of D-glucose. The final specific rotation is that of the equilibrium mixture.

α-D-glucose (36%)   D-glucose (0.02%) β-D-glucose (64%)

Note that the equilibrium mixture of the anomers of D-glucose contains a greater percentage of the β-anomer than of the α-anomer. The reason is that the β-anomer is the more stable of the two. From our discussion of conformational analysis in Chapter 4, this is the expected result. The hydroxyl group at carbon 1 is *equatorial* in the β-anomer, but *axial* in the α-anomer.

α-D-glucose β-D-glucose

Other monosaccharides also exhibit mutarotation. In water solution, the other aldoses with a 5-hydroxyl group also exist primarily in the pyranose forms. However, the percentages of the various species involved in the equilibrium may vary. For example, the equilibrium mixture of D-ribose in water is 56% β-pyranose, 20% α-pyranose, 18% β-furanose, 6% α-furanose, and a trace of the open-chain, aldehyde form.

Although the β-anomer of the pyranose rings is generally the more stable anomer, this is not always the case. For example, α-D-mannose is more stable than its β-anomer and predominates in an equilibrium mixture. This apparent anomaly, termed the **anomeric effect,** arises from interactions between the polar substituents on the ring.

β-D-ribopyranose (56%) α-D-ribopyranose (20%)

D-ribose

β-D-ribofuranose (18%) α-D-ribofuranose (6%)

Because of the facile conversion in water of the hemiacetal OH group between α and β, it often is not possible to specify the configuration at this carbon. For this reason, we will sometimes represent the hemiacetal OH bond with a squiggle, which means the structure may be α or β or a mixture.

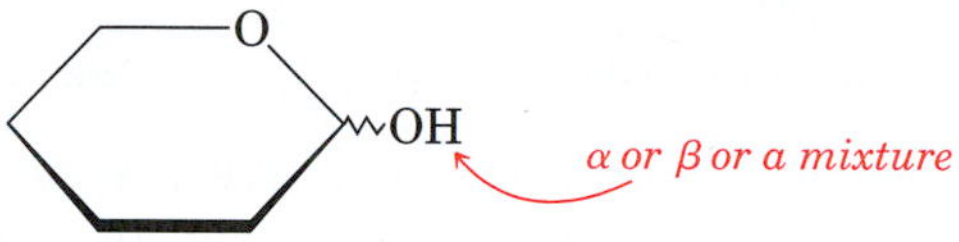

α or β or a mixture

Section 23.5
Glycosides

When a hemiacetal is treated with an alcohol, an acetal is formed (Section 13.4B). The acetals of monosaccharides are called **glycosides** and have names ending in **-oside.**

$$\underset{\text{a hemiacetal}}{\overset{\overset{\displaystyle OH}{|}}{RCHOR}} + R'OH \;\overset{H^+}{\rightleftharpoons}\; \underset{\text{an acetal}}{\overset{\overset{\displaystyle OR'}{|}}{RCHOR}} + H_2O$$

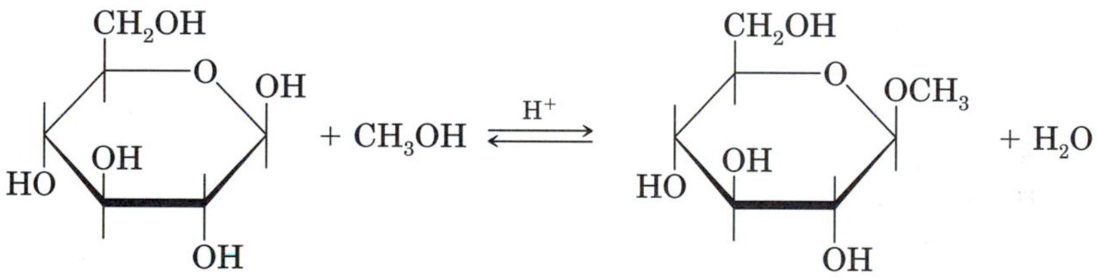

β-D-glucopyranose   methyl β-D-glucopyranoside

a glycoside

The glycoside carbon (carbon 1 in an aldose) is easy to recognize because it has two OR groups attached.

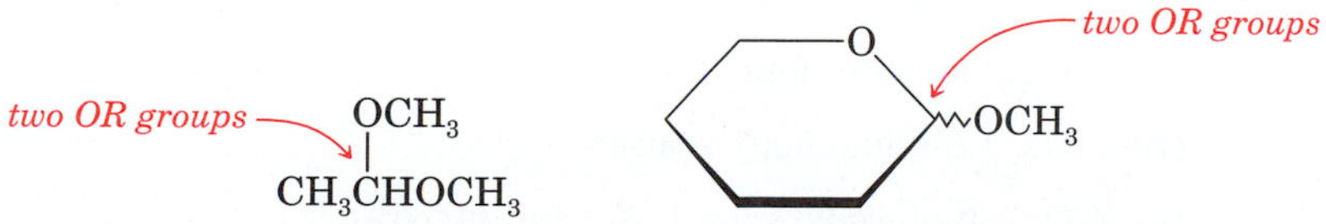

two OR groups — $\overset{\overset{\displaystyle OCH_3}{|}}{CH_3CHOCH_3}$

an acetal   *a glycoside*

two OR groups

Although a hemiacetal of a monosaccharide is in equilibrium with the open-chain form and with its anomer in water solution, an acetal is stable in neutral or alkaline solution. Therefore, a glycoside is not in equilibrium with the aldehyde or with its anomer in water solution. However, glycosides can be hydrolyzed to the hemiacetal (and aldehyde) forms by treatment with aqueous acid. This reaction is simply the reverse of glycoside formation.

methyl β-D-glucopyranoside   D-glucopyranose

Disaccharides and polysaccharides are glycosides; we will discuss these compounds later in this chapter. Other types of glycosides are also common in plants and animals. *Amygdalin* and *laetrile* (Section 13.4C) are glycosides found in the kernels of apricot pits and bitter almonds. *Vanillin* (used as vanilla flavoring) is another example of a structure found in nature as a glycoside, in this case, as a β-D-glucoside. In these types of glycosides, the non-sugar portion of the structure is called an **aglycone.** Vanillin is the aglycone in the following example.

vanillin   vanillin β-D-glucoside
                                         (glucovanillin)

STUDY PROBLEMS

23.8   Circle the hemiacetal (or acetal) carbon atom and its two oxygen atoms in the following monosaccharides and state if the anomer shown is α or β:

23.9   Draw Haworth formulas for the following names:

(a) α-D-galactopyranose

(b) methyl β-D-mannopyranoside

(c) 4-O-(α-D-glucopyranosyl)-β-D-mannopyranose

(d) β-D-arabinofuranosyl α-L-arabinofuranoside

Section 23.6
Oxidation of Monosaccharides

An aldehyde group is easily oxidized to a carboxyl group. Chemical tests for aldehydes depend upon this ease of oxidation (Section 13.7). Sugars that can be oxidized by such mild oxidizing agents as Tollens reagent, an alkaline solution of $Ag(NH_3)_2{}^+$, are called **reducing sugars** (because the inorganic oxidizing agent is *reduced* in the reaction). The cyclic hemiacetal forms of all aldoses are readily oxidized because they are in equilibrium with the open-chain aldehyde form.

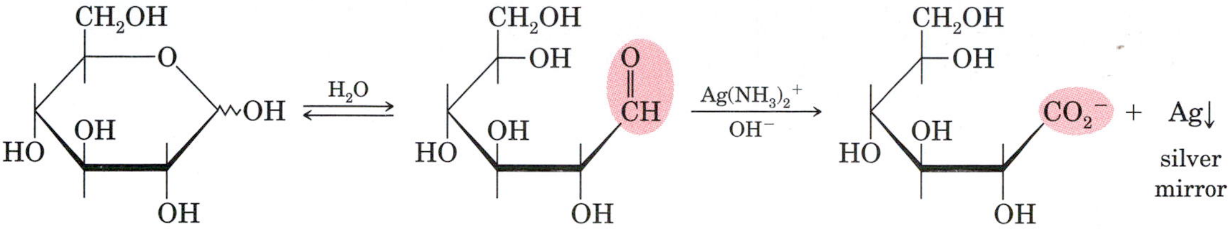

D-glucopyranose

a reducing sugar

Although fructose is a ketone, it is also a reducing sugar.

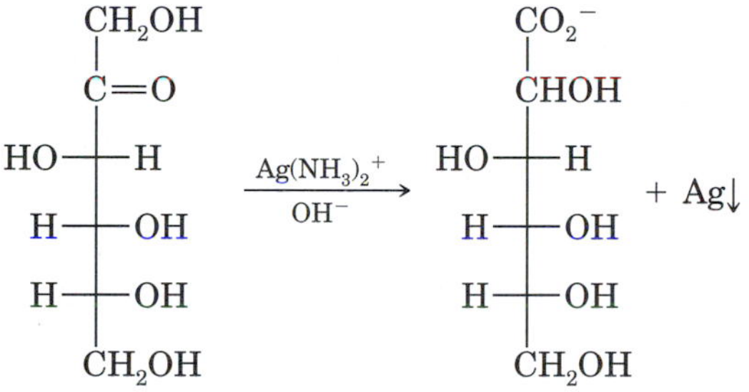

D-fructose

a reducing sugar

The reason that fructose can be oxidized so readily is that, in alkaline solution, fructose is in equilibrium with two diasteromeric aldehydes through an enediol tautomeric intermediate.

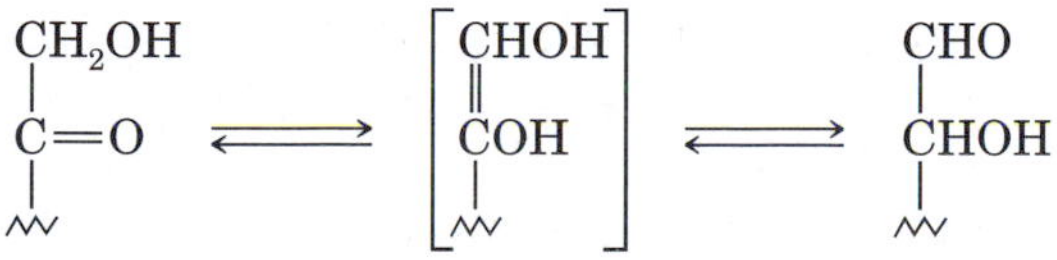

| *a ketose* | *an enediol intermediate* | *an aldose* |

In glycosides, the carbonyl group is blocked. Glycosides are **nonreducing sugars.**

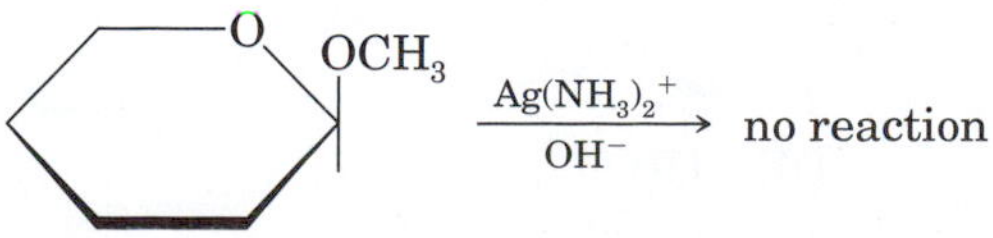

A. Aldonic Acids

The product of oxidation of the aldehyde group of an aldose is a polyhydroxy carboxylic acid called an **aldonic acid.** Although Tollens reagent can effect the conversion, a more convenient and less expensive reagent for the synthetic reaction is a buffered solution of bromine.

$$
\begin{array}{ccc}
\text{CHO} & & \text{CO}_2\text{H} \\
\text{H}\!-\!\text{OH} & & \text{H}\!-\!\text{OH} \\
\text{HO}\!-\!\text{H} & \xrightarrow[\text{pH 5–6}]{\text{Br}_2 + \text{H}_2\text{O}} & \text{HO}\!-\!\text{H} \\
\text{H}\!-\!\text{OH} & & \text{H}\!-\!\text{OH} \\
\text{H}\!-\!\text{OH} & & \text{H}\!-\!\text{OH} \\
\text{CH}_2\text{OH} & & \text{CH}_2\text{OH} \\
\text{D-glucose} & & \text{D-gluconic acid}
\end{array}
$$

an aldonic acid

In alkaline solution, the aldonic acids exist as open-chain carboxylate ions. Upon acidification, they form lactones (cyclic esters), just as any γ- or δ-hydroxy acid would (Section 15.6). Most aldonic acids have both γ and δ hydroxyl groups, and either a five- or a six-membered ring could be formed. The five-membered rings (γ-lactones) are favored in these cases.

$$
\begin{array}{ccc}
① \text{CO}_2\text{H} & & \\
② \text{H}\!-\!\text{OH} & & \\
③ \text{HO}\!-\!\text{H} & \underset{\longleftarrow}{\overset{-\text{H}_2\text{O}}{\longrightarrow}} & \\
④ \text{H}\!-\!\text{OH} & & \\
⑤ \text{H}\!-\!\text{OH} & & \\
⑥ \text{CH}_2\text{OH} & & \\
\text{D-gluconic acid} & & \text{a lactone}
\end{array}
$$

STUDY PROBLEM

23.10   Predict the product (if any) of bromine oxidation of each of the following compounds:

(a)
$$
\begin{array}{c}
\text{CHO} \\
\text{H}\!-\!\text{OH} \\
\text{CH}_2\text{OH}
\end{array}
$$

(b), (c), (d) structures

B. Aldaric Acids

Vigorous oxidizing agents oxidize the aldehyde group and also the terminal hydroxyl group (a primary alcohol) of a monosaccharide. The product is a

polyhydroxy dicarboxylic acid called an **aldaric acid.** Aldaric acids also form
lactones readily.

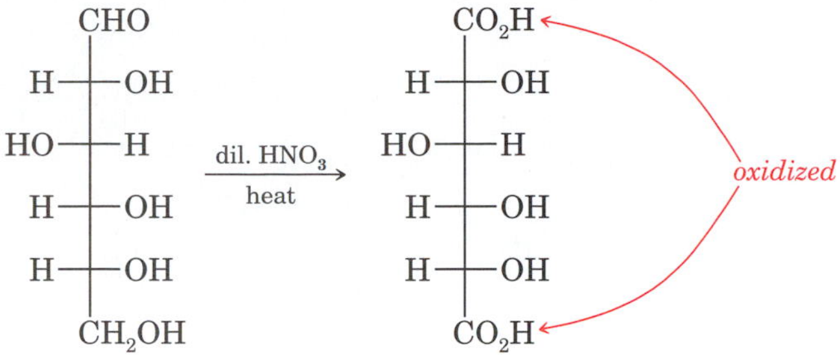

D-glucose → dil. HNO₃ / heat → D-glucaric acid

an aldaric acid

STUDY PROBLEM

23.11 Will each of the following monosaccharides yield (1) a *meso*-aldaric acid or
(2) an optically active aldaric acid upon reaction with HNO₃?

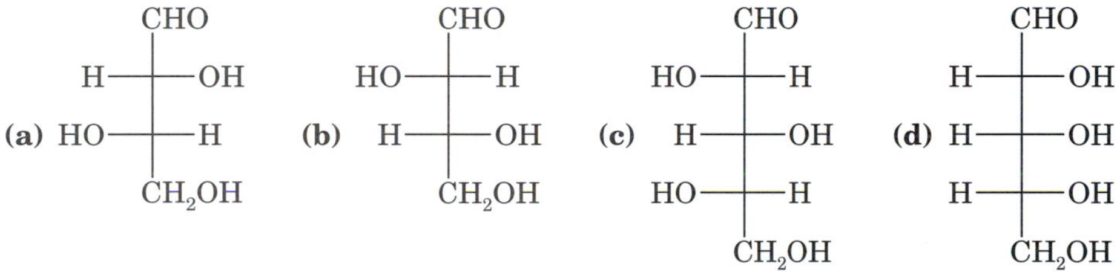

C. Uronic Acids

Although it is not easy to do in the laboratory, in biological systems the terminal CH_2OH group can be oxidized enzymatically without oxidation of the
aldehyde group. The product is called a **uronic acid.**

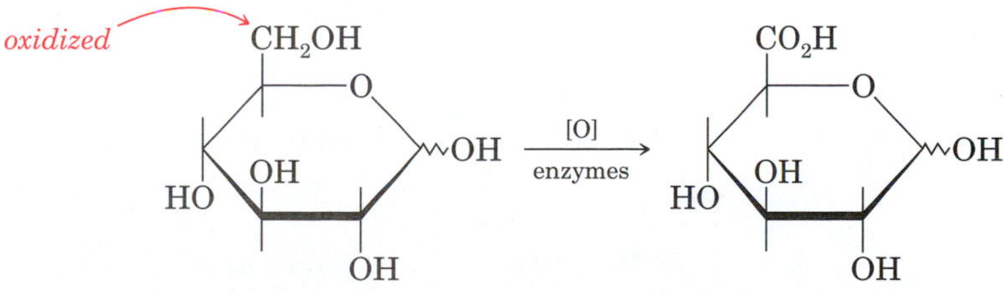

D-glucose → [O] enzymes → D-glucuronic acid

a uronic acid

Glucuronic acid is important in animal systems because many toxic substances are excreted in the urine as **glucuronides,** derivatives of this acid.
Also, in plant and animal systems, D-glucuronic acid can be converted to L-gulonic acid, which is used to biosynthesize L-ascorbic acid (vitamin C). (This
last conversion does not take place in primates or guinea pigs, which require
a dietary source of vitamin C.) The fact that a compound of the D-series be-

comes a compound of the L-series is not due to a biochemical change in configuration; rather, the change arises from the change in the numbering of the
carbons, as may be seen in the following equation.

D-glucuronic acid   L-gulonic acid

L-ascorbic acid
(vitamin C)

D. Periodic Acid Oxidation

The **periodic acid oxidation** is a test for 1,2-diols and for 1,2- or α-hydroxy
aldehydes and ketones. A compound containing such a grouping is oxidized
and cleaved by periodic acid (HIO_4). In the case of a simple 1,2-diol, the products are two aldehydes or ketones.

The periodic acid reaction goes through a cyclic intermediate, a fact that
explains why isolated hydroxyl groups are not oxidized.

STUDY PROBLEM

23.12   Predict the products of the periodic acid oxidation of the following compound:

$$\underset{CH_3CH-CH-CH_2}{\overset{OH \quad OH \quad OCH_3}{|\qquad|\qquad|}}$$

In the reaction of an α-hydroxy aldehyde or ketone, the carbonyl group is oxidized to a carboxyl group, while the hydroxyl group is again oxidized to an aldehyde or ketone.

to $-CO_2H$ *to* $-CHO$

$$\underset{RC-CHR}{\overset{O \; OH}{\|\;\;|}} \xrightarrow{HIO_4} \underset{RCOH}{\overset{O}{\|}} + \underset{HCR}{\overset{O}{\|}} + HIO_3$$

$$\underset{HC-CHR}{\overset{O \;\; OH}{\|\;\;\;|}} \xrightarrow{HIO_4} \underset{HCOH}{\overset{O}{\|}} + \underset{HCR}{\overset{O}{\|}} + HIO_3$$

The periodic acid oxidation is used in carbohydrate analysis. In carbohydrates, the oxidation of the interior hydroxyl groups proceeds further than it does with a simple 1,2-diol. For example, the products of the periodic acid oxidation of erythrose are formaldehyde and formic acid in the molar ratio of 1:3.

oxidized to
HCO_2H

oxidized to HCHO

$$\begin{array}{c} CHO \\ H-\!\!\!-OH \\ H-\!\!\!-OH \\ CH_2OH \end{array} \xrightarrow{HIO_4} \underset{HCH}{\overset{O}{\|}} + 3\,\underset{HCOH}{\overset{O}{\|}}$$

formaldehyde   formic acid

erythrose

This oxidation of the CHOH groups in erythrose to formic acid can be rationalized by considering the reaction to be stepwise.

Step 1:

oxidized

$$\begin{array}{c} CHO \\ H-\!\!\!-OH \\ H-\!\!\!-OH \\ CH_2OH \end{array} \xrightarrow{HIO_4} \begin{array}{c} CHO \\ H-\!\!\!-OH \\ CH_2OH \end{array} + \underset{HCOH}{\overset{O}{\|}}$$

formic acid

Step 2:

oxidized

$$\begin{array}{c} CHO \\ H-\!\!\!-OH \\ CH_2OH \end{array} \xrightarrow{HIO_4} \begin{array}{c} CHO \\ | \\ CH_2OH \end{array} + \underset{HCOH}{\overset{O}{\|}}$$

formic acid

Step 3:

$$\begin{array}{c} \text{CHO} \\ | \\ \text{CH}_2\text{OH} \end{array} \xrightarrow{\text{HIO}_4} \underset{\text{formaldehyde}}{\text{HCH}} + \underset{\text{formic acid}}{\text{HCOH}}$$

■ **Section 23.7**

Reduction of Monosaccharides

Both aldoses and ketoses can be reduced by carbonyl reducing agents, such as hydrogen and catalyst or a metal hydride, to polyalcohols called **alditols.** The suffix for the name of one of these polyalcohols is **-itol.** The product of reduction of D-glucose is called D-*glucitol,* or *sorbitol.*

D-glucose → D-glucitol (sorbitol)

Natural D-glucitol has been isolated from many fruits (for example, cherries, plums, apples, pears, and mountain ash berries) and from algae and seaweed. Synthetic D-glucitol is used as an artificial sweetener.

■ **Section 23.8**

Reactions at the Hydroxyl Groups

The hydroxyl group in carbohydrates behave in a manner similar to that of other alcohol groups. They can be esterified by either carboxylic acids or inorganic acids, and they can be converted to ethers. Carbohydrates can also act as diols and form cyclic acetals or ketals with aldehydes or ketones. These reactions were discussed in Chapters 7 and 13.

A. Acetate Formation

A common reagent for esterification of alcohols is acetic anhydride, with either sodium acetate or pyridine as an alkaline catalyst. If the reaction is carried out below 0°C, the acylation reaction is faster than the α–β anomeric interconversion. Under these conditions, either α- or β-D-glucose yields its corresponding pentaacetate. At higher temperatures, a mixture of the α- and β-pentaacetates is formed, with the β-pentaacetate predominating.

β-D-glucopyranose   penta-*O*-acetyl-β-D-glucopyranose

B. Ether Formation

Dimethyl sulfate is an inorganic ester with an excellent leaving group. This compound is used to form methyl ethers.

dimethyl sulfate   *a methyl ether*

When a monosaccharide is treated with an excess of dimethyl sulfate and NaOH, all the hydroxyl groups (including a hemiacetal or hemiketal OH group) are converted to methoxyl groups.

D-glucose   methyl tetra-*O*-methyl-D-glucopyranoside (55%)

In a typical Williamson ether synthesis ($RO^- + RX \rightarrow ROR + X^-$; Section 8.2B), the alkoxide must be prepared with a stronger base than NaOH. In the case of the carbohydrates, NaOH is a sufficiently strong base to yield alkoxide ions. (The inductive effect of the electronegative oxygens on adjacent carbons renders each hydroxyl group more acidic than a hydroxyl group in an ordinary alcohol.) Because the acetal bond is stable in base, the configuration at the anomeric carbon of a glycoside is not changed in this methylation reaction.

STUDY PROBLEM

> **23.13**   Give the structure of the product of the treatment of methyl 2-deoxy-α-D-ribofuranoside with: **(a)** acetic anhydride; **(b)** an alkaline solution of dimethyl sulfate.

C. Cyclic Acetal and Ketal Formation

Because carbohydrates contain numerous OH groups, it is sometimes desirable to block some of them so that selective reactions can be carried out on the other hydroxyl groups. Acetals and ketals are two common blocking groups (Section 13.4B). For example, an aldehyde, such as benzaldehyde, reacts with 1,3-diol groupings in sugar molecules. Other aldehydes and ketones can react preferentially at different diol groupings. (In some cases, the product is the furanose, rather than the pyranose ring.) The different products arise because of subtle (generally unpredictable) steric and electronic effects.

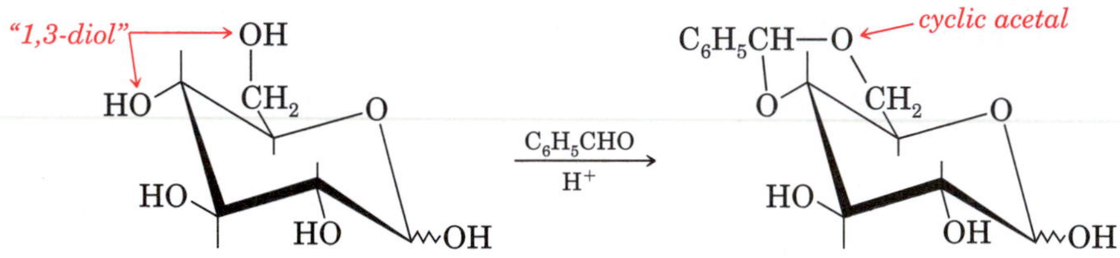

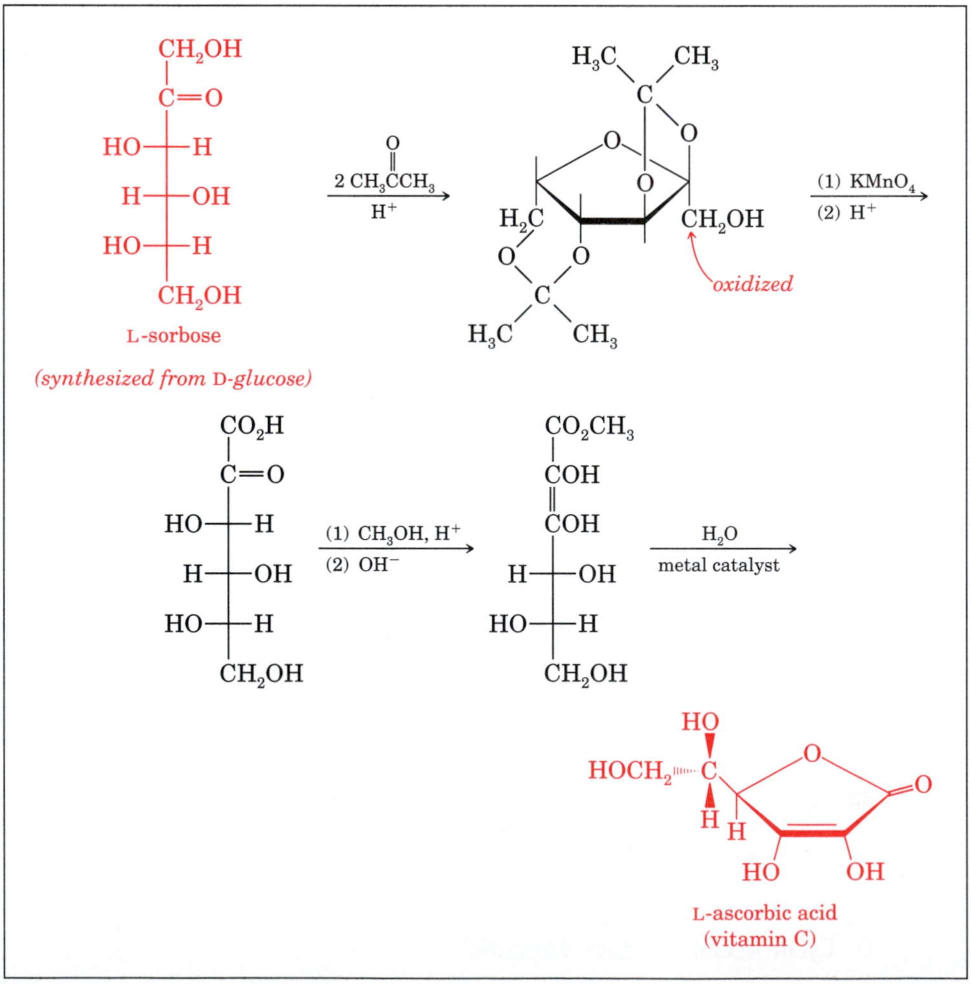

Figure 23.4 The conversion of L-sorbose to L-ascorbic acid (vitamin C).

In the commercial conversion of L-sorbose to vitamin C, acetone is used to block four hydroxyl groups so that a single CH_2OH group can be oxidized. This conversion is outlined in Figure 23.4.

Figure 23.5 summarizes the important reactions of monosaccharides.

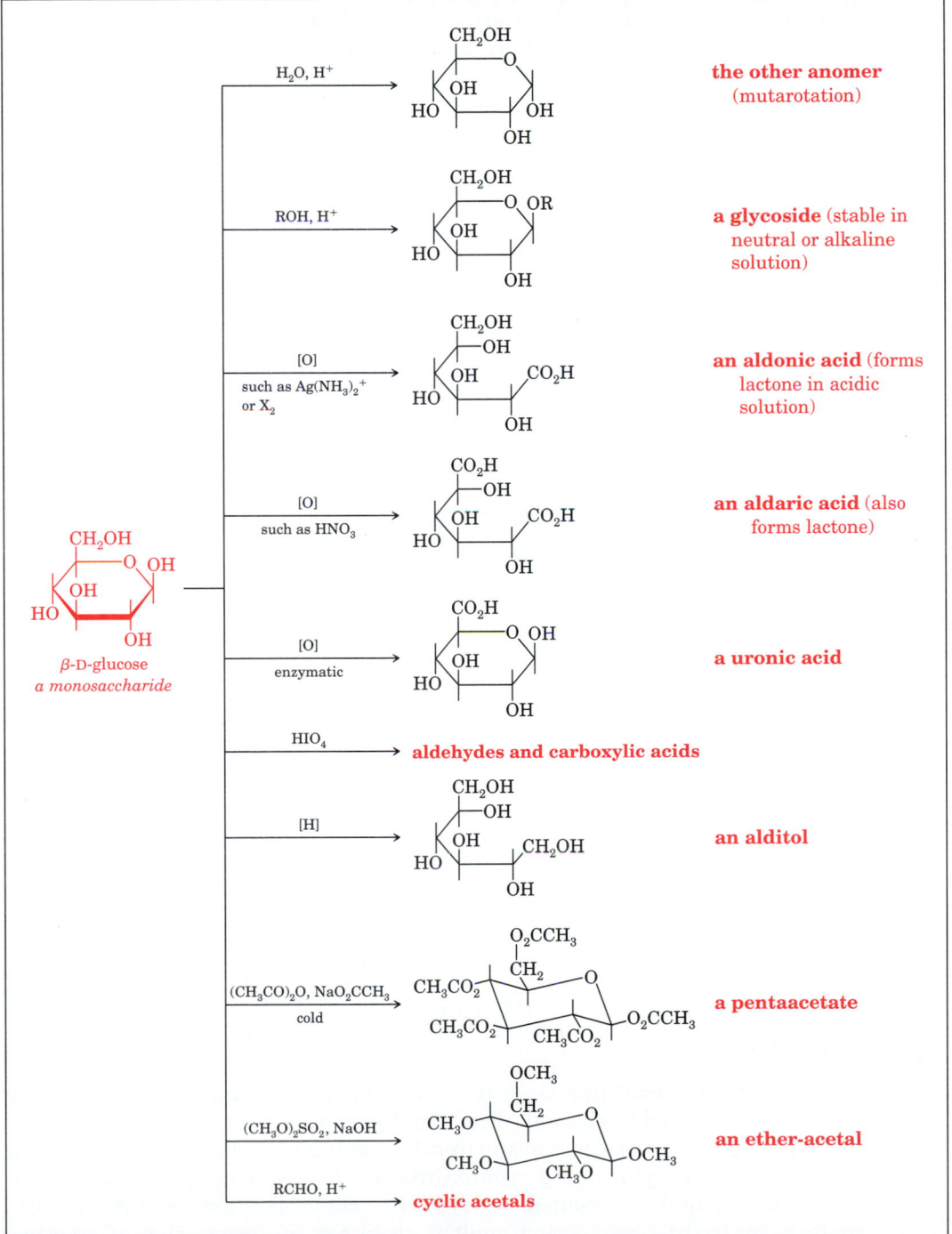

Figure 23.5   Summary of the important reactions of monosaccharides.

Section 23.9
Disaccharides

A **disaccharide** is a carbohydrate composed of two units of monosaccharide joined together by a glycoside link from carbon 1 of one unit to an OH of the other unit. A common mode of attachment is an α or β glycoside link from the first unit to the 4-hydroxyl group of the second unit. This link is called a 1,4'-α or a 1,4'-β link, depending on the stereochemistry at the glycoside carbon.

a 1,4'-β link
(conformational)   a 1,4'-β link
(Haworth)

Let us look at the preceding structures more closely. Unit 1 (the left-hand unit in each structure) has a β-glycoside link to unit 2. In aqueous solution, this glycoside link is fixed. It is not in equilibrium with the anomer. However, unit 2 (the right-hand unit in each structure) contains a hemiacetal group. In aqueous solution, this particular group is in equilibrium with the open-chain aldehyde form and with the other anomer.

A. Maltose

The disaccharide **maltose** is used in baby foods and malted milk. It is the principal disaccharide obtained from the hydrolysis of starch. Starch is broken down into maltose in an apparently random fashion by an enzyme in saliva called *α-1,4-glucan 4-glucanohydrolase*. The enzyme *α-1,4-glucan maltohydrolase,* found in sprouted barley *(malt),* converts starch specifically into maltose units. In beer-making, malt is used for the conversion of starches from corn or other sources into maltose. An enzyme in yeast *(α-glucosidase)* catalyzes the hydrolysis of the maltose into D-glucose, which is acted upon by other enzymes from the yeast to yield ethanol. One molecule of maltose yields two molecules of D-glucose, regardless of whether the hydrolysis takes place in a laboratory flask, in an organism, or in a fermentation vat.

$$\text{starch} \xrightarrow[\text{H}^+ \text{ or enzymes}]{\text{H}_2\text{O}} \text{maltose} \xrightarrow[\text{H}^+ \text{ or enzymes}]{\text{H}_2\text{O}} \text{D-glucose} \xrightarrow[\text{enzymes}]{} \text{CH}_3\text{CH}_2\text{OH}$$

ethanol

A molecule of maltose contains two units of D-glucopyranose. The first unit (shown on the left) is in the form of an α-glycoside. This unit is attached to the oxygen at carbon 4′ in the second unit by a 1,4′-α link.

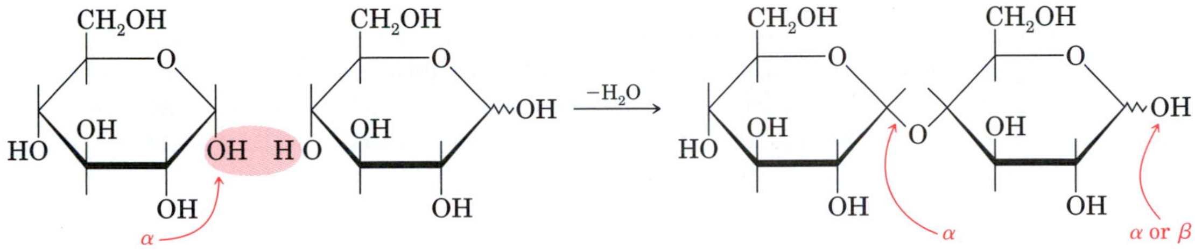

maltose

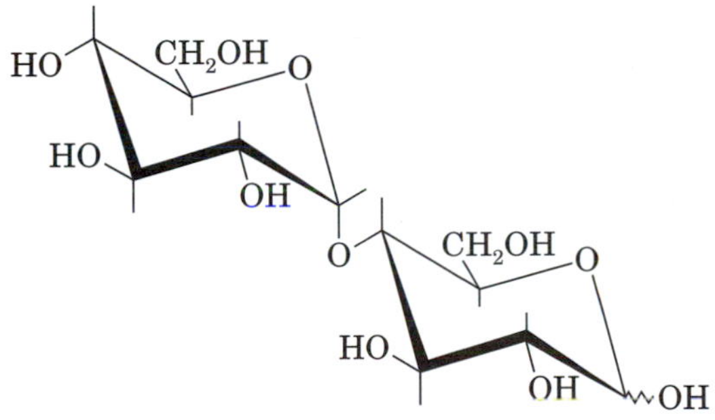

conformational formula for maltose
4-*O*-(α-D-glucopyranosyl)-D-glucopyranose

The anomeric carbon of the second unit of glucopyranose in maltose is part of a hemiacetal group. As a result, there are two forms of maltose (α- and β-maltose), which are in equilibrium with each other in solution. Maltose undergoes mutarotation, is a reducing sugar, and can be oxidized to the carboxylic acid **maltobionic acid** by a bromine–water solution.

STUDY PROBLEM

23.14   Give the structures of the products:

(a) α-maltose $\xrightarrow{\text{H}_2\text{O, H}^+}$ (b) β-maltose $\xrightarrow{\text{Br}_2, \text{H}_2\text{O}}$

B. Cellobiose

The disaccharide obtained from the partial hydrolysis of cellulose is called **cellobiose.** Like maltose, cellobiose is composed of two glucopyranose units joined together by a 1,4′-link. Cellobiose differs from maltose in that the 1,4-linkage is β rather than α.

cellobiose
4-*O*-(β-D-glucopyranosyl)-D-glucopyranose

Chemical hydrolysis of cellobiose with aqueous acid yields a mixture of α- and β-D-glucose, the same products that are obtained from maltose. Cellobiose can also be hydrolyzed with the enzyme *β-glucosidase* (also called *emulsin*), but not by *α-glucosidase,* which is specific for the α link (that is, the link in maltose).

STUDY PROBLEM

23.15   Give the structures of the products:

(a) α-cellobiose $\xrightarrow{\text{H}_2\text{O, H}^+}$ (b) β-cellobiose $\xrightarrow{\text{H}_2\text{O, H}^+}$

(c) α-cellobiose $\xrightarrow{\text{Br}_2,\text{H}_2\text{O}}$ (d) α-cellobiose $\xrightarrow{\text{Tollens reagent}}$

(e) α-cellobiose $\xrightarrow{\text{β-glucosidase}}$

C. Lactose

The disaccharide **lactose** (milk sugar) is different from maltose or cellobiose in that it is composed of two different monosaccharides, D-glucose and D-galactose.

lactose
4-*O*-(β-D-galactopyranosyl)-D-glucopyranose

Lactose is a naturally occurring disaccharide found only in mammals. Cow's milk and human milk contain about 5% lactose. Lactose is obtained commercially as a by-product in the manufacture of cheese.

In normal human metabolism, lactose is hydrolyzed enzymatically to D-galactose and D-glucose; then the galactose is converted to glucose, which can undergo metabolism. A condition called **galactosemia** that affects some infants is caused by lack of the enzyme used to convert galactose to glucose. Galactosemia is characterized by high levels of galactose in the blood and urine. Symptoms range from vomiting to mental and physical retardation and sometimes death. Treatment consists of removing milk and milk products from the diet. (An artificial milk made from soybeans may be substituted.)

D. Sucrose

The disaccharide **sucrose** is common table sugar. Sugar cane was grown domestically as early as 6000 B.C. in India. (The words "sugar" and "sucrose" come from the Sanskrit word *sarkara*.) Sucrose was encountered by the soldiers of Alexander the Great, who entered India in 325 B.C. In later centuries, the use of sucrose was spread by the Arabs and the Crusaders. Sugar cane was introduced into the New World by Columbus, who brought some to Santo Domingo in 1493. In the 1700s, it was discovered that certain beets also contain high levels of sucrose. The discovery meant that sugar could be obtained from plants grown in temperate climates as well as from sugar cane grown in the tropics.

sucrose
β-D-fructofuranosyl α-D-glucopyranoside

Whether it comes from beets or sugar cane, the chemical composition of sucrose is the same: one unit of fructose joined to one unit of glucose. The glycoside link joins the ketal and acetal carbons, β from fructose and α from glucose. Note the difference between sucrose and the other disaccharides we have discussed: in sucrose, *both* anomeric carbon atoms (not just one) are used in the glycoside link. In sucrose, neither fructose nor glucose has a hemiacetal group; therefore, sucrose in water is not in equilibrium with an aldehyde or keto form. Sucrose does not exhibit mutarotation and is not a reducing sugar.

Invert sugar is a mixture of D-glucose and D-fructose obtained by the acidic or enzymatic hydrolysis of sucrose. The enzymes that catalyze the hydrolysis of sucrose, called *invertases*, are specific for the β-D-fructofuranoside link and are found in yeast and in bees. (Honey is primarily invert sugar.) Because of the presence of free fructose (the sweetest sugar), invert sugar is sweeter than sucrose. A synthetic invert sugar called *Isomerose* is prepared by the enzymatic isomerization of glucose in corn syrup. It has commercial use in the preparation of ice cream, soft drinks, and candy.

The name "invert sugar" is derived from inversion in the sign of the specific rotation when sucrose is hydrolyzed. Sucrose has a specific rotation of +66.5°, a *positive* rotation. The mixture of products (glucose, $[\alpha] = +52.7°$, and fructose, $[\alpha] = -92.4°$) has a net *negative* rotation.

Section 23.10
Polysaccharides

A **polysaccharide** is a compound in which the molecules contain many units of monosaccharide joined together by glycoside links. Upon complete hydrolysis, a polysaccharide yields monosaccharides.

Polysaccharides serve three purposes in living systems: structural, nutritional, and as specific agents. Typical architectural polysaccharides are *cellulose,* which gives strength to the stems and branches of plants, and *chitin,* the structural component of the exoskeletons of insects. Typical nutritional polysaccharides are *starch* (as is found in wheat and potatoes) and *glycogen,* an animal's internal store of readily available carbohydrate. *Heparin,* an example of a specific agent, is a polysaccharide that prevents blood coagulation.

heparin

Polysaccharides can also be bonded to other types of molecules, as in *glycoproteins* (polysaccharide–protein complexes) and *glycolipids* (polysaccharide–lipid complexes).

A. Cellulose

Cellulose is the most abundant organic compound on earth. It has been estimated that about 10^{11} tons of cellulose are biosynthesized each year, and that cellulose accounts for about 50% of the bound carbon on earth! Dry leaves contain 10–20% cellulose; wood, 50%; and cotton, 90%. The most convenient laboratory source of pure cellulose is filter paper.

Cellulose forms the fibrous component of plant cell walls. The rigidity of cellulose arises from its overall structure. Cellulose molecules are chains, or microfibrils, of up to 14,000 units of D-glucose that occur in twisted rope-like bundles held together by hydrogen bonding.

A single molecule of cellulose is a linear polymer of 1,4'-β-D-glucose. Complete hydrolysis in 40% aqueous HCl yields only D-glucose. The disaccharide isolated from partially hydrolyzed cellulose is cellobiose, which can be further hydrolyzed to D-glucose with an acidic catalyst or with the enzyme emulsin. Cellulose itself has no hemiacetal carbon—it cannot undergo mutarotation or be oxidized by such test reagents as Tollens reagent. (There may be a hemiacetal at one end of each cellulose molecule. However, this is but a small portion of the whole and does not lead to observable reaction.)

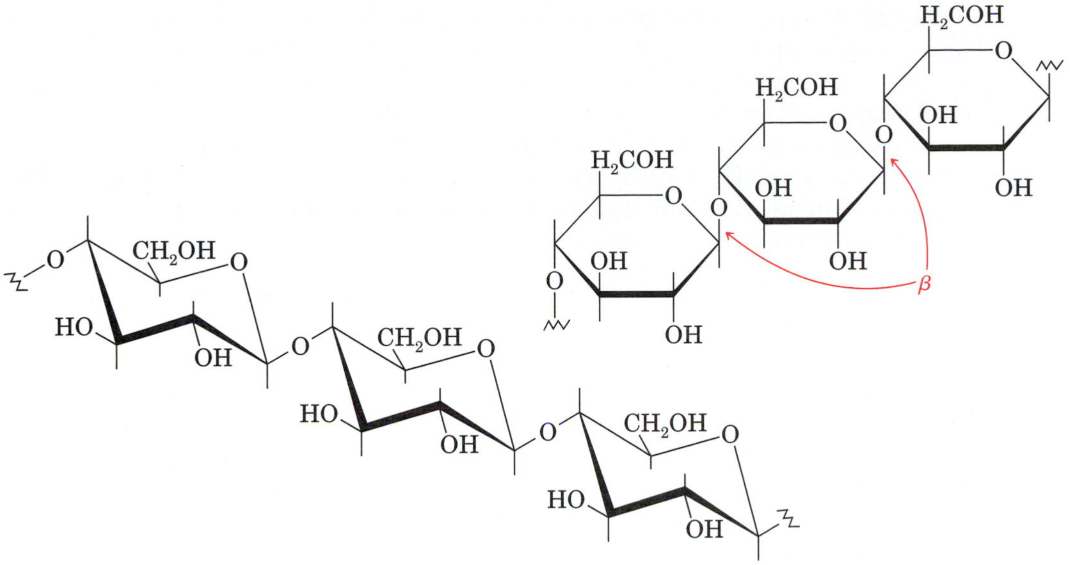

cellulose

Although mammals do not produce the proper enzymes for breaking down cellulose into glucose, certain bacteria and protozoa do have these enzymes. Grazing animals are capable of using cellulose as food only indirectly. Their stomachs and intestines support colonies of microorganisms that live and reproduce on cellulose. The animals use these microorganisms and their by-products as food.

B. Starch

Starch is the second most abundant polysaccharide. Starch can be separated into two principal fractions based upon solubility when triturated (pulverized) with hot water: about 20% of starch is **amylose** (soluble) and the remaining 80% is **amylopectin** (insoluble).

Amylose Complete hydrolysis of amylose yields only D-glucose; partial hydrolysis yields maltose as the only disaccharide. We conclude that amylose is a linear polymer of 1,4′-linked α-D-glucose. The difference between amylose and cellulose is the glycoside link: β in cellulose, α in amylose. This difference is responsible for the different properties of these two polysaccharides.

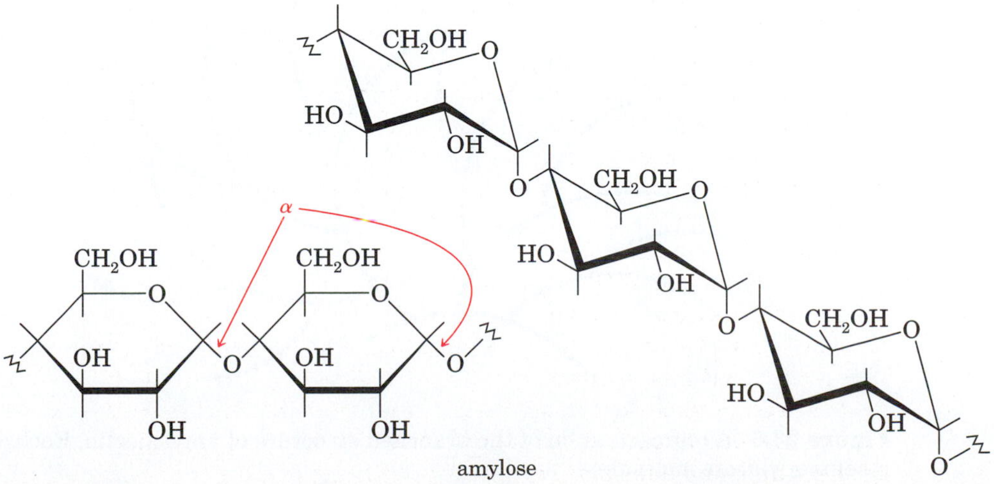

amylose

There are 250 or more glucose units per amylose molecule; the exact number depends upon the species of animal or plant. (Measurement of chain length is complicated by the fact that natural amylose degrades into smaller chains upon separation and purification.)

Amylose molecules form helices or coils around I_2 molecules. A deep blue color arises from electronic interactions between the two. This color is the basis of the **iodine test for starch,** in which a solution of iodine is added to an unknown as a test for the presence of starch.

Amylopectin A much larger polysaccharide than amylose, contains 1000 or more glucose units per molecule. Like the chain in amylose, the main chain of amylopectin contains 1,4'-α-D-glucose. Unlike amylose, amylopectin is *branched* so that there is a terminal glucose about every 25 glucose units (Figure 23.6). The bonding at the branch point is a 1,6'-α-glycosidic bond.

amylopectin

Complete hydrolysis of amylopectin yields only D-glucose. However, incomplete hydrolysis yields a mixture of the disaccharides maltose and isomaltose, the latter arising from the 1,6'-branching. The oligosaccharide mixture ob-

Figure 23.6 A representation of the branched structure of amylopectin. Each -•- represents a glucose molecule.

tained from the partial hydrolysis of amylopectin, referred to as **dextrins,** is used to make glue, paste, and fabric sizing.

$$\text{amylopectin} \xrightarrow{\text{H}_2\text{O}} \text{dextrins} \xrightarrow{\text{H}_2\text{O}} \text{maltose + isomaltose} \xrightarrow{\text{H}_2\text{O}} \text{D-glucose}$$

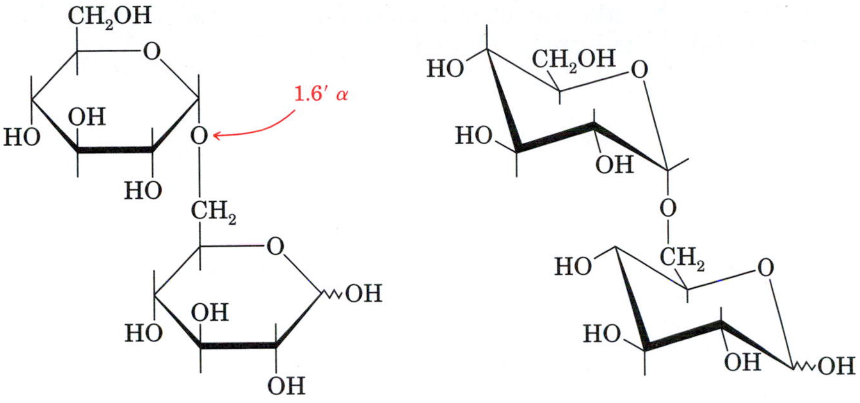

isomaltose
6-*O*-(α-D-glucopyranosyl)-D-glucopyranose

Glycogen is a polysaccharide that is used as a storehouse (primarily in the liver and muscles) for glucose in an animal system. Structurally, glycogen is related to amylopectin. It contains chains of 1,4'-α-linked glucose with branches (1,6'-α). The difference between glycogen and amylopectin is that glycogen is more branched than amylopectin.

C. Chitin

The principal structural polysaccharide of the arthropods (for example, crabs and insects) is **chitin.** It has been estimated that 10^9 tons of chitin are biosynthesized each year! Chitin is a linear polysaccharide consisting of β-linked *N*-acetyl-D-glucosamine. Upon hydrolysis, chitin yields 2-amino-2-deoxy-D-glucose. (The acetyl group is lost in the hydrolysis step.) In nature, chitins are bonded to nonpolysaccharide material (proteins and lipids).

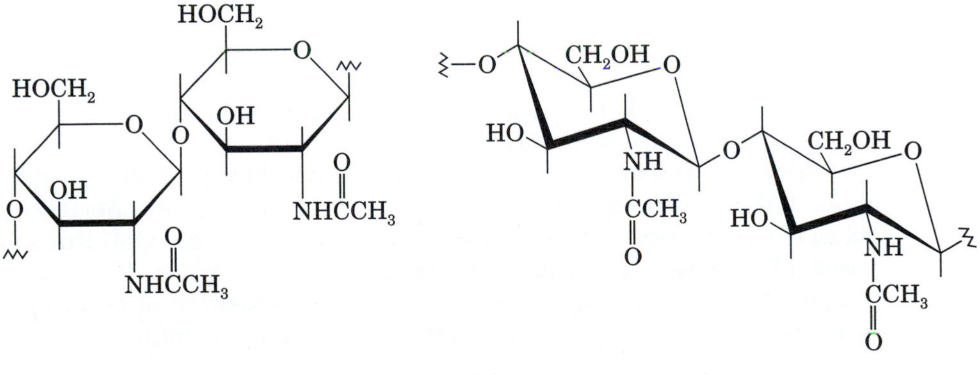

chitin

STUDY PROBLEM

23.16 Give the structures for the major organic products when chitin is treated with: **(a)** hot dilute aqueous HCl; **(b)** hot dilute aqueous NaOH.

Summary

Carbohydrates are polyhydroxy aldehydes and ketones or their derivatives. A **monosaccharide** is the smallest carbohydrate; it does not undergo hydrolysis to smaller units. Monosaccharides may be classified as to the number of carbons and to the principal functional group:

an aldotriose   a tetrulose

Epimers are diastereomers that differ in configuration at only one chiral carbon atom.

Natural monosaccharides generally belong to the **D-series.** Because of the presence of both hydroxyl and carbonyl groups, monosaccharides that can form **furanose** or **pyranose** hemiacetal or hemiketal rings undergo cyclization. The cyclization creates a new chiral carbon and therefore gives rise to a pair of diastereomers called **α and β anomers.** In solution, the anomers are in equilibrium with each other.

Because of the equilibrium, a monosaccharide undergoes reactions typical of aldehydes as well as reactions at the hydroxyl groups. The reactions of monosaccharides are summarized in Figure 23.5.

Disaccharides are composed of two monosaccharide units joined by a glycoside link from one unit to an OH group of the second unit. **Maltose** is composed of two D-glucopyranose units joined by a 1,4'-α link. **Cellobiose** is composed of two D-glucopyranose units joined by a 1,4'-β link. **Lactose** is composed of β-D-galactopyranose joined to the 4-position of D-glucopyranose. **Sucrose** is composed of α-D-glucopyranose and β-D-fructofuranose joined by a 1,2' link.

A **polysaccharide** is composed of many monosaccharide units joined by glycosidic links:

cellulose:	1,4'-β-D-glucopyranose
amylose:	1,4'-α-D-glucopyranose
amylopectin:	1,4'-α-D-glucopyranose with 1,6'-α-branching

Essay Problem for Chapter 23

Determination of the Ring Size of Cyclic Monosaccharides

The ring size of a cyclic monosaccharide can be determined by the following chemical procedure. Complete methylation of the saccharide is carried out using dimethyl sulfate under alkaline conditions. The resulting methylated methyl glycoside is ring-opened in acid solution and vigorously oxidized to yield several products. For example, the cyclic monosaccharide (**1**) when subjected to the above conditions yielded the products shown below.

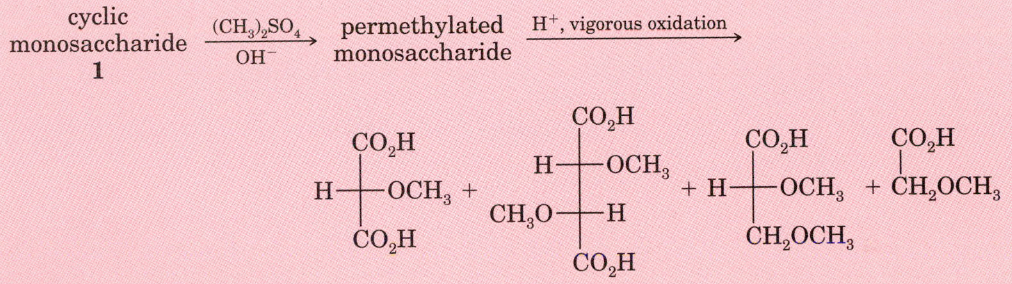

$$\text{cyclic monosaccharide } \mathbf{1} \xrightarrow[\text{OH}^-]{(CH_3)_2SO_4} \text{permethylated monosaccharide} \xrightarrow{\text{H}^+,\ \text{vigorous oxidation}}$$

Questions

1. What are the products of the vigorous oxidation of

$$CH_3CH_2\overset{\overset{\displaystyle OH}{|}}{CH}CH_2CH_2CH_3$$

   and

$$CH_3CH_2\overset{\overset{\displaystyle O}{\|}}{C}CH_2CH_2CH_3 ?$$

   Write equations to explain your answer.

2. Write the equation for the reaction of dimethyl sulfate with α-glucose. What are the two types of ether linkages in permethylated glucose?
3. What are the acid hydrolysis product(s) of permethylated glucose?
4. If the acid hydrolysis product(s) of permethylated glucose are vigorously oxidized, what products would be formed?
5. What is the ring size of the monosaccharide given in the essay?

Study Problems*

23.17 What does each of the indicated portions of the following name signify?

methyl 2,3,4,6-tetra-*O*-methyl-D-glucopyranoside

   (a) tetra-*O*-methyl   **(b)** gluco   **(c)** pyran   **(d)** oside

   (e) the methyl preceding the parent name as a separate word

---

* For information concerning the organization of the *Study Problems* and *Additional Problems*, see the *Note to student* on page 41.

23.18 Classify the following monosaccharides (as an aldohexose, for example):

(a)
$$\begin{array}{c} CHO \\ HO \overline{} H \\ H \overline{} OH \\ CH_2OH \end{array}$$

(b)
$$\begin{array}{c} CH_2OH \\ C{=}O \\ CH_2OH \end{array}$$

(c)
$$\begin{array}{c} CHO \\ CH_2 \\ HO \overline{} H \\ HO \overline{} H \\ CH_2OH \end{array}$$

23.19 State whether each of the monosaccharides in the preceding problem belongs to the D series or to the L series, if either.

23.20 Match each of the following classes of compound with a structure on the right:

(a) a hexulose

(1)

(b) a pentopyranose

(2)

(c) a pentofuranose

(3)

(d) a pentofuranoside

(4)

23.21 Write equations for cyclization reactions of the following monosaccharides showing the formation of the α and β six-membered-ring diastereomers. (Use Haworth formulas.)

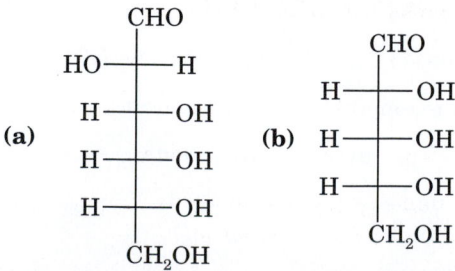

(a)
$$\begin{array}{c} CHO \\ HO \overline{} H \\ H \overline{} OH \\ H \overline{} OH \\ H \overline{} OH \\ CH_2OH \end{array}$$

(b)
$$\begin{array}{c} CHO \\ H \overline{} OH \\ H \overline{} OH \\ H \overline{} OH \\ CH_2OH \end{array}$$

23.22 Draw Haworth formulas for the following monosaccharides. [For (c) and (d), refer to Figure 23.3.]

(a)

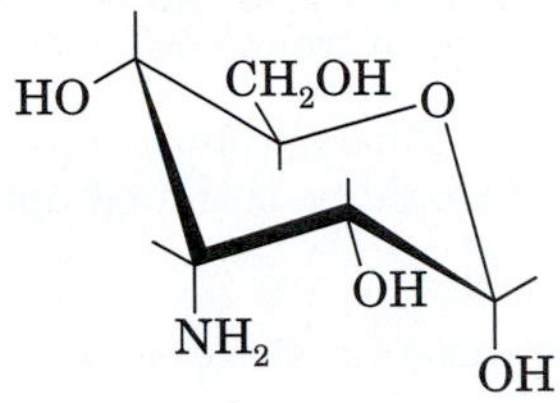

(b)

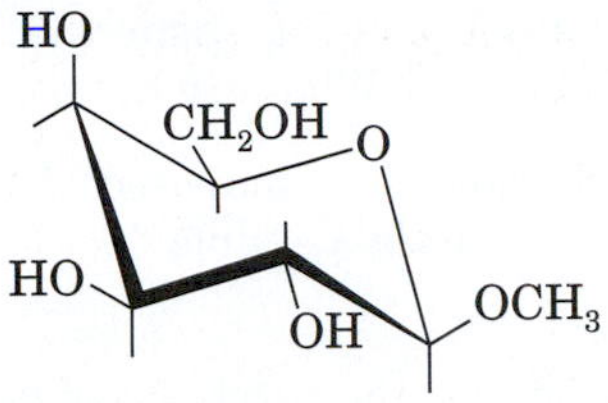

(c) β-D-altropyranose

(d) x-D-lyxofuranose

23.23 Write equations (using Haworth formulas) that illustrate:

(a) the mutarotation of pure β-D-arabinofuranose in water

(b) the conversion of β-D-fructofuranose to β-D-fructopyranose

(c) the mutarotation of β-maltose (Section 23.9A)

23.24 Give the Haworth formulas for the major organic products:

(a) D-glucose + $(CH_3)_2CHOH$ $\xrightarrow{H^+}$

(b) D-galactose + CH_3CH_2OH $\xrightarrow{H^+}$

(c) methyl α-D-ribofuranoside $\xrightarrow[\text{heat}]{H_2O, H^+}$

23.25 Draw and label the aldonic, aldaric, and uronic acids for each of the following mono-saccharides:

(a)

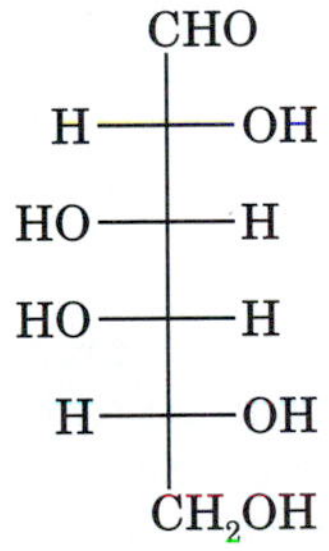

(b)

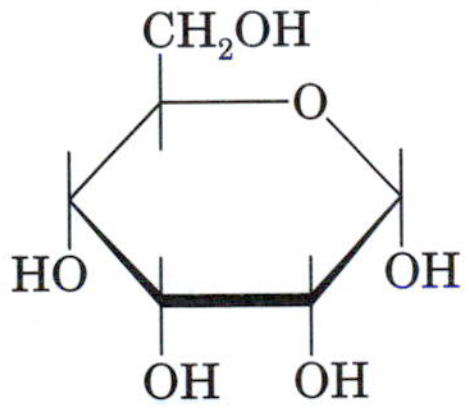

23.26 Three compounds (A, B, and C) were subjected to oxidation with HIO_4. The following products were obtained from each reaction, respectively. What are the structures of A, B, and C?

(a) $CH_3\overset{O}{\overset{\|}{C}}OH + H\overset{O}{\overset{\|}{C}}CH_2CH_3$

(b) $CH_3\overset{O}{\overset{\|}{C}}CH_2CH_2CH_2\overset{O}{\overset{\|}{C}}H$

(c) $H\overset{O}{\overset{\|}{C}}H + CH_3\overset{O}{\overset{\|}{C}}CH_3$

23.27 Write equations showing each step in the periodic acid oxidation of arabinose.

23.28 Write equations to represent the reaction of β-D-ribofuranose with **(a)** an excess of diethyl sulfate plus NaOH, and **(b)** an excess of acetic anhydride.

23.29 Methyl β-D-gulopyranoside is treated with: (1) excess dimethyl sulfate plus NaOH; (2) H_2O, H^+; and then (3) hot HNO_3. Write the equations that illustrate the steps in this reaction sequence.

23.30 *Trehalose* is a nonreducing sugar with the formula $C_{12}H_{22}O_{11}$. Upon hydrolysis, only D-glucose is obtained. What are the possible structures of trehalose?

23.31 If a polysaccharide were composed of 1,4'-α-linked D-glucopyranose with 1,3'-α-linked branches, what are the possible disaccharides that would be obtained upon partial hydrolysis?

23.32 Write equations for the reactions of D-galactopyranose (mixture of α and β) with the following reagents:

    (a) $Ag(NH_3)_2^+$, OH^- **(b)** Br_2, H_2O, pH 6   **(c)** HNO_3, heat

    (d) CH_3OH, H^+ **(e)** (1) $NaBH_4$; (2) H_2O, H^+

Additional Problems

23.33 The oxidation of D-fructose with Tollens reagent yields a mixture of anions of D-mannonic acid and D-gluconic acid. Explain.

23.34 A carbohydrate A ($C_{12}H_{22}O_{11}$) was treated with (1) CH_3OH, H^+, and (2) excess CH_3I and Ag_2O. The product B was hydrolyzed to 2,3,4,6-tetra-*O*-methyl-D-galactose and 2,3,6-tri-*O*-methyl-D-glucose. When A was treated with aqueous acid, the products were D-galactose and D-glucose in equal amounts. When A was treated with aqueous Br_2, a carboxylic acid C was isolated. Hydrolysis of C with aqueous HCl resulted in D-gluconic acid as the only acid. What are the structures of A, B, and C?

23.35 *Raffinose* is a trisaccharide found in beets. Complete hydrolysis of raffinose yields D-fructose, D-glucose, and D-galactose. Partial enzymatic hydrolysis of raffinose with invertase yields D-fructose and the disaccharide *melibiose*. Partial hydrolysis of raffinose with an α-glycosidase yields D-galactose and sucrose. Methylation of raffinose followed by hydrolysis yields 2,3,4,6-tetra-*O*-methylgalactose; 2,3,4-tri-*O*-methylglucose; and 1,3,4,6-tetra-*O*-methylfructose. What are the structures of raffinose and melibiose?

23.36 *Bleomycin* is an antitumor antibiotic active against squamous cell carcinomas. Part of the structure of this compound is a reducing disaccharide A with the formula $C_{12}H_{22}O_{11}$. Compound A is hydrolyzed in acidic solution to D-mannose and L-gulose. The hydrolysis is also catalyzed by enzymes specific for α-mannose glycosides. When A is subjected to treatment with (1) excess dimethyl sulfate + NaOH and then (2) acidic hydrolysis, the products are the 2,3,4,6-tetramethyl ether of D-mannose and the 3,4,6-trimethyl ether of L-gulose. What is the structure of A?

23.37 *Linustatin* ($C_{16}H_{27}NO_{11}$) is a disaccharide in defatted meal from flaxseed (*Linum usitatissimum*) that can protect livestock against the toxic effects of selenium compounds in forage plants. Acidic hydrolysis of linustatin yields D-glucose as the only monosaccharide. Enzymatic hydrolysis with β-glucosidase yields two compounds: D-glucose and a glycoside. Comparison of the ^{13}C NMR spectrum of linustatin with those of known compounds indicates the presence of two CH_3 groups, one quaternary C (no hydrogens bonded to it), and one unsubstituted —CH_2OH group, as well as other carbon atoms. The infrared spectrum of linustatin shows strong absorption at 3400 cm^{-1} (2.94 μm) and weak absorption at 2240 cm^{-1} (4.46 μm). No carbonyl absorption is apparent. What is the structure of linustatin?

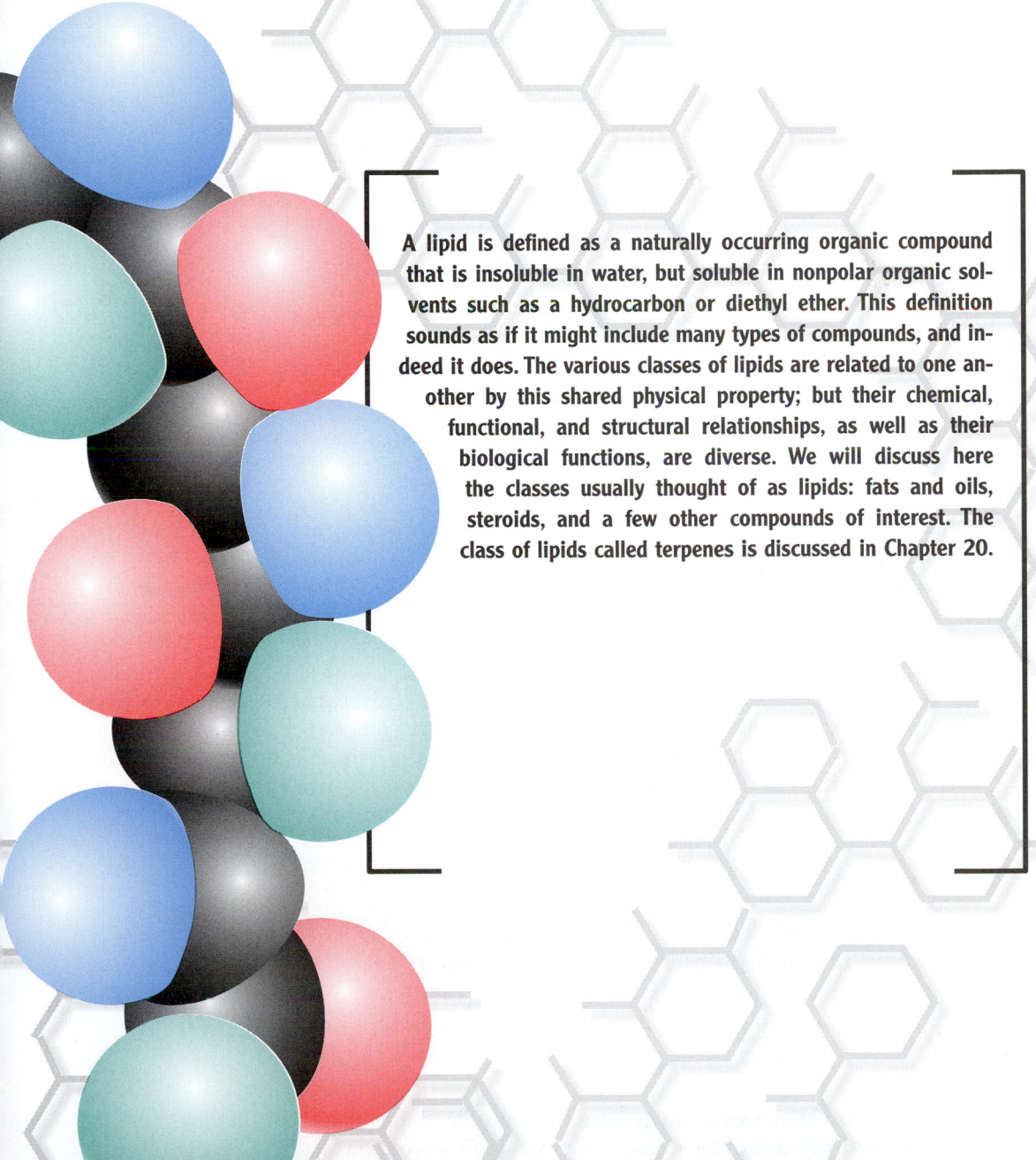

24 LIPIDS

A lipid is defined as a naturally occurring organic compound that is insoluble in water, but soluble in nonpolar organic solvents such as a hydrocarbon or diethyl ether. This definition sounds as if it might include many types of compounds, and indeed it does. The various classes of lipids are related to one another by this shared physical property; but their chemical, functional, and structural relationships, as well as their biological functions, are diverse. We will discuss here the classes usually thought of as lipids: fats and oils, steroids, and a few other compounds of interest. The class of lipids called terpenes is discussed in Chapter 20.

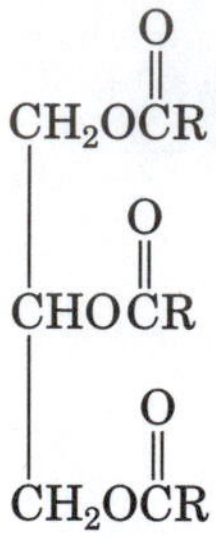

$$CH_2OCR$$ (with C=O above)

$$CHOCR$$ (with C=O above)

$$CH_2OCR$$ (with C=O above)

a fat:
a triglyceride,
or triacylglycerol

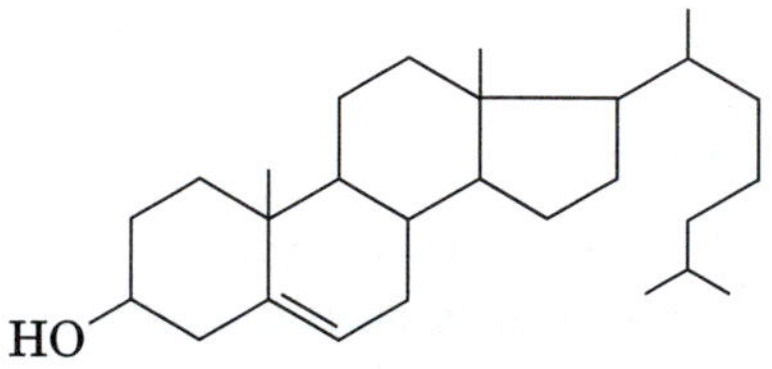

HO —

cholesterol

a steroid

Section 24.1

Fats and Oils

Fats and oils are **triglycerides,** or **triacylglycerols;** both terms mean triesters of glycerol. The distinction between a fat and an oil is arbitrary: at room temperature a fat is solid and an oil is liquid. Most glycerides in animals are fats, while those in plants tend to be oils; hence the terms *animal fats* (bacon fat, beef fat) and *vegetable oils* (corn oil, safflower oil).

The carboxylic acid obtained from the hydrolysis of a fat or oil, called a **fatty acid,** generally has a long, unbranched hydrocarbon chain. Fats and oils are often named as derivatives of these fatty acids. For example, the tristearate of glycerol is named tristearin, and the tripalmitate of glycerol is named tripalmitin. A fat or oil can be named as an ester in the usual manner, for example, glyceryl tristearate and glyceryl tripalmitate.

$$CH_2O_2C(CH_2)_{16}CH_3$$
$$CHO_2C(CH_2)_{16}CH_3 + 3\ H_2O \xrightarrow{\ H^+\ }$$
$$CH_2O_2C(CH_2)_{16}CH_3$$

glyceryl tristearate
(tristearin)

a typical fat

$$CH_2OH$$
$$CHOH \quad + 3\ CH_3(CH_2)_{16}CO_2H$$
$$CH_2OH$$

glycerol
(glycerine)

stearic acid

a fatty acid

Because they can be hydrolyzed to smaller molecules, fats and oils are said to be **complex lipids.** Lipids such as cholesterol that cannot be hydrolyzed are called **simple lipids.**

STUDY PROBLEM

24.1 Write the formula of the triacylglycerol that can be formed from each of the following sets of fatty acids:

(a) 3 palmitic acids   **(b)** 3 palmitoleic acids

(c) 2 palmitic acids and 1 butanoic acid

Fatty acids can also be obtained from **waxes,** such as beeswax. In these cases, the fatty acid is esterified with a simple long-chain alcohol.

$$C_{25}H_{51}CO_2C_{28}H_{57} \qquad C_{27}H_{55}CO_2C_{32}H_{65} \qquad C_{15}H_{31}CO_2C_{16}H_{33}$$

in beeswax   *in carnauba wax*   cetyl palmitate

in spermaceti

Most naturally occurring fats and oils are *mixed* triglycerides; that is, the three fatty-acid portions of the glyceride are not the same. Table 24.1 lists some representative fatty acids, and Table 24.2 shows the fatty-acid composition of some plant and animal triglycerides.

Almost all naturally occurring fatty acids have an *even* number of carbon atoms because they are biosynthesized from the two-carbon acetyl groups in acetylcoenzyme A.

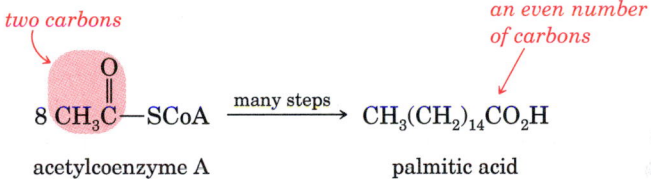

Acetylcoenzyme A reaction scheme showing: *two carbons* and *an even number of carbons* labels, $8\ CH_3C{-}SCoA$ (acetylcoenzyme A) $\xrightarrow{\text{many steps}}$ $CH_3(CH_2)_{14}CO_2H$ (palmitic acid).

Table 24.1   Selected fatty acids and their sources[a]

Name of Acid	Structure	Source
Saturated:		
butyric	$CH_3(CH_2)_2CO_2H$	milk fat
palmitic	$CH_3(CH_2)_{14}CO_2H$	animal and plant fat
stearic	$CH_3(CH_2)_{16}CO_2H$	animal and plant fat
Unsaturated:		
palmitoleic	$CH_3(CH_2)_5CH{=}CH(CH_2)_7CO_2H$	animal and plant fat
oleic	$CH_3(CH_2)_7CH{=}CH(CH_2)_7CO_2H$	animal and plant fat
linoleic[b]	$CH_3(CH_2)_4CH{=}CHCH_2CH{=}CH(CH_2)_7CO_2H$	plant oils
linolenic[b]	$CH_3CH_2CH{=}CHCH_2CH{=}CHCH_2CH{=}CH(CH_2)_7CO_2H$	linseed oil
arachidonic[b]	$CH_3(CH_2)_4(CH{=}CHCH_2)_4(CH_2)_2CO_2H$	plant oils

[a] The carbon–carbon double bonds in naturally occurring unsaturated fatty acids are *cis*.
[b] Essential fatty acids that must be present in the human diet and that are used for the synthesis of prostaglandins (see Section 24.4).

Table 24.2 Approximate fatty-acid composition of some common fats and oils

Source	Composition (%)[a]			
	Palmitic	Stearic	Oleic	Linoleic
corn oil	10	5	45	38
soybean oil	10	—	25	55
lard	30	15	45	5
butter	25	10	35	—
human fat	25	8	46	10

[a] Other fatty acids are also found in lesser amounts.

The hydrocarbon chain in a fatty acid may be saturated or it may contain double bonds. The most widely distributed fatty acid in nature, oleic acid, contains one double bond. Fatty acids with more than one double bond are not uncommon, particularly in vegetable oils. These oils are the so-called *polyunsaturates*.

The configuration around any double bond in a naturally occurring fatty acid is *cis*, a configuration that results in the low melting points of oils. A saturated fatty acid forms zigzag chains that can fit compactly together, resulting in high van der Waals attractions. Therefore, saturated fats are solids. If a few *cis* double bonds are present in the chains, the molecules cannot form neat, compact lattices, but tend to coil. Polyunsaturated triglycerides tend to be oils. Figure 24.1 shows models of the two types of chains.

A constant fluidity of cell membranes in different environments is maintained by the ratio of saturated to unsaturated fatty acids in the membranes. For example, bacteria grown at higher temperatures contain more saturated fatty acids in their cell membranes than do bacteria grown at lower temperatures.

STUDY PROBLEM

24.2 Complete the following equations:

(a) $cis\text{-}CH_3(CH_2)_7CH{=}CH(CH_2)_7CO_2H + KMnO_4 \xrightarrow[\text{cold}]{OH^-}$

(b) $(9Z, 12Z)\text{-}CH_3(CH_2)_4CH{=}CHCH_2CH{=}CH(CH_2)_7CO_2H \xrightarrow[\text{(2) } H_2O_2,\ H^+]{\text{(1) } O_3}$

(c) $CH_3(CH_2)_{14}CO_2H + Br_2 \xrightarrow{PBr_3 \text{ catalyst}}$

(d) $C_{25}H_{51}CO_2C_{28}H_{57} + NaOH \xrightarrow[\text{heat}]{H_2O,\ CH_3CH_2OH}$

(e) $C_{15}H_{31}CO_2C_{16}H_{33} \xrightarrow[\text{(2) } H_2O,\ H^+]{\text{(1) LiAlH}_4}$

(f) $\begin{array}{l} CH_2O_2CC_{15}H_{31} \\ | \\ CHO_2CC_{15}H_{31} \\ | \\ CH_2O_2CC_{15}H_{31} \end{array} \xrightarrow[\text{(2) } H_3O^+]{\text{(1) NaOH, } H_2O, \text{ heat}}$

A saturated triglyceride:

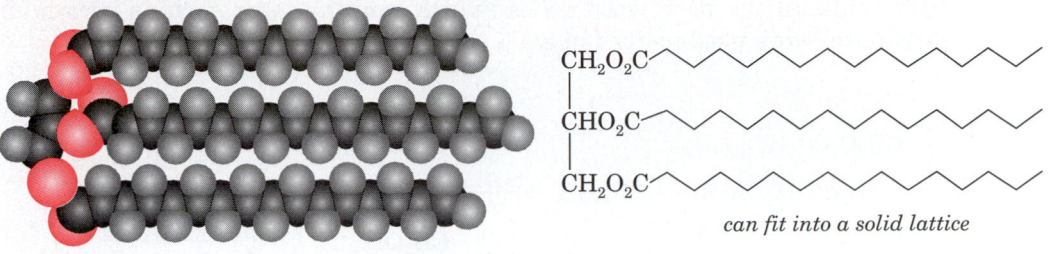

$$CH_2O_2C$$
$$CHO_2C$$
$$CH_2O_2C$$

can fit into a solid lattice

An unsaturated triglyceride:

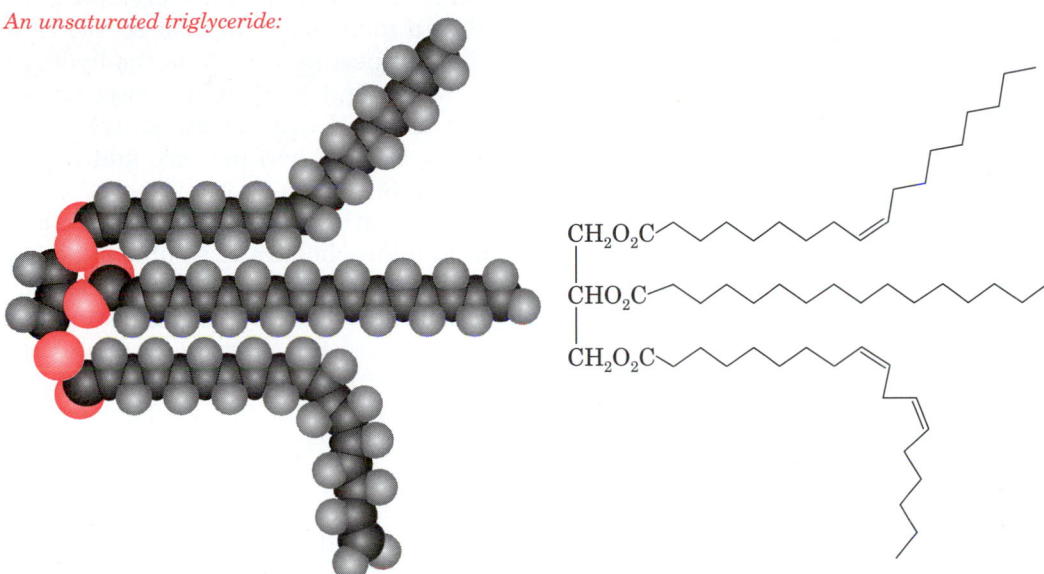

$$CH_2O_2C$$
$$CHO_2C$$
$$CH_2O_2C$$

cannot fit as well into a lattice

Figure 24.1   The shapes of saturated and unsaturated triglycerides. Adapted from William H. Brown, *Introduction to Organic and Biochemistry, 4th ed.* (Copyright 1987 by Wadsworth, Inc. Used by permission of Brooks/Cole Publishing Co.)

Section 24.2
Soaps and Detergents

Soaps are the alkali metal salts (usually sodium salts) of fatty acids. Soaps contain primarily C_{16} and C_{18} salts, but may also contain some lower-molecular-weight carboxylates.

Soaps were probably discovered by the ancient Egyptians several thousand years ago. Soap making by the Teutonic tribes was reported by Julius Caesar. The technique of soap making was lost in many parts of Europe during the Dark Ages, but was rediscovered during the Renaissance. The use of soap did not become widespread until the eighteenth century.

Soap is manufactured today by practically the same techniques as were used in ages past. Molten tallow (beef fat) or other fat is heated with lye (sodium hydroxide) and is thus saponified to glycerol and sodium salts of fatty acids. In the past, wood ashes (which contain bases such as potassium carbonate) were used instead of lye.

Saponification:

$$CH_2O_2C(CH_2)_{16}CH_3$$
$$|$$
$$CHO_2C(CH_2)_{16}CH_3 + 3\ NaOH \xrightarrow{heat} CHOH + 3\ CH_3(CH_2)_{16}CO_2{}^- Na^+$$
$$|$$
$$CH_2O_2C(CH_2)_{16}CH_3$$

$$CH_2OH$$
$$|$$
$$CHOH$$
$$|$$
$$CH_2OH$$

sodium stearate

glyceryl tristearate   glycerol   *a soap*

Once the saponification is complete, salt is added to help precipitate the soap, the water layer containing glycerol is drawn off, and the glycerol is recovered by distillation. Glycerol is used as a moisturizer in tobacco, pharmaceuticals, and cosmetics. (The moisturizing properties arise from the hydroxyl groups, which can hydrogen-bond with water and prevent its evaporation.) The soap is purified by boiling in fresh water to leach out excess lye, NaCl, and glycerol. Additives such as pumice, dyes, and perfume are added. The solid soap is then melted and poured into a mold.

A molecule of a soap contains a long hydrocarbon chain plus an ionic end. The hydrocarbon portion of the molecule is hydrophobic and soluble in nonpolar substances, while the ionic end is hydrophilic and water-soluble. Because of the hydrocarbon chain, a soap molecule as a whole is not truly soluble in water. However, soap is readily suspended in water because it forms **micelles,** clusters of 50–150 soap molecules with their hydrocarbon chains grouped together and with their ionic ends facing the water (see Figure 24.2).

The value of a soap is that it can emulsify oily dirt so that it can be rinsed away. This ability to act as an emulsifying agent arises from two properties of the soap. First, the hydrocarbon chain of a soap molecule dissolves in nonpolar substances, such as droplets of oil. Second, the anionic end of the soap molecule, which is attracted to water, is repelled by the anionic ends of soap molecules protruding from other drops of oil. Because of these repulsions between the soap–oil droplets, the oil cannot coalesce, but remains suspended.

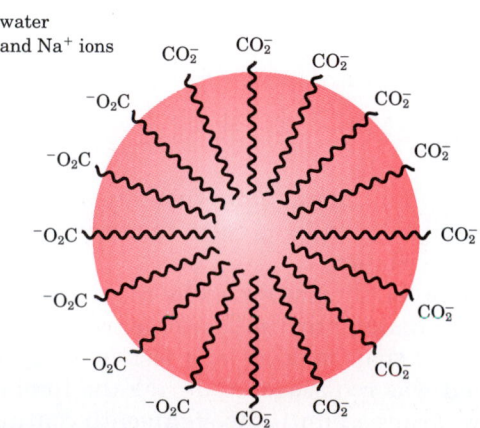

Figure 24.2 A micelle of the alkylcarboxylate ions of a soap.

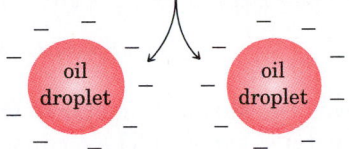

in soapy water, oil droplets repel each other because of similar charges of soap's carboxylate groups

The principal disadvantage of soaps is that they precipitate with hard water (water that contains Ca^{2+}, Mg^{2+}, Fe^{3+}, etc.) and leave a residue variously called scum, bathtub ring, or tattletale gray.

$$2\ RCO_2^- + Ca^{2+} \longrightarrow (RCO_2)_2Ca\downarrow$$

insoluble

After World War II, **synthetic detergents** were developed. Detergents are long-chain sulfonate or sulfate salts of sodium ($RSO_3^-\ Na^+$ and $ROSO_3^-\ Na^+$). Detergents have the advantage of not precipitating with the metal ions in hard water.

$$CH_3(CH_2)_{11}-\!\!\!\bigcirc\!\!\!-SO_3^-\ Na^+ \qquad CH_3(CH_2)_{11}OSO_3^-\ Na^+$$

sodium *p*-dodecylbenzenesulfonate   sodium dodecyl sulfate (sodium lauryl sulfate)

Soaps and detergents belong to a general class of compounds called **surfactants** (from *surface-active agents*), which are compounds that can lower the surface tension of water. Any surfactant molecule contains a hydrophobic end (one or more hydrocarbon chains) and a hydrophilic end (usually, but not necessarily, ionic). The hydrocarbon portion of a surfactant molecule must contain 12 or more carbon atoms to be effective.

General symbol for a surfactant:

hydrophobic tail ———   ———*hydrophilic head*

Surfactants can be classed as *anionic, cationic,* or *neutral,* depending on the nature of the hydrophilic group. Soaps, with their carboxylate groups, are anionic surfactants. The antibacterial "benzalkonium" chlorides (*N*-benzyl quaternary ammonium chlorides) are examples of cationic surfactants. A neutral surfactant contains nonionic, but polar, groups that can hydrogen bond with water. These are becoming increasingly popular because they are superior for cleaning synthetic fabrics.

$$\overset{\displaystyle O}{\underset{\displaystyle \|}{C_{11}H_{23}C}}O^-\ Na^+ \qquad C_{12}H_{25}OSO_3^-\ Na^+$$

anionic surfactants

$$\bigcirc\!\!\!-CH_2-\overset{\displaystyle CH_3}{\underset{\displaystyle CH_3}{\overset{\displaystyle |}{\underset{\displaystyle |}{N^+}}}}-C_{12}H_{25}\ Cl^-$$

a cationic surfactant (a benzalkonium chloride)

$$\overset{\overset{\displaystyle O}{\displaystyle \|}}{R}COCH_2CH_2O-(CH_2CH_2O)_2-CH_2CH_2OH$$

a neutral surfactant

Surfactants lower the surface tension of water by disrupting the hydrogen bonding on the water's surface. They cause this disruption by positioning their hydrophilic heads on the surface of the water and their hydrophobic tails extending away from the water's surface.

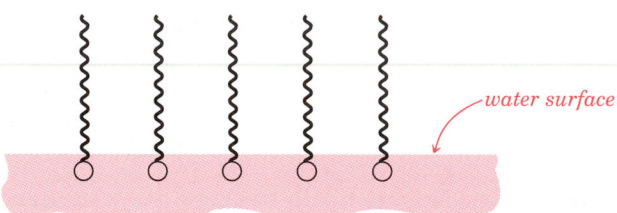

water surface

STUDY PROBLEMS

24.3 Write equations for a reaction sequence that would convert glyceryl trioleate into a sodium alkyl sulfate detergent.

24.4 Which of the following structures would you predict to have detergent properties?

(a) $CH_3(CH_2)_{11}CH_2\overset{\overset{\displaystyle O}{\displaystyle \|}}{C}OCH_3$ **(b)** $CH_3(CH_2)_{11}CH_2CH_2O$-glucose

(c) $CH_3(CH_2)_{11}CH_2CH_2OSO_3H$ **(d)** $CH_3(CH_2)_{11}CH_2CH_2\overset{+}{N}(CH_3)_3\ \ Cl^-$

(e)

[structure of a steroid-based compound with HO, CH₃, H₃C groups and $\overset{\overset{\displaystyle}{\displaystyle}}{C}NHCH_2CO_2^-\ \ Na^+$ side chain]

24.5 Using a 12-carbon continuous-chain alkyl group, draw a structure that illustrates each of the following terms:

(a) soap   **(b)** an anionic surfactant   **(c)** a cationic surfactant

(d) a neutral surfactant   **(e)** a *p*-alkylbenzenesulfonate detergent

Section 24.3

Phospholipids

Phospholipids are lipids that contain phosphate ester groups. **Phosphoglycerides,** one type of phospholipid, are closely related to the fats and oils.

These compounds usually contain fatty-acid esters at two positions of glycerol with a phosphate ester at the third position. Phosphoglycerides are distinctive because their molecules contain *two* long hydrophobic tails and a highly polar hydrophilic group—a dipolar-ion group. Phosphoglycerides are, therefore, neutral surfactants. They are excellent emulsifying agents. In mayonnaise, the phosphoglycerides of the egg yolk keep the oil emulsified in the vinegar.

Two types of phosphoglycerides:

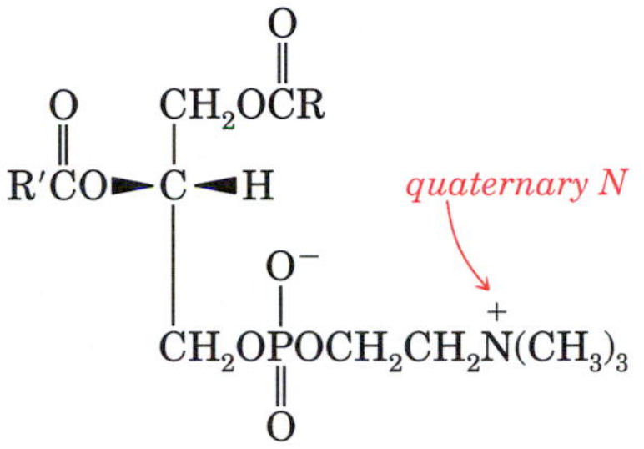

$$
\begin{array}{cc}
\underset{\text{R'CO} \blacktriangleright \text{C} \blacktriangleleft \text{H}}{\overset{\overset{O}{\|}}{}} \;\; \overset{\overset{O}{\|}}{\text{CH}_2\text{OCR}} & \text{quaternary N} \\
\text{CH}_2\text{OPOCH}_2\text{CH}_2\overset{+}{\text{N}}(\text{CH}_3)_3 &
\end{array}
$$

a lecithin,
or phosphatidylcholine

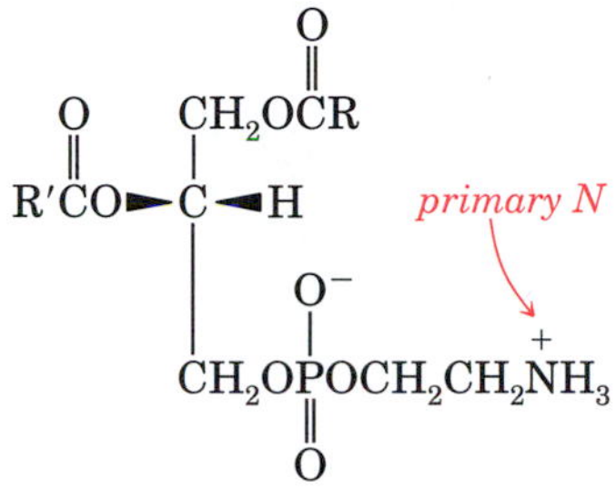

a cephalin,
or phosphatidylethanolamine

General symbol for a phosphoglyceride:

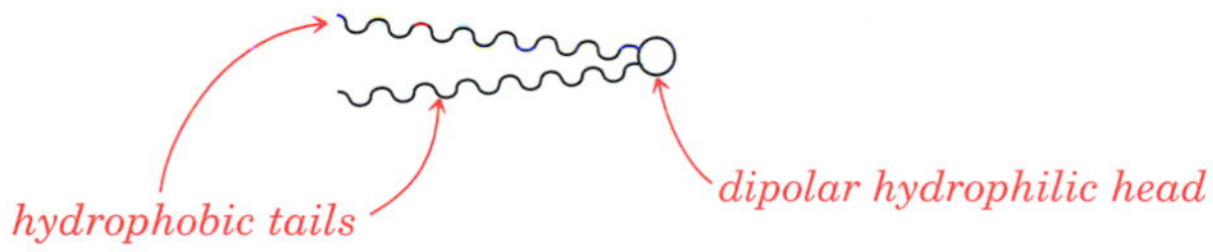

hydrophobic tails   *dipolar hydrophilic head*

Lecithins and **cephalins** are two types of phosphoglyceride that are found principally in the brain, nerve cells, and liver of animals and are also found in egg yolks, wheat germ, yeast, soybeans, and other foods. These two types of compounds are similar to each other in structure. Lecithins are derivatives of choline chloride, $HOCH_2CH_2N(CH_3)_3^+ \, Cl^-$, which is involved in the transmission of nerve impulses. Cephalins are derivatives of ethanolamine, $HOCH_2CH_2NH_2$.

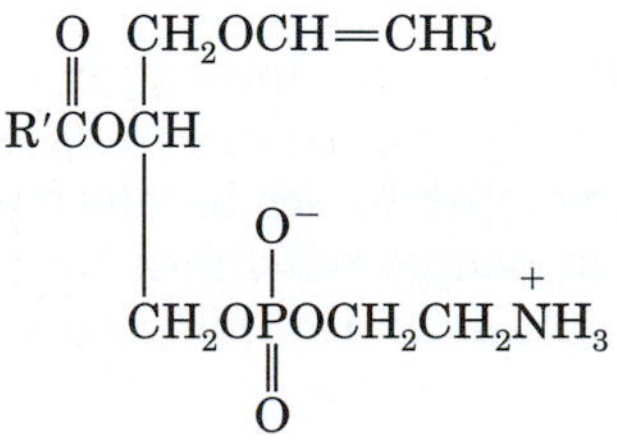

a plasmalogen

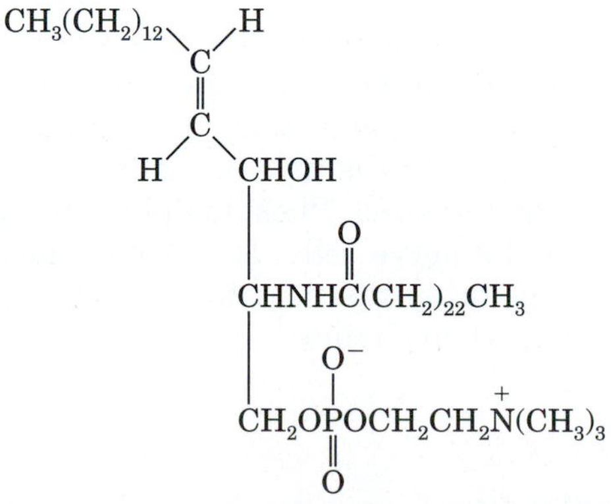

sphingomyelin

a sphingolipid

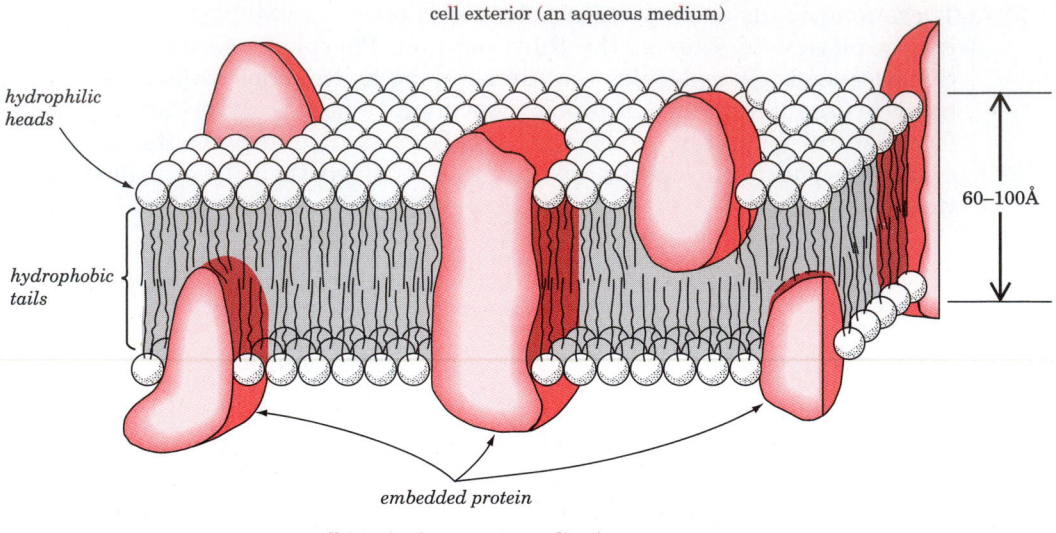

cell exterior (an aqueous medium)

hydrophilic heads

hydrophobic tails

60–100Å

embedded protein

cell interior (an aqueous medium)

Figure 24.3   The bilayer of phospholipids in a cell membrane.

Other classes of phospholipids are represented by **plasmalogens,** which have vinyl ether groups instead of ester groups at carbon 1 of glycerol, and **sphingolipids,** of which sphingomyelin is an example. Sphingomyelin is a phosphate ester, not of glycerol, but of a long-chain allylic alcohol with an amide side chain.

STUDY PROBLEM

> **24.6**   Hydrolysis of sphingomyelin yields phosphoric acid, choline, a 24-carbon fatty acid, and *sphingosine.* What is the structure of sphingosine?

Phospholipids are important in cell membranes. These membranes are formed primarily from proteins associated with a bilayer, or double layer, of phosphoglyceride molecules with their hydrophobic ends pointing inward and their hydrophilic ends pointing outward. The hydrocarbon portion of the membrane does not allow passage of water, ions, or polar molecules. The function of the embedded proteins is to pass water, ions, and other substances *selectively* in and out of the cell (see Figure 24.3).

It is thought that the sphingolipids such as sphingomyelin contribute strength to the myelin (nerve cell) sheath by the intertwining of their hydrocarbon chains. Phospholipids are also thought to act as electrical insulation for the nerve cells. The myelin sheaths of people with multiple sclerosis (and some other diseases that affect this membrane) are deficient in the long hydrocarbon chains.

Section 24.4

Prostaglandins

One of the more exciting areas of biochemical research today is that of **prostaglandins.** These compounds were isolated and studied by the Swedish scientists Sune Bergstrom and Bengt Samuelsson, who received the 1982 Nobel Prize for their work in this area. The prostaglandins were first discovered in semen, and it was recognized that they were synthesized in the prostate gland (hence the name). We now know that prostaglandins are found throughout the body and are also synthesized in the lungs, liver, uterus, and other organs and tissues.

Prostaglandins are hormone-like compounds that alter the activity of the cells where they are synthesized and the activity of adjoining cells. The nature of a prostaglandin's activities varies from one type of cell to another and from one type of prostaglandin to another. For example, administration of remarkably small doses of some prostaglandins stimulates uterine contractions and can cause abortions. Imbalances in prostaglandins can lead to nausea, diarrhea, inflammation, pain, fever, menstrual disorders, asthma, ulcers, hypertension, drowsiness, or blood clots. Aspirin inhibits prostaglandin synthesis by acetylating and thus deactivating *cyclooxygenase,* a key enzyme in the synthesis.

The prostaglandins are 20-carbon carboxylic acids that contain cyclopentane rings. They are biosynthesized from the 20-carbon unsaturated fatty acids.

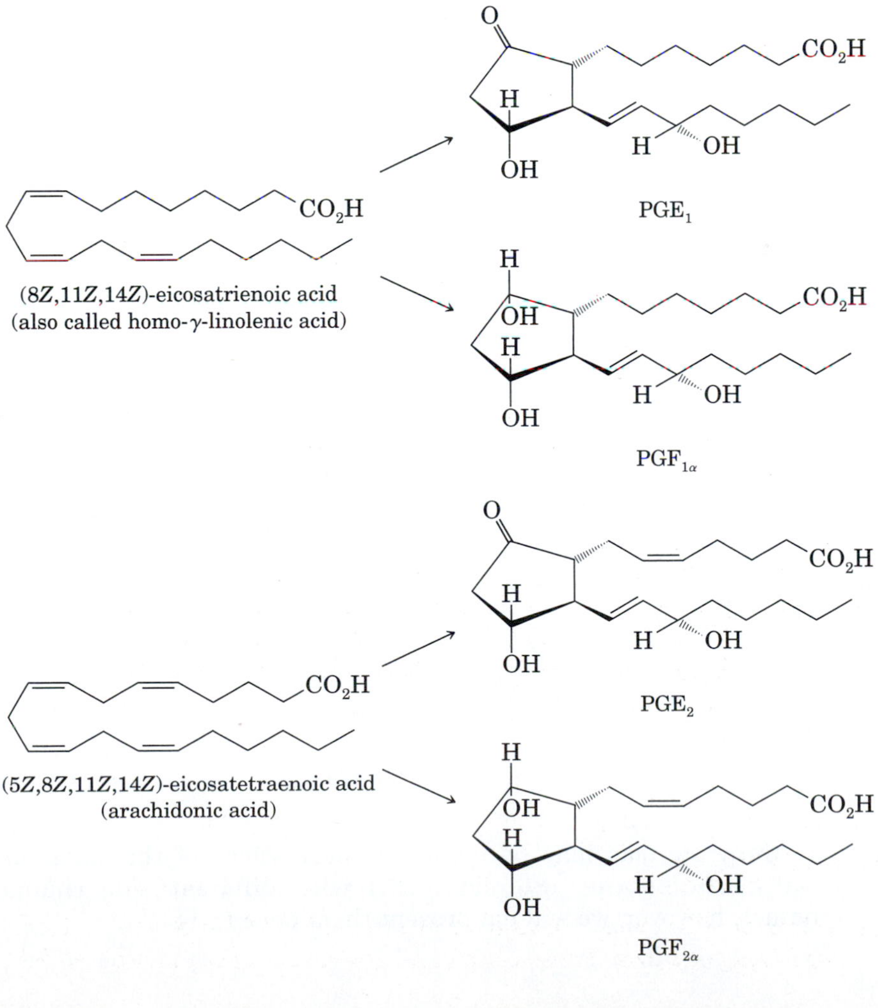

(8Z,11Z,14Z)-eicosatrienoic acid (also called homo-γ-linolenic acid)

PGE$_1$

PGF$_{1\alpha}$

(5Z,8Z,11Z,14Z)-eicosatetraenoic acid (arachidonic acid)

PGE$_2$

PGF$_{2\alpha}$

There are several known prostaglandins, but the four shown as products in the preceding equations are the most common ones. Although similar to each other in structure, the prostaglandins differ in (1) the number of double bonds, and (2) whether the cyclopentane portion is a diol or a keto alcohol. The terms PGE_1, $PGF_{1\alpha}$, PGE_2, and $PGF_{2\alpha}$ are symbols for these compounds. PG means *prostaglandin*, E means the *keto alcohol*, and F means the *diol*. The subscript numbers refer to the number of double bonds. (For example, the subscript 1 means *one* double bond.) The subscript α refers to the configuration of the OH at carbon 9 (*cis* to the carboxyl side chain).

STUDY PROBLEM

24.7   Show what functional groups must be inserted into each of the following compounds in order to yield the compound on the right.

(a) $(CH_2)_6CH_3$ $(CH_2)_7CH_3$ $\longrightarrow$ PGE_1

(b) $(CH_2)_6CH_3$ $(CH_2)_7CH_3$ $\longrightarrow$ PGE_2

Section 24.5
Steroids

We have mentioned steroids several times. Now let us take a more detailed look at these compounds. A **steroid** is a compound that contains the following ring system. The four rings are designated *A, B, C,* and *D.* The carbons are numbered as shown, starting with ring *A*, progressing to ring *D*, then to the angular (bridgehead) methyl groups, and finally to a side chain if it is present.

cholestane

Many steroids may be named as derivatives of this structure, which is called **cholestane.** (Steroid nuclei with different side chains also have names; however, we will not present them here.)

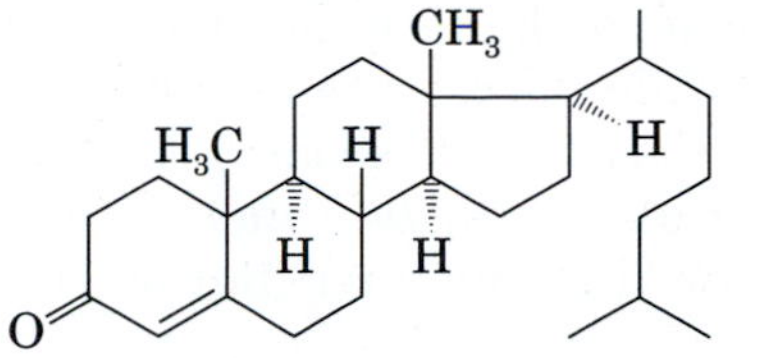

4-cholesten-3-one

a steroid with a cholestane skeleton

Steroids are found in almost all types of living systems. In animals, many steroids act as hormones. These, as well as synthetic steroids, find wide use as pharmaceuticals. In this section, we will briefly consider the conformation of the steroid ring system and then turn our attention to some important steroids. Finally, we will briefly discuss the biosynthesis of these compounds.

A. Conformation of Steroids

Recall from Section 4.5B that the more stable isomer of a 1,2-dialkylcyclo-hexane is the one with both substituents *equatorial*. A *trans*-1,2-disubstituted cyclohexane can exist in a preferred conformation in which both substituents are equatorial, while the *cis* isomer must have one axial substituent and one equatorial substituent. The *trans* isomer is therefore the more stable one.

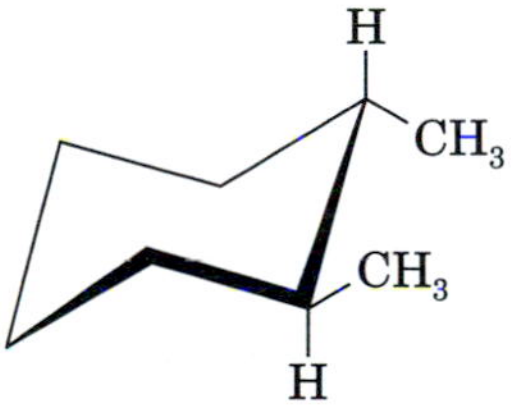

trans (e,e)   *more stable than*   *cis (a,e)*

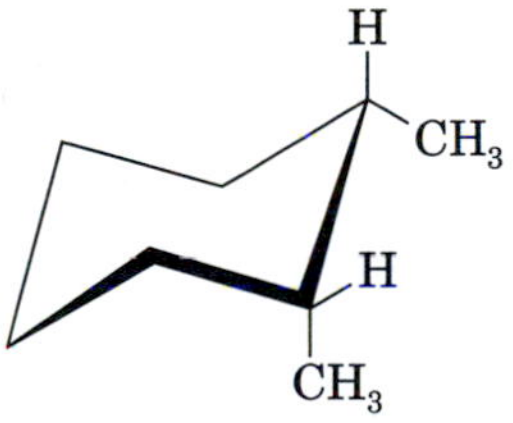

One ring fused to another in the 1,2-positions can be *cis* or *trans*. The following structures represent *cis*- and *trans*-decalin.

cis-decalin: or 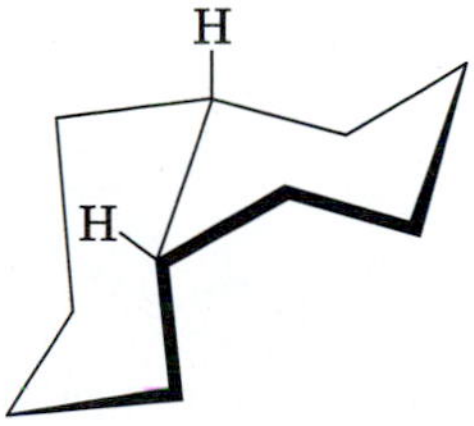   *Note cis H's as well as cis R groups.*

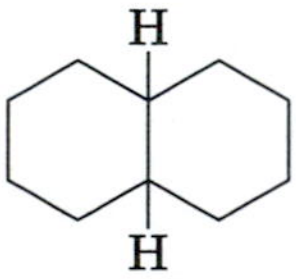

trans-decalin: or 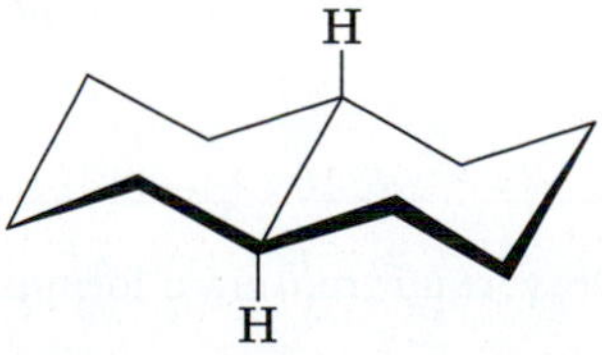   *Note trans H's.*

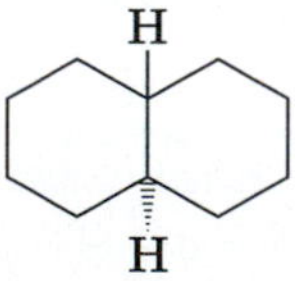

As can be seen in the formulas for *trans-* and *cis*-decalin, the *trans* ring juncture is *e,e,* while the *cis* ring juncture is *a,e.* The *trans* isomer is more stable than the *cis* isomer by about 3 kcal/mol. The steroid nucleus contains three ring junctures (*A/B*, *B/C*, and *C/D*). In nature, these are usually the more stable *trans* ring junctures (but we will encounter an exception later in this chapter).

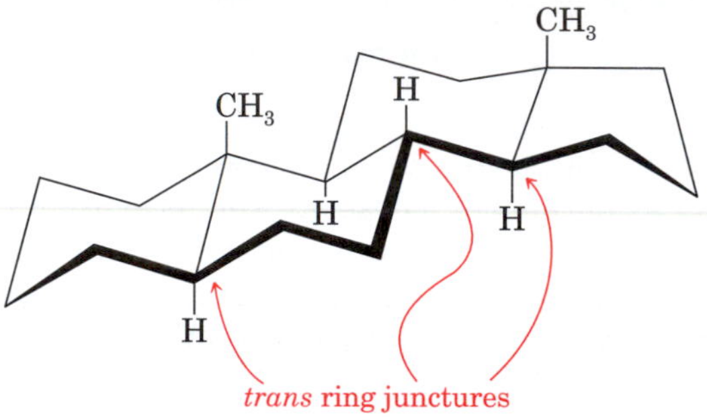

trans ring junctures

Groups substituted on a steroid ring system can be *below* or *above* the plane of the ring as the ring system is generally drawn. A group that is below the plane (*trans* to the angular methyl groups) is called an α group, while one that is above the plane (*cis* to the angular methyl groups) is called a β group.

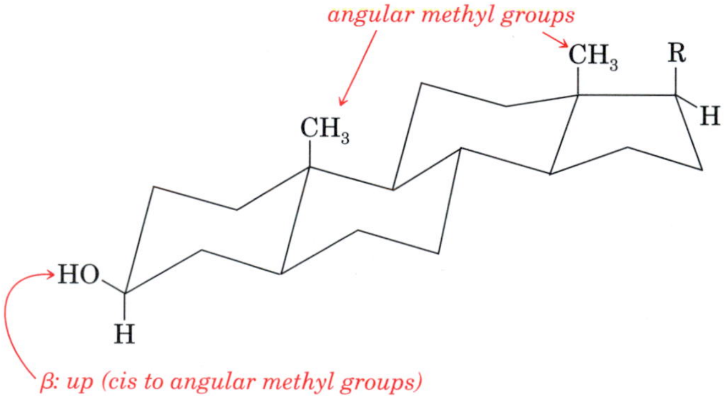

angular methyl groups

β: up (cis to angular methyl groups)

The terms α and β can be used in the names of steroids to designate the stereochemistry of substituents. In the name, α or β immediately follows the position number for the substituent.

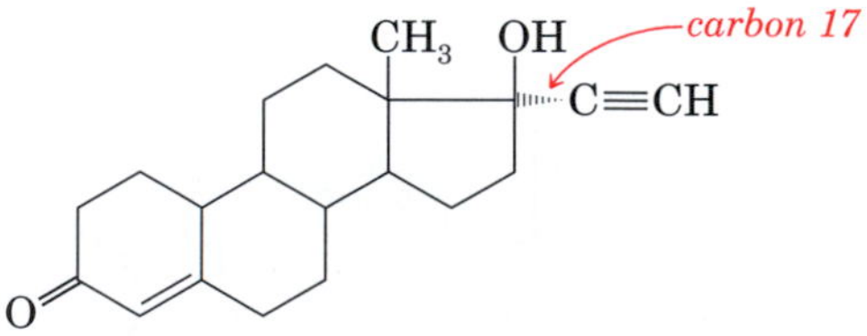

carbon 17

17 α-ethynyl-17β-hydroxy-4-estren-3-one

an oral contraceptive, usually called norethindrone

STUDY PROBLEMS

24.8   Draw conformational formulas for:

(a) cholestane   **(b)** 3α-cholestanol   **(c)** norethindrone

24.9   Designate each ring juncture as *cis, trans,* or *neither:*

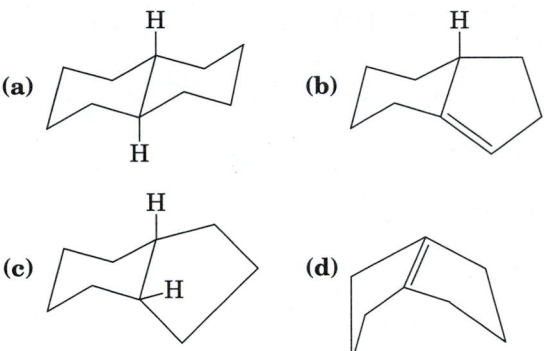

(a)   (b)   (c)   (d)

24.10   Tell whether each position shown is α, β, or neither:

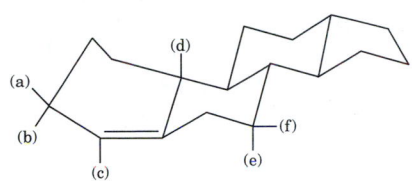

B. Some Important Steroids

Cholesterol, the most widespread animal steroid, is found in almost all animal tissues. Human gallstones and egg yolks are especially rich sources of this compound. Cholesterol is an important component of cell membranes of higher animals and is a necessary intermediate in the biosynthesis of the steroid hormones. However, because it can be synthesized from acetylcoenzyme A, it is not a dietary necessity. High levels of blood cholesterol are associated with arteriosclerosis (hardening of the arteries), a condition in which cholesterol and other lipids coat the insides of the arteries.

cholesterol
(5-cholesten-3β-ol)

A steroid related to cholesterol, *7-dehydrocholesterol,* which is found in the skin, is converted to *vitamin D* when irradiated with ultraviolet light. This reaction is discussed in Section 21.5.

Cortisone and **cortisol** (hydrocortisone) are two of 28 or more hormones secreted by the adrenal cortex. They are widely used to treat inflammation due to allergies or rheumatoid arthritis. Many related steroids with a carbonyl or hydroxyl group at carbon 11 have similar activity.

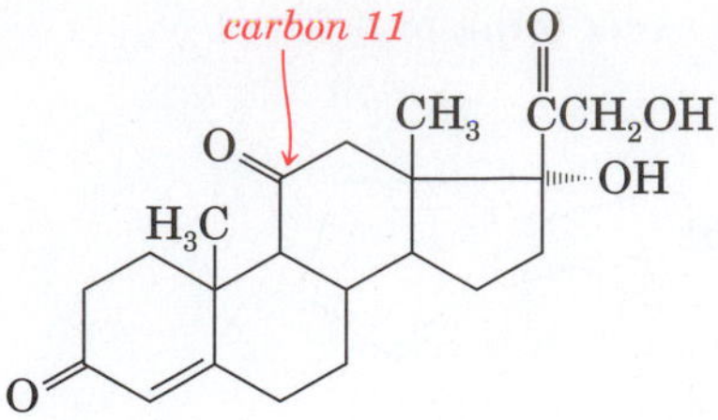

carbon 11

cortisone

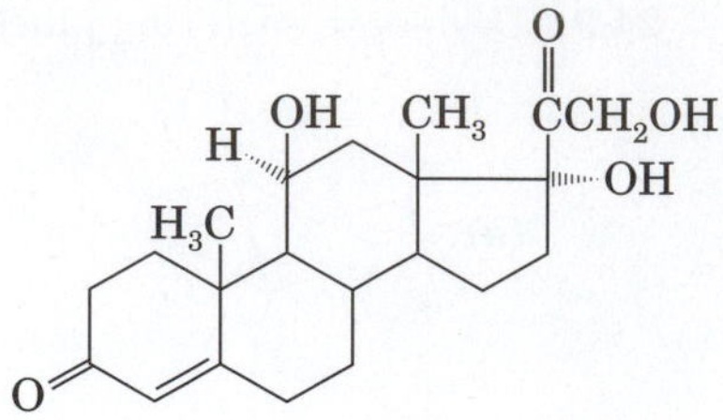

cortisol

Sex hormones are produced primarily in the testes or the ovaries. Their production is regulated by pituitary hormones. The sex hormones impart secondary sex characteristics and regulate the sexual and reproductive functions. Male hormones are collectively called **androgens;** female hormones, **estrogens;** and pregnancy hormones, **progestins.**

In pregnant females, the presence of progesterone suppresses ovulation and menstruation. Synthetic progestins, such as norethynodrel (Enovid), are used to suppress ovulation as a method of birth control.

Androgens:

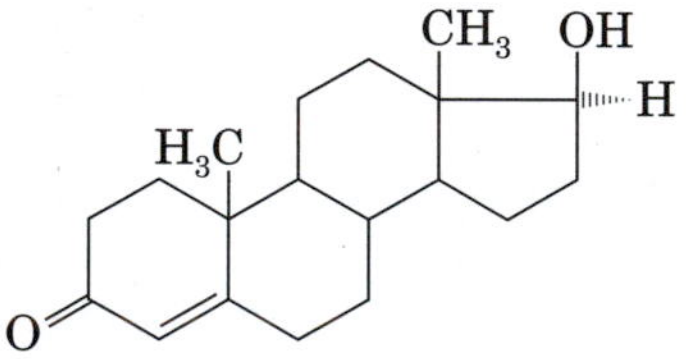

testosterone

the principal male hormone

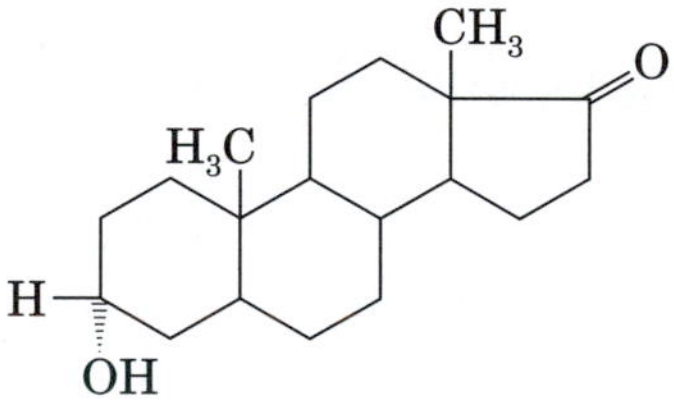

androsterone

a metabolized form of testosterone

Estrogens:

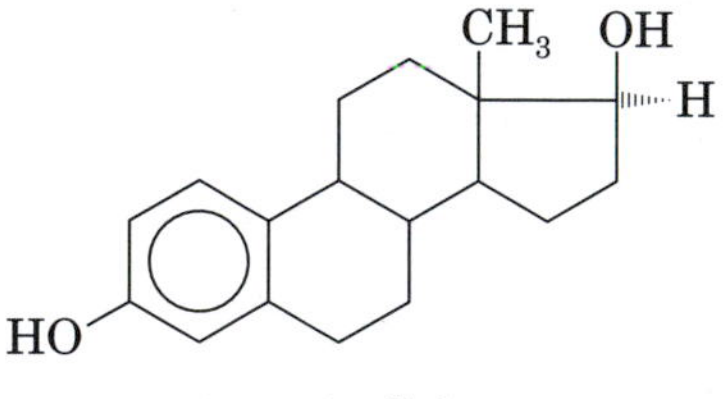

estradiol

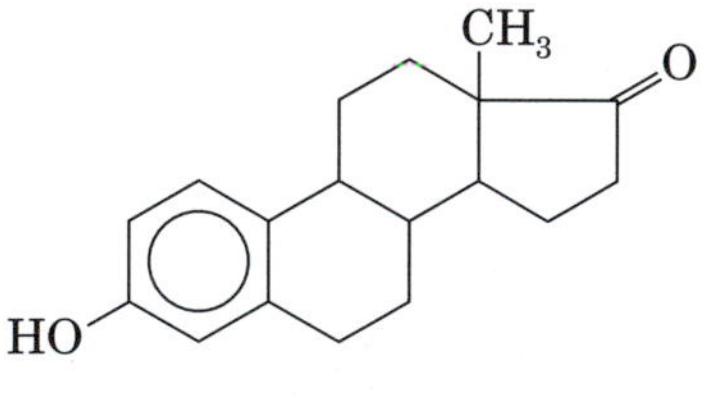

estrone

Progestins:

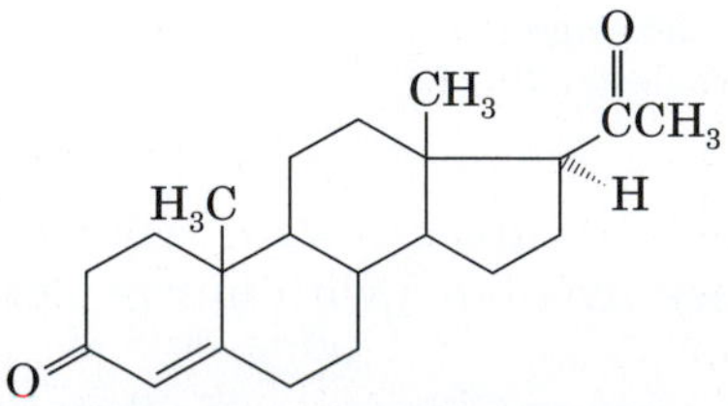

progesterone

suppresses ovulation

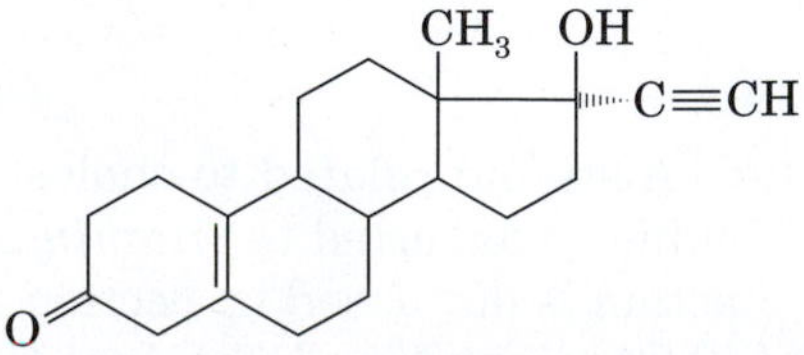

norethynodrel

a synthetic progestin

The tissue-building steroids are closely related to the androgens. Note the similarity in structure between methandrostenolone and testosterone.

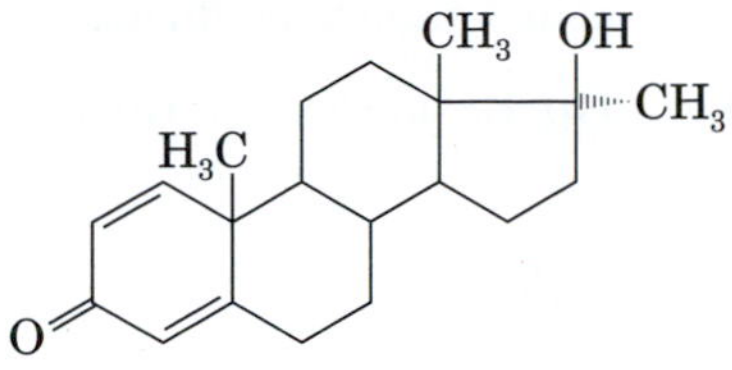

methandrostenolone

a tissue-building steroid

Bile acids are found in bile, which is produced in the liver and stored in the gall bladder. The structure of cholic acid, the most abundant bile acid, follows. Cholic acid, as well as other bile acids, has a *cis A/B* ring juncture instead of the usual *trans A/B* ring juncture.

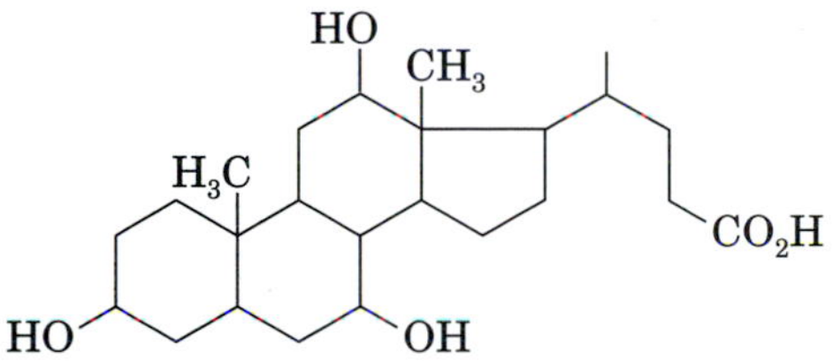

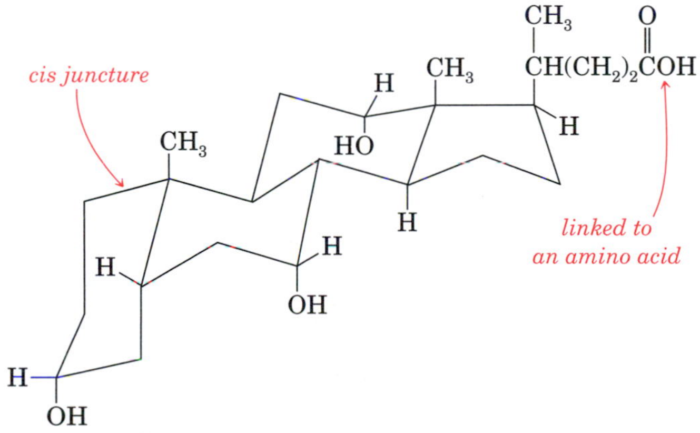

cholic acid

Bile acids are secreted into the intestines in combination with sodium salts of either glycine or taurine ($H_2NCH_2CH_2SO_3H$). The bile acid–amino acid link is an amide link between the carboxyl group of the bile acid and the amino group of the amino acid. In this combined form, the bile acid–amino acid acts to keep lipids emulsified in the intestines, thereby promoting their digestion.

$$RCO_2H + H_2NCH_2CO_2^- \ Na^+ \xrightarrow[-H_2O]{enzymes} R\overset{O}{\overset{\|}{C}}NHCH_2CO_2^- \ Na^+$$

cholic acid   sodium salt of glycine   *an emulsifying agent*

24.11 Write equations for the reactions of cholesterol with the following reagents, showing the principal organic products:

(a) H_2CrO_4 (b) Br_2 in CCl_4 (c) NBS, $h\nu$ (d) NaH

C. Biosynthesis of Steroids

The biosynthesis of steroids proceeds by way of *terpenes* (Section 20.1). An abbreviated biosynthetic path to terpenes is shown in Figure 24.4. The first step is an enzymatic ester condensation of the acetyl portions of acetylcoenzyme A. Intermediates in the formation of terpenes are the pyrophosphates (diphosphates) of mevalonic acid and a pair of isopentenyl alcohols. Figure 24.4 shows the formation of the sesquiterpene *farnesol*, which is used to synthesize the higher terpenes.

The precursor for steroids is the triterpene squalene, which arises from the condensation of two farnesol molecules. The sequence of reactions leading

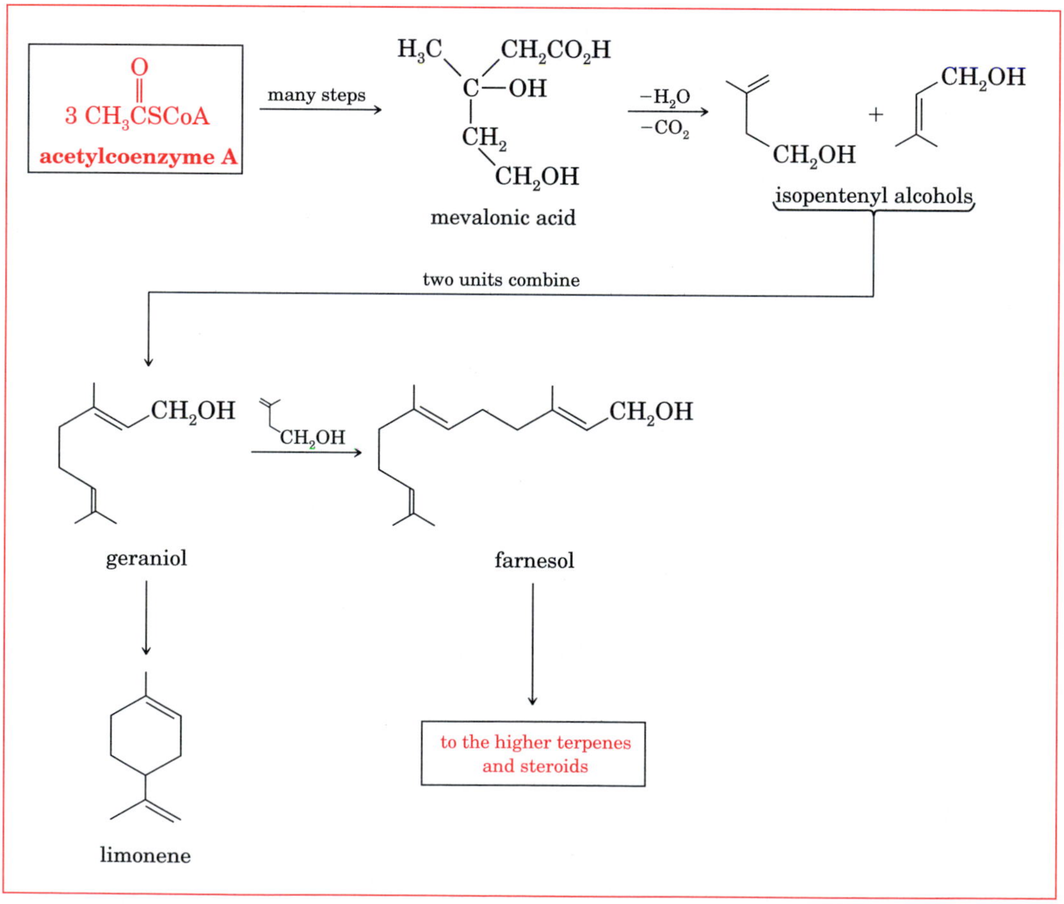

Figure 24.4 Generalized biosynthetic pathway leading to the terpenes. (To emphasize the terpene portions of the compounds, the pyrophosphate groups are not shown.)

from squalene to lanosterol, an intermediate that can be converted to cholesterol and other steroids, is shown in Figure 24.5. The first step is an enzymatic epoxide formation. Next, the epoxide is protonated and then a series of electron shifts results in the closure of four rings and the formation of the steroid ring system. The intermediate product of the ring-closure step is a carbocation. The final steps are a series of 1,2-shifts involving hydrides and methyl groups. Finally, a proton is lost and a stable alkene is generated. It is amazing that this whole sequence is catalyzed by only one enzyme, squalene oxidocyclase.

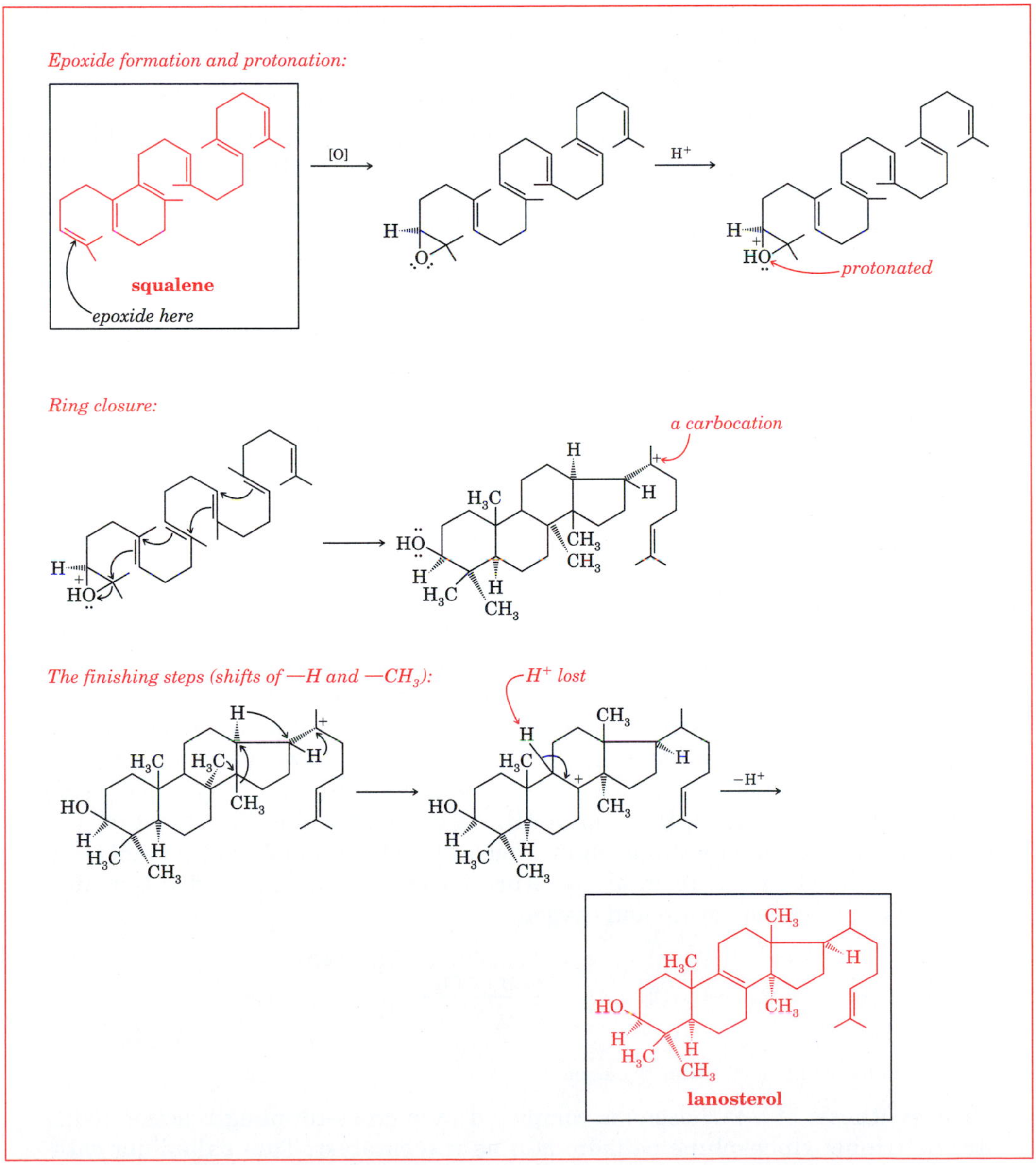

Figure 24.5 The conversion of squalene to lanosterol.

Summary

Fats and **oils** are triglycerides, or triacylglycerols; that is, triesters of glycerol and long-chain fatty acids. Generally, oils (liquid) contain more unsaturation than fats (solid). Fatty acids contain even numbers of carbon atoms and *cis* double bonds.

A **soap** is the alkali-metal salt of a fatty acid. A **detergent** is a salt of a sulfonate or sulfate that contains a long hydrocarbon chain. These compounds are **surfactants.** Naturally occurring waxes are esters of fatty acids and long-chain alcohols.

Phosphoglycerides, such as lecithins and cephalins, generally contain glycerol esterified with two fatty acids and a phosphatidylamine. Other phospholipids may differ in structure, but all contain long hydrocarbon chains plus a phosphate group and an amino group.

Prostaglandins are 20-carbon carboxylic acids containing a cyclopentane ring, hydroxyl groups, one or more carbon–carbon double bonds, and sometimes a keto group. They are biosynthesized from unsaturated 20-carbon fatty acids.

Cholesterol, cortisone, the sex hormones, and the bile acids are all **steroids,** compounds that contain the following ring system:

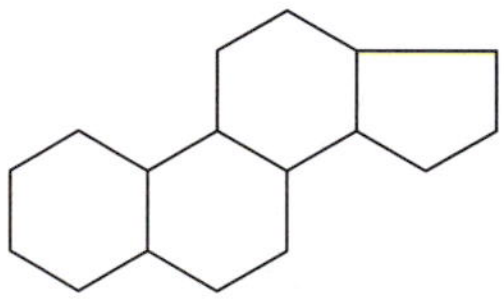

Essay Problem for Chapter 24

Synthesis of α-Farnesene, a Sesquiterpene

α-Farnesene [(3*E*,6*E*)-3,7,11-trimethyl-1,3,5,11-dodecatetraene] **(1)** is a naturally occurring sesquiterpene found in apple skins, in some species of ants, and in aphid alarm pheromone. The oxidation products of α-farnesene are believed to cause apple scald, a blanching then browning of apple skin. α-Farnesene was first synthesized in the 1920s by L. Ruzicka. A modern synthesis employs organometallic reagents. Reaction of 1-buten-3-yne with trimethylalane using dichlorobis(cyclopentadienyl)-zirconium(II) as catalyst yields (*E*)-(2-methyl-1,3-butadienyl)dimethylalane **(2)** as an intermediate. The alane **2** is reasonably stable and can be stored at room temperature for several days provided that it is protected from both moisture and oxygen.

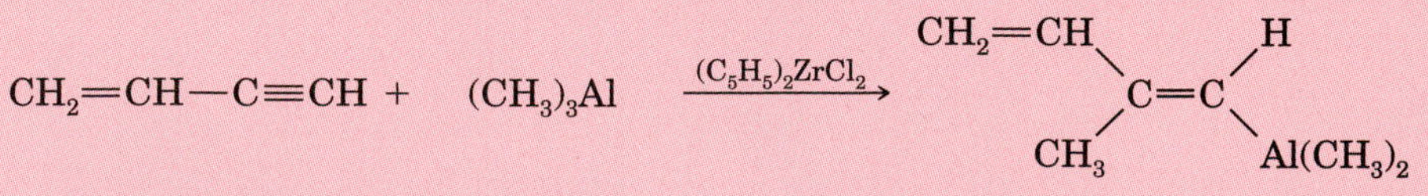

$$CH_2{=}CH{-}C{\equiv}CH \ + \ (CH_3)_3Al \xrightarrow{(C_5H_5)_2ZrCl_2}$$

1-buten-3-yne   trimethylalane   **2**

The synthesis of α-farnesene is completed by a cross-coupling reaction using tetrakis(triphenylphosphine)palladium(0) as the catalyst. This palladium catalyst allows stereo- and regiospecific coupling between allylic derivatives and alkenyl and aryl metals. To synthesize α-farnesene, a mixture of the palladium catalyst, geranyl chloride **(3)** and alane **2** is stirred for 6 hours at 25–30°C. Work-up consists of the addition of 3M HCl at 0° followed by extraction, chromatography, and distillation. All of the palladium compounds must be removed prior to distillation to prevent isomerization and polymerization of the alkene. (E. Negishi and H. Matsushita, *Org. Syn.* **1984**, *62*, 31.)

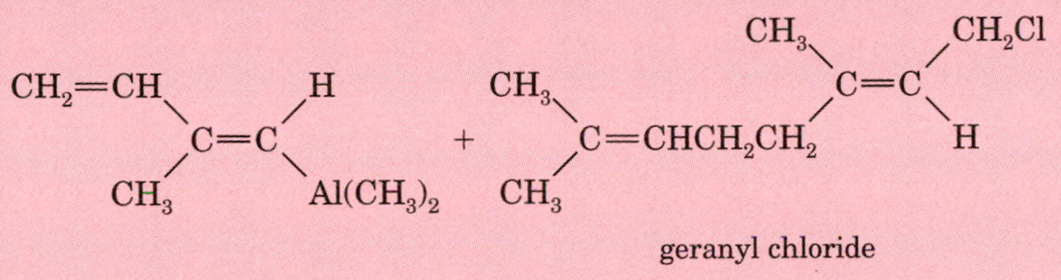

geranyl chloride

2   **3**

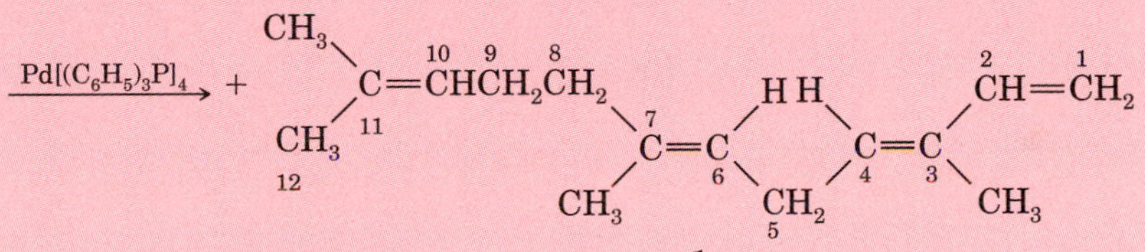

α-farnesene

1

Questions

1. Is the addition of the trimethylalane to the triple bond a *syn* or an *anti* addition reaction?

2. Suggest a procedure to obtain geranyl chloride from geraniol.

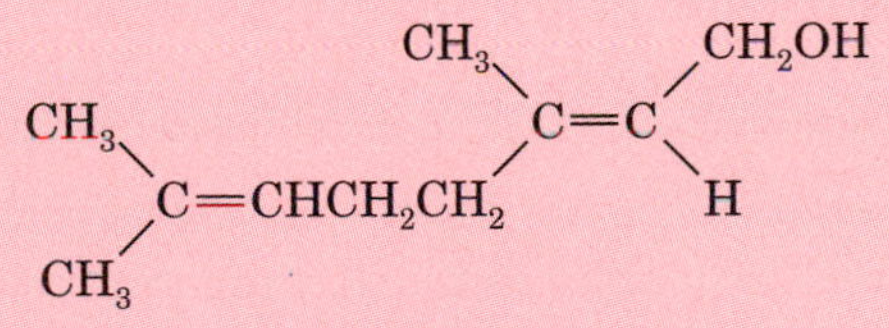

geraniol

3. Suggest a stereospecific synthesis of the 6-Z isomer of α-farnesene using the procedure discussed in the essay.

4. Although a chloro group is an excellent leaving group in palladium-catalyzed cross-coupling reactions, an acetate group can also be used. Design a synthesis of α-farnesene starting with oil of lemon grass (citral A; (E)-3,7-dimethyl-2,6-octadienal).

5. Is the method discussed in the essay suitable for the synthesis of β-farnesene? Explain your answer.

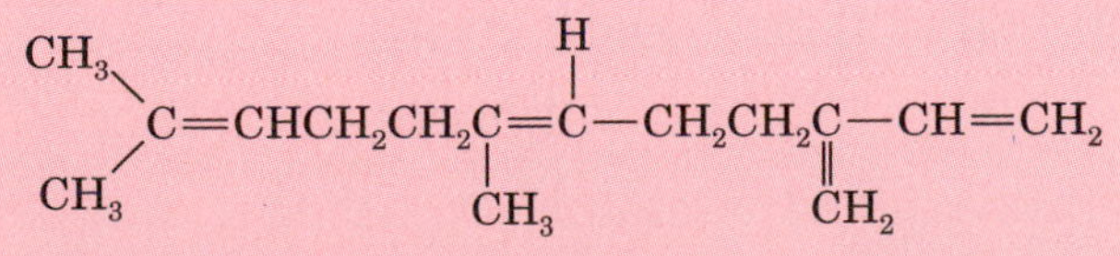

β-farnesene

24.12 Name the following compounds:

(a) $CH_3(CH_2)_{16}\overset{\displaystyle O}{\overset{\displaystyle \|}{C}}OCH(CH_3)_2$

(b) $\begin{array}{l} CH_2O_2C(CH_2)_7CH{=}CH(CH_2)_7CH_3 \\ | \\ CHO_2C(CH_2)_7CH{=}CH(CH_2)_7CH_3 \\ | \\ CH_2O_2C(CH_2)_7CH{=}CH(CH_2)_7CH_3 \end{array}$

(c) $Na^+\ {}^-O_3S{-}\langle \rangle{-}(CH_2)_9CH_3$

(d) $Na^+\ {}^-O_3SO(CH_2)_5CH_3$

(e)

24.13 Hydrolysis of *trimyristin,* a fat obtained from nutmeg, yields only one fatty acid, myristic acid. This same acid can be obtained from 1-bromododecane (the 12-carbon, continuous-chain alkyl bromide) and diethyl malonate in a malonic ester synthesis. What are the structures of myristic acid and trimyristin?

24.14 A fat of unknown structure is found to be optically active. Saponification, followed by acidification, yields two equivalents of palmitic acid and one equivalent of oleic acid. What is the structure of the fat? Write an equation for its saponification, followed by acidification.

24.15 Starting with glyceryl tristearate as the only organic reagent, show by equations how you would prepare a wax.

24.16 Write formulas for the products of complete hydrolysis of each of the following phosphoglycerides with dilute aqueous NaOH.

(a) $\begin{array}{l} \qquad\qquad\qquad CH_3(CH_2)_{16}CO_2CH_2 \\ CH_3(CH_2)_7CH{=}CH(CH_2)_7CO_2CH \quad O^- \\ \qquad\qquad\qquad\qquad\quad | \qquad\quad | \\ \qquad\qquad\qquad\qquad\quad CH_2OPOCH_2CH_2\overset{+}{N}(CH_3)_3\ Cl^- \\ \qquad\qquad\qquad\qquad\qquad\; \| \\ \qquad\qquad\qquad\qquad\qquad\; O \end{array}$

(b) $\begin{array}{l} \qquad\qquad\qquad CH_3(CH_2)_{16}CO_2CH_2 \\ CH_3(CH_2)_7CH{=}CH(CH_2)_7CO_2CH \quad O^- \\ \qquad\qquad\qquad\qquad\quad | \qquad\quad | \\ \qquad\qquad\qquad\qquad\quad CH_2OPOCH_2CH_2\overset{+}{N}H_3\ Cl^- \\ \qquad\qquad\qquad\qquad\qquad\; \| \\ \qquad\qquad\qquad\qquad\qquad\; O \end{array}$

---

* For information concerning the organization of the *Study Problems* and *Additional Problems,* see the *Note to student* on page 41.

24.17 *Cerebrosides* are glycosphingolipids that are found primarily in the sheaths of nerve cells. What fatty acid and sugar would be isolated from hydrolysis of the following cerebroside in acidic solution?

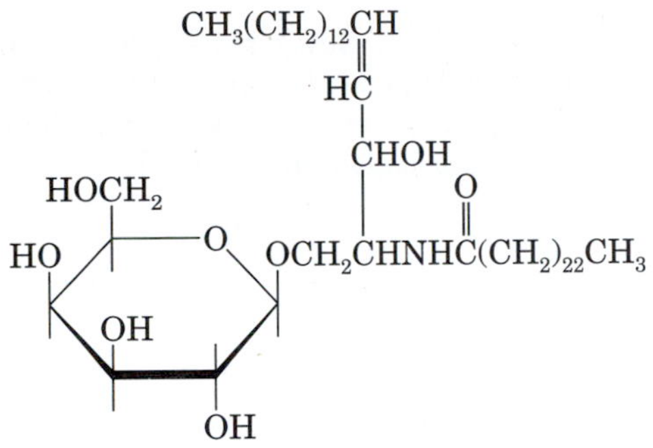

$$CH_3(CH_2)_{12}CH$$

24.18 Draw the structure of each of the following steroids.

(a) 5-cholesten-3α-ol

(b) 1-cholesten-3-one

(c) 4,6-cholestadien-3β-ol

(d) 3β-acetoxy-6α,7β-dibromocholestane

24.19 The structure of the bile acid *desoxycholic acid* is shown below. Redraw the structure showing the conformation of the ring system.

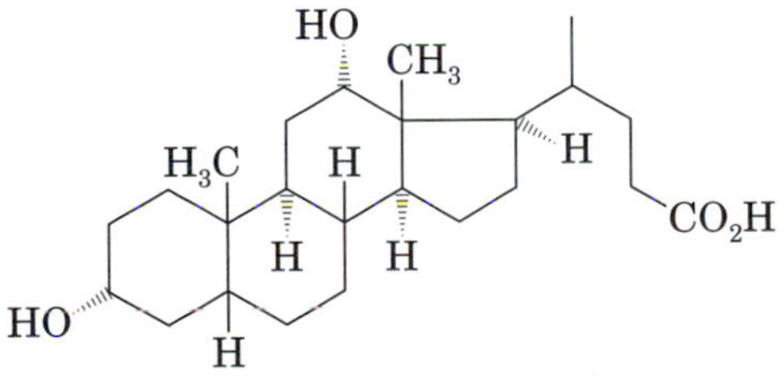

Additional Problems

24.20 Show how you could convert palmitoleic acid to: **(a)** ethyl palmitoleate; **(b)** ethyl palmitate; **(c)** nonanedioic acid; **(d)** 2-chloropalmitic acid; **(e)** heptanal; **(f)** 2-methyl-10-heptadecen-2-ol, $CH_3(CH_2)_5CH{=}CH(CH_2)_7C(OH)(CH_3)_2$.

24.21 A mixed triglyceride contains two units of stearic acid and one unit of palmitoleic acid. What are the major organic products when this triglyceride is treated with:

(a) an excess of dilute aqueous NaOH and heat?

(b) H_2, copper chromite catalyst, heat, and pressure?

(c) bromine in CCl_4?

24.22 What are the major organic products of ozonolysis, followed by oxidative work-up, of each of the following fatty acids? **(a)** palmitic acid; **(b)** palmitoleic acid; **(c)** linoleic acid; **(d)** linolenic acid.

24.23 How would you distinguish chemically between: **(a)** tripalmitin and tripalmitolein; **(b)** beeswax and beef fat; **(c)** beeswax and paraffin wax; **(d)** linoleic acid and linseed oil; **(e)** sodium palmitate and sodium *p*-decylbenzenesulfonate; **(f)** a vegetable oil and a motor oil?

24.24 How would you separate a mixture of estradiol and testosterone?

24.25 When cholic acid is treated with acetic anhydride, one monoacetate is formed preferentially. What is the structure of this monoacetate and why is it formed preferentially?

24.26 A biochemist prepared the following samples of mevalonic acid labeled with carbon-14. The three mevalonic acids were fed to a series of plants. Later, the isopentenyl alcohols from the plants were isolated. In which position would the carbon-14 be found in each of the product isopentenyl alcohols?

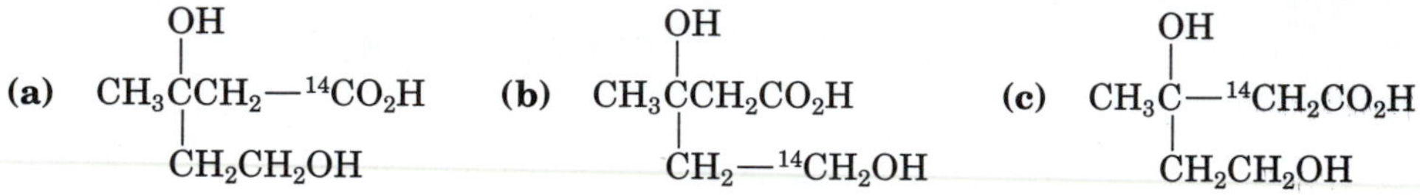
(a)
OH
CH3CCH2—14CO2H
CH2CH2OH
(b)
OH
CH3CCH2CO2H
CH2—14CH2OH
(c)
OH
CH3C—14CH2CO2H
CH2CH2OH

24.27 In a plant, two isopentenyl alcohol molecules combine to yield geraniol (see Figure 24.4). Identify the position of carbon-14 in geraniol arising from the labeled mevalonic acid in Problem 24.26(b).

24.28 *Prostacyclin,* also called prostaglandin I_2, dilates blood vessels and inhibits blood platelet aggregation (the start of the blood-clotting mechanism). This prostaglandin is unstable in aqueous acid and undergoes the following reaction. Suggest a mechanism.

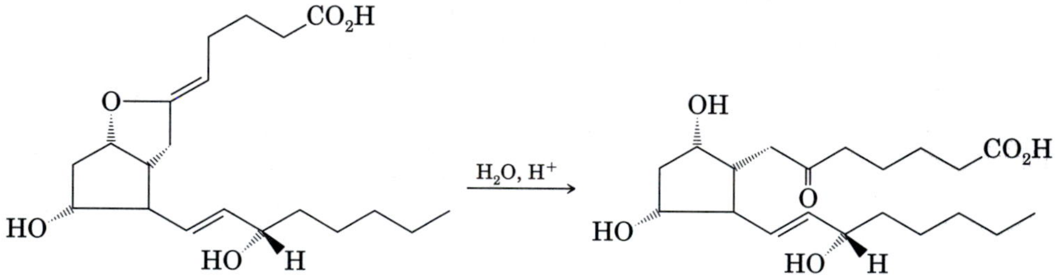
CO2H
O
HO
HO
H
H2O, H+
OH
CO2H
O
HO
HO
H

prostacyclin

AMINO ACIDS AND PROTEINS

Proteins are among the most important compounds in an animal organism. Appropriately, the word protein is derived from the Greek *proteios,* which means "primary." Proteins are *polyamides,* and hydrolysis of a protein yields *amino acids.*

$$\zeta-\underset{\underset{R}{|}}{NHCHC}\overset{\overset{O}{\|}}{}-\underset{\underset{R'}{|}}{NHCHC}\overset{\overset{O}{\|}}{}-\zeta \xrightarrow[\text{heat}]{\text{H}_2\text{O, H}^+}$$

a protein

$$\underset{\underset{R}{|}}{H_2NCHCO_2H} + \underset{\underset{R'}{|}}{H_2NCHCO_2H} \quad \text{etc.}$$

amino acids

Only twenty amino acids are commonly found in plant and animal proteins, yet these twenty amino acids can be combined in a variety of ways to form muscles, tendons, skin, fingernails, feathers, silk, hemoglobin, enzymes, antibodies, and many hormones. We will first consider the amino acids and then discuss how their combinations can lead to such diverse products.

Section 25.1

The Structures of Amino Acids

The amino acids found in proteins are **α-aminocarboxylic acids.** Variation in the structures of these monomers occurs in the side chain.

α-amino group ⟶ H_2N—C—H *variation in structure occurs in the side chain*

(with CO_2H above the central carbon and R below)

The simplest amino acid is aminoacetic acid ($H_2NCH_2CO_2H$), called *glycine,* which has no side chain and consequently does not contain a chiral carbon. All other amino acids have side chains, and therefore their α carbons are chiral. Amino acids from proteins belong to the L-series; that is, the groups around the α carbon have the same configuration as L-glyceraldehyde. (It is interesting that *racemic* α-amino acids have been detected in certain carbonaceous meteorites.)

CHO
HO►C◄H
CH_2OH

L-glyceraldehyde

CO_2H
H_2N►C◄H
R

an L-amino acid

STUDY PROBLEMS

25.1 While naturally occurring amino acids belong to the L-series, not all have an (*S*) configuration. Which L-amino acid has an (*R*) configuration? (See Table 25.1.)

25.2 Which of the following amino acids can exist as more than one pair of enantiomers? (See Table 25.1.)

(a) threonine   **(b)** histidine   **(c)** serine   **(d)** isoleucine

Table 25.1   The common amino acids found in proteins

Name	Abbreviation	Structure
alanine	ala	CH_3CHCO_2H | NH_2
arginine*	arg	$H_2NCNHCH_2CH_2CH_2CHCO_2H$ || | NH NH_2
asparagine	asn	O || $H_2NCCH_2CHCO_2H$ | NH_2
aspartic acid	asp	$HO_2CCH_2CHCO_2H$ | NH_2

(continued)

Table 25.1 *(continued)*   **The common amino acids found in proteins**

Name	Abbreviation	Structure
cysteine	cys	$HSCH_2CHCO_2H$ $\quad\quad\underset{NH_2}{\vert}$
glutamic acid	glu	$HO_2CCH_2CH_2CHCO_2H$ $\quad\quad\quad\quad\quad\underset{NH_2}{\vert}$
glutamine	gln	$\overset{\displaystyle O}{\overset{\|}{H_2NC}}CH_2CH_2CHCO_2H$ $\quad\quad\quad\quad\quad\underset{NH_2}{\vert}$
glycine	gly	CH_2CO_2H $\underset{NH_2}{\vert}$
histidine*	his	CH_2CHCO_2H (imidazole ring) $\quad\quad\underset{NH_2}{\vert}$
isoleucine*	ile	$\overset{\displaystyle CH_3}{\overset{\vert}{CH_3CH_2CHCHCO_2H}}$ $\quad\quad\quad\quad\underset{NH_2}{\vert}$
leucine*	leu	$(CH_3)_2CHCH_2CHCO_2H$ $\quad\quad\quad\quad\quad\underset{NH_2}{\vert}$
lysine*	lys	$H_2NCH_2CH_2CH_2CH_2CHCO_2H$ $\quad\quad\quad\quad\quad\quad\quad\underset{NH_2}{\vert}$
methionine*	met	$CH_3SCH_2CH_2CHCO_2H$ $\quad\quad\quad\quad\underset{NH_2}{\vert}$
phenylalanine*	phe	CH_2CHCO_2H (phenyl ring) $\quad\quad\underset{NH_2}{\vert}$
proline	pro	CO_2H (pyrrolidine ring, N–H)
serine	ser	$HOCH_2CHCO_2H$ $\quad\quad\quad\underset{NH_2}{\vert}$
threonine*	thr	$\overset{\displaystyle OH}{\overset{\vert}{CH_3CHCHCO_2H}}$ $\quad\quad\quad\underset{NH_2}{\vert}$
tryptophan*	trp	CH_2CHCO_2H (indole ring) $\quad\quad\underset{NH_2}{\vert}$
tyrosine	tyr	$HO-$(phenyl ring)$-CH_2CHCO_2H$ $\quad\quad\quad\quad\quad\quad\underset{NH_2}{\vert}$
valine*	val	$(CH_3)_2CHCHCO_2H$ $\quad\quad\quad\quad\underset{NH_2}{\vert}$

* Essential amino acid.

A. Dipolar Ion Formation

Amino acids do not always behave like organic compounds. For example, they have melting points of over 200°, whereas most organic compounds of similar molecular weight are liquids at room temperature. Amino acids are soluble in water and other polar solvents, but insoluble in nonpolar solvents such as diethyl ether or benzene. Amino acids have large dipole moments. Also, they are less acidic than most carboxylic acids and less basic than most amines.

$$RCO_2H \qquad RNH_2 \qquad \underset{\displaystyle R}{\overset{\displaystyle CO_2H}{H_2NCH}}$$

$$pK_a = \sim 5 \qquad pK_b = \sim 4$$

$$pK_a = \sim 10$$
$$pK_b = \sim 12$$

Why do amino acids exhibit such unusual properties? The reason is that an amino acid contains a basic amino group and an acidic carboxyl group in the same molecule. An amino acid undergoes an internal acid–base reaction to yield a **dipolar ion,** also called a **zwitterion** (from German *zwitter,* "hybrid"). Because of the resultant ionic charges, an amino acid has many properties of a salt. Furthermore, the pK_a of an amino acid is not the pK_a of a —CO_2H group, but that of an —NH_3^+ group. The pK_b is not that of a basic amino group, but that of the very weakly basic —CO_2^- group.

$$\underset{\displaystyle R}{\overset{\displaystyle CO_2H}{H_2\ddot{N}-\overset{|}{\underset{|}{C}}-H}} \quad \rightleftharpoons \quad \underset{\displaystyle R}{\overset{\displaystyle CO_2^-}{H_3\overset{+}{N}-\overset{|}{\underset{|}{C}}-H}}$$

a dipolar ion

Table 25.1 contains a complete list of the twenty amino acids commonly found in proteins. Even though amino acids exist as dipolar ions, their structures are commonly represented in the nonionic form, as shown in the table. Also included in this table are the three-letter abbreviations for the amino acid names. The use of these abbreviations will be discussed later in this chapter.

STUDY PROBLEM

25.3 Write an equation showing dipolar ion formation for each of the following amino acids. (See Table 25.1 for the structures of these amino acids.)

(a) threonine   **(b)** leucine

B. Importance of Side–Chain Structure

How can polymers composed of twenty similar amino acids have such a wide variety of properties? Part of the answer lies in the nature of the side chains in the amino acids. Note in Table 25.1 that some amino acids have side chains that contain carboxyl groups; these are classified as **acidic amino acids.** Amino acids containing side chains with amino groups are classified as **basic amino acids.** These acidic and basic side chains help determine the structure and reactivity of the proteins in which they are found. The rest of the amino acids are classified as **neutral amino acids.** The side chains of neutral amino acids are also important. For example, some of these side chains contain —OH, —SH, or other polar groups that can undergo hydrogen bonding, which we will find is an important feature in overall protein structure.

$$
\begin{array}{cccc}
\text{CO}_2\text{H} & \text{CO}_2\text{H} & \text{CO}_2\text{H} & \text{CO}_2\text{H} \\
| & | & | & | \\
\text{H}_2\text{NCH} & \text{H}_2\text{NCH} & \text{H}_2\text{NCH} & \text{H}_2\text{NCH} \\
| & | & | & | \\
\text{CH}_2\text{CH}_2\text{CO}_2\text{H} & (\text{CH}_2)_4\text{NH}_2 & \text{CH}(\text{CH}_3)_2 & \text{CH}_2\text{OH}
\end{array}
$$

glutamic acid	lysine	valine	serine
an acidic amino acid	*a basic amino acid*	*a neutral amino acid with a nonpolar side chain*	*a neutral amino acid with a polar side chain*

The characteristics of a protein are changed if a carboxylic acid group in a side chain is converted to an amide. Note the difference in the side chains of glutamic acid and glutamine.

$$
\begin{array}{cc}
\text{CO}_2\text{H} & \text{CO}_2\text{H} \\
| & | \\
\text{H}_2\text{NCH} \quad \textcolor{red}{\textit{acidic}} & \text{H}_2\text{NCH} \quad \overset{\text{O}}{\underset{\ \ }{}} \quad \textcolor{red}{\textit{neutral, polar}} \\
| & | \qquad \| \\
\text{CH}_2\text{CH}_2\text{CO}_2\text{H} & \text{CH}_2\text{CH}_2\text{CNH}_2
\end{array}
$$

glutamic acid	glutamine

The —SH group in cysteine plays a unique role in protein structure. Thiols can undergo oxidative coupling to yield disulfides (Section 8.6). This coupling between two units of cysteine yields a new amino acid called cystine and provides a means of cross-linking protein chains. Giving hair a "permanent wave" involves breaking some existing S—S cross-links by reduction and then reforming new S—S links in other positions of the protein chains.

The order in which amino acids are found in a protein molecule determines the relationship of the side chains to one another and consequently determines how the protein interacts with itself and with its environment. For example, a hormone or other water-soluble protein contains many amino acids with polar side chains, while an insoluble muscle protein contains a greater proportion of amino acids with nonpolar side chains.

The importance of the side chains of amino acids is illustrated by the condition known as **sickle-cell anemia.** The difference between normal hemoglobin and sickle-cell hemoglobin is that, in a protein molecule of 146 amino acid units, one single unit has been changed from glutamic acid (with an acidic side chain) to valine (with a nonpolar side chain). This one small error in the protein renders the affected hemoglobin less soluble and thus less able to perform its prescribed task of carrying oxygen to the cells of the body.

Section 25.2
Synthesis of Amino Acids

The common amino acids are relatively simple compounds, and racemic mixtures of most amino acids can be synthesized by standard techniques. The racemic mixtures can then be resolved to yield the pure enantiomeric amino acids.

The **Strecker synthesis** of amino acids, developed in 1850, is a two-step sequence. The first step is the reaction of an aldehyde with a mixture of ammonia and HCN to yield an aminonitrile. Hydrolysis of the aminonitrile results in the amino acid.

Step 1:

$$CH_3CH(O) \underset{}{\overset{NH_3}{\rightleftharpoons}} CH_3CHNH_2(OH) \overset{-H_2O}{\rightleftharpoons} CH_3CH=NH \overset{HCN}{\longrightarrow} CH_3CHCH(NH_2)$$

acetaldehyde   2-amino-propanenitrile

Step 2:

$$CH_3CHCN(NH_2) \xrightarrow[\text{(2) neutralize}]{\text{(1) } H_2O, H^+} CH_3CHCO_2H(NH_2)$$

(RS)-alanine (60%)

Another synthetic route to amino acids is by the **amination of an α-halo acid** with an excess of ammonia. (An excess of NH_3 must be used to neutralize the acid and to minimize overalkylation. See Section 18.4A.)

$$(CH_3)_2CHCHCO_2H(Br) \xrightarrow[\text{(2) neutralize}]{\text{(1) excess } NH_3} (CH_3)_2CHCHCO_2H(NH_2) + NH_4Br$$

(RS)-valine (48%)

The **Gabriel phthalimide synthesis** (Section 18.4A) is a more elegant route to amino acids. The advantage of this synthesis over direct amination is that overalkylation cannot occur.

Reductive amination of an α-keto acid is another procedure used to obtain racemic amino acids. (Remember, carboxyl groups are not easily reduced.)

Table 25.2 Summary of synthetic routes to α-amino acids	
Reaction	Section Reference

Substitution:

$$\underset{\text{X}}{\text{RCHCO}_2\text{H}} \xrightarrow[\text{(2) H}^+]{\text{(1) NH}_3} \underset{\text{NH}_2}{\text{RCHCO}_2\text{H}} \qquad\qquad\qquad 18.4\text{A}$$

$$\text{XCH(CO}_2\text{C}_2\text{H}_5)_2 \xrightarrow[\substack{\text{(2) NaOC}_2\text{H}_5,\,\text{RX} \\ \text{(3) H}_2\text{O, H}^+,\,\text{heat}}]{\text{(1) K}^+\,\text{phthalimide}} \underset{\text{NH}_2}{\text{RCHCO}_2\text{H}} \qquad 18.4\text{A}$$

Strecker Synthesis:

$$\underset{\text{RCH}}{\overset{\text{O}}{\|}} \xrightarrow[\substack{\text{(2) HCN} \\ \text{(3) H}_2\text{O, H}^+}]{\text{(1) NH}_3, -\text{H}_2\text{O}} \underset{\text{NH}_2}{\text{RCHCO}_2\text{H}} \qquad\qquad 25.2$$

Reductive Amination:

$$\underset{\text{RCCO}_2\text{H}}{\overset{\text{O}}{\|}} \xrightarrow{\text{H}_2,\,\text{NH}_3,\,\text{Pd}} \underset{\text{NH}_2}{\text{RCHCO}_2\text{H}} \qquad\qquad 13.6\text{D}$$

$$\underset{\text{CH}_3\text{CCO}_2\text{H}}{\overset{\text{O}}{\|}} \xrightarrow{\text{H}_2,\,\text{NH}_3,\,\text{Pd}} \underset{\text{NH}_2}{\text{CH}_3\text{CHCO}_2\text{H}}$$

an α-keto acid   *(RS)-alanine*

The reactions leading to α-amino acids are summarized in Table 25.2.

STUDY PROBLEM

25.4   Write flow equations showing four different ways to synthesize racemic phenyl-alanine.

Section 25.3

Reactions of Amino Acids

A. Amphoterism of Amino Acids

An amino acid contains both a carboxylate ion ($-\text{CO}_2^-$) and an ammonium ion ($-\text{NH}_3^+$) in the same molecule. Therefore, an amino acid is **amphoteric:** it can undergo reaction with either an acid or a base to yield a cation or an anion, respectively.

In acid:

$$\underset{\text{R}}{\overset{\text{CO}_2^-}{\underset{|}{\overset{|}{\text{H}_3\overset{+}{\text{N}}-\text{C}-\text{H}}}}} + \text{H}^+ \rightleftharpoons \underset{\text{R}}{\overset{\text{CO}_2\text{H}}{\underset{|}{\overset{|}{\text{H}_3\overset{+}{\text{N}}-\text{C}-\text{H}}}}}$$

a cation

In base:

$$\underset{\text{H}}{\overset{\text{CO}_2^-}{\underset{|}{\overset{|}{\text{H}_2\overset{+}{\text{N}}-\text{C}-\text{H}}}}} + \text{OH}^- \;\rightleftharpoons\; \underset{\text{R}}{\overset{\text{CO}_2^-}{\underset{|}{\overset{|}{\text{H}_2\ddot{\text{N}}-\text{C}-\text{H}}}}} + \text{H}_2\text{O}$$

an anion

STUDY PROBLEMS

25.5 When each of the following amino acids is dissolved in water, would the solution be acidic, basic, or near-neutral? (Refer to Table 25.1 for the structures.)

    (a) glutamic acid   **(b)** glutamine   **(c)** leucine

    (d) lysine   **(e)** serine

25.6 Monosodium glutamate (MSG) is widely used as a condiment. What is the most likely structure for this compound? (*Hint*: Which carboxyl group in glutamic acid is more acidic?)

25.7 Predict the product of reaction of **(a)** proline and **(b)** tyrosine with an excess of aqueous HCl and with an excess of aqueous NaOH.

Isoelectric points You might think that an aqueous solution of a so-called neutral amino acid would be neutral. However, aqueous solutions of neutral amino acids are slightly *acidic* because the $-\text{NH}_3^+$ group is a stronger acid than $-\text{CO}_2^-$ is a base. The result of the difference in acidity and basicity is that an aqueous solution of alanine contains more amino acid anions than cations. We can say that alanine carries a *net negative charge* in aqueous solution.

At pH 7, alanine carries a net negative charge:

weaker base ⟶

stronger acid ⟶

$$\underset{\text{CH}_3}{\overset{\text{CO}_2^-}{\underset{|}{\overset{|}{\text{H}_3\overset{+}{\text{N}}-\text{C}-\text{H}}}}} + \text{H}_2\text{O} \;\rightleftharpoons\; \underset{\text{CH}_3}{\overset{\text{CO}_2^-}{\underset{|}{\overset{|}{\text{H}_2\text{N}-\text{C}-\text{H}}}}} + \text{H}_3\text{O}^+$$

        no net charge   *net negative charge*

    If a small amount of HCl or other acid is added to an alanine solution, the acid–base equilibrium is shifted so that the net charge on the alanine ions becomes zero. The pH at which an amino acid carries no net ionic charge is defined as the **isoelectric point** of that amino acid. The isoelectric point of alanine is 6.0.

At pH 6, alanine carries no net charge:

$$\underset{\text{CH}_3}{\overset{\text{CO}_2^-}{\underset{|}{\overset{|}{\text{H}_2\text{N}-\text{C}-\text{H}}}}} + \text{H}_3\text{O}^+ \;\rightleftharpoons\; \underset{\text{CH}_3}{\overset{\text{CO}_2^-}{\underset{|}{\overset{|}{\text{H}_3\overset{+}{\text{N}}-\text{C}-\text{H}}}}}$$

Isoelectric points can be determined by **electrophoresis,** a process of measuring the migration of ions in an electric field. This is accomplished by placing an aqueous solution of an amino acid on an adsorbent between a pair of electrodes. In this cell, anions migrate toward the positive electrode and cations migrate toward the negative electrode. If alanine or another neutral amino acid is dissolved in plain water, there is a net migration of amino acid ions toward the positive electrode. At its isoelectric point, an amino acid exhibits no net migration toward either electrode in an electrophoresis cell.

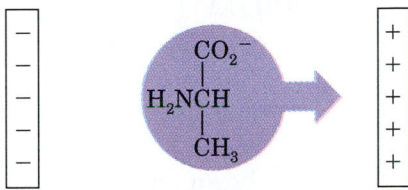

Isoelectric points can also be determined by titration. Figure 25.1 shows the titration curve for the titration of the cationic form of glycine, $H_3N^+—CH_2CO_2H$, with base. As base is added, the completely protonated ion is converted to the neutral dipolar ion, $H_3N^+—CH_2CO_2^-$. When one half of the cationic form has been neutralized, the pH equals the pK_1 for the reaction.

$$\overset{+}{H_3}NCH_2CO_2H \rightleftharpoons \overset{+}{H_3}NCH_2CO_2^- + H^+$$

$$K_1 = \frac{[\overset{+}{H_3}NCH_2CO_2^-][H^+]}{[\overset{+}{H_3}NCH_2CO_2H]}$$

When $[\overset{+}{H_3}NCH_2CO_2^-] = [\overset{+}{H_3}NCH_2CO_2H],$

$$K_1 = [H^+] \quad \text{and therefore} \quad pK_1 = pH$$

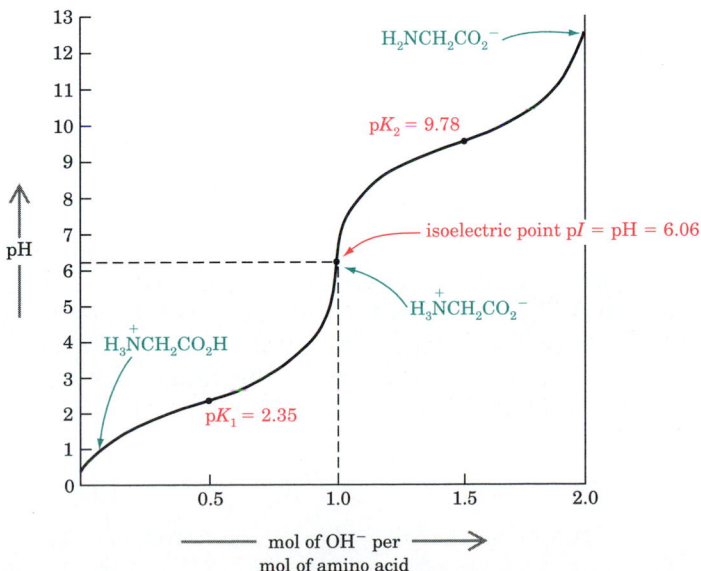

Figure 25.1 Titration curve for glycine hydrochloride.

As more base is added, all the cationic form is converted to the neutral dipolar ion. The pH at which this occurs is the isoelectric point. (Again, refer to Figure 25.1.) As yet more base is added, the dipolar ion is converted to an anion. At the halfway point, the pH equals the pK_2.

$$\overset{+}{H_3}NCH_2CO_2^- \;\rightleftharpoons\; H^+ + H_2NCH_2CO_2^-$$

$$K_2 = \frac{[H^+][H_2NCH_2CO_2^-]}{[\overset{+}{H_3}NCH_2CO_2^-]}$$

When $[H_2NCH_2CO_2^-] = [\overset{+}{H_3}NCH_2CO_2^-]$,

$$K_2 = [H^+] \quad \text{and therefore} \quad pK_2 = pH$$

The isoelectric point is shown on the titration curve. It can also be calculated as the average of pK_1 and pK_2.

$$\text{isoelectric point } pI = \frac{2.35 + 9.78}{2} = 6.06$$

The isoelectric point of a given amino acid is a physical constant. The value varies from amino acid to amino acid, but falls into one of three general ranges. For a *neutral amino acid,* the isoelectric point, which depends primarily on the relative pK_a and pK_b of the $-NH_3^+$ and $-CO_2^-$ groups, is around 5.5–6.0.

The second carboxyl group in an *acidic amino acid* means that there is another group that can interact with water. A water solution of an acidic amino acid is distinctly acidic, and the amino acid ion carries a net negative charge.

A *greater concentration of H^+* is required to bring an acidic amino acid to the isoelectric point than is needed for a neural amino acid. The isoelectric points for acidic amino acids are near pH 3.

A *basic amino acid* has a second amino group that undergoes reaction with water to form a positive ion. Hydroxide ions are needed to neutralize a basic amino acid and bring it to its isoelectric point. For basic amino acids, we would expect the isoelectric points to be above pH 7, and that indeed is the case. These isoelectric points are in the range of 9–10. Isoelectric points of some representative amino acids are found in Table 25.3.

Table 25.3	Isoelectric points for some amino acids	
Name	Structure	Isoelectric Point

Neutral:

alanine
$$CH_3CHCO_2H$$
$$\quad | $$
$$\quad NH_2$$
6.00

glutamine
$$\overset{O}{\overset{||}{H_2NCCH_2CH_2}}CHCO_2H$$
$$\qquad\qquad\quad |$$
$$\qquad\qquad\quad NH_2$$
5.65

Acidic:

glutamic acid
$$HO_2CCH_2CH_2CHCO_2H$$
$$\qquad\qquad\quad |$$
$$\qquad\qquad\quad NH_2$$
3.22

aspartic acid
$$HO_2CCH_2CHCO_2H$$
$$\qquad\quad |$$
$$\qquad\quad NH_2$$
2.77

Basic:

lysine
$$H_2N(CH_2)_4CHCO_2H$$
$$\qquad\qquad |$$
$$\qquad\qquad NH_2$$
9.74

arginine
$$H_2NCNH(CH_2)_3CHCO_2H$$
$$\quad ||\qquad\qquad |$$
$$\quad NH\qquad\quad NH_2$$
10.76

STUDY PROBLEMS

25.8 Suggest a reason for the fact that the isoelectric point of lysine is 9.74, but that for tryptophan (Table 25.1) is only 5.89. (*Hint:* Think of a reason why the heterocyclic N in tryptophan is not basic.)

25.9 Calculate the isoelectric points of the following amino acids:

(a) (RS)-isoleucine: $pK_1 = 1.92$, $pK_2 = 9.73$

(b) (RS)-phenylalanine: $pK_1 = 2.58$, $pK_2 = 9.24$

(c) (S)-proline: $pK_1 = 2.00$, $pK_2 = 10.60$

(d) (RS)-valine: $pK_1 = 2.29$, $pK_2 = 9.72$

B. Acylation

The amino group of an amino acid can be readily acylated with either an acid halide or an acid anhydride to yield an amide group. Because an amide nitrogen is not basic, an acylated amino acid does not form a dipolar ion. For this reason, acylated amino acids exhibit physical properties typical of organic compounds instead of those of salts. In the synthesis of peptides (small proteins), the N-acyl group is used as a blocking group (see Section 25.7).

$$CH_3COCCH_3 + \overset{+}{N}H_3CHCO_2^- \longrightarrow CH_3C-NHCHCO_2H + CH_3COH$$
$$\qquad\qquad\qquad CH_2CH(CH_3)_2 \qquad\qquad\qquad CH_2CH(CH_3)_2$$

acetic anhydride   leucine   N-acetylleucine (85%)

C. Reaction with Ninhydrin

α-Amino acids react with ninhydrin to form a blue-violet product called **Ruhemann's purple.** The reaction is used as a spot test to detect the presence of amino acids on chromatography paper. Because the reaction is quantitative, it also finds use in automated amino acid analyzers, instruments that determine the percentages of the various amino acids present in a sample.

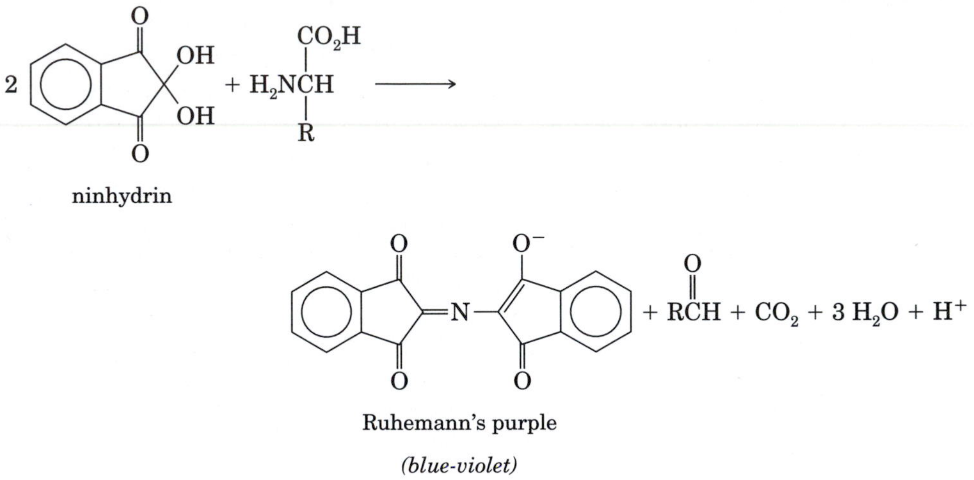

ninhydrin

Ruhemann's purple

(blue-violet)

Figure 25.2 summarizes the reactions of amino acids.

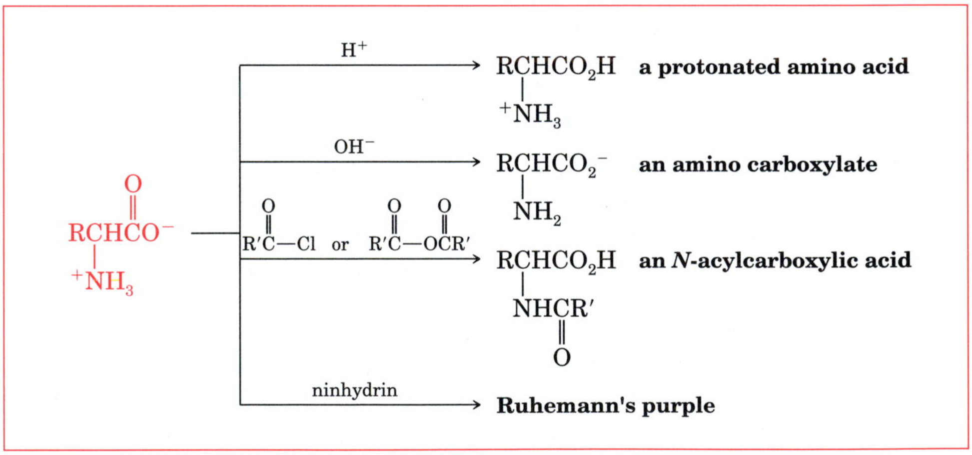

Figure 25.2 Summary of reactions of amino acids.

Section 25.4

Peptides

A **peptide** is an amide formed from two or more amino acids. The amide link between an α-amino group of one amino acid and the carboxyl group of another amino acid is called a **peptide bond.** The following example of a peptide formed from alanine and glycine, called *alanylglycine,* illustrates the formation of a peptide bond.

$$
\text{H}_2\text{NCHCOH} + \text{H}_2\text{NCH}_2\text{COH} \xrightarrow{-\text{H}_2\text{O}} \text{H}_2\text{NCHC}-\text{NHCH}_2\text{COH}
$$

— peptide bond

alanine	glycine	alanylglycine

a dipeptide

Each amino acid in a peptide is called a **unit,** or a **residue.** Alanylglycine has two residues: the alanine residue and the glycine residue. Depending upon the number of amino acid residues in the molecule, a peptide may be referred to as a **dipeptide** (two residues), a **tripeptide** (three residues), and so forth. A **polypeptide** is a peptide with a large number of amino acid residues. What is the difference between a polypeptide and a protein? None, really. Both are polyamides constructed from amino acids. By convention, a polyamide with fewer than 50 amino acid residues is classified as a polypeptide, while a larger polyamide is considered to be a protein.

In the dipeptide alanylglycine, the alanine residue has a free amino group and the glycine unit has a free carboxyl group. However, alanine and glycine could be joined another way to form *glycylalanine,* in which glycine has the free amino group and alanine has the free carboxyl group.

Two different dipeptides from alanine and glycine:

alanylglycine	glycylalanine

The greater the number of amino acid residues in a peptide, the greater is the number of structural possibilities. Glycine and alanine can be bonded together in two ways. In a tripeptide, three amino acids can be joined in six different ways. Ten different amino acids could lead to over four trillion decapeptides!

For purposes of discussion, we must represent peptides in a systematic manner. The amino acid with the free amino group is usually placed at the left end of the structure. This amino acid is called the **N-terminal amino acid.** The amino acid with the free carboxyl group is placed at the right and is called the **C-terminal amino acid.** The name of the peptide is constructed from the names of the amino acids as they appear left-to-right, starting with the N-terminal amino acid.

N-terminal on left   *C-terminal on right*

alanyltyrosylglycine

a tripeptide

For the sake of convenience and clarity, the names of the amino acids are often abbreviated. We have shown the abbreviations of the twenty common amino acids in Table 25.1. Using the abbreviated names, alanyltyrosylglycine becomes *ala–tyr–gly*.

SAMPLE PROBLEM

What is the structure of leu–lys–met?

Solution

$$\underset{\text{leu}}{\underset{\underset{\text{(CH}_3)_2\text{CHCH}_2}{|}}{\text{H}_2\text{NCH}\overset{\overset{\text{O}}{\|}}{\text{C}}}}\text{---}\underset{\text{lys}}{\underset{\underset{\text{(CH}_2)_4\text{NH}_2}{|}}{\text{NHCH}\overset{\overset{\text{O}}{\|}}{\text{C}}}}\text{---}\underset{\text{met}}{\underset{\underset{\text{CH}_2\text{CH}_2\text{SCH}_3}{|}}{\text{NHCHCO}_2\text{H}}}$$

STUDY PROBLEM

25.10 Write formulas for the following peptides:

(a) his-ala   **(b)** asp-trp-ala   **(c)** val-gly-cys-trp

Section 25.5

Bonding in Peptides

As we mentioned in Section 15.8C, an amide bond has some double-bond character due to partial overlap of the p orbitals of the carbonyl group with the unshared electrons of the nitrogen.

$$-\overset{\overset{\ddot{\text{O}}:}{\|}}{\text{C}}-\ddot{\text{N}}\big\langle \longleftrightarrow -\overset{\overset{^-:\ddot{\text{O}}:}{|}}{\text{C}}=\overset{+}{\text{N}}\big\langle$$

Evidence for the double-bond character of a peptide bond is found in bond lengths. The bond length of the peptide bond is shorter than that of the usual C—N single bond: 1.32 Å in the peptide bond versus 1.47 Å for a typical C—N single bond in an amine.

Because of the double-bond character of the peptide bond, rotation of groups around this bond is somewhat restricted, and the atoms attached to the carbonyl group and to the N all lie in the same plane. X-ray analysis shows that the amino acid side chains around the plane of the peptide bond are in a *trans* type of relationship. This stereochemistry minimizes steric hindrance between side chains.

Section 25.6
Determination of Peptide Structure

The determination of a peptide structure is not an easy task. Complete hydrolysis in acidic solution yields the individual amino acids. These may be separated and identified by such techniques as chromatography or electrophoresis. The molecular weight of the peptide can be determined by physical–chemical methods. With this information, the chemist can determine the number of amino acid residues, the identity of the amino acid residues, and the number of residues of each amino acid in the original peptide. But this information reveals nothing about the *sequence* of the amino acids in the peptide. Several techniques have been developed to determine this sequence. The first is **terminal-residue analysis.**

A. Terminal-Residue Analysis

The Edman reagent Analysis for the *N*-terminal can be accomplished by treating the peptide with phenyl isothiocyanate (the Edman reagent, named after the Swedish biochemist P. Edman, who developed it). The isothiocyanate undergoes reaction with the free amino group in a series of steps that eventually results in cleavage of the *N*-terminal amino acid from the peptide and in the formation of a phenylthiohydantoin (PTH), a derivative of an amino acid that can be isolated and characterized. The cleavage reaction is carried out by heating the intermediate adduct with acid in an anhydrous solvent such as nitromethane. (If water were present, the entire peptide would be hydrolyzed.)

phenyl isothiocyanate   *N-terminal*
(Edman reagent)   *residue*

a phenylthiohydantoin

Why cannot the chemist continue to treat the peptide stepwise with phenyl isothiocyanate? In each step the *N*-terminal residue could be broken off until the entire peptide was degraded and its order of amino acids determined. To a certain extent, this can be done; however, each cycle of thiourea formation and hydrolysis results in some internal hydrolysis of the remaining peptide. After about 20–30 cycles, the hydrolysis of the peptide is sufficient to produce many smaller peptides, each of which has an *N*-terminal residue. The phenylthiohydantoin is thus no longer that of a single end group, but a mixture from a variety of end groups. In recent years, automated peptide analyzers called *sequenators* have been used to degrade and analyze peptides of 20–60 amino acid residues.

The Sanger reagent Another reagent useful for determining the *N*-terminal residue is the Sanger reagent, 1-fluoro-2,4-dinitrobenzene. The fluoro group of the Sanger reagent can undergo aromatic nucleophilic substitution with amines. The substitution is facile because the intermediate carbanion is stabilized by the nitro groups (Section 12.4).

1-fluoro-2,4-dinitrobenzene

a 2,4-dinitrophenylamine

The Sanger reagent reacts readily with the *N*-terminal amino acid of a peptide and converts the amino group to an arylamino group. After complete hydrolysis of the treated peptide, the *N*-terminal amino acid remains bonded to the 2,4-dinitrophenyl group and thus can be separated from the other amino acids and identified. The principal disadvantage of using the Sanger reagent is that a peptide cannot be degraded one amino acid at a time, as in an Edman degradation.

from the N-terminal amino acid

The Sanger reagent was developed by Sir Frederick Sanger, who received the 1958 Nobel Prize for being the first to establish the complete amino acid sequence of a protein (insulin).

The *C*-terminal amino acid residue can be determined enzymatically. *Carboxypeptidase* is a pancreatic enzyme that specifically catalyzes hydrolysis of the *C*-terminal amino acid, but not that of other peptide bonds.

$$\text{rest of peptide} -\underset{\underset{R}{|}}{\text{NHCHCO}_2\text{H}} + \text{H}_2\text{O} \xrightarrow{\text{carboxypeptidase}} \text{rest of peptide} + \underset{\underset{R}{|}}{\text{H}_2\text{NCHCO}_2\text{H}}$$

C-terminal
amino acid

The principal difficulty with this method of analysis is that after the original *C*-terminal amino acid has been cleaved, the remaining portion of the peptide has another *C*-terminal residue, which in turn can undergo cleavage. Very careful experimental work is required to determine which amino acid in the resulting mixture was the *C*-terminal residue of the starting peptide.

B. Internal Sequence of Amino Acids

A large polypeptide must usually be hydrolyzed into smaller fragments for determination of the internal amino acid sequence. The hydrolysis mixture is separated and the order of amino acid residues in each fragment determined (by end-group analysis, for example). The structures of the fragments are then pieced together like a jigsaw puzzle to give the structure of the entire peptide.

SAMPLE PROBLEM

What is the structure of a pentapeptide that yields the following tripeptides when partially hydrolyzed?

gly–glu–arg,   glu–arg–gly,   arg–gly–phe

Solution
Fit the pieces together:

gly–glu–arg
glu–arg–gly
arg–gly–phe

The pentapeptide is gly–glu–arg–gly–phe.

STUDY PROBLEM

25.11   Complete hydrolysis of an octapeptide yielded the following amino acids:

2 ala, his, leu, lys, pro, thr, tyr. Partial hydrolysis yielded the following tripeptides: leu-ala-tyr; thr-pro-leu; lys-ala-his; and his-thr-pro. What is the structure of the peptide?

In theory, partial hydrolysis can be achieved by heating the polypeptide with water and acid or base. In practice, proteolytic (peptide-hydrolyzing) en-

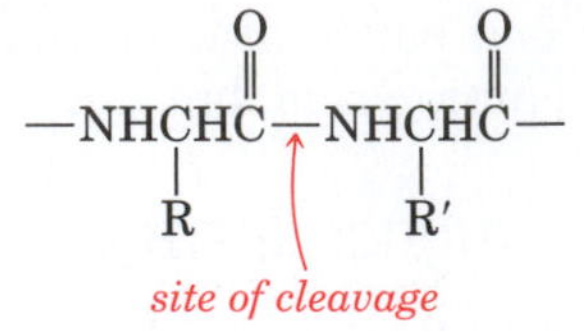

$$\begin{array}{ccc} & O & O \\ & \| & \| \\ -NHCHC & NHCHC- \\ | & | \\ R & R' \end{array}$$

site of cleavage

chymotrypsin: R = try, phe, trp
pepsin: R' = try, phe, trp
trypsin: R = lys, arg
cyanogen bromide (BrCN): R = met
thermolysin: R' = ile, leu, val

Figure 25.3 Specificities of proteolytic enzymes and reagents used to cleave internal peptide bonds.

zymes or chemical reagents are used. These enzymes and reagents have the advantage that they cleave the polypeptides at specific peptide bonds. Some examples are listed in Figure 25.3.

A new and indirect way of determining the sequence of amino acids in a protein molecule is by isolating the portion of a DNA molecule that is responsible for the biosynthesis of that protein and then determining the base sequence in the DNA fragment (Chapter 26).

SAMPLE PROBLEM

List the amino acids and peptides formed when the following peptide is treated with chymotrypsin:

tyr–ala–leu–tyr

Solution
Chymotrypsin cleaves the amide bond on the carbonyl side of tyrosine:

tyr–ala–leu–tyr $\longrightarrow$ tyr + ala–leu–tyr

STUDY PROBLEM

25.12 List the amino acids and peptides formed when the following polypeptide is treated with: **(a)** trypsin; **(b)** chymotrypsin; **(c)** pepsin:

trp–met–gly–gly–phe–arg–phe–trp–val–lys–ala–gly–ser

Section 25.7
Synthesis of Peptides

The first peptide was synthesized by Emil Fischer, who in 1902 also put forth the idea that proteins were polyamides.

The synthesis of ordinary amides from acid chlorides and amines is a straightforward reaction (Section 15.3C):

$$RCOCl + R'NH_2 \longrightarrow RCONHR'$$

However, the synthesis of peptides or proteins by this route is not straightforward. The principal problem is that there is more than one way in which the amino acids may join.

$$gly + ala \longrightarrow gly\text{–}ala \quad or \quad ala\text{–}gly \quad or \quad gly\text{–}gly \quad or \quad ala\text{–}ala$$

$$gly\text{–}ala + phe \longrightarrow phe\text{–}gly\text{–}ala \quad or \quad gly\text{–}ala\text{–}phe \quad or \quad gly\text{–}ala\text{–}gly\text{–}ala$$
$$or \quad phe\text{–}phe$$

To prevent unwanted reactions, every other reactive group, including reactive groups in side chains, must be blocked. By leaving only the desired amino group and carboxyl group free, the chemist can control the positions of reaction. (See Figure 25.4.)

The criteria for a good blocking group are (1) that it be inert to the reaction conditions needed for forming the desired amide link, and (2) that it be readily removable when the synthesis is complete. One such blocking group is a *carbamate group*—inert to the amide-formation reaction, but easily removed in a later step without disturbing the rest of the molecule. This approach to peptide synthesis was developed in 1932.

The carbamate group formed from benzyl chloroformate, as in the following equation, is called the *carbobenzoxy* or benzyloxycarbonyl group, frequently abbreviated CBz or simply Z.

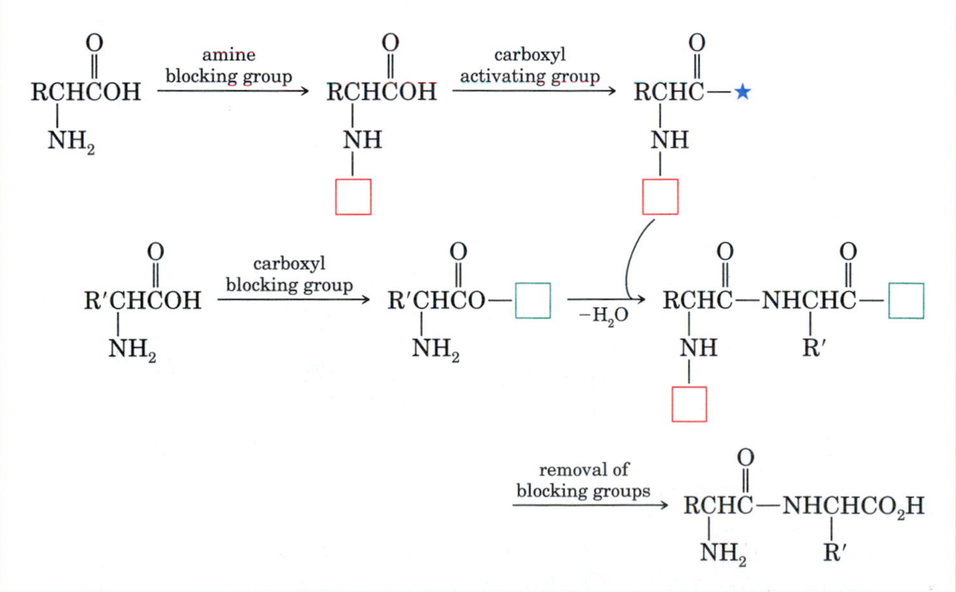

Figure 25.4 A general procedure for synthesizing peptides in the laboratory.

Preparation of carbamate to protect an amino group:

$$\text{benzyl chloroformate} + \text{glycine} \xrightarrow[\text{−HCl}]{\substack{\text{NaOH, H}_2\text{O} \\ 5°}} \text{benzyloxycarbonylglycine (80\%)}$$

(Z)

benzyl chloroformate   glycine

from $Cl_2C{=}O + C_6H_5CH_2OH$

carbamate group

benzyloxycarbonylglycine (80%)

The amino-blocked glycine could be treated with $SOCl_2$ to form the acid chloride and then treated with a new amino acid to form an amide. However, acid chlorides are highly reactive and unwanted side reactions may occur despite blocking. To circumvent this problem, the amino-blocked glycine is usually treated with ethyl chloroformate to yield an **activated ester.**

$$\text{Z—NHCH}_2\text{COH} + \text{ClCOC}_2\text{H}_5 \xrightarrow{\text{−HCl}} \text{Z—NHCH}_2\text{COCOC}_2\text{H}_5$$

amino-blocked   ethyl   *"activated ester group"*
glycine   chloroformate   *(an anhydride-carbonate)*

Like an acid chloride, this activated ester can undergo reaction with an amino group of another amino acid to give the desired dipeptide.

$$\text{Z—NHCH}_2\text{COCOC}_2\text{H}_5 + \text{H}_2\text{NCHCO}_2\text{H} \xrightarrow[\text{−C}_2\text{H}_5\text{OH}]{\text{−CO}_2} \text{Z—NHCH}_2\text{C—NHCHCO}_2\text{H}$$

$$\underset{\text{CH}_2\text{C}_6\text{H}_5}{} \qquad \underset{\text{CH}_2\text{C}_6\text{H}_5}{}$$

phenylalanine   amino-blocked gly–phe
or its ethyl ester

At this point, the sequence can be repeated to add a third amino acid. When the peptide synthesis is complete, the carbamate group is cleaved by reduction to yield the free peptide.

Removal of carbamate blocking group:

$$\text{CH}_2\text{—OC—NHCH}_2\text{C—NHCHCO}_2\text{H} \xrightarrow{\text{H}_2,\ \text{Pd}}$$

$$\underset{\text{CH}_2\text{C}_6\text{H}_5}{}$$

$$\text{CH}_3 + \text{CO}_2 + \text{H}_2\text{NCH}_2\text{C—NHCHCO}_2\text{H}$$

$$\underset{\text{CH}_2\text{C}_6\text{H}_5}{}$$

gly–phe

STUDY PROBLEM

25.13 Write the equations illustrating the addition of alanine to the amino-blocked gly–phe.

A. Solid-Phase Peptide Synthesis

New and better methods of synthesizing peptides are always under investigation. One relatively new technique is called **solid-phase peptide synthesis,** or the **Merrifield peptide synthesis** (after Bruce Merrifield at Rockefeller University, who developed the technique and was awarded a Nobel prize in 1984 for his work). In this type of synthesis, resins hold the C-terminal amino acid by the carboxyl group as the peptide is being synthesized. The resin is a polystyrene that contains about 1% p-(chloromethyl)styrene units.

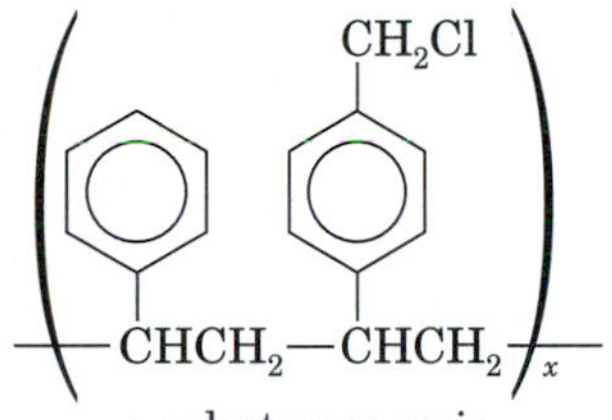

a polystyrene resin
containing p-(chloromethyl)styrene units

The amino group of the first amino acid is initially blocked, often as a *tert*-butoxycarbonyl ("Boc") group. This amino-blocked amino acid, as the carboxylate, reacts with the benzylic chloride groups of the resin to form ester groups (a typical substitution reaction between a carboxylate and a benzylic halide).

Step 1 (formation of Boc derivative):

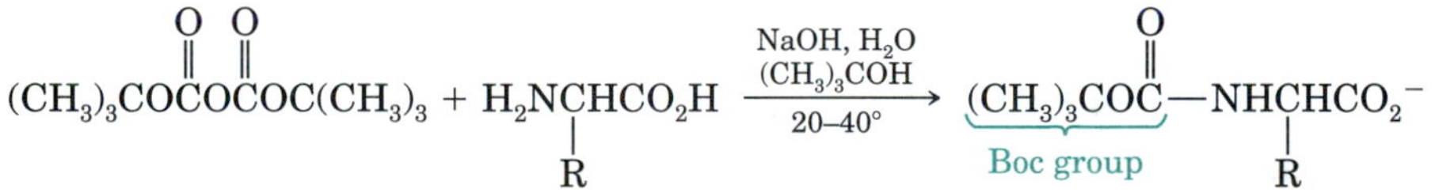

Step 2 (reaction with benzylic chloride group):

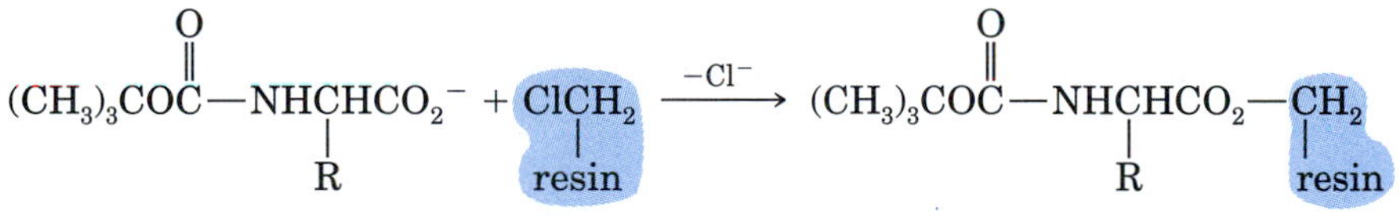

The amino-blocking group is removed by treatment with an anhydrous acid, such as HCl in acetic acid. Then, a second amino-blocked amino acid (with an activated carbonyl group) is added.

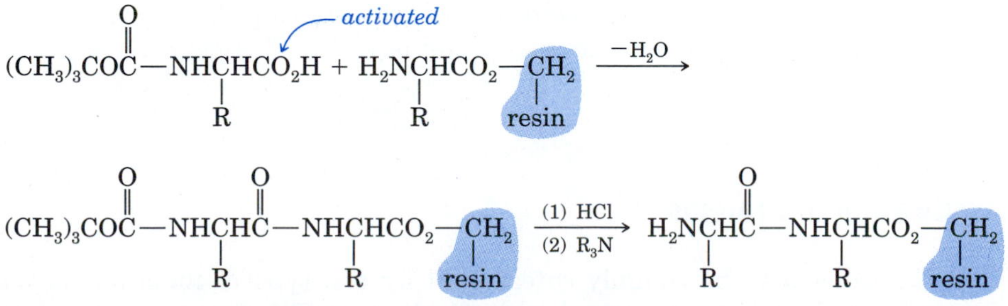

A common technique for activating the $—CO_2H$ group (so that it will undergo reaction with the amine) is by the addition of dicyclohexylcarbodiimide to the carboxylic acid. This compound reacts with the carboxylic acid to yield an intermediate with a leaving group that can be displaced by the amine in a typical nucleophilic acyl substitution reaction. The product is the amide.

dicyclohexylcarbodiimide (DCC)

good leaving group

the amide

dicyclohexylurea

In the solid-phase peptide synthesis, the product of this reaction is the amino-blocked dipeptide, still bonded to the resin at the carboxyl end. The amino-blocking group is removed, and the process is repeated until the peptide synthesis is complete. A controlled cleavage of the resin–ester bond with an acid such as anhydrous HF releases the peptide and removes the final amino-blocking group.

The classical type of peptide synthesis is tedious because intermediate peptides must be isolated and purified. However, in a solid-phase peptide synthesis, most impurities can simply be washed away from the resin after each step. This technique is successful enough that commercial **automated peptide synthesizers** have been developed.

Unfortunately, solid-phase peptide synthesis techniques are not practical for synthesizing large protein molecules in a pure state and in high yields. Poor-quality products may result because the resin tends to hold some impurities. However, recent advances in "gene-splicing" have provided routes to high yields of pure peptides and proteins.

Section 25.8
Classification of Proteins

Proteins may be roughly categorized by the type of function they perform. These classes are summarized in Table 25.4. **Fibrous proteins** (also called **structural proteins**), which form skin, muscles, the walls of arteries, and hair, are composed of long thread-like molecules that are tough and insoluble.

Table 25.4 **Classes of proteins**	
Class	**Comments**

Fibrous, or Structural (Insoluble):

collagens	form connective tissue; comprise 30% of mammalian protein; lack cysteine and tryptophan; rich in hydroxyproline
elastins	form tendons and arteries
keratins	form hair, quills, hoofs, nails; rich in cysteine and cystine

Globular (Soluble):

albumins	examples are egg albumin and serum albumin
globulins	an example is serum globulin
histones	occur in glandular tissue and with nucleic acids; rich in lysine and arginine
protamines	associated with nucleic acids; contain no cysteine, methionine, tyrosine, or tryptophan; rich in arginine

Conjugated (Combined with Other Substances):

nucleoproteins	combined with nucleic acids
mucoproteins	combined with $> 4\%$ carbohydrates
glycoproteins	combined with $< 4\%$ carbohydrates
lipoproteins	combined with lipids, such as phosphoglycerides or cholesterol

Another functional type of protein is the class of **globular proteins.** These are small proteins, somewhat spherical in shape because of folding of the protein chains upon themselves. Globular proteins are water-soluble and perform various functions in an organism. For example, *hemoglobin* transports oxygen to the cells. *Insulin* aids in carbohydrate metabolism. *Antibodies* render foreign protein inactive. *Fibrinogen* (soluble) can form insoluble fibers that result in blood clots. *Hormones* carry messages throughout the body.

Conjugated proteins, proteins connected to a nonprotein moiety such as a sugar, perform various functions throughout the body. A common mode of linkage between the protein and nonprotein is by a functional side chain of the protein. For example, an acidic side chain of the protein can form an ester with an —OH group of a sugar molecule.

Section 25.9
Higher Structures of Proteins

The sequence of amino acids in a protein molecule is called the **primary structure** of the protein. However, there is much more to protein structure than just the primary structure. Many of the properties of a protein are due to the orientation of the molecule as a whole. The shape (such as a helix) into which a protein molecule arranges its backbone is called the **secondary structure.** Further interactions, such as folding of the backbone upon itself to form a sphere, result in the **tertiary structure.** Interactions between certain protein subunits, such as between the globins in hemoglobin, result in the **quaternary structure.** The secondary, tertiary, and quaternary structures are collectively referred to as the **higher structure** of the protein.

One protein that has been well-studied in terms of its secondary structure is **keratin,** found in fur and feathers. Each protein molecule in keratin has the shape of a spiral, called a *right-handed α-helix* (Figure 25.5). "Right-

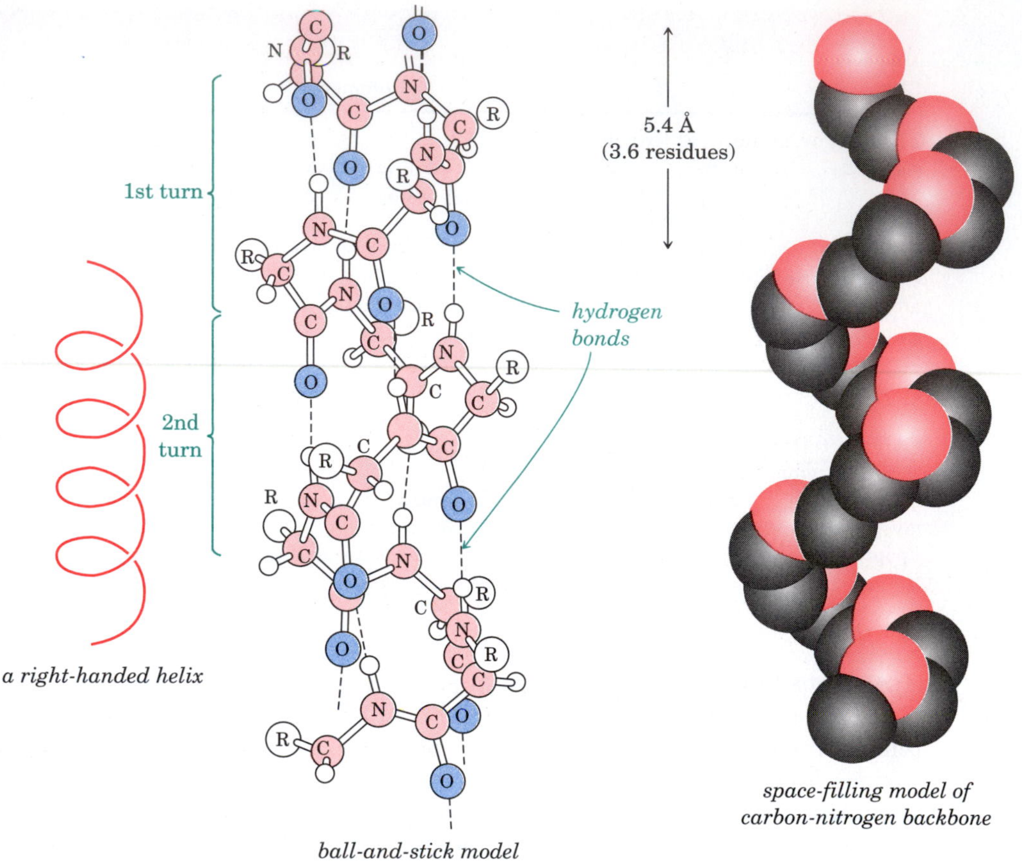

5.4 Å
(3.6 residues)

hydrogen bonds

1st turn

2nd turn

a right-handed helix

ball-and-stick model

space-filling model of carbon-nitrogen backbone

Figure 25.5 The protein chains in keratin form right-handed α-helices. The ball-and-stick model is adapted from C. B. Anfinsen, *The Molecular Basis of Evolution* (John Wiley and Sons, Inc., New York, 1964). The space-filling model is adapted from W. H. Brown, *Introduction to Organic and Biochemistry, 4th ed.* (Copyright 1987 by Wadsworth, Inc. Used by permission of Brooks/Cole Publishing Co.)

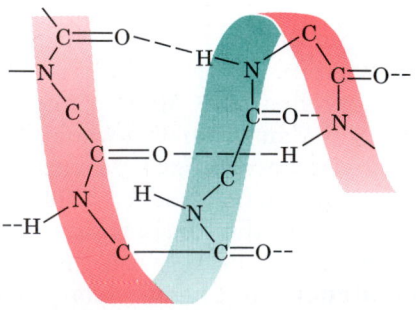

Figure 25.6 The helix in keratin is held in its shape by hydrogen bonds. (Nonparticipating R's and H's have been omitted.) From R. J. Fessenden and J. S. Fessenden, *Chemical Principles for the Life Sciences* (Allyn and Bacon, Inc., Boston, 1976).

handed" refers to the direction of the turns in the helix; the mirror image is a left-handed helix. In the mid-1930s, the term "α" was coined to differentiate the X-ray pattern of keratin from that of some other proteins.

In keratin, each turn of the helix contains 3.6 amino acid residues. The distance from one coil to the next is 5.4 Å. The helix is held in its shape primarily by hydrogen bonds between one amide–carbonyl group and an NH group that is 3.6 amino acid units away (Figure 25.6). The helical shape gives a strong, fibrous, flexible product.

Hydrogen bonding between an α-amino group and a carbonyl group is one contributor to the shape of a protein molecule. Other inter- and intramolecular interactions also contribute to the higher structure. Some of these interactions are hydrogen bonding between side chains, S—S cross-links, and *salt bridges* (ionic bonds such as RCO_2^- ^+H_3NR between side chains). The most stable higher structure is the one with the greatest number of stabilizing interactions. (Each hydrogen bond lends ~ 5 kcal/mol of stability.) Given a particular primary structure, a protein naturally assumes its most stable higher structure.

Let us look at some other types of protein structure. **Collagen** is a general classification of a tough, strong protein that forms cartilage, tendons, ligaments, and skin. Collagen derives its strength from its higher structure of "super-helices": three right-handed α-helical polypeptides entwined to form a triple left-handed helical chain. The entwined molecules are collectively called a *tropocollagen molecule.* One of the these tropocollagen molecules is about 15 Å in diameter and 2800 Å long. The tropocollagen triple helix, like a single helix, is held together by hydrogen bonding.

Gelatin is obtained by boiling collagen-containing animal material. However, gelatin is not the same type of protein as collagen. It has been found that the molecular weight of gelatin is only one-third that of collagen. Presumably, in gelatin formation, the tropocollagen molecule unravels and the single strands form hydrogen bonds with water, resulting in the characteristic gel-formation.

Helical structures are not the only type of secondary structure of proteins. Another type of structure, referred to as a β-*sheet,* or *pleated sheet,* is found in silk fibroin. The pleated sheet is an arrangement in which single protein molecules are lined up side by side and held there by hydrogen bonds between the chains (Figure 25.7).

The protein chains in silk fibroin are not simply stretched-out zigzag chains. Analysis by X-ray diffraction shows repetitive units every 7.0 Å. This repetition probably arises from a puckering (or "pleating") in the chains that alleviates steric hindrance. It is interesting that silk fibroin contains 46% glycine (no side chain) and 38% of a mixture of alanine and serine (small side chains. The lack of bulky R groups in these amino acids allows the side-by-side arrangement of protein chains in the fibroin structure.

7.0 Å per pleat

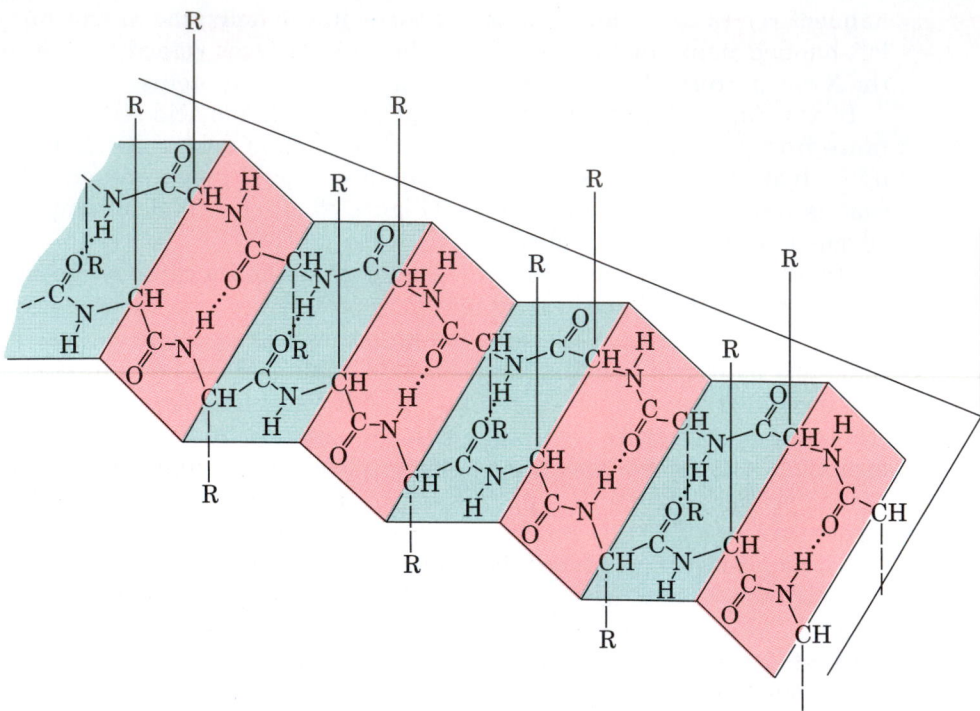

Figure 25.7 The pleated sheet structure of silk fibroin. From P. Karlson, *Introduction to Modern Biochemistry, 3rd ed.* (Georg Thieme Verlag, Stuttgart, 1968).

A *globular protein* depends on a tertiary structure to maintain its intricately folded, globular shape, which is necessary to maintain solubility. In a globular protein, polar hydrophilic side chains are situated on the outside of the sphere (to increase water-solubility) and nonpolar hydrophobic side chains are arranged on the interior surface, where they may be used to catalyze nonaqueous reactions. The unique surface of each globular protein enables it to "recognize" certain complementary organic molecules. This recognition allows enzymes to catalyze reactions of particular molecules, but not others.

Figure 25.8 lists definitions and examples of the higher structures of proteins.

primary: amino acid sequence
secondary: shape of backbone
    Examples: α-helix, β-pleated protein
tertiary: folding of helix
    Example: folded helix in a globular protein
quaternary: interactions between two or more protein molecules
    Example: the association of four globins in hemoglobin

Figure 25.8 Summary of the higher structures of proteins.

Summary

A **protein** is a polyamide. Hydrolysis yields α-amino acids of L-configuration at the α carbon. Amino acids undergo an internal acid–base reaction to yield **dipolar ions.**

$$\xi-NHCHC-\xi \xrightarrow[H^+]{H_2O} H_2N-C-H \rightleftharpoons H_3\overset{+}{N}-C-H$$

an α-amino acid   *a dipolar ion*

Acidic amino acids are those with a carboxyl group in the side chain (R in the preceding equation). **Basic amino acids** contain an amino group in the side chain. **Neutral amino acids** contain neither $-CO_2H$ nor $-NH_2$ in the side chain, but may contain OH, SH, or other polar groups. **Cross-linking** in proteins may be provided by the SH group in cysteine, which can link with another SH in an oxidation reaction: $2\ RSH \rightarrow RSSR + 2\ H$.

Racemic amino acids may be synthesized by a variety of routes, which are summarized in Table 25.2.

The **isoelectric point** of an amino acid is the pH at which the dipolar ion is electrically neutral and does not migrate toward an anode or cathode. The isoelectric point depends on the acidity or basicity of the side chain.

$$\underset{(CH_2)_4NH_3{}^+}{\overset{CO_2{}^-}{H_3\overset{+}{N}CH}} \xrightleftharpoons{OH^-} \underset{(CH_2)_4NH_3{}^+}{\overset{CO_2{}^-}{H_2NCH}}$$

in H_2O, a net positive charge   *at pH 9.74, no net charge*

Reactions of amino acids include reactions with acid and base, acylation, and conversion to Ruhemann's purple.

A **peptide** is a polyamide of fewer than 50 amino acid residues. The ***N*-terminal amino acid** is the amino acid with a free α-amino group, while the ***C*-terminal amino acid** has a free carboxyl group at carbon 1. **Terminal residue analysis** (using such reagents as the Edman reagent, the Sanger reagent, and hydrolytic enzymes) to determine the *N*- and *C*-terminals and **partial hydrolysis** to smaller peptides are two techniques for peptide structure determination.

In the synthesis of a peptide, reactive groups (except for the groups desired to undergo reaction) must be blocked. A **carbamate group** may be used to protect an amino group. A **solid-phase peptide synthesis** provides a blocking group for the *C*-terminal carboxyl group.

Proteins are polyamides of more than 50 amino acid residues. The order of side chains in a protein determines its **higher structures,** which arise from internal and external hydrogen bonding, van der Waals forces, and other interactions between side chains. The higher structures of proteins give them a variety of physical and chemical properties so that they may perform a variety of functions.

Synthesis of *N*-Blocked α-Amino Acids

In addition to the 20 α-amino acids coded for in the nucleic acids, a large number of biologically interesting α-amino acids occur in nature. A general synthetic technique for these compounds is desirable so that they and their analogues can be prepared and studied. One synthetic method employs the β-lactone of *N*-blocked serine (**4**). This lactone is prepared by dehydration of *N*-(benzyloxycarbonyl)-(*S*)-serine (**1**) with dimethyl azodicarboxylate (**2**) and triphenylphosphine (**3**). Triphenylphosphine oxide, as well as the β-lactone, is a product of the reaction.

β-Lactones contain a strained four-membered ring that readily undergoes nucleophilic ring opening. The products of these reactions are β-substituted propionic acids. The nucleophile attacks the β-carbon rather than the carbonyl carbon. For example, β-propiolactone reacts with various nucleophiles to give the β-substituted product.

The β-lactone of *N*-blocked serine (**4**) reacts with a nucleophile in a manner that is analogous to its parent β-propiolactone. Attack occurs at the β-carbon rather than at the carbonyl carbon thus yielding β-substituted alanines. (S. V. Pansare, G. Huyer, L. D. Arnold, and J. C. Vederas, *Org. Syn.*, **1992**, *70*, 1.)

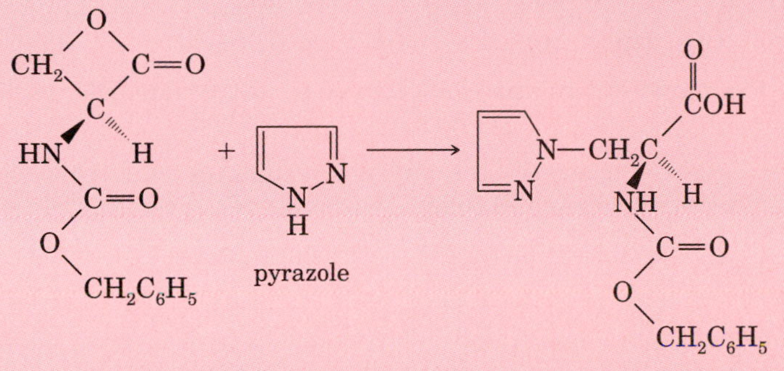

4   pyrazole   **5**

Questions

1. In the formation of the β-lactone **4**, the oxygen of the water that is lost is captured by the phosphorus of the triphenylphosphine. What happens to the two hydrogens of the water? Write an equation for this reaction.

2. Write an equation showing how N^α-benzyloxycarbonyl-β-(pyrazol-1-yl)-(S)-alanine **(5)** is formed from **4** and pyrazole. In your equation clearly identify the nitrogen of pyrazole that is nucleophilic and explain the reason for your choice.

3. In question 2, assume that the pyrazole attacks the carbonyl carbon of the β-lactone. What would be the product had this occurred?

4. How can the benzyloxycarbonyl blocking group be removed from a substituted alanine such as **5**? Using compound **5**, write an equation for this reaction.

5. Write equations showing how β-lactone **4** can be used to prepare the following compounds.

(a)

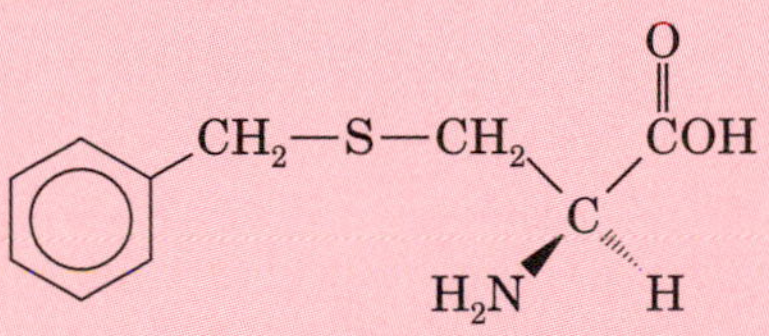

(b)

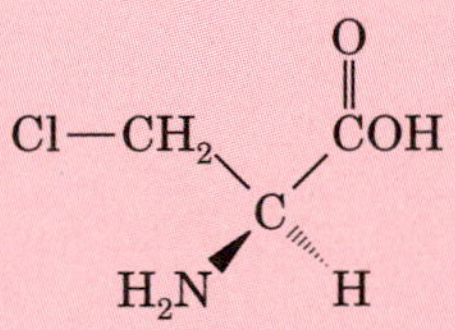

(c)

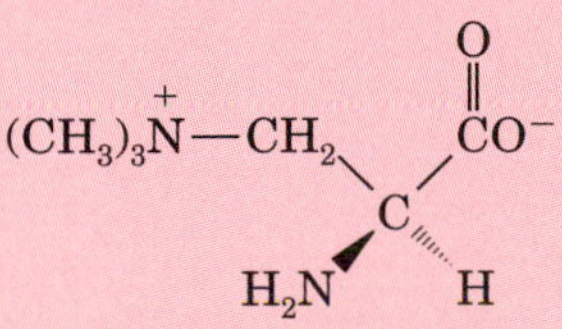

25.14 Write (1) the name of the amino acid and (2) the structure (as a dipolar ion) for each of the following abbreviations:

(a) gly   **(b)** asp   **(c)** ala   **(d)** trp

25.15 What is the structural relationship between D-glyceraldehyde, D-glucose, and L-alanine?

25.16 Outline a Strecker synthesis for: **(a)** phenylalanine, and **(b)** isoleucine. **(c)** What would be the configuration of the chiral carbons in the products?

25.17 Suggest a synthesis for valine from 3-methylbutanoic acid. Give the stereochemistry of the product.

25.18 Suggest a route to alanine from (R)-lactic acid (2-hydroxypropanoic acid) using a reductive amination. What is the stereochemistry of the product?

25.19 Predict the products of the reactions of alanine with: **(a)** dilute aqueous HCl; **(b)** dilute aqueous KOH; **(c)** methanol + H_2SO_4 with heat; **(d)** an equimolar quantity of acetic anhydride.

25.20 Write equations to represent: **(a)** dipolar-ion formation of histidine; **(b)** the equilibria between histidine and water; **(c)** the reaction of histidine with dilute aqueous HCl; **(d)** the reaction of histidine with dilute aqueous NaOH.

25.21 Explain the following observations:

(a) Although saturated carboxylic acids absorb at about 1720 cm^{-1} (5.81 μm) in the infrared region, amino acids do not absorb at this position.

(b) If a neutral solution of an amino acid is acidified, the infrared spectrum then shows absorption at 1720 cm^{-1}.

25.22 For each of the following amino acids, draw the structure that you would expect to be present at (1) pH 2, (2) pH 7, and (3) pH 11. Explain your answers.

(a) tyrosine   **(b)** glutamic acid

25.23 Complete the following equations:

(a) $\underset{\underset{+NH_3}{|}}{CH_3CHCO_2^-} + CH_3\overset{O}{\overset{||}{C}}\overset{O}{\overset{||}{C}}CH_3 \longrightarrow$

(b) ninhydrin + $\underset{\underset{+NH_3}{|}}{CH_3SCH_2CH_2CHCO_2^-} \longrightarrow$

(c) $CH_3(CH_2)_6\overset{O}{\overset{||}{C}}H \xrightarrow[\text{(3) } H_2O, H^+]{\substack{\text{(1) } NH_3 \\ \text{(2) } HCN}}$

(d) [indole structure with substituent $CH_2\overset{O}{\overset{||}{C}}CO_2H$] $\xrightarrow{H_2, NH_3, Pd}$

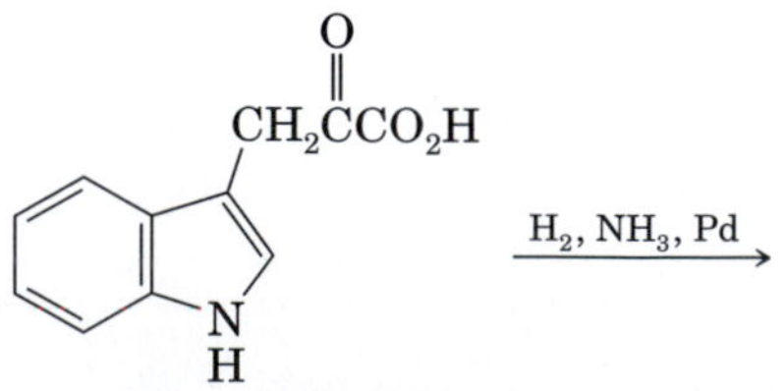

* For information concerning the organization of the *Study Problems* and *Additional Problems,* see the *Note to student* on page 41.

25.24 Unlike the other amino acids, proline does not form a blue-violet product when treated with ninhydrin. Explain.

25.25 In the following structure, label: **(a)** the peptide bond(s); **(b)** the *N*-terminal amino acid; **(c)** the *C*-terminal amino acid. **(d)** Is this structure a dipeptide, tripeptide, or tetrapeptide? **(e)** Would this peptide be considered acidic, basic, or neutral?

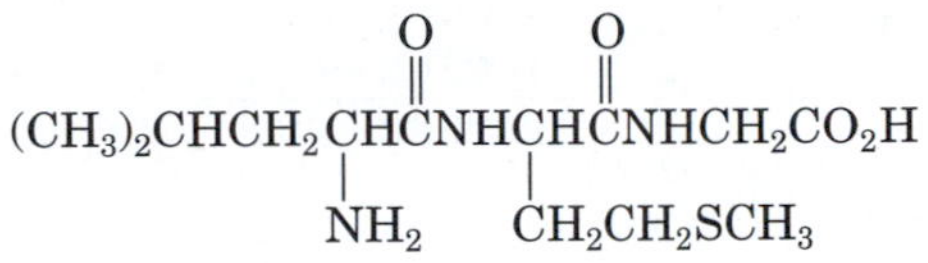

$$(CH_3)_2CHCH_2\underset{\underset{NH_2}{|}}{CH}\overset{\overset{O}{\|}}{C}NH\underset{\underset{CH_2CH_2SCH_3}{|}}{CH}\overset{\overset{O}{\|}}{C}NHCH_2CO_2H$$

25.26 Write the structures for **(a)** glycylglycine, and **(b)** alanylleucylmethionine.

25.27 Each of the following peptides is subjected to reaction with an alkaline solution of phenyl isothiocyanate, followed by acid hydrolysis. Give the structures of the products.

(a) gly–ala   **(b)** ala–gly   **(c)** ser–phe–met

25.28 Write the structures of the products of the reaction of 1-fluoro-2,4-dinitrobenzene, followed by hydrolysis, with: **(a)** valine; **(b)** alanylvaline; **(c)** glutamylglycine.

25.29 A pentapeptide obtained from treatment of a protein with trypsin contains arginine, aspartic acid, leucine, serine, and tyrosine. To determine the amino acid sequence, the peptide was treated with Edman reagent three times. The composition of the peptide remaining after each treatment was: (1) arginine, aspartic acid, leucine, serine; (2) arginine, aspartic acid, serine; and (3) arginine, serine. What is the amino acid sequence of the pentapeptide?

25.30 *Bradykinin* is a pain-causing nonapeptide that is released by globulins in blood plasma as a response to toxins in wasp stings. Partial hydrolysis of bradykinin results in the following tripeptides:

ser–pro–phe   gly–phe–ser   pro–phe–arg   arg–pro–pro

pro–gly–phe   pro–pro–gly   phe–ser–pro

What is the amino acid sequence in bradykinin?

25.31 Starting with monomeric amino acids, show how to prepare ala–gly and phe–val and then show how these dipeptides can be joined to yield ala–gly–phe–val. (Do not use a solid-phase synthesis.)

25.32 Complete the following equations:

(a) 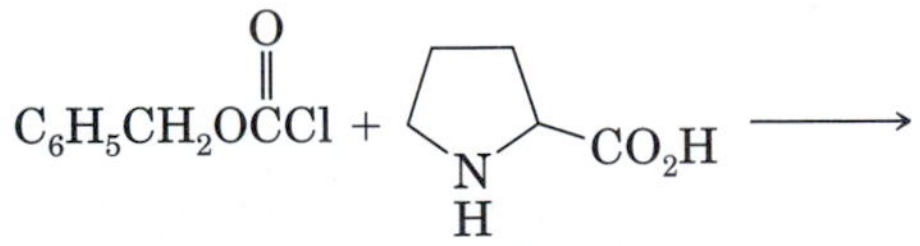 $C_6H_5CH_2O\overset{\overset{O}{\|}}{C}Cl$ + (pyrrolidine ring with N–H and CO_2H) $\longrightarrow$

(b) $C_6H_5CH_2O\overset{\overset{O}{\|}}{C}NH\underset{\underset{CH_2C_6H_5}{|}}{CH}\overset{\overset{O}{\|}}{C}OH$ + $ClCOC_2H_5$ $\longrightarrow$

(c) (C_6H_5)–N=C=N–(C_6H_5) + $CH_3CH_2\underset{\underset{NH_2}{|}}{\overset{\overset{CH_3}{|}}{CH}}CHCO_2H$ $\xrightarrow{H^+}$

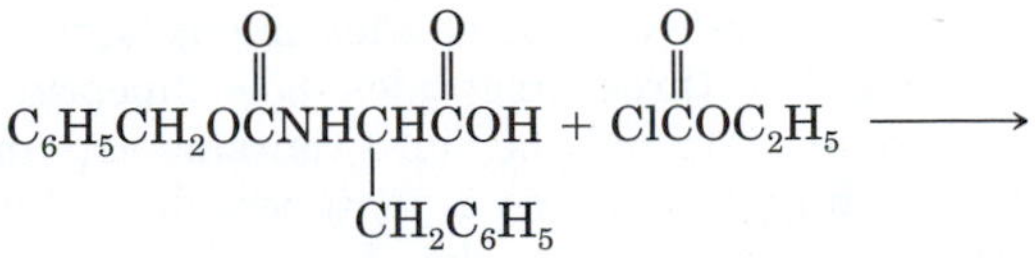

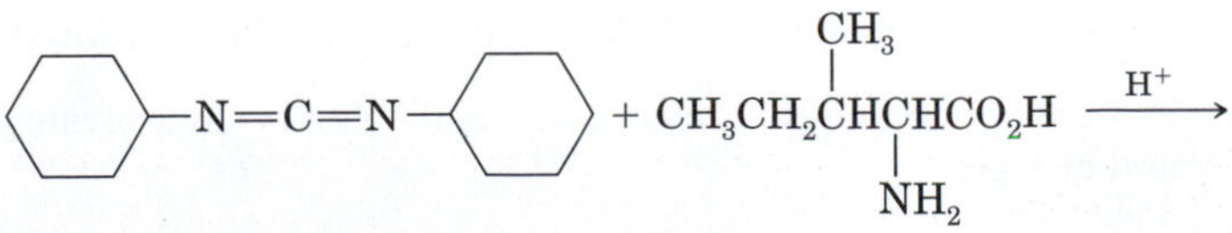

(d) $\underset{\underset{CH_3}{|}}{\underset{\overset{+}{N}H_3}{|}}$ C$_6$H$_5$CHCNHCHCO$_2^-$ $\xrightarrow[\text{(2) H}_2\text{O, H}^+\text{, heat}]{\text{(1) O}_2\text{N}\!-\!\!\!\langle\;\rangle\!\!-\!\text{F}}$

$$\text{C}_6\text{H}_5\overset{\overset{+}{N}\text{H}_3}{\underset{}{\text{CH}}}\text{C}\overset{\text{O}}{\|}\text{NHCH}\underset{\text{CH}_3}{\text{CO}_2^-}$$

(e) $\underset{\underset{CH_3}{|}}{\underset{\overset{+}{N}H_3}{|}}$ C$_6$H$_5$CHCNHCHCO$_2^-$ $\xrightarrow[\text{(2) CH}_3\text{NO}_2\text{,HCl}]{\text{(1) C}_6\text{H}_5\text{N}\!=\!\text{C}\!=\!\text{S}}$

Additional Problems

25.33   A new amino acid, γ-carboxyglutamic acid (gla) was discovered in 1974 as one of the amino acid residues in blood coagulation proteins and in proteins from calcified tissues. (The presence of this residue allows the protein to chelate metal ions.) The reason for the late discovery of this amino acid was that, in structure determinations, proteins are often hydrolyzed to smaller fragments by heating with acid.

(a)   Write an equation to show what happens to γ-carboxyglutamic acid when heated in aqueous acid.

(b)   Write an equation to show its probable reaction when heated in aqueous base.

$$(\text{HO}_2\text{C})_2\text{CHCH}_2\overset{\overset{\text{NH}_2}{|}}{\text{CHCO}_2\text{H}}$$

γ-carboxyglutamic acid (gla)

25.34   A peptide containing one equivalent each of tyr, ile, gly, arg, and cys had no *C*-terminal amino acid and no *N*-terminal amino acid. Explain.

25.35   Suggest a reason for the differences in the isoelectric points of **(a)** lysine (9.74) and histidine (7.59), and **(b)** lysine and arginine (10.76).

25.36   **(a)**   When the following peptide is treated with 1-fluoro-2,4-dinitrobenzene, followed by hydrolysis, *two* dinitrophenylamino acids are isolated. What are they?

(b)   Would the presence of the second product interfere with the determination of the peptide structure? Explain.

ala–lys–ala–gly

25.37   The complete hydrolysis of an acyclic nonapeptide yields a mixture of ala, asp, glu, gly, leu, lys, phe, tyr, and val. Terminal-residue analysis shows the *N*-terminal amino acid to be val and the *C*-terminal amino acid to be gly. Partial enzymatic hydrolysis with chymotrypsin yields a pentapeptide and a tetrapeptide, among other products. The tetrapeptide is partially hydrolyzed to three dipeptides. One dipeptide contains ala and gly; the second contains asp and tyr; and the third contains asp and ala. Partial hydrolysis of the nonapeptide with trypsin yields a pentapeptide and a tetrapeptide. The amino acid content of the tetrapeptide is glu, leu, lys, and val. What are the two possible structures of the nonapeptide? How could they be differentiated?

25.38   *Glutathione* is a tripeptide found in most living cells. Partial hydrolysis yields cys, glu, gly, glu–cys, and cys–gly.

(a) What is the amino acid sequence in glutathione?

(b) It has been discovered that glutamic acid forms a peptide link in glutathione with the *side-chain* carboxyl group, rather than the carboxyl group adjacent to the amino group. What is the structural formula of glutathione?

25.39 Using a procedure other than solid-phase peptide synthesis, write flow equations showing how the following peptides could be synthesized:

(a) ala–ile–phe

(b)
$$\overset{+}{H_3}NCH_2\overset{\overset{\displaystyle O}{\|}}{C}NHCH\overset{\overset{\displaystyle O}{\|}}{C}NHCH\overset{\overset{\displaystyle O}{\|}}{C}O^-$$

with CH_3 on the second α-carbon and $CH(CH_3)_2$ on the third α-carbon.

NUCLEIC ACIDS

The study of nucleic acids is a fascinating area of modern research. These polymers carry the genetic code and are responsible for its expression by protein synthesis. The two types of nucleic acids are **deoxyribonucleic acids (DNA)** and **ribonucleic acids (RNA)**. In all cells, DNA is the repository of the genetic code and RNA is involved in protein synthesis. Viruses, however, contain their genetic code in either DNA or RNA and use the host cells that they have infected for protein synthesis.

The component parts of nucleic acids can be obtained by direct or enzymatic hydrolysis (Figure 26.1). Complete hydrolysis yields a sugar (ribose from RNA or deoxyribose from DNA), purine and pyrimidine heterocyclic bases, and inorganic phosphate ions. Partial hydrolysis of a nucleic acid yields **nucleotides** (sugar bonded to a base and to a phosphate group). Nucleotides are the monomeric units of nucleic acids. Hydrolysis of the nucleotides results in **nucleosides** (sugar bonded to base).

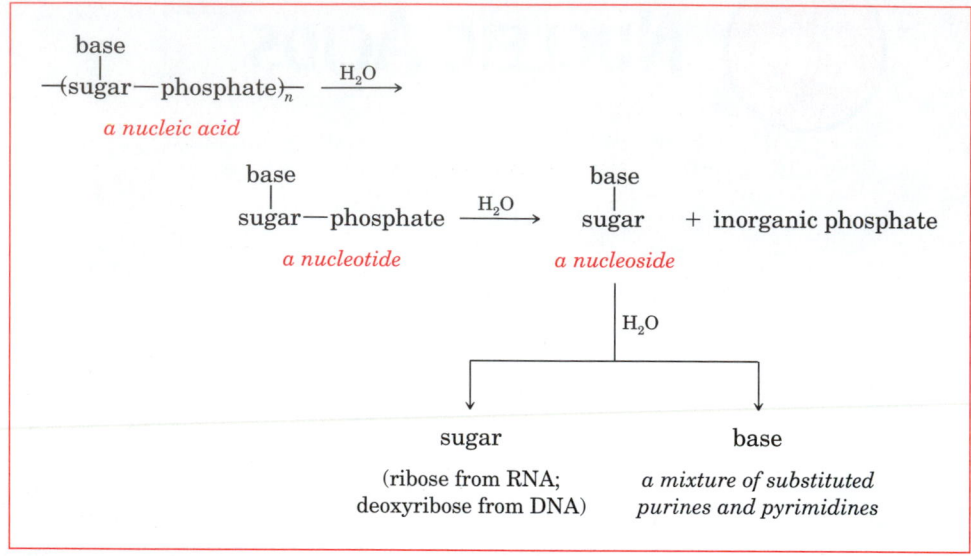

Figure 26.1 Hydrolysis of the nucleic acids.

In this chapter, we will first discuss the hydrolysis fragments. Next, we will consider DNA itself—its primary and higher structures—and some related topics. We will then consider RNA. We will conclude with an overview of the role of RNA in protein synthesis.

Section 26.1
Hydrolysis Products of the Nucleic Acids

In this section, we will discuss the complete hydrolysis products of the nucleic acids (the phosphate, the sugars, and the bases), then the nucleosides, and finally the nucleotides.

A. The Inorganic Phosphate

The inorganic phosphate generated by the complete hydrolysis of the nucleic acids is either $H_2PO_4^-$ or HPO_4^{2-}, depending on the pH and the medium used for hydrolysis.

In the nucleotides, the phosphate group is bonded to a hydroxyl group of the sugar. Because the protonated phosphate group is acidic, protonated nucleotides are also acidic.

B. The Sugars

The sugar in RNA, as the name ribonucleic acid implies, is β-D-ribose, which is in the furanose form. The sugar in DNA (deoxyribonucleic acid) is 2-deoxy-β-D-ribose, also in the furanose form. The prefix 2-deoxy in this name means that the structure lacks an oxygen at position 2.

$$
\begin{array}{l}
\text{CHO} \\
\text{H}\!-\!\!-\!\text{OH} \\
\text{H}\!-\!\!-\!\text{OH} \\
\text{H}\!-\!\!-\!\text{OH} \\
\text{CH}_2\text{OH}
\end{array}
$$

D-ribofuranose

$$
\begin{array}{l}
\text{CHO} \\
\text{CH}_2 \\
\text{H}\!-\!\!-\!\text{OH} \\
\text{H}\!-\!\!-\!\text{OH} \\
\text{CH}_2\text{OH}
\end{array}
$$

2-deoxy-D-ribofuranose

C. The Bases

Complete hydrolysis of nucleic acids yields several heterocyclic ring systems, usually called **bases.** These compounds are substituted pyrimidines and purines.

purine

the parent ring system
of adenine and guanine

pyrimidine

the parent ring system
of thymine, uracil, and cytosine

The four principal bases obtained from DNA are adenine, guanine, cytosine, and thymine. These bases are usually identified in abbreviated formulas by the first letters of their names (A, G, C, and T).

The four principal bases found in DNA:

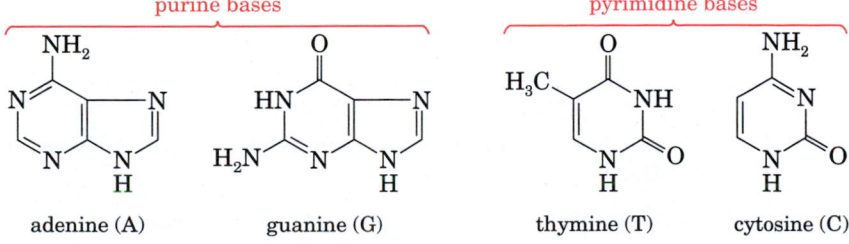

purine bases

adenine (A)   guanine (G)

pyrimidine bases

thymine (T)   cytosine (C)

The four principal bases obtained from RNA are adenine, guanine, cytosine, and uracil. Three of these bases are the same as those found in DNA, but uracil (U) in RNA replaces thymine (T) in DNA.

In RNA: *In DNA:*

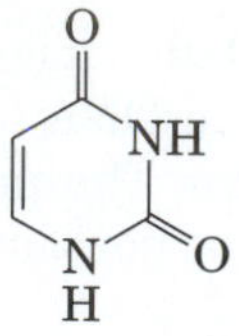

uracil (U)   thymine (T)

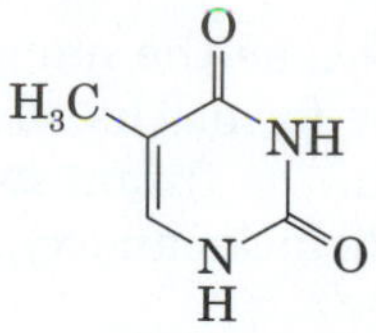

Although the bases we have described are the most common, a variety of so-called minor bases have also been isolated. In some cases, their concentrations are not really minor. For example, some plant DNA contains up to 7% 5-methylcytosine. These minor bases result from enzymatic modification of a principal base after the nucleic acid has been biosynthesized instead of being incorporated during the biosynthesis.

In Table 26.1, the structures of some of the minor bases in DNA and RNA are listed. Transfer RNA (tRNA) is a particularly rich source for these minor bases.

Tautomers   The pyrimidine and purine bases can undergo lactam–lactim tautomerization, a type of keto–enol tautomerization. In neutral and acidic media, the lactam form of these bases predominates.

Examples of lactam–lactim tautomerization:

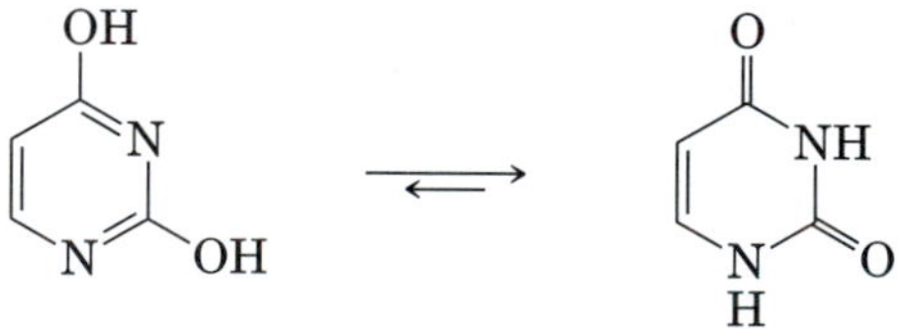

lactim form of uracil   *a lactam form of uracil*

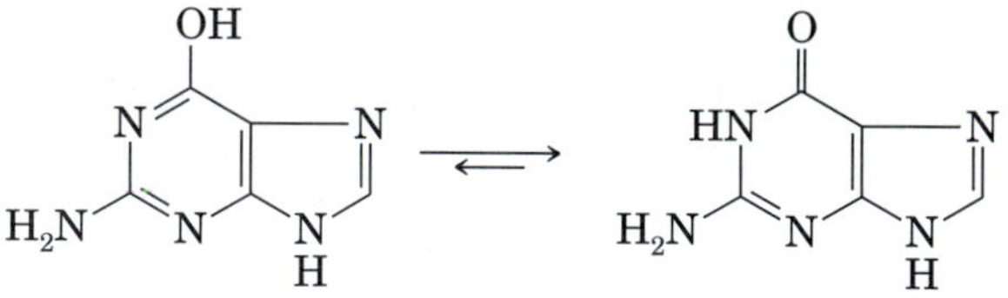

lactim form of guanine   *a lactam form of guanine*

STUDY PROBLEMS

26.1   There are two lactam–lactim tautomers of uracil in addition to those shown above. Draw their structures.

26.2   Show the possible lactam–lactim structures for thymine.

Acid–base reactions   Some of the bases from the nucleic acids are amphoteric. They can be protonated to form cationic salts or can lose a proton to form an anion. For example,

Acid–base reactions for uracil:

uracil

Table 26.1	**A few of the minor bases and nucleosides found in the nucleic acids**

Name	Formula	Where Found
5-methylcytosine		many types of DNA
N^6-methyladenine		many types of DNA and RNA
dihydrouracil		tRNA
1-methylguanine		tRNA
ribothymidine		tRNA
pseudouridine (ψ-uridine)		tRNA
1-methylinosine		tRNA

The ionic form of the bases depends on the pH of the solution. At physiological pH (near 7), all the bases are in a neutral nonionic form. The ionic forms of the bases, however, play a major role in the mechanism of denaturation of the nucleic acids by acids and bases (Section 26.2C).

Pyrimidine is a relatively weak base; its pK_b is 12.7. Uracil and thymine, both pyrimidine bases, are both far less basic than even pyrimidine. The reason for this low basicity is that the nitrogens in uracil and thymine are in amide groups in their lactam forms. The unshared pair of electrons of an amide nitrogen is delocalized by the carbonyl groups (Section 15.8C). The delocalized electrons are not available for bonding and, thus, these so-called bases are not basic.

STUDY PROBLEM

26.3   Using formulas in your answer, explain why cytosine is far more basic than uracil or thymine. (*Hint:* Consider the resonance stabilization of the cationic salt.)

A purine ring system contains imidazole and pyrimidine fused rings. The five-membered imidazole ring is basic, while the pyrimidine ring is only weakly basic.

purine
($pK_b = 11.7$)

Guanine, like the parent purine, is protonated at nitrogen 7. The two ring systems have similar pK_b values (11.0 for guanine and 11.7 for purine). Nitrogen 1 of guanine is an amide nitrogen and, therefore, not basic.

guanine

The amino group of adenine changes the position of protonation of the purine ring. The amino group stabilizes the adjacent positive charge at nitrogen 1 by resonance, but it cannot stabilize a positive charge at position nitrogen 7. Therefore, protonation of adenine occurs at nitrogen 1.

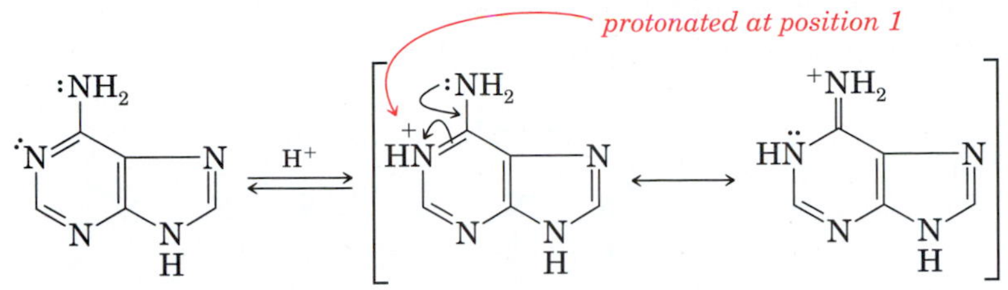

protonated at position 1

adenine

D. The Nucleosides

A nucleoside contains a purine or pyrmidine base bonded to a sugar by a β-glycosidic linkage.

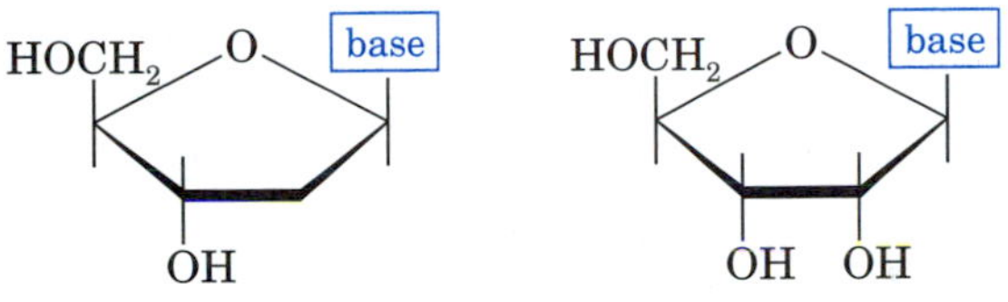

a nucleoside from DNA   *a nucleoside from RNA*

These glycosides are often called **N-glycosides** to differentiate them from the common carbohydrate *O*-glycosides. The chemical properties of the *N*- and *O*-glycosides are similar.

Because the bond is glycosidic, the anomeric carbon of the sugar is locked into one configuration (β for practically all known nucleosides).

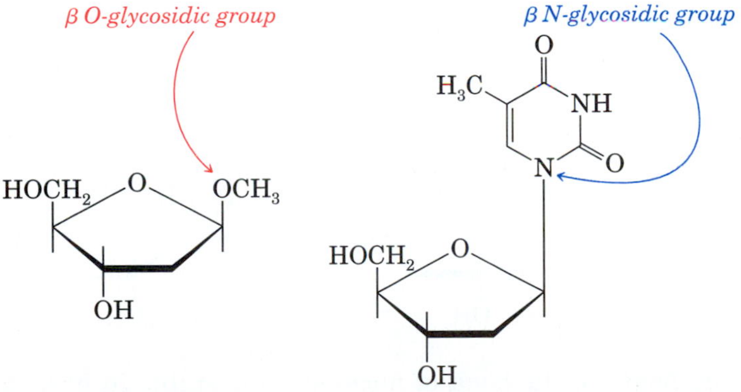

β O-glycosidic group   *β N-glycosidic group*

The structures of the four common nucleosides that can be isolated from DNA hydrolysis are shown in Figure 26.2. The principal nucleosides obtained from RNA are the same as those obtained from DNA with two exceptions: the sugar is ribose instead of deoxyribose, and uracil replaces thymine as a base.

When numbering the two ring systems of a nucleoside, we use primes for the sugar's positions.

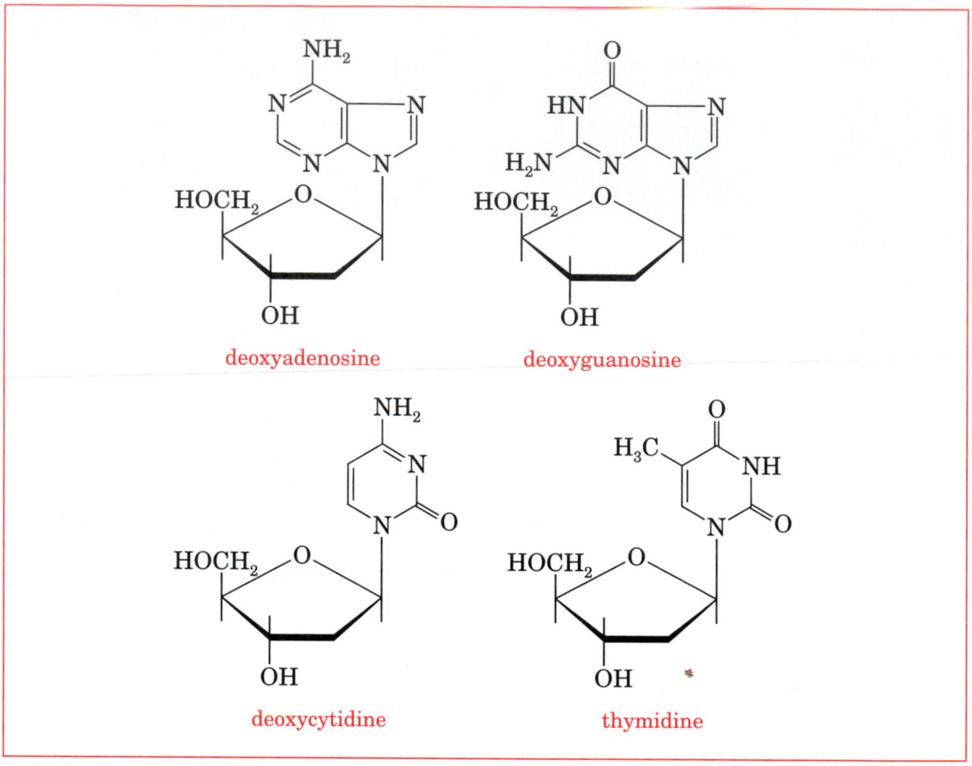

Figure 26.2 The principal nucleosides from the hydrolysis of DNA. (The deoxynucleoside of thymine is called thymidine because thymine is not a principal base in nucleic acids other than DNA.)

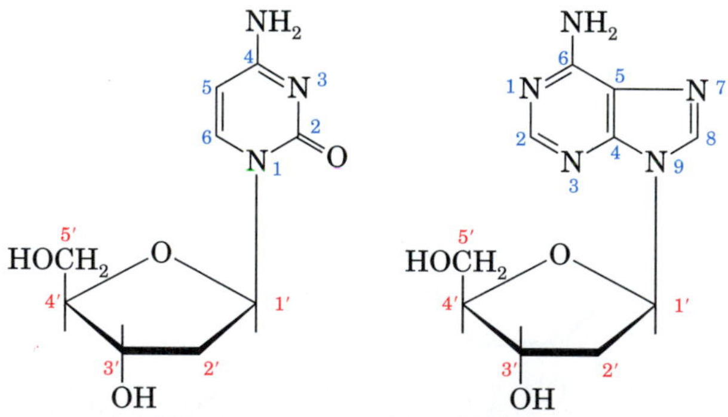

The N-glycosidic bond of nucleosides is stable in base, but readily cleaved in acid. The rate of acidic cleavage depends on the structure of the heterocyclic base. The purine nucleosides are rapidly hydrolyzed by aqueous acid to a heterocyclic base and a sugar. The pyrimidine nucleosides are cleaved only after prolonged treatment with concentrated acid.

E. The Nucleotides

A nucleotide contains a base, a sugar, and a phosphate group. The nucleotides are monomers for the nucleic acid polymers.

base
|
phosphate—[sugar]
a nucleotide

base
|
⅔—phosphate—[sugar]—⅔
the nucleic acids

Deoxyribonucleotides with a phosphate group at position 3′ or position 5′ are known. The 5′-phosphates are more common and we will emphasize them in our discussions. Figure 26.3 shows the structures of the principal deoxyribonucleotides.

A ribose residue in a nucleoside has *three* hydroxyl groups (at positions 2′, 3′, and 5′) that can form phosphate esters. Alkaline hydrolysis of RNA yields ribonucleotides with the phosphate group at position 2′ or 3′. Enzymatic hydrolysis of RNA yields nucleotides with either a 3′-phosphate or a 5′-phosphate depending on the enzyme used. The 3′- and 5′-phosphates are analogous to those obtained from DNA. The 2′-phosphate arises from a cyclic phosphate produced during the alkaline hydrolysis.

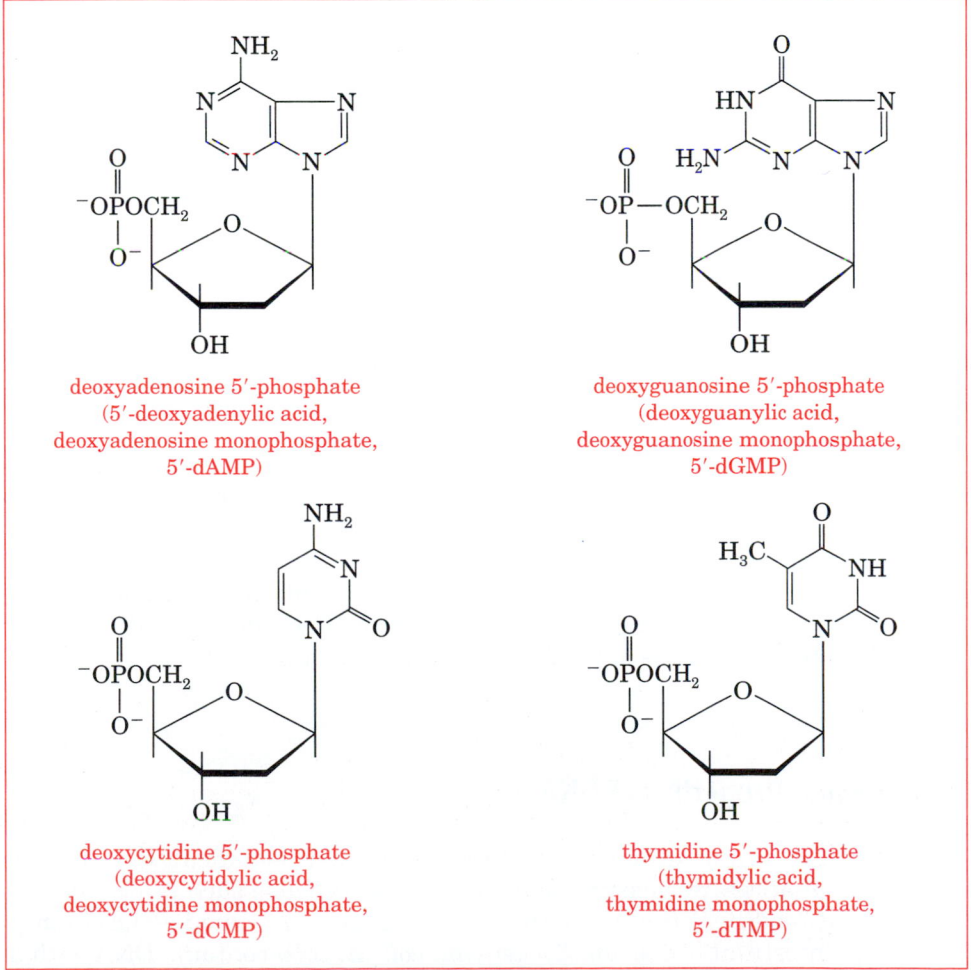

deoxyadenosine 5′-phosphate
(5′-deoxyadenylic acid,
deoxyadenosine monophosphate,
5′-dAMP)

deoxyguanosine 5′-phosphate
(deoxyguanylic acid,
deoxyguanosine monophosphate,
5′-dGMP)

deoxycytidine 5′-phosphate
(deoxycytidylic acid,
deoxycytidine monophosphate,
5′-dCMP)

thymidine 5′-phosphate
(thymidylic acid,
thymidine monophosphate,
5′-dTMP)

Figure 26.3 Formulas and names for the principal deoxyribonucleotides.

Table 26.2 pK_a values of the principal ribonucleotides			
Nucleotide	First Phosphate Proton	Second Phosphate Proton	Base
adenosine 5'-phosphate	0.9	6.1	3.8
uridine 5'-phosphate	1.0	6.4	9.5
cytidine 5'-phosphate	0.8	6.3	4.5
guanosine 5'-phosphate	0.7	6.1	2.4, 9.4

a cyclic phosphate

The structures of the principal ribonucleotides with a phosphate group at the 5' position are the same as the corresponding deoxyribonucleotides with two exceptions: ribose replaces deoxyribose as the sugar and uracil replaces thymine as a base.

The nucleotides are amphoteric. They contain a phosphoric acid ester group with two acidic protons as well as the basic heterocyclic ring. Table 26.2 lists the pK_a values of the principal ribonucleotides.

Section 26.2
Structure and Properties of DNA

DNA is a giant, even among macromolecules. It is difficult to isolate DNA and obtain accurate measurements of its mass because such large molecules fragment readily. Estimates of the size, however, can be made. The chromosome of the intestinal bacterium *Escherichia coli (E. coli)* contains DNA with a molecular weight of about 2.6×10^9. DNA molecules from higher animals are larger and more complex than those from bacteria.

A. Primary Structure of DNA

The DNA polymer consists of nucleoside residues linked by phosphate groups. The phosphate forms an inorganic ester link between the 3′ hydroxyl of one sugar residue and the 5′ hydroxyl of the next sugar (Figure 26.4). A linear DNA polymer, therefore, has a free 5′-hydroxyl group at one end and a free 3′-hydroxyl group at the other end.

A DNA molecule contains a backbone of sugar–phosphate groups bearing a series of bases (A, G, C, and T). The order in which the bases occur is called the **base sequence.** A number of shorthand formulas have been developed to represent the base sequence of a single DNA strand. The simplest method is to list the letters for the base sequence starting at the 5′-hydroxyl end of the chain (on the left) and proceeding toward the 3′-hydroxyl end (on the right)— for example, A—C—A—.

B. Secondary Structure of DNA: The Double Helix

How does the DNA polymer with its base sequence carry the genetic code? In 1953, J. D. Watson and F. H. C. Crick proposed a model for DNA that accounts for its behavior. In 1962, these two men, along with Maurice Wilkins,

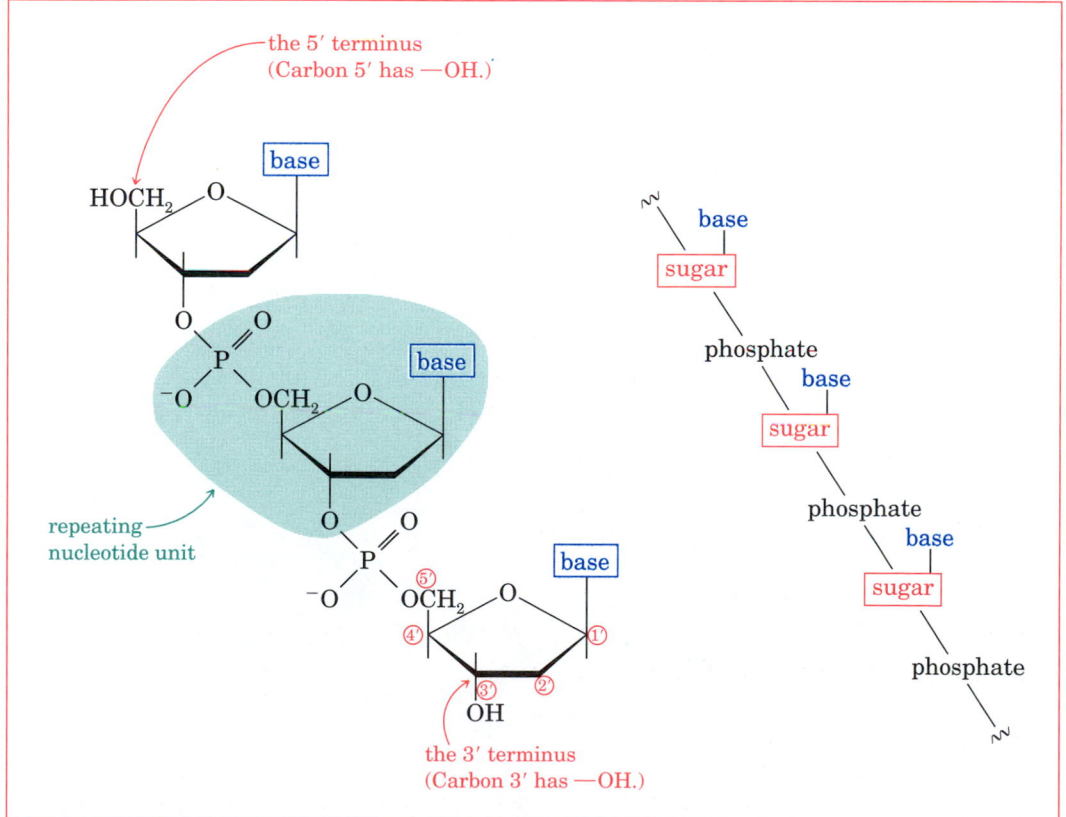

Figure 26.4 Primary structure of DNA. Each base is one of four heterocycles—cytosine, thymine, adenine, or guanine.

who provided evidence for the structure of the model by X-ray analysis, were awarded the Nobel prize for their work.

The Watson–Crick model of DNA is a double helix of two long antiparallel DNA molecules held together by hydrogen bonds. In this model, antiparallel means that two linear DNA molecules are parallel but aligned in opposite directions. Each end of the double helix thus consists of a 5′ terminus (from one molecule) and a 3′ terminus (from the other molecule).

5′ end ⟵ A—G—T—C—A—A—G—T—G—G—C—C ⟶ 3′ end

3′ end ⟵ T—C—A—G—T—T—C—A—C—C—G—G ⟶ 5′ end

The hydrogen bonds between the DNA strands are not random, but are specific between pairs of bases: guanine is hydrogen bonded to cytosine, and adenine to thymine.

Why are the hydrogen bonds specific—thymine to adenine and cytosine to guanine but not other combinations? Thymine and adenine can be joined by *two* hydrogen bonds (approximate total strength, 10 kcal/mol). Cystosine and guanine can be joined by *three* hydrogen bonds (approximate total strength, 17 kcal/mol). No other pairing of the four bases leads to such strong hydrogen bonding. Figure 26.5 shows the structures, dimensions, and hydrogen bonds of the bases.

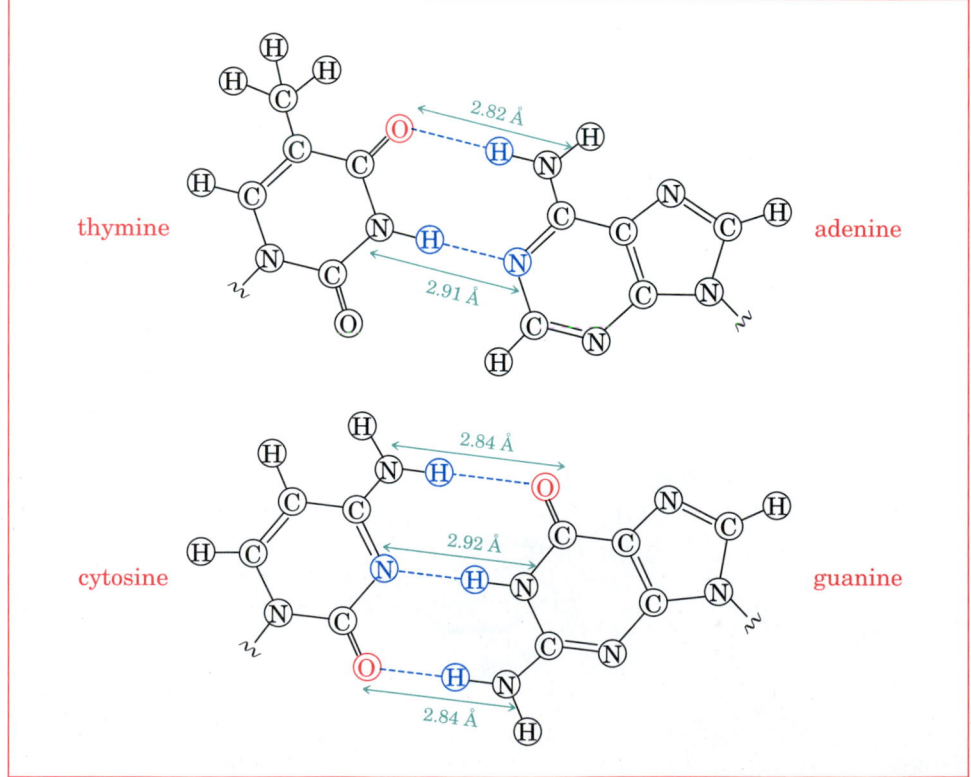

Figure 26.5 Hydrogen bonds and dimensions in base pairs of DNA: (a) thymine to adenine and (b) cytosine to guanine.

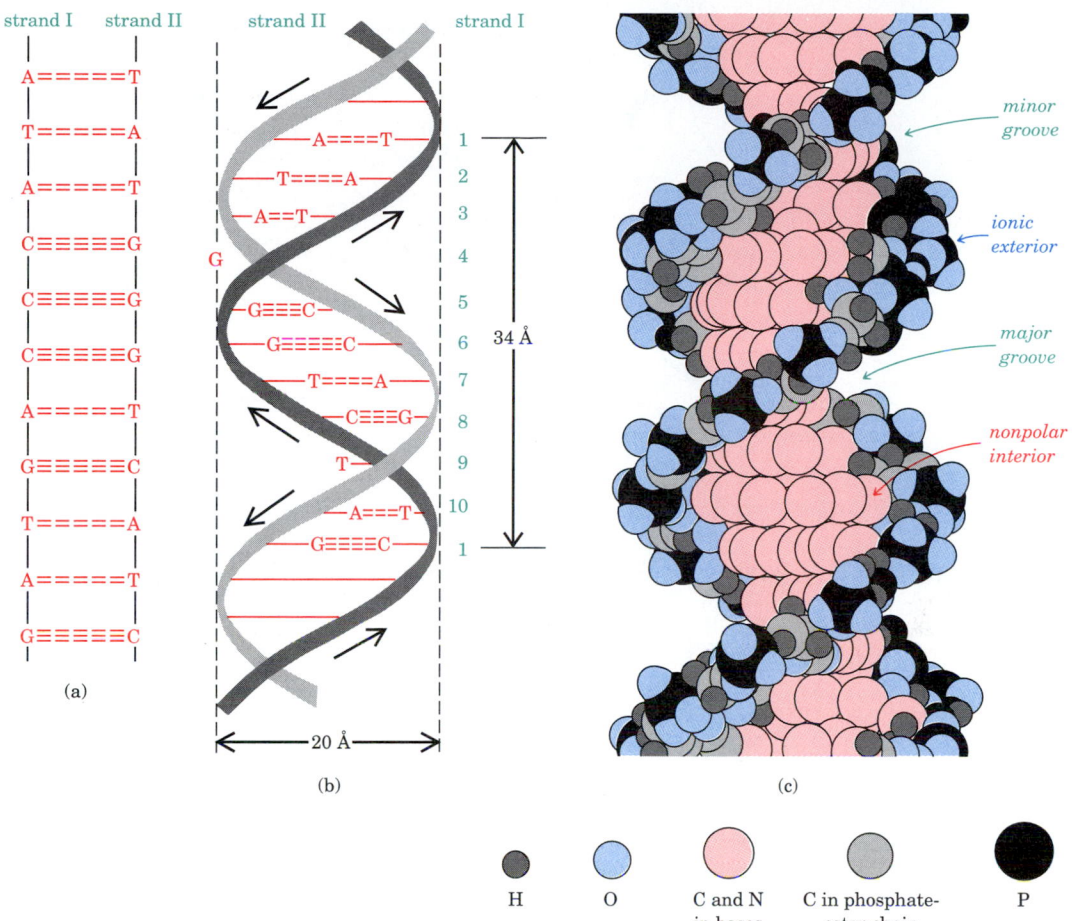

Figure 26.6 Three representations of the double-helical model of DNA. (a) Two un-coiled DNA molecules are joined by hydrogen bonds between complementary base pairs. (Not shown are the sugar and phosphate units.) (b) The DNA strands are coiled in a double helix, with ten base pairs for every complete turn of the helix. (c) A "space-filling" model of DNA. (Adapted from William H. Brown, *Introduction to Organic and Biochemistry, 4th ed.* Copyright 1987 by Wadsworth, Inc. Used by permission of Brooks/Cole Publishing Co.)

Picture the double-stranded helix of DNA held together by a series of par-ticular hydrogen-bonded pairs, as shown in Figure 26.6(a). Wherever an ade-nine (A) appears in one strand, a thymine (T) appears opposite it in the other strand. The two strands are completely complementary in this respect. Figure 26.6 shows the base-pairing and also shows a space-filling model of the ap-pearance of the DNA double helix.

In the DNA double helix, the purine and pyrimidine bases are on the *in-side* of the helix, and the deoxyribose and phosphate units are on the *outside*. In addition to hydrogen bonding between the stacked purines and pyrim-idines, other forces, such as induced dipole attractions, contribute to the maintenance of the two-stranded structure. The energy of these attractions is called the **stacking energy.** Figure 26.7 shows a representation of the DNA helix viewed down the axis.

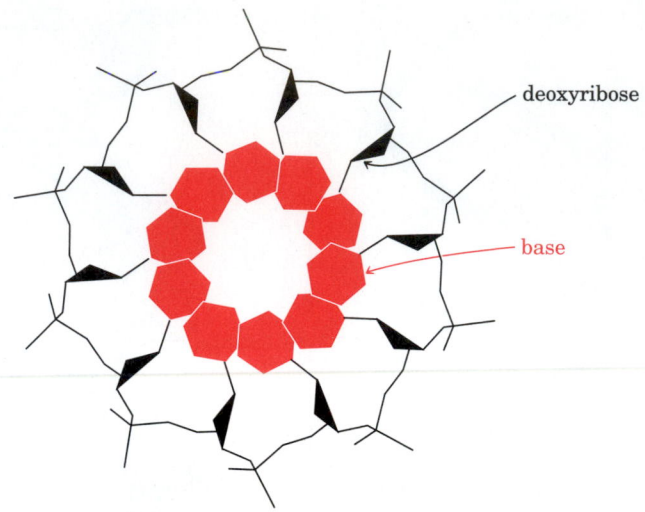

Figure 26.7 Diagram of DNA viewed down the helical axis.

26.4 With formulas, show why thymine and guanine cannot form as strong hydrogen bonds as thymine and adenine.

26.5 Answer the following questions concerning the representation of DNA below.

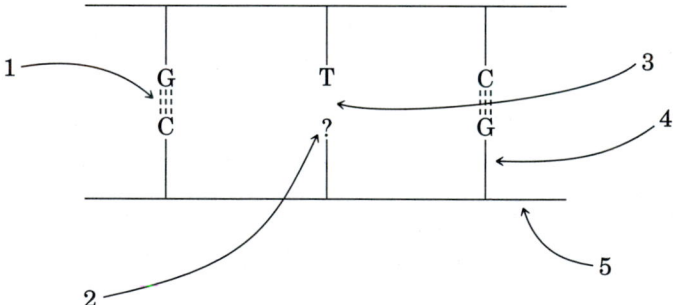

(a) What do the dashed lines indicated by 1 represent?

(b) What base should be at the position indicated by 2?

(c) How many dashed lines should be at the position indicated by 3?

(d) What does the line indicated by 4 represent?

(e) What does the line indicated by 5 represent?

C. Denaturation and Renaturation

Disruption of the secondary structure of a polymer is called *denaturation*. DNA, like proteins, can be denatured by changes in pH, increases in temperature, and some reagents. Loss of the base-pair hydrogen bonding causes the double-stranded helix to unwind and separate.

Strong acid or base causes the normally neutral heterocyclic bases in DNA to acquire ionic charges. These charges affect the bases' ability to form effec-

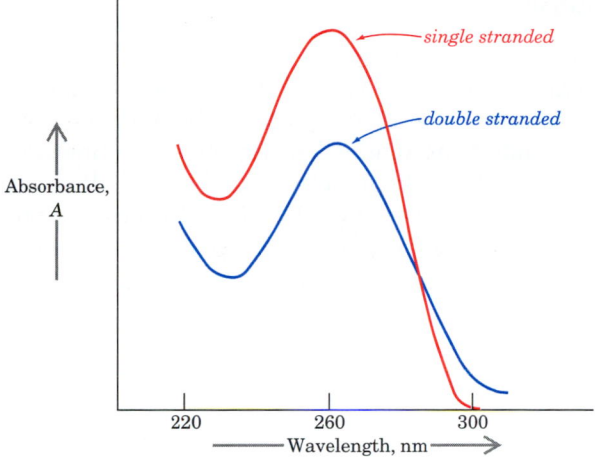

Figure 26.8 The absorbance of DNA at a wavelength of 260 nm increases when the double helix is unraveled into single strands.

tive hydrogen bonds. Changes in hydrogen bonding between the bases is easily detected by ultraviolet spectroscopy. Because of the π-electron systems, all the heterocyclic bases absorb ultraviolet radiation of characteristic wavelengths and with characteristic absorptivities. When the bases are strongly hydrogen bonded, the absorbance is diminished. As denaturation occurs, an increase in the absorbance occurs (Figure 26.8).

When denatured DNA is held at its **annealing temperature,** a temperature just below its denaturation temperature, or "melting point," the double-stranded helix can re-form. **Renaturation** is the reassembly of the original two strands that were separated.

Renaturation is a two-step process. The first step is **nucleation**—the joining of the two strands by a few base pairings. Nucleation is generally slow because the process is second order, but circular DNA nucleates rapidly because the process is first order. Once the two strands are loosely joined, the second step, a rapid **zippering** (the pairing of the remaining bases), occurs.

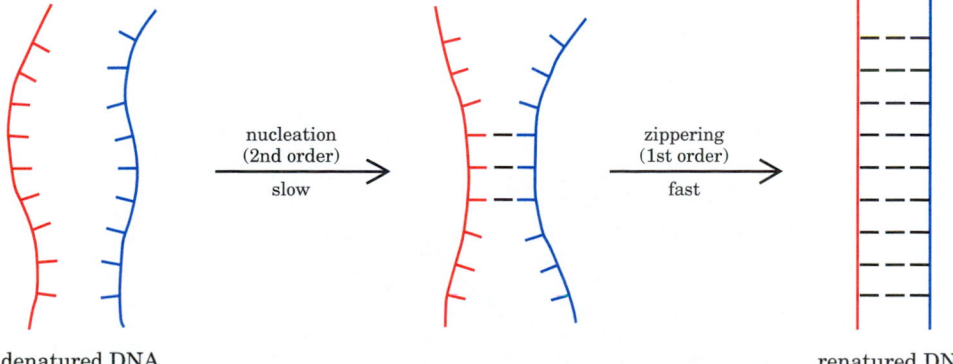

In addition to concentration, the rate of renaturation is dependent on the complexity of the DNA. The synthetic nucleic acids polyadenine and polyuracil undergo rapid renaturation because almost any collision between the two strands results in base pairing. Mammalian DNA, on the other hand, renatures very slowly because few collisions are fruitful.

D. Replication

In living cells, *the genetic code is contained by the DNA and consists of a particular sequence of bases that specifies the sequence of amino acids to be incorporated into any one protein* (a particular enzyme, for example). For passing the genetic code from one cell to its offspring (mother cell to daughter cell), the DNA duplicates itself exactly in a process called **replication.**

In replication, the double helix of DNA becomes enzymatically unraveled. Each strand serves as a template, or pattern, for the synthesis of a new complementary chain. When the syntheses are complete, two double-stranded helices exist where only one did before. Thus, after cell division, both cells contain identical DNA.

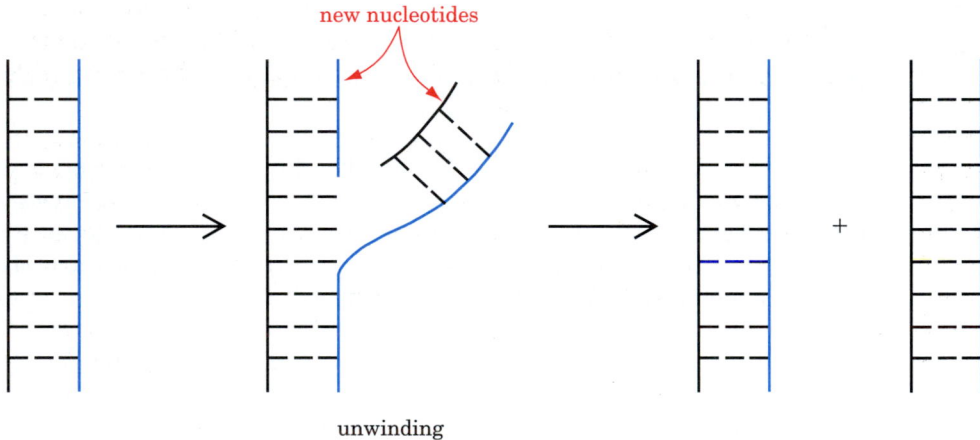

Replication is a complex series of reactions that requires many different enzymes. Addition of new nucleotides to the growing DNA chain, which is cat-

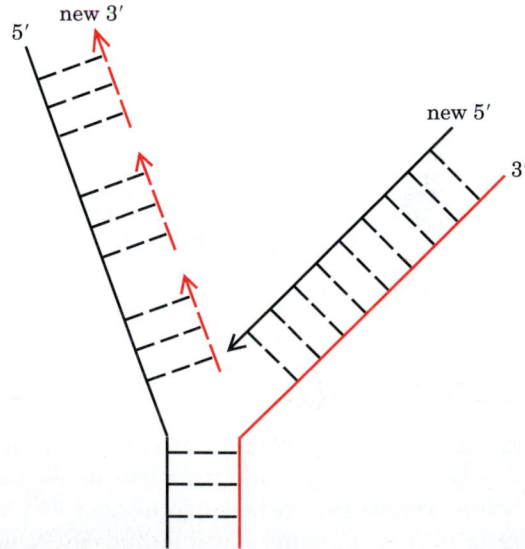

alyzed by the enzyme *DNA polymerase,* occurs by the substitution of a nucleotide 5'-triphosphate at the free 3'-hydroxyl group of the growing chain. A diphosphate ion is the leaving group; one monophosphate group becomes part of the chain. Figure 26.9 illustrates the addition of a nucleotide.

Both DNA strands are synthesized in the $5' \rightarrow 3'$ direction. Because the two existing chains are aligned in opposite directions, one strand must have a 3'-hydroxyl group near the **replication fork,** or site of unraveling, while the other strand must have a free 5'-hydroxyl group. The complement of the $3' \rightarrow 5'$ strand is synthesized as a continuous single chain, but the other new strand must be synthesized discontinuously, as a series of short chains. These pieces of this new DNA strand are then bonded together by other enzymes.

E. Conformers of DNA

Because of the chain of single bonds in its phosphodiester backbone, DNA can flex into several conformations. The three most important conformers are the

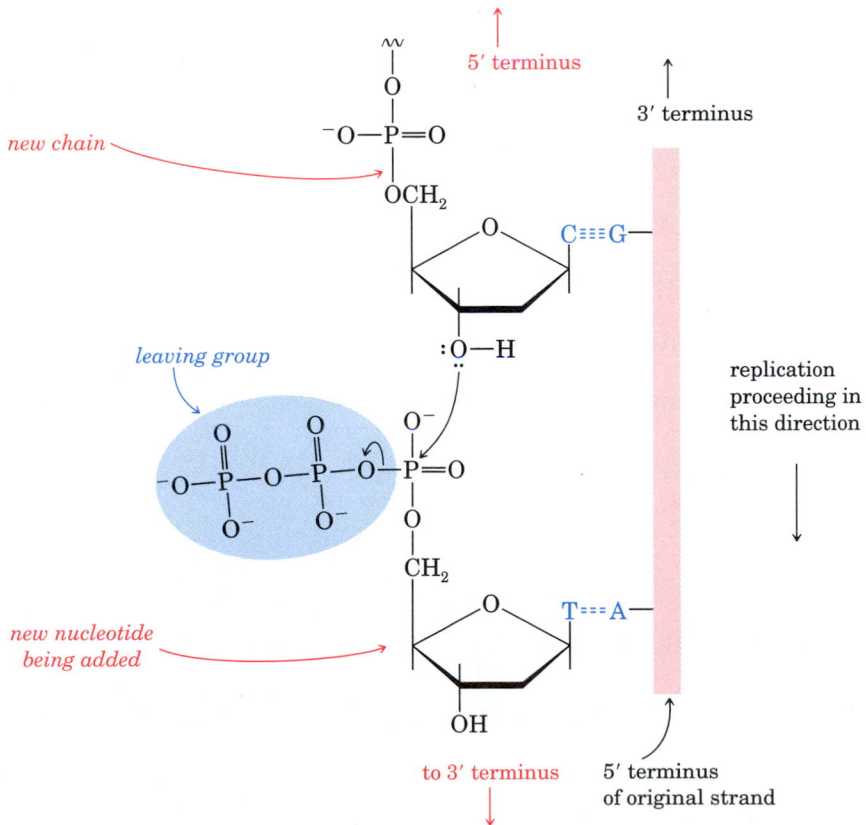

Figure 26.9 Addition of a nucleotide to a growing DNA chain. The nucleotide triphosphate loses a diphosphate ion in the enzyme-catalyzed reaction.

B-form, the A-form, and the Z-form. B-DNA, also called the Watson–Crick B-form, is the naturally occurring form of DNA. A-DNA is derived from the B-form by dehydration. The formation of Z-DNA occurs only when the DNA chains have specific base sequences.

The major structural features of the DNA molecule affected by conversion of one form to the other are (1) the conformation of the sugar, (2) the projection of the plane of the base around the glycosidic bond, and (3) the angle at which the bases are stacked in the helix in reference to the helical axis.

Sugar conformation   The two major conformations of 2-deoxyribose are the *2′-endo* conformer and the *3′-endo* conformer. The *2′-endo* conformer has carbon 2′ puckered toward the base whereas the *3′-endo* conformer has carbon 3′ puckered in that direction.

Haworth projection   *2′ endo*   *3′ endo*

conformational formulas

The 2′-endo conformation of deoxyribose is found in B-DNA, while the 3′-endo conformer occurs in A-DNA. The conformation of a Z-DNA sugar depends on the base to which it is bonded. If the base is cytosine, the sugar is in the 2′-endo conformation. If the base is guanine, the sugar is in the 3′-endo conformation.

Glycosidic-bond conformers   Rotation about the *N*-glycosidic bond allows the base to be oriented toward the sugar *(syn)* or away from it *(anti)*.

syn   *anti*

In B-DNA and A-DNA, all bases are in the *anti* orientation. In Z-DNA, the pyrimidine bases are *anti* and the purine bases are *syn*.

Stacking angle   In B-DNA, the planes of the bases are stacked nearly parallel to one another and at approximately right angles to the axis passing through the center of the helix. In A-DNA, the bases are tilted about 19° from the perpendicular to the helical axis. The tilt of the Z-DNA bases (9°) is intermediate between that in A-DNA and B-DNA.

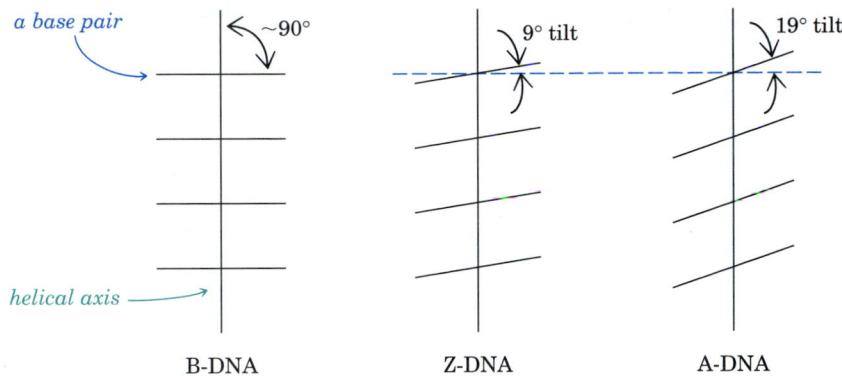

Properties of the conformers   **B-DNA** is the natural form of DNA. This form is characterized by a polar exterior (sugar–phosphate) and a relatively nonpolar interior (bases). The external spiral of anionic phosphate groups requires neutralization with cations, commonly protonated proteins. The nonpolar bases exclude most water molecules from the interior of the DNA and, therefore, they form hydrogen-bonded pairs with maximum strength.

The B-DNA helix is characterized by a major groove and a minor groove that spiral along the surface of the helix. (See Figure 26.6.) In spite of the nonregularity of bases in the sequence, the major groove is lined with carbon 6, carbon 8, and nitrogen 7 of the purines and carbons 4–6 of the pyrimidines. The narrower minor groove is paved with carbon 2 and nitrogen 3 of the purine bases and carbon 2 of the pyrimidine bases.

B-DNA is a right-handed helix. The number of bases per turn of the helix is 10–10.4 and the pitch (rise) per turn of the helix is 34 Å.

A-DNA is formed from B-DNA by dehydration. The double helix is compressed, resulting in a greater number of bases per turn (about 10.7–11). The bases are tilted, and the sugar is in the *3'-endo* conformation. Like B-DNA, A-DNA is a right-handed helix but the major groove is wider and deeper and the minor groove is narrower and shallower than in B-DNA.

Z-DNA is a left-handed helix. However, it is not a left-handed version of either A-DNA or B-DNA. Z-DNA can be formed only by DNA containing alternating pyrimidine and purine bases. Z-DNA can be formed from synthetic polynucleotides or from segments of natural B-DNA containing the required base sequence. The formation of a Z-DNA segment occurs when base pairs in the chain are flipped over, as shown in Figure 26.10. Methylation of cytosine enhances the tendency of a segment of B-DNA to form a Z-DNA segment.

F. Base Sequences in DNA

Exons and introns   Not all bases in DNA are involved in the informational portion of the genetic code. For example, many of the genes in eukaryotic DNA (DNA from cells containing well-defined nuclei) are segmented. That is, the

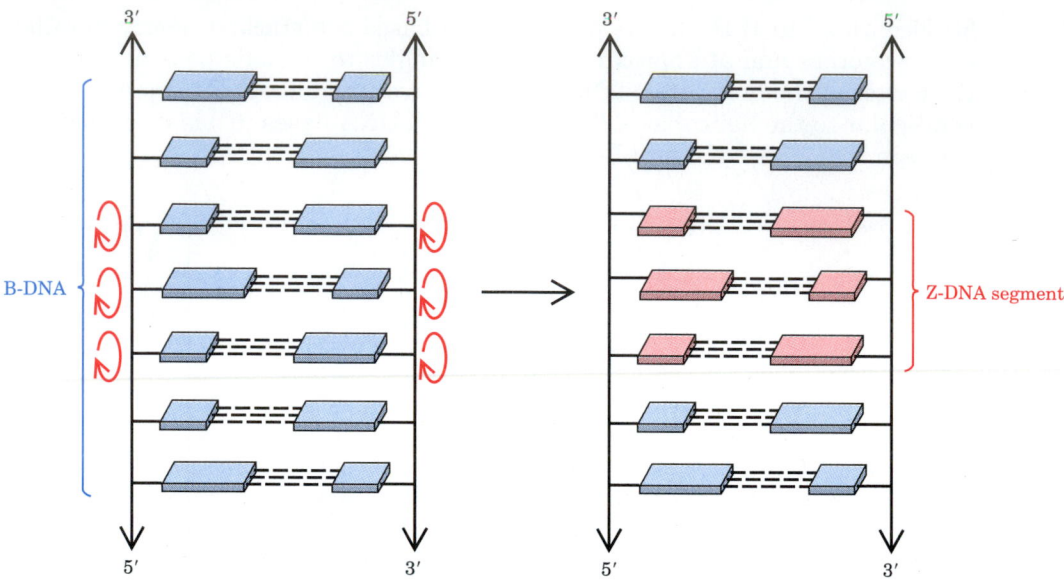

Figure 26.10   Diagram showing how bases can rotate in the conversion of a segment of B-DNA to Z-DNA.

gene has portions that contain information, called **exons,** and portions that do not contain genetic information, called **introns.** The function of the introns is not yet known.

Palindromes   Sequences of paired bases with the same order in both strands as we proceed from the 5′ direction to the 3′ end are called **palindromes,** a term describing a word or phrase that reads the same forward or backward (for example, radar).

A palindrome base sequence:

$$5' \longleftarrow \text{G—A—A—T—T—C} \longrightarrow 3'$$
$$3' \longleftarrow \text{C—T—T—A—A—G} \longrightarrow 5'$$

Restriction endonucleases   Often called **restriction enzymes,** restriction endonucleases catalyze the hydrolysis, or cutting, of DNA at certain locations. These enzymes, found in bacteria, cut only specific base sequences, commonly palindrome sequences. The function of restriction enzymes is the cleavage of foreign DNA that enters the bacterial cell. Associated with these enzymes are *modifying enzymes* that modify the bacteria's own DNA (by methylation of certain bases, for example) to prevent its cleavage by the restriction enzyme.

More than 500 restriction enzymes have been isolated. They are used to cleave DNA at specific locations in genetic engineering and in gene-mapping of chromosomes. These enzymes are also used in the fingerprinting of DNA, a technique used in criminology laboratories, for example, to establish the individual source of body fluids (blood, semen) or to determine parentage.

Cleavage of a palindromic sequence by a restriction enzyme obtained from a strain of E. coli:

cleave here

$$5' \longleftarrow \text{G}-\text{A}-\text{A}-\text{T}-\text{T}-\text{C} \longrightarrow 3'$$
$$3' \longleftarrow \text{C}-\text{T}-\text{T}-\text{A}-\text{A}-\text{G} \longrightarrow 5'$$

and cleave here

G. Species Variations in DNA

Because the base sequence of DNA determines the type of protein in an organism, the DNA of each species of plant or animal has a different base sequence. DNA molecules in different types of organisms also differ in size and shape. In general, the size of a DNA molecule is proportional to the amount of information it contains. Viruses have the smallest DNA molecules (about 5000 base pairs), and humans have the largest (about 2.9×10^9 base pairs). A DNA molecule of *E. coli* contains about 4.3×10^6 base pairs, which contain the code for 2000–3000 different proteins.

In eukaryotic cells, the long-chain DNA double-stranded helix is found in the cell nucleus, where it is complexed with proteins to form the chromosomes. The nucleus in each cell of an organism contains DNA of the same structure. Therefore, each cell contains all the genetic information required for the entire organism! Exceptions are cells that produce antibodies. The DNA molecules in these cells contain variable regions so that these cells can "custom-tailor" antibodies to fight specific antigens (invaders). Other exceptions are the germ cells produced by meiosis for the purpose of sexual reproduction.

In prokaryotic cells (cells that lack true nuclei, such as those of bacteria and blue-green algae), DNA forms circular, compactly folded molecules in the cytoplasm (cellular fluid). Some bacteria contain small circular double-stranded DNA molecules called **plasmids** in addition to their main DNA molecules. The genetic information in plasmids is independent of that in the principal bacterial DNA and is instrumental in the development of bacterial resistance to antibiotics. Because of their small size and relative lack of complexity, plasmids are used in genetic research and engineering.

Viruses are small particles composed of either DNA or RNA (not both) protected by a protein coating. Viruses can be simple in structure because they use cells of higher life forms, called *host cells,* to reproduce viral nucleic acids and proteins.

Section 26.3

Structure and Properties of RNA

RNA and DNA are similar in primary structure, but differ in a few key features. (1) The sugar in RNA is ribose while in DNA it is deoxyribose. (2) RNA contains uracil in place of thymine. (3) RNA molecules are single stranded (while DNA molecules are double stranded) and are in the A-form (Section 26.2E). Finally, (4) RNA molecules are generally smaller than DNA. For example, human DNA contains about 46 million base pairs, and the largest RNA contains only about 50,000.

The three major classes of RNA found in cells are:

1. **Messenger RNA (mRNA):** As its name implies, mRNA carries genetic information from DNA to the site of protein synthesis, a portion of the cell containing the ribosomes.

2. **Ribosomal RNA (rRNA):** Together with the ribosomal proteins, rRNA serves as the structural material for the ribosome.

3. **Transfer RNA (tRNA):** Molecules of tRNA carry the amino acids to the ribosome for incorporation into a growing polypeptide chain.

In addition to these three classes, a number of other types of RNA exist. Within the cell, for example, small RNA molecules serve specialized functions, such as mRNA processing. Also, the genetic codes for a number of viruses are contained in RNA instead of DNA. For brevity, we will limit our discussion to the principal RNA classes.

A. General Structure of RNA

The primary structure of RNA is similar to that of DNA—a series of sugar units (*ribose,* in this case) linked together by $5' \rightarrow 3'$ phosphodiester bonds. Each ribose residue is bonded to a base at the anomeric carbon. The principal RNA bases are *adenine, guanine, cytosine,* and *uracil* (replacing thymine).

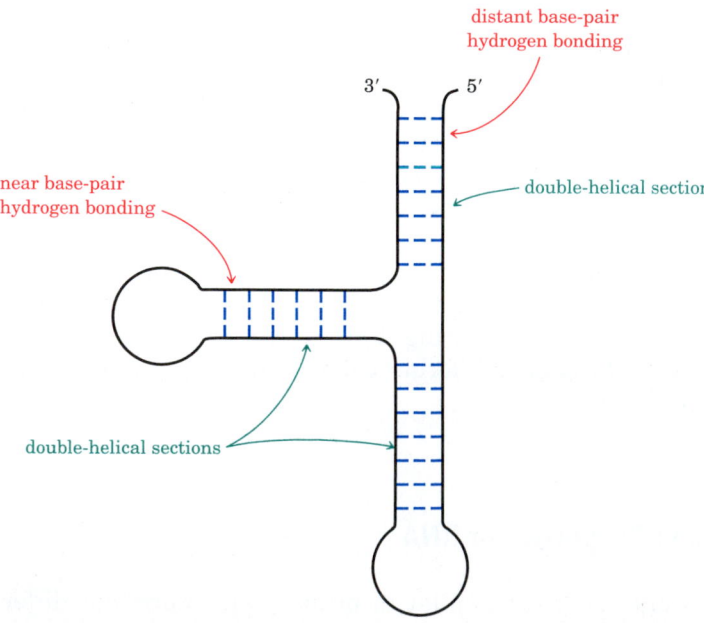

Figure 26.11 Typical structural features of RNA.

Uracil, like thymine, forms strong hydrogen bonds with adenine. These structural features were discussed in Section 26.1.

Although an RNA molecule is single stranded, it is neither linear nor a random coil. Its secondary structure is a helix with base pairing between distant bases of the same strand. In addition, base pairing between near bases of the same strand gives hairpin-loop structures (Figure 26.11). The tertiary structure of RNA involves folding and base stacking.

Although the major classes of RNA are single stranded, double-stranded RNA molecules do occur in certain viruses. In these double-stranded molecules, the 2′-hydroxyl group of the ribose forces the chain into a conformation similar to that of the A-form of DNA.

B. Messenger RNA (mRNA)

Messenger RNA is only a small percentage of the total RNA in the cell. The reason is, in part, that mRNA molecules have a rather short lifetime of 1–3 minutes in prokaryotic cells and 2–24 hours in eukaryotic cells.

Cells synthesize various types of protein from the genetic code in DNA. Protein synthesis is accomplished in two stages. First, a portion of the DNA unwinds. In eukaryotic (nucleated) cells, the sequence of bases in one **gene,** the part of the DNA that codes for one type of protein, is used as a template for the synthesis of messenger RNA (mRNA). Thus, the mRNA is the complement of the portion of the DNA strand that synthesized it. This synthesis of mRNA is called **transcription,** a term referring to the transcription of the genetic code from DNA to RNA.

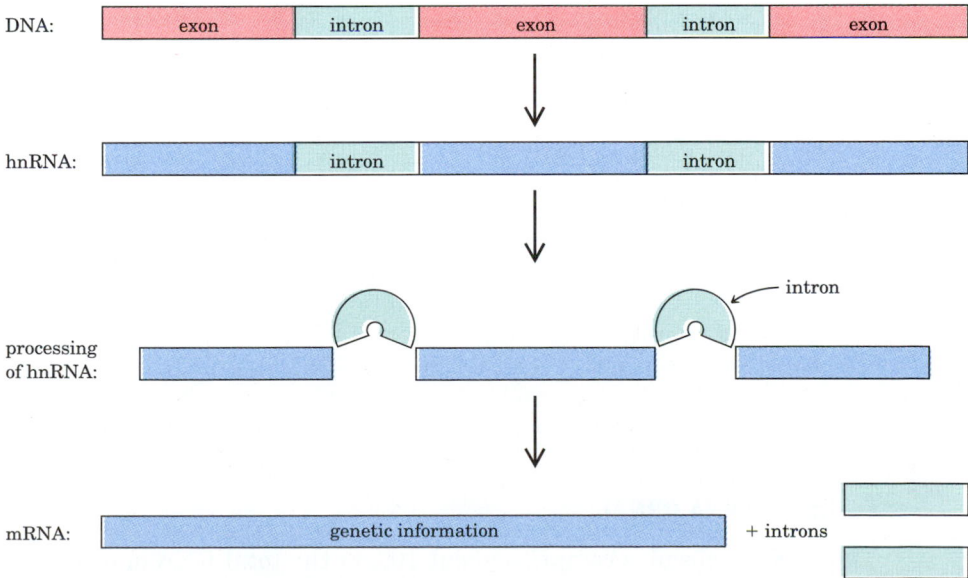

Figure 26.12 Diagram of the conversion of hnRNA to mRNA by the enzymatic removal of introns and the splicing of exons.

In mRNA, a particular series of three bases in a row, called a **codon,** specifies a particular amino acid. For example, C—C—U specifies proline. We will discuss codons in greater detail in Section 26.4A.

Because each type of RNA molecule carries information for one gene from DNA to the ribosomes, each mRNA contains a different base sequence. Consequently, mRNA is a heterogeneous group of compounds that is difficult to study.

In eukaryotic cells, the DNA base sequence for a single gene, consisting of both introns and exons (Section 26.2F), is first transcribed from one portion of one DNA strand onto a nascent mRNA, called **hnRNA** (heterogeneous nuclear RNA), which is then modified, or processed, in the nucleus to mRNA. The first steps in processing involve removing the introns and splicing the exons together (Figure 26.12).

Before the mRNA leaves the nucleus, the 5′ end of its chain is capped by the substitution of a 7-methylguanylyl group and the methylation of the 2′-hydroxyl groups of the first two ribose residues. The 3′ end of the RNA chain is tailed by the addition of up to 200 adenylyl residues.

Capping of the 5′ terminus:

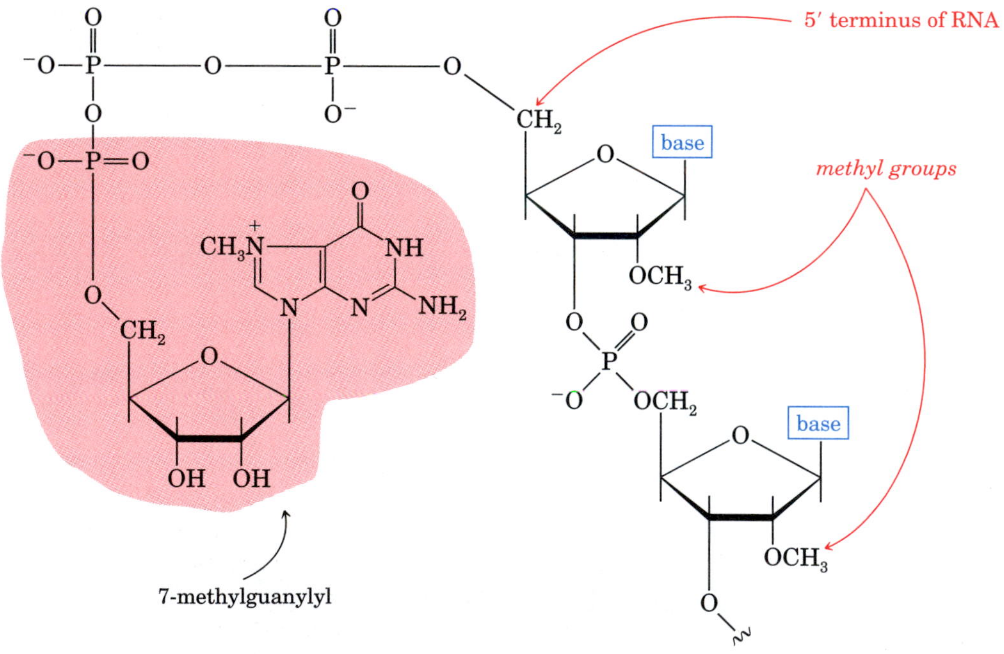

C. Transfer RNA (tRNA)

The tRNA molecules comprise about 10% of the total RNA of a cell. There are 50–60 different types of tRNA in a cell, with at least one tRNA for each amino acid codon (three-base sequence in mRNA). For convenience, the different types of tRNA are designated by the amino acid they carry (tRNA^{ala} is tRNA that carries alanine).

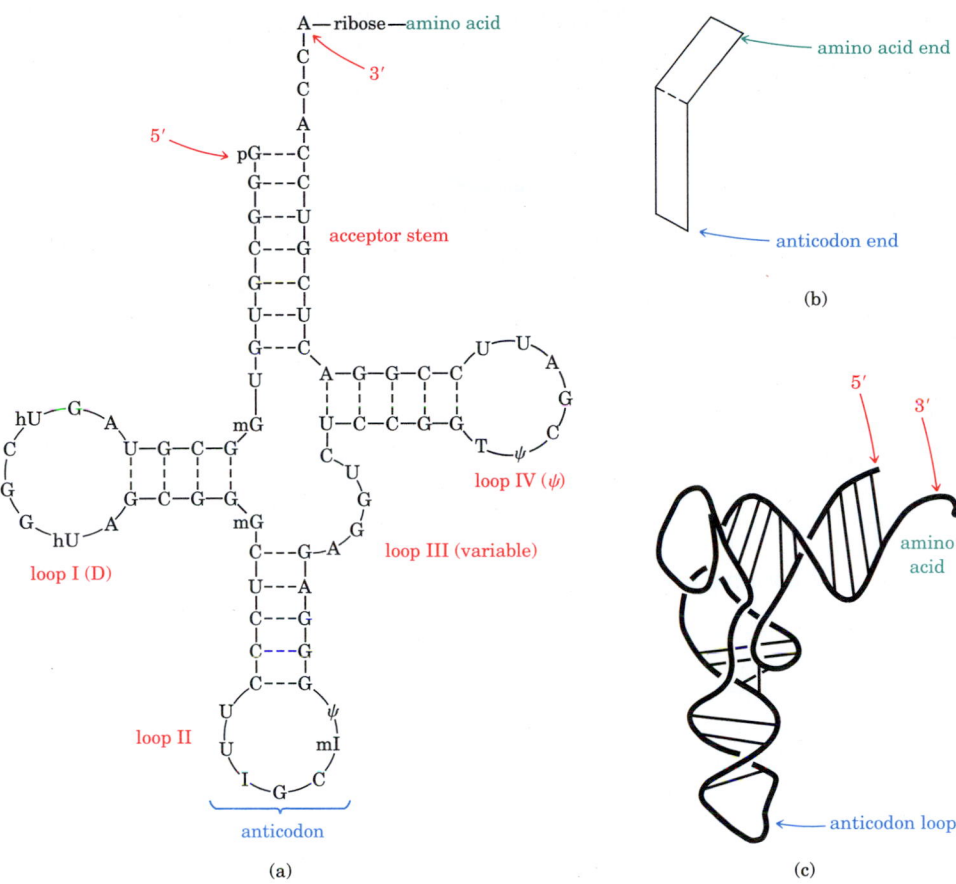

Figure 26.13 Representations of a typical tRNA molecule: (a) diagram showing cloverleaf structure; (b) diagram emphasizing the folded L shape; (c) three-dimensional representation.

All types of tRNA have similar chemical and physical properties. Each tRNA consists of a single RNA strand folded into a four-loop cloverleaf structure. The loops are labeled I through IV. Loop I is called the *dihydro-U loop* or D loop because it contains dihydrouridine (see Table 26.1). Loop II is called the *anticodon loop* because it contains a series of three bases (the **anticodon**) complementary to the codon in mRNA for its amino acid. Loop III is called the *variable loop* and contains 3–21 bases depending upon the tRNA. Loop IV is called the *pseudouridine* (Ψ) *loop* because it contains a pseudouridine base (Table 26.1).

In its tertiary structure, the tRNA molecule folds back upon itself to form an L-shaped structure with the amino acid at one end of the L and the anticodon at the other end of the L. This folding allows the bases at the 5′ and 3′ ends of the molecule to pair and gives rise to a region called the **acceptor stem** (see Figure 26.13). The last three bases at the 3′ end of all tRNA molecules are C—C—A; the amino acid is bonded to the 3′-hydroxyl group of the terminal ribose.

Bonding of the amino acid to the terminal adenosine nucleoside at the 3' end of the chain:

OCH$_2$ — O — adenine

R — CHCO — OH

NH$_2$ — O

amino acid ester at 3' hydroxyl of ribose

26.6 Draw a portion of a tRNA structure showing the bonding between D-ribofuranose and proline.

D. Ribosomal RNA (rRNA) and the Ribosomes

The most abundant ribonucleoproteins (ribonucleic acid–protein complexes) in the cytoplasm are the ribosomes. They are composed of ribosomal RNA (rRNA) and numerous proteins.

Ribosomes are isolated by the centrifugation of ruptured cells. They are identified by **Svedberg (S) units,** with 1S equal to 10^{-13} second, a unit used in determining the rate of sedimentation of compounds in an ultracentrifuge, a high-speed analytical centrifuge.

Prokaryotic cells contain 70S ribosomes, while eukaryotic cells contain larger 80S ribosomes. The 70S ribosome of *E. coli* has been extensively studied. It can be dissociated into two subunits, the 50S and 30S subunits. (Because they represent a physical property, not a structural feature, the S units are not additive.) The 50S subunit contains 34 proteins and two RNA molecules, 23S and 5S RNA. The 23S RNA contains 2904 nucleotides, and the 5S RNA contains 120 nucleotides. The 30S subunit has 21 proteins and one RNA, 16S RNA, which contains 1542 nucleotides. The 70S ribosome thus contains a total of 3 RNA molecules and 55 protein residues (Figure 26.14).

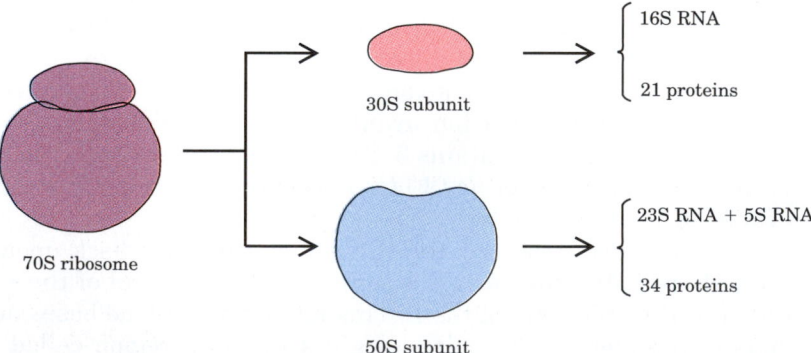

Figure 26.14 Dissociation of the prokaryotic 70S ribosome of *E. coli* into its components.

Section 26.4

The Role of RNA in Protein Biosynthesis

After transcription and processing, the mRNA is used as a template for the synthesis of the protein. This process, called **translation,** refers to the translation of the nucleotide code into a sequence of amino acids in a peptide molecule. Translation encompasses the integrated actions of mRNA, many types of transfer RNA (which deliver appropriate amino acids to mRNA), ribosomes (particles that provide the sites of the synthesis), and many enzymes.

replication

DNA $\xrightarrow{\text{transcription}}$ mRNA $\xrightarrow{\text{translation}}$ peptide or protein

tRNA

nucleotides   amino acids

A. Reading the Codons

When mRNA leaves the nucleus of a eukaryotic cell, ribosomes, which are also partly formed in the nucleus, become attached to the 5′ end of its chain. The ribosomes are the actual sites of protein synthesis.

Once the mRNA molecule is bonded to ribosomes, the bases are read *three at a time* as the ribosomes progress along the chain toward the 3′ end. Each set of three bases, a *codon,* signals the incorporation of a single amino acid into the growing peptide chain. The codons for the amino acids, which are the same in all known life forms, have been determined experimentally. They are listed in Table 26.3.

Bases are read three at a time;
each codon (set of three) designates one amino acid.

a segment of mRNA

5′ ⟵ ⌇—C—C—U—C—U—C—G—C—U—⌇⟶ 3′

codon for   *codon for*   *codon for*
proline   *leucine*   *alanine*

In Table 26.3, note that more than one codon can signal a particular amino acid to be incorporated into a protein. In addition, some codons serve special functions. For example, the codon AUG serves two functions: (1) as an initiator codon signaling for the start of synthesis of a peptide, and (2) as a codon for the incorporation of methionine into the growing peptide chain. Other special-purpose codons are UAA, UAG, and UGA, all of which signal STOP. When the ribosomal synthesis site encounters one of these stop codons, the peptide chain is released from the ribosome and assumes its secondary and tertiary structures.

Table 26.3 Codons for the amino acids	
Amino Acid	Codons[a]
phenylalanine (phe)	UUU, UUC
serine (ser)	UCU, UCC, UCA, UCG, AGU, AGC
tyrosine (tyr)	UAU, UAC
cysteine (cys)	UGU, UGC
tryptophan (trp)	UGG
leucine (leu)	CUU, CUC, CUA, CUG, UUA, UUG
proline (pro)	CCU, CCC, CCA, CCG
histidine (his)	CAU, CAC
glutamine (gln)	CAA, CAG
arginine (arg)	CGU, CGC, CGA, CGG, AGA, AGG
lysine (lys)	AAA, AAG
asparagine (asn)	AAU, AAC
isoleucine (ile)	AUU, AUC, AUA
methionine (met) or *N*-formylmethionine (fmet)	AUG
threonine (thr)	ACU, ACC, ACA, ACG
valine (val)	GUU, GUC, GUA, GUG
alanine (ala)	GCU, GCC, GCA, GCG
aspartic acid (asp)	GAU, GAC
glutamic acid (glu)	GAA, GAG
glycine (gly)	GGU, GGC, GGA, GGG

[a] When preceded by an initiator region, the codon AUG signals "Start a new peptide molecule beginning with *N*-formylmethionine, or fmet." The codons UAA, UAG, and UGA signal termination of the synthesis.

B. Anticodon of tRNA

The amino acids are carried to the mRNA–ribosome by tRNA molecules. One key feature of the tRNA structure is its anticodon, a series of three bases that is complementary to a particular codon. For example, if mRNA contains the codon GCC (for the amino acid alanine), the tRNA that carries alanine would contain the anticodon CGG. The anticodon allows a tRNA to recognize the correct location on the mRNA molecule.

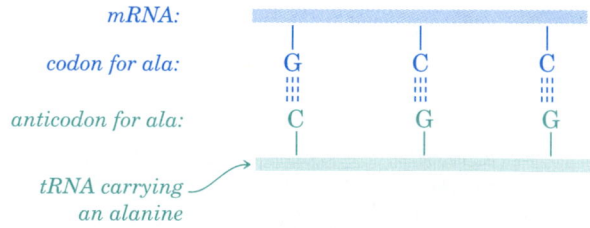

STUDY PROBLEM

26.7 Referring to Table 26.3, draw the structure of the peptide that would be formed if the following base sequences were found in mRNA:

(a) A C U U U U A A G G A U C U U G G U → 3′ end

(b) G A A U G C A G G G G G G C G A G G A G G G → 3′ end

C. Steps in Protein Biosynthesis

Figure 26.15 diagrams the overall process of protein biosynthesis. The three principal steps in this process are *initiation, chain elongation,* and *termination.*

Initiation In the presence of certain protein initiator factors, mRNA becomes bound to a ribosome by a leader sequence near the 5′ end of the mRNA. In bacterial cells, the START codon (AUG) after the leader sequence calls for *N*-formylmethionine (fmet) as the first amino acid of the peptide to be synthesized. Depending on the protein's structure, the formyl group or even the methionine residue may be hydrolyzed from the final protein. However, most proteins in a bacterium like *E. coli* have methionine as their *N*-terminal amino acid.

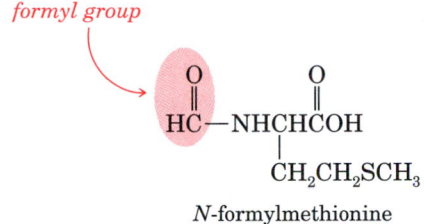

formyl group

N-formylmethionine

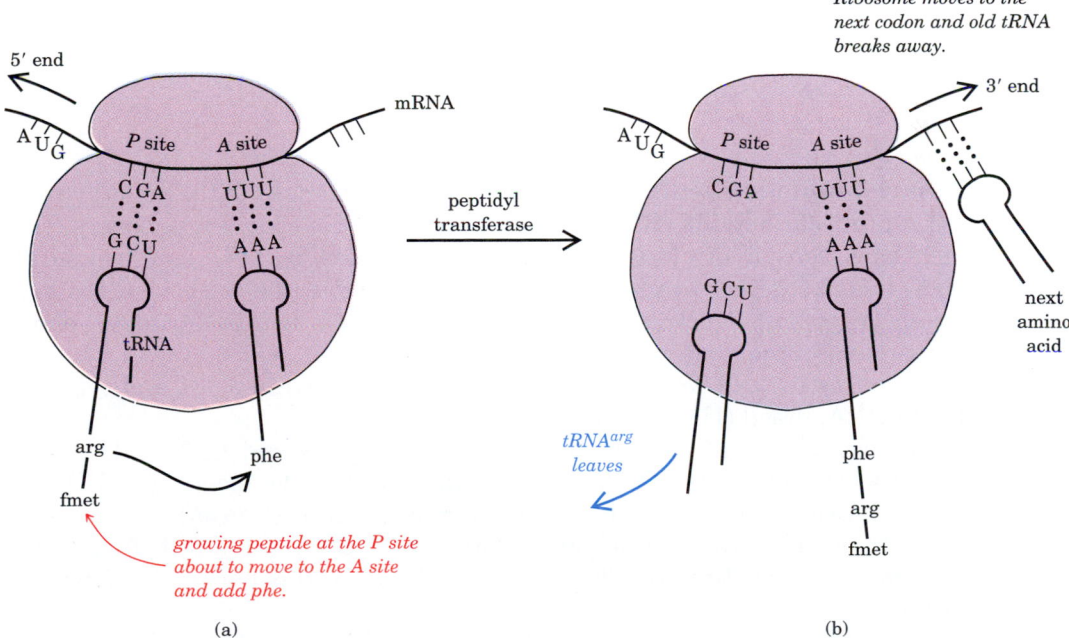

Figure 26.15 Biosynthesis of a peptide. (a) *fMet,* or *N*-formylmethionine (the initiating amino acid, with the codon AUG), has become bonded to *arg* (codon CGA). The next amino acid to be added will be *phe* (codon UUU). (b) The growing peptide chain has been transferred to phe. The ribosome moves along the mRNA chain (to the right) so that the peptide chain can be transferred to the next amino acid.

The codon AUG for "start with *N*-formylmethionine" is the same as the codon for the incorporation of methionine itself in the interior of the peptide chain. However, the tRNA that brings the *N*-formylmethionine to start a peptide chain differs from the tRNA that brings methionine itself. Initiation factors on the mRNA chain also help differentiate *N*-formylmethionine from methionine.

Chain elongation (1) The tRNA holding the second amino acid becomes hydrogen bonded to the mRNA codon and positions its amino acid at the **amino acid binding site (acyl site),** or **A site,** on the ribosome. (2) A peptide bond is formed between the *N*-formylmethionine and the incoming amino acid. (3) The dipeptide is translocated to the **peptide binding site (peptidyl site),** or **P site,** of the ribosome. (4) The ribosome moves along the mRNA chain so that the next tRNA with its amino acid can move in to the A site, and the process is repeated for the next amino acid.

Termination The peptide is complete when the ribosome encounters one of the STOP codons. These codons terminate the sequence because no tRNA molecules correspond to them. Protein-release factors catalyze the hydrolysis of the peptide so that it and the ribosomes are released from each other and from the mRNA.

Although one mRNA directs the synthesis of a specific protein, it can direct the synthesis of many molecules of this protein. As one ribosome, holding a growing peptide, moves along the mRNA chain, other ribosomes become successively attached to the 5′ end of the mRNA to follow along behind the first tRNA. Thus, several peptide molecules can be synthesized, one after the other, in assembly-line fashion.

Summary

Deoxyribonucleic acids (DNA), which are the carriers of the genetic code, and **ribonucleic acids (RNA),** which implement the code, are polymers. The backbones consist of sugar molecules (deoxyribose in DNA and ribose in RNA) linked together by phosphate units. Each sugar unit is also bonded to a heterocyclic base. Nucleic acids can be hydrolyzed to **nucleotides** (phosphate–sugar–base) and to **nucleosides** (sugar–base).

The bases in DNA are adenine (A), guanine (G), thymine (T), and cytosine (C). In RNA, uracil (U) replaces thymine.

Hydrogen bonding between specific base pairs (G and C, T and A) and stacking energy hold DNA in an antiparallel double-stranded helix (Watson–Crick model). DNA can undergo **replication** because the base pairs are specific. The three principal conformers of DNA are the **B-form,** the **A-form,** and the **Z-form.** The B-form of DNA is the naturally occurring form.

The portion of a DNA strand specific for the coding of the amino acid sequence of one protein is called a **gene.** This portion of DNA synthesizes a complementary **mRNA (transcription),** which leaves the nucleus to direct the synthesis of that protein molecule **(translation)** with the aid of **ribosomes.**

The bases of mRNA are read three at a time. Each three-base sequence (a **codon**) designates a particular amino acid to be incorporated into a growing peptide chain. The **tRNA** molecules carry amino acids to the site of protein synthesis and recognize their proper locations because of **anticodons.**

Essay Problem for Chapter 26

Synthetic Antiviral Agents

Though viral infections are estimated to account for 60% of illnesses in industrialized countries, only a few antiviral agents have been discovered. The lack of antiviral agents is directly related to the simplicity of viruses. While viruses carry all the information needed for their own reproduction, they differ from all other pathogens in that they use the reproductive machinery of their host cells for replication.

One group of antiviral agents is guanosine analogues. Typical of this group is Acyclovir [9-(2-hydroxyethoxymethyl)guanine] **(1)**, synthesized by Howard J. Schaeffer of Wellcome Research in 1976. (H. J. Schaeffer, L. Beauchamp, P. deMiranda, and G. B. Elion, *Nature,* **1978,** *272(13),* 583.)

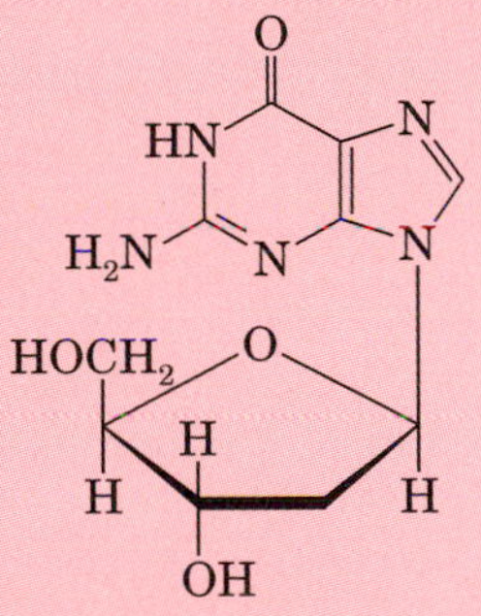

deoxyguanosine

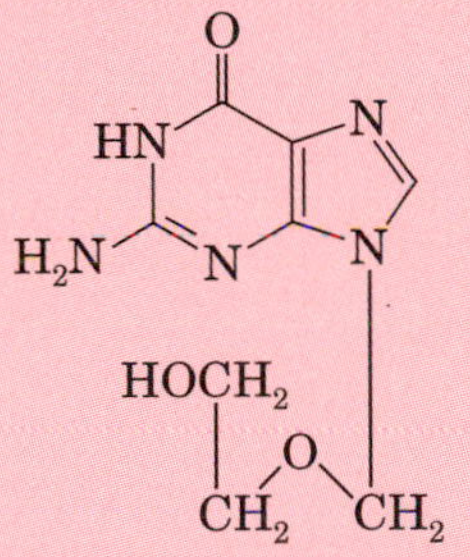

Acyclovir **(1)**

[9-(2-hydroxyethoxymethyl)guanine]

Acyclovir was synthesized as follows.

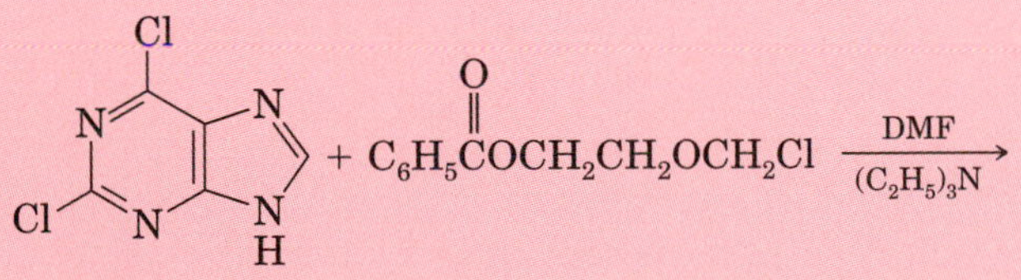

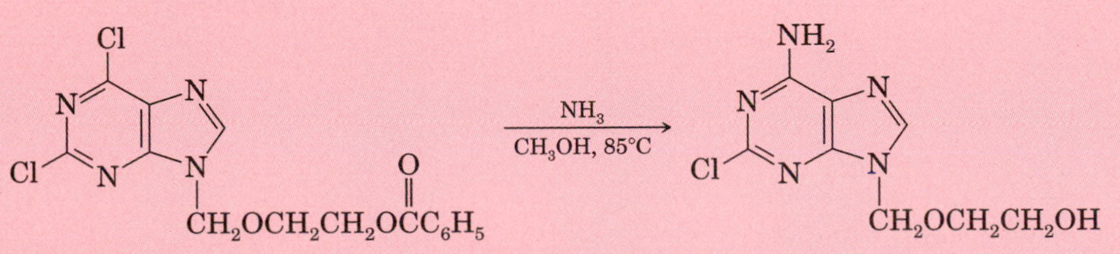

Acyclovir is converted to its monophosphate in herpes-infected cells. Because the conversion is catalyzed by a herpes virus, only a limited amount of the monophosphate is formed in herpes-free cells. The monophosphate is converted to the triphosphate by the cell's enzymes. The triphosphate inhibits the herpes DNA polymerase (but not the cellular polymerase) because Acyclovir triphosphate lacks a 3′-hydroxyl group, which is necessary for elongation of the growing viral nucleic acid.

Questions

1. What reaction mechanism is involved in conversion of 2,6-dichloropurine (2) to compound 3?
2. Why is the benzoate blocking group needed in the synthesis? How is it removed?
3. What are the intermediates when compound 4 is converted to compound 5 by nitrous acid?
4. DHPG [9-(1,3-dihydroxy-2-propoxymethyl)guanine] (6) is more soluble in water than Acyclovir and, consequently, it is more active against herpes infections. Why is DHPG more soluble than Acyclovir?

DHPG (6)

[9-(1,3-dihydroxy-2-propoxymethyl)guanine]

5. Suggest a synthesis for DHPG.

26.8 Draw a structure or structures that illustrate each of the following terms:

    (a)   a pyrimidine base found in RNA   **(b)**   a purine base found in DNA

    (c)   the backbone of a DNA polymer   **(d)**   a nucleotide

    (e)   a nucleoside

26.9 Draw formulas for **(a)** 2-deoxy-β-D-ribose 3-phosphate and **(b)** 2-deoxy-β-D-ribose 5-phosphate.

26.10 Draw the formula for a tautomer of uracil that shows its close relationship to pyrimidine.

26.11 Using uracil, β-D-ribose, and a phosphate group, draw formulas for **(a)** a nucleoside and **(b)** a nucleotide.

26.12 Write equations for the complete acid hydrolysis of 2′-deoxycytidine 5′-phosphate.

26.13 Adenosine 5′-triphosphate (ATP) is important in many biological schemes. What is its structure?

26.14 Why are guanine and cytosine paired in the DNA helix, but not adenine and cytosine? Use formulas to show the hydrogen bonding.

26.15 Write formulas showing the hydrogen bonds between nucleotides containing the following pairs:

    (a)   uracil and adenine

    (b)   uracil and guanine

    (c)   uracil and cytosine

26.16 In the preceding problem, state which pairs of bases are paired in nucleic acids. Explain why.

26.17 Why does the melting point (denaturation temperature) of DNA increase with increasing guanosine–cytosine content?

26.18 List the following compounds in order of increasing molecular weight (lowest first).

    (a)   mRNA   **(b)**   tRNA   **(c)**   DNA

26.19 Define the following terms:

    (a)   replication   **(b)**   transcription   **(c)**   translation

26.20 **(a)**   A portion of a DNA molecule has the following base sequence. What is the base sequence of the complementary strand?

$$5'\text{ end} \leftarrow \text{A T T C G G T A T} \rightarrow 3'\text{ end}$$

    (b)   What is the base sequence in an RNA molecule synthesized by the DNA shown above?

* For information concerning the organization of the *Study Problems* and *Additional Problems,* see the *Note to student* on page 41.

26.21 Three different codons of mRNA result in the incorporation of isoleucine in a protein: AUU, AUC, and AUA.

(a) What sequences of bases in DNA give rise to these codons?

(b) What are the corresponding anticodons in tRNA?

Additional Problems

26.22 (a) *Theophylline,* which occurs in tea and is also synthesized industrially, is a cardiac stimulant, diuretic (urine-secretion stimulant), and muscle relaxant. The structure of theophylline is a purine ring system with carbonyl groups at positions 2 and 6 and methyl groups at nitrogens 1 and 3. The protonated nitrogen is at position 7. What is the structure of theophylline?

(b) *Caffeine* has the same structure as theophylline with an additional *N*-methyl group at position 7. What is its structure?

26.23 What tripeptides would be formed by the sequence of codons in (a) and by the mutated sequence in (b)?

(a) mRNA: $5' \leftarrow$ UUU—UAU—AGU $\rightarrow 3'$

(b) mRNA: $5' \leftarrow$ UUU—UAC—AGU $\rightarrow 3'$

26.24 In sickle-cell anemia, valine replaces a glutamic acid unit of normal hemoglobin. Referring to Table 26.3, postulate the change occurring in the genetic code for hemoglobin that causes sickle-cell anemia.

26.25 Write letters for all possible base sequences in mRNA that would produce the dipeptide glu–arg.

26.26 Bradykinin, a peptide that lowers blood pressure, increases capillary permeability, and causes pain, has the following sequence of amino acids:

arg-pro-pro-gly-phe-ser-pro-phe-arg

What is the codon sequence in mRNA that is needed to produce this peptide?

26.27 An iodo group occupies about the same amount of space as does a methyl group. 5-Iodouridine is an antiviral drug that is incorporated into viral nucleic acid. To which nucleoside does this drug correspond?

APPENDIX
NOMENCLATURE OF
ORGANIC COMPOUNDS

A complete discussion of definitive rules of organic nomenclature would require more space than can be allotted in this text. We will survey some of the more common nomenclature rules, both IUPAC and trivial.* The IUPAC nomenclature rules comprise a number of nomenclature styles. Of these various styles, the substitutive (substituents modifying a parent hydrocarbon) and radicofunctional (functional-group class designation) are by far more common. The IUPAC rules are more permissive among the various nomenclature styles than many authors seem to appreciate. For example, isopropyl alcohol (radicofunctional) and 2-propanol (substitutive) are both IUPAC-approved names for the same compound. The substitutive nomenclature style can be applied to a greater variety of compounds than any of the other IUPAC-approved styles, and it is the style to which most chemists are referring when they use the terms "IUPAC nomenclature" and "the IUPAC name." The following references contain more detail:

IUPAC Nomenclature of Organic Chemistry, Sections A, B, C, D, E, F, and H. Pergamon Press, Oxford, 1979.

A Guide to IUPAC Nomenclature of Organic Compounds. Recommendations 1993. Prepared for publication by R. Panico, W. H. Powell, K.-C. Richter. Blackwell Scientific Publications, London, 1993.

Chemical Abstracts Service, *Naming and Indexing of Chemical Substances for Chemical Abstracts during the Ninth Collective Period (1972–1976) (January–June, 1972),* American Chemical Society, Columbus, Ohio, 1973.

Chemical Abstracts Service, *Combined Introductions to the Indexes to Volume 66 (January–June, 1967),* American Chemical Society, Columbus, Ohio, 1968.

A. M. Patterson, L. T. Capell, and D. F. Walker, *The Ring Index, 2nd Ed.,* American Chemical Society, Washington, D.C., 1960; *Supplement I (1957–1959),* 1963; *Supplement II (1960–1961),* 1964; *Supplement III (1962–1963),* 1965.

R. S. Cahn and O. C. Dermer, *Introduction to Chemical Nomenclature, 5th Ed.,* Butterworths, London, 1979.

---

* *Chemical Abstracts* has its own nomenclature system, which we will not cover here. To find a *Chemical Abstracts* index name, refer to one of the references listed or consult a recent *Index Guide.*

Alkanes

The names for the first thirty continuous-chain alkanes are listed in Table A1.

Branched alkanes In the name of an alkane with alkyl substituents, the longest continuous chain is considered the parent. The parent is numbered from one end to the other, the direction being chosen to give the lower numbers, *locants,* to the substituents. (When considering complex cases in which a question about numbering arises, the lower individual numbers up to the first point of difference are the correct locants. The locants with the lower sum may *not* be the correct ones. Therefore, the locants 1,7,7- would be preferred to 2,2,8-.) The entire name of the structure is composed of (1) the locants for the positions of the substituents; (2) the names of the substituents; and (3) the name of the parent.

Alkyl substituents The names of the alkyl substituents (also called *branches,* or *radicals*) are derived from the names of their corresponding alkanes with the ending changed from *-ane* to *-yl*. For example, CH_3CH_2— is ethyl (from ethane). Multiple substituents are placed in alphabetical order,

$$CH_2CH_3$$
$$|$$
$$CH_3CH_2CHCH_2CH_2CH_3$$
① ② ③ ④ ⑤ ⑥

3-ethylhexane

$$CH_3 \quad CH_2CH_3$$
$$| \qquad |$$
$$CH_3CHCH_2CHCH_2CHCH_2CH_3$$
① ② ③ ④ ⑤ ⑥ ⑦ ⑧
$$|$$
$$CH_3$$

4-ethyl-2,6-dimethyloctane

each preceded by its locant and like substituents grouped together. Some common branched alkyl substituents have trivial names (see Table A2).

Alkenes and Alkynes

Unbranched hydrocarbons having one double bond are named in the IUPAC system by replacing the ending *-ane* of the alkane name with *-ene*. With two

Table A1 Names of some continuous-chain alkanes

Molecular Formula	Name	Molecular Formula	Name
CH_4	methane	$C_{16}H_{34}$	hexadecane
C_2H_6	ethane	$C_{17}H_{36}$	heptadecane
C_3H_8	propane	$C_{18}H_{38}$	octadecane
C_4H_{10}	butane	$C_{19}H_{40}$	nonadecane
C_5H_{12}	pentane	$C_{20}H_{42}$	icosane
C_6H_{14}	hexane	$C_{21}H_{44}$	henicosane
C_7H_{16}	heptane	$C_{22}H_{46}$	docosane
C_8H_{18}	octane	$C_{23}H_{48}$	tricosane
C_9H_{20}	nonane	$C_{24}H_{50}$	tetracosane
$C_{10}H_{22}$	decane	$C_{25}H_{52}$	pentacosane
$C_{11}H_{24}$	undecane	$C_{26}H_{54}$	hexacosane
$C_{12}H_{26}$	dodecane	$C_{27}H_{56}$	heptacosane
$C_{13}H_{28}$	tridecane	$C_{28}H_{58}$	octacosane
$C_{14}H_{30}$	tetradecane	$C_{29}H_{60}$	nonacosane
$C_{15}H_{32}$	pentadecane	$C_{30}H_{62}$	triacontane

Table A2	Trivial names for some common alkyl groups

Structure	Name
$CH_3CH_2CH_2—$	propyl
$(CH_3)_2CH—$	isopropyl
$CH_3CH_2CH_2CH_2—$	butyl
$(CH_3)_2CHCH_2—$	isobutyl
$CH_3CH_2CH(CH_3)—$	*secondary-,* or *sec*-butyl
$(CH_3)_3C—$	*tertiary-, tert,* or *t*-butyl
$CH_3CH_2CH_2CH_2CH_2—$	pentyl (or amyl)
$(CH_3)_2CHCH_2CH_2—$	isopentyl (or isoamyl)
$(CH_3)_3CCH_2—$	neopentyl

or more double bonds, the ending is *-adiene, -atriene,* etc. The chain is numbered to give the lowest locants to the double bonds. (The *lower* number of the two carbons joined by the double bond is used to give the position.)

$$\overset{①}{C}H_3\overset{②}{C}H=\overset{③}{C}H\overset{④}{C}H_2\overset{⑤}{C}H_3 \qquad \overset{①}{C}H_2=\overset{②}{C}H\overset{③}{C}H=\overset{④}{C}H\overset{⑤}{C}H_3$$

2-pentene*   1,3-pentadiene*

Unbranched hydrocarbons having one triple bond are named by replacing the ending *-ane* of the alkane name with *-yne.* If there are two or more triple bonds, the ending is *-adiyne, -atriyne,* etc. The chain is numbered to give the lowest locants to the triple bonds. Again, the lower number is used to give the position. For example, $CH_3CH_2C{\equiv}CH$ is 1-butyne.

Unbranched hydrocarbons with both double and triple bonds are named by replacing the ending *-ane* of the alkane with the ending *-enyne.* When necessary, position numbers are used. The position numbers are chosen to be as low as possible. For example, $CH{\equiv}CCH=CHCH_3$ is 3-penten-1-yne, not 2-penten-4-yne. If there is a choice, the double bond receives the lower number. For example, $CH{\equiv}CCH=CH_2$ is 1-buten-3-yne, not 3-buten-1-yne. Trivial names for some alkenes and alkynes are listed in Table A3.

Table A3	Trivial names for some alkenes and alkynes

Structure	Name
$CH_2=CH_2$	ethylene
$CH{\equiv}CH$	acetylene
$CH_2=C=CH_2$	allene
$CH_2=CHCH_3$	propylene
$CH_3C{\equiv}CH$	methylacetylene
$(CH_3)_2C=CH_2$	isobutylene
$CH_2=C(CH_3)CH=CH_2$	isoprene

* The 1993 IUPAC recommendations on organic nomenclature specify an alternative positioning of locants in a name: locants may be placed immediately before the parts of the name to which they apply. The alternatives for these two names are: pent-2-ene and penta-1,3-diene.

Branched alkenes and alkynes In the IUPAC name of a branched alkene or alkyne, the parent chain is the longest chain that contains the maximum number of double or triple bonds. (This may or may not be the longest continuous chain in the structure.)

$$
\begin{array}{c}
\text{CH}_3\text{CH}_2\text{CH}_2 \\
| \qquad \textcircled{4} \qquad \textcircled{6}\ \textcircled{7} \\
\text{CH}_2\!\!=\!\!\text{CHC}\!\!=\!\!\text{CCH}\!\!=\!\!\text{CHCH}_3 \\
\textcircled{1} \quad \textcircled{2}\ \textcircled{3}\ |\textcircled{5} \\
\text{CH}_2\text{CH}_2\text{CH}_3
\end{array}
$$

3,4-dipropyl-1,3,5-heptatriene

Several unsaturated substituents have trivial names; some of these are listed in Table A4, along with their systematic names.

In complex compounds, the symbol Δ is sometimes used to denote double bonds.

is $\Delta^{4,6}$-cholestadiene

Geometric isomers There are two methods for naming geometric isomers. In one, the prefixes *cis-* (same side) and *trans-* (opposite sides) are used.

cis-2-butene   *trans*-2-butene

The other method uses (*E*) (groups with higher priority, according to the Cahn–Ingold–Prelog system, on opposite sides) or (*Z*) (groups with higher priority on the same side) for designation of geometric isomers. The priority rules are listed in Section 4.1B (page 124).

Table A4 IUPAC names for some unsaturated groups	
Structure	Name
$\text{CH}_2\!\!=$	methylene
$\text{CH}_2\!\!=\!\!\text{CH}\!-$	ethenyl (vinyl)[a]
$\text{CH}_3\text{CH}\!\!=$	ethylidene
$\text{CH}\!\equiv\!\text{C}\!-$	ethynyl
$\text{CH}_2\!\!=\!\!\text{CHCH}_2\!-$	2-propenyl (allyl)[a]
$\text{CH}_3\text{CH}\!\!=\!\!\text{CH}\!-$	1-propenyl
$(\text{CH}_3)_2\text{C}\!\!=$	isopropylidene
	2-cyclopenten-1-yl

[a] Names in parentheses are trivial.

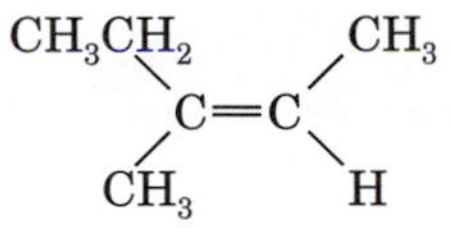
CH3CH2
CH3
C=C
CH3
H

(*Z*)-3-methyl-2-pentene

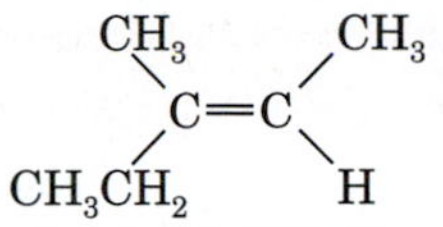
CH3
CH3
C=C
CH3CH2
H

(*E*)-3-methyl-2-pentene

ethyl has priority over methyl; methyl has priority over H

Cyclic Hydrocarbons

Cycloalkanes and cycloalkenes The names of saturated monocyclic hydrocarbons are formed by prefixing *cyclo-* to the name of the alkane with the same number of carbon atoms.

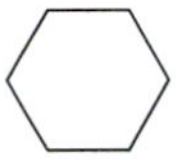

is cyclohexane

A ring is considered the parent unless there is a longer chain attached.

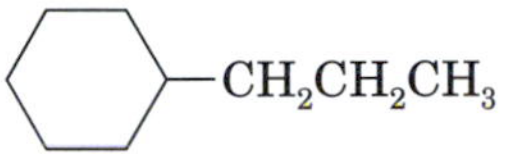
—CH2CH2CH3

propylcyclohexane

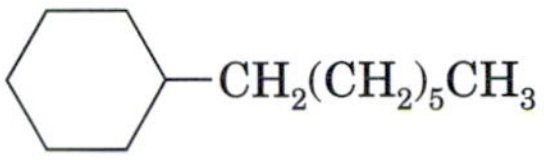
—CH2(CH2)5CH3

1-cyclohexylheptane

Alkyl substituents are named as prefixes and are given the lowest possible locants.

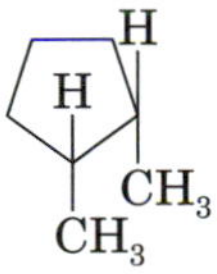
H
H
CH3
CH3

cis-1,2-dimethylcyclopentane

Unsaturated monocyclic hydrocarbons are named by changing the ending from *-ane* to *-ene* (*-adiene,* etc.). The ring is numbered to give the lowest locants possible to the double bonds. A ring with alkyl substituents and double bonds is numbered so that the double bonds receive the lowest possible locants.

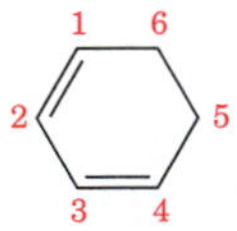
1
6
2
5
3
4

1,3-cyclohexadiene

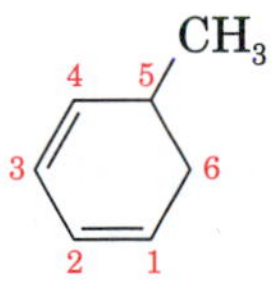
CH3
4
5
3
6
2
1

5-methyl-1,3-cyclohexadiene

Some common terpene ring structures have individual names.

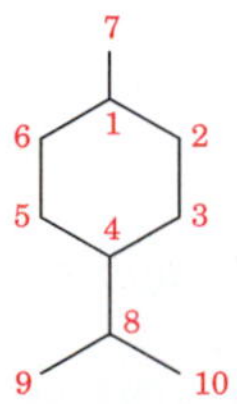

menthane

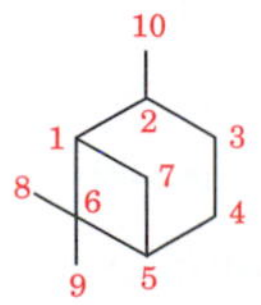

pinane

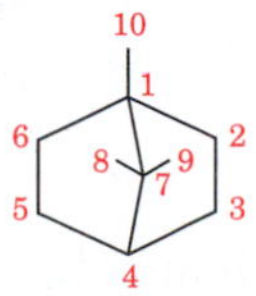

bornane

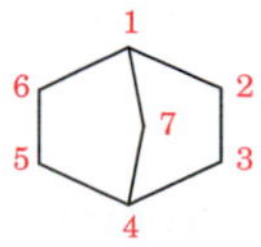

norbornane

Table A5	Names for some arenes and aryl groups		
Structure	Name	Structure	Name
Arenes:		**Aryl Groups:**	
	benzene	C_6H_5—	phenyl-
	naphthalene	$C_6H_5CH_2$—	benzyl-
	anthracene	CH_3——	p-tolyl-
	phenanthrene		1-naphthyl- (α-naphthyl-)
—CH_3	toluene	$C_6H_5CH{=}CHCH_2$—	cinnamyl-
—$CH(CH_3)_2$	cumene		
—$CH{=}CH_2$	styrene		
	o-xylene		
	mesitylene		

Aromatic hydrocarbons Many aromatic hydrocarbons are referred to by their trivial names (see Table A5). Systems composed of five or more linear fused benzene rings are named by a Greek-number prefix followed by *-cene*. (The prefix denotes the number of fused rings.)

pentacene

The aromatic system is the parent unless a longer chain is attached.

$C_6H_5CH_2CH_2CH_3$ $C_6H_5CH_2(CH_2)_5CH_3$

propylbenzene   1-phenylheptane

The positions of two substituents on a benzene ring are designated either by prefix numbers or by *o-, m-,* or *p-* (*ortho-, meta-,* or *para-*). For more than two substituents, numbers must be used.

o-dibromobenzene   *m*-dibromobenzene   *p*-dibromobenzene   1,2,3-triethylbenzene

The principal group, or a group that is part of the parent (for example, the CH_3 in toluene), is always considered to be bonded to position 1 on the ring. The numbers of substituents are chosen to be as low as possible.

m-chlorophenol   *p*-nitrotoluene   3,5-dimethylstyrene

Heterocycles

Some common heterocycles are listed in Table A6. In monocyclic heterocycles with only one heteroatom, that atom is considered position 1. Other ring systems are numbered by convention (see Table A6).

To name monocyclic compounds with one or more heteroatoms, prefixes may be used: *oxa-* (—O—), *aza-* (—NH—), *thia-* (—S—). For unsaturated rings, the ring size is designated by suffixes: *-ole* means five, and *-ine* means six (used only with rings that contain nitrogen). *Example:* An *oxazole* is a five-membered ring containing O and N. In numbering, O has priority over N.

1,3-oxazole   1,3,5-triazine

Heteroatoms in Chains

The *oxa-, aza-* system may also be used for naming aliphatic compounds. This method is called *replacement nomenclature.* All atoms in the chain are numbered so that the heteroatoms receive the lowest possible numbers. The parent is the alkane that has the same number of carbon atoms as the total number of atoms in the continuous chain (heteroatoms, but not hydrogens, included).

$CH_3OCH_2CH_2OCH_2CH_2OCH_2CH_3$

2,5,8-trioxadecane

Table A6   Names of some common heterocycles

Structure	Name	Structure	Name
furan	furan	pyrimidine	pyrimidine
4H-pyran	4H-pyran (γ-pyran)	morpholine	morpholine
pyrrole	pyrrole	thiophene	thiophene
pyrazole	pyrazole	indole	indole
imidazole	imidazole	carbazole	carbazole
pyridine	pyridine	purine	purine
piperidine	piperidine	quinoline	quinoline
piperazine	piperazine	isoquinoline	isoquinoline

Putting Names Together

The prefixes   In the IUPAC system, alkyl and aryl substituents and many functional groups are named as prefixes to the parent name (for example, iodomethane). Some common functional groups named as prefixes are listed in Table A7. Other prefix names are sometimes used for carbonyl groups, hydroxyl groups, etc. These are mentioned under their specific headings.

Like treatment of like things   Names should be as simple as possible and as consistent as possible. Two identical substituents should be treated alike, even though a few other rules may be broken. Although $C_6H_5CH_3$ would be

Table A7   Some common functional groups named as prefixes

Structure	Name	Structure	Name
—OR	alkoxy-[a]	—F	fluoro-
—NH$_2$	amino-	—H	hydro-[b]
—N=N—	azo-	—I	iodo-
—Br	bromo-	—NO$_2$	nitro-
—Cl	chloro-	—NO	nitroso-

[a] *methoxy, ethoxy-,* etc., depending upon R.
[b] *Hydro-* is a prefix used to designate a hydrogenated derivative of an unsaturated parent. *Perhydro-* means completely hydrogenated.

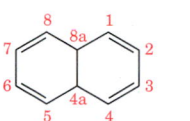 

4a,8a-dihydronaphthalene   perhydrophenanthrene

methylbenzene or toluene (rather than phenylmethane), (C$_6$H$_5$)$_2$CH$_2$ is called diphenylmethane.

Prefixes to designate like things   In simple compounds, the prefixes *di-, tri-, tetra-, penta-, hexa-,* etc. (not italicized) are used to indicate the number of times a substituent is found in the structure: e.g., dimethylamine for (CH$_3$)$_2$NH or dichloromethane for CH$_2$Cl$_2$.

In complex structures, the prefixes *bis-, tris-,* and *tetrakis-* (not italicized) are used: *bis-* means two of a kind; *tris-,* three of a kind; and *tetrakis-,* four of a kind.

[(CH$_3$)$_2$N]$_2$—   is bis(dimethylamino)- and not di(dimethylamino)-

The prefix *bi-* is used for (1) "double" molecules, and (2) bridged hydrocarbons.

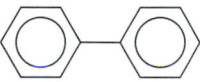

biphenyl   bicyclo[2.2.0]hexane

a double molecule   *a bridged hydrocarbon*

Order of prefixes   Prefixes are listed in *alphabetical order* (ethylmethyl-). In alphabetizing, a prefix denoting the number of times a substituent is found (di-, tri-, etc.) is disregarded. Ethyldimethyl- is the correct alphabetical order, even though *d* comes before *e*.

When to drop a vowel   In *conjunctive names* (names that are formed by combining two names), vowels are not elided, but are maintained. For example, the *e* in indoleacetic acid is retained (not indolacetic acid).

If there are two successive suffixes, the vowel at the end of the first suffix is dropped (propenoic acid, not propeneoic acid), unless it is followed by a consonant (propanediol, not propandiol).

Nomenclature priority of functional groups The various functional groups are ranked in priority as to which receives the suffix name and the lowest position number. A list of these priorities is given in Table A8. These are *not* the same priorities that are used for (E) and (Z) or (R) and (S).

$$\underset{\underset{\displaystyle NH_2}{|}}{CH_3\overset{\displaystyle \overset{O}{\|}}{C}CHCH_3}$$ is 3-amino-2-butanone (not 2-amino-3-butanone)

The principal functional group is indicated at the end of the name. A name with only one functional ending is preferred. *Example:* $HOCH_2CH_2CO_2H$ is 3-hydroxypropanoic acid. In names that must have two endings, the terminal ending refers to the principal functional group (see Table A8). *Example:* $CH_3CH{=}CHCO_2H$ is 2-butenoic acid. An exception to this rule is found in the naming of -enynes.

Numbering of the parent The parent is numbered so that the principal function receives the lowest number (see preceding section). Greek letters are reserved for trivial names and may not correspond directly to the IUPAC numbers. *Alpha* (α, the first letter of the Greek alphabet) means *on the nearest carbon,* which is often number 2 in the systematic name.

$$\underset{}{CH_3\overset{\overset{\displaystyle Br}{|}}{C}HCO_2H}$$

2-bromopropanoic acid (substitutive)
α-bromopropionic acid (trivial)

Table A8   Nomenclature priority^a

Structure	Name
—N(CH$_3$)$_3$$^+$ (as one example)	onium ion
—CO$_2$H	carboxylic acid
—SO$_3$H	sulfonic acid
—CO$_2$R	ester
—COX	acid halide
—CONR$_2$	amide
—CN	nitrile
—CHO	aldehyde
—CO—	ketone
ROH	alcohol
ArOH	phenol
—SH	thiol
—NR$_2$	amine
—O—O—	peroxide
—MgX (as one example)	organometallic
$\overset{\displaystyle \backslash \quad /}{\underset{\displaystyle / \quad \backslash}{C{=}C}}$	alkene
—C≡C—	alkyne
R—, X—, etc.	other substituents

^{a} Highest nomenclature priority is at the top.

Numbering of substituents and use of parentheses   If two numbering systems are required for identification of all atoms in the molecule, primes (') are often used for one of the systems to prevent confusion.

Alkyl substituents are numbered separately from the parent chain, beginning at the point of attachment. In these cases, a prime is not necessary provided that parentheses are used to enclose complex prefixes.

H_2N——CH_2CH_2Br

p-(2-bromoethyl)aniline

$CH_3CHClCH_2$
$CH_2=CHCH=CHC=CH_2$

2-(2-chloropropyl)-1,3,5-hexatriene

Configuration around a chiral carbon   Four different substituents around a chiral carbon can have either an (R) or (S) configuration. For designation of the configuration as (R) or (S), the structure is placed with the lowest-priority group in the rear. (See Section 4.1B for determination of priority, which is *not the same* as nomenclature priority.) The direction from the highest-priority group to the second-highest-priority group is then determined. If the direction is *clockwise,* the chiral carbon is (R); if it is *counterclockwise,* then the carbon is (S). If the structure has only one chiral carbon, (R) or (S) is used as the first prefix in the name. If the molecule has more than one chiral carbon, the designation of each chiral carbon and its position number is enclosed in parentheses in the prefix—e.g., $(2R,3R)$-dibromopentane.

H in rear

(R)-2-butanol   (S)-2-butanol

Carboxylic Acids

Four principal types of names for carboxylic acids are: (1) substitutive; (2) trivial; (3) carboxylic acid; and (4) conjunctive. In addition, it is sometimes necessary to name a carboxyl group with a prefix.

Substitutive names   Except for acids of one to five carbons and some fatty acids, aliphatic monocarboxylic acids are named by the substitutive system. The longest chain containing the —CO$_2$H group is chosen as the parent, and the chain is numbered starting with the carbon of the —CO$_2$H group as position 1. The name is taken from the name of the alkane with the same number of carbons, with the final -e replaced by -oic acid.

$$CH_3CH_2CH_2CH_2CH_2CO_2H$$

hexanoic acid

Substituents are designated by prefixes. A double bond is designated as a suffix preceding -oic acid.

$$\underset{CH_2=CHCH_2\overset{\displaystyle CH_3}{\overset{\displaystyle |}{CH}}CH_2CO_2H}{}$$

3-methyl-5-hexenoic acid or 3-methylhex-5-enoic acid

Trivial names Table A9 gives a list of commonly encountered trivial names for carboxylic acids.

Carboxylic acid names A carboxylic acid name is used when a —CO$_2$H group is attached to a ring. The name is a combination of the name of the ring system with the suffix -carboxylic acid. The carboxyl group is considered bonded to position 1 of the ring unless the ring system has its own unique numbering system. (The carbon of the —CO$_2$H group is not numbered as it is in a substitutive name.)

Table A9 Trivial names of some monocarboxylic acids

Structure	Name of Acid	Structure	Name of Acid
Saturated Chain:		**Other Functionality[b]:**	
HCO_2H	formic	$CH_3COCH_2CO_2H$	acetoacetic
CH_3CO_2H	acetic	$CH_3CH(OH)CO_2H$	lactic
$CH_3CH_2CO_2H$	propionic	CH_3COCO_2H	pyruvic
$CH_3(CH_2)_2CO_2H$	butyric	$CH_3COCH_2CH_2CO_2H$	levulinic
$CH_3(CH_2)_3CO_2H$	valeric	**On Rings:**	
$CH_3(CH_2)_{10}CO_2H$	lauric	$C_6H_5CO_2H$	benzoic
$CH_3(CH_2)_{12}CO_2H$	myristic		
$CH_3(CH_2)_{14}CO_2H$	palmitic		
$CH_3(CH_2)_{16}CO_2H$	stearic	salicylic	salicylic
Unsaturated Chain[a]:			2-naphthoic
$CH_2=CHCO_2H$	acrylic		
$CH_2=C(CH_3)CO_2H$	methacrylic		
$trans$-$CH_3CH=CHCO_2H$	crotonic		nicotinic

[a] For unsaturated fatty acids, see Table 24.1, page 963.
[b] For α-amino acids, see Table 25.1, page 986.

cyclohexanecarboxylic acid   2-pyridinecarboxylic acid

Conjunctive names Conjunctive names are combinations of two names: in the following examples, the name of the ring plus the name of the acid.

cyclohexaneacetic acid   indole-2-acetic acid

Diacids and polycarboxylic acids Diacids may be named systematically as *-dioic acids*. For example, $HO_2CCH_2CO_2H$ is propanedioic acid. Trivial names are commonly used. Some of these are listed in Table A10.

Table A10 Trivial names of some diacids	
Structure	Name of Acid
Aliphatic:	
HO_2CCO_2H	oxalic
$HO_2CCH_2CO_2H$	malonic
$HO_2C(CH_2)_2CO_2H$	succinic
$HO_2C(CH_2)_3CO_2H$	glutaric
$HO_2C(CH_2)_4CO_2H$	adipic
$HO_2C(CH_2)_5CO_2H$	pimelic
$HO_2C(CH_2)_6CO_2H$	suberic
$HO_2C(CH_2)_7CO_2H$	azelaic
$HO_2C(CH_2)_8CO_2H$	sebacic
cis-$HO_2CCH{=}CHCO_2H$	maleic
trans-$HO_2CCH{=}CHCO_2H$	fumaric
$HO_2CCH(OH)CH(OH)CO_2H$	tartaric
Aromatic:	
	phthalic
	isophthalic
	terephthalic

Aliphatic polycarboxylic acids containing more than two carboxyl groups are named by the carboxylic-acid system. The longest chain with the greatest number of carboxyl groups is chosen as the parent. If there is unsaturation, the double or triple bonds are included in the chain if possible.

③ ② ①
$$HO_2CCH_2CHCH_2CO_2H \quad \text{is 1,2,3-propanetricarboxylic acid}$$
$$| \\ CO_2H$$

Prefix names The following example shows the use of the *carboxy-* prefix in naming a compound impossible to name as a trioic acid.

$$HO_2CCH_2CHCH_2CO_2H \\ | \\ CH_2CO_2H$$

3-(carboxymethyl)pentanedioic acid

Sulfonic acids Sulfonic acids are named by adding the ending *-sulfonic acid* to the name of the rest of the structure.

$$CH_3CH_2SO_3H \qquad C_6H_5SO_3H$$

ethanesulfonic acid   benzenesulfonic acid

Acid Anhydrides

Acid anhydrides are named from the names of the component acid or acids with the word *acid* dropped and the word *anhydride* added. (Either substitutive or trivial acid names may be used.)

benzoic anhydride   acetic propionic anhydride

Cyclic anhydrides are named from the parent diacid.

maleic acid   maleic anhydride

Acid Halides

Acid halides are named by changing the ending of the carboxylic acid name from *-ic acid* to *-yl* plus the name of the halide. An ending of *-yl halide* in the name of a *diacid* implies that both carboxyl groups are acid-halide groups.

acetyl chloride   benzoyl bromide   succinyl chloride

Alcohols

The names of alcohols may be (1) substitutive; (2) radicofunctional; or, occasionally, (3) conjunctive. *Substitutive names* are taken from the name of the alkane with the final -*e* changed to -*ol*. For polyols, the prefix, *di-, tri-,* etc. is placed just before -*ol,* with the locants placed at the start of the name, if possible.

$$CH_3CH_2CH_2CH_2CH_2OH \qquad HO-\langle\ \rangle-OH$$

1-pentanol $\qquad\qquad$ 1,4-cyclohexanediol

In cases where confusion is possible, the locant precedes -*ol. Example:* $CH_3CH{=}CHCH_2OH$ is 2-buten-1-ol (or but-2-en-1-ol with alternative positioning of locants).

In a molecule containing a functional group with higher nomenclature priority and in complex molecules, the prefix *hydroxy-* is used. For example, $CH_3CH(OH)CH_2CO_2H$ is 3-hydroxybutanoic acid.

Radicofunctional names are generally composed of the name of the alkyl group plus the word *alcohol.*

$$(CH_3)_3COH \qquad C_6H_5CH_2OH$$

tert-butyl alcohol $\qquad$ benzyl alcohol

Conjunctive names are used principally with structures in which ring systems are bonded to an aliphatic alcohol.

pyridine-3-methanol

Polyols Structures with two OH groups on adjacent carbons (1,2-diols) are sometimes given trivial *glycol* names: the name of the *alkene* (not the alkane) from which the diol could be formed and the word *glycol. Glycerol* and *glycerin* are trivial names for 1,2,3-propanetriol.

$$\overset{\displaystyle OH}{\underset{\displaystyle |}{}}\qquad\qquad\qquad \overset{\displaystyle OH}{\underset{\displaystyle |}{}}$$
$$CH_3CHCH_2OH \qquad\qquad HOCH_2CHCH_2OH$$

1,2-propanediol (substitutive) $\qquad$ 1,2,3-propanetriol (substitutive)
propylene glycol (trivial) $\qquad\qquad$ glycerol (trivial)

Phenols

Phenols are compounds in which an OH is attached directly to an arene ring. In these cases, phenol (or naphthol, etc.) is considered the parent.

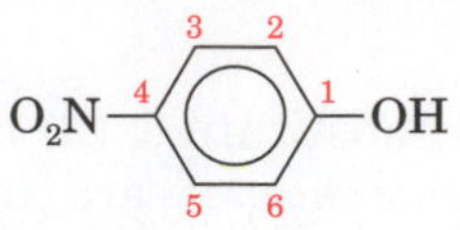

p-nitrophenol

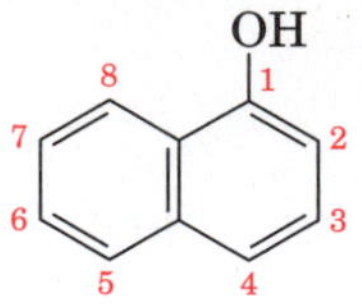

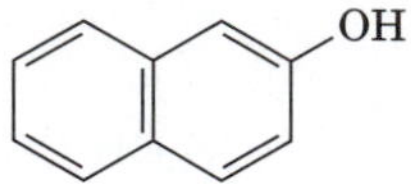

1-naphthol (substitutive)
α-naphthol (trivial)

2-naphthol (substitutive)
β-naphthol (trivial)

Many phenols and substituted phenols have trivial names.

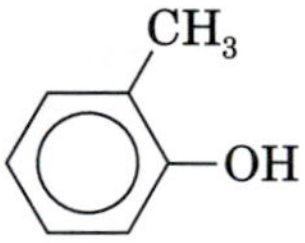

o-cresol 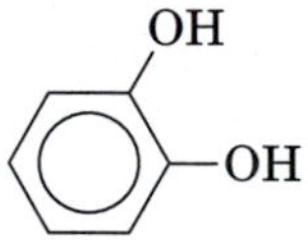   pyrocatechol 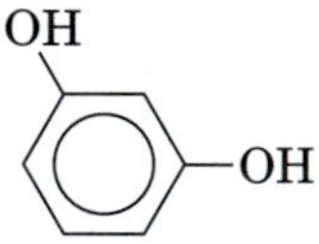   resorcinol    hydroquinone

Aldehydes

Aldehydes may be named by the substitutive system or by trivial aldehyde names. In the substitutive system, the *-oic acid* ending of the corresponding carboxylic acid is changed to *-al*.

$$CH_3CH_2CH_2CH_2CH_2CHO$$

hexanal

When a —CHO group is attached to a ring, the name is a combination of the name of the ring system with the suffix *-carbaldehyde*.

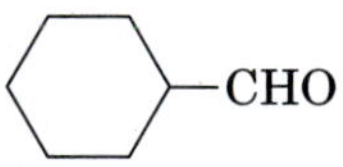

cyclohexanecarbaldehyde

In trivial names, the *-ic* (or *-oic*) *acid* ending is changed to *-aldehyde*.

$$CH_3CHO \qquad C_6H_5CHO$$

acetaldehyde   benzaldehyde

Some aldehydes have specific trivial names:

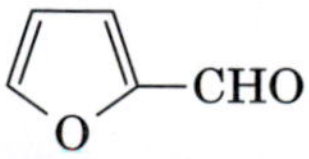

$HOCH_2CH(OH)CHO$

2-furaldehyde   glyceraldehyde
or furfural

Nitriles

The functional group of a nitrile is the *cyano group,* —C≡N. In the substitutive system, nitriles are named from the alkane parent that contains the same number of carbons (including the C in the —C≡N group) and the suffix *-nitrile* to denote ≡N.

carbon 1

3-carbons

$CH_3CH_2C≡N$ $(CH_3)_2CHCN$

propanenitrile   2-methylpropanenitrile

When the cyano group is bonded to a ring, the ending *-carbonitrile* (to denote the entire —C≡N group) is used.

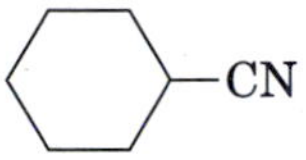

cyclohexanecarbonitrile 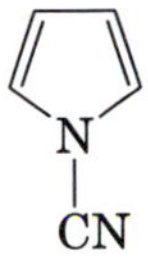   1-pyrrolecarbonitrile

If groups of higher priority are present, the prefix *cyano-* (to denote the entire —C≡N group) is used.

$$\overset{O}{\overset{\|}{HOCCH_2CH_2C}}≡N$$

3-cyanopropanoic acid

In trivial names, the *-ic acid* or *-oic acid* ending of the corresponding carboxylic acid name is replaced by *-onitrile.*

from benzoic acid

from acetic acid

$CH_3C≡N$ —C≡N

acetonitrile   benzonitrile

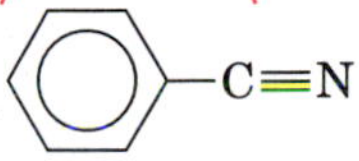

Occasionally, nitriles are named as cyanides, such as phenyl cyanide for C_6H_5—CN. These names are also trivial.

Amides

In both the substitutive and trivial systems, an amide is named by dropping the *-ic* (or *-oic*) *acid* ending of the corresponding acid name and adding *-amide.*

$$\overset{O}{\overset{\|}{CH_3(CH_2)_4CNH_2}}$$ $$\overset{O}{\overset{\|}{CH_3CNH_2}}$$

hexanamide (substitutive)   acetamide (trivial)

Substituents on the amide nitrogen are named as prefixes preceded by *N*- or *N,N*-. $C_6H_5CONHCH_3$ is *N*-methylbenzamide, and $C_6H_5CON(CH_3)_2$ is *N,N*-dimethylbenzamide. *N*-Phenylamides have trivial names of *anilides*.

$$\underset{\text{acetanilide}}{CH_3\overset{\overset{\textstyle O}{\|}}{C}NHC_6H_5} \qquad \underset{\text{benzanilide}}{C_6H_5\overset{\overset{\textstyle O}{\|}}{C}NHC_6H_5}$$

Sulfonamides are named by attaching the ending *-sulfonamide* to the name for the rest of the structure. *Example:* $C_6H_5SO_2NH_2$ is benzenesulfonamide.

p-aminobenzenesulfonamide
(sulfanilamide)

Amines

The preferred method for naming simple amines is by using the name of the hydrocarbon portion of the structure, omitting the final *-e* before another vowel, and adding the ending *-amine*. Note that the largest alkyl group bonded to the *N* is considered the parent alkane.

$$\underset{\text{ethanamine}}{CH_3CH_2NH_2} \qquad \underset{\text{1,2-ethanediamine}}{H_2NCH_2CH_2NH_2} \qquad \underset{\text{\textit{N}-methylethanamine}}{CH_3NHCH_2CH_3}$$

larger group

The more commonly encountered names, also recognized by IUPAC, are formed by listing the alkyl groups, not the alkanes, before the ending *-amine*.

$$\underset{\text{ethylamine}}{CH_3CH_2NH_2} \qquad \underset{\text{\textit{N}-methylethylamine}}{CH_3NHCH_2CH_3}$$

If functionality of higher nomenclature priority is present, the prefix *amino-* is used.

$$\underset{\text{2-amino-1-propanol}}{\overset{\overset{\textstyle NH_2}{|}}{CH_3CHCH_2OH}} \qquad \underset{\text{2-(methylamino)-1-propanol}}{\overset{\overset{\textstyle NHCH_3}{|}}{CH_3CHCH_2OH}}$$

Some arylamines have their own names.

$$\underset{\text{aniline}}{C_6H_5NH_2} \qquad \underset{\text{\textit{p}-toluidine}}{CH_3-\text{⬡}-NH_2}$$

The prefix *aza-* is sometimes used to identify a nitrogen in a chain or ring; see page 1059.

Amine salts Amine salts are named as *ammonium salts* or (in simple cases) as amine hydrochlorides, etc.

$$(CH_3)_3NH^+ \; Cl^-$$

trimethylammonium chloride
(trimethylamine hydrochloride)

Cyclic salts often are named as *-inium salts.*

$$C_6H_5NH_3^+ \; Br^-$$

anilinium bromide
(aniline hydrobromide)

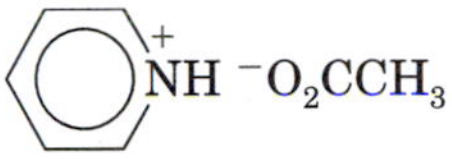

pyridinium acetate

Esters and Salts of Carboxylic Acids

Esters and salts of carboxylic acids are named with two words. The first word is the name of the substituent on the oxygen. The second word is derived from the name of the parent carboxylic acid with the ending changed from *-ic acid* to *-ate.*

$CH_3(CH_2)_4CO_2CH_3$ methyl hexanoate (substitutive)
$CH_3CO_2CH_2CH_3$ ethyl acetate (trivial)
$CH_3CH_2CO_2Na$ sodium propanoate (or sodium propionate)

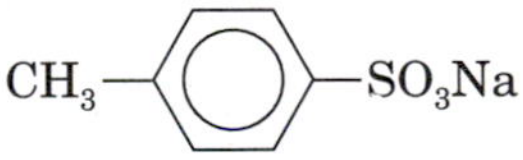

CH_3—〇—SO_3Na sodium *p*-toluenesulfonate (or sodium tosylate)

Ethers

Ethers are usually named with the names of the alkyl or aryl groups bonded to the O followed by the word *ether.* (These are radicofunctional names.)

$$CH_3CH_2OCH_2CH_3 \qquad CH_3O-\langle \rangle$$

diethyl ether

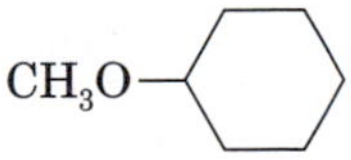

cyclohexyl methyl ether

In the names of more complex ethers, an *alkoxy-* prefix may be used. (This is the IUPAC preference.)

$$\overset{OCH_3}{\underset{|}{CH_3CH_2CH_2CHCH_2CO_2H}}$$

3-methoxyhexanoic acid

Sometimes the prefix *oxa-* is used; see page 1059.

Ketones

In the substitutive names for ketones, the *-e* of the parent alkane name is dropped and *-one* is added. A locant is used if necessary.

$$CH_3\overset{O}{\overset{\|}{C}}CH_2CH_2CH_3 \qquad CH_3\overset{O}{\overset{\|}{C}}CH=CHCH_3 \qquad CH_3\overset{O}{\overset{\|}{C}}CH_2\overset{O}{\overset{\|}{C}}CH_3$$

    2-pentanone   3-penten-2-one   2,4-pentanedione

e retained before consonant

In a complex structure, a keto group may be named in the substitutive system with the prefix *oxo-*. (The prefix *keto-* is also sometimes encountered.) Contrast the use of oxo- with that of oxa- (an ether).

$$CH_3\overset{O}{\overset{\|}{C}}CH_2CO_2H \qquad \text{is 3-oxobutanoic acid}$$

In radicofunctional names, the alkyl groups bonded to the C=O group are named, and the word *ketone* is added. Methyl ethyl ketone ($CH_3COCH_2CH_3$) is a common example. The CH_3CO— group is sometimes called the *aceto-*, or *acetyl,* group, while the C_6H_5CO— group is the *benzo-*, or *benzoyl,* group. Also encountered are *-phenone, -naphthone,* or *-acetone* endings, where one of these groups is an attachment to the ketone carbonyl group.

$$CH_3\overset{O}{\overset{\|}{C}}CH_2\overset{O}{\overset{\|}{C}}CH_3 \qquad CH_3\overset{O}{\overset{\|}{C}}C_6H_5 \qquad C_6H_5\overset{O}{\overset{\|}{C}}C_6H_5$$

   acetylacetone   acetophenone   benzophenone

Some ketones also have specific trivial names: CH_3COCH_3 is called acetone.

Organometallics and Metal Alkoxides

Organometallic compounds, those with C bonded to a metal, are named by the name of the alkyl or aryl groups plus the name of the metal.

$$(CH_3CH_2)_3Al \qquad\qquad C_6H_5MgBr$$

   triethylaluminum   phenylmagnesium bromide

Organosilicon or boron compounds are often named as derivatives of the metalloid hydrides. SiH_4 is silane; $(CH_3)_4Si$ is tetramethylsilane. BH_3 is borane; $(CH_3CH_2)_3B$ is triethylborane.

Sodium or potassium alkoxides are named as salts: the name of the cation plus the name of the alcohol with the *-anol* ending changed to *-oxide.* Salts of phenols are called *phenoxides.* (An alternate ending is *-olate.*)

$$CH_3CH_2ONa \qquad\qquad C_6H_5OK$$

   sodium ethoxide   potassium phenoxide
  (sodium ethanolate)   (potassium phenolate)

Glossary of some prefix symbols used in organic chemistry

(+)   dextrorotatory

(−)   levorotatory

(±)   racemic

α-   alpha: (1) on the adjacent carbon; (2) refers to configuration of carbon 1 in sugars; (3) refers to configuration of substituents on steroid ring systems

aldo-   aldehyde

allo-   closely related

andro-   relating to male

anhydro-   denoting abstraction of water

antho-   relating to flowers

anthra-   relating to coal or to anthracene

anti-   on opposite faces or sides

β-   beta: (1) opposite to that of α in configuration; (2) second carbon removed from a functional group or a heteroatom

bi-   twice or double

bisnor-   indicating removal of two carbons

chromo-   color or colored

cis-   on the same side of a double bond or ring

cyclo-   cyclic

Δ-   double bond

D-   on the right in the Fischer projection (see Section 23.3A)

d-   dextrorotatory; (+) is preferred

de-   removal of something, such as hydrogen *(dehydro-)* or oxygen *(deoxy-)*

dextro-   to the right, as in dextrorotatory

dl-   racemic; (±) is preferred

(*E*)-   on the opposite sides of a double bond

endo-   (1) on or in the ring and not on a side chain; (2) opposite the bridge side of a ring system:

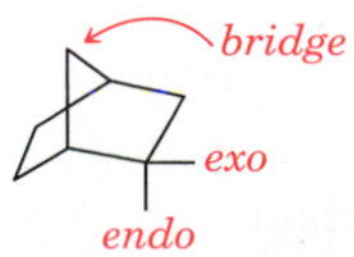

(3) attached as a bridge within a ring, as 1,4-*endo*-methyleneanthracene

(continued)

Glossary of some prefix symbols used in organic chemistry (continued)

epi-   (1) epimeric; (2) a bridge connection on a ring, as 9,10-epidioxyanthracene

erythro-   related in configuration to erythrose

exo-   (1) on a side chain attached to a ring; (2) on the bridge side of a ring system (see *endo-*)

gem-   attached to the same atom

hemi-   one-half

hydro-   (1) denotes presence of H; (2) sometimes relating to water

hypo-   indicating a low, lower, or lowest state of oxidation

i-, iso-   (1) methyl branch at the end of the chain; (2) isomeric; (3) occasionally *i-* is used for *inactive*

L-   on the left in the Fisher projection (see Section 23.3A)

l-   levorotatory; $(-)$ is preferred

leuco-   colorless or white

levo-   to the left, as in levorotatory

m-, meta-   (1) 1,3 on benzene; (2) closely related compound, a metaldehyde

meso-   (1) with a plane of symmetry and optically inactive; (2) middle position of certain cyclic organic compounds

n-   *normal:* continuous chain (archaic and not sanctioned by IUPAC)

neo-   one C connected to four other Cs

nor-   (1) removal of one or more Cs (with Hs); (2) structure isomeric to that of root name, as norleucine.

o-, ortho-   1,2 on benzene

oligo-   few (units of)

p-, para-   (1) 1,4 on benzene; (2) polymeric, as paraformaldehyde

per-   saturated with, as in *perhydro-,* or *peroxy-*

peri-   (1) 1,8 on naphthalene; (2) fusion of ring to two or more adjoining rings, as perixanthenoxanthene.

pheno-   relating to phenyl or benzene

poly-   many (units of)

Ψ-, pseudo-   has a resemblance to

pyro-   indicating formation by heat

(R)-   clockwise configuration around a chiral carbon

(RS)-   racemic

(continued)

Glossary of some prefix symbols used in organic chemistry (continued)

s-   abbreviation for *secondary-* or *symmetrical-*

(S)-   counterclockwise configuration around a chiral carbon

seco-   denoting ring cleavage

sec-   abbreviation for *secondary-*

sym-   symmetrical

syn-   on the same face or side

t-, tert-   abbreviation for *tertiary-*

threo-   related in configuration to threose

trans-   on opposite sides of a double bond or ring

uns-, unsym-   unsymmetrical

v-, vic-   vicinal: on adjacent Cs

(Z)-   on the same side of a double bond

ANSWERS TO PROBLEMS

The answers to the problems within the textual material are given here. For the answers to the chapter end problems, see the *Solutions Manual* that accompanies this text.

Chapter 1

1.1 **(a)** Mg $1s^2 2s^2 2p^6 3s^2$ **(b)** Si $1s^2 2s^2 2p^6 3s^2 3p^2$

1.2 **(a)** boron (B)   **(b)** oxygen (O)

1.3 **(a)**

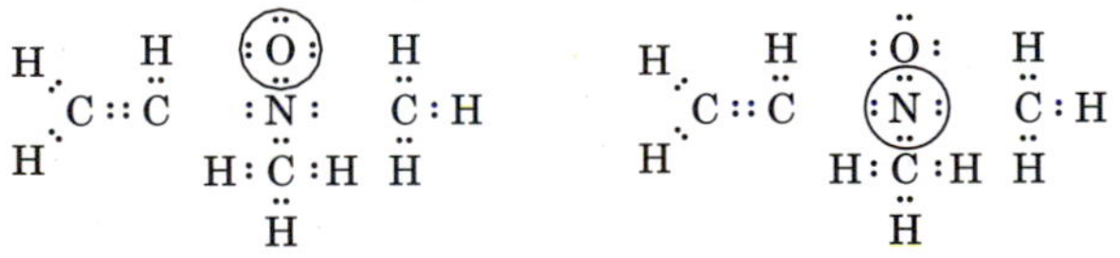

(Lewis structures)

(b)

(Lewis structures)

1.4 **(a)** 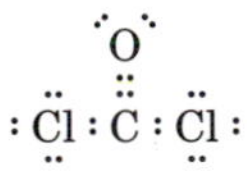 $:\!\ddot{C}l\!:\!\ddot{C}\!:\!\ddot{C}l\!:$ **(b)** (Lewis structure)   **(c)** (Lewis structure)

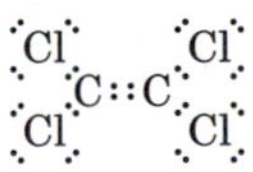

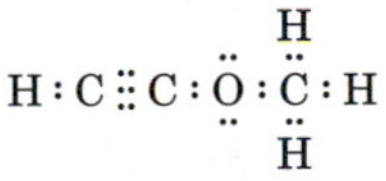

(There are other correct answers.)

(d) 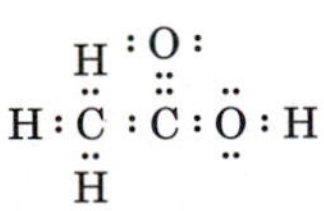 (Lewis structure)

(There are other correct answers.)

1.5 **(a)** neutral molecule, no formal charges   **(b)** ion: $\overset{0}{C}H_3\overset{+1}{N}H_3$

(c) neutral molecule: $:\!\overset{-1}{C}\!::\!\overset{+1}{O}\!:$

1.6 **(a)** all atoms 0;   **(b)** $CH_2\!=\!\overset{+1}{N}\!=\!\overset{-1}{N};$ **(c)** 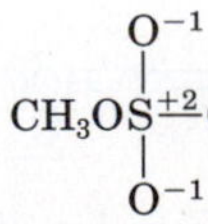 $CH_3O\overset{\pm2}{S}OH$, other atoms 0

$$\overset{O^{-1}}{\underset{O^{-1}}{\mid}}$$

1.7 (a) $\overset{\delta+}{CH_3CH_2}-\overset{\delta-}{Cl}$ (b) $CH_3-\overset{\overset{\displaystyle O^{\delta-}}{\|}}{\underset{\delta+}{C}}-CH_3$

1.8 (a) heterolytic: $CH_3-\overset{\overset{\displaystyle CH_3}{|}}{\underset{\underset{\displaystyle CH_3}{|}}{C}}\overset{\frown}{-Cl} \longrightarrow CH_3-\overset{\overset{\displaystyle CH_3}{|}}{\underset{\underset{\displaystyle CH_3}{|}}{C^+}} + Cl^-$

homolytic: $CH_3-\overset{\overset{\displaystyle CH_3}{|}}{\underset{\underset{\displaystyle CH_3}{|}}{C}}\overset{\frown}{-Cl} \longrightarrow CH_3-\overset{\overset{\displaystyle CH_3}{|}}{\underset{\underset{\displaystyle CH_3}{|}}{\overset{\cdot}{C}}} + Cl\cdot$

(b) heterolytic: $CH_3CH_2CH_2\overset{\frown}{-Cl} \longrightarrow CH_3CH_2\overset{+}{C}H_2 + Cl^-$

homolytic: $CH_3CH_2CH_2\overset{\frown}{-Cl} \longrightarrow CH_3CH_2\overset{\cdot}{C}H_2 + Cl\cdot$

1.9 (a) , $CH_2=CH_2$

(b) $H-\overset{\overset{\displaystyle H}{|}}{\underset{\underset{\displaystyle H}{|}}{C}}-\overset{\overset{\displaystyle Cl}{|}}{\underset{\underset{\displaystyle Cl}{|}}{C}}-H$, CH_3CHCl_2

1.10 (a) $(CH_3)_2CHCH_2Cl$; (b) CH_3CHCl_2; (c) $CH_3(CH_2)_3CHClCH_2Cl$;

(d) $(CH_3)_2C=C(CH_3)_2$; (e) $CH_2(C\equiv N)_2$

1.11 (a) (b) (c)

1.12 (a) (b) (c)

1.13 (a) $CH_3CH_2CH_2\overset{\overset{\displaystyle H}{|}}{\underset{\underset{\displaystyle H}{|}}{\ddot{N}H}}\text{---}:\overset{\overset{\displaystyle H}{|}}{N}CH_2CH_2CH_3$

(b) $CH_3\overset{\overset{\displaystyle H}{|}}{\ddot{O}H}\text{---}:\overset{\overset{\displaystyle H}{|}}{\ddot{O}}CH_3,\quad CH_3\overset{\overset{\displaystyle H}{|}}{\ddot{O}H}\text{---}:\overset{\overset{\displaystyle H}{|}}{O}H,\quad CH_3\overset{\overset{\displaystyle H}{|}}{\ddot{O}}:\text{---}H\ddot{O}:,\quad H_2\ddot{O}:\text{---}H\overset{\overset{\displaystyle H}{|}}{\ddot{O}}:$

(c) (d)

1.14 (a) base: $^-:\ddot{N}H_2 + H_2\ddot{O}: \;\rightleftharpoons\; :NH_3 + {}^-:\ddot{O}H$

(b) acid: $CH_3CH_2CH_2\overset{\overset{\displaystyle :\ddot{O}:}{\|}}{C}OH + H_2\ddot{O}: \;\rightleftharpoons\; CH_3CH_2CH_2\overset{\overset{\displaystyle :\ddot{O}:}{\|}}{C}\ddot{O}:^- + H_3O:^+$

(c) base: $^-:\ddot{O}CH_2CH_3 + H_2\ddot{O}: \;\rightleftharpoons\; H\ddot{O}CH_2CH_3 + {}^-:\ddot{O}H$

(d) acid:

(e) base: ⬡ :NH + H₂O: ⇌ ⬡ NH₂⁺ + ⁻:ÖH

1.15 (a) CH₃CH₂CO—H + HOH ⇌ CH₃CH₂CO:⁻ + HÖH⁺

(b) ⬡—CO—H + ⁻:ÖH → ⬡—CO:⁻ + HÖH

(c) (CH₃)₂NH + H—ÖH ⇌ (CH₃)₂NH₂⁺ + ⁻:ÖH

(d) ⬡N(H) + H—ÖCCH₃ ⇌ ⬡N⁺(H)(H) + ⁻:ÖCCH₃

1.16 (a) CH₃—⬡—CO⁻   **(b)** ⬡—O⁻   **(c)** CH₃OSO⁻ (with O)

(d) ⬡—S(=O)(=O)—NH₂   **(e)** ⬡—OH   **(f)** (phthalimide) C(=O)—NH—C(=O)

1.17 (a) HO—H is a stronger acid than H—CH₃ because O is more electronegative than C and thus is more able to carry a negative ionic charge.

$$HO—H + B:⁻ \rightleftharpoons HO:⁻ + HB$$

$$H—CH₃ + B:⁻ \rightleftharpoons ⁻:CH₃ + HB$$

(b) H—NH₂ is a stronger acid than H—CH₃ because N is more electronegative than C.

(c) HS—H is a stronger acid than HO—H because S is below O in the periodic table. Thus, S contains one more electron shell than O, is larger than O, and is better able to carry a negative charge.

$$HS—H + B:⁻ \rightleftharpoons HS:⁻ + HB$$

$$HO—H + B:⁻ \rightleftharpoons HO:⁻ + HB$$

1.18 (a) FCH₂COH (with O); Fluoro is more electronegative than iodo.

(b) CH₃C(Cl)(Cl)—COH (with O); Two chloro groups are better than one.

(c) CH₃C(Cl)(Cl)—COH (with O); Both chloro groups are closer to the carboxyl group.

1.19 (a) CH₃CCH₃ (with O), Lewis base; H⁺, Lewis acid

(b) (CH₃)₃C⁺, Lewis acid; Cl⁻, Lewis base

(c) $CH_3\overset{\overset{\textstyle O}{\|}}{C}OCH_3$, Lewis acid; $^-OCH_3$, Lewis base

1.20 $pH = pK_a + \log \dfrac{[A^-]}{[HA]}$

$[A^-] = [HA]$
therefore, $pH = pK_a = 2.85$

1.21 (a) < (b) < (c)

1.22 (c) < (b) < (a)

Chapter 2

2.1 (a) 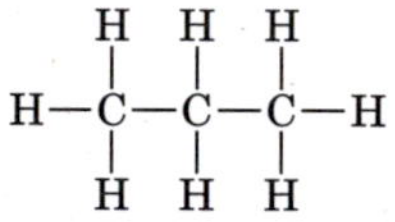 (b) 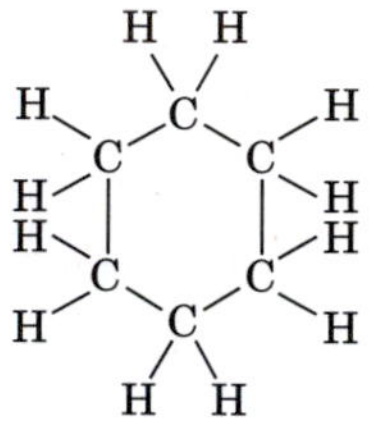 All C—C bonds are sp^3–sp^3 and all C—H bonds are sp^3–s.

2.2 (a) 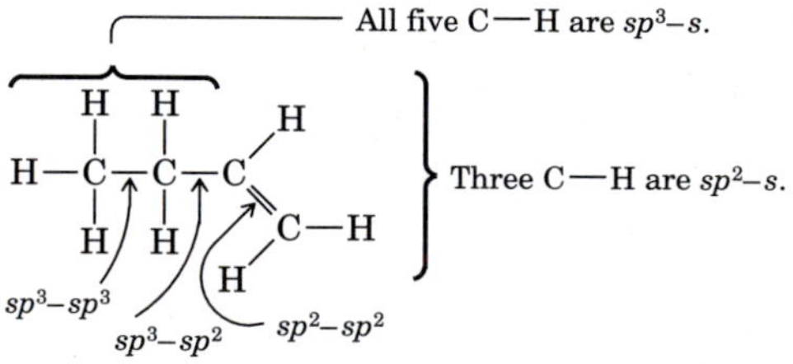 All five C—H are sp^3–s.

Three C—H are sp^2–s.

sp^3–sp^3 sp^3–sp^2 sp^2–sp^2

(b)

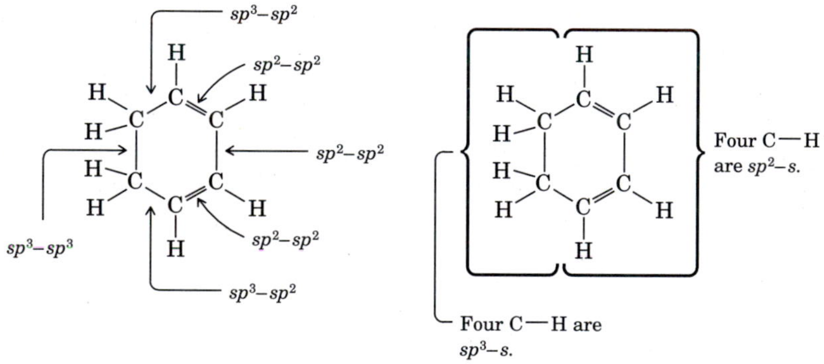

sp^3–sp^2
sp^2–sp^2
sp^2–sp^2
sp^3–sp^3
sp^2–sp^2
sp^3–sp^2

Four C—H are sp^2–s.

Four C—H are sp^3–s.

2.3

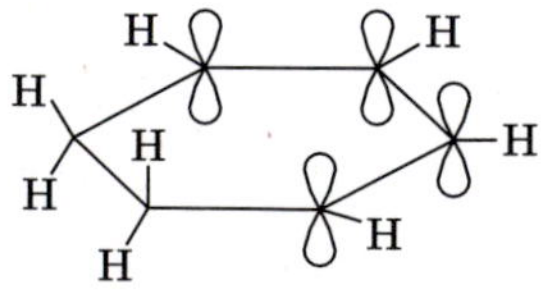

All C—C bonds lie in one plane. The hydrogens of the C—H groups that are bonded to the C=C also lie in the plane. The hydrogens of the other four C—H groups lie above and below the plane.

2.4 $H_3C\!-\!C\!\equiv\!CH$

two p–p and one sp–sp

sp^3–sp

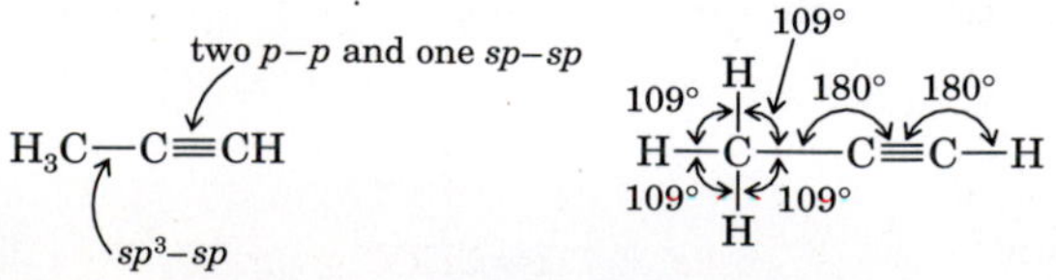

2.5 (a) (2), (1); (b) (2), (1); (c) (1), (2), (3)

2.6 (a) H_2 C=C HCH$_2$ CH ; (b) ⬡—NH$_2$; (c)

2.7 RCO$_2$H

2.8 (a) H—C—H (b) CH$_3$CH$_2$CH$_2$CH$_2$CH [The hydrocarbon can also be branched. For example,

(CH$_3$)$_2$CHCH$_2$CH.]

(c) ⬡—CH

2.9 (a)
H—C—N—C—N—H All C—H and N—H bonds are sp^3–s,
and all C—N bonds are sp^3–sp^3

(b)
sp^2–sp^2 and p–p
sp^2–s
sp^3–sp
H—N—C—N—C—C≡N: All unmarked C—H and N—H bonds are sp^3–s.
sp–sp and two p–p
sp^3–sp^2 sp^3–sp^3

(c)
sp^2–sp^3
sp^2–sp^2
:N   N—H
sp^2–s sp^2–sp^3

All C—C and C—N bonds not marked are sp^3–sp^3.

All C—H bonds and the N—H bond not marked are sp^3–s.

2.10 The following structures are not the only possible answers:

(a) CH$_3$CH$_2$CH$_2$CH (b) CH$_3$CCH$_2$CH$_3$ (c) CH$_3$CH$_2$CH$_2$COH (d) CH$_3$CCH$_2$COH

2.11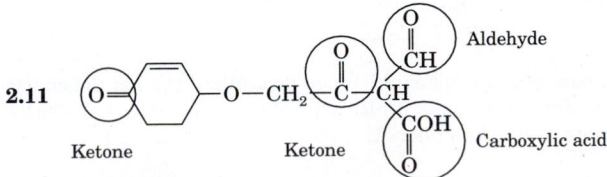

Ketone   Ketone   Aldehyde   Carboxylic acid

2.12 Five conjugated double bonds, no isolated double bonds.

2.13 (a) CH$_3$CH=CHCH=CHCH=CHCH$_3$ (b) CH$_2$=CHCH$_2$CH=CHCH=CHCH$_3$

(c) CH$_2$=CHCH$_2$CH=CHCH$_2$CH=CH$_2$

2.14 (a) 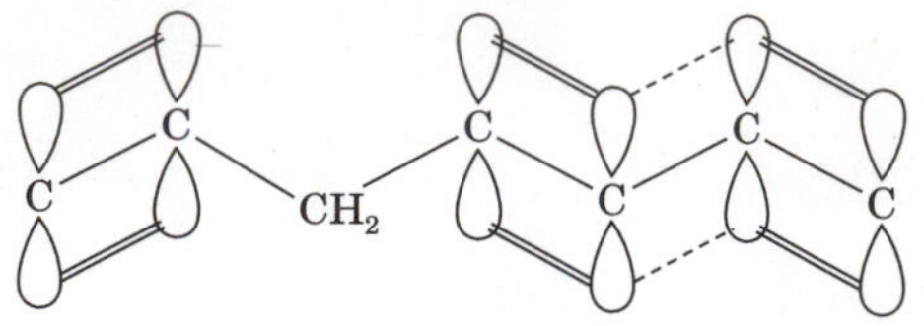 **(b)**

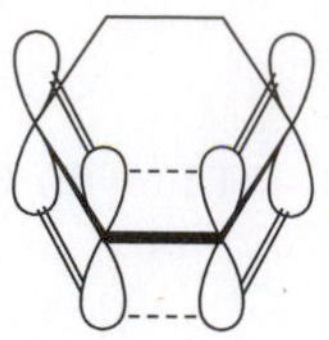

2.15 $CH_2{=}CHCCH_3$ (with O double-bonded to C)

2.16

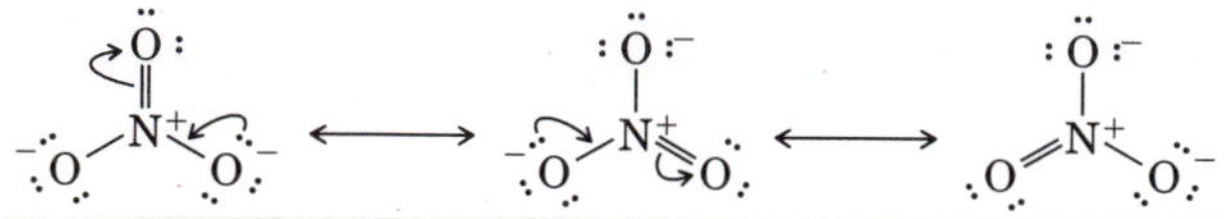

2.17 The right-hand structure is the major contributor because each atom has an octet.

2.18 (a) $CH_3CH{=}CH{-}\overset{+}{C}HCH_3 \longleftrightarrow CH_3\overset{+}{C}H{-}CH{=}CHCH_3$; *equivalent*

(b)

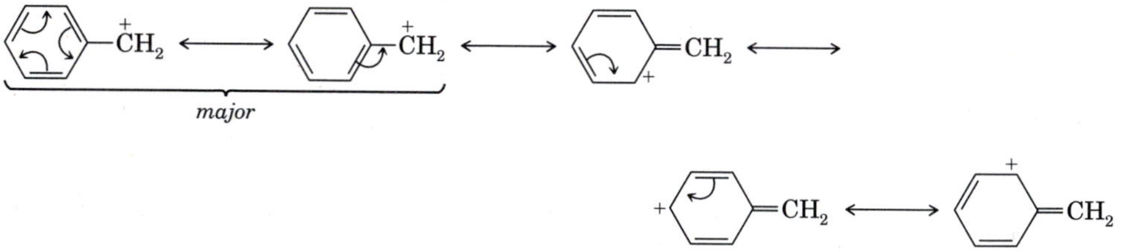

(c)

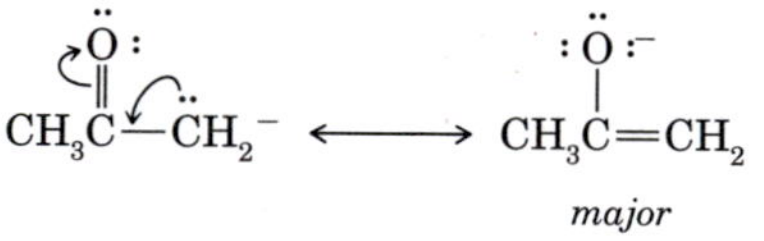

(d)

2.19 *p*-Nitrophenol is the stronger acid. The resonance structures for the conjugate base of *p*-nitrophenol follow. (The resonance structures for phenol are given in the text just before problem 2.19.)

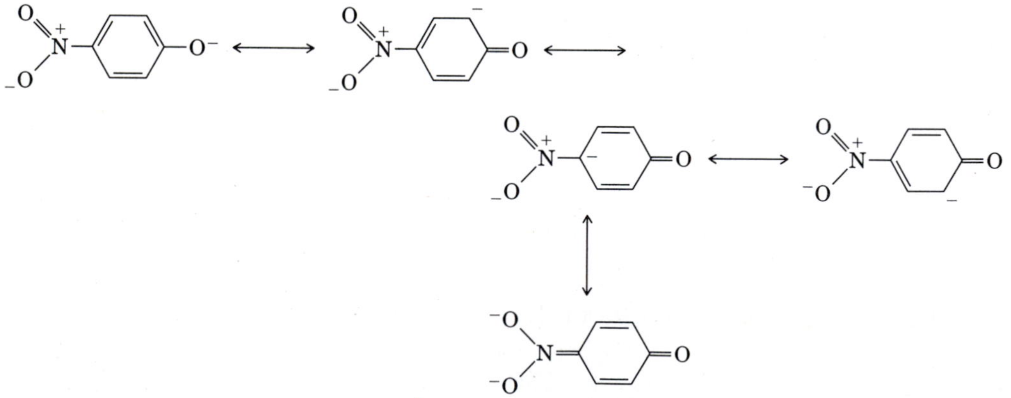

The negative charge of the conjugate base of *p*-nitrophenol is more delocalized than is the charge of the conjugate base of phenol. Consequently, the anion of *p*-nitrophenol is more stable than that of phenol and, therefore, *p*-nitrophenol is more acidic than phenol.

Chapter 3

3.1 **(a)** $CH_3CH_2CH_2\overset{\overset{\displaystyle O}{\|}}{C}H$ **(b)** $CH_3\underset{\underset{\displaystyle OH}{|}}{C}HCH_2CH_3$ **(c)**

3.2 **(a)** $HOCH_2\overset{\overset{\displaystyle O}{\|}}{C}H;$ **(b)** $-CH_2OH;$ **(c)** $CH_3CH_2CH_2CH_2\overset{\overset{\displaystyle O}{\|}}{C}H$

3.3 **(a)** 2   **(b)** 3

3.4 **(a)** $H-CO_2H$ CH_3CO_2H $CH_3CH_2CO_2H$
$CH_3CH_2CH_2CO_2H$ $CH_3CH_2CH_2CH_2CO_2H$

(b) $CH_3-\overset{\overset{\displaystyle O}{\|}}{C}CH_3$ $CH_3CH_2-\overset{\overset{\displaystyle O}{\|}}{C}CH_3$ $CH_3CH_2CH_2-\overset{\overset{\displaystyle O}{\|}}{C}CH_3$

$CH_3CH_2CH_2CH_2-\overset{\overset{\displaystyle O}{\|}}{C}CH_3$ $CH_3CH_2CH_2CH_2CH_2-\overset{\overset{\displaystyle O}{\|}}{C}CH_3$

(c) $H-CH=CH_2$ $CH_3CH=CH_2$ $CH_3CH_2CH=CH_2$
$CH_3CH_2CH_2CH=CH_2$ $CH_3CH_2CH_2CH_2CH=CH_2$

3.5 **(a)** 3-ethylhexane   **(b)** 2-methylpropane

3.6 **(a)** $CH_3CH_2\underset{\underset{\displaystyle CH_2CH_3}{|}}{C}HCH_2CH_2CH_3$

(b) $CH_3CH_2CH_2\underset{\underset{\displaystyle CH_2CH_3}{|}}{C}H\overset{\overset{\displaystyle CH_3}{|}}{C}HCH_2CH_2CH_3$

3.7 **(a)** $CH_3CH_2CH_2-$;   **(b)** $(CH_3)_2CHCH_2-$;   **(c)** $CH_3CH_2CH_2\overset{\overset{\displaystyle C(CH_3)_3}{|}}{C}HCH_2CH_2CH_2CH_3$

3.8 **(a)** $Cl_2CHCH_2Cl;$ **(b)** $Cl-$$-NO_2$

3.9 **(a)** 3-bromo-4,4-dichloro-3-methylheptane

(b) 3-bromo-5-isopropyl-1,3-dinitrocyclohexane

3.10 **(a)** 2-butene

(b) cyclopentene (A number is not necessary because there can be only one cyclopentene.)

(c) 1,2-dichloro-1-butene

3.11 **(a)**   **(b)** $HC\equiv C-C\equiv CCH_2CH_3$

(c) $CH_3\underset{\underset{\displaystyle CH_3}{|}}{C}H-$

(Note that the number 1 is implied for the carbon–carbon double bond.)

3.12 $CH_3CH_2CH_2CH_2\overset{\overset{\displaystyle O}{\|}}{C}H$; $CH_3\overset{\overset{\displaystyle O}{\|}}{C}CH_2CH_2CH_3$ or $CH_3CH_2\overset{\overset{\displaystyle O}{\|}}{C}CH_2CH_3$; $CH_3CH_2CH_2CH_2\overset{\overset{\displaystyle O}{\|}}{C}OH$

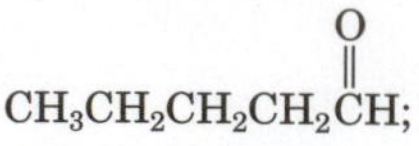

pentanal   2-pentanone   3-pentanone   pentanoic acid

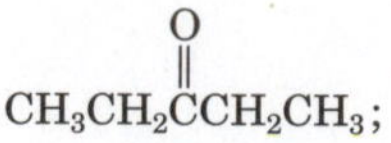

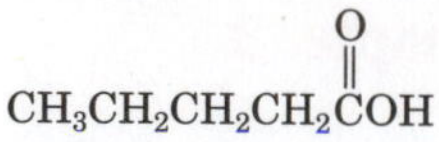

3.13   **(a)** 2,4-dichloro-3-pentanone

       (b) 3-ethyl-3-pentanol

       (c) 5-iodo-3-propylcyclohexene

       (d) 3-chlorocyclohexanone

       (e) 4-methyl-2-hexanone

       (f) 2-bromobutanoic acid

       (g) 1,3,5-tribromobenzene

       (h) phenylethanal

       (i) 5-*tert*-butyl-1,3-cyclohexadiene

3.14   The test fuel must contain 85% isooctane and 15% heptane.

Chapter 4

4.1   **1** and **2** represent the same compound.
3 is a structural isomer of **(1–2)** and of **4**.
4 is a geometrical isomer of **(1–2)**.

4.2   **(a)** 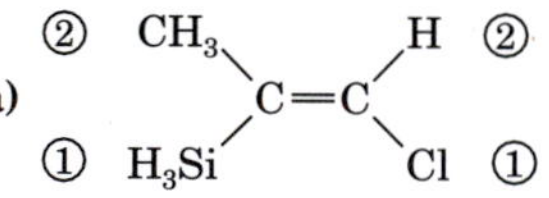 (Z)   **(b)** 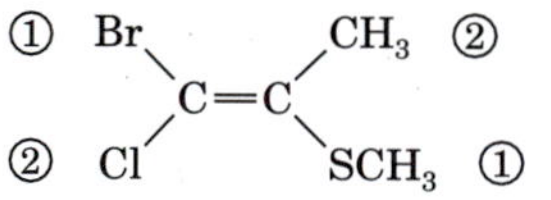 (E)

4.3 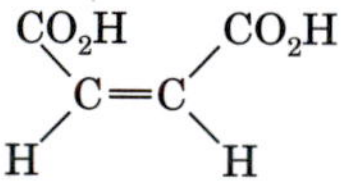   *maleic acid* 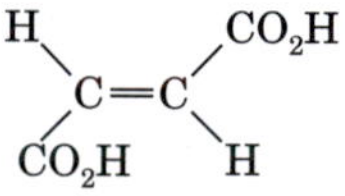   *fumaric acid*

4.4   **(a)** (E):   **(b)** (Z):   **(c)** (E):   **(d)** (Z)

4.5   **(a)** 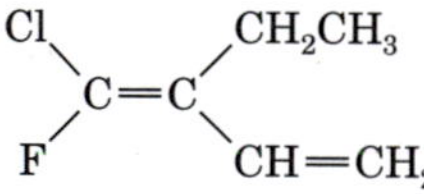   **(b)**

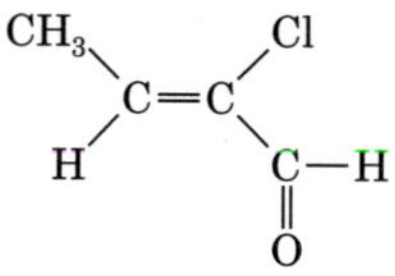

4.6

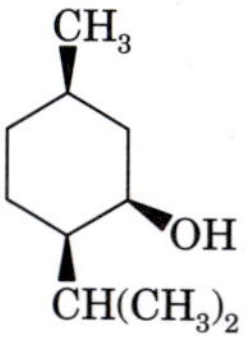

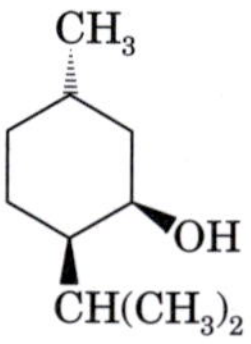

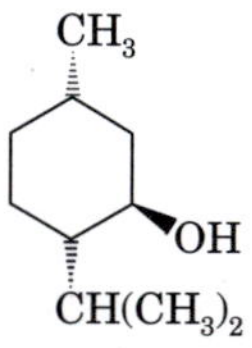

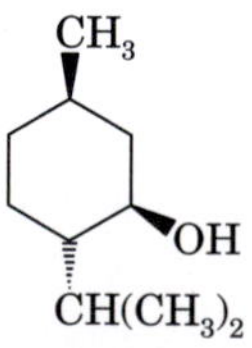

menthol

4.7   **(a)** 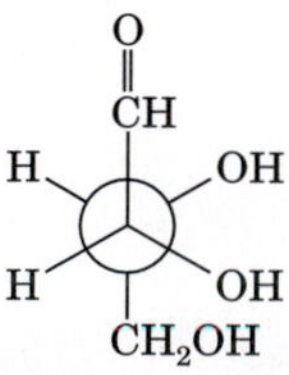   **(b)**

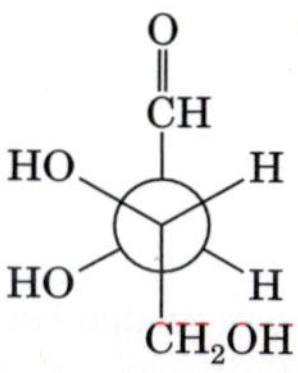

4.8 **(a)**

Br / H, H / H, H / Cl
anti

Br / H, Cl / H, H / H
gauche

(b)

CO₂H / H, H / H, H / OH
anti

CO₂H / H, OH / H, H / H
gauche

4.9 **(a)** *trans e,e;* **(b)** *cis a,e;* **(c)** *trans e,e;* **(d)** *trans e,e*

4.10 **(a)** $CH_3\overset{*}{C}H\overset{O}{\overset{\|}{C}}OH$ $\underset{NH_2}{|}$
(b) no chiral carbon
(c) $ClCH_2\overset{*}{C}HCH_2OH$ $\underset{CH_3}{|}$
(d) $HO\overset{*}{C}H$ with $\overset{O}{\overset{\|}{C}H}$ above and CH_2OH below

4.11 **(a)** $C_6H_5\blacktriangleright\overset{H}{\underset{CH_3}{C}}\blacktriangleleft Br$ $Br\blacktriangleright\overset{H}{\underset{CH_3}{C}}\blacktriangleleft C_6H_5$
dimensional
(b) $H\blacktriangleright\overset{CH_3}{\underset{CH_2CH_3}{C}}\blacktriangleleft OH$ $HO\blacktriangleright\overset{CH_3}{\underset{CH_2CH_3}{C}}\blacktriangleleft H$
dimensional

4.12 **(a)** $\underset{(CH_3)_2CH}{}\overset{CO_2H}{\underset{NH_2}{C}}\cdots H$ **(b)** $\underset{HO_2CCH_2}{}\overset{CO_2H}{\underset{NH_2}{C}}\cdots H$ (There are other correct dimensional formulas.)

4.13 **(a)** $H_2N-\overset{CO_2H}{\underset{CH_3}{|}}-H$ **(b)** $H_2N-\overset{CO_2H}{\underset{CH_2SH}{|}}-H$ (There are other correct Fischer projections.)

4.14 **(a)** Zero, because it is a racemic mixture.

(b) The solution is, in effect, half racemic and half (S)-enantiomer. The observed rotation is therefore half that of the pure (S)-enantiomer, or +8.0°.

4.15 **(a)** (R); **(b)** (S); **(c)** (R); **(d)** (S)

4.16 **(a)** $\underset{CH_3CH_2\ (CH_2)_3CH_3}{}\overset{Br}{C}\cdots H$ **(b)** $\underset{CH_3CH_2CH_2\ CH_3}{}\overset{OH}{C}\cdots H$

4.17 **(a)** four; **(b)** two; **(c)** eight

4.18

$H\blacktriangleright\overset{CH_3}{C}\blacktriangleleft Cl$
$H\blacktriangleright\overset{}{\underset{CH_3}{C}}\blacktriangleleft Cl$
meso

$Cl\blacktriangleright\overset{CH_3}{C}\blacktriangleleft H$
$H\blacktriangleright\overset{}{\underset{CH_3}{C}}\blacktriangleleft Cl$

$H\blacktriangleright\overset{CH_3}{C}\blacktriangleleft Cl$
$Cl\blacktriangleright\overset{}{\underset{CH_3}{C}}\blacktriangleleft H$

enantiomers

4.19   Make models of the (*R,R*) and (*S,S*) enantiomers to verify the mirror-image relationships of the conformers and to verify the nonsuperimposability of the enantiomers.

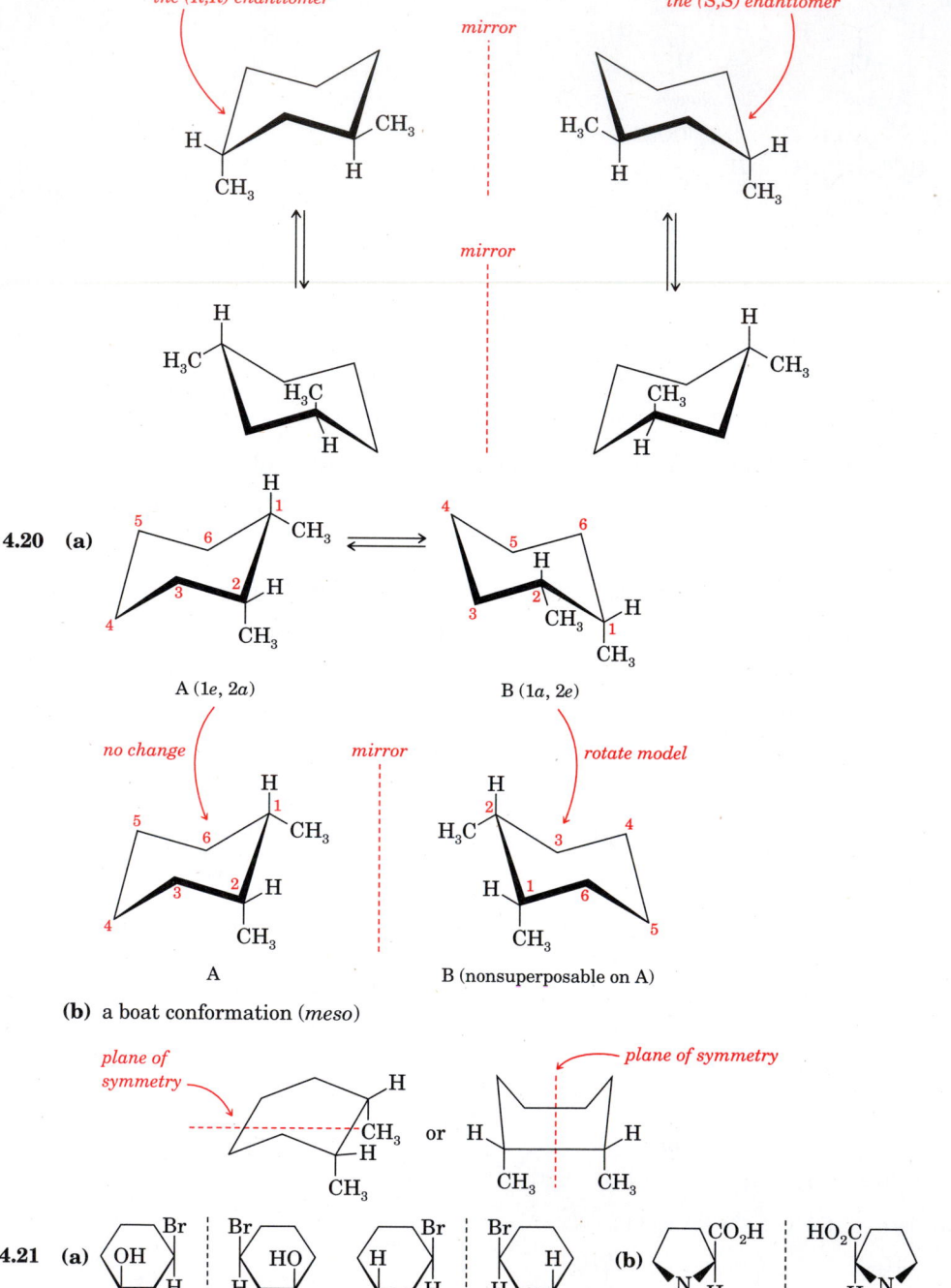

the (R,R) enantiomer   *mirror*   *the (S,S) enantiomer*

mirror

4.20   **(a)**

A (1*e*, 2*a*)   B (1*a*, 2*e*)

no change   *mirror*   *rotate model*

A   B (nonsuperposable on A)

(b) a boat conformation (*meso*)

plane of symmetry   *plane of symmetry*

or

4.21   **(a)**   **(b)**

the cis enantiomers   *the trans enantiomers*   *enantiomers*

Neither structure has a *meso* form.

4.22 **(a)**

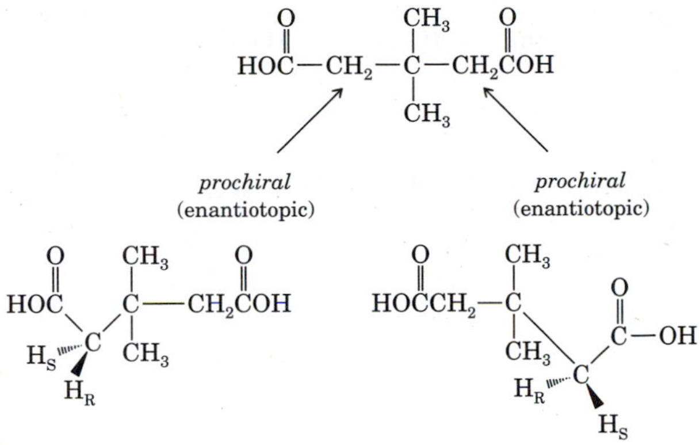

prochiral
(enantiotopic)

prochiral
(enantiotopic)

(b)

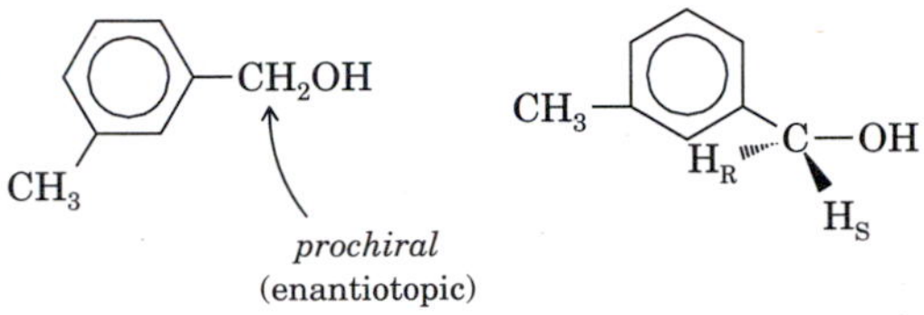

prochiral
(enantiotopic)

(c) $CH_3CH{-}CH_2CN$
         $\underset{OH}{|}$

prochiral
(diastereotopic)

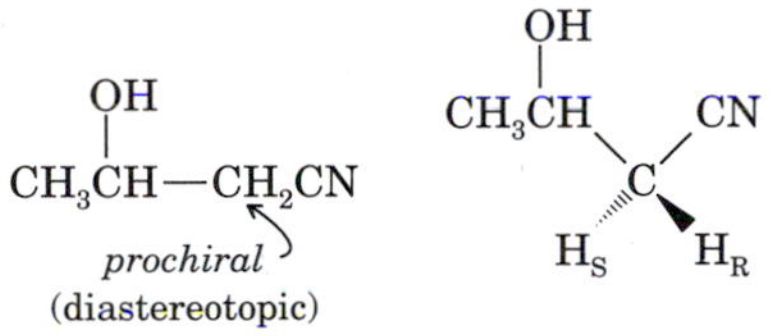

In each case, the hydrogen that is replaced by Y is H_R or H_S.

4.23 (1) Treat the optically active amine with the racemic carboxylic acid to obtain a mixture of diastereomeric salts.

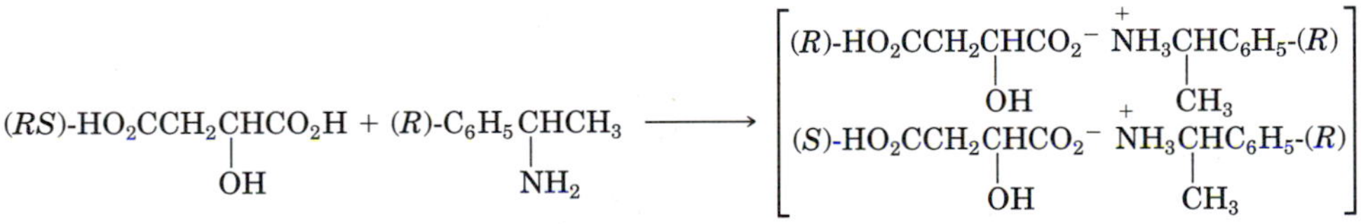

(2) Separate the mixture of diastereomers by a physical process, such as fractional crystallization.

$$\begin{bmatrix} (R,R)\text{-salt} \\ (S,R)\text{-salt} \end{bmatrix} \xrightarrow[\text{crystallization}]{\text{fractional}} (R,R)\text{-salt} + (S,R)\text{-salt}$$

(3) To obtain the protonated carboxylic acid free of the amine, treat each salt with aqueous mineral acid. The amine would remain as a water-soluble salt; thus, a water-insoluble carboxylic acid could be extracted by an immiscible organic solvent. The dicarboxylic acid in this problem would also be water soluble. Therefore, a continuous liquid–liquid extraction with an immiscible organic solvent would be required to isolate the carboxylic acid free from the amine salt.

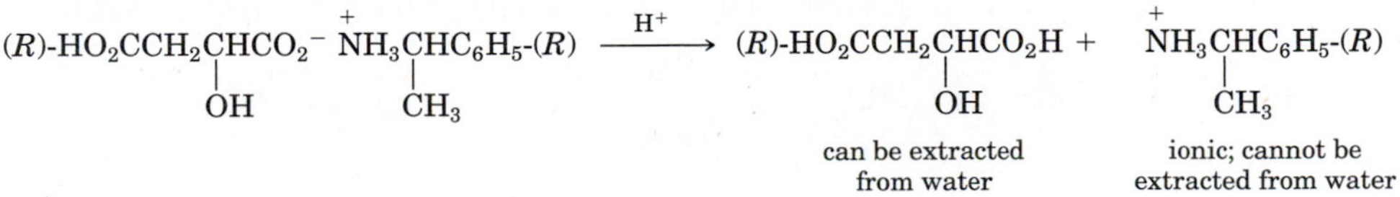

can be extracted
from water

ionic; cannot be
extracted from water

The same procedure can be used to obtain the (*S*)-dicarboxylic acid.

Chapter 5

5.1 **(a)** 3-bromopentane; 2°   **(b)** 3-chloro-2-ethylcyclohexanol; 2°

        (c) *cis*-1,4-dibromocyclohexane, both 2°   **(d)** 4-chloro-1-phenyl-2-butanol; 1°

5.2 **(a)** —Br

(b) $HOCH_2CH$—$CHCH_2CH_3$
with Cl and Cl below the two CH carbons

(c) $CH_3CH_2CHCCH_2CH_2CH_3$ with CH_3 above, and I and CH_2CH_3 below

(d) $CH_3CH=CHCH_2CHBr_2$

5.3 **(a)** phenyl—$CHCH_3$ (Br below) $+ ^-OH \longrightarrow$ phenyl—$CHCH_3$ (OH below) $+$ phenyl—$CH=CH_2 + H_2O$ $+$

(b) cyclohexyl—$Br + CH_3O^- \longrightarrow$ cyclohexyl—$OCH_3 +$ cyclohexene $+ CH_3OH$

(c) $CH_3CH_2CHCH_2CH_3$ (Br below) $+ ^-OH \longrightarrow CH_3CH_2CHCH_2CH_3$ (OH below) $+ CH_3CH=CHCH_2CH_3 + H_2O$

(d) cyclopentyl with Br $+ CH_3CH_2O^- \longrightarrow$ cyclopentyl with OCH_2CH_3 $+$ cyclopentene $+ CH_3CH_2OH$

5.4 $NC^- +$ CH_3CH_2, H (S) C—Br $\longrightarrow$ NC—C with H, CH_2CH_3 (R), CH_3 $+ Br^-$

5.5 **(a)** transition state with δ^- OCH_3, —H, Br, δ^- (chair cyclohexane)

(b) transition state with δ^- Br, H, CH_3, CH_3—, —H, CH_3, δ^- OH

transition state

5.6 **(a)** The 1-bromo-2-phenylethane will react more rapidly because bromobenzene (an aryl halide) does not undergo S_N2 reactions.

(b) $CH_3CH_2CH_2Cl$ will react more rapidly because it is a 1° alkyl halide and has a less sterically hindered S_N2 transition state.

(c) Iodocyclohexane will react more rapidly. An iodo substituent is a better leaving group than a chloro substituent because iodine is more polarizable.

(d) C_4H_9Cl will react with the nucleophile more rapidly because it is a 1° alkyl halide and has a less sterically hindered S_N2 transition state. Tertiary alkyl halides do not undergo S_N2 reactions.

5.7 **(a)** $(CH_3CH_2)_3COH + HI$ **(b)** cyclohexane with CH_3 and OCH_3 $+ HCl$

5.8 **(a)** $\xrightarrow{-Br^-}$ [indane cation with O_2N] $\xrightarrow[(2) -H^+]{(1) CH_3OH}$ indane with OCH_3 and O_2N

(b) $\xrightarrow{-Br^-}$ $\left[C_6H_5-\overset{+}{\underset{CH_3}{C}}\cdots H \right]$ $\xrightarrow[\text{(2) }-H^+]{\text{(1) } H\overset{O}{\overset{\|}{C}}OH}$ $(RS)-C_6H_5\overset{O\overset{\|}{C}H}{\underset{}{\overset{|}{C}H}}CH_3$

achiral   *racemic*

(c) $\xrightarrow{-Cl^-}$ $\left[(CH_3)_3C\cdots \overset{\overset{CH_3}{+}}{\bigcirc} \atop H \right]$ $\xrightarrow[\text{(2) }-H^+]{\text{(1) } H_2O}$

$(CH_3)_3C\cdots \overset{\overset{OH}{\overset{|}{\bigcirc}}CH_3}{} \atop H$ + $(CH_3)_3C\cdots \overset{\overset{CH_3}{\overset{|}{\bigcirc}}OH}{} \atop H$

5.9   **(b)** < **(a)** < **(c)**

5.10   **(a)** no rearranged carbocation

(b) $CH_3\overset{+}{C}CH_2CH_3$
            $\underset{CH_3}{|}$

(c) $\bigcirc\!\!\!+\!\!-CH_3$ **(d)** no rearranged carbocation

5.11   **(a)** $(CH_3)_2\overset{CH_3}{\underset{:\ddot{B}r:}{\overset{|}{C}}}-CHCH_2CH_3$ $\underset{\longrightarrow}{\overset{-:\ddot{B}r:^-}{\longleftarrow}}$ $\left[(CH_3)_2\overset{CH_3}{\overset{|}{C}}-\overset{+}{C}HCH_2CH_3 \right]$ $\longrightarrow$ $\left[(CH_3)_2\overset{+}{C}-\overset{CH_3}{\overset{|}{C}}HCH_2CH_3 \right]$

nonrearranged carbocation: $\left[(CH_3)_3C\overset{+}{C}HCH_2CH_3 \right]$

$\xrightarrow{H_2\ddot{O}:}$ $(CH_3)_3C\overset{\overset{+}{:}\overset{OH_2}{}}{\underset{}{C}}HCH_2CH_3$ $\xrightarrow{-H^+}$ $(CH_3)_3C\overset{OH}{\overset{|}{C}}HCH_2CH_3$

$\xrightarrow{CH_3CO_2H}$ $(CH_3)_3C\underset{\underset{+\overset{\|}{O}}{HO\overset{\cdot\cdot}{C}CH_3}}{\overset{|}{C}}HCH_2CH_3$ $\xrightarrow{-H^+}$ $(CH_3)_3C\underset{O_2CCH_3}{\overset{|}{C}}HCH_2CH_3$

rearranged carbocation: $\left[(CH_3)_2\overset{+}{C}CHCH_2CH_3 \right]$
                                                  $\underset{CH_3}{|}$

$\xrightarrow{H_2\ddot{O}:}$ $(CH_3)_2\overset{\overset{+}{:}\overset{OH_2}{}}{\underset{\underset{CH_3}{|}}{C}}CHCH_2CH_3$ $\xrightarrow{-H^+}$ $(CH_3)_2\overset{OH}{\underset{\underset{CH_3}{|}}{\overset{|}{C}}}CHCH_2CH_3$

$\xrightarrow{CH_3CO_2H}$ $(CH_3)_2\overset{\overset{+}{\overset{\cdot\cdot}{HO}}\overset{O}{\overset{\|}{C}}CH_3}{\underset{\underset{CH_3}{|}}{C}}CHCH_2CH_3$ $\xrightarrow{-H^+}$ $(CH_3)_2\overset{O_2CCH_3}{\underset{\underset{CH_3}{|}}{\overset{|}{C}}}CHCH_2CH_3$

(b)

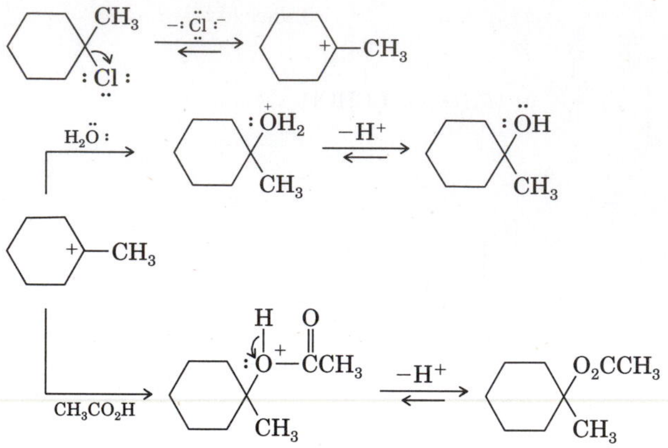

5.12 **(a)** allylic;   **(b)** vinylic;   **(c)** aryl;   **(d)** benzylic

5.13 **(a)**

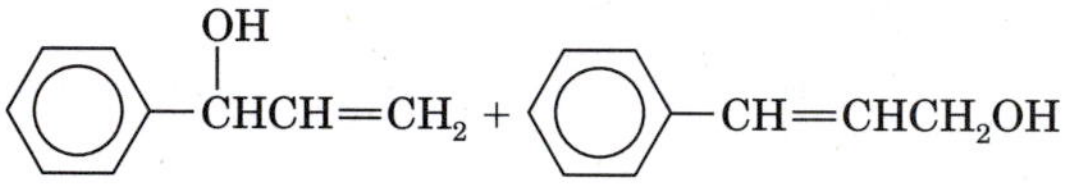

(b)

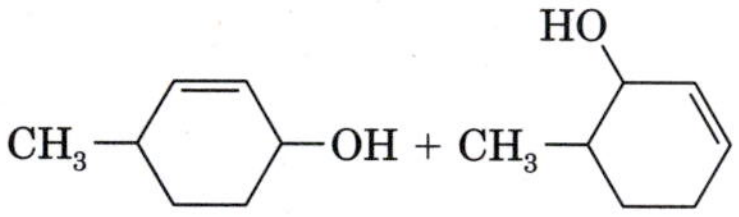

5.14

5.15 **(a)** (E)- and (Z)-$CH_3CH=CHCH_2CH_3$ + $CH_2=CHCH_2CH_2CH_3$

(b) (E)- and (Z)-$CH_3\underset{\underset{CH_2CH_3}{|}}{C}=CHCH_3$ + $CH_2=C(CH_2CH_3)_2$

5.16

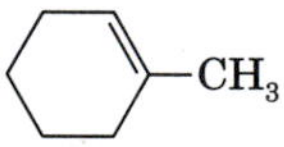

5.17

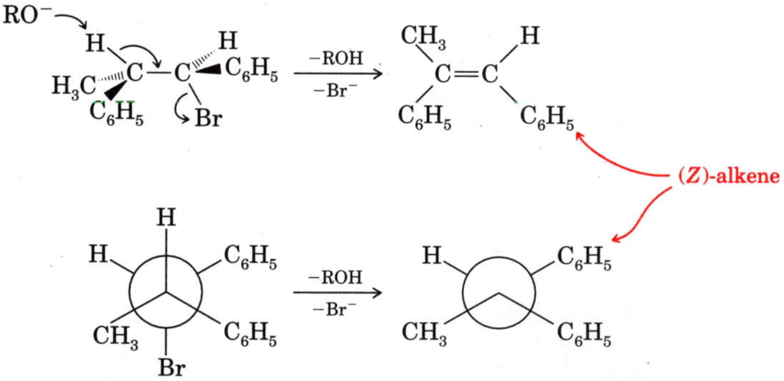

(Z)-alkene

5.18 The enantiomers yield (Z)-1-bromo-1,2-diphenylethene (*trans*-phenyls); the *meso* form yields the (E)-isomer (*cis*-phenyls).

5.19 **(a)** $(CH_3)_2CHCHBrCH_2CH_3 \xrightarrow{-HBr} (CH_3)_2C=CHCH_2CH_3$ + $(CH_3)_2CHCH=CHCH_3$

Saytzeff   *Hofmann*

(b)

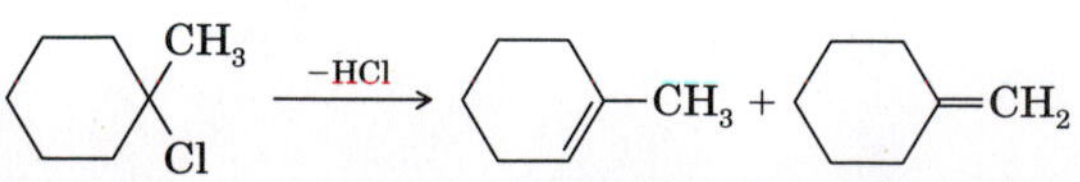

Saytzeff   *Hofmann*

5.20 CH_3S^- is the better nucleophile because S is larger and more polarizable than O.

5.21 **(a)** $CH_3(CH_2)_5\overset{\overset{\displaystyle OH}{|}}{C}HCH_3$ + $CH_3(CH_2)_5\overset{\overset{\displaystyle OCH_2CH_3}{|}}{C}HCH_3$ + *trans*-$CH_3(CH_2)_4CH{=}CHCH_3$

The two substitution products are formed by an S_N1 path, and the alkene by an E1 path. The alkyl halide is secondary. A weak nucleophile, but polar solvent, is used. These conditions favor carbocation reactions, while the higher temperature favors elimination.

(b) $HOCH_2\overset{\overset{\displaystyle OH}{|}}{C}HCH_2O{-}\langle\bigcirc\rangle$ + NaCl

The organic halide is primary. A reasonably strong nucleophile (but weak base) is used. We would expect reaction to occur by an S_N2 path.

(c) $CH_3CH_2OCH_2CH_2CN$ + NaCl

The halide is primary and ^-CN is a strong nucleophile. We would expect substitution by an S_N2 path.

(d) $CH_3\overset{\overset{\displaystyle OCH_2CH_3}{|}}{C}HCH{=}CH_2$ + $CH_3CH{=}CHCH_2OCH_2CH_3$ + HCl

The halide is allylic. The solvent is polar and weakly nucleophilic. We would expect substitution by an S_N1 path and the possibility of allylic rearrangement.

5.22 **(a)** ^-OH **(b)** $CH_3CO_2^-$ **(c)** CH_3S^- **(d)** $C_6H_5O^-$

5.23 **(a)** $\langle\bigcirc\rangle{-}O^-$ + $CH_3CH_2CH_2I \longrightarrow$

(b) $C_6H_5CH_2Br$ + CH_3S^- or $C_6H_5CH_2S^-$ + $CH_3I \longrightarrow$

(c) CH_3CH_2Br + $^-OCH_2CH_2CH_3$ or $CH_3CH_2CH_2Br$ + $^-OCH_2CH_3 \longrightarrow$

(d) $C_6H_5CO_2^-$ + $CH_2{=}CHCH_2Br \longrightarrow$

(e) $C_6H_5O^-$ + $CH_3CH_2Br \longrightarrow$

(f) $(CH_3)_2CHCH_2CHBrCH_3 \xrightarrow[CH_3OH]{^-OCH_3}$ or $(CH_3)_2CHCHBrCH_2CH_3 \xrightarrow[(CH_3)_3COH]{^-OC(CH_3)_3}$

Chapter 6

6.1 **(a)** $H{:}\overset{\cdot\cdot}{\underset{\cdot\cdot}{S}}{\cdot}$ **(b)** $H{:}\overset{\overset{\displaystyle H}{|}}{\underset{\underset{\displaystyle H}{|}}{C}}{:}\overset{\overset{\displaystyle H}{|}}{\underset{\underset{\displaystyle H}{|}}{C}}{:}\overset{\cdot\cdot}{\underset{\cdot\cdot}{O}}{\cdot}$ **(c)** $H{:}\overset{\overset{\displaystyle \cdot}{}}{\underset{\underset{\displaystyle H}{|}}{C}}{:}\overset{\cdot\cdot}{\underset{\cdot\cdot}{Cl}}{:}$

6.2 *initiation:* $Cl_2 \xrightarrow{h\nu} 2\ Cl{\cdot}$

propagation: $\langle\hexagon\rangle$ + Cl· $\longrightarrow$ $\langle\hexagon\rangle$·+ HCl

$\langle\hexagon\rangle$· + Cl_2 $\longrightarrow$ $\langle\hexagon\rangle{-}Cl$ + Cl·

termination: $\langle\hexagon\rangle$· + Cl· $\longrightarrow$ $\langle\hexagon\rangle{-}Cl$ or other combination of two radicals

6.3 $CHCl_3$ + Cl· $\longrightarrow$ ·CCl_3 + HCl

·CCl_3 + Cl_2 $\longrightarrow$ CCl_4 + Cl·

6.4 nine

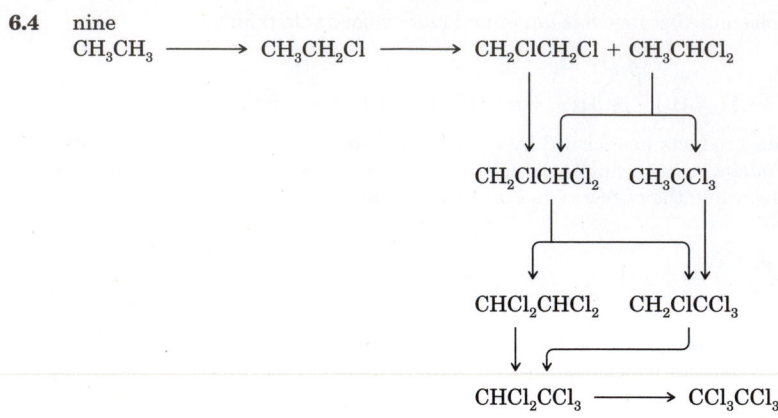

$$CH_3CH_3 \longrightarrow CH_3CH_2Cl \longrightarrow CH_2ClCH_2Cl + CH_3CHCl_2$$

$$CH_2ClCHCl_2 \quad CH_3CCl_3$$

$$CHCl_2CHCl_2 \quad CH_2ClCCl_3$$

$$CHCl_2CCl_3 \longrightarrow CCl_3CCl_3$$

6.5 **(a)**

	ΔH (kcal/mol)
$(CH_3)_3CH \longrightarrow (CH_3)_3C\cdot + H\cdot$	91
$Cl-Cl \longrightarrow 2Cl\cdot$	58
$(CH_3)_3C\cdot + Cl\cdot \longrightarrow (CH_3)_3CCl$	-83
$H\cdot + Cl\cdot \longrightarrow HCl$	$\underline{-103}$
	-37

(b)

	ΔH (kcal/mol)
$CH_4 \longrightarrow CH_3\cdot + H\cdot$	104
$Cl_2 \longrightarrow 2\,Cl\cdot$	58
$CH_3\cdot + Cl\cdot \longrightarrow CH_3Cl$	-83.5
$H\cdot + Cl\cdot \longrightarrow HCl$	$\underline{-103}$
	-24.5

Reaction **(a)** releases more energy.

6.6 not racemic,

$$\underset{(RS)}{CH_3}\overset{Cl}{\underset{|}{C}}H\underset{\underset{(R)}{|}}{\overset{}{C}}H CH_2Cl$$
$$CH_3$$

because the old chiral center remains (R). The newly generated chiral center is (R) and (S); therefore, a mixture of diastereomers is formed: (R,R) and (S,R).

6.7 **(a)** — 3° and benzylic

H_3C

$-C\textcircled{H}(CH_3)_2$

(b) — 2°, allylic, and benzylic

$-C\textcircled{H_2}CH=CH_2$

(c) — 2° and allylic

CH_3

$-\overset{|}{\underset{|}{C}}CH=CHC\textcircled{H_2}CH_3$

CH_3

(d)

CH_3

— 2° and allylic

— 3° and benzylic

CH_3

6.8   In each case, hydrogen (1) would be extracted at a faster rate than hydrogen (2).

6.9   **(a)** $CH_3CH\!=\!CH\!-\!CH_2 \longleftrightarrow CH_3\dot{C}H\!-\!CH\!=\!CH_2$

(b)

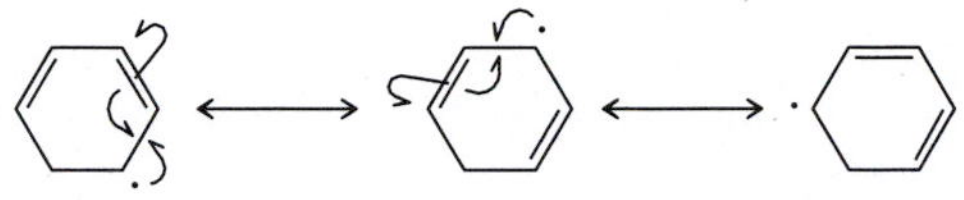

(c)

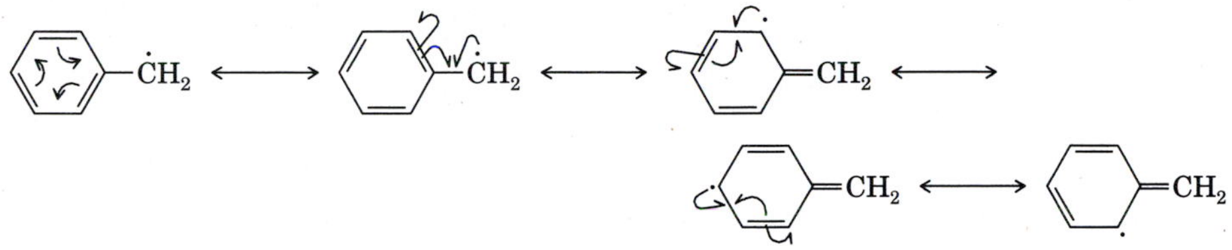

6.10   **(a)** $C_6H_5CHBrCH_2CH_2CH_3$;   **(b)** $C_6H_5CH\!=\!CHCH_2Br$

6.11

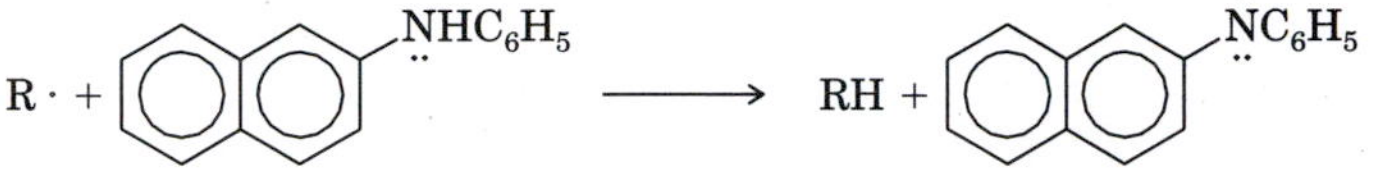

relatively stable free radical
(resonance-stabilized by both the
phenyl and naphthyl groups)

6.12 $-\!(CF_2CF_2)_x\!-$

6.13 $CH_2\!=\!CHCN$

Chapter 7

7.1   **(a)** 2,2-dimethylpropanol

(b) 4-*sec*-butyl-2,5-cyclohexadienol

(c) 1,2-propanediol or propylene glycol

7.2   **(a)**   **(b)**

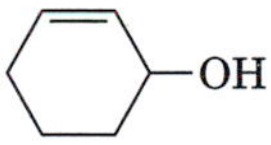

(c) 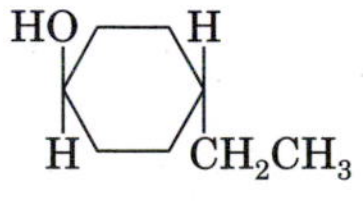   **(d)**

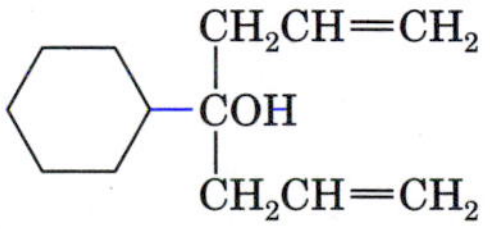

7.3   **(a)** 2°, benzylic;   **(b)** 3°, but *not* allylic;   **(c)** 2°, allylic

7.4   **(a)**

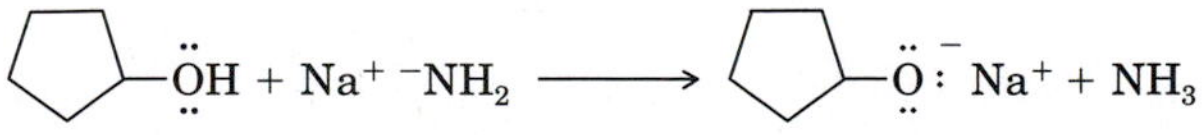

(b)

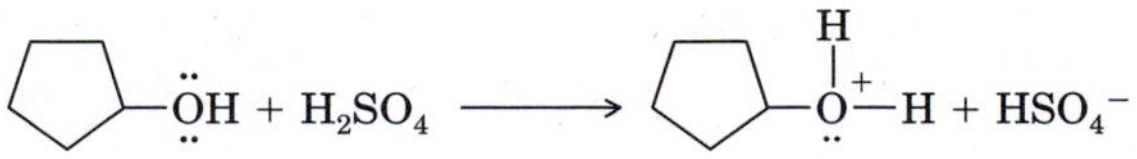

7.5   **(a)** $CH_3CH{=}CHCH_3$ $\xrightarrow{H_2O,\ H^+}$

$CH_3\overset{O}{\overset{\|}{C}}CH_2CH_3$ $\xrightarrow[\text{(2) } H_2O,\ H^+]{\text{(1) } NaBH_4}$ → $CH_3\overset{OH}{\underset{|}{C}}HCH_2CH_3$

(b) CH_3—⬠—CH_3 $\xrightarrow{H_2O,\ H^+}$

CH_3—⬠(O)—CH_3 $\xrightarrow[\text{(2) } H_2O,\ H^+]{\text{(1) } NaBH_4}$ → CH_3—⬠(OH)—CH_3

7.6   **(a)** $C_6H_5\overset{Br}{\underset{|}{C}}HCH_3$ **(b)** $CH_3CH_2\overset{Br}{\underset{|}{C}}HCH_2CH_3$

(c) $(R)\text{-}C_6H_5CH_2\overset{Br}{\underset{|}{C}}HCH_3$ **(d)** $(CH_3)_3C$⬡(H, H)Br

In each case, the chloro or iodo compound could also have been used.

7.7   **(a), (d),** and **(f),** which contain acidic hydrogens that would destroy any Grignard reagent formed.

7.8   **(a)** C_6H_5Br $\xrightarrow[\text{ether}]{Mg}$ C_6H_5MgBr $\xrightarrow[\text{(2) } H_2O,\ H^+]{\text{(1) } HCHO}$ $C_6H_5CH_2OH$

(b) $CH_3(CH_2)_3Cl$ $\xrightarrow{^-OH}$ $CH_3(CH_2)_3OH$

(c) CH_3CH_2Br $\xrightarrow[\text{ether}]{Mg}$ CH_3CH_2MgBr $\xrightarrow[\text{(2) } H_2O,\ H^+]{\text{(1) } C_6H_5CHO}$ $C_6H_5\overset{OH}{\underset{|}{C}}HCH_2CH_3$

7.9   **(a)** (1) ⬡—CH_2CH_2MgBr $\xrightarrow[\text{(2) } H_2O,\ H^+]{\text{(1) } CH_3\overset{O}{\overset{\|}{C}}CH_2CH_3}$

(2) CH_3MgBr $\xrightarrow[\text{(2) } H_2O,\ H^+]{\text{(1) } C_6H_5CH_2CH_2\overset{O}{\overset{\|}{C}}CH_2CH_3}$

(3) CH_3CH_2MgBr $\xrightarrow[\text{(2) } H_2O,\ H^+]{\text{(1) } C_6H_5CH_2CH_2\overset{O}{\overset{\|}{C}}CH_3}$

(b) (1) CH_3CH_2MgBr $\xrightarrow[\text{(2) } H_2O,\ H^+]{\text{(1) } C_6H_5\overset{O}{\overset{\|}{C}}CH_3}$

(2) C_6H_5MgBr $\xrightarrow[\text{(2) } H_2O,\ H^+]{\text{(1) } CH_3CH_2\overset{O}{\overset{\|}{C}}CH_3}$

(3) CH_3MgBr $\xrightarrow[\text{(2) } H_2O,\ H^+]{\text{(1) } CH_3CH_2\overset{O}{\overset{\|}{C}}C_6H_5}$

7.10   **(a)** (furan-ring)$-CH_2OH$ $+ PBr_3$ ⟶ (furan-ring)$-CH_2Br$ $+ H_3PO_3$

Although HBr could also be used for this conversion, some ring opening would occur (see Section 8.2).

(b) [furan ring structure]—CH$_2$OH + SOCl$_2$ $\xrightarrow{\text{pyridine}}$ [furan ring structure]—CH$_2$Cl + SO$_2$ + HCl

PCl$_3$ or HCl (ZnCl$_2$ catalyst) would also be suitable reagents.

7.11 [cyclopentane ring with H, OH, D] $\xrightarrow[-\text{HCl}]{\text{SOCl}_2}$ [cyclopentane ring with H, H, OSOCl, D] $\xrightarrow{\text{Cl}^-}$ [cyclopentane ring with H, Cl, H, D] + SO$_2$ + Cl$^-$

trans D and Cl

7.12 **(a)** CH$_3$CH$_2$CH=C(CH$_3$)$_2$ (most) + CH$_3$CH$_2$CH$_2$C=CH$_2$ with CH$_3$ substituent

(b) *cis*- and *trans*-CH$_3$CH=CHCH$_3$ (mostly *trans*) + CH$_3$CH$_2$CH=CH$_2$

(c) [bicyclohexylidene structure] (most) + [cyclohexyl-cyclohexene structure]

(d) (C$_6$H$_5$)$_2$C=CHCH$_3$

7.13 (CH$_3$)$_2$C—CHCH$_3$ with CH$_3$ and :OH $\xrightarrow{\text{H}^+}$ (CH$_3$)$_2$C—CHCH$_3$ with CH$_3$ and :OH$_2^+$ $\xrightarrow{-\text{H}_2\text{O}}$

(CH$_3$)$_2$C—CHCH$_3$ with CH$_3$ and + $\xrightarrow{\text{1,2 methyl shift}}$ (CH$_3$)$_2$C$^+$—CCH$_3$ with CH$_3$ and H $\xrightarrow{-\text{H}^+}$ (CH$_3$)$_2$C=C(CH$_3$)$_2$

7.14 CH$_3$S(=O)(=O)—O:$^-$ $\longleftrightarrow$ CH$_3$S(=O)=O: with :O:$^-$ $\longleftrightarrow$ CH$_3$S=O: with :O: and :O:$^-$

7.15 (S)-CH$_3$CH(CH$_2$)$_5$CH$_3$ with OH

7.16 **(b)** < **(c)** < **(a)**; **(a)** is the most oxidized of the three.

7.17 **(a)** oxidizing agent   **(b)** reducing agent   **(c)** neither

(d) reducing agent   **(e)** neither   **(f)** oxidizing agent

7.18 **(a)** [cyclopentane]=O;   **(b)** C$_6$H$_5$CO$_2$H

7.19 CH$_3$CH$_2$OH $\xrightarrow{\text{H}_2\text{CrO}_4}$ CH$_3$CO$_2$H $\xrightarrow[\text{heat}]{\text{CH}_3\text{CH}_2\text{OH, H}^+}$ CH$_3$CO$_2$CH$_2$CH$_3$

7.20 **(a)** (CH$_3$)$_2$CHCH$_2$Br $\xrightarrow[\text{ether}]{\text{Mg}}$ (CH$_3$)$_2$CHCH$_2$MgBr $\xrightarrow[\text{(2) H}_2\text{O, H}^+]{\text{(1) CH}_3\text{CCH}_2\text{CH}_3 \text{ (O)}}$ product

(b) [cyclohexyl]—Br $\xrightarrow[\text{ether}]{\text{Mg}}$ [cyclohexyl]—MgBr $\xrightarrow[\text{(2) H}_2\text{O, H}^+]{\text{(1) HCHO}}$ product

(c) CH_2=$CHCH_2Br$ $\xrightarrow[\text{ether}]{\text{Mg}}$ CH_2=$CHCH_2MgBr$

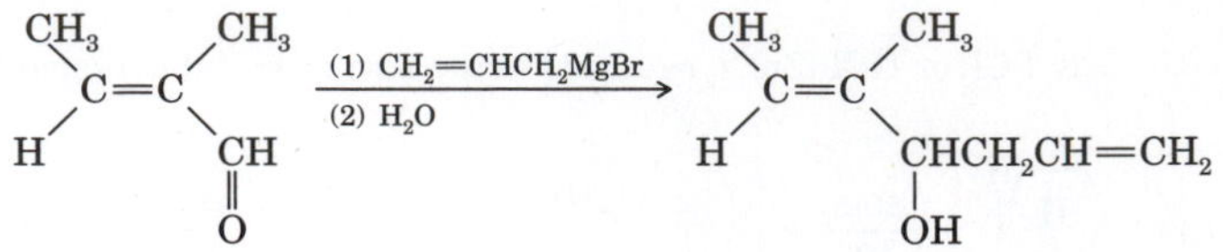

(*E*) isomer

7.21 **(a)** 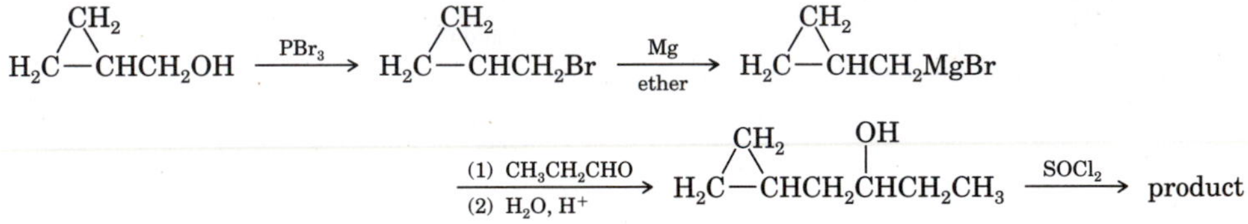 $\xrightarrow{\text{PBr}_3}$ $\xrightarrow[\text{ether}]{\text{Mg}}$

$\xrightarrow[\text{(2) } H_2O, H^+]{\text{(1) } CH_3CH_2CHO}$ $\xrightarrow{\text{SOCl}_2}$ product

(b) CH_3CH_2OH $\xrightarrow{\text{HBr}}$ CH_3CH_2Br $\xrightarrow[\text{ether}]{\text{Mg}}$ CH_3CH_2MgBr

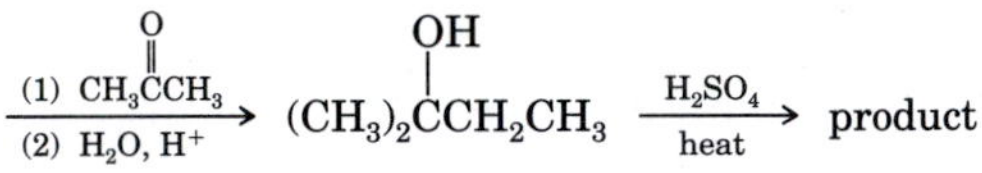

$\xrightarrow[\text{(2) } H_2O, H^+]{\text{(1) } CH_3CCH_3}$ $(CH_3)_2CCH_2CH_3$ (OH) $\xrightarrow[\text{heat}]{H_2SO_4}$ product

(c) $(CH_3)_2CHCH_2OH$ $\xrightarrow{\text{PBr}_3}$ $(CH_3)_2CHCH_2Br$ $\xrightarrow[\text{ether}]{\text{Mg}}$ $(CH_3)_2CHCH_2MgBr$

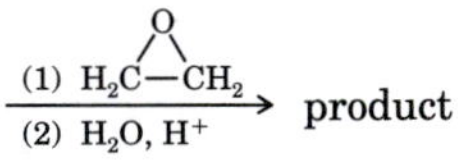

$\xrightarrow[\text{(2) } H_2O, H^+]{\text{(1) } H_2C-CH_2 \text{ (epoxide)}}$ product

(d) 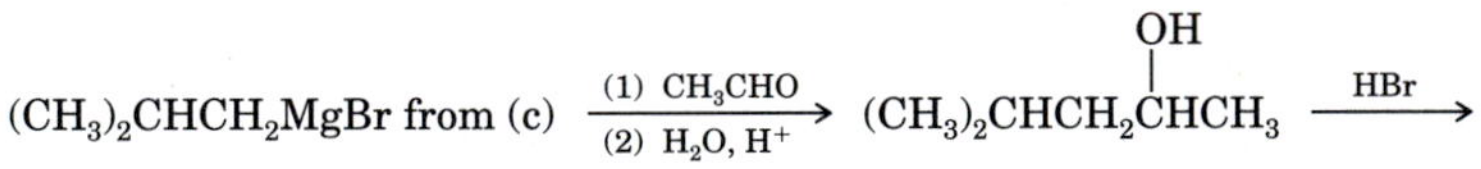 $(CH_3)_2CHCH_2MgBr$ from (c) $\xrightarrow[\text{(2) } H_2O, H^+]{\text{(1) } CH_3CHO}$ $(CH_3)_2CHCH_2CHCH_3$ (OH) $\xrightarrow{\text{HBr}}$

$(CH_3)_2CHCH_2CHBrCH_3$ $\xrightarrow{\text{CN}^-}$ product

Chapter 8

8.1 **(a)** 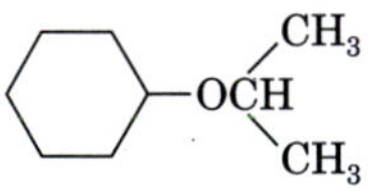 **(b)** 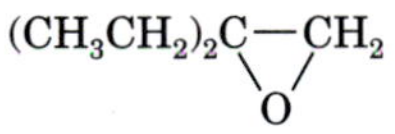 $(CH_3CH_2)_2C-CH_2$ (O) **(c)** 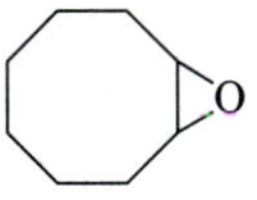 **(d)**

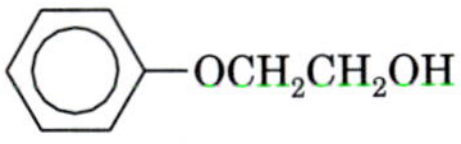

8.2 **(a)** 1,4-di-*tert*-butoxybenzene

(b) 1,2-epoxycyclohexane (cyclohexene oxide)

8.3 **(a)** $CH_3CH_2Br + {}^-OCH(CH_3)_2 \longrightarrow CH_3CH_2OCH(CH_3)_2 + Br^-$

(b) $CH_3CH_2CH_2Br + {}^-OC(CH_3)_3 \longrightarrow CH_3CH_2CH_2OC(CH_3)_3 + Br^-$

(c)

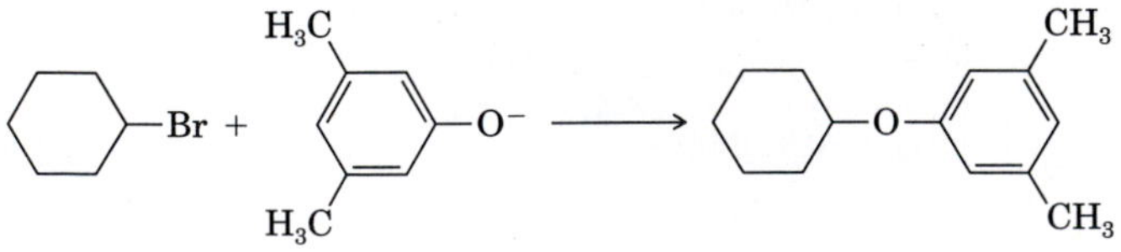

(d) $C_6H_5CH_2Br + C_6H_5O^- \longrightarrow C_6H_5CH_2OC_6H_5$

8.4

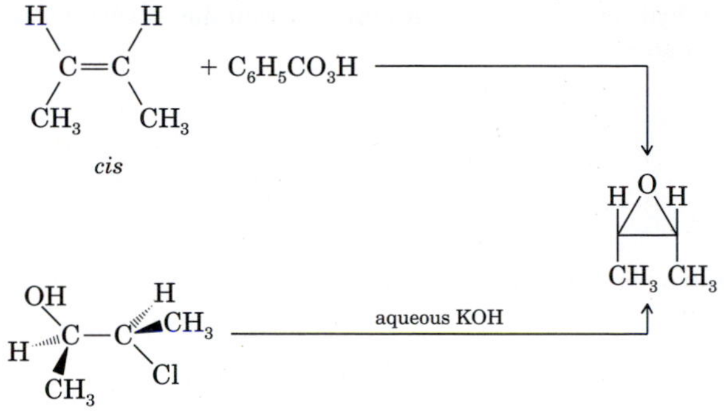

(1R,2R) or (1S,2S)

8.5 **(a)**

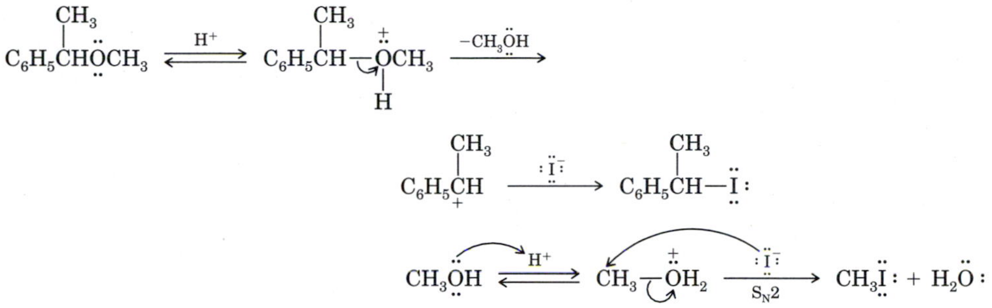

(b)

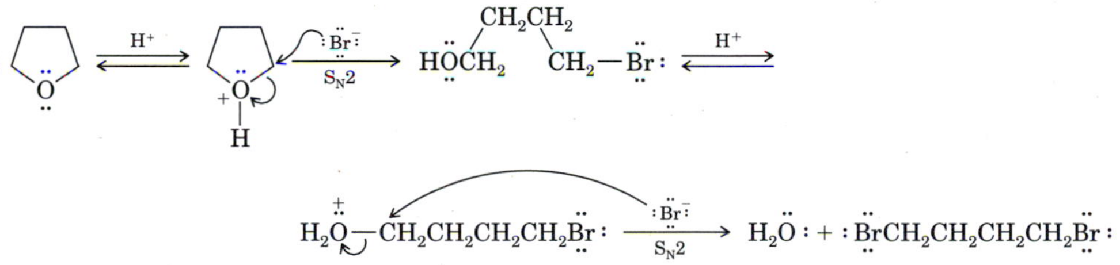

8.6 **(a)** $(CH_3)_3CCH_2CH_2OH \xrightarrow{\text{HBr}} (CH_3)_3CCH_2CH_2Br$

$$\xrightarrow{\text{Mg}} (CH_3)_3CCH_2CH_2MgBr \xrightarrow[\substack{(1) \ CH_2-CH_2 \\ \qquad \text{O} \\ (2) \ H_2O, \ H^+}]{} \text{product}$$

(b)

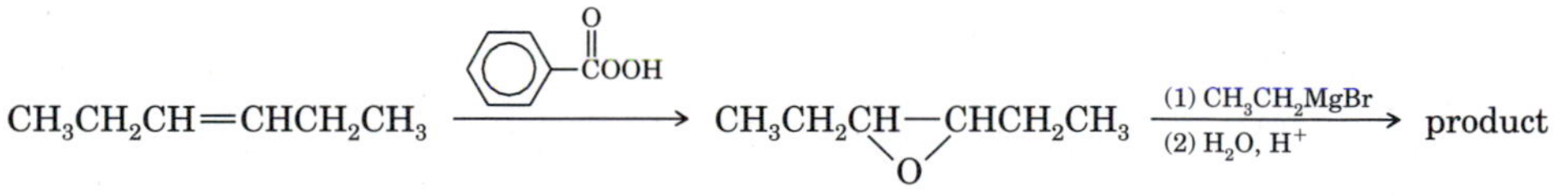

(c)

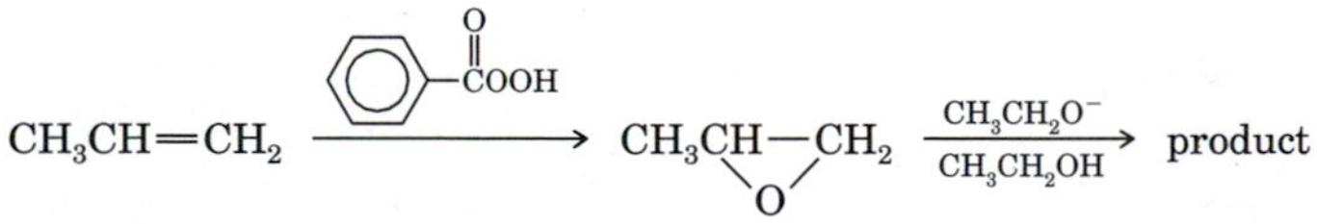

8.7 **(a)**

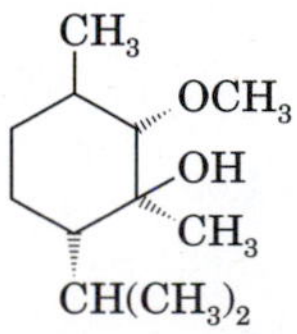

(b)

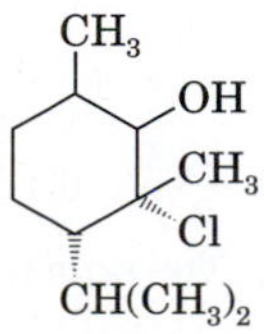

8.8   Dimethyl sulfoxide contains only two hydrophobic methyl groups and contains a very polar S=O group, which can hydrogen-bond with water.

$$CH_3-\underset{\delta+}{S}-CH_3$$

hydrogen bond

8.9   The electrons are priority (4). The ranking is O, 1; phenyl, 2; CH_3, 3; $2e^-$, 4.

Chapter 9

9.1   **(b)** < **(c)** < **(a).** The intensity of infrared absorption depends on the magnitude of the oscillating dipole. Compound **(b)** is a symmetrical alkene; therefore, there would be no change in dipole as the bond oscillates. (This double bond shows no absorption in the infrared spectrum.) Compound **(a)** has a permanent dipole because of the difference in electronegativity between oxygen and carbon. When the carbonyl group oscillates, there is a large change in dipole. Compound **(c)** is intermediate between these two extremes.

9.2   I **(d)**; II **(b)**

9.3   The structure of the compound is $CH_3CH_2OCH_2CH_3$.

$$CH_3CH_2OCH_2CH_3 + HI \longrightarrow CH_3CH_2I + CH_3CH_2OH$$

$$CH_3CH_2OH + HI \longrightarrow CH_3CH_2I + H_2O$$

Assignment of infrared absorption:

$2850 - 2975$ cm^{-1} ($3.4 - 3.5$ μm): sp^3 C—H stretch
1380 cm^{-1} (7.25 μm): sp^3 C—C stretch
1110 cm^{-1} (9.0 μm): C—O stretch

9.4   Compound I is an ester while compound II is a ketone. Although both spectra show C=O absorption, the spectrum of compound I also shows strong absorption in the 1230 cm^{-1} (8.1 μm) region (C—O stretch).

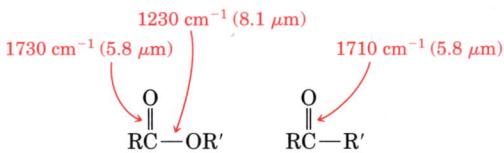

9.5   **(a)** 2   **(b)** 2   **(c)** 1

   (d) 1   **(e)** 1

9.6   **(a)** C(H₂)–CHCH₃
              | |
              CH₂—CH₂

   (b) CH₃OCH₂OCH₂OC(H₃)

   (c) C(H₃)CCH₂CHCH₂CC(H₃)
              with CH₃, OH, C(H₃) groups and C(H₃), C(H₃) below

9.7   **(a)** 3:2:2:2   **(b)** 6:4:1

9.8   **(a)** $\dfrac{14.0}{13.9} = 1.01$ $\dfrac{13.9}{13.9} = 1.00$ $\dfrac{125}{13.9} = 8.99$ **(b)** $\dfrac{50.6}{8.3} = 6.1$ $\dfrac{8.3}{8.3} = 1.00$ $\dfrac{50.2}{8.3} = 6.0$

   The ratio is: 1:1:9   The ratio is: 6:1:6

9.9   **(a)** six neighboring protons: 7 peaks   **(b)** nine neighboring protons: 10 peaks

   (c) one neighboring proton: 2 peaks   **(d)** one neighboring proton: 2 peaks

9.10   **(a), (c), (d), (e)**

9.11 **(a)** one at room temperature. (At low temperatures, NMR can distinguish between equatorial hydrogens and axial hydrogens.)

(b) one

9.12 **(a)** two for each

(b) two for the group arising from the CH_2 protons and three in the group arising from the other proton

(c) two in the group arising from the CH_3 protons and four in the group arising from the other proton

9.13 Cl—⟨◯⟩—$\overset{\overset{\displaystyle O}{\|}}{C}CH_3$

9.14 Two Cl atoms have a greater electron-withdrawing power than one Cl atom.

9.15 *triplet, 2.9* ↘ *triplet, 4.3* ↙
$C_6H_5CH_2CH_2O_2CCH_3$
singlet, 7.15 ↑   *singlet, 2.0* ↑

9.16 I(**e**); II(**d**); III(**a**)

9.17

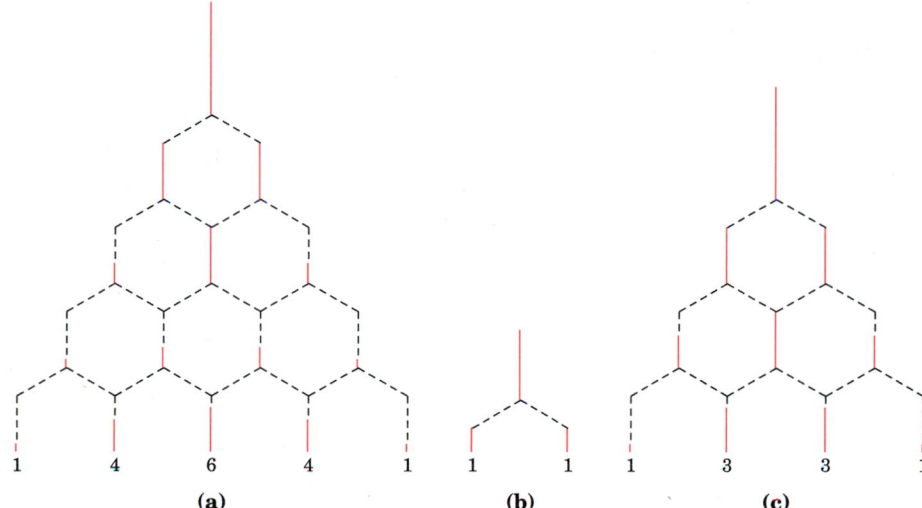

| 1 | 4 | 6 | 4 | 1 | | 1 | 1 | | 1 | 3 | 3 | 1 |
| (a) | | | | | | (b) | | | (c) | | | |

9.18

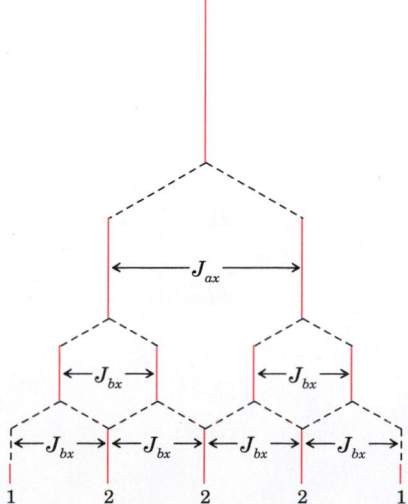

| 1 | 2 | 2 | 2 | 1 |

9.19   The compound contains an ester group. A partial structure can be written.

$$\overset{\displaystyle O}{\underset{\displaystyle \parallel}{-C}}\!O- \; + \; C_4H_7Br$$

Saturated noncyclic alkyl groups have the general formula C_nH_{2n+1} or, if containing Br, $C_nH_{2n}Br$. The C_4H_7Br grouping must thus contain one double bond or be part of a ring.

From the ^{13}C NMR spectra, we deduce the following groups, in order of their appearance in the spectra:

$$\overset{O}{\underset{\parallel}{-C-}} \qquad \overset{Br}{\underset{\mid}{CH=}} \qquad -CH= \qquad -OCH_3 \qquad -CH_2-$$

            not bonded   *not bonded to O*
            to O or Br

Putting the pieces together, we arrive at the structure:

$$BrCH{=}CHCH_2\overset{\displaystyle O}{\overset{\displaystyle \parallel}{C}}OCH_3 \quad \textit{(cis or trans)}$$

9.20 $(CH_3)_2CHOH$ **9.21** $C_6H_5CH_2OH$ **9.22** $CH_3CH_2CHBrCO_2H$

9.23 $\langle\!\!\bigcirc\!\!\rangle\!-\overset{\displaystyle O}{\overset{\displaystyle \parallel}{C}}CH_2CH_2CH_3$

Chapter 10

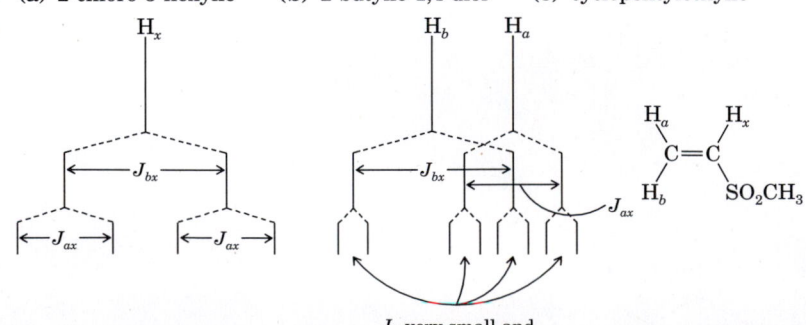

10.1 (a) ... (b) ... (c) ... (d) ...

10.2 (a) ... (b) ... (c) ...

10.3 (a) 2-chloro-3-hexyne   (b) 2-butyne-1,4-diol   (c) cyclopentylethyne

10.4

J_{ab}very small and peaks overlap

10.5

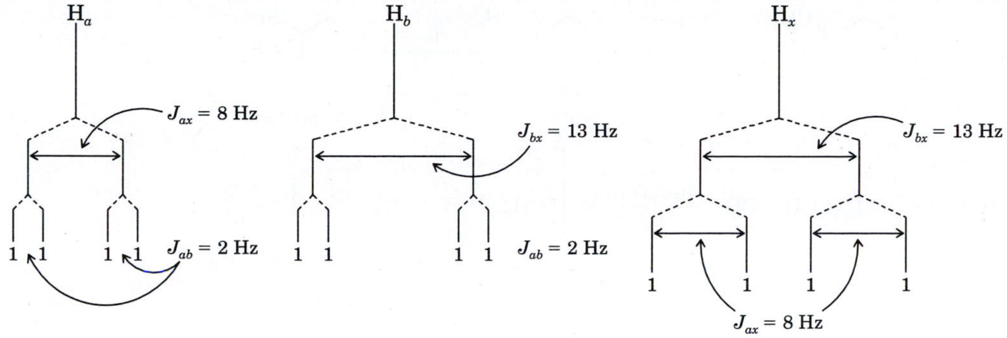

10.6 (a) $(CH_3)_2C=CHCH_3$ **(b)** (E)- and (Z)-$C_6H_5C=CHCH_3$
$\overset{|}{C}H_3$

(c)

$\xrightarrow{-HBr}$

$\underset{\substack{| \\ H}}{C_6H_5} \overset{\substack{CH_2CH_3 |}}{=} C \underset{\substack{| \\ CH_3}}{}$ (Z)

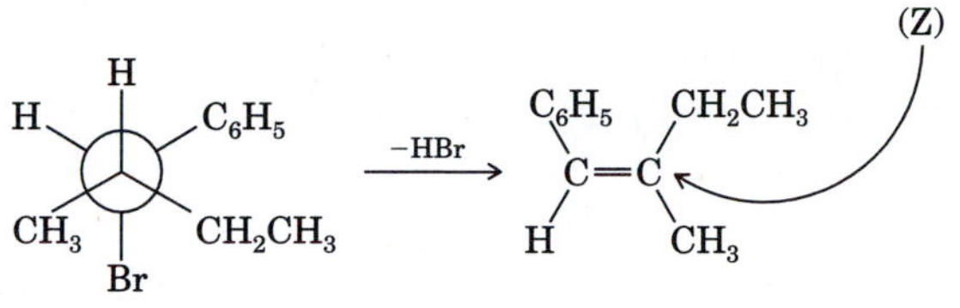

anti conformation

(d)

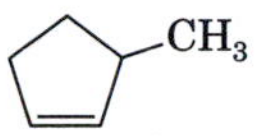

10.7 (a) $CH_3C\equiv CH \xrightarrow{NaNH_2} CH_3C\equiv C:^-Na^+ \xrightarrow{CH_3CH_2Br} CH_3CH_2C\equiv CCH_3$

(b) $CH\equiv CH \xrightarrow[-CH_3CH_3]{1\ CH_3CH_2MgBr} CH\equiv CMgBr \xrightarrow[\text{(2) } H_2O,\ H^+]{\text{(1) } CH_3CH_2\overset{O}{\overset{||}{C}}H} CH_3CH_2\overset{\overset{OH}{|}}{C}HC\equiv CH$

10.8 (a) $CH_3CH_2\overset{+}{C}HCH_3 \xrightarrow{Br^-} CH_3CH_2\overset{\overset{Br}{|}}{C}HCH_3$

(c) $(CH_3)_2\overset{+}{C}CH_2CH_3 \xrightarrow{Br^-} (CH_3)_2\overset{\overset{Br}{|}}{C}CH_2CH_3$

10.9 (a) $(CH_3)_3CCH=CH_2 \xrightarrow{HI} (CH_3)_3CCHICH_3 + (CH_3)_2CICH(CH_3)_2$

(b) $C_6H_5CH=CH_2 \xrightarrow{HCl} C_6H_5CHClCH_3$

(c) $(CH_3)_2CHCH=CH_2 \xrightarrow{HCl} (CH_3)_2CHCHClCH_3 + (CH_3)_2CClCH_2CH_3$

(d) $(CH_3)_3CCH=CHCH_3 \xrightarrow{HCl} (CH_3)_3CCH_2CHClCH_3$
$ + (CH_3)_3CCHClCH_2CH_3 + (CH_3)_2CClCHCH_2CH_3$
$\overset{|}{C}H_3$

10.10 (a) $ROOR \longrightarrow 2\ RO\cdot$
$RO\cdot + HBr \longrightarrow ROH + Br\cdot$

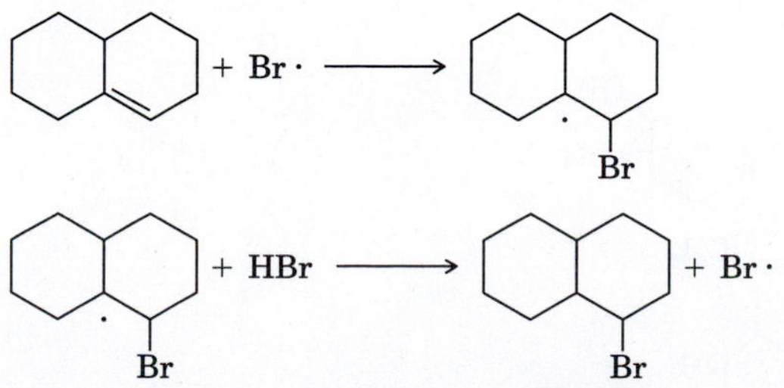

$+ Br\cdot \longrightarrow$

$+ HBr \longrightarrow$

$+ Br\cdot$

the product

Termination by coupling of any two radicals.

(b) 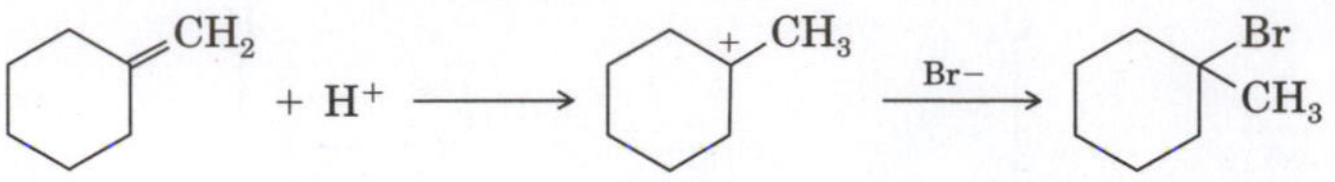 + H⁺ ⟶ $\xrightarrow{\text{Br}^-}$

the product

10.11 (a) 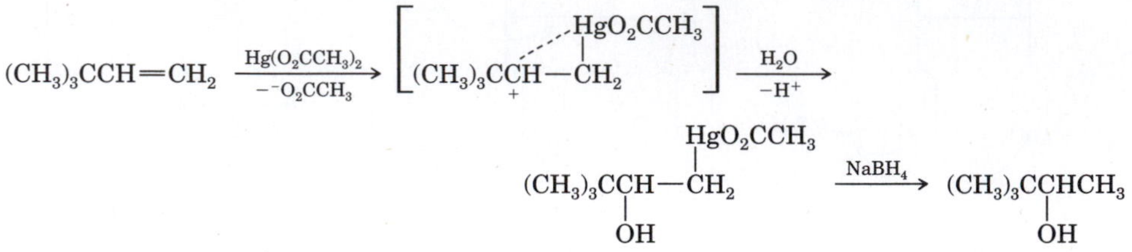 $(CH_3)_3CCH{=}CH_2 \xrightarrow[--O_2CCH_3]{Hg(O_2CCH_3)_2}$ $\left[(CH_3)_3CCH\overset{HgO_2CCH_3}{\underset{+}{-}}CH_2 \right] \xrightarrow[-H^+]{H_2O}$

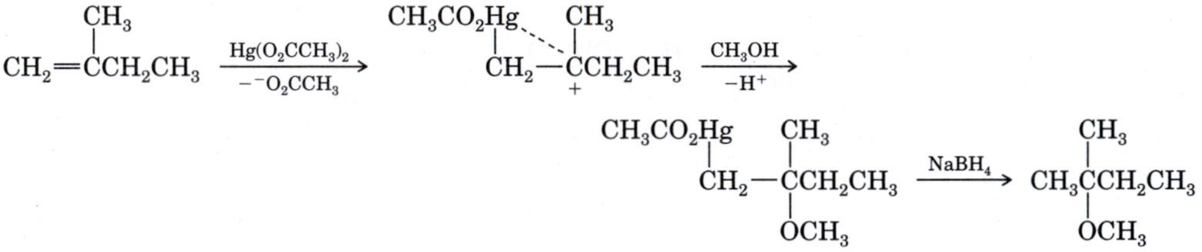

$(CH_3)_3CCH\overset{HgO_2CCH_3}{-}CH_2 \xrightarrow{\text{NaBH}_4} (CH_3)_3CCHCH_3$
$\qquad\quad |$
$\qquad\quad OH$ ... OH

(b) 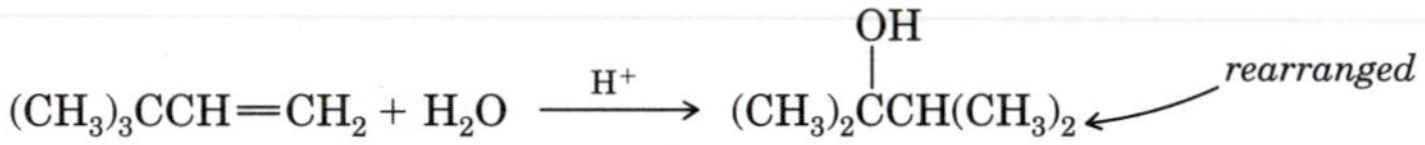 $(CH_3)_3CCH{=}CH_2 + H_2O \xrightarrow{H^+} (CH_3)_2\overset{OH}{\underset{|}{C}}CH(CH_3)_2 \xleftarrow{\textit{rearranged}}$

10.12 $CH_2{=}\overset{CH_3}{\underset{|}{C}}CH_2CH_3 \xrightarrow[--O_2CCH_3]{Hg(O_2CCH_3)_2}$ $CH_2\overset{CH_3CO_2Hg}{-}\overset{CH_3}{\underset{+}{C}}CH_2CH_3 \xrightarrow[-H^+]{CH_3OH}$

$CH_2\overset{CH_3CO_2Hg}{-}\overset{CH_3}{\underset{|}{C}}CH_2CH_3 \xrightarrow{\text{NaBH}_4} CH_3\overset{CH_3}{\underset{|}{C}}CH_2CH_3$
$\qquad\quad |$
$\qquad\quad OCH_3$ ... OCH_3

10.13 $\xrightarrow{^-OOH}$ ⟶

$\xrightarrow{^-OH}$

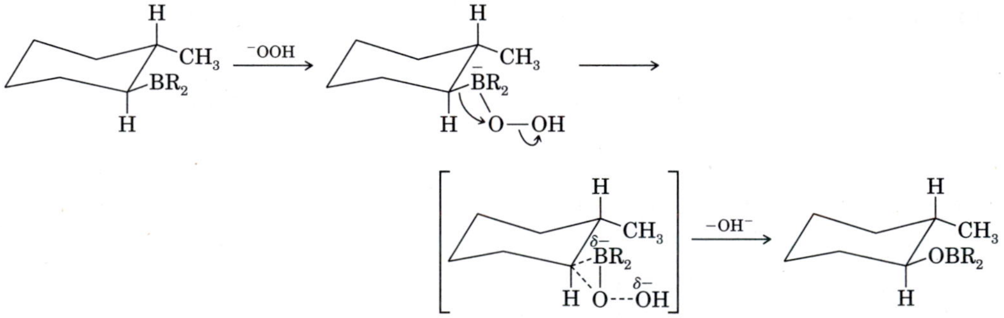

trans transition state
(cannot yield a *cis* alcohol)

10.14 (a)

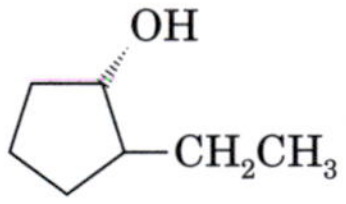

trans (and racemic)

(b)

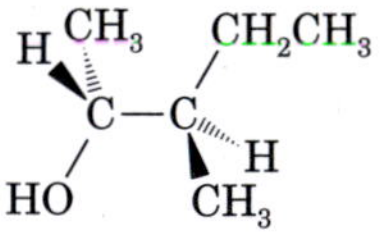

(2*R*,3*R*)-3-methyl-2-pentanol
and its (2*S*,3*S*) enantiomer

10.15 (a) 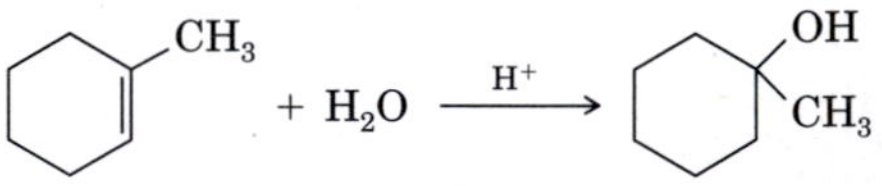 + H₂O $\xrightarrow{H^+}$

(b) 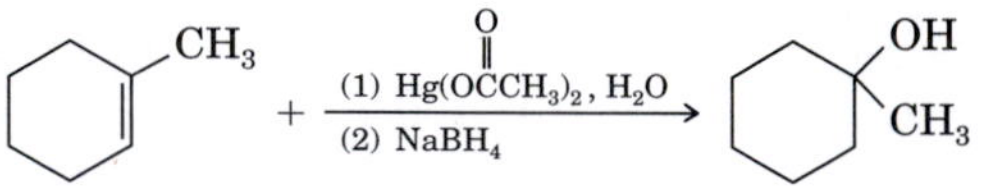 + $\xrightarrow[\text{(2) NaBH}_4]{\text{(1) Hg(OCCH}_3)_2, \text{ H}_2\text{O}}$

(c) 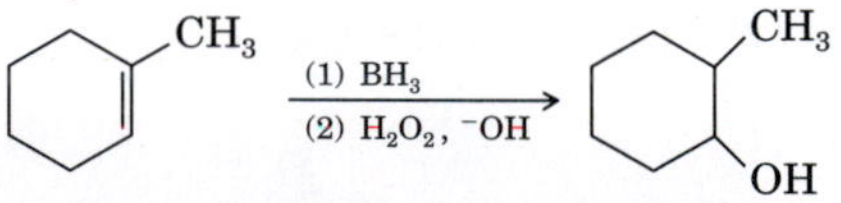 $\xrightarrow[\text{(2) H}_2\text{O}_2, \text{ }^-\text{OH}]{\text{(1) BH}_3}$

10.16

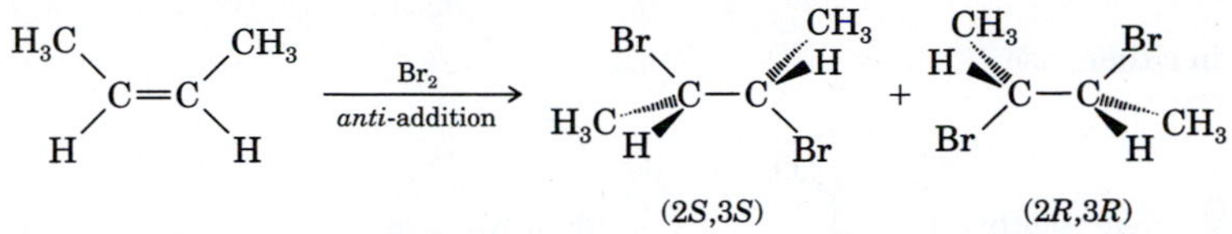

(2S,3S)   (2R,3R)

10.17 No, because the intermediate contains Br.

10.18

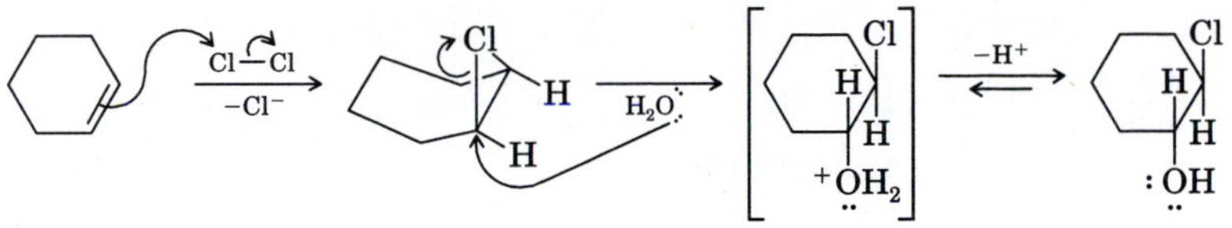

10.19 (a)

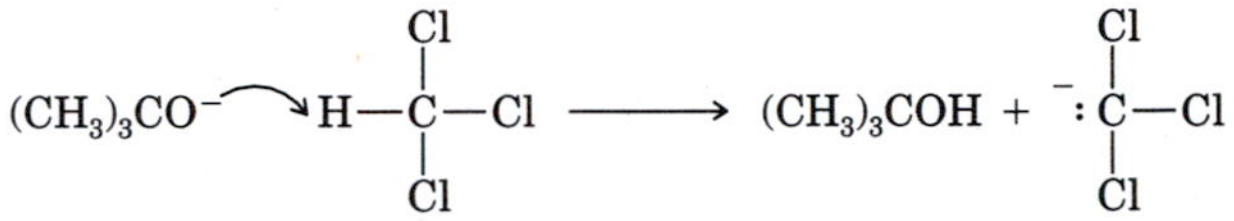

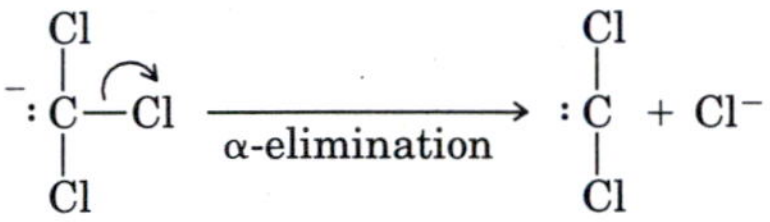

(b)

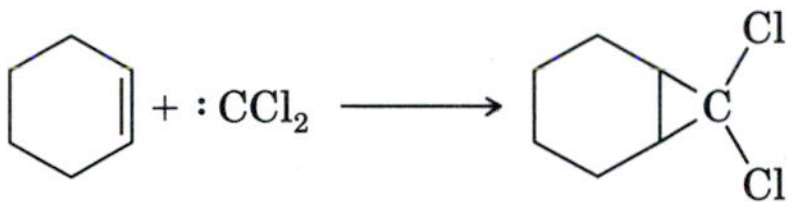

10.20 (a) $(CH_3)_2C=CH_2$ **(b)**

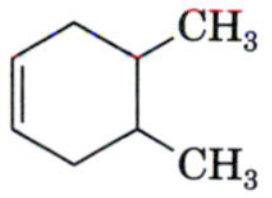

   (c) $(Z)\text{-}C_6H_5CH=CHCH_3$

In each case, the least stable member of the pair was selected.

10.21 (a) 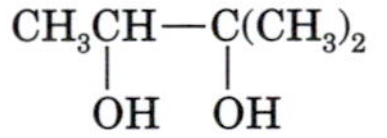 $CH_3CH-C(CH_3)_2$
           $\quad\quad\ \ |\quad\ \ |$
           $\quad\quad\ OH\ \ OH$

   (b)

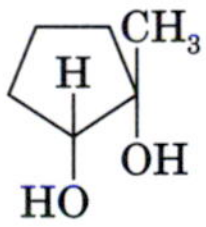

   Both products are racemic.

10.22 (a)

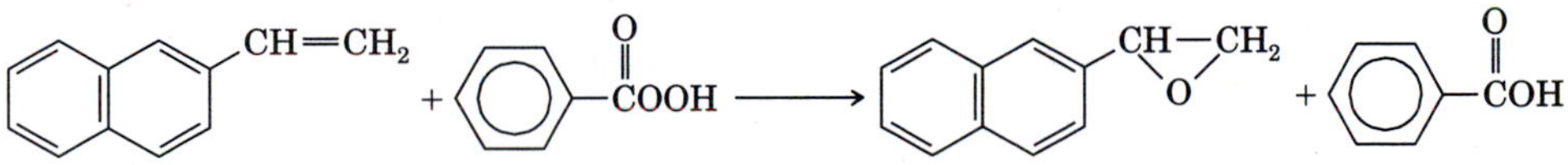

   (b)

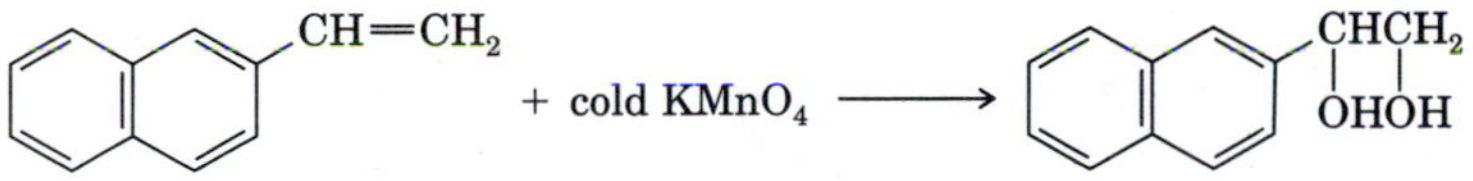

   (c)

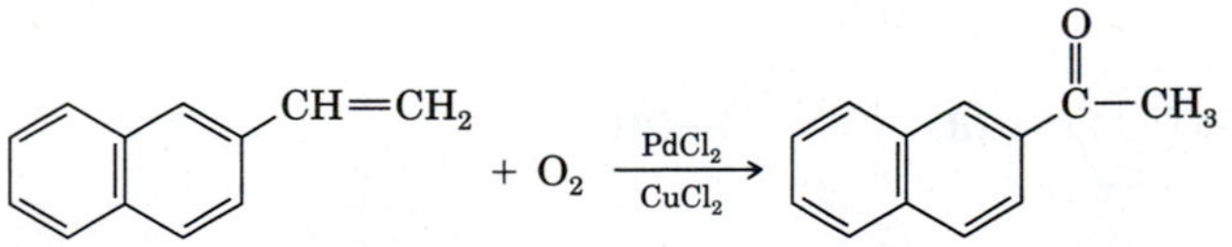

10.23 (a) $CH_3(CH_2)_7CH=CH-(CH_2)_7\overset{\displaystyle O}{\overset{\displaystyle \|}{C}}OH$

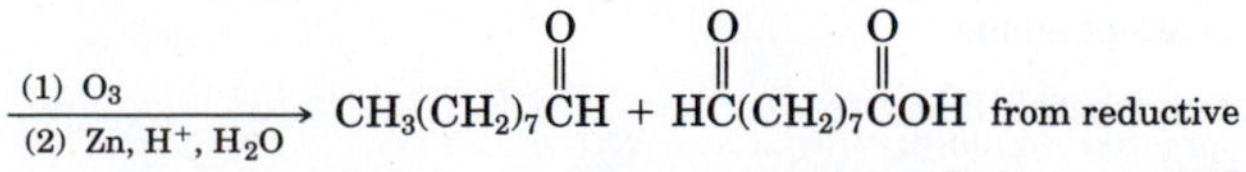

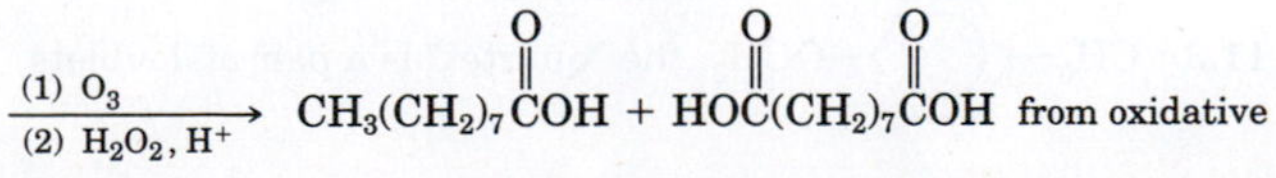

$\xrightarrow[\text{(2) Zn, H}^+,\text{ H}_2\text{O}]{\text{(1) O}_3}$ $CH_3(CH_2)_7\overset{O}{\overset{\|}{C}}H + H\overset{O}{\overset{\|}{C}}(CH_2)_7\overset{O}{\overset{\|}{C}}OH$ from reductive

$\xrightarrow[\text{(2) H}_2\text{O}_2,\text{ H}^+]{\text{(1) O}_3}$ $CH_3(CH_2)_7\overset{O}{\overset{\|}{C}}OH + HO\overset{O}{\overset{\|}{C}}(CH_2)_7\overset{O}{\overset{\|}{C}}OH$ from oxidative

(b) 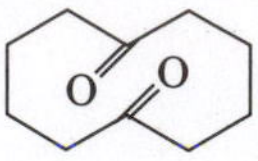 in either case

(c) 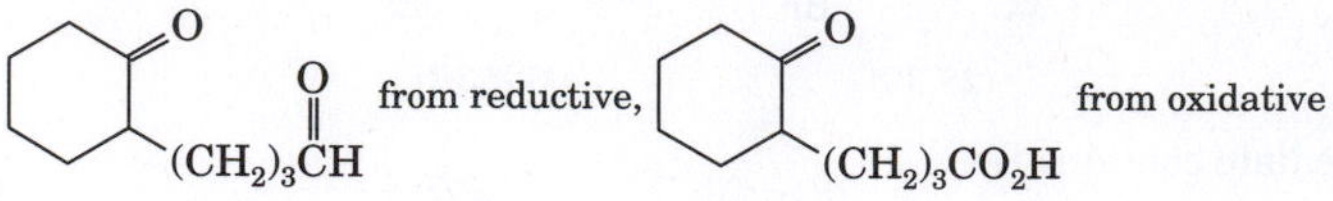 from reductive, from oxidative

10.24 (a) 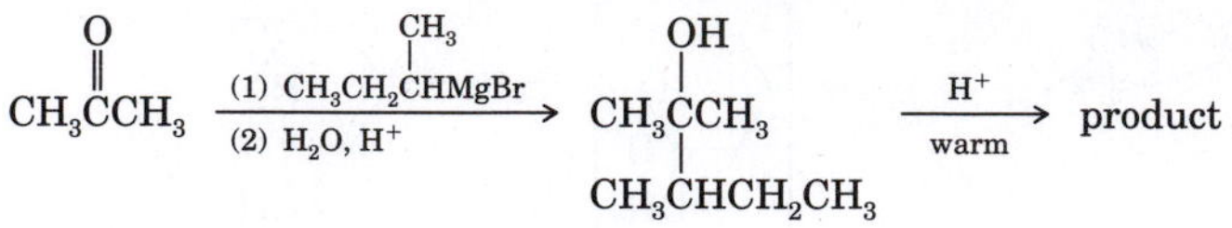 $CH_3\overset{O}{\overset{\|}{C}}CH_3$ $\xrightarrow[\text{(2) } H_2O, H^+]{\text{(1) } CH_3CH_2\overset{CH_3}{\overset{|}{C}}HMgBr}$ $CH_3\overset{OH}{\underset{CH_3CHCH_2CH_3}{\overset{|}{C}}}CH_3$ $\xrightarrow[\text{warm}]{H^+}$ product

(b) $CH_3CH_2CH{=}CH_2$ $\xrightarrow{Br_2}$ $CH_3CH_2CHBrCH_2Br$ $\xrightarrow[\text{(2) } H^+]{\text{(1) } NaNH_2}$ product

10.25 (a) 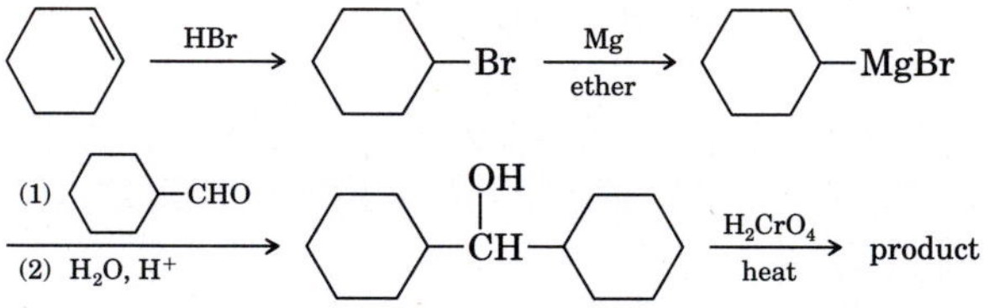 $\xrightarrow{HBr}$ $-Br$ $\xrightarrow[\text{ether}]{Mg}$ $-MgBr$

$\xrightarrow[\text{(2) } H_2O, H^+]{\text{(1)} \text{◯}-CHO}$ $\xrightarrow[\text{heat}]{H_2CrO_4}$ product

(b) 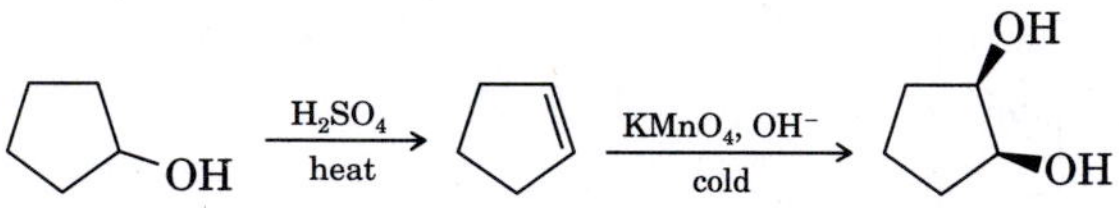 $\xrightarrow[\text{heat}]{H_2SO_4}$ $\xrightarrow[\text{cold}]{KMnO_4, \, OH^-}$

(c) 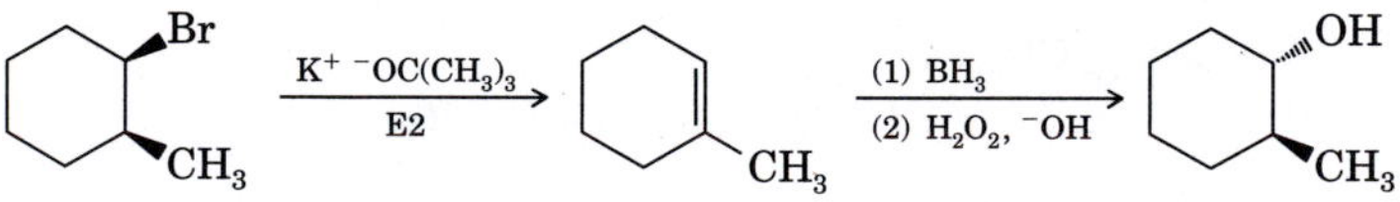 $\xrightarrow[\text{E2}]{K^+ \, {}^-OC(CH_3)_3}$ $\xrightarrow[\text{(2) } H_2O_2, \, {}^-OH]{\text{(1) } BH_3}$

(d) 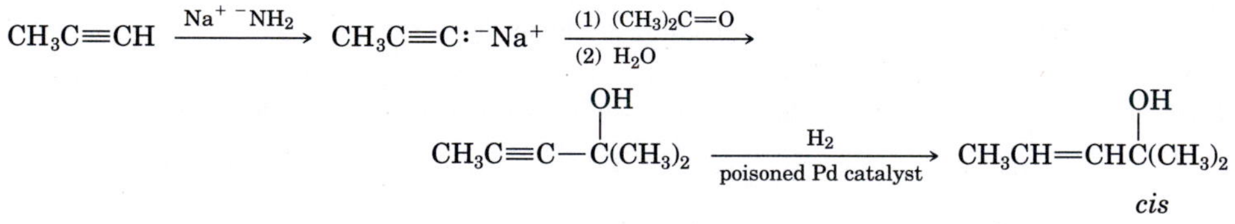 $CH_3C{\equiv}CH$ $\xrightarrow{Na^+ \, {}^-NH_2}$ $CH_3C{\equiv}C{:}^-Na^+$ $\xrightarrow[\text{(2) } H_2O]{\text{(1) } (CH_3)_2C{=}O}$

$CH_3C{\equiv}C{-}\overset{OH}{\overset{|}{C}}(CH_3)_2$ $\xrightarrow[\text{poisoned Pd catalyst}]{H_2}$ $CH_3CH{=}CH\overset{OH}{\overset{|}{C}}(CH_3)_2$

cis

Chapter 11

11.1 (a) 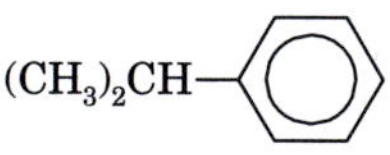 $(CH_3)_2CH-$

(b) 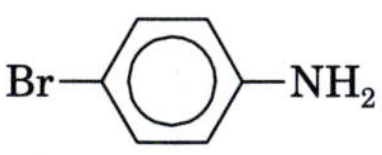 $Br-$$-NH_2$

(c) $\left(\text{◯}\right)_2 C{=}CH-$

(d) $-OH$ with O_2N substituents

11.2 (a) benzyl bromide

(b) 2-chloro-6-nitrotoluene

(c) *p*-iodoacetophenone

(d) *m*-isobutylacetanilide

11.3 $CH_3-$$-OCH_3$; the "quartet" is a pair of doublets.

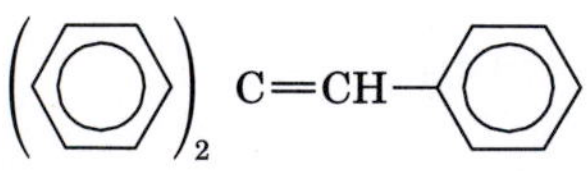

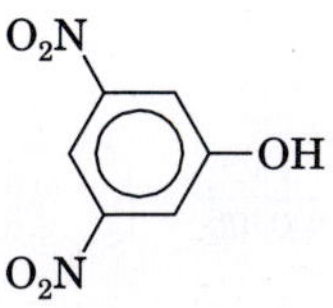

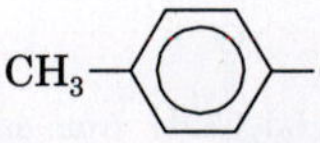

11.4 not aromatic

11.5 Compound **(a)** is not aromatic because one ring carbon is sp^3 hybridized. Ion **(b)** contains sp^2-hybridized ring atoms, but eight $(4n)$ pi electrons, and thus is not aromatic. Ion **(c)** contains sp^2-hybridized ring atoms and six $(4n + 2)$ pi electrons: the cycloheptatrienyl cation is aromatic.

11.6 **(a)** 6 pi electrons $(4n + 2)$; aromatic

(b) The compound has 6 pi electrons $(4n + 2)$ but it would not be aromatic because three of the carbon atoms in the ring system are sp^3 hybridized and, therefore, do not have a p orbital. These sp^3 carbon atoms prevent the three double bonds from being conjugated.

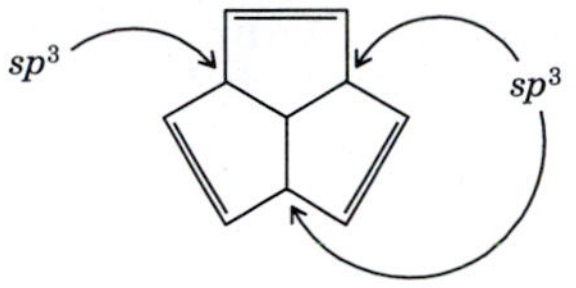

(c) 14 pi electrons $(4n + 2)$; aromatic, but only on the outside of the ring

11.7 **(a)** $C_6H_5\overset{\underset{\displaystyle |}{CH_3}}{CH}CH_2CH_3$; **(b)** $C_6H_5C(CH_3)_3$; **(c)** $C_6H_5\overset{\underset{\displaystyle |}{CH_2CH_3}}{C}(CH_3)_2$; **(d)** $(C_6H_5)_2CH_2$

11.8

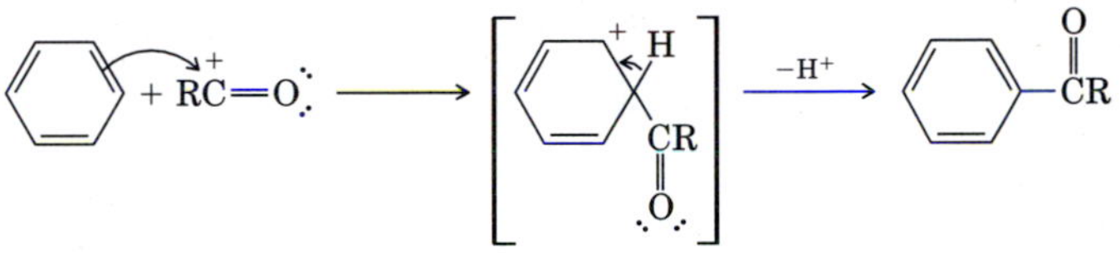

11.9 **(a)** $C_6H_5\overset{\underset{\displaystyle |}{\underset{\displaystyle CH_3}{}}}{\overset{\displaystyle CH_3}{C}}CH_2CH_3$ **(b)** $C_6H_5CH_2\overset{\underset{\displaystyle |}{}}{\overset{\displaystyle CH_3}{C}}HCH_2CH_3$

11.10 Two parts o-, two parts m-, and one part p-.

11.11 **(a)** 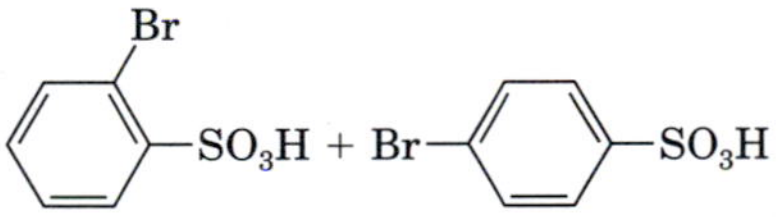 **(b)**

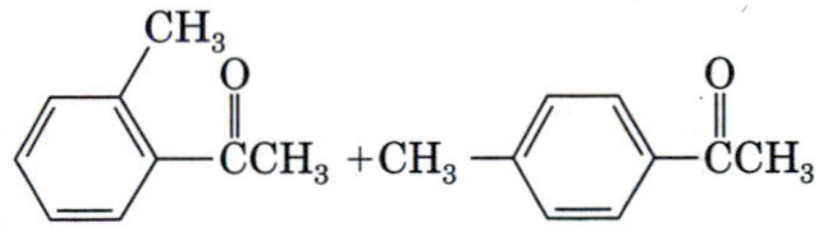

(c) 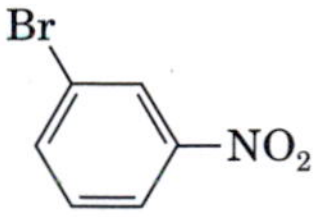 **(d)**

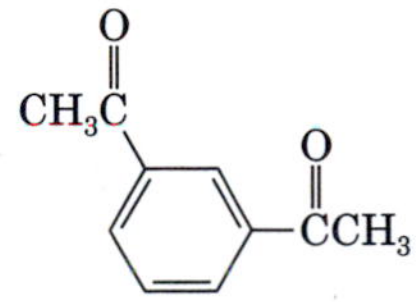

(e) 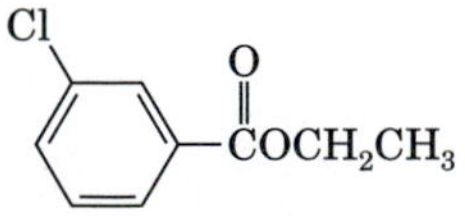 **(f)**

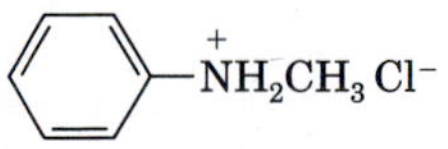

(g)

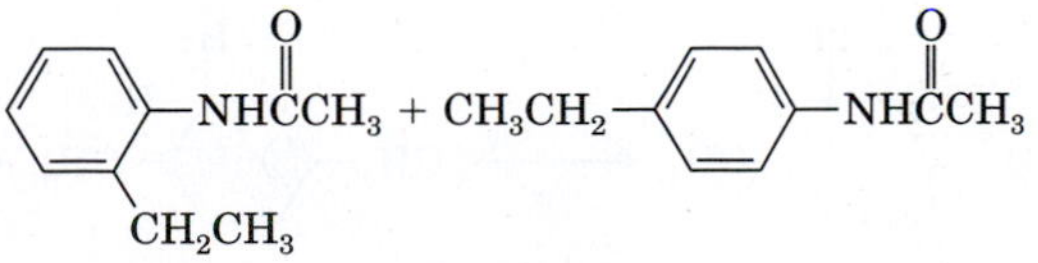

11.12

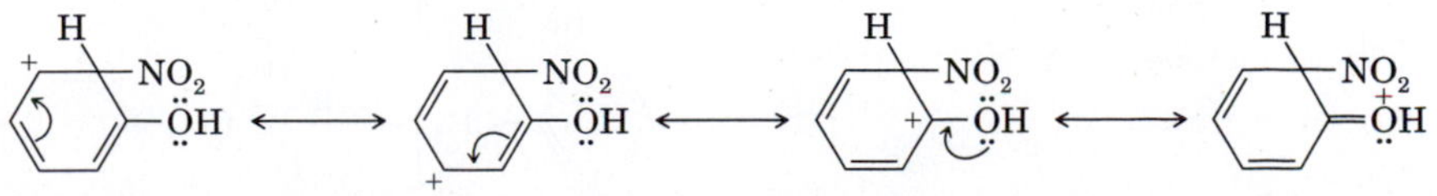

11.13 The unshared electrons of the nitrogen are delocalized by the carbonyl group and are less available for donation to the ring. Because the amide nitrogen is partially positive, it also exerts a greater electron-withdrawal than an amine nitrogen.

11.14 (a) $(CH_3)_2CH$—⟨ ⟩—Br + CH(CH$_3$)$_2$ substituted ⟨ ⟩—Br

(b) $(CH_3)_2CH$—⟨ ⟩—OH + CH(CH$_3$)$_2$ substituted ⟨ ⟩—OH

(c) $(CH_3)_2CH$—⟨ ⟩—CH$_3$ + CH(CH$_3$)$_2$ substituted ⟨ ⟩—CH$_3$

(d) ⟨ ⟩—CH(CH$_3$)$_2$.   Relative rates: (b) > (c) > (d) > (a).

11.15 (a) *o,p*   **(b)** *m*   **(c)** *o,p*   **(d)** *o,p*   **(e)** *m*   **(f)** *m*   **(g)** *m*

11.16 (a) Cl—⟨ ⟩—NH$_2$ + Cl substituted ⟨ ⟩—NH$_2$ with Cl

(b) ⟨CH$_3$, NO$_2$⟩—OCH$_3$ + CH$_3$—⟨NO$_2$⟩—OCH$_3$

(c) ⟨ ⟩—C(=O)NH—⟨NO$_2$⟩ + ⟨ ⟩—C(=O)NH—⟨ ⟩—NO$_2$

because the right-hand ring is activated and the left-hand ring is deactivated

Chapter 12

12.1 (a) CH$_3$—⟨ ⟩—CH$_3$ $\xrightarrow{Br^+}$ [CH$_3$—⟨Br,H⟩—CH$_3$ ⟷ CH$_3$—⟨Br,H⟩—CH$_3$ ⟷

CH$_3$—⟨Br,H⟩—CH$_3$] $\xrightarrow{-H^+}$ CH$_3$—⟨Br⟩—CH$_3$

(b) CH_3— ⟨ring⟩—$\overset{H}{\underset{}{C}H_2}$ $\xrightarrow[-HBr]{\cdot Br}$ [CH_3—⟨ring⟩—$\dot{C}H_2$ ⟷ CH_3—⟨ring⟩=CH_2 ⟷

CH_3—⟨ring⟩=CH_2 ⟷ CH_3—⟨ring⟩=CH_2] $\xrightarrow[-Br\cdot]{Br_2}$ CH_3—⟨ring⟩—CH_2Br

12.2 **(a)** ⟨bicyclic ring⟩ + hot $KMnO_4$ ⟶ ⟨benzene ring with⟩ $\overset{O}{\underset{}{\overset{\|}{C}OH}}$ and $\underset{O}{\overset{}{C-OH}}$ + CO_2

(b) ⟨bicyclic ring⟩ + Br_2 $\xrightarrow{light}$ ⟨bicyclic ring with Br⟩

(c) ⟨bicyclic ring⟩ + Br_2 $\xrightarrow{Fe}$ ⟨bicyclic ring with Br⟩ + ⟨bicyclic ring with Br⟩

12.3 **(a) and (b)** ⟨benzene ring⟩—$O^- Na^+$ + CH_3CH_2OH

12.4 **(a)** ⟨benzene ring with two OH⟩ $\xrightarrow{H_2CrO_4}$ ⟨ring with two O⟩

(b) C_6H_5OH $\xrightarrow[AlCl_3]{CH_3CH_2\overset{O}{\overset{\|}{C}}Cl}$ ⟨ring with $\overset{O}{\overset{\|}{C}CH_2CH_3}$ and OH⟩ $\xrightarrow{\text{separate from } p \text{ isomer}}$ $\xrightarrow[HCl]{Zn(Hg)}$ ⟨ring with $CH_2CH_2CH_3$ and OH⟩

(c) C_6H_5OH $\xrightarrow{OH^-}$ $C_6H_5O^-$ $\xrightarrow[heat]{CO_2}$ ⟨ring with CO_2^- and OH⟩ $\xrightarrow{H^+}$ ⟨ring with CO_2H and OH⟩

12.5 **(a)** CH_3—⟨ring⟩ $\xrightarrow[H_2SO_4]{HNO_3}$ CH_3—⟨ring⟩—NO_2 $\xrightarrow[(2) NaOH]{(1) Fe, HCl}$

CH_3—⟨ring⟩—NH_2 $\xrightarrow[0°]{\underset{HCl}{NaNO_2}}$ CH_3—⟨ring⟩—$\overset{+}{N_2}$ $\xrightarrow[warm]{H_2O}$ product

(b) ⟨benzene ring⟩ $\xrightarrow[H_2SO_4]{HNO_3}$ ⟨ring⟩—NO_2 $\xrightarrow[FeBr_3]{Br_2}$ ⟨ring with Br⟩—NO_2

$\xrightarrow[(2) NaOH]{(1) Fe, HCl}$ ⟨ring with Br⟩—NH_2 $\xrightarrow[0°]{\underset{HCl}{NaNO_2}}$ ⟨ring with Br⟩—$\overset{+}{N_2}$ $\xrightarrow{HBF_4}$ product

(c) CH_3—⟨ring⟩—$\overset{+}{N_2}$ from **(a)** $\xrightarrow[KCN]{CuCN}$ product

12.6 (a)

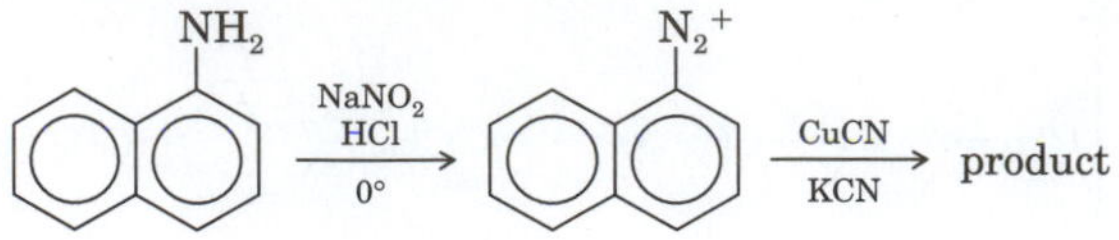

(b) CH_3—⟨⟩—NO_2 $\xrightarrow[\text{(2) OH}^-]{\text{(1) Fe, HCl}}$ CH_3—⟨⟩—NH_2 $\xrightarrow[0°]{\text{NaNO}_2 \atop \text{HCl}}$

CH_3—⟨⟩—N_2^+ Cl^- $\xrightarrow{\text{CuBr}}$ product

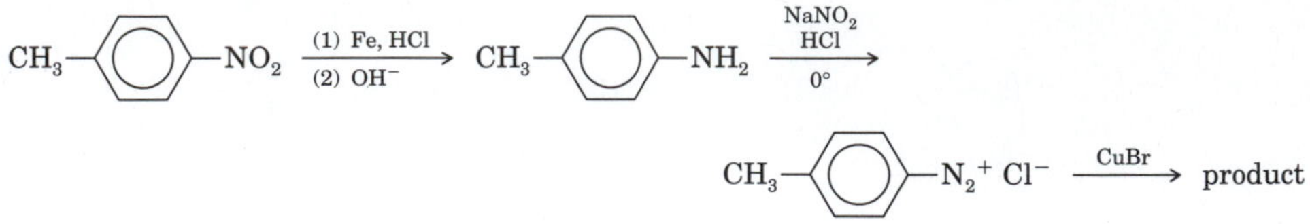

(c)

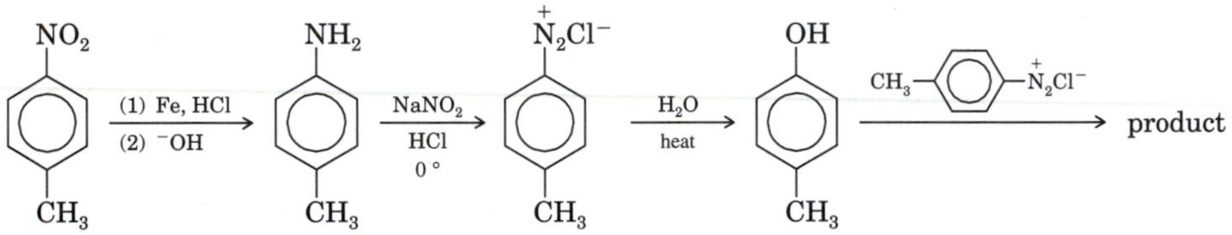

12.7

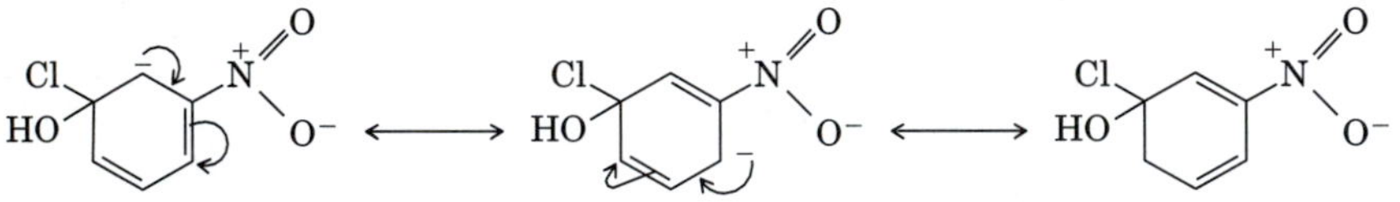

The negative charge never resides on the carbon bonded to the nitro group.

12.8

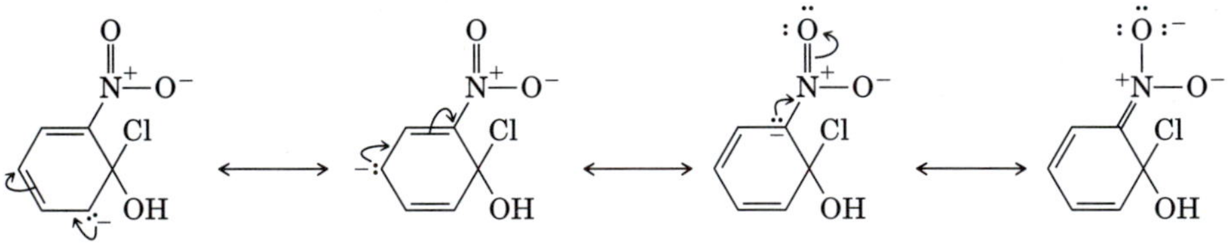

12.9 (a) C_6H_6 $\xrightarrow[\text{H}_2\text{SO}_4]{\text{HNO}_3}$ $C_6H_5NO_2$ $\xrightarrow[\text{FeCl}_3]{\text{Cl}_2}$

—NO_2 $\xrightarrow[\text{(2) OH}^-]{\text{(1) Fe, HCl}}$ —NH_2

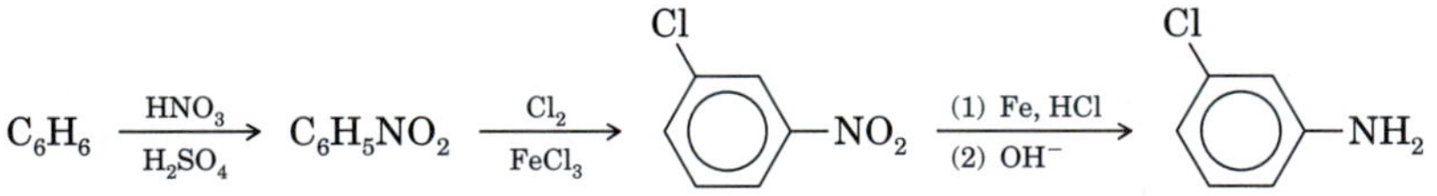

(b) $C_6H_6 + ClC(CH_2)_3CH_3$ $\xrightarrow{\text{AlCl}_3}$ $C_6H_5\overset{O}{\overset{\|}{C}}(CH_2)_3CH_3$ $\xrightarrow[\text{HCl}]{\text{Zn/Hg}}$ $C_6H_5(CH_2)_4CH_3$

(c) excess $C_6H_6 + ClCCH_2C_6H_5$ $\xrightarrow{\text{AlCl}_3}$ $C_6H_5\overset{O}{\overset{\|}{C}}CH_2C_6H_5$ $\xrightarrow[\text{HCl}]{\text{Zn/Hg}}$ $C_6H_5CH_2CH_2C_6H_5$

(d) excess $C_6H_6 + CHCl_3$ $\xrightarrow{\text{AlCl}_3}$ $(C_6H_5)_3CH$

(e) C_6H_6 $\xrightarrow[\text{H}_2\text{SO}_4]{\text{HNO}_3}$ $C_6H_5NO_2$ $\xrightarrow[\text{(2) OH}^-]{\text{(1) Fe, HCl}}$ $C_6H_5NH_2$

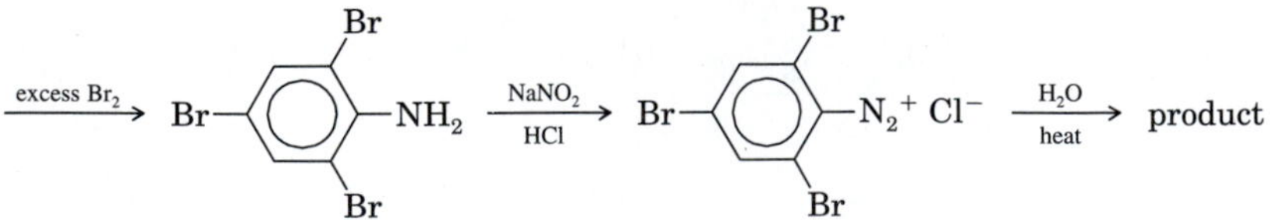

$\xrightarrow{\text{excess Br}_2}$ Br——NH_2 $\xrightarrow[\text{HCl}]{\text{NaNO}_2}$ Br——N_2^+ Cl^- $\xrightarrow[\text{heat}]{\text{H}_2\text{O}}$ product

Chapter 13

13.1 (a) $BrCH_2\overset{O}{\overset{\|}{C}}CH_3$ **(b)** $CH_3\overset{O}{\overset{\|}{C}}-\overset{O}{\overset{\|}{C}}CH_3$

(c) CH$_3$CCH$_2$CH$_2$—⟨benzene ring⟩—OH (with C=O)

(d) (CH$_3$)$_2$C=CHCH$_2$CH$_2$, CH$_3$ CH, C=C, with O and H

13.2 **(a)** 2-ethylhexanal

(b) 2-decanone

(c) 5-methyl-2-cyclohexene-1,4-dione

(d) hexanedial

(e) *p*-hydroxybenzaldehyde

(f) 3-methylcyclohexanone

13.3 **(a)** CH$_3$CCH$_2$CHO or any other structure with keto and aldehyde groups 1,3 to each other.

(b) H$_2$C=CHCCH$_3$ or any other structure with a carbon–carbon double bond and a keto group in conjugation.

(c) BrCH$_2$CHO or any other aldehyde with a bromine on the α carbon (carbon adjacent to the aldehyde carbon).

(d) CH$_3$CCH$_2$CH$_2$OH or any other 1,3-hydroxyketone.

13.4 **(a)** CH$_3$CH$_2$CH$_2$CH$_2$CH$_2$OH $\xrightarrow[25°]{\text{CrO}_3\cdot2\text{ pyridine}}$

(b) ⟨structure⟩ $\xrightarrow[\text{heat}]{\text{H}_2\text{CrO}_4}$

(c) ⟨structure⟩ $\xrightarrow[\text{heat}]{\text{H}_2\text{CrO}_4}$

(d) (CH$_3$)$_2$CHCH$_2$OH $\xrightarrow[25°]{\text{CrO}_3\cdot2\text{ pyridine}}$

13.5 **(b)**

13.6 ⟨benzene ring⟩—CH (with O double bond)

13.7 **(a)** Br$_2$CHCH + H$_2$O $\rightleftharpoons$ Br$_2$CHCHOH (with O and OH)

(b) Br$_2$CHCH$_2$CH + H$_2$O $\xleftarrow{\ \rightarrow\ }$ Br$_2$CHCH$_2$CHOH (with O and OH)

The hydrate in **(a)** would be more stable than the hydrate in **(b)**. In **(a)**, the bromine atoms are closer to the carbonyl group and, therefore, their electron-withdrawing inductive effect is greater.

13.8 **(a)** ⟨structure with OH, OCH$_2$, CH$_2$OH⟩
hemiketal

(b) ⟨structure with O, O⟩
ketal

(c) ⟨structure with OH, —OCH CH$_3$⟩
hemiacetal

13.9 **(a)** =O + HOCH₂CH₂OH   **(b)** =O + HOCH₂CH₂OH

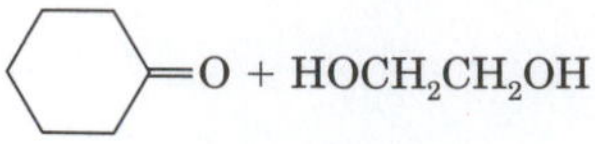

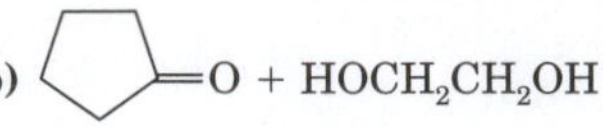

(c) —OH + $\overset{O}{\overset{\|}{H C}}$CH₃

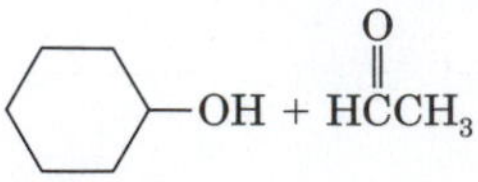

13.10 **(a)** C₆H₅MgBr;   **(b)** CH₃CH₂CH₂MgBr

13.11 **(a)** and **(b)** contain acidic protons relative to RMgX.

13.12 **(a)** —$\overset{O}{\overset{\|}{C}}$H + H₂NCH₂CH₃ $\overset{H^+}{\rightleftharpoons}$

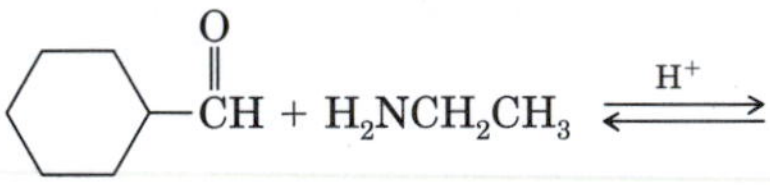

(b) CH₃——$\overset{O}{\overset{\|}{C}}$H + C₆H₅NH₂ $\overset{H^+}{\rightleftharpoons}$

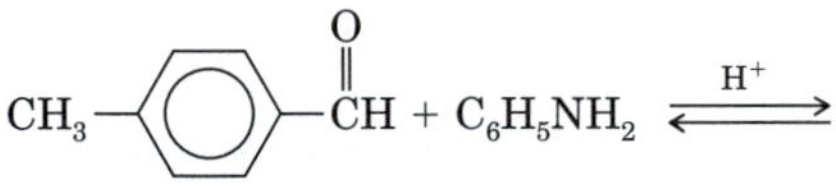

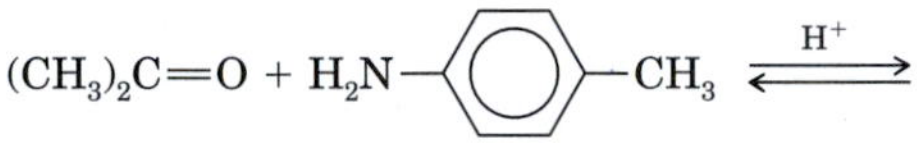

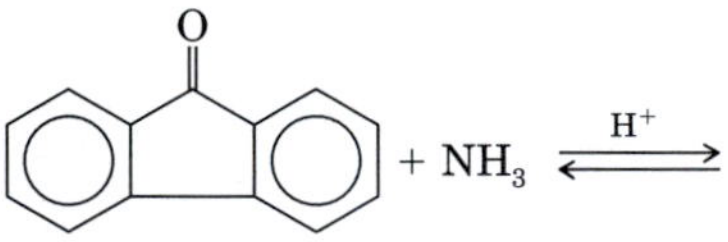

(c) (CH₃)₂C=O + H₂N——CH₃ $\overset{H^+}{\rightleftharpoons}$

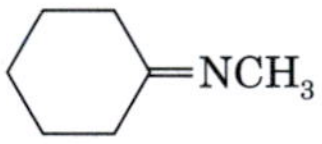

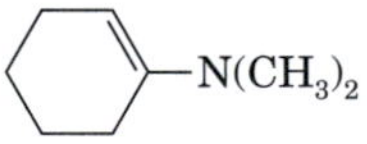

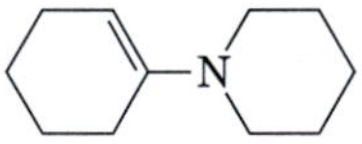

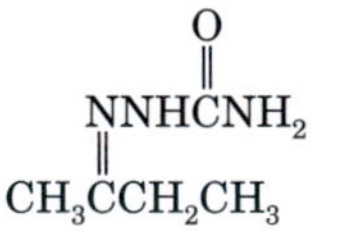

(d) + NH₃ $\overset{H^+}{\rightleftharpoons}$

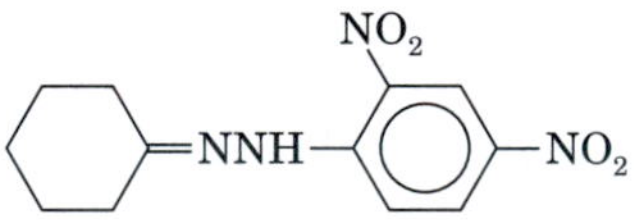

13.13 **(a)** =NCH₃   **(b)** —N(CH₃)₂   **(c)** —N

13.14 **(a)** CH₃$\overset{O}{\overset{\|}{C}}$CH₂CH₃ with NNHCNH₂ label   **(b)** =NNH— with NO₂ and NO₂   **(c)** C₆H₅$\overset{NNH_2}{\overset{\|}{C}}$CH₃

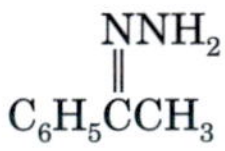

13.15 There are two possible Wittig routes to each product.

(a) (1) CH₃CH₂CH₂Br $\xrightarrow{(C_6H_5)_3P}$ (C₆H₅)₃P⁺—CH₂CH₂CH₃ Br⁻ $\xrightarrow{C_4H_9Li}$

(C₆H₅)₃P=CHCH₂CH₃ $\xrightarrow{C_6H_5CHO}$ C₆H₅CH=CHCH₂CH₃

(2) C₆H₅CH₂Br $\xrightarrow{(C_6H_5)_3P}$ (C₆H₅)₃P⁺—CH₂C₆H₅ Br⁻ $\xrightarrow{C_4H_9Li}$

(C₆H₅)₃P=CHC₆H₅ $\xrightarrow{CH_3CH_2CHO}$ C₆H₅CH=CHCH₂CH₃

(b) (1) $\xrightarrow{(C_6H_5)_3P}$ $\xrightarrow{C_4H_9Li}$

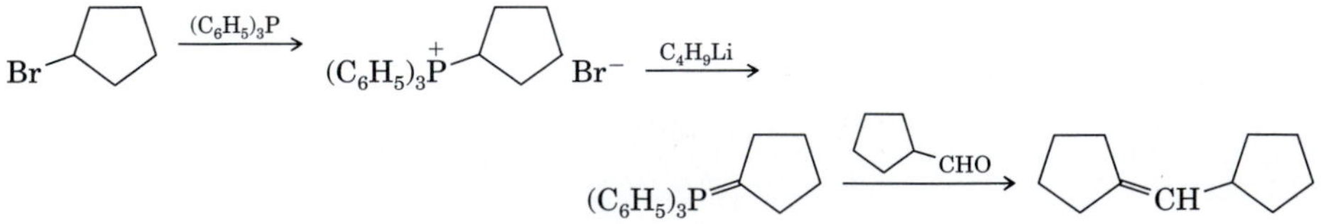

$\xrightarrow{CHO}$

(2) $\xrightarrow{(C_6H_5)_3P}$ $\xrightarrow{C_4H_9Li}$

$\xrightarrow{}$ product

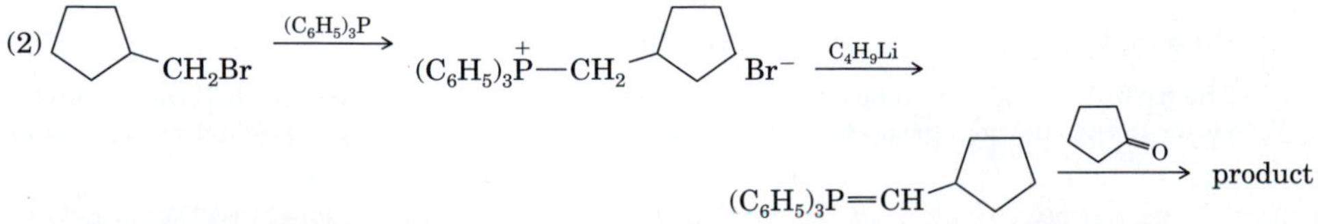

(c) (1) $BrCH_2CH_2CH_2CH_2Br$ $\xrightarrow{2\ (C_6H_5)_3P}$ $(C_6H_5)_3\overset{+}{P}—CH_2CH_2CH_2CH_2—\overset{+}{P}(C_6H_5)_3 + 2\ Br^-$ $\xrightarrow{2\ C_4H_9Li}$

$(C_6H_5)_3P{=}CHCH_2CH_2CH{=}P(C_6H_5)_3$ $\xrightarrow{}$ product

(with *o*-phthalaldehyde, CHO / CHO)

(2) (benzene ring with CH_2Br / CH_2Br) $\xrightarrow{2\ (C_6H_5)_3P}$ (benzene ring with $CH_2—\overset{+}{P}(C_6H_5)_3\ Br^-$ / $CH_2—\overset{+}{P}(C_6H_5)_3\ Br^-$) $\xrightarrow{2\ C_4H_9Li}$

(benzene ring with $CH{=}P(C_6H_5)_3$ / $CH{=}P(C_6H_5)_3$) $\xrightarrow{OHCCH_2CH_2CHO}$ product

13.16 $CH_3(CH_2)_4CHO$ $\xrightarrow[(2)\ H_2O,\ H^+]{(1)\ CH_3MgI}$

CH_3CHO $\xrightarrow[(2)\ H_2O,\ H^+]{(1)\ CH_3(CH_2)_3CH_2MgBr}$ $\longrightarrow$ $CH_3(CH_2)_4\overset{\displaystyle OH}{\underset{\displaystyle |}{C}}HCH_3$

$CH_3(CH_2)_4\overset{\displaystyle O}{\overset{\displaystyle ||}{C}}CH_3$ $\xrightarrow[\text{heat, pressure}]{H_2,\ Ni}$

13.17 (a) CH_3CH_2CHO $\xrightarrow[(2)\ H_2O,\ H^+]{(1)\ NaBH_4}$

(b) (cyclohexenone) $=O$ $\xrightarrow[(2)\ H_2O,\ H^+]{(1)\ NaBH_4}$ **(c)** (cyclohexane ring)$—\overset{\displaystyle O}{\overset{\displaystyle ||}{C}}CH_3$ $\xrightarrow[(2)\ H_2O,\ H^+]{(1)\ NaBH_4}$

13.18 (a) and **(c)**. Compound **(c)** is a hemiacetal, which is in equilibrium with the aldehyde in alkaline solution. Compound **(d)** is an acetal, which is not in equilibrium with the aldehyde under the alkaline conditions of the Tollens test.

13.19 (a) $C_6H_5\overset{\displaystyle \ddot O:}{\overset{\displaystyle ||}{C}}—\underset{}{\ddot C}H—\overset{\displaystyle :\ddot O}{\overset{\displaystyle ||}{C}}CH_3$ $\longleftrightarrow$ $C_6H_5\overset{\displaystyle \ddot O:}{\underset{}{C}}—CH{=}\overset{\displaystyle :\ddot O:^-}{C}CH_3$ $\longleftrightarrow$ $C_6H_5\overset{\displaystyle :\ddot O:^-}{C}{=}CH—\overset{\displaystyle :\ddot O}{\overset{\displaystyle ||}{C}}CH_3$

(b) $CH_3\overset{\displaystyle \ddot O:}{\overset{\displaystyle ||}{C}}—\underset{}{\ddot C}H—\overset{\displaystyle :\ddot O}{\overset{\displaystyle ||}{C}}CH_3$ $\longleftrightarrow$ $CH_3\overset{\displaystyle \ddot O:}{C}—CH{=}\overset{\displaystyle :\ddot O:^-}{C}CH_3$ $\longleftrightarrow$ $CH_3\overset{\displaystyle :\ddot O:^-}{C}{=}CHCCH_3$

13.20 Enol form of ethyl acetoacetate: $CH_3C\overset{}{\underset{}{=}}\underset{\displaystyle H}{C}\overset{}{\underset{}{—}}C\overset{}{—}OCH_2CH_3$ (with $O\cdots H\cdots \ddot O$ hydrogen bond)

13.21 (a) (benzene ring)$—\overset{\displaystyle O}{\overset{\displaystyle ||}{C}}CH_2Cl$ **(b)** (cyclohexanedione ring with two Br)

13.22 (a) (decalin ring with $\overset{}{\underset{\displaystyle ||}{C}}CH_3$, O) **(b)** $(CH_3)_2CHC\overset{\displaystyle O}{\overset{\displaystyle ||}{}}CH_3$ **(c)** $CH_3\overset{\displaystyle O}{\overset{\displaystyle ||}{C}}—$(benzene ring)$—\overset{\displaystyle O}{\overset{\displaystyle ||}{C}}CH_3$

13.23 (a) CH₃CCH₂CH₂OCH₃ (with =O above first C) $\xrightarrow[\text{(2) H}_2\text{O, H}^+]{\text{(1) C}_6\text{H}_5\text{MgBr}}$ CH₃CCH₂CH₂OCH₃ (with OH and C₆H₅ on central C) $\xrightarrow[-\text{H}_2\text{O}]{\text{H}^+,\text{ heat}}$

(E) and (Z)-CH₃C=CHCH₂OCH₃ (with C₆H₅ substituent)

(b) CH₃CCH₃ (with =O) $\xrightarrow[\text{(2) H}_2\text{O, H}^+]{\text{(1) NaBH}_4}$ CH₃CHCH₃ (with OH) $\xrightarrow[\text{heat}]{\text{H}_2\text{SO}_4}$ CH₃CH=CH₂

$\xrightarrow{\text{NBS}}$ BrCH₂CH=CH₂ $\xrightarrow[\text{(2) CH}_3(\text{CH}_2)_3\text{Li}]{\text{(1) (C}_6\text{H}_5)_3\text{P}}$ (C₆H₅)₃P=CHCH=CH₂

$\xrightarrow{(\text{CH}_3)_2\text{C}=\text{O}}$ H₂C=CHCH=C(CH₃)₂

(c) HCCH₂CH (with two =O groups) $\xrightarrow[\text{(2) H}_2\text{O, H}^+]{\text{(1) 2 CH}_3\text{CH}_2\text{MgBr}}$ CH₃CH₂CHCH₂CHCH₂CH₃ (with two OH groups)

(d) [cyclohexanone] $\xrightarrow[]{\text{Cl}_2,\text{ H}^+}$ [2-chlorocyclohexanone] $\xrightarrow[\text{heat}]{\text{K}^+\,{}^-\text{OC(CH}_3)_3}$

[cyclohexenone] $\xrightarrow[\text{H}_2\text{O}_2]{\text{OsO}_4}$ [1,2-dihydroxycyclohexanone, HO and OH]

(e) [cyclopentanol, ─OH] $\xrightarrow{\text{H}_2\text{CrO}_4}$ [cyclopentanone, =O]

$\downarrow$ HBr

[cyclopentyl bromide, ─Br] $\xrightarrow[\text{ether}]{\text{Mg}}$ [cyclopentyl─MgBr] $\xrightarrow[\text{(2) H}_2\text{O, H}^+]{\text{(1) [cyclopentanone]}}$ product

(f) [cyclopentanone with CH₃ and CH₂CH₂CO₂CH₃] $\xrightarrow{\text{Br}_2,\text{ H}^+}$ [brominated product with Br, CH₃, CH₂CH₂CO₂CH₃] $\xrightarrow[\text{heat}]{\text{K}^+\,{}^-\text{OC(CH}_3)_3}$ [cyclopentenone with CH₃ and CH₂CH₂CO₂CH₃]

The ketone contains only one position with alpha hydrogens; therefore, bromination occurs here. Acidic conditions are preferable to alkaline conditions (see Section 13.10). A sterically hindered base should be used in the elimination to minimize formation of the substitution product.

Chapter 14

14.1 (a) cyclopentanecarboxylic acid

(b) lactic acid, 2-hydroxypropanoic acid, α-hydroxypropionic acid

(c) potassium 2-methylpropanoate

(d) 3-ethyl-5-hydroxypentanoic acid

(e) 6,6-dimethylheptanoic acid

(f) 3-methylpentanedioic acid

14.2 (a) $CH_3(CH_2)_7\overset{O}{\overset{\|}{C}}OH$ (b) $K^{+-}O\overset{O}{\overset{\|}{C}}CH_2CH_2CH_2CHCH_2CH_2CH_2CH_3$ (c) $HO\overset{O}{\overset{\|}{C}}(CH_2)_6\overset{O}{\overset{\|}{C}}OH$
$\qquad\qquad\qquad\qquad\qquad\qquad\qquad\qquad\qquad\qquad |$
$\qquad\qquad\qquad\qquad\qquad\qquad\qquad\qquad\qquad CH_3$

(d) $H\overset{O}{\overset{\|}{C}}OH$ (e) $C_6H_5CH_2\overset{O}{\overset{\|}{C}}OH$ (f) $\bigcirc\!\!-CH_2CH_2CH_2CHCH_2CH_2\overset{O}{\overset{\|}{C}}OH$
$\qquad\qquad\qquad\qquad\qquad\qquad\qquad\qquad\qquad\qquad\qquad\qquad |$
$\qquad\qquad\qquad\qquad\qquad\qquad\qquad\qquad\qquad\qquad\qquad\quad CH_2C_6H_5$

$\qquad\qquad OCH_3 \qquad\qquad\qquad CH_3CH_2 \quad H$
$\qquad\qquad\quad |$
(g) $CH_3CHCH_2CO_2H$ (h) $\qquad C=C$
$\qquad\qquad\qquad\qquad\qquad\qquad\qquad\quad H \qquad\quad CO_2H$

14.3 (a) $CH_3CH_2Br \xrightarrow{Mg} CH_3CH_2MgBr \xrightarrow[\text{(2) H}^+]{\text{(1) CO}_2} CH_3CH_2\overset{O}{\overset{\|}{C}}OH$

(b) $CH_3CH_2CH_2OH \xrightarrow[H_2SO_4]{CrO_3} CH_3CH_2\overset{O}{\overset{\|}{C}}OH$

(c) $CH_3CH_2CH{=}CH_2 \xrightarrow[H_2SO_4]{CrO_3} CH_3CH_2\overset{O}{\overset{\|}{C}}OH + CO_2$

14.4 (a) $HOCH_2CH_2Cl \xrightarrow{KCN} HOCH_2CH_2CN \xrightarrow{H_2O, H^+} HOCH_2CH_2CO_2H$

(b) $\bigcirc\!\!\bigcirc\!\!-Br \xrightarrow{Mg} \bigcirc\!\!\bigcirc\!\!-MgBr \xrightarrow[\text{(2) H}^+, H_2O]{\text{(1) CO}_2} \bigcirc\!\!\bigcirc\!\!-\overset{O}{\overset{\|}{C}}OH$

14.5 (a) bromoacetic acid, because Br is more strongly electron withdrawing than a phenyl group

(b) dibromoacetic acid, because two Br are more strongly electron withdrawing than one is

(c) 2-iodopropanoic acid, because the I is one carbon closer to the carboxylate group and thus exhibits a stronger inductive effect.

14.6 (a) *p*-bromobenzoic acid   (b) *m*-bromobenzoic acid

(c) 3,5-dibromobenzoic acid

The reason for these relative acidities is that Br is electron withdrawing by the inductive effect and stabilizes the conjugate base relative to the acid. In (c), two Br atoms exert a stronger electron-withdrawing effect than one Br atom.

$\qquad\qquad\qquad\qquad\qquad\qquad\qquad\qquad\qquad\qquad$ stabilizes anion by electron-
$\qquad\qquad\qquad\qquad\qquad\qquad\qquad\qquad\qquad\qquad$ withdrawing inductive
$\qquad\qquad\qquad\qquad\qquad\qquad\qquad\qquad\qquad\qquad$ effect

Br $\qquad\qquad\qquad\qquad\qquad\qquad$ Br
$\bigcirc\!\!-CO_2H \rightleftharpoons \bigcirc\!\!-CO_2^- + H^+$

14.7 (a) $\bigcirc\overset{CO_2^- Na^+}{\underset{CO_2^- Na^+}{}}$ (b) $C_6H_5CH_2O_2CCH_2CH_3$

(c) $CH_3CO_2^- + CH_3OH$ (d) $CH_3CO_2^-\ CH_3NH_3^+$

(e) $CH_3CO_2H + ClCH_2CO_2^-$

14.8 **(a)** 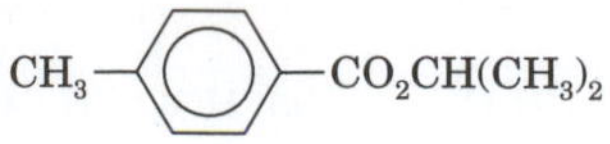 CH$_3$—⟨benzene⟩—CO$_2$CH(CH$_3$)$_2$ **(b)** CH$_3$CH$_2$O$_2$C—⟨benzene⟩—CO$_2$CH$_2$CH$_3$

(c) (R)-CH$_3$CO$_2$CHCH$_2$CH$_3$
         |
         CH$_3$

14.9

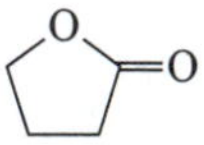

14.10 **(a)** 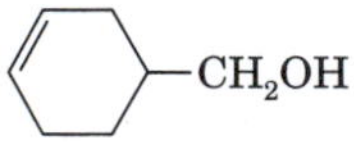 ⟨cyclohexene⟩—CH$_2$OH   **(b)** 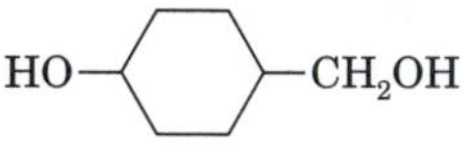 HO—⟨cyclohexane⟩—CH$_2$OH

cis and trans

14.11 **(a)** 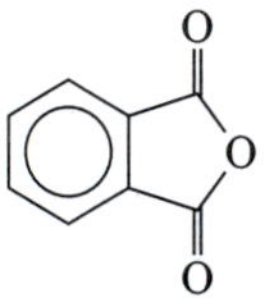   **(b)**

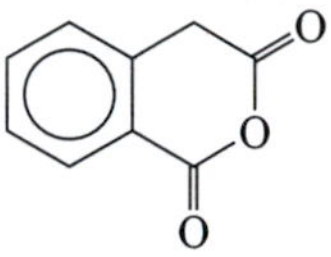

14.12

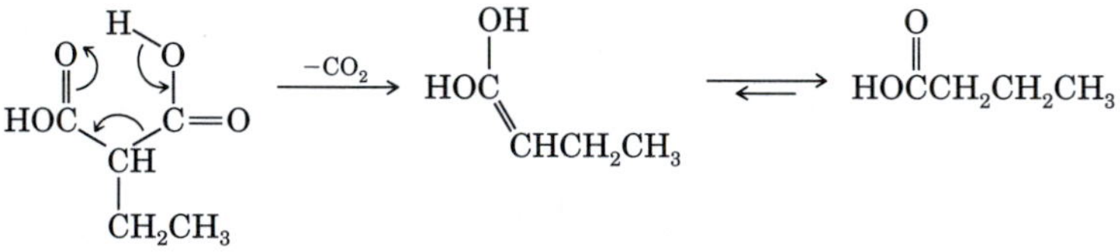

14.13 **(a)** CH$_3$CO$_2$H   **(b)** CH$_3$CCH$_2$CH$_3$ (with O double bond)   **(c)** no reaction

14.14 CH$_3$CH$_2$CH$_2$CO$_2$H + CH$_3$OH $\xrightarrow[\text{heat}]{\text{H}^+}$

CH$_3$CH$_2$CH$_2$CO$_2$H $\xrightarrow[\text{(2) CH}_3\text{I}]{\text{(1) OH}^-}$ → CH$_3$CH$_2$CH$_2$CO$_2$CH$_3$

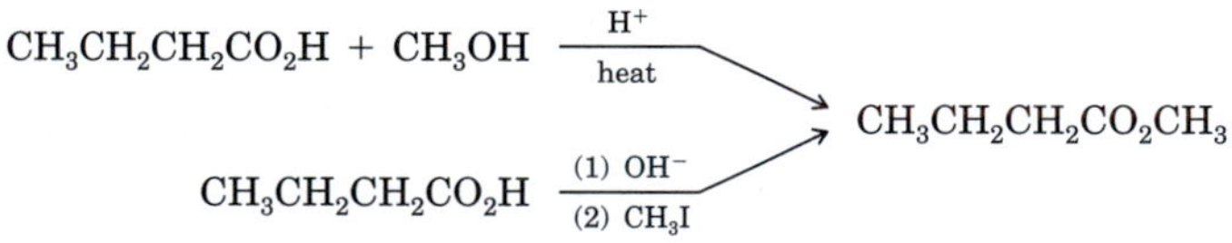

14.15 **(a)** CH$_3$CH$_2$CO$_2$H $\xrightarrow[\text{(2) H}_2\text{O}]{\text{(1) LiAlH}_4}$ CH$_3$CH$_2$CH$_2$OH $\xrightarrow{\text{HBr}}$ CH$_3$CH$_2$CH$_2$Br $\xrightarrow[\text{ether}]{\text{Mg}}$

CH$_3$CH$_2$CH$_2$MgBr $\xrightarrow[\text{(2) H}_2\text{O, H}^+]{\text{(1) CO}_2}$ product

(b) ClCH$_2$CH$_2$CO$_2$H $\xrightarrow[\text{(2) KCN}]{\text{(1) neutralize with OH}^-}$ NCCH$_2$CH$_2$CO$_2^-$ $\xrightarrow{\text{H}_2\text{O, H}^+}$ product

(c) 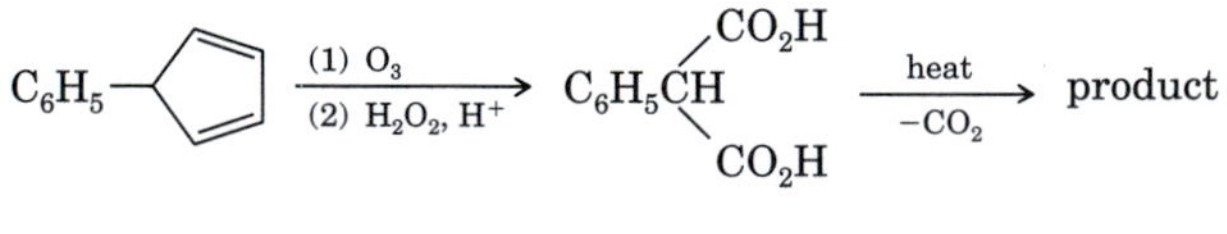 C$_6$H$_5$—⟨cyclopentadiene⟩ $\xrightarrow[\text{(2) H}_2\text{O}_2, \text{H}^+]{\text{(1) O}_3}$ C$_6$H$_5$CH(CO$_2$H)(CO$_2$H) $\xrightarrow[\text{-CO}_2]{\text{heat}}$ product

Chapter 15

15.1 **(d)** < **(a)** < **(b)** < **(c)**

15.2 CH$_3$OCCH$_2$CH$_2$CN (with O double bond on C)

15.3 **(a)** decanoyl chloride

(b) 3,5-dimethoxybenzoyl chloride

15.4 **(a)** $CH_3(CH_2)_4\overset{\overset{\displaystyle O}{\|}}{C}OH + SOCl_2 \longrightarrow$

(b)

naphthalene with $\overset{\overset{\displaystyle O}{\|}}{C}OH$ $+ SOCl_2 \longrightarrow$

(c) CH_3O— benzene ring with two more CH_3O groups —$\overset{\overset{\displaystyle O}{\|}}{C}OH + SOCl_2 \longrightarrow$

PCl_3, PCl_5, or $POCl_3$ (but not HCl) could be used in place of $SOCl_2$.

15.5 $CH_3CO_2^- Na^+ + Cl^- +$ unreacted CH_3NH_2

15.6 **(a)** CH_3—benzene ring—$\overset{\overset{\displaystyle O}{\|}}{C}Cl + HO\overset{\underset{\displaystyle CH_3}{|}}{C}HCH_3 \xrightarrow{\text{pyridine}}$

(b) $(CH_3)_2CH(CH_2)_8\overset{\overset{\displaystyle O}{\|}}{C}O^- Na^+ + CH_3\overset{\overset{\displaystyle O}{\|}}{C}Cl \longrightarrow$

(c) cyclohexyl—$\overset{\overset{\displaystyle O}{\|}}{C}Cl + 2\ HN\overset{\frown}{\underset{\smile}{\ \ }}O \longrightarrow$

(d) acenaphthylene with $\overset{\overset{\displaystyle O}{\|}}{C}Cl$ $+$ naphthalene with OH $\xrightarrow{\text{pyridine}}$

15.7 **(a)** $(CH_3)_2CHCl \xrightarrow[\text{(2) CuI}]{\text{(1) Li}} LiCu[CH(CH_3)_2]_2 \xrightarrow{C_6H_5\overset{\overset{\displaystyle O}{\|}}{C}Cl} \text{product}$

(b) $CH_3I \xrightarrow[\text{(2) CuI}]{\text{(1) Li}} LiCu(CH_3)_2 \xrightarrow{\text{cyclohexyl-}\overset{\overset{\displaystyle O}{\|}}{C}Cl} \text{product}$

or cyclohexyl—$I \xrightarrow[\text{(2) CuI}]{\text{(1) Li}} LiCu(C_6H_{11})_2 \xrightarrow{CH_3\overset{\overset{\displaystyle O}{\|}}{C}Cl} \text{product}$

15.8 **(a)** $C_6H_5CHBrCO_2H$ **(b)** $ClCH_2CHClCO_2H$

15.9 $(CH_3)_2CH\overset{\overset{\displaystyle O}{\|}}{C}OH + SOCl_2 \longrightarrow (CH_3)_2CH\overset{\overset{\displaystyle O}{\|}}{C}Cl + SO_2 + HCl$

(a) $(CH_3)_2CH\overset{\overset{\displaystyle O}{\|}}{C}Cl + CH_3CH_2OH \xrightarrow[-HCl]{\text{pyridine}} (CH_3)_2CH\overset{\overset{\displaystyle O}{\|}}{C}OCH_2CH_3$

(b) $(CH_3)_2CH\overset{\overset{\displaystyle O}{\|}}{C}Cl \xrightarrow[\text{(2) H}_2\text{O, H}^+]{\text{(1) LiAlH[OC(CH}_3)_3]_3} (CH_3)_2CH\overset{\overset{\displaystyle O}{\|}}{C}H$

(c) $(CH_3)_2CH\overset{\overset{\displaystyle O}{\|}}{C}Cl + (CH_3CH_2)_2CuLi \longrightarrow (CH_3)_2CH\overset{\overset{\displaystyle O}{\|}}{C}CH_2CH_3$

15.10 (a) pentanoic anhydride

(b) acetic benzoic anhydride or benzoic ethanoic anhydride

15.11 (a)

naphthalene with $OCCH_3$ (O double bond) ester group

(b)

benzene ring with $COCH_2CH_3$ (O above) and COH (O below)

15.12 (a)

$$\text{CH}_3\text{-C}_6\text{H}_4\text{-NH}_2 \xrightarrow{(CH_3CO)_2O} \text{CH}_3\text{-C}_6\text{H}_4\text{-NHCCH}_3 \xrightarrow{HNO_3}$$

$$\text{(ring, CH}_3, \text{-NHCCH}_3, \text{NO}_2) \xrightarrow{H_2O,\ OH^-} \text{(ring, CH}_3, \text{-NH}_2 + CH_3CO_2^-, \text{NO}_2)$$

(b)

$$\text{CH}_3\text{-C}_6\text{H}_4\text{-NHCCH}_3 \xrightarrow{HNO_3} O_2N\text{-C}_6\text{H}_3(\text{CH}_3)\text{-NHCCH}_3 \xrightarrow{H_2O,\ OH^-} O_2N\text{-C}_6\text{H}_3(\text{CH}_3)\text{-NH}_2$$

(c) The nitric acid would form a salt with the amine. The resulting $-^+NH_3$ group is strongly deactivating. (Then, oxidation of the $-^+NH_3$ group occurs.)

$$\text{CH}_3\text{-C}_6\text{H}_4\text{-NH}_2 + HNO_3 \longrightarrow \text{CH}_3\text{-C}_6\text{H}_4\text{-}^+NH_3\ NO_3^-$$

15.13 (a) $\text{CH}_3\text{-C(CH}_3)_2\text{-OCCH}_3$ (with O)

(b) $\text{C}_6\text{H}_5\text{-OCCH}_2\text{CH}_2\text{CH}_3$ (with O)

(c) $\text{C}_6\text{H}_5\text{-CH}_2\text{OC-C}_6\text{H}_5$ (with O)

(d) $\text{CH}_3\text{C(CH}_3)\text{-CH}=\text{CH}-\text{C(O)}-\text{OCH}=\text{CH}_2$

15.14 (a) $\text{CH}_3\text{CH}_2\text{CH}_2\text{CO}_2\text{H} + \text{CH}_3\text{OH} \underset{}{\overset{H^+}{\rightleftharpoons}} \text{CH}_3\text{CH}_2\text{CH}_2\text{CO}_2\text{CH}_3 + \text{H}_2\text{O}$

$$\text{CH}_3\text{CH}_2\text{CH}_2\text{CO}_2\text{H} \xrightarrow{SOCl_2} \text{CH}_3\text{CH}_2\text{CH}_2\overset{O}{\overset{\|}{\text{C}}}\text{Cl} \xrightarrow[\text{pyridine}]{CH_3OH} \text{CH}_3\text{CH}_2\text{CH}_2\text{CO}_2\text{CH}_3$$

(b) $\text{CH}_3\text{CO}_2\text{H} + \text{HOCH}_2\text{CH}=\text{C(CH}_3)_2 \underset{}{\overset{H^+}{\rightleftharpoons}} \text{CH}_3\text{CO}_2\text{CH}_2\text{CH}=\text{C(CH}_3)_2 + \text{H}_2\text{O}$

$$\text{CH}_3\overset{O}{\overset{\|}{\text{C}}}\text{O}\overset{O}{\overset{\|}{\text{C}}}\text{CH}_3 + \text{HOCH}_2\text{CH}=\text{C(CH}_3)_2 \longrightarrow \text{CH}_3\text{CO}_2\text{CH}_2\text{CH}=\text{C(CH}_3)_2 + \text{CH}_3\text{CO}_2\text{H}$$

(Acetyl chloride could also be used.)

15.15 (a) $\text{C}_6\text{H}_5\text{CO}_2\text{CH}_2\text{CH}_3 + \text{OH}^- \longrightarrow \text{C}_6\text{H}_5\text{CO}_2^- + \text{HOCH}_2\text{CH}_3$

(b) [structure] $+ 2\ OH^- \longrightarrow HOCH_2CH_2OH + {}^-\overset{\overset{O}{\|}}{O}C-\overset{\overset{O}{\|}}{C}O^-$

(c) [structure] $+ 2\ OH^- \longrightarrow$ [structure with $CH_2CH_2CO_2^-$ and O^-] $+ H_2O$

15.16 (a) $CH_3CO_2C_6H_5$; the C_6H_5O is more electron withdrawing and a better leaving group

(b) $CH_3CO_2CH_3$; less steric hindrance

(c) $CF_3CO_2CH_3$ because the CF_3 withdraws electron density from the $C=O$ carbon

(d) $CH_3CO_2CH_3$; less steric hindrance

15.17 (a) $CH_3\overset{\overset{:O:}{\|}}{C}OCH_2CH_3 \underset{}{\overset{H^+}{\rightleftharpoons}} \left[CH_3\overset{\overset{+\ddot{O}H}{\|}}{C}OCH_2CH_3 \right] \overset{CH_3\ddot{O}H}{\rightleftharpoons}$

$\left[CH_3\overset{\overset{:\ddot{O}H}{|}}{\underset{\underset{+}{CH_3\ddot{O}H}}{C}}OCH_2CH_3 \right] \overset{proton}{\underset{transfer}{\rightleftharpoons}} \left[CH_3\overset{\overset{:\ddot{O}}{}}{\underset{\underset{H}{CH_3\ddot{O}}}{C}\overset{H}{}}\ddot{O}CH_2CH_3 \right] \overset{-H^+}{\rightleftharpoons} CH_3\overset{\overset{O}{\|}}{C}OCH_3 + HOCH_2CH_3$

(b) $CH_3\overset{\overset{:\ddot{O}}{\|}}{C}OCH_2CH_3 \overset{:\ddot{O}CH_3}{\rightleftharpoons} \left[CH_3\overset{\overset{:\ddot{O}:^-}{}}{\underset{\underset{:OCH_3}{}}{C}}\ddot{O}CH_2CH_3 \right] \rightleftharpoons CH_3\overset{\overset{\ddot{O}:}{\|}}{C}OCH_3 + {}^-:\ddot{O}CH_2CH_3$

15.18 When the hydroxy acid is treated with a chlorinating agent, the hydroxyl group would be converted to a chloro substituent.

$$CH_3\underset{\underset{OH}{|}}{C}HCO_2H \overset{SOCl_2}{\longrightarrow} CH_3\underset{\underset{Cl}{|}}{C}H\overset{\overset{O}{\|}}{C}Cl$$

Treatment of a carboxylic ester of hydroxy acid with ammonia would yield the hydroxy amide because NH_3 does not replace OH^-.

$$CH_3\underset{\underset{OH}{|}}{C}HCO_2H \overset{CH_3OH,\ H^+}{\longrightarrow} CH_3\underset{\underset{OH}{|}}{C}HCO_2CH_3 \overset{NH_3}{\longrightarrow} CH_3\underset{\underset{OH}{|}}{C}H\overset{\overset{O}{\|}}{C}NH_2$$

15.19 (a) $CH_3CH_2CO_2CH_3 \overset{(1)\ 2\ CH_3MgI}{\underset{(2)\ H_2O,\ H^+}{\longrightarrow}} CH_3CH_2\underset{\underset{}{}}{\overset{\overset{OH}{|}}{C}}(CH_3)_2$

(b) $CH_3CH_2CO_2CH_3 \overset{(1)\ 2\ CH_3CH_2MgBr}{\underset{(2)\ H_2O,\ H^+}{\longrightarrow}} CH_3CH_2\overset{\overset{OH}{|}}{C}(CH_2CH_3)_2$

15.20 $H\overset{\overset{O}{\|}}{C}OCH_2CH_3 \overset{(1)\ 2\ CH_3CH_2MgBr}{\underset{(2)\ H_2O,\ H^+}{\longrightarrow}} HOCH(CH_2CH_3)_2$

15.21 (a) $HO_2CCH_2\underset{\underset{+NH_3\ Cl^-}{|}}{C}HCO_2H + C_6H_5CH_2\underset{\underset{+NH_3\ Cl^-}{|}}{C}HCO_2H + CH_3OH$

(b) Na^+ $^-O_2CCH_2\underset{\underset{NH_2}{|}}{C}HCO_2^-$ Na^+ + $C_6H_5CH_2\underset{\underset{NH_2}{|}}{C}HCO_2^-$ Na^+ + CH_3OH

15.22

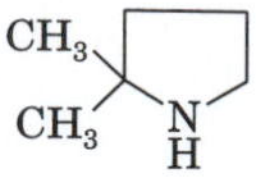

15.23

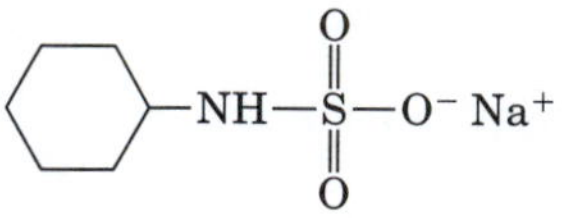

15.24 (a) CH_3CH_2CN; **(b)** $CH_3CH_2CH_2CN$

15.25 (a)

(b) $(CH_3)_2CHCH_2CO_2CH_3$ $\xrightarrow[\text{(2) } H_2O, H^+]{\text{(1) } 2\ CH_3CH_2CH_2MgBr}$ product

(c) $CH_3CH_2CH_2OH$ $\xrightarrow{H_2CrO_4}$ $CH_3CH_2CO_2H$ $\xrightarrow{SOCl_2}$ $CH_3CH_2\overset{O}{\overset{||}{C}}Cl$

$CH_3CH_2CH_2OH$ $\dashrightarrow^{HBr}$ $CH_3CH_2CH_2Br$ $\xrightarrow{KCN}$ $CH_3CH_2CH_2CN$

$\xrightarrow[\text{(2) } H_2O, H^+]{\text{(1) } LiAlH_4}$ $CH_3CH_2CH_2CH_2NH_2$ $\xrightarrow{CH_3CH_2\overset{O}{\overset{||}{C}}Cl}$ product

Chapter 16

16.1 $CH_3\underset{\underset{Br}{|}}{C}H\underset{\underset{Br}{|}}{C}HCH_3$; $CH_3\underset{\underset{Cl}{|}}{C}H\underset{\underset{Cl}{|}}{C}H\underset{\underset{Cl}{|}}{C}H\underset{\underset{Cl}{|}}{C}HCH_3$

16.2 (a) 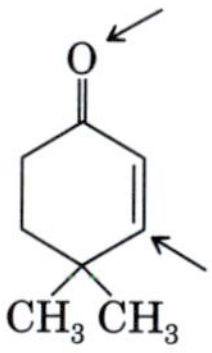 **(b)** 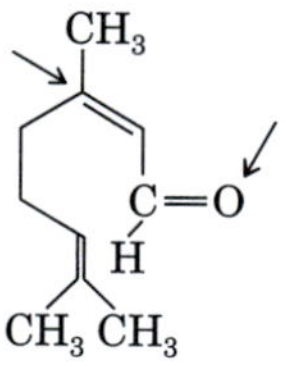 **(c)** 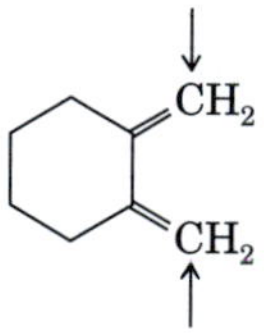 **(d)**

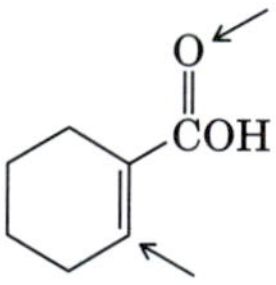

16.3 (a) $CH_3CH{=}CH{-}CH{=}CHCH_3$; equivalent   **(b)** 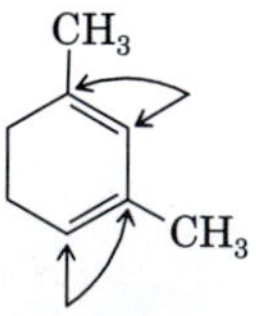 ; not equivalent

(c) 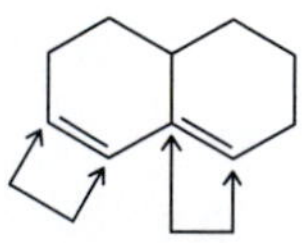 ; not equivalent   **(d)** 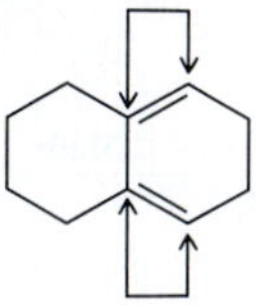 ; equivalent

16.4 π_6^* _____

π_5^* _____

π_4^* _____

π_3 ↑↓

π_2 ↑↓

π_1 ↑↓

16.5 **(a)** π_4^* _____   **(b)** π_4^* _____

π_3^* _____ π_3^* _____

π_2 ↑↓ π_2 ↑↓   ↑↓ n

π_1 ↑↓ π_1 ↑↓

16.6 $[CH_3CH_2\overset{+}{C}H\!-\!CH\!=\!CHCH_3 \longleftrightarrow CH_3CH_2CH\!=\!CH\overset{+}{C}HCH_3] \longrightarrow$

$CH_3CH_2CHICH\!=\!CHCH_3 + CH_3CH_2CH\!=\!CHCHICH_3$

16.7 **(a)**

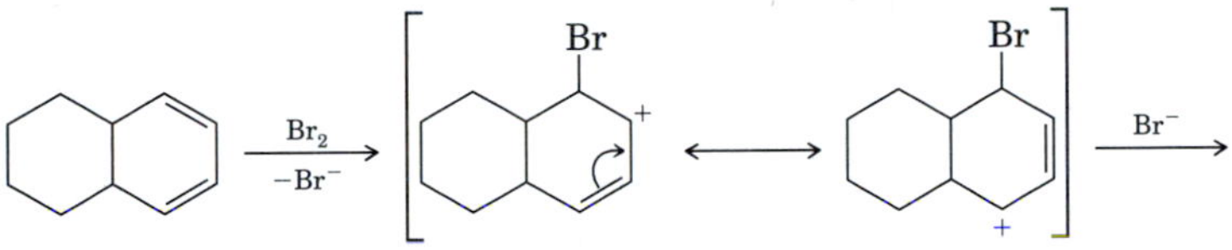

resonance structures for the intermediate

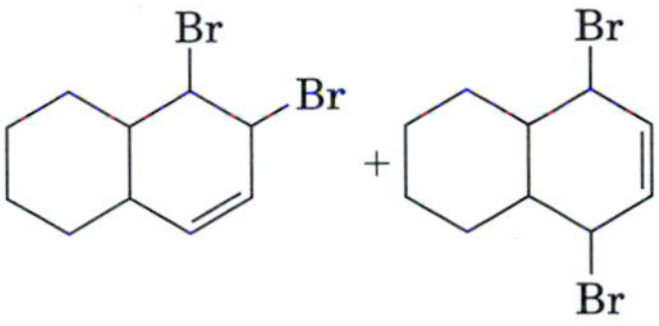

(b)

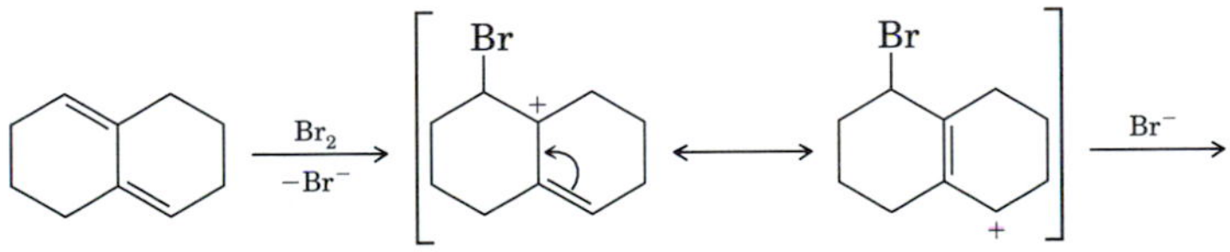

resonance structures for the intermediate

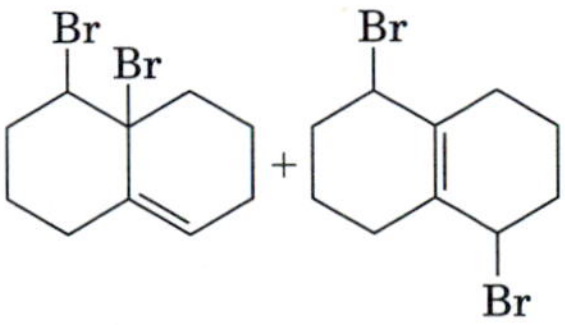

16.8 The *trans* isomer, because it is more stable.

16.9 A, $BrCH_2CH\!=\!CHCH_2Br$;   B, $BrCH_2\underset{\underset{Br}{|}}{CH}CH\!=\!CH_2$

16.10 $\left(\begin{array}{cc} Cl & CH_2 \\ \diagdown & \diagup \\ C\!=\!C \\ \diagup & \diagdown \\ CH_2 & H \end{array} \right)_x$

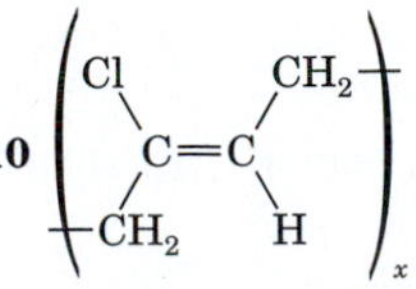

16.11 **(a)** s-*trans*   **(b)** s-*cis*

(c) s-*trans*

Only **(a)** is interconvertible

16.12 (a) 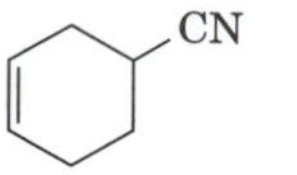 **(b)**

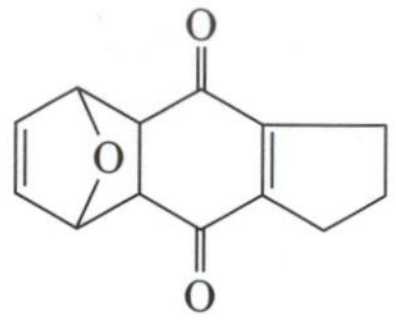

16.13  + CH$_3$CH$_2$O$_2$CC≡CCO$_2$CH$_2$CH$_3$ $\xrightarrow{\text{heat}}$

16.14

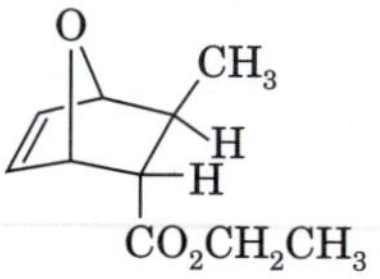

16.15 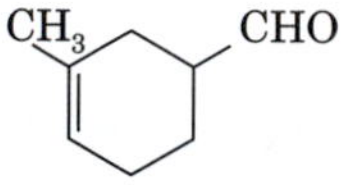 and

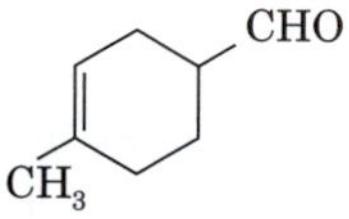

16.16 (a) 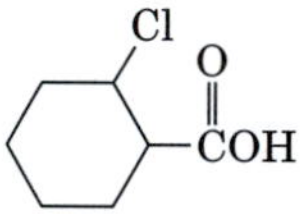 **(b)** 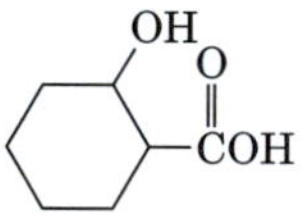 **(c)** CH$_3$CHCH$_2$CO$_2^-$ H$_3\overset{+}{\text{N}}$CH$_3$

cis and *trans*   *cis* and *trans* $\underset{\text{NHCH}_3}{|}$

16.17 (a) 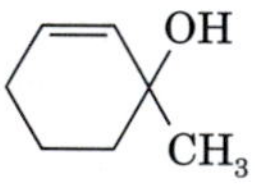 **(b)** 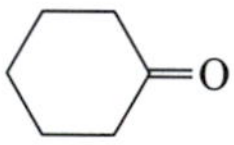 **(c)** 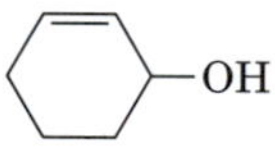 **(d)** 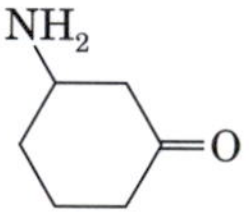 **(e)**

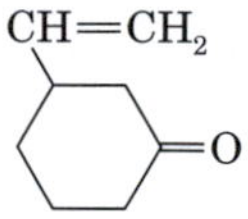

Chapter 17

17.1 (a) The hydrogens at position (2) because the negative charge of the conjugate base can be delocalized by both carbonyl groups.

(b) The hydrogens at position (1) because the negative charge of the conjugate base can be delocalized by the carbonyl group.

(c) The hydrogens at position (3) for the same reason given for part **(a)**.

17.2 (a) CH$_3$CH$_2$$\overset{\overset{\displaystyle O}{\|}}{\text{C}}OCH_2CH_3$ + Na$^+$ $^-$OCH$_2$CH$_3$ $\rightleftharpoons$ CH$_3\overset{-}{\underset{\overset{\displaystyle ..}{\text{Na}^+}}{\text{CH}}}$$\overset{\overset{\displaystyle O}{\|}}{\text{C}}OCH_2CH_3$ + CH$_3$CH$_2$OH

(b) CH$_3$CH$_2$O$\overset{\overset{\displaystyle O}{\|}}{\text{C}}CH_2$CN + Na$^+$ $^-$OCH$_2$CH$_3$ $\rightleftharpoons$ CH$_3$CH$_2$O$\overset{\overset{\displaystyle O}{\|}}{\text{C}}\overset{-}{\underset{\overset{\displaystyle ..}{\text{Na}^+}}{\text{CH}}}$CN + CH$_3CH_2$OH

(c) CH$_3$NO$_2$ + Na$^+$ $^-$OCH$_2$CH$_3$ $\rightleftharpoons$ Na$^+$ $^-$:CH$_2$NO$_2$ + CH$_3$CH$_2$OH

(d) 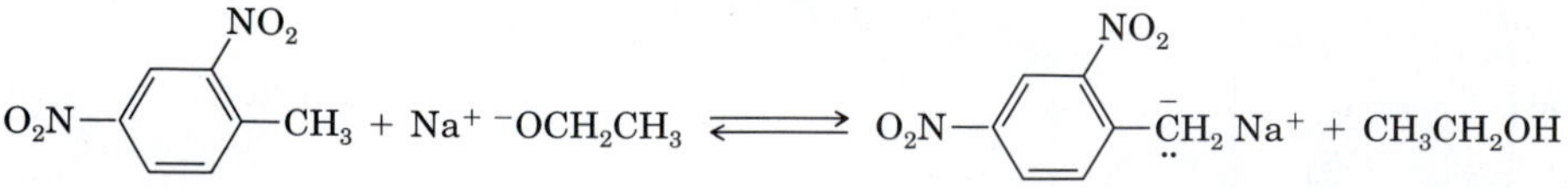 O$_2$N——CH$_3$ + Na$^+$ $^-$OCH$_2$CH$_3$ $\rightleftharpoons$ O$_2$N——$\overset{-}{\underset{..}{\text{CH}}}_2$ Na$^+$ + CH$_3$CH$_2$OH

(e) 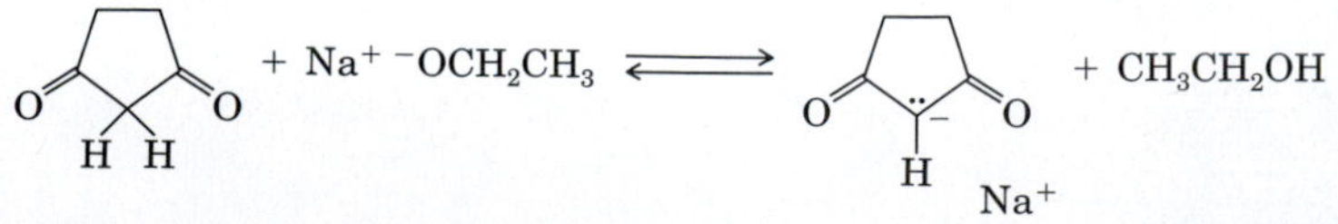 + Na$^+$ $^-$OCH$_2$CH$_3$ $\rightleftharpoons$ + CH$_3$CH$_2$OH

(f) $CH_3(CH_2)_4CH(CO_2C_2H_5)_2 + Na^+ \, {}^-OCH_2CH_3 \rightleftharpoons CH_3(CH_2)_4\overset{-}{C}(CO_2C_2H_5)_2 + CH_3CH_2OH$
$\phantom{CH_3(CH_2)_4CH(CO_2C_2H_5)_2 + Na^+ \, {}^-OCH_2CH_3 \rightleftharpoons CH_3(CH_2)_4}Na^+$

17.3 **(a)** $CH_3CH_2CH_2CH(CO_2C_2H_5)_2$ **(b)** $CH_3CH_2\underset{\underset{CH_3}{|}}{\overset{\overset{CH_3}{|}}{C}}(CO_2C_2H_5)_2$

17.4 **(a)** $CH_3CH_2CH_2CH(CO_2C_2H_5)_2 \xrightarrow[\text{heat}]{H_2O,\ OH^-}$

$ CH_3CH_2CH_2CH(CO_2{}^-)_2 \xrightarrow{H^+} CH_3CH_2CH_2CH(CO_2H)_2$

(b) $(CH_3)_2C(CO_2C_2H_5)_2 \xrightarrow[\text{heat}]{H_2O,\ H^+} [(CH_3)_2C(CO_2H)_2] \xrightarrow[-CO_2]{\text{heat}} (CH_3)_2CHCO_2H$

17.5

an enol

17.6 **(a)** the amide of a substituted acetic acid; start with diethyl malonate

(b) a substituted acetic acid; from diethyl malonate

(c) a substituted acetone; from ethyl acetoacetate

17.7 **(a)**

(b)

17.8 **(a)** $CH_2(CO_2C_2H_5)_2 \xrightarrow[\text{(2) } CH_3(CH_2)_3Br]{\text{(1) } NaOC_2H_5} CH_3(CH_2)_3CH(CO_2C_2H_5)_2 \xrightarrow[\text{(2) } H^+,\ \text{cold}]{\text{(1) } OH^-,\ H_2O,\ \text{heat}} \text{product}$

(b) $CH_2(CO_2C_2H_5)_2 \xrightarrow[\text{(2) } NaOC_2H_5,\ \text{then } CH_3I]{\text{(1) } NaOC_2H_5,\ \text{then } CH_3I} (CH_3)_2C(CO_2C_2H_5)_2 \xrightarrow[\text{heat}]{H_2O,\ H^+} \text{product}$

(c)

(d) $CH_2(CO_2C_2H_5)_2 \xrightarrow[\text{(2) } BrCH_2CO_2C_2H_5]{\text{(1) } NaOC_2H_5} \text{product}$

(e) $CH_2(CN)_2 \xrightarrow[\text{(2) } Br(CH_2)_4Br]{\text{(1) } NaOC_2H_5} Br(CH_2)_4CH(CN)_2 \xrightarrow{NaOC_2H_5} BrCH_2(CH_2)_3\overset{..}{\underset{..}{C}}(CN)_2 \longrightarrow \text{product}$

(f) $\xrightarrow[\text{(2) }(CH_3)_2CHCH_2Br]{\text{(1) }NaOC_2H_5}}$ product

17.9 (a) $\xrightarrow[\text{cold}]{Li^+ \ ^-N[CH(CH_3)_2]_2}$ $\xrightarrow[\text{allow to warm}]{C_6H_5CH_2Br}$ product

(b) $\xrightleftharpoons[\text{warm}]{Na^+ \ ^-OC_2H_5}$ $\xrightarrow{CH_3I}$ product

(a) is under kinetic control; **(b)** is under thermodynamic control.

17.10 (a) $CH_3CH_2\overset{O}{\overset{\|}{C}}CH_2CH_3 + HN$ $\xrightleftharpoons{H^+}$ $CH_3CH_2\overset{N}{\overset{|}{C}}=CHCH_3$ $\xrightarrow{CH_3I}$

$CH_3CH_2\overset{+N}{\overset{\|}{C}}CH(CH_3)_2$ $\xrightarrow{H_2O, \ H^+}$ $CH_3CH_2\overset{O}{\overset{\|}{C}}CH(CH_3)_2$

(b) $+ HN$ $\xrightleftharpoons{H^+}$ $\xrightarrow[\text{}]{BrCH_2\overset{O}{\overset{\|}{C}}CH_3}$

$\xrightarrow{H_2O, \ H^+}$

17.11 (c) and **(e)** because they contain α hydrogens

17.12 (a) $C_6H_5CH=CCHO$ with CH_2CH_3 **(b)** $C_6H_5CH=C\overset{O}{\overset{\|}{C}}CH_2CH_3$ with CH_3

17.13 (a) $C_6H_5CHO + CH_3CH_2CHO \xrightarrow{OH^-}$ product   **(b)** $CH_3\overset{O}{\overset{\|}{C}}(CH_2)_4\overset{O}{\overset{\|}{C}}CH_3 \xrightarrow[\text{(2) }H^+, \text{ heat}]{\text{(1) }OH^-}$ product

(c) $-CHO + CH_3CHO \xrightarrow{OH^-}$ product   **(d)** $OHC(CH_2)_5CHO \xrightarrow{OH^-}$ product

17.14 (a) $(CH_3CH_2)_2C=O + CH_2CN$ with $CO_2C_2H_5$ $\xrightarrow[\text{CH}_3CO_2H]{NH_4O_2CCH_3}$ (heat) product

(b) $=O + CH_2(CN)_2 \xrightarrow[\text{CH}_3CO_2H]{NH_4O_2CCH_3}$ (heat) product

17.15 (a) $C_6H_5CH_2\overset{\overset{\displaystyle O}{\|}}{C}CHCO_2C_2H_5$ **(b)** $C_6H_5CH_2\overset{\overset{\displaystyle O}{\|}}{C}CH_2C_6H_5$
    $\underset{\displaystyle C_6H_5}{|}$

17.16 (a) $\xrightarrow{NaOC_2H_5}$ $\xrightarrow[\text{heat}]{H_2O,\ OH^-}$

$\xrightarrow{H^+}$

(b) $\xrightarrow{NaOCH_3}$ $\xrightarrow{H^+}$

17.17 (a) $C_2H_5O_2C\overset{\overset{\displaystyle O}{\|}}{C}CH_2CO_2C_2H_5$ **(b)**

17.18 (a) $C_6H_5CO_2C_2H_5 + CH_3CO_2C_2H_5$ **(b)** $H\overset{\overset{\displaystyle O}{\|}}{C}OC_2H_5 + CH_3\overset{\overset{\displaystyle O}{\|}}{C}OC_2H_5$

(c) $2\ C_4H_9CO_2C_2H_5$ **(d)** $C_2H_5O\overset{\overset{\displaystyle O}{\|}}{C}-\overset{\overset{\displaystyle O}{\|}}{C}OC_2H_5 + CH_3\overset{\overset{\displaystyle O}{\|}}{C}OC_2H_5$

17.19 $CH_3\overset{\overset{\displaystyle O}{\|}}{C}SCoA \xrightarrow{-H^+}$ $^-\!:CH_2\overset{\overset{\displaystyle O}{\|}}{C}SCoA \xrightarrow[\text{ester condensation}]{CH_3CSCoA,\ -HSCoA}$ $CH_3\overset{\overset{\displaystyle O}{\|}}{C}-CH_2\overset{\overset{\displaystyle O}{\|}}{C}SCoA \xrightarrow[-HSCoA]{H_2O}$

an enolate

$CH_3\overset{\overset{\displaystyle O}{\|}}{C}-CH_2\overset{\overset{\displaystyle O}{\|}}{C}O^- \xrightarrow[-CO_2]{H^+} CH_3\overset{\overset{\displaystyle O}{\|}}{C}CH_3$

acetoacetate ion   acetone

17.20 (a) $CH_3CH(CO_2C_2H_5)_2 + CH_3CH_2CH{=}CHCO_2C_2H_5 \xrightarrow{NaOC_2H_5}$

$CH_3CH_2\underset{\underset{\displaystyle CH_3C(CO_2C_2H_5)_2}{|}}{C}HCH_2CO_2C_2H_5 \xrightarrow[\text{(2) H}^+]{\text{(1) H}_2O,\ OH^-,\ heat}}$ product

(b) $CH_2(CO_2C_2H_5)_2 + CH_3CH{=}CHCO_2C_2H_5 \xrightarrow[\text{(2) H}_2O,\ H^+]{\text{(1) NaOC}_2H_5}$

$CH_3\underset{\underset{\displaystyle CH(CO_2C_2H_5)_2}{|}}{C}HCH_2CO_2C_2H_5 \xrightarrow[\text{heat}]{H_2O,\ H^+}$ product

(c) $+\ CH_2(CO_2C_2H_5)_2 \xrightarrow[\text{(2) H}_2O,\ H^+]{\text{(1) NaOC}_2H_5}$ $\xrightarrow[\text{heat}]{H_2O,\ H^+}$ product

17.21 (a) $(CH_3)_3C\overset{\overset{\displaystyle O}{\|}}{C}OC_2H_5 + CH_3\overset{\overset{\displaystyle O}{\|}}{C}OC_2H_5 \xrightarrow[\text{(2) H}_2O,\ H^+]{\text{(1) NaOC}_2H_5}$ product

(b)

(c) $2\ CH_3CH_2\overset{\overset{\displaystyle O}{\|}}{C}H \xrightarrow{\ OH^-\ }$ product

(d) $(CH_3)_2C{=}O + CH_2(CN)_2 \xrightarrow{NH_4^+\ {}^-O_2CCH_3}$ product

(e) $C_6H_5\overset{\overset{\displaystyle O}{\|}}{C}CH_3 + C_2H_5O\overset{\overset{\displaystyle O}{\|}}{C}{-}\overset{\overset{\displaystyle O}{\|}}{C}OC_2H_5 \xrightarrow[\text{(2) }H_2O,\ H^+,\ \text{cold}]{\text{(1) NaOC}_2H_5}$ product

(f)

Chapter 18

18.1 **(a)** 1°; **(b)** 2°; **(c)** salt of 3°; **(d)** 1°; **(e)** quaternary; **(f)** 3°

18.2 **(a)** benzylamine **(b)** diisopropylamine **(c)** tetramethylammonium bromide
(d) allylamine **(e)** 1,10-decanediamine

18.3 **(a)** $CH_3(CH_2)_6NH_2$ **(b)** $CH_3CH_2\overset{\overset{\displaystyle NH_2}{|}}{C}HCH_2NH_2$

(c) $HOCH_2CH_2NH_2$ **(d)** $H_2N\overset{\overset{\displaystyle CH_2C_6H_5}{|}}{\underset{\underset{\displaystyle CH_3}{|}}{C}}H$ or

(e) $CH_3\overset{\overset{\displaystyle }{}}{\underset{\underset{\displaystyle NHCH_3}{|}}{C}}HCH_2CH_3$ **(f)** $(CH_3)_2\overset{+}{N}H_2\ {}^-O_2CCH_3$

18.4 **(c)** contains a chiral N, while **(d)** contains a chiral C. Each exists as a pair of enantiomers.

18.5 **(a)** (1) $CH_3CH_2CH_2Br$, (2) H_2O, OH^- **(b)** (1) $CH_2{=}CHCH_2Br$, (2) H_2O, OH^-
(c) (1) $C_6H_5CH_2Br$, (2) H_2O, OH^-

18.6

18.7 **(a)**

(b) $BrCH_2CH_2CH{=}CH_2 \xrightarrow{KCN} NCCH_2CH_2CH{=}CH_2 \xrightarrow[\text{(2) }H_2O,\ H^+]{\text{(1) LiAlH}_4}$ product

(c) $C_6H_5CO_2H \xrightarrow{SOCl_2} C_6H_5\overset{\overset{\displaystyle O}{\|}}{C}Cl \xrightarrow{2(CH_3)_2NH} C_6H_5\overset{\overset{\displaystyle O}{\|}}{C}N(CH_3)_2 \xrightarrow[\text{(2) }H_2O,\ H^+]{\text{(1) LiAlH}_4} C_6H_5CH_2N(CH_3)_2$

18.8 (a) $(R)\text{-}C_6H_5CH_2\overset{\overset{\displaystyle CH_3}{|}}{C}HNH_2$ (b) $H_2NCH_2CH_2CH_2NH_2$

18.9 (a) (1) $CH_3O\text{---}\langle\ \rangle\text{---}CH_2Br$ $\xrightarrow[-KBr]{}$

$CH_3O\text{---}\langle\ \rangle\text{---}CH_2\ddot{N}:$ (phthalimide) $\xrightarrow[\text{heat}]{NaOH, H_2O}$ product

(2) $CH_3O\text{---}\langle\ \rangle\text{---}CH_2\overset{\overset{\displaystyle O}{\|}}{C}NH_2$ $\xrightarrow[OH^-]{Br_2}$ product

(b) (1) $(R)\text{-}CH_3CH_2\overset{\overset{\displaystyle Br}{|}}{C}HCH_3$ $\xrightarrow[\substack{-KBr\\(S_N2)}]{}$ $(S)\text{-}CH_3CH_2\overset{\overset{\displaystyle N(phthalimide)}{|}}{C}HCH_3$ $\xrightarrow[\text{heat}]{NaOH, H_2O}$ product

(2) $(S)\text{-}CH_3CH_2\overset{\overset{\displaystyle O}{\|}}{\underset{\underset{\displaystyle CH_3}{|}}{C}}H\overset{}{C}NH_2$ $\xrightarrow[OH^-]{Br_2}$ product

18.10 *p*-Methylaniline is a stronger base than aniline because the methyl group is *electron-releasing*. *p*-Nitroaniline is a weaker base than aniline because the nitro group is *electron-withdrawing*.

18.11 $HO\overset{\overset{\displaystyle CH_2CO_2H}{|}}{\underset{\underset{\displaystyle CH_2CO_2H}{|}}{C}}CO_2^-$ $H_2\overset{+}{N}\langle\ \rangle NH$

18.12 $CH_3CH_2CH_2CH_2NH_2$

$\xrightarrow[\text{HCl}]{NaNO_2}$ (left) $\xrightarrow[\text{HCl}]{NaNO_2}$ (down)

$CH_3CH_2CH_2CH_2\text{---}\overset{+}{N}\equiv N$

$\xrightarrow{-N_2}$ $CH_3CH_2CH_2CH_2Cl$

$\xrightarrow{-N_2}$ $CH_3CH_2CH_2CH_2\overset{+}{O}H_2$ $\xrightarrow{-H^+}$ $CH_3CH_2CH_2CH_2OH$

Cl$^-$ or H$_2$O

$CH_3CH_2\overset{\overset{\displaystyle H}{|}}{C}HCH_2\text{---}\overset{+}{N}\equiv N$

$\xrightarrow{-N_2}$ $CH_3CH_2\overset{+}{C}HCH_3$

$\xrightarrow{-H^+}$ $CH_3CH=CHCH_3 + CH_3CH_2CH=CH_2$

$\xrightarrow{Cl^-}$ $CH_3CH_2CHClCH_3$

$\xrightarrow[-H^+]{H_2O}$ $CH_3CH_2\overset{\overset{}{}}{\underset{\underset{\displaystyle OH}{|}}{C}}HCH_3$

18.13 (a) $CH_3CH_2CH_2CH_2N(CH_3)_2 + CH_2{=}CH_2 + H_2O$

(b) $C_6H_5CH{=}CH_2 + (CH_3)_2NCH_2CH_3 + H_2O$

Chapter 19

19.1 (a) 2-amino-1-naphthol   **(b)** 1,2,6,7-tetramethylphenanthrene

(c) 9,10-dimethoxyanthracene   **(d)** 1-bromo-5-chloronaphthalene

19.2 Each compound has 18 pi electrons. Each compound should have about (6×18) or 108 kcal/mol of resonance energy.

19.3 The nitro group withdraws electron density by the inductive effect. Because the oxidizing agent attacks the other, less positive ring in Reaction 1, we would say that the attack is electrophilic.

    The amino group donates electrons to the ring by resonance. Therefore, its ring is more negative. Again, this is the ring attacked by the electrophilic agent.

19.4 (a)

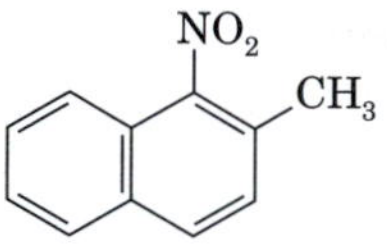

(b)

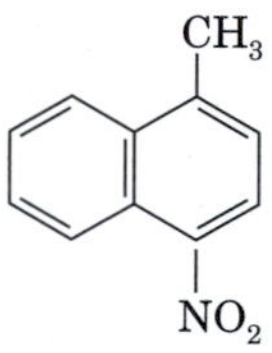

(c)

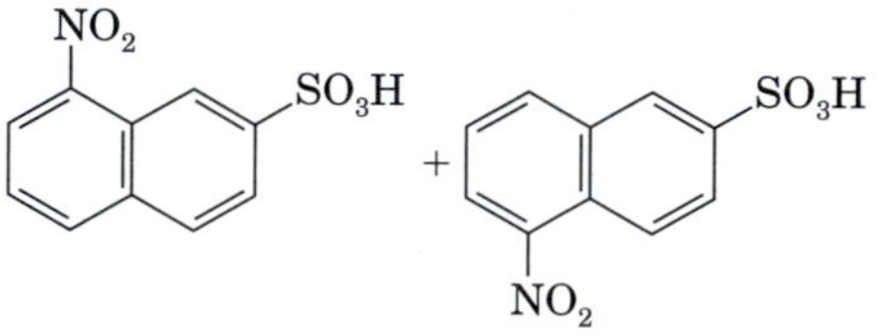

(d)

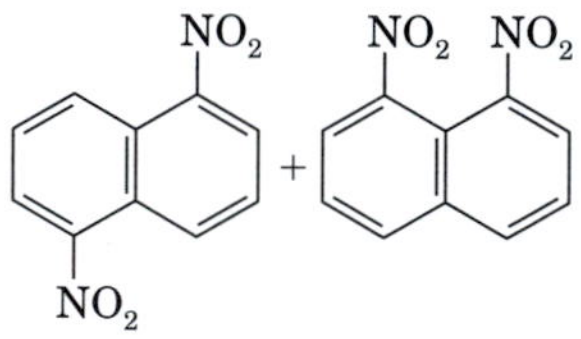

19.5 (a)

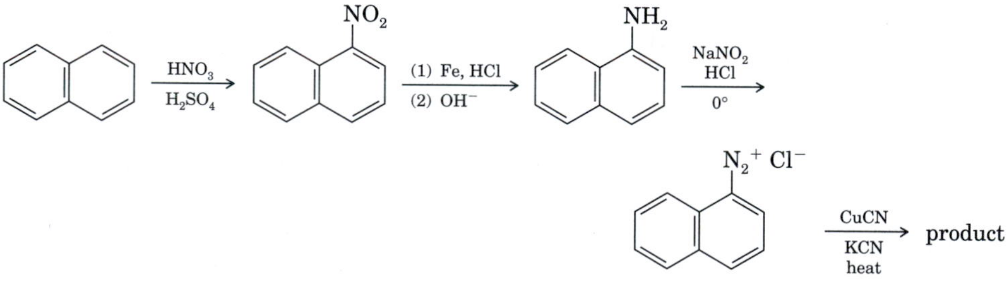

(b)

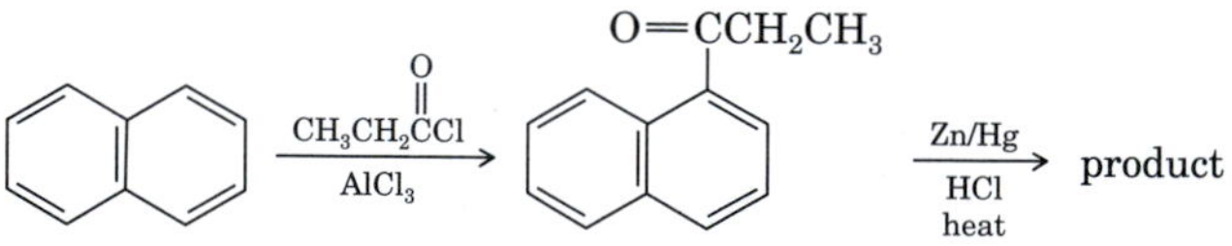

(c)

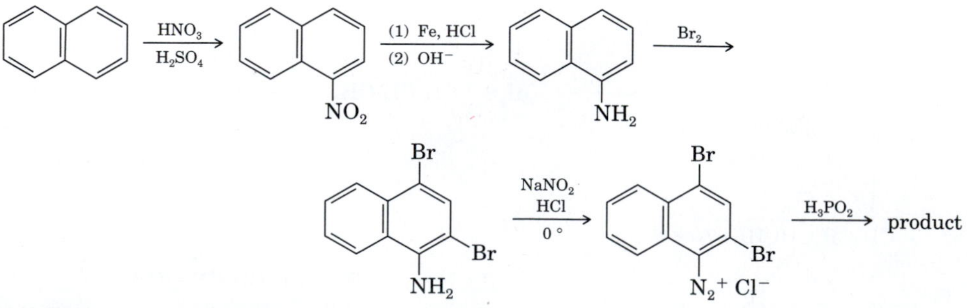

19.6

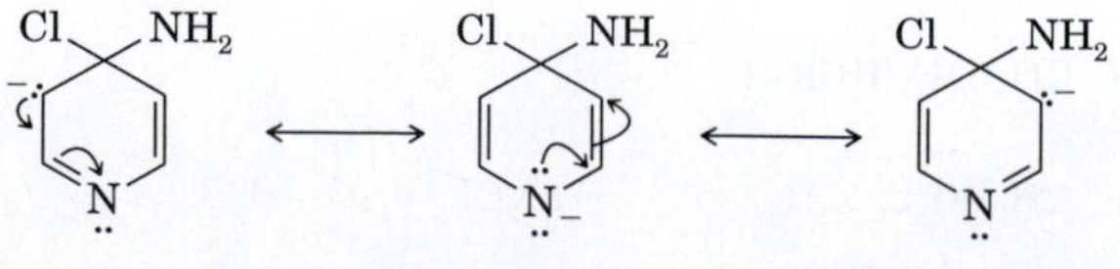

19.7

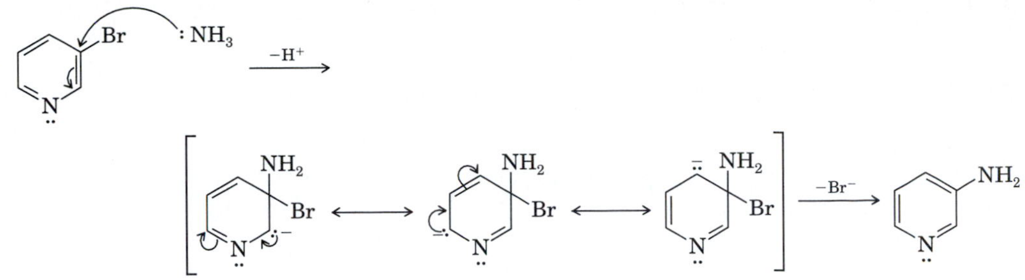

resonance structures for intermediate

19.8

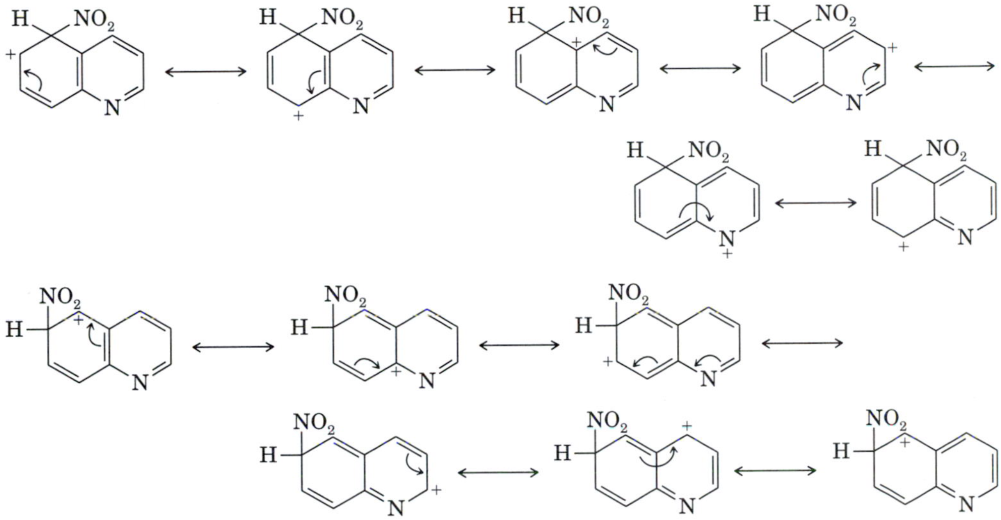

The 5-intermediate has three resonance structures with one aromatic ring, while the 6-intermediate has only two.

19.9 **(a)**

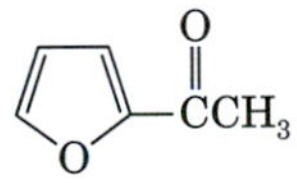

(b)

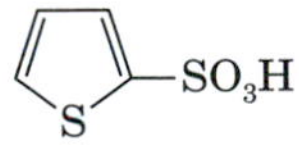

(c)

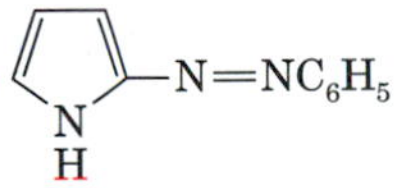

(d)

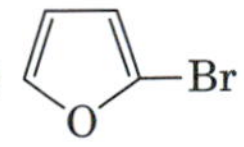

19.10 **(a)**

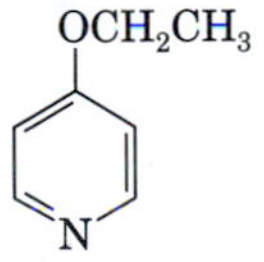

(b)

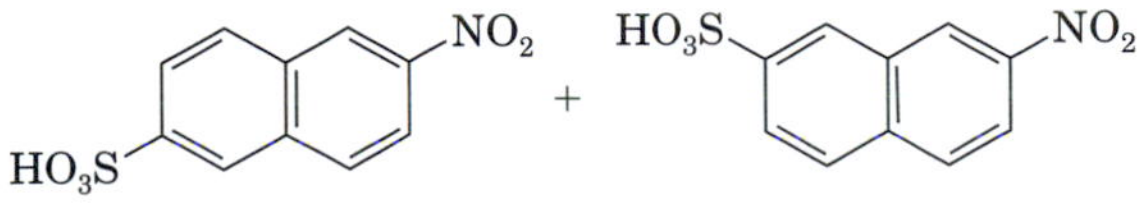

(c) 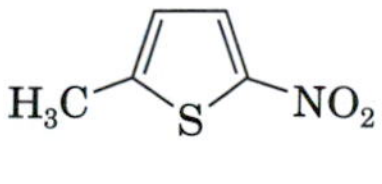 H_3C

NO_2

(d)

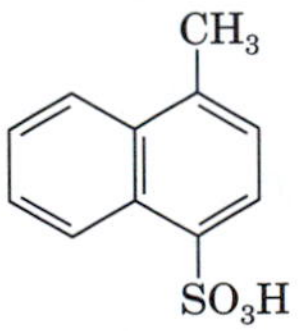

(e)

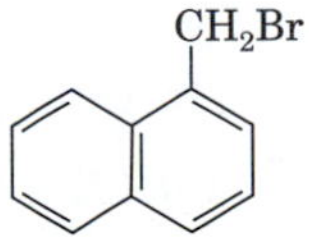

(f)

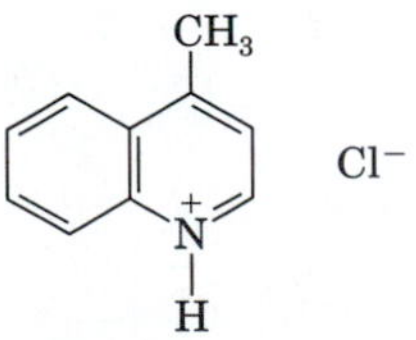

Cl^-

(g)

$+ CH_3CH_3$

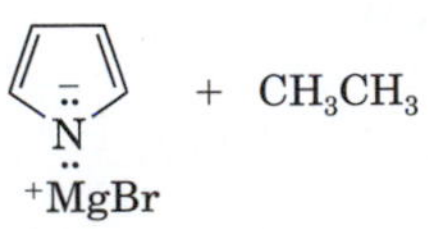

(h)

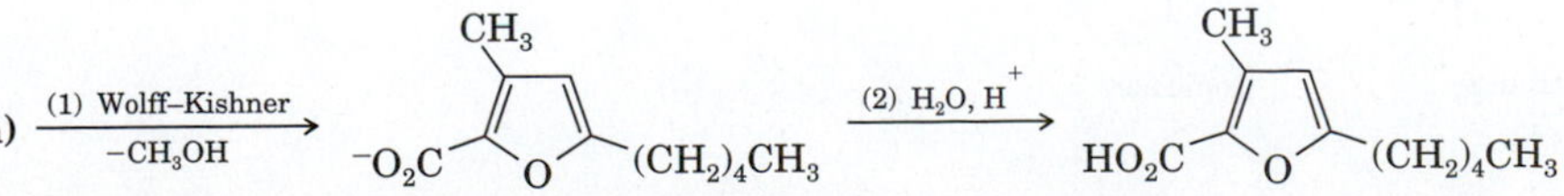

Chapter 20

20.1 **(a)**, **(c)**, and **(d)** are composed of isoprene units and therefore are classified as terpenes or terpenoids.

20.2 **(a)**

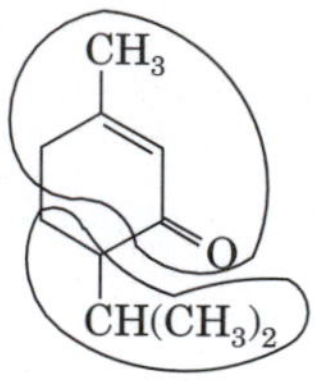

(b)

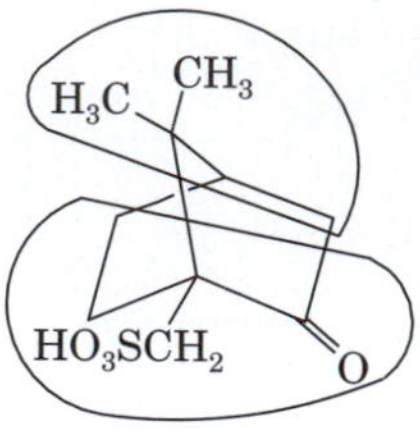

(c)

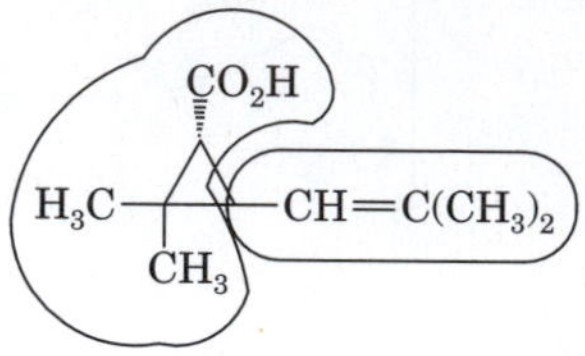

20.3

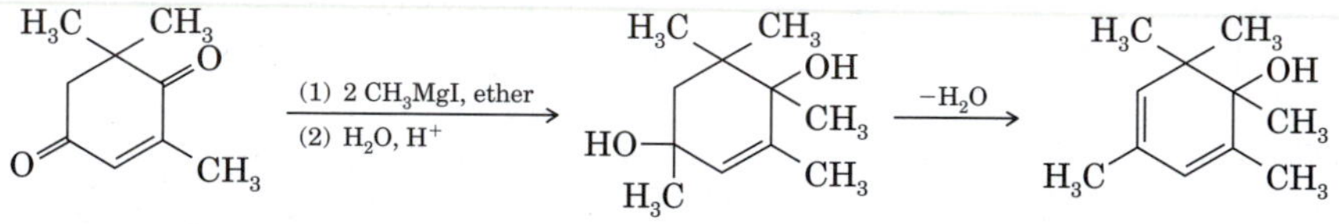

20.4

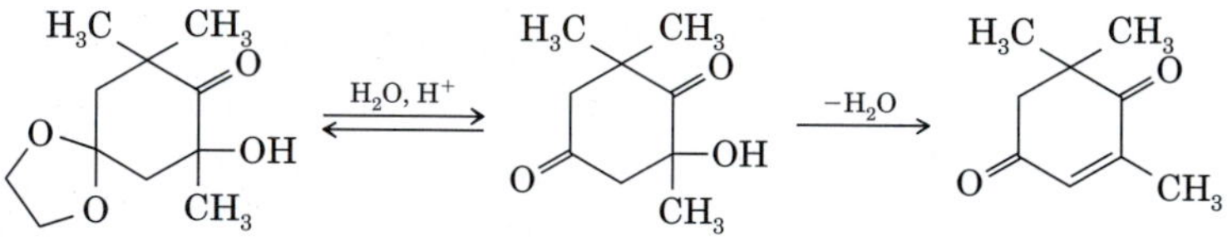

20.5 Under acidic conditions, dehydration and rearrangements could occur.

20.6 **(a)** $\dfrac{120 - 20}{120 + 20} \times 100\% = 71\%$

(b) $\dfrac{50 - 50}{50 + 50} \times 100\% = 0$

20.7

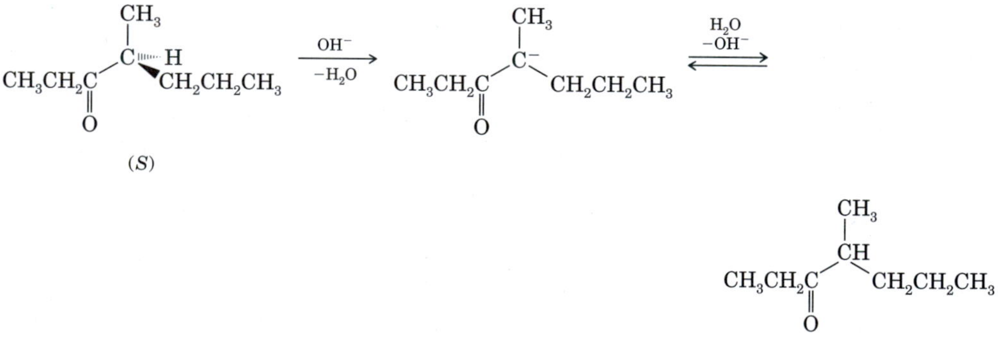

20.8 **(a)** $CH_3\overset{O^-}{\underset{\underset{CH_2CH_2CH_2CH_3}{|}}{C}}CH_2CH_3 + Li^+$　　**(b)** $\overset{OH}{\underset{|}{CH_2}}-CH_2CH_2CH_2\overset{O}{\overset{||}{C}}OCH_3$

(c) $CH_3\overset{NNHC_6H_5}{\overset{||}{C}}CH_3 + H_2O$　　**(d)** $^-CH_2\overset{NNHC_6H_5}{\overset{||}{C}}CH_3$

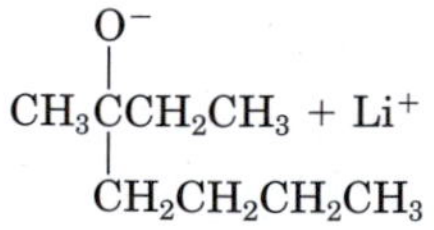

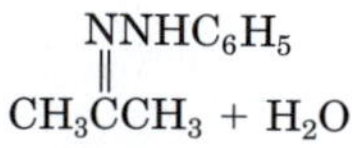

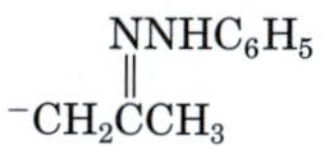

20.9

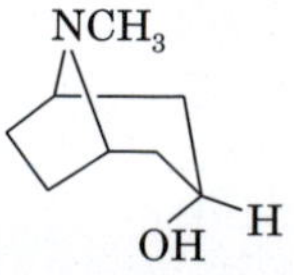

tropine 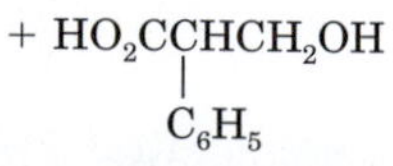　　　　　tropic acid

20.10

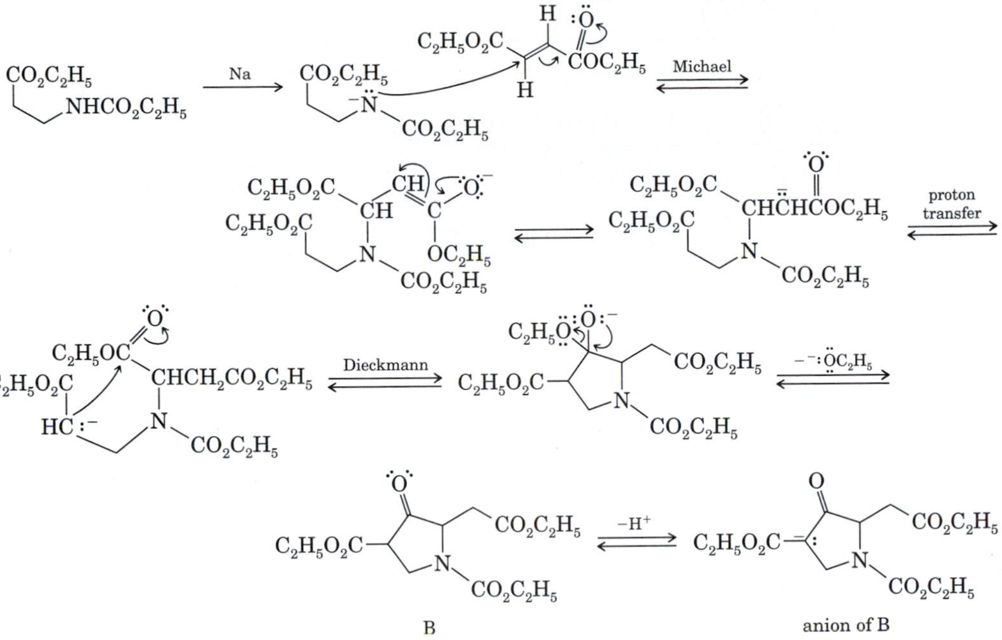

$CO_2C_2H_5$ $NHCO_2C_2H_5$ $\xrightarrow{\text{Na}}$

$C_2H_5O_2C$ ... N^- $CO_2C_2H_5$ ... $C_2H_5O_2C$ H H COC_2H_5 $\xrightarrow{\text{Michael}}$

$C_2H_5O_2C$ $C_2H_5O_2C$ CH CH C $\overset{..}{O}{}^{:-}$ N OC_2H_5 $CO_2C_2H_5$ $\rightleftharpoons$ $C_2H_5O_2C$ $C_2H_5O_2C$ CHCHCOC$_2$H$_5$ N $CO_2C_2H_5$ O $\xrightarrow{\text{proton transfer}}$

C_2H_5OC $C_2H_5O_2C$ $HC:^-$ CHCH$_2$CO$_2$C$_2$H$_5$ N $CO_2C_2H_5$ $\underset{\text{Dieckmann}}{\rightleftharpoons}$ $C_2H_5O_2C$ C_2H_5O O $CO_2C_2H_5$ N $CO_2C_2H_5$ $\xrightarrow{-:\overset{..}{O}C_2H_5}$

$C_2H_5O_2C$ O $CO_2C_2H_5$ N $CO_2C_2H_5$ $\underset{-H^+}{\rightleftharpoons}$ $C_2H_5O_2C=$ O $CO_2C_2H_5$ N $CO_2C_2H_5$

B   anion of B

Chapter 21

21.1 **(a)** [4 + 4] cycloaddition   **(b)** sigmatropic rearrangement

(c) [4 + 2] cycloaddition   **(d)** [2 + 2] cycloaddition

(e) electrocyclic reaction

21.2 π_6^* _____

π_5^* _____

π_4^* $\downarrow$

π_3 $\uparrow$

π_2 $\uparrow\downarrow$

π_1 $\uparrow\downarrow$

21.3 The double bonds in 1,4-pentadiene are not conjugated; thus, their π orbitals are independent of each other.

first C=C: π_2^* _____   second C=C: π_2^* _____

       π_1 $\uparrow\downarrow$ π_1 $\uparrow\downarrow$

21.4 **(a)** π_2^* _____ LUMO

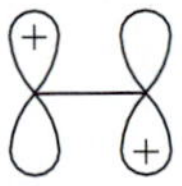

π_1 $\uparrow\downarrow$ HOMO

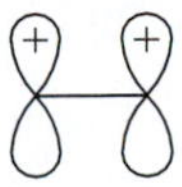

(b) π_3^* _____ LUMO

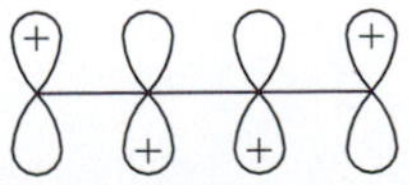

π_2 $\uparrow\downarrow$ HOMO

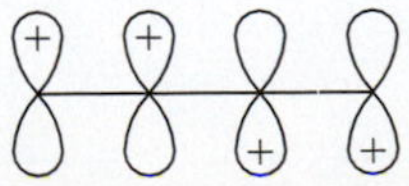

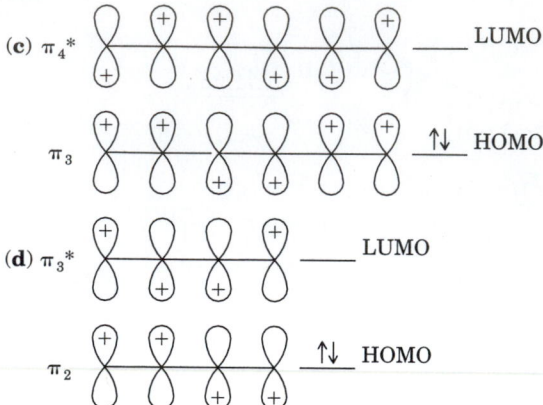

(c) $\pi_4{}^*$ _____ LUMO

π_3 ↑↓ HOMO

(d) $\pi_3{}^*$ _____ LUMO

π_2 ↑↓ HOMO

21.5 [6 + 4]

21.6 **(a)** $2\ C_6H_5CH{=}CHCO_2H \xrightarrow{h\nu}$ **(b)** $\begin{array}{l} CH_2CH{=}CH_2 \\ | \\ CH_2CH{=}CH_2 \end{array} \xrightarrow{h\nu}$

In **(a),** either _cis-_ or _trans_-cinnamic acid (3-phenylpropenoic acid) could be used. Other stereoisomers (and structural isomers) of the product might be obtained as by-products.

21.7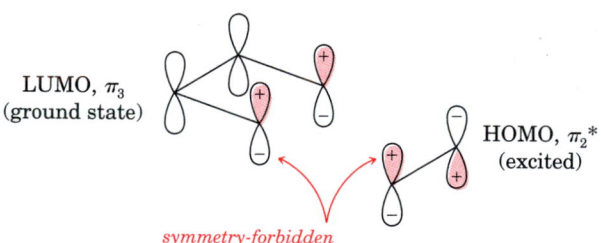

21.8 A photo-induced [4 + 2] cycloaddition cannot occur when _either_ the dienophile or the diene is excited.

LUMO, π_3 (ground state)   HOMO, $\pi_2{}^*$ (excited)

symmetry-forbidden

21.9 A [6 + 4] cycloaddition reaction is thermally induced because this reaction path is symmetry-allowed.

π_3, HOMO of the triene $\pi_3{}^*$, LUMO of the diene

21.10

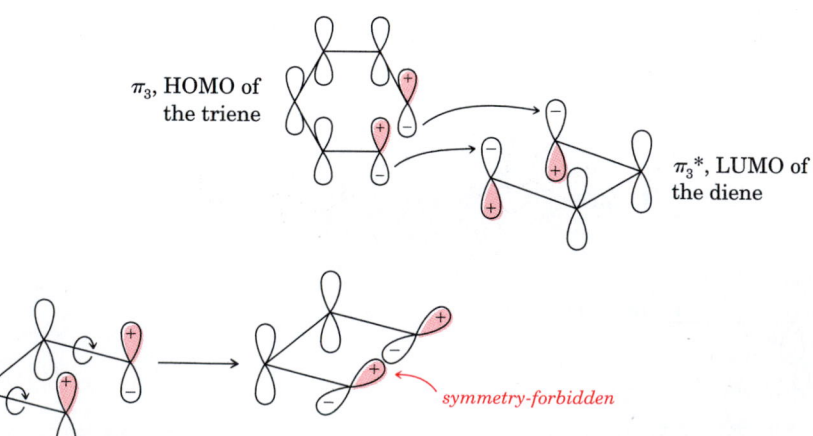

symmetry-forbidden

21.11

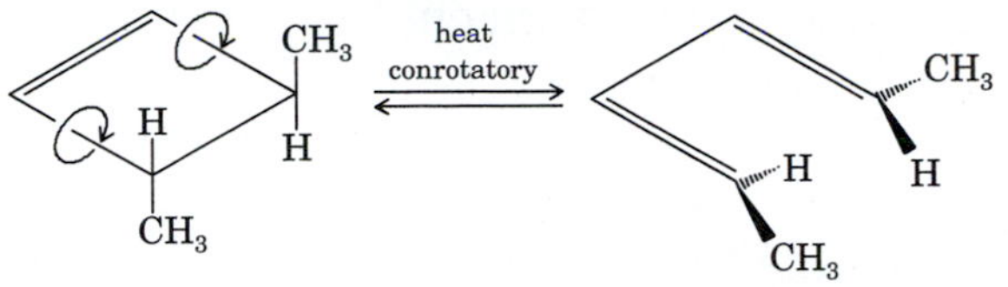

(2E, 4E)-2,4-hexadiene

21.12 (a)

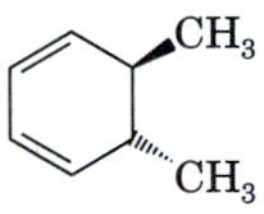

trans

(b)

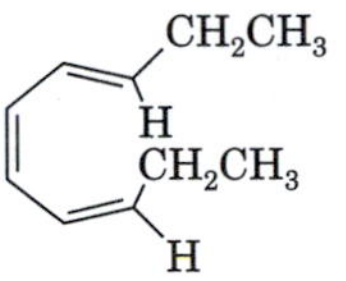

(3E,5Z,7Z)

21.13 (a) [3,3];   **(b)** [1,5]

21.14 (a) can proceed by a [1,5] sigmatropic shift, and **(b)** is a [3,3] sigmatropic shift. Both proceed readily.

Chapter 22

22.1   Tyrian purple

22.2 $\epsilon = \dfrac{A}{cl} = \dfrac{1.05}{(0.000100)(1.00)} = 10{,}500$

22.3   (c), (b), (a)

22.4   (b) 446 nm;   **(c)** 416 nm

22.5   In the following formulas, (1) the chromophores are circled and (2) the auxochromes are identified with arrows.

(a)

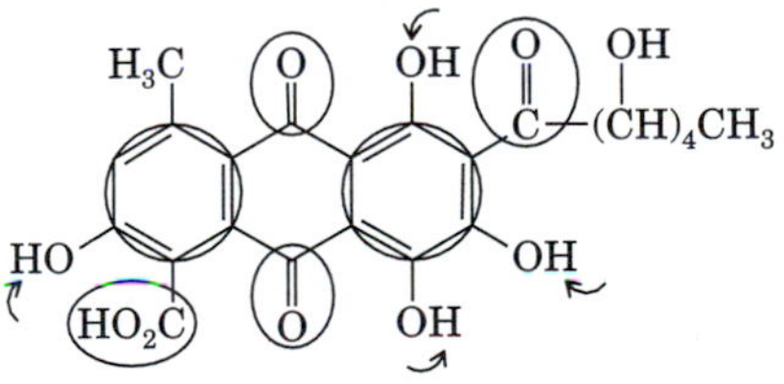

(b)

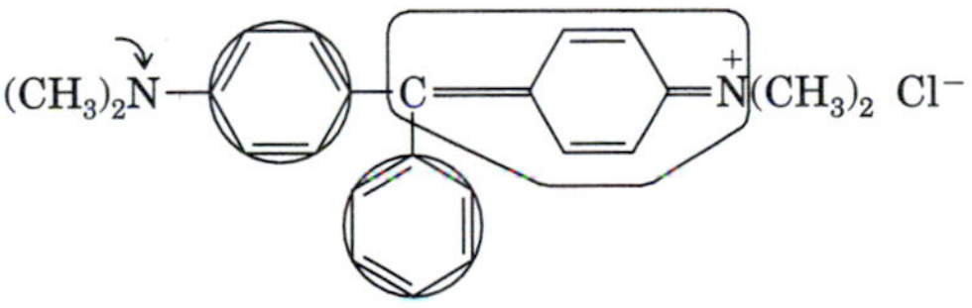

22.6   A compound that appears violet absorbs at a shorter wavelength (about 570 nm) than one that appears blue-green (about 650 nm), and therefore is the compound with less delocalization. The structure in **(b),** with only two —N(CH₃)₂ groups, has less delocalization and is the violet-colored compound. The structure in **(a),** therefore, is the blue-green compound.

22.7

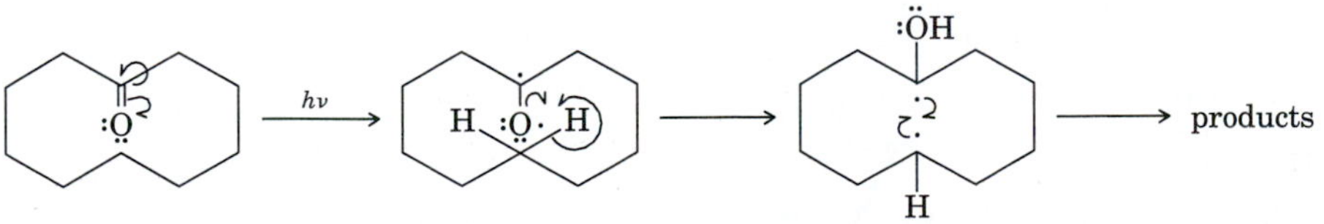

22.8   (a) $[CH_4]^{+}$, ion-radical;   **(b)** $[CH_3]^{+}$, cation;

(c) $[CH_2CH_2]^{+}$, ion-radical;   **(d)** $[CH_3CH_2]^{+}$, cation

22.9   (a) $[CH_4]^{+}$;   **(b)** $[CH_3CH_2CH_3]^{+}$

22.10 (a) 16;   **(b)** 43;   **(c)** 32;   **(d)** 18

22.11 (a) 30;   **(b)** 98;   **(c)** 46;   **(d)** 172

22.12 *B* contains Br, and *C* contains Cl.

22.13 (a) 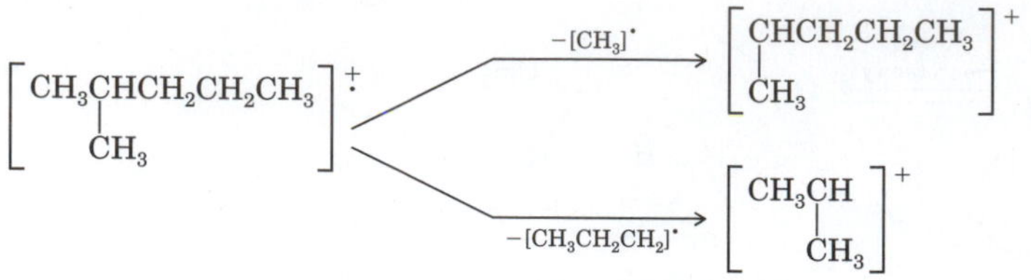 $\left[\begin{array}{c} CH_3CHCH_2CH_2CH_3 \\ | \\ CH_3 \end{array}\right]^{+\cdot}$

$\xrightarrow{-[CH_3]^{\cdot}} \left[\begin{array}{c} CHCH_2CH_2CH_3 \\ | \\ CH_3 \end{array}\right]^{+}$

$\xrightarrow{-[CH_3CH_2CH_2]^{\cdot}} \left[\begin{array}{c} CH_3CH \\ | \\ CH_3 \end{array}\right]^{+}$

$[CH_3]^+$ and $[CH_3CH_2CH_2]^+$ might also be observed.

(b) 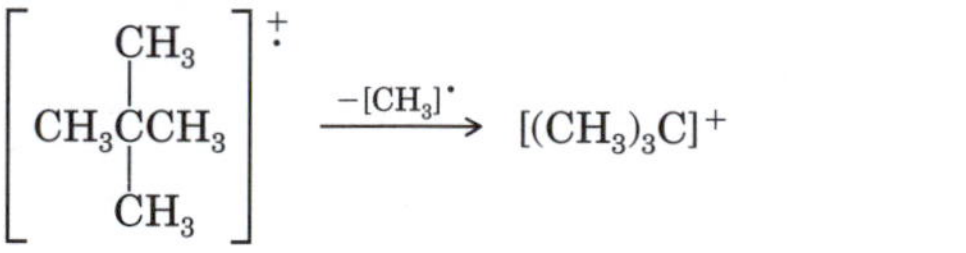 $\left[\begin{array}{c} CH_3 \\ | \\ CH_3CCH_3 \\ | \\ CH_3 \end{array}\right]^{+\cdot} \xrightarrow{-[CH_3]^{\cdot}} [(CH_3)_3C]^+$

(c) $[CH_2{=}CHCH_2CH_2CH_3]^{+\cdot} \xrightarrow{-[CH_3CH_2]^{\cdot}} [CH_2{=}CHCH_2]^+$

22.14 $CH_3CH_2OCH_2CH_3$

22.15 (a) 58; **(b)** 44; **(c)** 60; **(d)** 88 and 102

Chapter 23

23.1 (a) hexulose (ketohexose) **(b)** aldopentose **(c)** aldohexose

23.2 (a) D **(b)** L **(c)** L

23.3 L-(+)- and *meso*-tartaric acid

23.4

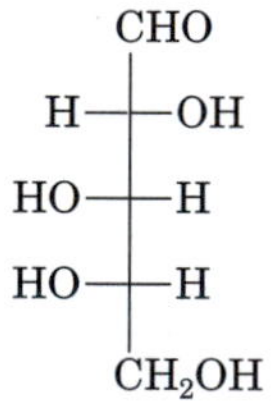

```
        CHO
    H ──── OH
   HO ──── H
   HO ──── H
       CH₂OH
```

L-arabinose

the enantiomer of D-*arabinose*

23.5 L-idose

23.6

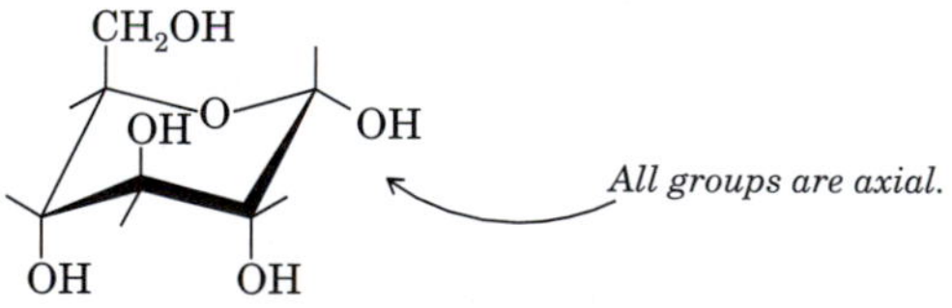

All groups are axial.

23.7 α:

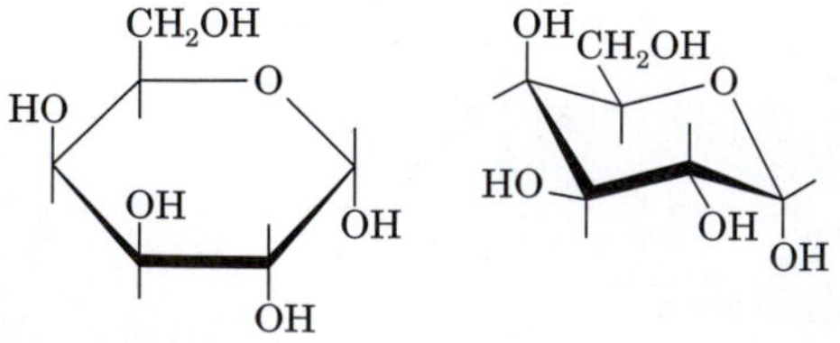

β:

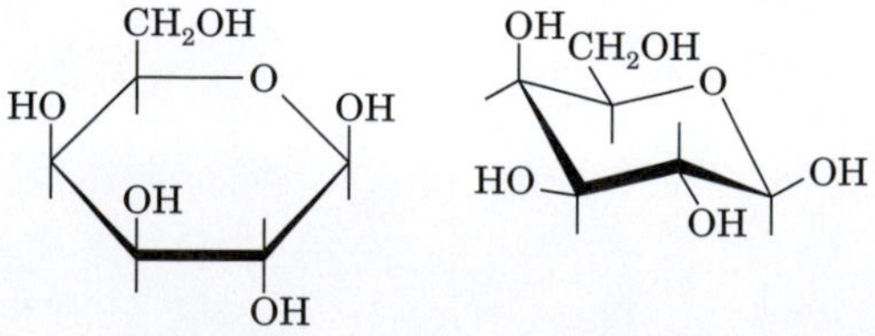

23.8 **(a)**

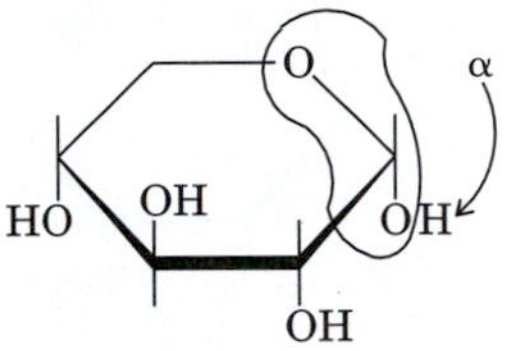

(b)

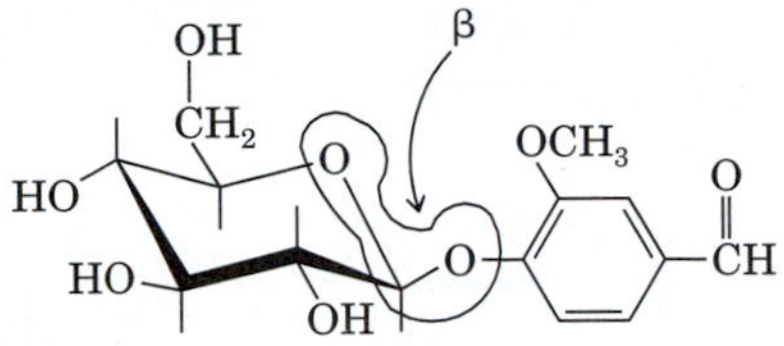

23.9 **(a)**

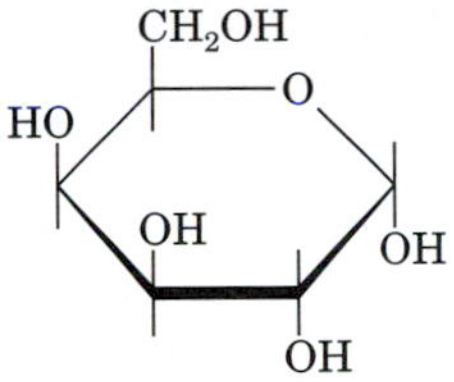

(b)

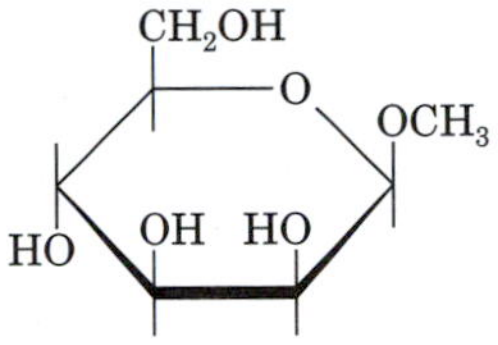

(c)

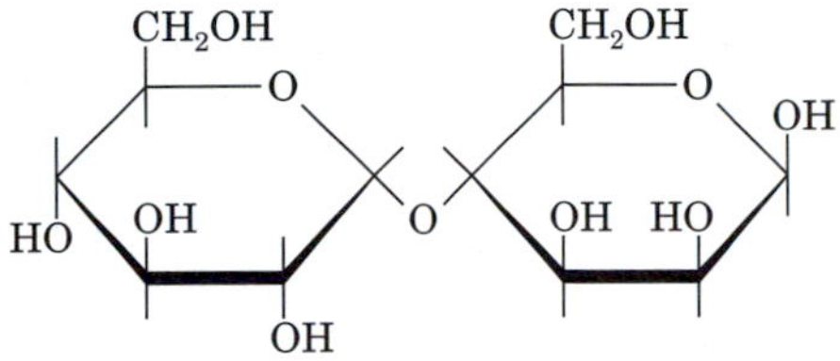

(d)

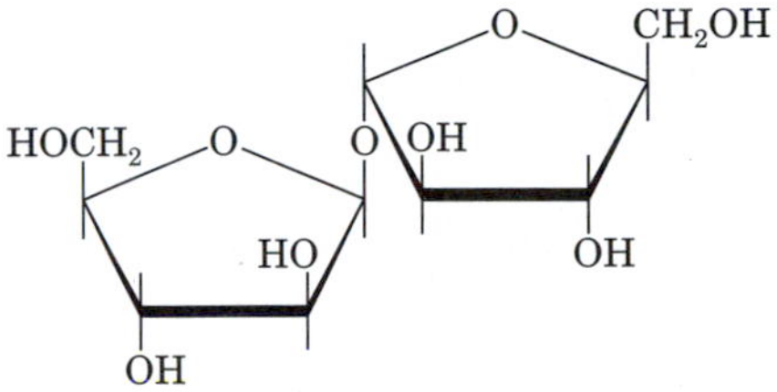

23.10 **(a)**

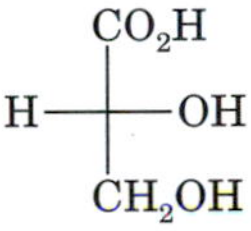

(b)

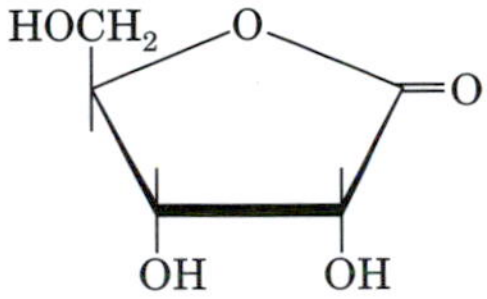

(c) none

(d)

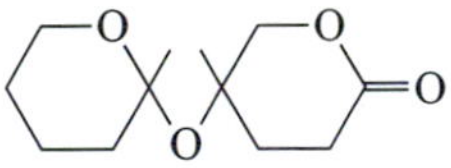

23.11 **(a)** optically active   **(b)** optically active   **(c)** *meso*   **(d)** *meso*

23.12 $CH_3\overset{O}{\overset{\|}{C}}H + H\overset{O}{\overset{\|}{C}}CH_2OCH_3$

23.13 **(a)**

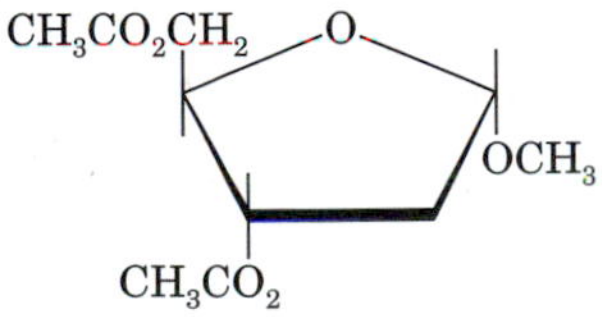

(b)

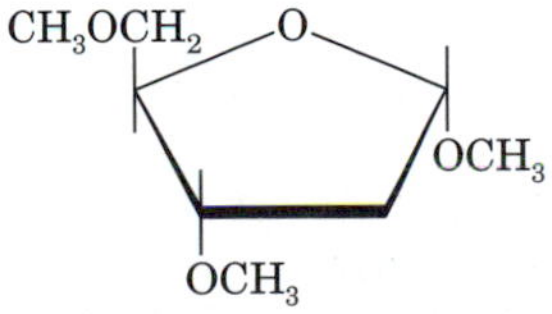

23.14 **(a)**

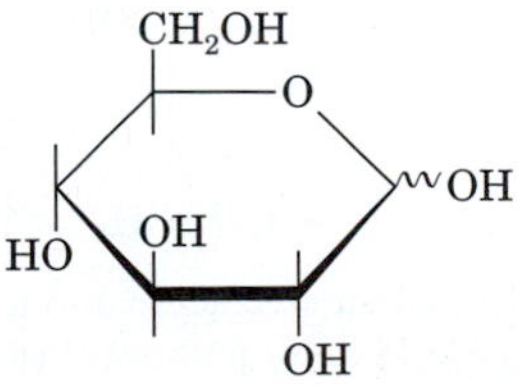

(b)

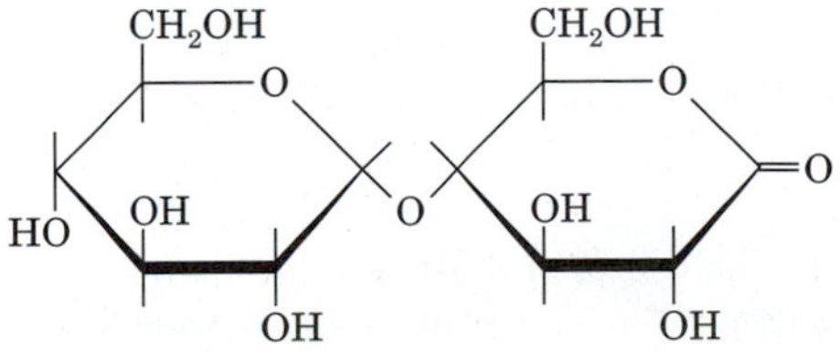

23.15 **(a)**, **(b)** and **(e)**:

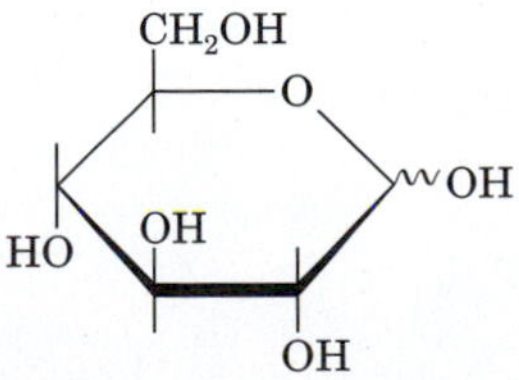

(c)

$$CH_2OH$$

HOCH$_2$ ... OH ... O ... CH$_2$OH ... O ... OH ... =O

HO OH OH

OH

(d)

HOCH$_2$ ... O ... CH$_2$OH — OH ... OH ... CO$_2^-$

HO OH OH

OH

23.16 (a)

$$CH_2OH$$

O

HO OH wOH + CH$_3$CO$_2$H

NH$_3^+$ Cl$^-$

(b)

$$CH_2OH$$

O

O OH + CH$_3$CO$_2^-$

NH$_2$

Chapter 24

24.1 (a)
$$CH_2O_2C(CH_2)_{14}CH_3$$
$$CHO_2C(CH_2)_{14}CH_3$$
$$CH_2O_2C(CH_2)_{14}CH_3$$

(b)
$$CH_2O_2C(CH_2)_7CH{=}CH(CH_2)_5CH_3$$
$$CHO_2C(CH_2)_7CH{=}CH(CH_2)_5CH_3$$
$$CH_2O_2C(CH_2)_7CH{=}CH(CH_2)_5CH_3$$

(c) (R) and (S)
$$CH_2O_2C(CH_2)_{14}CH_3$$
$$CHO_2C(CH_2)_{14}CH_3$$
$$CH_2O_2CCH_2CH_2CH_3$$

and

$$CH_2O_2C(CH_2)_{14}CH_3$$
$$CHO_2CCH_2CH_2CH_3$$
$$CH_2O_2C(CH_2)_{14}CH_3$$

24.2 (a)

HO OH
$$C{-}C$$
H CH$_3$(CH$_2$)$_6$CH$_2$ H CH$_2$(CH$_2$)$_6$CO$_2$H

or

H H
CH$_3$(CH$_2$)$_6$CH$_2$ $C{-}C$ CH$_2$(CH$_2$)$_6$CO$_2$H
HO OH

(b) $CH_3(CH_2)_4CO_2H$, $HO_2CCH_2CO_2H$, and $HO_2C(CH_2)_7CO_2H$

(c) $CH_3(CH_2)_{13}\underset{\underset{Br}{|}}{C}HCO_2H$ **(d)** $C_{25}H_{51}CO_2^-$ Na$^+$ + $C_{27}H_{55}CH_2OH$

(e) $C_{15}H_{31}CH_2OH + C_{15}H_{31}CH_2OH$ **(f)** $\underset{\underset{OH}{|}}{CH_2}{-}\underset{\underset{OH}{|}}{CH}{-}\underset{\underset{OH}{|}}{CH_2}$ + 3 $C_{15}H_{31}CO_2H$

24.3
$$CH_2O_2C(CH_2)_7CH{=}CH(CH_2)_7CH_3$$
$$CHO_2C(CH_2)_7CH{=}CH(CH_2)_7CH_3 \xrightarrow[\text{heat, pressure}]{\text{excess H}_2,\ \text{catalyst}}$$
$$CH_2O_2C(CH_2)_7CH{=}CH(CH_2)_7CH_3$$

$$\underset{\underset{CH_2OH}{|}}{\overset{\overset{CH_2OH}{|}}{CHOH}} + 3\ CH_3(CH_2)_{16}CH_2OH \xrightarrow{H_2SO_4}$$

$$CH_3(CH_2)_{16}CH_2OSO_3H \xrightarrow{\text{NaOH}} CH_3(CH_2)_{16}CH_2OSO_3Na$$

24.4 **(b)**, **(c)**, **(d)**, and **(e)**, all of which contain a large amount of hydrocarbon character and a highly polar or ionic group. The ester group in **(a)** is not sufficiently polar to lend detergent activity.

24.5 **(a)** $CH_3(CH_2)_{10}CO_2^-$ Na$^+$

(b) same as **(a)** or **(e)**   **(c)** $CH_3(CH_2)_{10}CH_2\overset{+}{N}(CH_3)_3$ Cl$^-$

(d) $CH_3(CH_2)_{10}CO_2CH_2CH_2O(CH_2CH_2O)_2CH_2CH_2OH$

(e) $CH_3(CH_2)_{10}CH_2{-}\bigcirc{-}SO_3^-$ Na$^+$

24.6 *trans*-CH$_3$(CH$_2$)$_{12}$CH=CHĊHCHCH$_2$OH with OH on the CH and NH$_2$ below

24.7 **(a)**

Five functional groups would have to be inserted.

(b)

Six functional groups would have to be inserted.

24.8 **(a)**

(b)

(c)

24.9 **(a)** *trans*   **(b)** neither   **(c)** *cis*   **(d)** neither

24.10 (a) beta   **(b)** alpha   **(c)** neither   **(d)** beta   **(e)** alpha   **(f)** beta

24.11 (a)

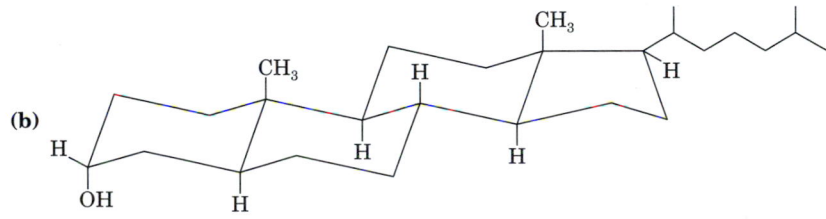

In **(a)**, you also might have oxidized the carbon–carbon double bond.

(b) cholesterol $\xrightarrow[\text{CCl}_4]{\text{Br}_2}$

(c) cholesterol $\xrightarrow{\text{NBS}}$

+ 7-bromocholesterol

(d) cholesterol $\xrightarrow{\text{NaH}}$

Chapter 25

25.1

L-cysteine

The —CH_2SH group has higher priority than does the —CO_2H group because S has a higher atomic number than does O.

25.2 Compounds **(a)** and **(d)** contain more than one chiral carbon and, therefore, can exist as more than one pair of enantiomers.

25.3 (a) $CH_3\overset{\displaystyle|}{\underset{\displaystyle OH}{CH}}-\overset{\displaystyle|}{\underset{\displaystyle NH_2}{CH}}-\overset{\displaystyle O}{\overset{\|}{C}}OH \rightleftharpoons CH_3\overset{\displaystyle|}{\underset{\displaystyle OH}{CH}}-\overset{\displaystyle|}{\underset{\displaystyle {}^+NH_3}{CH}}-\overset{\displaystyle O}{\overset{\|}{C}}O^-$

(b) $(CH_3)_2CHCH_2\overset{\displaystyle|}{\underset{\displaystyle NH_2}{CH}}\overset{\displaystyle O}{\overset{\|}{C}}OH \rightleftharpoons (CH_3)_2CHCH_2\overset{\displaystyle|}{\underset{\displaystyle {}^+NH_3}{CH}}-\overset{\displaystyle O}{\overset{\|}{C}}O^-$

25.4 (1) $C_6H_5CH_2\overset{\displaystyle O}{\overset{\|}{C}}H \xrightarrow[-H_2O]{NH_3} C_6H_5CH_2CH{=}NH \xrightarrow{HCN}$

$C_6H_5CH_2\overset{\displaystyle|}{\underset{\displaystyle NH_2}{CH}}CN \xrightarrow[\text{(2) neutralize}]{\text{(1) } H_2O,\ H^+,\ \text{heat}} C_6H_5CH_2\overset{\displaystyle|}{\underset{\displaystyle {}^+NH_3}{CH}}CO_2^-$

(2) $C_6H_5CH_2\overset{\displaystyle|}{\underset{\displaystyle Br}{CH}}CO_2H \xrightarrow[\text{(2) neutralize}]{\text{(1) excess } NH_3} \text{product}$ **(3)** $C_6H_5CH_2\overset{\displaystyle O}{\overset{\|}{C}}CO_2H \xrightarrow[Pd]{H_2,\ NH_3} \text{product}$

(4) $BrCH(CO_2C_2H_5)_2$

$\xrightarrow[\text{(2) } C_6H_5CH_2Br]{\text{(1) } NaOC_2H_5}$

$\xrightarrow[\text{(2) neutralize}]{\text{(1) } H_2O,\ H^+,\ \text{heat}} \text{product}$

25.5 **(a)** acidic;   **(b)** near-neutral;   **(c)** near-neutral;   **(d)** basic;   **(e)** near-neutral

25.6 $HO_2CCH_2CH_2\underset{\underset{\displaystyle NH_2}{|}}{C}HCO_2^- \, Na^+$

25.7   **(a)** [pyrrolidine ring]$-CO_2H \; Cl^-$,   [pyrrolidine ring]$-CO_2^- \; Na^+$

    (b) $HO-$[benzene ring]$-CH_2\underset{\underset{\displaystyle NH_3^+}{|}}{C}HCO_2H \; Cl^-$, $^-O-$[benzene ring]$-CH_2\underset{\underset{\displaystyle NH_2}{|}}{C}HCO_2^- \; 2 \, Na^+$

25.8   Like the nitrogen in pyrrole (Section 19.9), the nitrogen in tryptophan has no unshared bonding electrons; therefore, tryptophan is a neutral amino acid.

25.9   The isoelectric point is the average of the two pK_a values.

    (a) $pI = \dfrac{1.92 + 9.73}{2} = 5.82$ **(b)** 5.91   **(c)** 6.30   **(d)** 6.00

25.10 **(a)** [imidazole ring]$CH_2\underset{\underset{\displaystyle ^+NH_3}{|}}{C}H\overset{\overset{\displaystyle O}{\|}}{C}-NH\underset{\underset{\displaystyle CH_3}{|}}{C}H\overset{\overset{\displaystyle O}{\|}}{C}O^-$

    (b) $HO_2CCH_2\underset{\underset{\displaystyle ^+NH_3}{|}}{C}H\overset{\overset{\displaystyle O}{\|}}{C}-NH\underset{\underset{\displaystyle CH_2}{|}}{C}H\overset{\overset{\displaystyle O}{\|}}{C}-NH\underset{\underset{\displaystyle CH_3}{|}}{C}H\overset{\overset{\displaystyle O}{\|}}{C}O^-$ [indole ring]

    (b) $(CH_3)_2CH\underset{\underset{\displaystyle ^+NH_3}{|}}{C}H\overset{\overset{\displaystyle O}{\|}}{C}-NHCH_2\overset{\overset{\displaystyle O}{\|}}{C}-NH\underset{\underset{\displaystyle CH_2SH}{|}}{C}H\overset{\overset{\displaystyle O}{\|}}{C}-NH\underset{\underset{\displaystyle CH_2}{|}}{C}H\overset{\overset{\displaystyle O}{\|}}{C}O^-$ [indole ring]

25.11   leu-ala-tyr
         thr-pro-leu
      his-thr-pro
lys-ala-his
—————————————
lys-ala-his-thr-pro-leu-ala-tyr

25.12 **(a)** Trypsin cleaves the amide bond (to the right) of lys and arg.

    trp-met-gly-gly-phe-arg⁄phe-trp-val-lys⁄ala-gly-ser $\xrightarrow[H_2O]{trypsin}$

          trp-met-gly-gly-phe-arg + phe-trp-val-lys + ala-gly-ser

    (b) Chymotrypsin cleaves the amide bond (to the right) of tyr, phe, and trp.

    trp⁄met-gly-gly-phe⁄arg-phe⁄trp⁄val-lys-ala-gly-ser $\xrightarrow[H_2O]{chymotrypsin}$

          2 trp + met-gly-gly-phe + arg-phe + val-lys-ala-gly-ser

    (c) Pepsin cleaves the amide bond (to the left) of tyr, phe, and trp.

    trp-met-gly-gly⁄phe-arg⁄phe⁄trp-val-lys-ala-gly-ser $\xrightarrow[H_2O]{pepsin}$

          trp-met-gly-gly + phe-arg + phe + trp-val-lys-ala-gly-ser

25.13

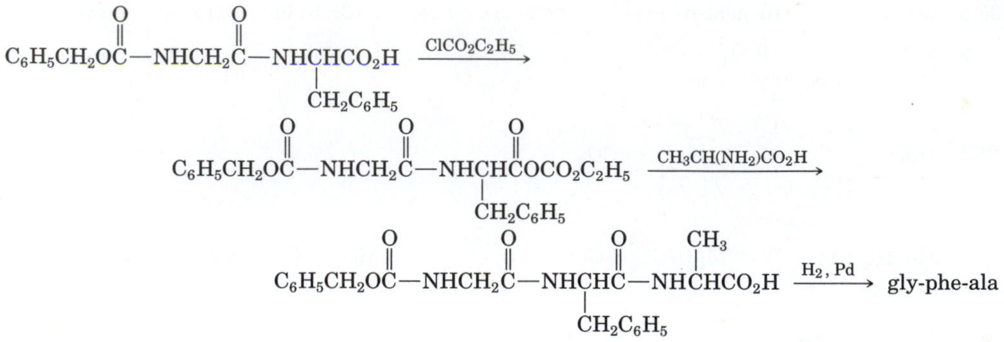

Chapter 26

26.1

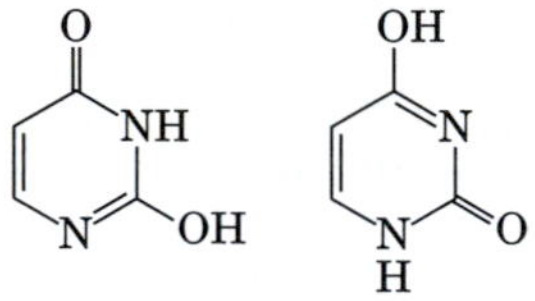

26.2

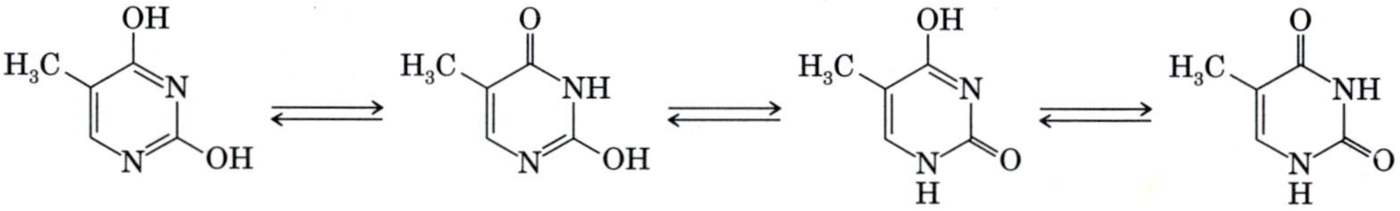

26.3 Cytosine is more basic because it contains a nitrogen with an electron pair that is not involved in resonance with a carbonyl group.

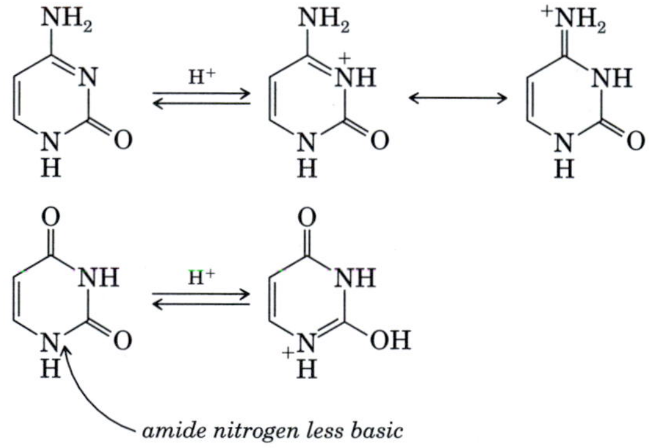

amide nitrogen less basic

26.4

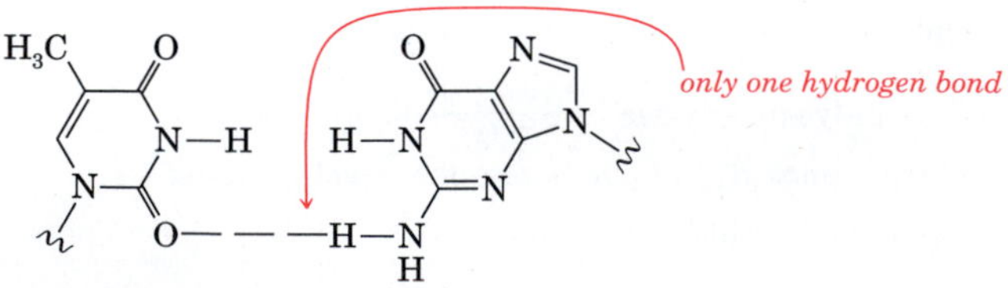

only one hydrogen bond

thymine   guanine

26.5 **(a)** hydrogen bonding between guanine and cytosine

(b) adenine (A)

(c) two, because there are two hydrogen bonds between thymine and adenine

(d) the covalent bond between the nitrogen at position 9 of guanine and the carbon at position 1 of a deoxyribose unit of the polymeric chain

(e) the deoxyribose–phosphate backbone of the DNA polymer

26.6

26.7 **(a)** thr-phe-lys-asp-leu-gly   **(b)** glu-cys-arg-gly-arg-gly-gly

INDEX